# Design of Steel Structures

## LIMIT STATES METHOD

N. Subramanian

*Consulting Engineer*
*Maryland*
*USA*

OXFORD
UNIVERSITY PRESS

OXFORD
UNIVERSITY PRESS

Oxford University Press is a department of the University of Oxford. It furthers the University's objective of excellence in research, scholarship, and education by publishing worldwide. Oxford is a registered trade mark of Oxford University Press in the UK and in certain other countries.

Published in India by
Oxford University Press
YMCA Library Building, 1 Jai Singh Road, New Delhi 110001, India



First published in 2016



ISBN-13: 978-0-19-946091-5
ISBN-10: 0-19-946091-4

Typeset in Times
by Cameo Corporate Services Limited, Chennai
Printed in India at Repro India Ltd

# PREFACE

'Design' in its broad sense covers planning of the building, choosing a configuration of the structure for the given requirements, structural analysis and design, detailing of various components, estimation of the material requirements, and the cost. The structural design of a structure simply means that the elements of the structure are to be proportioned and joined together in such a way that they will be able to withstand all the loads (load effects) that are likely to act on it during its service life, without excessive deformation; the structure should not collapse due to an abnormal loading such as that occurring during an earthquake. Structural design is often considered an art as well as a science. It must balance theoretical analysis with practical considerations such as the degree of certainty of loads and forces, the actual behaviour of the structure as distinguished from the idealized analytical and design model; the actual behaviour of the material, compared to assumed elastic behaviour; and the actual properties of materials used, compared to the assumed ones.

Steel is one of the major construction materials used all over the world. It has many advantages over other competing materials such as concrete and wood; these advantages include high strength to weight ratio, high ductility (hence its suitability for earthquake-resistant structures), uniformity, and recyclability (100% recycling is possible with steel). According to www.recycle-steel.org, in North America, more steel is recycled than paper, plastic, aluminium, and glass combined. Every ton of steel recycled conserves 1100 kg of iron ore, 650 kg of coal, and 55 kg of limestone. Presently, several grades and shapes of steel products exist.

Structural knowledge is increasing rapidly and continuously as techniques for analysis, design, fabrication, and erection of structures are being improved constantly and new types of structures are being introduced. Hence, designers need to have a sound knowledge of structural steel behaviour, including the material behaviour of steel, and the structural behaviour of individual elements and of the complete structure. Unless the structural engineer is abreast of the recent developments and understands the relationships between structural behaviour and design criteria implied by the rules of the design codes, he/she will be following the codal rules rigidly and blindly and may even apply them incorrectly in situations beyond their scope.

The latest Indian Standard code of practice for general construction using hot-rolled steel sections (IS 800:2007) was released in February 2008. This third revision of the code is based on the limit states method of design (the earlier versions of the code were based on the working or allowable stress method).

Another important change in the code is in the convention for the member axis. In order to be compatible with the results of computer programs, the code has chosen the *x-x* axis along the member, *z-z* axis as the major axis (parallel to the flanges in the case of I-sections), and *y-y* axis as the minor axis (perpendicular to the flanges in the case of I-sections). To explain the various clauses of this new revision of the code, I wrote the book titled *Design of Steel Structures*, which is the first book in India based on this code. This book is a sequel to the previous book and contains essentially similar material with enhanced features. This book also adopts the convention of member axis prescribed in IS 800 and uses SI units throughout the book.

Only extracts from the Indian Standard code of practice have been included in the text, and hence the readers are advised to refer to the latest code (IS 800:2007) published by the Bureau of Indian Standards, New Delhi for complete details. It is recommended that readers also refer to the latest version of the codes on design loads (IS 875 and IS 1893), dimension of sections (IS 808 or IS Handbook No. 1, IS 1161, IS 12778, IS 4923, and IS 811), specification of steel (IS 2062, IS 8500, IS 6639, and IS 3757), bolts (IS 1364 and IS 4000), and welding (IS 816).

## ABOUT THE BOOK

The objectives of writing this book are (a) to explain to the students the behaviour of various elements of structures and to provide the basis for the codal rules, (b) to compare the Indian code provisions with those of other countries, wherever necessary (in the global economy, many engineers are required to design using codes of other countries), (c) to provide ample examples so that students understand the concepts clearly, (d) to give information on latest developments, (e) to provide case studies on failures and important recent features, (f) to provide design aids, and (g) to provide sources for further reading so that interested readers may refer them for more details.

# FOREWORD

Steel structures are the landmarks of modern societies. Examples include the Golden Gate bridge in San Francisco, the Disney Concert Hall in Los Angeles, the Beijing Olympic Stadium, the Millenium bridge or the Orbit in London, Calatrava's train stations, and thousands more.

Using structural steel as a construction material allows ultimate flexibility, airy spaces, rapid construction, long spans, superior seismic performance, and much more. Leading architects worldwide have been quick to tap the opportunities offered by that material for freeform innovation, particularly in the last decade with the rapid growth of free-form architecture, and these attractive structures stand like testimonies to a society's commitments to a better future. Countries who aspire to participate in that future will inevitably find themselves as players in that arena. Structural engineers support that vision by their ability to design cost effective, safe, serviceable, and sustainable steel structures.

The textbook *Design of Steel Structures – Limit States Method* by Dr N. Subramanian, based on the Indian Steel Design code, provides Indian students and practising engineers more than an introduction to structural steel. At over 900 pages, it is a comprehensive presentation of the design methods and details relevant to steel structures (with material for the development of at least three semester-long courses). Its scope ranges from materials and specification, to design of seismic-resistant system, including coverage on structural analysis methods, loads, design of structural members and connections, and much more.

Most importantly, the textbook covers the important design requirements to prevent critical failure modes that are too often omitted in shorter-length steel design textbooks. For example, the issue of prying action in bolted connections, often overlooked by other authors, is covered here. Catastrophic connection failures could occur if prying action is accidentally overlooked by practising engineers.

Another valuable aspect of this textbook is the inclusion of case studies to put in perspective practical applications or to highlight the significance of design details—such as the description of the Hyatt Regency walkway collapse—which was a significant event that most senior structural engineers remember, but that is often unknown by younger generations.

Knowing first-hand the dedication and work it takes to write a textbook (having just completed the Second Edition of *Ductile Design of Steel Structures*, co-authored with Chia-Ming Uang and Rafael Sabelli), and recognizing the enormous investment of time it takes to do so, I wish Dr Subramanian much success with this textbook, and that it will be wholeheartedly and broadly embraced by academics and practising engineers across India, as the reference document on the topic.

**Michel Bruneau, Ph D, P.Eng.**
Professor, University at Buffalo, USA
Lead Author of *Ductile Design of Steel Structures*

## CONTENTS AND COVERAGE

The entire text consists of 18 chapters out of which 14 chapters completely cover the undergraduate curriculum. Four more chapters on *Loading and Load Combinations*, *Methods of Structural Analysis*, *Seismic-resistant Design*, and *Design Principles for Steel Enabling Structures* are meant for higher studies and for practising engineers. The chapters on *Corrosion Protection*, *Fire-resistant Design*, *Fatigue-resistant Design*, and *Design for Torsion* are provided in the online resource centre for those interested in these topics.

Each chapter comprises ample number of tables, figures, and examples. In addition, each chapter contains a *summary*, giving the salient points covered in that chapter and also *multiple-choice questions*, *exercises, review questions*, and *references* to help students assimilate the ideas presented in the chapters, to apply them to get a feel of the results obtained, and to gather additional material from literature. In addition, *case studies of failures* and some important aspects of structural design are interspersed throughout the text, to enhance the usefulness of the text.

*Chapter 1* provides a brief discussion on the historical developments, steel making processes, and metallurgy of steel. It also presents relevant information on the material properties of different types of steel.

*Chapter 2* introduces the design considerations and the role of structural design in the complete design process. Since every designer must have knowledge about the various types of structures, a brief discussion of the different types of steel structures is also presented. Various design philosophies are explained with their advantages and drawbacks. Many failures are attributed to the lack of determination of the loads acting on structures. Hence, the various loads that can act on a structure are discussed (with emphasis on wind and earthquake loads), as per the latest Indian codes in *Chapter 3*.

Several methods of analysis that accurately determine the forces and moments in various elements of structures have been developed. In fact, structural analysis is an integral part of the design process. Hence, these analysis methods are briefly described in *Chapter 4*.

The two methods used to connect the elements of steel structures, namely, bolted and welded connections, are discussed in *Chapters 5 and 6*. Discussions on simple bolted and welded joints and the use of adhesives have also been included in these chapters.

*Chapter 7* deals with the design of tension members. Plastic and local buckling behaviour of steel sections are covered in *Chapter 8*, as they will be useful in understanding the future chapters. *Chapter 9* covers the design of axially loaded compression members, which covers the multiple column curves adopted in IS 800 and compares it with the column curves of other codes. Different types of compression members including built-up compression members, sleeved columns, aluminium and stainless steel columns, cased columns, and steel arches are discussed. The effective length concept is also explained as per Indian and American codes.

The design of flexural members is covered in Chapters 10–12. *Chapter 10* deals with the design of laterally supported and laterally unsupported beams. It includes several aspects of beam design and the design of latticed beams, castellated beams, lintels, and purlins. Design of plate girders is explained in *Chapter 11*, which also includes discussions on plate girders with corrugated webs. *Chapter 12* describes the design of gantry girders and contains several practical aspects of their construction.

The design of beam-columns, which are subjected to both axial loads and bending moments are discussed in *Chapter 13*. Moment-resistant bolted and welded connections are discussed in *Chapter 14*. Design of slab bases, gusseted base plates, anchor bolts, shear connectors, grillage foundation, and beam bearing plates are coved in *Chapter 15*. With the information provided in Chapters 1–15, it is possible to design any type of structure consisting of tension members, compression members, flexural members, or beam-columns. To demonstrate this, the design roof trusses and industrial buildings are dealt with in *Chapter 16*.

*Chapter 17* on seismic-resistant design covers various aspects of planning (various plan and vertical irregularities to be avoided in seismic zones), different seismic force resisting systems (such as special moment frames, braced frames, and dual systems), seismic moment connections (as per American codes), steel plate shear walls, buckling-restrained braces, base isolation, and energy absorbing devices. *Chapter 18* deals with the design and practical aspects of steel enabling structures.

The appendices provide some useful data such as properties of a variety of structural steel sections, loads due to tsunami, properties of soils, bending moment, shear force, deflection of beams and frames, and computer programs (in Visual Basic) for the design of beams and beam-columns, design aids in the form of tables and charts, and the collection of design formulae.

We have taken care to present error-free material. I would like to specially thank the brilliant engineer *Kumar Abhishek Singh* of Kajima India Pvt. Ltd, Gurgaon, for going through the entire book and checking all the examples. In spite of our efforts, some errors might have crept in inadvertently; I would highly appreciate if these errors are brought to my attention. Any suggestions for improvement are also welcome.


## ACKNOWLEDGEMENTS

Sir Isaac Newton once said, *'If I have seen farther than others, it is because I have stood on the shoulders of giants.'* In the same way, I have been greatly influenced in the preparation of this book by the books, papers, and lectures of several great professors. I would like to apologize for any phrase or illustrations used in this book inadvertently without acknowledgement.

I am grateful to my teachers *Prof. P. Sabapathy* of Thiagarajar College of Engineering, *Prof. D. Venugoplan* of Anna University, and *Prof. C. Ganapathy* of Indian Institute of Technology, Madras because of whom I developed a great interest in steel structures. I also learnt much from *Prof. J. Lindner* and *Prof. Ch. Petersen*, while working in Germany as an Alexander von Humboldt fellow. My understanding of this subject was greatly influenced by books and publications of several authors. These books and authors are listed in the suggested reading section, at the end of the book.

While attending the presentations and discussions held in the sub-committee, which prepared the draft of IS 800:2007, I received valuable inputs from *Dr R. Narayanan* (former head of Education and Publication Division of Steel Construction Institute, UK), *Prof. V. Kalyanaraman*, *Dr S.R. Sathish Kumar*, and *Dr. Arul Jayachandran* (IIT Madras), *Dr T.K. Bandyopadhyay* (Former Joint Director General of Institute for Steel Development and Growth, Kolkata and now Senior Vice President, ThyssenKrupp Industries India Pvt. Ltd), *Dr C.N. Srinivasan* (M/s C.R. Narayana Rao Architects and Engineers), *Dr R.N. Krishna* (former Secretary General of Indian Concrete Institute), and other members of the sub-committee. I also learnt a lot from the discussions I had on several occasions with *Prof. A.R. Santhakumar* (former Dean of Anna University and Professor, IIT Madras).

I would like to wholeheartedly thank *Prof. Jayanta Banerjee* of Sabita Devi Education Trust, Braincare Group of Institutions, Barasat, West Bengal, *Prof. Mahendrakumar Madhavan* (IIT Hyderabad), *Prof. N. Umamaheswari* and *Prof. G. Augustine Maniraj Pandian* (SRM University, Chennai) for their extensive feedback and valuable suggestions.

I am grateful to the American Institute of Steel Construction Inc., Chicago, for giving permission to include numerous figures, photos, and extracts from their codes and papers published in Modern Steel Construction and Engineering Journal. I would also like to thank the following organizations/publishers for permitting to reproduce their material: American Society of Civil Engineers; Steel Construction Institute, Ascot, UK; Elsevier, UK; American Welding Society; Canadian Institute of Steel Construction; McGraw-Hill Education; The James F. Lincoln Arc Welding Foundation, Cleveland, USA; Taylor & Francis Group; John Wiley & Sons and Industrial Press, Inc., USA. Images/Tables/Information are also included in the book with the permission from INSDAG, New Delhi, Survey of India, National Information Centre of Earthquake Engineering (NICEE) at IIT Kanpur, Lehigh Preserve, Lehigh University, National Oceanic and Atmospheric Administration, USA, US Coast Guard, ASD Westok Ltd, UK, AFCONS Infrastructure Limited, and STRUCTURE Magazine of ASCE. Many thanks to Er Jayant Lakhlani of Lakhlani Associates, Rajkot, Gujarat, Er Karthik Muthusamy of California, Mr Tejas Mehta of Gaithersburg, MD, USA, Mr Binod Therat of Singapore, Dr Philip Oldfield of the University of Nottingham, UK (Thanks to Ar. Daniel Safarik, Editor of CTBUH Journal for arranging to get the permission), Mr William C. Gibb of Steel Cast Connections LLC, USA, Mr Henry Gallart of SidePlate Systems Inc., USA, Ms Kelly Luttrell of ConXTech, Pleasanton, CA, Mr Bill Cobb of SkylineScenes.com, Mr Emilio I. Panizo, Mr Uwe Tress, Mr Juan Pablo Ortiz Arechiga (the last three through Flickr), for providing photos of structures. I also wish to thank *Prof. R.B. Kulkarni* of Gogte Institute of Technology, Belgaum, Karnataka, for his permission to include the charts developed by him in Appendix F.

I am grateful to *Prof. Michel Bruneau*, Professor, University at Buffalo, USA, for writing the foreword for this book.

I would like to thank all those who assisted me in the preparation of this book. First and foremost I would like to thank *Dr S. Seetharaman*, Former Deputy Director, SERC, Madras and Presently Professor, S.A. Engg. College, Avadi, Chennai, for writing the chapters on Design of Beams and Design of Plate Girders, and *Er Vivek Abhyankar*, Senior Manager (Design), AFCONS Infrastructure limited, Mumbai for writing the chapter on Design Principles for Steel Enabling Structures. Thanks also are due to Dr D.S. Ramachandra Murthy, former Deputy Director, SERC, and Director (R&D) & Former Vice Chancellor, St Peter's Institute of Higher Education and Research, St Peter's University, Chennai, for reviewing and offering comments on the chapter on Fatigue; Prof. C. Ganapathy, Former Professor, IIT Madras and Emeritus Professor, St Xavier's Catholic College of Engineering, Chunkankadai, Nagercoil for going through the entire manuscript and offering comments; Ms R.M. Biruntha for her help in working out the examples in the chapters on Plate Girders and Gantry Girders; Mr D. Muthukumar for analysing the problems in Chapter 4; Dr Arul Jayachandran, Dr M. Sivakumar of SERC, Er Naseef Ummer of IIT, Delhi, and Ar. S. Sreevidya for their help in locating the literature; and Ms S. Mangalam for computer analysis and design of trusses in Chapter 16.

*Respectfully dedicated to Prof. P. Sabapathy,*

*Prof. C. Ganapathy, Prof. J. Lindner, and Prof. Ch. Petersen,*

*from whom I learnt the intricacies of steel design.*

I will be failing in my duty if I do not acknowledge the help and assistance I received from Ms S. Chithra at all the stages of this book writing project. Lastly, I acknowledge the excellent support and coordination provided by the editorial team at Oxford University Press, India.

I am thankful to all those students and teachers who sent emails or expressed happiness for the book on *Design of Steel Structures*. I am sure the students and teachers will welcome this book also, which now contains revised material and more examples in almost all the chapters (with two new chapters), design aids, as well as case studies. This book will serve the students as a guide to structural steel design as per IS 800:2007, not only in college, but also in their future professional life, as it contains a large amount of practical details.

**N. Subramanian**
**Gaithersburg, MD, USA**

# FEATURES OF

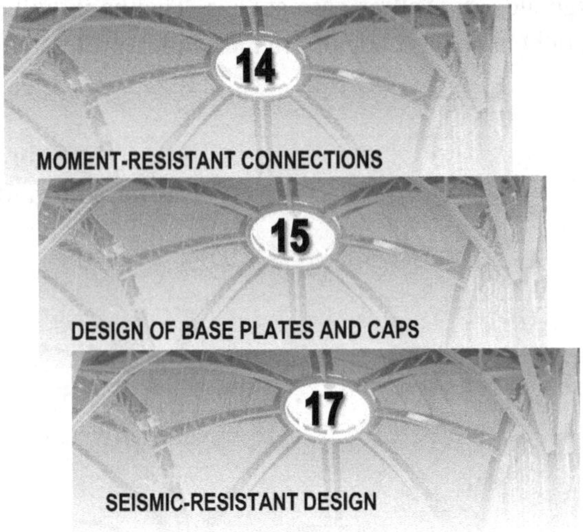


## COVERAGE

Based on the Limit State Method (IS 800:2007) of design, this book newly introduces 'Seismic Design' and 'Design Principles of Steel Enabling Structures'. It lays greater focus on the 'Design of Column Bases and Caps' and 'Moment-resistant Connections' through exclusive chapters. Advanced topics such as Fire-resistant Design, Fatigue-resistant Design, Design for Torsion and Corrosion Protection of Steel Structures have been covered in the form of chapters in the Online Resource Center (india.oup.com/orcs/9780199460915).

## CASE STUDIES

Several case studies that discuss and analyse structural failures, conceptual applications, and new developments in the industry are provided in each chapter

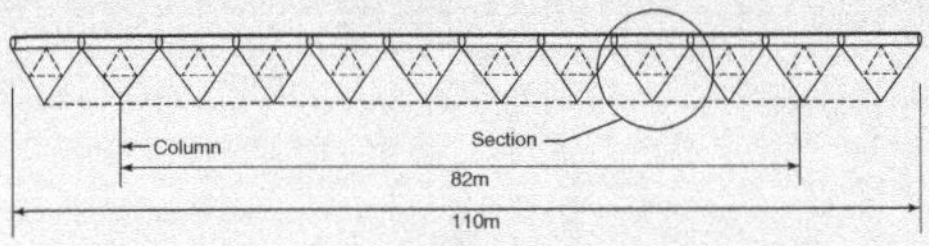

CASE STUDY

Hartford Civic Centre Roof Collapse

The Hartford Civic Centre Coliseum, Connecticut, USA, was completed in 1973. The space frame roof structure was 7.6 m high and covered 110 m by 91 m, with clear spans of 64 to 82 m.



(a) Elevation

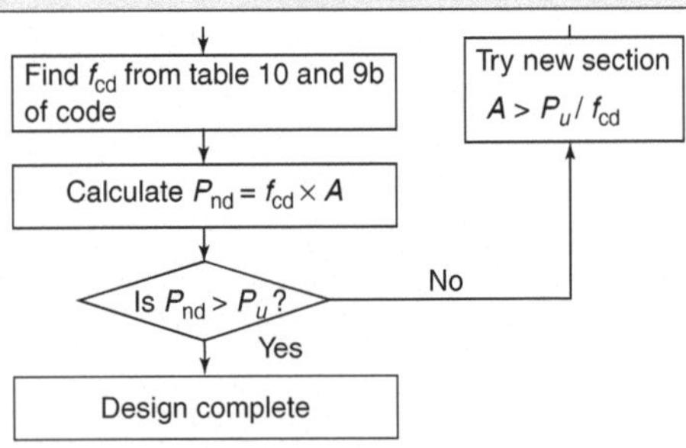


**FIG. 9.45** Flowchart for design of compression members

## FLOWCHARTS

Flowcharts are provided wherever necessary for systematic and methodical explanation of design procedures for tension members, compression members, beams, plate girders, and gantry girders

## ILLUSTRATIONS AND IMAGES

Numerous pictures from construction sites or of buildings that manifest the industrial usage of a given concept or structure are provided

Neatly labelled and shaded 3D diagrams that illustrate complex structures are included

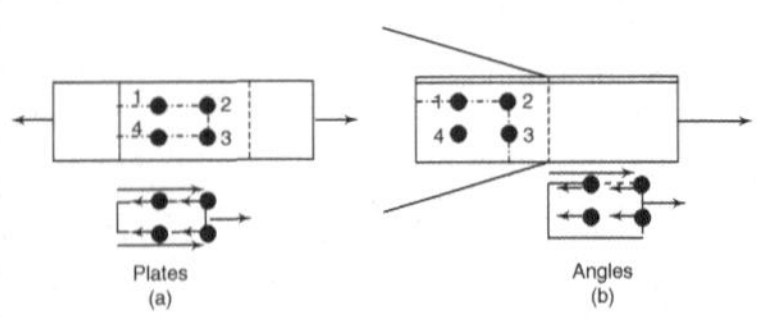


**FIG. 7.7** Block shear failure in plates and angles

**FIG. 7.9** Typical blockshear failure

# THE BOOK

## SUMMARY

with a general discussion of building materials
ıforced and prestressed concrete, structural steel,
›es on to present a historical review of iron and
ng process is briefly outlined. The metallurgical
which are relevant to structural steel designers,
:nted.
has several advantages over other competing
ıcrete and wood. The chemical composition and
:s of different kinds of steel are included. (It is
ecify the steel for the project at hand.) Properties

temperature. Of course the quality
fabrication are also equally important
fracture. One should avoid severe stre
stress flow. In critical cases, this is
selection.
Secondly, one must consider fati
the incidence of high levels of fluctua
consideration. In general, this has to be
structures, bridges, and structures sup
Residual stresses and stress concentrati

**SUMMARY**

The summary at the end of every chapter briefly recapitulates the topics covered in the chapter for a quick look-up

**MULTIPLE-CHOICE QUESTIONS AND EXAMPLES**

A variety of solved examples and MCQs help in further clarification of concepts and good practice of the theory taught in the chapter

## MULTIPLE-CHOICE QUESTIONS

1.1 Unit mass of steel is
(a) 2400 kg/m³ (c) 6850 kg/m³
(b) 7850 kg/m³ (d) 7500 kg/m³
1.2 Young's modulus of steel is
(a) $2 \times 10^5$ MPa (c) $2.1 \times 10^5$ MPa
(b) $2 \times 10^6$ MPa (d) 22000 MPa
1.3 Poisson's ratio of steel is
(a) 0.25 (c) 0.30

1.5 Structural steel (low-carbo
the range of
(a) 0.4 to 0.12%
(b) 0.02 to 0.08%
1.6 To produce one tonne of
required:
(a) 1.5 tonne of iron ore, 1
stone, and 1.5 tonne of

## EXAMPLES

**Example 3.1:**
An industrial building has been designed to resist a floor live load of 5 kN/m² [as per IS 875 (Part 2)]. However, later statistical readings were taken on similar slabs and the observed live loads in kilonewtons per square metre on various slabs are 13 × 3.2, 15 × 3.8, 35 × 4, 10 × 4.2, and 10 × 4.4. (13 × 3.2 indicates 13 slabs, each having a load of 3.2 kN/m².)
Determine the characteristic load if the accepted probability of load should not exceed 5%, assuming normal distribution.

$W_d = 0.432 \times$ span of purlin
$= 0.432 \times 4 = 1.728$ kN/m

*Snow load on purlin:*
Depth of snow load at Shimla, $d = 1.5$ m.
Ground snow load as per Eqn (3.12),

$s_0 = (3 - 2e^{-1.5(1.5)})1.5 = 4.1$

Note that, as per Eqn (3.13),

$s_0 = 1.97h^{1.36} = 1.97(1.5)^{1.36} =$

## EXERCISES

3.1 The measurement of loads on floor slabs of residential buildings are given below in kilonewtons per square metre.
8 × 0.90, 12 × 1.1, 15 × 1.2, 30 × 1.4, 40 × 1.5, 15 × 1.6, 5 × 1.7 (8 × 0.90 means eight samples of 0.9 kN/m² each)
Determine the characteristic load on the floors if the acceptable probability of the load is not to exceed 5% of the specified load.
3.2 An office building was designed to resist a floor load of 4 kN/m². After the building was constructed, measurements were taken on the various floors as 12 × 3.2, 10 × 3.8, 20 × 4, 15 × 4.2, and 8 × 4.4.

statistical table for normal distributio
book on statistics.
3.3 A roof having a span of 24 m and r
4 m apart. Assuming ACC sheeting,
live loads on the purlins, assumin
1.40 m.
3.4 A roof truss is of 16 m span and 2.67
Assuming GI sheeting, compute the ult
for which the purlins have to be desi
situated in Srinagar.

**EXERCISES**

The exercises given at the end of each chapter serve as a rich bank of unsolved questions for practice

**GLOSSARY**

A list of key terms that helps in revisiting important concepts and understanding the jargon has been given at the end of the book

## GLOSSARY

**A**

**Alloy steel** – Refers to steels made with deliberate additions of one or more alloying elements during steelmaking to enhance the properties of the steel. The most common alloying elements are Mn, Cr, Mo, and Ni.

**Accidental loads** – Loads that act occasionally and are of short duration, such as those due to impact of vehicles,

capacity of concrete is not exceeded. It is usually anchore into the concrete foundation by means of anchor bolts.

**Beam** – A straight or curved structural member, primaril supporting loads, applied at right angles to the longitudin axis/moments. These are members subjected primarily t bending and shear actions. Beams may be simply supporte continuous, cantilevered, or have fixed ends.

# COMPANION ONLINE RESOURCES

Visit india.oup.com/orcs/9780199460915 to access teaching and learning solutions online.

The following resources are available to support the students and faculty using this book.

**Students and Faculty:**

- Computer programs for design
- Additional chapters on advanced topics

**Faculty:**

- Solutions manual
- Chapter-wise PowerPoint presentations
- Image bank

## Steps to register and access Online Resources

Resources for instructors and students are developed to complement each text book and varies from book to book.

**Step 1: Getting Started**

- Go to india.oup.com

**Step 2: Browse quickly by**

- BASIC SEARCH
  - AUTHOR
  - TITLE
  - ISBN
- ADVANCED SEARCH
  - KEYWORDS
  - AUTHOR
  - TITLE
  - SUBTITLE
  - PUBLICATION DATE

**Step 3: Select title**

- Select Product
- Select Online Resources

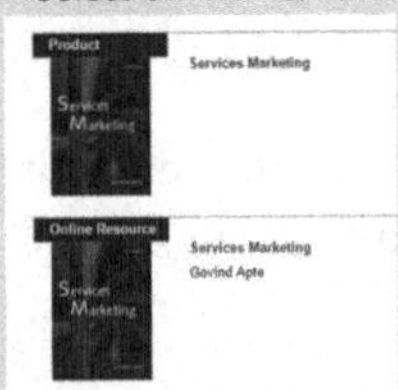

**Step 4: View Resources**

- Click on "View all resources"

**Step 5: Sign in with your Oxford ID**

**Step 6: if you do not have an Oxford ID, register with us**

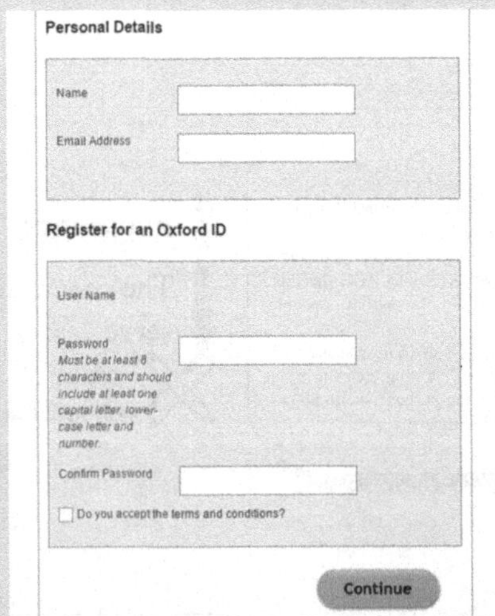


**Step 7: Fill in your details**

- Fill the detailed registration form with correct particulars.
- Fields marked with '*' in the form, are mandatory.
- Update

**Step 8: Validation**

- We shall revert to you within 48 hours after verifying the details provided by you. Once validated, please login using your username and the password and access the resources.

**Step 9: Confirmation**

- You will receive a confirmation on your email ID.

**Step 10: Visit us again**

- Go to india.oup.com
- Sign in with Oxford ID

**Step 11: Visit your licensed products**

- Go to "Resources" section

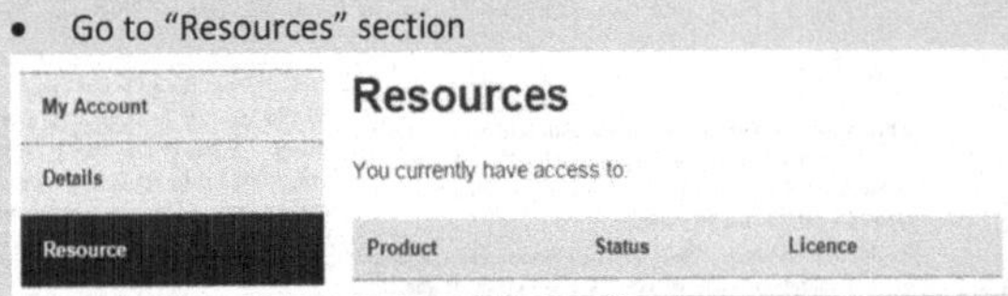


**Step 12: Download Resources**

- Click on the title
- View online resources
- Select resource type
- Download the resource you require.

For any further queries, please write to us at HEMarketing.in@oup.com with your mobile number.

# BRIEF CONTENTS

# DETAILED CONTENTS

# List of Symbols

$A$ area of cross section; surface area of cladding
$A_b$ area of bolt; gross area of horizontal boundary elements in SPSW
$A_{br}$ required bearing area
$A_c$ minor diameter area of the bolt; gross area of vertical boundary elements in SPSW
$A_d$ area of diagonal member
$A_e$ effective cross-sectional area; effective frontal area in wind
$A_F$ moment amplification factor
$A_f$ total flange area of the smaller connected column; floor area; required area of flange plates
$A_g$ gross cross-sectional area
$A_{gf}$ gross cross-sectional area of flange
$A_{go}$ gross cross sectional area of outstanding leg
$A_h$ design horizontal seismic coefficient
$A_k$ design horizontal acceleration spectrum value of mode $k$
$A_n$ net area of the total cross section
$A_{nb}$ net tensile cross sectional area of bolt
$A_{nc}$ net cross sectional area of the connected leg
$A_{ne}$ effective net area
$A_{nf}$ net cross sectional area of each flange
$A_{no}$ net sectional area of outstanding leg
$A_o$ initial cross-sectional area of tensile test coupon
$A_q$ cross-sectional area of the stiffener in contact with the flange
$A_s$ tensile stress area
$A_{sb}$ shank gross cross sectional area (nominal area) of a bolt
$A_{st}$ area of stiffener
$A_t$ total area of the compartment in fire
$A_{tg}$ gross sectional area in tension from the centre of the hole to the toe of the angle perpendicular to the line of force (block shear failure)
$A_{tn}$ net sectional area in tension from the centre of the hole to the toe of the angle perpendicular to the line of force (block shear failure)
$A_v$ shear area
$A_{vg}$ gross cross sectional area in shear along the line of transmitted force (block shear failure)
$A_{vn}$ net cross sectional area in shear along the line of transmitted force (block shear failure)
$A_w$ effective cross-sectional area of weld; window area; effective area of walls; area of web
$a, b$ larger and smaller projection of the slab base beyond the rectangle considering the column, respectively
$a_o$ peak acceleration
$a_1$ unsupported length of individual elements being laced between lacing points
$B$ breadth of flange of I-section; length of side of cap or base plate of a column
$B_s$ background factor
$b_o$ outstand/width of the element
$b_1$ stiff bearing length; stiffener bearing length
$b_e$ effective width of flange between pair of bolts
$b_{fc}$ width of column flange
$b_f$ breadth or width of the flange
$b_p$ panel zone width between column flanges at beam column junction
$b_s$ shear lag distance; stiffener width
$b_{sh}$ average breadth of the structure between heights $s$ and $h$
$b_{0h}$ average breadth of the structure between heights 0 and $h$
$b_t$ width of tension field
$b_w$ width of outstanding leg
$C$ centre-to-centre longitudinal distance of battens; coefficient related to thermal properties of wall, floor, etc.; spacing of transverse stiffener; moisture content of insulation
$C_1$ equivalent uniform moment factor
$C_{dyn}$ dynamic response factor
$C_e$ effective width of interior patch load
$C'_f$ frictional drag coefficient
$C_f$ force coefficient of the structure
$C_{fs}$ cross-wind force spectrum coefficient
$C_i$ specific heat of insulation material
$C_{Lh}$ lateral horizontal load for cranes
$C_m$ coefficient of thermal expansion; equivalent moment factor
$C_p$ cost of purlin
$C_{pe}$ external pressure coefficient
$C_{pi}$ internal pressure coefficient
$C_r$ cost of roof covering
$C_s$ specific heat of steel
$C_t$ cost of truss
$c_m$ moment reduction factor for lateral torsional buckling strength calculation
$D$ overall depth/diameter of the cross section
$D_o$ outer diameter
$d$ depth of web; nominal diameter; grain size of crystals; diagonal length; depth of snow; base dimension of the building
$d_2$ twice the clear distance from the compression flange angles, plates, or tongue plates to the neutral axis
$d_a$ depth of angle
$d_b$ beam depth; diameter of bolt

$d_c$ column depth
$d_g$ centre-to-centre of the outermost bolt of the end plate connection
$d_h$ diameter of the hole
$d_i$ thickness of insulation
$d_o$ nominal diameter of the pipe column or the dimensions of the column in the depth direction of the base plate
$d_p$ panel zone depth in the beam-column junction
$E$ modulus of elasticity for steel; energy released by earthquake
$E(T)$ modulus of elasticity of steel at T°C
$E(20)$ modulus of elasticity of steel at 20°C
$E_e$ equivalent elastic modulus of rope
$EL_x$ earthquake load in $x$ direction
$EL_y$ earthquake load in $y$ direction
$E_p$ modulus of elasticity of the panel material
$E_{sh}$ strain-hardening modulus
$E_t$ tangent modulus of elasticity
$E'_t$ reduced tangent modulus
$e$ eccentricity, head diagonal of bolt
$e_b$ edge distance of bolt
$F$ net wind force on cladding
$F'$ frictional drag force
$F_{br}$ strength of lateral bracing
$F_{cf}$ flange contribution factor
$F_o$ minimum bolt pretension
$F_{bsd}$ bearing strength of stiffener
$F_q$ stiffener force
$F_{qd}$ design buckling resistance of stiffener
$F_w$ protection material density factor; ultimate web crippling load
$F_{xd}$ design resistance of load carrying web stiffener
$F_x$ external load; force or reaction on stiffener
$F_z$ along-wind equivalent static load at height $Z$
$f$ actual normal stress range for the detail category, uniaxial stress, frequency of vortex shedding
$f_1, f_2, f_3$ principal stresses acting in three mutually perpendicular directions
$f_a$ stress amplitude
$f_b$ actual bending stress; bending stress at service load
$f_{bc}$ actual bending stress in compression
$f_{bd}$ design bending compressive stress corresponding to lateral buckling
$f_{bt}$ actual bending stress in tension at service load
$f_c$ average axial compressive stress
$f_{cc}, f_{cr}$ elastic buckling stress of a column or plate; Euler's buckling stress = $\pi^2E/(KL/r)^2$
$f_{ck}$ characteristic compressive strength of concrete
$f_{cr,b}$ extreme fibre compressive elastic lateral buckling stress
$f_{cd}$ design compressive stress
$f_d$ stress range at constant amplitude
$f_e$ equivalent stress
$f_f$ fatigue strength corresponding to $N_{sc}$ cycle of loading
$f_{fd}$ design normal fatigue strength
$f_{feq}$ equivalent constant amplitude stress range
$f_{fmax}$ highest normal stress range
$f_{fn}$ normal fatigue stress range for $5 \times 10^6$ cycles
$f_k$ characteristic strength
$f_l$ fatigue limit
$f_L$ stress range at cut-off limit
$f_m$ mean stress
$f_{max}$ maximum stress
$f_{min}$ minimum stress
$f_o$ proof stress
$f_0$ first mode natural frequency of vibration of a structure in the along-wind direction
$f_{0.2}$ 0.2% proof stress
$f_{pl}$ steel stress at proportional limit
$f_q$ shear stress
$f_R$ characteristic value of fatigue strength at loading cycle $N_R$
$f_{st}$ shear stress due to torsion
$f_{sw}$ warping shear stress
$f_t$ tension field strength
$f_u$ characteristic ultimate tensile stress
$f_{ub}$ characteristic ultimate tensile stress of the bolt
$f_{up}$ characteristic ultimate tensile stress of the connected plate
$f_{uw}$ ultimate tensile stress of the weld
$f_v$ yield strength in the panel utilizing tension field action
$f_{wd}$ design strength of weld
$f_y$ characteristic yield stress
$f_y(T)$ yield stress of steel at T°C
$f_y(20)$ yield stress of steel at 20°C
$f_{yb}$ characteristic yield stress of bolt
$f_{yf}$ characteristic yield stress of flange
$f_{yn}$ nominal yield strength
$f_{yst}$ design yield stress of stiffener
$f_{yp}$ characteristic yield stress of connected plate
$f_{yq}$ characteristic yield stress of stiffener material
$f_{yw}$ characteristic yield stress of the web material
$f_0$ yield strength of very large isolated crystals
$G$ shear modulus of rigidity for steel; thickness of grout
$G_v$ gust factor
$G^*$ design dead load
$g$ gauge length between the centre of the holes perpendicular to the load direction; acceleration due to gravity; gap for clearance and tolerance
$g_R$ peak factor for resonant response
$H$ height of section; transverse load
$H_p$ heated perimeter
$H_v$ calorific value of the $v$th combustible material
$H_w$ window height

$h$ depth of the section; storey height
$h_b$ total height from the base to the floor level concerned
$h_c$ height of the column
$h_e$ effective thickness
$h_{ef}$ embedment length of anchor bolt
$h_o$ distance between flange centroids of I-section
$h_i$ thickness of protection material; height of floor $i$
$h_L$ height of the lip
$h_s$ storey height
$h_y$ distance between shear centre of the two flanges of the cross section
$I$ moment of the inertia of the member about an axis perpendicular to the plane of the frame; impact factor fraction; importance factor
$I_b$ moment of inertia of brace
$I_f$ second moment of area of the foundation pad; insulation factor
$I_{fc}$ moment of inertia of the compression flange
$I_{ft}$ moment of inertia of the tension flange
$I_h$ turbulence intensity at height $h$
$I_p$ polar moment of inertia
$I_q$ moment of inertia of a pair of stiffener about the centre of the web, or a single stiffener about the face of the web
$I_s$ second moment of inertia of stiffener
$I_t$ St. Venant's torsional constant
$I_w$ warping constant
$I_y$ moment of inertia about the minor axis
$I_z$ moment of inertia about the major axis
IF interference factor
$K$ effective length factor
$K_a$ area averaging factor
$K_b$ effective stiffness of the beam and column; effective length factor for beams against lateral bending
$K_c$ combination factor; stiffness of column
$K_d$ wind directionality factor
$K_h$ reduction factor to account for the bolt holes in HSFG connection
KL effective length of the member
KL/$r$ appropriate effective slenderness ratio of the section
KL/$r_y$ effective slenderness ratio of the section about the minor axis
KL/$r_z$ effective slenderness ratio of the section about the major axis
(KL/$r$)$_o$ actual maximum effective slenderness ratio of the laced or battened column
(KL/$r$)$_e$ effective slenderness ratio of the laced column accounting for shear deformation
$K_m$ mode share correction factor for cross-wind acceleration
$K_y$, $K_z$ moment amplification factor about respective axes
$K_w$ warping restraint factor
$k$ regression coefficient; constant; mode shape power exponent for the fundamental mode of vibration
$k_1$ probability factor or risk coefficient
$k_2$ terrain, height and structure size factor
$k_3$ topography factor
$k_4$ importance factor for cyclonic region
$k_b$, $k_{br}$ stiffness of bracing
$k_s$ modulus of sub-grade reaction
$k_{sec}$ web distortional stiffness
$k_{sm}$ exposed surface area to mass ratio
$k_t$ brace stiffness excluding web distortion, torsion parameter
$k_{tb}$ stiffness of torsional bracing
$k_v$ shear buckling coefficient
$L$ actual length; unsupported length; centre-to-centre distance of the intersecting members; length of the end connection; cantilever length; land in weld
$L_b$ laterally unbraced length or distance between braces
$L_c$ length of end connection measured from the centre of the first bolt hole to the centre of the last bolt hole in the connection; distance between gantry girders
$L_{cf}$ clear distance between flanges of vertical boundary elements
$L_e$ effective horizontal crest length
$L_g$ gauge length of tensile test coupon
$L_h$ measure of the integral turbulence length scale at height $h$
$L_m$ maximum distance from the restraint at plastic hinge to an adjacent restraint (limiting distance)
$L_o$ length between points of zero moment (inflection) in the span
$L_w$ effective length of weld; length of wall
$l_a$ length of the angle
$l_e$ distance between prying force and bolt centre line
$l_g$ grip length of connection
$l_j$ length of the joint
$l_s$ length between points of lateral support
$l_t$ elongation due to temperature; length of top angle
$l_v$ distance from bolt centre line to the toe of the fillet weld or to half the root radius for a rolled section
$M$ bending moment; magnitude of earthquake
$M_{1sway}$ maximum first order end moment as a result of sway
$M_{br}$ required flexural strength of torsional bracing
$M_{cr}$ elastic critical moment corresponding to lateral torsional buckling
$M_d$ design flexural or bending strength
$M_{dv}$ design bending strength of the section under high shear
$M_{dy}$ design bending strength as governed by overall buckling about minor axis
$M_{dz}$ design bending strength as governed by overall buckling about major axis

$M_{eff}$ reduced effective moment
$M_{fr}$ reduced plastic moment capacity of the flange plate
$M_{fd}$ design plastic resistance of the flange alone
$M_{nd}$ design strength under combined axial force (uni-axial moment acting alone)
$M_{ndy}$, $M_{ndz}$ design strength under combined axial force and the respective uni-axial moment acting alone
$M_o$ cross-wind base overturning moment; first order elastic moment
$M_p$ plastic moment capacity of the section
$M_{pb}$ moment in the beam at the intersection of the beam and column centre lines
$M_{pc}$ moments in the column above and below the beam surfaces
$M_{pd}$ plastic design strength
$M_{pf}$ plastic design strength of flanges only
$M_{pr}$ reduced plastic moment capacity of the section due to axial force or shear
$M_q$ applied moment on the stiffener due to eccentric load
$M_{tf}$ moment resistance of tension flange
$M_u$ second order elastic moment; factored moment; required ultimate flexural strength of a section
$M_y$ factored applied moment about the minor axis of the cross section; yield moment capacity about minor axis
$M_{yq}$ yield moment capacity of the stiffener about an axis parallel to web
$M_z$ factored applied moment about the major axis of the cross section
$m$ mass; slope of the fatigue strength curve
$m_v$ mass of $v$th combustible material
$m^1$ non-dimensional moment parameter = $M_u/M_{bp}$
$N_d$ design strength in tension or in compression
$N_f$ axial force in the flange
$N_{sc}$ number of stress cycles
$n$ number of parallel planes of battens; mean probable design life of structure in years; reduced frequency; number of cycles to failure; factored applied axial force; number of bolts in the bolt group/critical section; number of stress cycles; number of storeys
$n_1$, $n_2$ dispersion length
$n_e$ number of effective interfaces offering frictional resistance to slip
$n_n$ number of shear planes with the threads intercepting the shear plane in a bolted connection
$n_s$ number of shear planes without threads intercepting the shear plane in a bolted connection
$n'$ number of rows of bolts
$P$ factored applied axial force; point load
$P_{bf}$ design strength of column web to resist the force transmitted by beam flange
$P_{cc}$ elastic buckling strength under axial compression
$P_{crip}$ crippling strength of web of I-section
$P_d$ design axial compressive strength
$P_{dy}$, $P_{dz}$ design compression strength as governed by flexural buckling about the respective axis
$P_{dw}$ design strength of fillet weld
$P_e$, $P_{cr}$ elastic Euler buckling load; $\pi^2EI/L^2$
$P_f$ probability of failure
$P_k$ modal participation factor for mode $k$
$P_{min}$ minimum required strength for each flange splice
$P_N$ probability that an event will be exceeded at least once in $N$ years
$P_n$ nominal axial strength
$P_{ny}$ axial strength of the member bent about its weak axis
$P_u$ maximum load in the column
$p$ pitch length between centres of holes parallel to the direction of the load; pitch of thread in bolt, pressure
$p_d$ design wind pressure
$p_s$, $p_1$, $p_2$ staggered pitch length along the direction of the load between lines of the bolt holes (Fig. 5.21)
$p_z$ wind pressure at height $Z$
$Q$ prying force; nominal imposed load; static moment of the cross section = $A\,\bar{y}$
$Q^*$ design imposed load
$Q_a$ accidental load (action)
$Q_c$ characteristic load (action)
$Q_d$ design load (action)
$Q_f$ fire load
$Q_i$ load effect $i$; design lateral force at floor $i$
$Q_m$ mean value of load
$Q_p$ permanent loads (action)
$Q_v$ variable loads (action)
$q_f$ fire load/unit floor area
$R$ ratio of the mean compressive stress in the web (equal to stress at mid depth) to yield stress of the web; reaction of the beam at support; stress ratio; response reduction factor; resultant force; root opening of weld; local radius of curvature of beam; return period
$R_d$ design strength of the member at room temperature
$R_i$ net shear in bolt group at bolt $i$
$R_k$ connection stiffness
$R_m$ mean value of resistance
$R_n$ nominal strength of resistance
$R_{nw}$ design strength of fillet weld per unit length
$R_{res}$ resultant force in the weld
$R_r$ response reduction factor
$R_{tf}$ resultant longitudinal shear in flange
$R_u$ ultimate strength of the member at room temperature; ultimate strength of joint panel
$r$ appropriate radius of gyration
$r_a$ root radius of angle
$r_b$ root radius of beam flange

$r_1$ minimum radius of gyration of the individual element being laced together
$r_f$ ratio of the design action on the member under fire to the design capacity
$r_{vv}$ radius of gyration about the minor axis ($v$-$v$)
$r_y$ radius of gyration about the minor axis
$r_z$ radius of gyration about the major axis
$S$ minimum transverse distance between the centroid of the rivet or bolt or weld group; strouhal number; size reduction factor; spacing of truss
$S_a$ spectral acceleration
$S_d$ spectral displacement
$S_p$ spring stiffness
$S_v$ spectral velocity
$S_1, S_2$ stability functions
$s$ design snow load; size of weld
$s_a$ actual stiffener spacing
$s_c$ anchorage length of tension field along the compression flange
$s_{ii}, s_{jj}$ stability functions
$s_{ii}^*, s_{jj}^*$ stability function for semi-rigid frames
$s_o$ ground snow load
$s_t$ anchorage length of tension field along the tension flange (distance between adjacent plastic hinges)
$T$ Temperature in °C; factored tension in bolt; natural period of vibration; applied torque
$T_a$ approximate fundamental natural period of vibration
$T_b$ applied tension in bolt
$T_d$ design strength under axial tension
$T_{db}$ design strength of bolt under axial tension; block shear strength of plate/angle
$T_{dg}$ yielding strength of gross section under axial tension
$T_{dn}$ rupture strength of net section under axial tension; design tension capacity
$T_{dw}$ design strength of weld in tension
$T_e$ externally applied tension
$T_{eq}$ equivalent fire rating time
$T_f$ factored tension force of friction type bolt; furnace temperature
$T_{f,max}$ maximum temperature reached in natural fire
$T_l$ limiting temperature of the steel
$T_{nb}$ nominal strength of bolt under axial tension
$T_{nd}$ design tension capacity
$T_{ndf}$ design tensile strength of friction type bolt
$T_{nf}$ nominal tensile strength of friction type bolt
$T_o$ ambient (room) temperature
$T_s$ steel temperature at time $t$
$T_t$ ambient gas temperature at time $t$
$T_u$ ultimate net section strength
$t$ thickness of element/angle, time in minutes
$t_a$ thickness of top angle
$t_b$ thickness of base plate
$t_{cw}$ thickness of column web
$t_e$ effective throat thickness of weld
$t_f$ thickness of flange; required fire rating time
$t_{fc}$ thickness of compression flange
$t_{fail}$ time to failure of the element in case of fire
$t_{fb}$ thickness of beam flange
$t_p$ thickness of plate/end plate
$t_{pkg}$ thickness of packing
$t_q$ thickness of stiffeners
$t_s$ thickness of web stiffener; duration of fire
$t_v$ time delay in minutes
$t_w$ thickness of web
$U$ shear lag factor
$V$ factored applied shear force; mean wind speed
$V_B$ total design seismic base shear; basic wind speed
$V_b$ shear in batten plate
$V_{bf}$ factored frictional shear force in HSFG connection
$V_{cr}$ critical shear strength corresponding to web buckling (without tension field action)
$V_d$ design shear strength; design mean wind velocity
$V_{db}$ shear capacity of outstanding leg of cleat
$V_{dw}$ design strength of weld in shear
$V_g$ gradient wind speed; gust speed
$V_h$ design wind speed at height $h$
$V_L$ longitudinal shear force
$V_R$ vector resultant shear in weld
$V_{nb}$ nominal shear strength of bolt
$V_{nbf}$ bearing capacity of bolt for friction type connection
$V_p$ plastic shear resistance under pure shear or shear strength of web
$V_n$ nominal shear strength or resistance
$V_{npb}$ nominal bearing strength of bolt
$V_{nsb}$ nominal shear capacity of a bolt
$V_{nsf}$ nominal shear capacity of a bolt as governed by slip or friction type connection
$V_{sb}$ factored shear force in the bolt
$V_{sd}$ design shear capacity
$V_{sdf}$ design shear strength of friction type bolt
$V_{sf}$ factored design shear force of friction bolts
$V_T$ applied transverse shear
$V_{tf}$ shear resistance in tension field
$\bar{V}$ average or mean velocity
$V_{yw}$ yield strength of web plate of I-section
$V_z$ mean or design wind speed at height z above the ground
$V'$ Instantaneous velocity fluctuation above the mean velocity
$W$ appropriate load; width; seismic weight; ventilation factor
$W_e$ equivalent cross-wind static force per unit-height
$w$ uniform pressure from below on the slab base due to axial compression under factored load; intensity of uniformly distributed load

$w_{tf}$ width of tension field
$X$ distance from a point to any other point
$x_t$ torsional index
$\bar{x}$ distance from centre of gravity in $x$ direction
$Y_s$ yield stress
$\bar{y}$ distance from centre of gravity in $y$ direction
$y_g$ distance between point of application of the load and shear centre of the cross section
$y_s$ coordinate of the shear centre with respect to centroid
$Z$ section modulus; height above ground; zone factor
$Z_e$ elastic section modulus
$Z_g$ depth of boundary layer
$Z_p$ plastic section modulus
$Z_{pr}$ plastic modulus of the shear area about the major axis; reduced plastic modulus
$\alpha$ coefficient of linear expansion; imperfection factor; power law coefficient; included angle in groove weld
$\alpha_{LT}$ imperfection factor
$\alpha_t$ coefficient of thermal expansion
$\beta$ reliability index; the ratio of structural damping to critical damping of a structure
$\beta_{lj}$ reduction factor for overloading of end bolt
$\beta_{lg}$ reduction factor for the effect of large grip length
$\beta_{pkg}$ reduction factor for the effect of packing plates
$\beta_M$ ratio of smaller to the larger bending moment at the ends of a beam column
$\beta_{My}$, $\beta_{Mz}$ equivalent uniform moment factor for flexural buckling for $y$-$y$ and $z$-$z$ axes, respectively
$\beta_{MLT}$ equivalent uniform moment factor for lateral torsional buckling
$\chi$ stress reduction factor due to buckling under compression
$\chi_m$ stress reduction factor $\chi$ at $f_{ym}$
$\chi_{LT}$ strength reduction factor for lateral torsional buckling of a beam
$\delta$, $\Delta$ storey deflection or drift; deflection
$\delta_b$ moment amplification factor for braced member
$\delta_L$ horizontal deflection of the bottom of storey due to combined gravity and notional load
$\delta_m$ moment amplification factor
$\delta_p$ load amplification factor
$\delta_s$ moment amplification factor for sway frame
$\delta_U$ horizontal deflection of the top of storey to combined gravity and notional load
$\varepsilon$ yield stress ratio; $(250/f_y)^{1/2}$; strain corresponding to stress $f$, resultant emissivity of surface
$\varepsilon_p$ plastic strain
$\varepsilon_{sh}$ strain hardening strain
$\varepsilon_u$, $\varepsilon_{br}$ ultimate strain
$\varepsilon_y$ yield strain
$\upsilon$ shape factor
$\phi$ strength or resistance reduction factor; cumulative distribution function; solidity ratio; inclination of the tension field stress in web; configuration factor; angle of twist
$\phi_{i,k}$ mode shape coefficient at floor $i$ in mode $k$
$\phi_s$ sway index
$\gamma$ unit weight of steel
$\gamma_f$, $\gamma_{fk}$, $\gamma_{lf}$ partial safety factor for load
$\gamma_{fft}$ partial safety factor for fatigue load
$\gamma_m$ partial safety factor for material
$\gamma_{m0}$ partial safety factor against yield stress and buckling
$\gamma_{ml}$ partial safety factor against ultimate stress
$\gamma_{mb}$ partial safety factor for bolted connection with bearing type bolts
$\gamma_{mf}$ partial safety factor for bolted connection with HSFG bolts
$\gamma_{mft}$ partial safety factor for fatigue strength
$\gamma_{mi}$ partial safety factor depending upon the type of failure as prescribed in IS: 800
$\gamma_{mw}$ partial safety factor for strength of weld
$\bar{\lambda}$ non-dimensional slenderness ratio

$$= \left(KL/r/\sqrt{\pi^2 E/f_y} = \sqrt{f_y/f_{cc}} = \sqrt{P_y/P_{cc}}\right)$$

$\lambda_{cr}$ elastic buckling load factor
$\lambda_e$ equivalent slenderness ratio
$\lambda_i$ effective thermal conductivity of insulation
$\lambda_{LT}$ non-dimensional slenderness ratio
$\lambda_y$, $\lambda_z$ non-dimensional slenderness ratio about respective axis
$\lambda_w$ non-dimensional web slenderness ratio for shear buckling
$\mu$ Poisson's ratio; shape coefficient or factor for snow load
$\mu_c$ correction factor; capacity reduction factor for fatigue
$\mu_f$ coefficient of friction (slip factor)
$\mu_r$ capacity reduction factor for non-redundant load path
$\theta^l$ non-dimensional rotation parameter = $\theta_r/\theta_p$
$\theta$ ratio of the rotation at the hinge point to the relative elastic rotation of the far end of the beam segment containing plastic hinge, upwind slope of ground
$\theta_p$ plastic rotation
$\theta_r$ rotation of semi-rigid joint
$\rho_s$ density of steel
$\rho_i$ dry density of insulation
$\rho_i'$ effective density of insulation
$s$ Stefon–Boltzmann constant
$\tau$ actual shear stress for the detail category
$\tau_b$ shear stress corresponding to buckling
$\tau_{cr}$ elastic critical shear buckling stress
$\tau_e$ equivalent shear stress
$\tau_f$ fatigue shear stress range for $N_{sc}$ cycle
$\tau_{fd}$ design fatigue shear strength

$\tau_{fmax}$ highest shear stress range
$\tau_{fn}$ fatigue shear stress range at $5 \times 10^6$ cycles for the detail category
$\tau_L$ shear stress range at cut-off limit
$\tau_o$ grout-concrete bond strength
$\tau_{vf}$ shear stress in the weld due to vertical force
$\tau_{vf1}$ shear stress in the weld due to bending moment
$\tau_w$ shear stress in weld throat; shear stress due to shear force
$\tau_y$ shear yield stress
$\xi_i$ reduction factor for geometric imperfection
$\psi$ ratio of the moment at the ends of the laterally unsupported length of a beam
$\omega$ circular natural frequency

*Note:* The subscripts *y* and *z* denotes the *y-y* and *z-z* axes of the section, respectively. For symmetrical sections, *y-y* denotes the minor principle axis whilst *z-z* denotes the major principal axis.

# 1

# MATERIALS, STRUCTURES, AND SPECIFICATIONS

## INTRODUCTION

In early societies, human beings lived in caves and almost certainly rested in the shade of trees. Gradually, they learnt to use naturally occurring materials such as stone, timber, mud, and biomass (leaves, grass, and natural fibres) to construct houses. Then followed brick making, rope making, glass, and metal work. From these early beginnings, the modern materials manufacturing industries developed. Today the iron and steel industry is the basic or key industry for any country. Iron and steel are considered as the basic raw material for several subsidiary industries such as engineering, automobiles, locomotives, machine tools, and ship building. The unique position of iron among the metals may be attributed to its abundance and to the wide range of properties that can be imparted to it by various treatments and by alloying it with various amounts of other elements.

The principal modern building materials are masonry, concrete (mass, reinforced, and prestressed), glass, plastic, timber, and structural steel (in rolled and fabricated sections). All the mentioned materials have particular advantages in a given situation and hence the construction of a particular building type may involve the use of various materials, e.g., a residential building may be constructed using load-bearing masonry, concrete frame or steel frame. The designer has to think about various possible alternatives and suggest a suitable material which will satisfy economic, aesthetic, and functional requirements.

We will now briefly discuss the use and advantages of the four basic materials which are employed extensively.

**Masonry** It is mainly used for load-bearing walls and walls taking in-plane or transverse loads. It is durable, fire-resistant, and aesthetically pleasing. It can be used for buildings of moderate height, i.e., of up to 20 storeys. (Unfortunately the masonry produced in India does not have uniform quality and that produced in South India has low strength. Hence buildings with load-bearing masonry are built only up to three to four floors.)

**Reinforced and prestressed concrete** Reinforced concrete framed or shear wall construction, if properly mixed, vibrated, and cured with water, is very durable and fire-resistant. Since reinforced concrete can be cast to any required shape, it is used for a variety of constructions including those of tall buildings and floors of all types of buildings. Prestressed concrete is used for floor construction of large-span structures and in buildings, bridges, and towers. In India, though concrete is used extensively in all types of construction, except by a small number of big companies, quality control is not exercised during the mixing of concrete. Moreover, concrete is not cured properly with water for the duration prescribed by the code. Also the steel reinforcements (especially the smaller-diameter rods) available in the market are produced by re-rollers and do not possess the required ductility and strength. Since concrete can be mixed and cast in to any required shape, it is misused by several small contractors, who do not give much importance to design or detailing. These factors have led to the deterioration of several concrete structures all over the country and also resulted in the failure of others in the recent earthquakes. Since prestressed concrete is used in major constructions and is used by major contracting companies, the quality of prestressed concrete in India is up to the required standards.

**Structural steel** Its main advantages are strength, speed of erection, prefabrication, and demountability. Structural steel is used in load-bearing frames in buildings, and as members in trusses, bridges, and space frames. Steel, however, requires fire and corrosion protection. In steel buildings, claddings and dividing walls are made up of masonry or other materials, and often a concrete foundation is provided. Steel is also used in conjunction with concrete in composite constructions and in combined frame and shear wall constructions. In many cases, the fabrication of steel members is done in the workshop and the members are then transported to the site and assembled. Tolerances specified for steel fabrication and erections are small compared to those for reinforced concrete structures. Moreover, welding, tightening of high-strength friction grip bolts, etc., require proper training. Due to these factors, steel structures are often handled by trained persons and assembled with proper care, resulting in structures with better quality. Steel offers better compressive and tensile strength than concrete and enables lighter constructions. Also, unlike masonry or reinforced concrete, steel can be easily recycled.

**Wood** Wood imparts natural, human warmth that steel and concrete lack. Due to this, wood has long been used for housing (up to three floors) and for historical structures in western countries such as the USA, the UK, Germany, France, and Japan, where there is cold climate. However, the development of wood composites—thin, pressed sheets—combined with joints and steel frames, has changed the scene. Glued laminated wood has been used in a number of large-span structures. Prominent wood composite structures are the Tacoma Dome and the North Michigan University stadium in the USA, and the Odate Jukai Dome in Japan. All these domes have diameters in the range of 160–180 m. Since wood is a natural product, it does not cause any environmental hazards, though the resins used in glued laminated wood may contain harmful chemicals. However, not all types of wood can be used for construction and quality wood is in short supply. In India, wood is used only for doors and windows. (Nowadays even doors and windows are made of aluminum, steel, ferrocement, or plastic.)

Some of the physical properties of structural steel, concrete, and wood are compared in Table 1.1. Structural steel has superior properties and many advantages over concrete and wood (see also Section 1.16).

In this chapter we will discuss the manufacture and those properties of structural steel, which are important in the selection of the material for a particular situation. We will also discuss the various types of steels, the available hot- and cold-rolled sections, and the various types of structures that can be built using these sections.

## 1.1 HISTORICAL DEVELOPMENT

Steel has been known since 3000 BC. Foam steel was used during 500–400 BC in China and then in Europe. The Ashokan pillar made with steel and the iron joints used in Puri temples are more than 1500 years old. They demonstrate that this know-how was available before the modern blast-furnace technology, which was developed in AD 1350 (Gupta 1998).

The large-scale use of iron for structural purposes started in Europe in the latter part of the eighteenth century. The first major application of cast iron was in the 30.4-m-span Coalbroakadale Arch Bridge by Darby in England, constructed in 1779 over the river Severn. The use of cast iron (which is weak in tension) as primary construction material was continued up to about 1840.

Until the end of the eighteenth century, cast iron was usually obtained from its ore by melting it in furnaces fired by charcoal. In 1740, Abraham Darby II found a way of converting coal into coke, which revolutionized the iron-making process. A later development in this process was the combination of limestone with the impurities in the ore and coke to form a slag, which could be run off independently of the iron. The iron so produced was very brittle and liable to crack under strain. These disadvantages were to a certain extent overcome by the invention of the reverberatory furnace in 1784 by Henry Cort. This method reduced the carbon content of the metal and the resulting product was named *wrought iron*, which was stronger, flexible, and had a higher tensile strength than cast iron. Cast iron had a carbon content of 2–4%, and wrought

**TABLE 1.1** Physical properties of major structural materials

| Item | Mild steel | Concrete[a] M20 grade | Wood |
|---|---|---|---|
| Unit mass (kg/m$^3$) | 7850(100) | 2400(31)[c] | 290–900(4–11) |
| Maximum stress (MPa) | | | |
| Compression | 250(100) | 20(8) | 5.2–23[b](2–9) |
| Tension | 250(100) | 3.13(1) | 2.5–13.8(1–5) |
| Shear | 144(100) | 2.8(1.9) | 0.6–2.6(0.4–1.8) |
| Young's modulus (MPa) | $2 \times 10^5$(100) | 22,360(11) | 4600–18000(2–9) |
| Coefficient of linear expansion (°C × $10^{-6}$) | 12 | 10–14 | 4.5 |
| Poisson's ratio | 0.3 | 0.2 | 0.2 |

[a]Characteristic compressive strength of 150–mm cubes at 28 days
[b]Parallel to grain
[c]Relative value as compared to steel

iron less than 0.15%. During 1826 wrought iron chains were used in the Menai Straits suspension bridge designed by Thomas Telford (the chains have since been replaced by steel chains). Robert Stephenson's Britannia Bridge was the first box girder wrought iron bridge. It was in use until around the nineteenth century.

Steel was first introduced in 1740, but was not available in large quantities until Sir Henry Bessemer of England invented and patented the process of making steel in 1855. In 1865, Siemens and Martin invented the open-hearth process and this was used extensively for the production of structural steel. In steel, the carbon content varies from 0.25% to 1.5%. The first major structure to use the new steel exclusively was Fowler and Baker's Railway Bridge at the Firth of Forth. A comparison of the properties of cast iron, wrought iron, and steel is provided in Table 1.2.

Companies such as Dorman Long started rolling steel I-sections by 1880. During 1879, Gilchrist and Thomas introduced the 'basic' lining into the Bessemer converter and open-hearth furnace. Using this lining made of magnasite or lignasite, it was possible to remove phosphorus from the locally available high-phosphorous iron ore. Riveting was used as a fastening method until around 1950 when it was superseded by welding. Bessemer steel production in Britain ended in 1974 and the last open-hearth furnace closed in 1980. The basic oxygen steel making (BOS) process using the CD converter was invented in Austria in 1953. In the latter part of the nineteenth century and the early twentieth century, newer

**TABLE 1.2** Comparison of cast iron, wrought iron, and steel

| Property | Cast iron | Wrought iron | Steel |
|---|---|---|---|
| Composition | It is a crude form of iron, containing 2.5–4.5% carbon. | It is the purest form of iron, containing up to 0.20% carbon. | It is midway between cast iron and wrought iron, containing 0.1–1.1% carbon. |
| Structure | It has a crystalline structure. | It has a fibrous structure with a silky lustre. | It has a granular structure. |
| Specific gravity | Its specific gravity varies from 7 to 7.5. | Its specific gravity is 7.80. | Its specific gravity is 7.85. |
| Melting point | Its melting point is about 1250 °C. It contracts on melting. | It melts at about 1500 °C. | Its melting point is between 1300 °C and 1400 °C. |
| Hardness | It is quite hard and can be hardened by heating and sudden cooling. | It cannot be hardened or tempered. | It can be hardened and tempered. |
| Strength | Its ultimate compressive strength is 600–700 MPa and ultimate tensile strength 120–150 MPa. | Its ultimate compressive strength is 200 MPa and ultimate tensile strength is about 400 MPa. | Its ultimate compressive strength is 180–350 MPa and ultimate tensile strength is 310–700 MPa. |
| Reaction to sudden shock | It does not absorb shocks. | It cannot stand sudden heavy shocks. | It absorbs shocks. |
| Magnetization | It cannot be magnetized. | It does not form permanent magnets but can be temporarily magnetized. | It can form permanent magnets. |
| Rusting | It does not rust easily. | It rusts more than cast iron. | It rusts easily. |
| Malleability and ductility | It is neither malleable nor ductile. | It is tough, malleable, ductile, and moderately elastic. | It is tough, malleable, and ductile. |
| Forging and welding | It is brittle and cannot be welded or rolled into sheets. | It can be easily forged or welded. | It can be rapidly forged or welded. |
| Uses | Because of its non- rusting property, it is used in the manufacture of parts most likely to rust, such as water pipes, sewers, and drain pipes. It is used for making parts of machines, which are not likely to be subjected to shocks or to tension. Lampposts, carriage wheels, rail chairs, and railings are usually made of cast iron. | As it can withstand sudden shocks without permanent injury, it is used to make chains, crane hooks, railway couplings, etc. | It is used as reinforcement in RCC and as structural members, bolts, rivets, and sheets (plain and corrugated). High-carbon steel is used for those parts of machinery where hard, tough, elastic, and durable material is required. It is used for making cutlery, files, and machine tools. |

technologies resulted in better and new grades of steel. Today we have several varieties of steel made with alloying elements such as carbon, manganese, silicon, chromium, nickel, and molybdenum (see Sections 1.5 and 1.6). The electric arc furnace is used to make special steels such as stainless steel. Further information on the history of steel can be found in Pannel 1964, Derry and Williams 1960, and Buchanan 1972.

## 1.2 PROCESSES USED FOR IRON AND STEEL MAKING

In this section, let us briefly discuss the different processes used to make iron and steel.

### 1.2.1 Iron Making

The important iron ores that are commonly used in the manufacturing process are haematite, limonite, magnetite, pyrite, and siderite. Iron production is a continuous process and consists of chemically reducing iron ore (iron ores are compounds of iron with non-metallic elements and contain impurities such as carbon, manganese, phosphorus, silicon, and sulphur) in a blast furnace using coke and crushed limestone.

The resulting material, called cast iron, contains carbon, sulphur, and phosphorus. The principle of iron making has not changed in the past 2000 years. However, the actual techniques employed as well as the scale of production have changed considerably. Nowadays, blast furnaces operate continuously over a period of several years, producing up to 8000 tonnes of molten iron every 24 hours.

### 1.2.2 Steel Making

Three main processes exist for the production of steel. The oldest of these is the open-hearth process. Since it was slow and uneconomical, it has been replaced largely by the *basic oxygen steel making* (BOS) process and the *electric arc method.* (The electric arc furnace is used mainly to make special steels such as stainless steel.) Steel production is basically a batch process and involves reducing the carbon, sulphur, and phosphorus levels and adding, when necessary, manganese, chromium, nickel or vanadium.

### *Integrated Steel Plants*

Today most structural steel is made in integrated steel plants using the BOS process shown in Fig. 1.1. Iron ore lumps, scrap steel (up to 30%), pellets, coke (made from cooking coal), and fluxes such as limestone and dolomite are used as the major raw materials. The main steps involved in the manufacturing process are as follows.

**Melting** Raw materials are charged in a blast furnace, where hot air is pumped to melt iron and fluxes at 1600 °C. The molten metal when cooled and solidified is called *pig iron*. Alternatively, it can be further refined to make steel. The excess carbon and other unwanted impurities are floated off as slag (this slag is blended with clinker to make blast furnace cement, which is used in high-performance concretes).

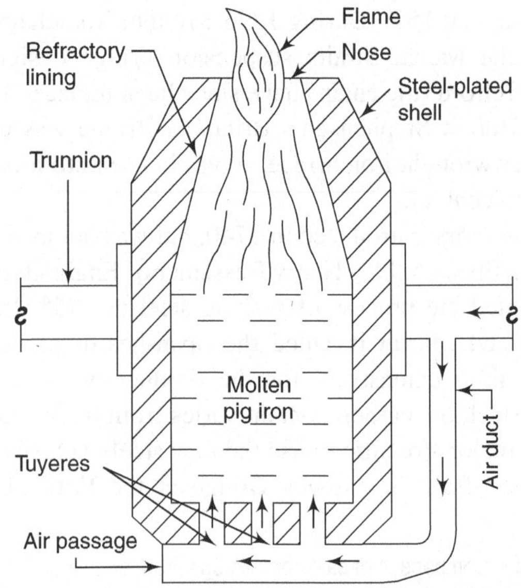


**FIG. 1.1** Basic oxygen steel making (BOS) process (Rangwala et al. 1997)

**Refining** Molten metal from the blast furnace is taken to the steel melting shop where the impurities are further reduced in a basic-oxygen furnace (LD converter) or an open-hearth furnace (see Fig. 1.2). The working of the LD converter is as follows. (This process was invented in Austria in 1953 and first adopted in two towns—Linz and Donawitz, and hence the name LD converter.)

1. The converter is tilted and is charged with molten pig iron from a Cupola furnace or sometimes directly from the blast furnace. (The converter is mounted on two horizontal trunnions as shown in Fig. 1.1, so that it can be tilted or rotated at any suitable angle.)

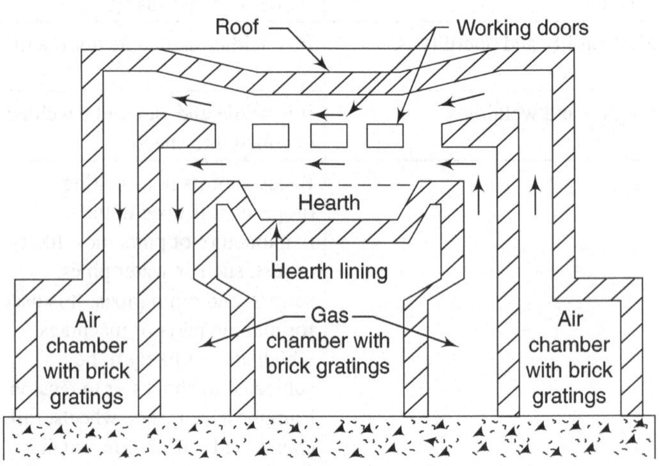


**FIG. 1.2** Open-hearth furnace (Rangwala et al. 1997)

2. The converter is brought to an upright position and a jet of pure oxygen is blown at extraordinary speed through the tuyeres. (There exist several variations—top blowing, bottom blowing, and a combination of both.)
3. The oxygen passes through the molten pig iron. A high temperature is developed and the excess elements present in pig iron, such as carbon, silicon, manganese, sulphur, and phosphorus, are oxidized. At this time, a reddish yellow flame is seen at the nose of the converter, accompanied by a roaring sound. The temperature and chemical composition are carefully monitored and samples are taken for chemical analysis and subsequent examination of physical properties; the results of these appear in the mill certificate given to the purchaser of the steel.
4. When the intensity of flame is considerably reduced, the oxygen supply is shut off. It has to be noted that the supply of oxygen should be carefully controlled to avoid the trapping of gas pockets in steel, especially when the steel is cast into ingots. These gas pockets may lead to defects in the final rolled steel product.
5. The converter is then tilted in the discharge position and this batch process typically produces 50–350 tonnes of steel, depending on the size of the furnace, every one hour to eight hours (compared with a minimum of 10 hours in the open-hearth process).

Deoxidizers, such as silicon and/or aluminum are used to control the dissolved oxygen content. Steel which has the highest degree of deoxidation {containing less than 30 parts per million (ppm) of oxygen} is termed *killed steel. Semi-killed steel* has an intermediate degree of deoxidation (about 30–150 ppm of oxygen). Steel containing the lowest degree of deoxidation is called *rimmed steel*. Rimmed steel may contain scattered blowholes throughout its structure. Such steel is most prone to brittle fracture. Structural steel sections are often produced using either killed steel or semi-killed steel, depending upon the intended use and the thickness. During continuous casting, only killed steel is used. Generally structural steel contains carbon (in the range of 0.10–0.25%) manganese (0.4–0.12%), sulphur (0.025–0.05%), and phosphorus (0.025–0.050%) depending on end use and specifications. The crude steel in liquid form is taken in a ladle for further refining/addition of ferro-alloys, etc.

**Casting** The liquid steel is taken out of the bottom as a continuous ribbon of steel. When sufficiently cooled, it is cut into semi-finished products, such as billets, blooms, and slabs. This method, called *continuous casting* (also known as the *concast method*), is different from the old method (still in use in older plants), where liquid steel is first solidified in large blocks called ingots (weighing about 5–40 tonnes) and then rolled into semi-finished products, involving higher energy and waste in reheating.

**Hot rolling** The semi-finished products, such as billets, blooms, and slabs, are heated at 1200 °C to make metal malleable and then rolled into finished products, such as plates, structural sections, bars, and strips. The greater the reduction in size, the greater the work hardening, which produces varying properties in a section. The variation in cooling rates for the different thicknesses introduces *residual stresses*, which may be relieved by the subsequent straightening process (see also Section 1.8). Further processing of steel can include cold rolling, pickling (to remove oxides and mill scale from the surface of the steel), and coating.

The schematic diagram showing the various stages of manufacturing of structural steel sections from the iron ore are shown in Fig. 1.3. Figure 1.4 shows the relative proportion of the semi-finished products.

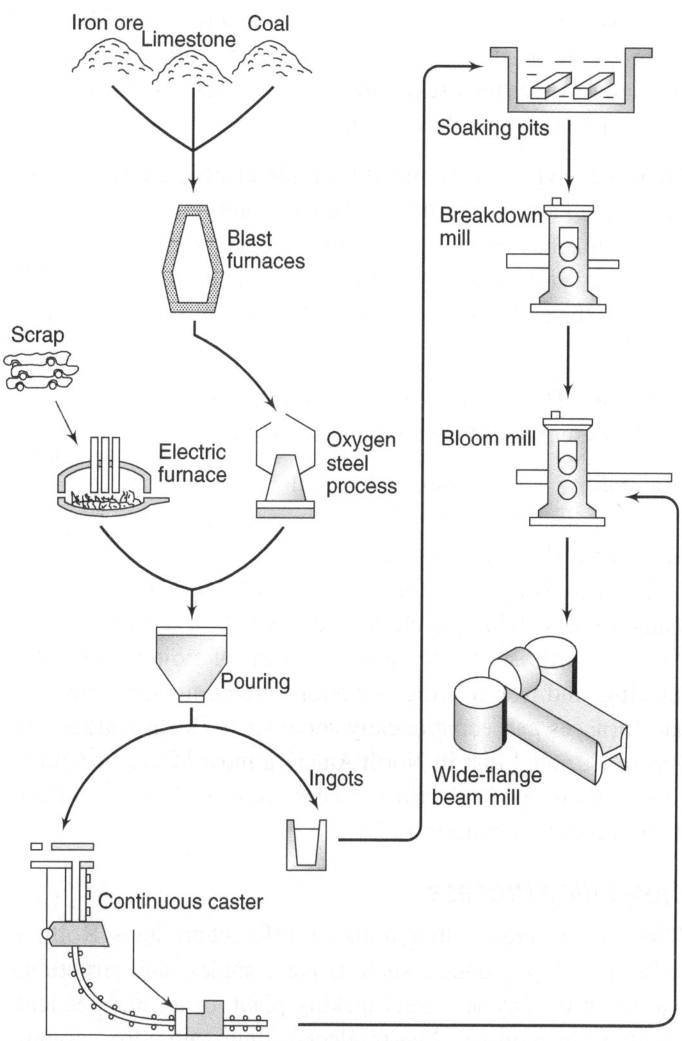


**FIG. 1.3** Schematic diagram showing the various stages of manufacturing structural steel sections from iron ore (Kulak & Grondin 2002)
*Courtesy*: Canadian Institute of Steel Construction

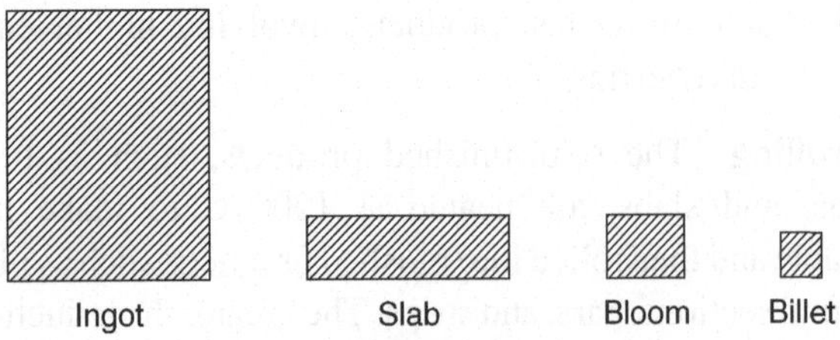

**FIG. 1.4** Relative proportions of semi-finished products

### *Mini Steel Plants*

Mini steel plants use raw materials such as scrap, fluxes, and ferro-alloys. Sponge iron can also be used to substitute scrap by up to 50%. The main steps in the manufacturing process in an electric arc furnace are as follows.

**Melting** Scrap or sponge iron, fluxes (minerals such as limestone, which are used to collect impurities), and ferro alloys are melted in an electric arc furnace, wherein electric current is passed through three large graphite electrodes, creating an arc between the electrodes and scrap steel. The heat so produced is sufficient to melt the scrap steel. An alternate method involves the use of induction furnaces, which are very small and are being phased out.

**Refining** The molten metal from the electric arc furnace is taken in a ladle, for refining. The metallurgy of steel in terms of carbon, phosphorus content, alloy elements, etc., is controlled at this stage. Manufacture of stainless steel requires the addition of nickel (7–8%) and chromium (15–18%).

**Casting** The liquid steel is cast into semi-finished products such as billets, blooms, and slabs.

**Hot rolling** The semis such as billets, blooms, and slabs are heated at 1200 °C to make metal malleable and then rolled into finished products.

Steel making using the electric arc furnace is a batch process with a cycle time of about two to three hours. Since the process uses scrap metal instead of molten iron, coke making, and iron making operations are eliminated. Electric arc furnaces can economically serve small, local markets. It has to be noted that in North America most of the structural steel is produced in electric furnaces (making steel the world's largest recycled material).

### *Hot-rolling Process*

There are different rolling mills for different products. Rolling mills for long products such as bars, angles, and structural parts can be part of a steel making plant, or an independent small-scale industry. Flat-product rolling mills are capital-intensive, as they have to meet strict quality parameters. Such rolling mills produce flat products such as HR plates, strips, or coils. A hot-rolled sheet's thickness can be further reduced by cold rolling, i.e., rolling in CR mills at room temperature. Cold-rolled products can be zinc coated in a galvanizing plant to make galvanized plates or coils.

Although the chemical composition of steel dictates its potential mechanical properties, its final mechanical properties are strongly influenced by the rolling process, finishing temperature, cooling rate, and also the heat treatment (if any).

The reheating, together with the actual mechanical working involved in rolling, modifies the steel in such a way that its tensile strength is considerably increased. The most common rolling practice involves squeezing the heated semi-finished products between a pair of rotating cylinders [see Fig. 1.5(a)] called *rollers* or *rolls*. The two rollers revolve at the same speed in opposite directions. Each pass, of which there may be up to 15, reduces the thickness of the ingot or slab by 50 mm. The steel leaves the mills in the form of 10-m-long semi-finished products, which are inspected both visually and ultrasonically for surface and internal defects such as cracks, blow holes, or slag inclusions. The billets coming out of bloom mills are reheated by using a series of furnaces and passed through grooved or profiled rolls [Fig. 1.5(b)], which operate on all four edges to turn the flat products into structural shapes such as angles, I-sections, and channels. The rolling process reduces it in cross section, elongates it, and shapes the steel into required shape by refining the grain size of the material.

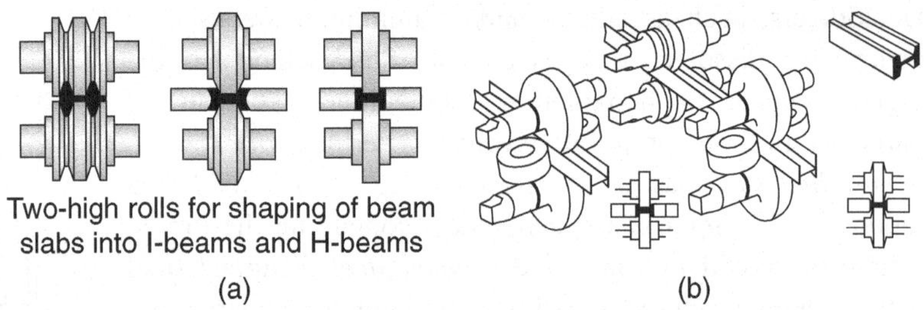

**FIG. 1.5** Rolls used for shaping I- and H-beams

As per the World Steel Association (www.worldsteel.org), the global production of crude steel during 2013 was estimated as 1606 mt and the consumption was 1481 mt. (China and India held the first and fourth ranks with 779 mt and 81.2 mt respectively.)

## 1.3 IRON AND STEEL IN INDIA

There are archaeological evidences of the use of iron in India during the Indus valley civilization. Two excellent examples are the iron pillar near Kutub Minar erected in the fifth century and the 14 m tall iron post in the Kodachadri village in Karnataka.

As early as in 1907, Jamsetji Nusserwanji Tata set up the first integrated steel manufacturing plant at Jamshedpur in Bihar. (It is significant to note that our country possesses enormous deposits of very rich iron ores, but they cannot be fully exploited because there are no coal mines in their vicinity.) During the first two five-year plans, Jawaharlal Nehru, the then Prime Minister of India, authorized the setting up of major steel plants at Bhilai, Rourkela, and Durgapur. In Karnataka, Sir Mokshakumdam Visweswarayya established the Bhadravathi steel plant. At the time of independence, India produced 1.3 million tonnes of steel per year. After the plants were set up at Durgapur, Bhilai, and Rourkela, the production of steel increased to 6.5 million tonnes in 1969–70 and to 21.3 million tonnes in 1995–96. The steel sector now consists of seven integrated plants (see Fig. 1.6) and about 180 mini steel plants and rerollers. (The mini steel plant requires less than one half man-hour per tonne of steel produced as against two man-hour per tonne in the most efficient integrated steel plant.) During the British rule several steel structures were constructed in India. The Howrah Bridge and second Hooghly cable-stayed bridge in Calcutta are examples of steel-intensive bridge construction. There are numerous bridges (built by the Railways) and industrial buildings exclusively using steel.

**FIG. 1.6** Map of India showing locations of integrated steel plants

In 2013, the annual production of steel was about 81.2 million tonnes and is likely to increase in the future. At present, India is the fourth largest producer of steel in the world. However, the per capita consumption of steel in India is low, about 60 kg/person/year as compared to 515 kg in China and 225 kg for the world. In India, a major part of steel is consumed in engineering applications, followed by automobiles and constructions.

The continuous casting method accounts for only 17% of the total steel produced. The remaining 83% is accounted for by the ingot route, which results in high energy costs. The productivity of our blast furnaces is around 50% of that of blast furnaces of world standard due to the low quality of indigenous coking coals and adoption of outdated technologies. Integrated steel plants in India (e.g., SAIL) consume 7.5 gigacalories per tonne of crude steel, whereas plants in other developed countries consume about 5 gigacalories or even less.

The National Mineral Development Corporation (NMDC), Kudremukh Iron Ore Co. (KIOCL), and Sesa Goa (Sesa) are the major merchant producers of iron ore. SAIL and Tata Steel have their captive iron ore mines. Pig iron is mainly produced by KIOCL, Sesa Goa, and Usha Ispat. In addition, there are many mini blast furnace pig iron producers. Even integrated steel plants such as SAIL and RINL produce significant amounts of pig iron. Gas-based sponge iron is produced by Essar Steel, Ispat Industries, and Vikram Ispat (a division of Grasim). SAIL, RINL, Tata Steel, and Jindal Vijayanagar (JVSL) are the largest primary steel producers. Essar steel, Ispat Industries, and Lloyds Steel are the largest secondary steel producers in India. In 2011–12, India produced 5.78 mt of pig iron, 20.37 mt of sponge iron, and 73.42 mt of finished steel. India imported 7.87 mt and exported 5.25 mt of finished steel.

To produce one tonne of crude steel nearly 1.75 tonnes of iron ore, 1.35 tonnes of coal, 0.5 tonne of limestone, and 0.5 tonne of other materials (dolomite, manganese, and other alloy materials) are required. The primary energy requirement to produce one tonne of crude steel in India is about 430–540 Kwh. As iron ore and coal are heavy and bulky materials, iron and steel plants should be located close to the raw material source. Generally, large-scale deposits of iron ore and coal do not occur in close proximity to the market centre. Hence, the location of iron and steel plants is determined based on the calculation of the cost of transporting coal, iron ore, and the finished product to the market centre. In places like the UK and Japan, where imported ore is used, coastal locations are preferred.

## 1.4 METALLURGY OF STEEL

The metallurgy of steel is outside the scope of this book. However, since a little knowledge of it will help the structural engineer understand the ductile behaviour, welding, corrosion, and fire protection aspects of steel, a brief discussion of the metallurgical composition of steel is included in this section.

### 1.4.1 Crystalline Structure of Metals

Solid metals such as iron consist of an aggregate of crystals or grains in which the atoms are arranged in a regular three-dimensional geometrical pattern called a space lattice. In an ideal close-packed lattice arrangement, the unit cell is a *face-centred cubic unit* (fcc) or a *hexagonal close-packed structural* cell. The face-centred cubic unit has atoms at each corner of the cube and in the centre of each face [Fig. 1.7(a)]. The hexagonal structure cell has atoms at each corner, the centre of the top and bottom faces, and also three atoms within the unit cell [Fig. 1.7(c)]. With planes of slightly less dense packing, we have the body-centred cubic structure (bcc structure), which has an atom at each corner and one in the centre of the cube [Fig. 1.7(b)]. These are the three basic lattices occurring in common metals.

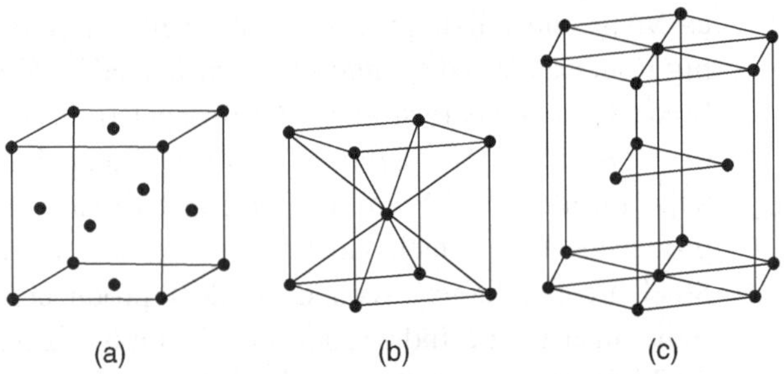


**FIG. 1.7** The three most common crystal structures in metals and alloys (a) Face-centred cubic (b) Body-centred cubic (c) Hexagonal close-packed

Pure iron, when heated to its melting point, undergoes several crystalline transformations. Up to a temperature of 910 °C, iron has a bcc crystalline structure and remains as 'ferrite' or '$\alpha$-iron'. (It has to be noted that the magnetic properties of iron decrease with increase in temperature and vanish at about 768 °C, called the 'Curie point'. The iron that exists between 768 °C and 910 °C and having a bcc structure is called '$\beta$-iron'.) Between 910 °C and 1400 °C, it transforms to 'austenite' or '$\gamma$-iron', which has fcc structure. From 1400 °C up to its melting temperature of about 1539 °C, iron reverts to the bcc structure and is called '$\delta$-ferrite'. On cooling the molten iron back to room temperature, the transformations are reversed almost at the same temperature when heated. These different phases of iron are summarized in Table 1.3.

**TABLE 1.3** Different phases of iron

| Stable temperature range °C | Form of matter | Phase | Identification symbol |
|---|---|---|---|
| >2740 | Gaseous | Gas | Gas |
| 1539–2740 | Liquid | Liquid | Liquid |
| 1400–1539 | Solid | bcc | $\delta$-ferrite |
| 910–1400 | Solid | fcc | $\gamma$-austenite |
| <910 | Solid | bcc | $\alpha$-ferrite |

### 1.4.2 Structure of Steel

When carbon in small quantities is added to iron, steel is obtained. The crystallization of such alloys during solidification is generally similar to but more complex than that of pure iron. The type of solid formed is governed by factors such as the chemical nature of the metals and the relative sizes of atoms. Different crystalline forms appearing in the structures are referred to as *phases* and appear to be homogenous although they may contain more than one type of atom. Solid structures may be described as substitutional (when atoms are within 14% of the same diameter of the base metal) or interstitial solid solutions. When iron combines to form a compound with elements such as carbon, the compound is interstitial due to the small size of the carbon atoms in comparison with that of the iron atoms. These small atoms enter the interstices or gaps between the atoms in the iron lattice. Figure 1.8 shows the equilibrium diagram for alloys of iron and carbon, containing 0 to 6.7% carbon.

Ferrite or $\alpha$-iron up to 910 °C dissolves carbon in an interstitial solid solution to a maximum of 0.025%. $\gamma$-iron is capable of dissolving carbon to a maximum of 2%. Cementite or iron carbide is an interstitial compound of iron and carbon, containing 6.7% carbon, with the chemical formula $Fe_3C$. It is extremely hard and brittle. At *E* (0.8% carbon), the austenite solid solution reacts to form a fine laminated mixture of ferrite and iron carbide. This eutectoid mixture, known as pearlite, is hard and has low ductility. (Ductility is the ability of a substance to undergo large plastic deformation.) These metallurgical terms of iron are summarized in Table 1.4.

It has to be noted that as carbon dissolves in the interstices, it distorts the original crystal lattice of iron and provides increased mechanical strength. However, the addition of more carbon reduces ductility and causes problems during the welding process.

Structural engineers are interested in the portion of the phase diagram shown in Fig. 1.8 that has up to 2.2% carbon. This portion can be divided into two parts, one to the left of the eutectoid point (0.8% carbon), called hypo-eutectoid steel, and the other to the right of this point, called hyper-eutectoid steel. Iron containing very low carbon (up to 0.0025%) is called

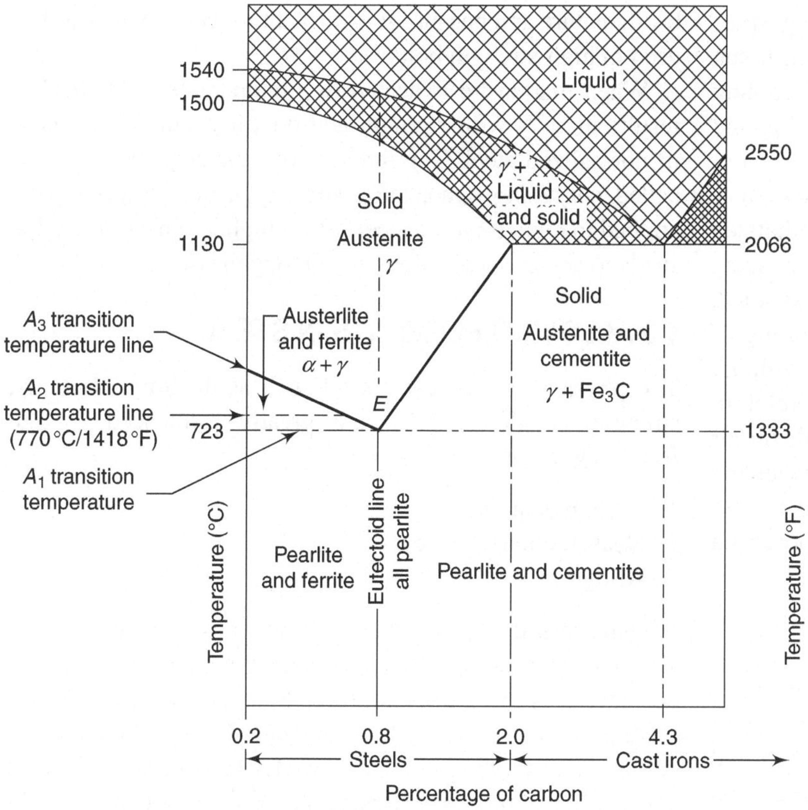


**FIG. 1.8** Equilibrium diagram for alloys of iron and carbon

**TABLE 1.4** Metallurgical terms involving iron

| Name | Metallurgical term | Percentage of carbon (max.) | Crystal structure |
|---|---|---|---|
| $\alpha$-iron | Ferrite | 0.02 | bcc |
| $Fe_3C$ | Cementite | 6.67 | — |
| Ferrite + cementite laminar mixture | Pearlite | 0.80 (overall) | — |
| $\gamma$-iron | Austenite | 2.0 (depends on temperature) | fcc |

very low carbon steel, and has a 100% ferrite microstructure. It is very soft, ductile, and has low mechanical strength.

During manufacture, when the steel is cooled very slowly from a higher temperature, ordinary structural steel with a pearlite ferrite microstructure is formed. As mentioned previously, ferrite is soft and ductile, and pearlite is hard and hence imparts mechanical strength to steel. The higher the carbon content, the higher will be the pearlite content and hence the higher will be the mechanical strength. The amount of pearlite for a given carbon content is calculated according to the following formula.

$$\text{Volume fraction of pearlite} = \frac{\text{percentage of carbon}}{0.8\% \text{ of carbon}} \tag{1.1}$$

However, as the pearlite content increases, ductility is reduced. Steel with more than 0.85% carbon is of no great significance in civil engineering, though it is used in the manufacture of cutting tools, where high hardness is required. More details about the microstructure of steel may be found in Llewellyn and Hudd (1998) and Theling (1984).

### 1.4.3 Heat Treatment of Steel

Since changes occur in the solid state, it is possible to modify the structure of steel and improve the properties by heat treatment. *Heat treatment* involves the heating and cooling of steel under controlled conditions to change its structural and physical properties. *Annealing* and *normalizing* are the processes used to refine the structure of steel. A material that has not been cold worked can be annealed. This should not be confused with annealing as used to remove the effects of cold work. In the annealing process, the steel is heated to a temperature just greater than the $A_3$ temperature (> 910 °C, see Fig. 1.8) and held at that temperature to achieve uniformity of composition and temperature prior to slow cooling, usually in the furnace. On heating, after the $A_1$ line is reached, pearlite changes to austenite, which has a very fine grain size. Cooling from $A_3$ to $A_1$ permits the separation of ferrite, which moves to the austenite grain boundaries because of the small grain size, and at the $A_1$ temperature, the remaining austenite changes to pearlite. Sufficient cooling time allows the carbon diffusion and transformation process to get completed. This refined pearlite + ferrite microstructure shows both increased strength and ductility. The yield strength of steel is related to the grain size by the following equation (Theling 1984)

$$f_y = f_0 + \frac{k}{\sqrt{d}} \tag{1.2}$$

where $f_y$ is the yield strength, $f_0$ is the yield strength of very large isolated crystals (for mild steel it can be taken as 5 MPa), $d$ is the grain size in millimetres, and $k$ is a constant (assumed as 38 for mild steel). Equation (1.2) clearly shows that decreasing grain size enhances the yield strength.

*Normalizing* is a process similar to annealing, except that in normalizing the steel is removed from the furnace and allowed to cool in still air. The changes occurring are the same as during annealing but less time at high temperature and the faster cooling rate give a slightly finer grain structure and finer laminations in the pearlite, as shown in Fig. 1.9. These finer structures result in slightly improved properties compared to those obtained as a result of annealing. Normalizing is cheaper than annealing since the steel is kept in the furnace for less time. However, it can only be used for fairly uniform sections, where air cooling is unlikely to cause distortion due to differential cooling and contraction.

Mild steel plates, structural sections, etc., show very good properties of strength with ductility in the normalized condition. Heat treatment is costly and hence for many purposes the exposure of normalizing can be avoided, provided the finishing temperature during the hot rolling of steel is appropriate. A finishing temperature for hot rolling, which is only slightly above the $A_3$ temperature (refer to Fig. 1.8), gives a very fine austenite grain size and, on air cooling, a microstructure very similar to that obtained by normalizing. It may be noted that in India the heat treated steels amount hardly to about 5% of the steel produced (Rangwala 1997).

### 1.4.4 Quenching and Tempering

When small sections of steel are water quenched from the $\gamma$-region, the cooling rate is too great to allow the separation of ferrite and formation of pearlite by the nucleation and growth process. The fcc austenite is unstable, however, and the change to a bcc structure similar to ferrite cannot be prevented. This structure, known as martensite, is extremely hard and brittle, owing to the distortion produced in the lattice by the carbon retained in the supersaturated solution. It is also seen from Fig. 1.9 that oil *quenching*, where the cooling rate is slightly slower, results in a mixture of martensite and pearlite. Martensite is not used in structural steel construction due to its brittle nature and because it is difficult to weld. However, it is used in high-strength bolts.

When martensite is heated to a temperature of 600 °C, it results in steel with strength and ductility superior to those in normalized steel. This process of reheating martensite is called *tempering*. Quenching and tempering are principally applied to higher-carbon steels to get high hardness or to alloy steels to achieve high strength and toughness.

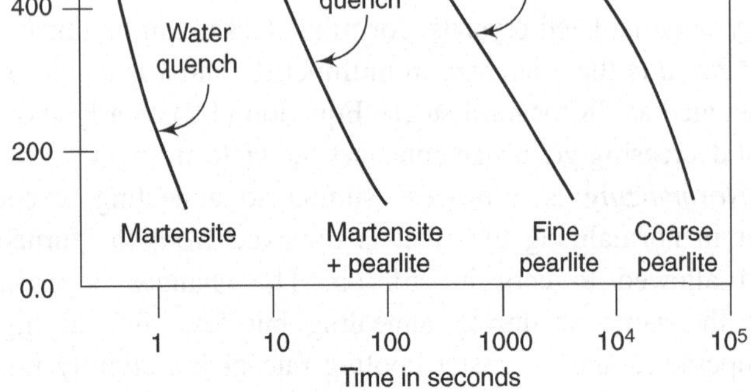


**FIG. 1.9** Change in microstructure as a function of cooling

*Source:* Teaching materials, INSDAG
*Courtesy:* INSDAG, New Delhi

## 1.5 ALLOYING ELEMENTS IN STEEL

The physical properties of steel such as ductility, elasticity, strength, toughness, etc., are greatly influenced by the following factors.

1. Carbon content
2. Heat-treatment process
3. Alloying elements

We have already discussed the first two factors in the previous sections. Depending upon the carbon content, the steel is designated as low-carbon steel (carbon content 0.10–0.25%), medium-carbon steel (carbon content 0.25–0.60%) and high-carbon steel (carbon content 0.60–1.10%). Table 1.5 shows various uses of steel of each category. Structural steels normally have a carbon content less than 0.25%. As already discussed, increasing the carbon content increases the hardness, yield, and tensile strength of steel. However, it decreases the ductility and toughness. Carbon also has greater influence on weldability. Mild steel is widely used for structural work and will be discussed in detail in the later sections of this chapter.

Manganese, silicon, sulphur, phosphorus, copper, vanadium, nickel, chromium, columbium, molybdenum, and aluminium

**TABLE 1.5** Uses of steel

| Type of steel | Carbon content | Uses |
|---|---|---|
| Mild steel | Up to 0.10% | Steel sections used in buildings, bridges, etc. |
| Medium-carbon steel | Up to 0.25% | Boiler plates, structural steel, etc. |
| | Up to 0.45% | Rails, tyres, etc. |
| | Up to 0.60% | Hammers, large stamping and pressing dies, etc. |
| High-carbon steel or hard steel | Up to 0.75% | Sledgehammers, springs, stamping dies, etc. |
| | Up to 0.90% | Miner's drills, smith's tools, stone mason's tools, etc. |
| | Up to 1.00% | Chisels, hammers, saws, wood working tools, etc. |
| | Up to 1.10% | Axes, cutlery, drills, knives, punches, etc. |

are some of the other elements that may be restricted in, or added to, structural steel. In recent years, microalloyed steels or high-strength low-alloy (HSLA) steels have been developed. They are basically carbon manganese steels in which small amounts of aluminium, vanadium, mobium, etc., are used to control the grain size. Molybdenum is also added (up to 0.5%) to refine the lamellar spacing in pearlite and to make it evenly distributed. Alloy steels are termed as low-alloy steels (total alloy content < 5%), medium-alloy steels (total alloy content 5–10%) and high-alloy steels (total alloy content > 10.0%). Based on manganese content, steels are also classified as carbon manganese steels (Mn > 1%) and carbon steels (Mn < 1%). (It has to be noted that the atomic diameter of manganese is larger than that of iron. Hence manganese exists as a substitutional solid solution in the ferrite crystal, by displacing the smaller iron atoms.)

If the silicon content is less than 0.2%, it has no appreciable effect on the physical properties of steel. If it is raised to about 0.30 to 0.40%, the elasticity and strength of steel are considerably increased without serious reduction in ductility. More than 2% of silicon causes brittleness.

If the sulphur content is between 0.02 and 0.10%, it has no appreciable effect on the ductility or strength of steel. The sulphur content, however, decreases malleability (*malleability* is the ability of materials, both in cold and hot states, to be bent or pressed to different shapes without fracture) and weldability of hot metals. A sulphur content of more than 0.10% decreases the strength and ductility of steel.

It is desirable to keep the phosphorus content of steel below 0.12%. It reduces the shock resistance, ductility, and strength of steel. If present in quantities between 0.30 and 1.00%, manganese helps to improve the strength and hardness of mild steel in more or less the same way as carbon; it also increases the toughness of steel. However, when its content exceeds 1.50%, it increases the formation of martensite and hence decreases ductility and toughness.

### 1.5.1 Weldability of Steel

In most cases, members of steel are welded during fabrication. Hence, steels must not only possess high strength but must also be suitable for welding. For good weldability, steel should not show high hardness in welded parts, but should have adequate elongation and notch toughness even in the *heat-affected zone* adjacent to a weld. Since weldability is affected by the kinds and amounts of alloying elements present in the steel, it is important to restrict both to the extent possible.

A major factor in weldability is the carbon equivalent, $C_{eq}$, of the chemical components in steel. The smaller this value, the better is the weldability. The carbon equivalent may be calculated by an equation such as that shown below, in which each symbol refers to the proportion of weight of that particular element in percentage (IS 2062 : 2011).

$$C_{eq} = \frac{C+Mn}{6} + \frac{(Cr+Mo+V)}{5} + \frac{(Ni+Cu)}{15} \tag{1.3}$$

where C is carbon, Mn is manganese, Cr is chromium, Mo is molybdenum, V is vanadium, Ni is nickel, and Cu is copper.

High-strength steels tend to have a high carbon equivalent. When the carbon equivalent exceeds a certain limit ($C_{eq}$ = 0.30–0.43), the loss of weldability is compensated by the reheating or post-heating of the weld zone. However, if the carbon content is less than 0.12%, then $C_{eq}$ can be tolerated up to 0.45%.

## 1.6 CHEMICAL COMPOSITION OF STEEL

Several varieties of steel are produced in India. The Bureau of Indian Standards (BIS) classifies structural steels into different categories based on the ultimate yield strength of the basic material and their use (see IS 7598). They are listed along with the appropriate codes of practice issued by BIS in Table 1.6.

**TABLE 1.6** Types of steel and their relevant IS standards

| Type of steel | Relevant IS standards |
|---|---|
| Structural steel | 2062, 1977, 3502, 5517, 8500 |
| Steel for rivets | 1148, 1149, 7557, 1929, 2155, 2998 |
| Steel for tubes and pipes | 1239, 1914, 806, 1161, 10748, 4923 |
| Steel for sheets and strips | 277, 1079, 12367, 513, 12313, 14246 |
| Steel for rerolling | 2830, 2831 |
| Steel for reinforced concrete | 432,1786, 2090 |
| Steel for bolts, nuts, and washers | 1363, 1364, 1367, 3640, 3757, 6623, 6639, 730, 4000, 5624, 6649, 8412, 10238, 12427 |
| Welding | 814, 1395, 816, 819, 1024, 1261, 1323 |
| Steel for filer rods/wires, electrodes | 1278, 1387, 7280, 6419, 6560, 2879, 4972 |
| Steel casting | 1030, 2708, 2644, 276 |

The chemical compositions of some typical steels specified by the Bureau of Indian Standards are listed in Table 1.7. For details of chemical composition of other steels refer IS 1977 [structural ordinary (low tensile) quality], IS 8500 (medium and high strength quality). A comparison of the chemical compositions of mild steels of different grades as given in international standards is shown in Table 1.8. This table will be helpful if the structural engineer has to design a structure using a different code of practice.

**TABLE 1.7** Chemical compositions (in percentage) of some typical structural steels

| Type of steel | Designation | IS code | C (max.) | Mn (max.) | S (max.) | P (max.) | Si (max.) | | Carbon equivalent |
|---|---|---|---|---|---|---|---|---|---|
| Standard structural steel | Fe 410 A[a] | 2062 | 0.23 | 1.5 | 0.050 | 0.050 | 0.4 | SK[b] | 0.42 |
| | Fe 410 B | 2062 | 0.22 | 1.5 | 0.045 | 0.045 | 0.4 | SK | 0.41 |
| | Fe 410 C | 2062 | 0.20 | 1.5 | 0.040 | 0.040 | 0.4 | K | 0.39 |
| Micro-alloyed medium-/high-strength steel | Fe 440 | 8500 | 0.20 | 1.3 | 0.050 | 0.050 | 0.45 | | 0.40 |
| | Fe 540 | 8500 | 0.20 | 1.6 | 0.045 | 0.045 | 0.45 | | 0.44 |
| | Fe 590 | 8500 | 0.22 | 1.8 | 0.045 | 0.045 | 0.45 | | 0.48 |

[a]Fe stands for steel and the number after Fe is the tensile strength in N/mm$^2$ or MPa
[b]K—killed steel, SK—semi-killed steel (explained in Section 1.2.2)
C = carbon, Mn = manganese, S = sulphur, P = phosphorus, Si = silicon

**TABLE 1.8** A comparison of the chemical compositions of different steel-grade designations (Madhu Mohan et al. 2002)

| Standard | Grade of steel | Maximum permissible limits (%) | | | | | |
|---|---|---|---|---|---|---|---|
| | | C | Mn | S | P | Si | Ni |
| IS 2062 | Fe410 WA | 0.23 (±0.02) | 1.50 (±0.05) | 0.05 (±0.005) | 0.05 (±0.005) | — | — |
| ISO | Fe 430 A | 0.24 | — | 0.05 | 0.06 | — | — |
| BS 4360 | 40 EE | 0.16 | 1.50 | 0.03 | 0.04 | 0.50 | |
| ASTM | A 36 | 0.15 | 0.60 | 0.06 | 0.04 | 0.30 | 1.0 |

## 1.7 TYPES OF STRUCTURAL STEEL

The structural designer is now in a position to select structural steel for a particular application from the following general categories.

**Carbon steel (IS 2062)** Carbon and manganese are the main strengthening elements. The specified minimum ultimate tensile strength for these steels varies from about 410 to 440 MPa and their specified minimum yield strength from about 230 to 300 MPa (see Table 1 of IS 800 : 2007).

**High-strength carbon steel** As discussed already, such steel has a high carbon content and hence shows reduced ductility, toughness, and weldability. This steel is specified for structures such as transmission lines and microwave towers, where relatively light members are joined by bolting. Such steels have a specified ultimate tensile strength, ranging from about 480–550 MPa, and a minimum yield strength of about 350–400 MPa.

**Medium- and high-strength microalloyed steel (IS 2062)** Such steel has a low carbon content but achieves high strength due to the addition of alloys such as niobium, vanadium, titanium, or boron (total microalloying elements restricted to less than 0.25%). Such steel has a specified ultimate tensile strength ranging from 440 to 590 MPa and a minimum yield strength of about 300–450 MPa.

**High-strength quenched and tempered steels** These steels are heat treated to develop high strength. Though they are tough and weldable, they require special welding techniques. They have a specified ultimate tensile strength between 700 and 950 MPa and a minimum yield strength between 550 and 700 MPa.

**Weathering steels** These are low-alloy atmospheric corrosion-resistant steels, which are often left unpainted. They have an ultimate tensile strength of about 480 MPa and a yield strength of about 350 MPa.

**Stainless steels** These are essentially low-carbon steels to which a minimum of 10.5% (maximum 20%) chromium and 0.50% nickel is added. More details about stainless steel are given in Section 1.15.

**Fire-resistant steels** Also called thermomechanically treated steels, they perform better than ordinary steel under fire.

## 1.8 MECHANICAL PROPERTIES OF STEEL

The mechanical properties of steel depend upon the following factors:

1. Chemical composition
2. Rolling methods
3. Rolling thickness

4. Heat treatment
5. Stress history

The important mechanical properties of steel are ultimate strength (also called tensile strength), yield stress (also called proof stress), ductility, weldability, toughness, corrosion resistance, and machinability.

The last four properties are often associated with the fabrication of steel structures and are important for the durability of the material.

### 1.8.1 Ultimate Strength or Tensile Strength

Ultimate strength, which is the minimum guaranteed *ultimate tensile strength* (UTS) at which the steel would fail, is obtained from a tensile test on a standard specimen, generally called a *coupon*. A typical specimen as per IS 1608 is shown in Fig. 1.10. In this test, the gauge length $L_g$ and the initial cross-sectional area $A_0$ are important parameters. The dimensions of the specimens are established to ensure that failure occurs within the designated gauge length. The test coupons are actually cut out from a specified portion of the member for which the tensile strength is required. The initial gauge length is taken as $5.65\sqrt{A_0}$ in the case of a specimen with a rectangular cross section and five times the diameter in the case of a circular specimen.

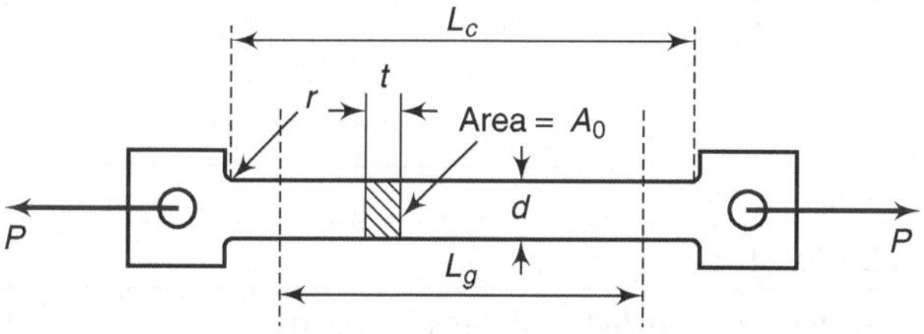


**FIG. 1.10** Standard tensile test specimen as per IS 1608

The coupon is fixed in a tensile testing machine, with specified distances between the grips, and tested under uniaxial tension. The loads are applied through the threaded ends. A typical stress–strain curve of ordinary and high-strength steel specimen subjected to a gradually increasing tensile load is shown in Fig. 1.11(a) and the stress–strain curve of mild steel specimen is shown in Fig. 1.11(b).

The ultimate tensile strength is the highest stress at which a tensile specimen fails by fracture and is gien by

$$\text{Ultimate tensile strength} = \frac{\text{ultimate tensile load}}{\text{original area of cross section}} \quad (1.4)$$

The ultimate tensile strength or stress is called the engineering ultimate tensile stress and is different from the true or absolute ultimate stress. As a specimen is subjected to tension, the area of cross section of the specimen decreases due to Poisson's effect, and the true stress is obtained by considering the actual

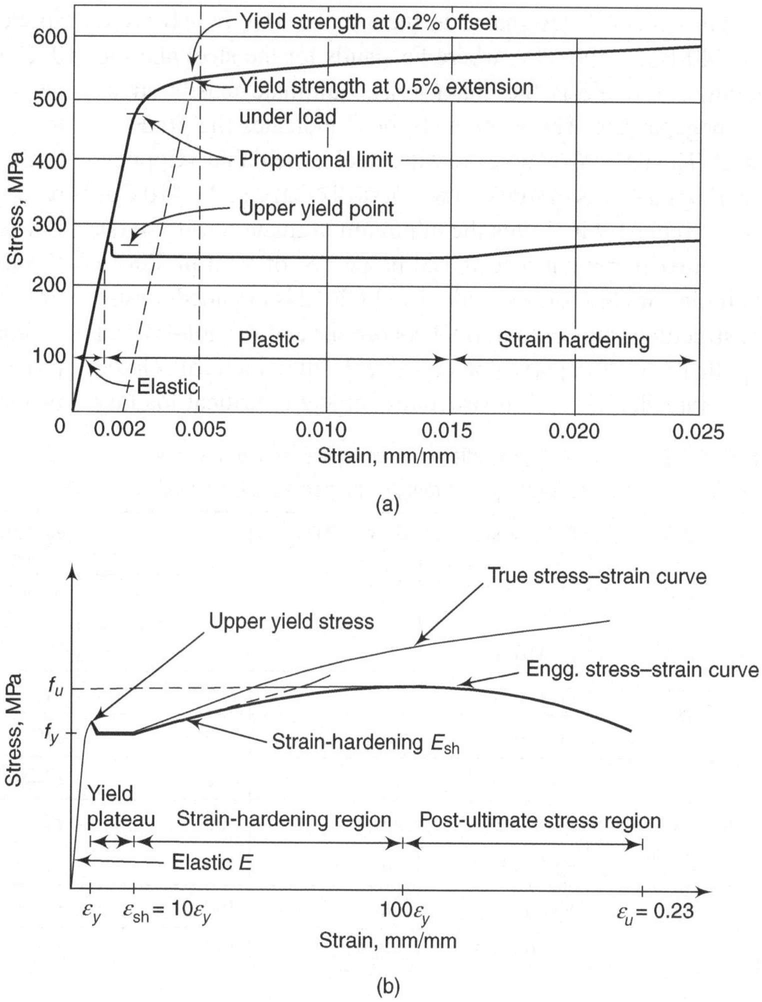


**FIG. 1.11** Typical stress–strain curves of mild steel—(a) Stress–Strain curves of ordinary and high-strength steel (b) Stress–Strain curve of a mild steel specimen

area of cross section at that instant instead of the original area of cross section. Thus it is calculated by the following formula.

$$\text{True stress} = \frac{\text{ultimate tensile load}}{\text{actual area of cross section}} \quad (1.5)$$

In most engineering structures, the stress in the members in service conditions does not cause any appreciable change in the cross section. Moreover, it is extremely difficult to determine the area of cross section at various stages of loading. Hence in all design calculations, only the original area of cross section is considered.

Note that steel is specified according to the guaranteed ultimate tensile stress designated in the code as the *characteristic ultimate tensile stress*, $f_u$ [which is defined as the minimum value of stress below which not more than a specified percentage (usually 5%) of corresponding stresses of samples tested are expected to occur]. See the next section for a discussion of characteristic strength. However, in some countries like the USA, steel is specified according to the (characteristic) yield strength.

Thus, steel is designated in India as Fe 310, Fe 410 WA, Fe 540 B, Fe 590, etc., where Fe stands for the steel and the number after Fe is the characteristic ultimate tensile stress in megapascals. The letter A, B, or C indicates the grade of steel. The letter W denotes that the steel is weldable. (Copper-bearing quality is designated with a suffix Cu, e.g., Fe 410 Cu-WA.) Table 1.9 indicates the minimum ultimate tensile stress and other important mechanical properties of steel produced in India. Grade A steel specified by IS 2062 is intended for use in structures subject to normal conditions and for non-critical applications (for parts not prone to brittle fracture). Grade B is intended for use in structures subject to critical loading applications, where service temperature does not fall below 0 °C. Grade B steel is generally specified for those structural parts which are prone to brittle fracture or are subjected to severe fluctuations of stress (for example members in bridges). Naturally, such steel is also specified for structural parts prone to both conditions. Grade C steel has guaranteed low temperature (up to –40 °C) and impact properties. Grade C steel is used in members or structures where the risk of brittle fracture requires consideration due to their design, size and/or service conditions. A comparison of the mechanical properties of mild steel as given in international codes is given in Table 1.10.

**TABLE 1.9** Mechanical properties of some typical structural steels
(a) Ultimate tensile strength, yield strength, and percentage elongation

| Type of steel | Designation | UTS (MPa) | Yield strength (MPa) Thickness (mm) | | | Min. percentage elongation (gauge length = $5.65\sqrt{A_0}$ ) | Charpy V-notch impact energy (min.) |
|---|---|---|---|---|---|---|---|
| | | | < 20 | 20–40 | > 40 | | |
| Standard structural steel (IS 2062) | Fe 410 A | 410 | 250 | 240 | 230 | 23 | — |
| | Fe 410 B | 410 | 250 | 240 | 230 | 23 | 27 |
| | Fe 410 C | 410 | 250 | 240 | 230 | 23 | 27 |
| | | | **<16** | **16–40** | **41–63** | | |
| Micro-alloyed medium-/high-strength steel (IS 8500) | Fe 440 B | 440 | 300 | 290 | 280 | 22 | 30 |
| | Fe 540 B | 540 | 410 | 390 | 380 | 20 | 25 |
| | Fe 490 B | 490 | 350 | 330 | 320 | 22 | 25 |
| | Fe 590 B/570 B | 590/570 | 450 | 430 | 420 | 20 | 20 |

**TABLE 1.9(b)** Other mechanical properties as per IS 800 : 2007

| Property | Value |
|---|---|
| Modulus of elasticity ($E$) | $2 \times 10^5$ MPa |
| Shear modulus ($G$) | $E/[2(1 + \mu)] = 0.769 \times 10^5$ MPa for $\mu = 0.3$ |
| Poisson's ratio ($\mu$)<br>(i) Elastic range<br>(ii) Plastic range | <br>0.3<br>0.5 |
| Unit mass of steel, $\rho$ | 7850 kg/m$^3$ |
| Coefficient of thermal expansion, $\alpha_t$ | $12 \times 10^{-6}$/°C |
| Brinell hardness number | 150–190 |
| Vickers hardness number | 157–190 |
| Approximate melting point | 1530°C |
| Thermal conductivity | 0.14 cal/cm$^2$ s/1 °C/cm |

**TABLE 1.10** Comparison of mechanical properties of different steel grade designations (Madhu Mohan et al. 2002)

| Standard | Grade of steel | Tensile strength ($f_u$), MPa | Yield strength, ($f_y$) MPa | | |
|---|---|---|---|---|---|
| | | | $t < 20$ mm | $20 < t < 40$ mm | $t > 40$ mm |
| IS 2062 | Fe 410W A | 410 | 250 | 240 | 230 |
| EN 10027-1 | S 235 | 360 | 235 | 235 | 215 |
| BS 4360 | 43 EE | 430 | 275 | 265 | 255 |
| ASTM | A 36 | 400 | 248 | — | — |

After reaching the ultimate tensile stress, a localized reduction in area, called *necking*, begins, and elongation continues with diminishing load until the specimen breaks. After failure, the fractured surface of the two pieces is found to form a cup-and-cone arrangement. Cup-and-cone fracture is considered as an indication of *ductile fracture*.

As shown in Fig. 1.11(a), initially a specimen of steel has a linear stress–strain curve whose slope equals Young's modulus of elasticity $E$. Thus,

$$\text{Modulus of elasticity} = \frac{\text{stress within the proportional limit}}{\text{strain}} \tag{1.6}$$

This can be expressed as

$$E = \frac{f}{\varepsilon} \tag{1.7}$$

where $f$ is the uniaxial stress below the proportional limit, and $\varepsilon$ is the strain corresponding to the stress $f$.

The values of $E$ vary in the range 200,000–210,000 MPa and an approximate value of 200,000 MPa is assumed in the code. The steel obeys *Hooke's law* in this linear range. That is, it remains elastic and recovers to the original shape perfectly on unloading. The limit of the elastic behaviour is often closely associated with the yield stress $f_y$ and the corresponding yield strain $\varepsilon_y = f_y/E$. Beyond this limit, the steel flows plastically without any increase in stress until the 'strain hardening' strain $\varepsilon_{sh}$ is reached. This *plastic range* is usually considerable, and accounts for the ductility of steel. The stress increases above the yield stress $f_y$, when the 'strain hardening' strain $\varepsilon_{sh}$ is exceeded, until the ultimate tensile stress $f_u$ is reached. As indicated earlier, at this stage, large local reductions in the cross section occur, and the load capacity decreases until tensile failure takes place.

The yield strain for mild steel is of the order of 0.00125 or 0.125%. Depending on the steel used, $\varepsilon_{sh}$ generally varies between 5 $\varepsilon_y$ and 15 $\varepsilon_y$. The average value of 10 $\varepsilon_y$ is taken as the yield plateau of structural steels. The value of $\varepsilon_u$ is taken as 100 $\varepsilon_y$ and that of $\varepsilon_{br}$ as 0.23 mm/mm. The initial slope of the strain-hardening part of the curve is termed the *strain-hardening modulus*, $E_{sh}$. It is much less steep than the elastic part, with $E_{sh}/E$ being typically between 1/30 and 1/100 (Alpsten 1973). The strain-hardening range is not consciously used in design, but some of the buckling limitations are conservatively derived to preclude buckling even at strains well beyond onset of strain hardening.

Yielding is sometimes accompanied by an abrupt decrease in load, as shown in Fig. 1.11(a), which results in upper and lower yield points. The upper yield point ($f_{yu}$) is influenced by the shape of the test specimen and by the testing machine itself, and is sometimes completely suppressed. The lower yield point ($f_{yl}$) is much less sensitive and is considered to be more representative. The stress–strain curve shown in Fig. 1.11(a) is typical of low-carbon (mild) steel. Note that the upper as well as lower yield points tend to increase with increase in speed of loading (strain rate). Typical values of the ratio $f_{yu}/f_{yl}$ for normal structural steel range from about 1.05 to 1.10. The term *yield stress* is commonly used to mean either yield point or yield strength when it is not necessary to make the distinction. Steel in compression has the same modulus of elasticity as in tension. The lower yield stress is also the same for tension and compression and there is about the same length of level yielding (contraction).

## Parameters That Influence Yield Stress

The strain rates used in tests to determine the yield stress of a particular steel type are significantly higher than the nearly static rates often encountered in actual structures (McGuire 1968 and Alpsten 1973). The increase in tensile and yield strength at higher strain rates is shown in Fig. 1.12. The lower yield stress at zero strain rates is called the *static yield level* and may be 10% to 15% lower than the yield stress reported in the rolling mill acceptance test (Gaylord et al. 1992).

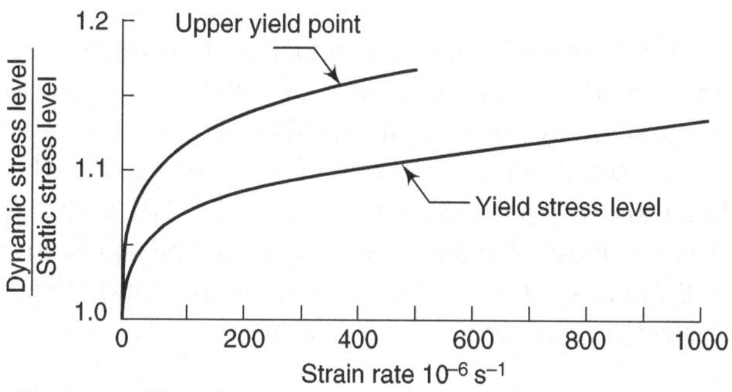


**FIG. 1.12** Effect of strain rate on upper yield point and yield stress of structural steel

*Courtesy*: Lehigh Preserve, Lehigh University Libraries, Lehigh University

However, since the values obtained from the majority of these mill tests are not more than 10% higher or lower than the static rate values, the net effect, when averaged over a complete design, may not be significant (Nethercot 2001). At higher temperatures, the reverse takes place (i.e., at higher strain rates there is reduction in yield strength). This fact needs to be considered only in blast-resistant design. This is of less practical importance in earthquake engineering applications since the strain rate will be well within the range. Mild steel and medium-strength steels have clear yield points and should not be stressed beyond the yield point as the deformation will be large and uncontrollable beyond yield. At strain rates characteristic of seismic response (0.01–0.10/s), steel exhibits a significant increase in yield strength (10–20%) above static test values. However, under cyclic straining, i.e., straining under cyclic loads, the effective strain rate decreases, minimizing this effect.

Yield stress may also be influenced by the position from which the test coupons are taken. For example, the webs of the I-section are thinner than those of the flanges and hence tend to possess a slightly finer grain structure as a result of faster cooling after rolling. Due to this the yield stress at the web will be higher than that at the flange (Alpsten 1973). It has to be noted that in most situations, the flanges of I-sections contribute most to their load-carrying capacity, since most of the area is concentrated in the flanges. Hence, structural designers must be careful in selecting the appropriate value for material strength for use in their calculations. In order to use plastic design or in earthquake-resistant structures, the steel should satisfy the following criteria.

1. The yield plateau should extend for at least six times the strain at first yield.
2. The ultimate/yield stress ratio must be greater than 1.25. (To develop an inelastic rotation capacity, a structural member needs adequate length of yield region along the axis of the member. The larger the ultimate to yield ratio, the longer is the yield region.)
3. The minimum elongation must be 15% on a gauge length of $5.65\sqrt{A_0}$.

It is also preferable that the actual yield strength based on the tensile test of steel does not exceed the specified yield strength by more than 120 MPa. Figure 1.13 shows the stress–strain curves of different types of steel produced in India and the permanent strain line. Fe 410 grade mild steel is the most commonly used in structural applications. Fe 370 grade steel is used in less important works. (Fe 310 mild steel is used primarily for furniture, doors, windows, etc.)

High-carbon steels do not usually have a pronounced yield point. Instead, after a range of linear elastic behaviour, which ends at a point called the *proportional limit*, the rate of increase in stress begins to drop till the tensile strength is reached (the upper curve of Fig. 1.13). In this case, yielding is arbitrarily defined by a yield strength which is usually taken to be that stress which leaves the specimen with a permanent set (plastic elongation) of 0.2% when the specimen is unloaded. It is

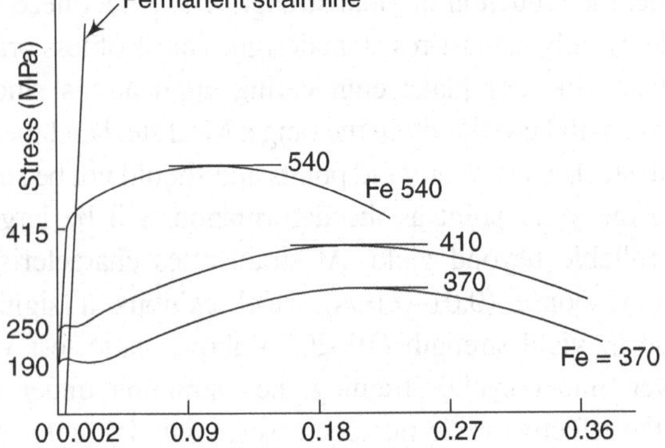


**FIG. 1.13** Stress–strain curves of different types of steel produced in India

obtained by drawing a line parallel to the elastic portion at 0.2% strain, which intercepts the stress–strain curve, as shown in Fig. 1.13. However some standards (e.g., ASTM specification, A370) define the yield stress as the stress corresponding to a 0.5% elongation under load.

The yield stress $f_y$ also varies significantly with the chemical constituents of the steel (e.g., the percentage of carbon and manganese), the heat treatment used, and with the amount of working which occurs during the rolling process. Thus, thinner plates which are more worked have higher yield stresses than thicker plates. The yield stress is also increased by cold working.

## 1.8.2 Inelastic Cyclic Response

The stress–strain response of most materials under cyclic loading is different from that under single (monotonic) loading. For fatigue analysis, it is necessary to consider the cyclic material behaviour for strength and life calculations.

When steel is subjected to cyclic loading in the inelastic range, the yield plateau is suppressed and the stress–strain curve exhibits the *Bauschinger effect*, in which non-linear response develops at a strain much lower than the yield strain, as shown in Fig. 1.14. As seen from this figure, as the amplitude of response increases, the stress level for a given strain also increases and can substantially exceed the stress indicated by the monotonic stress–strain curve.

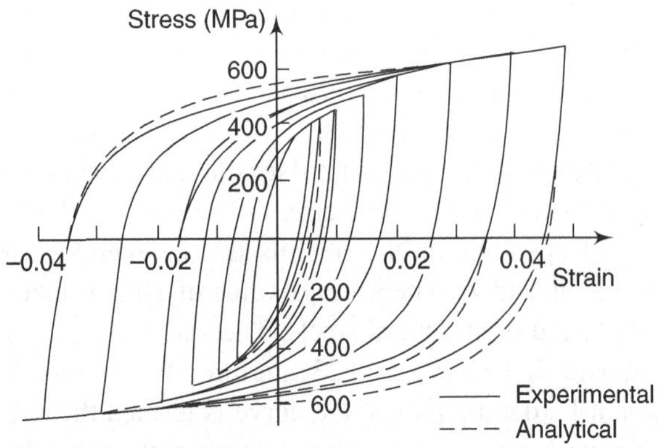


**FIG. 1.14** Stress–Strain curve of steel subjected to cyclic loading

## 1.8.3 Characteristic Strength

Variations in material properties (due to the non-uniform molecular structure of the material, and variations and inconsistencies in the manufacturing process, which depend on the degree of control, etc.) should be recognized and taken into consideration in the design process. The material properties that are of greatest importance in the design of structures using steel are yield strength, maximum percentage elongation, and Young's modulus. Other properties that are of less importance are hardness, impact resistance, and melting point.

If a number of samples are tested for a particular property (e.g., yield strength) and the number of specimens with the same strength (frequency) are plotted against the strength, then the results approximately fit a *normal distribution curve*, as shown in Fig. 1.15.

This curve can be mathematically expressed by the equation shown in Fig. 1.15, which can be used to define 'safe' values for design purposes. When defined, this safe value of yield strength, is called *characteristic strength*. If the characteristic strength is defined as the mean strength, then from Fig. 1.15, 50% of the material has a characteristic strength below this value and hence is not acceptable. Hence a characteristic value which has a particular chance (often 95%) of being exceeded in any standard tension test is chosen.

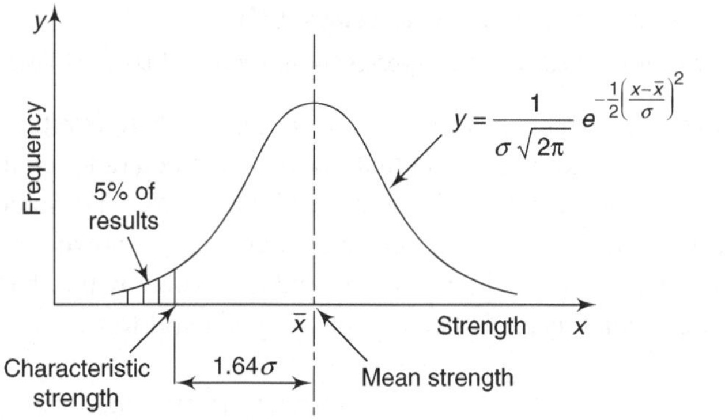


**FIG. 1.15** The normal distribution curve

Thus, the characteristic strength is calculated from the equation

$$f_k = f_{\text{mean}} - 1.64\sigma \tag{1.8}$$

where $\sigma$ is the standard derivaton for $n$ samples, and is given by

$$\sigma = \left[\frac{\sum(f_{\text{mean}} - f)^2}{(n-1)}\right]^{0.5} \tag{1.9}$$

The characteristic strength of steel is the value obtained from tests at the rolling mills, but by the time the steel becomes part of the finished structure, its strength might have been reduced (e.g., by corrosion or accidental damage). The strength to be used in design calculations is therefore the characteristic strength divided by a partial safety factor. The value of the partial safety factor adopted for steel is given in Table 5 of IS 800 : 2007 as 1.10 for yielding resistance.

### 1.8.4 Ductility

*Ductility* may be described as the ability of a material to change its shape without fracture. In other words, the ductility of a structure or its members is the capacity to undergo large inelastic deformations without significant loss of strength or stiffness. The stress–strain curve of a material also indicates the ductility. It is the amount of permanent strain, i.e. strain exceeding proportional limit up to the point of fracture. The ductility of the tension test specimen is measured by determining the percentage elongation (comparing the final and the original lengths over a specified gauge distance). The specified gauge length according to the code is as follows.

$$\text{Gauge length} = 5.65\sqrt{A_0} \tag{1.10}$$

Percentage elongation

$$= \frac{(\text{elongated length between gauge point} - \text{gauge length}) \times 100}{\text{gauge length}} \tag{1.11}$$

The measured elongation is influenced by the gauge length, strain rate of test, and failure within or outside the gauge length.

Values of 20% can be obtained for mild steel but are less for high-strength steel. By improper testing one may get percentage elongation values of 20–60 and hence the test houses should be careful with the testing process. A high value is advantageous because it allows the redistribution of stresses at ultimate load and the formation of plastic hinges. Design procedures based on inelastic behaviour require large ductility, particularly for the treatment of stresses near holes or abrupt changes in member shapes as well as for the design of connections. The minimum required percentage elongations of steel produced in India as per the code are given in Table 1.9(a) (also see Table 1 of IS 800 : 2007). Designers normally use specified properties, only rarely calling for their own material tests in the case of steel. For most standard mild steels, the values are greater than the minimum required. However, rerolled steel with impurities or improperly controlled steel, which may absorb more carbon, has higher strength but less percentage of elongation.

The minimum elongation requirement safeguards the structure in the following ways (Dayaratnam 1996).

1. In case of excessive or sudden loading, as, for instance, when there is an earthquake or a blast, the structure undergoes large deformations before collapsing. This gives adequate warning time to the occupants of the building. (Ductility is a very important property, especially when the structure is subjected to seismic loads and it has been found that ductile structures have performed well in such situations.)
2. It helps avoid fracture at a welded zone, which may be caused by differential cooling or heating.
3. It helps absorb local deformations and avoids cracking at stress concentration points.
4. It helps absorb deformations during transportation, handling, and fabrication.
5. It helps absorb bending or straightening of members and avoids cracking during fabrication.

Eurocode 3 sets the following requirements in order that the structures possess adequate ductility.

1. $\frac{f_u}{f_y} \geq 1.10$
2. Elongation at failure > 15%
3. $\varepsilon_u \geq 15\varepsilon_y$ (see Fig. 1.11)

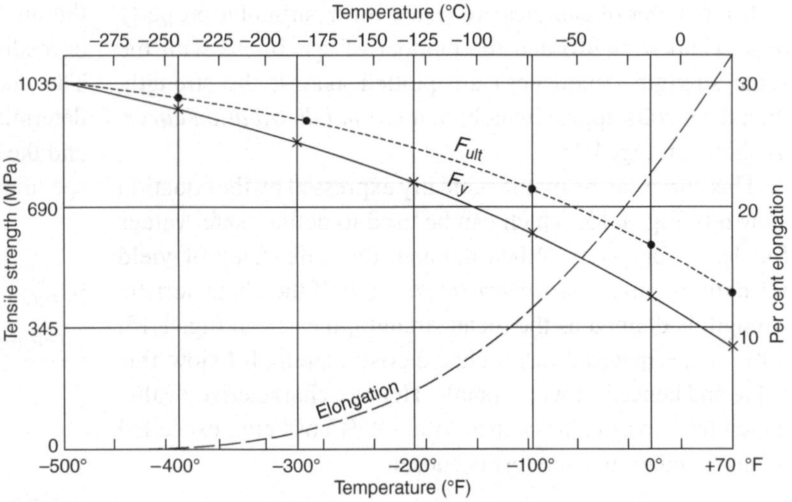


**FIG. 1.17** Effect of low temperature on steel (Bowles 1980)

*Source:* Bowles, Structural Steel Design © 1980. Reproduced with permission of McGraw-Hill Education

## 1.8.5 Low Temperature and Toughness (Brittle Fracture)

In structural steel design, toughness is a measure of the ability of steel to resist fracture under impact loading, i.e., the capacity to absorb large amounts of energy. *Toughness* can be an important design criterion, particularly for structures subject to impact loads (e.g., bridges) and for those subject to earthquake loads. The area under the stress–strain curve is a measure of toughness (see Fig. 1.16). The triangular area shown in Fig. 1.16 is the elastic strain energy. Hence both strength and ductility contribute to toughness. Thus, because its modulus of elasticity is only one-third that of steel, an aluminium structural member can absorb three times the energy for the same stress as a steel member of the same dimensions, provided the stress does not exceed the proportional limit.

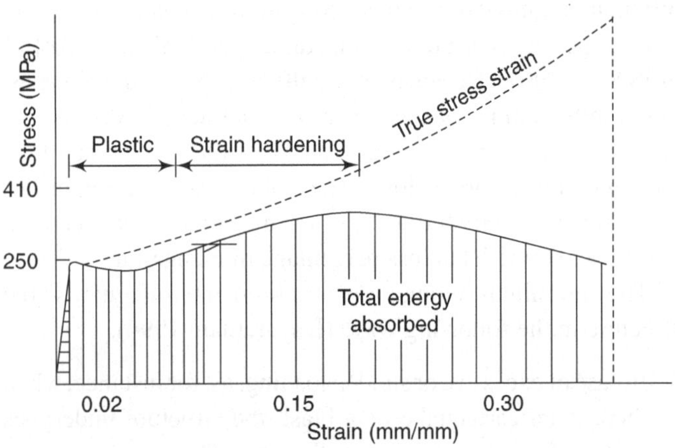


**FIG. 1.16** Toughness of steel

At room temperature, common structural steel is very tough and fails in a ductile manner. At temperatures below 0 °C, the yield strength of steel is only marginally affected, while there is substantial reduction in ductility and toughness (see Figs 1.17 and 1.19). (The ductility of a material is sometimes indicated by the ratio of the total energy absorbed by the material to the elastic energy.) Due to this phenomenon, at low temperature, steel structures sometimes fail suddenly and without warning. As we are aware, when the material elongates, there is a contraction in the lateral dimension owing to Poisson's ratio effect. If the lateral dimensions are fully (or even partially) restrained, the steel will pull apart without fully developing the yield potential. This type of failure is termed *brittle fracture*. Figure 1.18 illustrates the stress concentration effect that occurs at a notch (a notch may be any abrupt change in cross-sectional area). The notched section attempts to contract under tension, as shown by the dashed semicircles in this figure, and is resisted by the shaded zones, which is equivalent to applying a lateral force.

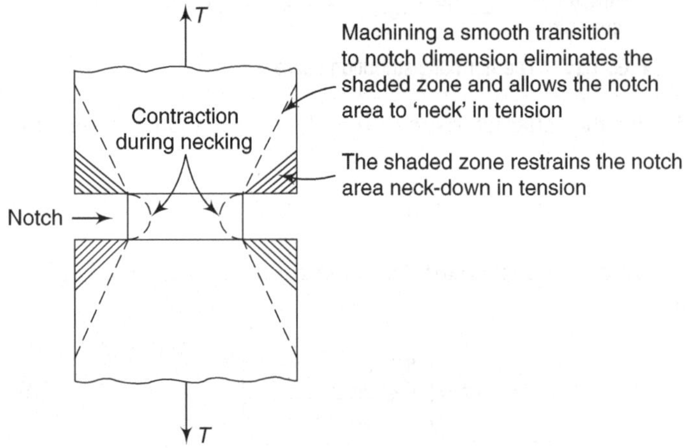


**FIG. 1.18** Stress concentration effect at a notch

*Source:* Bowles, Structural Steel Design © 1980. Reproduced with permission of McGraw-Hill Education

A right combination of low temperature, an abrupt change in section size (notch effect) or an imperfection, and the presence of tensile stress can initiate brittle fracture. This may begin as a crack, which may propagate and cause the member to fail. Most brittle fractures occur under static load at stress levels which are not excessive, but they may also be due to the dynamic application of a load or some overload.

The first failure due to brittle fracture was identified in 1886 (McGuire 1968). One of the best known brittle fractures occurred in Boston in a riveted steel tank of diameter 27.5 m and height 15 m, which contained 7.57 million litres of molasses. The tank failed in an explosive manner on January 15, 1919.

Similar sudden failures of steel water tanks, oil tanks, transmission lines, ships, plate girders, and bridges have occurred in the past (McGuire 1968). Most of these failures have occurred under normal service conditions, in welded structures, and at low temperatures. The subject of brittle fracture received greater attention when a large number of failures occurred in welded ships during World War II. Due to certain design changes that were implemented, the number of failures was reduced. However, during the cold winter of 1977, several spectacular failures occurred in bridge structures in Illinois, Minnesota and Pennsylvania in the USA. On January 22, 1988, several brittle failures occurred in a bridge in Providence, Rhode Island, the USA (Gaylord et al. 1992).

Extensive research on brittle fracture using the concepts of fracture mechanics has shown that material toughness, crack size, and stress are the primary factors which determine susceptibility to such fracture. The *Charpy V-notch test* (CVN test) is commonly used to evaluate the behaviour of metals as they are affected by an abrupt change in cross section (IS 1757). In this test, a standard rectangular simply supported beam having a V-notch at mid-length is subjected to an impact load by a swinging pendulum. The larger the impact value, the tougher is the material. The temperature of the specimen is varied, the energy absorbed by each specimen is recorded, and the energy–temperature curve is plotted as shown in Fig. 1.19. From this curve, a transition temperature corresponding to some level of energy absorption (usually 20 or 27 J) is selected. The transition temperature is the temperature below which fractures are mostly brittle and above which they are mostly ductile. A value of 20 J is generally used to define the lower transition temperature and 60 J has sometimes been specified for the higher one (Gaylord et al. 1992). If the temperature is above the ductility transition temperature, there is appreciable plastic flow at the root of a notch before cracking begins.

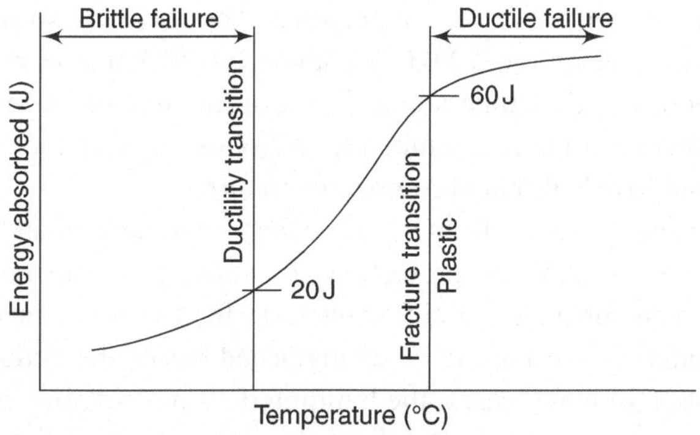


**FIG. 1.19** Energy–Temperature curve of steel

Though the CVN test has been a common means of determining notch toughness, other tests such as drop-weight, nil-ductility test (NDT), dynamic tear test, etc., have also been developed. Details of this may be found in Barsom and Rolfe 1987.

Structural steels vary greatly in toughness. Highly killed, fine grain steel with a suitable chemical composition or specially heat-treated steel exhibit considerable toughness. IS 2026 and IS 1757 codes allow the use of only those steels that exhibit a minimum energy absorption capacity at a predetermined temperature (e.g., 20 J at 23 ± 5 °C). The best combination of properties, a yield strength of 400–450 MPa, and a CVN impact value of 27 J down to –50 °C is produced in steels by treating them by a combination of niobium or vanadium, and aluminium. By lowering the carbon content to about 0.18%–0.16%, we may obtain CVN impact values of 27 J at –30 °C. In addition to the chemistry of steel, size of plates, residual stress, and cold work also affect toughness. (Thick plates, large residual stress, and cold work are detrimental.)

From the designer's point of view the most satisfactory way of dealing with brittle fracture is to reduce the likelihood of the occurrence by a sensible choice of material. In other words, the service temperature should be above the specified low temperature of the steel and the CVN impact value more than the one specified, to prevent brittle fracture. Brittle fracture may be controlled by using the following methods (Bowles 1980).

1. Members and their connections should be detailed to minimize stress concentrations.
2. Steels that have been especially alloyed for a low-temperature environment should be used.
3. If possible, the notch should be machined or ground into a smooth transition.
4. The fabrication and assembly sequence should be specified in order to minimize residual tensile stresses.
5. A pre-tension stress should be applied at elevated or normal temperature to the notched zone to cause a local yield before the service loads and temperatures are encountered.
6. The rate of tensile strain application or stress level should be reduced.
7. Using thick plates with mutually perpendicular welds stressed in the 'through thickness' direction should be avoided.
8. The welding electrodes or other welding consumables selected should have a specified Charpy impact value equal to or greater than that specified for the parent metal.

Clause 2.4.4 of BS 5950-1: 2000 and Tables 3.7 and 3.8 of Hong Kong steel code provide recommendations for the selection of thickness of each element such that brittle fracture could be avoided.

## 1.8.6 Lamellar Tearing

*Lamellar tearing* is a form of brittle fracture that may occur in certain welded joints. For example, a tear can occur if a large weld (or welds from both sides) is placed on a thick plate, since the shrinkage strains from the welding operation will be large and restrained (see Fig. 1.20). The restraint may be developed due to the weld on the far side, because of the member thickness or due to a combination of both the factors.

Generally I-sections are adequately ductile when loaded either parallel or transverse to the rolling direction (see Fig. 1.21). However, when the strain is localized in the 'through-thickness' direction (at one thick flange of the section), a restrained situation exists because the strain cannot redistribute from the flange through the web to the opposite flange. The large localized 'through-thickness' strain may exceed the yield strain and lead to lamellar tear (Salmon and Johnson 1996). A thin, stiffened column is also susceptible to lamellar tearing, since the flange stiffeners that are welded to the column flange produce a restraint. A large overmatch of electrode and base metal in a full penetration butt weld also tends to increase the possibility of tearing.

The use of fillet welds, a joint design that allows weld shrinkage to occur in the rolling direction (so that the shrinkage pulls on the fibres longitudinally in their strongest orientation), and the sequence of welding to minimize shrinkage strains are practical methods used to avoid lamellar tearing (ASCE paper, 1982 and Thornton, 1973). A number of joint alternatives (see Fig. 1.22) that may be used to reduce lamellar tearing are given in AISC, Engineering Journal (1973).

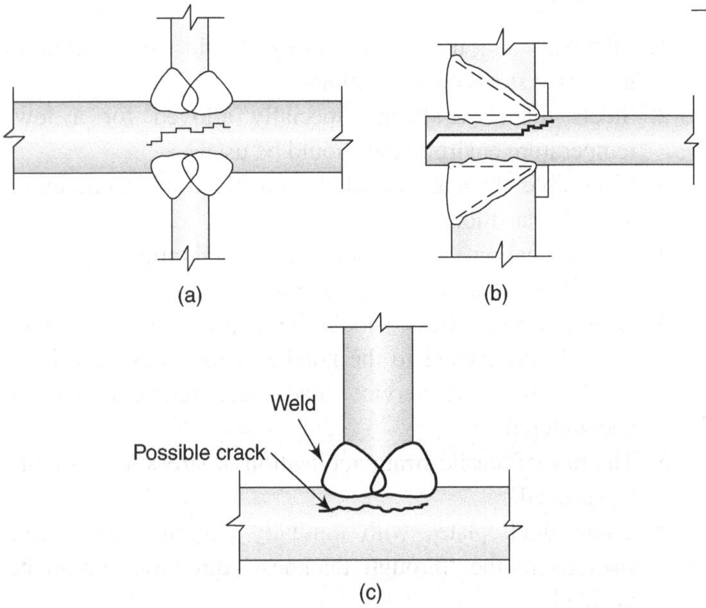

**FIG. 1.20** Lamellar tearing at welded joints

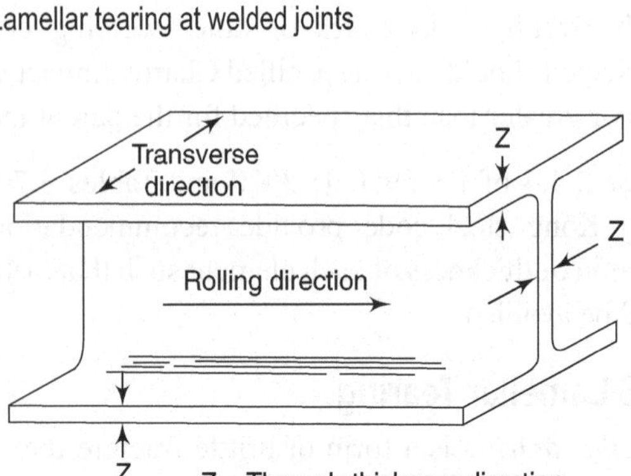

**FIG. 1.21** Definition of through-thickness direction

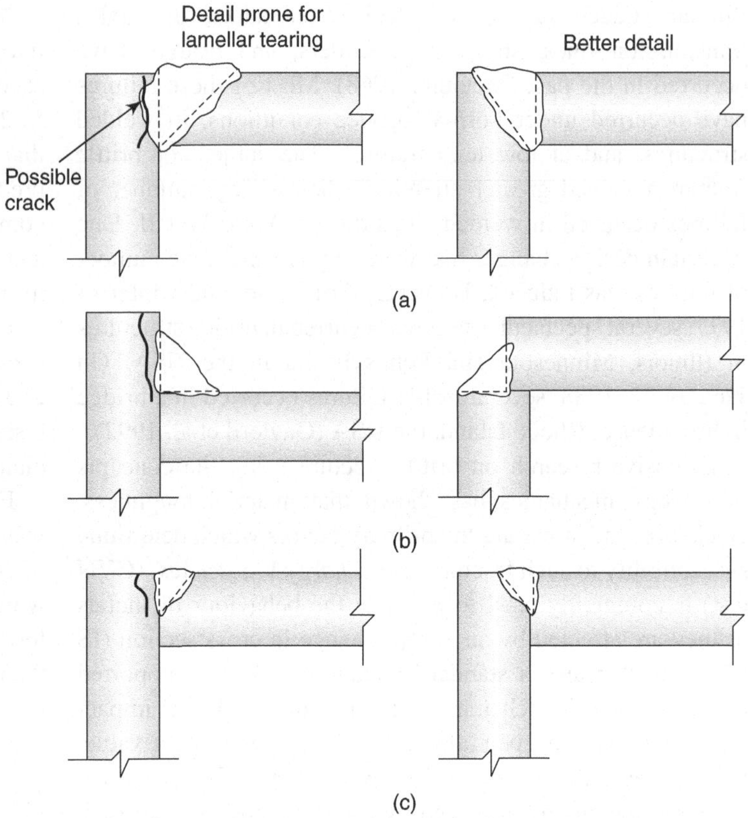

**FIG. 1.22** Joint alternatives to reduce Lamellar tearing

## 1.8.7 High-temperature Effects

Steel is not a flammable material. However, its strength reduces with rise in temperature. The yield as well as tensile strength at 500 °C are about 60–70% that at room (about 21 °C) temperature. The drop in strength is much higher at still higher temperatures (for example at 800 °C it is only 15% of that at room temperature).

Hence, steel frames enclosing materials that are flammable require fire protection, to control the temperature of steel members for a sufficient time for the occupants to seek safety or for the fire be extinguished before the building collapses. In many cases, the building does not collapse even at high temperatures. But the members are deformed beyond acceptable limits, and hence have to be replaced.

## 1.8.8 Resistance to Corrosion

Steel readily corrodes in moist air. Exposure to sea water, acid, or alkaline vapours hasten the process. It has been estimated that more than 0.075 mm of the thickness of steel members will be lost every year in an industrial environment in which sulphur dioxide is present. Hence, structural steel members should be protected effectively against corrosion.

The most common method of protecting a steel member involves the use of paint or metallic coating, or a plastic coat in the case of metallic sheeting. Steels with a copper content of 0.2–0.5% have improved resistance to atmospheric corrosion but still need to be protected. Paint systems consist of a zinc- or aluminium-based primary coat on which two or three layers of finishing coats are applied. Metallic coatings involve galvanizing and sheradizing (both of which use zinc), electroplating (usually applied to fasteners), and metal spraying using either zinc or aluminium. For the coating or painting to be effective, the surfaces of the steel members have to be cleaned effectively before treatment. Several methods are available and the usual one consists of blast cleaning the surface using small abrasive particles such as those of iron, which are directed to the surfaces of the members by using compressed air or an impeller. Instead of protective treatment, one can go in for special corrosion-resistant steel, which on exposure to weather forms a protective surface layer of oxide film. Such weathering steels contain a greater amount of phosphorus and some chromium and copper than normal steel. They cost 20% more than normal steel but this initial cost may be offset by savings in weight, protective treatment, and maintenance.

## 1.8.9 Hardness

*Hardness* is a measure of the resistance of the material to indentations and scratching. Several methods are available to determine the hardness of steel and other metals. In all these methods, an 'indentor' is forced on to the surface of the specimen. On removal, the size of the indentation is measured using a microscope. Based on the size of the indentation, the hardness of the specimen is determined. The various methods used to determine hardness are listed in Table 1.11 (also see Fig. 1.23). Rockwell hardness testing is not normally used for structural steel (Peterson 1990).

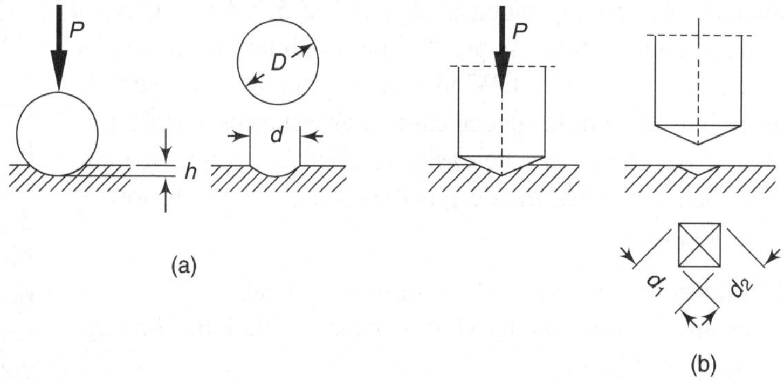


**FIG. 1.23** Illustration of methods to determine hardness

The Brinell hardness number (BHN) is given by the ratio of the applied load and the spherical area of the indentation. Mathematically,

$$\text{BHN} = \frac{0.204P}{\pi D\left[D - \sqrt{(D^2 - d^2)}\right]} \tag{1.12}$$

where $P$ is the load in Newtons, $D$ is the ball diameter in millimetres, and $d$ is the diameter of the indentation in millimetres.

The Vickers hardness number is given by

$$\text{VHN} = \frac{0.189P}{d^2} \tag{1.13}$$

where $P$ is the load in Newtons, and $d$ is the diagonal length of the indentation in millimetres = $(d_1 + d_2)/2$.

**TABLE 1.11(a)** Various testing methods used to determine the hardness of metals

| Testing methods | Indentor | Typical values for steel |
|---|---|---|
| Brinell hardness | Hard steel ball | 150–190 |
| Vickers hardness | Square based diamond pyramids of 135° included angle | 157–190 |
| Rockwell hardness | Diamond core with 120° included angle | 80–105 |

**TABLE 1.11(b)** Typical values of hardness for some metals and alloys

| Metal | Brinell Hardness (BHN) | Vickers Hardness (VHN) | Rockwell Hardness (RHN) |
|---|---|---|---|
| Fe 410 Structural steel | 122 | 126 | 70.5 |
| Fe 500 Structural steel | 149 | 157 | 80.8 |
| Aluminium | 20 | — | — |
| Hardened tool steel XT 72 W 18 Cr4 V1-Annealed (IS 7291:1981) | 255 | 265 | — |
| Grey iron castings (FG 150)-IS 210:1993 | 130–180 | 139–190 | 75–89 |

### 1.8.10 Fatigue Resistance

*Fatigue* is the term used in connection with the initiation and propagation of microscopic cracks into macrocracks by the repeated application of alternating stresses. The damage and failure of materials under cyclic loads is called *fatigue damage*. Fatigue need not be considered unless numerous significant fluctuations (usually taken as $2 \times 10^6$ to $5 \times 10^6$ cycles) of stress are anticipated. As per the code, ordinary reversals due to wind may be ignored. Wind-induced oscillations have to be taken into account in special cases. For instances, oscillations of $2 \times 10^6$ cycles can be easily reached in a lighting mast. The code states that the designer should check the following members for fatigue.

1. Members supporting lifting or moving loads,
2. Members subjected to wind-induced oscillations of a large number of cycles,
3. Members subjected to repeated stress cycles from vibrating machinery, and
4. Members subjected to crowd-induced oscillations.

Thus, fatigue effects are more likely to occur in bridges and cranes due to the cyclic nature of loading, which causes reversal of stresses. Welds are susceptible to a reduction in strength due to fatigue because of the presence of small cracks, local stress concentrations, and abrupt changes of geometry. It has to be noted that the incidence of fatigue and fatigue crack growth is independent of steel grade. Also, fatigue cracks are far more common than brittle fracture.

## 1.9 RESIDUAL STRESSES

Higher temperatures in the range of 600–700 °C are involved during the rolling of steel sections. Steel members are also subjected to high temperatures during fabrication by welding and also when material or members are cut by flame-cut (in these cases, heat is applied locally to selected parts of the cross sections). Cooling of these members or materials always takes place unevenly (even for a hot-rolled member placed on a cooling bed after rolling, some parts of the section, for example, the flange tips of an I-section, cool faster than the flange-to-web junctions, similarly the central portion of the web tends to cool faster than the junctions). Due to this uneven heating and cooling, structural members normally contain *residual stresses.* Residual stresses may also result from the cold straightening of bent members (Gaylord et al. 1992). Although it is possible to remove these stresses by subsequent reheating and slow cooling, this process is very expensive and hence not attempted in normal structural engineering applications (such a removal of residual stress is mainly used in pressure vessels).

As discussed earlier, in an I-section, the flange tips and web interiors cool faster than other parts of the cross section, resulting in residual tensile stresses in the region of junctions (which cool more slowly) and compressive stresses in the remainder of the cross section (which cool first). The typical distribution of residual stresses in a standard I-section is shown in Fig. 1.24(a). The stresses also vary across the thickness, and the pattern shown in Fig. 1.24(a) represents the average of the values across the thickness. Residual stresses tend to increase in magnitude with increase in size of the element. Both magnitude and distribution of thermal residual stress are influenced to a considerable degree by the geometry of the cross section, as shown in Fig. 1.24(b) (Gaylord et al. 1992). The magnitude of tensile residual stresses may reach up to $0.3 f_y$ and the compression residual stress up to $0.5 f_y$ in rolled I-sections (Trahair et al. 2001).

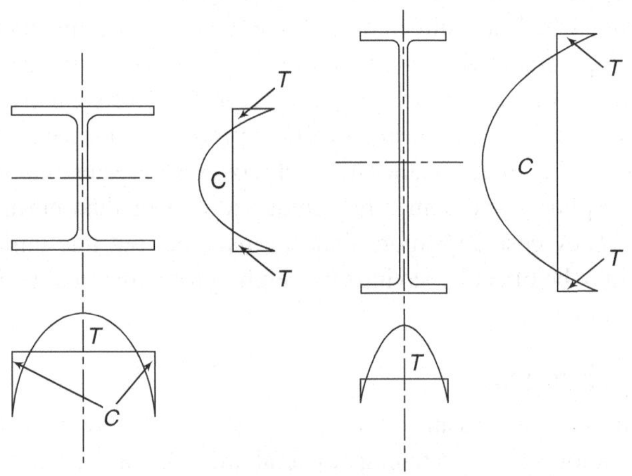


**FIG. 1.24** Residual stress distribution in standard I-sections

Because of the high concentration of heat, tensile residual stresses at the weld in welded members usually equal the yield strength of the weld metal itself; the compensating longitudinal compressive stresses are also quite high. Residual stresses in welded shapes are determined by the section geometry and the method of preparation of the components. Thus, a welded I-section fabricated from rolled plates has a different residual stress distribution from that of an I-section welded from plates flame-cut to width.

The typical distribution of residual stresses in welded sections made of plates with rolled edges is shown in Fig. 1.25. Large residual tensions develop at the corners of a welded box (Fig. 1.26). On the other hand, residual stresses in a hot-rolled square box are very low.

In steel sections which are made of quenched and tempered steels, the residual stresses are small. This is because such steels are partially stress-relieved. Thermal residual stresses

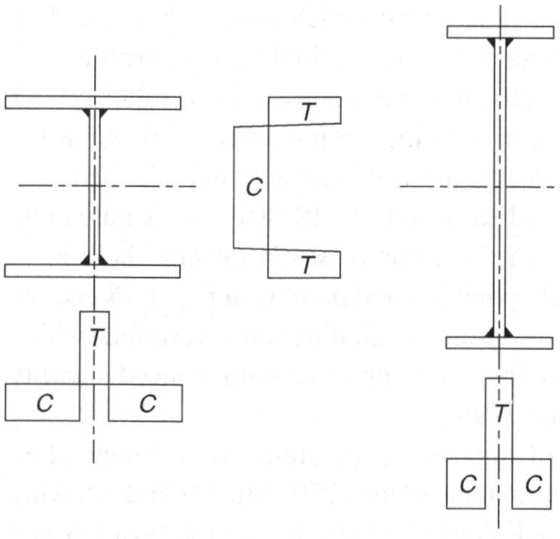


**FIG. 1.25** Residual stress distribution in welded I-beams

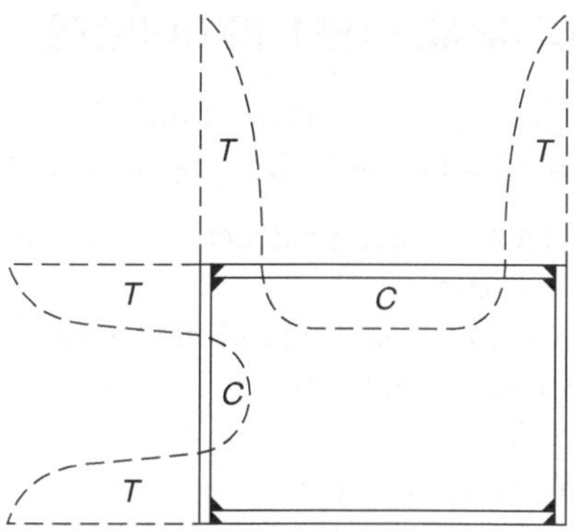


**FIG. 1.26** Residual stresses in a welded box section

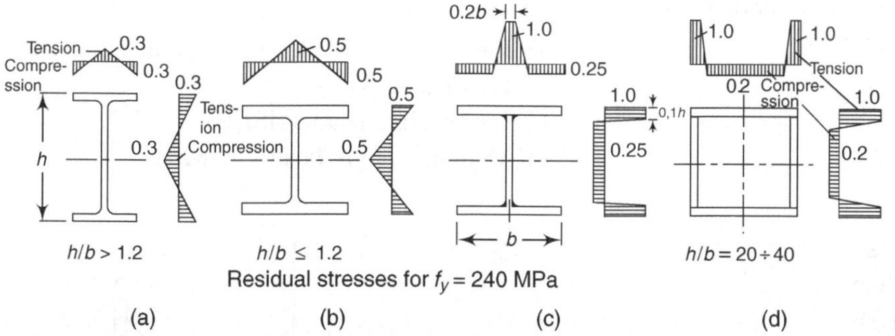


**FIG. 1.27** Idealized residual stress distribution as per Eurocode

extend almost the full length of the member. Figure 1.27 shows the idealized residual stress distributions in a rolled and welded cross sections as per Eurocode 3.

Fabricating operations such as cambering and straightening by cold bending also induce residual stresses. These residual stresses have the same magnitudes as those of thermal residual stresses but differ in distribution. They are superimposed on the thermal residual stresses. If a member is straightened by a continuous straightening process (rotorizing), the residual stress pattern is changed throughout the length of the member. On the other hand, if it is straightened at discrete points by gagging, thermal residual stresses are not altered over much of the length (Gaylord et al. 1992).

Because residual stresses must themselves be in equilibrium, their effect on structural behaviour is limited. Although such stresses do not affect the yield strength of the member, they do lower the proportional limit and increase the strain at the initiation of overall yielding. The most important consequence of this in a statically loaded structure is to cause the member to behave as if it has a non-uniform distribution of yield stress over its cross section. This factor is important in the design of compression members (columns), since under compressive stress, the regions that contain residual compressive stress yield earlier (at loads which produce an applied stress less than $f_y$). Similarly, members in bending (beams) also yield early and hence tend to deflect more (Tall 1974). Residual stresses are also to be considered in the design of members subject to fluctuating loads (fatigue), since they reduce the member's resistance to the growth of stable cracks. Such stresses also reduce the member's resistance to the growth of cracks in an unstable manner due to brittle fracture (see also Section 1.8.5).

## 1.10 STRESS CONCENTRATION

Steel structures often have connected elements which may have abrupt change in geometry and may contain holes for bolts. These features produce *stress concentrations*, which are localized stresses, greater than the average stress in the member (tensile stresses adjacent to a hole are often three times the average tensile stress). According to the theory of elasticity, the distribution of stress on the net section of an infinitely wide plate containing a hole at its centre line is given by (Timoshenko and Goodier 1970, and Roark and Young 1975)

$$f = f_1\left[1 + 0.5\left(\frac{r}{x}\right)^2 + 1.5\left(\frac{r}{x}\right)^4\right] \tag{1.14}$$

where $f_1$ is the stress that would exist in a member without a hole, $r$ is the radius of the hole, and $x$ is the distance from the centre of the hole to any point on the transverse section.

Since the stresses will be highly localized, the above equation can be applied with good accuracy to plates of finite width. Note that Eqn (1.14) is valid only if the stress at the edges of the holes does not exceed the proportional limit. After the stress reaches the proportional limit, the stress distribution will depend on the nature of the stress–strain curve of the steel (Gaylord et al. 1992).

The stress concentration at the holes is usually neglected in structural design and it is assumed that the stress is uniformly distributed over the net area of cross section. Since structural steel is sufficiently ductile to equalize the stress over the area, this assumption is justified in most cases. However, if the average stress in a member is high, the stress concentration effect should not be ignored. Stress concentration effects have been found to be critical in the webs of plate girders. Stress concentrations are also associated with fatigue and can also affect brittle fracture (see Section 1.8.5).

## 1.11 COLD WORK AND STRAIN HARDENING

Suppose a tensile specimen is loaded beyond the yield point, such as point *A*, as shown in Fig. 1.28. When it is unloaded it does not return to point *O*, but to point *B*. This means that a permanent set *OB* has occurred, and that the ductility capacity has been reduced from *OF* to *BF*. Upon being reloaded, the specimen behaved as if the stress–strain origin has shifted to point *B*. Thus the plastic zone prior to strain hardening has been reduced.

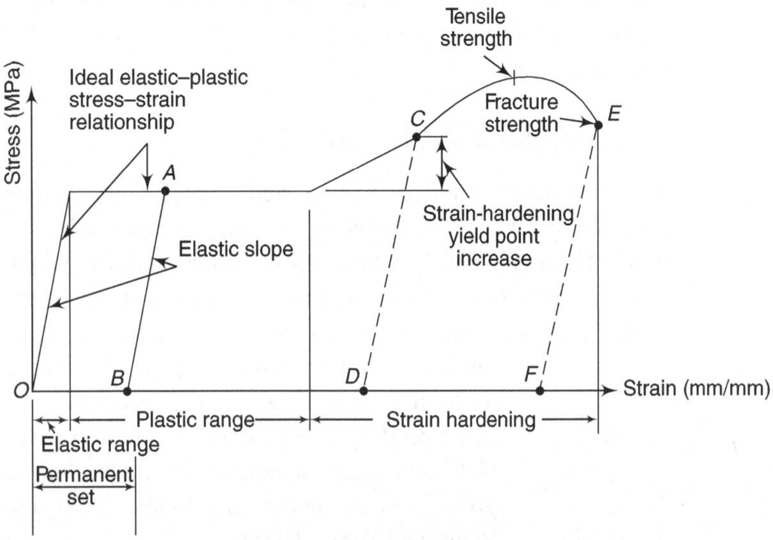


**FIG. 1.28** Effect of cold work and strain hardening

If the specimen is loaded up to point *C* in the stress–strain curve, the unloading follows the dashed line *CD*. In this case, the origin is shifted to point *D*; but the yield point has increased. This increase in yield point is known as *strain-hardening effect*. Now the available ductility is much reduced, from *OF* in the original state to only *DF* as shown in Fig. 1.28. This kind of loading beyond the elastic range to cause reduction in ductility, when done in atmospheric temperature, is referred to as *cold work*. (It may be noted that in real structures the state of stress will not be in uniaxial tension or compression, and hence the cold work effect is more complex.)

Inelastic deformation occurs at the bends of plates when they are made by *cold forming* [cold forming is a process by which sheets (hot rolled or cold rolled) are folded into the desired section profile by a series of forming rolls in a continuous train of roller sets or brakes]. Cold working into the strain-hardening range at the bend locations increases the yield strength and this increased strength is permitted in design by IS 801. Upon unloading and after a period of time, due to *strain ageing*, there is an increase in the yield point beyond point *C* in Fig. 1.28 and the original shape of the stress–strain diagram is restored (with a plastic zone and strain–hardening zone) with reduced ductility (Salmon & Johnson 1996).

The process of *annealing* [heating to a temperature beyond the transformation range (750–780 °C) and allowing the material to cool slowly] results in recrystallization and elimination of the effects of cold work.

## 1.12 STRUCTURAL STEEL PRODUCTS

Structural steel products of interest to designers can be divided into the following categories.

1. Flat hot-rolled products—plates, flat bars, sheets, and strips
2. Hot-rolled sections—rolled shapes and hollow structural sections
3. Bolts
4. Welding electrodes
5. Cold-rolled shapes

Hot-rolled sections and products consist of the following (see Fig. 1.29).

1. Rolled beams
   (a) Junior beams (ISJB, meaning Indian Standard Junior Beams)
   (b) Lightweight beams (ISLB)
   (c) Medium-weight beams (ISMB)

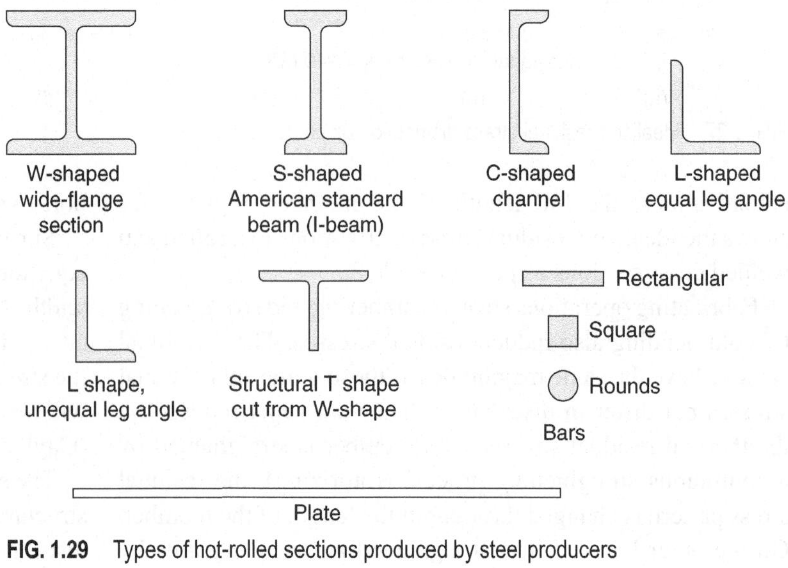


**FIG. 1.29** Types of hot-rolled sections produced by steel producers

   (d) Wide-flange beams (ISWB)
   (e) Heavyweight beams/columns (ISHB)
   (f) Column sections (ISSC)
2. Channels: Junior, light, and medium and parallel flange (ISJC, ISLC, ISMC, ISMCP)
3. Equal angles (ISEA or ISA)
4. Unequal angles (ISA)
5. T sections (ISJT, ISLT, ISST, ISNT and ISHT)
6. Rolled bars
   (a) Round (ISRO)
   (b) Square (ISSQ)
7. Tubular sections (ISLT, ISMT, ISHT)
8. Plates (ISPL)
9. Strips (ISST)
10. Flats (ISFl)

The standard rolled steel shapes and the nomenclature used in relation to them are mentioned in Table 1.12. Such nomenclature should be properly understood since it will be followed in the subsequent chapters.

Steel tubes are designated in terms of their nominal bore in millimetres and self-weight. Rolled-steel circular or square rods are designated, respectively, in terms of diameter or side (e.g., ISRO 10 mm or ISSQ 10 mm).

IS Handbook No. 1 and IS 808, published by the Bureau of Indian Standards, provide the dimensions, weights, and geometrical properties of steel beam, column, channel and angle sections (see also Appendix A). Since these publications do not contain the plastic modulus and shape factor for steel I-beams, they are provided in Annex H of IS 800 : 2007.

### 1.12.1 Choice of Section

The design of steel sections is governed by the cross-sectional area, *section modulus*, and *radius of gyration*. Though IS 808 and IS Handbook No.1 list the properties of various sections, due to the limitations of rolling mills only a few sections are available in the market. Therefore, design is governed by not only sectional properties but also the availability of the section. Another factor governing choice is the ease with which sections can be connected. In India ISMB beams are the most commonly produced. So are limited number of ISHB sections. Also, only medium channels are available. Only a limited number of unequal angles are available in the market. Also, not all the equal-angle sections are available readily in the market. Hence it will be a good idea to get a list of the available sections from steel producers like SAIL and plan the design accordingly. Appendix A gives the sectional properties of some of the sections manufactured in India.

The Indian Standard beam, commonly called the I-beam, may not be frequently used in future (because of excessive material in the web and relative lack of lateral stiffness due to the narrow flanges) and will be replaced by the parallel wide-flange beam, which is the most commonly used section for columns and beams in several countries. The channel and angle are commonly used either alone or with other sections. Channels are used mainly as

**TABLE 1.12** Rolled-steel structural shapes

| Rolled section | Indian standard designation | Remarks |
|---|---|---|
| 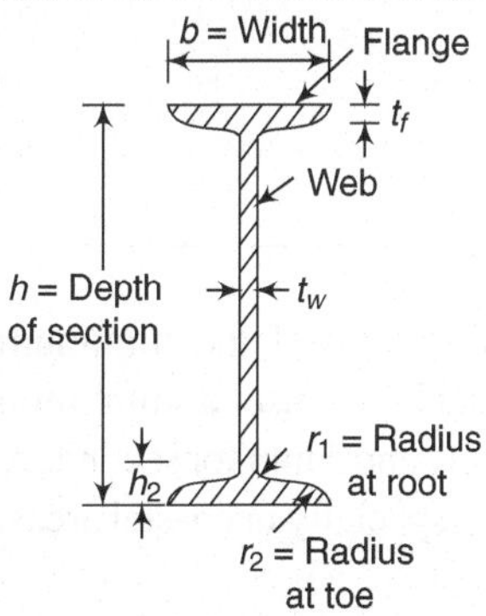 | ISJB<br>ISLB<br>ISMB<br>ISWB<br>ISHB | A beam section referred to as ISMB 400 at 0.616 kN/m is an ISMB with a depth of 400 mm and a weight of 0.616 kN per metre length. |
| 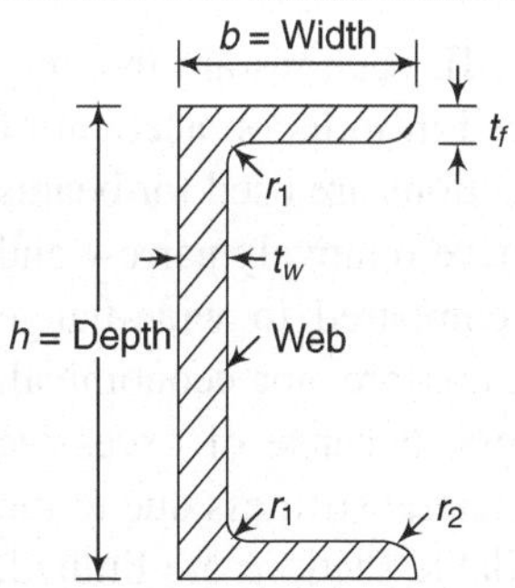 | ISJC<br>ISLC<br>ISMC | Channel sections are referred to, for example, as ISMC 200 at 0.221 kN/m. |

*(Contd)*

**TABLE 1.12** *(Contd)*

| Rolled section | Indian standard designation | Remarks |
|---|---|---|
| 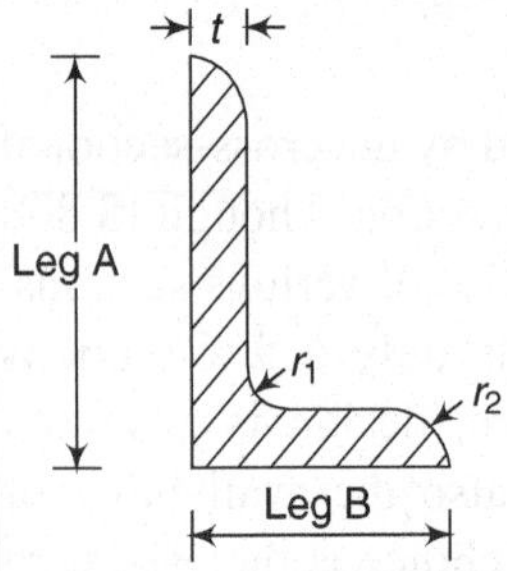 | ISA | Angles are equal or unequal. For equal angles, $A = B$. For unequal angles $A > B$. Angles are referred to, for example, as ISA 60 × 60 × 6, indicating equal angles with legs 60 mm each and thickness 6 mm. An example of an unequal angle is ISA 100 × 75 × 6. |
| 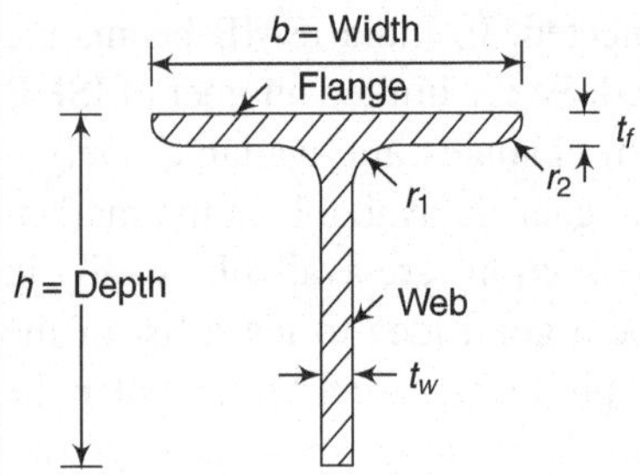 | ISNT<br>ISHT<br>ISST<br>ISLT<br>ISJT | Tee sections are referred to, for example, as ISNT 100 at 0.147 kN/m, indicating that the depth of the section is 100 mm. |
| Plates | $t \geq 5$ mm | Plates are referred to in terms of width × thickness, e.g., 900 × 10 indicates a plate 900 mm wide and 10 mm thick. |
| Strips | $t < 5$ mm | Strips are referred to in terms of width × thickness. |
| Flats | $t \geq 5$ mm<br>$b \leq 2250$ mm | Flats are referred to in terms of $b \times t$. |
| 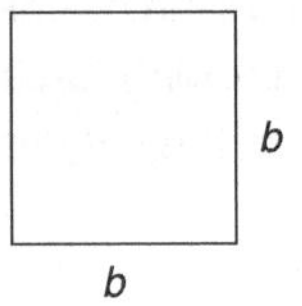 | | A square bar is referred to in terms of its sides, e.g., a 20 mm square bar. |
| 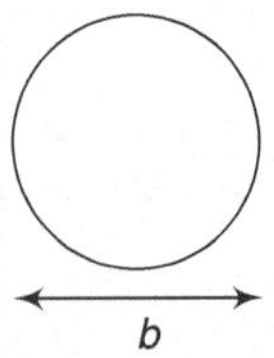 | | A round bar is referred to in terms of its diameter, e.g., a 20 mm diameter bar. |

purlins in industrial buildings and angles in trusses and towers. The structural tee is commonly used for chord members in trusses, though in India double angles are also used for the purpose.

It has to be noted that the rollers used to produce the structural shapes undergo wear, and coupled with the factors involved in the rolling process, only shapes of nominal dimensions (varying from theoretical or design values) can be produced. IS 1852 specifies allowable rolling tolerances, including amount of flange and web warping, and the deviation of web depth permitted for the section to be satisfactory. Designers should be careful about these tolerances, especially while using smaller sections which may be produced by small rerollers. The designed cross-sectional properties may not match with the actual cross section properties. Also, as mentioned previously, the rerolled sections may tend to have higher strength at the risk of reduced ductility.

Though IS 800 : 2007 code has removed the minimum thickness requirements, it is advisable to use a minimum thickness of 6 mm for the main members and 5 mm for secondary members exposed to the atmosphere, especially in coastal areas.

### 1.12.2 Wide-flange Sections

As mentioned in Section 1.12.1, ISMB sections are the only I-sections that are normally produced in India on account of the calibre rolling method. These sections are used for beams as well as columns. Such sections have relatively narrow and sloping flanges and a thick web compared to wide-flange sections (see Fig. 1.30). ISMB beams are not economical, especially for compression members, because of excessive material in the web and the lack of lateral stiffness due to the narrow flanges. Also, since the available sections are limited, when a section is slightly inadequate, the choice is limited to

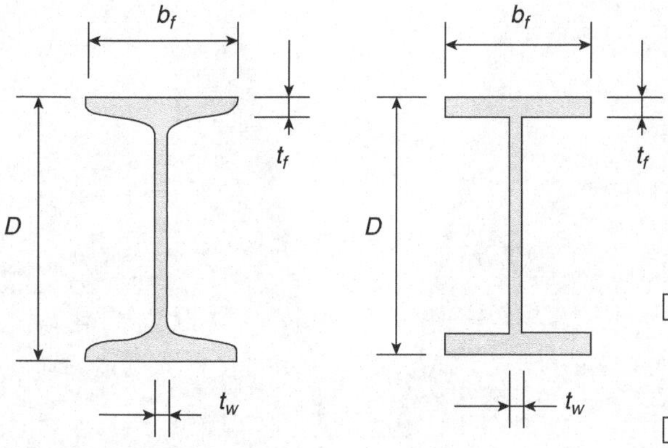


**FIG. 1.30** Standard and wide-flange sections

either the next available section (which may be 25–45% heavier in weight) or built-up sections through welding, using which involves time and cost.

The main features of wide-flange beams called W-beams in USA which make them more popular than Indian standard I-beams called S-beams in USA are the followings.

1. Wide-flange beams provide excellent sectional performance, with high bending and buckling resistance due to the H-shaped arrangement of flanges and the web.
2. The use of such beams reduces fabrication difficulties—since there is no taper in the H-beam flange, no tapered washer is necessary for bolting, and the gussets can be welded to the inner surface of the beam flange (See Fig.1.31). Unlike tapered-flange beams, H-beams can be readily butt welded, and a sound welding is assured.

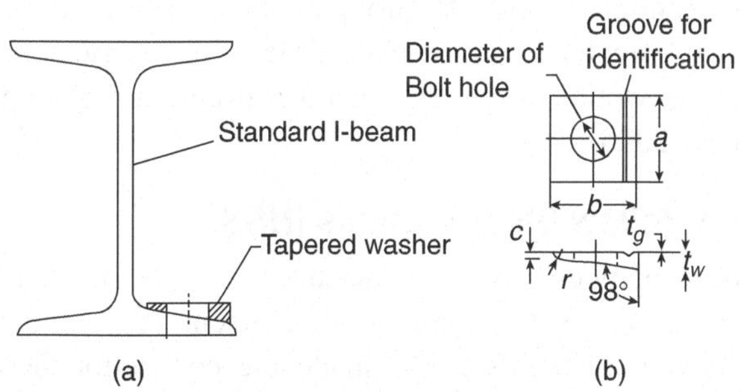


**FIG. 1.31** Tapered washer for standard I-beam (IS 5374:1975)

3. Since H-beams have a higher section modulus for the same weight, using them is economical. (A saving of the order of 10–24% can be achieved.)

Using a new manufacturing technology, it is now possible to have beams with the same depth but with different flange and web thickness, and also flange width. This facilitates simple design and improves fabrication efficiency. These types of beams are compared with conventional rolled wide-flange beams in Fig. 1.32, which are manufactured with 'specified depth' between flanges. This method also incorporates an advanced temperature-control technology in the rolling process to produce webs thinner than those of conventional I-beams. These thinner webs open the way to more economical design, since more material is needed only in the flanges (Subramanian 1982). These wide parallel-flange beams and columns are manufactured in India by M/s Jindal Steel and Power Limited (JSPL) at Raigarh, Chhattisgarh. The sectional properties of some of these beams are given in Appendix A. (Other sectional properties may be obtained from the manufacturers.)

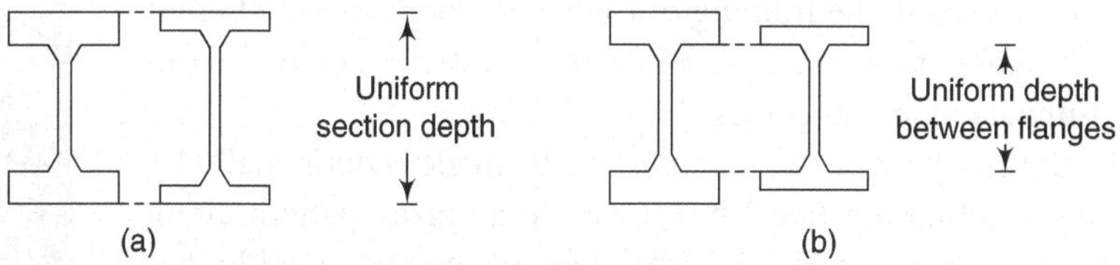


**FIG. 1.32** Wide-flange beams with specified depth between flanges (a) The Nittetsuhyper beam, and (b) Conventional rolled wide-flange beams

### 1.12.3 Welded and Hybrid Sections

Hot-rolled plates or flame-cut plates can be welded together to form I-sections or box girders as shown in Figs 1.25 and 1.26 (Such built-up sections can also be made by using riveting or bolting.) *Welded I-beams* with top and bottom plates and welded stiffeners are often used as plate girders. Welding makes it possible to combine any structural shape to get the desired properties. Figure 1.33 shows the different possibilities

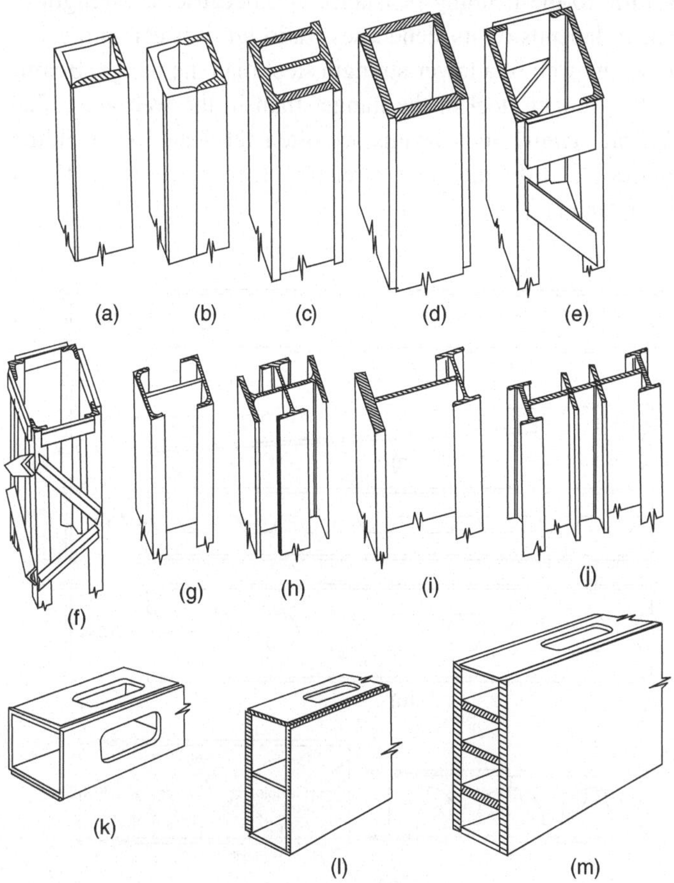


**FIG. 1.33** Built-up welded compression members

*Source:* Blodgett (1966). Reprinted with permission from The James F. Lincoln Arc Welding Foundation

that exist to develop built-up compression members using welding.

*Tapered girders* are fabricated either by welding two flange plates to a tapered web plate or by cutting a rolled I-beam lengthwise along its web at an angle, turning one half end for end, and then welding the two halves back together again along the web as shown in Fig. 1.34(a). Tapered girders are widely used in the framing of roofs over large areas, where it is desirable to minimize the number of interior columns or to eliminate them altogether.

Similarly, *castellated beams* can be made economically by flame-cutting a rolled I-beam web in a zigzag pattern along its centreline [see Fig. 1.34(b)]. One of the two equal halves is turned end for end and is welded to the other half. The result is a deeper beam, stronger and stiffer than the original. Castellated beams with circular holes can also be fabricated as shown in Fig. 1.34(c). Castellated beams have more section modulus and moment of inertia and result in greater economy.

Figure 1.35 shows a structure with castellated beams, that led to significant savings in construction costs. Properties of castellated beams made of ISMB beams and channels are given in Appendix A.

Since the web of a beam or a plate girder contributes only a little to the bending resistance and because its strength in shear depends on its slenderness ratio $h/t$, it is economical to have the web of a lower-strength steel than the flange. Beams with stronger steel in the flanges than in the web are called *hybrid beams*. Such beams are often fabricated by welding plates of different steel strengths (see Section 11.10 of Chapter 11)

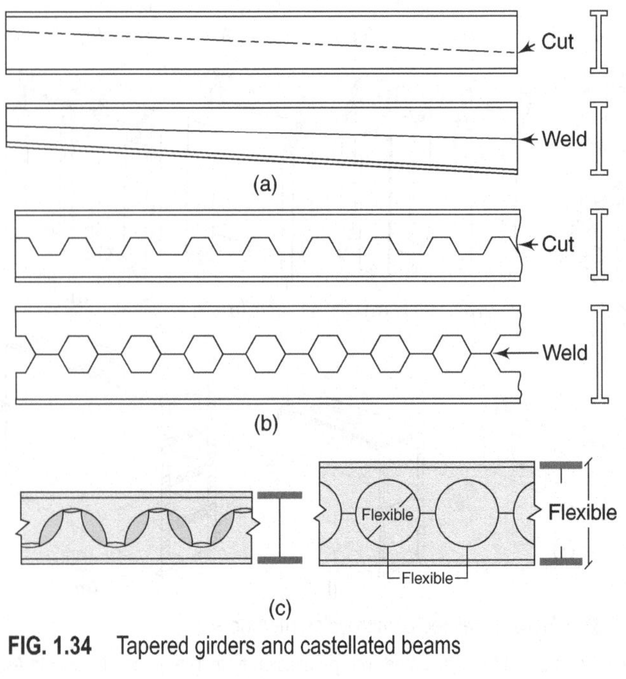


**FIG. 1.34** Tapered girders and castellated beams

**FIG. 1.35** Cellular beams over Rangers FC, an indoor training facility in Scotland
*Courtesy:* ASD Westok Ltd, UK

With the development of high-performance steels (see Section 1.14), it is now considered advantageous to replace the flat webs of conventional built-up I-beams with corrugated webs. These corrugated plates could be produced by cold-forming long, flat plates. The corrugations can have a trapezoidal shape or a sine-wave cross section (see Section 11.13 of Chapter 11). A Japanese company has developed welding equipment that automatically senses and adjusts the equipment position as it welds a corrugated plate to a flange plate. Corrugated webs allow deeper girders that are not susceptible to web stability problems, resulting in thinner webs and smaller flanges. A plate girder with a corrugated web, therefore, would weigh less than a conventional plate girder (Kulicki 2000).

## 1.12.4 Hollow Steel Sections (HSS)

Tubular members are being used extensively in plane and space trusses as tubes are more efficient in compression. The advent of welding has made the connection between tubular members using gusset plates possible and resulted in the widespread use of tubular sections (see Section 6.16 of Chapter 6). Welded connections without gusset plates require proper planning for profiling and welding the ends, and have to be executed carefully. A tube is an efficient section, having the same radius of gyration in all directions. Care should, however, be taken to see that the $d/t$ (diameter to thickness) ratio of the section is small enough to prevent failure by local buckling before the failure of the whole member. Recently, square and rectangular tubes have been introduced in India and these, of course, are much easier to connect because of their flat surfaces. With square or rectangular hollow sections, the smaller tube can be simply sawed with a single cut at the

## FLOW DRILL AND HOLLO BOLT SYSTEMS

The use of standard bolt is often impossible in steel hollow sections, as there is no access to the inside of the tube to allow for tightening. The use of gusset plates and brackets overcome this problem but are not aesthetically pleasing. Several techniques have been developed which permit bolt installation and tightening from one side of the connection only; they are called *blind bolts*. Commercially available examples include *Flowdrill* and *Hollo Bolt*.

*Flowdrilling* is basically a thermal drilling process which makes a hole through the wall of a SHS without removal of metal normally associated with a drilling process. The formed hole is then threaded by the use of a roll thread forming tool, leaving a threaded hole which will accept a standard fully threaded bolt. In flow drill system, the tungsten carbide bit of the Flowdrill tool is brought into contact with the RHS wall, where it generates sufficient heat to soften the steel. The bit is then advanced through the wall and when it is done, the metal 'flows' to form an internal bush. The tool is also fitted with means of removing any surplus material which may arise on the outside of RHS section. In the next stage, the flowdrill bush so made is threaded with a coldform flowtap and then the connection can be made with bolts. The complete cycle is shown in the figure below. Note that Flowdrill method can not be used with pre-galvanized sections.

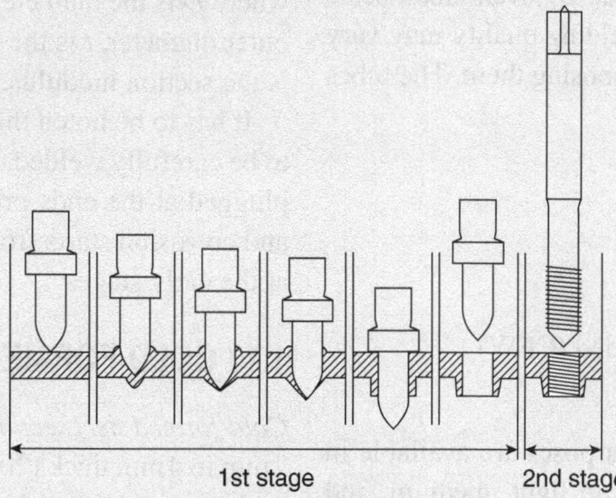


Flowdrill operations

In the hollo-bolt system, a pre-assembled unit is inserted through normal tolerance holes in both the RHS and attachment plate. As the bolt is tightened the cone is drawn into the body, spreading the legs, and forming secure fixing. After installation, the hollo-bolt head and collar only are visible as shown below.

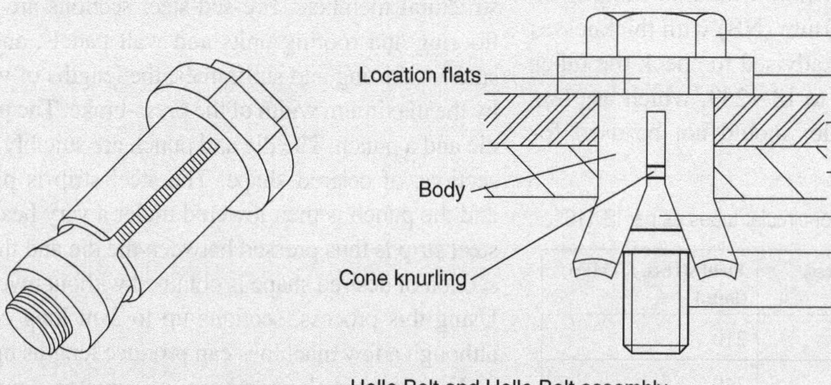


Hollo Bolt and Hollo Bolt assembly

required angle and welded to the bigger tube. Field erections of such rolled hollow sections (RHS) are simplified by the use of Saxe clips (Blodgett 1966).

The flow drill and hollo bolt provide a choice of methods to produce bolted joints with hollow sections, which offer the following benefits.

1. They produce bolted joints of structural capacity in RHS.
2. They minimize the change in the fabrication process by using connection details that are standard in the construc tion industry.
3. They reduce fabrication by removing the need to weld plates or other fittings onto the outside surface of the RHS.
4. They maintain aesthetics by producing a flush face on the RHS after fabrication.

In India, till recently only two grades of RHS, i.e., $f_y = 210$ MPa ($f_u = 330$ MPa) and $f_y = 240$ MPa ($f_u = 450$ MPa) were produced. Now, HSS as per the specifications given in IS 4923 with $f_y$ = 310 MPa ($f_u$ = 450 MPa) are also available. The properties of hollow sections are given in Appendix A. Rectangular hollow sections (RHS) are available in sizes 50 × 25 × 2 mm to 300 × 200 × 10 mm and square hollow sections (SHS) in sizes from 25 × 25 × 2 mm to 250 × 250 × 10 mm.

Basically there are two major classifications of hollow steel sections—welded and seamless tubes. Hot-rolled or cold-rolled steel sheets are rolled into tubes and welded along the length, resulting in a longitudinal welded joint. Since they are easy to manufacture, they are made by even small-scale industries. Hence the thickness and welding quality may vary and the designer has to be careful in choosing them. The tubes may also be classified as follows:

1. Hot-finished welded (HFW)
2. Hot-finished seamless (HFS)
3. Cold-drawn seamless (CDS)
4. Electric-resistance welded (ERW)
5. Cold-drawn electric-resistance welded (CEW)
6. Oxy-acetylene welded (OAW)

As per IS 1161, tubes for structural purposes are available in three grades of steel (see Table 1.13) in light, medium, and heavy categories. The yield stress of Yst 240 is the same as that of mild steel and this is often the grade (medium-section) available in the market.

YSt 310 Grade tubes are also now available ranging in diameter from 15 mm (NB) to 300 mm (NB) with thicknesses of 2 mm to 10 mm. Engineers are advised to check the tubes used at site, as tubes confirming to IS 1239, which are less expensive and are easily available, should not be used for structural purposes.

**TABLE 1.13** Strength of steel used for circular tubes as per IS 1161

| Grade | Ultimate tensile stress, $f_u$ (MPa) (min.) | Yield stress, $f_y$ (MPa) (min.) |
|---|---|---|
| Yst 210 | 330 | 210 |
| Yst 240 | 410 | 240 |
| Yst 310 | 450 | 310 |

Tubes are often manufactured in lengths ranging from 4 m to 7 m and should have the following tolerances.

Outside diameter: Up to 40 mm, nominal bore + 0.5 mm to −1.0 mm
Over 40 mm, nominal bore + 1.0 mm to −1.0 mm

Thickness: ±10%

The sectional properties of tubes and hollow sections are given in Appendix A. The properties of the tubes may be approximated by the following relations.

$$A = \pi D t \quad (1.15a)$$

$$r = 0.3535D \quad (1.15b)$$

$$Z = \frac{\pi D^2 t}{4} \quad (1.15c)$$

$$I = \frac{\pi D^3 t}{8} \quad (1.15d)$$

where $D$ is the diameter at mid-thickness = $(D_O - t)$, $D_O$ is the outer diameter, $t$ is the thickness, $r$ is the radius of gyration, $Z$ is the section modulus, and $I$ is the moment of inertia.

It has to be noted that tubes having less than 3.25 mm have to be carefully welded. If the tubes and hollow sections are not plugged at the ends properly, they retain the moisture inside and corrosion starts from the inside, which will not be noticed at the early stages.

## 1.13 COLD-FORMED STEEL SECTIONS

*Cold-formed steel sections* are made from light-gauge steel strips, 2 mm to 4 mm thick (20 B.G.–8 B.G.) and occasionally as thick as 5 mm, cold formed to shape in a rolling mill or press-brake, and are also known as cold-rolled and pressed-steel sections.

Considerable development has taken place in the production of floor and roof units and metal trim, and also in the field of structural members. Pressed-steel sections are largely used for flooring and roofing units and wall panels, and for metal trim such as skirting and subframes, the lengths of which are limited by the maximum width of the press-brake. The press consists of a die and a punch. The die and punch are suitably shaped to obtain sections of desired shape. The steel strip is placed on the die and the punch is then lowered under a very heavy pressure. The steel strip is thus pressed between the die and the punch, and the section of desired shape is obtained without involving any shock. Using this process, sections up to 3 m long can be produced; although a few machines can produce lengths up to nearly 6 m.

For structural members of greater length, cold-rolled sections are used. These are formed into the required shape by passing a metal strip between six to fifteen progressive sets of forming spindles or rollers, each pair of which adds successively to the shaping of the strip, the final pair producing the final section. As per IS 801, cold-formed steel sections are available in four grades of steel (see Table 1.14); the grade St 42 is similar to mild steel and is the only grade often available in the market. The basic sections rolled are plain angles and channels, lipped angles and channels, and zeds (see Fig. 1.36). Outwardly lipped channels are commonly called hat sections. The length is limited only by considerations of transport.

**TABLE 1.14** Strength of steel used for cold-formed steel sections

| Grade | Ultimate tensile stress, $f_u$ (MPa) (min.) | Yield stress, $f_y$ (MPa) (min.) |
|---|---|---|
| St 34 | 340 | 210 |
| St 42 | 420 | 240 |
| St 50 | 500 | 300 |
| St 52 | 520 | 360 |

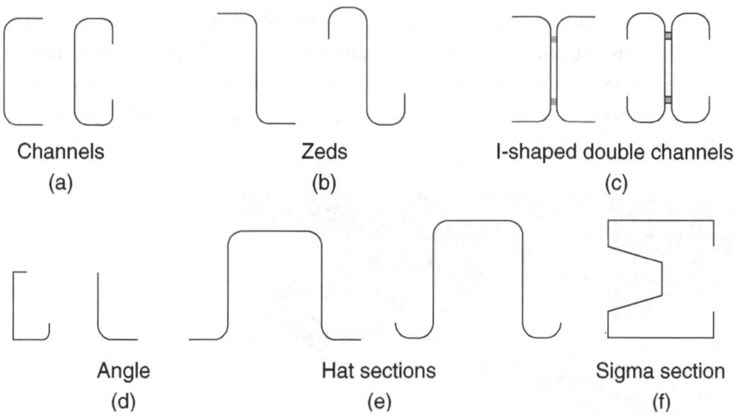


**FIG. 1.36** Various cold-formed steel sections

There is virtually no limit to the shape that can be rolled—the designer can choose a shape best suited to any particular purpose. Properties of Z- and C-sections, which are often used as purlins, are given in Appendix A.

Cold-rolled sections are most efficiently used with structures of moderate loads and span, in which circumstances they can be cheaper than hot-rolled members. Erection of the structure in such cases is often cheaper and easier because of its light weight and rigidity. Cold-rolled sections are used in India for purlins, girts, wind bracings, and roof trusses. In countries like the USA, they have been used in columns (with channels placed back to back), I-section beams formed by channels (for light loads over short spans), lattice beams, and rigid frames.

Connections are made by various types of welds, self-tapping screws, bolts, cold rivets, and hot rivets, and sections can be formed to push-fit into each other, thus avoiding the use of gusset plates. For example, top–hat flange sections can be used for beams into which the bracing members fit, so that the node connection is direct.

Structural considerations are similar to those in the design of light-alloy structures. The design of thin-wall structures requires special consideration, due to the possibility of local instability, and, as in aluminium sections, lips to the edges are provided to give increased stiffness to the section (Yu 2000). The need for protection from corrosion is important, because of the following factors.

1. Such structures are made up of very thin plates, of less than 6 mm thickness (mostly 2.0–4.0 mm thick)
2. It is difficult to reach inside the faces for painting.
3. The re-entrant corners may hold dust and moisture.

The structural steel or strip steel should conform to IS 1079 and the design is done as per IS 801. It should be noted that this code is still in the working-stress-method format and a draft code based on the limit states method is under consideration by BIS. Hence the provision of this code will not be discussed in this book.

## 1.14 HIGH-PERFORMANCE STEEL

A new grade of steel, called *high-performance steel*, HPS-485W was developed by a cooperative program between the Federal Highway Administration, the American Iron and Steel Institute, and the Department of the Navy in August 1994. Three grades of HPS are available: HPS 345W, HPS 485W, and HPS 690W, out of which HPS 485W (yield strength = 485 MPa) is widely used in the USA. The benefits related to HPS include enhancements in weldability, toughness, corrosion resistance, ductility, fatigue and fire resistance, formability, and strength. These factors combined led to construction elements of higher economic efficiency, ease of maintenance, and longer service life.

Although this new steel has the same strength levels as the currently available AASHTO N270 Grade 485W steel (yield strength = 485 MPa), its unique chemical and physical properties allow the use of more economical fabrication practices. The main differences between M270 steel and HPS are that the latter contains almost half as much carbon as the former, and one-tenth as much sulphur. Toughness values of over 200 J are commonly obtained, as compared to values of 30 to 50 joules for the current steel.

Because of the low carbon levels, minimum or no preheat would be required, which allows increased productivity of fabrication and reduced cost. HPS steel bridges can also be recycled to the extent of virtually 100%, which offers a great advantage when it comes to the environment.

### 1.14.1 Hydrogen-induced Cracking

A main thrust of the HPS Research Program was to develop bridge steels with significantly improved weldability. Improving weldability reduces the high cost of fabrication associated high preheat temperatures, heat input control, post-weld treatment, and other stringent controls, and to eliminate hydrogen-induced cracking in the weldment.

*Hydrogen-induced cracking*, also known as delayed cracking or cold cracking, has been one of the most common and serious problems encountered in steel weldments in bridges.

The common source of hydrogen is moisture. Grease, oxides, and other contaminants are also potential sources of hydrogen. Hydrogen from these sources can be introduced into the weld region through the welding electrode, shielding materials, base metal surface, and the atmosphere.

Hydrogen-induced cracking can occur in the weld *heat-affected zone* (HAZ) and in the *fusion zone* (FZ). While the

## CASE STUDY

### High-performance steel bridge in Tennessee, USA

The bridge in Tennessee is a two-span continuous structure located on state Route 53 over Martin Creek in Jackson County. The structure provides 8.5 m wide roadway over two spans, both 72 meters long. This bridge consists of three continuous welded plate girders, fabricated from HPS-485W, 2 m deep and spaced at 3.2 m centers. These plate girders act compositely with a cast-in-place concrete deck slab of thickness 212.5 mm. The bridge is jointless, having integral, pile supported abutments. The design was fully optimized for the 485-MPa steel, using the AASHTO Load and Resistance Factor (LRFD) Bridge Design Specifications. In the positive moment region, the top flanges are 450 mm wide and vary from 25 to 50 mm thickness in four increments. In the negative moment region, the top flanges vary in four increments of 28 mm by 450 mm to 62.5 mm by 750 mm. The webs vary from 11 mm by 1800 mm in the positive moment region to 12.5 mm by 1800 mm over the negative moment region.

Cost estimates prepared by the Tennessee Department of Transportation indicate that the steel weight was reduced by almost 25 percent compared to the original grade 345W design. Because HPS-485W costs slightly more than grade 345W steel, this resulted in a 16 percent reduction in the total cost to fabricate and erect steel for this bridge.

*Source*: www.fhwa.dot.gov

reasons for cracking are the same, controlling the factors that cause cracking can be different for the HAZ and FZ. For the HAZ, cracking is controlled in the modern steel-making processes themselves, which incorporate means to avoid susceptible microstructures and eliminate sources of hydrogen in the base metal (steel), and by using proper welding techniques, including preheat and heat input. For the FZ, control of susceptibility to hydrogen-induced cracking is achieved by adding alloying elements in the consumables, and using proper welding techniques, including preheat and heat input.

The most common and effective method of eliminating hydrogen-induced cracking is specifying a minimum preheat and interpass temperature for welding. In general, the higher the preheat, the less is the chance of formation of brittle microstructures and the greater is the time for the hydrogen to diffuse from the weld. However, preheating is time consuming and costly. One of the goals of developing high-performance steels is to reduce or eliminate preheat.

### 1.14.2 Weathering Characteristic of HPS

It was part of the initial research objective to develop HPS with 'weathering characteristic', meaning HPS should have the ability to perform without painting under normal atmospheric conditions. HPS steels have slightly better atmospheric corrosion resistance than the conventional grade 345W or 485W steels. For example, as measured in accordance with ASTM G101, the atmospheric corrosion resistance index (CI) for conventional Grade 485W is 6.0, while the index for HPS 485W is 6.5. Designers should follow the same guidelines and detailing practice as in conventional weathering grade steels to ensure successful applications of HPS steels in unpainted conditions.

HPS 485W is produced by quenching and tempering (Q&T) or thermal-mechanical controlled processing (TMCP). The chemical and mechanical properties of these steels as per the American Standards are shown in Tables 1.15 and 1.16.

Figure 1.37 shows the CVN transition curves for HPS 485W and conventional 345 W steel. The brittle ductile transition of

**TABLE 1.15** Chemical requirements of high performance steel (www.fhwa.dot.gov)

| Steel grade | | C | Mn | P | S | Si | Cu | Ni | Cr | Mo | V | AR | N |
|---|---|---|---|---|---|---|---|---|---|---|---|---|---|
| HPS | Min. | 0 | 1.1 | — | — | 0.3 | 0.25 | 0.25 | 0.45 | 0.02 | 0.04 | 0.01 | — |
| 485W and HPS 345W | Max. | 0.11 | 1.35 | 0.02 | 0.006 | 0.5 | 0.4 | 0.4 | 0.7 | 0.08 | 0.08 | 0.04 | 0.015 |

**TABLE 1.16** Mechanical properfies of HPS plates

| | HPS 345W up to 100 mm as rolled | HPS 485W up to 100 mm (Q&T), 50 mm (TMCP) |
|---|---|---|
| Yield strength (MPa) | 345 | 485 |
| Ultimate tensile strength (MPa) | 485 | 585–760 |
| CVN (min.) | 41 J at −120 °C | 48 J at −230 °C |

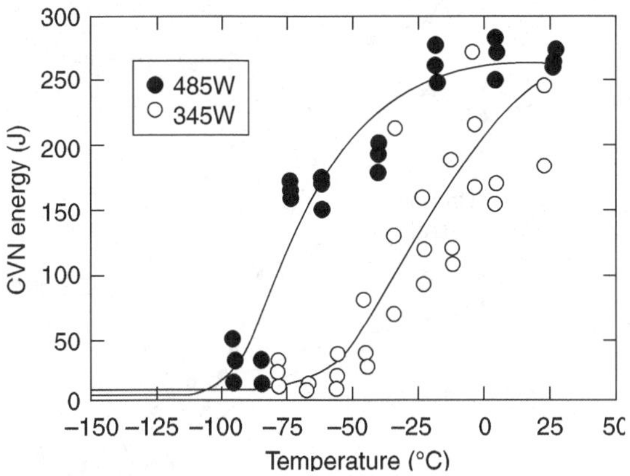


**FIG. 1.37** Charpy V-notch (CVN) transition curves for high-performance steels

HPS occurs at a much lower temperature than for conventional grade 345W steel. This means that HPS 485W remains fully ductile at a lower temperature, whereas conventional 345W steel begins to show brittle behaviour.

As the use of HPS 485W generally results in a smaller member and lighter structures, designers should pay attention to deformations, the global buckling of members, and the local buckling of components. The service limit states should be checked for deflection (limit of *L*/500), handling, shipping and construction procedures, and sequences. The AASHTO HPS guide encourages the use of hybrid girders, i.e., combining the use of HPS 485W and grade 345W steels. A hybrid combination of HPS 485W in the regular moments region, and grade 50W and HPS 345W in other areas, results in the optimum use of HPS and maximum economy.

The Nebraska Department of Transportation was the first to use HPS 485W in the design and construction of the 45-m simple span Snyder bridge, a welded plate girder (1.37-m) bridge, which opened to traffic in October 1997. The use of HPS resulted in a 24.2% reduction in steel cost and a 10.6% saving in total cost in Tennessee, USA in 1996. HPS is currently produced by Bethlehem-Lukens Plate, Oregon Steel Mills, and U.S. Steel.

Since the late 1990s more than 250 HPS steel bridges have been opened to traffic and 150 more bridges are under various stages of construction in the USA alone.

Though HPS is considered superior to normal mild steel, the following drawbacks have also been identified.

1. High yield-to-tensile ratios (less ductile behaviour, which is not favourable in earthquake zones),
2. Cost premium for base material, and
3. Currently, the AASHTO Manual for Design of Steel Bridges includes 50W limitations that prevent full utilization of the higher yield strength of HPS (Jamshidi et al. 1997).

The Building and Fire Research Laboratory, sponsored by the National Institute of Standards and Technology, USA is currently developing design methodologies and specifications for the design of HPS systems. More details about HPS may be found in Jamshidi et al. (1997), and Lwin (2002).

## 1.15 STAINLESS STEEL

Though structural steel offers many advantages, such as high strength to weight ratio, easy fabrication and erection, prefabrication, demountability and recycling, uniform properties, ductility, etc., its use in coastal areas is very limited due to fear of corrosion. Hence *stainless steel* was developed at the beginning of the twentieth century as a rust-resistant alloy steel. Since it was expensive, it was used in small quantities, for limited applications, for use in military supplies, and in the chemical industry. During the 1950s, it was possible to mass produce stainless steel at low cost due to the advancement in refining processes. This led to the greater use of stainless steel in kitchen products and other consumer durables, especially in Japan. Recent years have seen the growth of stainless steel applications in the construction industry, in roofing, interiors, and exteriors of buildings, and structural applications.

In India, stainless steel is produced in various grades (SS 301, SS 304, SS 304L, SS 310S, SS 316, SS 316L, SS 321, SS 409, SS 409M, SS 410S, SS 420, and SS 430) and forms but has found limited use in structural applications. Its use has been limited to those applications where aesthetics rather than structural performance is the sole criterion. However, though stainless steel production in India was only 20,000 tonnes in 1978, it increased to 3.4 million tonnes in 2015.

Stainless steel is essentially a low-carbon steel to which chromium has been added. It is this addition of chromium, in amounts greater than 10.5% by weight, that gives the steel its unique 'stainless' corrosion-resistant properties.

Like other steels, stainless steel has a high strength to weight ratio, is weldable and environment friendly, and its main advantages over the other steels are its aesthetically pleasing appearance, corrosion resistance, high tensile strength, high toughness, and impact and heat resistance. The most important difference from the structural designer's point of view is that it has a non-linear stress–strain relationship, similar to that of high-strength steels (Fig. 1.38). Thus, as compared to carbon steel, in the case of which there is a linear elastic behaviour up to the yield stress and a plateau before strain hardening is encountered, stainless steel has a more rounded response with no well-defined yield stress. Therefore, the yield strength of stainless steel is generally quoted in terms of a proof strength defined for a particular offset permanent strain (conventionally the 0.2% strain), as indicated in

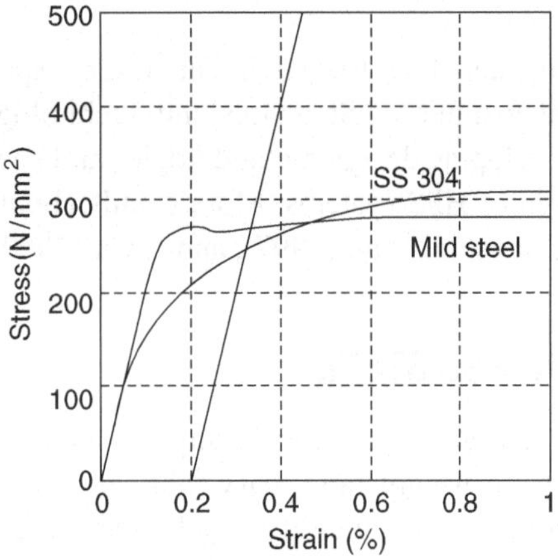


**FIG. 1.38** Stress–Strain curves of stainless steel and carbon steel

Fig. 1.38. The initial modulus is approximately the same as that for structural steel, but the proportional limit is generally quite low. For some stainless steels, the stress–strain curve may be different in tension and compression and the results may also be different in the longitudinal and transverse directions, depending upon the method by which they are obtained. Therefore, the guidelines used in designing conventional steel members are not applicable for stainless steel members. Hence several countries have developed separate guidelines for the design of stainless steel structural members (Euro Inox, 1994, ASCE/ANSI-8-90, ENV, 1993-1-4). However, Indian Code provisions for the design of stainless steel members are not available. This is one of the reasons why stainless steel members are not used in engineering structures in India.

Apart from being more resistant to corrosion, stainless steel is superior to carbon steel in such qualities as elongation and fire resistance. Figure 1.39 shows how yield strength and tensile strength depend on temperature. While the proof strength of ordinary steel begins to decline at a steel temperature of 300 °C to 500 °C, that of stainless steel has a small rate of decrease up to about 700 °C. This means that stainless steel has far superior fire resistance, making possible the construction of buildings using stainless steel without fire-proof insulation.

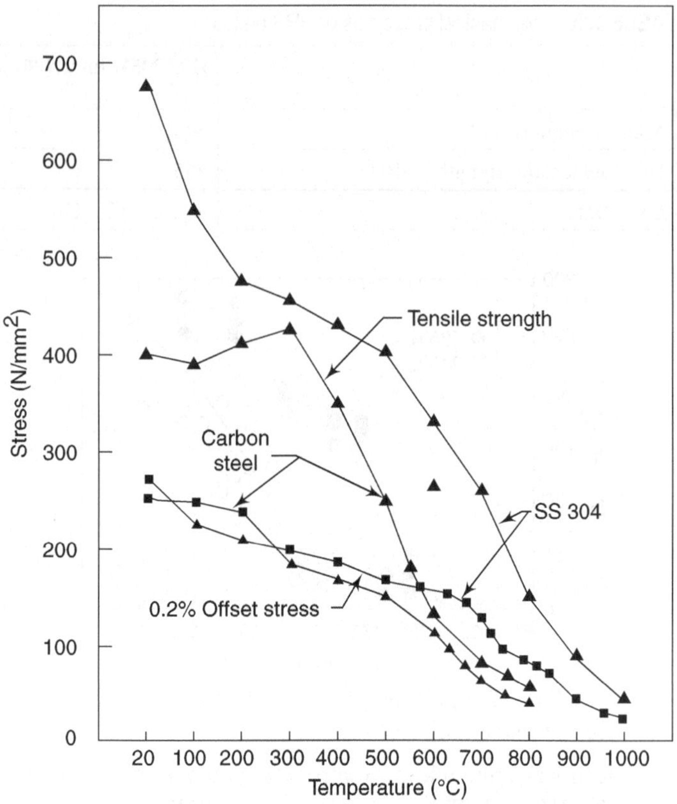


**FIG. 1.39** Stress–Temperature relation of stainless steel and carbon steel

Cloud Gate or The Bean, at Millennium Park, Chicago, Illinois, is a stainless steel 10 m × 13 m × 20 m sculpture by Indian-born British artist Anish Kapoor, which opened in 2006. The outer shell comprises 168 stainless steel panels, each 10 mm thick. It received the American Welding Society's Extraordinary Welding Award

Of the various available grades, SS 304, SS 304L, SS 306, SS 409, and SS 430 are suitable for structural applications. The chemical, mechanical, and physical properties of these grades are given in Tables 1.17, 1.18, and 1.19. (The properties for mild steel are also given for comparison.)

Most stainless steels fall into one of three main classes—martensitic, ferritic, and austenitic, the names taken from their metallurgical structures. Martensitic types are rarely used in

**TABLE 1.17** Chemical composition (in percentage) of stainless steels (Salem Steel 1998)

| Grade | C (max.) | Si (max.) | Mn (max.) | Cr | Ni | P (max.) | S (max.) |
|---|---|---|---|---|---|---|---|
| SS 316 | 0.08 | 0.75 | 2.00 | 16–18 | 10–14 | 0.045 | 0.03 |
| SS 304 | 0.08 | 0.75 | 2.00 | 18–20 | 8–10.5 | 0.045 | 0.03 |
| SS 304L | 0.03 | 0.75 | 2.00 | 18–20 | 8–12 | 0.045 | 0.03 |
| SS 409 | 0.08 | 1.00 | 1.00 | 10.50–11.75 | 0.50 (max.) | 0.045 | 0.03 |
| SS 430 | 0.12 | 1.00 | 1.00 | 16–18 | 0.75 (max.) | 0.040 | 0.03 |
| Mild steel (Fe 410) | 0.22 | 0.40 | 1.50 | — | — | 0.045 | 0.045 |

**TABLE 1.18** Mechanical properties (annealed condition) of cold-rolled stainless steel (IS 6603 and Salem Steel 1998)

| Grade | UTS (N/mm²) (min.) | 0.2% proof stress (N/mm²) (min.) | Min. percentage elongation on 50 mm GL | Hardness RH (max.) |
|---|---|---|---|---|
| SS 316 | 515 | 205 | 40 | 95 |
| SS 304 | 515 | 205 | 40 | 92 |
| SS 304 L | 485 | 170 | 40 | 92 |
| SS 409 | 380 | 205 | 20 | 88 |
| SS 430 | 450 | 205 | 22 | 89 |
| Mild steel (Fe 410) | 410 min | 250 min | 23 | 80–105 |

GL–Gauge length, RH–Rockwell hardness

**TABLE 1.19** Physical properties of CRSS (IS 6603 and Salem Steel 1998)

| Details | SS 316 | SS 304/304 L | SS 409 | SS 430 | Mild Steel |
|---|---|---|---|---|---|
| Density (kg/m³) | 77.5 | 77.5 | 75.5 | 75.5 | 78.5 |
| Modulus of elasticity (N/mm²) | 193,000 | 193,000 | 200,000 | 200,000 | 210,000 |
| Coefficient of thermal expansion (°C × $10^{-6}$, 0–500 °C) | 19.8 | 18.4 | 11.52 | 11.34 | 12.0 |
| Melting range (°C) | 1400–1420 | 1400–1455 | 1430–1510 | 1430–1510 | 1350–1450 |

buildings, only when high strength or hardness is required. Ferritic steels, which are magnetic, contain chromium alone most commonly to the extent of 12–30%. They are identified in the AISI 400 series (SS 409 and SS 430 grade). They are low-priced stainless steels useful in less aggressive environments, where environmental condition is of less importance (Mathur 2000).

These steels possess good resistance to corrosion and have good weldability. (However, their corrosion resistance is inferior to that of the 300 series.) They are highly recommended for components which need a paint coating, such as commercial garages, canopies, and road signs.

Austenitic steels are many in number and varied in composition. They contain 16–26% chromium and 6–22% nickel. They are in annealed condition, non-magnetic, and have excellent corrosion resistance. Although not hardenable on cold working, they develop high strength on such treatment. They have excellent weldability and are identified in the AISI 300 series (Euro Inox 1994). Normally used grades in structural applications are SS 304, SS 304L, and SS 316. Grade 304 should not be used for external elements near coastal areas. Grade 316 is the best grade for all external applications. (SS 316 is basically a 304 grade with the addition of 2–3% molybdenum.)

The outstanding plastic deformation capacity of stainless steel makes it an ideal material for space trusses, bracings, column bases, and base-isolated structures for earthquake-response control. Life-cycle cost analysis shows that stainless steel structures will be economical after 20 years. The current design philosophy of designing structures which should have a minimum life span of 100 years with materials which can be recycled makes stainless steel the ideal structural material of the future.

### 1.15.1 Surface Finish

Cold-rolled (thin gauge up to 3mm) stainless steels are available with the following finishes (Mathur 2000).

No. BA—A bright, smooth mirror finish

No. 2B—A bright, smooth, silvery grey, moderately reflective finish (most common finish for sheet material)

No. 2D—A matt, dull, silvery grey, non-reflective finish (gives better paint adhesion)

No. 3—A directional, uniform, polished finish, using 100 to 120 grit abrasive

No. 4—A bright, polished finish with a visible directional grain from a 120 to 180 grit abrasive.

**FIG. 1.40** Stainless steel double-layer grid roof at Hissar, Haryana, designed by the author

The above finishes are mill finishes. Other finishes such as various polished finishes, coloured BA finishes, regidized patterns, and decorative steel are also commercially available in stainless steel. A space frame roof made of stainless steel, designed by the author, and erected at the entrance of M/s Jindal Strips Limited at Hissar, Haryana is shown in Fig. 1.40 (Subramanian 2002).

## 1.16 ADVANTAGES OF STEEL AS A STRUCTURAL MATERIAL

As mentioned in the introduction, structural steel offers several advantages over other competing materials. These advantages are as follows.

**High strength** The high strength of steel per unit weight means that structures made of steel sections weigh less than those made of other materials. On the basis of the strength/weight ratio, steel is at least 3.5 times more efficient than concrete. To replace one unit area of steel in tension (with a yield stress of 450 MPa), we would need to use an equivalent concrete (M30) area of about 200 units. Concrete as a material is efficient when the load carried is essentially compressive. This fact is of great importance for long-span bridges, tall buildings, and structures with poor foundation conditions.

**High ductility** As we have seen already, the property of a material which enables it to withstand extensive deformation without failure under high tensile stresses is said to be its *ductility*. A material that does not have this property is probably hard and brittle and might break if subjected to a sudden shock. In structures built with structural steel, occasional human errors such as accidental overloading do not cause problems, due to the ductility of steel. A further advantage of structures made of steel, due to its ductility, is that when highly overloaded (for example due to earthquakes), the large deflections give visible evidence of impending failure. This aspect is very important for the safety of the occupants. In structural members under normal loads, high stress concentrations may develop at various points. The ductile nature of the usual structural steels enables them to yield locally at those points, thus preventing premature failures.

**Uniformity** The quality of steel-intensive construction is invariably superior, when compared with that of construction involving other materials. This is especially important in India, where quality control in construction sites is poor (resulting in poor performance, especially in concrete structures where water–cement ratio and curing are not controlled properly at site). Moreover, the properties of steel do not change appreciably with time as do those of reinforced concrete.

**Environment-friendly** Structural steel is recyclable and environment-friendly. Over 570 million tonnes of steel are recycled annually worldwide, which represents 50% of all steel produced. Steel is the world's most versatile material to be recycled. Due to this, steel can be used in one product after another without losing its quality. Steel from cans can be easily used, by recycling, in strong suspension cables. Recycling of steel saves energy and primary resources, and reduces waste. It also has scrap value even if it is not reusable in the present form. Another characteristic of a steel structure is that it can readily be disassembled at the end of its useful life. This has many environmental and economic advantages. It means that the steel components can be reused in future structures without the need for recycling, resulting in the saving of energy and avoidance of $CO_2$ emitted from the steel production processes. It has to be noted that concrete components cannot be recycled easily and that the ecofriendly disposal of demolished concrete waste is very difficult.

**Versatility** Using structural steel, it is possible to fasten different members together by simple connection techniques such as welding, bolting, and riveting. Steel members can also be rolled into a wide variety of sizes and shapes, as described in Section 1.12.

**Prefabrication** Often, steel components are manufactured at the factory (which means that they are produced using strict supervision and quality control), transported to the site, and erected using bolting and a minimum amount of welding. This kind of *prefabrication* is not easy in concrete structures. The prefabrication of steel structures results in the proper planning of construction, saving in time and money, speedy erection, and better quality of finished structures. Lighter steel members facilitate easy handling and erection.

**Permanence** Steel frames that are properly maintained last indefinitely. Several structures are available to testify to the durability of steel structures (e.g., the Eiffel Tower and the Railway Bridge across the Firth of Forth, both built in 1890). Under certain conditions, weathering steels do not require any painting or maintenance. Concrete is subjected to deterioration by the same environmental factors that steel is (namely, chloride contamination, alkali-silicate reactions, sulphate attack, etc.). However, steel is easily repairable at almost any stage of corrosion. But once the reinforcements in a concrete structure corrode, which they do frequently due to poor site control, it is extremely difficult to repair and restore them to the original condition. In Belgium and Japan, it has been found that steel bridges outlast prestressed concrete bridges by 15–26 years. This experience is confirmed by the premature failure of many prestressed concrete bridges in India.

**Additions to existing structures** The repair and retrofit of steel members and their strengthening at a future date (for example, to take into account enhanced loading) is simpler than in concrete members. Thus, new bays or even entire new wings, can be added to existing steel-frame buildings, and steel bridges may often be widened. Of course, special precautions have to be taken while welding on a member already carrying loads.

**Least disturbance to the community** Steel-intensive construction causes the least disturbance to the community in which the structure is located. Fast-track construction techniques developed in recent years have demonstrated that steel structures cause the least disruption to traffic and minimize financial losses to the community and business. Such construction also results in far less environmental pollution.

**Fracture toughness** Structural steels are tough (as we have discussed already, the ability of a material to absorb energy in large amounts is called toughness). Due to its toughness and ductility, steel members can be subjected to large deformations during fabrication and erection without fracture, thus allowing them to be bent, hammered, sheared, and have holes punched in them without visible damage.

**Elasticity** Steel behaves closer to design assumptions than most materials because it follows Hooke's law up to fairly high stresses. The moments of inertia of steel members can be definitely calculated while the values obtained for a reinforced-concrete structure are not accurate and depend on the extent of cracking (McCormac 1989).

## 1.17 DISADVANTAGES OF USING STEEL

Though steel has several advantages, it also has the following disadvantages.

**Maintenance costs** Most steels are susceptible to corrosion when freely exposed to air and water, and must therefore be periodically painted. However, weathering steels do not need to be painted. Steel members in the interior of buildings (not exposed to rain) do not corrode quickly.

**Fireproofing costs** Although structural members are incombustible, their strength is tremendously reduced at temperatures commonly reached in fires when the other materials in a building burn. Furthermore, since steel is an excellent heat conductor, non-fireproofed steel members may transmit enough heat from a burning section or compartment of a building to ignite materials which come into contact with them in adjoining sections of the building. Due to these factors, the steel frame of a building may have to be protected by materials with certain insulating characteristics. In addition, the building may have to include a sprinkler system, in order to meet the building code requirements.

**Susceptibility to buckling** The longer and more slender the compression members, the greater is the danger of buckling. Though steel has a high strength per unit of weight, steel columns have to be stiffened against buckling, resulting in uneconomical solutions sometimes.

**Fatigue** Another undesirable property of steel is that its strength may be reduced if it is subjected to a large number of stress reversals or several variations of tensile stress. (There are fatigue problems only when tension is involved.) Hence we often reduce the estimated strengths of such members if more than a prescribed number of cycles of stress variation are anticipated.

## 1.18 TYPES OF STEEL STRUCTURES

The structural engineer will be concerned with the design of a variety of structures, which may include the following.

**Buildings** These may include rigid, semirigid, or simple connected frames, load-bearing walls, cable-stayed and cantilevered structures. Buildings may be simple or multi-storeyed, with single or many spans. For multi-storeyed buildings, several lateral bracing systems have been developed, such as trussed, staggered truss, rigid central core, etc. Buildings are also classified according to use, such as residential, commercial, office, industrial, etc. These buildings may include a steel frame as shown in Fig. 1.41 or have a steel roof supported by load-bearing walls. The steel skeleton of the buildings may be rigid or pinned, a two- or three-hinged arch or a truss-on-column system. The truss also may be rigid or pin-connected, and may assume various shapes or have several bracing systems (see Chapter 16).

The building frame is actually a three-dimensional skeletal system, but in practice it is usually taken as rigid in only one plane. Some buildings are rigid in the *XY* as well as *YZ* planes, but this type of frame will not be considered in this text [see Subramanian (1999) for the details of three-dimensional frameworks]. The planar frame resulting from considering only the principal frame elements and/or the rigidity is termed as a *bent* and may have one or several stories. The spacing between bents in the third dimension is called the bay or *bay width. Spandrels* or floor beams are used to span the bays in multistoryed buildings with girders (usually heavier members than floor beams) spanning between columns or bents. Figure 1.42 illustrates additional terms or members used in one-storey industrial buildings, where lateral bracings are often provided only in selected bays.

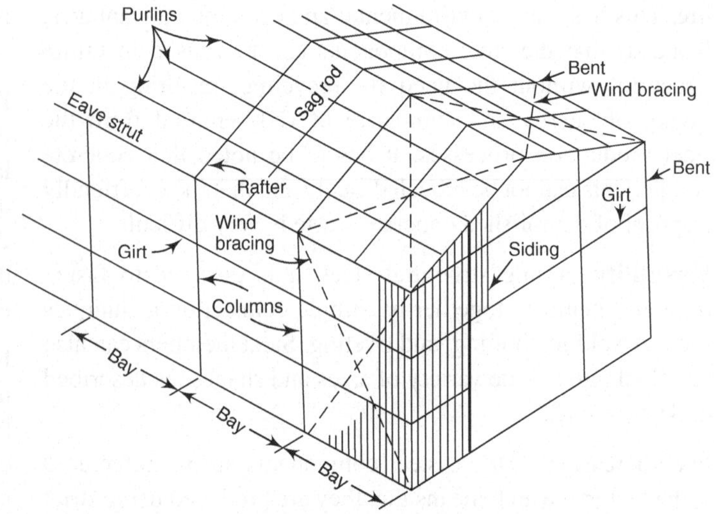


**FIG. 1.42** Different structural members in industrial buildings

The roof systems of all buildings consist of a framing, some kind of decking, and a waterproof covering. The main roof framing consists of the *rafters* or the *truss* in any bent. Spanning the bay width are *purlins* spaced at about 0.6–2m at centres, depending on the type of roof decking or sheeting used. *Sag rods* are provided as additional support for the purlins used on sloping roofs. Purlin design is rather complex for sloping roofs due to the unsymmetrical bending (see Section 10.18 of Chapter 10). The roof deck rests on purlins and may be a metal deck, precast concrete slabs, wood planking, gypsum sheets, GI sheets, polycarbonate sheets, fibre-glass sheets, or asbestos (which has been banned in several countries due to health hazards).

The siding may be made up of metal sheets, metal sandwich sheets (sometimes called curtain walls) consisting of two metal sheets with some insulating filler, brick, and precast or poured concrete, or asbestos sheets. A lightweight siding is carried by girts and spandrel beams in high-rise buildings and by *eaves struts* and *girts* in industrial buildings (see Fig. 1.42). Note that a spandrel beam is similar to a girt and is located at floor level (as the most exterior floor beam), and carries a proportion of live load. It also carries a part of the load acting on the siding. If the siding is heavy (e.g., brick or concrete blocks), built-up sections (using channels or angles depending on the load) can be used for the spandrel beam. It is necessary to establish the floor framing system very early in the design process, so that the load flow is identified and the sizes of different members are approximately fixed for computer analysis.

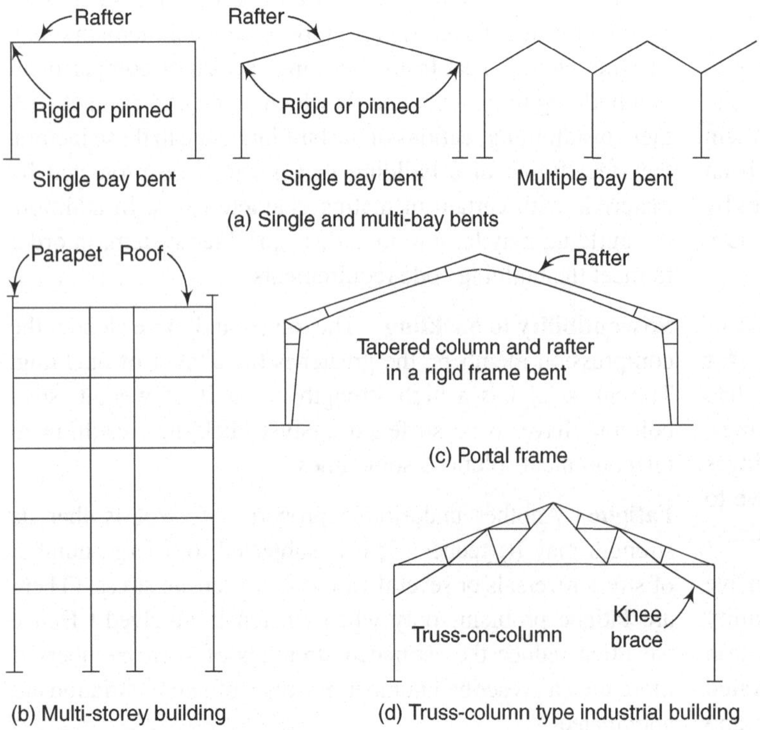


**FIG. 1.41** Types of steel structures

**Bridges** Bridges may be classified as truss, plate-girder, arch, cantilever, cable-stayed, or suspension (using cables as principal load-carrying members).

The truss and plate-girder bridges are commonly adopted for small to moderate spans and cable-stayed and suspension bridges for long spans. Many types of trusses like Pratt, Warren, Parker, Baltimore, K-type, Whipple truss, etc., are used in bridges. The truss types include both deck (traffic on top of truss) and through types (traffic passes between the trusses of the bridge). The deck type is preferred if the clearance below the truss is not a critical factor, because the piers could be made shorter. Many truss bridges combine both types. A combination of trusses (for long spans) and girders (for shorter or approach spans) is also adopted in practice. Bridges may also be classified as railroad, highway, or road and pedestrian bridges, depending upon use.

**Towers** Towers may be of different types, such as lighting towers, power transmission towers, observation towers, towers for radar and TV installation, telephone relay towers, and windmill towers. Towers may be self-supporting or cable-stayed. Most towers are made of steel angles or tubes, which are bolted at site.

**Storage tanks** They may be rectangular, circular, or spherical. They can be used to store oil or water. They may rest on the ground or be elevated on a staging.

**Other structures** Silos, bunkers, domes, folded plates, offshore platforms, chimneys, cooling towers.

In addition to these structures, the structural engineer may be called upon to design ships, parts of various machines and other mechanical equipment, and automobiles (bus and car bodies, chassis), etc. Figure 1.43 shows the photographs of some different types of steel structures.

(a) Beijing National Stadium also known as Bird's Nest, built for 2008 Summer Olympics in Beijing, China is the world's largest steel structure. This 91,000-seat stadium was designed by Swiss architects Herzog & de Meuron, ArupSport and China Architectural Design & Research Group. It is 330 m long, 220 m wide, 69.2 m tall, and contains 110,000 tonnes of steel

(b) Golden Gate Bridge, San Francisco, having a main span of 1280 m, is one of the longest suspension bridges in the world (Photo by Karthik Muthusamy)

(c) 411 m tall Empire State Building, New York, USA, built in 1931, which remained the world's tallest building for 40 years

(d) A typical pre-engineered building. They are adopted for low-rise buildings with eave height of up to 30 m and can be very economical and speedy

**FIG. 1.43** Photographs of some of the different types of steel structures (*Contd*)

(e) Transmission line towers are often made of steel angles and bolted at site (Photo by Binod Therat)

(f) Steel elevated water tanks are used to store water and provide pressure in the water distribution system. A tank elevated to 21 m creates about 0.207 MPa of discharge pressure, which is sufficient for most applications. The shapes of the tanks and their surface appearance can be varied according to the needs.

**FIG. 1.43** (*Contd*)

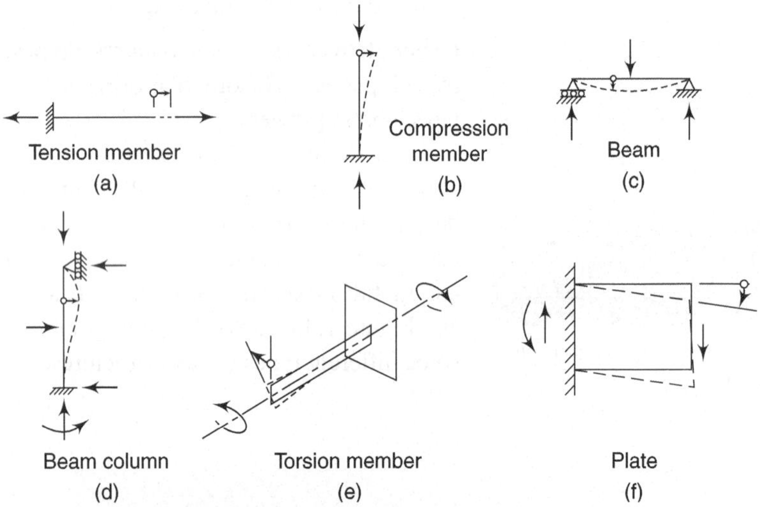


**FIG. 1.44** Various types of structural members

In this book, we will confine our attention to the design of elements which are encountered in buildings. The various structural members used in buildings can be classified into the following according to the method by which they transmit forces in the structures (see Fig. 1.44).

**Beams** *Beams* (also called *girders*) primarily carry bending stresses which are the largest at the extreme fibre. Thus, most of the material of the beam remains understressed, and is not fully utilized, especially if it is of rectangular section. Also, in simply supported beams the bending moment reduces towards the support and if the beam section is uniform, even the extreme fibres carry smaller stresses. Thus, the overall efficiency of the use of material in beams is low. Yet beams are used so commonly because they give flat floors and roofs, are readily available, and require less fabrication at site. The design of beams is covered in Chapters 10 to 12.

**Tension members or ties** *Tension members* are structural elements that are subjected to direct axial forces, which try to elongate the members. They occur as components of trusses, hangers and cables for floors or roofs, in bracing systems, as tie rods, and similar members. An axially loaded tension member carries uniform stress and hence would give 100% efficiency in the use of materials. But in practice the efficiency is below 100% because the effective area of a tension member is less than the gross area due to the presence of rivet/bolt holes, shear lag effect, etc. A suspension cable, however, has a structural efficiency of 100% since it is flexible and can carry only tension. However, the use of a cable may require strong anchorage, and a stiffening arrangement to check large deformation under load, vibration due to wind, and other effects. The design of tension members is covered in Chapter 7.

**Compression members (columns and struts)** *Compression members* are those members in a structure that are subjected to loads that tend to decrease their length. Such members are used as the vertical load-resisting elements of a building structure, called columns, as the posts that resist compressive components of a load in a truss, and as bridge piers. In a building structure, forces and moments are transmitted to the columns through beams at each floor or the roof level of the structure. If the column is required to resist a load acting concentric to the original longitudinal axis of the member, it is termed an *axially loaded column* or simply a *column*. Even axially loaded columns will not be able to use 100% of the material, due to the phenomenon of *buckling*. The design of compression members is discussed in Chapter 9 and that of beam columns (which carry bending moments in addition to compressive forces) in Chapter 13.

**Torsional members** Such members are often encountered in shafts of machines. Of course, torsional effects should be considered in spandrel beams also.

**Plates** Some built-up members and several light-gauge cold-formed members are made up of slender plate elements, which are liable to buckle locally (see Chapter 8).

Often, a combination of actions, such as bending and compression, bending and tension, compression bending and torsion, etc., occurs in frameworks where several members are joined together. In many cases, one action may be predominant and the other may be secondary. In the design, such secondary action is usually not taken into account. However, if the secondary action is considerable, the member has to be designed for the combined actions.

As already indicated, these members (beams, columns, tension members, etc) are joined together in a framework by means of welding, bolting, or riveting. These fastening techniques and designs of connections are covered in Chapters 5, 6, and 14.

## 1.19 ALUMINIUM ALLOYS

Because of their greater strength to unit weight ratio compared to that of steel, aluminium alloys are being increasingly used for structural purposes. Aluminium also has greater resistance to corrosion than does steel. Aluminium in its pure form is not very strong and has a tensile strength of only 89 MPa. However, by cold working, heat treatment, and alloying, its strength can be raised to about 550 MPa. The density of aluminium is one-third that of steel (2700–2800 kg/m$^3$). Its modulus of elasticity is also about one-third that of steel (0.62–0.73 × $10^5$ MPa) while its coefficient of thermal expansion is twice that of steel (about 23 × $10^{-6}$/°C). It does not have a well-defined yield point, which is therefore assumed to be the 0.2% proof stress.

Because of the lower modulus of elasticity, aluminium members are more susceptible to buckling and hence the advantage of reduced weight is often offset by the use of thicker members. However, the use of aluminium results in economical structures due to the additional advantage of reduced maintenance. The design concepts of aluminium structures are similar to those of steel structures. More details about the design aspects of aluminium may be found in IS 8147 and Arya and Ajmani (1989).

### 1.19.1 Mathematical Representation of Non-linear Stress–Strain Curves

The pronounced yield plateau of structural mild steel is non-existent in aluminium and stainless alloys. The non-linear stress–strain curves of these materials can be expressed mathematically using the Ramberg–Osgood expression,

$$\varepsilon = \frac{f}{E_0} + \varepsilon_p \left(\frac{f}{f_p}\right)^n \tag{1.16}$$

Equation (1.16) was originally developed for aluminium alloys but has proven suitable for other non-linear metals including high-strength steels and stainless-steel alloys. It involves the initial Young's modulus ($E_0$), the proof stress ($f_p$) corresponding to the plastic strain $\varepsilon_p$, and a parameter ($n$) which determines the sharpness of the knee of the stress–strain curve. In the design of aluminium and stainless–steel structures, it has become industry practice to use the 0.2% proof stress ($f_{0.2}$) as the equivalent yield stress. For this proof stress, the stress–strain relationship takes the form,

$$\varepsilon = \frac{f}{E_0} + 0.002\left(\frac{f}{f_{0.2}}\right)^n \tag{1.17}$$

It has also become standard practice to determine the parameter $n$ using the 0.01% and 0.2% proof stresses, which leads to the following expression.

$$n = \frac{\ln(20)}{\ln\left(\dfrac{f_{0.2}}{f_{0.01}}\right)} \tag{1.18}$$

Equation (1.18) ensures that the Ramberg–Osgood approximation matches exactly the measured stress–strain curve at the 0.01% and 0.2% proof stresses. It generally provides close approximations to measured stress–strain curves for stresses up to the 0.2% proof stress.

Rasmussen (2001) derived the full-range stress–strain curve for stainless steel.

## 1.20 FABRICATION AND ERECTION

According to the design of various members of a structure, the various required sections are procured and are fabricated at site or factory. Tolerances for the fabrication of steel structures should conform to IS 7215. The various activities in the fabrication shop include the following.

1. Exact cutting of length by sawing, shearing, cropping, thermal cutting, or machining, based on the fabrication drawing of the structure
2. Straightening of members
3. Cambering of beams
4. Drilling or punching of holes
5. Welding of gusset plates
6. Machining of butt joints, caps, and bases
7. Surface preparation, such as shot blasting
8. Painting or galvanizing after pickling in acids
9. Marking
10. Shop assembly and erection
11. Inspection and testing
12. Packing

The numbered parts are then transported to the site and the structure is erected following the erection tolerances specified in IS 12843. The normal tolerances after erection are given in Table 33 of IS 800:2007.

The straightness tolerances incorporated in the design rules are given in Table 34 of IS 800:2007. Where curvature exceeds these values, the effect of additional curvature on the design calculations should be reviewed. A tension member should not deviate from its correct position relative to the members to which it is connected by more than 3 mm along any setting axis.

After the structure is erected, the specified protective treatment should be applied on the surfaces of the steel members and joints. No painting should normally be used on the contact surfaces in the friction connection. More guidelines for the fabrication and erection of steel structures are given in Section 17 of IS 800 : 2007. Some recommendations for steel work tenders and contracts are also given in Annex G of the code.

### 1.20.1 Errors that Lead to Failures

To err is human but the consequences of an error in structural design can lead to loss of life and damage to property. Hence it is necessary to appreciate where errors can occur. Small errors can occur due to rounding of figures but these generally do not lead to failures. The majority of structural failures (whether it is the collapse of the structure or functional failure) are due to errors in design, construction, or operation (see Table 1.20). It has been reported that 85% of building failures occur due to human errors (Brown and Yin, 1988). Hence it is imperative for engineers and contractors to consciously avoid these errors.

**TABLE 1.20** Percentage of failures due to the type of activity (based on reported failures)

| Type of activity | Proportion of errors |
|---|---|
| Planning and design | 37–55% |
| Construction | 24–53% |
| Use and maintenance | 21–31% |
| Others | 20–23% |

The common errors that occur in the planning and design phase are due to the following (Martin and Purkiss 1992, and Subramanian 1984, 1989).

1. Ignorance of the physical behaviour of the structure under load, which leads to errors in basic assumptions used in theoretical analysis
2. Errors in selecting and estimating the loads, especially the erection forces
3. Numerical errors in calculations—these could be eliminated by proof-checking; however when speed is of paramount importance, checking of calculations is often neglected
4. Lack of consideration for certain effects such as fatigue, brittle fracture, residual stresses, etc.
5. Insufficient allowances for temperature strains, tolerances, etc.
6. Insufficient information about new materials, methods of analysis and design, detailing, erection procedures, etc.

Nowadays computer programs are being used as black boxes; that is, without knowing the limitations of these programs. Such usage leads to erroneous results. Errors that may occur during fabrication and erection are as follows.

1. Using the wrong grade of steel or wrong types of electrodes for welding,
2. Using the wrong weight of section (see also Section 1.12.1),
3. Errors in fabrication (holes not matching, oversized holes, lack of fit, improper welding, welding distortions, etc.), and
4. Errors due to improper quality control.

Errors can also occur during the life of the structure, which affect the safety of its occupants. Such errors include

1. overloading due to change of occupancy,
2. loading which is not expected during the design stage (an earthquake of greater magnitude, flood, tsunami, etc.),
3. alteration of the structural system (removal of the web of the flange to provide service ducts, addition of heavy partitions, balconies, etc.), and
4. poor maintenance.

A study of the various types of failure has been provided by Levy and Salvadori (1992), Kaminetzky (1991), and Feld and Carper (1997). Case studies of building failures provide opportunities to learn from previous mistakes (Subramanian 1999, 2000, 2014 a; Subramanian & Mangalam 1997, 1998; Kevin et al. 2000).

## 1.21 AESTHETICS OF STEEL STRUCTURES

Although architects pay attention to the aesthetic qualities of structures like theatres, stadia, office buildings, residential units, etc., industrial buildings and non-habitat structures, which are made of steel members, are not aesthetically designed. It is very difficult to assess the aesthetic qualities of any structure, because taste differs from person to person. Aesthetics may be viewed as a composite impression of all visual aspects of design. In a normal building, aesthetic qualities may be judged with reference to its site, the physical layout and environment, the view from within and from without, and the appearance of the building with reference to its surroundings.

In tower structures, the aesthetic quality may be judged in terms of the proportion and weights of the different elements, the simplicity of form of the bracing system, the sizes of the joints as compared to those of the members, and the harmony or contrast of the structure with the built or natural environment. Of course, these aesthetic factors should be considered in addition to functionalism, perfect manufacture, construction safety, stability, durability, disassembly, and reusability.

**FIG. 1.45** Guggenheim museum, Bilbao, Spain designed by Architect Frank O. Gehry
*Courtesy:* Emilio I. Panizo

Almost 2000 years ago, the Roman engineer-architect Vitruvius formulated certain attributes for assessing the aesthetic quality of any structure. According to him, order, symmetry, arrangement, eurhythmy, propriety, and economy are the basic characteristics that influence the aesthetic quality of a structure (Kavanagh 1975).

However, whether a person finds a structure beautiful depends upon whether it is able to evoke an emotion in him or her, which in itself is extremely complex in nature. The magnitude and combinations of the qualities listed previously obviously cannot be specified, since they evoke different responses in different individuals and are beyond the scope of our discussion. Nowadays, designers have moved away from notions of symmetry and order towards more expressive and dynamic structures, as seen in Fig. 1.45, which shows the Guggenheim museum, Bilbao, Spain. More details about aesthetic design may be found in Kavanagh (1975), Subramanian (1987, 2003, 2014 b), and Billington (1983).

## 1.22 COMPOSITE CONSTRUCTION

Except in certain structures such as towers, portal frames, industrial structures covered with trusses and purlins, cranes, etc., steelworks exist with reinforced concrete. In many places, the properties of reinforced concrete (strong in compression, greater rigidity) can be advantageously combined with those of structural steel. Such *composite construction* includes (a) concrete-encased steel columns, (b) concrete-filled steel columns, (c) concrete encased steel beams, and (d) steel beams supporting concrete slabs (see Fig. 1.46). The slab shown in Fig. 1.46(d) can be either a solid reinforced concrete slab or a concrete slab on a corrugated metal deck. In either case, a device called a *shear connector* is provided to prevent the slip that may occur at the interface between concrete and steel. In contrast to ordinary structural steel design, which considers only the strength of the steel, composite design assumes that the steel and concrete work together in resisting the loads. Composite construction has the following advantages.

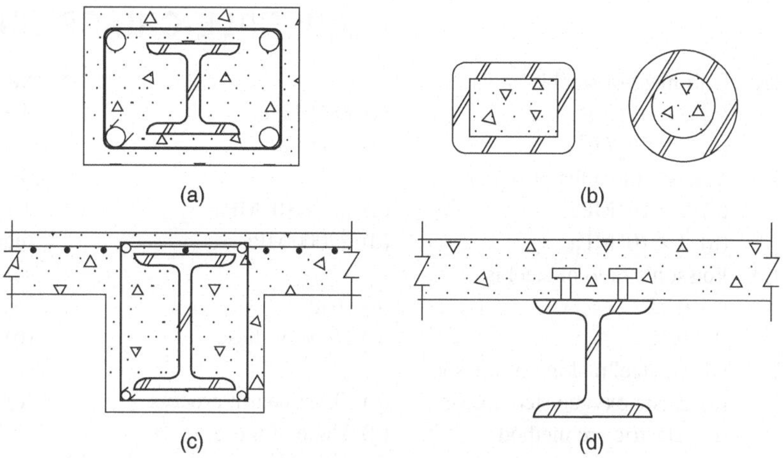


**FIG. 1.46** Various types of composite construction

1. In composite beams, the construction depth is reduced by approximately one-third.
2. The weight of steel beams is considerably reduced (by up to 50%).
3. The presence of concrete may be utilized to meet the necessary fire resistance.
4. The use of the propping system, to carry the load of wet concrete, may be eliminated (since the corrugated metal deck serves the purpose of permanent shuttering for the concrete slab and acts as tension reinforcement for the composite slab spanning between the main beams), and results in speedy construction (Subramanian 1992).
5. The use of concrete in columns (encased I-sections or filled hollow sections) results in a high load-carrying capacity.
6. Reduced sections result in a reduced load on the foundation and consequent economy in foundation design.

In India, the design of composite construction is done by using IS 11384 : 1985. Design of composite construction will not be considered in this book. More details may be found in Nethercot (2001), Martin and Purkiss (1992), and Kulak and Grondin (2002).

## SUMMARY

This chapter begins with a general discussion of building materials such as masonry, reinforced and prestressed concrete, structural steel, and wood. It then goes on to present a historical review of iron and steel. The steel making process is briefly outlined. The metallurgical properties of steel, which are relevant to structural steel designers, are also briefly presented.

Structural steel has several advantages over other competing materials such as concrete and wood. The chemical composition and mechanical properties of different kinds of steel are included. (It is quite important to specify the steel for the project at hand.) Properties of special steels such as stainless steel, high-performance steel, and cold-formed steel are also discussed. The important properties of structural steel are its ultimate and yield stress, ductility, toughness and weldability. We should remember that the yield stress, as measured by the tension coupons, is affected by several factors, such as the rate of loading and position from where the test coupons are taken. Variation in material properties can be incorporated in the design by the concept of characteristic strength. Ductility and toughness are very important when a steel structure is subjected to earthquake loads or impact loads.

In any design, the engineer's first task is to think about material selection. First, he or she should consider the working temperature range and form of the structure, and decide whether the components require thick material. A simple yardstick is that satisfactory performance against brittle fracture can be expected if the Charpy impact toughness exceeds 27 J at the lowest working temperature. Of course the quality of design, detailing, and fabrication are also equally important in the prevention of brittle fracture. One should avoid severe stress raisers to achieve smooth stress flow. In critical cases, this is as important as material selection.

Secondly, one must consider fatigue and ascertain whether the incidence of high levels of fluctuating stresses warrant further consideration. In general, this has to be checked in crane supporting structures, bridges, and structures supporting rotating machinery. Residual stresses and stress concentration effects should be properly accounted for in the design.

Though, mainly Fe 410 grade mild steel is used in India, it is prudent to check whether it is possible to use high-strength steel. Though the cost of high-strength steel may be 10–20% higher, its use may result in reduction in steel weight and subsequent foundation cost. Of course, while using high-strength steels, the deflection criteria may become critical due to the reduced section sizes. Other special types of steels are also available and may be specified, depending upon site conditions and cost considerations.

A brief introduction to corrosion, fire protection, and fatigue resistance of steel is also given. An overview of the different types of structures, types of sections (circular, I, angle, channel, hollow sections, etc.) and types of members (compression, tension, bending), is presented, which will be quite useful in understanding the concepts discussed in later chapters.

## MULTIPLE-CHOICE QUESTIONS

1.1 Unit mass of steel is
(a) 2400 kg/m$^3$ (c) 6850 kg/m$^3$
(b) 7850 kg/m$^3$ (d) 7500 kg/m$^3$

1.2 Young's modulus of steel is
(a) $2 \times 10^5$ MPa (c) $2.1 \times 10^5$ MPa
(b) $2 \times 10^6$ MPa (d) 22000 MPa

1.3 Poisson's ratio of steel is
(a) 0.25 (c) 0.30
(b) 0.22 (d) None of these

1.4 Oldest steel making process is
(a) Basic oxygen steel making (c) Open-hearth process
(b) Electric arc method (d) None of these

1.5 Structural steel (low-carbon steel) normally contains carbon in the range of
(a) 0.4 to 0.12% (c) 0.10 to 0.25%
(b) 0.02 to 0.08% (d) 0.5 to 1%

1.6 To produce one tonne of steel the following materials are required:
(a) 1.5 tonne of iron ore, 1.25 tonne of coal, 1.5 tonne of limestone, and 1.5 tonne of other materials
(b) 1.25 tonne of iron ore, 1.50 tonne of coal, 1.0 tonne of limestone, and 1.0 tonne of other materials
(c) 1.75 tonne of iron ore, 1.25 tonne of coal, 0.75 tonne of limestone, and 0.5 tonne of other materials

(d) 1.75 tonne of iron ore, 1.35 tonne of coal, 0.5 tonne of limestone, and 0.5 tonne of other materials

1.7 Up to what temperature does iron have a bcc crystalline structure and remain as ferrite?
(a) 900 °C (b) 950 °C (c) 910 °C (d) 1000 °C

1.8 Carbon equivalent is calculated using
(a) $\frac{C+Mn}{5}+\frac{Cr+Mo+V}{10}+\frac{Ni+Cu}{15}$
(b) $\frac{C+Mn}{6}+\frac{Cr+Mo+V}{5}+\frac{Ni+Cu}{15}$
(c) $\frac{C+Mn}{10}+\frac{Cr+Mo+V}{6}+\frac{Ni+Cu}{15}$
(d) None of these

1.9 Coefficient of thermal expansion of steel is
(a) $11 \times 10^{-6}$/°C (c) $13 \times 10^{-6}$/°C
(b) $12 \times 10^{-6}$/°C (d) None of these

1.10 Minimum elongation of steel specimen on a gauge length of $5.65\sqrt{A_0}$ is
(a) 10% (b) 15% (c) 20% (d) 12.5%

1.11 The energy absorption at lower and higher transition temperatures specified in Charpy V-notch test are
(a) 20 J and 60 J (c) 20 J and 50 J
(b) 15 J and 40 J (d) 25 J and 60 J

1.12 The magnitude of tensile and compressive residual temperatures in rolled sections may reach
(a) $0.2f_y$ and $0.8f_y$ (c) $0.3f_y$ and $0.5f_y$
(b) $0.4f_y$ and $0.5f_y$ (d) $0.3f_y$ and $0.6f_y$

1.13 Which of the following is true?
(a) Use of cold rolled sections result in saving of weight than hot rolled section
(b) Cold rolled sections are prone for local buckling
(c) Cold rolled sections offer wide variety of shapes
(d) The maximum thickness of cold rolled section may be 20 mm.

1.14 Stainless steel has chromium to the extent of
(a) less than 5.75% (c) about 8%
(b) greater than 10.5% (d) greater than 15%

1.15 State whether the following are true.
(a) Steel is a environment friendly material
(b) Steel is amenable for prefabrication
(c) Steel offers high ductility
(d) Steel is highly resistant to fire

## REVIEW QUESTIONS

1.1 Briefly state the advantages and use of the four basic materials: masonry, reinforced and prestressed concrete, structural steel, and wood.
1.2 Compare the properties of cast iron, wrought iron, and steel. Why is steel considered superior for structural applications compared to the other two materials?
1.3 Describe the production of structural steel in integrated steel plants.
1.4 What is the difference between killed and semi-killed steel?
1.5 What are the differences between the production of steel in integrated steel plants and mini steel plants?
1.6 What is the difference between hot rolling and cold rolling?
1.7 Describe the phase diagram of steel.
1.8 What are the heat treatments employed to improve the properties of steel?
1.9 How is normalizing different from annealing?
1.10 How does carbon content affect the properties of steel?
1.11 What are the factors that influence the ductility, elasticity, strength, and toughness of steel?
1.12 What is the relation between weldability and carbon equivalent?
1.13 What are the types of structural steel?
1.14 List the important mechanical properties of steel along with the factors that influence them.
1.15 Describe the test to predict the tensile strength of steel.
1.16 Give the values of yield strength, Young's modulus, coefficient of thermal expansion, and ultimate tensile strength for mild steel as per IS 800 : 2007.
1.17 Sketch the typical stress–strain curve of steel, indicating the three important regions.
1.18 What are the parameters that influence the yield stress of steel?
1.19 What is meant by characteristic strength?
1.20 What is ductility? Why is it important?
1.21 Under what circumstances does brittle failure of steel take place?
1.22 What is the Charpy V-notch test?
1.23 List the methods by which brittle fracture may be controlled.
1.24 Write short notes on
(a) Lameller tearing
(b) high-temperature effects on steel
(c) hardness
(d) corrosion resistance of steel
(e) fatigue resistance
1.25 Sketch the residual stresses in a typical rolled I-beam, welded I-beam, and welded box section. How are residual stresses induced in steel sections?
1.26 What is the effect of residual stress in the design of compression members?
1.27 Under what circumstances are stress concentration effects considered to be critical in design?
1.28 List the hot-rolled steel sections used in practice.
1.29 State the advantages of using wide-flange beams over narrow ISMB beams.
1.30 Write short notes on hollow steel sections and cold-formed steel sections.
1.31 What is hydrogen-induced cracking?
1.32 What are the characteristics of high-performance steel?
1.33 What are the advantages of stainless steel? What is the main difference in chemical composition between stainless steel and normal mild steel?

1.34 State the main advantages and disadvantages of steel as a structural material.

1.35 Discuss the following.
- (a) Fabrication and erection of steel structures
- (b) Non-linear stress–strain curves
- (c) Aluminium alloys
- (d) Various possibilities of structures using steel as a building material
- (e) Aesthetics of steel structures
- (f) Advantages of composite construction

## REFERENCES


Alpsten, G.A., *Variation in Mechanical and Cross-sectional Properties of Steel*, Swedish Institute of Steel Construction, Publication No.42, 1973.

Arya, A.S. and Ajmani, J.L., *Design of Steel Structures*, Fourth edition, Nem Chand and Bros, Roorkee, UP, 1989, pp. 587–609.

ASCE/ANSI – 8 – 90, *Specification for the Design of Cold-formed Stainless Steel Structural Members*, ASCE, 1991.

Barsom, J.M. and Rolfe, S.T., *Fracture and Fatigue Control in Structures—Applications of Fracture Mechanics*, Second edition, Prentice Hall, Inc., Englewood Cliffs, New Jersey, 1987.

Billington, D.P., *The Tower and the Bridge, The New Art of Structural Engineering*, Princeton University Press, Princeton, NJ, 1983.

Blodgett, O.W., *Design of Welded Structures*, The James P. Lincoln Arc Welding Foundation, Cleveland, Ohio, 1966.

Bradford, M.A. and Oehlers, D.J., *Elemental Behaviour of Composite Steel and Concrete Structural Elements* Butterworths - Heinemann, 2000, p. 280.

Brown, C.B. and Yin, X., 'Errors in Structural Engineering', *Journal of Structural Engineering*, ASCE, Vol. 114, No. 11, Nov 1998, pp. 2575–2593.

Buchanan, R.A., *Industrial Archeology in Britain*, Penguin, 1972.

Byfield, M.P., 'Material and Geometric Properties of Structural Steel for Use in Design', *The Structural Engineer*, Vol. 75, Jul–Dec 1997, pp. 363–367.

'Causes and Prevention of Lamellar Tearing', *Civil Engineering*, ASCE Vol. 52, No. 4, Apr 1982, pp. 74–75.

'Commentary on Highly-restrained Welded Connections', *Engineering Journal*, American Institute of Steel Construction, Vol.10, No. 3, Third quarter, 1973, pp. 61–73.

Dayaratnam, P., *Design of Steel Structures*, Second edition, Wheeler Publishing, Allahabad, 1996, p. 868.

Derry, T.K. and Williams, T., *A Short History of Technology*, Oxford University Press, 1960.

Dowling, P.J., Knowles, P., and Owens, Q.W., *Structural Steel Design*, The Steel Construction Institute and Butterworths, London, 1988, p. 399.

ENV 1993 – 1 – 4, *Eurocode 3 Design of Steel Structures*, Part 1.4, 'General Rules and Supplementary Rules for Stainless Steels', CEN, 1992.

Euro Inox, 'European Stainless Steel Development Group, *Design Manual for Structural Stainless Steel*, Nickel Development Institute, Toronto, Canada, 1994.

Feld, J. and Carper, K., *Construction Failures*, Second edition, Wiley, New York, 1997.

Gupta, T.N., ed., *Building Materials in India: 50 years—A Commemorative Volume*, Building Materials and Technology Promotion Council, New Delhi, 1998, p. 560.

Haseltine, B.A., 'Comparison of Energy Requirements for Building Materials and Structures', *The Structural Engineer*, Vol. 153, No. 9, 1975, pp. 357–365.

Hoshino, K., 'Market Development for Stainless Steel', *Steel Today and Tomorrow*, No. 155, Sep 2001, pp. 5–8.

IS:6603 – 1972, *Specification for Stainless Steel Bars and Flats*, Bureau of Indian Standards, New Delhi, 1972.

IS:1852-1985 – *Rolling and Cutting Tolerances for Hot Rolled Steel Products*, Fourth revision, Bureau of Indian Standards, New Delhi.

IS:1161-1979, *Specification for Steel Tube for Structural Purposes*, Third Edition, Bureau of Indian Standards, New Delhi.

IS:158147-1976 – *Code of Practice for Use of Aluminium Alloys in Structures*, Bureau of Indian Standards, New Delhi.

IS:2062-2011 *Hot Rolled Medium and High Tensile Structural Steel-Specification,* Seventh Revision, Bureau of Indian Standards, New Delhi

IS:4923-1997 – *Standard Specification for Hollow Steel Sections for Structural Use,* Second edition, Bureau of Indian Standards, New Delhi.

IS:801-1975 - *Code of Practice for Cold-formed Light Gauge Steel Structural Members in General Building Construction*, First Revision, Bureau of Indian Standards, New Delhi.

IS:806-1968 – *Code of Practice for Use of Steel Tubes in General Building Construction,* First Revision, Bureau of Indian Standards, New Delhi.

IS:8500-1991, *Structural Steel—Micro Alloyed (Medium and High-strength Qualities) – Specifications*, First revision, Bureau of Indian Standards, New Delhi.

Jamshidi, M., Fallaha, S., Azizinamini, A., Price, K., and Cress, M. 'Application of High-performance Steel in Steel Bridge Construction', *Building to Last Structures Congress Proceedings*, Vol. 1, 1997, ASCE, New York.

Kavanagh, T., 'Some Aesthetic Considerations in Steel Design', *Journal of the Structural Division*, ASCE, Vol. 101, No. 11, 1975, pp. 2257–2275.

Kaminetzky, D., *Design and Construction Failures: Lessons from Forensic Investigations*, McGraw Hill, New York, 1991.

Kevin, L.R., Clark, M., and Knott, A.W., 'Failure Analysis Case Information Disseminator', *Journal of Performance of Constructed Facilities*, ASCE, Vol. 16, No. 3, Aug 2000, pp. 127–131.

Kulak, G.L., and Grondin, G.Y., *Limit States Design in Structural Steel*, Seventh edition, Canadian Institute of Steel Construction, 2002.

Kulicki, J.M., 'The Shape of Things to Come', *Civil Engineering*, ASCE, Vol. 70, No. 2, Feb 2000, pp. 40–41.

Levy, M. and Selvadori, M., *Why Buildings Fall Down: How Structures Fail*, W.W. Norton, New York, 1992.

Llewellyn, D.T. and Hudd, R.C., *Steels: Metallurgy and Application*, Butterworths–Heineman, Third edition, 1998, p. 400.

Lwin, M.M., *High Performance Steel Designers Guide*, Second edition, US Department of Transportation, San Francisco, USA, 2002 (http://www.steel.org/infrastructure/bridges/high-performance/hps-designersguide/intro.htm), 30 Nov 2004.

Madhu Mohan, V., Devadas Manoharan, P., and Santhakumar, A.R., 'Towards a Consistent Design Criteria for the Transforming Indian Structural Steel Design Methodology', *All India Seminar on Modern Trends in Steel Structures*, 8–9 Feb 2002, Nagpur, pp. 92–101.

Martin, L.H., and Purkiss, J.A., *Structural Design for Steelwork to BS 5950*, Edward Arnold, London, 1992, p. 468.

Mathur, N.C., 'Stainless Steel in Architecture, Building and Construction', *Steel in Construction 2000*, Joint Plant Committee, Govt of India, Feb 2000, pp. 179–193.

McCormac, J.C. and Nelson, J.K., *Structural Steel Design: LRFD Method*, Third edition, Prentice Hall Inc., Englewood Cliffs, New Jersey, 2002, p. 736.

McGuire, W., *Steel Structures*, Prentice Hall, Englewood Cliffs, New Jersey, 1968.

Nethercot, D.A., *Limit States Design of Structural Steelwork*, Third edition, Spon Press, London, 2001, p. 270.

Pannel, J.P.M., *An Illustrated History of Civil Engineering*, Thames and Hudson, 1964.

Peterson, Ch., *Stahlbau–Grundlagen der Berechnung und baulichen Ausbuildung von Stahlbauten*, Second edition, Vieweg Verlag, Braunschweig, 1990, p. 1413.

Rangwala, S.C., Rangwala, K.S., and Rangwala, P.S., *Engineering Materials*, Twenty second edition, Charotar Publishing House, Anand, 1997.

Rasmussen, K. Jr., *Full Range Stress–Strain Curves for Stainless Steel Alloys*, Research Report No. R 811, Department of Civil Engineering, The University of Sydney, November 2001, p. 45.

Roark, J.R. and Young, W.C., *Formulae for Stress and Strain*, Fifth edition, McGraw Hill, 1975.

*Salem Steel—Users Guide*, Steel Authority of India, Salem Steel Plant, Salem, India, 1998, p. 48.

Salmon, C.G. and Johnson, J.E., *Steel Structures—Design and Behaviour*, Fourth edition, Harper Collins College Publishers, New York, 1996, p. 1024.

Sreevidya, S. and Subramanian, N., 'Aesthetic Appraisal of Antenna Towers', *Journal of Architectural Engineering*, ASCE, Vol. 9, No. 3, Sep 2003, pp. 103–108.

Subramanian, N., Aesthetics of Non-habitat Structures, *The Bridge and Structural Engineer*, Journal of ING/IABSE, Vol. 17, No. 4, 1987, pp. 75–100.

Subramanian, N., 'India's first Stainless Steel Space Frame Roof', *Fifth International Conference on Space Structures*, University of Surrey, UK, Aug 2002.

Subramanian, N., 'New Steel Beams Increase Efficiency and Reduce Costs', *Civil Engineering and Construction Review*, Vol. 5, No. 7, Jul 92, p. 29.

Subramanian, N., *Principles of Space Structures*, Second edition, Wheeler Publishing, Allahabad, 1999, p. 666.

Subramanian, N., 'Restoration of Cold Storage Factory Building', *Journal of Performance of Constructed Facilities*, ASCE, Vol. 14, No. 4, Nov. 2000, pp. 155–159.

Subramanian, N., 'Improper Detailing Results in Delayed Commissioning', *Journal of Performance of Constructed Facilities*, ASCE, Vol. 13, No. 1, Feb 1999, pp. 29–33.

Subramanian, N. and Mangalam, S., 'Restoration of a Factory Building', *Journal of the Institution of Engineers (India)*, Vol. 79, May 1998, pp. 14–17.

Subramanian, N. and Mangalam, S., 'Distress to Greenhouses due to Improper Execution', *Civil Engineering and Construction Review*, Vol. 10, No. 8, Aug 1997, pp. 45–48.

Subramanian, N., 'Composite Floors Speed Up Construction', *Civil Engineering and Construction Review*, Vol. 5, No. 1, Jan 1992, p. 18.

Subramanian, N., 'Diagnosis of the Causes of Failures', *The Bridge and Structural Engineer*, Vol. 19, No. 1, Mar 1989, pp. 24–42.

Subramanian, N., 'Why Buildings Collapse?' *Science Reporter*, Vol. 24, No. 5–6, May–June 1984, pp. 232–238.

Subramanian, N., 'The Failures that Changed the Perception of Our Designs', *The Bridge and Structural Engineer*, Journal of ING/IABSE, Vol. 44, No. 4, Dec 2014a, pp. 29–51.

Subramanian, N., 'Can Architecture be Absurd?', *The Master Builder*, Vol. 16, No. 11, Nov 2014b, pp. 1–7.

Tall, L., *Structural Steel Design*, Ronald Press, New York, 1974.

Theling, K.E., *Steel and its Heat Treatment*, Butterworths, 1984.

Thornton, C.H., 'Quality Control in Design and Supervision can Eliminate Lamellar Tearing, *Engineering Journal*, AISC, Vol. 10, No. 4 (Fourth quarter), pp. 112–116, 1973.

Timoshenko, S.P. and Goodier, J.N., *Theory of Elasticity*, Third edition, McGraw Hill, New York, 1970.

Trahair, N.S., Bradford, M.A., and Nethercot, D.A., *The Behaviour and Design of Steel Structures* to BS 5950, Third edition, Spon Press, London, 2001, p. 448.

Yu, W.W., *Cold-formed Steel Design*, Third edition, Inter Science, 2000, p. 848.

# 2

# BASIS OF STRUCTURAL DESIGN

## 2.1 DESIGN CONSIDERATIONS

Structural design, though reasonably scientific, is also a creative process. The aim of a structural designer is to design a structure in such a way that it fulfils its intended purpose during its intended lifetime and be adequately safe (in terms of strength, stability, and structural integrity), and have adequate serviceability (in terms of stiffness, durability, etc.). In addition, the structure should be economically viable (in terms of cost of construction and maintenance), aesthetically pleasing, and environment-friendly.

Safety is of paramount importance in any structure, and requires that the possibility of collapse of the structure (partial or total) is acceptably low not only under normal expected loads (service loads), but also under less frequent loads (such as due to earthquakes or extreme winds) and accidental loads (blasts, impacts, etc.). Collapse due to various possibilities such as exposure to a load exceeding the load-bearing capacity, overturning, sliding, buckling, fatigue fracture, etc. should be prevented.

Another aspect related to safety is stability—the structure as a whole should be stable under all conditions. (Even if a portion of it is affected or collapses, the remaining parts should be able to redistribute the loads.) In other words, *progressive failure* should be minimized.

*Serviceability* is related to the utility of the structure—the structure should perform satisfactorily under service loads, without discomfort to the user due to excessive deflection, cracking, vibration, etc. Other considerations of serviceability are durability, impermeability, acoustic and thermal insulation, etc. It may be noted that a design that adequately satisfies the safety requirement need not satisfy the serviceability requirement. For example, an I-beam at the roof level may have sufficient stiffness for applied loads but may result in excessive deflections, leading to cracking of the slab it is supporting, which will result in loss of permeability (leaking). Similarly, exposed steel is vulnerable to corrosion (thereby affecting durability).

Increasing the design margins of safety may enhance safety and serviceability, but increase the cost of the structure. For overall economy one should look into not only the initial cost but also the *life-cycle cost* and the long-term environmental effects. For example, using a very-high-strength steel to reduce weight often will not reduce cost because the increased unit price of high-strength steel will make the lighter design more costly. In bridges and buildings, the type of corrosion and fire protection selected by the designer will greatly influence the economy of the structure.

While selecting the material and system for the structure the designer has to consider the long-term environmental effects. Effects considered include maintenance, repair and retrofit, recycleability, environmental effects of the demolished structure, adoptability of fast track construction, demountability, and dismantling of the structure at a future date.

## 2.2 STEPS INVOLVED IN CONSTRUCTION

The construction of any structure involves many steps. Although the structural designer is not responsible for each of these steps, he should be involved in most of them so that the resulting structure is safe, stable, serviceable, durable, and is economically viable and aesthetically pleasing, and does not have an adverse impact on the environment. The necessary steps may be listed as follows.

1. A prospective owner identifies a location and arranges the necessary finance for construction. He also chooses the architect or project manager who, in turn, chooses the various consultants (structural, geotechnical, survey, etc.).
2. A land surveyor surveys the land and draws the contours.
3. An architect or engineer (project manager) studies the applicable bye-laws and draws a plan of the structure in

such a way that it meets the fire protection, health, and safety requirements.

4. The competent authority approves the plan, by checking it against the local bye-laws.
5. The geotechnical engineer investigates the site conditions, level of water table, nature of soil (whether expansive or not), etc., and gives a soil report.
6. The form, shape, and size of the structure are determined by the architect with the help of the structural engineer (based on preliminary design), such that the resulting structure is stable, economical, and resists the external loads efficiently.
7. Suitable materials of construction (steel, concrete, wood, brick, plastics, etc.) are selected after considering the required performance, cost, availability of labour, and cost of transportation to site. While choosing the materials, consideration should be given to design and detailing procedures, and procedures for shop fabrication and field construction.
8. The structural engineer estimates the probable loads (dead, live, wind, snow, earthquake, etc.) that are acting on the structure.
9. The structural engineer arrives at the structural system after comparing various possible systems. In a building, heating and air-conditioning requirements or other functional requirements may dictate the use of a structural system that is not the most efficient from a purely structural viewpoint, but which is the best, bearing the total building in mind. While choosing structural forms, layouts and details, the following should be considered.
   (a) The structure should have low sensitivity to hazardous conditions.
   (b) The structure should suffer only local damage even after serious damage to any one individual element by the hazard.
   (c) The structure should give ample warning before any collapse (should have various load paths and redistribution of loads).
10. A suitable structural analysis, mostly with the aid of computers, is done to determine the internal forces acting on various elements of the structural system, based on the various loads and their combinations.
11. Considering the critical loading conditions, the sizes of various elements are determined following the codal provision. The design should be done in such a way that
   (a) the structure will remain fit with adequate reliability and be able to sustain all actions (loads) and other influences experienced during construction and use,
   (b) the structures should have adequate durability under normal maintenance, and
   (c) the structure should not be seriously damaged or collapse under accidental events such as explosions, impacts, or due to consequences of human error.
12. The detailed structural drawings are then prepared once again following codal provisions and approved by the structural engineer.
13. The architect or project manager develops detailed architectural drawings and specifications.
14. The estimator arrives at the quantities involved and the initial cost of construction.
15. Based on these quantities, a tender for the building is floated.
16. Comparing the cost quoted by different tenders, the general contractor for the structure is chosen.
17. The contractor, based on the structural drawings, prepares the fabrication and erection drawings and a bill of quantity of materials (BOQ). The structural engineer again approves these drawings.
18. The contractor constructs the building based on the specifications given by the architect/project manager. During the construction, the contractor consults the architect/project manager if any changes have to be made due to the site conditions. The structural engineer must also convey to the fabricator and erector his or her concept of the structure and specific methods of execution (if any). For example, the 'push-up' system of erection of domes using the *Pantadome system* involves some specific erection sequences (Subramanian 1999).
19. The structural engineer, with the help of quality control inspectors, inspects the work of the fabricator and erector to ensure that the structure has been fabricated/erected in accordance with his or her designs and specifications. Similarly the architect/project manager also inspects the construction periodically to check whether it is built as per the specifications.
20. In some important buildings, 'as-built' drawings are prepared as a permanent record of the building.
21. After the structure is constructed and handed over to the owner, the owner, by appointing suitable consultants and contractors, maintains the building till its intended age.

From these steps, it may be clear that accurate calculations alone may not produce safe, serviceable, and durable structures. Suitable materials, quality control, adequate detailing, good supervision, and maintenance are also equally important.

These 21 steps briefly summarize the various activities involved in the construction of a structure. While executing the various steps, the structural engineer has to interact with the architect/project manager and also with others (electrical engineers, mechanical engineers, civil engineers, geotechnical engineers, surveyors, urban planners, estimators, etc.), and incorporate their requirements into the design (e.g., load due to mechanical and electrical systems). It has to be noted that steps 8 to 14, which are followed mainly in the design office, are not straightforward operations but are iterative, as shown in Figs 2.1 and 2.2. This book mainly covers only step 11—the design of structural elements to safely carry the expected loads and to ensure that the elements and the structure perform satisfactorily. Some guidelines and discussions are included about steps 7 and 10.

Compared to analysis (where all the parameters are known), design is a creative process. It involves the selection of the span, assessment of loads, and the choice of material, cross section, jointing method and systems, etc. Hence, there is no unique solution to a design problem. The designer has to make several decisions, which will affect the final construction and its cost. Therefore, the designer has to use his engineering judgment in order to reduce the cost and arrive at an efficient solution to the problem.

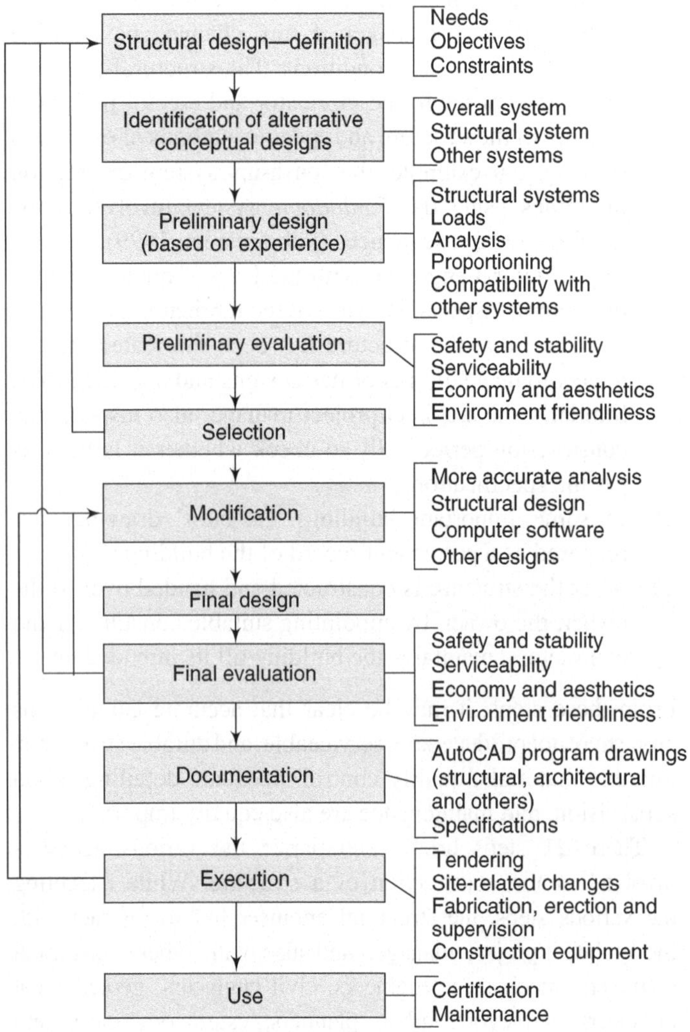


**FIG. 2.1** Overall iterative design process

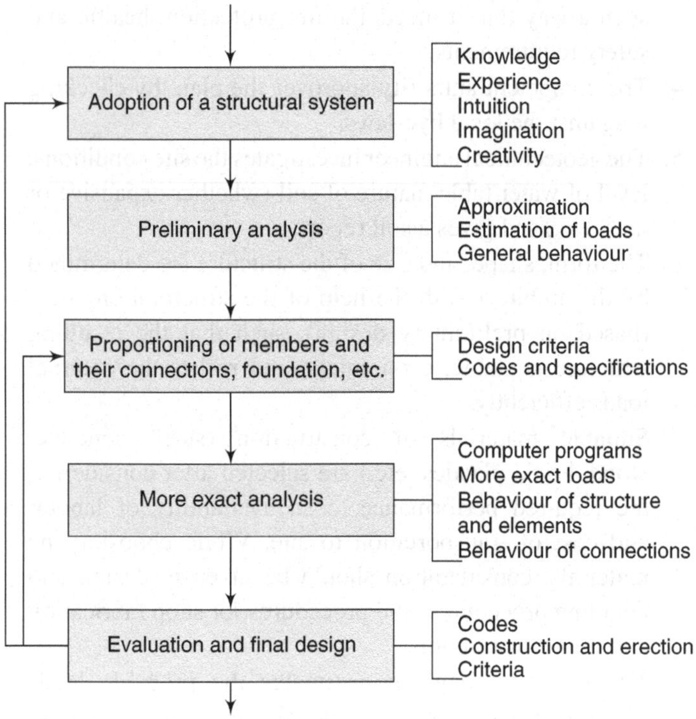


**FIG. 2.2** Iterative structural design process

## 2.3 ROLE AND RESPONSIBILITIES OF THE DESIGNER

The objective of design should be the achievement of an acceptable probability that the structure will perform satisfactorily for the intended purpose during the design life. With the appropriate degree of safety, the structure should sustain all the loads and deformations during construction and its designed life, and also have adequate resistance to accidental loads and fire.

As already discussed, the designer has to consider several factors while designing the structure. These factors include

1. material to be used,
2. arrangement of the structural system (e.g., bracing and flooring system),
3. type of jointing (welding, bolting, etc.),
4. method of fabrication and erection,
5. installation of services (lift, water supply, power, ventilation, heating and cooling, etc.),
6. type of cladding, roofing, etc.,
7. safety, economy, and aesthetics,
8. required corrosion protection,
9. required fire protection, and
10. operating/maintenance and life-cycle costs.

It is the structural designer's job to ensure that the best structural system is selected, within the scope of the imposed constraints. Today's structural engineer has several aids such as computer programs, handbooks, and charts, and hence should spend more time on thinking about design concepts and select the best structural system for the project at hand.

For most structures, the designer should specify a grade of structural steel that is readily available, keep the structural layout and structural details (e.g., connections) as simple as possible, use sections that are readily available, and use the maximum possible repetition of member sections and connection details. (It has to be noted that all the sections listed in IS 808 are not readily available in the market. However, some steel mills are ready to supply the required section, if a minimum prescribed quantity is ordered.)

The designer should also have some knowledge about the weld defects. [Many times, the detected weld 'defect' is not detrimental to the behaviour of the structure, while the attempted repair of the defect may introduce a condition that is more detrimental! Hence the designer, for the guidance of inspectors at site, must establish an acceptance level for defects (Nethercot 2001).] The designer should also be aware of the non-destructive testing techniques that are available for testing high-strength bolted joints and welded joints. Though it is highly impractical to test each and every joint, it may be prudent to test most critical joints, especially in important structures (e.g., a welded splice in a highly stressed bridge girder tension flange). It is preferable for the designer to have a knowledge of fabricating shop capabilities (e.g., size of available zinc baths for galvanization) and erection techniques.

It would be useful for the structural designer to study the descriptions of actual projects (available in several magazines and journals), discuss technical matters with practising engineers or attend technical meetings organized by professional bodies, visit fabrication shops and construction sites, and visualize and appreciate the structural behaviour of various structural systems (Nethercot 2001). In comparison with these aspects of actual design practice, the actual proportioning of members, detailing of connections, etc., are normally much more straightforward.

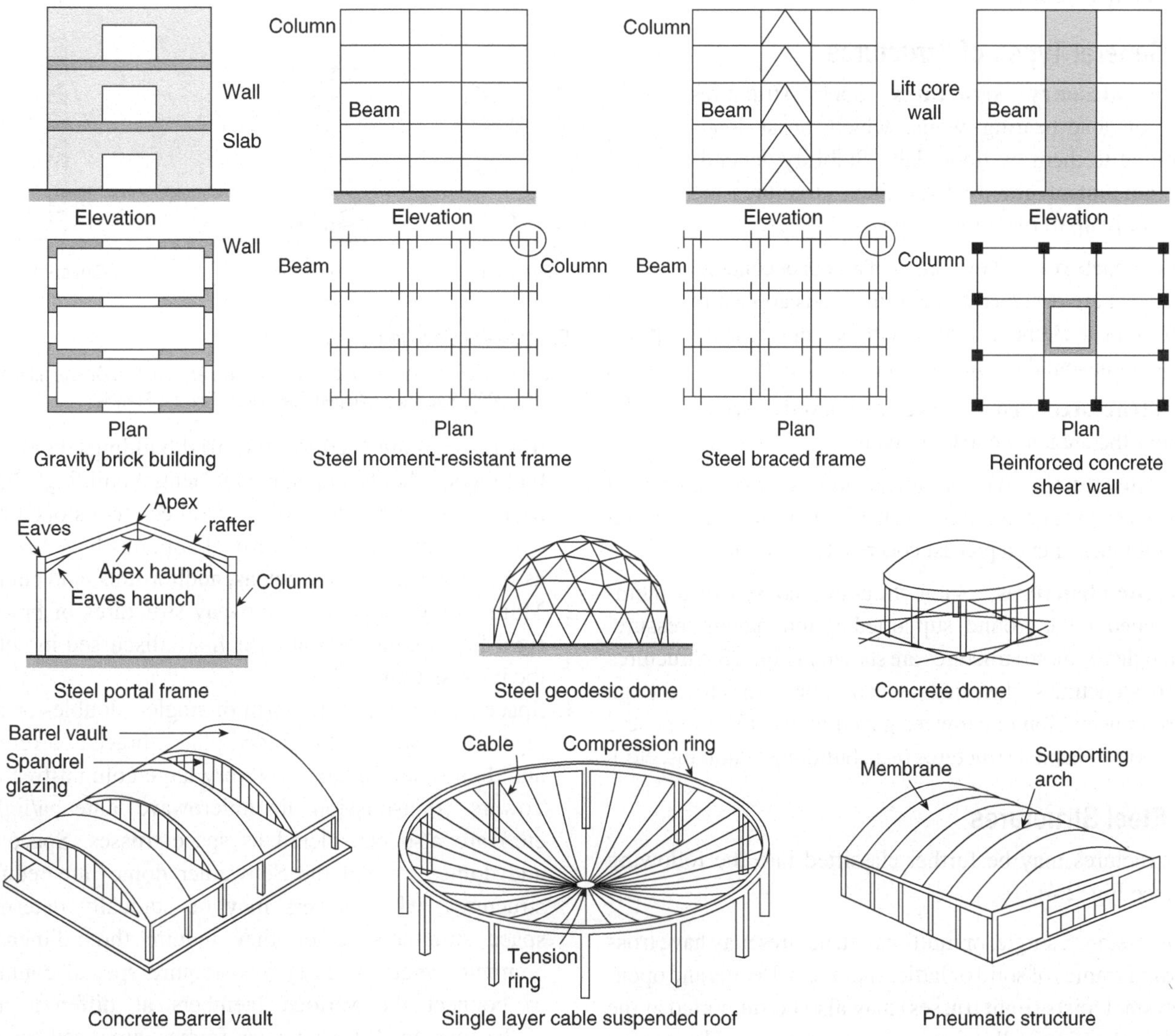


**FIG. 2.3** General types of structures

## 2.4 STRUCTURAL SYSTEMS

The art of structural design is manifested in the selection of the most suitable structural system for a given structure. The arrangement of beams/girders/joints or trusses, and columns to support the vertical (gravity) loads and the selection of a suitable bracing system or a column and beam/truss arrangement to resist the horizontal (lateral) loads poses a great challenge to the structural engineer, since they will determine the economy and functional suitability of the building. In bridge design, the choice of a continuous or simple span structure, plate girders, box girders or trusses, arches, cable suspension/cable-stayed girders, steel orthotropic deck (bridge floor) or concrete deck will determine not only economy but also the resulting aesthetics of the bridge (Subramanian 1987). The selection of a suitable system is made mainly based on previous data or experience.

Based on the way structures resist loads, they may be classified as follows.

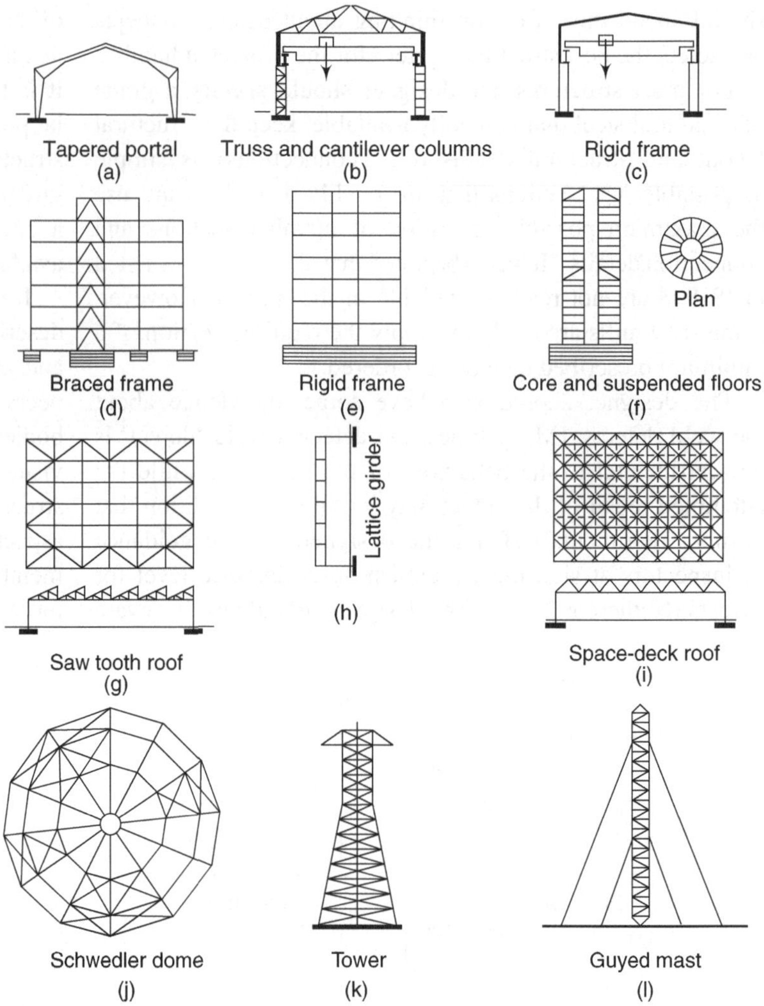


**FIG. 2.4** Examples of steel-frame structures

*Source:* Fig. 1.2 of Al Nageim and MacGinley, *Steel Structures, Practical Design Studies*, Third edition, CRC Press, 2005. Reprinted with permission from Taylor & Francis.

### 2.4.1 General Types of Structures

**Gravity masonry structures** Such structures consist of load-bearing walls, which resist loads transmitted to them by floor slabs. Stability depends on the amount of gravity load. Such structures are suitable only up to two or three floors.

**Framed structures** They consist of a steel or concrete skeleton that are designed to resist the loads applied to them from plate elements (concrete floors and masonry walls) and transmit these loads to the foundations.

**Shell structures** They have a curved surface enclosing the area and carrying loads.

**Tension structures** Tension structures consist of cables, which span between anchor structures (in the form of steel beams), and use membranes or even precast concrete to cover the area.

**Pneumatic structures** Such structures consist of a membrane sealed to the ground, supported by internal air pressure.

Examples of these structures are shown in Fig. 2.3. Structures are also sometimes classified as non-habitat structures (e.g., bridges, transmission line towers, guyed masts, TV towers, and chimneys) and habitat structures (e.g., buildings, stadiums, etc.).

### 2.4.2 Steel Structures

Steel structures may be further classified into the following types (see Fig. 2.4).

1. Single-storey, single, or multi-bay structures may have truss or rigid frames of solid or lattice members. Beams and open-web steel joists (light trusses) may also be supported at the ends by bearing walls of masonry construction. These types of structures are used for industrial buildings, commercial buildings, schools, and some residential buildings. Pitched roof portal frames consisting of rolled-steel sloped beams connected by welding to vertical columns have been used as industrial structures, arenas, auditoriums, and churches.
2. Multi-storey, single, or multi-bay structures of braced or rigid frame construction (which are discussed in detail in the next section).
3. Space structures, in the form of single-, double- or multi-layer grids, steel-frame folded plates, braced barrel walls, and domes, are required for very large column-free areas. Towers (transmission line/microwave antenna/lighting, etc.) are also considered as space trusses. Space grids and domes (except the Schwedler dome) are redundant structures, while towers may be statically determinate space structures. Often they require three-dimensional computer analysis. They also require special connectors to connect the various members at different angles (Subramanian 1999). Space frames are used to cover

**TABLE 2.1** Categories of main structural systems

| Type of structure | Use and example | Main considerations in design |
|---|---|---|
| Single-storey structures | Low rise, lightly loaded, open-web joist, trusses | Usually simple analysis and design |
| Steel frame | Braced or rigid frame with wide variety of types | Simple, rigid and semi-rigid connections will dictate the complexity of design |
| Long span | Large column-free areas; double- and multilayer grids, domes, etc. | Three-dimensional analysis and special connectors |
| Tension | Large spans, stressed skin domes, and folded plates | Proper connection between the skin of the roof and the main load-bearing members |
| High rise | Tall buildings having more than 20 storeys, framed tube, braced tube, tube in tube and bundled tube | Resistance to lateral forces |

large spans such as those occurring in large arenas, auditoriums, swimming pools, theatres, airport hangars, tennis or baseball grounds, ballrooms, etc.

4. Tension structures, tensegritic and cable-supported (cable-suspended or cable-stayed) roof structures (Subramanian 1999).
5. In a stressed skin structure, where the cladding is also designed as a load-bearing member, thus stabilizing the structure, special shear connections are necessary, in order that the sheeting acts integrally with the main frames of the structure (Subramanian 1999). It is possible to reduce the cost by considering the stiffening effect of the sheeting.
6. High-rise constructions: Tall buildings with more than 20 storeys are often considered in large cities, where land costs are very high. In the design of such structures, the designer should pay attention to the system resisting the lateral loads (MacGinley 1999). Several interesting systems have been developed, and a few are discussed in the next section.

Table 2.1 summarizes these structural systems and their design considerations.

It has to be noted that combinations with concrete (in the form of *shear walls* or floor slabs) are structurally important in many buildings. If adequate interconnection between the concrete slab and the steel beam is provided in buildings and bridges (in the form of *shear connectors*), the resulting system, called *composite construction*, is both structurally and economically advantageous. Braced, rigid frame, truss roof, and space-deck construction are shown in Fig. 2.4 for comparison. Only framed structures are discussed in detail in the book. Space structures, tension structures, stressed skin structures, and shell types, e.g., tanks, are not considered. Analysis, design, and construction aspects of space frames, tension structures, and stressed skin systems are available in Subramanian (1999). Details of the design of composite constructions are available in Kulak and Grondin (2002), Salmon and Johnson (1996), and Johnson (1994).

For framed structures, the main elements are the beam, column, beam-column, tie, and lattice member. Beams are those elements which are subjected to bending moments, and shear forces only and columns are subjected to axial loads (tensile or compressive). Beam-columns are subjected to axial loads as well as bending moments. In some special cases, beams are subjected to torsional moments. Beams and columns can be rolled or built-up H, box, or I. For long-span constructions, for example, to support the slab of a bridge, normal rolled sections may not have sufficient depth to act as beams. In such cases, deep welded plate girders, box girders, castellated girders, open-web joists or trusses may replace them. For very long spans, deep trusses or arches may be necessary.

## Steel-framed Buildings

Most steel structures belong to the category of braced and rigid frame construction. They are suitable for large industrial buildings, commercial buildings, schools, or for multi-storey buildings, which can be office buildings, hospitals, or residential buildings. They consist essentially of regularly spaced columns joined by beams or girders. Secondary beams span between these main beams and provide support to the concrete floor or roof sheeting. Depending on the type of beam-column connections employed, such systems may be classified as *simple construction* or as *continuous construction*.

**Simple construction** In simple construction (see clause 4.2 and F.4 of the code), the ends of beams and girders are connected to transmit transverse shear only and are free to rotate under load in the plane of bending. Hence hinged ends are assumed for the beams. Connections are usually made by welding plates or angles to a beam or column in the fabricator's shop and bolted at site to the connecting beam or column (see Fig. 2.5).

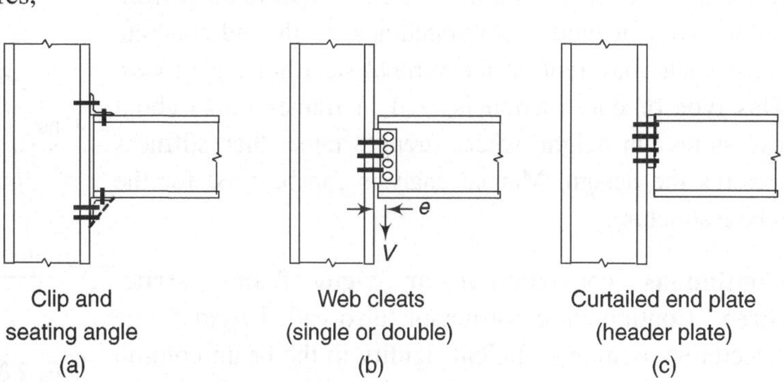


**FIG. 2.5** Simple beam to column connections

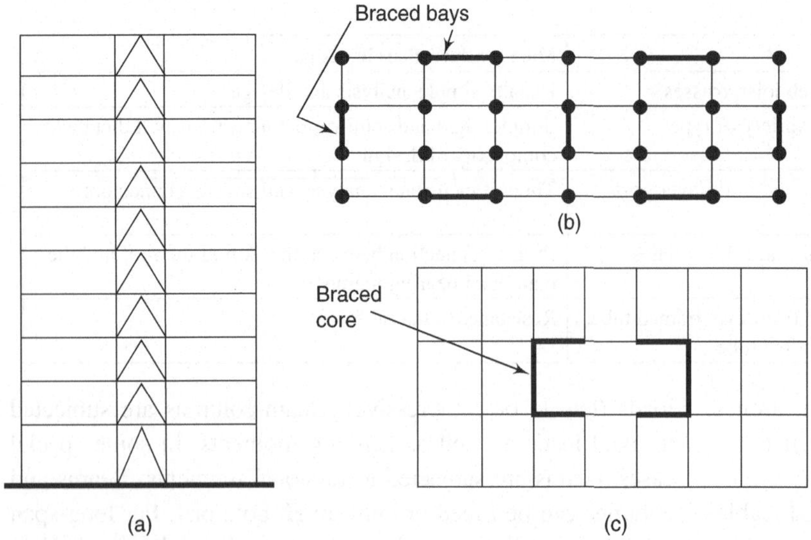


**FIG. 2.6** All-steel braced structures (a) Vertical bracing (b) Bracing on perimeter/interior walls (c) Bracing around core

*Source:* Fig. 7.1 of Al Nageim and MacGinley, *Steel Structures, Practical Design Studies*, Third edition, CRC Press, 2005. Reprinted with permission from Taylor & Francis

These simple constructions are statically determinate and hence the beams are designed as simply supported and the columns are designed for the axial loads (due to the reaction from the beams) and the moments produced by the eccentricity of the beam reactions as shown in Fig. 2.5(b). (A minimum distance of 100 mm from the face of the column is specified in the code clause 7.3.3.1.) In such frames lateral forces due to the wind or earthquake are generally resisted by bracings (usually made of angles), forming vertical or horizontal trusses as required.

The braced bays can be grouped around a central core, distributed around the perimeter of the building, or staggered through various elevations as shown in Fig. 2.6. The floors act as horizontal diaphragms to transmit load to the braced bays. *Bracing* must be provided in two directions and all connections are taken as pinned. The bracing should be arranged to be symmetrical with respect to the building plan, to avoid twisting. The unbraced portion of the building frame in effect 'leans' on the braced portion to keep from falling over. In multi-storey buildings, reinforced concrete shear walls may replace the vertical steel bracing trusses. This type of construction is used in frames up to about five storeys in height, where strength rather than stiffness governs the design. Manual analysis can be used for the whole structure.

**Continuous construction or rigid frame structures** Continuous construction (also called rigid frame structures) assume sufficient rigidity in the beam-column connections, such that under the action of loads the original angles between intersecting members are unchanged (see clause 4.2 and F4.2 of the code). Connections are usually made in the fabricator's shop as well as the site, by welding and bolting. The connections shown in Fig. 2.7 can be adopted in rigid frame construction, which transfer both shear and moment from beam to column. Fully welded connections can also be considered as rigid beam-to-column connections. Such connections naturally involve additional fabrication and higher erection costs. However, the greater rigidity produced in the structure due to its ability to develop flexural action results in reduced member sizes and the elimination of bracings. This form of construction is used for low-rise industrial buildings (Figs 2.4a and 2.4c) and for multi-storey buildings (Fig. 2.8a).

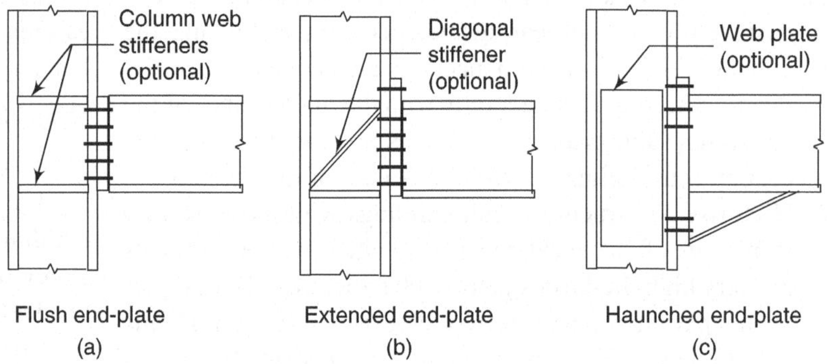


**FIG. 2.7** Rigid beam-to-column connections

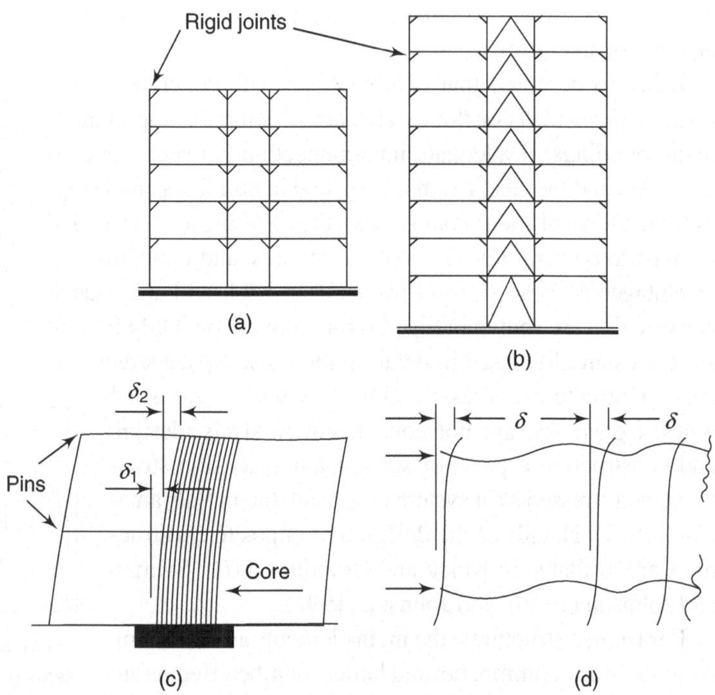


**FIG. 2.8** Continuous construction systems (a) Rigid frame (b) Rigid frame with bracing (c) Cantilever deflection (d) Rigid frame deflection

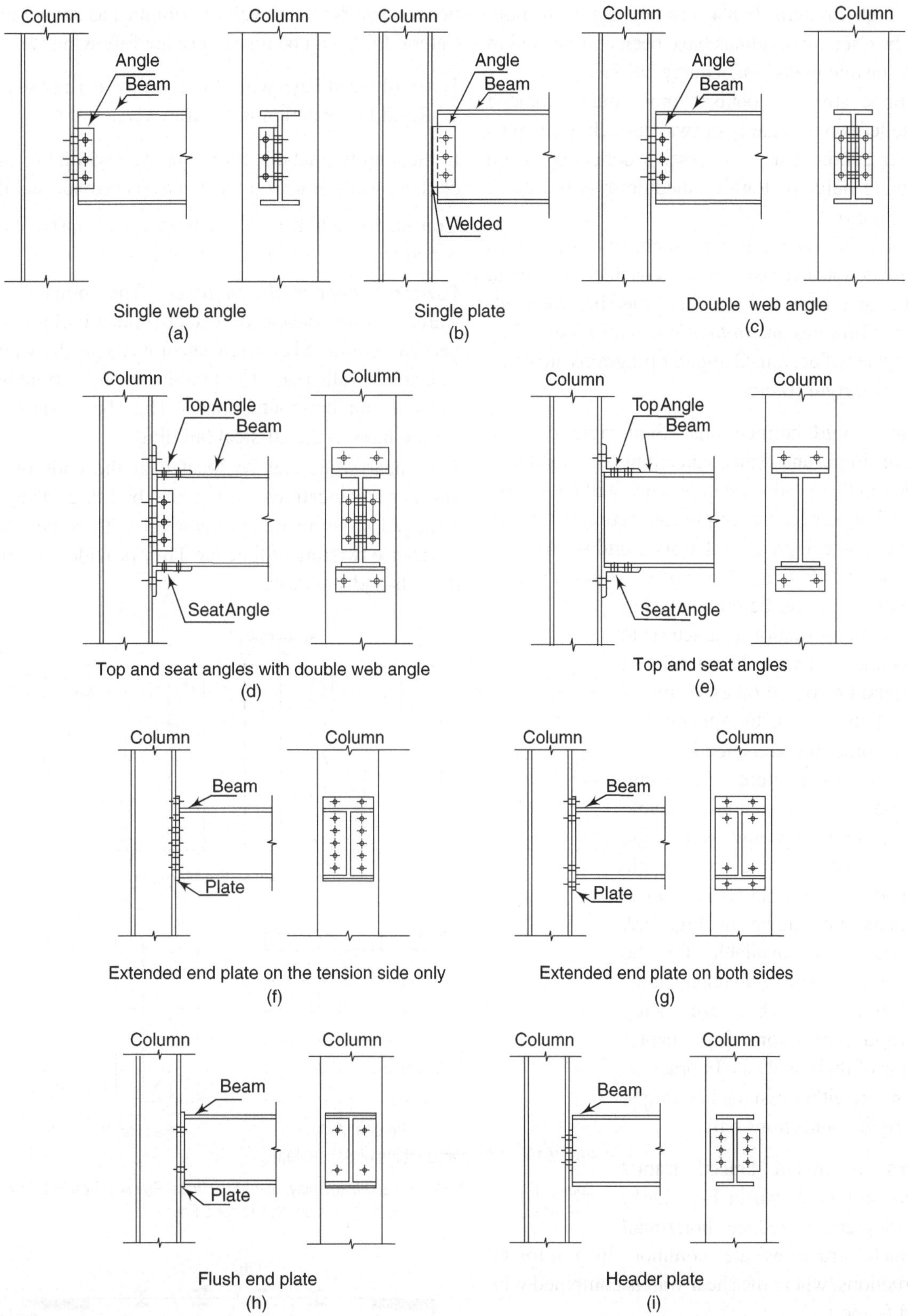


**FIG. 2.9** Examples of semi-rigid connections

In rigid frame structures, bending in beams and columns resists horizontal load. The columns, particularly in the upper storeys, must resist heavy moments. So sections will be much larger than in braced buildings.

The frame normally has H-section columns. It is rigid in one direction only, across the short span, and is braced longitudinally. As mentioned previously, the connections are expensive, being welded or made with beam ends with haunches and high-strength bolts. Frames rigid in both directions with box-section columns have been constructed in areas subject to seismic loads (MacGinley 1997).

The rigid frame structure deflects more than a braced structure. The deflection is made up of sway in each storey plus overall cantilever action. Due to excessive deflection, rigid frames are suitable only for low- or medium-rise buildings (up to about 15 floors).

The interaction between adjacent members present in continuous construction necessitates the consideration of at least a group of interconnected members together for analysis. Since such subframes are *statically indeterminate*, they require several cycles of design. Computer programs are often used to analyse such rigid frames.

**Frame with semi-rigid connections** Semi-rigid connections fall between simple and rigid connections. As a matter of fact, any connection that is adopted in practice will be a semi-rigid connection. This is because simple connections as shown in Fig. 2.5 do have some degree of rotational rigidity and rigid connections as shown in Fig. 2.7 also experience some degree of joint deformation. Hence, before analysing the frame, the moment-rotation characteristic of the adopted connection has to be established by a rational method or based on experiments (clause 4.2.1.2 of the code). In Appendix F, the code gives some recommendations for obtaining the moment-curvature relationship of single web-angle connections, double web-angle connections, top and seat angle connections (without double web-angle connections), and header plate connections. These connections are shown in Fig. 2.9. Computer programs are available for the analysis of frames with semi-rigid connections (Chen et al. 1995). Research is still being done on semi-rigid connections. See Chapter 4 for more details of their analysis. In practice, most connections are either assumed as simple connections or rigid connections only.

**Mixed systems** A mixed braced frame/rigid frame structure as shown in Fig. 2.8(b) can also be adopted to reduce horizontal deflections. (Such structures are common in reinforced concrete constructions, where the shear wall is combined with a concrete rigid frame.)

The different modes of deflection for the cantilever/core/braced bay and rigid frame sway are shown in the Figs 2.8(c) and (d).

**Staggered lattice girder system** This system, developed in the USA and shown in Fig. 2.10, is useful for long narrow buildings with a central corridor, e.g., hotels, classrooms, or offices. Storey-deep lattice girders, staggered on adjacent floors, span between wall or columns, as shown in the figure. Lateral loads can be resisted in the following two ways:

1. End-braced bays with floors acting as rigid diaphragms
2. Rigid frame action in the transverse frames

In the longitudinal direction, braced bays on the outside walls or shear walls at lift shafts/stairwells provide stability.

**Composite structures** Such structures can be of two kinds—concrete core structures, and concrete shear wall structures.

***Concrete shear wall structures*** The composite steel-shear wall structure consists of a steel-framed building braced with vertical reinforced concrete shear walls, as shown in Fig. 2.11. The shear walls placed in two directions at right angles carry vertical and horizontal loads. The shear walls replace the braced bays in the all-steel building.

The shear walls can be located at the ends or sides or in appropriate locations within the building. They should be arranged so as to be symmetrical with respect to the plan, otherwise twisting will occur. They provide fire-proof walls at the lifts and staircase.

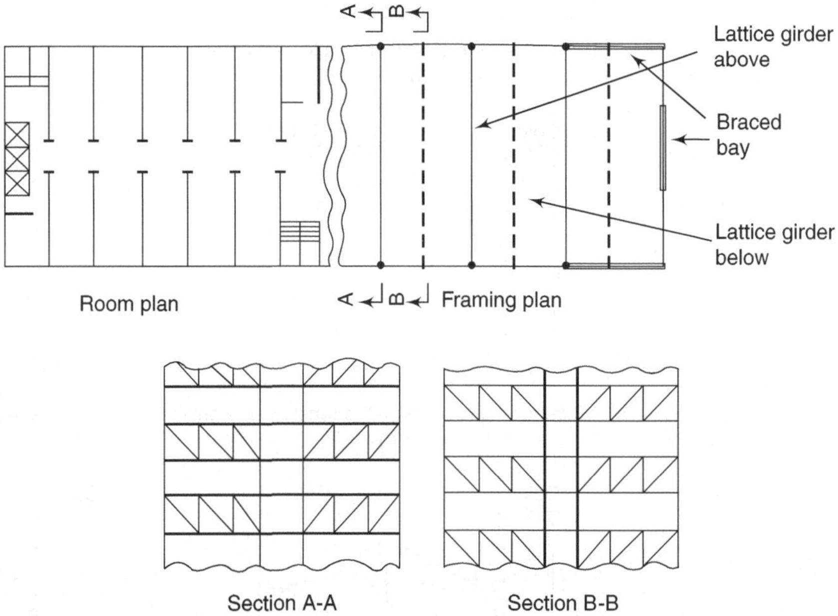


**FIG. 2.10** Staggered lattice girder system

*Source:* Fig. 7.2 of Al Nageim and MacGinley, *Steel Structures, Practical Design Studies*, Third edition, CRC Press, 2005. Reprinted with permission from Taylor & Francis.

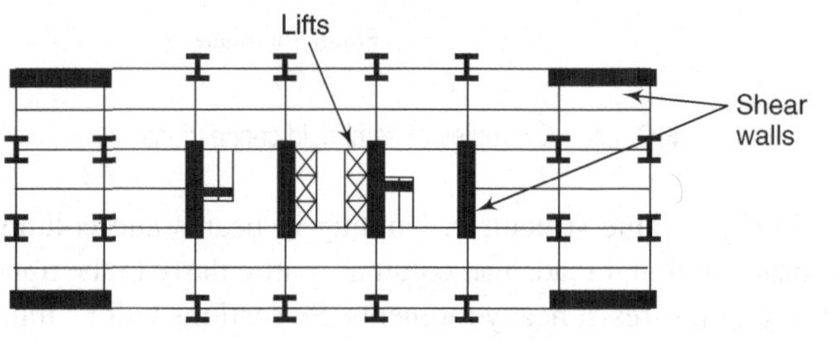


**FIG. 2.11** Steel building with concrete shear walls

***Concrete core structures (MacGinley 1997)*** The steel frame with a concrete core structure is very common, and adopted for offices (Fig. 2.12). The main advantages are as follows.

1. The space between the concrete core and the perimeter steel columns is column-free, resulting in maximum flexibility for division using lightweight partitions.
2. The concrete core provides a rigidly constructed fire-resistant shaft for lifts and staircases.
3. No bracing is required on the perimeter walls, resulting in uniform façade treatment on all faces.

The core is designed to resist all the wind loads on the building, the loads from lifts, stairs, etc., and part of the floor loads. The perimeter columns carry the remaining floor loads. The floor plan may be square, rectangular, or circular. The floor steel may be

1. supported on the core and perimeter columns,
2. cantilevered out from the core, and
3. suspended from an umbrella girder at the top of the core (Fig. 2.13).

Construction is rapid, using slip forming for the core, which is then used to erect the building. Construction can be carried out within the area of the building.

The core may be open or closed in form. Closed box or tubular cores are designed as vertical cantilevers. Open cores, generally of channel or H-section, are designed as connected cantilever shear walls. Cores may be of steel, reinforced concrete, composite steel and concrete, with steel columns at extremities or cased steel sections.

Where floor girders are cantilevered out from the core, two possible arrangements are as follows.

1. In a circular building, the girders can be supported from steel columns embedded in the core (Fig. 2.12a). The concrete core resists wind load.
2. In a square building, the cantilever girders on adjacent floors can be arranged to span at right angles to each other. This avoids the need for cantilevers to cross each other at right angles. Using hangers, the two edge beams on any floor (parallel to the cantilevers on that floor) are then suspended from the floor above (Fig. 2.12b).

**Suspended structures** In suspended structures, an umbrella girder is provided at the top of the core from which hangers for the outer ends of the floor beams are suspended (Fig. 2.13). In very tall buildings, additional umbrella girders can be introduced at intermediate locations along the height.

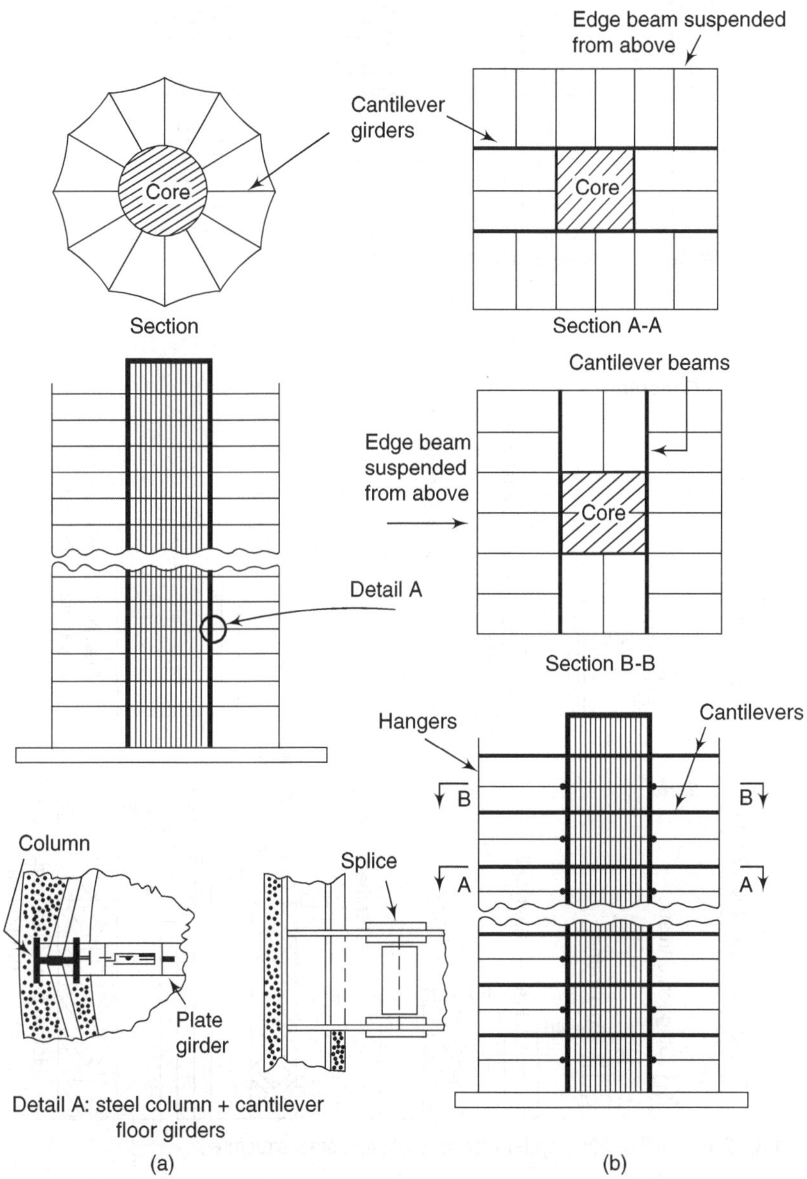


**FIG. 2.12** Buildings with cantilevered floors (a) Circular building (b) Square building

*Source:* Fig. 7.5 of Al Nageim and MacGinley, *Steel Structures, Practical Design Studies*, Third edition, CRC Press, 2005. Reprinted with permission from Taylor & Francis.

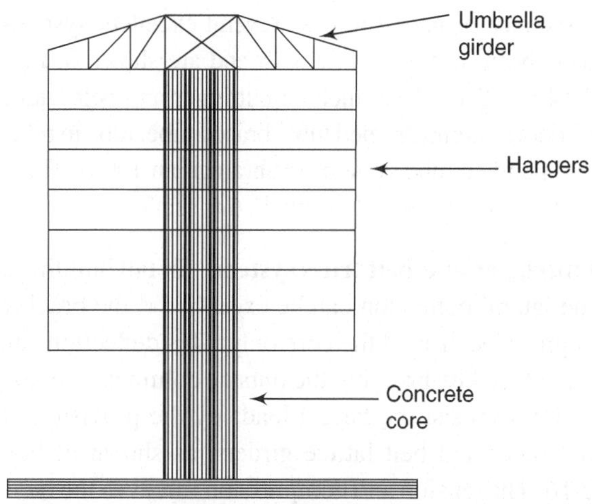


**FIG. 2.13** Suspended structure (may not be suitable for earthquake-prone areas)

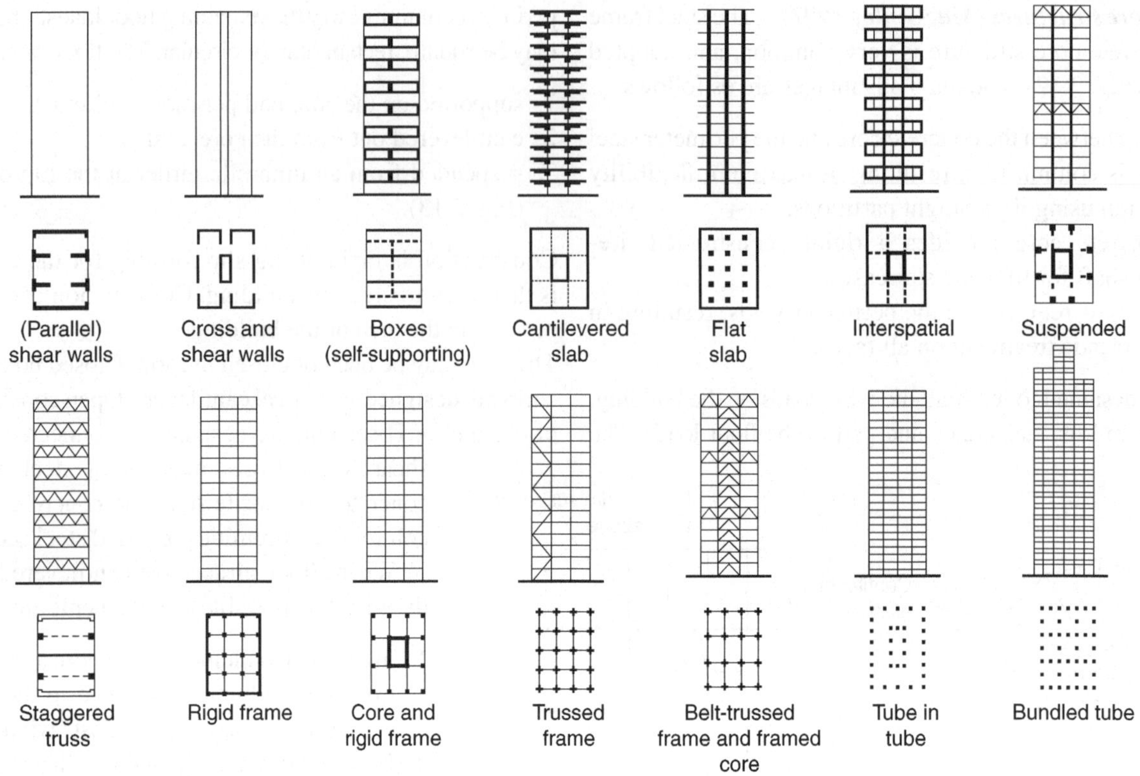


**FIG. 2.14** High-rise structural systems

All loads, both vertical and horizontal, are carried on the core in suspended structures. High-strength steel sections or cables are used for the suspension members. The time for erection can be shorter than for a conventional structure built upwards. The core is constructed first by slip forming from which the steelwork is suspended.

### 2.4.3 High-rise Structural Systems

As pointed out previously, to build tall buildings economically, the designer must pay attention to the resistance of lateral forces. Several excellent systems have been invented in the past and are shown in Figs 2.14 and 2.15. They include outrigger and belt lattice girders systems, framed tube, braced tube, tube in tube, and bundled tube systems (Subramanian 1995). These systems are discussed briefly in this section.

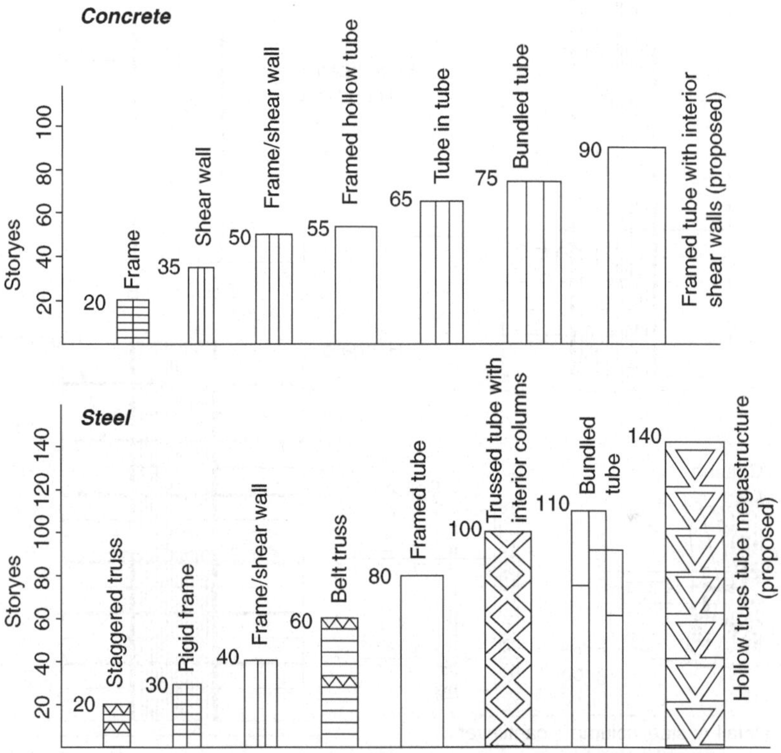


**FIG. 2.15** Tall building system for concrete and steel structures

**Outrigger and belt truss system** In tall buildings, the lateral deflection can be excessive if the bracing is provided around the core only. This deflection can be reduced by bringing the outside columns also into action to resist the lateral loads by the provision of outrigger and belt lattice girders, as shown in Fig. 2.16. The tension and compression forces in the outer columns apply a couple to the core, which acts against the cantilever bending under wind loads. The belt truss surrounding the building brings all external columns into action (MacGinley 1997). A single outrigger and belt lattice girder system at the top or additional systems at different heights of a very tall building can be provided.

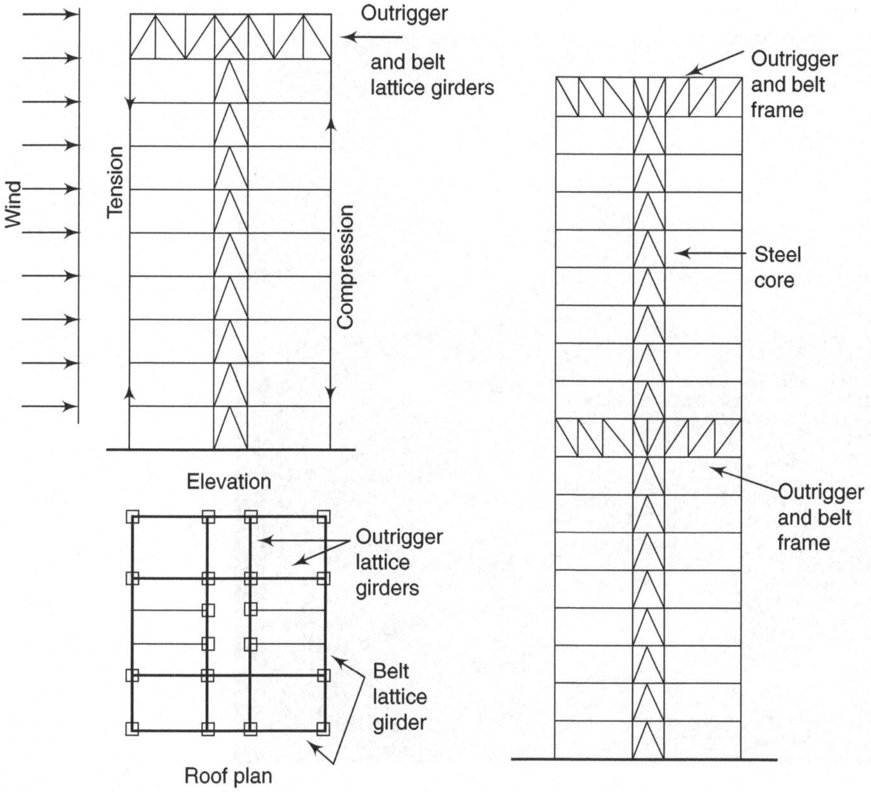


**FIG. 2.16** Outrigger and belt lattice girder system

*Source:* Fig. 7.3 of Al Nageim and MacGinley, *Steel Structures, Practical Design Studies*, Third edition, CRC Press, 2005. Reprinted with permission from Taylor & Francis.

**Tube structures** The tube type of structure was developed by Dr Fazlur Khan of the USA for very tall buildings, say over 80 storeys in height. If the core type of structure were used, the deflection at the top would be excessive. The tube system is very efficient with respect to structured material used and results in a considerable saving in material when compared with conventional designs.

In this system the perimeter walls are so constructed that they form one large rigid or braced tube, which acts as a unit to resist horizontal load. The framed tube shown in Figs 2.17(a) and (b) simulates the action of a perforated hollow tube by utilizing closely spaced exterior columns 1–3 m apart, tied at each floor level and interacting with deep spandrel (exterior wall) beams. The small perforations form spaces for windows, and the normal curtain walling is eliminated. This system was adopted for the 110-storey Twin World Trade Center Towers in New York, USA (which was destroyed by the terrorist attacks in 2001) and for the 72-storey First Bank Tower in Toronto, Canada.

In the single-tube structure, the perimeter walls carry the entire horizontal load and their share of the vertical load. Internal columns and/or an internal core, if provided, carry vertical loads only. In the tube-within-a tube system shown in Fig. 2.17(e), the internal tube can be designed to carry part of the horizontal load. Very tall stiff structures have been designed on the bundled tube system shown in Fig. 2.17(f), which consists of a number of tubes constructed together. This reduces the *shear lag* problem that is more serious if a single tube is used (MacGinley 1997). Shear lag is a phenomenon in which the stiffer (or more rigid) regions of the structure or structural component attract more stresses than the more flexible regions. Shear lag causes stresses to be unevenly distributed over the cross section of the structure or structural component.

## CASE STUDY

### Twin Towers of World Trade Centre

The twin 110-storey towers of World Trade Center in New York City, USA, were designed by Architect Minoru Yamasaki and Structural Engineers Leslie E. Robertson Associates in the early 1960s using a *framed tube* structural system. The North Tower was completed in December 1970 and the South Tower was finished in July 1971.

Both had a 63 m square plan, 350,000 m$^2$ of office space and the facades were sheathed in aluminum-alloy. The buildings were designed with narrow office windows 45 cm wide which also reflected Yamasaki's fear of heights as well as his desire to make building occupants feel secure.

The World Trade Center towers had high-strength, closely spaced load-bearing perimeter steel columns (acting like *Vierendeel trusses*) that supported all lateral loads like wind loads, and shared the gravity load with the core columns (see also Fig. 2.17a). The 59 perimeter columns (per side) were constructed using prefabricated modular pieces each consisting of three columns, three storeys tall, connected by spandrel plates. Adjacent modules were bolted together with the splices occurring at mid-span of the columns and spandrels. The spandrel plates were located at each floor, transmitting shear stress between columns, allowing them to work together in resisting lateral loads. The joints between modules were staggered vertically so that the column splices between adjacent modules were not at the same floor.

The core of the towers housed the elevator and utility shafts, restrooms, three stairwells, and other support spaces. The core of each tower—a combined steel and concrete structure—was a rectangular area 27 by 41 m and contained 47 steel columns running from the bedrock to the top of the tower. The large, column-free space between the perimeter and core was bridged by prefabricated

floor trusses. The floors trusses provided lateral stability to the exterior walls and distributed wind loads among the exterior walls. The trusses supported 100 mm thick lightweight concrete slabs that were laid on a fluted steel deck. The floors were connected to the perimeter spandrel plates with *viscoelastic dampers*, in order to reduce the sway of the buildings.

Outrigger truss systems were provided from the 107th floor to the top of the building. There were six such trusses along the long axis of the core and four along the short axis. This truss system allowed some load redistribution between the perimeter and core columns and also supported the tall communication antenna on the top of each building. When completed in 1972, the North Tower became the tallest building in the world for two years, surpassing the Empire State Building after a 40-year reign.

The World Trade Center experienced a fire on 13 February 1975 and a bombing on 26 February 1993. The blast opened a 30 m hole through five sublevels in the North tower in addition to killing six people. Following the bombing, floors that were blown out were repaired to restore the structural support they provided to columns.

On 11 September 2001, terrorists crashed a hijacked plane into the northern facade of the north tower, impacting between the 93rd and 99th floors. Seventeen minutes later, a second team of terrorists crashed another hijacked plane into the south tower, impacting between the 77th and 85th floors. After burning for 56 minutes, the south tower collapsed due to the plane impact and as a result of buckling of steel columns due to the ensuing fire. The north tower collapsed after burning for approximately 102 minutes, with the attacks on the World Trade Center resulting in 2,750 deaths.

On the northwest corner of the WTC site, a new 551 m tall Freedom Tower (now called One World Trade Center), designed by David M. Childs of Skidmore, Owings & Merrill, was built from April 2006 and opened in November 2014.

Ref: http://en.wikipedia.org/wiki/World_Trade_Center

The following videos show the failure of the twin towers due to terrorist attacks:

http://www.youtube.com/watch?v=E15GbdVaV-Q

http://www.youtube.com/watch?v=cjq-yK2ak_U&feature=related

http://www.youtube.com/watch?v=KDXBSym2R7c&feature=related

The analysis of a tube structure may be carried out using a space frame program. The main feature shown up in the analysis for horizontal load is the drop-off in load taken by the columns in the flange faces. This is caused by shear lag in the beam-column frame, as shown in Fig. 2.17(c). Simple beam theory would give uniform load in these columns.

The framed tube can be relatively flexible and bracing the tube, as shown in Fig. 2.17(d), provides additional stiffness. This helps reduce shear lag in the flange tube faces as the diagonal members make all exterior columns act together as a rigid tube. The 100-storey John Hancock Center in Chicago uses this type of braced (or trussed) tube system.

Dr F.Z. Khan carried the tube concept still further and constructed the 442 m high Sears Building (now called Willis Tower), Chicago as a bundled tube. In this system, a number of relatively small-framed tubes or diagonally-braced tubes are bundled together for great efficiency in resisting lateral forces (Fig. 2.17f).

## CASE STUDY

### John Hancock Centre

The 100 storey, 344 m tall John Hancock Center of Chicago, Illinois, was a collaborative effort of Skidmore, Owings and Merrill, structural engineer Fazlur Khan, and Bruce Graham. When completed in 1969, it was the tallest building in the world outside New York City.

The distinctive X-bracing at the exterior of the building, is a part of its 'trussed or braced tubular system'. The braced tube system, developed by engineer Fazlur Khan, helps the building stand upright during wind and earthquake loads as the X-bracing provides additional stiffness to the tube. The external bracings of the tube reduce the shear lag in the flange tube faces as shown in Fig. 2.17(c), as the bracings make all the exterior columns to act together as a rigid tube. In addition to higher performance against lateral loads, it also increases usable floor space. This original concept made the John Hancock Center an architectural icon, and won numerous awards.

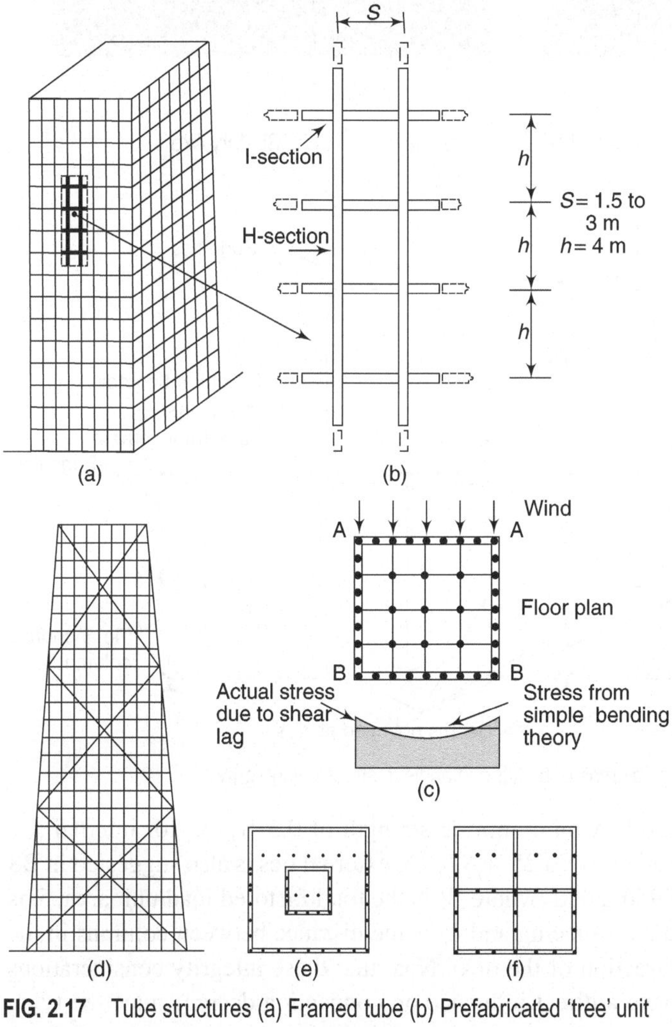


**FIG. 2.17** Tube structures (a) Framed tube (b) Prefabricated 'tree' unit (c) Stress distribution in walls AA, BB (d) Braced tube (e) Tube in tube (f) Bundled tube (Buildings may have more storeys than shown.)

*Source:* Fig. 7.7 of Al Nageim and MacGinley, *Steel Structures, Practical Design Studies*, Third edition, CRC Press, 2005. Reprinted with permission from Taylor & Francis.

**Diagrid systems** Recently, *diagrid systems*, which use perimeter diagonals for structural effectiveness and lattice-like aesthetics, have generated renewed interest in architectural and structural designers of tall buildings. The difference between braced tube structures and current diagrid structures is that, for diagrid structures, almost all the conventional vertical columns are eliminated (see Figs 2.18a and 2.18b). This is possible because the diagonal members in diagrid structural systems can carry gravity loads as well as lateral forces due to their triangulated configuration in a distributive and uniform manner (Ali and Moon 2007). Compared with conventional framed tubular structures without diagonals, diagrid structures are much more effective in minimizing shear deformation because they carry shear by axial action of the diagonal members, whereas conventional tubular structures carry shear by the bending of the vertical columns and horizontal spandrels (Moon et al. 2007). The optimum angle for the diagonals was found to be 63 degrees up to 40 to 50 storeys, and 69 degrees for 60 to 100 storeys (Moon et al. 2007). One of the serious issues in diagrid structures is the constructability of joints as it may become more complicated and more expensive to join different members at the joints. However, prefabrication of nodal elements may reduce the work at the site. Because of the triangular configuration of the diagrid structural system, it is not necessary to have rigid connections at the nodes; pin connections using bolts can be used, which will make connections at the site easy and convenient. This kind of bolting and prefabrication will solve the constructability problem of diagrid structures (Moon 2009). The 30 St Mary Axe in London (also known as The Gherkin) and the Hearst Headquarters in New York (See Fig. 2.18) both by Norman Foster (Structural engineers are Arup and WSP

**FIG. 2.18** Braced tube and diagrid structures (a) Braced tube (b) Diagrid (c) Hearst Headquarters, New York (d) 30 St Mary Axe

*Source:* CTBUH, *Photo Courtesy:* Dr. Philip Oldfield, University of Nottingham

Cantor Seinuk, respectively), are very good examples of major tall buildings incorporating diagrids. Other examples include the 438.6 m tall 103 storey Guangzhou International Financial Center designed by Wilkinson Eyre (Structural engineer: Arup) in Guangdong China, and the 123 floor, 556 m tall Lotte Super Tower designed by Kohn Paderson Fox (Structural engineer: Leslie E. Robertson Associates) in Seoul, South Korea.

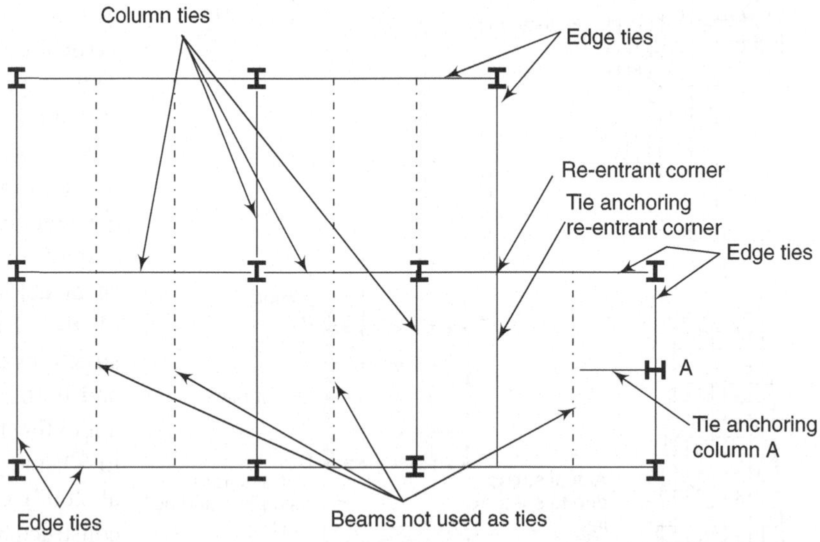


**FIG. 2.19** Tying columns of building to achieve structural integrity

## 2.5 STRUCTURAL INTEGRITY

To reduce the risk of localized damage spreading to all parts, buildings should be effectively tied together at each principal floor level. It is important to effectively hold each column in position by means of horizontal ties in two directions (preferably at right angles), at each principal floor level supported by the column. Horizontal ties are also required at roof level, except where the steel work supports only cladding weighing 0.7 kN/m$^2$ or less and carrying only imposed roof loads and wind loads. At re-entrant corners the tie member nearest to the edge should be anchored into the steel framework, as shown in Fig. 2.19. All these horizontal tie members should be capable of resisting a minimum factored tensile load (should not be considered as additive to other loads) of 75 kN at floor level and 40 kN at roof level. A minimum tie strength of 0.5 $W_f S_t L_a$ for internal ties and one of 0.25 $W_f S_t L_a$ for external ties is also suggested in BS 5950, 2000 (where $W_f$ is the total factored load/unit area, $S_t$ is the tie spacing and $L_a$ is the distance between columns in the direction of the ties). Note that these integrity considerations have a direct influence on connection design, since the tying action of beams requires the connection to possess adequate direct tension capacity. Experimental work done in the UK has established that end plates and web cleats of 8 mm thickness fastened to column flanges by two M20 grade 8.8 bolts will meet

**CASE STUDY**

### Ronan Point Collapse

Progressive collapse provisions were introduced in the British code as early as 1970, as a result of the Ronan Point collapse in 1968. This involved a 23-storey tower building in Newham, East London, which suffered a partial collapse when a gas explosion demolished a load-bearing wall, causing the collapse of one entire corner of the building. Four people were killed in the accident and 17 were injured. (Ronan Point was repaired after the incident. It was demolished in 1986 for a new low-rise housing development project.)

Due to the failure of Ronan Point apartment, many other similar large panel buildings were demolished. The Building Research Establishment, UK published a series of reports in the 1980s to advise councils and building owners what they should do to check the structural stability of their buildings. As a result of terrorist attacks on embassies abroad, along with the Murrah Federal Building in Oklahoma City, abnormal load requirements were introduced in the US codes. (Structural integrity requirements have been introduced in IS 800 only now.)

*Photo Source*: Ellingwood et al. 2007, NIST report

the requirement of ties resisting 75 kN factored tensile load (Nethercot 2001).

Each portion of a building between expansion joints should be treated as a separate building. By tying the structure together as shown in Fig. 2.19, alternative load paths, which enhance the safety, may be made available. To ensure sway resistance, no portion of the structure should be dependent on only one bracing system. All columns should be continuous vertically through the floors.

Precast concrete or other heavy floor or roof units must be properly anchored at both ends. At the edge of the structure, horizontal ties capable of resisting 1% of the maximum factored column loads should restrain columns. Key elements that would risk the collapse of greater area (greater than 15% of the floor area or 70 $m^2$, whichever is less) should be identified and designed for accidental loading (see clause 5.1.2 of IS 800 : 2007 for more on this).

## 2.6 ANALYSIS AND DESIGN

In the structural design process the term *analysis* refers to the determination of the axial forces, bending moments, shears, torsional moments, etc., acting on different members of a structure, due to the applied loads and their combinations (static or dynamic). In general, *design* may involve the development of the structural layout and system or the arrangement of the different members. But for the design engineer, design involves the selection of sizes of members to resist the forces and moments determined in the analysis phase, safely and economically. In the design phase, we will normally design not only the members but also their connections and the foundations, so that the loads are transmitted to the soil.

For statically determinate structures (simply supported beams, cantilevers, trusses, etc.), the analysis is relatively simple and the laws of statics can be used to determine the forces and moments on each member. The relative stiffness of intersecting members does not affect analysis. After the analysis is completed and the critical moments and forces in the different members are tabulated, the design of the member is a straightforward process using an appropriate method (e.g., the working stress method, the limit states method, etc.). For statically determinate structures, there is no need for re-analysis or redesign of the members.

However, for *statically indeterminate analysis*, the procedure is rather complex. A number of analytical methods have been developed, which include the slope deflection method, moment distribution method, Kani's method, portal method, cantilever method, matrix methods, etc. In these methods, assumptions are usually made regarding the distribution of applied load among the members according to the relative stiffness of connecting members, the response and behaviour of members and structures to the applied loads, the rigidity of joints, etc. Moreover, to perform the analysis, the sizes of various structural elements should be known in advance; for this a preliminary design is generally required. Thus, in these types of structures, analysis and design are interactive processes.

After the first cycle of analysis has been completed, the different members are designed as per the codal rules—it is usually necessary to re-analyse the structure to check the validity of original member sizes. For complex structures, several cycles of analysis and design may be required (many times, three cycles are found to be sufficient). The various methods of analysis of structures are discussed briefly in Chapter 4.

As already mentioned, a statically indeterminate analysis requires a preliminary member sizes, which are often assumed based on experience or using approximate methods. Handbooks often provide formulae and coefficients to simplify the preliminary design of continuous beams or simple rigid jointed frames such as portal frames (Davison and Ownes 2002, and Brockenbrough and Merritt 1999).

Various *computer programs* are available for the analysis and design of different types of structures. They include ABACUS, ADINA, ANSYS, ASKA, GT-STRUDL (www.gtstrudl.gatech.edu), SAP, and STRESS. This list is not exhaustive. Many of these packages were developed for use in Mainframe computers. Recently, a number of packages have been developed for use with IBM PC or compatible systems. Notable among them are SAP2000, STAAD III and STAAD PRO (www.reiworld.com), ETABS (www.csiberkeley.com), DAST, LARSA, STRAP (www.ATIR.com), RISA3D (www.risatech.com), SPACEGASS (www.spacegass.com), MIDAS (www.midasuser.com), ROBOTMillennium (www.robobat.com), STRUCAD * 3D, GT-STRUDL, and STRUDS. Windows versions of these packages are also available. These programs are quite general in terms of loading, geometric configurations, and support conditions.

With these programs, it is now possible to analyse any structure with any complicated geometry subjected to any pattern of loading (static or dynamic), and having any boundary condition or discontinuity.

Analytical methods and modelling techniques used by these computer programs offer various levels of sophistication and refinement. However, while using these programs, the designer should be aware of any assumptions used and the limitations of these programs. This is because any amount of mathematical precision cannot make up for the use of an analytical method that is not applicable to the structure being designed.

The code suggests the use of any one of the following methods of analysis. (See Chapter 4 for the details of these methods of analysis.)

1. Elastic (first-order or second-order) analysis,
2. Plastic analysis,
3. Advanced analysis (second-order plastic analysis or frame-instability analysis), and
4. Dynamic analysis.

It may be of interest to note that many of the steel buildings in India are analysed using first-order elastic analysis programs only (e.g., STAAD III), although non-linear analysis methods have been employed in a few important high-rise structures.

## 2.7 CODES AND SPECIFICATIONS

A structural engineer is often guided in his efforts by the *code of practice*. A code represents the consensus of opinion of experienced engineers and professionals. It may not cover in detail every situation a designer may encounter. Often the designer must exercise his own judgement in interpreting and applying the requirements of a code. It has to be noted that strict adherence to codes may often hamper the adoption of innovative designs. Codes are basically written for the purpose of protecting the public. They provide guidelines for the design and construction of structures. They are revised at regular intervals to reflect new developments (in research, materials, construction techniques, etc.), experience gained from past design practice, behaviour of existing structures, and failure of structures. Codes contain recommended loads for a given locality, and recommended fire and corrosion protection. They also contain rules governing the ways in which loads are to be applied, and design rules for steel, concrete, and other material. These rules may be in the form of detailed recommendations or by reference to other standards that provide specific design rules. The codes should be regarded as aids to design containing stress levels, design formulae, and recommendations for good practice, rather than as a manual or textbook on design.

The codes serve at least four distinct functions.

1. They ensure adequate structural safety, by specifying certain essential minimum requirements for the design.
2. They aid the designer in the design process. Often the results of sophisticated analysis are made available in the form of simple formulae or charts.
3. They ensure consistency among different engineers.
4. They protect the structural engineer from disputes, though codes in many cases do not provide legal protection.

*Project specifications*, along with design drawings, are given to the builder by the architect/project manager. These specifications and the way in which the drawings are prepared and presented vary from organization to organization. They include the following items.

1. Materials that must be used in the structure,
2. Sizes of structural members,
3. Joint details,
4. Expected quality, tolerance, and
5. Instructions on how the construction work is to be done.

Whoever writes the specification, the structural engineer should be involved in preparing or approving its technical contents.

In India, the *Bureau of Indian Standards* issues the codes and standard handbooks. Committees representing producers, designers, educators, fabricators, government bodies, and other interested bodies write them. The draft is circulated to a larger section of engineers, designer, and professionals. The committee considers their comments and finally the Bureau of Indian Standards prints the code. It is strongly advised that the reader should possess a copy of the following codes, which will be referred frequently in the book.

1. IS 800 : 2007—Code of practice for general construction in steel
2. IS 875 : 1987—Code of practice for design loads for buildings and structures (Part 1—Dead loads, Part 2—Imposed loads, Part 3—Wind loads, Part 4—Snow loads and Part 5—Special loads and load combinations)
3. IS 1893 (Part 1) : 2002—Criteria for earthquake-resistant design of structures
4. IS 808 : 1989—Dimensions for hot-rolled steel beams, columns, channels and angle sections.

In addition, the designer may need to refer to a number of other codes covering topics such as steel properties, welding of structural steel works, properties of fasteners (bolts), and also codes for the design of bridges, towers, silos, and off-shore structures.

The IS 800 : 2007 code also lists other codes published by the Bureau of Indian Standards that may be useful to the steel designer. It is to be noted that the codes IS 801, IS 806, and IS 4000 are still in the working stress method format and will be revised shortly in the limit states design format.

In certain cases, it may be useful for the designer to consult codes of other countries. Some of these that will be helpful are listed here.

1. Specification for structural steel buildings, American Institute of Steel Construction, Chicago, USA, (ANSI/AISC 360–2010)
2. Seismic provision for structural steel building, American Institute for Steel Construction, Chicago (ANSI/AISC 341–2010) 2016 versions of AISC 360 and AISC 341 will be available shortly at www.aisc.org
3. AS 4100-1998 Steel structures (second edn), Standards Australia, Standards Association of Australia, Homebush
4. BS 5950-2000 Structural use of steelwork in Buildings—Part I. Code of practice for design in simple and continuous construction, hot-rolled sections, British Standards Institution, London
5. CAN/CSA-S16.1–09 Limit states design of steel structures, Canadian Standards Association, Toronto, Canada
6. ENV 1993-1-1:2005 Eurocode 3: Design of steel structures Part 1-1 General rules and rules for buildings, European Committee for Standardization, Brussels

As discussed earlier, the loads acting on the structure as well as their strength (resistance) must be considered as variable. A thorough analysis of all uncertainties that might influence achieving a 'limit state' is not practical. It may be impossible to define the probability distributions of loads and resistances (strengths), since doing so may require hundreds of tests on samples of components. Hence the current approach is to not use a complicated reliability analysis but a simplified method.

For simplicity of explanation, most of the discussions and examples in this book deal with the design of elements in buildings. For other structures such as bridges, towers, etc., the design procedures are similar to those followed in the design of buildings. However in some structures such as bridges and crane girders, which are subjected to repeated loads, failure by fatigue should be considered in addition to static strength requirements. Guidelines for fatigue design are given in Section 13 of IS 800 : 2007.

## 2.8 DESIGN PHILOSOPHIES

Over the years, various design philosophies have evolved in different parts of the world, with regard to structural steel design.

The earliest codified design philosophy is the *working stress method of design* (WSM). This traditional method of design, based on linear elastic theory, is still surviving in some countries (including India and the USA), although it has now been replaced by the modern limit states design philosophy. In the recent revision of the code (IS 800 : 2007), the provisions relating to the WSM design procedure have been relegated from the main text of the code to a section (section 11) at the end of the code (with only a few pages devoted to it) 'so as to give greater emphasis to limit states design'.

The WSM was followed by the *ultimate strength design (USD) or plastic design,* which was developed in the 1950s. It was based on the ultimate strength of steel at ultimate loads. This method was introduced as an alternative to WSM in the IS 800 code in 1962.

In the mid 1960s, probabilistic concepts of design received a major impetus (Madsen et al. 1986 and Subramanian 1974). The philosophy was based on the theory that the various uncertainties in design could be handled more rationally in the mathematical framework of probability theory. The risk involved in design was quantified in terms of a probability of failure. Such probabilistic methods are known as *reliability-based methods*. However, this theory was not accepted in professional practice, mainly because it appeared to be complicated and intractable (mathematically and numerically).

For codification, the probabilistic 'reliability method' approach was simplified and reduced to a deterministic format involving multiple *(partial) safety factors* (rather than probability of failure). Based on the CEB-FIP recommendations, the philosophy of the *limit states method* (LSM) was introduced in the British Code CP 110(1972) (now BS 8110), and the Indian Concrete Code IS 456 : 1978. Limit states design was first adopted for steel structures in the Canadian Code in 1974, which was followed by the British Codes BS 5950 and BS 5400 (bridge code). In the USA, the American Institute of Steel Construction introduced the LSM in the form of *load and resistant factor design* (LRFD) in 1993, though the working stress method is still followed in the USA for the design of steel structures. A similar format was introduced in the American Concrete Code also in 1971 (ACI 318-71).

### 2.8.1 Working Stress Method

This was the traditional method of design not only for structural steel, but also for reinforced concrete and timber design. The conceptual basis of working stress method (WSM) is simple. The method basically assumes that the structural material behaves in a linear elastic manner, and that adequate safety can be ensured by suitably restricting the stresses in the material due to the expected *working loads* (service loads) on the structure.

The stresses under the working loads are obtained by applying the methods of 'strength of materials', such as the simple bending theory. The first attainment of yield stress of steel is taken to be the onset of failure. The limitations due to non-linearity (geometric as well as material) and buckling are neglected.

The stresses caused by the 'characteristic' loads are checked against the *permissible (allowable) stress*, which is a fraction of yield stress. Thus the permissible stress may be defined in terms of a *factor of safety*, which takes care of the overload or other unknown factors. Thus,

$$\text{Permissible (allowable) stress} = \frac{\text{yield stress}}{\text{factor of safety}} \qquad (2.1a)$$

Thus, in the working stress method

$$\text{Working stress} \leq \text{permissible stress} \qquad (2.1b)$$

Each member of the structure is checked for a number of different combinations of loadings. Usually, a factor of safety of 1.67 is adopted for tension members and beams. A value of 1.92 is used for long columns and 1.67 for short columns. A value of 2.5–3 is used for connections. (However using the WSM, the 'real' safety against 'failure' is unknown.) Since dead load, live load, and wind load are all unlikely to act on the structure simultaneously the stresses are checked as follows.

$$\text{Stress due to dead load} + \text{live load} < \text{permissible stress}$$

$$\text{Stress due to dead load} + \text{wind load} < \text{permissible stress}$$

$$\text{Stress due to dead load} + \text{live load} + \text{wind load} < 1.33\,(\text{permissible stress})$$

There are many limitations and shortcomings of the working stress method. First, the main assumption of linear elastic behaviour and the implied assumption that the stresses under working loads can be kept within the 'permissible stresses' are not found to be realistic. Many factors are responsible for this, such as the effects of stress concentrations, the long-term effects of creep and shrinkage, residual stresses, and other secondary effects. All such effects result in significant local increases in and redistribution of the calculated stresses. The WSM does not consider the consequences of material non-linearity and the non-linear behaviour of members in the post-buckled state. Moreover, steel components have the ability to tolerate high elastic stress by yielding locally and redistributing the loads. WSM does not provide a realistic measure of the actual factor of safety underlying a design. It also fails to discriminate between different types of load that act simultaneously, but have different degrees of uncertainty. This can, at times, result in very unconservative designs, particularly when two different loads (say, dead loads and wind loads) have counteracting effects.

In spite of these shortcomings, it may be stated that most structures designed in accordance with WSM have generally performed satisfactorily for many years. The size of tension members is about the same in both LRFD and WSM when the live load to dead load rate (LDR) is about 3. When the dead load becomes more predominant, there will be economy in using LRFD. With LDR greater than 3, WSM will be slightly (about 3%) more economical (Salmon and Johnson 1996). However, for other members, WSM results in relatively larger member sizes and hence in less deflections. There are instances where WSM results in considerable overdesign and where it is not safe (Allen 1972 and Gordon 1978). The WSM method is notable for its essential simplicity, in concept as well as application.

It has to be noted that serviceability requirements such as deflection limits are always investigated at service load conditions, even when the LSD design is used to satisfy safety requirements.

### 2.8.2 Ultimate Load Design

The shortcomings of WSM led to the development of *ultimate load design* (ULD). This method is also referred to as the *load factor method* or the *ultimate strength method*. When applied to steel structures, it is normally referred to as the *plastic design method*. The plastic design method was validated by research undertaken at Cambridge University, UK by Prof. Baker and his colleagues (Neal 1977). Plastic design is a special case of limit states design, wherein the limit state is attained when the members reach plastic moment strength $M_p$ and the structure is transformed into a mechanism (see Chapter 8). Plastic moment strength is the moment strength when all the fibres of the cross section of a member are at the yield stress $f_y$. (One side of the neutral axis is in tension and the other side in compression.)

The safety measure in the design is introduced by an appropriate choice of the *load factor*, defined as the ratio of the ultimate load (design load) to the working load. The ultimate strength design makes it possible for different types of loads to be assigned different load factors under combined loading conditions, thereby overcoming the related shortcomings of WSM. For example, IS 800 : 1984 suggested load factors as shown in Table 2.2. Thus in ULD,

$$\sum(\text{Working load} \times \text{LF}) \leq \text{ultimate load} \quad (2.2)$$

**TABLE 2.2** Load factors for plastic design

| Working loads | Load factor (IS 800:1984) (min.) | Load factor (BS 449:1969) |
|---|---|---|
| All loads and combinations except DL + IL + WL or EL | 1.7 | 1.75 |
| DL + IL + WL or EL | 1.3 | 1.40 |

*Note:* DL = Dead load, IL = Imposed load, WL = Wind load, EL = Earthquake load

It is to be noted that the satisfactory 'strength' performance at ultimate loads does not guarantee satisfactory 'serviceability' performance at normal service loads. Hence the codes (e.g., IS 800 : 1984) suggest that the deflections under working loads should be within the prescribed limits of the code. Plastic design does not permit using other limit states such as instability, fatigue, or brittle fracture (Neal 1977).

## 2.8.3 Limit States Design

There are various levels of reliability analysis to evaluate the risk involved and to obtain a probability-based assessment of structural safety. These uncertainties in design and reliability methods are discussed briefly in this section.

### *Uncertainties in Design*

To safeguard against the risk of failure (collapse or unserviceability), safety margins are normally provided in design. In WSM and USD, these safety margins were assigned (in terms of 'permissible stresses' in WSM and 'load factors' in USD) primarily on the basis of engineering judgment. Structures designed according to these methods were found, in general, to be safe and reliable. However, the safety margins provided in these methods lacked scientific basis. The science of reliability-based design was developed with the objective of providing a rational solution to the problem of 'adequate safety'.

The variables such as loads, material strength, and member dimensions (see Figs 2.20 and 2.21) are subject to varying degrees of uncertainty and randomness. Hence the structure must be designed to provide for the possibility of overload. Similarly the deviations in the dimensions of members or strength of material, even though within acceptable tolerance, can result in a member having less than the computed strength. Thus, the design should take into account the possibility

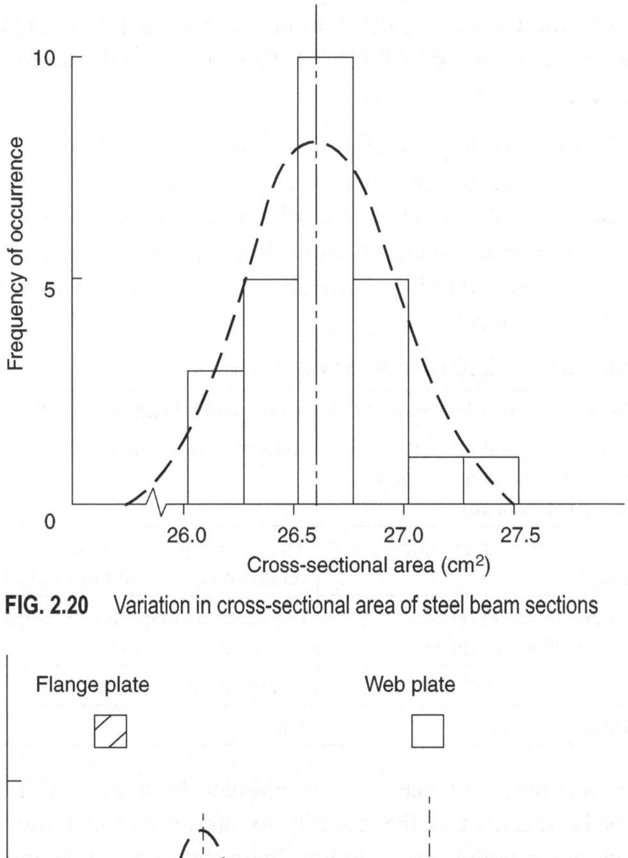


**FIG. 2.20** Variation in cross-sectional area of steel beam sections

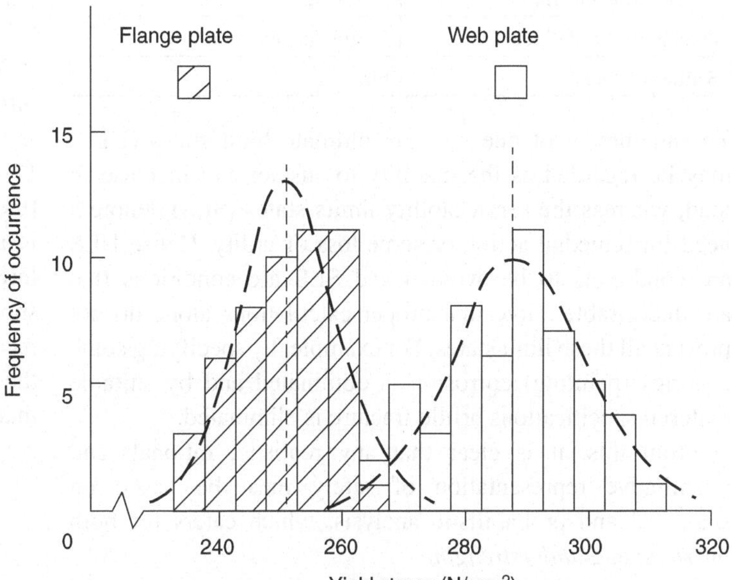


**FIG. 2.21** Variation in strengths of the material of steel I-sections

of understrength also. Further, there are idealization and simplifying assumptions used in the theories of structural analysis and design. There are also several unforeseen factors that influence the prediction of strength and serviceability. They include construction methods, workmanship and quality control, intended service life of the structure, human errors, possible future change of use, frequency of loading, etc. These uncertainties make it impossible for the designer to guarantee that the structure will be absolutely safe.

## Limit States

In limit states design, we prefer to use the term 'limit states', rather than 'failure'. Thus, a *limit state* is a state of impeding failure, beyond which a structure ceases to perform its intended function satisfactorily.

The limit states usually considered relevant for structural steel work are listed in Table 2.3. They are normally grouped into the following two types.

1. Ultimate (safety) limit states, which deal with strength, overturning, sliding, buckling, fatigue fracture, etc.
2. Serviceability limit states, which deal with discomfort to occupancy and/or malfunction, caused by excessive deflection, vibration, corrosion (and subsequent loss of durability), etc.

**TABLE 2.3** Limit states considered by the code

| Ultimate (safety) Limit State (ULS) | Serviceability Limit State (SLS) |
|---|---|
| Strength (including yielding, buckling, and transformation into a mechanism) | Deformation and deflection |
| Stability against overturning and sway | Vibration (e.g., wind-induced oscillations, floor vibration, etc.) |
| Failure due to excessive deformation or rupture | Repairable damage due to fatigue (cracking) |
| Fracture due to fatigue | Corrosion and durability |
| Brittle fracture | Fire |

The attainment of one or more ultimate limit states (ULS) may be regarded as the inability to sustain any increase in load, whereas the serviceability limits states (SLS) denote a need for remedial action or some loss of utility. Hence ULS are conditions to be avoided and SLS are conditions that are undesirable. Note that proper calculations alone do not prevent all these limit states. For example, by specifying some coatings (painting) corrosion is eliminated and by suitable material specifications brittle fracture is eliminated.

From this, it is clear that any realistic, rational, and quantitative representation of safety must be based on statistical and probabilistic analysis, which caters for both *overload* and *understrength.*

Design for the ultimate limit state may be conveniently explained with reference to the type of diagram shown in Fig. 2.22. This figure shows the hypothetical frequency distribution curves for the effect of loads on the structural element and the resistance (strength) of the structural element. When the two curves overlap, shown by the shaded area, the effect of the loads is greater than the resistance of the element and the element fails. Thus, the structure and its elements should be proportioned in such a way that the overlap of the two curves is small, which means that the probability of failure is within the acceptable range.

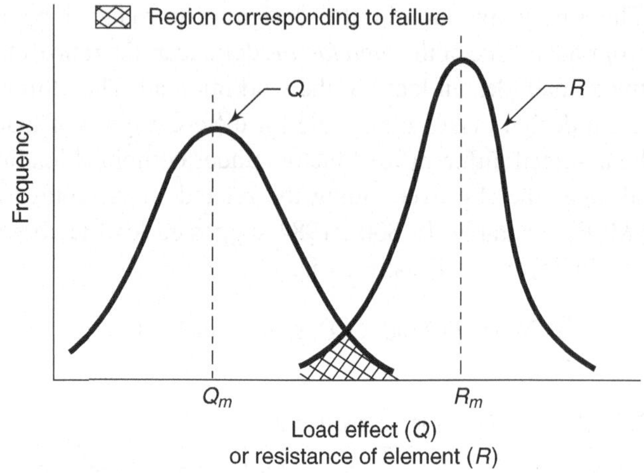


**FIG. 2.22** Frequency distribution curves

## Levels of Reliability Methods

There exist a number of levels of reliability analysis. These are differentiated by the extent of probabilistic information that is used (Unnikrishna Pillai and Menon 2003).

A full-scale probabilistic analysis is generally described as a *Level III reliability method*, which uses probability of failure, $P_f$, to evaluate the risk involved. It is highly advanced, mathematically difficult, and generally used at research level. It is clearly unsuitable for general use.

We may simplify the problem by limiting the probability information of the basic variables to their 'second moment statistics' (i.e., *mean* and *variance*). Such a method is called a *Level II reliability method.* In this method, the structural failure [the achievement of a limit state is examined by comparing resistance $R$ with load effect $Q$ in logarithmic form, observing $\ln(R/Q)$ as shown in Fig. 2.23]. In this figure, the hatched region shows the failure. The distance between the failure line and the mean value of the function [ln $(R/Q)$] is defined by $\beta$, called the *reliability index.* The larger the value of $\beta$, the greater is the margin of safety of the system. $\beta$ may be expressed as

$$\beta = \frac{\ln\left(\frac{R_m}{Q_m}\right)}{\sqrt{(V_R^2 + V_Q^2)}} \tag{2.3}$$

where $V_R$ is $\sigma_R/R_m$ and $V_Q$ is $\sigma_Q/R_m$.

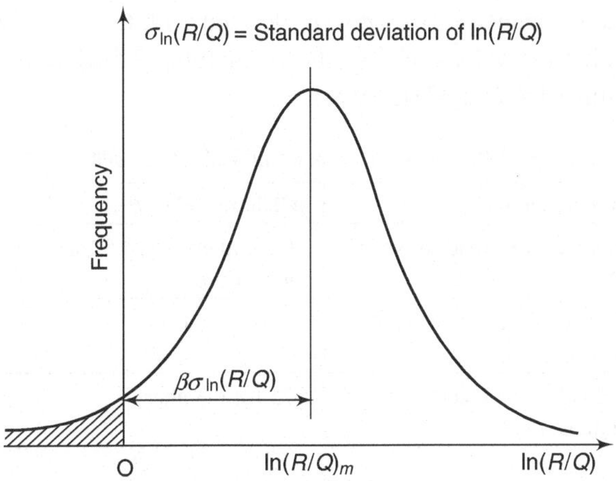


**FIG. 2.23** Level II reliability method

$R_m$ and $Q_m$ are the mean values of the resistance and load, and $\sigma_R$ and $\sigma_Q$ are the standard deviations of the resistance and load respectively.

However, even such a 'simplified method' is unsuitable for day-to-day use in a design office, as it requires the application of optimization techniques for the determination of $\beta$ (Ravindra and Galambos 1978; Galambos 1981; and Ranganathan 1990).

The values of the reliability index $\beta$ corresponding to various failure probabilities $P_f$ can be obtained from the standardized normal distribution function of the cumulative densities, and are given in Table 2.4.

**TABLE 2.4** Reliabilty index for various failure probabilities

| $\beta$ | 2.32 | 3.09 | 3.72 | 4.27 | 4.75 | 5.2 | 5.61 |
|---|---|---|---|---|---|---|---|
| $P_f = \phi(-\beta)$ | $10^{-2}$ | $10^{-3}$ | $10^{-4}$ | $10^{-5}$ | $10^{-6}$ | $10^{-7}$ | $10^{-8}$ |

It has to be noted that the values given in Table 2.4 are valid only if the safety margin is normally distributed.

For code use, the method must be as simple as possible, using deterministic rather than probabilistic data. Such a method, called *Level I reliability method* or first-order second moment reliability method, is used in our code to obtain a probability-based assessment of structural safety.

### *Characteristic Load and Characteristic Strength*

In normal design calculations, we generally use a single value for each load and for each material property, with a margin to take care of all uncertainties. Such a value is termed as *characteristic strength* (or resistance) or *characteristic load.*

The characteristic strength of a material (such as steel, concrete, or wood) is defined as that value of strength below which more than a prescribed percentage of test results will fall. This prescribed percentage is normally taken as 95. Thus the characteristic yield strength of steel is the value of yield strength below which not more than 5% of the test values may fall.

Similarly, the characteristic load, $Q_c$, is defined as that load which is not expected to be exceeded with more than 5% probability, during the life span of a structure. Thus, the characteristic load will not be exceeded 95% of the time.

The design values are derived from the characteristic values through the use of partial safety factors, both for material and for loads. The acceptable failure probability, $P_f$, for particular classes of structures is generally derived bearing in mind past practice, consequences of failure, and cost considerations. Having chosen $P_f$ and $\beta$, determination of partial safety factors is an iterative process (Galambos 1981).

## 2.9 LIMIT STATES METHOD

The philosophy of the *limit states method of design* (LSM) represents a definite advancement over the traditional design philosophies. Unlike WSM, which bases calculations on service load conditions alone, and unlike ULM, which bases calculations on ultimate load conditions alone, LSM aims for a comprehensive and rational solution to the design problem, by considering *safety* at ultimate loads and *serviceability* at working loads.

The LSM philosophy uses a multiple safety factor format that attempts to provide adequate safety at ultimate loads as well as adequate serviceability at service loads, by considering all possible 'limit states'. The selection of the various multiple safety factors is supposed to have a sound probabilistic basis, involving the separate consideration of different kinds of failure, types of materials, and types of loads (Kulak & Grondin 2002).

### 2.9.1 Limit State of Strength

The limit state of strength may be worked out according to the multiple safety factor design format, the load and resistance factor design format, or the partial safety factor format.

### *Multiple Safety Factor Format*

The objective of limit state design is to ensure that the probability of any limit state being reached is acceptably low. This is made possible by specifying appropriate *multiple safety factors* for each limit state (Level I reliability). Of course, in order to be meaningful, the specified values of safety factors should result in 'target reliability' (the reliability that will produce designs which will provide the required amount of safety and at the same time result in economic structures).

Most national codes introduced multiple safety factors in limit state design in the 1970s, primarily based on experience, tradition, and engineering judgement (Nethercot 2001; Kulak & Grondin 2002). Subsequently, code-making authorities were engaged in the process of code calibration, to determine the range of the reliability index $\beta$ (or its equivalent probability of failure, $P_f$) underlying the specified safety factors for different practical situations (Ravindra and Galambos 1978). With every code revision, conscious attempts are made to specify more rational reliability-based safety factors, in order to achieve practical designs that are satisfactory and consistent in terms of the degree of safety, reliability, and economy (Nethercot 2002; Galambos & Ravindra 1978).

## Load and Resistance Factor Design Format

Of the many multiple safety factor formats in vogue, perhaps the simplest to understand is the *load and resistance factor design* (LRFD) format, which is adopted by the American Institute of Steel Construction (AISC 1994). As per LRFD, the expression for structural safety is

$$\text{Design resistance } (\phi R_n) \geq \text{design load effect}\left(\sum \gamma_i Q_i\right) \quad (2.4)$$

where the left side of the Eqn (2.4) represents the strength (or resistance) of the system or component and the right side, the load expected to be carried by the system/component. The nominal strength $R_n$ is multiplied by a strength reduction (or resistance reduction) factor $\phi$ to obtain the *design strength*. Similarly, the various load effects $Q_i$ (such as dead load, live load, wind load, etc) are multiplied by their respective overload factors $\phi_i$ to obtain the sum $\Sigma \gamma_i Q_i$.

The *resistance factor* $\phi$ accounts for 'understrength' and is less than unity. This takes care of

1. the possibility of unfavourable deviation of material strength from the characteristic value,
2. the possibility of unfavourable reduction in member strength due to fabrication and tolerances,
3. the possibility of unfavourable variation of member sizes,
4. uncertainty in the theoretical assumptions, and
5. uncertainty in the calculation of strength of members.

On the contrary, the *load factors* $\gamma_i$, which account for overloading and the uncertainties associated with $Q_i$, are generally greater than unity. These load factors account for

1. the possibility of unfavourable deviation of the load from the characteristic value,
2. possibility of inaccurate assessment of the load,
3. uncertainty in the assessment of effects of the load, and
4. The uncertainty in the assessment of the limit state being considered.

The method used to develop LRFD uses the mean values $R_m$ and $Q_m$ (see Fig. 2.22), and the standard deviation $\sigma_R$ and $\sigma_Q$ of the resistance and the load respectively.

The resistance factors $\phi$ were derived with the objective of obtaining the values of $\beta$ as shown in Table 2.5 (Salmon and Johnson 1996 and AISC 1994).

**TABLE 2.5** Reliability index associated with different loads

| Load combination | Reliability index, $\beta$ |
|---|---|
| Dead load + live load (or snow load) | 3.0 for members, 4.5 for connection |
| Dead load + live load + wind load | 2.5 for members |
| Dead load + live load + earthquake load | 1.75 for members |

Because of the lower probability of wind or earthquake loads occurring simultaneously with the live load, a lower value of $\beta$ was chosen. Similarly, in keeping with the tradition of making connections stronger than members, a higher value of $\beta$ was chosen for the connection.

Equation (2.4) may be rearranged as

$$\sum Q_i < R_n\left(\frac{\gamma}{\phi}\right) \quad (2.5)$$

which is representative of the safety concept underlying WSM, $\gamma/\phi$ here denoting the 'factor of safety' applied to material strength, in order to arrive at the permisible stress for design.

Alternatively, Eqn (2.4) can be rearranged as

$$R_n \geq \frac{\gamma}{\phi}\sum Q_i \quad (2.6)$$

which is representative of the safety concept underlying ULM, $\gamma/\phi$ here denoting the so-called 'load factor' (ULM terminology) applied to the load in order to arrive at the ultimate load for design.

## Partial Safety Factor Format

The multiple safety factor adopted by the code is in the so-called *partial safety factor format*, which is expressed as

$$R_d \geq \sum \gamma_{\text{if}} Q_{\text{id}} \quad (2.7)$$

where $R_d$ is the design strength (or resistance) computed using the reduced material strength $R_u/\gamma_m$, where $R_u$ is the characteristic material strength and $\gamma_m$ is the partial safety factor for the material, and allows for uncertainties of element behaviour and possible strength reduction due to manufacturing tolerances and imperfections in the material. The values as suggested by the code are given in Table 2.6. It has to be noted that there is no statistical study based on Indian steel properties to guide the choice of partial safety factors. The partial safety factors suggested in Table 2.6 are based on the values given in Eurocode 3. $Q_{\text{id}}$ is the design action (load effect) computed for the enhanced loads ($\gamma_{\text{Df}} \cdot$ DL, $\gamma_{\text{Lf}} \cdot$ LL, and $\gamma_{\text{Qf}} \cdot$ EL)

**TABLE 2.6** Partial safety factor for materials, $\gamma_m$

| Definition | | Partial safety factor | |
|---|---|---|---|
| Resistance, governed by yielding, $\gamma_{m0}$ | | 1.10 | |
| Resistance of member to buckling, $\gamma_{m0}$ | | 1.10 | |
| Resistance, governed by ultimate stress, $\gamma_{m1}$ | | 1.25 | |
| Resistance of connection | Shop fabrications | | Field fabrications |
| (a) Bolts—friction type, $\gamma_{mf}$ | 1.25 | | 1.25 |
| (b) Bolts—bearing type, $\gamma_{mb}$ | 1.25 | | 1.25 |
| (c) Rivets | 1.25 | | 1.25 |
| (d) Welds, $\gamma_{mw}$ | 1.25 | | 1.50 |

involving separate partial load factors $\gamma_{Df}$ (for dead load), $\gamma_{Lf}$ (for live load), and $\gamma_{Qf}$ (for wind or earthquake load). The partial safety factors for loads $\gamma_f$ make allowances for possible deviation of loads and the reduced possibility of all loads acting together. The values suggested by the code are shown in Table 2.7. The partial safety factors for loads given in this Table are also based on Eurocode 3 with slight modifications. It is to be noted that when more than one imposed load acts simultaneously, the leading load is that causing larger action effect. In industrial structures, dead and live loads on roofs are usually small; crane loads combined with wind loads are critical in most cases. In such cases, crane load is the leading load which should be multiplied by the factor 1.5 (see Table 2.7) and roof live load will be treated as accompanying live load. When calculating the earthquake forces, only dead load of crane has to be considered since the suspended load is not likely to experience the interia force due to earthquake. The other terms involved are the *nominal* yield strength of steel, $f_y$, on the side of the resistance, and the nominal load effects DL, LL, and EL representing, respectively, dead loads, live loads, and wind/earthquake loads.

It may be noted that, whereas the multiplication factor $\phi$ is generally less than unity, the dividing factor $\gamma_m$ is greater than unity, giving the same effect. All the load factors are generally greater than unity, because overestimation usually results in improved safety. However, one notable exception to this rule is the dead load factor $\gamma_{Df}$, which is taken as 0.9 whenever dead load contributes to stability against overturning or sliding, or while considering reversal stresses when dead loads are combined with wind/earthquake loads. In such cases, underestimating the counter-acting effects of dead load results in greater safety.

Note that the load factors are reduced when different types of loads (DL, LL, and EL) are acting simultaneously at their peak values. (This is referred to sometimes as the *load combination effect*.) This is because of the reduced probability of all the loads acting concurrently (see also Section 3.15).

### 2.9.2 Structural Stability

The code deals with three forms of structural stability (clause 5.5). They are (a) general stability, (b) stability against overturning, and (c) sway stability.

**TABLE 2.7** Partial safety factors for loads, $\gamma_f$, for limit states

| Combination | Limit state of strength | | | | | Limit state of serviceability | | | |
|---|---|---|---|---|---|---|---|---|---|
| | | LL | | | | | LL | | |
| | DL | Leading | Accompanying (CL, SL, etc.) | WL/EL | AL | DL | Leading | Accompanying (CL, etc.) | WL/EL |
| DL + LL + CL | 1.5 | 1.5 | 1.05 | — | — | 1.0 | 1.0 | 1.0 | — |
| DL + LL + CL + WL/EL | 1.2 | 1.2 | 1.05 | 0.6 | — | 1.0 | 0.8 | 0.8 | 0.8 |
| | 1.2 | 1.2 | 0.53 | 1.2 | | | | | |
| DL + WL/EL | 1.5 (0.9)[a] | — | — | 1.5 | — | 1.0 | — | — | 1.0 |
| DL + ER | 1.2 (0.9) | 1.2 | — | — | — | — | — | — | — |
| DL + LL + AL | 1.0 | 0.35 | 0.35 | — | 1.0 | — | — | — | — |

[a]This value is to be considered whenever dead load contributes to stability against overturning or causes reduction in stress due to other loads such as wind uplift in roof trusses.

Abbreviations: DL = dead load, LL = imposed load (live load), WL = wind load, SL = snow load, CL = crane load (vertical/horizontal), AL = accidental load, ER = erection load, and EL = earthquake load.

*Note:* The effects of actions (loads) in terms of stresses or stress resultants may be obtained from an appropriate method of analysis.

### General Stability

It should be ensured that the structure as a whole and each of its elements remains stable from the commencement of erection until demolition. When two members are incapable of keeping themselves in equilibrium then sufficient external bracings should be provided for stability.

### Stability Against Overturning

The structure as a whole or any part of it should be designed to prevent instability due to overturning or sliding, while designing tall or cantilever structures. While checking for overturning, uplift or sliding, the loads should be multiplied by the relevant $\gamma_f$ factors given in Table 4 of the code. The code suggests the following.

1. The loads and forces should be divided into components aiding instability and those resisting instability.
2. The forces and loads causing instability should be combined using the appropriate load factors given in the code.
3. The permanent loads and effects causing resistance should be multiplied by a partial safety factor of 0.9 and added together with design resistance (after being multiplied by the appropriate partial safety factor). Note that a low factor is applied to those forces which provide restraint, to guard against them being overestimated.
4. The resistance effect should be greater than or equal to the destabilizing effect.

### Sway Stability

This condition imposes that there must not be excessive lateral deformation under applied loads (Fig. 2.24). In the following cases, the applied horizontal loads on structures may be small compared to the vertical loading.

1. When the structures are sheltered by an adjoining building,
2. When the structure is entirely enclosed, and
3. When the building is wide in relation to the applied load (a long multi-storey frame along its length will sometimes come in this category).

In order to prevent these situations from becoming critical, the code requires all structures to be checked for a minimum notional horizontal load (based on the vertical load at each level). The notional loads may also arise from practical imperfections, such as lack of fit.

These notional loads are specified in Section 4.3.6 of the code as 0.5% of the factored load (dead and imposed loads) (see Fig. 2.25).

This notional horizontal force should not be

1. applied while considering overturning,

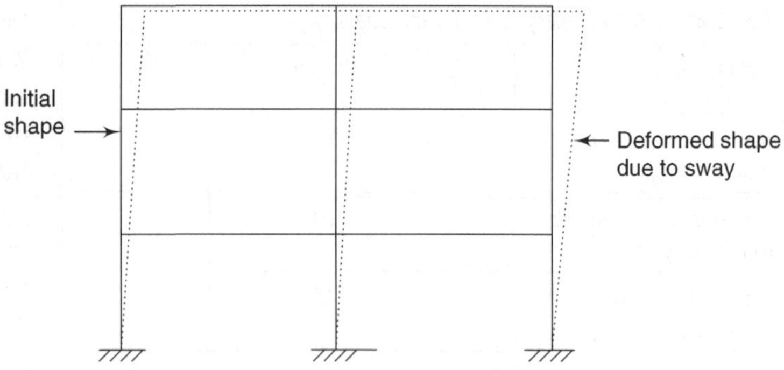


**FIG. 2.24** Lateral deformations under applied loads

2. combined with other horizontal loading such as wind or earthquake,
3. combined with temperature effects, and
4. taken to contribute to net shear on the foundation.

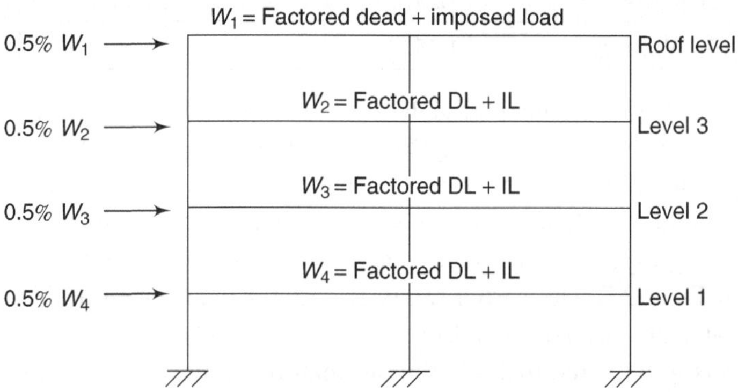


**FIG. 2.25** Notional horizontal forces

The notional horizontal load should be applied on the whole structure, in both orthogonal directions, in one direction at a time, and should be taken as acting simultaneously with a factored gravity load. Moreover when the ratio of the height to the lateral width of a building is less than unity, such notional loads need not be considered. More details about these notional loads are provided by Structural Engineering Institute/ASCE 1997.

## 2.9.3 Serviceability Limit States

In such limit states, the variable to be considered is a serviceability parameter $\Delta$ (representing deflection, vibration, etc.). A limit state or failure is considered to occur when a specified maximum limit of serviceability, $\Delta_{all}$ is exceeded (Fig. 2.26). In Fig. 2.26, $P_f$ is the probability of failure and $f_\Delta$ ($\Delta$) is the frequency distribution curve for $\Delta$. It may be noted that unlike the strength limit state shown in Fig. 2.22, the limit defining failure is deterministic and not probabilistic. Serviceability limit states relate to satisfactory performance and correspond to excessive deflection, vibration, local deformation, durability, and fire resistance.

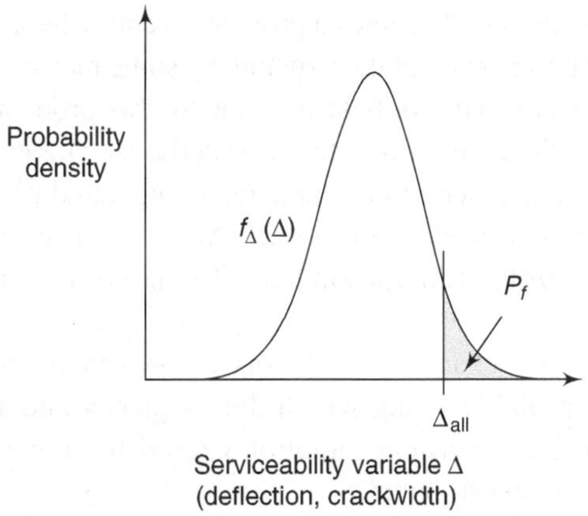


**FIG. 2.26** Reliability model for serviceability design

The load factor, $\gamma_f$, should be taken as unity for all serviceability limit state calculations, since they relate to the criteria governing normal use.

## Deflections

The maximum deflection affecting the strength and stability of the structure is controlled by the strength limit state. However, excessive deflection should not cause sagging, plaster cracking, or failure to align plant and machinery (e.g., in lifts). Excessive deflection of beams causes damage to supported non-structural elements such as partitions, excessive vibrations of floors, or impairs the usefulness of the structure (e.g., it may distort door frames so that doors do not open or close). On roofs, a major deflection-related concern is ponding of water. Excessive deflections are often indicative of excessive vibration and noise transmission, both serviceability problems. Deflections are to be calculated for all combinations of loads specified in the code, by using an elastic analysis and checked for the maximum values specified in the code. Some of the deflection limits specified by IS 800 : 2007 for industrial buildings are shown in Table 2.8. See also Table 6 of the code for other buildings.

Serviceability, instead of strength, may and often does control the design of beams. In order to avoid the sagging appearance due to deflection, we may pre-camber the beam, trusses, and girders. The code recommends that for spans greater than 25 m, a camber equal to the deflections due to dead load plus half the live load may be used. This deflection should be calculated without considering the effect of impact or dynamic loads. Guidelines for general serviceability limit states and especially for steel structures may be found in Galambos and Ellingwood (1986) and Ellingwood (1989).

The code states that very flexible roofs should be designed to withstand loads due to ponding of water, or accumulation of snow, but has not given any guidelines. According to AISC to prevent ponding of water accumulated on flat roofs, the support member should satisfy

$$\frac{L}{d} \leq \frac{4137}{f_b} \tag{2.8}$$

where $f_b$ is the computed service load bending stress in megapascals.

More discussions on deflection considerations are provided in Chapter 10.

## Vibration

With the development of lighter construction (using high-strength steels, cold-rolled sections, etc.) and use of longer spans, there is a higher risk of vibrations becoming critical

**TABLE 2.8** Deflection limits for industrial buildings

| Deflection | Design load | Member | Supporting element | Maximum deflection |
|---|---|---|---|---|
| Vertical | Live load/Wind load | Purlin and girts | Elastic cladding | Span/150 |
| | Live load/Wind load | Purlin and girts | Brittle cladding | Span/180 |
| | Live load | Simple span | Brittle cladding | Span/300 |
| | Live load | Cantilever | Brittle cladding | Span/150 |
| | Live load | Simple span | Elastic cladding | Span/240 |
| | Live load | Cantilever | Elastic cladding | Span/120 |
| | Live load/ Wind load | Rafter supporting | Profiled metal sheeting | Span/180 |
| | | | Plastered sheeting | Span/240 |
| Lateral | No cranes | Column | Elastic cladding | Height/150 |
| | No cranes | Column | Masonry/brittle cladding | Height/240 |

*Notes:*

1. Brittle claddings are asbestos sheeting, fibre concrete sheeting, un-reinforced masonry, etc., which crack easily under large deflections. Metal cladding is elastic cladding.
2. For vertical and lateral limits for buidings with cranes refer Table 6 of IS 800 : 2007.

in a number of situations. Vibration has to be checked when vibrating loads such as due to machinery, cranes, etc., are applied to plant platforms. Activities such as dancing, marching, and drilling and impact loads also produce vibration. In these cases, care must be taken to ensure that the structural response will not amplify the disturbing motion. The code recommends that flexible structures (with height to effective width ratio exceeding 5:1) should be investigated for lateral vibration under dynamic loads. Open-web steel joists (trusses) or steel beams supporting concrete decks may experience walking vibration problems. The code gives guidelines to estimate floor frequency, damping, and acceleration due to vibration, in Annex C. When vibration becomes a problem, we may have to change the natural frequency of the structure by some means. Changing the load factor will not help in reducing this problem. Design charts for floor vibration, design criteria for vibration due to walking, and other information regarding building vibration may be found in Hatfield (1992), Allen and Murray (1993), Murray (1991), Ellingwood and Tullin (1984), and Allen (1980).

Software packages (e.g., Floorvibe—www.floorvibe.com) are also available, using which the frequency and amplitude resulting from transient vibration caused by human activity can be quickly estimated.

## CASE STUDY

### Vibration of London Millennium Footbridge

The London Millennium Footbridge is a pedestrian-only steel suspension bridge crossing the River Thames in London, England, linking Bankside with the City. The bridge was designed by Arup, Foster and Partners, and Sir Anthony Caro. Due to height restrictions, and to improve the view, the bridge's suspension design had the supporting cables below the deck level, giving a very shallow profile. Construction of the bridge began in 1998, and it was opened on 10 June 2000.

Londoners nicknamed the bridge 'the Wobbly Bridge' after they felt an unexpected and uncomfortable swaying motion on the first two days after the bridge opened.

You may watch these vibrations at: http://www.youtube. com/watch? v=gQK21572oSU

The bridge's movements were caused by a 'positive feedback' phenomenon, known as *synchronous lateral excitation*. The natural sway motion of people walking caused small sideways oscillations in the bridge, which in turn caused people on the bridge to sway in step, increasing the amplitude of the bridge oscillations and continually reinforcing the effect.

Any bridge with lateral frequency modes of less than 1.3 Hz, and sufficiently low mass, could witness the same phenomenon with sufficient pedestrian loading. The greater the number of people, the greater will be the amplitude of the vibrations.

After extensive analysis by the engineers, the problem was fixed by the retrofitting of 37 fluid-viscous dampers (energy dissipating) to control horizontal movement and 52 tuned mass dampers (inertial) to control vertical movement. After a period of testing, the bridge was successfully reopened on 22 February 2002. The bridge has not been subject to significant vibration since then.

*Photo Courtesy*: Mr Uwe Tress

### Durability and Corrosion

A durable steel structure is one that performs satisfactorily the desired functions during its service life under the anticipated environment and exposure conditions. This implies that the member's designed cross-sectional area should not deteriorate and its strength should not be reduced mainly due to *corrosion*.

Several factors that may affect the durability of the steel construction may include

1. the environment,
2. the degree of exposure,
3. the shape of the member,
4. the structural detailing,
5. the protective coating employed, and
6. the ease of maintenance.

The methods of corrosion protection are governed by actual environmental conditions as per IS 9172. Section 15 of IS 800 gives the requirements for durability and also specifications for different coating systems under different atmospheric conditions.

Painting is the normal method of protection used to prevent corrosion. In order to ensure the durability of the painting system, the steel surface is first cleaned and then subjected to methods such as blast cleaning to remove any rust or other loose material. This type of process leaves the surface clean and slightly rough, enabling a good base to be formed. Good designers will also try to 'design for prevention' by avoiding traps for dirt and moisture (Subramanian 2000; British Steel Corporation 1996; and European Coal and Steel Community 1982).

It has to be noted that the code does not specify any minimum thickness for members (though the earlier revision of the code, IS 800 : 1984, specified a minimum thickness of 6 mm for members directly exposed to weather and fully accessible for cleaning and repainting and 8 mm for members directly exposed to weather but not accessible for cleaning and repainting). It is assumed that if the durability requirements as given in Section 15 are followed, it will ensure adequate protection for all thicknesses.

## CASE STUDY

### Painting of Golden Gate Bridge

Prior to 1965, coatings were generally oil- or alkyd-based and contained pigments using lead and/or chromium compounds as the corrosion inhibitors. In addition, they were often applied directly to steel surface that had been subjected to a surface preparation made by power tool cleaning (often using rotary impact tools). These old type of coatings were expected to last about 8 to 10 years. As a result many bridges have several layers of coating on them, which fall-off due to their own weight. Coatings with overall thickness of 6 mm or more have been encountered in the past.

The current standard for bridge coating entails the use of a three-layer coating system consisting of an inorganic zinc-rich primer, an epoxy mid-coat, and a urethane topcoat.

The primer is applied on blast-cleaned surface.

The 2737 m long Golden Gate Bridge in San Francisco, USA, when built during 1933 to 1937 was coated with lead-based paint. During 1965 the lead paint was removed and an inorganic zinc-rich paint system was applied. When the bridge was examined recently, it revealed that that the coating system is in excellent overall condition even after 44 years.

It has to be noted that the metallic zinc pigment in the zinc-rich paint will be able to provide galvanic protection to the steel surface, until the zinc itself is consumed. When the zinc is depleted, the steel will eventually rust. The time it takes the zinc to be consumed is dependent on the weather, duration of wetness, and the number of wet/dry cycles encountered. The additional two layers of coating protect the zinc by limiting the entry of moisture and oxygen. It is estimated that after about 30 to 55 years, another overcoat is possibly required. Thus, the new coating system may increase the life of steel structures to more than 100 years.

*Photo Courtesy*: Karthik Muthusamy

## CASE STUDY

### Richard J. Daley Center, Chicago

Weathering steel is a steel-alloy containing copper, chromium, nickel, and other alloying elements to enhance corrosion-resistance. When the steel rusts under normal atmospheric exposure, it forms a rusty-orange-brown protective impervious layer that precludes further corrosion. Minor damage to the coating is self-healing. Over time, as the thickness of coating increases, the texture roughens and the oxide coating changes from rusty-orange-brown to dark purple-brown patina. Thus weathering steel does not require any painting to protect it and provides significant life-cycle cost savings.

The Richard J. Daley Center, Chicago, built in the year 1965 was the first building to be constructed entirely with a type of weathering steel, called Cor-Ten. The Pablo Picasso sculpture outside the center is also made of weathering steel. As the picture shows, even after more than 40 years of existence, they look great due to proper maintenance of the building and the sculpture.

Of course weathering steel is not a material of choice in environments that are constantly wet or humid or for coastal structures. It has been used in several short span bridges such as the award-winning Box Elder Creek Bridge in Colorado. More details about weathering steel is available online at: http://www.aisc.org/content.aspx?id=17892

Richard J. Daley Center and Pablo Picasso Sculpture

### *Fatigue*

The fatigue limit state is important in structures where distress to the structure due to repeated loading is to be considered. Thus, fatigue is an important criterion for bridges, crane girders, platforms carrying vibrating machinery, etc. In such structures, failure may occur at stress levels well below normal yield. The code states that stress changes due to fluctuations in wind loading need not be considered as fatigue. Though fatigue failure is usually considered as an ultimate limit state, fatigue checks are carried out at working loads (with load factor $\gamma_f$ equal to unity), considering the stress variation, which occur in the normal working cycle. This apparent anomaly (i.e., checking the fatigue limit state at working load) exists because fatigue failure occurs due to the very large number of application of loads normally expected to act on the structure (i.e., the specified load). Other ultimate limit state failures, such as the failure of a column, could occur due to a single application of a greater load than that normally expected to act on the structure; thus a factored load is used in such design calculations.

Section 13 of the code gives guidelines for fatigue design but does not consider the effect of the following:

1. Corrosion fatigue
2. Low-cycle (high-stress) fatigue
3. Thermal fatigue
4. Stress corrosion cracking
5. Effects of high temperature (>150°C)
6. Effects of low temperature (< brittle transition temperature)

The code states that fatigue assessment is not normally required for building structures except in the case of the following members:

1. Those supporting lifting or rolling loads
2. Those subjected to repeated stress cycles from vibrating machinery
3. Those subjected to wind-induced oscillations of a large number of cycles in life
4. Those subjected to crowd-induced oscillations of a large number of cycles in life

The code also states that fatigue assessment is not necessary if any of the following conditions is satisfied.

1. The highest normal stress range,

$$f_{\text{fmax}} \leq \frac{27\mu_c}{\gamma_{\text{mft}}} \tag{2.9}$$

2. The highest shear stress range,

$$\tau_{\text{fmax}} \leq \frac{67\mu_c}{\gamma_{\text{mft}}} \tag{2.10}$$

3. The total number of actual stress cycles,

$$N_{SC} \leq 5\times 10^6 \left(\frac{27\mu_c}{\gamma_{\text{mft}} f_{\text{feq}}}\right)^3 \tag{2.11}$$

where,

$$f_{\text{feq}} = \left[\frac{\sum_{i=1}^{\gamma_s} n_i f_{\text{fi}}^3 + \sum_{j=\gamma_s}^{r} n_j f_{\text{fj}}^5}{n}\right]^{1/3} \tag{2.12}$$

where $f_{\text{feq}}$ is the equivalent constant amplitude stress range in megapascals, $\mu_c$ is the correction factor as given in the code, $\gamma_{\text{mft}}$ is the partial safety factor against fatigue failure as per the code, $n$ is the number of stress cycles, $n = \sum_{i=1}^{\gamma} n_i$, and $f_{\text{fi}}, f_{\text{fj}}$ are the stress ranges falling above and below $f_{\text{fn}}$, the stress range corresponding to the detail at $5 \times 10^6$ life cycles.

For the purpose of design against fatigue, the code classifies different details (of members and connections) under different fatigue classes. The design stress range corresponding to various numbers of cycles are given in the code for each fatigue class. The requirements of the code are to be satisfied at each critical location of the structure subjected to cyclic loading, considering the relevant number of cycles and magnitudes of stress range expected to be experienced during the life of the structure.

### Brittle Fracture

As with fatigue, brittle fracture will rarely occur in building construction. Such fracture is the sudden failure of the material under service condition, caused by low temperature or sudden change in stress. Since thick material is more prone to brittle fracture than thin material, limiting thicknesses are often prescribed by the codes for the various members. In addition to the thickness, the stress in the material, and the type of details employed in the construction also have significant effect on the risk of brittle fracture. As discussed in Chapter 1, the Charpy V-notch impact test is used to find out the temperature below which fractures are mostly brittle and above which fractures are ductile. Brittle fracture can be avoided by using a steel quality and welding electrodes that have a specified Charpy impact value.(IS 1757 : 1988 allows the use of only those steels that exhibit a minimum energy absorption capacity at a predetermined temperature, say 20 J at 23 ± 5°C.)

### Fire Resistance

Fire resistance of a steel member is a function of its mass, geometry, support conditions, type of fire and the fire protection methods adopted.

## 2.10 FAILURE CRITERIA FOR STEEL

The structural behaviour of a material close to failure may be classified as ductile or brittle. For example, cast iron exhibits a linear load displacement behaviour until fracture occurs suddenly with little or no plastic deformation. Such behaviour is termed *brittle behaviour*. In contrast, mild steel, though exhibiting a linear load-displacement relationship up to yield undergoes large plastic deformation before fracture. Thus, it is considered as a ductile material.

Thus, the minimum yield strength is the design strength in IS 800, and is an important failure criterion for steel. However, only when the load-carrying member is subject to uniaxial tensile stress, the properties obtained from the tension test will be identical with those of the structural member. In a real structure, the yielding will not be well defined, unlike as observed in the tension test. Hence in these cases, yielding is assumed to be achieved when any one component of stress reaches the uniaxial value $f_y$.

For all states of stress other than uniaxial, a definition of yielding is needed. These definitions are called *yield criteria* and are expressed as equations of interaction between the stresses acting.

The four generally accepted theoretical yield criteria are as follows (Martin and Purkiss 1992).

1. The maximum stress theory, which states that yield occurs when the maximum principal stress reaches the uniaxial tensile stress.
2. The maximum strain theory, which states that yield occurs when the maximum principal tensile strain reaches the uniaxial tensile strain at yield.
3. The maximum shear stress theory, which states that yield occurs when the maximum shear stress reaches half of the yield stress in uniaxial tension.
4. The distortion or shear strain energy theory, which states that yielding occurs when the shear strain energy reaches the shear strain energy in simple tension.

The most commonly accepted theory, developed by Huber, von Mises, and Hencky, developed in 1913, gives the uniaxial yield stress in terms of the principle stresses $f_1$, $f_2$, and $f_3$ as follows.

$$2Y_s^2 = (f_1 - f_2)^2 + (f_2 - f_3)^2 + (f_3 - f_1)^2 \tag{2.13}$$

In Eqn (2.13), $f_1$, $f_2$, and $f_3$ are the stresses that act in three mutually perpendicular planes of zero shear, and $Y_s$ is the 'yield stress' that may be compared with the uniaxial value, $f_y$.

For most structural design situations, one of the principal stresses is either zero or small enough to be neglected. Thus, Eqn (2.13) can be expressed in terms of direct stresses $f_b$, $f_{bc}$ and $f_{bt}$, and shear stress $f_q$ on two mutually perpendicular planes. It can be shown from Mohr's circle of stress that the principal stresses

$$f_1 = \frac{(f_b + f_{bc})}{2} - \left[\frac{(f_b - f_{bc})^2 + f_q^2}{4}\right]^{1/2} \tag{2.14}$$

and
$$f_2 = \frac{(f_b - f_{bc})}{2} + \left[\frac{(f_b - f_{bc})^2 + f_q^2}{4}\right]^{1/2} \tag{2.15}$$

If Eqns (2.14) and (2.15) are inserted in Eqn (2.13), with $f_3 = 0$, and $Y_s$ is equal to the design stress $f_y$, we get

$$f_y^2 = f_{bc}^2 + f_b^2 - f_{bc} f_b + 3f_q^2 \tag{2.16}$$

In monoaxial case, we get

$$f_y^2 = f_{bc}^2 + 3f_q^2 \tag{2.17}$$

### 2.10.1 Shear Yield Stress

The yield point for pure shear can be determined from a stress–strain curve with shear loading, or if the multiaxial yield criterion is known, it can be used. It can be shown that by using Eqn (2.13), the yield condition for shear stress acting alone is (Salmon and Johnson 1996)

$$\tau_y = \frac{f_y}{\sqrt{3}} = 0.58 f_y \tag{2.18}$$

### 2.10.2 Poisson's Ratio ($\mu$)

When stress is applied in one direction, strains are induced not only in that direction but also in the other two mutually perpendicular directions. Usually, the value of Poisson's ratio $\mu$ is obtained from the uniaxial stress condition, where it is the ratio of the transverse strain to the longitudinal strain under load. IS 800 in Clause 2.2.4.1 suggests a value of 0.3 for $\mu$, which is applicable in the elastic range, where the material is compressible. This value may reach 0.5 in the plastic range, where the material is incompressible (i.e., the resistance is constant, irrespective of the strain value).

### 2.10.3 Shear Modulus of Rigidity

If a specimen is loaded in pure shear, it produces a stress–strain curve, with a straight-line portion whose slope is the shear modulus of elasticity. Using the theory of elasticity, the shear modulus of elasticity is expressed as (Timoshenko and Goodier 1970)

$$G = \frac{E}{2(1+\mu)} \tag{2.19}$$

IS 800 suggests that the physical properties of structural steel, irrespective of its grade, which are given in Table 1.9(b).

## SUMMARY

Structural design is considered a science as well as an art. The aim of any structural designer should be to design a structure in such a way that it will fulfill its intended purpose during its intended lifetime and be adequately safe (in terms of strength, stability, and structural integrity), and have adequate serviceability (in terms of stiffness, durability, etc). In addition, the structure should be economically viable, (in terms of cost of construction and maintenance), aesthetically pleasing, and environment-friendly. The various steps involved in the construction activities are listed and the role and responsibilities of the designer while carrying out these activities are emphasized in the chapter.

Brief descriptions of the various structural systems that have evolved over the past have been given. Depending on the particular situation, and the requirement of the client, a structural system may be chosen. The concept of structural integrity is explained.

Analysis refers to the determination of the internal forces acting on different members of a structure, due to the application of external actions (forces). Analysis is followed by design, which involves the selection of the sizes of different members based on the criteria stipulated in the national code of practices. A brief description of the methods of analysis and the available computer programs are given. Further details about the available methods of analysis for steel structures are provided in Chapter 4. (From this chapter, it may be realized that the advanced methods of analysis integrate the analysis and design process into a single interactive activity.)

A discussion on codes and specifications which will guide the designer in the design process has been provided. The codes published by the Bureau of Indian Standards for the design of steel structures are listed and the reader is advised to obtain a copy of these codes, since they may be required to understand the material provided in the chapters to follow.

The various design philosophies that have been evolved in the past, namely, the working stress method, ultimate strength/load design (plastic design), and limit states design, are briefly discussed. The various limit states that have to be considered in design are explained. A brief description of the reliability of these methods of design is also provided.

The terms characteristic load and characteristic strength are explained. The basis of the limit states method is discussed along with the various partial safety factors adopted by the code for materials and loads. A brief introduction to stability is provided. The various serviceability limit states, related to the satisfactory performance of the structure (as opposed to ultimate/safety limit states, which are concerned with strength, stability, fatigue fracture, etc) are also briefly explained along with the deflection limits specified by the code. The failure criteria, which are required to define the yielding of the material, when subjected to multiaxial stresses, are also discussed.

As we have discussed in this chapter, though analysis and design are interactive processes, for convenience they are currently performed as separate activities, the design phase following the analysis phase. Once we finalize and decide upon the structural configuration for a particular design project, the next step is to calculate the various external forces that may act on the structure. The methods of calculating these loads are explained in the next chapter.

## MULTIPLE-CHOICE QUESTIONS

2.1 State whether the following is the job of a structural engineer or not.
(a) He/she has to survey the land and prepare a contour map.
(b) He/she has to estimate the probable loads that may act on the structure.
(c) He/she has to decide on the structural system.
(d) He/she has to do suitable structural analysis and then design the members.
(e) He/she has to float the tender and choose the general contractor for the building.

2.2 State true or false: The structural engineer
(a) should have knowledge about weld defects
(b) should be aware of non-destructive testing techniques
(c) should have a knowledge of fabricating shop capabilities
(d) should visualize the behaviour of structures as well as its members

2.3 Which of the following do not use steel extensively?
(a) Pneumatic structures
(b) Shell structures
(c) Framed structures
(d) Tension structures

2.4 Which of the following statements are wrong?
(a) Towers are considered as plane frames
(b) In simple construction, ends of beams are connected to transmit shear only
(c) Domes often require three dimensional analysis
(d) Cables are used in compression structures
(e) Composite construction use steel beams and concrete slab, which are connected by shear connectors

2.5 The horizontal tie members provided for structural integrity should be designed for a factored tensile load of
(a) 50 kN at floor level and 40 kN at roof level
(b) 100 kN at floor level and 50 kN at roof level
(c) 75 kN at floor level and 40 kN at roof level
(d) 85 kN at floor level and 85 kN at roof level

2.6 Which of the following is used for indeterminate structural analysis?
(a) Moment distribution method
(b) Slope deflection method
(c) Matrix methods
(d) None of these

2.7 Which are the following codes are necessary for the design of steel structures
(a) IS 800:2007
(b) IS 875:1987 (Part 1–5)
(c) IS 456:2000
(d) IS 808: 1989

2.8 State yes or no. In the working stress method, the following factors of safety are used.
(a) 1.92 for long columns
(b) 2.0 for tension members
(c) 2.5 to 3.0 for connections
(d) 2.5 for beams

2.9 The load factor for dead load + imposed load in ultimate load design is
(a) 1.5 (c) 2.0
(b) 1.7 (d) 1.6

2.10 Which of the following is considered as the ultimate limit state in limit states design?
(a) Stability against overturning
(b) Corrosion and durability
(c) Vibration
(d) Strength of member

2.11 Which of the following is considered as serviceability limit state in limit states design?
(a) Failure due to excessive deformation
(b) Brittle fracture
(c) Fire
(d) Deflection

2.12 Characteristic load is the value of load which will not be exceeded with more than
(a) 10% probability during the life span
(b) 5% probability during the life span
(c) 0% probability during the life span
(d) 20% probability during the life span

2.13 Partial safety factor for material resistance governed by yielding is taken in IS 800 as
(a) 1.0 (c) 1.3
(b) 1.2 (d) 1.1

2.14 Partial safety factor for material resistance governed by ultimate stress is taken in IS 800 as
(a) 1.25 (c) 1.20
(b) 1.5 (d) 1.10

2.15 Partial safety factor for resistance of bolted connection is taken in IS 800:2007 as
(a) 1.0 (c) 1.25
(b) 1.2 (d) 1.50

2.16 Partial safety factor for resistance of field welds is taken in IS 800 as
(a) 1.50 (c) 1.20
(b) 1.25 (d) 1.10

2.17 Partial safety factor for dead load is taken in IS 800:2007 as
(a) 1.60 (c) 1.50
(b) 1.40 (d) 2.00

2.18 Partial safety factor for imposed load is taken in IS 800:2007 as
(a) 1.60 (c) 1.50
(b) 1.40 (d) 1.20

2.19 Partial safety factor for earthquake load is taken in IS 800: 2007 as
(a) 1.60 (c) 1.50
(b) 1.40 (d) 1.20

2.20 Partial safety factor for dead load when it provides stability against overturning or sliding is taken in IS 800:2007 as
(a) 1.60 (c) 1.50
(b) 1.40 (d) 0.90

2.21 Notional loads are to be applied at each floor level along with factored gravity loads in one orthogonal direction at a time with an intensity of
(a) 1.0% of the factored live loads
(b) 0.5% of the factored dead and live loads
(c) 0.5% of the factored dead loads
(d) 1.5% of the factored dead and live loads

2.22 The vertical deflection limit specified for purlins in industrial buildings with elastic cladding under W.L./ I.L. is
(a) Span/150 (c) Span/180
(b) Span/240 (d) Span/300

2.23 The vertical deflection limit specified for simple beams in industrial buildings with brittle cladding under Live load is
(a) Span/150 (c) Span/180
(b) Span/240 (d) Span/300

2.24 The lateral deflection limit specified for columns in industrial buildings with no cranes but with elastic cladding is
(a) Height/150 (c) Height/180
(b) Height/240 (d) Height/300

2.25 There is no minimum thickness provision in IS 800 for durability. However, it is better to adopt a thickness of
(a) 5 mm for members exposed to weather
(b) 4 mm for members exposed to weather
(c) 6 mm for members exposed to weather
(d) 8 mm for members exposed to weather

## REVIEW QUESTIONS

2.1 What are the main objectives of a structural engineer while designing steel structures?
2.2 What are the various steps involved in the construction of steel structures?
2.3 What are the roles and responsibilities of the designer?
2.4 List the different types of steel structures.
2.5 Distinguish between simple and continuous construction.
2.6 Write short notes on the following:
(a) composite structures,
(b) suspended structures,
(c) frame with semi-rigid joints,
(d) tube structures, and
(e) Outriggers and belt truss systems.
2.7 How can the structural integrity of steel buildings be established?
2.8 Distinguish between analysis and design.
2.9 Why are codes of practices necessary and what is the function of codal specifications?
2.10 List a few Indian standard codes that are followed while designing structures made of steel.
2.11 Distinguish between the working stress method, ultimate strength design, and limit states design.
2.12 What are the limitations and shortcomings of the working stress method?
2.13 What are the possible limit states that are considered in the limit states method?
2.14 What are the features of the load and resistant factor design?
2.15 What are the partial safety factors (of resistance) for materials adopted by the code?
2.16 What are the partial load factors adopted for the following loading combinations? (a) DL (b) LL (c) WL (d) DL + LL
2.17 What are the three forms of structural stability considered by the code?
2.18 Write a brief note about the various serviceability limit states considered by the codes.
2.19 What are the four accepted yield criteria?
2.20 State the Huber, von Mises, and Hencky failure theory.

## REFERENCES

*AISC Manual of Steel Construction—Load and Resistance Factor Design*, Second edition, Vols 1 and 2, American Institute of Steel Construction, Chicago, 1994.

Al Nageim, H.K. and MacGinley, T.J., *Steel Structures, Practical Design Studies*, Third edition, CRC Press, 2005, p. 352.

Ali, M.M. and Moon, K.S., 'Structural Developments in Tall Buildings: Current Trends and Future Prospects', *Architectural Science Review*, Vol. 50, No. 3, 2007, pp. 205–223.

Allen, D.E. and Murray, T.M., 'Design Criteria for Vibration due to Walking', *Engineering Journal*, AISC, Vol. 30, No. 4, Fourth quarter 1993, pp. 117–129.

Allen, D.E., 'Building Vibration from Human Activities', *Concrete International*, Vol. 12, No. 6, Jun 1990, pp. 66–73.

Allen, D.E., *Structural Safety, Division of Building Research*, National Research Council of Canada (CBD 147) Ottawa, 1972.

British Steel, *The Prevention of Corrosion on Structural Steelwork*, British Steel Corporation, 1996.

Brockenbrough, R.L. and Merritt, F.S., ed., *Structural Steel Designers Hand Book*, Third edition, McGraw Hill Professional, New York, 1999.

BS 5950-2000, *Structural Use of Steelwork in Buildings: Part I – Code of Practice for Design in Simple and Continuous Construction, Hot Rolled Section*, British Standards Institution, London, 2000, p. 213.

Chen, W.F., Goto, Y., and Liew, J.Y.R., *Stability Design of Semi-Rigid Frames*, Wiley Inter Science, 1995, p. 488.

Davison, B. and Owens, G.W., ed., *Steel Designers Manual*, Sixth edition, Blackwell Publishers, London, 2002.

Ellingwood, B., 'Serviceability Guidelines for Steel Structures', *Engineering Journal*, AISC, Vol. 26, No. 1, First quarter 1989, pp. 1–8.

Ellingwood, B. and Tullin, A., 'Structural Serviceability: Floor Vibration', *Journal of Structural Engineering*, ASCE, Vol. 110, No. ST2, Feb 1984, pp. 401–418, Disc. V.111, No. ST5, May 1985, pp. 1158–1161.

Ellingwood, B.R., Smilowitz, R., Dusenbeery, D.O., Duthinh, D., Lew, H.S., and Carino, N.J., *Best Practices for Reducing the Potential for Progressive Collapse in Buildings*, NISTIR 7396, National Institute of Standards and Technology, Gaithersburg, Feb 2007, p. 216 (http://www.fire.nist.gov/bfrlpubs/build07/art008.html).

European Coal and Steel Community, *Durability of Steel Structures*, 1982.

Galambos, T.V. and Ravindra, M.K., 'Properties of Steel for Use in Load and Resistance Factor Design', *Journal of Structural Div.*, ASCE, Vol. 104, No. ST9, Sep 1978.

Galambos, T.V., 'Load and Resistance Factor Design', *Engineering Journal*, AISC, Third quarter 1981, Vol. 18, No. 3, pp. 74–82.

Galambos, T.V. and Ellingwood, B., 'Serviceability Limit States Deflections', *Journal of Structural Engineering*, ASCE, Vol. 112, No.3, Jan 1986, pp. 67–84

Gordon, J.E., *Structures (or Why Things Don't Fall Down)*, Penguin Books, New York, 1978.

Hatfield, F.J., 'Design Chart for Vibration of Office and Residential Floors', *Engineering Journal*, AISC, Vol. 29, No. 4, Fourth quarter 1992, pp. 141–144.

IS:800-1984 *Indian Standard Code for General Construction in Steel*, Second revision, Indian Standards Institution, New Delhi, Feb 1985.

IS:9172-1979 *Recommended Design Practice for Corrosion Protection of Steel Structures*, Bureau of Indian Standards, New Delhi.

Johnson, R.P., 'Composite Structures of Steel and Concrete', Vol. 1, *Beams, Columns, Frames and Applications in Buildings*, Blackwell Scientific Publications, London, 1994.

Kulak, G.L. and Grondin, G.Y., *Limit States Design in Structural Steel*, Seventh edition, Canadian Institute of Steel Construction, 2002.

Madsen, H.O., Krenk, S., and Lind, N.C., *Methods of Structural Safety*, Prentice Hall, Inc., Englewood Cliffs, NJ, 1986.

Martin, L.H. and Purkiss, J.A., *Structural Design of Steelwork to BS 5950*, Edward Arnold, London, 1992, p. 468.

Moon, K.S., Connor, J. J., and Fernandez, J. E., 'Diagrid Structural Systems for Tall Buildings: Characteristics and Methodology for Preliminary Design', *The Structural Design of Tall and Special Buildings*, Vol. 16, No.2, 2007, pp. 205–230.

Moon, K.S., 'Design and Construction of Steel Diagrid Structures', *Proceedings of the Nordic Steel Construction Conference*, Malmo, Sweden, 2–4 Sep 2009, pp. 398–405 (http://www.nordicsteel2009.se/pdf/72.pdf).

Murray, T.M., 'Building Floor Vibrations', *Engineering Journal*, AISC, Vol. 28, No. 3, Third quarter 1991, 102–109, Errata, Vol. 28, No. 4, Fourth quarter 1991, p. 176.

Neal, B.G., *The Plastic Methods of Structural Analysis*, Chapman and Hall, London, 1977.

Nethercot, D.A., *Limit States Design of Structural Steelwork*, Third edition, Spon Press, London, 2001.

Ranganathan, R., *Reliability Analysis and Design of Structures*, Tata-McGraw Hill Publishing Co.Ltd., New Delhi, 1990.

Ravindra, M.K. and Galambos, T.V., 'Load and Resistance Factor Design for Steel', *Journal of Structural Div*, ASCE, Vol. 104, Sep 1978, pp. 1337–1353.

Salmon, C.G., Johnson, J.E., and Malhas, F.A., *Steel Structures—Design and Behaviour*, Fifth edition, Harper Collins, College Publishers, New York, 2008, p. 888.

Smith, B.S. and Coull, A., *Tall Building Structures: Analysis and Design*, Inter Science, 1991.

Structural Engineering Institute, *Effective Length and Notional Load Approaches for Assessing Frame Stability: Implications for American Steel Design*, American Society of Civil Engineers, 1997, p. 442.

Subramanian, N., 'Corrosion Protection of Steel Structures', *The Master Builder*, Vol. 2, No. 1, Jan–Feb 2000, pp. 63–66 and No. 2, April–May 2000, pp. 14–17.

Subramanian, N., *Principles of Space Structures*, Wheeler Publishing, New Delhi, Second edition, 1999, pp. 604–605.

Subramanian, N., 'Aesthetics of Non-Habitat Structures', *The Bridge and Structural Engineer*, Journal of ING/IABSE, Vol. 17, No. 4, Dec 1987, pp. 75–100.

Subramanian, N., 'Computer Analysis and Design of Tall Structures', *Civil Engineering and Construction Review*, Vol. 8, No. 4, April 1995, pp. 42–46.

Subramanian, N., 'Safety in Engineering Structures', *Fourth National Conference on Industrial Safety*, College of Engineering, Guindy, Chennai, Nov 1974.

Timoshenko, S.P. and Goodier, J.N., *Theory of Elasticity*, McGraw Hill, New York, Third edition, 1970.

Unnikrishna Pillai, S. and Menon, P., *Reinforced Concrete Design*, Second edition, Tata-McGraw Hill Publishing Co. Ltd, New Delhi, 2003.

# 3

# LOADING AND LOAD COMBINATIONS

## INTRODUCTION

Before designing any structure or different elements such as beams and columns one has to determine the various natural and man-made loads acting on them. There may be different types of load and the designer must evaluate the magnitude of each of the loads acting on the structure. The designer has to consider the combination of loadings which will result in the most unfavourable loading with respect to the stresses in the members. The usual forms of loading include

- dead loads (DL), also called fixed loads,
- imposed loads (IL), previously called live loads (LL),
- wind loads (WL),
- earthquake loads (EL),
- snow loads (in some regions of India),
- ice loads (in some regions of India),
- crane loads (mainly in industrial buildings),
- temperature loads (loads due to contraction or expansion resulting from variation in temperature),
- erection and fabrication loads,
- dynamic loads (due to earthquake and wind),
- impact loads (in some structures such as bridges and cranes),
- longitudinal loads (caused by the acceleration or deceleration of moving loads),
- blast and explosion loads (due to terrorist activities or accidents),
- loading due to fire,
- foundation settlement (differential settlement of the structure as a whole or of its components),
- hydrostatic loads (to be considered in liquid-retaining structures or in hydraulic structures),
- wave and current loads (to be considered in marine and offshore structures),
- earth pressure (in basements, retaining walls, column footings, etc.),
- creep and shrinkage loads (not important for steel structures),
- fatigue effects,
- dust loads (in certain industrial buildings and buildings in certain locations, especially in deserts),
- prestressing loads,
- effect due to stress concentration and eccentric connections, rigidity of joints differing from design assumptions, etc., and
- construction loads, elastic shortening of members, $P$-$\Delta$ effects, etc. The second-order moments arising from the joint displacements are called *$P$-$\Delta$ effects*, while the second-order moments arising from member deflections are called *$P$-$\delta$ effects*.

Thus, a load on a structure may be due to the following:

1. Mass and gravitational effect ($mg$)—examples of such loads are dead loads, imposed loads, snow, ice, earth loads, hydraulic pressure, etc.,
2. Mass and its acceleration effect ($ma$)—examples of such loads are those caused by earthquakes, wind, impact, blasts, etc., and
3. Environmental effects: examples are loads due to temperature difference, settlement, shrinkage, etc. They are also termed *indirect loads*.

Based on these, loads may also be classified as 'gravity' and 'non-gravity' loads.

IS 800 : 2007 classifies actions (loads) as follows.

***Permanent actions* ($Q_p$)** Actions due to the self-weight of structural and non-structural components, fixed equipment and machinery, etc.

***Variable actions*** **($Q_v$)** Actions due to construction and service stage loads such as imposed (live) loads (crane loads and snow loads), wind loads, earthquake loads, etc.

***Accidental actions*** **($Q_a$)** Actions due to the impact of vehicles, due to explosions, fire, etc.

Though there are several types of loading, in most cases, we may have to consider only the major loads, i.e., DL, IL, WL, EL, and temperature loads. Many other loads such as those due to differential settlement, eccentric connections, etc., produce only secondary effects. However, in some cases they may become critical. The designer should use his judgment in considering whether to include these loads or not. Loads have to be chosen judiciously to ensure the safety of the structure. At the same time, overdesign and wastage of precious resources should be avoided.

While accessing the loads acting on a structure, it is customary to refer to the appropriate code of practice. In India, basic data on dead, live, and wind loads for buildings are given in IS 875, with more specialized information on matters such as load produced by cranes in industrial buildings provided by other codes (e.g., IS 807). Earthquake loads are specified in IS 1893. For towers and other forms of structures, the necessary loading data are provided in the code of practice appropriate to that type of structure (e.g., IS 802, IS 9178, and IS 6533). In the following sections, we will discuss the features given in the codes, the methods to calculate these loads, and their combinations to be applied on structures.

## 3.1 CHARACTERISTIC ACTIONS (LOADS)

Determining the loads for which a given structure has to be proportioned is one of the most difficult problems in design and is often not taught in course work. Decisions have to be made on the type of loads the structure may experience during its lifetime, combinations of loads, etc. The probability that a specific load will be exceeded during the life of the structure usually depends on the period of exposure (or life) of the structure and the magnitude of the design load. For example, the wind load acting on a structure varies from year to year based on the wind speed and the maximum that can be expected to occur at a given location only once in so many years. This period is called a (mean) *return period*. Thus, a return period is defined as the number of years, the reciprocal of which gives the probability of extreme wind exceeding a given wind speed in any one year. (In general, return period may be defined as the average time between consecutive occurrences of the same event. Return period is only an average duration and not the actual time between occurrences, which would be highly variable.) Such return periods can be determined from a statistical analysis of wind speed records. Of course, extremes of other natural phenomena such as snow, earthquake, or flood also occur in-frequently and return periods for specific extremes can be determined similarly.

Thus, if the return period $R$ of a wind speed of, say, 200 km/hr at a certain locality is 50 years, the probability that there will be a wind speed greater than 200 km/hr in any one year is $1/R = 1/50 = 0.02$. However, for design purposes we are not interested in the probability that it will be exceeded in any one year but rather the probability that it will be exceeded during the life of the structure. Note that if $1/R$ is the probability that the wind speed will be exceeded in any year, $(1 - 1/R)$ is the probability that it will not be exceeded. If we consider $N$ as the life of the structure in years, then the probability that it will not be exceeded during the life of the structure will be $(1 - 1/R)^N$. Therefore, the probability that it will be exceeded at least once in $N$ years is

$$P_N = 1 - (1 - 1/R)^N \tag{3.1}$$

Thus, the probability that a wind speed of 200 km/hr with a return period of 50 years will be exceeded at least once in 50 years (life of the structure) is

$$P_{50} = 1 - (1 - 0.02)^{50} = 0.635$$

That is, there is a 63.5 chance that the structure will be exposed to a wind exceeding 200 km/hr. The code [IS 875 (Part 3)] states that a value of $P_N = 0.63$ is normally considered sufficient for the design of buildings and structures against wind effects. Thus, if the acceptable risk is 0.63, it is sufficient to design the structure to resist the wind load from a 200 km/hr wind. It has to be remembered that there will be a margin of safety (due to the partial safety factors for loads and materials), and hence the structure may not collapse under a wind of this speed.

Ideally, loads applied to a structure during its life should be considered statistically and a characteristic load determined. Thus, the *characteristic load* may be defined as the load which will not be exceeded by a certain accepted or preassigned probability (usually 5%) during the life of the structure as shown in Fig. 3.1. For all practical purposes, the specified load in the codes is taken as the characteristic load in the absence of statistical data.

When the statistical data is available, the probability that a load may exceed the characteristic load is expressed as

$$p_f = \frac{\phi(Q_m - Q_c)}{\sigma} \tag{3.2}$$

where $p_f$ is the probability that the load exceeds the characteristic load $Q_c$, $Q_c$ is the characteristic load, $Q_m$ is the statistical mean of the observed maximum loads, $\sigma$ is the standard deviation of the loads, and $\phi$ is the cumulative distribution function.

The value of the cumulative distribution function depends on the statistical distribution of loads. The statistical distribution of dead loads and live loads is often assumed to be a normal distribution (as shown in Fig. 3.1), though some codes assume it to be of Weibull type distribution (Trahair et al. 2001). Equation (3.2) can be rearranged as

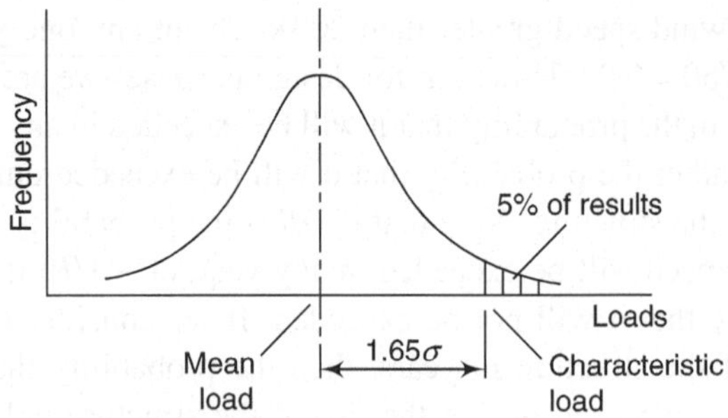


**FIG. 3.1** Definition of characteristic loads

$$\frac{Q_m - Q_c}{\sigma} = \phi^{-1}(pf) = -k \quad (3.3)$$

or

$$Q_c = Q_m + k\sigma \quad (3.4)$$

where

$$k = \phi^{-1}(p_f) \quad (3.5)$$

and $k$ is a coefficient associated with the preassigned probability of failure. For the characteristic load which will not be exceeded by a probability of 5%, we have $p_f = 5/100$ or

$$k = \phi^{-1}(0.05)$$

From standard tables for normal distribution, the value of $k$ for $p_f = 0.05$ is obtained as 1.65.

Thus, Eqn (3.4) can now be expressed as

$$Q_c = Q_m + 1.65\sigma \quad (3.6)$$

Example 3.1 illustrates the concept of characteristic load.

### Design Actions (Loads)

The design action $Q_d$ is arrived at by multiplying the characteristic actions (loads) by partial safety factors, $\gamma_f$. (These factors are often referred to as *load factors*.)

$$Q_d = \Sigma \gamma_f Q_{ck} \quad (3.7)$$

where $\gamma_f$ are the partial safety factors for loads given in Table 2.7, which are normally derived based on statistical analysis (Ravindra & Galambos 1978).

These partial safety factors are provided to take into account

1. the possibility of unfavourable deviation of the load from the characteristic value,
2. the possibility of inaccurate assessment of load,
3. variation in dimensional accuracy,
4. uncertainty in the assessment of effects of the load, and
5. uncertainty in the assessment of the limit state being considered.

When more than one imposed load can act simultaneously, leading load is considered as that causing the larger action effect. The load factor for water may be taken as 1.4 (as per BS: 8007). Though this value appears to be very conservative, if used for the design of a tank, for example, it allows for the tank overflowing, dimensional changes, and the possibility of the tank being filled with a denser liquid.

## 3.2 DEAD LOADS

A load fixed in magnitude and in position is called a *dead load.* Determining the dead load of the structure requires the estimation of the weight of the structure together with its associated 'non-structural' components. Thus, we have to calculate and include the weight of bare steelwork (including items such as bolts, nuts, and weld material), slabs, beams, walls, columns, partition walls, false ceiling, facades, cladding, water tanks, stairs, plaster finishes, and other services (cable ducts, water pipes, etc.). It has to be noted that the dead load of a steel structure is not known before it is designed. Normally, an initial value is assumed or estimated based on experience. For example, the weight of a lightly loaded truss may be taken as 250 N/m$^2$. After the design process, the assumed dead weight has to be compared with the actual dead load. If the difference between the two loads is significant, the assumed dead load should be revised and the structure redesigned. For the majority of steel buildings, the weight of structural steel will be less than 30% of the total dead load and hence, even large inaccuracies in the initial estimated value will not result in significant redesign. The dead load may be assessed from the knowledge of the specific weights or from that of the total weights of all the permanent items that constitute the building. The weights of frequently encountered structural and non-structural materials and also the architectural and roofing materials are given in Tables 3.1–3.4. Dead weights of other materials are provided in the code IS 875 (Part 1—Dead Loads).

**TABLE 3.1** Weights of structural materials

| Materials | Weight (kN/m$^3$) |
|---|---|
| Aggregate | |
| (a) Sand | 18.1 |
| (b) Stone | 16–18.5 |
| (c) Brick | 10–14 |
| (d) Slag | 7–8 |
| Brick | |
| (a) Common burnt | 16–18.5 |
| (b) Refractory | 17–20 |

(*Contd*)

**TABLE 3.1** *(Contd)*

| Materials | Weight (kN/m³) |
|---|---|
| (c) Heavy duty | 24–25 |
| (d) Brick masonry in CM 1:4 | 20 |
| Stone masonry | 20.4–26.5 |
| Brick ballast | 11.8 |
| Stone ballast | 16–19 |
| Mortar | |
| (a) Cement | 20.4 |
| (b) Lime | 17.3 |
| Cement concrete | |
| (a) Foam | 8–10 |
| (b) Light-weight | 10–14 |
| (c) Lean | 22 |
| (d) Plain | 24 |
| (e) Reinforced cement concrete | 25 |
| (f) Prestressed concrete | 25.5 |
| Steel | 77 |
| Aluminium | 22–26 |
| Cement | 14.1 |
| Timber | |
| (a) Teak | 6.13 |
| (b) White cedar | 7.22 |
| (c) Fir | 4.56 |
| (d) Sal | 7.85 |

**TABLE 3.2** Weights of architectural materials

| Materials | Weight (N/m²) |
|---|---|
| Aluminum sheets 1-mm thick | 27.5 |
| Fire insulation boards | 30–50 |
| Wood wool slab 25 mm thick | 150 |
| Fibre boards | 20–50 |
| Block board 25 mm thick | 120 |
| Gypsum panel and partition 75 mm thick | 430 |
| Gypsum plaster 13 mm thick | 215 |
| Plywood 10 mm thick | 70 |
| Glass fibre slab 25 mm thick | 20–40 |
| Linoleum 3 mm thick | 59 |
| Expanded metal | 40–60 |
| 5-mm glass | 125 |
| 10-mm mastic asphalt | 215 |
| 20-mm cement plaster | 450 |

**TABLE 3.3** Weight of roofing and flooring materials

| Materials | Weight (N/m²) |
|---|---|
| Roofing sheets | |
| (a) GI sheet 1.6 mm thick | 156 |
| (b) Steel sheet 1 mm thick | 77.5 |
| (c) AC sheet 6 mm thick | 160–170 |
| Mangalore tiles with battens | 639 |
| Asphalt roofing 10 mm thick | 220 |
| Plain cement concrete 40 mm | 850 |
| Floor finishes | 600–1200 |
| Terrazzo floor | 850 |
| Corrugated plastic sheeting | 45 |
| Glass sheet | |
| (a) 1.3 mm | 33.4 |
| (b) 2 mm | 52 |
| (c) 2.6 mm | 64.8 |
| Wooden floor | |
| (a) Hardwood, 22 mm | 157 |
| (b) Softwood, 22 mm | 201 |
| Gutters (asbestos) | |
| (a) Boundary wall gutters | 84–156 |
| (b) Valley gutters | 130–243 |
| (c) Half round gutters | 43–87 |
| Steelwork for roofing | 160–230 |

**TABLE 3.4** Weight of non-structural materials

| Materials | Weight (kN/m³) |
|---|---|
| Clay (undistributed) | 18 |
| Gravel rammed | 19–21 |
| Earth | |
| (a) Dry | 13–15 |
| (b) Moist, loose | 14–16 |
| (c) Moist, compact | 17–18 |
| Wet dense sand | 17–19 |
| Dry sand | 15–16 |
| Wet silt | 17–18 |
| Flour sacks | 3–6 |
| Grains, sugar, etc. | 7–8 |
| Coal | 7–12 |
| Granite | 26–28 |
| Marble | 27 |
| Sandstone | 22–27 |
| Stacked books | 8–9 |
| Dust | 5–8 |
| Textile bundles | 8–17 |
| Water | 10 |

It has to be noted that dead loads vary during construction, but thereafter remain constant, unless significant modifications are made to the structure or its permanent equipment.

The dimensions used to estimate the dead loads should be average, so that such loads can be estimated consistently. Overestimating of dead loads will result in uneconomical structures (dead loads are often overestimated); sometimes it may also result in unconservative design, especially when the dead load component is acting in the opposite sense of the resultant wind load. For example, the weights of tanks should be considered when full and empty. When wind loads are acting, empty tanks often cause the worst effect.

Gaylord et al. (1992) report that the dead load of the Quebec across the St Lawrence river was 20–30% larger than assumed in the design, though this discrepancy was not found to be the cause of the failure (the bridge collapsed during erection, killing more than 100 workmen). Bezant and Frangopal (2002) discuss the size effect hidden in excessive dead load factor.

The code (IS 800 : 2007) states that the self-weight computed on the basis of nominal dimensions and unit weights as given in IS 875 (Part 1) may be taken to represent the characteristic dead load.

## 3.3 IMPOSED LOADS

*Imposed loads* (previously referred to as live loads) are gravity loads other than dead loads and cover factors such as occupancy by people, movable equipment and furniture within buildings, stored materials such as books, machinery, and snow. Hence, they are different for different types of buildings—residential, offices, warehouses, etc. Thus, they vary often in space and in time.

Imposed loads may be subdivided into two groups: (a) those which are applied gradually, in which case the static equivalent can be used, and (b) those that are applied dynamically (for example, repeated loads and impact loads). Gradually applied imposed loads may vary slowly over time (Ravindra and Galambos 1978). However, in the past, codes specified only the extreme values.

Imposed loads are specified in the codes based on observation and measurement. They are generally expressed as static loads for convenience, although there may be minor dynamic forces involved. As per the Indian code [IS 875 (Part 2—Live loads), imposed loads are classified into the following groups:

1. Residential (dwelling, hotels, hostels, boarding and lodging houses, clubs, etc.)
2. Educational (classrooms, libraries, etc.)
3. Institutional (office rooms, bedrooms, kitchens, general storage, etc.)
4. Assembly halls (with and without fixed seating, etc.)
5. Office and business buildings (rooms with and without separate storage, computer rooms, filing/store rooms, strong rooms, dining rooms, etc.)
6. Mercantile buildings (shops)
7. Industrial (with light- /medium- /heavy-duty machinery/ equipment)
8. Storage buildings (warehouses, cold storages), etc.

Though live loads occur at random on any floor, they are often assumed to be uniformly distributed on the floor/roof area. The imposed (live) loads on floors as per IS 875 are given in Table 3.5, with some overlap of the sets. For complete guidance, the reader should refer to IS 875 (Part 2). These live loads are assumed to be present on the floors and have to be transmitted to the foundation through beams (cross-beams) and columns. The imposed floor loads given in Table 3.5 include a small allowance for impact and other dynamic effects that may occur during normal occupancy. However, they do not include forces resulting from the acceleration and braking of vehicles or the movement of crowds. When the mechanical stacking of materials is involved, or other abnormal loads are expected to be applied on the floor, then actual values of the loads should be used, and not those given in Table 3.5.

When the distribution of imposed loads over large areas is being considered, the maximum imposed loads specified in the codes, which correspond to rare events, are often reduced. This is done in order to make a small allowance for the decreased probability that the maximum imposed load will act on all areas at the same time. The code allows for a reduction in live load in the design of a single beam or girders by 5% for each 50 $m^2$ of floor area supported (for areas greater than 50 $m^2$), subject to a maximum reduction of 25%. However, this reduction should not be applied to floors of storage buildings. Similarly, in multi-storey buildings, the probability that all the floors will be simultaneously loaded with the maximum live loads as prescribed in the code is remote and hence reduction of column loads is allowed. Thus, the live loads given in Table 3.5 may be reduced in the design of columns, walls, and foundations of multi-storey buildings, as given in Table 3.6. Note that such reduction is not permissible, if we consider earthquake loads (see Section 3.12.3).

The imposed loads on various types of roofs as per IS 875 (Part 2) are given in Table 3.7. The horizontal loads acting on parapets, parapet walls, and balustrades as per IS 875 (Part 2) are given in Table 3.8. These loads will be assumed to act at handrail or coping level.

While designing a floor, it is not necessary, in many cases, to consider the concentrated loads and line loads (like partition walls), if the floor is capable of distributing the load. And for the design of supporting beams, the distributed load (e.g., 1 $kN/m^2$ for partition walls) is often used.

**Table 3.5** Imposed floor loads for different occupancies

| Type of building | UDL (kN/m$^2$) | Concentrated load (kN) |
|---|---|---|
| Residential (dwellings, houses, hotels, hostels, hospital, and educational) | | |
| (a) All rooms (living rooms, bedrooms, toilets, and bathrooms) | 2.0 | 1.8 |
| (b) Kitchen | 3 | 4.5 |
| (c) Corridors, staircases | 3–4 | 4.5 |
| (d) Balconies | 3.0 | 1.5 per m run at edge |
| (e) Restaurants, dining rooms, reading rooms without storage, boiler and plant | 4 | 2.7–4.5 |
| (f) Store rooms, boilers, projection | 5 | 4.5 |
| (g) Garage (vehicle up to 25 kN) | 2.5 | 9.0 over 750 mm square |
| (h) Garage (vehicle up to 40 kN) | 7.5 | 9.0 over 750 mm square |
| (i) Classrooms, dining halls, kitchens | 3 | 2.7 |
| (j) Libraries, archives (stacks) | 6 up to 2.2 m ht and +2 per m ht above 2.2 | 4.5 |
| Assembly buildings (theatres, waiting rooms, etc.) | | |
| (a) Hall with fixed seats, restaurants, corridors | 4 | 4.5 |
| (b) Without fixed seating, stages, and projection | 5 | 3.6 |
| (c) Office, kitchen | 3 | 4.5 |
| (d) Dressing room, lounge, toilets | 2 | 1.8–4.5 |
| (e) Boiler plants | 7.5 | 4.5 |
| (f) Crowded corridors, staircases | 5 | 4.5 |
| Offices, business, and mercantile | | |
| (a) With separate storage | 2.5 | 2.7 |
| (b) Without separate storage | 4 | 4.5 |
| (c) Store, vaults, and boiler | 5 | 4.5 |
| (d) Wholesale shop | 6 | 4.5 |
| (e) Retail shops, corridors, staircases, and stationery stores | 4 | 3.6–4.5 |
| (f) Halls, dining rooms, and kitchens | 3 | 2.7 |
| (g) Bathrooms and toilets | 2 | — |
| (h) Boiler and plant | 5 | 6.7 |
| Industrial and storage | | |
| (a) Work areas without machinery | 2.5 | 4.5 |
| (b) Machinery and work area | | |
| (i) Light duty | 5 | 4.5 |
| (ii) Medium | 7 | 4.5 |
| (iii) Heavy duty | 10 | 4.5 |
| (c) Boiler plant, corridors, and staircases | 5 | 4.5 |
| Storage buildings | | |
| (a) Storage rooms | 2.4/m height with min. 7.5 | 7 |
| (b) Cold storage | 5/m height with min. 15 | 9 |
| (c) Boiler and plant | 7.5 | 4.5 |
| Stairs landing and balconies | | |

(*Contd*)

**TABLE 3.5** *(Contd)*

| Type of building | UDL ($kN/m^2$) | Concentrated load (kN) |
|---|---|---|
| (a) Not liable to overcrowding | 4 | 4.5 and 1.3 kN at free edge (if cantilever) |
| (b) Liable to overcrowding | 5 | 4.5 and 1.3 kN at free edge (if cantilever) |

*Notes:*

1. Load due to partition: Increase the floor load by 33.3% per metre run of partition wall subject to a minimum of 1 $kN/m^2$. The total weight per metre run must be less than 4 kN/m.
2. There should be a reduction in live load to be carried by beams and columns (except in case of machinery).

**Table 3.6** Reduction in live load applicable to columns

| Floor measured from top | Percentage |
|---|---|
| 1 (top or roof) | 0 |
| 2 | 10 |
| 3 | 20 |
| 4 | 30 |
| 5–10 | 40 |
| 11 to ground floor | 50 |

**Table 3.7** Imposed loads on various types of roofs

| Type of roof | Uniformly distributed imposed load measured on plan area | Minimum imposed load measured on plan |
|---|---|---|
| Flat, sloping, or curved roof with slopes up to and including 10° | | 3.75 kN uniformly distributed over any span of 1 m width of the roof slab and 9 kN uniformly distributed over the span of any beam or truss or wall. |
| (a) Access provided | 1.5 $kN/m^2$ | |
| (b) Access not provided (except ladder for maintenance) | 0.75 $kN/m^2$ | Half of case (a) above |
| Roof with slope greater than 10° | For roof membrane sheets or purlins 0.75 $kN/m^2$; For every degree increase in slope over 10 degrees; reduce by 0.02 $kN/m^2$ | 0.4 $kN/m^2$ |
| Curved roof with slope of line obtained by joining springing point to the crown with the horizontal, greater than 10° | $0.75-0.52r^2$ $kN/m^2$, where $r = h/l$, $h$ is the height of the highest point of the structure measured from its springings, and $l$ is the chord width of the roof if the roof is singly curved and the shorter of the two sides if it is doubly curved. | 0.4 $kN/m^2$ |

*Note:*

1. The roof will be designed for imposed loads given above or for snow/rain load, whichever is greater.
2. All roof covering (other than glass) should be capable of carrying an incidental load of 900 N concentrated over an area of 125 $mm^2$.
3. Trusses, beams, columns, and girders excluding purlins can be designed for 2/3 of the live load on the roof.

**Table 3.8** Horizontal loads on parapets, parapet walls, and balustrades

| Usage Area | Intensity of horizontal load (kN/m run) |
|---|---|
| Light-access stairs, gangways, landings, balconies, and parapet walls (domestic) | |
| (a) not more than 600 mm wide | 0.25 |
| (b) more than 600 mm wide | 0.35 |
| All other stairways, landings, balconies, all parapets, and handrails to roofs (not subject to overcrowding) | 0.75 |
| Parapets and balustrades in areas of assembly in buildings likely to be overcrowded | 2.25 |

*Note:* In case of guard parapets on a floor of a multi-storeyed car park or crash barriers provided in certain buildings for fire escape, the value of imposed horizontal load (together with impact load) may be determined.

Srinivasa Rao and Krishnamurthy (1993) conducted a load survey on imposed loads acting on office buildings and found that the maximum imposed load is only of the order of 2.35 $kN/m^2$ in such buildings without separate provision for store rooms. (This value is much smaller than 4 $kN/m^2$ as specified in IS 875 (Part 2).) A similar survey was conducted by Sunil Kumar (2002), and Sunil Kumar and Kameswara Rao (1994) on residential buildings. The design imposed loads specified in the code are much higher than those estimated by the load survey. Ruiz and Trujillo (1997) and Ruiz and Soviano (1997) conducted a Monte Carlo analysis on the live load survey of office buildings and classrooms and found that the values specified in Mexican codes are quite conservative.

Similar load surveys have been conducted in the USA (Gaylord et al. 1992), UK, and Canada. However, all these surveys involve the observation of load intensity at an instant of time, i.e., at the time of the observation. But, as mentioned already, floor loads in a particular structure vary randomly with time, and it is important to know the magnitude of the lifetime loads. However, very little information of this kind is available. Hence the codes often specify only extreme values (which occur rarely and may be regarded as lifetime maximum loads). Thus the present imposed load specified in IS 875 (Part 2) may be considered to represent characteristic live loads, which have a 95% probability of not being exceeded over a 50-year period, based on Weibull type distribution (Trahair et al. 2001; Walpole & Myers 1978).

Examples 3.2 to 3.4 illustrate the methods of calculating dead and imposed loads.

### 3.3.1 Dynamic Imposed Loads

Dynamic imposed loads which act on structures include repeated loads as well as impact and blast loads. Repeated loads are significant in the context of fatigue problems in which the designer is concerned with the magnitude, range, and the number of repetitions of loads which are frequently applied. As opposed to repeated loads, impact loads (which may induce brittle fracture) are usually specified by values of extreme magnitude which correspond to rare events. Where static loads dominate, dynamic imposed loads are often replaced by the static force equivalent. However, when dynamic loads are significant, a proper dynamic analysis must be made and the response of the structure due to the dynamic loads should be obtained (Clough and Penzien 1993). We should also be careful to check whether the period of fundamental mode of vibration of the structure does not coincide with the frequency of pulses of the pulsating loads; otherwise resonance will occur.

Dynamic forces (vibration, shock, acceleration, retardation, and impact) are of importance in the design of overhead travelling cranes, or lifts, which are found in industrial buildings and tall buildings. The impact loads they cause, although not high, should be properly taken care of in the design. Impact due to vertical crane loads are converted empirically into equivalent static loads through an impact factor, which is normally a percentage of the crane load. Table 3.9 shows the impact factors as suggested by the IS 875 code for cranes and lifts. Thus, if the impact is 25%, the live load is multiplied in the calculation of the forces by 1.25.

It is important to note that the AISC code specifies that the live load on hangers supporting floor and balcony construction has to be increased by one-third for impact (Gaylord et al. 1992). Dynamic loads due to earthquakes and winds are discussed in Sections 3.12 and 3.13, respectively.

In the design of *crane runway girders* (see Fig. 3.2) and their connections, the horizontal forces caused by moving crane trolleys must be considered. The intensity of the horizontal load (also called lateral load) is a function of the weight of the trolley, the lifting load, and the acceleration of trolley. As per IS 875 (Part 2), the lateral load may be taken as $C_{Lh}$ = 10% of the weight of trolley and the lifted load in the case of electrically operated cranes (EOT) with a trolley having a rigid mast for the suspension of the lifted weight and $C_{Lh}$ = 5% of the weight of the trolley and the lifted load for all other EOT and hand-operated cranes.

**Table 3.9** Additional impact loads on buildings

| Structure | Impact allowance in percentage |
|---|---|
| Lifts and hoists | |
| (a) Frames | 100 |
| (b) Foundations | 40 |
| Reciprocating machinery—frames and foundations | 50 |
| Light machinery—structure and foundation | 20 |
| Electric overhead cranes | |
| (a) Girders | 25 |
| (b) Columns (class III and IV cranes) | 25 |
| (c) Columns (class I and II cranes) | 10 |
| (d) Foundations | 0 |
| Hand-operated cranes | |
| (a) Girders | 10 |
| (b) Columns and foundations | 0 |

*Note:* In addition to the impact allowance in the vertical direction, additional loads in lateral and longitudinal directions must be applied on beams and columns as a percentage of the static load.

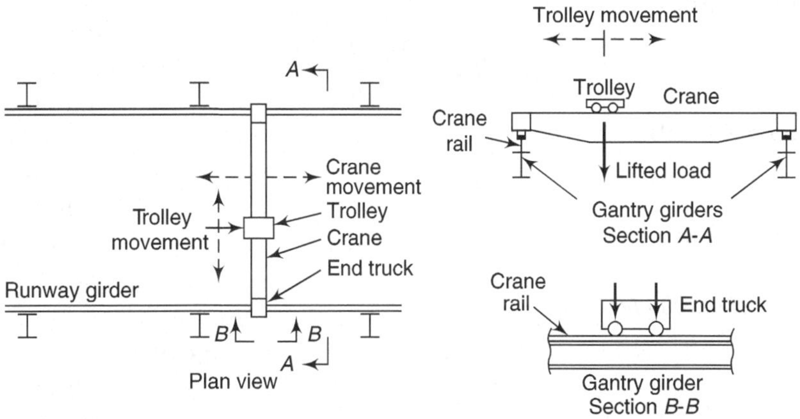


**FIG. 3.2** Loads due to crane movement

The above force should be applied at the top of the rail acting in either direction normal to the runway rails, and is distributed among all the wheels on one side of rail track. In addition, due to the acceleration and deceleration of the entire crane, a longitudinal tractive force is transmitted to the gantry girder through the friction of the end track wheels with the crane rail. IS 875 (Part 2) specifies that 5% of the maximum static wheel load of the crane is to be applied as longitudinal force, at the top of the rail.

The impact of moving live loads is a complex phenomenon and depends on the speed of the moving vehicle, its mass relative to the mass of the bridge, and the irregularities in the floor and in the wheels of the vehicle. The impact or dynamic forces due to moving vehicles are expressed as a function of the vehicle loads. Such loads are classified into class AA, class A, and class B. For class A or B loading, the impact loads are taken as

$$I = 0.54 \tag{3.8a}$$

for $L < 3.0$

$$I = \frac{9}{L + 13.5} \tag{3.8b}$$

for $3 < L < 45$ m, and

$$I = 0.154 \tag{3.8c}$$

for $L > 45$ m

where $I$ is the impact factor fraction and $L$ is the effective span in simply supported or continuous bridges.

For class AA loading and 70 *R* loading the impact load is specified by the IRC (Indian Road Congress) code as follows.

*Tracked vehicles*

$$I = 0.25 \tag{3.9a}$$

for $L < 5$ m.

$$I = 0.25 - \frac{0.15(L-5)}{4} \tag{3.9b}$$

for $5 \leq L \leq 9$ m.

$$I = 0.1 \tag{3.9c}$$

for $L \geq 9$ m.

*Wheeled vehicles*

$$I = 0.25 \tag{3.10a}$$

for $L \leq 23$ m.

$$I = \frac{9}{(L + 13.5)} \tag{3.10b}$$

for $23 \leq L \leq 45$ m.

$$I = 0.154 \tag{3.10c}$$

for $L > 45$ m.

For the details of class AA, class A, and class B loading refer to IRC code (2000). Wen and Yeo (2001) provide more details about the design of live loads for passenger car parking garages.

### 3.3.2 Snow Loads

*Snow loads* are to be considered in the mountainous (Himalayan) regions in the northern parts of India. Thus, the roofs in these regions should be designed for the actual load due to snow or for the imposed loads specified in IS 875 (Part 2), whichever is greater. The snow load on the roof is usually less than that on the ground. Since snow has a variable specific gravity, even if one knows the depth of snow considering which a design has to be made, the load per unit area of the roof is difficult to assess. Freshly fallen snow weighs up to 960 N/m$^3$ and packed snow 1600 N/m$^3$. The procedure for obtaining the snow load on a roof consists of multiplying the ground snow (corresponding to a 50-year mean return period) by a coefficient to take care of the effect of the roof slope, wind exposure, non-uniform accumulation of snow on pitched or curved roofs, multiple-series roofs or multilevel roofs, and roof areas adjacent to projections on a roof level (SEI/ASCE 7-10).

It is obvious that if the roof is steep, less snow will accumulate. In some cases, partial loading of snow should also be considered in addition to full loading, since in those cases partial loading may result in maximum effect. Wind may also act on a structure that is carrying snow load; though maximum snow and maximum wind load are not considered to act simultaneously.

However, it is important to consider drift formation due to wind, since a majority of snow-related roof damage is due to drifted snow. It has been reported that about $210,000,000 (1990 US dollars) in insured damages in the USA alone is attributed to snow drifting for the time period of 1977–1989 (O'Rourke and Auren 1997). Similar structural damages due to snow have also been reported in the UK (Walton 1990).

As per IS 875 (Part 4), snow load on a roof should be taken as

$$s = \mu s_0 \tag{3.11}$$

where $s$ is the design snow load in kilonewtons per square metre, $\mu$ is the shape factor or shape coefficient, and $s_0$ is the ground snow load in kilonewtons per metre square.

The ground snow load is measured according to the depth of snow on the ground. The code [IS 875 (Part 4)] does not give any specific value for the ground snow load. Dayaratnam (1996) suggests the following expression for determining the ground snow load

$$s_0 = (3 - 2e^{-1.5d})d \text{ kN/m}^2 \tag{3.12}$$

where $d$ is the depth of snow in metres.

Tobiasson and Greatorex (1996) suggest the following simple formula

$$s_0 = 1.97d^{1.36} \tag{3.13}$$

The depth of snow in some cities as given by Dayaratnam (1996) is given in Table 3.10.

**Table 3.10** Maximum depth of snow

| Place | Depth (mm) |
|---|---|
| Jammu and Kashmir | |
| Banihal pass | 2500 |
| Fatola (Ladakh) | 1500 |
| Gulmarg | 500 |
| Srinagar | 300 |
| Punjab and Himachal Pradesh | |
| Balani (high peak) | 6000 |
| Chamba | 2500 |
| Galog (Sirmur) | 2000 |
| Kilba (Kinnaur) | 1000 |
| Lahaul district | 1000 |
| Shimla | 1500 |
| Dhundhi | 580 |
| Patseo | 1810 |
| Uttar Pradesh | |
| Nandadevi peak | 4600 |
| Gukanda (Garhwal) | 1750 |
| Mussoorie | 1100 |
| Nainital | 350 |

Since sufficient data are not available to determine the shape coefficient on a statistical basis, the code [IS 875 (Part 4)] gives nominal values (assuming that the distribution of snow in the direction parallel to the eaves is uniform), as shown in Fig. 3.3. The value of the shape factor for curved roofs, multilevel roofs, complex multilevel roofs, and roofs with local projections and obstructions are given in IS 875 (Part 4).

Investigating the unbalanced or drifted snow loading on single-gable roofs using a database of 28 case histories, O'Rourke and Auren (1997) found that roof slope, roof width (i.e., ridge to eave distance), and ground snow load are the key parameters influencing drift size. They also found that substantial drift of snow (as shown in Fig. 3.4) is possible near the eaves of low-sloped gable roofs (slope less than 15°),

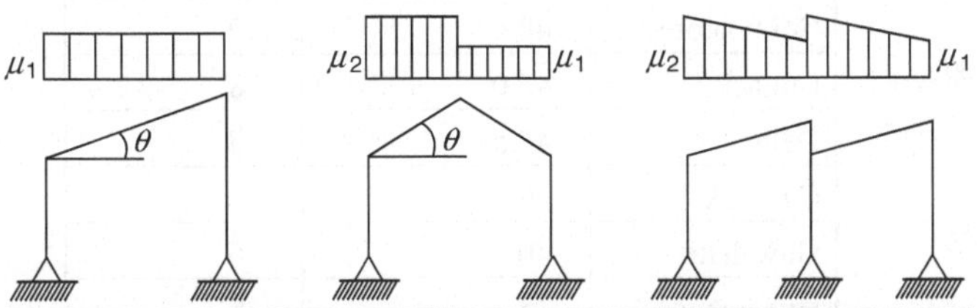


| θ° | $\mu_1$ | $\mu_1$ | $\mu_2$ | $\mu_1$ | $\mu_2$ |
|---|---|---|---|---|---|
| 0–15 | 0.8 | 0.8 | 0.8 | 0.8 | $0.8(1 + \theta/30)$ |
| 15–30 | 0.8 | 0.8 | $0.8 + 0.4\,(\theta/15 - 1)$ | | |
| 30–60 | $0.8\,\alpha$ | $0.8\,\alpha$ | $1.2\,\alpha$ | $0.8\,\alpha$ | 1.6 |
| > 60 | 0 | 0 | 0 | 0 | 1.6 |

where $\alpha = (60 - \theta)/30$, a reduction of 25% is allowed in case the buildings are not shielded by other buildings.

**FIG. 3.3** Snow load coefficients

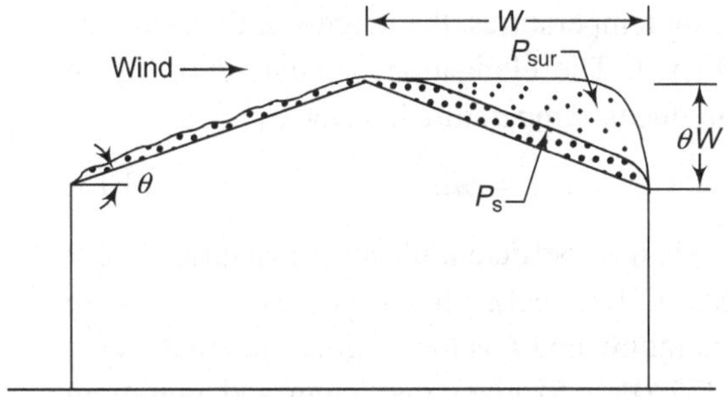


**FIG. 3.4** Maximum possible unbalanced snow distribution

particularly if the roof width is large and ground snow load is moderate. If $h_g$ is the ground snow depth, the surcharge snow load height at roof eave is $\beta h_g$. O'Rourke and Auren proposed the following relation for roofs that are symmetric about the ridge line.

$$\begin{aligned} \beta &= 0.5, \quad \frac{L}{W} \le 1 \\ &= 0.33 + 0.167L/W, \quad 1 < L/W < 4 \qquad (3.14) \\ &= 1.0, \quad \frac{L}{W} \ge 4 \end{aligned}$$

For gable roofs with larger slopes, larger than 16°, the case histories suggest that the peak drift load occurs away from the eave. For such cases, uniform load on the leeward side (with load intensity as an increasing function of $L/W$ ratio, reflecting wind direction effects) is suggested. For unsymmetrical roof geometries, O'Rourke and Auren recommend $\beta = 1.0$ (that is, 50% of the windward roof snow is applied to the leeward side as a triangular surcharge).

*Rain-on-snow surcharge*, which may be critical for certain roofs with low slopes and short eaves to ridge distance, is discussed by O'Rourke and Downey 2001. Snow loading is also discussed by O'Rourke and Wrenn 2004. Snow loads are not considered in the design of bridges, since heavy snow loads would make the bridge impossible to cross or else compel traffic to move at such a pace as to reduce the dynamic effect.

### 3.3.3 Ice Loads on Wires

*Ice loads* are required to be taken into account in the design of overhead electrical transmission and communication lines, overhead contact lines for electric traction, aerial masts, and similar structures in zones subject to ice formation. The thickness of ice deposit around wires may be taken as 3–10 mm, depending upon the location of the structure. The mass density of ice may be assumed to be equal to 900 kg/m$^3$. While considering the wind force on cables, the increase in diameters due to ice formation should be taken into account.

## 3.4 TEMPERATURE EFFECTS

Steel members expand or contract depending on increase or decrease of temperature. When a metal building is subjected to varying temperatures, the lengths of the elements get affected. The elongation or contraction of an element due to temperature is given by

$$l_t = \alpha t L \quad (3.15)$$

where $\alpha$ is the coefficient of linear expansion ($12 \times 10^{-6}$ per °C for steel), $t$ is the increase or decrease in temperature, and $L$ is the length of the member.

IS 875 (Part 5) gives maximum and minimum air temperature contours as shown in Figs 3.5 and 3.6. Table 3.11 gives the extreme temperatures of some selected cities. Note that structures such as microwave towers, transmission line towers, etc., experience more stresses due to the temperature differences than protected structures like industrial bents. For the portion of the structure below ground level, the variation of temperature is generally insignificant.

The structural analysis must take into account (a) change of the mean temperature to the initial temperature and (b) the temperature gradient through the section. Several finite element programs such as ANSYS and NASTRAN have the capability to analyse plane and space frames subjected to thermal loads.

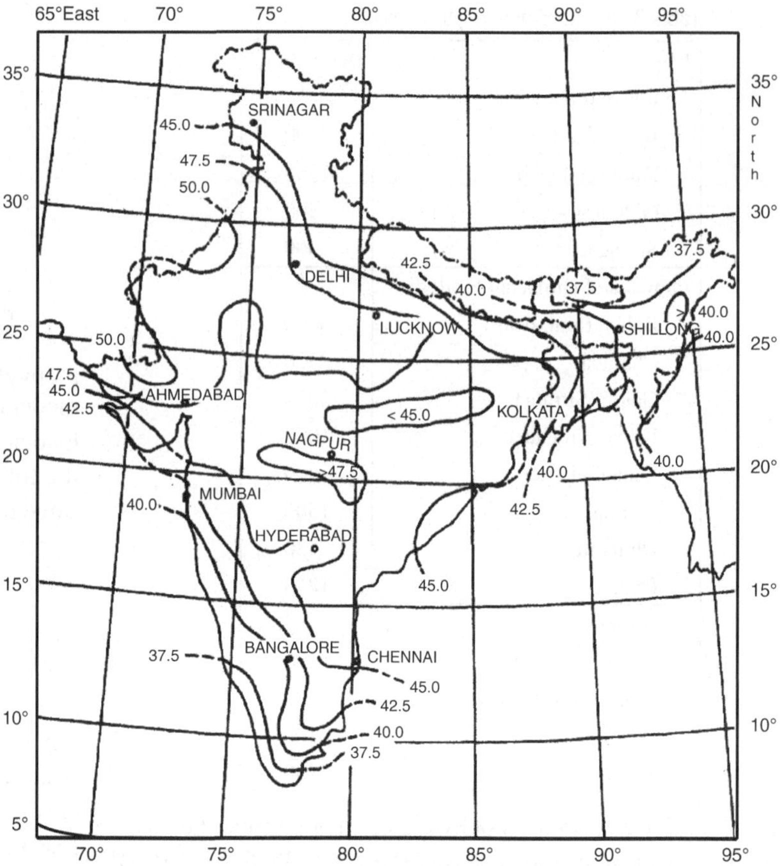


**FIG. 3.6** Maximum temperatures as per IS 875 (Part 5)

*Source:* Reprinted with permission from Survey of India, Government of India.

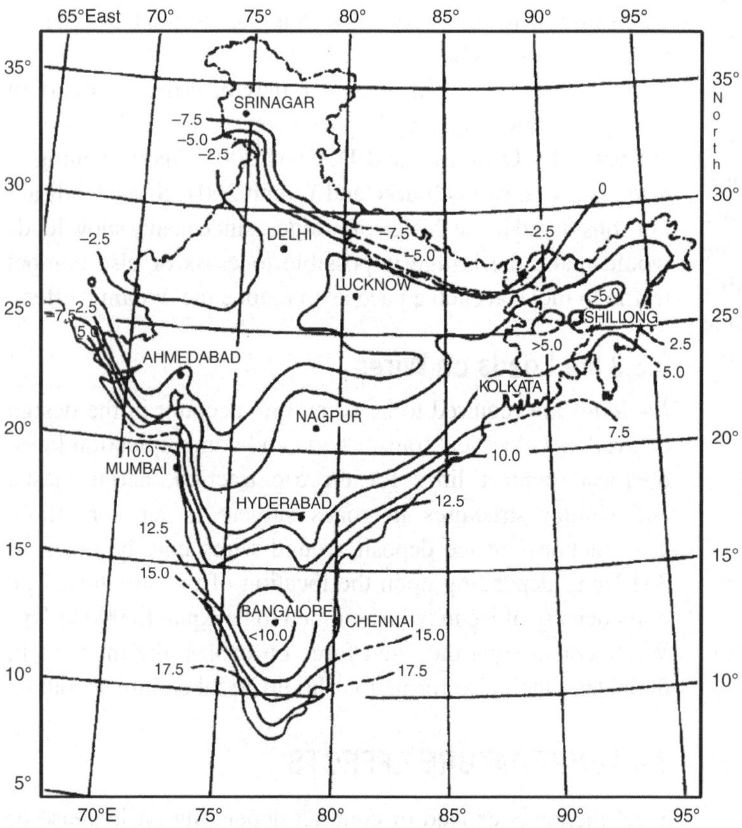


**FIG. 3.5** Minimum temperatures as per IS 875 (Part 5)

*Source:* Reprinted with permission from Survey of India, Government of India.

**Table 3.11** Extreme air temperatures in degrees celsius in the shade and basic wind velocities

| City | Highest (max.) | Lowest (min.) |
|---|---|---|
| Agra | 47.5 | −3 |
| Ahmedabad | 47.5 | 0 |
| Amritsar | 47.5 | −2.5 |
| Bangalore | 42 | 10 |
| Bhopal | 45 | 0 |
| Mumbai | 42 | 10 |
| Kolkata | 42 | 8 |
| Coimbatore | 40.4 | 12 |
| Cuttack | 45.0 | 8 |
| Delhi | 47.5 | −3 |
| Gangtok | 40 | 2 |
| Guwahati | 40 | 6 |
| Hyderabad | 45 | 8 |
| Kanpur | 47 | 0 |
| Lucknow | 47 | 0 |
| Chennai | 45 | 10 |
| Nagpur | 48 | 2 |
| Panjim, Goa | 34 | 10 |
| Srinagar | 40 | −8 |
| Vijayawada | 45 | 12 |

*Temperature effects* are critical only in places where the difference in temperature is very large in short intervals (for example in deserts where there is a large difference in day and night temperatures). Only if the structure is long, the length exceeding about 40 m, will the cumulative elongation of elements warrant the inclusion of thermal stresses to be considered along with stresses on applied loads. Hence, expansion joints are often provided at 40 m intervals to minimize temperature effects, and thermal stresses are not considered in such situations. Moreover, in many industrial buildings, the roofs are relieved of the thermal stresses by providing free movement at the top of the column in the transverse direction (by using rollers or slotted holes which permit free movement in the desired direction). In the longitudinal direction, longitudinal braces are often provided which take care of the temperature effects.

The gaps at the expansion joints should be such that the joints

1. accommodate the expected expansion/contraction due to seasonal variation of temperature and
2. avoid pounding of adjacent units under earthquake loads. The structure adjacent to the joint should be supported preferably on separate columns but not necessarily on separate foundations.

The code (IS 800 : 2007) provides the following guidelines for the provision of expansion joints.

1. If longitudinal bracings are provided at the central bay of the building, the length without expansion joints may be up to 180 m in case of covered buildings and 120 m in the case of open gantries (Fig. 3.7).
2. If longitudinal bracings are provided in two bays near the centre of the building as shown in Fig. 3.8, then the maximum centre line distance between the two braced bays may be restricted to 50 m for covered buildings and 30 m for open gantries. In addition, the maximum distance between the centres of the bracing to the nearest expansion joint or the end of the building should be restricted to 90 m in the case of covered buildings (60 m for open gantries). Thus the maximum length in this case is restricted to 230 m for covered buildings and 150 m for open gantries.

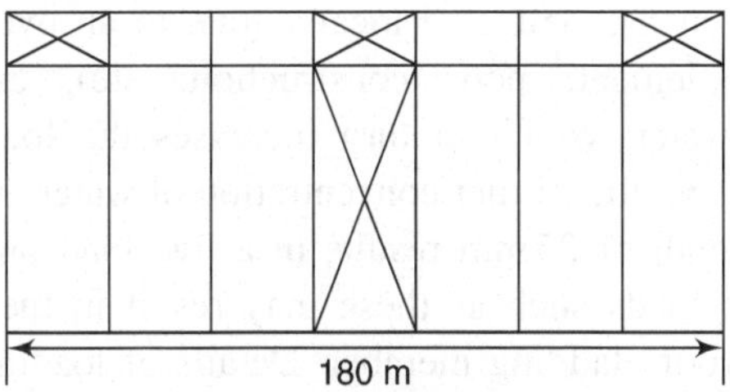


**FIG. 3.7** Maximum length of building with one bay of bracing

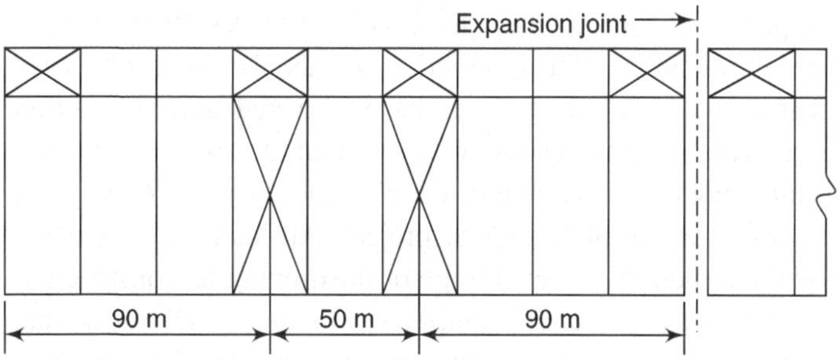


**FIG. 3.8** Maximum length of building/section with two bays of bracings

3. The maximum width of covered buildings should be restricted to 150 m, beyond which suitable expansion joints should be provided.

In practice, bracings are often provided at the end bays and middle bays of buildings to resist wind loads. Providing braces at the end bays is not desirable as far as temperature effects are concerned since they restrict horizontal movements of the system, resulting in cumulative effects, which cause more stresses. However, since wind loads normally dominate the bracing design, the provision of bracings in the end bays are not detrimental to the structure.

## 3.5 HYDROSTATIC AND SOIL PRESSURE

In the design of structures below ground level, e.g., basement walls, retaining walls, etc., the pressure exerted by soil or water, or both must be considered. Permissible bearing pressures on subsoil are discussed in Appendix C. Many established theories are available (Bowles 1990) to calculate the active earth pressures. While calculating the earth pressures, the possible surcharge due to moving loads or other stationary loads should be considered. When a portion or whole of the soil is below the free water table, the lateral earth pressure is evaluated by considering reduced weights of soil (due to buoyancy) and the full hydrostatic pressure. All foundation slabs and other footings below the water table should be designed to resist a uniformly distributed uplift equal to the full hydrostatic pressure.

Structures should be checked against overturning and horizontal sliding. Imposed loads having a favourable effect should be disregarded for this purpose. Due consideration should be given to the possibility of the soil being permanently or temporarily removed and for a submerged foundation only the buoyant weight of the foundation or soil should be considered in the calculation for overturning.

## 3.6 ERECTION LOADS

*Erection loads* are not directly stipulated in building codes. These loads may control the design of certain members of cantilever bridges, cable-supported structures, etc. The engineer responsible for erection may be held legally responsible for damages or loss of life resulting from a structural failure during erection. It has also been observed that many structural failures have occurred during erection rather than after construction. The quantification of erection loads may be difficult since erection methods and equipment tend to vary from project to project. All loads to be carried by the structure due to storage or positioning of construction materials and erection equipment (including all loads due to operation of such equipments) are considered as erection loads. Deepthi et al. (2002) studied the reliability of multi-storey buildings during construction and found that if the design process ignores construction loading on the slab during construction, then a dramatic loss of safety can occur (also see SEI/ASCE 37-02 for design loads on structures during construction). Several transmission line and microwave towers have been reported to have failed during construction (due to sudden gusty winds). Hence proper provisions should be made, including the use of temporary bracings to take care of the stresses due to erection loads. Dead load, wind load, and appropriate imposed load have to be considered on the structure during the period of erection, along with the erection loading. The structure with the temporary bracings should be capable of sustaining these loads within the limits for stresses and deflections.

## 3.7 ACCIDENTAL LOADS

*Accidental loads*, which in most cases are of short duration, occur due to (a) impacts and collisions, (b) explosions, and (c) fire.

Such loads may cause severe consequences, if special measures are not taken, but their probability of occurrence is small. The code IS 875 (Part 5) gives some guidelines for calculating the impact forces due to collisions between vehicles and structural elements (these forces should be considered in the design of safety railings), the horizontal load acting along the crane track (produced by impact of the crane on the buffer stop), the internal pressure due to the explosion of gas cylinders in closed rooms, vertical loads on air raid shelters, fire loads, etc.

In normal structures, these loads (except fire load) are not considered. However, in some situations, these loads have to be considered for the structural safety and integrity of the structure. For example, in May 1968, a natural gas explosion in a kitchen located in the 18th floor and in one of the four corners of a 23-storey precast concrete building in Ronan Point, London, triggered a progressive collapse of all the corner units above and below the unit (Levy and Salvadori 1992). The spectacular nature of the collapse created an enormous impact on the philosophy of structural design and resulted in important revisions of design codes, which made mandatory for design engineers to provide an alternative load path, in case of member failures due to any cause, to resist the loads previously borne by the failed parts. Other well-known progressive structural failures are the Commonwealth Avenue collapse in Boston and the Bailey's cross road collapse in Virginia, USA. Note that progressive failures are most common in (a) high-rise concrete flat-plate structures (during construction or earthquakes), (b) formwork for concrete structures, (c) high-rise structures constructed with precast concrete elements, and (d) lift—slab and hung ceiling systems. Another disaster due to explosion was the failure of the Oilrig Piper Alpha, 160 km off the east coast of Scotland in the North Sea on 6 July, 1988, in which 167 people were killed by the fire triggered by the explosion.

## 3.8 OTHER LOADS

A brief description of other loads, which may become critical in some situations, is given in this section.

**Settlements** Depending upon the soil condition at site, there may be a differential settlement of the foundation, which will cause additional forces in indeterminate structures (Subramanian and Venugopal 1976, and Venugopal and Subramanian 1977). Differential settlements can be minimized by choosing a foundation system (a raft or pile foundation, or by providing heavy tie beams connecting the individual footings—the tie beams should be designed to resist the moments produced by the differential settlements) or considering the specified differential settlement in the analysis itself (many computer programs are capable of considering support settlements). Settlement is a critical condition for single-storey buildings founded on expansive clays (many failures have been reported all over the world) and for buildings situated in a plot having different soil profiles.

**Wave and current loads** Such loads have to be considered for offshore structures. These are random loads and one has to consult specialist literature to determine them (Radhakrishnan et al. 1979 and Subramanian and Pranesh 1982). The force exerted by the water current should be considered in the design of bridge piers, abutments, and other waterfront structures (see Appendix B).

**Ponding** Ponding is a condition where water collects on a flat roof which has deflected locally (due to an overload, foundation settlement, poor construction, etc), causing collection of water, which in turn increases the load and deflection, causing the further concentration of water. Noting that a water depth of 25 mm results in a live load pressure of 0.25 kN/m$^2$, loads such as these may result in the local failure of the roof cladding member. Details of loading due to ponding may be found in Salmon and Johnson (1996) (see also Section 2.9.3).

**Lack-of-fit** Stresses may also result from improper fit, i.e., when a member of improper size (length) is forced into place during construction.

**Resonant loads** A company of soldiers marching in step exerts resonant load on a bridge, when their step is in rhythm with the oscillation of a bridge. Some old, small bridges are known to have collapsed under resonant loads (Wolmuth and Surtees 2004).

**Differential shortening of columns** Columns in tall buildings are subjected to large axial displacements because they accumulate loads from a large number of floors above the level being considered. A 60-storey interior column in a steel building may shorten by as much as 50–76 mm because of dead and live loads. If this shortening is not given due consideration, problems may develop in the performance of curtain walls and levelness of the floor system. The axial loads in all columns of a tall building are seldom the same, giving rise to the problem of differential shortening of columns. More details about differential shortening of columns and the corrective methods may be found in Taranath (1998).

More details about the various abnormal loadings on structures may be found in Garas et al. (2000).

## 3.9 NATURAL HAZARDS

Natural hazards such as earthquakes, tsunamis, windstorms, floods, avalanches, and landslides result in loss of life, damage to property, and loss of industrial production. In the last few decades, there has been an increasing trend in the occurrence of such natural calamities. (For example, the number of earthquakes measuring above 6.0 on the Richter scale in the 1960s was 51, in the 1970s it was 475, in the 1980s it was 1085, and in the 1990s it was 1492 and in the 2000s it reached a figure of 1611.) As a result of improved disaster management, mitigation, and construction methods, the number of lives lost has been reduced, especially in developed countries. However, economic losses continue to increase whenever such calamities strike densely populated areas. It can be seen from Table 3.12 that the Asia-Pacific region is the worst sufferer in terms of lives lost as well as economic losses. Since the 1960s economic losses from natural disasters on a global scale have tripled, while insured losses have quintupled.

**TABLE 3.12** Losses due to all disasters during 1985–2000 (Krishna et al. 2004)

| | Worldwide | Asia-Pacific | Percentage (Asia-Pacific) |
|---|---|---|---|
| Lives lost | 536,250 | 443,480 | 82.7 |
| Economic losses (US $ million) | 895,800 | 426,270 | 47.5 |

This data does not include the 2004 South Asia tsunami in which more than 230,000 people lost their lives, and which resulted in insured property losses of $2–10 billion.

*Source:* Reprinted with permission from NICEE

Table 3.13 lists the deadliest natural disasters and Table 3.14 gives the hazard-wise breakup of losses for the period 1900–1976.

**TABLE 3.13** Deadliest natural disasters during 1900–2015
(*Source: The World Almanac*, US Geological Survey)

| Year | Location | Event (magnitude) | Estimated death toll |
|---|---|---|---|
| 1931 | Huang He River, China | Flood | 3.7 million |
| 1970 | Bangladesh | Cyclone | 300,000 |
| 1976 | Tangshan, China | Earthquake (7.5) | 255,000 |
| 1920 | Gansu, China | Earthquake (8.6) | 200,000 |
| 1927 | Tsinghai, China | Earthquake (7.9) | 200,000 |
| 1923 | Kanto, Japan | Earthquake (7.9) | 143,000 |
| 1991 | Bangladesh | Cyclone | 132,000 |
| 1948 | Turkmenistan | Earthquake (7.3) | 110,000 |
| 1908 | Messina, Italy | Earthquake (7.2) | 70,000–100,000 |
| 1932 | Gansu, China | Earthquake (7.6) | 70,000 |
| 1970 | Peru | Earthquake (7.9) | 66,000 |
| 1935 | Quetta, Pakistan | Earthquake (7.5) | 30,000–60,000 |
| 1942 | Bengal, India | Hurricane | 40,000 |
| 1990 | Iran | Earthquake (7.7) | 40,000–50,000 |
| 2004 | South East Asia | Tsunami (9.0) | 230,000–290,000 |
| 2005 | Pakistan | Earthquake (7.6) | 75,000 |
| 2008 | China | Earthquake (7.9) | 87,000 |
| 2008 | Myanmar | Cyclone (category 3 to 4) | 138,300 |
| 2010 | Haiti | Earthquake (7.0) | 314,000 |
| 2011 | Japan | Earthquake/ Tsunami (9.0) | 19,000 |
| 2013 | Philippines | Typhoon Haiyan (category 5) | 7,500 (11.8 million were affected ) |

**TABLE 3.14** Hazard-wise losses due to all disasters for the period 1900–1976 (Krishna, P. et al. 2004)

| | Gross loss | Landslides, avalanches, and volcanic eruptions (%) | Cyclones and other windstorms (%) | Floods (%) | Earthquakes (%) |
|---|---|---|---|---|---|
| Death | 4.85 million | 2.93 | 10.83 | 28.10 | 58.14 |
| People rendered homeless | 232 million | — | 12.07 | 75.48 | 12.45 |
| Estimated economic losses | 131,200 million US$ | 7.62 | 36.43 | 18.37 | 37.58 |

*Source:* Reprinted with permission from NICEE

Probability estimates for the occurrence of earthquakes, hurricanes, flood, and tornado disasters with 1000 fatalities per event in the United States during 1- 10-, and 20-year exposure times and estimate of mean return period in years are shown in Table 3.15. Though it may not reflect the trend in other parts of the world, this table provides some data on the pattern of these disasters.

**TABLE 3.15** Probability estimates for the occurrence of natural disasters (www.coastal.er.usgs.gov/hurricane-forecast/barton4.html)

| Exposure time | 1000 fatalities per event | | | |
|---|---|---|---|---|
| Disaster | 1 year | 10 years | 20 years | Return period (in years) |
| Earthquakes | 0.01 | 0.14 | 0.26 | 67 |
| Hurricanes | 0.06 | 0.46 | 0.71 | 16 |
| Floods | 0.004 | 0.04 | 0.08 | 250 |
| Tornadoes | 0.006 | 0.06 | 0.11 | 167 |

It has been found that floods and tornadoes have a relatively shorter return period for small events, while earthquakes and hurricanes have relatively short return periods for larger events.

The natural hazards map of India (excluding earthquakes) is shown in Fig. 3.9. It clearly shows that the coastal states of Tamilnadu, Andhra Pradesh, Orissa, West Bengal, and Gujarat are more seriously affected by cyclonic storms. Moreover, these storms are more frequent in the east coast than in the west, and cause widespread damage. Since wind and earthquake loads are more important, we will discuss their characteristics in the next two sections. Guidelines on flood damage reduction is provided by FEMA-15 1981 and FEMA-55 2000.

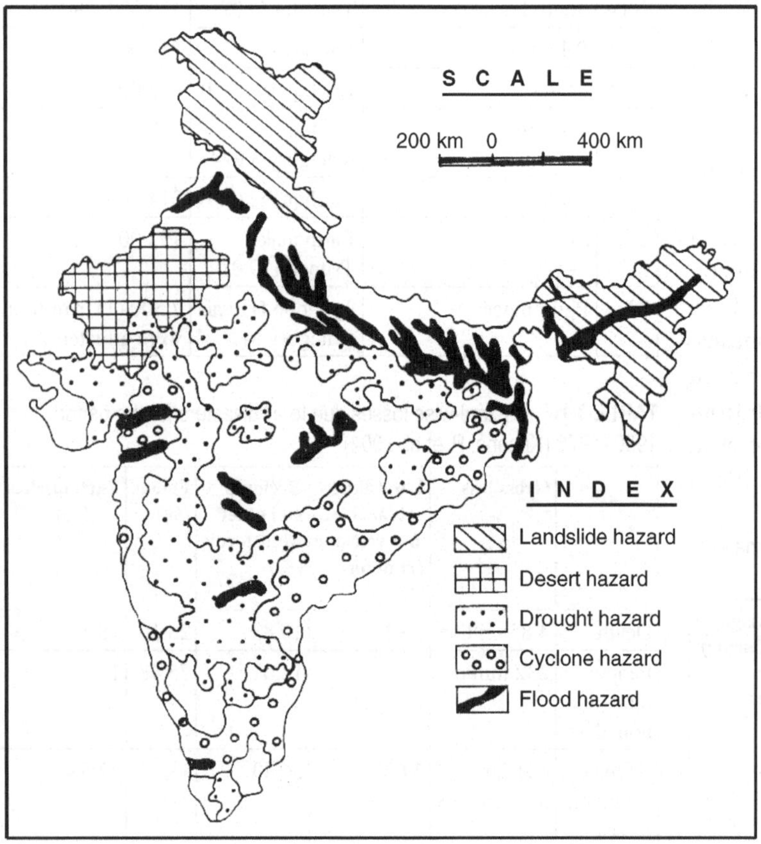


**FIG. 3.9** Estimated natural hazards map of India

## 3.10 EXTREME WINDS

Wind is the term used for air in motion and is usually applied to the natural horizontal motion of the atmosphere. Winds are produced by differences in atmospheric pressure, which are primarily due to differences in temperature. These temperature differences are caused by the differences in the thermal properties of land mass, ocean surfaces, and the unequal distribution of heat from the sun. When temperatures of adjacent regions become unequal, the warmer (lighter) air tends to rise and flow over the colder, heavier air. Winds initiated like this are further modified by the rotation of the earth.

The movement of air near the surface of the earth is three-dimensional, with the horizontal component much greater than the vertical component. This wind flow manifests itself into many forms, such as

1. gales,
2. cyclones/hurricanes/typhoons,
3. tornadoes,
4. thunderstorms, and
5. localized storms.

Some of these storms are shown in Fig. 3.10.

Friction from the surface of the earth creates a 'boundary layer' flow, but the characteristics of flow vary depending on the storm type. At a certain height above ground level (about 458 m), called the *gradient wind height*, the horizontal wind flow is no longer influenced by the ground or *boundary layer* effect.

Among the various types of storms, *cyclones* (see Fig. 3.11) result in extensive damage to structures, followed by tornadoes. The moving surface air produces six belts of wind around the earth, as shown in Fig. 3.12. There is no prevailing wind within the equatorial belt, which lies roughly between 5°S and 5°N. Therefore, near the equator and up to about 565 km on either side of it, there lies a low-pressure belt in which the air is hot and calm. The air in this region rises above the earth. Almost all cyclones form in the region of 5° to 30° above (in the clockwise direction during January to March) and below (in the counterclockwise direction during

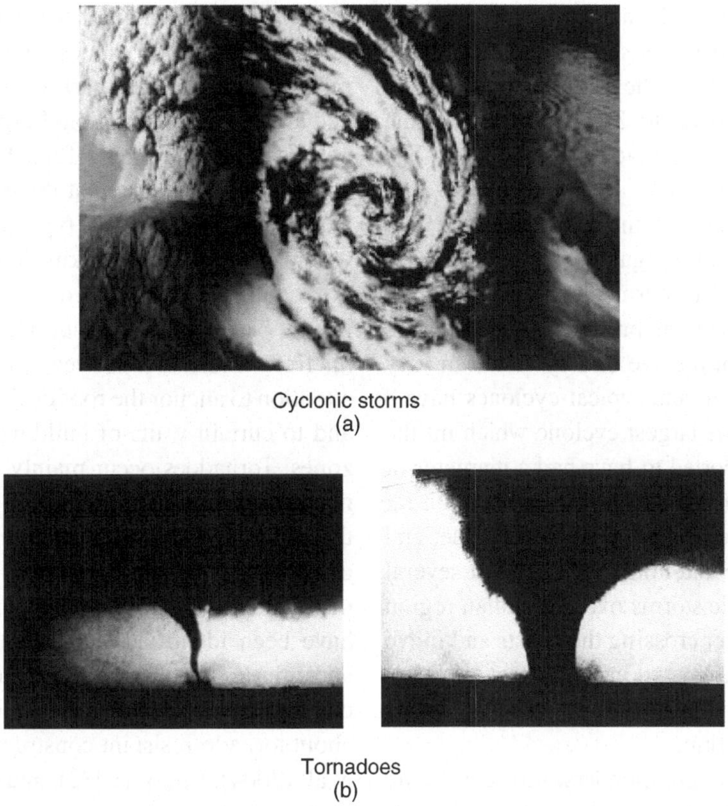


**FIG. 3.10** Various types of windstorms

*Source:* Bhandari et al. 2004, Reprinted with permission from NICEE.

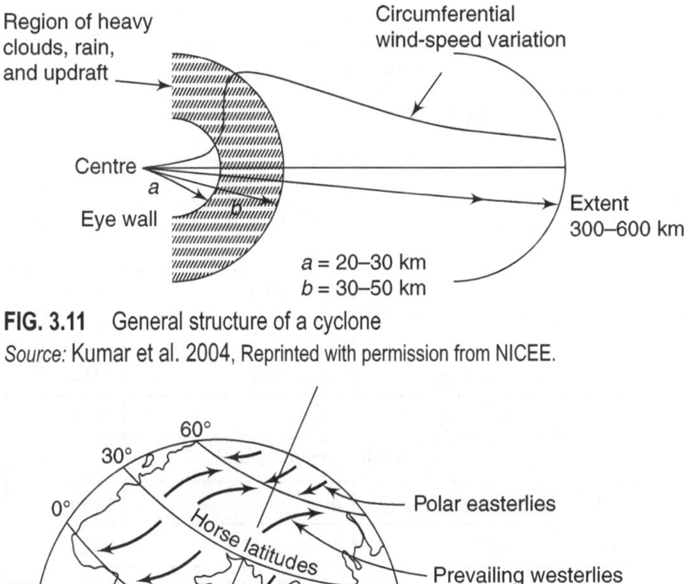


**FIG. 3.11** General structure of a cyclone

*Source:* Kumar et al. 2004, Reprinted with permission from NICEE.

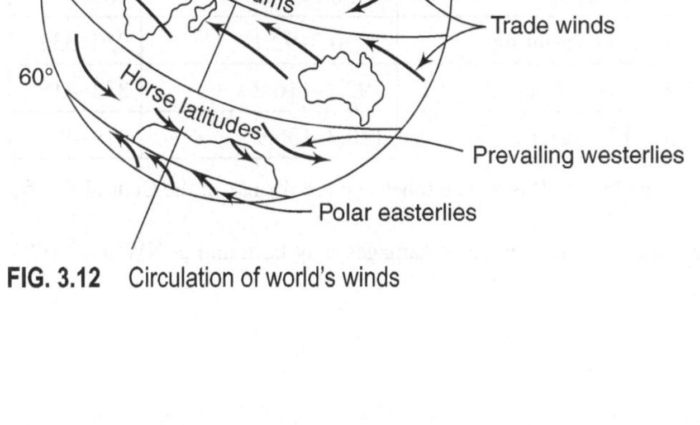


**FIG. 3.12** Circulation of world's winds

September to December). The total number of cyclones per year globally is around 80. The north-eastern hemisphere is the worst hit and experiences about 45% of the cyclones. The Indian region experiences about 6 cyclones/year mostly on the eastern coast. Compared to other countries of the world, the loss of life in India and Bangladesh due to tropical cyclones is much higher (see Table 3.13). Extreme winds such as thunderstorms, hurricanes, cyclones, tornadoes, and typhoons impose loads on structures that are many times greater than those normally assumed in their design. Some standards, such as the SEI/ASCE 7–05 and BS 6399 (Cook 2002), provide for hurricane wind speeds for a specified probability of occurrence and the draft Indian wind code suggests an increase in wind speed for certain parts of India, near sea coasts, which are identified to be affected by cyclonic storms. (See Section 3.15)

Wind speeds of 9–31m/s are typically reached in *thunderstorms* (which are common in temperate summer weather), and are often accompanied by swirling winds exerting high suction forces on roof and cladding elements. *Hurricanes/cyclones/typhoons* consist of high-velocity winds blowing circularly around a low-pressure centre known as the eye of the storm [See Fig. 3.10(a)]. From the edge of the storm towards its centre, the atmospheric pressure drops sharply, raising the wind velocity. The wind speeds of fully developed cyclones can reach up to 30–36 m/s and in the case of severe cyclones, they may reach up to 90 m/s. (The storms in northwestern Pacific Ocean are called *typhoons*, a name of Chinese origin, meaning 'wind which strikes'. The storms which occur in the Bay of Bengal and the seas of north Australia are called *cyclones*. The storms that originate in the tropical regions of the Atlantic Ocean or the Caribbean sea are called *hurricanes*.) The size of the eye of a cyclone can vary in diameter from 6 km to 40 km and typical cyclones have a diameter of about 600 km. (The largest cyclone which hit the island of Guam in 1979 is reported to have had a diameter of 2252 km and a wind speed of 85 m/s.) Cyclones in India far exceed the wind speed for the design given in the code, and are accompanied by sea surge and flooding. They last several hours. A feature of the cyclonic storms over the Indian region is that they rapidly weaken after crossing the coasts and move as depressions/lows inland. After striking the coast, a severe storm affects regions up to 60 km inland; sometimes this distance may extend up to 120 km.

*Tornadoes* develop within severe thunderstorms consisting of a rotating column of air, usually accompanied by a funnel-shaped downward extension of a dense cloud having a vortex of about 60–240 m diameter, whirling destructively at speeds of 75–135 m/s. Tornadoes often occur during the summer, mostly in the northern parts of India, mainly near Bangladesh, and are of limited duration. Large tornadoes may have several vortices within a single funnel. If the bottom of the funnel can be seen, it means that the tornado has touched down and begun its destruction. A typical tornado travels at 9–14 m/s, touches ground for 8–10 km, and has a funnel 90–150 m wide. Tornadoes are the most destructive of all wind forces, and it may be impractical to design buildings to sustain a direct hit from a tornado. However, the engineer should pay extra attention to anchor the roof deck (in case of wooden buildings) and to curtain walls of buildings located in tornado-affected zones. Tornadoes occur mainly in flat country. Wind alone is not the only damaging element of a tornado. The pressure at the centre of a tornado is extremely low. As the storm passes over a building, the pressure inside a building is far greater than that outside, causing the building to explode. (Tornadoes have been identified as one of the leading causes of death and injuries among natural disasters in the USA, and cause damages in excess of $100 million per year.) More details about tornado-resistant construction may be found in Whalen et al. (2004), Minor (1982), and Coulbourne et al. (2002).

The various scales for measuring windstorms are shown in Table 3.16.

**TABLE 3.16** Various scales for measuring windstorms

| **Beaufort scale** | | | | | **Saffir–Simpson hurricane scale[1]** | | | |
|---|---|---|---|---|---|---|---|---|
| **Bft Descriptive term** | | **Mean wind speed at 10 m above surface** | | **Wind pressure kg/m²** | **SS** | **Descriptive term** | **Mean wind speed** | |
| | | **m/s** | **km/h** | | | | **m/s** | **km/h** |
| 0 | Calm | 0–0.2 | 0–1 | 0 | 1 | Weak | 32.7–42.6 | 118–153 |
| 1 | Light air | 0.3–1.5 | 1–5 | 0–0.1 | 2 | Moderate | 42.7–49.5 | 154–177 |
| 2 | Light breeze | 1.6–3.3 | 6–11 | 2.0–0.6 | 3 | Strong | 49.6–58.5 | 178–209 |
| 3 | Gentle breeze | 3.4–5.4 | 12–19 | 0.7–1.8 | 4 | Very strong | 58.6–69.4 | 210–249 |
| 4 | Moderate breeze | 5.5–7.9 | 20–28 | 1.9–3.9 | 5 | Devastating | ≥69.5 | ≥250 |
| 5 | Fresh breeze | 8.0–10.7 | 29–38 | 4.0–7.2 | **Fujita tornado scale[2]** | | | |
| 6 | Strong breeze | 10.8–13.8 | 39–49 | 7.3–11.9 | **F** | **Descriptive term** | **m/s** | **km/h** |
| 7 | Near gale | 13.9–17.1 | 50–61 | 12.0–18.3 | 0 | Weak | 17.2–32.6 | 62–117 |
| 8 | Gale | 17.2–20.7 | 62–74 | 18.4–26.8 | 1 | Moderate | 32.7–50.1 | 118–180 |
| 9 | Strong gale | 20.8–24.4 | 75–88 | 26.9–37.3 | 2 | Strong | 50.2–70.2 | 181–253 |
| 10 | Storm | 24.5–28.4 | 89–102 | 37.4–50.5 | 3 | Devastating | 70.3–92.1 | 254–332 |
| 11 | Violent storm | 28.5–32.6 | 103–117 | 50.6–66.5 | 4 | Annihilating | 92.2–116.2 | 333–418 |
| 12 | Hurricane | > 32.7 | > 118 | > 66.6 | 5 | Disaster | 116.3–136.9 | 419–493 |

*Notes:* 1. There are three scales; the one given here is for use in the Atlantic and North-east Pacific Basin. The other two are for use in the Central Pacific Basin and western North Pacific (see NWSI 10-604, 2013).

2. Details of Enhanced Fujita scale (EF Scale), which has additional damage indicators and degree of damages may be found at NWSI 10-1605 (www.spc.noaa.gov/efscale/ef-ttu.pdf).

## CASE STUDY

### Hurricane Katrina

Hurricane Katrina struck the New Orleans area of USA on the early morning of 29 August 2005. It developed into a powerful category Five hurricane, on the Saffir–Simpson scale of hurricane intensity, with the highest wind speed of 280 km/h. It caused severe destruction along the Gulf coast from central Florida to Texas, much of it due to the storm surge. The storm surge breached Louisiana's levees at 53 different points, leaving 80 per cent of the city submerged, leaving several victims clinging to rooftops, and sending several others to shelters around the country. At least 1836 people lost their lives in the actual hurricane and in the subsequent floods, making it the deadliest hurricane in the US since the 1928 Okeechobee hurricane. NASA satellite images showed that the floods that buried up to 80 per cent of New Orleans subsided only by 15 September 2005. Katrina redistributed over one million people from the central Gulf coast elsewhere across the United States, which became the largest diaspora in the history of the United States.

(a)

(b)

(a) Satellite image of the 25 mile-wide eye of Hurricane Katrina (Photo: NASA) (b) Flooded I-10/I-610/West End Blvd interchange and surrounding area of northwest New Orleans and Metairie, Louisiana (Photo: US Coast Guard, Kyle Niemi)

The total damage from Katrina is estimated at $81.2 billion (2005 US dollars), nearly double the cost of the previously most expensive storm, Hurricane Andrew, when adjusted for inflation. Federal disaster declarations covered 233,000 km² of the United States, an area almost as large as the United Kingdom. The hurricane left an estimated three million people without electricity. The Superdome, which was sheltering many people who were not evacuated, sustained significant damage.

The damage was due to the breach that occurred in the I-wall on the eastern side of Inner Harbour Navigation Canal during Katrina. The failure of levees and floodwalls resulted in catastrophic flooding in the Greater New Orleans area, with floodwaters in many areas up to 2.5 m above the lowest floor of the building. Many buildings constructed with the first floor elevation above the base flood elevation (BFE) were severely damaged or destroyed when the floodwaters rose well above the first floor. The duration of the floodwaters in New Orleans contributed to further damages, with some areas remaining under water for several weeks. This long-duration flooding saturated some types of building materials beyond the point where they could be salvaged and contaminated the materials with chemical and biological substances in the floodwaters (FEMA 549, 2006).

It is to be noted that many of the broken levees have been reconstructed after Katrina damage. Instead of the I-wall structure in the old levee system, the new HSDRRS (Hurricane and Storm Damage Risk Reduction System) has a T-wall structure that is more stable especially in weak soil areas. The new T-wall is taller with concrete slope paving and has longer sheet pile. New H-piles were driven for additional support. Adoption of modern building codes, such as the IBC, International Residential Code (IRC), or National Fire Protection Association (NFPA) 5000, is recommended. These codes include up-to-date design and construction provisions that are consistent with the NFIP. The 2006 editions of the IBC and NFPA 5000 incorporate flood load (ASCE 7-05) and flood-resistant construction (ASCE 24-05) standards.

The following video shows the devastating effect of the Hurricane Katrina http://www.youtube.com/watch?v=R3V2fAVngEA

An animation of Hurricanes is provided by http://news.bbc.co.uk/2/hi/science/nature/4588149.stm

### 3.10.1 Characteristics of Wind

Wind is a phenomenon of great complexity because of many flow situations arising from the interaction of wind with structures. Wind loads on buildings and structures are functions of the wind flow and the effect of that flow on the structural system or structural or nonstructural component being considered.

Wind flow depends on (Schechter et al. 1995)

1. the basic wind speed,
2. the mean recurrence interval of the wind speed,
3. the characteristics of the terrain surrounding the building or structure,
4. the characteristic height above ground for the system being considered, and
5. the directional properties of the wind climate.

The effects of the wind flow on the structural system, or the structural or nonstructural component being considered, depend upon

1. the aerodynamics of the building or structure,
2. the position(s) of the area(s) acted on by the wind flow,
3. the magnitude(s) of the area(s) of interest,
4. the porosity of the building envelope,
5. the selection of the probability that the peak fluctuating wind load acting on the system or element will be exceeded during the windstorm considered in design, and
6. the susceptibility of the structural system under consideration to steady and time-dependent (dynamic) effects induced by the wind load.

Several wind engineering simplifications and assumptions are made to meaningfully predict wind behaviour.

### 3.10.2 Variation of Wind Velocity with Height

At the interface between the air flow and the surface of the earth, the surface friction reduces the air velocity adjacent to the earth's surface to almost zero. Above the surface, the velocity increase is curvilinear, varying from zero at the ground surface to a maximum at some distance from the ground. This height at which the velocity ceases to increase is called the *gradient height* (approximately 200–500 m from the ground level) and the corresponding velocity, the *gradient velocity*. The height through which wind velocity is affected by topography is called the *atmospheric boundary layer.* The variation of wind speed with height is also dependent upon the ground roughness (which is dependent on the density, size, and height of buildings, trees, vegetation, etc., on the ground) and is thus different for each terrain category. (See Fig. 3.13.) The wind load within this layer is in the domain of turbulent flow and can be mathematically predicted by a logarithmic equation. However, in practice, wind speed variation is given by a simpler power law expression of the form (Davenport 1960)

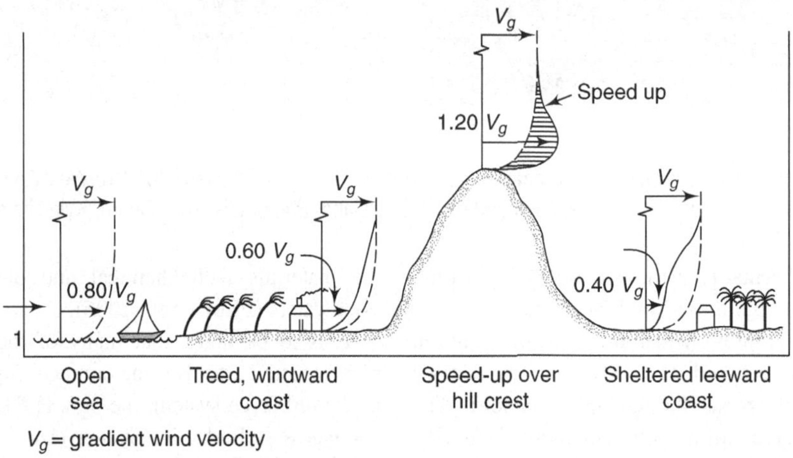


**FIG. 3.13** Typical influence of terrain and topography

$$V_z = V_g \left( \frac{Z}{Z_g} \right)^{\alpha} \tag{3.16}$$

where $V_z$ is the mean wind speed at height $Z$ above the ground, $V_g$ is the gradient wind speed assumed to be constant above the boundary layer, $Z_g$ is the depth of the boundary layer, and $\alpha$ is the power law coefficient.

Thus, by knowing the mean wind speed at the gradient height and the value of $\alpha$, the wind speed at a height $Z$ can be calculated by using Eqn 3.16. The exponent $\alpha$ and the depth of the boundary layer $Z_g$ vary with terrain roughness. The value of $\alpha$ ranges from a low of 0.125 for open country to about 0.4 for built-up urban areas, signifying that wind speed reaches its maximum value over a longer height in an urban terrain than in open country. (See Fig. 3.14.) Also, it takes quite a distance, called *fetch length*, for wind to travel over a typical terrain to fully develop the speed profile idealized for that terrain category. The pressure and suction generated by wind are functions of wind speed and in general increase with building height. The velocity at 10 m above ground level is normally used as the basic value for design purposes.

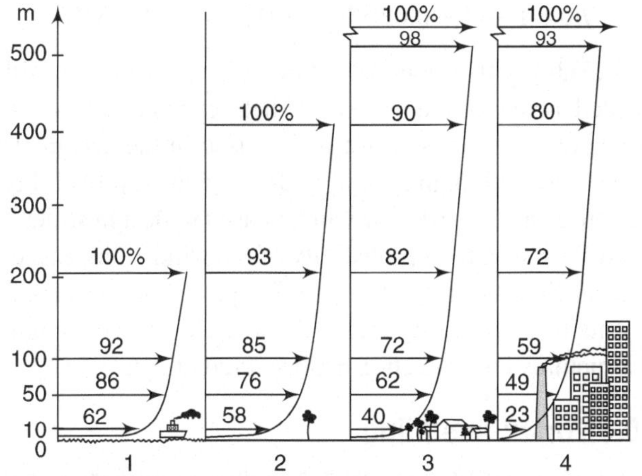


**FIG. 3.14** Boundary layer profile for different approach terrains

### 3.10.3 Turbulent Nature of Wind

Winds moving at speeds greater than 0.9–1.3 m/s are turbulent (since air has a very low viscosity, about one-sixteenth that of water), causing particles of air to move randomly in all directions. The typical variation of wind velocity with time, as measured by instruments called *anemometers* or *anemographs* (which are usually installed at meteorological observatories at heights generally varying from 10–30 m above ground level), is shown in Fig. 3.15(a).

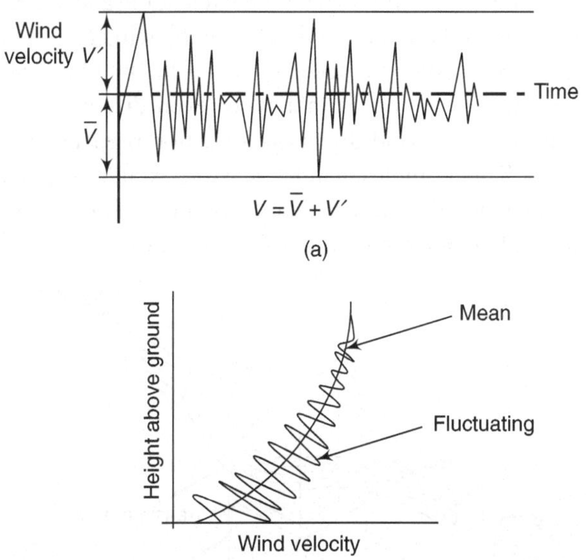


**FIG. 3.15** Variation of wind velocity with (a) time and (b) height

The velocity at any instant $V_t$ can be represented as the summation of the average velocity and the instantaneous value of velocity fluctuation about the mean value, as shown in Fig. 3.15(a). Thus,

$$V_t = \bar{V} + V' \tag{3.17}$$

where $V_t$ is the velocity at instant $t$, $\bar{V}$ is the average or mean velocity, and $V'$ is the instantaneous velocity fluctuation about the mean velocity $\bar{V}$.

The mean and fluctuating velocity along the height above the ground are shown in Fig. 3.15(b). Rapid bursts in the velocity of wind are called *gusts*. As already discussed, within the earth's boundary layer, both components (mean and fluctuating) vary not only with height but also depend upon the terrain and topography as already shown in Fig. 3.13. It is generally found that the size of the wind flow affects the size of the turbulence. Thus, the flow of a large mass of air has a larger overall turbulence than the corresponding flow of a small mass of air. Because of this random nature, the properties of wind are often studied statistically, by obtaining the required mean or the average. Because wind speed changes constantly, different averages are obtained by using different averaging times.

Spectral analysis techniques provide a convenient method for dealing with such random turbulence of wind. A complete treatment of the method is beyond the scope of this book. Interested readers may refer Simiu and Scanlan (1996) and Holmes (2001) for the details of these methods.

While dealing with rigid structures, the consideration of the 'equivalent static' wind is adequate. However, flexible and tall structures are sensitive to gusts that last about 1 s. Thus, for tall buildings, considerations of the wind energy spectrum, integral length scale, averaging time, and the frequencies of the structure are of utmost importance. As pointed out earlier, the averaging time may affect the wind velocity. The longest averaging time used in structural engineering applications is one hour. Averaging times of 2-3 s and 10 minutes have also been used. It is to be noted that, as the averaging time decreases, the maximum velocity of wind increases. For example, while a one-hour average of wind velocity may be 13.4 m/s, the same wind velocity averaged for one minute may be as high as 23.8 m/s. The influence of averaging time on velocity is given in Fig. 3.16, and may be used for adjusting non-3-s wind to 3-s gusts.

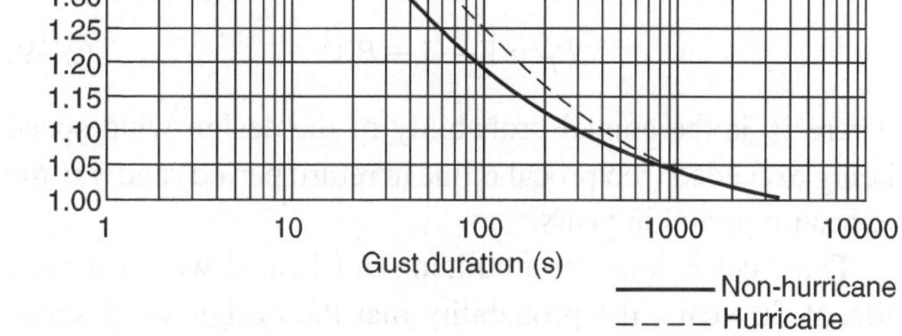


**FIG. 3.16** Ratio of probable maximum speed averaged over *t*-sec to hourly mean wind speed

**Example** *Given that a wind speed averaged over a 60 s interval is 25.0 m/s, determine the 3 s gust speed for a non-hurricane condition.*

***Solution***

From Fig. 3.16, the ratio

$$\frac{V_3}{V_{3600}} = V_{60}/V_{3600} = 1.25$$

The ratio

$$\frac{V_3}{V_{3600}} = 1.525$$

Therefore, the 3-s gust speed = (1.525/1.25) × 25 = 30.5 m/s for a non-hurricane condition.

As per IS 875, for flexible structures (with time period > 1.0 s), one must use the gust speed rather than mean wind speed. The gust speed can be obtained by multiplying the mean wind speed by the *gust factor* (also called *gust response factor*) $G_v$. Thus,

$$V_g = G_v V \quad (3.18)$$

where $V_g$ is the gust speed, $G_v$ is the gust factor, and $V$ is the mean wind speed.

The current version to be released by BIS shortly retains gust response factor and gives expressions for intensity of turbulence, mean hourly wind speed, background factor, energy ratio, and size reduction factor.

## 3.10.4 Return Period

In determining the wind load, a probabilistic approach is often used. Thus, the intensity of wind is considered as a function of *duration recurrence interval (return period).* The speed of wind is considered to vary with the return period. In the code, a mean return period of 50 years is used. A 50 year return period of wind of 55 m/s (200 km/hr) at Chennai means that, on the average, Chennai will experience a wind speed of 55 m/s within a period of 50 years. The return period of 50 years corresponds to a probability of occurrence of 1/50 = 0.02 or 2%. Thus, the chance that a wind exceeding 55 m/s will occur in Chennai within a given year is 2%. The probability that the design wind speed will be exceeded at least once during the exposed period of $n$ years is given by

$$P_N = 1 - (1 - P_a)^n \quad (3.19)$$

where $P_a$ is the annual probability of the design wind speed being exceeded (reciprocal of mean return period) and $n$ is the exposure period in years.

Thus, if we consider a building in Chennai with a service life of 50 years, the probability that the design wind speed will be exceeded at least once during the 50-year lifetime of the building is

$$P_N = 1 - (1 - 0.02)^{50} = 1 - 0.36 = 0.64 = 64\%$$

The probability that wind speed of a given magnitude will be exceeded increases or decreases with the exposure period of the building and the mean return period used in the design. Thus it is seen that although a wind with low annual probability of occurrence, such as 50-year wind, is used to design structures, there still exists a high probability of the wind being exceeded within the lifetime of the structure. The probability level of 0.64 is normally considered sufficient for the design of buildings and structures for protection against wind effects.

## 3.10.5 Vortex-shedding

In general, wind blowing past a body can be considered to be diverted in three mutually perpendicular directions, giving rise to forces and moments about the three axes. However, in structural engineering applications, the force and moment with respect to the vertical axes are not considered important (they are significant in aeronautical engineering applications). Thus the effects of wind are considered in two dimensions, as shown in Fig. 3.17. *Along-wind* or simply wind is the term used to refer to *drag forces* and *transverse wind* is used to refer to *cross-wind.* The cross-wind response, that is, motion in a plane perpendicular to the direction of wind, dominates over the along-wind response for many tall buildings (Taranath 1998). While the maximum lateral wind loading and deflection are generally in the direction parallel to the wind (along-wind direction), the maximum acceleration of the building (which may lead to perception of motion or even discomfort) may occur in the cross-wind direction (direction perpendicular to the wind). This complex nature of wake-excited response is due to the interaction of turbulences, building motion, and the dynamics of negative pressure wake formation.

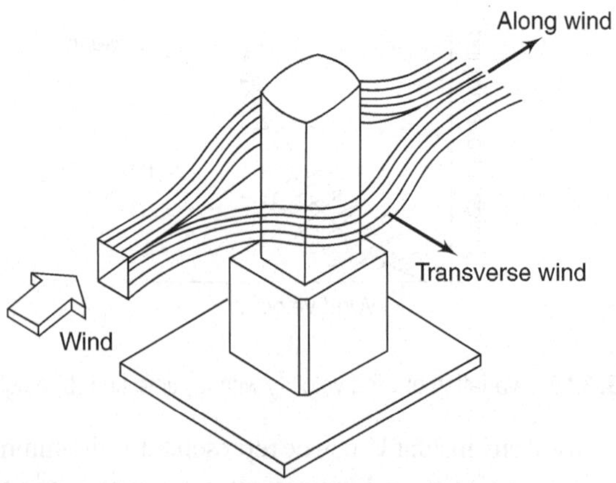


**FIG. 3.17** Simplified two-dimensional flow of wind

Consider a rectangular building subject to wind flow, as shown in Fig. 3.18. The originally parallel stream lines of the wind are now displaced at the boundary of the building. This results in spiral vortices being shed (i.e., they break away from the surface of the building) periodically from the sides of the building into the downstream flow of wind called the *wake*. Due to these vortices, an impulse is applied to the building in the transverse direction. At high wind speeds, the vortices are shed alternatively, first from one and then from the other side as shown schematically in Fig. 3.18. The transverse impulses are applied alternatively to the left and then to the right. The frequency of transverse impulse is half that of the along-wind impulse. This kind of shedding, which gives rise to vibration in the flow as well as transverse direction of the wind, is called *vortex shedding* or the *Karman vortex street.*

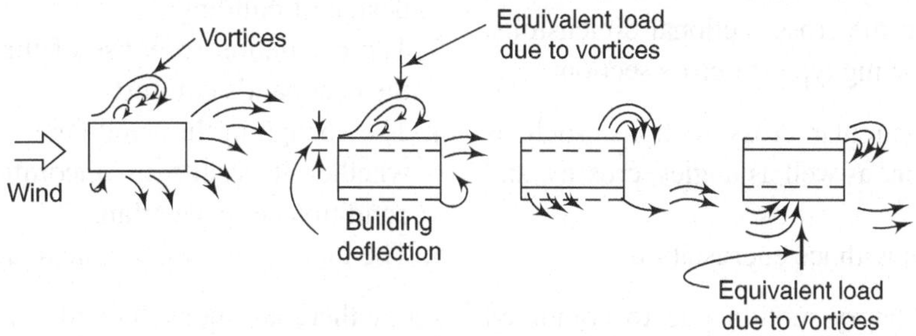


**FIG. 3.18** Vortex-shedding phenomenon

The shedding frequency $f$, caused by vortex shedding, can be calculated by using the following simple formula

$$f = \frac{SV_d}{b} \tag{3.20}$$

where $f$ is the frequency of vortex shedding in hertz, $S$ is the Strouhal number, $V_d$ is the design mean wind velocity at the top of the building (m/s), and $b$ is the breadth of the structure or structural members in the horizontal plane normal to the wind direction (m).

The Strouhal number is not a constant but varies irregularly with wind velocity. The code suggests the following values.

*Circular structures*

$S = 0.20$ for $bV_z$ not greater than 7, and $S = 0.25$ for $bV_z$ greater than 7.

*Rectangular structures*

$$S = 0.15 \text{ for all values of } bV_z.$$

When the wind velocity is such that the frequency of vortex shedding becomes approximately equal to the natural frequency of the building, a resonance condition is created, and hence vortex shedding becomes critical. According to IS 875 (Part 3), if the ratio of length to maximum transverse width is less than 2.0, vortex shedding need not be considered.

### 3.10.6 Interference Effect

A structure such as a building, tower, or a chimney, when placed in a flow of air, experiences pressure and forces. When one or more similar or dissimilar structures are placed downstream or upstream of the structure, the 'stand-alone' values of pressures and forces get altered. This is termed *interference effect*. Interferences occur irrespective of whether the structures involved are rigid or flexible. In the former, it is the 'wake' of one structure that affects the other, whereas in the latter, the deflections of the structure affect the wake itself (see Fig. 3.19). It is very difficult to quantify the interference effect. Systematic wind tunnel studies can be conducted to give some guidelines on such an effect (see also Section 3.13.9).

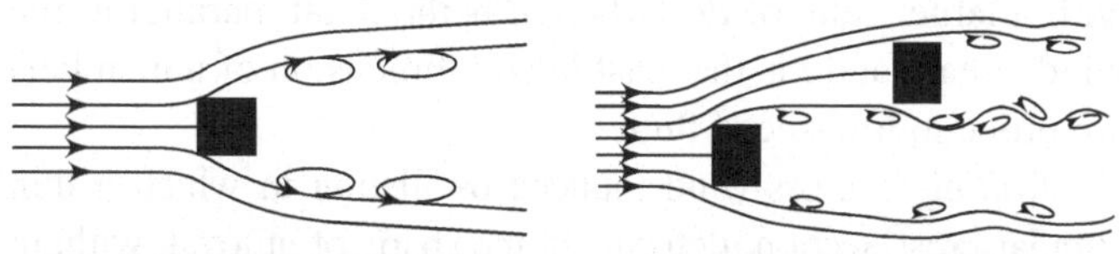


**FIG. 3.19** Typical wind flow around isolated and two nearby buildings (interference effect)

### 3.10.7 Dynamic Effects

The dynamic part of the wind sets up oscillations in a flexible structure, which may be defined as the one having a fundamental time period of vibration more than 1 s. Oscillations are thus caused in the along-wind response. Furthermore, flexible structures also respond in the cross-wind direction on account of vortex shedding. In the cross-wind direction, a flexible structure tends to oscillate due to the shedding of the eddies alternatively from either side of the structure at regular intervals. The frequency of eddy shedding is dependent on the size and shape of the structure, and wind speed, all grouped into a non-dimensional parameter, called the Strouhal number. Thus, the dynamic response is attributed to the following actions of wind.

1. Non-correlation of the fluctuating along-wind pressures over the height and width of the structure
2. Resonant vibrations of a structure
3. Vortex-shedding forces acting in a direction normal to the wind, causing a cross-wind as well as a torsional response

The interference effect of adjacent buildings also intensify the along-wind and cross-wind response. Motions in the direction of wind (known as *buffeting*) are caused by fluctuating wind forces associated with gusts. These excitations depend on the gust energy available at the resonant frequency.

The designer must be aware of the three forms of wind-induced motion, namely, galloping, flutter, and ovalling, which are characterized by increasing amplitudes of oscillation with increase of wind speed.

*Galloping* is movement-induced, self-excited, periodic oscillation in a direction perpendicular to the flow with amplitudes much larger than the cross sectional dimensions. It may be critical in the following types of cross section:

1. All structures with non-circular cross sections, such as triangular, square, polygon, as well as angles, crosses, and T-sections
2. Twisted cables and cables with ice encrustations

*Flutter* is an unstable oscillatory motion due to combined bending and torsion, which occurs in flexible plate like structures. It may be set up at wind speeds much less than those required for exciting individual modes of motion. Long-span suspension bridges and any other member of a structure with a large value of $d/t$ (where $d$ is the depth parallel to the wind stream and $t$ is the least lateral dimension of a member) are prone to low-speed flutter.

*Ovalling* is cross-wind induced oscillation in which a thin circular cross section, deforms in the shape of an oval, without significant bending in the longitudinal axis. Thin-walled structures with open ends at one or both the ends, such as oil-storage tanks, chimneys, natural-draught cooling towers, etc., where the ratio of the diameter of the minimum lateral dimension to the wall thickness is greater than 100, are prone to ovalling oscillations. Dynamic effects are further discussed in Section 3.13.10.

### 3.10.8 Measurement of Wind Loads

Gustav Eiffel was the first to conduct studies and measure displacements at the top of his 300 m tall Eiffel tower (in about 1889) under the action of winds. Similar studies were also conducted on the 16-storey Monadnock building at Chicago at about the same time. New York's Empire State Building was the first to be instrumented for wind load measurements. Concern with the effect of wind on buildings increased after the collapse of the *Tacoma Narrows bridge* in 1940. After 1960, several full-scale studies were conducted on buildings located in Canada, Australia, the USA, the UK, Holland, and Hong Kong. A similar study on a lattice tower was also conducted by the Structural Engineering Research Center (SERC), Chennai. However, many difficulties have been reported in the field measurement of wind loads.

Many researchers have made a comparison of structural wind loads for a number of tall buildings, determined from wind tunnel studies and code values. In the design of very tall buildings, it is recognized that the wind tunnel approach is a more refined method for arriving at design wind loads. For several important tall buildings, *wind tunnel tests* have been conducted to study their dynamic response.

Wind tunnel tests are done to get the following information (Taranath 1998).

1. The intensity and scale of pressure fluctuations on exterior panels and glass surfaces
2. The shear forces at various floor levels for the overall design of buildings
3. The oscillation response of the building in the context of the occupant's comfort
4. The change in the wind environment at ground level, i.e., whether it results in uncomfortable or even dangerous conditions to pedestrians
5. The interference effect due to adjoining buildings

Today there are more than 30 wind-tunnel facilities in North America alone, showing the importance of wind tunnel experiments on buildings and structures. In India also, several boundary-layer wind tunnel facilities exist, provided by NAL, Bangalore, IISc., Bangalore, SERC Chennai, IIT Bombay, IIT Roorkee, and IIT Delhi. A small wind tunnel test may cost approximately Rs 350,000.

More details about modelling in a tunnel, instrumentation, etc., may be found in Krishna and Gupta (1992) and Taranath (1998).

## 3.11 EARTHQUAKES

The crust of the earth is composed of about 13 large plates and several small ones, ranging in thickness from 32–240 km. The plates are in constant motion, riding on the molten mantle below and travelling at the rate of about a millimeter per week. The continental drift, the formation of mountains like the Himalayas, volcanic eruptions, and earthquakes are due to the movement of these plates (this theory is called the *platetectonics*). If plates carrying two continental masses, grind past each other or collide at their boundaries, they give rise to *earthquakes*. Therefore, regions close to the plate boundary are highly seismic and regions away from the boundaries exhibit less seismicity. The Indian subcontinent, which forms part of the Indo-Australian plate, is pushing against the Eurasian plate along the Himalayan belt. Hence, the Himalayan belt is highly seismic, whereas peninsular India, which is not traversed by any plate boundary, is relatively less seismic. Some consider that earthquakes may also be caused by other actions such as underground explosions due to the testing of nuclear bombs, or due to the construction of dams (e.g., the Koyna dam).

There are three great belts of the earth, where earthquakes are more frequent and violent. One of these belts almost encircles the Pacific Ocean, going through New Zealand, the Philippines, Japan and Alaska, and along the west coasts of North and South America. The second stretches across South Asia, encompassing Italy, Greece, Turkey, and parts of India into the Mediterranean region and the third is located from the south of Iceland through the middle of the Atlantic Ocean. Although most of the recorded earthquakes have occurred in these well-defined earthquake belts, an examination of the records reveals the truly global nature of this phenomenon. Thus, even in the seismically inactive parts of the world, some measure of earthquake resistance should be built into the design of all structures, since the failure of structures due to earthquakes will be catastrophic. For example, the Latur earthquake of 1993 occurred in a place which was previously considered to be the most stable region on the earth. Scientists estimate that over one million earthquakes occur every year! Some recent earthquakes and their effects are listed in Table 3.17. The recent data on past earthquakes in India (five in the past decade) is much higher than at any time in the past 200 years. 100,000 people have been killed by earthquakes in the Indian region in the past two centuries.

**TABLE 3.17** Some recent earthquakes and their effects

| Year | Location | Magnitude on Richter Scale | Effects |
|---|---|---|---|
| September 1985 | Mexico City | 8.1 | 300 buildings collapsed in downtown Mexico City. 8000 people killed. Property damage—$8 billion. Due to soft subsoil, buildings were subjected to an acceleration of 1.0g. 8- to 15-storey buildings, which had a natural frequency close to that of the soil, failed. Many liquefaction failures. |
| January 1994 | Northridge, 50 km northwest of central Los Angeles, USA | 6.7 | 50 deaths and 5000 injured, property damage of $50 billion–$100 billion, peak ground acceleration of 1.0g in several locations. More than 10,000 buildings damaged. |
| June 1990 | Gilan Province, Northern Iran | 7.7 | 40,000 deaths and 60,000 injured. 500,000 people were left homeless. 100,000 adobe houses sustained major damages or collapsed. |
| October 1989 | Loma Prieta, Santa Cruz Mountains, California | 7.1 | 60 buildings sustained varying degrees of damage with an estimated repair cost of $160 million. 41 deaths due to the falling of upper deck of interstate (Nimitz freeway) 880 in Oakland on to the lower roadway. Failure of Oakland Bay Bridge. Damage to many roads and bridges. |
| July 1993 | Japan Sea, off Southwest Hokkaido | 7.8 | Earthquake and subsequent tsunami in Northern Japan resulted in 196 deaths and losses worth $60 million. Majority of damage caused by tsunamis of height 30.6 m (Second highest ever recorded in Japan). Damage also due to liquefaction and landslides. |
| January 1995 | Kobe, Japan | 7.2 | This was the most damaging to strike Japan since the great Kanto earthquake of 1923 that killed 143,000 people. More than 5100 people died and 27,000 were injured. One-fifth of the city's population of 1.5 million left homeless. 105,000 buildings were destroyed with an estimated loss of $100 billion–$200 billion. |
| September 1993 | Latur, Maharashtra, India | 6.3 | 7600 people died, 16,000 injured, 30,000 houses destroyed. |
| January 2001 | Bhuj, Gujarat, India | 6.9 | About 20,000 people died and 166,000 were injured 600,000 people were homeless, 348,000 houses were destroyed and 844,000 damaged. Losses estimated to be $1.3 billion–$5 billion. 15.9 million people of the total population of 37.8 million were affected. |
| December 2004 | Indian Ocean, west coast of Northern Sumatra, Indonesia | 9.0 (fifth most powerful since 1900) | This was the second largest earthquake since the 9.2 magnitude Good Friday earthquake off Alaska, USA in 1964 and tied for the fourth largest since 1900. Caused damage and death as far as the east coast of Africa. Estimate of damage was in the range of $5 billion–$10 billion. |
| October 2005 | Pakistan | 7.6 | 100,000 people killed and more than 138,000 injured and extensive damage in northern Pakistan. At least 32,335 buildings collapsed. An additional 1360 people killed and 6266 injured in India. An estimated 4 million people were left homeless, several landslides also occurred. |
| May 2008 | Eastern Sichuan, China | 7.9 | At least 69,195 people killed, 374,643 injured, and 18,392 missing and presumed dead. More than 45.5 million provinces and regions were affected. At least 15 million people were evacuated from their homes and more than 5 million were left homeless. An estimated 5.36 million buildings collapsed and more than 21 million damaged. The total economic loss was estimated at US$ 86 billion. There were also several landslides. |
| January 2010 | Haiti | 7.0 | 316,000 people killed, 300,000 injured, and 1.3 million displaced. 188,383 buildings damaged in the Port-an-Prince area and in much of eastern Haiti. |

(*Contd*)

**TABLE 3.17** (*Contd*)

| Year | Location | Magnitude on Richter Scale | Effects |
|---|---|---|---|
| March 2011 | Tohoku, Japan | 9.0 | Largest ever recorded in Japan. Up to 40.5 m tall tsunami waves (highest ever recorded in Japan).<br>Costliest natural disaster in world history. One of the five most powerful earthquakes in the world since modern record. Killed an estimated 15,889 persons, 6152 injured, and 2611 people missing. 127,290 buildings totally collapsed, 272,788 half-collapsed, and 747,989 partially damaged. Extensive structural damage to Fukushima-Daiichi nuclear reactors. Economic losses estimated at US$ 235 billion. |

Earthquakes cause the ground to shake violently in all directions, lasting for a few seconds in a moderate earthquake or for a few minutes in very large earthquakes. [For example, the P (Primary) waves of the Kobe earthquake started at 05:47:15 and ended at 05:47:21. The S (Secondary) waves started at 05:47:22 and ended at 05:47:28 (see Section 3.11.1 for definition of P and S waves). Thus, in just 13 seconds, 5100 people died.] Earthquakes may trigger landslides, cause floods, make the ground heave and crack, and thereby result in large-scale destruction to life and property. Many earthquakes are also followed by fire due to the breakage of electric cables (see Scawthorn et al. 2005 for details on fires following earthquake). Many suffer due to dislocated or broken water supply and gas lines. Major earthquakes are also followed by numerous minor tremors.

## CASE STUDY

### The Indian Ocean Tsunami 2004

The Indian Ocean Tsunami was due to an under-sea earthquake (Sumatra–Andaman earthquake) that occurred on 26 December 2004, with an epicenter off the west coast of Sumatra, Indonesia. The quake, with a Richter magnitude of 9.1 to 9.3, is the second largest earthquake ever recorded on a seismograph. This earthquake had the longest duration of faulting ever observed, between 8.3 and 10 minutes.

2004 Indian Ocean tsunami hitting the shores of Ao Nang, Thailand (Photo: David Rydevik)

This earthquake caused an estimated 1600 km fault surface to slip under the ocean to of about 15 m, resulting in the earthquake (followed by the tsunami) to be felt simultaneously as far away as Bangladesh, Sri Lanka, India, Malaysia, Myanmar, Thailand, Singapore, and the Maldives. The slip did not happen instantaneously but took place in two phases over a period of several minutes. Due to the slip, the sea floor is estimated to have risen by several metres, displacing an estimated 30 km$^3$ of water and triggering devastating tsunami waves (with wave heights of 15m). Because of the distances involved, the tsunami took anywhere from fifteen minutes to seven hours to reach the various coastlines. In many places the waves reached as far as 2 km inland. The wave reached a height of 24 m when coming ashore along large stretches of the coastline, rising to 30 m in some areas when traveling inland.

An analysis by the United Nations found that a total of 229,866 people were lost, including 186,983 dead, and 42,883 missing, due to the tsunami in towns and villages along the coast of the Indian Ocean. The livelihoods of over 3 million survivors were destroyed. Beyond the heavy toll on human lives, it has caused an enormous environmental impact that will affect the region for many years to come.

A village near the coast of Sumatra lies in ruin after the tsunami that struck South East Asia

The following video shows the destructive power of this tsunami: http://news.bbc.co.uk/media/news_web/video/40673000/bb/40673025_bb_16x9.asx

http://www.youtube.com/watch?v=R-jlyfzGP-o

An animation of tsunami may be seen in this video: http://news.bbc.co.uk/2/hi/science/nature/7533972.stm

Guidelines for the design of structures against tsunami are scarce (it is generally not feasible or practical to design normal structures to withstand tsunami loads), but warning and evacuation systems have been developed (see FEMA 55 and FEMA P646 at www.fema.gov).

### 3.11.1 Characteristics of an Earthquake

The point beneath the surface, where the rocks break and move is called the *focus* or *hypocentre* of an earthquake. The amplitude of vibrations diminishes with distance from the *epicentre*, which is the point on the earth's surface immediately above the hypocentre.

Three types of waves are generated during an earthquake: (i) the *primary waves*, or P waves, which are the fastest compression waves (they travel at a speed of $V_P = 6–13$ km/s), pulling and pushing the rock particles parallel to the direction of propagation, (ii) The *secondary waves* or S waves, which are transverse shear waves, and are responsible for most of earthquake damage; they are slower than P waves (S waves travel at a speed of $V_s = 3.5 - 7.5$ km/s; $V_P/V_s \approx \sqrt{3}$) and cause rock mass to move in a direction at right angles to the direction of propagation, and (iii) *surface waves* which include *Rayleigh and Love waves* (R and L waves). L waves, travel from the epicentre along the surface of the earth transverse to the direction of propagation and R waves produce a circular motion, similar to the way waves travel in the ocean.

P waves and S waves may be grouped together as *body waves* and have a higher frequency range and attenuate more rapidly than surface waves. (Considering an earthquake centred about 8500 km from the observation point, the time between the start of a P wave and that of an S wave is 10 min.) The intensity of an earthquake reduces with the distance from the epicentre. The reduction of earthquake intensity, called *attenuation*, is at a faster rate for higher-frequency components than for low-frequency components. (Attenuation relationships are explained in detail by Paulay and Priestley 1992.) Hence structures with longer natural periods, such as high-rise buildings and bridges situated some distance away from the epicentre, may experience greater seismic loads than low-rise buildings which have a short natural period (Williams 2004). Therefore a 50-storey building may be at a greater risk than a single-storey building.

Earthquakes are recorded on a strong-motion *accelerograph* or *seismograph*, which is triggered only when the ground acceleration exceeds a preset value. Earthquake records can be made for three components of ground acceleration, two horizontal and one vertical. Displacement as well as acceleration seismographs can be used to record an earthquake.

Figure 3.20 shows the recorded (often cited) ground acceleration for the El Centro earthquake, California.

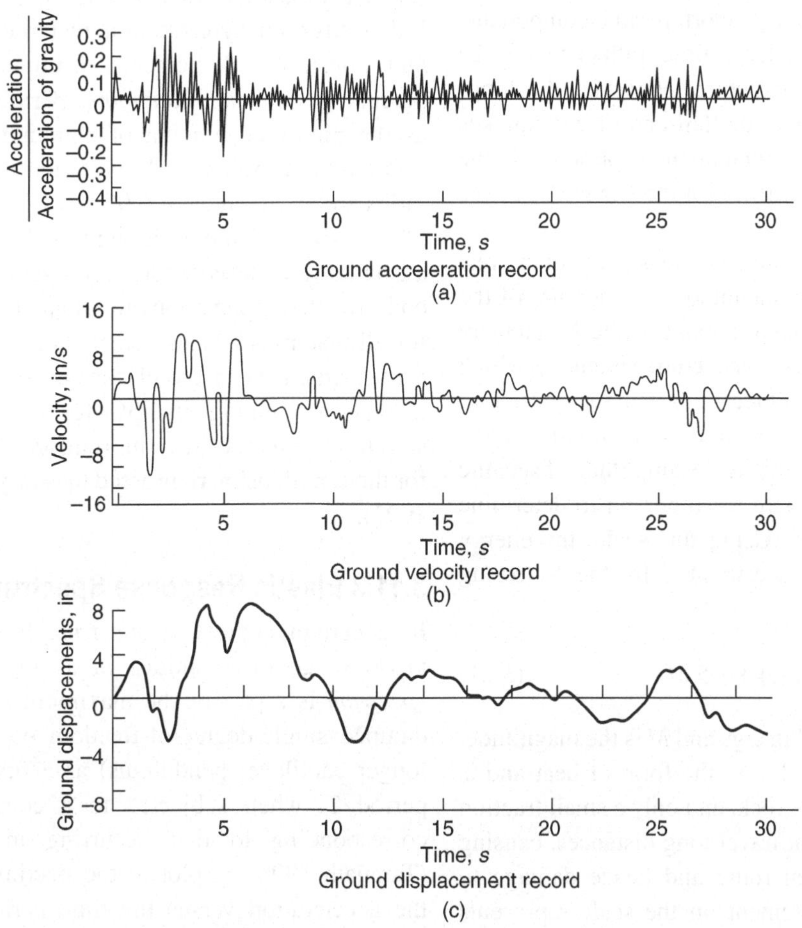


**FIG. 3.20** El Centro, California, earthquake record

*Source:* Taranath 1998, Reprinted with permission from McGraw-Hill Eduction.

It also shows the ground velocity and displacement plotted as functions of time. It is seen from Fig. 3.20(a), that during a short initial period, the intensity of ground acceleration increases to strong shaking, followed by a strong acceleration phase, which gradually decreases. The maximum recorded *peak ground acceleration* (PGA) is about $0.33g$ (where $g$ is the acceleration due to gravity, equal to $9.807\,\text{m/s}^2$), the maximum ground velocity is 0.348 m/s, and the maximum ground displacement from the initial position is about 211 mm. Note that in the 1994 Northridge earthquake, California, the peak values of acceleration, velocity, and displacement were $0.843g$, 1.28 m/s, and 325.5 mm. Usually the ground motion is recorded in two mutually perpendicular horizontal directions. Hence the PGA value can be different in different directions. The vertical PGA value is normally taken as two-thirds of the horizontal PGA. Thus the predominant direction of ground motion is usually horizontal. However, vertical acceleration should be considered in structures with long spans and in those in which stability is a criterion in design. Special attention should be paid to the effect of the vertical component of ground motion on cantilever beams, girders, and slabs. During the first few seconds of an earthquake, the acceleration of the ground reaches a peak value and is associated with relatively short period components of the range 0–0.5 s, which have little influence on the fundamental response of the structure. However, the long-period components that occur at the tail-end of earthquakes (with periods closer to the fundamental period of the building) have a significant influence on the behaviour of the structure.

The *earthquake's magnitude* and its intensity are of interest to the structural engineer. The magnitude is a measure of the amount of energy released by the earthquake, while the intensity is the apparent effect of the earthquake. Unlike intensity, which can vary with location, the magnitude is constant for a particular earthquake. The magnitude is measured on the *Richter scale*, which is a logarithmic scale, utilizing the amplitude of seismic vibrations recorded on a standard seismograph to determine the strength of an earthquake. Using this scale, the energy released, measured in ergs, is estimated by the following equation.

$$\text{Log}_{10}E = 11.4 + 1.5\,M \tag{3.21}$$

where $E$ is the energy released in ergs and $M$ is the magnitude.

Most of the energy released is in the form of heat and is accounted for by fracturing the rock, and only a small fraction of it by the seismic waves that travel long distances, causing the shaking of the ground en route and hence damage to the structures. A unit measurement on the scale represents a tenfold increase in amplitude and an increase in energy release of 31.6 times. Thus, an earthquake of magnitude 6 is 31.6 times more powerful than one measuring 5. The area of potential damage is considerably larger for M6 to M8 range (2000–10000 km$^2$). Earthquakes of Richter magnitude 6, 7, and 8, respectively, are categorized as moderate, major, and great.

Although an earthquake has only one magnitude, it has many different intensities. The intensity at a place is evaluated considering the three features of shaking—perception by people, performance of buildings, and changes to natural surroundings. Two commonly used intensity scales are the *Modified Mercalli Intensity (MMI) scale* and the *MSK scale*. Details of MSK scale are available in Appendix D of the draft IS 1893 code (Jain & Murty 2004). The MMI as well as the MSK scales are quite similar and range from I (least perceptive) to XII (most severe).

### 3.11.2 Damping

When a system is allowed to vibrate freely (with no external forces acting on the system), the amplitude of vibration decays with time due to loss of energy in a number of ways. This type of vibration is called *damped vibration*. In buildings, the extent of damping is due to internal friction, imperfect elasticity of material, slipping, sliding, etc. Thus, it depends on the construction materials, type of connections, and nonstructural elements. Damping is expressed as a percentage of *critical damping*. (Critical damping is defined as the minimum amount of damping necessary to prevent oscillation altogether.) Representative values of damping ratios are in the range of 0.05–0.10. A damping coefficient of 5% of critical damping is suggested in the IS 1893 code for all building materials. (See commentary C.7.5.3 of the draft code for the explanations to adopt the same damping value for all materials.) Measurements of the dynamic response of actual structures in the elastic range close to yield strength indicate a damping of 2-3% for steel structures, 5–7% for structures constructed using reinforced concrete, and 7–10% for those built using reinforced masonry (Paulay and Priestley, 1992).

### 3.11.3 Elastic Response Spectrum

In a certain context, a *spectrum* is a representation of a broad range of quantities in a single graph. A *response spectrum* is a plot of the maximum response of a range of damped single-degree-of-freedom oscillators (progressively longer cantilever pendulums) as a function of their natural period $T$, when subjected to a common ground motion, corresponding to that occurring in a given earthquake (Taranath 1998). A plot of the maximum response, such as the acceleration versus the time period of the pendulums, provides the *acceleration response spectrum*, as shown in Fig. 3.21.

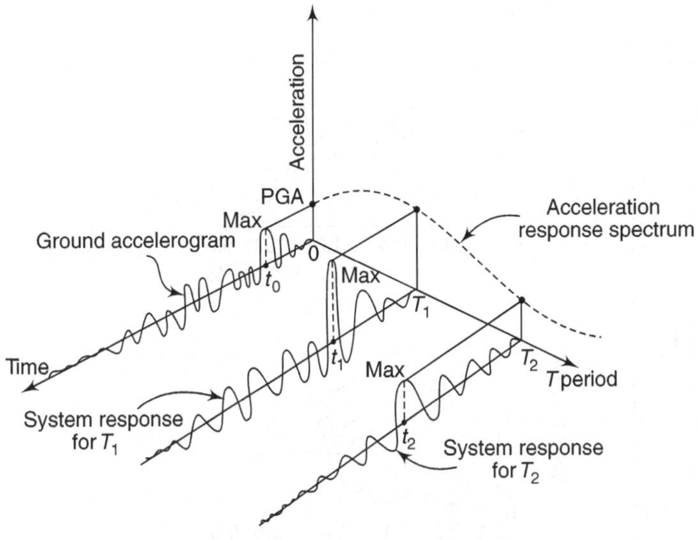


**FIG. 3.21** Graphical description of response spectrum

*Source:* Taranath 1998, Reprinted with permission from McGraw-Hill Education

The absolute value of the peak acceleration response occurring during the excitation for each pendulum is represented by a point on the acceleration spectrum curve. Response spectrum curves are generated by numerical integration (called Duhamel integration) of actual earthquake records to determine maximum values for each period of vibration. The damping ratio of the oscillator may be varied and the response is recorded for displacement, velocity, or acceleration. Due to resonance, the spectra have the tendency to amplify the ground motion in some range of period. The aim of generating these spectra is to emphasize this amplification. In many cases, the maximum response of the spectral acceleration $S_a$ will be desired for seismic-resistant design. The acceleration spectrum so obtained for the El Centro earthquake is shown in Fig. 3.22(a), and response spectra for different powerful earthquakes are shown in Fig. 3.22(b).

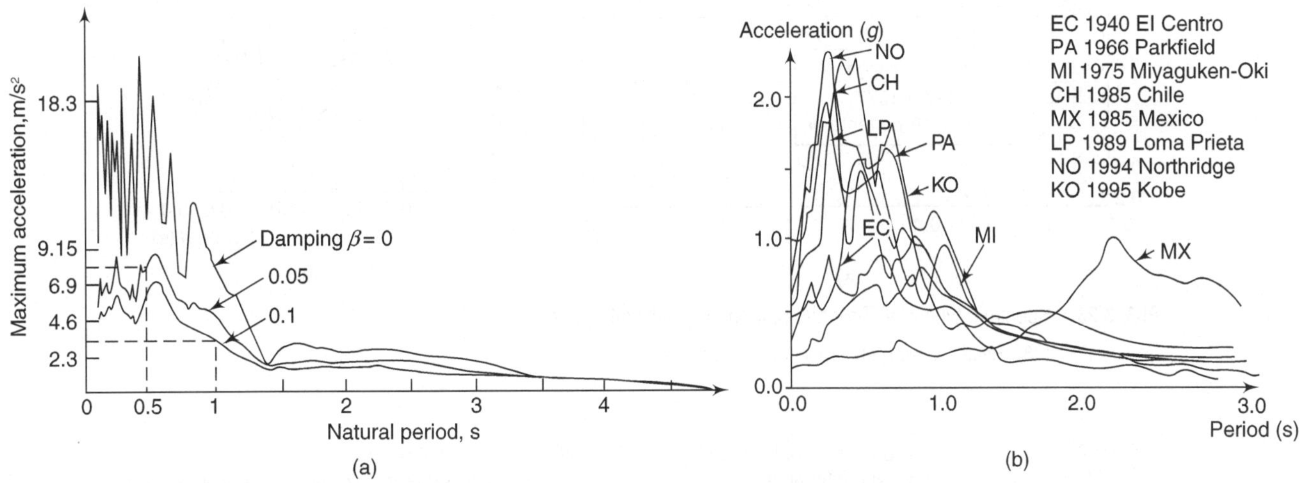


**FIG. 3.22** (a) Acceleration spectrum: El Centro earthquake (b) Response spectra for different strong earthquakes

*Sources:* (a) Taranath 1998, Reprinted with permission from McGraw-Hill Education.
(b) Gioncu and Mazzoloni 2003, Reprinted with permission from CRC Press.

There exists a relationship between the spectral velocity $S_v$, spectral acceleration $S_a$, and spectral displacement $S_d$, as given below.

$$S_a = \omega S_v = \omega^2 S_d \tag{3.22}$$

where

$$\omega^2 = \frac{k}{m} \quad [3.22(a)]$$

in which k is the stiffness of the system, and m is the mass of the system.

Because of this interrelationship, all three spectra may be plotted on the same graph, using tripartite axes and logarithmic scales, as shown Fig. 3.23, with different damping coefficients $\beta$.

The peaks and valleys in the response spectrum shown in Fig. 3.22(a) may not occur at the same values of natural periods for another earthquake ground motion even under similar site conditions. Moreover, most buildings in practice have at least some degree of damping and hence the peaks in response spectra are of little significance. Hence, as an alternative to the use of several earthquake spectra, a smooth spectrum representing an upper bound response to ground motions is normally used. A *smoothed elastic design acceleration response spectrum (SEDRS)*, as adopted by the Indian Code, is shown in Fig. 3.24. This gives the design response acceleration coefficient for 5% damping and for three different soil types.

The values given in Fig. 3.24 can be represented mathematically by the following equations.

1. For rocky or hard soil sites,

$$\frac{S_a}{g}\begin{cases} = 2.50,\ 0.0 \le T \le 0.10 \text{ (for fundamental mode)} \\ = 1 + 15T,\ 0.0 \le T \le 0.10 \text{ (for higher modes)} \\ = 2.50,\ 0.10 \le T \le 0.40 \\ = \dfrac{1.0}{T},\ 0.40 \le T \le 4.00 \\ = 0.25,\ T > 4.00 \end{cases} \tag{3.23}$$

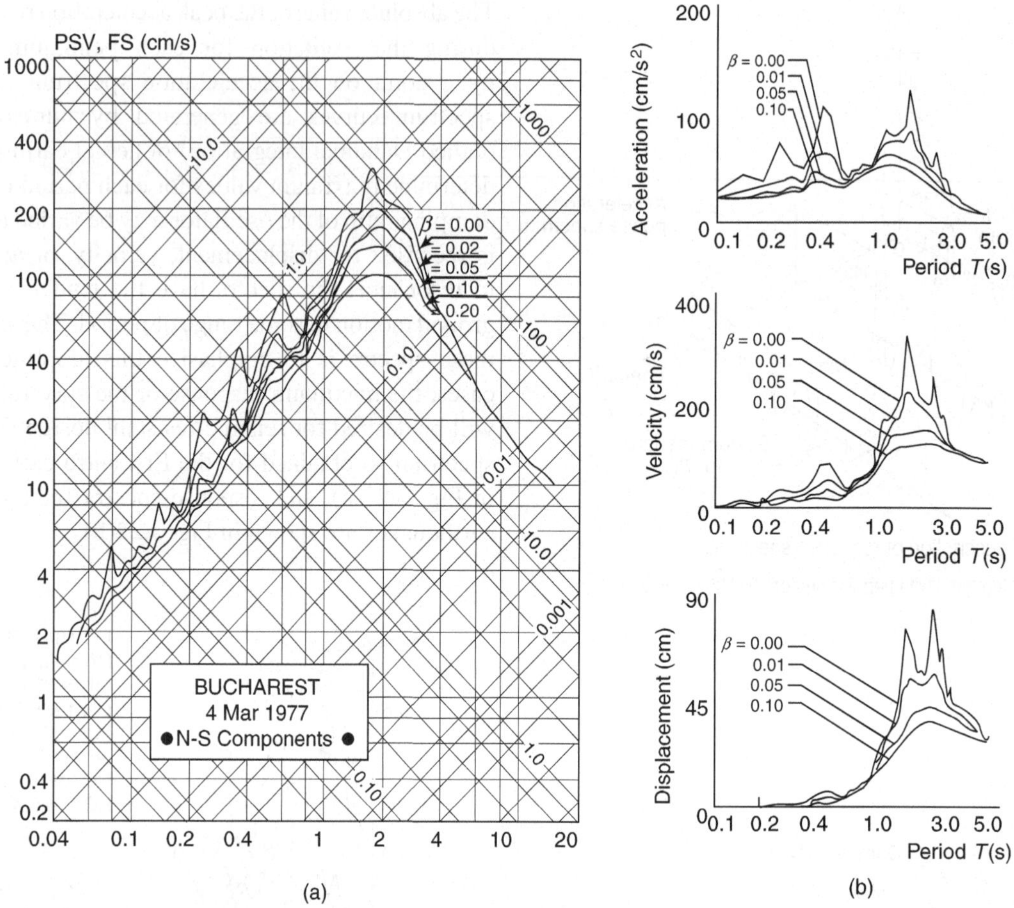


**FIG. 3.23** Spectrum types: (a) Tripartite plot and (b) Separate plots

*Source:* Ifrim 1984, Reprinted with permission from CRC press.

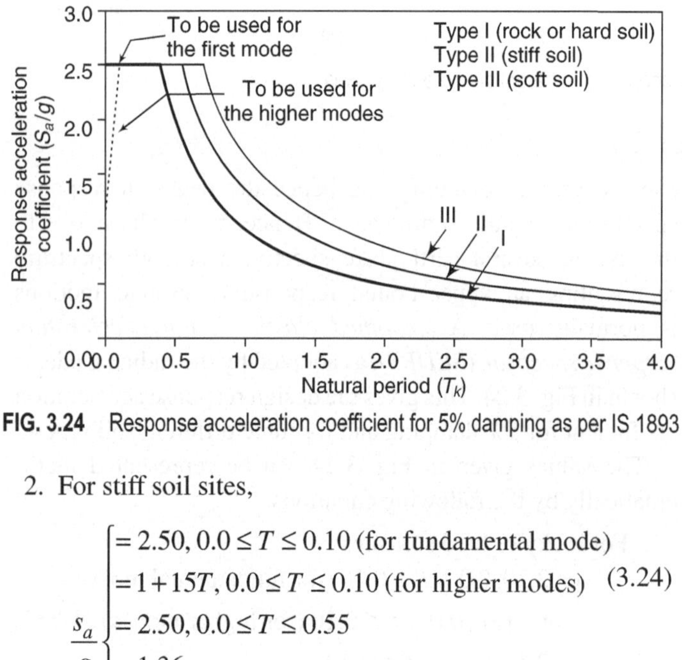


**FIG. 3.24** Response acceleration coefficient for 5% damping as per IS 1893

2. For stiff soil sites,

$$\frac{S_a}{g} \begin{cases} = 2.50, 0.0 \le T \le 0.10 \text{ (for fundamental mode)} \\ = 1 + 15T, 0.0 \le T \le 0.10 \text{ (for higher modes)} \\ = 2.50, 0.0 \le T \le 0.55 \\ = \dfrac{1.36}{T}, 0.55 \le T \le 4.0 \\ = 0.34, T > 4.0 \end{cases} \tag{3.24}$$

3. For soft soil sites,

$$\frac{S_a}{g} \begin{cases} = 2.50, 0.0 \le T \le 0.10 \text{ (for fundamental modes)} \\ = 1 + 15T, 0.0 \le T \le 0.10 \text{ (for higher modes)} \\ = 2.5, 0.0 \le T \le 0.67 \\ = \dfrac{1.67}{T}, 0.67 \le T \le 4.00 \\ = 0.42, T > 4.00 \end{cases} \tag{3.25}$$

The multiplying factors for obtaining $S_a/g$ values for other damping (these should not be applied to the point at zero

**TABLE 3.18** Multiplying factors for obtaining $S_a/g$ values for other damping (IS 1893)

| Damping percentage | Factors |
|---|---|
| 0 | 3.20 |
| 2 | 1.40 |
| 5 | 1.00 |
| 7 | 0.90 |
| 10 | 0.80 |
| 15 | 0.70 |

*Note:* For other damping percentages refer IS 1893-Part 1: 2002.

period) are given in Table 3.18. It is important to recognize that the response spectra approach does not consider the duration effects. Survivability of a structure depends not only on peak response levels but also on the duration of strong ground shaking and the number of cycles where response approaches the peak response levels. (This fact was realized during the 1985 Mexico earthquake.)

### *Unique Design Spectra*

For very important structures or where local soil conditions are not amenable to simple classification, the use of a recommended smooth spectrum curve will not yield safe design. In these cases, site-specific studies are performed by geotechnical consultants to determine more precisely the expected intensity and character of seismic motion. More details of developing such spectra are found in Taranath (1998).

### *Near-source Factor*

In regions subjected to large-magnitude earthquakes, locations close to the fault rupture may experience a ground acceleration up to twice that at a distance of 10 km from the source. Because in the past, the majority of ground motions were recorded in far-field regions, the concepts developed in the codes were based on such earthquakes only. There was a great damage during the Northridge, USA (1994) and Kobe, Japan (1995) earthquakes because these towns were situated in the near-field region. Also, in the near-field region, the ground motion has an impulse-type motion, similar to a blast, and the duration of the earthquake is very short. The velocities in near-field regions are very high (during the Northridge and Kobe earthquakes, velocities in the range of 150–200 cm/s were recorded compared to 30–40cm/s for far-field regions). Moreover, the vertical component in near-field regions may be greater than the horizontal component, due to the direct propagation of P waves. Some codes (e.g., the UBC code) introduce two near-source amplification factors—an acceleration-based factor for short-period structures, and a velocity-based factor for periods exceeding one second. For near-field earthquakes, a new methodology based on *displacement spectra* may yield better results than the usual acceleration-based design method (which is a force-based approach). It is also interesting to note that many steel structures were damaged by the fracture of connections during the Northridge and Kobe earthquakes without global collapses, giving rise to the idea that the local fracture of these connections can transform the original rigid structure into a structure with semirigid joints. More details about *near-field earthquakes* are provided by Gioncu and Mazzolani (2002).

Till now, we have discussed the characteristics of wind and earthquakes. In the next two sections we will discuss, respectively, the determination of earthquake and wind forces as per IS 1893 (Part 1) and IS 875 (Part 3).

## 3.12 DETERMINATION OF EARTHQUAKE LOADS

Earthquake load differs from other loads such as dead load, live load, etc., in many respects.

1. As discussed in the previous section, earthquake loading is uncertain and it is difficult to determine the amplitudes, duration, and frequency of earthquake motion.
2. In structures normally designed for gravity loading, the lateral loading caused by earthquake ground motion may cause severe damage unless special arrangements are made to resist them.
3. Earthquake loads are cyclic and induce reversal of stresses. Hence, axially loaded members have to resist both tension and compression, while beams have to be designed for positive as well as negative bending moments at the same section.
4. Earthquake loads are dynamic and produce different degrees of response in different structures. When the ground, under a structure having a mass, suddenly moves, the inertia of the mass tends to resist the movement, as shown in Fig. 3.25, and creates forces, called *inertia forces*, which are equal to the product of the mass of the structure and acceleration ($F = ma$). The mass is equal to the weight ($W$) divided by the acceleration due to gravity, i.e., $m = W/g$.

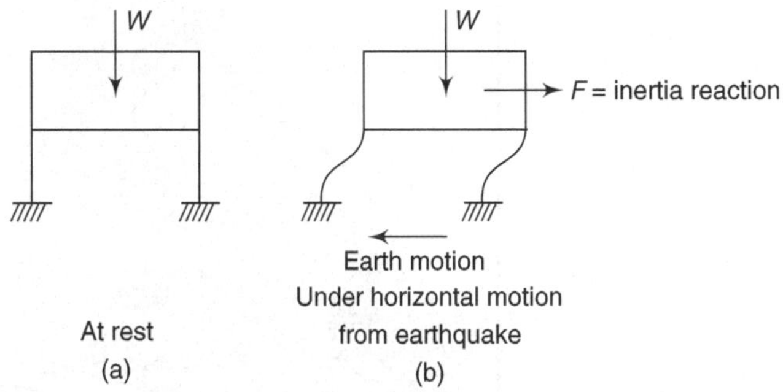


**FIG. 3.25** Force developed by earthquake

Large earthquakes have much lower frequency of occurrence than smaller ones. Hence it may be uneconomical to design a building (which will have a normal design life of 50–100 years), so that it remains undamaged during a large earthquake that may take place, say, once in 500 years. Hence in the code (IS 1893), the following *seismic design philosophy* is adopted.

1. Minor and frequent earthquakes should not cause any damage to the structure.
2. Moderate earthquakes should not cause significant structural damage but could have some non-structural damage (the structure should become operational once the damaged main members are repaired and strengthened).
3. Major and infrequent earthquakes should not cause collapse (the structure will become dysfunctional for further use,

but will stand so that people can be evacuated and property recovered).

Hence the structures are designed for much smaller forces than actual seismic loads during strong ground shaking. Note that this approach is different than that adopted in the case of wind, dead, live, and other loads, where the structure is designed for the anticipated loads.

For the purpose of determining seismic forces, the country is classified into four seismic zones, as shown in Fig. 3.26. It may be of interest to note that the Koyna earthquake of 1967 occurred in Seismic zone I (which triggered a major revision of the zonal map in 1970). Similarly, the Latur earthquake of 1993 occurred in Seismic zone I, which resulted in the revision of the seismic zone map in 2002). The shaking intensity associated with the different zones is given in Table 3.19.

**TABLE 3.19** MSK intensity related to different seismic zones in India

| Zone | Probable shaking intensity (MSK) |
|---|---|
| II | VI and lower |
| III | VII |
| IV | VIII |
| V | IX and higher |

It has to be noted that the values of the seismic zone factor ($Z$) given in the code are empirical and have been arrived at on the basis of engineering judgment rather than on any rational analysis of expected peak ground acceleration (PGA) and service life.

The code requires that the designer either (a) use a dynamic analysis of the structure, or (b) for usual rectangular medium-height buildings (regular buildings) use an empirical lateral *base shear force* (see Table 3.20). The dynamics of earthquake

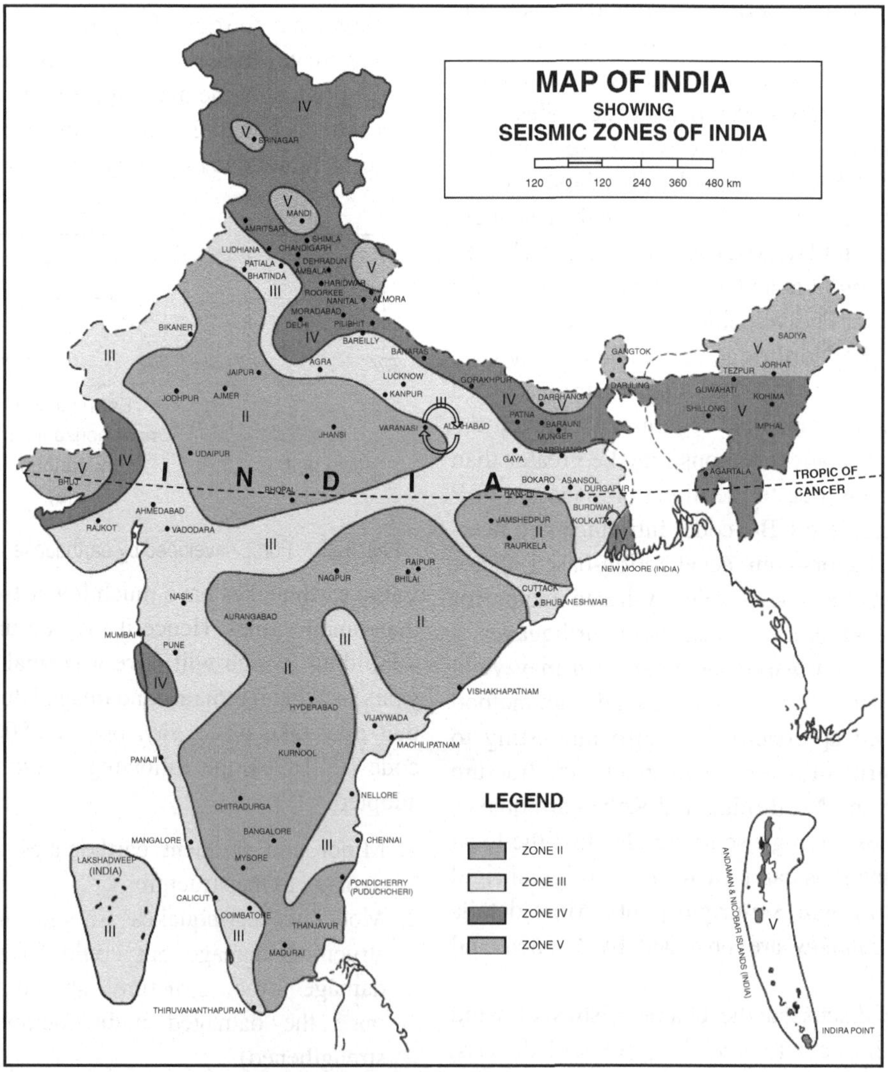


**FIG. 3.26** Seismic zones of India as per IS 1893

*Source:* Reprinted with permission from Survey of India, Government of India.

action on structures is outside the scope of this book, and the reader may refer to Chopra (2000), Clough and Penizien (1993), and Mazzolani and Piluso (1996) for the details of dynamic analysis methods.

**TABLE 3.20** Requirement of dynamic analysis as per IS 1893

| Seismic zone | Regular buildings | Irregular buildings |
|---|---|---|
| II and III | Height > 90 m | Height > 40 m |
| IV and V | Height > 40 m | Height > 12 m |

*Notes:* 1. Large-span industrial buildings may also require dynamic analysis.
2. Buildings with high level of torsion irregularity are prohibited in zones IV and V.

The code also cautions that engineers need to have an in-depth knowledge and understanding of dynamic analysis in order to be able to correctly model the structures and interpret the results. It also suggests that the results of computer analysis should be checked by using approximate methods such as Rayleigh's method and Dunkerlay's method (Clough and Penzien 1993 and Paz 1991). A comparison of various methods of dynamic analysis of multi-storey frames is provided by Radhakrishnan and Subramanian (1978). Similarly, one must be careful about the use of correct units while performing dynamic analysis since it is common for huge errors to occur just because units of mass and weight are mixed up (draft code, 2004)

The following are important factors that influence earthquake-resistant design:

1. Geographical location of the structure
2. Site soil and foundation condition
3. Importance of the structure
4. Dynamic characteristics of the structure, such as strength, stiffness, ductility, and energy dissipation capacity

These factors are considered directly or indirectly in all the methods of analysis.

## Equivalent Lateral Base Shear Force Procedure

The *equivalent lateral base shear force* procedure has traditionally been used by several building codes to simplify the design process. (This procedure is applied to regular structures only.) As per the code (IS 1893) the *design horizontal seismic coefficient* for a structure is

$$A_h = ZI\,(S_a/g)/2R \tag{3.26}$$

where

$Z$ = *zone factor*, given in Table 3.21, for the *maximum considered earthquake* (*MCE*).

**TABLE 3.21** Zone factor Z

| Seismic zone | II | III | IV | V |
|---|---|---|---|---|
| Z | 0.10 | 0.16 | 0.24 | 0.36 |

The factor of 2 in the denominator of Eqn (3.26) is used to reduce the MCE to the design basis earthquake (DBE).

$I$ = *importance factor*, depending on the functional use of the structure. For important service and community buildings such as hospitals, schools, emergency buildings, large community halls, and power stations, it is taken as 1.5, and for all other buildings (ordinary buildings), it is considered to be 1.0.

$R$ = *response reduction factor*, depending on the perceived seismic damage performance of the structure, characterized by ductile or brittle deformations. The values of $R$ for steel buildings are given in Table 3.22 (See also Section 17.6 of Chapter 17).

**TABLE 3.22** Response reduction factor $R$, for building systems

| Lateral load resisting system | $R$ |
|---|---|
| Braced frame systems | |
| (a) Ordinary concentrically braced frames (OCBF) | 4 |
| (b) Special concentrically braced frame (SCBF) | 4.5 |
| (c) Eccentrically braced frame (EBF) | 5 |
| Moment-resisting frame systems | |
| (a) Ordinary moment-resisting frame (OMRF) | 4 |
| (b) Special moment-resisting frame (SMRF) | 5 |

$S_a/g$ = response acceleration coefficient as given by Fig. 3.24 or Eqns (3.23) to (3.25), based on appropriate natural periods and the damping of the structure.

The $Z$ values given in the code represent a reasonable estimate of PGA in the respective zones. Thus a value of $Z$ equal to 0.36 in Zone V implies that a value of $0.36g$ is reasonably expected in Zone V. But it does not imply that acceleration in Zone V will not exceed $0.36g$. For example, during the 2001 Bhuj earthquake, a PGA of about $0.6g$ was recorded at Anjar, 44 km away from the epicentre (Draft code IS 1893).

### 3.12.1 Natural Frequencies

A structure with $N$ degrees of freedom has $N$ natural frequencies and $N$ mode shapes. The lowest of the natural frequencies of the structure is called its *fundamental natural frequency*, expressed in hertz. The associated natural period is called the *fundamental natural period*, which is the reciprocal of the natural frequency and is expressed in seconds. Where a number of modes are to be considered for dynamic analysis, the value of $A_h$ [see Eqn (3.26)] for each mode should be determined using the natural period of vibration of that mode. For underground structures and foundations at depths of 30 m or below, the design horizontal acceleration spectrum value should be taken as half the value obtained from Eqn (3.26). For structures and foundations between 30 m below ground level and the ground level, the value should be linearly interpolated between $A_h$ and $0.5A_h$. For vertical motions, the value should be taken as two-thirds of the design horizontal acceleration spectrum (IS 1893 : 2002).

### 3.12.2 Equivalent Static Method

The seismic weight of each floor is calculated as its full dead load plus the appropriate amount of imposed load (Table 3.23). While computing this weight, the weight of columns and walls in any storey should be appropriately apportioned to the floors above and below the storey. It has to be noted that buildings designed for storage purposes are likely to have large percentages of service load present at the time of earthquake. Other appropriate loads such as snow or permanent equipment should also be considered.

**TABLE 3.23** Percentage of imposed load to be considered while calculating seismic weight

| Imposed uniformly distributed floor load (kN/m$^2$) | Percentage of imposed load |
|---|---|
| Up to and including 3.0 | 25 |
| Above 3.0 | 50 |

*Note:* The imposed load on the roof need not be considered. No further reduction for large areas or for the number of storeys above the one under consideration [as envisaged in IS 875 (Part 2)] for static load cases is allowed (see also Section 3.3).

In the *equivalent static method* (also referred to as the *seismic coefficient method*), which accounts for the dynamics of the building in an approximate manner, the total *design seismic base shear* if determined by

$$V_B = A_h\, W \tag{3.27}$$

where $A_h$ is the design horizontal acceleration spectrum value as per Eqn (3.26), using the approximate fundamental natural period $T_a$, as given in Eqn (3.28) in the considered direction of vibration, and $W$ is the seismic weight of the building.

Note that the design lateral force should be computed for the building as a whole, i.e., when considering the building as consisting of two-dimensional frames, the seismic design force for each such frame should not be calculated separately based on the tributary mass shared by the frame (see Fig. 3.27). This is incorrect since only a fraction of the building mass is considered in such seismic load calculations.

The following assumptions are involved in the static procedure.

1. the fundamental mode of the building makes the most significant contribution to the base shear.
2. the total building mass is considered as against the modal mass that would be used in dynamic procedure. Both these assumptions are valid for low- and medium-rise buildings which are regular.

After the base shear force $V_B$ is determined, it should be distributed along the height of the building (to the various floor levels), using the following expression.

$$Q_i = V_B \left( \frac{W_i h_i^k}{\sum_{j=1}^{n} W_j h_j^k} \right) \tag{3.28}$$

where $Q_i$ is the design lateral force at floor $i$, $W_i$ is the seismic weight of floor $i$, $h_i$ is the height of floor $i$ measured from the base, $n$ is the number of storeys in the building or the number of levels at which the masses are located, and $k$ is a coefficient.

The value of $k$ equal to 2 is adopted in the Indian Code. It has to be noted that for tall buildings, the contribution of higher modes can be significant, though the first mode may still contribute to the maximum response. Hence, other codes (NEHRP or SEI/ASCE 7-02) adopt

$$k = 1 \text{ for } T \le 0.5\text{s}$$

and

$$k = 2 \text{ for } T \ge 2.5\text{s}$$

and a linearly varying value of $k$ for $T$ in the range of 0.5–2.5 s. However, the value of 2 adopted by IS 1893 is conservative (draft IS 1893).

The use of equivalent static method is explained in Example 3.11.

### 3.12.3 Approximate Fundamental Natural Period

The *approximate fundamental natural period* of vibration, in seconds, for a moment-resisting frame without brick infill panels is given by the code (FEMA 450 2001) as

$$T_a = 0.085h^{0.75} \tag{3.29a}$$

where $h$ is the height of the building in metres.

For all other buildings, including moment-resisting frame buildings with brick infills,

$$T_a = \frac{0.09h}{\sqrt{d}} \tag{3.29b}$$

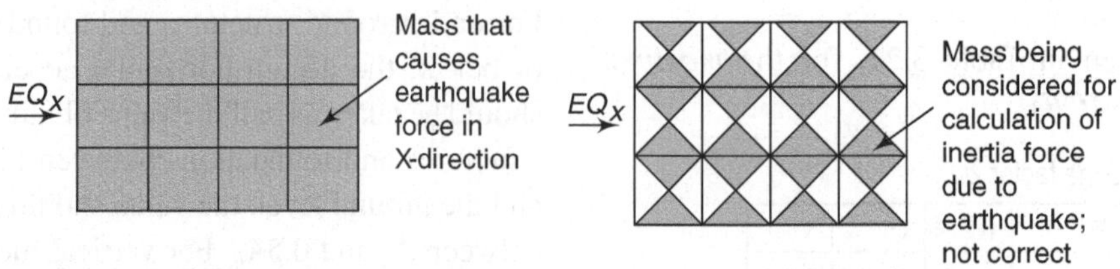


**FIG. 3.27** Calculation of design seismic force on the basis of tributary mass on two-dimensional frames leads to significant underdesign

where $d$ is the base dimension of the building at the plinth level, along the considered direction of the lateral force, in metres.

For buildings with concrete and masonry shear walls,

$$T_a = \frac{0.075h^{0.75}}{\sqrt{A_w}} \quad (3.29c)$$

where $A_w$ is the total effective area of the walls in the first storey of the building in square metres, and is calculated as

$$A_w = \sum A_{wi}[0.2 + (L_{wi}/h)]^2 \quad (3.29d)$$

$A_{wi}$ is the effective cross sectional area of the wall $i$ in the first storey of the building in square metres, $L_{wi}$ is the length of the shear wall $i$ in the first storey in the considered direction of the lateral forces, in metres. The value of $L_{wi}/h$ should be limited to below 0.9.

A comparison of the above equations with the observations of buildings during earthquakes is given in draft IS 1893. Based on this, it is suggested that the approximate natural period [given by Eqn (3.29)] be used in the calculations, even though one may obtain an exact value, especially for irregular structures (which may be more than this value) by using a dynamic analysis computer program (Jain 1995). This suggestion has been made to prevent the application of lower design seismic forces calculated using the large natural period obtained by the programs. (See Fig. 3.24.) For irregular buildings, dynamic analysis is used primarily for assessing the distribution of earthquake forces according to building height and to the different vertical elements, and not for assessing the magnitude of design seismic forces (Jain 1995) (see Example 3.12).

## 3.12.4 Dynamic Analysis Methods

In tall buildings, the contribution of higher modes may be important. In irregular buildings the *mode shape* may not be regular, and in industrial buildings with large spans and heights, the assumptions of the static procedure (the fundamental mode of vibration is the most dominant and mass and stiffness are evenly distributed) may not be valid. Hence, for these buildings, dynamic analysis methods are suggested by the code. Such methods are grouped into the *response spectrum method* (multi-storey buildings, irregular buildings, overhead water ranks, and bridge piers are often designed using this method) and *time-history response analysis* (most important structures such as nuclear reactors, large-span structures, or very tall buildings are designed using this method). The understanding of these methods requires some knowledge of structural dynamics and a brief description of these methods is given below.

### Response Spectrum Method

This is the most widely used method in seismic analysis. It can be performed by using the design spectrum shown in Fig. 3.24 or a site-specific design spectrum. A multi-storey structure as shown in Fig. 3.28 is idealized as a multi-storey shear building by assuming that the mass is lumped at the floor and roof diaphragm levels, that the diaphragms are infinitely rigid, and that the columns are axially inextensible but laterally flexible. The dynamic response of the system is represented by the lateral displacements of the lumped masses with the number of degrees of dynamic freedom (or modes of vibration $n$) being equal to the number of masses. An *undamped free vibration* analysis of the building can be performed as per established methods of mechanics using the appropriate masses and elastic stiffness of the structural system, to obtain the natural periods ($T$) and mode shapes ($\phi$) of its modes of vibration. Mode shapes depend on the distribution of mass and the stiffness of the building. (The second, third, fourth, and fifth mode periods are equal to approximately 0.327, 0.186, 0.121, and 0.083 times the fundamental mode.)

When the ground motion is applied to the base of the multimass system, the deflected shape of the system is the combination of all mode shapes, which can be obtained by the superposition of the vibrations of each lumped mass. A *modal analysis procedure* is utilized to determine the dynamic response of a multi-degree-of-freedom system (Paz 1991, Clough and Penzien 1993, and Chopra 2000). Modal analysis as suggested by IS 1893 is discussed in the next section. Each individual mode of vibration of a multimass system has its own period of vibration (with a unique mode shape formed by a line connecting the deflected masses, as shown in Fig. 3.28)

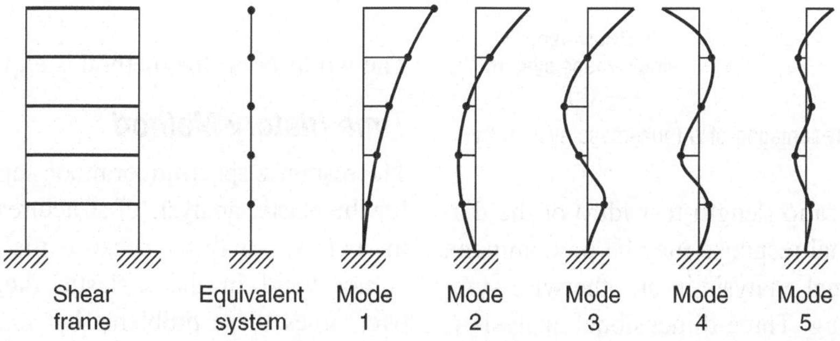


**FIG. 3.28** Mode shapes of multi-storey structure

and may be represented by a single-degree-of-freedom system of the same period, as shown in Fig. 3.29. This concept provides a computational basis for using response spectra based on single-mass systems for analysing multi-storey buildings. Given the period, mode shape, and mass distribution of a multi-storey building, we can use the response spectra of a single-degree-of-freedom system for computing the deflected shape, storey accelerations, forces, and moments. Each predominant mode is analysed separately and the results are combined statistically in order to obtain the total response (Taranath 1998). Methods such as *square-root-of-the sum-of-the squares method (SRSS)* or the *complete quadratic combination method (CQC)* can be used when natural periods of different modes are well separated (when they differ from each other by more than 10% of the lower frequency) and the damping ratio does not exceed 5%. The CQC method can account for modal coupling effects and is suggested by IS 1893. Many software programs can implement this method. However, when the natural periods are closely spaced, the response due to such modes is combined in the absolute sum and this resultant is combined with the responses of the remaining well separated modes. IS 1893 requires that as many modes should be included in the analysis as are necessary to ensure that at least 90% of the total seismic mass is excited by them. (Hence, a consideration of only the first three modes will underestimate the design force.) Modal analysis may be performed by matrix methods, numerical methods, or by iterative techniques such as those proposed by Rayleigh, Stodola, or Holzer (see Williams 2004). Several available commercial computer software packages such as STAAD Pro, SAP 2000, etc., have routines to perform response spectrum analysis.

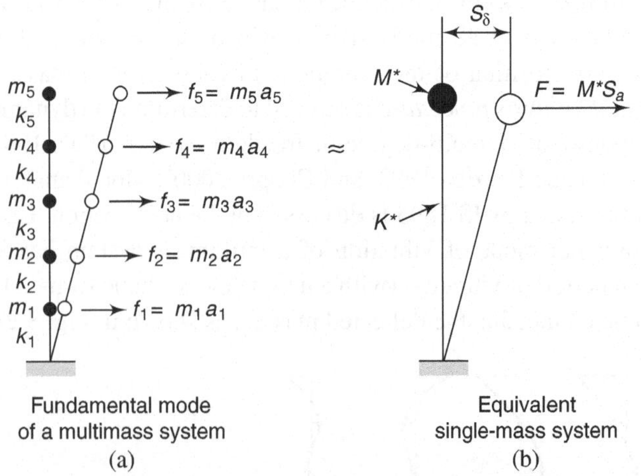


**FIG. 3.29** Mathematical representation of a multi-mass system by a single-mass system

Note that when the aspect ratio (length-to-width) of the diaphragm is large, the torsional response may be predominant, requiring a three-dimensional analysis in an otherwise symmetrical and regular building. Three-dimensional analysis is also required for irregular structures.

## Modal Analysis Method

As discussed in the above section, a building with a regular or nominally irregular plan configuration can be modelled as a system of masses lumped at the respective floor levels with each mass having one degree of freedom, that of lateral displacement in the direction under consideation.

The modal mass $M_k$ of mode $k$ is given by

$$M_k = \frac{\left[\sum_{i=1}^{n} W_i \phi_{ik}\right]^2}{g \sum_{i=1}^{n} W_i (\phi_{ik})^2} \tag{3.30}$$

where $g$ is the acceleration due to gravity, 9.807 m/s$^2$, $\phi_{ik}$ is the mode shape coefficient at floor $i$ in mode $k$, $W_i$ is the seismic weight of floor $i$, and $n$ is the number of storeys or masses.

The loads acting at any floor level $i$ due to the $k$th mode of vibration is given by the equation.

$$Q_{ik} = A_k \phi_{ik} P_k W_i \tag{3.31}$$

where $A_k$ is the design horizontal acceleration spectrum value, using the natural period of vibration ($T_k$) of mode $k$ [$A_k = ZI\,S_a/(2Rg)$], and

$P_k$ = the modal participation factor of mode $k$

$$= \frac{\sum_{i=1}^{n} W_i \phi_{ik}}{\sum_{i=1}^{n} W_i (\phi_{ik})^2} \tag{3.32}$$

The storey shear force in each mode in storey $i$ and in mode $k$ is given by

$$V_{ik} = \sum_{i=1}^{n} Q_{ik} \tag{3.33}$$

Storey shear forces due to all modes considered are obtained by using the CQC method given in the code or by the SRSS method.

The design lateral forces at roof and floor $i$ are, respectively, obtained as

$$F_{\text{roof}} = V_{\text{roof}} \tag{3.34}$$

and

$$F_i = V_i - V_{i+1} \tag{3.35}$$

The working of this method is explained in Example 3.12.

## Time History Method

The response spectrum or mode superposition method is useful for the elastic analysis of structures. It cannot be directly used in inelastic analysis because the principle of superposition is not valid in the inelastic range. Time history analysis overcomes these problems but requires larger computational effort. This method consists of a step-by-step integration in which the time domain is discretized into a number of small increments, and for each time interval the equations of motion are solved, with the displacements and velocities of the previous step serving as initial functions. Thus, ground acceleration time histories are required to conduct the time history method of analysis. For this, ground motions recorded under similar site conditions in past earthquakes may be used. Alternatively, synthetically generated ground motion may also be used. Such ground motion should be compatible with the spectrum given in IS 1893 (Fig. 3.24) or with the site-specific spectrum, whichever is applicable. This method is applicable to elastic as well as inelastic analyses. In elastic analysis, the stiffness characteristics of the structure are assumed to be constant throughout the duration of the earthquake. Whereas in inelastic analysis, for each incremental time, modifications to structural stiffness can be made, which may be caused by the formation of plastic hinges. More details of these methods may be found in Paz (1991), Chopra (2000), Taranath (1998), and Clough and Penzien (1993). The time history method is used only for specialized structures like nuclear power stations. Non-linear inelastic computer programs are available but are not used in normal design practice because of the following factors.

1. Their use requires a special background in structural dynamics.
2. The results obtained are difficult to interpret and apply to traditional design problems.
3. They require extensive computer hardware, which is expensive.

Hence, analyses used in practice are based on linear elastic analysis, such as the response spectrum method.

## 3.13 DETERMINATION OF WIND LOADS AS PER IS 875 (PART 3)

IS 875 (Part 3) : 1987 gives the basic wind speeds averaged over a short interval of 3 s and having a return period of 50 years at a height of 10 m above ground level in different parts of the country. The entire country is divided into six wind zones as shown in Fig. 3.30. The wind speed for some important cities/towns is given in Table 3.24.

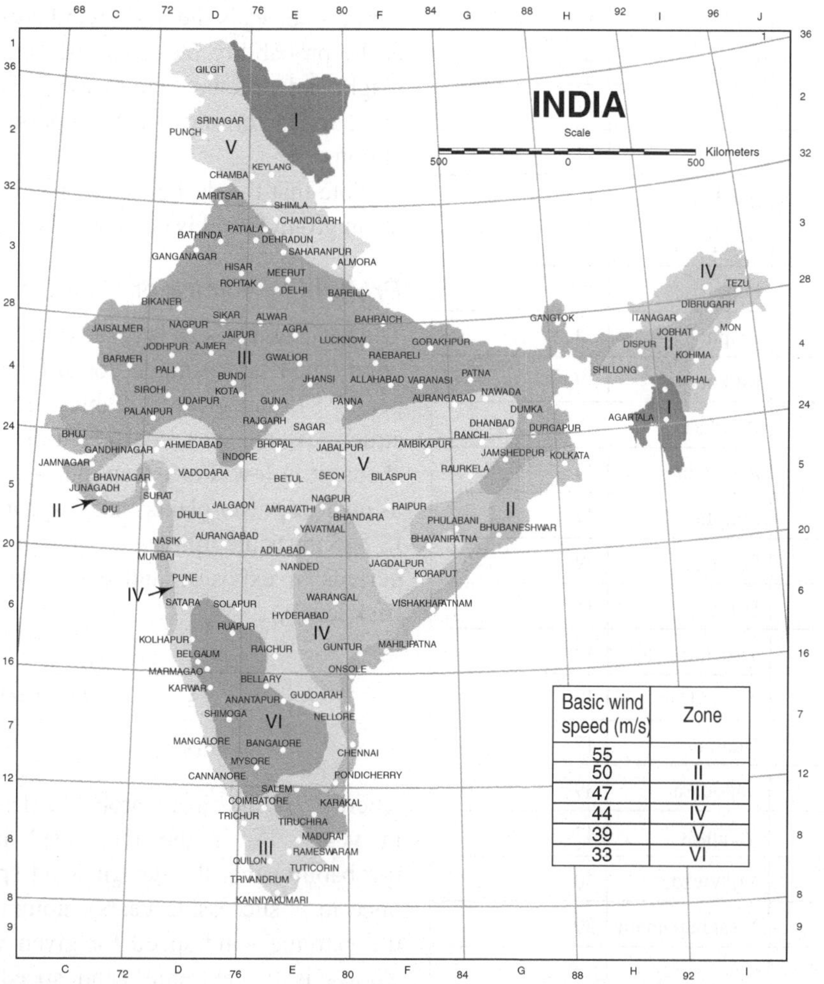


**FIG. 3.30** Basic wind speed in m/s (based on 50 year return period), as per IS 875 (Part 3)

*Source:* Reprinted with permission from Survey of India, Government of India.

**TABLE 3.24** Basic wind speed at 10 m height for some important cities/towns

| City/town | Basic wind speed (m/s)) | City/town | Basic wind speed (m/s)) |
|---|---|---|---|
| Agra | 47 | Jhansi | 47 |
| Ahmedabad | 39 | Jodhpur | 47 |
| Ajmer | 47 | Kanpur | 47 |
| Amritsar | 47 | Kohima | 44 |
| Aurangabad | 39 | Kurnool | 39 |
| Bangalore | 33 | Lakshadweep | 39 |
| Barauni | 47 | Lucknow | 47 |
| Bareilly | 47 | Ludhiana | 47 |
| Bhatinda | 47 | Madurai | 39 |
| Bhilai | 39 | Mangalore | 39 |
| Bhopal | 39 | Mysore | 33 |
| Bhubaneswar | 50 | Nagpur | 44 |
| Bhuj | 50 | Nainital | 47 |
| Bikaner | 47 | Nasik | 39 |
| Bokaro | 47 | Nellore | 50 |
| Mumbai | 44 | Panaji | 39 |
| Kolkata | 50 | Patiala | 47 |
| Calicut | 39 | Patna | 47 |
| Chandigarh | 47 | Pondicherry | 50 |
| Chennai | 50 | Port Blair | 44 |
| Coimbatore | 39 | Pune | 39 |
| Cuttack | 50 | Raipur | 39 |
| Darbhanga | 55 | Rajkot | 39 |
| Darjeeling | 47 | Roorkee | 39 |
| Dehradun | 47 | Rourkela | 39 |
| Delhi | 47 | Shimla | 44 |
| Durgapur | 47 | Srinagar | 39 |
| Gangtok | 47 | Surat | 44 |
| Guwahati | 50 | Tiruchirapalli | 47 |
| Gaya | 39 | Trivandrum | 39 |
| Gorakhpur | 47 | Udaipur | 47 |
| Hyderabad | 44 | Vadodara | 39 |
| Imphal | 47 | Varanasi | 47 |
| Jabalpur | 47 | Vijaywada | 50 |
| Jaipur | 47 | Visakhapatnam | 50 |
| Jamshedpur | 47 | | |

*Note:* For other cities see Appendix A of IS 875 (Part 3): 1987.

As we have already discussed, the wind pressure at any height of a structure depends on the following.

1. Velocity and density of the air
2. Height above ground level
3. Shape and aspect ratio of the building
4. Topography of the surrounding ground surface
5. Angle of wind attack
6. Solidity ratio or openings in the structure

Depending on the above factors, the wind can create positive pressure or negative pressure (suction) on the sides of the building.

### 3.13.1 Design Wind Speed

The *design wind speed* is obtained from the basic wind speed shown in Fig. 3.30, after modifying it to include risk level, terrain roughness, height and size of structure, and local topography as (see draft IS 875)

$$V_z = V_b\, k_1\, k_2\, k_3\, k_4 \tag{3.36}$$

where $V_z$ is the design wind speed at any height $z$ in metres per second, $V_b$ is the basic wind speed (given in Fig. 3.30, $k_1$ is the probability factor or risk coefficient, $k_2$ is the terrain height and structure size factor, $k_3$ is the topography (ground contours) factor, and $k_4$ is the importance factor for the cyclonic region.

Note that the design wind speed up to a 10 m height from mean ground level is considered constant.

#### Probability Factor or Risk Coefficient ($k_1$)

The basic wind speed is based on a 50 year return period of the wind. The risk coefficient for different classes of structures as per IS 875 are given in Table 3.25. Note that for important structures, the risk coefficient is high and for temporary sheds, it is low. For exceptionally important structures such as nuclear power plants, satellite communication towers, etc., IS 875 gives the following equation for $k_1$, based on different periods of exposure and chosen probability of exceedence (risk level).

$$k_1 = \frac{X_{N,P}}{X_{50,0.63}} = \frac{A - B[\text{In}\{-(1/N)\text{In}(1 - P_N)\}]}{A + 4B} \tag{3.37}$$

where $N$ is the mean probable design life of the structure in years, $P_N$ is the risk level in $N$ consecutive years (probability that the design wind speed is exceeded at least once in $N$ successive years), nominal value = 0.63, $X_{N,\,P}$ is the extreme wind speed for given values of $N$ and $P_N$, and $X_{50,0.63}$ is the extreme wind speed for $N$ = 50 years and $P_N = 0.63$.

**TABLE 3.25** Risk coefficients for different classes of structures in different wind-speed zones

| Class of structure | Mean probable design life of structure in years | $k_1$ factor for basic wind speed[a] (m/s) of | | | | | |
|---|---|---|---|---|---|---|---|
| | | 33 | 39 | 44 | 47 | 50 | 55 |
| All general buildings and structures | 50 | 1.0[a] | 1.0 | 1.0 | 1.0 | 1.0 | 1.0 |
| Buildings and structures presenting a low degree of hazard to life and property in the event of failure, such as isolated towers in wooded areas, buildings in farms, other than residential buildings | 25 | 0.94 | 0.92 | 0.91 | 0.90 | 0.90 | 0.89 |
| Important buildings and structures such as hospitals, buildings used for communication purposes, communication towers, power plant structures | 100 | 1.05 | 1.06 | 1.07 | 1.07 | 1.08 | 1.08 |

[a]$k_1 = 1$ corresponds to a probability of 0.02 (2%) that the design wind speed will be exceeded once within 50 years.

*Note:* For temporary sheds, formwork, and false work, see Table 1 of IS 875 (Part 3): 1987.

**TABLE 3.26** Values of coefficients A and B to be used in Eqn (3.39)

| Zone | A | B |
|---|---|---|
| 33 m/s | 83.2 | 9.2 |
| 39 m/s | 84.0 | 14.0 |
| 44 m/s | 88.0 | 18.0 |
| 47 m/s | 88.0 | 20.5 |
| 50 m/s | 88.8 | 22.8 |
| 55 m/s | 90.8 | 27.3 |

Values of coefficients $A$ and $B$ for different basic wind speed zones are given in Table 3.26.

### Terrain and Height Factor ($k_2$)

The following terrains are considered by the code (photographs to explain the terrain categories are also given in the draft IS 875 and Cook 1985).

**Category 1** Open terrain with few or no obstructions, which have any objects surrounding the structure with heights less than 1.5 m (e.g., open sea coasts, flat treeless plains, etc.).

**Category 2** Open terrain with scattered obstructions, which have heights in the range of 1.5–10 m above ground level (e.g., open land adjacent to a sea coast, undeveloped sparsely built outskirts of towns and suburbs, airfields, and open parklands).

**Category 3** Terrain with closely spaced buildings with heights up to 10 m with or without isolated tall buildings (e.g., well-wooded areas, towns, and industrial areas fully or partially developed).

**Category 4** Terrain with a large number of high and closely spaced buildings (e.g., city centres with heights above 25 m, fully developed areas).

While selecting the terrain category, the designer should consider the following.

(a) A structure may lie in two different types of terrain for two different wind directions.
(b) The future development of the surrounding area, which may alter the ground roughness and hence the terrain category.

IS 875 (Part 3) gives the values of $k_2$ (see Table 3.27) by which the basic wind speed has to be multiplied to obtain the wind speed at different heights in each terrain category. It has to be noted that the velocity profile for a given terrain category does not develop to full height immediately with the commencement of that terrain category but changes gradually in a terrain over a *fetch distance. Fetch* is the distance at which the wind profile stabilizes for a given terrain. The relation between fetch and developed height is given in Table 3.28. A method to calculate the velocity profile for structures exceeding the developed height given in Table 3.28 is given in Appendix B of the code.

**TABLE 3.27** $k_2$ factors to obtain design wind-speed variation with height in different terrains

| Height (m) | Terrain category 1 | Terrain category 2 | Terrain category 3 | Terrain category 4 |
|---|---|---|---|---|
| 10 | 1.05 | 1.00 | 0.91 | 0.80 |
| 20 | 1.12 | 1.07 | 1.01 | 0.80 |
| 50 | 1.20 | 1.17 | 1.12 | 1.10 |
| 100 | 1.26 | 1.24 | 1.20 | 1.20 |
| 200 | 1.32 | 1.30 | 1.27 | 1.27 |
| 300 | 1.35 | 1.34 | 1.31 | 1.30 |
| 400 | 1.38 | 1.37 | 1.34 | 1.32 |
| 500 | 1.40 | 1.39 | 1.36 | 1.34 |

*Notes:* 1. For intermediate values of height, linear interpolation may be used.
2. For other heights see Table 2 of IS875 (Part 3): 1987.

**TABLE 3.28** Fetch and developed height relationship

| Fetch ($x$) (km) | Developed height, $h_x$ (m) | | | |
|---|---|---|---|---|
| | Terrain category 1 | Terrain category 2 | Terrain category 3 | Terrain category 4 |
| 0.2 | 12 | 20 | 35 | 60 |
| 0.5 | 20 | 30 | 55 | 95 |
| 1 | 25 | 45 | 80 | 130 |
| 2 | 35 | 65 | 110 | 190 |
| 5 | 60 | 100 | 170 | 300 |
| 10 | 80 | 140 | 250 | 450 |
| 20 | 120 | 200 | 350 | 500 |
| 50 | 180 | 300 | 400 | 500 |

### Topography Factor ($k_3$)

The basic wind speed $V_b$, given in Fig. 3.30, does not take into account local topographic features such as hills, valleys, cliffs, escarpments, or ridges, which can significantly affect wind speed in their vicinity. (See Fig. 3.13.) The effect of topography is to accelerate wind near the summits of hills or crests of cliffs, escarpments, or ridges, and decelerate the wind in valleys or the near the foot of cliffs, steep escarpments, or ridges. The value of $k_3$ for level ground or when the upwind slope is less than 3° may be taken as 1.0. For slopes greater than 3°, it is confined in the range 1.0–1.36, and is given by

$$k_3 = 1 + Cs \tag{3.38a}$$

where $$C = 1.2\frac{Z}{L}$$

for $$\theta = 3° - 17° \tag{3.38b}$$
$$= 0.36 \text{ for } \theta > 17°,$$

$Z$ is the height of the crest or hill, $\theta$ is the upwind slope of the ground (see Fig. 3.31), $L$ is the length of the upward slope, and $s$ is a factor obtained from Figs 3.32 and 3.33.

The $s$ factors given in Figs 3.32 and 3.33 depend on the effective horizontal crest length, given by

$$L_e = L,\ 3° < \theta° < 17° \tag{3.39a}$$

$$L_e = \frac{Z}{0.3},\ \theta > 17° \tag{3.39b}$$

All structures lying in the affected zone ($1.5L_e$ on the windward side and $2.5L_e$ on the leeward side from the crest) should be designed using this $k_3$ factor. Note that the maximum effect is seen to occur at the crest of a cliff or escarpment and reduces gradually with distance from the crest. Also, locally, $k_3$ reduces from the base of a structure to its top.

### Importance Factor for Cyclonic Regions ($k_4$)

In order to ensure greater safety of structures located within regions 60 km wide of the east coast as well as on the Gujarat coast (where the wind speeds given in Fig. 3.30 are often exceeded (> 70 m/s during cyclones), the following values of $k_4$ are given in the draft code and IS 15498:2004.

1. For structures of post-cyclone importance (such as cyclone shelters, hospitals, schools, community buildings, communication towers, power plant structures, water tanks, etc.), $k_4 = 1.30$
2. For industrial structures, $k_4 = 1.15$
3. For all other structures, $k_4 = 1.0$
4. For non-cyclonic regions, $k_4 = 1.0$

## 3.13.2 Design Wind Pressure

Bernoulli's equation for streamline flow can be used to determine the local pressure at the stagnation point as a column of air strikes (at 90°) on an immovable body. Thus

$$p_z = 0.5\, \rho V_z^2 \tag{3.40}$$

where $p_z$ is the wind pressure in newtons per square metre at height $Z$, $\rho$ is the mass density of air (1.20 kg/m$^3$ at sea level and at 20°C, and $V_z$ is the design wind speed in metres per second at height $Z$.

Substituting the value for mass density of air, we get

$$p_z = 0.6\, V_z^2 \tag{3.40a}$$

The design wind speed is obtained as [draft IS 875 (Part 3)]

$$p_d = K_d\, K_a\, K_c\, p_z \tag{3.41}$$

where $K_d$ is the wind directionality factor, $K_a$ is the area averaging factor, and $K_c$ is the combination factor.

### Wind Directionality Factor ($K_d$)

Considering the randomness in the directionality of wind and recognizing that the pressure or force coefficients are determined for specific directions, this factor has been specified and accounts for a reduced probability of

1. maximum winds coming from any given direction, and
2. maximum pressure coefficient occurring for any given direction.

For circular, near-circular, and axisymmetric sections, which offer a uniform resistance, irrespective of the wind, the value of $K_d$ is taken as 1.0. For cyclone-affected regions also, the same value of $K_d$ is adopted. For all other buildings, solid or open signs, lattice frameworks, and trussed towers (triangular, square, and rectangular), $K_d$ is taken as 0.9.

### Area Averaging Factor ($K_a$)

Pressure coefficients are often obtained by averaging pressures over specified tributary areas. It is well known that the correlation decreases with increasing area. Hence an area averaging factor $K_a$, which is a reduction factor, is specified, as given in

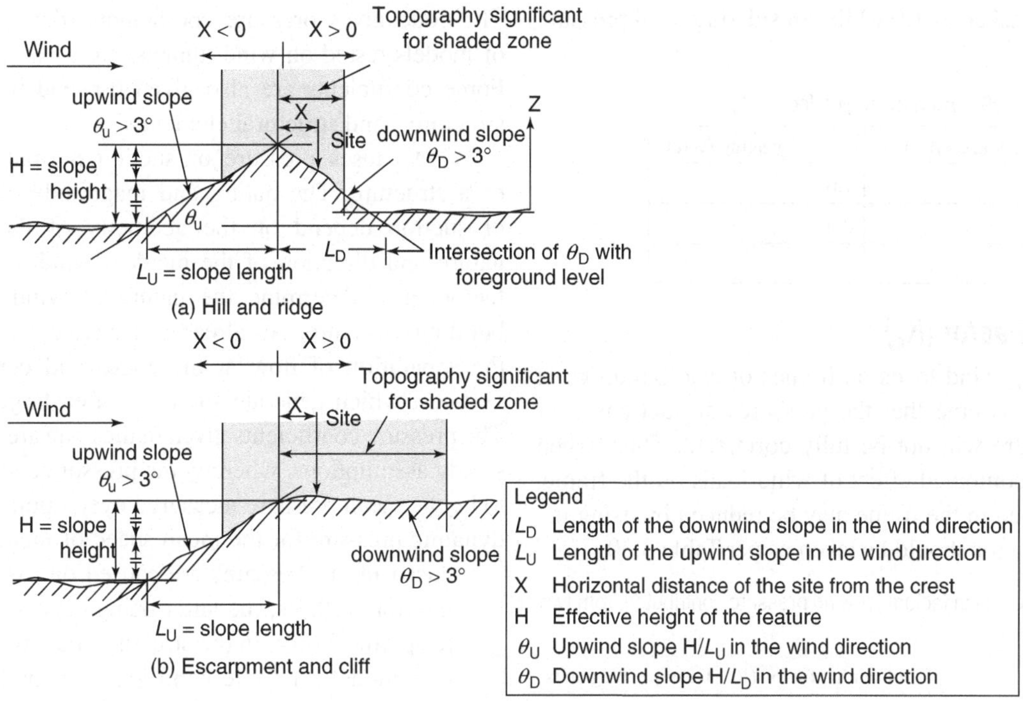


**FIG. 3.31** Definition of topography and topographic dimensions

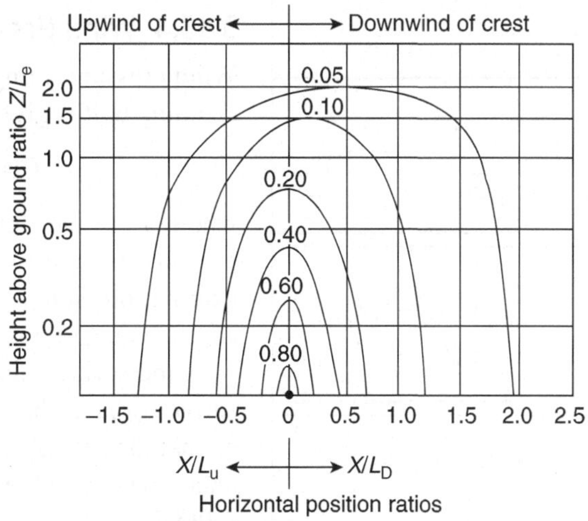


**FIG. 3.32** Topographic location factors *s* for hills and edges

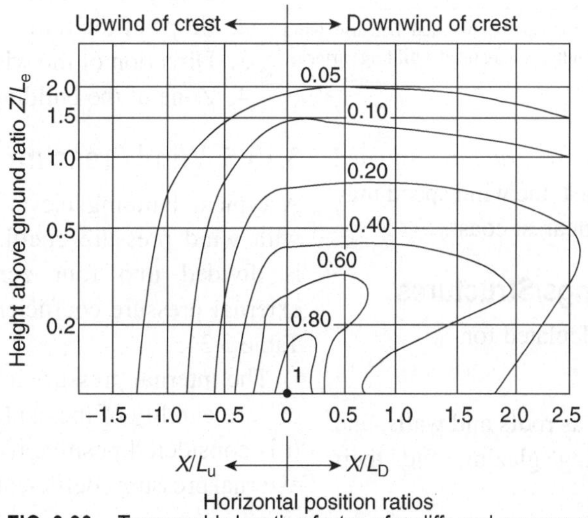


**FIG. 3.33** Topographic location factor *s* for cliffs and escarpments

Table 3.29. $K_a$ is taken as 1.0 while considering local pressure coefficients.

**TABLE 3.29** Area averaging factor, $K_a$

| Tributary area, $A$ (m$^2$) | Area averaging factor |
|---|---|
| ≤10 | 1.0 |
| 10–25 | 0.9 |
| ≥100 | 0.8 |

## *Combination Factor ($K_c$)*

While calculating wind loads on frames of clad buildings, it is reasonable to assume that the pressures or suctions over the entire structure will not be fully correlated. Thus, when considering the combined effect of wind loads on the frame, the forces obtained in the frame may be reduced by using the factor $K_a$. This factor should be taken as per Table 3.30.

**TABLE 3.30** Combination factors for wind pressure contributed from two or more building surfaces

| Design case | Combination factor ($K_c$) | Example diagrams |
|---|---|---|
| Where wind action from any single surface contributes 75% or more to an action effect | 1.0 | — |
| Pressures from windward and leeward walls in combination with positive or negative roof pressures | 0.8 | 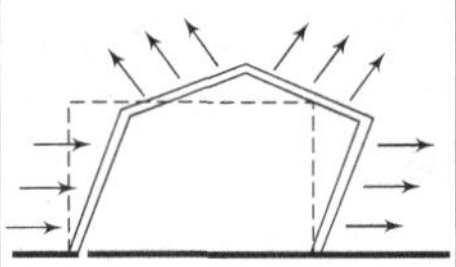 |
| Positive pressures on roofs in combination with negative internal pressures (from a wall opening) | 0.8 | 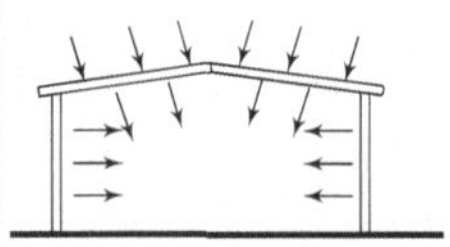 |
| Negative pressures on roofs or walls in combination with positive internal pressures | 0.95 | 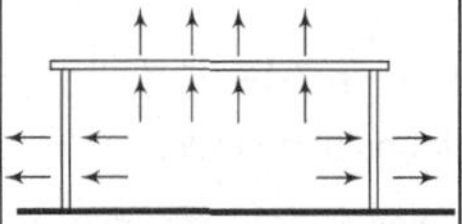 |
| All other cases | 1.0 | — |

*Note:* The action combination factors less than 1.0 can be justified because wind pressures are highly fluctuating and do not occur simultaneously on all building surfaces.

*Source:* Reprinted with permission from NICEE.

The code recommends that for *offshore structures* situated at a distance of up to 200 km off the coast, the wind speed may be taken as 1.5 times the value on the nearest coast.

### 3.13.3 Wind Pressure on Buildings/Structures

The wind load on a building can be calculated for

1. the building as a whole,
2. individual structural elements such as roofs and walls, and
3. individual cladding units including glazing and their fixings

The code gives pressure coefficients (derived on the basis of models tested on wind tunnels) for a variety of buildings. Force coefficients are also given for clad buildings, unclad structures, and structural elements.

Wind causes pressure or suction normal to the surface of a structure. The nature and magnitude of such pressure or suction depend on the geometry of the structure, the nature and direction of the incident wind and so on. These factors also determine the nature of wind flow around a building structure. As already discussed in Section 3.10.5, the separation of flow at the edges and corners results in vortices, which generate suctions, often large in magnitude. The pressure coefficients given in the code are based on quasi-steady assumptions, whereby the pressure coefficient is taken to be the mean pressure measured over a point divided by the dynamic pressure for the mean speed of incident wind (peak gust value). Pressures are caused on exterior as well as interior surfaces, the latter being dependent on openings (or permeability) in the structure, mostly in the walls. (A new model for static wind loads for buildings has been proposed recently by Chen and Kareem 2004.)

### 3.13.4 Wind Pressure on Roofs

Wind pressure acting normal to the individual element or cladding unit is given by

$$F = (C_{pe} - C_{pi})\,Ap_d \tag{3.42}$$

where $F$ is the net wind force on the element, $A$ is the surface area of the element or cladding, $C_{pe}$ is the external pressure coefficient, $C_{pi}$ is the internal pressure coefficient, and $p_d$ is the design wind pressure.

If the surface design pressure varies with height, the surface area of the structural element may be subdivided so that the specified pressures are taken over appropriate areas. The wind pressure coefficients depend on the following:

1. Shape of the building or the roof
2. Slope of the roof
3. Direction of the wind with respect to the building
4. Zone of the building

### 3.13.5 Wind Pressure on Industrial Buildings

A typical building elevation is shown in Fig. 3.34, along with wind pressure coefficients, $C_{pe}$ and $C_{pi}$. The building is divided into four zones and four local zones. The external pressure coefficients in different zones are given in Table 3.31.

The internal pressure in a building depends on the degree of permeability of the cladding and the direction of the wind. It is considered positive if acting from inside to outside. The external pressure coefficient is considered positive while acting

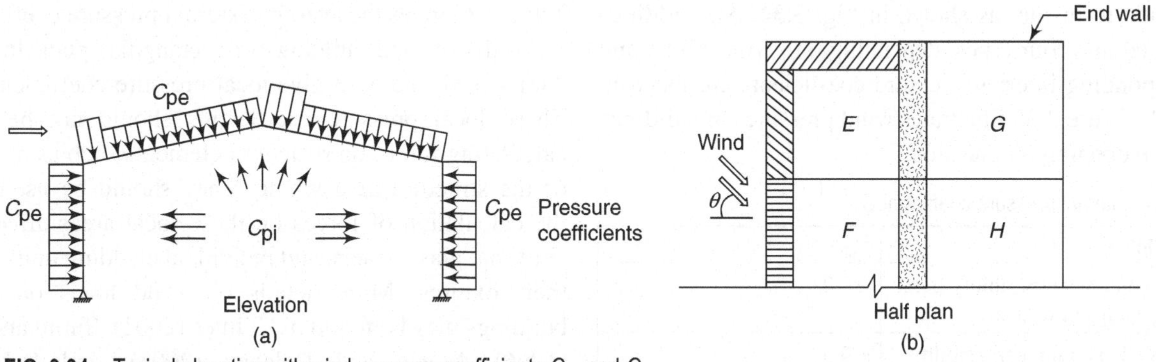


**FIG. 3.34** Typical elevation with wind pressure coefficients $C_{pe}$ and $C_{pi}$

**TABLE 3.31** External pressure coefficients ($C_{pe}$) for pitched roofs of rectangular clad buildings

| Building height ratio | Roof angle $\alpha$ degrees | Wind angle $\theta$ (0°) | | Wind angle $\theta$ (90°) | | Local coefficients | | | |
|---|---|---|---|---|---|---|---|---|---|
| | | EF | GH | EG | FH | | | | |
| | | (see key plan) | | | | | | | |
| $h/w \leq 1/2$ | 0 | –0.8 | –0.4 | –0.8 | –0.4 | –2.0 | –2.0 | –2.0 | — |
| | 5 | –0.9 | –0.4 | –0.8 | –0.4 | –1.4 | –1.2 | –1.2 | –1.0 |
| | 10 | –1.2 | –0.4 | –0.8 | –0.6 | –1.0 | –1.4 | | –1.2 |
| | 20 | –0.4 | –0.4 | –0.7 | –0.6 | –0.8 | | | –1.2 |
| | 30 | 0 | –0.4 | –0.7 | –0.6 | | | | –1.1 |
| | 45 | +0.3 | –0.5 | –0.7 | –0.6 | | | | –1.1 |
| | 60 | +0.7 | –0.6 | –0.7 | –0.6 | | | | |
| $1/2 < h/w \leq 3/2$ | 0 | –0.8 | –0.6 | –1.0 | –0.6 | –2.0 | –2.0 | –2.0 | — |
| | 5 | –0.9 | –0.6 | –0.9 | –0.6 | –2.0 | –2.0 | –1.5 | –1.0 |
| | 10 | –1.1 | –0.6 | –0.8 | –0.6 | –2.0 | –2.0 | –1.5 | –1.2 |
| | 20 | –0.7 | –0.5 | –0.8 | –0.6 | –1.5 | –1.5 | –1.5 | –1.0 |
| | 30 | –0.2 | –0.5 | –0.8 | –0.8 | –1.0 | | | –1.0 |
| | 45 | +0.2 | –0.5 | –0.8 | –0.8 | | | | |
| | 60 | +0.6 | –0.5 | –0.8 | –0.8 | | | | |

*Notes:* 1. $h$ is the height to the eaves or parapet and $w$ is the lesser horizontal dimension of a building.
2. Where no local coefficients are given, the overall coefficients apply.
3. For hipped roofs, the local coefficient for the hip ridge may be conservatively taken as the central ridge value.
4. For building height ratio 3/2 < h/w < 6 see Table 5 of IS 875 (Part 3): 1987.

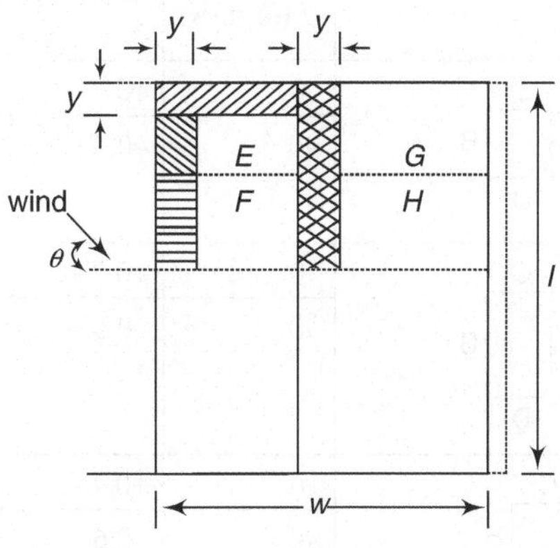


Key plan, $y = h$ or $0.15w$, whichever is less

from outside to inside, as shown in Fig. 3.34. All buildings are classified into four types, depending on permeability, and the corresponding internal pressure coefficients are listed in Table 3.32. Figure 3.35 illustrates wind pressures in buildings with a large opening on one side.

**TABLE 3.32** Internal pressure coefficient $C_{pi}$

| Type of building | $C_{pi}$ |
|---|---|
| Buildings with low permeability (less than 5% openings in wall area) | ± 0.2 |
| Buildings with medium permeability (5–20% openings in wall area) | ± 0.5 |
| Buildings with large permeability (openings in wall area > 20%) | ± 0.7 |
| Buildings with one side opening | See Fig. 3.35 |

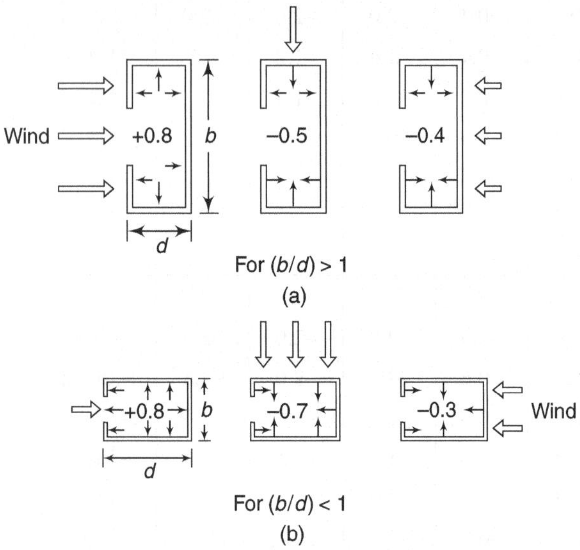


**FIG. 3.35** Internal wind pressure coefficient for buildings with one large side opening and top closed as per IS 875 (Part 3)

Table 3.33 gives the average external pressure coefficients for the walls of clad buildings of rectangular plan. In addition, Tables 3.31 and 3.33 give local pressure coefficient as well. These local pressure coefficients should not be used for calculating forces on structural elements such as roofs, walls, or the structure as a whole. They should be used only for the calculation of forces on these local areas affecting roof sheeting, glass panels, and individual cladding units including their fixtures. More details on wind loads on industrial buildings may be found in Holmes (2001), Simiu and Scanlan (1996), Mehta and Delahay (2003), and Subramanian (1990).

Note that wind directions other than 0° and 90° can give values higher than those at 0° and 90°. However, the code gives values for 0° and 90° only for simplicity in design. The code gives external coefficients for monoslope and hipped roofs (hipped roofs experience smaller suction as compared to pitched roofs of similar plan dimension and hence the code gives a relief of 20% in the pressure coefficients of hipped roofs), canopy roofs (e.g., those in open-air parking garages, railway platforms, stadiums, theatres, etc.), curved roofs, pitched and sawtooth roofs of multispan buildings, overhangs from roofs, cylindrical structures, roofs and bottoms of cylindrical structures, combined roofs, roofs with skylights, grandstands, and spherical structures.

### 3.13.6 Force Coefficients

The total wind load for a building as a whole is given by the code [IS 875 (Part 3)] as

$$F = C_f A_e p_d \tag{3.43}$$

where $F$ is the force acting in the direction specified, $C_f$ is the force coefficient of the structure, $A_e$ is the effective frontal area, and $p_d$ is the design wind pressure.

**TABLE 3.33** External pressure coefficients ($C_{pe}$) for walls of rectangular clad buildings

| Building height ratio | Building plan ratio | Elevation / Plan | Wind angle $\theta$ (degrees) | $C_{pe}$ for surface A | B | C | D | Local $C_{pe}$ |
|---|---|---|---|---|---|---|---|---|
| $h/w \le 1/2$ | $1 < l/w \le 3/2$ | 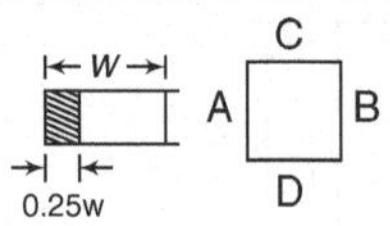 | 0 | +0.7 | −0.2 | −0.5 | −0.5 | −0.8 |
| | | | 90 | −0.5 | −0.5 | +0.7 | −0.2 | |
| | $3/2 < l/w < 4$ | 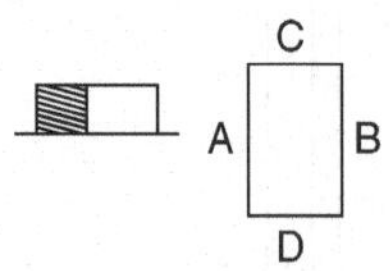 | 0 | +0.7 | −0.25 | −0.6 | −0.6 | −1.0 |
| | | | 90 | −0.5 | −0.5 | +0.7 | −0.1 | |
| $½ < h/w \le 3/2$ | $1 < l/w \le 3/2$ | 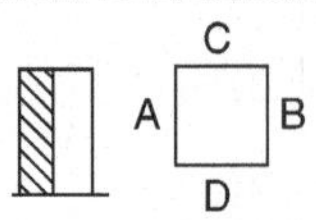 | 0 | +0.7 | −0.25 | −0.6 | −0.6 | −1.1 |
| | | | 90 | −0.6 | −0.6 | +0.7 | −0.25 | |

*(Contd)*

**TABLE 3.33** *(Contd)*

| Building height ratio | Building plan ratio | Elevation | Plan | Wind angle $\theta$ (degrees) | $C_{pe}$ for surface A | B | C | D | Local $C_{pe}$ |
|---|---|---|---|---|---|---|---|---|---|
| | $3/2 < l/w < 4$ | 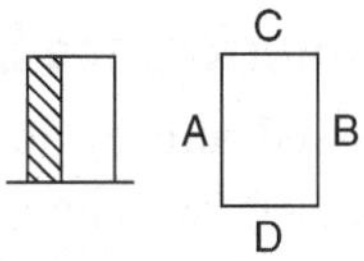 | | 0 | +0.7 | –0.3 | –0.7 | –0.7 | –1.1 |
| | | | | 90 | –0.5 | –0.5 | +0.7 | –0.1 | |
| $3/2 \le h/w < 6$ | $1 < l/w \le 3/2$ | 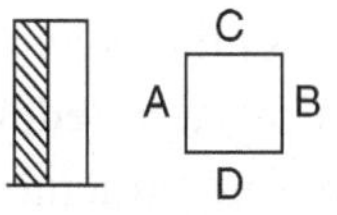 | | 0 | +0.8 | –0.25 | –0.8 | –0.8 | –1.2 |
| | | | | 90 | –0.8 | –0.8 | +0.8 | –0.25 | |
| | $3/2 < l/w < 4$ | 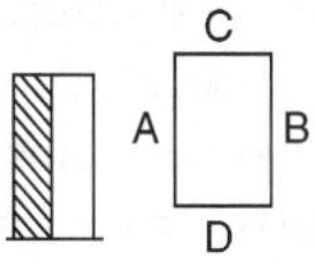 | | 0 | +0.7 | –0.4 | –0.7 | –0.7 | –1.2 |
| | | | | 90 | –0.5 | –0.5 | +0.8 | –0.1 | |

*Notes:* 1. $h$ is the height to the eaves or parapet, $l$ is the greater horizontal dimension, and $w$ is the lesser horizontal dimension of the building.
2. For building height ratio $h/w = \infty$ and $l/w = 1.5$, 1, and 2, see Table 4 of IS875 (Part 3): 1987

In order to determine the critical load, the total wind load should be calculated for each wind direction. If the surface design pressure varies with height, the surface area of the building should be subdivided so that specific pressures are taken over appropriate areas. A force taken to act in the direction of the wind is called *drag* and that in a direction perpendicular to it is called *lift*.

The code [IS 875 (Part 3)] gives force coefficients for rectangular clad buildings in uniform flow and for other clad buildings of uniform sections. Force coefficients for free-standing walls and hoardings, solid circular shapes mounted on a surface, and unclad buildings and frameworks (individual members of infinite length, flat-sided members, circular sections, wires and cables, single frames, multiple-frame buildings, and lattice towers) are also given.

## 3.13.7 Lattice Towers

*Force coefficients for latticed towers* of square or equivalent triangle section with flat sided members for wind blowing at any face are as given in Table 3.34. A comparison of the wind loads calculated as per different codes of practice and experimental results are provided by Subramanian (1991). The solidity ratio, $\phi$, is equal to the effective area (the projected area of all the individual members) of a tower (or section) normal to the wind direction divided by the area enclosed by the boundary (or section) of the tower normal to the wind direction. Such towers often taper from the base towards the top. The frontal area exposed to the wind and hence the solidity ratio changes from bottom to top of these towers. Thus, it may be necessary to divide the tower into several sections along the height and compute the forces on each part separately.

**TABLE 3.34** Overall force coefficients for towers composed of flat-sided members

| Solidity ratio $\phi$ | Force coefficient for Square towers | Equilateral triangular towers |
|---|---|---|
| 0.1 | 3.8 | 3.1 |
| 0.2 | 3.3 | 2.7 |
| 0.3 | 2.8 | 2.3 |
| 0.4 | 2.3 | 1.9 |
| 0.5 | 2.1 | 1.5 |

For square lattice towers, it is important to distinguish between 'wind onto face' and 'wind onto corner', since the former is critical for the design of bracings, and latter for that of leg members of the towers.

For square lattice towers with flat-sided members, the maximum load, which occurs when the wind blows into a corner should be taken as 1.2 times the load that occurs when the wind blows against a face. However, for equilateral-triangle lattice towers with flat-sided members, the load may be assumed to be constant for any inclination of the wind onto a face.

Force coefficients for lattice towers of square section with circular members (all in the same flow regime) are as given in Table 3.35. Force coefficients for lattice towers of equilateral-triangle section with circular members (all in the same flow regime) are as given in Table 3.36.

Note that the wind loading on tower appurtenances, such as ladders, conduits, lights, elevators, etc., should be calculated using appropriate net pressure coefficients for these elements after making allowances for shielding effect from other elements. These loads should be added to those acting on the

**TABLE 3.35** Overall force coefficients for square towers composed of rounded members

| Solidity ratio of front face, $\phi$ | Force coefficient for | | | |
|---|---|---|---|---|
| | Subcritical ($DV_d < 6$ m²/s) | | Supercritical ($DV_d \geq 6$ m²/s) | |
| | Flow onto face | Onto corner | Flow onto face | Onto corner |
| 0.05 | 2.4 | 2.5 | 1.1 | 1.2 |
| 0.1 | 2.2 | 2.3 | 1.2 | 1.3 |
| 0.2 | 1.9 | 2.1 | 1.3 | 1.6 |
| 0.3 | 1.7 | 1.9 | 1.4 | 1.6 |
| 0.4 | 1.6 | 1.9 | 1.4 | 1.6 |
| 0.5 | 1.4 | 1.9 | 1.4 | 1.6 |

$V_d$ = design wind velocity at height $Z$
$D$ = diameter of circular member

**TABLE 3.36** Overall force coefficients for equilateral triangular towers composed of rounded members

| Solidity ratio of front face, $\phi$ | Force coefficients for | |
|---|---|---|
| | Subcritical flow ($DV_d < 6$ m²/s) all wind directions | Supercritical flow ($DV_d \geq$ 6 m²/s) all wind directions |
| 0.05 | 1.8 | 0.8 |
| 0.1 | 1.7 | 0.8 |
| 0.2 | 1.6 | 1.1 |
| 0.3 | 1.5 | 1.1 |
| 0.4 | 1.5 | 1.1 |
| 0.5 | 1.4 | 1.2 |

tower body (calculated using Table 3.34, 3.35, or 3.36) and applied at the appropriate nodes during tower analysis.

## 3.13.8 Frictional Drag

In certain buildings of a special shape, a force due to frictional drag is taken into account in addition to those loads specified for rectangular clad buildings. Doing so is necessary only where the ratio $d/h$ or $d/b$ is greater than 4. The frictional drag force, $F'$, in the direction of the wind is given by the following formulae.

If $h \leq b$,

$$F' = C_f'(d-4h)\,bp_d + C_f'(d-4h)\,2\,hp_d \quad (3.44a)$$

If $h > b$,

$$F' = C_f'(d-4b)\,bp_d + C_f'(d-4b)\,2\,hp_d \quad (3.44b)$$

The first term in each case gives the drag on the roof and the second that on the walls. $C_f'$ is 0.01 for smooth surfaces without corrugations or ribs across the wind direction, 0.02 for surfaces with corrugations across the wind direction, and 0.04 for surfaces with ribs across the wind direction.

For other buildings, the frictional drag has been indicated, where necessary, in the tables of pressure coefficients and force coefficients.

## 3.13.9 Interference Effects

As discussed already in Section 3.10.6, when a structure is surrounded by others of similar size, the wind forces/ pressures acting on structures upstream/downstream are modified significantly, due to the high turbulence and the formation of large 'eddies'. The modifications in the flow affect the rigid structures of small height differently from the flexible tall structures. In general, the wind pressures get enhanced, though there can also be some shielding effect between two very closely spaced buildings/structures. In tall structures, whereas the mean along-wind response may often be reduced, the fluctuating along-wind response (dynamic component) and cross-wind response are enhanced. It is of interest to note that interfering structures sometimes bring the vortex shedding frequency close to the natural frequency of the structure, while in some other situations, they push it away from the natural frequency. In such cases, it is better to conduct detailed wind/ computational fluid dynamics (CPD) studies.

Interference is governed by a large number of geometrical parameters (such as shape, size, relative distance, and orientation between the structures involved) and flow parameters [wind direction, angle of attack, Strouhal number, or Reynolds number (in the case of rounded bodies)]. The draft IS 875 (Part 3) code gives a non-dimensional wind interference factor (IF), which should be used as a multiplying factor to be applied to the design wind pressure/force. The interference factor for roofs of low buildings is given in Fig. 3.36, which incorporates the building effect when two buildings are too close ($x < 0.5b$) and the maximum interference effect when the distance between them is $2b$ to $5b$, where $b$ is the minimum lateral dimension.

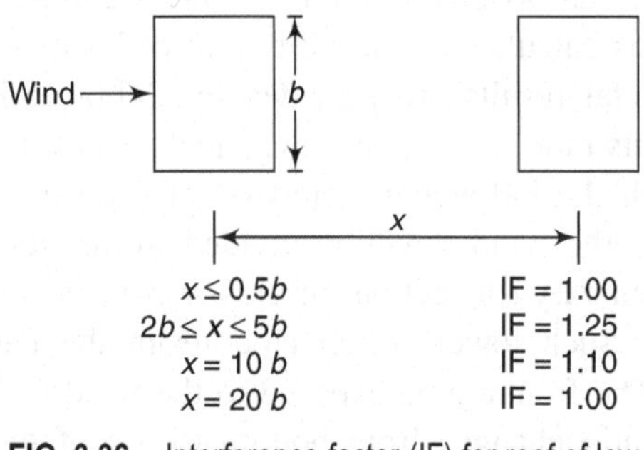


**FIG. 3.36** Interference factor (IF) for roof of low buildings

*Source:* Kumar et al. 2004, Reprinted with permission from NICEE

For tall structures, the complex phenomenon of interference has to be studied through wind tunnel studies only. However, for preliminary analysis, the interference factor (IF) can be assumed as given in Table 3.37.

**TABLE 3.37** Interference factor (IF) for tall buildings

| Zone | Z1 | Z2 | Z3 | Z4 |
|---|---|---|---|---|
| IF | 1.35 | 1.25 | 1.15 | 1.07 |

The interference effect due to buildings of height less than one-third of the height of the building under consideration may be neglected. The values given in Table 3.37 may be linearly interpolated to obtain the interference effects due to buildings having one-third to full height of the building under consideration.

Examples 3.5 to 3.9 illustrate the calculation of static wind loads on buildings and towers.

### 3.13.10 Dynamic Effects

The draft code [IS 875 (Part 3)] stipulates that the dynamic effects of winds (excitations along and across the direction of the wind) should be studied for the following types of buildings (flexible slender buildings) and closed structures:

1. Those with a height to minimum lateral dimension ratio of more than 5.0 ($h/b > 5.0$)
2. Those whose fundamental natural frequency (first mode) is less than 1.0 Hz

For these buildings, the calculated wind pressure at height $z$ should be multiplied by the dynamic response factor $C_{\text{dyn}}$ (instead of the gust factor $G$ as suggested in the 1987 version of the code). For structures whose natural frequency in the first mode is more than 1 Hz, the dynamic response factor $C_{\text{dyn}}$ may be taken as 0.9. It has also been found that the contribution from higher modes of vibration is rarely significant.

The code gives the following formulae to calculate the fundamental time period in the absence of a rational method of analysis. For moment-resistant frames without bracings or shear walls,

$$T = 0.1n \tag{3.45a}$$

where $n$ is the number of storeys, including basement storeys.

For all other structures,

$$T = \frac{0.09H}{\sqrt{d}} \tag{3.45b}$$

where $H$ is the total height of the structure in metres and $d$ is the maximum base dimension of the building in metres, in a direction parallel to the applied wind force.

Note that the formula given in Eqn. (3.45a) is different from that given in IS 1893 (see Section 3.12.3), which is more accurate and hence may be adopted for wind response also.

#### *Along-wind Response*

For the calculation of action effects (bending moments, shear forces, and member forces) at a height $s$ on the structure, the wind pressures on the structure at a height $z$ should be multiplied by the dynamic response factor, $C_{\text{dyn}}$. This factor is dependent on both $z$ and $s$, and $s < z < h$ (Fig. 3.37). For calculation of base bending moment, deflections, and acceleration at the top of the structure, a single value of $C_{\text{dyn}}$ should used with $s$ taken as zero.

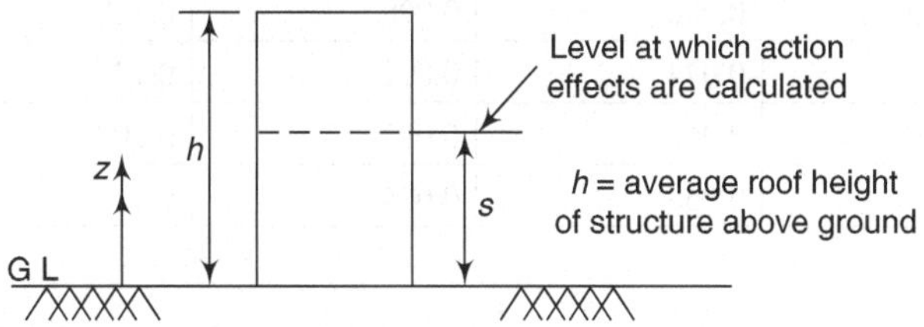


**FIG. 3.37** Notation for heights

The along-wind load on a structure on a strip area $A_e$ at any height $z$ is given by

$$F_z = C_f A_e p_z C_{\text{dyn}}$$

where $C_f$ is the force coefficient for the building, $A_e$ is the effective frontal (strip) area considered for the structure at height $z$, and $p_z$ is the wind pressure at height $z$ obtained as $0.6V_z^2$ (N/m$^2$), and $C_{\text{dyn}}$ is the dynamic response factor = total load/mean load, and is given by

$$C_{\text{dyn}} = \frac{1 + 2I_h[g_v^2 B_s + (H_s g_R^2 SE)/\beta]^{0.5}}{1 + 2g_v I_h} \tag{3.46}$$

where $I_h$ is the turbulence intensity, obtained from Table 3.38 by setting $z = h$, $g_v$ is the peak factor for the up-wind velocity fluctuations, which is taken as 3.5, and $B_s$ is the background factor, which is a measure of the slowly varying background component of the fluctuating response, caused by low-frequency wind and speed variations, and is calculated as

$$B_s = \frac{1}{1 + \dfrac{[36(h-s)^2 + 64b_{\text{sh}}^2]^{0.5}}{2L_h}} \tag{3.47}$$

$H_s$ is the height factor for the resonant response = $1 + (s/h)^2$. The peak factor for resonant response (1 hour period) is given by

$$g_R = \sqrt{[2\log_e(3600 f_0)]} \tag{3.48}$$

The size reduction factor is given by

**TABLE 3.38** Variation of turbulence intensity ($I_h$) with height for different terrain categories

| Height (z)(m) | Terrain category 1 | Terrain category 2 | Terrain category 3 | Terrain category 4 |
|---|---|---|---|---|
| 10 | 0.157 | 0.183 | 0.239 | 0.342 |
| 15 | 0.152 | 0.176 | 0.225 | 0.342 |
| 20 | 0.147 | 0.171 | 0.215 | 0.342 |
| 30 | 0.140 | 0.162 | 0.203 | 0.305 |
| 40 | 0.133 | 0.156 | 0.195 | 0.285 |
| 50 | 0.128 | 0.151 | 0.188 | 0.270 |
| 75 | 0.118 | 0.140 | 0.176 | 0.248 |
| 100 | 0.108 | 0.131 | 0.166 | 0.233 |
| 150 | 0.095 | 0.117 | 0.150 | 0.210 |
| 200 | 0.085 | 0.107 | 0.139 | 0.196 |
| 250 | 0.080 | 0.098 | 0.129 | 0.183 |
| 300 | 0.074 | 0.092 | 0.121 | 0.173 |
| 400 | 0.068 | 0.082 | 0.108 | 0.155 |
| 500 | 0.058 | 0.074 | 0.098 | 0.141 |

$$S = \frac{1}{\left[1+\dfrac{4f_0 h(1+g_v I_h)}{V_h}\right]\left[1+\dfrac{4f_0 b_{0h}(1+g_v I_h)}{V_h}\right]} \tag{3.49}$$

$E$ is ($\pi$/4) times the spectrum of turbulence in the approaching wind stream, given as follows:

$$E = \frac{\pi N}{(1+70N^2)^{5/6}} \tag{3.50}$$

$\beta$ is the ratio of structural damping to the critical damping of a structure, as given in Table 3.39, $b_{sh}$ is the average breadth of the structure between heights $s$ and $h$, $L_h$ is the measure of the integral turbulence length scale at height $h$ = 100 $(h/10)^{0.25}$, $f_0$ is the first mode natural frequency of vibration of a structure in the along-wind direction in hertz, and $b_{0h}$ = average breadth of the structure between heights 0 and $h$.

$$N = \text{reduced frequency} = \frac{f_0 L_h(1+g_v I_h)}{V_h} \tag{3.51}$$

where $V_h$ is the design wind speed at height $h$.

**TABLE 3.39** Damping coefficients for different kinds of structures

| Nature of structure | Damping coefficient ($\beta$) |
|---|---|
| Welded steel structures | 0.010 |
| Bolted steel structures | 0.020 |
| Reinforced or prestressed concrete | 0.020 |

## Cross-wind Response of Tall Buildings and Towers of Rectangular Cross section

Cross-wind response of tall structures are of three kinds: incident turbulence, cross-wind displacement, and wake excitation. Incident turbulence is important for canopy and similar roofs and bridge decks. Cross-wind displacements occur only in flexible structures with damping that is a fraction of 1% of the critical damping. Wake excitation is the most common type of cross-wind excitation and is caused by shedding of the vortices by a structure as explained in Section 3.10.5. In the following sections, methods for determining equivalent static wind force and base overturning moments in the cross-wind direction for tall buildings due to wake excitation are given as per the draft IS 875 (Part 3) code.

**Equivalent static wind force** The equivalent cross-wind static force per unit height as a function of z in newton per metre height is given by

$$W_e(z) = 0.6(V_h)^2\, d\, C_{dyn} \tag{3.52}$$

where $d$ is the lateral dimension of the structure parallel to the wind stream, and

$$C_{dyn} = \frac{1.5 g_R (b/d) K_m (z/h)^k \sqrt{\pi C_{fs}/\beta}}{(1+g_v I_n)^2} \tag{3.53}$$

where

$K_m$ = mode shape correction factor for cross-wind acceleration = 0.76 + 0.24 $k$

$k$ = mode shape power exponent for the fundamental mode of vibration
= 1.5 for uniform cantilever
= 0.5 for a slender-framed structure (moment-resistant)
= 1.0 for a building with a central core and moment-resisting façade
= 2.3 for a tower decreasing in stiffness with height, or with a large mass at the top

$C_{fs}$ = cross-wind force spectrum coefficient generalized for a linear mode shape (Figs 3.38–3.41)

$g_R$, $g_v$, $I_h$, and $\beta$ have already been defined in the previous section.

A comparative study of along-wind load effects of tall buildings as per international codes of practice is provided by Zhou et al. (2002).

**Cross-wind base overturning moment** The cross-wind base overturning moment in newton-metres is given by

$$M_0 = 0.5 g_R b\left[\frac{0.6(V_z)^2}{(1+g_y I_h)^2}\right] h^2 \left(\frac{3}{k+2}\right) K_m \sqrt{\frac{\pi C_{fs}}{\beta}} \tag{3.54}$$

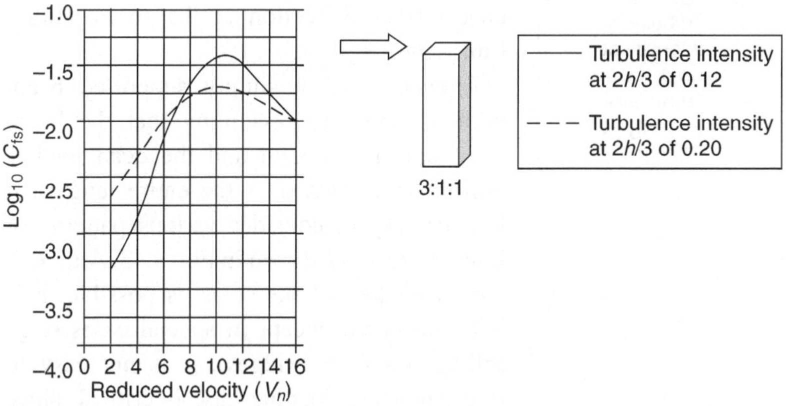


**FIG. 3.38** Cross-wind force spectrum coefficient for a 3:1:1 square section

*Source:* Kumar et al. 2004, Reprinted with permission from NICEE.

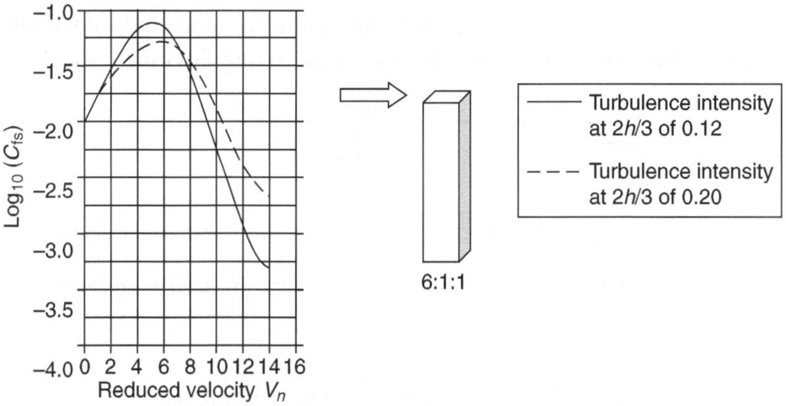


**FIG. 3.39** Cross-wind force spectrum coefficient for a 6:1:1 square section

*Source:* Kumar et al. 2004, Reprinted with permission from NICEE.

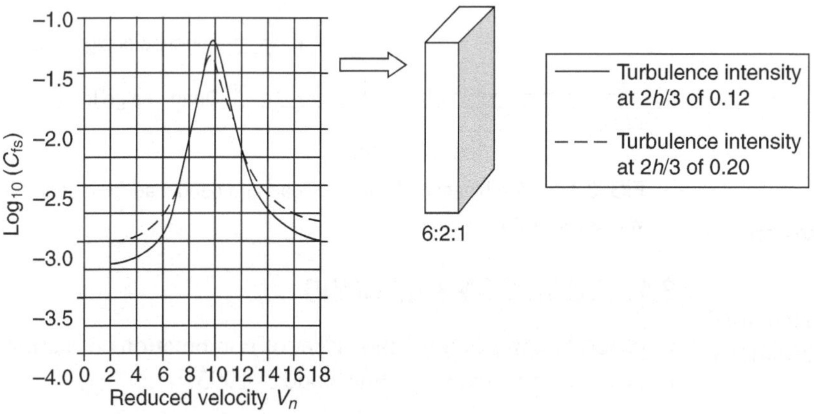


**FIG. 3.40** Cross-wind force spectrum coefficient for a 6:2:1 rectangular section

*Source:* Kumar et al. 2004, Reprinted with permission from NICEE.

where $[3/(k + 2)]K_m$ is the mode shape correction factor for cross-wind base overturning moment. Calculation of cross-wind response is not required for lattice towers.

Example 3.10 shows the calculation of wind load on a tall building using the gust factor approach.

## 3.14 PATTERN LOADING

All possible loads appropriate to a structure should be considered in design calculations. The loads should be considered separately and in realistic combinations to determine which is the most critical for the strength and stability of the structure.

For continuous structures, connected by rigid joints or continuous over supports, vertical loads should be arranged in the most unfavourable but realistic pattern for each element. The arrangement of live loads considered in the analysis will be limited to the following combinations:

1. When the loading pattern (i.e., live and dead load) is fixed, the arrangement concerned
2. Where the nominal imposed load ($Q$) is variable and does not exceed three-quarters of the nominal dead load ($G$), the design imposed load ($Q^*$) and design dead load on all spans (design imposed load is the characteristic imposed load multiplied by a appropriate partial safety factor)
3. Where the nominal live load ($Q$) is variable and exceeds three-quarters of the nominal dead load ($G$)
   (a) the design dead load ($G^*$) on all spans with full design imposed load ($Q^*$) on alternate spans [see Fig. 3.42(b)],
   (b) the design dead load ($G^*$) on all spans with full design imposed load ($Q^*$) on two adjacent spans [see Fig. 3.42(a)], and
   (c) the design dead load plus design imposed load on all spans.

It may be noted that arrangement 2 is permitted only for frames and not for continuous beams and one-way slabs. Also, it should be noted that the redistribution of moments cannot be applied in this case.

The loading arrangement 3(c) is required to find the maximum loads on columns and 3(a) and 3(b) are required to determine the maximum moments occurring in beams of rigid frames (as shown in Fig. 3.42).

The following points should be noted while considering pattern loading (Martin and Purkiss 1992).

1. Dead load factors need not be varied when considering pattern loading, but should be varied while considering stability against overturning (1.5, 1.2, and 0.9, as given in Table 2.7).

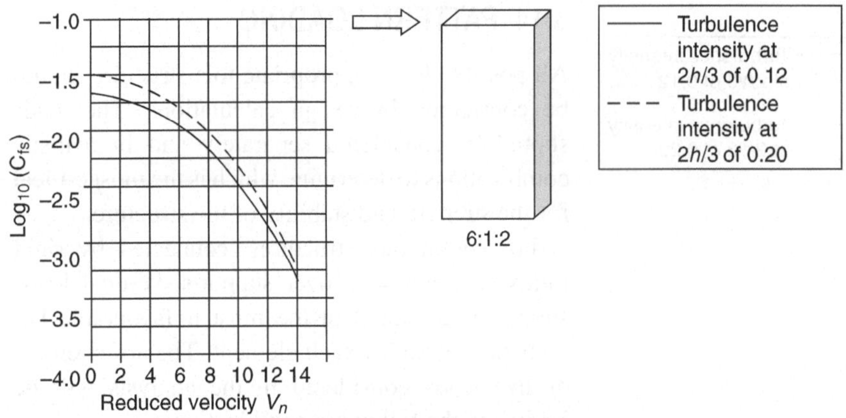


**FIG. 3.41** Cross-wind force spectrum coefficient for a 6:1:2 square section
*Source:* Kumar et al. 2004, Reprinted with permission from NICEE.

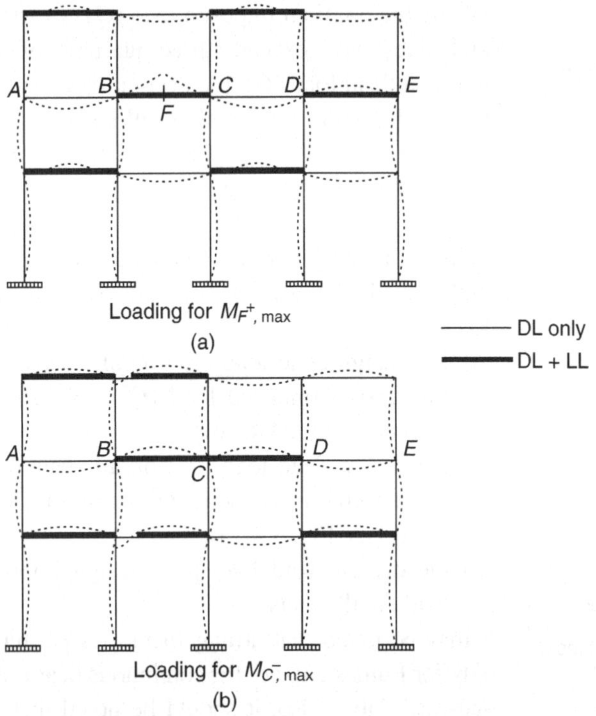


**FIG. 3.42** Gravity load patterns and the influence lines for bending moment

2. When horizontal loads are being considered (dead load + imposed load + wind or earthquake load combination), pattern loading need not be applied.
3. For dead load + wind load combination, the wind load should not be patterned except for the asymmetric loads as given in IS 875 (Parts 3).

From the above discussions, it is obvious that the critical loading condition for strength of a simply supported beam is when it supports the maximum design dead load and imposed load at the ultimate limit state. The size of the beam can be determined from the bending moment and shear derived from this loading condition, and the size of the beam should be checked for deflection at the serviceability limit state.

However, for a simply supported beam with a cantilever (assuming that the beam is of uniform section and the dead load is uniformly applied over the entire length), it is necessary to consider various patterns of imposed load, as shown in Fig. 3.43. Although partial loading of the spans is possible, it is not considered except in special cases (e.g., rolling loads on bridges). It is not easy to find out which loading case is critical, since it depends on the relative span dimensions and loads. While checking the overturning of a structure, the combination of minimum dead load, minimum live load, and maximum horizontal local (wind or earthquake) may become critical.

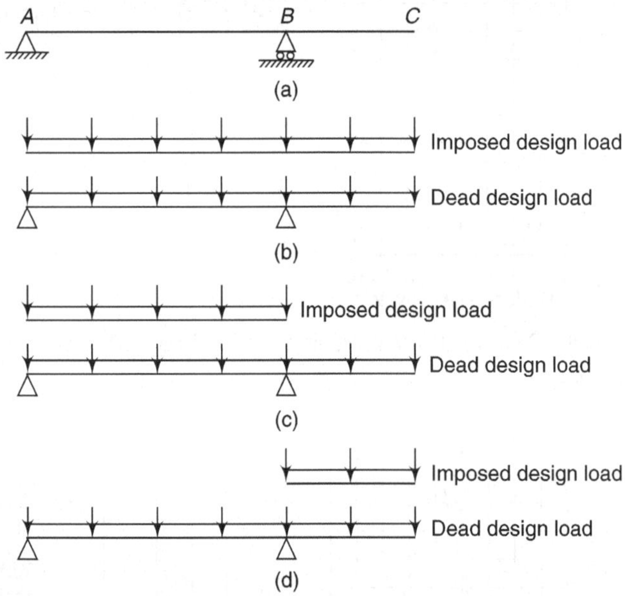


**FIG. 3.43** Combination of imposed load and dead load for a beam with a cantilever

## 3.15 LOAD COMBINATIONS

A general expression for the effect of combinations of action (load) effects is given by (Eurocode 3–1993)

$$\text{Load effect} = \sum \gamma_{G,j} G_{k,j} + \gamma_{Q,1} G_{k,1} + \sum_{i>1} \gamma_{Q,i} \psi_{0,i} Q_{k,i} \quad (3.55)$$

where $G_{k,j}$ are the characteristic values of the permanent actions, $Q_{k,1}$ is the characteristic value of one of the variable actions, $Q_{k,i}$ are the characteristic values of other variable actions, $\gamma_{G,j}$ are the partial safety factors for the permanent actions $G_{k,j}$, $\gamma_{Q,i}$ are partial safety factors for the variable actions $Q_{k,i}$, and $\psi_{0,i}$ are factors that are derived statistically using the theory of probability.

For accidental design situations,

$$\text{Load effect} = \sum_j \gamma_{\text{GA},j} G_{k,j} + A_d + \psi_{1,1} Q_{k,1} + \sum_{i>1} \psi_{2,i} Q_{k,i} \quad (3.56)$$

where $A_d$ are the design values for accidental design situations, $\gamma_{\text{GA},j}$ are partial safety factors for permanent actions along with an accidental design situation.

The basic concept is that, in any loading case, each of the actions is multiplied by the appropriate load factor, and a sum of the factored load is used in the design. When there is more than one variable action, the second variable action is multiplied by a factor, $\psi$, as in the combination formula above. This effectively reduces the loading to account for the fact that the probability of all the variable actions reaching their maximum value at the same time is very low. The IS code gives values for partial safety factors for various load combinations as shown in Table 2.7, combining $\gamma$ and $\psi$ factors together.

Thus, in general, we have the following load combinations.

1. 1.5 (DL + IL) + 1.05(CL or SL),
2. 1.2 (DL + IL) + 1.05(CL or SL) ± 0.6(WL or EL),
3. 1.2 (DL + IL ± WL or EL) + 0.53 (CL or SL),
4. 1.5 (DL ± WL or EL),
5. 0.9 DL ± 1.5 (WL or EL),
6. 1.2 (DL + ER),
7. 0.9DL + 1.2 ER, and
8. DL + 0.35(IL + CL or SL) + AL,

where DL = dead load, IL = imposed load (live load), WL = wind load, SL = snow load, CL = crane load (vertical/horizontal), AL = accidental load, ER = erection load, and EL = earthquake load. If we do not consider crane load, snow load, and accidental load we should consider the following 13 loading cases for a building in which horizontal (lateral) load is resisted by frames or walls oriented in two orthogonal directions, say *X* and *Y*. (See Fig. 3.44.)

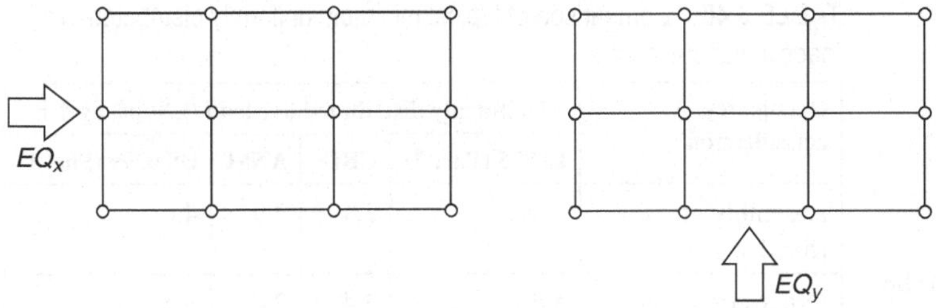


**FIG. 3.44** Earthquake loading in both directions

1. 1.5 (DL + IL)
2. 1.2 (DL + IL + $EL_x$)
3. 1.2 (DL + IL – $EL_x$)
4. 1.2 (DL + IL + $EL_y$)
5. 1.2 (DL + IL – $EL_y$)
6. 1.5 (DL + $EL_x$)
7. 1.5 (DL – $EL_x$)
8. 1.5 (DL + $EL_y$)
9. 1.5 (DL – $EL_y$)
10. 0.9 DL + 1.5$EL_x$
11. 0.9 DL – 1.5$EL_x$
12. 0.9 DL + 1.5 $EL_y$
13. 0.9 DL – 1.5$EL_y$

where $EL_x$ is the design earthquake load in the *x*-direction and $EL_y$ is the design earthquake load in the *y*-direction.

Since horizontal loads are reversible in direction, in some cases design is governed by the effect of horizontal load minus that of gravity loads. In such situations, a load factor greater than 1.0 for gravity loads will make the calculations unconservative. Hence a load factor of 0.9 is specified for gravity loads in load combinations 10–13. Many designs of footings, columns, and beams at the ends in framed structures are governed by these load combinations (Draft IS 1893 : 2005).

For microwave and other towers, where the live load is considerably small, the loading combination of 1.5 (DL + WL) governs the design. In industrial structures, dead and imposed loads on roofs are usually small, and crane loads combined with wind load are critical in most cases. In such cases, the crane load is the leading load, and has to be multiplied by 1.5, and the roof live load, etc., are treated as the accompanying live load. For the calculation of earthquake forces, only the dead load of the crane should be considered, since the suspended load is not likely to experience the inertia force due to earthquake.

Since dead loads can be estimated more precisely than live loads, several codes assume a smaller value of load factor for dead loads (e.g., 1.25 in the Canadian code, 1.4 in the American and British codes) and a higher value of load factor for imposed loads (e.g., 1.5 in the Canadian code 1.6 in AISC, and British codes). However, in the Indian code the same load factor of 1.5 is used for dead as well as imposed loads.

Since the code assumes that maximum earthquake and wind load will not occur simultaneously, in the above combinations, $EL_x$ or $EL_y$ may be replaced by $WL_x$ or $WL_y$ in places where the wind load is predominant. It may be noted that in many occasions, loading combinations 6, 7, 8, and 9 may govern the design.

When there is snow load on a roof, it should replace the imposed load on the roof in the above load combinations (also see Ellingwood and Rosowsky 1996). When imposed load is combined with earthquake load, the effect of the earthquake load should be calculated for the full dead load plus the percentage of the imposed load, as given in Table 3.23.

### 3.15.1 Load Combinations for Non-orthogonal Buildings

In structures with a non-orthogonal lateral load resisting system, the lateral load resisting elements may be oriented in a number of directions. In such buildings, considering *X*- and *Y*-direction loads acting separately, as discussed in the previous section, may be unconservative for elements not oriented in the *X*- and *Y*-directions.

A lateral load resisting system in the form of frames or walls offers maximum resistance when the load is in the direction of the element. But in non-orthogonal structures, it may be tedious to apply lateral loads in each of the directions in which the lateral load resisting elements are oriented. Hence, in such buildings, as shown in Fig. 3.45, IS 1893 suggests that the buildings be designed for the following:

1. 100% design earthquake load in the *X*-direction and 30% design earthquake load in the *Y*-direction, acting simultaneously, and
2. 100% design earthquake load in the *Y*-direction and 30% design earthquake load in the *X*-direction, acting simultaneously.

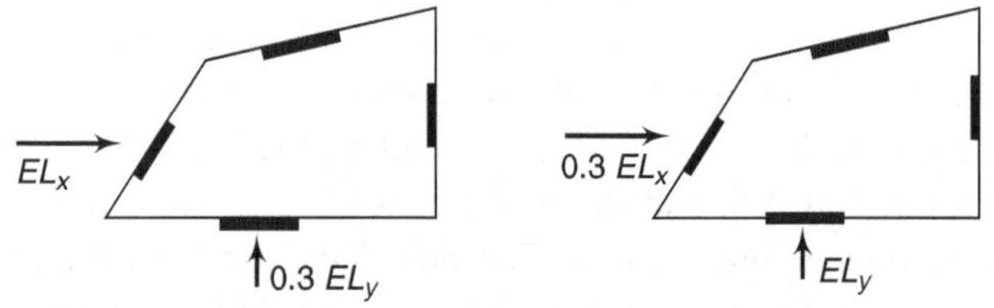


**FIG. 3.45** 100% + 30% rule for non-orthogonal lateral load resisting systems

IS 1893 suggests that the above rules should also be applied to a building which is torsionally unbalanced about both the orthogonal axes. Since the directions of earthquake forces are reversible, the following eight additional possibilities result.

1. $EL_x + 0.3\ EL_y$
2. $0.3EL_x + EL_y$
3. $EL_x - 0.3EL_y$
4. $0.3EL_x - EL_y$
5. $-(EL_x + 0.3EL_y)$
6. $(0.3EL_x + EL_y)$
7. $-(EL_x - 0.3EL_y)$
8. $-(0.3EL_x - EL_y)$

It is important to note that the 100% + 30% rule will govern the design of corner columns of buildings with orthogonal lateral load resisting systems. However, the code dispenses with this rule for orthogonal buildings to save the effort involved in design.

### 3.15.2 Combination of Three Earthquake Components

In complex structural systems (like nuclear reactors), we need to consider the effect of vertical earthquake loads also. When all the three components of an earthquake motion are considered, the responses due to each component may be combined using the assumption that when the maximum response from one component occurs, the responses from the other two components are 30% of their maximum. All possible combinations of the three components ($EL_x$, $EL_y$, and $EL_z$) including variations in signs (plus or minus) should be considered. This extension of the 100% + 30% rule results in the following 24 combinations.

1. $\pm EL_x \pm 0.3\ EL_y \pm 0.3\ EL_z$ (8 combinations)
2. $\pm EL_y \pm 0.3\ EL_x \pm 0.3\ EL_z$ (8 combinations)
3. $\pm EL_z \pm 0.3\ EL_x \pm 0.3\ EL_y$ (8 combinations)

where *X* and *Y* are two orthogonal directions and *Z* is the vertical direction. The code (IS 1893) also suggests the following SRSS method as an alternative to these combinations.

$$EL = \sqrt{(EL_x)^2 + (EL_y)^2 + (EL_z)^2}$$

However, when the SRSS method is used, the signs of the stress resultants (e.g., axial force, shear force, bending moment) of the members are lost. Hence, it may become difficult for the design engineer to assign the sign to the response quantities.

### 3.15.3 Load Combinations for Plastic Design

In the plastic design of steel structures, the following load combinations are considered.

(a) 1.7 (DL + IL),
(b) 1.7 (DL ± EL or WL),
(c) 1.3 (DL + IL ± EL or WL), and
(d) 0.9 DL ± 1.7 EL or WL.

A comparison of typical maximum uniformly distributed live loads given by different codes is given in Table 3.40.

**TABLE 3.40** Comparison of typical minimum uniformly distributed live loads in various codes

| Occupancy classification | Uniformly distributed live load (kN/m$^2$) as per | | | |
|---|---|---|---|---|
| | IS 875 (Part 2) | UBC | ANSI | BS 6399 (Part 1) |
| Assembly areas (fixed seats) | 4.0 | 2.4 | 2.9 | 4.0 |
| Offices (without separate storage) | 4.0 | 2.4 | 2.4 | 3.5 |
| Schools (classrooms) | 3.0 | 1.9 | 1.9 | 4.0 |

Since the overall factor of safety depends upon both loading and design rules, it is essential that the designer reconsiders the safety factors if the loading is taken from another code.

## EXAMPLES

**EXAMPLE 3.1:** An industrial building has been designed to resist a floor live load of 5 kN/m$^2$ [as per IS 875 (Part 2)]. However, later statistical readings were taken on similar slabs and the observed live loads in kilonewtons per square metre on various slabs are 13 × 3.2, 15 × 3.8, 35 × 4, 10 × 4.2, and 10 × 4.4. (13 × 3.2 indicates 13 slabs, each having a load of 3.2 kN/m$^2$.)

Determine the characteristic load if the accepted probability of load should not exceed 5%, assuming normal distribution.

**SOLUTION:**

The measured live loads are denoted by

$$Q_{vi} = 13 \times 3.2,\ 15 \times 3.8,\ 35 \times 4,\ 10 \times 4.2,\ \text{and } 10 \times 4.4$$

Total number of samples, $n = 13 + 15 + 35 + 10 + 10 = 83$

The mean value, $Q_{vm} = 3.91$

Standard deviation, (Eqn. 1.9) $\sigma = 0.3537$

Coefficient of variation, $C_v = \dfrac{\sigma}{Q_{vm}} = 0.09$

(The above calculations can be done using a scientific calculator.)

The characteristic live load, $Q_{vk} = Q_{vm} + 1.65\sigma$

$= 3.91 + 1.65 \times 0.3537 = 4.49\ \text{kN/m}^2$

Thus, the characteristic value that should have been used in the design is 4.49 kN/m$^2$ or 4.5 kN/m$^2$, which is less than that assumed in the design.

**EXAMPLE 3.2:** Roof trusses of 12 m span and 2.4 m rise are spaced at 4 m. The roof covering consists of ACC sheeting. Compute the ultimate bending moments for which the purlins have to be designed if the structure is situated in Shimla.

**SOLUTION:**

Length of the rafter = $\sqrt{(6^2 + 2.4^2)} = 6.462$ m

Maximum spacing of purlins = 1.4 m

(since we are using an ACC roofing).

No. of purlin spacings = $\dfrac{6.462}{1.4} = 4.61$

Provide 5 purlins, with a spacing of 6.462/5 = 1.29 m.

**Step 1:** *Dead load on purlin*

$$\text{Slope of roof} = \theta = \tan^{-1}\left(\frac{2h}{L}\right) = 21.8°$$

Weight of sheeting (as per Table 3.3) = 160 N/m$^2$

Assume the weight of fixtures, overlaps, etc., to be 50 N/m$^2$.

Total intensity $q_d = 210$ N/m$^2$

DL on purlin from sheeting, $w_s = q_d \times$ purlin spacing

$= 210 \times 1.29 = 0.271$ kN/m

Assuming ISMC 150,

Self-weight = 0.161 kN/m

Dead load, $w_d = 0.432$ kN/m

Total dead load on purlin,

$$W_d = 0.432 \times \text{span of purlin}$$
$$= 0.432 \times 4 = 1.728\ \text{kN/m}$$

**Step 2:** *Snow load on purlin*

Depth of snow load at Shimla, $d = 1.5$ m.

Ground snow load as per Eqn (3.12),

$$s_0 = (3 - 2e^{-1.5\,(1.5)})1.5 = 4.18\ \text{kN/m}^2$$

Note that, as per Eqn (3.13),

$$s_0 = 1.97h^{1.36} = 1.97\,(1.5)^{1.36} = 3.42\ \text{kN/m}^2$$

Take the value of ground snow load to be 4.18 kN/m$^2$.

Roof snow load intensity,

$$s = \mu\, s_0$$

From Fig. 3.3, for $\theta = 21.8°$,

$$\mu_1 = \mu_2 = 0.8$$

Therefore, $s = 0.8 \times 4.18 = 3.344$ kN/m$^2$

Hence, the snow load on a purlin is

$$w_s = 3.344 \times 1.29 = 4.314\ \text{kN/m}$$

The total snow load on the purlins,

$$W_s = 4.314 \times 4\cos 21.8°$$
$$= 16.02\ \text{kN}$$

The purlin may be assumed to be continuous and the negative BM at the support and positive BM at the middle of the end span are (snow load is treated as fixed load), respectively,

$$M^- = \gamma_f \frac{WL}{10} = 1.5(W_d + W_s)\, L/10$$
$$= 1.5(1.728 + 16.02) \times 4/10 = 10.65\ \text{kN/m}$$

$$M^+ = \gamma_f \frac{WL}{12} = 1.5(1.728 + 16.02) \times 4/12 = 8.874\ \text{kN/m}$$

**EXAMPLE 3.3:** A roof truss has a span of 12 m and a pitch of 2.4 m which is placed at 3.5 m c/c. Calculate the live load on the roof truss.

**SOLUTION:**

Span of the roof truss = 12 m

Pitch of the truss = 2.4 m

Spacing of truss = 3.5 m c/c

Pitch of truss, $\theta = \tan^{-1}(2.4/6) = 21.8°$

Live load for inaccessible roof = 0.75 kN/m$^2$

Live load deduction for slopes more than 10° $\Big\}$ $= (0.75 - 0.02 \times 11.8) \times \dfrac{2}{3}$

$= 0.34$ kN/m$^2$

However as per IS 875 (Part 2),

Minimum LL to be considered for design = 0.40 kN/m$^2$

Therefore, Design LL = 1.5 × 0.40 kN/m$^2$

$= 0.6$ kN/m$^2$

**EXAMPLE 3.4:** A commercial building, shown in Fig. 3.46, has seven storeys. The roof is accessible and all the floors are used as offices. Calculate the load on the interior column BC on the second floor and AB on the first floor, assuming the spacing of columns in the perpendicular direction as 4 m.

Live load on each floor = 2500 N/m$^2$

Live load on roof, with access = 1500 N/m$^2$

Assuming the thickness of the slab to be 125 mm,

Dead load = 3125 N/m$^2$

Add dead load of floor finish, etc., (say) = 1000 N/m$^2$

Total dead load = 4125 N/m$^2$

Height of each storey = 3 m

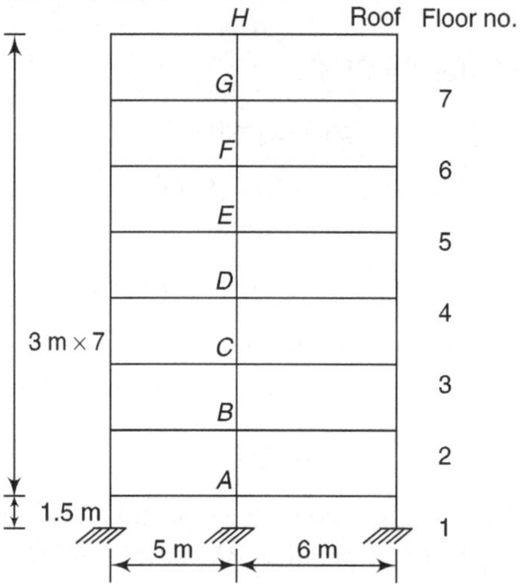


**FIG. 3.46** Live load reduction on a seven-storey building

**SOLUTION:**

Loads from the various floor levels are computed as shown in Table 3.41.

The live load has been reduced as per IS:875 (Part 2).

Area contributing load to each column = $4 \times \left(\frac{5+6}{2}\right) = 22 \text{ m}^2$

Live load on roof = 1500 N/m$^2$ = 1.5 kN/m$^2$

**TABLE 3.41**

| Column | Floor | Live load (kN) | Dead load (kN) | Load on column (kN) |
|---|---|---|---|---|
| GH | Roof | 1.5 × 22 = 33 | 4.125× 22 = 90.75 | 33 + 90.75 = 123.75 |
| FG | 7th floor | 2.5 × 22 = 55 | 4.125× 22 = 90.75 | (33 + 55) × 0.9 + 90.75 × 2 = 260.70 |
| EF | 6th floor | 2.5 × 22 = 55 | 4.125× 22 = 90.75 | (33 + 2 × 55) × 0.8 + 90.75 × 3 = 386.65 |
| DE | 5th floor | 2.5 × 22 = 55 | 4.125× 22 = 90.75 | (33 + 3 × 55) × 0.7 + 90.75 × 4 = 501.60 |
| CD | 4th floor | 2.5 × 22 = 55 | 4.125× 22 = 90.75 | (33 + 4 × 55) × 0.6 + 90.75 × 5 = 605.55 |
| BC | 3rd floor | 2.5 × 22 = 55 | 4.125× 22 = 90.75 | (33 + 5 × 55) × 0.6 + 90.75 × 6 = 729.30 |
| AB | 2nd floor | 2.5 × 22 = 55 | 4.125× 22 = 90.75 | (33 + 6 × 55) × 0.6 + 90.75 × 7 = 853.05 |

Live load on floors = 2500 N/m$^2$ = 2.5 kN/m$^2$

Design load on column BC = 1.5 × 729.30 = 1093.95 kN

Design load on column AB = 1.5 × 853.05 = 1279.58 kN

Note that if the live load reduction is not considered, the load on column AB will be 1.5(123.75 + (55 + 90.75) × 6) = 1497.375 kN

Thus there is an increase of 15% load. Also note that the dead load on the roof in a real structure may be more due to the type of weathering course adopted.

**EXAMPLE 3.5:** A rectangular industrial building situated in an industrial area is to be designed in Chennai. The building has a height of 4.5 m and a size of 10 m × 40 m. The walls of the building have 20 openings of size 1.2 m × 1.5 m. The building has a flat roof supported on load-bearing walls. [See Fig. 3.47(a).] Compute the design wind pressure and design forces on the walls and roofs of the building.

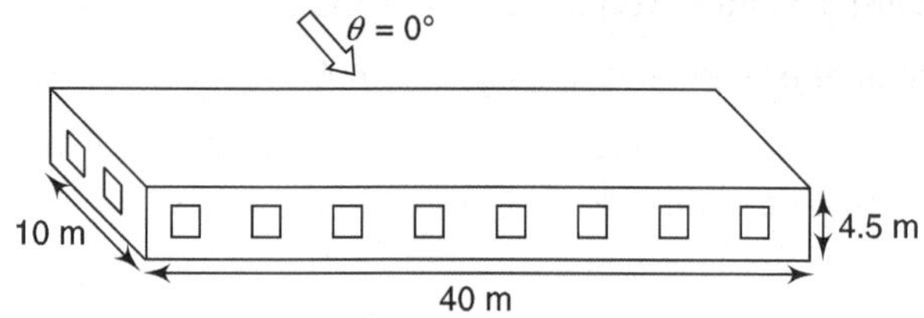


**FIG. 3.47 (a)**

**SOLUTION:**

The basic wind speed in Chennai (from wind zone map or Table 3.24),

$$V_b = 50 \text{ m/s}$$

An industrial building is to be designed for a 50-year life, so the risk coefficient from Table 3.25 is

$$k_1 = 1$$

Since the building is proposed to be erected in an industrial area in a city, it is considered to belong to category 3. The terrain factor, from Table 3.27 for a height of 4.5 m, is

$$k_2 = 0.91$$

The ground is assumed to be plain, and hence the topography factor is

$$k_3 = 1$$

Importance factor for cyclonic regions, for an industrial building,

$$k_4 = 1.15$$

Design wind speed, $V_z = V_b k_1 k_2 k_3 k_4$

$$= 50 \times 1 \times 0.91 \times 1 \times 1.15 = 52.325 \text{ m/s}$$

Wind pressure, $p_z = 0.6\ V_z^2 = 0.6(52.325)^2$

$$= 1642 \text{ N/m}^2 = 1.642 \text{ kN/m}^2$$

Wind directionality factor, $K_d = 1.0$

**Step 1:** *Area averaging factor* $K_a$

Tributary area of short walls = $10 \times 4.5 = 45\,\text{m}^2$, $K_{a1} = 0.873$

Tributary area of long walls = $40 \times 4.5 = 180\,\text{m}^2$, $K_{a2} = 0.8$

Tributary area of roof = $40 \times 10 = 400\,\text{m}^2$, $K_{a3} = 0.8$

**Step 2:** *Permeability of the building*

Area of the walls = $4.5(2 \times 10 + 2 \times 40) = 450\,\text{m}^2$

Area of all the openings = $20 \times 1.5 \times 1.2 = 36\,\text{m}^2$

% opening area = 8%, between 5% and 20%. Hence the building is of medium permeability.

Design wind pressure for short walls,

$$p_d = p_z\, k_d\, k_a = 1.642 \times 1 \times 0.873 = 1.433\,\text{kN/m}^2$$

Design wind pressure on long wall and roof

$$= 1.642 \times 1 \times 0.8 = 1.314\,\text{kN/m}^2$$

**Step 3:** *Wind load calculations*

$$F = (C_{pe} - C_{pi})\, A p_d$$

Internal pressure coefficient (Table 3.32)

$$C_{pi} = \pm 0.5$$

**Step 4:** *External pressure coefficients*

*On roof*: Using Table 3.31, with roof angle 0° without local coefficients, for $h/w$ = 0.45, the following coefficients are obtained.

**TABLE 3.42** External pressure coefficients

| Portion of roof | Wind incidence angle | |
|---|---|---|
| | 0° | 90° |
| E | −0.8 | −0.8 |
| F | −0.8 | −0.4 |
| G | −0.4 | −0.8 |
| H | −0.4 | −0.4 |

**Step 5:** *Design pressure coefficients for walls*

For $h/w = 0.45$ and $l/w = 4$, using Table 3.33, we get the values of external pressure coefficients for walls as shown in Table 3.43.

**TABLE 3.43** External pressure coefficients for walls

| Wall | Wind incidence angle | |
|---|---|---|
| | 0° | 90° |
| Wall A | +0.7 | −0.5 |
| Wall B | −0.25 | −0.5 |
| Wall C | −0.6 | +0.7 |
| Wall D | −0.6 | −0.1 |

Note that the pressure coefficients are given only for buildings with $l/w$ ratio up to 4. For longer buildings, i.e., for those with $l/w$ ratio >4.0, the values given in the table up to $l/w$ = 4.0 should be used.

The above values have to be combined with the internal pressure coefficients $C_{pi} = \pm 0.5$.

Thus,

$C_{p\,\text{net}}$ for walls A or B

$$= 0.7 - (-0.5) = +1.2 \quad \text{(pressure)}$$
$$= -0.5 - (+0.5) = -1.0 \quad \text{(suction)}$$

$C_{p\,\text{net}}$ for walls C or D [see Fig. 3.47(b)],

$$= 0.7 - (-0.5) = +1.2 \quad \text{(pressure)}$$
$$= -0.6 - (+0.5) = -1.1 \quad \text{(suction)}$$

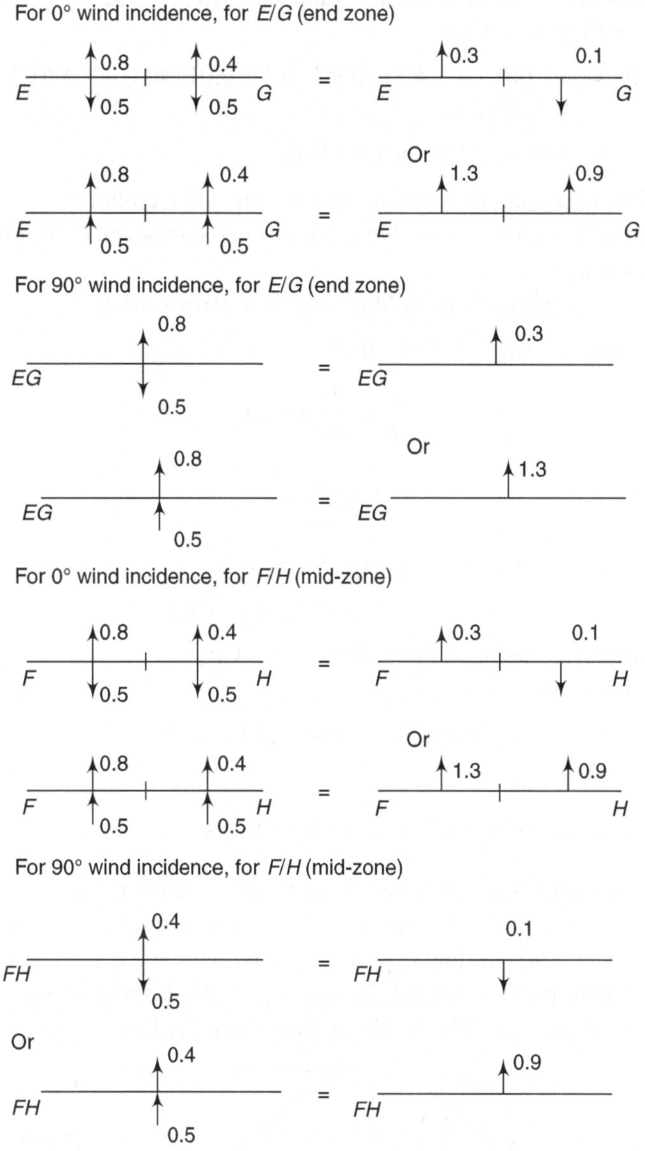


**FIG. 3.47(b)** Net roof pressure coefficients for different zones and combinations

**Step 6:** *Design pressure for walls*

For long walls,

$$F = C_{p\text{net}} \times p_d$$
$$= 1.2 \times 1.314 = 1.5768\,\text{kN/m}^2 \quad \text{(pressure)}$$
$$= -1 \times 1.314 = -1.314\,\text{kN/m}^2 \quad \text{(suction)}$$

For short walls,

$$F = 1.2 \times 1.433 = 1.7196\,\text{kN/m}^2 \quad \text{(pressure)}$$
$$= -1.1 \times 1.433 = -1.5763\,\text{kN/m}^2 \quad \text{(suction)}$$

For the roof,

$$F = 1.3 \times 1.314 = 1.7082\,\text{kN/m}^2 \quad \text{(pressure)}$$
$$= -0.1 \times 1.314 = -0.1314\,\text{kN/m}^2 \quad \text{(suction)}$$

*Calculation of force due to frictional drag:* Since 40/4.5 = 8.8 > 4.0 (even though 40/10 = 4.0), the frictional drag due to wind has to be considered. This will act in the longitudinal direction of the building along the wind. Here $h < b$, and hence Eqn (3.44a) is used.

$$F' = 0.01\,(40 - 4 \times 4.5)\,10 \times 1.314 + 0.01\,(40 - 4 \times 4.5)\,2 \times 4.5 \times 1.314$$

$$= 2.891 + 2.601 = 5.492\,\text{kN/m}^2$$

This frictional drag will act on the roof of the building

**Step 7:** *Alternate calculation using force coefficients given in the code*

Size of the building = 40 m × 10 m × 4.5 m

Therefore, $h/b = 4.5/10 = 0.45$

$$\frac{a}{b} = \frac{10}{40} = 0.25$$

and

$$\frac{b}{a} = \frac{40}{10} = 4$$

As per Fig. 4 of the code [IS 875 (Part 3)],

$$C_{f1} = 1.2,\ \ C_{f2} = 1.0$$

The force acting on the building = $C_f A_e p_a$

For 0° wind,

Force = 1.2 × (40 × 4.5) × 1.314 = 283.824 kN

For 90° wind,

Force = 1 × (10 × 4.5) × 1.314 = 59.13 kN

**EXAMPLE 3.6:** A factory is proposed to be built at Nellore on a hillock. The height of the hill is 150 m and the slope is 1 in 3. The building is proposed to be built at a distance of 100 m from the base of the hill (Fig. 3.48). Find the design wind pressure. The height of the building is 10 m.

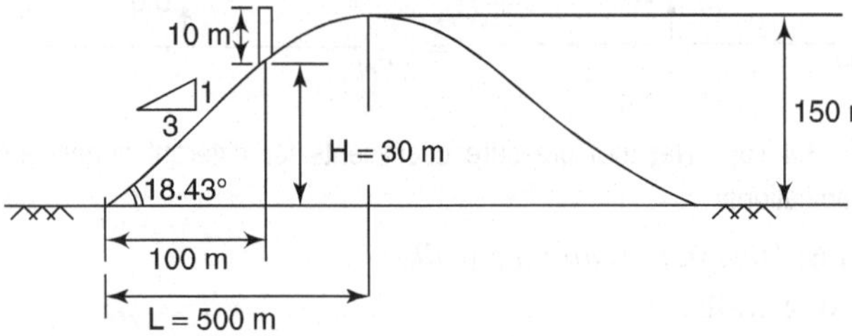


**FIG. 3.48**

**SOLUTION:**

Basic wind speed, $V_b = 50$ m/s

Risk coefficient, $k_1 = 1.00$

Terrain category 2 is assumed and the terrain coefficient, from Table 3.27 is

$$k_2 = 1.00$$

Topography factor, $k_3$:

$Z$ = effective height of the hill = 150 m

$\theta$ = 1 in 3 = $\tan^{-1}$ (1/3) = 18.43°

$L$ = length of upwind slope = 150(3) = 450 m

$L_e$ = effective horizontal crest length of the hill

= Z/0.3 for $\theta > 17°$

= 150/0.3 = 500 m

For $\theta = 18.43°$, $C = 0.36$

$X$ is the horizontal distance of the building from the crest measured considered positive towards the leeward side and negative towards the windward side.

$$X = -(L_e - 100) = -(500 - 100) = -400\text{ m}$$

$$X/L_e = -400/500 = -0.8$$

$$H = (150/500)100 = 30\text{ m}$$

$$H/L_e = 30/500 = 0.06$$

From Fig. 3.33,

$$X/L_e = -0.8 \text{ and } H/L_e = 0.06$$

$$s = 0.10$$

$$K_3 = 1 + Cs$$

$$= 1 + 0.36 \times 0.10 = 1.036$$

Design wind speed,

$$V_z = V_b k_1 k_2 k_3 k_4$$

$$= 1 \times 1.0 \times 1.036 \times 1 \times 50 = 51.8\text{ m/s}$$

Wind pressure, $p_z = 0.6\,V_z^2 = 0.6(51.8)^2 = 1610\,\text{N/m}^2$

Assuming the size of the building to be 10 m × 20 m,

$$K_a = 0.8 \text{ and } K_d = 0.9$$

Design wind pressure,

$$p_d = p_z K_a K_d$$

$$= 1610 \times 0.8 \times 0.9$$

$$= 1160\,\text{N/m}^2$$

**EXAMPLE 3.7:** An important community centre is proposed to be built on a hill top at Tirupathi. The size of the proposed building is 30 m × 60 m and its height is 5 m, as shown in Fig. 3.49(a). Calculate the design wind pressure on the building. The hill is 100 m high, with upwind and downwind slopes, respectively, of 18° and 10°. The building has 30 openings of 1.5 m × 1.5 m size. The roof is of metal sheeting and the roof angle is 15°. Calculate also the local wind pressures on the roof and wall cladding. The columns and trusses are at 5 m c/c longitudinally, the purlins are at 1.4 m c/c, and the columns at gable ends are at 5 m c/c.

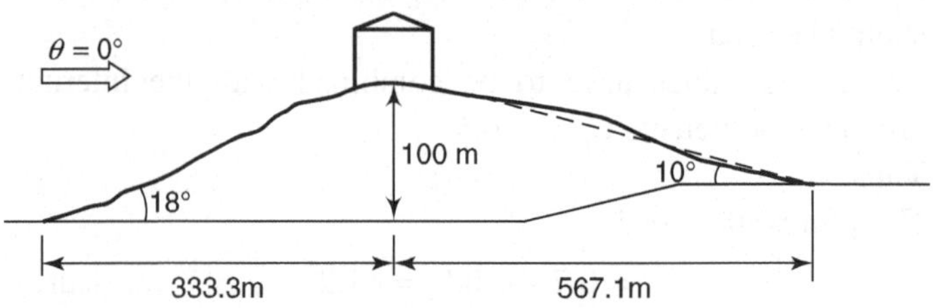


**FIG. 3.49 (a)** Size and height of proposed building

**SOLUTION:**

Basic wind velocity at Tirupathi = 39 m/s

$$k_1 = 1.06$$

A terrain category 2 can be assumed for a moderately developed area.

$$\text{Terrain factor } k_2 = 1.00$$

*Topography factor:* Let us consider the wind from the left and that from the right.

*Wind from the left:*

$$Z = 100\,\text{m}$$

$$L = 100/\tan 18° = 307.77\,\text{m}$$

$$k_3 = 1 + Cs$$

For $\theta = 18^{\circ}$, $C = 0.36$

$s$ is obtained from Fig. 3.33 for the crest position.

$$L_e = Z/0.3 = 100/0.3 = 333.3$$

$$H/L_e = 100/333.3 = 0.3$$

For $X/L = 0$ and $H/L_e = 0.3$, we get from Fig. 3.33,

$$s = 0.5$$

$$k_3 = 1 + 0.36 \times 0.5 = 1.18$$

*Wind from the right:*

$$L = 100/\tan 10° = 567.1\,\text{m}$$

$$C = 1.2\,Z/L = 1.2 \times 100/567.1 = 0.21$$

For $\theta = 10°$, $L_e = L = 567.1\,\text{m}$

$$H/L_e = 100/567.1 = 0.176$$

For $X/L = 0$ and $H/L_e = 0.176$, $s = 0.7$

$$k_3 = 1.0 + 0.21 \times 0.7 = 1.147$$

Thus, the critical value of $k_3$ is 1.18. The importance factor for cyclonic regions, $k_4$ is 1.0. Therefore, design wind speed,

$$V_z = V_b\,k_1k_2k_3k_4 = 39 \times 1.06 \times 1 \times 1.18 \times 1 = 48.78\,\text{m/s}$$

$$p_z = 0.6(V_z)^2 = 0.6(48.78)^2 = 1427.7\ \text{N/m}^2 = 1.428\,\text{kN/m}^2$$

Wind directionality factor, $K_d = 0.9$

*Area averaging factor, $K_a$:*

Tributary area for column = $5 \times 5 = 25\,\text{m}^2$, $K_{a1} = 0.9$

Tributary area for trusses = $2 \times 15.529 \times 5 = 155.3\,\text{m}^2$, $K_{a2} = 0.8$

Tributary area for purlins = $1.4 \times 5 = 7\text{m}^2$, $K_{a3} = 1.0$

Tributary area for short walls for the design of wind braces in plan = $30 \times 5 + 0.5 \times 30 \times 4.02 = 210.3\,\text{m}^2$, $K_{a4} = 0.8$

*Permeability of the building:*

Area of all the walls = $5(2 \times 30 + 2 \times 60) + 2 \times (0.5 \times 30 \times 4.02)$

$$= 1020.6\,\text{m}^2$$

$$\text{Area of all the openings} = 30 \times 1.5 \times 1.5 = 67.5\,\text{m}^2$$

$$\text{Percentage of openings} = (67.5/1020.6) \times 100 = 6.61$$

Hence the building is of medium permeability

$$p_d = p_z K_d K_a = 1.428 \times 0.9 \times K_a = 1.285\,K_a$$

For various members and components, use the appropriate value of $K_a$ as calculated above. As per the note under Section 5.3 of the draft code, since the height of the building is less than 10 m, the pressures may be reduced by 20% for calculations of stability and design of the framing. Internal pressure coefficient,

$$C_{pi} = \pm 0.5$$

*External pressure coefficients:* Using Table 3.31, $h/w = 5/30 = 0.16 < 0.5$, and $\alpha = 15°$, the pressure coefficients may be obtained as shown in Table 3.44.

**TABLE 3.44** External pressure coefficients

| Portion of roof | Wind incidence angle | |
|---|---|---|
| | 0° | 90° |
| E | −0.8 | −0.75 |
| F | −0.8 | −0.60 |
| G | −0.4 | −0.75 |
| H | −0.4 | −0.60 |

The net roof pressure coefficients are shown in Fig. 3.49(b).

*Design pressure coefficients for walls:* Referring to Table 3.33, for $h/w = 0.16 < 0.5$ and $l/w = 2.0$, $C_{pe}$ values for walls are as shown in Table 3.45.

**TABLE 3.45** $C_{pe}$ for walls

| $C_{pe}$ value for | Angle of incidence | |
|---|---|---|
| | 0° | 90° |
| Wall A | +0.7 | −0.5 |
| Wall B | −0.25 | −0.5 |
| Wall C | −0.6 | +0.7 |
| Wall D | −0.6 | −0.1 |

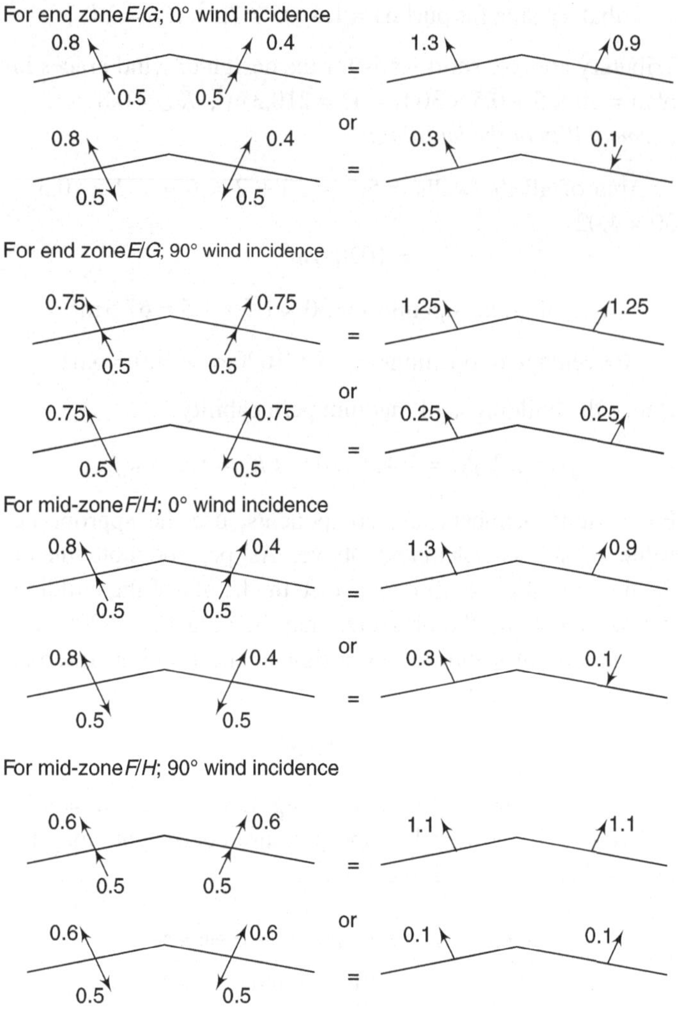


**FIG. 3.49 (b)** Net roof pressure coefficients for different zones and combination

These will have to be combined with the $C_{pe}$ value of ± 0.5. Thus,

$$C_{p\,\text{net}} \text{ for walls A or B} = 0.7 - (-0.5) = +1.2 \quad \text{(pressure)}$$
$$= 0.5 - (+0.5) = -1.0 \quad \text{(suction)}$$
$$C_{p\,\text{net}} \text{ for walls C or D} = 0.7 - (-0.5) = +1.2 \quad \text{(pressure)}$$
$$= -0.6 - (+0.5) = -1.1 \quad \text{(suction)}$$

Let us obtain the local pressure coefficients for the design of cladding and fasteners from Table 3.31 for a roof angle of 15°.

$$\left.\begin{array}{l}\text{Maximum local } C_{pe} \text{ for} \\ \text{roof at edges and ridge}\end{array}\right\} \begin{array}{l} = -1.2 - (+0.5) \\ = -1.7 \end{array}$$

To obtain the local pressure coefficients, the design pressure, $p_d$, is computed taking the values of $K_a = 1$. Therefore,

$$p_d = 1.428 \times 0.9 = 1.285\,\text{kN/m}^2$$

The zone of local coefficients is $0.15 \times 30 = 4.5$ m, or 5 m whichever is smaller. Therefore, we use the value of 4.5 m. Thus, for the edge regions as indicated in Table 3.31, for a distance of 4.5 m, the cladding and fasteners should be checked for a pressure of $1.7 \times 1.285 = 2.18\,\text{kN/m}^2$.

$$\frac{d}{h} = \frac{60}{5} = 12 \quad \text{and} \quad \frac{d}{b} = \frac{60}{30} = 2$$

Hence, frictional drag force as per Section 3.13.8 should also be considered in the design.

> **EXAMPLE 3.8:** A microwave tower of height 50 m is proposed to be built over a hill top [see Fig. 3.50]. The height of the hill is 450 m, and it has a gradient of 1 in 5. The horizontal approach distance is 1.5 km from level ground. The tower is proposed to be built at Chamba, Kashmir. Compute the design wind pressure.

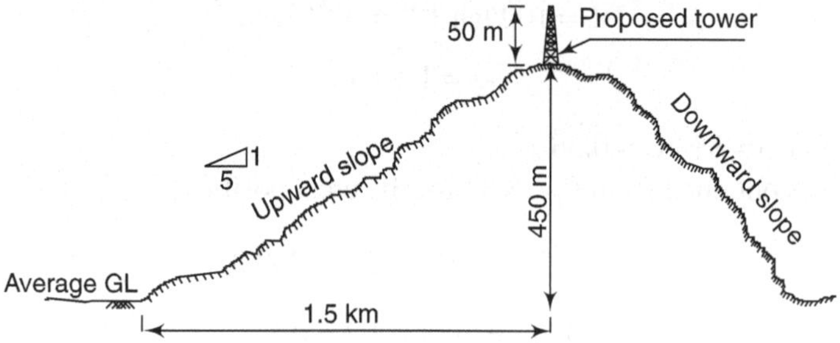


**FIG. 3.50** Wind pressure on tower on a hill

**SOLUTION:**

Basic wind speed at Chamba = 39 m/s

Risk coefficient for communication tower,

$$k_1 = 1.06$$

Assuming terrain category 1,

$$k_2 = 1.14$$

**Step 1:** *Topography factor*

$$Z = \text{effective height of hill} = 450\,\text{m}$$
$$\theta = \text{slope} = 1 \text{ in } 5 = \tan^{-1}(1/5) = 11.31°$$
$$L = \text{length of upward slope} = 1.5\,\text{km} = 1500\,\text{m}$$
$$L_e = \text{effective horizontal length of the hill}$$
$$= L \text{ for } \theta < 17° = 1500\,\text{m}$$

For $\theta < 17°$,

$$C = 1.2\,Z/L = 1.2 \times 450/1500 = 0.36$$
$$H = 500\,\text{m},\ X = 0$$
$$\frac{X}{L} = 0, \quad H/L_e = \frac{500}{1500} = 0.33$$

From Fig. 3.33, $s = 0.5$

Therefore, topography factor,

$$k_3 = 1 + 0.36 \times 0.5 = 1.18.$$

Importance factor for cyclonic regions,

$$k_4 = 1.0$$

The design wind speed is

$$V_z = k_1 k_2 k_3 V_b = 1.06 \times 1.14 \times 1.18 \times 1 \times 39 = 55.6\,\text{m/s}$$

Wind pressure, $p_z = 0.6(55.6)^2 = 1855\,\text{N/m}^2$
Design wind pressure,

$$p_d = p_z\, k_d\, k_a$$

$$K_d = 0.9$$

Assume a solidity ratio $\phi$ of 0.2.
Total area of tower = 168.75 m$^2$ (assumed)
Area on which wind impinges = $168.75 \times 0.2 = 33.75\,\text{m}^2$
Therefore, $K_a = 0.888$
Hence, $p_d = 0.9 \times 0.888 \times 1855 = 1483\,\text{N/m}^2$
For $\phi = 0.2$, for a square tower,

$$C_f = 3.3$$

from Table 3.34. Therefore,

$$\text{Wind load} = C_f A_e\, p_d$$
$$= 3.3 \times 1483 \times A_e$$
$$= 4892\, A_e\ \text{N}$$

where $A_e$ is the effective frontal area of tower. For $A_e = 33.75\,\text{m}^2$, wind load on tower = 165.12 kN

The above calculations have been made only for the purpose of guidance. For accurate calculations, the tower has to be divided into several equal parts and $A_e$ and $\phi$ have to be worked out for each of these parts. The loads so calculated have to be applied at the appropriate nodes.

**EXAMPLE 3.9:** Determine the wind pressure for a bridge (assume a life span of 100 years), of 60 m span, located at about 50 km from Madurai. Also determine the design wind force in terms of the width b of the bridge.

**SOLUTION:**
For Madurai, $V_b = 39\,\text{m/s}$

$$k_1 = 1.06$$

for a life span of 100 years.
Considering category 2, for the site located at 50 km from Madurai,

$$k_2 = 1.0$$
$$k_3 = 1.0 \text{ (level ground)}$$
$$k_4 = 1.0$$
$$V_2 = k_1 k_2 k_3 k_4 V_b$$
$$= 1.06 \times 1 \times 1 \times 1 \times 39 = 41.34\,\text{m/s}$$

Wind pressure,

$$p_z = 0.6 V_z^2 = 0.6 \times 41.34^2 = 1025\ \text{N/m}^2$$

Design wind pressure,

$$p_d = p_z\, K_a\, K_d$$
$$K_d = 0.9 \text{ (lattice framework)}$$
$$K_a = 0.9 \text{ (assumed)}$$

Design wind pressure = $1025 \times 0.9 \times 0.9 = 830\ \text{N/m}^2$
As per clause 6.3.3.2 of the IS 875 (Part 3) code, normal force,

$$F_n = C_{\text{fn}}\, p_d\, Klb$$

Transverse force,

$$F_t = C_{\text{ft}}\, p_d\, Klb$$

For $\theta = 0$ and assuming that the bridge is made of wide flanged I sections,

$$C_{\text{fn}} = 1.6 \text{ and } C_{\text{ft}} = 0$$

as per Table 26 of the code [(IS 875 (Part 3)]. Let $l/b$ = 10. Then from Table 25 of IS 875 (Part 3), $K = 0.69$. Therefore,
Wind force normal to the bridge = $1.6 \times 830 \times 0.69 \times 60 \times b/1000$
$= 54.98\, b\,\text{kN}$

Note that if the same bridge is made of rolled I sections, the coefficient $C_{\text{fn}}$ is 2.05 instead of 1.6 and hence the wind force will be increased by 28%.

**EXAMPLE 3.10:** Calculate design wind forces using the gust factor approach on a steel multi-storey building 60 m tall and of size 10 m × 20 m as shown in Fig. 3.51, to be constructed in Mumbai, about 500 m from the sea shore. Assume the average storey height to be 3.0 m. The frames are to be spaced at 5 m c/c in both directions. The building has its smaller dimension facing the sea.

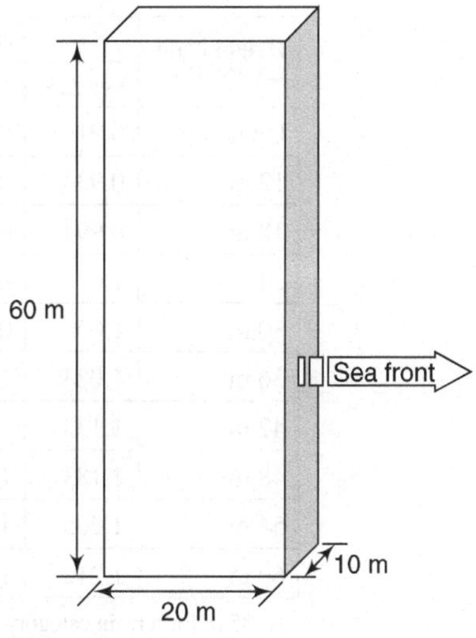


**FIG. 3.51**

**Solution:**
Height,

$$H = 60\,\text{m}$$

Least lateral dimension,

$$B = 10\,\text{m}$$

$$H/B = 60/10 = 6 > 5$$

Hence, dynamic analysis is required as per Clause 7.1 of the code. For Mumbai, the basic wind speed

$$V_b = 44\,\text{m/s}$$

One face of the building is exposed to terrain category 1, which transforms into terrain category 3 at a distance of 500 m. However, the other faces are exposed to terrain category 4, since they are located in a developed area having high-rise buildings. Hence, we have to calculate the combined wind profile as per Appendix B of IS 875 (Part 3) and clause 5.3.2.4. From Table 3.28, for fetch length $X = 0.5$ km, developed height in terrain category 3 is 35 m. Hence up to a height of 35 m, the $k_2$ factor will be as per terrain category 3 and above 35 m, it will be as per terrain category 1.

Risk coefficient $k_1 = 1.0$

The terrain factor $k_2$, is given in Table 3.46.

Topography factor $k_3 = 1.0$

Cyclonic region factor $k_4 = 1.0$

Wind directionality factor $K_d = 0.90$

Area averaging factor $K_a = 1.0$ (assumed) since the influence of the area is accounted for in the dynamic response factor.

Design wind speed $= 44 \times 1 \times k_2 \times 1 \times 1 = 44\,k_2$ m/s

$$p_z = 0.6V_z^2$$

$$p_d = p_z \times 0.9 \times 1.0 = 0.9\,p_z$$

Table 3.46 gives the values of $k_2$, $V_z$, and $p_z$.

*Wind-induced lateral forces on the structure:* These have to be calculated at every storey level for each wind direction for the three different cases in this structure.

$$F = C_f A_e p_d C_{dyn}$$

as per IS 875 (Part 3, Sec. 6.3).

*Force coefficient calculation:* Along the long side,

$$a/b = 20/10 = 2, \quad h/b = 60/10 = 6$$

Hence from Fig. 4 of IS 875 (Part 3), $C_f = 1.2$

Along the short direction,

$$a/b = 10/20 = 0.5, \quad h/b = 60/20 = 3, \; C_f = 1.35$$

*Dynamic response factor calculation* (*Along-wind*)

***(a) Wind onto wider face:***

$$H = 60\,\text{m}, \; b_{sh} = 20\,\text{m}, \; d = 10\,\text{m}$$

Assuming that the building has infilled walls, the fundamental time period,

$$T = 0.09H/\sqrt{d} = 0.09 \times 60/\sqrt{10} = 1.708\ s$$

$$f_0 = 1/T = 1/1.708 = 0.586\,\text{Hz}$$

From Table 3.39,

$$\beta = 0.02, \; L_h = 100(h/10)^{0.25} = 100(60/10)^{0.25} = 156.5\,\text{m}$$

*(i) For base floor (s = 0):*

$$B_s = \frac{1}{1 + (36 \times 60^2 + 64 \times 20^2)^{0.5}/(2 \times 156.5)} = 0.4427$$

$H_s = 1 + (0/60)^2 = 1.0,$

$V_h = 49.28$ m/s at $h = 60$ m.

**TABLE 3.46** Design wind speed and pressure

| Height from ground (m) | $k_2$ | | $V_z$ (m/s) | | $p_z$ (kN/m$^2$) | |
|---|---|---|---|---|---|---|
| | Sea face | Other faces | Sea face | Other faces | Sea face | Other faces |
| Up to 9 m | 0.91 | 0.8 | 40.04 | 35.2 | 0.962 | 0.743 |
| 12 m | 0.934 | 0.8 | 41.10 | 35.2 | 1.014 | 0.743 |
| 18 m | 0.994 | 0.8 | 43.74 | 35.2 | 1.148 | 0.743 |
| 24 m | 1.03 | 0.868 | 45.32 | 38.19 | 1.232 | 0.875 |
| 30 m | 1.06 | 0.97 | 46.64 | 42.68 | 1.305 | 1.093 |
| 36 m | 1.083* | 1.009 | 47.65 | 44.40 | 1.362 | 1.183 |
| 42 m | 1.133 | 1.048 | 49.85 | 46.11 | 1.491 | 1.276 |
| 48 m | 1.183 | 1.087 | 52.05 | 47.83 | 1.626 | 1.373 |
| 54 m | 1.205 | 1.108 | 53.02 | 48.75 | 1.687 | 1.426 |
| 60 m | 1.212 | 1.120 | 53.33 | 49.28 | 1.706 | 1.457 |

*At 35 m for terrain category 1, 1.06 + (1.12 – 1.06)35/20 = 1.075; At 36 m, 1.075 + (1.20 – 1.075) × 1/15 = 1.083

$$G_R = \sqrt{2\log_e(3600\times 0.586)} = 3.913$$

$$I_h = 0.2612$$

for terrain category 4 at height = 60 m from Table 3.38.

$$G_v = 3.5$$

$$S = \frac{1}{[1+4f_0h(1+g_vI_h)/V_h][1+4f_0b_{oh}(1+g_vI_h)/V_h]}$$

$$= 1/\{[1+4\times 0.586\times 60(1+3.5\times 0.2612)/49.28]$$
$$\times [1+4\times 0.586\times 20(1+3.5\times 0.2612)/49.28]\}$$

$$= \frac{1}{6.463\times 2.821} = 0.0548$$

$$N = \frac{f_0L_h(1+g_vI_h)}{V_h} = \frac{0.586\times 156.5(1+3.5\times 0.2612)}{49.28}$$

$$= 3.562$$

$$E = \frac{\pi N}{(1+70\mathrm{N})^{5/6}} = 0.039$$

$$C_{dyn} = \frac{1+2\times 0.2612[(3.5^2\times 0.4427)+(1\times 3.913^2 \times 0.0548\times 0.039/0.02)]^{0.5}}{1+2\times 3.5\times 0.2612}$$

$$= \frac{2.3880}{2.8284} = 0.844$$

*(ii) For floor level at mid height (s = 30 m):*

$$H_s = 1+(s/h)^2 = 1+\left(\frac{30}{60}\right)^2 = 1.25$$

$$B_s = \frac{1}{1+(36\times 30^2+64\times 20^2)^{0.5}/(2\times 156.5)} = 0.5652$$

$$C_{dyn} = 0.9067$$

*(iii) Top-floor level (s = 60 m):*

$$H_s = 2.0$$

$$B_s = \frac{1}{1+(64\times 20^2)^{0.5}/(2\times 156.5)} = 0.6617$$

$$C_{dyn} = 0.9766$$

***(b) Wind onto smaller face:***

$$H = 60\,\mathrm{m},\ d = 20\,\mathrm{m},\ b = 10\,\mathrm{m},\ L_h = 156.5\,\mathrm{m}$$

$$f_0 = \frac{1}{T} = \sqrt{20}/(0.09\times 60) = 0.828\ Hz$$

$$B = 0.02$$

*(i) For base floor (s = 0):*

$$B_s = \frac{1}{1+(36\times 60^2+64\times 10^2)^{0.5}/(2\times 156.5)} = 0.4591$$

$$H_s = 1+(0/60)^2 = 1.0,\ V_h = 53.33\,\mathrm{m/s}$$

$$g_R = \sqrt{2\log_e(3600\times 0.828)} = 4.0$$

$$I_h = 0.124$$

for terrain category 1 from Table 3.38.

$$g_v = 3.5$$

$$S = 1/([1+\{4\times 0.828\times 60(1+3.5\times 0.124)/53.33\}]$$
$$\times [1+\{4\times 0.828\times 10(1+3.5\times 0.124)/53.33\}])$$

$$= \frac{1}{6.3434\times 1.8906} = 0.0834$$

$$N = \frac{0.828\times 156.5(1+3.5\times 0.124)}{53.33} = 3.4844$$

$$E = \frac{\pi N}{(1+70\mathrm{N}^2)^{5/6}} = 0.0396$$

$$C_{dyn} = \frac{[1+2\times 0.124(3.5^2\times 0.4591+1\times 4^2\times 0.0834 \times 0.0396/0.02)^{0.5}]}{(1+2\times 3.5\times 0.124)}$$

$$= \frac{1.1713}{1.868} = 0.917$$

*(ii) Floor level at mid height (s = 30 m):*

$$H_s = 1.25,\ B_s = 0.6137,\ C_{dyn} = 0.972$$

*(iii) Top-floor level (s = 60 m):*

$$H_s = 2.0,\ B_s = 0.7964,\ C_{dyn} = 1.05$$

*For cross-wind response:* As per the Canadian code, the across-wind acceleration will exceed along-wind acceleration if $\sqrt{(WD/H)}$ is less than 1/3 where $W$ and $D$ are the dimensions of the building in the across-wind and along-wind directions. This code also gives acceptable acceleration limit for human perception as 0.5–3% of acceleration due to gravity—the lower limit for apartment buildings and the higher one for office buildings. In our case,

$$\sqrt{(WD/H)} = \sqrt{(20\times 10)/60} = 1.825 > 0.333$$

Hence, across-wind acceleration will not exceed along-wind acceleration. However, the calculation for $C_{dyn}$ is shown below for clarity of the equations presented in the draft code.

$$C_{dyn} = \frac{1.5g_R(b/d)K_m(z/h)^k\sqrt{(\pi C_{fs}/\beta)}}{(1+g_vI_h)^2}$$

$K_m = 0.76 + 0.24\,k$. For $h = 1.0$, $K_m = 1.0$.

*Wind on to wider face:*

For $Z = h$, Reduced velocity,

$$V_n = \frac{V_z}{[f_0b(1+g_vI_h)]} \quad \text{(as per Section 3.13.10)}$$

$$= \frac{49.28}{0.586\times 20(1+3.5\times 0.2612)} = 2.19\,\mathrm{m/s}$$

**TABLE 3.47** Variation of design wind pressure with height

| Height from ground (m) | $p_z$ (kN/m²) | | $p_d$ for buildings | | | $p_d$ for claddings |
|---|---|---|---|---|---|---|
| | Sea face | Other faces | Sea face, $G = 1.05$ | Opp. sea face $G = 1.05$ | Wide faces $G = 0.9766$ | |
| Up to 9 m | 0.962 | 0.743 | 1.010 | 0.780 | 0.726 | 1.010 |
| 12 m | 1.014 | 0.743 | 1.065 | 0.780 | 0.726 | 1.065 |
| 18 m | 1.148 | 0.743 | 1.205 | 0.780 | 0.726 | 1.205 |
| 24 m | 1.232 | 0.875 | 1.294 | 0.919 | 0.855 | 1.294 |
| 30 m | 1.305 | 1.093 | 1.370 | 1.148 | 1.067 | 1.370 |
| 36 m | 1.362 | 1.183 | 1.430 | 1.242 | 1.155 | 1.430 |
| 42 m | 1.491 | 1.276 | 1.566 | 1.340 | 1.246 | 1.566 |
| 48 m | 1.626 | 1.373 | 1.707 | 1.442 | 1.341 | 1.707 |
| 54 m | 1.687 | 1.426 | 1.771 | 1.497 | 1.393 | 1.771 |
| 60 m | 1.706 | 1.457 | 1.791 | 1.530 | 1.423 | 1.791 |

$I_h = 0.285$

at a height of $2h/3$.

Using Fig. 3.38, and referring to the bold line,

$\text{Log}_{10}\, C_{fs} = -3.24$, $C_{fs} = 5.75 \times 10^{-4}$

$$C_{dyn} = 1.5 \times 3.913(20/10) \times \left[\frac{1.0}{(1 + 3.5 \times 0.285)^2}\right](1)^{1.0}$$

$$\sqrt{(\pi \times 5.75 \times 10^{-4}/0.02)} = 0.883$$

*Floor-by-floor wind load:* This may be obtained by multiplying the exposed area tributary to the level by the corresponding value of design pressure at the floor height (from the above table) and by the $C_f$ values along the long side (1.2) and along the short side (1.35).

In Table 3.47, the higher wind speed has been used to calculate the pressure on the cladding on all the four faces. The designer may choose to vary it face to face. It is of interest to note that the SEI/ASCE7-10 code gives separate pressure coefficients for claddings, which are higher than those specified for the design of wind-resisting structural systems. (The building surface including the roof is divided into seven zones, and the values are given for these zones in terms of the tributary area of the cladding element exposed to the wind—the cladding elements near the boundary of the walls are subjected to more pressure.)

It has also to be noted the code does not give any guidelines for the torsional response of the building. Torsion may be introduced by (a) partial wind loading, (b) the eccentricity of the elastic centre with respect to the resultant wind load vector, and (c) the eccentricity of the elastic centre with respect to the centre of mass. Torsional effects are important for buildings taller than about 18 m. SEI/ASCE 7-10 gives more guidelines with respect to the torsional response of buildings due to wind loads.

**Example 3.11:** Consider the three-storey steel building shown in Fig. 3.52, located in Ludhiana (Seismic zone IV). Soil conditions are medium stiff and the entire building is supported on a raft foundation. The steel frames are infilled with unreinforced brick masonry. Determine the seismic load on the structure as per IS 1893. The seismic weights as shown in the figure have been calculated considering 50% of the live load lumped at the floors and no live load at roof.

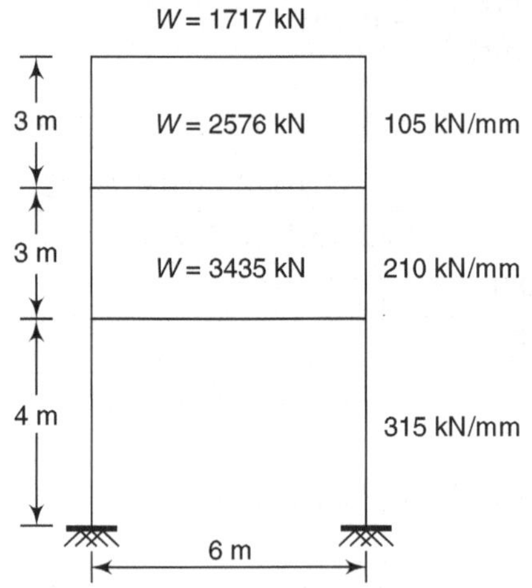


**FIG. 3.52**

**Solution:**

For seismic zone IV, the zone factor is 0.24 (Table 2 of IS 1893). Being an office building, the importance factor is 1.0 (Table 6 of IS 1893). The building has a special moment-resisting frame and hence $R = 5$.

Total seismic weight of the structure = $\Sigma W_i = 1717 + 2576 + 3435 = 7728$ kN

$$h = 4 + 3 + 3 = 10\text{m}$$

Assume the depth of building to be 15 m.

*Fundamental period:* The lateral load resistance is provided by moment-resisting frames infilled with brick masonry panels. Hence for EL in the $X$-direction,

$T = 0.09h/\sqrt{d}$ (Clause 7.6.2 of IS 1893)

$$= \frac{0.09 \times 10}{\sqrt{15}} = 0.23 \text{ s}$$

From Fig. 2 of IS 1893, for $T = 0.23$ s,

$$S_a/g = 2.5$$

$$A_h = \frac{ZI(S_a/g)}{2R} \quad \text{(Clause 6.4.2 of IS 1893)}$$

$$\frac{0.24 \times 1 \times 2.5}{2 \times 5} = 0.06$$

Design base shear, $V_B = A_h W$

$$= 0.06 \times 7728 = 463.68\,\text{kN}$$

*Force distribution with building height:* The design base shear is distributed with height as per clause 7.7.1 and the relevant calculated values are shown in Table 3.48.

**TABLE 3.48** Lateral load distribution as per static method

| Storey level | $W_i$(kN) | $h_i$(m) | $W_i h_i^2$ | $W_i h^2_i/\Sigma W_i h^2_i$ | Lateral force at *i*th level for EL in direction (kN) | |
|---|---|---|---|---|---|---|
| | | | | | *X* | *Y* |
| 3 | 1717 | 10 | 171,700 | 0.486 | 225.35 | 225.35 |
| 2 | 2576 | 7 | 126,224 | 0.358 | 166.00 | 166.00 |
| 1 | 3435 | 4 | 54,960 | 0.156 | 72.33 | 72.33 |
| Σ | 7728 | — | 352,884 | 1.000 | 463.68 | 463.68 |

*EL in Y-direction:*

$$T = \frac{0.09 \times 10}{\sqrt{6}} = 0.367\ \text{s}$$

Therefore, $\frac{S_a}{g} = 2.5$

and $A_h = 0.06$

Hence for this building, the design seismic force in the *Y*-direction is the same as that in the *X*-direction.

**EXAMPLE 3.12:** For the three-storey building shown in Fig. 3.53, the dynamic properties (natural periods and mode shapes of vibration) in the *X*-direction have been obtained by carrying out a free vibration analysis as shown in the figure. Obtain the design seismic force in the *X*-direction by the dynamic analysis (modal analysis) method, and distribute it along the building height. Assume that the building is in Ludhiana (Seismic zone IV) and is supported by medium soil. The building has a special moment-resisting frame, and hence R = 5.

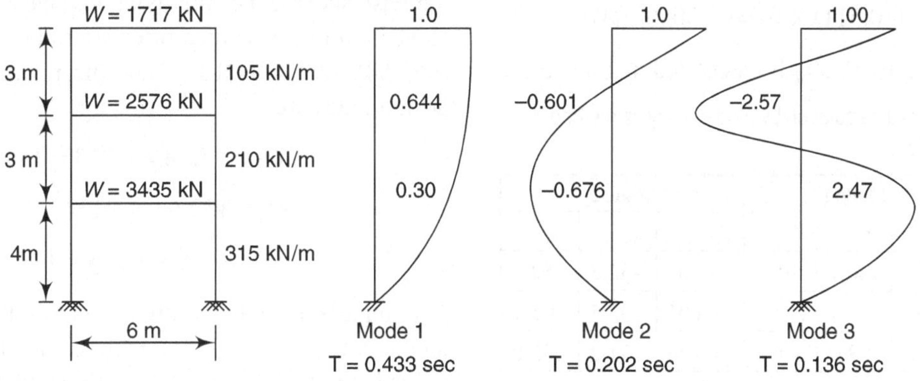


**FIG. 3.53**

**SOLUTION:**

Table 3.49 illustrates the calculation of modal mass and modal participation factors. It is seen that the first two modes exite 95.45% of the total mass, which is greater than the 90% specified by the code.

**TABLE 3.49** Calculation of modal mass and mode participation factor

| Storey level | Weight (kN) | Mode 1 | | | Mode 2 | | |
|---|---|---|---|---|---|---|---|
| | | $\phi_i$ | $W_i\phi_i$ | $W_i\phi_i^2$ | $\phi_i$ | $W_i\phi_i$ | $W_i\phi^2_i$ |
| 3 | 1717 | 1.00 | 1717 | 1717 | 1.00 | 1717 | 1717 |
| 2 | 2576 | 0.644 | 1659 | 1068 | −0.601 | −1548 | 930 |
| 1 | 3435 | 0.30 | 1031 | 309 | −0.676 | −2322 | 1570 |
| Σ | 7728 | — | 4407 | 3094 | — | −2153 | 4217 |
| $M_k = \frac{\sum(W_i\phi_i)^2}{g\sum W_i\phi_i^2}$ | | 6277.2 kN/g = 639,877 kg | | | 1099.2 kN/g = 112,051 kg | | |
| Percentage of total weight | | | 81.23 | | | 14.22 | |
| $p_k = \frac{\sum W_i\phi_i}{\sum W_i\phi_i^2}$ | | | 1.424 | | | −0.511 | |

The lateral load $Q_{ik}$ acting at $i$th floor in the $k$th mode is

$$Q_{ik} = A_h \phi_{ik} P_k W_i$$

Assuming medium soil, let us work out the values for the first two modes.

*Mode 1:*

$$T_1 = 0.433 \text{ s and } S_a/g = 2.50$$

according to clause 6.4.2. of IS 1893.

$$A_h = A_{h1} = \frac{ZI(S_a/g)}{2R} = 0.24 \times 1 \times \frac{2.5}{(2\times 5)} = 0.06$$

$$Q_{i1} = 0.06\phi_{ik} \times 1.424 \times W_i$$
$$= 0.08544\ \phi_{ik} W_i$$

*Mode 2:*

$$T_2 = 0.202 \text{ s}$$
$$S_a/g = 2.5$$

$$A_h = A_{h2} = \frac{ZI(S_a/g)}{2R} = \frac{0.24\times 1\times 2.5}{(2\times 5)} = 0.06$$

$$Q_{i2} = 0.06\phi_{ik}(-0.511) \times W_i = 0.03066\ \phi_{ik} W_i$$

The storey shear force $V_{ik} = \Sigma Q_{ik}$ is calculated in Table 3.50.

**TABLE 3.50** Lateral load calculation by modal analysis method (earthquake in *X*-direction)

| Floor level | Weight $W_i$, kN | Mode 1 | | | Mode 2 | | |
|---|---|---|---|---|---|---|---|
| | | $\phi_{i1}$ | $Q_{i1}$ | $V_{i1}$ | $\phi_{i2}$ | $Q_{i2}$ | $V_{i2}$ |
| 3 | 1717 | 1.00 | 146.7 | 146.7 | 1.00 | –52.6 | –52.6 |
| 2 | 2576 | 0.644 | 141.7 | 288.4 | –0.601 | 47.5 | –5.1 |
| 1 | 3435 | 0.30 | 88.0 | 376.4 | –0.676 | 71.2 | 66.1 |

Since the two modes are well separated (see clause 7.8.4.4 of IS 1893), the contribution of different modes can be combined by SRSS method.

$$V_3 = (146.7^2 + 52.6^2)^{0.5} = 155.8\,\text{kN}$$
$$V_2 = [288.4^2 + 5.1^2]^{0.5} = 288.4\,\text{kN and}$$
$$V_1 = [376.4^2 + 66.1^2]^{0.5} = 382.2\,\text{kN}$$

The externally applied design loads are then obtained by

$$Q_3 = V_3 = 155.8$$
$$Q_2 = V_2 - V_3 = 288.4 - 155.8 = 132.6 \text{ and}$$
$$Q_1 = V_1 - V_2 = 382.2 - 288.4 = 93.8$$

Therefore,

$$\Sigma Q = 382.2$$

Clause 7.8.2 requires that the base shear obtained by dynamic analysis ($V_B$ = 382.2 kN) be compared with that obtained from the empirical fundamental period as per Clause 7.6. If $V_B$ is less than that obtained from the empirical equation, the response quantities are to be scaled up.

The base shear for this frame as per the empirical fundamental period was calculated in Example 3.11 as 463.68 kN. Thus, the base shear as per the dynamic analysis of 382.2 is lower. Hence, all response quantities are to be scaled up in the ratio (463.68/382.2 = 1.213). Thus the revised values of applied design loads are

$$Q_3 = 155.8 \times 1.213 = 189$$
$$Q_2 = 132.6 \times 1.213 = 160.8 \text{ and}$$
$$Q_3 = 93.8 \times 1.213 = 113.8$$

A comparison of the design values for static and dynamic methods is given in Table 3.51. Notice that though the base shear is comparable for the static as well as dynamic cases, there is considerable difference in the lateral load distribution with the building height, which affects the storey moments.

**TABLE 3.51** Base shear at different levels

| Floor level | $Q$ (static) kN | $Q$ (dynamic scaled) (kN) | Storey shear, $V$ (static) (kN) | Storey shear (dynamic scaled) (kN) | Storey moment, $M$ (static) (kN m) | Storey moment, $M$ (dynamic) (kN m) |
|---|---|---|---|---|---|---|
| 3 | 225.35 | 189.10 | 225.35 | 189 | 676.05 | 567.3 |
| 2 | 166.0 | 160.8 | 391.35 | 349.8 | 1850.10 | 1616.4 |
| 1 | 72.33 | 113.8 | 463.68 | 463.6 | 3704.82 | 3471.8 |

**EXAMPLE 3.13:** Roof design loads include a dead load of 1.60 kN/m$^2$, a live load of 1.15 kN/m$^2$, and a wind pressure of 0.70 kN/m2 (upward or downward). Determine the governing loading.

**SOLUTION:**

The load combinations are

(a) 1.5 (DL + LL) = 1.5(1.6 + 1.15) = 4.125 kN/m$^2$

(b) 1.2 (DL + LL + WL) = 1.2(1.6 + 1.15 + 0.70) = 4.14 kN/m$^2$

(c) 1.2 (DL + LL – WL) = 1.2(1.6 + 1.15 – 0.70) = 2.46 kN/m$^2$

(d) 0.9DL + 1.5WL = 0.9 × 1.6 + 1.5 × 0.70 = 2.49 kN/m$^2$

(e) 0.9DL – 1.5WL = 0.9 × 1.6 – 1.5 × 0.70 = 0.39 kN/m$^2$

(f) 1.5 (DL + WL) = 1.5(1.6 + 0.70) = 3.45 kN/m$^2$

(g) 1.5 (DL – WL) = 1.5(1.6 – 0.70) = 1.35 kN/m$^2$

The second load combination governs. Hence the roof has to be designed for a total factored load of 4.14 kN/m$^2$. It may be noted that the fifth load case produces the minimum load. When the dead load is comparatively small, it will result in a negative value for the combination, which will be critical for the overturning or stability checks. Also, since it is a simple calculation, we are in a position to find the governing load combination. In a complex structural system, it may not be simple to evaluate the governing loading condition. Moreover, one loading combination may be critical for one set of members (say columns), and another combination may be critical for another set of members (say bracings). Hence, in these cases, a computer program will be quite useful to calculate the critical forces in any member due to any combination of loads.

## SUMMARY

The four main phases in a structural design process are (i) determining the structural system, (ii) calculating the various loads acting on the system, (iii) analysing the structural system for these loads, and (iv) designing the various members as per the codal provisions. Of these, determining the various loads is the most difficult and important phase, since the final design is based on them. One of the main reasons for several failures is the lack of consideration of the loads acting on the structures. Hence, in this chapter, we have briefly discussed all the loads that may act on any structure. Of the several natural and man-made loads, the following are considered important: (a) dead loads, (b) imposed loads (live and snow loads), (c) wind loads, and (d) earthquake loads. Though dead loads can be evaluated accurately, determining imposed, wind, and earthquake loads is difficult due to their uncertain probability of occurrence. Through continued research, we are now able to determine these loads fairly accurately. However, while calibrating the codal loads, some simplifying assumptions are often made. In order to ascertain the value of these loads, it is important to know their characteristics. Hence, some details about wind and earthquake loads are given, in addition to the provisions given in the Indian wind [IS 875 (Part 3)] and earthquake [IS 1893 (Part 1)] codes.

Some loads such as impact loads due to traffic on a bridge, crane loads, wind loads, and earthquake loads are dynamic in nature. However, most often they are converted to equivalent static loads. Dynamic analysis is resorted to only in the case of flexible structures, whose natural frequency in the first mode is less than 1.0 Hz or whose height to least lateral dimension ratio is more than about 5. Complicated structures should be avoided, especially in earthquake zones, since their analysis and modelling is difficult. For structures with complicated geometries, wind loading parameters should be derived from modal analysis in wind tunnels. It is very important to realize that the earthquake codes require the designer to design the structure only to a fraction of the load that may act on the building. Hence the designer has to detail the structure in such a way that, during a major earthquake, the structure may be damaged, but the occupants are able to escape on account of the ductility of the material and overstrength factors. The examples provided at the end of this chapter may clarify the concepts discussed and the references given at the end may be consulted for more details. More examples may be found in the IS explanatory handbooks on IS 1893 (Part 1), IS 875 (Part 3), and SP 64 (S & T): 2001.

Once the loads acting in the structure have been calculated, it may be analysed for the given loads. It is often analysed for the various loading combinations, using the partial load factors specified in the codes (which are again calibrated based on a large volume of data) and the stresses in various members for the worst combination of loads are determined. The next chapter discusses the various analytical methods adopted to accurately determine the axial force, shear forces, and bending moments acting in the various members of a structure.

## MULTIPLE-CHOICE QUESTIONS

3.1 Which of the following loads should be considered while designing buildings?
(a) Dead loads (c) Dynamic loads
(b) Wind loads (d) Wave loads

3.2 Which of the loads should be considered while designing bridges?
(a) Crane loads (c) Dust loads
(b) Impact loads (d) Imposed loads

3.3 Which of the loads should be considered while designing gantry girders?
(a) Snow loads (c) Longitudinal loads
(b) Impact loads (d) Blast loads

3.4 The wind load code (IS 875-Part 3) considers that the return period of wind as
(a) 50 years (c) 100 years
(b) 60 years (d) life of the structure

3.5 Partial safety factors are provided in IS 800 to take care of
(a) Variation in member sizes
(b) Inaccuracies in the assessment of loads
(c) Unfavorable deviation of loads from their characteristic values
(d) All of these

3.6 The dead load of steel is taken as
(a) 70 kN/m$^3$ (c) 85 kN/m$^3$
(b) 77 kN/m$^3$ (d) 100 kN/m$^3$

3.7 The imposed load for offices with separate storage as per IS 875-Part 2 is
(a) 2.0 kN/m$^2$ (c) 3.0 kN/m$^2$
(b) 2.5 kN/m$^2$ (d) 4.0 kN/m$^2$

3.8 The imposed load for offices without separate storage as per IS 875-Part 2 is
(a) 2.0 kN/m$^2$ (c) 3.0 kN/m$^2$
(b) 2.5 kN/m$^2$ (d) 4.0 kN/m$^2$

3.9 The imposed load for trusses with no access as per IS 875-Part 2 is
(a) 1.5 kN/m$^2$ (c) 0.75 kN/m$^2$
(b) 1.0 kN/m$^2$ (d) 0.80 kN/m$^2$

3.10 To consider the effect of partition walls, we may assume a distributed load on floor as

(a) 1.5 kN/m$^2$ (c) 0.75 kN/m$^2$
(b) 1.0 kN/m$^2$ (d) 0.80 kN/m$^2$

3.11 The effect of dynamic forces due to EOT cranes on girders may be taken as
(a) 10% of imposed load (c) 20% of imposed load
(b) 15% of imposed load (d) 25% of imposed load

3.12 Freshly fallen snow weighs up to
(a) 800 N/m$^3$ (c) 960 N/m$^3$
(b) 1000 N/m$^3$ (d) 900 N/m$^3$

3.13 When longitudinal braces are provided at the central bay, the length without expansion joints may be taken as
(a) 120 m in the case of covered buildings
(b) 150 m in the case of covered buildings
(c) 180 m in the case of covered buildings
(d) 200 m in the case of covered buildings

3.14 When longitudinal braces are provided in two bays near the center, the maximum length of building is restricted to
(a) 200 m in the case of covered buildings
(b) 210 m in the case of covered buildings
(c) 230 m in the case of covered buildings
(d) 250 m in the case of covered buildings

3.15 Windstorms are measured by using
(a) Beaufort scale
(b) Saffir-Simpson hurricane scale
(c) Hurricane severity index
(d) All of these

3.16 State whether the following are right or wrong:
(a) Wind speed variation with height is calculated using a power law.
(b) Wind velocity is measured by anemometers.
(c) As the averaging time decreases, the maximum velocity of wind increases.
(d) Maximum acceleration occurs in the along wind direction.

3.17 The crust of the earth is composed of
(a) 10 large plates and several small ones
(b) 11 large plates and several small ones
(c) 12 large plates and several small ones
(d) 13 large plates and several small ones

3.18 State whether the following are right or wrong
(a) Secondary waves cause most of the damage
(b) Earthquakes are recorded using accelerographs
(c) Magnitude of an earthquake is the measure of effects of earthquake
(d) Intensity is the measure of energy released during earthquake

3.19 For steel structures we may assume a damping of
(a) 5–7% (b) 7–10% (c) 2–3% (d) 3–5%

3.20 In IS 1893-Part 1, India is classified into
(a) 2 earthquake zones (c) 5 earthquake zones
(b) 4 earthquake zones (d) 3 earthquake zones

3.21 Percentage of imposed load considered while calculating seismic load when imposed load is less than 3.0 kN/m$^2$ is
(a) 25% (b) 30% (c) 40% (d) 50%

3.22 The approximate fundamental natural period of vibration for MRF without brick infill walls is taken in IS 1893 as (with $h$ and $d$ as height and base width of building in $m$)
(a) $T_a = 0.075h^{0.75}$ (c) $T_a = 0.09h/\sqrt{d}z$
(b) $T_a = 0.085h^{0.75}$ (d) $T_a = 0.075h^{0.85}$

3.23 The storey drift in any storey calculated with a partial load factor of 1.0, is restricted in IS 1893 to
(a) 0.004 times the storey height
(b) 0.005 times the storey height
(c) 0.003 times the storey height
(d) None of these

3.24 As per IS 15498:2004 additional wind load should be considered for structures located near the coast of
(a) 40 km wide region (c) 60 km wide region
(b) 50 km wide region (d) 80 km wide region

3.25 Dynamic wind effects should be considered for buildings with
(a) height to lateral dimension ratio (h/b) > 6 and fundamental frequency < 1.0 Hz
(b) height to lateral dimension ratio (h/b) > 6 and fundamental frequency > 1.0 Hz
(c) height to lateral dimension ratio (h/b) > 5 and fundamental frequency < 1.0 Hz
(d) height to lateral dimension ratio (h/b) > 5 and fundamental frequency > 1.0 Hz

## EXERCISES

3.1 The measurement of loads on floor slabs of residential buildings are given below in kilonewtons per square metre.
8 × 0.90, 12 × 1.1, 15 × 1.2, 30 × 1.4, 40 × 1.5, 15 × 1.6, 5 × 1.7 (8 × 0.90 means eight samples of 0.9 kN/m$^2$ each)
Determine the characteristic load on the floors if the acceptable probability of the load is not to exceed 5% of the specified load.

3.2 An office building was designed to resist a floor load of 4 kN/m$^2$. After the building was constructed, measurements were taken on the various floors as 12 × 3.2, 10 × 3.8, 20 × 4, 15 × 4.2, and 8 × 4.4.
Determine the probability of the loads exceeding the specified load of 4 kN/m$^2$.
Hint: After calculating $(Q_m - Q)/\sigma$, the probability of exceeding the specified load (Q) can be found out by referring to the statistical table for normal distribution, given in any standard book on statistics.

3.3 A roof having a span of 24 m and rise of 4 m is spaced at 4 m apart. Assuming ACC sheeting, estimate the dead and live loads on the purlins, assuming a purlin spacing of 1.40 m.

3.4 A roof truss is of 16 m span and 2.67 m rise is spaced at 5 m. Assuming GI sheeting, compute the ultimate bending moments for which the purlins have to be designed if the structure is situated in Srinagar.

3.5 A roof truss has a span of 20 m and a rise of 4 m, and is placed at 4 m c/c. Calculate the live load on the roof truss.

3.6 A six-storey building is to be used for residential purposes. Calculate the load on an interior column in the ground floor, assuming that the columns are placed in a grid of 6 m × 4 m. Consider live load reduction as per IS 875 (Part 2).

3.7 A tall building is proposed in Mumbai, where there are some existing tall buildings. Determine the design wind pressure. Use the following data:
(a) Level ground
(b) Design for a return period of 50 years
(c) Basic wind speed = 44 m/s
(d) Size of the building = 30 m × 40 m and height = 60 m
Estimate the risk, topography, terrain coefficients, and compute design wind speed and pressure.

3.8 Estimate the design wind pressure (in terms of the effective frontal area) for a 100 m high lattice tower located in the outskirts of Allahabad. Assume a solidity ratio of 0.2.

3.9 Determine the wind pressure for a bridge (assume a life of 100 years), of span 30 m located at about 30 km from Chennai. Also find the design wind force in terms of the width of the bridge.

3.10 A hospital is proposed to be built on a hill top near Bhuj. The height of the hillock above the surroundings is 400 m, and the slope is 1 in 6. The basic wind speed in Bhuj is 50 m/s. The size of the building is 60 m × 120 m, and the height is 12 m. Compute the design wind pressure on the building.

3.11 Compute the design wind pressure and design forces on the walls and roofs of a two-storey building which is 6.5 m high and has a size of 10 m × 30 m. Assume there are six openings on each floor of size 1.2 m × 1.2 m in the wall of length 30 m and two similar openings in each floor in the wall of length 10 m. The building has a flat roof and is supported on load-bearing walls.

3.12 We have the following data for a small shed.
Span of truss = 15 m (fink-type truss)
Rise of truss = 4 m
Eaves height = 7 m above GL
Spacing of truss = 4 m
Spacing of purlin = 4 m
The shed is situated in the industrial zone of Bangalore on plain land. Determine the following:
(a) the design wind speed,
(b) the design wind pressure on the building, and
(c) the wind load on one span of the purlin when the wind is parallel to the ridge, and normal to the ridge.

3.13 Calculate design wind forces using the gust-factor approach on a steel multi-storey building of size 10 m × 30 m and 51 m tall to be constructed at Chennai. Assume the average storey height as 3 m. The frames are to be spaced at 5 m c/c in both directions. The building is to be constructed about 1000 m from the sea shore and its smaller dimension is to face the sea.

3.14 Consider the four-storey office building shown below, located in Shillong (Seismic zone V). The soil conditions are medium stiff, and the entire building is supported on a raft foundation. The RC frames are infilled with brick masonry. The lumped weight due to dead loads is 12 kN/m$^2$ on the floors and 10 kN/m$^2$ on the roof. The floors carry a live load of 4 kN/m$^2$ and the roof 1.5 kN/m$^2$. Determine the design seismic load on the structure by the equivalent static method. Assume that the frames are steel moment-resisting frames with $R = 5$.

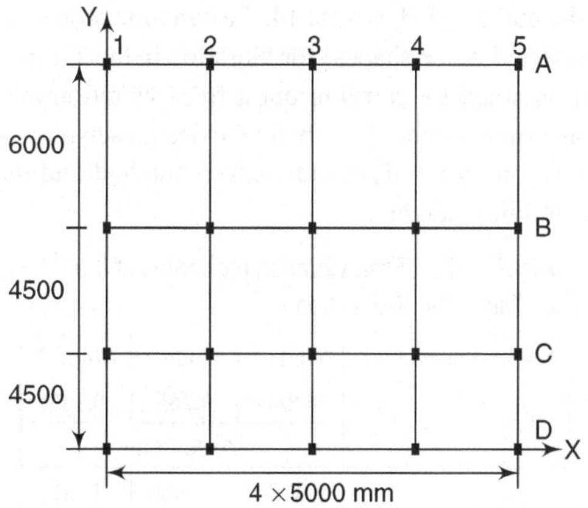

Plan

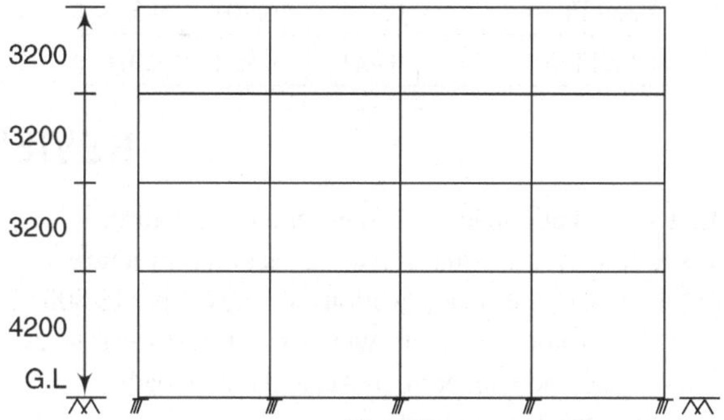

Elevation

[**Ans.** Design base shear = 1560 kN]

3.15. A three-storey building and the seismic weights acting on it are shown below. Assuming that the building is in Seismic zone IV and is supported by soft soil, determine the design seismic load on the structure by the
(a) equivalent static method and
(b) dynamic method.

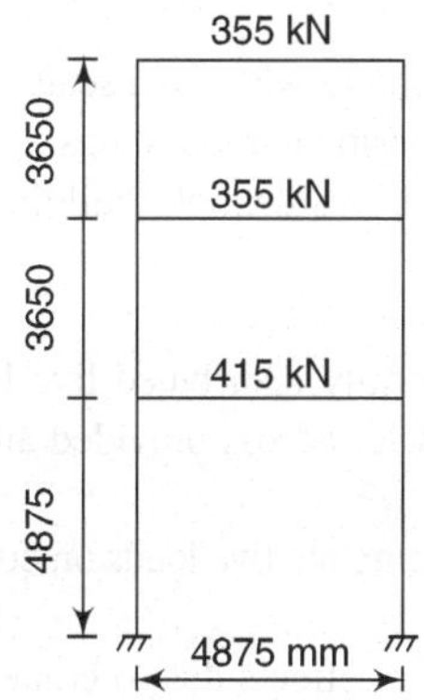


The free vibration properties of this building are shown in Table 3.52.

**TABLE 3.52** Free vibration properties

| Natural period (s) | Mode 1<br>0.114 | Mode 2<br>0.040 | Mode 3<br>0.021 |
|---|---|---|---|
| | Mode shapes | | |
| Roof | 1.000 | 1.000 | 1.000 |
| 2nd floor | 0.726 | −0.583 | −2.377 |
| 1st floor | 0.340 | −1.146 | 1.378 |

3.16 For the building in Exercise 14, the dynamic properties (natural periods and mode shapes) for vibration in the *X*-direction have been obtained by carrying out a free- vibration analysis and are shown in Table 3.53. Obtain the design seismic force in the *X*-direction by the dynamic-analysis method and distribute it with building height.

**TABLE 3.53** Free vibration properties of the building in the X-direction

| Natural period (s) | Mode 1 | Mode 2 | Mode 3 |
|---|---|---|---|
| | 0.860 | 0.265 | 0.145 |
| | Mode shape | | |
| Roof | 1.000 | 1.000 | 1.000 |
| 3rd Floor | 0.904 | 0.216 | −0.831 |
| 2nd Floor | 0.716 | −0.701 | −0.574 |
| 1st Floor | 0.441 | −0.921 | 1.016 |

(**Ans.** (a) As per dynamic analysis, $Q_1 = 185.8$ kN, $Q_2 = 285.7$ kN, $Q_3 = 369.0$ kN, $Q_4 = 314.1$ kN, and base shear = 1154.6 kN.
(b) After scaling the dynamic analysis results, $Q_1 = 233$ kN, $Q_2 = 358$ kN, $Q_3 = 463$ kN, $Q_4 = 394$ kN, and base shear = 1448 kN)

3.17 Roof design loads include a dead load of 1.75 kN/m$^2$, a live (or snow) load of 1.25 kN/m$^2$, and a wind pressure of 0.75 kN/m$^2$ (upward or downward). Determine the governing loading.

3.18 The axial forces on a building column are as follows: 250 kN dead load, 200 kN floor live load, 50 kN roof live load, and 175 kN wind load. Determine the required strength of the column.

## REVIEW QUESTIONS

3.1 List a few static loads that may act on a structure.
3.2 List a few dynamic loads that may act on a structure.
3.3 What are the three classifications of loads as per IS 800 : 2007?
3.4 List the IS code numbers which we have to use to evaluate wind, dead and live, earthquake, and crane loads.
3.5 Define characteristic load.
3.6 What is the use of partial safety factors? How are they different from the factor of safety used in the working stress method?
3.7 Is it safe to overestimate the dead load? If your answer is no, give your reasons.
3.8 What are the classifications under which imposed loads are grouped as per IS 875 (Part 2)?
3.9 Give the imposed floor loads for the following occupancies.
(a) Residential
(b) Assembly building with fixed seats
(c) Balconies in residential buildings
(d) Staircases in residential buildings and commercial buildings
(e) Cold storages
3.10 What is the uniformly distributed live load for a roof with a slope of 18° (a) with access provided and (b) with access not provided?
3.11 Why are we reducing the live loads on columns of multi-storey buildings?
3.12 How are the impact effects due to cranes normally accounted for in live loads?
3.13 How are the snow loads on roofs taken into account?
3.14 What are the parameters that influence drifted snow loads?
3.15 When do temperature effects become critical?
3.16 What are the guidelines as per IS 800 : 2007 for the provision of expansion joints?
3.17 Are loads occurring during erection critical? What are the erection loads that should be considered in design?
3.18 What kind of loads are considered as accidental loads?
3.19 List some of the abnormal loadings on structures.
3.20 Write short notes on cyclones and tornadoes.
3.21 What are the two scales adopted for measuring windstorms.
3.22 What are the parameters that affect a wind flow?
3.23 List the characteristics of a structural system that affect the behaviour of a system due to wind loads.
3.24 How is the variation of wind velocity with height expressed using the power law expression?
3.25 What are gusts? What is the averaging time used in the draft wind code?
3.26 What is the return period used in the wind code?
3.27 What is meant by vortex shedding? State the formula to calculate the shedding frequency.
3.28 What is meant by interference effect?
3.29 What are the factors that result in the dynamic response of the structure due to wind?
3.30 State the theory of platetectonics.
3.31 What are the three types of waves that are generated during an earthquake?
3.32 What is the equipment that is used to record earthquakes?
3.33 What is the Richter scale and how is the magnitude of an earthquake measured using it?
3.34 What are the intensity scales in use?
3.35 Indicate the typical damping values of steel, concrete, and masonry structures.
3.36 What is a response spectrum?
3.37 What is a tripartite spectrum?
3.38 Sketch the acceleration response spectrum as adopted in IS 1893 (Part 1).
3.39 How does the near source factor affect ground motion and ground acceleration?
3.40 What is the seismic design philosophy adopted by the IS 1893 code?
3.41 What are the important factors that influence earthquake-resistant design?
3.42 State the expression used in the code for horizontal seismic coefficient.

3.43 How is the base shear calculated using the equivalent static method?

3.44 What are the formulae given in the code for finding the fundamental natural period of vibration of steel buildings for (a) moment-resisting frames with brick infill panels, (b) without brick infills, and (c) with concrete and masonry shear walls?

3.45 Why does the dynamic analysis method yield better results than the equivalent static method?

3.46 What are the two methods used in the response spectrum analysis method when the natural periods of different nodes are well separated?

3.47 Write short notes on (a) risk coefficient ($k_1$), (b) terrain and height coefficient ($k_2$), (c) topography factor ($k_3$), and (d) importance factor for cyclonic regions ($k_4$).

3.48 What is the basis of wind directionality factor ($K_d$), area averaging factor ($K_q$), and combination factor ($K_a$).

3.49 What are the factors that influence wind pressure coefficient?

3.50 What are the local wind pressure coefficients? When are they used?

3.51 What is a solidity ratio?

3.52 When are frictional drag forces due to wind to be considered?

3.53 What are the factors that influence interference effect?

3.54 Under what conditions are the dynamic effects of wind to be considered?

3.55 What are the arrangements of live loads, as per the code, that produce maximum load effects?

3.56 List the 8 loading combinations to be considered and the 13 cases when considering only DL, LL, and EL.

3.57 What are the additional eight loading cases to be considered for non-orthogonal buildings?

3.58 What are the load combinations considered in plastic design?

## REFERENCES


Bazant, Z.P. and Frangopal, D.M., 'Size Effect Hidden in Excessive Dead Load Factor', *Journal of Structural Engineering*, Vol. 128, No. 1, Jan 2002, pp. 80–86.

Bhandari, N.M., Krishna, P., Kumar, K., and Gupta, A., *An Explanatory Hand book on IS: 875 (Part 3) Wind Loads on Buildings and Structures*, www.nicee.org/IITK - GSDMA/W06.pdf, Dec 2004, p. 116.

Bowels, J.E., *Foundation Analysis and Design*, 4th ed., McGraw Hill Book Co., New York, 1990.

Chen, X. and Kareem, A., 'Equivalent Static Wind Loads on Buildings: New Model', *Journal of Structural Engineering*, ASCE, Vol. 130, No. 10, Oct 2004, pp. 1425–1435.

Chopra, A.K., *Dynamics of Structures, Theory and Applications to Earthquake Engineering*, 2nd edition, Prentice Hall Inc., Englewood Cliffs, NJ, 2000, pp. 844.

Clough, R.W., and Penzien, J., *Dynamics of Structures*, McGraw Hill, New York, 1993.

Cook, N.J., *The Designer's Guide to Wind Loading of Building Structures- Part I*, Butterworths, London, 1985.

Cook, N., *Wind Loading, A Practical Guide to BS 6399-2*, Thomas Telford, 2002, pp. 150.

Coulbourne, W.L., Tezak, E.S., and McAllister, T.P., 'Design Guidelines for Community Shelters for Extreme Wind Events', *Journal of Architectural Engineering*, ASCE, Vol. 8, No. 2, June 2002, pp. 69–77.

Dayaratnam, P., *Design of Steel Structures*, Second edition, Wheeler Publishing, 1996, pp. 868.

Deepti, C.E., Stewart, M.G., and Rosowsky, D.V., 'Structural Reliability of Multi storey Buildings During Construction', *Journal of Structural Engineering*, Vol. 128, No. 2, Feb. 02, pp. 205–213.

Devenport, A.G., 'Wind Loads on Structures', *National Research Council of Canada, Division of Building Research,* Tech. Paper 88, 1960.

Durgesh, C.R., *Draft Code with Commentary on Seismic Evaluation and Strengthening of Existing Buildings*, www.nicee.org/IITK-GSDMA/EQ06.pdf.

Ellingwood B. and Rosowsky, D., 'Combining Snow and Earthquake Loads for Limit States Design', *Journal of Structural Engineering.*, Vol. 122, No. 11, Nov. 1996, pp. 1364 – 1368.

FEMA – 15, *Design Guildelines for Flood Damage Reduction*, Federal Emergency Management Agency, 1981, pp. 106, (www.fema.gov/hazards/floods/lib15.shtm).

FEMA – 353, *Recommended Specifications and Quality Assurance Guidelines for Steel Moment Frame Construction for Seismic Application* (www.fema.gov/hazards/earthquakes/fema353.shtm).

FEMA – 55, Coastal Construction Manual, Federal Emergency Management Agency, 2000, 500 pp. (www.fema.gov/hazards/floods/lib55.shtm), 15th Feb. 2005.

FEMA 450, NEHRP *Recommended Provisions for New Buildings and Other Structures Part 1 – Provisions, Part 2 – Commentary*, Building Seismic Safety Council, National Institute of Building Sciences, Washington D.C. 2003. (www.bssconline.org/NEHRP2003/comments/provisions/)

FEMA 549, Hurricane Katrina in the Gulf Coast-Mitigation Assessment Team Report- Building Performance Observations, Recommendations, and Technical Guidance, Federal Emergency Management Agency (FEMA), Washington D.C., 2006.

Garas, F., Virdi, K.S., Mathews, R., and Clarke, J.L., *Abnormal Loading on Structures*, Spon Press, London, 2000, pp. 384.

Gioncu, V. and Mazzolani, F.M., Ductility of Seismic Resistant Steel Structures, Spon Press, London and New York, 2002, pp. 687.

Holmes, J.D., *Wind Loading of Structures*, Spon Press, London and New York, 2001, pp. 356.

IRC 6 – 2014, *Standard Specifications and Code of Practice for Road Bridges*, Section II, Loads and Stresses, Revised edition, The Indian Roads Congress, New Delhi, 2014.

IS: 1893 (Part 1) – 2002 Criteria for earthquake resistant Design of Structures, Part1: General Provisions and buildings, (Fifth Revision) Bureau of Indian Standards, New Delhi.

IS: 4911-1968 Criteria for blast resistant design of structures for explosions above ground, Bureau of Indian Standards, New Delhi.

IS: 6533 – 1989 Code of Practice for design and Construction of Steel Chimneys Part 1, Mechanical Aspects, Part 2 – Structural Aspects, Bureau of Indian Standards, New Delhi.

IS: 802 (Part1, Sec. 1) – 1995 Use of Structural Steel in Overhead Transmission Line Towers - Code of Practice, Part 1 Material, Loads and Permissible Stresses, Sec.1.Materials and loads and Sec. 2 – Permissible Stresses, (1992), Bureau of Indian Standards, New Delhi.

IS: 807 – 1976, Code of Practice for Design, Manufacture, Erection and Testing (Structural portion) of Cranes and Hoists, Bureau of Indian Standards, New Delhi.

IS: 875 – 1987, Code of Practice for Design Loads (other than earthquake) for buildings and structures (Part 1 – Part 5), Bureau of Indian Standards, New Delhi.

IS: 9178 – 1979, Criteria for design of Steel bins for Storage of bulk Materials Part 1 – General requirements and assessment of Loads, Part 2 – Design Storage Criteria, Bureau of Indian Standards, New Delhi.

IS 15498:2004, Guidelines for Improving the Cyclonic Resistance of Low Rise Houses and Other Buildings/Structures, Bureau of Indian Standards, New Delhi.

Jain, S.K., 'A Proposed Draft of IS: 1893 Provisions on Seismic Design of Buildings – Part II. Commentary and Examples', *Journal of Structural Engineering*, SERC, Vol. 22, No. 2, Jul 1995, pp. 73–90.

Jain, S.K. and Murty, C.V.R., *Proposed Draft Provisions and Commentary on Indian Code IS: 1893 (Part 1)*, www.nicee.org/IITK-GSDMA/EQ05.pdf, 30 Dec 2004.

Krishna, P. and Gupta, R., 'Studies on Structures using Wind Tunnel', *The Indian Concrete Journal,* Vol. 66, No. 11, Nov 1992.

Krishna, P. Kumar, K., Bhandari, N.M., and Ahuja, A.K., *Review of Indian Wind Code, IS: 875 (Part3) – 1987*, www.nicee.org/IITK-GSDMA/W01.pdf.

Kumar, K., Bhandari, N.M., and Krishna, P., *IS: 875 (part 3), Wind Loads on Buildings and Structures – Proposed Draft and Commentary*, www.nicee.org/IITK-GSDMA/W02. pdf, p. 104, Dec 2004.

Levy, M., and Salvadori, M., *Why Buildings Fall Down*, W.W.Norton and Company, New York, 1992.

Martin, L.H. and Purkiss, J.A., *Structural Design of Steel Work to BS 5950*, Edward Arnold, London, 1992, p. 468.

Mazzolani, F.M. and Piluso, V., *Theory and Design of Seismic Resistant Steel Frames*, E & FN Spon, 1996, p. 544.

Mehta, K.C., and Delahay, J., *Guide to the Use of the Wind Load Provisions of ASCE 7-02,* ASCE Press, 2003.

Minor, J.E., 'Tornado Technology and Professional Practice', *Journal of Structural Div. ASCE,* Vol. 108, No.11, Nov 1982, pp. 2411–2422.

Murty, C.V.R., *IITK – BMTPC Earthquake Tips*, www.nicee.org/EQTips/IITK-BMTPC.htm

Newmark, N.M. and Rosenbleuth, E., *Fundamentals of Earthquake Engineering*, Prentice Hall, 1971.

O'Rourke, M. and Auren, M., 'Snow Loads on Gable Roofs', *Journal of Structural Engineering*, ASCE, Vol.123, No.12, Dec 1997, pp. 1645–1651.

O'Rourke, M. and Downey, C., 'Rain – on – Snow Surcharge for Roof Design', *Journal of Structural Engineering*, ASCE, Vol. 127, No. 1, Jan 01, pp. 74–79.

O'Rourke, M. and Wrenn, P.D., Snow loads: *A Guide to the Use and Understanding of Snow Load Provision of ASCE 7-02*, ASCE Press, 2004, p. 128.

Paulay, T. and Priestley, M.J.N., *Seismic Design of Reinforced Concrete and Masonry Buildings*, John Wiley & Sons, New York, 1992, p. 744.

Paz, M., *Structural Dynamics: Theory and Computations*, Third edition, Van Nostrand Reinhold, 1991.

Radhakrishnan, R. and Subramanian, N., 'Dynamic Analysis of Multi Storey Frames (A comparison of various methods)', Bulletin, *Indian Society of Earthquake Technology*, Vol.15, No. 3, Sep 1978, pp. 39–49.

Radhakrishnan, R. Subramanian, N., and Amalnarayanan, C.R., 'Dynamic Analysis of Offshore Structures', *Journal of Structural Engineering*, Vol. 6, No. 4, Jan 1979, pp. 213–218.

Ravindra, M.K. and Galambos, T.V., 'Load and Resistant Factor Design for Steel', *Journal of the Structural Div.*, ASCE, Vol. 104, No. ST9, Sep 1978, pp. 1337–1354.

Ruiz, S.E. and Soriano, J.A., 'Design Live Loads for Office Buildings in Mexico and USA', *Journal of Structural Engineering*, ASCE, Vol. 123, No.6, Jul 1997, pp. 816–822.

Ruiz, S.E. and Trujillo, A.S., 'Design Live Loads for Classrooms in United States and Mexico', *Journal of Structural Engineering*, ASCE, Vol. 123, No. 12, Dec. 1997, pp. 1652–1657.

Scawthorn, C., Eidinger, I., and Sehiff, A., Ed., *Fire Following Earthquake*, TCLEE Monograph 26, American Society of Civil Engineers, Reston, VA, 2005, p. 352.

Schechter, M.M., Schechter, E., and Simiu, E., 'Developmental computer-based version of ASCE 7-95 Standard Provision for Wind Loads', *NIST Tech. Note 1415,* National Institute of Standards and Technology, Gaithersburg, MD., Nov 1995.

SEI/ASCE, 24-98: Flood Resistant Design and Construction, *Structural Engineering Institute/American Society of Civil Engineering*, Reston, VA, 2000, p. 72.

SEI/ASCE 37-02, 'Design Loads on Structures During Construction', *Structural Engineering Institute/American Society of Civil Engineers*, Reston, VA, 2002.

SEI/ASCE 7 – 10, *Minimum Design Loads for Buildings and Other Structures*, ASCE Press, 2010.

Simiu, E. and Scanlan, R. H., *Wind Effects on Structures: Fundamentals and Applications to Design,* 3rd edition, Wiley Inter science, New York, 1996, p. 704.

SP 64 (S & T): 2001, Explanatory Handbook on Indian Standard Code of Practice for Design Loads, Bureau of Indian Standards, New Delhi, p. 116.

Srinivasa Rao, P. and Krishnamurthy, C.S., 'Imposed Loads on Floors of Office Buildings – Results of Load Survey', *Indian Concrete Journal*, Vol. 67, No.11, Nov 1993.

Subramanian, N., 'Wind Load of Steel Microwave Towers - A Comparison of Codes', *Engineering Design*, Vol. XX, No. 182, Jan –Jun 1991, pp. 18–22.

Subramanian, N., 'Wind Loads on Industrial buildings', *Civil Engineering and Construction Review*, Vol. 3, No.12, Dec 1990, pp. 44–50.

Subramanian, N. and Pranesh, M.R., 'Fixed Pile Supported Steel Offshore Platforms – Overview', *Journal of Institution of Engineers (India)*, Vol. 63, Pt. CI3, Nov. 1982, pp. 151–158.

Subramanian, N. and Venugopal, M.S., 'Partial Failures of Some Industrial Sheds-A Case Study', *The Indian Concrete Journal*, Vol. 50, No. 5, May 1976, pp. 143–144, 159.

Sunil Kumar. and Kameswara Rao, C.V.S., 'Live Loads in Residential Buildings in Urban and Rural Areas', *The Indian Concrete Journal*, Vol. 68, No.10, Oct 1994.

Sunil Kumar. 'Live Loads in Residential Buildings, Part II – Life Time Maximum Loads', *Indian Concrete Journal*, Vol. 76, No. 4, Apr 2002.

Sunil Kumar, 'Live Loads in Residential Buildings: Part I. The Results of the Survey', *Indian Concrete Journal*, Vol. 76, No. 3, Mar 2002.

Taranath, B.S., *Steel, Concrete and Composite Design of Tall Buildings*, Second edition, McGraw Hill, New York, 1998, p. 998.

Trahair, N.S., Bradford, M.A., and Nethercot, D.A., *The Behaviour and Design of Steel Structures to BS: 5950,* Third edition, Spon Press, London, 2001, p. 448.

Venugopal, M.S. and Subramanian, N., 'Differential Movements in Soils', *Seminar on Problems of Building Foundations*, Madras, Feb 20, 1977.

Vickery, P.J. and Skerlj, P.F., 'Hurricane Gust Factors Revisited', *Journal of Structural Engineering*, ASCE, Vol. 131, No. 5, May 2005, pp. 825–832.

Walpole, R.E. and Myers, R.H., *Probability and Statistics for Engineers and Scientists*, MacMillan, New York, 1978.

Walton, C., 'Cases of Structural damage attributed to snow on roofs in the UK, 1964 – 1984', *Building Research Establishment*, Note N 57/90, Watford, England.

Wen, Y.K. and Yeo, G.L., 'Design Live Loads for Passenger Cars Parking Garages', *Journal of Structural Engineering*, ASCE, Vol. 127, No. 3, Mar 01, pp. 280–289.

Whalen, T.M., Gopal, S., and Abraham, D.M., 'Cost – Benefit Model for the Construction of Tornado Shelters', *Journal of Construction Engineering and Management*, ASCE, Vol. 130, No. 6, Nov–Dec 2004, pp. 772–779.

Wolmuth, B. and Surtess, J., 'Crowd Related Failures of Bridges', *Civil Engineering Proceedings of ICE*, Vol.156, Issue 3, 2004, pp. 116–123.

Zhou, Y., Kijewski, T., and Kareem, A., 'Along Wind Load Effects on Tall Buildings: Comparative Study of Major International Codes and Standards', *Journal of Structural Engineering*, ASCE, Vol.128, No. 6, Jun 2002, pp. 788–796.

# 4

# METHODS OF STRUCTURAL ANALYSIS

## INTRODUCTION

The term design, as discussed in Section 2.6, means the selection of sizes of members to resist the forces and moments found in the analysis phase, so that the resulting structure is safe and economic. In a statically determinate structure (for example, simply supported beam, cantilever truss, etc.), the analysis is relatively simple. However, the analysis of *indeterminate structures* (such as portal frames and multi-storey bends), requires the various cross-sectional properties of members, which are not known initially. Hence, in these types of structures, analysis and design are interactive. In other words, a preliminary analysis is done using the assumed member sizes and then the members and the details of structural joints are designed as per the codal rules (member capacity formulae and local stress criteria for joint detailing). Reanalysis is often required if the assumed member sizes do not satisfy the strength, stability, deflection, or other criteria specified in the codes of practice.

Despite its popular use in the past and present, the two-stage process (finding the forces through analysis and checking the member sizes against the codal rules) has the following limitations.

1. It does not consider the interaction of strength and stability between the member and the structural system in a direct manner {in current engineering practice, this interaction effect is indirectly and approximately accounted for by the use of the *effective length factor* (K-factor) concept}. It is well recognized that the failure modes of the structural system often do not resemble the elastic buckling mode of the structural system, which is the basis for determining the K-factors.
2. For non-rectangular and irregular frames, it is difficult and complicated to find the K-factor by the elastic theory. This is particularly true for a frame with *leaning columns*. (In most steel structures, only selected frames are rigidly connected to resist lateral loads. The remaining frames are connected by 'simple' joints and can resist only gravity loads. Such an arrangement is more economical because simple connections are cheaper to fabricate than rigid connections. In such cases, the rigid moment frames must carry the total lateral load for the building. The simply connected members rely on the rigid moment frames for their stability. Thus, the simply supported portions of the structure—the sway effects of which are now carried by the rigid frame—are said to be 'leaning' on the rigid frame.) The determination of K-factors for these leaned columns is so complex that its solution is often difficult to obtain by the direct use of computer programs, since it involves a clear judgement of behaviour (Chen & Lui 1987).
3. The most serious limitation occurs when the limit states design procedure is used. Thus, *elastic (first-order) analysis* is used to determine the forces acting on each member of the structural system, whereas *plastic analysis* is used to determine the strength of each member treated in isolation. Hence there is no verification of the *compatibility* between the isolated member and the other members as a part of the frame. The strength equations specified in the design codes do not consider the system compatibility. There is no guarantee that all the members will sustain their design loads as members of the given framework.
4. Limit states specification such as those given in IS 800: 2007 are based directly on factored loads and limits of resistance. Though the structural systems and their members may behave linearly under working loads, they will behave non-linearly before their respective limit of resistance is reached.

There is an increasing awareness to use inelastic (second-order) analysis for the consideration of strength and stability effects in the design of the overall structural system. Such approaches can model the true member performance explicitly in the analysis, thus taking into account the compatibility between the system and the member. Due to these advantages, some design specifications, such as the Canadian standard for steel structures (CAN/CSA-S16.1 2001), have eliminated the use of the effective length concept, that is, K-factors (Kennedy et al. 1990). Checks for the in-plane and out-of-plane stability of beam columns are based on the actual member lengths, and frame stability is checked solely through the second-order analysis procedure. The Australian standard (AS-4100 1998) and Eurocode 3 follow a similar approach. In addition to this, CAN-S16.1 limits the use of elastic analysis to cases where the ratio of second-order effects to first-order effects is less than 1.4. A similar limit has also been proposed in IS 800 : 2007, clause 4.4.2.

Hence, in this chapter, the various analysis methods which can model the real behaviour of the structural system more accurately and thereby reduce reliance on semi-empirical methods (dealing with the interaction between member and system stability effects) are presented. For in-plane behaviour of rigidly connected frames, the theory and formulations for conducting accurate second-order inelastic analyses are fairly established (Chen & Toma 1994; Chen & Sohal 1995; Chen & Kim 1997; McGuire et al. 2000). However, research on accurately modelling the more complex aspects of the three-dimensional behaviour that involves inelastic torsional–flexural buckling is still progressing. Hence, currently the most practical approach for stability design still requires the separation of the in-plane frame and member behaviour from out-of-plane member stability checks.

Several researchers have studied the moment–rotation relationship of different types of connections discussed later in Chapters 5, 6, 14, and 15. The stiffness or flexibility of a connection is dependent on the geometric parameters, such as plate or angle thickness, bolt size, and connection type. The moment–rotation relationship is found to be non-linear for all types of connections over the entire range of loading and varies depending on connection flexibility (the more flexible the connection, the larger the relative rotation). A collection of large data of available experimental results on connections has been developed (Chen & Kishi 1989; Chen & Toma 1994). Several mathematical models have been proposed to fit the moment–rotation curves from the experimental data (Frye & Morris 1975; Bjorhovde et al. 1990; Kishi & Chen 1990; Eurocode 3). These models vary widely in their complexity, ranging from the simplest linear model to polynomial, exponential, and power models. These models along with the second-order analysis method as applicable to frames with semi-rigid joints are also briefly discussed in this chapter.

Most of the discussion in this chapter assumes access to modern frame analysis software. {The details of stiffness equations, their methods of solution, etc., are not included in this book and may be found from Goto & Chen (1987), Dhillon & O'Malley (1987), Chen & Toma (1994), Chen et al. (1995), and McGuire et al. (2000)}. But it is important to idealize these structures to get the desired results from the computer analysis and hence, a few points for idealization of members are also included.

## 4.1 METHODS OF ANALYSIS

An overview of the analysis methods to calculate the frame response is shown in Fig. 4.1 using the load–displacement curves of a statically loaded frame. It can be observed from this figure that it is very difficult to model all sources of non-linearity and portray the actual behaviour of a practical structure. Though the degree to which the methods can model true behaviour differs, each method can yield valuable information to the design engineer.

The various methods of analysis that can be employed, as shown in Fig. 4.1, are as follows:

1. First-order elastic analysis
2. Second-order elastic analysis (includes methods to consider the effect of connection flexibility)
3. Linear buckling analysis
4. Inelastic buckling analysis
5. First-order plastic analysis
6. Second-order inelastic analysis (includes advanced analysis methods)
   (a) Plastic zone method
   (b) Elastic plastic hinge method
   (c) Refined plastic hinge method
   (d) Notional load plastic hinge method
   (e) Quasi-plastic hinge method

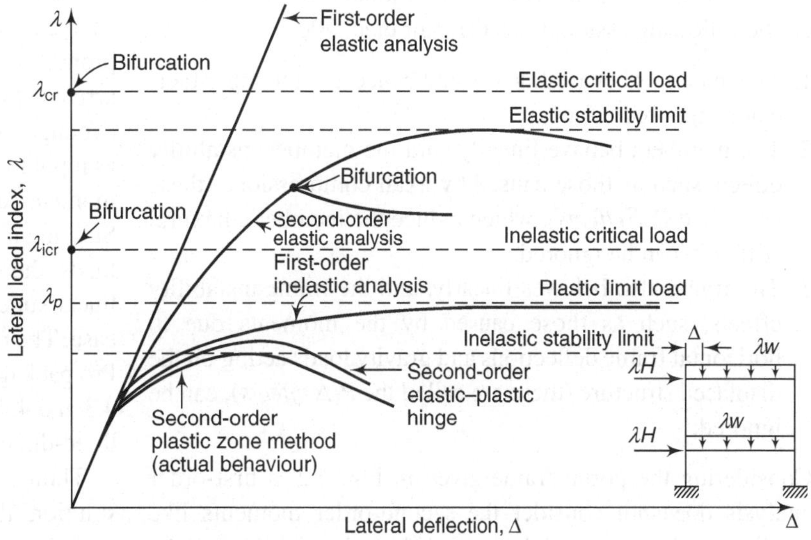


**FIG. 4.1** Load–displacement characteristics of different analysis methods

A brief description of the various methods of analysis is presented in the next section.

## 4.2 ELASTIC ANALYSIS

Elastic analysis assums linear-elastic material behaviour and may be done based on the equilibrium in the underformed geometry of the structure (first-order elastic analysis) or in the deformed geometry of the structure (second-order analysis). Both these methods are discussed briefly in the following sections.

### 4.2.1 First-order Elastic Analysis

This is the most basic method of analysis, in which the material is modelled as linear-elastic and the equilibrium is expressed in terms of the geometry of the undeformed structure. This assumption is valid when the elastic displacements are small compared to the dimensions of the structure. Thus, by definition, this method excludes non-linearity, but it generally represents conditions at service loads very well. Because of the assumption of linear-elastic behaviour, the principle of superposition may be used to combine load cases. A first-order elastic analysis is sufficient for normal framed structures, which are braced against sway. However, first-order elastic analysis will not yield sufficiently accurate results for some suspension systems, arches, tall buildings, and structures subject to early localized yielding or cracking. As mentioned earlier, all structures will exhibit significant non-linear response just prior to reaching their limit of resistance because of the yielding and buckling effects associated with the axial compressive forces. Hence, this non-linear behaviour is accounted for by the code formulae (that makes allowance for non-linearity in some empirical or semi-empirical manner) or by supplementary theoretical or experimental studies. Thus, the first-order elastic analysis of rigid jointed frames is based on the following assumptions (Trahair et al. 2001).

1. The materials behave linearly and hence all yielding effects can be ignored.
2. The members behave linearly, and the member instability effects such as those caused by axial compressions (these are called *P-δ effects*), which reduce the member's flexural stiffness, can be ignored.
3. The frame also behaves linearly, and the frame instability effects, such as those caused by the moments due to horizontal frame deflections and gravity loads acting on the displaced structure (these are called the *P-Δ effects*), can be ignored.

Considering the portal frame given in Fig. 4.2, a first-order analysis does not consider the second-order moments like $R_R(\delta + \Delta z/h)$ in the right-hand side column. Hence, the bending moment distribution given by the first-order elastic analysis is linear as shown in Fig. 4.2(a).

Though several manual methods are available for the first-order elastic analysis (such as the *slope deflection method* and

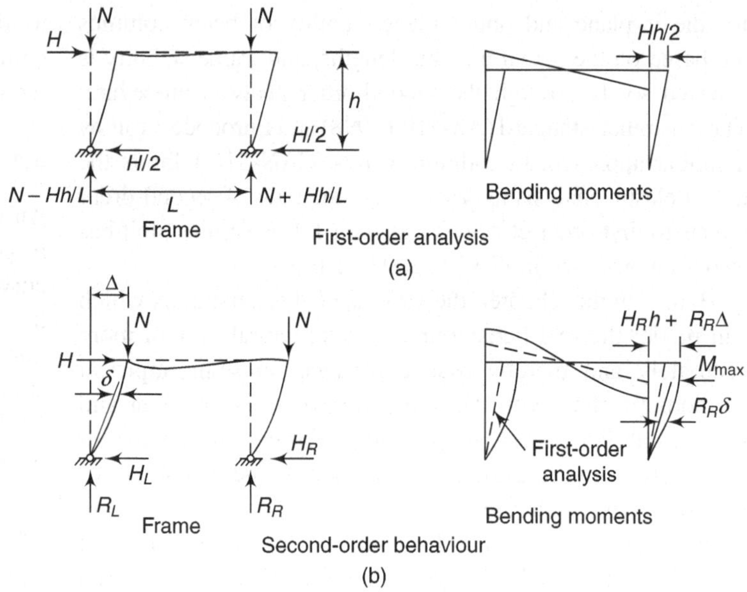


**FIG. 4.2** Simple portal frame

the *moment distribution method*), they are labour-intensive and error prone, especially for large frames. Hence, the designers in the past often utilized charts or available solutions for specific frames (Brockenbrough & Merritt 1999 and Davison & Owens 2002). However, the advent of computers and the development of matrix methods of analysis have resulted in the development of several software packages (see Section 2.6), which are now used extensively (Gere & Weaver 1969; Wang 1973; Livesley 1975; Subramanian 1983; Harrison 1990; Subramanian 1991, 1995). These first-order elastic analysis programs require the geometry of frame, member properties (such as area and moment of inertia, which are arrived at using some preliminary design), material properties (Young's modulus and yield stress), and boundary conditions as input to compute the joint displacements as well as member moments and forces for each specified set of loads (Pframe, Subramanian 1986). As these quantities are proportional to the loads, the results of individual analyses may be combined by linear superposition to get the results for a combined loading case. The results of first-order elastic analysis (using STAAD-Pro package) of a braced and unbraced frame are given in Figs 4.3 and 4.4, respectively. Some standard programs can handle three-dimensional frames as well.

Though the first-order elastic analysis provides an 'exact solution' that satisfies the requirements of compatibility and equilibrium of the undeformed structure, it does not provide

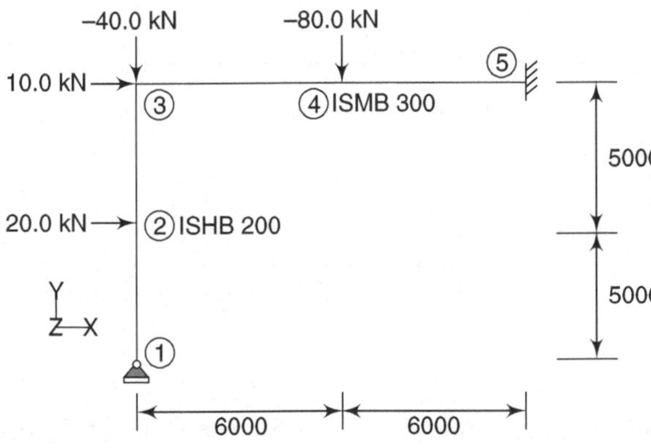


(a) Frame and loads

| Member | Section | $I_z$ (mm$^4$) | $A$ (mm$^2$) |
|---|---|---|---|
| 123 | ISHB 200 | $3608 \times 10^4$ | 4750 |
| 345 | ISMB 300 | $8604 \times 10^4$ | 5630 |

(b) Section properties

| Quantity | Units | First-order elastic | Second-order elastic | Elastic buckling | First-order plastic |
|---|---|---|---|---|---|
| $M_2$ | kNm | 20.04 | 20.04 | 0 | 17.2 |
| $M_3$ | kNm | −59.91 | −59.91 | 0 | −96.0 |
| $M_4$ | kNm | 134.98 | 134.98 | 0 | 170.2 |
| $M_5$ | kNm | −150.13 | −150.13 | 0 | −170.2 |
| $N_{123}$ | kN | −72.50 | −72.50 | −814.5 | −94.2 |
| $N_{345}$ | kN | −26.00 | −26.00 | −292.0 | −34.5 |
| $v_4$ | mm | 56.478 | 56.478 | (0.155) | — |
| $u_2$ | mm | 5.842 | 5.842 | (1.000) | — |
| Max horizontal deflection | mm | 8.076@ 3.333 | 8.076@ 3.333 | — | — |

(c) Analysis results

**FIG. 4.3** Analysis of a braced frame

any information about the influence of plasticity and stability on the behaviour of the structure. Hence, these influences are normally provided indirectly in member capacity checks through the use of column effective length and moment-amplification factors (see Chapter 13 and clause 9.3.2.2 of IS 800 : 2007). A brief description of these moment amplification factors are as follows.

## *P-δ* or Member Stability Effect

In braced frames, the joint displacements Δ are small, and only the *P-δ* effects are important. In unbraced frames, the *P*-Δ effects are important, though *P-δ* effects will also be present. Chen & Lui (1985) considering a beam column subjected to end moments and axial force (see Fig. 4.5) have shown that an estimate of the second-order elastic moment $M_u$ in lieu of a direct second-order elastic analysis can be obtained (with reasonable accuracy) by using the relation

$$M_u = \delta_b M_o \tag{4.1a}$$

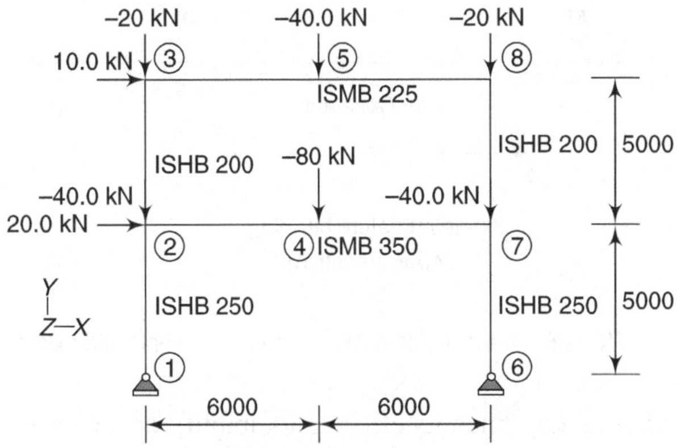


(a) Frame and loads

| Member | Section | $I$ (mm$^4$) | $A$ (mm$^2$) |
|---|---|---|---|
| 12,67 | ISHB 250 | $7737 \times 10^4$ | 6500 |
| 23,78 | ISHB 200 | $3608 \times 10^4$ | 4750 |
| 247 | ISMB 350 | $13630 \times 10^4$ | 6670 |
| 358 | ISMB 225 | $3442 \times 10^4$ | 3970 |

(b) Section properties

| Quantity | Units | First-order elastic | Second-order elastic | Elastic buckling | First-order plastic |
|---|---|---|---|---|---|
| $M_4$ | kN m | 143.93 | 143.93 | 0 | 222.7 |
| $M_5$ | kN m | 67.79 | 67.79 | 0 | 78.5 |
| $M_{76}$ | kN m | −123.24 | −132.72 | 0 | −158.2 |
| $M_{74}$ | kN m | −180.89 | −190.62 | 0 | −248.2 |
| $M_{78}$ | kN m | 57.65 | 57.87 | 0 | 90.5 |
| $M_8$ | kN m | −67.39 | −68.63 | 0 | −78.5 |
| $N_{12}$ | kN | −103.30 | −101.50 | −696.3 | −146.4 |
| $N_{23}$ | kN | −37.50 | −37.30 | −252.3 | −55.8 |
| $N_{67}$ | kN | −136.70 | −138.50 | −421.0 | −191.7 |
| $N_{78}$ | kN | −42.50 | −42.70 | −286.2 | −55.8 |
| $v_4$ | mm | 41.637 | 41.637 | (0.004) | — |
| $v_5$ | mm | 71.541 | 71.542 | (0.001) | — |
| $u_2$ | mm | 70.681 | 79.272 | (0.826) | — |
| $u_3$ | mm | 104.372 | 116.044 | (1.000) | — |
| Max. vertical deflection | mm | 42.241@ 5.0m | 42.578@ 5.0m | — | — |

(c) Analysis results

**FIG. 4.4** Analysis of an unbraced frame

and

$$\delta_b = C_m/[1 - (P_u/P_{cr})] \tag{4.1b}$$

where $M_u$ is the second-order elastic moment, $M_o$ is the first-order elastic moment, and $P_u$ is the maximum load in the column. $P_{cr}$ is the member buckling load (often called as critical load) defined by $\pi^2 EI/(KL)^2$, where $K$ is the effective length factor in the plane of bending based on the assumption that side-sway is prevented and $E$, $I$, and $L$ are member properties related to the axis of bending. For braced frame members having transverse loading between supports, a coefficient $C_m$ (*equivalent moment factor*) is used. The factor $\delta_b$ is called the

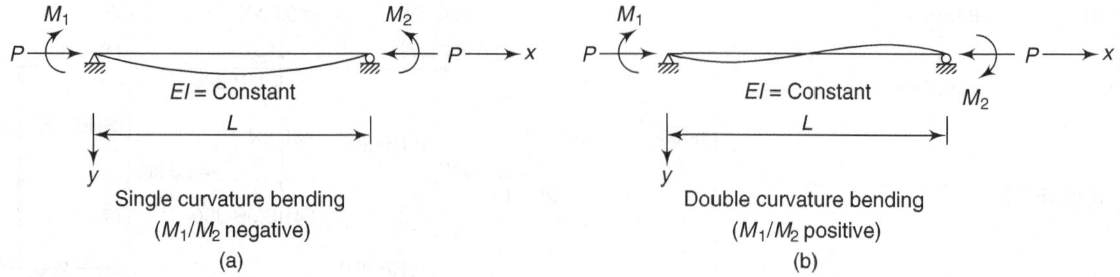


**FIG. 4.5** Beam-column subject to end moments and axial force

*amplification factor*, since its effect is to amplify or increase $M_o$. The value of $C_m$ may be determined by rational analysis (Chen & Lui 1986). For braced frame members subjected only to end moments and axial force, $C_m$ which converts the linearly varying moment to a uniform moment is given by $\psi = M_1/M_2$ (Chen & Kim 1997), where $M_1$ and $M_2$ are the numerically smaller and larger end moments obtained from the first-order analysis, respectively.

For $0 \leq X_c \leq L$

$$C_m = \sqrt{\frac{\{(\psi^2 + 2\psi \cos kL + 1}{[2(1 - \cos kL)]\}}} \quad (4.2a)$$

for $X_c < 0$ or $X_c > L$

$$C_m = \cos (kL/2) \quad (4.2b)$$

where $X_c$ is the location of the maximum moment along the length of the member and

$$k = \sqrt{(P/EI)}$$

The following simplified expressions are also available to replace Eqn (4.2a) (Salvadori 1955; Massonnet 1959; Austin 1961; Duan et al. 1989; Villettee et al. 2000):

for $-1.0 \leq \psi \leq 1.0$

$$C_m = 0.6 + 0.4\psi \geq 0.4 \quad (4.2c)$$

$$C_m = \sqrt{(0.3\psi^2 - 0.4\psi + 0.3)} \geq 0.4 \quad (4.2d)$$

$$C_m = 1/(1.75 + 1.05\psi + 0.3\psi^2) \geq 0.4 \quad (4.2e)$$

$$C_m = 1 + 0.25(P_u/P_{cr}) - 0.6(P_u/P_{cr})^{1/3}(\psi + 1) \quad (4.2f)$$

$$C_m = 0.79 + 0.21\psi - 0.36(\psi - 0.33)P_u/P_{cr} \quad (4.2g)$$

The first two simplified expressions were found to approximate Eqn (4.2a) adequately (Chen & Kim 1997). The last two expressions also consider the effect of axial load on $C_m$ and were found to give reasonable results.

The value of $M_1/M_2$ is positive when the member is bent in double (reverse) curvature and negative when it is bent in single curvature (see Fig. 4.5). The value of $C_m$ can be as low as 0.4 and it is provided by Chen & Lui (1985) for several different loading cases. The equations for approximate maximum moments in beam columns with transverse loads are provided by Trahair et al. (2001) and Chen & Atsuta (1977).

Based on the study of the inelastic behaviour of beam columns with unequal end moments, Trahair et al. (2007) proposed a design expression for $C_m$ in the form

$$C_m = (1 - \psi)/2 + [(1 + \psi)/2]^3(0.40 - 0.23P_u/P_{ny}) \quad (4.2h)$$

where $P_{ny}$ is the axial strength of the member bent about its weak-axis. Comparing this equation with Eqn (4.2c) shows that Eqn (4.2c) gives conservative results for most cases (Chen & Lui 1991).

### *P*-Δ or Frame stability effect

When lateral loads $\Sigma H$ act on a frame, the frame will deflect laterally until equilibrium is reached [Fig. 4.6(a)]. The corresponding lateral deflection can be calculated on the basis of the original configuration, and is referred to as first-order deflection, denoted by $\Delta_I$. Now if in addition to $\Sigma H$, vertical forces $\Sigma P$ act on the frame, then these forces will interact with the lateral displacement $\Delta_I$ caused by $\Sigma H$ to drift the frame further until a new equilibrium position, denoted by $\Delta$ is reached [Fig. 4.6(b)]. As indicated earlier, this phenomenon by which the vertical forces $\Sigma P$ interact with lateral displacement is called the $P$-$\Delta$ effect. The consequence of this effect is to increase the drift and the overturning moment. Since the additional deflection and overturning moment have detrimental effects on the stiffness and stability of the frame, they should be considered in design.

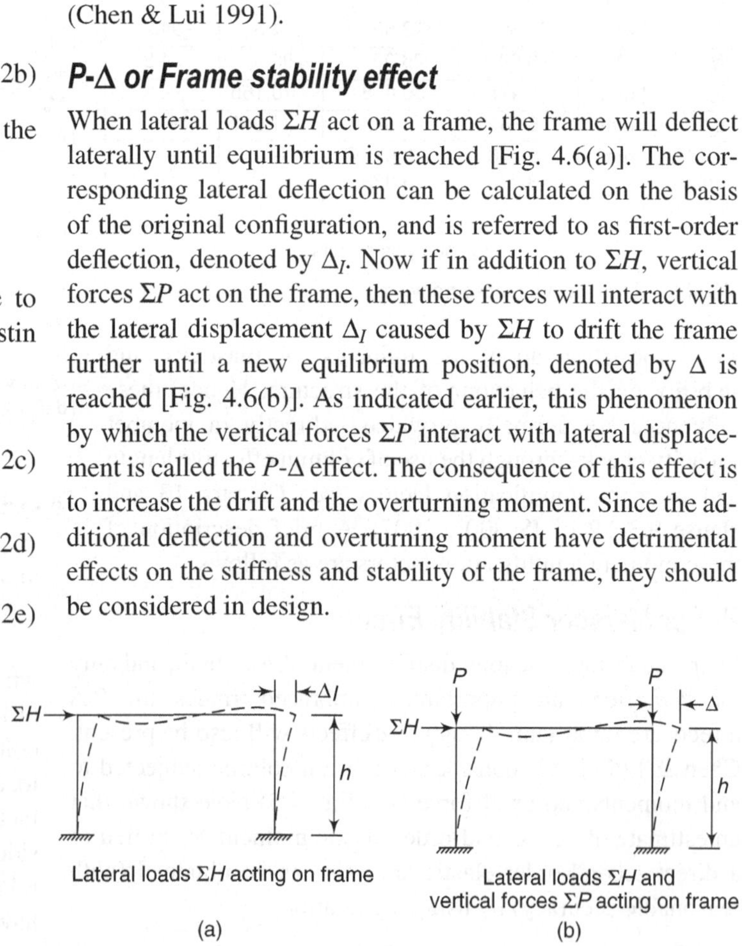


**FIG. 4.6** $P$-$\Delta$ effect

Two simplified approaches have been suggested to consider these $P$-$\Delta$ effects: storey-magnifier method and multiple-column magnifier method. By assuming that (i) each storey can behave independently of other storeys and (ii) the additional moments in the columns caused by the $P$-$\Delta$ effect is equivalent to that caused by the lateral force $\Sigma P\Delta/h$ (where $h$ is the storey height), the *storey-magnifier method* gives the value of $\Delta$ as given by Chen & Lui (1985)

$$\Delta = \Delta_I/[1 - (\Sigma P\Delta_I/\Sigma Hh)] \quad (4.3)$$

Thus, the maximum end-moment accounting for $P$-$\Delta$ effect can be written as

$$M = A_F M_{1\text{sway}} \quad (4.4a)$$

where $M_{1\text{sway}}$ is the maximum first-order end moment as a result of the swaying of the storey and $A_F$ is the moment amplification factor

$$= 1/[1 - (\Sigma P\Delta_I/\Sigma Hh)] \quad (4.4b)$$

Chen and Lui (1985) also give a procedure for calculating $M$, when the swaying of the frame is due to the asymmetry of frames and/or asymmetry of loading. This method works for frames with stiff beams in every storey so that a point of inflection occurs in every column in a storey.

*Multiple-column magnifier method* assumes that when instability occurs in a storey, all columns in that storey will become unstable simultaneously, and the method gives the following equation

$$M = A_F M_{1\text{sway}} \quad (4.5a)$$

where $A_F = 1/[1 - (\Sigma P/\Sigma P_e)]$ (4.5b)
and $P_e$ is the Euler buckling load.

### *Redistribution of Bending Moments*

The code in clause 4.4.3.4 allows for a redistribution of bending moments from the first-order elastic analysis, up to 15% of the peak calculated moment of the member under factored load, provided

1. the internal forces and moments in the member of the frame are in equilibrium with applied loads and
2. all the members, in which the moments are reduced, are either plastic or compact sections (see Chapter 8 and Table 2 of the code).

This redistribution is allowed because the frame will collapse only when sufficient number of plastic hinges are formed to convert the structure (or a part of it) into a 'mechanism'. Owing to the formation of plastic hinges at one location, moment redistributions take place in the 'inelastic' phase, due to the rotation capacity of the sections (Neal 1965). The 15% restriction is aimed at ensuring that plastic hinge formation does not take place under normal service loads. Details about moment redistribution are provided in Salmon & Johnson (1996). Such adjustments in the maximum bending moment attempts to bring the result close to plastic design.

## 4.2.2 Second-order Analysis

The basic objective of second-order analysis is to calculate the internal force equilibrium in the deformed geometry of the structure. This is explained with reference to the structure shown in Fig. 4.7. The bending moment due to both first-order and second-order elastic analyses are shown in Fig. 4.7(a). It is observed that the first-order analysis yields bending moments in the beam equal to those for a simply supported beam, whereas the second-order analysis results in larger bending moments in the beam and non-zero moments in the column. These additional moments are primarily due to the $P$-$\Delta$ effect, where $\Delta$ is the lateral drift of the frame and $P$ is the total vertical load.

For the frame shown in Fig. 4.7(b), side sway is restrained and hence the second-order moments in the beam and column are only due to the lateral deflection $\delta$, along the column. The additional moment in the column is equal to the column load $P$ multiplied by the deflection $\delta$ and hence it is termed as $P$-$\delta$ effect. The loss of rotational restraint of the column results in the reduction of the negative moment at the end of the column, which in turn increases the positive moment at the beam's mid-span. Based on this example, the following conclusions can be made (Galambos 1998).

1. Second-order behaviour affects moments in beams, columns, as well as connections. The axial forces and shears in all the members also change due to second-order effects.
2. The second-order moments need not have the same distribution as the first-order moments. They are not simply an amplification of the first-order moments. When second-order

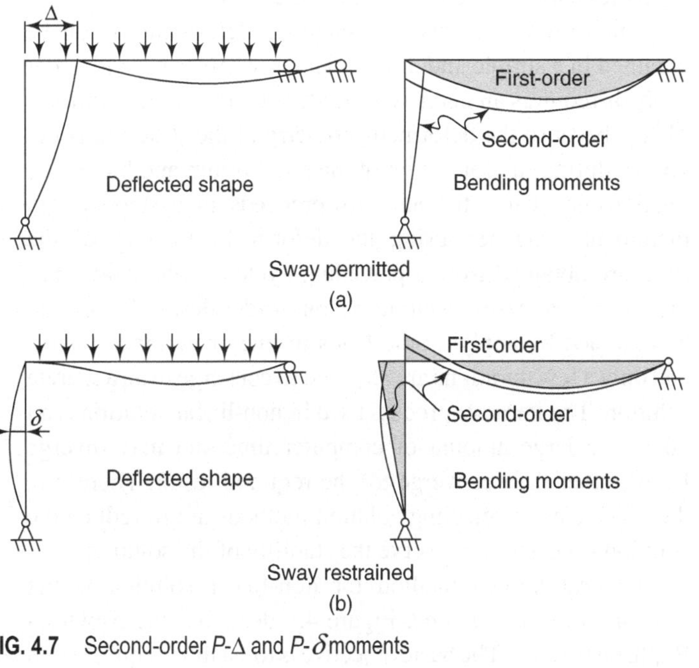


**FIG. 4.7** Second-order $P$-$\Delta$ and $P$-$\delta$ moments

moments are calculated as an amplification of first-order moments, it requires the decomposition of the first-order moments into sway and non-sway moments, calculation of separate amplifications factors for both cases, and combination of all the calculated terms. Moreover, these methods are complicated for calculating the maximum second-order moments in beams with distributed gravity loads.

3. Both braced and unbraced frames experience second-order effects (both $P$-$\Delta$ and $P$-$\delta$ effects). In the frames shown in Fig. 4.7 (which is an extreme case), one of the effects is dominant for sway and the other for non-sway condition. In Fig. 4.7(a), the second order $P$-$\Delta$ effect causes flexure of the left column, which may induce $P$-$\delta$ moments. Similarly, as shown in Fig. 4.7(b), slight joint translation due to axial shortening of the beam, may induce $P$-$\Delta$ effects.
4. Linear superposition of effects is not possible in second-order analysis. Thus, if we have lateral loads, they have to be combined with gravity effects and applied simultaneously in a single analysis. Moreover, since second-order effects are generally non-linear, all second-order analyses should be performed using the factored loads.

### 4.2.3 Second-order Elastic Analysis

In second-order elastic analysis also, the material is assumed to have a linear-elastic relationship. However, the equilibrium is calculated on the deformed geometry of the structure. A rigorous second-order analysis includes both the member curvature ($P$-$\delta$) and side-sway ($P$-$\Delta$) stability effects. (It may be of interest to note that the structural system becomes stiffer when its members are subjected to tension and it becomes softer.) The detrimental effects associated with second-order deformations due to the compressive forces are considered to be important in structures subjected to predominant gravity loads.

In first-order analysis, the unknown deformations can be obtained in a simple and direct manner, whereas second-order analysis requires an iterative procedure to obtain the solutions. This is because the deformed geometry of the structure is not known during the formation of the equilibrium and kinematic relationship. Thus, the analysis proceeds in a step-by-step incremental manner, using the deformed geometry of the structure obtained from a preceding cycle of calculation. For most practical cases, accurate second-order design forces can be obtained by applying the loads in one or two increments, and only a few iterations are required to converge to an accurate solution. The iterative process used in non-linear solutions can take up a large amount of computer time and may diverge from (rather than converge to) the required result. Therefore, the objectives in selecting solution methods are to reduce the solution time and to preserve the stability of the solution.

The best known method for non-linear solution is the *Newton–Raphson method.* Figure 4.8 describes the Newton–Raphson method. The basic objective is to establish equilibrium

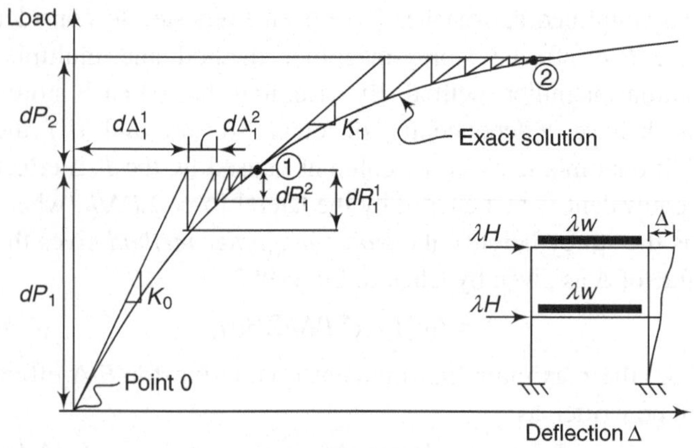


**FIG. 4.8** Newton–Raphson incremental-iterative solution method

at the end of each load increment. This is done by analysing the imbalance between the applied loads at the end of a linear step and the internal (element end) forces calculated from the results of that step. Thus, the imbalance between the linear approximation and the actual non-linear response is corrected in an iterative fashion. This scheme can be expressed in matrix form as follows (Galambos 1998):

$$\{dP\} - \{dR\} = [K_e]\{d\Delta\} \qquad (4.6)$$

where $\{dP\}$ is a vector of incremental applied nodal forces, $\{dR\}$ is a vector of imbalanced nodal forces (calculated as the difference between the resultant of the current internal member forces and the applied loads), $[K_e]$ is the linear stiffness matrix, and $\{d\Delta\}$ is the vector of incremental nodal displacements and rotations. Thus, the non-linear Eqn (4.6) is solved in a piecemeal linear fashion using the Newton–Raphson method, by applying successive increments of applied loads $\{dP_i\}$. If the imbalance $\left\{dR_i^1\right\}$ is significant (see Fig. 4.8), Eqn (4.6) is solved iteratively for successive values $\left\{dR_i^j\right\}$ and $\left\{d\Delta_i^j\right\}$, for $j$ = 2, 3, 4, ... until the equilibrium error represented by the unbalanced force vector is within acceptable limits. Conventional Newton–Raphson procedures pose some numerical difficulty when they approach the stability limit load and hence alternative solution schemes have been developed in the past. An overview of such solution procedures is provided by Chen and Lui (1991), Clarke in Chen and Toma (1994), and Yang and Quo (1994).

The use of first-order elastic stiffness matrix $[K_e]$ in Eqn (4.6), results in a large number of small increments $\{dP_i\}$. Moreover, the global equations will not be in a position to detect bifurcation points due to geometric instabilities. Hence, the use of stability functions is advocated in Chen & Lui (1991) and Galambos (1998).

#### Stability Functions

The *stability function* approach is developed directly from the differential equations for flexure, where the equilibrium equations are formulated on the deformed member, assuming small deformations (Livesley 1975). For a beam column subjected to end moments $M_A$ and $M_B$ and acted upon by an axial force $P$ as shown in Fig. 4.9, the incremental force–displacement relationship may be written as (Chen & Kim 1997).

$$\begin{Bmatrix} P \\ M_A \\ M_B \end{Bmatrix} = \frac{EI}{L}\begin{bmatrix} A/I & 0 & 0 \\ 0 & S_1 & S_2 \\ 0 & S_2 & S_1 \end{bmatrix}\begin{Bmatrix} \delta \\ \theta_A \\ \theta_B \end{Bmatrix} \tag{4.7}$$

where $P$ is the incremental axial force, $M_A$ and $M_B$ are the end moments, $\delta$ is the incremental axial displacement, $\theta_A$ and $\theta_B$ are the joint rotations, and $S_1$ and $S_2$ are the stability functions.

The stability functions may be derived as follows (Livesley 1975; Chen & Kim 1997; Subramanian 1999):

$$S_1 = s_{ii} = s_{jj} = \varepsilon(\sin\varepsilon - \varepsilon\cos\varepsilon)/[2(1 - \cos\varepsilon) - \varepsilon\sin\varepsilon] \tag{4.8a}$$

$$S_2 = s_{ij} = s_{ji} = \varepsilon(\varepsilon - \sin\varepsilon)/[2(1 - \cos\varepsilon) - \varepsilon\sin\varepsilon] \tag{4.8b}$$

where $\varepsilon^2 = PL^2/EI$ ($P$ negative). (4.9)

Similarly, when the axial force is tension ($P$ positive)

$$S_1' = \varepsilon(\sinh\varepsilon - \varepsilon\cosh\varepsilon)/[2(\cosh\varepsilon - 1) - \varepsilon\sinh\varepsilon] \tag{4.10a}$$

$$S_2' = \varepsilon(\varepsilon - \sinh\varepsilon)/[2(1 - \cosh\varepsilon) - \varepsilon\sinh\varepsilon] \tag{4.10b}$$

The shear induced at the ends of the member is given by

$$V = (EI/L^2)\{(S_1 + S_2)\theta_A + (S_1 + S_2)\theta_B - [2(S_1 + S_2) - \varepsilon^2](\Delta/L)\} \tag{4.11}$$

Note that when $P$ approaches zero, $\varepsilon^2$ also approaches zero. By using the L'Hospital's rule, it can be shown that $S_1$ reduces to 4 and $S_2$ reduces to 2, which are the same as in conventional slope-deflection equations. Also note that the equation for $S_1$ and $S_2$ are indeterminate when the axial force is zero. Hence in computer programs, power-series expansions are used to approximate the stability functions (Chen & Lui 1991; Subramanian 1999).

$$s_{ii} = 4 + \frac{2\varepsilon^2}{15} - \frac{11\varepsilon^4}{6{,}300} + \frac{\varepsilon^6}{27{,}000} - \frac{509\varepsilon^8}{582{,}120{,}000} + \cdots \tag{4.12a}$$

$$s_{ij} = 2 - \frac{\varepsilon^2}{30} + \frac{13\varepsilon^4}{12{,}6000} - \frac{11\varepsilon^6}{378{,}000} + \frac{907\varepsilon^8}{1164{,}240{,}000} - \cdots \tag{4.12b}$$

Equations (4.12a) and (4.12b) are applicable for members in tension ($P$ positive) and compression ($P$ negative). For most practical applications, Eqns (4.12a) and (4.12b) give an excellent correlation to the exact expressions given by Eqns (4.8) and (4.10). The tabulated values of stability functions can be found in Livesley & Chandler (1956) and Chen & Kim (1997), which also include a series expansion. Thus stability functions provide a simple means of accounting for the effect of axial force on the bending stiffness of the element. The advantage of stability functions is that they enable designers to use only one element per beam column and maintain accuracy in the element stiffness terms and in the recovery of element end forces for all ranges of axial loads.

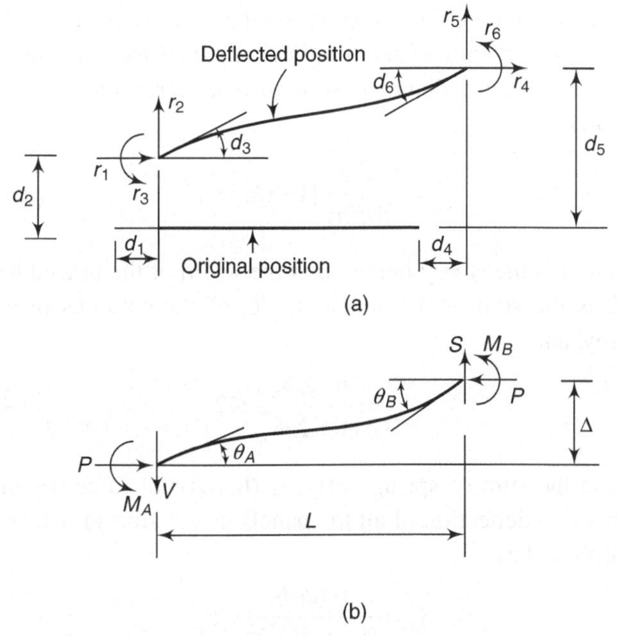


**FIG. 4.9** Typical beam column element

## Geometric Stiffness Approach

The finite element geometric stiffness approach as given here may lead to some error in the stiffness terms and in the force recovery process; hence while using the geometric stiffness approach, its range of applicability must be understood (White & Hajjar 1991). In the geometric stiffness approach, the stiffness matrix of a beam can be expressed as the sum of two matrices $[K_e]$ and $[K_g]$. $[K_e]$ is the *first-order frame stiffness* matrix and $[K_g]$ is called the *geometric stiffness matrix*. For an element $i$, the non-linear stiffness matrix is written as (Dhillon & O'Malley 1999):

$$[k]_i = [k_e]_i + [k_g]_i \tag{4.13}$$

where $[k_e]_i$ is expressed as follows.

$$[k_e] = \begin{bmatrix} \frac{AE}{L} & & & & & \\ 0 & \frac{12EI}{L^3} & & \text{Sym.} & & \\ 0 & \frac{6EI}{L^2} & \frac{4EI}{L} & & & \\ -\frac{AE}{L} & 0 & 0 & \frac{AE}{L} & & \\ 0 & -\frac{12EI}{L^3} & -\frac{6EI}{L^2} & 0 & \frac{12EI}{L^3} & \\ 0 & \frac{6EI}{L^2} & \frac{2EI}{L} & 0 & -\frac{6EI}{L^2} & \frac{4EI}{L} \end{bmatrix} \tag{4.14}$$

and

$$[k_g]_i = \frac{P}{L}\begin{bmatrix} 1 & & & & & \\ 0 & \frac{6}{5} & & & & \\ 0 & \frac{L}{10} & \frac{2L^2}{15} & \text{Sym.} & & \\ -1 & 0 & 0 & -1 & & \\ 0 & -\frac{6}{5} & -\frac{L}{10} & 0 & \frac{6}{5} & \\ 0 & \frac{L}{10} & -\frac{L^2}{30} & 0 & -\frac{L}{10} & \frac{2L^2}{15} \end{bmatrix} \quad (4.15)$$

where $L$ is the element length, $A$ is the area of cross section, $I$ is the moment of inertia, $E$ is the modulus of elasticity, and $P$ is the axial force in the element.

It is interesting to observe that these matrices can also be derived by retaining the first two terms in a Taylor series expansion for stability functions. Several textbooks contain general purpose computer programs that incorporate modules to do elastic second-order analysis (Wang 1973; Harrison 1990; Subramanian 1999).

Although the second-order elastic analysis can account for all the stability effects, it does not provide information on the actual inelastic strength of the structure. Therefore, in design, these combined stability and inelasticity effects must be accounted for in the code member-capacity checks. The load–displacement curve generated from a second-order analysis gradually approaches the horizontal straight line, which is the buckling load obtained from the elastic buckling analysis, as shown in Fig. 4.1.

Updated Langrangian procedure has also been used to derive geometric stiffness matrices, which represent $P$-$\Delta$ effects exactly, but in general, it only approximates the $P$-$\delta$ effects (White & Hajjar 1991; Yang & Quo 1994). Some approximate methods, such as *negative brace method* and *fictitious lateral load method* have also been developed, which could be used with linear analysis software, to model the second-order effects of frames, where deformations are usually not large. More details about various second-order elastic analysis methods may be found from Chen & Lui (1991), Liew et al. (1991), White & Hajjar (1991), and Galambos (1998). Second-order analysis programs that take into account the connection flexibility (see Section 4.5) are also available (Fyre & Morris 1975; Liew et al. 1993; Kishi et al. 1994; Chen et al. 1995).

It is interesting to note that many limit states code provisions prefer the direct second-order elastic analysis to frame design; most notably the Canadian limit states design specifications (CSA-S16.1 2001). In the Australian (AS-4100 1998) and the Indian code (IS 800 : 2007), the first-order elastic analysis with moment amplification is allowed when the moment amplification factors $\delta_s$ or $\delta_b$ ($K_{my}$ and $K_{mz}$ in the IS code—see clause 4.4.3 of the code) are not greater than 1.4. If they are greater than 1.4, a second-order analysis is required. The moment amplification factor $\delta_b$ for a braced member is given in Eqn (4.1b). The moment amplification factor $\delta_s$ for a sway member in rectangular frame is given by [this is similar to Eqn(4.3)]

$$\delta_s = \frac{1}{1-\Delta_s \sum P/(h_s \sum H)} \quad (4.16)$$

where $\Delta_s$ is the translational displacement of the top relative to the bottom in the storey height $h_s$ caused by the design horizontal storey shears $H$ at the column ends and $P$ is the design axial force in a column of the storey and the summations include all the columns of the storey.

Annex B of the IS code also states that in a frame where the elastic buckling load factor $\lambda_{cr}$, as determined below is greater than five, the changes in the effective stiffness of the members due to axial forces may be neglected.

$$\lambda_{cr} = 1/(200\phi_{smax}) \quad (4.17)$$

where $\phi_{smax}$ is the largest value of the sway index for any storey $\phi_s$ is given by

$$\phi_s = (\delta_U - \delta_L)/h_s \quad (4.18)$$

where $h_s$ is the storey height, $\delta_U$ is the horizontal deflection of the top of the storey due to the combined gravity and notional loads, and $\delta_L$ is the horizontal deflection of the bottom of the storey due to gravity and *notional load.*

The notional loads which are used to check the sway stability of the frame have been discussed in Section 2.9.2. The code (Annex B) also suggests that while calculating the sway, the *stiffening effect of infill wall* panels may be allowed by introducing a diagonal strut in that storey of area $A$ as follows.

$$A = \frac{k_3 \sum K_c}{h(h/b)}[1+(h/b)^2]^{3/2} \quad (4.19)$$

where $h$ is the storey height, $b$ is the width of the braced bay, $\Sigma K_c$ is the sum of the stiffness, $I/L$, of the columns in that storey, and

$$k_3 = \frac{h^2 \sum S_p}{80E \sum K_c} \le 2 \quad (4.20)$$

$\Sigma S_p$ is the sum of spring stiffness (horizontal force per unit horizontal deflection of all the panels in that storey), which is determined as

$$S_p = \frac{0.6h/b}{[1+(h/b)^2]^2} t_p E_p \quad (4.21)$$

where $t_p$ is the thickness of the wall panel and $E_p$ is the modulus of elasticity of the panel material.

The code also gives the following guidelines for considering a rigid jointed multi-storey building as a non-sway frame:

1. For a clad frame where the stiffening effect of cladding is not taken into account in the deflection calculation

$$\delta \leq h_s/2000 \qquad (4.22)$$

2. For an unclad frame or clad fame where the stiffening effect of cladding is taken into account in the deflection calculation

$$\delta \leq h_s/4000 \qquad (4.23)$$

When a rigid jointed frame does not comply with this criteria, it is classified as a sway frame according to the code, even if it is braced.

The code in Section 4.5.4 also gives some conditions for clad and un-clad frames, based on which the second-order effects of the loads acting on the structure in its deformed configuration may be neglected.

The results of second-order elastic analysis, using STAAD-Pro, of braced and unbraced frames are given in Figs 4.3 and 4.4. For the braced frame shown in Fig. 4.3, the difference between the second-order results and the first-order results is negligible. This is usually true for braced frames that have substantial bending effects and small axial compressions. For the unbraced frame shown in Fig. 4.4, the maximum second-order sway deflection is 11.2% larger than that of the first-order analysis, whereas the moment at the top of the right-hand lower storey column is about 5.4% larger than that of the first-order analysis. This shows that second-order effects are often significant in unbraced frames.

## 4.3 BUCKLING (CRITICAL LOAD) ANALYSIS

The elastic analysis discussed in the previous sections will not give the buckling behaviour of the structure. The buckling load of structures may be obtained by using a linear or inelastic buckling analysis. These two buckling analysis methods are discussed briefly in the following sections.

### 4.3.1 Linear Buckling Analysis

The elastic buckling load is calculated by linear buckling or bifurcation (or eigenvalue) analysis. The buckling loads are obtained from the solution of elastic frames subjected to idealized loads that do not produce direct bending of the structure. For example, in Fig. 4.10 the frame loads are replaced by a set of loads, which produce the same set of member axial forces without any bending. Before buckling, the displacements occur only in the directions of the applied loads. While buckling (when bifurcation occurs), the displacements begin and increase infinitely, assuming the linear theory of elastic stability, as shown in Fig. 4.1, as a horizontal straight line. The load at which these displacements occur is called as *buckling* or *bifurcation load.*

The results of an elastic buckling analysis may be used to approximate any second-order effect [see Figs 4.1 and 4.10(c)]. This analysis is also a common tool for calculating the effective lengths of columns. The effective length factor $K$ of columns can be calculated using the following expression (see Fig. 4.11):

$$K = \sqrt{(P_e/P_{cr})} = \sqrt{[(\pi^2 EI/L^2)/P_{cr}]} \qquad (4.24)$$

where $P_e$ is the Euler buckling load and $P_{\rm cr}$ is the axial load in the column at the incipient buckling of the frame.

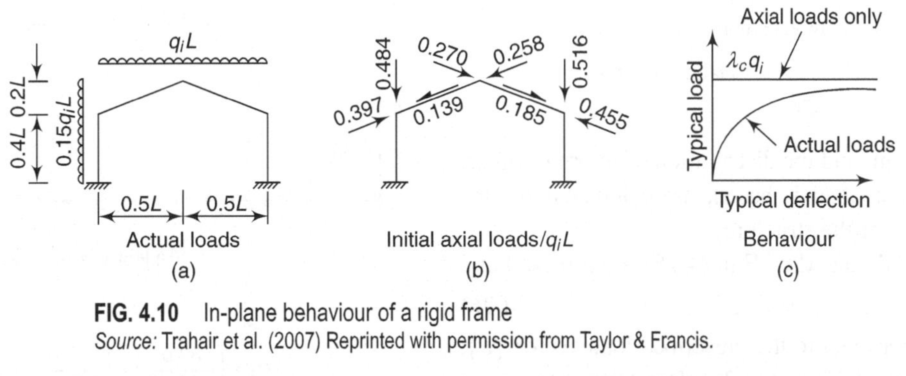


**FIG. 4.10** In-plane behaviour of a rigid frame
*Source:* Trahair et al. (2007) Reprinted with permission from Taylor & Francis.

Using a stiffness matrix formulation, the critical loads of frames and other structures may be calculated by using the following *eigenvalue problem.*

$$[K_e + \lambda_i K_g]\,\{\Delta_i\} = \{0\} \qquad (4.25)$$

where $[K_e]$ is the structural stiffness matrix, $[K_g]$ is the geometric stiffness matrix computed for a reference load $\{P_{\rm ref}\}$, $\lambda_i$ (an eigenvalue) is a load factor with respect to $\{P_{\rm ref}\}$, and $\{\Delta_i\}$ (an eigenvector) is the buckled mode shape. The lowest value of $\lambda$ that satisfies the equation for $\Delta \neq 0$ yields the elastic critical load vector $\lambda\{P_{\rm ref}\}$, and the corresponding $\{\Delta\}$ defines the buckled configuration. $[K_g]$ typically has the same form as the geometric stiffness matrix used in second-order analysis (see Section 4.2.3). However, in this case, the member forces that comprise the geometric stiffness are calculated by the first-order analysis. Thus, the resulting critical load is unique for the

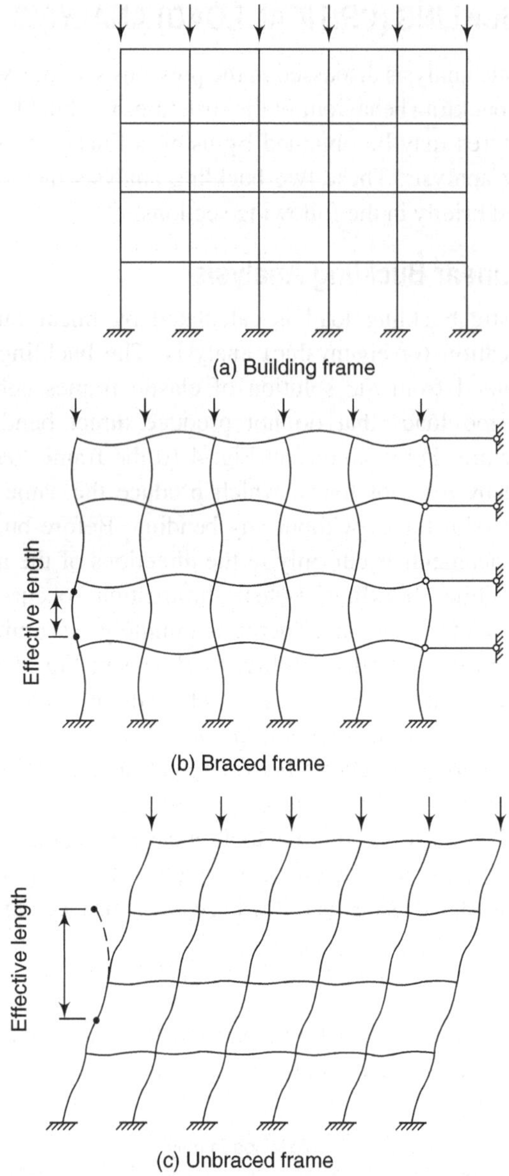


**FIG. 4.11** Buckling modes for rectangular frames and the *effective length concept*

given loading condition, and the distribution of internal forces, upon which the critical load is based, are calculated for the undeformed geometry of the structure.

The critical load calculated by Eqn (4.25) is expressed as

$$\{P_{cr,\,i}\} = \lambda_i\{P\} \tag{4.26}$$

where $\{P_{cr,\,i}\}$ is proportional to the prescribed load vector $\{P\}$. The subscript $i$ in Eqns (4.25) and (4.26) shows that there are as many buckling loads as there are number of degrees of freedom in Eqn (4.25). However, in many cases, the design engineers are only interested in the lowest critical value. The procedures adopted for solving Eqn (4.25) may be found in McGuire et al. (2000).

In addition to the rigorous computation method such as the one given earlier, many approximate and simplified methods for determining the critical loads of frames are also available (Horne 1975; ASCE 1997). However, since the *elastic critical loads* for most buildings are considerably higher than the inelastic limit loads (see Fig. 4.12), the linear buckling analysis is of limited practical value for the designers (Galambos 1998).

Alternatively, the effective length ratio $K_m$ of each compression member in a frame may be obtained by some simple charts, using the member end stiffness factors $k_1$ and $k_2$. The Indian code uses the charts developed by Wood (1973). It should be observed that these charts are applicable to the vertical column of regular rectangular frames with regular loading patterns in which each horizontal beam has zero axial force and all the columns buckle simultaneously in the same mode.

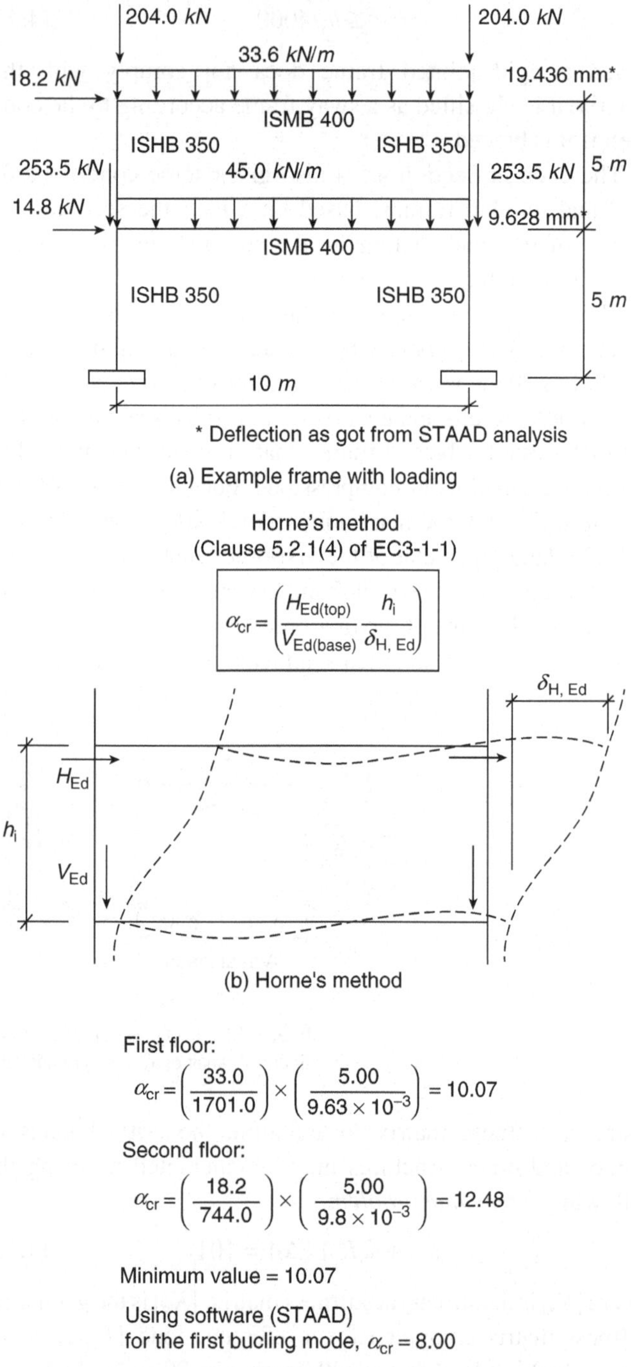


**FIG. 4.12** Example frame with buckling and collapse loads

A discussion about the various methods adopted to determine the elastic buckling loads of unbraced frames is provided by Trahair et al. (2007), though the computer programs based on Eqns (4.25) and (4.26) will also yield the required results.

### 4.3.2 Inelastic Buckling Analysis

As indicated earlier, the structural stiffness used in Eqn (4.25) is the elastic stiffness and the resulting critical load corresponds to elastic buckling. However, by modifying $[K_e]$ using a tangent modulus adjustment one can obtain critical loads that represent the actual limit state of the structure in a better manner (see Fig. 4.1). Thus, Eqn (4.25) can be adjusted to approximate the reduction in column stiffness due to axial compression and residual stresses (Ziemian et al. 1992). In this approach, the $[K]$ matrix in Eqn (4.25) is based on member stiffness, which is calculated using a tangent modulus for each member ($E_t$) consistent with the axial forces in that member at the critical load $\{P_{cr}\}$. The values of $E_t$ may be assumed based on some semi-empirical formulae derived from inelastic column curves (Chen & Kim 1997).

$$E_t = 4(f/f_y)(1 - f/f_y)E\xi_i \quad \text{for } P > 0.5P_y \tag{4.27}$$

$$E_t = E\xi_i \quad \text{for } P \leq 0.5P_y \tag{4.28}$$

where $E_t$ is the reduced tangent modulus, $\xi_i$ is the reduction factor for geometric imperfection, which may be taken as 0.85 for braced and unbraced frames, $f$ is the axial stress in the member, and $f_y$ is the yield stress of steel.

It is implicitly assumed that in Eqn (4.27), the proportional limit is half of the yield point. Other formulae that are used are as follows (Bowles 1980).

$$E_t = \lambda E \tag{4.29}$$

and

$$\lambda = (f_y - f_{cr}^2)/(f_y - f_{pl}^2) \tag{4.30}$$

where $f_y$ is the yield stress of steel, $f_{pl}$ is the steel stress at the proportional limit (which may be taken as $0.7f_y$ to $0.8f_y$), and $f_{cr}$ is the critical buckling stress ($= \pi^2 EI/L^2$). Petersen (1980) suggests the following formula for calculating the tangent modulus.

$$E_t = E[1 - \{(f - f_{pl})/(f_y - f_{pl})\}^2] \tag{4.31}$$

where $f$ is equal to $P/A$, and the other terms have already been defined.

Since the terms of the stiffness matrix $[K]$ are now dependent on the magnitude of the applied load vector $\{P_{cr}\}$, the adjustments in the tangent modulus result in an iterative solution procedure to solve Eqn (4.25). Computer programs are also available to calculate the inelastic critical loads of frames (Ganapathy & Subramanian 1977; Harrison 1990; Subramanian 1999). However, it is important to remember that though this tangent modulus adjustment procedure does not take into account the loss in stiffness due to the plastification of members under combined axial force and bending and the formation of plastic hinges; hence the critical loads calculated will not coincide with the inelastic limit point.

## 4.4 SOURCES OF NON-LINEARITY

Before discussing the various non-linear methods of analysis, it may be worthwhile to identify the different factors that influence steel frame design. In linear-elastic analysis, the material is assumed to be unyielding and the equations of equilibrium are formulated on the geometry of the undeformed configuration. Subsequent deformations are assumed to be very small such that they will not affect the equilibrium and mode of response of the system. Hence, it is possible to treat axial forces, bending moments, and torques as uncoupled actions and develop the stiffness equations for elements as discussed in Section 4.2. In reality, the behaviour of frames is non-linear due to the following effects (McGuire et al. 2000; Seetharaman et al. 2004).

### 4.4.1 Material Effects

The various material effects that influence the steel frame design are as follows.

1. *Gradual yielding due to bending*: Owing to bending, induced by external forces, the cross section will remain elastic, only when the bending is small. When the bending moment increases, the extreme fibres of beams yield gradually till the section becomes fully plastic (see Chapter 8). Due to this effect, the moment–curvature curve becomes non-linear (that is, a gradual transition from elastic to fully plastic). The gradual yielding also leads to the formation of plastic hinges in flexural members. In addition, the axial compression in beam columns also affects the moment-curvature relationship, especially after the initiation of yielding in the member. (A parabolic stiffness reduction function may be used in the non-linear analysis, to represent the gradual yielding of a section in flexure.) Similar inelastic interaction of axial-force, bending, shear, and torsion is also possible.
2. *Yielding due to residual stresses*: As mentioned earlier, the stress–strain curve of column members show a smooth transition from elastic to perfectly plastic due to the presence of residual stresses. As observed in Chapter 1 (Figs 1.24 to 1.27), these sections will have an initial stress distribution pattern due to residual stresses (residual stresses may result from manufacturing/fabrication processes, erection fit-up, construction sequencing, incidental thermal loads, or support settlements). This distribution will be altered by the presence of axial force, bending moment, etc., resulting in gradual yielding of the section (as the load is redistributed from the yielded region to fibres which are still elastic). This deterioration of the section due to plasticity is shown

in Fig. 4.13 and is similar to that occurring in a beam, where the cross-sectional stiffness is reduced to zero as a plastic hinge forms (this gradual degradation of stiffness may be modelled in the non-linear analysis by using the tangent modulus of the material).

3. *Plastic deformation of frames*: The plastification of cross sections will result in additional deformations.
4. Yielding in connection components and joint panels
5. Fracture of members and connections
6. Strain-hardening behaviour
7. Cyclic-plasticity effects
8. Load-path effects, shakedown, and incremental collapse

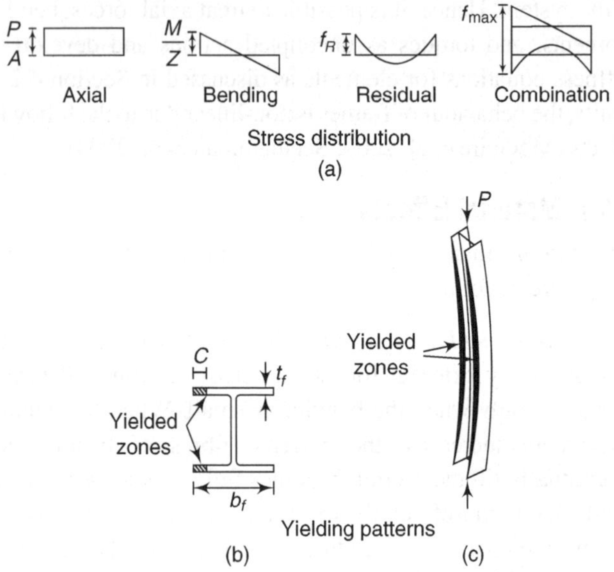


**FIG. 4.13** Partial yielding in a column due to residual stresses

### 4.4.2 Geometric Effects

The various geometric effects that influence the steel frame design are as follows.

1. Geometric effects include *initial imperfections* such as member out-of-straightness (or camber) and out-of-plumbness, resulting due to unavoidable errors during fabrication and erection. These imperfections induce additional moments in columns leading to further bending stiffness degradation (these could be considered in non-linear analysis either by modification of the tangent modulus or by representing them in the form of notional loads).
2. The *P-Δ effect*, a destabilizing moment equal to gravity load multiplied by the horizontal displacement, the frame undergoes as a result of the lateral load (see Section 4.2.1).
3. The *P-δ effect*, which is the influence of axial force on the flexural stiffness of an individual member. The P-Δ and P-δ effects often occur in combination (as discussed in Section 4.2.1) and may be considered in the category of *secondary bending moments*. Stability functions may be used to include these second-order effects.
4. Incidental connection and loading eccentricities.
5. Local buckling and cross section distortion (Subramanian 1995)
6 Finite rotation effects (three-dimensional structures)

### 4.4.3 Combined and Other Effects

Effects that act in combination and certain other effects that influence the steel frame design are listed as follows.

1. Plastic deformation plus P-Δ and/or P-δ effects
2. *Connection deformation*: The deformation behaviour of a typical beam column connection, when represented by non-linear moment–rotation relationship (discussed in detail in Section 4.5)
3. Panel zone deformations (see Section 4.5)
4. Contribution of infilling and secondary systems to strength and stiffness
5. Loading rate and duration
6. Finite member and joint size effects
7. Out-of-plane bracing elements
8. Foundation and support conditions
9. Shear connection to reinforced concrete slabs

Some of these factors are considered routinely in analysis, while some are included in design implicitly through specification provisions and established practice. Let us now discuss the methods which consider some of the important factors (it is impractical to consider all the effects in the structural analysis and hence the engineer should judge the magnitude of these effects before considering them).

## 4.5 JOINT AND CONNECTION EFFECTS

The three basic aspects of beam column joints that affect the frame behaviour are the finite size of the joint, shearing deformations of joint panels, and rotational deformations at the beam-to-joint connections. We will discuss these aspects separately.

### 4.5.1 Finite Size of Joint

The stiffening effect of the finite joint size often offsets the flexibility of joint panel and connection deformations. Hence, all these effects are usually ignored and the frame is modelled using the centreline dimensions and assuming rigid joints. Moreover, for fully welded or bolted moment connections with HSFG (high-strength friction grip) bolts, the relative rotation between the beams and columns is negligible. In frames with such connections, the combined effects of finite joint size and panel deformations may be approximately considered by modelling some fraction of the finite joint as rigid, using member end offsets (for example, normally an effective rigid joint size of half the true size of the joint is assumed as shown in Fig. 4.14. Figure 4.14(b) shows model A, which takes into account the finite width of the column shown in Fig. 4.14(a). Figure 4.14(c) shows the conventional model, model B, with the beam being flexible up to the column centrelines. Note that model A will be stiffer than model B, resulting in smaller deformation).

Figure 4.14(d) shows the beam with rigid ends under reverse bending; a common situation when lateral loads act on frames. The moment–rotation relationship for the beam can be expressed as (MacLeod 1990)

$$M = [6EI/\{L(1-\beta)^3\}]\theta \tag{4.32}$$

where $L$ is the centre to centre distance between the column, $C$ is the column width, and $\beta$ is equal to $C/L$.

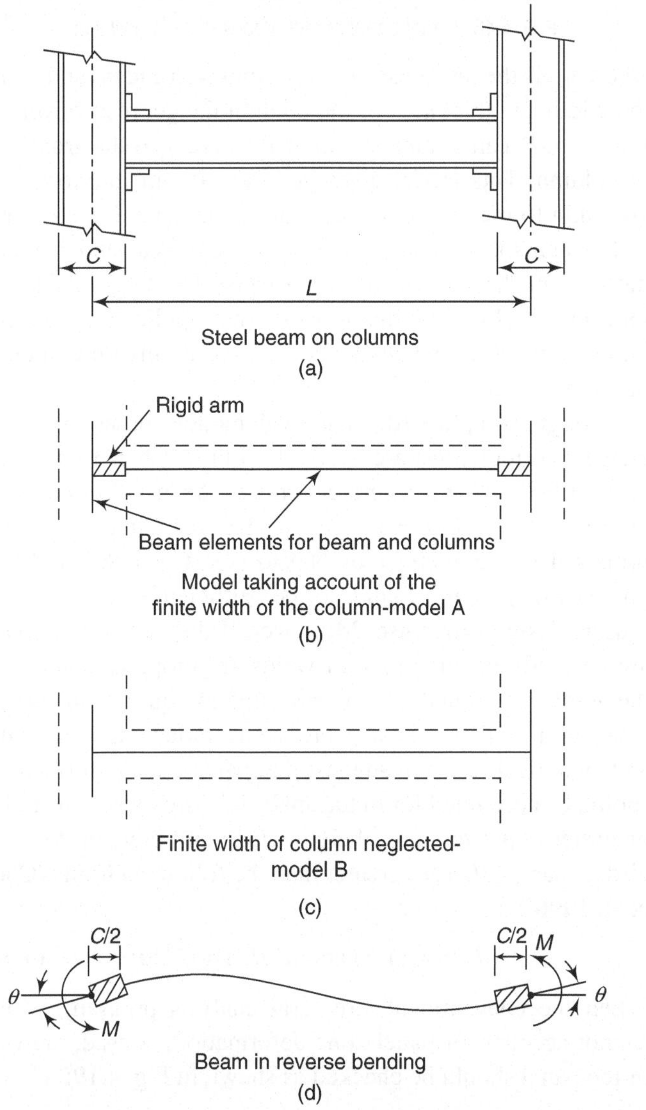


**FIG. 4.14** Modelling the effect of column width in analysis

The percentage increase in rotational stiffness, when the finite width of column is considered, may be expressed as (MacLeod 1990)

$$\alpha = 100[1/(1-\beta)^3] \tag{4.33}$$

A plot of $\alpha$ against $\beta$ is provided in Fig. 4.15, which shows the significant effect of rigid connection on the beam stiffness. The percentage difference in stiffness under a transverse load on the beam between model A and model B is also equal to $\alpha$. These arguments are also valid for the finite depth of beam in relation to the column stiffness.

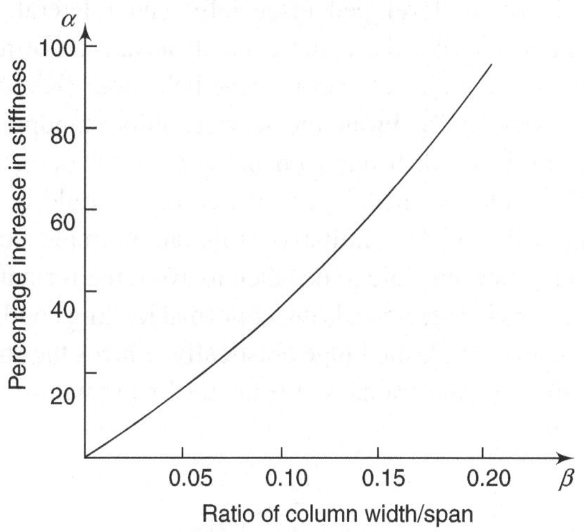


**FIG. 4.15** Effect of finite width of column on rotational stiffness

To evaluate frame stability correctly, joints and connections should be modelled directly in the analysis, wherever their effect is significant. Such an analysis is feasible through the combination of *rigid end links* (member end offsets) and *internal spring elements*, which are characterized by the relationship of the form:

$$P = k\delta \tag{4.34}$$

where $P$ is the direct force or moment, $k$ is spring stiffness, and $\delta$ is the movement in the direction of $P$. Figure 4.16 shows a joint element for in-plane conditions (up to three translational springs and three rotational springs can be included and various stiffness properties may be used). Both nodes would normally be in the same position but this is not a necessary condition. The details of frame analysis with member end offsets are given in Subramanian & Ganapathy (1980); Smith & Coull (1991).

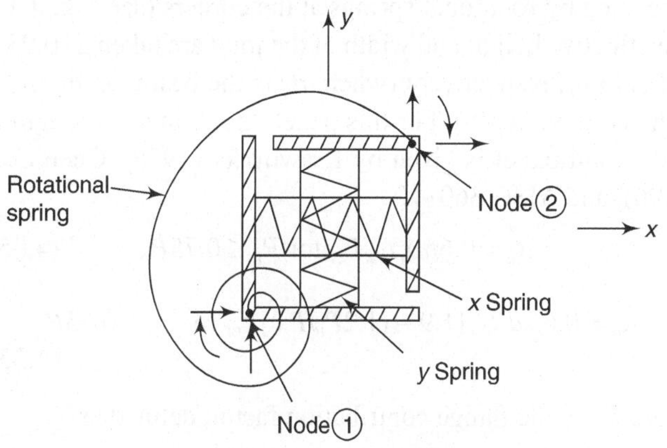


**FIG. 4.16** In-plane joint element with springs

### 4.5.2 Panel Zone Flexibility

A *panel zone* is the region of column web and flanges into which the beams are framed (Fig. 4.17). Although conventional analyses of frameworks are usually carried out without considering the effect of panel zone deformation, it has been shown analytically and experimentally that high shear forces are often developed (especially under lateral loads) in a joint panel, and the panel zone shear and deformations will have pronounced effect on frame behaviour (Krawinkler & Mohasseb 1987). From the serviceability standpoint, an increase in frame drift due to panel zone shear deformation is not desirable and hence proper measures should be taken to control this drift. Similarly, from an ultimate strength standpoint, measures are to be taken to avoid the formation of plastic hinges in these panels under normal loading conditions. The formation of plastic hinge drastically reduces the strength and stiffness of the frame and leads to the premature failure of the frame.

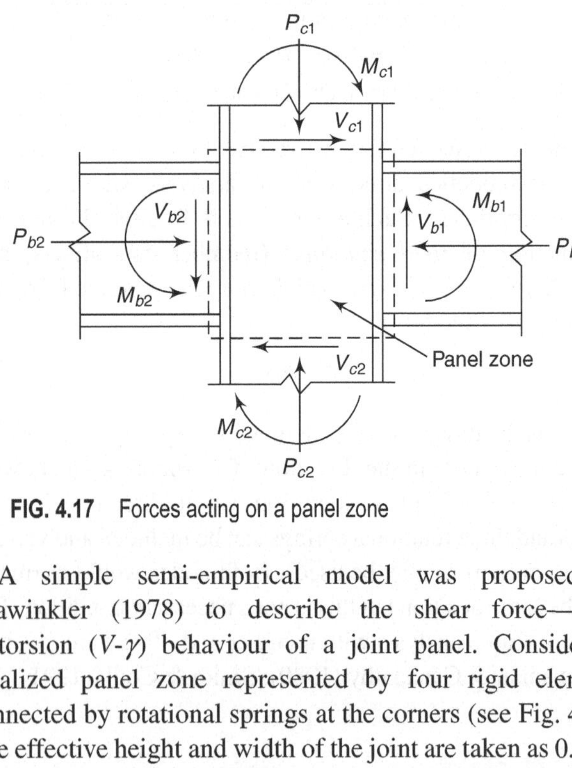


**FIG. 4.17** Forces acting on a panel zone

A simple semi-empirical model was proposed by Krawinkler (1978) to describe the shear force—shear distorsion ($V$-$\gamma$) behaviour of a joint panel. Consider an idealized panel zone represented by four rigid elements connected by rotational springs at the corners (see Fig. 4.18). The effective height and width of the joint are taken as 0.95 $d_b$ and 0.95 $d_c$, respectively (where $d_b$ is the beam depth and $d_c$ is the column depth). For this panel, the ultimate strength $R_u$ of the joint panel is given by Krawinkler (1978), Chen et al. (1996), and AISC 360–10.

$$R_u = 0.6 f_y d_c t_{cw} F_{cf} \text{ for } P_u \le 0.75 P_y \quad (4.35a)$$

$$R_u = 0.6 f_y d_c t_{cw} [1.9 - (1.2 P_u / P_y)] F_{cf} \text{ for } P_u > 0.75 P_y \quad (4.35b)$$

where $F_{cf}$ is the flange contribution factor, defined as

$$F_{cf} = 1 + 3 b_{fc} t^2_{fc} / (d_b d_c t_{cw}) \quad (4.35c)$$

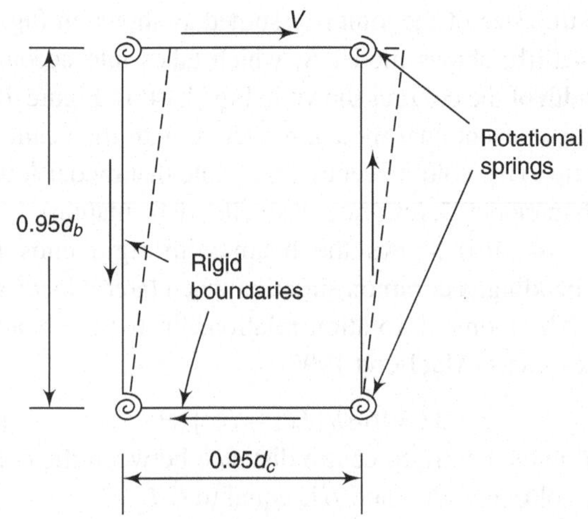


**FIG. 4.18** Joint panel model proposed by Krawinkler

where $t_{cw}$ is the thickness of the column web and $b_{fc}$ and $t_{fc}$ are the width and thickness of the column flange, respectively in Eqn 4.35, $P_y$ can be taken as $0.6 A_g f_y$ where $A_g$ is the gross area of column. This model gives good results for interior joints with thin to medium thickness column flanges. It also works well when the combined action of axial load and bending moment in the column does not cause yielding outside the joint panel. However, this model is not applicable to corner joints in which frame elements are present only on two faces of the panel zone.

Though complicated, analytical models based on finite element formulations are available (Lui & Chen 1986; Kato et al. 1988), which accurately model the panel zone shear deformation. It has been found that the simpler models proposed by Krawinkler or Nakao (see Chen & Lui 1991, pp. 310–12), were found to be sufficiently accurate for routine design office use. Moreover, if the panel-zone joints are carefully detailed and all welds are properly done, then the joints are found to possess superior quality in energy dissipation without causing premature failure. Experimental and analytical research suggest that panel-zone deformations should be accounted for in the analysis/design when the ratios of girder depth to storey height ($d_b/h$) and column depth to girder span ($d_c/L_b$) are greater than the following limits (Chen et al. 1996):

$$(d_b/h) = (1/8) \text{ and } (d_c/L_b) = (1/20) \quad (4.36)$$

When the conventional structural analysis program, which cannot account for panel-zone deformation, is used, the web of the panel should be checked as shown in Fig. 4.19:

$$\Sigma F = (M_{u1}/0.95 d_{b1}) + (M_{u2}/0.95 d_{b2}) - V_u > (R_u/\gamma_{mo}) \quad (4.37)$$

where $M_{u1}$ and $M_{u2}$ are the factored beam moments, $d_{b1}$ and $d_{b2}$ are the beam depths, $V_u$ is the factored column shear force, $R_u$ is the shear resistance of the panel zone according to Eqn (4.35), and $\gamma_{mo}$ is the material safety factor.

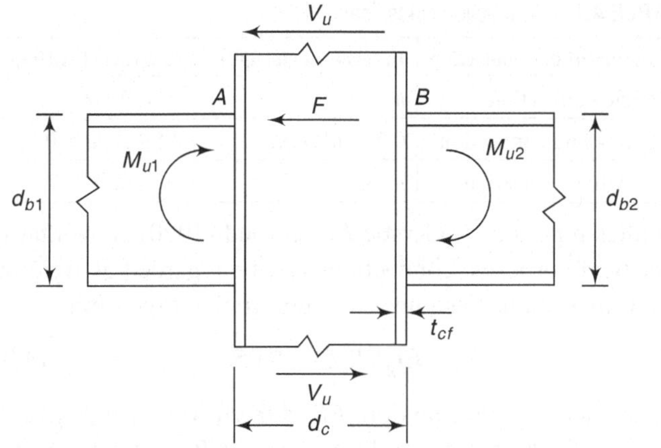


**FIG. 4.19** Equilibrium of forces acting on the panel-zone web

If $\Sigma F$ exceeds $(R_u/\gamma_{m0})$, the webs should be reinforced with *doubler plates*. This simplified approach is favoured in practical design because for the analysis program to take into account the panel-zone effects will involve extra degree of freedom or additional panel-zone elements in the stiffness formulations.

### 4.5.3 Modelling of Semi-rigid Connections

Conventional analysis and design of steel frames are usually carried out under the assumption that the connections joining the beams to columns are either fully rigid or ideally pinned. The rigid joint assumption implies that full slope continuity exists between the adjoining members and that the entire gravity moment is transferred from the beam to the column. On the other hand, an ideally pinned connection implies that the beams will behave as simply supported and the columns will not carry any gravity moment from the beams. However, experimental observations on different types of connection details have shown (see Fig. 4.20) that all connections used in current practice possess stiffness that fall between the fully

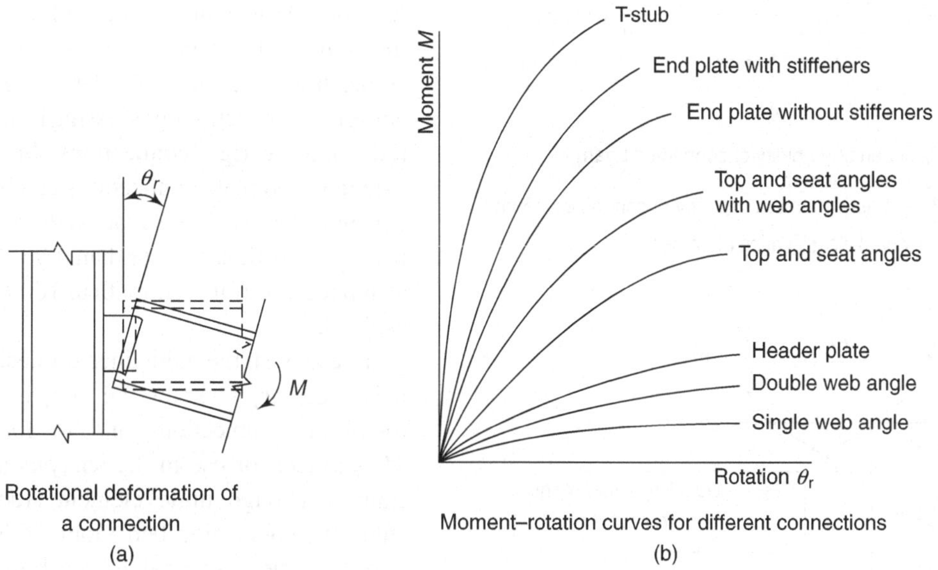


**FIG. 4.20** Different types of connections encountered in practice

rigid and ideally pinned extremes (Davison et al. 1987; Chen 1987, 1988; Chen & Kishi 1989; Arul Jayachandran et al. 2004).

The partially restrained connection not only changes the moment distribution among beams and columns, but also increases frame drift and accordingly increases the $P$-$\Delta$ effect in the frame analysis.

In engineering practice, it is important to know in what cases the connections can be assumed as rigid, semi-rigid, or flexible. The proposals for the *classification of connections* presented independently by Bjorhovde et al. (1990) and Eurocode 3 (1992) are shown in Figs 4.21 and 4.22. The classification system by Bjorhovde et al. is intended for the case in which prior knowledge of the member details is not available. This implies that the classification by Bjorhovde et al. does not consider the overall behaviour of frames. On the other hand, Eurocode 3 (EC3) classification is based on the overall strength of frames. Thus, EC3 classification is more rational if the layout and member details of the structural system are known in advance. However, ductility demand is not considered by the EC3 classification. It is also important to note that EC3 recognizes the fact that the semi-rigid action depends upon whether the frame is braced or unbraced. Hence, it has two different classification systems: one for the braced frame and the other for the unbraced frame.

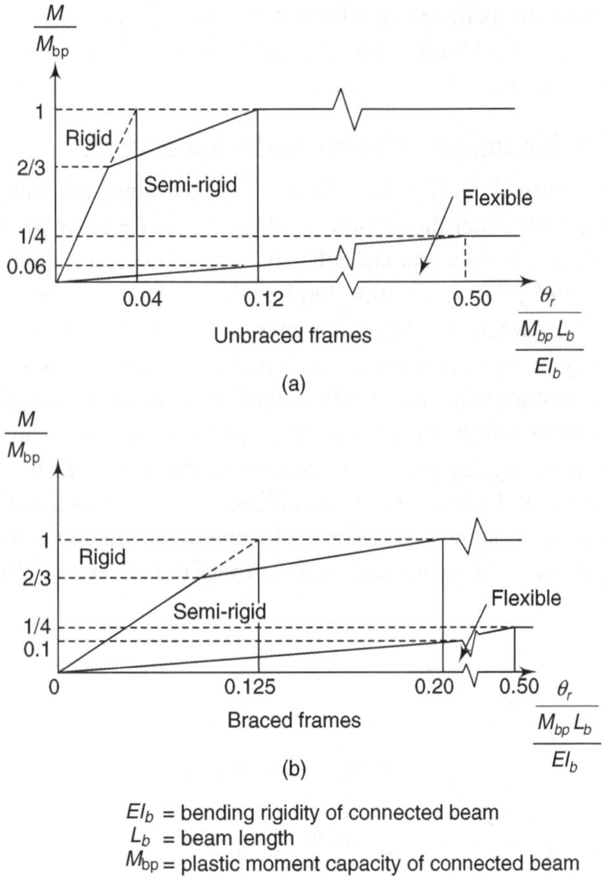


**FIG. 4.21** Eurocode 3 classification of frames

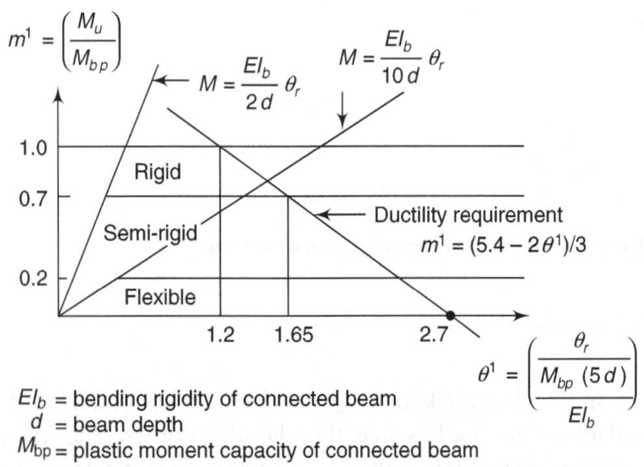


**FIG. 4.22** Classification system for frames proposed by Bjorhovde et al. (1990), Reprinted with permission from ASCE

Bjorhovde's classification is recommended by the code. As shown in Fig. 4.21 and Table 4.1, the connections are classified according to their ultimate strength or in terms of their initial elastic stiffness. It is based on the non-dimensional moment parameter ($m^1 = M_u/M_{bp}$) and non-dimensional rotation parameter ($\theta^1 = \theta_r/\theta_p$), where $\theta_p$ is the plastic rotation based on a reference length of beam equal to five times the depth of the beam.

**TABLE 4.1** Connection classification limits

| Nature of the connection | In terms of strength | In terms of stiffness |
|---|---|---|
| Rigid connection | $m^1 \geq 0.7$ | $m^1 \geq 2.5\theta^1$ |
| Semi-rigid connection | $0.7 > m^1 > 0.2$ | $2.5\theta^1 > m^1 > 0.5\theta^1$ |
| Flexible connection | $m^1 \leq 0.2$ | $m^1 \leq 0.5\theta^1$ |

Studies reported by Gerstle & Ackroud (1990) show that for practical purposes, connections can be regarded as rigid and the frame can be designed as a rigid frame if the limit

$$EI_g/(R_k L) < 0.05 \qquad (4.38)$$

is satisfied. In this equation, $EI_g$ is the flexural rigidity of the beam, $L$ is the length of the beam, and $R_k$ is the connection stiffness. Moreover, comparing the EC3 classification with a large database on moment–rotation characteristics of various connections, Kishi et al. (1997) have shown that single web angle and single plate connections, double web angle connections and header plate connections should be classified as flexible connections and top and seat angle connections with double web angle, top and seat angle connections, and flush-end-plate connections can be classified as semi-rigid connections. Hasan et al. (1997) have shown that extended end-plate connections possessing initial stiffness $\geq 10^{5.05}$ kNm/rad behave as rigid connections. All welded connections and extended end plate connections can be considered as rigid for design purposes. Arul Jayachandran et al. (2004) suggest that if the joint stiffness is less than 0.001 kNm, it may be assumed as hinged and if it is more than 1000 kNm it may be assumed as fixed.

To establish the guidelines for design of semi-rigid frames, it is necessary to know the $m$-$\theta_r$ behaviour of actual beam-to-column connections and to formulate the appropriate $M$-$\theta_r$ model for use in the analysis and design of semi-rigid frames. Though finite element studies have been used to study the connection behaviour (Krishnamurthy et al. 1979), at present the most used approach to describe the $M$-$\theta_r$ curve is to curve-fit the experimental data with simple expressions. Several analytical models have been developed to represent connection flexibility using available experimental test data: linear, bilinear, piece-wise linear (composed of a series of straight line segments), cubic B-spline (Jones et al. 1981), polynomial (Frye & Morris 1975), exponential function, and power function (Kishi & Chen 1990). The details of these models are summarized in Chen & Toma (1994) and Chen et al. (1995).

The code recommends the *Frye–Morris polynomial model*. The model has the form as shown in the following equation:

$$\theta_r = C_1(KM)^1 + C_2(KM)^3 + C_3(KM)^5 \qquad (4.39)$$

where $K$ is a standardization parameter dependent upon the connection type and geometry and $C_1$, $C_2$, and $C_3$ are curve fitting constants.

Table 4.2 shows the curve fitting constants $C_1$, $C_2$, and $C_3$ and the standardization constant $K$ for each connection type, while the size parameters for different types of connections are shown schematically in Fig. 4.23. Modified values of these constants are suggested in Table 44 of IS 800 : 2007.

**TABLE 4.2** Connection constants in Frye–Morris model

| Connection type | Curve-fitting constants | Standardization constants |
|---|---|---|
| Single web angle connection | $C_1 = 4.28 \times 10^{-3}$<br>$C_2 = 1.45 \times 10^{-9}$<br>$C_3 = 1.51 \times 10^{-16}$ | $K = 1.98 \times 10^{-6} d_a^{-2.4} t_a^{-1.81} g^{0.15}$ |
| Double web angle connection | $C_1 = 3.66 \times 10^{-4}$<br>$C_2 = 1.15 \times 10^{-6}$<br>$C_3 = 4.57 \times 10^{-8}$ | $K = 1.98 \times 10^{-6} d_a^{-2.4} t_a^{-1.81} g^{0.15}$ |
| Header plate connection | $C_1 = 5.10 \times 10^{-5}$<br>$C_2 = 6.20 \times 10^{-10}$<br>$C_3 = 2.40 \times 10^{-13}$ | $K = 1.17 \times 10^{-4} t_p^{-1.6} g^{1.6} d_p^{-2.3} t_w^{-0.5}$ |
| Top and seat angle with double web angle connection | $C_1 = 2.23 \times 10^{-5}$<br>$C_2 = 1.85 \times 10^{-8}$<br>$C_3 = 3.19 \times 10^{-12}$ | $K = 8.83 \times 10^{-4} d^{-1.287} t^{-1.128} t_c^{-0.415} l_a^{-0.694} g^{1.350}$ |
| Top and seat angle without double web angle connection | $C_1 = 8.46 \times 10^{-4}$<br>$C_2 = 1.01 \times 10^{-4}$<br>$C_3 = 1.24 \times 10^{-8}$ | $K = 1.28 \times 10^{-6} d^{-1.5} t_c^{-0.5} l_a^{-0.7} d_b^{-1.5}$ |
| End-plate connection without column stiffeners | $C_1 = 1.83 \times 10^{-3}$<br>$C_2 = -1.04 \times 10^{-4}$<br>$C_3 = 6.38 \times 10^{-6}$ | $K = 9.10 \times 10^{-7} d_g^{-2.4} t_p^{-0.4} d_b^{-1.5}$ |
| End-plate connection with column stiffeners | $C_1 = 1.79 \times 10^{-3}$<br>$C_2 = 1.76 \times 10^{-4}$<br>$C_3 = 2.04 \times 10^{-4}$ | $K = 6.10 \times 10^{-5} d_g^{-2.4} t_p^{-0.6}$ |
| T-stub connection | $C_1 = 2.1 \times 10^{-4}$<br>$C_2 = 6.2 \times 10^{-6}$<br>$C_3 = -7.6 \times 10^{-9}$ | $K = 4.59 \times 10^{-6} d^{-1.5} t^{-0.5} l_t^{-0.7} d_b^{-1.1}$ |

*Note:* Single web angle and double web angle can be regarded as flexible (simple) connections for drift check at service load and as semi-rigid for strength check at factored load (Kishi et al. 1997). Similarly, header-plate connections, top and seat angle connections (with and without double web angle) and end-plate connections without column stiffeners behave as semi-rigid connections (Kishi et al. 1997).

The code recommends the stiffness given Table 4.3, for preliminary analysis using a bilinear moment–curvature relationship. The values are based on secant stiffness at a rotation of 0.01 radian and typical dimension of connecting angle and other components as given in the table.

**TABLE 4.3** Secant stiffness

| Type of connection | Dimension (in mm) | Secant stiffness (in kN m/radian) |
|---|---|---|
| Single web connection angle | $d_a = 250$<br>$t_a = 10$<br>$g = 35$ | 1150 |
| Double web angle connection | $d_a = 250$<br>$t_a = 10$<br>$g = 77.5$ | 4450 |
| Top and seat angle connection without double web angle connection | $d_a = 300$<br>$t_a = 10$<br>$l_a = 140$<br>$d_b = 20$ | 2730 |
| Header plate | $d_p = 175$<br>$t_p = 10$<br>$g = 75$<br>$t_w = 7.5$ | 2300 |

(*Contd*)

The parameters given in Tables 4.2 and 4.3 are defined as follows (see Fig. 4.23 also):

$d_a$ = depth of the angle (in mm)
$t_a$ = thickness of the top angle (in mm)
$l_a$ = length of the angle (in mm)
$d_b$ = diameter of the bolt (in mm)
$d_g$ = centre to centre of the outermost bolt of the end-plate connection (in mm)
$t_p$ = thickness of end plate (in mm)
$t$ = thickness of column flange and stub connector (in mm)
$d$ = depth of the beam (in mm)
$l_t$ = length of the top angle (in mm)

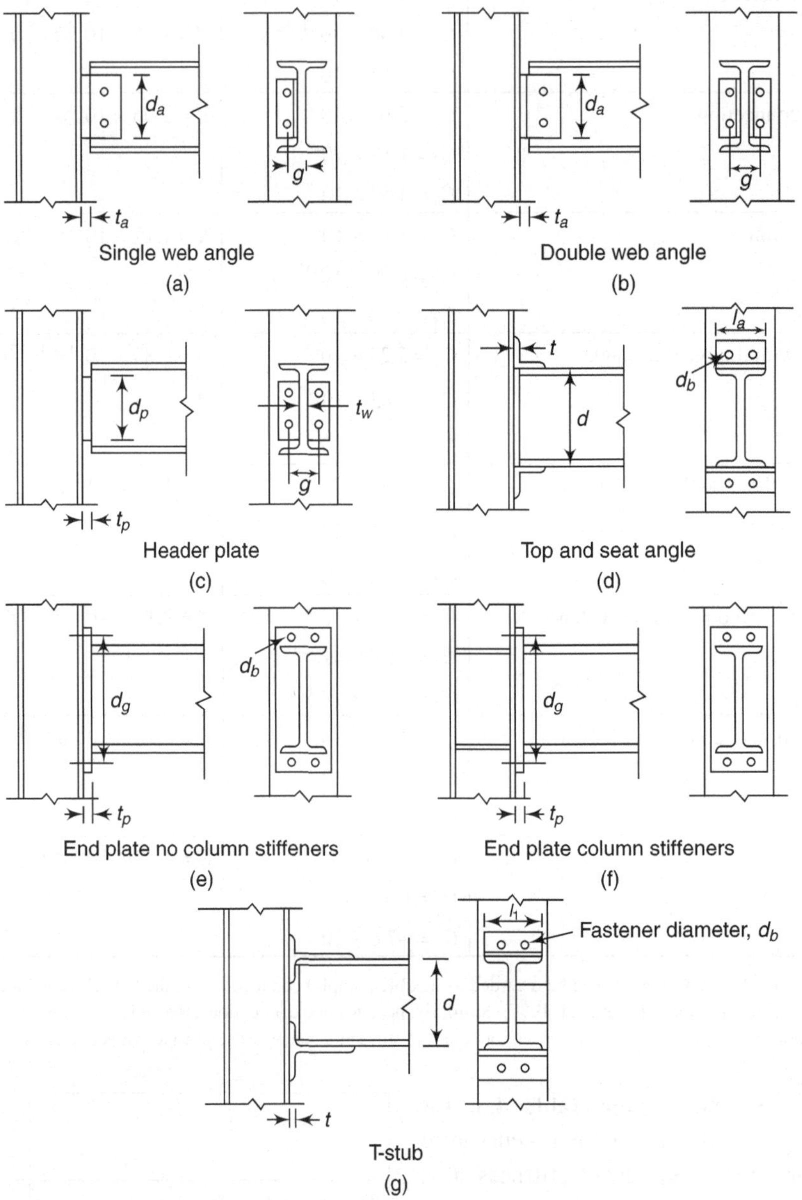


**FIG. 4.23** Common types of beam column connections and the size parameters

## 4.5.4 Analysis of Semi-rigid Frames

When a moment is applied to a connection in Fig. 4.20(a) the connection rotates according to the curves shown in Fig. 4.20(b); however, if the direction of the moment is reversed, the connection will unload and follow a different path which is almost linear with a slope equal to the initial slope of the $M$-$\theta_r$ curve. The typical moment–rotation behaviour of semi-rigid connections is shown in Fig. 4.24. Because of this behaviour, identical connections at the end of a beam may not always behave identically. To illustrate this, let us consider a portal frame subjected to gravity loads as shown in Fig. 4.25. In this case, the connection is subjected to a moment of $M_g$ on the $M$-$\theta_r$ curve [Fig. 4.25(a)]. At this point, if a lateral load is applied to the frame, the leeward connection will continue to hold higher loads, but the windward connection will unload [see Fig. 4.25(b)]. As a result of this, the apparent stiffness of the connections under the action of this lateral load will be different. For an extreme case (when the gravity load moment equals the ultimate moment capacity of the connections), the leeward connection will act like a pinned connection and the windward connection as a linear elastic connection (with a stiffness equal to its initial stiffness). This kind of loading/unloading of connection must be properly modelled to predict reliably the response of the frame (Chen & Kim 1997).

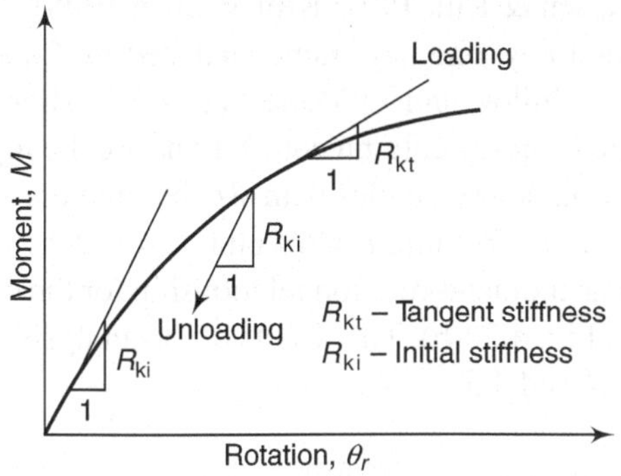


**FIG. 4.24** Typical moment–rotation behaviour of semi-rigid connections

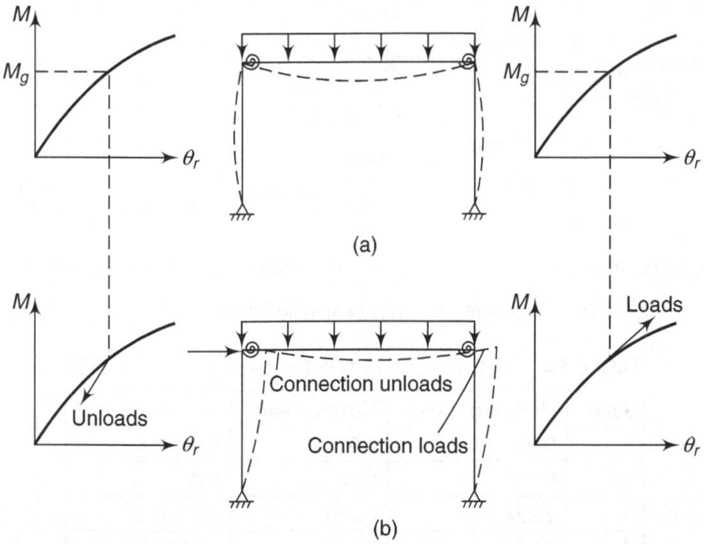


**FIG. 4.25** Effect of unloading/loading characteristics of connection in frames

To incorporate the effect of connection flexibility into the member stiffness relationship, the connection is modelled as a rotational spring with the moment–rotation relationship described by the Frye–Morris model given in Eqn (4.39). Consider a beam element subjected to end moments $M_A$ and $M_B$ and axial force $P$ with the connections attached at both the ends as shown in Fig. 4.26. The presence of the connections introduce relative incremental rotations $\theta_{rA}$ and $\theta_{rB}$ at the $A$th and $B$th end of the beam. Denoting $R_{kiA}$ and $R_{kiB}$ as the initial stiffness of connections $A$ and $B$, the relative rotation between the joint and the beam end (that is, rotational deformations of the connection) may be expressed as

$$\theta_{rA} = M_A/R_{kiA} \quad \text{and} \quad \theta_{rB} = M_B/R_{kiB} \tag{4.40}$$

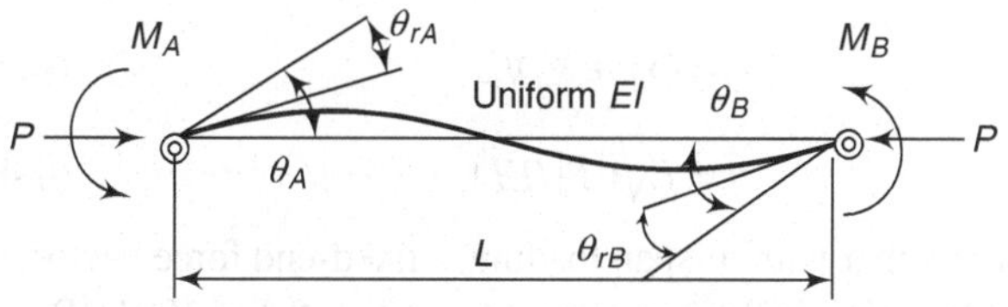


**FIG. 4.26** Beam-column element with semi-rigid connections

The initial stiffness $R_{ki}$ and $M_u$ can be obtained directly by generating the available database programs (Toma & Chen 1993).

If we denote $\theta_A$ and $\theta_B$ as the joint rotations at the $A$th and $B$th ends of the beam, respectively, then the slope deflection equations for the beam modified by the presence of flexible connections is given by

$$M_A = (EI/L)[s_{ii}(\theta_A - M_A/R_{kiA}) + s_{ij}(\theta_B - M_B/R_{kiB})] \tag{4.41a}$$

$$M_B = (EI/L)[s_{ij}(\theta_A - M_A/R_{kiA}) + s_{jj}(\theta_B - M_B/R_{kiB})] \tag{4.41b}$$

where $EI$ is the flexural rigidity of the beam, $L$ is the length of the beam, and $s_{ii}$ and $s_{ij}$ are the stability functions described in Section 4.2.3.

Solving the preceding equations for $M_A$ and $M_B$, we get

$$M_A = (EI/L)[s^*_{ii}\,\theta_A + s^*_{ij}\,\theta_B] \tag{4.42a}$$

$$M_B = EI/L\,[s^*_{ji}\,\theta_A + s^*_{jj}\,\theta_B] \tag{4.42b}$$

where $$s^*_{ii} = [s_{ii} + EIs^2_{ii}/(LR_{kiB}) - EIs^2_{ij}/(LR_{kiB})]/R^* \tag{4.43a}$$

$$s^*_{jj} = [s_{jj} + EIs^2_{ii}/(LR_{kiA}) - EIs^2_{ij}/(LR_{kiA})]/R^* \tag{4.43b}$$

$$s^*_{ij} = s^*_{ji} = s_{ij}/R^* \tag{4.43c}$$

and $R^* = [1 + EIs_{ii}/(LR_{kiA})]\,[1 + EIs_{ii}/(LR_{kiB})] - (EI/L)^2\,[s^2_{ij}/(R_{kiA}R_{kiB})]$ (4.43d)

The modified slope deflection equations can be written in the matrix form as

$$K = \frac{EI}{L}\begin{bmatrix} \frac{A}{I} & 0 & 0 & -\frac{A}{I} & 0 & 0 \\ & a & b & 0 & -a & c \\ & & s^*_{ij} & 0 & -b & s^*_{ij} \\ & \text{sym} & & \frac{A}{I} & 0 & 0 \\ & & & & a & -c \\ & & & & & s^*_{jj} \end{bmatrix} \tag{4.44}$$

where

$$a = (s_{ii}^{*} + 2s_{ij}^{*} + s_{jj}^{*} - u^2)/L^2 \quad (4.45a)$$

$$b = (s_{ii}^{*} + s_{ij}^{*})/L \quad (4.45b)$$

$$c = (s_{ij}^{*} + s_{jj}^{*})/L \quad (4.45c)$$

and

$$u = L\sqrt{(|P|/EI)} \quad (4.45d)$$

for a member with in-span loading, a fixed-end force vector must be added to the stiffness equation (see Chen & Lui 1991). Readers may refer to Chen & Lui (1991) (stability function approach) or Dhillon & O'Malley (1999) (geometric stiffness matrix approach) for more details about the derivation of these matrices.

Since the resulting equations are non-linear, the following iterative solution procedure is employed in the computer software (Dhillon & O'Malley 1999).

1. Divide the applied loads into a series of small increments.
2. Calculate the load increment vector $\{\Delta P\}$.
3. Calculate the member stiffness matrix $\{k\}_i$ for all members.
4. Assemble the member stiffness matrices to form the structure stiffness matrix $[K]$.
5. For $\{\Delta D\}$, solve the incremental stiffness equation $[K]\{\Delta D\} = \{\Delta P\}$.
6. Determine the member end actions.
7. Check convergences and if convergence is achieved, proceed to step 11.
8. Calculate connection secant stiffness (see Fig. 4.27).
9. Update the non-linear terms in the member stiffness, using the most recent connection stiffness, member forces, and structure geometry.
10. Repeat steps 3–9 until convergence is achieved.
11. At convergence, calculate accumulated displacements and member forces.
12. Continue till the analysis is complete for all load increments.

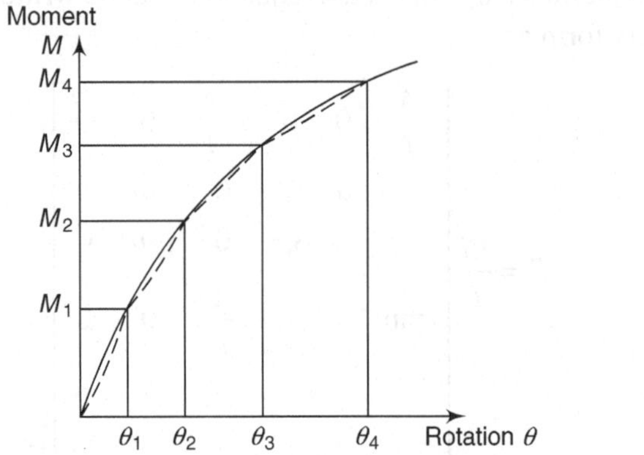


**FIG. 4.27** Connection secant stiffness through load increments

Software packages are available for the analysis of frames with semi-rigid connections (Chen et al. 1995; Kim & Chen 1998; Faella et al. 2000; Arthi et al. 2002). Details about large displacement second-order analysis, buckling analysis, and elastic–plastic analysis of semi-rigid frames are provided by Chen & Lui (1991) and Faella et al. (2000). Design aid for semi-rigid frame is provided by Kishi et al. (1995). Non-linear cyclic analysis of frames with semi-rigid connections has been dealt by Chen and Chui (2000) and seismic response of multi-storey steel buildings with flexible connections has been studied by Sivakumaran (1988). Benchmark problems to check the validity of results of computer programs are available (Chen & Kim 1997; Kim & Chen 1998).

The two-storey, one bay frame analysed by Chen and Lui (1991) and Dhillon and O'Malley (1999) is shown in Fig. 4.28(a). The beam-to-column connections are the top and seat angle with double web angles with size parameters as follows: $d = 204.6$ mm, $t = 7.9$ mm, $t_a = 6.4$ mm, $l_a = 152.4$ mm, and $g = 50.8$ mm. The moment–rotation relationship for the connection is shown in Fig. 4.28(b). The results of the analysis are shown in Tables 4.4 and 4.5.

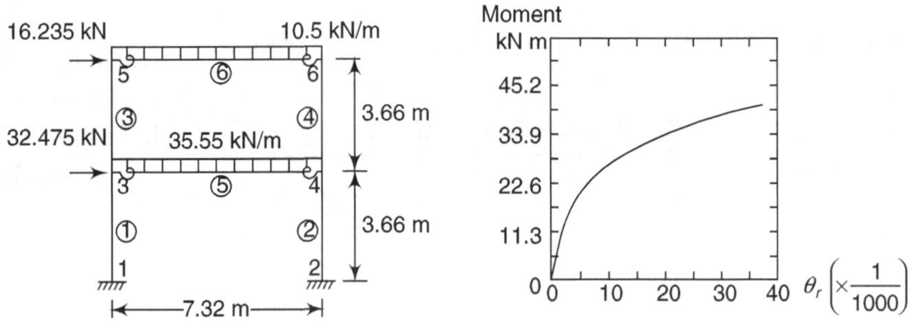


(a) Geometry of the frame (b) Moment–rotation curve of connections

**FIG. 4.28** Two-storey, one-bay semi-rigid frame

**TABLE 4.4** Joint displacements of the frame given in Fig. 4.28

| Joint | Horizontal (mm) | Vertical (mm) | Rotation (rad) |
|---|---|---|---|
| 1 | 0.00 | 0.00 | 0.00 |
| 2 | 0.00 | 0.00 | 0.00 |
| 3 | 25.58 | –0.54 | –0.01012 |
| 4 | 25.50 | –0.60 | –0.00976 |
| 5 | 62.26 | –0.80 | –0.00897 |
| 6 | 62.18 | –0.89 | –0.00859 |

**TABLE 4.5** Absolute member end forces for the frame given in Fig. 4.28

| Member | Axial force (kN) | Shear force (kN) | Moment (kN m) |
|---|---|---|---|
| 1 | 157.56 | 23.84 | 83.64 |
| 2 | 179.26 | 25.80 | 85.99 |
| 3 | 32.92 | 5.92 | 15.69 |
| 4 | 43.90 | 10.59 | 24.39 |
| 5 | 14.90 | 135.40 | 26.53 |
| 6 | 10.41 | 43.90 | 24.39 |

It has been found that flexible connections in comparison to rigid connections reduce beam-end moment and increase span moments, and result in higher moments in columns (Arul Jayachandran et al. 2004). Hence the semi-rigidity may reduce the beam sections and at the same time may govern the column design in frames. The semi-rigid connection results in lowering the overall weight of the frame mainly due to reduction in sizes. Dhillon and O'Malley (1999) found that the weight reduction in a typical semi-rigid frame is 7.2% compared to the rigid frame. The savings over the rigid frame will be more significant, if we consider the additional saving in connection cost. However, it is important to note that the drift increases in semi-rigid frame compared to the rigid frame by about 18.5%. Owing to this, in some cases the connection design may be controlled by drift limitations. Gerstle and Ackroyd (1990) provide the following prediction formulae, which can be used in the preliminary analysis for top and seat angles:

$$\Delta/H = W/[1315 + 2335\,(B/H)] \tag{4.46}$$

For flange plates,

$$\Delta/H = W/[1900 + 2335\,(B/H)] \tag{4.47}$$

where $\Delta$ is the lateral deflection at the top storey of the semi-rigid frame (in m), $H$ is the overall height of the frame (in m), $B$ is the overall width of the frame (in m), and $W$ is the lateral load intensity in $kN\,m^{-1}$ of vertical height.

Drift can be effectively reduced by increasing the connection stiffness, which implies that the designer must balance the overall building drift and connection stiffness in developing an economic framing system.

### 4.5.5 Modelling of Foundations

The modelling of the base of columns in ductile frames needs to be considered. The common assumption of full fixity at the column base may be valid only for columns supported by raft foundations or on individual foundation pads supported by short stiff piles or by basement walls. Foundation pads supported on deformable soil may have considerable rotational flexibility, resulting in column moments in the bottom storey quite different from those resulting from the assumption of a rigid base. The consequence can be unexpected column hinging at the top of lower storey columns. In such cases, the column base should be modelled by a rotational spring [Fig. 4.29(b)] of flexural stiffness (Paulay & Priestley 1992)

$$(M/\theta) = K_f = k_s I_f \tag{4.48}$$

where $k_s$ is the modulus of subgrade reaction in $kN/m^3$ and $I_f$ is the second moment of area of the foundation pad/soil interface corresponding with pad rotation, as shown in Fig. 4.29(a). The typical values of the modulus of subgrade reaction are given in Appendix C. For computer programs that do not have the facility to directly input the characteristics of rotational springs, the foundation rotational flexibility may be modelled by the use of a fictitious prismatic column number, as shown in Fig. 4.29(c) of length $l$ and flexural rigidity $EI$, such that

$$K = 4EI/l = k_s I_f \tag{4.49}$$

Alternatively, modelling as shown in Fig. 4.29(d) may also be used (Paulay & Priestley 1992).

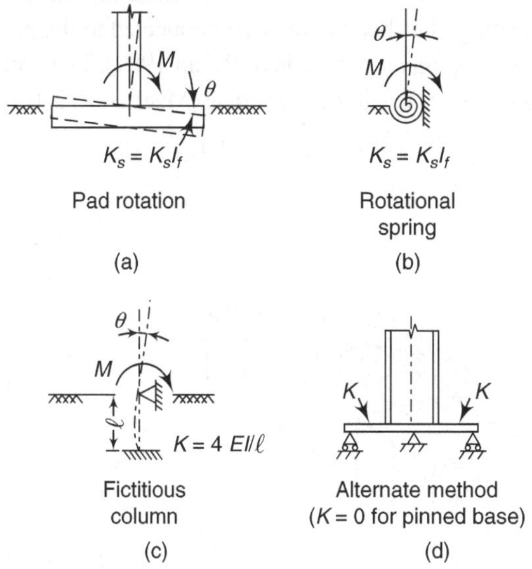


**FIG. 4.29** Modelling of column base rotational stiffness

IS 800 : 2007 stipulates that if the pedestal and foundation stiffness is not known, then the following may be assumed (see also clause 4.3.4).

1. When the column is rigidly connected to a suitable foundation, the stiffness of the pedestal will be taken as the stiffness of the column above base plate.
2. When the column is nominally connected to the foundation, 10% of the column stiffness may be assumed as the pedestal stiffness.
3. When an actual pin or rocker is provided in the connection between the steel column and pedestal, the base stiffness will be taken as zero and the column will be assumed as hinged at base.

### 4.5.6 Infilled Frames

Brickwork or blockwork which 'infills' the area between beams and columns in a frame alters the intended structural response, even if they are relatively weak. These infills, stiffen the frames and as a consequence the natural period of the structure will decrease, resulting in increased seismic forces. These infills are assumed to act as a diagonal compression strut (connected by pins to the frame corners) under the action of lateral loads. Paulay & Priestley (1992) recommend that the area of such a strut will be taken as equal to one fourth of the diagonal length of the panel multiplied by the wall thickness.

The provisions as given in the IS code for the consideration of infilled frames have been discussed in Section 4.2.3.2 [Eqns (4.19) to (4.21)].

### 4.5.7 Stressed Skin Design

In pitched portal structures [Fig. 4.30(a)], the shear stiffness of the sheeting can be modelled by representing each sheeting panel by a diagonal tension member. In order to utilize the shear stiffness of the sheeting, the sheeting must be connected to the purlins by proper *shear connectors* (Davis & Bryan 1982). The area of the diagonal member $A_d$ can be derived as (MacLeod 1990):

$$A_d = b/(E\sin^3 \theta c) \tag{4.50}$$

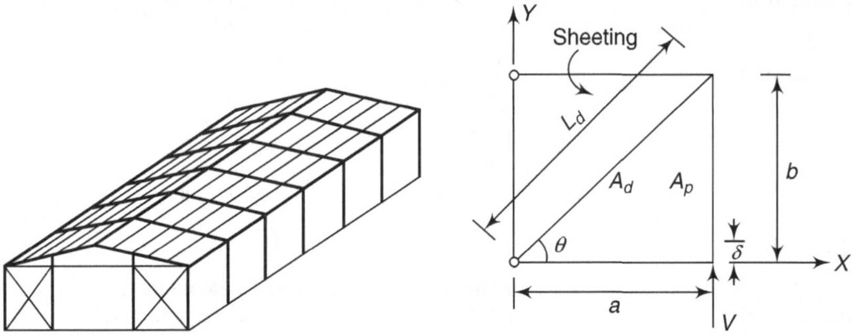


Pitched portal structure

(a)

Shear flexibility of sheeting represented by a diagonal member

(b)

**FIG. 4.30** Pitched portal structure

where $\theta = \tan^{-1}(b/a)$, $b$ is the length of the panel parallel to the corrugation of steel sheeting (the length in the direction of the shear load $V$), $a$ is the length of the panel perpendicular to the corrugations, $E$ is the Young's modulus, $c$ is the flexibility factor = $\delta/V$, and $\delta$ is the deflection of the panel in the direction of shear force $V$ [see Fig. 4.30(b)].

Davies & Bryan (1982) give the values of *flexibility factor*. The calculations for stressed skin design are also available in Errera & Apparao (1976) and Subramanian (1999). More details of modelling of steel structures for computer analysis may be found from MacLeod (1990) and Brown (1995).

## 4.6 FIRST-ORDER PLASTIC ANALYSIS

In the *first-order plastic analysis* method, all instability effects are ignored and the collapse strength of the frame is determined by using the rigid-plastic assumption (Neal 1977). The traditional method of calculating the plastic collapse load of a steel structure is the *limit analysis approach* where possible collapse mechanisms are searched to find the one that gives the minimum collapse load, called the *plastic limit load* (see Fig. 4.1). This procedure requires that the loads are assumed to be proportionate, that is, the vertical load and the lateral load increase together in a fixed ratio. The major advantage of the load stepping approach for non-linear modelling of frames is that one can observe the development of plastic hinges and the degradation of stiffness as the load is applied. When the collapse load is obtained one does not need to worry about the possibility of having missed a collapse mode provided the load steps are sufficiently small (MacLeod 1990). All members must be ductile so that the plastic moment capacity can be maintained at each hinge location over a range of hinge rotations sufficient to allow the plastic collapse mechanism to develop. Ductility is usually ensured by the following: (a) restricting the steel type to one which has sufficient yield plateau (extend for at least six times the yield strain) and significant strain hardening (see Sections 1.8.4 and 1.11) and (b) preventing reductions in rotation capacity by (i) local buckling—by restricting the dimensions of the different elements of the cross section, (ii) out-of-plane buckling effects—by restricting the unbraced lengths and providing torsional restraint against lateral buckling, and (iii) in-plane buckling effects—by restricting the member in-plane slenderness and axial compression forces. In addition to these, the members should not be subjected to impact loading. These restrictions are specified in clause 4.5.2 of the IS code.

The methods used for plastic analysis of frames are extensions of those discussed in Chapter 8, which incorporate reductions in the plastic moment capacity to account for the presence of axial forces. The manual application of these methods for frames is not simple as it is for beams, but computer programs are available, which may be used to arrive at the plastic limit load of frames (Subramanian 1974; Hancock et al. 1995; Chen and Sohal 1995). The details of these methods are available in Neal (1977), Horne and Morris (1981), Chen and Sohal (1995), and Davies and Brown (1996).

## 4.7 SECOND-ORDER INELASTIC ANALYSES

*Inelastic analysis* refers to any method of analysis in which the effects of material yielding are accounted for. The different types of inelastic methods may be generalized into the following three main groups:

1. Plastic zone or distributed plasticity method
2. Elastic–plastic hinge method
3. Refined or modified plastic hinge method

The plastic hinge methods are also referred to as concentrated plasticity methods. This generalization is based on the degree of refinement in the representation of yielding effects. The elastic–plastic hinge method is the simplest approach, whereas the plastic-zone method exhibits greatest refinement. The refined plastic-hinge method is an improvement over the elastic–plastic hinge method and hence requires less computational effort and is less costly (in terms of computer time) than the plastic-zone method. The concept of the spread of plasticity for the various inelastic analysis methods is shown in Fig. 4.31.

The second-order inelastic methods are often referred to as *advance analysis methods.* By this it means that they take into account the relevant material properties, residual stresses, geometric imperfections, second-order effects, three-dimensional effects, erection procedures, and interaction

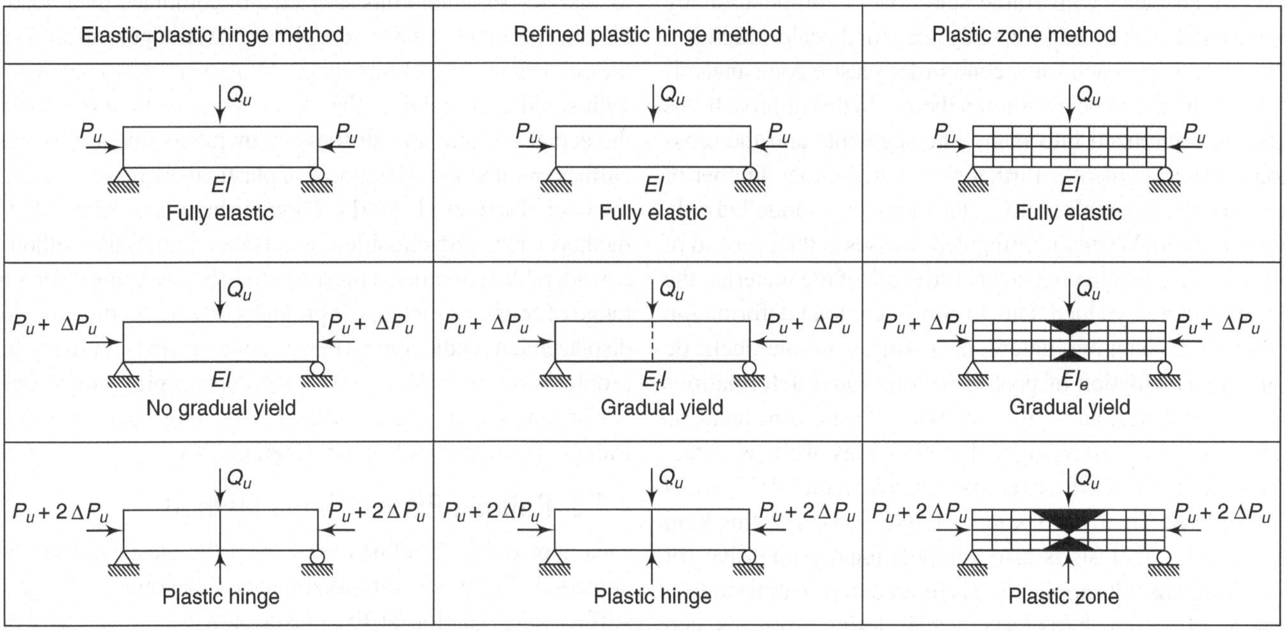


**FIG. 4.31** Concept of spread of plasticity for various advanced analysis methods

*Source:* Chen & Kim (1991) Reprinted with permission from Taylor & Francis.

with the foundations. Thus, the advance analysis methods incorporate both strength and stability behaviour in such a way that a separate member design is not required. They directly assess the strength and stability of the overall system, including the interaction of the member strength and stability. In addition to IS 800 : 2007, Eurocode 3 (EC3 1992), Canadian Code (CSA–S16.1 2009), and Australian Code (AS-4100 1998) permit the use of advanced analysis methods, which eliminate the tedious and sometimes confusing member capacity checks in the conventional limit states method. This often leads to significant savings in design. Let us now briefly discuss the various second-order inelastic methods of analysis.

## 4.7.1 Plastic Zone Method

In the plastic zone method, the frame members are discretized into finite elements, and the cross sections of these members are again subdivided into many sub-elements as shown in Fig. 4.32. The deflection at each division point along the member is obtained by numerical integration. The incremental load-deflection response at each loading step, which updates the geometry, captures the second-order effects (Chen & Kim 1997). Since the cross section is discretized into very small sub-elements, the residual stresses are considered constant within each sub-element. The stress state at each sub-element can be traced explicitly and hence the gradual spread of yielding can be predicted (see Fig. 4.31). Since it explicitly accounts for the second-order effects, spread of plasticity, and residual stresses, the plastic zone method eliminates the need for separate member capacity checks. Hence the plastic zone method has been accepted to provide the exact solution (see Fig. 4.1).

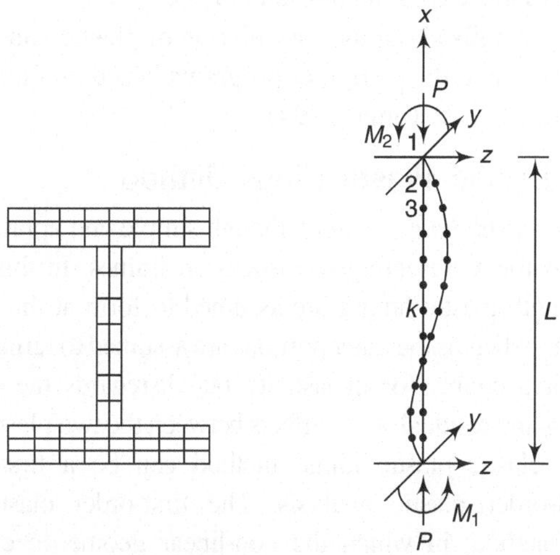


**FIG. 4.32** Plastic zone analysis model

There are two types of plastic zone analyses. The first approach involves the modelling of the structure with large number of three-dimensional shell elements and numerical integration of the elastic–plastic stiffness matrix. The constitutive matrix, which is elastic before yielding is replaced by an elastic–plastic constitutive matrix once yielding is detected. Based on the deformation theory of plasticity, the effects of combined normal and shear stresses may be accounted for. The three-dimensional spread of plasticity analysis, when combined with second-order theory (which deals with frame stability) is computationally intensive and hence is used for analysing small-scale structures.

The second approach for second-order plastic zone analysis is based on the use of beam column theory. In this approach, the member is discretized into many line segments and the cross section of each segment is further subdivided into a number of finite elements. The inelasticity in the member is modelled only for normal stress. When the computed stresses at the centroid of any fibres reach the uniaxial normal strength of the material, the fibre is considered yielded. Due to non-linear load deformation response and changing nature of stiffness in the inelastic regions, the calculation of post-yield forces and deformations require iterative trial-and-error processes. Plastic zone analysis methods have been developed for planar as well as three-dimensional structures (for example, Chen & Atsuta 1977; Vogel 1985; White & Chen 1990; Clarke 1994). However, this kind of plastic zone analysis is also computationally intensive for routine design use. Its application is limited to research studies, generation of design charts, very special design problems, and for generating benchmark solutions (which can be used to verify the accuracy of other less sophisticated analyses). As a matter of fact, the beam column interaction equations in some limit states codes were developed (for example, AISC–LRFD beam column equations) in part based on the curve-fit to the 'exact' strength curves obtained from plastic zone analysis similar to those by Kanchanalai (1977).

More details about the formulation of plastic zone analysis techniques and the computer programs based on them may be found in Chen & Toma (1994).

### 4.7.2 Elastic–Plastic Hinge Method

*Elastic–plastic hinge method*, though simple and approximate is efficient for representing inelasticity in frames. In this method, zero-length plastic hinges are assumed to form at the ends of a member, whereas the other portions are assumed to remain elastic. Thus, it accounts for inelasticity but disregards the spread of yielding and residual stress effects between the two plastic hinges.

The elastic–plastic hinge method can be a first-order or second-order plastic analysis. The first-order elastic–plastic hinge method, in which the non-linear geometric effects are neglected, predicts the same ultimate load as conventional rigid-plastic analysis (see Fig. 4.1). In second-order elastic–plastic hinge analysis, the deformed structural geometry is considered for formulating the stiffness equations. The geometric non-linearity is incorporated by using the stability functions, which facilitates the use of only one beam column element for a frame member. Hence, it provides a definite advantage over plastic zone method, as the discretization of the element and cross section into several sub-elements are avoided, resulting in quick and economic solution. This method is found to provide good results for frames comprising slender members and subjected predominantly to bending. However, when this method is used to analyse beam columns subjected to combined axial load and bending moment, it may overestimate the strength and stiffness of the element in the inelastic range. Moreover, for stocky members with significant yielding, the plastic hinge method over-predicts the actual strength and stiffness of members due to the gradual stiffness reduction as the spread of plasticity increases in an actual member (Liew et al. 1991). Though there is a debate that this method cannot be classified as advanced analysis method and considerable refinements must be made before using it for a wide range of framed structures, it provides essentially the same load–displacement predictions as plastic zone methods for many frame problems (Chen & Kim 1997). A computer program capable of performing second-order elastic–plastic hinge analysis of planar frames is available in Toma & Chen (1994).

### 4.7.3 Refined Plastic Hinge Method

This approach is a refined version of the elastic–plastic hinge approach. Two modifications are made to account for the gradual stiffness degradation at the plastic hinge locations as well as gradual member stiffness degradation between two plastic hinges. A section stiffness (two-surface) degradation function is adopted to reflect the gradual yielding effect in forming plastic hinges (see Fig. 4.33). The member inelastic behaviour is modelled in terms of member force rather than stress as used in the plastic zone approach. A tangent modulus concept is used to account for the residual stress effect along the member. Even the connection flexibility can be accounted for by modelling the connections as rotational springs tied to the end of the elements and by enforcing equilibrium and compatibility conditions between the connection element and the beam column element.

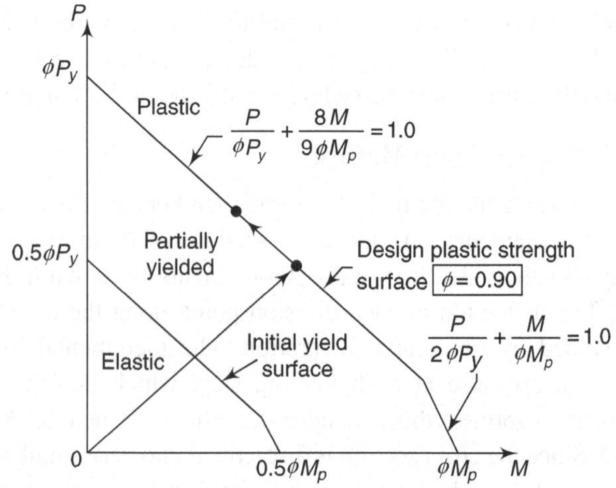


**FIG. 4.33** Two-surface stiffness degradation model for refined plastic hinge analysis

*Source:* Chen & Toma (1994) Reprinted with permission from Taylor & Francis.

Liew et al. (1993) used the Column Research Council [now the Structural Stability Research Council (SSRC)] tangent modulus and LRFD tangent modulus to account for both the effects of residual stresses and geometric imperfections. It has been found that by using the CRC tangent modulus model, different members with different residual stresses can be incorporated and the effect of geometric imperfections can be considered separately. A reduced tangent modulus approach has been suggested by Chen and Kim to consider the effect of geometric imperfections. Since the refined plastic hinge method strikes a balance between the requirements of realistic representation of frame behaviour (see Fig. 4.1) and ease of use, it is considered suitable for future design office use. More details about this method and the computer software developed based on this method are available in Chen and Toma (1994), Chen et al. (1995), and Chen and Kim (1997). Figures 4.34 and 4.35 show the results of a frame analysed by Chen and Kim (1997) using a software based on the refined plastic hinge (reduced tangent modulus) method. After determining the force distribution at collapse, the only design check that is required is that the member forces do not exceed the plastic strength of the cross section, where the plastic hinge forms. In the American code this check is specified by the following equation.

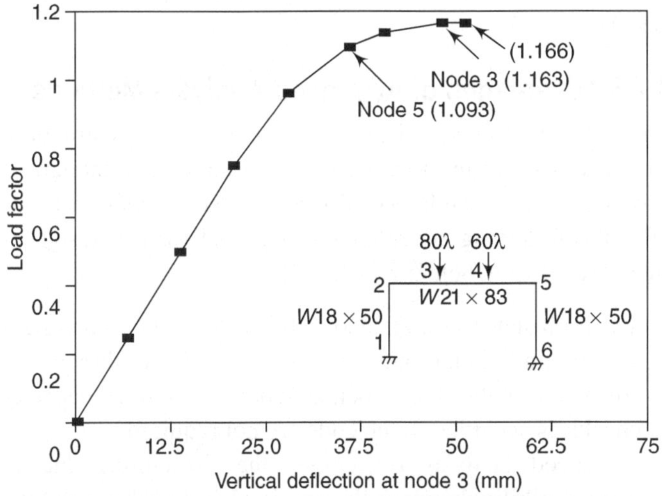


**FIG. 4.34** Load–displacement curves for the portal frame

*Source:* Chen & Kim (1997) Reprinted with permission from Taylor & Francis.

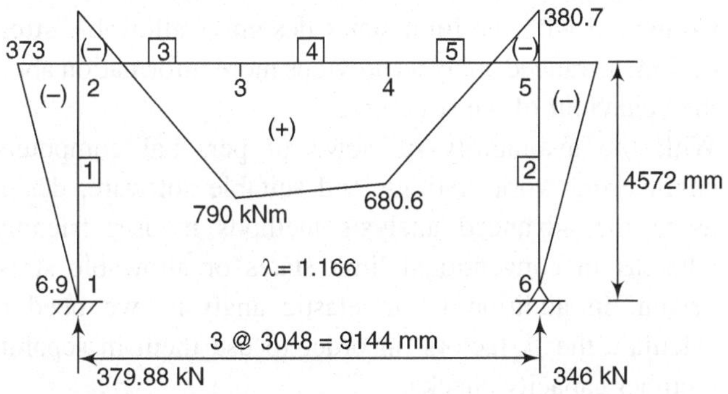


**FIG. 4.35** Force distribution at collapse for the example frame

$$P/P_y + 8M/(9M_p) = 1.0 \quad \text{for } P/P_y \geq 0.2 \qquad (4.54a)$$

$$P/(2P_y) + M/M_p = 1.0 \quad \text{for } P/P_y < 0.2 \qquad (4.54b)$$

## 4.7.4 Notional Load Plastic Hinge Method

One approach to refine the second-order elastic-plastic hinge analysis for frame design is to specify artificially large values of frame imperfections (that is, initial out-of-plumbness). These large imperfections are intended to account for the effects of residual stresses, frame imperfections, and distributed plasticity that are not considered in the analysis. This is the approach adopted by the Eurocode 3, where a maximum out-of-plumbness defined by $\psi_0 = L/200$ (for an unbraced frame), where $L$ is the column length, is used (see Fig. 4.36). Eurocode 3 also allows the initial sway imperfection to be replaced by a closed system of equivalent horizontal forces. It has to be cautioned that using the wrong direction of geometric imperfection in braced frames with rigid joints may increase the bending stiffness of the column rather than reduce it (see Fig. 4.37).

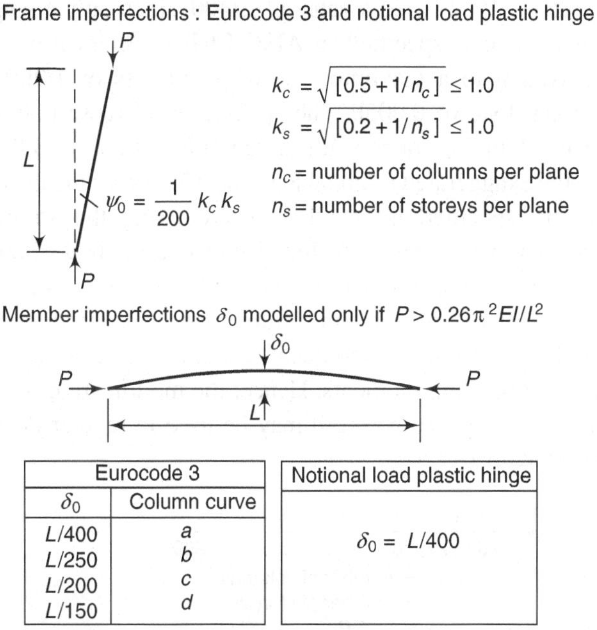


**FIG. 4.36** Comparison between the notional load plastic hinge method with Eurocode 3 method

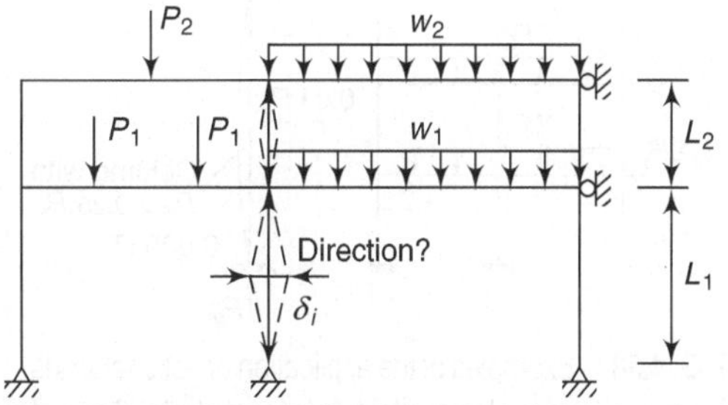


**FIG. 4.37** Difficulty in determining the direction of geometric imperfection

Liew et al. (1994) developed an analysis, which utilizes a set of equivalent notional lateral loads to account approximately for the influence of member imperfections and distributed plasticity effects on the overall system strength. These notional lateral loads, which are expressed as a fraction of the storey gravity loads or member axial force, are chosen to allow adequately for frame and member instability when they are combined with the applied lateral loads for the global frame analysis (see Section 2.9.2). For sway frames subjected to combined gravity and lateral loads, the notional loads should be added to the lateral loads (Chen & Kim 1997). The notional load approach has been adopted in the Canadian and Australian codes also. The notional load method also introduces the difficulty of determining the direction in which the notional loads should be applied to the frame.

Chen and Kim (1997) suggest the use of 'further reduced' tangent modulus model to avoid the difficulties of determining the direction of notional loads or geometric imperfections.

Eurocode specifies different maximum out-of-straightness values ranging from L/150 to L/400 depending on the column curves. However, in the notional load plastic hinge approach Liew et al. (1994) used a value of L/400, since only a single column curve is specified in AISC-LRFD specifications. It is equivalent to a maximum out-of-plumbness of 1/200 or a notional load of 0.005P. The application of these notional loads to a building frame is illustrated in Fig. 4.38. In IS 800 : 2007, the exaggerated notional loads of 0.5% times the gravity loads are specified to avoid over predicting the strength of the member. It has been found that this method, though very simple, under-predicts the strength by more than 20% in the various leaning column frames and over-predicts the strength by 10% in isolated beam columns subjected to axial forces and bending moments. Hence, the modification of this approach is required before it may be used in regular design (Chen & Kim 1997).

$$\phi = \frac{1}{200}\left(\frac{1}{2}+\frac{1}{n_c}\right)^{0.5}\left(\frac{1}{5}+\frac{1}{n_s}\right)^{0.5} \leq \frac{1}{200}$$

$n_c$ = number of columns
$n_s$ = number of storeys

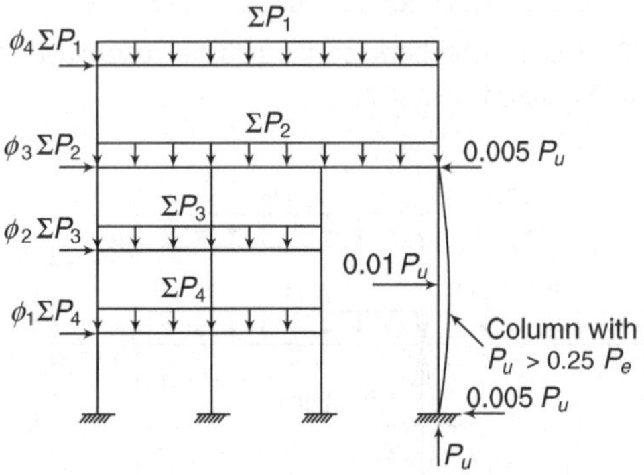


**FIG. 4.38** Examples of the application of notional loads for second-order elastic–plastic hinge analysis (Chen et al. 1995), Reprinted with permission from John Wiley

### 4.7.5 Quasi-plastic Hinge Method

This approach is an intermediate approach between the plastic zone and the elastic–plastic hinge method. It requires less computation but predicts strengths of frames within 5% accuracy, as compared to the plastic zone method.

Quasi-plastic hinge method uses an element specially developed for considering gradual plastification under combined axial force and bending. Initial yield and full plastification surface are used to analytically represent the gradual yielding of the cross section. A residual stress pattern (with a maximum value at flange tips equal to $0.3f_y$, and a linear transition of stress from flange tips to the web joint and constant tensile stress through the web) is used to determine an initial yield surface. A fully plastic surface is generated by calibration to a plastic zone solution. The parameters of the full plastification equation are determined by a curve-fit procedure. The inelastic stiffness matrix is derived from the incremental flexibility relationships, which in turn is obtained by successive integration along the member length. As the method is based on flexibility relationships, it cannot be extended to 3D analysis. Though this model eliminates the necessity of the refined model through the cross section, it requires many elements along the length of the members. More details of this model are provided by Attala et al. (1994).

### 4.7.6 Advantages of Advanced Analysis Methods

The code offers advanced analysis as another tool for structural engineers for use in the design of steel structures, though its adoption is not mandatory. The advanced analysis methods offer the following advantages when used for the design of steel structures (Chen & Kim 1997).

1. The limit states strength and stability of a structural system and its individual members are considered directly in advanced analysis and hence separate member capacity checks as specified in the code are not required.
2. Advanced analysis overcomes the difficulties due to incompatibility between the elastic global analysis and the limit state member design in the conventional limit states or LRFD method.
3. Compared with the limit states design or allowable stress design, advanced analysis provides more information about the behaviour of the structure.
4. With the availability of powerful personal computers, engineering work stations, and suitable software, design using the advanced analysis methods is user friendly, whereas in conventional limit states or allowable stress design, in addition to an elastic analysis, we need to calculate the '*k*-factors' in order to use them in separate member capacity checks.

5. Through advanced analysis, it is possible to redistribute the internal forces throughout the structural system. This results in economy, especially for highly indeterminate steel frames.
6. Since the analysis and design are integrated, advance analysis offers a time-effective solution, as it eliminates tedious and often confused member capacity checks including the calculation of *k*-factors (as is done in limit states/allowable stress method or LRFD).

However, the following points should be borne in mind while using the advanced analysis method.

1. Some advance analyses may require the use of at least two elements per member to determine the elastic buckling loads, whereas in linear analysis we use only one element per member.
2. In some cases, where a higher accuracy is required for the analysis of some elements, or where the location of plastic hinge is not at the two ends (for example, propped cantilevers), more than two elements may be needed.
3. Unlike linear analysis, an advanced analysis involves many iterations and factorizations of the tangent stiffness matrix.
4. It is a normal practice for the designer to first carry out a linear analysis for checking the proper modelling of frames before undertaking an advance analysis. Hence, it is better to choose an advance analysis which uses the same discretization of the frame (without further member division into more elements).
5. When a greater number of elements are used, the stiffness matrix is large, which leads to greater truncation errors and a higher possibility of divergence during iterations, especially for very large structures.
6. The three important considerations while dealing with the design of steel structures, namely, accuracy, simplicity, and efficiency, should be satisfied by the chosen analysis.

## 4.7.7 Direct Analysis Method of Design

Direct analysis method (DAM) of design for stability was first introduced in the AISC 360-05 code. The direct analysis method allows a more accurate determination of the load effects in the structure through the inclusion of the effects of geometric imperfections and stiffness reductions directly within the structural analysis. This also allows the use of $K = 1.0$ in calculating the in-plane column strength within the beam–column interaction equations. This is a significant simplification in the design of steel moment frames and combined systems. This method essentially consists of the following four steps:

1. A second-order analysis
2. Use of reduced stiffness in the analysis
3. Application of a *notional load* in the analysis under certain circumstances
4. Strength check of members using $K = 1$ for compressive strength

### Second-order Analysis

The *second-order analysis* must be conducted for all factored load combinations. (It has to be noted that the principle of superposition does not apply; hence results for different loads cannot be superimposed.) In the second-order analysis both $P$-$\Delta$ effects (the effects of loads acting on the displaced locations of joints or nodes in the structure) and $P$-$\delta$ effects (the effects of loads acting on the deformed shapes of individual members) must be considered. It should be noted that when the following conditions are satisfied, the effect of $P$-$\delta$ on the response of the structure can be neglected and a $P$-$\Delta$-only second-order analysis is permitted: (a) the structure supports gravity loads primarily; (b) the ratio of the second-order drift to the first-order drift is less than 1.7; and (c) less than one-third of the total gravity load on the structure is supported by columns that are part of moment–moment-resisting frames in the direction of translation being considered. The second-order analysis must be conducted for factored load combinations. (Note that the principle of superposition does not apply; results for different loads cannot be superimposed.)

Instead of doing a rigorous second-order analysis, the designer still has the option of using the familiar '$B_1$ and $B_2$' procedure, in which the results of a first-order analysis are amplified by factors $B_1$ and $B_2$ to account for second-order effects. Factor $B_1$, calculated for each beam-column and each direction of bending of the beam-column, accounts for $P$-$\delta$ effects; $B_2$, calculated for each story of the building and each direction of lateral translation of the story, accounts for $P$-$\Delta$ effects (See Appendix 8 of AISC 360-10).

### Reduced Stiffness in Analysis

A factor of 0.8 has to be applied to the stiffness of all members in the analysis. An additional factor, $\tau_b$, has to be applied to the flexural rigidity of framed columns in which the factored axial force is greater than half the yield capacity. Instead of applying the additional reduction factor $\tau_b$ (if there are indeed columns to which the additional factor is applicable), the designer may apply at each floor an additional lateral load of 0.001 times the vertical load applied on that floor.

### Application of Notional Loads

A *notional load* (which is an additional lateral load intended to simulate the effects of initial out-of-plumbness of the building) must be applied under certain conditions (see also Section 4.7.4). When required, the magnitude of the notional load on each floor is 0.002 times the vertical load applied on that floor (based on a nominal initial story out-of-plumbness ratio of 1/500). If the ratio of the second-order drift to the first-order drift (a ratio reasonably approximated by the $B_2$ multiplier) is greater than 1.5, notional loads must be included in all load combinations. If the ratio is less than 1.5, notional loads need to be applied only in gravity-only load combinations; they need not be applied in combination with other lateral loads.

### Member Strength Checks

When the forces and moments in members have been determined from the analysis outlined in the previous subsections (second-order reduced stiffness notional loads when applicable), member capacities can be checked using an effective length factor, $K$, of unity for members subject to compression.

A comparison of the three stability design methods prescribed in AISC 360-10 is provided in Table 4.6.

**TABLE 4.6** Comparison of the stability design methods prescribed in AISC 360-10

| Method | Type of analysis | Member stiffness in analysis | Inclusion of notional lateral loads | Column effective length in strength check |
|---|---|---|---|---|
| Direct analysis method | Second-order | Reduced EI and EA ($0.8\tau_b$ EI and 0.8 EA) | In some cases | $K = 1$ |
| Effective length method | Second-order | Nominal EI and EA | In some cases | Sideways buckling analysis |
| First-order analysis method | First-order | Nominal EI and EA | Yes, always | $K = 1$ |

*Note:* When $\frac{P_u}{P_y} \le 0.5$, $\tau_b = 1.0$; when $\frac{P_u}{P_y} > 0.5$, $\tau_b = 4\left(\frac{P_u}{P_y}\right)\left(1-\frac{P_u}{P_y}\right)$, with $P_y = A_g f_y$

### 4.7.8 Serviceability Design

As the service loads are usually substantially less than the factored loads used for strength design, the behaviour of a frame under service loads is usually closely approximated by the predictions of a first-order analysis (Section 4.2.1). Due to this, the serviceability design of a frame is usually carried out using the results of a first-order elastic analysis. The serviceability design of a frame is mostly based on the requirement that the service load deflections must not exceed the specified values. The member sizes are systematically changed in the analysis, until this requirement is met.

Frame deflections are usually measured in terms of a non-dimensional parameter called *drift*. *Inter-storey drift* is the ratio of inter-storey deflection to storey height. *Frame drift* is the ratio of the lateral deflection at the top of the frame to the frame height. IS 800 : 2007 limits the frame drift to 1/300 times the building height $H$ and storey drift to 1/300 times the storey height (see Section 2.9.3 and Table 2.8). However, for some multi-storey frames a deflection limit of $H/500$ may be more suitable, when the cladding cannot accommodate larger movements. (According to the ASCE Ad hoc committee report (1986), the accepted range of drift limits by first-order analyses is H/750 to H/250, with a typical value of H/400. Typical limits on the inter-storey drift are 1/500 to 1/200 times the storey height). It is important to note that a steel building with standard dimensions and designed to a first order drift limit of H/500, may experience additional second-order deflections in the range of 10% to 30% of the first-order ones (Chen & Kim 1997). Hence, if second-order analysis is used to access the serviceability of a frame, then the limits for lateral and inter-storey drift should be relaxed. Sometimes, it may also be necessary to check the other limit states serviceability conditions such as (a) vibration, (b) thermal expansion and contraction, (c) connection slip, (d) camber, and (e) corrosion.

## 4.8 PRACTICAL OUT-OF-PLANE ADVANCED ANALYSIS

The advanced methods of analysis discussed till now are applicable to 2D frames and 2D loading which fail in-plane by yielding. These methods are also limited to frames in which local and lateral buckling are prevented, so that failure is in-plane by excessive yielding caused by the combination of primary actions (first-order moments and axial forces) and second-order moment components resulting from in-plane instability effects. There have been recent attempts to develop advanced analyses either for the out-of-plane failure of 2D frames with in-plane loading, or for the biaxial bending and torsion of 3D frames under 3D loading. The 3D plastic zone analysis developed by Jiang et al. (2002) considers warping torsion and member out-of-plane buckling. Trahair & Chan (2003), presented a number of suggestions for the development of practical methods of advanced analysis for the out-of-plane behaviour of 2D frames under 2D loadings. These include the development of an element, which will be sufficiently accurate so that only one element per member will be required, and includes allowances for residual stresses and initial crookedness and twists and for the reductions in the out-of-plane stiffness caused by yielding. It was also suggested that the element be simplified by ignoring distorsion and pre- and post-buckling effects. Practical advance analysis of out-of-plane behaviour should be able to account for the influences of moment and axial force distributions, load heights, and end restraints on elastic and inelastic lateral buckling. It should also be consistent with the code formulations of beam and column out-of-plane strengths (Trahair & Chan 2003).

## SUMMARY

In the conventional limit states method, including LRFD, first- or second-order elastic analysis is used to analyse a structural system and separate member design is performed using the equations presented in the codes of practices. The effective length factor is used to account for the interaction between a structural system and its component members. However, it has been realized that elastic frame analysis and limit state member strength design are not compatible. The design equations become more complicated to make up the simplified assumptions of elastic behaviour in analysis and to consider the second-order effects.

In this chapter, the various analysis and design concepts related to first-order elastic, first-order plastic, second-order elastic, buckling, second-order inelastic, and advanced analysis methods have been provided. Joint and connection effects are also discussed and the means of considering the semi-rigid joints are also provided.

Realizing the incompatibilities of using the limit states design with the linear elastic analysis, modern codes such as IS 800 : 2007, Canadian, and Australian and Eurocode 3 allow the use of advanced analysis methods. While the limit state method accounts for the real behaviour of frames implicitly in its column strength and beam column interaction equations, the advance analysis method accounts for these effects explicitly through stability functions, stiffness degradation functions, and geometric imperfections.

Among the various advanced analysis methods, including plastic zone, quasi-plastic zone, elastic–plastic hinge, notional load plastic hinge, and refined plastic hinge methods, the refined plastic hinge method is found to be the most practical method for design office use. It is because it retains the efficiency of plastic zone method and provides simplicity of computation and sufficient accuracy for practical use. The method is developed by imposing simple modifications to the elastic–plastic hinge method. They include a modification to account for the gradual sectional stiffness degradation at the plastic hinge locations and to include the gradual member stiffness degradation between two plastic hinge locations.

Since the advanced analysis captures the limit state of strength and stability, the engineer need not perform the tedious and often confusing member capacity checks specified in the codes. Another important advantage of using advanced analysis is that it accurately predicts the behaviour of the system as well as the member, which is not provided by the conventional design methods. The rapid development of computer hardware and software has made it possible to adopt the advanced analysis techniques for design office use.

However, for a second-order inelastic analysis to be used in actual design practice, the capability of different analysis approaches for representing beam column and connection performance must be clearly identified. Also, any limit states that are not properly modelled in the analysis must be well understood such that appropriate specification checks can be performed.

Once the forces in the various members are evaluated using a linear elastic or second-order elastic analysis, the design of members in the conventional limit states analysis can be carried out by using the specification checks. The design of various members based on these checks is discussed in the next few chapters.

## MULTIPLE-CHOICE QUESTIONS

4.1 State whether the following are true or false with respect to the first-order elastic analysis:
   (a) Material is modeled as linear-elastic
   (b) Equilibrium is established in the undeformed geometry of the structure
   (c) It is concerned with service loads
   (d) None of these

4.2 Clause 4.4.3.4 of IS 800 allows for the redistribution of bending moments obtained from first order elastic analysis up to an extent of
   (a) 10% (b) 15% (c) 20% (d) 25%

4.3 Second-order analysis
   (a) results in larger bending moments due to P-delta effects
   (b) considers equilibrium in the deformed geometry of the structure
   (c) results can be linearly superposed when using factored loads
   (d) none of these is correct

4.4 Stability functions
   (a) are a simple means of accounting for the effect of axial force on moments
   (b) can be approximated by power-series expansions
   (c) allows the use of only one element per beam-column in the matrix analysis
   (d) all of these are correct

4.5 Clause 4.4.3 of IS 800 allows first-order analysis with moment amplifications when moment amplification factors are less than
   (a) 1.4 (b) 1.3 (c) 1.2 (d) 1.1

4.6 Annex B of IS 800 states that the changes in the effective stiffness of the member due to axial force may be neglected when the elastic buckling load factor is greater than
   (a) 4 (b) 5 (c) 6 (d) 3

4.7 A rigid jointed clad building, where stiffening effect of cladding in not taken into account, can be considered as non-sway frame when deflection of individual storey, $\delta$
   (a) $\delta \leq h_s/3000$, $h_s$ is the height of the storey
   (b) $\delta \leq h_s/4000$, $h_s$ is the height of the storey
   (c) $\delta \leq h_s/2000$, $h_s$ is the height of the storey
   (d) $\delta \leq h_s/5000$, $h_s$ is the height of the storey

4.8 The sources of non-linearity are due to
   (a) gradual yielding due to bending
   (b) yielding due to residual stresses
   (c) the $P$-$\Delta$ effect
   (d) none of these

4.9 Panel-zone deformations have to be considered in the analysis when
   (a) girder depth to storey height ratio is greater than 1/8
   (b) column depth to girder span ratio is greater than 1/20
   (c) both (a) and (b)
   (d) none of these

4.10 In IS 800 Annex F, simple connections are to be assumed when
(a) the connection transfers only shear at some nominal eccentricity
(b) the frame has about 5 stories in height
(c) stiffness rather than strength governs the design
(d) all of these

4.11 Fully welded connections can be assumed as
(a) simple connections
(b) rigid connections
(c) semi-rigid connections

4.12 Full fixity of columns may be assumed when columns are supported on
(a) rafts
(b) pile foundation
(c) basement walls
(d) individual footings

## EXERCISES

4.1 Assuming some initial member sizes analyse the frames given in Figs 4.3(a) and 4.4(a) by using computer programs based on:
(a) Linear-elastic analysis
(b) Second-order elastic analysis
and compare the results.

4.2 If you have access to computer programs that can do the following analysis:
(a) Elastic buckling analysis
(b) Inelastic buckling analysis
Compute the buckling loads for the frames given in Figs 4.4(a) and 4.12(a) and compare the results.

4.3 For the frame given in Fig. 4.28(a), consider that the beams are connected to the columns, by using top and seat angle without double web angle connections. Using Table 4.2 and Eqn (4.39) define the $M$-$\theta_r$ curve. If you have access to a computer program, which can take care of connection flexibility, analyse the frame and compare the results with those obtained using the assumption of rigid joints.

4.4 Assume that the column base is modelled as a rotational spring [see Eqn (4.48)], assuming clay soil and analyse the frame using a software having facility to analyse connection flexibility. Compare the results with fixed support condition.

4.5 Consider the effect of infilled brick wall of 230-mm thickness by using a diagonal compression strut having an area as calculated using Eqn (4.19). Analyse the frame given in Fig. 4.4(a) with the infill and compare the results obtained without infill.

4.6 Calculate the plastic collapse loads of frames given in Figs 4.4(a) and 4.12(a), using a suitable computer program.

4.7 If you have access to any advanced analysis software program, analyse frames in Figs 4.4(a) and 4.12(a) and interactively design the members. Compare the analysis results and designed member section with the results obtained using linear-elastic analysis and code checks.

## REVIEW QUESTIONS

4.1 Under what situation will the analysis and design become interactive?

4.2 What are the limitations of the two-stage design process (finding the forces in various members through a linear analysis and checking the member sizes against the codal rules)?

4.3 What is the main advantage of using inelastic analysis for the direct design of members?

4.4 List the different methods that can be adopted for the analysis of 2D steel frames.

4.5 State the assumptions involved in the first-order elastic analysis.

4.6 Though the first-order analysis provides an 'exact solution', the results do not reflect the true-behaviour of a structure. Why?

4.7 What are the main advantages of first-order elastic analysis?

4.8 How are the influences of plasticity and stability incorporated while designing members using first-order analysis methods?

4.9 Write short notes on:
(a) $P$-$\delta$ or member stability effect
(b) $P$-$\Delta$ or frame stability effect

4.10 What is meant by amplification factor?

4.11 Under what circumstances is redistribution of bending moments allowed in the code?

4.12 What is the basic difference between first-order and second-order analyses?

4.13 List some of the differences in results between first- and second-order analyses.

4.14 How are the results in a second-order elastic analysis obtained (since they are based on the equilibrium in the deformed geometry of the structure)?

4.15 Write a short note on the Newton–Raphson method.

4.16 What are stability functions?

4.17 The stability functions will become indeterminate when the axial force in the members approach zero. How is this difficulty overcome in computer programs?

4.18 What is the main difficulty in using second-order elastic analysis results? How are these difficulties overcome in the design?

4.19 Under what circumstances can the change in the effective stiffness of the members due to axial force be neglected as per the code?

4.20 How can the stiffening effect of a cladding or in-fill be taken into account in the analysis?

4.21 What is the guideline given in the code for considering a frame as a sway frame?

4.22 What is the main use of a linear buckling analysis?

4.23 How are the critical loads of frames determined in the stiffness matrix approach?

4.24 The elastic critical load of frames is of limited value to the designers. Why?

4.25 What is the main adjustment made in the stiffness matrix to arrive at the inelastic critical load?

4.26 List the main sources of non-linearity.

4.27 What are the three basic aspects of beam column joint that affect the member behaviour?

4.28 How can we approximately consider the finite size of the joint in the routine stiffness analysis program?
4.29 What are the semi-empirical equations for predicting the ultimate strength of a joint, as proposed by Krawinkler?
4.30 What are the limiting ratios of panel zone, above which the panel zone deformations have to be considered in the analysis?
4.31 When the panel zone deformation is high, what is the remedy adopted in the joint panel webs?
4.32 Write short notes on the following classification of connections:
(a) Eurocode 3 classification
(b) Classification by Bjorhovde et al.
4.33 Compare the Eurocode and Bjorhovde et al. classification system for connections.
4.34 What is the simple rule suggested by Gerstle and Ackroyd for regarding a joint as rigid?
4.35 What are the different analytical models developed to respresent connection flexibility using experimental data?
4.36 State the polynomial model as suggested by Frye–Morris and adopted by the code.
4.37 Why do the identical semi-rigid connections at the end of the beam of a rigid frame not behave identically?
4.38 List the various steps in a solution procedure employed in normal computer codes incorporating semi-rigid joints.
4.39 How can the foundation be modelled to take into account the flexibility at the base of a frame?
4.40 What are the IS code provisions for considering foundation stiffness?
4.41 How can the shear stiffness of the sheeting in a pitched portal frame be considered in the analysis?
4.42 How is the plastic collapse load of a structure determined in the first-order plastic analysis?
4.43 What are the conditions that should be imposed such that ductility is ensured and plastic collapse mechanism is allowed to develop in the plastic analysis?
4.44 What are the three main groups of inelastic analysis methods?
4.45 Write short notes on:
(a) Plastic zone method
(b) Elastic–Plastic hinge method
(c) Refined Plastic hinge method
(d) Notional load plastic hinge method
(e) Quasi-plastic hinge method
4.46 Why is the plastic zone method used only in limited research studies and for generating benchmark solutions?
4.47 What is the advantage of elastic–plastic hinge method over the plastic zone method?
4.48 What are the two modifications that are done in the elastic–plastic hinge method so that it may be called as refined plastic hinge method?
4.49 What are the practical problems encountered in notional plastic hinge method while modelling structures in the computer?
4.50 What are the drawbacks of the quasi-plastic hinge methods?
4.51 What are the advantages offered by the advanced analysis methods?
4.52 Why is a linear-elastic method sufficient for the serviceability limit states design?
4.53 What is the difference between frame drift and inter-storey drift? What are the limits prescribed by the IS code for these two drifts?
4.54 How do you differentiate an out-of-plane advanced analysis from the ordinary advanced analysis method?

## REFERENCES


'Ad Hoc Committee on Serviceability: A Critical Appraisal and Research Needs', *Journal of Structural Engineering*, ASCE, Vol. 112, No. 12, Dec 1986, pp. 2646–2664.

Allen, H.G. and Bulson, P.S., *Background to Buckling*, McGraw Hill, London, 1980.

Arthi, N., Arul Jayachandran, S., and Seetharaman, S., 'Numerical Analysis of Semi-rigid Sway Frames using Gauss-Seidel Iterations', *All India Seminar on Modern Trends in Steel Structures,* organized by ACCE(I) and IE(I), Nagpur, 8–9th Feb 2002, pp. 57–65.

Arul Jayachandran, S., Seetharaman, S., and Kishore, I.V.K., 'Semi-rigid Connections in Steel Frames—A Note for the Designers of Structural Steelwork', *Steel in Construction*, INSDAG, Vol. 5, No. 1, Jan 2004, pp. 57–63.

*ASCE Technical Committee on LRFD: Effective Length and Notional Load Approaches for Assessing Frame Stability: Implications for American Steel Design*, American Society of Civil Engineers, New York, 1997.

Attala, M.N., Deierlein, G.G., and McGuire, W., 'Spread of Plasticity: Quasi-plastic Hinge Approach', *Journal of Structural Engineering*, ASCE, Vol. 120, No. 8, Aug 1994, pp. 2451–2473.

Austin, W.J., 'Strength and Design of Metal Beam-columns', *Journal of Structural Div.,* ASCE, Vol. 87, No. ST4, Apr 1961, pp. 1–32.

Barakat, M. and Chen, W.F., Practical Analysis of Semi-rigid Frames, *Engineering Journal,* AISC, Vol. 27, No. 2, 1990, pp. 54–68.

Bjorhovde, R., Colson, A., and Brozzetti, J., 'Classification System for Beam-to-column Connections', *Journal of Structural Engineering*, ASCE, Vol. 116, No. 11, Nov 1990, pp. 3059–3076.

Brockenbrough, R.L. and Merritt, F. S., ed., *Structural Steel Designers Handbook*, Third edition, McGraw Hill Professional, New York, 1999.

Brown, D.G., *Modeling of Steel Structures for Computer Analysis*, The Steel Construction Institute, Ascot, UK 1995, p. 52.

Cabrero, J.M. and Bayo, E., 'Development of Practical Design Methods for Steel Structures with Semi-rigid Connections', *Engineering Structures*, Vol. 27, No. 8, Jul 2005, pp.1125–1302.

Chan, S.L. and Chui, P.P.T., *Nonlinear Static and Cyclic Analysis of Steel Frames with Semi-rigid Connections,* Elsevier, UK, 2000.

Chen, W.F. and Lui, E.M., 'Stability Design Criteria for Steel Members and Frames in the United States', *Journal of Constructional Steel Research*, Vol. 5, 1985, pp. 31–74, Discussion by Vandepitte, Vol. 6, No. 1, 1986, pp. 76–79.

Chen, W.F. and Atsuta, T., *Theory of Beam Columns, Vol. II: Space Behaviour and Design,* McGraw Hill, New York, 1977.

Chen, W.F. and Kishi, N., 'Semi-rigid Steel Beam-to-column Connections: Data Base and Modeling', *Journal of structural Engineering*, ASCE, Vol. 115, No. 1, Jan 1989, pp. 105–119.

Chen, W.F. and Lui, E.M., *Stability Design of Steel Frames*, CRC Press, Boca Raton, Florida, 1991, p. 380.

Chen, W.F. and Lui, E.M., *Structural Stability: Theory and Implementations*, Elsevier, New York, 1987.

Chen, W.F. and Sohal, I.S., *Plastic Design and Second-order Analysis of Steel Frames*, Springer-Verlag, 1995.

Chen, W.F. and Toma, S., ed., *Advanced Analysis of Steel Frames: Theory, Software and Applications*, CRC Press, Boca Raton and New York, 1994, p. 400.

Chen, W.F., ed., *Joint Flexibility in Steel Frames,* Journal of Constructional Steel Research, Vol. 8, Special Issue, 1987.

Chen, W.F., ed., *Steel Beam-to-Column Building Connections*, Journal of Constructional Steel Research, Vol. 10, Special Issue, 1988.

Chen, W.F. and Kim, S.E., *LRFD Steel Design Using Advanced Analysis*, CRC Press, Boca Raton and New York, 1997, p. 441.

Chen,W.F., Goto,Y., and Liew, J.Y.R., *Stability Design of Semi-rigid Frames*, John Wiley & Sons, New York, 1995, p. 488.

Clarke, M.J., *Plastic-zone Analysis of Frames*, Chapter 6 of Advance Analysis of Steel Frames: Theory, Software and Applications, W.F. Chen and S. Toma, eds, CRC Press, Boca Raton, Florida, 1994, pp. 259–319.

Davies, J.M. and Brown, B.A., *Plastic Design to BS 5950*, Blackwell Science Ltd., Oxford, 1996.

Davies, J.M., and Bryan, E.R., *Manual of Stressed Skin Diaphragm Design*, Granada, 1982.

Davison, B. and Owens, G.W., *Steel Designers Manual*, Sixth edition, Blackwell Publishers, Oxford and The Steel Construction Institute, Ascot, 2002.

Davison, J.B., Kirby, P.A., and Nethercot, D.A., 'Rotational Stiffness Characteristics of Steel Beam-to-column Connections', *Journal of Constructional Steel Research*, Vol. 8, 1987, pp.17–54.

Dhillon, B.S. and O'Malley III, J.W., 'Interactive Design of Semi-rigid Steel Frames', *Journal of Structural Engineering*, ASCE, Vol. 125, No. 5, May 1999, pp. 556–564.

Errera, S.J. and Apparao, T.V.S.R., 'Design of I-shaped Beams with Diaphragm Bracing', *Journal of Structural Division*, ASCE, Vol. 102, No. ST4, Apr 1976, pp.769–781.

Faella, C., Piluso, V., and Rizzano, G., *Structural Steel Semi-rigid Connections: Theory, Design, and Software*, CRC Press, FL, 2000.

Frye, M.J. and Morris, G.A., 'Analysis of Flexibly Connected Steel Frames', *Canadian Journal of Civil Engineering*, Vol. 2, No. 3, 1975, pp. 280–291.

Galambos, T.V., *Guide to Stability Design Criteria for Metal Structures*, Fifth edition, John Wiley, New York, 1998, p. 911.

Ganapathy, C. and Subramanian, N., 'A Computer Program for the Buckling Analysis of Space Frames', *Computer Society of India, Annual Convention*, Pune, Jan 1977.

Gere, J.A. and Weaver, W., *Analysis of Framed Structures*, Van Nostrand Company Ltd, 1969.

Gerstle, K.H. and Ackroyd, M.H., 'Behaviour and Design of Flexibility-connected Building Frames', *Engineering Journal*, AISC, Vol. 27, No. 1, 1990, pp. 22–29.

Goto, Y. and Chen, W.F., Second-order elastic analysis for frame design, *Journal of Structural Div.,* ASCE, Vol. 113, No. 7, Jul 1987, pp. 1501–1519.

Hancock, G.J., Papangelis, J.P., and Clarke, M.J., *Plane Rigid Frame Structural Analysis (PRFSA) User Manual*, Centre for Advanced Structural Engineering, University of Sydney, Australia, 1995, (www.civil.usyd.edu.au/case/prfsa.php).

Harrison, H.B., *Structural Analysis and Design, Parts 1 and 2*, Second edition, Pergamon Press, Oxford, 1990.

Hasan, R., Kishi, N., Chen, W.F., Komuro, M., 'Evaluation of Extended End-plate Connections', *Journal of Structural Engineering*, ASCE, Vol. 123, No. 12, Dec 1997, pp. 1595–1602.

Horne, M.R., 'An Approximate Method for Calculating the Elastic Critical Loads of Multi-storey Plane Frames', *The Structural Engineer*, Vol. 15, No. 6, Jun 1975, pp. 242–248.

Horne, M.R., and Morris, L.J., *Plastic Design of Low-Rise Frames*, Granada, London, 1981.

Jiang, X-M, Chen, H., Liew, J.Y.R., 'Spread of Plasticity Analysis of Three-dimensional Steel Frames', *Journal of Constructional Steel Research*, Vol. 58, 2002, pp. 193–212.

Jones, S.W., Kirby, P.A., and Nethercot, D.A., 'Modeling of Semi-rigid Connection Behaviour and its Influence on Steel Column Behaviour', in *Joints in Structural Steel Work,* Howlett, J.H., Jenkins. W.M., and Stainby, R., eds, Pentech Press, pp. 5.73–5.78.

Kanchanalai, T., *The Design and Behaviour of Beam-columns in Unbraced Steel Frames*, AISI Project No. 189, Report No.2, Civil Engineering/Structures Research Lab., University of Texas at Austin, 1977, p. 300.

Kato, B., Chen, W.F., and Nakao, M., 'Effects of Joint-panel Shear Deformation on Frames', *Journal of Constructional Steel Research*, Vol. 10, 1988, pp. 269–320.

Kennedy, D.J.L., Picard, A., and Beaulieu, D., 'New Canadian Provisions for the Design of Steel Beam-columns', *Canadian Journal of Civil Engineering*, Vol. 17, 1990, pp. 873–893.

Kim, Y. and Chen, W.F., 'Practical Analysis for Partially Restrained Frame Design', *Journal of Structural Engineering*, ASCE, Vol. 124, No. 7, Jul 1998, pp. 736–749.

Kishi, N. and Chen, W.F., 'Moment-rotation Relation of Semi-rigid Connections with Angles', *Journal of Structural Engineering*, ASCE, Vol. 116, No. 7, Jul 1990, pp. 1813–1834.

Kishi, N., Chen, W.F., Goto, Y., and Matsuoka, K.G., 'Analysis Program for the Design of Flexibly Jointed Frames', *Computers and Structures*, Vol. 49, No. 4, 1994, pp. 705–713.

Kishi, N., Chen, W.F., Goto, Y., and Matsuoka, K.G., 'Design Aid of Semi-rigid Frame Design', *Engineering Journal*, AISC, Vol. 32, No. 3, 1995, pp. 90–107.

Kishi, N., Hasan, R., Chen, W.F., and Goto, Y., 'Study of Eurocode 3 Steel Connection Classification', *Engineering Structures*, Vol. 19, No. 9, 1997, pp. 772–779.

Krawinkler, H. and Mohasseb, S., 'Effect of Panel Zone Deformation on Seismic Response', *Journal of Constructional Steel Research*, Special Issue on Joint Flexibility in Steel Frames (W.F. Chen, ed.), Vol. 8, 1987, pp. 233–250.

Krawinkler, H., 'Shear in Beam-column Joints in Seismic Design of Steel Frames', *Engineering Journal*, AISC, Vol. 15, No. 3, 1978, pp. 82–91.

Krishnamurthy, N., Huang, H.T., Jefferey, P.K., and Avery, L.K., 'Analytical M-$\theta$ Curves for End-plate Connections', *Journal of Structural Div.*, ASCE, Vol. 105, No. ST1, Jan 1979, pp. 133–145.

Liew, J.Y.R., White, D.W., and Chen, W.F., Beam-column Design in Steel Frameworks—Insights on Current Methods and Trends, *Journal of Constructional Steel Research* Vol. 18, No. 4, 1991, pp. 259–308.

Liew, J.Y.R., White, D.W., and Chen, W.F., 'Notional Load Plastic Hinge Method for Frame Design', *Journal of Structural Engineering*, ASCE, Vol. 120, No. 5, May 1994, pp. 1434–1454.

Liew, T.Y.R., White, D.W., and Chen, W.F., 'Second-order Refined Plastic Hinge Analysis for Frame Design', Parts I and II, *Journal of Structural Engineering*, ASCE, Vol. 119, No. 11, Nov 1993, pp. 3196–3237.

Livesley, R.K. and Chandler, D.B., *Stability Functions for Structural Frameworks*, Manchester University Press, 1956.

Livesley, R.K., *Matrix Methods of Structural Analysis*, Second edition, Pergamon Press, 1975.

Lui, E.M. and Chen, W.F., 'Frame Analysis with Panel Zone Deformation', *International Journal of Solids and Structures*, Vol. 22, No. 12, Dec 1986, pp. 1599–1627.

MacLeod, I.A., *Analytical Modeling of Structural Systems: An Entirely New Approach with Emphasis on Behaviour of Building Structures*, Ellis Horwood, New York, 1990, p. 222.

Massonet, C., 'Stability Considerations in the Design of Steel Columns', *Journal of Structural Div.*, ASCE, Vol. 85, 1959, pp. 75–111.

McGuire, W., Gallagher, R.H., and Ziemian, R.D., *Matrix Structural Analysis*, Second edition, John Wiley, New York, 2000, p. 460.

Neal, B.G., *The Plastic Methods of Structural Analysis*, Third edition, Chapman & Hall, London, 1977, p. 358.

*P-Frame-User Manual for Plane Frame*, Computer Design Consultants, Chennai, India, 1986.

Paulay, T. and Priestley, M.J.N., *Seismic Design of Reinforced Concrete and Masonry Buildings,* John Wiley & Sons, New York, 1992, p. 744.

Petersen, C., *Statik und Stabilität der Baukonstruktionen*, Friedr. Vieweg & Sohn, Braunschweig/Wiesbaden, Germany, 1980, p. 960.

Salvadori, M.G., 'Lateral Buckling of I-beams', *Transactions of the American Society of Civil Engineers*, Vol. 120, 1955, pp. 1165–1177.

Seetharaman, S., Pandian, N., and Arul Jayachandran, '5 Advanced Analysis Techniques for Steel Frames—An Overview', *Proceeding of the Structural Engg. Convention*, IIT, Kharagpur, Dec 2003, pp. 85–94.

Sivakumaran, K.S., 'Seismic Response of Multi-storey Steel Building with Flexible Connections', *Engineering Structures*, Vol. 10, 1988, pp. 239–248.

Smith, B.S. and Coull, A., *Tall Building Structures: Analysis and Design*, Inter Science, 1991.

Subramanian, N., 'Computer Analysis and Design of Tall Structures', *Civil Engineering and Construction Review,* Vol. 8, No. 4, Apr 1995, pp. 42–46.

Subramanian, N. and Ganapathy, C., 'Computer Analysis of Space Frames with Offset Members', *Computers and Structures*, Vol. 11, Apr 80, pp. 297–303.

Subramanian, N., 'Distortional Buckling of Lipped Channel Beam', *International Conference on Stability of Structures,* PSG College of Tech., Coimbatore, India, 7–9 Jun 1995.

Subramanian, N., ed., *First National Seminar on Computers and Civil Engineering,* Organized by Computer Design Consultants, Madras, 26 Nov 1983.

Subramanian, N., ed., *National Seminar on Computers in Civil Engineering*, Organized by the Association of Consulting Civil Engineers, Chennai, Apr 1991.

Subramanian, N., *Plastic Design of Minimum Weight Frames by Moment Distribution*, M Sc. Thesis, University of Madras, May 1974, p. 164.

Subramanian, N., *Principles of Space Structures*, Second edition, Wheeler Publishing, Allahabad, 1999, p. 660.

Trahair, N.S. and Chan, S.L., 'Out-of-plane Advanced Analysis of Steel Structures', *Engineering Structures*, Vol. 25, 2003, pp. 1627–1637.

Trahair, N.S., Bradford, M.A., Nethercot, D.A., and Gardner, L., *The Behaviour and Design of Steel Structures to EC3*, Fourth edition, CRC Press, London and New York, 2007.

Villettee, M., Boissonnade, N., Muzeau, J-P, Jaspart, J-P, *Development of Comprehensive Formula for the Design of Beam-columns*, Internal report, Baudin–Chateauneuf, LERMES-CUST, University of Liege, Netherlands, 2000.

Vogel, U., 'Calibrating Frames', *Der Stahlbau*, Vol. 10, 1985, pp. 1–7.

Wang, C.K., *Computer Methods in Advanced Structural Analysis*, Intext Educational Publishers, New York, 1973.

White, D.W. and Chen, W.F., 'Second-order Inelastic Analysis for Frame Design', *Proc. of the National Symposium on Advances on Steel Structures*, V. Kalyanaraman, ed., IIT Madras, Tata-McGraw Hill, 7–9 Feb 1990.

White, D.W. and Hajjar, J.F., 'Application of Second-order Elastic Analysis in LRFD: Research to Practice', *Engineering Journal*, AISC, Vol. 24, No. 4, 1991, pp. 133–148.

Wood. R.H., 'Rigid-jointed Multi-storey Steel Frame Design: A State-of-the-art Report', Current Paper CP 25/73, *Building Research Establishment*, Watford, Sep 1973.

Yang, Y.B. and McGuire, W., 'Stiffness Matrix for Geometric Nonlinear Analysis', *Journal of Structural Div.*, ASCE, Vol. 112, No. 4, Apr 1986.

Yang, Y.B. and Quo, S.R., *Theory and Analysis of Nonlinear Framed Structures*, Prentice Hall, Upper Saddle River, NJ, 1994.

Yu, C.H. and Shanmugam, N.E., 'Stability of Frames with Semi-rigid Joints', *Computers and Structures*, Vol. 23, No. 5, 1986, pp. 639–648.

Ziemian, R.D., McGuire, W., and Deierlein, G.G., 'Limit States Steel Design using Second-order Inelastic Analysis: Planar Frame Studies and 3-D High-rise Frame Study', *Journal of Structural Engineering*, Vol. 118, No. 8, Sep 1992, pp. 2532–2567.

# 5

# BOLTED CONNECTIONS

## INTRODUCTION

Any steel structure is an assemblage of different members such as beam, columns, and tension members, which are fastened or connected to one another, usually at the member ends. Many members in a steel structure may themselves be made of different components such as plates, angles, I-beams, or channels. These different components have to be connected properly by means of fasteners, so that they will act together as a single composite unit. Connections between different members of a steel framework not only facilitate the flow of forces and moments from one member to another but also allow the transfer of forces up to the foundation level. Connections are also required for extending the lengths of different types of members, connecting steel columns to footings, and for joining two parts of a structure during erection. A structure is only as strong as its weakest link. Unless properly designed and detailed, the connections may become weaker than the members being joined. It is desirable to avoid connection failure before member failure due to the following reasons.

1. A connection failure may lead to a catastrophic failure of the whole structure.
2. Normally, a connection failure is not as ductile as that of a steel member failure.
3. For achieving an economical design, it is important that connectors develop full or a little extra strength of the members it is joining.

Connection failure may be avoided by adopting a higher safety factor for the joints than the members.

The behaviour of joints or connections is very complex due to the various factors which influence them, such as geometric imperfections, lack of fit, residual stresses, connection flexibility, geometric complexity, slipping, and non-linear load deformation characteristics (Owens & Cheal 1989). In bolted or riveted joints, a variety of components such as angle cleats, end plates, stiffeners, and bolts or rivets are used to transfer and disperse loads from one member to the other. Due to the use of bolts or rivets, discrete rather than continuous load paths are employed to transfer the loads. The bolts or rivets themselves have complex behaviour. Hence, the design of bolted or riveted joints is much more complicated and involved than those of the members. Complex design procedures have been developed in the past to produce more economical and safe connections. But due to the variety of connections, possible configurations, variability of behaviour, and practical limitations in fabrication and erections, complex design procedures have not been found suitable. Hence, based on experimental results, past performance (with respect to static and dynamic loads), and the ductility of steel, many approximations and assumptions are made in the design of these connections. Moreover, some standard connections types have been evolved in the past, which have shown satisfactory performance, for which simplified analytical procedures can be used (see Figs 5.1 and 5.2). The behaviour and design of these connections are discussed in this chapter.

The basic goal of connection design is to produce a joint that is safe, economical, and simple (so that it can be manufactured and assembled at site without any difficulty). It is also important to standardize the connections in a structure and to detail it in such a way that it allows sufficient clearance and adjustment to accommodate any lack of fit, resists corrosion, is easy to maintain, and provides reasonable appearance.

Connections (or structural joints) may be classified according to the following parameters:

1. Method of fastening such as rivets, bolts, and welding—connections using bolts are further classified as *bearing* or *friction type connections*

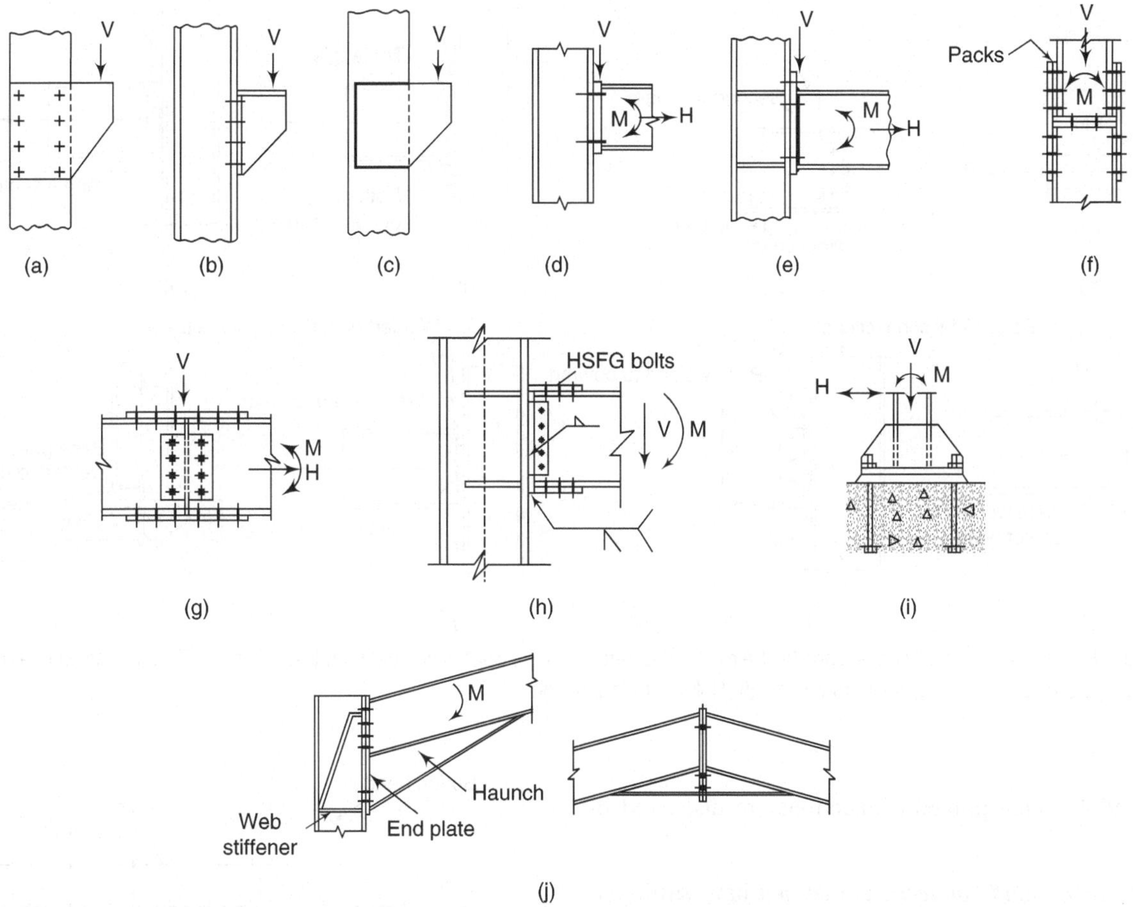


**FIG. 5.1** Examples of 'rigid' connections (a) Eccentrically loaded bolted connection (b) Bolted bracket (c) Eccentric weled bracket (d) Flush end plate (e) Extended end plate (f) Column splices (g) Bolted beam splice (h) Cover plate connection (i) Column-base plate connection (j) Eaves and apex connection in portal frames

2. Connection rigidity—simple, rigid (so that the forces produced in the members may be obtained by using an indeterminate structural analysis), or semi-rigid
3. Joint resistance—bearing connections and friction connections (these are explained in subsequent sections)
4. Fabrication location—shop or field connections
5. Joint location—beam column, beam-to-beam, and column-to-foundation
6. Connection geometry—single web angle, single plate, double web angle, top and seat angles (with and without stiffeners), end plates, header plate, welded connections using plates and angles, etc. (most of these connections are shown in Fig. 2.9 of Chapter 2).
7. Type of force transferred across the structural connection—shear connections, shear and moment connection or simply moment connection, tension or compression, and tension or compression with shear.

Structural connections transmit forces which result in linear and rotational movements. The linear movements at a joint are generally small but the rotational movement depends on the stiffness of the type of connection.

According to the IS code, based on connection rigidity, the joints can be defined as follows:

**Rigid** *Rigid connections* develop the full moment capacity of connecting members and retain the original angle between the members under any joint rotation, that is, rotational movement of the joint will be very small on these connections. Examples of rigid connections are shown in Fig. 5.1.

**Simple** In *simple connections* no moment transfer is assumed between the connected parts and hence are assumed as hinged (pinned). The rotational movement of the joint will be large in this case. Actually, a small amount of moment will be developed but is normally ignored in the design. Any joint eccentricity less than about 60 mm is neglected. Examples of hinged (pinned) connections are shown in Fig. 5.2. Some simple connections, for example, tie bars, are connected by real pins as shown in Fig. 5.2(a). If the pins are not corroded or blocked with debris, they will act as pin joints. Tie bars are rarely used now since the safety depends on a single pin and also due to the cost of manufacture and malfunctioning of pins due to corrosion or debris (Martin

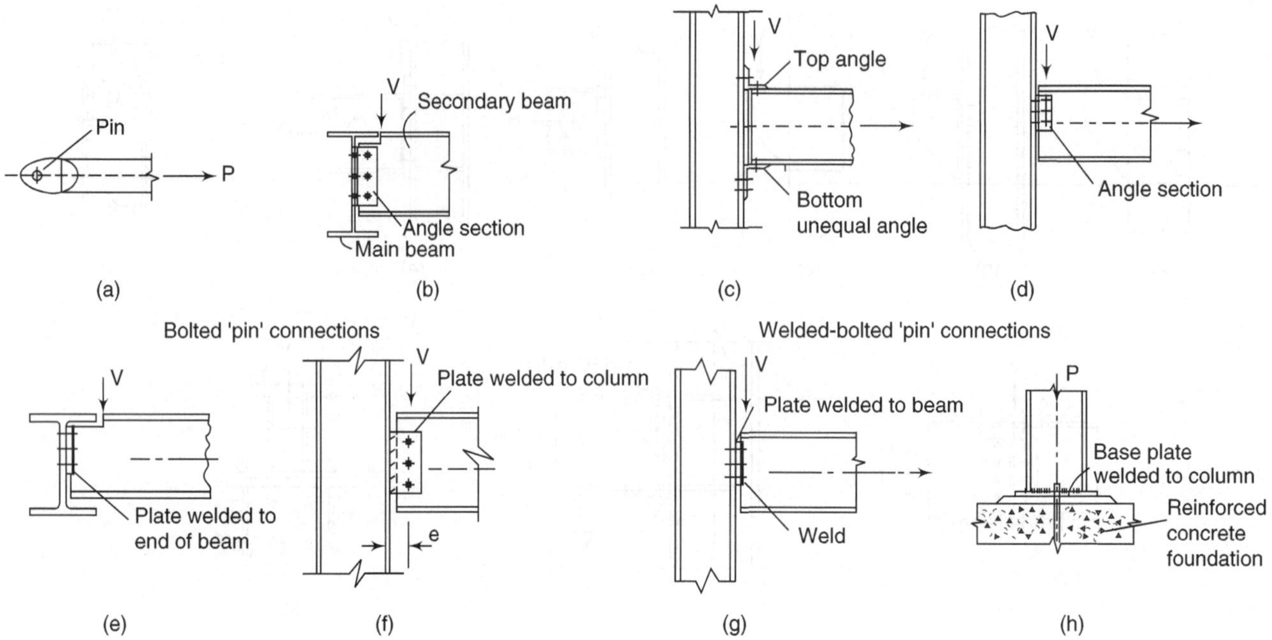


**FIG. 5.2** Examples of 'pinned' connections (a) Tie rod (b) Beam-to-beam (c) Seated angle (beam-to-column) (d) Beam-to-column (e) Beam-to-beam (f) Beam-to-column (g) Beam-to-column (h) Column-to-foundation

& Purkiss 1992). Such pinned connections are discussed in Section 5.6.

**Semi-rigid** *Semi-rigid connections* may not have sufficient rigidity to hold the original angles between the members and develop less than the full moment capacity of the connected members. The design of these connections requires determining the amount of moment capacity (or moment–rotation relationship of the connection) based on test results or rational methods (say 20%, 30%, or 75% of moment capacity).

In reality, all the connections will be semi-rigid. However, for convenience we assume some of them as rigid and some as hinged. We will discuss simple bolted connections in this chapter. Simple welded connections are discussed in the next chapter. Moment-resistant bolted and welded connections are discussed in Chapter 14.

## 5.1 RIVETS AND RIVETED CONNECTIONS

For many years rivets were the sole practical means of producing safe and serviceable steel connections. A rivet is made up of a round ductile steel bar (mild or high tensile steel as per IS 1929 and IS 2155) called *shank*, with a head at one end (see Fig. 5.3). The head may have different shapes as shown in Fig. 5.3. The snap and pan heads form a projection beyond the plate face, whereas the counter sunk head will be flush with the surface of the plate face.

The length of the rivet to be selected should be longer than the grip of the rivet (see Fig. 5.3), sufficient to form the second head. The installation of a rivet requires the heating of the rivet

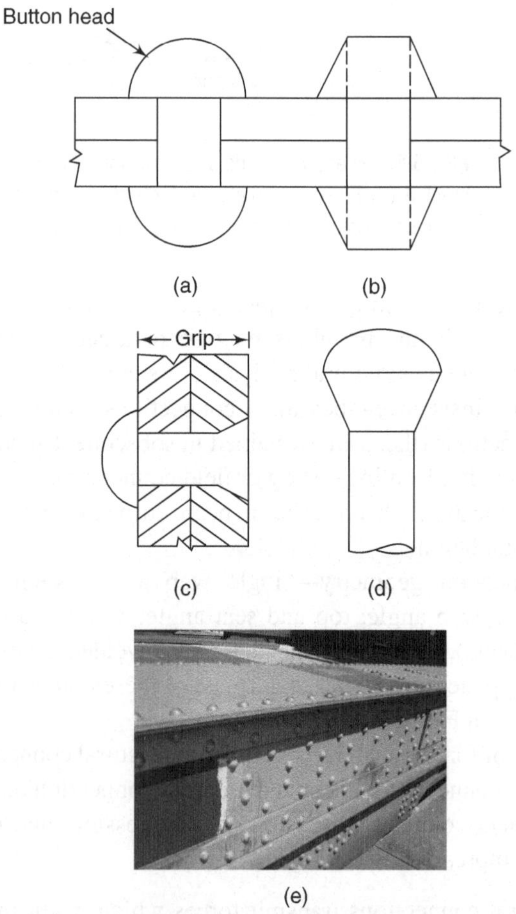


**FIG. 5.3** Types of rivets (a) Snap (b) Pan (c) Flat countersunk (d) Round countersunk (e) Riveted road bridge in Downtown Chicago

to a cherry red colour (approximately 980°C), inserting it into an oversize hole (approximately 1.5 mm more than the size of the rivet), applying pressure to the preformed head while at the same time squeezing the plain end of the rivet using a pneumatic driver to form a round head. During this process, the shank of the rivet completely or nearly fills the hole into which it had been inserted. Upon cooling, the rivet shrinks, thereby producing a clamping force. Owing to this, a riveted joint is intermediate between a friction and a bearing type connection. Since the amount of clamping produced is not dependable, (rivets are often inspected after installation, wherein loose rivets are detected and replaced) a bearing type connection is commonly assumed. The riveted joint has had a long history of success under fatigue stresses as in the several railway bridges throughout the world. However, riveting operations require at least four persons—one to heat and toss the rivet to the driving crew, one to catch the hot rivet and insert it in the hole, one to handle the backup bar, and one to drive the rivet with a pneumatic hammer (cold formed rivets eliminate the heating, but may not be tight if not carefully driven. Also since there is no shrinking, the clamping force is less. Such cold formed rivets are used in aircraft fabrication and in handles for cutlery).

Riveting is no longer used in engineering structures for the following reasons:

1. The necessity of pre-heating the rivets prior to driving
2. The labour costs associated with large riveting crews
3. The cost involved in careful inspection and removal of poorly installed rivets
4. The high level of noise associated with driving rivets

At present, shop welding and friction bolting (the loosening of nuts of ordinary bolts under continued dynamic loads has been eliminated by the introduction of high-strength friction grip bolts) have completely replaced riveting. Hence, in the rest of this chapter, we will focus on bolted connections. Readers should be aware that the design of riveted connection is similar to the design of bolted connection, except that the diameter of the rivet is taken as the diameter of the hole (diameter of rivet + clearance) in place of the nominal diameter of the bolt.

## 5.2 BOLTED CONNECTIONS

There are several types of bolts used to connect structural members. Some of them are listed as follows:

1. Unfinished bolts or black bolts or C grade bolts (IS 1363 : 2002)
2. Turned bolts
   (a) Precision bolts or A grade bolts (IS 1364 : 2002)
   (b) Semi-precision bolts or B grade bolts (IS 3640 : 1982, IS 1364 : 2002)
3. Ribbed bolts
4. High strength bolts (IS 3757 : 1985 and IS 4000 : 1992)

### 5.2.1 Black Bolts

Black bolts are also referred to as ordinary, unfinished, rough, or common bolts. They are the least expensive bolts. However, they may not produce the least expensive connection since the connection may require a large number of such bolts. They are primarily used in light structures under static loads such as small trusses, purlins, girts, bracings, and platforms. They are also used as temporary fasteners during erection where HSFG bolts or welding are used as permanent fasteners. They are not recommended for connections subjected to impact, fatigue, or dynamic loads. These bolts are made from mild steel rods with a square or hexagonal head and nuts as shown in Fig. 5.4, conforming to IS 1363.

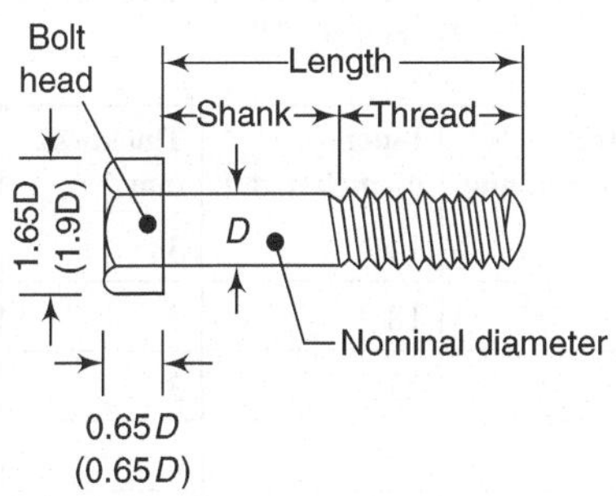


(a)

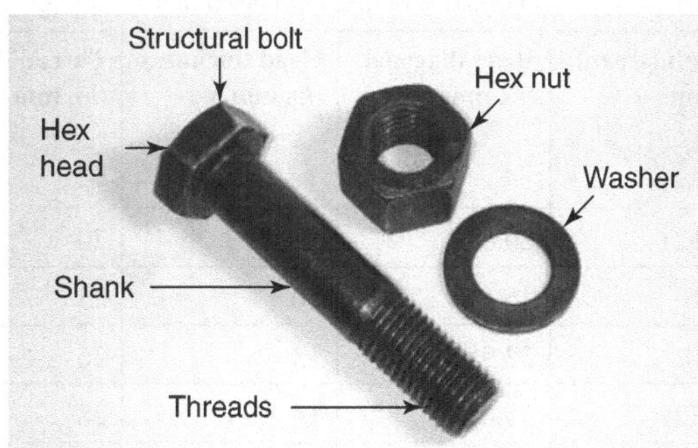


(b)

**FIG. 5.4** (a) Hexagonal head black bolt and nut. Figures in brackets are for high-strength bolts and nuts, (b) Structural steel fastener—bolt, nut, and washer (Copyright© American Institute of Steel Construction, Inc. Reprinted with permission. All rights reserved.)

In steel construction, generally, bolts of property class 4.6 are used. In property class 4.6, the number 4 indicates 1/100$^{th}$ of the nominal ultimate tensile strength in N/mm$^2$ and the number 6 indicates the ratio of yield stress to ultimate stress, expressed as a percentage. Thus, the ultimate tensile strength of a class 4.6 bolt is 400 N/mm$^2$ and yield strength is 0.6 times 400, which is 240 N/mm.$^2$ The tensile properties of fasteners used in steel construction are listed in Table 5.1. Due to the high percentage elongation of these bolts, they are more ductile. For bolts of property class 4.6, nuts of property class 4 are used and for bolts of property class 8.8, nuts of property class 8 or 10 are used.

**TABLE 5.1** Tensile properties of fasteners used in steel construction

| Specification | Grade/ classification | Properties | | |
|---|---|---|---|---|
| | | Yield stress, MPa (Min) | Ultimate tensile stress, MPa (Min) | Elongation percentage (Min) |
| IS 1367 (Part 3) (ISO 898) Specifications of fasteners-threaded steel for technical supply conditions | 3.6 | 180 | 330 | 25 |
| | 4.6 | 240 | 400 | 22 |
| | 4.8 | 320 | 420 | 14 |
| | 5.6 | 300 | 500 | 20 |
| | 5.8 | 400 | 520 | 10 |
| | 6.8 | 480 | 600 | 8 |
| | 8.8 ($d$ < 16 mm) | 640 | 800 | 12 |
| | 9.8 | 720 | 900 | 10 |
| | 10.9 | 940 | 1040 | 9 |
| | 12.9 | 1100 | 1220 | 8 |
| IS 7557 | Annealed condition | 160 | 330–410 | 30 |
| Specification for steel wire (up to 20 mm) for the manufacture of cold forged rivets | As-drawn condition | 190 | 410–490 | 20 |

Though square heads cost less, hexagonal heads give better appearance, are easier to hold by wrenches, and require less turning space. Most of the connections with black bolts are made by inserting them in clearance holes of about 1.5 mm to 2 mm more than the bolt diameter and by tightening them with nuts. They are produced in metric sizes ranging from 5–36 mm and designated as M5 to M36. In structural steelwork, M16, M20, M24, and M30 bolts are often used. The bolts used in steelwork have a course pitch of thread, that is, 2, 2.5, 3, and 3.5 mm for 16, 20, 24, and 30 mm diameter bolts, respectively. The ratio of net tensile area at threads to nominal plain shank area of the bolt is 0.78 as per IS 1367 (Part 1). The other dimensions of the bolt are as given in Table 5.2. These dimensions are so chosen that the bolt head does not fail unless the shank fails.

**TABLE 5.2** Dimensions of hexagon head black bolts (grade 4.6) as per IS 1364 (Part 1)

| Bolt size ($d$), mm | Head diagonal ($e$), mm | Head thickness ($k$), mm | Thread* length ($b$), mm | Pitch of thread, mm | Washer (IS 5370) | | |
|---|---|---|---|---|---|---|---|
| | | | | | Outer diameter, mm | Inner diameter, mm | Thickness, mm |
| (12) | 20.88 | 8 | 20 | 1.75 | 24 | 14 | 3 |
| 16 | 26.17 | 10 | 23 | 2.0 | 30 | 18 | 3 |
| 20 | 32.95 | 13 | 26 | 2.5 | 37 | 22 | 3 |
| (22) | 35.03 | 14 | 28 | 2.5 | 39 | 24 | 3 |
| 24 | 39.55 | 15 | 30 | 3.0 | 44 | 26 | 4 |
| (27) | 45.20 | 17 | 33 | 3.0 | 50 | 30 | 4 |
| 30 | 50.85 | 19 | 35 | 3.5 | 56 | 33 | 4 |
| 36 | 60.79 | 23 | 40 | 4.0 | 66 | 39 | 5 |

*For length $l < 125$ mm. For $125 < l < 200$, $b$ is 6 mm more and for $l > 200$, $b$ is 19 mm more. Sizes in brackets not preferred.

When the nuts on the bolts are tightened with spud wrenches, little tension is produced and hence there will not be any clamping force induced in the joint. Sometimes a hole is drilled in the bolt and a cotter pin with a castellated nut is used to prevent the nut from turning off the bolt (when subjected to fatigue loads).

### 5.2.2 Turned Bolts (Close Tolerance Bolts)

These are similar to unfinished bolts, with the difference that the shanks of these bolts are formed from a hexagonal rod. The surface of these bolts are prepared and machined carefully to fit in the hole. Tolerances allowed are about 0.15 mm to 0.5 mm. Since the tolerance available is small, these bolts are expensive. The small tolerance necessitates the use of special methods to ensure that all the holes align correctly. These bolts (precision and semi-precision) are used when no slippage is permitted between connected parts and where accurate alignment of components is required. They are mainly used in special jobs (in some machines and where there are dynamic loads).

### 5.2.3 Ribbed Bolts (Fluted Bolts)

These bolts have a round head similar to a rivet and raised ribs parallel to the shank. The actual diameter of a ribbed bolt is slightly larger than the hole into which it is driven. While driving the bolt into the hole, it cuts into the edges around the hole, producing a relatively tight fit. These are used in bearing type connections and in connections where there are stress reversals.

A variation of this type of bolt is called an *interference-body bolt*, which has serrations instead of longitudinal ribs. These serrations are around the shank as well as parallel to the shank. Ribbed bolts and interference-body bolts are difficult to drive when several layers of plates are required to be connected. These types of bolts are rarely used in ordinary steel structures.

### 5.2.4 High-strength Bolts

In the 1950s, the concept of using *high-strength bolts* and a summary of research and behaviour of such bolts was presented to practicing engineers and the steel fabrication industry (Munse 1956). The principal factor that delayed the immediate acceptance of high-strength bolts was the high cost of the bolt including two hardened washers (these washers serve two purposes: (a) to distribute the clamping pressure to a larger area of the softer metal of the fastened parts, and to prevent the nut or bolt head from damaging the component or member and (b) to prevent the threaded portion of the bolt from bearing on the connected members). This requirement was partially caused by the nut and head tending to dig (called *galling*) into the soft steel ($f_y = 230$ MPa) available at that time. After the washers could be reduced to one (located under the element, head or nut, being turned) or eliminated (except in special situations) and the greater strength of a bolt over that of rivet could be utilized in design, high-strength bolts became economical over rivets. Thus, in the 1960s the bearing type connection was recognized as an acceptable substitute for riveted connection. The turn-of-the-nut method was introduced in the 1960s as an alternative to the torque wrench method. In the 1980s, the philosophy of design, now called *slip-critical connections* was introduced (Kulak et al. 1987).

High-strength bolts are made from bars of medium carbon steel. The bolts of property class 8.8 and 10.9 are commonly used in steel construction. These bolts should conform to IS 3757 and their tensile properties are given in Table 5.1. As discussed in Chapter 1, their high strength is achieved through quenching and tempering process or by alloying steel. Hence, they are less ductile than black bolts. The material of the bolts do not have a well-defined yield point. Instead of using yield stress, a so-called *proof load* is used. The proof load is obtained by multiplying the tensile stress area (may be taken as the area corresponding to root diameter at the thread and is approximately equal to 0.8 times the shank area of bolt) with the proof stress. In IS 800, the proof stress is taken as 0.7 times the ultimate tensile stress of the bolt. (In other codes such as the American code, the proof stress is taken as the yield stress, established by the 0.2% offset strain.) This bolt tension $0.7f_u$ gives adequate reserve strength, should this bolt be somewhat over stressed (e.g., 3/4 turn instead of 1/2 turn in the turn-of-the-nut method). Note that grade 10.9 bolts have lower ductility than grade 8.8 bolts and the margin between the 0.2% yield strength and the ultimate strength is also lower.

The high-strength friction grip (HSFG) bolt, nut, and washer dimensions are shown in Table 5.3 (also see Fig. 5.4 for approximate bolt dimensions). Although the sizes of bolts from M16 to M36 are available, bolts of sizes M16, M20, M24, and M30 are commonly used in practice. These bolts are identified by the manufacturer's identification symbol and the property class identification symbol 8S, 8.8S, 10S, or 10.9S, which will be embossed on the heads of these bolts. The suffix letter S denotes a high-strength structural bolt with a large series hexagon head. High-strength nuts and washers (8S or 10S nut property class with 8.8S bolt and 10S nut property class with 10.9S bolt) should be used with these bolts. The nut and head of the bolts are sufficiently large to provide an adequate bearing area. Since, these bolts have a tensile strength much higher than the ordinary black bolts, the number of bolts required at a joint is considerably reduced. The

**TABLE 5.3** High-strength friction grip bolts as per IS 3757 : 1985

| Diameter, *d* mm | M16 | M20 | (M22) | M24 | (M27) | M30 | M36 |
|---|---|---|---|---|---|---|---|
| Head diagonal, *e* mm | 29.56 | 37.29 | 39.55 | 45.20 | 50.85 | 55.37 | 66.44 |
| Head thickness, *k* mm | 10 | 12.5 | 14 | 15 | 17 | 18.7 | 22.5 |

*(Contd)*

**TABLE 5.3** *(Contd)*

| | | | | | | | |
|---|---|---|---|---|---|---|---|
| Nut diagonal, mm | 29.56 | 37.29 | 39.55 | 45.20 | 50.85 | 55.37 | 66.44 |
| Nut thickness, mm | 13 | 16 | 18 | 19 | 22 | 24 | 26 |
| Washer outer diameter,**D* mm | 30 | 37 | 39 | 44 | 50 | 56 | 60 |
| Washer inner diameter, *d* mm | 18 | 22 | 24 | 26 | 30 | 33 | 36 |
| Washer thickness, heavy, mm | 4 | 4 | 4 | 4 | 5 | 5 | 5 |
| Thread length, ***b* mm | | | | | | | |
| <100 | 31 | 36 | 38 | 41 | 44 | 49 | 56 |
| >100 | 38 | 43 | 45 | 48 | 51 | 56 | 63 |
| Pitch of thread, *p* mm | 2 | 2.5 | 2.5 | 3 | 3 | 3.5 | 4 |

* The outside diameter of a washer is an important dimension when detailing, for example, to avoid overlapping an adjacent weld.

**The thread length depends on the length of the bolt, which is calculated as grip length plus the allowance for grip as given in Table 5.4.

vibration and impact resistance of the joint are also improved considerably.

**TABLE 5.4** Allowance for bolt length

| Nominal size of the bolt | Allowance for Grip*, in mm |
|---|---|
| 16 | 26 |
| 20 | 31 |
| 24 | 36 |
| 30 | 42 |
| 36 | 48 |

*The allowance includes thickness of one nut and one washer only. If additional washers are used or where threads are excluded (in bearing type joints) from shear plane, high allowance may be required.

The percentage elongation of 12% at failure of these bolts is less than the black bolts, but is still acceptable for design purposes. Special techniques are used for tightening the nuts to induce a specified initial tension in the bolt, which causes sufficient friction between the faying faces. These bolts with induced initial tension are called *high-strength friction grip (HSFG) bolts*. Due to this friction, the slip in the joint (which is associated with black bolts) is eliminated and hence the joints with HSFG bolts are called non-slip connections or *friction type connections* (as opposed to the *bearing type connections* of ordinary black bolts). The induced initial tension in the bolt is called the *proof-load* of the bolt and the coefficient of friction between the bolt head and the faying surfaces is called the *slip factor*. The bolts of property class 8.8 can be hot-dip galvanized [as per IS 1367 (Part 13)] whereas bolt of class 10.9 and above should be considered carefully before they are hot-dip galvanized since this may cause hydrogen embrittlement (IS 3757). When the grade 8.8 bolts are galvanized, the bolts and mating nuts should be provided with a suitable lubricant so that seizure will not take place in assembly. The effect of such lubricants has been studied by Subramanian (1984) and Subramanian & Petersen (1987).

### 5.2.5 Bolt Tightening Methods

When slip resistant connections are not required and when bolts are not subjected to tension, high-strength bolts are tightened to a 'snug-tight' condition to ensure that the load transmitting plies are brought into effective contact (this may be achieved as a result of a few impacts of an impact wrench or the full effort of a man using an ordinary spud wrench).

However, the reliability of HSFG bolts in a non-slip or friction type connection depends on the method of tightening of the bolt, which will ensure whether the required proof load (pre-tension) is obtained. The three methods that may be used in practice are *the turn-of-the-nut tightening* (part-turn method), *direct tension indicator tightening, and calibrated wrench tightening* (torque control method).

**Turn-of-the-nut tightening method** Turn-of-the-nut tightening method, also known as part-turn method, is the simplest and most common method. Developed in the 1950s and 1960s, the specified pre-tension in the bolt is considered to be obtained by a specified rotation of the nut from the 'snug-tight' condition. Although the snugness or the initial tightness can vary due to the surface condition of the pieces being tightened, this variation does not affect the clamping force significantly (Kulak et al. 1987). In this method, after the bolts are snug-tight, permanent marks (these permanent marks may be used in a subsequent inspection) are made on bolts and nuts to identify the relative position of the bolt and nut and to control the final nut rotation. Each nut is then tightened by a specified turn of the nut from the snug-tight position depending on the length of the bolt as prescribed in IS 4000 (see Table 5.5 which also gives the minimum tension which should be available in the bolt after tightening). This is shown graphically in Fig. 5.5. It has to be noted that some codes do not allow this method of tightening for higher grade bolts with parallel shanks because of the concern over their limited ductility (Owens & Cheal 1989).

**TABLE 5.5** Minimum bolt tension and nut rotation from snug-tight condition (IS 4000 : 1992).

| Nominal size of bolt | Length of bolt,* mm | | Minimum bolt tension in kN for bolts of property class | |
|---|---|---|---|---|
| | Nut rotation 1/2 turn | Nut rotation 3/4 turn | 8.8 | 10.9 |
| M16 | <120 | >120 < 240 | 94.5 | 130 |
| M20 | <120 | >120 < 240 | 147 | 203 |
| M24 | <160 | >160 < 350 | 212 | 293 |
| M30 | <160 | >160 < 350 | 337 | 466 |
| M36 | <160 | > 160 < 350 | 490 | 678 |

*Length is measured from the underside of the head to the extreme end of the shank.

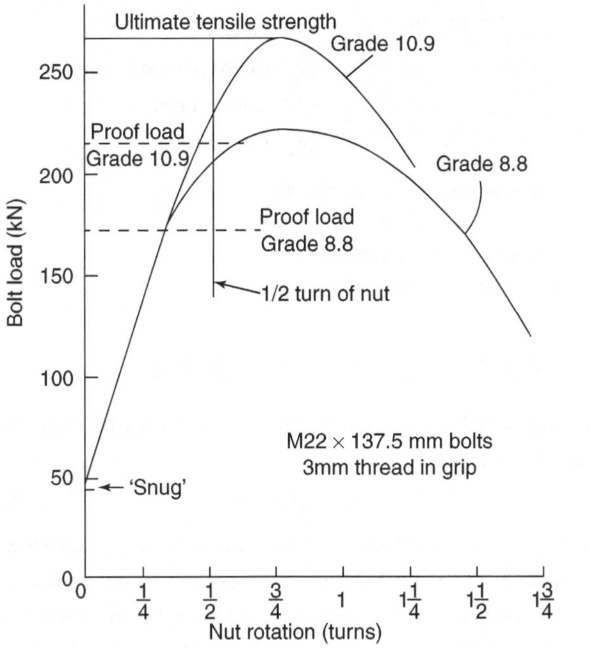


**FIG. 5.5** Typical induced tension/nut rotation relationship for parallel shank HSFG bolts

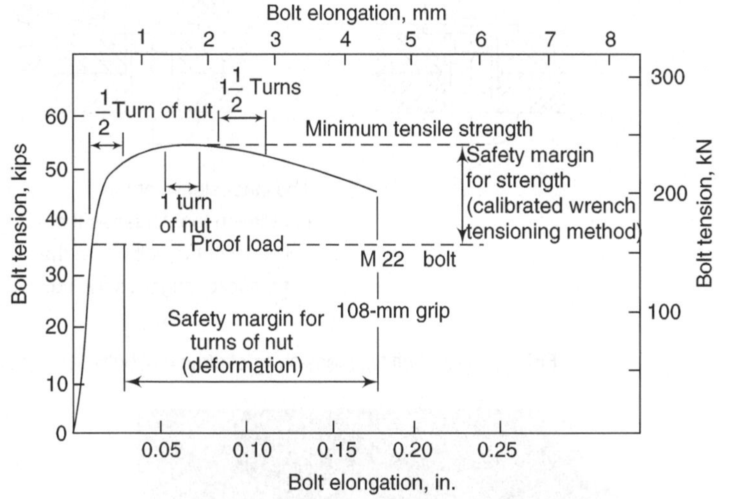


**FIG. 5.6** Behaviour of bolt while using turn-of-the-nut method

*Source:* Rump & Fisher 1963. Reprinted with permission from ASCE. All rights reserved.

It generally takes 2.25–2.5 turns from the snug-tight position to fracture the bolt (Ragupathi 1995). In the American practice, 1/3 turn of the nut is prescribed for bolt length less than $4d$, 1/2 turn of the nut for bolt lengths from $4d$ to $8d$ or 200 mm, and 2/3 turn of the nut for bolt lengths greater than $8d$ or 200 mm, where $d$ is the diameter of bolt (Salmon & Johnson 1996). When the turn-of-the-nut method is used, the bolts are tensioned using 1/8 turn increments. In this method, the bolt deformation is a critical factor as shown in Fig. 5.6.

**Direct tension indicator tightening** There are two types of proprietary load indication devices. The first type of device indicates the load by producing a measurable change in gap between the nut and the gripped material (Struik et al. 1973). In this method, a hardened washer containing a number of protrusions on one face is used (see Fig. 5.7). This washer is inserted between the head or nut and the gripped material with the protrusions bearing against the underside of the head or nut, leaving a gap maintained by the protrusions. When the bolts are tightened, the protrusions are compressed and the gap is reduced. The bolt tension is determined by measuring the remaining gap using a gauge (this gap should be less than 0.38 mm). The specified gap corresponds to a specified tension in the bolt.

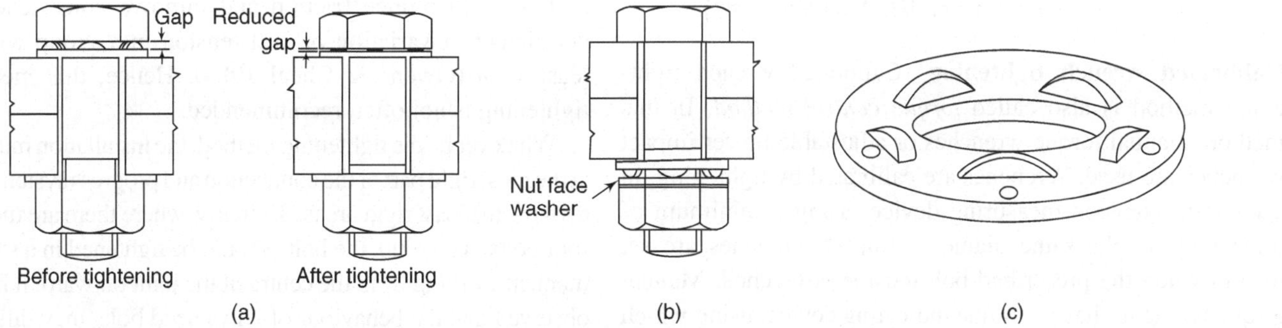


**FIG. 5.7** Proper use and orientation of direct-tension indicator (DTI) (a) DTI under bolt head, nut turned (bolt head could also be turned) (b) DTI under nut, nut turned (bolt head could also be turned) (c) Typical load indicating washer

The second type uses proprietary fasteners designed to indirectly indicate the bolt tension. In this type, the bolt is tightened by turning a nut, which has a protruding nib; the tightening is continued till the nib shears off (Fig. 5.8). Both these methods are irreversible and only show that the bolts were once tensioned to their proof load. Hence, they should not be used if there is an interaction between neighbouring bolts, since these methods will not be in a position to indicate the current bolt load (Owens & Cheal 1989). These kinds of bolts are not yet available in India.

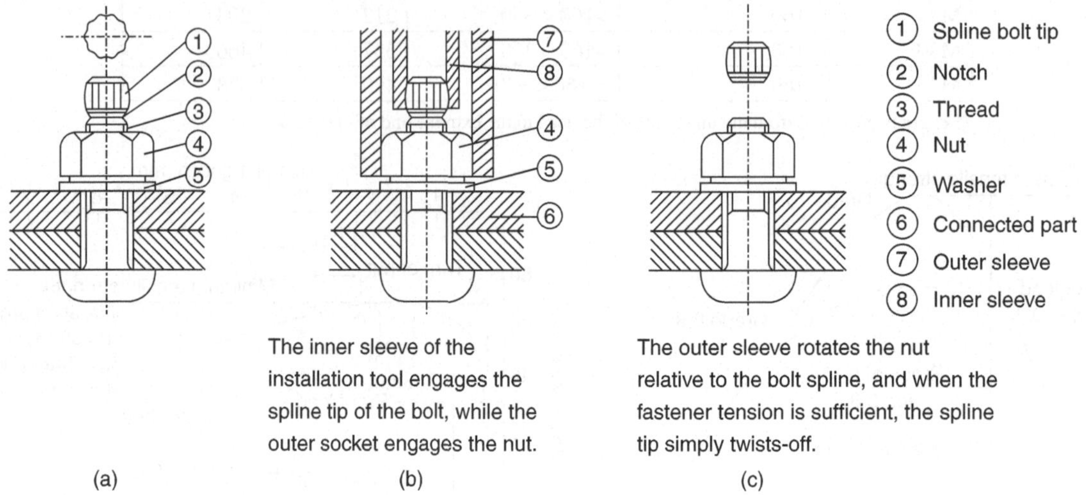


**FIG 5.8** Twist off type tensioning of structural bolts (a) Bolt before clamping (b) During clamping (c) After clamping

Bolt tightening using impact wrench (*Source:* Four Story Office Building Case Study, *AISC Teaching Aid.* Reprinted with permission from AISC. All rights reserved.)

**Calibrated wrench tightening** Calibrated wrench tightening method is also called *torque control method.* In this method, manual torque wrenches or adjustable power impact wrenches are used. Wrenches are calibrated by tightening, in a hydraulic tension-measuring device, using a minimum of three bolts of the same diameter. Impact wrenches are set to stall when the prescribed bolt tension is reached. Manual torque wrenches have a torque indicating device, using which the torque required to produce the initial tension is measured. The relationship between torque and tension may be expressed as (Ragupathi 1995)

$$T = kPd \tag{5.1}$$

where $T$ is the applied torque, $P$ is the required bolt tension, $d$ is the diameter of the bolt shank, and $k$ is the torque coefficient. The torque coefficient $k$ is taken as 0.15 for highly lubricated bolts and as 0.25 for rusty bolts. The calibration should be checked at least once a day or the wrench may be used on a bolt of different grade or size. A hardened washer must be used under the nut or the head, whichever is turned for tightening the bolt. Without a hardened washer, the results will not be reliable. The torque wrenches are adjusted to produce 1.1 times the proof load of the bolts, since most of the torque effort goes to overcome friction in the threads.

The possibility of over-torquing the bolts with power wrenches should not cause any concern, since such over-torquing usually fractures the bolt and hence the bolt may be replaced. However, even with careful execution (lubrication of bolts will reduce friction and improve consistency); the coefficient of variation of bolt tension in a group will be at least 15% (Owens & Cheal 1989). Hence, this method of tightening is not often recommended.

Whatever be the tightening method, the installation must begin at the most rigid part of the connection and progress systematically towards the least rigid areas. Similarly, where there are more than four bolts in a group, the bolts should be tightened in a staggered manner, working from the centre of the joint outward. It has been observed that the behaviour of galvanized bolts may differ from the behaviour of normal, uncoated high-strength bolts.

The turn-of-the-nut method is the cheapest and it is also more reliable. Hence, it is often preferred over the calibrated wrench method. The direct tension indicator tightening method using the protruded washer is less accurate than torque spanner according to Bahia & Martin (1981). The torque wrench and turn-of-the-nut method produce a proof load which is higher than the specified value (Kulak et al. 1987). However, since the turn-of-the-nut method often induces a bolt tension that may exceed the elastic limit of the threaded portion, repeated tightening of high-strength bolts may be undesirable.

### 5.2.6 Advantages of Bolted Connections

The black bolts offer the following advantages over riveted or welded connections:

1. Use of unskilled labour and simple tools
2. Noiseless and quick fabrication
3. No special equipment/process needed for installation
4. Fast progress of work
5. Accommodates minor discrepancies in dimensions
6. The connection supports loads as soon as the bolts are tightened (in welds and rivets, cooling period is involved).

The main drawback of the black bolt is the slip of the joint when subjected to loading. When large forces are to be resisted, the space required for the joint is extensive. Also, precautions such as the provision of special locking devices or the use of pre-loaded high-strength bolts are required in situations involving fluctuating loads.

Though the material cost of HSFG bolts are about 50% higher than black bolts and require special workmanship for installation, they provide the following advantages.

1. HSFG bolts do not allow any slip between the elements connected, especially in close tolerance holes (see Fig. 5.9), thus providing rigid connections.

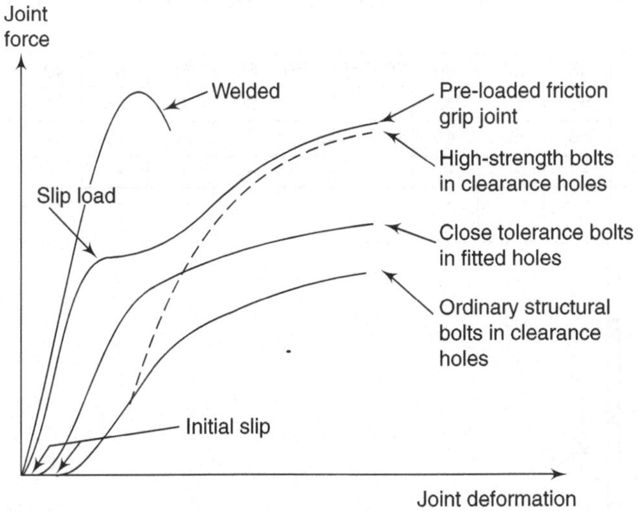


**FIG. 5.9** Typical load–deformation behaviour of different types of fasteners

2. Due to the clamping action, load is transmitted by friction only and the bolts are not subjected to shear and bearing.
3. Due to the smaller number of bolts, the gusset plate sizes are reduced.
4. Deformation is minimized.
5. Since HSFG bolts under working loads do not rely on resistance from bearing, holes larger than usual can be provided to ease erection and take care of lack of fit. Thus the holes may be standard, extra large, or short/long slotted. However, the type of hole will govern the strength of the connection.
6. Noiseless fabrication, since the bolts are tightened with wrenches.
7. The possibility of failure at the net section under the working loads is eliminated.
8. Since the loads causing fatigue will be within proof load, the nuts are prevented from loosening and the fatigue strength of the joint will be greater and better than welded and riveted joints. Moreover, since the load is transferred by friction, there is no stress concentration in the holes.
9. Unlike riveted joints, few persons are required for making the connections.
10. No heating is required and no danger of tossing of bolt. Thus, the safety of the workers is enhanced.
11. Alterations, if any (e.g., replacement of the defective bolt) are done easily than in welded or riveted connections.

However, bolting usually involves a significant fabrication effort to produce the bolt holes and associated plates or cleats. In addition, special procedures are required to ensure that the clamping actions required for pre-loaded friction-grip joints are achieved. The connections with HSFG bolts may not be as rigid as a welded connection.

### 5.2.7 Bolt Holes

Bolt holes are usually drilled. *Punched holes* (punched full size or punched undersize and reamed) are preferred by steel fabricators because it is simple and saves time and cost. However, punching can reduce ductility and toughness and may lead to brittle fracture. Hence, punched holes should not be used where plastic tensile straining can occur (Owens et al. 1981). IS 800 allows punched holes only in materials whose yield stress $f_y$ does not exceed 360 MPa and where thickness does not exceed $(5600/f_y)$ mm. It also disallows punched holes in cyclically loaded details. Holes should not be formed by gas cutting, since they affect the local properties of steel, though plasma cutting is allowed in the code for statically loaded members (clause 17.2.4.5).

Bolt holes are made larger than the bolt diameter to facilitate erection and to allow for inaccuracies in fabrication.

Table 5.6 shows the standard values of holes for different bolt sizes (the clearance is 1.0 mm for bolts less than 14 mm and 2 mm for bolts between 16 mm and 24 mm and 3 mm for bolts exceeding 24 mm).

**TABLE 5.6** Bolt diameter, pitch, and edge distances as per IS 800

| Nominal diameter of bolt, mm | 12 | 14 | 16 | 18 | 20 | 22 | 24 | 27 | 30 | Above 36 |
|---|---|---|---|---|---|---|---|---|---|---|
| Diameter of hole, mm | 13.0 | 15.0 | 18.0 | 20.0 | 22.0 | 24.0 | 26.0 | 30 | 33 | Bolt diameter + 3 mm |
| Minimum edge distance,* mm | | | | | | | | | | |
| (a) for sheared or rough edge | 22 | 26 | 31 | 34 | 37 | 41 | 44 | 51 | 56 | 1.7 × hole diameter |
| (b) for rolled, sawn, or planed edge | 18 | 23 | 27 | 30 | 33 | 36 | 39 | 45 | 50 | 1.5 × hole diameter |

*The edge distances in this table, which are for standard holes, must be increased if oversize or slotted holes are used (see Fig. 5.10).

Max. edge distance = $12t\varepsilon$ where $\varepsilon = (250/f_y)^{0.5}$ Pitch (min.) 2.5 × nominal diameter of bolt

Pitch (max.) $32t$ or 300 mm (a) parts in tension $16t$ or 200 mm, whichever is less

(b) parts in compression $12t$ or 200 mm, whichever is less (c) tacking fasteners $32t$ or 300 mm, whichever is less and $16t$ or 200 mm, whichever is less for plates exposed to weather, where $t$ is the thickness of the thinner outside plate or angle.

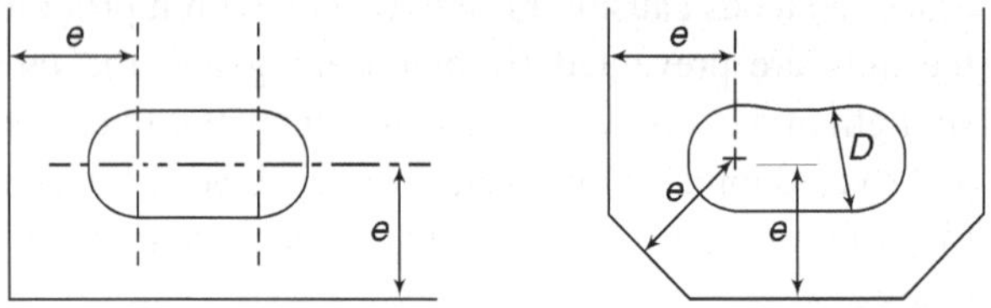


e = end or edge distance e = end or edge distance

**FIG. 5.10** Slotted holes

Oversize holes [should not exceed 1.25$d$ or ($d$ + 8) mm in diameter, where $d$ is the nominal bolt diameter in mm] and slotted holes are allowable but should not be used often. A *slotted hole* should not exceed the appropriate hole size in width and 1.33$d$ in length, for short slotted hole and 2.5$d$ in length, for long slotted hole (See Fig. 5.10). As discussed in Section 3.4 slotted holes are used to accommodate movements in a structure. However, if holes are longer than 2.5$d$, shear transfer in the direction of the slot is not admissible even in a friction type connection (see also Section 10.2.1 of code).

Bolt holes reduce the gross cross sectional area of sections (plates, angles, etc.). The net value is used in the calculations, when the element is in tension (see Chapter 7). As already discussed, bolt holes produce stress concentration, but this is offset by the fact that yield at highly stressed cross section will work-harden before fracture, resulting in the yield of adjacent cross section also. Whereas, if the member is in compression, then the gross cross section of the member is used in the calculation, because at yield the bolt hole deforms, transferring part of the load to the shank of the bolt or be resisted by bearing.

### 5.2.8 Spacing and Edge Distance of Bolt Holes

The centre-to-centre distance between individual fasteners in a line, in the direction of load/stress is called the *pitch*. The distance between any two consecutive fasteners in a zigzag pattern of bolts, measured parallel to the direction of load/stress is called a *staggered pitch*. A minimum spacing of 2.5 times the nominal diameter of the fastener is specified in the

| Leg size, mm | 200 | 150 | 130 | 125 | 115 | 110 | 100 | 95 | 90 | 80 | 70–75 | 65–60 | 55 | 50 | 45 | 40 |
|---|---|---|---|---|---|---|---|---|---|---|---|---|---|---|---|---|
| $g_1$, mm | 115 | 90 | 80 | 75 | 70 | 65 | 60 | 55 | 50 | 45 | 40 | 35 | 30 | 28 | 25 | 21 |
| $g_2$, mm | 75 | 55 | 50 | 45 | 45 | 45 | 40 | 55 | 50 | 45 | 40 | 35 | 30 | 28 | 25 | 21 |
| $g_3$, mm | 85 | 65 | 55 | 55 | 50 | 45 | 40 | – | – | – | – | – | – | – | – | – |
| Maximum bolt size for double row of bolts | 27 | 22 | 20 | 20 | 12 | 12 | 12 | – | – | – | – | – | – | – | – | – |

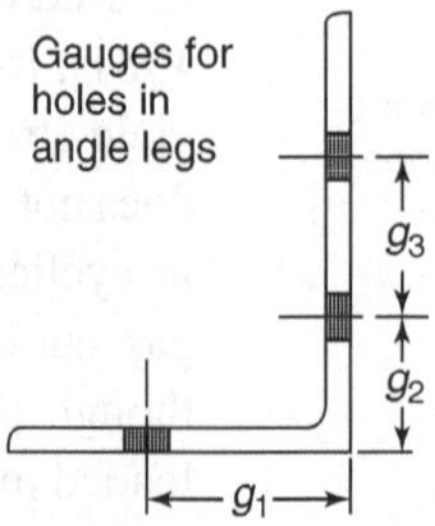


*Note:* For other size of angles, refer Table XXXI of ISI handbook on Structural Steel Sections SP-6(1): 1964

**FIG. 5.11** Workable gauge distance in angle legs

code to ensure that there is sufficient space to tighten the bolts, prevent overlapping of the washers, and provide adequate resistance to tear-out of the bolts. It also limits any adverse interaction between high bearing stresses due to neighbouring bolts. Similarly, the code specifies maximum pitch values, as given in Table 5.6. These values are specified to prevent buckling of plates in compression between the bolts, to ensure that the bolts act together as a group to resist the loads and to avoid corrosion by ensuring adequate bridging of the paint film between plates. The spacing between adjacent parallel line of fasteners, transverse to the direction of load/stress is called *gauge distance*. Figure 5.11 gives the gauge distance as specified in SP-1, published by the Bureau of Indian Standards.

The distance from the centre of a fastener hole to the edge of an element (measured at right angles to the direction of the load) is called the *end* or *edge distance.* The edge distance should be sufficient for bearing capacity and to provide space for the bolt head, washer, and nut. Hence, minimum edge distances are specified in the code and are given in Table 5.6. The maximum edge distance to the nearest line of fasteners from an edge of any unstiffened part should not exceed $12t\varepsilon$ where $\varepsilon = (250/f_y)^{0.5}$ and $t$ is the thickness of the thinner outer plate. (This rule is not applicable to fasteners interconnecting the components of back-to-back tension members.) In corrosive environment, the maximum edge distance should not exceed 40 mm plus $4t$.

## 5.3 BEHAVIOUR OF BOLTED JOINTS

Loads are transferred from one member to another by means of the connections between them. A few typical bolted connections are given in Fig. 5.12.

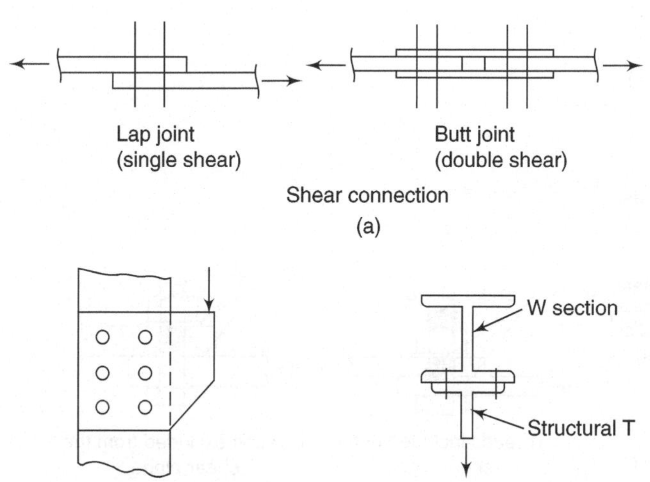


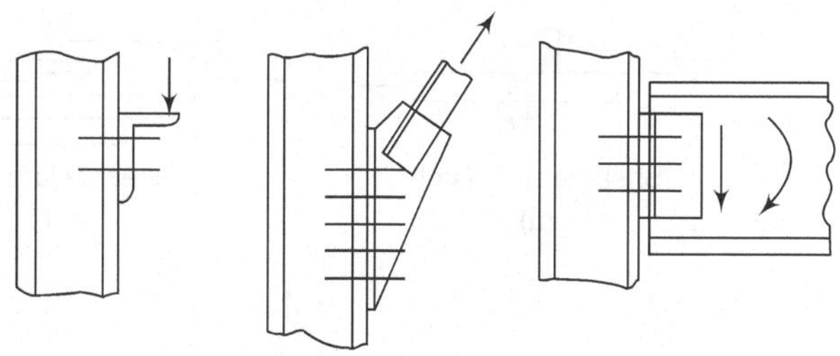


**FIG. 5.12** Typical bolted connections

The possible 'limit states' or failure modes that may control the strength of a bolted connection are shown in Fig. 5.13. Thus, any joint may fail in any one of the following modes:

1. Shear failure of bolt
2. Shear failure of plate
3. Bearing failure of bolt
4. Bearing failure of plate
5. Tensile failure of bolts
6. Bending of bolts
7. Tensile failure of plate

In bearing type connections using black bolts or high-strength bolts, as soon as the applied load overcomes the very small amount of friction at the interface, slip will occur and the force is transferred from one element to another element through bearing of bolts (see Fig. 5.9). Note that for a single bolt in normal clearance holes, slip may be twice the bolt clearance in extreme circumstances, i.e., up to 3 mm to 4 mm, depending on the bolt size (Owens & Cheal 1989). However, in practice there will be multiple number of bolts in a connection and the movement due to mismatch of the bolts will be less.

Once the bolts are in bearing, the connection will behave linearly, until yielding takes place at one or more of the following positions (Owens & Cheal, 1989):

1. At the net section of the plate(s), under combined tension and flexure
2. On the bolt shear plane(s)
3. In bearing between the bolt and the side of the hole

The forces acting on the bolt are shown in Fig. 5.14(a).

The response of the connection becomes non-linear after yielding takes place, as plasticity spreads in the presence of strain hardening and failure takes place at one of the critical sections/locations listed above (see Fig. 5.13). The mode of failure and the point of initiation of yielding depends upon the proportions and relative material strength of the components.

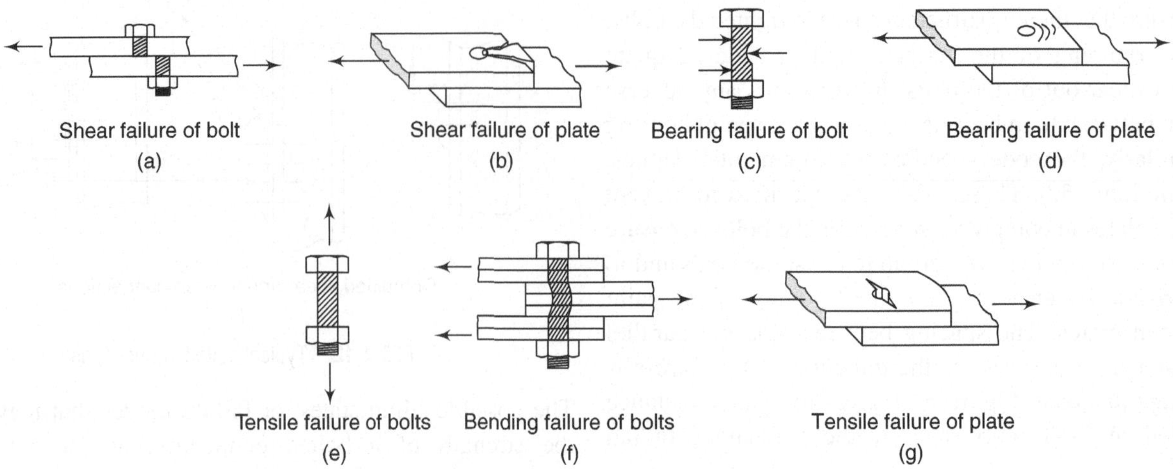


**FIG. 5.13** Possible modes of failure of bolted connections

Bolt tension fracture limit state

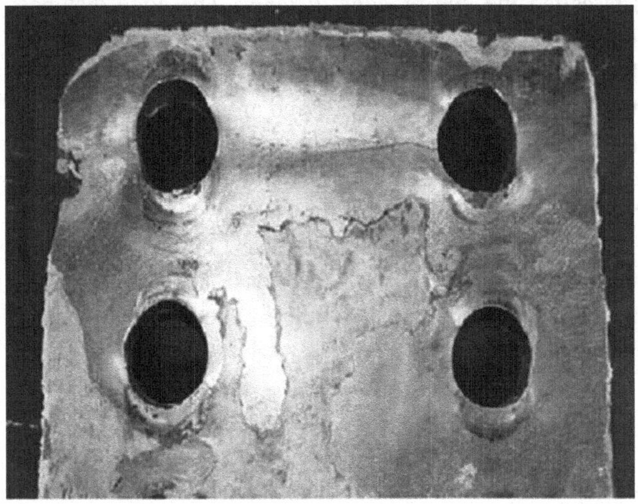

Bolt bearing limit state

Photos by J.A. Swanson and R. Leon of Georgia Institute of Technology (*Source: AISC Connections Teaching Toolkit—A Teaching Guide for Structural Steel Connections.* Reprinted with permission from AISC. All rights reserved.)

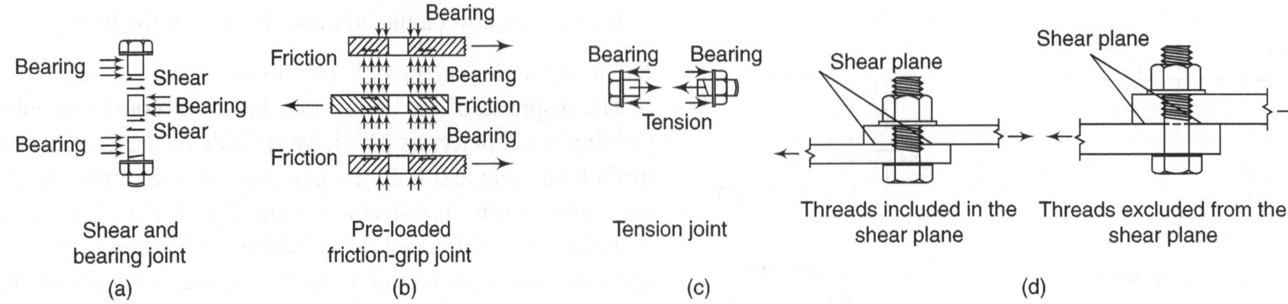


**FIG. 5.14** Force transmission through bolts (Trahair et al. 2001)

In a multibolt connection, the behaviour is similar except that the more highly loaded bolt starts to yield first, and the connection will become less stiff. At a later stage, due to redistribution of forces, each bolt is loaded to its maximum capacity. However, it is generally assumed that equal size bolts share equally in transferring the external force as shown in Fig. 5.15(b), even during service loads. However, in a long bolted connection, the shear force is not evenly distributed among the bolts, and consequently the bolts at the end of a joint resist the highest amount of shear force, as shown in Fig. 5.15(c). In such joints, the end bolt forces may be so high that it may lead to a progressive joint failure called 'unbuttoning'. If the joint is short, the forces in the bolts will be redistributed by plastic action, and hence the bolts will share the shear force equally.

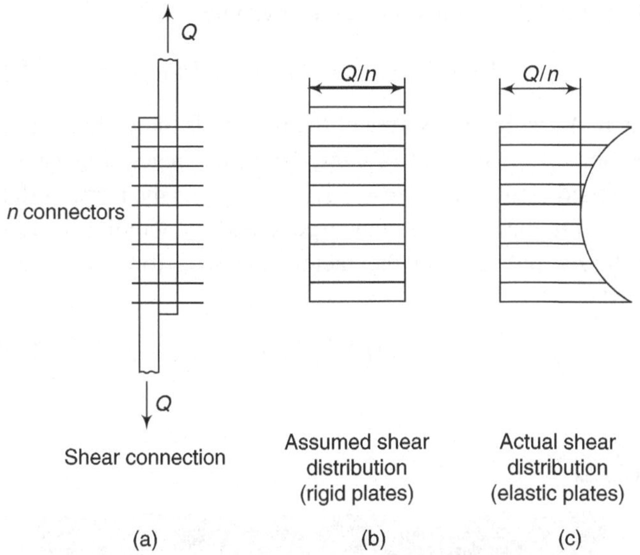


**FIG. 5.15** Shear distribution in a long bolted connection

Shear and bearing connections using close tolerance bolts in fitted holes behave in a similar manner to connections with clearance holes, except that the bolt slip is considerably small (see Fig. 5.9). As mentioned earlier, close tolerance bolts are rarely used.

In the case of HSFG bolts, the slip in the bolt will not occur immediately but at a load which overcomes the frictional resistance provided by the pre-load of the bolt (see Fig. 5.9). After slip occurs, the behaviour of the bolt is similar to the normal bolts. In this case also, it is commonly assumed that equal size bolts share the loads equally in transferring the external force.

The flexibility of a connection is determined from the sum of the flexibilities of its different components (bolts, plates, and cleats used in the connection). Note that plates (gusset plates) are comparatively stiff when loaded in their own plane and are considered as rigid. However, when bent out of their plane, they are comparatively flexible. The overall behaviour of any connection should be carefully assessed by determining the force flow through the connection and by synthesizing the responses of the elements to their individual loads (Trahair et al. 2001).

## 5.4 DESIGN STRENGTH OF ORDINARY BLACK BOLTS

Expressions for design strength of ordinary black bolts subjected to shear, tension, and bearing forces are given in this section. In addition, when bolts are subjected to tension, there may be additional forces due to flexibility of connections, which are called *prying forces*. Methods to calculate prying forces and interaction equation for bolts subjected to combined shear and tension forces are also covered. Tension capacity of plates and efficiency of joints are also discussed.

### 5.4.1 Bearing Bolts in Shear

The nominal capacity of a bolt in shear $V_{nsb}$ [Figs 5.14(a), 5.14(d), and 5.13(a)] depends on the ultimate tensile strength $f_u$ of the bolt, the number of shear planes $n$ ($n = n_n + n_s$), and the areas $A_{sb}$ (nominal shank area) and $A_{nb}$ (net shear area at the threads) of the bolt in each shear plane. It is expressed in the code as

$$V_{nsb} = (f_u/\sqrt{3})(n_n A_{nb} + n_s A_{sb})\beta_{lj}\beta_{lg}\beta_{pkg} \tag{5.2}$$

where $n_n$ is the number of shear planes with threads intercepting the shear plane, $n_s$ is the number of shear planes without threads intercepting the shear plane [See Fig. 5.14(d)], $\beta_{lj}$ is the reduction factor which allows for the overloading of end bolts that occur in long connections, $\beta_{lg}$ is the reduction factor that allows for the effect of large grip length, and $\beta_{pkg}$ is the reduction factor to account for packing plates in excess of 6 mm.

IS 800 stipulates that the factored shear force $V_{sb}$ should satisfy

$$V_{sb} \le V_{nsb}/\gamma_{mb} \tag{5.3}$$

where $\gamma_{mb}$ is the partial safety factor for the material of the bolt, taken as 1.25 in the code.

When the net shear area at the threads is not given, it may be taken, for ISO thread profile, as

$$A_{nb} = (\pi/4)(d - 0.9382p)^2 \tag{5.4}$$

where $d$ is the shank or nominal diameter of bolt in mm and $p$ is the pitch of the thread in mm.

The net tensile stress area will be approximately 78 – 80% of the gross area. It has been found that the shear strength is controlled by joint length, grip length, and the thickness of the packing plates. However, the pitch (distance between the centres of adjacent fasteners along the line of principal stress) is not an important variable.

**Reduction factor for long joints** When the joint length $l_j$, of a splice or end connection, in tension or compression, exceeds $15d$ in the direction of load (the joint length is taken as the distance between the first and last rows of the bolts in a joint, measured in the direction of the load transfer), the nominal shear capacity $V_{nsb}$ is multiplied by a reduction factor $\beta_{lj}$ as shown in Eqn (5.2). This reduction factor is given by

$$\beta_{lj} = 1.075 - l_j/(200\ d) \quad \text{for } 0.75 \le \beta_{lj} \le 1.0 \qquad (5.5)$$

This reduction factor should not be applied when the distribution of shear over the length of joint is uniform as in the connection of web of a section to the flanges.

**Reduction factor for large grip lengths** When the total thickness of connected plates or plies (grip length) $l_g$ exceeds five times the nominal diameter of the bolts, the nominal shear capacity $V_{nsb}$ is multiplied by a reduction factor $\beta_{lg}$ as shown in Eqn (5.2). This reduction factor is given by

$$\beta_{lg} = 8d/(3d + l_g) \qquad (5.6)$$

The value of $\beta_{lg}$ calculated using Eqn (5.6) should not be more than $\beta_{lj}$ given in Eqn (5.5) and the grip length $l_g$ is also restricted to $8d$ by the code.

The reason for the above reduction in strength is that as the grip length increases, the bolt is subjected to greater bending moments due to the shear forces acting on them [see Fig. 5.13(f)].

**Reduction factor for packing plates** Similar to the grip length, the thickness of packing plates also influence the nominal shear capacity $V_{nsb}$. Thus, when the packing plates are greater than 6 mm, the shear capacity is multiplied by the reduction factor $\beta_{pkg}$. This reduction factor is given by

$$\beta_{pkg} = (1 - 0.0125 t_{pkg}) \qquad (5.7)$$

where $t_{pkg}$ is the thickness of the thicker packing plate in mm.

### 5.4.2 Bolts in Tension

The nominal capacity of a bolt in tension $T_{nb}$ [Figs 5.14(c) and 5.13(e)] depends on the ultimate tensile strength $f_{ub}$ of the bolt and net tensile stress area $A_n$ (taken as the area at the bottom of the threads) of the bolt, and is given by

$$T_{nb} = 0.90 f_{ub} A_n < f_{yb} A_{sb} (\gamma_{m1}/\gamma_{m0}) \qquad (5.8)$$

where $A_{sb}$ is the shank area of the bolt, $f_{yb}$ is the yield stress of the bolt, $\gamma_{m1}$ is the partial safety factor for material resistance governed by ultimate stress = 1.25, and $\gamma_{m0}$ is the partial safety factor for material resistance governed by yielding = 1.10. IS 800 stipulates that the factored tension force $T_b$ should satisfy

$$T_b \le T_{nb}/\gamma_{mb} \qquad (5.9)$$

## CASE STUDY

### Kemper Arena Collapse

Designed by the German architect Helmut Jahn, the Kemper Arena's large flat roof was suspended on hangers from 3 large triangulated space frame trusses, spanning the shorter dimension of the roof (see photo). The three trusses, each 16.5 m wide, were spaced 30 m apart, and were made of 1.2 m diameter pipe sections. This 17,600-seat capacity arena was of size 97 by 108 m and opened for public in 1974.

The secondary steel plane trusses were supported by the space frame by pipe hangers at 42 different panel points. Each of these hangers carried 622 kN in tension. The space frames were bolted together by ASTM A-490 high strength bolts, loaded in shear. The hangers were attached to the space frames by a welded gusset plate with a pin-connection, and to the top-chords of the secondary steel trusses by a steel plate welded to the hangers, and then bolted with four A-490 bolts, vertically loaded in tension (see figure on next page).

In order to reduce the storm water runoff into the city storm sewers, the roof was designed to hold water as a temporary reservoir. The roof only had eight 130 mm diameter drains. The local code actually required 55 drains. The water was designed to pour out over scuppers, when the water depth exceeded 50 mm. This feature resulted in ponding loads. As the water accumulates,

Kemper Memorial Arena
*Courtesy:* Mr Bill Cobb of skylinescenes.com

the middle of the roof begins to sag or deflect. This causes more water to accumulate and thus more of a deflection, continuing until the roof becomes unstable.

On 4 June 4 1979 at about 7:10 p.m., a severe wind (with 110 km/h speed) and rain storm (dumping 108 mm of rain per hour) caused a portion of Kemper Arena's roof to collapse. A portion of the roof approximately 61 by 66 m collapsed into the arena. Since the arena was not in use at the time, no one was injured.

The investigations done by Exponent Engineering and Scientific consulting, revealed the following:

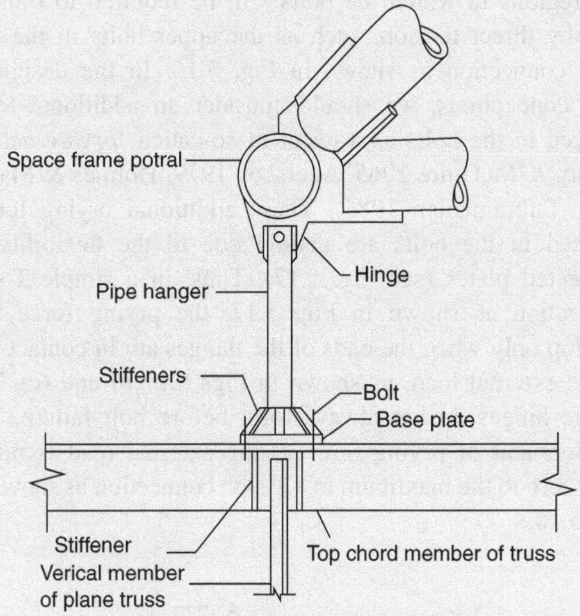


Details of the kemper arena hanger assembly

1. Wind loading caused many cycles of loading and unloading of the bolts in the tension connection. Analysis demonstrated that failure of just one of four bolts per hanger resulted in a large prying load on the three remaining bolts. Once they failed, the adjacent hangers were overloaded, resulting in a cascading failure. The bolt failure due to the dynamic effects of wind and the ponding load were determined to be the most probable causes of failure.
2. The rain load due to ponding was not properly estimated, while designing the roof. Commonly accepted but sometimes erroneous assumption of uniform rainwater loading, may not be true under high wind conditions.
3. The lack of redundancy in the roof support system (ability to transfer loads after partial system failure) and even distribution of connection load to individual bolts were also shortcomings of the design.

Based on the investigation, the different issues were addressed, the roof repaired and the arena reopened within a year. The important lesson to be learnt by this failure is that high-strength bolts, which are relatively brittle, should not be used in joints subjected to fatigue loads. HSFG bolts, with pre-tensioning, are suitable for such connections.

where $\gamma_{mb}$ is the partial safety factor for material resistance of connection = 1.25.

If any of the connecting plates is sufficiently flexible, then additional prying forces may be induced in the bolt (see Section 5.4.4 for the details).

### 5.4.3 Bolts in Bearing

When an ordinary bolt is subjected to shear forces, it comes into contact with the plates, after the slip occurs. The bearing limit state relates to deformation around a bolt hole, as shown in Fig. 5.13(d) (bearing failure for a large end distance). A shear tear-out failure (also called end bearing failure) as shown in Fig. 5.13(b) occurs when the end distance is small. Bearing failure in bolts (see Fig. 5.13(c)) is possible only by using low strength bolt with very high grade plates, which will not occur in practice.

The nominal bearing strength of the bolt $V_{npb}$ is given by

$$V_{npb} = 2.5k_b d t f_u \quad (5.10)$$

where $f_u$ is the ultimate tensile stress of the plate in MPa, $d$ is the nominal diameter of the bolt in mm, and $t$ is the summation of the thicknesses of the connected plates experiencing bearing stress in the same direction. (If the bolts are countersunk, then $t$ is equal to the thickness of the plate minus one half of the depth of counter sinking.)

$k_b$ is smaller of $e/(3d_o)$, $p/(3d_o) - 0.25$, $f_{ub}/f_u$ and 1.0 where $f_{ub}$ is the ultimate tensile stress of the bolt, $e$ is the end distance, $p$ is the pitch of the fastener along bearing direction, and $d_o$ is the diameter of bolt hole. $V_{npb}$ should be multiplied by a factor 0.7 for over size or short slotted holes and by 0.5 for long slotted holes. The factor $k_b$ takes into account inadequate edge distance or pitch and also prevents bearing failure of bolts. If we adopt a minimum edge distance of 1.5 × bolt hole diameter and a minimum pitch of 2.5 × diameter of bolt, the value of $k_b$ can be taken as 0.50.

IS 800 stipulates that the bolt bearing on any plate subjected to a factored shear force $V_{sb}$, should satisfy

$$V_{sb} \leq V_{npb}/\gamma_{mb} \quad (5.11)$$

where $\gamma_{mb}$ is the partial safety factor for the material resistance of connection = 1.25.

In plates, the bearing strength is approximately a linear function of the end distance (Owens & Cheal 1989). Both bolts and plates are subject to significant triaxial containment. Due to this, bearing behaviour of plates is influenced by the proximity of neighbouring holes or boundary (edge distance). Away from holes or boundaries, significant hole elongations commence at a nominal stress of $2f_u$ but failure will occur only at about $3f_u$. Though the presence of threads in the bearing zone increase the flexibility, they do not reduce the bearing strength. Similarly, bolt material often sustains bearing stresses in excess of twice the ultimate tensile strength (Owens & Cheal 1989). Hence, it is not generally necessary to consider bolt bearing in design. Bearing in the thinner plate will control for plate thicknesses up to about one half of the bolt diameter.

The bearing stresses will be maximum at the centre line of the bolt and minimum at the edge of the bolts. Similar distribution will be there in the elevation also. High bearing stresses will lead to excessive deformation of connections (Bahia & Martin 1981).

Equations (5.2), (5.8), and (5.10), which express the design shear, tensile strength, and bearing strength of a bolt, respectively, can be presented in the form of tables to avoid repeating these calculations. Tables 5.7, 5.8, and 5.9 will aid the designer while designing joints using ordinary bolts.

**TABLE 5.7** Strength of bolts in clearance holes

| | Bolt grade 4.6 N/mm$^2$ | Bolts grade 8.8 N/mm$^2$ | Other grades of bolts N/mm$^2$ |
|---|---|---|---|
| Shear strength $v_{nsb}$ | 185 | 370 | $f_u/(\sqrt{3} \times 1.25)$ |
| Bearing strength $v_{npb}$ | 400 | 800 | $2.5k_b f_u/1.25^*$ |
| Tension strength $t_{nb}$ | 272 | 576 | $0.9f_{ub}/1.25$ but not greater than $f_{yb}(1.25/1.1)$ |

*assuming $k_b = 0.5$.

**TABLE 5.8** Bearing strength $v_{npb}$ of connected parts for ordinary bolts in clearance holes in n/mm$^2$

| Grade 410 | Grade 540 | Grade 570 | Other grades |
|---|---|---|---|
| 820 | 540 | 570 | $2.5k_b f_u/1.25$ |

**TABLE 5.9** Design capacity of ordinary bolts based on net tensile area (Grade 4.6)

| Bolt size, $d$(mm) | Tensile stress area (mm$^2$) | Shank area (mm$^2$) | Tension capacity, $T_b$(kN) $t_{nb}$ = 272 MPa | Single shear capacity, $V_{sb}$(kN) $v_{nsb}$= 185 MPa | Minimum thickness of ply for bolt bearing $v_{nsb}$ = 400 MPa $t_{bb} = t_e$, mm |
|---|---|---|---|---|---|
| (12) | 84.3 | 113 | 24.6 | 15.6 | 3.2 |
| 16 | 157 | 201 | 43.8 | 29.0 | 4.5 |
| 20 | 245 | 314 | 68.5 | 45.3 | 5.6 |
| (22) | 303 | 380 | 82.9 | 56.0 | 6.3 |
| 24 | 353 | 452 | 98.7 | 65.3 | 6.8 |
| (27) | 459 | 573 | 124.9 | 84.9 | 7.8 |
| 30 | 561 | 707 | 154.2 | 103.8 | 8.6 |
| 36 | 817 | 1018 | 222.0 | 151.1 | 10.5 |

$V_{sb} = A_{nb}v_{nsb}$; $T_b = A_{sb}f_{yb}$ (1.25/1.1) or 0.9 $A_n f_{ub}$; $t_{bb} = V_{npb}/(dv_{npb})$; Sizes in brackets are not preferred

### 5.4.4 Prying Forces

Moment-resisting beam-to-column connections often contain regions in which the bolts will be required to transfer load by direct tension, such as the upper bolts in the end plate connection as shown in Fig. 5.16. In the design of such connections, we should consider an additional force induced in the bolts as a result of so-called '*prying action*' (Douty & McGuire 1965, Agerskov 1979, Holmes & Martin 1983, Subramanian 1984). These additional prying forces induced in the bolts are mainly due to the flexibility of connected plates (see Fig. 5.17). Thus, in a simple T-stub connection as shown in Fig. 5.17, the prying force will develop only when the ends of the flanges are in contact due to the external load, as shown in Figs 5.17(b) and (c). The plastic hinges do not always form before bolt failure. The development of prying force as the external load is raised from zero to the maximum in a T-stub connection as shown in Fig. 5.18.

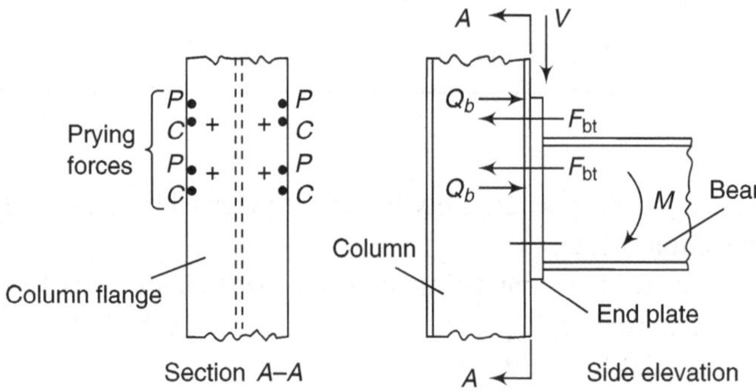


**FIG. 5.16** Prying forces in a beam-to-column connection

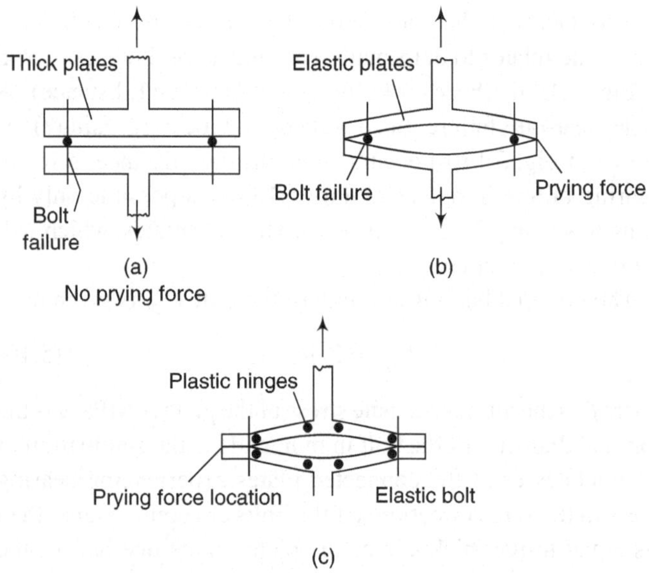


**FIG. 5.17** Failure modes due to prying forces

Prying action
Photo by J.A.Swanson and R.Leon of Georgia Institute of Technology (*Source: AISC Connections Teaching Toolkit—A Teaching Guide for Structural Steel Connections.* Reprinted with permission from AISC. All rights reserved)

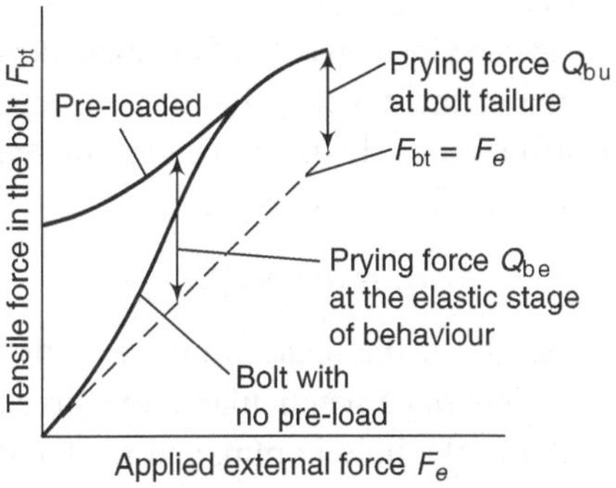


**FIG. 5.18** Relationship between external force and bolt force for a T-stub

Several researchers have studied this problem and proposed equations to calculate the prying force developed in the bolt (Astanesh 1985; Kulak et al. 1987; Owens & Cheal 1989; Thornton 1985, 1992). IS 800 has adopted the equation proposed by Owens & Cheal (1989) and the additional force $Q$ in the bolt due to prying action (see Fig. 5.19)

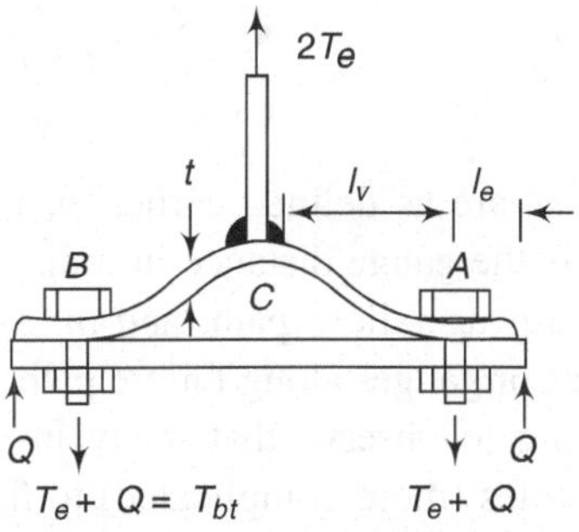


**FIG. 5.19** Forces acting in the elastic stage for prying force theory

$$Q = \{l_v/2l_e\}\,[T_e - \beta\gamma f_o\, b_e t^4/(27 l_e l_v^2)] \qquad (5.12)$$

where $l_v$ is the distance from the bolt centre line to the toe of the fillet weld or to half the root radius of a rolled section in mm and $l_e$ is the distance between prying force and bolt centre line in mm.

This distance is taken as the minimum of either the end distance or the value given by

$$l_e = 1.1t\sqrt{\beta\, f_o/f_y} \qquad (5.13)$$

where $\beta = 2$ for non-tensioned bolt and 1 for pre-tensioned bolt, $\gamma = 1.5$, $b_e$ is the effective width of flange per pair of bolts in mm, $f_o$ is the proof stress which must be consistent with the units of $T_e$ (kN/mm$^2$ and kN), and $t$ is the thickness of the end plate in mm.

The second term in Eqn (5.12) is usually relatively small and hence may be neglected to yield the formula

$$Q = T_e l_v/(2l_e) \qquad (5.14)$$

This formula is obtained if plastic hinges are assumed at the bolt line and the root, that is, when minimum flange thickness is used in design.

It is interesting to observe that the British code BS 5950 allows prying to be treated in a simpler way by requiring that the bolt tensile strengths as given in that code be reduced by 20%.

When the distance from the bolt line to the edge of the flange is relatively large, the prying force $Q$ will act at some intermediate point rather than at the edge of the flange. Owens & Cheal (1989) proposed the following formula for minimum prying force and for minimum flange thickness design

$$l_e = 1.1t\sqrt{\beta\, f_o/f_y} \qquad (5.15)$$

where $f_y$ is the yield stress of the plate and $f_0 = 0.70\,f_{ub}$. Other quantities have been defined already.

For practical design using HSFG or grade 8.8 bolts this formula leads to the following values of $l_e$:

For pre-loaded bolts $l_e = 1.5t$

For non-pre-loaded bolts $l_e = 2.0t$

The thickness of end plate to avoid yielding of the plate is obtained by equating the moment in the plate at the bolt centre line and a distance $l_v$ from it [see Fig. 5.17(c) and 5.19] to the plastic moment capacity of the plate $M_p$. Thus we have,

$$M_A = Ql_e \text{ and } M_c = Tl_v - Ql_e \qquad (5.16)$$

$$M_A = M_c = Tl_v/2 = M_p \qquad (5.17)$$

Taking
$$M_p = (f_y/1.10)\,(b_e t^2/4) \qquad (5.18)$$

The minimum thickness for the end plate can be obtained as

$$t_{min} = \sqrt{4.40 M_p/(f_y b_e)} \qquad (5.19)$$

The corresponding prying force can be obtained as $Q = M_p/l_e$. If the total force in the bolt $(T + Q)$ exceeds the tensile capacity of the bolt, then the thickness of the end plate has to be increased. Example 5.2 illustrates the effect of prying forces.

### 5.4.5 Bolts with Shear and Tension

Where bolts are subjected to both tension and shear, as in the connections shown in Fig. 5.20, then their combined effect may be conveniently assessed from a suitable interaction diagram. Tests on bolts under shear and tension showed elliptical or circular interaction curves for the ultimate strength of bolts (Chesson et al. 1965, Khalil & Ho 1979). The following equation for the circular interaction curve has been proposed in the code.

$$(V/V_{sd})^2 + (T_e/T_{nd})^2 < 1.0 \quad (5.20)$$

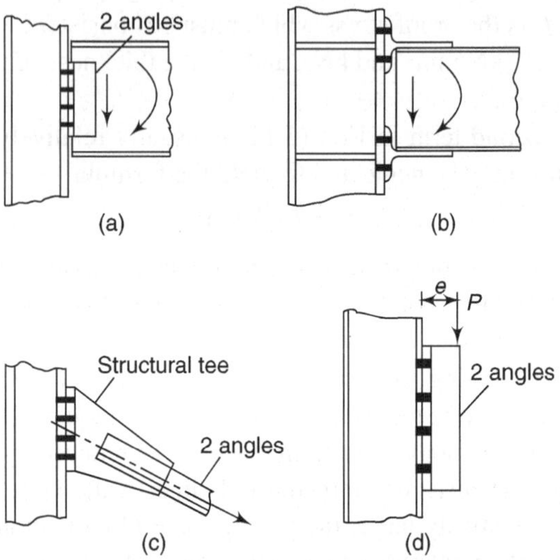


**FIG. 5.20** Typical combined shear and tension connections

where $V$ is the applied factored shear, $V_{sd}$ is the design shear capacity, $T_e$ is the externally applied factored tension, and $T_{nd}$ is the design tension capacity.

This interaction curve is graphically shown in Fig. 5.21. Example 5.12 illustrates the use of Eqn (5.20).

### 5.4.6 Efficiency of a Joint

Holes are drilled in the plates for the connection with bolts, hence the original strength of the full section is reduced. The joint which causes minimum reduction in strength is said to be more efficient. Thus, for better efficiency, a section should have the least number of holes at the critical section. The efficiency, expressed in percentage, is the ratio of the actual strength of the connection to the gross strength of the connected members. It can also be expressed as

Efficiency = (Strength of joint per pitch length/Strength of solid plate per pitch length) × 100 (5.21)

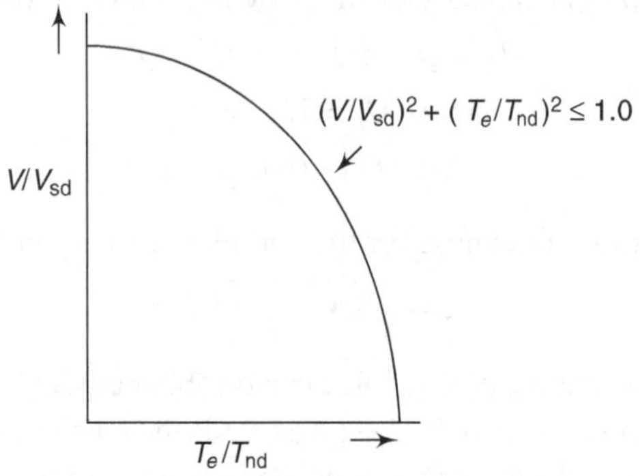


**FIG. 5.21** Interaction curve for bolts

### 5.4.7 Tension Capacity of Plate

The plate in a joint may fail in tension through the weakest section due to the holes. The holes may be arranged in the longitudinal direction of the plate, so that the number of holes is equal in all the rows across the width [see Fig. 5.22(a)], or staggered so that the number of holes across the width is reduced. In the first case, the plate will fail across the weakest section, whereas in the second case the failure is along a zigzag pattern. The tension capacity $T_{nd}$ of the plate is expressed as

$$T_{nd} = 0.9 f_u A_n / \gamma_{m1} \quad (5.22a)$$

where $f_u$ is the ultimate stress of material in MPa, $A_n$ is the net effective area of the plate in mm$^2$, and $\gamma_{m1}$ is the partial safety factor taken as 1.25 in the code.

Thus, the load carrying capacity of the plate depends on the net effective area of the plate, which in turn depends on the arrangement of the holes. If the holes are not-staggered, the net area $A_n$ can be easily computed as

$$A_n = (b - nd_h)t \quad (5.22b)$$

where $b$ is the width of the plate in mm, $n$ is the number of holes along the width $b$, perpendicular to the direction of load, $d_h$ is the diameter of the hole in mm, and $t$ is the thickness of the plate in mm.

Whenever there is more than one hole and the holes are not lined up transverse to the loading direction, more than one potential failure line may exist [see Fig. 5.22(b)]. The controlling failure line is one which gives the largest stress on an effective net area. In many cases, the critical failure path is also the path that has the minimum net area. Based on experimental evidence, a simplified empirical relationship has been proposed by Cochran (1922) for staggered rows of holes as

$$A_n = \left[ b - nd_h + \sum_{i=1}^{m} p_{si}^2/(4g_i) \right] t \quad (5.22c)$$

where $A_n$, $b$, $t$, $d_h$ are as defined earlier, $p_s$ is the staggered pitch in mm, $g$ is the gauge distance in mm, $n$ is the number of holes in the zigzag failure path, and $m$ is the number of staggered pitches or gauges along failure path.

It is interesting to observe that many investigators have proposed other rules (more complicated) to find the net area of plates with staggered holes, but none of them were found to give better results (Munse & Chesson 1963; Kulak et al.

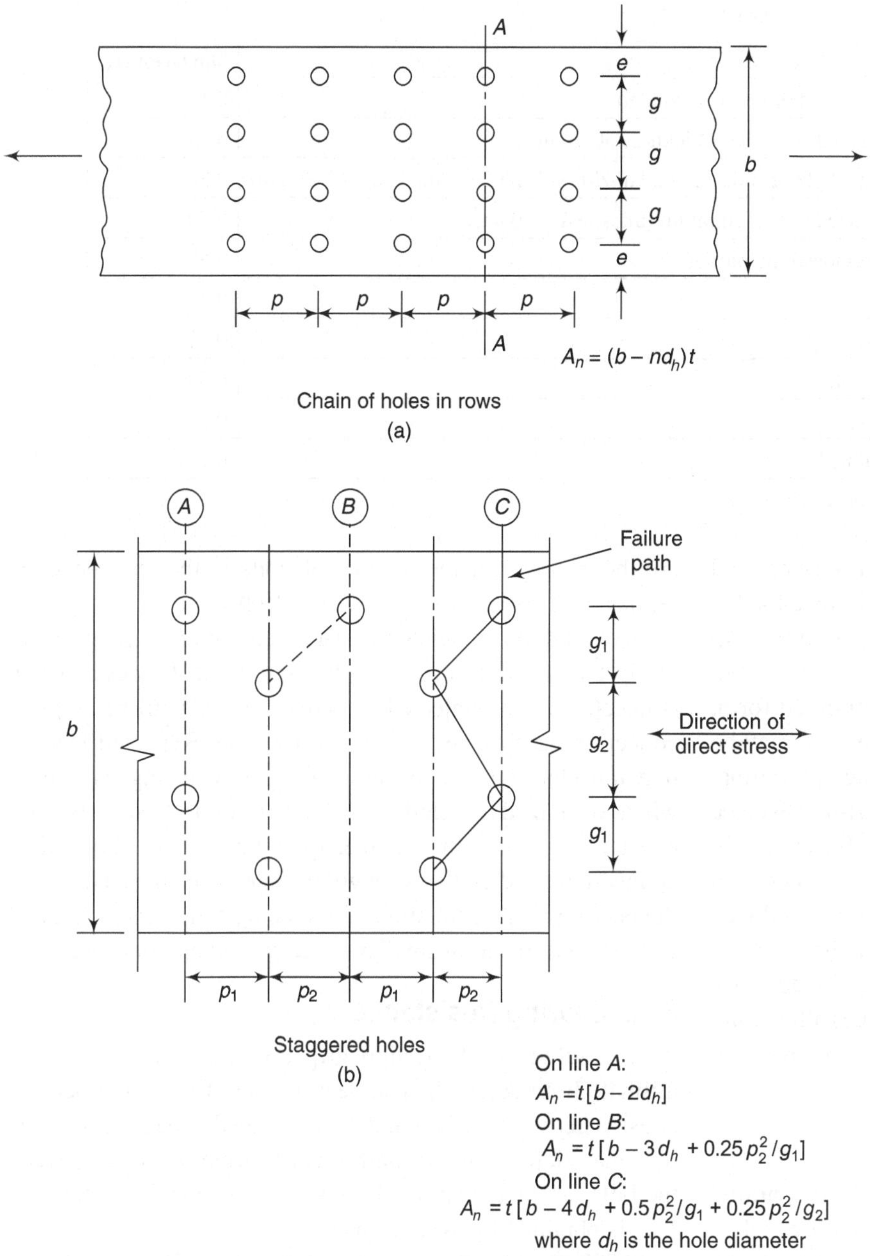


**FIG. 5.22** Tension capacity of plates

1987). More discussion on net section through staggered lines of fasteners is provided by McGuire (1968).

All possible failure paths (straight as well as zigzag) are to be tried as shown in Fig. 5.22(b) and the corresponding net areas are to be computed as per Eqns (5.22b) and (5.22c), to find the minimum net area of the plate. If the tensile load on the plate is more than its tensile strength, the plate fails in tension through rupture.

## 5.5 DESIGN STRENGTH OF HIGH STRENGTH FRICTION GRIP BOLTS

As we have seen already, HSFG bolts are used when forces are large, where space for the connection is limited, where erection cost can be reduced by using fewer bolts or where the structures are subjected to dynamic loads. Thus, they provide 'rigid' fatigue resistant joints.

It may be noted that HSFG bolts may be subdivided into parallel shank and waisted shank types. A parallel shank bolt, which is the most commonly used (and discussed in this section), is designed not to slip at serviceability load; but slips into bearing at ultimate load. Thus, only when the externally applied load exceeds the frictional resistance between the plates, the plates slip and the bolts bear against the bolt holes. A waisted shank bolt has higher strength and is designed not to slip both at service and ultimate load. Hence, waisted shank HSFG bolts are more rigid at ultimate load and need not be checked for bearing or long joint capacity (BS 5950, Martin & Purkiss 1992).

### 5.5.1 Slip Resistance

As mentioned earlier, the initial pretension in bolt develops clamping forces at the interface of elements being joined [see Fig. 5.14(b)]. The frictional resistance to slip between the plate surfaces subjected to clamping force, opposes slip due to externally applied shear.

The design slip resistance or nominal shear capacity of a bolt $V_{nsf}$ of the parallel shank and waisted shank friction grip bolts is given by the code as

$$V_{nsf} = \mu_f n_e K_h F_o \tag{5.23}$$

where $\mu_f$ is the *coefficient of friction* (called as *slip factor*) as specified in Table 5.10 ($\mu_f \le 0.55$); $n_e$ is the number of effective interfaces offering frictional resistance to slip; $K_h = 1.0$ for fasteners in clearance holes, 0.85 for fasteners in oversized and short slotted holes and for fasteners in long slotted holes loaded perpendicular to the slot, and 0.7 for fasteners in long slotted holes loaded parallel to the slot; $F_o$ is the minimum bolt tension (proof load) at installation and may be taken as $A_{nb} f_o$, $A_{nb}$ is the net area of the bolt at the threads; $f_o$ is the proof stress, taken as $0.7 f_{ub}$; and $f_{ub}$ is the ultimate tensile stress of bolt.

IS 800 stipulates that for a bolt subjected only to a factored design force $V_{sf}$, in the interface of connections at which slip cannot be tolerated, should satisfy the following

$$V_{sf} \le V_{nsf}/\gamma_{mf} \tag{5.24}$$

**TABLE 5.10** Slip factor, $\mu_f$, for preloaded bolts (see also IS 800:2007)

| S. no. | Preparation | Treatment of surface | Slip factor ($\mu_f$) |
|---|---|---|---|
| 1. | Blasted with shot or grit | Hot dip galvanized or red lead painted | 0.10 |
| | | Spray-metallized with zinc (thickness 50-70 μm) | 0.25 |
| | | Painted with ethylzinc silicate or alkalizinc silicate coat (thickness 60-80 μm) | 0.30 |
| | | Spray-metallized with aluminium (thickness > 50 μm) | 0.50 |
| | | Loose rust removed, no pitting | 0.50 |
| 2. | Sand blasted | Untreated | 0.48 |
| | | After mild rusting | 0.52 |
| 3. | Wire brushed | Clean mill scale | 0.33 |
| 4. | Untreated | Untreated | 0.2 |
| 5. | Untreated | Red lead painted | 0.1 |

*Note:* As per IS 4000:1992 contact surfaces should not be sand blasted

where $\gamma_{mf} = 1.10$ if slip resistance is designed at service load and $\gamma_{mf} = 1.25$ if slip resistance is designed at ultimate load.

It may be noted that the resistance of a friction grip connection to slip in service is a serviceability criterion, but for ease of use it is presented in the code in a modified form, suitable for checking under factored loads. Thus, $V_{nsf}$ may be evaluated at a service load or ultimate load using appropriate partial safety factors, depending upon whether slip resistance is required at service load or ultimate load. Because slip resistance certainly must relate to the contact area between pieces, the $K_h$ values used for standard holes are reduced when larger holes reduce the contact areas. Note that IS 800 : 2007 allows a value of 0.7 for fasteners in long slotted holes parallel to load though British and American codes allow only 0.6. (The region of annular ring of high contact pressure will extend to 0.5–0.7$d$ from the edge of the hole.)

## 5.5.2 Long Joints

Similar to black bolts, the design slip resistance $V_{nsf}$ for parallel shank friction grip bolts is reduced for long joints (i.e., when the distance between the first and last rows of bolts in the joint, measured in the direction of the load transfer exceeds 15$d$ in the direction of the load) by a factor $\beta_{lj}$ given by

$$\beta_{lj} = 1.075 - l_j/(200d) \text{ but } 0.75 < \beta_{lj} \leq 1.0 \quad (5.25)$$

where $d$ is the nominal shank diameter of the bolt.

For waisted shank friction grip bolts, design slip resistance need not be reduced for long joints. Note that the IS code does not consider waisted shank friction grip bolts. The British Code (BS 5950 part 1) specifies a factor $k$ to be used with Eqn (5.25), which is equal to 1.1 for parallel shank bolts and 0.9 for a waisted shank bolt.

As may be noted, the slip factor (coefficient of friction) is taken as 0.48–0.52 for treated (sand or shot or grit blasted) and uncoated surfaces and taken as 0.1–0.33 for untreated or coated surfaces. These values have been obtained by tests and are the average for the number of bolts in the test specimen (Kulak et al. 1987; Owens & Cheal 1989).

It should be understood that overcoming slip does not imply that a failure mode has been reached. However, when connections are subjected to stress reversal, there is great concern regarding any slip at service load. Repeated loading may introduce fatigue concerns, if slip is occurring, especially when oversized or slotted holes are used. Generally, slip resistance governs the number of bolts used in slip-critical connections, rather than strength in bearing or shear. However, the bearing related equations for spacing of fasteners and end distance will result in smaller spacings and end distances.

## 5.5.3 Bearing Resistance

As a parallel shank friction grip bolt slips into bearing at ultimate limit state when subject to shear forces, the bearing stresses between the bolt and the plate need to be checked. The bolt may deform due to high local bearing stresses between the bolt and plate and the design bearing capacity of the bolt $V_{npb}$ is obtained by using Eqn (5.10).

IS 800 stipulates that the factored shear force $V_{sf}$ should satisfy

$$V_{sf} \leq V_{npb}/\gamma_{mb} \quad (5.26)$$

where $\gamma_{mb}$ is the partial safety factor for the material of the bolt, taken as 1.25 in the code.

Note that while checking black bolts, the ultimate tensile capacity of the bolt or the plate, whichever is smaller is used. Since the bearing strength of HSFG bolts will be greater than the plates, no check on bearing strength of bolt is necessary.

An alternative mode of failure is that of the bolt shearing through the end of the plate as shown in Fig. 5.13(b). This may be controlled by specifying the end distances and pitches as given in Table 5.6 and by the $k_b$ factor of Eqn (5.10). The block shear resistance of the edge distance due to bearing force should also be checked for the connection (see Section 7.6.3 of Chapter 7 for details).

### 5.5.4 Tension Resistance

The design tensile strength of parallel shank and waisted shank friction grip bolts is similar to that of black bolts (see Section 5.4.2) and is given by

$$T_{nf} = 0.9 f_u A_{nb} \leq f_{yb} A_{sb} (\gamma_{m1} \gamma_{m0}) \quad (5.27)$$

where $f_u$ is the ultimate tensile stress of the bolt and $A_n$ is the net tensile area of the bolt and may be taken as the area at the root of the threads. $A_{sb}$ is the shank area of the bolt, $\gamma_{m1}$ is the partial safety factor for material resistance governed by ultimate stress = 1.25, $\gamma_{m0}$ is the partial safety factor for material resistance governed by yielding = 1.10, and $f_{yb}$ is the yield stress of bolt.

As per IS 800, the HSFG bolt subjected to a factored tension force $T_f$ should satisfy the following equation:

$$T_f \leq T_{nf}/\gamma_{mb} \quad (5.28)$$

where $\gamma_{mb}$ is the partial safety factor for the material of the bolt and is taken as 1.25 in the code.

It is important to note that the British code specifies the design tensile strength of HSFG bolts under factored load as

$$\begin{aligned} T_{nf} &= 0.9F_o = 0.9(0.8A_{sb})(0.7f_{ub}) \\ &= 0.504A_{sb}f_{ub} \\ &= 0.504(A_{nb}/0.8)f_{ub} \\ &= 0.63A_{nb}f_{ub} \quad (5.29) \end{aligned}$$

Comparing Eqns (5.27), (5.28), and (5.29) it is seen that the design tensile strength calculated as per IS 800 : 2007 will be higher than that calculated using the British code.

The effect of the prying force $Q$ has been shown in Fig. 5.18. When the external load is applied, part of the load (approximately 10%) of the load is equilibrated by the increase in bolt force. The balance of the force is equilibrated by the reduction in contact between the plates. This process continues and the contact between the plates is maintained until the contact force due to pre-tensioning is reduced to zero by the externally applied load. Normally, the design is done such that the externally applied tension does not exceed this level. After the external force exceeds this level, the behaviour of the bolt under tension is exactly similar to that of a bearing type of bolt. Eqns (5.12), (5.13), (5.15), and (5.19) may be used in the calculation of HSFG bolts subjected to prying forces.

### 5.5.5 Combined Shear and Tension

The interaction curve suggested for combined shear and tension for HSFG bolts (for which slip in the serviceability limit state is limited) is similar to that of black bolts (see Fig. 5.21) and is given below

$$(V_{sf}/V_{sdf})^2 + (T_f/T_{ndf})^2 \leq 1.0 \quad (5.30)$$

where $V_{sf}$ is the applied shear at service load, $V_{sdf}$ is the design shear strength, $T_f$ is the externally applied tension at service load, and $T_{ndf}$ is the design tension strength.

Since slip resistance is a service load consideration, the numerator terms of Eqn (5.30) are service loads $T$ and $V$ (tension and shear per bolt). It should be observed that any external tension will produce a corresponding reduction in the clamping force between the plies. Until the external load on a bolt exceeds the pre-compression force between the pieces, the tension force in the bolt will not change significantly from its initial tension (see Fig. 5.18).

Though the British and American codes suggest a simple linear interaction formula (due to external tension, there will be linear reduction in friction capacity of the connection), IS 800 suggests a circular interaction formula. Example 5.13 illustrates the use of Eqn (5.30).

Equations (5.23), (5.24), and (5.26)–(5.28) express the design shear, bearing, and tensile resistance of HSFG bolts and can be presented in the form of a table, as shown in Table 5.11, to avoid repeated calculations. Figure 5.23 shows a bracing member connected to the other members of a bridge structure using guest plates and HSFG bolts.

**TABLE 5.11** Capacities of high-strength friction grip bolts in clearance holes ($K_h = 1$ and $\mu = 0.48$)

| Bolt diameter (mm) | Stress area of bolt (mm$^2$) | | Proof load of bolt (kN) Property class | | Tensile capacity (kN) Property class | | Slip resistance in single shear ($n_e = 1$, $\gamma_{mf} = 1.25$) (kN) Property class | |
|---|---|---|---|---|---|---|---|---|
| | Thread | Shank | 8.8 | 10.9 | 8.8 | 10.9 | 8.8 | 10.9 |
| (12) | 84.3 | 113 | 47.2 | 61.3 | 48.5 | 63 | 18.1 | 23.5 |
| 16 | 157 | 201 | 87 | 114 | 90.4 | 117 | 33.7 | 43.8 |
| 20 | 245 | 314 | 137 | 178 | 141 | 183 | 52.6 | 68.4 |
| (22) | 303 | 380 | 169 | 220 | 174 | 226 | 65.1 | 84.7 |
| 24 | 353 | 452 | 197 | 256 | 203 | 264 | 75.9 | 98.6 |
| (27) | 459 | 572 | 257 | 334 | 264 | 343 | 98.7 | 128 |
| 30 | 561 | 706 | 314 | 408 | 323 | 420 | 120 | 156 |
| 36 | 817 | 1017 | 457 | 595 | 470 | 611 | 175 | 228 |

Tension capacity = $0.9f_u A_{nb}/1.25$

Slip resistance = $\mu_f n_e K_h F_o/\gamma_{mf}$

where $\gamma_{mf}$ = 1.1(at service load) and 1.25 (at ultimate load), $F_0$ is the minimum bolt tension (proof load) = $A_{nb}f_o$; $f_o = 0.7f_{ub}$, and $n_e$ is the number of effective interfaces offering frictional resistance to slip.

FIG. 5.23 Example of a connection using HSFG bolts

### 5.5.6 Block Shear Failure

As a result of some research work carried out in USA, it was found that angle, gusset plate, and coped beams connections may fail as a result of block shear (Kulak & Grondin 2000). Failure occurs in shear at a row of bolt holes parallel to the applied loads, accompanied by tensile rupture along a perpendicular face. This type of failure results in a block of material being torn out by the applied shear force as shown in Fig. 5.24. The block shear strength $T_{db}$ of a connection is taken as the smaller of

$$T_{db1} = [A_{vg}f_y/(\sqrt{3}\,\gamma_{m0}) + 0.9f_uA_{tn}/\gamma_{m1}] \quad (5.31a)$$

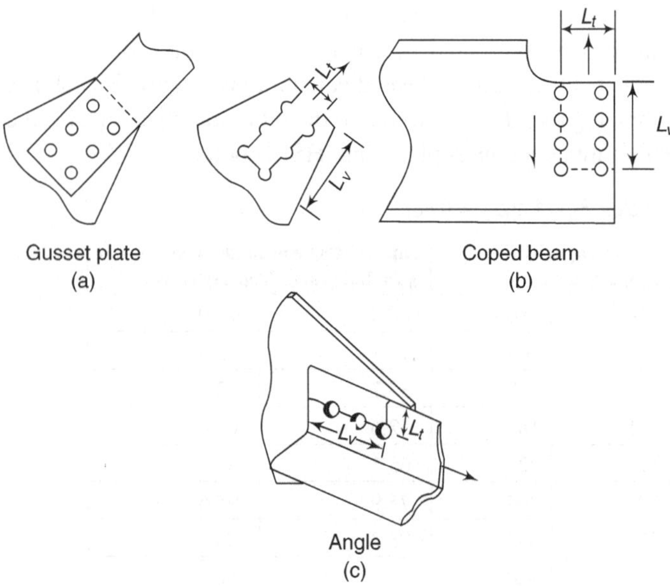


FIG. 5.24 Examples of block shear failure

and

$$T_{db2} = [0.9f_uA_{vn}/(\sqrt{3}\,\gamma_{m1}) + f_yA_{tg}/\gamma_{m0}] \quad (5.31b)$$

where $A_{vg}$ and $A_{vn}$ are the minimum gross and net area, respectively, in shear along a line of transmitted force (along $L_v$ in Fig. 5.24); $A_{tg}$ and $A_{tn}$ are the minimum gross and net area, respectively, in tension from the hole to the toe of the angle or next last row of bolt in gusset plates (along $L_t$ in Fig. 5.24); $f_u$ and $f_y$ are the ultimate and yield stress of the material, respectively; and $\gamma_{m0}$ and $\gamma_{m1}$ are the partial safety factor for material resistance governed by yielding (1.10) and governed by ultimate stress (1.25), respectively.

The use of Eqn (5.31) is illustrated in Example 5.14.

## 5.6 PIN CONNECTIONS

When two structural members are connected by means of a cylindrical pin, the connection is called a pin connection. Pins are manufactured from steel bars of diameter 9 mm to 300 mm. Pin connections are provided when zero moment and free rotation is required. For satisfactory working of the connection, the friction between the pin and the connected members should be minimized. Hence in practice pins are machined to a polished surface and the holes in the members are also machined to make it smooth and frictionless. Pins are often used in chain link cables in suspension bridges, diagonal bracing connections in beams and columns, hinged arches, and truss bridge girders. Figure 5.25 shows a pin connection between the foundation and structure of a bridge.

FIG. 5.25 Pin connection between the foundation and the structure of a bridge

To make a pin connection, one end of the bar is forged like a fork and a hole is drilled. The end of the other bar is also forged and drilled to form an eye. The eye bar is inserted in the jaws of the fork and then the pin is inserted through the hole in the fork and the eye of the other bar. Usually, a cotter pin or screw is provided so that the pin will not move out of the position as shown in Fig. 5.26.

### 5.6.1 Design of Pins

Provisions for the design of pins is not available in IS 800. The following provisions are available in BS 5950 (Part 1).

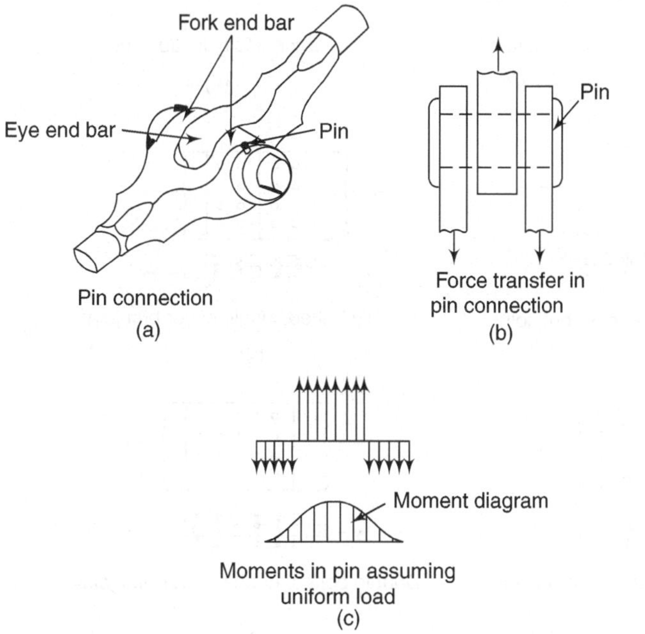


**FIG. 5.26** Pin connection and moment diagram in a pin

**Shear capacity** The shear capacity of a pin may be taken as follows:

1. If rotation is not required and the pin is not intended to be removed: $0.6f_{yp}A$
2. If rotation is required or if the pin is intended to be removed: $0.5f_{yp}A$

where, $A$ is the cross sectional area of the pin and $f_{yp}$ is the yield strength of the pin.

**Bearing capacity** The bearing capacity of the pin may be taken as follows:

1. If rotation is not required and the pin is not intended to be removed: $1.5f_y dt$
2. If rotation is required or if the pin is intended to be removed: $0.8f_y dt$

where $d$ is the diameter of the pin, $f_y$ is the lower of the yield strength of the pin and the connected part, and $t$ is the thickness of the connected part.

**Bending** The moments in a pin should be calculated on the basis that the connected parts form simple supports. It is generally assumed that the reactions between the pin and the connected part are uniformly distributed along the length in contact of each part.

The moment capacity of the pin should be taken as follows:

1. If rotation is not required and pin is not intended to be removed: $1.5f_{yp}Z$
2. If rotation is required or if the pin is intended to be removed: $1.0f_{yp}Z$

where $f_{yp}$ is the yield strength of the pin and $Z$ is the section modulus of the pin.

Assuming that the pin is a cylindrically shaped beam and is not intended to be removable

$$M_u = 1.5f_{yp}Z$$

or

$$M_u = 1.5f_{yp}(\pi d^3/32)$$

or

$$d = [21.33M_u/(\pi f_{yp})]^{1/3} \tag{5.32}$$

where $M_u$ is the ultimate bending moment in Nmm and $f_{yp}$ is the yield strength of the pin.

The design of a pin connection is shown in Example 5.3.

## 5.7 SIMPLE CONNECTIONS

In many cases, a connection is required to transmit a force only and there may not be any moment acting on the group of connectors, even though the connection may be capable of transmitting some amount of moment. Such a connection is referred to as a *simple*, *force*, *pinned*, or *flexible connection*.

As already shown in Fig. 5.14, two types of load transfers occur in these connections. In the first, the force acts in the connection plane (formed by the interface between the two connected plates) and the fasteners between these plates act in shear [Fig. 5.14(a)]. In the second type, the force acts out of the plane of the connection and the fasteners act in tension as shown in Fig. 5.14(c). In practice there will always be some eccentricity and the moment due to this small eccentricity is ignored. The different types of simple connections found in steel structures may be classified as follows:

1. Lap and butt joints
2. Truss joint connections
3. Connections at beam column junctions
   (a) Seat angle connection
   (b) Web angle connection
   (c) Stiffened seat angle connection
   (d) Header plate connection
4. Tension and flange splices

Let us now discuss these connections briefly.

### 5.7.1 Lap and Butt Joints

Lap and butt joints are often used to connect plates or members composed of plate elements. Though lap joints are the simplest, they result in eccentricity of the applied loads. Butt joints on the other hand eliminate eccentricity at the connection.

#### *Lap Joints*

When two members which are to be connected are simply overlapped and connected together by means of bolts or welds, the joint is called a lap joint [see Figs 5.27(a)–(d)]. A single

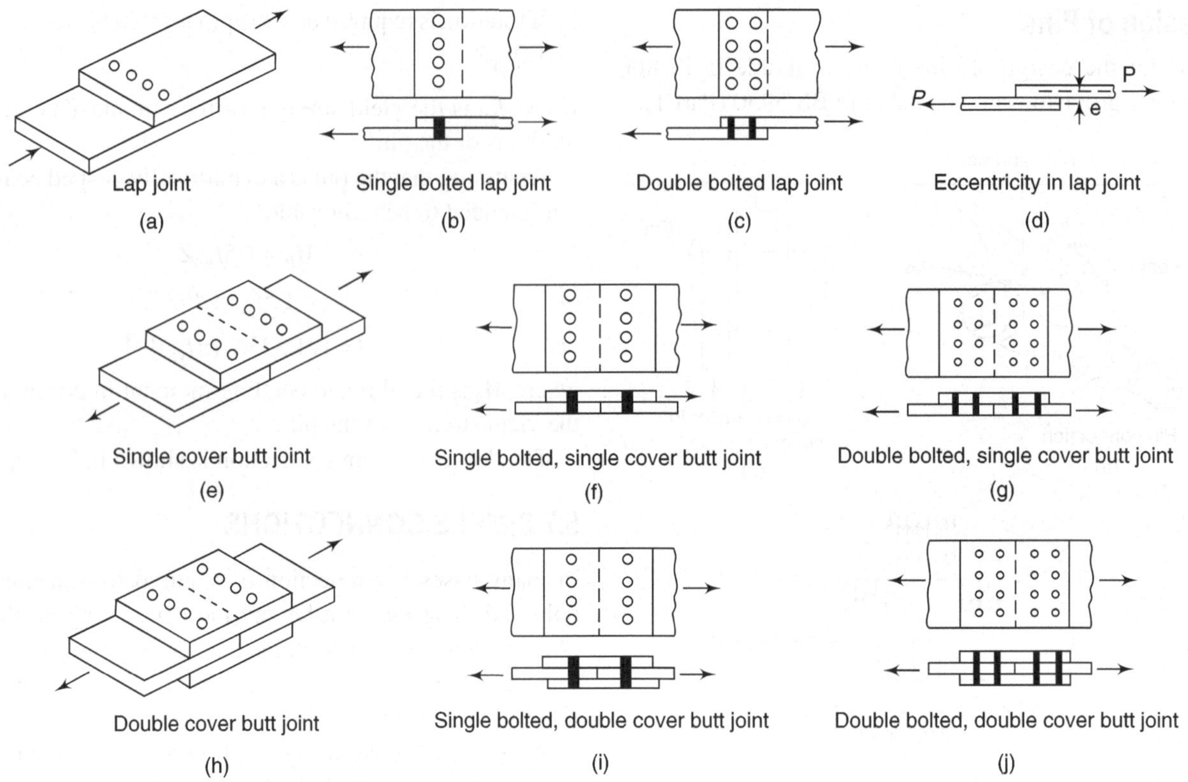


**FIG. 5.27** Lap and butt joints

bolted lap joint and a double bolted lap joint are shown in Figs 5.27(b) and 5.27(c), respectively. The drawback of such a lap joint is that the centre of gravity of load in one member and the centre of gravity of load in the second member do not coincide and hence an eccentricity as shown in Fig. 5.27(d) is created. Due to this a couple $P \times e$ is formed, which causes undesirable bending in the connection leading to failure of bolts in tension. To minimize the effect of bending in a lap joint, at least two bolts in a line should be provided. Moreover, due to the eccentricity, the stresses are distributed unevenly across the contact area between the bolts and members to be connected. Hence, the use of lap joints is not often recommended. The design of lap joint is illustrated in Examples 5.1, 5.4, 5.5, and 5.6.

### Butt Joints

In butt joints, the members to be connected are placed against each other and are bolted or welded together through the use of additional plates, called *cover plates*. The cover plates may be provided on either one or both sides of the connection as shown in Figs 5.27(e)–(j). If the cover plate is provided on one side of the joint only it is called as a *single cover butt joint* [see Figs 5.27(e)–(g)] and when provided on both sides of the joint, it is called as a *double cover butt joint* [see Figs 5.27(i) and (j)]. As in the case of lap joints, in the case of single cover butt joint also, some eccentricity is involved. Double cover butt joints are preferred due to the following reasons.

(a) In double cover butt joints, the eccentricity of load is eliminated and hence there is no bending in the bolts.

(b) In double cover butt joints, the total shear force to be transmitted is split into two parts and hence the bolts are in double shear. Whereas in the case of lap joint or single cover butt joint, the force acts in one plane of the bolt and hence the bolts are said to be in single shear. Thus, the shear carrying capacity of bolts in a double cover butt joint is double that of a bolt in a lap joint or single cover butt joint. The design of butt joint is illustrated in Examples 5.7 – 5.11.

(c) The thickness of the cover plate is generally taken as (i) 1.125 times the thickness of the main plate for single cover butt joint and (ii) 0.625 times the thickness of the main plate for double cover butt joint.

## 5.7.2 Truss Joint Connections

Even though truss joints are assumed to be hinged, the detailing using gusset plates and multiple fasteners or welding does not represent hinged condition. However, in practice the secondary moment associated with such a joint is disregarded, unless the loading is cyclic.

Typical truss joints are shown in Fig. 5.28, which use gusset plates in order to accommodate the bolts. These gusset plates are connected to the members of the truss by means of bolts or welds using lap or butt joints depending on whether one or two angles are used for the member. Note that in Fig. 5.28(c), end plates are used to facilitate the connection of shop welded parts of a truss at the site using bolts. As stated earlier, the lateral dimensions of a gusset plate are determined principally

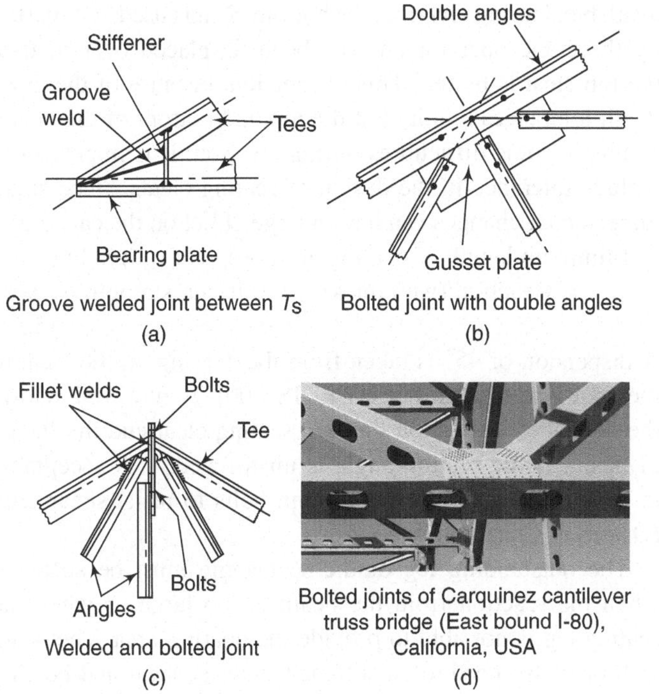


Groove welded joint between $T_S$ (a) | Bolted joint with double angles (b) | Welded and bolted joint (c) | Bolted joints of Carquinez cantilever truss bridge (East bound I-80), California, USA (d)

**FIG. 5.28** Typical joints in trusses

by the fastener requirements of the members. However, the determination of the thickness of the gusset plate poses problems. The gusset plates are not easy to analyse because of their odd shape and the discrete introduction of forces through fasteners. Moreover, the elementary formula based on beam theory are valid only for beams whose span is more than twice the depth and at cross sections not closer to concentrated loads than about half the depth. The ordinary gusset plates fall considerably short of these requirements and hence the results obtained by the application of beam formulae are questionable and may be misleading (Gaylord et al. 1992).

Based on the experimental study on gusset plates commonly found in Warren type bridges, Whitmore (1952) proposed that the maximum direct stress in a gusset plate from an individual member could be estimated adequately by ensuring that the member force is distributed uniformly over an effective area given by 30° dispersion from the outer row of fasteners as shown in Fig. 5.29(a) [similar dispersion can be assumed for welded connections and are shown in Figs 5.29(b) and 5.29(c)]. Whitmore's proposal is valid for stresses in gusset plates loaded in the elastic range.

The behaviour of gusset plates connections in the inelastic range was studied by Hardash & Bjorhovde (1985), who proposed a block shear model [see Eqn (5.31)] to predict the ultimate capacity of gusset plate connections in tension. Based on the monotonic compression test results, Thornton (1984) proposed a 45° dispersion angle to evaluate the effective width, instead of the Whitmore's 30° to account for the load redistribution in the

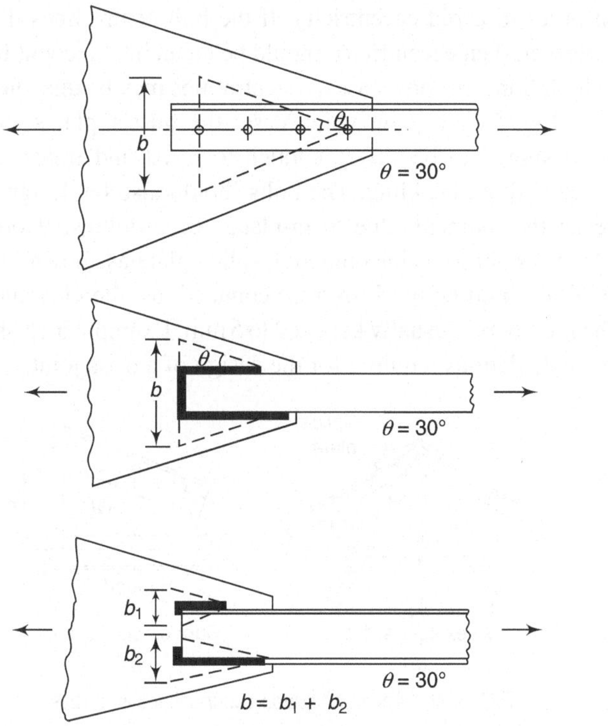


**FIG. 5.29** Effective width of gusset plates as per Whitmore (1952)

gusset plate. This method resulted in good agreement with the test results. A review of the recent design and behaviour of gusset plate connections is provided by Cheng et al. (2000). Compressive stresses may develop parallel to and at the edge of the gusset plates. Therefore, the width of the top edge must not be too large, compared with the thickness, or the plate may buckle (Gaylord et al. 1992; Cheng et al. 2000). This kind of buckling, called *local buckling*, may be prevented if the unsupported edge of a gusset plate is restricted to 42ε times the thickness (Gaylord et al. 1992), where $\varepsilon = (250/f_y)^{0.5}$.

In the absence of any computation for gussets, the thickness of gussets as shown in Table 5.12 is recommended by Ragupathi (1995).

Usually, a gusset plate thickness slightly higher than the thickness of the angle is chosen to ensure safety.

Trusses may be single plane trusses in which the members are connected on the same side of the gusset plate or double plane trusses in which the members are connected to both sides of the gusset plate.

The centre line of all the members connected at a joint, should meet at a point as shown in Fig. 5.28 and 5.30(a),

**TABLE 5.12** Thickness of gussets used in trusses

| **Maximum design force in diagonal (kN)** | Up to 200 | 200–450 | 450–750 | 750–1650 | 1650–2250 |
|---|---|---|---|---|---|
| **Thickness of gusset in mm** | 8 | 10 | 12 | 16 | 18 |

in order to avoid eccentricity. If the bolt centre lines do not intersect, then eccentricity should be taken into account in the calculations. In some cases, eccentricities may be unavoidable [see Fig. 5.30(b)], in which case, the gusset plates should be designed to be strong enough to resist and transmit the force without buckling. The bolts should also be designed to resist the moments due to in-plane eccentricities. If out-of-plane instability is encountered, splice plate as shown in Fig. 5.30(a) could be used. In truss connections, the clearance of the members is usually kept at 2 to 5 mm. Example 5.15 shows the calculations required for the design of a truss joint.

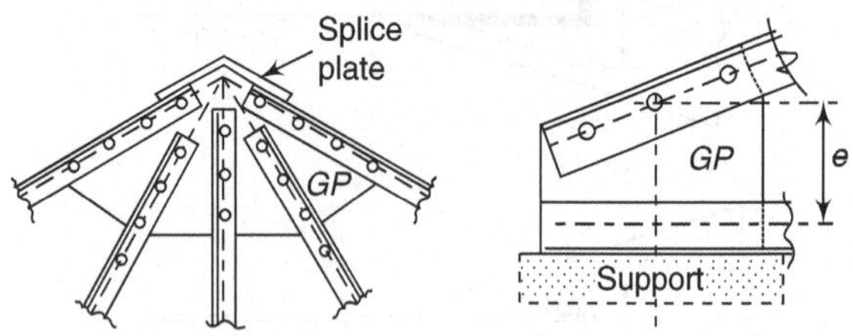


**FIG. 5.30** Apex and support connections in a truss

## 5.7.3 Clip and Seating Angle Connections

*Clip and seating angle connection* also called *angle seat connection* or *seat angle connection* [Figs 5.2(b) and 5.31] transfers reaction from the beam to the column through the angle seat. The top cleat (called *clip angle*) is provided for lateral or torsional restraint (support) to the top flange of the beam and is bolted to the top flange. This is important for lateral stability. The angle seat may be bolted or fillet welded to the column and the angles are assumed to be rigid compared to the bolts.

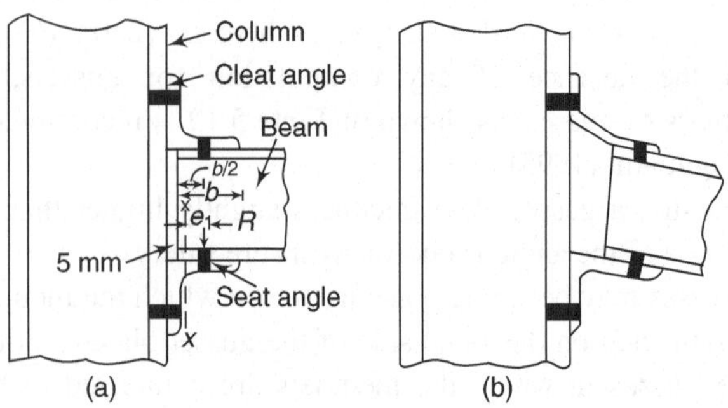


**FIG. 5.31** Angle seat connections

The beam reaction is transferred by bearing, shear, and bending of the horizontal leg of the bottom angle, by vertical shear through the fasteners, and by horizontal force in the fasteners between the vertical leg and the column. The beam is designed as simply supported and the column is designed for the eccentric beam reaction. The clearance between the end of the beam and flanges of the column should be minimum (between 2 and 5 mm) so that maximum bearing length is available. During erection, the bottom cleat is used as a marker by the crane operator and the beam is placed over it, then the top cleat is bolted. This connection, because of the close depth tolerances required at the top and bottom of the beam, results in some difficulties during erection. It is important to include tolerance in the design of seating cleats, since small dimensional changes can have a large effect on the capacity.

Minimum length of bearing at the edge of root radius

$$= \text{Reaction/(web thickness} \times \text{design strength of web)} \tag{5.33}$$

A dispersion of 45° is taken from the bearing on the cleat to the root line (clause 8.7.1.3 of IS 800). Figure 5.32 shows the details of dispersion through seating cleat and the flange of beam. If the seating angle is unstiffened, it is acceptable to design the seating cleat/column bolts to resist shear only (Owens & Cheal 1989).

The outstanding leg of the seat angle must be stiffened when the reaction from the beam is too large or when the seating leg is not able to provide the bearing area. The beam web may also need to be stiffened to resist shear and bearing (see Chapter 10). When large reactions from the beam occur, more number of bolts may have to be provided to connect the seat angle to the column (normally four bolts are provided). In such a case, an additional angle called stiffener angle is provided. Packings equal to the thickness of seat angle are used between the column flange and the stiffener angle (see Fig. 5.33). In general, a pair of stiffener angles is provided and the outstanding legs are tack bolted. Note that when the seating angle is stiffened, the horizontal leg of the angle can no longer flex, and the bolts connecting the column flange to the stiffener angles should be designed to resist the corresponding bending moment (arising due to the eccentricity between the centre of the bearing length and the column face) as well as shear.

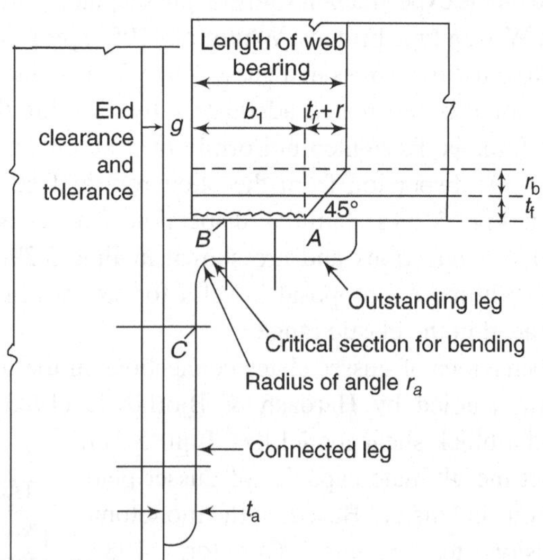


**FIG. 5.32** Details of dispersion through seating angle and flange of beam

The eccentricity should be taken as the outstand of the horizontal leg.

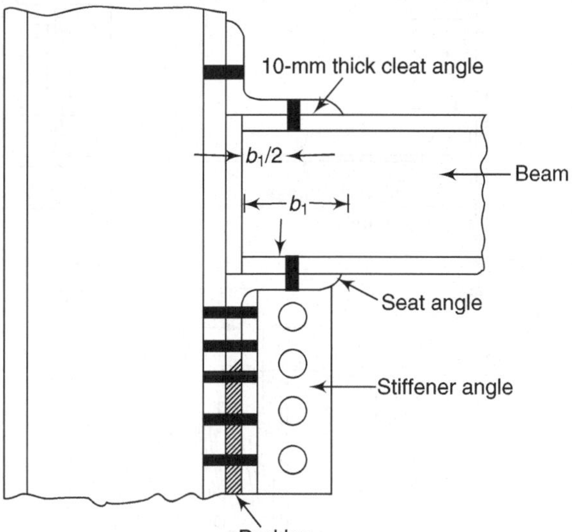


**FIG. 5.33** Stiffened seat connection

The clip angle at the top of the beam can only resist shear by flexure of the legs [see Fig. 5.31(b)]. This is much more flexible than the bottom seating angle and hence should not be assumed to contribute to the shear resistance of the connection. However, Yang and Jeon (2009), from their experimental investigations found that initial rotational stiffness and plastic moment capacity increased with increasing top angle thickness rather than that of the seat angle thickness. They also proposed analytical models to predict initial stiffness and plastic moment capacity. Babu and Sreekumar(2012), based on their experimental and FEM studies, found that the ductile behaviour of a beam-column connection is improved by increasing the number of bolts on the top and seat angles and the thickness of top and seat angles.

## 5.7.4 Design of Unstiffened and Stiffened Seat Connection

The various steps involved in the design of unstiffened and stiffened seat connections are explained in this section.

**Unstiffened seat connection** The design of unstiffened seating angle connection, consists of the following steps:

1. The selection of seat angle having a length equal to the width of the beam.
2. The length of the outstanding leg of the seat angle is calculated on the basis of web crippling of the beam. The seat leg length is kept more than the calculated bearing length given by

$$b = [R/t_w\,(f_{yw}/\gamma_{m0})] \tag{5.34}$$

   where $R$ is the reaction from the beam, $t_w$ is the thickness of the web of the beam, $f_{yw}$ is the yield strength of the web of the beam, and $\gamma_{m0}$ is the partial safety factor for material = 1.10.
3. A dispersion of 45° is taken from the bearing on the cleat to the root line [Note: The dispersion at a slope of 1 in 2.5 as specified in IS 800 clause 8.7.4 is not applicable for this purpose (Owens & Cheal 1989).] Thus, the length of bearing on the cleat

$$b_1 = b - (t_f + r_b) \tag{5.35}$$

   where $t_f$ is the thickness of beam flange and $r_b$ is the root radius of beam flange. The distance of end bearing on cleat to root angle

$$b_2 = b_1 + g - (t_a + r_a) \tag{5.36}$$

   where $t_a$ is the thickness of angle, $r_a$ is the root radius of angle, and $g$ is the clearance and tolerance.
4. The connected leg is so chosen that at least two horizontal rows of bolts can be accommodated. Hence, the assumed angle should have a leg length of 100 mm or more.
5. The reaction is assumed to be uniformly distributed over bearing length $b_1$. The thickness of seat angle is so chosen that the outstanding leg does not fail in bending on a section at the toe of the fillet. The bending moment

$$M_u = R\,(b_2/b_1)\,(b_2/2) \tag{5.37}$$

   This is equated against the moment capacity

$$M_d = 1.2Z\,(f_y/\gamma_{m0}) \tag{5.38}$$

   If the assumed angle thickness is not sufficient (i.e., when $M_d < M_u$), then the section is revised.
6. The shear capacity of the outstanding leg of cleat is calculated as

$$V_{dp} = wtf_y/(\sqrt{3}\,\gamma_{m0}) \tag{5.39}$$

   The above value should be more than the reaction from the beam.
7. The number of bolts is calculated as below

$$n = \text{Reaction/Strength of bolt} \tag{5.40}$$

8. A cleat angle of nominal size is provided on the top flange of the beam, connected by two bolts on each of its legs.

The design of a seat angle connection is presented in Example 5.16.

### *Stiffened Seat Connection*

As discussed in Section 5.7.3, the behaviour of a stiffened seat connection is different than that of an unstiffened seat connection. Hence, the design of a stiffened seat connection consists of the following steps.

1. Assume the size of seat angle on the basis of bearing length as described in unstiffened seat connection.
2. The outstanding leg of the stiffener angle must provide the required bearing area. The outstanding leg must not exceed 14ε times the thickness, where $\varepsilon = (250/f_y)^{0.5}$ to avoid local buckling (clause 8.7.1.2). The required bearing area is calculated as

$$A_{br} = R/(f_y/\gamma_{m0}) \tag{5.41}$$

where $R$ is the reaction from beam, $f_y$ is the yield strength, and $\gamma_{m0}$ is the partial safety factor for material, taken as 1.10. The thickness of the stiffener angle should be more than the thickness of the web of the beam being supported.

3. Due to the stiffener, the seat angle is not flexible and the bolts in the connecting leg are subjected to moments in addition to shear. This connection behaves like a bracket connection (see Section 14.1.2). It can be assumed that the reaction from the beam acts at the middle of the outstanding leg of the angle.
4. From step 3, the eccentricity, the bending moment, and the tension acting in the critical bolts are computed, similar to the bracket connection.
5. Using the interaction formula [Eqn (5.20)], the critical bolt is checked.
6. A nominal angle is provided at the top of the beam and is connected with two nominal size bolts on each leg of the cleat angle.

The calculations for a typical stiffened seat connection are provided in Example 14.1 of Chapter 14. Note that the stiffened seat angle connection is not a simple connection and has been included here for convenience.

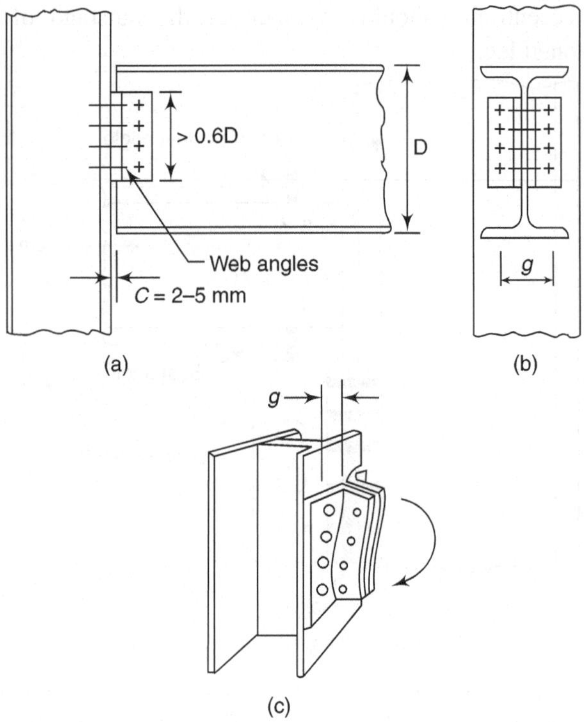


**FIG. 5.34** Web angle connection and flexibility of web angle

### 5.7.5 Web Angle Connection

A *web cleat connection*, also known as *web cleat* or *angle cleat connection* [see Fig. 5.2(c) and 5.2 (d)] is used to transfer beam reaction through web angles either to the flange or to the web of the supporting member. One or two angle cleats may be used, and bolted to the beam web and to the supporting members. Double web cleat connection is preferred over single sided web cleat connection. The flanges may be notched or coped [see Fig. 5.2(d)], if required. The clearance between the beam and the supporting member is kept at 2-5 mm to provide sufficient end distance in the web. This connection also has very little moment capacity, as there is significant flexibility in the angle legs connected to the supporting member and the bolted connection to the beam web (see Fig. 5.34).

The beam reaction is transferred by shear and bearing from the web of the beam to the web bolts and to the angle cleats. These are then transferred by the cleat angle to the bolts at the junction of supporting member, and then to the supporting member mainly by shear and also by tension and compression. The beam is designed as a simply supported beam and the supporting member is designed for the eccentric beam reaction.

The length of the web angle is decided based on the number of bolts and the pitch and is normally not less than 0.6 times the depth of the beam. (The maximum length is the clear depth of the web between the fillets, which may be approximately taken as 0.75 times the depth of the beam.) If the depth of the web cleat is less than about 0.6 times the depth of beam web, then the bolts may be designed for shear forces only. Otherwise, assuming pure shear transfer at the column face, the bolts connecting the cleats to the beam web should be designed for the moment due to eccentricity. Normally, the connection is made in the compression zone of the beam in order to provide lateral restraint to the compression flange. A minimum thickness of 8 mm is chosen for beams having a depth of up to 450 mm. Note that thicker angles will decrease the flexibility of the joint and may introduce end moments. The joint may be made flexible by making the gauge distance (see Fig. 5.34) on the column flange as large as possible, which is usually kept in the range of 100 mm to 140 mm. The web angle connection may be combined with seat angle connection, when the reaction from the beam is heavy.

It is preferable to use friction grip fasteners for the cleat-to-column connection. If ordinary bolts are used, the combined effect of the eccentric load and the clearances in the bolt holes could lead to an unacceptable twist at the end of the beam (Owens & Cheal 1989). The design calculations for a typical web angle connection are shown in Example 5.17.

### 5.7.6 Flexible End Plate Connection

*Flexible end plate connection*, also called as *header plate connection* [Fig. 5.2 (e)] consists of an end plate connected to the web of the beam, at the beam ends, by fillet welds. This plate is, in turn, connected to the column flange or web by means of bolts. This connection behaviour is similar to the legs of web angles connected to the column flange. However, it is difficult to ensure that the connection has sufficient flexibility/ductility to accommodate the end rotation of the beam. In practice, the flexibility may be achieved by limiting the thickness of the plate ($t < d/3$ for grade 4.6 bolts and $t < d/2$ for grade 8.8 bolts) and by positioning the bolts not too close to the web and flange of the beam, and by suitably selecting pitch and gauge length of bolts. An empirical thickness of 8 mm is used for beams having a depth of 450 mm and 10 mm for beams having greater heights. The end plate depth is kept to a minimum to reduce end moments. The length '$a$' in Fig. 5.35 is kept less than $30t$ (where $t$ is the thickness of end plate) so that the web of the beam does not come into contact with the column. The flanges of the beam may be notched or coped, if required. In these cases, the length of the cut portion is limited to $20t_w$ to prevent buckling of the web, where $t_w$ is the thickness of the web.

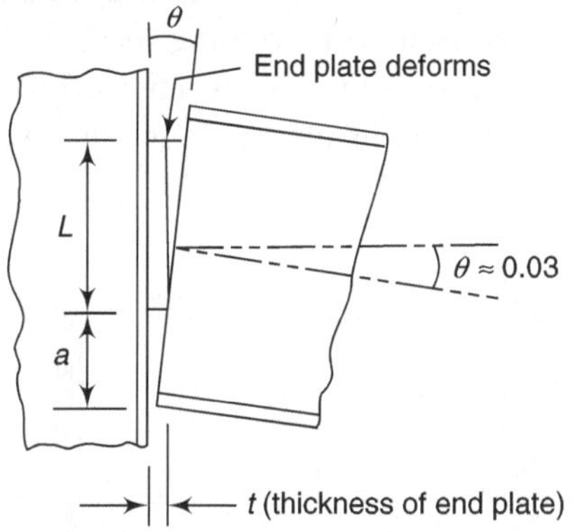


**FIG. 5.35** Header plate connection (connecting bolts are not shown for clarity)

The beam is designed for zero end moment, while the end plate augments the web shear and bearing capacity of the beams. The supporting member (e.g., column) has to be designed for the eccentric beam reaction. The reaction from the beam is transferred by weld shear to the end plate, by shear and bearing to the bolts, and by shear and bearing to the supporting member.

The weld is always referred by its leg size and the design strength of a shop-site welded fillet weld may be found from Table 6.5. If $V$ is the reaction, $L_w$ is the length of weld, $s$ is the size of weld, and $f_{wd}$ is the design strength in weld material, then the equation for the size of the weld is given by

$$s = V/(L_w f_{wd}) \geq 6\,\text{mm} \tag{5.42}$$

A typical design of flexible end plate connection is shown in Example 5.18.

### 5.7.7 Web Side Plate Connection

A *web side plate connection* also called a *fin plate connection* or *single plate shear connection* consists of a fin plate that is fillet welded to the supporting member, and bolted to the beam web [Fig. 5.2(f)]. The flanges may be notched or coped, when required.

The problem with the design of this kind of connection is that the 'bending' flexibility, which was available in properly positioned angle cleat and end plate connection, is not available. Hence, the movement required to accommodate the end rotation of a simply supported beam has to be achieved by the hole clearances and by the elongation of the bolt holes. Hence in this type of connection, to achieve the above, the following must be considered (Owens & Cheal 1989).

1. Only ordinary bolts should be used.
2. The bolts should be designed to fail by bearing of the connected plies and not by shear of the bolt.
3. The edge distances must be greater than two times the bolt diameter.
4. The resistance of the welds attaching the fin plate to the support is made greater than the moment that can be applied by the bolts. To achieve ductility, a minimum size of weld, relative to the web thickness is selected.

However, as per the IS code provisions, it is not possible to make the bolts to fail in bearing.

Since this connection has limited flexibility, it may be classified as *semi-rigid*. As discussed above, it can be used for joints in simple constructions, provided that the degree of interaction between the members can be established. The beam reaction is transferred by shear through the bolts to the fin plate, by in-plane bending and shear to the welds, and by vertical and horizontal shear to the supporting member (Trahair et al. 2001). The beam and supporting member are designed for the reaction and moment.

To limit the effect of lateral eccentricity of the beam-web from the fin plate, the following guidelines are used (Owens & Cheal 1989):

1. Depth of cleat > 0.75 × depth of beam
2. Thickness of cleat > (5 × Reaction)/($f_y$ × depth of fin plate)

The moment that could be applied by the bolts is checked with the moment capacity of plate. Since fillet welds loaded transversely have limited ductility, the following is checked:

1. Minimum weld size is arrived at by using a weld size of ($0.6t/0.7$) where $t$ is the thickness of the fin plate.
2. The welds are checked for the moment that could be developed by the bolts.

The moment developed by the bolt is given by

$$M = \text{capacity of bolt} \times \Sigma r_i^2/r_i \tag{5.43}$$

The moment capacity of the plate = $1.2Zf_y/\gamma_{m0}$ (5.44)

The horizontal shear in weld = Moment/$Z$ of weld group

Vertical shear = Reaction/Total weld length (5.45)

The design of fin plate to resist reaction and applied moment is provided by Trahair et al. (2001) and a typical design is given in Example 5.19. More details about single plate shear connections may be found in Astaneh et al. (1989) and Muir and Hewitt (2009).

## 5.8 BEAM-TO-BEAM CONNECTIONS

The simple connections such as clip and seating angle connection, web angle connection, and flexible end plate connections, discussed in the earlier sections for connecting beam-to-columns, can be adopted for beam-to-beam connections also. If the supporting beam and the supporting girder are required to have their top flange at the same level (for fixing floor chequered plates or floor slabs), the top flange of the beam is cut away (coped) to provide space for the supporting girder.

In practice most of the secondary beams are connected to main beams by web cleats and bolts since the web of the main beam may not be strong enough to support a seating angle. (Moreover, the seat connection requires more space in the vertical direction.) When the beam ends are coped, the web of the secondary beam has to be checked for block shear. In addition, the loss of top flange will also reduce the bending strength. To reduce the effect of notched end and to increase the robustness of the element, stiffeners may be welded just below the coping [see Figs 5.36(c) and (d)], which are required only on one side of the web. If two secondary beams on either side of the main supporting girder are on the same grid line, same bolts are used to connect both beams to the supporting girder (see Fig. 5.36). The arrangement used for bolts and cleats, is suitable where larger (left hand) beam is erected before the smaller one [Fig. 5.36(c)]. The top bolts through main girder webs will be in double shear after the secondary beam is attached. The larger beam can be held in position by the bottom pair of bolts while the smaller one is erected. If the smaller beams are to be erected first, they are propped and supported in some way until the larger one is erected.

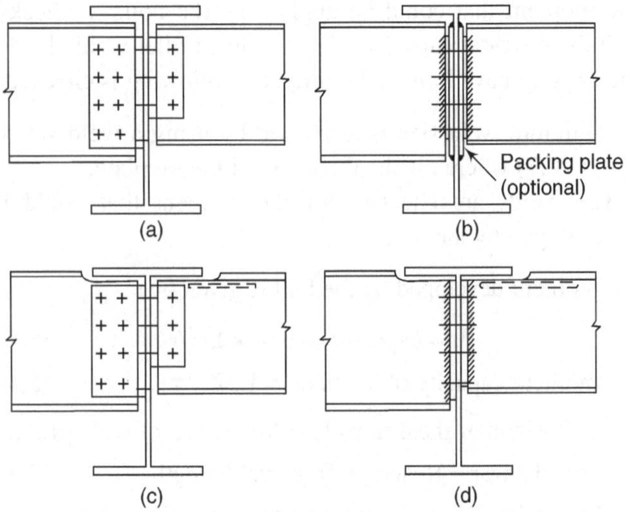


**FIG. 5.36** Simple beam-to-beam connections

Sometimes rigid connections may be provided for moment continuity between secondary beams, which may result in economy in beam section. Rigid connections are also necessary if a cantilever beam is supported by a girder. However, in this case, the end of the cantilever will transfer shear force and bending moment—this bending moment will be transferred to the primary beam as a torsion. Hence, the primary beam has to be designed for bending, shear, and torsion. If the primary beam is torsionally flexible, then torsion transferred to it may be neglected. Figure 5.37 shows some typical details of moment-resisting beam connections. To ease erection, packing plates are used while connecting short, wide secondary beams (Owens & Cheal 1989). Note that when top flanges are required to be at the same level, a combined splice plate/end plate connection with coped ends, as shown in Fig. 5.37(b) is used. For avoiding local rotation, HSFG bolts may be utilized in these connections. More details on beam-to-beam connections may be found in Owens & Cheal (1989).

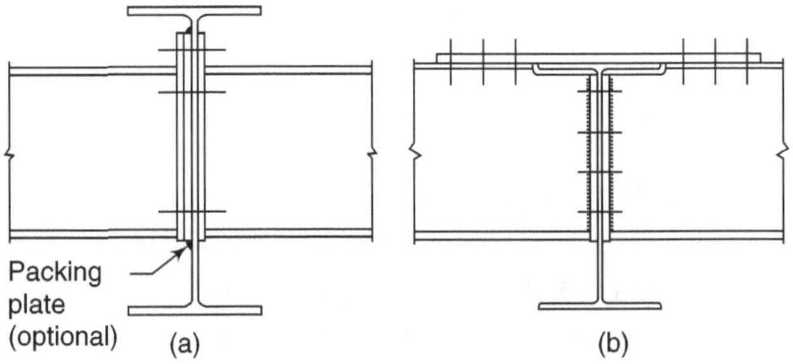


**FIG. 5.37** Moment-resisting beam-to-beam connections

The design of a beam connection using bolted web cleats is shown in Example 5.20.

## 5.9 BEAM AND COLUMN SPLICES

Rolled beams, plate girders, and columns are spliced due to several reasons, which include the following:

1. The full length may not be available from the mill
2. Often the size of section which can be transported is based on the size of the trucks. Hence, the maximum length is often restricted to 6 m. Thus, the fabricator may find it economical to splice even when full length could be obtained
3. The designer may desire to use splice points to camber the beam
4. The designer may desire a change in section to fit the variation in strength required along the span of a beam

### 5.9.1 Beam Splices

Many field splices for beam and plate girders use laping splice plates and high-strength bolts as connectors (see

Fig. 5.38). Splices are designed for the moment *M* and the shear *V* occurring at the spliced section, or they are designed for some specification prescribed higher values. On each side of the joint, one must provide enough bolts to transfer the shear and bending moment.

**FIG. 5.38** Bolted beam splice (Copyright© American Institute of Steel Construction, Inc. Reprinted with permission. All rights reserved.)

For beam splices, each element of the splice is designed to do the work the sections underlying the splice plates could do, if uncut. Plates on the flange should be designed to do the work of the flange (area of splice plate should not be less than 5% in excess of the area of flange element splices) and plates over the web should be designed to do the work of the web. The simplified partition of resisting all the moments by the flange splices and all shear by the web splices should not be used for deep plate girders with slender webs. These web splices should be designed to resist both shear and its share of the moment. Also double row of bolts should be used on each side of such web splices [where tension field action (see chapter 11 on plate girders) is predominant]. In web splices, the moment about the centroid of the bolt group on either side of the splice should be designed for moment due to eccentricity. IS code stipulates that in no case should the strength of the spliced portion be less than 50% of the effective strength of the material spliced.

In accordance with the principles used for designing bolted connections, the forces designed should be those acting at the centre of gravity of the bolt group (Salmon & Johnson 1996). Most splices are located near a point of effective restraint; if possible, splices should not be located in critical sections. Thus, they are often located where either the shear or the bending moment is low, that is, at locations where minimum strength is required.

Figure 5.39 illustrates two basic forms of beam splice, both of which may have several variants (Nethercot 2001). Flush end plates are used when the bending moments to be resisted are modest. Singly or doubly extended end plates are used when resisting high moments of one sign or full reversal, respectively. When end plate thickness is greater than 25 mm, the susceptibility of lamellar tearing should be checked. Moreover, end plate connections have no allowance for length corrections. For the end plate arrangement, design is similar to that discussed for the beam-to-column end plate, that is, the bolts are designed to share the shear force equally and a group of bolts are designed to take the tension due to the moment (T-stub action).

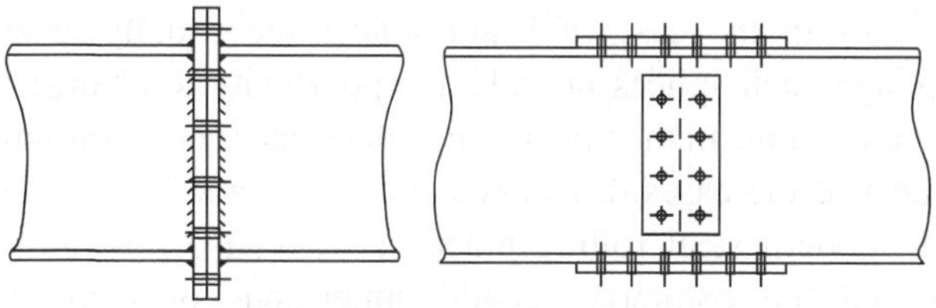

**FIG. 5.39** Types of beam splices

Reasonable care should be given to the flatness of the end plate, which will tend to distort in the presence of heavy welding on one face. When a change in size between the two sections of the beam occurs at an end plate splice, it can be easily accommodated by introducing longitudinal stiffeners to the larger beam as shown in Fig. 5.40. The flange cover plates and bolts (after deducting the bolt hole area) in Fig. 5.39(b), should be designed to take the moment, while the web plates and bolts are designed to resist the shear force plus the secondary moment due to eccentricity. For large beams and plate girders, flange plates are provided on both faces (top and bottom) of the beam flanges. The use of HSFG bolts will result in reasonable number of bolts, thus reducing the splice length.

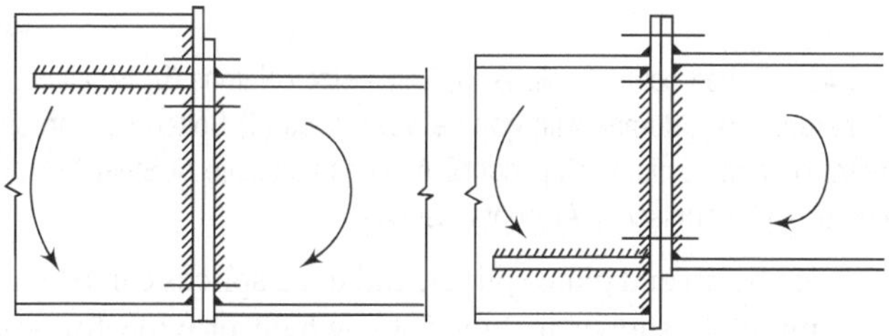

Coplanar compression flange
(a)

Coplanar tension flange
(b)

**FIG. 5.40** End plate connections between beams of different sizes

If there is a small difference in the height of the two sections of beams that are being jointed, the first line of bolts near the junction in the cover plate may be ignored when calculating the slip resistance of the connection, using HSFG bolts (Owens & Cheal 1989). If the difference in thickness is relatively large, shims or packing plates may be used. The design of a bolted beam splice is worked out in Example 5.21.

## 5.9.2 Column Splices

The connections as discussed for beams may be adopted for column splices also. Design of column splices does not pose much difficulty, since the high tensile forces that caused problems in the case of beam splices are usually absent. Although such splices provide an opportunity to change the size of columns in the upper floors, the same size is continued in most of the cases due to economic and practical reasons. With modern steel rolling practices, it is also possible for retaining the common outside dimensions over the full column height and reduce only the thickness of flanges/web in the upper floors (see Fig. 1.32).

When the columns carry predominantly axial forces, either direct end bearing arrangement or leaving a gap between the ends may be used (see Fig. 5.41). In the first case, the load is transferred through the contact area (the ends of both columns are machined or cut precisely) and the splices are designed only to resist any accidental tension due to some uplift loading or internal explosion in the building. Whereas in the second case [Fig. 5.41(c)], the whole load is transmitted by means of the splice plates. In these cases it is preferable to use HSFG bolts in the connection. Changes in the size of the columns can be accommodated by using packing plates as shown in Figs 5.41(b) and 5.41(c).

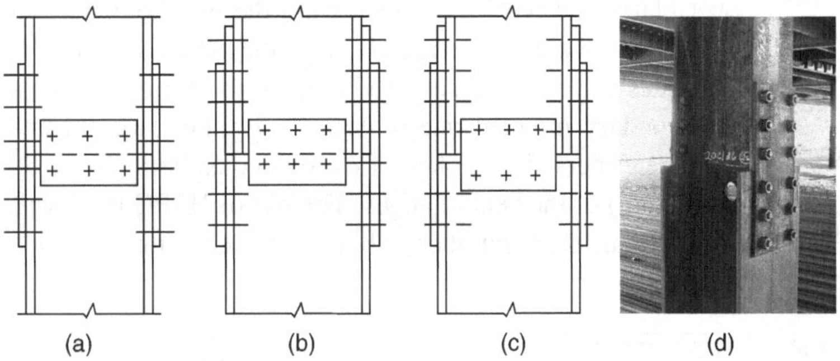


**FIG. 5.41** Bolted column splices (a) Same size columns (b) Small change in size (c) Different size columns with gap between ends (d) Bolted column splice of two different column sections (Copyright© American Institute of Steel Construction Inc. Reprinted with permission. All rights reserved.)

Similar to beam end plate splices, end plate splices can be used for columns as shown in Fig. 5.42, which provide for load reversals in columns in a convenient way. Short end plates are used for moderate moments and extended end plates for heavy moments. It is normal practice to position splices just above floor level so that the effects of flexing of the column may be neglected. However, in regions of seismic activity, the splices should be placed near mid-height of the columns, where bending moments will be minimum. The design of end plate connections can be treated in a similar way to end plate beam splices. Changes in column sizes can be easily tackled with the use of appropriate stiffeners, as shown in Fig. 5.42(c). IS code also stipulates that the centroidal axis of the splices should coincide with the centroidal axis of members jointed; if this condition is not satisfied, the design should cater for the resulting stress considering eccentricity. Examples 5.22 and 5.23 show the calculations required for the design of column splices.

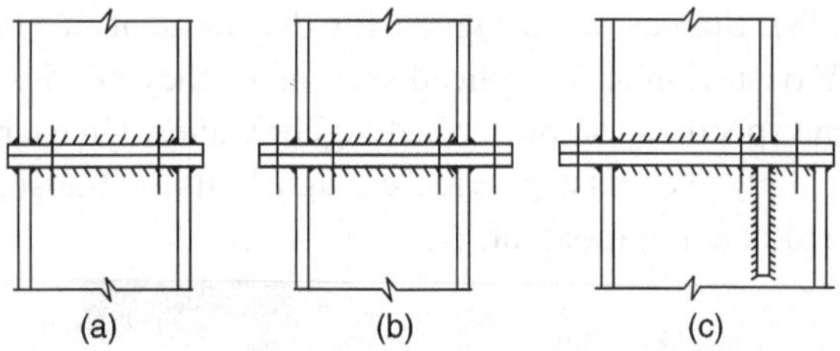


**FIG. 5.42** Column splices using end plates (a) Same size columns (b) Same size columns with high moment (c) Different size columns with high moment

## 5.10 SEMI-RIGID CONNECTIONS

As discussed in Section 4.5, connections can be classified as rigid, semi-rigid, and flexible for the purpose of analysis and design. Simple (flexible) connections are assumed to transfer only shear at some nominal eccentricity and may be used in frames up to approximately five storeys in height, where strength rather than stiffness governs the design. In such frames, a separate lateral load resisting system, in the form of bracings or shear walls should be provided. Examples of simple beam-to-column connections are those using clip and seating angle, web cleats, and header plate.

Rigid connections transfer significant moments to the columns and are assumed to undergo negligible deformations at the joint. These characteristics are necessary in sway frames for stability and also to resist lateral loads. Flush plate or extended end plate connections, haunched end plate connections, and fully welded connections are considered as rigid beam-to-column connections. They are used in high rise and slender structures where stiffness requirements control the design.

Semi-rigid connections fall between the above two categories. In fact simple connections do have some degree of rotational restraint and 'rigid' connections do experience some degree of joint deformation as shown in Fig. 4.20(b) (Dhillon & O'Malley III 1999).

In order to take into account the rotational restraint provided by these connections, they have to be tested in the laboratory. The simplest method of analysis will be to idealize these connections as equivalent rotational springs with either bilinear or non-linear moment rotation characteristics. The classifications proposed by Bjorhovde and the Frye–Morris model are given in the code, and discussed in Section 4.5.3 (Frye & Morris 1975 and Bjorhovde et al. 1990). However, due to the difficulty of evaluating and establishing the degree of restraint and analysing the frames with such joints (several general purpose computer programs have provisions to give the rotational restraints at the ends of the beams), the effect of semi-rigid connections is not considered in practice. Their effect is also not considered in the plastic design method.

Most of the designers consider these connections as rigid or simple in the analysis. The semi-rigid connections (top and seat angles, top and seat angles with web angles, web side plate, etc.,) cause a shift in moments from the beam ends to spans; and as a result, the overall maximum bending moments decrease in beams, but increase in some columns as compared to rigid frames (see Fig. 5.43). However, the softening of the connections results in a large increase in the building drift. Economic designs may be produced by balancing positive and negative moments in the beams. However, the drift may become critical in unbraced steel frames and may control the design of connections. Hence, an economic framing system can only be achieved by balancing the building drift with the stiffness of the connection. Due to these problems, the semi-rigid joints are treated in practice as simple connections only and used in frames which are braced adequately. Refer Chapter 4 for the details of analysis of the frame with semi-rigid joints. Once the end moments and shears are found out from such an analysis, the connections could be designed to transfer these moments and shears.

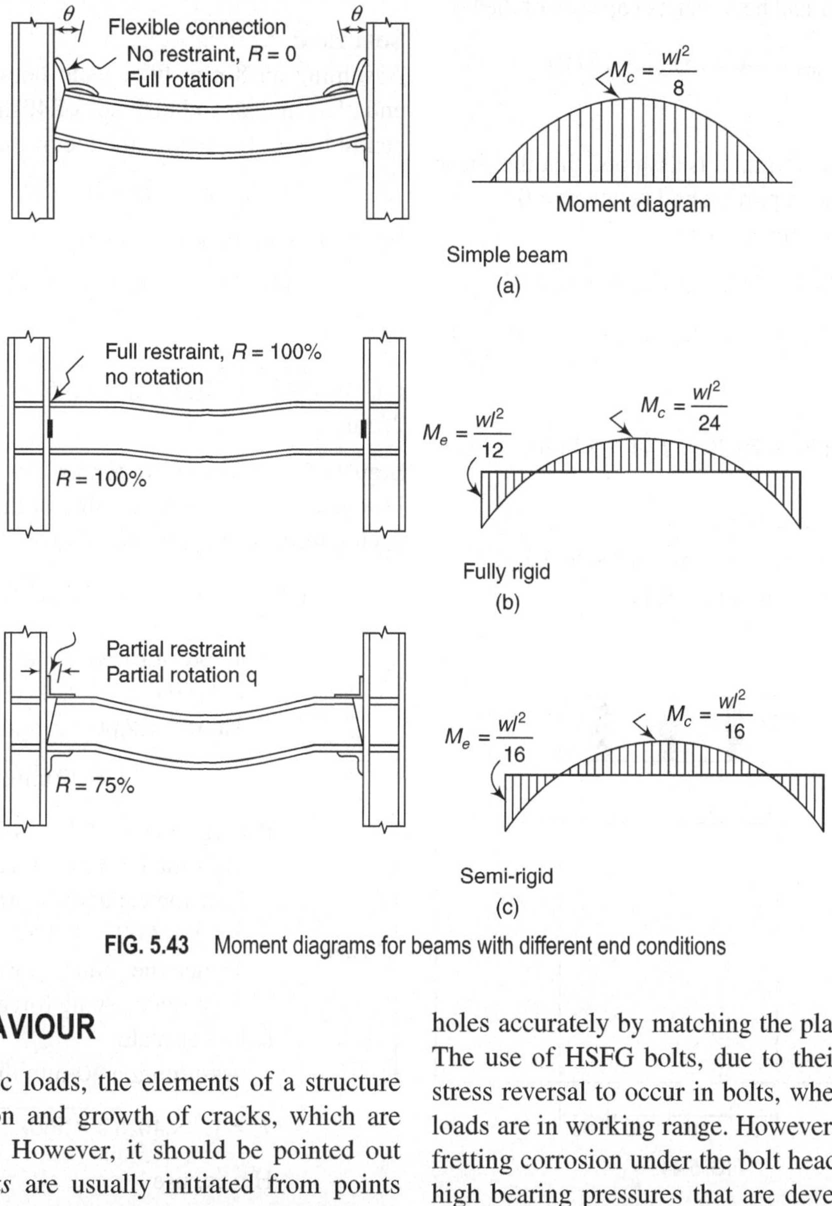


**FIG. 5.43** Moment diagrams for beams with different end conditions

## 5.11 FATIGUE BEHAVIOUR

Owing to repeated cyclic loads, the elements of a structure may experience initiation and growth of cracks, which are called as fatigue cracks. However, it should be pointed out that these *fatigue cracks* are usually initiated from points of stress concentration, which occur near the bolt holes. In order to minimize the effect of stress concentration, it is important to use HSFG bolts instead of the normal black bolts in structures likely to be subjected to fatigue loads. Other measures to reduce fatigue effects include using proper shape of gusset plates (which ensure smooth flow of forces), drilling holes accurately by matching the plates to be connected, etc. The use of HSFG bolts, due to their pre-tensioning, avoids stress reversal to occur in bolts, when the fluctuations in the loads are in working range. However, these bolts may exhibit fretting corrosion under the bolt head on the plies, due to the high bearing pressures that are developed below them. It is important to note that tension connections using HSFG bolts perform extremely well under fatigue. Shear connections using HSFG bolts are much better than most welded connections, whereas shear connections using black bolts are inferior to welded connections and hence should not be used in structures subjected to fatigue loads.

## EXAMPLES

**EXAMPLE 5.1:** Design a lap joint between two plates as shown in Fig. 5.44 so as to transmit a factored load of 70 kN using M16 bolts of grade 4.6 and grade 410 plates.

**SOLUTION:**
Strength calculation

Nominal diameter of the bolt $d = 16$ mm

Hole diameter = 16 + 2 = 18 mm

Bolts are in single shear and hence shear capacity of the bolt

$= (f_u/\sqrt{3})\,(n_n A_{nb} + n_s A_{sb})/\gamma_{mb} = (400/\sqrt{3})\,(1 \times 157)/1.25$

= 29.0 kN

Since the top plate is only 12 mm, it is assumed that the shear plane is through the threaded portion and hence $n_s = 0$.

Bearing capacity of the thinner plate

$= 2.5 k_b d t f_u / \gamma_{mb} = (2.5 \times 0.49 \times 410/1.25) \times 16 \times 12 = 77.15$ N

$k_b$ is smaller of 30/(3 × 18), 40/(3 × 18) – 0.25, 400/410, 1.0.

Hence, $k_b = 0.49$,

Bolt value = 29 kN

Required number of bolts = 70/29 = 2.41 ≃ 3 bolts

Detailing

Minimum pitch = 2.5 × $d$ = 40 mm

Minimum edge distance = 29 mm (as per Table 5.6)

Provide three bolts as shown in Fig. 5.44.

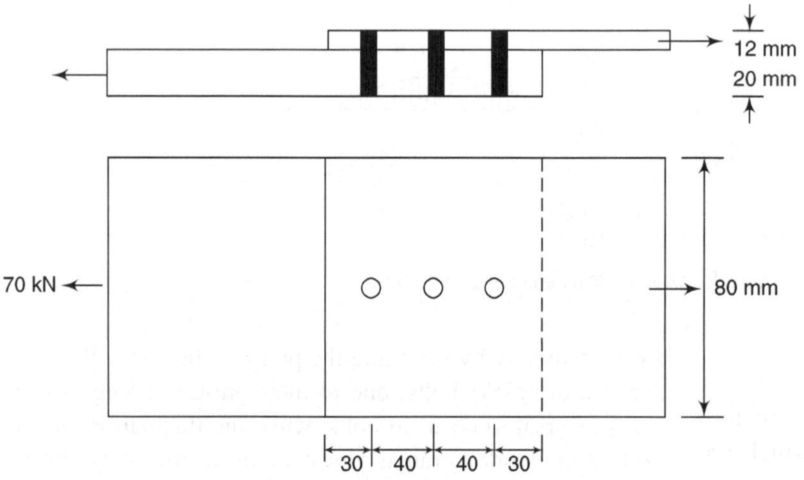


FIG. 5.44

**EXAMPLE 5.2:** Design a hanger joint as shown in Fig. 5.45 to carry a factored load of 300 kN. Use an end plate of size 250 mm × 150 mm and appropriate thickness, M24 HSFG bolts (2nos) and Fe 410 steel for end plate ($f_y = 250$ MPa).

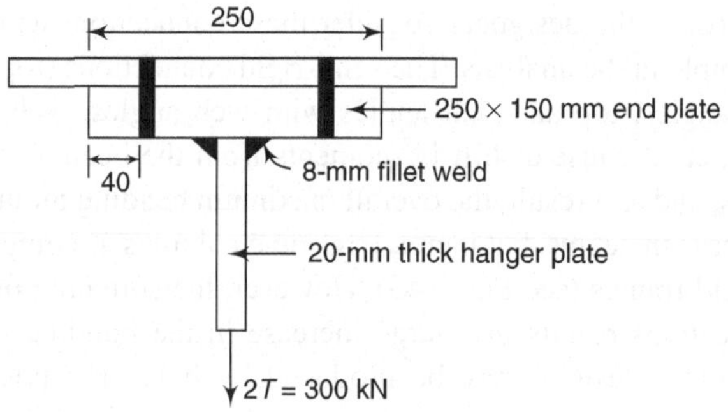


FIG. 5.45

**SOLUTION:**
Assuming an 8-mm fillet weld between the hanger and the end plate and an end distance of 40 mm, the distance from the centre line of the bolt to the toe of the fillet weld,

$$l_v = (250/2) - 10 - 8 - 40 = 67 \text{ mm}$$

For minimum thickness design,

$$M = T l_v / 2 = (300/2) \times 67/2 = 5025 \text{ kNmm}$$

and

$t_{\min} = \sqrt{[(4.4 \times 5025 \times 10^3)/(250 \times 150)]} = 24.28$ mm, say 25 mm.

**Step 1:** *Check for prying forces*

Distance $l_e$ from the centre line of the bolt to the point where prying force acts is the minimum of edge distance (40 mm) or

$$l_e = 1.1 t \sqrt{\beta f_o / f_y} = 1.1 \times 25 \sqrt{(2 \times 560/250)} = 58 \text{ mm}$$

Note proof stress of the $M24$ bolt = 0.7 × 800 = 560 MPa

Hence, adopt

$$l_e = 40 \text{ mm}$$

Prying force = $M/l_e$ = 5025/40 = 125.6 kN

Bolt load = 300/2 + 125.6 = 276 kN

Tension capacity of $M24$ bolt = $0.9 f_u A_{nb}/1.25$ = 0.9 × 800 × 353/1.25 = 203.3 kN < 276 kN

Hence the joint is unsafe.

To reduce prying force and hence bolt load, use a thicker plate

Assuming a 40-mm thick plate,

$$l_e = 1.1 \times 40 \sqrt{(2 \times 560/250)} = 93\, mm$$

Hence, use

$$l_e = 40 \text{ mm (edge distance)}$$

**Step 2:** *Allowable prying force*

$$Q = 203.3 - (300/2) = 53.3 \text{ kN}$$

Allowable moment at the toe of the weld = $T l_v - Q l_e$ = (300/2) × 67 – 53.3 × 40 = 7918 Nm

Moment of capacity = $(f_y/1.10)(b_e t^2/4)$

= $(250/1.10) \times (150 \times 40^2/4) \times 10^{-3}$

= 13636 Nm > 7918 Nm

Hence the joint is safe

Proof stress = $0.7 \times 800 = 560$ MPa = 0.560 kN/mm$^2$

Prying force as per Eqn (5.12)

= $67/(2 \times 40)[150 - 2 \times 1.5 \times 0.560 \times 150 \times 40^4/(27 \times 40 \times 67^2)]$

= 0.8375[150 − 133] = 14.24 kN < 53.3 kN

Hence the joint is safe.

Thus, it is observed that an end plate thickness of 40 mm is required to avoid the development of significant prying forces in the bolts. Note that the second term in the above calculation is not small. It will be small only when the thickness of the plate is small.

**Example 5.3:** Design a pin for the tension link shown in Fig. 5.46. Use plates of grade Fe 410 ($f_y$ = 250 MPa).

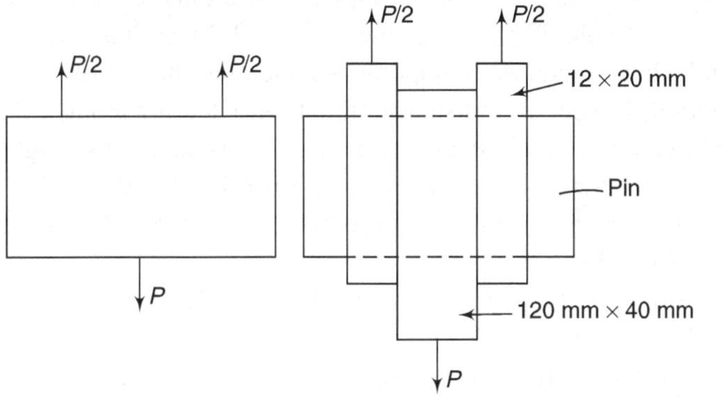


**FIG. 5.46**

**Solution:**

Ultimate pull transmitted by each 20-mm plate

$$P' = 250 \times 120 \times 20 \times 10^{-3} = 600 \text{ kN}$$

$$P = 2 \times 600 = 1200 \text{ kN}$$

Maximum bending moment = Pl/4 = 1200 × (40 + 10 + 10)/4

= 18,000 kNmm

$d = [21.33M_u/(\pi f_{yp})]^{1/3} = [(21.33 \times 18000 \times 10^3)/(\pi \times 250)]^{1/3}$ = 78.76 mm

Hence, provide an 80-mm-diameter pin.

*Check*

Shear force in the pin = 1200 kN

Shear capacity of the pin = $0.6 \times 250 \times \pi \times 80^2/4 \times 10^{-3}$ = 754 kN < 1200 kN

Hence, increase the size of the pin to 105 mm. Now,

Shear capacity = 1298 kN > 1200 kN. Hence the joint is safe in shear.

Bearing force = 600 kN

Bearing capacity = $1.5 \times 250 \times 105 \times 20 \times 10^{-3}$ = 787.5 kN > 600 kN

Hence the joint is safe.

**Example 5.4:** A member of a truss consists of two angles ISA 75 × 75 × 6 placed back to back. It carries an ultimate tensile load of 150 kN and is connected to a gusset plate 8-mm thick placed in between the two connected legs. Determine the number of 16-mm-diameter 4.6 grade ordinary bolts required for the joint. Assume $f_u$ of plate as 410 MPa.

**Solution:**

The arrangement of joint is as shown in Fig. 5.47. The bolts are in double shear. They bear against 8-mm gusset and two 6-mm angles, the former controlling the value in bearing as shown below.

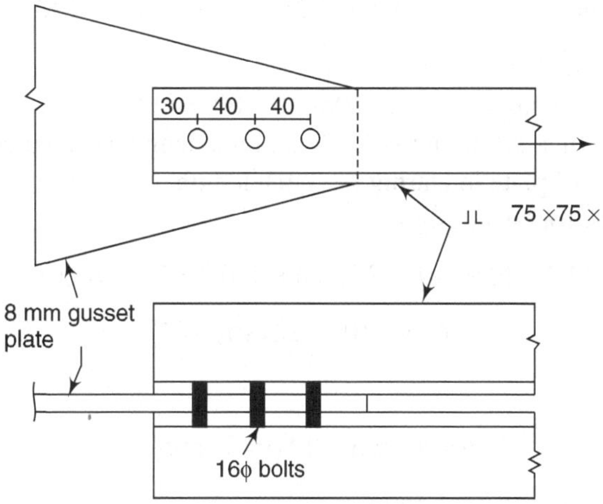


**FIG. 5.47**

From Table 5.9,

Strength in double shear for 16-mm-diameter 4.6 grade bolts = 2 × 29 = 58 kN

Strength in bearing on 8-mm plate = (2.5 × 0.49 × 16 × 8 × 410)/(1.25 × 1000)

= 51.4 kN

(Note: with $e$ = 30 and $p$ = 40, $k_b$ = 0.49 from Example 5.1)

Strength in bearing on two 6-mm thick angles

= [2.5 × 0.49 × (2 × 6) × 16 × 400]/(1.25 × 1000) = 75.2 kN

Thus,

Strength of the bolt = 51.4 kN

Required number of bolts = 150/51.4 = 2.92

Therefore, provide three bolts as shown in Fig. 5.47.

**Example 5.5:** The plates of a 6-mm thick tank are connected by a single bolted lap joint with 20-mm diameter bolts at 60-mm pitch. Calculate the efficiency of the joint. Take $f_u$ of plate as 410 MPa and assume 4.6 grade bolts.

**SOLUTION:**

Strength of the bolt in single shear (from Table 5.7) = (245 × 185)/1000 = 45.3 kN

The same value may be obtained directly from Table 5.9.

Assuming $e = 35$, $k_b$ should be least of 35/(3 × 22), 60/(3 × 22) – 0.25, 400/410, 1.0. Thus $k_b = 0.53$. Design strength in bearing on 6 mm plate = 2.5 × 0.53 × 20 × 6 × 410/(1.25 × 1000) = 52.15 kN.

Strength of the net section per pitch length (see Chapter 7) $= 0.9A_n f_u/\gamma_{m1}$

$$= 0.9(60 - 22) \times 6 \times 410/(1.25 \times 1000) = 67.3\text{ kN}$$

Strength of the bolt in single shear is 45.3 kN, being the smallest of 45.3, 52.15, and 67.3, determines the strength of the joint. Thus,

Strength of the joint per pitch length = 45.3 kN

Original strength of the plate per pitch length $= 0.9f_u A/\gamma_{m1}$

$$= 0.9 \times 410 \times 60 \times 6/(1.25 \times 1000) = 106.3\text{ kN}$$

Therefore,

Joint efficiency = (45.3/106.3) × 100 = 42.6%

Note that the strength of bolt per pitch length is equal to the strength of plate in tearing per pitch length.

Thus, for best design

$$45.3 = [0.9 \times (p - 22) \times 6 \times 410]/(1.25 \times 1000)$$

$$56.625 \times 10^3 = 2214(p - 22)$$

or

$$p = 47\text{ mm} < 2.5\phi = 50\text{ mm}$$

Hence, provide $p = 50$ mm

Now,

Original strength of the plate = (0.9 × 410 × 6 × 50)/(1.25 × 1000) = 88.56 kN

Efficiency = (45.3/88.56) × 100 = 51.2%

Note that we have adopted a bolt diameter of 20 mm for a 6-mm plate. As per *Unwin's formula* we can use a 16-mm bolt. If we rework with a 16-mm bolt, and spacing of 40 mm

Bolt strength in single shear = 29 kN

Original strength of the plate = (0.9 × 410 × 6 × 40)/(1.25 × 1000) = 70.85 kN

Therefore,

Efficiency = (29/70.85) × 100 = 40.9%

Thus, the use of a bolt with a large diameter slightly increases the efficiency.

*Note:* As per Unwin's formula, when $t \geq 8$mm, $d = 6\sqrt{t} > t$

**EXAMPLE 5.6:** Calculate the efficiency of a zigzag double bolted lap joint as shown in Fig. 5.48. Assume Fe 410 grade plate and grade 4.6 bolts of diameter 20 mm and 8-mm thick plates.

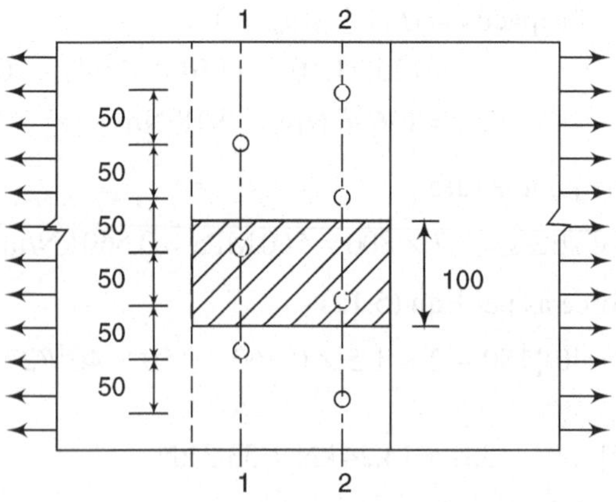


FIG. 5.48

**SOLUTION:**

Strength of a 20-mm-diameter bolt

(a) in single shear (from Table 5.9) = 45.3 kN
(b) in bearing on a 8-mm plate, (with $k_b = 0.5$) = (2.5 × 0.5 × 20 × 8 × 410)/(1.25 × 1000) = 65.6 kN

Consider the 100-mm length of the joint shaded as shown in Fig. 5.48, which represents typical conditions. Line 1-1 has one bolt hole 22 mm in diameter. Line 2-2 also has one bolt hole, but if plate-*A* tears along 2-2, the bolt line 1-1 must fall with it. Hence, the resistance of plate-*A* in tearing along 2-2 = Strength of the net section of plate 2-2 plus one bolt strength.

Hence for plate-*A*, net section 1-1 is critical. Here,

Net strength of the plate = [0.9(100 – 22) × 8 × 410]/(1.25 × 1000) = 184 kN

Strength of two bolts (Table 5.9) = 2 × 45.3 = 90.6 kN

Therefore,

Strength of the joint = 90.6 kN (smaller of 184 kN and 90.6 kN)

Gross strength of the plate = (0.9 × 100 × 8 × 410)/(1.25 × 1000) = 236 kN

Efficiency = (90.6/236) × 100 = 38.4%

**EXAMPLE 5.7:** A single-bolted double-cover butt joint is used to connect two plates 6-mm thick (see Fig. 5.49). Assuming the bolts of 20-mm diameter at 60-mm pitch, calculate the efficiency of the joint. Use 410 MPa plates and 4.6 grade bolts.

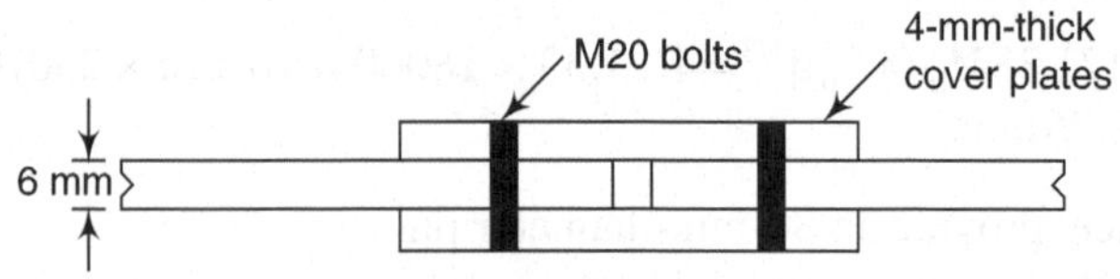


FIG. 5.49

**SOLUTION:**

Strength of a 20-mm-diameter bolt in double shear (from Table 5.9)

$$= 45.3 \times 2 = 90.6\text{ kN}$$

Thickness of the cover plate > 5/8*t* = 5/8 × 6 = 3.75 m

Therefore, provide 4-mm thick cover plates.

The thickness to be considered in the calculations is the thickness of the main plate or the sum of the cover plates, whichever is less. Hence,

Strength of the bolt in bearing (assuming $k_b = 0.5$) = (2.5 × 0.5 × 20 × 6 × 400)/(1.25 × 1000) = 48 kN

Strength of the plate in tearing = $0.9(p - d) \times tf_u/1.25$

$$= 0.9(60 - 22) \times 6 \times 410/(1.25 \times 1000)$$
$$= 67.3\,\text{kN}$$

The strength of the plate in bearing is the minimum. Hence,

Strength of the joint = 48 kN

Strength of the solid plate per pitch length = (0.9 × 60 × 6 × 410)/(1.25 × 1000)

$$= 106.3\,\text{kN}$$

Joint efficiency = (48/106.3) × 100 = 45.17%

Compared with Example 5.5, it is found that the efficiency of butt joint is 1.06 times greater than the lap joint with the same configuration.

**EXAMPLE 5.8:** Two plates 10-mm and 18-mm thick are to be jointed by a double-cover butt joint. Assuming cover plates of 8-mm thickness, design the joint to transmit a factored load of 500 kN. Assume Fe 410 plate and grade 4.6 bolt.

**SOLUTION:**

Assume 20-mm diameter bolts.

Bolt strength in double shear (from Table 5.9) = 2 × 45.3 = 90.6 kN

Since the joint has a packing plate greater than 6 mm (see Fig. 5.50), the bolt strength has to be reduced by

$$\beta_{pkg} = (1 - 0.0125 t_{pkg})$$
$$= (1 - 0.0125 \times 8) = 0.9$$

Hence,

Bolt strength in shear = 90.6 × 0.9 = 81.54 kN

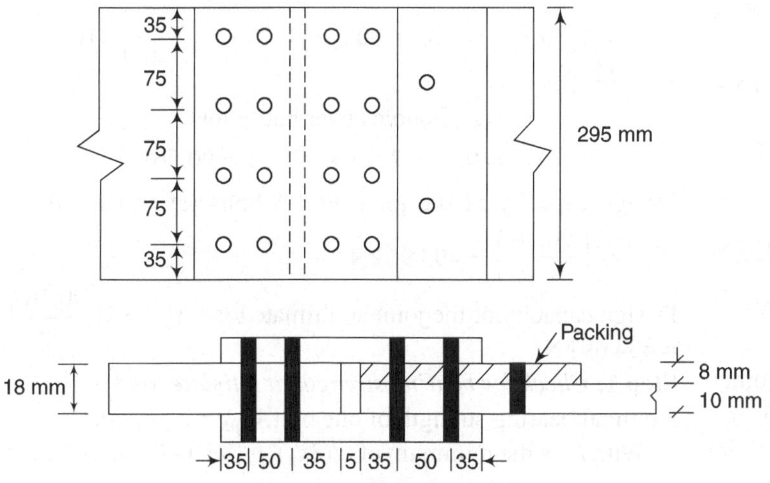


**FIG. 5.50**

Bolt strength in bearing (assuming $k_b = 0.5$) = (2.5 × 0.5 × 20 × 10 × 410/(1.25 × 1000) = 82 kN

Note: Strength of the joint is governed by the strength of the inner plate at the weakest section.

Hence,

Bolt strength = 81.54 kN

Required number of bolts = 500/81.54 = 6.13

Provide eight bolts, four bolts in two rows.

Strength of bolt/pitch length = 2 × 81.54 = 163.08 kN

Equating it to the strength/pitch length of the plate in tearing

$$163.08 = 0.9 \times (p - 22) \times 10 \times 410/(1.25 \times 1000)$$

$$203.85 \times 10^3 = 3690(p - 22)$$

$p = 77.24\,\text{mm} > 2.5 \times 20 = 50\,\text{mm}$

Minimum spacing of bolts = 2.5 × 20 = 50 mm

Minimum edge distance = 1.5 × 22 = 33mm

Provide edge distance of 35 mm

Hence, provide $p$ = 75 mm. Provide additional two bolts of 20-mm diameter on the packing plate as shown in Fig. 5.50.

Design strength of plate

$$= \frac{0.9 A_n f_u}{\gamma_m} = \frac{0.9 \times (295 - 4 \times 22) \times 10 \times 410}{1.25 \times 1000}$$
$$= 611\,\text{kN} > 500\,\text{kN}$$

**EXAMPLE 5.9:** Consider the joint in Example 5.8, but subjected to a tensile factored load of 700 kN. Assume 16 mm diameter and grade 4.6 bolts and design the joint.

**SOLUTION:**

Bolt strength of 16 mm bolt in double shear (Table 5.9) = 2 × 29 = 58 kN

Required number of bolts = 700/58 = 12

Provide 15 bolts, in five rows at a spacing of say 65 mm, as shown in Fig. 5.51.

Minimum edge distance = 27 mm (Table 5.6)

Minimum pitch = 2.5 × 16 = 40 mm.

Adopt an edge distance of 40 mm and gauge (cross-pitch) as 110 mm

**Step 1:** *Strength of bolts in shear*

Since the joint has a packing plate greater than 6 mm

Reduction factor for packing plate

$$\beta_{pk} = (1 - 0.0125\, t_{pk})$$
$$= (1 - 0.0125 \times 8) = 0.9$$

Length of the joint = 4 × 65 = 260 mm > 15 $d_b$ = 15 × 16 = 240 mm

Hence, Reduction factor for long joint

$\beta_{lj} = 1.075 - 0.005\,(l_j/d_b) = 1.075 - 0.005\,(260/16) = 0.994$

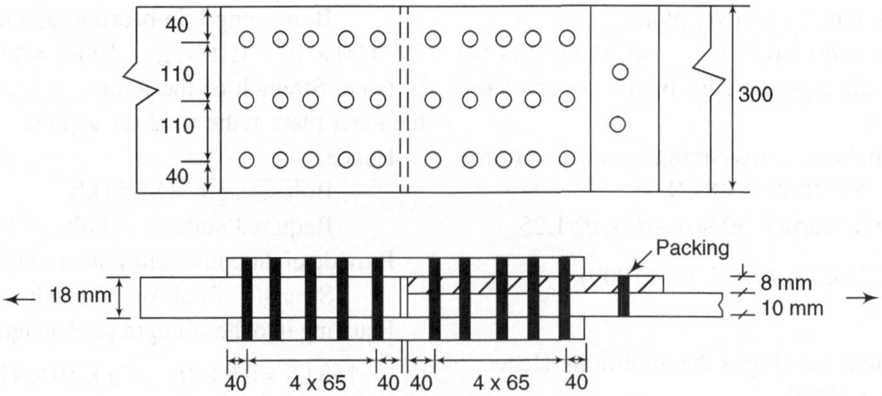


FIG. 5.51

Assuming that one shear plane is in the shank and another in the thread of the bolts, nominal shear strength of the bolt in double shear

$$=\beta_{pk}\,\beta_{lj}\,f_{ub}\,(n_n A_{nb}+n_s A_{sb})/\sqrt{3}$$

$$=0.9\times 0.994\times 400\left(1\times 0.78\times\pi\times\frac{16^2}{4}+1+\pi\times\frac{16^2}{4}\right)$$

= 73,940 N = 73.94 kN

Design shear strength of the bolt = 73.94/1.25 = 59.15 kN

Design shear strength of 15 bolts in the joint = 15 × 59.15 = 887.25 kN > 700 kN

**Step 2:** *Strength of bolts in bearing*

$k_b$ is the minimum of e/(3d$_o$), p/(3d$_o$) – 0.25, $f_{ub}/f_u$, 1.0

Hence it is the minimum of 40/(3 × 18), 65/(3 × 18) – 0.25, 400/410, 1.0

Hence $k_b$ = 0.74

Hence nominal strength of one bolt in bearing = 2.5$k_b$dt$f_u$

= 2.5 × 0.74 × 16 × 10 × 410 / 1000 = 121.36 kN

*Note:* Thickness of the thinner plate will govern the bearing strength.

Design strength of bolt in bearing = 121.36 / 1.25 = 97.09 kN

Design strength of 15 bolts in bearing = 97.09 × 15 = 1456.35 kN > 700 kN

Hence strength of the bolts in the connection = 887.25 kN

Design strength of the plate

$$\frac{0.9A_n f_n}{\gamma_{m0}}=\frac{0.9\times(300-3\times 18)10\times 410}{1.25\times 1000}=726.2\,\text{kN}>700\,\text{kN}$$

Hence the joint is safe to take the applied load of 700 kN.

**Example 5.10:** Determine the capacity of the joint shown in Fig. 5.52, if 20 mm HSFG bolts are used on clean mill scale surface.

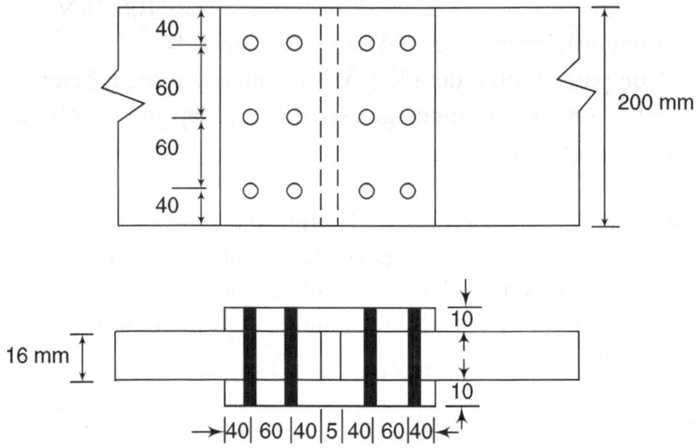


FIG. 5.52

**Solution:**

For HSFG grade 8.8 bolt, $f_{ub}$ = 800 N/mm$^2$

Assuming fasteners in clearance holes, $K_h$ = 1.0

Coefficient of friction, $\mu_f$ (Table 5.10) = 0.33 (clean mill scale)

Nominal shear capacity of bolt

$$V_{nsf}=\mu_f n_e K_h F_o$$

$$F_o=A_{nb}(0.7f_{ub})=(0.7\times 800)\times\left(0.78\times\pi\times\frac{20^2}{4}\right)\times 10^{-3}=137.22\,kN$$

$n_e$ = 2 (Double cover butt joint)

$V_{nsf}$ = 0.33 × 2 × 1 × 137.22 = 90.56 kN

Design capacity of the joint with 6 bolts (at service load),

$$V_{dsf}=6\left(\frac{90.56}{1.1}\right)=493.96\,\text{kN}$$

Design capacity of the joint at ultimate load, $V_{dsf}=6\left(\frac{90.56}{1.25}\right)$ = 434.69 kN

**Step 1:** *Check for bearing strength at ultimate load*

Nominal bearing strength of one bolt, $V_{npb}$ = 2.5$k_b$ $dtf_u$

With $k_b$ is the minimum of $e/(3d_o)$, $p/(3d_o)$ - 0.25, $f_{ub}/f_u$, 1.0

Hence it is the minimum of 40/(3 × 22), 60/(3 × 22) – 0.25, 400/410, 1.0

Hence $k_b = 0.606$

$V_{npb} = 2.5 \times 0.606 \times 20 \times 10 \times 410 \times 10^{-3} = 124.23\,kN$

Design strength of 6 bolts in bearing = 124.23 × 6/1.25 = 596.3 > $V_{dsf}$

Tensile strength of the plate

$$\frac{0.9A_n f_u}{\gamma_{mo}} = \frac{0.9 \times (200 - 3 \times 22)16 \times 410}{1.25 \times 1000} = 632.9\,kN > V_{dsf}$$

Hence design capacity = 434.69 kN

**Example 5.11:** Design a butt joint to connect two plates 175 × 10 mm (Fe 410 grade) using M20 bolts. Arrange the bolts to give maximum efficiency.

**Solution:**

Thickness of the cover plate = (5/8) × 10 = 6.25 mm

Provide double cover plates each having a thickness of 8 mm.

The tensile force the main plate can carry (assuming two bolts in a line)

= 0.9 × (175 – 2 × 22) × 10 × 410/(1.25 × 1000) = 386.7 kN

Shear strength of the bolt = 2 × 45.3 = 90.6 kN

Bearing strength of the bolt (Assuming $k_b = 0.5$) = (2.5 × 0.5 × 20 × 10 × 410)/(1.25 × 1000) = 82 kN

Hence,

Strength of the bolt = 82 kN

Required number of bolts = 386.7/82 = 4.7

Hence, provide five bolts of 20-mm-diameter as shown in Fig. 5.53.

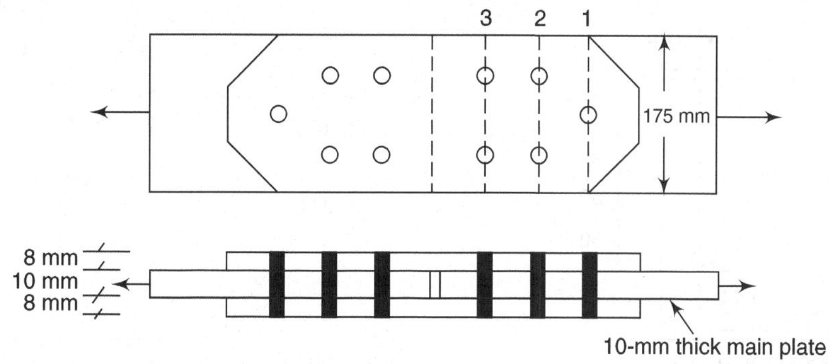


FIG. 5.53

**Example 5.12:** Determine the adequacy of the fasteners in Fig. 5.54, when 20-mm-diameter grade 4.6 bolts are used. Assume that the strength of the column flange and the structural T (ST) sections do not govern the design. Neglect prying action.

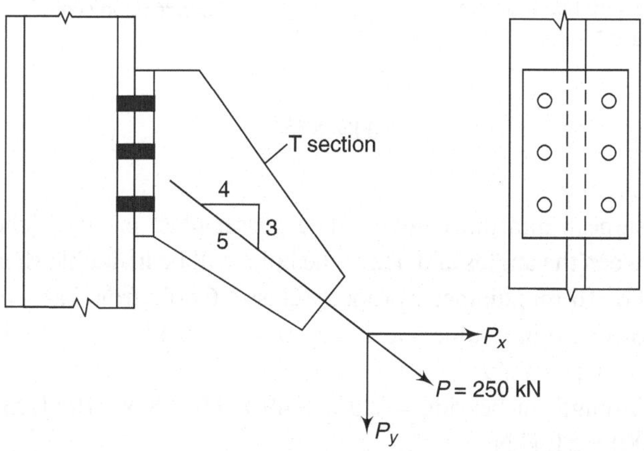


FIG. 5.54

**Solution:**

Tension component $P_{ux} = 250 \times 4/5 = 200$ kN

Shear component $P_{uy} = 250 \times 3/5 = 150$ kN

The factored loads $T_u$ and $V_u$ per bolt

Tension $T_u = 200/6 = 33.33$ kN/bolt

Shear $V_u = 150/6 = 25$ kN/bolt

For a 20-mm-diameter bolt,

Strength in single shear (from Table 5.9) = 45.3 kN

Strength in tension (from Table 5.9) = 68.5 kN

As per Eqn (5.20),

$(V/V_{sd})^2 + (T_e/T_{nd})^2 \leq 1.0$

$(25/45.3)^2 + (33.33/68.5)^2 = 0.305 + 0.238 = 0.54 < 1.0$

Hence, the six grade 4.6 bolts of 20-mm-diameter are sufficient to carry the load of 250 kN applied at the joint.

**Example 5.13:** Repeat Example 5.12 assuming the use of M20 parallel-shank HSFG grade 8.8 bolts in clearance holes and a factored load of 500 kN.

**Solution:**

Taking $\mu = 0.48$,

Tensile capacity of one bolt (Table 5.11) = 141 kN

Slip resistance of one bolt = 52.6

Tensile load per bolt = 400/6 = 66.67 kN

Shear load per bolt = 300/6 = 50 kN

Checking the interaction Eqn (5.30),

$(50/52.6)^2 + (66.67/141)^2 = 0.903 + 0.224 = 1.13 > 1.0.$

Hence unsafe.

Note that if the connection is only required to function as non-slip under service loads, the interaction equation has to be checked at service stage as follows:

Tensile load per bolt (assuming load factor of 1.5) = 400/(1.5 × 6) = 44.44 kN

Shear load per bolt = 300/(1.5 × 6) = 33.33 kN

Thus, at service load,

$(33.33/52.6)^2 + (44.44/141)^2 = 0.402 + 0.100 = 0.502 < 1.00$

Hence the connection is safe.

**EXAMPLE 5.14:** An ISMB 600 is connected to a column by web cleats with a single row of bolts. If the reaction is 350 kN and there are four 20-mm-diameter bolts through the web, as in Fig. 5.55, check if the section is adequate for block shear failure.

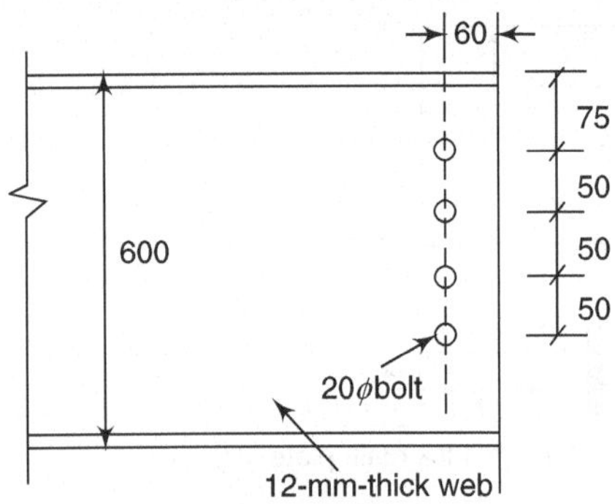


FIG. 5.55

**SOLUTION:**

**Step 1:** *Block shear capacity*

$$T_{db1} = A_{vg} f_y/(\sqrt{3} \times 1.1) + 0.9 f_u A_{tn}/1.25$$

$$T_{db2} = 0.9 f_u A_{vn}/(\sqrt{3} \times 1.25) + f_y A_{tg}/1.1$$

Design strength of the web = 250 N/mm²

Plate or web thickness = 12 mm

Net length of shear face = (3 × 50 + 75) – 3.5 × 22 = 148 mm

Net length of tension face = 60 – 0.5 × 22 = 49 mm

$$A_{vg} = 12 \times (3 \times 50 + 75) = 2700\,\text{mm}^2$$

$$A_{vn} = 12 \times 148 = 1776\,\text{mm}^2$$

$$A_{tg} = 12 \times 60 = 720\,\text{mm}^2$$

$$A_{tn} = 12 \times 49 = 588\,\text{mm}^2$$

Therefore,

$$T_{db1} = [2700 \times 250/(\sqrt{3} \times 1.1) + 0.9 \times 410 \times 588/1.25]/1000$$
$$= 354.283 + 173.58 = 527.86\,\text{kN}$$

or

$$T_{db2} = [0.9 \times 410 \times 1776/(\sqrt{3} \times 1.25) + 250 \times 720/1.1]/1000$$
$$= 302.69 + 163.636 = 466.32\,\text{kN}$$

Therefore,

$$T_{db} = 466.32\,\text{kN}$$

The value of $T_{db}$ = 466.32 kN is much higher than the applied reaction of 350 kN and hence there will not be any block shear failure in this case.

**EXAMPLE 5.15:** Design a connection of a truss joint as shown in Fig. 5.56, using M16 black bolts of property class 4.6 and grade 410 steel. Assume that the members shown are capable of resisting the loads.

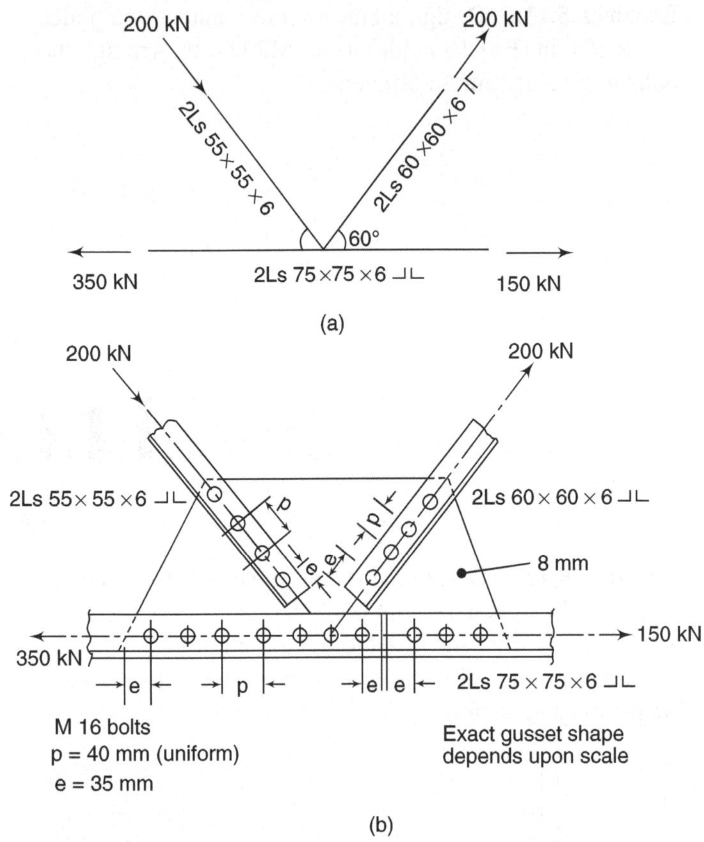


FIG. 5.56

**SOLUTION:**

Assume 8-mm thick gusset. The gusset plate is sandwiched between the angles and hence the bolts will be in double shear.

For 16-mm diameter property class 4.6 bolt, from Table 5.9

Strength in double shear = 29.0 × 2 = 58 kN; $k_b$ = 40/(3 × 18) – 0.25 = 0.49

Strength in bearing = 2.5 × 0.49 × 16 × 8 × 410/(1.25 × 1000) = 51.4 kN

Hence,

Strength of bolt = 51.4 kN

Bolts for 200 kN = 200/51.4 = 3.9 (hence provide four bolts)

Bolts for 150 kN = 150/51.4 = 2.9 (hence provide three bolts)

Bolts for 350 kN = 350/51.4 = 6.8 (hence provide seven bolts)

Provide edge distance = 2 × 16 = 32, say 35 mm

Pitch = 2.5 × 16 = 40 mm

**Step 1:** *Check for gusset plate*

Distance from first bolt to last bolt in member carrying 200 kN

$$= 3p = 3 \times 40 = 120\,\text{mm}$$

'Whitmore' effective width = 2tan30° × 120 = 138.56 mm

Capacity of plate = 0.9 × 138.56 × 8 × 410/(1.25 × 1000) = 327.22 kN > 200 kN

Hence the connection is safe. Note that the connection should be checked for block shear failure as shown in Example 5.14. The required calculations are shown as follows:

Net length of shear face = (3 × 40 + 35) – 3.5 × 18 = 92 mm

Net length of tension face = (60 – 35) – 0.5 × 18 = 16 mm (as per Fig. 5.10)

$$A_{vg} = 6 \times (3 \times 40 + 35) = 930\,\text{mm}^2$$

$$A_{vn} = 6 \times 92 = 552\,\text{mm}^2$$

$$A_{tg} = 6 \times 25 = 150\,\text{mm}^2$$

$$A_{tn} = 6 \times 16 = 96\,\text{mm}^2$$

$T_{db1}$ = [930 × 250/(√3 × 1.1) + 0.9 × 410 × 96/1.25]/1000 = 150.37 kN

and

$T_{db2}$ = [0.9 × 410 × 552/(√3 × 1.25) + 250 × 150/1.1]/1000 = 128.17 kN

Thus,

$$T_{db} = 128.17\,\text{kN} > 200/2 = 100\,\text{kN}$$

Hence the connection is safe. Adopt the structural details as shown in Fig. 5.56(b).

**Example 5.16:** Design a seat angle connection between a beam MB 300 and column SC 200 for a reaction of beam of 100 kN, using M20 bolts of property class 4.6. Take Fe 410 grade steel ($f_y$ = 250 MPa). See Fig. 5.57.

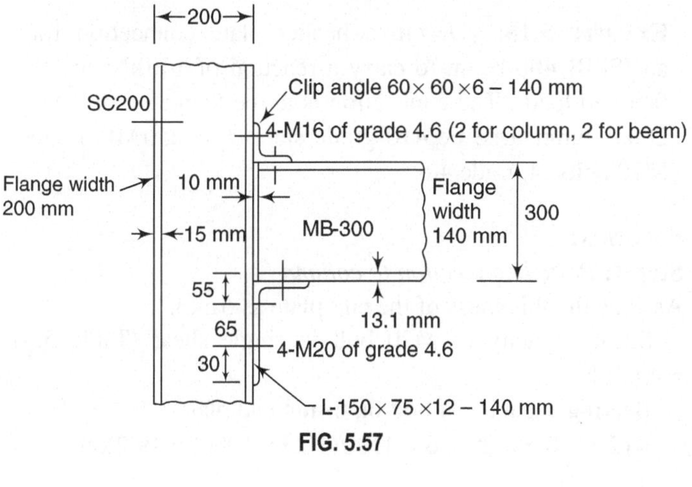


**FIG. 5.57**

**Solution:**

Assuming 150 × 75 × 12 angle

For M20 bolt,

Thickness of column flange (ISSC 200) = 15 mm

Strength in single shear (from Table 5.9) = 45.3 kN

Strength in bearing = 2.5 × 0.5 × 20 × 12 × 400/(1.25 × 1000) = 96 kN

Note: Least thickness in bearing is used. Since bolt ultimate tensile stress is 400 MPa compared to the plate of 410 MPa, the lower value is used in the calculation.

Hence,

Strength of bolt = 45.3 kN

Required number of bolts = 100/45.3 = 2.21

Provide four M20 bolts.

**Step 1:** *Seating angle*

Width of MB 300 = 140 mm

Hence,

Length of angle = 140 mm

Length of bearing required at root line of beam

$= R/(t_w f_{yw}/\gamma_{m0})$

= (100 × 1000)/(7.7 × 250/1.1) = 57.15 mm

Assume end clearance of beam from face of column as 5 mm and tolerance of 5 mm.

Required length of outstanding leg = 57.15 + 10 = 67.15 < 75 mm. Hence, the length is as required.

Length of bearing on cleat = $b_1 = 57.15 - (t_f + r)$

= 57.15 – (13.1 + 14) = 30.05 mm

For 150 × 75 × 12 angle, distance from the end of bearing on cleat to root angle (*A* to *B*) in Fig. 5.32,

$$b_2 = b_1 + 5 + 5 - (t + r_a) \text{ of angle}$$

$$= 30.05 + 10 - (12 + 10) = 18.05\,\text{mm}$$

Assume uniformly distributed load over bearing length $b_1$.

Moment at root of angle (point *B*) due to load to right of *B* = (100 × 18.05/30.05) × (18.05/2) = 542 KNmm

Moment capacity = $1.2 Z f_y/\gamma_{m0}$

= 1.2 × (250/1.1) × 140 × $12^2$/6 × $10^{-3}$

= 916 KNmm < 542 KNmm

Hence, the connection is safe.

Therefore, provide 150 × 75 × 12 mm seating angle

Shear capacity of the outstanding leg of cleat

= $wtf_y/(\sqrt{3} \times 1.10)$ = 140 × 12 × 250/(√3 × 1.10 × 1000)

= 220 kN > 100 kN

Hence, the shear capacity is as required.

Shear strength of beam

$V_d = A_v f_{yw}/(\sqrt{3} \times 1.10)$ = (300 × 7.7) × 250/(√3 × 1.10 × 1000)

= 303 kN > 100 kN

Hence, web does not need any stiffener at support.

Hence, provide a seating angle of 150 × 75 × 12 mm with four M20 bolts and provide a top clip angle of 60 × 60 × 6 with four M16 bolts of grade 4.6 as shown in Fig. 5.57.

**Example 5.17:** Design a bolted web angle connection for an ISMB 400 beam, to carry a reaction of 140 kN due to factored loads. The connection is to the flange of a column ISSC 200 in grade Fe 410 steel (see Fig. 5.58).

**Solution:**

For ISMB 400, $b_f = 140$ mm, $t_w = 8.9$ mm; using M20 HSFG bolts of grade 8.8,

Shear capacity of M20 bolt in single shear (from Table 5.11) = 52.6 kN

Bearing capacity on web of beam (web thickness of ISMB 400 = 8.9 mm)

= 2.5 × 0.5 × 20 × 8.9 × 410/(1.25 × 1000) = 72.98 kN

Strength of the bolt = 52.6 kN

Using web cleats of ISA 90 × 90 × 8, with a gauge distance $g = 50 + 50 + 8.9 = 108.9$ mm

Use four M20 bolts with a vertical pitch of 70 mm as shown in Fig. 5.58. Assuming the shear to be acting on the face of the column, due to the eccentricity, some horizontal shear forces will be acting on the bolt group in addition to the shear due to reaction.

$e = 50$ mm

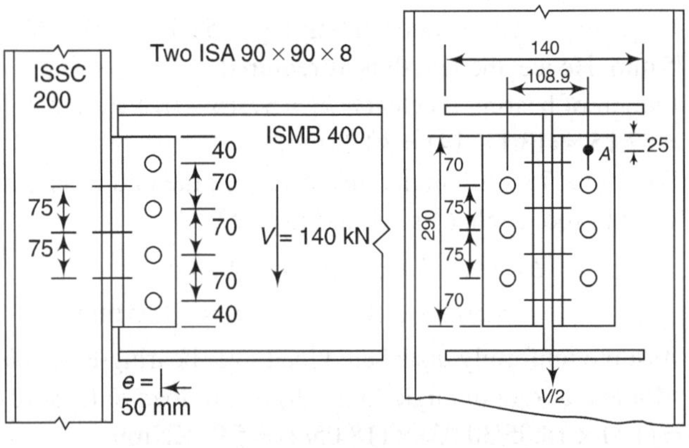


**FIG. 5.58**

Horizontal shear force on bolt due to moment due to eccentricity

$$= V_x e_x r_i / \sum r_i^2$$

$= 140 \times 50 \times 105/[2\,(35^2 + 105^2)] = 30$ kN

Vertical shear force per bolt = 140/4 = 35 kN

Resultant shear force = $\sqrt{(35^2 + 30^2)} = 46$ kN < 52.6 kN

Hence, the connection is safe.

**Step 1:** *Connection to column flange*

Assuming $\mu = 0.48$, slip resistance per bolt

= 0.48 × 1 × 1 × (245 × 0.7 × 800)/(1.25 × 1000) = 52.6 kN

Bearing resistance on 8 mm cleat per bolt

= 2.5 × 0.5 × 8 × 20 × 410/(1.25 × 1000) = 65.6 kN

Hence,

Bolt strength = 52.6 kN

Try six bolts as shown in Fig. 5.58 with a vertical pitch of 75 mm, 54.45 mm from the centre line of beam web. Similar to the previous case, the shear transfer between the beam web and the angle cleats may be assumed to take place on the face of the beam web.

Horizontal shear force on bolt due to moment due to eccentricity = $V_x e_x\, r_i / \Sigma\, r_i^2$

$= (140/2) \times 54.45 \times 75/(2 \times 75^2) = 25.41$ kN

Alternatively, if the centre of pressure is assumed 25 mm below the top cleat, then

Horizontal shear force = $(140/2) \times 54.45 \times 195/(45^2 + 120^2 + 195^2)$

= 13.65 kN < 25.41 kN

Vertical shear per bolt = 140/6 = 23.33

Resultant shear force = $\sqrt{25.41^2 + 23.33^2}$ = 34.5 < bolt capacity. Hence the connection is safe.

**Step 2:** *Cleat angle*

Take two angles ISA 90 × 90 × 8 of length 290 mm. Check bending at bolt line of connection to column flange.

Bending moment = 140/2 × 54.45 = 3811.5 Nm

Moment capacity =

$$\frac{1.2 f_y Z}{\gamma_{mo}} = \left[1.2 \times 250 \times \left(8 \times \frac{290^2}{6}\right) / (1.1 \times 10^3)\right]$$

= 33,640 Nm > 3,811.5 Nm

Hence chosen cleat angle size is adequate.

**Step 3:** *Check for shear strength of cleat and beam*

$$\text{Shear strength of single cleat} = \frac{f_y A_v}{(\sqrt{3} \times 1.1)} = \frac{250 \times (290 \times 8)}{(\sqrt{3} \times 1.1)}$$

$$\times 10^{-3} = 304\text{ kN} > \frac{140}{2} = 70\text{ kN}$$

$$\text{Local shear strength of beam} = \frac{f_y A_v}{(\sqrt{3} \times 1.1)} = \frac{250 \times (400 \times 8.9)}{(\sqrt{3} \times 1.1)}$$

$\times 10^{-3} = 467$ kN > 140 kN.

Hence it is safe to carry the shear.

**Example 5.18:** Design a header plate connection for an ISMB 400 beam to carry a reaction of 140 kN due to factored loads. The connection is to the flange of an ISSC 200 column. Use Fe 410 grade steel ($f_y$ = 250 MPa) and M20 bolts of grade 4.6.

**Solution:**

**Step 1:** *Bolted connection to column*

Assume the thickness of the end plate as 6 mm.

Shear capacity of M20 bolt in single shear (Table 5.9) = 45.3 kN

Bearing capacity of bolt on 6-mm end plate

= (2.5 × 0.5 × 20 × 6 × 410)/(1.25 × 1000) = 49.2 kN

Strength of bolt = 45.3 kN

Required number of bolts = 140/45.3 = 3.09

Provide four M20 bolts of grade 4.6.

**Step 2:** *End plate*

Length of end plate = 60 + 2 × 45 = 150 mm

Provide $a$ = 150 < 30 × 6 = 180 mm

Use a 160 × 150 × 6-mm end plate

Length of the fillet weld connecting end plate to beam web = 150 – 2 × 6 = 138

Assuming site welded (see Chapter 6)

Size of the weld = $(140 \times 10^3)/(2 \times 138 \times 158) = 3.21$

Provide a 6-mm fillet weld.

Shear stress on the web of the beam at the end plate $= (140 \times 10^3)/(8.9 \times 150) = 104.9\,\text{MPa} < 250/(\sqrt{3} \times 1.10) = 131.2\,\text{MPa}$

Hence, the connection is safe.

**Step 3:** *Check for shear strength of end plate*

$$\text{Shear strength of end plate} = \frac{f_y A_v}{(\sqrt{3} \times 1.1)} = \frac{250 \times 636}{(\sqrt{3} \times 1.1)} \times 10^{-3}$$

= 83.5 kN<140 kN

*Note:* $A_v = t_p\,(l_p - 2d_h) = 6(150 - 2 \times 22) = 636\,\text{mm}^2$

Hence increase the thickness of end plate to 10 mm

Adopt the details as shown in Fig. 5.59.

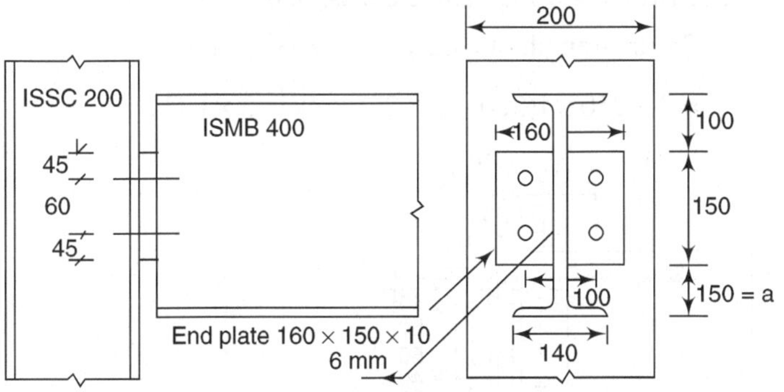


FIG. 5.59

**EXAMPLE 5.19:** Design a web side plate connection (welded to the column and site bolted to the beam) for an ISMB 400 in Fe 410 grade steel and to carry a reaction of 140 kN, due to factored loads. The connection is to the flange of an ISSC 200 column.

**SOLUTION:**

**Step 1:** *Connection to web of beam*

Assume M20 grade 8.8 bolts and a 10-mm thick cleat plate.

Shear capacity of bolt in single shear

$= (245 \times 800)/1.25 \times \sqrt{3} \times 1000) = 90.53\,\text{kN}$

Bearing capacity of connected ply (assuming $k_b = 0.5$)

$= (2.5 \times 0.5 \times 20 \times 8.9 \times 410)/(1.25 \times 1000) = 72.98\,\text{kN}$

Note that as per IS provisions, bearing capacity will be greater than the shear capacity and hence it is not possible to design the bolt to fail by bearing.

Required number of bolts = 140/72.98 = 1.9

Provide three M20 bolts of grade 8.8.

**Step 2:** *Web side plate*

Provide a 100 × 10 mm plate, 300-mm long, with bolt line 60 mm from the face of the column.

**Step 3:** *Check for eccentricity effects due to the offset of beam web*

Minimum thickness = (5 × reaction)/($f_y$ × depth of plate)

= (5 × 140 × 1000)/(250 × 300)

= 9.33 < 10 mm,

hence the chosen thickness is fine.

Moment at the face of column due to the eccentricity of load

= 140 × 60 = 8400 Nm

Moment that could be developed by bolts

$= (2 \times 90.5 \times 100^2)/100 = 18{,}100\,\text{Nm}$

Design for a moment of 18,100 Nm

Moment capacity of plate $= 1.2 \times (250/1.1) \times (10 \times 300^2/6) \times 10^{-3}$

= 40, 909 Nm > 18,100 Nm, hence safe.

Effective length of 6-mm fillet weld = 300 – 2 × 6 = 288 mm

Horizontal shear at the end of weld due to moment $= (18{,}100 \times 6)/(2 \times 288^2) = 0.655\,\text{kN/mm}$

Vertical shear = 140/(2 × 288) = 0.243 kN/mm

Resultant shear $= \sqrt{(0.655^2 + 0.243^2)} = 0.699\,\text{kN/mm}$

Approximate minimum weld size for ductility (Owens & Cheal 1989) = $0.6t/0.7$

= 0.6 × 10/0.7 = 8.57

Use a 8-mm fillet weld, which has a strength of 1.058 kN/mm.

An end clearance of 20 mm is provided so that the fillet weld will have ample clearance (see Fig. 5.60).

Shear capacity of web side plate $= f_y A_v/(1.1 \times \sqrt{3})$

$A_v = t_p\,(l_p - 3d_h) = 10\,(300 - 3 \times 22) = 2340\,\text{mm}^2$

Thus shear capacity = $250 \times 2340/(1.1 \times \sqrt{3}) = 307\,\text{kN} > 140\,\text{kN}$

Hence the shear capacity is adequate.

Richard et al. (1982) have found from experiments that some eccentricity is developed at the connection and should be accounted for in the design of the plate weldment and its supports. They suggested that for beams with uniformly distributed load, the eccentricity could be calculated by using the formula

$$E = nhl/(384d)$$

where $E$ is the eccentricity as measured from the bolt line to the point of inflection in the beam [see Fig. 5.60], $n$ is the number of bolts used, $h$ is the height of the bolt pattern, $l$ is the length of the beam, and $d$ is the depth of the beam. The test results have shown that this formula is valid for bolt pitch up

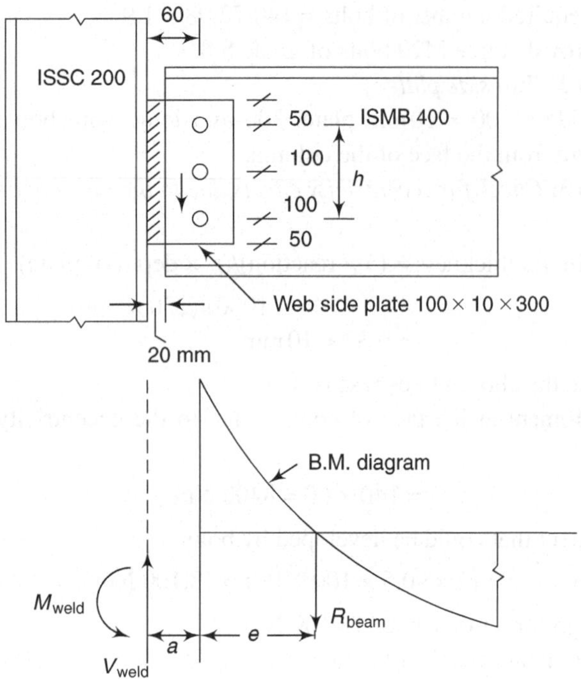


FIG. 5.60

to 150 mm [for concentrated loads the eccentricity coefficients are available in Richard et al. (1980) or may be taken from the point of inflection of the bending moment (BM) diagram].

Assuming a span of beam = 6 m,

$$E = (3 \times 200 \times 6000)/(384 \times 400) = 23.44\,\text{mm}$$

Hence, the moment acting on the weld line

$$M = 140(23.44 + 60) = 11{,}682\ \text{Nm} < 18{,}100\ \text{Nm}$$

(assumed in the calculations)

Hence safe.

**EXAMPLE 5.20:** Design a bolted web cleat connection for an ISMB 600 and two coped beams of size ISMB 400 (300 kN reaction due to factored loads) and ISMB 250 (75 kN reaction due to factored loads) using grade 8.8 bolts of 20 mm diameter (see Fig. 5.61).

**SOLUTION:**

For M20 grade 8.8 bolts:

Shear capacity in single shear = $245 \times 800/(\sqrt{3} \times 1.25 \times 1000) = 90.5\,\text{kN}$

Shear capacity in double shear = $2 \times 90.5 = 181\,\text{kN}$

**Step 1:** *Left hand side beam* ISMB 400

Web thickness = 8.9 mm

Bearing capacity of bolt against the web; with $k_b = 0.6$

$= 2.5 \times 0.6 \times 20 \times 8.9 \times 410/(1.25 \times 1000) = 87.6\,\text{kN}$

Try five bolts at 60 mm spacing and 40-mm edge distance.

Horizontal shear force on bolt due to eccentricity

$= 300 \times (50 + 0.5 \times 12) \times 120/[2\,(60^2 + 120^2)]$

$= 56\,\text{kN}$

Vertical shear force per bolt = 300/5 = 60 kN

Resultant shear force = $\sqrt{(56^2 + 60^2)} = 82.07\,\text{kN}$ > bolt capacity. Hence the connection is safe.

**Step 2:** *Connection to web of supporting beam*

Assuming a cleat of 2L ISA 90 × 90 × 8 mm

Bearing capacity on 8 mm cleat = $2.5 \times 0.6 \times 20 \times 8 \times 410/(1.25 \times 1000) = 78.72\,\text{kN}$

Bearing capacity of 78.72 kN governs

Try eight bolts, two columns at a horizontal gauge length of 120 mm with a vertical pitch of 50 mm for the first three bolts and 140-mm for the last bolt with an edge distance of 40 mm in the cleat. Assuming centre of pressure at 25 mm below top of cleat,

Horizontal shear force on bottom bolt due to moment due to eccentricity

$= [300 \times 0.5(120 - 8.9) \times 255]/[2(15^2 + 65^2 + 115^2 + 255^2)] = 25.69\,\text{kN}$

Vertical shear force per bolt = 300/8 = 37.5 kN

Resultant shear force = $\sqrt{(37.5^2 + 25.69^2)} = 45.46\,\text{kN}$ < bolt capacity

Hence the connection is ok.

Use two 90 × 90 × 8 angle cleats, 320 mm long.

Shear strength of cleats = $f_y A_v/(\sqrt{3} \times 1.1)$

$= 2 \times 250 \times [(320 - 5 \times 22) \times 8]/(\sqrt{3} \times 1.1)$

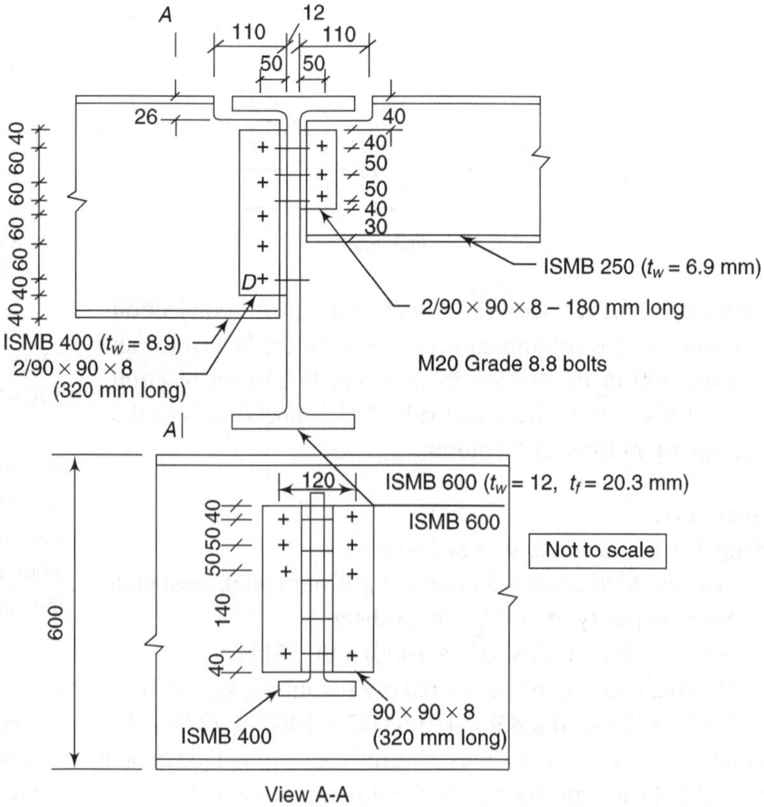


FIG. 5.61

$= 440\,kN > 300\,kN$

**Step 3:** *Right hand side beam*

ISMB 250 with $t_w = 6.9\,mm$

Bearing capacity of web of beam $= 2.5 \times 0.5 \times 20 \times 6.9 \times 410/(1.25 \times 1000) = 56.58\,kN$

Try three bolts at 50-mm vertical pitch and 40-mm edge distance. Horizontal shear force on bolt due to moment (due to eccentricity)

$= [75 \times (50 + 0.5 \times 12) \times 50]/(2 \times 50^2) = 42\,kN$

Vertical shear per bolt $= 75/3 = 25\,kN$

Resultant shear force $= \sqrt{42^2 + 25^2} = 48.9\,kN <$ bolt capacity

**Step 4:** *Connection to web of supporting beam*

As seen in Fig. 5.61, this is connected to the web of supporting beam by six bolts, in two columns at a horizontal gauge of 120 mm c/c and a pitch of 50 mm. Assuming centre of pressure 25 mm below top of cleat, horizontal shear force on bottom bolt due to moment (due to eccentricity)

$= [75 \times 0.5(120 - 6.9) \times 115]/[2(15^2 + 65^2 + 115^2)] = 13.8\,kN$

Vertical shear force per bolt $= 75/6 = 12.5\,kN$

Resultant shear force per bolt $= \sqrt{(13.8^2 + 12.5^2)} = 18.6\,kN$ < bolt capacity in double shear. Hence the connection is safe.

Use two $90 \times 90 \times 8$ angle cleats 180 mm long.

Shear strength of cleats $= 2 \times 250 \times [(180 - 3 \times 22) \times 8]/(\sqrt{3} \times 1.1)$

$= 239\,kN > 75\,kN$

**Step 5:** *Web of supporting beam*

Check for combined load for left and right hand beam.

Bearing capacity on 12 mm web $= 2.5 \times 0.5 \times 20 \times 12 \times 410/(1.25 \times 1000)$

$= 98.4\,kN$

Total horizontal shear force due to moment due to eccentricity (on third row)

$= [(25.69 \times 115/255] + 13.8 = 25.38\,kN$

Total vertical shear force $= 37.5 + 12.5 = 50\,kN$

Resultant shear force $= \sqrt{(50^2 + 25.38^2)} = 56.1\,kN <$ bearing capacity.

Hence, the connection is safe.

**Step 6:** *Left hand beam*

At the end of the notch, depth of section $= 400 - 26 = 374\,mm$

Shear capacity $= (374 \times 8.9) \times (250/\sqrt{3})/(1.1 \times 1000)$
$= 436.76\,kN >$ shear force $= 300\,kN$

**Step 7:** *Shear capacity through bolt holes*

Block shear capacity:

$T_{db1} = [250/(\sqrt{3} \times 1.1) \times 8.9 \times (54 + 4 \times 60) + (0.9 \times 410/1.25) \times 8.9 \times (50 - 22/2)]/1000 = 445.8\,kN$

$T_{db2} = [(0.9 \times 410/(\sqrt{3} \times 1.25) \times 8.9 \times (54 + 4 \times 60 - 4.5 \times 22) + 50 \times 8.9 \times (250/1.10)]/1000 = 396.92\,kN$

Hence,

$T_{db} = 396.92\,kN$

Shear force = 300 kN < shear capacity. Hence the connection is safe.

**Step 8:** *Right hand side beam*

At the end of the notch, depth of section $= 250 - 26 = 224$

Shear capacity $= (224) \times 6.9 \times (250/\sqrt{3})/(1.1 \times 1000)$
$= 202.80\,kN >$ shear force $= 75\,kN$

Block shear capacity of web

$T_{db1} = [250/(\sqrt{3} \times 1.1) \times 6.9 \times (54 + 2 \times 50) + (0.9 \times 410/1.25) \times 6.9\,(50 - 22/2)]/1000 = 218.86\,kN$

$T_{db2} = [0.9 \times 410/(\sqrt{3} \times 1.25) \times 6.9 \times (54 + 2 \times 50 - 2.5 \times 22) + 50 \times 6.9 \times (250/1.1)]/1000 = 194.83\,kN$

$T_{db} = 194.83\,kN$

Shear force = 75 kN < Shear capacity

Hence, the connection is safe.

**Example 5.21:** Design a bolted splice for an ISMB 400 section to transfer a factored bending moment of 120 kNm and a factored shear of 80 kN. Assume that the flange splice carries all the moment and that the web splice carries only the shear (see Fig. 5.62).

**Solution:**

For ISMB 400, $t_f = 16.0\,mm$ and $t_w = 8.9\,mm$

**Step 1:** *Flange splice*

Force in the flanges $= BM/(D - t_f) = 120 \times 10^3/(400 - 16) = 312.5\,kN$

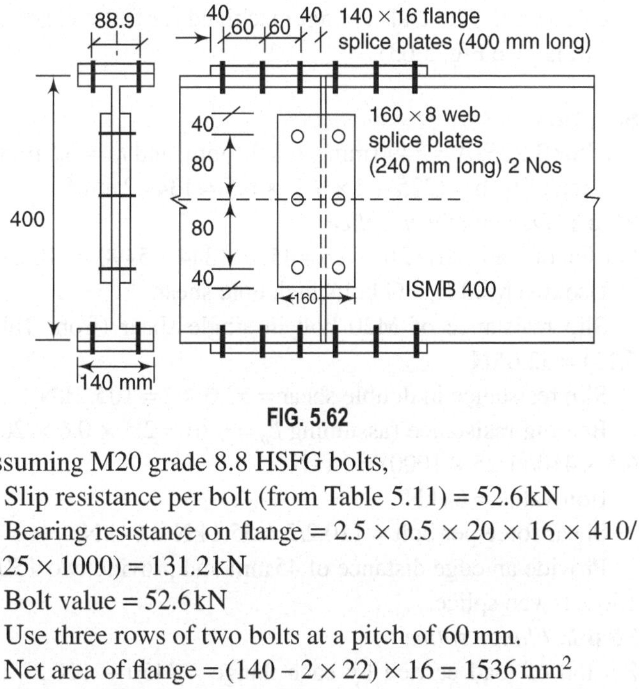


**FIG. 5.62**

Assuming M20 grade 8.8 HSFG bolts,

Slip resistance per bolt (from Table 5.11) = 52.6 kN

Bearing resistance on flange $= 2.5 \times 0.5 \times 20 \times 16 \times 410/(1.25 \times 1000) = 131.2\,kN$

Bolt value = 52.6 kN

Use three rows of two bolts at a pitch of 60 mm.

Net area of flange $= (140 - 2 \times 22) \times 16 = 1536\,mm^2$

Flange capacity $= (250/1.1) \times 1536/1000 = 349\,kN >$ flange force $= 312.5\,kN$

Hence the connection is safe.

Use a 140-mm wide splice plate.

Required thickness of plate = $312.5 \times 10^3/[250(140 - 2 \times 22)/1.10]$

$$= 14.32\,\text{mm}$$

Use two flange splice plates of size 400 × 140 × 16 mm

**Step 2:** *Web splice*

For M20 HSFG bolt in grade 8.8 in double shear

Slip resistance per bolt (from Table 5.11) = 2 × 52.6 = 105.2 kN

Try 8-mm thick web splice plate on both sides of the web.

Bearing resistance per bolt = 2.5 × 0.5 × 20 × 8 × 410/(1.25 × 1000) = 65.6 kN

Bolt value = 65.6 kN

Try three bolts at a vertical pitch of 80 mm and 40 mm from the centre of the joint

Horizontal shear force on bolt due to moment (due to eccentricity)

$$= 80 \times 40 \times 80/(2 \times 80^2) = 20\,\text{kN}$$

Vertical shear force per bolt = 80/3 = 26.67 kN

Resultant shear force = $\sqrt{(26.67^2 + 20^2)}$ = 33.34 kN < 65.6 kN

Use two web splice plates of size 240 × 160 × 8 mm. The designed splice details are shown in Fig. 5.62.

**EXAMPLE 5.22:** Design a bolted cover plate splice for an ISHB 225 column connected to an ISHB 225, to transfer a factored axial load of 450 kN, both columns are of grade Fe 410 steel. The ends are not machined for full contact in bearing (see Fig. 5.63).

**SOLUTION:**

For ISHB 225, $A = 5490\,\text{mm}^2$, $t_f = 9.1$ mm, and $t_w = 6.5$ mm

Area of web = $(225 - 2 \times 9.1) \times 6.5 = 1344.2\,\text{mm}^2$

**Step 1:** *Design of web splice*

Portion of load carried by web = 450 × 1344.2/5490 = 110.2 kN

Use two M20 HSFG bolts in double shear.

Slip resistance of M20 bolt in single shear (from Table 5.11) = 52.6 kN

Slip resistance in double shear = 52.6 × 2 = 105.2 kN

Bearing resistance (assuming $k_b = 0.6$) = 2.5 × 0.6 × 20 × 6.5 × 410/(1.25 × 1000) = 64 kN

Bolt value = 64 kN

Shear force per bolt = 110.2/2 = 55.1 kN < 64 kN

Provide an edge distance of 45 mm and provide two 150 × 150 × 8 web splice.

**Step 2:** *Flange splice*

Portion of load carried by each flange = (450 – 110.2)/2 = 169.9 kN

For four M20 HSFG bolts of grade 8.8 in two rows in single shear.

Total slip resistance = 4 × 52.6 = 210.4 kN

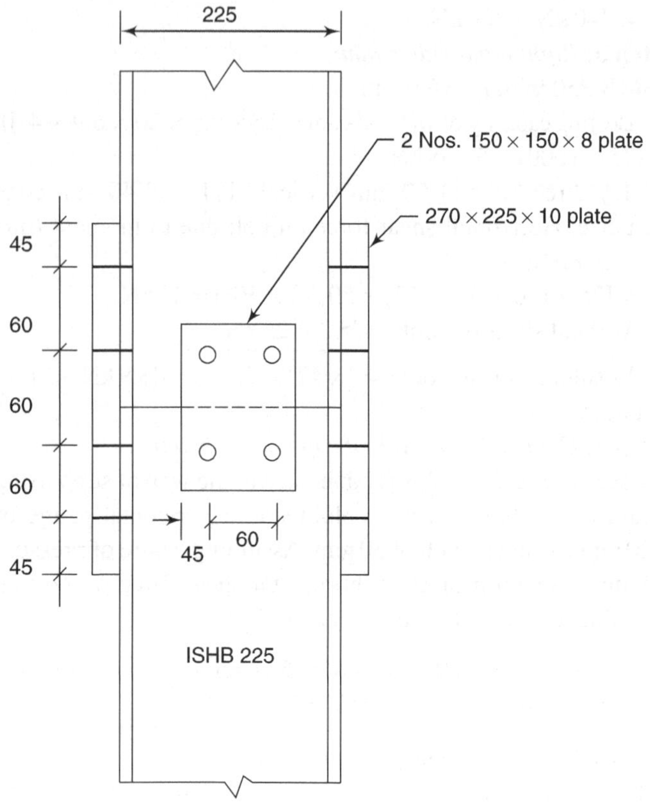


**FIG. 5.63**

Assuming $k_b = 0.5$

Total bearing resistance = 4 × 2.5 × 0.5 × 20 × 9.1 × 410/(1.25 × 1000)

$$= 298.5\,\text{kN}$$

Total bolt strength = 210.4 kN > 169.9 kN

Hence the connection is safe.

Provide an edge distance of 45 mm and a vertical pitch of 60 mm. Provide a 225-mm wide splice plate.

Thickness of flange splice plate = $169.9 \times 10^3/[(250/1.1) \times (225 - 2 \times 22)]$

$$= 4.13\,\text{mm}$$

However, provide 10-mm thick flange plate (thickness should be equal to the flange thickness of the column).

Adopt two splice plate of size 270 × 225 × 10 mm. The designed splice details are shown in Fig. 5.63.

**EXAMPLE 5.23:** Check the ability of the column splice illustrated in Fig. 5.64 to transfer a combination of the following: compression of 400 kN, moment of 100 kNm, and shear of 20 kN. Assume that the splice is designed for direct bearing and that M20 bolts are used. The steel is Fe 410 and the bolts are grade 8.8.

**SOLUTION:**

The bending moment is assumed to be carried by the flange cover plates.

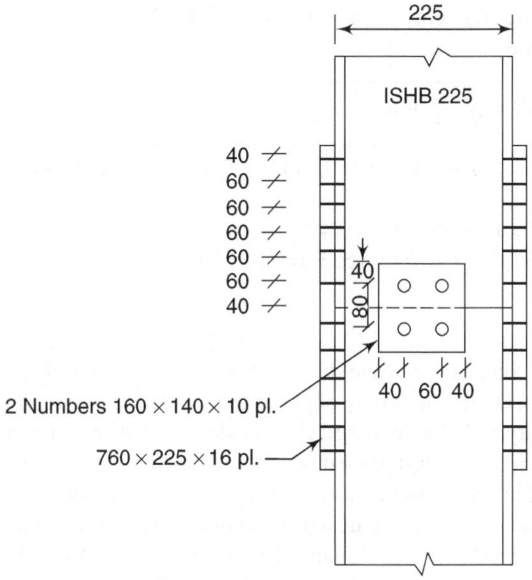


**FIG. 5.64**

The force in the cover plate = ± 400/2 + 100 × 10^3/225
= 644.44 kN and 244.44 kN

Adopt a cover plate of width = 225 mm

Rotation will take place about an axis near the outer edge of the flange of the upper column section (Martin & Purkiss 1996). Thickness of the flange plate, from the moment of forces about the axis of rotation (assuming the width of cover plate as 225 mm)

$$T_p = [M \pm P(D_u/2)]/[(B_p - 2d_b)\,(f_y/\gamma_{m0})D_u]$$

$$= [100 \times 10^6 + 400 \times 10^3\ (225/2)]/[(225 - 2 \times 22) \times (250/1.1)\ 225]$$

$$= 15.67\ \text{mm}$$

Provide a 16-mm thick plate.

Number of M20 bolts (slip resistance $V_{sb}$ for an M20 bolt = 52.6 kN as per Table 5.11)

$$n_b = [M \pm P(D_u/2)]/(V_{sb}D_u)$$
$$= [100 \times 10^6 + 400 \times 10^3\ (225/2)]/(52.6 \times 10^3 \times 225)$$
$$= 12$$

Hence, use 12 bolts in two rows with a vertical pitch of 60 mm and an edge distance of 40 mm. Use a flange plate of size 760 × 225 × 16 mm. Where the ends of the column are machined and in contact, the horizontal shear force on the column will be resisted by the friction force, in part of whole, at the point of contact (at the axis of rotation).

It may be assumed that the machined surfaces will have $\mu = 0.15$. Thus,

$$\text{Frictional resistance} = \mu(M/D_u + P/2)$$
$$= 0.15\ (100 \times 10^3/225 + 400/2)$$
$$= 96.67\ \text{kN} > 20\ \text{kN (applied shear)}$$

Therefore, theoretically no shear connection is required. However, in practice a web plate with nominal bolts is generally provided. Provide a web plate of size 160 × 140 × 10 mm with two bolts of M20 on either side of the splice. (see Fig. 5.64 for the details of the splice connection)

**Example 5.24:** Calculate the strength of the splice shown in Fig. 5.65 with respect to shear and moment. All the bolts are 20 mm diameter HSFG bolts.

**Solution:**

**Step 1:** *Moment connection*

Flange splice plates and bolts resist moment.

Slip resistance of M20 grade 8.8 HSFG bolt (Table 5.11) = 52.6 kN

Moment of resistance = 10 × 52.6 × 550 × 10 $^{-3}$ = 289.3 kNm

**Step 2:** *Shear connection*

Web splice resist all shear, say F kN

Bolt group

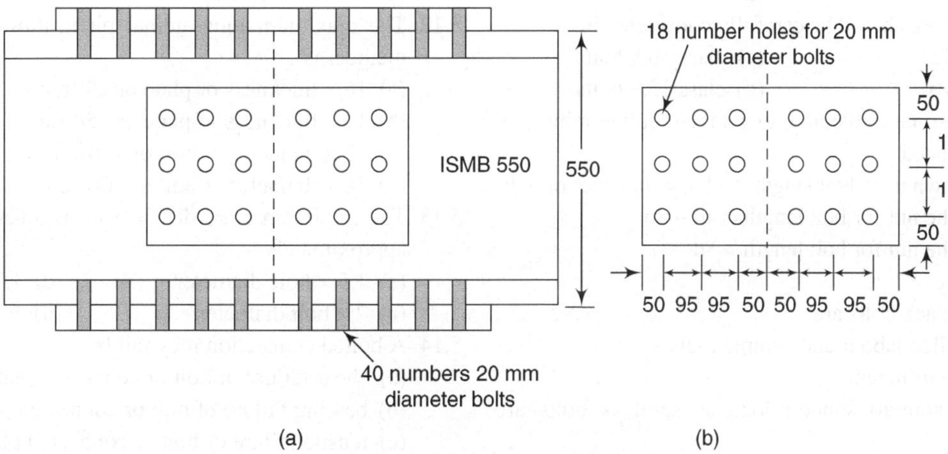


**FIG.5.65** (a) Joint (b) Web splice plate

$$\sum x^2 + \sum y^2 = (6 \times 95^2) + (6 \times 130^2) = 15.5 \times 10^4 \text{ mm}^2$$

$$R_x = \frac{My}{\Sigma x^2 + \Sigma y^2} = \frac{(145F)130}{15.5 \times 10^4} = 0.1216 \text{ F kN}$$

$$R_y = \frac{Mx}{\Sigma x^2 + \Sigma y^2} = \frac{(145F)95}{15.5 \times 10^4} = 0.0889 \text{ F kN}$$

Load due to shear, $R_v = F/9 = 0.1111$ F kN

Resultant,

$$R = \sqrt{(R_v + R_y)^2 + R_x^2}$$

$$= \sqrt{(0.1111 + 0.0889)^2 + 0.1216^2}\ F = 0.234F$$

Equating it to the bolt force, 0.234F = 52.6 kN

Thus, shear resistance F = 224.78 kN

## SUMMARY

Different components in a steel structure such as beams, columns, and tension members have to be connected by means of fasteners, so that all these members will act together to resist the applied forces. Riveted, bolted, and welded connections are normally used in steel structures. Out of these, the riveted connections have become obsolete, due to the difficulties encountered while making these connections. Hence, in this chapter, the various bolts and their tightening methods have been discussed. The design expressions for black and HSFG bolts were presented along with pin connections, which are mainly used in tension members. The various types of simple connections have been covered along with their design methods. Due to the variety of possible connection configurations, the variability of behaviour, and the practical limitations in fabrication and erection, the design methods presented are based on the rational philosophy of simple analysis, using higher load factors of safety, rather than on complex design procedures. Several examples have been worked out to explain the concepts presented. Moment-resistant bolted connections are discussed in Chapter 14. The various types of simple welded connections are discussed in the next chapter.

## MULTIPLE-CHOICE QUESTIONS

5.1 According to IS 800, based on connection rigidity, joints are considered as
(a) rigid (c) semi rigid
(b) pinned (d) all of these

5.2 Rivets are no longer used in steel construction because
(a) they need pre-heating
(b) they require larger crew
(c) of the high level of noise involved
(d) all of these

5.3 In property class 6.8 of bolts the numbers 6 and 8 indicate ultimate tensile stress of __________ and yield stress of __________.
(a) 800 MPa / 600 MPa (c) 800 MPa / 480 MPa
(b) 600 MPa / 480 MPa (d) none of these

5.4 The ratio of net tensile area at threads to nominal shank of the area of bolt is taken in IS 1367 as
(a) 0.78 (c) 0.785
(b) 0.80 (d) 0.85

5.5 Hot dip galvanizing should be carefully considered in
(a) class 4.6 bolts (c) class 10.9 bolts
(b) class 8.8 bolts (d) class 12.9 bolts

5.6 Using turn-of-the-nut tightening, to get the required minimum bolt tension we need
(a) 1/3 turn of the nut for bolt length < 4d (d = diameter of bolt)
(b) 1/2 turn of the nut for bolt length = 4d – 8d
(c) 2/3 turn of the nut for bolt length > 8d
(d) none of these

5.7 Advantages of black bolts are
(a) use of unskilled labour and simple tools
(b) noiseless environment
(c) connection able to support load as soon as bolts are tightened
(d) all of these

5.8 Advantages of HSFG bolts are
(a) they are cheaper than black bolts
(b) do not allow any slip between the connected elements
(c) suitable in connections subjected to fatigue
(d) none of these

5.9 The diameter of bolt hole can exceed the diameter of the bolt by about
(a) 1.5 to 2 mm (c) 1 to 3 mm
(b) 2 to 5 mm (d) > 5 mm

5.10 The minimum pitch allowed in the code is
(a) 1.5 × diameter of bolt (c) 3 × diameter of bolt
(b) 2.5 × diameter of bolt (d) 2.7 × diameter of bolt

5.11 The maximum longitudinal pitch allowed in bolted compression member is
(a) 16 × thickness of plate or 250 mm
(b) 12 × thickness of plate or 200 mm
(c) 12 × diameter of bolt or 200 mm
(d) 16 × diameter of bolt or 300 mm

5.12 The maximum longitudinal pitch allowed in bolted tension member is
(a) 16 × thickness of plate or 200 mm
(b) 12 × thickness of plate or 150 mm
(c) 12 × diameter of bolt or 150 mm
(d) 16 × diameter of bolt or 200 mm

5.13 The minimum edge distance in member with rolled edge is approximately
(a) 1.5 × hole diameter (c) 1.7 × hole diameter
(b) 2 × bolt diameter (d) 1.7 × bolt diameter

5.14 A bolted connection may fail by
(a) shear failure of bolt or connected plate
(b) bearing failure of bolt or connected plate
(c) tensile failure of bolt or connected plate
(d) none of these

5.15 Prying effect will be more pronounced in
(a) thin plates
(b) thick plates
(c) intermediate thickness plate
(d) large diameter bolts

5.16 Tension capacity of connected plates is given by
(a) $T_{nd} = 0.8 f_u A_n / 1.25$ (c) $T_{nd} = 0.9 f_u A_g / 1.25$
(b) $T_{nd} = 0.9 f_y A_g / 1.25$ (d) $T_{nd} = 0.9 f_u A_n / 1.25$

5.17 The design bearing strength of bolt is given by
(a) $2.5\, k_b\, dtf_u / 1.25$ (c) $2.5\, k_b\, dtf_u / 1.10$
(b) $2.0\, k_b\, dtf_u / 1.25$ (d) $2.0\, k_b\, dtf_u / 1.25$

5.18 The design shear capacity of HSFG bolt at ultimate load is given by
(a) $\mu_f n_e K_n F_o / 1.10$ (c) $0.9 f_u A_g / 1.25$
(b) $\mu_f n_e K_n F_o / 1.25$ (d) $0.9 f_y A_g / 1.25$

5.19 The coefficient of friction for surface blasted with grit and painted with alkalizinc silicate of thickness 70 μm can be assumed as
(a) 0.30 (c) 0.50
(b) 0.40 (d) 0.33

5.20 Whitmore suggested an effective area from the outer row of fasteners with
(a) 25° dispersion
(b) 30° dispersion
(c) 35° dispersion
(d) 22.5° dispersion

## EXERCISES

5.1 Design a lap joint between two plates of size 100 × 16 mm thick and 100 × 10 mm thick so as to transmit a factored load of 100 kN using a single row of M16 bolts of grade 4.6 and grade 410 plates.

5.2 Design a hanger joint as shown in Fig. 5.45 to carry a factored load of 200 kN. Use end plate of size 250 mm × 100 mm and appropriate thickness, two M20 HSFG bolts and Fe 410 steel for end plate ($f_y$ = 250 MPa). Assume 8-mm fillet weld between the hanger and the end plate.

5.3 Design a pin for the tension link shown in Fig. 5.46. Use plates of grade Fe 410 ($f_y$ = 250 MPa) and size two 160 × 16 mm on one side and 160 × 32 mm as the other side.

5.4 A member of a truss consists of two angles ISA 65 × 65 × 6 placed back to back. It carries an ultimate tensile load of 125 kN and is connected to a gusset plate 8-mm thick placed in between the two connected legs. Determine the number of 16-mm-diameter grade 4.6 ordinary bolts required for the joint. Assume $f_u$ of plate as 410 MPa.

5.5 The plates of a tank 8-mm thick are connected by a single bolted lap joint with 20-mm-diameter bolts at 50-mm pitch. Calculate the efficiency of the joint. Take $f_u$ of plate as 410 MPa and assume grade 4.6 bolts.

5.6 A single bolted double cover butt joint is used to connect two plates 8-mm thick (see Fig. 5.49). Assuming the bolts of 20-mm diameter at 50-mm pitch, calculate the efficiency of the joint. Use 410 MPa plates and grade 4.6 bolts.

5.7 Two plates 10-mm and 14-mm thick are to be jointed by double cover butt joint. Assuming cover plates of 8-mm thickness, design the joint to transmit factored load of 300 kN. Assume Fe 410 plate and 16-mm-diameter grade 4.6 bolt.

5.8 Design a butt joint to connect two plates 200 × 8 mm (Fe 410 grade) using M16 bolts. Arrange the bolts to give maximum efficiency.

5.9 Determine the adequacy of the fasteners in Fig. 5.54, to carry a factored load of 160 kN when 16-mm-diameter grade 4.6 bolts are used. Assume that the strength of the column flange and the structural T (ST) sections do not govern the design. Neglect prying action.

5.10 Repeat the above exercise assuming the use of M16 parallels-shank HSFG 8.8 grade bolts in clearance holes and a load of 320 kN.

5.11 An ISMB 400 is connected to a column by web cleats with a single row of bolts. If the reaction is 220 kN and there are four 16-mm-diameter bolts through the web, as in Fig. 5.55, check if the section is adequate for block shear failure.

5.12 Design a connection of a truss joint as shown below, using M16 black bolts of property class 4.6 and grade 410 steel. Assume that the members shown are capable of resisting the loads and use 12-mm thick gusset plates.

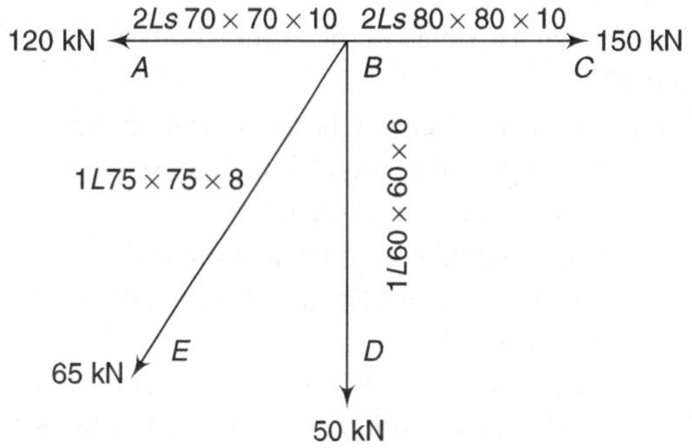


5.13 Design a seat angle connection between a beam MB 250 and column HB 200 for a reaction of beam 90 kN, using M16 bolts of property class 4.6. Take Fe 410 grade steel with $f_y$ = 250 MPa. (see Fig. 5.57 for the arrangement of the connection)

5.14 Design a bolted web angle connection for an ISMB 350 beam, to carry a reaction of 90 kN due to factored loads. The connection is to the flange of a column ISHB 200 in grade Fe 410 steel. (see Fig. 5.58 for the arrangement use *M*16 bolts).

5.15 Design a header plate connection for an ISMB 300 beam to carry a reaction of 90 kN due to factored loads. The connection is to the flange of a ISHB 200 column. Use Fe 410 grade steel ($f_y$ = 250 MPa) and *M*16 bolts of grade 4.6.

5.16 Design a web side plate connection (welded to the column and site bolted to the beam) for ISMB 300 in Fe 410 grade steel and to carry a reaction of 90 kN, due to factored loads. The connection is to the flange of an ISHB 200 column. Use *M*16 bolts.

5.17. Design a bolted web cleat connection for an ISMB 400 and a coped beam of size ISMB 300 (180 kN reaction due to factored loads) using bolts of 16-mm diameter and grade 8.8.

5.18. Design a bolted splice for an ISMB 300 section to transfer a factored bending moment of 100 kNm and a factored shear of 60 kN. Assume that the flange splice carries all the moment and that the web splice carries only the shear. Use M16 bolts.

5.19. Design a bolted cover plate splice for an ISHB 200 column connected to an ISHB 200, to transfer a factored axial load of 370 kN, both columns are of grade Fe 410 steel. The ends are not machined for full contact in bearing.

## REVIEW QUESTIONS

5.1 What are the factors that influence the behaviour of joints or connections?

5.2 What is the basic goal of a connection design?

5.3 How are connections classified?

5.4 Write short notes on rigid, simple, and semi-rigid joints.

5.5 Describe the installation procedure of a rivet.

5.6 What are the reasons for riveting to become obsolete?

5.7 List some of the bolts that are used in structural connections.

5.8 Under what circumstances are black bolts not recommended?

5.9 In class 4.6 bolts, what do the numbers 4 and 6 indicate?

5.10 What are the normal bolt sizes used in structural applications?

5.11 What is the difference between
(a) black bolts and turned bolts (close tolerance bolts)?
(b) black bolts and ribbed bolts?

5.12 What is the main purpose of a washer in HSFG bolts?

5.13 Why are connections using HSFG bolts called slip-critical connections?

5.14 What are the two classes of HSFG bolts that are used in practice?

5.15 What is the main difference between bearing type connection and friction type connection? Which type of connection is preferable under alternating loads?

5.16 Describe the different bolt tightening methods.

5.17 What are the advantages of bolted connections over riveted or welded connections?

5.18 What are the advantages of HSFG bolts over black bolts?

5.19 Why are drilled holes preferable than punched holes?

5.20 Where are slotted holes used?

5.21 What is the difference between a pitch and a staggered pitch?

5.22 Why are minimum pitch values specified in the code?

5.23 List the failure modes that may control the strength of a bolted joint.

5.24 What is the maximum slip, a bolt in normal clearance hole may experience?

5.25 Describe the behaviour of a bolted connection using black bolts under increasing loads.

5.26 What is the major difference in behaviour between a joint with black bolts and a joint with HSFG bolts?

5.27 Write the expression for calculating the nominal capacity of black bolts.

5.28 Write the expression for reduction factors for joints with black bolts and with (a) long joints (b) large grip length (c) thick packing plates

5.29 Write the expression for calculating the nominal capacity of bolt in tension.

5.30 Write the expression for calculating the nominal bearing strength of the bolt.

5.31 Why does the bolt bearing capacity not often control the design?

5.32 What do you mean by prying forces?

5.33 Write the interaction equation specified by the code for black bolts.

5.34 Write the expression for calculating the tensile capacity of plate.

5.35 Define efficiency of a joint.

5.36 Write the empirical expression proposed by Cochrane to calculate the net area of the plate when staggered rows of holes are present.

5.37 Write the expression given in the code for calculating the design slip resistance of HSFG bolts.

5.38 Write the value of coefficient of friction (slip factor) for
(a) surfaces blasted with shot or grit and hot-dip galvanized
(b) sand blasted surfaces
(c) red lead painted surfaces
(d) surfaces blasted with shot or grit and painted with ethylzinc silicate coat of 45 $\mu$m thickness

5.39 What is the expression used to reduce the slip resistance of HSFG bolted joints for long joints?

5.40 What is the expression used to calculate the bearing resistance and tension resistance of HSFG bolted joint?

5.41 Write the expression used in the code for the interaction of tension and shear in HSFG bolted connection.

5.42 Write a short note on pin connection.

5.43 List some of the simple (pinned) connections used in steel structures.

5.44 Write short notes on:
(a) Lap joint (b) Butt joint

5.45 What are the advantages of butt joints over lap joints?

5.46 Describe the Whitmore's (30°) method of designing gusset plate?

5.47 What is the difference between unstiffened and stiffened seat connection?

5.48 Write short notes on:
(a) Unstiffened seat angle connection
(b) Stiffened seat angle connection
(c) Web angle connection
(d) Flexible end plate connection
(e) Web side plate connection

5.49 Write short notes on:
(a) Beam-to-beam connection
(b) Beam splices
(c) Column splices
(d) Semi-rigid connections

5.50 Why are HSFG bolts preferred in joints subjected to fatigue?

# REFERENCES


Agerskov, H. 'High-strength Bolted Connection Subject to Prying', *Journal of Structural Div.,* ASCE, Vol. 102, No. ST1, Jan 1976, pp. 161–175.

Astaneh, A., 'Procedure for Design and Analysis of Hanger Type Connections', *Engineering Journal*, AISC, Vol. 22, Second quarter, 1985, pp. 63–66.

Astaneh, A., Call, S.M., and McMullin, K.M., 'Design of Single Plate Shear Connections', *Engineering Journal*, AISC, First quarter, 1989, pp. 21–32.

Astaneh-Asl, A., 'Seismic Behaviour and Design of Gusset Plates', *Steel Tips*, Structural Steel Educational Council, Moraga, CA, Dec 1998, p. 34.

Astaneh-Asl, A., 'Gusset Plates in Steel Bridges—Design and Evaluation', *Steel Tips*, Structural Steel Educational Council, Moraga, CA, Apr 2010, p. 158.

Babu, S.S. and Sreekumar, S., 'A Study on the Ductility of Bolted Beam-Column Connections', *International Journal of Modern Engineering Research*, Vol. 2, No. 5, Sep–Oct 2012, pp. 3517–3521.

Bahia, C.S. and Martin, L.H., 'Experiments on Stressed and Unstressed Bolt Groups Subject to Torsion and Shear', *Proc. Conference of Joints in Structural Steelwork*, Teeside Polytechnic, UK, 1981.

Bickford, J.H., *An Introduction to the Design and Behavior of Bolted Joints*, Third edition, CRC Press, 1995, p. 992.

Butler, L.J., Pal, S., and Kulak, G.L., 'Eccentrically Loaded Welded Connections', *Journal of the Structural Div.*, ASCE, Vol. 98, No. ST5, May 1972, pp. 989–1005.

Cheng, J.J.R., Grondin, G.Y., and Yam, M.C.H., 'Design and Behaviour of Gusset Plate Connections', *Fourth International Workshop on Connections in Steel Structures,* 22–25 Oct, 2000, Roanoke, VA, USA, pp. 307–317.

Chesson, E. Jr, Faustino, N.L., and Munse, W.H., 'High-strength Bolts Subject to Tension and Shear', *Journal of the Structural Div.*, ASCE, Vol. 91, No. ST5, Oct 1965, pp. 155–180.

Cochrane, V.H., 'Rules for Rivet Hole Deductions in Tension Members', *Engineering News record*, Vol. 89, Nov. 16, 1922, pp. 847–848.

Douty, R.T. and McGuire, W., 'High-strength Bolted Moment Connections', *Journal of the Structural Div.*, ASCE, Vol. 91, No. ST2, Mar–Apr 1965, pp. 101–128.

Gaylord, E.H., Gaylord, C.N., and Stallmeyer, J.E., *Design of Steel Structures*, Third edition, McGraw Hill, New York, 1992.

Holmes, M. and Martin, L.H., *Analysis and Design of Structural Connections*, Ellies Horwood Ltd, 1983.

http://www.exponent.com/kemper_arena/, last accessed on 20 July 2015.

IS:1149-1982 High tensile steel rivet bars for structural purposes (third revision)

IS:1363-1992 Hexagon head bolts, screws and nuts of product grade C:
Part 1- Hexagon head bolts (size range M5 to M64) (third revision),
Part 2 - Hexagon head screws (size range M5 to M64) (third revision),
Part 3- Hexagon nuts (size range M5 to M64) (third revision)

IS:1364-1992 Hexagon head bolts screws and nuts of product grades A and B:
Part 1-Hexagon head bolts (size range M1.6 to M64) (third revision)
Part 2-Hexagon head screws (size range M1.6 to M64) (third revision)
Part 3-Hexagon nuts (size range M1.6 to M64) (third revision)
Part 4-Hexagon thin nuts (chamfered) (size range M1.6 to M64) (third revision)
Part 5-Hexagon thin nuts (unchamfered) (size range M1.6 to M10) (third revision)

IS:1367-1992 (Parts 1 to 18) Technical supply conditions for threaded steel fasteners

IS:1929-1982 Specification for hot forged steel rivets for hot closing (12 to 36mm diameter) (first revision)

IS:2155-1982 Specification for cold forged solid steel rivets for hot closing (6 to 16mm diameter) (first revision)

IS:3640-1982 Specification for hexagon fit bolts (first revision)

IS:3757-1985 Specification for high strength structural bolts (second revision)

IS:4000-1992 Code of practice for high strength bolts in steel structures (first revision)

IS:5369-1975 General requirements for plain washers and lock washers (first revision)

IS:5370-1969 Specification for plain washers with outside diameter 3 × inside diameter

IS:5372-1975 Taper washer for channels (ISMC) (first revision)

IS:5374-1975 Taper washers for I-beams (ISMB) (first revision)

IS:6610-1972 Specifications for heavy washers for steel structures

IS:6623-1985 Specifications for high strength structural nuts (first revision)

IS:6639-1972 Specifications for hexagonal bolts for steel structures

IS:6649-1985 Specification for hardened and tempered washers for high strength structural bolts and nuts (first revision)

Khalil, S.H. and Ho, C.H., 'Black Bolts under Combined Tension and Shear', *The Structural Engineer*, Vol. 57B, Dec 1979, pp. 69–76.

Kulak, G.L., Fisher, J.W., and Struik, J.H.A, *Guide to Design Criteria for Bolted and Riveted Joints*, Second edition, John Wiley, New York, 1987, p. 309.

Kulak,G.L. and Grondin, G.Y., 'Block Shear Failure in Steel Members—A Review of Design Practice', *Fourth International Workshop on Connection in Steel Structures*, Roanoke, VA, 22–25 Oct, 2000, pp. 329–339.

Lehman, D.E., Roeder, C.W., Herman, D., Johnson, S., and Kotulka, B., 'Improved Seismic Performance of Gusset Plate Connections', *Journal of Structural Engineering*, ASCE, Vol. 134, No.6, Jun 2008, pp. 890–901.

McGuire, W., *Steel Structures*, Prentice Hall, Inc., Englewood Cliffs, NJ, 1968, pp. 310–328.

Muir, L.S. and Hewitt, C.M., 'Design of Unstiffened Single-plate Shear Connection, *Engineering Journal*, AISC, Second quarter 2009, pp. 67–79.

Munse, W.H. and Chesson, E.Jr., 'Riveted and Bolted Joints: Net Section Design', *Journal of the Structural Div.*, ASCE, Vol. 89, No. ST1, Jan–Feb 1963, pp. 107–126.

Munse, W.H., Research on bolted connections, *Transactions, ASCE*, Vol. 121, 1956, pp. 1255-1266.

Narayanan, R., et al., *Teaching Material On Structural Steel Design* (45 chapters based on Limit State Design of Steel Structures), INSDAG, New Delhi, 2001, http://www.steel-insdag.org/TM_Contents.asp, last accessed on 20 July 2015.

Owens, G.W., Driver. P.J., and Kriege, G.J., 'Punched Holes in Structural Steel-work', *Journal of Constructional Steel Research*, Vol. 1, No. 3, 1981.

Owens, G.W. and Cheal, B.D., *Structural Steelwork Connections*, Butterworths, London, 1989, p. 330.

Ragupathi, M., *Design of Steel Structures*, Tata-McGraw Hill Publishing Co., New Delhi, 1995, p. 727.

Richard, R.M., Gillett, P.E., Kriegh, J.D., and Lewis, A., 'The Analysis and Design of Single Plate Framing Connections', *Engineering Journal*, AISC, Vol. 17, No. 2, 1980, pp. 38–51.

Richard, R.M., Kriegh, J.D., and Hormby, D.E., 'Design of Single Plate Framing Connections with A307 Bolts', *Engineering Journal*, AISC, Vol. 19, No. 4, 1982, pp. 209–213.

Rumpf, J.L., and Fisher, J.W., 'Calibration of A325 Bolts', *Journal of the Structural Div.*, ASCE, Vol. 89, No. ST6, Dec 1963, pp. 215–234.

Salmon, C.G., Johnson, J.E., and Malhas, F.A., *Steel Structures, Design and Behavior*, Fifth edition, Harper Collins College Publishers, New York, 2008, p. 888.

Struik, J.H.A., Oyeledun, A.O., and Fisher, J.W., 'Bolt Tension Control with a Direct Tension Indicator', *Engineering Journal*, AISC, Vol. 10, No. 1 (First quarter), pp. 1–5.

Subramanian, N. and Petersen, C., 'Behaviour of HSFG Bolts in Chimney Joints', *Third International Conference on Steel Structures*, Singapore, Mar 1987.

Subramanian, N., *Versuche an geschraubten Flanchverbindungen in L-Form*, Unpublished report, Lehrstuhl fuer Stahlbau, Der Universitaet Bundeswehr, Munich, Germany, 1984.

Tamboli, A.R. *Handbook of Structural Steel Connection Design and Details*, Second edition, McGraw Hill, New York, 2009, p. 656.

The SKF Group, *Bolt-tightening Handbook*, Catalogue No. TSI 1101 AE, Apr 2001, France, p.44.

Thornton, W.A., 'Prying Action—A General Treatment', *Engineering Journal*, AISC, Vol. 22, Second quarter, 1985, pp. 67–75.

Thornton, W.A., Strength and Serviceability of Hanger Connections, *Engineering Journal*, AISC, Vol. 29, Fourth quarter, 1992, pp. 145–149.

Trahair, N.S., Bradford, M.A., and Nethercot, D.A., *The Behaviour and Design of Steel Structures to BS 5950*, Third edition, Spon Press, London and New York, 2001, p. 448.

Whitmore, R.E., *Experimental Investigation of Stresses in Gusset Plates*, Bulletin No. 16, Engineering Experiment Station, University of Tennessee, USA, 1952.

# 6

# WELDED CONNECTIONS

## INTRODUCTION

*Welding* is a method of connecting two pieces of metal by heating to a plastic or fluid state (with or without pressure), so that fusion occurs. Welding is one of the oldest and reliable methods of jointing. Little progress in welding technology, as is known now, was made until 1877, though welding processes such as forge welding and brazing were known for at least 3000 years. Although there had been initial work on arc welding in the 1700s using carbon electrodes powered by batteries, development work intensified between 1880 and 1900 with the availability of electric generators to replace batteries. Professor Elihu Thompson was the first to patent the first resistance groove welding machine in 1885. Charles Coffin invented the metal arc process and patented it in USA in 1892, though Zerner introduced the carbon arc welding process in 1885. The concept of coated metal electrodes, which eliminated the problems associated with the use of bare electrodes, was introduced in 1889 by A.P. Strohmeyer. The metal arc process was first used (in 1889) in Russia by using uncoated, bare electrodes (Salmon & Johnson 1996).

Oxyacetylene welding and cutting was employed after 1903, due to the development of acetylene torches by Fouche and Picard. By the early 1900s, Lincoln Electric offered the first arc welding machine and by 1912 covered electrodes were patented. During World War I (1914–1918), welding techniques were used for repairing damaged ships. During the period 1930–1950, several improvements and techniques such as the use of granular flux to protect the weld and submerged arc welding were developed.

Today there are several welding processes available to join various metals and their alloys. Those processes, which will be used in welding structural steel, are discussed in the next section. The types of welds and welded joints and the advantages of using welding over bolts or rivets are also discussed. The behaviour and design of various simple welded connections are also outlined. A brief review of the methods of joining tubular connections is given. Several examples are given to illustrate the design procedures adopted for welded connections. This chapter concludes with the joints made using adhesives.

## 6.1 WELDING PROCESSES

Structural welding is nearly all electric; though some gas welding may also be used ['gas' denotes the use of a gas (usually acetylene/oxygen mixture) to produce a very hot flame to heat the parts and the weld filler material]. However, gas is used primarily for cutting pieces to shape. It is now possible to cut metals using mechanically controlled gas cutting equipment in fabrication shop, which results in smooth cuts similar to sawed cuts. Though gas welding is simple and inexpensive, it is slow and hence it is generally used for repair and maintenance work only.

In the most common welding processes of welding structural steel, electric energy is used as the heat source. Electric welding involves passing either direct or alternating current (mostly direct current is used) through an electrode (commonly the electrode is the anode and the operation uses 'reversed polarity'). By holding the electrode at a very short distance from the base metal, which is connected to one side of the circuit, an arc forms as the circuit is essentially 'shorted' (see Fig. 6.1). With this 'shorting' of the circuit, a very large current flow takes place, which melts the electrode's tip (at the arc) and the base metal in the vicinity of the arc. A temperature of about 3300–5000°C is produced in the arc. The electron flow making the circuit 'carries the molten electrode metal' to the base metal to build up the joint. The parameters that control the quality of weld are the electrode size and the current that produces sufficient heat to melt the base metal and minimizes electrode splatter.

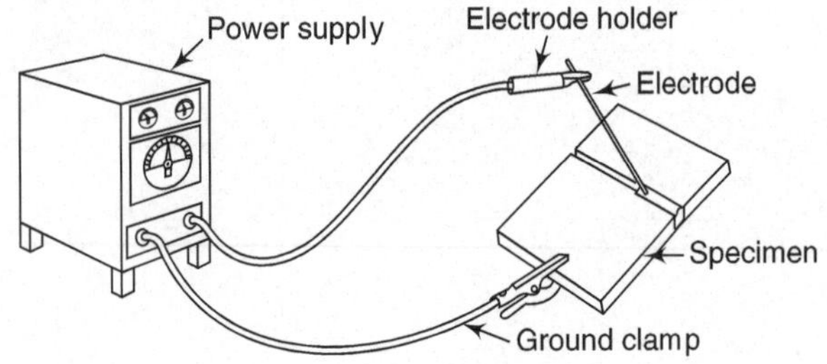


**FIG. 6.1** Shielded metal arc welding outfit

The different processes of arc welding that are used in structural steel applications are as follows:

1. Shielded metal arc welding (SMAW)
2. Submerged arc welding (SAW)
3. Gas-shielded metal arc welding (GMAW)
4. Flux core arc welding (FCAW)
5. Electroslag welding (ESW)
6. Stud welding (SW)

These processes are discussed briefly in the following sections.

### 6.1.1 Shielded Metal Arc Welding

*Shielded metal arc welding* (*SMAW*) also called 'stick' welding is a manual process and is the most common method of welding used in structural connections owing to low capital cost and flexibility. However, for long continuous welds automatic processes are preferred due to the consistent quality. The SMAW processes require the following set up (see Fig. 6.1):

1. Constant-current (CC) welding power supply
2. Electrode holder, lead, and its terminals
3. Ground clamp, lead, and its terminals
4. Welding electrodes

As stated earlier, the electrons flowing through the gap between the electrode and the metal produce an arc that furnishes the heat to melt both the electrode metal and the base metal. Temperatures within the arc exceed 3300°C. The arc heats both the electrode and the metal beneath it. Tiny globules of metal form at the tip of the electrode and are transferred to the molten weld pool on the base metal. As the electrode moves away from the molten pool, the molten mixture of electrode and base metal solidifies and the weld is completed (see Fig. 6.2).

Generally the electrode is stronger than the parent metal. For example, an E41 electrode, which would be used to weld grade 410 steel, gives a weld deposit which has a maximum yield strength of 330 MPa with a tensile strength in the range of 410–510 MPa (see Table 6.1). For *manual metal arc welding (MMA)*, the electrodes should comply with IS 2879, IS 1395, and IS 814. The electrode is coated with a flux. Heat from the electric current causes the combustion and

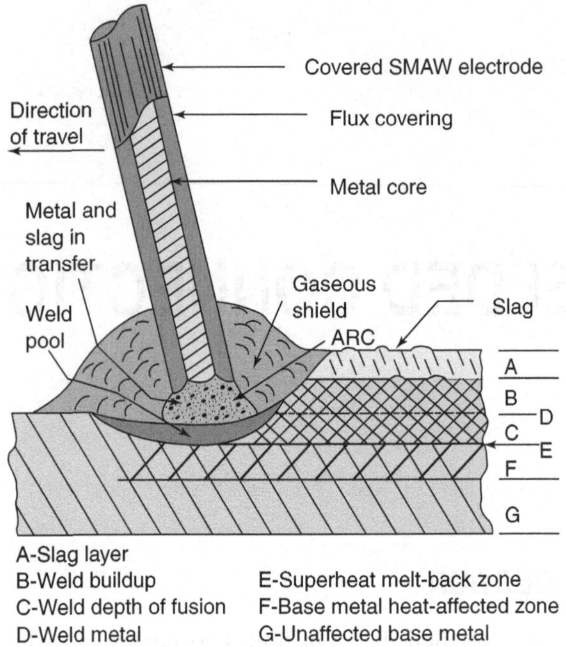


**FIG. 6.2** Shielded metal arc process

*Source:* AWS Welding Handbook, 2001. Reprinted with permission from © American Welding Society.

decomposition of the flux. This creates a gaseous shield to protect the electrode tip, the metal, and the molten pool from atmospheric contamination due to oxidation. The flux contains materials that coat the molten steel droplets as they transfer to the weld and become slag after cooling. This slag being lighter than the metal floats on the weld puddle's surface, which can be brushed away after the weld is solidified.

The flux coating on some electrodes contains a metal powder to provide additional heat and filler to increase the

**TABLE 6.1** Tensile properties of electrodes

| Specification | Grade/ Classification | Yield stress, MPa (Min) | Properties | |
|---|---|---|---|---|
| | | | Ultimate tensile stress, MPa, (Min) | Elongation percentage (Min) |
| IS 814 : 2004 | Ex40xx | 330 | 410–510 | 16 |
| | Ex41xx | 330 | 410–510 | 20 |
| | Ex42xx | 330 | 410–510 | 22 |
| | Ex43xx | 330 | 410–510 | 24 |
| | Ex44xx | 330 | 410–510 | 24 |
| Specification for covered electrodes for manual metal arc welding of carbon and carbon manganese steel | Ex50xx | 360 | 510–610 | 16 |
| | Ex51xx | 360 | 510–610 | 18 |
| | Ex52xx | 360 | 510–610 | 18 |
| | Ex53xx | 360 | 510–610 | 20 |
| | Ex54xx | 360 | 510–610 | 20 |
| | Ex55xx | 360 | 510–610 | 20 |
| | Ex56xx | 360 | 510–610 | 20 |

deposition rate (Galvery & Marlow 2001). The electrode flux and metal filler electrode determine the chemical, electrical, mechanical, and metallurgical properties of the weld as well as the electrode handling characteristics. On subsequent passes (necessary to weld thick parts), it is necessary to brush the earlier passes to remove any slag, dirt, or other foreign material, whose presence might cause a flaw in the weld. The electrodes are available in lengths of 225–450 mm, and diameters ranging from 3.2 to 6 mm. The maximum size of weld produced in one pass is about 8 mm (Bowles 1980).

SMAW can be both a low hydrogen and non-low hydrogen process (hydrogen causes the weld to crack), depending on the type of electrode used. The low hydrogen electrodes are identified by either 5, 6, or 8 as the last number in their designation. All others are non-low hydrogen (see Section 6.2). The strength of electrodes can be in the range of 415–760 MPa (Julicher 1981). Low hydrogen electrodes can be used for any structural steel application including welding of concrete reinforcing steel. The use of non-low hydrogen electrodes are limited to lower strength steels (Julicher 1981).

Using SMAW, welds from 3 mm to unlimited thicknesses are possible. Thicknesses less than 3 mm can be joined but require much greater skill, while those over 18 mm are more economically done by other methods like FCAW.

The advantages of SMAW process are (Galvery & Marlow 2001):

1. Low cost equipment
2. Welds many different metals including the most commonly used metals and alloys
3. Relatively portable and useful in confined spaces
4. Welds can be performed in any position
5. Process is less affected by wind and drafts than gas-shielded processes
6. No upper limit on the thickness of the metal to be welded
7. Can be performed under most weather conditions

The disadvantages of SMAW process are as follows:

1. Not suitable for metal sheets under 1.5-mm thickness
2. Operator duty cycle and overall deposition rate are usually lower with SMAW than with wire-fed processes, since the SMAW process must be stopped when the electrode is consumed and needs to be changed
3. The entire electrode cannot be used; the remaining stub in the electrode handle must be discarded wasting 25 mm to 50 mm of electrode
4. Frequent stops and starts during electrode changes can increase the chances of weld defects

SMAW is not permitted in rain, snow, blowing sand, and when the base metal is below –18°C. This is the most commonly used field method using DC welding machine.

### 6.1.2 Submerged Arc Welding

*Submerged arc welding* (SAW) also has a great range of electrode strengths and is a low hydrogen process. Submerged arc welding uses the arc from a consumable electrode while the arc and base metal are buried in a granular flux. The power supply may be either constant-current or constant-voltage. There are hand-held semi-automatic SAW machines, but the typical SAW process uses a motor-driven carriage. The machine automatically strikes an arc, feeds the electrode from large coils, distributes the flux ahead of the weld area, and also has provisions to gather and reuse the unmelted flux (see Fig. 6.3). SAW typically uses 2- to 6-mm diameter electrodes with a current of about 400–1000 amperes per electrode. Two electrodes may be used in side-by-side vertical, leading-and-following vertical or V-shape across the line of weld making the process very fast (Galvery & Marlow 2001). Some machines use three electrodes. Single pass welds on butt joints up to 75-mm thick are possible. Some flux granules melt in the arc forming an easily removable slag, which protects the weld from the atmosphere and slows cooling. The slag also shields the operator from the arc. The slag formed on the welds is removed by brushing. SAW is usually done in the flat welding position. Deposition rates approaching 90 kg/h are possible with two-electrode automatic machines (Galvery & Marlow 2001). This process is ideal for ship and bridge building where long straight welds are common.

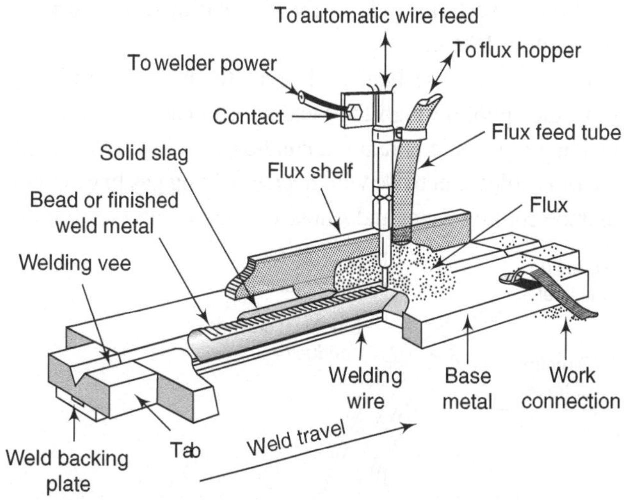


**FIG. 6.3** Submerged arc welding

*Source:* Galvery and Marlow (2001), Reprinted with permission from Industrial Press, Inc., NY, USA.

SAW offers the following advantages:

1. High deposition rate
2. High quality welds with good ductility, high impact strength, high density, and good corrosion resistance
3. Deeper penetration than SMAW process (backing strip is usually used for groove welds)
4. Mechanized process, but some equipment is hand-held

5. Good for welding long joints (in excess of 1 m)
6. High operator appeal as no eye protection needed

However, SAW has the following drawbacks:

1. Flat or horizontal fillets only
2. Long set-up time
3. Limited portability of equipment
4. Equipment must be positioned
5. More difficult to monitor the arc, which is out of view
6. Slag removal needed

Due to these limitations, SAW is usually restricted to shop welding.

### 6.1.3 Gas-shielded Metal Arc Welding

*Gas-shielded metal arc welding* (GMAW) is also referred to as *metal-active gas welding* (MAG) or *metal-inert gas* (MIG) welding. GMAW is also a low hydrogen and manually operated process with a great range of electrode strengths. In this process, the welder positions the electrode wire coming from the center of the contact tip to where the weld is to begin, actually touching the welding wire to the base metal. When the welder squeezes the trigger on the welding gun, the following three events happen simultaneously (Jeffus 2002).

1. The trigger turns on the welding power supply and the electrode wire strikes an arc. In the flat position currents up to 400 A (spray transfer) can be used to give high welding speeds. In vertical and overhead positions, the current is kept below 180 A.
2. The wire feed mechanism begins feeding electrode wire from the spool through the welding gun cable. Note that the welding gun cable is really a bundle of cables: power cables, control cables, electrode wire liner, shielding gas line, and may include cooling water and smoke extractor lines (see Fig. 6.4).

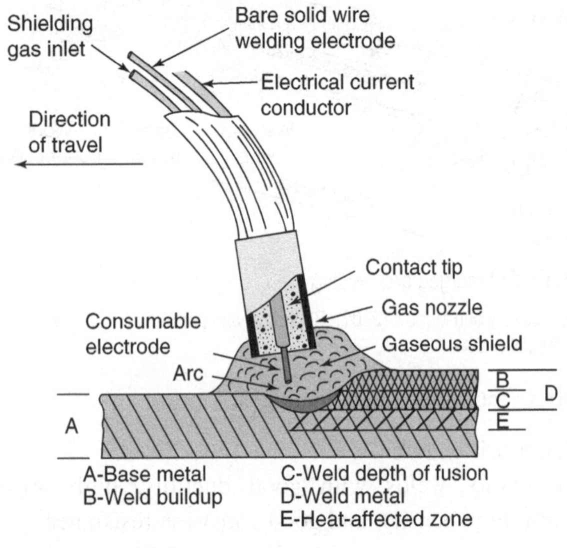


**FIG. 6.4** GMAW process

*Source:* AWS Welding Handbook, 2001. Reprinted with permission from © American Welding Society.

3. An electrically operated valve (solenoid) opens and feeds shielding gas from the regulator/flow-meter to surround the electrode and weld pool, shielding them from the atmosphere, particularly oxygen and nitrogen. The shielding gas may be carbon dioxide or a mixture of argon and carbon dioxide.

Welding begins as the section of electrode wire between the tip and the base metal is heated and deposited into the weld. As the wire is consumed, the feed mechanism supplies more electrode wire at a proper rate to maintain a steady arc. The welder manipulates the gun and lays down the weld in the desired pattern. To stop the process when the weld is completed, the welder releases the trigger shutting off the welding current, wire feed, and shielding gas. GMAW can weld metals such as aluminium, carbon steel, copper, low-alloy steel, magnesium, nickel, stainless steel, and titanium. It is mainly used for fillet welded joints (e.g., beam-to-column and stiffener-to-panel connections). Bare wires of diameter 0.9–1.6 mm or flux core wires of diameter 2.4–3.2 mm can be used in this process.

Some of the advantages of GMAW are as follows (Galvery & Marlow 2001).

1. GMAW is the only consumable process that can weld all commercial alloys.
2. GMAW, with its continuous wire electrode, overcomes the start-and-stop cycle of SMAW and leads to fewer discontinuities and higher deposition rates.
3. Since there is no need to stop welding to change consumed electrodes, long continuous welds can be done manually.
4. GMAW can easily be adapted to fully automate the welding processes.
5. All welding positions can be used.
6. There is significantly higher utilization of filler metal than SMAW. There is no end loss as with upper, unconsumed end of an SMAW electrode.
7. Deeper penetration than SMAW process may permit smaller fillets of the same strength.
8. Metals as thin as 0.58 mm (24 gauge) may be welded.
9. Easier to learn than most other welding processes.
10. There are no practical thickness limitations.
11. There is very little spatter and no slag with properly adjusted equipment and many manufacturers paint or plate over GMAW welds with little or no additional surface preparation.

The disadvantages of GMAW are as follows (Galvery & Marlow 2001).

1. Welding equipment used in this process can be more expensive, more complicated, and slightly less portable than SMAW.
2. GMAW is more difficult to use in tight quarters as the gun is large and the gun cable is somewhat stiff and inflexible.

## CASE STUDY

### Space Needle, Seattle

The Space Needle is an observation tower in Seattle, Washington, USA. It was built during the 1962 World's Fair, and is 184 m tall, 42 m wide, and weighs 9550 tons. It is built to withstand winds of up to 89 m/s (322 km/h) and earthquakes of up to 9.1 magnitude (in 2001 an earthquake of magnitude 6.8 jolted the Space Needle). It also has 25 lightning rods. The Space Needle sways only 16 mm per 10 km/h of wind speed. Unlike other similar structures, such as the CN Tower in Toronto, the Space Needle is not used for broadcasting purposes. It has an observation deck at 160 m and the rotating SkyCity restaurant at 150 m. The architects are John Graham & Company and the structural engineers are John K. Minasian and Victor Steinbrueck.

The structure is bolted to the foundation with 72 bolts, each 9.1 m long. Automatic submerged-arc welding was used extensively while fabricating the unique and complex built-up columns in the shop. Approximately 50% of all shop welding was with the submerged-arc process; 25% with self shielding cored wire, semi-automatically; and the remainder manual stick electrode (Blodgett 1966). The structure required 3400 tons of structural steel.

Two views of Space Needle, Seattle, USA

3. The large size of the GMAW gun combined with the 12- to 25-mm length of the electrode wire that sticks out, makes it harder to see the arc and achieve quality welds.
4. GMAW outdoor use is limited to very calm days or to where shielding screens can be used to prevent the shielding gas from being blown away.
5. GMAW cannot be performed outdoors in greater than 8 km/hr breeze.

### 6.1.4 Flux Core Arc Welding

The structure and chemical composition of the flux core arc welding (FCAW) electrode wire is the differentiating factor between the GMAW and FCAW processes. Unlike GMAW, FCAW electrode wire is not solid and consists of a thin-walled metal tube filled with flux. The powered flux provides alloying elements, arc stabilizers, deoxidizers, slag formers, and shielding gas generating chemicals. The volume and forcefulness of the shielding gas eliminates the need for external shielding gas. The slag produced also shields the weld pool from atmospheric oxygen and nitrogen and retards its cooling rate to reduce martensite formation. Working in tandem, the shielding gas and slag allow FCAW to be used successfully in field conditions (see Fig. 6.5). Most of the time no physical shielding is needed to protect the weld from the wind. FCAW is a low hydrogen process with a great range of electrode strengths and can be used in all positions.

Most FCAW is semi-automatic, that is, the welder manually positions the torch and the wire feeder continuously feeds the electrode wire into the weld. However, the process has been made fully automatic with a computer-driven robot guiding the torch along a preset path. Very complex parts can be made in this way.

The advantages of FCAW are as follows (Galvery & Marlow 2001).

1. Owing to its excellent penetration, FCAW can use groove angles as narrow as 30°, saving as much as 50% of the filler metal needed for SMAW (which requires larger groove angles) and in some cases can avoid joint beveling in metal up to 13 mm thick.
2. Different thickness material can be welded with the same electrode by power supply adjustment.
3. FCAW is easier to carry out than SMAW.
4. It has excellent weld pool control.

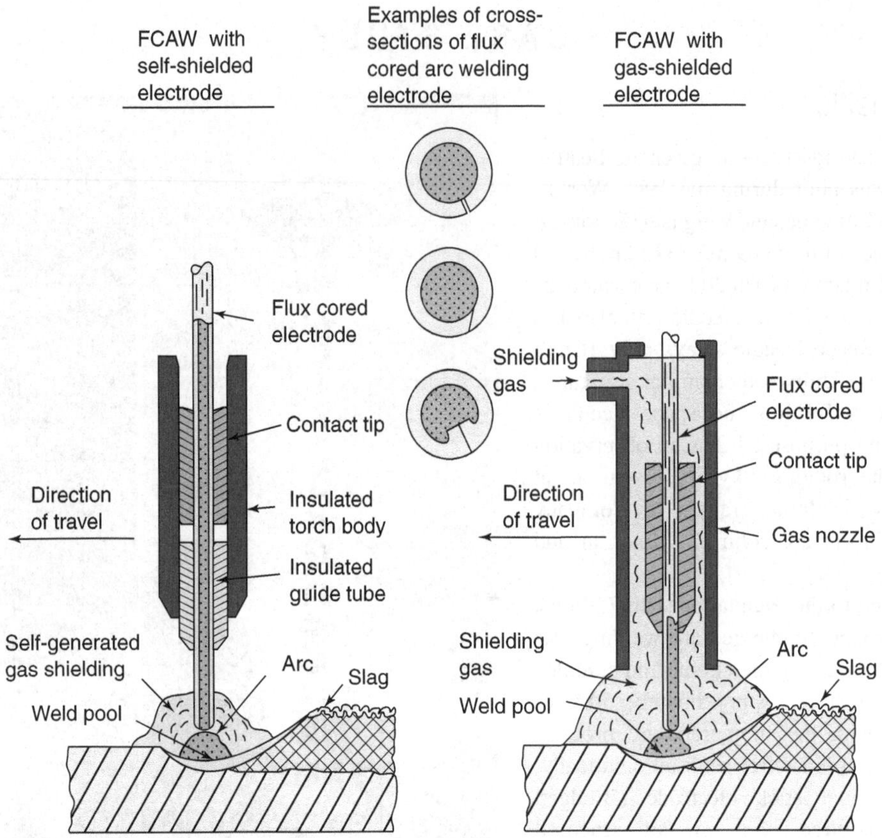


**FIG. 6.5** FCAW process

*Source:* AWS Welding Handbook, 2001. Reprinted with permission from © American Welding Society.

5. FCAW has better welder visibility of the weld pool than GMAW because the gas diffuser nozzle is not needed and can be removed.
6. FCAW oxidizers and fluxing agents permit excellent quality welds to be made on metals having some surface oxides and mill scale. Often metal from flame cutting can be welded without further preparation, thus resulting in considerable saving.
7. High deposition rates of more than 11.3 kg/h are possible compared to 4.5 kg/h for SMAW with 6-mm diameter electrodes.
8. Self-shielded FCAW electrodes work better in windy field conditions than GMAW electrodes.
9. There is no stub loss as in SMAW, where the average stub loss is 11% of the electrode; in FCAW nearly all the electrode is used.
10. Unlimited thicknesses can be joined with multiple passes.
11. Welds in all positions can be made.
12. Wire feeder and welder must be close to the point of welding.

However FCAW has the following drawbacks.

1. Slag removal requires an additional production step, similar to SMAW.
2. FCAW generates large volumes of fumes and smoke, requiring additional ventilation indoors and reducing welder visibility during the process.

## 6.1.5 Electroslag Welding

*Electroslag welding (ESW)* is not really an arc welding process since an arc is only used to start it (Galvery & Marlow 2001; Jeffus 2002). The process begins by striking an arc between the wire electrodes and the bottom of the joint. Powered flux, added to the joint, melts from the arc heat into a liquid slag floating above the arc. Once there is adequate slag, the arc is stopped and the process continues with the heat generated by the flow of electrode current through the molten slag. Wire feeders add metal to the weld pool and an oscillator mechanism moves the electrodes over the joint to provide an even filler metal distribution. Water-cooled copper shoes on the joint faces contain the slag and the weld metal (see Fig. 6.6). They move up the side of the joint as the weld progresses. This process was developed to join thick sections. Plates with thicknesses from 25–450 mm can be welded in one pass (Bowles 1980). Multiple solid or flux-cored electrodes may be used to speed up the process. Excellent penetration is achieved with extreme heat from the slag.

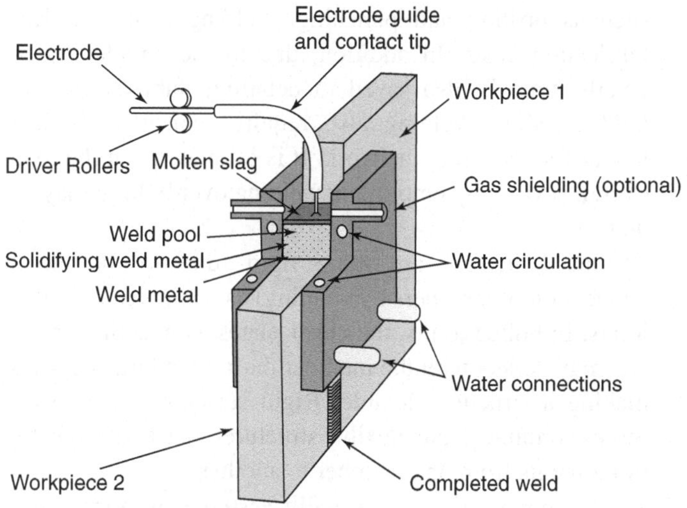


**FIG. 6.6** Electrogas welding

*Source:* AWS Welding Handbook, 2001. Reprinted with permission from © American Welding Society.

The advantages of ESW are as follows:

1. Joins thick materials
2. Has high deposition rate
3. Multiple electrodes may be used
4. Minimum joint preparation
5. Low distortion

The drawbacks of ESW are as follows:

1. Welds flat or vertical joints only
2. Very complicated set-up
3. Cooling water is needed

Electroslag welding is used to shop weld components (in a vertical position) used in bridge work such as girder and floor plates. A major difference between electrogas welding (EGW), which is shown in Fig. 6.6, and its cousin electroslag welding (ESW) is that the arc in EGW is not extinguished, instead it remains struck throughout the welding process.

### 6.1.6 Stud Welding

*Arc stud welding (SW)* permits rapid attachment of studs (shear connectors used in composite construction) and other fasteners to steel beams without piercing the structure metal. The power source may be DC constant-current supply or a capacitor-discharge supply. The SW process begins by inserting a stud into the stud gun, placing a ceramic cup or ferrule on the end of the stud, and positioning the gun perpendicular to the surface of the structure (see Fig. 6.7). This break-away ceramic cup helps contain the arc heat and shields the weld metal from the atmosphere. When the gun trigger is squeezed, an arc forms between the stud and the structure metal. The control electronics determines how long this arc has to run to ensure molten metal on both the stud and the structure. It then releases the stud using spring pressure in the gun to force the molten surfaces together. When stud welding in aluminium, a shielding gas is needed.

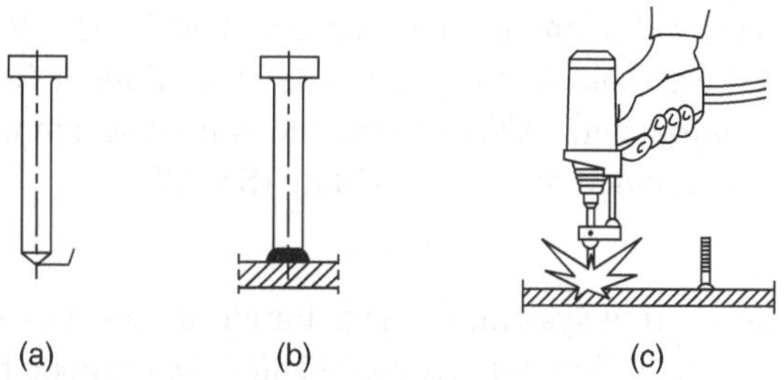


**FIG. 6.7** Stud welding

The advantages of SW are as follows:

1. Simple
2. Fast
3. Can be automated

The drawbacks of SW are as follows:

1. Needs clean surfaces
2. Equipment sensitive to adjustment

### 6.1.7 Choice of Welding Process

One of the aforementioned processes is selected for a particular application, based on the following parameters.

***Location of the welding operation*** If welding is done in a fabrication shop, SAW, GMAW, FCAW, and ESW can be used. For field applications SMAW is preferred.

***Accuracy of setting up*** SAW, spray transfer GMAW, and ESW require accurate set-up.

***Penetration of weld*** Penetration of FCAW and SAW is better than SMAW.

***Volume of weld to be deposited*** FCAW, GMAW, and ESW have high deposition rates.

***Position of welding*** SAW and ESW are not suitable for overhead positions. FCAW and GMAW can be used in all positions. SMAW is probably the best for overhead works, especially at site.

***Access to joint*** In easily accessible joints SAW and GMAW are used. In cramped joints SMAW is used.

***Steel composition*** GMAW and SAW are less likely to lead to heat-affected zone (HAZ) cracking.

***Thickness of connecting parts***

***Comparative cost***

## 6.2 WELDING ELECTRODES

As stated earlier, a variety of *electrodes* are available so that a proper match of base metal strength and metallurgical properties to the weld metal can be chosen. As discussed already, if the molten metal is not shielded properly it will absorb hydrogen and other gases from the atmosphere, and due to oxidation will result in weld defects such as porosity, brittleness, and susceptibility to corrosion. Hence, only coated electrodes are used in structural welding. Welding electrodes are classified (by the American Welding Society, in cooperation with ANSI) using the following numbering system for shielded metal arc welding (SMAW):

Exxxbc

In this numbering system, E stands for electrodes. xxx stands for two or three-digit number establishing the ultimate tensile strength of the weld metal. As per IS 814 the following values are available: 40, 41, 42, 43, 44, 50, 51, 52, 53, 54, 55, 56 kg/cm$^2$. The value of b indicates the suitability of welding positions, which may be flat, horizontal, vertical, and overhead, that is, b = 1 denotes suitability for all positions, 2 denotes suitability for flat positioning of work, 4 denotes suitability for flat, horizontal, overhead, and vertical down. 'c' stands for coating and operating characteristics. The value of c equal to 5, 6, 8 indicates low hydrogen.

The various grades of electrodes as per IS 814 and their tensile properties are shown in Table 6.1. Electrode classification for other processes is provided by Tamboli (1999). Further information on compositions and choice of electrodes is available in Gourd (1980) and Owens and Cheal (1989).

## 6.3 ADVANTAGES AND DISADVANTAGES OF WELDING

Welding offers the following advantages over bolting or riveting.

1. Welded connections eliminate the need for making holes in the members, except for a few employed for erection purposes. Since the holes at the ends govern the design of bolted connections (edge distance), a welded connection results in a member with a smaller gross section. This has a greater influence in the case of tension members, since the calculation of net section is eliminated.
2. Welding offers airtight and watertight jointing of plates and hence is employed in the construction of water/oil storage tanks, ships, etc.
3. Welded joints are economical, since they enable direct transfer of stresses between the members. Moreover, the splice plates and bolt material are eliminated. The required size of gusset plates is also smaller, because of reduced connection length. Due to the elimination of operations such as drilling and punching, welding results in less fabrication costs. In addition, due to the simple design details, time is also saved in detailing, fabrication, and field erection. Welding also requires considerably less labour for executing the work. It is estimated that the total overall savings by employing welding over bolting may be up to 15%.
4. Welded structures are more rigid (due to the direct connection of members by welding) as compared to bolted joints. In bolted joints, the cover plates, connecting angle, etc. may deflect with the member during load transfer thus making a structure flexible. Rigid structures are always more economical than flexible structures, due to the transfer of moments from one member to another.
5. Welded connections are usually aesthetic in appearance and appear less cluttered in contrast to bolted connections. This is evident from Fig. 6.8, which shows a bolted and welded plate girder.
6. Welding offers more freedom to the designer in choosing sections. The designer is not bound by the available rolled sections, but may build up any cross section, which may be economical and advantageous. Welding has resulted in the innovation of open web joists, castellated beams, tapered beams, vierendeel trusses, composite construction, tubular trusses, and offshore platforms.
7. Welding is practicable even for complicated shapes of joints. For example, connections with tubular sections can be made easily by welding, whereas it is nearly impossible to make them using bolting. Tubular sections are structurally economical as compression members and their use in trusses is feasible due to welding.
8. Alterations can be made with less expense in case of welding as compared to bolting. It is also easy to correct mistakes in fabrication during erection, whereas a mismatch of holes in a bolted connection is very difficult to correct. Also members can be shortened by cutting and rejoined by suitable welding. In the same way, members can be lengthened by splicing a piece of the same cross-section.
9. A truly continuous structure is formed by the process of fusing the members together. This gives the appearance of a one-piece integrated structure. Usually, the strength of a

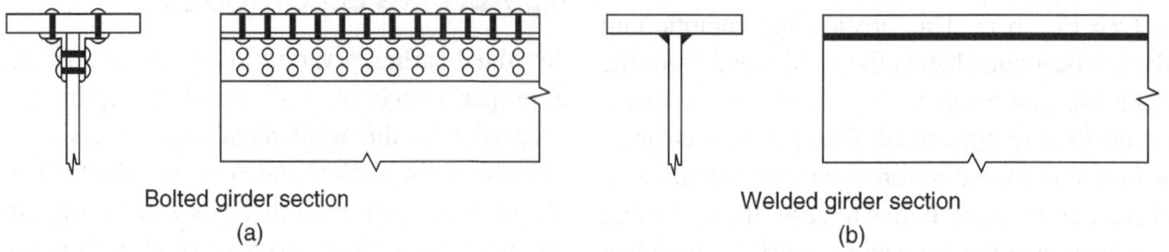


**FIG. 6.8** Appearance of bolted and welded plate girders

welded joint is as strong as or stronger than the base metal, thus there are no restrictions in the placement of joints.
10. The efficiency of a welded joint is more than a bolted joint. In fact 100% efficiency can be obtained using welding.
11. Due to the elimination of holes, stress concentration effect is considerably less in welded connections.
12. The process of welding is relatively silent compared to riveting and bolting (drilling holes) and requires less safety precautions.

However, welding has the following disadvantages.

1. Welding requires highly skilled human resources.
2. The inspection of welded joints is difficult and expensive, whereas inspection of bolted joints is simple. Moreover, non-destructive testing is required in important structures.
3. Members jointed by welding may distort, unless proper precautions are taken. Welded joints have large residual stresses.
4. Costly equipment is necessary to make welded connections.
5. Welded connections are prone to cracking under fatigue loading.
6. Proper welding may not be done in field conditions, especially in vertical and overhead positions.
7. The possibility of brittle fracture is more in the case of welded joints than in bolted connections.
8. The welding performed in the field is expensive than performed in the shop.
9. Welding at the site may not be feasible due to lack of power supply.

Several factors influence the welding cost, which include the following (Salmon & Johnson 1996):

1. Cost of preparing the edges to be welded (in case of groove welds)
2. Amount of weld material required
3. Ratio of the actual arc time to overall welding time
4. The handling required (cranes and special equipment needed during erection)
5. General over head costs
6. Cost of pre-heating, if any

More details about welding costs may be found in Blodgett (1966) and Donnelly (1968).

## 6.4 TYPES AND PROPERTIES OF WELDS

The welds may be grouped into four types as follows:

1. Groove welds
2. Fillet welds
3. Slot welds
4. Plug welds

These are shown in Fig. 6.9. Each type of weld has its own advantage and may be selected depending on the situation. It has been found that fillet welds are used extensively (about 80%) followed by groove welds (15%). Slot and plug welds are used rarely (less then 5%) in structural engineering applications. Fillet welds are suitable for lap and T-joints (see Section 6.5) and groove welds are suitable for butt, corner, and edge joints. Each of these four types of welds are discussed further in the following sections.

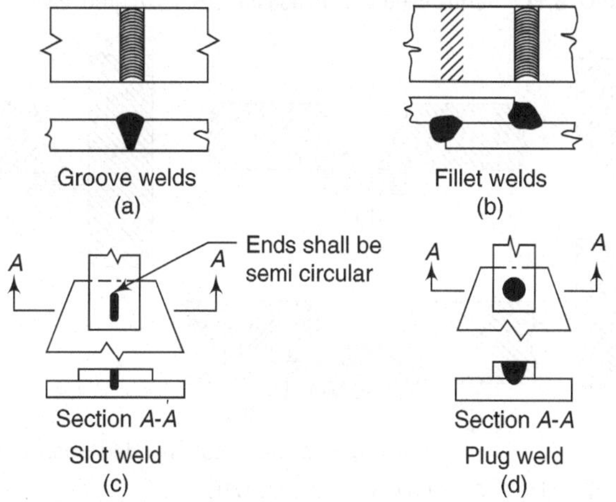


**FIG. 6.9** Types of welds

### 6.4.1 Groove Welds

*Groove welds* are used to connect structural members that are aligned in the same plane and often used in butt joints. Groove welds may also be used in T-connections. The grooves have a slope of 30°–60°. Edge preparation becomes necessary for plates over 10-mm thick for manual arc welding, and over 16-mm thick for automatic welding. Various types of groove welds are shown in Fig. 6.10. The square groove weld is used to connect plates up to 8-mm thickness. The terms that are associated with a completed groove weld are shown in Fig. 6.11. Partial penetration groove welds should not be used especially in fatigue situations.

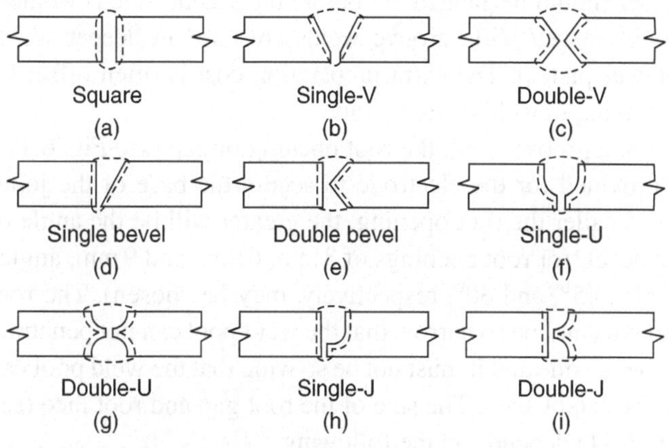


**FIG. 6.10** Types of groove welds

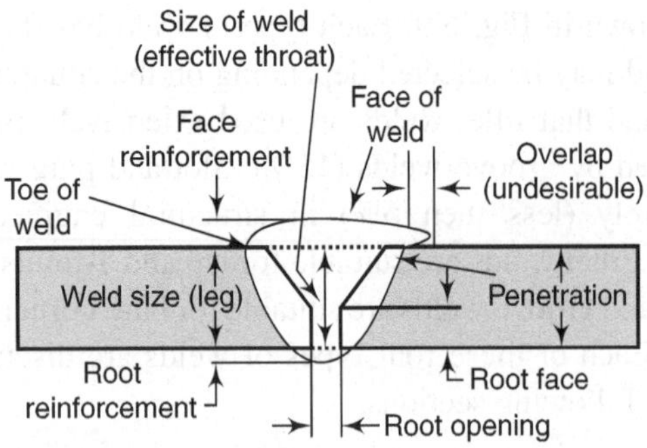


**FIG. 6.11** Terms used to describe the parts of a groove weld

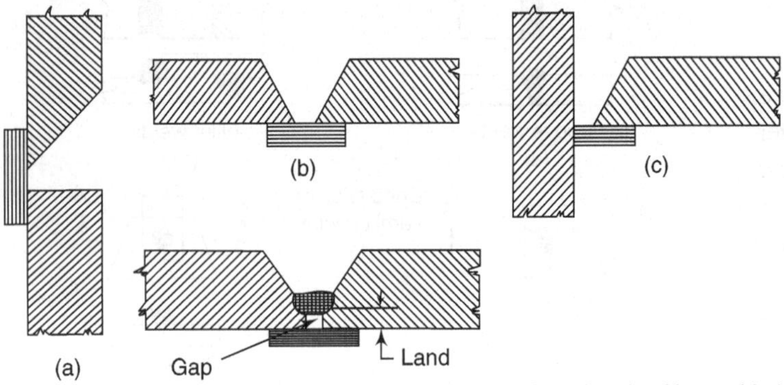


(d) Back-up plate not recommended when land is provided

**FIG. 6.12** Use of back-up plate in groove weld

*Source:* Blodgett (1966). Reprinted with permission from The James F. Lincoln Arc Welding Foundation, USA.

To ensure full penetration and a sound weld, a back-up strip is provided at the bottom of single-V/bevel/J or U grooves. Thus, the back-up strips are commonly used when all welding is done from one side or when the root opening is excessive (see Fig. 6.12). The back-up strip introduces a crevice into the weld geometry and prevents the problem of burn-through. The back-up strip can be left in place or removed after welding the pieces. The welder should not use a pack-up strip when root face (also called as land) is provided, since it will result in the formation of a gas pocket as shown in Fig. 6.12(d), thus preventing a full penetration of weld (Salmon & Johnson 1996). In the case of double-bevel/V/J/U groove weld, a spacer may be provided at the centre of the plate to prevent burning through; however the spacer should be gauged out before the second side is welded. Double-bevel/V/J/U groove welds are used in deeper welds (thicker plates). The extra preparation cost is often offset by the saving in weld metal volume.

For a groove weld, the root opening or gap (see Fig. 6.11), is provided for the electrode to access the base of the joint. The smaller the root opening, the greater will be the angle of the bevel (for root openings of 3 mm, 6 mm, and 9 mm, angles of 60°, 45°, and 30°, respectively, may be chosen). The root gap must not be so narrow that the weld pool can not penetrate to the far side and it must not be so wide that the weld pool can not be established. The size of the root gap and root face (see Fig. 6.11) depends on the following:

1. Type of welding process
2. Welding position
3. Volume of deposited weld material
4. Cost of preparing edges
5. Access for arc and electrode
6. Shrinkage and distortion

For semi-automatic or manual metal arc welding, typical values of root gap and root face are 2–3 mm and 1–2 mm, respectively. Since greater penetration can be achieved with submerged arc welding, values of 0–2 mm and 4–6 mm can be used for root gap and root face, respectively.

The choice between single or double penetration depends on access on both sides, the thickness of the plate, the type of welding equipment, the position of the weld, and the means by which the distortion is controlled.

Since weld metal is expensive compared to the base metal, the groove is made of double-bevel or double-V for plates of thickness more than 12 mm, and made of double-U or double-J for plates of thickness more than 40 mm. For plates between 12–40 mm, single-J and single-U grooves may be used (for details of edge preparation refer to AWS D1.1:2000 code).

Since groove welds will transmit the full load of the members they join, they should have the same strength as the members they join. Hence, only full penetration groove welds are often used.

## 6.4.2 Fillet Welds

*Fillet welds* are most widely used due to their economy, ease of fabrication, and adoptability at site. They are approximately triangular in cross section and a few examples of application of fillet weld are shown in Fig. 6.13. Unlike groove welds, they require less precision in 'fitting up' two sections, due to the overlapping of pieces. Hence, they are adopted in field as well as shop welding. Since they do not require any edge preparation (edge conditions resulting from flame cutting or shear cutting procedures are generally adequate), they are cheaper than groove welds.

In connections, members generally intersect at right angles, but intersection angles between 60° and 120° can be used, provided the correct throat size is used in design calculations (see Section 6.9.2). Fillet welds are assumed to fail in shear.

## 6.4.3 Slot and Plug Welds

*Slot and plug welds* are not used exclusively in steel construction. When it becomes impossible to use fillet welds or when the length of the fillet weld is limited, slot and plug welds are used to supplement the fillet welds. Plug welds are also occasionally used to fill up holes in connections, such as beam-to-column seat angles where temporary erection bolts have

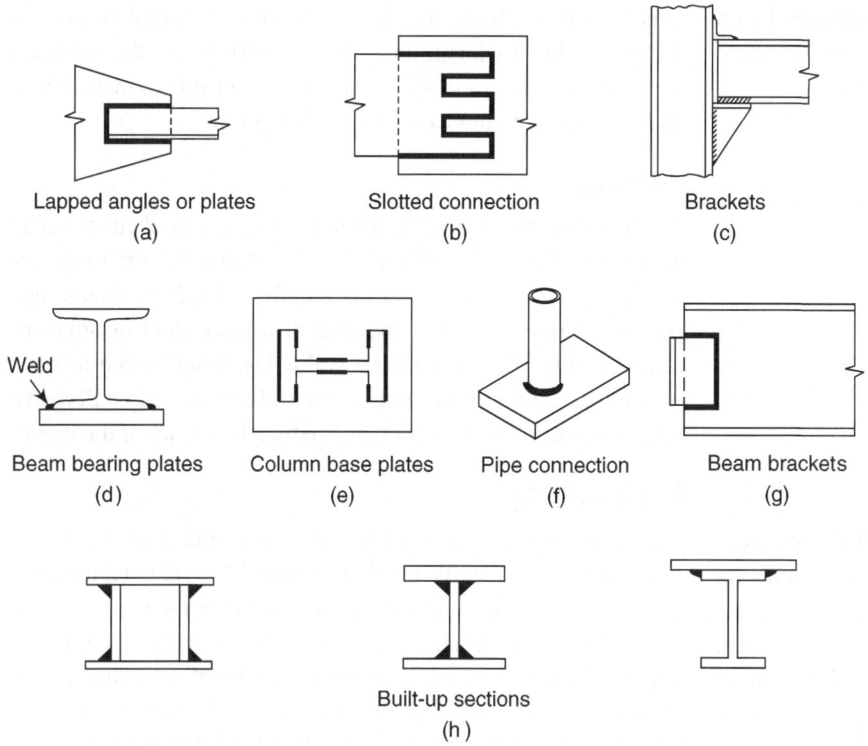


**FIG. 6.13** Typical uses of fillet welds

been placed to align members prior to welding. They are also assumed to fail in shear. Thus, their design strength is similar to that of fillet welds. The penetration of these welds into base metal is difficult to ascertain. Moreover, the inspection of these welds is difficult. Hence, they are normally not used to connect members carrying tensile loads. Slot and plug welds are useful in preventing overlapping parts from buckling.

## 6.4.4 Structure and Properties of Weld Metal

The weld metal is a mixture of parent metal and steel melted from the electrode. The solidified weld metal has properties characteristic of cast steel. Hence, it has higher yield to ultimate ratio but lower ductility compared to structural steel. When the weld pool is cooling and solidifying, the parent metal along side the joint is subjected to heating and cooling cycles and the metallurgical structure of this steel in this region will be changed (see Fig. 6.14). This region is called the *heat-affected zone* (HAZ). The structure of HAZ will be governed by

1. the composition of the steel (carbon equivalent) and
2. the cooling rate of HAZ.

The cooling rate of HAZ is affected by the following factors (Dowling et al. 1988):

1. Arc energy (heat input)
2. Thickness of steel
3. Temperature of steel pieces during welding (pre-heat)
4. Type of joint

The change in structure in HAZ should be considered in the design stage by selecting a suitable Charpy V-impact value for the (see Section 1.8.5) welding electrode (its Charpy impact value should be equal to or greater than that specified for thc parent metal), corrosion resistance, etc. Pre-heating of joints will also help to reduce HAZ cracks. However, pre-heating increases the cost of welding.

Pre-heating of joints is done for the following reasons:

1. To disperse hydrogen away from the weld pool and HAZ
2. To remove any surface moisture in highly humid conditions
3. To bring the steel to ambient temperature (20°C), in cold climates
4. To reduce shrinkage stresses in welds and adjacent base metal (important in highly restrained joints)
5. To lower the transition temperature of the weld and adjacent base metal
6. To increase the notch toughness in the weld zone

Charts for finding the required pre-heat temperature are provided by Blodgett (1966). In critical cases, 'pre-heat' is maintained for a considerable period of time after welding. High currents usually give low impact properties. (*Note:* The electrode diameter is chosen to match the current being used.)

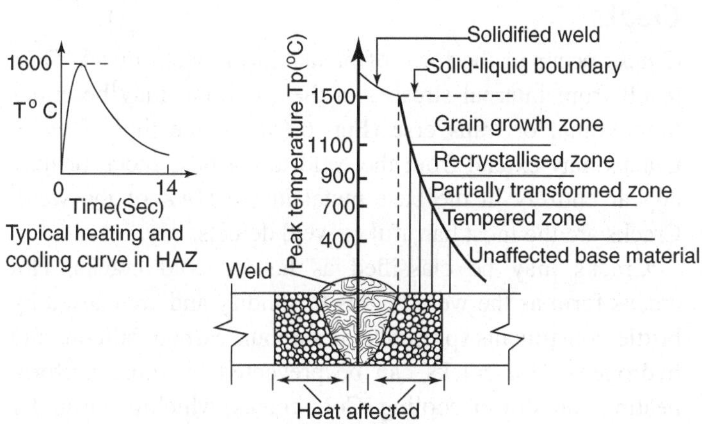


**FIG. 6.14** Heat-affected zones (HAZ) in welded parts

## 6.4.5 Weld Defects

The production of sound welds is governed by the type of joint, its preparation and fit-up, the root opening, etc. In addition to this, the choice of electrode, the welding position, the welding current and voltage, the arc length, and the rate of travel also affect the quality of weld (Gaylord et al. 1992). Accessibility of the welding operation is also important, since the quality of weld is determined to a considerable extent by the positioning of the electrode. Some of the common defects in the welds are as follows:

1. Incomplete fusion
2. Incomplete penetration
3. Porosity
4. Inclusion of slag
5. Cracks
6. Undercutting
7. Lamellar tearing

These defects are shown schematically in Fig. 6.15 and discussed briefly.

### Incomplete Fusion

*Incomplete fusion* (see Fig. 6.15a) may result if the surfaces to be jointed have not been cleaned properly and are coated with oxides, mill scale, and other foreign materials. Insufficient current supplied by the welding equipment (which results in the base metal not reaching its melting point) and high rate of welding will also result in incomplete fusion.

### Incomplete Penetration

This defect (see Fig. 6.15b) is due to the failure of weld metal to penetrate the complete depth of the joint, where full penetration has been specified. This defect, found in groove welds, is due to unsuitable groove design for the selected welding process, use of large size electrodes, insufficient welding current, and excessive welding rates.

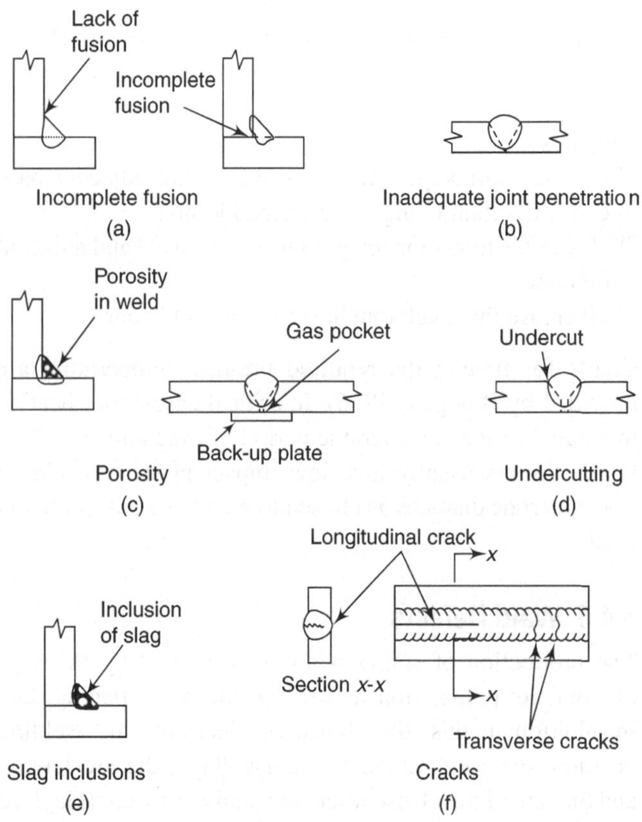


**FIG. 6.15** Possible weld defects

### Porosity

*Porosity* (see Fig. 6.15c) is formed when gas pockets or voids are draped during the cooling process. It may be due to excessively high current or longer arc length. Sometimes a large gas pocket is concentrated at the root of a fillet weld or adjacent to a back-up plate in a groove weld. The latter is due to poor welding procedure and careless use of back-up strips. Porosity results in stress concentration and reduced ductility of the metal.

### Undercutting

*Undercutting* (see Fig. 6.15d) is due to a local decrease of the thickness of the parent metal at the weld toe. This results in a groove in the base metal adjacent to the toe of the weld and left unfilled by weld metal during the welding process. Undercut may result in loss of the gross section and will act as a notch or stress riser. This defect may be due to excessive current or a very long arc. This defect can be easily detected by visual inspection and corrected by depositing additional weld material.

### Slag Inclusion

*Slag* is formed during the welding process in the form of metal oxides as a result of chemical reactions of the melted electrode coating. Normally the slag floats over the surface (since it has lower density than the molten weld metal) and is removed after the weld is cooled. However, if the cooling is rapid, the slag may get trapped in the weld before it could rise to the surface. Also, when several passes of welds are made to achieve the desired weld thickness, the slag that forms after each successive pass should be removed. The non-removal of slag after each pass in a multipass weld is a common cause of slag inclusion. Overhead welds are also more prone to slag inclusions [Fig. 6.15(e)] and hence adequate care should be taken in such occasions.

### Cracks

*Cracks* occur in the form of breaks in the weld metal. They result from internal stresses in the weld and may be either longitudinal or transverse (Fig. 6.15f) to the line of weld. Cracks may extend from the weld to the base metal or may appear entirely in the base metal in the HAZ of the weld. Cracks are the most harmful of weld defects.

Cracks may be classified as hot or cold cracks. Hot cracks form as the weld begins to solidify and are caused by brittle constituents (presence of sulphur, carbon, silicon, and hydrogen). Hot cracks can be prevented by more uniform heating and slower cooling. Cold cracks, which occur under room temperature, run parallel to but under the weld in the base metal. Use of low hydrogen electrodes along with proper pre-heating and post-heating may reduce the possibility of cold cracks (see Section 1.14.1 also).

### Lamellar Tearing

*Lamellar tearing* is a type of cracking (brittle fracture) that occurs in the base metal beneath the weld (see Fig. 1.20). Lamellar tearing is caused by the combined effect of large relative strains in the base metal due to localized stresses, strain restraint in the base metal, and loading perpendicular to the mill rolling direction of the member being welded. If a large weld or welds from both sides are imposed on a thick piece of metal, a lamellar tearing can occur. A large overmatch of the electrode and base metal in a full penetration groove weld tends to increase the possibility of lamellar tearing. Restrained corner and T-joints are most susceptible to lamellar tearing. Flange stiffeners welded to thin column flanges produce a restraint, thus increasing the possibility of lamellar tearing in these flanges. The presence of hydrogen increases the vulnerability of lamellar tearing. Groove welds rarely experience lamellar tearing. Lamellar tearing can be reduced by adopting the following:

1. Using plate material with improved through-thickness properties
2. Using joint alternatives that reduce lamellar tearing (see Fig. 1.22)
3. Using low-strength and low hydrogen welding consumables
4. Pre-heating to reduce the tendency of lamellar tearing
5. Using some order of welding which will minimize shrinkage strains

Lamellar tearing is also discussed in Section 1.8.6 of Chapter 1.

## 6.4.6 Welding Quality Control

Since a small error in a weld may lead to a catastrophic collapse, the following checks are to be made before, during, and after welding (Blodgett 1966):

1. Checks before welding
   (a) Proper included angle
   (b) Proper root opening
   (c) Proper root face
   (d) Proper alignment
   (e) Cleanliness of the joint
2. Checks during welding
   (a) Proper type and size of electrode
   (b) Proper welding current and polarity
   (c) Proper tack welds
   (d) Good fusion
   (e) Proper pre-heat and interpass temperature
   (f) Proper sequencing of passes
   (g) Proper travel speed
   (h) Absence of overlap
   (i) In vertical welding, tilt of crater
   (j) Slight reinforcement on groove weld
   (k) Filled craters
3. Checks after welding
   (a) Absence of excessive undercut
   (b) Filled craters
   (c) Slight reinforcement on groove weld
   (d) Full size on fillet welds
   (e) Absence of cracks

In addition, a qualified welder, who knows the weld qualification procedures, should be employed to execute the job. Visual inspection (which is dependent on the competence of the observer) and non-destructive tests (may be employed for important structures) should be used to determine the type and distribution of weld defects (Gaylord et al. 1992). Any poor or suspicious weld should be cut and replaced. A welding gauge may be used to rapidly check the size of the fillet welds.

### Non-Destructive Tests On Welds

The *non-destructive tests* usually employed on welds are as follows.

***Liquid penetrant inspection*** In the liquid or dye penetrant inspection, a dye is brushed or sprayed on the surface of the weld. It penetrates into the surface cracks of the weld. Then a developer is sprayed on, which is stained red by the dye, which rises from the surface defects by capillary action. A fluorescent liquid may also be used (instead of the aforementioned procedure), which detects surface cracks upon exposure to black light.

***Magnetic particle inspection*** Magnetic particle inspection (magnaflux) is based on inducing a strong magnetic field in a short section of the ferromagnetic material. Poles develop where there are leakages in the field. Thus, the cracks would cause north and south poles to form on each side of the cracks. This will be detected by the iron powder sprinkled on the test area, which migrates to the poles. The resulting pattern is used to identify the location, size, and shape of the cracks.

***Radiographic inspection*** Radiographic inspection uses shortwave radiations, such as x-rays or gamma rays to discover the surface and subsurface flaws in the weld. When a beam of x-rays or gamma rays are beamed on the weld surface using a portable machine they encounter less resistance at a defect. This method is expensive and hence used on important structures only. Although it gives reliable results for groove welds, the interpretation of pictures for fillet weld is difficult. Moreover, while using this method, proper care should be taken to protect the workers from radiation of x-rays.

***Ultrasonic inspection*** Ultrasonic inspection is also effective in detecting surface as well as subsurface defects. In this method, using ultrasonic equipment, sound waves are sent into

## CASE STUDY

### Failure of the King's Bridge Due to Faulty Welding

Dowling et al. (1988) report the failure of King's bridge in Melbourne, Australia, which failed, while in service on 10th July 1962 (Melbourne's winter) due to brittle fracture, when a 45-ton vehicle was passing over it. This plate girder bridge consisted of four plate girders, spanning 30 m and topped with reinforced concrete deck slab. Each plate girder's bottom flange was made of high-strength 400 × 19 mm plate, supplemented in the region of high bending moment by cover plates of size 300 × 19 mm or 360 × 12 mm. The cover plates were attached to the flange by a 5-mm fillet weld all round as shown in Fig. 6.16.

The longitudinal welds connecting the cover plates were made before the short 80 mm transverse welds at the ends. They provided complete restraint against contraction, when the transverse welds were placed, resulting in transverse crack in flange plates. The transverse welds were made in three passes. In some instances, the cracks caused in the main flange plate by the first run were covered up by a subsequent pass. In other cases, the cracks caused by the last run were covered up with priming paint. The penetration of paint coats into the cracks showed later that the cracks passed through the full thickness of the flange even before the girders left the factory. In the span that failed, cracks existed in the main flange plate under seven of the eight transverse fillet welds. Thus, the most likely and most dangerous cracks were regularly missed by the inspectors, who however repaired several less harmful longitudinal cracks. All the seven cracks developed into complete flange failure, partly by brittle fracture and partly by fatigue, under a load that was well within the design load of the bridge.

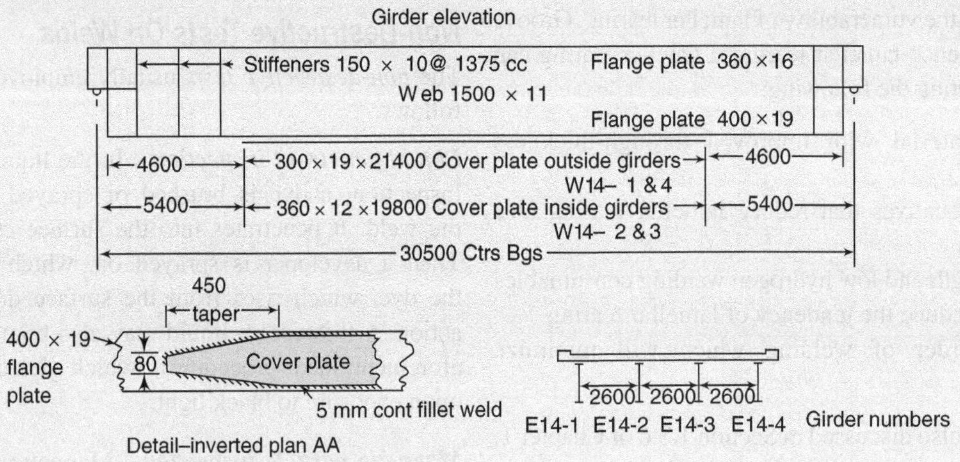


**FIG. 6.16** Details of plate girder and welding in King's Bridge, Melbourne, Australia (Dowling et al. 1988)

the surface of the welds and the reflected waves are monitored by a receiver. This data is displayed as visual patterns on an oscilloscope screen.

In interpreting the results of any inspection procedure, care should be taken to access the severity of defects. Some defects may be relatively unimportant, while others may be critical in specific service requirement.

## 6.5 TYPES OF WELDED JOINTS

The *types of welded joints* depend on the following factors:

1. The size and shape of the members connected at the joint
2. The type of loading
3. The area available for welding at the joint
4. Relative cost of various types of welds

The five basic types of welded joints which can be made in four different welding positions such as flat, horizontal, vertical, and overhead are as follows (see Fig. 6.17):

1. Butt joint
2. Lap joint
3. T-joint
4. Corner joint
5. Edge joint

These joints are discussed briefly in the following sections.

### 6.5.1 Butt Joints

A *butt joint* is used to join the ends of flat plates of nearly equal thickness. This type of joint eliminates the eccentricity developed using a lap joint (see Fig. 6.17). The butt joint obtained from a full penetration groove weld has 100% efficiency (i.e. the weld is considered as strong as the parent plate). As mentioned previously, the groove welds used in butt joints can be full penetration or partial penetration depending upon whether the penetration is complete or partial through the thickness (see Section 6.4.1 for the discussion on groove welds). Such butt joints also minimize

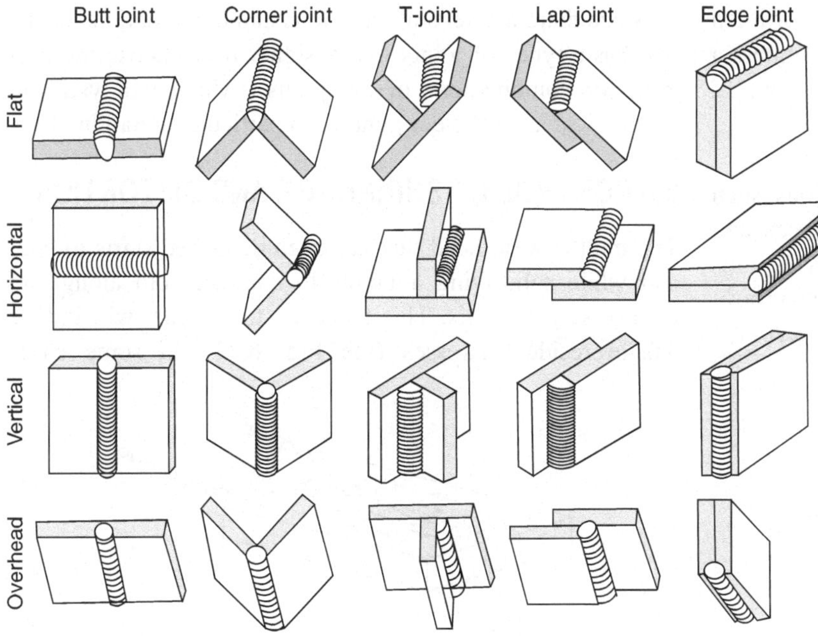


FIG. 6.17 Five basic weld joints may be made in four different welding positions

Reinforcement makes the butt joint stronger under static load and the flow of forces is generally smooth. However, when in the case of fatigue loads, stress concentration develops in the reinforcement, leading to cracking and early failure. Hence, under these circumstances, the reinforcement should be either removed by machining or kept within limits (normally within 0.75–3 mm) to avoid stress concentration. Similarly, when plates of two different thicknesses and/or widths are joined, the wider or thicker part should be reduced at the butt joint to make the width or thickness equal to the smaller part, the slope being not steeper than one in five [see Fig. 6.18(b)]. Where the reduction of the dimension of the thicker part is impracticable, and/or where dynamic/alternating forces are not involved, the weld metal shall be built up at the junction with the thicker part to dimensions at least 25% greater than those of the thinner part, or alternatively to the dimensions of the thicker member [see Fig. 6.18(c)]. Their main drawback is that the edges of the plates which are to be connected must usually be specially prepared and very carefully aligned before welding. They also result in high residual stresses.

the size of the connection and are aesthetically pleasing than lap joints. Face reinforcement (see Fig. 6.11) is the extra weld metal that makes the throat dimension greater than the thickness of the welded material. The provision of reinforcement increases the efficiency of the joint and ensures that the depth of the weld is at least equal to the thickness of the plate. Reinforcement is normally provided, since it is difficult for the welder to make the weld flush with the parent metal.

Due to the accurate placement of parts before welding, butt joints are often made in shops, where it is possible to control the welding process. Field butt joints are rarely used.

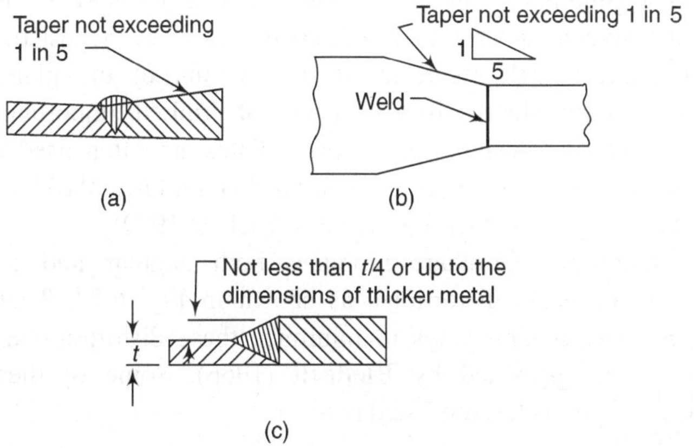


FIG. 6.18 Butt welding of parts of unequal thickness and/or unequal width

## 6.5.2 Lap Joints

*Lap joints* are most commonly used because they offer ease of fitting and ease of jointing. Thus, they do not require any special preparation (even flame cut or sheared edges can be used) and can accommodate minor errors in fabrication or minor adjustment in length. Lap joints utilize *fillet welds* (see Section 6.4.2) and hence are well suited for shop as well as field welding. Some examples of lap joints are shown in Fig. 6.19. The connections using lap joints may require a small number of erection bolts, which may either be removed after welding or left in place. The additional advantage of lap joints is that plates with different thicknesses can be joined without any difficulty. Hence, it is often preferred in truss joints as

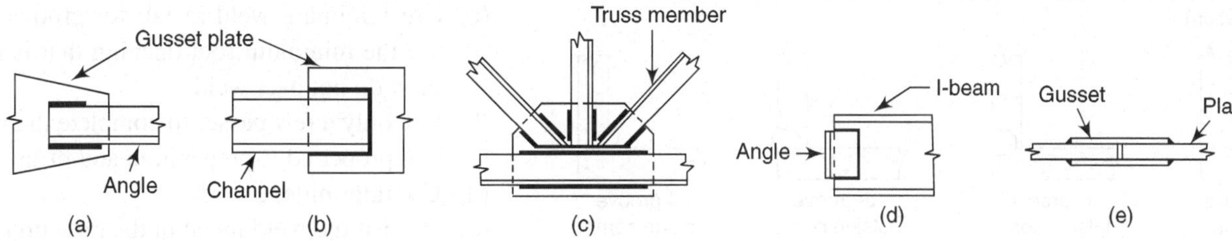


FIG. 6.19 Some examples of welded lap joints

shown in Fig. 6.19(c). However, the main drawback of a lap joint is that it introduces some eccentricity of loads, unless a double lap joint is used as in Fig. 6.19(e).

### 6.5.3 Tee Joints

*T-joints* are often used to fabricate built-up sections such as T-shapes, I-shapes, plate girders, hangers, brackets, and stiffeners, where two plates are joined at right angles. T-joints can be made by using either fillet or groove welds. The groove weld edge shapes used on T-joints are shown in Fig. 6.20.

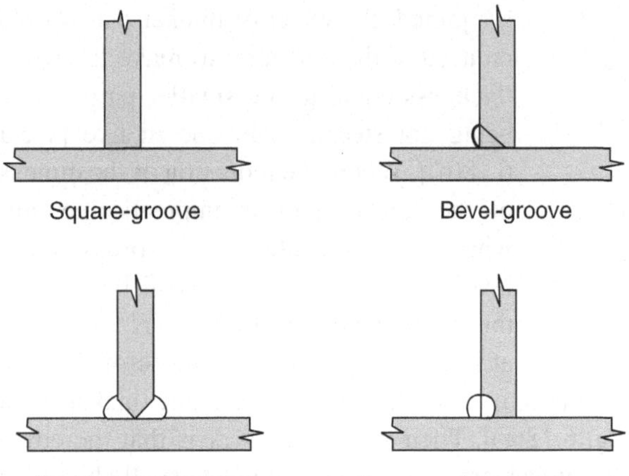


**FIG. 6.20** Edge shapes used on T-joints (Double bevel/J groove are often used with thick plates)

### 6.5.4 Corner Joints

*Corner joints* are used to form built-up rectangular box sections, which may be used as columns or beams to resist high torsional forces. The groove weld edge shapes for corner joints are shown in Fig. 6.21.

### 6.5.5 Edge Joints

*Edge joints* are not used in structural engineering applications. They are used to keep two or more plates in a given plane (see Fig. 6.17).

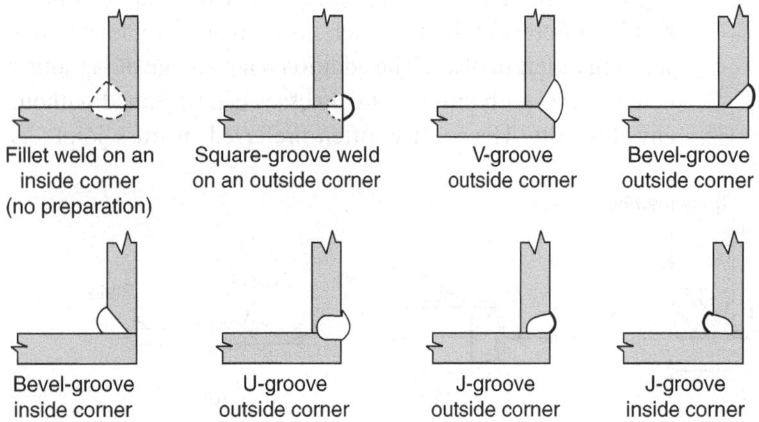


**FIG. 6.21** Corner joint edge shapes

Since there are several variations and combinations of the five basic types of joints, the designer may choose the best joint (or combinations of the joints), which will yield an economical and efficient joint, for a particular situation.

## 6.6 CONTROL OF SHRINKAGE AND DISTORTION

The molten weld bead that has been deposited starts to cool and while solidifying attempts to contract both along and transverse to its axis. This tendency to contract will induce tensile residual stresses (see Fig. 6.22), in some cases

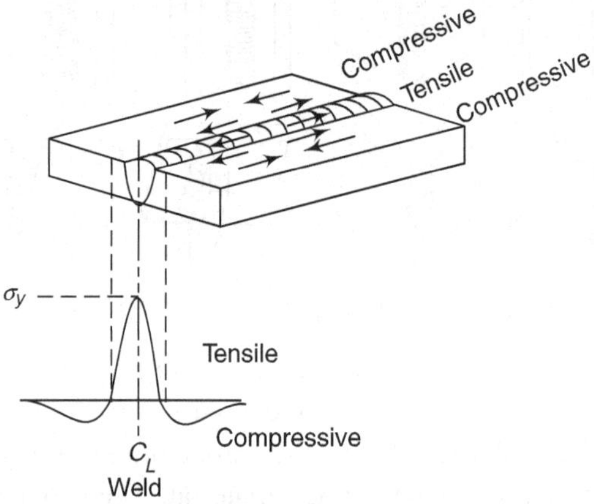


**FIG. 6.22** Residual stress in welds

accompanied by tensile yielding, where it is resisted by surrounding structure. When the surrounding structure is less than fully rigid (which is often the case in practice), it results in distortions (see Fig. 6.23).

Longitudinal shrinkage can cause slender plates to buckle. If the weld is eccentric to the effective centroid of the resisting structure, it will result in overall bowing of the plates. These longitudinal distortions are of limited importance in connection design, since stocky plates are often used in connections, which sustain overall longitudinal shrinkage without any local buckling (Owens & Cheal 1989).

Transverse shrinkage produces both angular and out of plane distortions, as shown in Fig. 6.24. There are several ways to minimize these distortions and are provided by Blodgett (1966). Some of these approaches are listed here.

1. Reduce the shrinkage forces by incorporating the following:
   - (a) Use minimum weld metal; for groove welds use the minimum root opening that is necessary; do not over weld
   - (b) Use only a few passes to complete the weld
   - (c) Use proper edge preparation and fit-up
   - (d) Use intermittent welds
   - (e) Deposit the weld metal in the direction opposite to the progress of welding the joint

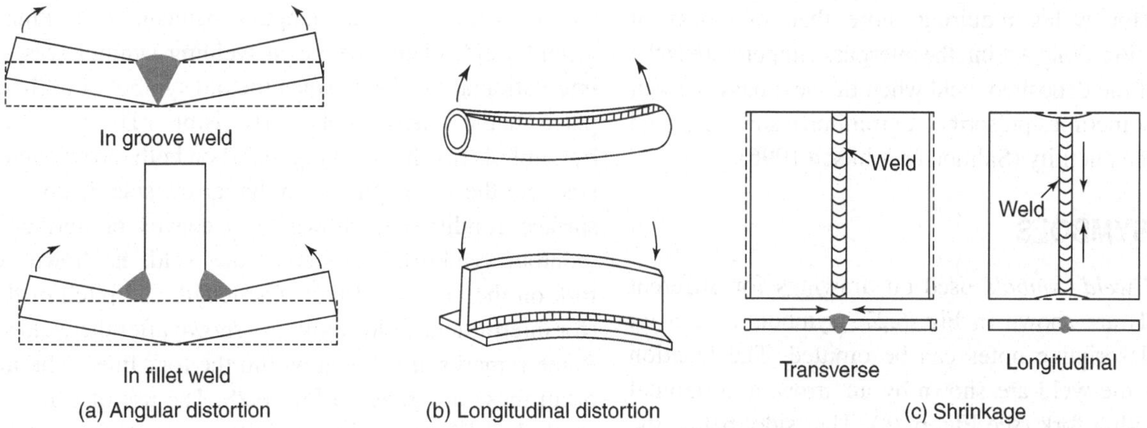


Shrinkage $S = k\dfrac{A}{t}$ — $k$ = empirical factor (0.1 to 0.17), $A$ = Cross section of the weld, $t$ = thickness of plate

**FIG. 6.23** Distortion due to welding

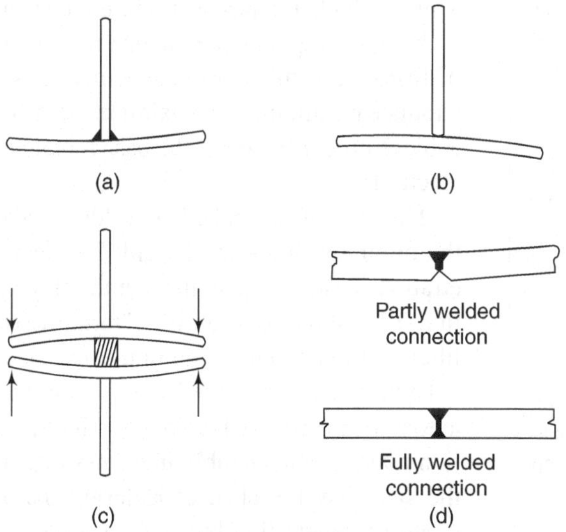


**FIG. 6.24** Welding distortion and its control (a) Unrestrained distortion due to fillet welds (b) Presetting to eliminate distortion in completed structure (c) Restraint and flexure to eliminate distortion (d) Counteracting distortions

2. Allow the shrinkage to occur freely as follows.
   (a) Prior to welding, pre-set the elements so that after welding and its associated distortions, the geometry is correct. As an example, the end plate shown in Fig. 6.24(b) is pre-dished (either by pressing or by spot heating) and the web plate is set a few degrees off the right angle to the end plate.
   (b) Use pre-bending of pieces as shown in Fig. 6.24(c).
3. Balance shrinkage forces by incorporating the following.
   (a) Use symmetry in welding (as shown in Fig. 6.24(d) for groove welds); Fillet welds on each side of a piece contribute to the counteracting effects (Fig. 6.25).

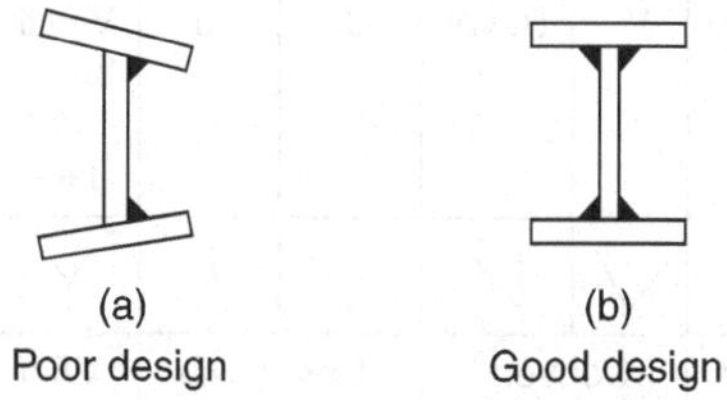


**FIG. 6.25** Symmetry of welding to minimize distortion

   (b) Use scattered and intermittent weld segments. As shown in Fig. 6.26, initially the intermittent welds may be placed in a staggered fashion and then the remaining portions may be filled. For many structures such as plate girders, this kind of technique is adopted. First short segments of weld are placed at strategic locations so that the different pieces are held together and then the remaining portions are filled.
   (c) Use *peening* (i.e., stretching the metal by a series of blows, using a hammer).
   (d) Use clamps, jigs, etc., to force the weld metal to stretch as it cools.

In practice, more than one method may be used at the same time for a particular situation. Minimum pre-heat and *interpass*

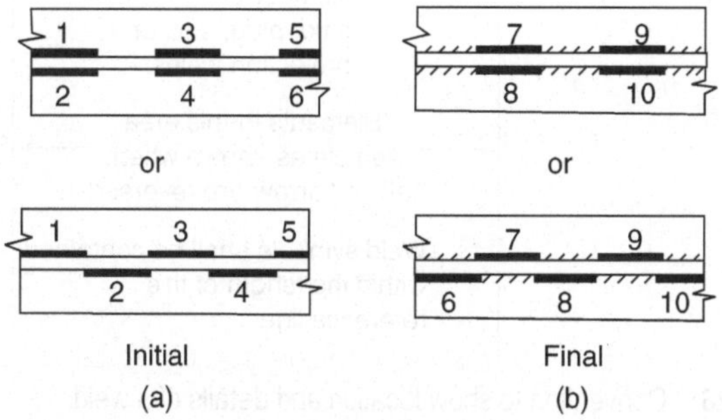


**FIG. 6.26** Sequences for intermittent welds

*temperature* (for welds requiring more than one pass of welding operation along a joint, the interpass temperature is the temperature of the deposited weld when the next pass is about to begin) are sometimes prescribed to minimize shrinkage and ensure adequate ductility (Salmon & Johnson 1996).

## 6.7 WELD SYMBOLS

The *standard weld symbols* used on drawings for different types of welds are shown in Fig. 6.27. Symbols save a lot of space as descriptive notes can be omitted. The location and details of the weld are shown by an arrow, a horizontal line ending with a fork (see Fig. 6.28). The 'side' below the arrow is called the arrow side and the 'side' above is called the other side. A circle at the kink indicates a weld all round and a vertical line and triangular pennant at the kink shows a field weld. (This change in welding symbol was adopted internationally in 1976, since the old symbol of a filled in dot placed at the kink was not clearly visible.) The weld is denoted by symbols (as shown in Fig. 6.28) on both 'arrow' and 'other' side, but the weld symbol on the 'arrow side' is inverted. The surface condition is shown by a convex or horizontal line (contour symbol). The size of the welds is shown near the fork on the horizontal line. The length and pitch of the weld (for intermittent welds only) are shown after the weld symbol. Some remarks are shown within the fork lines. The meaning of all these are given in Fig. 6.28. The use of some of these symbols is illustrated in Fig. 6.29.

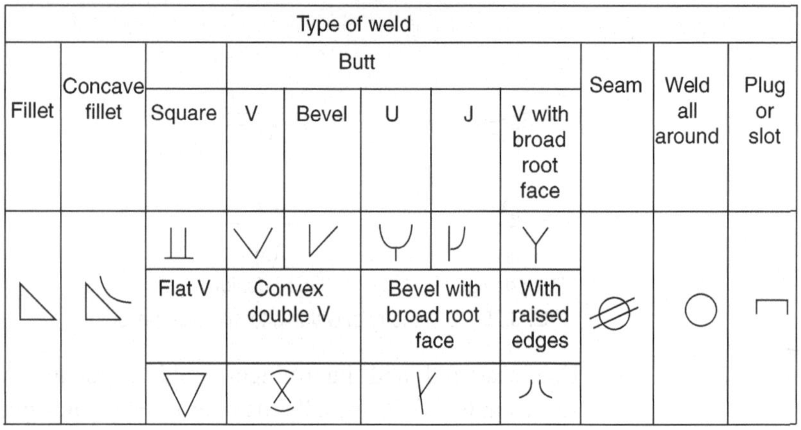


**FIG. 6.27** Basic weld symbols

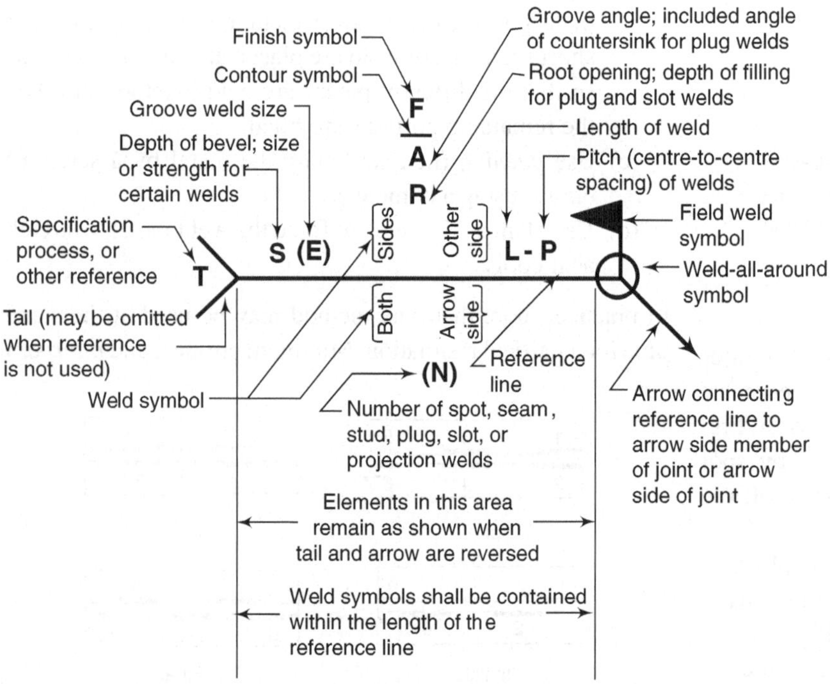


**FIG. 6.28** Convention to show location and details of a weld

*Source:* Galvery and Marlow (2001). Reprinted with permission from Industrial Press, Inc., NY, USA.

## 6.8 WELD SPECIFICATIONS

Thicker plates dissipate the heat due to arc welding vertically as well as horizontally while thinner plates dissipate heat only horizontally. Thus, in thicker plates heat is removed from the welding area quickly and hence results in lack of fusion. For this reason, specifications often stipulate minimum and maximum weld sizes to achieve proper fusion of the base metal and the electrode.

The size of normal fillet welds is taken as the minimum leg size of weld. For deep penetration welds, where the depth of penetration beyond the root run is 2.4 mm, the size of fillet weld is taken as minimum leg size plus 2.4 mm. For fillets made by semi-automatic or automatic process, where the depth of penetration may be considerably in excess of 2.4 mm, the size may be taken considering the actual depth of penetration but subject to the agreement between the purchaser and contractor (IS 800 : 2005).

### 6.8.1 Minimum Weld Size

To ensure fusion, minimize distortion, and to avoid the risk of cracking IS 800 : 2007 and IS 816 provide for a minimum size weld based on the thickness of the pieces being joined. The size of fillet weld should not be less than 3 mm nor more than the thickness of the thinner part joined. The minimum size of the first run or of a single run fillet weld should be as per Table 6.2. Usually the weld size closer to the minimum size is selected. Large size welds require more than one run of welding, which means that after the first run, chipping and cleaning of the weld is required to remove the slag. This will increase

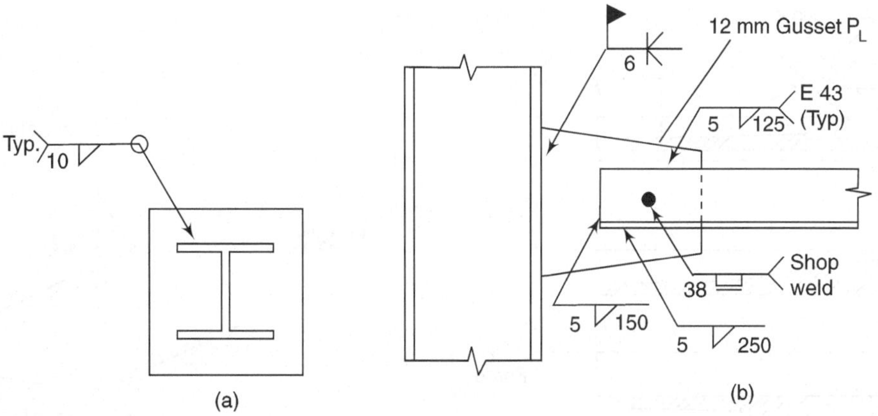


**FIG. 6.29** Illustration of some of the welding symbols

**TABLE 6.2** Minimum size of a single run fillet weld

| Thickness of thicker part, mm | Minimum size, (IS 800:2007) mm | Minimum size (AISC 360-2010) |
|---|---|---|
| Up to 6 | – | 3 |
| Up to 10 | 3 | – |
| 6 to 13 | – | 5 |
| 10 to 20 | 5 | – |
| 13 to 19 | – | 6 |
| 20 to 32 | 6 | – |
| >19 | – | 8 |
| 32 to 50 mm(see notes below) | 8 for first run 10 for minimum size of weld | – |

*Notes:* 1. When the minimum size of the fillet weld given in the table is greater than the thickness of the thinner part, the minimum size of the weld should be equal to the thickness of the thinner part. The thicker part shall be adequately preheated to prevent cracking of the weld.
2. Where the thicker part is more than 50 mm thick, special precautions like pre-heating should be taken.

the cost of welding. Also note that a smaller size weld will be cheaper than a larger one for the same strength, considering the volume of welding. For example, a 300-mm-long 5-mm weld will have the same strength (198.3 kN) compared to a 150-mm-long 10-mm size weld. However, the volume of a 10-mm weld (7500 mm$^3$) is twice that of a 5-mm weld (3750 mm$^3$).

## 6.8.2 Maximum Fillet Weld Size Along Edges

The maximum size of fillet weld used along the edges of pieces being jointed is limited to prevent the melting of the base material at the location where the fillet would meet the corner of the plate, if the fillet were made to the full plate thickness. The maximum permitted size is as follows.

1. Along the edge of the plate, less than 6-mm thick, the maximum size is equal to the thickness of the plate (see Fig. 6.30).

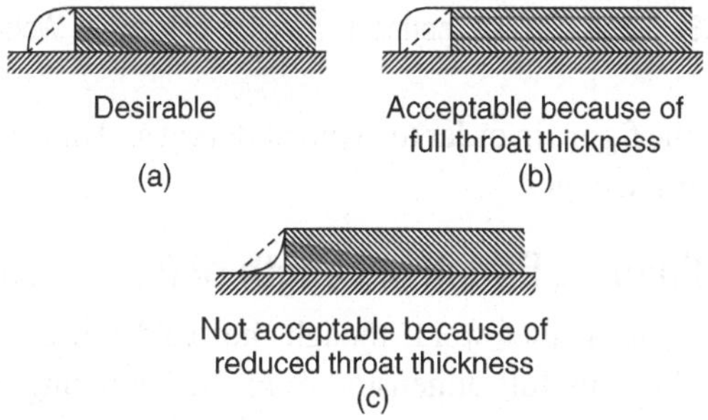


**FIG. 6.30** Limitations of the size of fillet weld applied along the edge of a plate

2. Where the fillet weld is applied to the square edge of a plate of thickness greater than 6 mm, size of the weld should be at least 1.5 mm less than the edge thickness [see Fig. 6.31(a)]. This limit is specified such that the total strength may be developed without overstressing the adjacent metal.
3. Where the fillet weld is applied to the rounded toe of the rolled section, the size of the weld should not exceed 3/4 of the thickness of the section at the toe [see Fig. 6.31(b)].

End fillet weld normal to the direction of force should be of unequal size with the throat thickness of the weld not less than $0.5t$, where $t$ is the thickness of the part as shown in Fig. 6.32. The thickness of the welds should be negotiated at a uniform slope.

Often, the lap joint lends itself to welds being applied on either side of the joint, as shown in Fig. 6.32(b). In such situations, the welds have to be interrupted at the corners, as

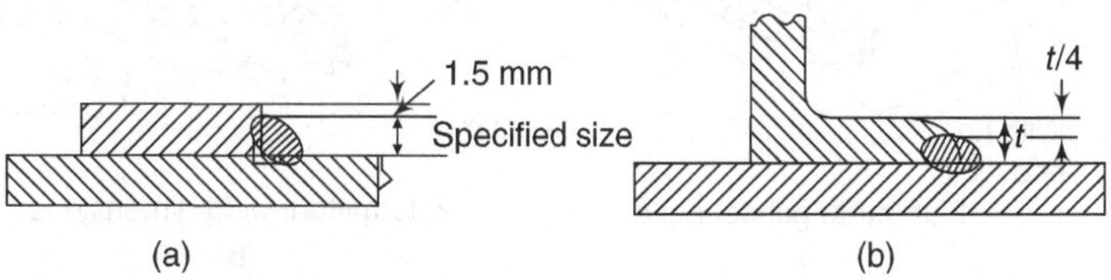


**FIG. 6.31** Size of fillet welds

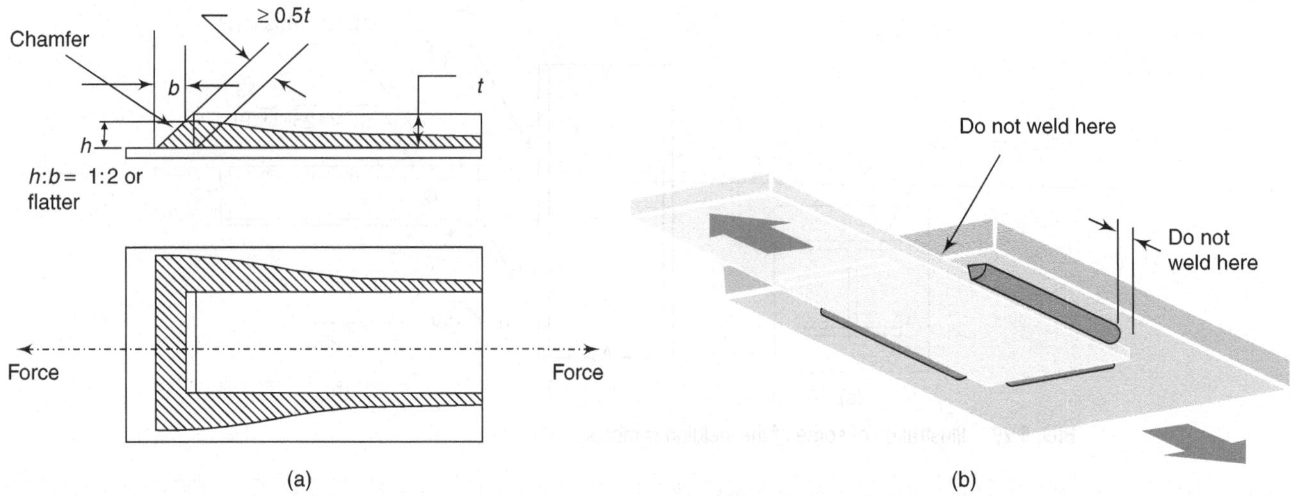


**FIG. 6.32** End fillet weld normal to direction of force (a) Welding along three edges (b) Welding along all edges

shown in the figure in order to avoid undercut and unacceptable melting of the edges.

## 6.8.3 Minimum Effective Length of Fillet Weld

When placing a fillet weld, though the welder tries to build up the weld to its full dimension from the beginning, there is always a slight tapering off where the weld starts and where it ends. Therefore, a minimum length of four times the size of the weld is specified (see Fig. 6.33b). If this requirement is not met, the size of the weld should be one fourth of the effective length.

For these reasons, the effective length is taken equal to its overall length minus twice the size of weld. *End returns* as shown in Fig. 6.33(d) are made equal to twice the size of the weld to relieve the high stress concentration at the ends. Most designers neglect the end returns in the effective length calculation of the welds. End returns must be provided for welded joints that are subject to eccentricity, stress reversals, or impact loads.

The stress distribution in welded joints is complex and non-uniform. Figure 6.34 shows the typical stress distribution at service load for longitudinal fillet welds in a lap joint. The actual variation of shear stress in the weld from points *A* to *B* depends on the length of the weld and also the ratio of the width of the plates being joined. Thus it is seen that there is a serious stress concentration at the edge of the plate. In order to control it, the length of the longitudinal (side) fillets should not be less than the width of the plate (see Fig. 6.33a). The uneven stress distribution increases as the width of the plate increases. For this reason, the perpendicular distance between longitudinal fillet welds is limited to 16 times the thickness of the thinner plate jointed. If the plate is wider than this limit, slot or plug welds may be introduced, which tend to improve the distribution of stress in plate. Figure 6.35 shows the typical shear variation for fillet welds loaded transverse to the weld axis, which is also non-uniform.

## 6.8.4 Overlap

The overlap of plates to be fillet welded in a lap joint should not be less than 4 times the thickness of the thinner part (see Fig. 6.33c).

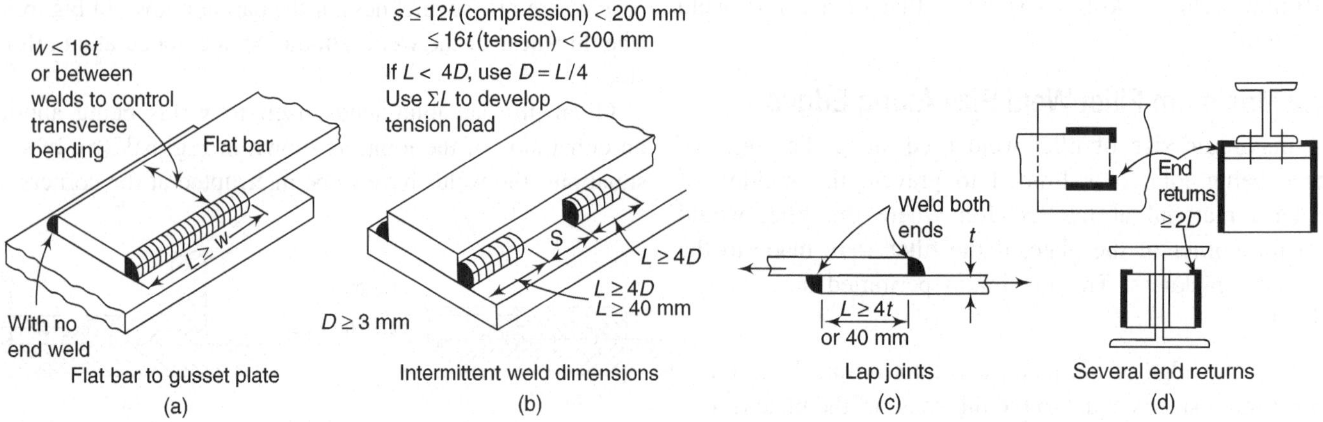


**FIG. 6.33** Several welded connection specifications

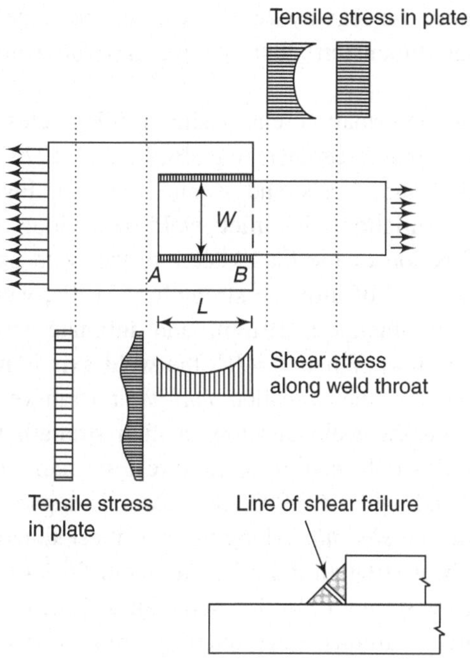


**FIG. 6.34** Behaviour of side fillet welds

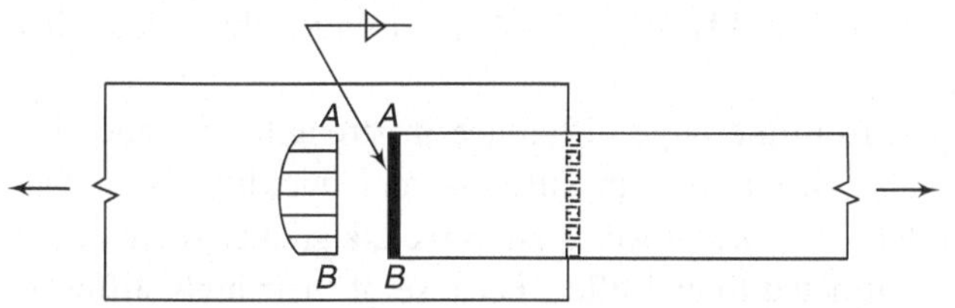


**FIG. 6.35** Typical stress distribution in a lap joint with transverse fillet welds

## 6.8.5 Effective Length of Groove Welds

The effective length of groove welds in butt joints is taken as the length of continuous full size weld, but it should not be less than four times the size of the weld.

## 6.8.6 Effective Length of Intermittent Welds

The intermittent fillet welds should have an effective length not less than four times the weld size, with a minimum of 40 mm, as already shown in Fig. 6.33(b). The clear spacing between the effective lengths of intermittent welds should not exceed 12 and 16 times the thickness of thinner plate jointed for compression and the tension joint respectively, and should never be more than 200 mm. The intermittent groove weld in butt joints should have an effective length of not less than four times the weld size and the longitudinal space between the effective lengths of the intermittent welds should not be more than 16 times the thickness of the thinner part joined. The IS code prohibits the use of intermittent welds in joints subjected to dynamic, repetitive, and alternate stresses.

## 6.9 EFFECTIVE AREA OF WELDS

The effective areas of a groove or fillet weld is the product of the effective throat dimension ($t_e$) multiplied by the effective length of the weld. The effective throat dimension of a groove weld or a fillet weld depends on the minimum width of expected failure plane and is explained in the next section.

### 6.9.1 Groove Weld

The effective throat thickness of a complete penetration groove weld is taken as the thickness of the thinner part joined (see Figs 6.36a and 6.36b). The effective throat thickness of T- or L-joints are taken as the thickness of the abutting part. Reinforcement (see Fig. 6.11), which is provided to ensure full cross-sectional area, is not considered as part of the effective throat thickness.

The effective throat thickness of a partial penetration joint weld is taken as the minimum thickness of the weld metal common to the parts joined, excluding reinforcement (see Fig. 6.36). In unsealed single groove welds of V-, U-, J-, and bevel-types and groove welds welded from one side only, the throat thickness should be at least 7/8th of the thickness of the thinner part joined. However, for the purpose of stress calculation, the effective throat thickness of 5/8th thickness of the thinner member only should be used (IS 816 : 1969). The unwelded portion in incomplete penetration welds, welded from both sides, should not be greater than 0.25 times the thickness of the thinner part joined, and should be central in the depth of the weld (Fig. 6.36d). In this case also, a reduced effective throat thickness of 5/8th of the thickness of the thinner part should only be used in the calculations. Groove welds used in butt joints, where the penetration is less than those specified here, due to non-accessibility, should be considered as non-load carrying for the purposes of design calculations.

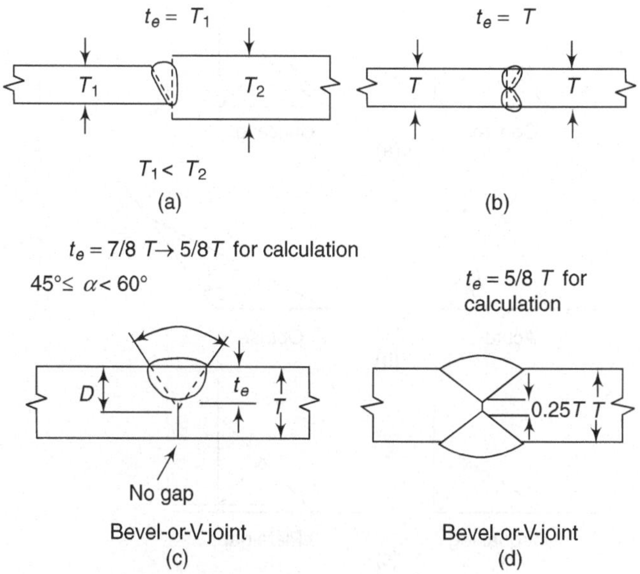


**FIG. 6.36** Effective throat dimensions for groove welds

### 6.9.2 Fillet Weld

The effective throat dimension of a fillet weld is the shortest distance from the root of the face of the weld, as shown in Fig. 6.37. The effective throat thickness of a fillet weld should not be less than 3 mm and should not exceed $0.7a$ ($1.0a$, under special circumstances), where $a$ is the size of the weld in mm. Thus, if the fillet weld is having unequal lengths (which is a rare situation), as shown in Fig. 6.37(b), the value of $t_e$ should be computed from the diagrammatic shape of the weld. In general, the effective throat thickness is the height of the triangle inscribed within the weld which has its base along the free boundary of the weld (see Fig. 6.38). Unequal-legged fillet welds are used when the vertical leg of the weld cannot be increased, for example when attaching a channel (used as a shear connector in composite construction) over a beam flange. Here the vertical leg is limited to the flange thickness at the outer edge of the channel. Additional strength may be obtained by increasing the horizontal leg of the fillet. However, note that by increasing the horizontal dimension $b$ of the weld to twice that of the vertical dimension $a$, the throat dimension can be increased by 26.5%, while the area increases by 100%. Whereas, if we

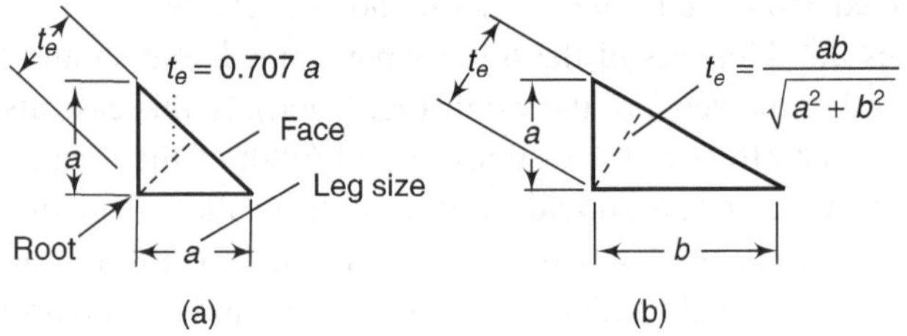


**FIG. 6.37** Effective throat dimensions for fillet welds

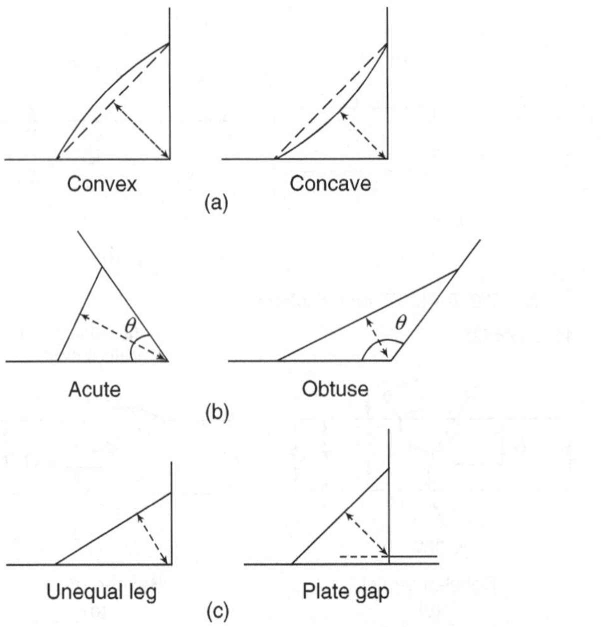


**FIG. 6.38** Effective throat thickness for various fillet weld profiles

increase both the legs by 26.5%, we get the same strength (and throat dimension), but the increase in weld area is only 60%.

The load–deformation relationship of fillet welds has been studied by several researchers (e.g., Butler et al. 1972; Swannel 1981; Neis 1985; Lesik and Kennedy 1990). Figure 6.39 shows the variations in fillet weld behaviour with the relative direction of the load vector to the weld axis (for a 8-mm fillet weld of ultimate strength 565 MPa, weld length 50 mm, plate thickness 19 mm, and ultimate strength of plate = 511 MPa). When $\theta = 0°$, the weld axis is normal to the load vector, the so-called *end fillet* (transverse fillet) situation, and the weld develops a high strength with less ductility (with deformation at rupture less than 1 mm). On the other hand, when $\theta = 90°$, the weld axis is parallel to the load vector, the *side-fillet* (longitudinal fillet) situations, and the weld shear strength is limited to about 56% of the weld metal tensile strength (Butler and Kulak 1968). However, the side fillet exhibits more ductility (rupture occurring at over 2-mm deformation). Intermediate orientations show intermediate values of both strength and ductility. Thus, the end fillet welds are 30%–40% stronger than side fillet welds.

Apart from the large difference in strength, end and side fillets also differ in both stiffness and ductility. Note that when the two types of welds are mixed, a greater share of the load is attracted to end fillets because of their high stiffness, which may result in the side fillets not developing to their full capacity. The code stipulates that the effective throat thickness in fillet welds joining faces inclined to each other should be taken as follows:

$$\text{Effective throat thickness} = K \times \text{size of weld}$$

where $K$ is a constant depending upon the angle between fusion faces (see Fig. 6.39), as given in Table 6.3.

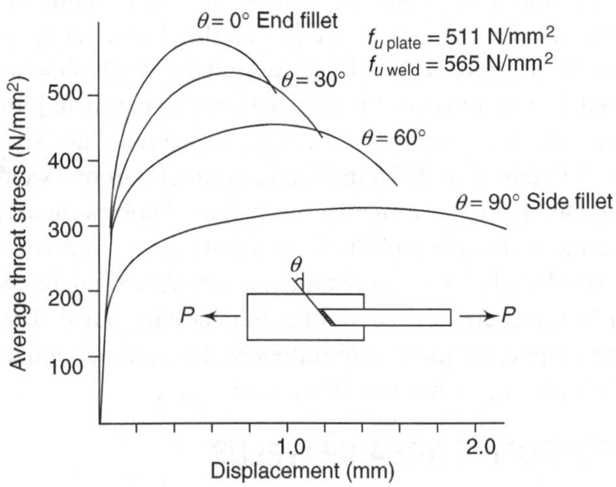


**FIG. 6.39** Load-deformation curves for an 8-mm leg fillet weld at varying angles $\theta$ to load vector

**TABLE 6.3** Throat size reduction factor *K* for different angles between fusion faces

| Angle $\psi$ between fusion faces in degrees | Reduction factor *K* (BS EN 1011–2:2001) | Reduction factor *K* (IS 800:2007) | |
|---|---|---|---|
| 60 to 90 | 0.70 | 0.70 | 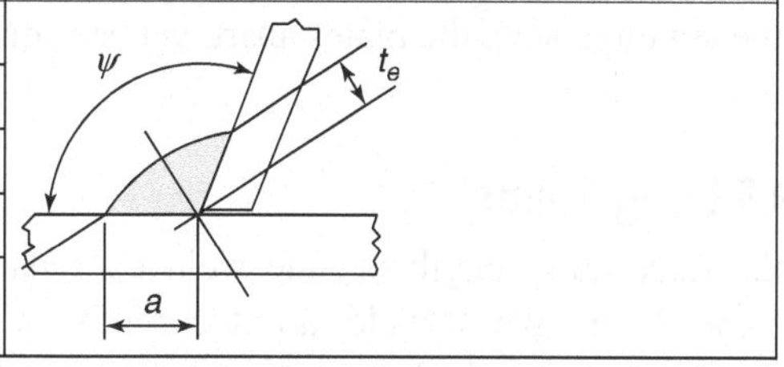 |
| 91 to 100 | 0.65 | 0.65 | |
| 101 to 106 | 0.60 | 0.60 | |
| 107 to 113 | 0.55 | 0.55 | |
| 114 to 120 | 0.50 | 0.50 | |

*Note:* Fillet welds are normally used for connecting parts whose fusion faces form angles between 60° and 120°. If the angle between fusion faces is greater than 120°, for example at the toe of an inclined bracing, then a full penetration butt weld should be used. If the angle is less than 30°, then adequacy of the weld must be shown.

## 6.9.3 Plug Weld

Effective area of plug welds should be taken as the nominal area of the hole in the plane of the faying surface. IS code stipulates that they should not be designed to carry any stresses.

## 6.9.4 Concave and Convex Welds

By appearance, the concave fillet weld shown in Fig. 6.40(a) seems to be larger than the convex weld of Fig. 6.40(b). However, it can be easily seen by inspecting both the cross-sections that the concave weld has less penetration and smaller throat than the convex weld. Thus, the convex weld will be stronger even though it may have lesser deposited weld metal.

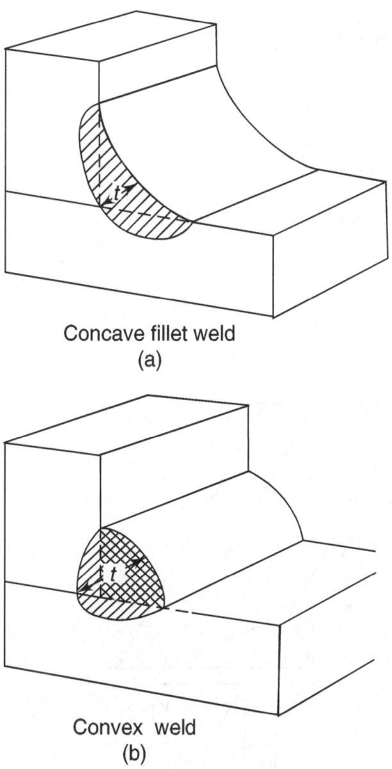


**FIG. 6.40** Concave and convex welds

*Source:* Blodgett (1966). Reprinted with permission from The James F. Lincoln Arc Welding Foundation, USA.

Designers originally favoured the concave fillet weld because they thought that it offers a smooth flow of stresses. However, from experience, it has been found that single pass fillet welds of this shape have a great tendency to crack upon cooling, which outweighs the effect of improved stress distribution. When a concave fillet weld cools and shrinks, its outer surface is stressed in tension. This is because, freezing of the molten weld due to the quenching effect of the plates commences along the sides of the joint. Due to this, the last material to freeze lies in the plane along the centerline of the weld (see Fig. 6.41). Whereas in a convex fillet, the weld shrinks while cooling, but without stressing the outer face in tension and hence does not crack (Blodgett 1966).

For multi-pass fillet welds, the convex bead shape usually applies only to the first pass. Hence, when concave welds are desired for special design considerations, such as stress flow, they should be made in more passes—the first slightly convex and the rest can be build up to form a concave fillet weld (Blodgett 1966).

To obtain a crack free fillet weld it is also desirable to provide a small gap between the two plates that are welded, so that the plates can 'move in' slightly as the weld shrinks. This reduces the transverse stress in the weld. The small gap may be obtained as follows (Blodgett 1966).

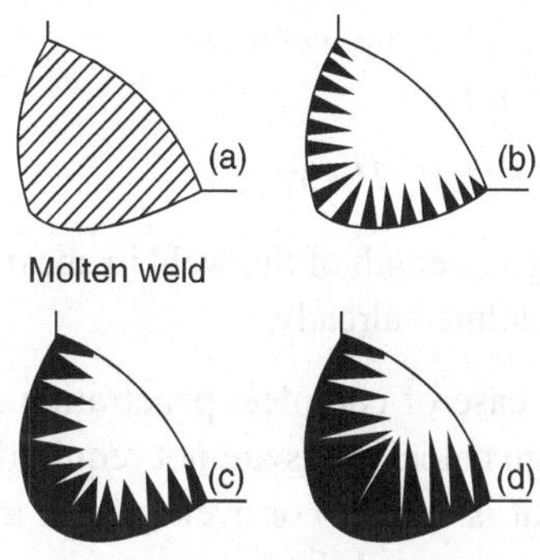


**FIG. 6.41** Progressive cooling of a weld bead (the dark portions show the frozen molten weld)

*Source:* Blodgett (1966). Reprinted with permission from The James F. Lincoln Arc Welding Foundation, USA.

1. Inserting spacers made of soft steel wire between the plates—the soft wire will flatten out as the weld shrinks.
2. A deliberately rough flame cut edge—the small peaks of the cut edge keep the plates apart, yet can squash out as the weld shrinks.

### 6.9.5 Long Joints

If the maximum length of side weld exceeds $150t_e$, where $t_e$ is the throat size of weld, a reduction factor as per clause 10.5.7.3 of the code has to be applied to the calculated strength.

## 6.10 DESIGN OF WELDS

The following assumptions are usually made in the analysis of welded joints.

1. The welds connecting the various parts are homogenous, isotropic, and elastic.
2. The parts connected by the welds are rigid and their deformation is, therefore, neglected.
3. Only stresses due to external forces are considered. The effects of residual stresses, stress concentrations, and the shape of the weld are neglected.

### 6.10.1 Groove Welds

As per IS 800 : 2007, the groove welds in butt joints will be treated as parent metal with a thickness equal to the throat thickness and the stresses shall not exceed those permitted in the parent metal.

1. For tension or compression normal to effective area and tension and compression parallel to the axis of the weld

$$T_{dw} = f_y L_w t_e / \gamma_{mw} \tag{6.1}$$

where $T_{dw}$ is the design strength of the weld in tension, $f_y$ is the smaller of yield stress of the weld and the parent metal in MPa, $t_e$ is the effective throat thickness of the weld (see Section 6.9.1) in mm, $L_w$ is the effective length of the weld in mm, and $\gamma_{mw}$ is the partial safety factor taken as 1.25 for shop welding and as 1.5 for site welding.

2. For shear on effective area

$$V_{dw} = L_w t_e f_{yw} / (\sqrt{3}\gamma_{mw}) \tag{6.2}$$

where $V_{dw}$ is the design strength of the weld in shear. Other quantities have been defined already.

As stated earlier, in the case of complete penetration groove weld in butt joints, design calculations are not required as the weld strength of the joint is equal to or even greater than the strength of the member connected. In the case of incomplete penetration groove weld in butt joints, the effective throat thickness is computed and the required effective length is determined and checked whether the strength of the weld is equal to or greater than the strength of the member connected or the applied external force.

### 6.10.2 Fillet Welds

The actual distribution of stress in a fillet weld is very complex. A rigorous analysis of weld behaviour has not been possible so far. Multi-axial stress state, variation in yield stress, residual stresses, and strain hardening effects are some of the factors, which complicate the analysis. In many cases, it is possible to use the simplified approach of average stresses in the weld throat given as follows.

Consider each of the fillet welds connecting the two plates as shown in Fig. 6.42 of length $L$ and throat thickness $a$, which is inclined at $\alpha = \tan^{-1}(s_2/s_1)$. Each welds transfer a longitudinal shear $V_L$ and transverse forces of shears $V_{Tx}$ and $V_{Ty}$ between the plates. The average normal and shear stresses $f_w$ and $\tau_w$ on the weld throat may be expressed in terms of forces as (Trahair et al. 2001)

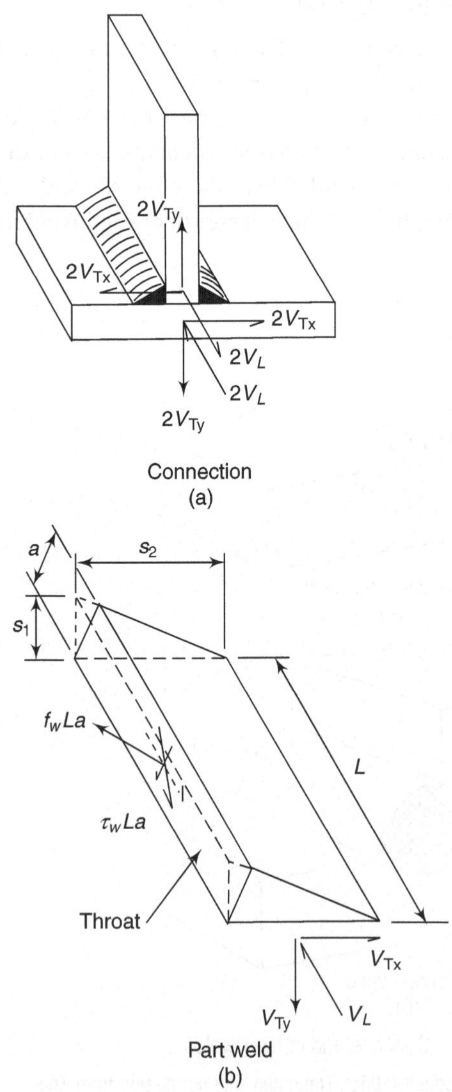


**FIG. 6.42** Forces in a fillet welded joint (Trahair et al. 2001)

$$f_w La = V_{Tx}\sin\alpha + V_{Ty}\cos\alpha \quad (6.3)$$

$$\tau_w La = [V_{Tx}\cos\alpha - V_{Ty}\sin\alpha)^2 + V_L{}^2]^{0.5} \quad (6.4)$$

In addition to these there are local stresses in the weld arising from bending effects, shear lag, and stress concentrations, together with longitudinal normal stresses induced by compatibility between the weld and end plate, which are often neglected in the design.

Thus, it is a usual practice to assume that the static strength of the weld is determined by the average throat stresses $f_w$ and $\tau_w$ alone, and that the ultimate strength of the weld is reached when (Owens & Cheal 1989)

$$f_{UW} = \beta\sqrt{(f_w^2 + \gamma\tau_w^2)} \quad (6.5)$$

where $f_{uw}$ is the ultimate tensile stress of the weld. It has been found that the parameter $\gamma$ varies between 1.8 and 3.0. However, it was decided to adopt a value of $\gamma = 3.0$, because of the consequent similarity with von Mises Yield criterion (see Section 2.10 of Chapter 2). (Note that Eqn 6.5 relates to interactive resistance to rupture and not to onset of yielding.)

The factor $\beta$ is a function of the parent metal and weld metal strength. If the design strength $f_{uw}$ is related to the parent metal strength, then the lower the parent metal strength, the lower could be the value of $\beta$. This is because, the ratio of the weld metal strength to the parent metal strength increases with increasing parent metal strength. Typical values of $\beta$ are (Owens & Cheal 1989)

$$\beta = 0.7 \text{ for } f_y < 240 \quad (6.5a)$$

$$\beta = 0.8 \text{ for } 240 < f_y < 280 \quad (6.5b)$$

$$\beta = 0.85 \text{ for } 280 < f_y < 340 \quad (6.5c)$$

$$\beta = 1.0 \text{ for } 340 < f_y < 400 \quad (6.5d)$$

Disregarding the factor $\beta$, Eqn (6.5) can be written as

$$f_{UW} = \sqrt{(f_w^2 + 3\tau_w^2)} \quad (6.6)$$

Substituting Eqns (6.3) and (6.4) into Eqn (6.6), and rearranging we get,

$$3(V_{Tx}^2 + V_{Ty}^2 + V_L^2) - 2(V_{Tx}\sin\alpha + V_{Ty}\cos\alpha)^2 = (f_{uw}La)^2 \quad (6.7)$$

This equation is often simplified conservatively to (by ignoring the second term on the left hand side of equation)

$$V_R/La = f_{uw}/\sqrt{3} \quad (6.8)$$

where $V_R$ is the vector resultant of the applied forces $V_{Tx}$, $V_{Ty}$, and $V_L$ and is written as

$$V_R = \sqrt{(V_{Tx}^2 + V_{Ty}^2 + V_L^2)} \quad (6.9)$$

IS 800 has replaced the weld shear strength $f_{uw}/\sqrt{3}$ in Eqn (6.8) by the design strength $f_u/(\sqrt{3}\gamma_{mw})$ and considers $f_u$ as the smaller of the ultimate stress of the weld and parent metal. (Note that the weld metal always has a higher strength; hence we should use the parent metal strength only in the equation as per IS code.) It is interesting to note that other codes (e.g. British code) consider the weld metal strength also and hence as per BS 5950, the weld shear strength $f_{uw}/\sqrt{3}$ in Eqn (6.8) is replaced by the lesser of $0.5U_e$ and $0.55U_s$, where $U_e$ and $U_s$ are the minimum ultimate tensile strengths of weld electrode and parent metal, respectively.

Hence as per IS 800 : 2007 the design strength is given by

$$P_{dw} = L_w t_e f_u/(\sqrt{3}\gamma_{mw}) \quad (6.10a)$$

or

$$P_{dw} = L_w K s f_u/(\sqrt{3}\gamma_{mw}) \quad (6.10b)$$

where $P_{dw}$ is the design strength of the fillet weld and $s$ is the size of the weld. Other terms have been defined earlier.

Tables 6.4 and 6.5 have been prepared to simplify the calculation while using Eqn (6.10).

**TABLE 6.4** Design strength $f_{wd}$ in weld material for $f_u = 410$ MPa

| Type of stress | Design stress, N/mm$^2$ |
|---|---|
| Direct tension or compression (shop welding) | 189 |
| Direct tension or compression (site welding) | 158 |

For other electrodes $f_{wd} = f_u/(\sqrt{3} \times \gamma_{mW})$ and $\gamma_{mw} = 1.25$ (shop welding) = 1.5 (site welding).
Capacity in kN for fillet weld (shop welded) = 0.7 × size × 189 × effective length/1000

**TABLE 6.5** Design capacity of fillet welds

| Leg length, s (mm) | Design capacity per unit length, $R_{nw}$ (kN/mm) | | | |
|---|---|---|---|---|
| | Fe 410 with E43 and E51 electrodes | | Fe 540, shop welded | |
| | Shop welded, $f_{wd}$ = 189 MPa | Site welded, $f_{wd}$ = 158 MPa | With E43, $f_{wd}$ = 189 MPa | With E51, $f_{wd}$ = 235 MPa |
| 4 | 0.529 | 0.442 | 0.529 | 0.658 |
| 5 | 0.661 | 0.553 | 0.661 | 0.822 |
| 6 | 0.793 | 0.663 | 0.793 | 0.987 |
| 8 | 1.058 | 0.884 | 1.058 | 1.316 |
| 10 | 1.323 | 1.106 | 1.323 | 1.645 |
| 12 | 1.587 | 1.327 | 1.587 | 1.974 |
| 15 | 1.984 | 1.659 | 1.984 | 2.467 |
| 18 | 2.381 | 1.990 | 2.381 | 2.961 |
| 20 | 2.646 | 2.212 | 2.646 | 3.290 |
| 22 | 2.910 | 2.433 | 2.910 | 3.619 |
| 25 | 3.307 | 2.765 | 3.307 | 4.112 |

$R_{nw} = 0.7 \times s \times f_{wd}/1000$

The design procedure is as follows.

1. Assume the size of the weld based on the thickness of the members to be joined.
2. By equating the design strength of the weld to the external factored load, the effective length of the weld to be provided is calculated. The length may be provided either as longitudinal fillet welds (parallel to the load axis) or as transverse fillet welds (perpendicular to the load axis) along with longitudinal fillet welds. It is a common practice to treat both the welds as if they are stressed equally. If the length exceeds $150t_e$, reduce design capacity as per code.
3. If only the longitudinal fillet weld is provided, a check is made to see if the length of each longitudinal fillet weld is more than the perpendicular distance between them.
4. End returns of length equal to twice the size of the weld are provided at each end of the longitudinal fillet weld.

### 6.10.3 Intermittent Fillet Welds

Intermittent fillet welds are provided to transfer calculated stress across a joint, when the strength required is less than that developed by a continuous fillet weld of the smallest practical size. Such intermittent welds are often found in the connection of stiffeners to the web of plate girders. The fillet weld length required is computed as a continuous fillet weld. A chain of intermittent fillet welds of total length equal to the computed length, is provided as shown in Fig. 6.43. Intermittent fillet welds shown in Fig. 6.43(a) are structurally better than those shown in Fig. 6.43(b), since they reduce the distortion due to the balancing nature of the welds.

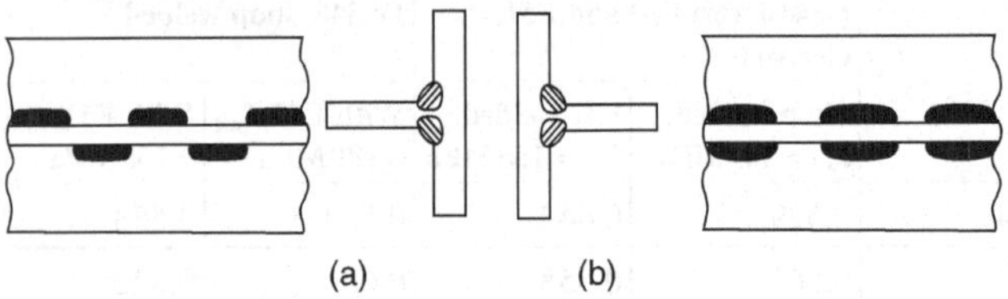


**FIG. 6.43** Intermittent fillet weld

When providing intermittent fillet welds, the following points should be kept in mind.

1. Intermittent fillet welds are not economical, unless the weld is of minimum size. It is because, a smaller size weld of a longer length is more economical for the same strength. (Though the strength of the fillet weld increases proportionally with the size, the weight of the weld metal increases as the square of the size.)
2. Intermittent welding is difficult to make, if automatic welding is adopted.
3. For exposed conditions, continuous welding offers better life and ease of maintenance. However, if intermittent welds are used in such conditions, the welds should be returned around the corners.
4. Intermittent welding should not be used when a structure is subjected to dynamic or repetitive loads.
5. Though intermittent groove welds may be used in butt joints to resist shear forces only, they are not generally recommended.

In the design of intermittent welds, the following procedure is adopted (IS 816 : 1969).

1. Assume the size of weld and compute the total length of required intermittent weld.
2. The minimum effective length (four times the size of weld or 40 mm) and clear spacing ($12t$ for compressions and $16t$ for tension and should not be less than 200 mm, where $t$ is the thickness of the thinner plate joined) clauses of IS codes should be followed.
3. At the ends, the longitudinal intermittent fillet weld should be of length not less than the width of the member, otherwise transverse welds should be provided. If transverse welds are provided along with longitudinal intermittent welds, the total weld length at the ends should not be less than twice the width of the member.

### 6.10.4 Slot or Plug Welds

Longitudinal fillet welds in slots should be considered as having the same strength as ordinary longitudinal fillet weld. In addition, the following points should be considered (IS 816 : 1969).

1. Width or diameter of the weld should be less than three times the thickness or 25 mm, whichever is greater.
2. Corners at the enclosed ends or slots should be rounded with a radius not less than 1.5 times the thickness or 12 mm, whichever is greater.
3. The distance between the edge of the part and the edge of the slot or hole or between adjacent slots or holes, should be not less than twice the thickness and not less than 25 mm for holes.
4. A combination of plug weld and other types of welds is permissible and the strength of the joint is the sum of the individual capacities of the welds.

Slots if provided should be set back behind the beginning of the welds on the edges of the plate to ensure that the effective area of the plate is not reduced and that the deposition of slots does not lead to serious stress concentration. Thus, the joint shown in Fig. 6.44(b), may appear to have higher strength than the one in Fig. 6.44(a); however it fails at a smaller load due to the high stress concentration developed at the re-entrant angles of the notches.

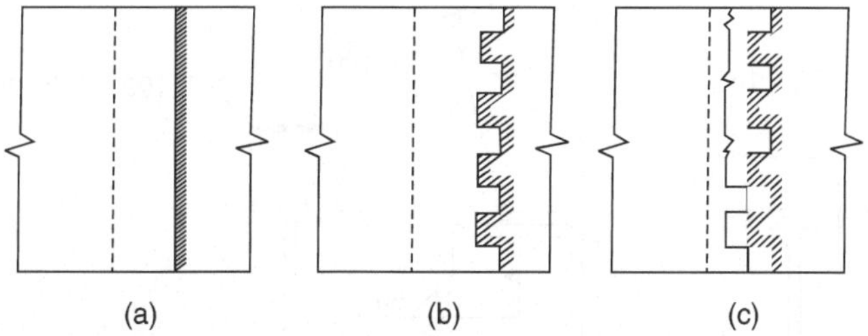


**FIG. 6.44** Wrong method of increasing length of fillet welds by means of square slots (a) Joint with fillet weld (b) Joint with slot and fillet weld (c) Failure of the joint shown in (b)

## 6.11 SIMPLE WELDED JOINTS

In this section we will discuss the design of some simple welded joints such as truss member connections, angle seat connections, web angle and end seat connections, and end plate connections (see Fig. 6.45).

### 6.11.1 Design of Fillet Welds for Truss Members

In the design of welds connecting tension or compression members, the welds should be at least as strong as the members they connect and the connection should not result in significant eccentricity of loading. Truss members often consist of single or double angles, and occasionally T-shapes and channels. Consider the angle tension member shown in Fig. 6.46, with two longitudinal welds (on the two sides parallel to the axis of the load) and a transverse weld (perpendicular to the axis of the load). The axial force $T$ in the member will act along the centroid of the member. The force $T$ has to be resisted by the forces $P_1$, $P_2$, and $P_3$ developed by the weld lines. The forces $P_1$ and $P_2$ are assumed to act at the edges of the angle and the force $P_3$ at the centroid of the weld length, located at $d/2$. Taking moments about point A located at the bottom edge of the member and considering clockwise moments as positive, we get

$$\Sigma M_A = -P_1 d - P_2 d/2 + Ty = 0 \tag{6.11}$$

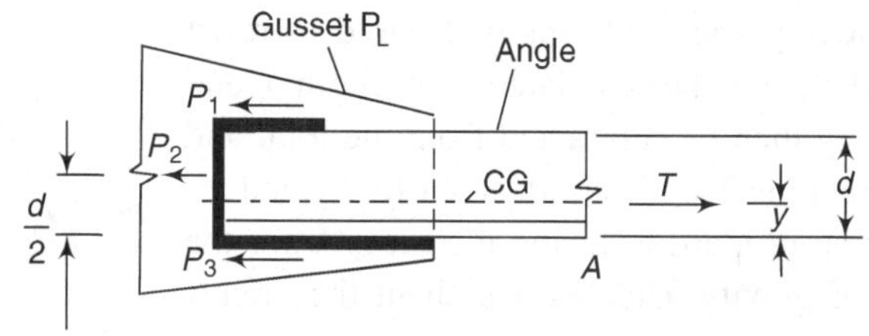


**FIG. 6.46** Balancing the welds on a tension member connection

Hence,

$$P_1 = Ty/d - P_2/2 \tag{6.12}$$

The force $P_2$ is equal to the resistance $R_w$ of the weld per mm multiplied by the length $L_w$ of the weld.

$$P_2 = R_w L_{w2} \tag{6.13}$$

Considering the horizontal equilibrium, we get

$$\Sigma F_H = T - P_1 - P_2 - P_3 = 0 \tag{6.14}$$

Solving Eqns (6.12) and (6.14) simultaneously, we get

$$P_3 = T(1 - y/d) - P_2/2 \tag{6.15}$$

Designing the connection shown in Fig. 6.46, to eliminate the eccentricity caused by the unsymmetrical welds is called *balancing the weld.* The procedure adopted for balancing the weld is as follows.

1. After selecting the proper weld size and electrode, compute the force resisted by the end weld $P_2$ (if any) using Eqn (6.13).
2. Compute $P_1$ using Eqn (6.12).
3. Compute $P_3$ using Eqn (6.15) or

$$P_3 = T - P_1 - P_2 \tag{6.16}$$

4. Compute the lengths $L_{w1}$ and $L_{w3}$ on the basis of

$$L_{w1} = P_1/R_w \text{ and } L_{w3} = P_3/R_w \tag{6.17}$$

**FIG. 6.45** Some simple welded connections (a) Column splice with CJP grove weld (b) Welded double angle connection (c) Unstiffened seat connection (Copyright © American Institute of Steel Construction Inc., Reprinted with permission. All rights reserved.)

Alternatively, the total length required to resist the load, $L_w$ may be calculated. The length of end weld may then be subtracted from the total and the remaining length is allocated to $P_1$ and $P_2$ in inverse proportion to the distances from the centre of gravity. Discussions about the effects of balanced connections on angle compression members have been provided by Temple and Sakla (1994).

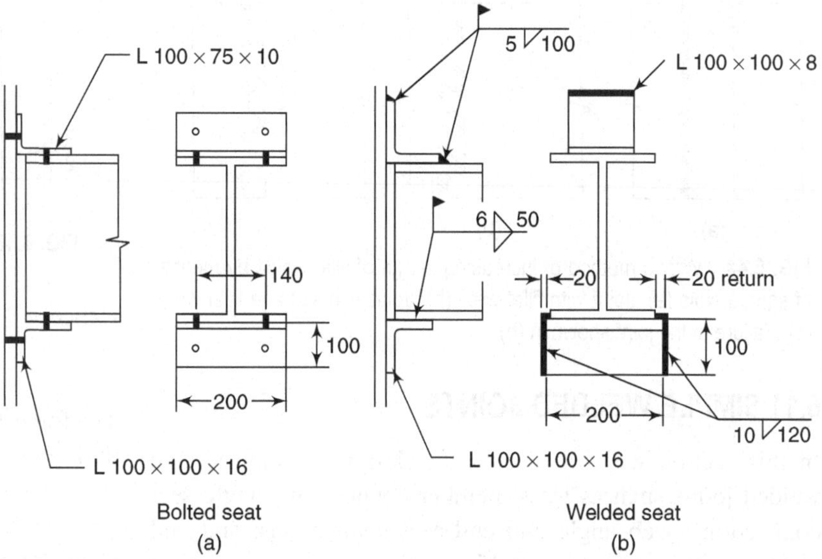


**FIG. 6.48** Welded seat angle connection (for comparison bolted seat angle connection is also shown; The angle sizes and weld sizes may change depending on load)

## *Single Sided Welds in Tension*

Examples of unsatisfactory and satisfactory welds for tension connections are given in Fig. 6.47. In the examples shown in Fig. 6.47(a), the eccentricity between the line of action of the load and the throat centroid creates a moment on the weld throat. Hence, this should be avoided in practice. In the symmetric arrangements shown in Fig. 6.47(b), though there is a small variation in stress across the weld throats, with little ductility, this variation is redistributed, resulting in uniform stress fields on the weld throats.

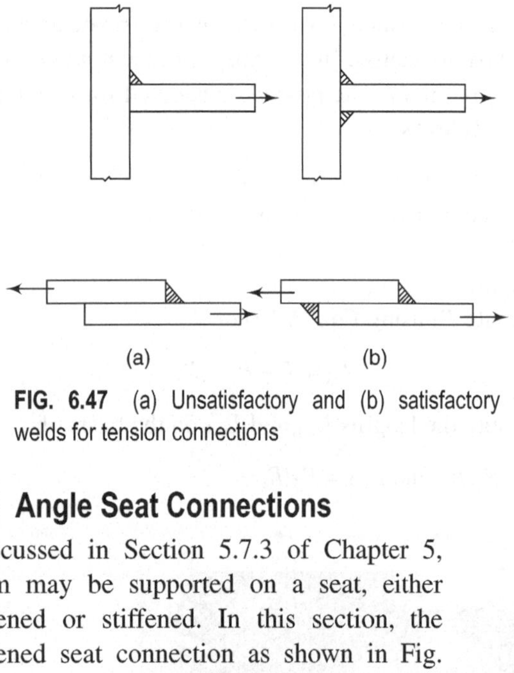


**FIG. 6.47** (a) Unsatisfactory and (b) satisfactory welds for tension connections

## 6.11.2 Angle Seat Connections

As discussed in Section 5.7.3 of Chapter 5, a beam may be supported on a seat, either unstiffened or stiffened. In this section, the unstiffened seat connection as shown in Fig. 6.48 is discussed, where an angle is designed to carry the entire reaction. This type of connection uses a top clip angle, whose intended function is to provide lateral support to the compression flange. The seated connection is designed to transfer only the vertical reaction and should not give significant restraining moment at the end of the beam. Hence, the seat and top angle are selected in such a way that they are relatively flexible. The behaviour of welded seat angle connections has been studied by Roeder & Dailey (1989).

The thickness of seat angle is determined by the flexural strength at a critical section of the angle as shown in Fig. 6.49 (the bolted seat angle is shown for comparison). When the beam is attached to the seat with bolts (which is often the case), as shown in Fig. 6.49(b), the rotation of the beam at the end creates a force that tends to restrain the pull away from the column. Thus, the critical section for flexure is near the base of the fillet on the outstanding leg. For welded seat, since the weld along the end holds the angle tight against the column, the critical section is the same (whether or not the beam is attached to the seat) as that for the previous case of bolted beam, connected to seat. For practical purposes, the distance $c$ may be taken at 9.5 mm from the face of the angle. The bending moments on the critical section of the angle and on the connection to the column are determined by multiplying the beam reaction with the distance to the critical section.

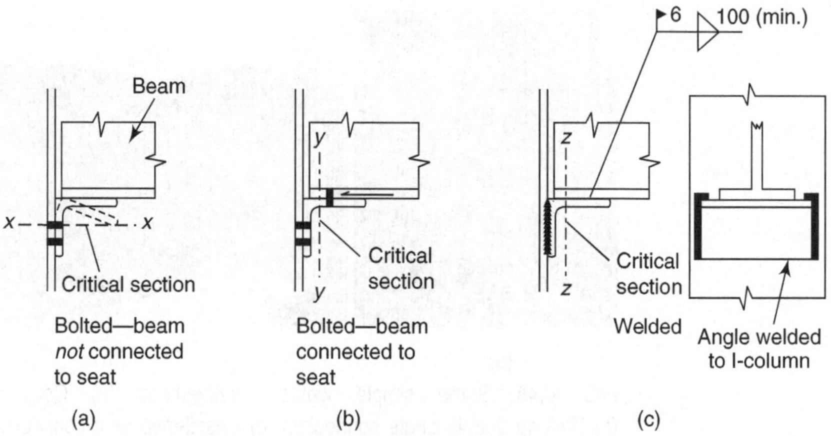


**FIG. 6.49** Critical section for flexure on seats

The code does not state how the computation of this bending moment is to be made. The conservative approach of assuming the reaction at the center of the full contact width (see Fig. 6.50a) will result in excessively thick angles. The less conservative approach of assuming the reaction at the center of the required bearing length $b$ (called as stiff bearing length in clause 8.7.1.3 of code) measured from the end of the beam (see Fig. 6.50b), has been suggested by Blodgett (1966). Another rational distribution is a triangular distribution as shown in Figs. 6.50(c) and (d) for flexible angle and thick angle, respectively (Salmon & Johnson 1996).

The design of the unstiffened angle seat involves the following steps.

1. Selection of seat angle having a length equal to width of the beam.
2. Length of the outstanding leg of the seat angle is calculated on the basis of web crippling of the beam.

$$b = R/[t_w(f_{yw}/\gamma_{m0})] \tag{6.18}$$

   where $R$ is the reaction of the beam, $t_w$ is the thickness of the web of the beam, $f_{yw}$ is the yield strength of the web, and $\gamma_{m0}$ is the partial safety factor for material = 1.10.
3. Determine the length of the bearing on cleat

$$b_1 = b - (t_f + r_b) \tag{6.19}$$

   where $t_f$ is the thickness of the flange of the beam and $r_b$ is the root radius of the beam flange.
4. Determine the distance from the end of the bearing on cleat to the root of the angle

$$b_2 = b_1 + g - (t_a + r_a) \tag{6.20}$$

   where $g$ = erection clearance + tolerance, $t_a$ is the thickness of the angle, and $r_a$ is the root radius of the angle.
5. The bending moment at the critical section may be calculated by assuming that the reaction from beam is uniformly distributed over bearing length $b_2$

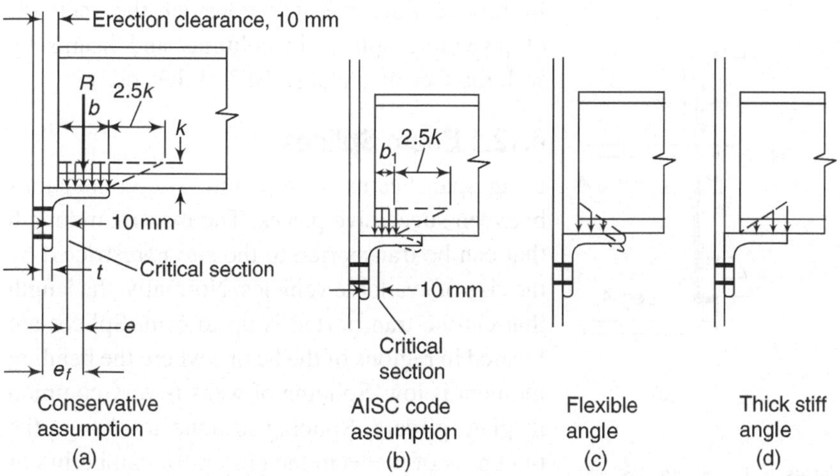


**FIG. 6.50** Bearing stress assumptions for seated connections

$$M_u = R \times (b_2/b_1) \times b_2/2 \tag{6.21}$$

   By equating it to the strength of solid rectangular section (angle leg), bent about its weak axis, the thickness of the seat angle may be determined.
6. Determine the required weld size.
   (a) Without taking eccentricity, the length of the weld on each side can be found out by using

$$L_w = R/(2 \times R_w) \tag{6.22}$$

   where $R_w$ is the strength of weld per mm.
   (b) If the eccentricity is considered, the resultant force in the weld due to shear and bending (Salmon & Johnson 1996) is given by

$$R_{\text{res}} = [R/(2L_w^2)]\sqrt{[L_w^2 + 20.25(b_2/2)^2]} \tag{6.23}$$

### 6.11.3 Web Angle and End Plate Connections

The field-welded shear connection using web angles is shown in Fig. 6.51. The intension of such a connection is that the angles are as flexible as possible so that the beams are capable of rotating at the ends. [The single plate framing connection discussed in Section 5.7.7 is a modification of web angle connection where a single plate instead of the pair of angles is welded to the column flange (either in shop or in field) and then bolted to the beam web at site (Astanesh et al. 1989).] They are assumed to provide simply supported end condition and designed to transmit shear only. The pair of angles are shop welded to the beam and field welded or connected to the column by means of HSFG bolts at site. The angles (called clip angles) project out of the beam web by a distance of about 12 mm (called set back), so that the beam can be fitted with acceptable tolerances. When beams intersect and have the same depth, the flanges are coped (cut away) as shown in Fig. 6.51(c), resulting in some loss of shear strength. The coping of beams will result in block shear failure and are susceptible to local web buckling (Gupta 1984; Yura et al. 1982).

Erection bolts are used to erect these beams and then the angles are welded at site. These bolts are often provided at the bottom of the angle. Generally 100-mm size legs are used for connecting the beam and the leg size at the column or girder side is kept a little longer. The length of the angle is kept equal to the distance between fillets of the beam, so that sufficient length is available for welding. Usually a weld size 2–3 mm smaller than the web angle thickness is chosen.

The connection, though assumed to transfer only shear forces, due to the eccentricity of connection, is also subjected to a bending moment (Blodgett 1966). Due to the rotation effect, the field welds cause web angles to press against the beam web at the top and tear

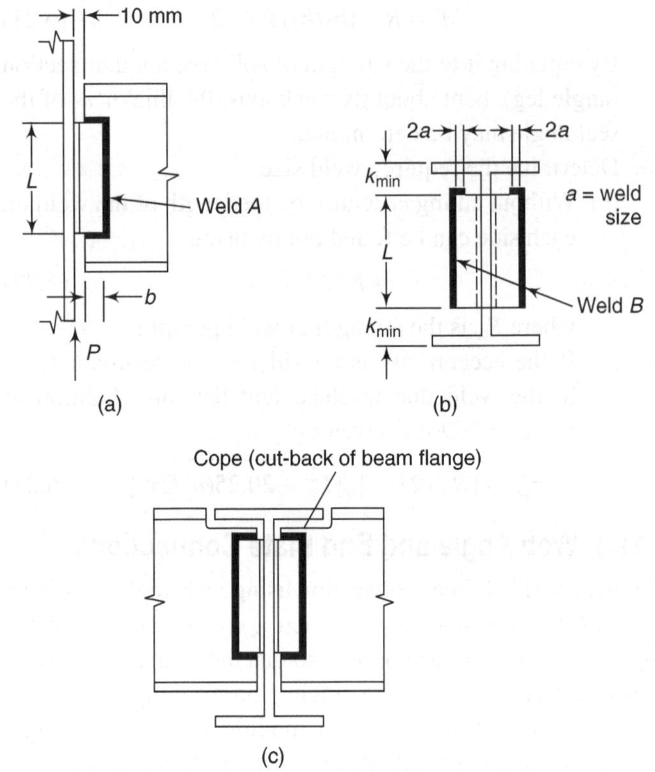


**FIG. 6.51** Simple shear double-angle connections

apart from the bottom, thus indicating horizontal shear in the fillet weld. It is assumed that the neutral axis is at a distance of $L/6$ from the top of the angle. The horizontal shear is taken as zero at this point and maximum at the bottom of the angle (see Fig. 6.52). Neglecting the effects of the returns at top, the horizontal component $R_x$ can be obtained from moment equilibrium (Blodgett 1966).

Applied moment from load = Resisting moment of weld

$$(P/2)e_2 = (2/3)RL \tag{6.24}$$

where $L$ is the length of weld

Hence,

$$R = 0.75Pe_2/L$$

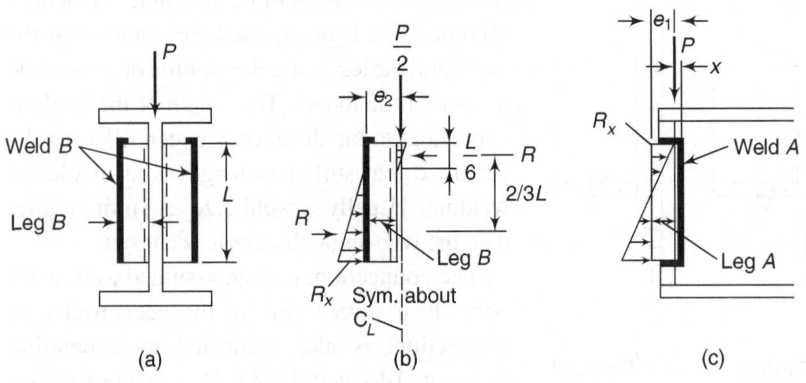


**FIG. 6.52** Eccentric shear in the plane of the web

*Source:* Blodgett (1966). Reprinted with permission from The James F. Lincoln Arc Welding Foundation, USA.

From force triangle, we get

$$R = 0.5R_x \times (5/6)L$$

From these two equations, we get

$$R_x = 9Pe_2/(5L^2) \tag{6.25}$$

Vertical force on weld

$$R_v = P/(2L) \tag{6.26}$$

Resultant force on weld

$$R_{res} = \sqrt{\{[9Pe_2/(5L^2)]^2 + [P/(2L)]^2\}} \tag{6.27}$$

or

$$R_{res} = [P/(2L^2)]\sqrt{(L^2 + 12.96e_2^2)}\,\text{N/mm} \tag{6.28}$$

This equation neglects eccentricity $e_1$, which tends to cause tension at the top of the weld lines [see Fig. 6.52(c)]. The flexural tension component $R_x$ at the top of the weld $B$ is

$$R_x = My/I = Pe_1(L/2)/[2L^3/12] = 3Pe_1/L^2 \tag{6.29}$$

Thus,

$$R_{res} = \sqrt{[P/(2L^2)] + (3Pe_1/L^2)^2} \tag{6.30}$$

or

$$R_{res} = P/(2L^2)\sqrt{(L^2 + 36e_1^2)}\,\text{N/mm} \tag{6.31}$$

Note that these equations disregard the weld returns, which have the greatest effect if $L$ is short. Considering the returns to be equal to $L/12$, Salmon and Johnson (1996), derived the following equation

$$R_{res} = P/(2L^2)\sqrt{(L^2 + 20.25e_1^2)}\,\text{N/mm} \tag{6.32}$$

## 6.12 BEAM AND COLUMN SPLICES

As already mentioned in Chapter 5, sections used for beams and columns may have to be spliced due to practical reasons. In this section, we will consider the methods of providing splices in columns and beams by welding (see also clause 10.7 of the code).

### 6.12.1 Beam Splices

Long span beams may require site connections between successive pieces. The maximum length that can be transported to the site is restricted by the size of available vehicles. Normally, the length that can be transported is up to 6 m. Splices are located in regions of the beams where the bending moment is low. Splicing of webs is very common in plate girders. Splicing is done to change the thickness of plates in the girder, for cambering or when the required length of plate is not available.

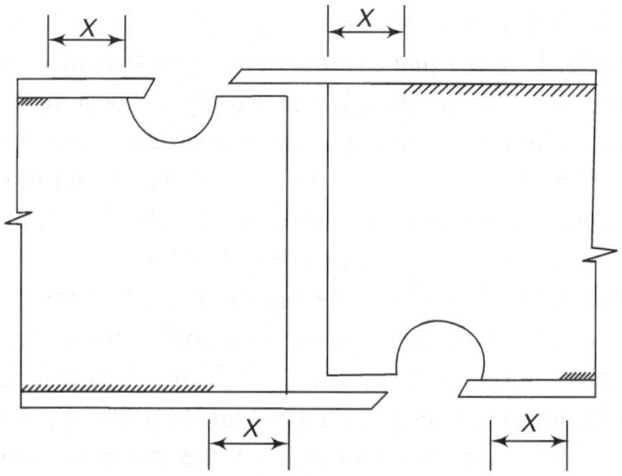


**FIG. 6.53** Staggered form of arrangement for groove welded beam splice

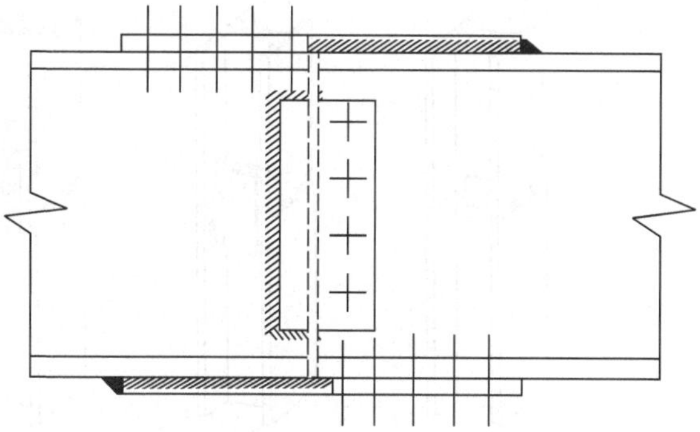

**FIG. 6.55** A hybrid splice connection

Beams may be spliced using groove welds and using a staggered form of arrangement as shown in Fig. 6.53. Alignment of the flange may be facilitated by omitting the flange-to-web welds for a short distance *X* on either side of the connection before assembly, as shown in this figure. In order to minimize distortion, the flange-to-web welds are completed after completing the web-to-web welds. The cope holes in the web adjacent to the flange welds will improve the access for welding the flanges. Filling of the cope holes may be necessary in box girders, for corrosion protection. This kind of splicing requires temporary support during welding and is often expensive (though providing aesthetic appearance).

The alignment problems of groove welded splices can be overcome by using splice plates, which can be fillet welded (see Fig. 6.54). Hybrid connections consist of splice plates attached to one beam section by welding (done in fabrication shop), which are connected to the other beam section by HSFG bolts (see Fig. 6.55). In this case, the final tightening of the bolts has to be carried out after the welding is complete. Since both welding and drilling of holes are necessary, this type of hybrid connection is used rarely. Note that when welds are combined with fasteners of bearing type in a joint, the welds must be designed to transfer all the load, and the capacity of bolts are usually ignored.

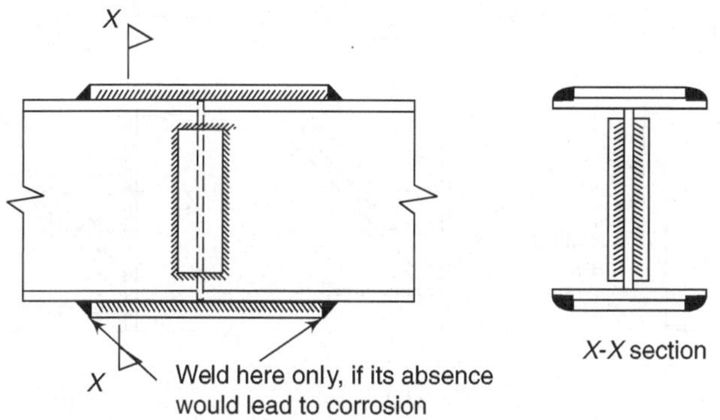


**FIG. 6.54** Fillet welded beam splice

## 6.12.2 Column Splices

When column sections are to be spliced by welding (see Fig. 6.56), the ends are first milled for a square bearing surface. Then the splice angles are shop welded [see Figs 6.56(a) and (c)]. The outstanding legs of these angles are provided with holes for erection bolts to engage the outstanding legs of the other two angles that are shop welded to the upper column section. Since they do not project beyond the ends of the columns, they are not damaged during transit. Once these two sections are brought to the site, the splice angles are bolted and the two column sections are welded using full penetration groove welds or partial penetration welds. Instead of splice angles, splice plates can be used [see Figs 6.56(b), (d), and (e)]. In Fig. 6.56(e), notice the filler plates, which are used when two different sizes of columns are connected. It is also possible to use welded end plates for making column splices. Both faces of these end plates are machined for bearing and normally have a thickness of more than 20 mm [see Figs 6.56(f), (g), (h), and (i)]. When two different sizes of columns are connected, a web stiffener [Fig. 6.56(i)] is used, which assists in diffusing the load into the lower column. This web stiffener should have similar proportions to the upper column flange. Though column splices are normally placed just above floor level, in earthquake regions they must be located in regions of near zero bending moments.

# 6.13 BRACING CONNECTIONS

Bracings are often used in framed construction to enhance the lateral stiffness (see Section 2.4.2 in Chapter 2). Thus in these braced frames, diagonal members are added to the basic rectangular layout in the vertical and/or horizontal planes. Such addition requires modifications to the beam-to-connections as shown in Fig. 6.57. When the arrangement maintains concentric member center lines as in Fig. 6.57(a), then the additional forces for which the individual connection components must be designed can be determined easily.

Resolving the tensile force F in the upper diagonal into vertical and horizontal components, the forces at the various member interfaces are determined as follows:

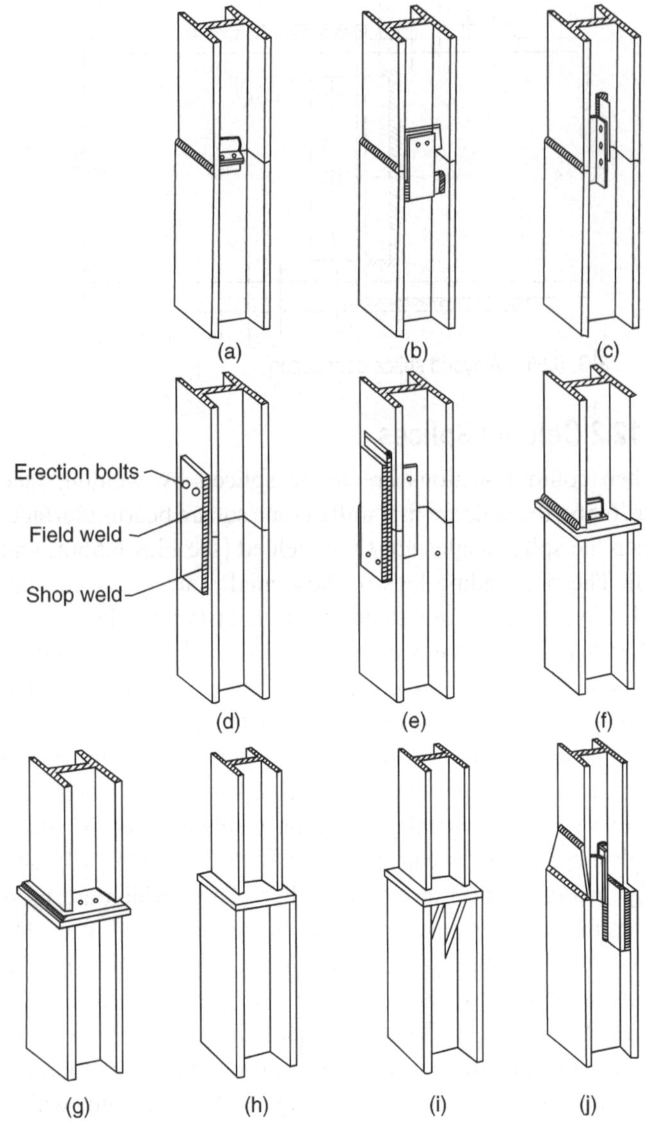


**FIG. 6.56** Welded column splices

*Source:* Blodgett (1966). Reprinted with permission from The James F. Lincoln Arc Welding Foundation, USA.

Welds between gusset and beam = Horizontal shear = F cos $\alpha$

Welds between gusset and column = Vertical shear = F sin $\alpha$

Bolts between gusset and column = Vertical shear + End reaction due to beam loads = F cos $\alpha$

This assumption neglects the effect of the eccentricity of half the column width which produces a moment on the column flange bolts equal to the product of the total vertical shear times the eccentricity.

Shifting the point of intersection of the member axes to the column face would eliminate this effect; however the same effect has to be transferred to the column. For example, a more compact gusset can be achieved if the axes meet at the column-beam flange intersections, as shown in Fig. 6.57(b). Since the eccentricity is in the plane of the column and beam webs, the eccentric moment produced can usually be designed to be resisted easily by the bending strength of the column and or beam about their major axes. The gusset welds and the bolts are in pure shear and hence need not be checked for shear/tension as would be required for Fig. 6.57(a).

Figure 6.57(c) shows a bracing connected to the column only. This results in even greater simplicity since far fewer bolts are required. Furthermore, the beam does not require any holes for connecting the gusset and can thus be detailed as a standard beam. The end plate on the gusset is, however, subject to prying action and hence needs to be of adequate thickness. The bolts would also need to have sufficient tensile strength and it is better to use HSFG bolts.

For bracings with lighter loads the detail shown in Fig. 6.57(d) may be used. This type of connection not only has a compact gusset, but also results in an economical solution. The plate is shop-welded to the column and needs to be located accurately. This can be done by marking the plate as shown and providing a corresponding mark on the column flange; these marks should be included on the detailed drawing.

In cases where the beam forms part of the bracing system and acts as a tie or a strut, the horizontal component of the force in the inclined bracing member has to be transmitted into it. In details shown in Figs 6.57(a) and (b), this is done through the beam flange, but in details shown in Figs 6.57(c) and (d), the force is transmitted from the bracing to the column and then from the column into the beam end. When the resulting force in the beam is tensile, the end plate and its bolts have to be checked for this force, in addition to the end shear in the considered beam.

As mentioned already, the effect of the eccentricity in connections in Figs 6.57(b), (c), and (d) is usually not

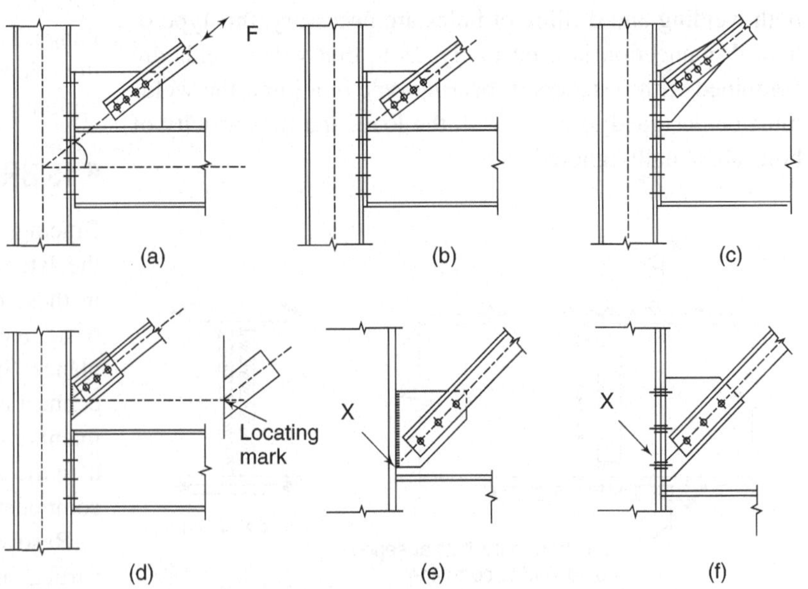


**FIG. 6.57** Different types of bracing end connections

significant, but should be checked nevertheless, especially in cases in Figs 6.57(c) and (d) where it is more pronounced.

When a gusset as shown in Fig. 6.57(c) is used, the line of action of the bracing force should pass through the mid-depth of the flange of the cleat as shown. However, in details shown in Figs 6.57(e) and (f), although the bracings have been brought closer to the beam flange thus reducing the bracing-beam-column eccentricity, the eccentricity on the flange bolts and the welding will cause a serious overstress at points X because of the concentration of force at these points.

An explanation to the various possible approaches to the design of bracing connections including seismic design of steel braced frame systems may be found in Sabelli et al. (2013) and Thornton and Muir (2009).

## 6.14 TUBULAR CONNECTIONS

*Tubular members* and hollow sections are being used extensively in plane and space frames, because they are more efficient in resisting compression (due to the large radius of gyration). However, connecting two tubular members is possible only through welding and requires extensive end preparation of the tubes. Using automatic oxyacetylene tube cutting machines, it is now possible to cut the ends to any required shapes [see Fig. 6.58(a)]. If such machines are not employed, connections have to be done using gusset plates [Fig. 6.58(b)]. In such connections, the tubes are slit, the gusset plate is inserted and welded. The ends of the branch tubes are either capped with plates or quarter sphere caps and welded to the gusset plates of the main tube [see Fig. 6.58(c)]. Gusset plates provide additional length for fillet welding to the tube. However, they tend to stiffen the tube resulting in concentration of stress in the tubes at end of the gusset plate. This effect may be reduced by tapering the gusset plates at their ends. For this reason, gusset plates should not be used in connections subjected to fatigue loading. Welding of the tubular members may be done by using complete penetration groove weld, fillet weld, or a combination of both. Special patented connectors are used in space trusses. Figure 6.58(d) shows such a connection (MERO) developed in Germany. Figure 6.59 shows a *double layer space grid* built using such a connector and situated in Vatican City. More details of such connections may be found in Subramanian (2006).

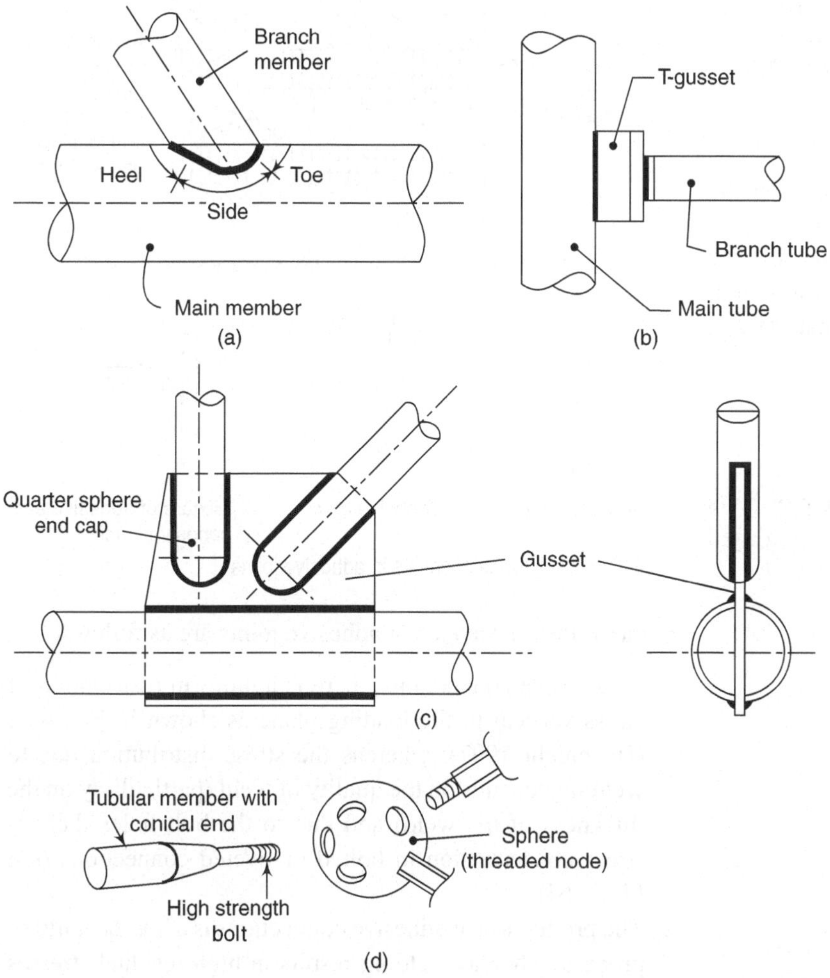


**FIG. 6.58** Tubular connections

The strength of connection welded to the face of a *hollow structural section* (HSS) is primarily a function of the geometric parameters of the connection and cannot be increased by simply increasing the weld strength (Kloiber 2001). Moreover, unlike open sections such as wide flange shapes, it is not possible to add a stiffener to the back of the connection to provide a more uniform load transfer. Instead, the weld must be designed to ensure that yielding and redistribution will occur in the base metal, thereby allowing the weld to develop the required design load before reaching its ultimate breaking strength (Kloiber 2001).

The most common direct welded connections are axially loaded truss type members forming joints in a T, Y, or K configurations. The main member or chord should always be checked to make sure the wall thickness is adequate for all of the connection forces delivered by branch members. The joint configuration for K-type truss connections can be classified as gapped or overlapped and as stepped or matched as shown in Fig. 6.60.

*Gapped connections* are more economical to fabricate, as they require only a single bevel cut and the gap allows fit up tolerance. Stepped connections are more economical to fabricate than matched connections, because they permit the use of fillet welds. Matched connections require a flare-bevel groove weld, which are relatively inefficient.

FIG. 6.59 A double layer space grid in Vatican City

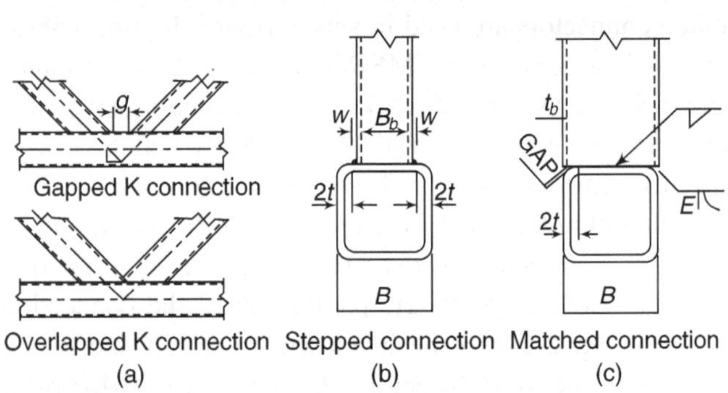


FIG. 6.60 Connections in hollow structural sections

Single plate framing connections (see Fig. 6.61) can be used to connect wide flange beam to hollow section columns and beams. The design of tubular connections is outside the scope of this book and the details may be found in Marshall (1991), Packer & Henderson (1997), AISC Manual (1997), Zhao et al. (2001), Kosteski & Packer (2003).

## 6.15 APPLICATION OF ADHESIVES

Normally steel members are connected with the help of bolts (in the past by rivets) or welds. Adhesives are now being used as an alternative to bolts and welds. Until now, adhesives have not been used extensively due to the lack of knowledge about the mechanism, design calculations, and long-term resistance as well as the working of the sticking surfaces. However, in the 1960s, some bridges were built with adhesive connections, with some additional bolts for added safety. Since there was no knowledge of the methods of calculation, the construction and design of these bridges were based on experiments. These bridges are still performing well, testifying the long-lasting behaviour of these adhesives (Brockmann & Neeb 2001). The quality of these adhesives has been greatly improved and a variety of adhesives have been developed.

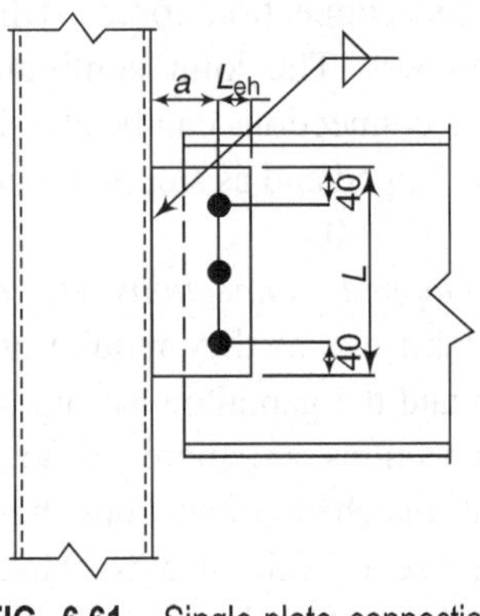


FIG. 6.61 Single-plate connection in hollow sections

The adhesive technique is a positive jointing of substances. The active cohesive and adhesive forces in an adhesive layer are responsible for its strength (Fig. 6.62). If one or both of these forces are exceeded due to stress, the adhesive connection fails. The various failure modes are the cohesive crack (fracture of adhesive layer when the internal forces are exceeded), adhesive crack (due to separation of materials), or a combination of both. These are shown in Fig. 6.63.

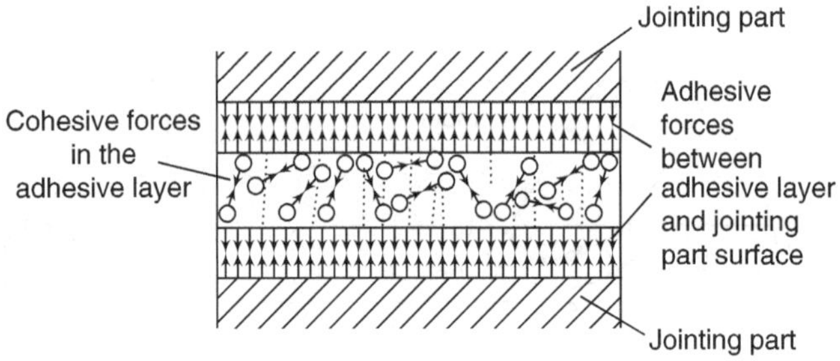


FIG. 6.62 Adhesive and cohesive forces in an adhesive sealing

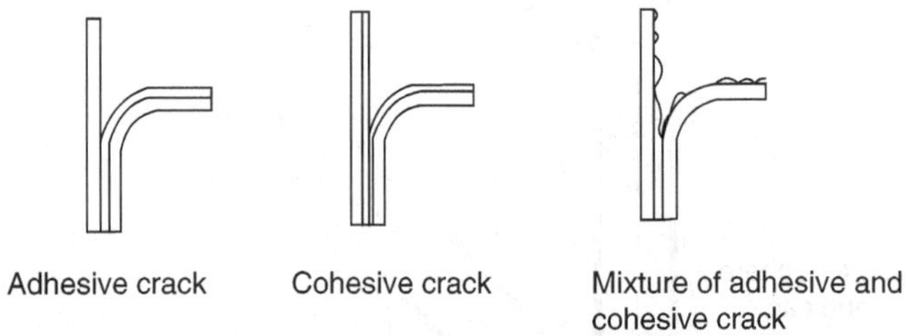


FIG. 6.63 Types of cracks in adhesive joints

Some of the advantages of adhesive joints are as follows.

1. In adhesive connections, there is a uniform distribution of stress vertical to the loading plane as shown in Fig. 6.64 (Habenicht 1997), whereas the stress distribution due to welding depends on the quality of weld (particularly on the thickness of the weld) and due to the bolt holes there is stress concentration in bolted or riveted connections (see Fig. 6.64).
2. The production of adhesive connections is a low-heat fitting process, whereas welding results in high residual stresses in the material.

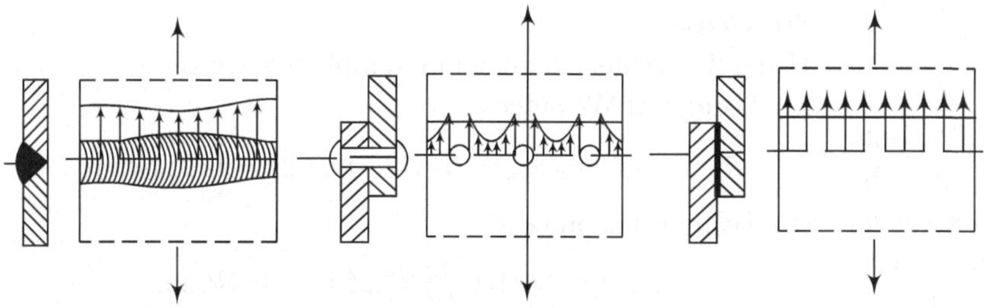

FIG. 6.64 Stress distribution of different types of joints

3. In joints subjected to cyclic loads, adhesive functions like a viscous damper because of the lower elastic modules.

However, the adhesive joints have the following drawbacks.

1. Adhesive joints require the preparation of jointing surfaces and need certain time for the curing of the adhesive to acquire sufficient strength.
2. Adhesives are affected by high temperatures (>200°C).
3. The strength and deformation characteristics of the jointing parts do not match with the adhesive sealing.
4. The deformation of the adhesive sealing is influenced by the jointing parts. Adhesives have elastic–plastic/viscous-elastic deformation behaviour, whereas steel has a linear-elastic behaviour.

By conducting different tests on adhesive joints, Pasternak et al. (2004) have found that adhesive connections show higher strength and stiffness compared with bolted connections. Preloading of the bolts has no influence on the carrying capacity and the stiffness of a combined connection with bolts and adhesives (Pasternak et al. 2004). The properties of structural adhesives are provided by Hartshorn (1986) and Habenicht (1997). We hope that in future the calculation methods will be refined and many connections will be made with adhesives.

## EXAMPLES

**EXAMPLE 6.1:** Two plates of thickness 14 mm and 12 mm are to be jointed by a groove weld as shown in Fig. 6.65. The joint is subjected to a factored tensile force of 350 kN. Assuming an effective length of 150 mm, check the safety of the joint for
*Case (i)* Single-V groove weld joint
*Case (ii)* double-V groove weld joint

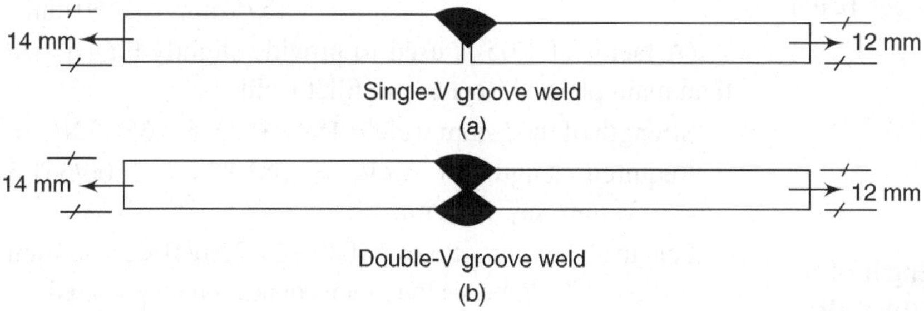


FIG. 6.65

Assume that Fe 410 grade steel plates are used and that the welds are shop welded.

**SOLUTION:**
*Case (i)* Single-V groove weld: In this case, incomplete penetration results due to single-V groove weld.

Hence, throat thickness
$t_e = 5/8t = 5/8 \times 12 = 7.5$ mm

Effective length of weld

$$L_w = 150 \text{ mm}$$

$$\text{Strength of weld} = L_w t_e f_y/\gamma_{mw} = 7.5 \times 150 \times (250/1.25)/1000 = 225 \text{ kN} < 350 \text{ kN}$$

Hence the joint is not safe.
*Case (ii)* In the case of double-V groove weld, complete penetration takes place

Throat thickness = thickness of thinner plate = 12 mm
Strength of weld = 12 × 150 × (250/1.25)/1000 = 360 kN > 350 kN

Hence, the joint is safe.

**EXAMPLE 6.2:** A web of a plate girder consists of 1000 × 16-mm plate of grade Fe 410 and is to be provided with a splice at a section where the factored shear and bending moment to be resisted by the web are $V = 1000$ kN and $M = 350$ kNm. The flange plate thickness = 50 mm each. Design a web splice.

**SOLUTION:**
Size of the web plate = 1000 × 16 mm
Thus,

$$d = 1000 \text{ mm}; \ t_w = 16 \text{ mm}$$

Overall depth of girder = $h = 1000 + 50 + 50 = 1100$ mm
Factored shear force

$$V_u = 1000 \text{ kN}$$

Factored moment

$$M_u = 350 \text{ kNm}$$

Since it is a web plate, double-V groove weld could be used.
Effective thickness of the weld

$$t_e = t_w = 16 \text{ mm}$$

Shear capacity of the weld

$$f_v = 16 \times 250/(\sqrt{3} \times 1.25) = 1847 \text{ N/mm}$$

Length of weld required to resist shear

$= V_u/f_v = 1000 \times 1000/1847 = 541.3$ mm $< d = 1000$ mm

Design bending strength of the web = $1.2 Z_e f_y/\gamma_{m0}$ or $\beta_b Z_p f_y/\gamma_{m0}$

$$= 1.2(16 \times 1000^2/6) \times 250/(1.1 \times 10^6)$$
$$= 727.3\,\text{kNm} < 350\,\text{kNm}$$

*Note:* Though the yield stress of electrodes is 330 MPa, only the yield stress in the plate is used in the calculations. Since the mechanical properties of the weld metal are always assumed to be stronger than the connecting parts, full penetration groove welds are adequately strong and hence no checking is required. These calculations have been given just to show the strength of the welded joint.

**Example 6.3:** The tie member of a truss is made of ISA 65 × 65 × 6 and is subjected to a factored tension load of 90 kN. The length of the angle is not enough to go from end to end and hence a splice has to be provided. Design a groove welded joint.

**Solution:**

Provide a single-V groove weld.

The effective throat thickness = 5/8*t* = 5/8 × 6 = 3.75 mm

Perimeter length of angle available for welding = 65 + 65 = 130 mm

Strength of the weld = 3.75 × 130 × (250/1.25)/1000 = 97.5 kN > 90 kN

Area of the angle = 744 mm$^2$

Design strength of the member = 744 × 250/1.1 = 169 kN > 90 kN

*Note:* Though the strength of the weld is adequate, partial penetration welds should be avoided in direct tension. If not, the secondary bending stresses arising from the eccentricity of the throat area have to be considered.

**Example 6.4:** Determine the effective throat dimension of a 10-mm fillet weld made by (a) shielded metal arc welding (SMAW) and (b) submerged arc welding (SAW).

**Solution:**

(a) Using SMAW process

Effective throat thickness

$$t_e = 0.7a = 0.7 \times 10 = 7\,\text{mm}$$

(b) Using SAW process

Using SAW process, we are in a position to get better penetration than SMAW process. Thus,

$$a = 10 + 2.4 = 12.4\,\text{mm}$$

Hence,

$$t_e = 0.7a = 0.7 \times 12.4 = 8.68\,\text{mm}$$

**Example 6.5:** Determine the design shear strength of a 10-mm fillet weld produced by (a) SMAW and (b) SAW, assuming E 51 electrodes, Fe 410 material, and shop welding.

**Solution:**

Using the results obtained in Example 6.4, we get

(a) Using SMAW process

$$R_w = t_e(f_u/\sqrt{3})/\gamma_{mw} = 7(410/\sqrt{3})/1.25 = 1325\,\text{N/mm}$$

(b) Using SAW process

$$R_w = 8.68(410/\sqrt{3})/1.25 = 1643\,\text{N/mm}$$

Thus, the strength of the weld produced by using the SAW process is about 1.25 times higher than the strength of the weld produced by using the SMAW process.

**Example 6.6:** Design a connection to joint two plates of size 250 × 12 mm of grade Fe 410, to mobilize full plate tensile strength using shop fillet welds, if (a) a lap joint is used (b) a double cover butt joint is used.

**Solution:**

*(a) Connection using lap joint*

Plate strength = $f_y A_g / \gamma_{m0}$

= 250(250 × 12)/(1.10 × 1000) = 681.82 kN

Size of the weld:

Minimum = 5 mm (Table 6.3)

Maximum = 12 – 1.5 = 10.5 (clause 10.5.8.1 of IS 800)

Though the minimum size is preferable, let us take the size of the weld as 8 mm in order to reduce the length of the connection.

Strength of the weld = $[f_u/(\sqrt{3} \times 1.25)]\ 0.7 \times \text{size}$

$= [410/(\sqrt{3} \times 1.25)] \times 0.7 \times 8$

$= 189 \times 0.7 \times 8 = 1058\,\text{N/mm}$

Required length of weld = 681.82 × 1000/1058 = 644 mm, say 650 mm

Adopting a welding scheme as shown in Fig. 6.66(a).

Weld length available at end = 150 mm

Length of weld on one side = (650 – 150)/2 = 250 mm < 150 × 0.7 × 8. Hence, long joint correction not required.

*(b) Connection using Butt joint*

Assume width of cover plate = 250 – 2 × 15 = 220 mm

Use an 8-mm thick cover plate (>5/8 × 12 = 7.5 mm)

Area of cover plate = 220 × 8 = 1760 mm$^2$

Required area of cover plate = 1.05 × 250 × 12/2 = 1575 mm$^2$ < 1760 mm$^2$

(A factor of 1.05 is used to provide slightly bigger size than main plate.) Use a 5-mm fillet weld.

Strength of the 5-mm weld = 189 × 0.7 × 5 = 661.5 N/mm

Required length of weld = 681.82 × 1000/661.5 = 1031 mm, say 1040 mm

Length of the connection = [(1040 – 2 × 220)/4] × 2 = 300 mm < 150 × 0.7 × 8, hence long joint connection not required.

Hence, provide two cover plates of size 300 × 220 mm as shown in Fig. 6.66(b).

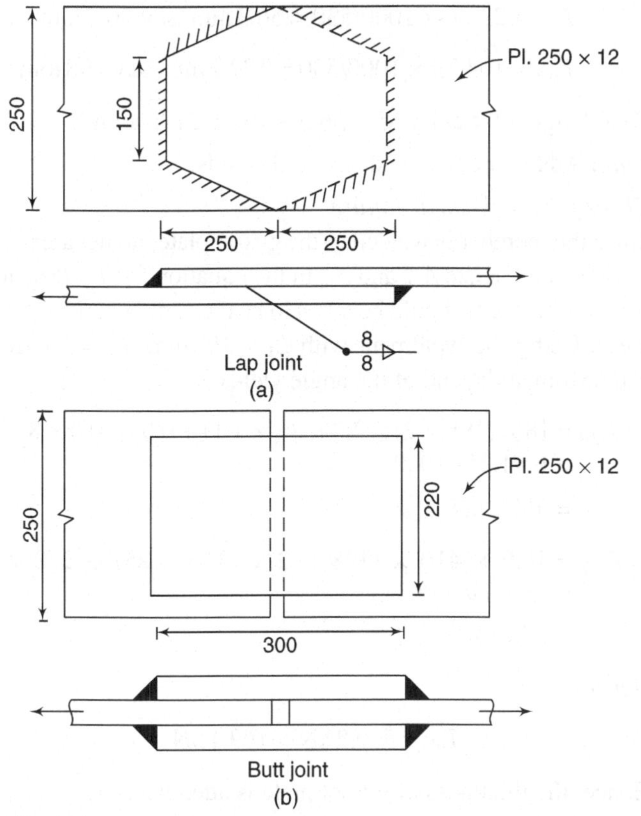


**FIG. 6.66**

**EXAMPLE 6.7:** Determine the size and length of the fillet weld for the lap joint to transmit a factored load of 120 kN shown in Fig. 6.67, assuming site welds, Fe 410 steel, and E 41 electrode. Assume width of the plate as 75 mm.

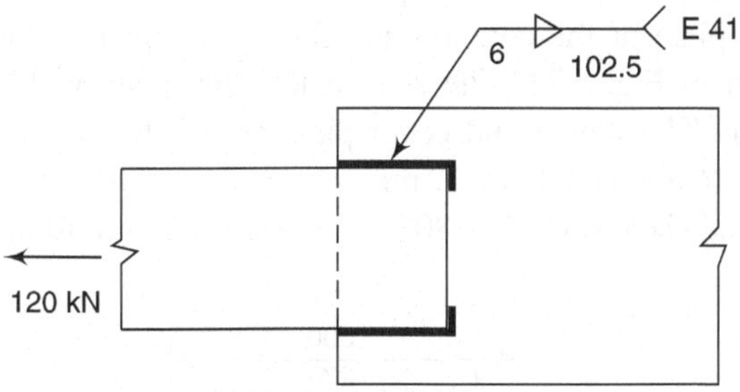


**FIG. 6.67**

**SOLUTION:**

Minimum size of weld for a 8-mm thick section = 3 mm

Maximum size of weld = 8 − 1.5 = 6.5 mm

Choose the size of weld as 6 mm.

Effective throat thickness = 0.7 × 6 = 4.2 mm

Strength of weld = $4.2 \times 410/(\sqrt{3} \times 1.5) = 662.7$ N/mm

Assuming that there are only two longitudinal (side) welds,

Required length of weld = $120 \times 10^3/662.7$

= 181 mm

Length to be provided on each side = 181/2 = 90.5 mm > 75 mm

Hence, provide 90.5-mm weld on each side with an end return of 2 × 6 = 12 mm.

Therefore, the overall length of the weld provided = 2 × (90.5 + 2 × 6) = 205 mm

**EXAMPLE 6.8:** Rework Example 6.7 using 3 mm welds.

**SOLUTION:**

Effective throat thickness = 0.7 × 3 = 2.1 mm

Strength of weld = $2.1 \times 410/(\sqrt{3} \times 1.5) = 331$ N/mm

Required length of the weld = $120 \times 10^3/331$

= 363 mm

Two possible solutions are shown in Fig. 6.68.

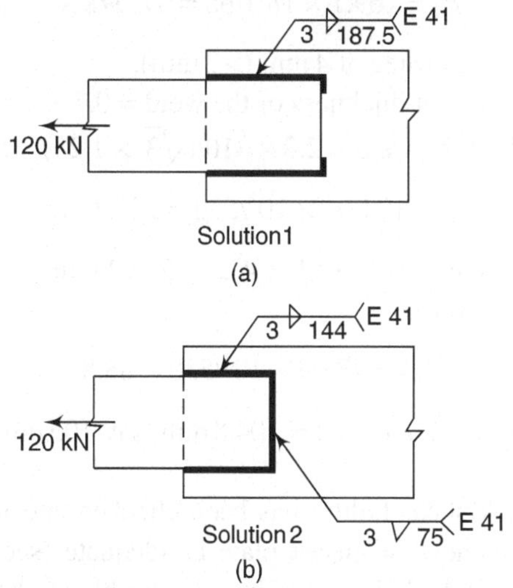


**FIG. 6.68**

(a) If only longitudinal welds are provided,

Length of each side weld = 363/2 = 181.5 mm

Total length on each side including end return = 181.5 + 2 × 3 = 187.5 mm

(b) If welds are provided on three sides

Length of each side weld = (363 − 75)/2 = 144 mm > 75 mm (width of the plate)

The solution as shown in Fig. 6.68(b) is preferred since it is more compact, reduces the overall length of the connection, and provides better stress distribution.

**EXAMPLE 6.9:** A tie member of a truss consisting of an angle section ISA 65 × 65 × 6 of Fe 410 grade, is welded to an 8-mm gusset plate. Design a weld to transmit a load equal to the full strength of the member. Assume shop welding. See Fig. 6.69.

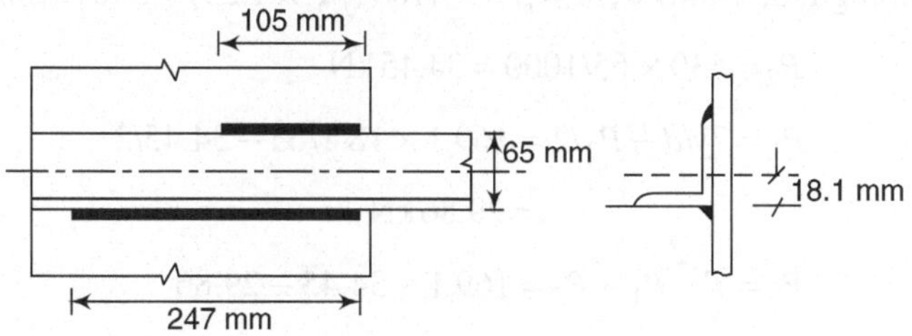


**FIG. 6.69**

**SOLUTION:**
From IS handbook no.1 or IS 808, the properties of ISA 65 × 65 × 6 are

$$A = 744\,\text{mm}^2$$

$$C_z = 18.1\,\text{mm}$$

Tensile capacity of the member = 744 × 250/1.1 = 169.1 kN

The force resisted by the weld at the lower side of the angle

$$P_1 = 169.1 \times (65 - 18.1)/65 = 122.01\,\text{kN}$$

Force to be resisted by the upper side of the angle

$$P_2 = 169.1 \times 18.1/65 = 47.09\,\text{kN}$$

Assuming a weld size of 4 mm (> 3 mm),

Effective throat thickness of the weld = 0.7 × 4 = 2.8 mm

Strength of the weld = $2.8 \times 410/(\sqrt{3} \times 1.25) = 530$ N/mm

$$L_{w1} = 122.01 \times 10^3/530 = 230.1\,\text{mm}$$

Hence, provide 230.1 + (2 × 4)2 = 246.1 mm, say 247 mm length at the bottom

$$L_{w2} = 47.09 \times 10^3/530 = 88.8$$

Provide 88.8 + (2 × 4) × 2 = 104.8 mm, say 105 mm length at the top.

The block shear failure has been checked and it is found that the thickness of gusset plate is adequate (see Example 6.10 for the calculations involved in the block shear failure check).

**EXAMPLE 6.10:** Design a joint according to the instructions given in Example 6.9, if the welding is done on three sides of the angle as shown in Fig. 6.70.

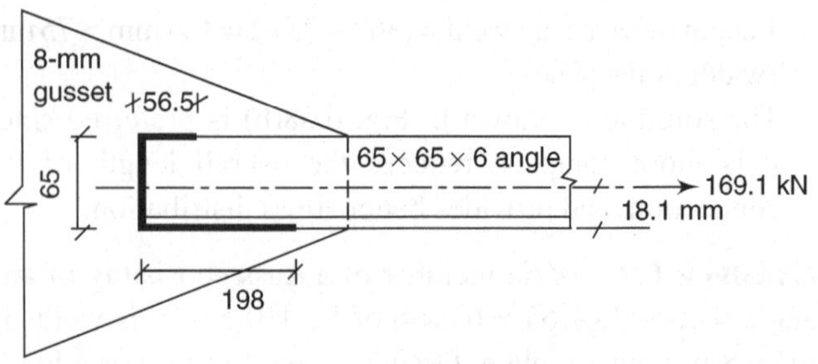


**FIG. 6.70**

**SOLUTION:**
Strength of 4-mm weld = $2.8 \times 410/(\sqrt{3} \times 1.25) = 530$ N/mm

$$P_2 = 530 \times 65/1000 = 34.45\,\text{kN}$$

$$P_1 = Ty/d - P_2/2 = 169.1 \times 18.1/65 - 34.45/2$$

$$= 29.86\,\text{kN}$$

$$P_3 = T - P_1 - P_2 = 169.1 - 34.45 - 29.86$$

$$= 104.79\,\text{kN}$$

$$L_{w1} = 29.86 \times 1000/530 = 56.3\,\text{mm, say } 56.5\,\text{mm}$$

$$L_{w3} = 104.79 \times 1000/530 = 197.7\,\text{mm, say } 198\,\text{mm}$$

Total length of weld = 65 + 56.5 + 198 = 319.5 mm

***Note*:** Add twice the weld size at the ends.

*Check for block-shear failure*

Since the member is welded to the gusset plate, no net areas are involved and hence $A_{vn}$ and $A_{tn}$ in the equation for $T_{db}$ (Section 6.4.1 of the code) should be taken to be the corresponding gross areas. Using the weldment with $L_1$ = 198 mm, $L_2$ = 56.5 mm and 65 mm at the end of the angle yields

$$T_{db1} = [8 \times (198 \times 2) \times 250/(\sqrt{3} \times 1.1) + 0.9 \times 410 \times 8 \times 65/1.25]/1000$$

$$= 569.2\,\text{kN}$$

$$T_{db2} = [0.9 \times 410 \times (198 \times 2)/(\sqrt{3} \times 1.25) + 250 \times 8 \times 65/1.1]/1000$$

$$= 658.1\,\text{kN}$$

Hence,

$$T_{db} = 569.2\,\text{kN} > 169.1\,\text{kN}$$

Hence, the thickness of gusset plate is adequate.

***Note:*** $L_2$ does not enter into this calculation because a shear rupture of the gusset plate along the toe of the angle runs for the full length of the contact with the toe, 198 mm, instead of only the length $L_2$.

**EXAMPLE 6.11:** Design a suitable fillet weld to connect the web plate to the flange plate and the flange plate to the cover plate of the compression flange of a plate girder as shown in Fig. 6.71. The size of the web plate is 1000 × 10 mm. The flange and cover plate are of the dimensions 300 × 16 mm and 275 × 12 mm, respectively. The maximum factored shear force $V$ = 800 kN. Assume shop welding.

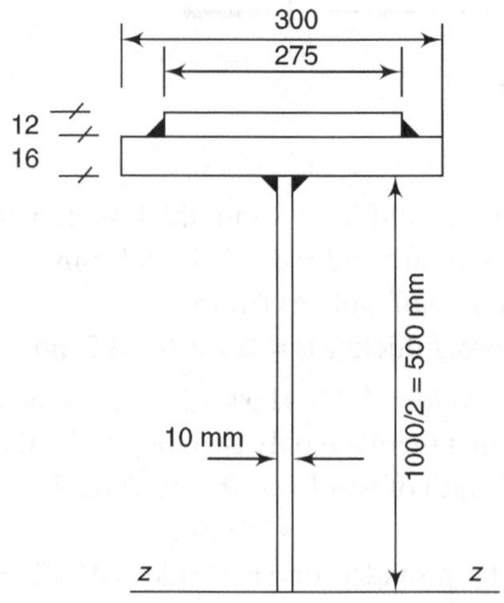


**FIG. 6.71**

**SOLUTION:**

*Connection of web and flange plate*

Minimum size of weld = 3 mm

Maximum size of weld = 10 – 1.5 = 8.5 mm

Adopt a 6-mm size weld

$$\tau_{vf,\,cal} = VA_f\bar{Y}_f/(I_{zz}\,\Sigma t)$$

$$t = 0.7 \times 6 = 4.2\,\text{mm}$$

$$\Sigma t = 2 \times 4.2 = 8.4\,\text{mm}$$

$$A_f\bar{Y}_f = 300 \times 16(500 + 8) + 275 \times 12(500 + 16 + 6)$$

$$= 243.84 \times 10^4 + 172.26 \times 10^4$$

$$= 416.1 \times 10^4$$

$$I_{zz} = 2 \times [275 \times 12^3/12 + 275 \times 12 \times 522^2 + 300 \times 16^3/12 + 300 \times 16 \times 508^2] + 10 \times 1000^3/12$$

$$= 5.109 \times 10^9\,\text{mm}^4$$

$$\tau_{vf,\,cal} = 800 \times 10^3 \times 416.1 \times 10^4/(5.109 \times 10^9 \times 8.4)$$

$$= 77.56\,\text{N/mm}^2 < 410/(\sqrt{3} \times 1.25) = 189\,\text{N/mm}^2$$

Hence, the weld is safe.

*Connection of flange plate to cover plate*

Adopt a 6-mm fillet weld.

$$t_e = 6 \times 0.7 = 4.2\,\text{mm}$$

$$\Sigma t = 2 \times 4.2 = 8.4\,\text{mm}$$

$$A\bar{Y} = 275 \times 12 \times (500 + 16 + 6) = 172.26 \times 10^4$$

$$\tau_{vf,\,cal} = 800 \times 10^3 \times 172.26 \times 10^4/(5.109 \times 10^9 \times 8.4)$$

$$= 32.11\,\text{N/mm}^2 < 189\,\text{N/mm}^2$$

Hence the weld is safe.

**EXAMPLE 6.12:** A circular pipe of 1.5 m diameter is fabricated using a 16 mm thick plate in Fe 410 grade steel by lapping over and welding it using 8 mm shop fillet weld on the inside and outside of the pipe as shown in Fig. 6.72. Determine the safe internal pressure that can be allowed inside the pipe.

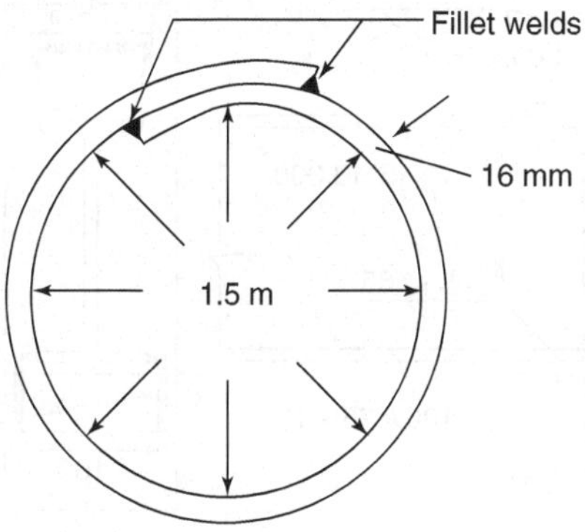


**FIG. 6.72**

**SOLUTION:**

For shop welding, $\gamma_{mw} = 1.25$

Size of weld = 8 mm

Effective throat thickness, $t_e = 0.7 \times 8 = 5.6\,\text{mm}$

Internal hoop force causing bursting of pipe = $p_u\,d/2$

Resistance offered by the weld per unit length = $\dfrac{2t_ef_u}{\sqrt{3}\gamma_{mw}}$

Hence, $\dfrac{p_ud}{2} = \dfrac{2t_ef_u}{\sqrt{3}\gamma_{mw}}$

Or $p_u = \dfrac{4t_ef_u}{d\sqrt{3}\gamma_{mw}} = \dfrac{4 \times 5.6 \times 410}{1500 \times \sqrt{3} \times 1.25} = 2.828\,\text{N/mm}^2$

Safe internal pressure = 2.828/1.5 = 1.885 N/mm$^2$

**EXAMPLE 6.13:** Determine the service load permitted on the connection shown in Fig. 6.73. Assume field welding and Fe 410 steel.

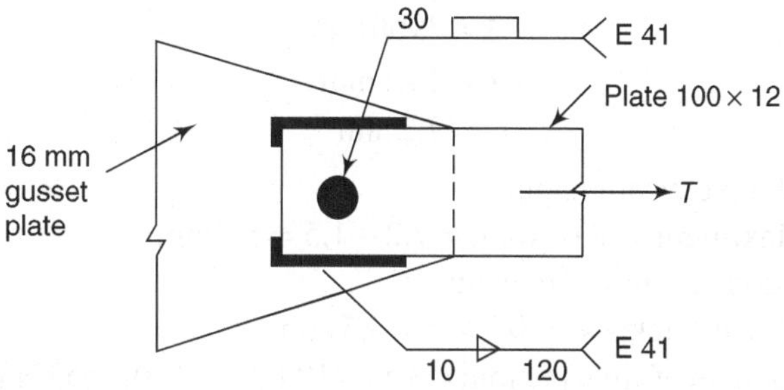


**FIG. 6.73**

**SOLUTION:**

The design strength per mm supplied by a 10-mm fillet weld,

$$R_{nw} = t_ef_y/(\sqrt{3} \times \gamma_{mw})$$

$$= 0.7 \times 10 \times 410/(\sqrt{3} \times 1.5) = 1104\,\text{N/mm}$$

Strength provided by the fillet welds = 120 × 1104/1000 = 132.48 kN

The design strength of a plug weld of diameter 30 mm

$$= \pi \times 30^2/4 \times 410/(\sqrt{3} \times 1.5 \times 1000) = 111.55\,\text{kN}$$

Design strength based on the weld = 132.48 + 111.55 = 244.03 kN

Tensile capacity of the plate

$f_yA_g/1.1 = 250 \times (12 \times 100)/(1.1 \times 1000) = 272.7\,\text{kN} > 244.03\,\text{kN}$

Service load carrying capacity

$$244.03 = 1.5T$$

Therefore,

$$T = 244.03/1.5 = 162.68\,\text{kN}$$

**EXAMPLE 6.14:** An ISMC 250 is used to transmit a factored force of 700 kN. The channel section is connected to a gusset plate 10 mm thick as shown in Fig. 6.74. Design a fillet weld if the overlap is limited to 300 mm. Use slot welds if required.

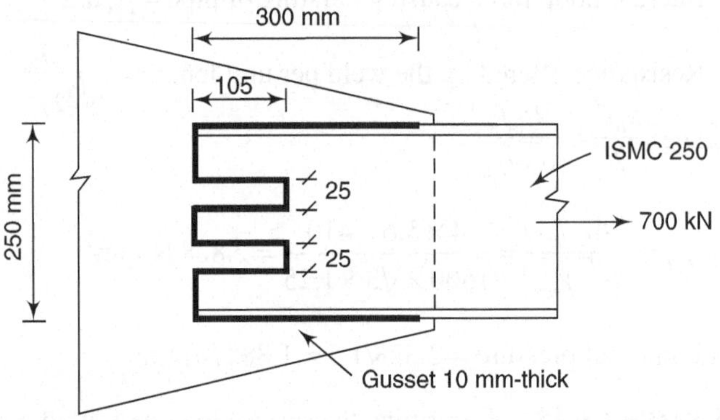


FIG. 6.74

**SOLUTION:**

The properties of ISMC 250 are

$$A = 3900\,\text{mm}^2$$
$$t_f = 14.1\,\text{mm}$$
$$t_w = 7.2\,\text{mm}$$

As slot welds are required

Maximum size of weld = 7.2 – 1.5 = 5.7 mm

Adopt a 5-mm size weld.

Throat thickness = 0.7 × 5 = 3.5 mm

Strength of weld per mm = $3.5 \times 410/(\sqrt{3} \times 1.5) = 552$ N/mm

Required length of weld = 700 × 1000/552 = 1268 mm, say 1270 mm

The maximum length of weld that can be provided in the channel

$$= 300 \times 2 + 250 = 850\,\text{mm} < 1270\,\text{mm}$$

Hence, use two slot welds of width 25 mm ($3t = 3 \times 7.2 = 21.6$ or 25 mm, whichever is greater)

Assume the length of slot weld as $x$ mm, then

$$1270 = 2 \times 300 + 250 + 4x$$

Hence,

$$x = 105\,\text{mm}$$

Therefore, provide 105-mm long fillet welds in slots as shown in Fig. 6.74.

**EXAMPLE 6.15:** Design a welded seat angle connection between a beam MB 300 and column HB 200 for a reaction of beam 100 kN, assuming Fe 410 grade steel ($f_y = 250$ MPa) and site welding.

**SOLUTION:**

For MB 300, $t_f = 13.1$ mm, $r_b = 14.0$ mm, and $t_w = 7.7$ mm

1. Width of flange MB 300 = 140 mm. Since the bottom seat angle has to be welded to the flange, assume length of angle = 140 + 2 × 10 = 160 mm.

2. Length of bearing required at the root line of the beam

$$b = R/[t_w(f_{yw}/\gamma_{m0})]$$
$$= 100 \times 1000/(7.7 \times 250/1.1)$$
$$= 57.15\,\text{mm}$$

Length of bearing on cleat = $b_1 = b - (t_f + r_b)$

$$= 57.15 - (13.1 + 14.0)$$
$$= 30.05\,\text{mm}$$

Assume end clearance = 5 mm, tolerance = 3.5 mm, and therefore

$$g = 5 + 3.5 = 8.5\,\text{mm}$$

Required length of outstanding angle = 57.15 + 8.5 = 65.65

Distance from the end of bearing on cleat to the root of the angle (assuming an angle of 100 × 75 × 10 mm) with $r_a = 8.5$ mm

$$b_2 = b_1 + g - (t_a + r_a)$$
$$= 30.05 + 8.5 - (8.5 + 10) = 20.05$$

Assume a uniformly distributed load over the bearing length $b_1$. Moment at root of angle due to load,

$$M_u = 100(20.05/30.05) \times 20.05/2 = 668\text{ Nm}$$

Moment capacity of the leg of the angle

$$= 1.2 \times (250/1.1) \times (160 \times 10^2/6) \times 10^{-3} = 727\text{ Nm} > 668\text{ Nm}$$

*Size of weld*

Maximum weld size = 10 – 1.5 = 8.5 mm

Minimum weld size = 5 mm

Use a 5-mm weld.

Strength of the 5-mm weld = $(5 \times 0.7) \times 410/(\sqrt{3} \times 1.5)$ = 552 N/mm

Length of the weld per side = $100 \times 10^3/(2 \times 552) = 90.5$ mm

Take the length of the weld per side = 90.5 + 2 × 5 = 100.5 mm, say 110 mm.

Use a clip angle of size 60 × 60 × 6 with a 3 mm weld.

The connection details are shown in Fig. 6.75.

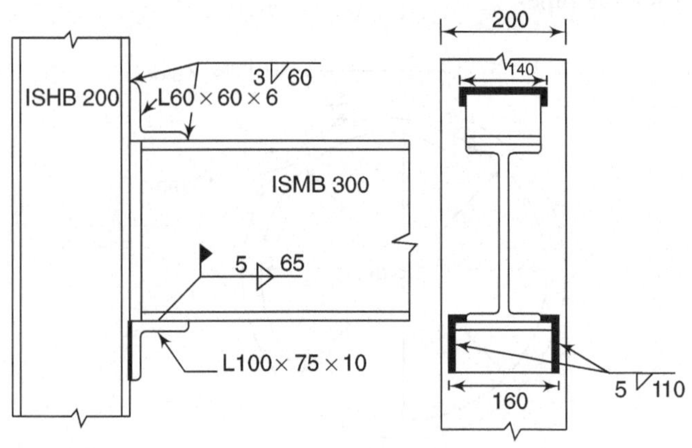


FIG. 6.75

$L_w = 110 -$ end returns $= 100$ mm

Taking eccentricity into account,

$$R_{es} = [R/(2L_w^2)]\sqrt{[L_w^2 + 20.25 \times (b_2/2)^2]}$$

$$= [100 \times 1000/(2 \times 100^2)]\sqrt{[100^2 + 20.25 \times (20.05/2)^2]}$$

$$= 549 \text{ N/mm}$$

Capacity of a 5-mm weld = 552 N/mm > 549 N/mm

Hence the connection is safe.

**EXAMPLE 6.16:** Design a welded end plate connection for a ISMB 400 in grade Fe 410 steel to carry a reaction of 120 kN due to factored loads. The connection has to be made to the flanges of ISHB 300.

**SOLUTION:**

For MB 400, $b_{fb} = 140$ mm, $t_{fb} = 16$ mm, and $r_b = 14$ mm

For HB 300, $b_{fc} = 250$ mm

***Bolted connection to column***

For M20 grade 4.6 bolts and a 6-mm thick end plate,

Shear capacity of the bolt in single shear (from Table 5.9) = 45.3 kN

Bearing capacity of the bolt on a 6-mm thick end plate

$$= 2.5 \times 0.5 \times 20 \times 6 \times 410/1.25 = 49.2 \text{ kN}$$

Required no. of bolts = 120/45.3 = 2.64

Hence, provide four M20 as shown in Fig. 6.76.

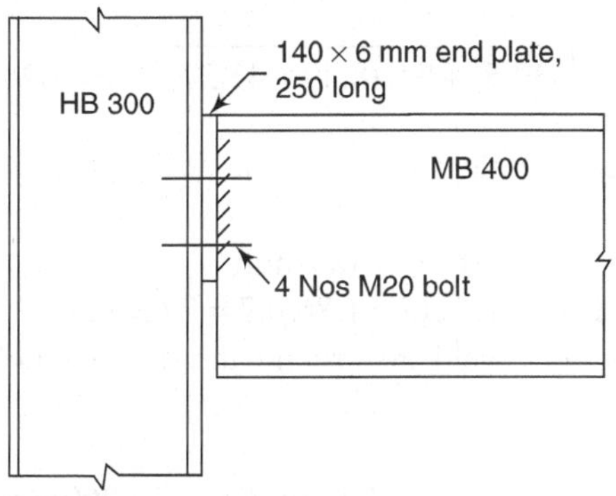


**FIG. 6.76**

***End plate***

Use a 140 × 6 mm plate, 250 mm deep.

Length of fillet welds connecting the end plate to the beam web

$$= 250 - 16 - 14 = 220 \text{ mm on each side}$$

Shear on weld $= 120 \times 10^3/(2 \times 220) = 272$ N/mm

Use a 6 mm fillet weld.

Strength of the weld $= 6 \times 0.7 \times 410/(\sqrt{3} \times 1.5) = 662$ N/mm > 272 N/mm

shear strength of end plate

$$\frac{f_y A_v}{(\sqrt{3} \times \gamma_m)} = \frac{250\,[(250 - 2 \times 22) \times 6]}{(\sqrt{3} \times 1.1)} = 162 \text{ kN} > 120 \text{ kN}$$

Hence the connection is safe.

**EXAMPLE 6.17:** Design a welded splice for an ISMB 400 section to transfer a factored bending moment of 120 kNm and a factored shear of 80 kN. Assume that the flange splice carries all the moment and that the web splice carries only the shear.

**SOLUTION:**

ISMB 400

$$t_f = 16 \text{ mm}, t_w = 8.9 \text{ mm, and } b_f = 140 \text{ mm}$$

Force in the flange $= \text{BM}/(D - t_f)$

$$= 120 \times 10^3/(400 - 16) = 312.5$$

Assuming 8-mm weld,

Capacity $= 8 \times 0.7 \times 410/(\sqrt{3} \times 1.5) = 883.7$ N/mm

Required length of the weld $= 312.5 \times 1000/883.7 = 354$ mm

Hence, adopt a cover plate of length = 400 mm

Use a 100-mm splice plate.

Required thickness of plate $= 312.5 \times 10^3/(250 \times 100/1.1)$

$= 13.75$ mm

Provide two flange splice plates of size 400 × 100 × 16 mm.

***Web splice***

Try 8-mm thick web splice plates on both sides of the web

Adopt a 6-mm fillet weld,

Capacity of the weld = 662.7 N/mm

Required length of the weld $= 80 \times 1000/662.7 = 120$ mm

Since the welds are on both sides, provide a 60-mm weld on each side and provide a web splice plate of size 200 × 60 × 8 mm. See Fig. 6.77 for the details.

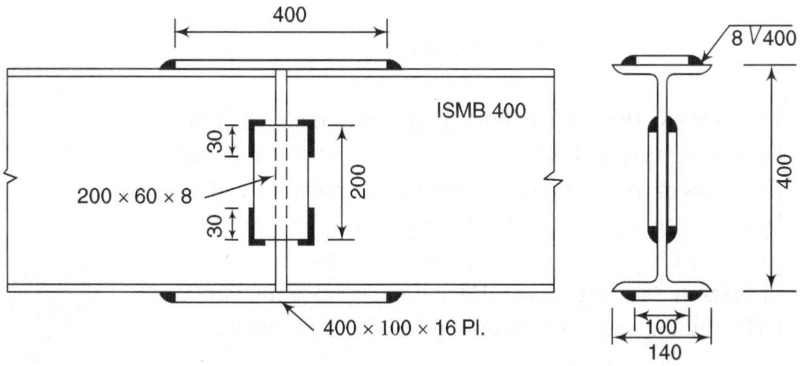


**FIG. 6.77**

**EXAMPLE 6.18:** Determine the factored load capacity $P_u$ of weld $B$ shown in Fig. 6.52, if a 6-mm weld is used, $L =$ 340 mm, and 100 × 75 × 8 angles are used.

**SOLUTION:**

(a) Using Eqn (6.28)

$$R_{res} = P_u/(2L^2)\sqrt{(L^2 + 12.96e_2^2)}$$

$$e_2 = 100 \text{ mm; leg size } L = 340 \text{ mm}$$

$$R_{res} = [P_u/(2 \times 340^2)]\sqrt{(340^2 + 12.96 \times 100^2)}$$

$$= 2.14P_u \times 10^{-3}$$

Strength of the 6-mm weld

$= 6 \times 0.7 \times 410/(\sqrt{3} \times 1.25) = 795.35\,\text{N/mm}$

Therefore, $2.14P_u \times 10^{-3} = 795.35$

or

$$P_u = 371.66\,\text{kN}$$

(b) Using Eqn (6.32)

$$R_{res} = [P_u/(2L^2)]\sqrt{(L^2 + 20.25\,e_1^2)}$$

Assuming that $e_1$ equals the other leg size = 75 mm

$$R_{res} = P_u/(2 \times 340^2)\sqrt{(340^2 + 20.25 \times 75^2)}$$

$$= 2.072P_u \times 10^{-3}$$

Thus,

$$2.072P_u \times 10^{-3} = 795.35$$

or

$$P_u = 383.84\,\text{kN}$$

(c) Neglecting the end returns of the weld

$$R_{res} = P_u/(2 \times 340^2)\sqrt{(340^2 + 36 \times 75^2)}$$

$$= 2.439 \times 10^{-3}P_u$$

Thus,

$$2.439 \times 10^{-3}P_u = 795.35$$

or

$$P_u = 326.04\,\text{kN}$$

**Example 6.19:** A 1.5 m long cantilever made of circular hollow section (CHS) has a diameter of 150 mm and a thickness of 6 mm. It is connected to a 12mm thick end plate by shop welding. This cantilever supports a vertical factored load of 4 kN at its free end and at a horizontal eccentricity of 400 mm (see Fig. 6.78). Design the connection between CHS and the end plate assuming Fe 410 grade steel.

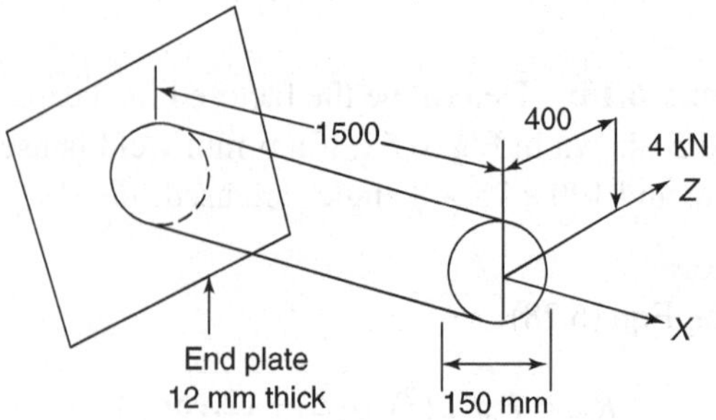


FIG. 6.78

**Solution:**

Assume that the CHS is connected to the end plate by fillet weld of thickness $t_e$. Then,

$$A_w = \pi D t_e = 471.2t_e\,\text{mm}^2$$

$$Z_w = \pi R^2 t_e = \pi \times 75^2 t_e = 17{,}671.45t_e\,\text{mm}^3$$

$$I_{pw} = 2\pi R^3 t_e = 2 \times \pi \times 75^3 t_e = 2{,}650{,}719t_e\,\text{mm}^4$$

***Design forces***

Factored shear force, $V_u = 4\,\text{kN}$

Factored bending moment, $M_u = 4 \times 1.5 = 6\,\text{kNm}$

Factored torsion, $T_u = 4 \times 0.4 = 1.6\,\text{kNm}$

Design strength of weld,

$$f_{d,w} = \frac{f_{u,w}}{\sqrt{3}\gamma_{mw}} = \frac{410}{\sqrt{3} \times 1.25} = 189.37\ \text{Mpa}$$

***Note***: for shop welding $\gamma_{mw} = 1.25$

***Stress in the weld due to applied forces***

Shear stress, $q = \dfrac{V_u}{A_w} = \dfrac{4 \times 10^3}{471.2t_e} = \dfrac{8.49}{t_e}$ MPa

Bending stress, $f_b = \dfrac{M_u}{Z_w} = \dfrac{6 \times 10^6}{17{,}671.45t_e} = \dfrac{339.5}{t_e}$ MPa

Stress due to torsion, $f_t = \dfrac{T_u R}{I_{pw}} = \dfrac{1.6 \times 10^6 \times 75}{2{,}650{,}719t_e} = \dfrac{45.27}{t_e}$ MPa

Equivalent shear stress in the weld (Clause 10.5.10.2.2 of IS 800)

$$f_e = \sqrt{f_b^2 + 3q^2} = \sqrt{339.5^2 + 3 \times (8.49 + 45.27)^2}/t_e = 352/t_e$$

The resultant stress should not exceed the design strength of the weld.

Hence, $352/t_e = 189.37$ or $t_e = 1.86\,\text{mm}$

Hence required size of weld = 1.86/0.7 = 2.65 mm

Provide 3 mm fillet weld all around the circular hollow section.

**Example 6.20:** A circular plate of diameter 200 mm is welded to anotherflat plate using 4 mm fillet weld as shown in Fig. 6.79. Calculate the torque that can be resisted by the weld, assuming Fe 410 steel and shop welding.

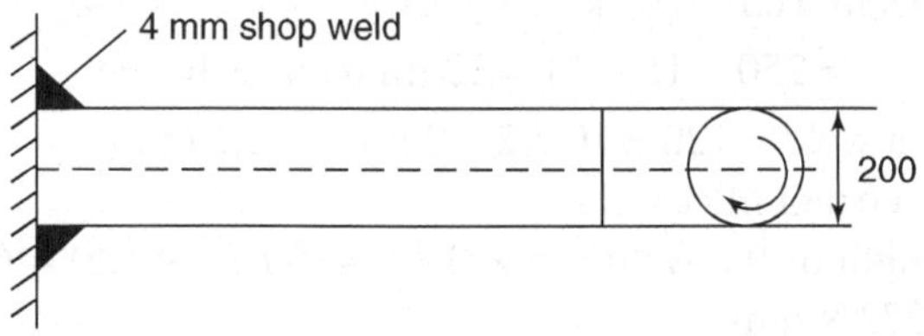


FIG. 6.79

**Solution:**

For Fe 410 steel, $f_u = 410\,\text{MPa}$

For shop welding, $\gamma_{mw} = 1.25$

Size of weld = 4 mm

Effective throat thickness, $t_e = 0.7 \times 4 = 2.8$ mm

Design strength of weld per mm length,

$$R_w = 1 \times t_e \times \frac{f_{u,w}}{\sqrt{3}\gamma_{mw}} = 1 \times 2.8 \times \frac{410}{\sqrt{3} \times 1.25} = 530.2 \text{ N/mm}$$

Total length of the weld, $L = \pi d = \pi \times 200 = 628.3$ mm

$$\text{Torque} = \frac{LR_w d}{2} = \frac{628.3 \times 530.2 \times 200}{2} \times 10^{-6} = 33.3 \text{kNm}$$

## SUMMARY

Welding is employed in several connections due to its aesthetic appearance. Moreover by using welding any difficult geometry can be jointed, whereas bolting requires some edge distances and also prior drilling. Several welding processes are available, each having some advantages and disadvantages. Users should choose a particular process based on their requirements. Welding, due to high heating, introduces distortion of connected parts. Some techniques have been given to minimize these distortions and also to reduce welding defects.

The various types of welds (groove, fillet, slot, and plug) and welded joints (butt, lap, T, and corner) have been discussed. The specifications for welded joints are also given. Most widely used welding symbols are explained. The concepts and design procedures of simple joints such as butt, lap, truss joints, angle seat, and web angle, connections have been discussed. Since it is impossible to consider all the combinations of connections, only those that are frequently encountered in practice are discussed. Splicing of beams and columns, which may be necessary due to the difficulty of transporting long length of members, are also included. The details of these improvements are also included. A brief description of the emerging jointing technique using adhesives has been included. The concepts presented have been explained with several numerical examples. Moment-resistant welded connections are covered in Chapter 14. Interested readers may find more details about the testing, analysis, or design of welded connections from the references cited in this chapter.

## MULTIPLE-CHOICE QUESTIONS

6.1 The most common manual method of welding used in structural connections is
(a) GMAW (b) ESW (c) SMAW (d) SAW

6.2 In the classification Exxxbc of SMAW electrodes, the xxx stands for
(a) suitability of welding position
(b) type of coating and operating characteristics
(c) ultimate tensile strength of weld metal
(d) none of these

6.3 Using welding, we can obtain up to
(a) 75% efficiency
(b) 85% efficiency
(c) 95% efficiency
(d) 100% efficiency

6.4 The type of welding used for connecting the member without eccentricity is
(a) groove welding
(b) fillet welding
(c) groove or fillet welding
(d) slot welding

6.5 If the thickness of groove weld is equal to that of the main plates, then the weld strength is
(a) equal to that of the main members
(b) equal or even more than that of the main members
(c) more than that of the main member
(d) about 95% of that of the main member

6.6 When plates of two different thicknesses are welded using groove welds, the effective throat thickness in stress calculations is taken as __________ of the thickness of thinner part.
(a) 1/4th (b) 5/8th (c) 6/8th (d) 7/8th

6.7 The members meeting at an angle can be welded using
(a) groove welds only
(b) fillet welds only
(c) groove or fillet welding
(d) plug welds

6.8 Which is considered as the mode of failure of a fillet weld?
(a) Shear (b) Tension (c) Bearing (d) Bending

6.9 The size of fillet weld is taken as
(a) side of the triangle of the fillet
(b) throat of the fillet
(c) size of the plate
(d) length of the fillet weld

6.10 What type of welding crack does the following image show?

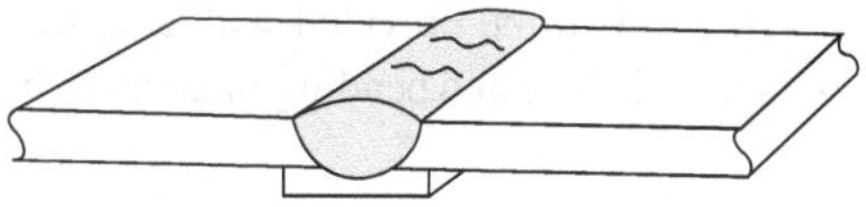

(a) Transverse (c) HAZ
(b) Centerline (d) Lamellar tearing

6.11 What type of welding crack does the following image show?

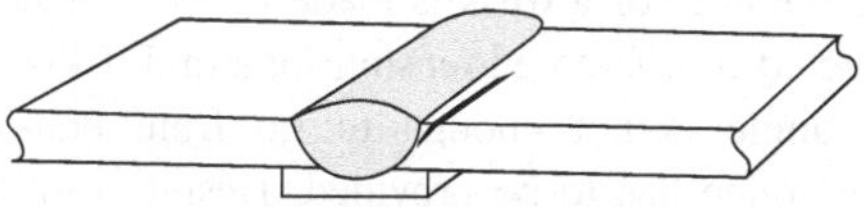

(a) Transverse (c) HAZ
(b) Centerline (d) Lamellar tearing

6.12 The ratio of the effective throat of fillet weld to its size is
(a) 0.707
(b) less than or equal to 0.707
(c) less than 1.0
(d) more than 0.707

6.13 The maximum size of fillet weld applied to the square edge of a plate of thickness greater than 6mm is
(a) not more than the thickness of plate
(b) 1.0 mm less than the thickness of plate

(c) 1.5mm less than the thickness of plate
(d) half the thickness of plate

6.14 Minimum length of fillet weld should be ______________ times the size of weld.
(a) 4 (b) 2 (c) 2 (d) 3

6.15 In order to reduce the effect of stress concentration at the edge of a plate, the length of longitudinal (side) fillets should be
(a) not less than 100mm
(b) not less than the width of the plate
(c) 16 times the thickness of the thinner plate joined
(d) 12 times the thickness of the thinner plate joined

6.16 The effective length of fillet weld may be taken as
(a) total length – 2 × weld size
(b) total length – 2 × throat size
(c) 0.9 times the total length
(d) total length – $\sqrt{2}$ × weld size

6.17 The clear spacing between the effective lengths of intermittent welds in plates subject to compression should be
(a) greater than 16 times the thickness of thinner plate joined
(b) between 12 to 16 times the thickness of thinner plate
(c) 12 times the thickness of thinner plate
(d) 16 times the thickness of thicker plate

6.18 The design strength of fillet weld of length $L_w$ and size $s$ is calculated as
(a) $L_w s f_u /(\sqrt{3}\gamma_{mw})$ (c) $L_w s f_y /(\sqrt{3}\gamma_{mw})$
(b) $L_w t_e f_y /(\sqrt{3}\gamma_{mw})$ (d) $L_w t_e f_u /(\sqrt{3}\gamma_{mw})$

6.19 The design capacity of a weld is reduced when the length of welded joint is greater than
(a) 100 $t_e$ (b) 120 $t_e$ (c) 150 $t_e$ (d) 200 $t_e$

6.20 Slot welding is used when two plates are placed
(a) one butting against the other
(b) at right angles to each other
(c) one next to other
(d) one above the other

6.21 For fillet welds subjected to normal ($f_w$) and shear ($\tau_w$) stresses, the equivalent stress is given by
(a) $\sqrt{f_w^2 + 3\tau_w^2}$ (c) $\sqrt{3f_w^2 + \tau_w^2}$
(b) $\sqrt{f_w^2 + \tau_w^2}$ (d) $\sqrt{0.5f_w^2 + 3\tau_w^2}$

## EXERCISES

6.1 Two plates of thickness 12mm and 10mm are to be jointed by a groove weld. The joint is subjected to a factored tensile force of 275kN. Assuming an effective length of 150mm, check the safety of the joint for
(a) single-V groove weld joint and
(b) double-V groove weld joint.
Assume Fe 410 grade steel plates and that the welds are shop welded.

6.2 A web of a plate girder consists of a (800 × 12)-mm plate of grade Fe 410 and is to be provided with a splice at a section where the factored shear and bending moment to be resisted by the web are

$$V = 800\text{kN and } M = 250\text{kNm}$$

The flange plate thickness = 2 × 20 = 40mm each. Design a web splice.

6.3 The tie member of a truss is made of ISA 75 × 75 × 6 and is subjected to a factored tension load of 110kN. The length of the angle is not enough to go from end-to-end, and hence a splice has to be provided. Design a groove welded joint.

6.4 Determine the effective throat dimension of 8-mm fillet weld made by (a) shielded metal arc welding (SMAW) and (b) submerged arc welding (SAW).

6.5 Determine the design shear strength of a 8-mm fillet weld produced by (a) SMAW and (b) SAW. Assume E 51 electrodes, Fe 410 material, and shop welding.

6.6 Design a connection to joint two plates of size 200 × 10mm of grade Fe 410 to mobilize full plate tensile strength using shop fillet welds, if (i) a lap joint is used, (ii) a double cover butt joint is used.

6.7 Determine the size and length of the fillet weld for the lap joint to transmit a factored load of 100kN shown in Fig. 6.67, assuming site welds, Fe 410 steel, and E 41 electrode.

6.8 Rework the Exercise 7 using 4-mm welds.

6.9 A tie member of a truss consisting of an angle section ISA 75 × 75 × 8 of Fe 410 grade, is welded to a 10-mm gusset plate. Design a weld to transmit a load equal to the full strength of the member. Assume shop welding (see also Fig. 6.69).

6.10 Redesign the joint in Exercise 6.9, if the welding is done on three sides of the angle as shown in Fig. 6.70.

6.11 Design a suitable fillet weld to connect the web plate to the flange plate and flange plate to the cover plate of the compression flange of a plate girder as shown in Fig. 6.71. The size of the web plate is 800 × 8mm. The flange and cover plate are of the size 250 × 12mm and 200 × 10mm, respectively. The maximum factored shear force $V$ = 650kN. Assume shop welding.

6.12 Determine the service load permitted on the connection similar to that shown in Fig. 6.73. The size of plate is 100 × 10mm and the size of the gusset plate is12mm. Assume a plug weld of 20mm diameter and an 8-mm fillet weld of length 100mm on each face of the plate. Assume field welding and Fe 410 grade steel.

6.13 An ISMC 300 is used to transmit a factored force of 800kN. The channel section is connected to a gusset plate 10-mm thick as shown in Fig. 6.74. Design a fillet weld if the overlap is limited to 350mm. Use slot welds if required.

6.14 Design a welded seat angle connection between a beam MB 400 and column HB 200 for a reaction of beam 120kN, assuming Fe 410 grade steel ($f_y$ = 250MPa) and site welding.

6.15 Design a welded header plate connection for a ISMB 300 in Fe 410 grade steel to carry a reaction of 80 kN due to factored loads. The connection has to be made to the flanges of ISHB 200.

6.16 Design a welded splice for an ISMB 300 section to transfer a factored bending moment of 80 kNm and a factored shear of 60 kN. Assume that the flange splice carries all the moment and that the web splice carries only the shear.

6.17 Determine the factored load capacity $P_u$ of the weld B shown in Fig. 6.80, if a 6-mm weld is used, $L = 250$ mm, and $125 \times 75 \times 8$ angles are used.

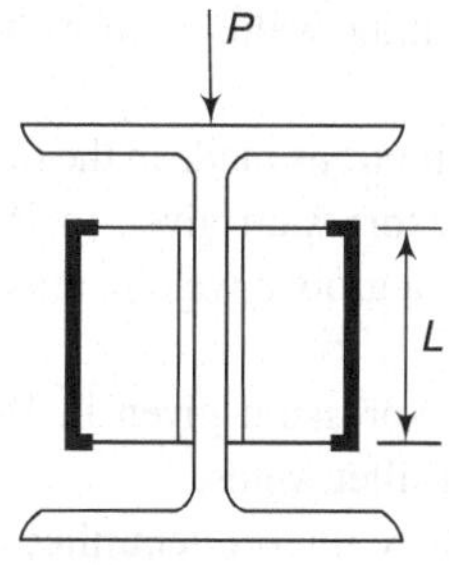


**FIG. 6.80**

## REVIEW QUESTIONS

6.1 List the different processes of arc welding used in structural steel applications.

6.2 Write the expansions for the following abbreviations:
(a) SMAW (b) SAW (c) GMAW (d) FCAW
(e) ESW (f) SW

6.3 Why is SMAW chosen as the common method of welding in structural steel applications?

6.4 What are the different components required in the SMAW process?

6.5 What is the maximum yield stress and ultimate tensile strength of E41 electrode?

6.6 What is the use of coating the electrode with a flux?

6.7 Why is it important to remove the slag before depositing material each time in a multi-pass welding?

6.8 What are the advantages and disadvantages of the SMAW process?

6.9 Why are low hydrogen electrodes preferred?

6.10 What is the main feature of the SAW process?

6.11 List the advantages and disadvantages of the SAW process.

6.12 Describe the GMAW process.

6.13 What are the advantages and disadvantages of the GMAW process?

6.14 What is the distinguishing feature of FCAW over GMAW?

6.15 List the advantages and disadvantages of the FCAW process?

6.16 Describe the ESW process.

6.17 List the advantages and disadvantages of the ESW process.

6.18 Where is the SW process typically used?

6.19 List the parameters based on which a particular welding process is selected.

6.20 In the numbering system Exxxbc, what does each component signify?

6.21 List some of the important advantages of welding over bolting?

6.22 Are there any disadvantages of using welding? What are they?

6.23 List the four types of welds.

6.24 When are backup strips used in groove welds? Under what condition should it not be used?

6.25 What are the various types of groove welds?

6.26 What do you mean by partial penetration and full penetration groove weld? Which one is preferred and why?

6.27 When is a 'spacer' used in groove welds? Can it be left behind in a welded joint?

6.28 What are the parameters that dictate the size of the root gap and root face of a groove weld?

6.29 What are the parameters that affect the choice of single and double penetration welds?

6.30 Why are fillet welds often adopted at site?

6.31 Is edge preparation necessary in fillet welds?

6.32 Under what circumstances are slot and plug welds used?

6.33 Why are slot and plug welds not preferred?

6.34 What are the factors that affect the heat affected zone (HAZ) of the weld?

6.35 Why is the preheating of a joint done?

6.36 List some of the common defects associated with welds.

6.37 What are the factors that affect the quality of a weld?

6.38 Write short notes on the following weld defects:
(a) Under cutting (b) Porosity
(c) Slag inclusion (d) Incomplete penetration
(e) Incomplete fusion (f) Cracks
(g) Lamellar tearing

6.39 How can the occurrence of lamellar tearing be reduced?

6.40 List the checks to be made before, during, and after welding to maintain quality control.

6.41 Write short notes on the following non-destructive testing methods of welds:
(a) Liquid penetration inspection
(b) Magnetic particle inspection
(c) Radiographic inspection
(d) Ultrasonic inspection

6.42 What are the five basic types of welded joints?

6.43 What is the main drawback of a butt joint?

6.44 What is the type of weld used in a butt joint?

6.45 What are lap joints? What is their main advantage? What welds are used in lap joints?

6.46 What types of welds are used in T-joints?

6.47 How can distortions in welded joints be minimized?

6.48 Write short notes on weld symbols.

6.49 What is the advantage of using weld symbols?

6.50 What is the use of end returns? What is the minimum length of an end return?

6.51 What is the minimum thickness of the groove weld to be provided in U, V, J, and bevel types? What is the effective throat thickness taken in the calculation?

6.52 When are unequal legged fillet welds used?

6.53 What effective throat thickness of fillet welds is taken into calculations, when joining faces are inclined to each other?

6.54 Which fillet weld is stronger—side or end?

6.55 Why are convex fillet welds used rather than concave fillet welds?
6.56 What are the assumptions made in the design of welded joints?
6.57 Write down the expression given in IS 800 to calculate the design strength of a groove weld in tension and a groove weld in shear.
6.58 Write down the expression given in IS 800 to calculate the design strength of fillet welds.
6.59 Cite a location where intermittent fillet welds are used.
6.60 In what locations intermittent welds are not to be used?
6.61 What are the points to be borne in mind while using slot or plug welds?
6.62 Write short notes on
(a) Fillet welds in truss members
(b) Web angle and end plate connections
(c) Angle seat connections
(d) Beam and column splices
(e) Tubular connections
6.63 What are the failure modes of joints made using adhesives?
6.64 What are the advantages of adhesive joints over joints made by welding or bolting?
6.65 What are the drawbacks of adhesive joints?

## REFERENCES


*AISC—Seismic Provisions for Structural Steel Buildings*, American Institute of Steel Construction, Chicago, 1997.

Althouse A.D., Turnquist, C.H., Bowditch, W., and Bowtich, K.E., *Modern Welding*, Ninth edition, The Good Heart-Willcox Company Inc., Illinois, 2000, p. 779.

Astanesh, A., Call, S.M., and McMullin, K.M., 'Design of Single Plate Shear Connections', *Engineering Journal*, AISC, Vol. 26, No, 1, First quarter 1989, pp. 21–32. Also see discussion by Richard, R.M., Vol. 27, No. 3, Third Quarter, 1990, pp. 121–126.

Ballio, G. and Mazzolani, F.M., *Theory and Design of Steel Structures*, Chapman and Hall, London, 1983.

Bangash, M.Y.H., *Structural Detailing in Steel*, Thomas Telford, London, 2000, p. 240.

Blodgett, O.W., *Design of Welded Structures*, The James F. Lincoln Arc Welding Foundation, Cleveland, 1966.

Brockmann, W. and Neeb, T., Stand der Klebtechnik im Stahlbau, *Der Stahlbau*, Vol. 70, No. 2, Feb 2001, pp. 106–115.

Butler, L.J. and Kulak, G.L. 'Strength of Fillet Welds as a Function of Direction of Load', Welding Research Supplement, *Welding Journal*, 1968, pp. 321–345.

Donnelly, J.A., 'Determining the Cost of Welded Joints', *Engineering Journal*, AISC, Vol. 5, No. 4, Oct 1968, pp. 146–147.

Dowling, P.J., Knowles, P., and Owens, G.W., *Structural Steel Design*, The Steel Construction Institute and Butterworths, London, 1988, p. 399.

Engelhardt, M.D. and Sabol, T.A., 'Reinforcing of Steel Moment Connections with Cover Plates: Benefits and Limitations', *Engineering Structures*, Vol. 20, Nos 4–6, Apr–Jun 1998, p. 510.

Galvery, Jr. W.L., Marlow, F.M., *Welding Essentials; Questions and Answers*, Expanded edition, Industrial Press, 2001, p. 480.

Ghosh, A., 'Modern Approach of Design of Welded Components Subjected to Fatigue Loading', *Journal of Structural Engineering*, ASCE, Vol. 130, No. 5, May 2004, pp. 812–820.

Gourd, L.M., *Principles of Welding Technology*, Edward Arnold, London, 1980.

Gupta, A.K., 'Buckling of Coped Steel Beams', *Journal of the Structural Engineering*, ASCE, Vol. 110, No. 9, Sep 1984, pp. 1977–1987, also see discussion by Cheng, J.J. and Yura, Y.A., Vol. 112, No. 1, Jan 86, pp. 201–204.

Gurney, T.R., *Fatigue Design Rules for Welded Steel Joints*, The Welding Institute, 1976.

Habenicht, G., *Kleben-Grundlagen, Technologie, Anwendungen*, Third edition, Springer-Verlag, Berlin, 1997.

*Handbook in Structural Steel Detailing*, Institute of Steel Development and Growth, Kolkata, Publication No. INS / PUB 018.

Hartshorn, ed., *Structural Adhesives: Chemistry and Technology*, Plenum Publ. Corp., 1986.

Hicks, J., *Welded Joint Design*, Third edition, Industrial Press, 2000, p. 141.

*Hollow Structural Section Connections Manual*, AISC, 1997.

IS:814-1991 Covered Electrodes for Manual Metal Arc Welding of Carbon and Carbon Manganese Steel (Fifth revision).

IS:816-1969 Code of Practice for Use of Metal Arc Welding for General Construction in Mild Steel (First revision).

IS:819-1957 Code of Practice for Resistance Spot Welding for Light Assemblies in Mild Steel.

IS:1024-1999 Code of Practice for Use of Welding in Bridges and Structures Subject to Dynamic Loading (Second revision).

IS:1261-1959 Code of Practice for Seam Welding in Mild Steel.

IS:1278-1972 Specification for Filler Rods and Wires for Gas Welding (Second revision).

IS:1323-1982 Code of Practice for Oxy-acetylene Welding for Structural Work in Mild Steels (Second revision).

IS:1395-1982 Low and Medium Alloy Steel Covered Electrodes for Manual Metal Arc Welding (Third revision).

IS:6419-1996 Welding Rods and Bare Electrodes for Gas Shielded Arc Welding of Structural Steel (First revision).

IS:7280-1974 Specifications for Bare Wire Electrodes for Submerged Arc Welding of Structural Steels.

Iwankiw, N.R., Design for Eccentric and Inclined Loads on Bolt and Weld Groups, *AISC Engineering Journal*, Fourth quarter, 1987.

Jeffus, L., *Welding: Principles and Applications*, Fifth edition, Thomson Delmar Learning, 2002, p. 944.

*Journal of Structural Engineering*, ASCE, Vol. 128, No. 4, April 02, Special Issue on Steel Moment Frames after Northridge, pp. 413–555.

Kloiber, L., 'Connections for Hollow Structural Sections', *Structure*, SEI/ASCE, Jul–Aug 2001, pp. 40–43.

Kosteski, N. and Packer, J.A., Longitudinal Plate and Through Plate-to-hollow Structural Section Welded Connections, *Journal of Structural Engineering*, ASCE, Vol. 129, No. 4, Apr 2003, pp. 478–486.

Lesik, D.F. and Kennedy, D.J.L., 'Ultimate Strength of Fillet Welded Connections Loaded in Plane', *Canadian Journal of Civil Engineering*, Vol. 17, No. 1, 1990, pp. 55–67.

Manuel, T.J. and Kulak, G.L., Strength of Joints that Combine Bolts and Welds, *Journal of Structural Engg*., ASCE, Vol. 126, No. 3, Mar 2000, pp. 279–287.

Marshall, P.W., *Design of Welded Tubular Connections—Basis and Use of AWS Code Provisions*, Elsevier, 1991, p. 412.

Martin, L.H. and Purkiss, J.A., *Structural Design of Steelwork to BS 5950*, Edward Arnold, London, 1992, p. 468.

Mazzolani, ed., *Moment Resistant Connections of Steel Frames in Seismic Areas: Design and Reliability*, E & FN Spon, London, 2000, p. 644.

Neis, V.V., 'New Constitutive Law for Equal Leg Fillet Welds', *Journal of the Structural Engineering*, ASCE, Vol. 111, No. 8, Aug 1985, pp. 1747–1759.

Packer, J.A. and Henderson, J.E., *Hollow Structural Section Connections and Trusses—A Design Guide,* Second edition, Canadian Institute of Steel Construction, Toronto, Canada, Jun 1997, p. 447.

Pasternak, H., Schwarzlos, A., and Schimmack, N., 'The Application of Adhesives to Connect Steel Members', *Journal of Constructional Steel Research,* Vol. 60, 2004, pp. 649–658.

Roeder, C.W. and Dailey, R.H., 'The Results of Experiments on Seated Beam Connections', *Engineering Journal*, AISC, Vol. 26, No. 3, Third quarter 1989, pp. 90–95.

Sabelli, R., C.W. Roeder, and J.F. Hajjar, 'Seismic Design of Steel Special Concentrically Braced Frame Systems: A Guide for Practicing Engineers,' *NEHRP Seismic Design Technical Brief No. 8*, National Institute of Standards and Technology, Gaithersburg, MD, NIST GCR 13-917-24, 2013, p. 31.

Salmon, C.G., Johnson, J.E., and Malhas, F.A., *Steel Structures, Design and Behavior*, Fifth edition, Harper Collins College Publishers, New York, 2008, p. 888.

Subramanian, N., *Space Structures: Principles and Practice*, Multi-Science Publishing Co. Ltd., Essex, UK, 2006.

Swannel, 'Rational Design of Fillet Weld Groups', *Journal of the Structural Div.*, ASCE, Vol. 107, N0.ST5, May 1981, pp. 789–802.

Temple, M.E. and Sakla, S.S.S., Balanced and Unbalanced Welds for Angle Compression Members, *Canadian Journal of Civil Engg*., Vol. 21, 1994, pp. 396–403.

Thornton, W.A., 'Connections—Art, Science and Information into the Quest for Economy and Safety', *Engineering Journal*, AISC, Vol. 32, No. 4, Fourth quarter, 1995, pp. 132–144.

Thornton W.A. and L.S. Muir, 'Design of Vertical Bracing Connections for High seismic Drift', *Modern Steel Construction*, AISC, Vol. 49, No. 3, Mar 2009, pp. 61–65.

www.engineersedge.com/weld_design_menu.shtml (Provides weld joint design equations and calculations), last accessed on 20 July 2015.

http://www.weldguru.com/ (Provides information about different types of welding, welding electrodes, welding codes, and latest developments), last accessed on 20 July 2015.

Yura, J.A., Birkemoe, P.C., and Ricles, J.M., 'Beam Web Shear Connections: An Experimental Study', *Journal of the Structural Div*., ASCE, Vol. 108, No. ST2, Feb 1982, pp. 311–325.

Zhao, X.L., et al., *Design Guide for Circular and Rectangular Hollow Section Welded Joints under Fatigue Loading*, CIDECT and Verlag TÜV Rheiland GMBH, Köln, Germany, 2001, p. 121.

# 7

# DESIGN OF TENSION MEMBERS

## INTRODUCTION

Steel tension members are probably the most common and efficient members in the structural applications. They are those structural elements that are subjected to direct axial tensile loads, which tend to elongate the members. A member in pure tension can be stressed up to and beyond the yield limit and does not buckle locally or overall. Hence, their design is not affected by the classification of sections, for example, compact, semi-compact, etc., as described in Chapter 8. The design stress $f_y$ as determined from Table 1.9, is therefore not reduced.

Wind bracings in X configuration are used in industrial buildings and multi-storey structures, where the members are so flexible that 'buckling' takes place under compressive stresses when lateral loads act in one direction. But they function as tension member when the direction of lateral load is changed. Tension members are used as hangers for stairwells, elevator cables, hanger support for curtain walls, and parts of the hoisting equipment (Bowles 1980). As stated earlier, tension members carry loads most efficiently, since the entire cross section is subjected to uniform stress.

Tension members occur as components of trusses (bottom chord of roof trusses), bridges, transmission line and communication towers, and wind bracing systems in multi-storey buildings (see Fig. 7.1). Some truss web members and members in towers may carry tension under certain loading cases and may be subjected to compression for other loading cases. Steel cables used in suspension bridges and in cable supported roofs are also examples of tension members. Such cables are also used in guyed towers as well as power line poles where alignment changes occur.

Tension members in the form of high-strength steel rods are often used to strengthen existing structures by attaching them to the bottom flange or to the chord of the structure. They are pre-tensioned to induce a compressive stress in the member, which must then be overcome when a load is applied (Bowles 1980). Sag rods are used to provide intermediate support to purlins and wall grit systems.

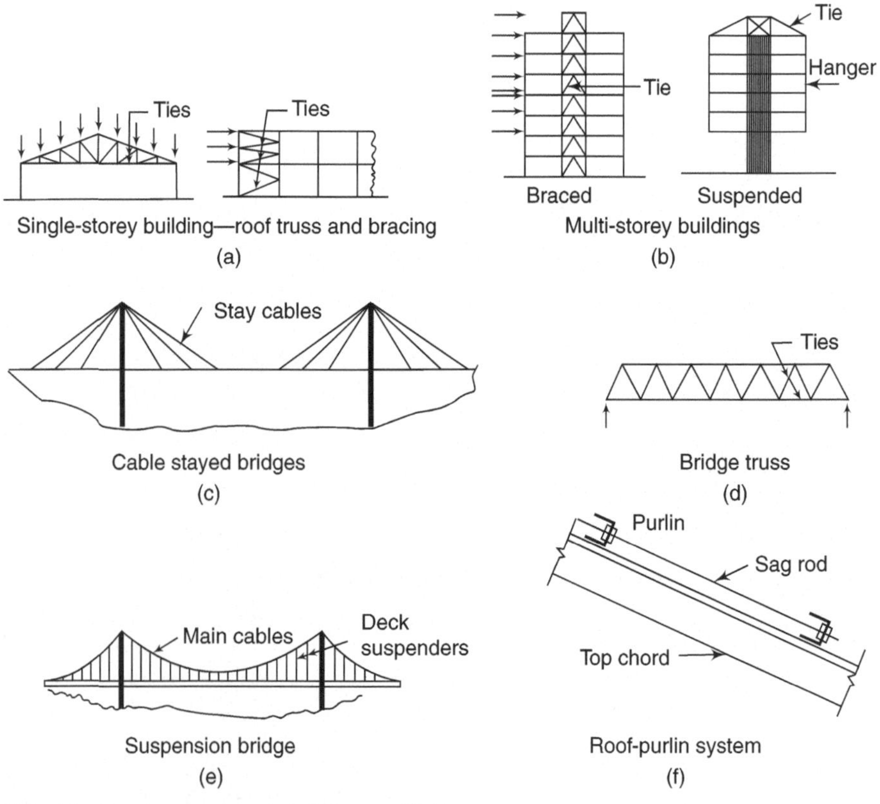


**FIG. 7.1** Tension members in buildings and bridges

The strength of these members is influenced by several factors such as the length of connection, size and spacing of fasteners, net area of cross section, type of fabrication, connection eccentricity, and shear lag at the end connection. To simplify the design procedure of tension members, considerable amount of research has been carried out (Salmon & Johnson 1996; Kulak & Wu 1997). This chapter discusses the effects of these parameters and the design of tension members as per IS 800.

## 7.1 TYPES OF TENSION MEMBERS

Tension members may consist of single structural shape or they may be built using a number of structural shapes. The cross section of some typical tension members are shown in Fig. 7.2. In most of the situations, single angle members may be found economical. However, a single angle section with bolted connection will produce eccentricity about both the planes, whereas a single angle with welded end connections may produce eccentricity about one plane only. Single angle members are used extensively in towers and as web members in trusses. Double angle sections are used as chord members in light roof trusses or in situations where some rigidity is required and where the members are subjected to reversal of stresses. The tension members in bridge trusses are made of single or built-up channels and I-sections. When two elements such as two angles are used as a single member, they should be interconnected at reasonable intervals to enable them to act together as one member. These two separate elements are often placed parallel to each other with a gap of about 6–10 mm between them. They should also have *spacer plates* placed at regular intervals between them, which are connected to these individual elements by tack welds or bolts as shown in Fig. 7.3. Though in Fig. 7.3 only welding is shown, similar rules apply for a bolted spacer (also called as *stitch plates*). Single angle or double angles are either bolted to a single gusset plate at each end or may be welded directly to the webs or flanges of T- or I-chord members.

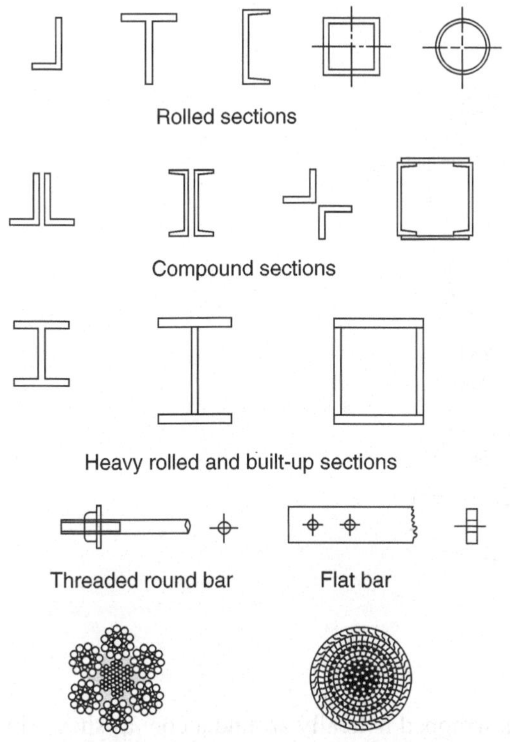


**FIG. 7.2** Cross section of typical tension members

Structural T-sections may be used as chord members of lightly loaded trusses, instead of the back-to-back two angle sections. The stem of the T-sections may be used to connect the single or double angle web members, thus eliminating the use of any gusset plate, especially in welded connections. Tubular members are also used in roof trusses as tension members.

I-sections, channel sections, and built-up sections using angles, channels, etc., are used when greater rigidity is required and hence are often used in bridge structures. The four angle sections are usually connected to two gusset plates. Two channels may be arranged as in Fig. 7.4 to connect to a single gusset plate, which fits between the webs or to connect to a gusset plate at each flange. Various shapes may be made by welding plates together (e.g., three plates welded to form I-sections).

Rods and bars are used as tension members in the bracing systems. As mentioned earlier, sag rods are used to support purlins, to support girt in industrial buildings, or as longitudinal ties. They are either welded to the gusset plates or threaded and bolted to the main members directly using nuts. When rods are used as wind bracings, they are pre-tensioned to reduce the effect of sway. Pre-tensioning these bars may be achieved in two different ways: (a) they may be cut a little short of their required length and pulled and bolted in place to provide the initial tension and (b) rods may be made in two sections, and joined with a turnbuckle, which may be adjusted to tighten the member.

Eyebars are members with rectangular cross section with enlarged heads at each end which are bored for pin connection. They were used extensively in early bridge trusses and are still used occasionally as hangers.

A wire is a single continuous element cold drawn from hot-rolled rods; they may be black wires or zinc coated hot dip galvanized wires. Wire rope is used for guy wires, floor suspenders in suspension bridges, hoisting lines, etc. Bridge strands and ropes are formed by the helical winding of wires and they elongate when subjected to tension. Bridge strands are used for small suspension bridges, suspended roofs, and similar structures. Most ropes are made of high tensile steel with proof stress in the range of 1000–1400 MPa. The proof stress of high tensile wires are measured at 0.7% residual strain. Ropes and cables are attached to other members or to the anchorages by various kinds of sockets (Subramanian

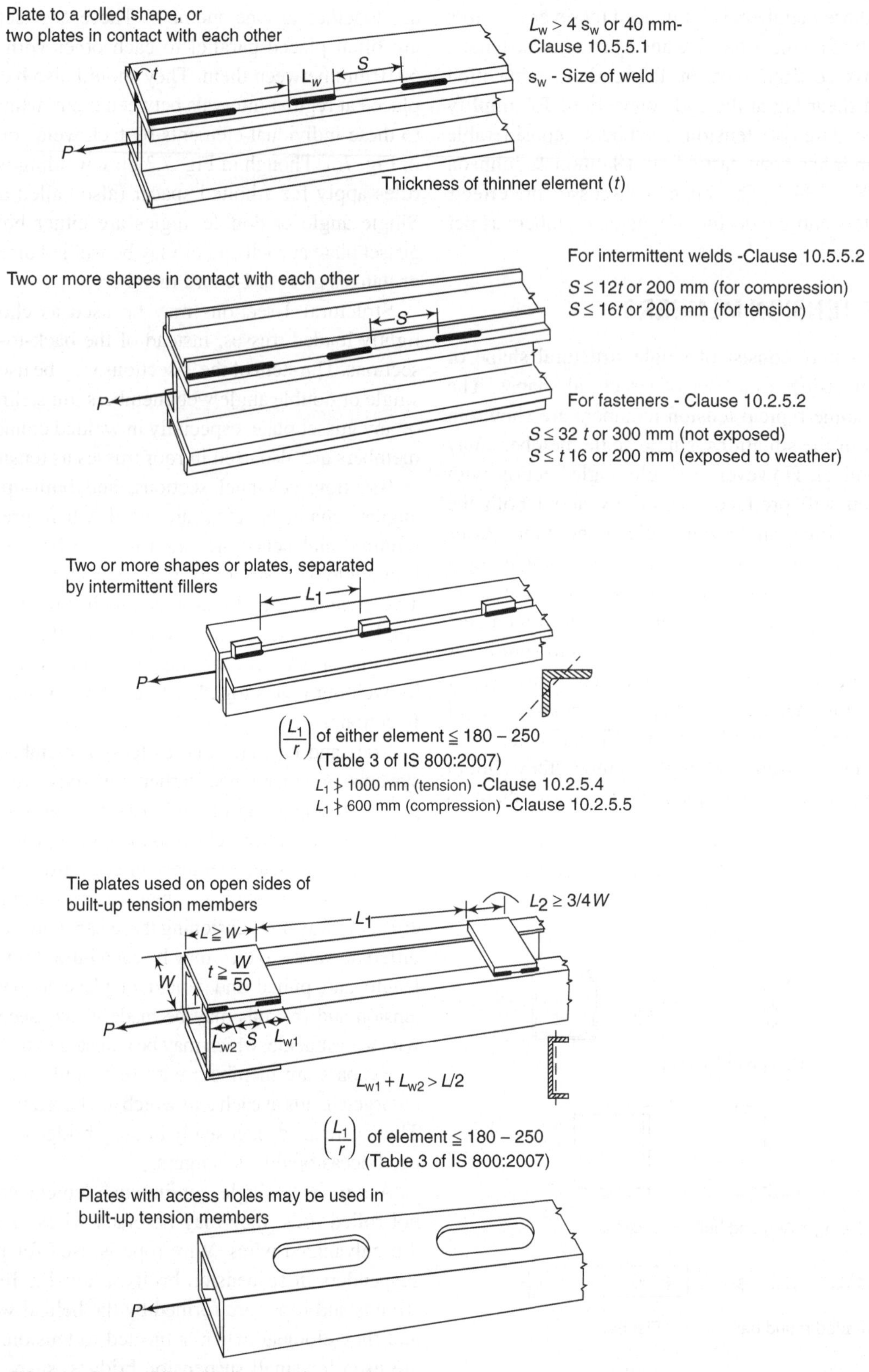


**FIG. 7.3** Tack welding of built-up tension members

2006). Main cables of large suspension bridges consist of parallel wires which are strung individually, squeezed together, and wrapped; a rope is composed of a plurality of strands wrapped helically around a core as shown in Fig. 7.2. For better protection against corrosion, the strands are locked as shown in Fig. 7.2. These strands are smooth surfaced.

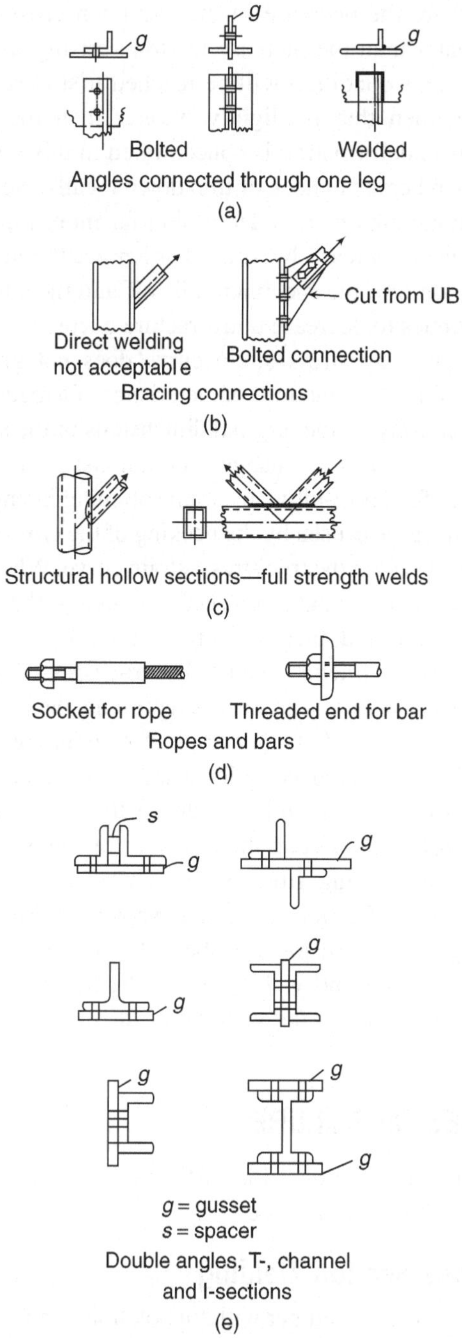


**FIG. 7.4** Member end connections

## 7.2 DESIGN OF STRANDS

As discussed in Section 7.1, strands are made of twisting wires and ropes. The diameter of the rope is called the nominal diameter and the area of cross section is

$$A_s = 0.6D^2 \tag{7.1}$$

The elongation of ropes consists of rigid body stretching and elastic elongation. The equivalent modulus of elasticity is given by

$$E_e = 0.65E \tag{7.2}$$

where $E$ is the Young's modulus of steel = $2 \times 10^5$ MPa and $E_e$ is the equivalent elastic modulus of the rope

Ropes are usually rated for their capacity and elongation by the manufacturers (see www.ushamartin.com). Typical minimum breaking loads are 2400 kN for a 50-mm diameter rope and 5300 kN for a 75-mm diameter rope. Working loads of up to 45% of minimum breaking strength are commonly used.

As the rope is subjected to tension, the wire with the shortest length will yield first. Thus, the capacity of the rope is always less than the total capacity of all the wires put together. Hence, it may be necessary to test load the ropes in order to provide a suitable factor of safety. Wire ropes are often pre-stressed through turnbuckles or mechanical or hydraulic jacks to avoid excessive elongations. The pre-tension could be as much as 0.5% to 1% of the length. The properties of wire ropes are given in Table 7.1.

**TABLE 7.1** Properties of wire ropes

| Property | Value |
|---|---|
| Ultimate stress | 1400 – 2000 MPa |
| Proof stress (0.7% residual strain) | 1000 – 1400 MPa |
| Percentage elongation | 3 to 4% |
| Area of cross section | 0.4 – 0.6 $D^2$ |
| Approximate modulus of elasticity | $1.3 \times 10^5$ MPa |

*Source*: Dayaratnam (1996)

## 7.3 SLENDERNESS RATIO

Although stiffness is not required for the strength of a tension member, a minimum stiffness is stipulated by limiting the maximum slenderness ratio of the tension member. The *slenderness ratio* of a tension member is defined as the ratio of its unsupported length ($L$) to its least radius of gyration. This limiting slenderness ratio is required in order to prevent undesirable lateral movement or excessive vibration. (As stated already, stability is of little concern in tension members.) The slenderness limits specified in the code for tension members are given in Table 7.2.

**TABLE 7.2** Maximum values of effective slenderness ratios as per IS 800

| Member | Maximum effective slenderness ratio ($L/r$) |
|---|---|
| A tension member in which a reversal of direct stress occurs due to loads other than wind or seismic forces | 180 |
| A member subjected to compressive forces resulting only from a combination of wind/earthquake actions, provided the deformation of such a member does not adversely affect the stresses in any part of the structure | 250 |

(*Contd*)

**TABLE 7.2** *(Contd)*

| Member | Maximum effective slenderness ratio ($L/r$) |
|---|---|
| A member normally acting as a tie in a roof truss or a bracing member, which is not considered effective when subject to reversal of stress resulting from the action of wind or earthquake forces | 350 |
| Members always in tension (other than pre-tensioned members) | 400 |

## 7.4 DISPLACEMENT OF TENSION MEMBERS

The increase in the length of a member due to axial tension under service loads is

$$\Delta = PL/(EA_g) \tag{7.3}$$

where $\Delta$ is the axial elongation of the member (mm), $P$ is the axial tensile force (un-factored) in the member (N), $L$ is the length of the member (mm), and $E$ is the modulus of elasticity of steel = $2.0 \times 10^5$ MPa. Note that displacement is a serviceability limit state criterion and hence is checked under service loads and not under factored loads.

## 7.5 BEHAVIOUR OF TENSION MEMBERS

The load–deformation behaviour of an axially loaded tension member is similar to the basic material stress–strain behaviour (see Fig. 1.11 of Chapter 1). When a member is subjected to tension, the area of cross section and the gauge length continuously change due to the Poisson effect and longitudinal strain, respectively (see also Section 1.8.1 of Chapter 1). Stresses and strains may be calculated using the initial area of cross section and the initial gauge length, which is referred to as the engineering stress and engineering strain or using the current area of cross section and the current gauge length, which is referred to as the true stress and true strain.

The engineering stress–strain curve is typically represented by four regions, namely the linear elastic region, the yield plateau, the strain hardening region and the strain softening (unloading) region. In the elastic region, stress is linearly proportional to strain. When the load exceeds a value corresponding to the yield strength, the specimen undergoes gross plastic deformation. There is a residual permanent deformation if the load is released to zero. The volume of the specimen remains constant during plastic deformation and as the specimen elongates, the cross section decreases uniformly along the gauge length. After some stage the stress to produce continued plastic deformation increases with increasing plastic strain, that is, the metal strain hardens. Initially, the strain hardening effect more than compensates for this decrease in area and the engineering stress continues to rise with the increasing strain. Eventually a point is reached where the decrease in the specimen cross-sectional area is greater than the increase in strain arising from strain hardening. This condition will be reached first at some point in the specimen that is slightly weaker than the rest. All further plastic deformation is concentrated in this region, and the specimen begins to neck or thin down locally. Because the cross-sectional area now is decreasing far more rapidly than the local stress increase by strain hardening, the actual load required to deform the specimen falls off and the engineering stress continues to decrease until fracture occurs.

The engineering stress–strain curve does not give a true indication of the deformation characteristics of a metal because it is based entirely on the original dimensions of the specimen, and these dimensions change continuously as the load increases. In fact, post-ultimate strain softening in engineering stress–strain curve caused by the necking of the cross section is completely absent in the true stress–strain curve. When the true stress based on the actual cross-sectional area of the specimen is used, it is found that the stress–strain curve increases continuously until fracture occurs. The true stress–strain curve is also known as *flow curve* since it represents the basic plastic flow characteristics of the material. Any point on the flow curve can be considered as the local stress for a metal strained in tension by the magnitude shown on the curve. However, since it is difficult to obtain the ordinates of true stress–strain curve, the engineering stress–strain curve is often utilized. As discussed in Section 1.8.1 and shown in Fig. 1.11(a), high-strength steel tension members do not exhibit a well-defined yield point and yield plateau. Hence the 0.2% offset load is usually taken as the yield point for such high-strength steel.

## 7.6 MODES OF FAILURE

In the following sections, the different modes of failure of tension members are discussed.

### 7.6.1 Gross Section Yielding

Generally a tension member without bolt holes can resist loads up to the ultimate load without failure. But such a member will deform in the longitudinal direction considerably (nearly 10%–15% of its original length) before fracture. At such a large deformation a structure becomes unserviceable. Hence, one of the limiting values in design strength is the one corresponding to the yielding of the gross cross section, calculated as follows

$$T_{dg} = f_y A_g / \gamma_{m0} \tag{7.4}$$

where $A_g$ is the gross area of cross section in mm$^2$, $f_y$ is the yield strength of the material (in MPa), and $\gamma_{m0}$ is the partial safety factor for failure in tension by yielding ($\gamma_{m0}$ = 1.10 as per IS 800).

### 7.6.2 Net Section Rupture

A tension member is often connected to the main or other members by bolts or welds. When connected using bolts, tension members have holes and hence reduced cross section, being referred to as the *net area*. Holes in the members cause stress concentration at service loads, as shown in Fig. 7.5(a). From the theory of elasticity, we know that the tensile stress adjacent to a hole will be about two to three times the average stress on the net area, depending upon the ratio of the diameter of the hole to the width of the plate normal to the direction of stress. The ratio of the maximum elastic stress to the average stress ($f_{max}/f_{avg}$) is referred to as the stress-concentration factor and a plot of the stress-concentration factor with the ratio of the hole diameter $2r$ to the net width $b_n$ is shown in Fig. 7.5(c). The effects of changes in the cross section of plates in tension are shown in the lower curves of Fig. 7.5(c). Stress concentration becomes very significant when repeated applications of load may lead to fatigue failure or when there is a possibility of a brittle fracture of a tension member under dynamic loads (see Section 1.8.5 of Chapter 1). Stress concentration may be minimized by providing suitable joint and member details; when it is not possible to do it, an estimate of the magnitude of local stresses can be made using Fig. 7.5(c).

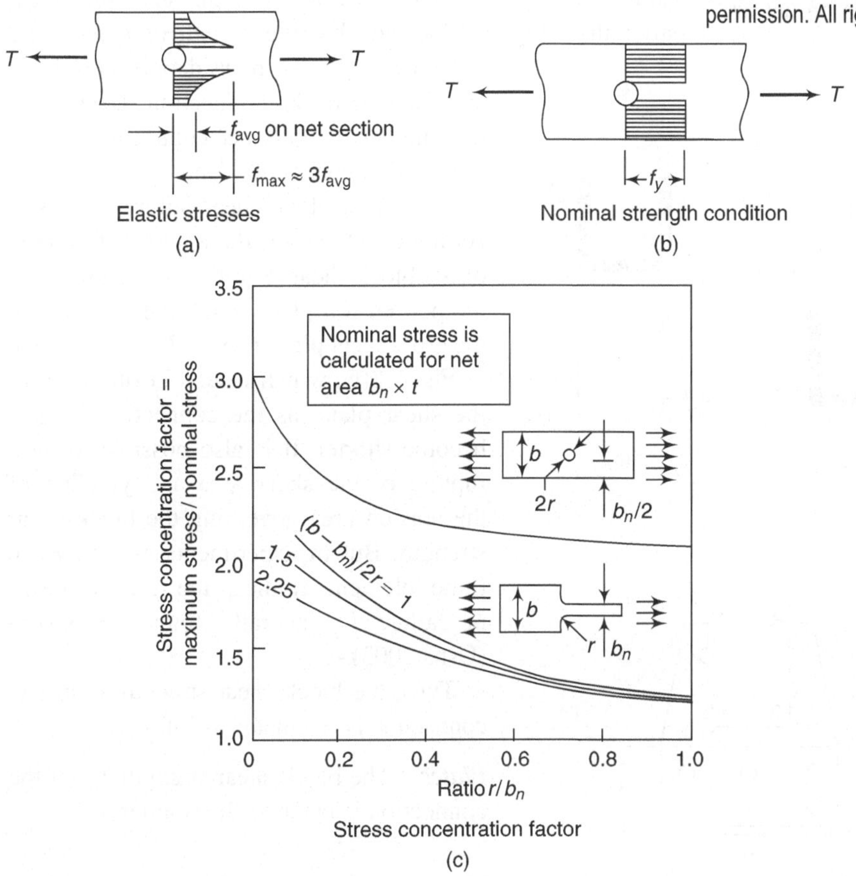


**FIG. 7.5** Stress concentration due to holes

When a tension member with a hole is loaded statically, the point adjacent to the hole reaches the yield stress $f_y$ first. On further loading, the stress at that point remains constant at yield stress and each fibre away from the hole progressively reaches the yield stress $f_y$ [see Fig. 7.5(b)]. Deformations continue with increasing load until finally rupture (tension failure) of the member occurs when the entire net cross section of the member reaches the ultimate stress $f_u$ (see Fig 7.6). The *net section rupture* is calculated as follows:

$$T_{dn} = 0.9 f_u A_n / \gamma_{m1} \tag{7.5}$$

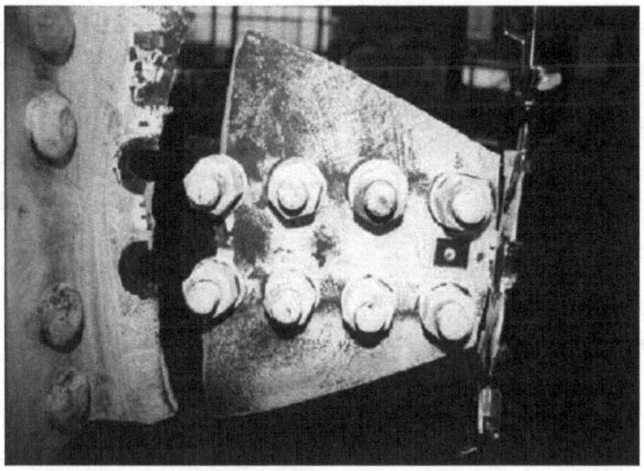

**FIG. 7.6** Tension rupture limit state

Photo by J.A. Swanson and R. Leon of Georgia Institute of Technology 

where $A_n$ is the net effective area of the cross section in mm$^2$, $f_u$ is the ultimate strength of the material in MPa, and $\gamma_{m1}$ is the partial safety factor for failure due to rupture of cross section ($\gamma_{m1}$ = 1.25 as per IS 800). Because of strain hardening, the actual strength of a ductile member may exceed that indicated by Eqn (7.4). However, since there is no reserve of any kind beyond the ultimate resistance an additional multiplier of 0.90 has been introduced in Eqn (7.5). Such a high margin of safety has been traditionally used in design when considering the fracture limit state than for the yielding limit state (Salmon & Johnson 1996). The 0.9 factor was included in the strength equation of Eqn (7.5), based on a statistical evaluation of a large number of test results for net section failure of plates.

Similarly, threaded rods subjected to tension could fail by rupture at the root of the threaded region. Thus, the design strength of the threaded rods in tension is given by Eqn (7.5) where $A_n$ is the net root area at the threaded sections. The various factors, which affect the net effective section, are discussed in Section 7.7.

### 7.6.3 Block Shear Failure

Originally observed in bolted shear connections at coped beam ends, block shear is now recognized as a potential failure mode at the ends of axially loaded tension members also. In this failure mode, the failure of the member occurs along a path involving tension on one plane and shear on a perpendicular plane along the fasteners as shown in Fig. 7.7. Other examples of *block shear failures* including failures in welded connections are given in Fig. 7.8.

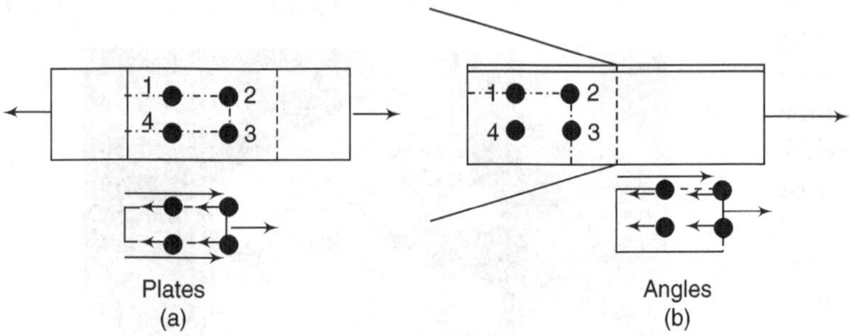


**FIG. 7.7** Block shear failure in plates and angles

It can be observed as shown in Fig. 7.8(a) that the gusset plate may fail in tension on the net area of section *a-a*, and in Fig. 7.8(c) it may fail on the gross area of section *a-a*. The angle member in Fig. 7.8(a) may also separate from the gusset plate by shear on net area 1-2 combined with tension on net area 2-2 as shown in Fig. 7.8(b). A similar fracture of the welded connection of Fig. 7.8(c) is shown in Fig. 7.8(d). The fracture of a gusset plate for a double angle member or of one of the gusset plates for an I-Section [Fig. 7.8(e)] is shown in Fig. 7.8(f). The gusset plate in Fig. 7.8(d) may also fail on the net section a-a. All these failures [Figs 7.8(b), (d), and (f)] are called block shear failures (see Fig. 7.9).

**FIG. 7.9** Typical blockshear failure

Photo by J.A. Swanson and R. Leon of Georgia Institute of Technology (Copyright© American Institute of Steel Construction, Inc., Reprinted with permission. All rights reserved.)

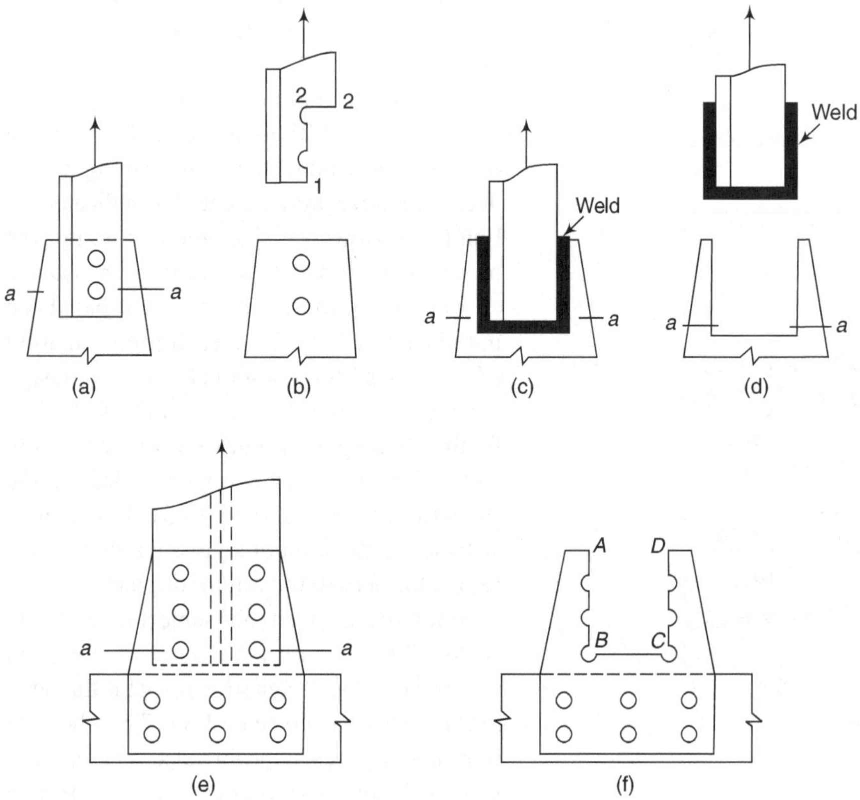


**FIG. 7.8** Examples of block shear failure

The block shear phenomenon becomes a possible mode of failure when the material bearing strength and bolt shear strength are higher. When high bearing strength of material and high-strength bolts are used, only a few bolts are required in the connection. Decreasing the number of bolts per connection results in a smaller connection length; however, the possibility of block shear failure increases. As indicated earlier, the appropriate model of the block shear failure is the rupturing of the net tension plane (BC) and yielding on the gross shear plane (AB and CD), as shown in Fig. 7.8(f), which results in rupturing of the shear plane as the connection lengths become shorter. It is also possible to have rupture of the shear area and yielding of the tension area governing the block shear strength. But in experiments as well as in finite element studies, the former mode of failure is generally found to govern (Usha 2003).

Thus, the block shear strength at an end connection is calculated as follows:

***Plates*** The block shear strength $T_{db}$ of the connection is taken as the smaller of

$$T_{db1} = A_{vg} f_y / (\sqrt{3}\,\gamma_{m0}) + 0.9 f_u A_{tn} / \gamma_{m1} \quad (7.6a)$$

$$T_{db2} = 0.9 f_u A_{vn}/(\sqrt{3}\,\gamma_{m1}) + f_y A_{tg}/\gamma_{m0} \quad (7.6b)$$

where $A_{vg}$ and $A_{vn}$ are the minimum gross and net area in shear along a line of transmitted force, respectively [1-2 and 4-3 as shown in Fig. 7.7(a) and 1-2 as shown in Fig. 7.7(b); $A_{tg}$ and $A_{tn}$ are the minimum gross and net area in tension from the hole to the toe of the angle or next last row of bolt in plates, perpendicular to the line of force, respectively [2-3 as shown in Figs 7.7(a) and (b)]; and $f_u$ and $f_y$ are the ultimate and yield stress of the material, respectively.

***Angles*** Strength as governed by block shear failure in angle end connection is calculated using Eqn (7.6) and appropriate areas in shear and tension as shown in Fig. 7.7(b).

The lower values of the design tension capacities, as given by Eqns (7.4) to (7.6), govern the design strength of plates or members with hole and should be greater than the factored design tension. Note that no net areas are involved in the failures of welded connections [see Fig. 7.8(c)]. Therefore, in applying Eqn (7.6) to this case in the second term of Eqn (7.6a), use $A_{tg}$ (instead of $A_{tn}$) and in the first term of Eqn (7.6b), use $A_{vg}$ (instead of $A_{vn}$).

Experimental work on block shear failure of angles in tension has been reported by Epstein (1992). Epstein & Thacker (1990) have studied the effect of bolt stagger on block shear failure of angles. Finite element studies on block shear failures have been conducted by Epstein & Gulia (1992); Epstein & Chamrajanagar (1994); Epstein & McGinnis (2000). Block shear tests in high-strength steel angles were conducted by Gross et al. (1995). A review of block shear failure of steel angle members is provided by Kulak & Grondin (2000) and an assessment of predictions using American code provisions by Orbison & Ziemian (1995).

Recently, Driver et al. (2006) proposed a unified equation for block shear failure to predict the capacities of angles, tees, gusset plates, and coped beams.

Based on this work, the 2009 version of the Canadian code has adopted the following equation

$$T_{db} = 0.75\left[U_t A_{tn} f_u + \frac{0.6 A_{vg}(f_y + f_u)}{2}\right] \quad (7.6c)$$

Where $U_t$ is the efficiency factor and equals 1.0 for flange-connected tees and for symmetric failure patterns and concentric loading; 0.6 for angles connected by one leg and stem connected tees; 0.9 for coped beams with one bolt line; and 0.3 for coped beams with two bolt lines. For $f_y > 485$ MPa $(f_y + f_u)/2$ should be replaced by $f_y$. Other terms are defined earlier.

## 7.7 FACTORS AFFECTING THE STRENGTH OF TENSION MEMBERS

As discussed already, the yielding of the gross section of tension member causes excessive elongation and hence the load corresponding to the yielding of gross section is taken as one limit state. However, the net section through the bolt holes at the ends of the member may be subjected to tensile stresses well in excess of the yield stress to as high as ultimate stress without the member suffering excessive elongation. Hence, the rupture strength of the net section through the bolt holes at the ends is considered another limit state. Several factors affect the rupture strength of the net section of tension members. They are briefly described below.

### 7.7.1 Effect of Bolt Holes

In order to make connections, tension members are often bolted to adjacent members directly or by using gusset plates. These bolt holes reduce the area of cross section available to carry tension and hence affect the strength as discussed in the following section.

#### *Methods of Fabrication*

There are generally two methods of making holes to receive bolts, namely punching and drilling. Due to punching, the material around the holes is deformed in shear beyond ultimate strength to punch out the hole. It has been found that the strength of members with *punched holes* may be 10%–15% less than members with *drilled holes* (Munse & Chesson 1963; Barth et al. 2002). This is due to the strain hardening effect of the material around the punched holes and the consequent loss of ductility.

Under cyclic loading, the material around the punched holes present the greatest scope for crack initiation due to stress concentration, and hence punched hole is not allowed under fatigue environment. In order to avoid the effects of strain hardening due to punching, either drilling the holes or sub punching the holes to a diameter 2 mm less than the hole size and reaming to the correct size is recommended (Salmon & Johnson 1996).

To account for the effects of punched holes, Munse & Chesson (1963) proposed a fabrication factor of 0.85 to be multiplied to the predicted capacities of members having punched holes. Presently in many specifications, the punching effect upon the net section strength is accounted for by taking the hole diameter as 2 mm larger than the actual hole size when computing the net area (see clause 3.6.1 of IS 800).

#### *Net Area of Cross Section*

The presence of a hole tends to reduce the strength of a tension member. When more than one bolt hole is present, the failure paths may occur along sections normal to the axis of the member, or they may include zigzag sections, if the fasteners are staggered (Fig. 7.10). *Staggering holes* improves the load carrying capacity of the member for a given row of bolts. When the bolts are arranged in a zigzag fashion with a pitch $p$ and gauge $g$, the net effective area of the plate with a width $B$ and thickness $t$ is given by

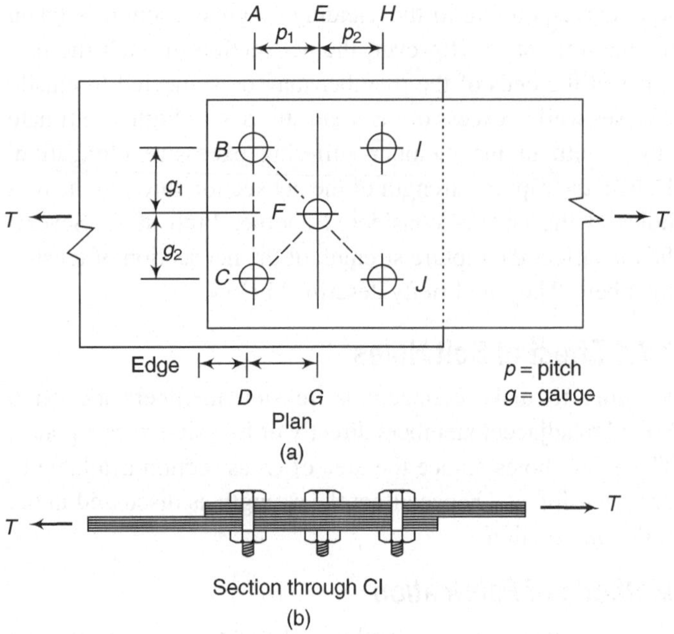


**FIG. 7.10** Staggered bolt holes

$$A_n = [B - nd_h + \Sigma(p^2/4g)]t \quad (7.7)$$

where $n$ is the number of bolt holes in the critical section considered, the summation is over all the paths of the critical sections normal to the direction of the tensile force, and $d_h$ is the diameter of the bolt hole. The above empirical relation was proposed by Cochrane in 1922 based on experimental evidence (see also Section 5.4.7 of Chapter 5). All possible failure paths (straight as well as zigzag) are to be considered and the corresponding net areas are to be computed as per Eqn (7.7) to find the minimum net area of the plate.

### *Effect of Bearing Stress*

When slip takes place between plates being joined by bolts, one or more fasteners come into bearing against the side of the hole. Consequently bearing stress is developed in the material adjacent to the hole and in the fastener. Initially this stress is concentrated at the point of contact. An increase in load causes local yielding and a larger area of contact resulting in a more uniform bearing stress distribution. The actual failure mode in bearing depends on the end distance, the bolt diameter, and the thickness of the connected material. Either the fastener splits out through the end of the plate because of the insufficient end distance or excessive deformations are developed in the material adjacent to the hole and the elongation of the hole takes place as shown in Fig. 7.11. Often a combination of these failure modes will occur (see also Section 5.3 of Chapter 5).

However, research (Munse & Chesson 1963) shows that as long as the bearing stress is less than 2.25 times the tensile stress, the effect of bearing stress can be neglected.

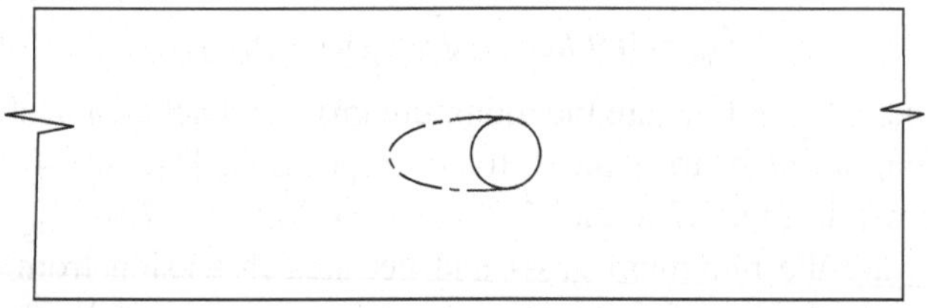

**FIG. 7.11** Elongation of bolt hole due to local yielding under bearing stress

## 7.7.2 Effect of Shear Lag

The force is transferred to a tension member (angles, channels, or T-sections) by a gusset or the adjacent member connected to one of the legs either by bolting or welding. The force thus transferred to one leg by the end connection locally gets transferred as tensile stress over the entire cross section by shear. Hence, the tensile stress on the section from the first bolt hole up to the last bolt hole will not be uniform. The connected leg will have higher stresses at failure even of the order of ultimate stress while the outstanding leg stresses may be even below yield stress. However, at sections away from the end connection, the stress distribution becomes more uniform.

Similarly consider an I-section as shown in Fig. 7.12(a), which is connected to the other members of a structure with two gusset plates attached to the flanges. This is the common connection found in bridge trusses as it is not practical to connect both webs and the flanges. It is obvious that the web is not fully effective in the region of the connection. As shown in Fig. 7.12(b), only a distance away from the connection, there will be uniform stress throughout the section. Because the internal transfer of forces from the flange region into the web region will be by shear (in the case of angles, transfer of force from one leg to the other will be by shear as indicated later in Fig. 7.16) and because one part 'lags' behind the other, the phenomenon is referred to as *shear lag*. The shear lag reduces the effectiveness of the component plates of a tension member that are not connected directly to a gusset plate. The efficiency of a tension member can be increased by reducing the area of such components not directly connected at ends.

The shear lag effect reduces with increase in the connection length. In a longer connection, consisting of a number of bolts, a major portion of the load is already distributed more uniformly over the cross section by the force transferred by other bolts, before it reaches the cross section through the innermost bolt, at which section the total force is maximum. Tests results indicate that as the length of the connection increases the outstanding leg is able to develop an average stress equal to the yield stress while the average stress in the connected leg reaches the ultimate strength of material. However, for a member with shorter connection, the average stress in the outstanding leg may not reach the yield strength (Kulak et al. 1987).

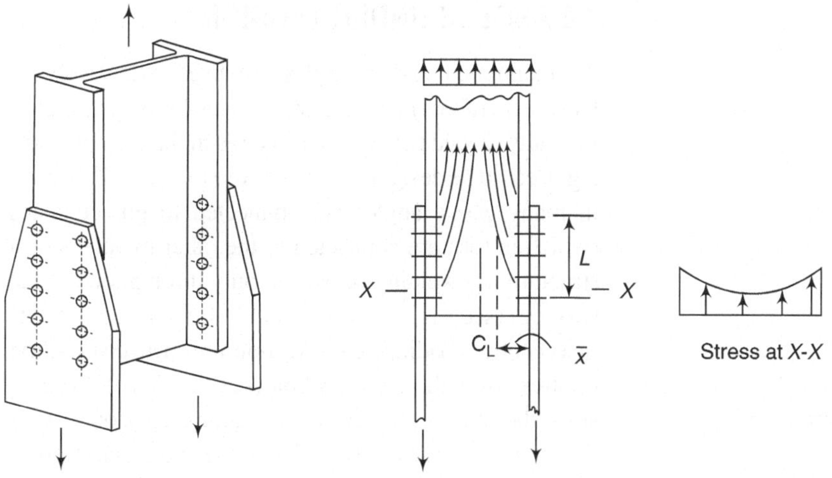


**FIG. 7.12** Shear lag in tension member

In addition to its effect on shear lag, an increase in the connection length of a specimen also allows for a larger restoring moment at the eccentric connection. The connection length affects the lever arm of the restoring moment. A longer connection length permits, therefore, a larger moment to develop, thus increasing the net section efficiency. A review of the shear lag research for both bolted and welded connections, along with design recommendation is provided by Easterling & Giroux (1993). The effects of shear lag need to be considered only when all the elements are not connected directly. Different design codes deal with it in different ways.

To account for shear lag effect, Munse & Chesson (1963) suggested a shear lag factor of $U = 1.0 - \bar{x}/L$ in the calculation of the net section capacity, where $\bar{x}$ refers to the distance from the face of the gusset plate to the centroid of the area tributory to that gusset plate [see Fig. 7.12(b)] and $L$ is the length of the connection. For a member with all of its elements transferring load, $U$ is taken as 1.0. The study conducted by Kulak & Wu (1997) revealed that the net section efficiency increases when the number of bolts in a line is increased up to four and after that there is no appreciable increase in the efficiency.

### 7.7.3 Geometry Factor

Tests on bolted joints show that the net section is more efficient if the ratio of the gauge length $g$ to the diameter $d$ is small (Kulak et al. 1987). The increase in the efficiency due to a smaller $g/d$ ratio is due to the suppression of contraction at the net section.

The reduction in area due to the yielding of the metal on the net section between holes is restrained by the metal on the adjacent parallel gross cross section. Therefore, a condition of biaxial stress develops, and there is an increase in the tensile strength. As the gauge is increased the efficiency decreases. This is because the concentrated forces applied to the member by the widely spaced fasteners produce non-uniform straining of such a nature that fracture begins locally before stress can be equalized over the net section. To account for the effect of gauge or $g/d$ ratio, Munse and Chesson (1963) proposed a geometry factor, $K_3$, given by Eqn (7.8), which is multiplied with the net section to account for this effect.

$$K_3 = 1.60 - 0.70(A_{ne}/A_g) \tag{7.8}$$

The value of $K_3$ generally varies in the range of 0.9 to 1.14.

### 7.7.4 Ductility Factor

The reduction in ductility tends to reduce the strength of the member. An increase in ductility tends to increase the net section strength by allowing a better plastic redistribution of stress concentration over the cross section thus distributing stresses more evenly when the steel is in the plastic state. Tension members with bolt holes made from ductile steels have proved to be as much as one-fifth to one-sixth times stronger than similar members made from less ductile steels having the same strengths (Kulak et al. 1987). To account for this effect, Munse and Chesson (1963) proposed a reduction factor

$$K_1 = 0.82 + 0.0032R_a \leq 1.0 \tag{7.9}$$

where $R_a$ is the area reduction ratio before rupture. In case of commonly used structural steels exhibiting minimum prescribed ductility, the ductility factor $k_1$ equals 1.0.

### 7.7.5 Spacing of Fasteners

The closer spacing of fasteners relative to their diameter may sometimes lead to block shear failure at the ends as discussed in Section 7.6.3, which has to be accounted for as a limit state.

### 7.7.6 Effect of Residual Stresses and Initial Crookedness

The stress–strain behaviour as discussed in Section 7.1 is for an ideal tension member and is based on the tensile-test coupon used to determine the basic material properties. However, member behaviour may differ from coupon behaviour for various reasons, such as slip in bolted and riveted connections, non-linear behaviour of the connections, and residual stresses in the member. As we have seen in Section 1.9 of Chapter 1, residual stresses result principally from non-uniform cooling of hot-rolled or welded members and from cold straightening of bent members. The residual stress distributions have been shown in Figs 1.24–1.27 of Chapter 1.

Let us consider a rectangular cross section with residual stress distribution as shown in Fig. 7.13. Now as the tensile load is increased from zero, local yielding commences before the general yield load $N_y$ is reached [Fig. 7.13(c)], and hence the range over which the load–extension behaviour is linear, decreases.

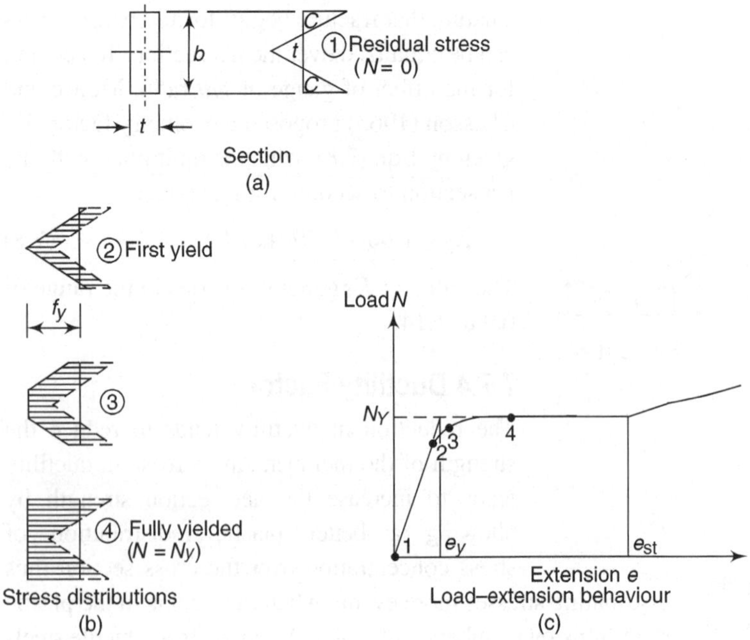


**FIG. 7.13** Effect of residual stresses on load–extension behaviour

However, the general yield load $N_y$ at which the whole cross section is yielded can still be reached due to the redistribution of the stresses. The residual stresses also result in local early strain hardening and reduce the plastic range of the member [see Fig. 7.13(c)]; however the member behaviour can be still be regarded as ductile. Though the residual stresses have no consequences with respect to the static strength of the member, they can be important if fatigue is involved. (The residual stresses can have a pronounced effect on the strength of compression members, as is shown in Chapter 9.)

If the tension member has an initial crookedness $v_0$ [see Fig. 7.14(a)] the axial load causes it to bend and deflect laterally as $v$ shown in Fig. 7.14(b). Due to this action, the member is partially straightened and the bending action in the central region is reduced. The bending induces additional axial stresses, which cause local yielding and strain hardening in the member similar to residual stresses [Fig. 7.13(c)]. Normally, the initial crookedness will be small, and hence the resulting reductions in the linear and plastic ranges will be comparatively small. (A tolerance of 3 mm has been specified in clause 17.13 3.1 of IS 800:2007) Hence, the behaviour of the member may be considered as ductile.

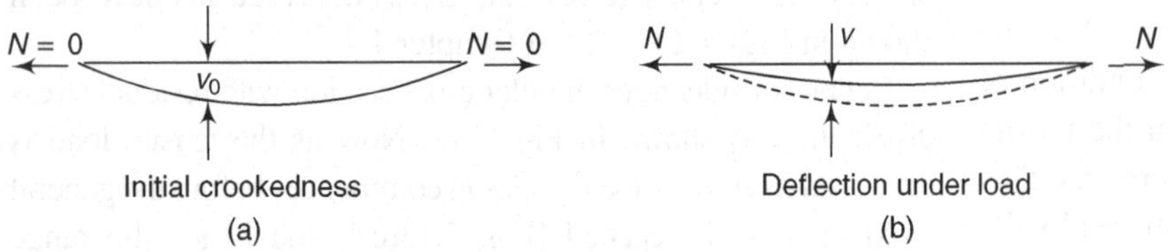


**FIG. 7.14** Tesnsion member with initial crookedness

## 7.8 ANGLES UNDER TENSION

As mentioned earlier angles are used extensively as tension members in towers, trusses, and bracings. Angles, if axially loaded through centroid (as in the case of tower legs), could be designed as in the case of plates. However, in many cases, angles are connected to gusset plates (which in turn are connected to the other members of a structure) by welding or bolting only through one of the two legs (see Fig. 7.15). This kind of connection results in eccentric loading, causing non-uniform distribution of stress over the cross section (see Fig. 7.16). Further, since the load is applied by connecting only one leg of the member, there is a shear lag at the end connections.

When the angles are connected to other angles through the centroid and when the holes are staggered on two legs of an angle (see Fig. 7.17), the gauge length $g$ for use in Eqn (7.7) is obtained by developing the cross section into an equivalent flat plate [see Fig. 7.17(b)] by revolving about the centre lines of the component parts. The critical net section can then be established by the procedure described for plates. Thus referring to Fig. 7.17(b), the gauge distance $g^*$ is obtained as

$$g^* = g_a - (t/2) + (g_b - t/2) = g_a + g_b - t \qquad (7.10)$$

An illustration of the calculations involved is given in Example 7.3. Examples of net section calculations are also provided in Chapter 5.

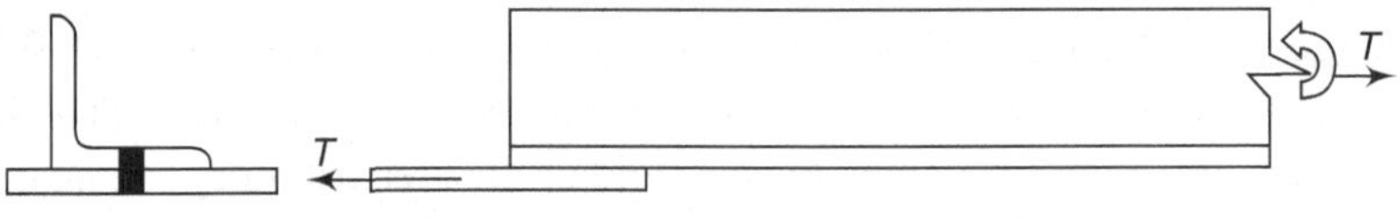


**FIG. 7.15** Angles eccentrically loaded through gussets

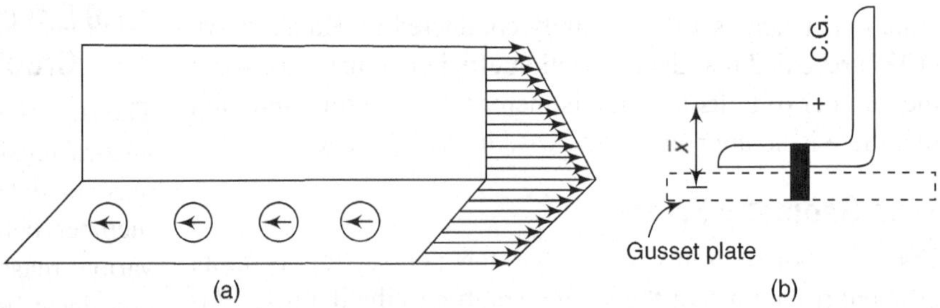


**FIG. 7.16** Distribution of stresses in an angle

### 7.8.1 Net Section Design

It was shown in Section 7.7 that the strength of a tension member with bolted or riveted connections can be predicted with good accuracy by taking into account the various factors affecting the strength of the net section. Hence, the following procedure has

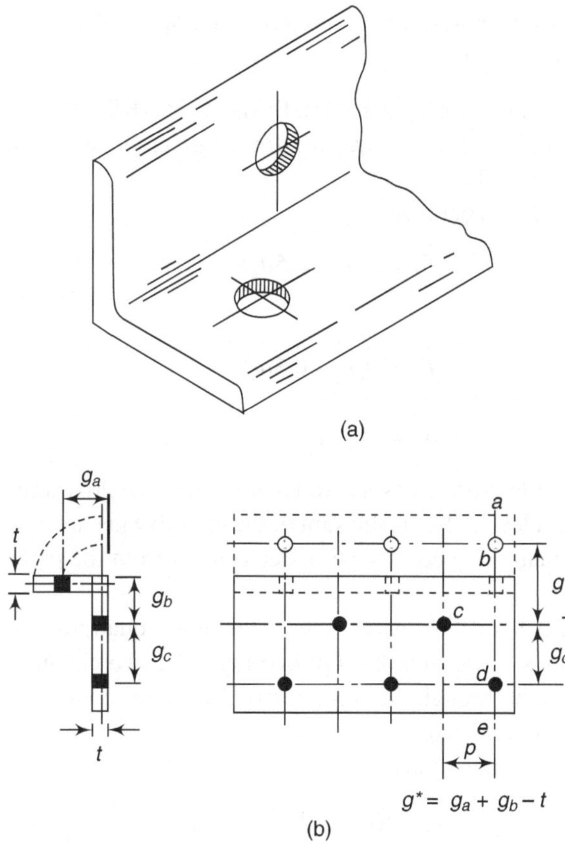


**FIG. 7.17** Fasteners in more than one plane

been suggested for the design of such members (Gaylord et al. 1992; Munse & Chesson 1963). To provide for necessary margin of safety against fracture, the capacity of the member should be determined by multiplying the effective net cross-sectional area with the specified minimum tensile strength $f_u$ divided by the partial safety factor $\gamma_{m1}$. Thus, the effective net area is defined by (Gaylord et al. 1992).

$$A_{ne} = K_1K_2K_3K_4A_n \quad (7.11a)$$

where $A_n$ is the net area of the cross section obtained by $p^2/4g$ rule [see Eqn (7.7)] and $A_{ne}$ is the effective area of the cross section modified by the other terms in the equation. These terms represent a ductility factor ($K_1$), a factor for the method of hole forming ($K_2$), a geometry factor reflecting hole spacing ($K_3$), and a shear leg factor ($K_4$).

It has been found that for more practical cases the ductility factor ($K_1$) and the geometric factor ($K_3$) can be taken as unity (Kulak & Wu 1997; Gaylord et al. 1992). As mentioned previously, the factor $K_2$ can be taken as 0.85 for punched holes or 1.0 for drilled holes. However in many specifications, including IS 800, rather than using a value of 0.85 for $K_2$ for the punching effect, it is accommodated by taking the hole diameter as 2-mm larger than the actual hole size when computing the net area. Hence, Eqn (7.11a) is simplified to

$$A_{ne} = K_4A_n \quad (7.11b)$$

Considering the recommendation of Munse & Chesson (1963), for $K_4$, the net section tension capacity of an angle connected by one of its legs to the gusset is given by

$$T_{dn} = (1 - \bar{x}/L)A_nf_u/\gamma_{m1} \quad (7.12)$$

where $L$ is the connection length (between extreme fasteners) and $\bar{x}$ is the distance from the face of the gusset plate to the centroid of the area tributory to that gusset plate (see Fig. 7.18).

The direct use of Eqn (7.12) presents a problem for the designer because the length of the connection $L$ must be known (or assumed) before it can be applied. Thus, it results in an iterative solution. The American code (ANSI/AISC 360–05) provision is based on this equation, with an upper limit of 0.9 for the term $(1 - \bar{x}/L)$. However, the specification suggests that 0.85 can be used for the $(1 - \bar{x}/L)$ term, if there are three or more bolts in line (and when $b_f < 2/3d$) and that 0.70 should be used when there are four or more fasteners. Similar values have been proposed for welded connections (1.00 for $L_w \geq 2w$, 0.87 for $2w > L_w \geq 1.5w$ and 0.75 for $1.5w > L_w \geq w$, where

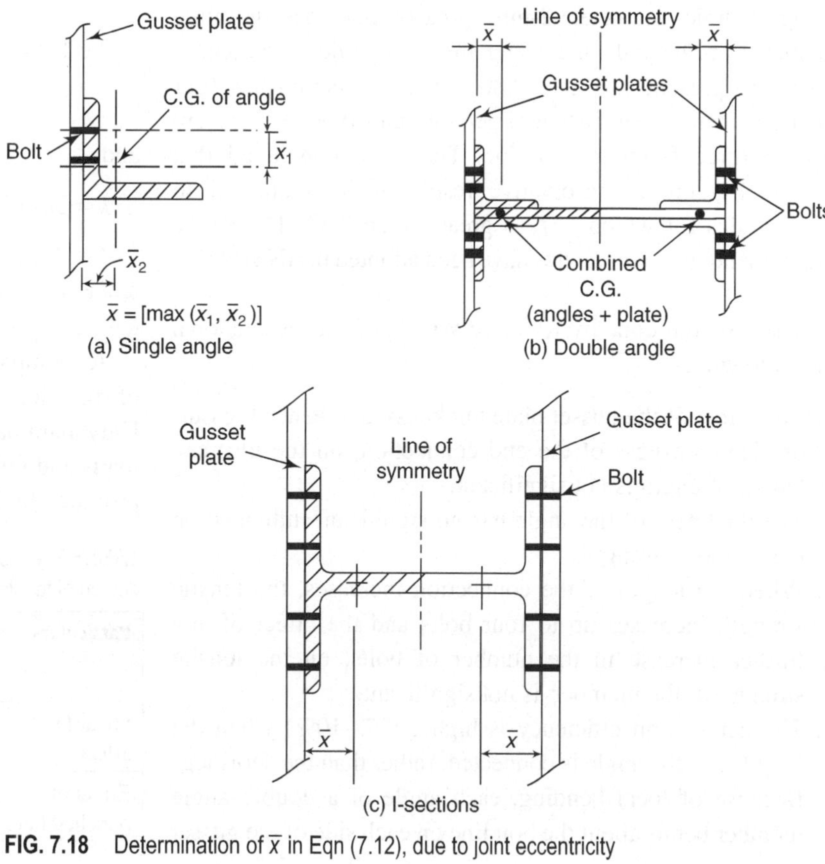


**FIG. 7.18** Determination of $\bar{x}$ in Eqn (7.12), due to joint eccentricity

$L_w$ is the length of the weld along one side of the plate and $w$ is the distance between longitudinal welds i.e., plate width).

Kulak & Wu (1997) conducted several tests on bolted angle tension members and studied the effects of various factors such as out-of-plane stiffness of gusset plates, angle thickness, connection by long leg and short leg, and connection length. They proposed the following net section strength formula based on these tests and the finite element analysis done by them (without the partial safety factor for material)

$$T_u = f_u A_n + \beta f_y A_o \quad (7.13)$$

where $A_n$ is the net area of the connected leg at the critical cross section, $A_o$ is the gross area of the outstanding leg, and $\beta = 1.0$ for members with four or more fasteners per line in the connection, or $\beta = 0.5$ for members with two or three fasteners per line in the connection. Kulak & Wu compared the ultimate load predicted by Eqns (7.12) and (7.13) and found that Eqn (7.13) provides conservative results and falls in a narrower scatter band of results than those predicted by Eqn (7.12). They also proposed the following equation for single and double angles connected by only one of the legs and made of Fe 410 steel.

$$T_u = U A_n f_u \quad (7.14)$$

where $A_n$ is the net area of the critical cross section calculated using a hole diameter 2-mm greater than the nominal diameter of the bolt and using the $p^2/4g$ rule if staggered holes are present and $U = 0.80$ if the connection has four or more member of fasteners per line and 0.60 if there are two or three fasteners per line. They also concluded that Eqn (7.14) gives conservative results if it is applied to steel grades for which $f_y/f_u$ is greater than 0.62. The above recommendations for angles have been adopted in ANSI/AISC code.

The research done by Kulak & Wu (1997) has also shown the following.

1. The effect of the gusset plate thickness, and hence the out-of-plane stiffness of the end connection, on the ultimate tensile strength is not significant.
2. The thickness of the angle has no significant influence on the member strength.
3. When the length of the connection increases, the tensile strength increases up to four bolts and the effect of any further increase in the number of bolts, on the tensile strength of the member, is not significant.
4. The net section efficiency is higher (7%–10%) when the long leg of the angle is connected, rather than the short leg.
5. Because of local bending, each angle of a double angle member bends about the bolt line on each side of the gusset plate; thus the double angles seem to act individually rather than as a rigid unit.

It may be of interest to note that the British Code (BS: 5950–1, 2000) gives the following equation for single angle, channel, or T-section members.

For bolted connections:

$$T_u = f_y(A_e - 0.5a_2) \quad (7.15)$$

For welded connections:

$$T_u = f_y(A_g - 0.3a_2) \quad (7.16)$$

and $$a_2 = A_g - a_1$$

where $A_g$ is the gross cross-sectional area, $a_1$ is the gross area of connected leg, and $A_e$ is the sum of the effective net areas of all the elements of a cross section determined from the $p^2/4g$ rule.

BS 5950-1 (2000) gives the following equations for double angles connected through one leg only, two channels connected only through the web, or two T-sections connected only through the flange.

For bolted connections:

$$T_u = f_y(A_e - 0.25a_2) \quad (7.17)$$

For welded connections:

$$T_u = f_y(A_g - 0.15a_2) \quad (7.18)$$

In the 1984 version of IS 800, the effective net area is given as

$$A_e = a_1 + a_2 k \quad (7.19a)$$

and

$$k = 3a_1/(3a_1 + a_2) \text{ for single angles} \quad (7.19b)$$

$$k = 5a_1/(5a_1 + a_2) \text{ for double angles or T-sections} \quad (7.19c)$$

where $a_1$ is the effective cross-sectional area of the connected leg and $a_2$ is the gross area of the unconnected leg.

A comparison of the provision of the various codes of practices is given in Table 7.3 (Usha 2003; Usha & Kalyanaraman 2002). A comparison of experimental failure loads and failure loads according to different specifications is provided by Madugula and Mohan (1998).

**TABLE 7.3** Comparison of factors considered by different codes for the design of tension members

| Particulars | IS 800: 1984 | BS-5950 1990 | Eurocode 3 1992 | AISC 1999 | CAN 1994 | AS 4100 | ASCE 1988 |
|---|---|---|---|---|---|---|---|
| Shear lag affect | Yes | Yes | Yes | Yes | Yes | Yes | Yes |
| Effect of punched holes | No | No | No | Yes | Yes | No | No |

(*Contd*)

**TABLE 7.3** *(Contd)*

| Particulars | IS 800: 1984 | BS-5950 1990 | Eurocode 3 1992 | AISC 1999 | CAN 1994 | AS 4100 | ASCE 1988 |
|---|---|---|---|---|---|---|---|
| Effect of ($f_y$/ $f_u$) ratio | No | No | No | No | No | No | No |
| Connection length | No | No | Yes | Yes | Yes | No | No |
| Block shear mode of failure | No | No | Yes | Yes | Yes | No | Yes |

*Source*: Usha (2002)

## 7.8.2 Indian Code (IS 800 : 2007) Provisions for Angle Tension Members

The net section strength of single and double angle tension members (either bolted or welded) and connected through one leg (including the shear lag effect) given in the code is based on the research done at IIT, Madras (Usha 2003; Usha & Kalyanaraman 2002). The design strength as governed by the tearing of the net section is given by the code as

$$T_{dn} = 0.9 f_u A_{nc}/\gamma_{m1} + \beta A_{go} f_y/\gamma_{m0} \quad (7.20)$$

and

$$\beta = 1.4 - 0.076[(b_s/L_c)(w/t)(f_y/f_u)] \le 0.9\, f_u \gamma_{m0}/(f_y \gamma_{m1}) \ge 0.7 \quad (7.21)$$

where $f_u$ and $f_y$ are the ultimate and yield stress of the material, $w$ and $t$ are the size and thickness of the outstanding leg, respectively, $b_s$ is the shear distance from the edge of the outstanding leg to the nearest line of fasteners, measured along the centre line of the legs in the cross section (see Fig. 7.19), $L_c$ is the length of the end connection measured from the center of first bolt hole to the centre of the last bolt hole in the end connection, $A_{nc}$ is the net area of the connected leg at critical cross section, computed after deducting the diameter of hole (the diameter of the holes should be taken as 2-mm larger than the nominal size in the case of punched holes), $A_{go}$ is the gross area of the outstanding leg, and $\gamma_{m0}$ and $\gamma_{m1}$ are the partial safety factor for the material ($\gamma_{m0} = 1.10$ and $\gamma_{m1} = 1.25$).

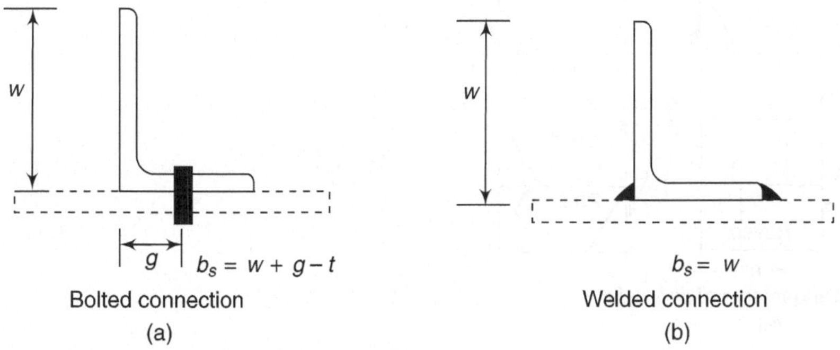


**FIG. 7.19** Definition of $b_s$ as per the code

Alternatively, the IS code suggests the use of an equation, for preliminary design, similar to Eqn (7.14), with the partial factor of safety for material as

$$T_{dn} = \alpha A_n f_u/\gamma_{m1} \quad (7.22)$$

with $\alpha = 0.6$ for one or two bolts, 0.7 for three bolts, and 0.8 for four or more bolts in the end connections or equivalent weld length.

It is important to observe that in the case of welds, the determination of the value of $\alpha$ is difficult since the welds may be transverse, longitudinal, or combined. Designers have to use their judgement to arrive at an equivalent number of bolts.

## 7.8.3 Welded Tension Members

For welded tension members other than flat plates or bars, when the load is transmitted by longitudinal welds only or a combination of longitudinal and transverse welds through just one leg of the angle, the effective net area $A_e$ will be $A_g U$. When a load is transmitted by transverse weld through just one leg of the angle, $A_e$ is the area of the connected leg and $U$ will be equal to one.

When flat plates or bars, connected by longitudinal welds, are used as tension members, they may fail prematurely by shear lag at the corners if the welds are too far apart. Therefore, the ANSI/AISC code states that when such situations are encountered, the length of the weld may not be less than the width of the plates or bars. For such situations, the value of $U$ corresponding to different values of $L_w$ are as follows:

$$L_w \ge 2w \quad U = 1.0 \quad (7.23a)$$

$$2w > L_w \ge 1.5w \quad U = 0.87 \quad (7.23b)$$

$$1.5w > L_w \ge w \quad U = 0.75 \quad (7.23c)$$

where $L_w$ is the length of the weld and $w$ is the width of the plate.

## 7.9 OTHER SECTIONS

The tearing strength $T_{dn}$ of double angles, channels, I-sections, and other rolled steel sections, connected by one or more elements to an end gusset is also governed by shear lag effects (see Section 7.7.2). The code suggests that the design tensile strength of such sections, as governed by the tearing of the net section, may also be calculated by using Eqn (7.20) to (7.22), where $\beta$ is calculated based on the shear leg distance $b_s$, taken from the farthest edge of the outstanding leg to the nearest bolt/weld line in the connected leg of the cross section. The net effective area of a single channel section connected through the web may be treated as for double angles connected by one leg each to the gusset. When a rolled or built-up channel or I-sections are connected through flanges, the web is found to be partially ineffective in resisting the tensile load. In such cases, the net area may be taken as the total area minus half the web area (Duggal 2000).

Based on Eqn (7.12), AISC specification gives a simplified equation for $\bar{x}$ for rectangular hollow structural sections (HSS) connected with a single concentric gusset plate, based on the outside dimensions

$$\bar{x} = (B^2 + 2BH)/[4(B + H)] \tag{7.24}$$

where $B$ and $H$ are the breadth and width of the HSS sections. Dowswell & Barber (2005) compared the above AISC equation with available test data and results of finite element studies and found them to be reasonably accurate and generally conservative.

## 7.10 TENSION RODS

A common and simple tension member is the threaded rod. Such rods are usually found as secondary members, where the required strength is small. Some examples of tension rods are as follows.

1. Sag rods are used to help support purlins in industrial buildings. Various arrangements of sag rods are shown in Fig. 7.20. Sag rods reduce the bending moment about the minor axis (channels which are used as purlins are weak about the minor axis), resulting in economy. These rods are threaded at the ends and bolted to purlins. Since individual sag rods are placed between successive pairs of purlins, they may be designed individually, each to carry a tangential component from all the purlins below it. Thus, sag rods just below and suspended from the ridge purlin will be subjected to maximum force.
2. Vertical ties to help support girts in the walls of industrial buildings [Fig. 7.21(a)].
3. Hangers, such as tie rods supporting a balcony [Fig. 7.21(b)]. When providing such hangers, proper detailing should be adopted and communicated properly to the fabricator. Otherwise failures such as those witnessed in the Hyatt Regency Hotel at Kansas city may result.
4. Tie rods to resist the thrust of an arch.

As mentioned previously, tie rods are also often used with an initial tension as diagonal wind bracings in walls, roofs, and towers. The initial tension reduces the deflection and vibration and also increases the stiffness of these rods.

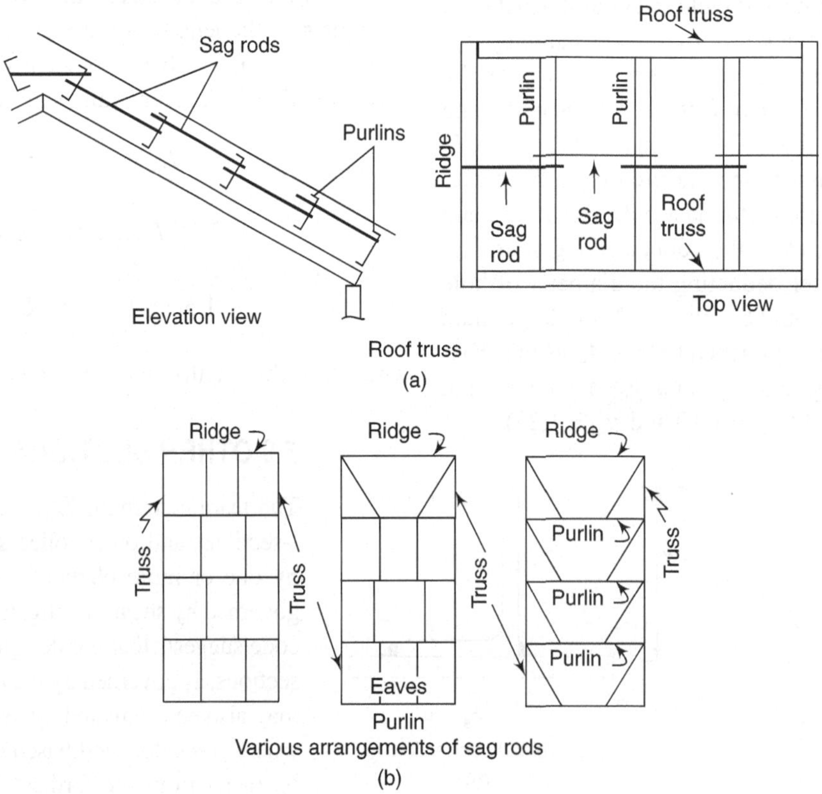


**FIG. 7.20** Use of sag rods to support purlins

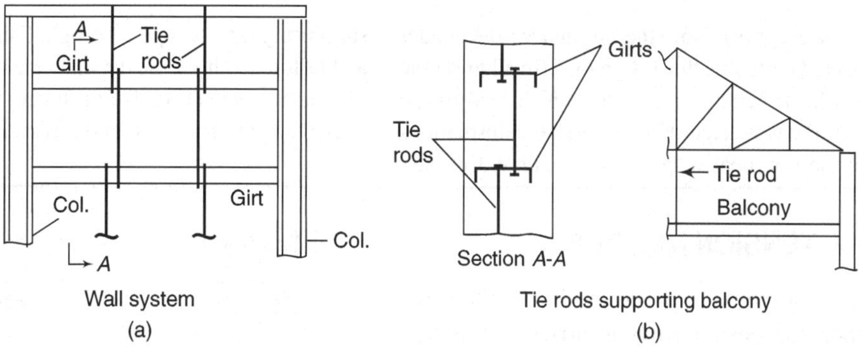


**FIG. 7.21** Uses of tension rods

## CASE STUDY

### Hyatt Regency Walkway Collapse

The 40-story Hyatt Regency hotel in Kansa City, Missouri, USA, was opened on 1 July 1980. The lobby of the hotel featured a multistory atrium, which had suspended concrete walkways on the second, third, and fourth levels. These three separate pedestrian walkways connected the north and south buildings. The fourth and second floor walkways hung one above the other and the third floor walkway hung offset to one side. These walkways all connected to steel trusses that hung from the atrium ceiling.

The two walkways were suspended from a set of steel tension rods of 32 mm, with the second floor walkway hanging directly underneath the fourth floor walkway. The walkway platform was supported on 3 cross-beams suspended by steel rods retained by nuts. The cross-beams were box beams made from C-channels welded toe-to-toe. The original design called for three pairs of rods running from the second floor all the way to the ceiling.

On 17 July 1981, when a party was going on, the 4th floor walkway failed and fell on the lower walkway, both walkways crashing into the floor three stories below killing 114 people and injuring 185. The third floor walkway was not involved in the collapse.

The cause of the failure is that the contractor replaced the one vertical suspension rod specified by the original designer, by two shorter rods; one from the upper support to the first walkway, and another from the bottom beam of the first walkway down to

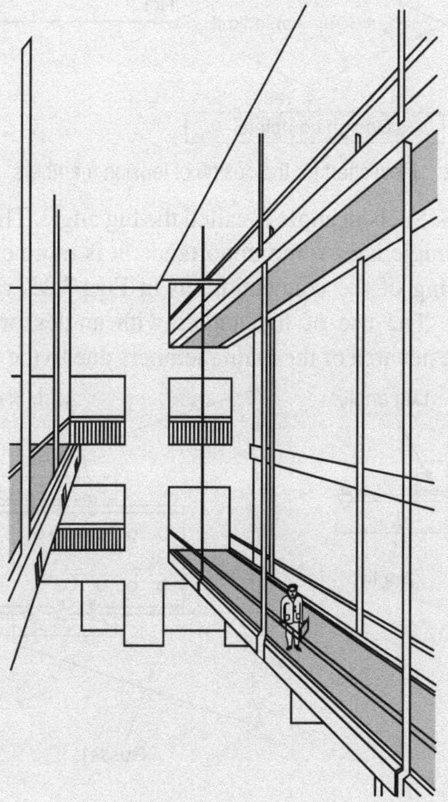

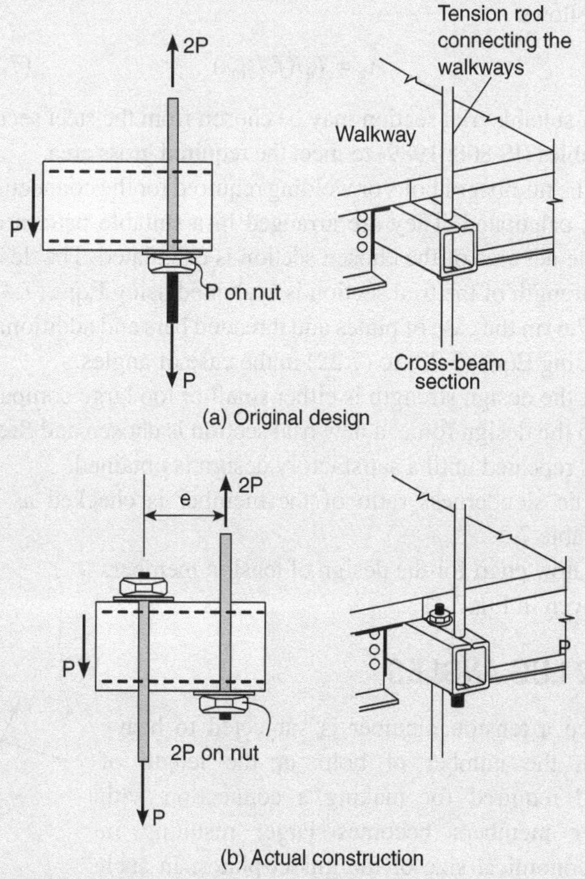


Difference between the design and construction of the walkway support system

the second walkway (see figure). Now the nut and washer under the upper rod is subjected to double the design load (in addition the eccentricity created a local bending moment), which lead to the failure; photographs of the wreckage show excessive deformations of the cross section; the box beams split at the weld and the nut supporting them slipped through. Failure to communicate properly and failure to check the details were cited as the main problems for the faulty connection detail; the connection that failed was never shown on any drawings, and it was not designed.

## 7.11 DESIGN OF A TENSION MEMBER

In the design of a tension member, based on the tensile force acting on the member, the designer has to arrive at the type and size of the member. The type of member is chosen based on the type of the structure and location of the member (e.g., double angles at the bottom chord or a rafter of roof trusses, angles or pipes for web members of roof trusses, etc.).

The design is iterative, involving a choice of a trial section and an analysis of its capacity. The various steps are as follows:

1. The net area required $A_n$ to carry the design load $T$ is obtained by the equation

$$A_n = T_u/(f_u/\gamma_{m1}) \tag{7.25}$$

2. From the required net area, the gross area may be computed by increasing the net area by about 25% to 40%. The required gross area may also be checked against that required from the yield strength of the gross section as follows

$$A_g = T_u/(f_y/\gamma_{m0}) \tag{7.26}$$

   A suitable trial section may be chosen from the steel section tables (IS 808: 1989) to meet the required gross area.
3. The number of bolts or welding required for the connections is calculated. They are arranged in a suitable pattern and the net area of the chosen section is calculated. The design strength of the trial section is evaluated using Eqns (7.4) to (7.6) in the case of plates and threaded bars and additionally using Eqns (7.20) to (7.22) in the case of angles.
4. If the design strength is either small or too large compared to the design force, a new trial section is chosen and Step 3 is repeated until a satisfactory design is obtained.
5. The slenderness ratio of the member is checked as per Table 7.2.

The flow chart for the design of tension members is given in Fig.7.22.

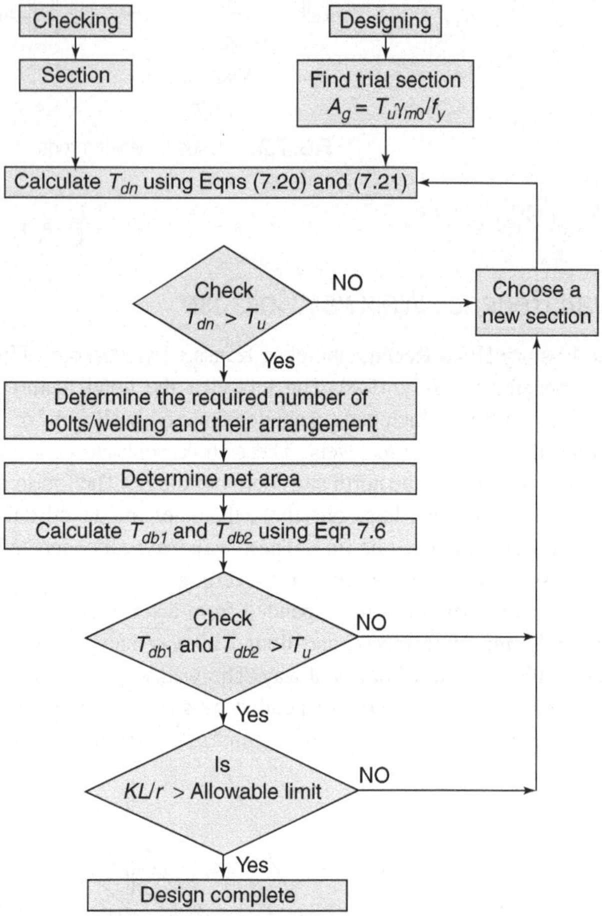


**FIG. 7.22** Flow chart for the design of tension members

## 7.12 LUG ANGLES

When a tension member is subjected to heavy load, the number of bolts or the length of weld required for making a connection with other members becomes large; resulting in uneconomical size of the gusset plates. In such situations, an additional short angle may be used to reduce the joint length and shear lag as shown in Fig. 7.23. Such an angle is called the lug angle. The location of the lug angle is of some importance; it is more effective at the beginning of the connection, as in Fig. 7.23, rather than at the end. The use of lug angles with angles or channels reduces the net area of the main members due to the additional

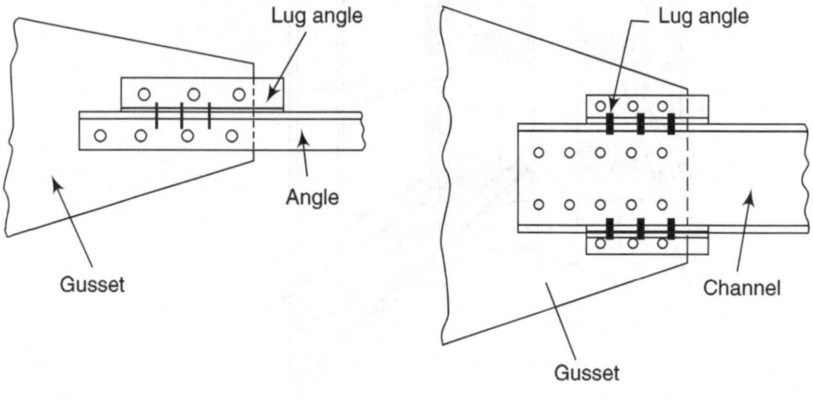


**FIG. 7.23** Lug angles

bolt holes in projected members. This reduction in the net area of the member should not be excessive. In the connections of the lug angles to the member or the gusset plate more than two bolts are used. Since both legs of the angles or channels are connected to the lug angles, the net area of the members should be calculated simply as gross area minus the area of the holes.

Lug angles may be eliminated by providing unequal angle sections with the wider leg as the connected leg and using two rows of staggered bolts. In many cases, the cost of providing the lug angles (including their connection and the extra fabrication required to make the holes) may be found to be expensive than providing extra length and thickness of gusset plate. Hence, they are not used in practice.

In the case of angle members, the lug angles and their connection to the gusset or other supporting member should be capable of developing a strength of not less than 20% in excess of the force in the outstanding leg of the angle, and the connection of the lug angle to the angle member should be capable of developing 40% in excess of the force (clause 10.12.2).

In channel members, however, the lug angles and their connections to the gusset or other supporting member should be capable of developing a strength of not less than 10% in excess of the force in the flange of the channel and the attachment of the lug angle to the member should have a strength not less than 20% in excess of that force.

### 7.12.1 Compensating for Reduction in Cross-sectional Area

Ends of tension members may be strengthened to offset reduction in the cross-sectional area due to threads, fastener holes, etc. To compensate for the loss of area due to threads, round or square bars may be 'upset' to form an increased cross section into which the threads may be cut. Upset bars may have an area at the root of the thread at least 15% greater than the area of the bar.

Similarly, tension members which are bolted to gusset plates may be strengthened at the ends to make-up for the loss of area due to bolt holes. In the case of welded or rolled I-sections, this may be done by welding to the flanges thicker plates of the same steel or plates of the same thickness but of high-strength steel. Gaylord et al. (1992) report that the tension members of the Carquinez straits bridge in California were increased in area at the ends by groove welding thicker plates as shown in Fig. 7.24. They also report that in the Benicia–Martinez bridge in California the end stubs were of high-strength steel with the thickness of flange and web same as that of the member.

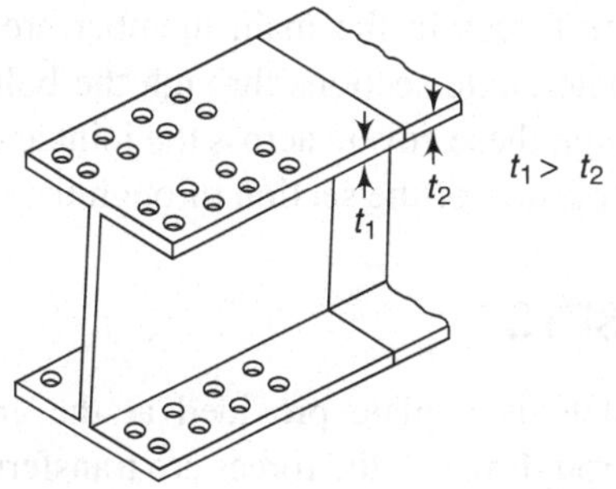


**FIG. 7.24** Compensating for reduction in cross-sectional area

## 7.13 SPLICES

When the available length is less than the required length of a tension member, splices are provided. The various types of splices that can be provided are shown in Figs 7.25(a) to (c). If the sections are not of the same thickness, packings are introduced, as shown in Fig. 7.25(d). Moreover, reduction in the capacity of bolt has to be considered for long joints or if the packing thickness is greater than 6 mm (see Section 5.4 of Chapter 5 for these calculations).

In the design of a tension splice, the effect of eccentricity is neglected; as far as possible it should be avoided. Thus Fig. 7.25(e) shows an angle section spliced on one leg of the angle only by a plate. Such an arrangement causes eccentricity and introduces bending moments. To overcome this, both the legs of the angle should be spliced, as shown in Fig. 7.25(a). The splice as shown in Fig. 7.25(b) is used in the legs of transmission line or communication towers and aids transfer of tensile loads, without any eccentricity.

The splice cover plates or angles and its connections should be designed to develop the net tensile strength of the main

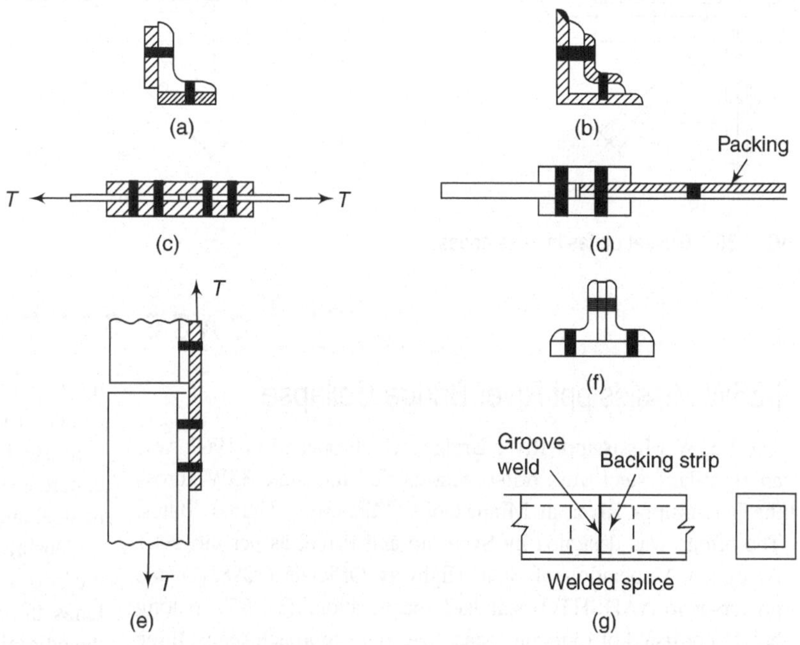


**FIG. 7.25** Splices in tension members

member. The forces in the main member are transferred to the cover plate angle sections through the bolts/welding and carried through these covers across the joint and is transferred to the other portion of the section through the fasteners.

## 7.14 GUSSETS

A gusset plate is a plate provided at the ends of tension members through which the forces are transferred to the main member. Gusset plates may be used to join more than one member at a joint as shown in Fig. 7.26. The lines of action of truss members meeting at a joint should coincide as shown in Fig. 7.26. If they do not coincide, secondary bending moments and stresses are created, which should be considered in the design.

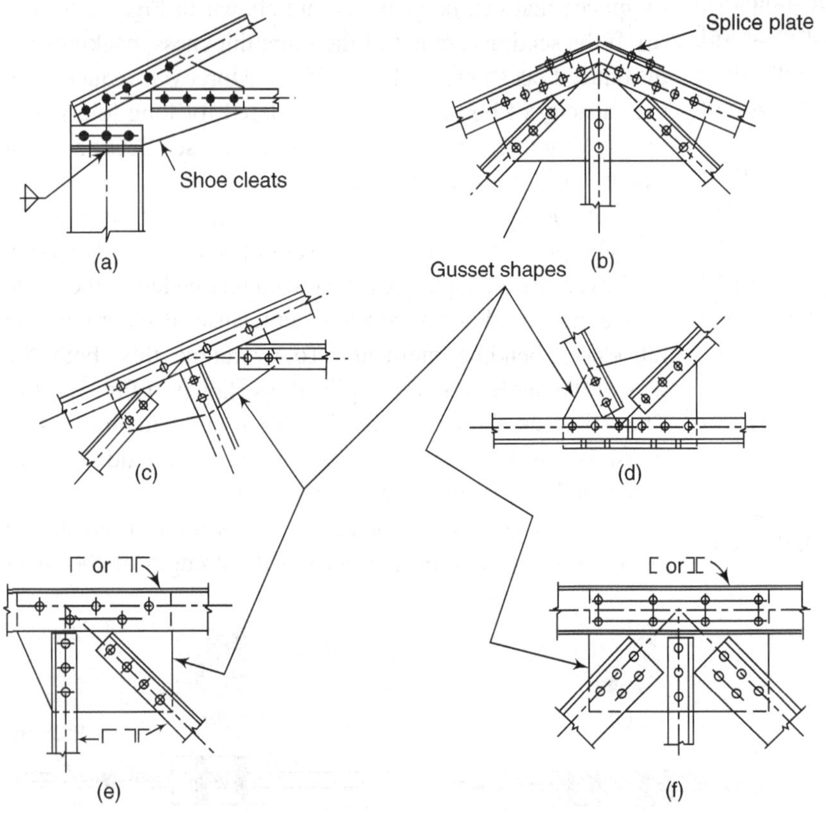


**FIG. 7.26** Gusset plates in truss nodes

The size and shape of the gusset plates are decided based on the direction of various members meeting at a joint. The plate outlines are fixed so as to meet the minimum edge distances specified for the bolts that are used to connect the various members at a particular joint. The shape of the gusset plate should be such that it should give an aesthetic appearance, in addition to meeting the edge distances of bolts, as mentioned earlier.

The stresses developed at the gusset plates are complex and hence the elementary beam formulae are not applicable (since these formulae are valid only for beams whose span is more than twice the depth and at cross sections not closer to concentrated loads than about half the depth; most of the gussets do not satisfy these requirements).

Compressive stresses may develop parallel to and at the edge of gusset plates. This is because deflection of a truss tends to change the angles between its members (Gaylord et al. 1992). Thus, the lower part of the gusset plate in Fig. 7.26(e) will be compressed if the angle between the adjacent members decreases. Therefore, the width of that edge should not be too large, compared with the thickness; else the plate will buckle. A bending of this kind, called *local buckling*, is discussed in Chapter 8. Hence, the edge of the plate should be less than 42 $\varepsilon t$ where $\varepsilon t = (250 f_y)^{0.5}$ and $t$ is the thickness of plate.

It is tedious to analyse the gusset plate for shear stresses, direct stresses, and bending stresses, and hence empirical methods have been used in the past to arrive at the thickness of the gusset plate (e.g., Whitmore method). More details of these methods are discussed in Section 5.7.2 of Chapter 5. The block shear model (see Section 7.6.3) could also be used to find the thickness of the gusset plate. It is a usual practice to provide the thickness of the gusset plate equal to or slightly higher than the members that are connected by the gusset plate. It is interesting to note that the failure of the I-35W bridge at Minneapolis, USA in August 2007, was due to the inadequate thickness of the gusset plate.

## CASE STUDY

### I-35W Mississippi River Bridge Collapse

The I-35W Mississippi River Bridge, which opened in 1967, was an eight-lane steel truss bridge that carried Interstate 35W across the Mississippi River in Minneapolis, Minnesota, United States. The bridge was designed by Sverdrup and Parcel as per the 1961 American Association of State Highway Officials (AASHO, the precursor to AAHSHTO) standard specification. The 579 m long bridge consisted of fourteen spans: five south approach spans, three main spans, and six north approach spans. The three main spans were of deck truss construction, while all but two of the approach spans were steel multi-girder construction, and the remaining two were of concrete slab construction. The centre span consisted of a single 140 m steel arched truss over the 119 m wide navigation channel.

During the weekday evening rush hour on 1 August 2007, with the bridge traffic moving very slowly through the limited number of lanes, the central span of the bridge suddenly collapsed, followed by the adjoining spans. The structure and deck collapsed into the river,

35 m below the deck. Approximately 100 vehicles were involved, with 35 people dead and 100 injured. At the time of collapse, the PSA construction company crew was working on the deck replacing the expansion joints, and removing and replacing the concrete.

The bridge has been inspected annually since 1993 and every other year prior to that. The State reports from 2001 and 2005 indicated the presence of fatigue cracks in the bridge's trusses, and that the bridge had no secondary system to bear the weight of traffic in the event of an unexpected failure. In 1990 the federal government, citing significant corrosion in its bearings, designated it as 'structurally deficient' (this term does not indicate any lack of safety, as over 75,000 bridges in USA have this rating). Some repair work was carried out in the bridge to arrest the fatigue cracks.

I-35W Mississippi River Bridge (before and after the failure)

The National Transportation Safety Board (NTSB), which investigated the bridge failure, has concluded that the gusset plates, which connected steel members of the bridge, were roughly half the thickness than they should have been and hence contributed to the failure of the bridge. NTSB investigators found 16 fractured gusset plates from the bridge's center span. The AASHO specification-1961, based on which this bridge was designed, had a provision which stated that the ratio of the free length of the gusset plate edge divided by the thickness should not exceed 48. However, the failed gusset plate had a ratio of 60, far more than that allowed by the code. Stiffeners in the form of angles or plates are to be added to the edges of gusset plates to prevent the edge buckling, in such cases. Note that edge buckling is one of the major failure modes of gusset plates in braced frames and trusses.

The NTSB investigation found no evidence that cracking, corrosion, or other causes 'played any role in the collapse of the bridge'. They also found no flaws in the steel and concrete material used in the bridge.

This bridge was replaced by the I-35W Saint Anthony Falls Bridge, which was constructed at the same location at an accelerated schedule, and opened to public on Sept. 18, 2008. It is a high performance concrete box-girder bridge, providing superior durability and multiple levels of structural redundancy, with in-built state-of-the-art sensor and monitoring system.

## 7.15 FATIGUE EFFECTS

When a member is subjected to either a large number of stress reversals or cyclic stresses, the result is application of a stress range to the member. The allowable stresses should be reduced to avoid a fatigue failure if the number of stress cycles and/or the stress range is large. It is usual to assume that fatigue can be neglected in normal building construction. However, in the design of bridge tension members, the fatigue effects should be considered. In such situation, it is advisable to use HSFG bolts to connect the member to the joint. An illustrative design of a bridge tension member for fatigue may also be found in Bowles (1980).

## EXAMPLES

**Example 7.1:** What is the net area $A_n$ for the tension member shown in Fig. 7.27, in case of (a) drilled holes and (b) punched holes.

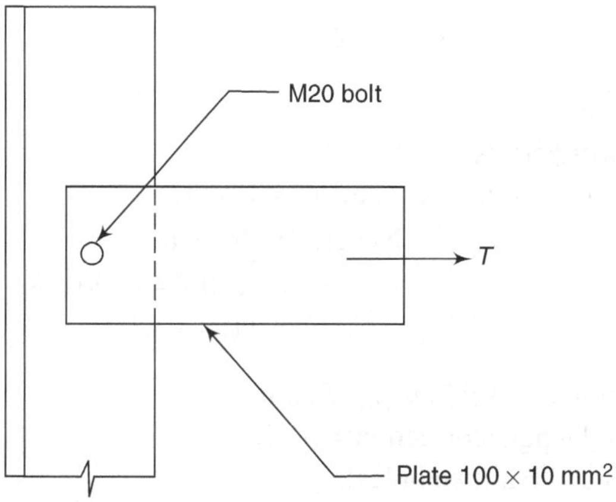


**FIG. 7.27**

**Solution:**

$$A_g = 100 \times 10 = 1000\,\text{mm}^2$$

(a) ***Net area of tension member for a drilled hole***

Hole diameter for an M20 bolt = 22 mm

$$A_n = A_g - (\text{hole diameter} \times \text{thickness of plate})$$
$$= 1000 - 22 \times 10 = 780\,\text{mm}^2$$

(b) ***Net area of tension member for a punched hole***

Hole diameter = 22 + 2 = 24 mm

$$A_n = 1000 - 24 \times 10 = 760\,\text{mm}^2$$

**EXAMPLE 7.2:** Determine the minimum net area of the plates as shown in Figs 7.28(a) and (b) with a plate of size of 210 × 8 mm and 16-mm bolts.

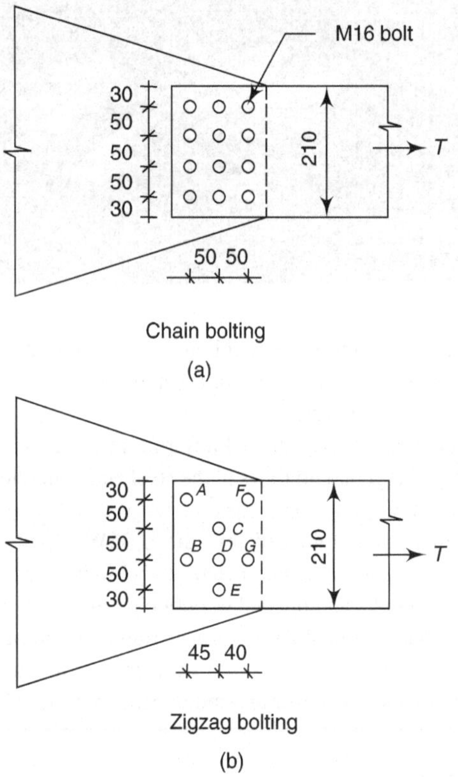


**FIG. 7.28**

**SOLUTION:**

(a) ***Chain bolting***

For a 16-mm bolt, hole diameter = 18 mm

$$\text{Net area} = (b - nd)t$$
$$= (210 - 4 \times 18) \times 8$$
$$= 1104\,\text{mm}^2$$

(b) ***Zigzag bolting [See fig. 7.28(b)]***

Staggered length correction = $p_i^2/4g_i$

Path AB and FG (two holes):

Net area = (210 − 2 × 18) × 8 = 1392 mm²

Path CDE (three holes):

Net area = (210 − 3 × 18) × 8 = 1248 mm²

Path ACDE (four holes and one stagger):

Net area = [210 − 4 × 18 + 45²/(4 × 50)]8 = 1185 mm²

Path FCDE (four holes and one stagger):

Net area = [210 − 4 × 18 + 40²/(4 × 50)]8 = 1168 mm²

Path ACG or FCB (three holes and two staggers):

Net area = [210 − 3 × 18 + 45²/(4 × 50) + 40²/(4 × 50)]8 = 1393 mm²

Path FCG (three holes and two staggers):

Net area = 210 − 3 × 18 + 2 × 40²/(4 × 50)]8 = 1376 mm²

The minimum net area is for path FCDE = 1168 mm². Note that the minimum net area occurs at a path which has the maximum number of holes and minimum number of staggers.

**EXAMPLE 7.3:** Determine the net area $A_n$ for the 125 × 75 × 6 angle with M20 bolt holes shown in Fig. 7.29.

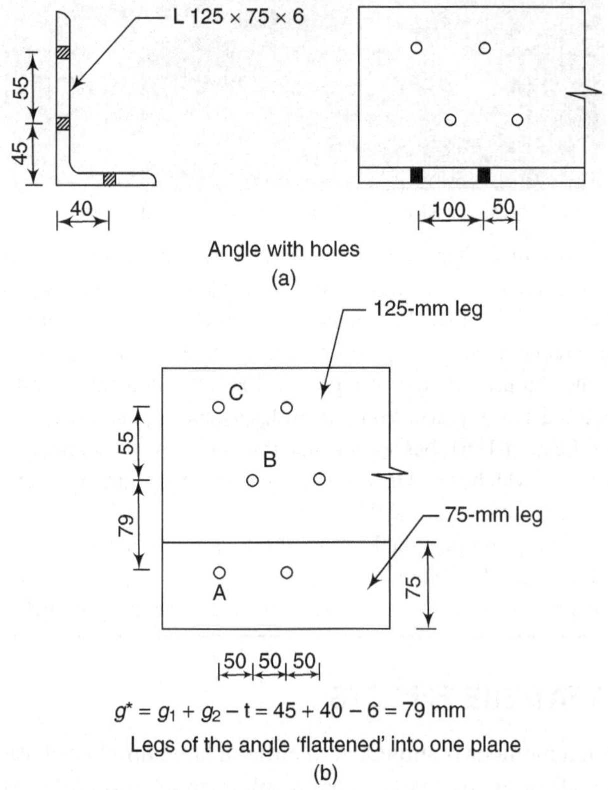


**FIG. 7.29** Effective net area of angle

**SOLUTION:**

For an M20 bolt, $d_h = 22$ mm

For net area calculation, the angle may be visualized as being flattened into a plate as shown in Fig. 7.29(b).

$$g^* = g_1 + g_2 - t = 45 + 40 - 6 = 79\,\text{mm}$$
$$A_n = A_g - \Sigma d_h t + \Sigma(p^2/4g)t$$

Path AC:

Net area = 1170 − 2 × 22 × 6 = 906 mm²

Path ABC:

Net area = 1170 − 3 × 22 × 6 + [50²/(4 × 55) + 50²/(4 × 79)] × 6 = 889.65 mm²

Since the smallest net area is 889.65 mm² for path ABC, therefore, that value holds.

> **Example 7.4:** Determine the design tensile strength of plate (160 × 8 mm) connected to 10-mm thick gusset using 16-mm bolts, as shown in Fig. 7.30, if the yield and the ultimate stress of the steel used are 250 MPa and 410 MPa, respectively.

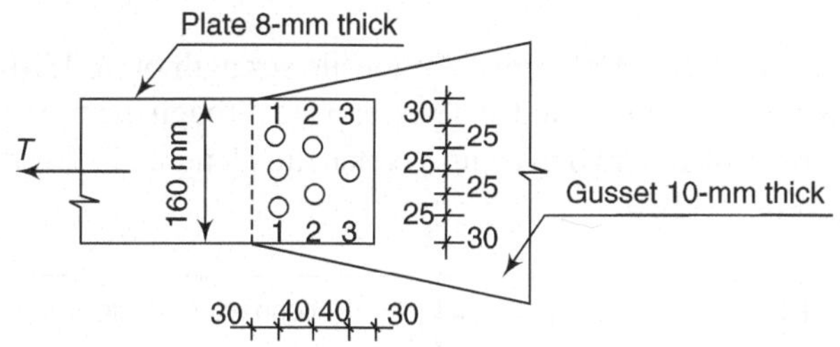


FIG. 7.30

**Solution:**

$$f_y = 250\,\text{MPa}$$

$$f_u = 410\,\text{MPa}$$

**Step 1:** *Calculation of net area*

$A_n$ (Path 11) $= (160 - 3 \times 18) \times 8 = 848\,\text{mm}^2$

$A_n$ (Path 1221) $= [(160 - 4 \times 18) + (2 \times 40^2)/(4 \times 25)] \times 8 = 960\,\text{mm}^2$

$A_n$ (Path 12321) $= [(160 - 5 \times 18) + (4 \times 40^2)/(4 \times 25)] \times 8 = 1072\,\text{mm}^2$

Factored design tension in member by

(a) Yielding of gross section,

$$T_{dg} = [f_y \times A_g / \gamma_{m0}]$$
$$= [250 \times (160 \times 8)/1.10] \times 10^{-3} = 290.9\,\text{kN}$$

(b) Rupture of net section

$$T_{dn} = (0.9 \times f_u \times A_n / \gamma_{ml})$$
$$= (0.9 \times 410 \times 848/1.25) \times 10^{-3} = 250.33\,\text{kN}$$

Therefore, the design tensile strength of the plate = 250.33 kN

**Step 2:** *Check for minimum edge distance*
Provided edge and end distance = 30 mm > 1.5 × 18 = 27 mm.
Hence, the edge distance is as required.

> **Example 7.5:** A single unequal angle 100 × 75 × 6 is connected to a 10-mm thick gusset plate at the ends with six 16-mm-diameter bolts to transfer tension as shown in Fig. 7.31. Determine the design tensile strength of the angle assuming that the yield and the ultimate stress of steel used are 250 MPa and 410 MPa:
> (i) if the gusset is connected to the 100-mm leg
> (ii) if the gusset is connected to the 75-mm leg

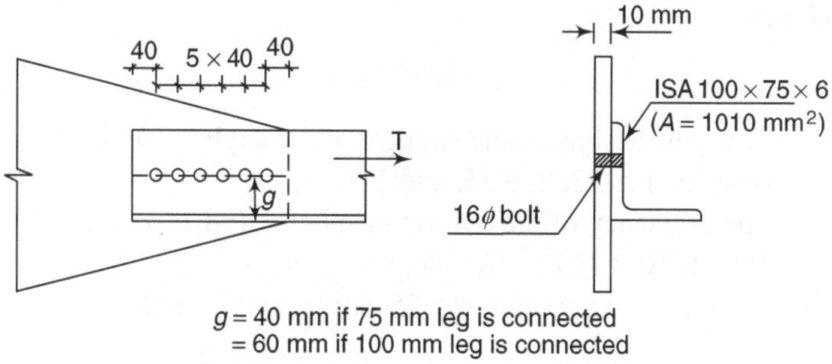


FIG. 7.31

**Solution:**

(i) *Gusset is connected to the 100-mm leg of the angle*

$$A_{nc} = (100 - 6/2 - 18) \times 6 = 474\,\text{mm}^2$$
$$A_{go} = (75 - 6/2) \times 6 = 432\,\text{mm}^2$$
$$A_g = 1010\,\text{mm}^2$$

(a) ***Strength governed by yielding of gross section***

$$T_{dg} = A_g f_y / \gamma_{mo} = (1010 \times 250/1.10) \times 10^{-3} = 229.55\,\text{kN}$$

(b) ***Strength governed by rupture of critical section***

$$T_{dn} = 0.9 f_u A_{nc} / \gamma_{m1} + \beta A_{go} f_y / \gamma_{m0}$$
$$\beta = 1.4 - 0.076(w/t)(f_y/f_u)(b_s/L_c)$$
$$= 1.4 - 0.076[(75-3)/6](250/410)[(72+60)/(5\times 40)]$$
$$= 1.4 - 0.367 = 1.033 > 0.7$$
$$\text{and} < 1.30[(0.9 \times 410/250)(1.1/1.25)]$$
$$T_{dn} = [0.9 \times 410 \times 474/1.25 + 1.033 \times 432 \times 250/1.10] \times 10^{-3}$$
$$= 139.92 + 101.42 = 241.34\,\text{kN}$$

Alternatively,

$$T_{dn} = \alpha A_n f_u / \gamma_{m1} = 0.8 \times [(474 + 432) \times 410/1.25] \times 10^{-3} = 237.73\,\text{kN}$$

Hence, take

$$T_{dn} = 241.34\,\text{kN}$$

(c) ***Strength governed by block shear***

$$A_{vg} = 6 \times (5 \times 40 + 40) = 1440\,\text{mm}^2$$
$$A_{vn} = 6 \times [(5 \times 40 + 40) - 5.5 \times 18] = 846\,\text{mm}^2$$
$$A_{tg} = 6 \times 40 = 240\,\text{mm}^2$$
$$A_{tn} = 6 \times (40 - 0.5 \times 18) = 186\,\text{mm}^2$$
$$T_{db1} = A_{vg} f_y / (\sqrt{3}\,\gamma_{m0}) + 0.9 f_u A_{tn} / \gamma_{m1}$$
$$= [1440 \times 250/(\sqrt{3} \times 1.1) + 0.9 \times 410 \times 186/1.25] \times 10^{-3} = 243.85\,\text{kN}$$
$$T_{db2} = 0.9 f_u A_{vn} / (\sqrt{3}\,\gamma_{m1}) + f_y A_{tg} / \gamma_{m0}$$
$$= [0.9 \times 410 \times 846/(\sqrt{3} \times 1.25) + 250 \times 240/1.10] \times 10^{-3} = 198.73\,\text{kN}$$

Hence,

$$T_{db} = 198.73\,\text{kN}$$

Thus, the design tensile strength of the angle = 198.73 kN (least of 198.73, 229.55, and 241.34).
The efficiency of the tension member = 198.73 × 1000 × 100/(1010 × 250/1.10) = 86.57%

(ii) *Gusset is connected to the 75-mm leg of the angle*

$$A_{nc} = (75 - 6/2 - 18) \times 6 = 324\,\text{mm}^2$$

$$A_{go} = (100 - 6/2) \times 6 = 582\,\text{mm}^2$$

$$A_g = 1010\,\text{mm}^2$$

(a) ***Strength as governed by yielding of gross-section***

$$T_{dg} = A_g f_y/\gamma_{m0} = 229.55\,\text{kN}$$

(b) ***Strength governed by tearing of net section***

$$T_{dn} = 0.9 f_u A_{nc}/\gamma_{m1} + \beta A_{go} f_y/\gamma_{m0}$$

$$\beta = 1.4 - 0.076(w/t)(f_y/f_u)(b_s/L_c)$$

$$= 1.4 - 0.076[(100 - 3)/6](250/410)[(97 + 40)/(5 \times 40)]$$

$$= 0.8868 > 0.7$$

$$T_{dn} = [0.9 \times 410 \times 324/1.25 + 0.8868 \times 582 \times 250/1.1] \times 10^{-3}$$

$$= 212.94\,\text{kN}$$

Alternatively,

$$T_{dn} = \alpha A_n f_u/\gamma_{m1}$$

$$= [0.8 \times (324 + 582) \times 410/1.25] \times 10^{-3}$$

$$= 237.73\,\text{kN}$$

Hence, take $T_{dn} = 212.94$ kN

(c) ***Strength governed by block shear***

$$A_{vg} = 6 \times (5 \times 40 + 40) = 1440\,\text{mm}^2$$

$$A_{vn} = 6 \times (5 \times 40 + 40 - 5.5 \times 18) = 846\,\text{mm}^2$$

$$A_{tg} = 6 \times 35 = 210\,\text{mm}^2$$

$$A_{tn} = 6 \times (35 - 0.5 \times 18) = 156\,\text{mm}^2$$

$$T_{db1} = A_{vg} f_y/(\sqrt{3}\,\gamma_{m0}) + 0.9 f_u A_{tn}/\gamma_{m1}$$

$$= [1440 \times 250/(\sqrt{3} \times 1.1) + 0.9 \times 410 \times 156/1.25] \times 10^{-3}$$

$$= 235\,\text{kN}$$

$$T_{db2} = 0.9 f_u A_{vn}/(\sqrt{3}\,\gamma_{m1}) + f_y A_{tg}/\gamma_{m0}$$

$$= [0.9 \times 410 \times 846/(\sqrt{3} \times 1.25) + 250 \times 210/1.10] \times 10^{-3}$$

$$= 191.91\,\text{kN}$$

Hence,
$T_{db} = 191.91$ kN

Thus, the design tensile strength of the angle = 191.91 kN (least of 229.55, 212.94 and, 191.91)
The efficiency of the tension member = 191.91 × 1000 × 100/(1010 × 250/1.10) = 83.6%
Hence, in this case, by connecting the short leg, the efficiency is reduced by about 3%.

**EXAMPLE 7.6:** Determine the tensile strength of an ISMC 175 when it is connected to a gusset plate through the web by two rows of 16 mm bolts with a connection length of 100 mm (see Fig. 7.32).

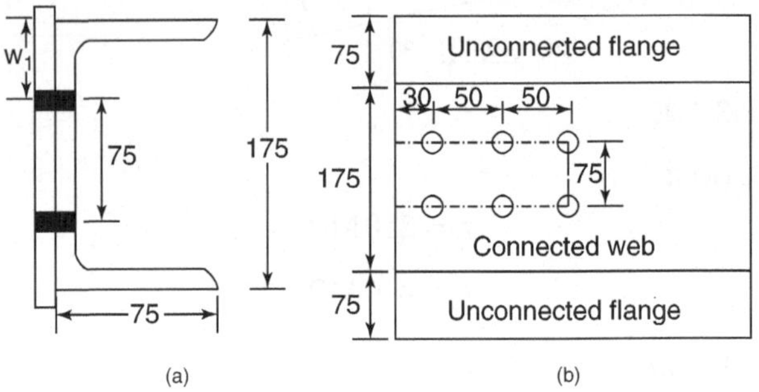


**FIG. 7.32** (a) Channel connected through web, (b) Equivalent flat plate

**SOLUTION:**
Properties of ISMC 175: Area, $A$ = 2490 mm², $h$ = 175 mm, $b_f$ = 75 mm, $t_f$ = 10.2 mm, and $t_w$ = 6 mm.
Minimum edge distance (Table 5.6) = 1.5 ×18 = 27 mm
Adopt an edge distance of 30 mm
Adopt a pitch of 50 mm > 2.5 ×16 = 40 mm
Area of unconnected portion, $A_{go}$ = 2490 − { [175 − 2(10.2/2)] − 2 × 18} 6 = 1717.2 mm²
Net area of connected portion, $A_{nc}$ = (175 − 10.2 − 2 × 18) × 6 = 772.8 mm²

(a) ***Strength governed by yielding of cross section***

$$T_{dg} = \frac{A_g f_y}{\gamma_{m0}} = \frac{2490 \times 250}{1.1 \times 1000} = 565.9\text{ kN}$$

(b) ***Strength governed by rupture of the critical section***

$w$ = 75 mm, $w_1$ = 50 mm,
$b_s = w + w_1 - t = 75 + 50 - (10.2 + 6)/2 = 116.9$ mm

$$\beta = 1.40 - 0.076\left(\frac{w}{t}\right)\left(\frac{f_y}{f_u}\right)\left(\frac{b_s}{L_c}\right)$$

$$= 1.40 - 0.076\left(\frac{75}{10.2}\right)\left(\frac{250}{410}\right)\left(\frac{116.9}{100}\right)$$

$$= 1.002 \le \frac{0.9 f_u \gamma_{m0}}{f_y \gamma_{m1}} = \frac{0.9 \times 410 \times 1.1}{250 \times 1.25} = 1.3$$

$$T_{dn} = \frac{0.9 f_u A_{nc}}{\gamma_{m1}} + \frac{\beta f_y A_{go}}{\gamma_{m0}}$$

$$= \left[\frac{0.9 \times 410 \times 772.8}{1.25} + \frac{1.002 \times 250 \times 1717.2}{1.1}\right] \times 10^{-3}$$

$$= 228.13 + 391.05 = 619.18\,\text{kN}$$

(c) ***Strength governed by block shear***

Assuming the connection as shown in Fig. 7.32

$$A_{vg} = 6 \times [(100 + 30) \times 2] = 1560 \text{ mm}^2$$

$$A_{vn} = 6 \times (100 + 30 - 2.5 \times 18) = 1020 \text{ mm}^2$$

$$A_{tg} = 6 \times 75 = 450 \text{ mm}^2$$

$$A_{tn} = 6 \times (75 - 18) = 342 \text{ mm}^2$$

$$T_{db1} = \frac{A_{vg} f_y}{\sqrt{3}\gamma_{m0}} + \frac{0.9 f_u A_{tn}}{\gamma_{m1}}$$

$$= \left[\frac{1560 \times 250}{\sqrt{3} \times 1.1} + \frac{0.9 \times 410 \times 342}{1.25}\right] \times 10^{-3}$$

$$= 305.65\,\text{kN}$$

$$T_{db2} = \frac{0.9 f_u A_{vn}}{\sqrt{3}\gamma_{m1}} + \frac{f_y A_{tg}}{\gamma_{m0}}$$

$$= \left[\frac{0.9 \times 410 \times 1020}{\sqrt{3} \times 1.25} + \frac{250 \times 450}{1.1}\right] \times 10^{-3}$$

$$= 276.12\,\text{kN}$$

Hence the tensile strength of the section is 276.12 kN (the least of 565.9 kN, 619.18 kN, 305.65 kN, 276.12kN).

**EXAMPLE 7.7:** Determine the tensile strength of an ISMC 175 when it is connected to gusset plates through the two flanges by two rows of 16 mm bolts with a connection length of 200 mm (see Fig. 7.33).

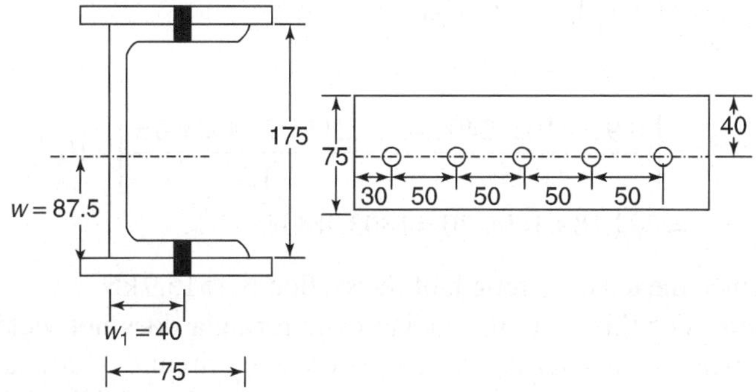


**FIG. 7.33**

**SOLUTION:**

Properties of ISMC 175: Area, $A$ = 2490 mm$^2$, $h$ = 175 mm, $b_f$ = 75 mm, $t_f$ = 10.2 mm, and $t_w$ = 6 mm.

Area of unconnected portion, $A_{go}$ = (175 – 10.2) 6 = 988.8 mm$^2$

Net area of connected portion,

$$A_{nc} = 2\left(75 - \frac{6}{2} - 18\right) \times 10.2 = 1101.6\,\text{mm}^2$$

(a) ***Strength governed by yielding of cross section***

$$T_{dg} = \frac{A_g f_y}{\gamma_{m0}} = \frac{2490 \times 250}{1.1 \times 1000} = 565.9 \text{ kN}$$

(b) ***Strength governed by rupture of the critical section***

$w$ = 175/2 = 87.5 mm, $w_1$ = 40 mm, $b_s = w + w_1 - t$ = 87.5 + 40 – (6 + 10.2)/2 = 119.4 mm

$$\beta = 1.40 - 0.076\left(\frac{w}{t}\right)\left(\frac{f_y}{f_u}\right)\left(\frac{b_s}{L_c}\right)$$

$$= 1.40 - 0.076\left(\frac{87.5}{6}\right)\left(\frac{250}{410}\right)\left(\frac{119.4}{200}\right) = 0.9965$$

$$T_{dn} = \frac{0.9 f_u A_{nc}}{\gamma_{m1}} + \frac{\beta f_y A_{go}}{\gamma_{m0}}$$

$$= \left[\frac{0.9 \times 410 \times 1101.6}{1.25} + \frac{0.9965 \times 250 \times 988.8}{1.1}\right] \times 10^{-3}$$

$$= 325.2 + 223.9 = 549.1 \text{ kN}$$

(c) ***Strength governed by block shear***

Assuming the connection as shown in Fig. 7.33

$$A_{vg} = 2 \times 10.2 \times (200 + 30) = 4692 \text{ mm}^2$$

$$A_{vn} = 2 \times 10.2 \times (200 + 30 - 4.5 \times 18) = 3039.6 \text{ mm}^2$$

$$A_{tg} = 2 \times 10.2 \times 35 = 714 \text{ mm}^2$$

$$A_{tn} = 2 \times 10.2 \times (35 - 0.5 \times 18) = 530.4 \text{ mm}^2$$

$$T_{db1} = \frac{A_{vg} f_y}{\sqrt{3}\gamma_{m0}} + \frac{0.9 f_u A_{tn}}{\gamma_{m1}}$$

$$= \left[\frac{4692 \times 250}{\sqrt{3} \times 1.1} + \frac{0.9 \times 410 \times 530.4}{1.25}\right] \times 10^{-3}$$

$$= 772.24\,\text{kN}$$

$$T_{db2} = \frac{0.9 f_u A_{vn}}{\sqrt{3}\gamma_{m1}} + \frac{\beta f_y A_{tg}}{\gamma_{m0}}$$

$$= \left[\frac{0.9 \times 410 \times 3039.6}{\sqrt{3} \times 1.25} + \frac{250 \times 714}{1.1}\right] \times 10^{-3}$$

$$= 680.32\,\text{kN}$$

Hence the tensile strength of the section is 549.1 kN (the least of 565.9 kN, 549.1 kN, 680.32 kN, 772.24kN).

***Note:*** When connected to the web the strength as in Example 7.6, the tensile strength was only 50 per cent, i.e., 276.12 kN.

**EXAMPLE 7.8:** Determine the tensile strength of an ISMB 350 section when it is connected to gusset plates through the two flanges as shown in Fig. 7.34, using four rows of 16 mm bolts with a connection length of 150 mm.

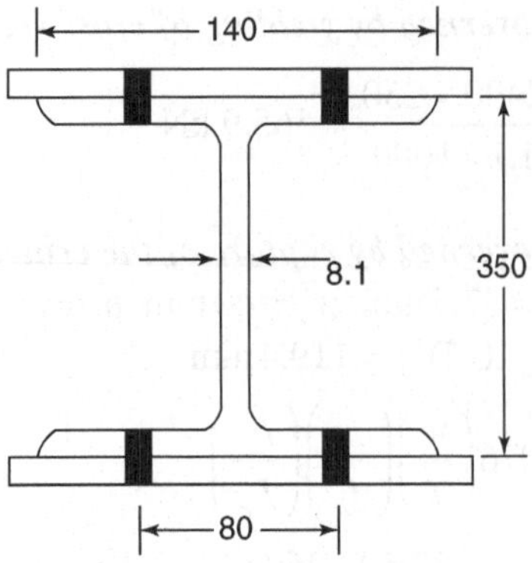


**FIG. 7.34**

**SOLUTION:**

Properties of ISMB 350: Area, $A = 6670\,\text{mm}^2$, $h = 350\,\text{mm}$, $b_f = 140\,\text{mm}$, $t_f = 14.2\,\text{mm}$, and $t_w = 8.1\,\text{mm}$.

Area of unconnected portion, $A_{go} = (350 - 14.2/2 - 14.2/2) \times 8.1 = 2719.98\,\text{mm}^2$

Net area of connected portion, $A_{nc} = 6670 - (350 - 14.2) \times 8.1 - 4 \times 18 \times 14.2 = 2927.62\,\text{mm}^2$

(a) ***Strength governed by yielding of cross section***

$$T_{dg} = \frac{A_g f_y}{\gamma_{m0}} = \frac{6670 \times 250}{1.1 \times 1000} = 1515.9 \text{ kN}$$

(b) ***Strength governed by rupture of the critical section***

$w = 350/2 = 175\,\text{mm}$, $w_1 = 80/2 = 40\,\text{mm}$, $b_s = w + w_1 - t$
$= 175 + 40 - (14.2 + 8.1)/2 = 203.85\,\text{mm}$

$$\beta = 1.40 - 0.076\left(\frac{w}{t}\right)\left(\frac{f_y}{f_u}\right)\left(\frac{b_s}{L_c}\right)$$

$$= 1.40 - 0.076\left(\frac{175}{8.1}\right)\left(\frac{250}{410}\right)\left(\frac{203.85}{150}\right) = +0.04 < 0.7$$

Hence take $\beta = 0.7$

$$T_{dn} = \frac{0.9 f_u A_{nc}}{\gamma_{m1}} + \frac{\beta f_y A_{go}}{\gamma_{m0}}$$

$$= \left[\frac{0.9 \times 410 \times 2927.62}{1.25} + \frac{0.7 \times 250 \times 2719.98}{1.1}\right] \times 10^{-3}$$

$$= 864.23 + 432.72 = 1296.95 \text{ kN}$$

Hence the tensile strength of the section is 1296.95 kN

***Note:*** Strength governed by block shear failure should also be checked as shown in the previous examples, which is left as an exercise for the reader.

**EXAMPLE 7.9:** Determine the tensile strength of an ISMB 350 section when it is connected through its web using two gusset plates as shown in Fig. 7.35, using two rows of 16 mm bolts with a connection length of 150 mm.

**SOLUTION:**

Properties of ISMB 350: Area, $A = 6670\,\text{mm}^2$, $h = 350\,\text{mm}$, $b_f = 140\,\text{mm}$, $t_f = 14.2\,\text{mm}$, and $t_w = 8.1\,\text{mm}$.

Area of unconnected portion, $A_{go} = 2 \times 140 \times 14.2 = 3976\,\text{mm}^2$

Net area of connected portion,

$A_{nc} = 6670 - 2 \times 140 \times 14.2 - 2 \times 18 \times 8.1 = 2402.4\ \text{mm}^2$

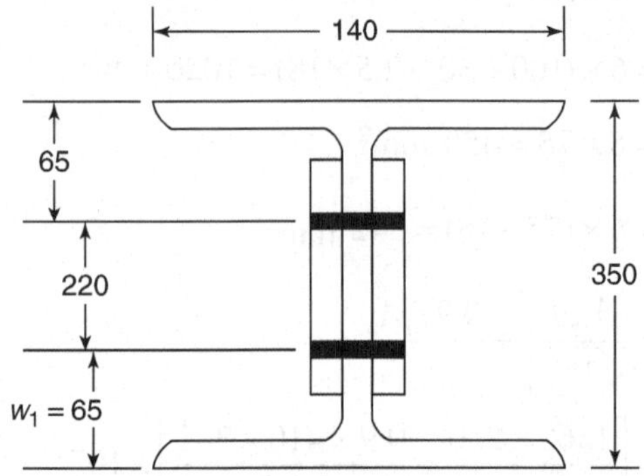


**FIG. 7.35**

(a) ***Strength governed by yielding of cross section***

$$T_{dg} = \frac{A_g f_y}{\gamma_{m0}} = \frac{6670 \times 250}{1.1 \times 1000} = 1515.9 \text{ kN}$$

(b) ***Strength governed by rupture of the critical section***

$w = 140/2 = 70\,\text{mm}$, $w_1 = 65\,\text{mm}$, $b_s = w + w_1 - t$
$= 70 + 65 - (14.2 + 8.1)/2 = 123.85\,\text{mm}$

$$\beta = 1.40 - 0.076\left(\frac{w}{t}\right)\left(\frac{f_y}{f_u}\right)\left(\frac{b_s}{L_c}\right)$$

$$= 1.40 - 0.076\left(\frac{70}{14.2}\right)\left(\frac{250}{410}\right)\left(\frac{123.85}{150}\right) = 1.211$$

$$T_{dn} = \frac{0.9 f_u A_{nc}}{\gamma_{m1}} + \frac{\beta f_y A_{go}}{\gamma_{m0}}$$

$$= \left[\frac{0.9 \times 410 \times 2402.4}{1.25} + \frac{1.211 \times 250 \times 3976}{1.1}\right] \times 10^{-3}$$

$$= 709.18 + 1094.30 = 1803.48\,\text{kN}.$$

Hence the tensile strength of the section is 1515.9 kN

***Note:*** For this case, the Indian code formula does not yield satisfactory results as strength governed by rupture of critical section is greater than strength governed by yielding of cross section! It has to be noted that the strength governed by block shear failure should also be checked as shown in the previous examples, which is left as an exercise for the reader.

**EXAMPLE 7.10:** Determine the tensile strength of a roof truss diagonal 100 × 75 × 6 mm ($f_y = 250\,\text{MPa}$) connected to the gusset plate by 4-mm welds as shown in Fig. 7.36.

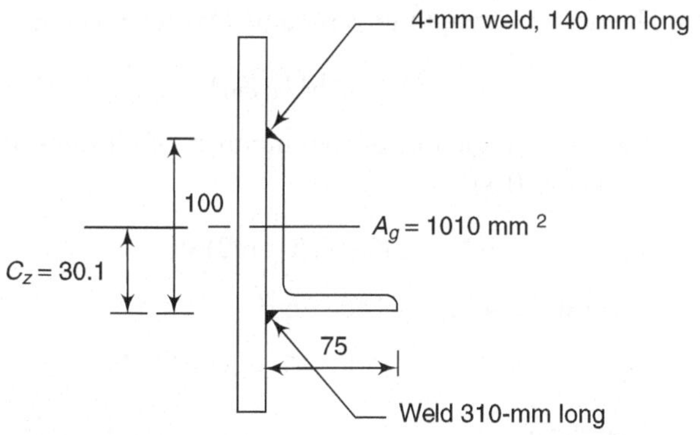


**FIG. 7.36**

**SOLUTION:**

Area of the connected leg = (100 – 6/2) × 6 = 582 $mm^2$
Area of the outstanding leg = (75 – 6/2) × 6 = 432 $mm^2$

$$A_g = 1010\,mm^2$$

(a) ***Strength governed by yielding of cross section***

$$T_{dg} = A_g f_y/\gamma_{mo} = (1010 \times 250/1.10) \times 10^{-3} = 229.55\,kN$$

(b) ***Strength governed by rupture of critical section***

$$T_{dn} = 0.9 f_u A_{nc}/\gamma_{m1} + \beta A_{go} f_y/\gamma_{m0}$$

Assuming average length of weld $L_w$ = 225 mm

$$\beta = 1.4 - 0.076(w/t)(f_y/f_u)(b_s/L_w)$$
$$= 1.4 - 0.076[(75 - 3)/6](250/410)(75/225)$$
$$= 1.215$$

Hence,

$$T_{dn} = [0.9 \times 410 \times 582/1.25 + 1.215 \times 432 \times 250/1.10] \times 10^{-3}$$
$$= 291.1\,kN$$

Alternatively,

$$T_{dn} = \alpha A_n f_u/\gamma_{m1} = [0.8(1010) \times 410/1.25] \times 10^{-3}$$
$$= 265.02\,kN$$

Hence, $T_{dn}$ = 291.1 kN

(c) ***Strength governed by block shear***
Since the member is welded to the gusset plate, no net areas are involved and hence $A_{vn}$ and $A_{tn}$ in the equation for $T_{db}$ (Section 6.3.1 of the code) should be taken as the corresponding gross areas (Gaylord et al. 1992). Assuming average length of the weld on each side as 225 mm and the gusset plate thickness as 8 mm,

$$T_{db1} = [8 \times (225 \times 2) \times 250/(\sqrt{3} \times 1.1) + 0.9 \times 410 \times 8 \times 100/1.25] \times 10^{-3}$$
$$= 708.53\,kN$$

$$T_{db2} = [0.9 \times 410 \times 8 \times (225 \times 2)/(\sqrt{3} \times 1.25) + 250 \times 8 \times 100/1.1] \times 10^{-3}$$
$$= 798.38\,kN$$

Hence, $T_{db}$ = 708.53 kN

Thus, tensile strength = 229.55 kN (least of 229.55, 291.1, and 708.53)
The efficiency of the tension member = 229.55 × 1000 × 100/(1010 × 250/1.10) = 100%
It is clear that since there is no reduction in the area in the welded connection, the efficiency of the tension member is not reduced.

Note that in the calculation, we have assumed the average length of weld as 225 mm on each side. However, the welding should be proportioned based on the position of the neutral axis.

Thus, for the tensile capacity = 229.55 kN, with capacity of 4-mm weld = 0.530 kN/mm

Length of the weld at the upper side of the angle

= (229.55 × 30.1/100)/0.530 = 130 mm, say 140 mm

Length of the weld at the bottom side of the angle

= [229.55 × (100 – 30.1)/100]/0.530 = 302 mm, say 310 mm

**EXAMPLE 7.11:** Select a suitable angle section to carry a factored tensile force of 170 kN assuming a single row of M20 bolts and assuming design strength as $f_y$ = 250 N/$mm^2$.

**SOLUTION:**

Approximate required area = 1.1 × 170 × $10^3$/250 = 748 $mm^2$
Choose 65 × 65 × 8 angle with $A$ = 976 $mm^2$
Strength governed by yielding = [976 × 250/1.1] × $10^{-3}$ = 221.82 kN

$$A_{nc} = \text{area of connected leg} = (65 - 4 - 22) \times 8 = 312\,mm^2$$
$$A_{go} = (65 - 4) \times 8 = 488\,mm^2$$

Required number of M20 bolts (Table 5.9) = 170/45.3 = 3.75
Provide four bolts at a pitch of 60 mm.

**Step 1:** *Strength governed by rupture of critical section*

$$T_{dn} = 0.9 f_u A_{nc}/\gamma_{m1} + \beta A_{go} f_y/\gamma_{m0}$$
$$\beta = 1.4 - 0.076 \times (65/8)(250/410)(61 + 35)/(3 \times 60)$$
$$= 1.199$$
$$T_{dn} = [0.9 \times 410 \times 312/1.25 + 1.199 \times 488 \times 250/1.10] \times 10^{-3}$$
$$= 225.08\,kN$$

Alternatively,

$$T_{dn} = \alpha A_n f_u/\gamma_{m1} = [0.8 \times (312 + 488) \times 410/1.25] \times 10^{-3}$$
$$= 209.92\,kN$$

**Step 2:** *Strength governed by block shear*
Assuming an edge distance of 40 mm,

$$A_{vg} = 8 \times (3 \times 60 + 40) = 1760\,mm^2$$
$$A_{vn} = 8 \times (3 \times 60 + 40 - 3.5 \times 22) = 1144\,mm^2$$

$A_{tg} = 8 \times 30 = 240\,mm^2$

$A_{tn} = 8 \times (30 - 0.5 \times 22) = 152\,mm^2$

$T_{db1} = [1760 \times 250/(\sqrt{3} \times 1.1) + 0.9 \times 410 \times 152/1.25] \times 10^{-3}$
$= 275.81\,kN$

$T_{db2} = [0.9 \times 410 \times 1144/(\sqrt{3} \times 1.25) + 250 \times 240/1.10] \times 10^{-3} = 249.52\,kN$

Tension capacity of the angle = 209.92 kN > 170 kN

Hence the angle is safe.

**EXAMPLE 7.12:** A tie member in a bracing system consists of two angles 75 × 75 × 6 bolted to a 10-mm gusset, one on each side using a single row of bolts [See Fig. 7.37(a)] and tack bolted. Determine the tensile capacity of the member and the number of bolts required to develop full capacity of the member. What will be the capacity if the angles are connected on the same side of the gusset plate and tack bolted [Fig. 7.37(b)]? What is the effect on tensile strength if the members are not tack bolted?

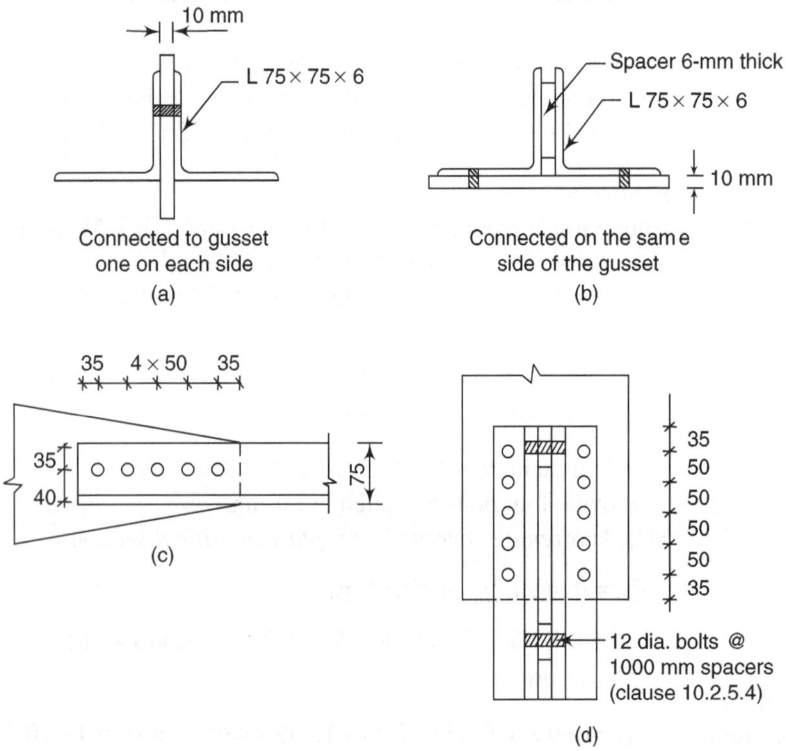


FIG. 7.37

**SOLUTION:**

(a) ***Two angles connected to opposite side of the gusset as in Fig. 7.37(a)***

(i) Design strength due to yielding of gross section $T_{dg} = f_y(A_g/\gamma_{m0})$

$A_g = 866\,mm^2$ (for a single angle)

$T_{dg} = 250 \times 2 \times (866/1.10) \times 10^{-3}$

$T_{dg} = 393.64\,kN$

(ii) The design strength governed by tearing at net section

$$T_{dn} = \alpha A_n(f_u/\gamma_{m1})$$

Assume a single line of four numbers of 20-mm-diameter bolts ($\alpha = 0.8$)

$A_n = [(75 - 6/2 - 22)6 + (75 - 6/2)6]2$

$A_n = (300 + 432)2 = 1464\,mm^2$

$T_{dn} = (0.8 \times 1464 \times 410/1.25) = 384.15\,kN$

Therefore,

Tensile capacity = 384.15 kN

**Step 1:** *Design of bolts*

Choose edge distance = 35 mm

Capacity of bolt in double shear (Table 5.9)

$= 2 \times 45.3 = 90.6\,kN$

Bearing capacity of the bolt does not govern as per Table 5.9.

Hence,

Strength of a single bolt = 90.6 kN

Provide five bolts. Then,

Total strength of the bolts = 5 × 90.6 = 453 kN > 384.15 kN

Hence the connection is safe.

Minimum spacing = $2.5t = 2.5 \times 20 = 50\,mm$

Hence, provide a spacing of 50 mm.

The arrangements of bolts are shown in Fig. 7.37(c).

**Step 2:** *Check for block shear strength (clause 6.4)*

Block shear strength $T_{db}$ of connection will be taken as

$$T_{db1} = [A_{vg}f_y/\sqrt{3}\,\gamma_{m0}) + (0.9A_{tn}f_u/\gamma_{m1})]$$

or

$$T_{db2} = 0.9f_uA_{vn}/\sqrt{3}\,\gamma_{m1} + (f_yA_{tg}/\gamma_{m0})$$

whichever is smaller.

$A_{vg} = (4 \times 50 + 35)6 = 1410\,mm^2$

$A_{vn} = (4 \times 50 + 35 - 4.5 \times 22)6 = 816\,mm^2$

$A_{tn} = (35.0 - 22/2)6 = 144\,mm^2$

$A_{tg} = (35 \times 6) = 210\,mm^2$

$T_{db1} = \{[(1410 \times 250)/(\sqrt{3} \times 1.10)] + [(0.9 \times 144 \times 410)/1.25]\} \times 10^{-3}$

$= 227.5\,kN$

$T_{db2} = \{[(0.9 \times 410 \times 816)/(\sqrt{3} \times 1.25)] + [(250 \times 210)/1.10]\} \times 10^{-3}$

$= 186.8\,kN$

For double angle,
block shear strength = 2 × 186.8 = 373.6 kN
Therefore,
Tensile capacity = 373.6 kN (least of 393.64 kN, 384.15 kN, and 373.6 kN)

(b) ***Two angles connected to the same side of the gusset plate [Fig. 7.37(b)]***

(i) Design strength due to yielding of the gross section = 393.64 kN

(ii) Design strength governed by tearing at the net section = 384.15 kN

Assuming ten bolts of 20 mm diameter, five bolts in each connected leg
Capacity of an M20 bolt in single shear = 45.3 kN
Total strength of bolts = 10 × 45.3 = 453 kN > 393.64 kN
Hence the connection is safe.

The arrangement of bolts is shown in Fig. 7.37(d). Since it is similar to the arrangement in Fig. 7.37(c), the block shear strength will be the same, i.e., 373.6 kN.

Hence, the tensile capacity = 373.6 kN

The tensile capacity of both the arrangements (angles connected on the same side and connected to the opposite side of gusset) are same, as per the code though the load application is eccentric in this case. Moreover, the number of bolts are ten whereas in case (a) we used only five bolts since the bolts were in double shear.

(c) ***If the angles are not tack bolted, they behave as single angles connected to gusset plate.***

In this case also the tensile capacity will be the same and we have to use ten M20 bolts. This fact is confirmed by the test and FEM results of Usha (2003), stating that 'The net section strength of double angles on opposite sides of the gusset and tack connected adequately over the length is nearly the same as that of two single angles acting individually. Current design provisions indicating greater efficiency of such double angles are not supported by the test and FEM results.'

**EXAMPLE 7.13:** It was shown in Example 7.12 that the capacity of 2 angles 75 × 75 × 6 mm is 373.6 kN. Design a single angle to carry 350 kN. Assume that the length of the member is 3 m.

**SOLUTION:**

The weight of 2 L 75 × 75 × 6 mm is 2 × 6.8 = 13.6 kg/m. Let us choose an unequal angle of size 150 × 75 × 8 mm with a weight of 13.7 kg/m and area = 1750 mm², $r_{vv}$ = 16.2 mm.

**Step 1:** *Design strength due to yielding of cross section*

$$T_{dg} = f_y A_g / \gamma_{m0} = 250 \times 1750/1.10 = 397.7\,\text{kN} > 350\,\text{kN}$$

**Step 2:** *Design strength governed by tearing of net section*

$$T_{dn} = \alpha A_n f_u / \gamma_{m1}$$

Assuming nine M20 bolts, with strength = 9 × 45.3 = 407.7 kN

$$A_n = (150 - 4 - 2 \times 22)8 + (75 - 4)8 = 1384\,\text{mm}^2$$

$$T_{dn} = (0.8 \times 1384 \times 410/1.25) \times 10^{-3} = 363.1\,\text{kN} > 350\,\text{kN}$$

**Step 3:** *Assuming a staggered bolting and block shear failure as shown in Fig. 7.38*

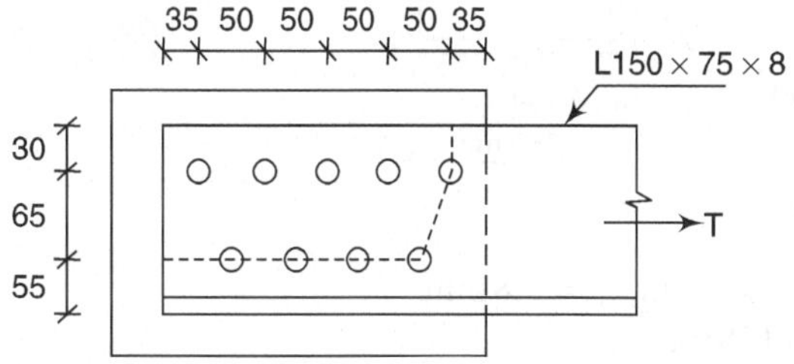


**FIG. 7.38**

$$A_{vg} = (3 \times 50 + 25 + 35) \times 8 = 1680\,\text{mm}^2$$

$$A_{vn} = (3 \times 50 + 25 + 35 - 3.5 \times 22) \times 8 = 1064\,\text{mm}^2$$

$$A_{tn} = (65 + 30 + 25^2/(4 \times 65) - 1.5 \times 22)8 = 515\,\text{mm}^2$$

$$A_{tg} = [65 + 30 + 25^2/(4 \times 65)] \times 8 = 779\,\text{mm}^2$$

Block shear capacity

$$T_{db1} = A_{vg} f_y / (\sqrt{3}\,\gamma_{m0}) + 0.9 A_{tn} f_u / \gamma_{m1}$$
$$= [1680 \times 250/(\sqrt{3} \times 1.1) + 0.9 \times 515 \times 410/1.25] \times 10^{-3}$$
$$= 220.44 + 152.03 = 372.47\,\text{kN}$$

$$T_{db2} = 0.9 A_{vn} f_u / (\sqrt{3}\,\gamma_{m1}) + A_{tg} f_y / \gamma_{m0}$$
$$= [0.9 \times 1064 \times 410/(\sqrt{3} \times 1.25) + (779 \times 250/1.1)] \times 10^{-3}$$
$$= 358.38\,\text{kN}$$

Hence,
$T_{db}$ = 358.38 kN > 350 kN
Check for stiffness (Table 3 of code)
$L/r$ = 3000/16.2 = 185 < 250
Hence the section is safe.

It is seen from Examples 7.12 and 7.13 that unequal angle section with its long leg connected has a carrying capacity similar to that of two equal angles of the same weight connected on the same side or opposite side of gusset plate. Hence wherever possible unequal angles (with its long leg connected) should be used. But unfortunately unequal angles are not freely available in the market.

**EXAMPLE 7.14:** A tension member in a bridge structure 10-m long is subjected to an axial tensile (factored) load of 1800 kN. Design the section with channels facing each other (see Fig. 7.39). Assume $f_u$ = 410 MPa and $f_y$ = 250 MPa.

**SOLUTION:**

Required area = 1.1 × 1800 × 1000/250 = 7200 mm²
Select two channels ISMC 300 (36.3 kg/m) each with the following properties

$A$ = 4630 mm²; $B$ = 90 mm, $g$ = 50 mm
$I_{zz}$ = 6420 × 10⁴ mm⁴
$I_{yy}$ = 313 × 10⁴ mm⁴

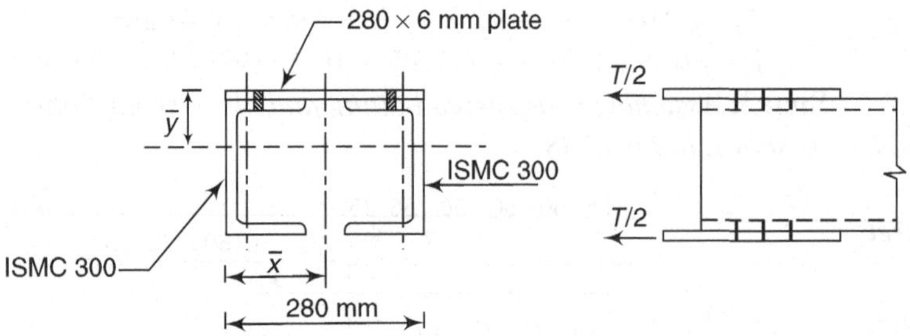


**FIG. 7.39**

$t_f = 13.6$ mm; $t_w = 7.8$ mm
$r_{zz} = 118$ mm
$C_y = 23.5$ mm

Assuming 16-mm-diameter bolts for the connection,

Diameter of hole = 16 + 2 = 18 mm

$$A_g = 2 \times 4630 = 9260\,\text{mm}^2$$

Assuming staggered bolts, deduction of two holes will be made in the calculation of the net area. Net area is calculated as per Section 7.9.

Net area provided = gross area − area of bolt holes − (0.5 × web area)

$= 9260 - 2 \times 18 \times 13.6 - 0.5\,[2 \times (300 - 2 \times 13.6) \times 7.8]$
$= 8770.4 - 2127.84 = 6642.56\,\text{mm}^2 < 7200\,\text{mm}^2$

As the net area provided by the two channels is less than the net area required, we have to provide additional cover plates.

Required net area of cover plates = 7200 − 6642.56 = 557.44 mm²

Let us provide an additional plate of size 280 × 6 mm with area 1680 mm² and arrange two channels as shown in Fig. 7.39.

$$\bar{y} = \frac{2 \times 4630(150+6) + 280 \times 6 \times 3}{2 \times 4630 + 280 \times 6} = 132.5\text{mm}$$

$I_{zz} = 2 \times 6420 \times 10^4 + 280 \times 6^3/12 + 280 \times 6(132.5 - 6/2)^2 + 2 \times 4630 \times (156 - 132.5)^2$
$= 161.7 \times 10^6\,\text{mm}^4$

$A = 2 \times 4630 + 280 \times 6 = 10940\,\text{mm}^2$

$I_{yy} = 2[313 \times 10^4 + 4630(280/2 - 23.5)^2] + 6 \times 280^3/12$
$= 142.91 \times 10^6\,\text{mm}^4$

$$r_{min} = \sqrt{(142.91 \times 10^6/10940)} = 114.3\text{ m}$$

$\lambda = l/r = 10 \times 1000/114.3 = 87.49 < 180$

Hence $\lambda$ is within allowable limit.

Net area $= 2 \times 4630 + 280 \times 6 - 2[18 \times 13.6 + 18 \times 6]$
$- 0.5[2 \times (300 - 2 \times 13.6) \times 7.8]$
$= 8106.56\,\text{mm}^2$

Tensile strength of the member (assuming more than four bolts)

$= (0.8 \times 8106.56 \times 410/1.25) \times 10^{-3} = 2127.2\text{ kN} > 1800\text{ kN}$

Hence, the tensile strength of the member is as required.

The two channels should be tied effectively at regular intervals with tie plates so as to function as a single unit (see clause 7.7.2.3 of the code)

Effective depth of tie plate = distance between centroids of the main components

$= 280 - 2 \times 23.5 = 233$ mm

Assuming 16-mm-diameter bolts and with edge distance = 30 mm

Overall depth = 233 + 2 × 30 = 293 mm ≈ 300 mm

Length of the plate = width of the member = 280 mm

Thickness of tie plate = (1/50)(distance between innermost connecting line of bolts)

$= (1/50)(280 - 2 \times 50) = 3.6$ mm

Provide (300 × 280 × 6)-mm tie plates.

**EXAMPLE 7.15:** Design sag rods for consecutive purlins near the supported end of a roof truss system as shown in Fig. 7.40. The purlins are supported at one-third points by sag rods. Also design the ridge rod between ridge purlins. The data given are c/c Spacing of truss = 6 m, spacing of purlin = 1.4 m, self weight of roofing = 200 N/m², intensity of wind pressure = 1500 N/m², slope of the roof truss = 25°, and no access is provided to the roof.

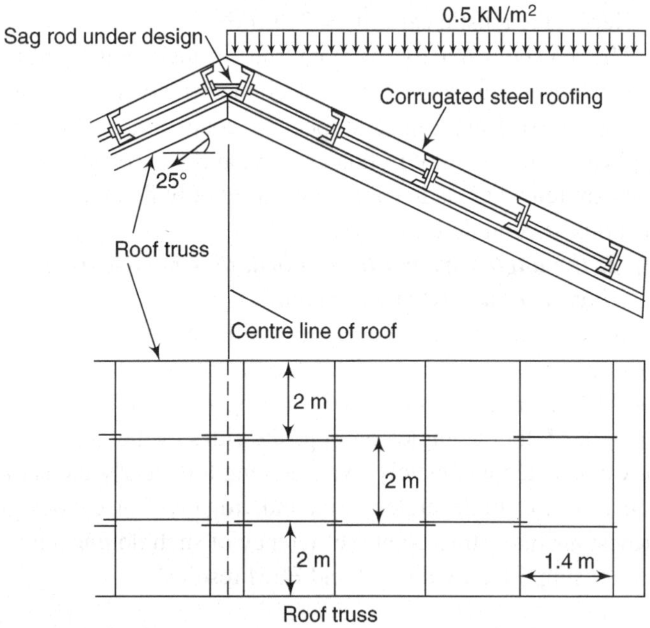


**FIG. 7.40** Roof and sag rods for Example 7.15

**SOLUTION:**

Dead load from roofing = 200 × 1.4 = 280 N/m

Self weight of purlin = 100 N/m (assumed)

Live load = (0.75 − 0.02 × 15) × 1000 = 450 N/m² > 0.40kN/m²

Live load on purlin = 450 × 1.4 = 630 N/m

Total gravity load = 630 + 280 + 100 = 1010 N/m

Wind load = 1500 × 1.4 = 2100 N/m (normal to roof)

Component of gravity load parallel to roof

$= 1010 \times \sin 25° = 426.8$ N/m

As the sag rods are placed at one-third points on the purlin,

Pull on sag rod = 426.8 × 6/3 = 853.6 N

Factored load = 1.5 × 853.6 = 1280.4 N

Required net area $= T_{dn} \times \gamma_{m1}/(0.9 f_u)$

$= 1280.4 \times 1.25/(0.9 \times 410) = 4.34\,\text{mm}^2$

Provide a 16-mm-diameter sag rod with a threaded area of 157 $mm^2$ between purlins. (Note that the provided rod should not be less than 16-mm diameter.)
Tie rod between ridge purlins
Pull in the tie rod = 4 × 853.6 × sec 25° = 3767 N
Factored load = 3767 × 1.5 = 5650 N
Required net area = $T_{dn}\gamma_{m1}/(0.9f_u)$
= 5650 × 1.25/(0.9 × 410) = 19.14 $mm^2$
Hence, provide 16-mm diameter tie rods between the ridge purlins.

**Example 7.16:** A diagonal member of a roof carries an axial tension of 450 kN. Design the section and its connection with a gusset plate and lug angle. Use $f_y$ = 250 MPa and $f_u$ = 410 MPa.

**Solution:**
Factored tensile load = 450 kN
Required net area of section = $T_u\gamma_{m1}/(0.9f_u)$
= 450 × 1000 × 1.25/(0.9 × 410)
= 1524 $mm^2$
Choose ISA 150 × 75 × 10 with A = 2160 $mm^2$, $r_{vv}$ = 16.1 mm
Providing 20-mm-diameter bolts; strength of a bolt in single shear = 45.3 kN
(Strength in bearing will not govern.)
Required number of bolts = 450/45.3 ≈ 10
Using a pitch of 2.5 × 20 = 50 mm and an edge distance of 30 mm
Length of gusset plate = 9 × 50 + 2 × 30 = 510 mm
Area of connected leg $A_{nc}$ = [150 – 22 – (10/2)] × 10 = 1230 $mm^2$
Area of outstanding leg $A_{go}$ = [75 – (10/2)] × 10 = 700 $mm^2$
$A_n$ = 1230 + 700 = 1930 $mm^2$ > 1524 $mm^2$
Tearing strength of the net section
$T_{dn} = \alpha A_n f_u/\gamma_{m1}$ = 0.8 × 1930 × 410/1.25
= 506.4 kN > 450 kN

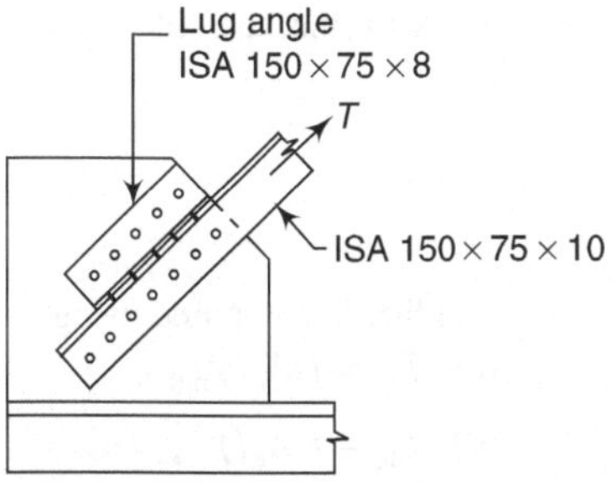

FIG. 7.41

Hence safe.
Without lug angle, the length of the gusset plate is 510 mm. If the bolts are staggered and arranged in two rows, the length of the gusset plate may be reduced. We will now provide a lug angle (see Fig. 7.41).

**Step 1:** *Design of lug angle*
Total factored tensile load = 450 kN
Gross area of the connected leg = [150 – (10/2)] × 10
= 1450 $mm^2$
Gross area of outstanding leg = [75 – (10/2)] × 10 = 700 $mm^2$
In an unequal angle, the load gets distributed in the ratio of the gross area of connected and outstanding legs.
Load shared by outstanding leg of main angle
= 450 × 700/(1450 + 700) = 146.5 kN
Load on lug angle = 1.2 × 146.5 = 175.8 (clause 10.12.2)

Required net area for lug angle = 175.8 × $10^3$ × 1.25/(0.9 × 410)
= 596 $mm^2$
Use ISA 150 × 75 × 8 angle with $A$ = 1750 $mm^2$
Assuming that the section is weakened by one row of 20-mm-diameter bolt
Net area = 1750 – 22 × 8 = 1574 $mm^2$ > 596 $mm^2$ (For simplicity the effect of $\Sigma p^2/4g$ is not considered here but should ideally be considered as explained in previous examples.)
The lug angle is also kept with its 75-mm long leg as outstanding leg
Number of bolts to connect 150-mm leg of lug angle with gusset plate
= 175.8/45.3 ≈ 4
Provide five bolts of 20 mm diameter to connect lug angle with gusset plate.
***Check***
Load on connected leg = 450 × 1450/(1450 + 700) = 303.5 kN
Required number of bolts = 303.5/45.3 ≈ 7
Hence provide seven 20-mm-diameter bolts to connect the diagonal tension member with the gusset.
Required number of bolts to connect outstanding legs of the two angles (clause 10.12.2)
= 1.4 × 146.5/45.3 ≈ 5
Hence, provide five bolts of 20 mm diameter.
Required length of gusset plate = 6 × 50 + 2 × 30 = 360 mm (compared with 510 mm without lug angle).

**Example 7.17:** Four guy cables are attached to a 250-m TV antenna tower at 100 m and 200 m as shown in Fig. 7.42. Each cable is subjected to a force as shown in Fig. 7.42. Design the cable given the following (Bowles 1980)

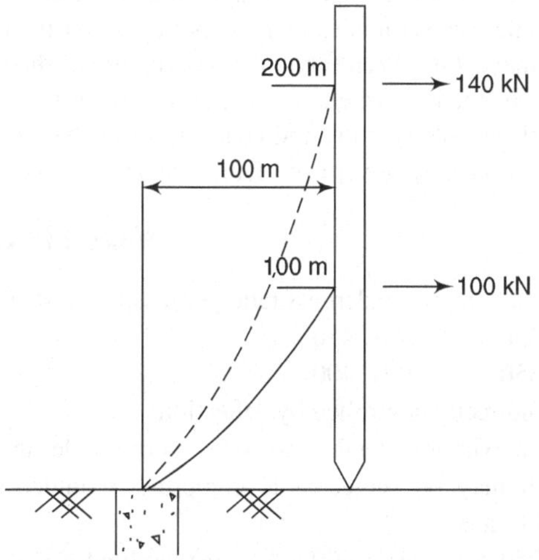

FIG. 7.42

$$H = wL^2/(8h)$$

where $w$ is the self weight of the cable, $H$ is the horizontal reaction, $L$ is the span length, and $h$ is the mid span sag.

$$\text{Cable tension } T = H\left[1+16\left(\frac{h}{L}\right)^2 + \tan^2\theta - 8\left(\frac{h}{L}\right)\tan\theta\right]^{0.5}$$

Limit the elongation to 100 mm.

**Solution:**
For the 100 m level, we obtain

$$\frac{wL^2}{8h} = 100\,\text{kN};\; w/h = (100 \times 8)/100^2 = 0.08$$

With $\theta = 45°$ and $\tan\theta = 1.0$, we will use a trial and error procedure to find the tension in the cable. With a safety factor = 2.5

| Trial | $H = w/0.08$ | $T$ | $T_u$ | Cable weight, $w$ kN/m | Diameter, mm |
|---|---|---|---|---|---|
| 1 | 0 (assumed) | 141.42 | 353.5 | 0.024 | 25 |
| 2 | 0.3 | 140.6 | 351.4 | Close enough to initial | |

Tension in cable = 351.4 kN

$$\text{Length of cable} = L[1 + \frac{8}{3}(h/L)^2 - (32/5)(h/L)^5]$$

$$= 100[1 + (8/3)(0.3/100)^2 - (32/5)(0.3/100)^5]$$

$$= 100.002\text{ m}$$

Hence it may be taken as 100 m.

Required area of cable $= 351.4 \times 10^3/1400 = 251\,\text{mm}^2$

Nominal diameter $= \sqrt{(A/0.6)} = \sqrt{(251/0.6)} = 20.5\,\text{mm}$

Select a rope of nominal diameter 25 mm.

***Check for elongation***

Equivalent modulus of elasticity $= 0.65 \times 2 \times 10^5$ MPa
$= 1.3 \times 10^5$ MPa

Area of cross section of cable $= 0.6D^2$
$= 0.6 \times 25^2 = 375\,\text{mm}^2$

Elongation at working load
$= PL/AE = 140.6 \times 10^3 \times 100 \times 1000/(375 \times 1.3 \times 10^5)$
$= 288.4\,\text{mm}$

Limiting elongation = 100 mm

Hence, the cables have to be prestressed.

Pre-stressing required $= P_i$

and

$P_iL/AE = (288.4 - 100)$

Hence,

$P_i = 188.4 \times 375 \times 1.3 \times 10^5/(100 \times 10^3)$
$= 91{,}845\text{ N} = 91.85\text{ kN}$

## SUMMARY

Steel tension members are the most common and efficient members in structural applications. These members are connected to other members of a structure using gusset plates and by bolts or welds. A variety of cross sections can be used for tension members. The behaviour of tension members with and without residual stresses and initial crookedness has been discussed. A brief discussion about the various factors such as the net area of cross section, method of fabrication, effect of bearing stress, and effect of shear lag have been provided. The various modes of failures of tension members have been identified. It has been found that the rupture of the net section at the end connections where the tensile stresses are the largest or the block shear failure at the end connections or the yield strength of gross section may govern the failure of tension members. The yielding of the gross section may be the governing failure mode of tension members connected by welding at the ends, whereas the other two failure modes will govern when the members are connected at the ends by bolts. The effect of connecting the end gusset plate to only one element of the cross section has been empirically accounted for by reducing the effectiveness of the outstanding legs, while calculating the net effective area. The methods for accounting for these factors in the design of tension members are discussed with emphasis to the formulae given in the Indian code. The iterative method of design of tension members is presented. The concepts presented are explained with the use of several examples.

## MULTIPLE-CHOICE QUESTIONS

7.1 The maximum slenderness ratio permissible in steel ties is
(a) 250 (c) 450
(b) 350 (d) 400
(e) indirectly controlled by deflection

7.2 The maximum slenderness ratio permissible in steel ties which may be subjected to compression under wind load condition is
(a) 250 (c) 400 (e) no limit
(b) 350 (d) 180

7.3 Which of the following is not a mode of failure in a tension member?
(a) Gross-section yielding
(b) Net section rupture
(c) Local buckling
(d) Block shear failure

7.4 Design strength in gross-section yielding is computed using
(a) $T_{dg} = f_uA_g/\gamma_{m0}$ (c) $T_{dg} = f_yA_g/\gamma_{m0}$
(b) $T_{dg} = f_uA_g/\gamma_{m1}$ (d) $T_{dg} = f_yA_g/\gamma_{m1}$

7.5 Tensile strength adjacent to a hole during the elastic stage of behaviour will approximately be __________ the average stress in the member
(a) three times
(b) 1.5 times
(c) less than two times

7.6 Design strength in net section rupture is computed using
(a) $T_{dn} = f_uA_g/\gamma_{m0}$ (c) $T_{dn} = 0.9f_yA_n/\gamma_{m0}$
(b) $T_{dn} = 0.9f_uA_n/\gamma_{m1}$ (d) $T_{dn} = 0.9f_uA_g/\gamma_{m1}$

7.7 What is the net width of the plate shown in the following figure, for calculating tension capacity?

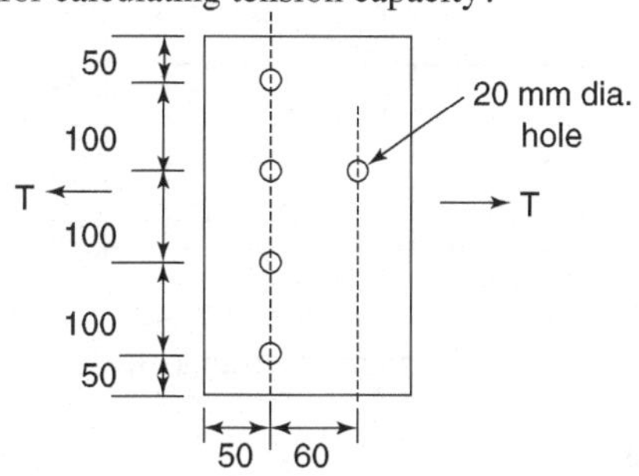


(a) 340 mm (c) 260 mm
(b) 320 mm (d) none of these

7.8 Which of the following statements are incorrect?
(a) Block shear is possible when material bearing strength and bolt shear strength are high
(b) In block shear, failure occurs along a path involving tension in one plane and shear on a perpendicular plane
(c) Block shear failure will not occur in welded connections
(d) Smaller connection length is susceptible for block shear failure

7.9 Which of the following equations is used for predicting block shear failure?

(a) $T_{\text{db}} = \dfrac{f_y A_{\text{vg}}}{\sqrt{3}\gamma_{\text{m0}}} + 0.9 f_u A_{\text{tn}}/\gamma_{\text{m1}}$

(b) $T_{\text{db}} = \dfrac{0.9 f_y A_{\text{vg}}}{\sqrt{3}\gamma_{\text{m0}}} + f_u A_{\text{tn}}/\gamma_{\text{m1}}$

(c) $T_{\text{db}} = \dfrac{f_y A_{\text{vg}}}{\gamma_{\text{m0}}} + 0.9 f_u A_{\text{tn}}/(\sqrt{3}\gamma_{\text{m1}})$

(d) $T_{\text{db}} = \dfrac{0.9 f_u A_{\text{vn}}}{\sqrt{3}\gamma_{\text{m1}}} + f_y A_{\text{tg}}/\gamma_{\text{m0}}$

7.10 For calculating the net area of flats with staggered holes, the area to be subtracted from the gross area is

(a) $\left(nd_n + \sum \dfrac{p^2}{4g}\right)t$ (c) $\left(nd_n - \sum \dfrac{p^2}{4g}\right)t$

(b) $\left(\sum \dfrac{p^2}{4g}\right)t$ (d) $(nd_n)t$

7.11 Which one of the following is an incorrect statement?
(a) The shear lag effect reduces with increase in the connection length
(b) Shear lag effect can be reduced by connecting longer leg of an unequal angle
(c) When an angle is connected through one leg only, at the connection there will be uniform stress.
(d) Shear lag effect need not be considered when all the elements of a section are connected directly.

7.12 The efficiency of a single unequal angle tie member will be better if
(a) it is connected through longer leg
(b) it is connected through shorter leg
(c) the shorter leg is welded
(d) the shorter leg is bolted

7.13 In angle sections, the lug angles and their connection to the gusset should be capable of withstanding $x$% more force in the outstanding leg of the angle, where $x$ is
(a) 10% (b) 20% (c) 30% (d) 40%

7.14 In angle sections, the lug angles and their connection to the gusset should be capable of withstanding $x$% more force in the outstanding leg of the angle, where $x$ is
(a) 10% (b) 20% (c) 30% (d) 40%

7.15 In channel sections, the attachment of the lug angles to the main member should be capable of withstanding $x$% more force in the member, where $x$ is
(a) 10% (b) 20% (c) 30% (d) 40%

7.16 The length of un-stiffened top edge of a gusset plate should be limited to
(a) $12\varepsilon t$ (b) $23\varepsilon t$ (c) $30\varepsilon t$ (d) $42\varepsilon t$

7.17 A gusset plate is subjected to
(a) compressive/tensile stress
(b) shear stress
(c) bending stress
(d) all of these

7.18 Which of the following members will be affected by shear lag?

(a)

(c)
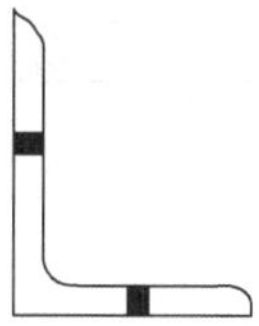
(b)
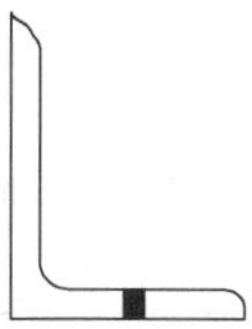
(d)

7.19 A more efficient tension member section will be
(a) bolted single angle section
(b) welded single angle section
(c) two angles connected to opposite sides of gusset plate
(d) single channel section

7.20 Cables can be used effectively as tension members in which of the following structures?
(a) Suspended roofs
(c) Guyed towers
(b) Cable stayed bridges
(d) All of these

## EXERCISES

7.1 What is the net area $A_n$ for the tension member shown in Fig. 7.43, when (a) the holes are made by drilling and (b) holes are made by punching. Assume M20 bolts.

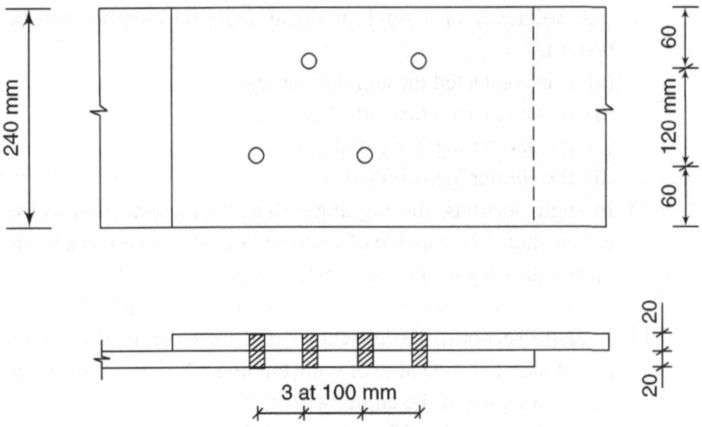


**FIG. 7.43**

7.2 Determine the minimum net area of the plates as shown in Fig. 7.44 with 16-mm-diameter bolts.

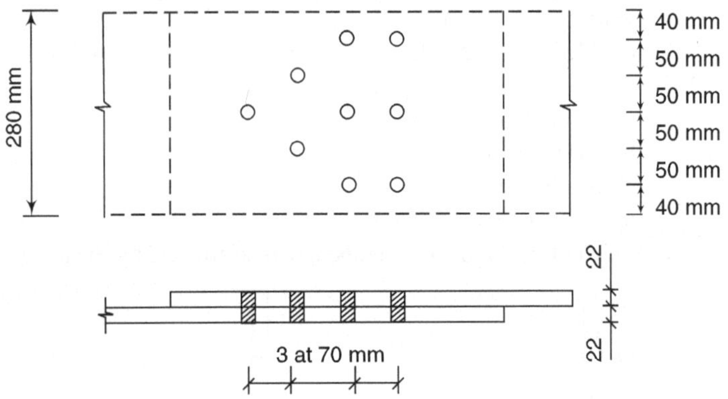


**FIG. 7.44**

7.3 Determine the net area $A_n$ for the $150 \times 75 \times 8$ angle with M20 bolt holes as shown in Fig. 7.45.

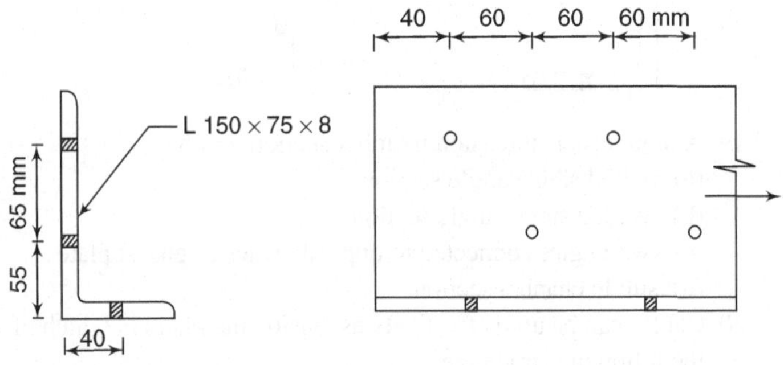


**FIG. 7.45**

7.4 Determine the design tensile strength of a plate (160 × 6 mm) connected to an 8-mm thick gusset using 16-mm-diameter bolts as shown in Fig. 7.46, if the yield and the ultimate stress of the steel used are 250 MPa and 410 MPa, respectively.

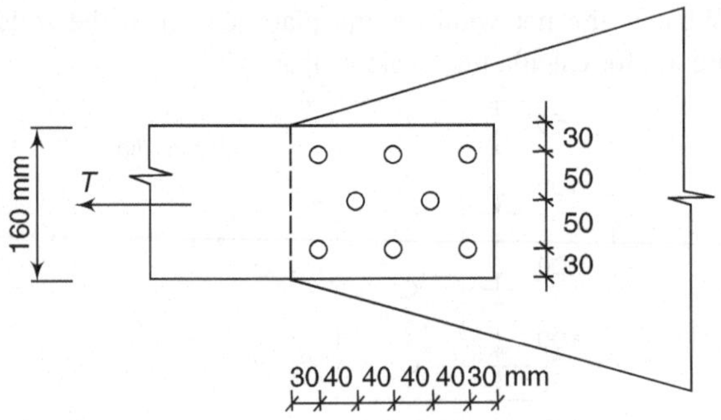


**FIG. 7.46**

7.5 A single angle $75 \times 75 \times 6$ is connected to a 8-mm thick gusset plate at the ends with four 20-mm-diameter bolts to transfer tension as shown in Fig. 7.47. Determine the design tensile strength of the angle. Assume that the yield and the ultimate stress of steel used are 250 MPa and 410 MPa, respectively.

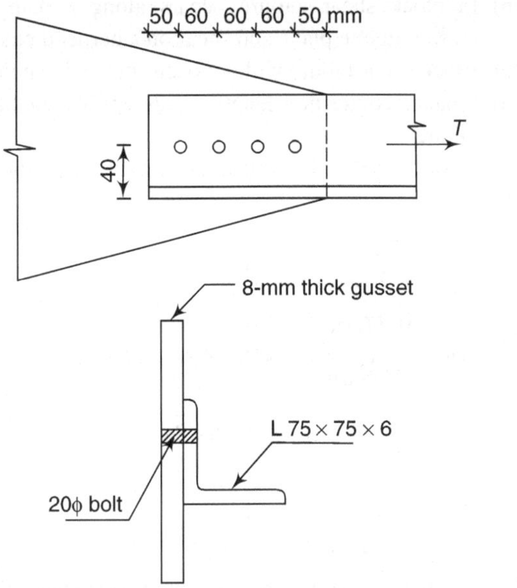


**FIG. 7.47**

7.6 Determine the tensile strength of a roof truss diagonal 200 × 100 × 10 mm ($f_y$ = 250 MPa) connected to the gusset plate by 6-mm welds as shown in Fig. 7.48.

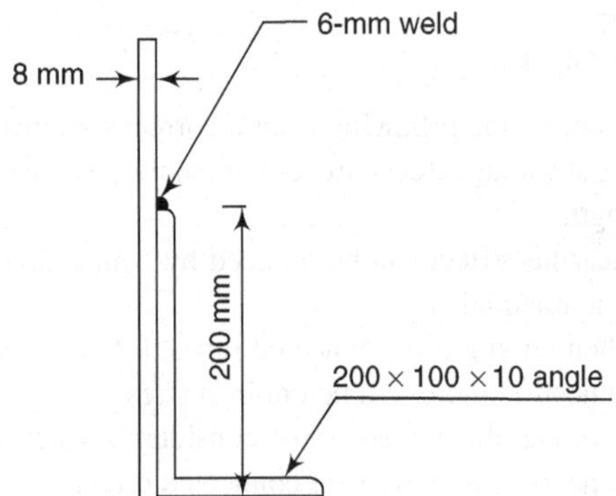


**FIG. 7.48**

7.7 Select a suitable angle section to carry a factored tensile force of 150 kN assuming (a) a single row of M16 bolts, (b) welded end connection. Assume design strength as $f_y = 250\,\text{N/mm}^2$.

7.8 Determine the tension capacity of 150 × 90 × 8 mm angles in Fe 410 steel assuming
(a) connection through the longer leg by two rows of M20 bolts and
(b) connection through shorter leg by a single row of M24 bolts.

7.9 A tie member in a bracing system consists of two angles 90 × 90 × 10 bolted to a 12-mm gusset, one on each side using a single row of bolts (See Fig. 7.49a) and tack bolted. Determine the tensile capacity of the member and the number of bolts required to develop full capacity of the member. What will be the capacity if the angles are connected on the same side of the gusset plate and tack bolted (Fig. 7.49b)?

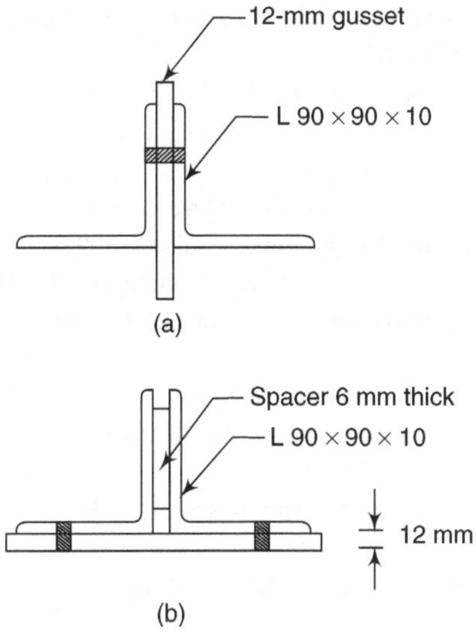


**FIG. 7.49**

7.10 Design a single angle to carry a tensile load of 500 kN. Assume that the length of the member is 3 m.

7.11 A tension member in a bridge structure having a length of 8 m is subjected to an axial tensile (factored) load of 2000 kN. Design the section with channels facing each other (see Fig. 7.50). Assume $f_u = 410\,\text{MPa}$ and $f_y = 250\,\text{MPa}$.

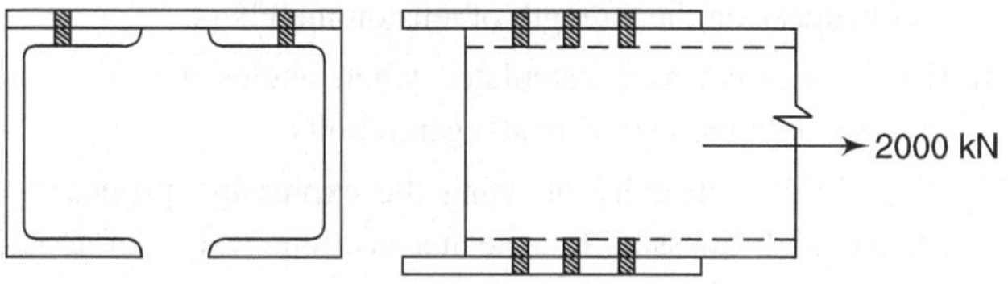


**FIG. 7.50**

7.12 Design sag rods for consecutive purlins near the supported end of a roof truss system as shown in Fig. 7.40. The purlins are supported at one-third points by sag rods. Also design the ridge rod between ridge purlins. The data given are as follows:
C/C Spacing of truss = 8 m
Spacing of purlin = 1.4 m
Self weight of roofing = 160 N/m$^2$
Intensity of wind pressure = 1000 N/m$^2$
Slope of the roof truss = 30°
No access is provided to the roof.

7.13 A diagonal member of a roof carries an axial tension of 500 kN. Design the section and its connection with a gusset plate and lug angle. Use $f_y = 250\,\text{MPa}$ and $f_u = 410\,\text{MPa}$.

## REVIEW QUESTIONS

7.1 List some of the tension members used in buildings and bridges.

7.2 What are the factors that influence the strength of tension members?

7.3 List the type of cross section that can be used as tension members and their use in typical structures.

7.4 How can the rods, which are used as tension members, be pre-tensioned?

7.5 Why are rods, which are used as tension members, required to be pre-tensioned?

7.6 What is the use of spacer plates or stitch plates? At what spacing are they connected to the members?

7.7 Write a short note on the use of ropes and strands in bridges.

7.8 Why is the capacity of a rope always less than the total capacity of wires put together?

7.9 What is the equivalent modulus of elasticity of ropes?

7.10 What is meant by slenderness ratio?

7.11 Write down the expression for the axial elongation of the member subjected to a tensile force?

7.12 What is the main difference between true and engineering stress–strain curve of a tension member?

7.13 How is the yield point of a high-strength steel tension member determined?

7.14 List the different modes of failures of a tension member.

7.15 What is the design stress of a tension member based on
(a) gross-section yielding (b) net section rupture

7.16 Write short notes on block shear failure in plates and angles.

7.17 Under what circumstances will the block shear failure be predominant?

7.18 Why are drilled holes preferred over punched holes?

7.19 By what methods can the effect of punched holes be considered in the design calculation?

7.20 Write down the expression to calculate the net area of cross section of a plate of width $B$, thickness $t$, and having staggered holes of pitch $p$ and gauge $g$.

7.21 What is the effect of high bearing stress due to bolts bearing against the holes?

7.22 What is meant by shear lag?

7.23 How can the effects of shear lag be considered in the design calculation?

7.24 Do the geometry and ductility factors affect the design strength of tension members considerably?

7.25 Write short notes on the effect of residual stress and initial crookedness on the strength of tension members.

7.26 How is the net area calculated when angles are connected through both the legs with staggered bolts?

7.27 What is the difficulty of using the expression proposed by Munse and Chesson for the net section design of tension members?

7.28 What is the expression suggested by Kulak and Wu for calculating the net section design of tension members?

7.29 Write down the expression given in IS 800 : 2007 for the net section design of angle tension members.

7.30 List the use of tension rods in building structures.

7.31 When rolled I-sections, channels, etc., are connected only through flanges to gusset plates, how is the net area calculated erring on the safer side?

7.32 List the various steps involved in the design of a tension member.

7.33 What is a lug angle? Why is it not used in practice?

7.34 Write short notes on splices to tension members.

7.35 How are the sizes of gussets determined?

## REFERENCES


AISC-LRFD, *Load and Resistance Factor Design Specification for Structural Steel Buildings*, American Institute of Steel Construction, Chicago, Ill, 1999.

AS4100, *Australian Standards—Steel Structures*, Standards Association of Australia, 1 The Crescent, Homebush, NSW 2140, 1998.

ASCE Manual No. 52, *Guide for Design of Steel Transmission Towers*, American Society of Civil Engineers, 1988.

Barth, K.E., Orbison, J.G., and Nukala, R., 'Behaviour of Steel Tension Members Subjected to Uniaxial Loading', *Journal of Constructional Steel Resea*rch, Vol. 58, 2002, pp. 1103–1120.

Bathon, L., Mueller III, W.H., and Kempner, Jr L.K., 'Ultimate Load Capacity of Single Steel Angles', *Journal of Structural Engineering*, ASCE, Vol. 119, No. 1, Jan 1993, pp. 279–300.

Bowles, J.E., *Structural Steel Design*, McGraw–Hill International Book Co., New York, pp. 217–253.

BS-5950 – Part 1, *Code of Practice for Design in Simple and Continuous Construction: Hot Rolled Sections*, British Standards Institute, London, 2000.

CAN/CSA-*S16.1-M94, Limit States Design of Steel Structures*, Canadian Standards Assoc., Rexdale (Toronto), Ontorio, Canada, M9W 1R3, 1994.

Dayaratnam, P., *Design of Steel Structures*, Second edition, Wheeler Publishing, 1996, p. 868.

Dowswell, B. and Barber, S., 'Shear Lag in Rectangular Hollow Structural Sections Tension Members: Comparison of Design Equations to Test Data', *Practice Periodical on Structural Design and Construction*, ASCE, Vol. 10, No. 3, Aug 2005, pp. 195–199.

Duggal, S.K., *Design of Steel Structures,* Second edition, Tata McGraw-Hill, 2000, p. 821.

Easterling, W.S. and Giroux, L.G., 'Shear Lag Effects in Steel Tension Members', *AISC Engineering Journal*, Third quarter, 1993, pp. 77–84.

Epstein, H.I, 'An Experimental Study of Block Shear Failure of Angles in Tension', *AISC Engineering Jo*urnal, Vol. 29, 1992, pp. 75–84.

Epstein, H.I. and Chamarajanagar, R., 'Finite Element Studies for Bolt Shear Tests', *Computers and Structures*, Vol. 61, No. 5, 1994, pp. 967–974.

Epstein, H.I. and Gulia, F.S., 'Finite Element Studies of Bolt Stagger Effects in Tension Members', *Computers and Structures*, Vol. 48, 1992, pp. 1153–1156.

Epstein, H.I. and McGinnis, M.J., 'Finite Element Modeling of Block Shear in Structural Tees', *Computers and Structures*, Vol. 77, 2000, pp. 571–582.

Epstein, H.I. and Thacker, B.H., 'The Effect of Bolt Stagger for Block Shear Tension Failures in Angles', *Computers and Structures*, Vol. 39, No. 5, 1990, pp. 571–576.

Eurocode3: *ENV 1993-1-1: Design of Steel Structures*, Part 1.1 General Rules and Rules for Buildings, Brussels, 1992.

Gaylord, E.H., Jr, Gaylord, C.N., and Stallmeyer, J.E., *Design of Steel Structures*, Third edition, McGraw-Hill Inc., New York, NY, 1992, pp. 115–182.

Gross, J.M., Orbison, J.G., and Ziemian, R.D, Block Shear Tests in High-strength Steel Angles, *AISC Engineering Journal*, Third quarter, 1995, pp. 117–122.

http://en.wikipedia.org/wiki/Hyatt_Regency_walkway_collapse, last accessed on 20 July 2015.

http://ethics.tamu.edu/ethics/hyatt/hyatt2.htm, last accessed on 20 July 2015.

http://failurebydesign.info/, last accessed on 20 July 2015.

Kulak, G.L. and Grondin, G.Y., 'Block Shear Failure in Steel Members—A Review of Design Practice', *Fourth International Workshop on Connections in Steel Structures*, 22–25 Oct 2000, Roanoke, VA.

Kulak, G.L. and Wu, E.Y., 'Shear Lag in Bolted Angle Tension Members', ASCE, *Journal of Structural Engineering*, Vol. 123, No. 9, Sept 1997, pp. 1144–1152.

Kulak, G.L., Fisher, J.W., and Struik, J.H.A., *Guide to Design Criteria for Bolted and Riveted Joints*, Second edition, John Wiley and Sons, New York, 1987, p. 309.

Madugula, M.K.S. and Mohan, S., 'Angles in Eccentric Tension', *Journal of Structural Engineering,* ASCE, Vol. 114, No. 10, 1988, pp. 2387–2396.

Munse and Chesson, 'Riveted and Bolted Joints: Net Section Design', *Journal of Structural Engineering*, ASCE, Vol. 89, No. 1, 1963, pp. 107–126.

Orbison, J.G. and Ziemian, R.D., 'Assessment of American Block Shear Load Capacity Predictions', *Journal of Constructional Steel Research*, Vol. 35, 1995, pp. 323–338.

Ragupathi, M., *Design of Steel Structures*, Tata McGraw-Hill, New Delhi, 1995.

Regan and Salter, 'Tests on Welded-angle Tension Members', *The Structural Engineer*, Vol. 62 B, No. 2, 1984, pp. 25–30.

Salmon, C.G. and Johnson, J.E., *Steel Structures, Design and Behavior*, Fourth edition, Harper Collins College Publishers, New York, 1996.

Subramanian, N., 'Rehabilitation of Steel Structures—Some Case Studies', *Advance course on Design of Steel Skeletal Structures*, 16–18 Dec 1998, SERC, Chennai.

Subramanian, N., *Space Structures: Principles and Practice*, Multi-Science Publishing Co. Ltd., Essex, U.K., 2006.

Subramanian, N., 'I-35W Mississippi River Bridge Failure—Is it a Wake-up Call?', *The Indian Concrete Journal*, Vol. 82, No. 2, Feb 2008, pp. 29–38

Usha and Kalyanaraman, V., 'Strength of Steel Tension Members', *Proc. All India Seminar on Modern Trends in Steel Structures*, Nagpur, 8–9 Feb 2002, pp. 21–29.

Usha, *Analytical Study of Nonlinear Behaviour of Steel Angle Tension Members*, Thesis submitted for M.S., (by Research), IIT, Madras, Sept 2003.

www.dot.state.mn.us/i35wbridge/ (it also includes a video of the collapse), last accessed on 20 July 2015.

www.ushamartin.com, last accessed on 20 July 2015.

# 8

# PLASTIC AND LOCAL BUCKLING BEHAVIOUR OF STRUCTURAL STEEL

## INTRODUCTION

As noted in Chapter 7, any structural design must satisfy the following two basic criteria:

1. Strength—the structure or member shall be able to satisfactorily carry the imposed loading
2. Stiffness—the structure or member shall be sufficiently stiff, such that it will not deform or distort more than the permissible limit

To satisfy these requirements, designers rely heavily on their analyses. Using the *elastic design* method (also termed the *working stress* or *allowable stress* method) the stresses and deformations are computed at working loads using the elastic properties of steel. As mentioned in Chapter 2, this method of design limits the structural usefulness of steel up to an allowable stress level, which is well below the elastic limit. Hence the structures designed using this method may be heavier than those designed by the plastic method, which uses the ultimate load rather than the yield stress as the design criterion. (The plastic design method, which utilizes the strength of steel beyond the yield stress, is also called the ultimate load design or load factor design method.) The plastic method is often used in the analysis of statically indeterminate structures and provides a rapid and rational approach for the analysis of such structures. However, both the elastic and the plastic methods neglect the influence of stability, which may significantly affect the load carrying capacity of structures or elements which are slender and subjected to compressive stresses.

The two aspects of structural behaviour under stress which are of particular importance and have considerable influence on the design of steel members are the following:

1. Behaviour of steel in the plastic region of the stress–strain curve [see Fig. 1.11(a)].
2. The tendency of unsupported compression members to 'buckle' and become unstable. *Buckling* may be defined as a structural behaviour in which a deformation develops in a direction or plane perpendicular to that of the load which produced it; this deformation changes rapidly with variations in the magnitude of applied load. Buckling occurs mainly in members that are subjected to compression (Dowling et al. 1988). Its effect is to decrease the load carrying capacity of a structure (i.e., reduce the strength) and also to increase the deformation (i.e., reduce the stiffness).

These two behavioural aspects will be briefly considered in this chapter to provide the necessary theoretical background material for the later design-oriented chapters on columns and beams.

## 8.1 PLASTIC THEORY

### 8.1.1 Basis of Plastic Theory

Tests on actual buildings carried out in the mid 1930s in USA and UK made it clear that a steel frame behaves differently from the assumptions made in conventional simple elastic theory. These tests showed that the real factor of safety of structural elements was very different from what has been assumed. This is because in the elastic design method, the member capacity is based on the attainment of yield stress. However, as discussed in Section 1.8.4 of Chapter 1, steel has a unique property called ductility, because of which it is able to absorb large deformations beyond the elastic limit without fracture. Due to this property, steel possesses reserve strength beyond its yield strength. The method which utilizes this reserve strength is called the *plastic method* of analysis.

The plastic theory makes the design process more rational, since the level of safety is related to the collapse load of the structure and not to the apparent failure at one point. The concept of design based on ultimate load was first developed in Hungary in 1914 by Dr Gabor Kazinczy. He carried out tests on fixed-ended beams and came to the conclusion that failure took place only when yielding occurred at the three cross sections at which the hinging action takes place. The German engineer Maier-Leibnitz showed that the ultimate capacity of continuous beams is not affected by the settlement of supports. Lord J.F. Baker of Cambridge University along with Prof. Horne and Prof. Heyman evolved the method of the simple plastic theory, which resulted in collapse loads being considered as the design criteria (Horne 1979). In USA, Van den Broek wrote the first published paper on plastic theory in 1939. Subsequent work at Lehigh University resulted in several publications (e.g., Beedle 1958; ASCE 1971). It has been found that plastic methods are easier to apply than elastic methods in some cases. However, the frames and their components must be checked rigorously for overall and local stability because of the high strains that might have to be endured.

Although the entire section yields in tension and the compression members yield under the ultimate load, plastification of the material across the depth as well as along the length of the member takes place in the beam under the ultimate transverse load, and the members can behave elasto-plastically. A plastic hinge concept was also was evolved to simplify computations (Davison & Owens 2004). The behaviour of laterally supported beams is discussed in the next section.

To explain the concepts of plastic analysis, consider an I-beam subjected to a steadily increasing bending moment $M$ as shown in Fig. 8.1. In the service load range the section is elastic as shown in Fig. 8.1(a) and the elastic condition exists until the stress at the extreme fibre reaches the yield stress $f_y$ [Fig. 8.1(b)]. Once the strain $\varepsilon$ reaches $\varepsilon_y$ (see Fig. 8.2), increasing the strain does not induce any increase in stress. This elastic–plastic stress–strain behaviour is the accepted idealization for structural steel having yield stresses up to about $f_y$ = 450 MPa.

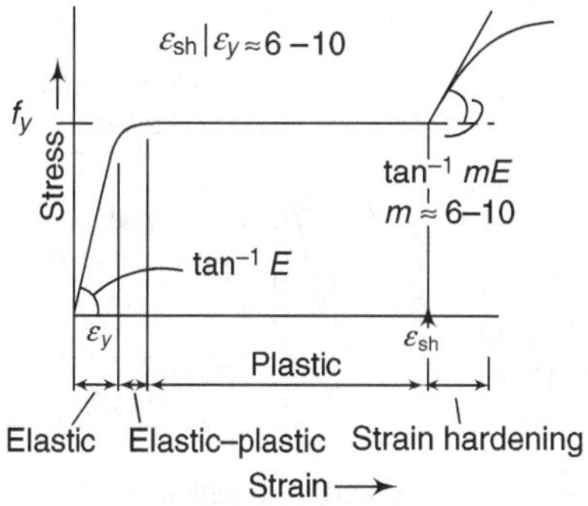


**FIG. 8.2** Idealized tensile stress–strain diagram for steel

When the yield stress reaches the extreme fibre as shown in Fig. 8.1(b), the nominal moment strength $M_n$ of the beam is referred to as the *yield moment* $M_y$ and is given by

$$M_n = M_y = Z_e f_y \tag{8.1}$$

where $Z_e$ is the elastic section modulus.

A further increase in the bending moment causes the yield to spread inwards from the lower and upper surfaces of the beam as shown in Fig. 8.1(c). The spread of yielding in an I-section with residual stresses is also shown in the left hand side of Fig. 8.1. This stage of partial plasticity occurs because of the yielding of the outer fibres without increase of stresses, as shown by the horizontal line of the idealized stress–strain diagram shown in Fig. 8.2. Upon increasing the bending moment further, the whole section yields as shown in Fig. 8.1(d). When this condition is reached, every fibre has a strain equal to or greater than $\varepsilon_y = f_y/E_s$. The nominal moment strength $M_n$ at this stage is referred to as the *plastic moment* $M_p$ and is given by

$$M_p = f_y \int_A y \, dA = f_y Z_p \tag{8.2}$$

where $Z_p = \int y \, dA$ is the *plastic section modulus.*

Any further increase in the bending moment results only in rotation, since no greater resisting moment than the fully plastic moment can be developed until strain hardening occurs. The corresponding moment–curvature relationship of the beam at various stages of loading is shown in Fig. 8.3(b); curvature may be defined as the reciprocal of radius of curvature [see Fig. 8.3(a)]. The portion DE of the curve in Fig. 8.3(b), a rising curve, is due to strain hardening. In the simple plastic theory it is conservatively assumed that the maximum moment $M_p$ is reached at a point where the curvature can increase indefinitely (point C on the curve shown in Fig. 8.3), i.e., neglecting the benefits of strain hardening. The maximum moment $M_p$ is called the *plastic moment of resistance*; the portion of the member where $M_p$ occurs,

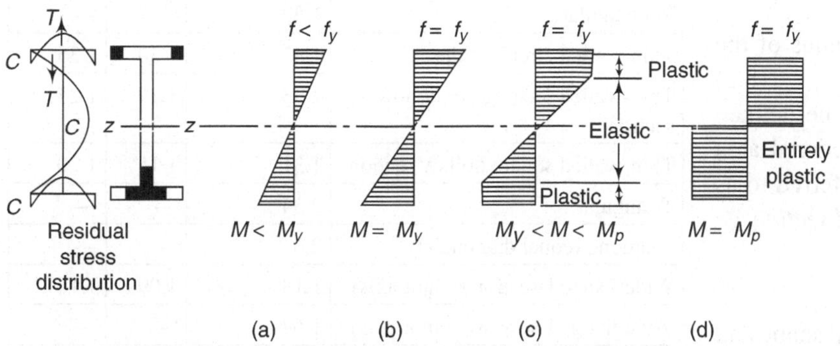


*Note*: Shaded portion indicates portion that have reached $f_y$ due to residual stress of stage *b*.

**FIG. 8.1** Stress distribution at different stages of loading

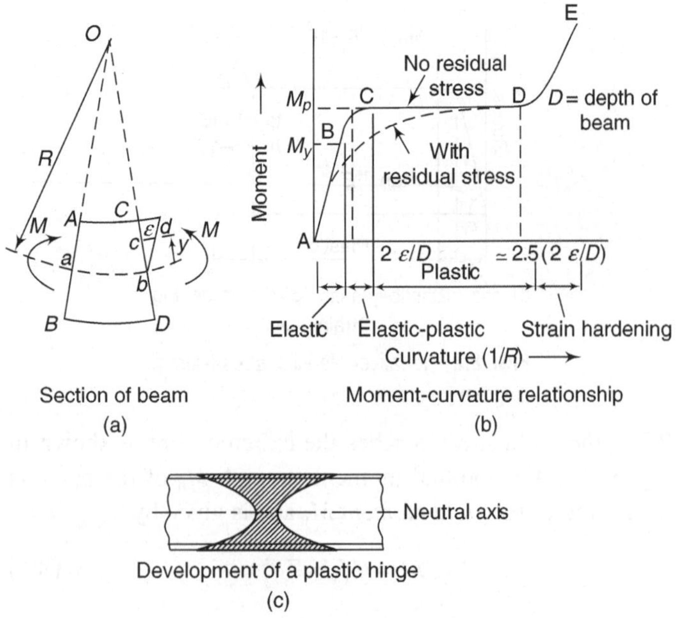


**FIG. 8.3** Moment-curvature relationship at various stages of loading

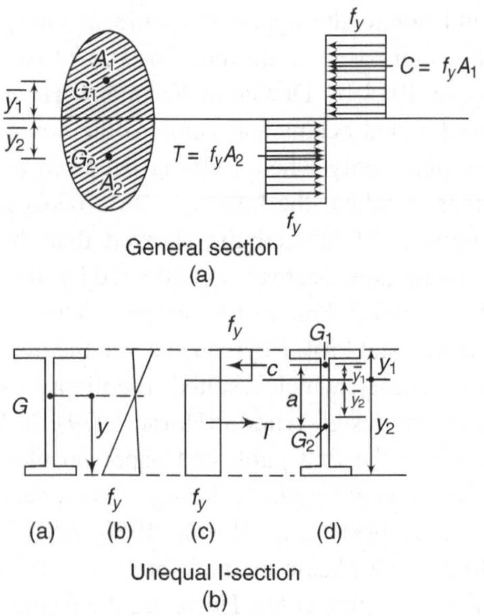


**FIG. 8.4** Force equilibrium of a section

termed a *plastic hinge*, can sustain large local increases of curvature [a plastic hinge is shown in Fig. 8.3(c)].

For equilibrium of normal forces, the tensile and compressive forces must be equal. In the elastic stage, when the bending stress varies from zero at the neutral axis to a maximum at the extreme fibres, this condition is achieved when the neutral axis passes through the centroid of the section. In the fully plastic stage, because the stress is uniformly equal to the yield stress, equilibrium is achieved when the neutral axis divides the section into two equal areas. Thus, considering the general cross section shown in Fig. 8.4 and equating the compressive and tensile forces, we get

$$f_y A_1 = f_y A_2$$

Since $A_1 = A_2$ and $A = A_1 + A_2$,

$$A_1 = A_2 = A/2$$

Plastic moment of resistance

$$M_p = f_y A_1 \bar{y}_1 + f_y A_2 \bar{y}_2$$

$$= f_y A/2 (\bar{y}_1 + \bar{y}_2)$$

Thus, $M_p = f_y Z_p$

where $Z_p = (A/2)(\bar{y}_1 + \bar{y}_2)$ is the plastic modulus of the section. (8.3)

Thus, the *plastic modulus* of the section may be defined as the combined statical moment of the cross-sectional area above and below the equal-area axis. It is also referred to as the *resisting modulus of the completely plasticized section.*

## 8.1.2 Shape Factor

The ratio $M_p/M_y$ is a property of the cross-sectional shape and is independent of the material properties. This ratio is known as the *shape factor* $\nu$ and is given by

$$\nu = M_p/M_y = Z_p/Z_e \tag{8.4}$$

For wide flange I-sections in flexure about the strong axis (*z-z*), the shape factor ranges from 1.09 to about 1.18, with the average value being about 1.14. One may conservatively take the plastic moment strength $M_p$ of I-sections bent about their strong axis to be at least 15% greater than the strength $M_y$ when the extreme fibre alone reaches the yield stress $f_y$. On the other hand, the shape factor for I-sections bent about their minor axis is about the same as for a rectangle, i.e., about 1.5. The shape factors of different cross sections are given in Table 8.1. Note that when the material at the centre of the section is increased, the value of $\nu$ increases. Examples 8.1 to 8.7 given at the end of this chapter explain the methods of determining the shape factors and plastic section moduli of different sections.

**TABLE 8.1** Shape factor of different cross sections

| Cross section | Shape factor $\nu$ | | |
|---|---|---|---|
| | Max. | Min. | Average |
| Circular | 16/3π = 1.70 | — | — |
| Rectangular | 1.50 | — | — |
| Thin-walled circular tube | 1.47 | 1.30 | 1.35 |
| Thin-walled rectangular hollow section | 1.33 | 1.19 | 1.25 |
| Thin-walled square hollow section | 1.31 | 1.15 | 1.20 |
| Triangular | 2.34 | — | — |
| Diamond (equal diagonals) | 2.0 | — | — |
| Wide flange I-section (major axis) | 1.18 | 1.09 | 1.14 |
| Wide flange I-section (minor axis) | 1.67 | — | — |
| Equal angles | 1.84 | 1.81 | 1.82 |

(*Contd*)

**TABLE 8.1** *(Contd)*

| Cross section | Shape factor $\nu$ | | |
|---|---|---|---|
| | Max. | Min. | Average |
| Unequal angles | 1.83 | 1.75 | 1.80 |
| Channels (major axis) | 1.22 | 1.16 | 1.18 |
| Channels (minor axis) | 1.8 | — | — |

For the theoretically ideal section in bending, i.e., two flange plates connected by a web of insignificant thickness, the value of $\nu$ will be 1. A value of shape factor nearly equal to one shows that the section is efficient in resisting bending. The moment–curvature behaviour of various cross sections is shown schematically in Fig. 8.5.

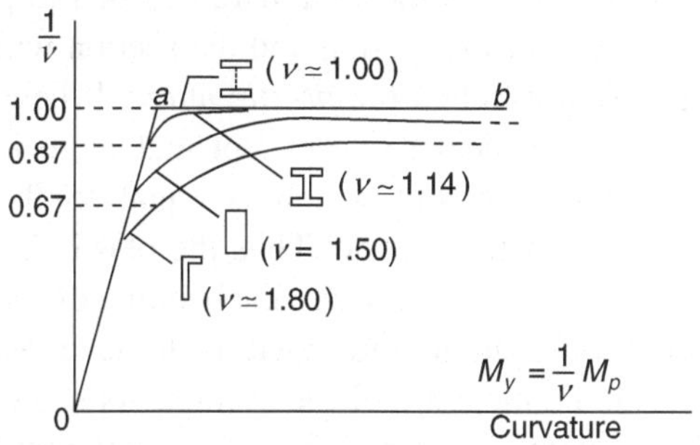


**FIG. 8.5** Moment–curvature relationship of various cross sections

## 8.2 GENERAL REQUIREMENTS FOR PLASTIC DESIGN

The following assumptions are made in plastic design to simplify computations.

1. The material obeys Hooke's law till the stress reaches $f_y$.
2. The yield stress and modulus of elasticity have the same value in compression and tension (see Fig. 8.6).

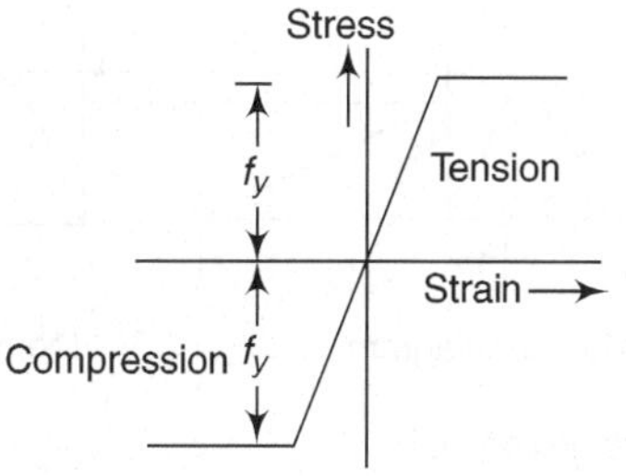


**FIG. 8.6** Stress–strain curve for perfectly plastic materials

3. The material is homogenous and isotropic in both the elastic and plastic states.
4. The material is assumed to be sufficiently ductile to permit large rotations of the section to take place.
5. Plastic hinge rotations are large when compared with the elastic deformations, so that all the rotations are concentrated at the plastic hinges. The segments between the plastic hinges are rigid.
6. The magnitude of bending moment caused by the external loads will at the most be equal to the plastic moment capacity of the section. A plastic hinge forms when the moment reaches the capacity of the section.
7. The influence of normal and shear forces on plastic moments is not considered.
8. Plane sections remain plane even after bending and the effect of shear is neglected.
9. The equilibrium of forces at the time of collapse is considered for the undeformed state of the structure.
10. No instability occurs in any member of the structure up to collapse.

Several codes (such as IS 800 and BS: 5950) allow the use of plastic design only when the loading is static and fatigue is not a design criterion. IS 800 stipulates that the following conditions should be satisfied in order to use the plastic method of analysis, unless adequate ductility of the structure and the plastic rotation capacity of its members are established for the design loading conditions by other means of evaluation (see clause 4.5.2 of IS 800).

1. The yield stress of steel used should not be greater than 450 MPa.
2. The stress–strain characteristics of the steel used should obey the following conditions, in order to ensure plastic moment redistribution.
   (a) The yield plateau (horizontal portion of the stress–strain curve) should be greater than 6 times the yield strain.
   (b) The ratio of the ultimate tensile stress to the yield stress should be more than 1.2.
   (c) The elongation on the standard gauge length should be more than 15%.
   (d) The steel should exhibit strain-hardening capacity.

In addition, the following conditions should also be satisfied.

3. The members shall be hot-rolled or fabricated using hot-rolled plates.
4. The cross sections of the members not containing plastic hinges should be 'compact' (see Section 8.13 for the discussion on compact section) and those of members containing plastic hinges should be 'plastic'.
5. The cross section should be symmetrical about its axis perpendicular to the axis of the plastic hinge rotation.

These limitations are intended to ensure that there is a sufficiently long plastic plateau to enable a hinge to form and that the steel will not experience premature strain hardening.

The fundamental concept of the plastic analysis method is that sufficient plastic hinges are formed at critical sections in the structure, which transforms the structure into a collapse mechanism. Extensive testing has adequately verified this concept (ASCE 1971).

### 8.2.1 Advantages and Disadvantages of Plastic Design

Plastic design methods offer the following advantages (Horne 1979).

1. Realization of uniform and realistic factor of safety for all parts of the structure (in contrast to elastic methods, where the safety factor varies).
2. Simplified analytical procedure and rapidity of obtaining design moments, since there is no need to satisfy elastic strain compatibility conditions.
3. Saving of material over elastic methods resulting in lighter structures.
4. No effect due to temperature change, settlement of supports, imperfections, erection methods, etc. (because their only effect is to change the amount of rotation required). This is in contrast to the elastic method, where extra calculations are required. However, calculations for instability and elastic deflections require careful considerations in plastic methods. The plastic design method is very popular for the design of some structures, e.g., beams and portal frames.
5. Gives some idea of the collapse mode and strength of the structure.
6. In the elastic method of design, the design process is repeated several times to obtain an optimum solution, whereas the plastic method of design produces a balanced section in a single attempt.

The disadvantages of plastic design method are the following.

1. Obtaining collapse load is difficult if the structure is reasonably complicated.
2. There is little saving in column design.
3. Difficult to design for fatigue.
4. Lateral bracing requirements are more stringent than for elastic design.
5. Calculations for elastic deformations require careful considerations.
6. When more than one loading condition occurs, it is necessary to perform separate calculations, one for each loading condition; the section requiring the largest plastic moment is selected. Unlike the elastic method of design, wherein the moments produced by different loading conditions can be added together, the plastic moments obtained by different loading conditions cannot be combined (i.e., the plastic moment calculated for a given set of loads is valid only for that loading condition). This is because the 'principle of superposition' becomes invalid when certain parts of the structure have yielded.

## 8.3 PLASTIC HINGE CONCEPT

In order to understand how a beam can fail, it is desirable to understand the concept of plastic hinges. We have discussed in Section 8.1.1 that as the external load and the bending moment increase, the rotation at a section increases proportionally up to the yield point. Any further increase in the moment will generate a non-linear moment–curvature relation, and the curvature will increase rapidly to reach an unbounded value as the moment tends to reach the value of $M_p$. The part on the moment-rotation curve where there is plasticity (shown by the non-linear curve) may be projected onto the moment diagram. This can then be projected onto the section and the area that has been plasticized can be determined (see Fig. 8.7). At the centre of the area there will be full plasticity over the full depth of the section and the section will behave like a hinge. This is called the *plastic hinge*. It behaves like a frictionless mechanical hinge except that there is always a fixed moment constraint which is equal to the plastic moment capacity of the section. Thus the plastic hinge may be defined as a *yielded zone*, where the bending of a structural member can cause an infinite rotation to take place at a constant plastic moment $M_p$ of the section. As shown above, plastic hinges form in a member at the maximum bending moment locations. However, at the intersections of two members, where the bending moment is the same, a hinge forms in the weaker member. Generally, hinges are located at restrained ends, intersections of members, and point loads. The hinges may not form simultaneously as the loading increases, but this is not important for calculating the final collapse load.

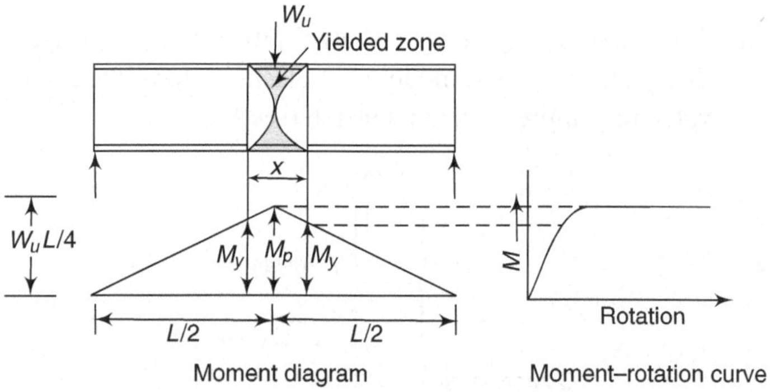


**FIG. 8.7** Plastic hinge concept

### 8.3.1 Length of Plastic Hinge

The value of moment adjacent to the yield zone is more than the yield moment up to a certain length $x$ of the structural member (see Fig. 8.7). This length is known as the *hinge length*. The hinge length depends upon the type of loading and the geometry of the cross section of the structural member.

Let us consider a simply supported rectangular beam with a central concentrated load $W_u$ as shown in Fig. 8.7. For this beam,

$$M_p = W_u L/4 = f_y\, bh^2/4 \text{ (since } Z_p = bh^2/4)$$

$$M_y = f_y\, bh^2/6 = (f_y\, bh^2/4)\,(2/3) = (2/3)M_p$$

Let $x$ be the length of plasticity zone. From the bending moment diagram shown in Fig. 8.7,

$$M_y / (L/2 - x/2) = M_p / (L/2)$$

or $$M_y / M_p = [(L - x)/2] / (L/2) = 1 - x/L$$

Thus, $(L - x)\, M_p = LM_y$

Using the notation for shape factor, $\upsilon = M_p/M_y$, and rearranging we get,

$$x = L\left(1 - \frac{1}{\upsilon}\right)$$

Substituting the value of $\upsilon = 1.5$, for rectangular sections, we get

$$x = L/3$$

Hence in this case, the hinge length equals to one-third of the span. For I-sections, $\upsilon$ is approximately equal to 9/8 and the plastic hinge length will be $L/9$.

For a uniformly loaded fixed-end beam Bowles (1980) showed that the hinge length at the ends (which is half the mid-span hinge length) is given by

$$x = \left(\frac{L}{2.83}\right)\sqrt{\left[1 - \left(\frac{1}{\nu}\right)\right]} \tag{8.5}$$

For an I-section, the length of the plastic hinge in the centre of the beam is (with $\upsilon = 9/8$)

$L_{\text{hinge}} = 2x = 2L / 8.49$, that is, 23% of span length.

Generally the number of plastic hinges $n = r + 1$, where $r$ is the number of redundancies or the indeterminacy. (However, there are exceptions, e.g., the partial collapse of a beam in a structure.)

## 8.4 PLASTIC-COLLAPSE LOAD

The load at which a sufficient number of plastic hinges are formed in a structure such that a collapse mechanism is created is called the *plastic-collapse load* or *plastic limit load*. The feature of plastic design which distinguishes it from elastic design is that it takes into account the favourable redistribution of bending moment which takes place in the indeterminate structures after the first hinge forms at the point of maximum bending moment. This redistribution may be considerable and the final load at which the collapse mechanism forms may be significantly higher than the load at which the first hinge forms. Thus, for example, in a fixed beam having a concentrated load at the one-third point, the first hinge forms at one of the supports; as the load is increased further, the moment at this hinge remains constant at $M_p$, while the moments at the other support and the load point increase until a second hinge is formed. When the load is increased further, the moments at these two hinges remain constant at $M_p$, until a third and final hinge is formed to make the beam a mechanism. In this beam, the final ultimate load will be 33% higher than the first hinge load.

The working loads are multiplied with a factor called the *load factor* (see Section 8.7.1) to obtain the plastic limit loads. Depending on the combination of loads and their probability of acting at the same time, different load factors are used (see Section 2.8.2 and Table 2.2). Though the structure is designed based on plastic design, the serviceability conditions at working load are valid even for plastic design, since the structure generally behaves elastically under working loads during its lifespan.

A number of examples of plastic-collapse mechanisms in cantilevers and single- and multiple-span beams are shown in Fig. 8.8. Cantilevers and overhanging beams generally collapse as single-bar mechanisms with a single plastic hinge at one of the supports [Fig. 8.8(a)]. Note the formation of the plastic hinge in the weaker section when there is a reduction in section capacity. Single-span beams collapse as two-bar mechanisms with a hinge (plastic or frictionless) at each support and a plastic hinge within the span. Note that when there is a uniform moment region, general plasticity may occur [Fig. 8.8(b)]. Multi-span beams generally collapse in one span only as a local two-bar mechanism, with a plastic hinge within the span and a hinge (plastic or frictionless) at each support. It is also possible to have a three-bar mechanism, where two adjacent spans combine to form a mechanism [Fig. 8.8(c)]. A similar mechanism may occur in an overhanging beam.

As noted already, plastic hinges normally occur at supports, points of concentrated load, and points where cross sections change. The location of a plastic hinge in a beam with a uniformly distributed load is, however, not well defined.

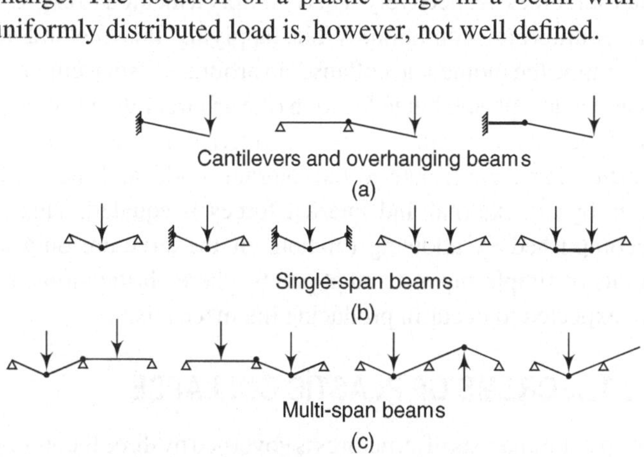


**FIG. 8.8** Plastic-collapse mechanisms in beams

**TABLE 8.2** Conditions to be satisfied in elastic and plastic analysis

| Elastic analysis | Plastic analysis |
|---|---|
| *Equilibrium:* The summation of the forces and moments acting on any free body must be equal to zero, i.e., $\Sigma F_x = 0$, $\Sigma F_y = 0$, $\Sigma M_y = 0$ | *Equilibrium:* The bending moment distribution defined by the assumed plastic hinges must be in static equilibrium with the applied loads and reactions. |
| *Compatibility:* Strain in one section is assumed to be equal to that in the adjoining section. | *Mechanism:* There must be sufficient number of plastic and frictionless hinges for the beam/structure to form a mechanism. The ultimate or collapse load is reached when a mechanism is formed. |
| *Moment–curvature relations:* $M/I = f/y = E/R$ (Navier's theorem)<br>where $f$ = stress at any layer,<br>$E$ = Young's modulus of elasticity,<br>$R$ = local radius of curvature of beam,<br>$M$ = maximum bending moment,<br>$I$ = moment of inertia of the cross section of beam, and<br>$y$ = the distance of the layer under stress from the neutral axis. | *Plasticity:* The bending moments in any section of a structure must be less than the plastic moment of the section $-M_p \leq M \leq M_p$ |

## 8.5 CONDITIONS OF PLASTIC ANALYSIS

The basic conditions that are to be satisfied for any structure in elastic and plastic analysis are shown in Table 8.2.

If all the three conditions are satisfied, the lowest plastic limit load (a unique value) is obtained. If only the equilibrium and mechanism conditions are satisfied (this forms the basis for the mechanism method of plastic analysis), an upper bound solution for the true ultimate load is obtained. A lower bound solution for the true ultimate load is obtained when the equilibrium and plasticity conditions only are satisfied (statical method of plastic analysis). See Section 8.7 for the mechanism and statical methods of analysis.

### 8.5.1 Principle of Virtual Work

The principle of virtual work may be stated as follows: If a system of forces in equilibrium is subjected to a virtual displacement, then the work done by the external forces equals the work done by the internal forces, i.e.,

$$W_E = W_I \tag{8.6}$$

where $W_E$ and $W_I$ are the work done by the external and internal forces, respectively. This is simply a method to express the equilibrium condition. While applying this method to determine the moment at collapse, an arbitrary displacement is assumed at a plastic hinge location (the arbitrary displacement must be one for which only the internal moments at the plastic hinges contribute to the internal work) and the work done by the external and internal forces is equated. This is accomplished by allowing rotations of the structure only at points of simple support and at points where plastic moments are expected to occur in producing the mechanism.

## 8.6 THEOREMS OF PLASTIC COLLAPSE

The plastic analysis of structures is governed by three theorems, which are as follows: The *static* or *lower bound theorem* states that a load computed on the basis of an assumed equilibrium moment diagram, in which the moments are nowhere greater than the plastic moment, is less than, or at the best equal to, the correct collapse load. Hence the static method represents the lower limit to the true ultimate load and has a maximum factor of safety. The static theorem was first suggested by Kist and its proof was given by Gvozder, Greenberg, and Horne (Horne 1979).

The *kinematic* or *upper bound theorem* states that a load computed on the basis of an assumed mechanism will always be greater than, or at the best equal to, the correct collapse load. Hence the kinematic method represents an upper limit to the true ultimate load and has a smaller factor of safety compared to the static method. A proof of this theorem was provided by Gvozder, Greenberg, and Prager (Horne 1979).

The upper and lower bound theorems can be combined to produce the *uniqueness theorem*, which states that the load that satisfies both the theorems at the same time is the correct collapse load.

## 8.7 METHODS OF PLASTIC ANALYSIS

Using the principle of virtual work and the upper and lower bound theorems, a structure can be analysed for its ultimate load by any of the following methods: (1) static method and (2) kinematic method.

In the mechanism or kinematic method, a mechanism is assumed and virtual work equations are used to determine the collapse load. The number of independent mechanisms ($n$) is related to the number of possible plastic hinge locations ($h$) and the number of degrees of redundancy ($r$) of the frame by the equation

$$n = h - r \tag{8.7}$$

In the statical or equilibrium method, an equilibrium moment diagram is obtained such that the moment at any section is less than or equal to the plastic moment capacity. Even though the equilibrium method gives a lower bound solution, the virtual work method is often used due to its simplicity of application in comparison with the equilibrium method.

If the upper and lower bound solutions obtained by the mechanism and statical methods coincide or are sufficiently close, then the assumed plastic hinge locations are correct. If, however, these bounds are not precise enough, then the location of the assumed hinge should be modified (an indication of this will be provided by the bending moments determined in the statical analysis) and the analysis repeated. The use of the mechanism and equilibrium methods of plastic analysis in single-span beams is explained in Examples 8.8–8.15.

### 8.7.1 Load Factors

Let us suppose that $W_c$ represents the collapse load of a beam which would yield the full plastic moment $M_p$. If a collection of beams having different end conditions (free or fixed) and the working load $W$ were first designed elastically and then plastically, the ratio $W_c/W$ for each of these beams will not be identical. Only the beams that are simply supported will produce a constant ratio of $W_c/W$ and for these cases the values of $W_c/W$ will be the lowest.

From a practical point of view, a minimum acceptable and constant load factor is required, and that found for a simply supported beam may be regarded satisfactory. Since the ratio $W_c/W$ represents the load factor, its actual value may be found by considering a simply supported beam with a central point load.

Elastically,

$$M = WL/4 = f_{bc} Z_e$$

Plastically,

$$M_p = W_c L/4 = f_y Z_p = f_y \, \nu Z_e$$

Hence,

$$M_p/M = W_c/W = (f_y/f_{bc})\nu$$

If we consider $f_{bc} = 0.66 f_y$ and a mean value of $\nu = 1.14$ (for I-sections), we get

$$\text{Load factor} = M_p/M = W_c/W = (1/0.66) \times 1.14 = 1.727 \quad (8.8)$$

Hence the load factor is normally taken as 1.7 (see also Section 2.8.2 of Chapter 2).

## 8.8 PLASTIC DESIGN OF CONTINUOUS BEAMS

The application of plastic design to continuous beams involves two separate steps.

1. Selection of a suitable section based on considerations of the plastic-collapse mechanism;
2. Ensuring that no other form of failure (local buckling) prevents the attainment of this collapse mechanism (see Section 8.13).

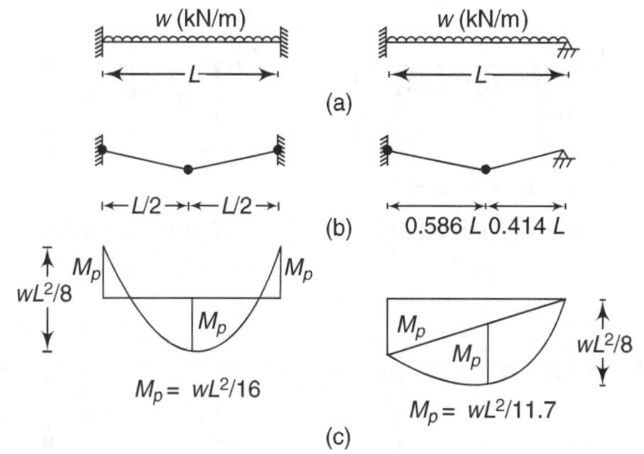


**FIG. 8.9** Plastic collapse of internal and end spans of continuous beams (a) Support conditions (b) Plastic collapse mechanisms (c) Bending moments at collapse

The first step requires the use of one of the standard techniques for plastic analysis, such as the mechanism method or equilibrium method. Plastic collapse can occur in an internal span or in the end span of a continuous beam. Therefore it is necessary to consider two cases: a fixed-end beam and a propped cantilever. In both cases, a mechanism will form after a sufficient number of plastic hinges have formed. These mechanisms and the bending moments at collapse are shown in Fig. 8.9. The use of these results is illustrated in Examples 8.16–8.18, given at the end of this chapter.

## 8.9 PLASTIC DESIGN OF PORTAL FRAMES

Plastic design is used extensively for the design of single-storey portal frame structures. An in-depth coverage of all the important aspects of design has been provided by Davies and Brown (1996) and Horne and Morris (1981). The actual application of plastic design involves the same two steps described in Section 8.8, namely, adequate strength and avoidance of secondary failures. In addition, the effect of axial load on vertical members should be considered (see Section 8.10).

The design of portals is a little complicated because there can be more than one mechanism of failure, and the mechanism giving the least collapse load has to be chosen. Single-bay portal frames with fixed bases have three redundancies and require four hinges to produce a mechanism.

Furthermore, as the number of redundancies increases, so does the number of possible modes of collapse. The possible mechanisms are classified into two groups: elementary and combined mechanisms. Whereas the former are independent of each other, the latter are linear combinations of elementary mechanisms. The elementary and combined mechanisms of a single-storey portal frame are shown in Fig. 8.10. The *beam mechanism* results when three plastic hinges form within any one of the elements of the frame [see Fig. 8.10(b)].

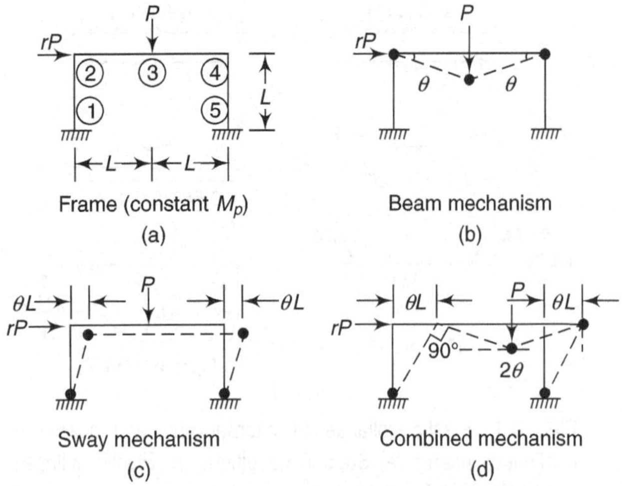


**FIG. 8.10** Collapse mechanisms of a single-bay, fixed-base portal frame

A set of columns in a storey may develop plastic hinges at the top and bottom of each column so as to generate a simple *sway mechanism* of the structure [see Fig. 8.10(c)]. The sway and beam mechanisms may be combined to form a *combined mechanism* as shown in Fig. 8.10(d). Thus, a structure may collapse partially or as a whole. As mentioned in Section 8.3, if the order of indeterminacy is $r$, then the minimum number of plastic hinges required for total collapse is $r + 1$, whereas the number of hinges required for partial collapse may be less. Examples given at the end of this chapter illustrate the determination of collapse loads for simple single-storey portal frames.

Figure 8.11 illustrates the different collapse mechanisms for a normal pitched portal frame, which are dependent on the values of vertical ($P$) and horizontal ($H$) loads. $P$ is assumed to represent any form of vertical loading such as dead or imposed, while $H$ represents wind loading. Note that in the figure, $P$ and $H$ have been idealized into two concentrated loads, the resulting modes of failures being indicative of the forms of collapse typical for such frames. For most portals, failure occurs by mode 1 [Fig. 8.11(b)]. Only when the frame is tall or the wind pressure is high relative to the vertical loading, failure by mode 2 will occur [Fig. 8.11(c)]. Failure by mode 3 is possible only when the wind load is very high and the maximum imposed load is very low. A similar behaviour is observed for frames with pinned bases.

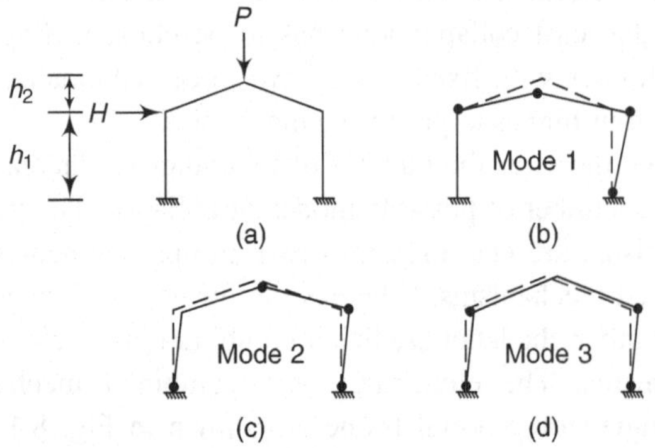


**FIG. 8.11** Collapse mechanisms of a single-bay fixed-base pitched portal frame

When the mechanism method is applied to structures with sloping members, the determination of displacements in the direction of applied forces is required. This difficulty may be resolved by the application of the *centre-of-rotation technique* (also called as *method of instantaneous centres*). The description of this method may be found in Horne and Morris (1981) and Davies and Brown (1996). Horne and Morris (1981) also provide charts for the direct selection of a factored value of $M_p$ for pinned- and fixed-base frames, subjected to a uniform vertical load plus a single horizontal eaves load. When the gravity load case governs the design, a good estimate of the required section (assuming uniform frame) may be obtained by using the following formulae (Horne & Morris 1981):

For pinned-base frames:

$$M_p = \gamma_L (wL^2/8)[1 + k + (1 + k)^{0.5}] \quad (8.9a)$$

For fixed-base frames:

$$M_p = \gamma_L (wL^2/8)[1/[1 + 0.5\,k + (1 + k)^{0.5}] \quad (8.9b)$$

where $\gamma_L = 1.7$ is the global load factor (see Sections 2.8.2 and 8.7.1), $k = h_2/h_1$, and $h_1$ and $h_2$ are defined in Fig. 8.11.

For single-bay frames having other geometries or frames having other than the same section throughout, the reader should consult Davies and Brown (1996) and Horne and Morris (1981). The design of portal frames using haunches, either at the eaves or at the apex (which improve the strength and stiffness of the frames), are also discussed in these references. Martin and Purkiss (1992) provide the complete design of a haunched portal frame as per BS 5950. Haunches shift the eaves plastic hinges to the column top immediately below the lower end of the haunch for a fixed-based frame (Morris 1981). Using haunches, about 5 to 10 per cent can be saved on rafter size, though it involves additional fabrication cost. Examples 8.19–8.21 illustrate the methods used to determine the collapse load of simple portal frames.

## 8.10 EFFECT OF AXIAL AND SHEAR FORCE ON PLASTIC MOMENT CAPACITY

The effect of axial force, apart from introducing buckling, is to reduce the plastic moment capacity of a section. The presence of axial force implies that the sum of tension and compression forces in the section is not zero and hence we have the following equation:

$$\int_A f_y dA - P = 0 \quad (8.10)$$

where $P$ is the axial force. This also means that the neutral axis moves away from the equal-area axis providing an additional area in tension or compression depending on the type of axial load.

Consider a rectangular section of width $b$ and depth $d$ subjected to an axial compressive force $N$ together with a moment $M$ as shown in Fig. 8.12. (Note that we have used the notation $N$ to denote the axial force, instead of $P$, since the force can be tensile or compressive.) The values of $M$ and $N$ are increased at a constant value $M/N$ until the fully plastic stage is attained. The stress distribution at this stage will be as shown in Fig. 8.12(a). This may be split up into two portions as shown in Figs 8.12(b) and (c) for convenience. The stress distribution shown in Fig. 8.12(b) may be assumed to balance the axial load and that shown in Fig. 8.12(c) to balance the bending moment. From these figures,

$$N = 2f_y b y_0 \tag{8.11}$$

and

$$M_{pr} = 2\int_{y_0}^{d/2} yb\,dy f_y = b[(d/2)^2 - y_0^2]f_y$$
$$= 0.25 f_y b(d^2 - 4y_0^2) \tag{8.12}$$

where $y_0$ is the distance from the neutral axis to the point of stress change for $M_p$ without axial force (see Fig. 8.12), $M_{pr}$ is the reduced plastic moment capacity due to axial force, and $f_y$ is the yield stress. Note that when the moment and axial force are applied alone, we get the moment without axial force

$$M_p = f_y b d^2/4 \tag{8.13a}$$

Thus,

$$M_{pr} = M_p(1 - 4y_0^2/d^2) \tag{8.13b}$$

For very low axial forces, say, less than 10–15% of the axial yield load, the reduction in $M_p$ may be ignored, since $y_0$ will be very small and $y_0^2$ will be smaller still. For example, if

$2y_0/d = 0.15$, $4y_0^2/d^2 = 0.0225$, and $(1 - 4y_0^2/d^2) = 0.9775$

then the reduction is about 2% only. The axial force alone (without any moment),

$$N_p = f_y b d \tag{8.13c}$$

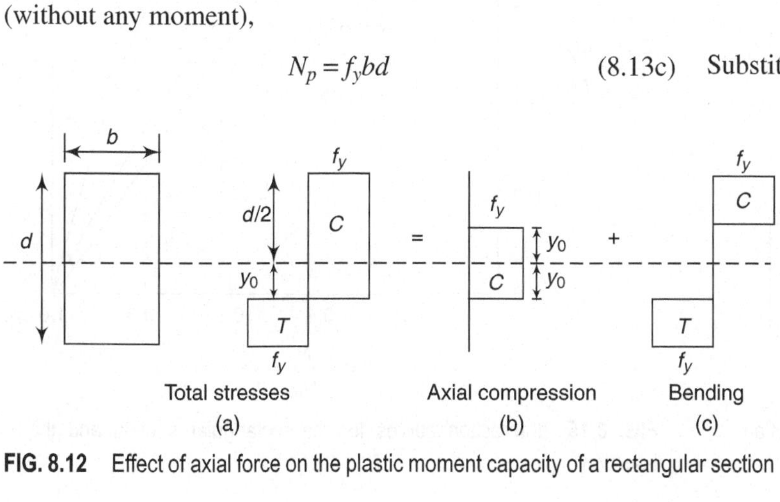


**FIG. 8.12** Effect of axial force on the plastic moment capacity of a rectangular section

If the ratio $N/N_p$ (the squash load ratio) is denoted by $n$, then from Eqn (8.11),

$$y_0 = N/(2f_y b) = nN_p/(2f_y b) = nd/2 \tag{8.14}$$

Substituting this value of $y_0$ in Eqn (8.13b) and putting $M_{pr} = Z_{pr}f_y$ and $M_p = Z_p f_y$, where $Z_{pr}$ is the reduced plastic modulus, we get

$$Z_{pr} = Z_p(1 - n^2) \tag{8.15}$$

Substituting $y_0 = N/(2f_y b)$ and using Eqns (8.13a) and (8.13c) in Eqn (8.12), we get the following interaction equation for rectangular sections:

$$(M_{pr}/M_p) + (N/N_p)^2 = 1 \tag{8.16}$$

## 8.10.1 Interaction Formula for I-sections

In the case of I-sections, two cases are to be considered: one when the neutral axis is within the web and the other when the neutral axis is within the flange (see Fig. 8.13). Dealing first with the case of Fig. 8.13(b),

$$N = 2y_0 t f_y \tag{8.17}$$

where $t$ is the thickness of the web. The reduced plastic moment $M_{pr}$ may be written as

$$M_{pr} = M_p - t[(2y_0)^2/4]f_y = M_p - ty_0^2 f_y \tag{8.18}$$

The above value of $M_{pr}$ is obtained by noting that the reduction of the plastic moment below the value $M_p$ (when no axial force is present) is represented by the loss of the plastic moment of the web area.

As in the case of a rectangular section, if the section becomes plastic under an axial thrust only, the value of the thrust would be $N_p = Af_y$, where $A$ is the area of cross section. This thrust $N_p$ is known as the *squash load* representing the load that would have been sustained by a very short member failing by pure axial compressive deformation. If the ratio of $N/N_p$ (the *squash load ratio*) is denoted by $n$, then from Eqn (8.17),

$$y_0 = N/(2tf_y) = nN_p/(2tf_y) = nA/(2t) \tag{8.19}$$

Substituting this value of $y_0$ into Eqn (8.18) and putting $M_{pr} = Z_{pr}f_y$ and $M_p = Z_p f_y$, we get (Annex I of BS 5950-1:2000)

$$Z_{pr} = Z_p - (A^2/4t)n^2 \tag{8.20}$$

This formula applies until the axial thrust becomes so high that the neutral axis ceases to lie within the web. The limiting thrust at which it happens is given by putting $y_0 = D/2 - T$. Since $y_0 = nA/(2t)$, this means that the formula given in Eqn (8.20) applies until $n$ reaches the value of $n_1$ where

$$n_1 = (D - 2T)t/A \tag{8.21}$$

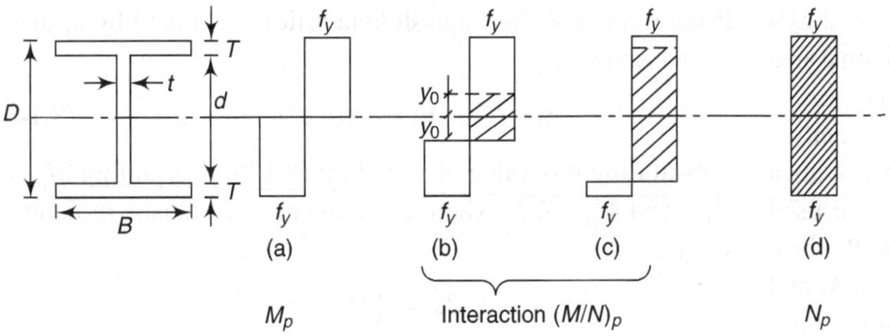


**FIG. 8.13** Effect of axial force on the plastic moment capacity of an I-section

Now let us consider the case when the neutral axis is within the flange [Fig. 8.13(c)]. It is seen that the core area is the area left after removing, from the original section of area $A$, two flange areas each of width $B$ and thickness $(D/2 - y_0)$.

Hence,

$$N = [A - 2(D/2 - y_0)B]f_y \quad (8.22)$$

The moment of resistance $M_{pr}$ may be derived from the two remaining flange areas, the centroids of which are at a distance $[y_0 + 0.5(D/2 - y_0)]$, i.e., $0.5(D/2 + y_0)$, from the centroidal axis. Hence,

$$M_{pr} = \text{area of part flange} \times \text{lever arm}$$

$$= 2[(D/2 - y_0)Bf_y]\,[0.5(D/2 + y_0)] \quad (8.23)$$

Substituting $N = nAf_y$ in Eqn (8.22) gives

$$y_0 = 0.5[D - (1 - n)A/B] \quad (8.24)$$

Hence, from Eqn (8.23), putting $M_{pr} = Z_{pr}f_y$, we get

$$Z_{pr} = (A^2/4B)\,(1 - n)(2BD/A - 1 + n) \quad (8.25)$$

The reduced plastic moduli of the I-section about the minor axis may be obtained in a similar manner, with the neutral axis either inside or outside the web [see Figs 8.14(a) and (b), respectively). When the neutral axis is inside the web, we get (BS 5950-1: 2000)

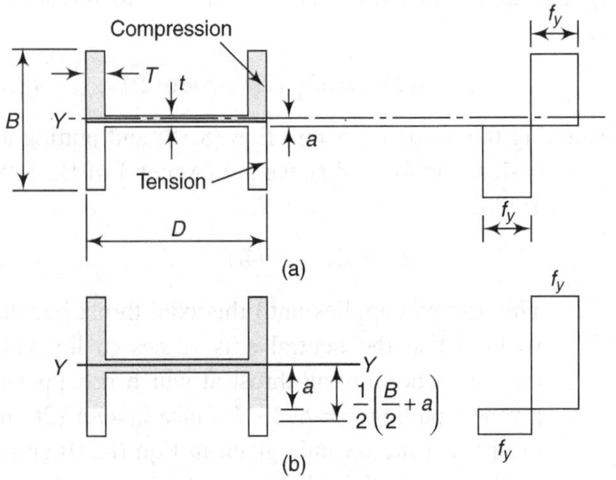


**FIG. 8.14** Effect of axial force on the plastic moment capacity of an I-section about the minor axis

$$Z_{pry} = Z_{py} - (A^2/4D)n^2 \quad (8.26)$$

This applies until $y_0 = t/2$, that is, until $n$ reaches the value $n_2 = Dt/A$. When neutral axis is outside the web [Fig. 8.14(b)], we get

$$Z_{pry} = (A^2/8T)(1 - n)\,[(4TB/A) - 1 + n] \quad (8.27)$$

Equation (8.27) applies when $n$ is greater than $n_2$.

The interaction curves of Eqns (8.16), (8.25), and (8.27) are shown in Fig. 8.15. An approximate straight line equation, which lies on the safer side for typical rolled I-sections bent about the major axis, can be written as (Petersen 1990; Salmon & Johnson 1996)

$$(N/N_p) + (1/1.18)(M/M_p) \le 1 \text{ for } N/N_p > 0.15 \quad (8.28a)$$

$$M = M_p \quad \text{for } N/N_p < 0.15 \quad (8.28b)$$

The formulae given in Eqns (8.16), (8.25), and (8.27) are found to be correct for sections made from three plates and are sufficiently accurate for rolled beams and columns with parallel flanges. (In section property tables, approximate allowances are made for the effects of the root fillets and hence those values may differ slightly from those predicted by these equations.) Taking, for example, ISMB 400 with an axial force of 610 kN,

$$T = 16 \text{ mm}, t = 8.9 \text{ mm}, B = 140 \text{ mm}, D = 400 \text{ mm}$$

$$N_p = A_g f_y = 7840 \times 250 \times 10^{-3} = 1960 \text{ kN}$$

$$Z_p = BT(D - T) + (D - 2T)^2 t/4 = 140 \times 16(400 - 16) + (400 - 32)^2 \times 8.9/4$$

$$= 1161.5 \times 10^3 \text{ mm}^3$$

$$A_g = 2BT + (D - 2T)t = 2 \times 140 \times 16 + (400 - 32) \times 8.9 = 7755.2 \text{ mm}^2$$

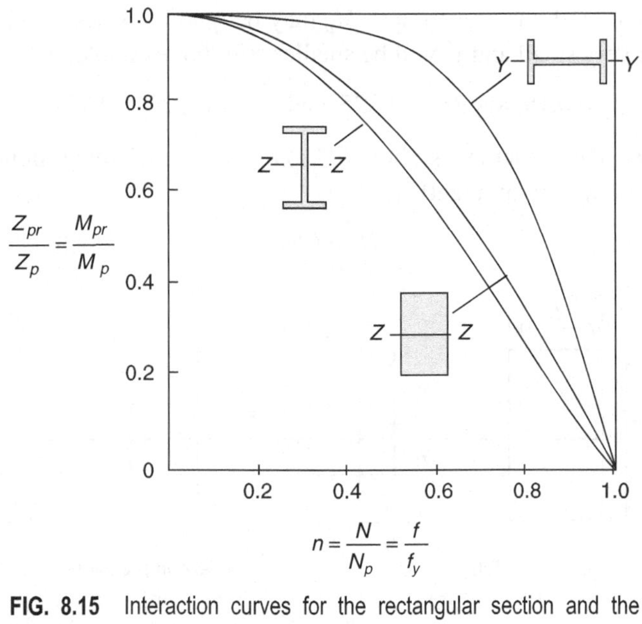


**FIG. 8.15** Interaction curves for the rectangular section and the I-section

As per the section table,

$$Z_p = 1176.18 \times 10^3 \text{ mm}^3$$

$$A = 7840 \text{ mm}^2$$

The values as per the section tables are higher, since they allow for root fillets.

$$n = 610/1960 = 0.3112$$

Change point for the value of $n_1 = (D - 2T)t/A_g$
$= (400 - 2 \times 16)8.9/7840$
$= 0.4178 > 0.3112$

Hence,

$Z_{p,\text{red}} = Z_p - (A^2/4t)n^2 = 1176.18 \times 10^3 - 7840^2/(4 \times 8.9) \times 0.3112^2$
$= 1176.18 \times 10^3 - 167.209 \times 10^3$
$= 1008.97 \times 10^3$ mm$^3$ (14% reduction)

If $n = 0.15$, $Z_p$ is reduced by $38.848 \times 10^3$ mm$^3$ only (3.3% reduction). Let us now consider $n = 0.60$. Since $n_1 = 0.4178 < n$,

$Z_{pr} = (A^2/4B)(1 - n)(2BD/A - 1 + n)$

$= 7840^2/(4 \times 140)(1 - n)(2 \times 140 \times 400/7840 - 1 + n)$

$= 109{,}760 (1 - n)(14.28 - 1 + n) = 609.64 \times 10^3$ mm$^3$ (48% reduction)

In single-storey portal frames, the value of $n$ will usually be less than 0.15 and hence the effect of axial force will be small (unless the frame supports heavy loads like cranes). However, in multi-storey frames, an allowance for the effect of axial loads should be made for columns, especially in the lower storeys.

### 8.10.2 Effect of Shear Force on Plastic Moment Capacity

When a member is subjected to uniaxial tensile or compressive stress $f$ in the presence of shear stress $\tau$, yield occurs [according to the von Mises criterion for material failure (see Section 2.10)] when

$$f_y^2 = f^2 + 3\tau^2 \quad (8.29)$$

Yield in pure shear (with $f = 0$) occurs when

$$\tau_y = f_y/\sqrt{3} = 0.58 f_y \quad (8.30)$$

Since most of the shear force acting on an I-section is taken by the web, the maximum capacity in shear $V_p$ is taken as

$$V_p = \tau_y A_w = 0.58 f_y A_w \quad (8.31)$$

where $A_w$ is the area of the web. The shear force in an I-section cannot exceed this value (though the strain-hardening effect may allow a slightly higher capacity). Maximum shear force and maximum moment can occur together only at the end of a beam, where plastic deformation is affected by the nature of

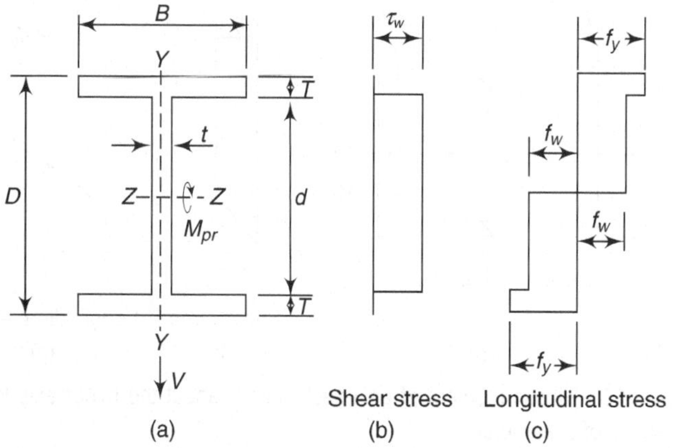


**FIG. 8.16** Plastic moment of an I-section bent about the major axis in the presence of shear force

the connection between the beam and column. Hence plastic moment capacity in the presence of shear force depends on the detailed geometry and loading of the member concerned.

Consider the symmetrical I-section shown in Fig. 8.16, subjected to a bending moment $M_{pr}$ and shear force $V$. The maximum shear capacity of the section is $V_p = dt\,\tau_y$ and the plastic moment capacity in the absence of shear is $M_p = Z_{pz} f_y$, where $Z_{pz}$ is the plastic modulus. Let us assume that the actual applied shear force $V$ is resisted by a uniform shear stress $\tau_w$. Hence

$$\tau_w = V/dt \quad (8.32)$$

If $M_{pr}$ and $\tau_w$ produce full plasticity, the distribution of stress will be as shown in Fig. 8.16(c). Thus the flanges will be stressed to the full yield values $\pm f_y$ and the webs to $f_w$. From Eqn (8.29), with $f = f_w$ and $\tau = \tau_w$ and Eqn (8.30), we get

$$f_w = \sqrt{(f_y^2 - 3\tau_w^2)} = f_y\sqrt{[1 - (\tau_w/\tau_y)^2]} \quad (8.33)$$

The reduced moment capacity can now be written as

$$M_{pr} = M_p - \text{loss of web capacity}$$

where the loss of web capacity is the plastic modulus of the web ($Z_{pw} = td^2/4$) multiplied by the difference between the full and reduced yield stresses, i.e., $(f_y - f_w)$. Therefore,

$$\text{Loss of web capacity} = Z_{pw}(f_y - f_w)$$

$$= z_{pw}\left\{f_y - f_y\sqrt{[1 - (\tau_w/\tau_y)^2]}\right\}$$

$$= z_{pw} f_y\left\{1 - \sqrt{[1 - (\tau_w/\tau_y)^2]}\right\} \quad (8.34)$$

Hence,

$$M_{pr} = M_P - Z_{PW} f_y\left\{1 - \sqrt{[1 - (\tau_w/\tau_y)^2]}\right\} \quad (8.35)$$

Alternatively, if we use $M_p = Z_{pz} f_y$ and $M_{pr} = Z_{prz} f_y$, where $Z_{prz}$ is the reduced plastic modulus in the presence of shear, we get

$$Z_{prz} = Z_{pz} - Z_{pw}\left[1 - \sqrt{(1 - q)^2}\right] \quad (8.36)$$

where $q = V/V_p$ is the ratio of shear force $V$ to capacity shear force $V_p$.

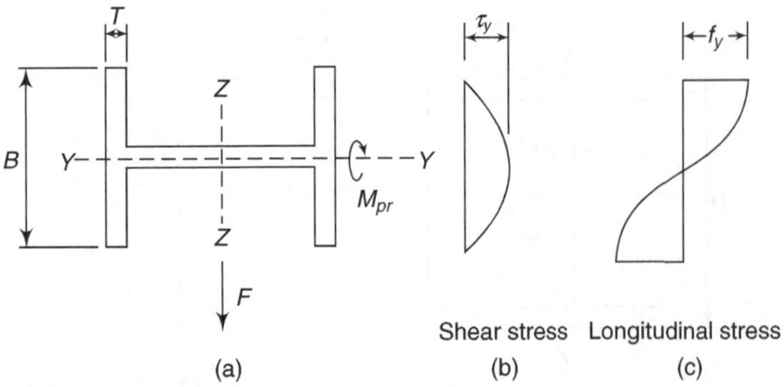


**FIG. 8.17** Plastic moment of an I-section bent about the minor axis in the presence of shear force

When an I-section is bent about its minor axis (see Fig. 8.17), the accompanying shear forces perpendicular to the web are taken by the flanges. Now the shear stress is distributed approximately parabolically as shown in Fig. 8.17(b) and the longitudinal stresses are also non-uniform [see Fig. 8.17(c)]. Horne (1951) has shown that the reduced plastic moment capacity in this case can be taken approximately as

$$M_{pr} = M_p[1 - 0.45(\tau_w/\tau_y)^2] \tag{8.37}$$

where $\tau_w$ is the shear stress calculated on the area of the flange. This formula may be applied up to $\tau_w = 0.8\ \tau_y$. Equation (8.37) can also be expressed in terms of the plastic modulus as

$$Z_{pry} = Z_{py}\,(1 - 0.45q^2) \tag{8.38}$$

where $Z_{py}$ and $Z_{pry}$ are the plastic modulus and reduced plastic modulus, respectively, about the minor axis and $q = \tau_w/\tau_y = V/V_p$, where $V_p = 2BT\ \tau_y = 1.155BTf_y$. It may be noted that the reduction of plastic capacity in bending about the minor axis of an I-section due to shear is very small and may be neglected.

To show the reduction of plastic section modulus due to shear, let us consider a beam ISMB 400 subjected to a shear force of 310 kN:

$$V/V_p = (310 \times 1000)/(0.58 \times 250 \times 400 \times 8.9) = 0.6$$

$$Z_{prz} = Z_{pz} - Z_{pw}\left[1 - \sqrt{(1-q)^2}\right]$$

$= 1176.18 \times 10^3 - 8\,.9 \times 368^2/4\left[1 - \sqrt{(1 - 0.6^2)}\right] = 1115.9 \times 10^3\ \text{mm}^3$

Percentage reduction in the plastic modulus of the whole section

$= 100\,(1176.18 \times 10^3 - 1115.9 \times 10^3)/(1176.18 \times 10^3) = 5.13\%$

Although the ratio $V/V_p = 0.6$ is relatively high, the reduction in the value of the plastic modulus of the whole section is not high.

## 8.11 SPECIAL CONSIDERATIONS

There are a number of special considerations while attempting a plastic design. These considerations are as follows:

1. Local buckling (concerned with the width-to-thickness ratios for the flange and web of a section)
2. Flange stability
3. Column action (effect of axial force on plastic moment capacity)
4. Deflection
5. Connection.

Local buckling is discussed in Section 8.12, and the effect of axial force on plastic moment capacity in Section 8.10. Other considerations/limitations are discussed briefly in the following.

### 8.11.1 Flange Stability

One of the major assumptions in plastic theory and design is that the beam is supported continuously laterally to prevent the failure of the compression flange by lateral buckling. Therefore this condition should be translated in practice for the design to be valid. Note that the plastic hinges require a certain amount of ductility in addition to their strength requirement—*rotation capacity* is a measure of this ductility (rotation capacity may be defined as that capacity which a given cross-sectional shape can accept at the plastic moment without premature failure occurring). Thus a section should ideally exhibit a rotation capacity that corresponds to a strain equal to the strain-hardening strain $\varepsilon_s$, similar to curve *A* shown in Fig. 8.18. Lateral instability can prevent a member from maintaining its nominal $M_p$ value during hinge rotation, since yielding dramatically reduces the resistance of a member to lateral buckling; the moment sustained by the section reduces with increased rotation instead of remaining constant as shown in curve *B* of Fig. 8.18. To prevent this, members should be adequately braced against both lateral and torsional displacements at the positions at which hinges are assumed in the failure mechanism.

The code (IS 800) suggests that torsional restraints against lateral buckling should be provided at all plastic hinge locations. BS: 5950-1:2000 suggests that both flanges should have lateral restraints at each plastic hinge location to resist a force equal to 2.5 per cent of the force in the compression

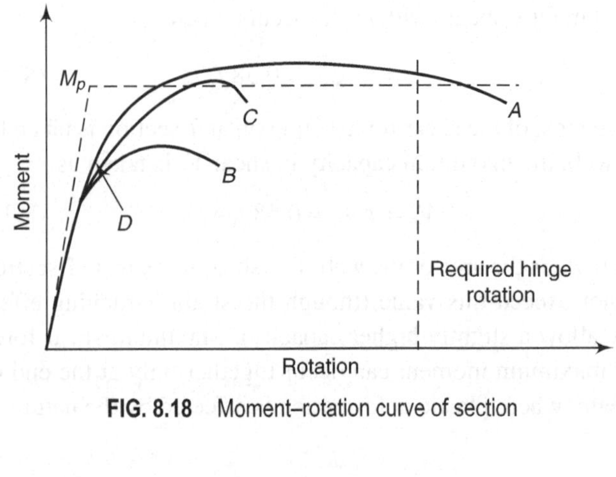


**FIG. 8.18** Moment–rotation curve of section

flange. When it is not practicable to provide such restraint directly at the hinge location, it should be provided within a distance of $D/2$ of the plastic hinge location, where $D$ is the total depth of the section. However, for the hinge that forms last, this requirement is not required since it needs only just to attain the $M_p$ value for the mechanism to be considered complete. Hence the code (IS 800) states that the torsional restraint requirement need not be met at the last hinge to form, provided it can be clearly identified.

Within a member containing a plastic hinge, the maximum distance from the restraint at the plastic hinge to an adjacent restraint should be limited to a length $L_m$. IS 800, based on BS 5950-1:2000, gives a conservative approximation for $L_{m1}$:

$$L_{m1} \le 38r_y/\{[(f_c/130) + (f_y/250)^2 (x_t/40)^2]^{0.5}\} \quad (8.39)$$

where

$r_y = \sqrt{(I_y/A)}$ is the radius of gyration about the minor axis (mm),

$f_c$ is the compressive stress on the cross section due to axial load (MPa),

$f_y$ is the yield stress (MPa),

$x_t$ is the torsional index

$= 1.132(AI_w/I_yI_t)^{0.5}$ (may be taken as $D/t_f$ as per BS: 5950-1:2000),

$A$ is the area of cross section,

$I_w$, $I_y$, $I_t$ are the warping constant, second moment of area of the cross section about the minor axis, and St.Venant's torsion constant, respectively,

$D$ is the depth of the section,

When the member has unequal flanges, $r_y$ should be taken as the lower of the values for the compression flange only or for the whole section. Where the cross section of the member varies within the length between restraints, the maximum value of $r_y$ and the minimum value of $x_t$ should be used. Example 8.22 illustrates the use of Eqn (8.39) in the design of a continuous beam.

The British code allows higher values for $L_m$, allowing for the effects of moment gradient for I-sections with uniform cross section with equal flanges when $D/B \ge 1.2$ (where $D$ is the overall depth and $B$ is the flange width) and $f_c$ does not exceed 80 MPa, as shown below (Horne 1964):

$$L_m = \xi L_{m1} \quad (8.40a)$$

where the factor $\xi$ is given by the following.

- For $1 \ge \psi \ge \psi_u$: $\xi = 1$ (8.40b)
- For $\psi_u > \psi > 0$: $\xi = 1 - (1 - KK_0)(\psi_u - \psi)/\psi_u$ (8.40c)
- For $0 \ge \psi > -0.75$: $\xi = K[(K_0 - 4(1 - K_0)\psi/3]$ (8.40d)
- For $\psi \le -0.75$: $\xi = K$ (8.40e)

$\psi$ is the end moment ratio (see Table 42 of code for the sign convention). The limiting value of $\psi_u$ for *Fe* 410 grade is

$$\psi_u = 0.44 + (x_t/270) - (f_c/200) \quad (8.41)$$

The coefficients $K_0$ and $K$ may be obtained as follow:

$$K_0 = (180 + x_t)/300 \quad (8.42a)$$

For $20 \le x_t \le 30$: $K = 2.3 + 0.03\,x_t - x_t f_c/3000$ (8.42b)

For $30 \le x_t \le 50$: $K = 0.8 + 0.08\,x_t - (x_t - 10) f_c/2000$ (8.42c)

The spacing of restraints in member lengths not containing plastic hinges should satisfy the requirements on lateral buckling strength of beams (see Chapter 10). A simpler equation is given for $L_m$ in the Canadian code, CAN/CSA-S16.1-09, as below:

$$L_m = 550\, r_y/\sqrt{f_y} \quad \text{for } \psi < -0.5 \quad (8.43a)$$

and

$$L_m = 980\, r_y/\sqrt{f_y} \quad \text{for } \psi \ge -0.5 \quad (8.43b)$$

## 8.11.2 Stiffeners at Plastic Hinge Locations

Web stiffeners should be provided at points where a concentrated load is applied within $D/2$ of a plastic hinge location, which exceeds 10% of the shear capacity of the member (checked as per the provisions of beams; see Chapter 10). The stiffener should be provided within a distance of half the depth of the member, on either side of the hinge location, and designed to carry the applied loads (see Chapter 11 for the design of stiffeners). If the stiffeners consist of plates, then the outstanding width-to-thickness ratio $b/t$ should not exceed the values given for plastic section (see Section 8.12). When other sections are used, the ratio $(I_{so}/I_t)^{0.5}$ should not exceed the values given for the plastic section, where $I_{so}$ is the second moment of area of the stiffener about the face of the element perpendicular to the web and $I_t$ is St. Venants' torsion constant of the stiffener.

## 8.11.3 Fabrication Restriction

Within a length equal to the member depth, on either side of the plastic hinge location, the following restrictions should be applied to the tension flange.

1. Holes, if provided, should be drilled. If punched, they should be punched 2 mm undersize and reamed.
2. All sheared or hand flame cut edges should be finished smooth by grinding, clipping, or planing.

## 8.11.4 Second-order Effects

Any second-order effect of the loads acting on the structure in its deformed configuration may be neglected, where the elastic buckling load factor ($\lambda_{cr}$) satisfies the condition $\lambda_{cr}/\lambda_p \ge 10$. See also Section 4.5.4 of IS 800 : 2007.

1. When, $4.6 \le \dfrac{\lambda_{cr}}{\lambda_p} < 10$, the second-order effects may be neglected, provided the design load effects are amplified by a factor

$$\delta_p = 0.9/\{1 - (1/\lambda_{cr})\} \quad (8.44)$$

2. When $\lambda_{cr}/\lambda_p < 4.6$, a second-order elasto-plastic analysis should be carried out.

The elastic buckling load factor ($\lambda_{cr}$) shall be the ratio of the elastic buckling load set of the frame to the design load set of the frame, determined by using

1. approximate methods (approximate methods are suggested for sway and non-sway frames in Section 4.6.2 of IS 800)
2. rational buckling analysis of the whole frame (see Chapter 4 for the details of this analysis)

These provisions in IS 800 are similar to those found in the Australian code AS 4100-1998.

### 8.11.5 Connections

It has been shown experimentally that in an actual structure plastic hinges do not form at the intersection of member centre-lines but rather at the interface of the beam and the column members, i.e., at the connections. Hence, the connections in a continuous beam or a portal should be such that they permit the development of the plastic moment of the connected members, without the connection itself yielding. The only exception is the connections that are not located at the points of maximum bending moment; in these cases, they may be designed to carry the appropriate moment and shear at that section. (See Section 4.5.3 of IS 800 : 2007).

Due to this, it is necessary to take the finite size of joints into account in the analysis. If the depth of the beam on either side of a column member is different, then the average depth can be used in the calculations to determine the position of the column hinge. In portal frames, the finite size of haunches are normally taken into account in the analysis; thus the hinge will not form within the length of the haunch. One way to avoid the plastic hinge forming at the connections is to make the connection stronger than the plastic moment capacity of beams or columns, so that the hinges are displaced and form at the end of adjacent beams (see Section 17.9 of Chapter 17).

### 8.11.6 Deflection

As can be noticed, even if we use plastic design, the member under working loads, is in the elastic range of behaviour. Hence deformations and deflections under load would still be in the elastic range and the normal limitations of deflections must be applied for structures designed using plastic theory.

### 8.11.7 Strain Hardening and Residual Stress

The concept of a plastic hinge has been derived from the basic assumption that infinite strain in tension or compression takes place at the yield stress. However, after plastic strains of the order of 6–10 times the elastic strain at yield have occurred, structural steel show an increase in stress (with the initial slope of the stress–strain curve being about 5% of that during the elastic range) referred to as strain hardening (see Fig. 1.11). This small increase of stress may prevent the 'infinite curvature' (assumed in plastic analysis) from developing at sections where the theoretical plastic moment capacity is developed. Due to this, the moment of resistance may rise above the full plastic value in simply supported beams with a central concentrated load. However, in portal frames, under uniform vertical loading, the bending moments increase above the theoretical plastic hinge value in regions of high moment gradient and remain below that value in regions of low moment gradient. This does not cause any major problem, provided the joints have a sufficient margin of strength above the appropriate plastic hinge values to prevent joint failure, or alternatively the joints themselves provide sufficient rotation capacity (Horne & Morris 1981).

Due to the self-equilibrating residual stresses (see Section 1.9 and Figs 1.24 and 1.25 of Chapter 1), the moment–curvature relationship may depart earlier from elastic linearity (see Fig. 8.3), thus causing yielding to start at lower loads. (This also results in a consequent increase in the deflections, which occurs at all subsequent load levels.) Nevertheless, it returns close to that for an initially stress-free beam by the time the curvature reaches about 2.5 times the curvature at first yield in a similar but initially stress-free beam, as shown in Fig. 8.3 (Horne & Morris 1981). Since in plastic-collapse mechanisms such curvatures are reached at and near plastic hinges before the attainment of the theoretical plastic-collapse load, the residual stresses do not have any effect on collapse load, especially when we use plastic sections (see Section 8.13).

## 8.12 LOCAL BUCKLING OF PLATES

The concept of stability as it applies to structural systems may be understood best by considering the conditions of equilibrium. If a structural system that is in equilibrium is disturbed by a force, it has two basic alternatives when the disturbing force is removed.

1. It will return to its original position, in which case we refer to the system as being *stable*.
2. It will continue to deform and as a consequence be incapable of supporting the load it supported before the disturbance occurred, in which case the system is called *unstable*.

Instability is thus characterized as a change in geometry, which results in the loss of the ability to support load. Stability, specifically the loss of ability to support load, is an extremely important consideration in the development of the limit states design criterion. Thus *buckling* may be defined as a structural behaviour in which a mode of deformation develops in a direction or plane perpendicular to that of the loading which produces it; such a deformation changes rapidly with increase in the magnitude of the applied loading. It occurs mainly in members or elements that are subjected to compressive forces.

While using plastic design, it is assumed that plastic deformation can take place without the geometry of the structure changing to such an extent that the conditions of equilibrium are significantly modified. Such changes in geometry can arise at three levels, namely (Horne & Morris 1981),

1. deformation within the cross section of a member (resulting from *local buckling* in the plate elements constituting the web or the flange),
2. displacements within the length of the member relative to straight lines drawn between the corresponding points of the end sections (due to the bending and/or twisting of the member), and
3. overall change of geometry of the structure, causing the joints to displace relative to each other (e.g., the sway deformation in multi-storey frames).

These three levels of deformation are thus associated, respectively, with local, member, and frame instability. The problem of member stability is discussed in Chapter 9. However, in plastic design it is assumed that plastic deformations, leading to some redistribution of stresses and bending moments, can take place before instability sets in and this means that the theoretical elastic instability load should be significantly above the plastic limit load. We will discuss local buckling and its consequences in this section.

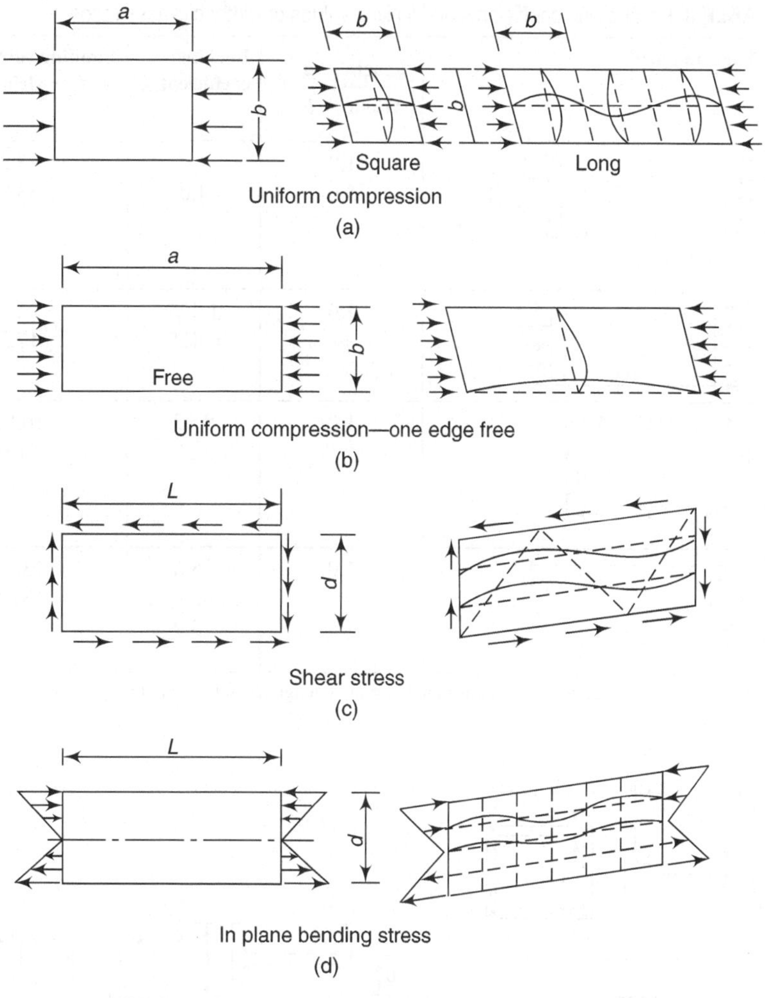


**FIG. 8.20** Elastic buckling of plates with different loading and boundary conditions

## 8.12.1 Elastic Buckling of Plates

Local stability of the compressed elements of a section without transverse stiffeners can be studied by reference to the elastic stability of an infinite plate having width $b$ and thickness $t$, as shown in Fig. 8.19. Assume that the plate is loaded by compressive forces acting on the simply supported sides having width $b$. The critical stress of this plate is given by (Timoshenko & Gere, 1961)

$$f_{cr} = (k\,\pi^2 E)/[12\,(1 - \mu^2)(b/t)^2] \tag{8.45}$$

where $\mu$ is Poisson's ratio of the material, $b/t$ is the width-to-thickness ratio of the plate, $k$ is the buckling coefficient, and $E$ is Young's modulus of rigidity of the material.

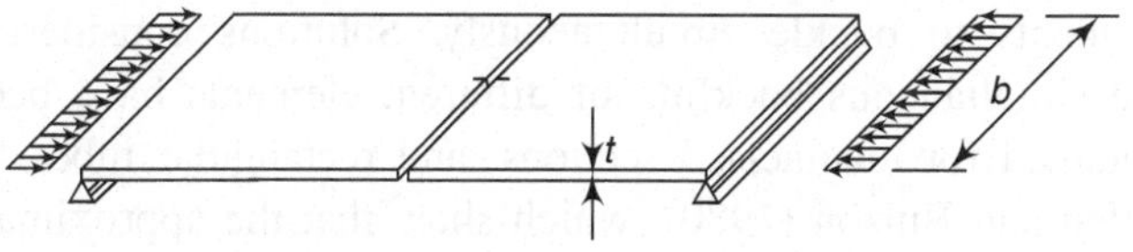


**FIG. 8.19** Infinite plate subjected to compressive forces

The value of the coefficient $k$ depends on the constraints along the non-loaded edges of the plate. The way in which the plates buckle and also the value of their critical buckling stress depend on the edge conditions, dimensions, and loading. The buckled plate patterns of four different plates with different loading and boundary conditions are shown in Fig. 8.20. Some values of $k$ for the four plates are given in Table 8.3. It has to be noted that the edge conditions affect not only the critical buckling stress but also the post buckling behaviour.

On the basis of the boundary conditions, plate elements in structural members can be divided into two categories: unstiffened elements and stiffened elements. *Unstiffened elements* are supported along one edge parallel to the axial stress (e.g., legs of single angles, flanges of beams, and stems of T-sections). *Stiffened elements* are supported along both the edges parallel to the axial stress (e.g., flanges of square and rectangular hollow sections, perforated cover plates, and webs of I-sections and channel sections (see Fig. 8.21).

**TABLE 8.3** Buckling coefficients and limiting values of width/thickness ratios.

| Plate and load | $\frac{\text{Length}}{\text{Width}} = \frac{a}{b}$ | Bucking coefficient, $k$ | Limiting value of $b/t$ for $f_{cr}$ = yield stress |
|---|---|---|---|
| 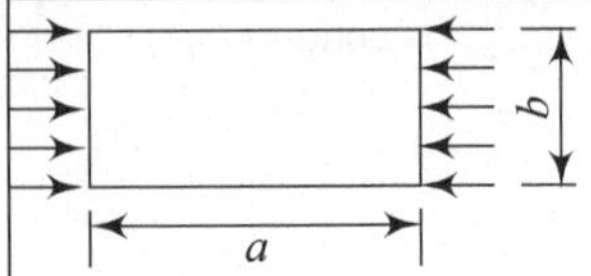 | 1.0<br>5.0 | 4.0<br>4.0 | 53.8<br>53.8 |
| 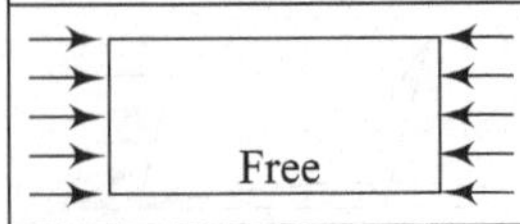 | 1.0<br>∞ | 1.425<br>0.425 | 32.1<br>17.5 |
| 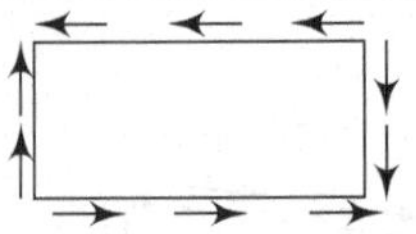 | 1.0<br>∞ | 9.35<br>5.35 | 108.2<br>81.9 |
| 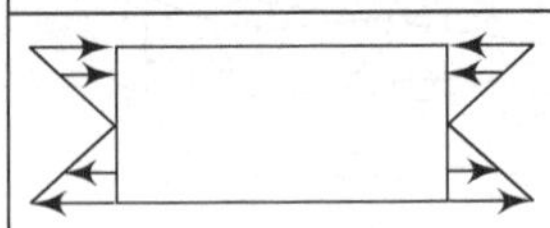 | 1.0<br>∞ | 25.6<br>minimum<br>23.9 | 136.1<br>131.4 |

All edges simply supported except as noted. $a$ = plate length/stiffener spacing on a plate girder.

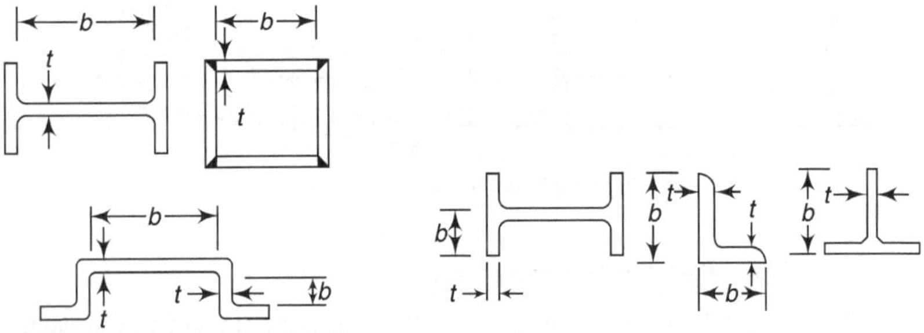

Elements supported along two edges (stiffened elements)
(a)

Elements supported along one edge only (unstiffened elements)
(b)

**FIG. 8.21** Stiffened and unstiffened compression elements

## 8.12.2 Simply Supported Plates

The lowest value of buckling coefficient $k$ for simply supported plates is 4.0 (Timoshenko & Gere 1961). The buckling stress $f_{cr}$ varies inversely as the square of the plate slenderness or width-to-thickness ratio $b/t$, when the dimensionless buckling stress $f_{cr}/f_y$ is plotted against a modified plae slenderness ratio (Trahair et al. 2001)

$$\sqrt{(f_y/f_{cr})} = b/t\sqrt{\{(f_y/E)[12(1-\mu^2)/(\pi^2 k)]\}} \quad (8.46)$$

If the material ceases to be linear elastic at the yield stress $f_y$, the above result is valid for $\sqrt{(f_y/f_{cr})} \geq 1$. This limit is equivalent to a width-to-thickness ratio $b/t$ given by (with $E = 2.0 \times 10^5$ MPa and $\mu = 0.3$)

$$b/t\sqrt{(f_y/250)} = 53.8 \quad (8.47)$$

## 8.12.3 Plates Free Along One Longitudinal Edge

The solution for the elastic buckling stress $f_{cr}$ of a thin flat plate simply supported along both transverse edges and one longitudinal edge and free along the other longitudinal edge [see Fig. 8.20(b)] can still be expressed in the general form of Eqn (8.45). The buckling coefficient $k$ is now approximated by

$$k = 0.425 + (b/a)^2 \quad (8.48)$$

For long plate elements, which are used as flange outstands in many structural steel members, the buckling coefficient $k$ is close to the minimum value of 0.425. In this case the elastic buckling stress (for a steel with $E = 2 \times 10^5$ MPa and $\mu = 0.3$ is equal to the yield stress $f_y$ when

$$(b/t)\sqrt{(f_y/250)} = 17.5 \quad (8.49)$$

## 8.12.4 Plates with Other Boundary Conditions

The edges of flat plates may be fixed or elastically restrained, instead of being simply supported or free. The elastic buckling loads of flat plates with various support conditions have been determined, and many values of buckling coefficient $k$ to be used with Eqn (8.45) are given in Timoshenko and Gere (1961), Bleich (1952), and Allen and Bulson (1980). Some of these coefficients are shown in Fig. 8.22.

Many structural steel compression members are assemblies of flat plate elements which are rigidly connected together along their common boundaries. The local buckling of such an assembly (e.g., welded box girder, plate girder) can be analysed approximately by assuming that the plate elements are simply supported along their common boundaries and free along any unconnected boundary. Thus, the buckling stress of each plate element can be determined using Eqn (8.45) with $k$ = 4.0 or 0.425 as appropriate. This approximation is conservative, as the rigidity of the joints between the plate elements causes all connected plate elements to buckle simultaneously. Solutions considering the simultaneous buckling of different elements have been obtained for channels, I-sections, and rectangular tubes by Allen and Bulson (1980), which show that the approximate solutions as discussed above underestimate the true value of $k$.

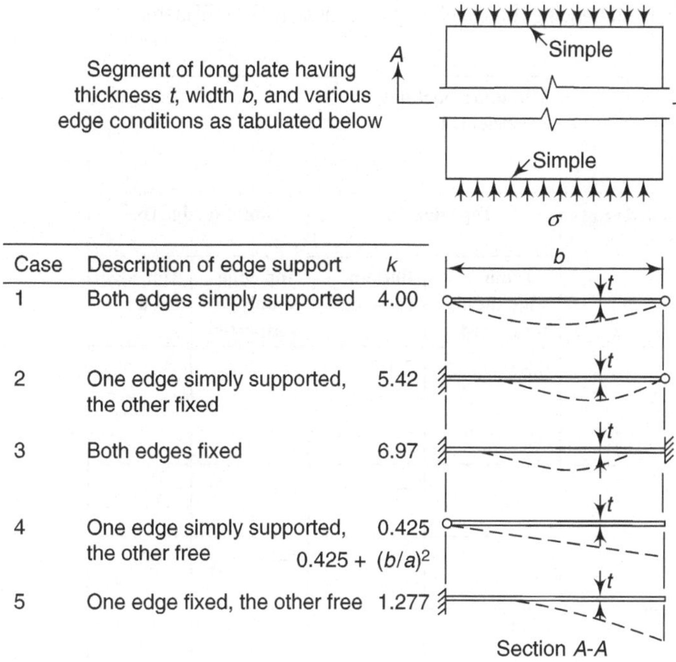


| Case | Description of edge support | k |
|---|---|---|
| 1 | Both edges simply supported | 4.00 |
| 2 | One edge simply supported, the other fixed | 5.42 |
| 3 | Both edges fixed | 6.97 |
| 4 | One edge simply supported, the other free | 0.425<br>0.425 + (b/a)² |
| 5 | One edge fixed, the other free | 1.277 |

**FIG. 8.22** Coefficients $k$ for Eqn (8.45)

## 8.12.5 Plate Elements in Shear

The elastic buckling stress $f_{cr}$ of a thin flat plate of length $L$, depth $d$, and thickness $t$ as shown in Fig. 8.20(c), simply supported along all four edges and loaded by shear stresses distributed uniformly along its edges can be written as (Timoshenko & Gere 1961)

$$f_{cr} = \{\pi^2 E/[12(1-\mu^2)]\}[k/(d/t)^2] \quad (8.50)$$

in which the buckling coefficient $k$ is approximated by

$$k = 5.35 + 4\,(d/L)^2 \quad (8.51)$$

when $L \geq d$, and

$$k = 5.35\,(d/L)^2 + 4 \quad (8.52)$$

when $L \leq d$.

When all edges are fixed, $k = 8.98 + 5.6(d/L)^2$ for $L \geq d$.

The shear stresses in many structural members are transmitted by unstiffened webs and hence the aspect ratio $L/d$ will be large. For these cases, the buckling coefficient approaches a value of 5.35 and the buckling stress has the value

$$f_{cr} = 5.35\pi^2 E/[12(1-\mu^2)(d/t)^2] \quad (8.53)$$

This elastic buckling stress is equal to the yield stress in shear $\tau_y = f_y/\sqrt{3}$ (see Section 2.10.1 of Chapter 2), when (with $E = 2.0 \times 10^5$ MPa and $\mu = 0.3$)

$$(d/t)\sqrt{(f_y/250)} = 81.9 \quad (8.54)$$

It has to be noted that the elastic buckling stress may be significantly increased by using intermediate transverse stiffeners (which will decrease the aspect ratio $L/d$, thus increasing the value of the buckling coefficient), or by using longitudinal stiffeners to decrease the depth–thickness ratio. The stiffener spacing which reduces the aspect ratio to between 0.5 and 2 is found to give best results, since that will result in only one buckle forming in each panel.

## 8.12.6 Plate Elements in Bending

For a thin flat plate of length $L$, width $d$, and thickness $t$ as shown in Fig. 8.20(d), simply supported along all four edges and loaded by bending stress distribution, which varies linearly across its width, the elastic buckling stress can be calculated as (Trahair et al. 2001)

$$f_{cr} = k\pi^2 E/[12(1-\mu^2)\,(d/t)^2] \quad (8.55)$$

where the buckling coefficient $k$ depends on the aspect ratio $L/d$ and the number of buckles along the plate. For long plates, the value of $k$ may be approximated as

$$k = 23.9 \quad (8.56)$$

for which the length of each buckle is approximately $2d/3$. Using this value of $k$, the limiting $d/t$ ratio may obtained as

$$(d/t)\sqrt{(f_y/250)} = 131.4 \quad (8.57)$$

This type of edge force distribution is associated with bending and the results are applicable to the thin webs of plate girders bent in this way. Buckling coefficients for long flat plates under compression and bending for various boundary conditions are given in Table 8.4 (see also Tables 34 to 36 of Bleich, 1952).

It has to be noted that the $b/t$ or $d/t$ limits given in Eqns (8.47), (8.49), (8.54), and (8.57) are applicable for perfectly flat plates and do not account for residual stresses and various imperfections, which lower the proportional limit. Therefore smaller values of $(b/t)$ or $(d/t)$ have to be used in design methods to avoid local buckling of plate elements in sections. However, it is not easy to determine the values to be used in design, since there can be considerable variations in out-of-flatness, residual stress, etc. Hence design codes suggest the limiting values based on judgement and experience.

For plates without residual stress (see Fig. 8.23), three regions must be considered for establishing strength: elastic buckling (Euler hyperbola), yielding (segments $AB$, $A'B$, and $A''B$), and strain hardening (Haaijer & Thürlimann, 1960). If $f_{cr}/f_y$ is defined as $1/\lambda_c^2$, Eqn (8.45) becomes

$$\lambda_c = (b/t)\sqrt{[12 f_y(1-\mu^2)/(\pi^2 Ek)]} \quad (8.58)$$

**TABLE 8.4** Buckling coefficients for long flat plates under compression and bending (Brockenbrough & Johnston 1974; Bijlaard 1957)

| Loading | Ratio of bending stress to uniform compression stress $f_{cb}/f_c$ | Minimum buckling coefficient, $k_c$ | | | | | |
|---|---|---|---|---|---|---|---|
| | | Unloaded edges simply supported | Unloaded edges fixed | Top edge free | | Bottom edge free | |
| | | | | Bottom edge simply supported | Bottom edge fixed | Top edge simply supported | Top edge fixed |
| $a$; $b$; $f_1$, $f_2$; $f_2 = -f_1$ | ∞ (Pure bending) | 23.9 | 39.6 | 0.85 | 2.15 | — | — |
| $f_2 = -2/3 f_1$ | 5.00 | 15.7 | — | — | — | — | — |
| $f_2 = -1/3 f_1$ | 2.00 | 11.0 | — | — | — | — | — |
| $f_2 = 0$ | 1.00 | 7.8 | 13.6 | 0.57 | 1.61 | 1.70 | 5.93 |
| $f_2 = 1/3 f_1$ | 0.50 | 5.8 | — | — | — | — | — |
| $f_1$, $f_2$; $f_2 = f_1$ | 0.0 (Pure compression) | 4.0 | 6.97 | 0.42 | 1.33 | 0.42 | 1.33 |

*Values given are based on plates having loaded edges simply supported and are conservative for plates having loaded edges fixed

It is seen from Fig. 8.23 that plates (curves b and c) achieve the strain-hardening condition at values of $\lambda$ relatively higher than those of columns. The values of $\lambda_c$ at which strain hardening commences (denoted by $\lambda_0$) for Fe 410 grade steel and as given by Haaijer and Thürlimann (1960) are shown in Fig. 8.23 and are as follow:

For columns: $\lambda_0 = 0.173$ ($KL/r = 15.7$)

For fixed flanges: $\lambda_0 = 0.461$ ($b/t = 14.3$)

For fixed webs: $\lambda_0 = 0.579$ ($b/t = 42.0$)

They found that the degree of rotational restraint along the loaded edge (simply supported or fixed) had no effect on $\lambda_0$. Since plates as well as columns contain residual stresses, the true strength should be represented by a transition curve, shown by dotted lines in Fig. 8.23, between the Euler curve and the point at which strain hardening commences. Thus the strength of plates under edge compression may be governed by the following five regions of $\lambda_c$: (1) strain hardening at low values of $\lambda_c$, (2) yielding (at $\lambda_c = 0.46 - 0.58$), (3) inelastic buckling (some fibres yielded and some remaining elastic) in the region represented by the transition curve ($\lambda_c = 0.58 - \sqrt{2}$), (4) elastic bucking represented by the Euler hyperbola at $\lambda_c \approx \sqrt{2}$, and (5) post-buckling strength with stress distribution and large deformation at $\lambda_c > 1.5$. For design purposes, the acceptable range of $\lambda_c$ values is specified by the codes.

Table 8.5 compares some typical limiting values of $b/t$ provided in the Indian (IS 800), British (BS: 5950), American (AISC-360), and Canadian codes (CAN/CSA-S16.1). These

| | | |
|---|---|---|
| Columns | $\lambda_0 = 0.173$ | ($KL/r = 15.7$) |
| Long hinged flanges | $\lambda_0 = 0.455$ | ($b/t = 8.15$) |
| Fixed flanges | $\lambda_0 = 0.461$ | ($b/t = 14.3$) |
| Hinged webs | $\lambda_0 = 0.588$ | ($b/t = 32.3$) |
| Fixed webs | $\lambda_0 = 0.579$ | ($b/t = 42.0$) |

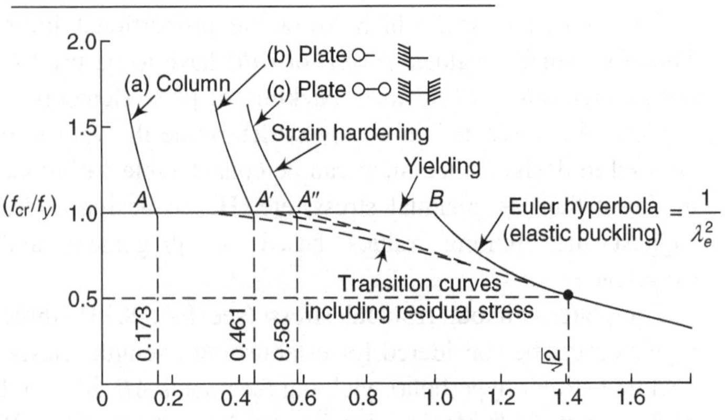


$$\lambda_c = \sqrt{\frac{f_y}{f_{cr}}} = \frac{b}{t}\sqrt{\frac{f_y\,(12)(1-\mu^2)}{\pi^2 Ek}}$$

**FIG. 8.23** Plate buckling compared with column buckling (Haaijer and Thürlimann 1960), Reprinted with permission ASCE

values are based on yield stress and thus form the basis for the semi-compact classification given in Table 8.6. Note that the IS 800 values are derived from BS: 5950 and hence are identical to the BS:5950 values.

**TABLE 8.5** Typical limiting values of $b/t$ for plates in compression for $f_y$ = 250 MPa

| Type of element | Specification | | | | Buckling coefficient $k$ |
|---|---|---|---|---|---|
| | IS 800 | BS: 5950 | AISC-360 | CAN/CSA-S16.1 | |
| Supported on one unloaded edge: | | | | | |
| (a) Single angle | 15.7 | 15.7 | 13 | 12.65 | 0.425 |
| (b) Outstand of compression flange (rolled section) | 15.7 | 15.7 | 16 | 12.65 | 0.70* |
| (c) Stem of T-section | 18.9 | 18.9 | 21 | 21.5 | 1.277 |
| Supported on both unloaded edges: | | | | | |
| (a) Webs | 42 | 42 | 42 | ≈ 42 | 5.0* |
| (b) Cover plates | 42 | 42 | 42 | 42 | 5.0 |
| (c) Perforated plates | — | — | 53 | 53 | 6.97 |

*See Salmon and Johnson (1996) for the details of how these values are obtained.

**TABLE 8.6** Table 8.4 Cross-sectional limits necessary to prevent local buckling in members (see also Table 2 of IS 800:2007)

| Type of element | Method of manufacture | Ratio | Limiting proportions for sections | | |
|---|---|---|---|---|---|
| | | | Plastic | Compact | Semi-compact |
| Outstand element of compression flange | Welded | $b/t_f$ | 8.4 | 9.4 | 13.6 |
| | Rolled | $b/t_f$ | 9.4 | 10.5 | 15.7 |
| Internal element of compression flange | Welded or rolled | $b/t_f$ | 29.3 | 33.5 | 42 |
| Web of an I-, H-, or box section | Welded or rolled: Neutral axis at mid depth | $d/t_w$ | 84 | 105 | 126 |
| | Generally if $r_1$ is negative | $d/t_w$ | 84/ $(1+r_1)$ but ≥ 42 | 105/ $(1+r_1)$ but ≥ 42 | 126/ $(1+2r_2)$ but ≥ 42 |
| | If $r_1$ is positive | $d/t_w$ | | 105/ $(1+1.5r_1)$ but ≥ 42 | |
| Web of channel | | $d/t_w$ | 42 | 42 | 42 |
| Angle (both criteria should be satisfied) | | $b/t$ and $d/t$ | 9.4 | 10.5 | 15.7 |
| Stem of a T-section (rolled or cut from a rolled I- or H-section) | | $d/t$ | 8.4 | 9.4 | 18.9 |

*Notes*:
1. These values should be multiplied by $\varepsilon = (250/f_y)^{0.5}$, if $f_y$ is not equal to 250 MPa
2. The stress ratios $r_1$ and $r_2$ are defined as follows:
   $r_1$ = average axial compressive stress / design compressive stress of web alone
   $r_2$ = average axial compressive stress / design compressive stress of overall section

$r_1$ and $r_2$ are negative if axial stress is tensile.

## 8.12.7 Stress Ratios for Classification

The stress ratios $r_1$ and $r_2$ used in Table 8.6 as well as later in Table 8.7 should be determined as follows (BS 5950-1:2000):

1. For I- or H-sections with equal flanges

$$r_1 = \frac{P_c}{d_w t_w f_{yw}} \quad \text{but } -1 < r_1 \le 1$$

$$r_2 = \frac{P_c}{A_g f_{yw}}$$

2. For I- or H-sections with unequal flanges

$$r_1 = \frac{P_c}{d_w t_w f_{yw}} + \frac{(b_{ft}t_{ft} - b_{fc}t_{fc}) f_{yf}}{d_w t_w f_{yw}} \quad \text{but } -1 < r_1 \le 1$$

$$r_2 = \frac{f_1 + f_2}{2 f_{yw}}$$

3. For RHS or welded box sections with equal flanges

$$r_1 = \frac{P_c}{2 d_w t_w f_{yw}} \quad \text{but } -1 < r_1 \le 1$$

$$r_2 = \frac{P_c}{A_g f_{yw}}$$

where, $A_g$ is the gross cross-sectional area, $b_{fc}$ is the width of compression flange, $b_{ft}$ is the width of tension flange, $d_w$ is the depth of web, $P_c$ is the axial compression (negative for tension), $f_1$ is the maximum compressive stress in the web (see Fig. 8.24), $f_2$ is the minimum compressive stress in the web (negative for tension), $f_{yf}$ is the design strength of the flanges, $f_{yw}$ is the design strength of the web (but $f_{yw} \le f_{yf}$), $t_{fc}$ is the thickness of compression flange, $t_{ft}$ is the thickness of tension flange, and $t_w$ is the thickness of web.

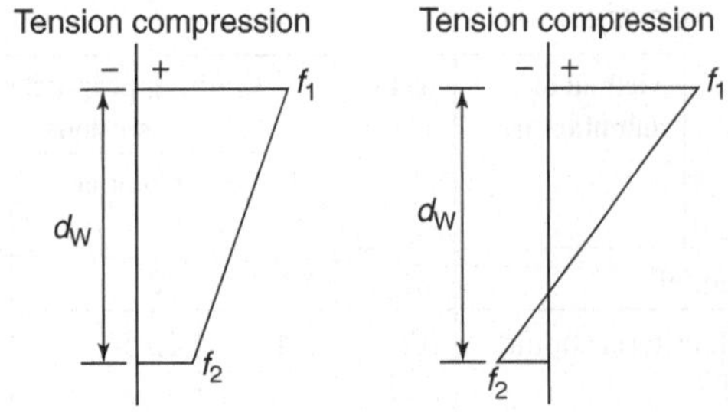


**FIG. 8.24** Stress ratios for semi-compact web

## 8.13 CROSS SECTION CLASSIFICATION

Determining the resistance (strength) of structural steel components requires the designer to consider first the cross-sectional behaviour and second the overall member behaviour. Whether in the elastic or inelastic material range, cross-sectional resistance and rotation capacity are limited by the effects of local buckling.

In the code (IS 800), cross sections are placed into four behavioural classes depending upon the material yield strength, the width-to-thickness ratios of the individual components (e.g., webs and flanges) within the cross section, and the loading arrangement. The four classes of sections are defined as follows.

**Plastic or class 1** Cross sections which can develop plastic hinges and have the rotation capacity required for the failure of the structure by the formation of a plastic mechanism (only these sections are used in plastic analysis and design).

**Compact or class 2** Cross sections which can develop their plastic moment resistance, but have inadequate plastic hinge rotation capacity because of local buckling.

**Semi-compact or class 3** Cross sections in which the elastically calculated stress in the extreme compression fibre of the steel member, assuming an elastic distribution of stresses, can reach the yield strength, but local buckling is liable to prevent the development of the plastic moment resistance.

**Slender or class 4** Cross sections in which local buckling will occur even before the attainment of yield stress in one or more parts of the cross section. In such cases, the effective sections for design are calculated by deducting the width of the compression plate element in excess of the semi-compact section limit.

The moment–rotation characteristics of these four classes of cross sections are shown in Fig. 8.25. As seen from this figure, class 1 (plastic) cross sections are fully effective under pure compression, and capable of reaching and maintaining their full plastic moment in bending and hence used in plastic design. These sections will exhibit sufficient ductility ($\theta_2 > 6\,\theta_1$, where $\theta_1$ is the rotation at the onset of plasticity and $\theta_2$ is the lower limit of rotation for treatment as a plastic section).

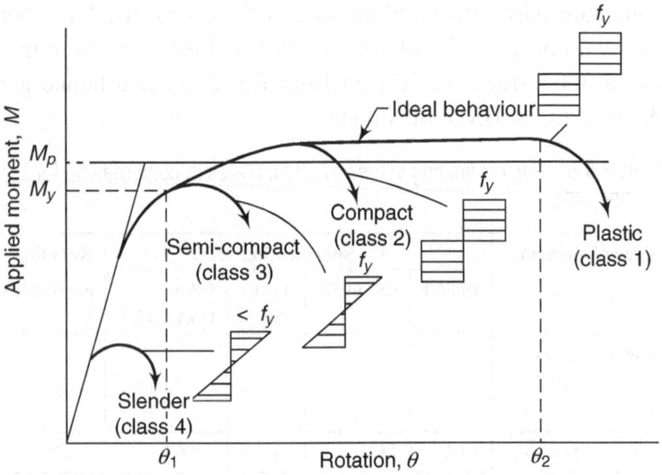


**FIG. 8.25** Moment–rotation behaviour of the four classes of cross sections as defined by IS 800

Class 2 (compact) cross sections have lower deformation capacity, but are also fully effective in pure compression and are capable of reaching their full plastic moment in bending (they have ductility in the range $\theta_1 < \theta_2 < 6\theta_1$). Class 3 (semi-compact) cross sections are fully effective in pure compression but local buckling prevents the attainment of the full plastic moment in bending; bending moment resistance in these cross sections is limited to the (elastic) yield moment only. For class 4 (slender) cross sections, the local or lateral buckling of the member occurs in the elastic range. An effective cross section is therefore defined based on the width-to-thickness ratios of the individual plate elements and this is used to determine the resistance of the cross section. The majority of the hot-rolled cross sections belong to class 1, 2, or 3, and hence their resistances may be based on the gross cross section properties obtained from section tables (IS 808).

For cold-formed cross sections, which predominantly have open cross sections (e.g., channel) and light gauge materials (with thickness less than 6 mm), the design will seldom be based on the gross cross section properties. The class 4 cross sections contain slender elements that are susceptible to local buckling in the elastic range of material behaviour. For the design of slender compression elements considering the strength beyond the elastic local buckling of elements (such as those of cold-formed sections), the reader should refer to IS 801 (see also Section 8.14.2). The effective cross sections are to be calculated in the case of slender welded plate girders (see Example 8.23).

The design moment capacity $M_d$ of each of the four classes of sections defined above may be calculated as follows (of course a partial factor of safety for materials as per IS 800 should also be applied to the design moment capacity):

$$\text{Plastic:} \quad M_d = Z_p f_y \tag{8.59a}$$

$$\text{Compact:} \quad M_d = Z_p f_y \tag{8.59b}$$

Semi-compact: $M_d = Z_e f_y$ (8.59c)

Slender: $M_d < Z_e f_y$ (8.59d)

where $Z_p$ and $Z_e$ are the plastic and elastic section moduli, respectively.

Each compressed (or partially compressed) element is assessed individually against the limiting width-to-thickness ratios for the class 1, 2, and 3 elements defined in Table 8.6. If an element fails to meet the class 3 limits, it should be considered as class 4. The different elements of a cross section can belong to different classes. In such cases, the section is classified based on the least favourable classification. A member with one or more slender elements has its effective area reduced to

$$A_{\text{eff}} = \Sigma(b_{\text{eff}} t) \quad (8.60)$$

in which the effective width of any slender plate element is determined from

$$B_{\text{eff}} = \lambda_{L3} t \varepsilon \leq b \quad (8.61)$$

where $\lambda_{L3}$ is the limiting width-to-thickness ratio of the semi-compact section. The limiting width-to-thickness ratios are modified by a factor ε, which takes into account the yield strength of the material (for a circular hollow section, the width-to-thickness ratios are modified by $\varepsilon^2$). The factor $\varepsilon$ is defined by

$$\varepsilon = (250/f_y)^{0.5} \quad (8.62)$$

where $f_y$ is the nominal yield strength of the steel. Note that increasing the nominal yield strength results in stricter classification limits.

The terms internal element, outstanding elements, etc., as used in Table 8.6, are described as follows.

**Internal elements** These are elements attached along both longitudinal edges to other elements or to longitudinal stiffeners connected at suitable intervals to transverse stiffeners (e.g., a web of I-sections and a flange and a web of box sections).

**Outstanding elements or outstands** These are elements attached along only one of the longitudinal edges to an adjacent element, the other edge being free to displace out of the plane (e.g., the flange overhang of an I-section, stem of a T-section, and leg of an angle section).

**Tapered elements** These elements may be treated as flat elements having the average thickness defined in IS 808.

## Compound Elements in Built-up Section

In the case of a compound element consisting of two or more elements bolted or welded together as shown in Fig. 8.26, the following width-to-thickness ratios should be considered:

1. Outstand width of compound element $b_e$ to its own thickness
2. The internal width of each added plate between the lines of welds or fasteners connecting it to the original section compared to its own thickness $(b_i/t_i)$
3. Any outstand of the added plates beyond the line of welds or fasteners connecting it to original section compared to its own thickness.

Note that stricter limits are imposed for welded elements, in recognition of the weakening effect of the more severe residual stresses present.

## Hollow Section Members

A circular hollow section member has no buckling effects when its diameter-to-thickness ratio $D/t$ satisfies

$$D/(t\varepsilon^2) \leq 88 \quad (8.63)$$

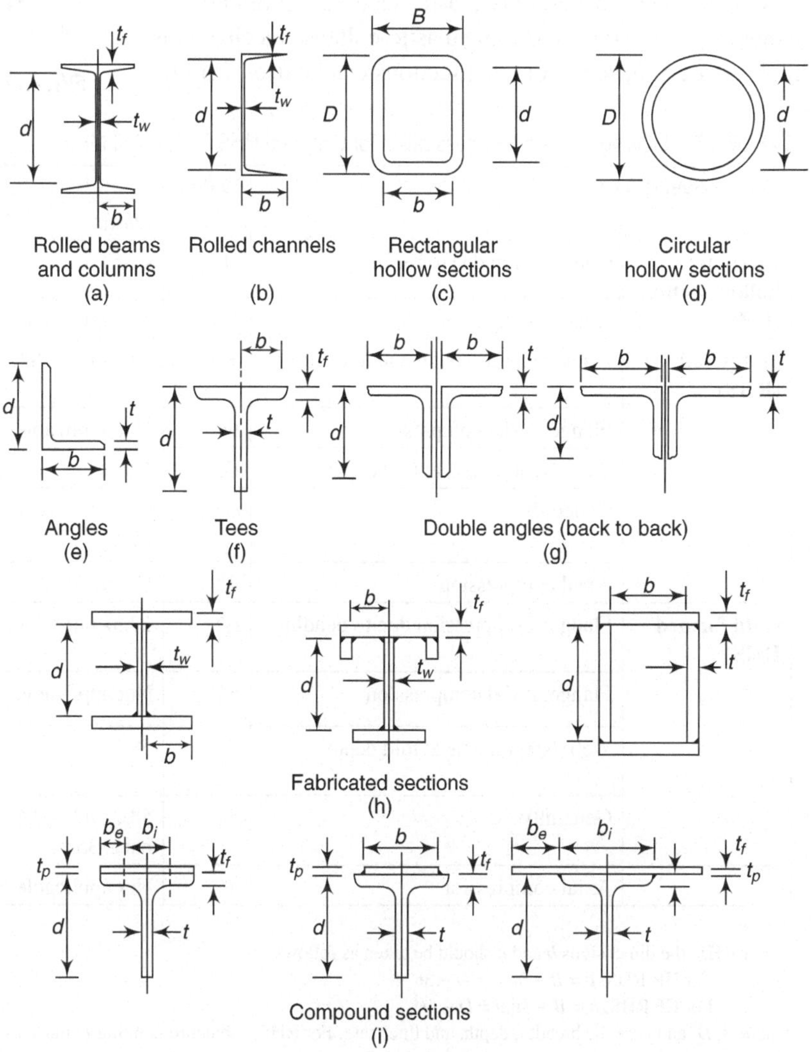


**FIG. 8.26** Definition of $b$, $d$, $t_f$, and $t_w$ as used in Table 8.6

A circular hollow section member which does not satisfy Eqn (8.63) is considered a slender member. Its effective area is reduced to

$$A_{\text{eff}} = A_g\{88/[D/(t\varepsilon^2)]\}^{0.5} \tag{8.64}$$

The limiting width-to-thickness ratios of hollow rectangular sections are not specified in IS 800. However, for these members, the values shown in Table 8.7, may be used which is based on BS: 5950.

The relationship between the design moment capacity $M_d$ and the compression flange slenderness $b/t$ indicating the $\lambda$ limits is shown in Fig. 8.27. In this figure, the value of $M_d$ for the semi-compact section is conservatively taken as $M_y$. Note that in the above classifications, it is assumed that the web slenderness $d/t$ is such that its buckling before yielding is prevented. When a beam is subjected to bending, the entire web may not be in uniform compression, and if the neutral axis lies at mid-depth, half of the web will actually experience tension. In such cases, the slenderness limits for webs are somewhat relaxed.

As mentioned earlier, only plastic sections should be used in indeterminate frames forming plastic-collapse mechanisms. In elastic design, semi-compact sections can be used with the understanding that the maximum stress reached will be $M_y$. Slender sections also have stiffness problems and are not preferable for hot-rolled structural steelwork. However, they are used extensively in cold-formed members. Plate girders are usually designed based on the tension field approach (which takes into account the post-buckling behaviour of plates) to achieve economy (see Chapter 11 for the details).

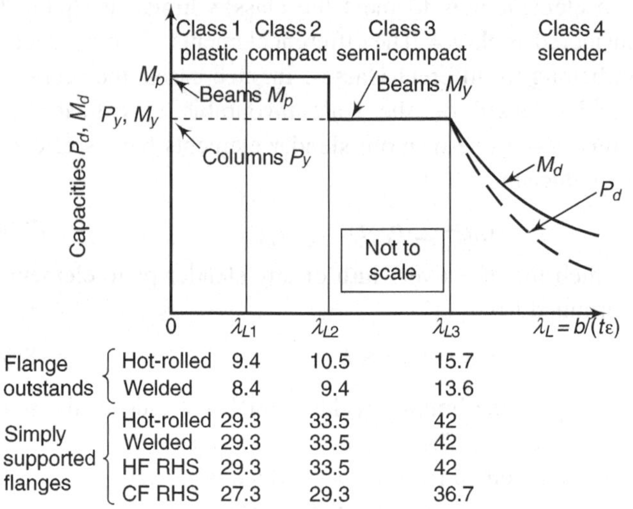


**FIG. 8.27** Moment capacities of fully braced beams and columns

**TABLE 8.7** Limiting width-to-thickness ratios for CHS and RHS for $f_y$ = 250 MPa

| Compression element | | Ratio | Limiting value | | |
|---|---|---|---|---|---|
| | | | Plastic | Compact | Semi-compact |
| **Circular hollow section (CHS)** | Compression due to bending | $D/t$ | $42\varepsilon^2$ | $52\varepsilon^2$ | $146\varepsilon^2$ |
| | Axial compression | $D/t$ | Not applicable | | $88\varepsilon^2$ |
| **Hot rolled (RHS)** | Flange: Compression due to bending | $b/t$ | $29.3\varepsilon$ but $\le 84\varepsilon - d/t$ | $33.5\varepsilon$ but $\le 65\varepsilon - 0.5d/t$ | $42\varepsilon$ |
| | Flange: Axial compression | $b/t$ | Not applicable | | |
| | Web: Neutral axis at mid depth | $d/t$ | $67\varepsilon$ | $84\varepsilon$ | $126\varepsilon$ |
| | Generally | $d/t$ | $67\varepsilon/(1 + 0.6r_1)$ but $\ge 42\varepsilon$ | $84\varepsilon/(1 + r_1)$ but $\ge 42\varepsilon$ | $126\varepsilon/(1 + 2r_2)$ but $\ge 42\varepsilon$ |
| | Axial compression | $d/t$ | Not applicable | | |
| **Cold formed RHS** | Flange: Compression due to bending | $b/t$ | $27.3\varepsilon$ but $\le 75.6\varepsilon - d/t$ | $29.3\varepsilon$ but $\le 56.7\varepsilon - 0.5d/t$ | $36.7\varepsilon$ |
| | Flange: Axial compression | $b/t$ | Not applicable | | |
| | Web: Neutral axis at mid depth | $d/t$ | $58.7\varepsilon$ | $73.4\varepsilon$ | $110.1\varepsilon$ |
| | Generally | $d/t$ | $58.7\varepsilon/(1 + 0.6r_1)$ but $\ge 35\varepsilon$ | $73.4\varepsilon/(1 + r_1)$ but $\ge 35\varepsilon$ | $110.1\varepsilon/(1 + 2r_2)$ but $\ge 35\varepsilon$ |
| | Axial compression | $d/t$ | Not applicable | | |

*Notes*:

1. For RHS, the dimensions $b$ and $d$ should be taken as follows:
   For HR RHS: $b = B - 3t$; $d = D - 3t$
   For CF RHS: $b = B - 5t$; $d = D - 5t$

where $B$, $D$, and $t$ are the breadth, depth, and thickness. For RHS, subject to bending $D$ must always be the web dimension

2. $\varepsilon = (250/f_y)^{0.5}$

Since the classification into plastic, compact, etc., is based on bending, it cannot be used for a compression member. In compression members, a criterion is used to determine whether the member is slender or not (see Chapter 9). However, in practice, compact or plastic sections are used for compression members, since they have more stiffness than semi-compact or slender members.

## 8.14 BEHAVIOUR AND ULTIMATE STRENGTH OF PLATES

Unlike columns, plates continue to support loads even after buckling. The post buckling strength has been considered in most of the codes, by using an effective width concept. This difference in behaviour of plates is explained in this section by comparing the behaviour of plates with that of columns.

### 8.14.1 Behaviour of Plates

Though the behaviour of columns is discussed in the next chapter, it may be interesting to compare the behaviour of a column and a plate. In the case of an ideal column, as the axial load is increased, the lateral displacement remains zero until the attainment of the critical buckling load called the *Euler load* (see Chapter 9 for more information on column buckling). If the axial load versus lateral displacement is plotted, we will get a line along the load axis up to $P = P_{cr}$ (Fig. 8.28). This is called the 'fundamental path'. When the axial load reaches the Euler buckling load, the lateral displacement increases indefinitely at constant load. This is the 'secondary path', which bifurcates from the fundamental path at the buckling load. The secondary path for columns represents *neutral equilibrium*. A smooth transition from the stable path to the neutral equilibrium path will occur for practical columns, having initial imperfections [as shown by the dashed curves in Fig. 8.28(a)].

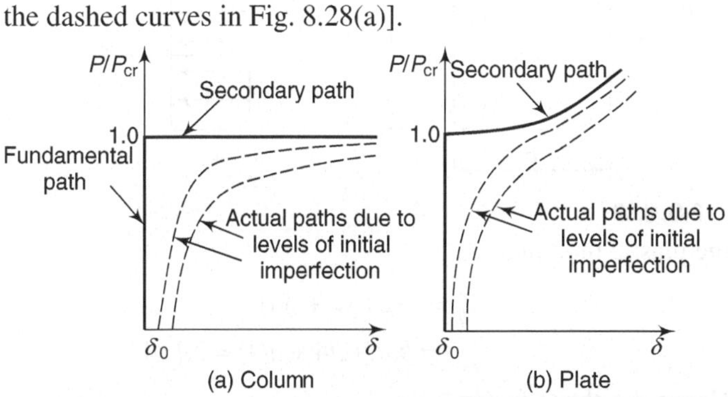


**FIG. 8.28** Load versus out-of-plane displacement curves

The fundamental path for a perfectly flat plate is similar to that for an ideal column. At the critical buckling load, this path bifurcates into a secondary path as shown in Fig. 8.28(b). However in this case, the secondary path shows that the plate can carry loads higher than the elastic critical load. Unlike columns, the secondary path for a plate is stable. Therefore, the elastic buckling of a plate need not be considered a collapse. However, plates having one edge free and other edges (outstands) simply supported have very little post-buckling strength.

The actual failure load of columns and plates is reached when the yielding spreads from the supported edges, triggering collapse; thereafter, the unloading occurs.

Local buckling causes loss of stiffness and redistribution of the stresses. Uniform edge compression in the longitudinal direction results in non-uniform stress distribution after buckling as shown in Fig. 8.29, and the buckled plate derives most of its stiffness from the longitudinal edge supports.

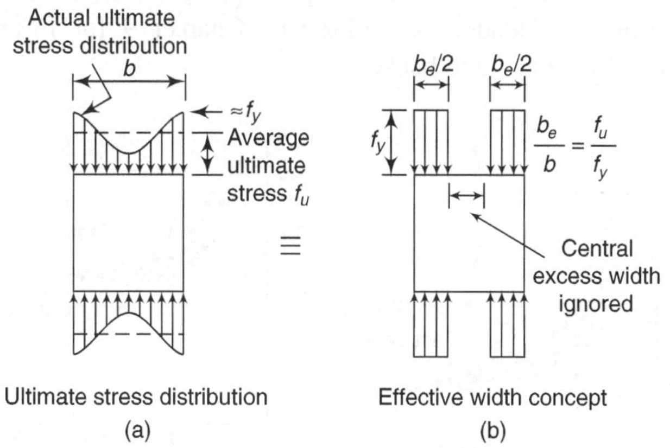


**FIG. 8.29** Effective width concept for plates with simply supported edges

Elastic post-buckling stiffness is measured in terms of the apparent modulus of elasticity, $E^*$ (the ratio of the average stress carried by the plate to the average strain). Figure 8.30 shows the post-buckling stiffness of plates having simply supported edges. The values of $E^*$ for long plates ($a \gg b$) for some typical longitudinal edge conditions have been given by Allen and Bulson (1980). In most of the cases, the value of $E^*$ is in the range 0.408–0.5$E$ and may be taken to be approximately equal to 0.5$E$ (Allen & Bulson 1980; Galambos 1998). In the case of practical plates, there is a gradual loss of stiffness as shown by the dashed curves in Fig. 8.30.

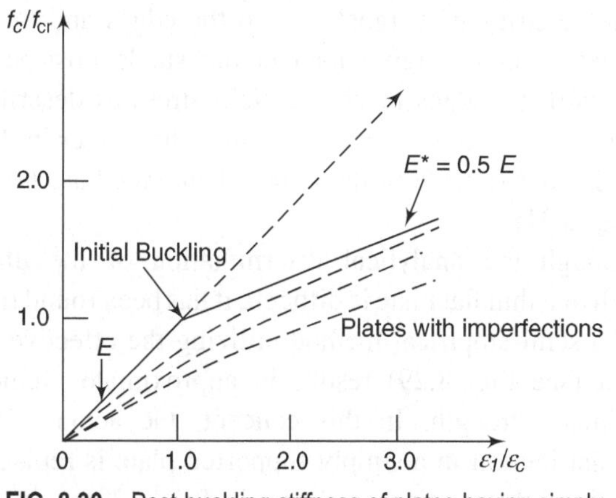


**FIG. 8.30** Post-buckling stiffness of plates having simply supported edges

## 8.14.2 Post-buckling Strength of Plates

Plate strength curves can be constructed in the same way as column strength curves (Fig. 8.31). In the case of ideal columns with low slenderness (i.e., stocky columns) failure is due to 'squashing' at the yield stress. However, for slender columns, failure occurs by buckling at or near the Euler load. Tests on practical columns indicate that failure always occurs below the failure load of an ideal column of the same slenderness (Galambos 1998). For stocky columns, the yield stress provides an upper bound and for slender columns the buckling stress provides an upper bound. Also the scatter in the test results is considerable particularly in the range of intermediate slenderness ratios (see Chapter 9 for further discussion on column buckling).

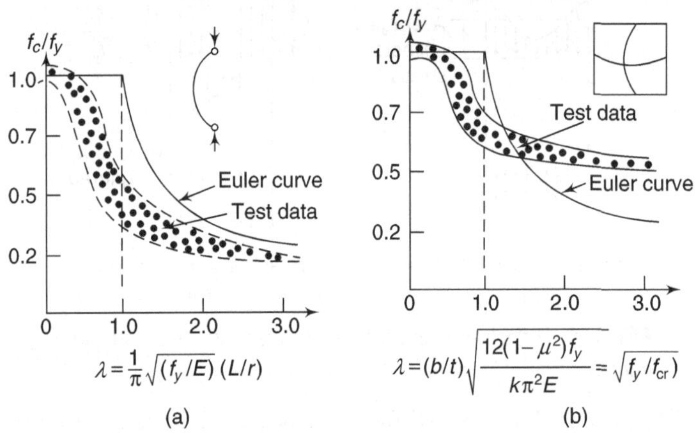


**FIG. 8.31** Column and plate strength curves

In the case of a flat plate simply supported on all four sides, we can expect failure by squasing if the $b/t$ ratio is less than the limiting value given by Eqn (8.55). Similarly, for $b/t$ ratios larger than the limiting value, failure of the plate after buckling at the critical buckling stress can be expected. However, tests on practical plates indicate that for large $b/t$ ratios, the failure stress is substantially greater than the critical buckling stress. This is due to the *post-buckling behaviour*, which is unique to plates. Thus, after initial buckling, the load on a plate from the middle strips gets transferred to the edges and the plate continues to carry higher load in the stable post-buckling range, until the edges reach the yield stress as described in the previous sections. As with columns, the scatter in the test results is considerable in the range of intermediate $b/t$ ratios (see Fig. 8.31).

Although the analytical determination of the ultimate strength of a thin flat plate is difficult, it has been found that the use of a semi-empirical method utilizing the effective width concept (see Fig. 8.29) results in approximate predictions of ultimate strength. In this concept, the actual ultimate stress distribution in a simply supported plate is replaced by a simplified distribution as shown in Fig. 8.29(b), in which the central portion of the plate is ignored and the remaining effective width $b_e$ is assumed to carry the yield stress $f_y$. From experiments on real plates with initial curvatures and residual stresses, it has been found that the effective width for hot-rolled and welded plates could be obtained from the following equation:

$$b_e/b = \alpha\sqrt{(f_{cr}/f_y)} \qquad (8.65)$$

where $\alpha$ is the parameter that indicates the inclusions of the influence of initial curvatures and residual stresses. A value of 0.65 has been suggested for $\alpha$ (Trahair et al. 2001).

Winter (1947) suggested the following formula for effective width based on tests on cold-formed steel sections:

$$(b_e/b) = (f_{cr}/f_y)[1 - 0.22\sqrt{(f_{cr}/f_y)}] \qquad (8.66)$$

This formula was first adopted in the AISC specification for light gauge cold-formed steel sections in 1946 and subsequently in the specifications of many other countries with minor modifications.

## EXAMPLES

**EXAMPLE 8.1:** Determine the shape factor for a rectangular beam of width $b$ and depth $d$.

**SOLUTION:**

For a rectangular section (Fig. 8.32), the elastic modulus

$$Z_e = (bh^3/12)/(d/2) = bd^2/6$$

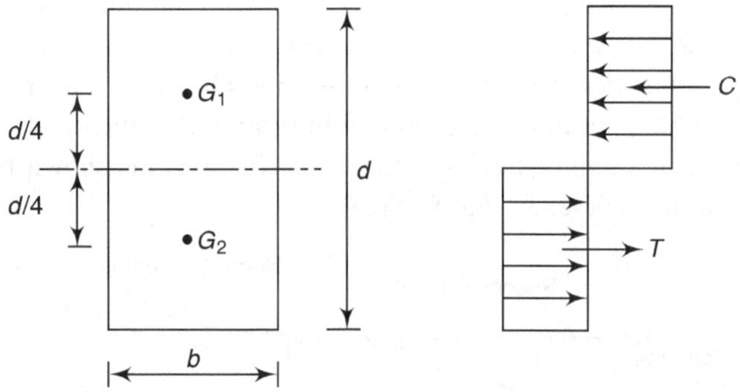


**FIG. 8.32**

the plastic modulus

$$Z_p = A/2\,(\bar{y}_1 + \bar{y}_2)$$
$$= bd/2(d/4 + d/4) = bd^2/4$$

Hence the shape factor

$$v = Z_p/Z_e = (bd^2/4)/(bd^2/6) = 1.5$$

**EXAMPLE 8.2:** Determine the plastic moment capacity and shape factor of the I-section shown in Fig. 8.33. This section is ISMB 400 with the root radius omitted. Assume $f_y = 250$ MPa.

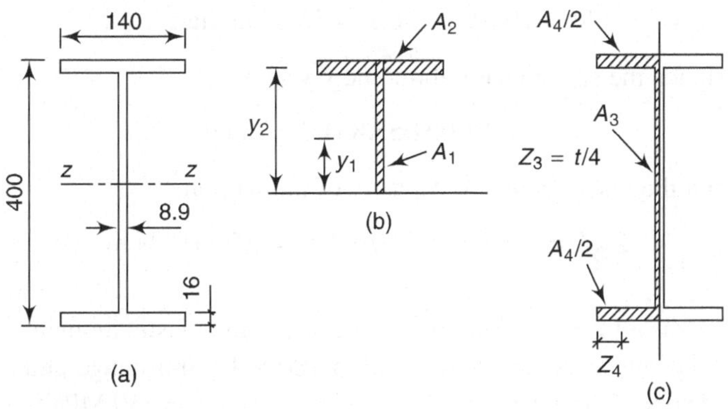


**FIG. 8.33**

**SOLUTION:**

**Step 1:** *Plastic section modulus about major axes*

To determine the plastic section modulus about the *z-z* axis, divide the section into areas $A_1$ and $A_2$ as shown in Fig. 8.33, where

$$A_1 = (D/2)t = (400/2) \times 8.9 = 1780 \text{ mm}^2$$
$$A_2 = (B - t)T = (140 - 8.9) \times 16 = 2097.6 \text{ mm}^2$$

and

$$\bar{y}_1 = D/4 = 400/4 = 100 \text{ mm}$$

$$\bar{y}_2 = (D/2 - T/2) = (400/2 - 16/2) = 192 \text{ mm}$$

The plastic section modulus

$$Z_p = 2(A_1 \bar{y}_1 + A_2 \bar{y}_2)$$
$$= 2(1780 \times 100 + 2097.6 \times 192)$$
$$= 1.1615 \times 10^6 \text{ mm}^3$$

The value given in Annex H of the code is $1176.18 \times 10^3$ mm$^3$, which is slightly greater (1.27%) because of the additional material at the root radius.

$I_{zz} = 2(140 \times 16^3/12) + 2 \times 140 \times 16 \times (200 - 8)^2 + (400 - 16 \times 2)^3 \times 8.9/12$

$= 202.20 \times 10^6$mm$^4$

$Z_e = 202.20 \times 10^6/200 = 1011.04 \times 10^3$ mm$^3$(compared to $1020.0 \times 10^3$mm$^3$in Annex H of the code)

Hence the shape factor

$Z_p/Z_e = 1.1615 \times 10^6/(1011.04 \times 10^3)$

$= 1.1488$ (compared to 1.1498 given in Annex H)

Note that this value together with the stress diagrams shown in Fig. 8.1 explain the use of the limit for moment capacity of $M_d \leq 1.2\ Z_e f_y/\gamma_{m0}$, which prevents plasticity at service load (clause 8.2.1.2)

**Step2:** *Plastic section modulus about minor axis*

Similarly, to determine the plastic section modulus about the *y-y* axis, divide the section into areas $A_3$ and $A_4$ as shown in Fig. 8.33(c), where

$$A_3 = (D - 2T)t/2 = (400 - 2 \times 16)8.9/2 = 1637.6 \text{ mm}^2$$
$$A_4 = 2\,(B/2)T = 2 \times 140/2 \times 16 = 2240 \text{ mm}^2$$

and

$$\bar{z}_3 = t/4 = 8.9/4 = 2.225 \text{ mm}$$
$$\bar{z}_4 = B/4 = 140/4 = 35 \text{ mm}$$

The plastic section modulus

$$Z_{py} = 2(A_3\,\bar{z}_3 + A_4\,\bar{z}_4)$$
$$= 2(1637.6 \times 2.225 + 2240 \times 35) = 0.1641 \times 10^6 \text{ mm}^3$$

$I_{yy} = 2 \times 16 \times 140^3/12 + (400 - 32) \times 8.9^3/12 = 7338.95 \times 10^3$ mm$^4$

$Z_{ey} = 7338.95 \times 10^3/70 = 104.842 \times 10^3$ mm$^3$ (compared to $88.9 \times 10^3$mm$^3$ as given in IS 808-1989)

Hence the shape factor about the *y-y* axis

$$\nu = Z_{py}/Z_{ey} = 164.1 \times 10^3/104.842 \times 10^3$$
$$= 1.565$$

and the plastic moment capacity

$$Z_{py}f_y = 164.1 \times 10^3 \times 250 \times 10^{-6} = 41.025 \text{ kNm}$$

**EXAMPLE 8.3:** Determine the plastic section moduli about the *z-z* and *y-y* axes and the plastic moment capacity of the channel section shown in Fig. 8.34(*a*). Assume $f_y = 250$ MPa.

**SOLUTION:**

**Step 1:** *Plastic section modulus about major axis*

To determine the plastic section modulus about the *z-z* axis, divide the section into areas $A_1$ and $A_2$ as shown in Fig. 8.34(b).

$$A_1 = (D/2)t = 300/2 \times 7.6 = 1140 \text{ mm}^2$$
$$A_2 = (B - t)T = (90 - 7.6) \times 13.6 = 1120.64 \text{ mm}^2$$
$$\bar{y}_1 = D/4 = 300/4 = 75 \text{ mm}$$

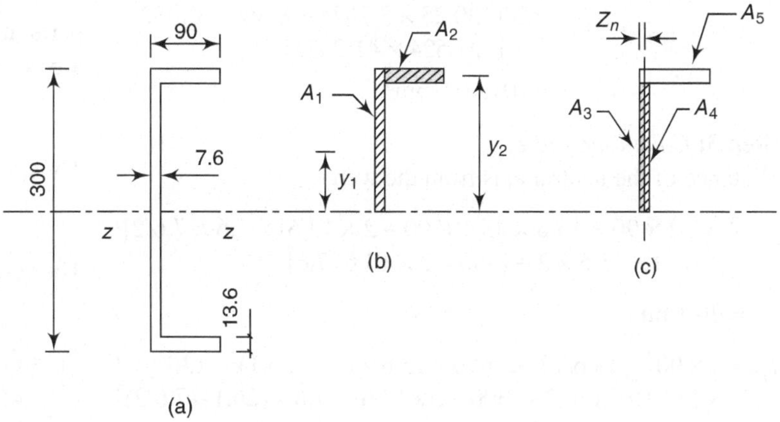


**FIG. 8.34**

$$\bar{y}_2 = D/2 - T/2 = 300/2 - 13.6/2 = 143.2 \text{ mm}$$

The plastic section modulus

$$Z_{pz} = 2(A_1\ \bar{y}_1 + A_2\ \bar{y}_2)$$

$$= 2(1140 \times 75 + 1120.64 \times 143.2) = 491.95 \times 10^3 \text{ mm}^3$$

$$I_{zz} = (300 - 2 \times 13.6)^3 \times 7.6/12 + 2 \times 90 \times 13.6^3/12 + 2 \times 90 \times 13.6 \times (150 - 13.6/2)^2$$

$$= 63094.77 \times 10^3 \text{ mm}^4$$

$$Z_{ez} = 63094.77 \times 10^3/150 = 420.631 \times 10^3 \text{ mm}^3$$

Shape factor about the $z$-$z$ axis

$$= Z_{pz}/Z_{ez}$$

$$= 491.95 \times 10^3/(420.631 \times 10^3) = 1.169$$

**Step 2:** *Plastic section modulus about minor axis*
Similarly, to determine the plastic section modulus of the $y$-$y$ axis, divide the section into areas $A_3$, $A_4$, and $A_5$ as shown in Fig. 8.34(c). Since the neutral axis divides the total area into two equal parts,

$$z_n = (A_1 + A_2)/D = (1140 + 1120.64)/300 = 7.535 \text{ mm}$$

Now the areas

$$A_3 = (D/2)z_n = (300/2)\ 7.535 = 1130.25 \text{ mm}^2$$

$$A_4 = (D/2 - T)\ (t - z_n) = (300/2 - 13.6)\ (7.6 - 7.535) = 8.866 \text{ mm}^2$$

$$A_5 = (B - z_n)T = (90 - 7.535) \times 13.6 = 1121.524 \text{ mm}^2$$

The distances

$$\bar{z}_3 = z_n/2 = 7.535/2 = 3.7675 \text{ mm}$$

$$\bar{z}_4 = (t - z_n)/2 = (7.6 - 7.535)/2 = 0.0325 \text{ mm}$$

$$\bar{z}_5 = (B - z_n)/2 = (90 - 7.535)/2 = 41.2325 \text{ mm}$$

The plastic section modulus about the $y$-$y$ axis,

$$Z_{py} = 2(A_3\ \bar{z}_3 + A_4\ \bar{z}_4 + A_5\bar{z}_5)$$

$$= 2(1130.25 \times 3.7675 + 8.866 \times 0.0325 + 1121.524 \times 41.2325)$$

$$= 101{,}004 \text{ mm}^3$$

**Step 3:** *Calculation of $Z_{pe}$*
Distance of the neutral axis from the web,

$$\bar{z} = [2 \times 90 \times 13.6 \times 45 + (300 - 2 \times 13.6) \times 7.6 \times 7.6/2]/[90 \times 13.6 \times 2 + (300 - 2 \times 13.6)\ 7.6]$$

$$= 26.1 \text{ mm}$$

$$I_{yy} = 2 \times 90^3 \times 13.6/12 + 2 \times 90 \times 13.6 \times (45 - 26.1)^2 + (300 - 2 \times 13.6) \times 7.6^3/12 + (300 - 2 \times 13.6) \times 7.6 \times (26.1 - 7.6/2)^2$$

$$= 3567.85 \times 10^3 \text{ mm}^4$$

$$Z_{ey} = 3567.85 \times 10^3/(90 - 26.1) = 55{,}834.9 \text{ mm}^3$$

Hence the shape factor about the $y$-$y$ axis

$$= 101{,}004/55{,}834.9 = 1.809$$

and the plastic moment capacity of the section,

$$Z_{py}f_y = 101{,}004 \times 250 \times 10^{-6} = 25.251 \text{ kN m}$$

**EXAMPLE 8.4:** Find the shape factor and plastic moment capacity of a tee section with a 100 × 12 mm flange and 180 × 10 mm web (see Fig. 8.35). Assume $f_y$ = 250 MPa.

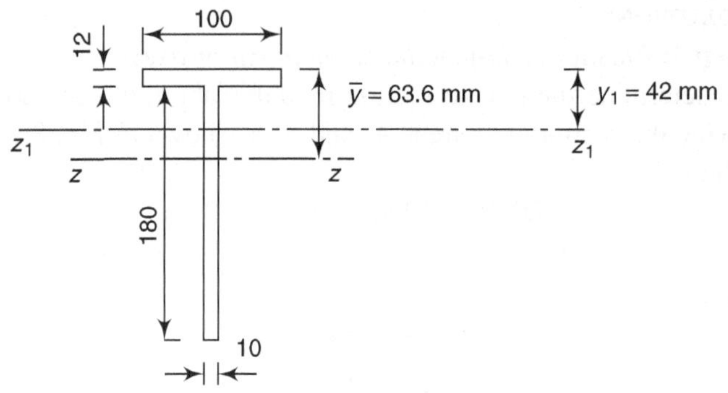


**FIG. 8.35**

**SOLUTION:**
**Step 1:** Let the distance of the neutral axis from the top be

$$\bar{y} = [100 \times 12 \times 12/2 + 180 \times 10 \times (90 + 12)]/(100 \times 12 + 180 \times 10) = 63.6 \text{ mm}$$

**Step 2:** Moment of inertia about the $z$-$z$ axis,

$$I_{zz} = 100 \times 12^3/12 + 100 \times 12 \times (63.6 - 6)^2 + 10 \times 180^3/12 + 10 \times 180 \times (90 - 63.6)^2$$

$$= 10.11 \times 10^6 \text{ mm}^4$$

**Step 3:** Elastic section modulus,

$$Z_{zz} = I_{zz}/y_{max} = 10.11 \times 10^6/(192 - 63.6) = 78{,}740 \text{ mm}^3$$

Let $z_1$-$z_1$ be the axis that divides the total area into two equal parts at a distance $y_1$ from the top (see Fig. 8.35). The total area of the section

$$A = 100 \times 12 + 180 \times 10 = 3000 \text{ mm}^2$$

Thus,

$$100 \times 12 + (y_1 - 12) \times 10 = 1500$$

Therefore

$$y_1 = 42 \text{ mm}$$

$\bar{y}_1$ = the distance of the CG of the area above the equal-area axis

$$= [100 \times 12 \times (42 - 12/2) + 10 \times 30/2]/(100 \times 12 + 30 \times 10)$$

$= 31.8$ mm

$\bar{y}_2$ = the distance of the CG of the area below the equal-area axis

$= [(180 + 12) - 42]/2 = 75$ mm

The plastic section modulus

$$Z_p = (3000/2)(31.8 + 75) = 160{,}200 \text{ mm}^3$$

The shape factor

$$Z_p/Z_e = 160{,}200/78{,}740 = 2.0345$$

The plastic moment capacity

$$Z_p f_y = 160{,}200 \times 250 \times 10^{-6} = 40.05 \text{ kN m}$$

**EXAMPLE 8.5:** Determine the shape factor for a triangular section of base $b$ and height $h$ as shown in Fig. 8.36.

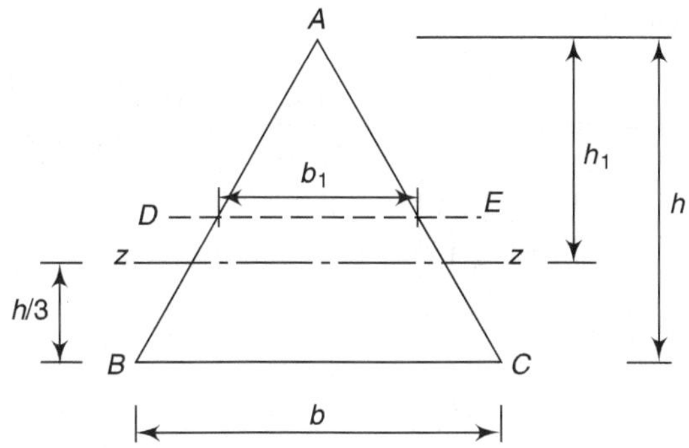


**FIG. 8.36**

**SOLUTION:**

Moment of inertia

$$I_{zz} = bh^3/36$$

Elastic section modulus

$$Z_e = (bh^3/36)/(2h/3)$$
$$= bh^2/24$$

Let $DE$ be the equal-area axis (see Fig. 8.36). Then,

$$0.5\, b_1 h_1 = 0.5(bh/2) \text{ or } b_1 h_1 = bh/2 \quad (1)$$

From similar triangles $ADE$ and $ABC$

$$h_1/b_1 = h/b \quad (2)$$

From Eqns (1) and (2), we get

$$h_1 = h/\sqrt{2} \text{ and } b_1 = b/\sqrt{2}$$

$$A = bh/2$$

$$\bar{y}_1 = h_1/3 = h/(3\sqrt{2})$$

$$\bar{y}_2 = [(h - h_1)/3]\,[(b_1 + 2b)/(b_1 + b)]$$
$$= [(h - h/\sqrt{2})/3]\,[2b + b/\sqrt{2})\,(b + b/\sqrt{2})]$$
$$= [h\,(\sqrt{2}-1)/3\sqrt{2})]\,(2\sqrt{2} + 1)/(\sqrt{2} + 1)$$
$$= 0.1548\, h$$

$$Z_p = A/2(\bar{y}_1 + \bar{y}_2) = bh/4\,[h/(3\sqrt{2}) + 0.1548h]$$
$$= 0.0976\, bh^2$$

The shape factor

$$\nu = Z_p/Z_e = 0.0976\, bh^2/(bh^2/24) = 2.343$$

The shape factor about the $y$-$y$ axis will be different and the calculation of the same is left as an exercise to the reader.

**EXAMPLE 8.6:** Determine the shape factor for a hollow tubular section with internal diameter as $d$ and external diameter as $D$ as shown in Fig. 8.37.

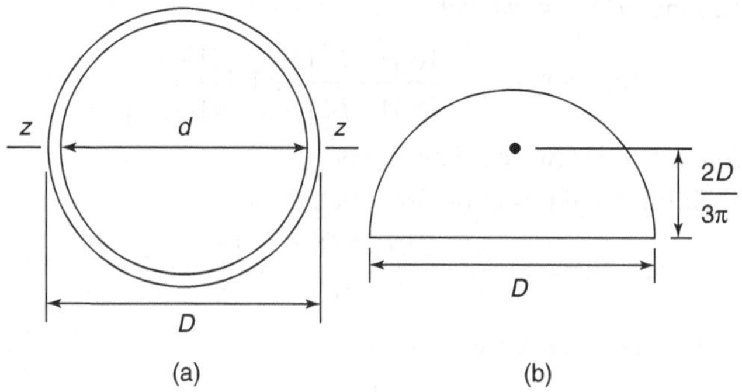


**FIG. 8.37**

**SOLUTION:**

Moment of inertia of the section about $z$-$z$ axis

$$I_z = \frac{\pi}{64}(D^4 - d^4)$$

Elastic modulus of the section, $Z_{ez} = \dfrac{I_z}{D/2} = \dfrac{\pi}{32D}(D^4 - d^4)$

Due to the symmetry of the cross section, the plastic neutral axis coincides with the elastic neutral axis. That is, it is located at the centre of the section.

Plastic section modulus, $Z_{pz} = \dfrac{A}{2}(\bar{y}_1 + \bar{y}_1)$

where $A$ is the area of the section $= (\pi/4)(D^2 - d^2)$

Hence the compression and tension areas are $A_c = A_t = (\pi/8)(D^2 - d^2)$

The distance of centre of gravity of a semicircular area from its diagonal chord is $2D/3\pi$ (see Fig. 8.37b). Hence the centre of gravity of the compression and tension areas from the plastic neutral axis is

$$y = \bar{y}_1 = \bar{y}_2 = \frac{\dfrac{\pi D^2}{8}\left(\dfrac{2D}{3\pi}\right) - \dfrac{\pi d^2}{8}\left(\dfrac{2d}{3\pi}\right)}{\pi(D^2 - d^2)/8} = \frac{2}{3\pi}\frac{\left(D^3 - d^3\right)}{\left(D^2 - d^2\right)}$$

$$Z_{pz} = \frac{A}{2}\left(\bar{y}_1 + \bar{y}_1\right) = A\bar{y} = \frac{\pi(D^2 - d^2)}{4}\left[\frac{2}{3\pi}\frac{\left(D^3 - d^3\right)}{\left(D^2 - d^2\right)}\right]$$

$$= \frac{\left(D^3 - d^3\right)}{6}$$

Shape factor,

$$v = \frac{Z_{pz}}{Z_{ez}} = \frac{\left(D^3 - d^3\right)}{6}\frac{32D}{\pi(D^4 - d^4)} = \frac{32D}{6\pi}\frac{\left(D^3 - d^3\right)}{(D^4 - d^4)}$$

$$= \frac{16}{3\pi}\frac{\left[1 - \left(\dfrac{d}{D}\right)^3\right]}{\left[1 - \left(\dfrac{d}{D}\right)^4\right]}$$

Denoting $d/D$ as $k$, we get

$$\text{Shape factor} = \frac{16}{3\pi}\frac{[1 - k^3]}{[1 - k^4]} = 1.7\frac{[1 - k^3]}{[1 - k^4]}$$

Let us consider two extreme cases.

(a) Solid circular section for which $d = 0$

$$\text{Hence, shape factor} = \frac{16}{3\pi}\frac{D(D^3)}{D^4} = \frac{16}{3\pi} = 1.7$$

(b) Thin circular section $d \to D$

$$\text{Shape factor} = \frac{16}{3\pi} \times \frac{3}{4} = \frac{4}{\pi} = 1.27$$

**Example 8.7:** Determine the shape factor for a hollow square section with internal side as $b$ and external side as $B$ as shown in Fig. 8.38.

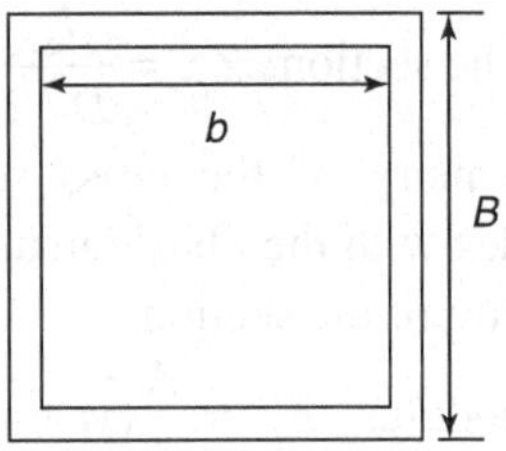


**FIG. 8.38**

**Solution:**

Due to the symmetry of the cross section, the plastic neutral axis coincides with the elastic neutral axis. That is, it is located at the centre of the cross section.

Elastic modulus of the section, $Z_{ez} = \dfrac{I_z}{B/2} = \left(\dfrac{B^4}{12} - \dfrac{b^4}{12}\right)\dfrac{2}{B}$

$$= \frac{B^4 - b^4}{6B}$$

Plastic section modulus, $Z_{pz} = \dfrac{A}{2}(\bar{y}_1 + \bar{y}_2)$

where $A$ is the area of the section $= (B^2 - b^2)$

Hence the compression and tension areas

$$A_c = A_t = (B^2 - b^2)\ /2$$

Hence the centre of gravity of the compression and tension areas from the plastic neutral axis is

$$y = \bar{y}_1 = \bar{y}_2 = \frac{\dfrac{B^2}{2}\left(\dfrac{B}{4}\right) - \dfrac{b^2}{2}\left(\dfrac{b}{4}\right)}{(B^2 - b^2)/2} = \frac{1}{4}\frac{\left(B^3 - b^3\right)}{\left(B^2 - b^2\right)}$$

$$Z_{pz} = \frac{A}{2}\left(\bar{y}_1 + \bar{y}_1\right) = A\bar{y} = (B^2 - b^2)\left[\frac{1}{4}\frac{\left(B^3 - b^3\right)}{\left(B^2 - b^2\right)}\right] = \frac{\left(B^3 - b^3\right)}{4}$$

Shape factor,

$$v = \frac{Z_{pz}}{Z_{ez}} = \frac{\left(B^3 - b^3\right)}{4}\frac{6B}{(B^4 - b^4)} = \frac{3B}{2}\frac{\left(B^3 - b^3\right)}{(B^4 - b^4)}$$

$$= \frac{3}{2}\frac{\left[1 - \left(\dfrac{b}{B}\right)^3\right]}{\left[1 - \left(\dfrac{b}{B}\right)^4\right]}$$

Denoting $b/B$ as $k$, we get

$$\text{Shape factor} = \frac{3}{2}\frac{[1 - k^3]}{[1 - k^4]} = 1.5\frac{[1 - k^3]}{[1 - k^4]}$$

Let us consider two extreme cases.

(a) Solid square section for which $b = 0$
Hence, shape factor = 1.5

(b) Thin hollow square section $b \to B$

$$\text{Shape factor} = \frac{3}{2} \times \frac{3}{4} = \frac{9}{8} = 1.125$$

**Example 8.8:** Find out the collapse load of a simply supported beam subjected to a concentrated load at mid-span as shown in Fig. 8.39.

**Solution:**

**Step 1:** *Kinematic method*

Since it is a determinate structure, one hinge is sufficient to make the beam a mechanism. Let $\Delta$ be the vertical virtual deflection at the load point. The slope of the beam at the supports

$$\theta = \Delta/(L/2) \text{ or } \Delta = \theta L/2$$

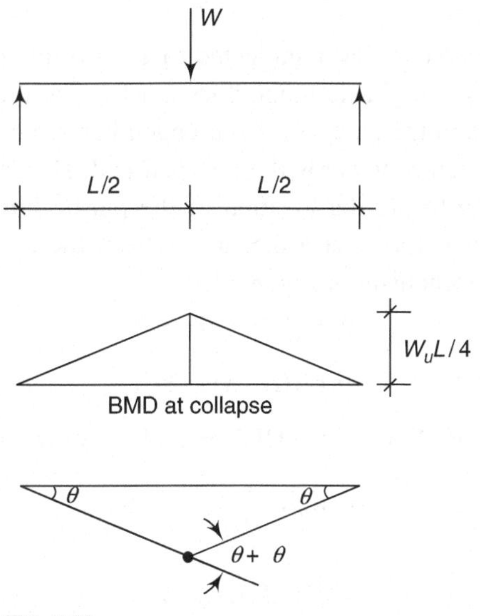


FIG. 8.39

Work done by the plastic hinge = $2M_p\theta$; work done by the load = $W_u\theta L/2$. At collapse, work by hinge = work by load

$\therefore$ $$2M_p\theta = W_u\theta L/2$$

$\therefore$ $$W_u = 4M_p/L$$

**Step 2:** *Static method*

$$M_p = W_uL/4 \text{ or } W_u = 4M_p/L$$

**Example 8.9:** Determine the collapse load of a fixed beam with a concentrated load at mid-span as shown in Fig. 8.40.

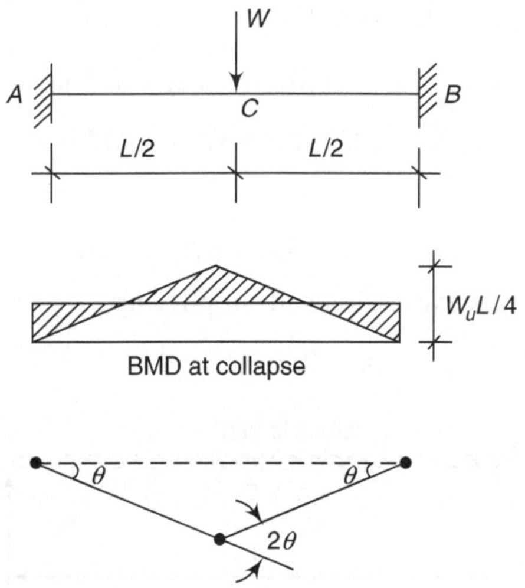


FIG. 8.40

**Solution:**

Note that three hinges are to be formed at $A$, $B$, and $C$ for the beam to be converted into a mechanism.

**Step 1:** *Kinematic method*

Work done by the plastic hinge = $M_p\theta + M_p(2\theta) + M_p\theta = 4M_p\theta$; work done by the displacement of the load = $W_u(\theta L/2)$. Equating these expressions, we get

$$4M_p\theta = W_u(\theta L/2)$$

or

$$W_u = 8M_p/L$$

**Step 2:** *Static method*

$$W_uL/4 = 2M_p \text{ or } W_u = 8M_p/L$$

**Example 8.10:** Determine the collapse load of the fixed beam as shown in Fig. 8.41.

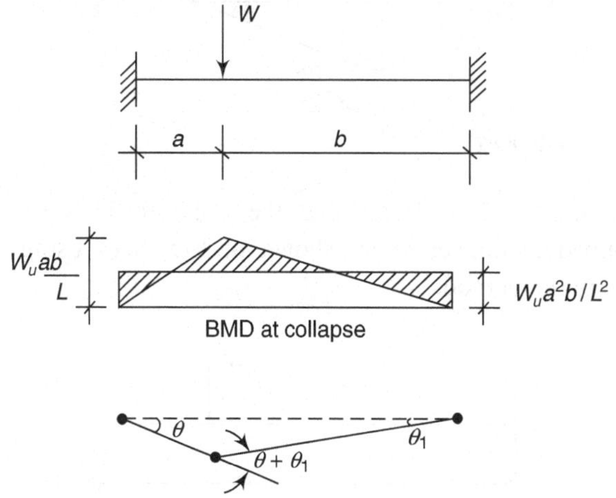


FIG. 8.41

**Solution:**

**Step 1:** *Kinematic method*

Work done by the plastic hinge

$$= M_p\theta + M_p\theta_1 + M_p(\theta + \theta_1)$$
$$= M_p\theta + M_p(a/b)\theta + M_p[\theta + (a/b)\theta]$$
$$= 2M_p\theta(1 + a/b)$$
$$= 2M_p\theta(a + b)/b = 2M_pL\theta/b$$

Work done by the displacement of the load = $W_ua\theta$. Equating the two expressions,
we get

$$W_ua\theta = 2M_pL\theta/b \text{ or } W_u = 2M_pL/ab$$

**Step 2:** *Static method*

$$W_uab/L = 2M_p \text{ or } W_u = 2M_pL/ab$$

**Example 8.11:** Determine the collapse load of the fixed beam shown in Fig. 8.42 using the kinematic method.

**Solution:**

*Kinematic method*

Work done by the plastic hinges = $M_p\theta + M_p\theta + M_p(2\theta) = 4M_p\theta$; work done by the displacement of the load = $W_u(1/2)(L/2\theta) = W_uL\theta/4$. Equating the two, we get

$$4M_p\theta = W_uL\theta/4 \text{ or } W_u = 16M_p/L$$

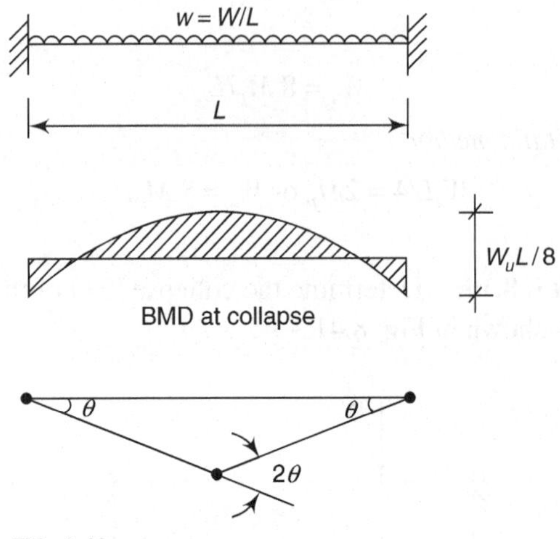


FIG. 8.42

**EXAMPLE 8.12:** Determine the collapse load of the propped cantilever beam shown in Fig. 8.43 using the kinematic method.

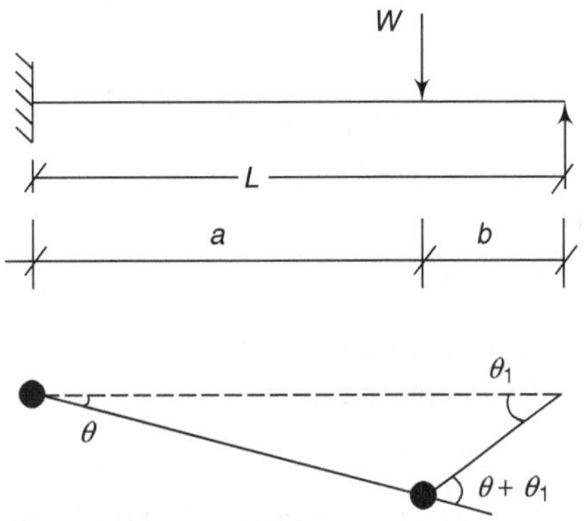


FIG. 8.43

**SOLUTION:**

Since the order of determinacy of this beam is 1, two hinges are required to generate a mechanism. Work done by the plastic hinges

$$= M_p\theta + M_p(\theta + \theta_1)$$
$$= M_p\theta + M_p[\theta + (a/b)\theta]$$
$$= M_p\theta(1 + 1 + a/b)$$
$$= M_p\theta[b + (a + b)]/b$$
$$= M_p\theta(L + b)/b$$

Work done by the displacement of two loads = $W_u a\theta$. Equating the two, we get

$$W_u a\theta = (L + b)M_p\theta/b$$
$$W_u = M_p(L + b)/ab$$

**EXAMPLE 8.13:** Determine the collapse load of a propped cantilever beam carrying uniformly distributed load as shown in Fig. 8.44.

**SOLUTION:**

For a propped cantilever subjected to a uniformly distributed load (UDL), the plastic hinge does not form at the centre of the span, as in the case of a fixed-ended beam, but at a point *C* near the centre towards the propped end. Hence, it is first necessary to locate the location of the plastic hinge. Let us assume that it forms at a distance *x* from the propped end. From the mechanism of Fig. 8.44,

$$\Delta = (L - x)\ \theta = x\theta_1$$
$$\theta_1 = [(L - x)/x]\theta$$

External work done by the UDL = load × average deflection

$$= (w_uL)\ \frac{1}{2}\ (L - x)\theta$$
$$= \frac{1}{2} w_uL(L - x)\theta$$

Internal work done by the hinges

$$= M_p\theta + M_p(\theta + \theta_1)$$
$$= M_p\theta + M_p\{\theta + [(L - x)/x]\theta\}$$
$$= M_p\theta[1 + 1 + (L - x)/x]$$
$$= M_p\theta(L + x)/x$$

By the principle of virtual work, external work done = internal work done:

$$\frac{1}{2}\ w_uL(L - x)\theta = M_p\theta(L + x)/x$$
$$M_p = (w_uL/2)x(L + x)/(L + x) \qquad (1)$$

For the maximum value of $M_p$, $dM_p/dx = 0$. Thus,

$$(w_uL/2)\ [(L + x)\ (L - 2x) - x(L - x)]/(L + x)^2 = 0$$

or

$$(L + x)\ (L - 2x) - x(L - x) = 0$$

Expanding this expression and simplifying,

$$x^2 + 2xL - L^2 = 0$$

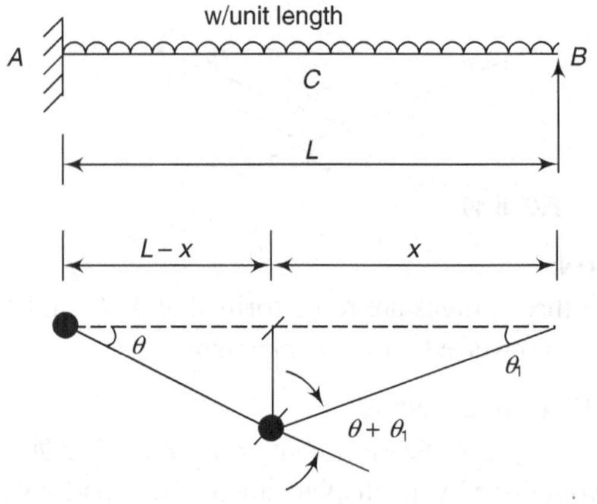


FIG. 8.44

Hence

$$x = \left[-2L \pm \sqrt{(2L^2 + 4L^2)}\right]/2 = -L \pm \sqrt{2}L = (\sqrt{2} - 1)L$$

$$= 0.414L \quad (2)$$

Substituting this value of $x$ in Eqn (1), we get

$$M_p = (w_uL/2) \times 0.414L(L - 0.414L)/(L + 0.414L)$$

$$= (w_uL/2)\ (0.414L \times 0.586/1.414)$$

$$= w_uL^2/11.656$$

Hence

$$w_u = 11.656M_p/L^2$$

If $W_u$ is the total uniformly distributed load = $w_uL$, then

$$W_u = 11.656M_p/L$$

**EXAMPLE 8.14:** A beam with fixed ends is subjected to a uniformly distributed load of $w$ on half of its length as shown in Fig. 8.45. Determine the collapse load if the beam has uniform cross section.

**SOLUTION:**

The beam shown in Fig. 8.45 may develop plastic hinges at points A, C, and D at a distance $x$ from support C.

From the mechanism shown in Fig. 8.44,

$$\Delta = (L - x)\theta = x\theta_1$$

Hence $\theta_1 = \dfrac{L - x}{x}\theta$

External work done = load × area of mechanism diagram under the load

$$= \frac{w_u}{L/2}\left[\left(\frac{x\theta_1 + \theta L/2}{2}\right)\left(\frac{L}{2} - x\right) + \left(\frac{x\theta_1 + 0}{2}\right)x\right]$$

$$= \frac{2w_u}{L}\left[\frac{1}{2}\left(x\frac{L - x}{x} + \frac{L}{2}\right)\left(\frac{L}{2} - x\right)\theta + \frac{x^2}{2}\left(\frac{L - x}{x}\right)\theta\right]$$

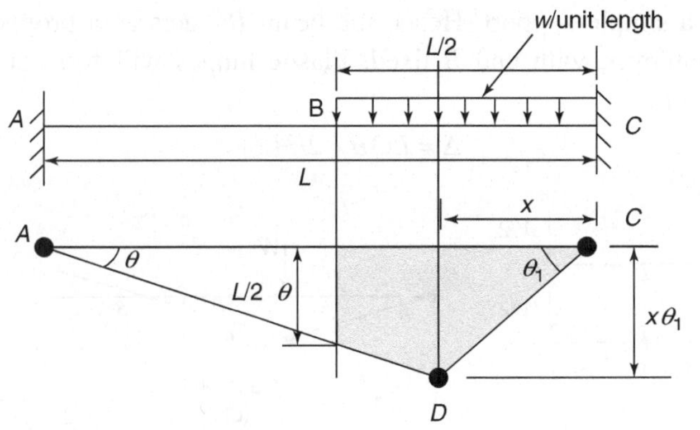


FIG. 8.45

$$= \frac{w_u}{L}\left[\left(L - x + \frac{L}{2}\right)\left(\frac{L}{2} - x\right) + x(L - x)\right]\theta$$

$$= \frac{w_u}{L}\left[\frac{3}{4}L^2 - \frac{3}{2}Lx - \frac{Lx}{2} + x^2 + Lx - x^2\right]\theta$$

$$= \frac{w_u}{L}\left[\frac{3}{4}L^2 - Lx\right]\theta = w_u\left[\frac{3}{4}L - x\right]\theta$$

Internal work done = $M_p\theta + M_p(\theta + \theta_1) + M_p\theta_1$

$$= 2M_p(\theta + \theta_1) = 2M_p\left(1 + \frac{L - x}{x}\right)\theta = 2M_p\left(\frac{L}{x}\right)\theta$$

By the principle of virtual work,

External work done = internal work done

$$w_u\left[\frac{3}{4}L - x\right]\theta = 2M_p\left(\frac{L}{x}\right)\theta$$

Hence, $M_p = \dfrac{w_u}{2L}\left[\dfrac{3}{4}Lx - x^2\right]$

For the maximum value of $M_p$, $dM_p/dx = 0$

Thus, $\dfrac{w_u}{2L}\left[\dfrac{3}{4}L - 2x\right] = 0$

Solving, $x = 3/8\text{L} = 0.375\text{L}$

Hence, $M_p = \dfrac{w_u}{2L}\left[\dfrac{3}{4}L \times 0.375L - (0.375L)^2\right] = 0.703\ w_u/L$

or $w_u = 14.22\ M_p/L$

**EXAMPLE 8.15:** Determine the collapse load of a fixed-end beam subjected to a load of $W$ at one-third span as shown in Fig. 8.46.

**SOLUTION:**

To convert the beam into a mechanism, three plastic hinges are required. A plastic hinge can form at the point where the cross section changes. Hence we have two possible mechanisms.

**Case 1:** Two plastic hinges at the supports and one plastic hinge at the point where the cross section changes [see Fig. 8.46(b)]:

External work done = load × deflection

$$= W_u(L/3)\theta$$

Internal work done = moment × rotation

$$= 1.5\ M_p\theta + M_p(\theta + \theta) + M_p\theta$$

$$= 4.5M_p\theta$$

By the principle of virtual work,

$$4.5M_p\theta = W_u(L/3)\theta$$

or

$$W_u = 13.5M_p/L$$

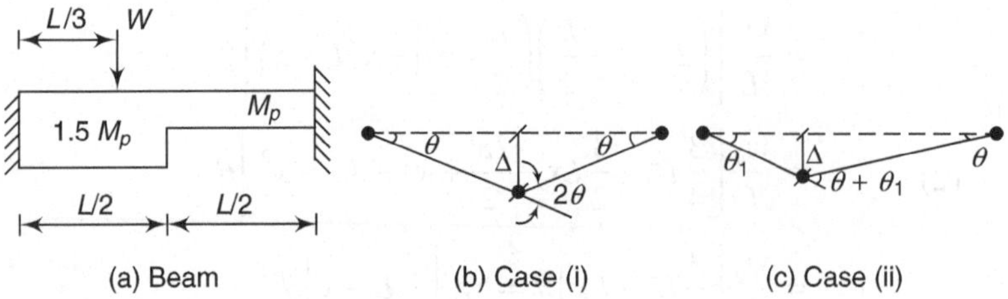


**FIG. 8.46**

**Case 2:** Two plastic hinges at the supports and one just below the concentrated load:

$$\Delta = (L/3)\ \theta_1 = (2/3)L\theta \text{ or } \theta_1 = 2\theta$$

External work done $= W_u(L/3)\theta_1$

$= W_u(L/3)2\theta = (2/3)W_uL\theta$

Internal work done $= 1.5M_p\theta_1 + 1.5\ M_p(\theta + \theta_1) + M_p\theta$

$= 1.5M_p(2\theta) + 1.5\ M_p(\theta + 2\theta) + M_p\theta$

$= 8.5M_p\theta$

By the principle of virtual work,

$$(2/3W_u\theta) = 8.5M_p\theta$$

Hence

$$W_u = 12.75M_p/L$$

The collapse load will be the smaller of the above two values. Hence the collapse load of the beam is $12.75M_p/L$.

**EXAMPLE 8.16:** Find the collapse load of the beam of uniform cross section shown in Fig. 8.47.

**SOLUTION:**

*Span CD*

Since the end $D$ is free, a plastic hinge will form at $C$.

External work done = load × deflection

$= (W/2) \times (L/3)\theta$

$= (WL/6)\theta$

Internal work done $= M_p\theta$

By the principle of virtual work,

$$(WL/6)\theta = M_p\theta$$

or

$$W_u = 6\ M_p/L$$

*Span AC*

Since the end $C$ is propped, hinges will form at $A$ and $B$ only. The end $D$ will be lifted up when a mechanism is formed and negative work is done by the load acting at $D$. Thus,

External work done $= W(L/2)\theta + (-W/2)(L/3)\theta$

$= WL\theta[(1/2) - (1/6)] = (WL/3)\theta$

Internal work done $= M_p\theta + M_p(\theta + \theta) = 3M_p\theta$

By the principle of virtual work,

$$(WL/3)\ \theta = 3M_p\theta$$

or

$$W_u = 9M_p/L$$

Hence the collapse load for the beam is $6M_p/L$.

**EXAMPLE 8.17:** Determine the collapse load for a continuous beam having a uniform cross section as shown in Fig. 8.48.

**SOLUTION:**

In the case of a continuous beam, the collapse load due to the beam mechanism forming in each span is calculated and the minimum collapse load is taken as the collapse load of the beam.

*Span AB*

The end A is a simple support and the end $B$ is an intermediate support. Hence the beam $AB$ will act as a propped cantilever with $B$ as the fixed end. Plastic hinges will be formed at $B$ and $D$. External work done $= 2W_uL/2\theta = W_uL\theta$; Internal work done $= M_p(\theta + \theta) + M_p\theta = 3M_p\theta$. By the principle of virtual work,

$$W_uL\theta = 3M_p\theta$$

Hence

$$W_u = 3M_p/L$$

*Span BC*

Here, the end $B$ is an intermediate support and the end $C$ is a simple support. Hence the beam $BC$ acts as a propped cantilever, with end $B$ fixed. Plastic hinges will form at $B$ and $E$.

$$\Delta = L/3\theta = 2/3L\theta_1$$

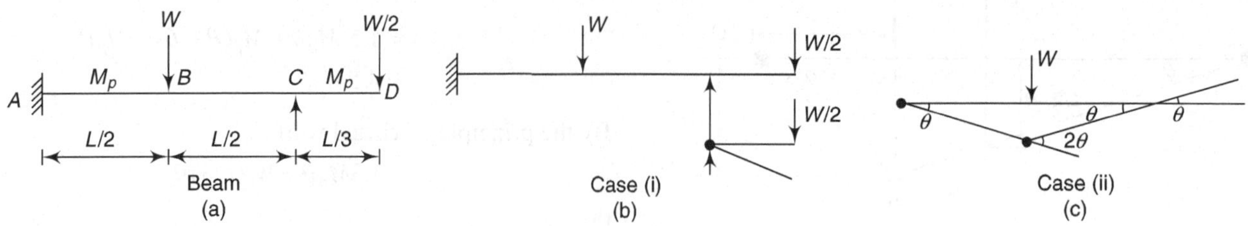


**FIG. 8.47**

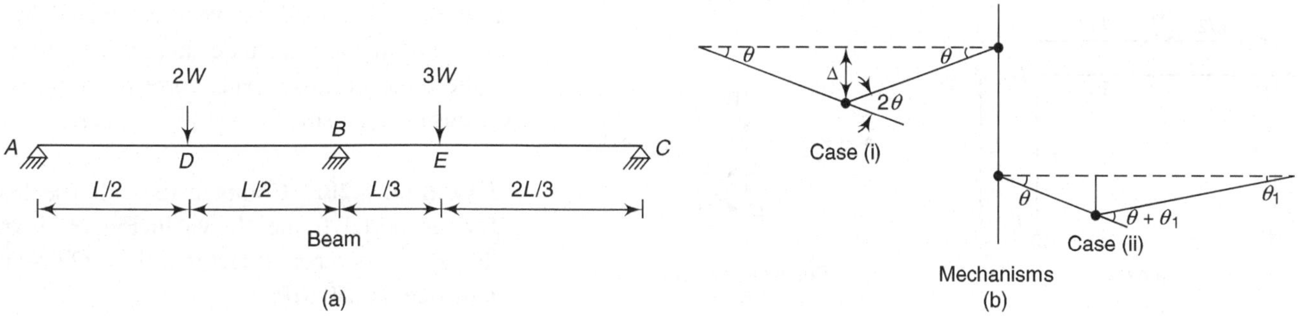


**FIG. 8.48**

Hence

$$\theta_1 = \theta/2$$

External work done = $3W_u 1/3L\theta = W_u L\theta$

Internal work done = $M_p\theta + M_p(\theta + \theta_1)$

$$= M_p\theta + M_p(\theta + \theta/2) = 2.5\,M_p\theta$$

By the principle of virtual work,

$$W_u L\theta = 2.5 M_p\theta$$

Hence

$$W_u = 2.5\,M_p/L$$

Therefore, the collapse load of the continuous beam is 2.5 $M_p/L$.

**EXAMPLE 8.18:** The three-span continuous beam shown in Fig. 8.49 is subjected to a total design load of 30 kN/m over its whole length. Check whether an ISMB 225 in Fe 410 steel would be a suitable section. It may be assumed that the beam is adequately braced against lateral deflection and twist over its whole length.

**SOLUTION:**

The beams *AB* and *CD* act as propped cantilever. The possible locations for the plastic hinges are at the intermediate supports *B* and *C* and at a point 0.414*L* from the supports *A* and *D*.

$$\Delta = 0.414L\theta_1 = 0.586L\theta \text{ or } \theta_1 = (0.586/0.414)\theta$$

External work done = load × average deflection

$$= (w_u)L \times (1/2) \times 0.586L\theta$$

$$= (0.586/2)w_u L^2\theta$$

Internal work done = moment × rotation

$$= M_p\theta + M_p(\theta + \theta_1)$$

$$= M_p\theta + M_p[\theta + (0.586/0.414)\theta]$$

$$= 3.415 M_p\theta$$

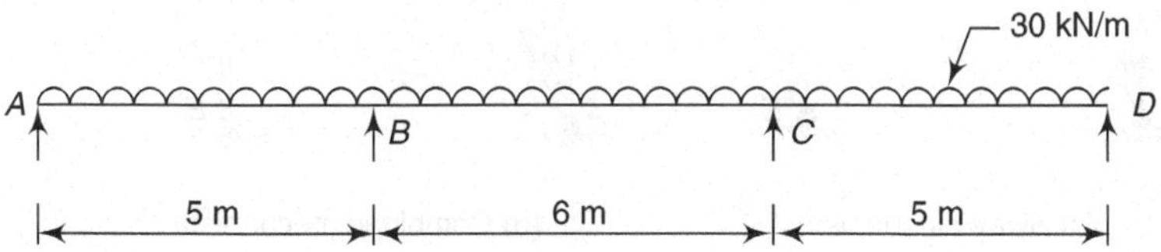


**FIG. 8.49**

By the principle of virtual work,

$$(0.586/2)w_u L^2\theta = 3.415 M_p\theta$$

$$M_p = w_u L^2/11.657 = 30 \times 5^2/11.657 = 64.34 \text{ kN m}$$

Beam *BC* will act as a beam with fixed ends. Hence,

$$M_p = w_u L^2/16 = 30 \times 6^2/16 = 67.5 \text{ kN m}$$

To obtain a uniform section throughout the continuous beam, the highest values amongst the plastic moments should be selected for design. To achieve economy, the beam cross section may be designed for the minimum plastic moment found in the above spans. Cover plates may be provided for the additional plastic moment required in the other spans. Hence, the design plastic moment

$$M_p = 67.5 \text{ kN m}$$

The required plastic section modulus,

$$Z_p = M_p/f_y = 67.5 \times 10^6/250$$

$$= 270 \times 10^3 \text{ mm}^3$$

From Annex H of IS 800, select a beam section ISMB 225 with $Z_p = 348.27 \times 10^3 \text{ mm}^3 > 270 \times 10^3 \text{ mm}^3$.

**EXAMPLE 8.19:** Determine the collapse load for the portal frame shown in Fig. 8.50, assuming that the beams and columns have the same cross section.

**SOLUTION:**

In the case of portal frames, the first step is to find the number of independent mechanisms and then all the possible combined mechanisms. The possible locations for the plastic hinges are *A*, *B*, *C*, and *D*. Hence the number of possible plastic hinges, $N = 4$; the degree of redundancy, $r = 5 - 3 = 2$; and the number of possible independent mechanisms, $n = N - r = 4 - 2 = 2$. The two independent mechanisms are the (a) beam mechanism and (b) sway mechanism. These two independent mechanisms may be combined to form other combined mechanisms.

*Beam mechanism*

Assuming that plastic hinges form at *B*, *C*, and *D* to give a beam mechanism,

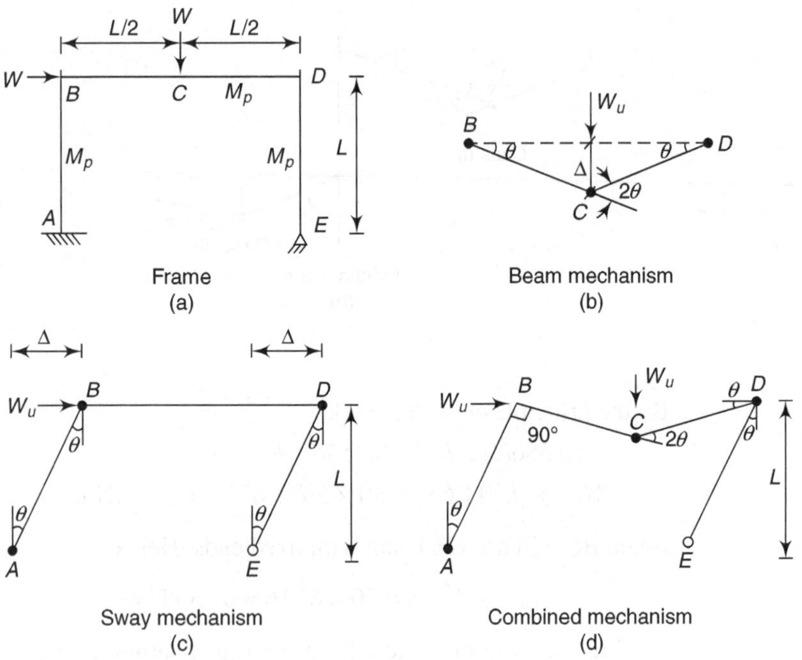


**FIG. 8.50**

$$\text{External work done} = W_u(L/2)\theta$$

$$\text{Internal work done} = M_p\theta + M_p(\theta + \theta) + M_p\theta = 4M_p\theta$$

By the principle of virtual work,

$$W_u(L/2)\theta = 4M_p\theta$$

Hence

$$W_u = 8M_p/L$$

*Sway mechanism*

With a plastic hinge forming at *A*, *B*, and *D* (*E* is a mechanical hinge with zero moment and hence no plastic hinge will develop at *E*), the sway mechanism shown in Fig. 8.50(c) will result.

$$\text{External work done} = W_uL\theta$$

$$\text{Internal work done} = M_p\theta + M_p\theta + M_p\theta = 3\,M_p\theta$$

By the principle of virtual work,

$$W_uL\theta = 3\,M_p\theta$$

Hence

$$W_u = 3\,M_p/L$$

*Combined mechanism*

A combined mechanism with plastic hinges at *A*, *C*, and *D* is shown in Fig. 8.50(d).

External work done = load × deflection

$$= W_uL\theta + W_u(L/2)\theta = 1.5\,W_uL\theta$$

Internal work done = moment × rotation

$$= M_p\theta + M_p\,(2\theta) + M_p\,(2\theta) = 5\,M_p\theta$$

By the principle of virtual work,

$$1.5W_uL\theta = 5M_p\theta$$

or

$$W_u = (10/3)M_p/L$$

As the collapse load has been calculated by the kinematic method in this example, it is the least of the three ultimate loads corresponding to the various mechanisms. Hence $W_u = 3M_p/L$.

**EXAMPLE 8.20:** Compute the collapse load for the portal frame shown in Fig. 8.51 and design the members if factored $W_u = 72$ kN and $f_y$ of steel is 250 MPa.

**SOLUTION:**

The possible locations of the plastic hinges are *A*, *B*, *C*, *D*, and *E*.

The number of possible plastic hinges, $N = 5$; degree of redundancy, $r = 6 - 3 = 3$; number of possible independent mechanisms, $n = N - r = 5 - 3 = 2$. As stated in the previous example, the beam and sway mechanisms can be combined to form combined mechanisms.

*Beam mechanism*

This mechanism is generated with three plastic hinges at *B*, *C*, and *D* and is associated with partial collapse [see Fig. 8.51(b)].

External work done = load × deflection

$$= 2W_u \times 2\theta = 4W_u\theta$$

Internal work done = moment × rotation

$$= M_p\theta + 2M_p(\theta + \theta) + M_p\theta = 6M_p\theta$$

By the principle of virtual work, $4W_u\theta = 6M_p\theta$. Hence $W_u = 1.5M_p$.

*Sway mechanism*

The sway mechanism is shown in Fig. 8.51(c), with plastic hinges in the columns at *A*, *B*, *D*, and *E*.

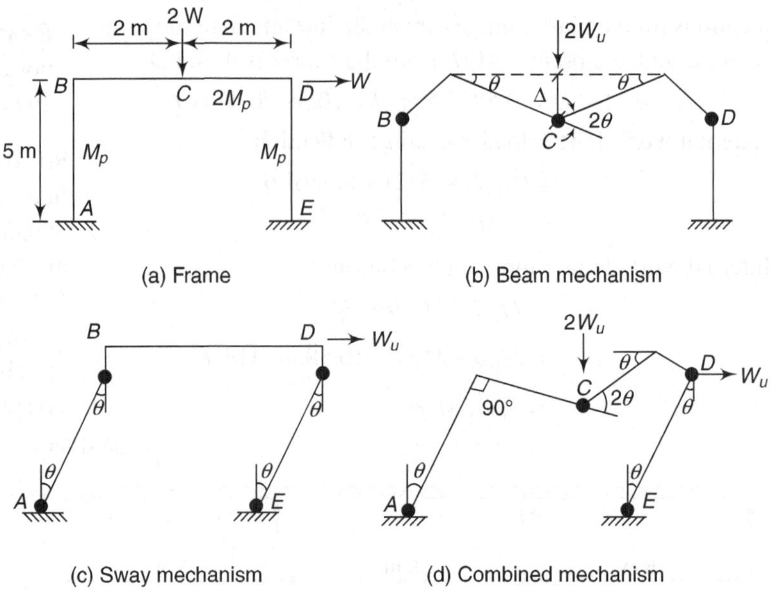


**FIG. 8.51**

External work done = $W_u L\theta = W_u 5\theta$

Internal work done = $M_p\theta + M_p\theta + M_p\theta + M_p\theta = 4M_p\theta$

By the principle of virtual work, $5W_u\theta = 4M_p\theta$. Hence $W_u = (4/5)M_p$.

*Combined mechanism*

The combined mechanism is shown in Fig. 8.51(d) with hinges forming at *A*, *C*, *D*, and *E*.

External work done = $2W_u \times 2\theta + W_u 5\theta = 9W_u\theta$

Internal work done = $M_p\theta + 2M_p(2\theta) + M_p\theta + M_p\theta + M_p\theta = 8M_p\theta$

By the principle of virtual work, $9W_u\theta = 8M_p\theta$. Hence $W_u = (8/9)M_p$.

Therefore, the collapse load for the frame is $(4/5)M_p$ [least of $1.5M_p$, $(4/5)M_p$, and $(8/9)M_p$].

*Design*

The section must be so designed that it resists the maximum value of $M_p$. From the preceding calculations,

$$M_p = (9/8)W_u = (9/8) \times 72 = 80 \text{ kN m}$$

Required $Z_p$ for column = $80 \times 10^6/250 = 320 \times 10^3$ mm$^3$

Required $Z_p$ for beam = $2 \times 320 \times 10^3 = 640 \times 10^3$ mm$^3$

Hence, ISMB 225 (with $Z_p$ = 348.27 mm$^3$) is required for the columns and ISMB 300 (with $Z_p$ = 651.74 mm$^3$) is required for the beam of the portal frame.

**EXAMPLE 8.21:** Find the collapse load for the frame of uniform cross section shown in Fig. 8.52(a) under the applied factored loads. Also find the minimum section required to resist the applied loads.

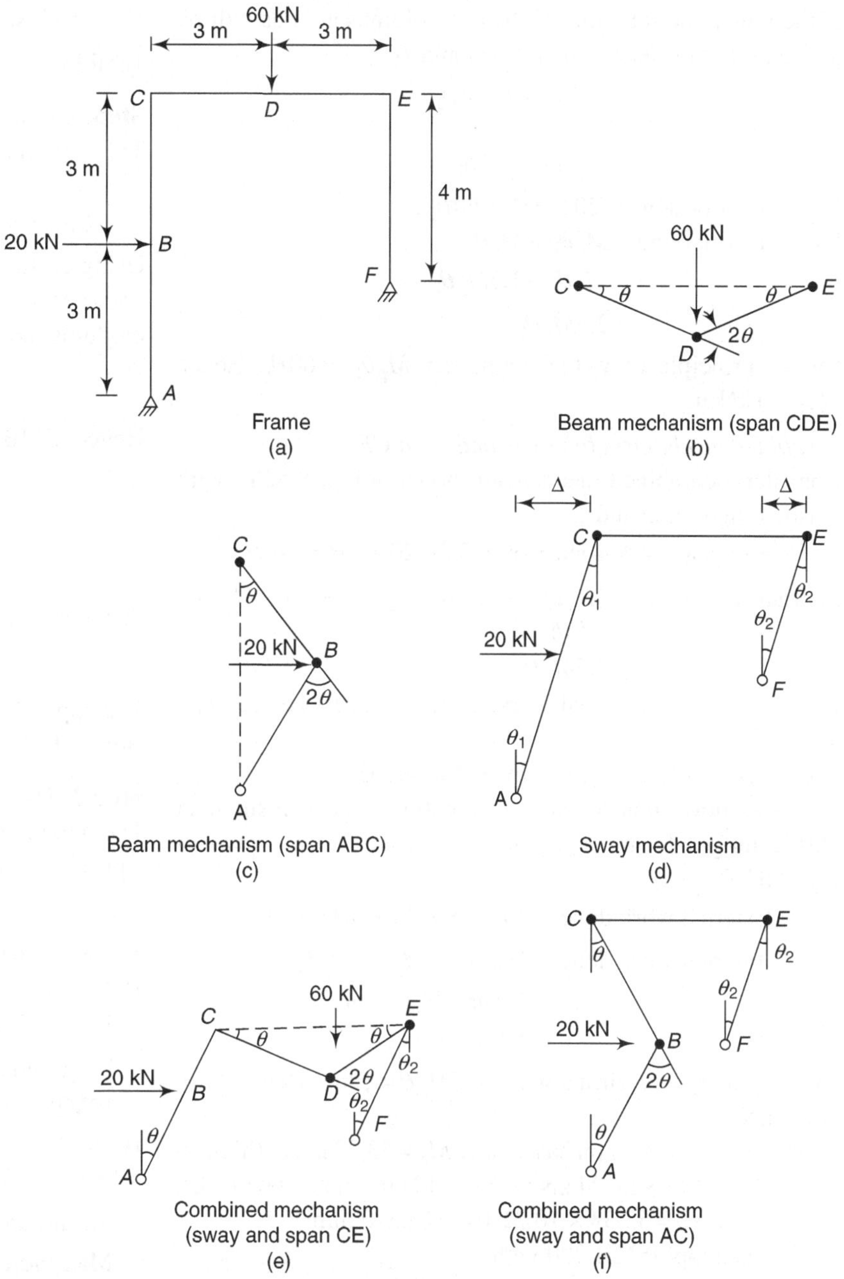


**FIG. 8.52**

**SOLUTION:**

The possible locations of plastic hinges are *B*, *C*, *D*, and *E*. The number of possible plastic hinges, $N = 4$; degree of redundancy, $r = 4 - 3 = 1$; number of possible independent mechanisms, $n = N - r = 4 - 1 = 3$.

*Beam mechanism for span CE*

For the beam mechanism shown in Fig. 8.52(b),

External work done = $60 \times 3\theta = 180\theta$

Internal work done = $M_p\theta + M_p(2\theta) + M_p\theta = 4M_p\theta$

By the principle of virtual work, $4M_p\theta = 180\theta$. Hence $M_p$ = 45 kN m.

*Beam mechanism for span AC*

Since the end *A* has a mechanical hinge, no plastic hinge will develop there. The beam mechanism will form with two plastic hinges at *C* and *B* as shown in Fig. 8.52(c).

External work done = $20 \times 3\theta = 60\theta$

Internal work done = $M_p\theta + M_p(2\theta) = 3M_p\theta$

By the principle of virtual work, $60\theta = 3M_p\theta$. Hence $M_p$ = 20 kN m.

*Sway mechanism*

A sway mechanism will be formed with plastic hinges at *C* and *E* as shown in Fig. 8.52(d). Now note that the columns *AC* and *FE* are of different lengths. To have the same deflection at the joints, the rotations of the two columns at ends will be different. Let us denote them by $\theta_1$ and $\theta_2$.

$$\Delta = 6\theta_1 = 4\theta_2$$

Hence,

$$\theta_2 = 1.5\theta_1$$

External work done = $20 \times 3\theta_1 = 60\theta_1$

Internal work done = $M_p\theta_1 + M_p\theta_2$

$= M_p\theta_1 + 1.5M_p\theta_1$

$= 2.5M_p\theta_1$

By the principle of virtual work, 2.5 $M_p\theta_1$ = $60\theta_1$. Hence $M_p$ = 24 kN m.

*Combined mechanism for sway and span CE*

Consider a combined mechanism shown in Fig. 8.52(e) with plastic hinges at *D* and *E*.

External work done = $60 \times 3\theta + 20 \times 3\theta = 240\theta$

Internal work done = $M_p(2\theta) + M_p(\theta + \theta_2) = 2M_p\theta + M_p(\theta + 1.5\theta)$

$= 4.5M_p\theta$

By the principle of virtual work, $4.5M_p\theta = 240\theta$. Hence $M_p$ = 53.33 kN m.

*Combined mechanism for sway and span AC*

Let us now consider another combined mechanism with plastic hinges forming at joints *B*, *C*, and *E* as shown in Fig. 8.52(f).

External work done = $20 \times 3\theta + 20 \times 3\theta = 120\theta$

Internal work done = $M_p(2\theta) + M_p\theta + M_p\theta_2$

$= 3M_p\theta + M_p(1.5\theta)$

$= 4.5M_p\theta$

By the principle of virtual work, $4.5M_p\theta = 120\theta$. Hence $M_p$ = 26.66 kN m.

Therefore, the section has to have $M_p$ = 53.33 kN m (highest of 45 kN m, 20 kN m, 24 kN m, 53.33 kN m, and 26.66 kN m).

Required $Z_p = 53.33 \times 10^6/250 = 213{,}320$ mm$^3$

Hence, adopt ISMB 200 with

$$Z_p = 253.86 \times 10^3 \text{ mm}^3$$

**EXAMPLE 8.22:** The spans and loading of a two span beam are shown in Fig. 8.53. Select a suitable I-section in Fe 410 grade steel using (a) elastic design procedure and (b) plastic design.

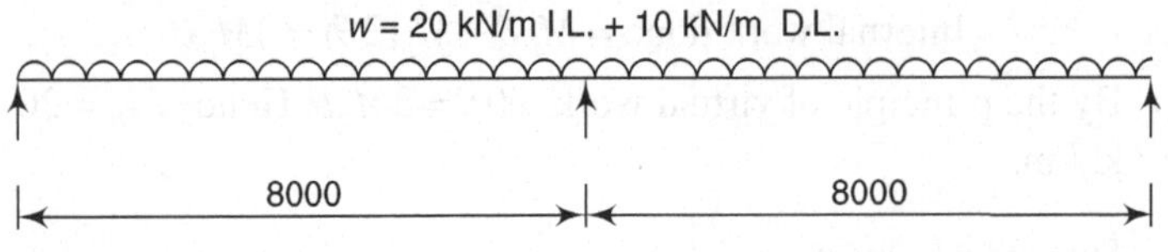


**FIG. 8.53**

**SOLUTION:**

*I.L* = 20 kN/m = 160 kN per span

*D.L* = 10 kN/m = 80 kN per span

Total load = 30 kN/m = 240 kN per span

**Step 1:** *Using elastic analysis*

The maximum bending moment at the centre support

$= WL/8 = 240 \times 8/8 = 240$ kN m

Shear = $1.25 \times 240 = 300$ kN

Using elastic design and assigning a stress of $0.66f_y = 0.66 \times 250 = 165$ MPa for Fe 410 grade steel, the minimum section modulus required

$$Z_e = 240 \times 10^6/165 = 1454.5 \times 10^3 \text{ mm}^3$$

Hence ISMB 500@ 86.9 kg/m is required with

$Z_e = 1810 \times 10^3$ mm$^3$

$t_w = 10.2$ mm

$D = 500$ mm

Shear capacity of the section = $(0.4 \times 250) \times 500 \times 10.2 \times 10^{-3}$

= 510 kN

The applied shear (300 kN) is less than 510 kN.

Hence the beam is safe.

**Step 2:** *Using plastic analysis*

The maximum bending moment (by considering each span as a propped cantilever)

$$M_p = W_uL^2/11.66$$

The load factor to be considered in plastic analysis = 1.7 (see Table 2.2 of Chapter 2)

Required $M_p = (30 \times 1.7) \times 8^2/11.66 = 279.93$ kN m

Required $Z_p = 279.93 \times 10^6/250 = 1{,}119{,}720$ mm$^3$

ISMB 400 @ 61.6 kg/m is required with $Z_p = 1176.18 \times 10^3$ mm$^3$, $D = 400$, $t_w = 8.9$ mm, $r_y = 28.2$ mm, $t_f = 16$ mm, $B = 140$ mm, $A = 7840$ mm$^2$, and $I_y = 622 \times 10^4$ mm$^4$

Shear capacity $250/\sqrt{3} \times 400 \times 8.9 \times 10^{-3} = 513.84$ kN

Maximum shear force at support = $279.93/8 + (240 \times 1.7)/2$ = 238.97 kN

This is less than 0.6 times the shear capacity ($0.6 \times 513.84$ = 308.3 kN) of the section, and hence there is no need to reduce the moment capacity of the section for shear.

Moment capacity = $250 \times 1176.18/10^{-3} = 294.045$ kN m

Both flanges must be restrained at hinge positions and at a point located at a maximum distance $L_m$ from the hinge position (clause 4.5.2.1 of the code)

$$L_m \le \frac{38r_y}{[(f_c/130) + (f_y/250)^2(x_t/40)^2]^{0.5}}$$

$x_t = 1.132(AI_w/I_yI_t)^{0.5}$

$h = 400 - 16 = 384$ mm

$I_w = I_f h^2/2 = t_f B^3 h^2/24 = 16 \times 140^3 \times 384^2/24 = 2.697 \times 10^{11}$ mm$^6$

$$I_t = \left(\frac{1}{3}\right)\left[2Bt_f^3 + ht_w^3\right] = \left(\frac{1}{3}\right)[2 \times 140 \times 16^3 + 384 \times 8.9^3]$$

$= 472{,}529$ mm$^3$

$x_t = 1.132 \times [7840 \times 2.697 \times 10^{11}/(622 \times 10^4 \times 472529)]^{0.5}$

$= 30.362$

$$L_m = \frac{38 \times 28.2}{[(0/130) + (250/250)^2 (30.362/40)^2]^{0.5}}$$

$= 1411.76$ mm

Although the weight of ISMB 400 (61.6 kg/m) as per plastic design is about 40.6% (103.7 kg/m) less than the weight as per elastic design, the additional cost of providing stiffeners and restraints in the plastic design may reduce the margin gained due to weight.

**EXAMPLE 8.23:** Check whether the moment capacity of a welded plate girder comprising two 650- × 25-mm flange plates and one 1500- × 15-mm web plate will be affected by flange local buckling assuming (a) Fe 410 steel with a design strength of $f_y = 250$ MPa (b) Fe 540 steel with a design strength of $f_y = 410$ MPa (see Fig. 8.54).

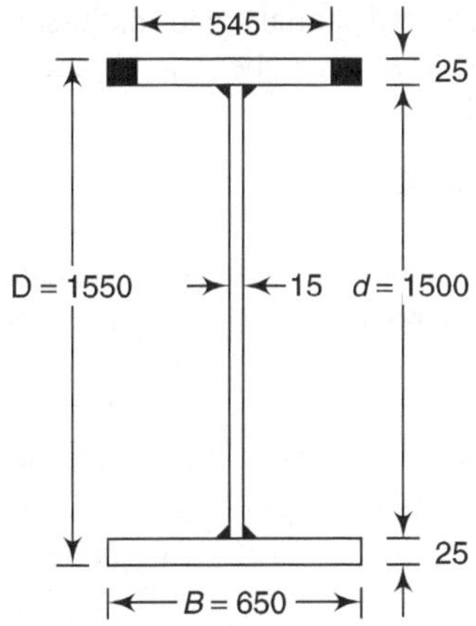


**FIG. 8.54**

**SOLUTION:**

(a) For $f_y = 250$ MPa, from Table 8.6, (Table 2 of IS 800), maximum outstand $b/t$ for the flange to be compact $= 9.4$.

Actual $b/t$, using Fig. 8.26 $= (325 - 15/2)/25 = 12.7 > 9.4$

Maximum $b/t$ for the flange to be semi-compact $= 13.6$

Hence the section is semi-compact and $M_d = Z_e f_y$

$I_z = BD^3/12 - (B - t_w)d^3/12$

$= 650 \times 1550^3/12 - (650 - 15) \times 1500^3/12$

$= 23116.15 \times 10^6$ mm$^4$

$Z_{ez} = 23116.15 \times 10^6/(1500/2 + 25) = 29827.3 \times 10^3$ mm$^3$

$M_d = 250 \times 29827.3 \times 10^3/10^6 = 7456.8$ kNm

$Z_p = 2Bt_f(D - t_f)/2 + t_w d^2/4$

$= 2 \times 650 \times 25\ (1550 - 25)/2 + 15 \times 1500^2/4$

$= 33{,}218.75 \times 10^3$ mm$^3$

$M_p = 33{,}218.75 \times 10^3 \times 250/10^6 = 8304.7$ kNm

Hence reduction in capacity from that corresponding to compact behaviour

$= [(8304.7 - 7456.8)/8304.7] \times 100 = 10.21\%$

(b) For $f_y = 410$ MPa, maximum $b/t$ for the flange to be semi-compact

$= 13.6 \times (250/410)^{0.5} = 10.6$

Therefore the section is slender. Limit for the effective flange width for semi-compact behaviour

$= 10.6 \times 25 = 265$ mm

Hence effective top width of the flange $= 265 \times 2 + 15 = 545$ mm (see Fig. 8.54)

*Location of neutral axis*

Taking moments about the top edge, the distance of the neutral axis from the top edge,

$\bar{y} = [545 \times 25 \times 25/2 + 1500 \times 15 \times (1500/2 + 25) + 650 \times 25 \times (1525 + 25/2)]/(545 \times 25 + 1500 \times 15 + 650 \times 25)$

$= 813.2$ mm

$I_z = (545 \times 25)\ (813.2 - 12.5)^2 + (15 \times 1500^3/12) + (15 \times 1500) \times (813.2 - 775)^2 + 650 \times 25\ (736.8 - 12.5)^2 + 545 \times 25^3/12 + 650 \times 25^3/12$

$= 2.1513 \times 10^{10}$ mm$^4$

$Z_z$ (top flange) $= 2.1513 \times 10^{10}/813.2 = 26454 \times 10^3$ mm$^3$

$M_d = 410 \times 26454 \times 10^3/10^6 = 10{,}846.2$ kNm

Capacity, if the whole section were as effective as the semi-compact section,

$M_{d1} = 410 \times 29827.3 \times 10^3/10^6 = 12{,}229.2$ kNm

Hence reduction in capacity from that corresponding to semi-compact section behaviour

$= [(12{,}229.2 - 10{,}846.2)/12{,}229.2] \times 100 = 11.3\%$

Capacity, if the whole section is as effective as the plastic section,

$M_p = 33218.75 \times 10^3 \times 410/10^6 = 13{,}619.7$ kNm

Hence reduction in capacity due to slenderness

$= [(13619.7 - 10{,}846.2)/13619.7] \times 100 = 20.4\%$

Thus, 20.4% of the capacity of the cross section could not be utilized, due to the slenderness of the cross section.

## SUMMARY

The basic concepts of plastic analysis have been introduced in this chapter. The section modulus, either elastic or plastic depending on the design philosophy adopted, is the most appropriate property to be considered while selecting a beam section. Methods for calculating the plastic modulus and shape factor (ratio of $M_P/M_y$) of any cross section have been explained with a number of examples. The requirements for plastic design as per IS 800 have been discussed. The advantages and disadvantages of plastic design have been listed. The plastic hinge concept, conditions to be satisfied for elastic/plastic analysis, and the methods of plastic analysis (mechanism and statical method) have been discussed. The theorems of plastic collapse and alternative patterns of hinge formation triggering plastic collapse have been discussed. The plastic designs of simply supported beams, fixed beams, propped cantilevers, continuous beams, and portal frames have been explained with the use of worked examples. The effect of axial force and shear force on plastic moment capacity has been discussed and interactive equations derived. A number of special considerations, such as local buckling, flange stability, deflection, connection, etc., that should be satisfied while attempting a plastic design have also been explained.

Most structural sections are assemblages of plates of slender proportions, which are prone to local buckling. Local buckling has the effect of reducing the overall load carrying capacity of columns and beams due to the reduction in the stiffness and strength of the locally buckled elements. Therefore it is desirable to avoid local buckling before the yielding of the member. The local buckling action can be analysed approximately by considering each plate of a cross section in isolation. The critical stress formulae of plates subjected to different loading and boundary conditions have been given is this chapter. From these formulae it is found that the buckling capacity of a plate is inversely proportional to the square of its width-to-thickness ratio ($b/t$ or $d/t$). One way of avoiding local buckling is to ensure that the width-to-thickness ratio of each component plate does not exceed a certain value. In IS 800, a comprehensive system has been introduced for classifying the cross section of members that are subjected to compression due to moment and/or axial loading. The classification depends primarily on the geometry of the cross section and has the following four classes: plastic (class 1), compact (class 2), semi-compact (class 3), and slender (class 4). These four classes have been described for various cross sections in this chapter.

Plastic and compact sections are preferable if limit state design is used, and only plastic sections can be used in mechanism-forming indeterminate frames (plastic design). Slender sections are to be avoided even in elastic design. Most hot-rolled steel sections have individual elements of sufficient thickness (may be classified as plastic, compact, or semi-compact) and hence local buckling is avoided before the yielding of the member. However, fabricated (welded) sections and thin-walled cold-formed steel members usually experience the local buckling of plate elements before the yield stress is reached. For such sections, only the effective area and moment of inertia should be used for calculating the capacity.

Substantial reserve strength exists in plates beyond the point of elastic buckling. The utilization of this reserve capacity may also be a design objective. Post-buckling reserve strength is normally taken into account in the design of cold-formed sections, approximately by using the empirical effective width concept.

Further details on plastic analysis, plastic design of frames (especially pitched roof portals and multi-storey frames), and buckling of plate elements may be found in the references given at the end of the book.

## MULTIPLE-CHOICE QUESTIONS

8.1 The shape factor depends on
(a) yield stress of the material
(b) geometry of the section
(c) redistribution of the moments
(d) none of these

8.2 Which of the following has maximum value of shape factor?
(a) Triangular section (c) Circular section
(b) Rectangular section (d) I-section

8.3 The conditions to be satisfied in plastic analysis are
(a) equilibrium (c) compatibility
(b) mechanism (d) all of these

8.4 Load factor in the plastic design is normally taken as
(a) 1.5 (b) 1.6 (c) 1.7 (d) 2.0

8.5 The collapse load $W_u$ of a simply supported beam of span $L$ and uniform cross section, loaded with a uniformly distributed load of $w = W/L$ is
(a) $12M_p/L$ (c) $8M_p/L$
(b) $4M_p/L$ (d) $16M_p/L$

8.6 The collapse load of a propped cantilever beam of span L and uniform cross section, loaded with a concentrated load at mid span is
(a) $11.77\,M_p/L$ (c) $4M_p/L$
(b) $6M_p/L$ (d) $8M_p/L$

8.7 For a simply supported beam of rectangular section loaded with a central point load, the plastic hinge length is given by
(a) $L/3$ (b) $L/9$ (c) $L/2.5$ (d) $2L/8.49$

8.8 The plastic section modulus $Z_{pz}$ of a rectangular section with depth $d$ and breadth $b$ is given by
(a) $bd^2/6$ (c) $bd^2/4$
(b) $db^2/6$ (d) $db^2/4$

8.9 If the order of indeterminacy is $r$, then the maximum number of plastic hinges required for total collapse is
(a) $r$ (b) $r+1$ (c) $r+2$ (d) $r-1$

8.10 The interaction formula for rectangular sections subjected to bending moment and axial force is

(a) $\left(M_{pr}/M_p\right)^2 + \left(N/N_p\right)^2 = 1$

(b) $\left(M_{pr}/M_p\right) + \left(N/N_p\right)^2 = 1$

(c) $\left(M_{pr}/M_p\right)^{2/3} + \left(N/N_p\right)^{2/3} = 1$

(d) $\left(M_{pr}/M_p\right) + \left(N/N_p\right) = 1$

8.11 The interaction formula for I-sections, bent about major axis and subjected to bending moment and axial force is

(a) $\left(\frac{1}{1.18}\right)\left(M/M_p\right)^2+\left(N/N_p\right)^2\le 1$

(b) $(1/1.18)\left(M/M_p\right)+\left(N/N_p\right)^2\le 1$

(c) $\left(\frac{1}{1.18}\right)\left(M/M_p\right)^{2/3}+\left(N/N_p\right)^{2/3}\le 1$

(d) $\left(\frac{1}{1.18}\right)\left(M/M_p\right)+\left(N/N_p\right)\le 1$

8.12 Where there is a concentrated load, web stiffeners are to be provided, within a distance equal to
(a) full depth of the member on either side of plastic hinge location
(b) half the depth of the member on either side of plastic hinge location
(c) twice the depth of the member on either side of plastic hinge location
(d) none of these

8.13 A semi-compact section is one which
(a) can develop their plastic moment resistance and has required rotation capacity for the formation of plastic collapse mechanism
(b) can develop their plastic moment resistance but does not have the required rotation capacity for the formation of plastic collapse mechanism
(c) can reach the yield strength, but cannot develop plastic moment resistance due to local buckling
(d) cannot reach even the yield strength due to local buckling

8.14 The limiting $b/t_f$ ratio of outstand element of rolled compression flange to be considered as plastic is, with $f_y = 250$ MPa
(a) 9.4 (b) 10.5 (c) 13.6 (d) 15.7

8.15 The limiting $b/t_f$ ratio of outstand element of welded compression flange to be considered as compact is, with $f_y = 250$ MPa
(a) 9.4 (b) 13.6 (c) 15.7 (d) 29.3

8.16 The limiting $d/t_w$ ratio of web of rolled I-section to be considered as plastic is, with $f_y = 250$ MPa
(a) 84 (b) 105 (c) 126 (d) 42

8.17 The limiting $d/t_w$ ratio of web of rolled channel to be considered as semi-compact is, with $f_y = 250$ MPa
(a) 42 (b) 84 (c) 105 (d) 126

8.18 The limiting $b/t_f$ ratio of circular hollow tube to be considered as compact is, with $f_y = 250$ MPa
(a) 52 (b) 146 (c) 52 (d) 88

8.19 The design plastic moment capacity $M_d$ of semi-compact section is calculated as
(a) $Z_e f_u$ (b) $Z_e f_y$ (c) $Z_p f_y$ (d) $< Z_e f_y$

## EXERCISES

8.1 Find the shape factor for the following sections:
(a) square of side $a$ with its diagonal parallel to $z$-$z$ axis as shown in Fig. 8.55(a) [Ans: $\nu = 2$]
(b) A diamond section with unequal diagonal, the longer being $d$ and shorter $b$ with the shorter diagonal placed parallel to the $z-z$ axis shown in Fig. 8.55(b) [Ans: $\nu = 2$]
(c) Circular section of radius $r$ as shown in Fig. 8.55(c) [Ans: $\nu = 1.7$]

8.2 Determine the plastic moment capacity of the sections shown in Figs 8.56(a) to (c), assuming $f_y = 250$ MPa

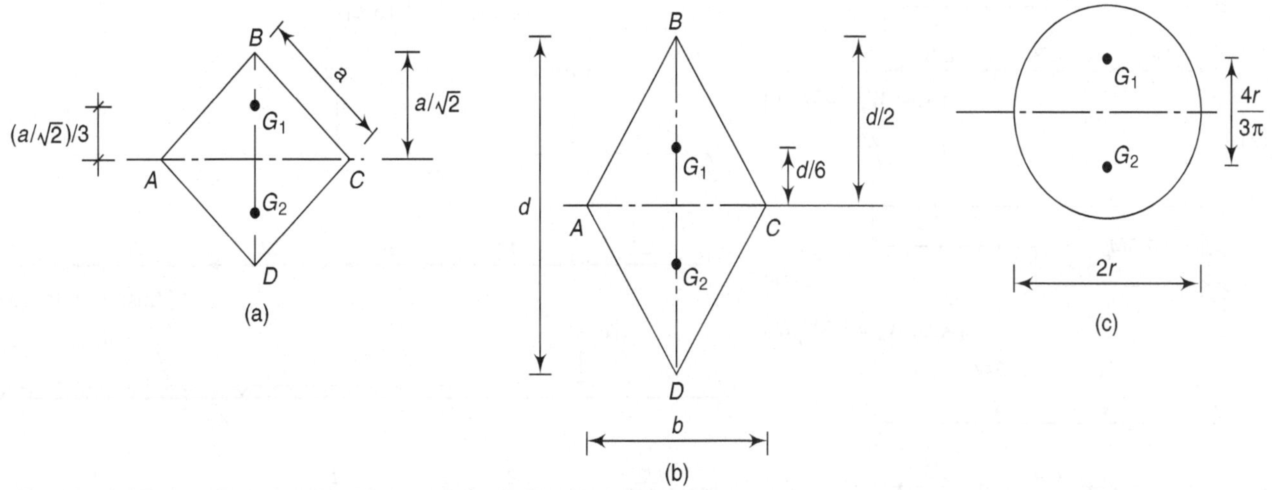


FIG. 8.55

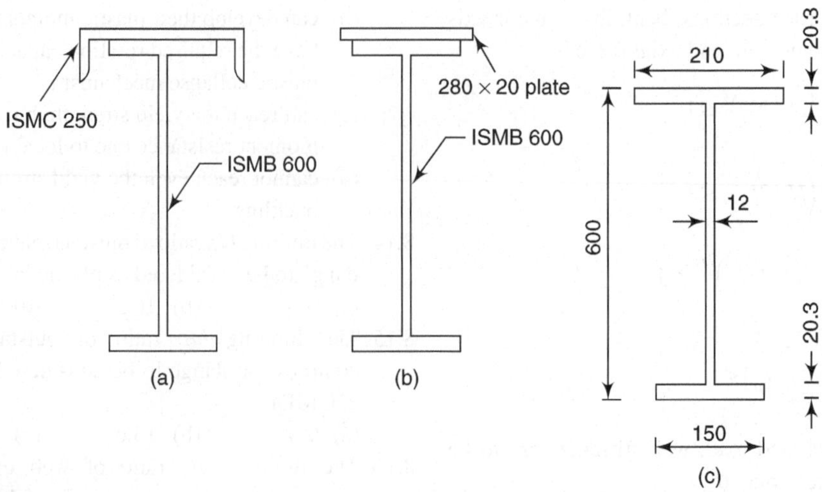


**FIG. 8.56**

8.3 Find out the collapse load of the following beams (assume that the beam is of uniform cross section, unless otherwise shown).

(a)

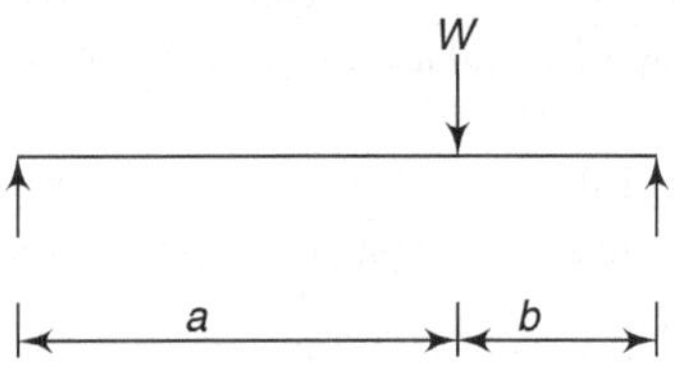


[Ans: $W_u = M_pL/ab$]

(b)

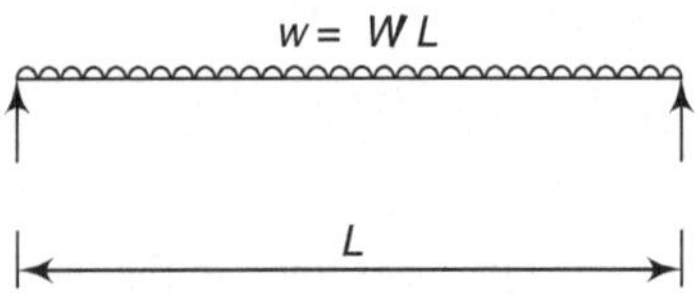


[Ans: $W_u = 8M_p/L$]

(c)

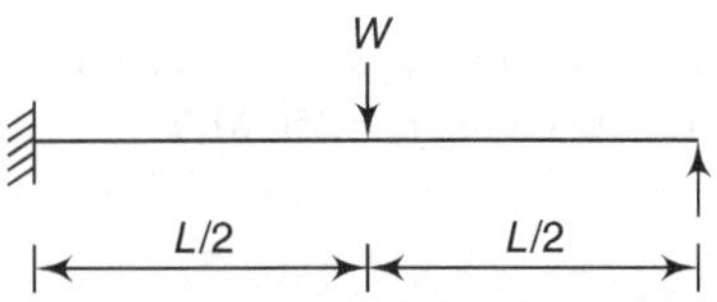


[Ans: $W_u = 6M_p/L$]

(d)

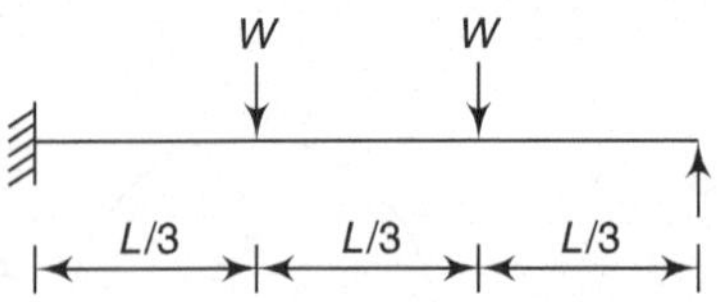


[Ans: $W_u = 4M_p/L$]

(e)

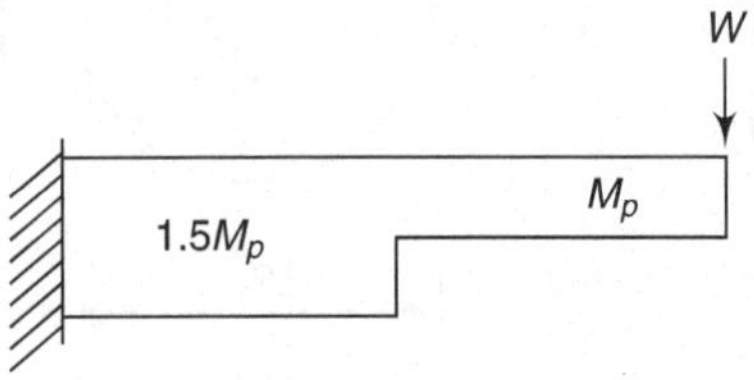


[Ans: $W_u = 1.5M_p/L$]

(f)

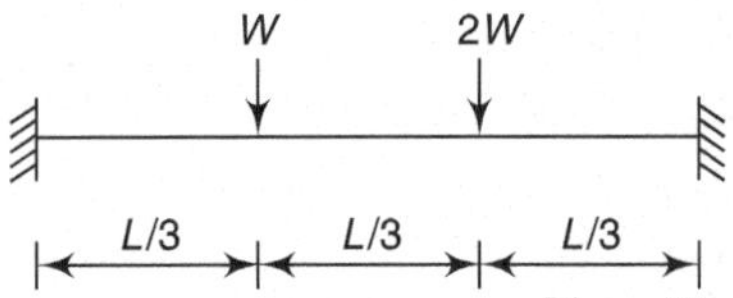


[Ans: $W_u = 3.6M_p/L$]

(g)

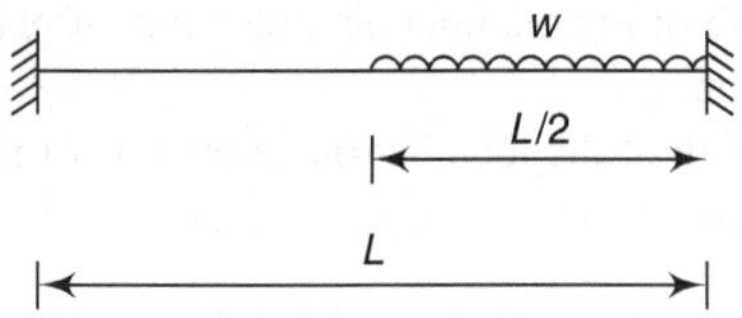


[Ans: $w_u = 14.22M_p/L$ and the plastic hinge will form at $0.375L$ from right support]

(h)

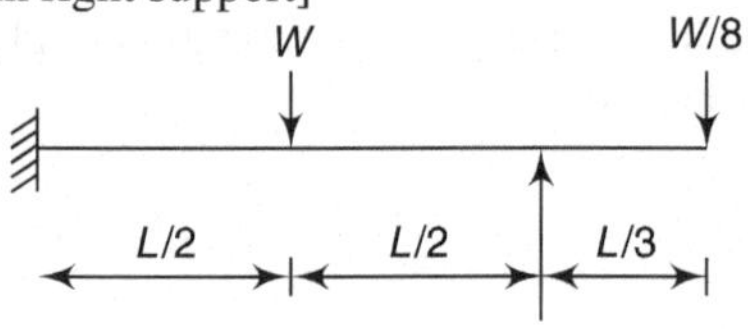


[Ans: $W_u = (72/11)M_p/L$]

(i)

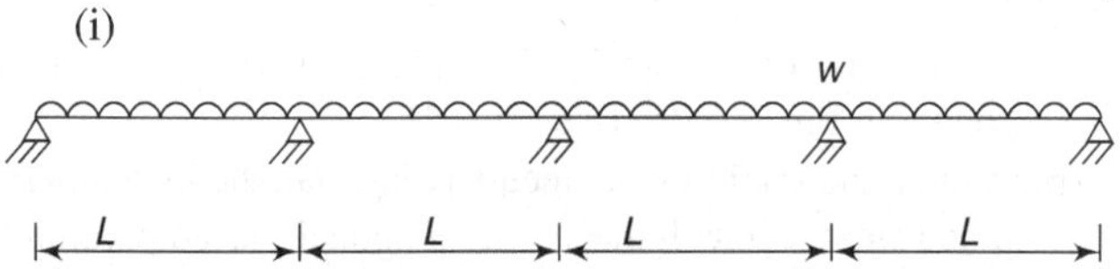


[Ans: $W_u = 12M_p/L^2$]

(j) Determine the span ratios of a continuous beam as shown in (i), such that all the spans collapse at the same time.
[Ans: If $L_1$ and $L_2$ are the length of exterior and interior spans, $L_2 = 1.15\,L_1$]

(k)

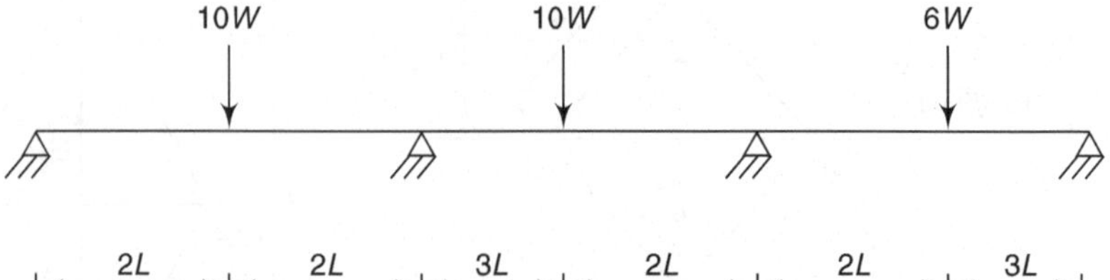


[Ans: $W_u = 0.15M_p/L$]

(l)

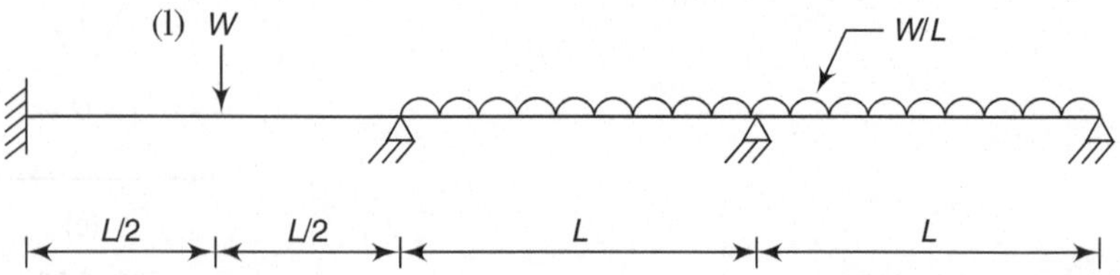


[Ans: $W_u = 8M_p/L$]

8.4 Design a suitable section for the beam shown in Fig. 8.57.

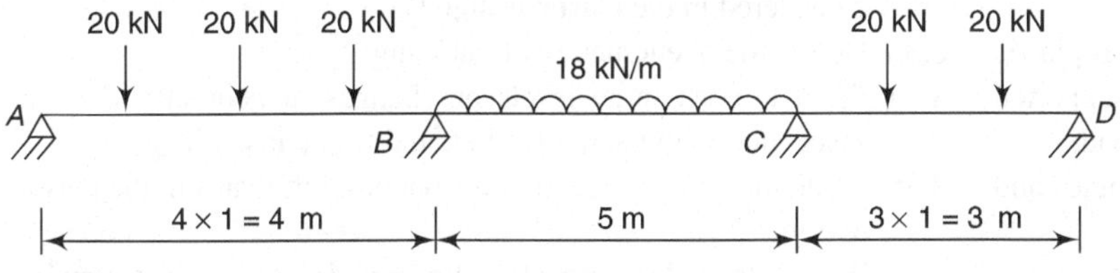


FIG 8.57

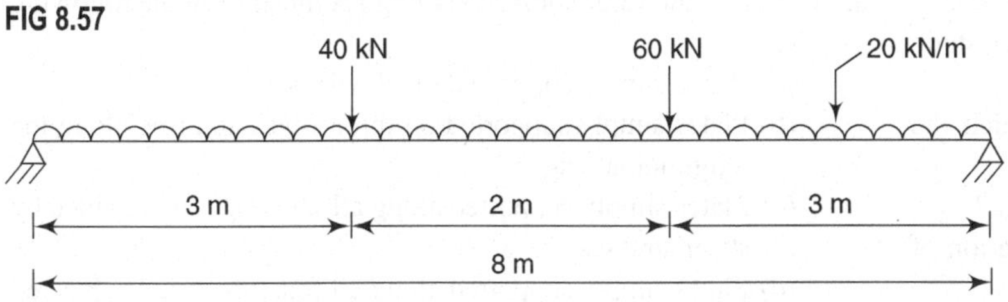


FIG 8.58

8.5 A simply supported beam is subjected to loads as shown in Fig. 8.58. Assume a load factor of 1.7 and design its section.

8.6 Find out the collapse load for a portal frame of uniform cross section as shown in Figs 8.59(a), (b), (c), and (d).

8.7 Design a suitable section for a two-span continuous beam, each having a span of 10 m and supporting a dead load of 15 kN/m and a live load of 25 kN/m by (a) elastic design procedure and (b) plastic design procedure.

8.8 Check whether the moment capacity of a welded plate girder consisting of two 600- × 20-mm flange plates and one 1000- × 15-mm web plate will be affected by flange buckling, assuming (a) Fe 410 grade steel ($f_y$ = 250 MPa), (b) Fe 540 grade steel ($f_y$ = 410 MPa).

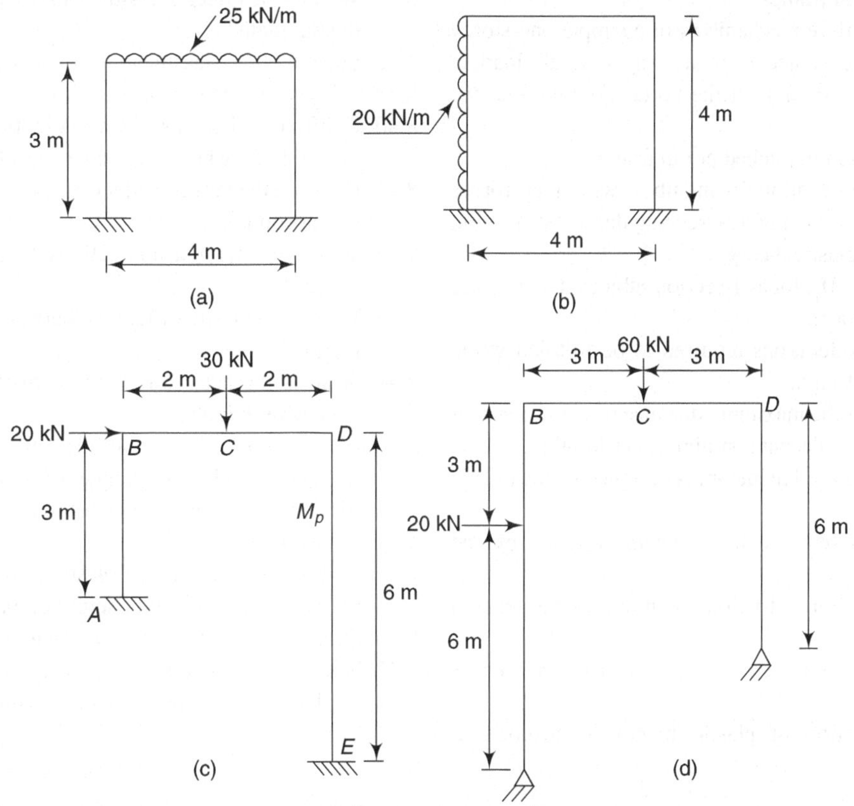


FIG 8.59

## REVIEW QUESTIONS

8.1 What are the two basic criteria any structural design must satisfy?

8.2 In what way plastic design methods are different from elastic design methods?

8.3 Distinguish between the elastic modulus and plastic modulus of a section.

8.4 State the assumptions made in plastic design to simplify computations.

8.5 State the condition to be satisfied in order to use plastic methods of analysis as per IS 800.

8.6 Define shape factor.

8.7 Will the shape factor of rectangular and circular cross sections be higher than that of I-sections? If yes, state the reason.

8.8 What are the advantages and disadvantages offered by plastic design?

8.9 What is a plastic hinge? In what way is it different from an ordinary hinge?

8.10 What are the points at which a plastic hinge is likely to form?

8.11 Describe the collapse of a fixed-ended beam, as assumed in plastic design.

8.12 What are the basic conditions to be satisfied in an elastic analysis and a plastic analysis?
8.13 Illustrate the difference between the yield moment and plastic moment of resistance by taking an example of a simply supported rectangular beam subjected to a central concentrated load.
8.14 What is the difference in collapse mechanisms of single- and multi-span beams?
8.15 State the principle of virtual work.
8.16 State the following theorems of plastic collapse: (a) static theorem (b) kinematic theorem and (c) uniqueness theorem.
8.17 What are the two methods of plastic analysis by which the collapse load can be determined?
8.18 What is the load factor usually adopted in plastic design?
8.19 What are the two steps that are involved in the application of plastic design to continuous beams?
8.20 What are the two groups of mechanisms considered in the plastic design of portal frames?
8.21 Draw the possible failure mechanisms of a simple one-storey one-bay portal frame subjected to a sway (lateral) load at the top of one of the columns of the portal. Assume that the supports are fixed.
8.22 Why are haunches used in pitched portal frames?
8.23 Write the interaction equation for members with axial forces and bending moment having (a) rectangular cross section (b) I-sections as per plastic theory.
8.24 Write the equation for $M_{pr}$ for an I-section subjected to bending moments and shear force.
8.25 List the special considerations required to be fulfilled while attempting a plastic design.
8.26 Write the equation for the maximum distance from the restraint at plastic hinge to the adjacent restraint as per IS 800.
8.27 Why are restraints required in members designed by the plastic analysis?
8.28 Where are stiffeners required in members designed by the plastic analysis?
8.29 What are the fabrication restrictions required for the tension flange of members?
8.30 When should second-order effects be considered as per IS 800?
8.31 How can the formation of plastic hinges be avoided at connections?
8.32 Under what conditions are residual stresses not required to be considered in the plastic design?
8.33 Define the phenomenon of buckling.
8.34 What are the three levels of changes in geometry that are associated with local, member, and frame instability?
8.35 Write the general equation for buckling of plates in the elastic region.
8.36 What is the value for the buckling coefficient $k$ in the following cases:
 (a) Plates simply supported on all edges
 (b) Plates simply supported on three edges and free along the longitudinal edge
 (c) Plates simply supported along all four edges and loaded by shear stresses
 (d) Plates simply supported along all four edges and subjected to bending
8.37 What is the effect of residual stress on the buckling strength of the flat plate?
8.38 State the four classifications of sections as per IS 800.
8.39 Is it possible to use slender section for beams?
8.40 If different elements of a cross section are of different classes, how will the whole section be classified?
8.41 How can the reduced effectiveness of slender sections be taken into account?
8.42 Why are only plastic sections to be used in indeterminate frames?
8.43 What is the basic difference between a plastic and a compact section?
8.44 What is the basic difference between a semi-compact and a compact section?
8.45 Give the $b/t$ or $d/t$ ratio for the following cases:
 (a) Outstand element (flange) of rolled beam of plastic section
 (b) Outstand element (flange) of rolled beam of semi-compact section
 (c) Web of an I-beam of plastic cross section
 (d) Web of an I-beam of semi-compact section
8.46 Describe the difference in behaviour of columns and plates.
8.47 What do you understand by post-buckling strength of plates?
8.48 Explain the concept of effective width. Where is it used?

## REFERENCES


Allen, H.G. and Bulson, P.S., *Background to Buckling*, McGraw-Hill Book Company (UK) Ltd, London, 1980, p. 582.

Beedle, L.S., *Plastic Design of Steel Frames*, John Wiley & Sons, Inc., 1958.

Bleich, H., *Buckling Strength of Metal Structures*, McGraw–Hill, New York, 1952.

Davies, J.M. and Brown, B.A., *Plastic Design to BS 5950*, Blackwell Science, 1996.

Davison, B. and Owens, G.W., *Steel Designers Manual*, Sixth edition, The Steel Construction Institute, Ascot and Blackwell Publishers, London, 2004.

Dowling, P.J., Knowles, P., and Owens, G.W., *Structural Steel Design*, The Steel Construction Institute and Butterworths, London, 1988, p. 399.

Galambos, T.V., ed., *Guide to Stability Criteria for Metal Structures*, Fifth edition, John Wiley & Sons, Inc., New York, 1998, p. 911.

Haaijer, G. and Thurlimann, B., 'On Inelastic Buckling in Steel', *Transactions of ASCE*, Vol. 125, 1960, pp. 308–344.

Horne, M.R., 'The Plastic Theory of Bending of Mild Steel Beams with Particular Reference to the Effect of Shear Forces', *Proc. Royal Society*, Vol. 207 (A), No. 216, 1951.

Horne, M.R., 'Safe Loads on I-section Columns in Structures Designed by Plastic Theory', Paper no. 6794, *Proc. of the Institution of Civil Engineers*, Vol. 29, 1964, pp. 137–150.

Horne, M.R., *Plastic Theory of Structures*, Pergamon Press Ltd, Oxford, 1979.

Horne, M.R. and Morris, L.J., *Plastic Design of Low-rise Frames*, Granada, London, 1981.

Joint Committee of Welding Research Council and the American Society of Civil Engineers, *Commentary on Plastic Design in Steel*, Second edition, ASCE Manual and Reports on Practice, No. 41, New York, 1971.

Kalyanaraman, V., Pekoz, T., and Winter, G., 'Unstiffened Compression Elements', *Journal of the Structural Div.*, ASCE, Vol. 103, No. ST 9, Sep 1977, pp. 1833–1848.

Martin, L.H. and Purkiss, J.A., *Structural Design of Steelwork to BS 5950*, Edward Arnold, London, 1992.

Morris, L.J., 'A Commentary on Portal Frame Design', *The Structural Engineer*, Vol. 59A, No. 19, 1981, pp. 394–402.

Petersen, C., *Stahlbau-Grundlagen der Berechnung und Baulichen Ausbuildung von Stahlbauten*, Second edition, Vieweg Verlag, Braunschweig, 1990.

Salmon, C.G. and Johnson, J.E., *Steel Structures: Design and Behavior*, Fourth edition, Harper Collins College Publishers, New York, 1996.

Timoshenko, S.P. and Gere, J.M., *Theory of Elastic Stability*, Second edition, McGraw-Hill, New York, 1961.

Trahair, N.S., Bradford, M.A., and Nethercot, D.A., *The Behaviour and Design of Steel Structures to BS 5950*, Third edition, Spon Press, London, 2001, p. 448.

Winter, G., 'Strength of Thin Steel Compression Flanges', *Transactions of ASCE*, Vol. 112, 1947, p. 527.

9

# DESIGN OF COMPRESSION MEMBERS

## INTRODUCTION

A structural member which is subjected to compressive forces along its axis is called a *compression member*. Thus, compression members are subjected to loads that tend to decrease their lengths. Except in pin-jointed trusses, such members (in any plane or space structure), under external loads, experience bending moments and shear forces. If the net end moments are zero, the compression member is required to resist load acting concentric to the original longitudinal axis of the member and is termed *axially loaded column,* or simply *column.* If the net end moments are not zero, the member will be subjected to an axial load and bending moments along its length. Such members are called *beam-columns* and are treated in Chapter 13.

Multi-storey steel building under construction in Maryland, USA

Close-up view of the column of the above building (Note the composite slab)

Let us consider an example of an axially loaded column shown in Fig. 9.1. For this column, the axial load $P$, to be resisted by it, is the sum of the beam shears $V_1 + V_2$. For the column shown in Fig. 9.1, the net end moment is assumed to be zero; this is true if the end moments and shears developed by the two beams are equal. Such situations arise in many interior columns of buildings having equal column spacing. Where the beam is not connected rigidly to the column, the beams will not develop significant end moments and in such situations also the column has to resist only the difference in end shears. In several interior columns, the net moment will be small and the member is designed as an axially loaded column.

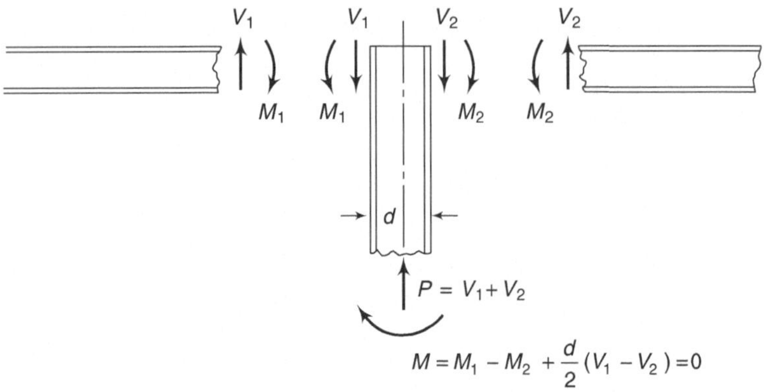


**FIG. 9.1** Axially loaded column

Different terms are used for a compression member depending upon its position in a structure. The vertical compression members in a building supporting floors or girders are normally called as columns (referred sometimes as *stanchions* in UK). They are subjected to heavy loads. Sometimes vertical compression members are called *posts*. The compression members used in roof trusses and bracings are called *struts*. They may be vertical or inclined and normally have small lengths. The top chord members of a roof truss are called the *principal rafter.* The principal compression member in a crane is called the *boom.* Short compression members at the junction of columns and roof trusses or beams are called *knee braces.* Some of these compression members are shown in Fig. 9.2.

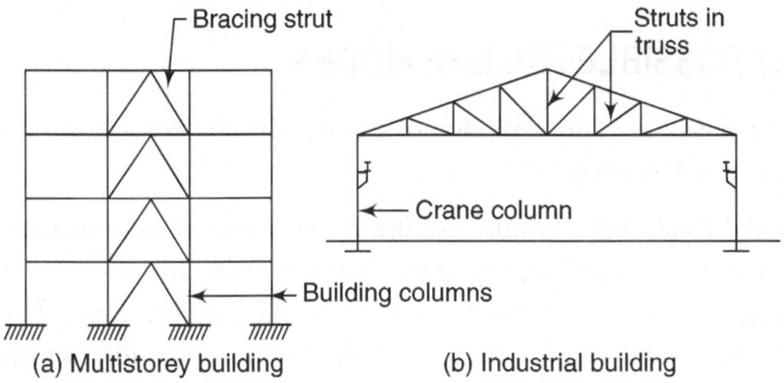


**FIG. 9.2** Types of compression members

It is well known from basic mechanics of materials that only very short columns can be loaded up to their yield stress. For long columns (see Section 9.4.1 also), *buckling* (deformation in the direction normal to the load axis) occurs prior to developing the full material strength of the member. Hence, a sound knowledge of stability theory is necessary for designing compression members in structural steel.

Since compression members have to resist buckling, they tend to be stocky and 'square' and circular tubes are found to be the ideal sections, since their radius of gyration is same in the two axes (see also Section 9.8). This situation is in contrast to the slender and more compact tension members and deep beam sections. Unlike the member subjected to tension, a compression member is designed on the assumption that its gross cross-sectional area will be effective in resisting the applied loads. Bolts may be used to connect columns to adjacent members. As the load is applied, the member will contract. It is assumed that the action of bolts is such that they will replace the material removed for holes. Thus, the bolt holes are often ignored in the design. Since compression members comprise of thin plates, they also experience local buckling, as discussed in Chapter 8.

## 9.1 CONSTRUCTION DETAILS

Columns in buildings are connected to other members and foundation through the following:

1. Beam-to-column connections
2. Column cap connections
3. Column base plates

Column splices are used to increase the height of columns or to connect different column cross sections.

Typical beam-to-column connections are discussed in Chapters 5, and 14 (see Figs 5.1, 5.2, 14.27, and 14.28) and column cap connections are shown in Fig. 9.3. Splices in compression members are discussed in Sections 5.9 and 6.12.2 (see Figs 5.41, 5.42, and 6.56). Column base plates which are used to connect columns with concrete foundations are discussed in Chapter 15.

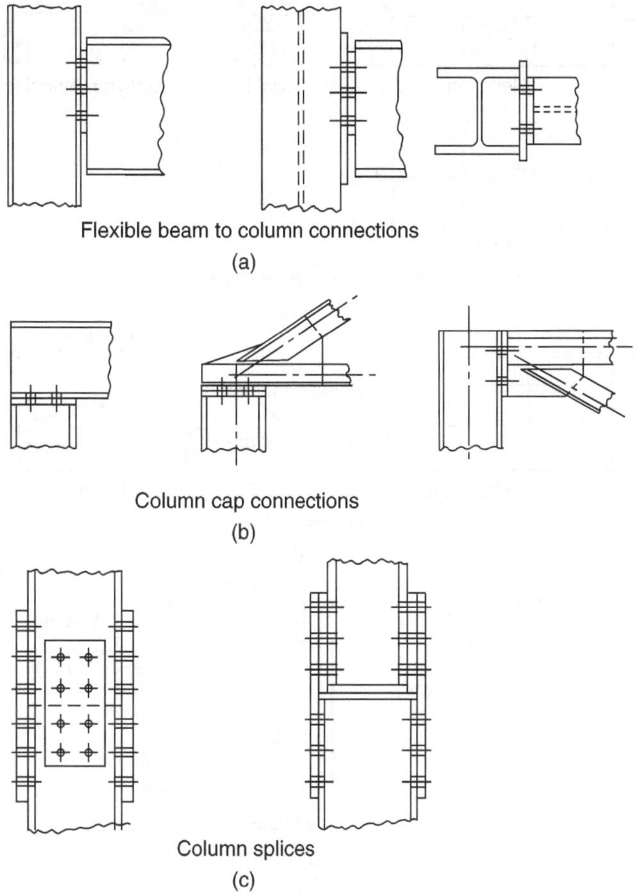


**FIG. 9.3** Typical column connections

### 9.1.1 Loads on Compression Members

As discussed previously, axial loading on columns in buildings is due to loads from roofs, floors, and walls transmitted to the column through beams and also due to its own self-weight, as shown in Fig. 9.4(a). Floor beam reactions are often eccentric to the column axis [see Fig. 9.4(a)] and if either the beam arrangement or the loading is asymmetrical will result in moments to be transmitted to the column. Wind loads in multi-storey buildings are usually applied at the respective floor levels and are assumed to be resisted by the bracings and hence in braced buildings do not cause large moments. However, in unbraced rigid framed buildings, the moments due to wind loads should also be taken into account in the design of columns [see Fig. 9.4(b)].

Similarly in industrial buildings, loads from crane and wind cause moments in columns as shown in Fig. 9.4(c). In such cases, wind load is applied to the column through the sheeting rails and may be taken as uniformly distributed throughout the length of the column. Design of columns subjected to axial load and bending moment is discussed in Chapter 13.

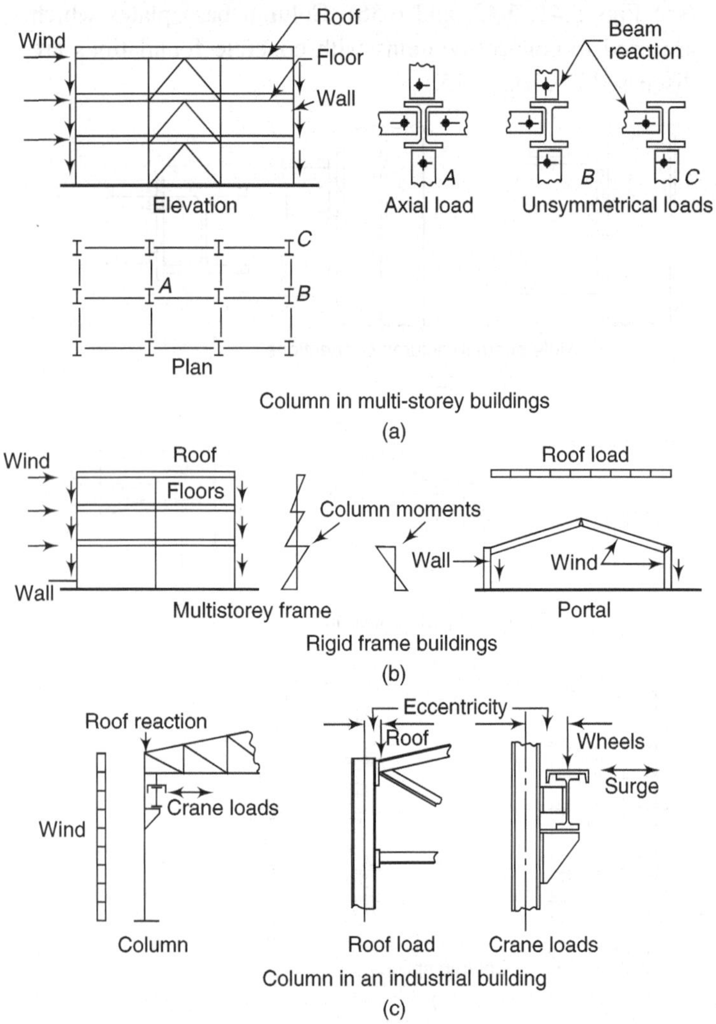


**FIG. 9.4** Loads and moments on compression members

The strength of a column depends on the following parameters:

1. Material of the column
2. Cross-sectional configuration
3. Length of the column
4. Support conditions at the ends (called restraint conditions)
5. Residual stresses
6. Imperfections

The imperfections include the following:

1. The material not being isotropic and homogenous
2. Geometric variations of columns
3. Eccentricity of load

It is difficult to assess the residual stress acting on each column cross section and also to assess the degree of support condition offered by a variety of connection details adopted in practice. Due to the large number of variables, which influence the strength of columns and beam-columns in the elastic and inelastic ranges, several researchers throughout the world have done extensive experimental and theoretical investigations on the behaviour of columns due to these variables.

In this chapter, the effect of these parameters on the strength and stability of columns are briefly discussed and the design, as per the Indian code of practice IS 800, is given. The method of design of open web columns (latticed, battened) are also provided.

## 9.2 POSSIBLE FAILURE MODES

The possible failure modes of an axially loaded column are discussed as follows:

***Local buckling*** Failure occurs by buckling of one or more individual plate elements, e.g., flange or web, with no overall deflection in the direction normal to the applied load. This failure mode may be prevented by selecting suitable width-to-thickness ratios of component plates. Alternatively, when slender plates are used, the design strength may be reduced (see Section 9.3).

***Squashing*** When the length is relatively small (stocky column) and its component plate elements are prevented from local buckling, then the column will be able to attain its full strength or 'squash load' (yield stress × area of cross section). See Section 9.4 also.

***Overall flexural buckling*** This mode of failure normally controls the design of most compression members. In this mode, failure of the member occurs by excessive deflection in the plane of the weaker principal axis. An increase in the length of the column, results in the column resisting progressively less loads (see Section 9.5).

***Torsional and flexural–torsional buckling*** Torsional buckling failure occurs by twisting about the shear centre in the longitudinal axis. A combination of flexure and twisting, called *flexural–torsional buckling* is also possible. Torsional buckling is a possible mode of failure for point symmetric sections. Flexural torsional buckling must be checked for open sections that are singly symmetric and for sections that have no symmetry. Note that open sections that are doubly symmetric or point symmetric are not subjected to flexural-torsional buckling, since their shear centre and centroid coincide. Closed sections are also immune to flexural–torsional buckling. Flexural–torsional buckling is discussed in Section 9.10.

In addition to these failure modes, in compound members (two or more shapes joined together to form a lattice cross section), failure of a component member may occur, if the joints between members are sparsely placed. Codes and specifications usually have rules to prevent such failures (see Section 9.15).

## 9.3 CLASSIFICATION OF CROSS SECTION

If individual plate elements which make up the cross section of a compression member, (for example, the web and two flanges in the case of an I-section), are thin, local buckling of the type shown in Fig. 9.5 may occur. For columns, it is frequently possible to eliminate this problem by limiting the proportions of component plates, such that local buckling will not influence the strength of the cross sections.

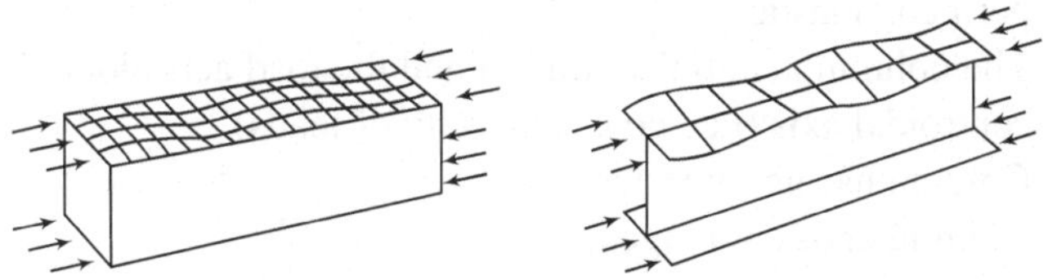


**FIG. 9.5** Local buckling in box- or I-section columns (deformation of a single flange only is shown for clarity)

The same classification that was set out for beams, in Section 8.13, is used for compression members. That is, to prevent local buckling, the width-to-thickness ratio of the different elements (web, flange, etc.,) of the compression member should be limited. Upper limits for the limiting width-to-thickness ratios for elements subjected to compression are given in Table 9.1. These limits are applicable to semi-compact sections. When more slender plates, having width-to-thickness ratios higher than the limits given in Table 9.1 are to be used, the strength of the section should be suitably reduced. However, it is a good practice to use plastic or compact sections for compression members, since they provide more stiffness than semi-compact or slender sections.

**TABLE 9.1** Limiting width-to-thickness ratios for axial compression elements

| Element | Ratio | Upper limit for semi-compact section |
|---|---|---|
| Outstand element of compression flange (I, H, or C) | | |
| (a) Rolled section | $b/t_f$ | $15.7\varepsilon$ |
| (b) Welded section | $b/t_f$ | $13.6\varepsilon$ |
| Internal element of compression flange (box) | $b/t_f$ | $42\varepsilon$ |
| Web of an I-H or box-section | $d/t_w$ | $126/(1+2r_2) \geq 42\varepsilon$ |
| Single angle or double angle with the components separated (all three criteria should be satisfied) | $b/t$<br>$d/t$<br>$(b+d)/t$ | $15.7\varepsilon$<br>$15.7\varepsilon$<br>$25\varepsilon$ |
| Circular hollow section | $D/t$ | $88\varepsilon^2$ |
| Hot rolled RHS—Flange | $b/t$ | $42\varepsilon$ |
| Web | $d/t$ | $126/(1+2r_2) \geq 42\varepsilon$ |
| Cold formed RHS—Flange | $b/t$ | $36.7\varepsilon$ |
| Web | $d/t$ | $110/(1+2r_2) \geq 37\varepsilon$ |

(*Contd*)

**TABLE 9.1** (*Contd*)

$\varepsilon = (250/f_y)^{0.5}$
For HR RHS: $b = B - 3t$; $d = D - 3t$
For CF RHS: $b = B - 5t$; $d = D - 5t$

## 9.4 BEHAVIOUR OF COMPRESSION MEMBERS

Before discussing the behaviour of compression members, it may be useful to know about the classification of compression members based on their length.

### 9.4.1 Long, Short, and Intermediate Compression Members

Compression members are sometimes classified as being long, short, or intermediate. A brief discussion about this classification is as follows:

***Short compression members*** For very short compression members the failure stress will equal the yield stress and no buckling will occur. Note that for a compression member to fall into this classification, it has to be so short (for an initially straight column $L \leq 88.85r$, for $f_y = 250$ MPa) that it will not have any practical application.

***Long compression members*** For these compression members, the Euler formula [see Section 9.5 and Eqn (9.9)] predicts the strength very well, where the axial buckling stress remains below proportional limit. Such compression members will buckle elastically.

***Intermediate length compression members*** For intermediate length compression members, some fibres would have yielded and some fibres will still be elastic. These compression members will fail both by yielding and buckling and their behaviour is said to be 'inelastic'. For the Euler formula Eqn (9.9) to be applicable to these compression members, it should be modified according to the reduced modulus concept or the tangent modulus concept (as is done in AISC code formula) to account for the presence of residual stresses. Now let us discuss the behaviour of short and slender columns.

### 9.4.2 Short Compression Members

Consider an axially compressed member of short length which is initially straight and made of material having the ideal rigid–plastic stress–strain relationship as shown in Fig. 9.6. At low values of external load $P$, there will be no visible deformation—neither lateral nor axial. Since $P$ is applied at the centroid of the section, apart from possible localized effects at the ends of the member, all parts of the member will experience the same value

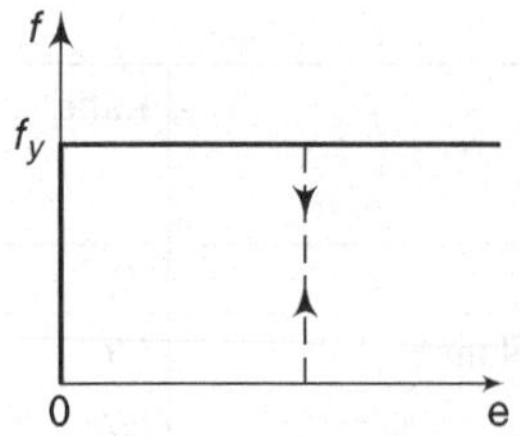


FIG. 9.6 Idealized material behaviour

of compressive stress $f_c = P/A$. Large deformation is possible only when $f_c$ reaches the yield stress $f_y$. At this stage, the member deforms axially. The value of the axial force at which this happens is termed as the '*squash load*' and is given by

$$P_y = f_y A \tag{9.1}$$

Typical laboratory compression test on the short length of the rolled section is often referred to as the *stub column test.* Since rolled steel sections have residual stresses, there will be non-uniform yielding of the cross section of the stub column. Thus, those fibres which contain residual compression have their effective yield point reduced; while those containing residual tension have theirs increased. However, in both cases, the squash load of the stub column can be achieved provided there is no local buckling. (Nethercot 2001).

### 9.4.3 Slender Compression Members

As mentioned earlier, the strength of the compression member decreases as its length increases, in contrast to the axially loaded tension member whose strength is independent of its length. Thus, the compressive strength of a very slender member may be much less than its tensile strength, as shown in Fig. 9.7 (also see Fig. 9.9). This decrease in strength is due to the following parameters, which are often grouped under the heading of imperfections: the initial lack of straightness, accidental eccentricities of loading, residual stresses, and variation of material properties over the cross section. In order to understand the effect of these parameters, we should study the behaviour of ideal, straight pin-ended columns, which is discussed in the next section.

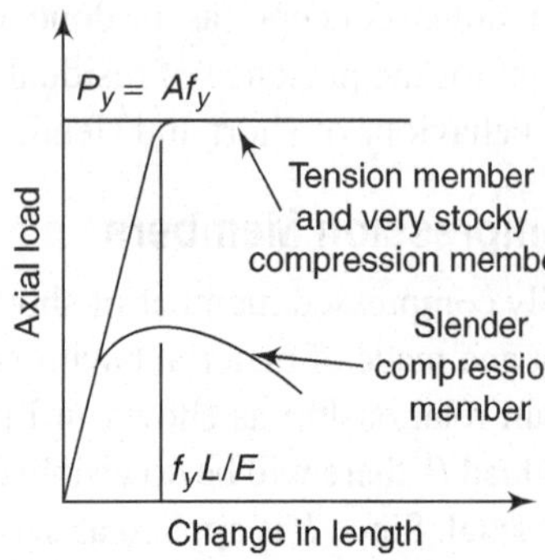


FIG. 9.7 Strengths of axially loaded members

## 9.5 ELASTIC BUCKLING OF SLENDER COMPRESSION MEMBERS

Though the first qualitative remarks on column strength and stability were given by Erone of Alexandria (75 BC) and similar descriptions of buckled columns by Leonardo Da Vinci (1452–1519), the theory of column buckling was first originated by Euler during 1744–1759 (Timoshenko 1953; Euler 1759). Euler considered an ideal column with the following attributes.

1. Material is isotropic and homogenous and is assumed to be perfectly elastic.
2. The column is initially straight and the load acts along the centroidal axis (i.e., no eccentricity of loads).
3. Column has no imperfections.
4. Column ends are hinged.

Such a column is also known as an *Euler column* and is shown in Fig. 9.8. A concentric load $P$ is applied to the upper end of the member, which remains straight until buckling occurs, when it is slightly bent as shown in Fig. 9.8.

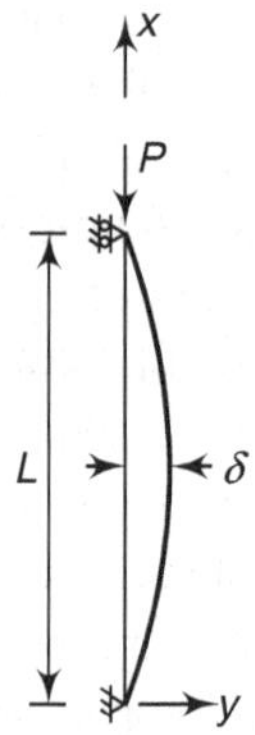


FIG. 9.8 Behaviour of a perfectly straight elastic pin-ended column

At any location $x$, the bending moment $M$ on the member bent slightly about the principal axis is

$$M = Py \tag{9.2}$$

and since

$$d^2y/dx^2 = -(M/EI) \tag{9.3}$$

where $E$ is Young's modulus and $I$ is the moment of inertia; the differential equation for the member becomes

$$(d^2y/dx^2) + (P/EI)y = 0 \tag{9.4}$$

After letting $k^2 = P/EI$, the solution of this second order differential equation may be expressed as

$$y = A_1 \sin kx + A_2 \cos kx \tag{9.5}$$

where $A_1$ and $A_2$ are unknown coefficients. Applying the boundary conditions (a) $y = 0$ at $x = 0$ and (b) $y = 0$ at $x = L$, one may obtain for condition (a),

$$A_2 = 0$$

and for condition (b),

$$0 = A_1 \sin kx \tag{9.6}$$

This equation is satisfied if $kL = N\pi$. Thus we get,

Buckling of a pin-ended column in a laboratory test
(Courtesy: Late Dr W. Kurth of Germany)

$$P = \pi^2EI/L^2,\ 4\pi^2EI/L^2, \ldots,\ n^2\pi^2EI/L^2 \tag{9.7}$$

The fundamental buckling mode, with a single curvature deflection ($y = A_1 \sin \pi x/L$) will occur when $N = 1$. Thus, the Euler critical load for a column with both ends hinged is

$$P_{cr} = \pi^2EI/L^2 \tag{9.8}$$

Or in terms of average critical stress, using $I = A_g r^2$

$$f_{cr} = P_{cr}/A_g = \pi^2E/\lambda^2 \tag{9.9}$$

where $\lambda$ is the slenderness ratio defined by

$$\lambda = L/r \tag{9.10}$$

Note that the buckling phenomenon is associated with the stiffness of the member. A member with low stiffness will buckle early than one with high stiffness. Increasing member lengths causes reduction in stiffness. The stiffness of the member is strongly influenced by the amount and distribution of the material in the cross section of the column; the value of $r$ reflects the way in which the material is distributed. Also note that any member will tend to buckle about the weak axis (associated with lesser ability to resist buckling).

Critical loads and critical stresses can be similarly found for struts having other end-restraint conditions. For example, if both ends are prevented from rotation as well as from lateral movement (fixed ends), the critical load may be $4\pi^2EI/L^2$. In such cases, the slenderness ratio is defined as $KL/r$, where $KL$ is the *effective length* of the member. This concept is discussed in Section 9.9.

Inspection of Eqn (9.9) indicates that the critical stress is inversely proportional to the slenderness ratio of the column and very large values of $f_{cr}$ can be obtained by using $L/r \to 0$. However, as per the differential equation for bending, Eqn (9.4), stress should be proportional to strain. Thus, the upper limit of validity is the proportional limit, which is often taken as $f_{cr} \to f_y$, though short columns may be loaded even into the strain-hardening range. Thus, one may write,

$$f_y = \pi^2E/\lambda_p^2 \tag{9.11}$$

Hence, $\lambda_p = \pi\sqrt{(E/f_y)}$ ; for $f_y = 250$ MPa, $\lambda_p = 88.85$ (9.12)

Thus, the changeover from yielding to buckling failure occurs at point $C$ (see Fig. 9.9), defined by the slenderness ratio $\lambda_p$. Figure 9.9(a) shows the strength curve for an axially loaded initially straight pin-ended column indicating the plastic yield (stocky columns) and elastic buckling (long column) behaviour. Figure 9.9(a) is often presented in a non-dimensional form as shown in Fig. 9.9(b).

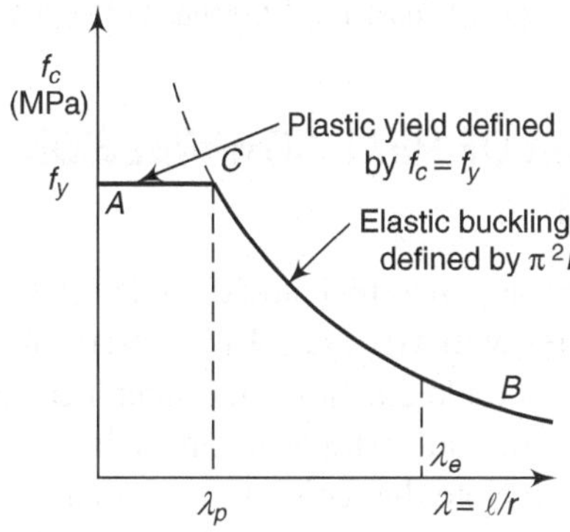


Strength curve
(a)

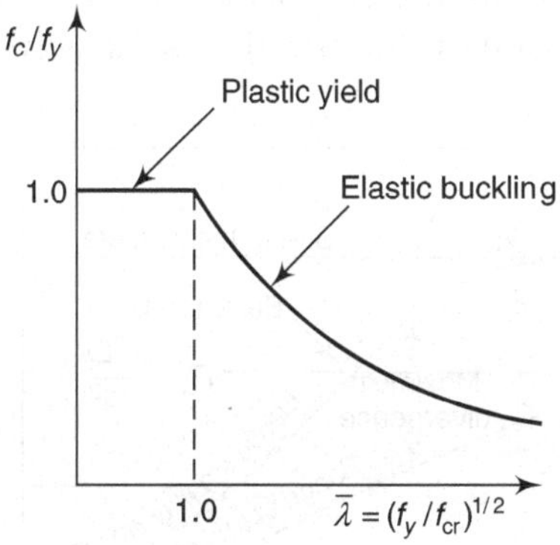


Strength curve in a non-dimensional form
(b)

**FIG. 9.9** Strength curve for an axially loaded initially straight pin-ended column

## 9.5.1 Large Deflection Behaviour: Post Buckling

The Euler analysis referred to above is applicable only if the deflections are small and the curvature can be expressed in the approximate form—$d^2y/dx^2$. The exact expression applicable to large as well as small deflections is (Chen & Lui 1987)

$$-\{(d^2y/dx^2)/[1 + (dy/dx)^2]^{3/2}\} \quad (9.13)$$

This expression leads to a more complex analysis, known as the theory of Elastica. The buckling behaviour which it predicts is slightly different from that given by Euler's theory [Fig. 9.10(a)] and is shown in Fig. 9.10(b).

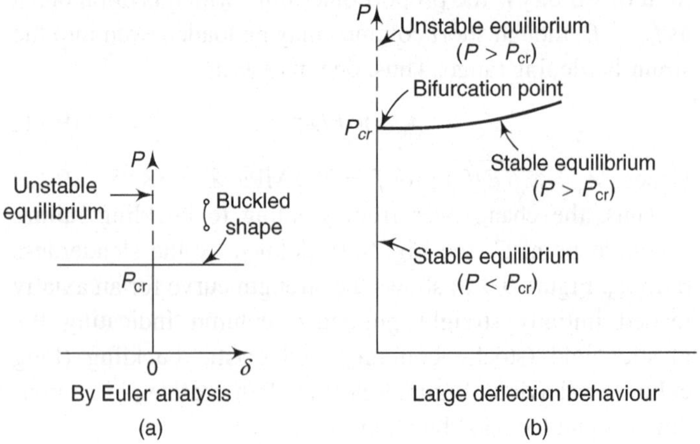


**FIG. 9.10** Load-deflection behaviour of a linear-elastic, initially straight, pin-ended column

## 9.6 BEHAVIOUR OF REAL COMPRESSION MEMBERS

Euler's approach was generally ignored for design because test results did not agree with it (see Fig. 9.11). Test results included effects of initial crookedness of the member, accidental eccentricity of load, end restraint, local or lateral buckling, and residual stress. Note that columns are usually an integral part of a structure and as such cannot behave independently. Due to the imperfections, the load deflection curve of a real column will be much different than that of an ideal column. Now let us consider these imperfections separately.

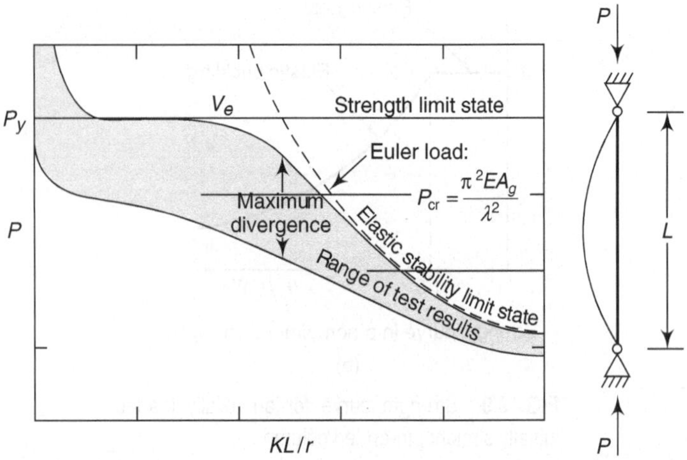


**FIG. 9.11** Typical experimental results showing column strength vs slenderness ratio

### 9.6.1 Initial Out-of-straightness

A pin-ended strut having an initial out-of-straightness, as shown in Figs 9.12(a) and 9.12(b), shows an initially straight,

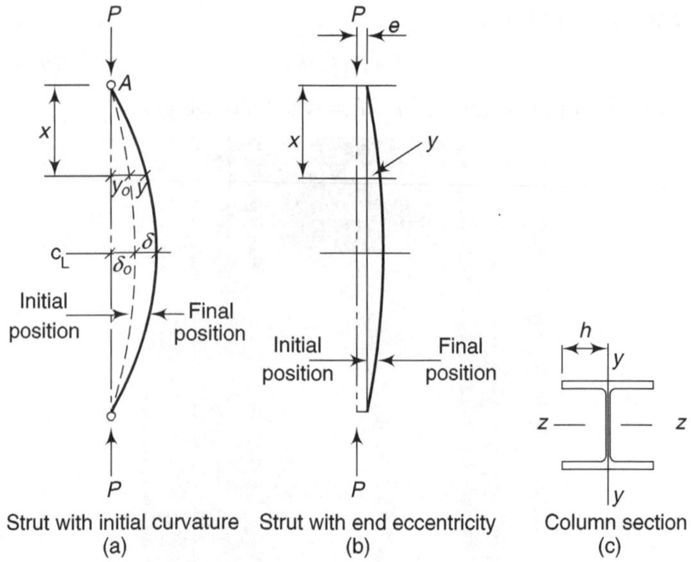


**FIG. 9.12** Pin-ended struts having imperfect geometry

pin-ended column in which loading is applied with an eccentricity. Both these cases lead to the superimposition of the bending stress on the applied axial stress.

Now consider a strut having initial out-of-straightness

$$y_o = \delta_0 \sin\pi x/L \quad (9.14)$$

where $\delta_0$ is the amplitude (initial bow or deflection at mid-height), which is known or can be found by measurement. (It has to be noted that a straightness tolerance of $0.001L$ generally, and $0.002L$ for members with hollow cross section, has been incorporated in the design rules of columns, as per Table 34 of IS 800:2007.) The initial curvature causes it to bend from the commencement of the application of the axial load, and this increases the maximum stress in the member. If the deflection of the centroidal axis of the strut, after the application of load $P$ at $x$ from $A$ is $y$, then the equilibrium equation is

$$EI(d^2y/dx^2) = -P(y + y_o) \quad (9.15)$$

The solution of Eqn (9.15) which satisfies the boundary conditions $y = 0$ at $x = 0$ and $y = 0$ at $x = L$ is the deflected shape

$$y = \delta \sin(\pi x/L) \quad (9.16)$$

where $\delta$ is the central deflection after the application of the axial load $P$. It is possible to show that, provided the material remains elastic, the central deflection is given by (Knowles 1987)

$$\delta = \delta_0/[1 - (P/P_{cr})] \quad (9.17)$$

Thus, the effect of the axial thrust $P$ is to increase the amplitude of the initial deflection by an *amplification factor* as follows

$$A.F. = 1/[1 - (P/P_{cr})] \quad (9.18)$$

In theory, therefore, as the applied load $P$ approaches the buckling load $P_{cr}$, the deflection tends to become infinity as shown in Fig. 9.13 (curve O′AB).

When the load increases, the deflection and also the bending moment increases. The resulting stress ($M/Z$) on the concave face of the column is compressive and hence adds to the average axial compressive stress ($f_c = P/A$). As the external load $P$ is increased further, first yield occurs when the stress on the concave face reaches the yield stress $f_y$. This load causing the first yield is designated as $P_y$ in Fig. 9.13(a). The corresponding stress distribution across the column is shown in Fig. 9.13(b). With further increase in the external load $P$, the zone of yielding spreads, both along the member and across the column cross section, which results in the deterioration of the bending stiffness of the column. Eventually, the column collapses when the failure load $P_f$ is reached [point D in the curve O′CDE of Fig. 9.13(a)]. The stress distribution across the cross section of the column at the collapse load is also shown in Fig. 9.13(b). Note that the extent of the post-first yield load increase (C to D in the curve) and the section plastification depend upon the slenderness ratio of the column.

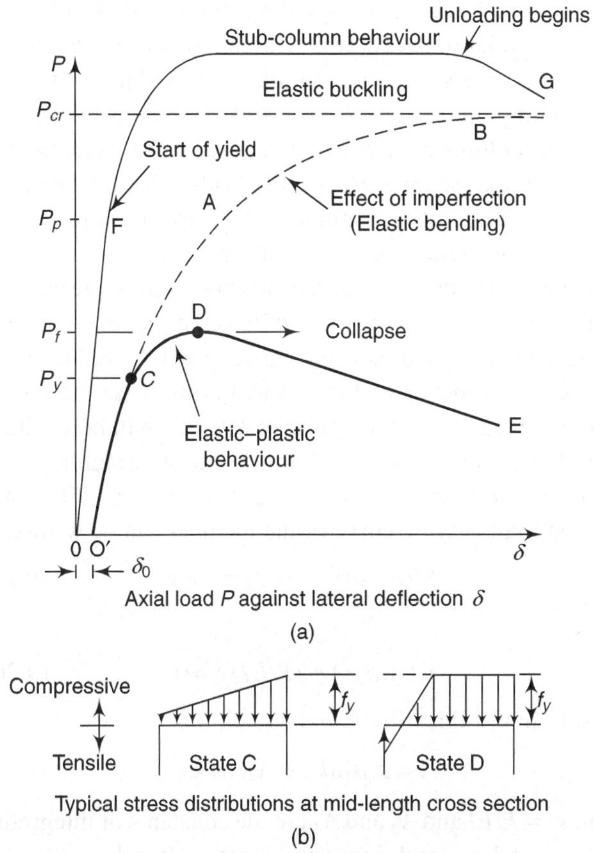


**FIG. 9.13** Load-deflection behaviour of struts with initial out-of-straightness

In Fig. 9.13(a), the stub column behaviour of the column (curve OFG) is also shown. From Fig. 9.13(a), it is clearly seen that $P_{cr}$ and $P_p$ are upper bounds to the loads $P_y$ and $P_f$, at first yield and at collapse. If the initial imperfection $\delta_0$ is small, $P_y$ may be close to $P_f$ and $P_p$. However, if the column is short and stocky, $P_{cr}$ will be very large, but $P_p$ can be expected to be close to $P_y$ (Dowling et al. 1988). In very slender columns, collapse will be by elastic buckling.

If a large number of columns with initial out-of-straightness are tested to failure and the data points representing the value of mean stress at failure are plotted against the slenderness ratio, the resulting lower bound curve to the test data will be as shown in Fig. 9.14(a). Thus, for short and stocky columns ($\lambda < \lambda_p$), the initial out-of-straightness will have negligible effect (since it is more a function of length) and hence they will attain the squash load $P_y$ ($P_y = Af_y$). For very slender columns ($\lambda > \lambda_e$), $P_y$ is much greater than $P_{cr}$ and hence the failure load will be close to the elastic critical load. However, for intermediate values of slenderness ratio ($\lambda_e < \lambda > \lambda_p$), the behaviour will be inelastic and the effect of initial out-of-straightness will be very pronounced (the greatest loss of strength will be approximately at $\bar{\lambda} = 1.0$ or $\lambda = \lambda_p$).

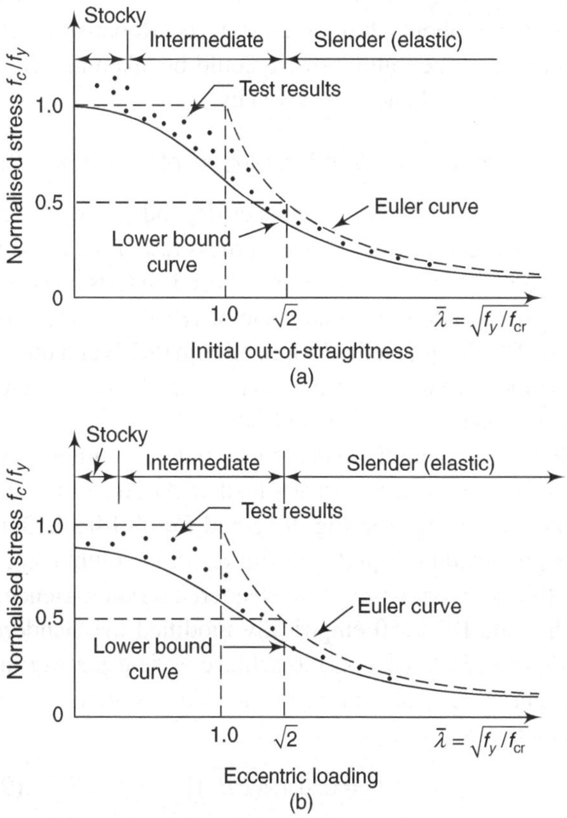


**FIG. 9.14** Typical strength curves for struts with initial out-of-straightness

The maximum stress at mid-height of the strut shown in Fig. 9.12(a) is given by

$$f_{max} = (P/A) + P\delta(h/I_y) \tag{9.19}$$

where $h$ is the distance from the centroid of the section to the extreme fibre on the concave side, $I_y$ is the moment of inertia about the weak axis.

The load at which $f_{max}$ reaches $f_y$ may be taken as the lower bound of the collapse load. Writing $I_y = Ar_y^2$, where $r_y$ is the radius of gyration about $y$-$y$ axis, letting $f_c = P/A$ (average axial stress), and $f_{cr} = P_{cr}/A$ (Euler stress), Eqn (9.19) may be written as

$$f_y = f_c + f_c\,(\delta_0 h/r_y^2)\,\{1/[1 - (f_c/f_{cr})]\} \tag{9.20}$$

Putting $\eta = \delta_0(h/r_y^2)$, which is called the *imperfection factor* or *Perry factor* and rearranging, we get the *Perry formula* (Ayrton & Perry 1886) as follows

$$(f_{cr} - f_c)\,(f_y - f_c) = \eta f_{cr} f_c \tag{9.21}$$

The compressive strength is given by the smaller root of the quadratic Eqn (9.21) and is given as follows (Allen & Bulson 1980)

$$f_c = \phi - [\phi^2 - f_{cr} f_y]^{0.5} \tag{9.22a}$$

where $\phi = [f_y + (\eta + 1)f_{cr}]/2$ (9.22b)

Based on about 200 column tests Robertson (1925) concluded that the initial bow $\delta_0$ could be taken as length of column/1000 and hence $\eta$ is given by

$$h = \delta_0 h/r_y^2 = (L/r_y)(0.001h/r_y) = \alpha(L/r_y) = \alpha\lambda \tag{9.23}$$

where $\alpha = (0.001h/r_y)$, a parameter dependent on the shape of the cross section, and $\lambda$ = slenderness ratio $L/r_y$. Robertson also evaluated the mean values of $\alpha$ based on tests on columns and suggested a lower bound value of $\alpha = 0.003$, for column designs. The design method based on Eqn (9.22) and the value of $\alpha$ mentioned earlier is termed as *Perry-Robertson Approach* and was adopted by the British Code.

It has been found that very stocky (stub) columns resisted loads in excess of their squash load ($f_yA$) due to the effect of strain hardening [see Fig. 9.11 and Fig. 9.14(a)]. But the values predicted by Eqn (9.22) will result in column strength lower than $f_y$, even for very low slenderness ratios. Hence, the British Code BS-5950 empirically modified the slenderness $\lambda$ in Eqn (9.23) to $(\lambda - \lambda_p)$ to achieve a yield plateau in the design curve. The value of $\lambda_p$ was arrived at to fit the observed test data as

$$\lambda_p = 0.2[\pi\sqrt{(E/f_y)}] \tag{9.24}$$

Thus, for calculating the elastic critical loads for $\lambda > \lambda_p$, the following modified formula was used.

$$f_{cr} = \pi^2 E/(\lambda - \lambda_p)^2 \text{ for } \lambda > \lambda_p \tag{9.25}$$

A comparison of this modification with the Robertson's curve and the test data are shown in Fig. 9.15.

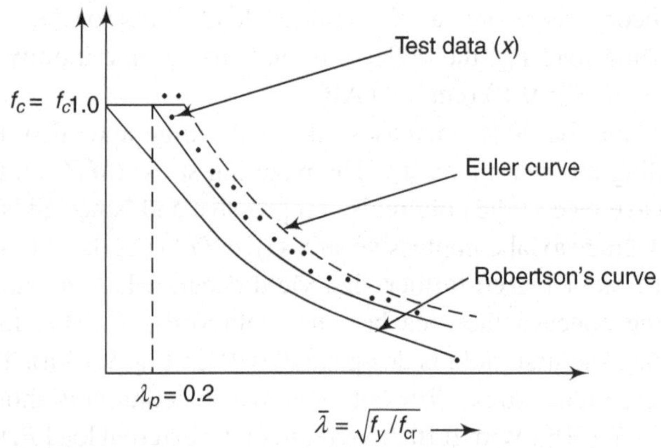


**FIG. 9.15** Design strut curve with a plateau at low slenderness values

## 9.6.2 Eccentricity of Applied Load

An initially straight pin-ended column, to which the loading is applied with an eccentricity $e$, is shown in Fig. 9.12(b). [Note that in real columns it is impossible to ensure that the load is applied at the exact centroid of the column (see Fig. 9.4)]. The applied load induces a bending moment, equal to $P \times e$, in the member, which will also cause the column to deflect laterally in a way similar to the column with initial out-of-straightness, as described in Section 9.6.1. Similar to the column with initial out-of-straightness, the greatest stress occurs at the concave face of the column at mid-height, leading to the yielding of the member at this location. Hence, the load-deflection behaviour in this case also will be similar to that shown in Fig. 9.13(a), except that the deflection is zero at zero load.

In practical situations, the eccentricity $e$ will be proportional to the size of the column and will not be a function of its length. Thus, the load carrying capacity will be reduced even for stocky members [see Fig. 9.14(b)]; otherwise the form of the strength curve obtained from test results will be similar to that obtained for columns with initial out-of-straightness.

For the column shown in Fig. 9.12(b), the differential equation to the elastic curve of the member can be written as

$$EI(d^2y/dx^2) = -M = -Py \tag{9.26a}$$

Hence,

$$(d^2y/dx^2) + (P/EI)y = 0 \tag{9.26b}$$

The solution of this differential equation is

$$y = A_3 \sin kx + A_4 \cos kx \tag{9.27a}$$

where $k^2 = P/EI$ and $A_3$ and $A_4$ are the constants of integration.

The boundary conditions are $y = e$ at $x = 0$ and $y = e$ at $x = L$. Substituting these conditions, and evaluating the constants, we finally get

$$y_{max} = e[\{(1 - \cos kL)/\sin kL\}\,\sin kx + \cos kx] \tag{9.27b}$$

The maximum deflection at mid-length ($x = L/2$) is obtained as

$$y_{max} = e[\{(1 - \cos kL)/\sin kL\}\sin(kL/2) + \cos(kL/2)] \quad (9.27c)$$

Noting that

$$\sin kL = 2\sin(kL/2)\cos(kL/2) \text{ and } 1 - \cos kL = 2\sin^2(kL/2)$$

Equation 9.27(c) could be simplified to

$$y_{max} = e\sec(kL/2) \quad (9.27d)$$

Thus, the maximum bending moment is

$$M_{max} = Pe\sec(kL/2) \quad (9.28)$$

The maximum stress becomes

$$f_{max} = (P/A) + (M_{max}/I)c \quad (9.29a)$$

where $c$ is the distance of the compression fibre from the centroidal axis. Substituting Eqn (9.27d) into Eqn (9.29a) we get,

$$f_{max} = (P/A) + (Pce/I)\sec(kL/2) \quad (9.29b)$$

Putting $I = Ar^2$ and $k = \sqrt{(P/EI)}$, the expression Eqn (9.29b) becomes

$$f_{max} = (P/A)\,[1 + (ec/r^2)\sec\{(L/2r)\sqrt{(P/EA)}\}] \quad (9.29c)$$

Denoting the average stress $(P/A)$ by $f_c$ and taking the value of $f_{max}$ as $f_y$, we get

$$f_c = f_y/[1 + (ec/r^2)\sec\{(L/2r)\sqrt{(P/EA)}\}] \quad (9.30)$$

This is called the *secant formula*. It may be of interest to note that the Indian Standard IS 800-1962 adopted the secant formula as the basis for predicting the average axial stress in the member with a factor of safety of 1.68 and adopting a value of $ec/r^2 = 0.20$, which was found to give the average value for a large number of column tests (Arya & Ajmani 1989).

## 9.6.3 Effect of Residual Stress

As we have discussed in Section 1.9 of Chapter 1, structural members normally contain residual stresses, due to the uneven cooling of different parts of the cross section. It can also be observed from Figs 1.24 to 1.26 of Chapter 1 that the pattern of residual stress distribution is different for rolled and welded cross sections and that the residual tensile stresses can sometimes be very high and may even reach up to yield stress (see Fig. 1.27).

When a column is subjected to axial compression, the uniform applied stresses $f_c$ is added to the residual stress $f_r$, leading to a non-uniform stress distribution. Assuming that the residual stress distribution is as per Fig. 1.27(a), the greatest stress occurs at the flange tips as shown in Fig. 9.16(a). (The stress distribution in the flange only is shown for clarity.) When the total stress $(f_c + f_r)$ is below the yield stress $f_y$, the member deforms elastically. However, when $(f_c + f_r)$ reaches $f_y$, yielding begins at the flange tips as shown in Fig. 9.16(b). Increasing the load further will result in the spread of the yield in the flanges and also in the web (see Fig. 9.17).

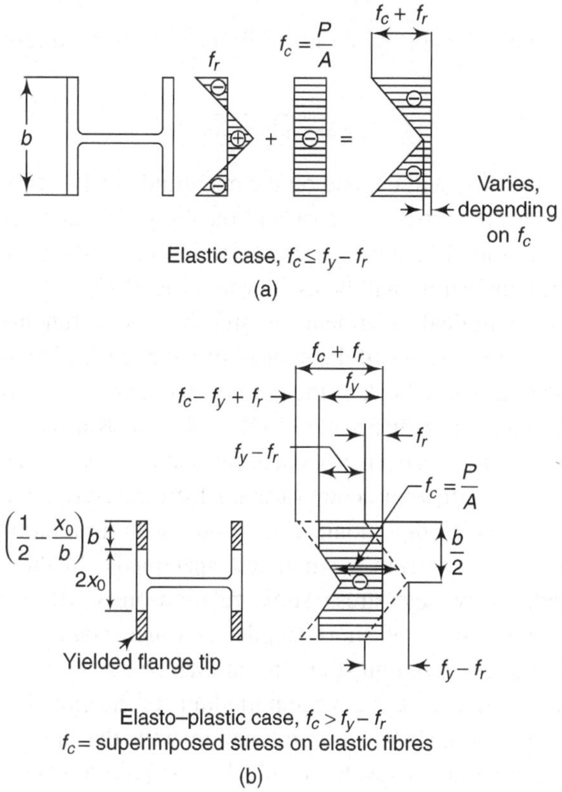


**FIG. 9.16** Stress distribution in a column with linear residual stress ($f_r$ may be assumed as $f_y/3$)

When the ends of the flange start yielding, we should only use the effective elastic area and effective moment of inertia to calculate the stiffness. Thus, the stiffness and moment curvature relationship will be continuously changing as the axial load is increased beyond the proportional limit $f_{pr}$. Thus,

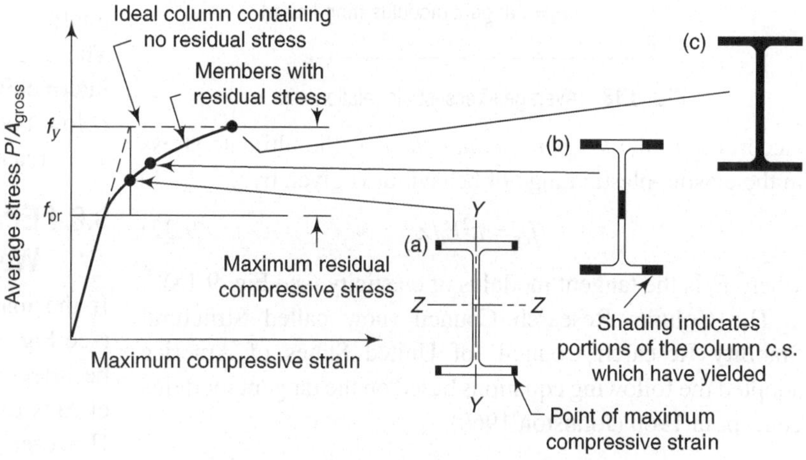


**FIG. 9.17** Influence of residual stress on average stress–strain curve

when parts of the cross section starts yielding, the resisting moment will be given by Eqn (9.9) but with the moment of inertia replaced by the moment of inertia of the elastic portion of the cross section $I_e$. Hence, the critical buckling stress is now given by [compare with Eqn (9.9)]

$$f_{cr} = \pi^2 E/(L/r_e)^2 \tag{9.31}$$

where the radius of gyration $r$ in Eqn (9.9) has been replaced by

$$r_e = \sqrt{(I_e/A_e)}$$

The value of $r_e$ will depend on the magnitude and distribution of the residual stress pattern and on the proportions of the cross section. The average stress–strain curve for a column with residual stress will be as shown in Fig. 9.17.

The analytical treatment of stiffness as a function of induced stress has been the subject of a considerable amount of research dating back to the late nineteenth century (Young 1807; Engesser 1889; Shanley 1946, 1947; von Karman 1908). Two methods have been proposed, one using the residual stress distribution (either the actual variation from measurements or a mathematical model) along with the stress–strain relationship of the material (from a small test specimen). In the other method, an average stress–strain relationship is determined from experiments of short lengths of rolled shapes having residual stress. Column strength can then be determined from the test results using the tangent modulus of the experimental curves (see Fig. 9.18) in combination with the appropriate slenderness ratio. Thus, the residual stress pattern is not used in the second approach (Yu & Tall 1971).

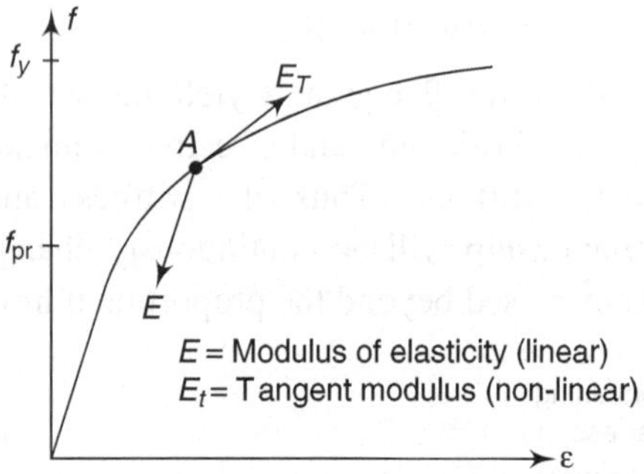


**FIG. 9.18** Average stress–strain relationship

According to the *tangent modulus theory*, the ultimate stress in the elastic–plastic range of behaviour is given by

$$f_{cr} = \pi^2 E_t/\lambda^2 \tag{9.32}$$

where $E_t$ is the tangent modulus of elasticity (see Fig. 9.18).

The Column Research Council (now called Structural Stability Research Council) of United States of America adopted the following equations based on the tangent modulus concept in 1960 (Johnston 1960)

$$(f_c/f_y) = 1 - \bar{\lambda}^2/4 \text{ for } \bar{\lambda} \le \sqrt{2} \tag{9.33a}$$

$$= 1/\bar{\lambda}^2 \text{ for } \bar{\lambda} \le \sqrt{2} \tag{9.33b}$$

where $\bar{\lambda} = \sqrt{(f_y/f_{cr})} = (KL/r)(1/\pi)\sqrt{(f_y/E)}$ (9.33c)

Note that at $\bar{\lambda} = \sqrt{2}$, the preceding parabolic equation and the Euler hyperbola become tangent to each other (see Fig. 9.19). This figure also shows the difference in behaviour between strong and weak axis of H-column. The preceding equations are slightly modified in the current ANSI/AISC 360-10 specifications to provide a constant factor of safety for all columns and for an initial out-of-straightness of about $L/1500$ as (Salmon & Johnson 1996)

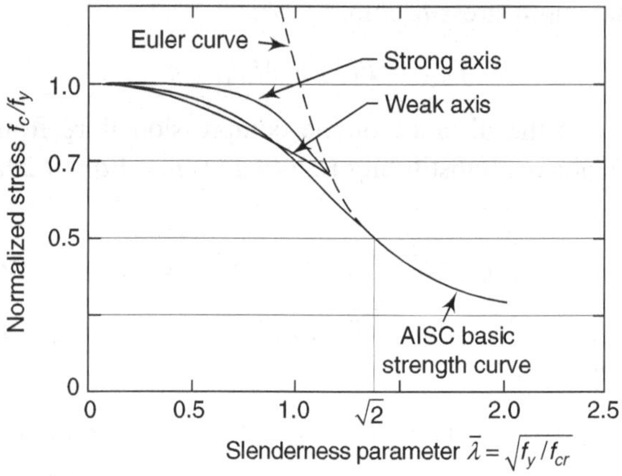


**FIG. 9.19** Column strength curves as per AISC code for H-section

$$P_n = A_g f_c \tag{9. 34a}$$

where $f_c = (0.658^{f_y/f_{cr}})f_y$ for $(KL/r) \le 4.71\sqrt{(E/f_y)}$ or $f_{cr} \ge 0.44 f_y$ (9.34b)

$$f_c = 0.877 f_{cr} \text{ for } (KL/r) > 4.71\sqrt{(E/f_y)} \text{ or } f_{cr} < 0.44 f_y \tag{9.34c}$$

and $$f_{cr} = \pi^2 E/\lambda^2 \tag{9.34d}$$

Note that for very stocky columns, the collapse load is not affected by residual stress. Very slender columns are also unaffected by residual stresses as they buckle elastically. The effect of residual stress is much felt in columns having intermediate slenderness, which buckle inelastically, at loads below both elastic critical load and the plastic squash load. The greatest reduction occurs at $\bar{\lambda} = 1.0$.

### 9.6.4 Effect of Strain Hardening and the Absence of Well-defined Yield Point

If the material experiences strain-hardening at high strains (see Fig. 1.11 of Chapter 1), the onset of first yield may not be affected but there may be increase in collapse load [as already discussed in Section 9.6.1 and shown in Fig. 9.14(a)]. However, this increase in collapse load is often ignored in designs, leading to higher margin of safety for short columns.

High-strength columns may have stress–strain curves, which will show non-linear material behaviour above the limit of proportionality $f_{pr}$ [see Fig. 1.11(a) of Chapter 1]. Moreover, high-strength steels have a yield stress, beyond which the stress–strain curve becomes more or less horizontal. Some steels and aluminium do not have a plastic plateau, but experience strain hardening throughout the inelastic range (similar to the stress–strain curve shown in Fig. 9.18). In such cases, the yield stress is often taken as the 0.2% proof stress, for design purposes [see Fig. 1.11(a)]. The strength curve for struts having these types of material behaviour is similar to the struts of ideal elastic–plastic material having residual stress (Fig. 9.19).

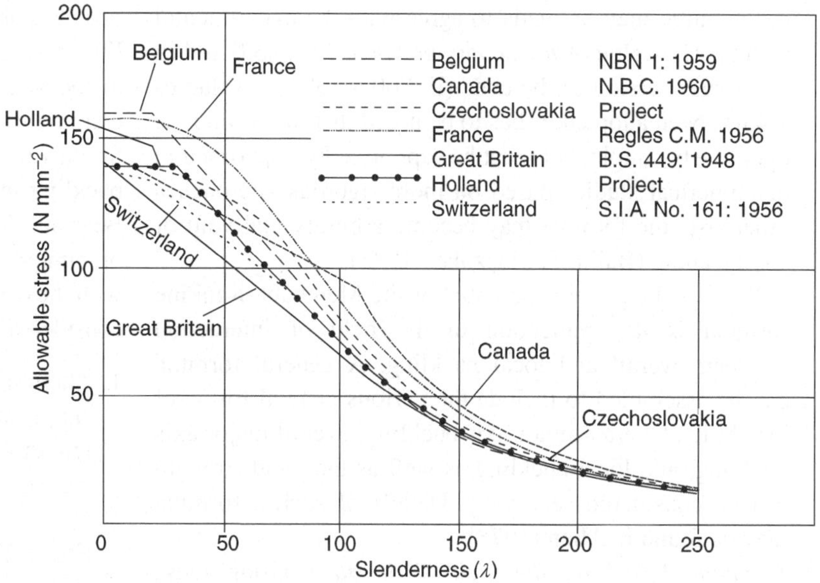


**FIG. 9.20** Comparison of allowable stress formulae adopted in different codes (Ganapathy 1975)

## 9.7 DEVELOPMENT OF MULTIPLE COLUMN CURVES

In real columns, all the effects mentioned in Section 9.6 occur simultaneously, i.e., out-of-straightness, eccentricity of loading, residual stresses, also depending on the material, lack of clearly defined yield point, and strain hardening. Note that only strain hardening tends to raise the column strength (that too only for short stocky columns), while the other effects lower the column strength for all or part of the slenderness ratio range.

Thus, the determination of the maximum strength of columns is a complicated process, often involving numerical integration, especially when initial imperfections and material non-linearities or residual stresses must be considered (see Allen & Bulson 1980—for the details of numerical integration techniques). These procedures are not suitable for design office use since they require lengthy calculations. Hence, simplified design formulae were developed and used in various design codes (Johnson 1976; Ganapathy 1975). These formulae (for the allowable stress method) are compared in Fig. 9.20. The establishment of an acceptable single formula for the design stress had been a subject of controversy, since, as we have discussed, there are several parameters that influence the strength.

In the past, four basic methods were used to establish column design formulae, curves or charts (Galambos 1998).

1. *Empirical formulae based on the results of column tests*: Such formulae are applicable only to the material and cross section on which the tests were performed. The earliest column formulae are of this type. However, a few formulae have been suggested during 1980s, based on computerized data banks, containing large amounts of column tests reported in the literature (e.g., Fukumoto et al. 1983). Since they are entirely based on tests and do not account for end restraints in a rational way, they are not accepted in modern codes. One such formula, based on Merchant–Rankine approach was adopted in the 1984 version of IS 800 and is discussed as follows.

***Merchant–Rankine approach***

It is possible to suggest a formula to correct the Euler load given in Eqn (9.31) in order that the predicted load complies with the following characteristics: For low values of $L/r$, the failure stress $f_c$ tends to become the yield stress of steel and for large value of $L/r$, it becomes the Euler's critical stress $f_{cr}$. Such a formula, as adopted in the earlier version of the code (IS 800 – 1984), is called the *Merchant–Rankine formula,* which is written in the following form:

$$(1/f_c^n) = (1/f_{cr}^n) + (1/f_y^n) \tag{9.35a}$$

where $f_c$ is the stress at failure, $f_{cr}$ is the elastic critical compressive stress (Euler's stress), and $f_y$ is the yield stress of steel.

The formula given in Eqn (9.35a) can be written as

$$f_c^n = (f_y^n f_{cr}^n)/(f_{cr}^n + f_y^n);$$

$$\text{or } f_c = (f_y f_{cr})/(f_{cr}^n + f_y^n)^{1/n} = f_y/[1 + (f_y/f_{cr})^n]^{1/n} \tag{9.35b}$$

The 1984 version of the Indian code (IS 800) adopted this formula with a factor of safety of 1.67, thus

$$f_{ac} = 0.6 f_y\ [1/[1 + (f_y/f_{cr})^n]^{1/n}] \tag{9.35c}$$

For zero slenderness ratio, the column fails though yielding, and the allowable compressive stress $f_{ac} = 0.6f_y$, with the multiplication factor becoming unity, since $f_{cr}$ becomes infinity. Thus, in the preceding formula, the buckling effect of the column is considered by the multiplying factor. By varying the exponential coefficient

$n$, the curve may be made to agree with the experimental results. The value of $n$ can vary between 1.0 to 3.0 and in the earlier version of the code (IS 800 – 1984), a value of 1.4 has been adopted. It has to be noted that the formula is operatively simple, but must be supported by experimental or numerical studies based on more rigorous approaches otherwise, the formula may become arbitrary and void of any meaning (Ballio & Mazzolani 1983).

Perhaps the greatest potential of the Merchant–Rankine formula is its application to the field of interaction between overall and local buckling. A general formula can be assembled to include the various general forms of instability. Overall minor axis buckling, overall major axis buckling, and local buckling as well as the yield strength can be considered separately. Details of such a formula may be found in Allen (1978).

2. *Formulae based on the yield limit state*: Using these formulae, the strength of the column is expressed in terms of the axial load that will give an elastic stress for an initially imperfect column which is then equated to the yield stress (as discussed in Section 9.6.1). An example of this type of formula is the Perry–Robertson formula given in the current IS 800 code and the British code BS: 5950 – Part1-2000. As discussed in Section 9.6.1, empirical factors are included to account for the initial imperfections of geometry and loading, but these formulae do not directly consider the inelastic column strength and account for end restraint rationally (Galambos 1998).
3. *Formulae based on tangent modulus theory*: These formulae rationally account for the bifurcation load (but not the maximum strength) of perfectly straight columns. The effects of imperfections are indirectly accounted for by considering that the maximum strength is reduced to the tangent-modulus strength. The CRC column strength curve as discussed in Section 9.6.3 is an example of this type of formula. It has been found that these curves can also account for the effect of end restraints.
4. *Formulae based on maximum strength*: The column design curves used in modern codes are basically obtained from numerical fit of curves obtained from maximum strength analysis of representative geometrically imperfect columns containing residual stress and are discussed as follows.

In order to develop a rational basis for predicting the strength of real columns (which have initial imperfections, end restraints, eccentricity of loading, and residual stresses), the committee 8 on instability (headed by Prof. Beer from Austria) of the European Convention for Constructional Steel work (founded in the year 1955 and shortly known as ECCS) launched an extremely large series of experiments on axially loaded columns. Seven European countries were involved and about thousand buckling tests were performed (Sfintesco 1970). In addition, Beer and Schulz (1970) developed strength curves by means of numerical simulation program and correlated the results with the experimental results. In the numerical simulation, they considered the following.

1. The initial configuration of columns is having a sinusoidal shape with amplitude $\delta_0$ equal to 1/1000th of the column length (which was justified by the actual test results and statistical evaluation of measured values).
2. Conventional distribution of residual stresses as shown in Fig. 1.27 of Chapter 1 that schematically represent those experimentally determined.

A statistical control of the numerical simulation method was also performed according to the Monte Carlo method. Based on these results, the ECCS committee concluded that in order to represent the real strength of columns, five different curves have to be used, depending on different cross sections of columns. These curves, called the *European multiple column curves*, are shown in Fig. 9.21.

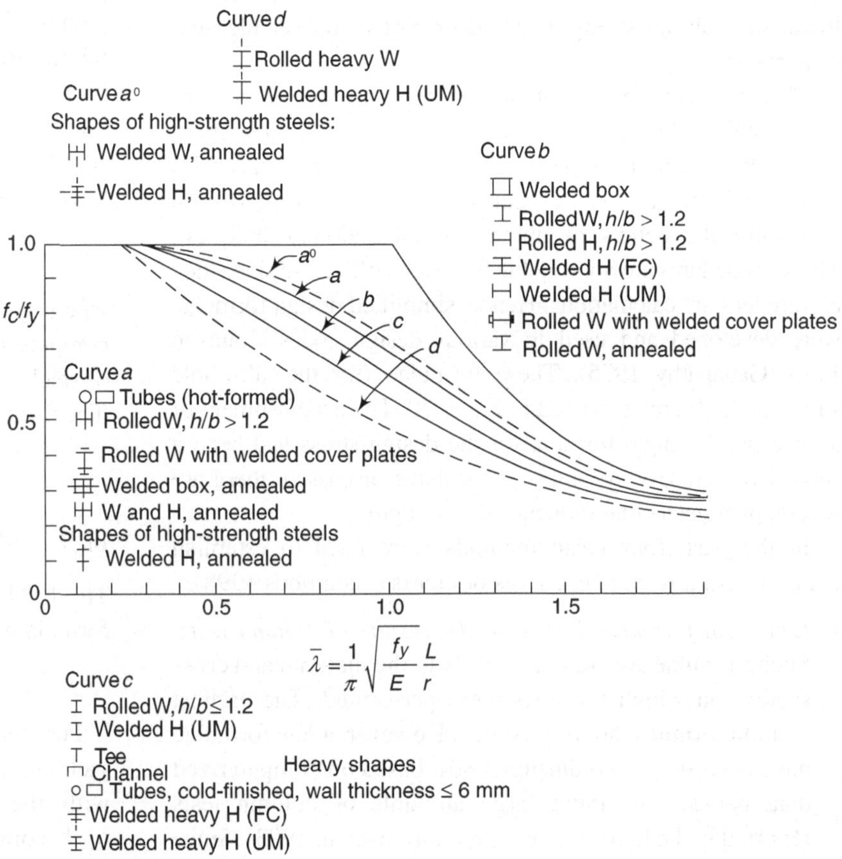


**FIG. 9.21** European multiple column curves (ECCS 1978) based on initial out-of-straightness, $\delta_0 = 0.001L$)

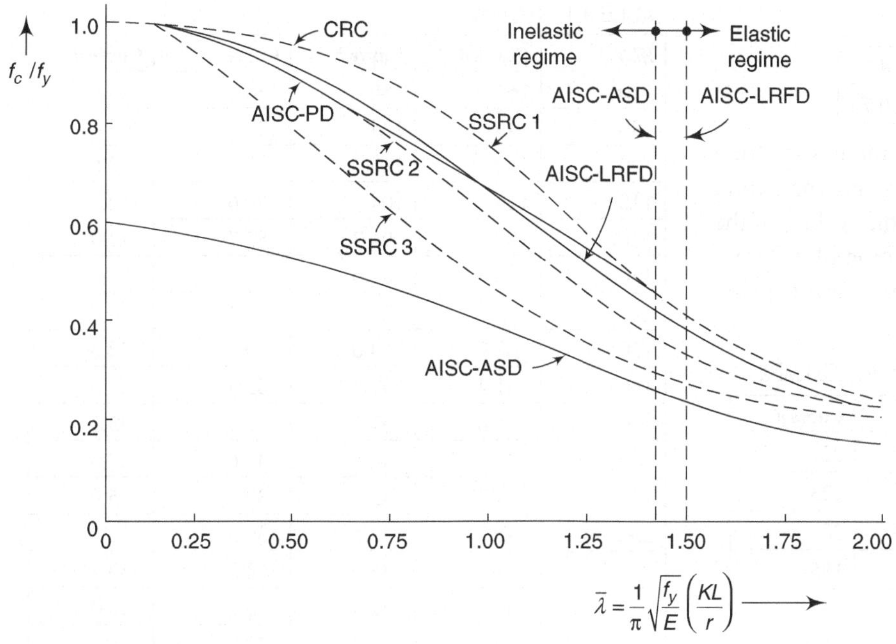


**FIG. 9.22** SSRC column strength curves developed by Bjorhovde

Rondal and Maquoi (1978) proposed two alternative equations to represent the ECCS curves and one such equation is as follows:

$$f_u/f_y = \beta - \sqrt{[\beta^2 - (1/\lambda^{-2})]} \qquad (9.36a)$$

where $\beta = [\{1 + \alpha(\bar{\lambda}^2 - 0.20) + \bar{\lambda}^2\}/(2\,\bar{\lambda}^2)]$ (9.36b)

The value of $a$ in this equation for curves $a_0$, $a$, $b$, $c$, and $d$ are taken as 0.125, 0.206, 0.339, 0.489, and 0.756, respectively. Lindner (1978) also proposed equations to represent ECCS curves.

Based on an extensive numerical study, Bjorhovde (1972) generated a set of 112 column curves for members for which measured residual stress distribution were available, assuming that the initial crookedness was of a sinusoidal shape having maximum amplitude of 1/1000[th] of the column length and that the end restraint is zero. These curves were reduced to a set of three column curves and adopted by SSRC. These curves known as *SSRC column strength curves* are shown in Fig. 9.22. Curve 1 is used for stress-relieved shapes, circular, and extruded box shapes and I-shaped columns of high-strength steel. Curve 2 is used for hot-rolled light and medium shapes as well as for welded built-up H-shapes bent about the major axis. Curve 3 is used for heavy rolled shapes and for low-grade welded shapes. Figure 9.22 also shows the AISC—allowable stress design (ASD), load and resistance factor design (LRFD), and plastic design (PD) curves, for comparison.

Algebraic representation of the three column strength curves, obtained by curve fitting are also available (Galambos 1998; Chen & Lui 1991). Rondal and Maquoi (1979) proposed the following equation to represent the SSRC curves.

$$\frac{f_u}{f_y} = \beta_1 - \sqrt{\left[\beta_1^2 - \left(\frac{1}{\bar{\lambda}}\right)^2\right]} \qquad (9.37a)$$

where $\beta_1 = (1 + \eta + \bar{\lambda}^2)/(2\bar{\lambda}^2)$ (9.37b)

$$\eta = \alpha(\bar{\lambda} - 0.15) \qquad (9.37c)$$

$\alpha = 0.103$, 0.293, and 0.622 for curves 1, 2, and 3, respectively.

Several European countries have adopted the ECCS curves. Eurocode 3 has adopted a modified form of ECCS curves. The Canadian code has adopted the SSRC column curves 1 and 2 and uses the following equation (Galambos 1998).

$$(f_u/f_y) = 0.9\text{A}\,(1 + \bar{\lambda}^{2n})^{-(1/n)} \qquad (9.37d)$$

with $n = 2.24$ (welded members) and 1.34 (hot-rolled members) for curves 1 and 2 respectively.

### 9.7.1 Multiple Column Curves in the IS Code

The Indian Code (IS 800) has adopted the multiple column curves, as shown in Fig. 9.23, which are similar to the curves given in the British Code BS 5950(Part 1)-2000, which is based on the Perry–Robertson approach. The design compressive strength is given by

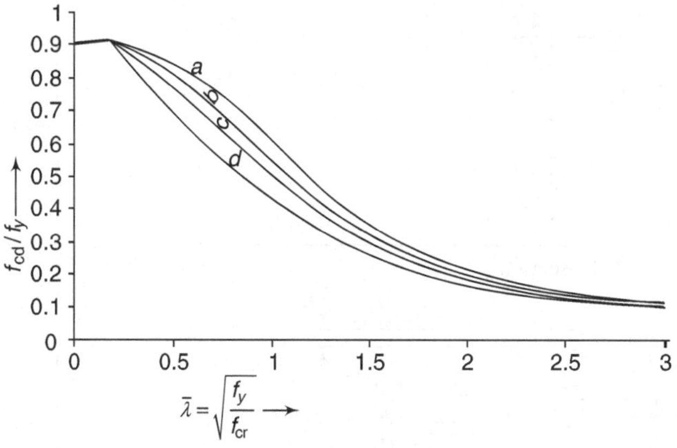


**FIG. 9.23** Column buckling curves as per IS 800-2007

$$P_d = A_e f_{cd} \qquad (9.38)$$

where $A_e$ is the effective sectional area.

The design stress in compression $f_{cd}$ is given by

$$f_{cd} = [(f_y/\gamma_{m0})/\{\phi + (\phi^2 - \bar{\lambda}^2)^{0.5}\}] = \chi f_y/\gamma_{m0} \le f_y/\gamma_{m0} \qquad (9.39a)$$

where $\phi = 0.5[1 + \alpha(\bar{\lambda} - 0.2) + \bar{\lambda}^2]$ (9.39b)

and $\bar{\lambda}$ is the non-dmensional effective slenderness ratio = $\sqrt{(f_y/f_{cr})}$, $f_{cr}$ is the Euler buckling stress = $\pi^2E/(KL/r)^2$, $KL/r$ is the effective slenderness ratio, $\alpha$ is the imperfection factor (see Table 9.2), $\chi$ is the stress reduction factor, and $\gamma_{m0}$ is the partial safety factor for material strength = 1.10.

**TABLE 9.2** Imperfection factor $\alpha$

| Buckling curve | *a* | *b* | *c* | *d* |
|---|---|---|---|---|
| $\alpha$ | 0.21 | 0.34 | 0.49 | 0.76 |

The stress reduction factor $\chi$ and the design compressive stress $f_{cd}$ for different buckling curves and effective slenderness ratio are given in Table 9.3 for $f_y = 250$ MPa. For other values of the yield stress, refer IS 800:2007. The classification of different sections under different buckling curves *a*, *b*, *c*, or *d* is given in Table 9.4.

**TABLE 9.3** Design compressive stress $f_{cd}$ (MPa) for $fy = 250$ MPa

| $KL/r\downarrow$ | Curve *a* | Curve *b* | Curve *c* | Curve *d* |
|---|---|---|---|---|
| 10 | 227 | 227 | 227 | 227 |
| 20 | 226 | 225 | 224 | 223 |
| 30 | 220 | 216 | 211 | 204 |
| 40 | 213 | 206 | 198 | 185 |
| 50 | 205 | 194 | 183 | 167 |
| 60 | 195 | 181 | 168 | 150 |
| 70 | 182 | 166 | 152 | 133 |
| 80 | 167 | 150 | 136 | 118 |
| 90 | 149 | 134 | 121 | 105 |
| 100 | 132 | 118 | 107 | 92.6 |
| 110 | 115 | 104 | 94.6 | 82.1 |
| 120 | 101 | 91.7 | 83.7 | 73.0 |
| 130 | 88.3 | 81.0 | 74.3 | 65.2 |
| 140 | 77.8 | 71.8 | 66.2 | 58.4 |
| 150 | 68.9 | 64.0 | 59.2 | 52.6 |
| 160 | 61.4 | 57.3 | 53.3 | 47.5 |
| 170 | 55.0 | 51.5 | 48.1 | 43.1 |
| 180 | 49.5 | 46.5 | 43.6 | 39.3 |
| 190 | 44.7 | 42.2 | 39.7 | 35.9 |
| 200 | 40.7 | 38.5 | 36.3 | 33.0 |
| 210 | 37.1 | 35.2 | 33.3 | 30.4 |
| 220 | 34.0 | 32.3 | 30.6 | 28.0 |
| 230 | 31.2 | 29.8 | 28.3 | 26.0 |
| 240 | 28.8 | 27.5 | 26.2 | 24.1 |
| 250 | 26.6 | 25.5 | 24.3 | 22.5 |

**TABLE 9.4** Selection of buckling curve depending on cross section

| Cross section | Limits | | Buckling about axis | Buckling curve |
|---|---|---|---|---|
| Rolled I-sections 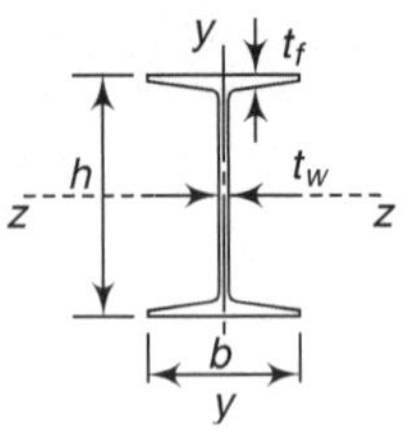 | $h/b > 1.2$: | $t_f \le 40$ mm | *z-z* | *a* |
| | | | *y-y* | *b* |
| | | 40 mm $< t_f \le 100$ mm | *z-z* | *b* |
| | | | *y-y* | *c* |
| | $h/b \le 1.2$: | $t_f \le 100$ mm | *z-z* | *b* |
| | | | *y-y* | *c* |
| | | $t_f > 100$ mm | *z-z* | *d* |
| | | | *y-y* | *d* |
| Welded I-section 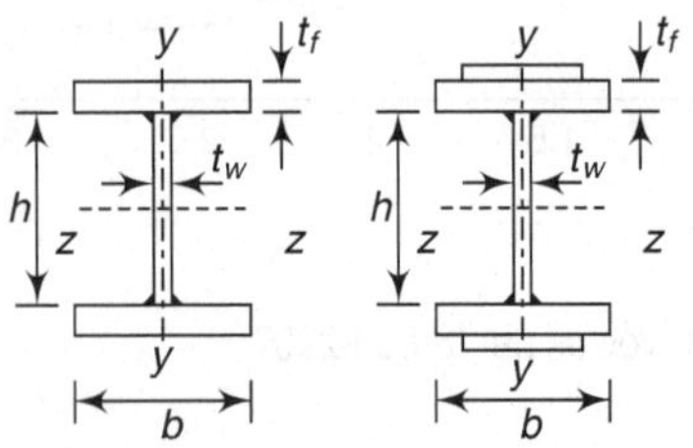 | | $t_f \le 40$ mm | *z-z* | *b* |
| | | | *y-y* | *c* |
| | | $t_f > 40$ mm | *z-z* | *c* |
| | | | *y-y* | *d* |
| Hollow section 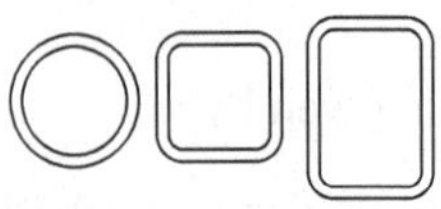 | Hot rolled | Any | | *a* |
| | Cold formed | Any | | *b* |
| Welded box section | Generally (except as below) | | Any | *b* |
| 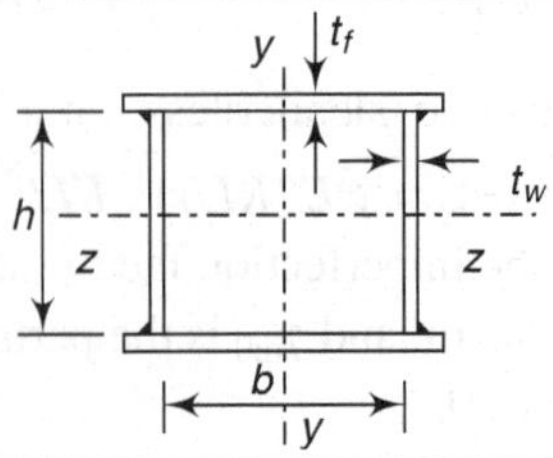 | Thick welds and | $b/t_f < 30$ | *z-z* | *c* |
| | | $h/t_w < 30$ | *y-y* | *c* |

(*Contd*)

**TABLE 9.4** *(Contd)*

| Cross section | Limits | | Buckling about axis | Buckling curve |
|---|---|---|---|---|
| Channel, angle, T- and solid sections 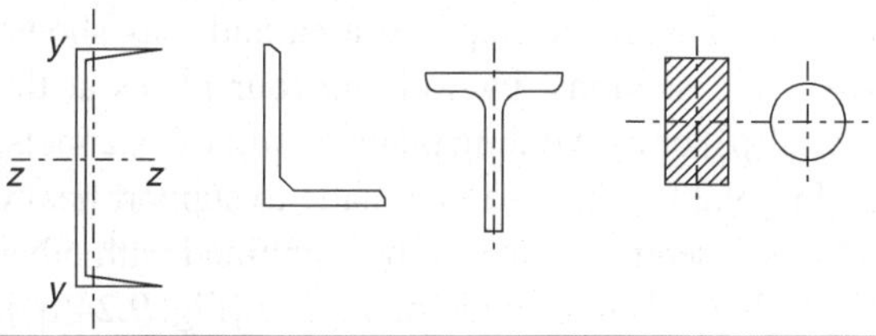 | | | Any | *c* |
| Built-up member 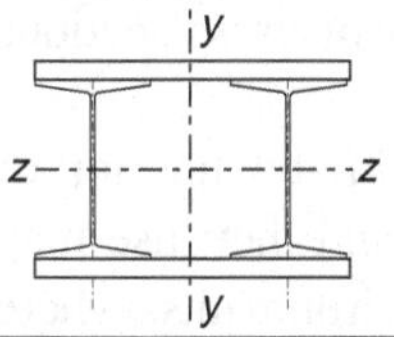 | | | Any | *c* |

## 9.8 SECTIONS USED FOR COMPRESSION MEMBERS

Though numerous sections may be selected to safely resist a compressive load in a given structure, from a practical viewpoint, the possible solutions are limited mainly due to the considerations of availability of sections, connection problems, and the type of structure in which they will be used. Figure 9.24 shows the possible configurations that may be used as compression members. From this figure it can be observed that the sections used as compression members are similar to tension members with some exceptions. These exceptions are caused by the fact that the strength of compression members vary inversely to the slenderness ratio [see Section 9.5 and Eqn (9.9)]. Thus, rods, bars, and plates are too slender to be used as compression members.

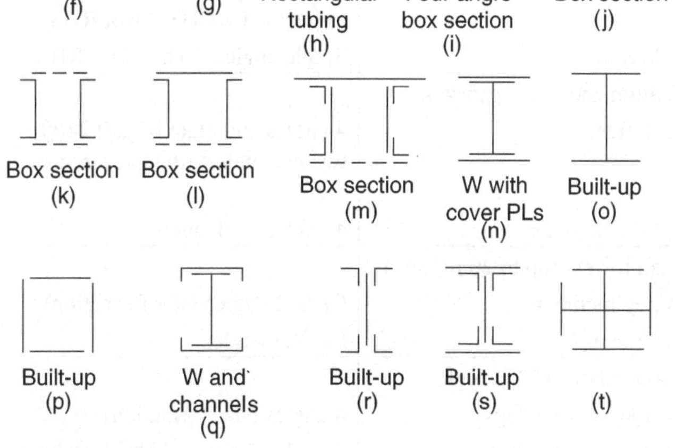


**FIG. 9.24** Shapes of compression members

Single angle members are used as compression and bracing members in lightly loaded trusses and towers. Equal-leg angles may be economical than unequal leg angles, because their least *r* values are greater for the same area of steel. Pair of angles bolted or welded back-to-back, using stitch plates at some constant intervals/at the joints (necessary for connections to other members), are normally used as top and bottom chord members of trusses. In this case, it may be desirable to use unequal-leg angles with the long legs back-to-back to give a better balance between the *r* values, about the *y* and *z* axes. In some structures, the outstanding legs of the double angle member may be bolted or welded directly to the connecting member. However, this results in non-axial loading, and hence either the eccentricity should be taken into account or reduced design stresses should be used in the design. Two-angle sections can also be used in the form of *star* angles [see Fig. 9.62(c)]. This cruciform arrangement of double angles is most effective because of its approximately equal radii of gyration in two directions. Such *star* angles are mostly used in bracings or legs of transmission line or microwave towers. Though it is possible to form a box section using two angles with welding, it is not used due to the difficulty of connecting with other members.

If roof trusses are welded, structural T-sections may be used for the top chord members. Such members will eliminate the gusset plates in lightly loaded trusses since the web members can be directly welded to the stems of the Ts. Single channels are rarely used due to the negligible *r* values about their web axis; hence are used only when some lateral support is provided in the weak direction. I-sections are often used as columns in buildings and as compression members in highway/railway bridges. Though the $r$ values of I-sections about the two axes are not same, they are better than those of channels. Since I-sections have thick flanges (which avoid the problem of local buckling) and are amenable for easy connections, they are often used as compression members in buildings.

Circular hollow sections (CHS) represent the optimum theoretical shape for compression members. There is no weak bending axis for CHS and $r \approx 0.35D$ ($D$ = diameter of CHS). The use of tubes as compression members was limited in the past due to the difficulty of connecting other members and manufacturing costs. However, after the introduction of welding, their use is increasing. They are being used as compression and bracing members in tubular trusses, space frames, and in the legs and other members of offshore oil platforms. It may be of interest to note that several famous bridges constructed during the nineteenth century (e.g., Firth or Forth bridge in Scotland, Ead's bridge in St Louis, USA) made extensive use of tube-shaped members.

Circular columns have the advantage of being equally rigid in all directions and are usually very economical unless moments are large. Tubes also have more torsional resistance. Moreover, round tubes when subjected to wind load, result in less force than flat sections. However, due to the shaping of the tubes at the welded connected ends, the connection cost may be high. Such connection problems are eliminated by the use of square (SHS) and rectangular hollow sections (RHS). Such SHS and RHS members are being used increasingly in different types of structures. In many structural applications, the weight of hollow sections required would be only 50% of that required for open profiles such as I- or channel sections.

When the area provided by rolled sections (I-, channel, tube, RHS, etc.,) is not sufficient to support the applied load, built-up sections as shown in Fig. 9.24(i)-(s) (also see Fig. 1.33 of Chapter 1) can be used by welding together angles, channels, I-sections, plates, etc., such that they act as a single compression member. The built-up section shown in Fig. 9.24(r) with four angles and a plate may be used when the load is moderate. Four angles arranged as shown in Fig. 9.24(i) are common in crane booms. Such arrangement is used for moderate loads and large length.

Two channels can be placed face-to-face [Fig. 9.24(j)] or back-to-back [Fig. 9.24(k)]. Two channels placed face-to-face provide a larger value of the radius of gyration than two channels placed back-to-back separated apart by the same distance. Hence, two channels placed face-to-face are ideal for use as a compression member than two channels placed back-to-back. However, in the case of two channels placed back-to-back the area of lacing is minimized and hence may be used in situations where a large radius of gyration is not required. The dotted lines in Fig. 9.24 represent connecting elements such as lacing or battens whereas the solid lines represent plates.

When two laced channels or I-sections are not sufficient, plates (called cover or flange plates) may be attached to the flanges as shown in Fig. 9.24(n), or three plates may be welded to form an I with the required area and $r$ as shown in Fig. 9.24(o). Box sections, by welding four plates at the corners [Fig. 9.24(p)] or by welding plates to angles, channels, or I-sections [Fig. 9.24(i)-(m)] can be made to support heavy loads in buildings. I-sections can also be combined with other sections such as channels to provide more area [Fig. 9.24(q)]. Even two I-sections can be built-up as shown in Fig. 9.24(t). Approximate radii of gyration of a variety of cross sections are given in Appendix A (Table A.15).

When the built-up section is located in the interior of a building, the cover plates may be solid and their use may reduce the fabrication cost. In exterior environments, where corrosion is a problem, it is necessary to have access to the interior of the section for maintenance and inspection. In such situations, sections with lacings and patterns may be used—which allow access to the interior of the section for cleaning and painting—without the careful fabrication required to completely seal the interior for ingress of moisture. In USA, perforated cover plates are preferred over lacing, since automatic gas cutting methods allow rapid cutting of the plate openings.

Thus, the choice of the sections in a particular situation depends on the following:

1. Functional aspects of the structure
2. Functional aspects of the member
3. Easy connectivity to other members
4. High radius of gyration

Table 9.5 gives some recommendations for these aspects.

**TABLE 9.5** Recommended structural shapes

| Structure/member | Type of cross section |
|---|---|
| Small span roof trusses (up to 20 m span) | |
| (a) Rafters and bottom ties | Double angles (connected back-to-back), Ts, CHS, SHS, RHS |
| (b) Web members | Single angles, CHS, SHS, RHS |
| Medium and long span roofs | |
| (a) Rafters | 4 angles and plate [Fig. 9.24(r)], two channels with plates, box-section |
| (b) Web members | 4 angles, 2 channels |
| Small towers (up to 20 m high) | |
| (a) Leg members | CHS, 2 angles (star formation) |
| (b) Bracings | CHS or angles |
| Towers about 150 m high | |
| (a) Bottom 40 m legs | 4 angles (star formation) |
| (b) Bracings and horizontal members | 2 angles (connected back-to-back) |

*(Contd)*

**TABLE 9.4** *(Contd)*

| Structure/member | Type of cross section |
|---|---|
| Members of space frames | CHS, SHS |
| Columns of multi-storey buildings | I-sections, built-up I-sections, box sections (for heavy loads) |
| Columns of industrial buildings | I-sections, built-up channels or I-sections |

## 9.9 EFFECTIVE LENGTH OF COMPRESSION MEMBERS

The effect of end restraints on column strength is usually incorporated in the design by the concept of *effective length*. (It must be emphasized that the concept of effective length is only a simple and convenient short cut and the designer is not obliged to use it, if he/she knows a better, yet sufficiently simple solution.) Hence, in this section, the effective length concept is first explained for idealized boundary conditions and then extended to cases where intermediate restraints are present. The Julian and Lawrence alignment charts (used in American and Canadian codes) and Wood's Curves (used in British and Indian codes), which are used to determine the effective length of columns of multi-storey buildings are also explained. The Indian code recommendations for the effective length factors of truss members are also discussed.

### 9.9.1 Effective Length for Idealized Boundary Conditions

Till now, the discussion in this chapter has been centered around pin-ended columns. Under ideal conditions, the boundary conditions of a column may be idealized in one of the following ways:

1. Both ends pinned (as already discussed in Section 9.5)
2. Both ends fixed
3. One end fixed and the other end pinned
4. One end fixed and the other end free

For all these conditions, the differential equations as given in Eqn (9.4) can be set up and the appropriate boundary conditions applied to get the following critical loads (Timoshenko & Gere 1961; Allen & Bulson 1980).

1. For column with both ends fixed:

$$P_{cr} = 4\pi^2 EI/L^2 = 4\pi^2 E/(L/r)^2 \quad (9.40)$$

2. For column with one end fixed and the other end pinned:

$$P_{cr} = 2\pi^2 E/(L/r)^2 \quad (9.41)$$

3. For columns with one end fixed and the other end free:

$$P_{cr} = \pi^2 E/[4(L/r)^2] \quad (9.42)$$

Using the length of the pin-ended column, as the basis for comparison, the critical load in the three cases mentioned earlier can be obtained by employing the concept of *effective length KL* where *K* is called the *effective length ratio* or effective length coefficient. (It is the ratio of the effective length to the unsupported length of the columns.) The value of *K* depends on the degree of rotational and translational restraints at the column ends. The *unsupported length L* is taken as the distance between lateral connections, or actual length in the case of a cantilever. In a conventional framed construction, *L* is taken as the clear distance between the floor and the shallower beam framing into the columns in each direction at the next higher floor level. In the case of a *flat slab* construction, *L* is taken as the clear distance between the floor and the lower extremity of the capital, the drop panel or slab, whichever is the least.

In other words, the *effective length* of a column in a given plane may be defined as the distance between the points of inflection (zero moment) in the buckled configuration of the column in that plane (see Fig. 9.25). Note that when there is relative translation at the ends of the column, the points of inflection may not lie within the member. In such a case, they may be located by extending the deflection curve beyond the column end(s) and by applying conditions of symmetry, as shown in Fig. 9.26. Thus, the effective length can also be defined as the length of an equivalent pin-ended column, having the same load-carrying capacity as the member under consideration. The smaller the effective length of a particular column, the smaller is its danger of lateral buckling and the greater its load carrying capacity. It must be recognized that it is very difficult to get perfectly fixed or perfectly hinged end conditions in practice.

Using *KL*, Eqns (9.40), (9.41), and (9.42) can be written as

$$P_{cr} = \pi^2 E/(KL/r)^2 \quad (9.43)$$

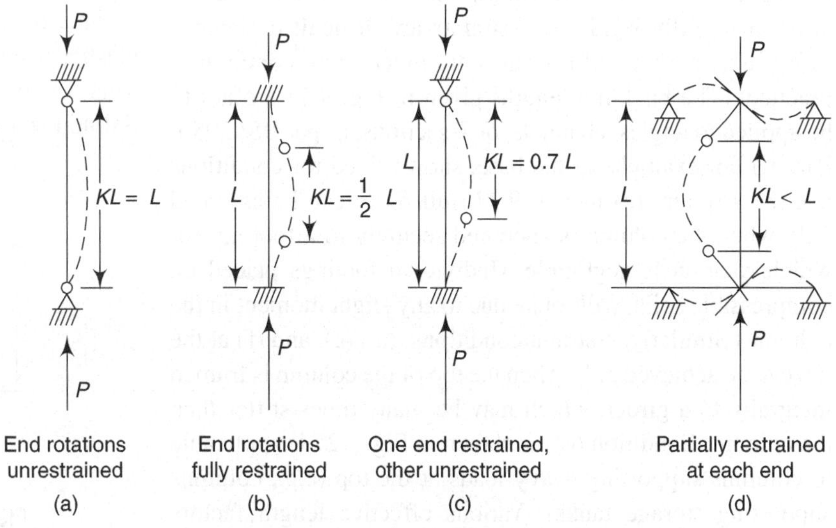


**FIG. 9.25** Effective length *KL* when there is no joint translation at the ends

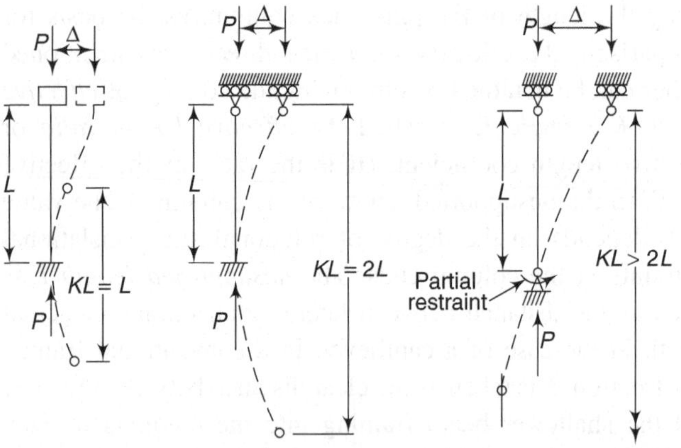


End rotations fully restrained (a) | One end rotation fully restrained, other unrestrained (b) | One end rotation partially restrained, other end unrestrained (c)

**FIG. 9.26** Effective length *KL* when there is joint translation at the ends

where K = 1.0 for columns with both ends pinned,

K = 0.5 for columns with both ends fixed,

K = 0.707 for columns with one end fixed and the other end pinned,

K = 2.0 for columns with one end fixed and the other end free,

K ≤ 1.0 for columns partially restrained at each end, and

K ≥ 2.0 for columns with one end unstrained and the other end rotation partially restrained.

Thus, the effective length is important in design calculations because the buckling load is inversely proportional to the square of the effective length.

Approximate values for effective lengths, which can be used in design, are given in IS 800 and are shown in Fig. 9.27 along with the theoretical *K* values [as given by Eqn (9.43)]. Note that the values given in the code are slightly more than the theoretical values given in (a), (b), and (c) of Fig. 9.27. It is because, fully rigid end restraints are difficult to achieve in practice; partial end restraints are much more common in practice. (The buckling lengths given in Fig. 9.27 are not to be applied to angles, channels, or T-sections, as per BS: 5950-Part 1.) For example, at the base, shown fixed for conditions (a), (b), (c), and (e) in Fig. 9.27, full fixity can be assumed only when the column is anchored securely to a footing, for which rotation is negligible. (Individual footings placed on compressible soils, will rotate due to any slight moment in the column.) Similarly, restraint conditions, (a), (c), and (f) at the top can be achieved only when the top of the column is framed integrally to a girder, which may be many times stiffer than the column. Condition (c), as shown in Fig. 9.27 is applicable to columns supporting heavy loads at the top (e.g., columns supporting storage tanks). Various effective length factors suggested by different codes of practice are given in Table 9.6.

| | (a) | (b) | (c) | (d) | (e) | (f) |
|---|---|---|---|---|---|---|
| Buckled shape of column is shown by dashed line | 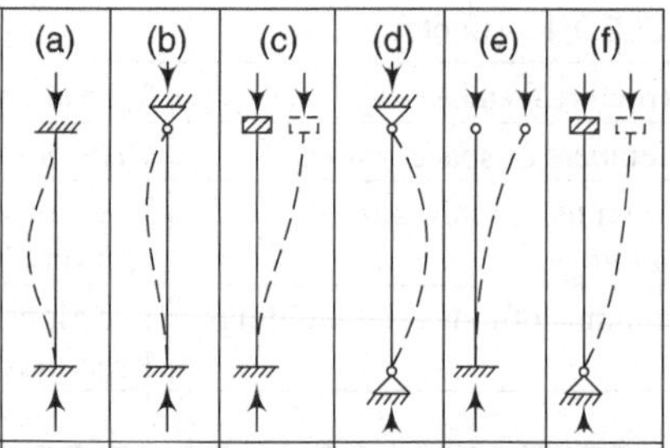 | | | | | |
| Theoretical *K* value | 0.5 | 0.7 | 1.0 | 1.0 | 2.0 | 2.0 |
| Recommended *K* value when ideal conditions are approximated | 0.65 | 0.80 | 1.2 | 1.0 | 2.0 | 2.0 |
| End condition code | Rotation fixed, translation fixed; Rotation free, translation fixed; Rotation fixed, translation free; Rotation free, translation free | | | | | |

**FIG. 9.27** Effective-length factors *K* for centrally loaded columns with various end conditions

**TABLE 9.6** Effective length factors for columns in various codes of practice

| End condition (see Fig. 9.27) | (a) | (b) | (c) | (d) | (e) | (f) |
|---|---|---|---|---|---|---|
| IS 800-2007 | 0.65 | 0.80 | 1.20 | 1.0 | 2.0 | 2.0 |
| BS: 5950-Part1: 2000 | 0.70 | 0.85 | 1.20 | 1.0 | 2.0 | — |
| AS 4100-1998 | 0.70 | 0.85 | 1.20 | 1.0 | 2.2 | 2.2 |
| CAN/CSA-S16.1-2001 | 0.65 | 0.80 | 1.20 | 1.0 | 2.0 | 2.0 |
| ANSI/AISC 360-2005 | 0.65 | 0.80 | 1.20 | 1.0 | 2.1 | 2.0 |
| AI J, 1970 | 0.65 | 0.80 | 1.20 | 1.0 | 2.1 | — |

## 9.9.2 Intermediate Restraints and Effective Lengths in Different Planes

In the previous sections, it was assumed that the compression member was supported only at its ends as shown in Fig. 9.27. If the member has an additional lateral support (bracing) which prevents it from deflecting at its centre, as shown in Fig. 9.28, the elastic buckling load is increased by a factor of four (i.e., the value of $P_{cr}$ will be $4\pi^2 EI/L^2$).

This restraint need not be completely rigid, but may be elastic, provided its stiffness exceeds a certain minimum value. This value of minimum stiffness $k_b$ has been derived by Salmon & Johnson (1996) as

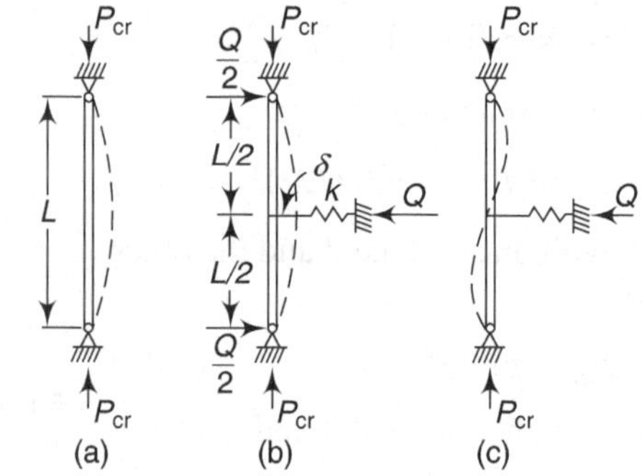


**FIG. 9.28** Compression member with an elastic intermediate restraint

$$k_b = \beta P_{cr}/L \quad (9.44a)$$

where, the value of $\beta$ varies from one (for one intermediate support) to four (infinite intermediate supports).

In addition to stiffness, a brace must provide adequate strength. The strength required of an ideal brace is given by (Salmon and Johnson, 1996)

$$Q = k_b\delta \quad (9.44b)$$

Note that until buckling of the column takes place, $\delta$ is zero and hence there will not be any force in the bracing. As discussed already, real compression members have initial crookedness, and if we consider an initial crookedness of $\delta_0$, the strength requirement will be

$$k_{b,\text{req}} = k_b(1 + \delta_0/\delta) \quad (9.44c)$$

$$Q = k_b(\delta + \delta_0) \quad (9.44d)$$

If we consider $\delta = \delta_0 = L/500$, we get the required stiffness and nominal strength of bracing as (Salmon & Johnson 1996)

$$k_{b,\text{req}} = 2\beta P_{cr}/L \quad (9.44e)$$

$$Q_n = k_b(0.004L) = 0.004\beta P_{cr} \quad (9.44f)$$

Calculations based on the preceding equations suggest that any member with a slenderness ratio of 200 will be adequate for bracing purposes. Hence many specifications and codes suggest a rule-of-thumb of using a bracing having a strength of 2 to 2.5% of the compressive strength of the compression elements being braced. This seems to be a conservative alternative to a rigorous analytical study (Lay & Galambos 1966). Such a clause is also available in the code (IS 800 clause 7.6.6.1).

When such restraints are provided, the buckling behaviour will be different about the two column axes. For example, consider a pin-ended column as shown in Fig. 9.29, which is braced about the minor axis against lateral movement (but not rotationally restrained) at a spacing of $L/4$. Now the minor axis buckling mode will be as shown in Fig. 9.29 with $K = 1/4$. If there is no bracing in the major axis, the effective length for buckling about major axis will be $L$. Hence, the slenderness ratio about the major and minor axis will be $L/r_z$ and $L/4r_y$ respectively. For ISHB sections $r_z < 4r_y$ and hence the major axis slenderness ratio will be greater, giving lower value of critical load and failure will occur by major axis buckling. If this is not the case, checks have to be carried out about both the axes. Figure 9.29(b) shows the ineffective use of bracing, since such a bracing provided in the major axis will not increase the load carrying capacity of the column. Note that for symmetric cross sections (square and circular solid or hollow sections), bracings must be such as to reduce the slenderness ratios in all directions (by providing bracing in more than one plane). Thus, it is seen that by providing sufficient number of brace points, it is possible to completely eliminate buckling and the members could be sized for direct stresses only [i.e., $A = P_u/(f_y \times \gamma_{m0})$] The minimum number of such brace points thus corresponds to making column lengths between brace points exactly equal to the transition length between short- and long-column behaviour. (Putting more brace points than this will not accomplish anything because the column size is already governed by direct stresses.)

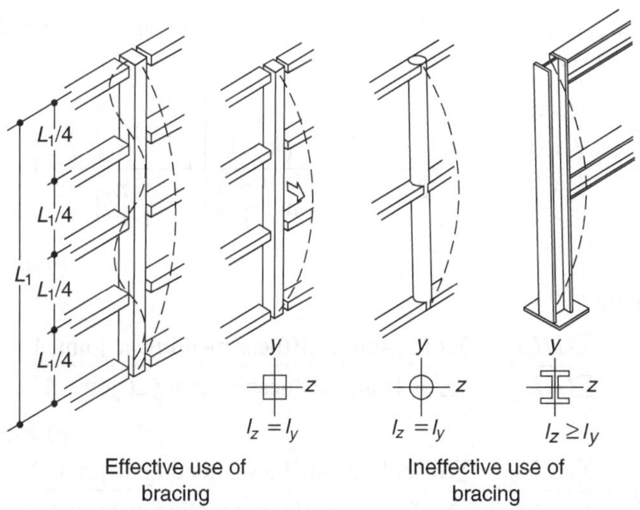


**FIG. 9.29** Column braced by a member in one plane

In many situations, it may not be possible to use bracing in more than one plane. For example, columns are often used with walls, where these walls can serve as lateral bracing in one plane; in other planes for functional reasons it may not be possible to provide bracings. In such situations, non-symmetric members could be employed advantageously with their strong axis oriented in the out-of-braced-plane buckling mode and their weak axis in the in-plane mode. A great level of efficiency may be obtained by keeping the buckling load about one axis exactly equal to the buckling load about the other axis. The number of braces required to achieve this may be found out by the relationship $r_z/r_y = L_z/L_y$. (This relationship may also be useful in the design of built-up columns—see Example 9.18.)

## 9.9.3 Columns in Multi-storey or Framed Buildings

As mentioned already, isolated columns are rare and they normally form a part of any framework. Moreover, their end conditions are influenced by the members to which they are connected. The more accurate determination of $K$ for such a compression member as part of any framework requires the application of methods of indeterminate structural analysis, modified to take into account the effects of axial load and inelastic behaviour on the rigidity of the members (see Chapter 4 for the details of such analysis). These procedures are not directly applicable to routine design and hence simple models are often used to determine the effective length factor for framed members. In such simplified approaches, a distinction is always made about sway (unbraced) and non-sway (braced) frames because of their distinct buckling modes (see

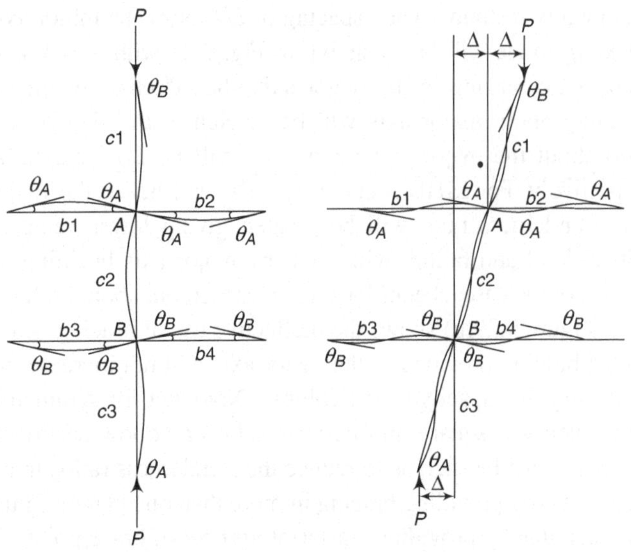


**FIG. 9.30** Sub-assemblage model for braced and unbraced frames

Fig. 9.30). In non-sway (braced) frames, the columns buckle in single curvature and hence their effective length factor will always be less than unity; whereas the columns in sway frames buckle in double curvature and hence their effective length factor will always be greater than unity.

## Braced Frames

A braced frame is considered to be a frame in which lateral stability is provided by diagonal bracing, shear walls, in-fill walls, or by any other equivalent means. It is not necessary that every bay of a multi-storey building has to be braced, since the unbraced bays are restrained by those that are braced. The code (IS 800) gives criteria for considering the frame as a non-sway frame in clause 4.1.2. [see Section 4.2.3.2, Eqns (4.22) and (4.23) of Chapter 4]. The ACI code (ACI 318-2014) suggests the calculation of the stability index $Q$, which is given by

$$Q = [(\Sigma P_u)/h_s](\Delta_u/H_u) \tag{9.45}$$

where $\Sigma P_u$ is the sum of axial loads on all columns in the storey, $h_s$ is the height of the storey, $\Delta_u$ is the elastic first-order lateral deflection of the storey, and $H_u$ is the total lateral force acting on the storey. If the value of $Q$ is less than 0.05, the column may be assumed as braced.

In the absence of bracing elements, the *lateral flexibility* measure of the storey $\Delta_u/H_u$ (storey drift per unit storey shear) may be taken for a typical intermediate storey as (Taranath 1988)

$$(\Delta_u/H_u) = h_s^2/[(12E\,\Sigma(I_c/h_s)] + h_s^2/(12E\Sigma I_b/L_b) \tag{9.46}$$

where $\Sigma I_c/h_s$ is the sum of ratios of the second moment of area to the height of all columns in the storey in the plane under consideration, $\Sigma I_b/L_b$ is the sum of ratios of the second moment of the area to the span of all floor beams in the storey in the plane under consideration, and $E$ is Young's modulus of elasticity of steel.

The application of this concept is demonstrated in Example 9.6. Note that Eqn (9.46) does not consider the effect of infills, bracings, or shear walls. However, it gives an idea of the lateral flexibility of the storey under consideration.

Various investigators have provided charts to permit easy determination of effective lengths for commonly encountered situations. Two simplified approaches are often recommended in the codes of practices. The American and Canadian codes follow the alignment charts originally developed by Julian and Lawrence in 1959 (Kavanagh 1962). The British and Indian codes follow the charts originally developed by Wood (1974). The Australian code also gives a chart similar to that of the Wood's chart.

**Julian and Lawrence alignment chart** The model used for the determination of $K$ for a column braced against side sway is shown in Fig. 9.30(a). The column under consideration is denoted as $c2$ in the figure. The following assumptions are made in the derivation of the $K$ factor.

1. All members are prismatic and behave elastically.
2. The axial forces in the beams are negligible.
3. All columns in a storey buckle simultaneously.
4. At a joint, the restraining moment provided by the beams is distributed among the columns in proportion to their stiffness ($I/L$)
5. At buckling, the rotations at the near and far ends of the beam are equal and opposite (i.e., the girders are bent in single curvature).
6. The frame is subjected to vertical loads, applied only at the joints.

Using the ordinary slope-deflection equations for beams $b1$, $b2$, $b3$, and $b4$ and the modified (for effect of axial load) slope deflection equations for column $c2$, a *transcendental equation for the effective length ratio K* can be established as (Chen & Lui1991)

$$\frac{G_A G_B}{4}\left(\frac{\pi}{K}\right)^2 + \left(\frac{G_A + G_B}{2}\right)\left[1 - \frac{(\pi/K)}{\tan\left(\dfrac{\pi}{K}\right)}\right] + \frac{2\tan\left(\dfrac{\pi}{2K}\right)}{\left(\dfrac{\pi}{K}\right)} - 1 = 0 \tag{9.47a}$$

where

$$G_A = \frac{\Sigma(I/L)_c}{\Sigma(I/L)_b} = \frac{\Sigma \text{ of column stiffness meeting at joint } A}{\Sigma \text{ of beam stiffness meeting at joint } A} \tag{9.47b}$$

$$G_B = \frac{\Sigma(I/L)_c}{\Sigma(I/L)_b} = \frac{\Sigma \text{ of column stiffness meeting at joint } B}{\Sigma \text{ of beam stiffness meeting at joint } B} \tag{9.47c}$$

The solution to Eqn (9.47) is expressed in a nomograph form (alignment chart) in Fig. 9.31(a). Values of $G_A$ and $G_B$ are given on two outside scales and for $K$ on the middle scale. The line joining $G_A$ and $G_B$ intersects the middle scale and will fetch the required value of $K$ [Fig. 9.31(b)]. Note that the range of $K$ is $0.5 \le K \le 1.0$. The lower bound ($K = 0.5$) corresponds to the case when the beams are sufficiently stiff compared to the column such that the end condition of the column to be designed approaches the fixed condition. The upper bound value ($K = 1.0$) corresponds to the case of flexible beams, such that the end condition of the column under consideration approaches the pinned condition.

Equation (9.47) was obtained on the assumption that all the members (beams and columns) bend according to a symmetrical slope having single curvature [see Fig. 9.30(a) and Fig. 9.31(c)]. If the beam ends opposite those meeting in the joint with the column are joined to a constraint allowing rotation [Fig. 9.31(d)] or preventing it [Fig. 9.31(e)], the resulting buckling mode shapes will be different and hence the stiffness $K_i$ for the beam can be expressed as (Chen & Lui 1991)

$$K_i = \gamma_i I_i/L_i \tag{9.47d}$$

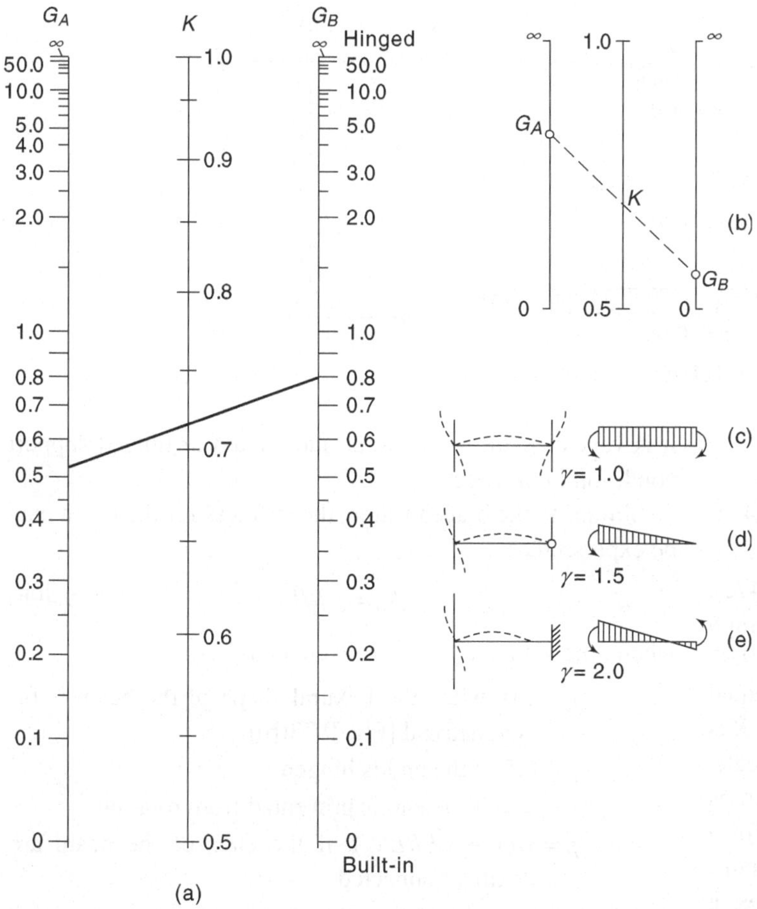


**FIG. 9.31** Nomograph effective length of column in braced frames, Julian and Lawrence (1959)

where $I_i$ is the moment of inertia in the bending plane, $L_i$ is the length, $\gamma_i$ is a coefficient, which is dependent on the flexural shape of the beam, and $\gamma_i$ is 1 for beams bending in single curvature, 1.5, if the other end is hinged, 2.0, if the other end is built-in (fixed), and $1/(1 + 2EI/LR_k)$, if the ends of the beam are flexibly connected. $R_k$ is the connection stiffness.

Similarly, if the column is hinged, $G_B$ is taken as $\infty$ and if it is built-in or fixed, it is taken as zero.

**Wood's curves for braced frames** Wood (1974) suggested a different type of approach to the problem. Considering a sub-element shown in Fig. 9.32(a) and using stability functions (see Section 4.2.3.1 of Chapter 4) and a stiffness distribution method (which is similar to the moment distribution process), he derived equations, which were then presented in the form of graphical charts as shown in Fig. 9.32(b).

In general, if a column is also connected to other columns as shown in Fig. 9.32(f), the coefficients $\beta_1$ and $\beta_2$ can be modified as

$$\beta_i = \Sigma K_c/[\Sigma K_c + \Sigma K_b] \tag{9.48a}$$

where $K_c$ is the stiffness of the column meeting in the joint under consideration. Note that in the case of a column built-in at its base, $\Sigma K_b = \infty$ and hence $\beta_2 = 0.0$. Conversely, for a hinged column $\Sigma K_b = 0$ and hence $\beta_2 = 1.0$.

Note that the curves in Fig. 9.32(b) were obtained on the assumption that beam ends are built-in and therefore their flexural shape has a point of contraflexure at 1/3 of the span. Beam stiffness is thus expressed by the following relation.

$$K_{b,i} = \gamma_i I_i/L_i \tag{9.48b}$$

where $\gamma_i$ is a coefficient, for which the following values must be assumed

$\gamma_i = 1$, if the ends are built in [Fig. 9.32(c)], $\gamma_i = 0.75$, if the end is hinged [Fig. 9.32(d)], and $\gamma_i = 0.50$, if the flexural shape has a single curvature [Fig. 9.32(e)].

## Frames in which Side Sway is not Prevented (Unbraced Frames)—Julian and Lawrence's Alignment Chart

The model for a column in a frame subjected to side sway is shown in Fig. 9.30(b). The column under consideration is denoted by $c2$ in the figure. The assumptions used for this model are the same as those used for the model of the braced frame, except assumption 5, which is assumed as follows: At buckling, the rotations at the near and far ends of the beams are equal in magnitude and direction (i.e., the beams are bent in double curvature).

Now the equation for determining $K$ is expressed as (Chen & Lu 1991)

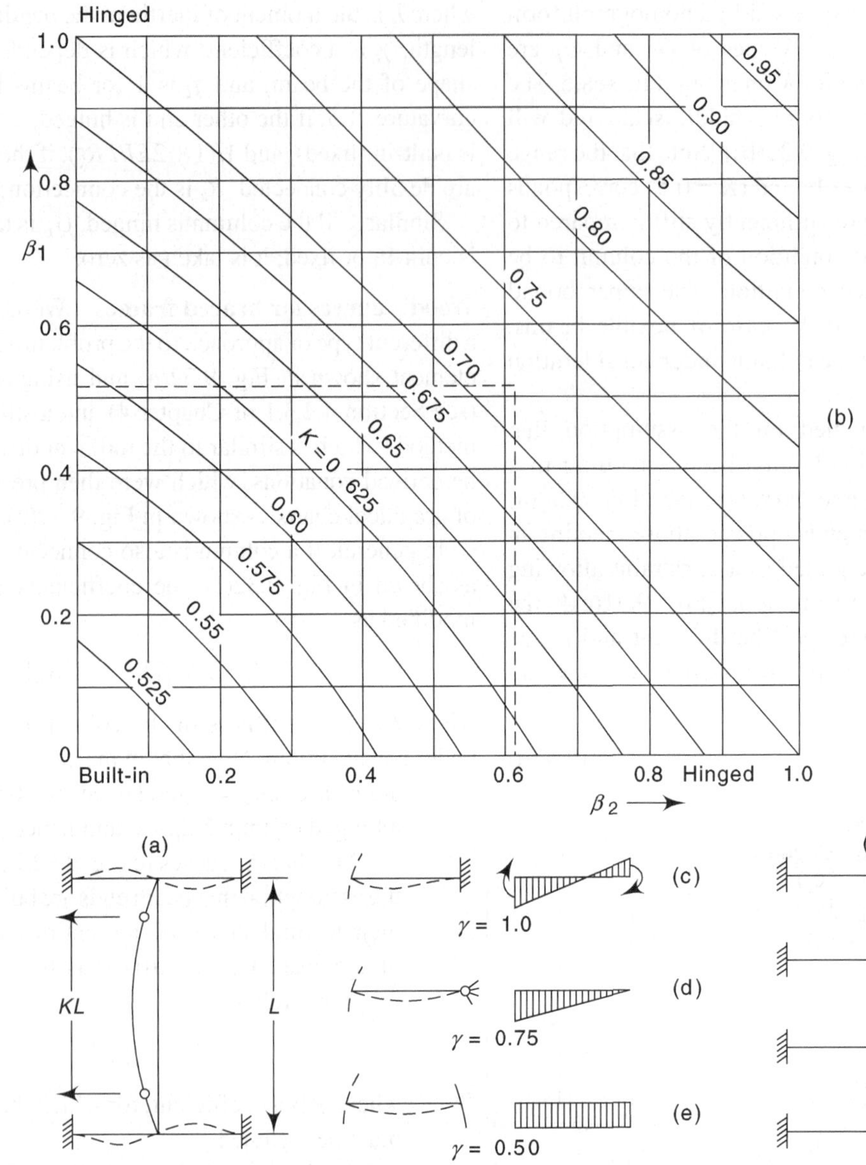


**FIG. 9.32** Wood's curves for the determination of effective length of braced frames

$$\frac{G_A G_B (\pi/K)^2 - 36}{6(G_A + G_B)} - \frac{(\pi/K)}{\tan\left(\dfrac{\pi}{K}\right)} = 0 \tag{9.49a}$$

where $G_A$ and $G_B$ are as defined in Eqns (9.47b) and (9.47c), respectively. Equation (9.49a) is expressed in a nomograph form by Julian and Lawrence (1959), as shown in Fig. 9.33 (Kavanagh 1962). Note that $K$ is always greater than or equal to unity. As explained in the previous section, the value of $K$ is obtained at the intersection of line $G_A - G_B$ on the central scale.

Theoretically, $G$ is equal to zero when the column is connected to the footing by a fixed support. Similarly for a column, which is connected to the footing by a hinged support, the theoretical value of $G$ is ∞. But for design purposes, a value of $G$ equal to 1 is used for fixed support and a value of $G$ equal to 10 for hinged support is recommended, since it is very difficult to obtain an ideal fixed or hinged support condition in practice.

Similar to the braced frame, the stiffness for the beam may be expressed as

$$K_i = \gamma_i I_i / L_i \tag{9.49b}$$

where, the value of $\gamma_i$ can be taken as follows:

- $\gamma_i = 1.0$, when the flexural shape of the beam is bi-symmetrical [Fig. 9.33(b)]
- $\gamma_i = 0.5$, if the end is hinged
- $\gamma_i = 2/3$, if the end is prevented from rotating
- $\gamma_i = 1/(1 + 6EI/LR_k)$, if the ends of the beam are flexibly connected

For semi-rigid connections, the values as shown in Table 9.7 should be used (Kishi et al. 1997).

**TABLE 9.7** Coefficient $\gamma_i$ for beams in unbraced frames (Kishi et al. 1997)

| End conditions | | $\gamma_i$ |
|---|---|---|
| *A* th end | *B* th end | |
| Rigid | Semi-rigid | $[1 + (2EI_b)/(L_bR_{kB})]/[1 + (4EI_b)/(L_bR_{kB})]$ |
| Semi-rigid | Rigid | $1/[1 + (4EI_b)/(L_bR_{kA})]$ |
| Semi-rigid | Pinned | $0.5/[1 + (3EI_b)/(L_bR_{kA})]$ |
| Semi-rigid | Semi-rigid | $[1 + (2EI_b)/(L_bR_{kB})]/R^*$ |

*Note:* $R^* = [1 + (4EI_b)/(L_bR_{kA})]\ [1 + (4EI_b)/(L_bR_{kB})] - [(EI_b)/Lb]^2\ [4/(R_{kA}R_{kB})]$
$R_{kA}$ and $R_{kB}$ are the spring constants at the end of semi-rigid connections

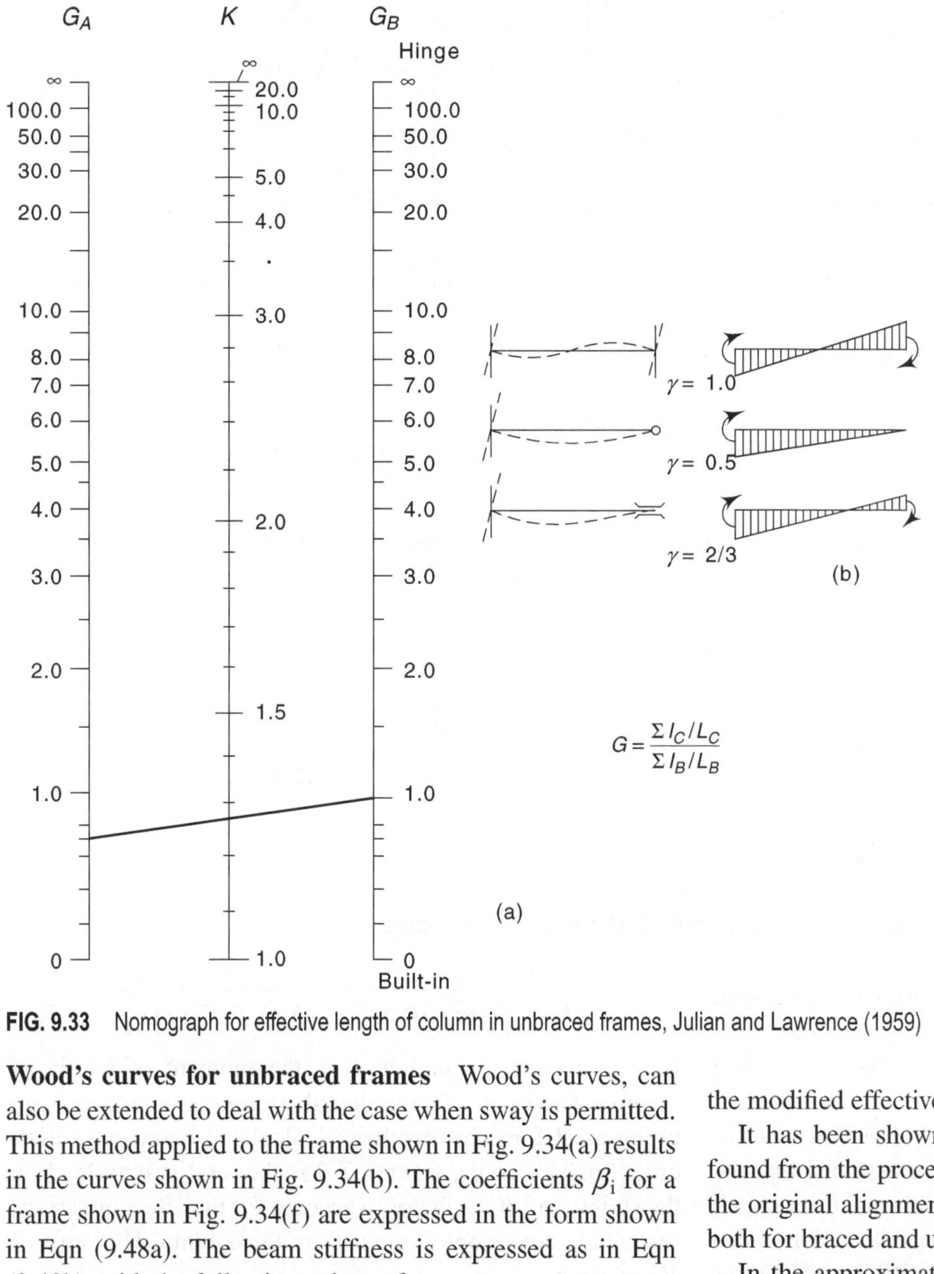


**FIG. 9.33** Nomograph for effective length of column in unbraced frames, Julian and Lawrence (1959)

**Wood's curves for unbraced frames** Wood's curves, can also be extended to deal with the case when sway is permitted. This method applied to the frame shown in Fig. 9.34(a) results in the curves shown in Fig. 9.34(b). The coefficients $\beta_i$ for a frame shown in Fig. 9.34(f) are expressed in the form shown in Eqn (9.48a). The beam stiffness is expressed as in Eqn (9.48b), with the following values of $\gamma_i$.

$\gamma_i = 1$ for columns,

$\gamma_i = 1$ for beams having one end prevented from rotating [Fig. 9.34(c)]

$\gamma_i = 0.75$ for beams having one end free to rotate [Fig. 9.34(d)]

$\gamma_i = 1.5$ for beams with a bisymmetrical beam shape, and therefore for beams connected to other columns [Fig. 9.34(e)]

It should be cautioned that the results for frames with sway permitted must be considered even more approximate than those given for frames with sway prevented. There have been many studies aimed at checking their reliability, and in many cases, the results were found to be unsafe or wrong when compared to a rigorous elastic stability analysis (Kuhn & Lundgren 1977). It is mainly due to the destabilizing effects of the vertical loads and the simplifying assumptions adopted to derive the curves. Hence, approximate methods like Wood's curves are acceptable when sway is permitted, only if the frames are substantially regular, i.e., when heights, moments of inertia, and axial forces in the columns do not differ considerably between themselves.

### Modifications to Nomographs for Different Column End Conditions

In developing the nomograph or alignment charts shown in Figs 9.31 and 9.33, it was assumed that the far ends of the columns (columns $C_1$ and $C_3$ as shown in Fig. 9.30) which were connected to the column under consideration (column $C_2$) were rigidly connected to other members of the frame. A procedure that can be used to account for far ends of these columns being hinged or fixed was proposed by Duan & Chen (1988, 1989). Using their method, the same alignment chart shown in Figs 9.31 and 9.33 can be used with additional calculation for stiffness ratio $G^*$ and $G_C$ to arrive at the modified effective length factor $K^*$.

It has been shown that the modified effective length $K^*$ found from the procedure suggested by Duan and Chen using the original alignment charts results in more accurate values, both for braced and unbraced frames (Chen & Lui 1991).

In the approximate charts (Figs 9.31 to 9.34) it has been assumed that the column sizes and loads are identical for all members in a storey. For cases when column sizes or column loads are not identical, adjustments to the nomographs or curves (Figs 9.31 to 9.34) are necessary to obtain the *K* factors

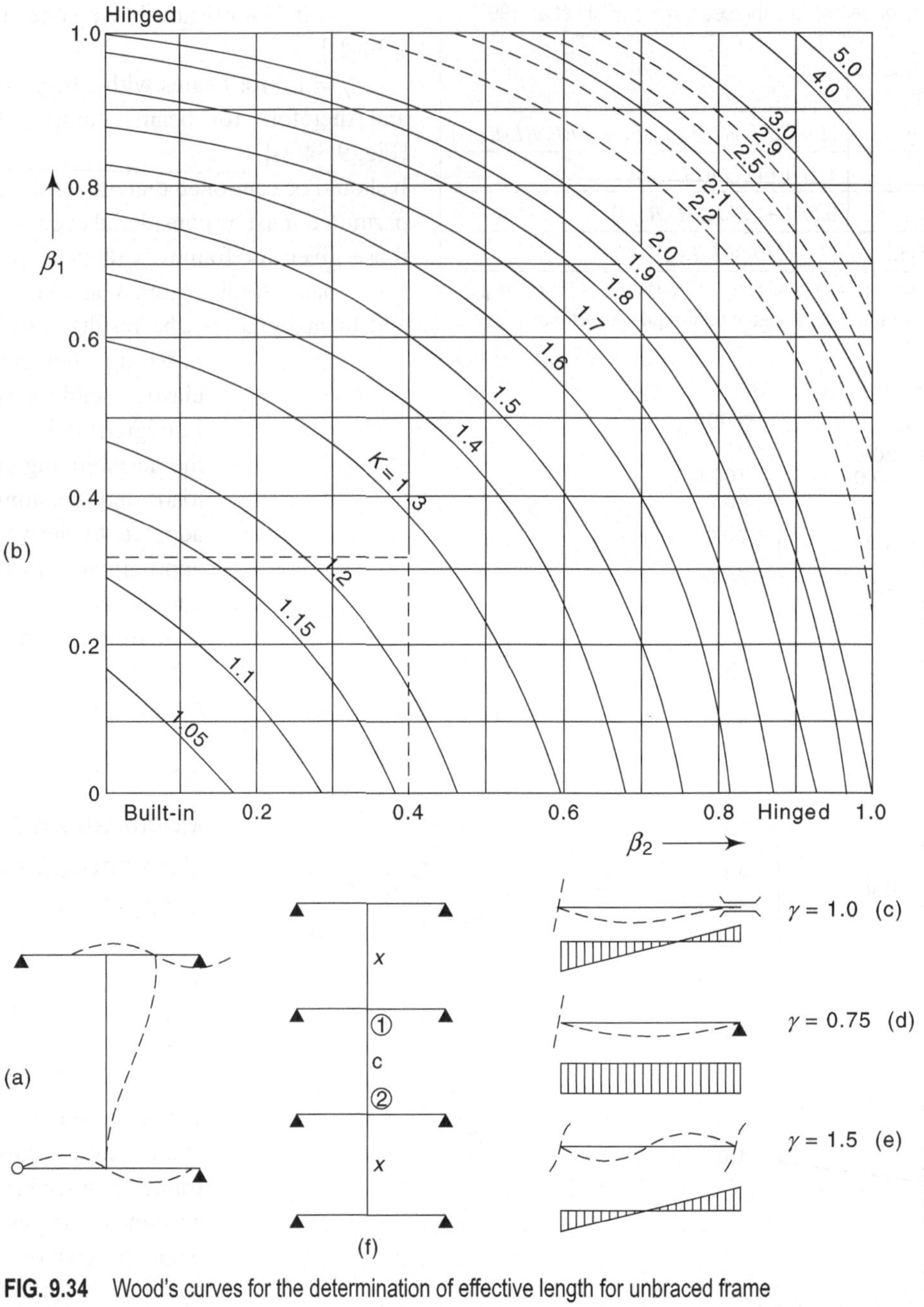


**FIG. 9.34** Wood's curves for the determination of effective length for unbraced frame

for the columns. Such correction curves have been developed by Chen and Chow (1969). More discussions about these curves, including worked examples are provided by Johnston (1976).

In the models for the braced frame in Fig. 9.30(a) and for the unbraced frame in Fig. 9.30(b), it was assumed that the axial forces are constant for all columns and axial forces in all beams are negligible. This condition can be relaxed for braced frames, using the procedure developed by Bridge and Fraser (1987).

The following equation was proposed by LeMessurier (1972) for calculating the effective length factor of member $i$ in a fram.

$$K_i = \sqrt{\{(\pi^2 I_i/P_i)[(\Sigma P + \Sigma C_L P)/\Sigma(\beta I)]\}} \quad (9.50a)$$

In the preceding equation, $I_i$ is the moment of inertia of member $i$ (whose effective length factor $K_i$ is required), $P_i$ is the axial force in member $i$ (calculated based on a first-order analysis of the frame), $\Sigma P$ is the total gravity load on the storey, and $\beta$ is a factor to account for the effect of various restraining conditions at the ends of the column. The value of $\beta$ is given by

$$\beta = \frac{[6(G_A + G_B) + 36]}{[2(G_A + G_B) + G_A G_B + 3]} \quad (9.50b)$$

$G_A$ and $G_B$ have been defined already in Eqns (9.47b) and (9.47c). $C_L$ is a factor to account for the reduction in column stiffness due to the presence of an axial force and is given by

$$C_L = (\beta K^2/\pi^2) - 1 \tag{9.50c}$$

where $K$ is calculated from Eqn (9.50a) or obtained from the nomograph shown in Fig. 9.33. Equation (9.50a) may be simplified and written as

$$K_i = \sqrt{\{[\pi^2 EI_i/(P_i L_i^2)](\Sigma P/\Sigma P_{cr})\}} > \sqrt{(5/8)}K \tag{9.50d}$$

where $\Sigma P_{cr}$ is the sum of the elastic critical loads of all columns in a storey evaluated using the $K$ factor obtained from the nomograph shown in Fig. 9.33. The limit on the right-hand side of Eqn (9.50d), is the minimum value of $K$ that accounts for the interaction between sidesway and nonsidesway buckling (ASCE Task committee, 1997). It has been found that the advantage of LeMessurier's methods is that it gives a more accurate value of $K$ for a sway frame when the $P$-$\Delta$ effect is significant. In addition, the method is applicable to frames with *leaned columns* (columns hinged at both ends which contribute only axial stiffness but not lateral stiffness to a frame), since the $K$ factor expression was derived for an entire storey of the frame (Chen & Lui 1991).

The effect of column yielding and end restraint on the effective length factor of a compression member has been investigated by a number of researchers (e.g., Yura 1971; Disque 1973; Christopher & Bjorhovde 1999), who used the tangent modules approach to arrive at inelastic $K$-factors. However, it was found that the $K$ factor obtained from inelastic analysis is usually smaller than that obtained from elastic analysis. This indicates that as long as inelasticity is confined to the columns and all restraining beams remain elastic, the use of the elastic $K$ factor will result in a conservative design (Chen & Lui 1991).

There will be a significant difference in strength between the columns of sway and non-sway frames. [The column strength depends largely on $1/(KL)^2$. Assuming $K$ as 1.5 for unbraced frames and 0.85 for braced frames, the strength of the columns in a sway frame is only one third that of the column in a non-sway frame.] This has led to several failures during construction, when a column designed for a no-sway condition has been temporarily left unbraced. Sway instability failures occur suddenly and hence result in injury or loss of life. The collapse of scaffoldings inadequately tied to the facade of a building is a common example of sway instability (Dowling et al. 1988). An improved $K$-factor formula which is applicable to column in fully braced, partially braced, and unbraced frames has been derived by Cheong–Siat–Moy (1999).

It is important to note that column design using K-factors can be tedious and confusing for complex building structures containing leaning columns and/or combined framing systems, particularly when inelasticity is considered. This confusion can be avoided if the advanced analysis methods described in Chapter 4 are used.

### *Effective Length (As per IS 800)*

The code (IS 800), gives the following equations for the effective length factor $K$, based on Wood's curves (see Annex D of the code):

For non-sway frames (braced frames):

$$K = \frac{[1 + 0.145(\beta_1 + \beta_2) - 0.265\beta_1\beta_2]}{[2 - 0.364(\beta_1 + \beta_2) - 0.247\beta_1\beta_2]} \tag{9.51a}$$

For sway frames (moment-resisting frames):

$$K = \left\{\frac{[1 - 0.2(\beta_1 + \beta_2) - 0.12\beta_1\beta_2]}{[1 - 0.8(\beta_1 + \beta_2) + 0.6\beta_1\beta_2]}\right\}^{0.5} \tag{9.51b}$$

with $\beta_i = \Sigma K_c/(\Sigma K_c + \Sigma K_b)$ (9.51c)

where $K_c$ and $K_b$ are the effective flexural stiffness of the columns or beams meeting at the joint at the ends of the columns and rigidly connected at the joints.

$$K_c \text{ or } K_b = C(I/L)$$

where $I$ is the moment of inertia about an axis perpendicular to the plan of the frame, $L$ is the length of the member, taken as centre-to-centre distance of the intersecting member, and $C$ is the connection factor as shown in Table 9.8.

**TABLE 9.8** Connection factor $C$

| Fixity conditions at far end | Connection factor $C$ | |
|---|---|---|
| | Braced frame | Unbraced frame |
| Pinned | $1.5(1 - \bar{n})$ | $0.5(1 - \bar{n})$ |
| Rigidly connected column | $1.0(1 - \bar{n})$ | $1.0(1 - 0.2\bar{n})$ |
| Fixed | $2.0(1 - 0.4\bar{n})$ | $0.67(1 - 0.4\bar{n})$ |

*Note:* $\bar{n} = P/P_{cr}$, where $P$ is the applied load and $P_{cr}$ is the elastic buckling load $= \pi^2 EI/(KL)^2$

Note that for calculating $C$ we need the effective length and hence the determination of effective length is an interactive process. Initially, we can assume $K = 1$ for calculating the value of $C$.

A review of the IS code provisions for effective length of columns in frames is provided by Dafedar et al. (2001).

## 9.9.4 Compression Members in Trusses

In trusses (triangulated framework), the loads are usually applied at the joints. Thus, if the joints are truly pinned, the members are axially loaded. Even though the compression members in a truss may be subjected to secondary moments, because of local yielding of extreme fibres of the members near the joints, these secondary moments gradually dissipate and hence may be neglected in the buckling analysis.

The effective length ratio $K$ of compression members in trusses does not exceed unity, because trusses are usually loaded only at the joints so that joint displacement at one end of the member relative to the other end is small. However, $K$ should not be taken less than unity, even in trusses with bolted, riveted, or welded joints, if the truss supports a fixed system of loads and is designed in such a way that all its member's fail simultaneously. It is because, in this case, the rotational restraint at the joints will be very minimum (Gaylord et al. 1992). In actual practice, a truss is never optimal with reference to its weight, as profiles must be unified to achieve fabrication economy; hence the maximum load capacity is attained only in one member. Due to this, the less loaded struts meeting at a joint will provide a degree of rotational restraint to the most loaded ones.

When designing for moving load system on trusses, $K$ can be reduced to 0.85 because conditions causing maximum stress in the member under consideration will not cause maximum stress in the members framing in to it, thus providing rotational restraints.

When designing the compression chord of the truss, the designer has to decide whether the purlins support the chord against buckling in the plane of the roof (Goyet et al. 1981). If it is not considered to support the chord, the distance between diagonal wind bracing connections (see Fig. 1.42 of Chapter 1) must be taken as the unsupported length for buckling in the plane of the roof and the distance between truss panel points as the unsupported length for the buckling in the plane of the truss. We will assume in the examples of this book that purlins support the chord members against buckling in the plane of the roof. If the chord member is of the same cross section for two or more panels (which is usually the case in practical situations), there will be rotational restraints at the joint, if the chord forces are unequal. In this case, the effective length will be less than unity and Galambos (1998) suggest a value of $K = 0.85$.

When the magnitude of force in the compressive chord changes at a sub-panel point that is not braced normal to the plane of the main truss [see Fig. 9.35(a)], the approximate value of the effective length factor for chord buckling normal to the plane of the truss, based on the two compressive forces in the chord members $F_1$ and $F_2$ is given by Galambos (1998) as follows:

$$K = 0.75 + 0.25\,(F_1/F_2), \text{ where } F_2 < F_1 \qquad (9.52)$$

If $F_2$ is a tensile force it should be given a minus sign and hence the preceding formula leads to a substantial reduction of effective length. However, the assumed value of $K$ should not be less than 0.5, even if the preceding formula gives lower values. The formula in Eqn (9.52) is also applicable to the vertical web member of a $K$-braced truss shown in Fig. 9.35(b).

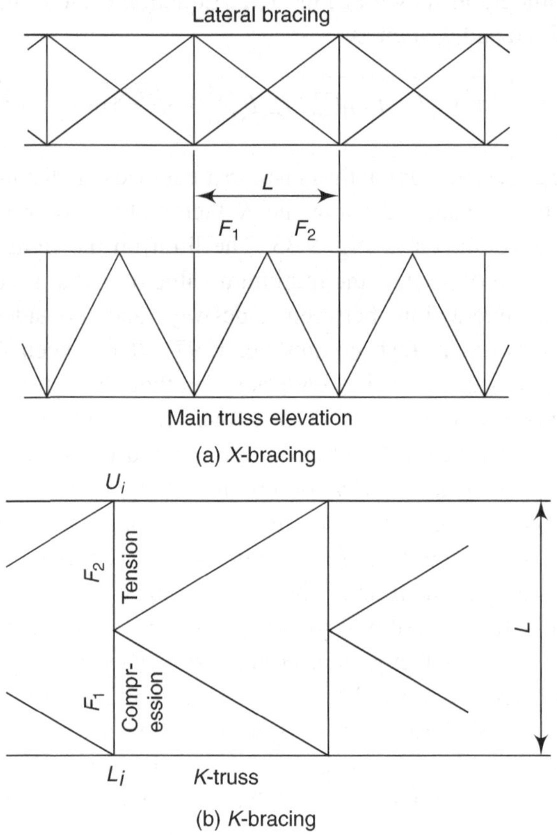


**FIG. 9.35** Effective length for truss members

Simple rules for the effective length of web members which are rotationally restrained by chord members are difficult to obtain. Bleich (1952) recommends a value of $K$ in the range of 0.57 to 1.0. If the truss carries moving loads, as discussed earlier, the rotational restraint may be appreciable and hence Galambos (1998) recommends a value of 0.85. For buckling normal to the plane of the main truss, the web compression member should be designed for $K = 1$, unless the details of cross frames (perpendicular to the truss) are available. If the cross bracing is of $K$ type [Fig. 9.35(b)] Galambos (1998) suggests a value of 0.8 and a value of 0.7 for an $X$ bracing [Fig. 9.35(a)], provided the top and bottom bracings are adequate to prevent joint translation. In the case of redundant trusses, there is reserve strength above the load at initial buckling of any compression member. Ballio and Mazzolani (1983) provide guidelines (based on the DIN Code) for determining the effective length factor for the legs of towers (see Fig. 9.36).

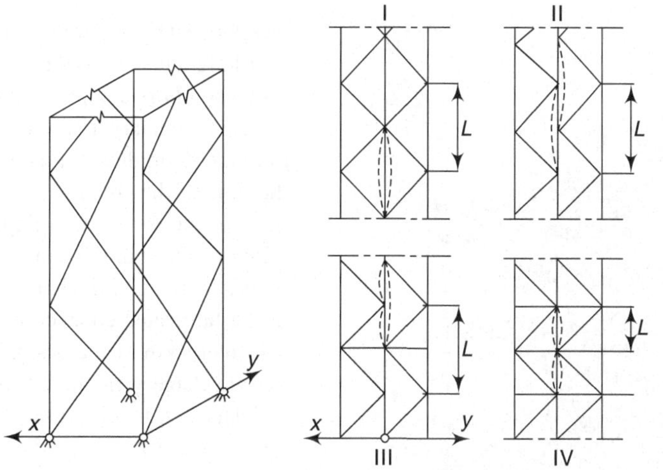


| | Section | Scheme | | | |
|---|---|---|---|---|---|
| | | I | II | III | IV |
| $\lambda = KL/r_{min}$ |  | $K = 1$ | $K = 0.7$ (0.8) | $K = 0.60$ (0.70) | $K = 1$ |
| | 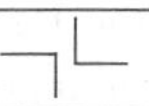 | $K = 1$ | $K = 0.85$ (0.90) | $K = 0.80$ (0.85) | $K = 1$ |
| $\lambda_x = K_x L/r_x$ |  | $K_x = 1$ | $K_x = 1$ | $K_x = 1$ | $K_x = 1$ |
| $\lambda_y = K_y L/r_y$ | 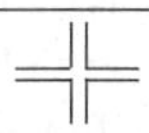 | $K_y = 1$ | $K_y = 1$ | $K_y = 1$ | $K_y = 1$ |

**FIG. 9.36** Effective length for the legs of towers (Ballio & Mazzolani 1983)

## CASE STUDY

### Hartford Civic Centre Roof Collapse

The Hartford Civic Centre Coliseum, Connecticut, USA, was completed in 1973. The space frame roof structure was 7.6 m high and covered 110 m by 91 m, with clear spans of 64 to 82 m.

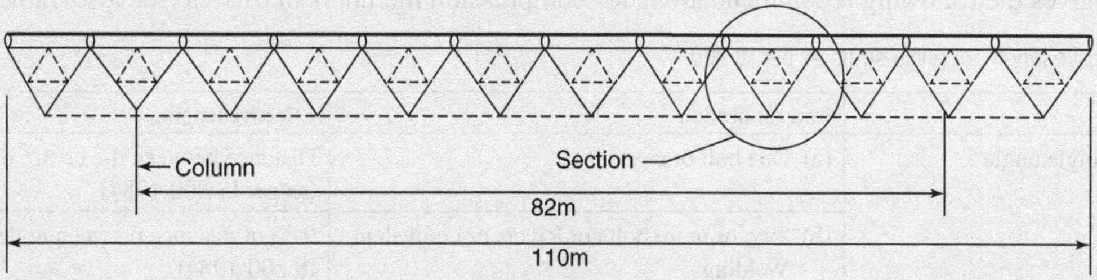


(a) Elevation

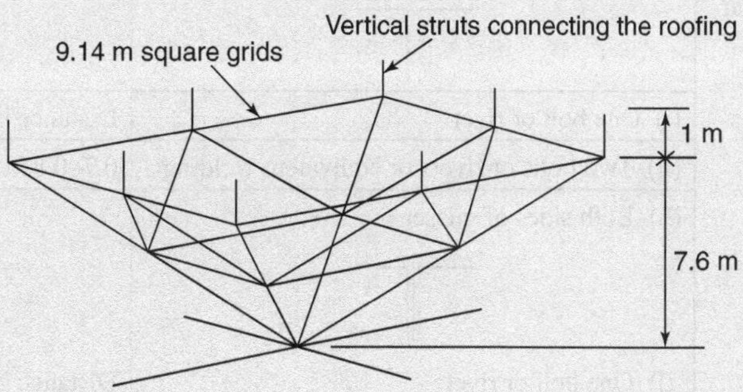


(b) Section

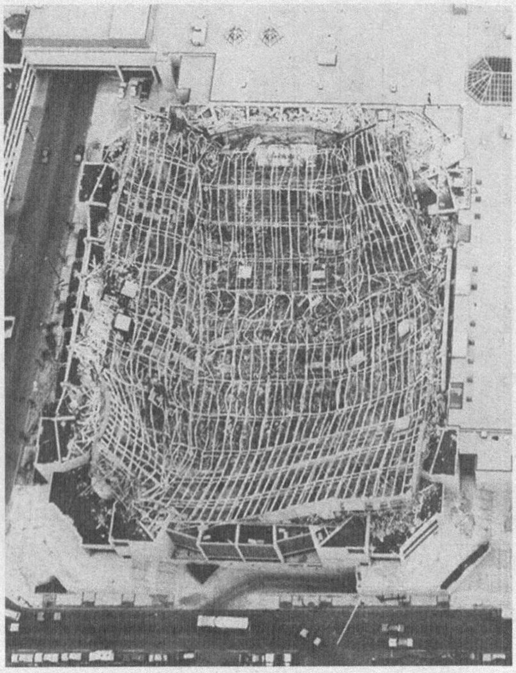

(c) View of the collapsed roof
Space frame roof of Hartford Civic Centre Coliseum roof
(*Source:* Lev Zetlin Associates)

The space frame consisted of Warren trusses with triangular bracing between the top and bottom chords (see figure). The two main layers were arranged in 9.14m by 9.14m grids composed of horizontal steel bars 6.4m apart. The 9.14m diagonal bars connected the nodes of the upper and lower layers, and in turn, were braced by horizontal bars at a middle layer (see figure). Struts 1 m long attached to the space truss were used to support the 75 mm wood fibre composition roofing.

On 17 January 1978, at 4:15 a.m. the roof crashed down 25.2 m to the floor, due to a large snow storm. Luckily it was empty at the time of the collapse, and no one was hurt. Though there were several causes for the collapse, the main cause was the relatively minor changes in the connections between steel components, that is, the fabrication deviating from design. A few centimeters shift of the fabricated connection, cut down the axial force capacity to less than tenth of the design value! Some angle sections found at the wreckage were found to have failed in block shear. Epstein and Thacker, in 1991 used finite element analysis and found that block shear was the mode of failure for these angles. This study also established the difference in behaviour of coped beams (where the load is applied to the connection in the plane of the web, which also is the block shear plane) and angles (where the load is applied eccentric to the failure plane).

In addition, the Hartford Civic Centre Coliseum roof design was extremely susceptible to torsional buckling of compression members which, as a mode of failure, was not considered by the computer analysis used by the designers. Had the designers chosen tubular or even I-sections, instead of the cruciform section adopted in the roof members, the failure might have been averted (the four steel angles forming the cruciform cross section has much smaller radius of gyration than tubes or I-sections, and hence not efficient in resisting compressive loads). The failure also showed that computer software should be used only as a software tool, and not as a substitute for sound engineering experience and judgment.

## *IS 800 Provisions for Trusses*

The code (IS 800) gives the following recommendations for compression members in trusses (see also Table 9.9).

**TABLE 9.9** Effective length for angle struts as per IS 800

| **Type** | **End connection** | **Effective length** |
|---|---|---|
| Discontinuous single angle | (a) One bolt or rivet | Distance between the centre of end fastening (as per IS 800-1984) |
| | (b) Two or more bolts or Rivets or equivalent Welding | 0.85 of distance between node points (as per IS 800-1984) |
| Discontinuous double angle, stitched together by bolts or welding at regular intervals | (a) Same side of gusset 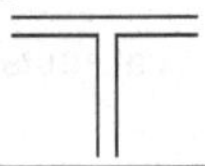 | |
| | (i) One bolt or rivet | Distance between centre of end fastenings |
| | (ii) Two bolts or rivets or equivalent welding | 0.7–0.85 of distance between nodes |
| | (b) Both sides of gusset 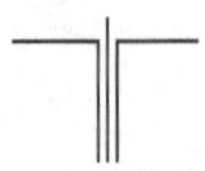 | |
| | (i) One bolt or rivet | Distance between centre of end fastenings |
| | (ii) Two bolts or rivets or equivalent welding | 0.7-0.85 of distance between nodes |

(*Contd*)

**TABLE 9.9** *(Contd)*

| Type | End connection | Effective length |
|---|---|---|
| | In a plane perpendicular to that of end gusset —for both (a) and (b) | Distance between centre of end fastening |
| Continuous angles (e.g., top and bottom chords of trusses, tower legs) | (a) Continuous | 0.7–1.0 of distance between nodes |
| | (b) In a plane perpendicular to the plane of truss | Distance between centres of nodes |

1. In the case of bolted, riveted, or welded trusses, the effective length *KL* of the compression member can be taken as 0.7 to 1.0 times the distance between the centres of connections, depending on the degree of end restraint provided.
2. In the case of members of trusses, for buckling in the plane perpendicular to the plane of the truss, the effective length may be taken as the distance between the centres of intersection.
3. In the case of double angle discontinuous struts, connected back-to-back on opposite sides of the gusset [see Fig. 9.24(b)] by not less than two bolts/rivets in line along the length of the angle, at each end or by equivalent welding, the value of *K* may be taken as 0.7 to 0.85, depending on the restraint provided. In the plane perpendicular to the truss, a value of $K = 1.0$ is recommended.
4. For double angle discontinuous struts connected back-to-back to one side of a gusset or other members by one or more bolts/rivets in each angle or by equivalent weld length, the same *K* values as above are recommended (i.e., $K = 0.7–0.85$ for in-plane and $K = 1.0$, for out-of-plane buckling).
5. For double angle continuous struts such as those forming the flanges, chords, or ties of trusses or trussed girders, or the legs of towers, values of *K* from 0.7 to 1.0 may be chosen depending on the end restraint provided.

For single angle discontinuous struts, the code does not give any recommendation for the effective length factor. Hence, the effective length factor of single angle discontinuous struts connected by single or more bolts/rivets or equivalent welding may be taken as 1.0 and the design be carried out as discussed in Section 9.13.1.

## 9.9.5 Effective Lengths of Columns in Simple Structures

The effective lengths of columns for single storey buildings of simple design may be determined by reference to the typical cases illustrated in Figs 9.37 to 9.39, provided that the following conditions are satisfied (BS 5950-1:2000).

1. In the *Y*-*Y* plane, the columns act as cantilevers tied together by the roof trusses, but in this plane, the tops of columns are not otherwise held in position or restrained in direction.

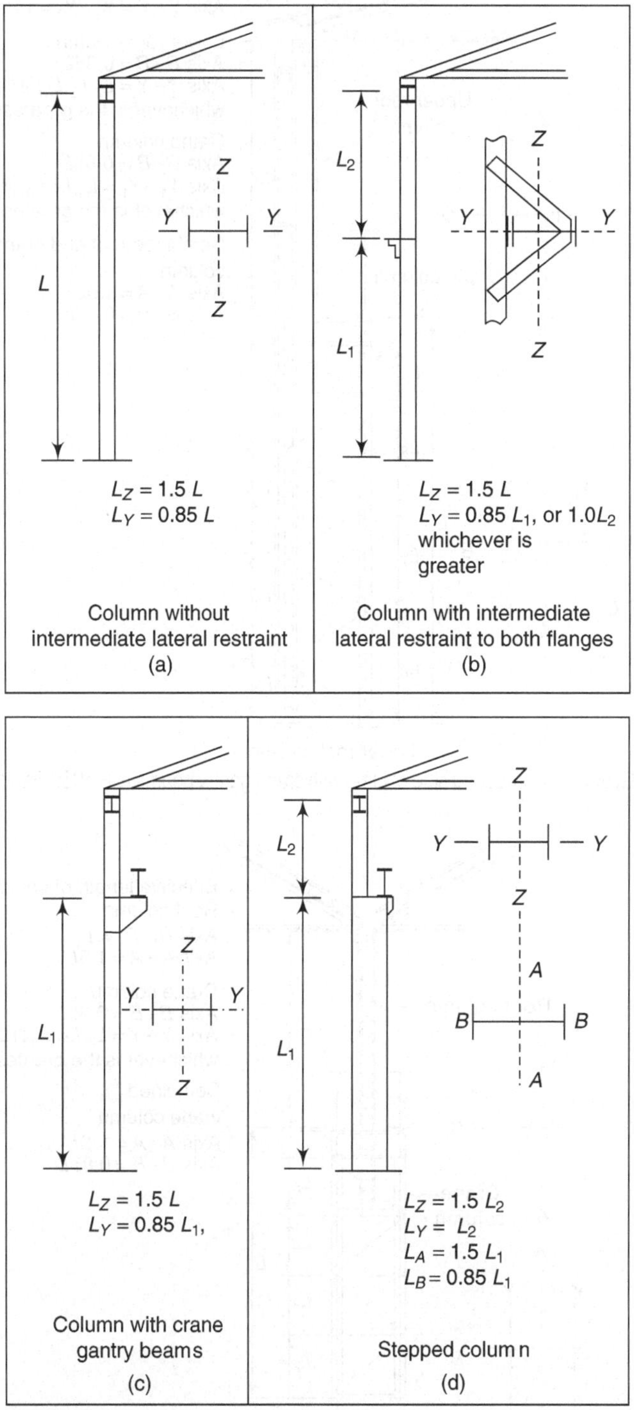


**FIG. 9.37** Effective lengths for side stanchions in single storey buildings

2. In the perpendicular plane (*Z-Z*), the tops of the columns are effectively held in position by members connecting them to a braced bay, or by other suitable means. In Figs 9.38 and 9.39, it is assumed that the braced bay also holds columns in position at crane girder level.

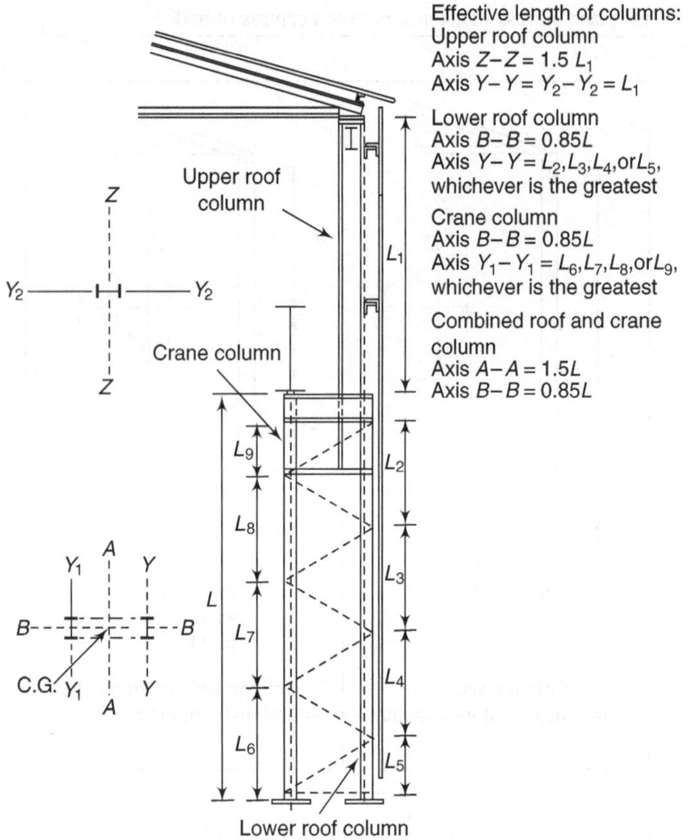


**FIG. 9.38** Compound side column with crane gantry beams (BS 5950 Part 1)

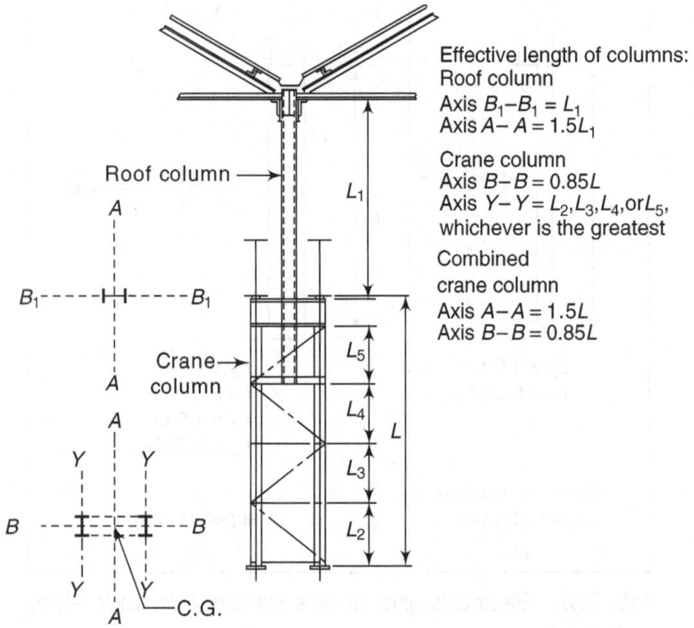


**FIG. 9.39** Compound valley column with crane gantry beams (BS 5950 Part 1)

3. The bases of columns are effectively held in position and restrained in direction in both the planes.
4. The foundations provide restraint equal to that provided by the bases of columns.

If there are different end conditions, then the following modifications are suggested by the British code.

1. If the tops of the columns are effectively held in position in the *Y-Y* plane by horizontal bracing or other suitable means, the effective length may be obtained from Fig. 9.27.
2. If the roof truss or other roof member is connected to the columns in such a way that they transfer appreciable moment to the column, the effective length may be obtained by using the Wood's curves (Fig. 9.32 or 9.34).
3. If in the *Z-Z* plane, the base of the column is not effectively restrained in direction, the effective lengths $0.85L$ or $0.85L_1$ in Figs 9.37 to 9.39 should be increased to $1.0L$ or $1.0L_1$ respectively.

## 9.9.6 Struts having Variable Section

*Tapered columns* (i.e., struts having gradually varying section) are no longer frequently adopted, since the expense of additional fabrication does not always result in economy in weight. Their use is often justified by aesthetic or functional considerations.

For a linearly tapered column, the depth at any distance $x$ from the smaller end can be expressed as (Fig. 9.40)

$$d_2 = d_0[1 + (x/L)\gamma] \tag{9.53}$$

where $d_o$ is the smallest depth at $x = 0$ and $\gamma$ is the tapering ratio, for a member whose depth at the large end is three times that of its smaller end $\gamma = 2$.

Based on the research done by Lee et al., Galambos (1998), they suggest that the buckling load for a tapered column of length $L$ can be considered equivalent to that of a prismatic (uniform) column having a cross section equal to that of the

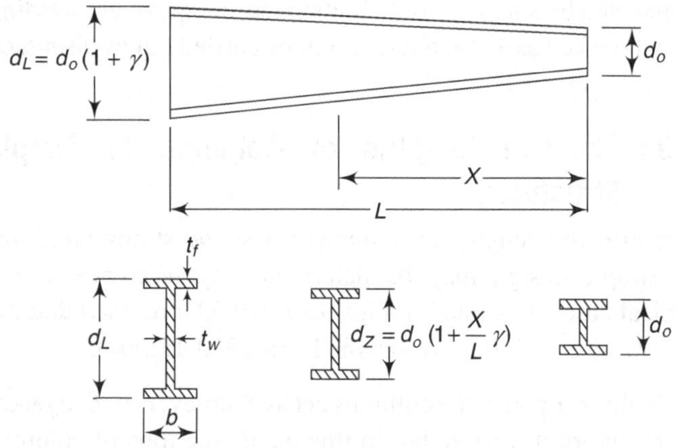


**FIG. 9.40** Geometry of linearly tapered I-column

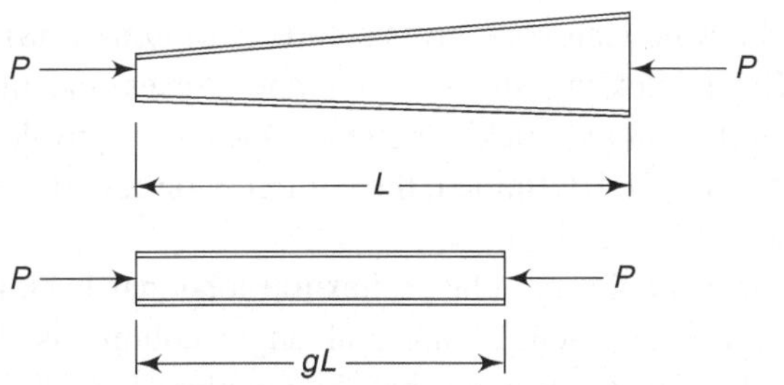


**FIG. 9.41** Tapered column and equivalent prismatic column

smaller end of the tapered column and a length equal to '$gL$' (see Fig. 9.41). Thus, for the column braced against weak-axis buckling, the buckling will occur about the strong axis, with

$$(P_{\mathrm{cr},z})_{\gamma} = \pi^2 EI_{Z0}/(gL)^2 \quad (9.54)$$

Since the variation of the weak-axis radius of gyration along the length of a web-tapered member (flange width is constant) is small, the buckling load about weak axis is given by

$$(P_{\mathrm{cr},y})_{\gamma} = (\pi^2 EI_{y0}/L^2) \quad (9.55)$$

Using a curve fitting technique, the value of $g$ was found to be (Galambos 1998)

$$g = 1.0 - 0.375\gamma + 0.080\gamma^2\,(1.0 - 0.0775\gamma) \quad (9.56)$$

The preceding elastic buckling formulae are applicable to long columns. For intermediate length columns, buckling will occur in the inelastic range and the formula will over predict the critical load. Since there is no available inelastic solution, one may use the multiple column curves [Section 9.7.1, Eqns (9.38) and (9.39)] given in the code with an effective length '$gL$'. Since welding is involved in producing these tapered sections, curves $b$ and $c$ may be chosen if $t_f \leq 40$ mm for major and minor axis buckling, respectively and curves $c$ and $d$ if $t_f >$ 40 mm. It may be of interest to note that a similar approach is suggested by the German code (Ballio and Mazzolani 1983). If the column is restrained at either or both ends, the buckling load as discussed earlier will be different. A restrained column of length $L$ may be considered as a pin-ended column of length $K_{\gamma}L$, where $K_{\gamma}$ is the effective length factor. Curves relating end restraint to effective length factor $K_{\gamma}$ are provided by Petersen (1980) and Galambos (1998) based on the work of Lee et al.

*Stepped columns* (formed by two or more parts differing in geometry and loading) are more common; they are typical of industrial buildings with heavy cranes. The equivalent length of a uniform column having the same buckling characteristics as that of the stepped column is needed to determine the axial compressive resistance and the Euler buckling load [for potential modes of failure by buckling or bending about the major ($z$-$z$) axis]. Tables to determine effective length factors are provided in IS 800 : 2007, Annex D2, for the cases when the column base is fixed and top is pinned, fixed, free, or held against rotation but not held in position. Tables and charts for the effective length factors for stepped columns are also available in Timoshenko and Gere (1961), Petersen (1980), Anderson and Woodward (1972), Agarval and Stafiej (1980), and Young and Budynas (2001).

## 9.10 FLEXURAL-TORSIONAL BUCKLING

Columns for which shear centre and the centroid coincide, fail in one of the following three independent buckling modes:

1. Buckling by lateral bending about the $z$-axis (major axis)
2. Buckling by lateral bending about the $y$-axis (minor axis)
3. Twisting about the shear centre

This is not the case with cross sections with only one axis of symmetry, such as channels, angles, or Ts. For such sections with one axis of symmetry there is a coupling of flexural and torsional deformations, which complicates the behaviour considerably. Hence, the column shown in Fig. 9.42 may fail in one of two buckling modes: (a) flexural buckling in the $z$-axis (major axis); however, if the flanges are so wide that $I_{zz} > I_{yy}$, it may buckle in the $y$-direction and (b) a combination of twisting and bending in the $y$-axis. When the member ends are subjected to a uniform compressive stress, the resultant axial force $P$ acts at the centroid $C$. Suppose as a consequence of flexural buckling in the $y$-direction, the section deflects by an amount $v$. A bending moment equal to $Pv$ is then induced in this section. However, this bending moment and the associated shear forces are acting in the centroidal plane $y$-$y$ and not in the shear centre plane $y'$-$y'$. Hence they will induce bending deformations and also cause the section to twist. Thus, purely flexural buckling in this direction is impossible and the bending and torsional deformations are coupled. This form of buckling is termed *flexural-torsional buckling* (Dowling et al. 1988).

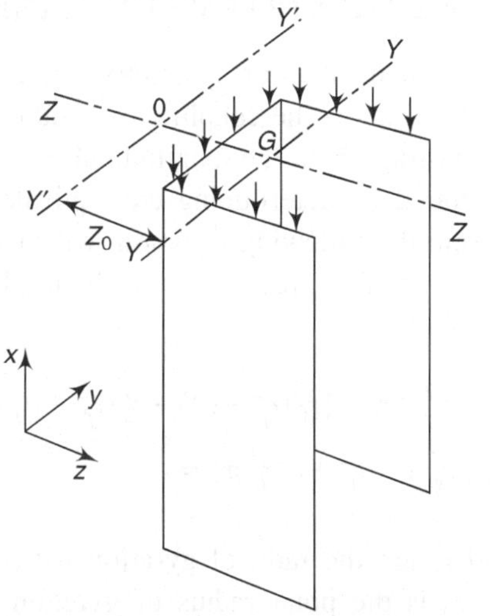


**FIG. 9.42** Compression member having a channel cross section

Galambos et al. (1992) suggest that the critical load for the coupled buckling mode can be found by determining the equivalent radius of gyration $r_{tb}$ given by

$$(1 - z_0^2/r_p^2) r_{tb}^4 - (r_z^2 + r_t)^2 r_{tb}^2 + r_z^2 r_t^2 = 0 \qquad (9.57a)$$

where $r_p$ is the polar radius of gyration about the shear centre $= \sqrt{(I_p/A)}$ and $r_t$ is the equivalent radius of gyration for torsional buckling given by Eqn (9.57b) . The solution of this equation produces two values of $r_{tb}$, the smaller of which will always be less than $r_z$.

$$r_t^2 = \frac{I_w + 0.04 I_t (KL)^2}{I_p} \qquad (9.57b)$$

where $I_w$ is the warping constant, $I_t$ is the torsional constant, and $I_p$ is the polar moment of inertia.

To determine the critical load for a column whose cross section has one axis of symmetry ($z$-axis) we compute $r_{tb}$ from Eqn (9.57) and compare it with $r_y$. The smaller of the two identifies the buckling mode and determines the corresponding critical stress. The effective length ratio $K$ can be used in evaluating the buckling load associated with $r_e$ and $r_y$, provided the boundary conditions for $\phi$ and $v$ (displacement in $y$-direction) are the same (Gaylord et al. 1992).

The solution of differential equations of equilibrium can also be presented in terms of $f_{crz}$, $f_{cry}$, and $f_{crx}$, where $f_{crz}$ and $f_{cry}$ are the Euler buckling stresses for the $z$ and $y$ axes, respectively, and $f_{crx}$ is the elastic torsional buckling stress. Such solutions are given in the American code.

After calculating the value of $r_{tb}$, the slenderness ratio ($KL/r_{tb}$) can be calculated and the buckling stress can be computed using Tables 9.3 and 9.4 and Fig. 9.23 (Tables 7.4, 7.2, and Fig. 7.1 of the code).

## 9.11 COLUMNS WITH NO AXIS OF SYMMETRY

For columns with no axis of symmetry, there is only one buckling mode which is the coupling of bending about axes $z$ and $y$ and twisting about the longitudinal axis. The critical load can be found by determining the equivalent radius of gyration $r_e$ from the following equation, which is a cubic in $r_e^2$. The smallest of the three roots must be used (Gaylord et al. 1992).

$$(r_p^2 - z_0^2 - y_0^2)^2 r_e^6 - [r_y^2 (r_p^2 - z_0^2) + r_z^2 (r_p^2 - y_0^2) + r_p^2 r_t^2]$$
$$r_e^4 + r_p^2 (r_z^2 r_y^2 + r_z^2 r_t^2 + r_y^2 r_t^2) r_e^2 = 1 \qquad (9.58)$$

where $r_z$ and $r_y$ are the radii of gyration for $z$ and $y$ axes, respectively, $r_p$ is the polar radius of gyration about shear centre $= \sqrt{(I_{ps}/A)}$, and $r_t$ is the equivalent radius of gyration for torsional buckling given by Eqn 9.57b.

The American code gives solutions to Eqn (9.58) in terms of Euler buckling stresses in $z$ and $y$ axes and the elastic flexural-torsional buckling stress. Once the equivalent radius of gyration is determined, the compression resistance may be found from Fig. 9.23.

For example, the elastic flexural-torsional buckling load, $P_{cr,\,ft}$ for a pin-ended unequal angle column is shown in Fig. 9.43. It is observed that the buckling load $P_{cr,\,ft}$ is less than any of $P_{cr,\,x}$, $P_{cr,\,y}$, and $P_{cr,\,z}$.

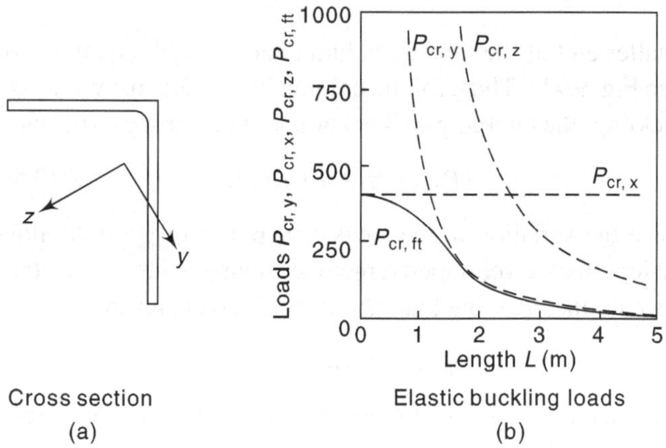


**FIG. 9.43** Elastic buckling of a typical simply supported angle section column (Trahair et al. 2001)

## 9.12 PREVENTION OF BUCKLING FAILURE

The buckling resistance of a member may be increased by increasing its cross-sectional area or by choosing a section having large value of radius of gyration $r$. A more effective approach will be to modify the buckling mode. This can be done by (a) preventing rotations at the end of a strut (b) restraining the midpoint of a pin-ended column using a bracing, which allows it to buckle in a mode having two half sine waves. This increases the critical load by a factor of four. (The stiffness required for the bracing for modifying the buckling behaviour of the compression member is very small—see Section 9.9.2.)

Similar possibilities exist for compressed plates. Stiffening the edge of an outstand (thin flange) by a lip for example, increases the buckling resistance of the plate. Analogous to a central bracing of a strut, a longitudinal stiffener may be attached along the middle of a simply supported plate to double the number of buckles and to quadruple the elastic critical stress. Similarly, transverse and/or longitudinal stiffeners may be attached to the webs of plate or box girders to increase their resistance to buckling under shear stresses.

Beams may be restrained against lateral and/or torsional displacements to eliminate lateral-torsional buckling and to develop their full plastic moment capacity. Overall buckling behaviour of frames can be improved by appropriate diagonal bracing, thick lift/service shaft walls or shear walls. A summary of buckling phenomena and parameters is provided in Table 9.10 (Dowling et al. 1988).

**TABLE 9.10** Summary of buckling phenomena and parameters (Dowling et al. 1988)

| Buckling phenomenon | Member type | Loading | Stiffness parameters | Slenderness ratio | Generalized slenderness |
|---|---|---|---|---|---|
| Flexural | Struts and columns | Axial compression | $EI_z$ or $E_y$ | $L/r_z$ or $L/r_y$ | $\sqrt{(f_y/f_{cr})}$ |
| Torsional | Struts* | Axial compression | $GI_t, EI_w$ | — | $\sqrt{(f_y/f_{cr})}$ |
| Flexural-torsional | Struts** | Axial compression | $GI_t$, $EI_w$, $EI_z$, $EI_y$ | — | $\sqrt{(f_y/f_{cr})}$ |
| Local | Thin flat plates or elements | Compression or shear | $Et^3\{12(1-\mu^2)\}$ | $b/t$ | $\sqrt{(f_y/f_{cr})}$ |
| Lateral-torsional | Beams | Major axis bending | $GI_t$, $EI_w$, $EI_y$ | — | $\sqrt{(M_p/M_{cr})}$ |

*open, thin-walled section with double or point symmetry
**open thin-walled section without double or point symmetry

## 9.13 SINGLE ANGLE STRUTS

Angles are perhaps the most basic and widely used for all rolled structural steel shapes. There is a wide range of sizes available and end conditions are relatively simple. Single angles are commonly used in many applications such as web members in steel joists and trusses, members of latticed transmission towers and communication structures, elements of built-up columns, and bracing members. In roof trusses, the single angle web members are often connected by one leg (thus introducing eccentricity with respect to the centroid of the cross sections) on one side of the chords and sometimes alternatively on opposite sides of T-sections as shown in Fig. 9.44. In towers, the bracing members meeting at a joint are connected to the opposite sides of the leg member, in order to reduce eccentricity as well as to reduce (or eliminate) gusset plates. The ease with which connections can be made contributes to the popularity of their use.

Due to the asymmetry of the angle cross section, the determination of the compression capacity under eccentric loading along with end restraints is complex. Many researchers have studied the load carrying capacity of single angle members (e.g., Madugula & Kennedy 1985; Woolcock & Kitipornchai 1986; Kitipornchai & Lee 1986; Bathon et al. 1993; Adluri & Madugula, 1996). A review of experimental and analytical research on angle members is provided by Galambos (1998).

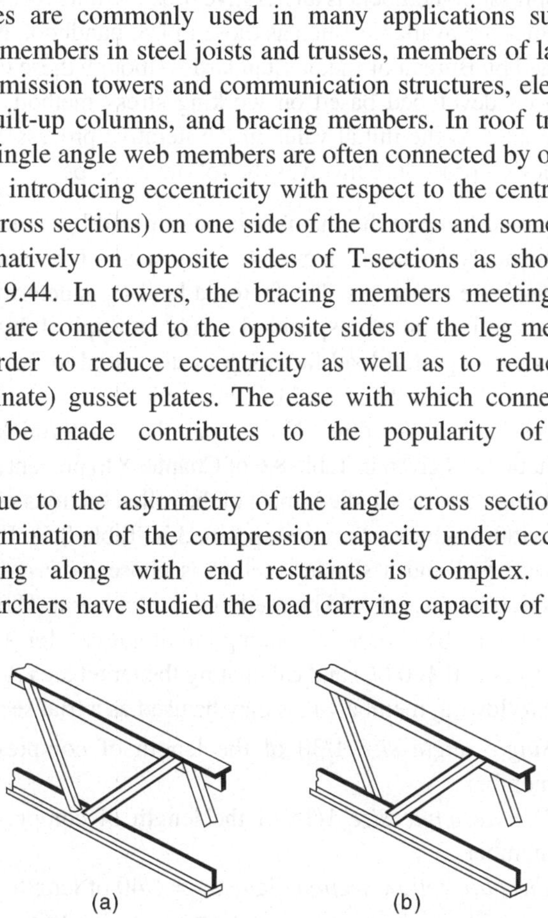


**FIG. 9.44** Web members in trusses (a) Same sides (b) Opposite sides

Angle members loaded through the centroid by a compressive axial force will buckle in flexural buckling about the minor principal axis of the cross section or in a torsional-flexural mode. When the width-to-thickness ratio of the legs of the angle are larger, the greater will be the possibility of torsional or flexural-torsional buckling being the controlling limit states. Note that the end restraint can significantly increase the ultimate strength of single angles of higher slenderness ratios, whereas it may weaken the ultimate strength in the lower slenderness ratio ranges. One of the most difficult tasks for designers is to judge or determine the end restraint and eccentricity condition for their specific application. Failure to consider the end restraints may lead to uneconomical designs whereas ignoring the end eccentricity may result in an unsafe design.

An empirical approach is adopted in ASCE Standard 10-1992, ASCE Manual 52-1988, ECCS recommendations-1985, and BS 5950-2000, to include the effects of end restraints and joint eccentricities by modifying the member's effective slenderness ratio. For example, in BS-5950, for single-angle struts connected to a gusset or directly to another member at each end by two or more fasteners in line along the angle or by equivalent welded connection, the slenderness ratio $L/r$ should not be less than

$$0.85L/r_v \quad \text{or} \quad 0.7L/r_a + 30 \tag{9.59a}$$

where $r_v$ is the radius of gyration about the minor principal axis and $r_a$ is the radius of gyration about the axis parallel to the plane of the gusset or the supporting member. Consideration of the rotational restraint permits the use of an effective length factor 0.85 while the end eccentricity effect is taken into account in the out-of-plane buckling. If a single fastener is used at each ends, the $L/r$ should be taken as not less than

$$L/r_v \quad \text{or} \quad 0.7L/r_a + 30 \tag{9.59b}$$

Thus, the rotational restraint effect is ignored for this connection.

The complete treatment of single angles is outside the scope of this book and the interested reader may refer to Galambos (1998). A separate publication is available from AISC for the design of single angle members subjected to compression bending and beam-column action (AISC-1993), though these provisions have been integrated in the current version of the code (AISC-360, 2010). Based on an inelastic analysis of steel angle columns with residual stresses Adluri and Madugula (1996) developed an average column curve for steel angles (similar to curve $c$ of Fig. 9.23). This curve could be represented by the following algebraic equations:

$$P_d = Af_y(1.0 - 0.18\lambda^2),\ 0 \le \lambda \le 0.73 \tag{9.60a}$$

$$P_d = Af_y(0.990 + 0.150\lambda - 0.360\lambda^2 - 0.020\lambda^3),\ 0.73 \le \lambda \le 1.10 \tag{9.60b}$$

$$P_d = Af_y(1.0 + 0.121\lambda - 0.367\lambda^2),\ 1.10 \le \lambda \le 1.18 \tag{9.60c}$$

$$P_d = Af_y(0.05 + 0.807\,\lambda^{-2}),\ 1.18 \le \lambda < 1.80 \tag{9.60d}$$

$$P_d = Af_y(0.008 + 0.938\lambda^{-2}),\ 1.80 \le \lambda \le 2.80 \tag{9.60e}$$

$$P_d = Af_y\lambda^{-2},\ 2.80 < \lambda \tag{9.60f}$$

where slenderness parameter $\lambda = (KL/r\pi)\sqrt{f_y/E}$ (9.60g)

## 9.13.1 Indian Code Provisions

The Indian code (IS 800) suggests that when the single angle is loaded concentrically in compression, the design strength may be evaluated using Fig. 9.23, choosing class $c$ curve. When the single angle is loaded eccentrically through one of its legs, the flexural torsional buckling strength may be evaluated using an equivalent slenderness ratio $\lambda_e$ given by

$$\lambda_e = \sqrt{\left[k_1 + k_2\lambda_{vv}^2 + k_3\lambda_\phi^2\right]} \tag{9.61a}$$

where $k_1$, $k_2$, $k_3$ are constants depending on end conditions as shown in Table 9.11.

$$\lambda_{vv} = (L/r_{vv}) / \left\{\varepsilon\sqrt{\pi^2 E/250}\right\} \tag{9.61b}$$

and

$$\lambda_\phi = (b_1 + b_2)/\left\{\varepsilon\sqrt{(\pi^2 E/250)} \times 2t\right\} \tag{9.61c}$$

where $L$ is the unsupported length of the member, $r_{vv}$ is the radius of gyration about the minor axis, $b_1$ and $b_2$ are the widths of the two legs of the angle, $t$ is the thickness of the leg, and $\varepsilon$ is the yield stress ratio $(250/f_y)^{0.5}$.

**TABLE 9.11** Constants $k_1$, $k_2$, and $k_3$

| No. of bolts at the end of member | Gusset/connecting member fixity* | $k_1$ | $k_2$ | $k_3$ |
|---|---|---|---|---|
| ≥ 2 | Fixed | 0.20 | 0.35 | 20 |
| | Hinged | 0.70 | 0.60 | 5 |
| 1 | Fixed | 0.75 | 0.35 | 20 |
| | Hinged | 1.25 | 0.50 | 60 |

*Stiffness of in-plane rotational restraint provided to the gusset/connecting member. For partial restraint, the $\lambda_e$ value can be interpolated between the $\lambda_e$ values for fixed and hinged cases.

Equations (9.61a) to (9.61c) are based on the research conducted at the Indian Institute of Technology, Madras.

## 9.14 DESIGN OF COMPRESSION MEMBERS

The strength of a compression member is based on its gross area $A_g$ (for slender cross section, $A_{eff}$ should be used). The strength is always a function of the effective slenderness ratio $KL/r$, and for short columns the yield stress $f_y$ of the steel. Since the radius of gyration $r$ depends on the section selected, the design of compression members is an iterative process, unless column load tables are available (such as those in I.S. handbook No. 1 published by Bureau of Indian Standards—though these tables have been developed based on working stress method, they may be used as the initial value in the iterative process). The usual design procedure involves the following steps.

1. The axial force in the member is determined by a rational frame analysis, as in Chapter 4, or by statics for statically determinate structures. The factored load $P_u$ is determined by summing up the specified loads multiplied by the appropriate partial load factors $\gamma_f$ as discussed in Sections 2.9.1 of Chapter 2 and 3.15 of Chapter 3.
2. Select a trial section. Note that the width/thickness limitations as given in Table 8.6 of Chapter 8 to prevent local buckling must be satisfied (most of the rolled sections satisfy the width-to-thickness ratios specified in Table 8.6). If it is not satisfied and a slender section is chosen, the reduced effective area $A_{eff}$ should be used in the calculation. The trial section may be chosen by making initial guesses for $A_{eff}/A$, and $f_{cd}$ (say 0.4–0.6$f_y$) and calculating the target area $A$.
   The following member sizes may be used as a trial section:
   1. *Single angle size* 1/30 of the length of compression member
   2. *Double angle size* 1/35 of the length of compression member
   3. *Circular hollow section* diameter = 1/40 of length

   The slenderness ratios as given in Table 9.12 will help the designer to choose the trial sections.

**TABLE 9.12** Slenderness ratios to be assumed while selecting the trial sections

| Type of member | Slenderness ratio ($L/r$) |
|---|---|
| Single angles | 100–150 |
| CHS, SHS, RHS | 90–110 |
| Single channels | 90–150 |
| Double angles | 80–120 |
| Double channels | 40–80 |
| Single I-section | 80–150 |
| Double I-sections | 30–60 |

3. Compute $KL/r$ for the section selected. The computed value of $KL/r$ should be within the maximum limiting value given in Table 9.13. Using Fig. 9.23 and Tables 9.3 and 9.4 compute $f_{cd}$ and the design strength $P_d = Af_{cd}$.

**TABLE 9.13** Maximum slenderness ratio of compression members

| Type of Member | $KL/r$ |
|---|---|
| Carrying loads resulting from dead loads and superimposed loads | 180 |
| Carrying loads resulting from wind and seismic loads only, provided the deformation of such a member does not adversely affect the stress in any part of the structure | 250 |
| Normally acting as a tie in a roof truss or a bracing system but subject to possible reversal of stress resulting from the action of wind or seismic forces | 350 |
| Lacing bars in columns | 145 |
| Elements (components) in built-up sections | 50 |

4. Compare $P_d$ with $P_u$. When the strength provided does not exceed the strength required by more than a few per cent, the design would be acceptable; otherwise repeat steps 2 through 4.

The flowchart for the design of compression members is given in Fig 9.45.

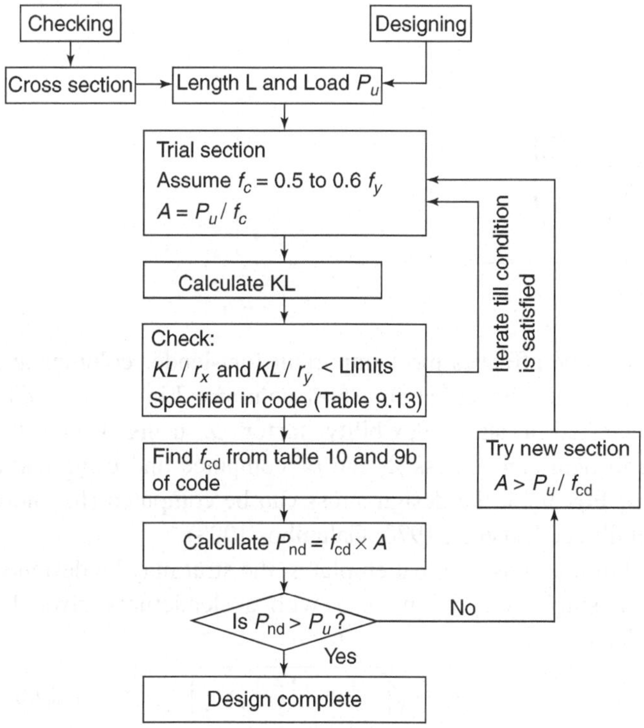


**FIG. 9.45** Flowchart for design of compression members

## 9.14.1 Limiting Slenderness Ratio

The maximum slenderness ratio of compression members under axial load is limited by the code as shown in Table 9.13.

The limit of 180 may be taken as normally applicable to primary members in compression, such as columns, compound column sections, etc. The limit of 250 may be applied to secondary members, which themselves do not support any structure, but form a part of one. The limit of 350 may be applied to bracing systems, which lend rigidity to a structure, but by themselves carry only nominal loads.

It may be of interest to note that the latest version of AISC code (AISC: 360-2010) does not specify any upper limit on the slenderness ratio, though the commentary to this code recommends an upper limit of 200. The upper limits provided are based on professional judgement and practical considerations such as economics, ease of handling, and care required to minimize any inadvertent damage during fabrication, transport, and erection.

# 9.15 BUILT-UP COMPRESSION MEMBERS

For large loads and for efficient use of material, *built-up columns* (also called as *combined columns* or *open-web columns*) are often used. They are generally made up of two or more individual sections such as angles, channels, or I-sections and properly connected along their length by *lacing* or *battening* so that they act together as a single unit (see Fig. 1.33 of chapter 1 and Fig. 9.46). Such laced combined compression members are often used in bridge trusses. According to the type of connection between the chords, built-up members may be classified as follows:

1. Laced members [Fig. 9.46(a)]
2. Struts with batten plates [Fig. 9.46(b)]
3. Battened struts [Fig. 9.46(c)]
4. Members with perforated cover plates [Fig. 9.46(f)]

In general, such struts can be considered either as simple or a built-up struts depending on the plane of bending. A strut having a cross section as shown in Fig. 9.46(d), for example, must be considered as a simple strut if it bends about the *z*-axis and considered as a built-up strut if it bends about the *y*-axis (Ballio & Mazzolani 1983). Struts having cross sections of the type shown in Fig. 9.46(e) behave as built-up struts both in *z* and *y* directions.

The effects of shear in built-up columns sets apart the design of these members from that of other columns. The importance of designing the elements connecting the main longitudinal members for shear was tragically demonstrated by the failure of the first Quebec Bridge in Canada during construction in 1907 (Galambos 1998). It has been found that about three fourths of the early failure of laced columns resulted from local rather than general column failure. Moreover, the critical

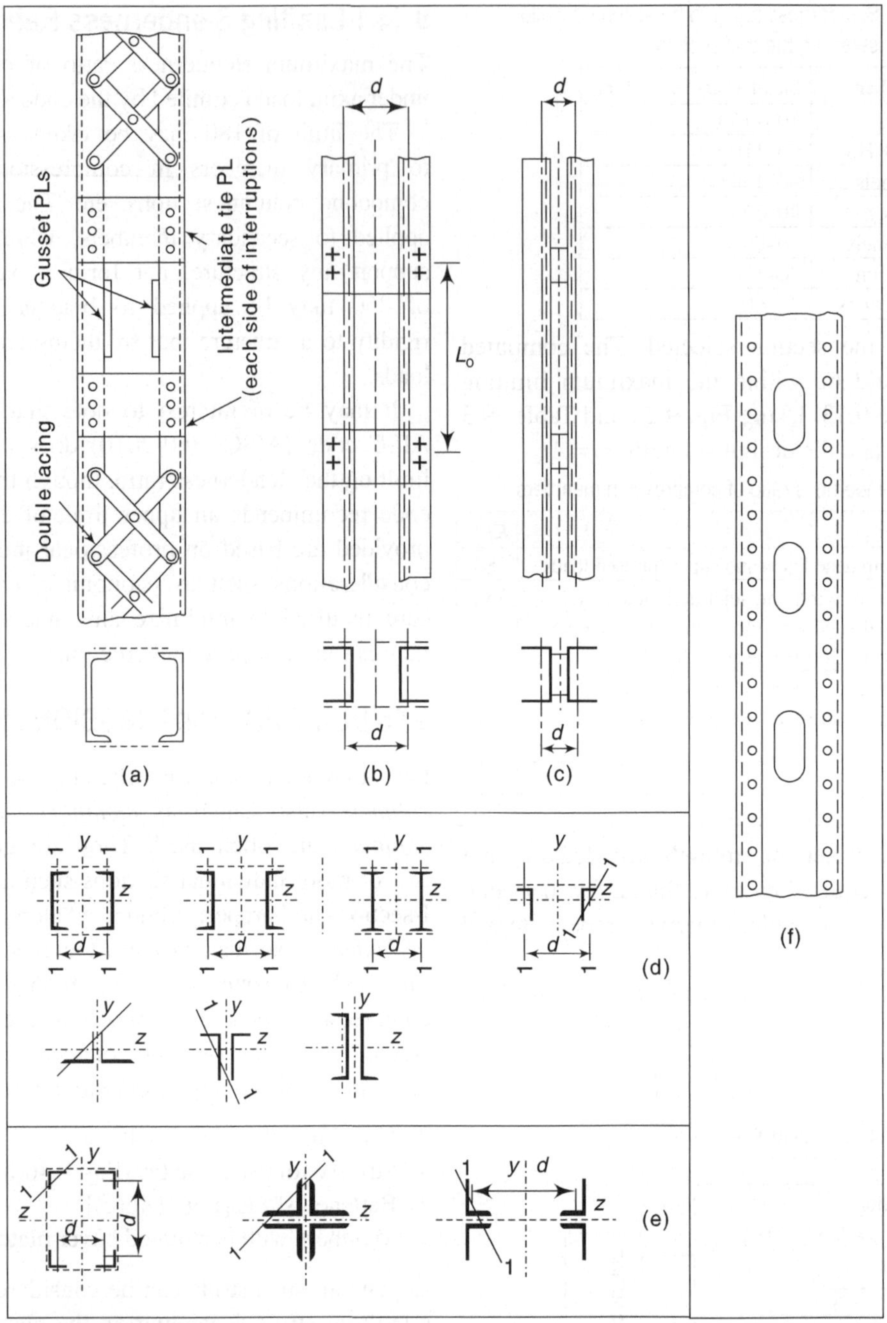


**FIG. 9.46** Built-up columns

load for a built-up column is less than that of a comparable solid column because the effect of shear on deflections is much greater for the former (Galambos 1998). The shear in column may be due to the following:

1. Lateral loads from wind, earthquake, gravity, or other causes
2. The slope of column with respect to the line of thrust due both to unintentional initial curvature and the increased curvature during buckling
3. The end eccentricity of the load due to either end connections or fabrication imperfections

The slope effect is most important for slender columns and the eccentricity effect for short columns. Lin et al. (1970) suggested a shear flexibility factor $\mu$, using which the equivalent slenderness $\lambda_e$ can be computed and using curve *c* of Fig. 9.23, the design stress can be computed (for more details see Lin et al. 1970; Galambos 1998).

For columns with batten plates, the strut may be designed as a single integral member with a slenderness given by (Bleich 1952)

$$\lambda_e = \sqrt{\left[\lambda_m^2 + \lambda_e^2\left(\pi^2/12\right)\right]} \quad (9.62)$$

With the limitations $\lambda_e > 50$ and $\lambda_e < 1.4\ \lambda_c$ where $\lambda_m = L/r_{\min}$ = strut slenderness, $\lambda_c = L_o/r_o$ = local chord slenderness between one batten plate and the next, and $L_o$ is the centre-to-centre distance of batten plate.

This equation was derived strictly for hinged-end columns. Bleich estimated that the buckling strength of a steel column having $L/r$ of 110 is reduced by about 10% when $\lambda_c$ = 40 and by greater amounts for larger values of $\lambda_c$. Aslani and Goel (1991) proposed the following equation based on their theoretical and experimental investigations, which is applicable to all built-up columns with general end conditions

$$\lambda_e = \sqrt{\left[\lambda_m^2 + 0.82\left(\frac{\alpha^2}{1+\alpha^2}\right)\lambda_c^2\right]} \tag{9.62a}$$

Where $\alpha = d/(2r_{ib})$ and $r_{ib}$ is the radius of gyration of individual components relative to its centroidal axis parallel to member axis of buckling and $d$ is the distance between centroids of individual components. Equation 9.62(a) was adopted in the AISC 360 code (from 1993 to 2005) for intermediate connectors that are welded or having pre-tensioned bolts. However, the AISC 360-2010 code specifies the following equations based on the statistical evaluations by Sato and Uang (2007).

1. When intermediate connectors are bolted snug tight

$$\lambda_e = \sqrt{\left[\lambda_m^2 + \lambda_c^2\right]} \tag{9.62b}$$

2. When intermediate connectors are welded or connected by means of pre-tensioned bolts
   (a) When $\lambda_c \le 40$

$$\lambda_e = \lambda_m \tag{9.62c}$$

   (b) When $\lambda_c > 40$

$$\lambda_e = \sqrt{\left[\lambda_m^2 + (K_i\lambda_c)^2\right]} \tag{9.62d}$$

where $K_i$ = 0.5 for angles back-to-back
= 0.75 for channels back-to-back
= 0.86 for all other cases

The most commonly adopted lacing systems are shown in Fig. 9.47. The simplest form of lacing consists of single bars connecting the two components [Fig. 9.47(b)]. The double lacing [Fig. 9.47(d)] is sometimes considered preferable,

Built-up column in a portal frame, Mumbai (Note the stubs in the column, which carry a small crane)

(a) General view under construction

(b) View of the lacings

4 ×15 m span continuous *lattice portal* frame for Kirloskar small engines division, Rajkot (*Note* how the lacings are welded inside the angles and the staggering on perpendicular faces)

*Courtesy*: Er Jayant Lakhlani

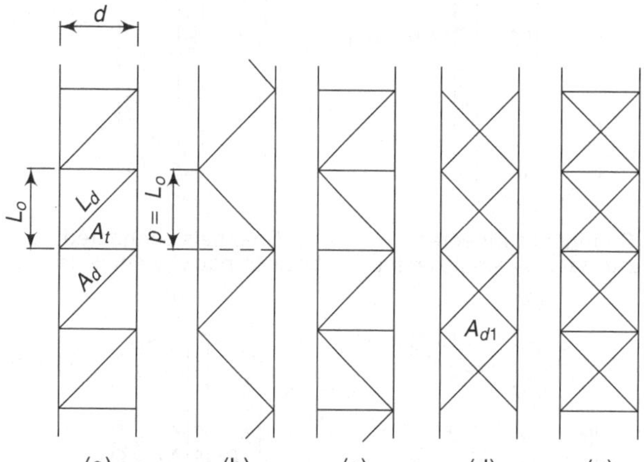


**FIG. 9.47** Commonly adopted lacing systems [(c) and (e) are not recommended]

Struts with *perforated* cover plates used in the Carquinez cantilever truss bridge (East bound I-80), California, USA

*Photo*: Sreemathi Subramanian

although a well-designed single lacing is equally effective. In single or double lacing systems, cross members perpendicular to the longitudinal axis of the strut should not be used [see Figs 9.47(c) and (e)]. The 'accordion' like action of the lacing system without cross members permits the lateral-expansion of the column. The introduction of cross members prevents the lateral expansion and thus forces the lacing bar to share the axial load on the strut. Note that the *lacing bars* and *batten plates* are not designed as load carrying elements. Their function is primarily to hold the main component members of the built-up column in their relative position and equalize the stress distribution in them. At the ends and at intermediate points where it is necessary to interrupt the lacing (for example, to admit gusset plates), the open sides are connected with *tie plates* (also called batten plates or stay plates). Tie plates are also provided at the top and bottom of the column (see Fig. 9.48).

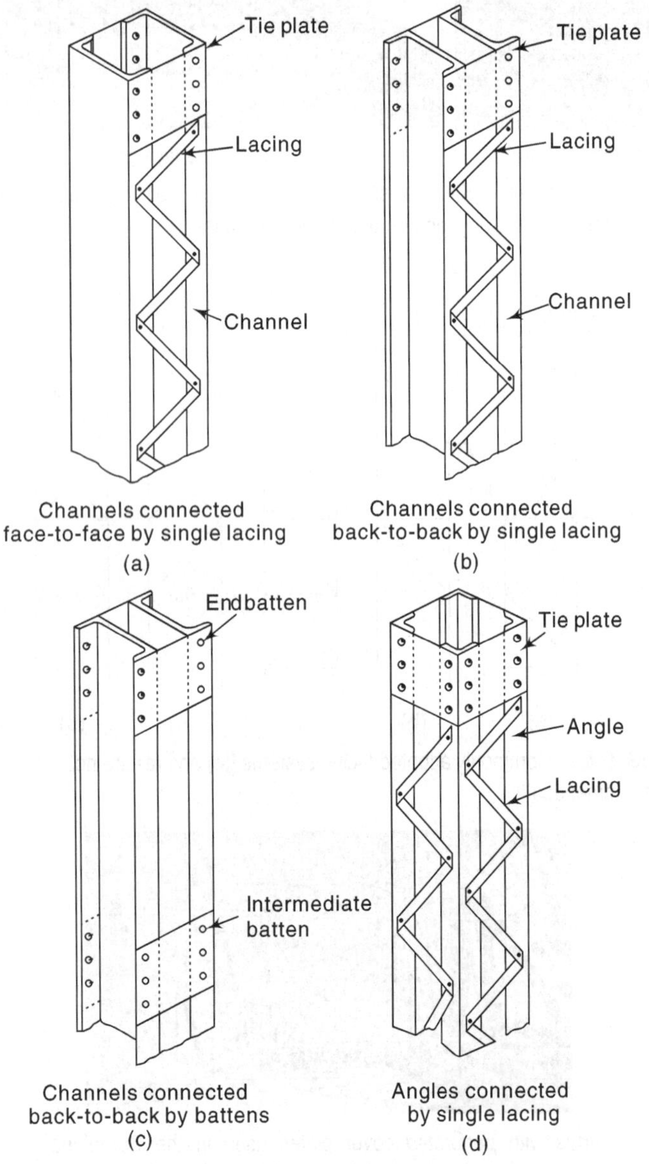


**FIG. 9.48** Isometric view of built-up columns

Saboo and Rai (2004), based on their experimental work, found that the column with the batten spacing as per IS 800 may not perform satisfactorily when subjected to earthquakes. They suggest the spacing of battens has to be reduced to half in the plastic hinge zones.

It should be noted that the battened columns have the least resistance to shear compared to columns with lacings and perforated plates, and may experience an appreciable reduction in strength. Hence, they are not generally used in the United States. Columns with perforated plates require no special considerations for shear effects. These cover plates contain perforations spaced axially, which afford access for welding or painting. After the advent of automatic cutting machines, the production of such perforated plates have become simpler. Hence, they are used extensively in USA. They result in reduction of fabrication and maintenance cost and offer superior stiffness and straightness. At present, they are not used in India and interested readers may refer to Salmon and Johnson (1996) for the design of built-up columns with perforated plates.

The effective slenderness of laced struts of the types shown in Figs 9.47(c), (d), and (e) (with two chords connected by lacings) may be obtained by using the following equation (Ballio & Mazzolani 1983)

$$\lambda_{eq} = \sqrt{\left[\lambda^2 + \pi^2 (A/A_d) L_d^3/(L_o d^2)\right]} \qquad (9.63)$$

where $\lambda$ is the strut slenderness = $L/r_{\min}$, $A$ is the overall cross-sectional area = $2A_1$, $A_1$ is the area of the individual chord, $A_d$ is the cross-sectional area of the diagonal lacing {= $2A_{d1}$ for Figs 9.47(d) and (e)}, $L_d$ is the length of the diagonal lacing, $d$ is the distance between the centroid of the chords, and $L_o$ is the chord length between the two successive joints.

Equation (9.63) is also applicable to the laced strut type shown in Fig. 9.47(b), provided $L_o$ is replaced by $p$.

If two I-sections of area $A$, moments of inertia ($I_z$ and $I_y$), and radii of gyration ($r_z$ and $r_y$) about the axes $Z$ and $Y$ respectively, and are kept side by side at a centre-to-centre distance of $x$ as shown in Fig. 9.48(a), their combined moment of inertia will be

$I_{zz} = 2I_z$ Hence $r_{zz} = r_z$

$$I_{yy} = 2\left[I_y + A\left(\frac{x}{2}\right)^2\right]$$

Equating $I_{zz}$ and $I_{yy}$ for maximum efficiency, we get

$$2I_z = 2[I_y + A\left(\frac{x}{2}\right)^2]$$

or

$$\left(\frac{x}{2}\right)^2 = \frac{I_z - I_y}{A} = r_{zz}^2 - r_{yy}^2$$

Hence $x \approx 2r_{zz}$ as $r_{yy}$ is much smaller than $r_{zz}$.

This formula can be used to find the optimum spacing of two I-sections in built-up columns.

### 9.15.1 Rules Specified in the Indian Code

In most of the codes, latticed members are designed and proportioned according to detailed empirical rules, most of which are related to local buckling requirements. In the Indian code (IS 800 : 2007), the following rules are given.

#### Lacings

1. The radius of gyration of the combined column about the axis perpendicular to the plane of lacing should be greater than the radius of gyration about the axis parallel to the plane of lacing [see Fig. 9.49(a)].

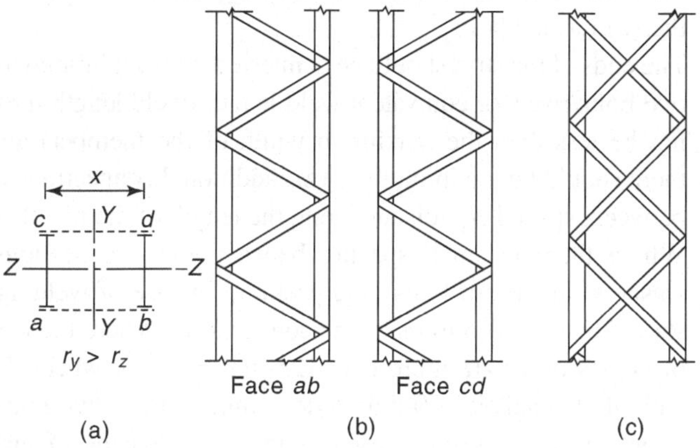


**FIG. 9.49** Lacing systems—(a) In plan (b) Single lacing (c) Double lacing

2. Lacing system should be uniform throughout the length of the column.
3. Single [Fig. 9.49(b)] and double laced [Fig. 9.49(c)] systems should not be provided on the opposite sides of the same member. Similarly, lacings and battens should not be provided on opposite sides of the same member.
4. Single laced system on opposite sides of the main component shall be in the same direction viewed from either side so that one is the shadow of the other. Thus, the lacing as shown in Fig. 9.49(b), for face *cd*, is not recommended.
5. The lacing shall be designed to resist a total transverse shear $V_t$ at any point in the member, equal to 2.5% of the axial force in the member; and this shear shall be divided among the lacing systems in parallel planes.
6. The lacings in addition should be designed to resist any shear due to bending moment or lateral load on the member.
7. The slenderness ratio of lacing shall not exceed 145.
8. The effective length shall be taken as the length between inner end bolts/rivets of the bar for single lacings and 0.7 times the length for double lacings effectively connected at intersections. For welded bars, the effective length is taken as 0.7 times the distance between the inner ends of the welds connecting the single bars to the members.
9. The minimum width of the lacing bar shall not be less than approximately three times the diameter of the connecting bolt/rivet; the thickness shall not be less than 1/40$^{th}$ of the effective length for single lacing and 1/60$^{th}$ for double lacing.
10. The spacing of lacing bars shall be such that the maximum slenderness ratio of the components of the main member between two consecutive lacing connections is not greater than 50 or 0.7 times the most unfavourable slenderness ratio of the combined column.
11. When welded lacing bars overlap the main members, the amount of lap should be not less than four times the thickness of the bar and the welding is to be provided along each side of the bar for the full length of lap. Where lacing bars are fitted between main members, they should be connected by fillet welds on each side or by full penetration butt weld.
12. Where lacing bars are not lapped to form the connection to the components of members, they shall be so connected that there is no appreciable interruption in the triangulated system.
13. Plates shall be provided at the ends of laced compression members and shall be designed as battens.
14. Flats, angles (normally adopted in practice), channels, or tubes may be used as lacings.
15. Lacing bars, whether in double or single shear shall be inclined at an angle of 40° to 70° to the axis of the built-up member.
16. The effective slenderness ratio $(KL/r)_e$ of the laced column shall be taken as 1.05 times $(KL/r)_o$, where $(KL/r)_o$ is the maximum actual slenderness ratio of the column, to account for shear deformation effects.
17. The required sections of lacing bars as compression/tension members may be determined using the appropriate design stresses $f_{cd}$ as given in Section 9.7.1 and $T_d$ in Sections 7.6 and 7.8.2 of Chapter 7.

#### Battens

The rules for the design of battens shall be the same as for lacings except for the following conditions (see Fig. 9.50).

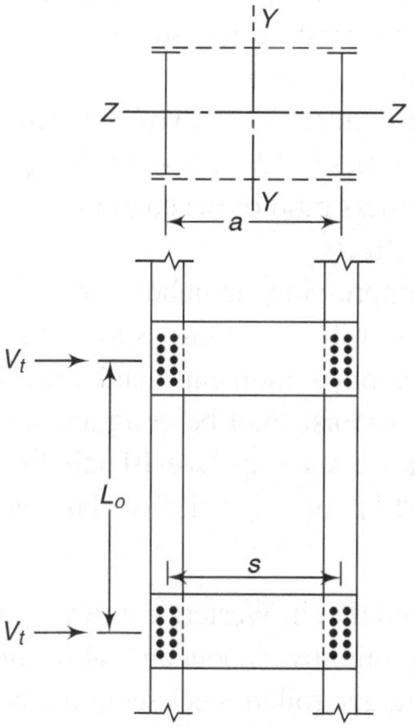


**FIG. 9.50** Battening system

1. The number of battens shall be such that the member is divided into not less than three bays.
2. Battens shall be designed to resist simultaneously

$$\text{Longitudinal shear } V_b = V_t L_o / ns \tag{9.64}$$

and

$$\text{Moment } M = V_t L_o / 2n \tag{9.65}$$

where $V_t$ is the transverse shear force, $L_o$ is the distance between centre-to-centre of battens, longitudinally, $n$ is the number of parallel planes of battens, and $s$ is the minimum transverse distance between the centroids of the bolt/rivet group/welding connecting the batten to the main member.
3. When plates are used for battens, the effective depth between the end bolts/rivets or welds shall not be less than twice the width of one member in the plane of battens; nor less than three quarters of the perpendicular distance between centroids of the main members for intermediate battens; and not less than the perpendicular distance between the centroids of main members for end battens.
4. The thickness of batten plates shall not be less than 1/50$^{th}$ of the distance between the innermost connecting transverse bolts/rivets or welds.
5. The requirement of size and thickness does not apply when other rolled sections are used for battens with their legs or flanges perpendicular to the main member.
6. When connected to main members by welds, the length of the weld connecting each end of the batten shall not be less than half the depth of the plate; at least one third of its length should be placed at each end of the edge; in addition the weld shall be returned along the other two edges for a length not less than the minimum lap (i.e., not less than four times the thickness of the plate). The length of the weld and depth of batten shall be measured along the longitudinal axis of the main member.
7. The effective slenderness ratio of battened column shall be taken as 1.10 times $(KL/r)_o$, where $(KL/r)_o$ is the maximum actual slenderness ratio of the column, to account for shear deformation effects.
8. Battened compression members, not complying with the preceding rules or those subjected to eccentricity of loading, applied moments, and lateral forces in the plane of the battens, shall be designed according to exact theory of elastic stability (see Bleich 1952; Timoshenko & Gere 1961) or empirically but verified by test results.

It should be noted that in Western countries such as USA and UK, due to the prohibitive labour and fabrication costs and the availability of larger rolled steel sections, built-up columns are seldom used nowadays.

## 9.16 COMPRESSION MEMBERS COMPOSED OF TWO COMPONENTS BACK-TO-BACK

Compression members may also be composed of two angles, channels, or Ts back-to-back in contact or separated by a small distance and connected together by bolting, rivetting, or welding [Fig. 9.46(d)]. In such a case, clause 7.8 of IS 800 : 2007 gives the following specifications.

1. The slenderness ratio of each member between the connections should not be greater than 40 or 0.6 times (0.75 times in AISC 360) the minimum slenderness ratio of the strut as a whole.
2. The ends of the strut should be connected with a minimum of two bolts/rivets or equivalent weld length (weld length must not be less than the maximum width of the member) and there should be a minimum of two additional connections in between, spaced equidistant along the length of the member. Where there is small spacing between the two sections, washers (in case of bolts) and packing (in case of welding) should be provided to make the connections. Where the legs of angles or Ts are more than 125-mm wide, or where the web of channels is 150-mm wide, a minimum of two bolts/rivets should be used in each connection. Spacing of tack bolts or welds should be less than 600 mm (Clause 10.2.5.5). If bolts are used, they should be spaced longitudinally at less than 4.5 times the bolt diameter and the connection should extend to at least 1.5 times the width of the member (Clause 10.2.3.2).
3. The bolts/rivets should be 16 mm or more in diameter for a member $\leq$ 10 mm thick and 20 mm in diameter for a member $\leq$ 16 mm thick, and 22 mm in diameter for members greater than 16 mm thick.
4. Such members connected by bolts/welding should not be subjected to transverse loading in a plane perpendicular to the riveted, bolted, or welded surfaces.
5. When placed back-to-back, the spacing of bolts/rivets should not exceed $12t$ or 200 mm and the longitudinal spacing between intermittent welds should not be more than $16t$, where $t$ is the thickness of the thinner section.

## 9.17 COMPRESSION MEMBERS WITH OTHER MATERIALS AND SHAPES

Recently, columns made of stainless steel, aluminium alloy, and composite construction are increasingly being used. Moreover, the buckling behaviour of columns may be improved by the concept of sleeved columns. Arches may also be considered as compression members. These compression members with other materials and shapes are discussed in this section.

### 9.17.1 Aluminium and Stainless Steel Compression Members

Similar to high-strength steel, aluminium and stainless steel have a rounded stress–strain behaviour without possessing a well-defined yield point as shown in Figs 1.11 and 1.38 of Chapter 1 and hence the yield stress is assumed to be 0.2% proof stress. The tangent modulus approach is used for the design of aluminium columns, simplified to a straight line in the inelastic range. Details of these formula and design methods are provided by Galambos (1998) and Gaylord et al. (1992). The current American recommendations for stainless steel columns use the tangent-modulus approach (ASCE-1991). The Eurocode approach is based largely on assumed analogies with the elastic perfectly plastic behaviour of carbon steel (www.euro-inox.org). Gardner and Nethercot (2004) have recently done extensive research on stainless steel structural members and proposed a new approach to the design.

### 9.17.2 Sleeved Column Concept

The behaviour of compression members may be improved by adopting the sleeved column concept developed by Sridhara (1991). In this concept a solid core is loosely placed inside a sleeve and the load is applied to the core only and no part of the applied load is imparted on to the sleeve. For this reason, the core is made slightly longer than the sleeve. The core, under the action of the load, bends and presses against the sleeve, thus causing primarily bending stresses in the sleeve. If the bending stress in the sleeve is kept within the yield strength of the sleeve material, it is possible to stress the core to extremely high stress levels, much beyond the yield strength of the core material. For example, a mild steel sleeve (tube) having an outer diameter of 33 mm and thickness 3.15 mm, when tested was able to take an axial concentric load of 90 kN before failure by yielding. Similarly, a solid mild steel core (rod) of diameter 25 mm and length 330 mm failed at 140 kN by yielding. The same rod with a diameter of 33 mm and length 330 mm was able to sustain a load of 240 kN. Whereas, when the 330-mm long core (rod) was placed inside the 300-mm long and 33-mm diameter sleeve and loaded through the core, it was able to sustain a load of 440 kN (Sridhara 1991). The stress in the mild steel core during the test reached an extremely high value of 896 MPa, whereas its actual yield strength is only 285 MPa. When the core is loaded beyond its yield strength or Euler's buckling load, it first bends in the gap provided and further buckling of the core is prevented by the sleeve. The core is thus subjected to compressive loads only and the sleeve to bending and shear stresses. The tests conducted by Sridhara confirmed that this concept can be utilized if the slenderness ratio of the sleeve is less than 25. Recently, this concept was successfully applied to water tank staging (Mahadevan et al. 1995). This concept can be utilized in latticed sleeved columns, crane booms, transmission towers, space frames, etc., provided suitable connections to other members are made (see also section 17.13 of Chapter 17).

### 9.17.3 Steel Arches

An *arch* is a curved structural member that spans between two points (supports) and subjected mainly to axial compression. It is the most efficient structural form that is both striking in appearance and aesthetic in character. Many types of rigid arches exist and are often characterized by their support conditions (fixed, two-hinged, or three-hinged). For primary uniformly distributed loading condition, the shape of a rigid arch is ideally parabolic, since it will be under pure axial compression. For a parabolic arch, the maximum forces are developed at the springing. The basic issue in arch design is whether or not to absorb the horizontal thrusts by some interior element, such as a tie member or by heavy foundations. When it is functionally possible, tie members are very often used, which are tension elements and they take up the outward arch thrusts in a highly efficient way.

The primary design consideration is how to cope with the behaviour of an arch in the lateral direction. It is obvious that a typical arch, which lies in one vertical plane must be prevented from simply toppling over sideways [see Fig. 9.51(a)]. Two methods are used to prevent this. One method is to use fixed base connections. Fixed arches, however, develop bending moments near supports, which are difficult to calculate. Moreover, for very large structures, the use of fixed-base connections for lateral stability may require massive foundations to prevent overturning. Another most commonly used method for achieving lateral stability is by laterally bracing arches with other elements, as shown in

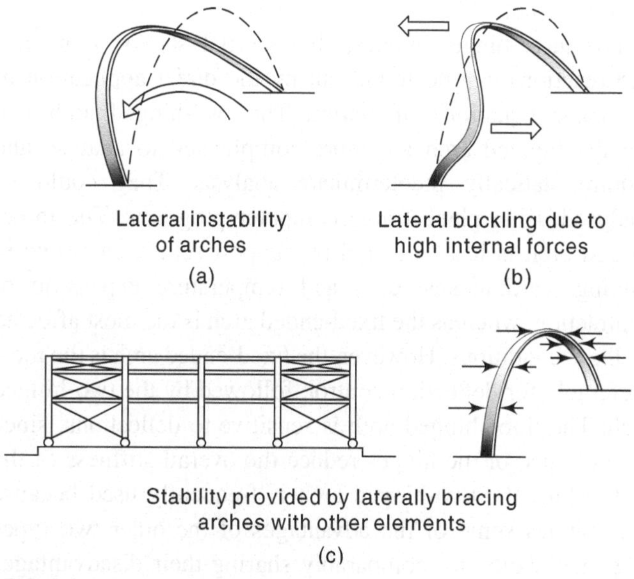


**FIG. 9.51** Arch design considerations

Arch bridge connecting USA and Canada near Niagara Falls, USA (note the bracings between the two arches)

Fig. 9.51(c). Thus, a pair of arches at either end of a structure may be stabilized through the use of diagonal elements. Interior arches may be stabilized by connecting them to the end arches by transverse members. Twin arch ribs with a lateral bracing system often occur in arch bridges.

The second major problem with respect to the behaviour of arches in the lateral direction is that of lateral buckling. Since internal forces are often fairly low, the elements of a steel arch can be fairly slender. Due to this, an out-of-plane buckling, as shown in Fig. 9.51(b) can occur. This problem may be solved either by increasing the stiffness of the arch in the lateral direction (by increasing the lateral dimension) or by bracing the arch periodically along its length with transverse members. Thus, the same system that is used to stabilize the arch from overturning laterally [Fig. 9.51(c)] also provides lateral bracing for arch members and prevents lateral buckling. If non-uniform loading or point loads are applied to arches, bending is developed in arch members in addition to axial forces. The higher the bending, the larger must be the member used.

For three-hinged arches, the internal shears, moments, and reactions can be found out by the direct application of the basic equations of statics. The two-hinged arch and the fixed-ended arch are more complicated to analyse and require statically indeterminate analysis. They could be analysed using plane frame computer programs. The three-hinged arch is least affected by support settlements (due to sinking, earthquakes, etc.) and temperature expansion or contraction, whereas the fixed-ended arch is the most affected by these conditions. However, the fixed-ended arch is the most preferable for deflection control, followed by the two-hinged arch. The three-hinged arch is sensitive to deflections, since the presence of the hinges reduce the overall stiffness of the arch. Thus, the two-hinged arch is frequently used because it combines some of the advantages of the other two types of arches while not comparably sharing their disadvantages (Schodek 2001).

The analysis and design of arches may be broadly classified according to their response to load and their in-plane mode of failure. These characteristics depend on the type of arch, such as the following (Galambos 1998):

Arch bridge in Germany

1. Slender arches, generally of solid web with rolled or built-up sections, subjected primarily to axial forces in the arch rib (similar to an axially loaded column)
2. Slender arches subjected to significant bending and deformation due to asymmetric or point loading (similar to beam-column problems as discussed in Chapter 13)
3. Stocky arches, with truss-like configuration, which are used when heavy bending occurs. In these arches, the failure will be primarily due to chord or flange failure arising from axial load and excessive bending (similar to any truss subjected to axial force and bending moment)
4. Arches composed of arch ribs and deck stiffening girders

The in-plane and out-of-plane (linear and non-linear) buckling behaviour is outside the scope of this book and interested readers may refer to Galambos (1998). Design procedures for arches based on ultimate inelastic strength studies, and using interaction type formula similar to beam-column formulas have been proposed by several authors. (e.g., Sakimoto & Komatsu 1983) and a review of these methods is provided by Galambos (1998).

As discussed earlier, most arches in practice are braced against lateral movement either continuously or at regularly spaced intervals. When pairs of arch ribs are braced closely with stiff transverse bars, the critical load per one arch rib of a set of braced arches can be increased up to 250% of that for the identical isolated single arch, so far as elastic buckling is concerned. For actual bridge arches, this increase may be limited by yielding of the material. The strength of braced arches in the plane of bracing is similar to the buckling strength of columns with lacing bars and battens (Timoshenko & Gere 1961). Simple approximate methods for determining strength of braced on unbraced steel arches which fail by lateral instability are provided by Sakimoto & Komatsu (1982). They suggested the use of column curve *c* of the ECCS multiple column curves or similar curves. Practical applications for actual steel bridges are also presented (e.g., Sakimoto & Sakata 1990). Flexural-torsional buckling of arches has also been studied (e.g., Bradford & Pi 2006; Papangelis & Trahair 1987).

It is interesting to note that the arch bridges are amongst the oldest man-made bridges. The first use of cast-iron for a sizable structure resulted in the construction of the 30-m span Coalbrookdale arch bridge over the river Severn in the year 1779 in England. Several steel railroad and highway arch bridges have been constructed all over the world (e.g., the 503-m long Sydney Harbour bridge, Sydney, Australia, built in 1932 and designed by Freeman Fox, the 255-m long continuous tied arch bridge in Duisburg, Germany built in 1950, the 228-m long Fort Pitt, double deck tied arch bridge, Pittsburg, constructed in the 1950s, the New River Gorge Bridge, West Virginia, USA with a central span of 518 m (world's longest arch bridge built in 1977) and the 216-m long Changjiang river (railway) bridge at Juijiang, China built in the year 1992). The introduction of curved structural sections has allowed more creative bridge designs to be built. Steel arches could span about 20 to 90 m. For greater span (240 – 300 m), the arches may be made up of truss work. Even pre-stressed arch steel bridges have been built (Troitsky 1990; Nazir 2003). The advent of cable stayed bridges has somewhat reduced the use of arch bridges, since the segments of arch bridges have to be erected carefully with the use of temporary supports.

### 9.17.4 Composite/Cased Columns

Steel-concrete composite columns are being used increasingly in building structures, since they advantageously combine the properties of concrete (strong in compression, greater rigidity) with those of structural steel. They may be broadly classified as those with rolled and built-up shapes encased in concrete (see Fig. 1.46a of Chapter 1) and those with concrete placed inside a steel tube or SHS/RHS (see Fig. 1.46b of Chapter 1). Steel columns are sometimes cased in concrete for fire protection. Concrete encased columns can be either square or rectangular in cross section. A composite column with an I-section embedded in concrete and provided with regular reinforcement at the corners is shown in Fig. 1.46(a) in Chapter 1. The vertical reinforcement should be provided with lateral ties at regular intervals as per the concrete code (IS: 456-2000). The use of such composite columns results in high load-carrying capacity. In the design of composite columns, it is assumed that steel and concrete work together in resisting the loads. The steel section must comprise at least 4% of the total cross-sectional area, otherwise the columns must be designed as ordinary reinforced concrete columns.

Unfortunately, both IS 800:2007 and IS: 11384-1985 do not have provisions for the design of composite columns. The design of cased columns discussed here is based on BS 5950. The following conditions must be fulfilled while using the empirical method suggested by BS 5950-1:2000 (see also Fig. 9.52).

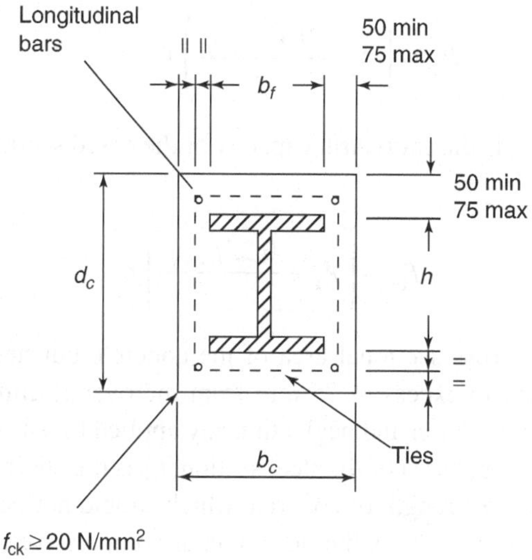


**FIG. 9.52** Cased column—conditions specified in BS 5950-1:2000

1. The steel section should be either single rolled or fabricated I- or H- section with equal flanges. A pair of rolled channels in contact back-to-back or separated back-to-back by not less than 20 mm nor more than half their depth.
2. The overall dimensions of the steel section should not exceed 1000 mm (depth) × 500 mm (width).
3. The enclosing concrete should have a minimum characteristic compressive strength of $f_{ck}$ = 20 MPa and should have a cover of more than 50 mm. The concrete should be thoroughly compacted, especially below cleats, cap plates, and beam soffits.
4. The ties around longitudinal bars should be more than 5 mm in diameter and their spacing should not exceed 200 mm.

The design procedure for axially loaded cased column is done as follows.

1. The effective length $L_e$ is taken as the lesser of
   (a) $40\, b_c$
   (b) $\dfrac{100 b_c^2}{d_c}$
   (c) $250\, r$

   where $b_c$ and $d_c$ are the breadth and depth of the cased column (see Fig. 9.52), $r$ is the minimum radius of gyration of the steel section alone, that is, $r_y$
2. The radius of gyration of the cased column about minor $y$-$y$ axis, $r_y$, is taken as $0.2 b_c$, but not more than $0.2\,(b_f + 150)$ mm, and not less than that of the steel section alone.
3. The radius of gyration of the cased column about major $z$-$z$ axis, $r_z$, is taken as the radius of gyration of the steel section alone.

4. The compression resistance of the cased column, $P_{cd}$, is determined as

$$P_{cd} = \left(A_g + \frac{0.45 f_{ck} A_c}{f_y}\right) f_{cd} < P_{cs} \quad (9.66a)$$

Where $P_{cs}$ is the short strut capacity of the cased section given by

$$P_{cs} = \left(A_g + \frac{0.25 f_{ck} A_c}{f_y}\right) f_y \quad (9.66b)$$

$A_c$ is the gross sectional area of the concrete but neglecting any casing in excess of 75 mm from the overall dimensions of the steel section and neglecting any applied finish, $A_g$ is the gross sectional area of the steel section, $f_{ck}$ is the characteristic compressive strength of concrete which should not be greater than 40 N/mm$^2$, $f_{cd}$ is the design axial compressive stress of steel section, determined as discussed in Section 9.7.1, and $f_y$ is the yield strength of steel section.

More details about the design and construction of composite sections may be found in the works of Oehlers and Bradford 1995; Johnston 1976; Roik and Bergmann 1992; Viest, et al. 1996; and Johnson 1994.

## 9.18 DISPLACEMENT

The decrease in length of a member due to axial compression under service loads is given by

$$\Delta = PL/(A_g E) \quad (9.67)$$

where $\Delta$ is the axial shortening of the member (in mm), $P$ is the service (unfactored) axial compressive force in the member (in N); $L$ is the length of the member (in mm), $E$ is the Young's modulus of rigidity (in N/mm$^2$), and $A_g$ is the gross area of cross section (in mm$^2$).

It should be noted that columns in tall buildings are subjected to large axial displacements because they accumulate loads from a large number of floors and are also relatively long. For example, a 60-storey interior column in a steel building may shorten as much as 50 to 75 mm because of dead and live loads. Due to this shortening, problems may develop in the performance of curtain walls and levelness of floor system, certain wall panels may bow or even pop off the building. Frame shortening may also force the mechanical and plumbing lines that are attached rigidly to the structure, to act as structural column, resulting in their distress. Also since the axial loads in all columns in a tall building are seldom the same, there will be *differential-shortening* of columns. Hence, it may be necessary to appropriately correct column lengths (say, at every sixth floor level). Certain columns may be made slightly longer than their nominal length to account for axial shortening. More details about procedures to calculate the corrections and making the corrections at construction site may be found in Taranath (1998).

## EXAMPLES

**EXAMPLE 9.1:** An ISHB 300 is to be used as a short column carrying axial load. Its compressive strength likely to be affected by local buckling assuming (a) Fe 410 steel with $f_y = 250$ MPa (b) Fe 540 steel with a design strength of $f_y = 410$ MPa.

**SOLUTION:**
From section tables, $b_f = 250$ mm, $t_f = 10.6$ mm, $D = 300$ mm, $R = 11.0$ mm, $t_w = 7.6$ mm, and $A = 7480$ mm$^2$.

(a) For $f_y = 250$ MPa, from Table 2 of the code, IS 800 : 2007,
$(b/t)$ allowable = 15.7
Actual $(b/t)$ for flange = (250/2)/10.6 = 11.79 < 15.7
$d/t$ for web = [300 − 2 (10.6 + 11)]/7.6 = 33.79
From Table 2 of code, $(d/t)$ max = 42
Thus full cross section is effective and design strength = 250 × 7480/(1.1 × 1000) = 1700 kN

(b) For $f_y = 410$ MPa,
Flange limit is $15.7\sqrt{(250/410)} = 12.26 > 5.9$
Web limit = $42\sqrt{(250/410)} = 32.79 < 33.79$

Therefore, cross section is slender on account of proportions of the web.

*Effective cross section*

$$d = 300 - 2(10.6 + 11) = 256.8$$

$$A_{\text{eff}} = 7480 - \{256.8 - 7.6 \times 32.79\}\, 7.6$$

$$= 7480 - (256.8 - 249.2)\, 7.6$$

$$= 7480 - 57.76 = 7422.24 \text{ mm}^2$$

Design strength = 7422.24 × 410/(1.1 × 1000) = 2766.5 kN

Note that in this case, the local buckling reduces the compression strength by about 1%. Since the web proportions control and most of the section's area is concentrated in the (semi-compact) flanges, the reduced effectiveness of the web results in much smaller loss of design capacity.

**EXAMPLE 9.2:** Determine the design axial load on the column section ISMB 350, given that the height of column is 3.0 m and that it is pin-ended. Also assume the following: $f_y = 250$ N/mm$^2$, $f_u = 410$ N/mm$^2$; E = $2 \times 10^5$ N/mm$^2$.

**SOLUTION:**
Unless the axis about which buckling will occur is obvious, all possibilities must be checked. For ISMB section, $r_y$ is normally between one fourth to one fifth of $r_z$ and hence the likely mode of failure is by buckling about the minor axis.

However, in case where different effective lengths apply for the two planes, both possibilities should normally be checked. Here both the possibilities are shown just for illustration.

*Cross section properties*

Flange thickness, $t_f = 14.2$ mm

Thickness of web, $t_w = 8.1$ mm

Flange width, $b = 140$ mm

Self weight, $w = 524$ N/m

Cross-sectional area, $A = 6670\ mm^2$

$$r_z = 143 \text{ mm}$$

$$r_y = 28.4 \text{ mm}$$

(a) *Buckling curve classification*

$$h/b = 350/140 = 2.5 > 1.2$$

$$t_f = 14.2 \text{ mm} < 40 \text{ mm}$$

Hence, we should use buckling curve '*a*' about *z-z* axis and '*b*' about *y-y* axis.

(b) *Effective length*

Since both ends are pinned effective length, $KL_y = KL_z = 3.0$ m

(c) *Non-dimensional slenderness ratio*

About *z-z* axs: $\alpha = 0.21$

$$\lambda_z = \sqrt{\left[f_y(KL/r)^2/(\pi^2 E)\right]}$$

$$= \sqrt{[250 \times (3000/143)^2/(\pi^2 \times 2 \times 10^5)]} = 0.2361$$

$$\phi = 0.5\ [1 + \alpha(\lambda - 0.2) + \lambda^2]$$

$$= 0.5[1 + 0.21\ (0.2361 - 0.2) + 0.2361^2] = 0.532$$

$$f_{cd} = (f_y/\gamma_{mo})/\{\phi + [\phi^2 - \lambda^2]^{0.5}\}$$

$$= (250/1.10)/\{0.532 + [0.532^2 - 0.2361^2]^{0.5}\}$$

$$= 225.3\ \text{N/mm}^2$$

About *y-y* axis: $\alpha = 0.34$

$$\lambda_y = \sqrt{\left[f_y(KL/r)^2/(\pi^2 E)\right]}$$

$$= \sqrt{[250 \times (3000/28.4)^2/(\pi^2 \times 2 \times 10^5)]} = 1.189$$

$$\phi = 0.5[1 + \alpha(\lambda - 0.2) + \lambda^2]$$

$$= 0.5[1 + 0.34(1.189 - 0.2) + 1.189^2] = 1.375$$

$$f_{cd} = (f_y/\gamma_{mo})/\{\phi + [\phi^2 - \lambda^2]^{0.5}\}$$

$$= (250/1.10)/\{1.375 + [1.375^2 - 1.189^2]^{0.5}\}$$

$$= 110\ \text{N/mm}^2$$

The same result may be obtained by using Table 9b of IS 800 : 2007. Thus, for $KL/r = 105.63$ and $f_y = 250$ MPa, from Table 9b, we get

$$f_{cd} = 110\ \text{N/mm}^2$$

(d) *Design stresses*

In *z*-direction,

$$f_{cd} = 225.3 \text{ MPa}$$

In *y*-direction,

$$f_{cd} = 110.0 \text{ MPa}$$

Hence design axial compressive stress,

$$f_{cd} = 110 \text{ MPa}$$

The design strength,

$$P_d = 6670 \times 110/1000$$

$$= 733.7 \text{ KN}$$

**EXAMPLE 9.3:** Determine the design axial load on the column section ISMB 350. The height of the column is 6.0 m as shown in Fig. 9.53. It is effectively restrained at mid-height by a bracing member in the *z-z* direction, but is free to move in the *y-y* direction and both the ends of the column are pinned. Also assume $f_y = 250\ \text{N/mm}^2$; $f_u = 410\ \text{N/mm}^2$, and $E = 2 \times 10^5\ \text{N/mm}^2$.

**SOLUTION:**

Properties of column section are same as in Example 9.2.

(a) *Buckling curve classification*

$h/b = 350/140 = 2.5 > 1.2$

$t_f = 14.2$ mm < 40 mm

Hence we should use buckling curve '*a*' about *z-z* axis and '*b*' about *y-y* axis.

(b) *Effective length*

Effective length in *z*-direction, $l_z = 6.0$ mm.

Since effectively restrained in *z*-direction at mid-height, effective length $l_y = 3.0$ mm.

(c) *Non-dimensional slenderness ratio*

About *z-z* axis: $\alpha = 0.21$

$$\lambda_z = \sqrt{[f_y(KL/r)^2/(\pi^2 E)]}$$

$$= \sqrt{[250(6000/143)^2/(\pi^2 \times 2 \times 10^5)]}$$

$$= 0.472$$

$$\phi = 0.5\ [1 + \alpha(\lambda - 0.2) + \lambda^2]$$

$$= 0.5[1 + 0.21(0.472 - 0.2) + 0.472^2]$$

$$= 0.640$$

$$f_{cd} = (f_y/\gamma_{mo})/\{\phi + [\phi^2 - \lambda^2]^{0.5}\}$$

$$= (250/1.10)/\{0.640 + [0.640^2 - 0.472^2]^{0.5}\}$$

$$= 212\ \text{N/mm}^2$$

About *y-y* axis: $\alpha = 0.34$

$$\lambda_y = \sqrt{[f_y(KL/r)^2/(\pi^2 E)]}$$

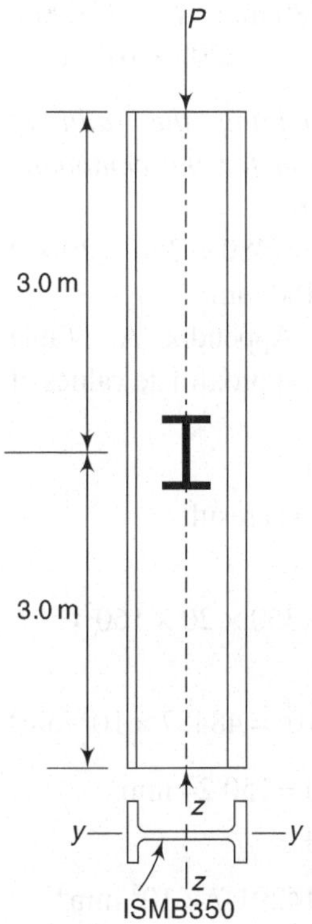

**FIG. 9.53**

$= \sqrt{[250(3000/28.4)^2/(\pi^2 \times 2 \times 10^5)]}$

$= 1.189$

$\phi = 0.5[1 + \alpha(\lambda - 0.2) + \lambda^2]$

$= 0.5[1 + 0.34(1.189 - 0.2) + 1.189^2]$

$= 1.375$

$f_{cd} = (f_y/\gamma_{mo})/\{\phi + [\phi^2 - \lambda^2]^{0.5}\}$

$= (250/1.10)/\{1.375 + [1.375^2 - 1.189^2]^{0.5}\}$

$= 110$ N/mm$^2$

The same result may be obtained from the Table 9b of the code for $f_y$ = 250 MPa and $KL/r$ = 3000/28.4 = 105.63, $f_{cd}$ = 110 N/mm$^2$

(d) *Design stresses*

In *z*-direction = $f_{cd,\,z}$ = 212 MPa

In *y*-direction, $f_{cd,\,y}$ = 110 MPa

Hence design axial compressive stress,

$f_{cd}$ = 110 MPa

Design strength = 6670 × 110/1000 = 733.7 KN

**EXAMPLE 9.4:** Calculate the compressive resistance of a compound column consisting of ISHB 300 with one cover plate of 350 × 20 mm on each flange (see Fig. 9.54) and having a length of 5 m. Assume that the bottom of the column is hinged and top is rotation fixed, translation free and $f_y$ = 250 MPa.

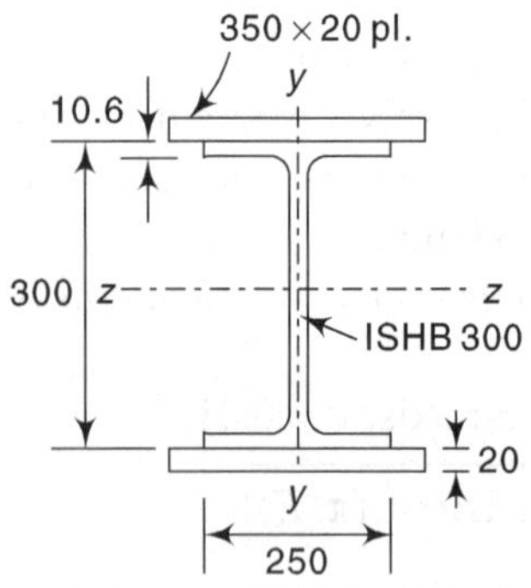


FIG. 9.54

**SOLUTION:**

From section tables for ISHB 300

Area = 7480 mm$^2$; $I_{zz}$ = 12600 × 10$^4$ mm$^4$; $I_{yy}$ = 2200 × 10$^4$ mm$^4$

(a) *Determining the radii of gyration for the compound section*

Area = 7480 + 350 × 20 × 2 = 21,480 mm$^2$

From Appendix A (Table A.15), approximate values of

$r_z = 0.40h = 0.40 \times 300 = 123$ mm

$r_y = 0.21b = 0.21 \times 350 = 73.5$ mm

Let us do the calculation to verify this result.

$I_{zz}$ for plates $= 2[I + A\,y_1^2]$

$= 2 \times [350 \times 20^3/12 + 350 \times 20 \times 160^2]$

$= 35887 \times 10^4$ mm$^3$

Total $I_{zz} = 12600 \times 10^4 + 35887 \times 10^4 = 48487 \times 10^4$ mm$^4$

Hence, $r_z = \sqrt{(48487 \times 10^4/21480)} = 150.24$ mm

$I_{yy}$ of ISHB 300 = 2200 × 10$^4$ mm$^4$

$I_{yy}$ of plates = 2 × 20 × 350$^3$/12 = 14291.7 × 10$^4$ mm$^4$

Total $I_{yy}$ = 2200 × 10$^4$ + 14291.7 × 10$^4$ = 16491.7 10$^4$ mm$^4$

$r_y = \sqrt{(16491.7 \times 10^4/21480)} = 87.62$ mm

(b) *Buckling curve classification*

From Table 10 of the code for buckling about any axis use curve c.

(c) Since $r_y$ is small, the buckling will be about the *y-y* axis

Effective length = 2 × 5000 = 10,000 mm (see Fig. 9.27)

$\lambda = KL/r_y$ = 10,000/87.62 = 114.13 < 180

From Table 9c of the code, for

$f_y$ = 250 MPa and $\lambda$ = 114.13,

$f_{cd}$ = 90.1 N/mm$^2$

Hence design strength = 21480 × 90.1/1000 = 1935.35 kN

**EXAMPLE 9.5:** A heavy column is required to support a gantry girder and a special H-section is to be fabricated. The trial section is shown in Fig. 9.55(a). Check its suitability to support a factored load of 11,000 kN, assuming both ends are pinned and a length of 8 m. Steel of design strength 250 N/mm$^2$ is to be used. Could a rolled section be suitably reinforced (by welding cover plates to its flanges) so as to provide an alternate solution?

**SOLUTION:**

(a) For the welded section shown in Fig. 9.55(a)

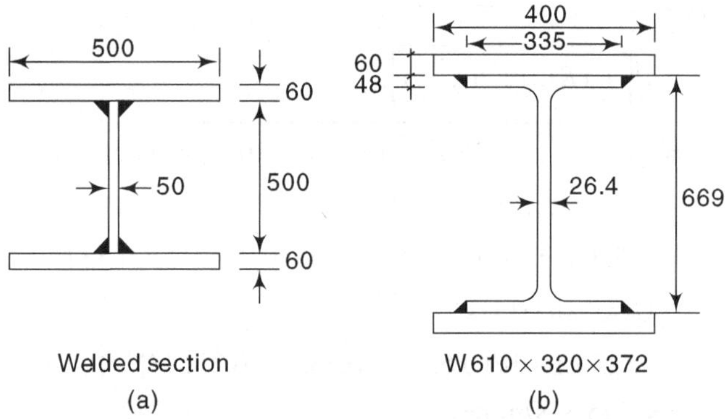


FIG. 9.55

$A = (500 \times 60) \times 2 + 500 \times 50 = 85,000$ mm$^2$

$I_y = 2(60 \times 500^3)/12 + 500 \times 50^3/12 = 125520.8 \times 10^4$ mm$^4$

$r_y = \sqrt{(I_y/A)} = \sqrt{125520.8 \times 10^4/85,000} = 121.5$ mm

$\lambda = KL/r_y = 1 \times 8000/121.5 = 65.83$

(i) *Section classification* (Table 2 of IS 800)

Flange $b/t_f$ = (500/2)/60 = 4.1 < 13.6

Web $d/t_w$ = 500/50 = 10 < 42

Hence, the section is not slender.

(ii) *Buckling curve classification*

From Table 10 of the code, $t_f$ > 40 mm

For *z-z* axis use curve '*c*'

For *y-y* axis use curve '*d*'

(iii) *Design strength*

From Table 9d, for $KL/r$ = 65.83, and $f_y$ = 230 N/mm$^2$, (As per Table 1 of the code)

$f_{cd} = 133.25 \text{ N/mm}^2$

Hence design strength = $f_{cd} \times A = 133.25 \times 85{,}000/1000 = 11{,}326$ kN > 11,000 kN

Hence, the section is suitable.

(b) The heaviest rolled section is wide flange W 610 × 320 with the following properties:

$A = 47630 \text{ mm}^2$, $I_y = 30200 \times 10^4 \text{ mm}^4$, $r_y = 79.6$ mm

Hence we have to provide substantial cover plates to make the area about 85,000 mm². As a first trial use 400 × 60 mm plates on both flanges as shown in Fig. 9.55(b).

$A = 47630 + 2\,(400 \times 60) = 95630 \text{ mm}^2$

$I_y = 30200 \times 10^4 + 2\,(60 \times 400^3)/12 = 94200 \times 10^4 \text{ mm}^4$

$r_y = \sqrt{(94200 \times 10^4/95{,}630)} = 99.25$ mm

From Table 10 of the code, since $t_f > 40$ mm use curve '*d*'.

$\lambda = KL/r_y = 1 \times 8000/99.25 = 80.60$

From Table 9d, for

$f_y = 230 \text{ N/mm}^2$ and $\lambda = 80.6$,
$f_{cd} = 112.28 \text{ N/mm}^2$

Capacity of the section = 112.28 × 95630/1000
= 10737 kN ≈ 11,000 kN

Hence the section is suitable.

**EXAMPLE 9.6:** The framing plan of a multi-storey building is given in Fig. 9.56. Assume that all the columns have a size of ISHB 400; the longitudinal beams (z-direction) have a size of ISMB 600 and the transverse beams (y-direction) have a size of ISMB 400. The storey height is 3.5 m, and the columns are assumed to be fixed at the base. For a column in a typical lower floor of the building, determine the effective length $KL_z$ and $KL_y$. For the purpose of estimating the total axial loads on the columns in the storey, assume a total distributed load of 35 kN/m² from all the floors above (combined). Assume Fe 410 grade steel.

**SOLUTION:**

Unsupported length of the column, in *y*-direction,

$L = 3500 - 400 = 3100$ mm

In *z*-direction, $L = 3500 - 600 = 2900$ mm

(a) *Relative stiffness of columns and beams*

ISHB 400: $I_z$ of column = $28100 \times 10^4 \text{ mm}^4$; $I_y = 2730 \times 10^4 \text{ mm}^3$;

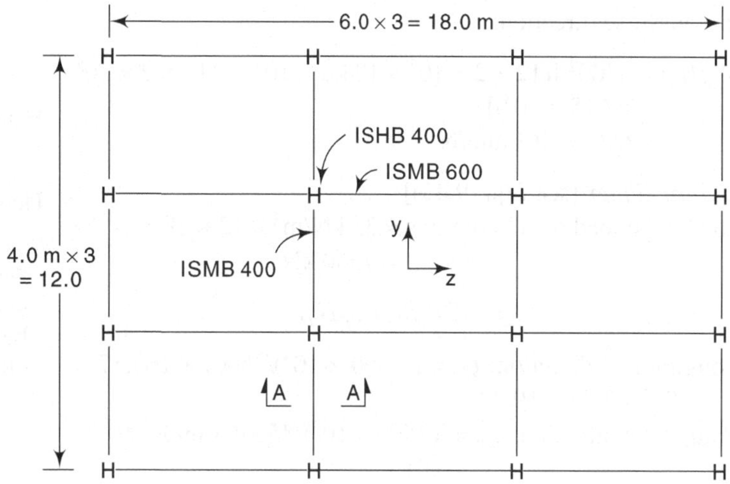


Typical framing plan
(a)

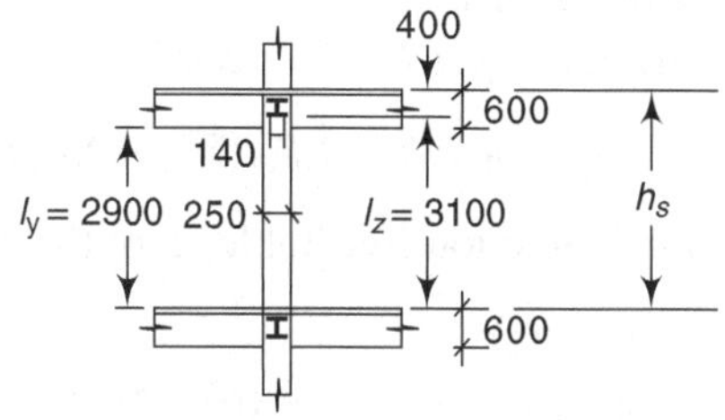


Section A–A
(b)

**FIG. 9.56**

$r_z = 169$ mm; $r_y = 52.6$ mm

ISMB 400: $I_z = 20500 \times 10^4 \text{ mm}^4$; $I_y = 622 \times 10^4 \text{ mm}^4$
ISMB 600: $I_z = 91800 \times 10^4 \text{ mm}^4$; $I_y = 2650 \times 10^4 \text{ mm}^4$

- Columns, 16 in number, ISHB 400, $h_s = 3500$ mm
  In *z*-direction

  $\Sigma I_c/h_s = 16 \times 28100 \times 10^4/3500 = 1284.5 \times 10^3 \text{ mm}^3$

  In *y*-direction = $16 \times 2730 \times 10^4/3500 = 124.8 \times 10^3 \text{ mm}^3$
- Longitudinal beams, 12 Nos, ISMB 600, $L_b = 6000$ mm

  $\Sigma(I_b/L_b)_z = 12 \times 91800 \times 10^4/6000 = 1836 \times 10^3 \text{ mm}^3$
- Transverse beams

  $\Sigma(I_b/L_b)_y = 12 \times 20500 \times 10^4/4000$
  $= 615 \times 10^3 \text{ mm}^3$

(b) *Determination of whether column is braced or unbraced*
Ignoring the contribution of in-fill walls, as per Eqn (9.46)

$\Delta_u/H_u = h_s^2/[12E\Sigma(I_c/h_s)] + 12E\Sigma(I_b/L_b)]$

In longitudinal direction:

$(\Delta_u/H_u) = 3500^2/[(12 \times 2 \times 10^5 \times 1284.5 \times 10^3) + (12 \times 2 \times 10^5 \times 1836 \times 10^3)]$
$= 163.57 \times 10^{-8}$ mm/N

In transverse direction:

$(\Delta_u/H_u) = 3500^2/[(12 \times 2 \times 10^5 \times 124.8 \times 10^3) + (12 \times 2 \times 10^5 \times 615 \times 10^3)]$

$= 690 \times 10^{-8}$ mm/N

*Stability index* [see Eqn (9.45)]

Total axial load on all columns $\approx 35$ kN/m$^2 \times 12 \times 18$

$= 7560$ kN

$$Q = (\Sigma P_u/h_s)(\Delta_u/H_u)$$

Longitudinal direction: $Q_z = [(7560 \times 10^3)/3500] \times 163.57 \times 10^{-8} = 3.53 \times 10^{-3}$

Transverse direction: $Q_y = [(7560 \times 10^3)/3500] \times 690 \times 10^{-8}$

$= 0.015$

Since both $Q_z$ and $Q_y$ are less than 0.05, the column in the storey can be considered braced in both longitudinal and transverse directions.

*Effective length as per IS 800 : 2007*

$$\beta_1 = \beta_2 = \Sigma K_c/(\Sigma K_c + \Sigma K_b)$$

Longitudinal direction (see Table 35 of the code)

$\Sigma K_c = C(I_c/h_s)$

$P_{cr} = \pi^2 EI/L^2 = \pi^2 \times 2 \times 10^5 \times 28100 \times 10^4/3500^2$

$= 45279 \times 10^3$ N

$\bar{n} = P/P_{cr} = 7560/45279 = 0.167$

$C = 2.0(1 - 0.4\bar{n}) = 2(1 - 0.4 \times 0.167) = 1.867$

$\Sigma K_c = 2[1.867 \times 28100 \times 10^4/3500] = 149.9 \times 10^3$ mm$^3$

$\Sigma K_b = 2 \times [91800 \times 10^4/6000] = 306 \times 10^3$ mm$^3$

$\beta_2 = 0.0$ (assumed fixed at base)

$\beta_1 = 149.9 \times 10^3/(149.9 \times 10^3 + 306 \times 10^3) = 0.328$

Hence,

$$K = \{[1 + 0.145(\beta_1 + \beta_2) - 0.265\beta_1\beta_2]\}/[2 - 0.364(\beta_1 + \beta_2) - 0.247\beta_1\beta_2)] \quad (9.51a)$$

Substituting the values of $\beta_1$ and $\beta_2$, we get the effective length factor

$$K = 0.557$$

Thus,

$$KL_z = 0.557 \times 2900 = 1615.3 \text{ mm}$$

*Transverse direction*

$P_{cr} = \pi^2 EI/L^2 = \pi^2 \times 2 \times 10^5 \times 2730 \times 10^4/3500^2$

$= 4399 \times 10^3$N

$\bar{n} = P/P_{cr} = 7560/4399 = 1.719$

$c = 2(1 - 0.4\bar{n}) = 2(1 - 0.4 \times 1.719) = 0.625$

*Note:* The column size has to be revised as $P_{cr}$ smaller than P. The following calculations are shown only for completeness.

$\Sigma K_c = 2(0.625 \times 2730 \times 10^4/3500) = 4875 \times 10^3$ mm$^3$

$\Sigma K_b = 2 \times (20500 \times 10^4/4000) = 102.5 \times 10^3$ mm$^3$

$\beta_1 = 4875/(4875 + 102.5 \times 10^3) = 0.045$

$\beta_2 = 0.0$ (fixed at base)

Hence $K = \{[1 + 0.145 \times 0.045 - 0.265 \times 0]/[2 - 0.364 \times 0.045 - 0.247 \times 0]\} = 0.51$

Hence $KL_y = 0.51 \times 3100 = 1581$ mm

**EXAMPLE 9.7:** For the braced frame shown in Fig. 9.57, calculate the effective length factor $K$ for the columns between the third and fourth floor in lines 1 and 2 using alignment chart and Wood's curves and compare the results.

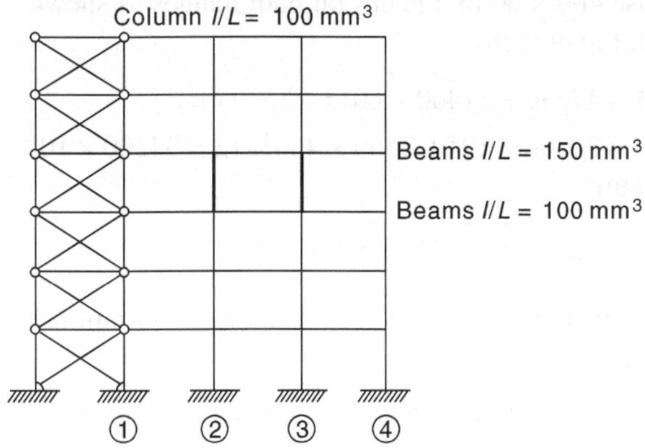


**FIG. 9.57**

**SOLUTION:**

(a) For columns in line 1, the beams give no contribution and hence $K = 1$.

(b) For columns in line 2, one beam has its other end connected to another column and hence may be regarded as having single curvature, whereas the other beam has its end connected to a hinge. Hence,

$G_A = [(100 + 100)/(150 + 1.5 \times 150)] = 0.533$

$G_B = [(100 + 100)/(100 + 1.5 \times 100)] = 0.80$

From Fig. 9.31(a), we get $K = 0.72$

By using Wood's curves,

$\beta_1 = [(100 + 100)/(100 + 100 + 0.5 \times 150 + 0.75 \times 150)] = 0.516$

$\beta_2 = [(100 + 100)/(100 + 100 + 0.5 \times 100 + 0.75 \times 100)] = 0.615$

From Fig. 9.32, $K \approx 0.72$

The preceding result may be obtained by using the following equations given in the code

$K = [1 + 0.145(\beta_1 + \beta_2) - 0.265\beta_1\beta_2]/[2 - 0.364(\beta_1 + \beta_2) - 0.247\beta_1\beta_2]$

$= [1 + 0.145 \times (0.516 + 0.615) - 0.265 \times 0.516 \times 0.615]/[2 - 0.364(0.516 + 0.615) - 0.247 \times 0.516 \times 0.615]$

$= 1.0799/1.5099 = 0.715$

Thus, both the methods give an effective length factor value of $K = 0.72$.

**EXAMPLE 9.8:** For the unbraced frame shown in Fig. 9.58, calculate the effective length factor K for the columns in lines 2 and 3 and between second and third floor, using alignment chart and Wood's curve and compare the results.

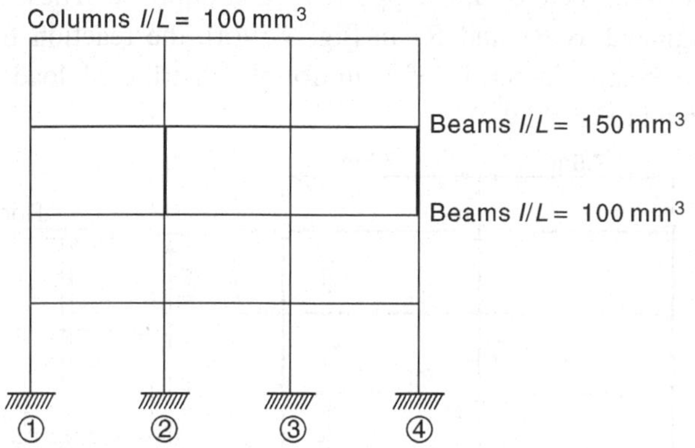


**FIG. 9.58**

**SOLUTION:**

(a) Using Julian and Lawrence Nomograph

For the columns of lines 2 and 3, we have

$G_A = [(100 + 100)/(150 + 150)] = 0.67$

$G_B = [(100 + 100)/(100 + 100)] = 1$

From Fig. 9.33, $K \approx 1.28$

(b) Using Wood's Curves

$\beta_1 = \{(100 + 100)/[100 + 100 + 1.5(150 + 150)]\} = 0.308$

$\beta_2 = \{(100 + 100)/[100 + 100 + 1.5(100 + 100)]\} = 0.4$

From Fig. 9.34, $K = 1.28$

The preceding value can be obtained from the formula given in the code as

$K = \{[1 - 0.2\ (\beta_1 + \beta_2) - 0.12\beta_1\beta_2]/[1 - 0.8(\beta_1 + \beta_2) + 0.6\beta_1\beta_2]\}^{0.5}$

$= [(1 - 0.2 \times 0.708 - 0.12 \times 0.308 \times 0.4)/(1 - 0.8 \times 0.708 + 0.6 \times 0.308 \times 0.4)]^{0.5}$

$= 1.289$

Thus, by using either method, we get the same result.

**EXAMPLE 9.9:** Calculate the compressive resistance of the leg of a transmission line tower consisting of a 200 × 200 × 20 angle section of height 3.0 m. Assume that the conditions at both ends of the *z-z* and *y-y* planes are such as to provide simple support. Take the design strength of steel as 250 N/mm$^2$ and assume that the load is concentrically applied to the angle.

**SOLUTION:**

Since the principal axes for an angle section do not coincide with the rectangular *y-y* and *z-z* axes, the buckling strength about the minor principal axis *v-v* should normally be checked. From Table 9.4, the curve *c* is appropriate for buckling about any axis. Therefore, only the axis about which the slenderness is greatest needs to be considered.

*Check for limiting thickness*

$$d/t = b/t = 200/20 = 10 < 15.7$$

$$(b + d)/t = 400/20 = 20 < 25.$$

Hence the section is not slender.

From section tables,

$$\text{area} = 7640 \text{ mm}^2$$

$$r_{xx} = r_{yy} = 61.4 \text{ mm},\ r_{vv} = 39.3 \text{ mm},$$

Maximum $\lambda = 3000/39.3 = 76.34$

From Table 9c of the code, for

$\lambda = 76.34$ and $f_y = 240$ MPa, ($f_y$ reduced as per Table 1 of the code)

$$f_{cd} = 138.49 \text{ N/mm}^2$$

Hence design strength = 138.49 × 7640/1000 = 1058 kN

**EXAMPLE 9.10:** Calculate the compressive resistance of a 200 × 200 × 20 angle assuming that the angle is loaded through only one leg, when

(a) it is connected by two bolts at the ends
(b) it is connected by one bolt at each end
(c) it is welded at each end

Assume that the member has a length of 3 m and $f_y = 250$ MPa.

**SOLUTION:**

(a) Connected by two bolts at the ends

(i) For two bolts at each end, for fixed condition,

$k_1 = 0.20$, $k_2 = 0.35$, and $k_3 = 20$ (from Table 12 of the code)

$\varepsilon = (250/f_y)^{0.5} = 1.0$

$\lambda_{vv} = (3000/39.3)/\sqrt{(\pi^2 E/250)} = 0.859$

$\lambda_\phi = (200 + 200)/\left[(2 \times 20)\sqrt{\pi^2 \times 2 \times 10^5/250}\right] = 0.1125$

$\lambda_e = [k_1 + k_2\lambda_{vv}^2 + k_3\lambda_\phi^2]$

$= \sqrt{[0.20 + 0.35 \times 0.859^2 + 20 \times 0.1125^2]} = 0.843$

From Table 10 of the code, select buckling curve 'c' from Table 7 of the code, $\alpha = 0.49$.

$f_{cd} = (f_y/\gamma_{mo})/[\phi + (\phi^2 - \lambda_e^2)^{0.5}]$

$\phi = 0.5\ [1 + 0.49(0.843 - 0.2) + 0.843^2] = 1.013$

$f_{cd} = (240/1.1)/[1.013 + (1.013^2 - 0.843^2)^{0.5}]$

$= 138.56 \text{ N/mm}^2$

$P_d = 138.56 \times 7640/1000 = 1058.6$ kN

(ii) If we consider the fixity condition as hinged (From Table 12 of the code)

$k_1 = 0.70$, $k_2 = 0.60$, and $k_3 = 5$

Hence $\lambda_e = \sqrt{(0.70 + 0.60 \times 0.859^2 + 5 \times 0.1125^2)}$

$= 1.098$

$\phi = 0.5\ [1 + 0.49(1.098 - 0.2) + 1.098^2] = 1.322$

$f_{cd} = (240/1.1)/[1.322 + (1.322^2 - 1.098^2)^{0.5}]$

$= 218.18/2.058 = 106\ \text{N/mm}^2$

$P_d = 106 \times 7640/1000 = 809.8\ \text{kN}$

(b) Connected by only one bolt at the ends and the fixity condition taken as hinged,

$$k_1 = 1.25,\ k_2 = 0.50,\ \text{and}\ k_3 = 60$$

Hence $\lambda_e = \sqrt{(1.25 + 0.50 \times 0.859^2 + 60 \times 0.1125^2)}$

$= 1.542$

$\phi = 0.5[1 + 0.49(1.542 - 0.2) + 1.542^2] = 2.018$

$f_{cd} = (240/1.1)/[2.018 + (2.018^2 - 1.542^2)^{0.5}]$

$= 218.18/3.32 = 65.72\ \text{N/mm}^2$

$P_d = 65.72 \times 7640/1000 = 502\ \text{kN}$

(c) When the strut is welded at each end, it is similar to case (a) and the strength will be equal to 1058.6 kN.

Note that when two bolts are provided at the ends, depending on the assumed fixity condition we get the capacity as 1058.6 kN or 809.8 kN. The value of 1058.6 kN is very close to the concentric loading case of Example 9.9, i.e., 1058 kN.

Let us cross-check this value with the Eqn (9.60).

$\lambda = (KL/r\pi)\sqrt{(f_y/E)}$

$= [3000/(39.3 \times \pi)]\ \sqrt{(250/2 \times 10^5)}$

$= 0.859$

$P_d = Af_y\ (0.990 + 0.150\lambda - 0.360\lambda^2 - 0.020\lambda^3)/\gamma_{m0}$

$= 250 \times 7640(0.990 + 0.15 \times 0.859 - 0.360 \times 0.859^2 - 0.02 \times 0.859^3)/(1.1 \times 1000)$

$= 191.03 \times 7640/1000 = 1459.5\ \text{kN}$

This shows that the value adopted by the Indian code is very conservative.

**Example 9.11:** A part of an office floor and the elevation of internal column stack A are shown in Figs 9.59(a) and (b). The roof and floor loads are given as follows:

Roof: Dead load (total) = 5 kN/m²

Imposed load = 1.5 kN/m²

Floors: Dead load (total) = 5 kN/m²

Imposed load = 4 kN/m²

Design column A for axial load only. The self weight of the column, including fire protection, may be taken as 1 kN/m. The roof and floor steel have the same layout. Use Fe 410 grade steel. The slabs for the floor and roof are pre-cast one way spanning slabs.

**Solution:**

When calculating the load on column lengths, the imposed load may be reduced as discussed in Chapter 3.

**Step 1:** *Loading*

Four floor beams are supported at column A. These are designated as *B*1 and *B*2 in Fig. 9.59(a), the reaction from these beams in terms of a uniformly distributed load are shown in Fig. 9.59(c).

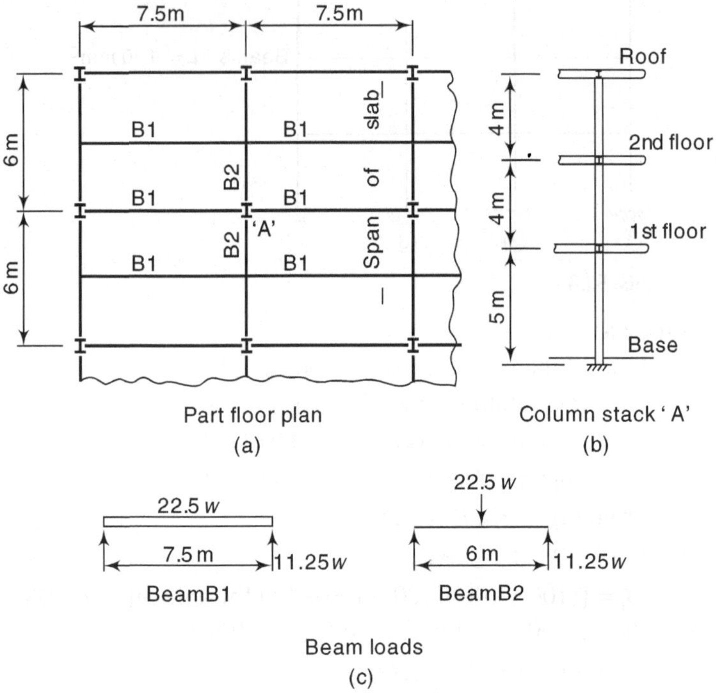


**FIG. 9.59**

Load on beam $B1 = 7.5 \times 3 \times w = 22.5\ w$ kN

where *w* is the uniformly distributed load. The dead and imposed loads are calculated separately to apply the reduction in imposed loads as suggested by IS 875. (This type of calculation may be useful when different load factors are applied to dead and imposed loads.) The self-weight of beam *B*2 is included in the reaction from beam *B*1. The loading calculations are shown in a tabular form as follows.

| Dead load, kN | Imposed load, kN, Reduction | | | Total design load, kN |
|---|---|---|---|---|
| | 0% | 10% | 20% | |
| (a) *Column between roof and 2nd floor* | | | | |
| *w* = 5 kN/m² | *w* = 1.5 kN/m² | | | |
| Two nos *B*1: 112.5 | 33.75 | | | |
| Two nos *B*2: 112.5 | 33.75 | | | |
| Self wt of | | | | |
| Column: 4.0 | | | | |
| Total: 229.0 | 67.5 | | | |

(*Contd*)

(*contd*)

| Dead load, kN | Imposed load, kN, Reduction | | | Total design load, kN |
|---|---|---|---|---|
| | 0% | 10% | 20% | |
| Design load = 1.5 (229.0 + 67.5) | | | | 444.75 |
| (b) *Column between first floor and second floor* | | | | |
| $w$ = 5 kN/m$^2$ | $w$ = 4 kN/m$^2$ | | | |
| Two nos B1:112.5 | 90.0 | 81.0 | | |
| Two nos B2: 112.5 | 90.0 | 81.0 | | |
| Self weight: 4.0 | | | | |
| Total: 229.0 | | 162.0 | | |
| Design load = 1.5 (229 + 162) = 586.5 + 444.75 | | | | 1031.25 |
| (c) *Column between ground floor and first floor* | | | | |
| $w$ = 5 kN/m$^2$ | $w$ = 4 kN/m$^2$ | | | |
| Two nos B1: 112.5 | 90.0 | | 72.0 | |
| Two nos B2: 112.5 | 90.0 | | 72.0 | |
| Self weight: 5.0 | | | | |
| Total: 230.0 | | | 144.0 | |
| Design load = 1.5(230 + 144) = 561 + 1031.25 | | | | 1592.25 |

**Step 2:** *Column design*

(a) Column between 2nd floor and roof
Design load = 444.75
Since most of the rolled sections are not slender, the local buckling check on $b/t$ ratio is not shown.
Try ISHB 150 @ 306 N/m
$A$ = 3900 mm$^2$, $r_y$ = 34.4 mm, $t_f$ = 9.0 mm, $h$ = 150 mm, $b$ = 150 mm
Assuming that the beam connections are simple shear type and the frame is prevented from sway using bracings/shear walls, from Table 11 of code,
$K$ = 1.0, since the shear type connections will permit end rotations but will restrain the translation.

$$L_e = 1.0 \times 4000 = 4000 \text{ mm}$$

Slenderness ratio $\lambda$ = 4000/34.4 = 116.28 < 180
Hence ok.
From Table 10 of the code, since $h/b$ = 1.0 and $t_f$ < 40 mm, for buckling about $y$-$y$ axis, use curve $c$. Thus from Table 9c of the code,

$$f_{cd} = 87.75 \text{ N/mm}^2$$

Compressive strength = 87.75 × 3900/1000 = 342.2 kN < 444.75 kN
Hence choose ISHB 200, with

$$A = 4750 \text{ mm}^2,\ r_y = 45.1 \text{ mm}$$

Now $\lambda$ = 4000/45.1 = 88.69
and $f_{cd}$ = 122.66 N/mm$^2$
Compressive strength = 4750 × 122.66/1000 = 582.6 kN > 444.75 kN
Hence, the selected section is safe.

(b) *Column between first floor and second floor*
Design load = 1031.25 kN
Assuming $f_{cd}$ = 125 N/mm$^2$, required area = 1031.25 × 1000/125 = 8250 mm$^2$
Try ISHB 300: A = 7480 mm$^2$, $r_y$ = 54.1 mm, $h$ = 300 mm, $b$ = 250 mm

$$\lambda = 4000/54.1 = 73.93 < 180$$

Hence ok.
Since $h/b$ > 1.2, $t_f$ < 40 mm, from Table 10 of the code, we can use curve $b$
From Table 9b, for

$$KL/r = 73.93 \text{ and } f_y = 250 \text{ MPa}$$
$$f_{cd} = 159.7 \text{ MPa}$$

Compressive strength = 7480 × 159.7/1000 = 1194 kN > 1031.25
Hence, the selected section is safe.

(c) *Column between the base and first floor*
Design load = 1592.25 kN
Let us assume that the base is fixed (foundation is raft slab or pile foundation).
For fixed at bottom and top hinged (Table 11 of the code)

$$K = 0.8$$

Hence,
effective length = 0.8 × 5000 = 4000 mm
Try ISHB 450: $A$ = 11100 mm$^2$, $r_y$ = 51.8 mm

$$\lambda = 4000/51.8 = 77.22 < 180. \text{ Hence ok.}$$

From Table 10 and Table 9b,

$$f_{cd} = 154.45 \text{ N/mm}^2$$

Compressive strength = 154.45 × 11100/1000 = 1714.4 kN > 1592.25 kN
Hence, the section is satisfactory.

**Example 9.12:** Design a column to support a factored axial load of 1100 kN. The column has an effective length of 6 m with respect to the $z$-axis and 4 m with respect to the $y$-axis. Use Fe 410 grade steel.

**Solution:**

Assume the design compressive stress as $f_{cd}$ = 0.6 $f_y$ = 150 MPa.

Required area = 1100 × 10$^3$/150 = 7333 mm$^2$

Try ISHB 300 at 588 N/m. The relevant properties of the section are as follows (IS 808-1989):

$A = 7480$ mm$^2$; $r_z = 130$ mm; $r_y = 54.1$ mm; $h/b = 300/250 = 1.2$, $t_f = 10.6$ m

$$L/r_z = 6000/130 = 46.15 < 180. \text{ Hence ok.}$$

$$L/r_y = 4000/54.1 = 73.94 < 180. \text{ Hence ok.}$$

From Tables 10 and 9b of the code, for

$$L/r_y = 73.94 \text{ and } f_y = 250 \text{ MPa}$$

$$f_{cd} = 159.7 \text{ N/mm}^2$$

Therefore, strength of the column = 159.7 × 7480/1000 = 1194.5 kN > 1100 kN

Hence, the chosen section is safe.

*Note:* If the beams in this example are rigidly connected to the columns, then the Wood's curves should be used to determine the effective length factor $K$.

**EXAMPLE 9.13:** The strut of a space frame member having a length of 3 m has to carry a factored load of 230 kN. Assuming $f_y = 220$ N/mm$^2$, design a circular tube to carry the load. Assume that the ends are simply supported.

**SOLUTION:**

Assume medium class pipe of outside diameter 114.3 mm with thickness 4.5 mm, weight = 121 N/m, $r = 38.9$ mm, and area = 1550 mm$^2$.

(a) *Check for limiting width-to-thickness ratio (see Table 8.7)*

$$D/t = 114.3/4.5 = 25.4 < 88\varepsilon^2 \; [88 \times (250/220)^2 = 93.8]$$

Hence, the tube is not slender.

(b) *Buckling curve classification*

From Table 10 of the code, for hot-rolled tubes, use buckling curve '$a$'.

(c) *Resistance to flexural buckling*

$$KL/r = 1 \times 3000/38.9 = 77.12$$

From Table 9a of the code,
for $f_y = 220$ N/mm$^2$ and $KL/r = 77.12$,

$$f_{cd} = 157.45 \text{ N/mm}^2$$

Design strength = $A \times f_{cd}$ = 1550 × 157.45/1000 = 244 kN > 230 kN

Hence, the assumed section is safe.

**EXAMPLE 9.14:** Repeat the problem given in Example 9.13 using a square tube, with $f_y = 240$ N/mm$^2$.

**SOLUTION:**

Assume a SHS of size 91.5 × 91.5 mm with a thickness of 4.5 mm, area = 1514 mm$^2$, weight = 118.8 N/m, $r = 35.2$ mm, and $\varepsilon = (250/f_y)^{0.5} = (250/240)^{0.5} = 1.042$

(a) *Check for limiting width-to-thickness ratio (see Table 8.7)*

$$D/t = 91.5/4.5 = 20.33 < 42\,\varepsilon (42 \times 1.042 = 43.7)$$

Hence, the section is not slender.

(b) *Buckling curve classification*

From Table 10 of the code, for hot-rolled tubes, use buckling curve '$a$'.

(c) *Resistance to flexural buckling*

$$KL/r = 1 \times 3000/35.2 = 85.3$$

From Table 9a of the code, for

$$f_y = 240 \text{ N/mm}^2 \text{ and } KL/r = 85.3$$

$$f_{cd} = 153.99 \text{ N/mm}^2$$

Hence design strength = 153.99 × 1514/1000 = 233.14 kN > 230 kN

Hence the assumed section is safe.

**EXAMPLE 9.15:** Design a double angle discontinuous strut to carry a factored load of 175 kN. The length of the strut is 3.0 m between intersections. The two angles are placed back-to-back and are tack bolted. Consider the following cases.

(a) Angles are placed on opposite sides of the gusset plate.
(b) Angles are placed on the same side of the gusset plate.
(c) Two angles in star formation.

Assume grade Fe 410 steel with $f_y = 250$ MPa.

**SOLUTION:**

Let us assume the design compressive stress is $0.4 f_y = 100$ MPa.

Required area = 175 × 1000/100 = 1750 mm$^2$

(a) *Angles placed on opposite sides of the gusset plate*

Let us try 2 ISA 70 × 70 × 8 mm at 83 N/m with $A = 2 \times 1060$ mm$^2$,
$I_y = 47.4 \times 10^4$ mm$^4$, $r_z = 21.2$ mm, $c_y = 20.2$ mm

$$A^* = 2 \times 1060 = 2120 \text{ mm}^2$$

Assuming a 10-mm thick gusset plate

$$I_{yy}^* = 2[I_y + A(c_y + t_g/2)^2]$$
$$= 2[47.4 \times 10^4 + 1060(20.2 + 5)^2]$$
$$= 229.42 \times 10^4 \text{ mm}^2$$

$$r_y^* = \sqrt{(229.42 \times 10^4/2120)} = 32.90 \text{ mm}$$

$$r_z^* = r_z = 21.2 \text{ mm}$$

*Note*: The asterisk sign * is used for the properties of double angle strut.

Minimum radius of gyration

$$r^* = r_z = 21.2 \text{ mm}$$

Effective length factor could be between 0.7 and 0.85 (clause 7.5.2.1)

Assume $K = 0.85$.

Hence, effective length = 0.85 × 3000 = 2550 mm

$L/r$ = 2550/21.2 = 120.28 < 180. Hence, the selected section is ok.

For $L/r$ = 120.28 and $f_y$ = 250 MPa, using Tables 10 and 9c of the code,

$$f_{cd} = 83.44 \text{ N/mm}^2$$

Hence strength of the member = 83.44 × 2120/1000 = 176.89 kN > 175 kN

*Note*: ISA 70 × 70 × 8 is not commonly rolled and hence the designer may choose 75 × 75 × 8 (89 N/m) or 80 × 80 × 6 (73 N/m), $r$ = 24.6 mm

For 80 × 80 × 6, $L/r$ = 103.65, $f_{cd}$ = 102.47 MPa

Hence capacity = 102.47 × 2 × 929/1000 = 190.4 kN.

Hence, provide 2 ISA 80 × 80 × 6 as shown in Fig. 9.60(a).

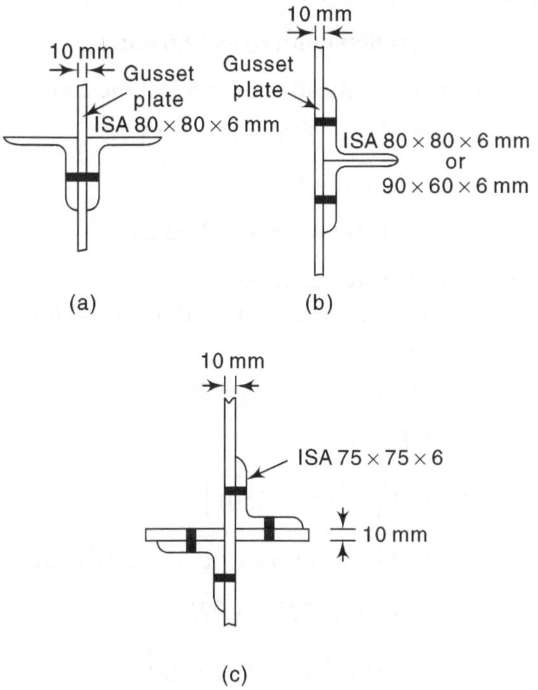


**FIG. 9.60**

(b) *Angles placed on the same side of the gusset plate*

Try 2$L$ 75 × 75 × 6 mm at 68 N/m with $A$ = 866 mm²; $r_z$ = 23 mm; $c_y$ = 20.6 mm;

$$I_z = I_y = 45.7 \times 10^4 \text{ mm}^4$$

$r_z$ for two angles combined is the same as for one angle, i.e., 23 mm

$$A^* = 2 \times 866 = 1732 \text{ mm}^2$$

$$I^*_{yy} = 2 \times 45.7 \times 10^4 + 2 \times 866 \times 20.6^2 = 164.9 \times 10^4 \text{ mm}^4$$

$$r^*_y = \sqrt{(I_{yy}/A)} = \sqrt{[164.9 \times 10^4/(2 \times 866)]} = 30.8 \text{ mm}$$

Minimum radius of gyration $r^* = r_z$ = 23 mm

Effective length = 0.85 × 3000 = 2550 mm

$L/r$ = 2550/23 = 110.87 < 180. Hence, the chosen section is safe.

For $L/r$ = 110.87 and $f_y$ = 250 MPa, from Table 10 and Table 9c of the code,

$$f_{cd} = 93.65 \text{ N/mm}^2$$

Hence capacity = 93.65 × 1732/1000 = 162.2 kN < 175 kN.

Hence try 80 × 80 × 6 at 73 N/m, with $A$ = 929 mm² and $r_z$ = 24.6 mm.

Now, $L/r$ = 2550/24.6 = 103.66.

From Table 10 and Table 9c of the code, for $L/r$ = 103.66 and $f_y$ = 250 MPa,

$$f_{cd} = 102.47 \text{ MPa}$$

Capacity = 102.47 × 929 × 2 = 190.4 kN > 175 kN. Hence the assumed angle is safe.

Let us try an unequal angle of size 90 × 60 × 6 at 68 N/m, with 10 mm gap,

with $A$ = 865 mm², $r_z$ = 28.6 mm, $r_y$ = 17.1 mm, $I_y$ = 25.2 × 10⁴ mm⁴,

$$c_y = 13.9 \text{ mm}.$$

Radius of gyration about $z$-$z$ axis for the combination is same as for one angle

$$r^*_z = 28.6 \text{ mm}$$

Radius of gyration of $r_y$ is obtained by finding the moment of inertia about $y$-$y$ axis

$$I^*_{yy} = 2 \times 25.2 \times 10^4 + 2 \times 865\ (13.9 + 10/2)^2$$
$$= 50.4 \times 10^4 + 61.79 \times 10^4$$
$$= 112.19 \times 10^4 \text{ mm}^4$$
$$r^*_y = \sqrt{[112.19 \times 10^4/(2 \times 865)]} = 25.46 \text{ mm}$$

Hence $r^*_{min}$ = 25.46 mm

$$L/r = 2550/25.46 = 100.13$$

From Tables 10 and 9c of the code for $L/r$ = 100.13 and $f_y$ = 250 MPa,

$$f_{cd} = 106.84 \text{ MPa}$$

Capacity = 106.84 × 2 × 865/1000 = 184.83 kN > 175 kN

Hence, the section is safe.

*Note*: From these calculations, it is seen that the designers' task is to design an angle that will be available in the market and also to choose a section that will provide an optimum section. It is clear that two unequal angles with the longer leg connected to the gusset plate provides the economic solution. Two angles on the opposite side of the gusset plate will be economical than two angles on the same side of the gusset plate. Tack welding at a spacing of 550 mm (using stitch plates) along the length has to be provided to avoid local buckling of the individual angles [see calculation in part (c)].

(c) *Star angle*

Choose two angles 75 × 75 × 6 mm at 68 N/m in a star configuration as shown in Fig. 9.60(c). The relevant properties of single 75 × 75 × 6 angle at 68 N/m are $A = 866$ mm$^2$; $r_z = r_y = 23$ mm; $c_z = c_y = 20.6$ mm; $r_u = 29.1$ mm; $r_v = 14.6$ mm

Assuming the gusset plate thickness as 10 mm,

$$A^* = 2 \times 866 = 1732 \text{ mm}^2$$

$$r_z^* = r_y^* \sqrt{[r^2 + (c_y + 10/2)^2]} = \sqrt{[23^2 + (20.6 + 10/2)^2]}$$

$$= 34.41 \text{ mm}$$

$$r_v^* = \sqrt{[r_v^2 + 2(c_y + 10/2)^2]} = \sqrt{[14.62^2 + 2(20.6 + 10/2)^2]}$$

$$= 39.04 \text{ mm}$$

$$r_u^* = 29.1 \text{ mm}$$

The least radius of gyration

$$r_{min}^* = 29.1 \text{ mm} = r_u \text{ of single angle}$$

Hence $\lambda = L/r = 2550/29.1 = 87.62 < 180$

For $\lambda = 87.62$ and $f_y = 250$ MPa from Tables 10 and 9c of the code,

$$f_{cd} = 124.57 \text{ N/mm}^2$$

Capacity of the section = 124.57 × 1732/1000
= 215.75 kN > 175 kN.

The chosen section has a capacity much higher than the required strength and hence, a lower available angle section may be chosen. The reader may do it as an exercise and select a suitable angle.

Note that the star type configuration results in a smaller section compared to the other two angle configurations in (a) and (b). Such star angles are often used for the legs of transmission or communication towers (see Subramanian 1992). They are also used as bracings in industrial buildings.

*Tack Welding*

Tack welding along the length should be provided to avoid local buckling of each of the elements. The spacing of the welds to be adjusted based on (clause 7.8.1 of the code)

$$\lambda_e \leq 0.6\lambda = 0.6 \times 87.62 = 52.05 \text{ or } \leq 40$$

Selecting the lowest value,

$$\lambda_e = 40$$

The spacing of the tack weld is given by

$$\lambda_e = S/r_v \leq 40$$

Hence $S = 40r_v = 40 \times 14.6 = 584$ mm

The welding should be designed to resist a transverse load of 2.5% of the axial load, i.e., 175 × 1000 × 2.5/100 = 4375 N.

Assuming a 3 mm fillet weld (as per Table 6.3),

Design strength of the weld (Table 6.5) = 158 N/mm$^2$

Hence length of weld = 4375/(0.7 × 3 × 158) = 13.2 mm

Hence provide a 3 mm tack welding of 15-mm length at 550 mm spacing (see Fig. 7.3 given in Chapter 7 for the details of stitch plates).

**EXAMPLE 9.16:** Design a single angle discontinuous strut to carry a factored load of 65 kN. Assume that the distance between its joints is 2.5 m and is loaded through one leg. Use $f_y = 250$ MPa.

**SOLUTION:**

Take effective length = $L$ = 2500 mm

Assume a slenderness ratio of 120 and the corresponding $f_{cd} = 83.7$ MPa (from Table 9.3)

$$\text{Area required} = 65 \times 1000/83.7 = 777 \text{ mm}^2$$

Choose ISA 75 × 75 × 6 at 68 N/m with

$$A = 866 \text{ mm}^2,\ r_{vv} = 14.6 \text{ mm}$$

Check for section classification (Table 2 of the code)

$$d/t = b/t = 75/6 = 12.5 < 15.7\varepsilon$$

Hence the section is semi-compact

$$(b + d)/t = 150/6 = 25 \leq 25\varepsilon$$

Hence take effective area as gross area.

Assuming two bolts at each end and fixed condition (from Table 12 of the code)

$$k_1 = 0.2,\ k_2 = 0.35, \text{ and } k_3 = 20$$

$$\varepsilon = (250/f_y)^{0.5} = 1.0$$

$$\lambda_{vv} = (L/r_{vv})/\left[\varepsilon\sqrt{(\pi^2 E/250)}\right]$$

$$= (2500/14.6)/\left[1.0 \times \sqrt{(\pi^2 \times 2 \times 10^5/250)}\right]$$

$$= 171.23 \times 0.01125 = 1.926$$

$$\lambda_\phi = (b_1 + b_2)/\left[\varepsilon\sqrt{(\pi^2 E/250)} \times 2t\right]$$

$$= 150 \times 0.01125/(2 \times 6) = 0.1406$$

$$\lambda_e = \sqrt{[k_1 + k_2\lambda_{vv}^2 + k_3\lambda^2]}$$

$$= \sqrt{[0.2 + 0.35 \times 1.926^2 + 20 \times 0.1406^2]} = 1.376$$

$$f_{cd} = (f_y/\gamma_{mo})/[\phi + (\phi^2 - \lambda_e^2)^{0.5}]$$

$$\phi = 0.5[1 + 0.49(1.376 - 0.2) + 1.376^2] = 1.735$$

$$f_{cd} = (250/1.1)/[1.735 + (1.735^2 - 1.376^2)^{0.5}]$$

$$= 81.4 \text{ N/mm}^2$$

$$P_d = 81.4 \times 866/1000 = 70.497 \text{ kN} > 65 \text{ kN}$$

Hence, the assumed section is safe.

*Note:* If the ends are hinged, the capacity is 36 kN only.

**EXAMPLE 9.17:** Design the single angle section of Example 9.16 assuming that it is concentrically loaded.

**SOLUTION:**

From Example 9.16, Choose ISA 75 × 75 × 6 with A = 866 mm$^2$, $r_{vv}$ = 14.6 mm

As per Example 9.16, the section is semi-compact and hence effective area may be taken as gross-area. As the angle is concentrically loaded, the design strength may be evaluated using Clause 7.1.2 of the code using Class C curve (Clause 7.5.1.1 of the code).

Assuming $K = 0.85$, $KL/r_{vv} = 0.85 \times 2500/14.6 = 145.5$

From Table 9c of the code, for $KL/r$ = 145.5 and $f_y$ = 250 MPa, $f_{cd}$ = 62.35 N/mm$^2$

Design compressive strength, $P_d = f_{cd}A = 62.35 \times 866/1000 = 54$ kN < 65 kN

Hence choose ISA 80 × 80 × 6 with $A$ = 929 mm$^2$, $r_{vv}$ = 15.6 mm

Hence $KL/r_{vv} = 0.85 \times 2500/15.6 = 136.2$

From Table 9c of the code, for $KL/r$ = 136.2 and $f_y$ = 250 MPa, $f_{cd}$ = 69.28 N/mm$^2$

Design compressive strength, $P_d = f_{cd}A = 69.28 \times 929/1000 = 64.3$ kN ≈ 65 kN

Hence the section is adequate.

**EXAMPLE 9.18:** Design a laced column 10-m long to carry a factored axial load of 1100 kN. The column is restrained in position but not in direction at both ends. Provide single lacing system with bolted connection.

(a) Design the column with two channels back-to-back
(b) Design the column with two channels placed toe-to-toe
(c) Design the lacing system with welded connections for channels back-to-back.

**SOLUTION:**

*Design of column*

$$P = 1100 \times 10^3 \text{ N}$$
$$L = 1.0 \times 10 = 10 \text{ m}$$

Assume a design stress of 125 MPa,

Required area = 1100 × 10$^3$/125 = 8800 mm$^2$

Select two ISMC 300 at 363 N/m. The relevant properties of ISMC 300 are (IS 808-1989)

$$A = 4630 \text{ mm}^2, r_{zz} = 118.0 \text{ mm}, r_{yy} = 26.0 \text{ mm}, t_f = 13.6 \text{ mm}$$
$$c_{yy} = 23.5 \text{ mm}, I_{zz} = 6420 \times 10^4 \text{ mm}^4, I_{yy} = 313 \times 10^4 \text{ mm}^4$$

Area available = 2 × 4630 = 9260 mm$^2$

Built-up sections will be economical, when the radius of gyration of the *y-y* axis is increased in such a way that it is more or less equal to the radius of gyration about the *z-z* axis. This is achieved by spacing the sections in such a way that $r_{zz}$ becomes $r_{min}$. Let us first check the safety of the section and then workout the required spacing between the two channels.

$$L/r_{zz} = 10 \times 10^3/118.0 = 84.74$$

The $L/r$ of the built-up column should be taken as 1.05 × ($L/r_{zz}$) = 1.05 × 84.74 = 88.98

For $L/r_{zz}$ = 88.98 and $f_y$ = 250 MPa, using Table 9c of the code, $f_{cd}$ = 122.53 MPa.

Load carrying capacity = $A_e f_{cd}$ = 9260 × 122.53/1000
= 1135 kN > 1100 kN

Hence the column is safe.

(a) Let us provide two channels back-to-back and connect them by lacing and denote $S$ as the spacing between two channels [see Fig. 9.61(a)].

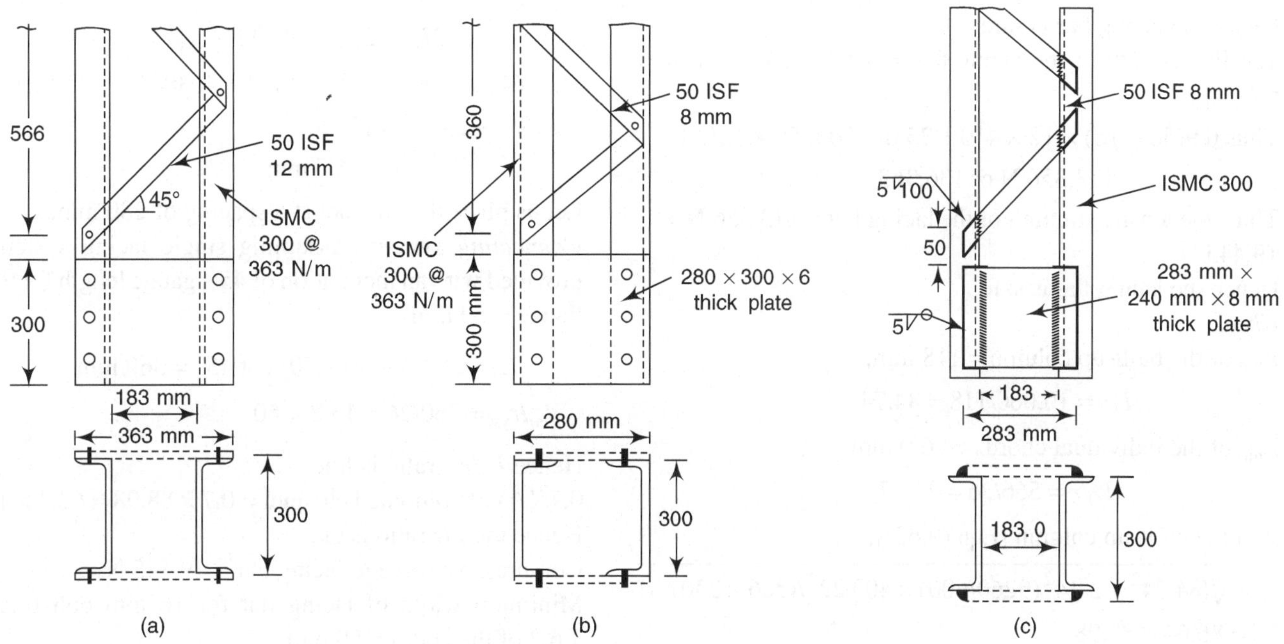


**FIG. 9.61**

*Spacing of channels*

$$2I_{zz} = 2[I_{yy} + A(S/2 + c_{yy})^2]$$

Thus, $2 \times 6420 \times 10^4 = 2[313 \times 10^4 + 4630(S/2 + 23.5)^2]$
or $(S/2 + 23.5)^2 = 13190$

$$S = 182.70 \text{ mm}$$

Let us keep the channels at a spacing of 183 mm.
*Lacing system* Using a single lacing system with the inclination of lacing bar = 45° (gauge length for a 90 mm flange = 50 mm)
Spacing of lacing bars, $L_o = 2(183 + 50 + 50) \cot 45°$
$= 2 \times 283 \times 1 = 566$ mm

$L_o/r_{yy}$ should be $< 0.7 \times L/r$ of whole column
$21.77 < 0.7 \times 88.98 = 62.3$
Hence safe.
Maximum shear = $(2.5/100) \times 1100 \times 10^3 = 27{,}500$ N
Transverse shear in each panel = $V/N = 27{,}500/2 = 13{,}750$ N
Compressive force in the lacing bar = $(V/N)$ cosec 45°
$= 13750 \times 1.414 = 19445$ N
Assuming 16-mm diameter bolts,
Minimum width of lacing flat (clause 7.6.2 of the code) = $3 \times 16$, say 50 mm
Minimum thickness = $(1/40)(183 + 50 + 50)$ cosec 45° = 10.01 mm
Provide 12-mm thick plate with a width of 50 mm
Minimum $r = t/\sqrt{12} = 12/\sqrt{12} = 3.464$ mm
$L/r$ of the lacing bar = $283 \times$ cosec $45°/3.464 = 115.5 < 145$
Hence safe.
For $L/r = 115.5$ and $f_y = 250$ MPa, using Table 9c of the code

$$f_{cd} = 88.6 \text{ MPa}$$

Load carrying capacity = $88.6 \times 50 \times 12 = 53{,}163$ N > 19,445 N
Hence the lacing bar is safe.
Tensile strength of the lacing flat = $0.9(B - d)tf_u/\gamma_{m1}$ or $f_yA_g/\gamma_{m0}$
Thus $0.9(50 - 18) \times 12 \times 410/1.25$ or $250 \times 50 \times 12/1.1$

113,356 N or 136,363

Thus, the tensile strength of the lacing flat = 113,356 N > 19,445 N
Hence the lacing flat is safe.
*Check*
$r_{min}$ of the built-up column = 118 mm,

$$L/r = 10{,}000/118 = 84.74$$

$r_{min}$ of the individual chords = 26.0 mm,

$$L_0/r = 566/26 = 21.77$$

$\lambda$ of the built-up column [Eqn (9.63)],

$$\lambda_e = \sqrt{\{84.74^2 + 3.14^2(9260/600) \times 400.22^3/(566 \times 230)^2)\}}$$
$$= 86.64 < 88.98$$

Hence, the column is safe.

*Connection:* Assuming that the 16 mm bolts of grade 4.6 are connecting both the lacing flats with the channel at one point and that the shear plane will not pass through the threaded portion of bolt.
Strength of bolt in double shear = $2 \times A_{sb} (f_u/\sqrt{3})/\gamma_{mb}$
$= 2 \times \pi \times 16^2/4 \times (400/\sqrt{3})/1.25 = 74{,}293$ N

Strength in bearing = $2.5 k_b dtf_u/\gamma_m$ (with $k_b = 0.49$) = $2.5 \times 0.49 \times 16 \times 12 \times 410/1.25$
$= 77{,}145$ N

Hence, strength of bolt = 74,293 N > 19,445 N
Hence one 16-mm diameter bolt of grade 4.6 is required.
*Tie plates:* Tie plates must be provided at the ends of the laced column.
Effective depth = $183 + 2 \times c_{yy} > 2 \times b_f$
$= 183 + 2 \times 23.5 = 230 \text{ mm} > 2 \times 90$
$= 180$ mm

Hence
Required overall depth of tie plate = $230 + 2 \times 25 = 280$ mm
(edge distance of 16-mm diameter bolts = 25 mm)
Provide a tie plate of 300-mm depth.
Length of tie plate = $183 + 2 \times 90 = 363$ mm
Required thickness of tie plate = $1/50(183 + 2g)$ = $1/50(183 + 2 \times 50) = 5.66$ mm
(where $g$ = gauge distance—see Fig. 5.10 of Chapter 5)
Hence, provide a tie plate of 6-mm thickness.
Provide a tie plate of size $363 \times 300 \times 6$ mm at both ends with six 16-mm diameter bolts.

(b) Consider the case of laced column with the two channels provided toe-to-toe.
*Spacing (see Fig. 9.61b):*

$$2I_{zz} = 2[I_{yy} + A(S/2 - c_{yy})^2]$$
$$2 \times 6420 \times 10^4 = 2[313 \times 10^4 + 4630(S/2 - 23.5)^2]$$

or $(S/2 - 23.5)^2 = 13190$

$$S = 276.7 \text{ mm}$$

Let us place the channel at a spacing of 280 mm.
*Connecting system:* Assuming single lacing system is provided with an inclination of 45°; gauge length for 90 m flange = 50 mm

$$L_o = 2(280 - 50 - 50) \cot 45° = 360 \text{ mm}$$
$$L_o/r_{yy} = 360/26 = 13.8 < 50$$

Hence $L_o/r_{yy}$ ratio is fine.
$0.7(L/r)$ of combined channel = $0.7 \times 88.98 = 62.3 > 13.8$
Hence the $L/r$ ratio is ok.
Compressive force in lacing bar = 19,445 N
Minimum width of lacing flat for 16 mm bolt (clause 7.6.2 of the code) = 50 mm
Minimum thickness = $1/40(280 - 50 - 50) \times$ cosec 45°

$= 6.36$ mm

Hence, provide a 50 × 8 mm flat.

*Check* $r_{min} = t/\sqrt{12} = 8/\sqrt{12} = 2.309$ mm

$$L/r = 180 \times \text{cosec } 45°/2.309 = 110.2 < 145$$

Hence the chosen flat is safe.

For $L/r = 110.2$ and $f_y = 250$ MPa, from Table 9c of the code,

$$f_{cd} = 94.4 \text{ N/mm}^2$$

Capacity of the lacing flat = 94.4 × 50 × 8
= 37,760 N > 19,445 N

Tensile strength of the lacing flat = $0.9(B - d)tf_u/\gamma_{m1}$ or $f_y A_g/\gamma_{m0}$
= 0.9(50 – 18) × 8 × 410/1.25 or 250 × 50 × 8/1.1
= 75,571 N or 90,909 N both > 19,445 N

Hence the lacing flat is safe.

*Connection:*

Strength of bolt in double shear {from (a)} = 74,293 N

Strength in bearing = $2.5k_b dtf_u/\gamma_{mb}$ = 2.5 × 0.49 × 16 × 8 × 410/1.25 = 51,430 N

Hence, strength of bolt = 51,430 N > 19,445 N

Therefore, provide one 16-mm diameter bolt of grade 4.6.

*Tie plates:*

Effective depth of tie plate = $S - 2c_{yy}$
= 280 – 2 × 23.5 = 233 mm > 2 × 90 = 180 mm

Required overall depth = 230 + 2 × 25 = 280 mm (edge distance of 16-mm diameter bolt = 25 mm)

Provide a 300 mm plate

Length of tie plate = 280 mm

Thickness of tie plate = 1/50(280 – 2 × 50) = 3.6 mm

Provide 6 mm.

Provide a tie plate of size 280 × 300 × 6 mm and use six bolts of 16-mm diameter and grade 4.6 to connect it to the channels. The arrangement is shown in Fig. 9.61(b). It is seen that by providing channels toe-to-toe, the lacing size and the tie plate size are reduced.

(c) From part (a) of this example,

Spacing of the channels = 183 mm

Compressive force in the lacing = 19,445 N

Effective length of lacing flat (welded)
= 0.7 × 183 × cosec 45° = 181.16 mm

Minimum thickness of flat = 1/40 × (183 × cosec 45°)
= 6.47 mm

Provide 50 × 8 mm lacing flat.

Minimum radius of gyration, $r = t/\sqrt{12} = 8/\sqrt{12} = 2.31$ mm

$$L/r = 181.16/2.31 = 78.4 < 145$$

Hence the $L/r$ ratio is ok.

For $L/r = 78.4$ and $f_y = 250$ MPa, using Table 9c of the code

$$f_{cd} = 138.56 \text{ N/mm}^2$$

Capacity of lacing bar = 138.56 × 50 × 8 = 55,424 N > 19,445 N

Hence the lacing bar is safe.

Overlap of lacing flat = 50 mm > 4 × 8 = 32 mm

Hence the lacing flat is safe.

*Connection*:

Thickness of flange of ISMC 300 = 13.6 mm

Minimum size of weld = 5 mm (Table 6.3)

Strength of weld/unit length = $0.7 \times 5 \times 410/(\sqrt{3} \times 1.5)$
= 552 N/mm

Required length of weld = 19445/552 = 35.2 mm

Adding extra length for ends, the weld length to be provided

$$= 36 + 2(2 \times 5) = 56 \text{ mm}$$

Provide 100 mm weld length at both ends.

*Tie plate*:

Overall depth of plate = $183 \times 2 \times c_{yy}$
= 183 + 2 × 23.5
= 230 mm > 2 × 90 mm

Let length of tie plate = 183 + 2 × 50 = 283 mm

Thickness of tie plate = 1/50(183 + 2 × 50) = 5.66 mm

Provide a 8 mm plate to accommodate a 5 mm weld.

Provide a tie plate of 283 × 240 × 8 mm size and connect it with 5 mm welds as shown in Fig. 9.61(c).

**EXAMPLE 9.19:** Design a batten system for the column in Example 9.18. Assume that the two channels are kept back-to-back. (See Fig. 9.62)

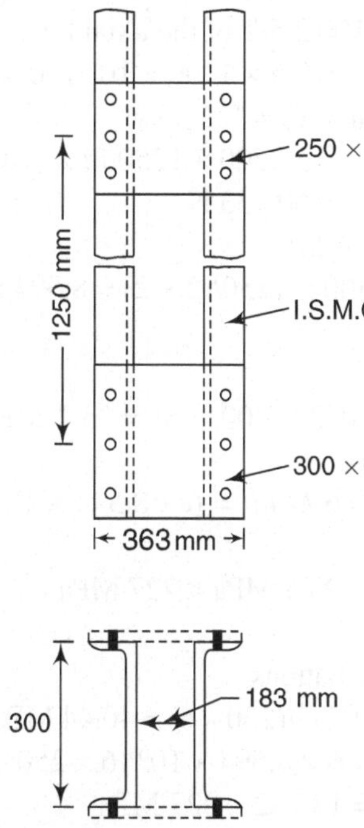


FIG. 9.62

**SOLUTION:**

From the previous example: with two ISMC 300 channels, Spacing of channel = 183 mm, $L/r = 84.74$

$L/r$ of the battened column = $1.1 \times 84.74 = 93.21$

For $L/r = 93.21$ and $f_y = 250$ MPa, from Table 9c of the code,

$$f_{cd} = 116.5 \text{ MPa}$$

Capacity of the built-up column = $2 \times 4630 \times 116.5/1000$ = 1078.8 kN ≈1100 kN

Hence the column is safe.

*Spacing of battens:* ($L_o/r_{yy}$) should be less than 0.7 times the slenderness of the built-up column.

i.e., $L_o/r_{yy} < 0.7\ (L/r)$

$$L_o < 0.7 r_{yy}\ (L/r) = 0.7 \times 26 \times 93.21 = 1696.4 \text{ mm}$$

Also, $L_o/r_{yy} < 50$

or $L_o < 50 \times 26 = 1300$ mm

Hence provide battens at a spacing of 1250 mm.

*Size of end battens:* Provide 20 mm bolts.

Edge distance (Table 5.6) = 32 mm

Effective depth = $183 + 2 \times 23.5 = 230$ mm $> 2 \times 90$ mm

Hence, chosen effective depth is safe.

Overall depth = $230 + 2 \times 32 = 294$ mm

Required thickness of batten = $1/50(183 + 2 \times 50) = 5.66$ mm

Length of the batten = $183 + 2 \times 90 = 363$ mm

Provide $363 \times 300 \times 6$ mm end batten plates.

*Size of intermediate batten:*

Effective depth = $3/4 \times 230 = 172.5$ mm $> 2 \times 90$ mm

Hence adopt an effective depth of 180 mm

Overall depth = $180 + 2 \times 32 = 244$ mm

Therefore, provide a $363 \times 250 \times 6$ mm batten plate.

*Design forces:*

Transverse shear $V$ = 2.5% of the axial load

$= 2.5 \times 1100 \times 10^3/100 = 27{,}500$ N

Longitudinal shear = $V_b = V_t L_o/(ns)$

$= 27{,}500 \times 1250/[2 \times (183 + 2 \times 50)]$

$= 60{,}733$ N

Moment $M = V_t L_o/(2n)$

$= 27{,}500 \times 1250/(2 \times 2) = 8.594 \times 10^6$ Nmm

*Check*

(a) For end battens

Shear stress = $60733/(300 \times 6) = 33.7 \text{ MPa} < (250/\sqrt{3})/1.1 = 131.2$ MPa

Bending stress = $6\ M/td^2 = 6 \times 8.594 \times 10^6/(6 \times 300^2) < (250/1.1)$

$= 95.5 \text{ MPa} < 227$ MPa

Hence safe.

(b) For intermediate battens

Shear stress = $60{,}733/(250 \times 6) = 40.44 \text{ MPa} < 131.2$ MPa

Bending stress = $6 \times 8.594 \times 10^6/(6 \times 250^2)$

$= 137.52 < 227$ MPa

Hence safe.

*Check*

$L/r$ of the built-up column = 84.74

$r_{min}$ of the individual chords = 26.0 mm

$$L/r = 1250/26 = 48.08$$

$\lambda$ of the built-up column [Eqn (9.62)],

$$\lambda_e = \sqrt{\{84.74^2 + (3.14^2/12) \times 48.08^2\}}$$

$$= 95.3 > 93.21\ (1.1 \times 84.74)$$

However, the difference is only 2.2% and hence the built-up column is safe.

*Connection:* The connections should be designed to transmit both shear and bending moment.

Assuming 20-mm diameter bolts.

Strength of bolt in bearing will not govern. Hence, strength of bolt = 45.3 kN

Required number of bolts = $60.733/45.3 = 1.34$

Let us provide three bolts to take into account the stresses due to bending moments.

*Check*

Force in each bolt due to shear = $60733/3 = 20244$ N

Adopt a pitch of 120 mm.

Force due to moment = $Mr/\Sigma r^2$

$= 8.594 \times 10^6 \times 120/(120^2 + 120^2)$

$= 35{,}808$ N

Resultant force = $\sqrt{(20244^2 + 35{,}808^2)}$

$= 41{,}134 \text{ N} < 45{,}300$ N

Hence safe.

Number of batten plates on one face = (length of column/spacing) + 1

$$= (10 \times 10^3)/1250 + 1 = 9 > 3$$

Hence the number chosen is ok.

**EXAMPLE 9.20:** Design a column having an effective length of 6 m and subjected to a factored axial load of 2400 kN. Provide the channels back-to-back connected by welded battens. Assume Fe 410 grade steel.

**SOLUTION:**

Required area of cross section = $2400 \times 10^3/(0.6 \times 250)$ = 16,000 mm$^2$

Let us assume two channels ISMC 400 at 501 N/m. Relevant properties of ISMC 400 as per IS 808 – 1989 are

$A = 6380$ mm$^2$; $b_f = 100$ mm; $t_f = 15.3$ mm; $r_z = 154$ mm, $r_y = 28.2$ mm;

$I_{zz} = 15{,}200 \times 10^4$ mm$^4$; $I_{yy} = 508 \times 10^4$ mm$^4$; $c_{yy} = 24.2$ mm

Area provided = $2 \times 6380 = 12{,}760$ mm$^2$

$$\lambda = L/r = 1.1 \times 6.0 \times 10^3/154 = 42.86$$

For $L/r = 42.86$ and $f_y = 250$ MPa, using Table 9c of the code,

$$f_{cd} = 193.71 \text{ MPa}$$

Capacity of the built-up column = 193.71 × 2 × 6380/1000 = 2471 kN > 2400 kN

Hence the column is safe.

*Spacing of channels:*

$$2I_{zz} = 2[I_{yy} + A(S/2 + c_{yy})^2]$$

$$2 \times 15200 \times 10^4 = 2[508 \times 10^4 + 6380\,(S/2 + 24.2)^2]$$

$$(S/2 + 24.2)^2 = 23028$$

Hence $S$ = 255.1 mm. Therefore, provide two ISMC 400 at a spacing of 256 mm back-to-back as shown in Fig. 9.63.

*Spacing of battens:*

$L_o/r_{yy}$ should be less than 0.7 × slenderness ratio of the built-up column. Hence,

$$L_o/r_{yy} < 0.7(L/r)$$

$$L_o < 0.7(L/r)r_{yy}$$

$$< 0.7 \times 42.86 \times 28.2$$

$$< 846 \text{ mm}$$

Also, $L_o/r_{yy}$ should be less than 50. Therefore,

$$L_o/r_{yy} < 50$$

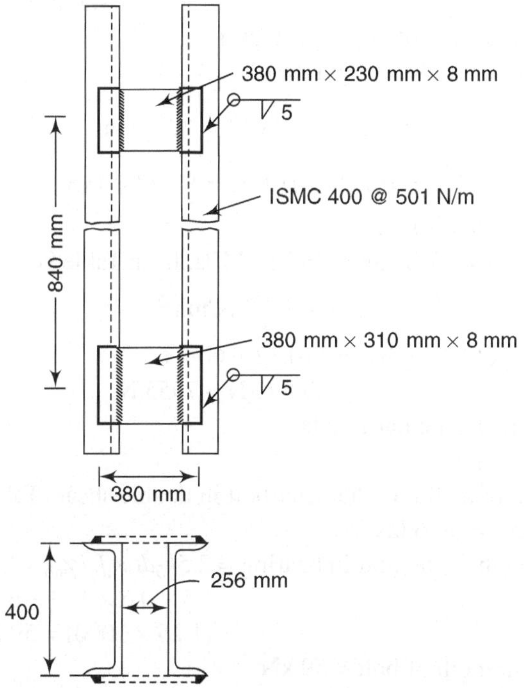


**FIG. 9.63**

i.e., $L_o < 50 \times 28.2 = 1210$ mm

Provide the batten at a spacing of 840 mm.

*Size of end battens*

Overall depth of batten = 256 + 2 × $c_{yy}$

= 256 + 2 × 24.2 = 304.4 mm

Provide a 62 mm overlap of batten on channel flange for welding.

Length of batten = 256 + 2 × 62 = 380 mm

Thickness of batten = 1/50 × 380 = 7.60 mm

Provide 380 × 310 × 8 mm end batten plate.

*Size of intermediate battens*

Overall depth = 3/4 × 304.4 = 228.3 mm

Provide 380 × 230 × 8 mm intermediate battens.

*Design forces:*

Transverse shear = 2.5% of axial load

= 2.5 × 2400 × 10³/100 = 60,000 N

Longitudinal shear $V_b = V_t L_o/ns$

= 60,000 × 840/[2(256 + 2 × 62/2)]

= 79,245 N

Moment $M = V_t L_o/2n$

= 60,000 × 840/(2 × 2) = 12.6 × 10⁶ N mm

*Check*

(a) For end battens,

Shear stress = 79,245/(310 × 8)

$= 31.9 \text{ MPa} < 250/(\sqrt{3} \times 1.1) = 131.2 \text{ MPa}$

Bending stress = $12.6 \times 10^6 \times 6/(8 \times 310^2)$

= 98.3 MPa < 250/1.1 = 227 MPa

(b) For intermediate battens,

Shear stress = 79,245/(230 × 8)

= 43.1 MPa < 131.2 MPa

Bending stress = $12.6 \times 10^6 \times 6/(8 \times 230^2)$

= 178.6 MPa < 227 MPa

Hence the battens are safe.

*Design of weld*:

Let the welding be done on all the four sides at each end of batten plate as shown in Fig. 9.63.

Let $t$ be the throat thickness of weld.

$I_{zz} = 2[(62 \times t^3/12) + (62 \times t)(230/2)^2] + 2 \times t \times 230^3/12$

Neglecting $62 \times t^3/12$, which will be insignificant, we get

$$I_{zz} = 366.77 \times 10^4 t \text{ mm}^4$$

$$I_{yy} = 2[t \times 62^3/12] + 2 \times 230 \times t^3/12 + 2 \times 230 \times t \times 31^2$$

Again neglecting $2 \times 230 \times t^3/12$, which will be very small, we get

$$I_{yy} = 48.18 \times 10^4\, t \text{ mm}^4$$

$$I_p = I_{zz} + I_{yy} = t(366.77 \times 10^4 + 48.18 \times 10^4)$$

$$= 414.95 \times 10^4 t \text{ mm}^4$$

The same result may be obtained by using Table 6.7.

$$I_p = t\frac{(b+d)^3}{6} = t\frac{(62+230)^3}{6}$$

$$= 414.95 \times 10^4 t \text{ mm}^4$$

$$r = \sqrt{[(230/2)^2 + (62/2)^2]} = 119.1 \text{ mm}$$

$$\cos\theta = 31/119.1 = 0.260$$

Direct shear stress = $79245/(2 \times 62 + 2 \times 230)t$
$= 135.7/t$ N/mm$^2$

Shear stress due to bending moment
$= 12.6 \times 10^6 \times 119.1/(414.95 \times 10^4 \times t)$
$= 361.65/t$ N/mm$^2$

Combined stress [clause 10.5.10 of the code]
$[3(135.7/t)^2 + (361.65/t)^2]^{0.5}$
$= 431.2/t < 410/(\sqrt{3} \times 1.25) = 189.4$
or $t = 431.2/189.4 = 2.28$ mm

Size of weld = 2.28/0.7 = 3.26 mm

The size of weld should not be less than 5 mm for 15.3 mm flange. Hence, provide a 5 mm weld to make the connections.

**EXAMPLE 9.21:** Design a built-up laced column with four angles to support an axial load of 900 kN. The column is 12-m long and both the ends are held in position and restrained against rotation. Assume Fe 410 grade steel.

**SOLUTION:**

Required area of the column = $900 \times 10^3/(0.6 \times 250) = 6000$ mm$^2$

Provide four angles ISA 100 × 100 × 8 mm. In this case we do not have any restriction of size (except for any architectural constraint). Hence, we will work backwards on the spacing from the assumed area and design compressive stress. The relevant properties are

$$A = 1540 \text{ mm}^2$$
$$c_{zz} = c_{yy} = 27.6 \text{ mm}$$
$$r_{zz} = r_{yy} = 30.7 \text{ mm}$$
$$I_{zz} = I_{yy} = 145 \times 10^4 \text{ mm}^4$$

Area provided = 4 × 1540 = 6160 mm$^2$

For 6280 mm$^2$, the required $f_{cd} = 900 \times 10^3/6160 = 146.1$ MPa

From Table 9c, allowable $L/r$ (for $f_{cd}$ = 146.1 MPa) ≈ 75

From Table 11 of the code, for the fixed condition, $K = 0.65$

*As it is a laced column*

$L = 1.05 \times [0.65 \times 12 \times 10^3] = 8190$ mm

Required $r = 8190/75 = 109.2$ mm

Moment of inertia of required section $I = Ar^2$
$= 6160 \times 109.2^2 = 73.46 \times 10^6$ mm$^4$

Equating required and provided moment of inertia,

$$73.46 \times 10^6 = 4 \times 145 \times 10^4 + 6160\ \bar{y}^2$$
$$\bar{y} = 104.8 \text{ mm}$$

Spacing of angle $S = 2 \times (104.8 + 27.6) = 264.8$ mm

Therefore, provide $S = 265$ mm

Now, $I_{zz} = I_{yy} = 4 \times 145 \times 10^4 + 6160(265/2 - 27.6)^2 = 73.58 \times 10^6$ mm$^4$

$$r = \sqrt{(73.58 \times 10^6/6160} = 109.3$$
$$L/r = 8190/109.3 = 74.93$$

From Table 9c, for $L/r = 74.93$ and $f_y = 250$ MPa,

$$f_{cd} = 144.11 \text{ MPa}$$

Capacity of the built-up column = 6160 × 144.11/1000 = 887.72 kN ≈ 900 kN

Hence the column is safe. If necessary, the spacing may be increased to 270 mm.

*Connecting system:*

Let us provide a double lacing system with the lacing flats inclined at 45°. Both are provided at the centre of the leg of angle.

Spacing of lacing bar, $L_o = (270 - 50 - 50)\cot 45°$
$= 170$ mm

$L_o/r_{yy} = 170/30.7 = 5.5 < 50$

It should also be less than $0.7 \times 74.9 = 52.43 > 5.5$

Shear force, $V = (2.5/100) \times 900 \times 10^3 = 22{,}500$ N

Transverse shear in each panel = $V/N = 22{,}500/2 = 11250$ N

As double lacing is provided,

Compressive force in lacing bar = $(V/2N)\operatorname{cosec}\theta$
$= 11250/2 \times \operatorname{cosec} 45°$
$= 7955$ N

*Section of lacing flat:*

Assuming 20 mm bolts, width of the flat (clause 7.6.2) = 60 mm

Length of lacing = (270 – 50 – 50) cosec 45° = 240.4

Minimum thickness of the lacing flat = (1/60) × 240.4 (Since double lacing)
= 4.0 mm

Provide a flat of size 60 × 6 mm.

Minimum radius of gyration,

$$r = t/\sqrt{12} = 6/\sqrt{12} = 1.73 \text{ mm}$$
$$L_1/r = 0.7(240.4)/1.73 = 97.27 < 145$$

Hence the flat is safe.

For $L_1/r = 97.27$ and $f_y = 250$ MPa, from Table 9c,

$$f_{cd} = 110.82 \text{ N/mm}^2$$

Capacity of lacing bar = 110.82 × 60 × 6
= 39,895 N > 7955 N

Hence the lacing bar is safe.

*Connections:*

Strength of a 20 mm diameter bolt in double shear (Table 5.9) = 2 × 45.3 = 90.6 kN

Strength of the bolt in bearing = $2.5k_b dt \times f_u/\gamma_{mb}$
$= 2.5 \times 0.6 \times 120 \times 6 \times 410/(1.25 \times 1000) = 59$ kN

Hence strength of bolt = 59 kN

Number of bolts = $2 \times 7955 \times \cot 45°/(59 \times 10^3) = 0.27$

Provide one 20-mm diameter bolt.

*Note:* The size of the lacing plate may be reduced to 50 × 6 m with a 16-mm diameter bolt.

*Tie plate*

Tie plates are to be provided at each end of the built-up column.

Effective depth = 270 – 2 × 27.6 = 214.8 > 2 × 100 mm

Overall depth of the tie plate = 200 + 2 × 25 = 250 mm

(Minimum edge distance of a 16-mm diameter bolt = 25 mm)

Thickness of tie plate = 1/50(270 – 50 – 50) = 3.4 mm

Therefore, provide a 270 × 250 × 6 mm tie plate and connect it with three 16-mm diameter bolts as shown in Fig. 9.64.

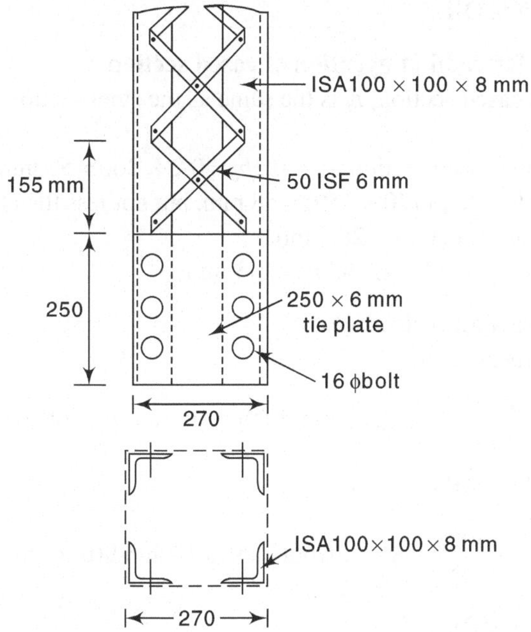


**FIG. 9.64**

**EXAMPLE 9.22:** A circular hollow section column has to be designed for an external application. The column is simply supported at its ends. The height of the column is 3.5 m and the applied factored axial load is 180 kN. Design a stainless steel column.

**SOLUTION:**

Since the application is external, choose grade 316 stainless steel with the following properties (see Tables 1.18 and 1.19):

$$E = 193{,}000 \text{ N/mm}^2$$

$$f_y = 0.2\ \% \text{ of the proof stress} = 205 \text{ N/mm}^2$$

Try 100 mm nominal bore medium tube with the following properties:

Diameter = 114.3 mm; $t$ = 4.50 mm; weight/m = 119 N/m; $A_g$ = 1550 mm$^2$; $I$ = 234.4 × 10$^4$ mm$^4$; $r$ = 38.9 mm

*Classification of the cross section (Table 2 of IS 800 : 2007)*

$$\varepsilon = (250/f_y)^{0.5} = (250/205)^{0.5} = 1.104$$

$$D/t = 100/4.5 = 22.2 < 88\varepsilon^2 = 88 \times 1.104^2 = 107.2$$

Hence, the section is non-slender.

*Resistance to flexural buckling*

$$KL/r = 1 \times 3500/38.9 = 89.97 \approx 90$$

From Table 10 of the code, we have to use curve 'b' (assuming the section is cold formed). From Table 9b, for $L/r$ = 90 and $f_y$ = 205 N/mm$^2$

$$f_{cd} = 121.5 \text{ N/mm}^2$$

(Interpolating between $f_y$ = 210 N/mm$^2$ and $f_y$ = 200 N/mm$^2$)

Capacity = 121.5 × 1550/1000 = 188 kN > 180 kN

Hence the member is safe. For more examples on stainless steel member design (using Eurocode) see www.euro-inox.org.

**EXAMPLE 9.23:** Design a circular hollow section column of grade YSt 240. Assume that the column has hinged ends, a height of 3.2 m, and supports a factored axial load of 125 kN.

**SOLUTION:**

For YSt 240 grade circular tube (as per IS 1161)

$$f_u = 410 \text{ MPa and } f_y = 240 \text{ MPa}$$

Assuming the design compressive stress $f_{cd}$ as $0.5f_y$ = 120 MPa

Required area = 125 × 1000/ 120 = 1042 mm$^2$

Try medium class tube of nominal bore 80 mm

A = 1070 mm$^2$, $r$ = 30 mm, and $t$ = 4 mm

*Classification of the cross section*

$$\varepsilon = \left(\frac{250}{f_y}\right)^{0.5} = \left(\frac{250}{240}\right)^{0.5} = 1.02$$

$$D/t = 80/4 = 20 < 88\varepsilon^2$$

Hence the section is non-slender.

*Calculate design stress*

$$KL/r = \frac{1 \times 3200}{30} = 106.67 < 180$$

Assuming hot finished seamless (HFS) type, from Table 10 of the code, we need to use buckling curve *a*.

From Table 9a of the code, for $L/r$ = 106.67 and $f_y$ = 240 MPa, $f_{cd}$ = 119.3 N/mm$^2$

Design capacity = 119.3 × 1070/1000 = 127.6 kN > 125 kN

Hence use 80 mm nominal bore medium class tube.

*Caution*: Tubes confirming to IS 1161 are not readily available in the market. Hence the designers should carefully check whether the correct tubes have been used.

**EXAMPLE 9.24:** Design a square hollow section column of grade YSt 310 to support a factored axial load of 135kN. Assume that the column is fixed at the bottom and free at the top and has a height of 2.5 m.

**SOLUTION:**

For YSt 310 grade circular tube (as per IS 1161)

$$f_u = 450 \text{ MPa } f_y = 310 \text{ MPa}$$

Assuming the design compressive stress $f_{cd}$ as 0.5 $f_y$ = 0.4 × 310 = 124 MPa

Required area = 135 × 1000/124 = 1089 mm$^2$

Since length is large assume a bigger size; try SHS of size 91.5 mm × 91.5 mm with thickness of 5.4 mm.

$A = 1785$ mm$^2$, $r = 34.8$ mm, and $t = 5.4$ mm

*Classification of the cross section:*

$$\varepsilon = \left(\frac{250}{f_y}\right)^{0.5} = \left(\frac{250}{310}\right)^{0.5} = 0.898$$

$$b = B - 3t = 91.5 - 3 \times 5.4 = 75.3 \text{ mm}$$

$$b/t = 75.3/5.4 = 13.94 < 42\varepsilon = 42 \times 0.898 = 37.7$$

Hence the section is non-slender.

*Calculate design stress:*

From Table 11 of the code, K = 2.0

$$KL/r = \frac{2 \times 2500}{34.8} = 143.7 < 180$$

Assuming hot rolled type, from Table 10 of the code, we need to use buckling curve *a*.

From Table 9a of the code, for $L/r = 143.7$ and $f_y = 310$ MPa, $f_{cd} = 76.5$ N/mm$^2$

Design capacity = 76.5 × 1785/1000 = 136.5 kN > 135 kN

Hence use SHS of size 91.5 mm × 91.5 mm medium class tube of thickness 5.4 mm.

**EXAMPLE 9.25:** Calculate the compressive resistance of the pin-ended ISMB 350 column of Example 9.2, having a length of 3 m, if it is encased in concrete of compressive strength 20 N/mm$^2$, as shown in Fig. 9.65.

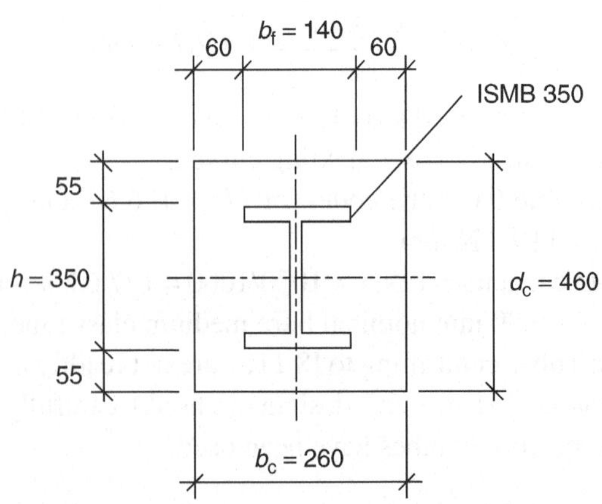


**FIG. 9.65**

**SOLUTION:**

Properties of ISMB 350

Flange width, $b_f = 140$ mm

Depth of section, $h = 350$ mm

Cross section area, $A_g = 6670$ mm$^2$

Radius of gyration, $r_z = 143$ mm, $r_y = 28.4$ mm

Area of concrete = $A_c = b_c \times d_c = 260 \times 460 = 119{,}600$ mm$^2$

**Check for effective length**

The effective length of column (3000 mm) should not exceed the least of:

1. $40b_c = 40 \times 260 = 10{,}400$ mm

2. $\frac{100b_c^2}{d_c} = \frac{100 \times 260^2}{460} = 14{,}696$ mm

3. $250\, r_y = 250 \times 28.4 = 7{,}100$ mm

Hence it is OK.

**Check for radii of gyration of cased section**

For the cased section, $r_z$ is the same as the steel section = 143 mm

For the cased section, $r_y = 0.2b_c = 0.2 \times 260 = 52$ mm $< 0.2\,(b_f + 150) = 0.2\,(140 + 150) = 58$ mm, but not less than that of the uncased section = 28.4 mm.

Hence $r_y = 52$ mm and $r_z = 143$ mm.

**Compression resistance**

Slenderness ratio

$\frac{KL_z}{r_z} = \frac{3000}{143} = 20.98$, from Tables 10 and 9(a) of the code, $f_{cd} = 225.4$ MPa

$\frac{KL_y}{r_y} = \frac{3000}{52} = 57.7$, from Tables 10 and 9(b) of the code, $f_{cd} = 184$ MPa

Hence design axial compressive stress, , $f_{cd} = 184$ MPa

$$f_{ck} = 20 \text{ MPa}, f_y = 250 \text{ MPa}$$

The compression resistance of the cased column, $P_{cd}$, is determined as

$$P_{cd} = \left(A_g + \frac{0.45 f_{ck} A_c}{f_y}\right) f_{cd}$$

$$= \left(6670 + \frac{0.45 \times 20 \times 119{,}600}{250}\right) 184 \times 10^{-3}$$

$$= 2019.5 \text{ kN}$$

which should not be greater than the short strut capacity, $P_{cs}$, given by

$$P_{cs} = \left(A_g + \frac{0.25 f_{ck} A_c}{f_y}\right) f_{cd}$$

$$= \left(6670 + \frac{0.25 \times 20 \times 119{,}600}{250}\right) 250 \times 10^{-3}$$

$$= 2265.5 \text{ kN}$$

Hence the compression resistance of the cased section is 2019.5 kN. Comparing it with the compressive resistance of the uncased column (Example 9.2) shows that the load capacity of the column has been increased from 733.7 kN to 2019.5 kN due to the casing by concrete.

## SUMMARY

A structural member subjected to compressive forces along its axis is termed as a compression member. Compression members in buildings are called as columns and those found in trusses are called struts or rafters. Although the structural column is a simple structural member, the interaction between the responses and characteristics of the material, the cross section, the method of fabrication, the imperfections and other geometric factors, and end conditions (restraints) make the column one of the most complex individual structural members.

The following parameters have an effect on the column strength:

1. Material properties (stress–strain relationship, yield stress)
2. Manufacturing method (hot-rolled/welded and cold-straightened shape resulting in different residual stress patterns)
3. Shape of the cross section (area, cross section geometry, and bending axis)
4. Length
5. Initial out-of-straightness (maximum value and shape of the distribution) and other imperfections (see also Tables 33 and 34 of IS 800: 2007)
6. End support conditions (pinned-with or without sway and restrained ends, with or without sway)

All these parameters that affect the strength of columns are discussed. The possible failure modes of columns are identified. The behaviour of compression members differ based on their length. Short or stocky columns can be loaded up to their yield stress and can attain their squash loads, provided the elements that make up the cross section are prevented from local buckling. Long compression members buckle elastically and hence their strength may be predicted by the Euler's formula. Intermediate length compression members fail both by yielding and buckling and hence their behaviour is inelastic. The behaviour of these slender columns is affected by the various parameters as listed before. To understand their behaviour, the elastic buckling of ideal pin-ended column is considered and Euler formula is derived. Then the effects of initial out-of-straightness, residual stresses, and material properties on the strength of compression members are discussed. Since there is a wide scatter in the strength of columns with different cross sections, many international codes have gone in for multiple column curves. Multiple column curves were first suggested by ECCS, which were followed by the American SSRC column strength curves. The Indian code has adopted the column curves similar to those given in the British code. The various individual and built-up or combined cross sections which can be used as compression members are discussed.

The effect of end supports (end restraints) on column strength is usually incorporated in the design by the concept of effective length. The effective length of columns with idealized boundary conditions are discussed and also the effect of intermediate restraints and the calculation of effective lengths of compressive members in different planes. The effective length of columns in multi-storey buildings is difficult to calculate, since the end restraints depend on the stiffness of the beams meeting at column ends and also on whether the frames are braced or unbraced. Columns in sway frames buckle in double curvature (hence the effective length factor will always be greater than unity) and the columns in braced (non-sway) frames, buckle in single curvature (hence the effective length factor will be less than unity).

The two widely used charts (alignment chart developed by Julian and Lawrence and Wood's curves) for the determination of effective length factor of columns in both braced and unbraced multi-storey buildings are illustrated. Since these charts are based on some idealized conditions, some modifications have been proposed by various researchers. The IS code provisions on effective length are based on the Wood's curves, with slight modifications.

The compression members in trusses have been considered separately since the loads are usually applied at the joints and the joints are often considered as pinned. Similarly, single-angle struts should also be treated separately, since they will be connected in most of the cases through one of their legs to the other members. The IS code provisions for such continuous and discontinuous angle members are summarized. Discussions on struts with variable cross section (tapered columns) and stepped columns are also included.

Columns for which the shear centre and centroid do not coincide will buckle in a coupled flexural and torsional mode. Sometimes we may also encounter columns with no axis of symmetry. Hence, a brief discussion about these columns and methods to prevent buckling failures is included. IS code provisions for the design of built-up columns with lacings and battens are also provided. The various steps involved in the design of compression members are also given. Some specifications for the compression members composed of two components back-to-back are also included. A brief description of compression members with other material properties and shapes, including steel arches is provided. Most of the concepts are explained with illustrative examples.

## MULTIPLE-CHOICE QUESTIONS

9.1 A vertical member primarily subjected to compression is called
(a) principal rafter (c) beam
(b) strut (d) stanchion

9.2 The failure mode in which an axially loaded compression member may fail is by
(a) local buckling (c) squashing
(b) overall flexural buckling (d) all of these

9.3 Which of the following sections are preferred as columns?
(a) ISLB (b) ISMB (c) ISHB (d) ISWB

9.4 It is a good practice to use the following section as a compression member.
(a) Plastic or compact section
(b) Semi-compact section
(c) Slender section
(d) Compact or semi-compact section

9.5 The effective width of outstand in a rolled I-section under compression is equal to ________
(a) the web depth
(b) flange width
(c) half of flange width
(d) half of (flange width – thickness of web)

9.6 The effective width of an outstand in a rolled angle under compression is
(a) equal to the width of the leg
(b) width of leg – thickness of leg
(c) depends on slenderness ratio
(d) equal to width of both the legs

9.7 The design compressive strength of an axially loaded compression member as per IS 800:2007 is based on
(a) Marchant–Rankine formula
(b) Perry–Robertson formula
(c) Rankine formula
(d) Secant formula

9.8 The design stress in compression, as per IS 800, is given by

$$f_{cd} = \chi f_y / \gamma_{m0}$$

where $\chi$ is the stress reduction factor given by
(a) $1/\phi + (\phi^2 - \lambda^2)^{0.5}$
(b) $1/[\phi + (\phi^2 - \lambda^2)^{0.5}]$
(c) $\phi + (\phi^2 - \lambda^2)^{0.5}$
(d) $1/[\phi + (\lambda^2 - \phi^2)^{0.5}]$

9.9 For rolled I-section the following curve is used for major axis buckling
(a) Curve A
(b) Curve B
(c) Curve C
(d) Curve D

9.10 For rolled angle section the following curve is used for major axis buckling
(a) Curve A
(b) Curve B
(c) Curve C
(d) Curve D

9.11 The optimum theoretical shape for compression members is ________
(a) double angles placed back to back
(b) cruciform arrangement of double angles
(c) tubular sections
(d) T-sections

9.12 Circular hollow sections have the following advantages:
(a) Are subjected to less wind loads than other flat sections
(b) Have high local buckling strength
(c) Have more torsional resistance
(d) All of these

9.13 Which of the following built-up sections are ideal for use as columns when the spacing between the individual member is equal in all the cases?

(a) 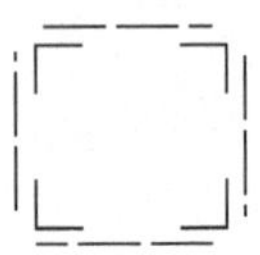
(c) 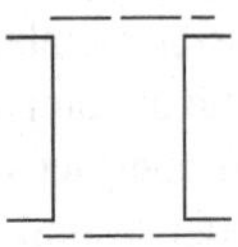
(b) 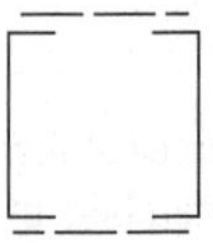
(d) 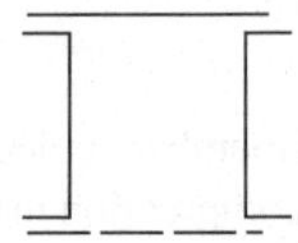

9.14 As per the code, the effective length of a column fixed at both ends is ________ ($L$ = Length of the column)
(a) $0.50L$
(b) $0.65L$
(c) $0.85L$
(d) $L$

9.15 As per the code, the effective length of a cantilever column is ________
(a) $1.2L$
(b) $0.65L$
(c) $2L$
(d) $3L$

9.16 The effective buckling length of a strut made of double angles back-to-back connected at the ends by at least two bolts is ________
(a) $1.2L$
(b) $0.7 - 0.85L$
(c) $0.85L$
(d) $0.7L$

9.17 The buckling resistance of a column may be increased by________
(a) having a section with large value of radius of gyration $r$
(b) increasing the size of the member
(c) providing bracings along the length of the column
(d) all of these

9.18 The best double angle compression member is ________
(a) equal angles on same side of gusset plate
(b) unequal angles with short legs back-to-back
(c) unequal angles with long legs back-to-back
(d) equal angles on opposite side of gusset plate

9.19 The slenderness of lacing should not exceed ________
(a) 180
(b) 145
(c) 200
(d) 140

9.20 The slenderness ratio of single angle bolted strut should be less than ________
(a) 180
(b) 250
(c) 145
(d) 350

9.21 The slenderness ratio of tie member in a truss subjected to possible reversal of stress should be less than ________
(a) 180
(b) 250
(c) 145
(d) 350

9.22 For the same load, unsupported length and end conditions, a laced column as compared to battened column:
(a) is stronger
(b) is weaker
(c) is having equal strength
(d) their strength cannot be compared

9.23 The slenderness ratio of components of laced column between consecutive lacing connections should be less than ________
(a) 180
(b) 250
(c) 145
(d) 50

9.24 The lacing bars in built-up columns should be designed to resist a total transverse shear equal to ________
(a) 2.5% of the column load
(b) 1.5% of the column load
(c) 3.5% of the column load
(d) 2.5% of the column load + additional shear due to B.M. or lateral load

9.25 The thickness of the lacing bar for single lacing should not be less than _________
(a) 1/30th of effective length
(b) 1/40th of effective length
(c) 1/50th of effective length
(d) 1/60th of effective length

9.26 For connecting lacing bars to column sections with 16 mm bolts, the width of lacing bar should not be less than _________
(a) 40 mm (b) 45 mm (c) 48 mm (d) 35 mm

9.27 The effective length of a battened column is _________
(a) increased by 10%
(b) increased by 5%
(c) 0.8 times the actual length of column
(d) 0.9 times the actual length of column

9.28 The effective depth of end batten should be more than _________
(a) twice the flange width of component column
(b) perpendicular distance between the centroids of main members
(c) longitudinal distance between innermost fasteners
(d) both (a) and (b)

9.29 The maximum lateral deflection allowed in a compression member in industrial buildings with elastic cladding, as per IS:800 is _________
(a) $L/250$ (b) $L/325$ (c) $L/150$ (d) $L/240$

9.30 The maximum lateral deflection allowed in a compression member in industrial buildings, supporting cab operated gantry girder is _________
(a) $L/150$ (b) $L/100$ (c) $L/240$ (d) $L/360$

## EXERCISES

*Note:* Use grade Fe 410 steel ($f_y = 250$ MPa) and $E = 2.0 \times 10^5$ MPa, if not specified in the problem.

9.1 Check whether an ISHB 400 in Fe 540 steel would be affected by local buckling effects when used as a column.

9.2 Determine the axial load capacity of a short length of square box column in Fe 540 steel fabricated by welding together four 600 × 14 mm plates.

9.3 A column consisting of ISHB 400 at 774 N/m has an unsupported length of 3.8 m. It is effectively held in position at both ends, restrained against rotation at one end. Calculate the axial load this column can carry.

9.4 Determine the design axial load of a column section ISHB 300. The column is having a height of 9 m and is effectively restrained by two bracings in the *y-y* direction at 3 m and 6 m and by one bracing member in the *z-z* direction at mid-height. Assume pinned condition at both ends of the column.

9.5 Determine the capacity of a column of size (wide flange section) W 250 × 250 × 101 (with weight = 1010 N/m, $r_z$ = 112.7 mm, $r_y$ = 65.6 mm, area = 12900 mm$^2$) with bracing and end condition as shown in Fig. 9.66.

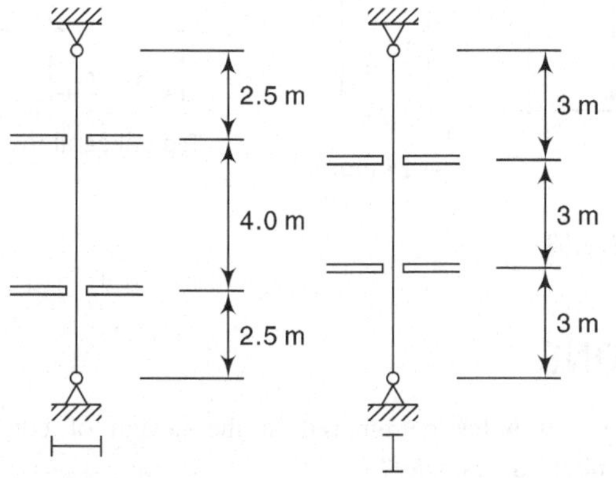


**FIG. 9.66**

9.6 For the same bracing and end condition as shown in Fig. 9.66 determine the capacity of ISMB 550 (weight = 1040 N/m) and compare the capacity with the W250 × 250 × 101 section of the previous example. (*Note:* Both the sections are having more or less equal weight.)

9.7 Determine the axial load capacity of the following section, assuming that the column has pinned ends and has a length of 3 m.
(a) ISMB 400 with one cover plate of size 250 × 12 mm on each flange [see Fig. 9.24(n)]
(b) ISMB 300 with one channel ISMC 300 welded to each flange of ISMB 300 [see Fig. 9.24(q)]
(c) Two ISMC 300 back-to-back with a distance of 200 mm between them [see Fig. 9.24(k)]
(d) Four angles ISA 90 × 90 × 6 with a plate 300 × 12 mm welded to form a column as shown in Fig. 9.24(r).

9.8 A single angle discontinuous strut ISA 65 × 65 × 6 is 1.2-m long. It is connected by one bolt at each end. Calculate the safe load this strut can carry.

9.9 A double angle discontinuous strut consists of two ISA 75 × 75 × 6 connected back-to-back to both sides of the 10-mm thick gusset plate with two bolts. The length of the strut is 4.5 m. Calculate the safe load carrying capacity of the section.

9.10 A double angle discontinuous strut 2.25 m long consists of two ISA 60 × 60 × 5 connected to the same side of the 8-mm thick gusset plate by one bolt. Calculate the load that this strut can carry.

9.11 An ISHB 400 column is an interior column with strong axis buckling in the plane of the frame as shown in Fig. 9.67. The columns above and below are also ISHB 400. The beams framing at the top are ISMB 400 and those at the bottom are ISMB 450. The columns are 4-m high and the beam span is 7 m. Determine the effective length of this column (*AB* in Fig. 9.67) as per the alignment chart as well as Wood's curves of IS 800 : 2007. Assuming that the column carries a factored load of 1200 kN, determine whether the column is safe to carry this load. Assume that the frame is braced by shear walls.

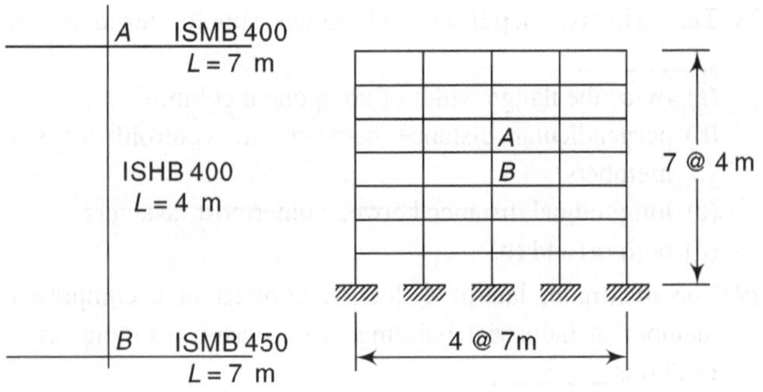


FIG. 9.67

9.12 For the same building shown in Fig. 9.67, determine the effective length and load carrying capacity of column *AB*, if the frame is a sway frame. Compare the results with those of Exercise 9.11.

9.13 A part of an office building and the elevator of the internal column stack *A* are similar to those shown in Fig. 9.67, except that the distance between the columns in *X*- and *Y*-direction is 6 m. Assuming the following loads, design an interior ground floor column for axial loads only.

Roof: Dead load (total) = 3 kN/m$^2$

Imposed load = 1.5 kN/m$^2$

Floors: Dead load (total) = 4 kN/m$^2$

Imposed load = 3 kN/m$^2$

Take the self-weight of the column including the fire protection as 0.8 kN/m. Also assume that the roof and floor have the same layout.

9.14 Design a column to support an axial load of 2500 kN, assuming that the column has an effective length of 8 m with respect to the *z*-axis and 4 m with respect to the *y*-axis. Since the maximum available rolled section (ISHB 450) may not be sufficient to support this load, design a built-up I-section with two flange plates.

9.15 Design a hollow circular tubular strut of a truss having an effective length of 1.8 m and to carry a factored load of 150 kN.

9.16 For the problem in Exercise 9.15, design a suitable hollow rectangular section and compare the weights.

9.17 Design a suitable section for a column 3.85-m long which is effectively held in position and restrained against rotation at both ends, in order to carry a factored load of 450 kN.

9.18 Design a single angle strut carrying a factored compressive load of 88 kN with length between centre-to-centre of intersection as 2.2 m. Design also the bolted end connections.

9.19 Design a single angle section having 3.25 m length to carry a factored compressive load of 14.5 kN due to wind. Design also the welded end connections for it.

9.20 Design a double angle section to act as a compression member in a truss having 2.1 m length and to carry a factored load of 150 kN.

(a) Connected back-to-back to both sides of a 12-mm thick gusset plate.

(b) Connected back-to-back to one side of the 12-mm thick gusset.

(c) Angles in star formation

9.21 Design a built-up column with two channels back-to-back having an effective length of 6.5 m and carrying a factored load of 1000 kN. Also design the lacings.

9.22 Design a built-up column with four angles laced together, having an effective length of 7.0 m and supporting a factored load of 1200 kN.

9.23 Design a built-up column with two channels toe-to-toe to carry a factored load of 2000 kN. Take the effective length as 5.5 m.

(a) Design it as a laced column and also design the lacings.

(b) Design it as a battened column and also design the battens.

9.24 Design a column of length 8 m to carry a factored axial load of 3000 kN, the ends having fixed in both position and direction. Design a suitable built-up column section using I-sections and batten plates.

9.25. Design a circular stainless steel column to carry a factored load of 200 kN and having an effective length of 5 m. Assume S 316 grade stainless steel.

9.26 A steel tower supports a water tank of size 3 m × 3 m × 2.5 m. The self-weight of the tank is 45 kN, when the tank is empty. The arrangement of the structure is shown in Fig. 9.68. Take unit weight of water as 9.81 kN/m$^3$. Take design wind pressure as per IS 875-Part 3, assuming that the tank is situated in New Delhi. Design the tower structure for the water tank.

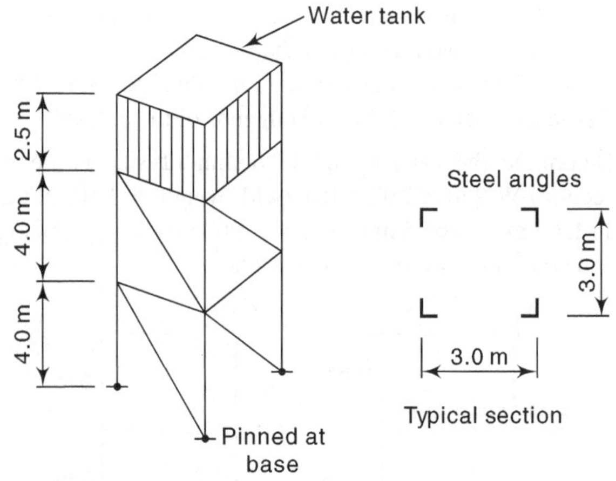


FIG. 9.68

## REVIEW QUESTIONS

9.1 Distinguish between column and beam column.

9.2 Cite the instances when a column may be regarded as an axially loaded column.

9.3 What is the basic difference in behaviour between tension and compression members, while resisting the loads?

9.4 Can bolt holes be ignored in the design of compression member? If yes, why?

9.5 State the parameters that affect the strength of compression members.

9.6 State the possible failure modes of an axially loaded column.

9.7 How can local buckling be eliminated in the compression member?
9.8 Why is it better to choose plastic or compact sections for columns?
9.9 How does the behaviour of a compression member differ based on its length?
9.10 Will a short column attain its squash load? Under what conditions?
9.11 What is the difference in behaviour of a long column and an intermediate length column?
9.12 What do you understand by inelastic buckling?
9.13 Derive the Euler's formula.
9.14 What are the basic assumptions that are made while deriving the Euler's formula?
9.15 Is buckling a strength problem or a stiffness problem?
9.16 What is the critical load of a fixed-fixed column according to Euler's theory?
9.17 Why is Euler's formula not adopted for practical designs?
9.18 What is the amplification factor?
9.19 Describe the behaviour of a steel column with an initial bow in the form of a sine curve.
9.20 Draw the $f_c$ versus $L/r$ curve for columns with initial out-of-straightness and show how it differs with an ideal straight column.
9.21 State the Perry formula.
9.22 How was the Perry-Robertson formula adjusted empirically by the British code in order to achieve a yield plateau for short columns?
9.23 What is the main difference in behaviour between a column with initial out-of-straightness and a column with eccentricity of applied load?
9.24 Describe how the residual stresses affect the column strength.
9.25 What is the tangent modulus? How can it be used to arrive at the inelastic buckling strength of columns?
9.26 What is the region of slenderness that is mostly affected by residual stress?
9.27 What is the effect of strain hardening on column strength?
9.28 What are the four basic ways that were used to establish column design formula?
9.29 What are multiple column curves? Why are several curves necessary to determine the strength of compression members?
9.30 What is the amplitude of initial out-of-straightness considered in developing the multiple column curves?
9.31 State the formula adopted in the code for predicting the strength of compression members.
9.32 Why a curve that will predict less strength is used for the welded I-section?
9.33 State the curve to be used for the following sections:
(a) Hot rolled CHS
(b) Cold rolled RHS
(c) Major axis buckling of welded box section
(d) Built-up section
(e) Channel, angle, *T*-, and solid sections
9.34 Can you guess why a lower strength curve is used for built-up as well as angle or channel section?
9.35 State the various cross sections that can be used as compression members.
9.36 Why are CHS/SHS better in resisting compression forces?
9.37 When are angles in star formation used?
9.38 What are the considerations for selecting a particular cross section for a compression member?
9.39 What are the boundary conditions for which closed formed solution for buckling loads could be obtained?
9.40 What is the necessity of using the effective length factor?
9.41 Define effective length of a column.
9.42 What is the minimum stiffness of a bracing that will effectively transform the first buckling mode of a column into the second mode?
9.43 Is it possible to prevent buckling of a column using sufficient number of bracings? How many bracings are required to do so?
9.44 What are the assumptions made in the derivation of *K* factor charts/curves?
9.45 Why separate curves/monograms are presented for braced and unbraced frames?
9.46 State the equation for finding the stability index.
9.47 Why have modifications been proposed by researchers to the basic alignment charts?
9.48 What is the method used in the Indian code for calculating the effective length of multi-storey frames?
9.49 Write short notes on
(a) Effective length of tapered columns
(b) Effective length of stepped columns
(c) Flexural-torsional buckling
(d) Buckling of column with no-axis of symmetry
(e) Methods to prevent buckling failure
9.50 Why is a different formula suggested for single angle struts in the Indian code?
9.51 What are the steps involved in the design of a compression member?
9.52 What are the classifications of built-up members?
9.53 What is the main purpose of lacings and battens?
9.54 State some of the important empirical rules that are to be followed while designing
(a) Lacing bars
(b) End battens
(c) Intermediate battens
9.55 Which built-up column is stronger (a) one with lacings (b) one with battens?
9.56 Why are double lacings with horizontal members not advisable?
9.57 State some of the empirical design rules given in the code for designing compression members composed of the components back-to-back.
9.58 Why separate treatment is required for the design of aluminium or stainless steel column?
9.59 Write short notes on:
(a) Three-hinged arches
(b) Fixed arches
(c) Two-hinged arches
9.60 Write the expression to calculate the decrease in length of a compression member.

# REFERENCES


Adluri S.M.R. and Madugula, M.K.S., 'Development of Column Curves for Single Angles', *Journal of Structural Engineering*, Vol. 122, No. 3, March 1996, pp. 318–325.

Agarval, K.M. and Stafiej, A.P., 'Calculation of Effective Length of Stepped Columns', *Engineering Journal*, AISC, 1980, No. 4, pp. 96–105.

ANSI/AISC 360-10, *Specification for Structural Steel Buildings*, American Institute of Steel Construction, Chicago, IL, 2010 (www.aisc.org).

AISC, *Specification for Load and Resistance Factor Design of Single-angle Members*, American Institute of Steel Construction, Chicago, USA, Dec 1993.

Allen, D., 'Merchant–Rankine Approach to Member Stability', *Journal of Structural Div.*, ASCE, Vol. 104, No. ST12, Dec 1978, pp. 1904–1914.

Allen, H.G. and Bulson, P.S., *Background to Buckling*, McGraw-Hill Book Company (UK) Ltd, London, 1980.

Anderson, J.P. and Woodward, J.H., 'Calculation of Effective Lengths and Effective Slenderness Ratios of Stepped Columns', *Engineering Journal*, AISC, 1972, No. 4, pp. 157–166.

ANSI/ASCE, 10-90, *Design of Latticed Steel Transmission Structures*, ASCE, New York, 1992.

ANSI-ASCE Standard No. ASCE-8-90, *Specification for the Design of Cold-formed Stainless Steel Structural Members*, ASCE, New York, 1991.

Arya, A.S. and Ajmani, J.L., *Design of Steel Structures*, Fourth edition, Nem Chand & Bros, Roorkee, 1989.

*ASCE Task Committee on Effective Length—Effective Length and Notional Load Approaches for Assessing Frame Stability: Implications for American Steel Design*, American Society of Civil Engineers, New York, 1997.

Aslani, F. and S.C. Goel, 'An Analytical Criterion for Buckling Strength of Built-up Compression Members', *Engineering Journal*, AISC, Vol. 28, No. 4, Fourth quarter 1991, pp. 159–168.

Ayrton, W.E. and Perry, J., 'On Struts', *The Engineer*, Vol. 62, 1866, p. 464.

Ballio, G. and Mazzolani, F.M., *Theory and Design of Steel Structures*, Chapman and Hall, London, 1983, p. 632.

Ballio, G. and Mazzolani, F.M., *Theory and Design of Steel Structures*, Chapman and Hall, London, 1983, pp. 505–506.

Bathon, L., Mueller III, W.H., and Kempner, Jr., L., 'Ultimate Load Capacity of Single Steel Angles', *Journal of Structural Eng*ineering, ASCE, Vol. 119, No. 1, Jan 1993, pp. 279–300.

Beer, H. and Schultz, G., 'Theoretical Basis of the European Column Curves', *Construction Metallique*, No. 3, Sep 1970, pp. 37–57.

Bjorhovde, R. and Tall, L., 'The Probabilistic Characteristics of Maximum Column Strength', *Proc. of the International Colloquium on Column Strength*, Paris, 23–34 Nov 1971.

Bjorhovde, R., 'Deterministic and Probabilistic Approaches to the Strength of Steel Columns', PhD Dissertation, Lehigh Univ., Bethlehem, PA, May 1972.

Bjorhovde, R., 'Effect of End Restraint on Column Strength—Practical Applications', *Engineering Journal*, AISC, Vol. 21, No. 1, 1985, pp. 42–45.

Bleich, F., *Buckling Strength of Metal Structures*, McGraw-Hill, New York, 1952.

Bradford, M.A. and Pi Y-L., 'Flexural–Torsional Buckling of Fixed Arches under Uniform Bending', *Journal of Constructional Steel Research*, Vol. 62, No. 1–2, Jan–Feb 2006, pp. 20–26.

Bridge, R.Q. and Fraser, D.J., 'Improved G-factor Method for Evaluating Effective Lengths of Columns', *Journal of Structural Engineering*, ASCE, Vol. 113, No. 6, Jun 1987, pp. 1341–1356.

BS 5950-1: 2000 *Structural Use of Steelwork in Building – Part 1: Code of Practice for Design—Rolled and Welded Sections*, British Standard Institution, London, Aug 2001.

Chen, W.F. and Lui, E.M., *Stability Design of Steel Frames*, CRC Press, Boca Raton, Florida, 1991.

Chen, W.F. and Lui, E.M., *Structural Stability—Theory and Implementation*, Elsevier, New York, 1987.

Cheong – Siat – Moy, F., 'An Improved K-factor Formula', *Journal of Structural Engineering*, Vol. 125, No. 2, Feb 1999, pp. 169–174.

Christopher, J.E. and Bjorhovde, R., 'Semi-rigid Frame Design Methods for Practicing Engineers', *Engineering Journal*, AISC, Vol. 36, First quarter 1999, pp. 12–28.

Chu, K.H. and Chow, H.L., Effective Column Length in Unsymmetrical Frames, *Publication of IABSE*, 1969, pp. 29–31.

Dafedar, J.B., Desai, Y.M., and Shiyekar, M.R., 'Review of Code Provisions for Effective Length of Framed Columns', *The Indian Concrete Journal*, Vol. 75, No. 6, Jun 2001, pp. 402–407.

DIN 18800, Teil 2, *Stahlbauten: Stabilitätsfälle, Knicken von Stäben und Stabwerben,* Deutsches Institut fur Normung e.v., Berlin, Dec 1980.

Disque, R.O., 'Inelastic K-factor for Column Design', *Engineering Journal*, AISC, Vol. 10, No. 2, Second quarter 1973, pp. 33–35.

Dowling P.J., Knowles, P., and Owens, G., (ed.) *Structural Steel Design*, Butterworths, London and The Steel Construction Institute, Ascot, 1988.

Driscoll, G.C., 'Effective Length of Columns with Semi-rigid Connections', *Engineering Journal*, AISC, Vol. 13, No. 4, Fourth quarter 1976, pp. 109–115.

Duan, L. and Chen, W.F., 'Effective Length Factor Columns in Unbraced Frames', *Journal of Structural Engineering*, ASCE, Vol. 115, No. 1, Jan 1989, pp. 149–165.

Duan, L. and Chen, W.F., Effective Length Factor for Columns in Braced Frames, *Journal of Structural Engineering*, ASCE, Vol. 114, No. 10, Oct 1988, pp. 2357–2370.

Engesser, F., Über die knickfestigkeit gerader stäbe, *Zeitschrift des Architekten-und Ingenieur-Vereins zu Hannover*, Vol. 35, 1889, pp. 455–462.

Epstein, H.I. and Thacker, B., "Effect of bolt stagger for block shear tension failures in angles", *Computers and Structures*, Vol.39, No.5, 1991, pp.571-76.

Epstein, H.I. and Aleksiewicz , L.J., 'Block Shear Equations Revisited Again', *Engineering Journal*, AISC, Vol. 45, No. 1, First quarter 2008, pp. 5–12.

Euler, L., Sur le Force de Colonnes, *Memoires de l'Academie Royale des Sciences et Belles, Lettres*, Vol. 13, Berlin, 1959 (English translation by J.A.Van den Broek, *American Journal of Physics*, Vol. 15, 1947, pp. 309).

Fukumoto, Y., Nethercot, D.A., and Galambos, T.V., 'Experimental Data for the Buckling of Steel Structures—NDSS Stability of Metal Structures', *Proceedings, Third International Colloquium*, SSRC, Toronto, Canada, May 1983, pp. 609–630.

Galambos, T.V., ed., *Guide to Stability Design Criteria for Metal Structures*, Fifth edition, John Wiley & Sons, 1998.

Ganapathy Chettiar, C., Permissible Stresses in Axial Compression in IS:800–1962, *ISI Bulletin*, Vol. 27, No. 7, Jul 1975, pp. 241–245.

Gardner, L. and Nethercot, D.A., Structural Stainless Steel Design—A New Approach, *The Structural Engineer*, Vol. 82, No. 21, 2004, pp. 21–28.

Gaylord, E.H. Jr., Gaylord, C.N., and Stallmeyer J.E., *Design of Steel Structures*, Third edition, McGraw-Hill International Edition, New York, 1992.

Goyet, V.V., Frey, Fr., and Massonnet, Ch., Ultimate Load of Trusses Buckling in their Plane, *IABSE Proceedings*, P-47/81, Nov 1981, pp. 141–164.

'Guide for Design of Steel Transmission Towers', *ASCE Manual and Reports on Engineering Practice*, No. 52, Second edition, ASCE, New York, 1988.

Johnson, B.G., ed., *Guide to Design Criteria for Metal Compression Members*, Column Research Council and John Wiley & Sons, New York, 1960.

Johnson, R.P., 'Composite Structures in Steel and Concrete', Vol. 1, *Beams, Columns, Frames and Applications in Buildings*, Blackwell Scientific Publications, London, 1994.

Johnston, B.G., ed., *Guide to Stability Design Criteria for Metal Structures*, Third edition, Wiley, New York, 1976.

Kavanagh, T.C., 'Effective Length of Framed Columns', *Transactions*, ASCE, Vol. 127, 1962, Part II, pp. 81–101.

Kirby, P.A. and Nethercot, D.A., *Design for Structural Stability*, Granada Publishing, London, 1979.

Kishi, N., Chen, W.F., and Goto Y., 'Effective Length Factor of Columns in Semi-rigid and Unbraced Frames', *Journal of Structural Engineering*, ASCE, Vol. 123, No. 3, Mar 1997, pp. 313–320.

Kitipornchai, S. and Lee, H.W., 'Inelastic Buckling of Single Angle, Tee and Double Angle Struts', *Journal of Constructional Steel Research*, London, Vol. 6, No. 1, 1986, pp. 3–20 also, Vol. 6, No. 3, 1986, pp. 219–236.

Knowles, P., *Design of Structural Steelwork*, Second edition, Surrey University Press, 1987.

Kuhn, G. and Lundgren, H.R., 'An Appraisal of the Effective Length Alignment Charts', *International Colloquium on Stability of Structures under Static and Dynamic Loads,* Washington DC, ASCE, 1977, pp. 212–242.

Lay, M.G. and Galambos, T.V., 'Bracing Requirements for Inelastic Steel Beams', *Journal of the Structural Div*., ASCE, Vol. 92, No. ST2, Apr 1966, pp. 207–228.

LeMessurier, W.J., 'A Practical Method of Second Order Analysis, Part 2—Rigid Frames', *Engineering Journal*, AISC, Vol. 14, No. 2, Second quarter 1972, pp. 49–67.

Lindner, J., Näherungen fur die Europäischen Knickspannungskurven, *Die Bautechnik*, No. 10, Oct 1978, pp. 344–347.

Lui, E.M., 'A Novel Approach to K-factor Determination', *Engineering Journal*, AISC, Vol. 29, No. 3, Third quarter 1992, pp. 150–159.

Madugula, M.K.S. and Kennedy, J.B., *Single and Compound Angle Members—Structural Analysis and Design*, Elsevier Applied Science, London, England, 1985.

Mahadevan, K., Sridhara, B.N., and Kalyanaraman, V., 'Buckling Strength of Core Loaded Sleeved Column Systems in Water Tank Staging', *International Conference on Stability of Structures, ICSS 95*, Jun 1995, Coimbatore, India, pp. 441–451.

Maquoi, R. and Rondal, J., Mise en equation des nouvelles courbes europeennes de flambement, *Construction Metallique*, Paris, No. 1, 1978.

Narayanan, R., ed., *Axially Compressed Structures: Stability and Strength*, Elsevier Applied Science Publishers, Essex, 1982.

Nazir, C.P., 'Prestressed Steel Arch Bridge', *Journal of the Institution of Engineers (India)*, Civil Engg., Div., Vol. 84, May 2003, pp. 72–76.

Nethercot, D.A., *Limit States Design of Structural Steelwork*, Third edition, Spon Press, London, 2001.

Oehlers, D.J. and Bradford, M.A., *Composite Steel and Concrete Structural Members, Fundamental Behaviour*, Pergamon Press, Elmsord, NY, 1995.

Papangelis, J.P. and Trahair, N.S., 'Flexural Torsional Buckling of Arches', *Journal of Structural Engg*., ASCE, Vol. 113, No. ST7, Jul 1987, pp. 1433–1443.

Petersen, C., Statik und Stabilität der Baukonstruktionen, *Friedr. Vieweg & Sohn*, Braunschweig/Wiesbaden, Germany, 1980, p. 960.

Robertson, A., 'The Strength of Struts', *ICE Selected Engineering paper* 28, 1925.

Roik, K. and Bergmann, R., *Composite Columns (in Constructional Steel Design),* P.J. Dowling, et al, (eds), Elsevier Science Publishing, New York, 1992, pp. 443–470.

Rondal, J. and Maquoi, R., 'Single Equation for SSRC Column-strength Curves', *Journal of the Structural Div.*, ASCE, Vol. 105, No. ST1, Jan 1979, pp. 247–250.

Sakimoto, T. and Komatsu, S., 'Ultimate Strength of Arches with Bracing Systems', *Journal of Structural Div*., ASCE, Vol. 109, No. 3, Mar 1983, pp. 613–627.

Sakimoto, T. and Sakata, T., 'The Out-of-plane Buckling Strength of Through-type Arch Bridges', *Journal of Constructional Steel Research*, Vol. 16, 1990, pp. 307–318.

Sakimoto, T. and Sakata, T., 'Ultimate Strength of Arches with Bracing Systems', *Journal of Structural Div.,* ASCE, Vol. 108, No. ST5, May 1982, pp. 1064–1076.

Salmon, C.G., Johnson, J.E., *Steel Structures: Design and Behaviour*, Fourth edition, Harper & Row Publishers, New York, 1996.

Sato, A. and C.-M. Uang, 'Modified Slenderness Ratio for Built-up Members', *Engineering Journal*, AISC, Vol. 44, No. 3, Third quarter 2007, pp. 269–280.

Schodek, D.L., *Structures*, Fourth edition, Prentice Hall, New Jersey, 2001.

Sfintesco, D., 'Experimental Basis of the European Column Curves', *Construction Metallique*, No. 3, Sep 1970, p. 5.

Shanley, F.R., 'Inelastic Column Theory', *Journal of the Aeronautical Science*, Vol. 14, No. 5, May 1947, pp. 261–264.

Shanley, F.R., 'The Column Paradox', *Journal of the Aeronautical Sciences*, Vol. 13, No. 12, Dec 1946, p. 678.

Smith, E. and Epstein. H., 'Hartford Coliseum Roof Collapse: Structural Collapse and Lessons Learned', *Civil Engineering*, ASCE, Apr 1980, pp. 59–62.

Sridhara, B.N., 'Design of Optimum Weight Compression Members', *International Conference on Steel and Aluminium Structures*, Singapore, May 1991, pp. 298–307.

Subramanian, N., 'Design of Antenna Towers', *The Bridge and Structural Engineer*, Vol. XXII, No. 2, Jul 1992, pp. 15–32.

Taranath B.S., *Structural Analysis and Design of Tall Buildings*, McGraw-Hill International edition, 1988.

Taranath, B.S., Steel, *Concrete and Composite Design of Tall Buildings*, Second edition, McGraw-Hill, New York, 1998.

Timoshenko S.P. and Gere, J.M., *Theory of Elastic Stability*, Second edition, McGraw-Hill International Edition, New York, 1961.

Timoshenko, S.P., *History of the Strength of Materials*, McGraw Hill, New York, 1953.

Trahair, N.S., Bradford M.A., and Nethercot, D.A., *The Behaviour and Design of Steel Structures to BS 5990,* Third edition, Spon Press, London, 2001.

Troitsky, M.S., *Prestressed Steel Bridges: Theory and Design*, Von Nostrand Reinhold Co., New York, 1990.

Viest, I.M., Cotaco, J.P., Furlong, R.W., Griffis, L.G., Leon, R.T. and Wyllie, L.A., *Composite Construction: Design for Buildings*, ASCE, McGraw-Hill, New York, 1996.

von Karman, T., Die knickfestigkeit gerader Stäbe, *Physicalische zeitschrift*, Vol. 9, 1908, p. 136.

Wood, R.H., 'Effective Lengths of Columns in Multistory Frames', *The Structural Engineer*, Vol. 52, No. 7, Jul 1974, pp. 235–244, No. 8, Aug. 1974, pp. 295–302, No. 9, Sep 1974, pp. 341–346.

Woolcock, S.T. and Kitipornchai, S., 'Design of Single Angle Web Struts in Trusses', *Journal of Structural Engineering*, ASCE, Vol. 113, No. 9, Sep 1987, pp. 2102–2107.

www.euro-inox.org, last accessed on 20 July 2015.

Young, T., *A Course of Lectures on Natural Philosophy and Mechanical Arts*, J. Johnson, London, England, 1807.

Young, W.C. and Budynas, R., *Roark's Formulas for Stress and Strain*, Seventh edition, McGraw-Hill Professional, New York, 2001, p. 832.

Yu, C.K. and Tall, L., 'Significance and Application of Stub Column Test Results', *Journal of the Structural Div*., ASCE, Vol. 97, No. ST7, Jul 1971, pp. 1841–1861.

Yura, J.A., 'The Effective Length of Columns in Unbraced Frames', *Engineering Journal*, AISC, Vol. 8, No: 2, Apr 1971, pp. 37–42, Discussions Vol. 9, No. 1, Jan 1972, pp. 40–41, 46, Oct 1972, pp. 167–168.

# 10

# DESIGN OF BEAMS

## INTRODUCTION

Beams are structural members that support loads which are applied transverse to their longitudinal axes. They are assumed to be placed horizontally and subjected to vertical loads. Beams have a far more complex load-carrying action than other structural elements such as trusses and cables. Little was known about the exact mechanism by which beams carry loads until recently, relative to the length of time, when such elements were discovered to have been used in early and recent civilizations. A sketch by Leonardo da Vinci indicates that he tried to describe beam action in a rational way. Galileo Galilei was the first to begin investigating the bending problem in a systematic way in 1638 (the term bending problem refers to the study of stresses and deformations in a beam bowed by the action of forces, applied perpendicular to the axis of the beam, so that fibres on one face of the beam are elongated and those on the opposite face are contracted). Although Galilei was not successful in solving the problem, his initial formulations helped future investigators such as Hooke, Mariotte, Leibnitz, Navier, and Coulomb. Auguste Coulomb (1736–1806) is credited with the final solution of the bending problem. Later investigators refined the solution and tackled other problems commonly associated with beams such as the effect of torsion and lateral torsional buckling.

The load transfer by a beam is primarily by bending and shear. Any structural member could be considered as a beam if the loads cause bending of the member. If a substantial amount of axial load is also present, the member is referred to as a beam-column (beam-columns are covered in Chapter 13). Though some amount of axial effects will be present in any structural member, in several practical situations, the axial effect can be neglected and the member can be treated as a beam.

If we consider a normal building frame, the beams that span between adjacent columns are called *main* or *primary beams/girders*. The beams that transfer the floor loading to the main beams are called *secondary beams/joists*. Beams may also be classified as follows:

***Floor beams*** A major beam of a floor system usually supporting joists in buildings; a transverse beam in bridge floors.

***Girder*** In buildings, girders are the same as floor beams; also a major beam in any structure. Floor beams are often referred to as girders.

***Girt*** A horizontal member fastened to and spanning between peripheral columns of an industrial building; used to support wall cladding like corrugated metal sheeting.

***Joist*** A beam supporting floor construction but not a major beam.

***Lintels*** Beam members used to carry wall loads over wall openings for doors, windows, etc.

***Purlin*** A roof beam, usually supported by roof trusses.

***Rafter*** A roof beam, usually supporting purlins.

***Spandrels*** Exterior beams at the floor level of buildings, which carry part of the floor load and the exterior wall.

***Stringers*** Members used in bridges parallel to the traffic to carry the deck slab; they will be connected by transverse floor beams.

Beams may be termed as *simple beams* when the end conditions do not carry any end moments from any continuity developed by the connection. A beam is called a *continuous beam*, when it extends continuously across more than two supports. A *fixed beam* has its ends rigidly connected to other members, so that the moments can be carried across the connection. In steel frames, the term 'fixed end' is somewhat a misnomer, since the ends of rigid connections are not fixed but allow some joint rotation.

Beam loadings will consist of both dead and live loads. A part of the beam dead load is its own weight. When long, heavy members are used, the self-weight may be a significant part of the total load. Hence, the beam section should be checked for adequacy for the applied loads, including the beam weight. For the usual structural frame of a steel building, bending effects are predominant in beams with negligible torsional effects.

In order to resist the bending effects in the most optimum way, much of the material in a beam has to be placed as far away as possible from the neutral axis. The web area of the beam has to be adequate for resisting the shear. Usually, maximum bending moment and maximum shear occur at different cross sections. However, in continuous beams, they may occur at the same cross section near the interior supports. Commonly used cross-sectional shapes are the rolled I-sections and the built-up I-sections or box sections, though channel sections are also often used as purlins (Fig. 10.1). Doubly symmetric shapes such as the standard rolled I-shapes are the most efficient.

Generally a beam may be subjected to simple, unsymmetrical, or biaxial bending. Simple bending takes place if the loading plane coincides with one of the principal planes of a doubly symmetric section such as an I-section or, in the case of a singly symmetric open section (C-section), the loading passes through the *shear centre* and is parallel to the principal plane. Unsymmetrical bending occurs if the plane of loading does not pass through the shear centre. In this case, bending is coupled with torsion. If simple bending takes place about both the principal planes without torsion, the beam is said to undergo biaxial bending. A typical example for biaxial bending in a beam is the roof purlin.

Several complications may arise in a beam design due to a variety of reasons, such as complex stress conditions, tendency of the member to buckle, and conflicting design requirements. Complex stress states may arise when loads are inclined to the principal axes, when unsymmetrical sections are used or where large values of shear and bending moment occur at a section. Buckling may take place in many ways: (a) lateral buckling of the whole beam between supports, (b) local buckling of flanges, and (c) longitudinal buckling of the web and buckling in the depth direction under concentrated loads or near the supports. As far as design is concerned, the depth of the beam is a very important parameter. While a large depth is desirable for moment resistance, the resulting thin web may not be able to resist lateral or web buckling.

In the design of a beam, two aspects are of primary consideration: (a) strength requirement, that is, the beam has adequate strength to resist the applied bending moments and accompanying shear forces and (b) stability consideration, that is, the member is safe against buckling. Another requirement is adequate bending stiffness, which is often satisfied by members meeting the requirement of strength. Beams develop higher stresses compared to axial members for the same loads and bending deflections are considerably high. Bending effects rather than shear govern the proportions of a rolled beam section, except in cases where shear-to-moment ratio is very large. Occasionally, bending deflections may form the primary design consideration. Shear deflections are usually very small and hence neglected except in the case of deep beams with very short spans.

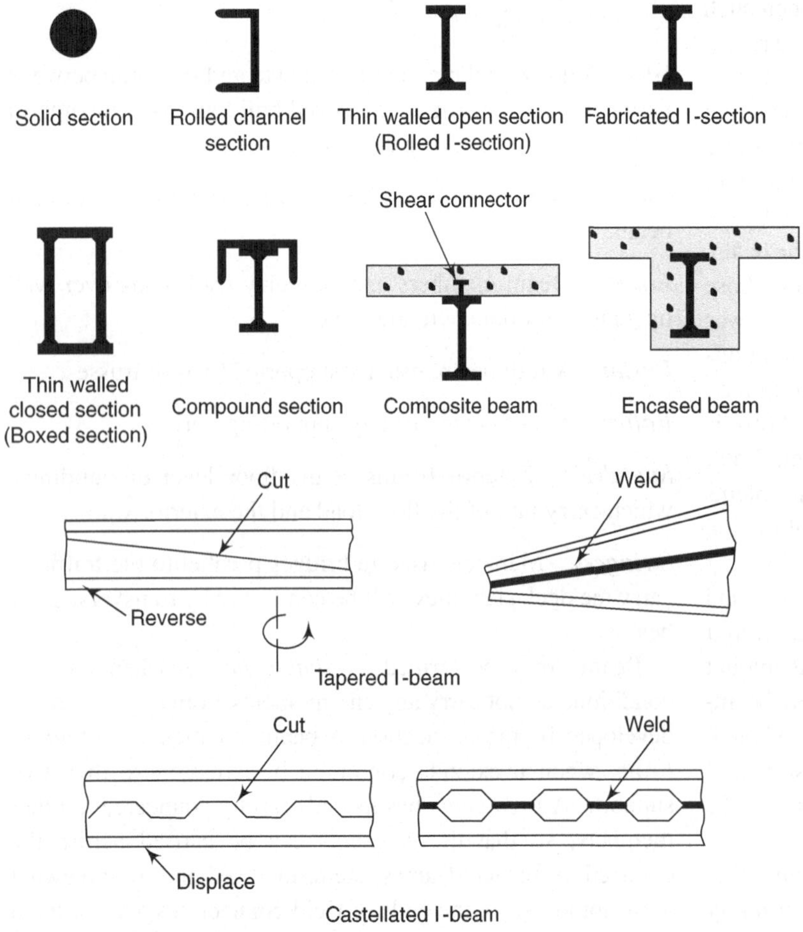


**FIG. 10.1** Types of steel beams

## 10.1 BEAM TYPES

There are various forms of beam cross sections used in practice, as shown in Fig. 10.1. The selection of a section would depend upon the use for which it is intended and on the overall economy. The beam chosen should possess required strength and it should not deflect beyond a limit. Generally thin-walled open sections are used as beams (e.g., rolled sections). This is mainly due to their capability for forming easy and convenient connections to adjacent members. If suitable rolled sections are not available, they may be fabricated out of rolled plates. Such members are called *plate girders*, which are dealt in detail in the next chapter. Compound sections may also be fabricated out of rolled sections for special purposes. Similarly, in *hybrid sections*, the flanges of higher yield stress compared to the web can also be formed. Certain specific features, such as the passage of services below the floor of a building, may necessitate the use of sections with openings in the web. Other economical types of beams that can be fabricated out of rolled sections are the *castellated* and *tapered beams*. *Composite steel beams* and concrete-encased steel beams, which act in concert with concrete, derive added strength, economy, and in the latter case fire protection also. Table 10.1 gives the important types of steel beams used in practice with their optimum span range. (Note that open web joists and trusses do not strictly come under the classification of beams.)

**TABLE 10.1** Beam types

| Type of beam | Optimum span range (m) | Application |
|---|---|---|
| Angles | 3–6 | For lightly loaded beams such as roof purlin and sheeting rail |
| Rolled I-sections | 1–30 | Most frequently used as a beam |
| Castellated beams | 6–60 | Long spans and light loads |
| Plate girders | 10–100 | Long spans with heavy loads such as bridge girders |
| Box girders | 15–200 | Long spans and heavy loads such as bridge girders |
| Open web joist (see Fig. 10.37) | 4–40 | Fabricated for large spans, using angles or tubes as chords and round bars for web diagonals |
| Trusses | 10–100 | Long spans and moderate loads such as industrial roofing |

## 10.2 SECTION CLASSIFICATION

The bending strength of a section depends on how well the section performs in bending. When a stocky beam (thick-walled beam) (see Fig. 8.1 of Chapter 8) is subjected to bending, extreme fibres in the maximum moment region reach the yield stress [see Fig. 8.1(b)]. Upon further loading, more and more fibres reach the yield stress and the stress distribution is as shown in Fig. 8.1(c). The bending moment that causes the whole cross section to reach yield stress [Fig. 8.1(d)], is termed as the plastic moment of the cross section $M_p$. The cross section is incapable of resisting any additional moment, but may maintain the plastic moment, acting as a plastic hinge, for some more amount of rotation. If the section is thin-walled (slender), it may fail by local buckling, even before reaching the yield stress. The ratio of ultimate rotation to yield rotation is called the *rotation capacity*. Four classes of section have been identified: (a) plastic (class 1), (b) compact (class 2), (c) semi compact (class 3), and (d) slender (class 4) based on their yield and plastic moments along with their rotation capacities. Refer to Section 8.13 and Fig. 8.25 of Chapter 8 for more details of this classification.

Local buckling can be prevented by limiting the width–thickness ratio. In the case of rolled sections, higher thickness of the plate is adopted to prevent local buckling. For built-up and cold-formed sections, longitudinal stiffeners are provided to reduce the width into smaller sizes. It is evident that only plastic sections can be used in indeterminate frames which form collapse mechanisms. Compact sections can be used in simply supported beams which fail after reaching $M_p$ at one section. Semi-compact sections can be used for elastic design, where the section fails after reaching $M_y$ at the extreme fibres. Slender sections are not preferred in hot-rolled structural steelwork; but they are extensively used in cold-formed members.

## 10.3 LATERAL STABILITY OF BEAMS

The failure modes of a beam member are presented in Table 10.2. A beam, as in category 1 in Table 10.2, subjected to transverse loads would undergo bending and deflect in the plane of the load. On increasing the transverse loads, the beam would attain its full plastic moment capacity. Two important assumptions have to be made to achieve this ideal beam behaviour. They are as follows:

1. The compression flange of the beam is restrained from moving laterally
2. Any form of local buckling is prevented

Thus, a beam loaded predominantly in flexure would attain its full moment capacity if the local and lateral instabilities of the beam are prevented. To ensure the first condition, the cross sections of flanges and web of the beam chosen must be *plastic* or *compact*. If significant ductility is required, sections must invariably be plastic. If the laterally unrestrained length of the compression flange of the beam is relatively long as in category 2 of Table 10.2, then a phenomenon, known as *lateral buckling* or *lateral torsional buckling* of the beam may take place. The beam would fail well before it attains its full moment capacity. This phenomenon has a close similarity to the Euler buckling of columns, where collapse is triggered before the column attains its squash load (full compressive yield load).

In steel structures, especially in buildings, beams are restrained laterally by the floor decks, which are placed

**TABLE 10.2** Main failure modes of hot-rolled beams

| Mode | | Comments |
|---|---|---|
| Excessive bending triggering collapse<br>Category 1 | 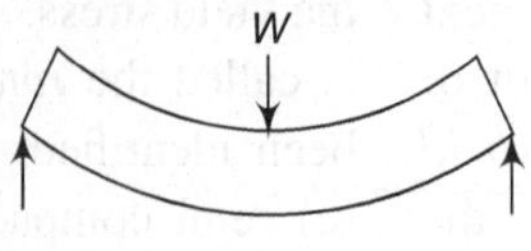  | This is the basic failure mode provided (a) the beam is prevented from buckling laterally and (b) the component elements are at least compact, so that they do not buckle locally. Such 'stocky' beams will collapse by plastic hinge formation. |
| Lateral torsional buckling of long beams which are not suitably braced in the lateral direction (i.e., 'unrestrained' beams)<br>Category 2 | 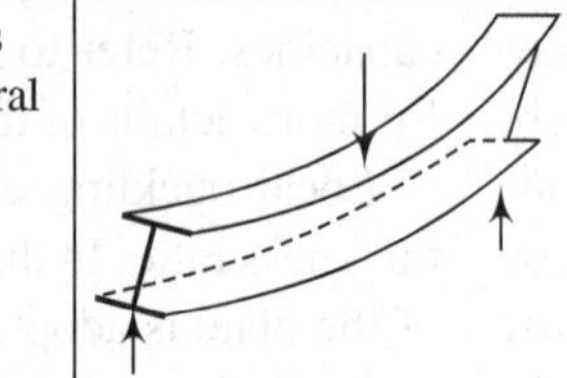 | Failure occurs by a combination of lateral deflection and twist. The proportions of the beam, support conditions, and the way the load is applied are all factors, which affect failure by *lateral torsional buckling*. |
| Failure by local buckling of a flange in compression or web due to shear or web under compression due to concentrated loads<br>Category 3 | 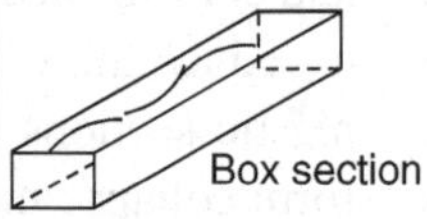  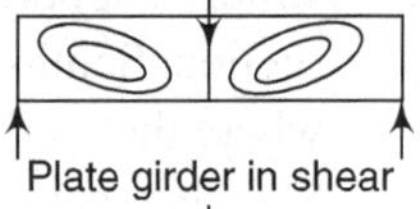  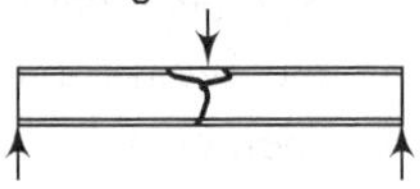 | Unlikely for hot-rolled sections, which are generally stocky. Fabricated box sections may require flange stiffening to prevent premature collapse.<br>Web stiffening may be required for plate girders to prevent shear buckling.<br>Load bearing stiffeners are sometimes needed under point loads to resist web buckling. |
| Local failure by (a) shear yield of web, (b) local crushing of web, and (c) buckling of thin flanges.<br>Category 4 |   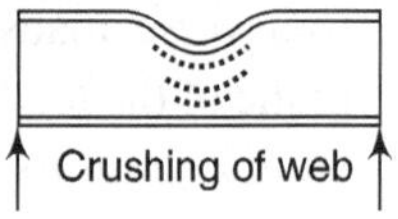    | Shear yield can only occur in very short spans and suitable web stiffeners will have to be designed.<br>Local crushing is possible when concentrated loads act on unstiffened thin webs. Suitable stiffeners can be designed.<br>This is a problem only when very wide flanges are employed. Welding of additional flange plates will reduce the plate *b*/*t* ratio and thus flange buckling failure can be avoided. |

on top of them. During the construction stage, before the floor decks are in place, if the beams are not adequately supported laterally, they may be susceptible to lateral buckling. Therefore, during construction stage, they may need special attention with regard to their lateral stability. If adequate lateral restraints are not provided to beams in the plane of their compression flanges, the beams would buckle laterally resulting in a reduction of their maximum moment capacity. Lateral buckling can be prevented, if adequate restraints are provided to the beam in the plane of the compression flange; such beams are called *laterally restrained beams*.

Beams of categories 3 and 4 given in Table 10.2 fail by local buckling, which should be prevented by adequate design measures, in order to achieve their capacities. It is interesting to explain the phenomenon of lateral buckling in more detail. Consider a simply supported, but laterally unsupported (except at the ends of the beam) beam of 'short-span' subjected to incremental transverse loads at its mid-section as shown in Fig. 10.2(a). The beam will deflect downwards, i.e., in the direction of the load [Fig. 10.2(b)].

The direction of the load and the direction of movement of the beam are the same. This is similar to a short column under axial compression. A 'long-span' beam [see Fig. 10.3(a)], when incrementally loaded will first deflect downwards, and when the load exceeds a particular value,

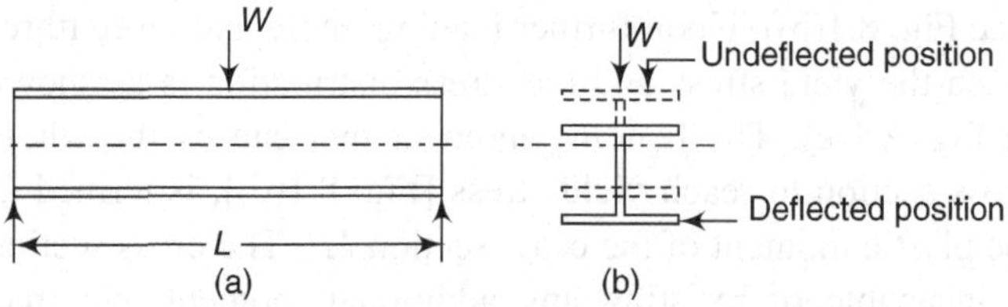


**FIG. 10.2** (a) Short span beam (b) Vertical deflection of the beam

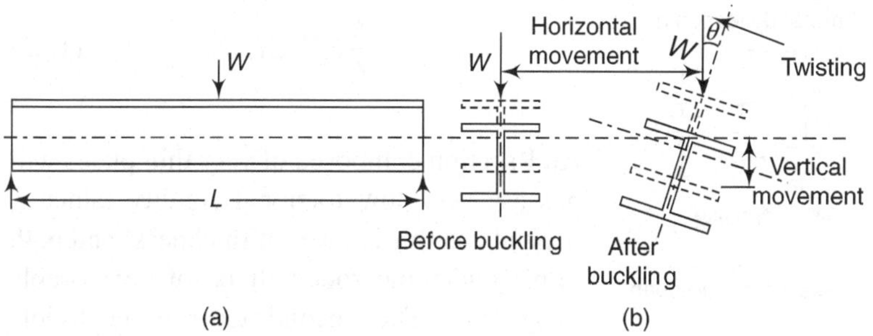


**FIG. 10.3** (a) Long span beam (b) Laterally deflected shape of the beam

it will tilt sideways due to instability of the compression flange and rotate about the longitudinal axis [Fig. 10.3(b)]. The three positions of the beam cross section shown in Fig. 10.3(b) illustrate the displacement and rotation that take place as the mid-section of the beam undergoes lateral torsional buckling. The characteristic feature of lateral buckling is that the entire cross section rotates as a rigid body without any cross-sectional distortion. This behaviour is very similar to an axially compressed long column, which after initial shortening in the axial direction, deflects laterally when it buckles.

In the case of axially loaded columns, the deflection takes place sideways and the column is said to buckle in a pure flexural mode. A beam, under transverse loads, has a part of its cross section in compression and the other in tension. The part under compression becomes unstable while the tensile stresses elsewhere tend to stabilize the beam and keep it straight. Thus, beams when loaded exactly in the plane of the web, at a particular load, will fail suddenly by deflecting sideways and then twisting about its longitudinal axis (Fig. 10.4). This form of instability is more complex (compared to column instability) since the *lateral buckling* problem is three-dimensional in nature. It involves coupled lateral deflection and twist, i.e., when the beam deflects laterally, the applied moment exerts a torque about the deflected longitudinal axis, which causes the beam to twist. The bending moment at which a beam fails by lateral buckling when subjected to a uniform end moment is called its *elastic critical moment* ($M_{cr}$). In the case of lateral buckling of beams, the elastic buckling load provides a close upper limit to the load carrying capacity of the beam. It is clear that lateral instability occurs only if the following two conditions are satisfied.

1. The section possesses different stiffness in the two principal planes.
2. The applied loading induces bending in the stiffer plane (about the major axis).

Similar to columns, the lateral buckling of unrestrained beams is also a function of its slenderness. Several factors influence the buckling capacity of beams. Some of them are cross-sectional shape of the beam, type of support, type of loading, and position of applied load.

## 10.3.1 Lateral Torsional Buckling of Symmetric Sections

As explained earlier, when a beam fails by lateral torsional buckling, it buckles about its weak axis, even though it is loaded in the strong plane. The beam bends about its strong axis up to the critical load at which it buckles laterally [see Figs 10.5(a) and (b)].

For the purpose of this discussion, the lateral torsional buckling of an I-section is considered with the following assumptions.

1. The beam is initially undistorted.
2. Its behaviour is elastic (no yielding).
3. It is loaded by equal and opposite end moments in the plane of the web.
4. The loads act in the plane of the web only (there are no externally applied lateral or torsional loads).
5. The beam does not have residual stresses.
6. Its ends are simply supported vertically and laterally.

Obviously, in practice, these ideal conditions are seldom met. For example, rolled sections invariably contain residual stresses. The effects of the deviations from the ideal case are discussed in a later section.

The differential equation for the angle of twist for the beam shown in Fig. 10.5 is

$$EI_w \frac{d^4\phi}{dx^4} - GI_t \frac{d^2\phi}{dx^2} - \frac{M^2}{EI_y}\phi = 0$$

The solution to this differential equation is given by (Bleich 1952; Timoshenko & Gere 1961; Salmon & Johnson 1996)

$$M_{cr} = (\pi/L)\sqrt{EI_y GI_t + \left(\frac{\pi E}{L}\right)^2 I_w I_y} \quad (10.1a)$$

where $EI_y$ is the minor axis flexural rigidity, $GI_t$ is the torsional rigidity, and $EI_w$ is the warping rigidity. Equation (10.1a) is the elastic

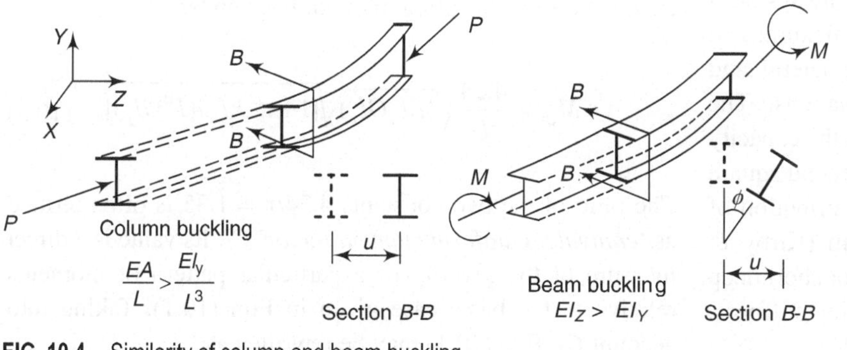


**FIG. 10.4** Similarity of column and beam buckling

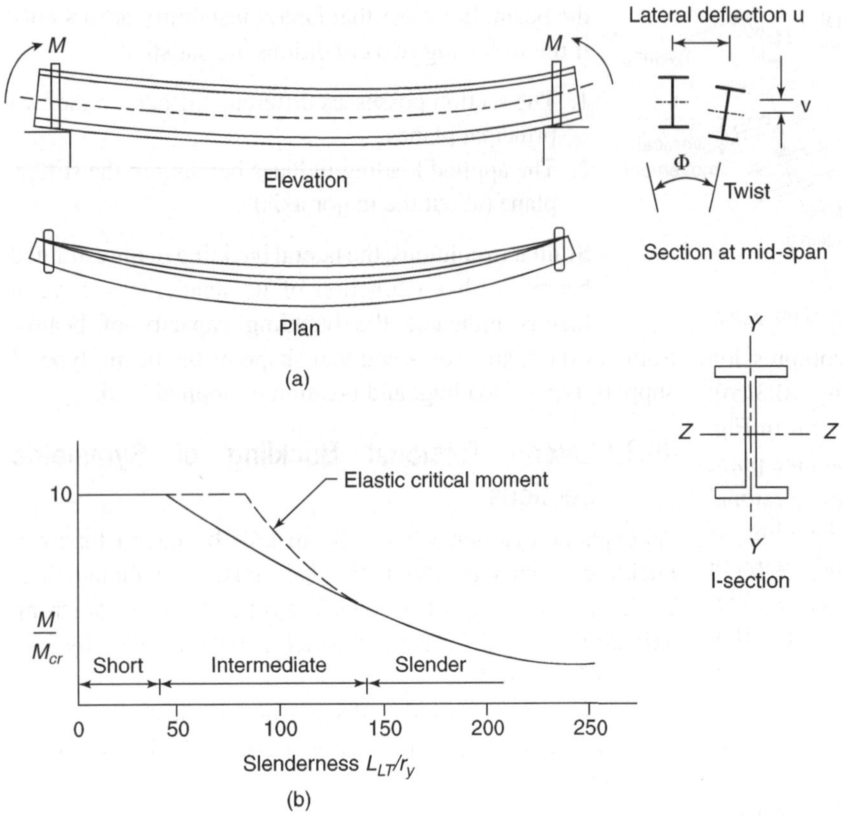


**FIG. 10.5** (a) Lateral torsional buckling of I-section beams (b) Buckling behaviour

lateral-torsional buckling strength of an *I*-shaped section under the action of constant moment in the plane of the web, over the laterally unbraced length *L*.

Thus, critical bending moment capacity attained by a symmetric I-beam subjected to equal end moments undergoing lateral torsional buckling between points of lateral or torsional support is a function of two torsional characteristics of the specific cross section: the pure torsional resistance under uniform torsion and the warping torsional resistance.

Equation (10.1a) may be rewritten as

$$M_{cr} = \frac{\pi}{L}(EI_y GI_t)^{0.5}\left[1 + \frac{\pi^2 EI_w}{L^2 GI_t}\right]^{0.5} \qquad (10.1b)$$

The torsion that accompanies lateral buckling is always non-uniform. It is evident from Eqn (10.1a) that the flexural and torsional stiffness of the member relate to the lateral and torsional components of the buckling deformations. The magnitude of the first term in Eqn (10.1b) relates the capacity of the shape to resist St Venant torsion and the second square root term in Eqn (10.1b) is a measure of the contribution of warping to the torsional resistance of the beam (Kirby & Nethercot 1979). In practice, this value is large for short deep girders. For long shallow girders with low warping stiffness, $I_w \approx 0$ and Eqn (10.1b) reduces to

$$M_{cr} = \frac{\pi}{L}(EI_y GI_t)^{0.5} \qquad (10.2)$$

An I-section composed of very thin plates will possess very low torsional rigidity (since $I_t$ depends on third power of thickness) and both terms under the root will be of comparable magnitude. The second term is negligible compared to the first for the majority of hot-rolled sections. But light gauge sections derive most of the resistance to torsional deformation from the warping action. The beam length also has considerable influence upon the relative magnitudes of the two terms as shown in the term $\pi^2 EI_w/L^2 GI_t$. Shorter and deep beams ($\pi^2 EI_w/L^2 GI_t$ term will be large) demonstrate more warping resistance, whereas, the term will be small for long and shallow beams (Kirby & Nethercot 1979).

It is also found that lateral instability is possible if

1. the section possesses different stiffness in two principal planes
2. the applied loading causes bending in the stiffer plane

Thus, the cross-sectional shape is a particularly important parameter in assessing the lateral buckling capacity of a beam. Conversely, the problem of lateral instability can be minimized or even eliminated by a judicious choice of the section as shown in Table 10.3. Although the five cross sections shown are having the same cross-sectional area, the values of their flexural and torsional properties relative to those of the square cross section show considerable variation. It is observed from Table 10.3 that the flat and deep I-sections, which have largest in-plane bending stiffness, have less buckling strength compared to the box section (Kirby & Nethercot 1979). The effect of cross section is further discussed in Section 10.4.1.

For a beam with central load acting at the level of centroidal axis, the critical buckling moment is given by

$$M_{cr} = \frac{4.24}{L}\sqrt{(EI_y GI_t)}\sqrt{[1 + \pi^2 EI_w/(L^2 GI_t)]} \qquad (10.3)$$

The ratio of the two constants $4.24/\pi = 1.35$ is often termed as '*equivalent uniform moment factor*' $C_1$. Its value is a direct measure of the severity of a particular pattern of moments relative to the basic case given in Eqn (10.1). Taking into account $C_1$, Eqn (10.1) may be rewritten as

**TABLE 10.3** Types of cross-section and their relative values of section properties (Kirby & Nethercot 1979)

| Section type | Square | Flat | Shallow beam (typical) | Deep beam (typical) | RHS (typical) |
|---|---|---|---|---|---|
| Section properties | | | | | |
| $A$ | 1 | 1 | 1 | 1 | 1 |
| $I_z$ | 1 | 25 | 12.45 | 45.59 | 16.94 |
| $I_y$ | 1 | 0.04 | 3.20 | 3.20 | 8.10 |
| $I_t$ | 1 | 0.04 | 0.034 | 0.033 | 4.731 |

$$M_{cr} = C_1(EI_y GI_t)^{1/2}\left[\frac{\pi}{L}\left(1+\frac{\pi^2}{B^2}\right)^{1/2}\right] \tag{10.4}$$

where $B^2 = L^2\, G\, I_t/EI_w$ or

$$M_{cr} = C_1(EI_y\, GI_t)^{1/2}\, \gamma \tag{10.5}$$

where $\gamma = \pi/\ell\,(1+\pi^2/B^2)^{1/2}$

Equation (10.5) is a product of three terms: the first term $C_1$ varies with the loading and support conditions; the second term varies with the material properties and the shape of the beam; and the third term $\gamma$ varies with the length of the beam. This equation is regarded as the basic equation for lateral torsional buckling of beams. The influence of the three terms mentioned earlier is discussed in the following section.

## CASE STUDY

### Lateral Torsional Buckling Failure of Marcy Bridge

A pedestrian footbridge near Marcy, New York, collapsed during construction on 12 October 2002, killing one and injuring nine workers. The bridge was designed as a single span composite box girder bridge consisting of a trapezoidal steel tub girder with stay-in-place forms and concrete deck. The Marcy Bridge was designed to span approximately 52 m across an extension of the Utica-Rome expressway (NYS Route 49). At the time of collapse, the concrete deck was being placed, starting at the north end and moving south, using an automatic screeding machine to distribute the concrete. Eyewitnesses reported that the bridge had been noticeably 'bouncy' while the concrete was placed, and when the concrete placement had progressed to about mid-span, the bridge suddenly twisted and rolled to the east, tipping off the abutments.

Prior to and during the placement of the concrete deck, the tub cross-section was closed only by stay-in-place forms, which were nominally connected to closure angles on each top flange and to each other by using self-tapping screws (see figure). The stay-in-place forms do not provide sufficient torsional strength or stiffness—thus the section is open and torsionally flexible. Computer analysis conducted by Exponent Engineering and Scientific Consulting confirmed that the tub girder did not have sufficient strength to carry the wet concrete without collapse. The mode of failure was considered as lateral torsional buckling of the entire girder cross-section. Such bridges are particularly susceptible to this type of failure, because they are very flexible in a twisting mode before the concrete deck hardens. (After curing, the concrete deck would have provided a closed section with increased torsional stiffness and would have prevented the lateral torsional buckling failure.)

The standard practice at that time, as recorded in bridge design specifications and national design standards, did not recognize this failure mode, and there were no requirements in those codes or standards to provide bracing, falsework, or other measures to prevent such a collapse. After the collapse, and a very similar accident in Sweden that occurred just months before this, bridge design standards in New York and the American Association of State Highway and Transportation Officials (AASHTO) were updated with provisions intended to prevent further accidents. Now these codes explicitly require diagonal bracing between the top-flanges in order to suppress this buckling mode.

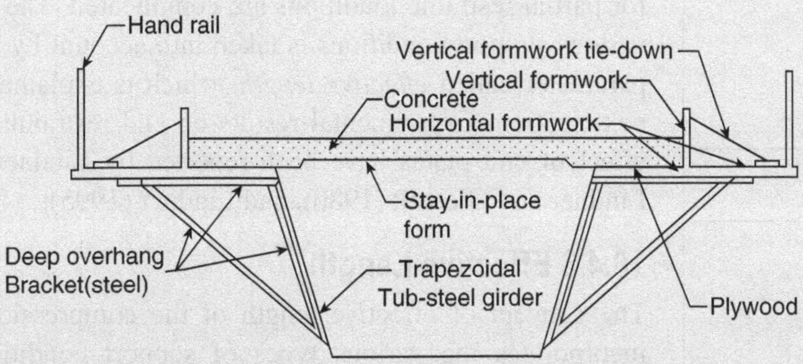

Cross-section of Marcy Bridge and the failure due to lateral torsional buckling (*Source*: Corr, et al. 2009, Peraza 2008.)
*Courtesy*: Structure magazine ASCE

## 10.4 FACTORS AFFECTING LATERAL STABILITY

Equation (10.1) is subject to the same limitations as is the Euler load $P_{cr}$ for struts, i.e., it represents the elastic critical moment for a perfect member. Thus, the elastic critical moment $M_{cr}$ as obtained in the previous section, is applicable only to a beam of I-section which is simply supported and subjected to end moments. In practical situations, support conditions, beam cross section, loading, etc., vary from the basic case and affect the lateral torsional buckling behaviour (Subramanian 1983). The following sections elaborate on these variations and make the necessary modifications to the basic case for design purposes.

### 10.4.1 Influence of Cross-sectional Shape on Lateral Torsional Buckling

Structural sections are generally made up of either open or closed sections. Examples of open and closed sections are shown in Fig. 10.6.

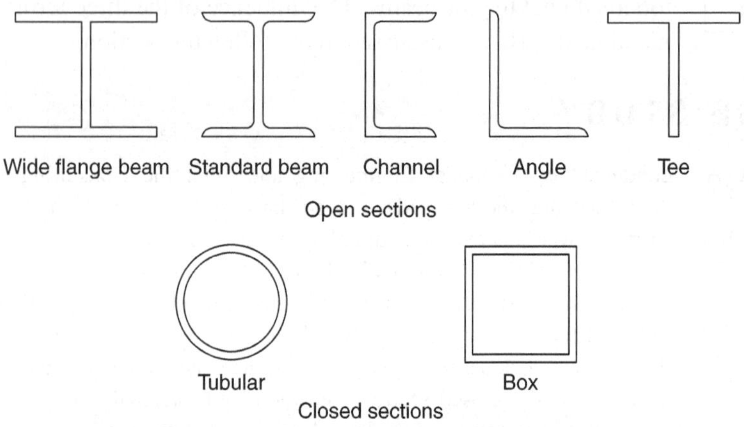


**FIG. 10.6** Cross-sectional shape which may affect lateral torsional buckling

The shape of the cross section of a beam is a very important parameter while evaluating its lateral buckling capacity. In other words, lateral instability can be reduced or even avoided by choosing appropriate sections. The effect of cross-sectional shape on lateral instability is illustrated in Table 10.3 and Fig. 10.7 for different types of section with same cross-sectional area. The figure shows that the I-section with the larger in-plane bending stiffness does not have matching stability. In contrast,

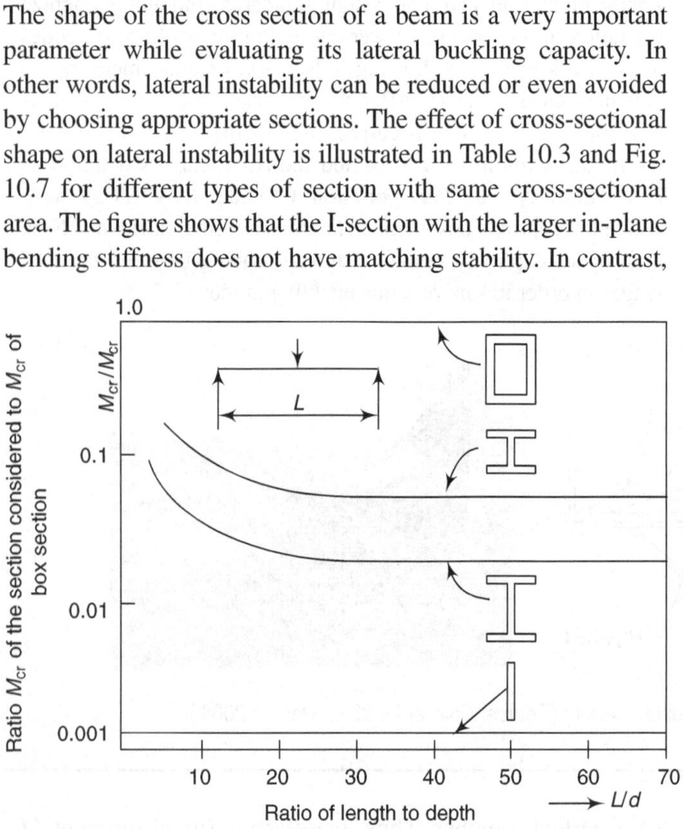


**FIG. 10.7** Effect of the type of cross section on buckling strength (Kirby & Nethercot 1979)

closed sections such as tubes, boxes, and solid shafts have high torsional stiffness, often as high as 100 times that of an open section. The hollow circular tube is the most efficient shape for torsional resistance, but is rarely employed as a beam element because of the difficulties encountered in connecting it to other members and lesser efficiency as a flexural member. I-sections are commonly used due to their easy availability and ease of connections. Box sections may be advantageously used as crane girders where the laterally unsupported length may be large.

In open cross sections (I and channel), a major part of the material is distributed in the flanges, i.e., away from their centroids, to improve their resistance to in-plane bending stresses. It is also convenient to connect such sections to adjacent members. In an ideal case, where the beams are assumed to be restrained laterally, their major axis bending strength is the principal design consideration. Though they possess high major axis bending strength, they are relatively weak in their minor axis bending and twisting. Thus, the use of open sections implies the acceptance of low torsional resistance inherent in them. From a conceptual point of view, the beam with an open cross section has to be regarded as an element having an enhanced tendency to fall over on its weak axis.

### 10.4.2 Support Conditions

The lateral restraint provided by the simply supported conditions assumed in the basic case is the lowest and therefore $M_{cr}$ is also the lowest. It is possible, by other restraint conditions, to obtain higher values of $M_{cr}$, for the same structural section, which would result in better utilization of the section and thus save material. As lateral buckling involves three kinds of deformations, namely *lateral bending*, *twisting*, and *warping*, it is feasible to think of various types of end conditions. But, the supports should either completely prevent or offer no resistance to each type of deformation. Solutions for partial restraint conditions are complicated. The effect of various support conditions is taken into account by way of a parameter called *effective length,* which is explained, in the next section. Experimental results on end restraints and the effect of end plates have been reported by Lindner (1980), Lindner and Gietzelt (1980), and Lindner (1995).

### 10.4.3 Effective Length

The concept of effective length of the compression flange incorporates the various types of support conditions. For the beam with simply supported end conditions and no intermediate lateral restraint, the effective length is equal to the actual length between the supports. When a greater amount of lateral and torsional restraints is provided at the supports, the effective length is less than the actual length and alternatively, the length becomes more when there is less restraint.

The effective length factor would indirectly account for the increased lateral and torsional rigidities provided by the restraints. Thus, there are two $K$ factors, $K_b$ (restraint against lateral bending) and $K_w$ (restraint against warping). The values of $K_b$ and $K_w$ are 1.0 for free bending and free warping (simply supported) and 0.5 for bending and warping prevented (fixed) support condition (Kirby & Nethercot 1979). The values of $K_b$ and $K_w$ vary with the proportions of beams and the accurate assessment of the degree of restraint provided by practical forms of connection is difficult. Hence in the code, this problem is treated in an approximate way by considering $K_b = K_w$ and treating the effective length as $L = KL$ (Kirby & Nethercot 1979). The effective lengths $KL$ of the compression flange for different end restraints according to IS 800 : 2007 are given in Table 10.4 (see also Table 15 of the code, which is based on BS 5950–1 : 2000).

**TABLE 10.4** Effective length of compression flanges

| **Effective length *KL* for beams, between supports** | |
|---|---|
| **Condition at supports** | **Effective length, *KL*** |
| Compression flange at the ends unrestrained against lateral bending (free to rotate in plan) | $L$* |
| Compression flange partially restrained against lateral bending (Partially free to rotate in plane at the bearings) | $0.85L$* |
| Compression flange restrained fully against lateral bending (rotation fully restrained in plan) | $0.75\,L$* |

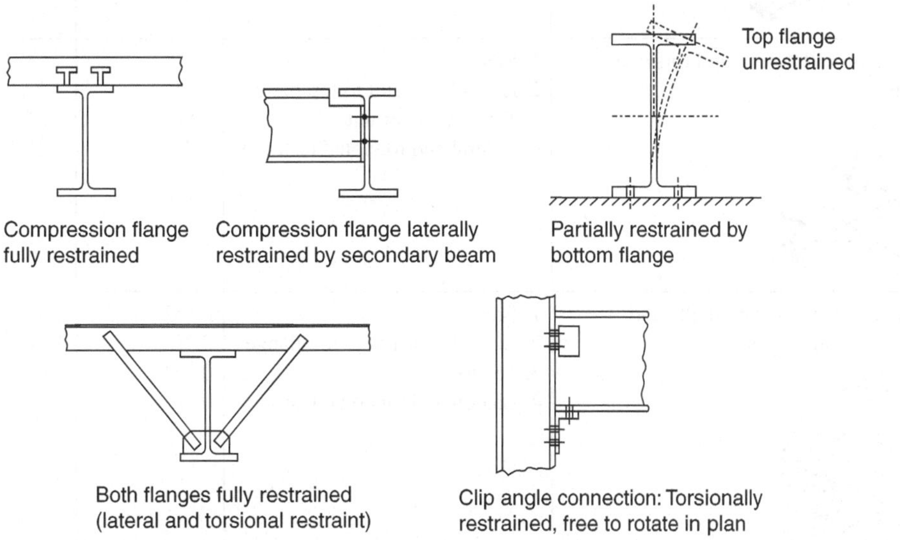

Restrained conditions at supports–see also Table 15 of the code

*When the ends of the beam are not restrained against torsion, or where the loading condition is destabilizing or when flanges are free to move laterally, these values have to be increased as per Table 15 of the code.

Assume a simply supported condition for beams and girders with span length $L$ and no lateral restraint of the compression flanges along the length, except torsional restraint at the ends of the beam. Restraint against torsion can be provided by (a) web or flange cleats, (b) bearing stiffeners acting together with the bearing of the beam, (c) lateral end frames or external supports to the ends of the compression flanges, and (d) the beam built into the supporting walls (Ojalvo & Chambers 1977).

## *Cantilever Beams*

Solutions in the form of tables, charts, and approximate formulae for the lateral-torsional buckling of cantilevers, including transverse loads applied at the level of either flange are available (Timoshenko & Gere 1961; Nethercot 1973b; Nethercot 1983; Trahair et al. 2001). Though the upper flange of the cantilever is in tension, it deforms more in the buckled shape and the lateral stability is further reduced, when the load is applied to the top flange of a cantilever, possessing relatively low vales of $L^2GI_t/EI_w$. For cantilevers, the most severe loading condition is the point load acting at the tip. Nethercot (1983) has shown that for most applications, the simple effective length method (similar to that used for struts) is satisfactory and the critical buckling moment is given by

$$M_{cr} = \frac{\pi}{KL}\sqrt{EI_y GI_t}\sqrt{\left[\left(1+\frac{\pi^2 EI_w}{(KL)^2 GI_t}\right)\right]} \qquad (10.6)$$

The effective length factors for various restraint conditions at the tip and at the root of the cantilever are given in Table 10.5.

**TABLE 10.5** Effective length $L_E$ for cantilevers without intermediate restraint (adapted from BS 5950-1:2000)

| Restraint conditions | | Loading conditions | |
|---|---|---|---|
| **At support** | **At tip** | **Normal** | **Destabilizing (top flange loading)** |
| (a) Continuous, with lateral restraint to top flange 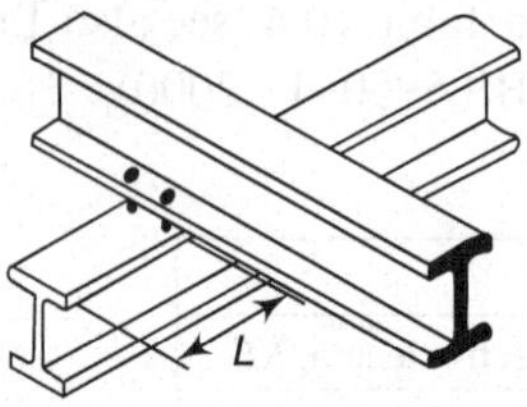 | 1. Free<br>2. Lateral restraint to top flange<br>3. Torsional restraint<br>4. Lateral and torsional restraint | 3.0*L*<br>2.7*L*<br>2.4*L*<br>2.1*L* | 7.5*L*<br>7.5*L*<br>4.5*L*<br>3.6*L* |
| (b) Continuous, with partial torsional restraint 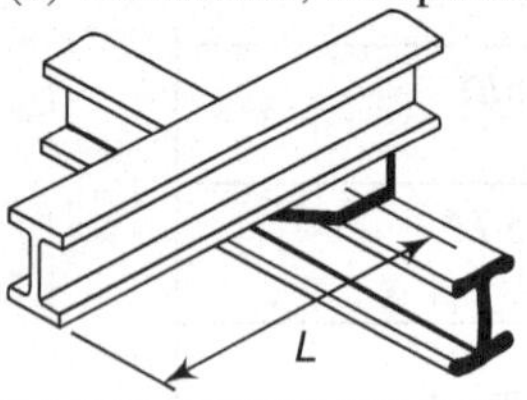 | 1. Free<br>2. Lateral restraint to top flange<br>3. Torsional restraint<br>4. Lateral and torsional restraint | 2.0*L*<br>1.8*L*<br>1.6*L*<br>1.4*L* | 5.0*L*<br>5.0*L*<br>3.0*L*<br>2.4*L* |
| (c) Continuous, with lateral and torsional restraint 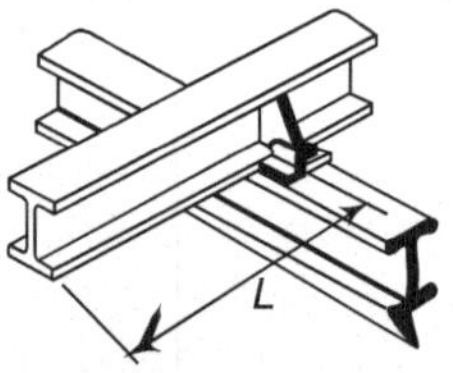 | 1. Free<br>2. Lateral restraint to top flange<br>3. Torsional restraint<br>4. Lateral and torsional restraint | 1.0*L*<br>0.9*L*<br>0.8*L*<br>0.7*L* | 2.5*L*<br>2.5*L*<br>1.5*L*<br>1.2*L* |
| (d) Restrained laterally, torsionally and against rotation on plane 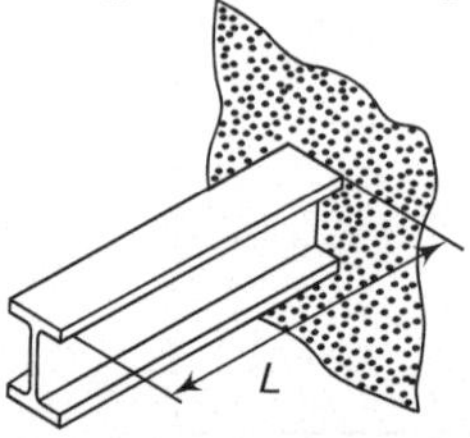 | 1. Free<br>2. Lateral restraint to top flange<br>3. Torsional restraint<br>4. Lateral and torsional restraint | 0.85*L*<br>0.7*L*<br>0.75*L*<br>0.5*L* | 1.4*L*<br>1.4*L*<br>0.6*L*<br>0.5*L* |

| Tip restraint conditions | | |
|---|---|---|
| 1. Free 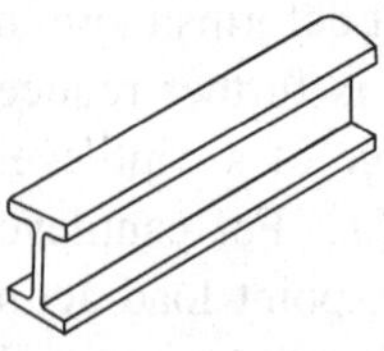 (not braced on plan) | 2. Lateral restraint to top flange 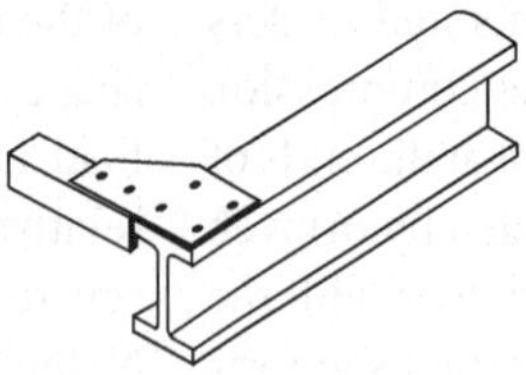 (braced on plan in at least one bay) | 3. Torsional restraint 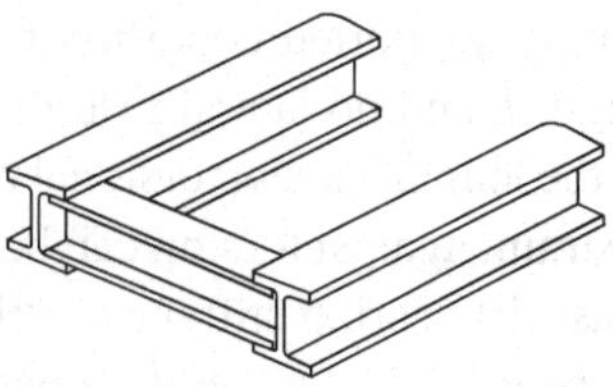 (not braced on plan) |
| 4. Lateral and torsional restraint 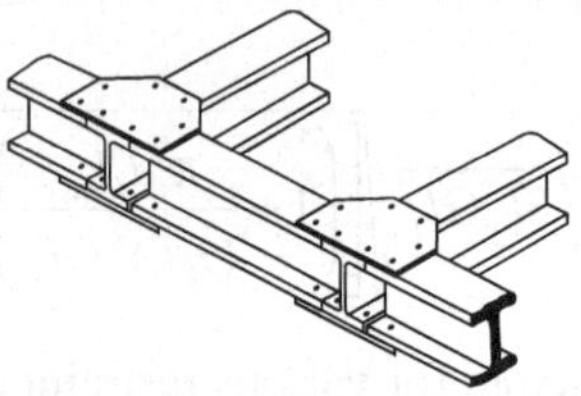 (braced on plan in at least one bay) | | |

Note that in IS 800 : 2007, in case (d), for free end and for torsional restraint, the values are given as $0.80L$ and $0.60L$ as against the values of $0.85L$ and $0.75L$ proposed by BS 5950–1 : 2000. These $K$ values include the effect of uniform loading, point loading, varying ratios of back span length to cantilever length, varying support conditions for the 'far' end of the supported portion of the beam, etc.; the values for the worst case only are presented in Table 10.5. Though other more exact methods are available (e.g., Bo Dowswell 2004), they are a bit too complex and do not cover enough cases.

Kitipornchai et al. (1984) studied the efficiency of bracing types and location along the span. They found that for discrete bracing, the most effective location is the top (tension) flange and as close to the cantilever tip as possible. They also found that for cantilever fully restrained at support, lateral restraints placed within 40% of the span are practically useless.

### *Intermediate Braces*

Provision of intermediate lateral supports (bracings) can increase the lateral stability of a beam. For the bracings to be effective, the designer should make certain that the braces themselves are prevented from moving in their axial direction. For example, for the beam shown in Fig. 10.8(a), the lateral buckling of the beams will not be prevented by the braces, and hence the unbraced length should be taken as $L$. To prevent this type of system-buckling, the bracings should be either anchored into a wall [Fig. 10.8(b)] or provided with diagonal bracings, which effectively transfer the loads to the columns

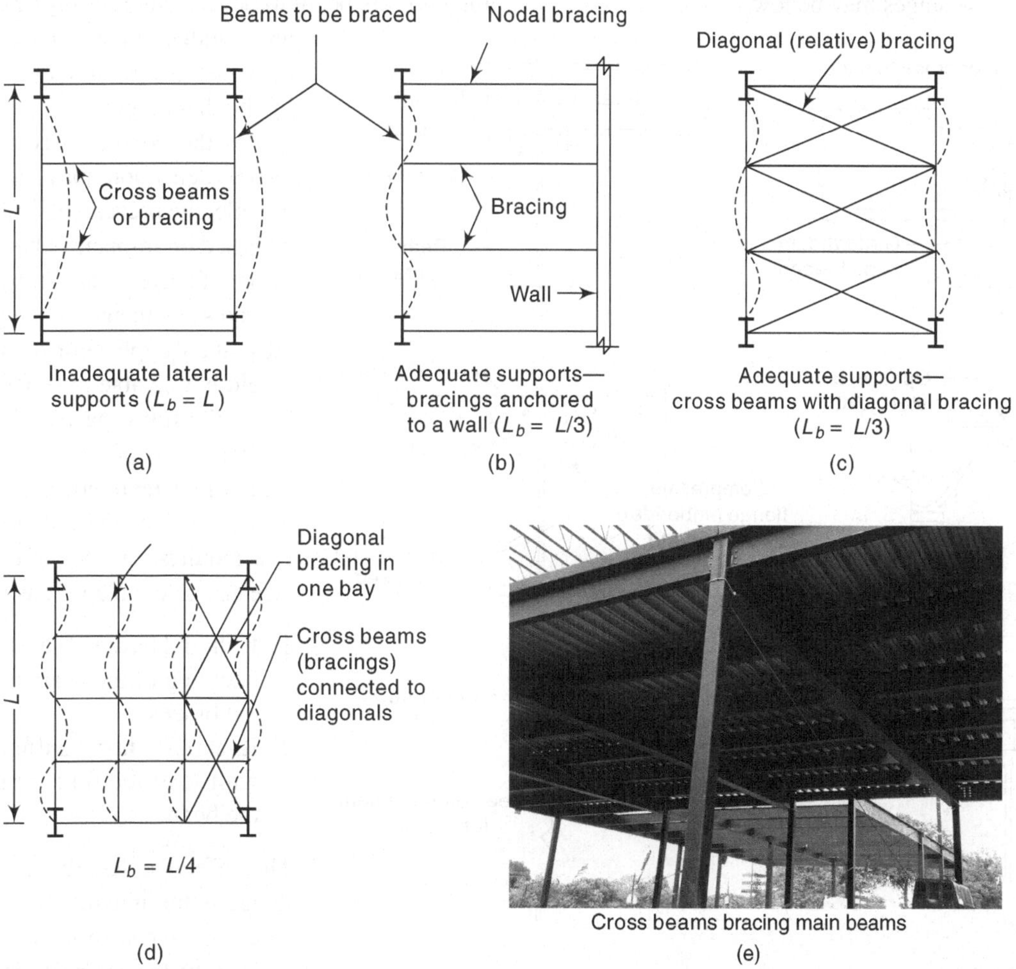


**FIG. 10.8** Lateral bracing systems for roof and floor bems

[Fig. 10.8(c)]. It has to be noted that the diagonal bracing need not be provided in all the bays [see Fig. 10.8(d)]. Generally, even a light bracing has the ability to provide substantial increase in stability. Lateral bracing may be either discrete (e.g., cross beams) or continuous [(e.g., beam encased in concrete floors) (see Fig. 10.9)].

The design engineer has to exercise his/her judgement to decide what constitutes a satisfactory lateral support (bracing) for a steel beam. For the girder in Fig. 10.9(a), the bracing having connections of appreciable depth and having sufficient stiffness may be adequate to prevent the twisting of the beam. Note that the connection should be near the compression flange of the beam/girder, to be effective. Lateral bracings may also be effective, when bracings are resting on the top flange of the beam and connected to the beam by bolts [see Fig. 10.9(b)] or when open web joists are welded or bolted to the top flange [see Fig. 10.9(c)]. Concrete slab incorporated in the compression flange [see Fig. 10.9(d)] or a concrete slab connected by means of shear connectors [see Fig. 10.9(e)] provides continuous and effective lateral bracing to the beam. However, when a concrete slab rests on the top flange of a beam, the engineer must study the situation carefully before counting on friction to provide full lateral support. Such friction may not be available in some situations, for example due to traffic in bridges and vibrating machinery in buildings. It is also important to check the adequacy of beams, during the intermediate stages of construction, as the amount of bracing provided during these stages may be low.

The corrugated or trapezoidal sheet-metal roofs, which are usually connected to the purlin with metal strips or by puddle welds at intervals may furnish only partial lateral support [see Fig. 10.9(f)]. Using *diaphragm action* of the adequately attached metal deck can considerably reduce the stiffness requirement for point bracing. Metal deck with about 60-mm thick concrete fill will provide many times the required stiffness. It has been found that the diaphragm action of the metal deck without the concrete slab may provide more than 1300 kN/m. More details about considering the diaphragm bracing action of metal decking is provided by Nethercot and Trahair (1975), Errera and Apparao (1976a, 1976b), Davies (1976), and Easley (1977). Lindner (1972) and Nethercot and Trahair (1975) have proposed techniques for checking whether a diaphragm is capable of providing full bracing and allowing the beam to develop its full in-plane capacity. Vogel and Lindner (1980) have given tables to calculate the rotational restraint provided by the diaphragm.

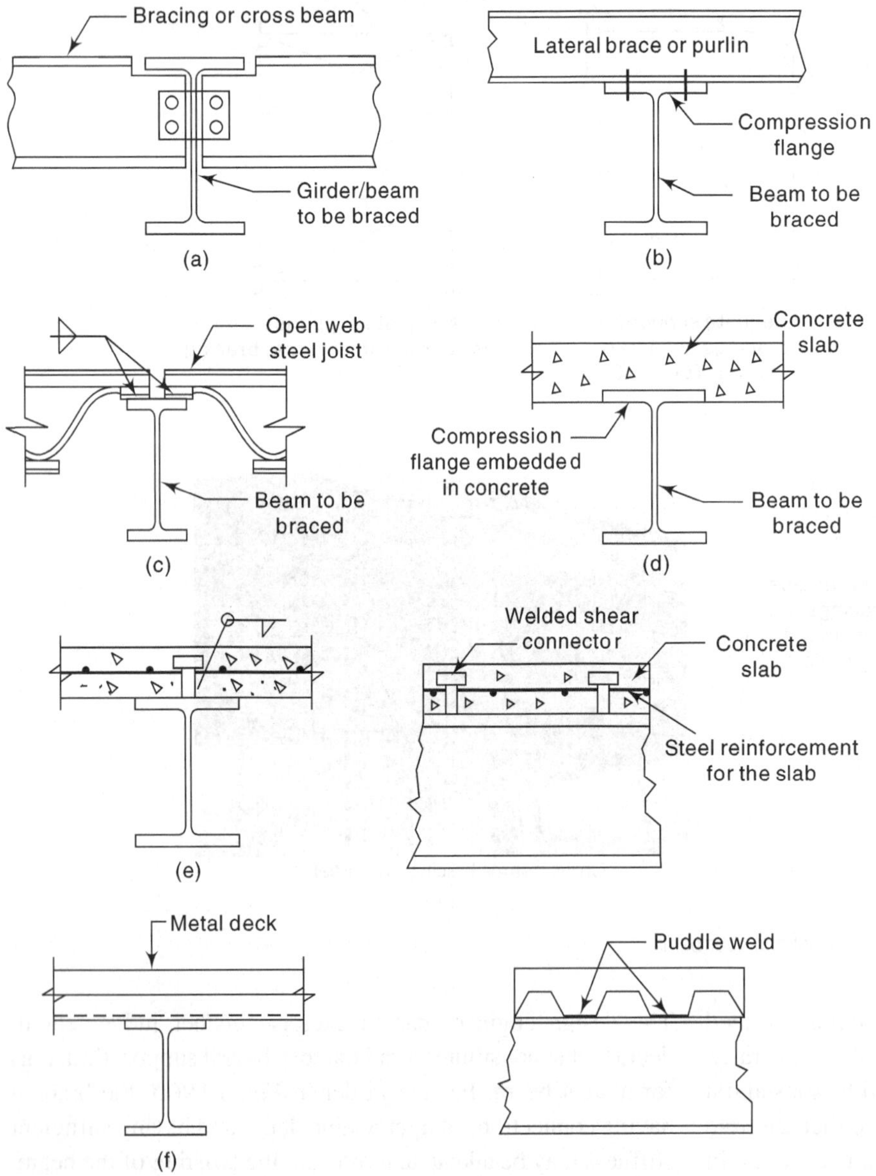


**FIG. 10.9** Examples of discrete and continuous lateral bracing of beams

For the continuously restrained beams, lateral deflection may be assumed to be completely prevented, and the design can be based on in-plane behaviour. For bracing to be effective, it has to be provided in the compression flange. Bracing provided below the point of application of the transverse load would not be able to resist the twisting and hence the full capacity of the beam cannot be achieved (Kirby & Nethercot 1979). Similarly lateral bracing system, attached near the beam centroid are ineffective. The two requirements for effective bracing are (see also Section 9.9.2) as follows.

1. It should have sufficient stiffness so that buckling of the beam occurs in between the braces.
2. It should have sufficient strength to withstand the force transferred to it by the beam.

The effective length of the compression flange is the maximum distance centre-to-centre of the restrained members. In the code only the strength aspect is covered. Thus it states that the effective lateral restraints for a beam should be capable of resisting 2.5% of the maximum force in the compression flange taken as divided equally between the numbers of points at which the restraint members occur.

Based on the research done by Winter (1960) and Yura (2001), the

American code (ANSI/AISC: 360-2010) suggests the strength and stiffness of lateral bracing as given in Table 10.6.

**TABLE 10.6** Bracing requirements

| Requirement | Relative | (Nodal) Discrete |
|---|---|---|
| Stiffness $k_{br}$, N/mm | $\left(\frac{1}{0.75}\right)\left(\frac{4M_uC_d}{L_bh_o}\right)$ | $\left(\frac{1}{0.75}\right)\left(\frac{10M_uC_d}{L_bh_o}\right)$ |
| Strength $F_{br}$, N | $\frac{0.008M_uC_d}{h_o}$ | $\frac{0.02M_uC_d}{h_o}$ |

$M_u$ = Required flexural strength, Nmm
$h_o$ = Distance between flange centroids, mm
$C_d$ = 1.0 for single curvature, 2.0 for double curvature
$L_b$ = Laterally unbraced length, mm

Lateral bracing should be connected to both flanges at the brace point nearest the inflection point for beams subjected to double curvature bending along the length to be braced. In members subjected to double curvature bending, the inflection point should not be considered as brace point (Galambos 1998). Two general types of bracings, as mentioned in Table 10.6 are considered. A *relative brace* controls the movement of the brace point with respect to adjacent braced points (truss system of bracing). A nodal or *discrete brace* controls the movement at the braced point without direct interaction with adjacent braced points. For the nodal/discrete brace to be effective, it should be anchored in a wall or other stiff member [see Fig. 10.8(b)].

**Torsional bracings** Torsional bracing may be provided in the form of cross frames or diaphragms at discrete locations or continuous bracing in the form of metal decks and slabs. The factors that have significant effect on lateral bracing, such as the number of bracings, top flange loading and brace location on the cross section, were found to be relatively unimportant while sizing a torsional brace (Yura 2001). However, the position of the braces on the cross section has an effect on the stiffness of the brace itself (see Fig. 10.10). The torsional brace attached near the bottom flange will bend in single curvature and the flexural stiffness will be $2EI_b/S$, while the brace attached to the top flange will bend in reverse curvature and the flexural stiffness in this case will be $6EI_b/S$, where $I_b$

Torsional bracings provided for girders in a *bridge at Gaithersburg, MD, USA*

is the moment of inertia of brace and $S$ is the spacing between the girders. The torsional brace stiffness requirements as given in the American code (ANSI/AISC: 360-2005), based on the work of Yura (2001), are as follows:

$$k_{tb} = k_t/[1 - (k_t/k_{sec})] \qquad (10.7)$$

where $k_{tb}$ is the required brace system stiffness, $h_o$ is the distance between flange centroids (in mm), $L$ is the span length (in mm), $E$ is Young's modulus of rigidity (in N/mm²), $n$ is the number of nodal braced points within the span, $I_y$ is the out-of-plane moment of inertia (in mm⁴), $C_1$ is the equivalent uniform moment factor (see Section 10.4.5), $t_w$ is the web thickness of the beam (in mm), $t_s$ is the web stiffener thickness (in mm), $b_s$ is the stiffener width for one sided stiffener (twice the individual stiffener width is taken for a pair of stiffeners) (in mm), and $M_u$ is the required flexural strength. The values of $k_t$ and $k_{sec}$ are as follows:

$k_t$ = brace stiffness excluding web distortion (see Fig. 10.11) in N mm/radian

$$= (1/0.75)\,[2.4LM_u^2/(nEI_yC_1^2)] \qquad (10.8)$$

$k_{sec}$ = web distortional stiffness, including the effect of web transverse stiffeners, if any, N mm/radian

$$= (3.3E/h_o)\,[1.5h_o\,t_w^3/12 + t_sb_s^3/12] \qquad (10.9)$$

Equation (10.7) will result in a negative value when $k_{sec} < k_t$, indicating that torsional beam bracing will not be effective due to web distortional stiffness. For rolled-sections ($h/t_w < 60$), cross-sectional distortion will not be significant if the diaphragm connection extends at least one-half the web depth (Galambos 1998). Note that the web stiffener need not extend from one flange to the other flange of the beam and may be stopped short by a distance equal to $4t_w$ from any beam flange that is not directly attached to the torsional brace.

The required flexural strength of the torsional bracing is given by

$$M_{br} = (0.024M_uL)/(nC_1L_b) \qquad (10.10)$$

where $L_b$ is the distance between braces in mm.

**Continuous torsional bracing** For continuous torsional bracings, Eqn (10.10) may be used with $L/n$ as 1.0. The distortional stiffness of an unstiffened web is given by

$$k_{sec} = 3.3Et_w^3/(12h_o) \qquad (10.11)$$

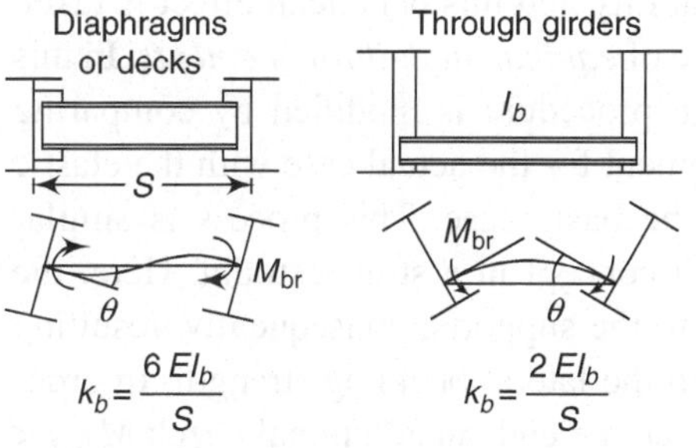


**FIG. 10.10** Diaphragm stiffness

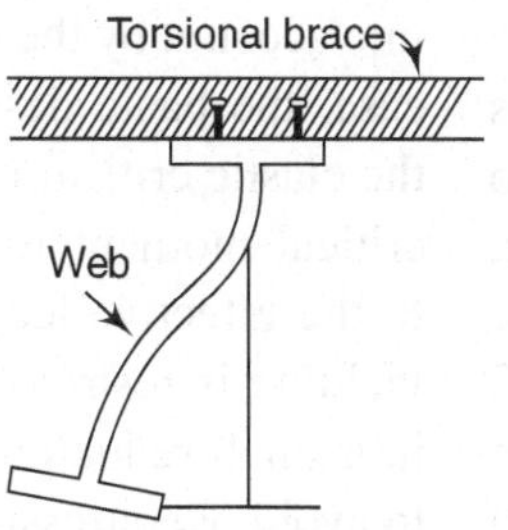


**FIG. 10.11** Web distortion

Equations (10.7) to (10.11) give strength and stiffness requirements for doubly symmetric beams. For singly symmetric beams, these equations will be conservative (ANSI/AISC: 360-2010). The $k_{sec}$ term in these equations accounts for cross-sectional distortion. A web stiffener at the brace point reduces cross-sectional distortion and improves the effectiveness of torsional brace.

Parallel chord trusses with both chords extended to the end of the span and attached to supports can be treated as beams. The value of $M_u$ in these equations may be taken as the maximum chord force times the depth of the truss, to determine the brace strength and stiffness (ANSI/AISC: 360-2010).

### 10.4.4 Level of Application of Transverse Loads

The lateral stability of a transversely loaded beam is dependent on the arrangement of the loads as well as the level of application of the loads with respect to the centroid of the cross section. Figure 10.12 shows a centrally loaded beam experiencing either destabilising or restoring effect when the cross section is twisted.

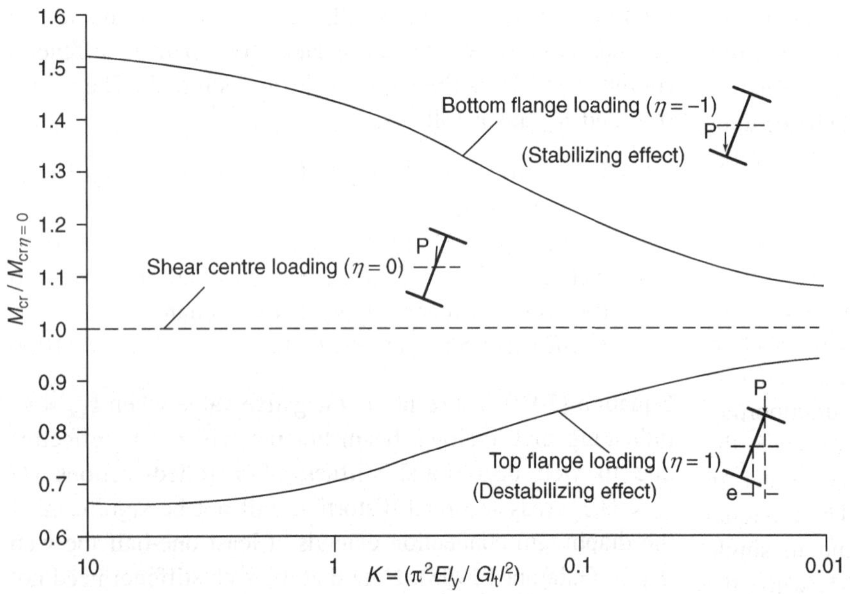


**FIG. 10.12** Effect of level of loading on beam stability

A load applied above the centroid of the cross section causes an additional overturning moment and becomes more critical than the case when the load is applied at the centroid. On the other hand, if the load is applied below the centroid, it produces a stabilising effect. The location of the load application has no effect if a restraint is provided at the load point. For example, IS 800 : 2007 takes into account the destabilising effect of the bottom flange loading by using a notional effective length of 1.2 times the actual span to be used in the calculation of effective length (see Table 15 of the code). Nethercot and Rockey (1972) and Nethercot (1983) provide an equation of the form

**TABLE 10.7** Coefficient to consider effect of level of load application

| Loading | $M$ | $A$ | $B$ |
|---|---|---|---|
| Central load, $P$ | $PL/4$ | 1.35 | $1 - 0.180W^2 + 0.649W$ |
| Uniformly distributed load, $w$ | $wL^2/8$ | 1.12 | $1 - 0.154W^2 + 0.535W$ |
| Point loads at $L_1$ from left and right supports, $L_2$ being distance between loads | $PL_1$ | $1+\left(\dfrac{L_1}{2L_1+L_2}\right)^2$ | $1 - 0.465W^2 + 1.636W$ |

$$W = \frac{\pi}{L}\sqrt{\frac{EI_w}{GI_t}}$$

$$C_{11} = AB^{2y/h} \tag{10.12}$$

to consider the effect of the level of application of the load. The value of $A$ and $B$ are taken from Table 10.7, $h$ is the beam depth and $y$ is the distance from mid-height to the point of load application. The distance $y$ has to be taken as positive for loading applied below mid-height and negative for loads applied above mid-height.

The value of $AB$ may be approximately taken as 1.4 and $C_{11}$ may be multiplied with $C_1$ values given in Section 10.4.5 to obtain critical loads which consider the effect of non-uniform bending and level of application of load. Thus, the load carrying capacity is lower for a beam loaded at the upper flange compared to the case where it is loaded at the lower flange.

### 10.4.5 Influence of Type of Loading

So far, only the basic case of beams loaded with equal and opposite end moments has been considered. But, in reality, loading patterns would vary widely from the basic case. Two reasons for studying the basic case in detail are (a) it is analytically amenable and (b) the loading condition is regarded as the most severe (see Fig. 10.13). Cases of moment gradient, where the end moments are unequal, are less prone to instability and this beneficial effect is taken into account by the use of *equivalent uniform moments*. In this case, the basic design procedure is modified by comparing the elastic critical moment for the actual case with the elastic critical moment for the basic case. This process is similar to the effective length concept in a strut restraint. Here, the yielding is restricted to the supports; consequently, resulting in a small reduction in the lateral buckling strength. In order to avoid overstressing at one end, an additional check $M_{max} < M_p$ should also be satisfied. In certain situations, maximum

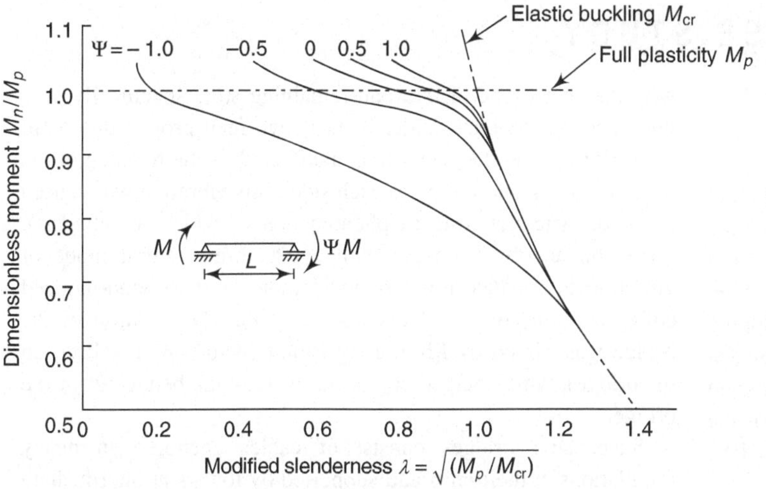


**FIG. 10.13** Inelastic buckling of beams with unequal end moments

moment occurs within the span of the beam. The reduction in stiffness due to yielding would result in a smaller lateral buckling strength.

To take into account the non-uniform bending, a simple modifier is applied to Eqn (10.1) in many codes (Salvadori 1956) as given below

$$M_{cr} = C_1 M_{ocr} \quad (10.13a)$$

where $M_{ocr}$ is the critical moment obtained from Eqn (10.1) and $C_1$ is the *equivalent uniform moment factor*. Various lower-bound formulae have been proposed for $C_1$. Kirby and Nethercot (1979) presented an equation for $C_1$ which can be applied to any shape of moment diagram curve between the brace points. This equation is slightly modified to take into account the cross-section monosymmetry by the ANSI/AISC: 360-05 code as (see Fig. 10.14)

$$C_1 = (12.5M_{max})/(2.5M_{max} + 3M_A + 4M_B + 3M_C]R_m \leq 3.0 \quad (10.13b)$$

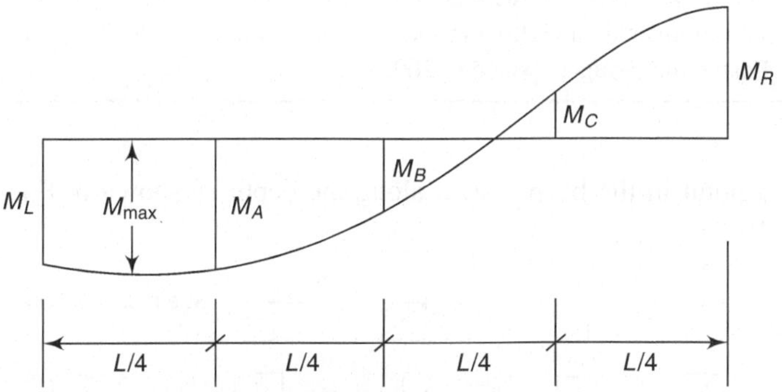


**FIG. 10.14** Shape of bending moment diagram

where $M_{max}$ is the absolute value of maximum moment in the unbraced segment in Nmm, $M_A$ is the absolute value of the moment at quarter point of the unbraced segment in Nmm, $M_B$ is the absolute value of the moment at centre line of the unbraced segment in Nmm, $M_C$ is the absolute value of moment at three quarter point of the unbraced segment in Nmm, $R_m$ is the cross-section monosymmetry parameter = 1.0 for doubly symmetric members, = 1.0 for singly symmetric members subjected to single curvature bending, and = $0.5 + 2(I_{yc}/I_y)^2$, for singly symmetric members subjected to reverse curvature bending. $I_{yc}$ is the moment of inertia about *y*-axis with respect to the compression flange, or for reverse curvature bending, with respect to the smaller flange in mm$^4$ and $I_y$ is the moment of inertia about the principal *y*-axis in mm$^4$.

The value of $C_1$ may be conservatively taken as 1.0 (leading to less economical designs) for all the cases and for cantilevers where the free end is unbraced. It should be noted that the ANSI/AISC code formula does not consider the possibility of $K_b$ being different from $K_w$ (see Section 10.4.3). Moreover, the values of $C_1$ for any support condition are the same regardless of the value of lateral bending and warping coefficients $K$ and $K_w$, respectively (Serna et al. 2006).

The British code (BS: 5950-1-2000) has adopted the following equation (it is called $M_{LT}$ in the code, which is the inverse of the factor $C_1$)

$$C_1 = [M_{max}/(0.2M_{max} + 0.15M_A + 0.5M_B + 0.15M_C)] \leq 2.273 \quad (10.13c)$$

The Australian code (AS: 4100-1998) has adopted an equation as follows:

$$C_1 = (1.7M_{max})/\sqrt{(M_A^2 + M_B^2 + M_C^2)} \leq 2.5 \quad (10.13d)$$

Though this equation is similar to that given in the ANSI/AISC code, it differs in the weightage given to the mid-span and quarter point moments. Recently, Serna et al. (2006) have proposed the following equation

$$C_1 = \sqrt{[35M_{max}^2/(M_{max}^2 + 9M_A^2 + 16M_B^2 + 9M_C^2)]} \quad (10.13e)$$

It has been shown that Eqn (10.13e) improves the results given by the American and British codes for some loading cases. Serna et al. (2006) also provide a general expression, which takes into account the end support conditions.

The Indian code gives various values of $C_1$ in a table in Annex E for different moment diagrams. This table is based on the Eurocode 3. The values of $C_1$ given in Annex E of the code for end moment loading may be approximated by the following equation (Gardner & Nethercot 2005) when $k = 1$

$$C_1 = 1.88 - 1.40\psi + 0.52\psi^2 \text{ but } C_1 \leq 2.70 \quad (10.13f)$$

where $\psi$ is the ratio of the end moments defined in Table 42 of the code. It should be noted that the Table 42 given in the Indian code considers the end rotational restraint against lateral bending $K$ and suggests that $K_w$ values be taken as equal to unity.

## CASE STUDY

### Wind-induced Collapse of Tacoma Narrows Bridge

The original Tacoma Narrows Bridge opened 1 on July 1940, and dramatically collapsed into Puget Sound on November 7 of the same year. This suspension bridge spanned the Tacoma Narrows strait between Tacoma and the Kitsap Peninsula, in the United States. At the time of its construction (and its destruction) Galloping Gertie (nick named due to its oscillations) was the third longest suspension bridge in the world, after the Golden Gate Bridge and the George Washington Bridge. 'Longest' is a comparison of the main spans in suspension bridges—it had a total length of 1810.2 m with a central (longest) span of 853.4 m.

Collapse of the Tacoma Narrows Bridge, Washington state, 1940

The failure of the bridge occurred due to the twisting of the bridge deck in mild winds of about 64 km/h. This failure mode is termed as torsional vibration mode (which is different from the transversal or longitudinal vibration mode). Thus, when the left side of the roadway went down, the right side would rise, and vice versa, with the centre line of the road remaining still. Specifically, it is the 'second' torsional mode. In fact, two men proved this point by walking along the center line, unaffected by the flapping of the roadway rising and falling to each side. This vibration was caused by aeroelastic fluttering (a phenomenon in which aerodynamic forces on an object couple with a structure's natural mode of vibration to produce rapid periodic motion). The wind-induced collapse occurred on 7 November 7 1940. The collapse of the bridge was filmed by Mr Barney Elliott (who was travelling on the bridge), which helped engineers to study the behaviour of the bridge.

Suspension bridges consist of cables anchored in heavy foundations at their ends and supported by towers at intermediate points. The deck is suspended from these cables. Thus they are more flexible than other types of bridges, and require bracing to reduce vertical and torsional motions. In the Tacoma Narrows bridge, instead of the usual deep open trusses, narrow and shallow solid I-beams were used in the decks, which resulted in the build-up of wind loads.

Solid I-beams in the deck of the original bridge and trussed deck of the reconstructed bridge

The bridge collapse boosted research in the field of bridge aerodynamics which resulted in better designs. After the collapse, two bridges were constructed in the same general location. The first one, now called the Tacoma Westbound Bridge, is 1822 m long, 12 m longer than Galloping Gertie. The second one, the Tacoma Eastbound Bridge, opened in 2007.

## 10.5 BUCKLING OF REAL BEAMS

The theoretical assumptions made in Section 10.3 are generally not realised in practice (see Table 10.8). In this section, the behaviour of real beams (which do not meet all the assumptions of the buckling theory) is explained. Effects of plasticity, residual stresses, and imperfections are described in the following sections.

### 10.5.1 Plasticity Effects

Initially, the case where buckling is not elastic is considered. All other assumptions hold good. As the beam undergoes bending under applied loads, the axial strain distribution at a point in the beam varies along the depth as shown in Fig. 10.15.

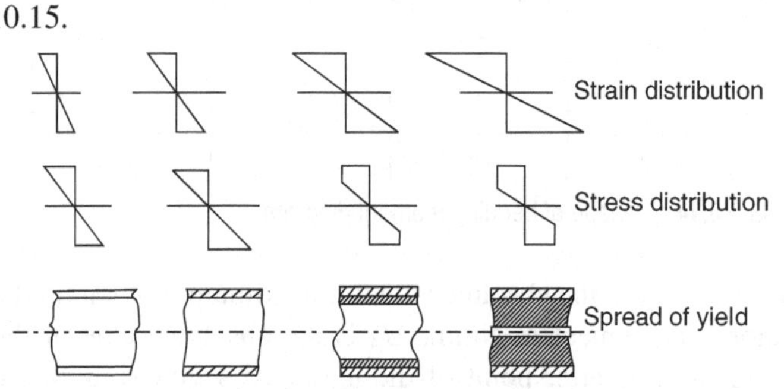


**FIG. 10.15** Plasticity distribution on cross section

**TABLE 10.8** Imperfections present in real beams (Nethercot 1974)

| Theoretical assumptions | Type of imperfection | Conditions in real beams | Comments |
|---|---|---|---|
| Initially perfectly straight undeformed beam | Geometrical | Initially deformed, initial bow* and initial twist*, deflections in plane of applied load | Neglecting this is unsafe, but its effect is slight in most cases |
| Loading applied in the plane of major axis only | Loading | Non-vertically applied loads*, eccentrically applied loads* | |
| Initially stress free, perfectly elastic, uniform cross section | Structural | Residual stresses present*, section may be partially yielded, variations of yield stress along and across section | Typically lower yield stress for flange than web, also different in tension and compression; effect is usually to err on the safe side |
| | | Variations in cross-sectional dimensions | Has only a slight effect, if the variation is within 2% to 5% |

*Considerable variations in these quantities

With the increase in loading, yielding of the section is initiated at the outer surfaces of the top and bottom flanges. If the $M_{cr}$ of the section as calculated by Eqn (10.1) is less than $M_y$, then the beam buckles elastically. In the case, where $M_{cr}$ is greater than $M_y$, some amount of plasticity is experienced at the outer edges before buckling is initiated. If the beam is sufficiently stocky, the beam section attains its full plastic moment capacity $M_p$. The interaction between instability and plasticity is shown in Fig. 10.16.

There are three distinct regions in the curve as follows.

1. Beams with high slenderness $\left(\sqrt{M_p/M_{cr}} > 1.2\right)$, where the failure of the beam is by elastic lateral buckling at $M_{cr}$.
2. Beams of intermediate slenderness $\left(0.4 < \sqrt{M_p/M_{cr}} < 1.2\right)$, where failure occurs by inelastic lateral buckling at loads below $M_p$ and above $M_{cr}$.
3. Stocky beams $\left(\sqrt{M_p/M_{cr}} < 0.4\right)$, which attain $M_p$ without buckling.

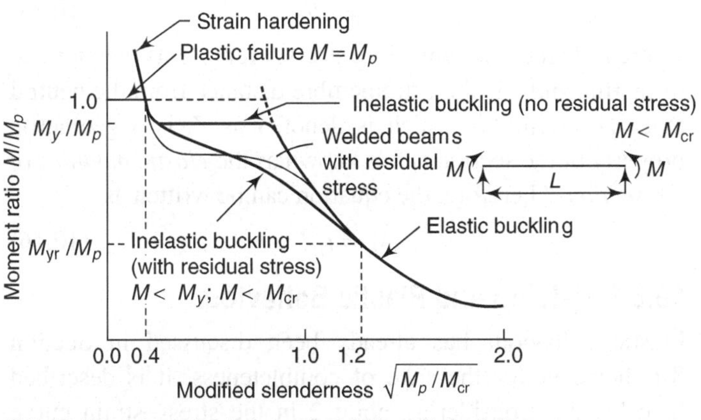


**FIG. 10.16** Interaction between instability and plasticity

## 10.5.2 Residual Stresses

It is normally assumed that a structural section in the unloaded condition is free from stress and strain. In reality, this is not true. During the process of manufacture of steel sections, they are subjected to large thermal expansions resulting in yield level strains in the sections. As the subsequent cooling is not uniform throughout the section, self-equilibrating patterns of stresses are formed. These stresses are known as *residual stresses* (see Section 1.9 of Chapter 1 for more details). Similar effects can also occur at the fabrication stage during welding or flame cutting of sections. A typical residual stress distribution in a hot-rolled steel beam section is shown in Fig. 1.24 of Chapter 1.

Due to the presence of residual stresses, yielding of the section starts at lower moments. Then, with the increase in moment, yielding spreads through the cross section. The inelastic range, which starts at $M_{yr}$ increases instead of the elastic range. The plastic moment value $M_p$ is not influenced by the presence of residual stresses.

## 10.5.3 Imperfections

The initial distortion or lack of straightness in beams may be in the form of a lateral bow or twist. In addition, the applied loading may be eccentric inducing more twist to the beam. It is clear that these initial imperfections correspond to the two types of deformations that the beam undergoes during lateral buckling. Assuming $M_{cr} < M_y$, the lateral deflection and twist increase continuously from the initial stage of loading assuming large proportion as $M_{cr}$ is reached. The additional stresses (lateral bending stresses and warping stresses) thus produced, would cause failure of the beam as the maximum stress in the flange tips reaches the yield stress. This form of failure by limiting the stress to yield magnitude is shown in Fig. 10.17. In the case of beams of intermediate slenderness, a small amount of stress redistribution takes place after yielding and the prediction by the limiting stress approach will be conservative. If residual stresses were also included, the failure load prediction would be conservative even for slender beams (Kirby & Nethercot 1979). For more discussions on initial geometric imperfection, readers may refer to Vinnakota (1977) and Nethercot (1983).

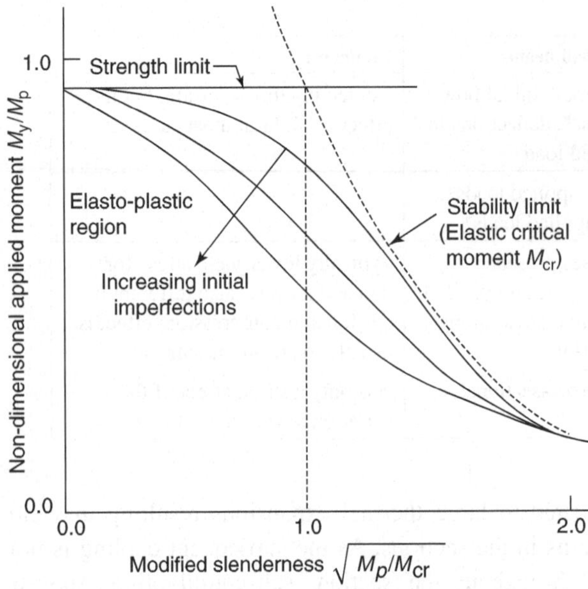


**FIG. 10.17** Beam failure curves for increasing imperfections

Thus, in order to understand the behaviour of real beams, it is necessary to consider the combined effects of instability and plasticity and also the role of factors such as residual stress and geometrical imperfections. When all these effects are considered, it may be found that slender beams fail more or less elastically by excessive deformation at loads that are close to $M_{cr}$, beams of intermediate slenderness fail inelastically by excessive lateral deformation and stocky beams will attain $M_p$ with negligible lateral deformation. This kind of behaviour is shown in Fig. 10.16.

## 10.6 BEHAVIOUR OF BEAMS IN BENDING

The various forms of response of a beam subjected to in-plane bending are given in Table 10.2. Which particular response would govern in a given problem is dependent upon the section properties of the beam, the form of the applied loading, type of support provided, and the yield stress of the material $f_y$. A laterally restrained and compact steel beam may fail by (a) flexure, (b) shears, or (c) bearing. These three failure conditions form the collapse limit states for such steel beams. Steel beams may also become unserviceable due to excessive deflection and this is considered as the limit state of serviceability. These limit states are explained in the following sections.

### 10.6.1 Steel Beams in Flexure

If a beam member, laterally restrained and compact, is loaded progressively, it deflects and its bending curvature varies along its length. The beam behaviour is linear elastic during the initial stages of loading. Linear elastic response is assumed by neglecting the effect of residual stresses, which are always present in structural sections.

With the assumption that plane sections which are normal to the longitudinal axis before bending, remains plane and normal even after bending, the strain at any fibre across the depth of the section is proportional to its distance from the neutral axis. Therefore, the variation of strain along the depth of the member can be obtained as (see Fig. 8.3 of Chapter 8)

$$\varepsilon = y/R \qquad (10.14)$$

In the linear elastic region, since stress is proportional to strain, one can get the corresponding stress $f$ for each strain $\varepsilon$ from the idealised stress–strain curve as shown in Fig. 10.18. It may be observed here that in an idealized stress–strain curve, the strain hardening effects are neglected. We choose several points on the stress–strain curve for detailed discussion on the behaviour of beams.

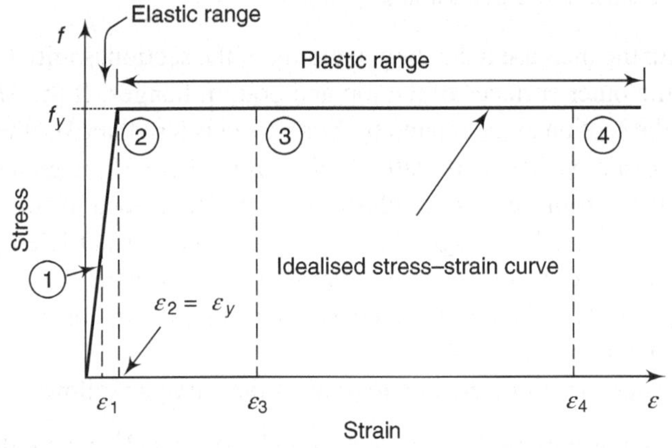


**FIG. 10.18** Idealised elasto-plastic stress–strain curve for steel

### 10.6.2 Elastic Behaviour

Consider a point 1 in the linear portion of the stress–strain curve. At this point the strain $\varepsilon_1$ is less than $\varepsilon_y$, the yield strain of the material. In this linear portion, the stress is directly proportional to the strain and from simple strength of materials approach, it can be shown that the moment of resistance is given by

$$M_c = f_1 I/c \qquad (10.15)$$

where $f_1$ denotes extreme fibre stress, $I$ represents the moment of inertia, and $c$ is the extreme fibre distance from the neutral axis. The term $I/c$, which is denoted as $Z$, is a geometric property of the section and is known as the *elastic modulus of the section*. Therefore, the equation can be written as

$$M_c = f_1 Z \qquad (10.16)$$

### 10.6.3 Yielding and Plastic Behaviour

Plastic behaviour has already been discussed in Section 8.1, however for the sake of completeness, it is described here briefly. Considering point 2 in the stress–strain curve in Fig. 10.18, it can be seen that the extreme fibre strain has reached yield value $\varepsilon_y$. The corresponding stress at point 2, $f_2$ becomes equal to $f_y$. The stress and strain are proportional up to this point as the extreme fibre is within the elastic range. The moment corresponding to this point, known as the *first yield moment*, causes the extreme fibres to yield and is given by

$$M_y = f_y Z \qquad (10.17)$$

It is seen from Fig. 10.19 that the extreme fibres after attaining yield stress do not take any more stresses. When loading is further increased, the outermost fibre strain, $\varepsilon_{max}$ at the maximum bending moment location, i.e., the midpoint of the beam, reaches a value $\varepsilon_3 > \varepsilon_y$ (denoted by point 3 in Fig. 10.18). At point 3, the strain is plastic; however, the fibre stresses in the plastified portion are still maintained at yield stress level $f_y$. It is also seen that the stresses are redistributed inwardly towards the neutral axis of the section and the interior fibres gradually attain yield stress value [see Fig. 10.19(a)]. Thus, the outer fibres are plastified while the inner fibres near to the neutral axis are elastic. At this stage, the moment capacity of the beam is the sum of the contributions from the plastic and elastic portions.

$$M_3 = \underbrace{f_y (b_f t_f + d_0 t_w)(z_c + z_t)}_{\text{(plastic)}} + \underbrace{f_y t_w d^2/6}_{\text{(elastic)}} \qquad (10.18)$$

where $b_f$ and $t_f$ are breadth and thickness of flange (which has been plastified), $d_0$ is the depth up to which the web is also plastified, $t_w$ is the thickness of web, $d$ is the depth of web which is not plastified, and $z_c$ and $z_t$ are the lever arm distances from the neutral axis [see Fig. 10.19(a)].

Upon further loading, the outer fibre strain increases rapidly and attains the stage shown as point 4 in Fig. 10.18. At this point, the elastic core near the neutral axis region becomes very small due to the inward spread of plasticity. It is seen from Fig. 10.20(c) that when the bending moment exceeds the yield moment value $M_y$, the curvature $\phi$ of the beam (which was proportional to bending moment earlier) increases far more rapidly compared to the previous rate of increase. When the whole cross section of the beam gets plastified, the curvature becomes infinity as shown in Fig. 10.20(c). Figure 10.20(b) shows such a fully plastified cross section. This is also shown in Fig. 10.19(b) where the two yield zones have merged at the neutral axis.

When the entire beam cross section becomes plastic, it resists any further rotation under a constant moment. At this stage, the beam is said to have formed a *plastic hinge*.

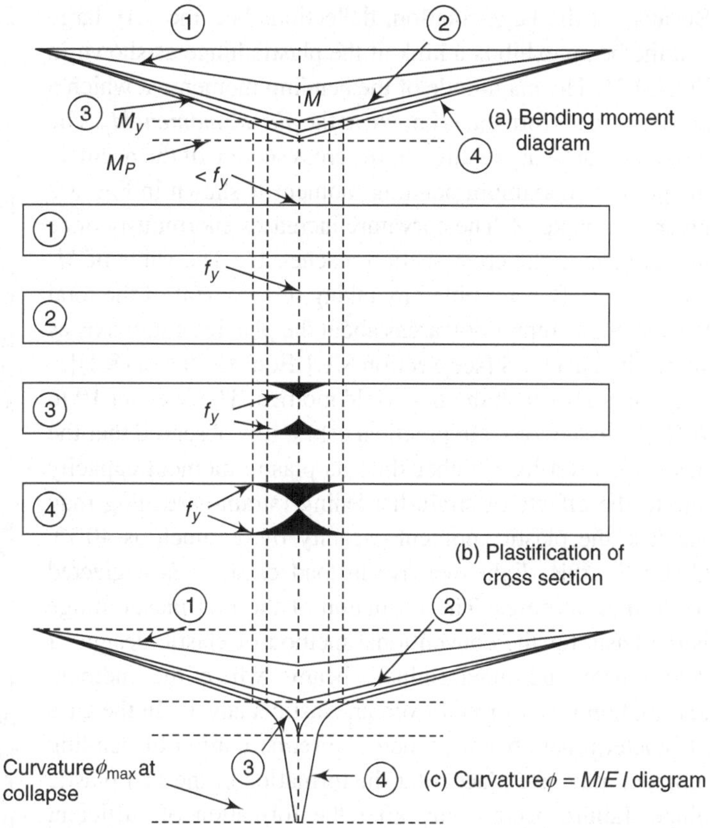


**FIG. 10.20** BM diagram and spread of plasticity across the thickness of the beam

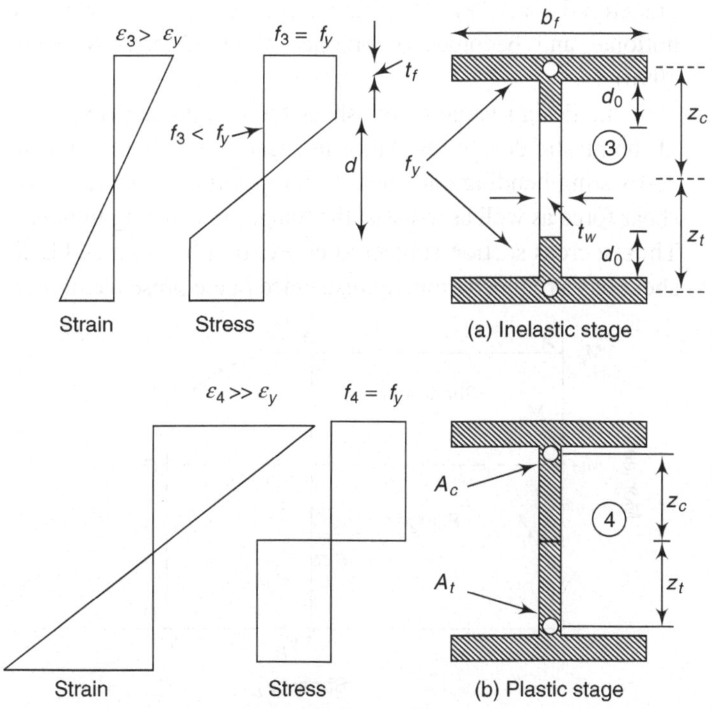


**FIG. 10.19** Strain and stress distributions in plastic range

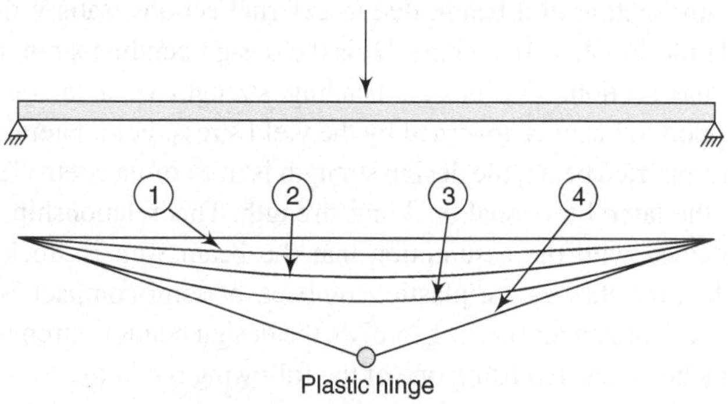


**FIG. 10.21** Simply supported beam and its deflection at various stages of loading

Because of the large rotation, deflections become very large and the beam exhibits a kink at the plastic hinge as shown in Fig. 10.21. The magnitude of the bending moment, at which a plastic hinge is formed, is known as the *plastic moment* $M_p$. The moment–curvature relation of the cross section of the beam, at the point of maximum bending moment is shown in Fig. 8.3 given in Chapter 8. The curvature increases enormously once the moment at the cross section reaches $M_p$. The value of $M_p$ could be easily determined by taking the moment of the total tension and compression areas about the plastic neutral axis as shown in Chapter 8 [see Section 8.1.1 Eqns (8.2) and (8.3)].

$M_p$ is higher than the first yield moment $M_y$ by about 10 to 20% (for I-sections). In practical cases, it is observed that the beam carries a load higher than its plastic moment capacity due to the effects of strain-hardening (strain-hardening may increase the plastic moment capacity by as much as 40%). Generally, this slight increase in load capacity is neglected for design purposes. The formation of the first plastic hinge is the basis for the conventional method of elastic design of determinate and indeterminate beams where the moment distribution is performed from an elastic analysis. In the case of indeterminate beams, where the redistribution of bending moment may take place after the formation of the first plastic hinge, failure occurs only after the formation of sufficient number of plastic hinges causing the beam to become a mechanism (see Chapter 8). Plastic design is based on the ultimate load which causes the formation of a mechanism in an indeterminate beam. Residual stresses, which are invariably present in structural sections, cause yielding to initiate at a lower load which results in increased deflections. This in no way affects the value of $M_p$ since the residual strains in any structural section are always self-equilibrating. Local buckling arising from plastic deformations of the compression flanges generally produces a reduction of moment capacity in the later stages of rotation.

## 10.7 DESIGN STRENGTH OF LATERALLY SUPPORTED BEAMS IN BENDING

For laterally supported beams, the factored design moment $M$ at any section in a beam, due to external actions, satisfy the relationship $M < M_d$, where $M_d$ is the design bending strength of the section. The design bending strength of a laterally supported beam is governed by the yield stress. For a laterally unsupported beam, the design strength is most often controlled by the lateral torsional buckling strength. This relationship is obtained with the assumption that the beam web is stocky. When the flanges are plastic, compact, or semi-compact but the web is slender (i.e., $d/t_w > 67\varepsilon$), the design bending strength may be calculated using one of the following methods.

1. The flanges resist the bending moment and the axial force acting on the section and the web resists only the shear.
2. The whole section resists the bending moment and the axial force and therefore the web has to be designed for combined shear and its share of normal stresses.

This is done by using simple elastic theory in the case of semi-compact webs and simple plastic theory in the case of compact and plastic webs.

Shear force does not have any influence on the bending moment for values of shear up to $0.6V_d$ (called the low shear load), where $V_d$ is the design shear strength. The provision for determining $V_d$ is explained a little later. When the design shear force $V$ is less than $0.6V_d$, the design bending strength $M_d$ will be taken as (clause 8.2.1.2 of code)

$$M_d = \beta_b Z_p f_y/\gamma_{m0} \le 1.2 Z_e f_y/\gamma_{m0} \le 1.5\ Z_e f_y/\gamma_{m0} \text{ for cantilevers} \quad (10.19a)$$

where $\beta_b = 1.0$, for plastic and compact sections and $\beta_b = Z_e/Z_p$, for semi-compact sections.

Thus, for class 3 semi-compact sections,

$$M_d = f_y Z_e/\gamma_{m0} \quad (10.19b)$$

For class 4 slender sections, $M_d = f_y' Z_e$, where $f_y'$ is the reduced design strength for slender sections. $Z_p$ and $Z_e$ are the plastic and elastic section moduli of the cross section, respectively, $f_y$ is the yield stress of the material, and $\gamma_{m0}$ is the material partial safety factor = 1.10 as per IS 800 : 2007.

The additional check ($M_d < 1.2 Z_e f_y/\gamma_{m0}$) is provided to prevent the onset of plasticity under unfactored dead, imposed, and wind loads. For most of the I-beams and channels given in IS 808, $Z_{pz}/Z_{ez}$ is less than 1.2 and hence the plastic moment capacity governs the design. For sections where $Z_{pz}/Z_{ez} > 1.2$, the constant 1.2 may be replaced by the ratio of factored load/unfactored load ($\gamma_f$). Thus, the limitation $1.2 Z_e f_y$ is purely notional and becomes in practice $\gamma_f Z_e f_y$ (Morris & Plum 1996).

If the design value of the shear force is greater than 60% of the plastic design resistance in shear, a member subject to co-existing bending and shear has to use its web to resist the shear force as well as to assist the flanges in resisting moment. Thus, a cross section subject to co-existing bending and high shear has a reduced moment resistance in the presence of high

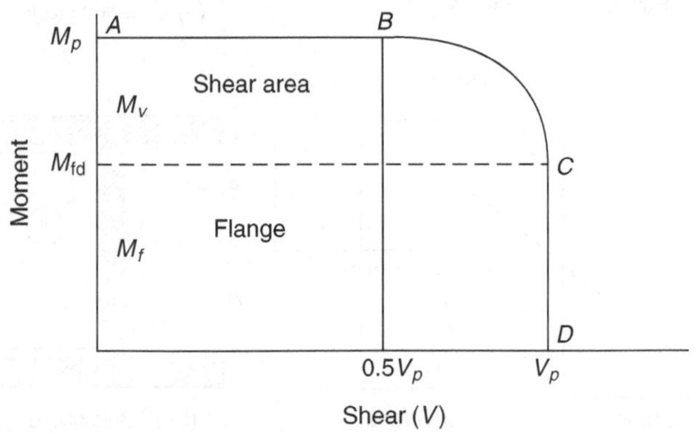


**FIG. 10.22** Interaction between moment and shear

shear. The interaction diagram between moment and shear is shown in Fig. 10.22 (Horne & Morris 1981).

When the design shear force (factored) $V$ exceeds $0.6V_d$ (called the high shear load), where $V_d$ is the design shear strength of the cross section, the design bending strength $M_d$ will be taken as (clause 8.2.1.3 of code).

$$M_d = M_{dv}$$

where $M_{dv}$ is the design bending strength under high shear and it is calculated as follows:

(a) *Plastic or Compact Section*

As the shear force $V$ is increased from zero, no reduction in plastic moment is assumed below the value of $M_p$ until $V$ reaches $0.5V_p$, where $V_p$ is the shear strength of the web given by $A_v f_v$, where $A_v$ is the shear area taken as $Dt_w$ for rolled sections and $dt_w$ for built-up sections ($d = D - 2t_f$). The design shear strength $f_v$ is taken as $0.6f_y$, where $f_y$ is the design strength in tension or compression, i.e., slightly greater than the true von Mises value of $f_y/\sqrt{3}$. When the full capacity in shear $V_p$ is reached, the shear area is assumed to be completely ineffective in resisting the moment, and hence the reduced plastic moment $M_{dv}$ becomes $M_{fd}$, where $M_{fd} = M_p - (D^2t_w/4) f_y$ for rolled sections and $M_{fd} = M_p - (d^2t_w/4)f_y$ for built-up sections. Between $V = 0.5V_p$ and $V_p$, $M_{dv}$ is assumed to reduce parabolically according to the curve BC given in Fig. 10.22 (clause 9.2.2 of code.)

$$M_{dv} = M_p - \beta(M_p - M_{fd}) \le 1.2\, Z_e f_y/\gamma_{m0} \quad (10.20a)$$

where $$\beta = (2V/V_p - 1)^2 \quad (10.20b)$$

In Eqn (10.20) $M_p$ is the design plastic moment of the whole section disregarding high shear force effect, but considering web buckling effects, $V$ is the factored applied shear force, $V_p$ is the design shear strength as governed by web yielding or web buckling = $0.6Dt_w f_y$, and $M_{fd}$ is the plastic design strength of the area of the cross section excluding the shear area, considering partial safety factor $\gamma_{m0}$.

This equation may also be written as (Horne & Morris 1981)

$$Z_{pr,z} = Z_{p,z} - Z_{pv}\,[(2\tau_w/f_v) - 1]^2 \quad (10.20c)$$

where $Z_{pv}$ is the plastic modulus of the shear area about the major axis of the cross section = $D^2t_w/4$, $\tau_w$ is the shear stress due to $V$ calculated on the shear area, $f_v$ is yield stress of the web, and $Z_{pr,z}$ is the reduced plastic modulus about the major axis.

When bending is about the minor axis of the I-section and the accompanying shear force is in a plane parallel to the flange, the shear stress $\tau_w$ is calculated on the combined area of the flanges. In this case, the reduced plastic modulus about minor axis $Z_{pr,y}$ is equal to the plastic modulus $Z_{py}$, when $\tau_w \le 0.5f_v$. When $\tau_w > 0.5f_v$, the plastic modulus decreases parabolically reaching zero when $\tau_w = f_v$. Thus, we get (Horne & Morris 1981)

$$Z_{pr,y} = Z_{p,y}\frac{4\tau_w}{f_v}(1 - \tau_w/f_v) \quad (10.20d)$$

(b) *Semi-compact section*

$$M_{dv} = Z_e f_y/\gamma_{m0} \quad (10.20e)$$

where $Z_e$ is the elastic section modulus of the whole section.

### 10.7.1 Holes in the Tension Zone

The effect of holes in the tension flange, on the design bending strength need not be considered if (clause 8.2.1.4 of the code)

$$(A_{nf}/A_{gf}) \ge (f_y/f_u)\,(\gamma_{m1}/\gamma_{m0})/0.9 \quad (10.21)$$

where $A_{nf}/A_{gf}$ is the ratio of net and gross area of the flange, $f_y/f_u$ is the ratio of yield and ultimate strength of the material, $\gamma_{m1}/\gamma_{m0}$ is the ratio of partial safety factors against ultimate to yield stress = 1.25/1.1 = 1.136 as per IS 800 : 2007.

When $A_{nf}/A_{gf}$ does not satisfy Eqn (10.21), the reduced flange area $A_{nf}$ satisfying Eqn (10.21) may be taken as the effective flange area in tension (Dexter and Altstadt 2003).

The effect of holes in the tension region of the web on the design flexural strength need not be considered, if the limit given in Eqn (10.21) is satisfied for the complete tension zone of the cross section, comprising the tension flange and tension region of the web. Fastener holes in the compression zone of the cross section need not be considered in the design bending strength calculation, except for oversize and slotted holes, or holes without any fastener.

### 10.7.2 Shear Lag Effects

The simple theory of bending is based on the assumption that plane sections remain plane after bending. In reality, shear strains cause the sections to warp. The effect in the flange is to modify the bending stresses obtained by simple bending theory. Thus, higher stresses are produced near the junction of a web and lower stresses at points away from the web as shown in Fig. 10.23. This phenomenon is known as *shear lag*. It results in a non-uniform stress distribution across the width of the

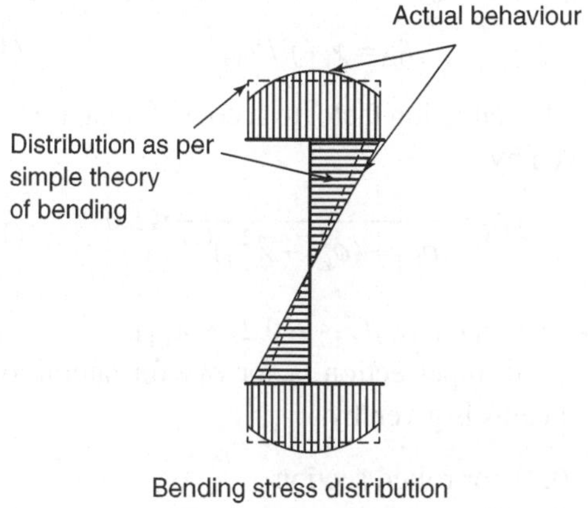


**FIG. 10.23** Shear lag effects in wide-flange beams

flange. The shear lag effects are minimal in rolled sections, which have relatively narrow and thick flanges. For normal dimensions of the flanges, the effects are negligible. But if the flanges are unusually wide, compared to their length (as in plate girders or box girders), these shear strains influence the normal bending stresses in the flanges.

The shear lag effects in flanges may be disregarded provided, for outstand elements (supported along one edge), $b_o \leq L_0/20$ and for internal elements (supported along two edges), $b_i \leq L_0/10$ where $L_0$ is the length between points of zero moment (inflection) in the span, $b_o$ is the width of the flange with outstand, and $b_i$ is the width of an internal element (clause 8.2.1.5 of the code).

When these limits are exceeded, the effective width of the flange for design strength may be taken conservatively as the values satisfying the limits given earlier.

## 10.8 DESIGN STRENGTH OF LATERALLY UNSUPPORTED BEAMS

Beam experiencing bending about major axis and when its compression flange is not restrained against lateral buckling, may fail by lateral torsional buckling before attaining its bending strength. The effect of lateral torsional buckling on flexural strength need not be considered when $\lambda_{LT} \leq 0.4$ (see Fig. 10.16) where $\lambda_{LT}$ is the non-dimensional slenderness ratio for lateral torsional buckling. The design bending strength of a laterally unsupported beam as governed by lateral torsional buckling as per clause 8.2.2 of the Indian code (IS 800 : 2007) is given by

$$M_d = \beta_b Z_p f_{bd} \tag{10.22}$$

with

$\beta_b = 1.0$ for plastic and compact sections
$= Z_e/Z_p$ for semi-compact sections

where $Z_p$ and $Z_e$ are the plastic section modulus and elastic section modulus with respect to extreme compression fibre and $f_{bd}$ is the design bending compressive stress.

The design bending compressive stress is given by

$$f_{bd} = \chi_{LT} f_y / \gamma_{m0} \tag{10.23a}$$

where $\chi_{LT}$ is the reduction factor to account for lateral torsional buckling given by

$$\chi_{LT} = \frac{1}{\phi_{LT} + (\phi_{LT}^2 - \lambda_{LT}^2)^{0.5}} \leq 1.0 \tag{10.23b}$$

in which $\phi_{LT} = 0.5[1 + \alpha_{LT}(\lambda_{LT} - 0.2) + \lambda_{LT}^2]$.

The values of imperfection factor $\alpha_{LT}$ for lateral torsional buckling of beams is given by

$\alpha_{LT} = 0.21$ for rolled section
$\alpha_{LT} = 0.49$ for welded section

$\gamma_{m0}$ = partial safety factor for material = 1.10

The non-dimensional slenderness ratio $\lambda_{LT}$, is given by

$$\lambda_{LT} = \sqrt{\beta_b Z_p f_y / M_{cr}} \leq \sqrt{1.2 Z_e f_y / M_{cr}} = \sqrt{\frac{f_y}{f_{cr,b}}} \tag{10.24}$$

where $M_{cr}$ is the elastic critical moment and $f_{cr,\,b}$ is the extreme fibre compressive elastic lateral buckling stress.

Note that Eqns (10.22) to (10.24) have been adopted from Eurocode 3. However, Eurocode gives multiple beam curves (similar to multiple column curves). Thus, four lateral-torsional buckling curves are provided in the Eurocode (selected on the basis of the overall height-to-width ratio of the cross section, the type of cross section, and whether the cross section is rolled or welded), whereas the Indian code offers only two curves (only making a distinction between rolled and welded sections). Eurocode 3 also defines lateral torsional buckling curves for two cases: (a) the general case and (b) rolled sections or equivalent welded sections.

The elastic lateral buckling moment $M_{cr}$ is given by (clause 8.2.2.1 of the code)

$$M_{cr} = \sqrt{\left\{\left(\frac{\pi^2 EI_y}{(KL)^2}\right)\left[GI_t + \frac{\pi^2 EI_w}{(KL)^2}\right]\right\}} \tag{10.25a}$$

Kerensky et al. (1956) simplified this expression by introducing the following approximations for doubly symmetric sections.

$I_y = b_f^3 t_f/6$, $I_w = I_y h^2/4$, $I_t = 0.9\, b_f t_f^3$, $b_f = 4.2 r_y$ and $E = 2.6G$

Thus, Eqn (10.25a) is reduced to

$$M_{cr} = \frac{\beta_{LT}\pi^2 EI_y h}{2(KL)^2}\left[1 + \frac{1}{20}\left[\frac{KL/r_y}{h/t_f}\right]^2\right]^{0.5} = \beta_b Z_p f_{cr,b} \tag{10.25b}$$

where $I_t$ is the torsional constant, $I_w$ is the warping constant, $I_y$ is the moment of inertia about the weak axis, $r_y$ is the radius of gyration of the section about the weak axis, $KL$ is the effective laterally unsupported length of the member (see Section 10.4.3 and clause 8.3.1 of the code), $h$ is the overall depth of the section, $t_f$ is the thickness of the flange, and $\beta_{LT} = 1.20$ for plastic and compact sections with $t_f/t_w \leq 2.0$ and 1.00 for semi-compact sections or sections with $t_f/t_w > 2.0$.

Using the same approximations, the extreme fibre compressive elastic buckling stress may be obtained as (with $E = 2.0 \times 10^5$ MPa and $Z_z = 1.1 b_f d_f h$)

$$f_{cr,b} = \left(\frac{1625}{KL/r_y}\right)^2\left[1 + \frac{1}{20}\left[\frac{KL/r_y}{h/t_f}\right]^2\right]^{0.5} \tag{10.26}$$

**TABLE 10.9** Critical stress $f_{cr,b}$ (in N/mm$^2$)

| | $h/t_f$ | | | | | | | | | | | | | |
|---|---|---|---|---|---|---|---|---|---|---|---|---|---|---|
| $KL/r_y$ | 8 | 10 | 12 | 14 | 16 | 18 | 20 | 25 | 30 | 35 | 40 | 50 | 60 | 80 | 100 |
| 10 | 22545.42 | 22249.41 | 22086.95 | 21988.41 | 21924.22 | 21880.10 | 21848.49 | 21799.88 | 21773.43 | 21757.47 | 21747.10 | 21734.90 | 21728.27 | 21721.68 | 21718.63 |
| 20 | 6218.90 | 5946.41 | 5793.01 | 5698.53 | 5636.36 | 5593.33 | 5562.35 | 5514.47 | 5488.28 | 5472.43 | 5462.12 | 5449.97 | 5443.36 | 5436.78 | 5433.73 |
| 30 | 3148.51 | 2905.13 | 2763.96 | 2675.23 | 2616.04 | 2574.67 | 2544.67 | 2497.92 | 2472.16 | 2456.49 | 2446.27 | 2434.19 | 2427.61 | 2421.04 | 2418.00 |
| 40 | 2035.61 | 1820.71 | 1692.57 | 1610.39 | 1554.72 | 1515.38 | 1486.60 | 1441.31 | 1416.11 | 1400.69 | 1390.59 | 1378.62 | 1372.07 | 1365.53 | 1362.49 |
| 50 | 1492.54 | 1302.79 | 1187.08 | 1111.50 | 1059.56 | 1022.43 | 995.02 | 951.42 | 926.88 | 911.76 | 901.82 | 889.98 | 883.48 | 876.97 | 873.94 |
| 60 | 1177.68 | 1009.25 | 904.72 | 835.39 | 787.13 | 752.25 | 726.28 | 684.51 | 660.71 | 645.94 | 636.17 | 624.48 | 618.04 | 611.57 | 608.55 |
| 70 | 973.68 | 823.07 | 728.32 | 664.69 | 619.91 | 587.23 | 562.70 | 522.81 | 499.81 | 485.42 | 475.85 | 464.33 | 457.96 | 451.53 | 448.52 |
| 80 | 831.04 | 695.29 | 609.01 | 550.48 | 508.90 | 478.32 | 455.18 | 417.18 | 395.00 | 381.01 | 371.65 | 360.33 | 354.03 | 347.65 | 344.65 |
| 90 | 725.66 | 602.40 | 523.41 | 469.41 | 430.74 | 402.10 | 380.28 | 344.13 | 322.79 | 309.22 | 300.08 | 288.96 | 282.74 | 276.42 | 273.44 |
| 100 | 644.57 | 531.86 | 459.18 | 409.17 | 373.13 | 346.27 | 325.70 | 291.31 | 270.81 | 257.66 | 248.76 | 237.86 | 231.72 | 225.45 | 222.49 |
| 110 | 580.18 | 476.47 | 409.26 | 362.77 | 329.09 | 303.86 | 284.44 | 251.74 | 232.05 | 219.33 | 210.66 | 199.99 | 193.94 | 187.74 | 184.80 |
| 120 | 527.75 | 431.79 | 369.35 | 325.97 | 294.42 | 270.67 | 252.31 | 221.20 | 202.30 | 190.00 | 181.57 | 171.13 | 165.18 | 159.04 | 156.12 |
| 130 | 484.20 | 394.96 | 336.71 | 296.10 | 266.45 | 244.05 | 226.67 | 197.04 | 178.90 | 167.01 | 158.82 | 148.62 | 142.77 | 136.70 | 133.80 |
| 140 | 447.43 | 364.07 | 309.51 | 271.36 | 243.42 | 222.25 | 205.77 | 177.53 | 160.11 | 148.63 | 140.68 | 130.70 | 124.95 | 118.96 | 116.08 |
| 150 | 415.95 | 337.76 | 286.48 | 250.53 | 224.14 | 204.08 | 188.43 | 161.48 | 144.75 | 133.66 | 125.94 | 116.21 | 110.56 | 104.64 | 101.79 |
| 160 | 388.68 | 315.08 | 266.72 | 232.76 | 207.76 | 188.72 | 173.82 | 148.08 | 132.01 | 121.29 | 113.79 | 104.29 | 98.75 | 92.91 | 90.08 |
| 170 | 364.82 | 295.32 | 249.58 | 217.40 | 193.67 | 175.56 | 161.36 | 136.73 | 121.28 | 110.92 | 103.65 | 94.38 | 88.94 | 83.18 | 80.38 |
| 180 | 343.76 | 277.93 | 234.56 | 203.99 | 181.42 | 164.16 | 150.60 | 127.01 | 112.14 | 102.13 | 95.07 | 86.03 | 80.70 | 75.02 | 72.24 |
| 190 | 325.04 | 262.52 | 221.28 | 192.18 | 170.66 | 154.18 | 141.22 | 118.60 | 104.27 | 94.60 | 87.74 | 78.93 | 73.70 | 68.10 | 65.35 |
| 200 | 308.27 | 248.76 | 209.46 | 181.70 | 161.14 | 145.38 | 132.97 | 111.25 | 97.44 | 88.08 | 81.42 | 72.83 | 67.70 | 62.19 | 59.46 |
| 210 | 293.17 | 236.39 | 198.86 | 172.33 | 152.66 | 137.56 | 125.65 | 104.77 | 91.45 | 82.39 | 75.93 | 67.55 | 62.52 | 57.09 | 54.39 |
| 220 | 279.49 | 225.21 | 189.30 | 163.90 | 145.04 | 130.56 | 119.12 | 99.02 | 86.16 | 77.39 | 71.11 | 62.93 | 58.01 | 52.67 | 50.00 |
| 230 | 267.04 | 215.05 | 180.64 | 156.27 | 138.17 | 124.25 | 113.25 | 93.89 | 81.46 | 72.96 | 66.86 | 58.88 | 54.06 | 48.80 | 46.16 |
| 240 | 255.67 | 205.78 | 172.75 | 149.34 | 131.94 | 118.54 | 107.95 | 89.27 | 77.25 | 69.01 | 63.08 | 55.30 | 50.58 | 45.39 | 42.78 |
| 250 | 245.23 | 197.29 | 165.53 | 143.00 | 126.25 | 113.35 | 103.13 | 85.10 | 73.47 | 65.47 | 59.70 | 52.11 | 47.48 | 42.38 | 39.80 |

IS 800 : 2007 uses a similar expression but the coefficient of 1625 is replaced by 1473.5. Table 10.9 gives the value of $f_{cr,\,b}$ for various values of $KL/r_y$ and $h/t_f$ based on IS 800 : 2007. Table 10.10 gives the values of design bending compressive strength corresponding to lateral torsional buckling [based on Eqn (10.23a)] for $\alpha_{LT} = 0.21$ and $\alpha_{LT} = 0.49$ for $f_y = 250$ MPa. Intermediate values may be obtained by interpolating the values given in these tables.

The elastic critical moment is affected by four components.

1. Moment gradient in the unsupported length
2. Boundary conditions at the lateral support points
3. Non-symmetric and non-prismatic nature of the member
4. Location of transverse load with respect to shear centre

The boundary conditions at the lateral supports have two components.

1. Torsional restraint—where the cross section is prevented from rotation about the shear centre
2. Warping restraint—where the flanges are prevented from rotating in their own plane about an axis perpendicular to the flange

The elastic critical moment corresponding lateral torsional buckling of a doubly symmetric prismatic beam subjected to uniform moment in the unsupported length and torsionally restraining lateral supports is given by [also see Eqns (10.1) and (10.5)].

$$M_{cr} = C_1 \frac{\pi^2 EI_y}{(KL)^2}\left[\frac{I_w}{I_y} + \frac{GI_t (KL)^2}{\pi^2 EI_y}\right]^{0.5} \quad (10.27)$$

where $I_y$, $I_w$, and $I_t$ are the moment of inertia about the minor axis, warping constant, and St Venant's torsion constant of the cross section, respectively, $G$ is the shear modulus of rigidity, $KL$ is the effective length between the points of restraint against lateral torsional buckling, and $C_1$ is the equivalent uniform moment factor.

More accurate calculation of the elastic critical moment for general case can be obtained from specialist literature (e.g., Roik et al. 1972; Trahair 1993) or by using computer programs or equations given in the next section, which are less approximate than that given earlier, which may be beneficially used to calculate elastic lateral buckling moment of more general cases of laterally unsupported beams.

**TABLE 10.10** Design bending stress $f_{bd}$ for $f_y = 250$ MPa (in N/mm²)

| $f_{cr}$ N/mm² | $\lambda_{LT}$ | $\alpha = 0.21$ | | | $\alpha = 0.49$ | | |
|---|---|---|---|---|---|---|---|
| | | $\phi_{LT}$ | $\chi_{LT}$ | $f_{bd}$ N/mm² | $\phi_{LT}$ | $\chi_{LT}$ | $f_{bd}$ N/mm² |
| 10000 | 0.1581 | 0.508 | 1.000 | 227.27 | 0.502 | 1.000 | 227.27 |
| 9000 | 0.1667 | 0.510 | 1.000 | 227.27 | 0.506 | 1.000 | 227.27 |
| 8000 | 0.1768 | 0.513 | 1.000 | 227.27 | 0.510 | 1.000 | 227.27 |
| 7000 | 0.1890 | 0.517 | 1.000 | 227.27 | 0.515 | 1.000 | 227.27 |
| 6000 | 0.2041 | 0.521 | 0.999 | 227.07 | 0.522 | 0.998 | 226.79 |
| 5000 | 0.2236 | 0.527 | 0.995 | 226.09 | 0.531 | 0.988 | 224.54 |
| 4000 | 0.2500 | 0.537 | 0.989 | 224.76 | 0.544 | 0.975 | 221.49 |
| 3000 | 0.2887 | 0.551 | 0.980 | 222.76 | 0.563 | 0.955 | 217.03 |
| 2000 | 0.3536 | 0.579 | 0.965 | 219.23 | 0.600 | 0.922 | 209.46 |
| 1000 | 0.5000 | 0.657 | 0.924 | 210.06 | 0.699 | 0.843 | 191.59 |
| 900 | 0.5270 | 0.673 | 0.916 | 208.10 | 0.719 | 0.828 | 188.12 |
| 800 | 0.5590 | 0.694 | 0.905 | 205.65 | 0.744 | 0.809 | 183.95 |
| 700 | 0.5976 | 0.720 | 0.891 | 202.48 | 0.776 | 0.787 | 178.82 |
| 600 | 0.6455 | 0.755 | 0.872 | 198.16 | 0.817 | 0.758 | 172.30 |
| 500 | 0.7071 | 0.803 | 0.844 | 191.90 | 0.874 | 0.720 | 163.70 |
| 450 | 0.7454 | 0.835 | 0.825 | 187.59 | 0.911 | 0.696 | 158.28 |
| 300 | 0.9129 | 0.992 | 0.725 | 164.87 | 1.091 | 0.592 | 134.53 |
| 150 | 1.2910 | 1.448 | 0.475 | 108.05 | 1.601 | 0.393 | 89.24 |
| 90 | 1.6667 | 2.043 | 0.310 | 70.49 | 2.248 | 0.266 | 60.49 |
| 80 | 1.7678 | 2.227 | 0.279 | 63.45 | 2.447 | 0.242 | 54.92 |

*(Contd)*

**TABLE 10.10** *(Contd)*

| $f_{cr}$ N/mm$^2$ | $\lambda_{LT}$ | $\alpha = 0.21$ | | | $\alpha = 0.49$ | | |
|---|---|---|---|---|---|---|---|
| | | $\phi_{LT}$ | $\chi_{LT}$ | $f_{bd}$ N/mm$^2$ | $\phi_{LT}$ | $\chi_{LT}$ | $f_{bd}$ N/mm$^2$ |
| 70 | 1.8898 | 2.463 | 0.247 | 56.22 | 2.700 | 0.216 | 49.11 |
| 60 | 2.0412 | 2.777 | 0.215 | 48.78 | 3.034 | 0.189 | 43.05 |
| 50 | 2.2361 | 3.214 | 0.181 | 41.16 | 3.499 | 0.162 | 36.72 |
| 40 | 2.5000 | 3.867 | 0.147 | 33.34 | 4.189 | 0.132 | 30.11 |
| 30 | 2.8868 | 4.949 | 0.112 | 25.34 | 5.325 | 0.102 | 23.19 |
| 20 | 3.5355 | 7.100 | 0.075 | 17.14 | 7.567 | 0.070 | 15.94 |
| 10 | 5.0000 | 13.504 | 0.038 | 8.73 | 14.176 | 0.036 | 8.28 |

### 10.8.1 Elastic Critical Moment of a Section Symmetrical About Minor Axis

In the case of a beam which is symmetrical only about the minor axis (see Fig. 10.24) with bending about major axis, the elastic critical moment for lateral torsional buckling is given by the general equation (clause E – 1.2 of the code)

$$M_{cr} = C_1 \frac{\pi^2 EI_y}{(KL)^2} \left\{ \left[ \left( \frac{K}{K_w} \right)^2 \frac{I_w}{I_y} + \frac{GI_t (KL)^2}{\pi^2 EI_y} + (c_2 y_g - c_3 z_j)^2 \right]^{0.5} - (c_2 y_g - c_3 z_j) \right\} \quad (10.28)$$

where $c_1$, $c_2$, and $c_3$ are factors depending upon the loading and end restraint conditions (see Table 10.11), and $K$ and $K_w$ are the effective length factors of the unsupported length accounting for boundary conditions at the end lateral supports.

The effective length factors $K$ and $K_w$ vary from 0.5 for full fixity (against warping) to 1.0 for free (to warp) case and 0.7 for the case of one end fixed and other end free. It is analogous to the effective length factors for compression members with end rotational restraint.

The $K_w$ factor refers to the warping restraint. Unless special provisions to restrain warping of the section at the end lateral supports are made, $K_w$ should be taken as 1.0. In Eqn (10.28), $y_g$ is the $y$ distance between the point of application of the load and the shear centre of the cross section and is positive when the load is acting below the shear centre, and is otherwise negative.

**TABLE 10.11** Constants $c_1$, $c_2$, $c_3$

| Loading and support conditions | Bending moment diagram | Value of $K$ | Constants | | |
|---|---|---|---|---|---|
| | | | $c_1$ | $c_2$ | $c_3$ |
| | $\psi = +1$ 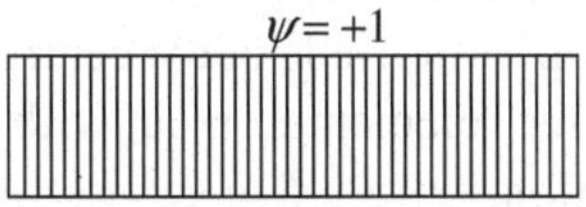 | 1.0<br>0.7<br>0.5 | 1.000<br>1.000<br>1.000 | — | 1.000<br>1.113<br>1.144 |
| | $\psi = +3/4$ 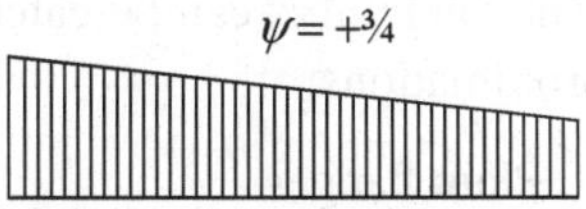 | 1.0<br>0.7<br>0.5 | 1.141<br>1.270<br>1.305 | — | 0.998<br>1.565<br>2.283 |
| | $\psi = +1/2$ 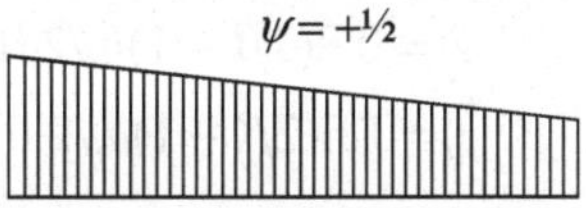 | 1.0<br>0.7<br>0.5 | 1.323<br>1.473<br>1.514 | — | 0.992<br>1.556<br>2.271 |
| | $\psi = +1/4$ 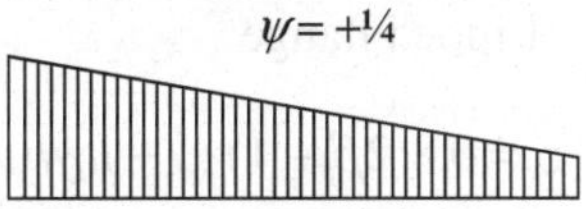 | 1.0<br>0.7<br>0.5 | 1.563<br>1.739<br>1.788 | — | 0.977<br>1.531<br>2.235 |
| | $\psi = 0$ 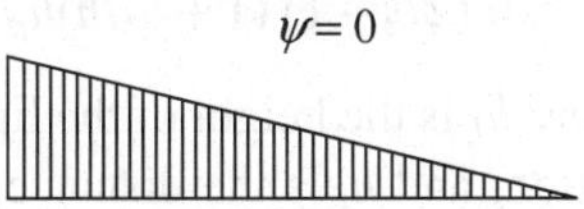 | 1.0<br>0.7<br>0.5 | 1.879<br>2.092<br>2.150 | — | 0.939<br>1.473<br>2.150 |

*(Contd)*

**TABLE 10.11** (*Contd*)

| Loading and support conditions | Bending moment diagram | Value of $K$ | Constants | | |
|---|---|---|---|---|---|
| | | | $c_1$ | $c_2$ | $c_3$ |
| 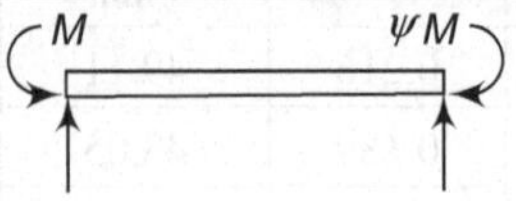 | $\psi = -1/4$ 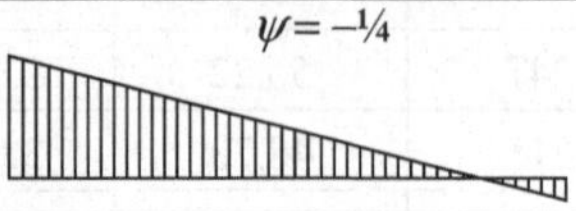 | 1.0<br>0.7<br>0.5 | 2.281<br>2.538<br>2.609 | —| 0.855<br>1.340<br>1.957 |
| | $\psi = -1/2$ 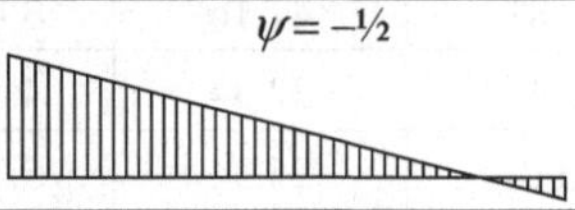 | 1.0<br>0.7<br>0.5 | 2.704<br>3.009<br>3.093 | — | 0.676<br>1.059<br>1.546 |
| | $\psi = -3/4$ 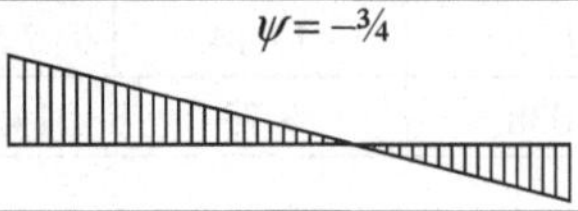 | 1.0<br>0.7<br>0.5 | 2.927<br>3.009<br>3.093 | — | 0.366<br>0.575<br>0.837 |
| | $\psi = -1$ 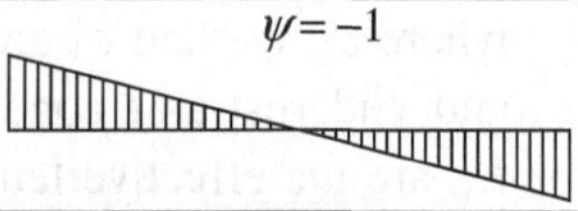 | 1.0<br>0.7<br>0.5 | 2.752<br>3.063<br>3.149 | — | 0.000<br>0.000<br>0.000 |
| 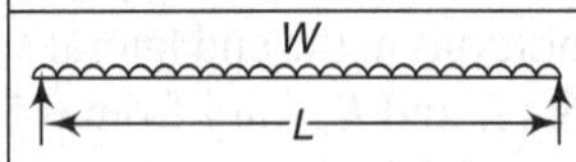 | 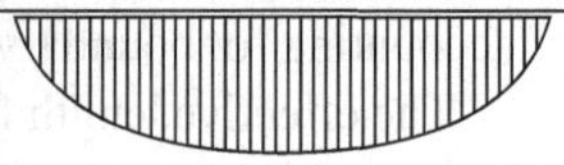 | 1.0<br>0.5 | 1.132<br>0.972 | 0.459<br>0.304 | 0.525<br>0.980 |
| 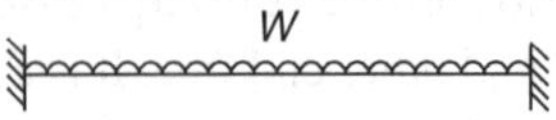 | 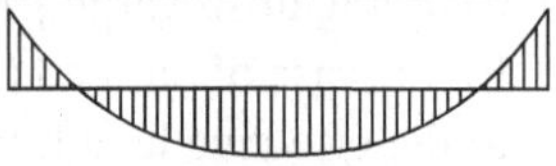 | 1.0<br>0.5 | 1.285<br>0.712 | 1.562<br>0.652 | 0.753<br>1.070 |
| 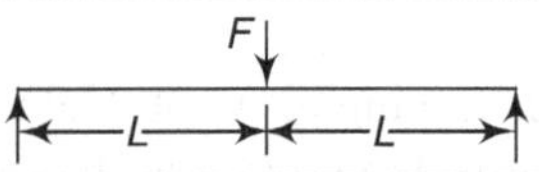 | 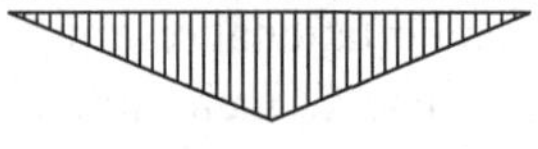 | 1.0<br>0.5 | 1.365<br>1.070 | 0.553<br>0.432 | 1.730<br>3.050 |
| 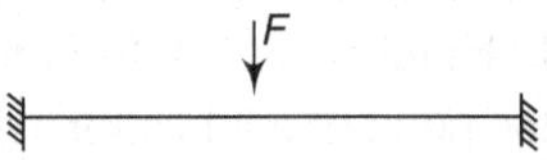 | 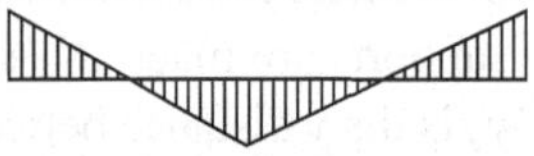 | 1.0<br>0.5 | 1.565<br>0.938 | 1.267<br>0.715 | 2.640<br>4.800 |
| 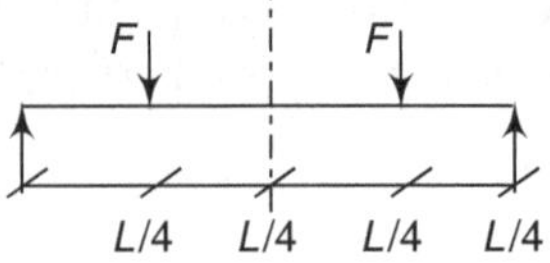 |  | 1.0<br>0.5 | 1.046<br>1.010 | 0.430<br>0.410 | 1.120<br>1.890 |

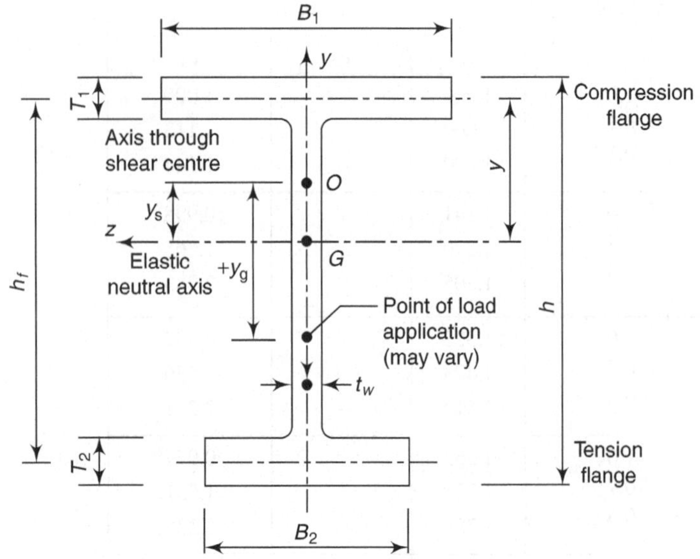


$y_s = h_f/[1 + (B_1/B_2)^3(T_1/T_2)] - \bar{y}$; $h_f = h - (T_1+T_2)/2$, $\bar{y} = \text{ABS}(y)$

**FIG. 10.24** Monosymmetric I-beams

$$z_j = y_s - 0.5 \int_A (z^2 + y^2)y dA/I_z \qquad (10.29a)$$

where $y_s$ is the coordinate of the shear centre with respect to the centroid, positive when the shear centre is on the compression side of the centroid, $y$ and $z$ are the coordinates of the elemental area with respect to centroid of the section.

The value of $z_j$ can be calculated by using the following approximations:

(a) Plain flanges

$$z_j = 0.8(2\beta_f - 1)h_f/2.0 \quad (\text{when } \beta_f > 0.5) \qquad (10.29b)$$

$$z_j = 1.0(2\beta_f - 1)h_f/2.0 \quad (\text{when } \beta_f \le 0.5) \qquad (10.29c)$$

(b) Lipped flange

$$z_j = 0.8(2\beta_f - 1)(1 + h_L/h)h_f/2 \quad (\text{when } \beta_f > 0.5) \qquad (10.29d)$$

$$z_j = (2\beta_f - 1)(1 + h_L/h)h_f/2 \quad (\text{when } \beta_f \le 0.5) \qquad (10.29e)$$

where $h_L$ is the height of the lip, $h$ is the overall height of the section, and $h_f$ is the distance between the shear centres of the two flanges of the cross section.

The torsion constant $I_t$ is given by

$$I_t = \sum b_i t_i^3/3 \text{ for open sections} \quad (10.30a)$$

$$= \frac{4A_e^2}{\sum(b/t)} \text{ for hollow sections} \quad (10.30b)$$

where $A_e$ is the area enclosed by the section and $b$ and $t$ are the breadth and thickness of the elements of the section, respectively.

The warping constant $I_w$, for I-sections mono-symmetric about the weak axis, is given by

$$I_w = (1 - \beta_f)\beta_f I_y h_f^2 \quad (10.31)$$

= 0 for angle, T-, narrow rectangle sections, and approximately for hollow sections

$$\beta_f = I_{fc}/(I_{fc} + I_{ft}) \quad (10.32)$$

where $I_{fc}$ and $I_{ft}$ are the moment of inertia of the compression and tension flanges, respectively, about the minor axis of the entire section. Note that for equal flange beams $\beta_f = 0.5$.

Trahair et al. (2001) showed that the buckling resistance for a *T*-beam with the flange in compression ($\beta_f$ = 1.0) is significantly higher than for an equal flanged I-beam ($\beta_f$ = 0.5) with the same value of $K_m$ [= $\pi^2 EI_y h_f^2/(4GI_t L^2)$], but the resistance is greatly reduced for a *T*-beam with the flange in tension ($\beta_f$ = 0.0). More discussions on the buckling of mono-symmetric I-beams may be found from Kitipornchai and Trahair (1980), Nethercot (1983), Wang and Kitipornchai (1986), and Kitipornchai et al. (1986). Solutions for tapered doubly symmetric and non-symmetric beams are provided by Nethercot (1973a), Bradford (1988), Bradford and Cuk (1988), Bradford (1989), and Petersen (1990). It may be conservative to use $M_{cr}$ for doubly symmetric I-shapes for singly symmetric I-shapes having compressive flange larger than tension flange.

### 10.8.2 Beams with Other Cross Sections

When an I-shaped beam is bent about its weak axis (*y*-axis), making the *y*-axis the netural axis, lateral-torsional buckling will not occur (Salmon, et al. 2008).

**Channels** Unless the loads pass through the shear centre, a channel is subjected to combined bending and torsion. Usual loadings through the centroid or in the plane of the web give rise to such combined stress. For loads in the plane parallel to the web, lateral buckling must be considered (Salmon & Johnson 1996).

For design purposes, lateral torsional buckling equations of symmetrical I-shaped sections may be applied, which are found to err on the unsafe side by about 6% only in the extreme cases (Hill 1954).

**Zees** For Zees, the loading in the plane of the web causes unsymmetrical bending (see Section 10.19). The effect of biaxial bending on *Z*-section was found to reduce the critical moment $M_z$ to about 90% of the value given by Eqn (10.1) (Hill 1954). Salmon and Johnson (1996) suggest using one half of the values of $M_{cr}$ obtained by using Eqn (10.1).

**Tee and double angle sections** The T-section may be thought of as a mono-symmetric I-shaped section with moment of inertia $I_y$ of one flange equal to zero. Rolled T-sections will rarely have strength controlled by the lateral torsional buckling limit state. Ellifritt et al. (1992) provide some information about flexural strength of WT sections. The lateral torsional buckling moment for Tees and double angles is given by the American code (ANSI/AISC:360-10) as

$$M_{cr} = [\pi\sqrt{(EI_y GI_t)}/L][B + \sqrt{(1 + B^2)}] \quad (10.33)$$

where $B = \pm\ 2.3\ (d/L)\ \sqrt{(I_y/I_t)}$ and $d$ is the depth of the T-section.

The plus sign for $B$ applies when stem is in tension and the minus sign when it is in compression. If the tip of the stem is in compression anywhere along the unbraced length, the negative value of $B$ should be used.

The American code ANSI/AISC: 360-10 gives guidance for calculating the lateral-torsional buckling strength of channels, circular, rectangular, and square hollow sections, Tees and double angles, single angles, and other unsymmetrical shapes. Trahair (2001) provides an extensive coverage on the moment capacities of angle sections.

### 10.8.3 Strength Formula Considering Inelastic Buckling

Inelastic buckling is usually taken into account by specifying some type of empirical curve which gives essentially the elastic solution for long beams and which terminates at $M_{cr} = M_p$, the plastic moment for short beams (Galambos 1998).

It is assumed that beams and columns behave in a similar manner in the elastic range. Thus, the lateral torsional buckling strength is determined by using the following equivalent slenderness parameter

$$\lambda_{LT} = \sqrt{(M_p/M_{cr})} \quad (10.34)$$

and by using an empirical beam formula.

In the German code, the buckling resisting moment is given by

$$M_d = \chi_{LT} M_{rd} \quad (10.35a)$$

The value of $M_{rd}$ should be determined in conjunction with the section classification and material design strength $f_y$. The factored design moment $M_u \leq M_d$. Based on the numerous tests conducted in Europe, the German code DIN: 18800, gives the value of $\chi_{LT}$ as

$$\chi_{LT} = [1/(1 + \lambda_{LT}^{2n})]^{1/n} \quad (10.35b)$$

where $n$ is a coefficient, which can be varied to get a number of buckling curves to take into account the different cross sections. For simplicity, the value of $n = 2.5$ has been chosen in the German code (Lindner & Subramanian 1982, Lindner et al. 1994).

The AISC code (ANSI/AISC: 360-10) uses a straight line transition from elastic buckling curve at $M_{cr} = M_r$, $L = L_r$ to $M_{cr} = M_p$, $L = L_p$, where

$$M_r = 0.7 f_y Z_z \quad (10.36a)$$

$Z_z$ is the elastic section modulus and $f_y$ is the yield stress. The lateral-torsional buckling stress is then determined for simply supported beams under uniform moment by

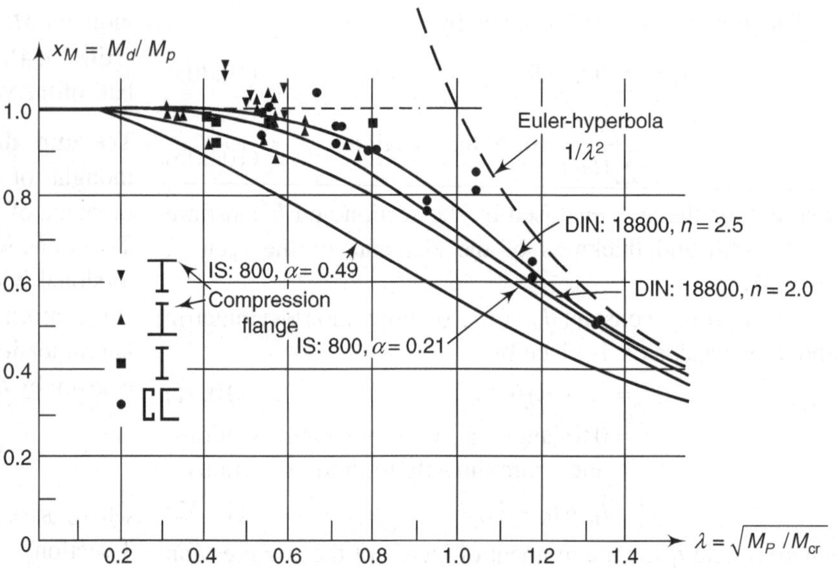


**FIG. 10.25** Comparison of Indian and German code formulae with experimental results

$$M_d = \begin{cases} M_p \text{ for } L \le L_p \\ C_1\,[M_p - (M_p - M_r)\,(L - L_p)/(L_r - L_p)] \le M_p \\ \quad \text{for } L_p \le L \le L_r \\ M_{cr} \text{ for } L \ge L_r \end{cases} \quad (10.36b)$$

where $L$ is the length between points that are braced against lateral displacement of compression flange or braced against the twist of the cross secion.

$$L_p = 1.76\, r_y \sqrt{E/f_y} = 790\, r_y / \sqrt{f_y} \quad (10.36c)$$

and $L_r$ is determined from Eqn (10.1) for $M_{cr} = M_r$ and $C_1 = 1.0$, and may be taken conservatively as $L_r = \pi r_{ts} \sqrt{E/(0.7 f_y)}$ [see Eqn (F2-6) of AISC 360:2010 for the exact expression for $L_r$]. For doubly symmetric section $r^2_{ts} = I_y h_o / (2 Z_z)$. The value of $r_{ts}$ may also be approximated accurately and conservatively as the radius of gyration of compression flange plus one sixth of the web (i.e., the portion of the beam under compresion) as

$$r_{ts} = b_f / \sqrt{[12(1 + h t_w / (6 b_f t_f))]} \quad (10.37)$$

In Eurocode 3, buckling curves similar to those of columns (see Section 9.7.1) are used. Indian code has also adopted the Eurocode 3 formulae (see Section 10.8). Fukumoto and his co-workers (Fukumoto et al. 1980) have also suggested formula based on statistical data of experimental results, which has been used by Yura et al. (1978) to arrive at resistance factors for beam design according to the AISC code formula.

A comparison of the DIN 18800 code formula with the Indian code formula and the experimental results are shown in Fig. 10.25. It has been taken to ensure that the resulting design curve corresponds to a mean value rather than lower bound to the experimental results. This decision was justified on the bases of the neglect of certain favourable effects, e.g., increased yield point, non-uniform moment, restraint from adjacent members, etc. (Lindner & Subramanian 1982). The value of $n = 2.0$ was suggested for unsymmetrical beams, when the smaller flange is compressed (Lindner & Gietzelt 1980). It is also seen from this figure that the Indian code (also Eurocode 3) formula provides a lower bound to the experimental results.

### 10.8.4 Compound/Built-up Beams

**Section classification** Compound sections are classified into plastic, compact, and semi-compact in the same way as discussed for rolled beams in Section 10.2. However, compound beams are treated as a section built-up by welding in the British code. (The Indian code has not specified this clearly.) The limiting width-to-thickness ratios have to be checked as follows (see Fig. 10.26).

1. Whole flange consisting of flange plate and rolled beam flange is checked using $b_1/t_f$, where $b_1$ is the total outstand of the compound beam flange and $t_f$ is the thickness of the flange of the rolled section.
2. The outstand $b_2$ of the flange plate from the rolled beam flange is checked using $b_2/t_p$, where $t_p$ is the thickness of the flange plate.
3. The width-to-thickness ratio of the flange plate between welds $b_3/t_p$ is checked, where $b_3$ is the width of the flange of rolled sections.
4. The rolled beam flange itself and the web must be checked as already discussed in Section 8.13 (Table 8.6 of Chapter 8).

**Moment capacity** The area of flange plates to be added to a given rolled sections to increase the strength by the required amount may be determined as follows, for a laterally restrained beam.

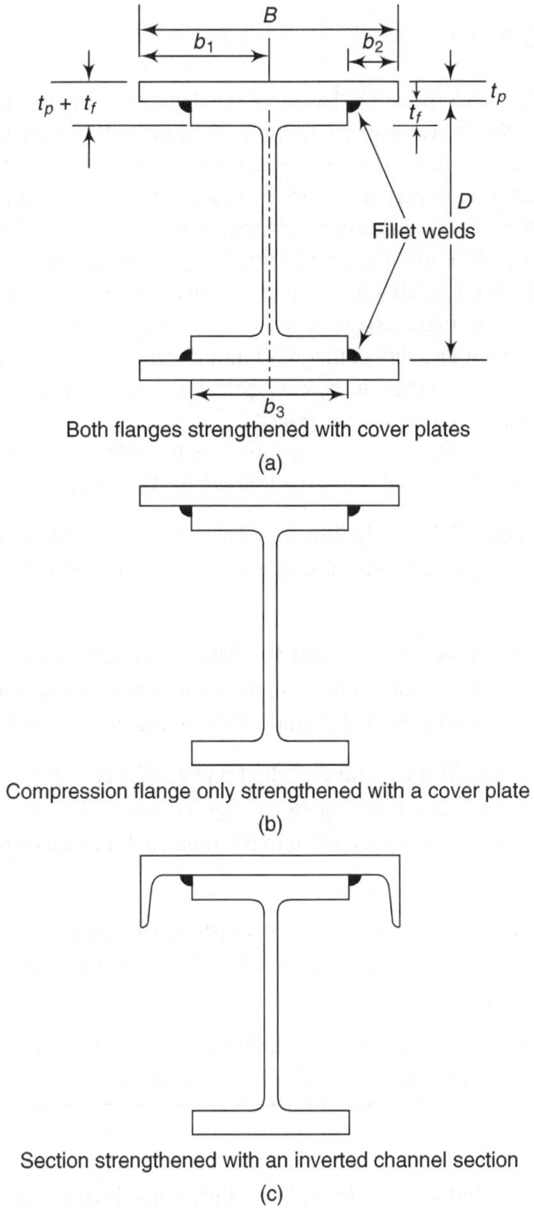


Both flanges strengthened with cover plates

(a)

Compression flange only strengthened with a cover plate

(b)

Section strengthened with an inverted channel section

(c)

**FIG. 10.26** Typical compound beam cross sections

Total plastic modulus required

$$Z_{p,z} = M/f_y \tag{10.38}$$

where $M$ is the applied factored moment. If $Z_{p,\text{rb}}$ is the plastic modulus of the rolled beam, the additional plastic modulus required is

$$\begin{aligned} Z_{p,\text{az}} &= Z_{p,z} - Z_{p,\text{rb}} \\ &= 2\,Bt_p(D + t_p)/2 \end{aligned} \tag{10.39}$$

where $Bt_p$ is the area of the flange plate and $D$ is the depth of the rolled beam. Using Eqn (10.39), dimensions of the flange plates can be quickly obtained. If the beam is not restrained, successive trials are required.

**Web** The web of the beam should be checked for shear, web buckling, and crippling at support and at points of concentrated loads.

**Welds connecting flange plates and beam flange** The fillet welds between flange plates and rolled beams should be designed to resist horizontal shear using elastic theory [Fig. 10.27(b)].

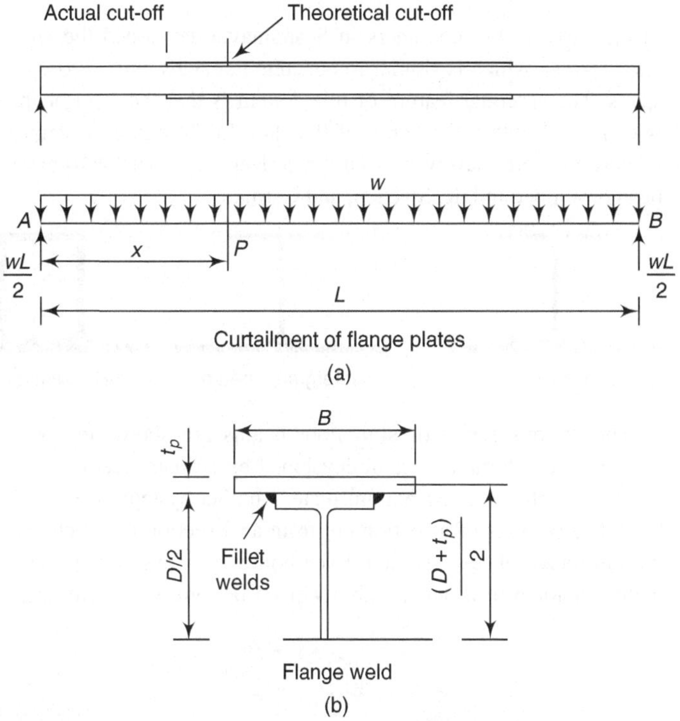


Curtailment of flange plates

(a)

Flange weld

(b)

**FIG. 10.27** Compound beam design

Horizontal shear in each fillet weld $= V_u B t_p (D - t_p)/4I_Z$ (10.40)

where $V_u$ is the factored shear force, $I_Z$ is the moment of inertia about the $Z$-$Z$ axis. The other terms have been defined earlier. The leg length can be selected from Table 6.6 using the minimum recommended size from Table 6.3 of Chapter 6. Intermittent welds may be specified, but continuous automatic welding considerably reduces the likelihood of failure due to fatigue or brittle fracture (MacGinley and Ang 1992).

## 10.8.5 Curtailment of Flange Plates

For a restrained beam with a uniformly distributed load, the theoretical cut-off points for the flange plates can be determined as follows [see Fig. 10.27(a)]. The moment capacity of the rolled beam

$$M_d = f_y Z_p \beta_b / \gamma_{m0} \le 1.2 Z_e f_y / \gamma_{m0} \tag{10.41}$$

Equate $M_d$ to the moment at $P$ at a distance $x$ from the support

$$wLx/2 - wx^2/2 = M_d$$

where $w$ is the factored uniform load and $L$ is the span of the beam. Solving for $x$, we will get the theoretical cut-off point. The flange plate should be continued beyond this point so that the weld on the extension can develop the load in the plate at the cut-off point.

## CASE STUDY

### Slim Floor Construction

In the early 1990s, engineers in Scandinavia developed the *slim floor system*, which is similar to concrete flat slabs, with 5 to 9 m spans. The essential feature of this system is that the steel beam is contained within the depth of the slab. In the earlier systems, precast concrete slabs were used. Later deep composite slabs have been developed in UK, Sweden, and Germany.

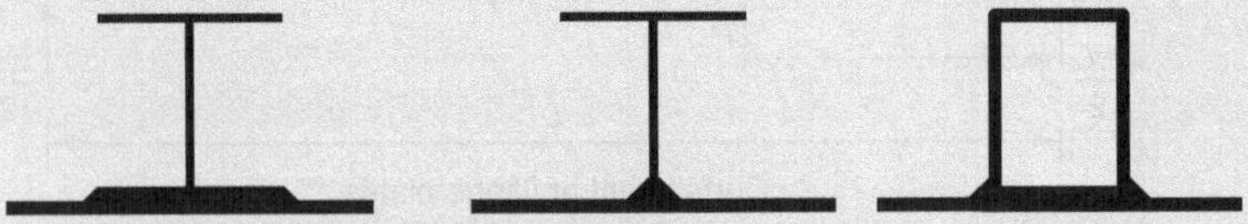


The various forms of slim floor beams are shown in these figures. The 'Slimflor' system developed by British Steel uses an I-section with a welded bottom plate. Another system developed by Arbed consists of a section cut from an I-section to which the bottom flange plate is welded. Even hollow sections, which have better torsional resistance, with a welded bottom plate have been used. Corus Corporation has developed another system called SLIMDEK®, which uses an asymmetric beam, is shown here.

Almost the whole steel section is protected from the fire by the floor slab. Hence periods of fire resistance up to 60 minutes are achievable without protection to the exposed bottom plate. Service ducts up to 160 mm deep × 320 mm long can be accommodated within the depth of Slimdek, by penetrating the web with circular or elongated openings. Extensive tests conducted at City University, London has shown that a 30 mm concrete cover to the top flange provides sufficient shear bond and hence welded shear studs are not necessary.

The slim floor system optimizes the effective volume of the building and offers the following advantages:

**Reduced floor thickness** This can be advantageous in tall buildings, where extra floors can be added for the same total height.

**Easy installation of equipment** The integrated beam makes it easier to build under-floor equipment (air-conditioning, piping, electrical networks, etc.) and simplify the fitting of false ceilings.

**Increased fire resistance** It eliminates the need for passive fire protection, resulting in savings in cost and time. It has built-in fire resistance of up to 60 minutes, for beams without web-opening.

**Economy** The amount of steel per square metre of floor is relatively low (in the range of 15 to 25 kg/m$^2$ for beam spans from 5 to 7.5 m)

More details and method of design of these slim floors may be found in Lawson, et al. 1997 and www.corusconstruction.com

### 10.8.6 Continuous Beams

The safe value for the critical load of a laterally continuous beam may be obtained by ignoring lateral continuity but taking account of the in-plane bending moment distribution in the beam. The effective length of each section is taken equal to the actual distance between supports and the critical moment is calculated for each segment. The lowest moments are associated with the critical segment (see Fig. 10.28). This segment may then be examined in detail, by calculating the stiffness ratios between it and the adjacent elements and thus determining the effective length. This effective length may then be used in a rework of the original calculations. This method gives accurate conservative results except when the moment gradients and restraint stiffness are unusually high (Nethercot & Trahair 1976).

The behaviour of beams, which are continuous in both transverse and lateral planes [Fig. 10.28(b)], is essentially similar to that of the laterally continuous beams, the only difference being more complex patterns of bending moments are present in the former case [see Fig. 10.28(b)]. It should be noted that the distances between points of zero moments should not be taken as effective lengths for checking the lateral buckling strengths, since the whole beam will be involved in the buckling. Hence the length of the beam between lateral restraints only should be used for the buckling calculations (Kirby & Nethercot 1979).

## 10.9 SHEAR STRENGTH OF STEEL BEAMS

Since shear force generally exists with bending moments, the maximum shear stress in a beam is to be compared with the shear yield stress. Though bending will govern the design in most steel beams, shear forces may control in the cases where the beams are short and carry heavy concentrated loads. The pattern of shear stress distribution in I-section is

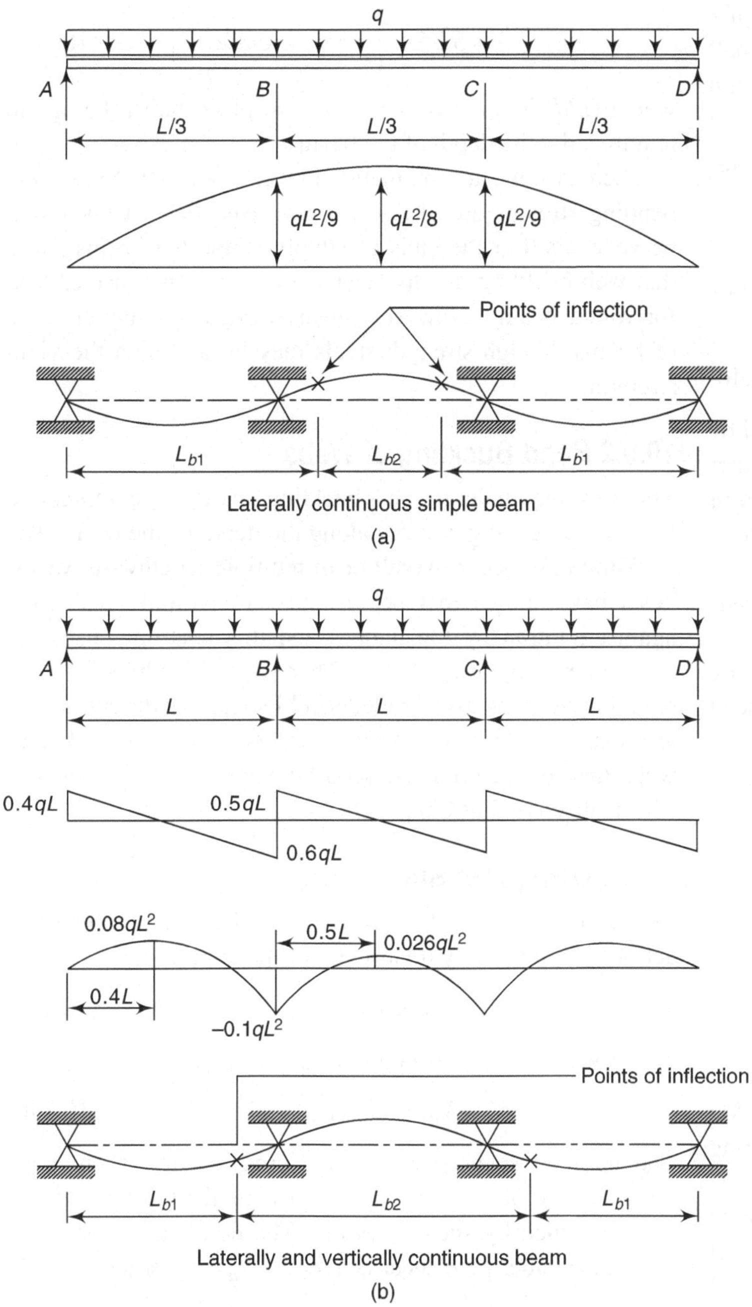


**FIG. 10.28** Laterally and vertically continuous beams

shown in Fig. 10.29. It may be seen that shear stress varies parabolically with depth, with the maximum occurring at the neutral axis.

Let us take the case of an I-beam subjected to the maximum shear force (at the support of a simply supported beam). The external shear $V$ varies along the longitudinal axis $x$ of the beam with bending moment as $V = dM/dx$. For beams of open cross section subjected to no twisting, the internal shear stresses $\tau$ which resist the external shear can be written as,

$$\tau = \frac{VQ}{I_z t} \tag{10.42}$$

where $V$ is the shear force at the section, $I_z$ is the moment of inertia of the entire cross section about the neutral axis, $Q\ (= A\bar{y})$ is the static moment of the cross section (above the location at which the stress is being determined) about the neutral axis and $t$ is the thickness of the portion at which $\tau$ is calculated. Using this equation, the maximum shear stress at the centroidal axis can be evaluated. These values for some simple profiles are given here.

Rectangular section $\tau_{max} = 3V/(2A)$ (10.43a)

Solid circular section $\tau_{max} = 4V/(3A)$ (10.43b)

Thin walled circular hollow section $\tau_{max} = 2V/A$ (10.43c)

$$\text{I-section } \tau_{max} = \frac{V}{I_z t_w}\left[\frac{b_f t_f (d + t_f)}{2} + \frac{t_w d^2}{8}\right] \tag{10.43d}$$

where $d = h - 2t_f$, $h$ is the height of the beam, $b_f$ and $t_f$ are the breadth and thickness of the flange, $t_w$ is the thickness of the web, and $I_z$ is the moment of inertia of beam about the $z$-$z$ axis.

Equation (10.43d) is plotted in Fig. 10.29(a), which represents shear stresses in the elastic range. It is observed from Fig. 10.29(a) that a significant proportion of shear force is carried by the web and the shear stress distribution over the web area is nearly uniform. Hence, for the purpose of design, we can assume without much error, the average shear stress for most commonly adopted sections (such as I, channel, T, etc.) which is given by

$$\tau_{av} = \frac{V}{t_w d_w} \tag{10.44}$$

where $t_w$ is the thickness of the web and $d_w$ is the depth of the web. Whenever there are bolt holes in the web, this stress is multiplied by the ratio of gross web area/net web area. The nominal shear yielding strength of webs is based on the von Mises yield criterion (see Section 2.10

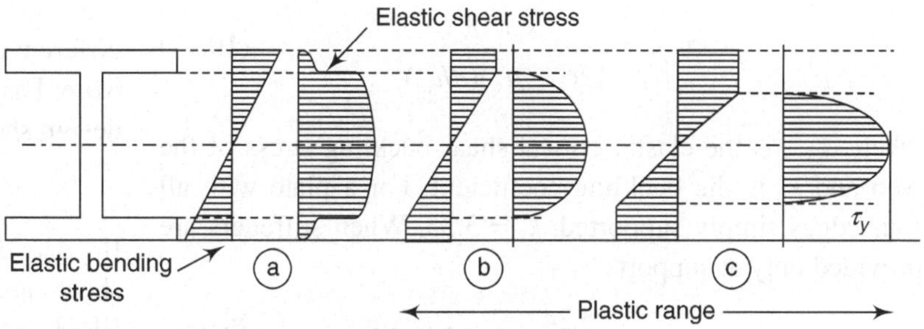


**FIG. 10.29** Combined bending and shear in beams

of Chapter 2), which states that for an unreinforced web of a beam, whose width to thickness ratio is comparatively small (so that web-buckling failure is avoided), the shear strength may be taken as

$$\tau_y = \frac{f_y}{\sqrt{3}} = 0.58f_y \qquad (10.45)$$

where $f_y$ is the yield stress.

Taking the shear yield stress as 60% of the tensile yields stress, $V_p$ can be written as

$$V_p \approx 0.6f_y t_w d_w \qquad (10.46)$$

This expression gives the nominal shear strength provided by the web when there is no shear buckling of the web. Whether that occurs will depend on $d_w/t_w$, the depth-thickness ratio of the web. If this ratio is too large, i.e., the web is too slender, the web may buckle in shear either inelastically or elastically.

When the shear capacity of the beam is exceeded, *shear failure* occurs by excessive shear yielding of the gross area of the webs as shown in Fig. 10.30. Shear yielding is very rare in rolled steel beams. Shear is rarely a problem in rolled steel beams, the usual practice being to design the beam for flexure and then to check it for shear.

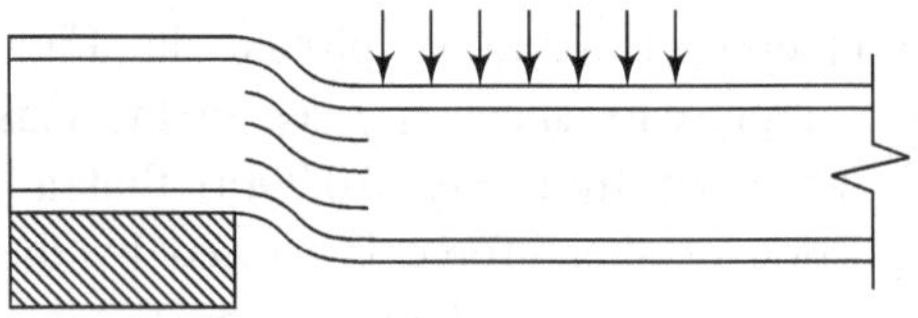

**FIG. 10.30** Shear yielding near support

### 10.9.1 Shear Buckling of Beam Webs

Since the web of an I-beam is essentially a plate, it may buckle due to shearing stresses which are less than the shearing yield strength of steel. In a plate subjected to pure shear, the shear stresses are equivalent to principal stresses of the same magnitude, one tension and another compression, acting at 45° to the shear stresses. This is shown in the Fig. 8.20(c) given in Chapter 8. Buckling takes place in the form of waves or wrinkles inclined at around 45°.

As discussed in Chapter 8, the shear stress at which buckling of a perfect plate takes place is given by

$$\tau_{cr,\,v} = \frac{k_v \pi^2 E}{12(1-\mu^2)(d/t_w)^2} \qquad (10.47)$$

where $\tau_{cr,\,v}$ is the elastic critical shear buckling stress of the web and $k_v$ is the buckling coefficient. For a plate with all four edges simply supported, $k_v$ = 5.35. When stiffeners are provided only at supports

$$k_v = 4 + \frac{5.34}{(L/d)^2} \qquad L/d \le 1 \qquad (10.48a)$$

$$= 5.35 + \frac{4}{(L/d)^2} \qquad L/d \ge 1 \qquad (10.48b)$$

where $(L/d)$ is the aspect ratio of the plate with $L$ being the length and $d$ the depth of the beam.

There is some approximation in using Eqn (10.48) as some bending stresses are always present. But these stresses will be very small at the ends of simply supported beams. Note that web buckling due to shear is not a design consideration for rolled beams. However, shear strength of thinner webs of beams of high-strength steels may be less than the yield strength.

### 10.9.2 Bend Buckling of Webs

The web may undergo local buckling due to the compressive part of the bending stresses along the depth of the beam. The buckling may occur overall or in multiple lengthwise waves. Tests have shown that beam webs are partially restrained against rotation by the flanges and this restraint raises the critical stress by at least 30%. There is no likelihood of bend buckling of the webs of rolled steel I shapes as the $d/t$ ratio is less than 67. However, in plate girders having much thinner webs than rolled sections, bend buckling deserves attention. This is discussed in Chapter 11.

### 10.9.3 Design for Shear

The factored design shear force $V$ in a beam due to external actions should satisfy (clause 8.4 of the code)

$$V \le V_d \qquad (10.49a)$$

where $V_d$, the design strength, is given by

$$V_d = V_n/\gamma_{m0} \qquad (10.49b)$$

The nominal shear strength of a cross section $V_n$ may be governed by plastic shear resistance or the strength of the web governed by shear buckling. The nominal plastic shear resistance under pure shear is gven by $V_n = V_p$, where

$$V_p = \frac{A_v f_{yw}}{\sqrt{3}} \qquad (10.50)$$

$A_v$ is the shear area and $f_{yw}$ is the yield strength of the web. The shear area may be calculated as given in Table 10.12 for different cross sections (Nethercot 1991).

*Note*: Fastener holes need not be accounted for in the plastic design shear strength calculation provided that

$$A_{vn} \ge (f_y/f_u)(\gamma_{m1}/\gamma_{m0})A_v/0.9$$

If $A_{vn}$ does not satisfy this condition, the effective shear area may be taken as that satisfying this limit. Block shear failure criteria may be verified at the end connections.

**TABLE 10.12** Shear areas

| Section | Shear Area $A_v$ |
|---|---|
| 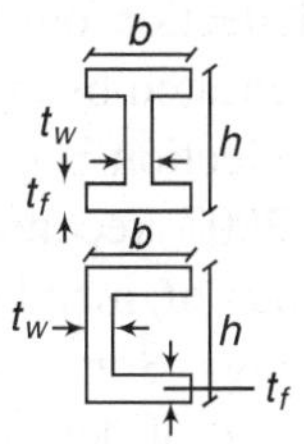  Hot-rolled (a) | $ht_w$<br>$ht_w$ |
|   Built-up (b) | $dt_w$ |
| 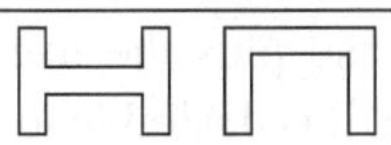 Built-up sections or rolled (minor axis bending) (c) | $2bt_f$ |
| 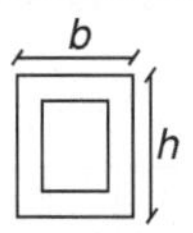  Rectangular hollow section (d) | $Ah/(b + h)$ |
| 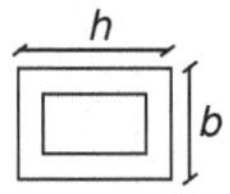  Rectangular hollow section (loaded parallel to width) (e) | $Ab/(b + h)$ |
| 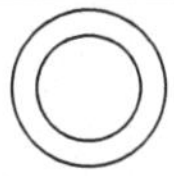 Circular hollow section (f) | $2A/\pi$ |
| 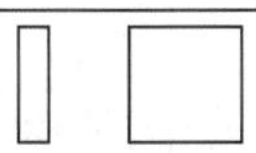 Plates and solid bar (g) | A |

## 10.10 MAXIMUM DEFLECTION

Although excessive vibration and excessive deflection are included in the 'limit state of serviceability', the codes usually limit only the deflection. A beam designed to have adequate strength may become unsuitable if it deflects excessively under the loads. Beams that deflect too much may not normally lead to a structural failure, but nevertheless endanger the functioning of the structure. For example, excessive deflection in a floor not only gives a feeling of insecurity, but also damages the non-structural components (cracking of plaster ceilings) attached to it. Excessive deflections in industrial structures often cause misalignment of the supporting machinery and cause excessive vibration. Similarly, high deflections in purlins may cause damage to the roofing material. Excessive deflection in the case of a flat roof results in accumulation of water during rainstorms called *ponding*.

Ponding has led to a large number of flat roof failures (which occur every year) in the United States. The best method of preventing ponding is to provide appreciable slope on the flat roof (20 mm/m or more) together with good drainage facilities. It has been estimated that about two-thirds of the flat roofs in the United States have slopes less than this minimum recommended value. It may cost about 3–6% more to construct a roof with this desired slope than building with no slope. When a large quantity of water is present on the roof, the strong winds that accompany rain storms, will push the water to one side of the roof, creating a large load on the roof beams (Ruddy 1986). Ponding failures can be prevented if the roof system (consisting of roof deck and supporting beams and girders) has sufficient stiffness. If this minimum stiffness is not provided, it is necessary to do calculations to make sure that a ponding failure is not possible. Theoretical calculations for ponding are complicated and finite element solutions are also available (Colombi 2006). The American code (ANSI/AISC: 360-10) provides some guidelines based on the work by Marino (1966).

Crane misalignment is another consequence of excessive deflection. Since these affect the performance of the structure in its working conditions, the serviceability check is done at working load levels. The deflection in beams is restricted by codes of practice by specifying deflection limitations which are usually in terms of deflection to span ratio (see Table 2.8). Note that the limits shown in Table 2.8 of Chapter 2 are for live load deflection only, as dead load deflection is assumed to be compensated by *cambering* of steel beams. Hence, only live load deflection is of concern in the completed structure. IS 800 : 2007 recommends maximum deflection values for a number of member types in industrial buildings and other buildings. Some specifications may include a limiting allowable ratio of depth of section to span. For example, the allowable stress design method of the American code suggests a limit of $L/d \leq 5500/f_y$ where $f_y$ is the yield stress in MPa, for floor beams and girders (fully stressed) and $L/d \leq 20$ when they are subjected to shock and vibration. The deflection limit is specified mainly to prevent the cracking of plaster.

The code also recommends that when the deflection, due to dead load plus the live load combination, is likely to be excessive, it is desirable to pre-camber the beams, trusses, and girders. For spans greater than 25 m, camber approximately equal to the deflection due to dead load plus half the live load may be used. It is usually more economical to select heavier

| Beam and Load | Maximum Moment | Deflection at Centre |
|---|---|---|
| 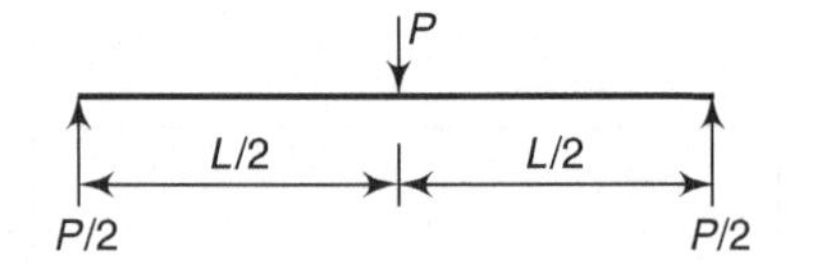 | $PL/4$ | $\frac{PL^3}{48EI}$ |
| 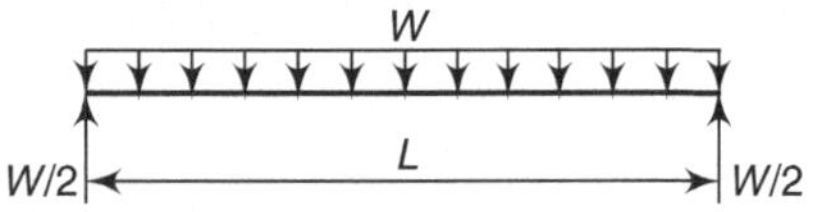 | $WL/8$ | $\frac{5WL^3}{384EI}$ |
| 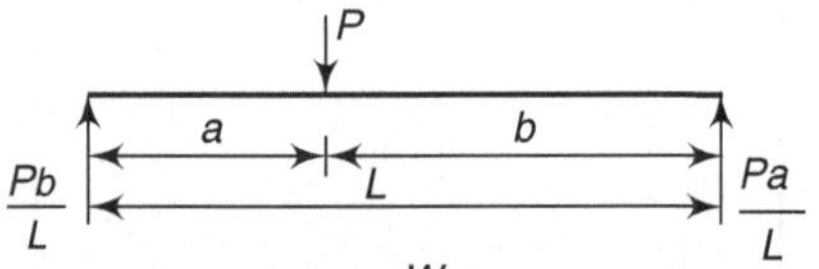 | $Pab/L$ | $\frac{PL^3}{48EI}\left[\frac{3a}{L} - 4\left(\frac{a}{L}\right)^3\right]$ |
| 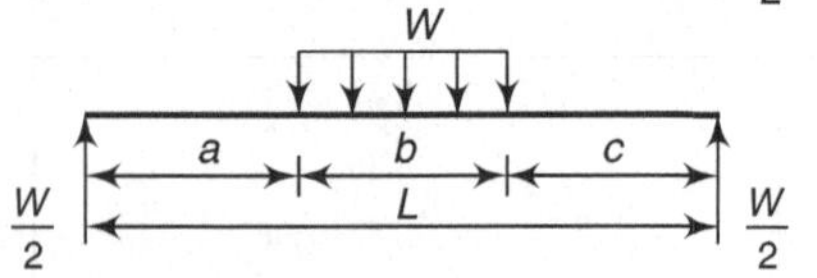 | $W\left(\frac{a}{2} + \frac{b}{8}\right)$ | $\frac{W}{384EI}[8L^3 - 4Lb^2 + b^3]$ |
| 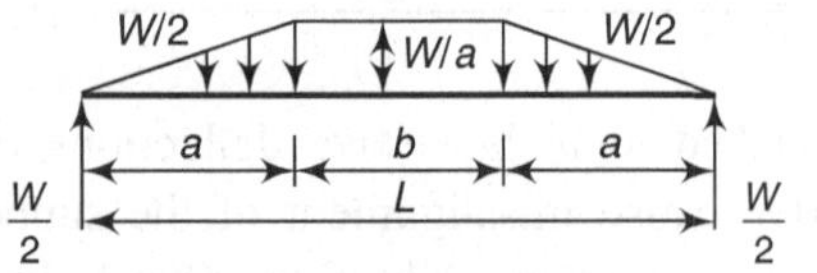 | $Wa/3$ | $\frac{W}{120EI}[16a^2 + 20ab + 5b^2]$ |
| 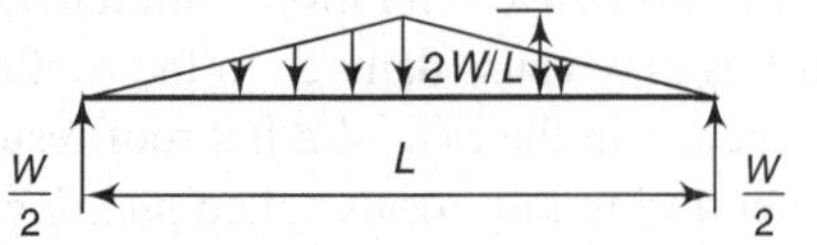 | $WL/6$ | $\frac{WL^3}{60EI}$ |
| 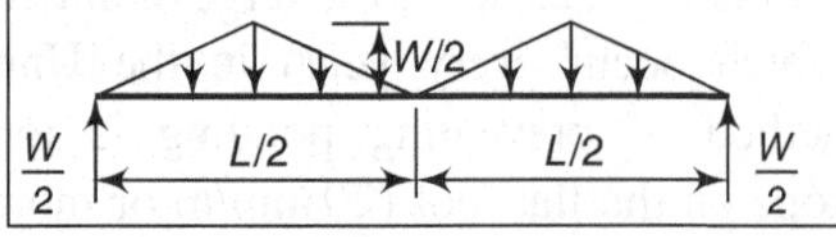 | $WL/8$ | $\frac{WL^3}{73.14EI}$ |

**FIG. 10.31** Simply supported beams: maximum moments and deflections

beams with larger moments of inertia to limit deflection than it is to camber sections. Because cambering requires exact dimensioning of members for proper fitting, it will result in extra cost for fabrication and erection. Hence often larger beam sections are used instead of cambering. Also if a higher strength steel is used, it may probably be economical to use Fe 410 grade steel for the members that deflect excessively. More discussions on deflections, camber, and other serviceability limit states are provided by Galambos and Ellingwood (1986), Ellingwood (1989), Ricker (1989), Murray et al. (1997), and West et al. (2003). Deflections for some common load cases for simply supported beams together with the maximum moments are given in Fig. 10.31. For general load cases, deflections can be calculated by moment area, conjugate beam, or virtual work procedures (Thandavamoorthy 2005). Some guidelines about design against floor vibration are given in Annex C of the code.

## 10.11 WEB BUCKLING AND WEB CRIPPLING

A heavy load or reaction concentrated on a short length produces a region of high compressive stresses in the vertical elements of the web either under the load or at the support. Away from the loading point, stresses are spread widely. But just under a load or above a reaction point, they may cause web failures such as web buckling as shown in the Fig. 10.32(b) and web crippling or web crushing as shown in Figs. 10.32 (a) and (c). Web buckling or vertical buckling occurs when the intensity of vertical compressive stress near the centre of the section becomes greater than the critical buckling stress for the web acting as a column. Tests indicate that for rolled steel beams, the initial failure is by web crippling rather than by buckling. But for built-up beams having greater ratios of depth to thickness of web, failure by vertical buckling may be more probable than failure by web crippling. The suggested solutions to the problem include spreading of the load over a longer portion of the flange, provision of web stiffeners at points of load, and reaction or thickening of the web. Since web crippling occurs only at few sections, it is economical to provide stiffeners at these sections.

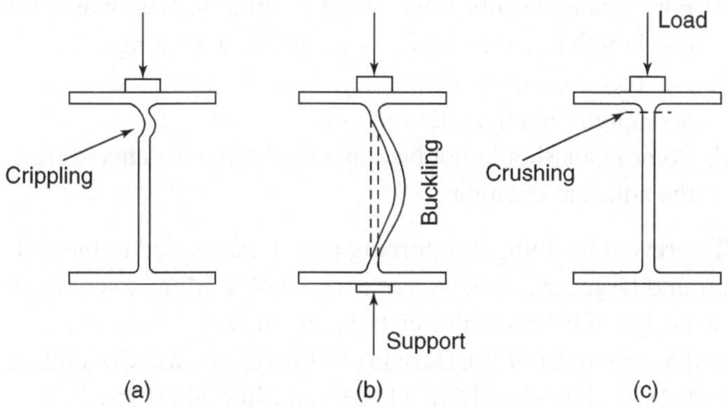


**FIG. 10.32** Local buckling of the web

In the case of web buckling, the web may be considered as a strut restrained by the beam flanges. Such *idealised struts* should be considered at the points of application of concentrated load or reactions at the supports as shown in Fig. 10.33 and Fig. 10.34.

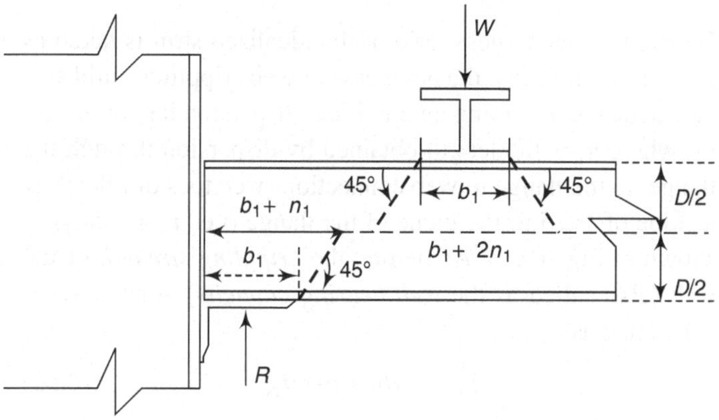


**FIG. 10.33** Dispersion of concentrated loads and reactions for evaluating web buckling

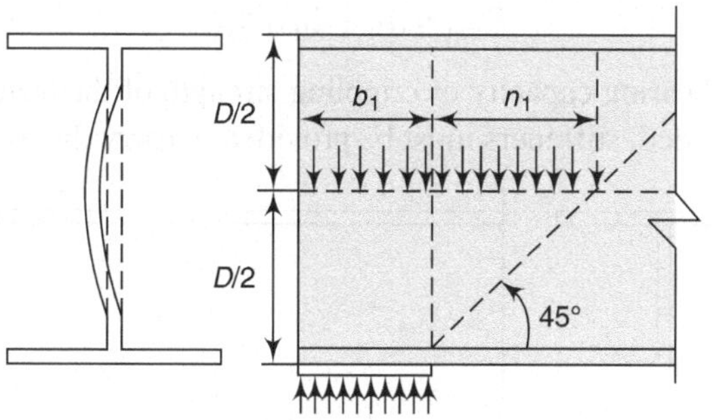


**FIG. 10.34** Effective width for web buckling

In both the cases, the load is spread out over a finite length of the web as shown in Fig. 10.33. This is known as the *dispersion length* and its theoretical treatment is complex. Hence empirical formulae based on experiments are used. One such assumption is that the dispersion length is taken as $(b_1 + n_1)$ where $b_1$ is the stiff bearing length and $n_1$ is the dispersion of 45° line at the mid-depth of the section as shown in Fig. 10.34. Hence the *web buckling strength* at the support is given by

$$F_{wb} = (b_1 + n_1)tf_c \tag{10.51}$$

where $t$ is the web thickness and $f_c$ is the allowable compressive stress corresponding to the assumed *web strut*. The effective length of the strut is taken as $L_E = 0.7d$ where $d$ is the depth of the *strut* in between the flanges. The slenderness ratio of the idealised web strut could be written as

$$\lambda = \frac{L_E}{r_y} = \frac{0.7d}{r_y}$$

Since, $$r_y = \sqrt{\frac{I_y}{A}} = \sqrt{\frac{t^3}{12t}} = \frac{t}{2\sqrt{3}} \quad (10.52)$$

$$\frac{L_E}{r_y} = 0.7d\frac{2\sqrt{3}}{t} \approx 2.5\frac{d}{t}$$

Hence, the slenderness ratio of the idealized strut is taken as $\lambda = 2.5d/t$. Similarly, the latter case of web crippling could also be calculated assuming an empirical dispersion length of $b_1 + n_2$, where $n_2$ is the length obtained by dispersion through the flange, to the flange to web connection (web toes of fillets), at a slope of 1:2.5 to the plane of the flange (i.e., $n_2 = 2.5d_1$) as shown in Fig. 10.35. As before, the *crippling strength* of the web (also called as the *web bearing capacity*) at supports is calculated as

$$F_{crip} = (b_1 + n_2)\, tf_{yw} \quad (10.53)$$

where $f_{yw}$ is the design yield strength of the web. At an interior point where concentrated load is acting, the crippling strength is given by

$$F_{crip} = (b_1 + 2n_2)\, tf_{yw} \quad (10.54)$$

If this bearing capacity or crippling strength of the beam web is exceeded, stiffeners must be provided to carry the load.

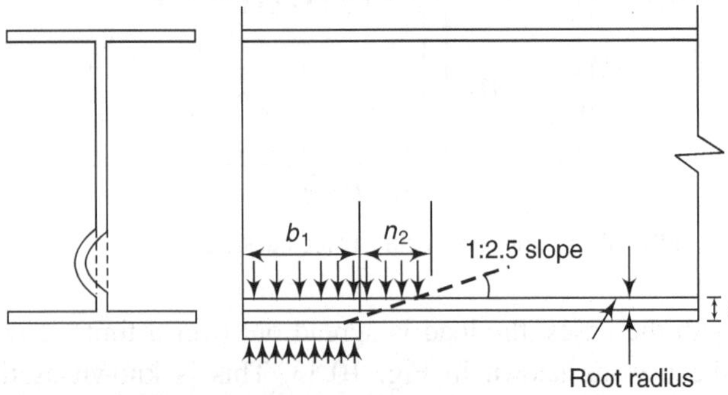


**FIG. 10.35** Effective width of web bearing

## 10.12 HOLES IN BEAMS

If beam connections are made with bolts, holes will have to be provided in the beam either in the web or flanges. These holes are relatively small and their effect will be small for flexure. The codal provisions for such cases are presented in Section 10.7.1. The strength of a steel beam depends on the strength of the compression flange. Thus, an open hole in the compression flange affects the strength of a steel beam more than a hole in the tension flange. A hole in the flange of a beam causes an increase in stress. If the hole in the compression flange contains a rivet or bolt, the strength reduction is lessened as the fastener can transmit compression. If the hole is located near the region of highest flexural stress, it may affect the ultimate strength of the beam. The increase in stress is of two types: (a) increase caused by redistribution of flexural stress to a smaller area and (b) concentration of stress near the edge of the hole. Analysis of stresses in a beam with holes depends on the effect of the holes on the position of the neutral axis. In general, if the holes do not remove more than 15% of the gross area of that flange, they are neglected provided there is no question of fatigue.

Holes in beam webs have less effect on flexural strength than holes in the flanges. Larger holes are sometimes cut in beam webs to provide space for utilities. Ideally these holes should be placed in the web only at sections of low shear. In the flanges, holes should be cut at points of low bending moment. If this is not possible, the effect of the holes should be accounted for. In some cases, the beam must be reinforced in the vicinity of the hole by welding doubler plates or by stiffening angles. The introduction of an opening in the web of the beam alters the stress distribution within the member and also influences its collapse behaviour. The shape of the web openings is usually regular (circular or rectangular). These openings decrease the stiffness of the beams resulting in larger deflections. The strength of the beams with openings may be governed by the plastic deformations due to both moment and shear at the openings. The strength realised will depend upon the interaction between the moment and shear. The reduction in moment capacity at the openings is small while the reduction in shear capacity may be significant. Figure 10.36 shows common types of web openings in beams.

General guidelines for design of beams with openings are as follows:

1. The hole should be centrally placed in the web and eccentricity should be avoided
2. The best location for the opening is within the middle one-third of the span
3. Web opening should be away from the support by twice the beam depth
4. Clear spacing between openings should be more than the beam depth $D$
5. The diameter of circular openings should be restricted to $0.5D$
6. For rectangular openings, depth should be less than $0.5D$ and length less than $1.5D$ for unstiffened openings
7. For rectangular openings, depth should be less than $0.7D$ and length less than $2D$ for stiffened openings
8. Point loads should not be applied within a distance $d$ from the adjacent opening

To prevent buckling, reinforcing plates are welded to the web around large size clearance or access holes. More discussions on design of beams with web holes are provided by Narayanan and Avanessian (1985), Darwin (1990), ASCE Task Committee (1992), Redwood and Cho (1993), and Burton (1995).

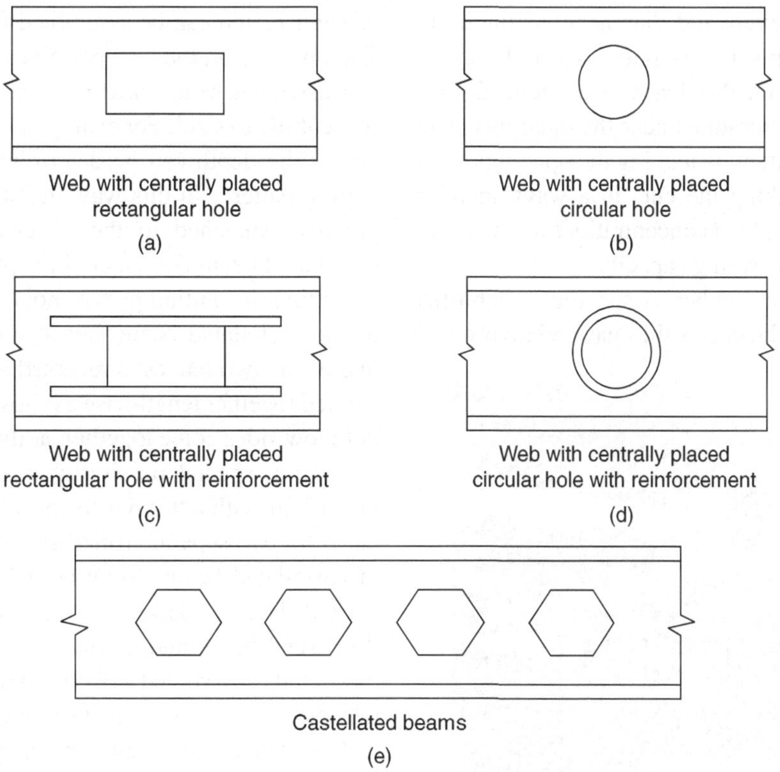


**FIG. 10.36** Holes in beams

## 10.13 LATTICED BEAMS

*Latticed beams* or *open-web steel joists* are generally prefabricated trusses as shown in Fig. 10.37. They are used to directly support floor and roof systems in buildings. For smaller joists, the web members may be continuous circular bar (most often found in Indian railway stations) which are welded to the flange members (angles or Ts) and they are referred to as *bar joists*. For the same span, an open-web joist will be lighter than a rolled shape and the openings in the web give an added advantage of allowing routing of utilities. Depending upon span, an open web joist may be more economical than rolled shapes.

Invariably, all open-web steel joists are designed as simply supported trusses with uniformly distributed loads on the top chord. Due to this loading, the top chord is subjected to bending as well as axial compression for which the member should be designed as a beam-column. In order to ensure lateral stability of the top chord, the floor or roof deck must be attached to the top chord such that it forms a continuous lateral support. The bottom chord, which is always in tension, is designed as a tension member. The web members are designed for vertical shear forces determined from the loading.

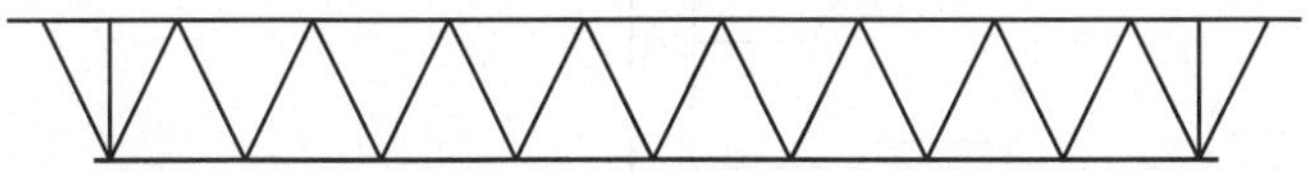

**FIG. 10.37** Latticed beam

Lateral stability of open-web girders needs careful consideration. Lateral stability is inversely proportional to the depth of the girder. For a given flange section, the lateral buckling load reduces in proportion with the increase in the depth of the girder. The spacing of web elements has negligible influence on the lateral buckling strength of chord members. The most efficient inclination for the web rods is found to be 45°.

Floors and roof decks usually consist of cast-in-place or precast concrete slabs. Open-web joists are available in standard depths and load capacities from manufacturers. Some open-web joists are designed to function as floor and roof joists, while others may function as girders supporting concentrated reactions from joists.

## 10.14 CASTELLATED BEAMS

A beam with a number of regular openings in its web is called a castellated beam. A castellated beam is formed by flame cutting a single rolled wide flange beam in a definite pattern and then rejoining the segments in such a way as to form a beam with a number of regular openings in its web. The new section with holes will have a depth at least 50% more and its section modulus increased by 2.25 times the original section. This allows the beams to span further than the parent rolled section, with the advantage that services may be accommodated within the depth of the section. The applied forces are carried by a combination of pure flexure and Vierendeel action in the webs caused by shear within the discontinuous web panels. The castellated beams have the disadvantage of reduced shear capacity due to stress concentrations near the openings. This can be alleviated to some extent by making the openings close to the neutral axis and making the cuts in a wavy manner. Stiffening can be provided at load concentrations and reaction points to improve its shear carrying capacity.

Castellated beams have a number of specific possibilities and advantages. They are light, cheap, they have relatively high resistance, and can be assembled fast at the construction site. The openings in the web permit easy installation of utilities.

Castellated beam in Danbury Garage, Connecticut (note the perpendicular bracings, which increase lateral stability)
*Photo:* Mr Tejas Mehta

Castellated beams offer a designer all kinds of opportunities for 'cutting to size'. For example, in simple straight castellated beams the depth can be determined at will by changing the cutting pattern. In this way, the strength of the beam can be precisely matched to the occurring loads. Additionally, a castellated beam that tapers in depth can quite easily be made by setting the cutting pattern not exactly parallel to the length of the castellated beam, but at a slight angle. After cutting, one of the two halves is reversed and the two halves are then welded together lengthwise as shown in Fig. 10.38. At one end both low sides come together, at the other end both high sides.

A third special possibility is the combining of a lighter upper half with a heavier lower half. Obviously, these will be cut from two separate rolled sections. As long as the number of castellated beams to be built is even, no material will be wasted. This last design is attractive, for example, when a sub-floor steel beam and a concrete floor are combined. In that case, studs are welded to the top flange of the castellated beam in order to ensure firm anchoring of the concrete to the beam.

Castellated beams can very easily be cambered and cranked. Steeply curved castellated beams are constructed from halves cut from separate rolled sections. The tooth widths and pitches of the inner and outer halves are set differently, to ensure that after cambering they will perfectly match one another. Curved

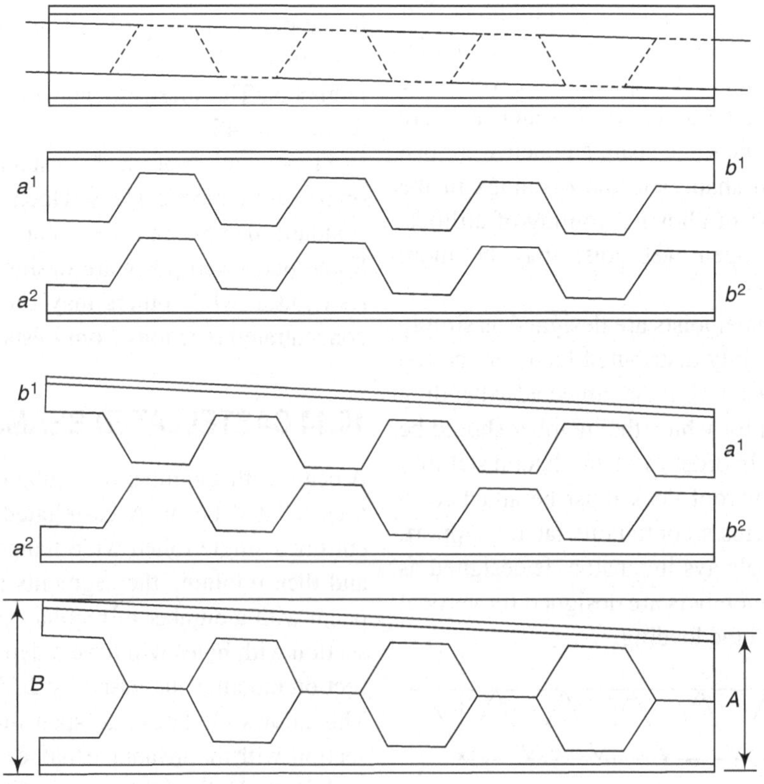


**FIG. 10.38** Castellated beams

castellated beams can be fabricated in a wide range of types; for example, circular, elliptical, and parabolic, to name a few.

In certain instances the choice of castellated beams may not be desirable; for example, when a beam is subjected to substantial concentrated loads, or when the castellated beam is used as a continuous beam across several supports. When very high requirements are set for fire resistance, castellated beams could be less attractive since the fire resistant coating has to be around 20% thicker than for rolled sections in order to obtain the same fire resistance.

There are a number of possible failure modes for castellated beams, which are as follows (Kerdal & Nethercot 1984; Zirakian & Showkati 2006).

***Vierendeel mechanism*** This occurs due to excessive deformation across one of the openings in the web and formation of hinges in the corners of the castellation.

***Lateral torsional buckling of the web*** This is caused by large shear at the welded joint.

***Rupture of welded joint in the web*** This arises due to excessive horizontal shear at the welded joint in the web.

***Lateral torsional buckling of the overall beam*** Castellated beams undergo lateral torsional buckling of the whole span.

***Plastic hinge formation*** This mode of failure occurs when lateral torsional buckling is prevented. Areas of plasticity are formed in the top and bottom flanges caused by a combination of stresses due to primary flexure and Vierendeel action. The hinge may form away from point of maximum bending moment due to the presence of shear and axial forces.

***Web buckling*** This is caused by heavy loading and short span of the beam. This may be avoided at a support by filling the first castellation by welding a plate in the hole.

The design of a castellated beam has to be carried out in such a way that all these failure modes are considered (Knowles 1985; Martin & Purkiss 1992).

**Design for flexure** There will be high primary bending stresses in a castellated beam. In addition, there will be bending and shear stresses due to Vierendeel action. The presence of high shear and axial forces may produce a reduction in the plastic moment capacity of the section. Consequently, the design process becomes iterative as the critical point for the design may not be the point of maximum moment. In the literature, a series of design interaction graphs are given to make the design more convenient (Knowles 1985). It is also suggested that a beam may be over designed by 5% for flexure. The moment capacity of a castellated beam is given by

$$M = f_y Z_p \tag{10.55}$$

where $Z_p$ is calculated at the centre line of a castellation. The design of a laterally unrestrained beam is similar to a plate girder, but considering the cross section properties on the centre line of the castellation (Nethercot & Kerdal 1982), the following should be done.

*Shear*: The shear stress is limited to $0.7f_y$ and is calculated by elastic method.

*Local buckling*: This is checked based on codal provisions.

*Web buckling*: While considering web buckling, the width of the web has to be taken at mid-height of the beam.

*Deflection*: The deflection of a castellated beam is caused both by primary flexure as well as by the Vierendeel action.

$$\delta_t = \delta_b + \delta_s \tag{10.56}$$

The usual procedure may be followed for calculating the deflection due to primary flexure, considering that the value of second moment of area is calculated at the centre of the castellation. In the design of castellated beams, rules given here are followed to get an economic and reliable design. The angle of cutting in the centre is kept as 45°. The provided depth of the beam at the minimum cross section is not less than one fourth the depth of the original beam.

## 10.15 LINTELS

Above openings in buildings such as for doors and windows, beams are provided to support the masonry that comes in between the opening and the slab above. Such beams are known as *lintels*. A lintel is preferably built flush with the sides of the wall. Flat and plate sections are used for small openings. For openings of moderate dimensions, back-to-back angles and inverted T-sections are best options. Channels and I-sections are the ideal choice for large openings.

The design of the lintels is carried out for a small portion of the masonry load above the opening. Two cases of loading arise depending upon the height above the lintel.

*Case 1*: If an equilateral triangle can be formed on the lintel and the slab is above the apex of the triangle, then the load of the masonry in the triangle is assumed to be acting over the beam. The maximum moment in this case is taken as $WL/6$.

*Case 2*: If the apex of the equilateral triangle formed on the lintel falls above the slab, then the load of the masonry in the rectangle formed by the lintel and the vertical height between the lintel and the slab acts on the lintel.

The lintel may be designed either as a laterally supported beam or as a laterally unsupported beam.

## 10.16 PURLINS

Purlins are beams used on trusses to support the sloping roof system between the adjacent trusses. Channels, angle sections, and cold formed C- or Z-sections are widely used as purlins. They are placed in an inclined position over the main rafters of the trusses. To avoid bending in the top chords of roof trusses, it is theoretically desirable to place purlins only at panel points. For larger trusses, however, it is more economical to space purlins at closer intervals. In India, where asbestos cement (AC) sheets are used, the maximum spacing of purlins is also restricted by the length of these sheets. AC sheets (though banned in many countries for the risk of lung cancer while working with them) provide better insulation to sun's heat (compared to GI sheets), which can be further improved by painting them white on the top surface. The maximum permissible span for these sheets is 1.68 m. A longitudinal overlap of not less than 150 mm is provided for AC sheets. The purlin spacing is so adjusted with lengths of the sheets that the longitudinal overlaps fall on the purlins to which they are directly bolted. Spacing of purlins should be so fixed that the cutting of sheets is avoided. Hence in practice when AC sheets are used, the purlin spacing is kept between 1.35 to 1.40 m. But in general, purlins are spaced from 0.6 m to about 2 m and their most desirable depth to span ratio is about 1/24. While the dead loads act through the centre of gravity of the purlin section, the wind loads act normal to the roof trusses. Thus, the purlin section is subjected to bending and twisting resulting in unsymmetrical bending.

Purlins may be designed as simple, continuous, and cantilever beams. The simple beam design yields the largest moments and deflections. For simply supported purlins the maximum moment will be $WL^2/8$ and if they are assumed as continuous, the moment will be $WL^2/10$. While erecting angle, channel- or I-section purlins, it is desirable that they are erected over the rafter with their flange facing up slope [see Fig. 10.39(a)]. In this position, the twisting moment does not cause instability. If the purlins are kept in such a way that the flanges face the downward slope, then the twisting moment will cause instability [Fig. 10.39(b)].

As we know, channels are very weak about their web axes and have a tendency to sag in the direction of the sloping roof and often sag rods are provided mid-way or at one-third points between the roof trusses to take up the sag. If sag rods are used they will provide lateral support with respect to y-axis bending. Consequently, the moment $M_{yy}$ is reduced and thereby the

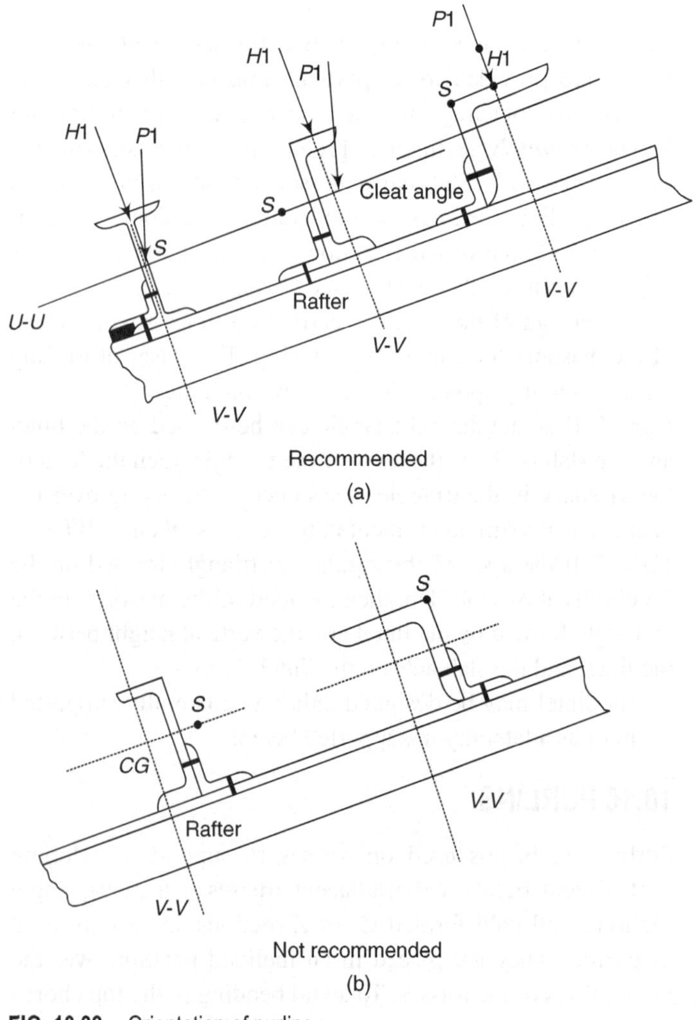


FIG. 10.39 Orientation of purlins

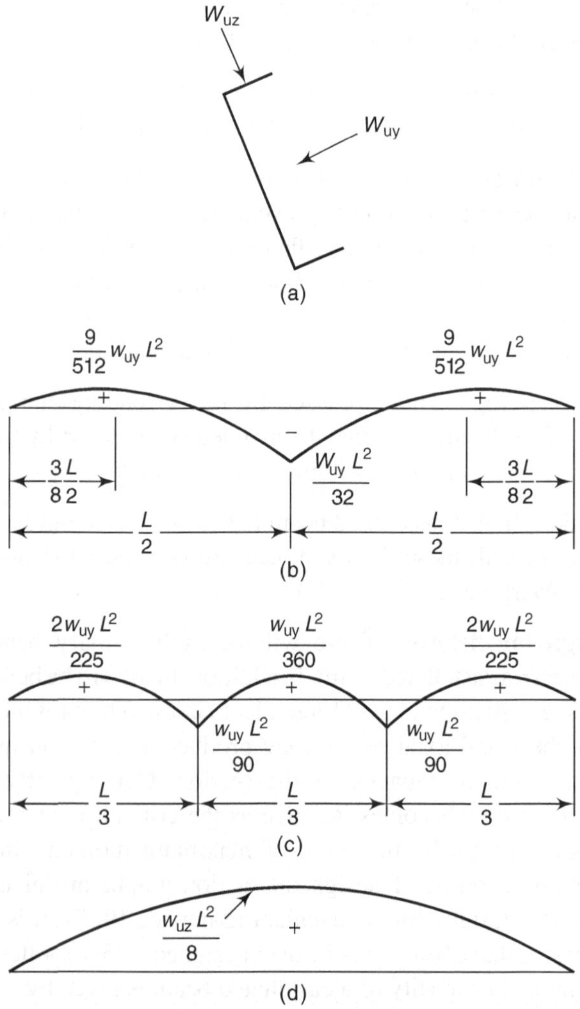


FIG. 10.40 (a) Channel purlin, (b) moment about web axis of purlins with sag rods at midspan, (c) moment about web axis of purlins with sag rods at one-third points, (d) moment about z-axis of purlin

required purlin section is smaller (see Fig. 10.40). The moment diagrams shown in Fig. 10.40 are developed on the assumption that the changes in length of the sag rods are negligible. It is further assumed that the purlins are simply supported at the trusses. This is a conservative assumption, since purlins are often continuous over two or more trusses and appreciable continuity will be achieved at their splices. The code also permits to take advantage of the continuity of purlins over supports (clause 8.9.1). Note that if sag rods are not used, the maximum moment about the web axis would be $w_{uy}L^2/8$. Thus, when one sag rod is used the moments are reduced by 75% and two sag rods are used as one-third points, the moments are reduced by 91%. In addition to providing lateral supports to purlins, sag rods also help to keep the purlins in proper alignment during erection until the roof deck is installed and connected to the purlins. Sag rods are often used with channel and I-section purlins, but are very rarely used with angle purlins.

The design of a purlin is a trial and error process. The span of the purlin is the centre-to-centre distances of the adjacent trusses. The gravity loads $P$, due to sheeting, self-weight of purlins, the live load (including snow load, if any), and the wind load $H$ are computed. The gravity loads act vertically downward and the wind loads act normal to the sheeting either inward or outward depending on the pressure or suction. Thus, the gravity loads are normal to the major axis of the purlin but are at an inclination. The components of the loads along *uu* and *vv* axes are determined. Taking advantage of the continuity of the purlin over the supports, the maximum bending moments about the two axes $M_u$ and $M_v$ are calculated separately and checked according to the biaxial bending requirements. This is applicable to channel and I-sections. The recommended bending moments are

$$M_u = PL/10 \quad \text{and} \quad M_v = HL/10$$

where $P$ acts along *v-v* axis, $H$ acts along *u-u* axis, and $L$ is the span of the purlin. It has to be noted that the purlins at the edges or end spans be designed considering local wind effects (see Tables 3.31 and 3.33 given in Chapter 3).

Various linear and non-linear formulae have been suggested for different types of cross sections (see Fig. 10.41). Horne and Morris (1981), recommend the following formulae which are safe and sufficiently accurate.

(a) for I-sections

$$[M_z/(Z_{pz}f_y)]^2 + M_y/(Z_{py}f_y) \le 1.0 \qquad (10.57a)$$

(b) For all solid sections and closed hollow sections

$$[M_z/(Z_{pz}f_y)]^{5/3} + [M_y/(Z_{py}f_y)]^{5/3} \le 1.0 \qquad (10.57b)$$

(c) For all other sections

$$M_z/(Z_{pz}f_y) + M_y/(Z_{py}f_y) \le 1.0 \qquad (10.57c)$$

However, the following equation may be used conservatively for biaxial moment of channel and I-sectios.

$$\frac{M_u}{M_{du}} + \frac{M_v}{M_{dv}} \le 1.0 \qquad (10.57d)$$

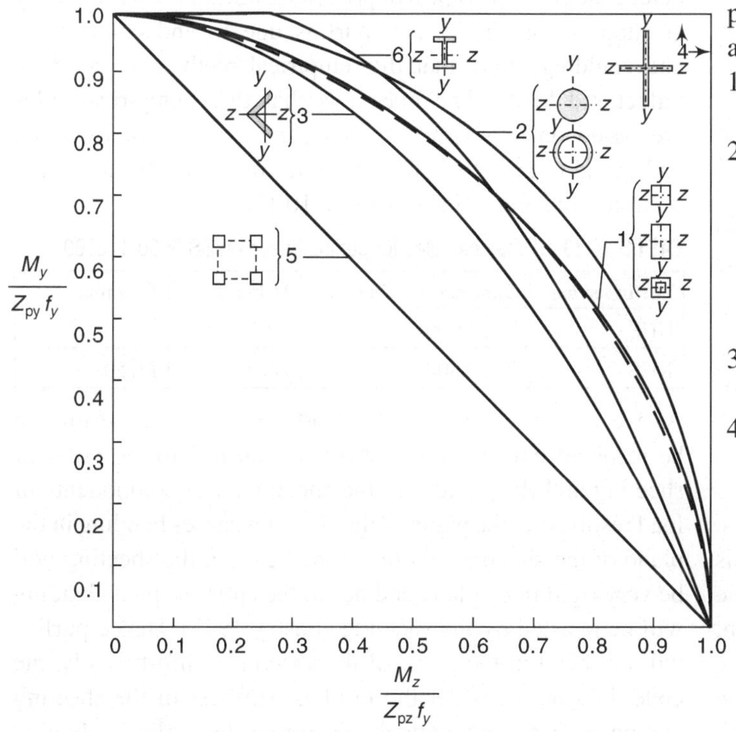


**FIG. 10.41** Interaction curves for biaxial bending (Horne & Morris 1981)

## 10.16.1 Design Procedure for Channel/I-section Purlins

As mentioned earlier, the design of purlins is a trial and error procedure and the various steps involved in the design are as follows:

1. The span of the purlin is taken as the centre-to-centre distance of adjacent trusses.
2. The gravity loads $P$, due to sheeting and live load, and the load $H$ due to wind are computed. The components of these loads in the direction perpendicular and parallel to the sheeting are determined. These loads are multiplied with partial safety factors $\gamma_f$ for loads (see Table 4 of the code) for the various load combinations.
3. The maximum bending moments ($M_z$ and $M_y$) and shear forces ($F_z$ and $F_y$) using the factored loads are determined.
4. The required value of plastic section modulus of the section may be determined by using the following equations (see Section 10.17)

$$Z_{pz} = M_z\, \gamma_{m0}/f_y + 2.5(d/b)[M_y\, \gamma_{m0}/f_y] \qquad (10.58)$$

where $\gamma_{m0}$ is the partial safety factor for material = 1.1, $d$ is the depth of the trial section, $b$ is the breadth of the trial section, $M_z$ and $M_y$ are the factored bending moments about the $Z$ and $Y$ axes, respectively, and $f_y$ is the yield stress of steel.

Since Eqn (10.58) involves $b$ and $d$ of a section, we must use a trial section and from this equation find out whether the chosen section is adequate or not.

5. Check for the section classification (Table 2 of the code).
6. Check for the shear capacity of the section for both the $z$ and $y$ axes (for purlins shear capacity will always be high and may not govern the design. The shear capacity in $z$ and $y$ axes is taken as (Morris & Plum 1996)

$$V_{dz} = f_y/(\sqrt{3}\,\gamma_{m0})A_{vz}$$

$$V_{dy} = f_y/(\sqrt{3}\,\gamma_{m0})A_{vy}$$

and

$$A_{vz} = ht_w$$

$$A_{vy} = 2b_f t_f$$

where $h$ is the height, $t_w$ is the thickness of the web, $b_f$ is the breadth of the flange, and $t_f$ is the thickness of the flange of I-channel section, respectively.

7. Compute the design capacity of the section in both the axes.

$$M_{dz} = Z_{pz}f_y/\gamma_{m0} \le 1.2Z_{ez}f_y/\gamma_{m0}$$

$$M_{dy} = Z_{py}f_y/\gamma_{m0} \le \gamma_f Z_{ey}f_y/\gamma_{m0}$$

Note that in the second equation 1.2 is replaced by $\gamma_f$. It is because, in the $y$-direction, the shape factor $Z_p/Z_e$ will be greater than 1.2 and hence if we use the factor as 1.2 we cannot prevent the onset of yielding under unfactored loads.

8. Check for local capacity by using the interaction equation

$$(M_z/M_{dz}) + (M_y/M_{dy}) \le 1.0$$

9. Check whether the deflection is under permissible limits (Table 6 of the code).
10. Under wind suction (combined with dead load), the bottom flange of the purlin, which is laterally unsupported will be under compression. Hence, under this loading case, the lateral-torsional buckling capacity of the section has to be calculated using Eqns (10.23) to (10.28) and an overall member buckling check is to be made by using the interaction equation

$$(M_z/M_{dz}) + (M_y/M_{dy}) \le 1.0$$

where $M_{dz}$ is the lateral-torsional buckling strength of the member.

## 10.16.2 Empirical Design of Angle Purlins

An angle purlin section is unsymmetrical about both axes. Angle purlin may be used wherever the slope of the roof is less than 30°. The vertical and normal loads acting on the purlin are determined and the maximum bending moment is calculated as $PL/10$ and $HL/10$, where $P$ and $H$ are the respective vertical and horizontal loads. A trial angle section may be assumed with depth as 1/45 of the span and width as 1/60 of the span. The trial depth and width are arrived to ensure that the deflection is within limits. Equation (10.57d) is again used to satisfy the adequacy of the section.

BS 5950–1:2000 gives the following empirical design method for purlins with the following general requirements.

1. Unfactored loads are used in the design.
2. The span should not exceed 6.5 m.
3. If the purlin spans one bay only, it should be connected at the ends by at least two bolts.
4. If the purlins are continuous over two or three spans, with staggered joints in adjacent lines of members, at least one end of any single bay member should be connected by not less than two bolts.

The rules for empirical design of angle purlins are as follows:

1. The roof slope should not exceed 30°.
2. The loading should be substantially uniformly distributed. Not more than 10% of the total load should be due to other type of load.
3. The elastic modulus about the axis parallel to the plane of cladding should not be less than WpL/1800 mm3 where Wp is the total unfactored load on one span (N) due to dead and imposed load or dead minus wind load and L is the span of purlin in mm.
4. Dimension D perpendicular to the plane of the cladding should not be less than $L/45$ and dimension $B$ parallel to the plane of cladding should not be less than $L/60$.

When sag rods are used, the sag rod spacing may be used to determine $B$. This empirical procedure allows for partial fixity at supports of single-span purlins due to end connections and cladding. Note that this empirical method is based on unfactored loads. The reason is that these empirical rules are based on experience from earlier versions of the code and found to be satisfactory. Similar rules for CHS and RHS sections are summarized in Table 10.13.

**TABLE 10.13** Empirical rules for purlins based on BS 5950-1: 2000

| Purlin section | $Z_p$(minimum), in mm$^3$ | D, in mm | B, in mm |
|---|---|---|---|
| CHS | $W_pL/2000$ | $L/65$ | — |
| RHS | $W_pL/1800$ | $L/70$ | $L/150$ |

As we have seen the gravity load acting on the purlin can be resolved into two components (a) normal to the rafter or sheeting and (b) parallel to the sheeting. The components of load resolved in the plane of the sheeting causes bending in the plane of the sheeting. As mentioned earlier, the sheeting will be very rigid in its plane and hence the entire in-plane loading will be resisted by the sheeting (Easley 1977). Hence purlins will not bend in the plane of the sheeting. Unfortunately, the code does not consider the in-plane stiffness of the sheeting and hence it recommends the design of the purlin for biaxial bending (clause 8.9 of the code). Hence, we shall design the purlins for biaxial bending, as per the conservative approach suggested by the code.

## 10.17 BIAXIAL BENDING

When a beam is subjected to a loading condition that produces bending about both the major (strong) and the minor (weak) axes, the beam undergoes biaxial bending. Such a loading case is shown in Fig. 10.42(a). As we discussed in Section 10.16, purlins are also subjected to biaxial bending. The load acts normal to the longitudinal axis but is inclined with respect to the principal axes of the cross section. Here, the load passes through the shear centre of the cross section. The *shear centre* is defined as the point through which the loads must act if there is to be no twisting or torsion of the beam. The location of the shear centre can be determined by equating the internal resisting torsional moment derived from the shear flow on the cross section to the external torque. A few of them are shown in Fig. 10.42(b).

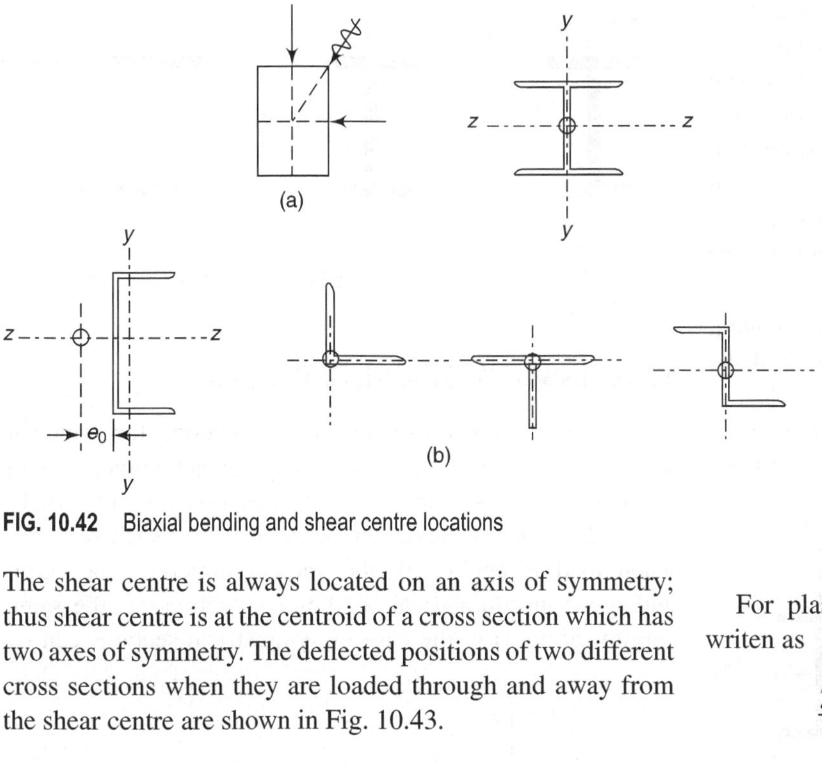


**FIG. 10.42** Biaxial bending and shear centre locations

The shear centre is always located on an axis of symmetry; thus shear centre is at the centroid of a cross section which has two axes of symmetry. The deflected positions of two different cross sections when they are loaded through and away from the shear centre are shown in Fig. 10.43.

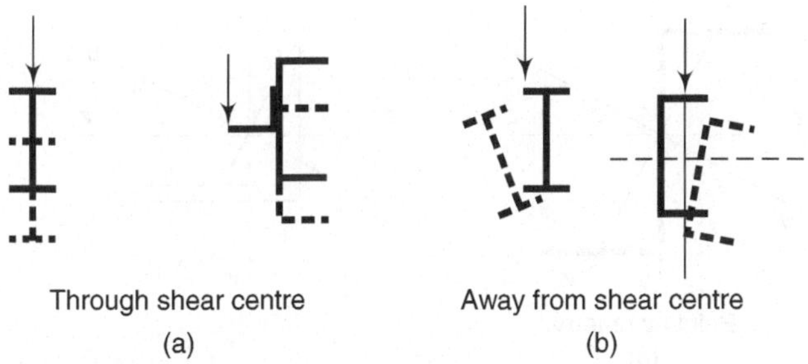


**FIG. 10.43** Deflected positions of sections loaded through shear centre and away from shear centre

When loads act through the shear centre, the problem is simple bending in two perpendicular directions. The load can be resolved in the perpendicular *z*- and *y*- directions, each producing bending about the other axis. The code (IS 800 : 2007) deals with combined loading through the use of interaction formulae which account for relative importance of each load effect with respect to the strength corresponding to that effect. To illustrate this, consider bending is about z-axis only.

$$M_z/M_{dz} \leq 1.0 \qquad (10.59a)$$

where $M_z$ is the factored bending moment about *z-z* axis and $M_{dz}$ is the design moment strength about the *z-z* axis.

Similarly, if bending were about the *y-y* axis only, the condition will be

$$M_y/M_{dy} \leq 1.0 \qquad (10.59b)$$

where $M_y$ is the factored bending moment about the *y-y* axis and $M_{dy}$ is the design moment strength about the *y-y* axis.

When bending occurs about both *z-z* and *y-y* axes, an interaction type of formula requires that the sum of the two ratios is less than 1.0 (clause 9.3.1 of the code); that is,

$$M_z/M_{dz} + M_y/M_{dy} \leq 1.0 \qquad (10.60)$$

This requirement gives the designer freedom to use the section effectively by allocating to one direction what has not been used up in the other direction. It may be noted here that any shape bent about its weak axis cannot buckle in the other direction; so lateral buckling is not a limit state for weak axis bending.

Beams subjected to biaxial bending can be designed only by trial and error. Lindner and Gietzelt (1985) provide a design method for beams subjected to biaxial bending and Sarma and Lakshmanan (1990) present a method to evaluate channel and tube purlins under biaxial loading.

For plastic and compact sections, Eqn (10.60) may be writen as

$$\frac{\gamma_{m0} M_z}{Z_{p,z} f_y} + \frac{M_y \gamma_{m0}}{Z_{p,z} f_y} \leq 1.0 \quad \text{with } \gamma_{m0} = 1.1 \qquad (10.61)$$

For design use, let us multiply Eqn (10.61) with $Z_{p,\,z}$ to get

$$Z_{p,z} \geq \frac{M_z \gamma_{m0}}{f_y} + \frac{M_y \gamma_{m0}}{f_y}\left(\frac{Z_{p,z}}{Z_{p,y}}\right)$$

If an approximate range of $(Z_{p,\,z}/Z_{p,\,y})$ is known, a fairly close design can often be obtained. The quantity $(Z_{p,\,z}/Z_{p,\,y})$ may be computed approximately using the beam properties neglecting the web effect. Thus, (with $b_f$ = flange width and $d$ = depth) for *I*-shaped beam, we get

$$\frac{Z_{p,z}}{Z_{p,y}} = \frac{I_{p,z}(b_f/2)}{(d/2)I_{p,y}} = \frac{2b_f t(d/2)^2(b_f/2)}{(d/2)(2)(1/12)tb_f^3} = \frac{3d}{b_f} \quad (10.62)$$

The error in this approximation, due to the omission of the moment of inertia of the web, is offset closely by increasing the coefficient of $d/b_f$ from 3 to 3.5. Gaylord et al. (1992) suggest using a factor of 2.5 for limit states method. Thus the required

$$Z_z = \frac{M_z\gamma_{m0}}{f_y} + \frac{M_y\gamma_{m0}}{f_y} \times 2.5 \times \frac{d}{b_f} \quad (10.63)$$

Note that Eqn (10.63) makes no allowance for lateral-torsional buckling and hence some trials may be necessary.

## 10.18 UNSYMMETRICAL BENDING

If the applied loads do not pass through the shear centre of the cross section, the beam will be subjected to flexure and torsion. An example of loading which gives rise to torsion is shown in Fig. 10.44(a). The load is applied to the centre of the top flange, but its line of action does not pass through the shear centre of the cross section. The equivalent system, as shown in Fig. 10.44(b), consists of the given force acting through the shear centre together with a twisting moment. A simplified way of treating the problems is shown in Fig. 10.45, in which the top flange is assumed to provide the entire resistance to the horizontal component of the load. In the second case, the two flanges resist equally the torsion produced by the external moment. Hence, the problem is reduced to the case of bending about two axes by considering the load acting through the shear centre.

There are two mutually perpendicular axes of bending for a beam cross section. If the loading takes place along these two axes, the beam is said to bend symmetrically and the axes are referred to as principal axes. When the bending takes place about either principal axis, the plane of loading coincides with the plane of bending. If the loading occurs in any other direction, unsymmetrical bending occurs. The approach for such problems is to resolve the load into two components perpendicular to the two principal axes shown in Fig. 10.45 and then to determine component bending moments about the two principal axes.

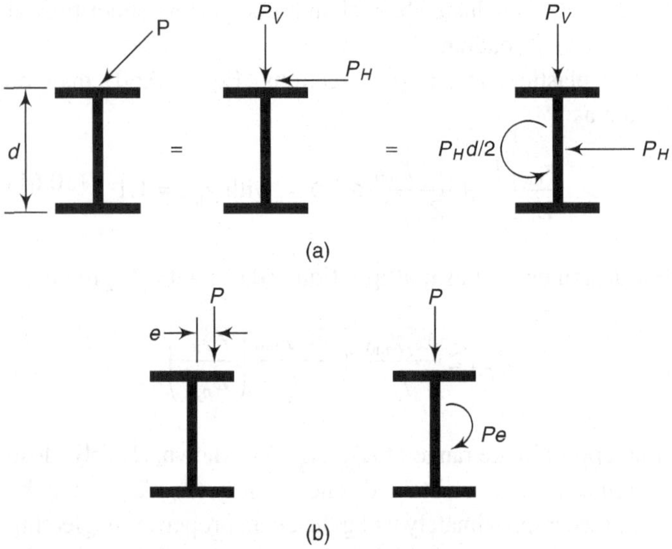


FIG. 10.44 Load not applied through shear centre

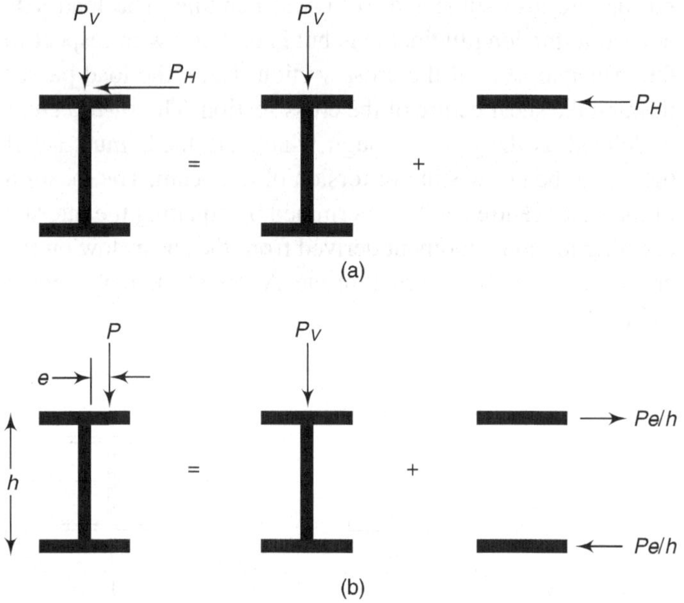


FIG. 10.45 Simplified way of treating cases with loads applied not through the shear centre

## 10.19 UNSYMMETRICAL SECTIONS

In steel sections such as angles and Z-sections, the principal axes are *u-u* and *v-v* axes and they do not coincide with the horizontal (*z-z*) and vertical axes (*y-y*). Figure 10.46 shows examples of angle and Z-sections. For such sections, when load is applied in the vertical direction, the beam undergoes unsymmetrical bending. In such cases, the same simplification as in the case of symmetrical sections can be

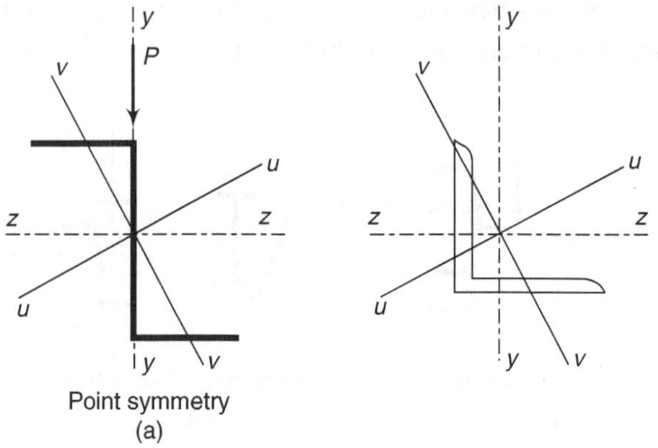


FIG. 10.46 Sections prone to unsymmetrical bending

used. Thus, the bending stresses can be determined about the two perpendicular principal axes. However the points of maximum stresses, $f_{z,\max}$ and $f_{y,\max}$, may not occur at the same point. Hence, the stresses must be calculated at various points and the maximum value of stress has to be used in the design.

## 10.20 DESIGN OF BEAMS

### *Laterally Supported Beam*

The design of laterally supported beam consists of selecting a section on the basis of the modulus of the section and checking it for shear, deflection, and web crippling. The steps to be followed are as follows.

1. The loads that may be acting on the beam are ascertained. The limiting strength design loads are obtained by suming up the loads multiplied by the appropriate partial load factors as given in Table 4 of the code.
2. A trial beam section is assumed and the distribution of the bending moment along the length of the beam is determined by an elastic analysis (if the beam is statically indeterminate) or by statics (if the beam is statically determinate). The maximum bending moment and shear force are calculated.
3. The required section modulus may be determined using Eqn (10.19).
4. From section tables (IS 808–1989), a suitable section is selected, which has a section modulus equal to or more than the calculated modulus. ISMB sections may be preferred as they are readily available in the market.
5. The beam is checked for shear as per Eqn (10.49).
6. The beam is checked for deflection, as per Table 6 of the code.
7. The beam is checked for web crippling as per Eqns (10.53) and (10.54).

### *Design of Laterally Unsupported Beams*

As discussed in this chapter, when the compression flange of the beam is laterally unsupported, lateral torsional buckling may take place, leading to failure at or below the elastic critical moment.

The design of laterally unsupported beams with equal flanges (e.g., I-sections and channel sections) or mono symmetric beams is essentially a trial and error process, since the section dimensions are not initially known. Hence we have to assume a section in order to compute the strength of that section. The design procedure is essentially the same as described for a laterally supported section, except that in step 3, the design strength $M_d$ is computed based on Eqn (10.22) or Eqn (10.28) (for monosymmetric section) and compared with the factored design moment $M$. If it is not equal to or greater than the factored design moment, next higher section is chosen and the process is repeated till a section which satisfies the condition $M_u \le M_d$, is found.

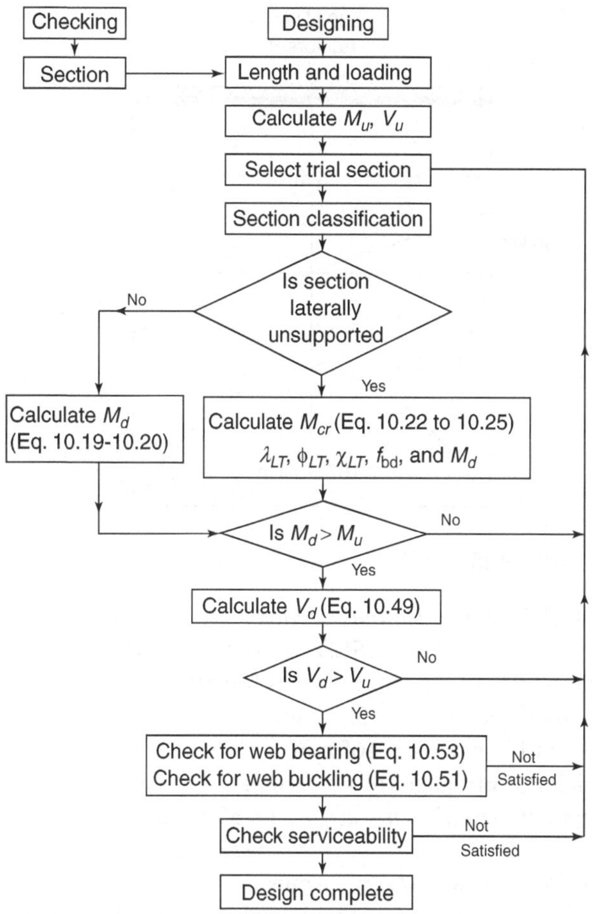


**FIG. 10.47** Flowchart for design of beams

The various steps involved in the design of beams are shown in Fig. 10.47.

## EXAMPLES

**EXAMPLE 10.1:** Design a simply supported beam of span 5 m carrying a reinforced concrete floor capable of providing lateral restraint to the top compression flange. The uniformly distributed load is made up of 20 kN/m imposed load and 20 kN/m dead load (section is stiff against bearing). Assume Fe 410 grade steel.

**SOLUTION:**

**Step 1:** *Calculation of factored loads*

Dead load = 1.5 × 20 = 30 kN/m

Live load = 1.5 × 20 = 30 kN/m

Total factored load on the beam = 60 kN/m (see Fig. 10.48)

**Step 2:** *Calculation of maximum bending moment and shear force*

Maximum bending moment = $60 \times 5^2/8$ = 187.5 kN m

**Step 3:** *Section modulus required ($Z_p$)*

$$Z_p\,(\text{required}) = \frac{M \times \gamma_{m0}}{f_y} = \frac{187.5 \times 10^6 \times 1.1}{250} = 825 \times 10^3\,\text{mm}^3$$

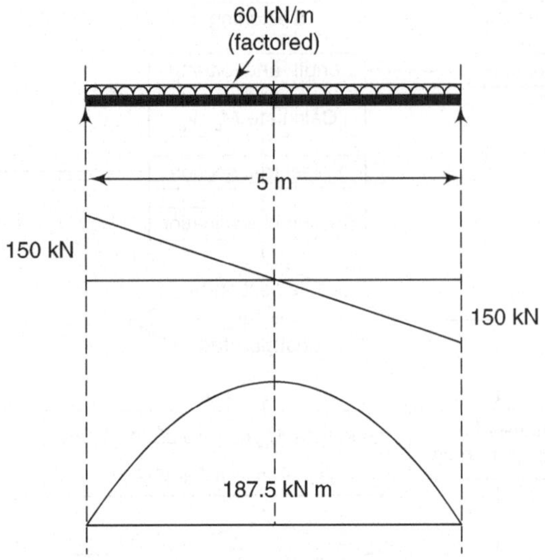


FIG. 10.48

**Step 4:** *Selection of suitable section*

Choose a trial section of ISLB 350 @ 0.486 kN/m

The properties of the section are as follows:

Depth of section ($h$) = 350 mm

Width of flange ($b$) = 165 mm

Thickness of flange ($t_f$) = 11.4 mm

Depth of web ($d$) = $h - 2(t_f + R) = 350 - 2(11.4 + 16)$ = 295.2 mm

Thickness of web ($t_w$) = 7.4 mm

Moment of inertia about major axis $I_{zz} = 13200 \times 10^4\,\text{mm}^4$

Elastic section modulus ($Z_e$) = $751.9 \times 10^3\,\text{mm}^3$

Plastic section modulus ($Z_p$) = $851.11 \times 10^3\,\text{mm}^3$

*Section classifiaton*

$$\varepsilon = \sqrt{\frac{250}{f_y}} = \sqrt{\frac{250}{250}} = 1$$

$$\frac{b}{t_f} = \frac{82.5}{11.4} = 7.23 < 9.4$$

$$\frac{d}{t_w} = \frac{295.2}{7.4} = 39.9 < 84$$

Hence the section is classified as a plastic section.

**Step 5:** *Adequacy of the section including self-weight of the beam*

Factored self weight of the beam = 1.5 × 0.486 = 0.73 kN/m

Total load acting on the beam = 60.73 kN/m

Maximum bending moment = 190 kN m

$$\text{Plastic section modulus equired} = \frac{190 \times 10^6 \times 1.1}{250}$$

$$= 836 \times 10^3\,\text{mm}^3$$

Since it is less than $851.11 \times 10^3\,\text{mm}^3$, hence, the chosen section is adequate.

**Step 6:** *Calculation of design shear force*

$$\text{Design shear force } V = \frac{wl}{2} = \frac{60.73 \times 5}{2} = 151.83\text{ kN}$$

**Step 7:** *Design shear strength of the section*

Design shear strength

$$V_d = \frac{f_y}{\gamma_{m0} \times \sqrt{3}} \times h \times t_w = \frac{250}{1.1 \times \sqrt{3}} \times 350 \times 7.4$$

$$= 340\text{ kN} > 151.83\text{ kN}$$

Also,

$0.6V_d = 204$

Therefore, the design shear force $V < 0.6V_d$

**Step 8:** *Check for design capacity of the section*

$\frac{d}{t_w} = 39.9$ (which is less than $67\varepsilon$)

Hence,

$$M = \beta_b Z_p \times \frac{f_y}{\gamma_{m0}}$$

$\beta_b = 1.0$, since the section is plastic section.

Thereore,

$$M_d = \frac{1.0 \times 851.11 \times 10^3 \times 250}{1.1} = 193.43\text{ kN m}$$

$$193.43\text{ kN m} \le \frac{1.2 \times Z_e \times f_y}{\gamma_{m0}} = \frac{1.2 \times 751.9 \times 10^3 \times 250}{1.1}$$

$$= 205\text{ kN m}$$

Hence the design capacity of the member is more than maximum bending moment $M_d$ (193.43 kN m > 190 kN m).

**Step 9:** *Check for deflection*

Deflection (which is a serviceability limit state) must be calculated on the basis of the unfactored imposed loads (assuming that the beam is cambered to compensate for dead load deflection).

$$\delta = \frac{5wl^4}{384EI} = \frac{5 \times 20 \times 5000^4}{384 \times 2 \times 10^5 \times 13200 \times 10^4} = 6.165\text{ mm}$$

$= 6.165$ mm

$$\text{Allowable maximum deflection } \Delta_{max} = \frac{L}{300} = \frac{5000}{300}$$

$$= 16.67\text{ mm}$$

The deflection is less than the allowable maximum deflection.

**Example 10.2:** Design a simply supported beam of 7 m span carrying a reinforced concrete floor capable of providing lateral restraint to the top compression flange. The total udl is made up of 100 kN dead load plus 150 kN imposed load. In addition, the beam carries a point load at mid-span made up of 50 kN dead load and 50 kN imposed load (assuming a stiff bearing length of 75 mm).

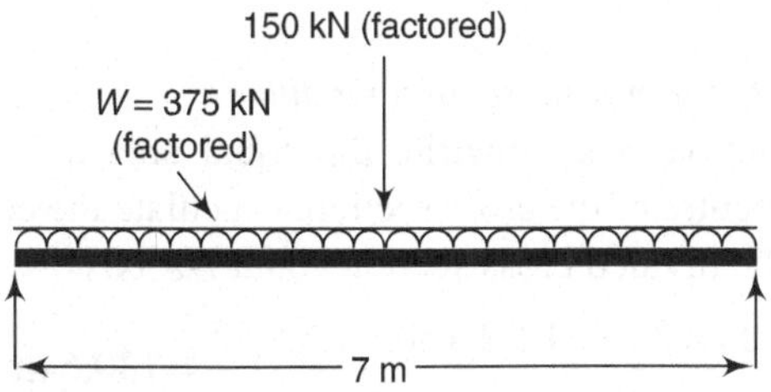


FIG. 10.49

**SOLUTION:**

**Step 1:** *Calculation of factored load*

Udl:

Dead load = 1.5 × 100 = 150 kN

Live load = 1.5 × 150 = 225 kN

Total udl acting on the beam = 375 kN

Concentrated load:

Dead load = 1.5 × 50 = 75 kN

Live load = 1.5 × 50 = 75 kN

Total concentrated load acting on the beam = 150 kN (see Fig. 10.49)

**Step 2:** *Calculation of maximum bending moment*

$$\text{Bending moment} = \frac{WL}{8} + \frac{WL}{4} = \frac{375 \times 7}{8} + \frac{150 \times 7}{4}$$

$$= 590.63\,\text{kN m}$$

**Step 3:** *Calculation of required plastic section moulus*

$$Z_p = \frac{M \times \gamma_{m0}}{f_y} = \frac{590.63 \times 10^6\ 1.10}{250} = 2600 \times 10^3\,\text{mm}^3$$

**Step 4:** *Selection of suitable section*

Choose a trial section of ISMB 550 @ 1.037 kN/m.

The properties of the section are as follows:

Overall depth ($h$) = 550 mm

Width of flange ($b$) = 190 mm

Thickness of flange ($t_f$) = 19.3 mm

Depth of web ($d$) = $h - 2(t_f + R) = 550 - 2(19.3 + 18)$ = 475.4 mm

Thickness of web ($t_w$) = 11.2 mm

Moment of inertia about major axis $I_{zz} = 64900 \times 10^4\,\text{mm}^4$

Elastic section modulus ($Z_e$) = $2359.8 \times 10^3\,\text{mm}^3$

Plastic section modulus ($Z_p$) = $2711.98 \times 10^3\,\text{mm}^3$

*Section classifiaton*

$$\varepsilon = \sqrt{\frac{250}{f_y}} = \sqrt{\frac{250}{250}} = 1$$

$$\frac{b}{t_f} = \frac{95}{19.3} = 4.922 < 9.4\varepsilon$$

$$\frac{d}{t_w} = \frac{475.4}{11.2} = 42.45 < 84\varepsilon$$

Hence the section is classified as a plastic section.

**Step 5:** *Adequacy of the section including self-weight of the beam*

Self-weight of the section (factored) = 1.5 × 1.037

= 1.5555 kN/m

$$\text{Maximum bending moment } 590.63 + \frac{1.5 \times 1.037 \times 7^2}{8}$$

$$= 600.16\,\text{kN m}$$

Required plastic section modulus

$$\frac{600.16 \times 10^6 \times 1.1}{250} = 2640.7 \times 10^3\,\text{mm}^3 < 2711.98 \times 10^3\,\text{mm}^3$$

Hence the chosen section is adequate.

**Step 6:** *Calculation of design shear force*

$$\text{Shearforce } V = \frac{1.555 \times 7}{2} + \frac{375}{2} + \frac{150}{2} = 268\ \text{kN}$$

**Step 7:** *Design shear strength of the section*

Design shear strength

$$V_d = \frac{f_y \times h \times t_w}{\gamma_{m0} \times \sqrt{3}} = \frac{250}{1.1 \times \sqrt{3}} \times 550 \times 11.2 = 808.29\,\text{kN}$$

Also,

$0.6V_d = 485\,\text{kN}$

The design shear force $V < 0.6V_d$

**Step 8:** *Check for design capacity of the section*

$$\frac{d}{t_w} = 42.45 \text{ which is less than } 67\varepsilon$$

$$M = \beta_b Z_p \times \frac{f_y}{\gamma_{m0}} < \frac{1.2 \times Z_e \times f_y}{1.10}$$

$\beta_b = 1.0$ since the section is plastic section

$$M_d = \frac{1.0 \times 2711.98 \times 10^3 \times 250}{1.1} = 616.36\,\text{kN m}$$

$$\leq \frac{1.2 \times Z_e \times f_y}{\gamma_{m0}} = \frac{1.2 \times 2359.8 \times 10^3 \times 250}{1.1} = 643.58\,\text{kN m}$$

Hence the design capacity of the member is more than the maximum bending moment $M_d$ (616.36 kN m > 600.16 kN m).

**Step 9:** *Check for deflection*

Actual deflection $\delta = \delta_{udl} + \delta_{pl}$

$$\delta = \frac{5Wl^3}{384EI} + \frac{Wl^3}{48EI}$$

$$= \frac{5 \times 150 \times 10^3 \times 7000^3}{384 \times 2 \times 10^5 \times 64900 \times 10^4} + \frac{50 \times 10^3 \times 7000^3}{48 \times 2 \times 10^5 \times 64900 \times 10^4}$$

= 7.91 mm

$$\text{Allowable maximum deflecton } \delta_{max} = \frac{L}{300} = \frac{7000}{300} = 23.33\,\text{mm}$$

Actual deflection is less than the allowable maximum deflection.

**Step 10:** *Check for web buckling at support*

Assume stiff bearing length $b_1 = 75$ mm

Depth of web $d = 475.4$ mm

$$I_{eff}\ \text{web} = \frac{b_1 \times t_w^3}{12} = \frac{75 \times 11.2^3}{12} = 8780.8\,\text{mm}^4$$

$$A_{eff}\ \text{web} = b_1 \times t_w = 75 \times 11.2 = 840\,\text{mm}^2$$

$$r = \sqrt{\frac{I_{eff}web}{A_{eff}web}} = \sqrt{\frac{8780.8}{840}} = 3.233\,\text{mm}$$

Effective length of web $d_{eff}$ = 0.7 times the depth of the web ($d$)

$$\lambda = \frac{d_{eff}}{r} = \frac{0.7 \times 475.4}{3.233} = 102.9$$

Therefore, $f_{cd} = 103.4\,\text{N/mm}^2$ (From Table 9c of the code)

$n_1 = 550/2 = 275\,\text{mm}$

$b_1 + n_1 = 75 + 275 = 350\,\text{mm}$

$A_b = 350 \times 11.2 = 3920\,\text{mm}^2$

Buckling resistance $= f_{cd} \times A_b$

$= 103.4 \times 3920/1000$

$= 405.3\,\text{kN} > 268\,\text{kN}$

Hence the web is safe.

**Step 11:** *Check for web bearing*

$$F_w = \frac{(b_1 + n_2) \times t_w \times f_y}{\gamma_{m0}}$$

$b_1 = 75\,\text{mm}$

$n_2 = 2.5 \times$ (Root radius + flange thickness)

$= 2.5\,(18 + 19.3)$

$= 93.25\,\text{mm}$ (since angle of dispersion is 1:2.5)

$t_w = 11.2\,\text{mm}$

$f_y = 250\,/\text{mm}^2$

$$F_w = \frac{(75 + 93.25) \times 11.2 \times 250}{1.10}$$

$= 428.27\,\text{kN} > 268\,\text{kN}$

Hence the web is safe.

**Example 10.3:** For the built-up section shown in Fig. 10.50, calculate the plastic section modulus and the shape factor.

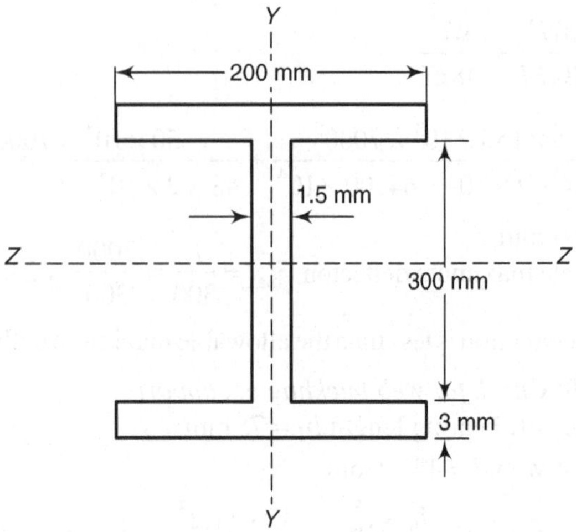


FIG. 10.50

**Solution:**

*Calculation of the plastic section modulus*

Since the section is symmetric, the equal area axis passes through the centre of the cross section. Calculate the centre of gravity for the divided cross section about *z-z* axis

$$y_{top} = \frac{200 \times 3 \times 151.5 + 150 \times 1.5 \times 75}{(200 \times 3) + (150 \times 1.5)} = 130.64\text{mm}$$

Also $y_{bottom} = 130.64\,\text{mm}$, since the section is symmetrical

$$Z_{pz} = \frac{A}{2}\,(y_{top} + y_{bottom})$$

$$= \frac{1650}{2}\,(130.64 + 130.64) = 215.56 \times 10^3\,\text{mm}^3$$

*Calculation of $Z_{py}$*

Divide the section into two equivalent portions along *y-y* axis and calculate the centre of gravity for the divided cross section

$$Z_{left} = \frac{[2 \times (100 \times 3 \times 50)] + (0.75 \times 300 \times 0.75/2)}{(2 \times 100 \times 3) + (0.75 \times 300)}$$

$= 36.47\,\text{mm}$

The section is also symmetric along *y-y* axis. Hence

$z_{right} = 36.47\,\text{mm}$

Therefore

$$Z_{py} = \frac{A}{2}\,(z_{left} + z_{right})$$

$$= \frac{1650}{2}\,(36.47 + 36.47) = 60.18 \times 10^3\,\text{mm}^3$$

*Shape factor calculation*

$$I_Z = 2 \times \left[\left(\frac{200 \times 3^3}{12}\right) + \left\{(200 \times 3) \times 151.5^2\right\}\right] + \frac{1.5 \times 300^3}{12}$$

$= 30.92 \times 10^6\,\text{mm}^4$

$$Z_e = \frac{30.92 \times 10^6}{153} = 202.08 \times 10^3\,\text{mm}^3$$

$$\therefore \text{ Shape actor } S = \frac{Z_{pz}}{Z_e} = \frac{215.56 \times 10^3}{202.08 \times 10^3} = 1.067$$

**Example 10.4:** A proposed cantilever beam is built into a concrete wall. It supports a dead load of 20 kN/m and a live load of 10 kN/m. The length of beam is 5 m. Select a suitable section with necessary checks. Assume stiff bearing length of 100 mm and the beam is laterally restrained (see also Example 10.7).

**Solution:**

**Step 1:** *Calculation of load*

Dead load $= 1.5 \times 20 = 30\,\text{kN/m}$

Live load $= 1.5 \times 10 = 15\,\text{kN/m}$

Total load $= 45\,\text{kN/m}$

**Step 2:** *Calculation of bending moment and shear force*

$$BM = \frac{wl^2}{2} = \frac{45 \times 5^2}{2} = 562.5\,\text{kNm}$$

$SF = wl = 45 \times 5 = 225\,\text{kN}$

**Step 3:** *Choosing a trial setion*

$$Z_p = \frac{M \times \gamma_{m0}}{f_y} = \frac{562.5 \times 10^6 \times 1.10}{250} = 2475 \times 10^3\,\text{mm}^3$$

Choose a trial section of ISMB 550 @ 1.037 kN/m.

Overall depth $(h) = 550\,\text{mm}$

Width of flange $(b) = 190\,\text{mm}$

Depth of web $(d) = 475.4\,\text{mm}$

Thickness of flange $(t_f) = 19.3\,\text{mm}$

Thickness of web $(t_w) = 11.2\,\text{mm}$

Moment of inertia about major axis, $I_z = 64900 \times 10^4\,\text{mm}^4$

Elastic section modulus $(Z_e) = 2359.8 \times 10^3\,\text{mm}^3$

Plastic section modulus $(Z_p) = 2711.98 \times 10^3\,\text{mm}^3$

*Section classification*

$$\frac{b}{t_f} = \frac{95}{19.3} = 4.92 < 9.4$$

$$\frac{d}{t_w} = \frac{475.4}{11.2} = 42.45 < 84$$

Hence the section is classified as plastic.

**Step 4:** *Calculation of shear capacity of the setion*

$$v_d = \frac{f_y}{\gamma_{m0} \times \sqrt{3}} \times h \times t_w = \frac{250}{1.1 \times \sqrt{3}} \times 550 \times 11.2 = 808.3\text{kN}$$

$0.6V_d = 485\,\text{kN} > 225\,\text{kN}$

**Step 5:** *Design capacity of the section*

$$M_d = \frac{Z_p \times f_y}{\gamma_{m0}} = \frac{2711.98 \times 10^3 \times 250}{1.10} = 616\text{kN}$$

$$\leq \frac{1.5 \times Z_e \times f_y}{\gamma_{m0}} = \frac{1.5 \times 2359.8 \times 10^3 \times 250}{1.1} = 804.47\text{KN m}$$

Hence the section is safe.

**Step 6:** *Check for deflection (assuming beam is cambered for dead load deflection)*

$$\delta = \frac{wl^4}{8EI} = \frac{10 \times 5000^4}{8 \times 2 \times 10^5 \times 64900 \times 10^4} = 6.02\text{mm}$$

$$\text{Allowable delection} = \frac{L}{150} = \frac{5000}{150} = 33.33\text{ mm}$$

**Step 7:** *Check for web buckling*

Cross-sectional area of web for buckling $A_b = (b_1 + n_1)t_w$

$b_1 = 100\,\text{mm}$

$n_1 = 550/2 = 275\,\text{mm}$

$A_b = (100 + 275) \times 11.2 = 4200\,\text{mm}^2$

Effective length of the web $= 0.7 \times d = 0.7 \times 475.4 = 332.8\,\text{mm}$

$$I = \frac{b \times t_w^3}{12} = \frac{100 \times 11.2^3}{12} = 11.707 \times 10^3\,\text{mm}^4$$

$A = 100 \times 11.2 = 1120\,\text{mm}^2$

$$r_{min} = \sqrt{\frac{11.707 \times 10^3}{1120}} = 3.23$$

$$\therefore \quad \lambda = \frac{l_{eff}}{r_{min}} = \frac{332.8}{3.23} = 103$$

Allowable stress $= 103.4\,\text{N/mm}^2$ (Table 9c of the code)

Capacity of the section $= 103.4 \times 4200$

$= 434.2\,\text{kN} > 225\,\text{kN}$

Hence the section is safe against web buckling.

**Step 8:** *Check for web bearing*

$$F_w = \frac{(b_1 + n_2) \times t_w \times f_y}{\gamma_{m0}}$$

$n_2 = 2.5(R + t_f) = 2.5\,(18 + 19.3) = 93.25\,\text{mm}$

$$F_w = \frac{(100 + 93.25) \times 11.2 \times 250}{1.10} = 491.91\text{kN} > 225\text{kN}$$

Hence the section is safe against bearing.

**EXAMPLE 10.5:** Design a continuous beam of spans 4.9 m, 6 m, and 4.9 m carrying a uniformly distributed load of 32.5 kN/m and the beam is laterally supported.

**SOLUTION:**

**Step 1:** *Factored load calculation*

Factored uniformly distributed load $= 1.5 \times 32.5 = 48.75\,\text{kN/m}$

**Step 2:** The bending moment and shear force distribution are shown in Fig. 10.51.

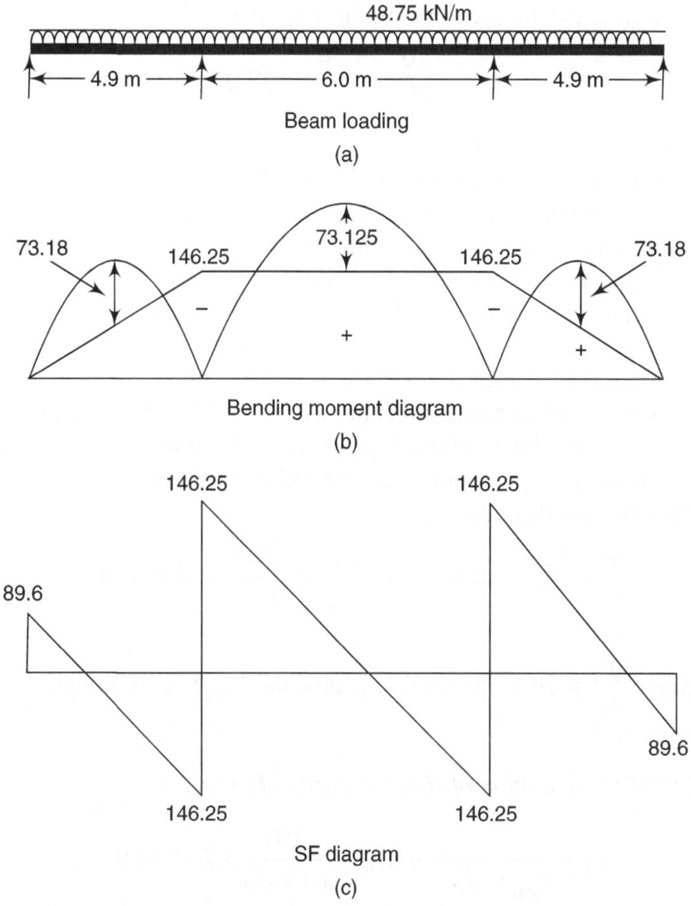


FIG. 10.51

Maximum bending moment = 146.25 kN m
Maximum shear force = 146.25 kN

**Step 3:** *Plastic section modulus reqired*

$$Z_p = \frac{M \times \gamma_{m0}}{f_y} = \frac{146.25 \times 10^6 \times 1.10}{250} = 643.5 \times 10^3 \text{ mm}^3$$

**Step 4:** *Selection of suitable section*
Choose a trial section of ISMB 300
Overall depth ($h$) = 300 mm
Width of flange ($b$) = 140 mm
Thickness of flange ($t_f$) = 12.4 mm
Depth of web ($d$) = $h - 2(t_f + R)$ = 300 – 2(12.4 + 14.0)
= 247.2 mm
Thickness of web ($t_w$) = 7.5 mm
Moment of inertia about major axis $I_z = 8603.6 \times 10^4$ mm$^4$
Elastic section modulus ($Z_e$) = $573.6 \times 10^3$ mm$^3$
Plastic section modulus ($Z_p$) = $651.74 \times 10^3$ mm$^3$

**Step 5:** *Check for adequacy of the section to include self-weight of the beam*
Self-weight of the beam = 0.442 kN/m

$$\text{Total bending moment} = 146.25 + \left(\frac{1.5 \times 0.442 \times 6^2}{12}\right)$$
$$= 148.2 \text{ kN m}$$

Therefore plastic section modulus required

$$Z_p = \frac{M \times \gamma_{m0}}{f_y} = \frac{148.2 \times 10^6 \times 1.10}{250} = 652.08 \times 10^3 \text{ mm}^3$$

> the provided section
Hence choose another trial section of ISHB 250.
Overall depth ($h$) = 250 mm
Width of flange ($b$) = 250 mm
Thickness of flange ($t_f$) = 9.7 mm
Depth of web ($d$) = $h - 2(t_f + R)$ = 250 – 2(9.7 + 10)
= 210.6 mm
Thickness of web ($t_w$) = 6.9 mm
Moment of inertia about major axis ($I_z$) = $7736.5 \times 10^4$ mm$^4$
Elastic section modulus ($Z_e$) = $618.9 \times 10^3$ mm$^3$
Plastic section modulus ($Z_p$) = $678.73 \times 10^3$ mm$^3$

*Section classification*

$$\frac{b}{t_f} = \frac{125}{9.7} = 12.88 < 15.7 \qquad \frac{d}{t_w} = \frac{210.6}{6.9} = 30.5 < 84$$

Since $\frac{b}{t_f} > 10.5$, the section is classified as a semi-compact section.

**Step 6:** *Calculation of shear capacity of the section*

$$V_d = \frac{f_y}{\gamma_{m0} \times \sqrt{3}} \times h \times t_w = \frac{250}{1.1 \times \sqrt{3}} \times 250 \times 6.9$$
$$= 226 \text{kN} > 146.25 \text{kN}$$

(For maximum shear, the load has to be very close to the support and not at the support.)
Hence the provided section is adequate. To explain the use of clause 9.2.2 of IS 800, let us assume that the shear force is 292.5 kN and assume another section of ISLB 350 → 0.495 kN/m.
Overall depth ($h$) = 350 mm
Width of flange ($b$) = 165 mm
Thickness of flange ($t_f$) = 11.4 mm
Depth of web ($d$) = $h - 2(t_f + R)$ = 350 – 2(11.4 + 16)
= 295.2 mm
Thickness of web ($t_w$) = 7.4 mm
Moment of inertia about major axis ($I_z$) = $13158.3 \times 10^4$ mm$^4$
Elastic section modulus ($Z_e$) = $751.9 \times 10^3$ mm$^3$
Plastic section modulus ($Z_p$) = $851.11 \times 10^3$ mm$^3$

*Section classification*

$$\frac{b}{t_f} = \frac{82.5}{11.4} = 7.23 < 9.4$$

$$\frac{d}{t_w} = \frac{295.2}{7.4} = 39.9 < 84$$

Hence the section is plastic.
Check for shear capacity of section

$$V_d = \frac{f_y}{\gamma_{m0} \times \sqrt{3}} \times h \times t_w = \frac{250}{1.1 \times \sqrt{3}} \times 350 \times 7.4 = 340 \text{ kN}$$

0.6 $V_d$ = 204 kN < 292.5 kN
This shows a high shear condition.
Check for moment capacity of the section [Eqn 10.20(a)]

$$M_{dv} = M_d - \beta\,(M_d - M_{fd}) \le \frac{1.2 \times Z_e \times f_y}{\gamma_{m0}}$$

where $M_{fd}$ is the plastic design strength of the area of cross section excluding the shear area.

$$\beta = \left[2 \times \left(\frac{V}{V_d}\right) - 1\right]^2$$
$$= \left[2 \times \left(\frac{292.5}{340}\right) - 1\right]^2 = 0.52$$

Calculation of section modulus of flange

$$Z_{fd} = Z_p - A_w y_w$$
$$= 851.1 \times 10^3 - \left(350 \times 7.4 \times \frac{350}{4}\right)$$
$$= 624.485 \times 10^3 \text{ mm}^3$$

$$\therefore \quad M_{fd} = \frac{Z_{fd} \times f_y}{\gamma_{m0}}$$

$$= \frac{624.485 \times 10^3 \times 250}{1.10} = 141.93 \text{kNm}$$

Moment capacity of the secion

$$M_d = \frac{Z_p \times f_y}{\gamma_{m0}} = \frac{851.11 \times 10^3 \times 250}{1.10}$$
$$= 193.43 \text{ kN m}$$
$$\therefore M_{dv} = 193.43 - 0.52(193.43 - 141.93)$$
$$= 165.65 \text{ kN m} < \frac{1.2 \times Z_e \times f_y}{\gamma_{m0}}$$
$$= \frac{1.2 \times 751.9 \times 10^3 \times 250}{1.10}$$
$$= 205.06 \text{ kN m} > 146.25 \text{ kN m}$$

Hence the section is safe.

**EXAMPLE 10.6:** Calculate the moment carrying capacity of a laterally unrestrained ISMB 400 member of length 3 m.

**SOLUTION:**

*Section properties:*

Overall depth $(h) = 400$ mm
Width of flange $(b) = 140$ mm
Thickness of flange $(t_f) = 16$ mm
Height of web $(d) = h - 2(t_f + R) = 400 - 2(16 + 14)$
$= 340$ mm
Thickness of web $(t_w) = 8.9$ mm
Moment of inertia about major axis $(I_z) = 20458.4 \times 10^4 \text{ mm}^4$
Moment of inertia about minor axis $(I_y) = 622.1 \times 10^4 \text{ mm}^4$
Elastic section modulus $(Z_e) = 1020 \times 10^3 \text{ mm}^3$
Plastic section modulus $(Z_p) = 1175.2 \times 10^3 \text{ mm}^3$

As the beam is laterally unsupported, the design bending strength is found by calculating the lateral torsional buckling moment:

$$M_{cr} = \sqrt{\frac{\pi^2 EI_y}{(KL)^2}\left(GI_t + \frac{\pi^2 EI_w}{(KL)^2}\right)}$$

$$G = \frac{E}{2(1+\mu)} = \frac{2 \times 10^5}{2 \times (1 + 0.3)} = 76.923 \times 10^3 \text{ N/mm}^2$$

$$I_t = \sum \frac{b_i t_i^3}{3} = \left[\frac{2 \times 140 \times 16^3}{3} + \frac{(400 - 16) \times 8.9^3}{3}\right]$$
$$= 4.725 \times 10^5 \text{ mm}^4$$

$KL = 3000$ mm

$h_f = 400 - 16 = 384$ mm

$I_w = (1 - \beta_f)\beta_f I_y h_f^2$

$$\beta_f = \frac{I_{fc}}{I_{fc} + I_{ft}} = 0.5$$

$I_w = (1 - 0.5) \times 0.5 \times 622.1 \times 10^4 \times 384^2$
$= 2.29 \times 10^{11} \text{ mm}^6$

$$M_{cr} = \sqrt{\frac{\pi^2 \times 2 \times 10^5 \times 622.1 \times 10^4}{3000^2}\left(76.923 \times 10^3 \times 4.725 \times 10^5 + \frac{\pi^2 \times 2 \times 10^5 \times 2.29 \times 10^{11}}{3000^2}\right)}$$
$$= 343.68 \text{ kN m}$$

*Section classification*

Outstanding element of compression flange $= b/t_f = 70/16 = 4.38 < 9.4$

Web with neutral axis at mid-section $= d/t_w = 340/8.9 = 38.2 < 84\varepsilon$

Hence the section is plastic.

*Calculation of moment carrying capacity*

$$\lambda_{LT} = \sqrt{\frac{\beta_b Z_p f_y}{M_{cr}}}$$
$$= \sqrt{\frac{1.0 \times 1175.2 \times 10^3 \times 250}{343.68 \times 10^6}} = 0.925$$

$\phi_{LT} = 0.5\,[1 + \alpha_{LT}(\lambda_{LT} - 0.2) + \lambda_{LT}^2]$

$\alpha_{LT} = 0.21$

$\phi_{LT} = 0.5[1 + 0.21\,(0.925 - 0.2) + 0.925^2] = 1.004$

$$\chi_{LT} = \frac{1}{\phi_{LT} + [\phi_{LT}^2 - \lambda_{LT}^2]^{0.5}} \le 1.0$$
$$= \frac{1}{1.004 + [1.004^2 - 0.925^2]^{0.5}} = 0.717 \le 1.0$$

$$f_{bd} = \frac{\chi_{LT} f_y}{\gamma_{m0}}$$
$$= \frac{0.717 \times 250}{1.10} = 163 \text{ N/mm}^2$$

$M_d = \beta_b Z_p f_{bd}$
$= 1.0 \times 1175.2 \times 103 \times 163$
$= 191.56$ kN m

The moment carrying capacity of the section = 191.56 kN m.

**EXAMPLE 10.7:** A proposed cantilever beam is built into a concrete wall and free at its end. It supports dead load of 20 kN/m and a live load of 10 kN/m. The length of the beam is 5 m. Select an available section with necessary checks. Assume bearing length of 100 mm.

**SOLUTION:**

**Step 1:** *Calculation of load*

Dead load = 1.5 × 20 = 30 kN/m
Live load = 1.5 × 10 = 15 kN/m
Total load = 45 kN/m

**Step 2:** *Calculation of bending moment and shear force*

$$BM = \frac{wl^2}{2} = \frac{45 \times 5^2}{2} = 562.5 \text{ kNm}$$

$SF = wl = 45 \times 5 = 225$ kN

**Step 3:** *Selection of initial section*

Assume $\lambda = 80$; $h/t_f = 20$

$$f_{cr,\,b} = \left(\frac{1473.5}{\lambda}\right)^2\left[1+\frac{1}{20}\left(\frac{\lambda}{h/t}\right)^2\right]^{0.5}$$

$$= \left(\frac{1473.5}{80}\right)^2\left[1+\frac{1}{20}\left(\frac{80}{20}\right)^2\right]^{0.5}$$

$$= 455.1\,\text{N/mm}^2$$

$$\lambda_{LT} = \sqrt{\frac{f_y}{f_{crb}}} = \sqrt{\frac{250}{455.1}} = 0.741$$

$$\phi_{LT} = 0.5[1 + \alpha_{LT}(\lambda_{LT} - 0.2) + \lambda^2_{LT}]$$
$$= 0.5[1 + 0.21(0.741 - 0.2) + 0.741^2]$$
$$= 0.831$$

$$\chi_{LT} = \frac{1}{\phi_{LT} + [\phi_{LT}^2 - \lambda_{LT}^2]^{0.5}} \le 1.0$$

$$= \frac{1}{0.831 + [0.831^2 - 0.741^2]^{0.5}} \le 1.0 = 0.828$$

$$f_{bd} = \frac{\chi_{LT} f_y}{\gamma_{m0}} = \frac{0.828\times 250}{1.10} = 188.27\text{N/mm}^2$$

$$\text{Therefore required section modulus} = \frac{562.5\times10^6}{188.27}$$
$$= 2987.6\times 10^3\,\text{mm}^3$$

Choose a section of ISWB 600 @ 1.337 kN/m

Overall depth ($h$) = 600 mm

Width of flange ($b$) = 250 mm

Thickness of flange ($t_f$) = 21.3 mm

Thickness of web ($t_w$) = 11.2 mm

Depth of web ($d$) = $h - 2(t_f + R) = 600 - 2(21.3 + 17)$ = 523.4 mm

Moment of inertia about major axis ($I_z$) = $106199\times10^4$ mm$^4$

Moment of inertia about minor axis ($I_y$) = $4702.5\times10^4$ mm$^4$

Elastic section modulus ($Z_e$) = $3540\times10^3$ mm$^3$

Plastic section modulus ($Z_p$) = $3986.6\times10^3$ mm$^3$

*Section classification*

Outstand of compression flange = 125/21.3 = 5.86 < $9.4\varepsilon$

Web with N.A at mid-depth = 523.4/11.2 = 46.74 < $83.9\varepsilon$

Therefore, the section is plastic.

Since flange thickness is > 20 mm, we should use $f_y$ = 240 MPa, as per Table 1 of IS 800.

**Step 4:** *Calculation of lateral torsional buckling moment*

$$M_{cr} = \sqrt{\frac{\pi^2 EI_y}{(KL)^2}\left(GI_t + \frac{\pi^2 EI_w}{(KL)^2}\right)}$$

$KL = 0.85L = 0.85\times5000 = 4250$ mm

$$G = \frac{E}{2(1+\mu)} = \frac{2\times10^5}{2\times(1+0.3)}$$
$$= 76.923\times10^3\,\text{N/mm}^2$$

$$I_t = \sum\frac{b_i t_i^3}{3} = \left[\frac{2\times250\times21.3^3}{3} + \frac{(600-21.3)\times11.2^3}{3}\right]$$
$$= 1.88\times10^6\,\text{N/mm}^2$$

$$I_w = (1-\beta_f)\beta_f I_y h_f^2$$

$$\beta_f = \frac{I_{fc}}{I_{fc} + I_{ft}} = 0.5$$

$$I_w = (1-0.5)\times0.5\times4702.5\times10^4\times578.7^2 = 3.94\times10^{12}\,\text{mm}^6$$

$$M_{cr} = \sqrt{\frac{\pi^2\times2\times10^5\times4702.5\times10^4}{4250^2}}$$
$$\sqrt{\left(76.923\times10^3\times1.88\times10^6 + \frac{\pi^2\times2\times10^5\times3.94\times10^{12}}{4250^2}\right)}$$
$$= 1719\,\text{kN m}$$

**Step 5:** *Calculation of moment carrying capacity of a section*

$$\lambda_{LT} = \sqrt{\frac{\beta_b Z_p f_y}{M_{cr}}}$$
$$= \sqrt{\frac{1.0\times3986.6\times10^3\times240}{1719\times10^6}} = 0.746$$

$$\phi_{LT} = 0.5[1 + \alpha_{LT}(\lambda_{LT} - 0.2) + \lambda_{LT}^2]$$
$$\alpha_{LT} = 0.21$$
$$\phi_{LT} = 0.5[1 + 0.21(0.746 - 0.2) + 0.746^2]$$
$$= 0.8356$$

$$\chi_{LT} = \frac{1}{\phi_{LT} + [\phi_{LT}^2 - \lambda_{LT}^2]^{0.5}} \le 1.0$$
$$= \frac{1}{0.8356 + [0.8356^2 - 0.746^2]^{0.5}} \le 1.0$$
$$= 0.825 \le 1.0$$

$$f_{bd} = \frac{\chi_{LT} f_y}{\gamma_{m0}}$$
$$= \frac{0.825\times240}{1.10} = 180\text{N/mm}^2$$

$$M_d = \beta_b Z_p f_{bd}$$
$$= 1.0\times3986.6\times10^3\times180$$
$$= 717.6\,\text{kN m} > 562.5\,\text{kN m}$$

**Step 6:** *Calculation of shear capacity of section*

$$V_d = \frac{f_y}{\gamma_{m0}\sqrt{3}}\times h\times t_w$$
$$= \frac{250}{1.1\times\sqrt{3}}\times600\times11.2$$
$$= 881.7\,\text{kN}$$

$0.6\ V_d = 529\,\text{kN} > 225\,\text{kN}$

**Step 7:** *Calculation of deflection (assume that there is no cambering)*

$$\delta_b = \frac{wl^4}{8EI}$$

$$w = 30\ \text{N/m}$$

$$\delta_b = \frac{30 \times 5000^4}{8 \times 2 \times 10^5 \times 106199 \times 10^4} = 11.03\text{m}$$

$$\text{Allowable deflection} = \frac{l}{150} = \frac{5000}{150} = 33.33\,\text{mm}$$

Hence the section is safe against deflection.

**Step 8:** *Check for web buckling*

$A_b = (b_1 + n_1)t_w$

$b_1 = 100\,\text{mm}$

$n_1 = 600/2 = 300\,\text{mm}$

$\therefore A_b = (100 + 300) \times 11.2 = 4480\,\text{mm}^2$

Effective length $= 0.7d = 0.7 \times 523.4 = 366\,\text{mm}$

$$I = \frac{bt^3}{12} = \frac{100 \times 11.2^3}{12} = 11.707 \times 10^3\,\text{mm}^4$$

$A = 100 \times 11.2 = 1120\,\text{mm}^2$

$$r_{\min} = \sqrt{\frac{I}{A}} = \sqrt{\frac{11.707 \times 10^3}{1120}} = 3.23\,\text{mm}$$

$$\lambda = \frac{l_{\text{eff}}}{r_{\min}} = \frac{366}{3.23} = 113.31$$

$\therefore f_{cd} = 91\,\text{N/mm}^2$ (From Table 9c of the code), with $f_y = 240\,\text{MPa}$

Strength of the section against web buckling $= 91 \times 4480 = 407.68\,\text{kN} > 225\,\text{kN}$

Hence the section is safe against web buckling.

**Step 9:** *Check for web bearing*

$F_w = (b_1 + n_2)t_w f_y / \gamma_{m0}$

$b_1 = 100\,\text{mm}$

$n_2 = 2.5(R + t_f) = 2.5(17.0 + 21.3) = 95.75\,\text{mm}$

$F_w = (100 + 95.75) \times 11.2 \times 250/1.10$

$= 498.27\,\text{kN} > 225\,\text{kN}$

Hence safe.

**Example 10.8:** Design a continuous beam of span 4.9 m, 6 m and 4.9 m carrying a total uniformly distributed load of 32.5 kN/m and laterally unrestrained with a bearing length of 100 mm.

**Solution:**

**Step 1:** *Load calculation*

Factored load $= 1.5 \times 32.5$

$= 48.75\,\text{kN/m}$

For the bending moment and shear force diagram see Fig. 10.51.

Maximum bending moment $= 146.25\,\text{kNm}$

Maximum reaction $= 146.25 + 146.25 = 292.5\,\text{kNm}$

**Step 2:** *Selection of initial section*

Assume $\lambda = 100$ and $h/t_f = 25$. Therefore $f_{\text{cr,b}} = 291.4\,\text{N/mm}^2$ (from Table 14 of the code).

$$\lambda_{\text{LT}} = \sqrt{\frac{f_y}{f_{\text{cr,b}}}} = \sqrt{\frac{250}{291.4}} = 0.926$$

$$\phi_{\text{LT}} = 0.5[1 + \alpha_{\text{LT}}(\lambda_{\text{LT}} - 0.2) + \lambda_{\text{LT}}^2]$$

$$= 0.5[1 + 0.21(0.926 - 0.2) + 0.926^2]$$

$$= 1.005$$

$$\chi_{\text{LT}} = \frac{1}{\phi_{\text{LT}} + [\phi_{\text{LT}}^2 - \lambda_{\text{LT}}^2]^{0.5}} \le 1.0$$

$$= \frac{1}{1.005 + [1.005^2 - 0.926^2]^{0.5}} \le 1.0$$

$$= 0.716 \le 1$$

$$f_{\text{bd}} = \frac{\chi_{\text{LT}} f_y}{\gamma_{\text{m0}}} = \frac{0.716 \times 250}{1.10} = 162.7\,\text{N/mm}^2$$

$$\text{Therefore, required section} = \frac{146.25 \times 10^6}{162.7} = 898.7 \times 10^3\,\text{mm}^3$$

Choose a section of ISLB 400 @ 0.57 kN/m

Overall depth $(D) = 400\,\text{mm}$

Width of flange $(b) = 165\,\text{mm}$

Thickness of flange $(t_f) = 12.5\,\text{mm}$

Thickness of web $(t_w) = 8\,\text{mm}$

Depth of web $(d) = h - 2(t_f + R) = 400 - 2(12.5 + 16)$

$= 343\,\text{mm}$

Moment of inertia about major axis $I_{zz} = 19281.5 \times 10^4\,\text{mm}^4$

Moment of inertia about minor axis $I_{yy} = 716 \times 10^4\,\text{mm}^4$

Elastic section modulus $(Z_{ez}) = 964.1 \times 10^3\,\text{mm}^3$

Plastic section modulus $(Z_{py}) = 1098.2 \times 10^3\,\text{mm}^3$

*Section classification*

Outstand of compression flange $= b/t_f = 82.5/12.5$

$= 6.6 < 9.4$

Web with N.A at mid-depth $= d/t_w = 343/8 = 42.8 < 84$

Therefore, the section is plastic.

**Step 3:** *Calculation of lateral torsional buckling moment*

$$M_{\text{cr}} = \sqrt{\frac{\pi^2 EI_y}{(KL)^2}\left(GI_t + \frac{\pi^2 EI_w}{(KL)^2}\right)}$$

$$G = \frac{E}{2(1+\mu)} = \frac{2 \times 10^5}{2 \times (1 + 0.3)} = 76.923 \times 10^3\,\text{N/mm}^2$$

$$I_t = \sum \frac{b_i t_i^3}{3} = \left[\frac{2 \times 165\ 12.5^3}{3} + \frac{(400 - 12.5) \times 8^3}{3}\right]$$

$= 2.81 \times 10^5 \text{ mm}^4$

$I_w = (1 - \beta_f)\beta_f I_y h_f^2$

$h_f = D - t_f = 400 - 12.5 = 387.5 \text{ mm}$

$$\beta_f = \frac{I_{fc}}{I_{fc} + I_{ft}} = 0.5$$

$I_w = (1 - 0.5) \times 0.5 \times 716 \times 10^4 \times 387.5^2 = 2.69 \times 10^{11} \text{ mm}^6$

$$M_{cr} = \sqrt{\frac{\pi^2 \times 2\times10^5 \times 716\times10^4}{6000^2}\left(\begin{array}{c}76.923\times10^3\times2.81\times10^5 \\ +\dfrac{\pi^2\times2\times10^5\times2.69\times10^{11}}{6000^2}\end{array}\right)}$$

$= 119.48 \text{ kN m}$

$$\lambda_{LT} = \sqrt{\frac{\beta_b Z_p f_y}{M_{cr}}} = \sqrt{\frac{1.0\times1098.2\times10^3\times250}{119.48\times10^6}} = 1.516$$

$\phi_{LT} = 0.5[1 + \alpha_{LT}(\lambda_{LT} - 0.2) + \lambda_{LT}^2]$

$= 0.5[1 + 0.21(1.516 - 0.2) + 1.516^2] = 1.79$

$$\chi_{LT} = \frac{1}{\phi_{LT} + [\phi_{LT}^2 - \lambda_{LT}^2]^{0.5}} \le 1.0$$

$$= \frac{1}{1.79 + [1.79^2 - 1.516^2]^{0.5}} = 0.365 \le 1$$

$$f_{bd} = \frac{\chi_{LT} f_y}{\gamma_{m0}} = \frac{0.365\times250}{1.10} = 83 \text{ N/mm}^2$$

$$\text{Therefore required section} = \frac{146.25\times10^6}{83} = 1762\times10^3 \text{ mm}^3$$

Choose a section of ISLB 500 @ 0.75 kN/m.

Overall depth ($D$) = 500 mm

Width of flange ($B$) = 180 mm

Thickness of flange ($t_f$) = 14.1 mm

Depth of the web ($d$) = $D - 2(t_f + R) = 500 - 2(14.1 + 17)$ = 437.8 mm

Thickness of web ($t_w$) = 9.2 mm

Moment of inertia about major axis $I_z = 38579 \times 10^4 \text{ mm}^4$

Moment of inertia about minor axis $I_y = 1060 \times 10^4 \text{ mm}^4$

Elastic section modulus ($Z_{ez}$) = $1543 \times 10^3 \text{ mm}^3$

Plastic section modulus ($Z_{pz}$) = $1770.8 \times 10^3 \text{ mm}^3$

Radius of gyration $r_y = 33.4$ mm

*Section classification*

Outstand of compression flange = $b/t_f = 90/14.1 = 6.38 < 9.4$

Web with N.A at mid-depth = $d/t_w = 437.8/9.2 = 47.58 < 83.9$

Therefore the section is plastic.

**Step 4:** *Calculation of moment carrying capacity of the section*

$$M_{cr} = \sqrt{\frac{\pi^2 EI_y}{(KL)^2}\left(GI_t + \frac{\pi^2 EI_w}{(KL)^2}\right)}$$

$$G = \frac{E}{2(1+\mu)} = \frac{2\times10^5}{2\times(1+0.3)} = 76.923\times10^3 \text{ N/mm}^2$$

$$I_t = \sum\frac{b_i t_i^3}{3} = \left[\frac{2\times180\times14.1^3}{3} + \frac{485.8\times9.2^3}{3}\right]$$

$= 4.63 \times 10^5 \text{ mm}^4$

$I_w = (1 - \beta_f)\beta_f I_y h_f^2$

$h_f = D - t_f = 500 - 14.1 = 485.9$

$$\beta_f = \frac{I_{fc}}{I_{fc} + I_{ft}} = 0.5$$

$I_w = (1 - 0.5) \times 0.5 \times 1060 \times 10^4 \times 485.9^2$

$= 6.256 \times 10^{11} \text{ mm}^6$

$$M_{cr} = \sqrt{\frac{\pi^2\times2\times10^5\times1060\times10^4}{6000^2}\left(\begin{array}{c}76.923\times10^3\times4.63\times10^5 \\ +\dfrac{\pi^2\times2\times10^5\times6.256\times10^{11}}{6000^2}\end{array}\right)}$$

$= 201.6 \text{ kN m}$

Conservatively,

$$M_{cr} = \frac{\pi^2 EI_y h}{2(KL)^2}\left[1 + \frac{1}{20}\left(\frac{KL/r_y}{h/t_f}\right)^2\right]^{0.5}$$

$$\frac{KL}{r_y} = \frac{6000}{33.4} = 179.64$$

$$\frac{h}{t_f} = \frac{500}{14.1} = 35.46$$

$$M_{cr} = \frac{\pi^2\times2\times10^5\times1060\times10^4\times500}{2\times6000^2}\left[1 + \frac{1}{20}\left(\frac{179.64}{35.46}\right)^2\right]^{0.5}$$

$= 219.55 \text{ kN m}$

From Table 10.9, for $KL/r_y = 179.64$ and $h/t_f = 35.46$, $f_{cr,\,b} = 101.66 \text{ N/mm}^2$

Therefore $M_{cr} = 101.66 \times 1770.8 \times 10^3 = 180.02 \text{ kN m}$.

*Note:* The variation in the values obtained by the approximate and accurate method is about 8.9 to 10.7%.

$$\lambda_{LT} = \sqrt{\frac{\beta_b Z_p f_y}{M_{cr}}}$$

$$= \sqrt{\frac{1.0\times1770.8\times10^3\times250}{201.6\times10^6}} = 1.48$$

$\phi_{LT} = 0.5[1 + \alpha_{LT}(\lambda_{LT} - 0.2) + \lambda_{LT}^2]$

$\alpha_{LT} = 0.21$

$\phi_{LT} = 0.5[1 + 0.21(1.48 - 0.2) + 1.48^2] = 1.7296$

$$\chi_{LT} = \frac{1}{\phi_{LT} + [\phi_{LT}^2 - \lambda_{LT}^2]^{0.5}} \le 1.0$$

$$= \frac{1}{1.7296 + [1.7296^2 - 1.48^2]^{0.5}} = 0.381 \le 1.0$$

$$f_{bd} = \frac{\chi_{LT} f_y}{\gamma_{m0}} = \frac{0.381 \times 250}{1.10} = 86.6\,\text{N/mm}^2$$

$$M_d = Z_p f_{bd}$$

$$= 1770.8 \times 10^3 \times 86.6 = 153.34\,\text{kN m} > 146.25\,\text{kN m}$$

**Step 5:** *Calculation of shear capacity of section*

$$V_d = \frac{f_y}{\gamma_{m0}\sqrt{3}} \times D \times t_w = \frac{250}{1.1 \times \sqrt{3}} \times 500 \times 9.2 = 603.59\,\text{kN}$$

$0.6\ V_d = 362.15\,\text{kN} > 292.5\,\text{kN}$

**Step 6:** *Calculation of deflection*

$$\delta_b = \frac{5wl^4}{384EI}$$

$w = 32.5\,\text{kN/m}$

$$\delta_b = \frac{5 \times 32.5 \times 6000^4}{384 \times 2 \times 10^5 \times 38579 \times 10^4} = 7.10\,\text{mm}$$

$$\text{Allowable deflection} = \frac{l}{300} = \frac{6000}{300} = 20\,\text{mm}$$

Hence safe against deflection.

**Step 7:** *Check for web buckling*

$A_b = (b_1 + n_1)t_w$, $b_1 = 100\,\text{mm}$, $n_1 = D/2 = 500/2 = 250\,\text{mm}$

$\therefore A_b = (100 + 250) \times 9.2 = 3220\,\text{mm}^2$

Effective length $= 0.7d = 0.7 \times 437.8 = 306.46\,\text{mm}$

$$I = \frac{bt^3}{12} = \frac{100 \times 9.2^3}{12} = 6489\,\text{mm}^4$$

$$A = 100 \times 9.2 = 920\,\text{mm}^2$$

$$r_{min} = \sqrt{\frac{I}{A}} = \sqrt{(6489 / 920)} = 2.66\,\text{mm}$$

$$\lambda = \frac{L_{eff}}{r_{min}} = \frac{306.46}{2.66} = 115.21$$

$\therefore$ From Table 9c of the code, $f_{cd} = 88.9\,\text{N/mm}^2$

Strength of the section against web buckling $= 88.9 \times 3220 = 286.26\,\text{kN} \approx 292.5\,\text{kN}$

Hence the section is safe against web buckling.

**Step 8:** *Check for web bearing*

$F_w = (b_1 + n_2)t_w f_y / \gamma_{m0}$

$b_1 = 100\,\text{mm}$

$n_2 = 2.5(t_f + R) = 2.5 \times (14.1 + 17) = 77.75$

$F_w = (100 + 77.75) \times 9.2 \times 250/1.10 = 371.65\,\text{kN} > 292.5\,\text{kN}$

Hence the section is safe against web bearing.

**Example 10.9:** Design a laterally unrestrained beam to carry a uniformly distributed load of 50 kN/m. The beam is unsupported for a length of 1.5 m and is simply placed on longitudinal beams at its ends.

**Solution:**

**Step 1:** *Calculation of load*

Factored load $= 1.5 \times 50 = 75\,\text{kN/m}$

**Step 2:** *Calculation of bending moment and shear force*

$$BM = \frac{wl^2}{8} = \frac{75 \times 1.5^2}{8} = 21.09\,\text{kN m}$$

$$SF = \frac{wl}{2} = \frac{75 \times 1.5}{2} = 56.25\,\text{kN}$$

**Step 3:** *Choosing an initial section*

Assume $\lambda = 100$; $h/t_f = 25$ and hence from Table 10.9, $f_{cr,b} = 291.31\,\text{N/mm}^2$

$$\lambda_{LT} = \sqrt{\frac{f_y}{f_{cr,b}}} = \sqrt{\frac{250}{291.31}} = 0.926$$

$$\phi_{LT} = 0.5[1 + \alpha_{LT}(\lambda_{LT} - 0.2) + \lambda_{LT}^2]$$

$$= 0.5[1 + 0.21(0.926 - 0.2) + 0.926^2]$$

$$= 1.005$$

$$\chi_{LT} = \frac{1}{\phi_{LT} + [\phi_{LT}^2 - \lambda_{LT}^2]^{0.5}} \le 1.0$$

$$= \frac{1}{1.005 + [1.005^2 - 0.926^2]^{0.5}} = 0.716 \le 1.0$$

$$f_{bd} = \frac{\chi_{LT} f_y}{\gamma_{m0}} = \frac{0.716 \times 250}{1.10} = 162.7\,\text{N/mm}^2$$

$$\text{Therefore required section} = \frac{21.09 \times 10^6}{162.7} = 129.6 \times 10^3\,\text{mm}^3$$

Choose a section of ISMB 175 @ 0.19 kN/m.

Overall depth $(D) = 175\,\text{mm}$

Width of flange $(B) = 90\,\text{mm}$

Thickness of flange $(t_f) = 8.6\,\text{mm}$

Thickness of web $(t_w) = 5.5\,\text{mm}$

Depth of web $(d) = D - 2(t_f + R) = 175 - 2(8.6 + 10) = 137.8\,\text{mm}$

Moment of inertia about major axis $I_z = 1270.6 \times 10^4\,\text{mm}^4$

Moment of inertia about minor axis $I_y = 85.1 \times 10^4\,\text{mm}^4$

Elastic section modulus $(Z_{ez}) = 145.2 \times 10^3\,\text{mm}^3$

Plastic section modulus $(Z_{pz}) = 161.65 \times 10^3\,\text{mm}^3$

Minimum radius of gyration $(r_y) = 18.6\,\text{mm}$

*Section classification*

Outstand of compression flange $= 45/8.6 = 5.23 < 9.4$

Web with N.A at mid depth $= 137.8/5.5 = 25.05 < 83.9$

Therefore the section is plastic.

**Step 4:** *Calculation of lateral-torsional buckling moment*

$$M_{cr} = \sqrt{\frac{\pi^2 EI_y}{(KL)^2}\left(GI_t + \frac{\pi^2 EI_w}{(KL)^2}\right)}$$

$$G = \frac{E}{2(1+\mu)} = \frac{2\times10^5}{2\times(1+0.3)} = 76.923\times10^3\,\text{N/mm}^2$$

$$I_t = \sum\frac{b_i t_i^3}{3} = \left[\frac{2\times90\times8.6^3}{3} + \frac{(175-8.6)\times5.5^3}{3}\right]$$

$= 47.39 \times 10^3\,\text{mm}^3$

$$I_w = (1-\beta_f)\beta_f I_y h_f^2$$

$$\beta_f = \frac{I_{fc}}{I_{fc}+I_{ft}} = 0.5$$

$h_f = 175 - 8.6 = 166.4\,\text{mm}$

$I_w = (1-0.5)\times0.5\times85.1\times10^4\times166.4^2$

$= 5.89\times10^9\,\text{mm}^6$

$$M_{cr} = \sqrt{\frac{\pi^2\times2\times10^5\times85.1\times10^4}{1500^2}\left(76.923\times10^3\times47.39\times10^3 + \frac{\pi^2\times2\times10^5\times5.89\times10^9}{1500^2}\right)}$$

$= 81.11\,\text{kNm}$

*Check* As per the approximate equation given in the code

$$M_{cr} = \frac{\pi^2 E I_y h}{2(KL)^2}\left[1+\frac{1}{20}\left(\frac{KL/r_y}{h/t_f}\right)^2\right]^{0.5}$$

$$\frac{KL}{r_y} = \frac{1500}{18.6} = 80.65;\quad \frac{h}{t_f} = \frac{175}{8.6} = 20.34$$

$$M_{cr} = \frac{\pi^2\times2\times10^5\times85.1\times10^4\times175}{2\times1500^2}\left[1+\frac{1}{20}\left(\frac{80.65}{20.34}\right)^2\right]^{0.5}$$

$= 87.30\,\text{kN m}$

This value is comparable to 81.11 kN m obtained by the exact method (7.6% difference)

$$\lambda_{LT} = \sqrt{\frac{Z_p f_y}{M_{cr}}} = \sqrt{\frac{161.65\times10^3\times250}{81.11\times10^6}} = 0.7$$

$\phi_{LT} = 0.5[1 + \alpha_{LT}(\lambda_{LT} - 0.2) + \lambda_{LT}^2]$

$\alpha_{LT} = 0.21$

$\phi_{LT} = 0.5[1 + 0.21(0.7 - 0.2) + 0.7^2] = 0.7975$

$$\chi_{LT} = \frac{1}{\phi_{LT} + [\phi_{LT}^2 - \lambda_{LT}^2]^{0.5}}$$

$$= \frac{1}{0.7975 + [0.7975^2 - 0.7^2]^{0.5}} = 0.848 \le 1.0$$

$$f_{bd} = \frac{\chi_{LT} f_y}{\gamma_{m0}} = \frac{0.848\times250}{1.10} = 192.67\,\text{N/mm}^2$$

$M_d = Z_p f_{bd} = 161.65\times10^3\times192.67 = 31.14\,\text{kN m} > 21.09\,\text{kN m}$

**Step 5:** *Calculation of shear capacity of section*

$$V_d = \frac{f_y}{\gamma_{m0}\sqrt{3}}\times D\times t_w = \frac{250}{1.1\times\sqrt{3}}\times175\times5.5 = 126.3\,\text{kN}$$

$0.6\,V_d = 75.78\,\text{kN} > 56.25\,\text{kN}$

**Step 6:** *Calculation of deflection*

$$\delta_b = \frac{5wl^4}{384EI}$$

$w = 50\,\text{kN/m}$

$$\delta_b = \frac{5\times50\times1500^4}{384\times2\times10^5\times1270.6\times10^4} = 1.3\,\text{mm}$$

$$\text{Allowable deflection} = \frac{l}{300} = \frac{1500}{300} = 5\,\text{mm}$$

Hence the section is safe against deflection.

**Step 7:** *Check for web buckling*

Assuming that longitudinal beams are of the same size,

$A_b = (b_1 + n_1)t_w$

$b_1 = (b_f - t_w)/2 = (90 - 5.5)/2 = 42.25\,\text{mm}$

$n_1 = D/2 = 175/2 = 87.5\,\text{mm}$

$\therefore A_b = (42.25 + 87.5)\times5.5 = 713.6\,\text{mm}^2$

$$I = \frac{bt_w^3}{12} = \frac{42.25\times5.5^3}{12} = 585.8\,\text{mm}^4$$

$A = 42.25\times5.5 = 232.38\,\text{mm}^2$

$$r_{min} = \sqrt{\frac{I}{A}} = \sqrt{\frac{585.8}{232.38}} = 1.59\,\text{mm}$$

$$\lambda = \frac{l_{eff}}{r_{min}} = \frac{0.7\times137.8}{1.59} = 60.67$$

$\therefore f_{cd} = 167\,\text{N/mm}^2$ (From Table 9c of the code)

Strength of the section against web buckling $= 167 \times 713.6 = 119.17\,\text{kN} > 56.25\,\text{kN}$

Hence the section is safe against web buckling.

**Step 8:** *Check for web bearing*

$F_w = (b_1 + n_2)t_w f_y/\gamma_{m0}$

$b_1 = 42.25\,\text{mm};$

$n_2 = 2.5(t_f + R) = 2.5(8.6 + 10) = 46.5\,\text{mm}$

$F_w = (42.25 + 46.5)\times5.5\times250/1.10 = 110.9\,\text{kN} > 56.25\,\text{kN}$

Hence the section is safe against web bearing.

**Example 10.10:** Design a steel beam having a clear span of 9 m and resting on 150-mm wide end bearings. The beam's spacing is 3 m and the beam carries a dead load of 5 kN/m², including the weight of the section. The imposed load on the beam is 15 kN/m². The beam depth is restricted to 575 mm and the yield strength of the steel is 250 N/mm².

**Solution:**

*Factored loads*

Dead load $= 1.5\times5 = 7.5\,\text{kN/m}^2$

Imposed load $= 1.5\times15 = 22.5\,\text{kN/m}^2$

The beams are spaced at 3 m intervals, therefore the load per meter

Dead load = 7.5 × 3 = 22.5 kN/m

Imposed load = 22.5 × 3 = 67.5 kN/m

Total load = 90 kN/m

Effective span = clear span + 2 × half the width of the end bearing

$= 9 + 2 \times 0.075 = 9.15\text{ m}$

$$\text{Reactions at support} = \frac{90 \times 9.15}{2} = 411.75\text{ kN}$$

$$\text{Mid-span moment} = \frac{wl^2}{8} = \frac{90 \times 9.15^2}{8} = 941.878\text{ kN m}$$

*Selection of the section*

Plastic section modulus required

$$Z_p = \frac{M \times \gamma_{m0}}{f_y} = \frac{941.878 \times 10^6 \times 1.10}{250} = 414.43 \times 10^4\text{ mm}^3$$

The section with the largest plastic modulus under 575 mm depth restriction is ISWB 550 @ 1.125 kN/m. The plastic section modulus of this section is $306.6 \times 10^4\text{ mm}^3$ which is less than the required value. The section must be strengthened with additional plates to provide the required plastic section modulus. The stiffness required to be provided to restrict the deflection problem can be calculated as follows:

Maximum deflection = Effective span/360 = 9150/360 = 25.42 mm.

The moment of inertia of the required beam section that can satisfy the deflection requirement mentioned earlier due to unfactored imposed load

$$I_z = \frac{5}{384} \times \frac{45 \times 9150^4}{2 \times 10^5 \times 25.42} = 80784 \times 10^4\text{ mm}^4$$

Hence provide ISWB 550 with 10 mm cover plates which gives $I_z = 106{,}266.1 \times 10^4\text{ mm}^4$.

*Check for shear*

Shear capacity of the section

$$V_d = \frac{f_y}{\gamma_{m0} \times \sqrt{3}} \times h \times t_w = \frac{250}{1.1 \times \sqrt{3}} \times 550 \times 10.5 = 757.7\text{ kN}$$

$0.6V_d = 0.6 \times 757.7 = 454.663$

The maximum shear force is 411.75 kN and it is less than $0.6V_d$.

*Check for plastic modulus*

The plastic modulus required for the section is $414.43 \times 10^4\text{ mm}^3$. ISWB 550 provides $306.6 \times 10^4\text{ mm}^3$. Assumed thickness of the plate is 10 mm and the total depth of the beam is 570 mm. Distance between the c/c of the plates is 560 mm.

Additional plastic section modulus to be provided by the plate = $107.83 \times 10^4\text{ mm}^3$

$$\text{Required area of the plate} = \frac{107.83 \times 10^4}{560} = 1925.535\text{ mm}^2$$

The provided area of the plate = 2000 mm². The plastic modulus $Z_p$ of the compound section = $306.6 \times 10^4 + 2000 \times (550 + 10) = 418.6 \times 10^4\text{ mm}^3$, which is more than the required plastic section modulus.

*Check for deflection*

Minimum $I_z$ required is $80784 \times 10^4\text{ mm}^4$

$I_z$ provided by ISWB 550 = $74906.1 \times 10^4\text{ mm}^4$

$I_z$ to be provided by the plates = $5878 \times 10^4\text{ mm}^4$

$I_z$ provided by the plates = $2 \times 200 \times 10 \times 280^2$

$= 31360 \times 10^4\text{ mm}^4$.

Total $I_z$ provided = $106266.1 \times 10^4\text{ mm}^4$ greater than the $I_z$ required.

*Curtailment of flange plates*

In the case of plated beams, the flange plates may be curtailed near the supports of the beam to save some steel. The theoretical cut-off points may be calculated using the algebraic method or by a geometrical construction. For this problem the theoretical cut-off points are computed using the algebraic method. The bending moment capacity for the beam ISWB 550

$$M = 306.6 \times 10^4 \times 250 = 766.5\text{ kN m}$$

At any point, distance $x$ from the support

$M_z = Rx - \dfrac{wx^2}{2}$, where $M_z = 766.5$ kN m, $R = 411.75$ kN, and $w = 90$ kN/m

Substituting the values

$$766.5 \times 10^6 = 411.75 \times 10^3 \times x - \frac{90 \times x^2}{2}$$

Solving this equation $x$ = 6500 mm or 2600 mm. Hence the theoretical cut-off point is 2600 from either side of the support.

**EXAMPLE 10.11:** Design a purlin on a sloping roof truss with the dead load of 0.15 kN/m² (cladding and insulation), a live load of 2 kN/m² and wind load of 0.5 kN/m² (suction). The purlins are 2 m centre to centre and of span 4 m, simply supported on a rafter at a slope of 20 degrees (see Fig. 10.52).

(a) Provide channel section purlin

(b) Provide channel purlin with a sag rod at mid-span

(c) Provide angle purlin

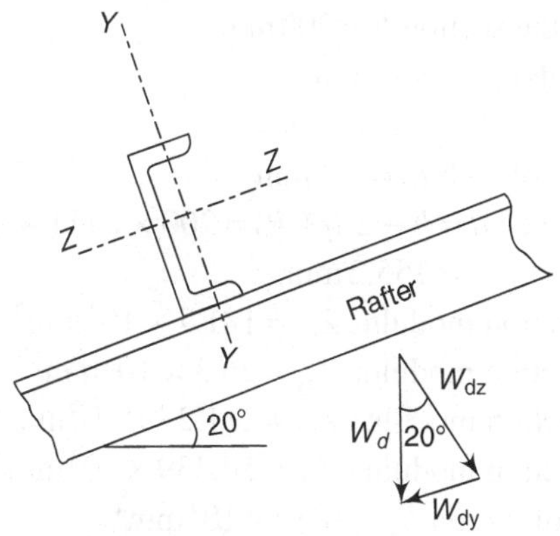


**FIG. 10.52**

**SOLUTION:**

*Load calculation*

Dead load = 0.15 × 2 = 0.3 kN/m

Live load = 2 × 2 = 4 kN/m

Wind load = 0.5 × 2 = 1 kN/m (suction)

$W_{dz} = 0.3 \times \cos 20° = 0.282$ kN/m

$W_{iz} = 4 \times \cos 20° = 3.76$ kN/m

$W_{wz} = -1$ kN/m

$W_{dy} = 0.3 \times \sin 20° = 0.103$ kN/m

$W_{iy} = 4 \times \sin 20° = 1.37$ kN/m

Note that $W_{wy}$ is zero as wind pressure is perpendicular to the surface on which it acts, i.e., normal to the rafter.

*Factored load combination*

Z-direction:

$WL + DL + LL = (1.2 \times 1.0) + (1.2 \times 0.282) + (1.2 \times 3.76) = 6.0552$ kN/m

$DL + LL = (1.5 \times 0.282) + (1.5 \times 3.76) = 6.063$ kN/m

Y-direction:

$DL + LL = (1.5 \times 0.103) + (1.5 \times 1.37) = 2.21$ kN/m

Bending moment and shear force calculation:

$M_z = 6.063 \times 4^2/8 = 12.126$ kN m

$M_y = 2.21 \times 4^2/8 = 4.42$ kN m

$F_z = 6.063 \times 4/2 = 12.126$ kN

$F_y = 2.21 \times 4/2 = 4.42$ kN

*(a) Channel section purlin*

Assume an ISMC 200 channel.

Plastic section modulus required

$$= \frac{M_z \times \gamma_{m0}}{f_y} + 2.5\frac{d}{b} \times \frac{M_y \times \gamma_{m0}}{f_y}$$

$$= \frac{12.126 \times 10^6 \times 1.1}{250} + 2.5 \times \frac{200}{75} \times \frac{4.42 \times 10^6 \times 1.1}{250}$$

$= 183 \times 10^3$ mm$^3$

Choose a channel section ISMC 200 @ 0.22 kN/m with plastic section modulus of

$Z_{pz} = 211.25 \times 10^3$ mm$^3$ and $Z_{py} = 56.339 \times 10^3$ mm$^3$.

*Section properties*

Cross sectional area $A = 2821$ mm$^2$

Depth of the section $h = 200$ mm

Width of flange $b = 75$ mm

Thickness of flange $t_f = 11.4$ mm

Thickness of web $t_w = 6.1$ mm

Depth of web $d = h - 2(t_f + R) = 200 \times 2\,(11.4 + 11) = 155.2$ mm

Elastic section modulus $Z_{ez} = 181.7 \times 10^3$ mm$^3$

Elastic section modulus $Z_{ey} = 26.3 \times 10^3$ mm$^3$

Plastic section modulus $Z_{pz} = 211.25 \times 10^3$ mm$^3$

Plastic section modulus $Z_{py} = 56.339 \times 10^3$ mm$^3$

Moment of inertia $I_{zz} = 1830 \times 10^4$ mm$^4$

Moment of inertia $I_{yy} = 141 \times 10^4$ mm$^4$

*Section classification*

$$\frac{b}{t_f} = \frac{75}{11.4} = 6.58 < 9.4$$

$$\frac{d}{t_w} = \frac{155.2}{6.1} = 25.44 < 42$$

Hence the section is plastic.

*Calculation of shear capacity of the section*

Z-direction

$$V_d = \frac{f_y}{\gamma_{m0} \times \sqrt{3}} \times h \times t_w = \frac{250}{1.1 \times \sqrt{3}} \times 200 \times 6.1 = 160.18\text{ kN}$$

$0.6V_d = 96$ kN > 12.126 kN

Y-direction

$$\text{Shear capacity} = \frac{250}{1.1 \times \sqrt{3}} \times 2 \times 75 \times \frac{11.4}{10^3} = 224.4\text{ kN} > 4.42\text{ kN}$$

Note that in purlin design, the shear capacity is usually high relative to the shear force.

*Design capacity of the section*

$$M_{dz} = \frac{Z_{pz} \times f_y}{\gamma_{m0}} = \frac{211.25 \times 10^3 \times 250}{1.1 \times 10^6} = 48\text{ kN m}$$

$$\le \frac{1.2 \times Z_{ez} \times f_y}{\gamma_{m0}} = \frac{1.2 \times 181.7 \times 10^3 \times 250}{1.1 \times 10^6} = 49.55\text{ kN m}$$

Hence $M_{dz} = 48$ kN m > 12.126 kN m

$$M_{dy} = \frac{Z_{py} \times f_y}{\gamma_{m0}} = \frac{56.339 \times 10^3 \times 250}{1.10} = 12.8\text{ kN m}$$

$$\le \frac{\gamma_f \times Z_{ey} \times f_y}{\gamma_{m0}} = \frac{1.5 \times 26.3 \times 10^3 \times 250}{1.1} = 8.96\text{ kN m}$$

Since the ratio $Z_p/Z_e$ is greater than 1.2, the constant in the preceding equation is replaced by the ratio of $\gamma_f = 1.5$. Hence $M_{dy} = 8.96$ kN m > 4.42 kN m.

*Overall member strength (local capacity)*

To ascertain the overall member strength, the following interaction equation should be satisfied.

$$\frac{M_z}{M_{dz}} + \frac{M_y}{M_{dy}} \le 1$$

$$\frac{12.126}{48} + \frac{4.42}{8.96} = 0.75 \le 1$$

Hence the overall member strength is satisfactory.

*Check for deflection due to live loads*

$$\delta = \frac{5wl^4}{384EI} = \frac{5 \times 3.76 \times 4000^4}{384 \times 2 \times 10^5 \times 1830 \times 10^4} = 3.42\text{ mm}$$

$$\text{Allowable deflection} = \frac{l}{180} = \frac{4000}{180} = 22.22\text{ mm}$$

Hence the section is safe.

*Check for wind suction*

The effect of wind suction has not been considered till now; it can become critical in some situations. It has to be combined with dead load.

Factored wind load $W_z = 0.9 \times 0.282 - 1.5 \times 1 = -1.246\,\text{kN/m}$

$W_y = 0.9 \times 0.103 = 0.0927\,\text{kN/m}$

*Buckling resistance of section*

Equivalent length $l_e = 4\,\text{m}$

Moment $= M_z = wl^2/8 = -1.246 \times 4^2/8 = -2.492\,\text{kN m}$

$M_y = 0.0927 \times 4^2/8 = 0.1854\,\text{kN m}$

The value of $M_z$ is much lower than the value 12.126 kN m earlier, but the negative sign indicates that the lower flange of the channel is in compression and this flange is unrestrained. Hence the buckling resistance of the channel must be found.

$$M_{cr} = \sqrt{\frac{\pi^2 EI_y}{(KL)^2}\left(GI_t + \frac{\pi^2 EI_w}{(KL)^2}\right)}$$

$$G = \frac{E}{2(1+\mu)} = \frac{2\times10^5}{2\times(1+0.3)} = 76.923 \times 10^3\,\text{N/mm}^2$$

$$I_t = \sum \frac{b_i t_i^3}{3} = \left[\frac{2\times75\times11.4^3}{3} + \frac{(200-11.4)\times6.1^3}{3}\right]$$

$= 88346.7\,\text{mm}^4$

$I_w = (1 - \beta_f)\beta_f I_y h_f^2$

$h_f = 200 - 11.4 = 188.6\,\text{mm}$

$$\beta_f = \frac{I_{fc}}{I_{fc} + I_{ft}} = 0.5$$

$I_w = (1 - 0.5) \times 0.5 \times 141 \times 10^4 \times 188.6^2$

$= 1.2538 \times 10^{10}\,\text{mm}^6$

$$M_{cr} = \sqrt{\frac{\pi^2\times2\times10^5\times141\times10^4}{4000^2}\left(76.923\times10^3\times88346.7 + \frac{\pi^2\times2\times10^5\times1.2538\times10^{10}}{4000^2}\right)}$$

$= 38.09\,\text{kN m}$

$$\lambda_{LT} = \sqrt{\frac{\beta_b Z_p f_y}{M_{cr}}}$$

$$= \sqrt{\frac{1.0\times211.25\times10^3\times250}{38.09\times10^6}} = 1.1775$$

$\phi_{LT} = 0.5[1 + \alpha_{LT}(\lambda_{LT} - 0.2) + \lambda_{LT}^2]$

$= 0.5[1 + 0.21(1.1775 - 0.2) + 1.1775^2]$

$= 1.296$

$$\chi_{LT} = \frac{1}{\phi_{LT} + [\phi_{LT}^2 - \lambda_{LT}^2]^{0.5}} \le 1.0$$

$$= \frac{1}{1.296 + [1.296^2 - 1.1775^2]^{0.5}} \le 1.0$$

$= 0.544 \le 1$

$$f_{bd} = \frac{\chi_{LT} f_y}{\gamma_{m0}} = \frac{0.544\times250}{1.10} = 123.71\,\text{N/mm}^2$$

$M_{dz} = Z_p f_{bd}$

$= 211.25 \times 10^3 \times 123.71$

$= 26.13\,\text{kN m} > 2.492\,\text{kN m}$

The buckling resistance $M_{dy}$ of the section need not be found out, because the purlin is restrained by the cladding in the z-plane and hence instability is not considered for a moment about the minor axis.

*Overall member strength*

To ascertain the overall member buckling strength, the following interaction should be satisfied.

$$\frac{M_z}{M_{dz}} + \frac{M_y}{M_{dy}} \le 1$$

$$\frac{2.492}{26.13} + \frac{0.1854}{8.96} = 0.097 < 1$$

Hence the overall member strength is satisfactory.

It has to be noted that the maximum bending moment occurs at the centre of the beam and the maximum shear force at the supports. Hence it is not necessary to check the moment capacity in the presence of shear force. Also purlins are not normally checked for web bearing and crippling as the applied concentrated loads are low (note the low value of shear force in this example).

*(b) Channel section purlin with one sag rod at mid-span*

Since the channel section purlin is provided with a sag rod at mid-span, the bending moment in the y-direction will be reduced considerably.

$M_y = 2.21 \times 4^2/32 = 1.105\,\text{kNm}$

$M_z = 12.126\,\text{kN m}$

Required section modulus $= (M_z \times \gamma_{m0}/f_y) + 2.5(d/b)(M_y\gamma_{m0}/f_y)$

Assuming ISMC 100 with $d = 100\,\text{mm}$ and $b = 50\,\text{mm}$,

Required $Z = (12.126 \times 10^6 \times 1.1/250) + 2.5 \times (100/50) \times (1.105 \times 10^6 \times 1.1/250)$

$= 77.66 \times 10^3\,\text{mm}^3$

Provide ISMC 150 with the following section properties.

Depth of section $h = 150\,\text{mm}$; $r_y = 22\,\text{mm}$

Width of flange $b = 75\,\text{mm}$

Thickness of flange $t_f = 9.0\,\text{mm}$

Thickness of web $t_w = 5.7\,\text{mm}$

Elastic section modulus $Z_{ez} = 105 \times 10^3\,\text{mm}^3$

Elastic section modulus $Z_{ey} = 19.5 \times 10^3\,\text{mm}^3$

Plastic section modulus $Z_{pz} = 119.82 \times 10^3\,\text{mm}^3 > 77.66 \times 10^3\,\text{mm}^3$

Moment of inertia $I_z = 788 \times 10^4\,\text{mm}^3$

*Section classification*

$b/t_f = 75/9.0 = 8.33 < 9.4$

$d/t_w = [150 - 2\,(9.0 + 10)]/5.7 = 19.65 < 42$

Hence the section is plastic. Shear capacity is not being checked since the shear force is small and hence the section will be adequate.

*Design capacity of the section*

$M_{dz} = (Z_{pz} \times f_y/\gamma_{m0}) = (119.82 \times 10^3 \times 250/1.1 \times 10^6)$
$= 27.23\,kN\,m$
$\leq (1.2 \times Z_{ez} f_y/\gamma_{m0}) = [(1.2 \times 105 \times 10^3 \times 250)/(1.1 \times 10^6)] = 28.63$

$Z_{py} = 2t_f b_f^2/4 + (h - 2t_f)\, t_w^2/4 = 2 \times 9.0 \times 75^2/4 + (150 - 2 \times 9.0) \times 5.7^2/4$
$= 26384.6\,mm^3$

$M_{dy} = (Z_{py} f_y/\gamma_{m0}) = 26384.6 \times 250/(1.1 \times 10^6) = 6.0\,kN\,m$
$\leq 1.5\, Z_{ey} f_y/\gamma_{m0} = 1.5 \times 19.5 \times 10^3 \times 250/(1.1 \times 10^6)$
$= 6.6\,kN\,m$

Hence the section is safe.

*Overall member strength due to live load*

For overall member strength, the following interaction equation must be satisfied.

$$(M_z/M_{dz}) + (M_y/M_{dy}) \leq 1.0$$
$$(12.126/27.23) + (1.105/6.0) = 0.629 < 1.0$$

Hence the member strength is satisfactory.

*Check for deflection due to live load*

$\delta = (5wl^4/384EI) = (5 \times 3.76 \times 4000^4)/(384 \times 2 \times 10^5 \times 788 \times 10^4)$
$= 7.95\,mm < 22.22\,mm$

Hence the section is safe.

*Check for wind suction*

From Example 10.11(a)

$M_z = 2.492\,kN\,m$
$M_y = 0.0927 \times 4^2/32 = 0.0464\,kN\,m$
$f_{cr} = [1473.5/(KL/r_y)]^2\, \{1 + (1/20)\, [(KL/r_y)/(h/t_f)]^2\}^{0.5}$
$KL/r_y = 4000/22 = 181.8$
$h/t_f = 150/9.0 = 16.67$

Thus, $f_{cr} = (1473.5/181.8)^2\, \{1 + (1/20)\, [181.8/16.67]^2\}^{0.5}$
$= 173.1\,N/mm^2$

$f_{bd} = 120.0\,N/mm^2$ (from Table 13a of the code)
$M_{dz} = Z_{pz} f_{bd} = 119.82 \times 10^3 \times 120.0/10^6 = 14.38\,kN\,m$

*Overall buckling strength*

For overall buckling strength, the following interaction equation should be satisfied.

$(M_z/M_{dz}) + (M_y/M_{dy}) = (2.492/14.38) + (0.0464/6.0)$
$= 0.18 < 1.0$

Hence the overall buckling strength is satisfactory.

Hence by using one sag rod, it was possible to reduce the section from ISMC 200 to ISMC 150 (about 25% reduction in weight).

*(c) Angle Section Purlin (as per BS 5950-1:2000)*

From Example 10.11(a) $M_z = 12.126\,kN\,m$; $W_p = (1.0 + 0.282 + 3.76) \times 4$
$= 20.168\,kN$

Moment at working load = 12.126/1.5 = 8.084 kN m

Let us assume that bending about *z-z* axis resists the vertical loads and the horizontal component is resisted by the sheeting.

Design strength $f_y = 250\,MPa$

Applied moment = moment capacity of single angle

$8.084 \times 10^6 = 250 \times Z_{ez}$

Required $Z_{ez} = 8.084 \times 10^6/250 = 32.33 \times 10^3\,mm^3$

Provide ISA 150 × 75 × 10 angle @ 0.17 kN/m,

with $Z_{ez} = 51.9 \times 10^3\,mm^3 > 20.168 \times 4 \times 10^6/1800$
$= 44.817 \times 10^3\,mm^3$

$d/t = 150/10 = 15.0 > 10.5$ but $< 15.7$

The section is *semi-compact.*

Leg length perpendicular to plane of cladding
$= 4000/45 = 88.88\,mm < 150\,mm$

Leg length parallel to the plane of cladding
$= 4000/60 = 66.66\,mm < 75\,mm$

Deflection need not be checked in this case.

**EXAMPLE 10.12:** Design the diagonal braces to provide lateral support to five ISMB 600 beams positioned as shown in Fig. 10.53. Assume that the braces are 2250 mm long, attached near the compression flange, and located along the supporting beams at the one third points. The girders have 14.4 m span. Use Fe 410 grade steel and assume a maximum factored bending moment of 472 kN m.

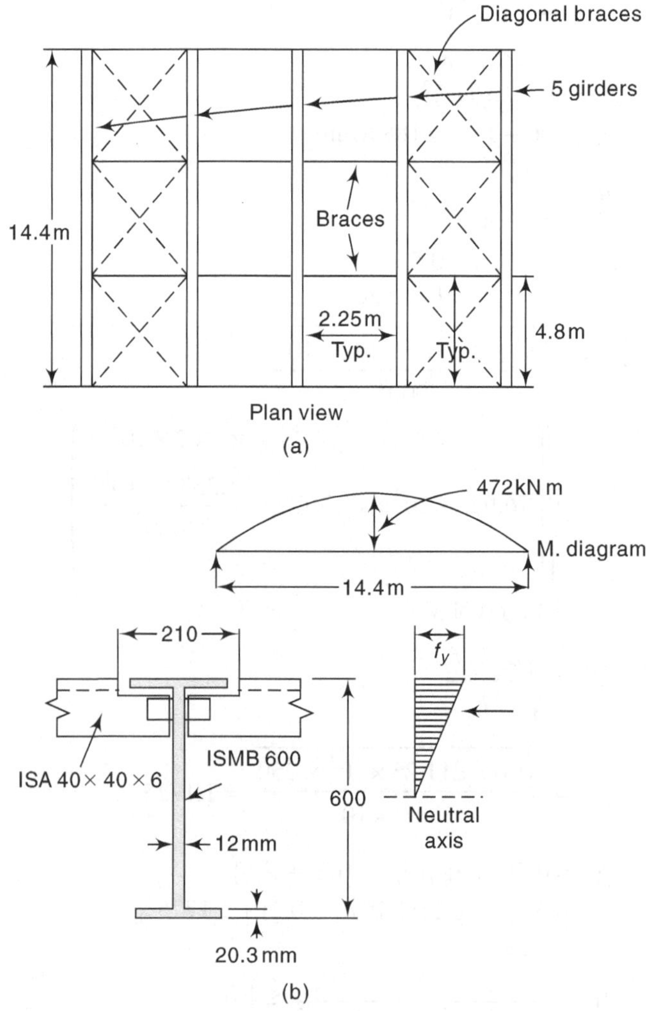


FIG. 10.53

**SOLUTION:**

Stiffness: $k_{br} = (4M_uC_d)/(0.75L_bh_o) = (4 \times 472 \times 10^6 \times 1)/[0.75 \times 4.8 \times 10^3 \times (600 - 20.3)] = 904.68$ N/mm for each girder

For 2.5 girders $= 904.68 \times 2.5 = 2261.7$ N/mm

Length of diagonal $= \sqrt{(2.25^2 + 4.8^2)} = 5.3$ m

$(AE/L)_b \cos^2\theta = (A_b \times 2 \times 10^5)/(5.3 \times 10^3) \times (2.25/5.3)^2 = 2261.7$

Hence $A_b = 332.6$ mm²

$F_{br} = 0.008M_uC_d/h_o = 0.008 \times 472 \times 10^6/(600 - 20.3) = 6514$ N

$A_b = 6514 \times (5.3/2.25) \times 1.1/250 = 67.5$ mm²

The stiffness requirement governs.

Hence provide ISA $L$ 40 × 40 × 6 with $A = 447$ mm².

**EXAMPLE 10.13:** For the same system in Example 10.12, instead of diagonal braces, design a diaphragm system with $M_u = 472$ kN m, $C_1 = 1.0$, four braces, $f_y = 250$ MPa, and $L = 14.4$ m.

**SOLUTION:**

The properties of the girder are as follows:

$h_o = 600 - 20.3 = 579.7$ mm, $t_w = 12$ mm

$I_y = 2650 \times 10^4$ mm⁴

Assume thickness of stiffener as 10 mm. The required strength is given by

$M_{br} = (0.024M_uL)/(nC_1L_b)$

$= (0.024 \times 472 \times 10^6 \times 14.4 \times 10^3)/(2 \times 1 \times 4.8 \times 10^3)$

$= 16.992 \times 10^6$ N mm

$Z_{p,\,req} = 16.992 \times 10^6 \times 1.1/250 = 74{,}765$ mm³

The stiffness of the diaphragm on the exterior girder is $6EI_{br}/S$. Since there are diaphragms on both sides of each interior girder, the stiffness is $2 \times 6EI_{br}/S$. The average stiffness available to each girder is $(2 \times 6 + 3 \times 12)/5 = 9.6EI_{br}/S$

$k_t = 2.4LM_u^2/(0.75nEI_yC_1^2)$

$= 2.4 \times 14.4 \times 10^3 \times (472 \times 10^6)^2/(0.75 \times 2 \times 2 \times 10^5 \times 2650 \times 10^4 \times 1^2)$

$= 968.5 \times 10^6$ N mm/rad

$k_{sec} = (3.3E/h_o)\,[1.5h_ot_w^3/12 + t_sb^3_s/12]$

$= (3.3 \times 2 \times 10^5/579.7)\,[1.5 \times 579.7 \times 12^3/12 + 10 \times 99^3/12]$

$= 1063.1 \times 10^6$ N mm/rad

Required $k_{tb} = k_t/[1 - (k_t/k_{sec})]$

$= 968.5 \times 10^6/[1 - 968.5/1063.1)]$

$= 10{,}884 \times 10^6$ N mm/rad

Min $I_{br} = k_ts/(9.6E) = 968.5 \times 10^6 \times 2250/(9.6 \times 2 \times 10^5)$

$= 113.49 \times 10^4$ mm⁴

Try ISMC 150 with $Z_p = 119.82 \times 10^3$ mm³ $> 74.765 \times 10^3$ mm³

and $I_z = 788 \times 10^4$ mm⁴

Thus available

$k_{tb} = 9.6 \times 2 \times 10^5 \times 788 \times 10^4/2250$

$= 6724.26 \times 10^6$ N mm/rad $< 10{,}884 \times 10^6$ N mm/rad

Hence provide ISMC 200 with $I_z = 1830 \times 10^4$ mm⁴.

$k_{tb} = 9.6 \times 2 \times 10^5 \times 1830 \times 10^4/2250$

$= 15{,}615 \times 10^6 > 10{,}884 \times 10^6$ N mm/rad

Hence the section is safe. Provide the channel near the bottom flange of the I-beam, as shown in Fig. 10.54.

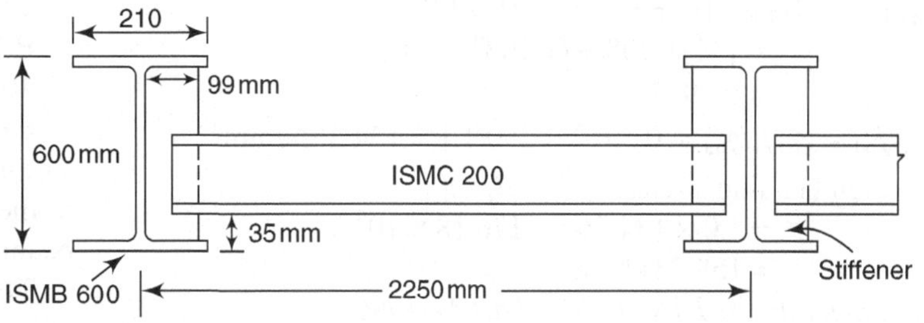


**FIG. 10.54**

**EXAMPLE 10.14:** Compute the design capacity of a simply supported ISMB 400 beam under uniform moment with $L = 3.80$ m, $f_y = 250$ MPa, and $E = 2 \times 10^5$ MPa. $E/G = 2.7$, as per
(1) IS 800 : 2007, (2) DIN: 18800-1990, (3) ANSI/AISC: 360-10.

**SOLUTION:**

The relevant properties of ISMB 400 are as follows:

$I_y = 622 \times 10^4$ mm⁴; $Z_{pz} = 1176.18 \times 10^3$ mm³; $Z_z = 1020 \times 10^3$ mm³; $r_y = 28.2$ mm; $t_w = 8.9$ mm; $t_f = 16$ mm; $h = 400 - 16 = 384$ mm; $b_f = 140$ mm;

$I_w = h_o^2I_y/4 = 384^2/4 \times 622 \times 10^4 = 2.293 \times 10^{11}$ mm⁶

$I_t = (1/3)(2b_ft_f^3 + ht_w^3) = (1/3)\,(2 \times 140 \times 16^3 + 384 \times 8.9^3)$

$= 0.4725 \times 10^6$ mm⁴

$$M_{cr} = (\pi/L)\sqrt{(EI_yGI_t)}\sqrt{[1 + (\pi^2 EI_w)/(GI_tL^2)]}$$

$$= (\pi/L)\sqrt{[(2 \times 10^5)^2 \times 622 \times 10^4 \times 0.4725 \times 10^6/2.6]}$$

$$\sqrt{1 + [(\pi^2 \times 2.6 \times 2.293 \times 10^{11})/(0.4725 \times 10^6 \times 3800^2)]}$$

$$= (\pi/3800) \times 2 \times 10^5 \times 10^4 \times 106.32\sqrt{(1 + 0.8624)}$$

$$= 239.9 \times 10^6 \text{ N mm}$$

$$f_{cr,b} = \frac{M_{cr}}{Z_z} = \frac{239.9 \times 10^6}{1020 \times 10^3} = 235.2\,MPa$$

As per the approximate formula [Eqn (10.26)]

$f_{cr,\,b} = [1473.5/(KL/r_y)]^2\,[1 + (1/20)\,\{(KL/r_y)/(h/t_f)\}^2]^{0.5}$

$KL/r_y = 3800/28.2 = 134.75$

$h/t_f = 400/16 = 25$

$f_{cr,\,b} = (1473.5/134.75)^2\,[1 + (1/20)\,(134.75/25)^2]^{0.5}$

$= 187.25$ MPa

Thus by using the approximate formula we get a value that is 20% less than the accurate value.

*Capacity as per IS 800 : 2007*

Non-dimensional slenderness ratio

$$\lambda_{LT} = \sqrt{(\beta_b Z_p f_y / M_{cr})}$$

$$= \sqrt{1 \times 1176.18 \times 10^3 \times 250/(239.9 \times 10^6)}$$

$= 1.107$

$\phi_{LT} = 0.5[1 + \alpha_{LT} (\lambda_{LT} - 0.2) + \lambda_{LT}^2]$
$\alpha_{LT} = 0.21$ (for rolled section)
$\phi_{LT} = 0.5[1 + 0.21(1.107 - 0.2) + 1.107^2]$
$= 1.208$
$\chi_{LT} = \{1/(\phi_{LT} + [\phi_{LT}^2 - \lambda_{LT}^2]^{0.5})\} \leq 1.0$
$= \{1/ [1.208 + (1.208^2 - 1.107^2)^{0.5}]\}$
$= 0.5912$
$f_{bd} = \chi_{LT} f_y/\gamma_{m0} = 0.5912 \times 250/1.1 = 134.36 \text{ N/mm}^2$

Design bending strength $M_d = \beta_b Z_p f_{bd}$
$= 1.0 \times 134.36 \times 1176.18 \times 10^3$
$= 158.03$ kN m

*Capacity as per DIN: 18800 (Part 2)–1990*

$M_p = 250 \times 1020 \times 10^3 = 255 \times 10^6 \text{ N mm} = 255$ kN m
$\chi_{LT} = [1/(1 + \lambda_{LT}^5)]^{0.4} = [1/(1 + 1.107^5)]^{0.4} = 0.676$
$M_d = 0.676 \times 255/1.1 = 156.69$ kN m

*Capacity as per ANSI/AISC 306–2005*

$M_r = 0.7 f_y Z_z = 0.7 \times 250 \times 1020 \times 10^3 = 178.5 \times 10^6$ N mm

Equating $M_r = (\pi/L_r)\sqrt{(EI_y GI_t)}\sqrt{[1 + (\pi^2 EI_w)/(GI_t L^2)]}$, we get

$L_r = \pi/(178.5 \times 10^6) \times 212.64 \times 10^9 \sqrt{1.8624} = 5107 \text{ mm} > L = 3800$ mm

$L_p = 790 r_y/\sqrt{f_y} = 790 \times 28.2/\sqrt{250} = 1409 \text{ mm} < L$
$= 3800$ mm

Hence $M_d = \phi C_1 [M_p - (M_p - M_r) [(L - L_p)/(L_r - L_p)] \leq M_p$
$= 0.9 \times 1.0 \times [255 - \{(255 - 178.5) \times (3800 - 1409)/(5107 - 1409)\}]$
$= 184.98$ kN m

It is observed from the preceding calculations that the Indian code provisions are conservative compared to the German and American provisions and that the American provisions give high value of moment capacity for the chosen section (about 13% higher than the Indian code provisions). It is mainly due to the straight line transition adopted by the American code in the inelastic range.

**EXAMPLE 10.15:** Design a lintel over an opening of 3.5 m. The lintel is supporting a wall of thickness 230 mm. The lintel has to support a uniform load of 50 kN in addition to the masonry. The weight of the masonry may be assumed as 19.2 kN/m$^3$ and the height of the wall above the lintel as 3 m. Assume the yield strength of steel as 250 MPa.

**SOLUTION:**

Assume 100 m bearing on either side of the lintel.

Effective span of the lintel = 3500 + 2 × 100
= 3700 mm

The lintel will support the weight of brickwork in the form of an equilateral triangle over the opening. The weight of this brick work

$$W = \frac{1}{2} \times 3.70 \times 3.70 \times \sin 60° \times 0.23 \times 19200$$

= 26,177.8 N = 26.178 kN

Uniform load over lintel = 50 kN

Assumed self-weight of lintel = 200 N/m

B.M. due to load $W = WL/6 = 26.178 \times 3.7/6 = 16.14$ kN m

Total uniform load = $(50 \times 1000 + 200 \times 3.7)/10^3 = 50.74$ kN

B.M. due to uniform load = $50.74 \times 3.7/8 = 23.467$ kN m

Total BM = 16.14 + 23.467 = 39.61 kN m

Factored BM = 1.5 × 39.61 = 59.415 kN m

The masonry wall is assumed not to provide any lateral support to the lintel and hence is designed as a laterally unsupported beam. Assume $\lambda = 80$ and $h/t_f = 20$.

Thus, from Example 10.7, we have

$f_{cr} = 455.1 \text{ N/mm}^2$

and

$f_{bd} = 188.27 \text{ N/mm}^2$

Required section modulus = $59.415 \times 10^6/188.27$
$= 315 \times 10^3 \text{ mm}^3$

Choose a section of ISMB 250 @ 0.373 kN/m with $Z_p = 465.71 \times 10^3 \text{ mm}^3$

The relevant properties of ISMB 250 are as follows:

$h$ = 250 mm; $b_f$ = 125 mm; $t_f$ = 12.5 mm; $t_w$ = 6.9 mm; $R$ = 13.0 mm; $r_y$ = 26.5 mm

Depth of web = $h - 2(t_f + R) = 250 - 2\,(12.5 + 13)$
= 199 mm

$I_z = 5130 \times 10^4 \text{ mm}^4$; $I_y = 335 \times 10^4 \text{ mm}^4$; $Z_{ez} = 410 \times 10^3 \text{ mm}^3$

*Section classification*

Outstand of compression flange = $62.5/12.5 = 5 < 9.4\varepsilon$

Web with N.A at mid-depth = $199/6.9 = 28.8 < 84\varepsilon$

Hence section is *plastic.*

*Calculation of lateral torsional buckling moment*

$$f_{cr} = [1473.5/(KL/r_y)]^2 \{1 + (1/20) [(KL/r_y)/(h/t_f)]^2\}^{0.5}$$

Since the lintel is embedded in masonry at the ends, it may not restrain the beam against torsion; hence the effective length of the compression flange of the beam is taken as (see Table 15 of IS 800 : 2007)

$KL/r_y = (1.2 \times 3700 + 2 \times 250)/26.5 = 186.4$
$h/t_f = 250/12.5 = 20$
$f_{cr} = (1473.5/186.4)^2 \{1 + (1/20) [186.4/20]^2\}^{0.5}$
$= 144.4 \text{ N/mm}^2$

$f_{bd} = 103.5 \text{ N/mm}^2$ (from Table 13a of the code)

$M_{dz} = Z_{pz} f_{bd} = 465.71 \times 10^3 \times 103.5/10^6$
$= 48.2 \text{ kN m} < 59.415$ kN m

This section is not safe. Provide ISMB 300.

*Check for shear*

Maximum shear force = (26.178 + 50.74) × 1.5/2 = 57.69 kN

Shear capacity of the section = $[250/(\sqrt{3} \times 1.1)] \times 300 \times 7.7/10^3 = 303.1 \text{ kN} > 57.69$ kN

Hence the section is safe.

*Check for deflection*

$\delta_b = [(WL^3)/(60EI)] + (5/384)\,(WL^3/EI)$

$= (26178 \times 3700^3)/(60 \times 2 \times 10^5 \times 8990 \times 10^4)$
$+ (5/384)(50740 \times 3700^3)/(2 \times 10^5 \times 8990 \times 10^4)$
$= 1.229 + 1.861 = 3.09$ mm
$\delta_{all} = L/180 = 3700/180 = 20.55$ mm $> 3.09$ mm
Hence the section is safe.

**EXAMPLE 10.16:** Design a castellated beam in grade Fe 410 steel to carry an imposed load of 4 kN/m and dead load of 4 kN/m over a simply supported span of 16 m. Assume that the compression flange is fully restrained.

**SOLUTION:**
Assume self-weight as 0.75 kN/m.
Take factored load $= 1.5(4 + 0.75 + 4) = 13.125$ kN/m
Maximum moment $= 13.125 \times 16^2/8 = 420$ kN m
Maximum shear force $= 13.125 \times 16/2 = 105$ kN

The capacity of castellated beam will be reduced due to the secondary effects of shear. Hence $Z$ required has to be increased to find a suitable trial section. As a first approximation, it may be taken as 1.05 times the value based on the net plastic section.

Hence $Z_{p,\,req} = 1.05M/(f_y/1.1) = 1.05 \times 420 \times 10^6/(250 \times 1.1) = 160 \times 10^6\ \text{mm}^3$

Select NCB 450 (0.724 kN/m) castellated from ISMB 450
$Z_z = 2113.27 \times 10^3\ \text{mm}^3$ (app. $Z_p = 1.2 \times 2113.27 \times 10^3\ \text{mm}^3$)

**1. Calculation of properties of the section (see Fig. 10.55)**

Area of T chord
$= 9.4(109 - 17.4) + 17.4 \times 150$
$= 3471.04\ \text{mm}^2$

Position of centroid of castellated T-section
$C_z = [91.6 \times 9.4 \times (17.4 + 91.6/2) + 17.4 \times 150 \times 17.4/2]/3471.04$
$= 22.22$ mm

Distance between centres of T chords
$h = D - 2C_z = 675 - 2 \times 22.2 = 630.6$ mm
Length of top of opening $= 114.25$ mm
$A_{T,\,web} = 9.4(109 - 17.4) = 861.04\ \text{mm}^2$
$I_z = 71{,}693 \times 10^4\ \text{mm}^4$
$I_T = 17.4 \times 150 \times (22.2 - 17.4/2)^2 + 91.6 \times 9.4\,(91.6/2 - 22.2)^2 + 91.6^3 \times 9.4/12 + 150 \times 17.4^3/12$
$= 1.6231 \times 10^6\ \text{mm}^4$
$Z_T = 1.6231 \times 10^6/86.8 = 18699\ \text{mm}^3$

As per Martin and Purkiss (1992),
$1/A_{fict} = 4.3207 D_s/(h^2 t_w) + 235.56 \times 10^{-6} D_s^2/I_T + 116.3 \times 10^{-3}/A_{T,\,web}$
where $D_s$ = Depth of the original section = 450 mm. Substituting the values of $h$, $t$, $I_T$, and $A_{T,\,web}$,

$1/A_{fict} = 4.3207 \times 450/(630.6^2 \times 9.4) + 235.56 \times 10^{-6} \times 450^2/(1.6231 \times 10^6) + 116.3 \times 10^{-3}/861.04$
$= 7.018 \times 10^{-4}\ \text{mm}^{-2}$.

**2. Section classification with $\varepsilon = 1$**

Flange outstand $b/t_f = 75/17.4 = 4.3 < 9.4\varepsilon$
Web slenderness $d/t = [675 - 2(17.4+15)]/9.4 = 64.91 < 84\varepsilon$
Hence the section is *plastic*.
Stem of the T-section: $(d/t)_{limit} < 18.9\varepsilon$ or $d_{limit} = 18.9 \times 9.4 = 177.66$ mm $> 91.6$ mm
Hence the section is plastic.

**3. Check for shear**

The elastic shear stress $= VQ/(I_z t)$, where $V = 105$ kN.
The maximum shear will occur at section $z$-$z$ shown in Fig. 10.55(b),
$Q = A\bar{y} = 150 \times 17.4\,(675/2 - 17.4/2) + 9.4 \times 91.6\,(457/2 + 91.6/2)$
$= 1094.3 \times 10^3\ \text{mm}^3$

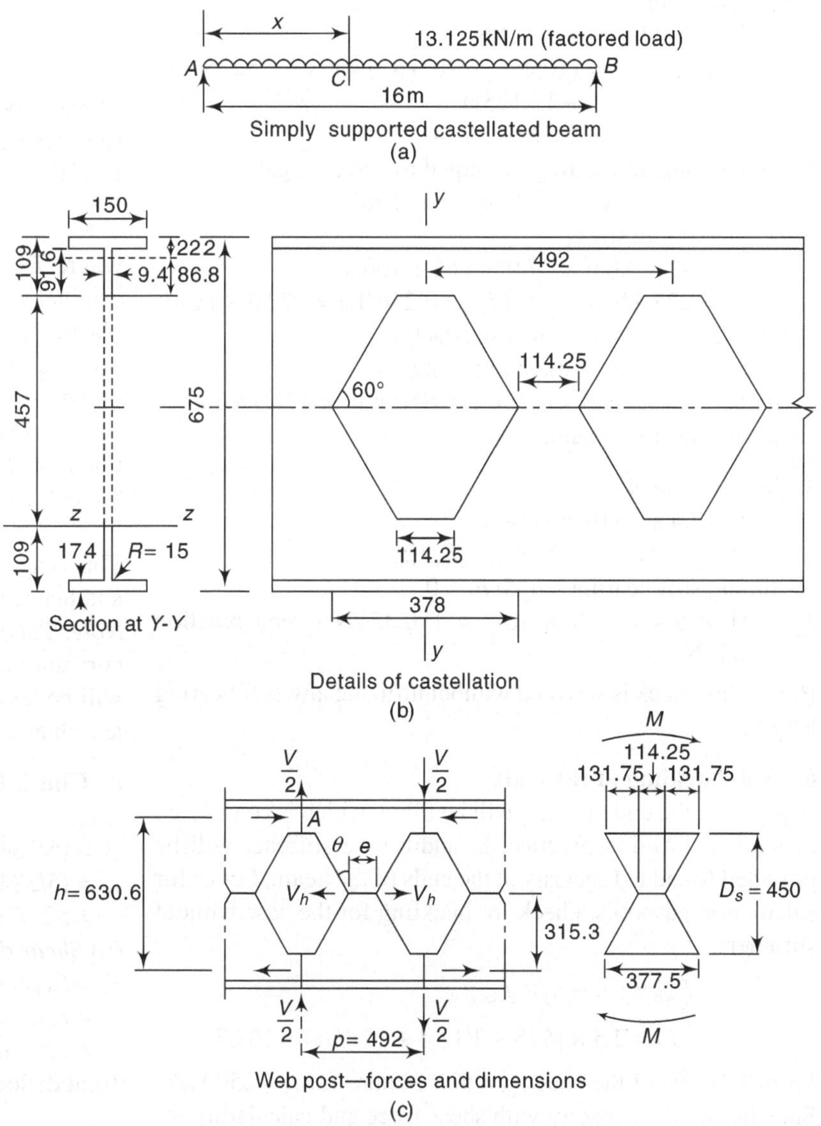


FIG. 10.55

$I_z = 71{,}693 \times 10^4\,\text{mm}^4$; $t = 9.4\,\text{mm}$
Hence,
Shear stress $= 105 \times 10^3 \times 1094.3 \times 10^3/(71{,}693 \times 10^4 \times 9.4) = 17.05\,\text{MPa} < 0.7 f_y = 0.7 \times 250 = 175\,\text{MPa}$
Hence the section is safe.

**4. Check for moment and shear**

Consider a section at a distance $x$ from the support. The maximum moment and shear effects are combined to give the worst combination at this point.

Moment at $C = 105x - 13.125x^2/2 = 105x - 6.56x^2$
Shear at $C = 105 - 13.125x$

Stress due to bending taking the T-section as beams,
$= M/(\text{area of } T \times \text{distance between centroids})$
$= M \times 10^6/(3471.04 \times 630.6) = 0.4568M\,\text{N/mm}^2$
where moment $M$ is in kN m.
Direct stress due to shear taking bending on the T-sections
$= (\text{shear}/2) \times (\text{width across top of opening}/2)Z_T$
$= (\text{shear}/2) \times (114.25/2) \times 10^3/18699$
$= 1.5275V\,\text{N/mm}^2$
where $V$ is the shear force in kN.

Combined stress $= (105x - 6.56x^2) \times 0.4568 + 1.5275(105 - 13.125x)$
$= -3x^2 + 27.9x + 160.4$

Differentiating and setting this equal to zero, we get

$$6x = 27.9 \text{ or } x = 4.65\,\text{m}$$

Hence maximum stress
$= -3 \times 4.65^2 + 27.9 \times 4.65 + 160.4$
$= 225.3\,\text{N/mm}^2 < (f_y/\gamma_{m0}) = 250/1.1 = 227.3\,\text{N/mm}^2$
Hence the selected section is satisfactory.
At mid-span the stress due to bending
$= 0.4658M = 0.4658 \times 420 = 195.6\,\text{N/mm}^2 < 227.3\,\text{N/mm}^2$
Hence the section is safe.

**5. Bearing check**

$n_2 = 2.5\,(17.4 + 15.0) = 81\,\text{mm}$
Bearing capacity $F_w = (b_1 + n_2)t_w f_{yw}/\gamma_{m0}$
Assuming stiff bearing length $b_1 = 0$
$F_w = 81 \times 9.4 \times 250 \times 10^{-3} = 190.35\,\text{kN} <$ end reaction $= 105\,\text{kN}$

Hence this check is satisfied without utilizing any stiff bearing length.

**6. Web buckling at the ends**

In practice, the end opening will be filled, which is considered as a good detailing practice. In addition, a stiffener will be provided for added security at the ends of the beam. Let us for calculation purposes, check for buckling for the case without stiffeners.

$$F_{wb} = (b_1 + n_1)t_w f_{cd}/\gamma_{m0}$$

$$d/t = 2.5 \times [675 - 2(17.4 + 15)]/9.4 = 162.3$$

From Table 9c of the code $f_{cd} = 52.1\,\text{N/mm}^2$ for $f_y = 250\,\text{MPa}$. Substituting the capacity with shear force and calculating for bearing length

$F_{wb} = 105 \times 10^3 = (b_1 + 675/2) \times 9.4 \times 52.1$
Therefore, $b_1 = -123\,\text{mm}$
This result indicates that no stiff bearing is required. However, in practice a nominal stiff bearing will be provided.

**7. Check for the buckling of web post of the castellation**

The bending stress in the web post is given by (Blodgett 1964)

$$f_{max} = 3V_h \tan\theta/4t_w e\theta^2$$

where $V_h$ is the shear force on the weld, $\theta$ is the angle of the side of the web post, $t_w$ is the web thickness, and $e$ is the minimum width.

For $\theta = 30°$, $f_{max} = 1.58V_h/(t_w e)$

$V_h$ may be calculated by taking moments about point $A$ (see Fig. 10.55)

$$V_h = \frac{V}{2}\frac{p}{(h/2)} = \frac{105}{2} \times \frac{492}{315.3} = 81.92\,\text{kN}$$

Shear stress due to horizontal shear $= 81.92 \times 10^3/(9.4 \times 114.25)$
$= 76.27\,\text{N/mm}^2 < f_{yw}/(\sqrt{3}\,\gamma_{mo}) = 131.2\,\text{N/mm}^2$
$f_{max} = 1.58 \times 81.92 \times 103/(9.4 \times 114.25)$
$= 120.5\,\text{N/mm}^2 < 250/\gamma_{m0} = 227.3\,\text{N/mm}^2$

The web post may fail by buckling due to bending. Equivalent slenderness of plate is given by (BS 5950-1:2000, Annex B2.7)

$$\lambda_{LT} = (2.8\beta L_e d / t^2)^{0.5}$$

where $L_e$ is the effective length, taken as the depth of the original section (i.e., 450 mm in this example), $\beta = 1$ for the plastic section, $d$ is the depth of the web post, taken as the average $(377.75 + 114.25)/2 = 246\,\text{mm}$, and $t$ is the thickness of the web = 9.4 mm.

Therefore, $\lambda_T = 2.8 \times (450 \times 246/9.4^2)^{0.5} = 99.1$

For $f_y = 250\,\text{N/mm}^2$ and $\lambda_{LT} = 99.1$, from Table 16 of BS 5950-1:2000,

$$f_b = 120\,\text{N/mm}^2$$

This is approximately equal to the actual stress ($120.5\,\text{N/mm}^2$) and hence it is safe.

*Note:* The moment applied to this web post puts it into double curvature and the value of moment gradient in this situation will be less than one. Hence the actual bending stress will be less than $120.5\,\text{N/mm}^2$.

**8. Check for deflection**

(a) *Bending deflection (Considering live load + dead load)*
$\delta_b = (5/384)wL^4/EI$
$= (5/384) \times 8.75 \times (16 \times 1000)^4/(2.0 \times 10^5 \times 71693 \times 10^4)$
$= 52.07\,\text{mm}$
(b) *Shear deflection*
$\delta_s = (wL^2/8)/GA_{fict}$
$= 8.75 \times (16 \times 1000)^2 \times 7.018 \times 10^{-4}/(8 \times 76923)$
$= 2.55\,\text{mm}$
Total deflection $= 52.07 + 2.55 = 54.62\,\text{mm}$
$< 16000/240 = 66.67\,\text{mm}$
Hence the beam is safe in deflection.

## SUMMARY

Beams are structural members that support loads which are applied transverse to their longitudinal axis. They resist the load primarily by bending and shear. Beams may be classified as lintels, spandrels, stringers, floor or roof beams, and purlins. Several types of beams exist, viz., angles, rolled I-sections, compound beams, castellated beams, plate girders, box girders, and joists. As in the case of columns, the sections are classified as plastic, compact, semi-compact, and slender depending on the width-to-thickness ratios of the individual plate elements. Local buckling can be prevented by limiting the width-to-thickness ratios. In many situations, the compression flanges of the beams are restrained by roofing elements in the form of trapezoidal sheeting or by reinforced concrete decking. In such cases, the beams attain their full plastic moment capacity, provided the beams chosen are plastic or compact.

When the laterally unrestrained length of the compression flange of the beam is long, then lateral buckling (also called lateral-torsional buckling) of the beam takes place. This phenomenon is similar to that of Euler buckling of columns. However, this form of instability is more complex compared to column instability, since the lateral buckling problem is three-dimensional is nature. The elastic lateral torsional buckling moment of symmetrical sections is derived based on several simplifying assumptions. It is observed that box sections provide greater resistance to lateral torsional buckling compared to I-sections, which are normally used as beams.

Several factors affect the lateral buckling behaviour, which include the type of cross section, type of loading, support conditions, restraint from other members, level of application of transverse load, effects of plasticity, residual stresses, and imperfections. The effect of these parameters is briefly discussed. We have to consider two effective length factors in the case of lateral torsional buckling of beams (one concerned with the restraint against lateral buckling and the other with the restraint against warping). Since this is complicated, in the code a single effective length factor only is specified for restraints against lateral bending, which has to be increased by 20% when restraints are not provided for warping. Intermediate restraints (in the form of bracings) and continuous restraints (in the form of decking) to the compression flange of the beams may be advantageously used to alter the lateral torsional buckling behaviour of beams. The stiffness and strength required for the bracings are also provided based on the American code provisions. Discussions about cantilever beams and continuous beams are also included. The effect of type of loading can be included by specifying an equivalent uniform moment factor. This factor given in various codes have been compared.

The inelastic buckling is usually taken into account by specifying some type of empirical curve which give essentially the elastic solution for long beams and which terminate at $M_{cr} = M_p$, the plastic moment for short beams. Similar to the empirical multiple column curves, multiple beam curves could be generated by using an equivalent slenderness ratio $\lambda_{LT} = \sqrt{M_p/M_{cr}}$. A brief description of the provisions in the German, American, and Eurocode are given. The provisions in the Indian code for the design strength of laterally supported and laterally unsupported beams are presented for both bending and shear. The provisions in the code for monosymmetric beams are also given. It has to be noted that the code gives expressions for the elastic critical moment, $M_{cr}$ for symmetric and monosymmetric beams only. Hence a brief description about the ways of determining $M_{cr}$ for other shapes of beams (viz., channels, Zs, Ts, double angles, and compound beams) are also given. Once the elastic critical moments of beams of any cross section are determined, the capacity of the beam may be found out by using Eqns (10.22) and (10.23).

The moment capacity of laterally continuous beams may be obtained by calculating all segmental critical loads individually (the distances between cross-beams may be considered as segments) and choosing the lowest value, assuming each segment is simply supported at ends. Shear forces may control short beams which carry heavy concentrated loads. Hence methods to consider shear strength of beams are given based on the provisions of the Indian code.

The beams should not excessively deflect or vibrate during the service life of the structure (limit state of serviceability). Hence some expressions and discussions are provided to check the beams for deflection due to dead and live loads and ponding due to accumulation of rain water. Information about and expressions for web buckling, web crippling, web and flange holes, lattice beams, castellated beams, lintels, purlins, biaxial bending, and unsymmetrical bending are also included.

## MULTIPLE-CHOICE QUESTIONS

10.1 Beams should be designed and checked for
- (a) strength requirements (bending moment, shear, and torsional moments)
- (b) stability requirements (local buckling, lateral torsional buckling)
- (c) stiffness requirements (deflection, vibration)
- (d) all of these

10.2 In indeterminate structures, which form collapse mechanisms, beams should be
- (a) plastic
- (b) semi-compact
- (c) compact
- (d) slender

10.3 The factors that influence the lateral torsional buckling capacity of beams are
- (a) cross-sectional shape of the beam
- (b) length of beam
- (c) type of support condition
- (d) position of applied load
- (e) all of these

10.4 The general equation for critical buckling moment is

(a) $M_{cr} = \frac{\pi}{L}\left(EI_y GI_t\right)^{0.5}\left[1+\frac{\pi^2 EI_w}{L^2 GI_t}\right]$

(b) $M_{cr} = \frac{\pi}{L}\left(EI_y GI_t\right)^{0.5}\left[1+\frac{\pi^2 EI_w}{L^2 GI_t}\right]^{0.5}$

(c) $M_{cr} = \frac{\pi}{L}\left(EI_y GI_t\right)^{2}\left[1+\frac{\pi^2 EI_w}{L^2 GI_t}\right]$

(d) $M_{cr} = \frac{\pi}{L}\left(EI_y GI_t\right)^{2}\left[1+\frac{\pi^2 EI_w}{L^2 GI_t}\right]^{0.5}$

10.5 The most efficient section for torsional resistance is
(a) I-section
(b) hollow circular tube
(c) wide-flanged I-section
(d) channel section

10.6 Restraint against torsion at the supports may be provided by
(a) web cleats
(b) bearing stiffeners
(c) beam built into the supporting wall
(d) any of these

10.7 The effective length of compression flange of a simply supported beam not restrained against torsion and subjected to destabilizing loads is taken as $KL$, where $K$ is
(a) 0.70 (b) 1.2 (c) 1.00 (d) 0.85

10.8 The effective length of compression flange of a simply supported beam with both flanges restrained against warping and ends fully restrained against torsion and subjected to normal loads is taken as $KL$, where $K$ is
(a) 0.70 (b) 1.2 (c) 1.00 (d) 0.85

10.9 The effective length of compression flange of a simply supported beam with warping not restrained in both flanges but ends fully restrained against torsion and subjected to normal loads is taken as $KL$, where $K$ is
(a) 0.70 (b) 1.2 (c) 1.00 (d) 0.85

10.10 Lateral support in the case of steel beams may be achieved by
(a) concrete slab connected by means of shear connectors to the compression flange
(b) concrete slab rests on the compression flange
(c) providing cross beams
(d) all of these

10.11 The effective lateral restraint for a beam should be capable of resisting $x$ per cent of the maximum force in the compression flange, where $x$ is
(a) 2 (b) 2.5 (c) 3.5 (d) 5

10.12 The requirement for effective bracing is:
(a) It should have sufficient strength
(b) It should be connected nearer to the compression flange
(c) It should have sufficient stiffness
(d) All of these

10.13 Stabilizing effect is produced in a beam, when the load is applied
(a) above the centroid of cross-section
(b) below the centroid of cross-section
(c) at the centroid of cross-section
(d) at the top flange

10.14 The design bending strength of laterally supported beam is given by $M_d = \beta_b Z_p f_y / \gamma_{m0}$. For plastic and compact sections $\beta$ is
(a) $Z_e/Z_p$ (b) 1.0 (c) 1.2 (d) 1.5

10.15 The shear lag effect in beam flanges are disregarded when the outstand of the beam is less than
(a) $L_0/10$ (b) $L_0/15$ (c) $L_0/20$ (d) $L_0/4$

10.16 The design bending strength of laterally unsupported beam is given by
(a) $M_d = \beta_b Z_p f_{bd}$
(b) $M_d = \beta_b Z_p f_y / \gamma_{m0}$
(c) $M_d = \beta_b Z_p f_{bd} / \gamma_{mo}$
(d) $M_d = \beta_b Z_e f_y / \gamma_{m0}$

10.17 The elastic lateral torsional buckling moment $M_{cr}$ for a laterally unsupported beam is given by
(a) $\beta_b Z_p f_{cr,b}$
(b) $\beta_b Z_p f_y / \gamma_{m0}$
(c) $\beta_b Z_p f_{bd}$
(d) $\beta_b Z_e f_{cr,b}$

10.18 The imperfection factor for welded steel section is
(a) 0.21 (b) 0.49 (c) 0.50 (d) 0.40

10.19 The beam is considered to be subjected to high shear when the factored shear force is greater than
(a) $0.4V_d$ (b) $0.6V_d$ (c) $0.8V_d$ (d) $0.5V_d$

10.20 Torsional constant, It, for open sections is given by
(a) $\sum b_i t_i^3$
(b) $\sum t_i b_i^3 / 3$
(c) $\sum b_i t_i^3 / 3$
(d) $\sum t_i b_i^3$

10.21 The warping constant, $I_w$, for equal flange I-beams is given by
(a) $0.25 I_y h_f^2$
(b) $0.5 I_y h_f^2$
(c) $0.25 I_y h_f^3$
(d) $0.25 I_z h_f^2$

10.22 The design shear strength of beam, $V_d$, is given by
(a) $A_v f_{yw} / (\gamma_{m1} \sqrt{3})$
(b) $A_v f_{yw} / (\gamma_{m0} \sqrt{3})$
(c) $A_v f_{uw} / (\gamma_{m1} \sqrt{3})$
(d) $A_v f_{uw} / (\gamma_{m0} \sqrt{3})$

10.23 Web crippling in steel beams occurs due to
(a) excessive bending moment
(b) failure of the web under concentrated load
(c) formation of plastic hinges
(d) none of these

10.24 Web crippling strength $F_{crip}$ at support is calculated using
(a) $(b_1 + n_2) t_f f_{yf}$
(b) $(b_1 + 2n_2) t_w f_{yw}$
(c) $(b_1 + n_2) t_w f_{yf}$
(d) $(b_1 + 2.5 d_1) t_w f_{yw}$

10.25 For an I-section beam, flange buckling is more critical than web buckling because
(a) flanges are thicker than webs
(b) web can develop post-buckling strength
(c) edge conditions for flanges are more favourable
(d) none of these

10.26 A limit on deflection is specified in the code because excessive deflections may cause
(a) problems for roof drainage
(b) a feeling of insecurity to the occupants of a building
(c) damage to non-structural components
(d) all of these

10.27 The deflection of steel beams in buildings other than industrial buildings, and supporting elements not susceptible to cracking is limited in IS 800 to span divided by
(a) 360 (b) 150 (c) 300 (d) 180

10.28 The deflection of steel beams in buildings other than industrial buildings, and supporting elements susceptible to cracking is limited in IS 800 to span divided by

(a) 360 (b) 150 (c) 300 (d) 180

10.29 IS 800 suggests that the purlins are designed as

(a) cantilever beams (c) continuous beams

(b) simply supported beams (d) tension members

10.30 Sag rods are designed as

(a) compression members

(b) tension members

(c) laterally supported beams

(d) laterally unsupported beams

## EXERCISES

10.1 Determine the buckling moment of an ISMB 400 section with $f_y = 250$ MPa when it is simply supported over a span of 3.5 m.

10.2 Select a rolled I-section to safely carry end moments of 545 kN m and 115 kN m over a laterally supported span of 5.5 m assuming the moments produce single curvature bending.

10.3 A simply supported beam spanning 5 m carries a udl of 3 kN/m including its self-weight. Floor construction restrains it against lateral buckling. What size of beam is required with $f_y = 250$ MPa?

10.4 Calculate the maximum value of a mid-span concentrated load $W$ that can be safely applied on a laterally unsupported beam of length 3 m. The beam is simply supported at either end and is free to rotate in plane but restrained torsionally at their ends. The beam section is ISMB 250 and the load is applied at the top flange at mid-span. $f_y$ of steel is 250 MPa. See Fig. 10.56.

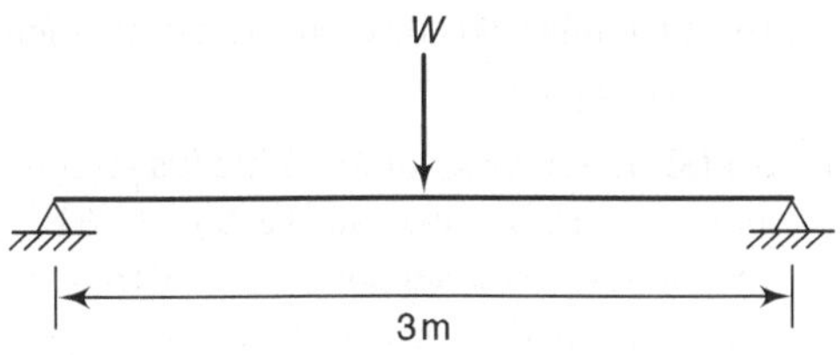


**FIG. 10.56**

10.5 For a built-up section, determine the following:

(a) Elastic section modulus $Z_e$ and yield moment $M_y$

(b) Plastic section modulus $Z_p$ and plastic moment $M_p$

Bending is about *z-z* axis and steel with $f_y = 250$ MPa is used. See Fig. 10.57.

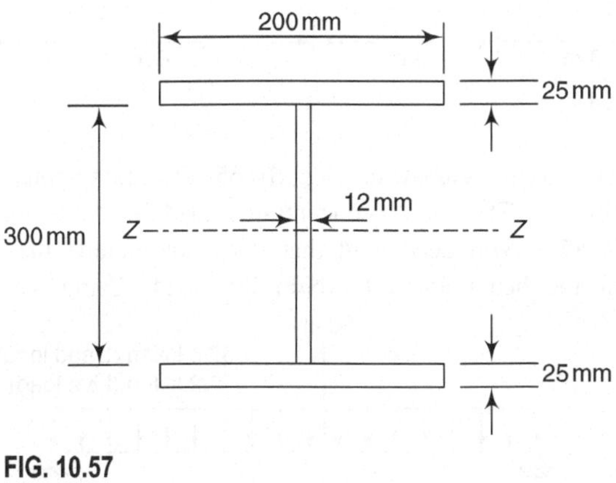


**FIG. 10.57**

10.6 Select a standard hot-rolled section of steel with $f_y$ = 250 MPa for the beam shown in Fig. 10.58. The beam has continuous lateral support. The maximum permissible live load deflection is $L$/360.

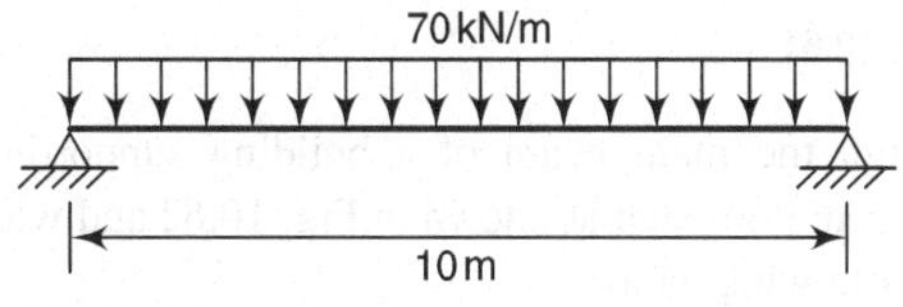


**FIG. 10.58**

10.7 The beam shown in Fig. 10.59 must support two concentrated live loads of 90 kN each at the quarter points. The maximum deflection must not exceed $L$/240. Lateral support is provided at the ends of the beam. Use $f_y$ = 250 MPa steel and select a rolled section.

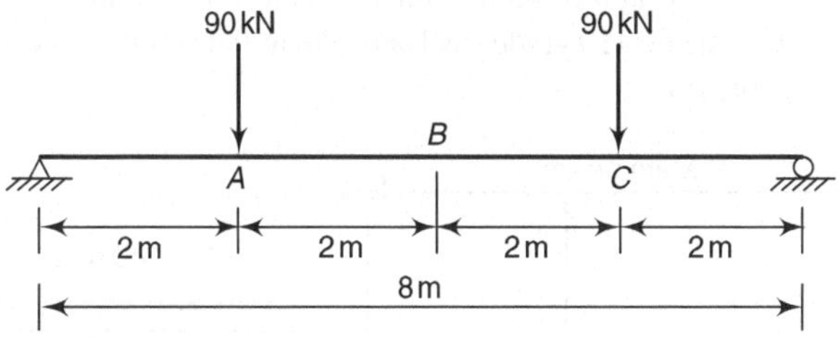


**FIG. 10.59**

10.8 Using steel with $f_y = 250$ MPa, select a rolled I-section for the beam shown in Fig. 10.60. The concentrated load is a service live load and uniform load is 30% dead load and 70% live load. Lateral bracing is provided at the ends and at mid-span. There is no restriction on deflection.

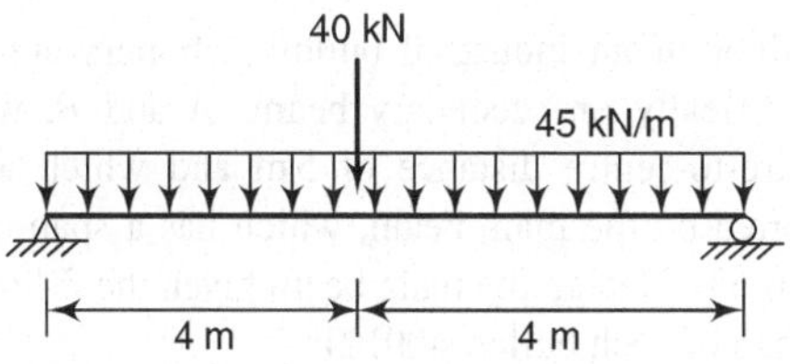


**FIG. 10.60**

10.9 Compute the moment carrying capacity and shear strength of a laterally unrestrained ISMB 300 member of length 3 m with $f_y$ = 250 MPa.

10.10 A hall in a building is 15 m × 18 m. A reinforced concrete slab 150-mm thick is supported by beams as shown in Fig. 10.61. Design the beams, assuming the unit weight of RCC as 25 kN/m$^3$ and using Fe 410 grade steel.

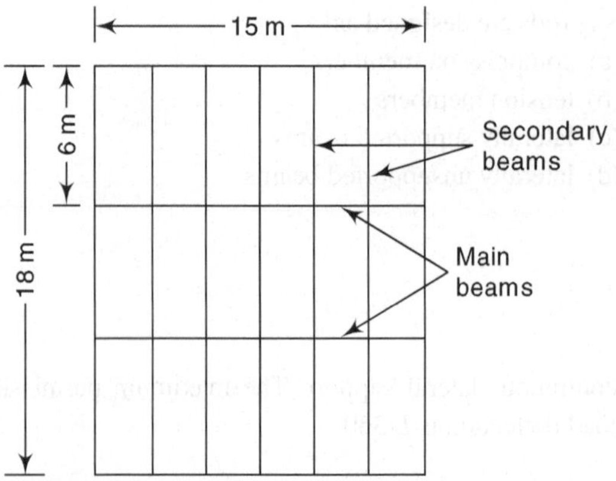


**FIG. 10.61**

10.11 Design the main beam of a building supporting concrete floor slab as shown in Fig. 10.62 and with the following data:

(a) Beam centres: 6 m
(b) Span (simply supported): 7.4 m
(c) Concrete slab (spanning in two directions): 240-mm thick
(d) Finished screed: 40-mm thick
(e) Imposed load: 4 kN/m$^2$
(f) Take weight of concrete slab as 24 kN/m$^3$ and total weight of 40-mm thick screed as 1.0 kN/m$^2$

Assume Fe 410 grade steel and take initial weight of beam as 1.0 kN/m.

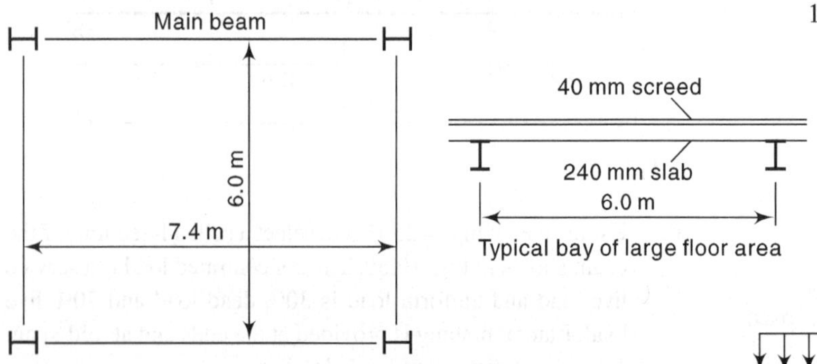


**FIG. 10.62**

10.12 In a floor of an industrial building, boilers are supported symmetrically on secondary beams *A* and *B*, which have a centre-to-centre distance of 5 m and which are in turn supported by the main beam, which has a span of 9 m (see Fig. 10.63). Design the main beam given the following data:

(a) Load of each boiler: 400 kN
(b) Open steel flooring (carried on beams A and B): 0.4 kN/m$^2$
(c) Imposed load (outside boiler area): 4.0 kN/m$^2$

Assuming the self-weight of secondary beam as 4 kN, design the main beam with $f_y = 250$ MPa.

10.13 A compound beam is to carry a uniformly distributed dead load of 400 kN and an imposed load of 600 kN. The beam is simply supported and has a span of 10 m. Assume 30 kN for the self-weight of the beam. The overall depth must not

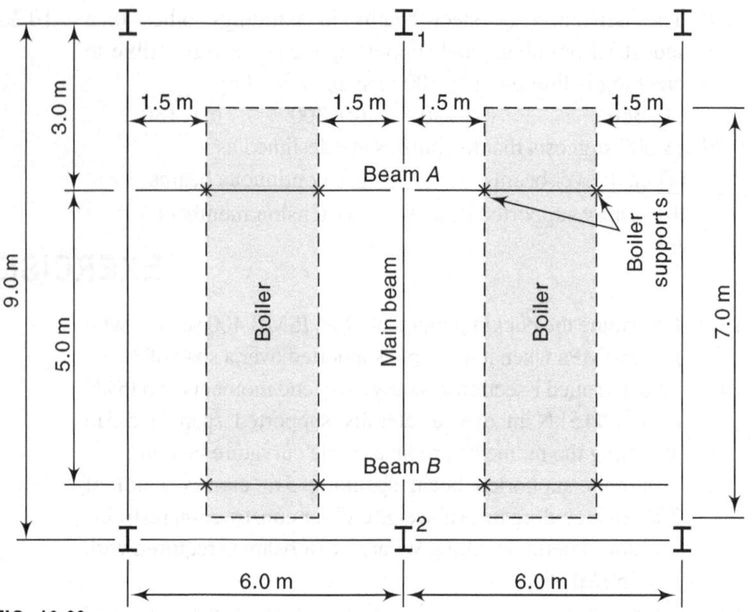


**FIG. 10.63**

exceed 700 mm. The length of stiff bearing at the ends is 215 mm. Full lateral support is provided for the compression flange. Use Fe 410 grade steel.

(a) Design the beam section and check deflection assuming uniform section throughout.
(b) Determine the theoretical cut-off points for the flange plates.

10.14 An ISMB 300 beam of $f_y = 250$ MPa steel is subjected to a loading as shown in the Fig. 10.64. There is a continuous lateral support. The concentrated loads *P* are service live loads and $w = P/10$ is a service dead load. Determine the maximum values of *P* and *w* that can be supported.

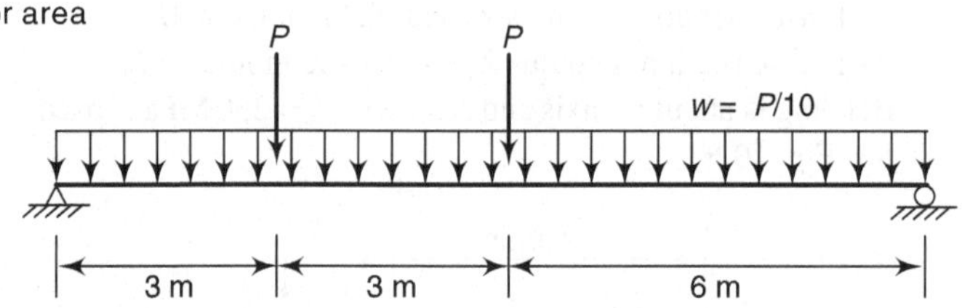


**FIG. 10.64**

10.15 The beam shown in Fig. 10.65 has continuous lateral support. The 90 kN concentrated load is a live load, udl is 87.6 kN/m dead load that does not include the weight of the beam and 29.2 kN/m live load. Using steel with

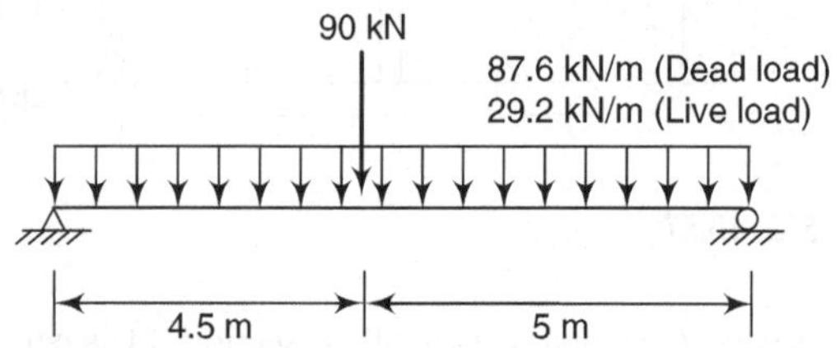


**FIG. 10.65**

$f_y$ = 250 MPa, select a rolled section. There is no limit on the deflection.

10.16 The cantilever beam shown in Fig. 10.66 is laterally supported only at the fixed end. The concentrated load is service live load. Use steel with $f_y$ = 250 MPa and select a rolled section. There is no limit on deflection.

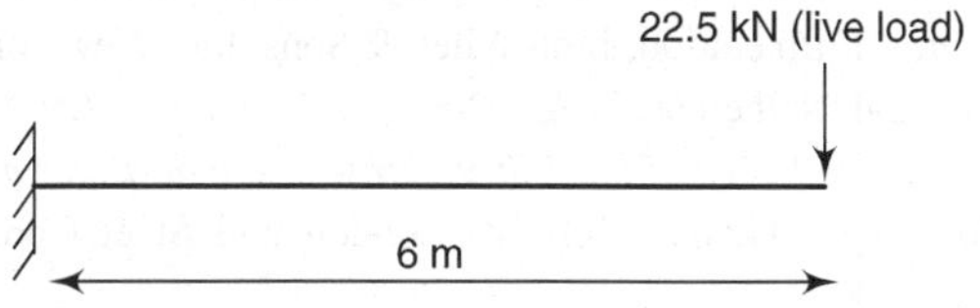


**FIG. 10.66**

10.17 Design an I-section purlin for an industrial building situated in Cuttack to support a 18 gauge galvanized corrugated iron sheet roof (with a dead weight of 131 N/m$^2$) with the following data:

Spacing of the truss c/c = 5.0 m
Span of truss = 15 m
Spacing of purlin c/c = 1.5 m
Intensity of wind pressure = 2.3 kN/m$^2$
Yield stress of steel = 250 MPa

10.18 Design a purlin for a roof truss having the following data:

Span of truss = 12 m
Spacing of truss = 4.0 m centre-to-centre
Inclination of the roof = 30°
Spacing of purlin = 1.6 m centre-to-centre
Wind suction = 1.4 kN/m$^2$
Roof covering is AC sheets (assume a weight of 1.68 kN/m$^2$). Take live load as 0.75 kN/m$^2$ on plan area less 0.02 kN/m$^2$ for every degree increase in slope over 10°, subject to a minimum 0.40 kN/m$^2$.

(a) Provide channel purlin
(b) Provide channel purlin with a sag rod at mid-span
(c) Provide angle purlin and design by the empirical method.

10.19 Design a lintel over an opening of 4.0 m supporting brickwork of thickness 230 mm. The lintel has to support a uniform load of 80 kN in addition to the masonry. Assume the weight of masonry as 19.2 kN/m$^3$ and height of the masonry above the lintel as 5 m. Use Fe 410 grade steel.

10.20 Design a castellated beam for a span of 14 m. The dead weight of the roofing is 750 N/m$^2$ and the live load is 2 kN/m$^2$. Cut the selected I-section at 60° and restrict the overall height to 800 mm. Use Fe 410 grade steel.

## REVIEW QUESTIONS

10.1 Why are rolled I-sections widely used as beam members?
10.2 What are the checks to be performed for beam member design?
10.3 What are castellated sections and under what conditions are they used?
10.4 What is the difference between bending and buckling of a beam member?
10.5 What is meant by lateral torsional buckling of a beam member?
10.6 Under what conditions can lateral buckling occur?
10.7 Under what conditions can a beam member be assumed as laterally restrained?
10.8 What is the difference between column buckling and beam buckling?
10.9 An I-beam loaded perpendicular to the minor axis does not buckle. Why?
10.10 What is local buckling of a beam member?
10.11 Differentiate between local and lateral buckling of beams.
10.12 How can the lateral buckling behaviour be improved in a beam member?
10.13 Write short notes on classification of members.
10.14 What is the value of strain percentage in a beam section at the fully plastic condition?
10.15 Under what conditions should a beam be checked for shear?
10.16 What is meant by plastic modulus of a section?
10.17 What is meant by shape factor of a section?
10.18 Which section performs best in torsion and why?
10.19 Why is an I-section commonly used as a beam member even though its lateral buckling strength is low?
10.20 What is meant by shear lag and how is it accounted for in design?
10.21 What is the effect of high shear force on plastic moment resistance and when should it be considered?
10.22 What is meant by effective length of a beam member?
10.23 What are destabilizing loads?
10.24 What are some of the factors that cause buckling of the compression flange?
10.25 A beam subjected to transverse loads èxperiences two types of shear. What are they?
10.26 Where does the largest value of horizontal shear occur in a beam?
10.27 Write down common situations where shear might become critical.
10.28 What is meant by web crippling?
10.29 What are the reasons for specifying deflection limitations?
10.30 What is meant by camber and why is it provided?
10.31 When does unsymmetrical bending occur in a beam?
10.32 What are some of the commonly used beam members that are subjected to unsymmetrical bending?
10.33 What is meant by shear centre and identify shear centre for *I*-, channel, and angle sections?
10.34 When does torsion occur in a beam?
10.35 What is meant by St Venant torsion and warping torsion?

## REFERENCES

'ASCE Task Committee on Design Criteria for Composite Structures in Steel and Concrete, Proposed Specification for Structural Steel Beams with Web Openings', *Journal of Structural Engineering*, ASCE, Vol. 118, No. ST12, Dec 1992, pp. 3315–3324 and Commentary, pp. 3325–3349.

Bleich, F., *Buckling Strength of Metal Structures*, McGraw–Hill, New York, 1952.

Bo Dowswell, 'Lateral–Torsional Buckling of Wide-flange Cantilever Beams', *Engineering Journal, AISC,* Vol. 41, No. 3, 2004.

Bradford, M.A. and Cuk, P.E., 'Elastic Buckling of Tapered Monosymmetric I-beams', *Journal of Structural Engineering, ASCE*, Vol. 114, No. 5, May 1988, pp. 977–996.

Bradford, M.A. and Trahair, N.S., 'Distortional Buckling of *I*-beams', *Journal of the Structural Div*., ASCE, Vol. 107, No. ST2, Feb 1981, pp. 355–370.

Bradford, M.A., 'Inelastic Buckling of Tapered Mono-symmetric *I*-beams', *Engineering Structures,* Vol. 11, No. 2, 1989, pp. 119–126.

Bradford, M.A., 'Lateral-Distortional Buckling of Steel I-section Members', *Journal of Constructional Steel Researc*h, Vol. 23, Nos 1–3, 1992, pp. 97–116.

Bradford, M.A., 'Stability of Tapered I-beams', *Journal of Constructional Steel Research*, Vol. 9, 1988, pp. 195–216.

Burton, H.S., 'Proposed Specifications for Structural Steel Beams with Web Openings', *Journal of Structural Engineering,* ASCE, Vol. 121, No. 3, Mar 1995, pp. 589–593 and Commentary, pp. 594–597.

Clark, J.W. and Hill, H.N., Lateral Buckling of Beams, *Journal of the Structural Div*., ASCE, Vol. 86, No. ST7, Jul 1960, pp. 175–196.

Colombi, P., 'The Ponding Problem on Flat Steel Roof Grids', *Journal of the Constructional Steel Research*, Vol. 62, No. 7, Jul 2006, pp. 647–655.

Corr, D.J., McCann, D.M., and McDonald, B.M., 'Lessons Learned from Marcy Bridge Collapse', *Forensic Engineering*, 2009, pp. 395–403.

Darwin, D., 'Design of Steel and Composite Beams with Web Openings (with companion computer program WEBOPEN)', *AISC Steel Design Guide*, No. 2, American Institute of Steel Construction, Chicago, IL, 1990.

Davies, J.M., 'Calculation of Steel Diaphragm Behaviour', *Journal of the Structural Div*., ASCE, Vol. 102, No. ST7, Jul 1976, pp. 1411–1430.

Dexter, R.J. and Altstadt, S.A., 'Strength and Ductility of Tension Flanges in Girders', in *Recent Developments in Bridge Engineering*, Mahmoud, K.M., ed. A.A. Balkema/Sweto & Zeitlinger, Lisse, The Netherlands, 2003, pp. 67–81.

Dowling, P.J., Knowles, P.R., and Owens, G.W., *Structural Steel Design*, The Steel Construction Institute, Ascot, and Butterworths, London, 1988.

Easley, J.T., 'Strength and Stiffness of Corrugated Metal Shear Diaphragms', *Journal of the Structural Div.*, ASCE, Vol. 103, No. ST1, Jan 1977, pp. 169–180.

Ellifritt,D.S., Wine,G., Sputo, T., and Samuel, S., 'Flexural Strength of WT Sections', *Engineering Journal*, AISC, Vol. 29, No. 2, pp. 67–74.

Ellingwood, B., 'Serviceability Guidelines for Steel Structures', *Engineering Journal, AISC*, Vol. 26, No. 1, 1989, pp. 1–8.

Errera, S.J. and Apparao, T.V.S.R., 'Design of I-shaped Beams with Diaphragm Bracing', *Journal of the Structural Div., ASCE*, Vol. 102, No. ST4, Apr 1976, pp. 769–781.

Errera, S.J. and Apparao, T.V.S.R., 'Design of I-shaped Columns with Diaphragm Bracing', *Journal of the Structural Div*., ASCE, Vol. 102, No. ST9, Sep 1976, pp. 1685–1701.

Fukumoto, Y., Itoh, Y., and Kubo, M., 'Strength Variation of Laterally Unsupported Beams', *Journal of the Structural Div*., ASCE, Vol. 106, No. ST1, Jan 1980, pp. 165–182.

Galambos, T.V. and Ellingwood, B., 'Serviceability Limit States: Deflection', *Journal of Structural Engineering, ASCE*, Vol. 112, No. 1, Jan 1986, pp. 67–84.

Galambos, T.V., ed., *Guide to Stability Design Criteria for Metal Structures*, Fifth edition, John Wiley & Sons, Inc., New York, 1998.

Gardner, L. and Nethercot, D.A., *Designers' Guide to EN 1993-1-1, Euro code 3: Design of Steel Structures – General Rules and Rules for Buildings*, Thomas Telford, London and Steel Construction Institute, 2005.

Gaylord, E.H. Jr, Gaylord, C.N., and Stallmeyer, J.E., *Design of Steel Structures*, Third edition, McGraw-Hill Inc., New York, 1992, pp. 367–372.

Hill, H.N., 'Lateral Buckling of Channels and Z-beams', *Transactions*, ASCE, Vol. 119, 1954, pp. 829–841.

Horne, M.R. and Morris, L.J, *Plastic Design of Low-rise Frames*, Granada, London, 1981.

http://en.wikipedia.org/wiki/Tacoma_Narrows_Bridge_%281940%29, last accessed on 20 July 2015.

Kerdal, D. and Nethercot, D.A., 'Failure Modes of Castellated Beams', *Journal of Constructional Steel Research,* Vol. 4, 1984, pp. 295–315.

Kerensky, O.A., Flint, A.R., Brown, W.C., 'The Basis for Design Of beams and Plate Girders in the Revised British Standard 153', *Proceedings of the Institution of Civil Engineers* (London), Vol. 5, No. Pt.III, Aug 1956, pp. 396–461.

Kindmann, R. and Muszkiewicz, R., Biegedrillknickmomente und eigenformen von biegeträgern unter berücksichtigung der drehbettung, *Stahlbau*, Vol. 73, No. 2, Feb 2004, pp. 98–106.

Kirby, P.A. and Nethercot, D.A., *Design for Structural Stability*, Granada Publishing, London, 1979.

Kitipornchai, S., Dux, P.F., and Richter, N.J., 'Buckling and Bracing of Cantilevers', *Journal of Structural Engineering*, ASCE, Vol. 110, No. 9, Sep 1984, pp. 2250–2262.

Kitipornchai, S., Wang, C.M., and Trahair, N.S., Buckling of Monosymmetric *I*-beams under Moment Gradient, *Journal of the Structural Div*., ASCE, Vol. 112, No. 4, Apr 1986, pp. 781–799.

Kittipornchai, S. and Trahair, N.S., 'Buckling Properties of Monosymmetric *I*-beams', *Journal of the Structural Div*., ASCE, Vol. 106, No. ST5, May 1980, pp. 941–957.

Knowles, P.R., *Design of Castellated Beams*, Steel Construction Institute, Ascot, U.K., 1985.

Lawson R.M., Mullett, D.L., and Rackham, J.W., *Design of Asymmetric Slimflor Beams using Deep Composite Decking*, The Steel Construction Institute, UK, SCI-P-175-1997.

Lindner, J. and Gietzelt, R., Zwiachsige Biegung und Längskraft-ein ergänzter Bemessungsvorschlag, *Stahlbau*, Vol. 54, 1985, pp. 265–271.

Lindner, J. and Subramanian, N., 'Allowable Stress Provisions of IS: 800-1962', *Journal of the Institution of Engineers (India),* Vol. 62, Pt.CI4, Jan 1982, pp. 246–248.

Lindner, J., Biegedrillknicken in Theorie,Versuch und Praxis, Berischte am Forschung und Entwicklung, Deutscher Ansscu für Stahlbau (DASt), 9-1980, *Stahabau Verlag*, Köln, pp.7–19.

Lindner, J., Gietzelt, R., Biegedrillknicken – Erläuterungen, Versuche, Beispiele, DASt, 10/1980, *Stahlbau – Verlag*, Köln, 1980.

Lindner, J., Influence of Constructional Detailing on the Lateral–Torsional Buckling Capacity of Beam Columns', in *Structural Stability and Design*, A.A.Balkema, Rotterdam, 1995, pp. 61–66.

Lindner, J., Mindeststeifigkeiten für den kippsicherheitsnachweis beim traglastverfahren, *Der Bauingenieur*, Vol. 47, H.7, 1972, pp. 238–240.

Lindner, J., Scheer, J., and Schmidt, H., Stahlbauten, Erläuterungen zu DIN 18800, Teil 1 bis Teil 4, Second edition, *Beuth und Ernst & Sohn*, Berlin, 1994.

MacGinley, T.J. and Ang, P.T.C., *Structural Steelwork – Design to Limit State Theory*, Butterworths/Heinemann, Oxford, 1992.

Marino, F.J., 'Ponding of Two-way Roof Systems', *Engineering Journal*, AISC, Vol. 3, No. 3, 1966, pp. 93–100.

Martin, L.H. and Purkiss, J.A., *Structural Design of Steelwork to BS 5950,* Edward Arnold, London, 1992.

Morris, L.J. and Plum, D.R., *Structural Steelwork Design to BS5950*, Second edition, Addison Wesley Longman, Essex, 1996.

Murray, T.M., Allen, D.E., and Ungar, EE., 'Floor Vibrations due to Human Activity', *Steel Design Guide*, No. 11, American Institute of Steel Construction, Chicago, IL, 1997.

Narayanan, R. and Avanessian, N.G-V, 'Design of Slender Webs having Rectangular Holes', *Journal of Structural Engineering*, ASCE, Vol. 111, No. 4, Apr 1985, pp. 777–787.

Narayanan, R., ed., *Beams and Beam Columns: Stability and Strength*, Applied Science Publishers, Barking, Essex, England, 1983.

Nethercot, D.A. and Trahair, N.S., 'Lateral Buckling Approximations for Elastic Beams', *The Structural Engineer*, Vol. 56, No. 6, Jun 1976, pp. 197–204.

Nethercot, D.A. and Kerdal, D., 'Lateral–Torsional Buckling of Castellated Beams', *The Structural Engineer*, Vol. 60B, No. 3, 1982, pp. 53–61.

Nethercot, D.A. and Rockey, K.C., 'A Unified Approach to the Elastic Lateral Buckling of Beams', *Engineering Journal*, AISC, Vol. 9, No. 3, Jul 1972, pp. 96–107.

Nethercot, D.A. and Trahair, N.S., 'Design of Diaphragm-braced *I*-beams', *Journal of the Structural Div.*, ASCE, Vol. 101, No. ST10, Oct 1975, pp. 2045–2061.

Nethercot, D.A., *Beams in Constructional Steel Design – An International Guide*, Dowling, P.J., Harding, J.E., and Bjorhovde, R., eds, Elsevier Applied Science, London/New York, 1992, pp. 45–65.

Nethercot, D.A., 'Elastic Lateral buckling of Beams', in *Beams and Beam Columns: Stability and Strength*, R.Narayanan, ed., Applied Science Publishers, Essex, England, 1983.

Nethercot, D.A., 'Imperfections and the Design of Steel Beams', *Proc., Institution of Civil Engineers*, Vol. 57, Jun 1974, pp. 291–306.

Nethercot, D.A., 'Lateral Buckling of Tapered Beams', *Publications, IABSE*, Vol. 33, -II, 1973a, pp. 173–192.

Nethercot, D.A., 'The Effective Lengths of Cantilevers as Governed by Lateral Buckling', *The Structural Engineer,* Vol. 51, 1973b, pp. 161–168.

Ojalvo, M. and Chambers, R.S., 'Effect of Warping Restraints on *I*-beam Buckling', *Journal of the Structural Div.*, ASCE, Vol. 103, No. ST12, Dec 1977, pp. 2351–2360.

Petersen, C., Statik und Stabilität der Baukonstruktionen, *Friedr. Vieweg & Sohn*, Braunschweig/Wiesbaden, 1980.

Redwood, R., and Cho, S.H., 'Design of Steel and Composite Beams with Web Openings', *Journal of Constructional Steel Research*, Vol. 25, 1993, pp. 23–41.

Ricker, D.T., 'Cambering Steel Beams', *Engineering Journal, AISC*, Vol. 26, No. 4, 1989, pp. 136–142.

Roik, K., Carl, J., and Lindner, J., Biege Torsions Probleme Gerader Dünnwandiger Stäbe, *Verlag von Wilhelm Ernst & Sohn*, Berlin, 1972.

Ruddy, J.L., 'Ponding of Concrete Deck Floors', *Engineering Journal*, AISC, Vol. 23, No. 3, 1986, pp. 107–115.

Salmon, C.G. and Johnson, J.E., *Steel Structures: Design and Behavior*, Fourth edition, Harper Collins College Publishers, New York, 1996.

Salvadori, M.G., 'Lateral Buckling of Eccentrically Loaded *I*-columns', *Transactions*, ASCE, Vol. 121, 1956, p. 1163.

Sarma, B.S. and Lakshmanan, N., 'Evaluation of Channel and Tube Purlins under Biaxial Loading', *Journal of Institution of Engineers, (India), Civil Engg. Div.*, Vol. 70, Jan. 1990, pp. 153–159.

Serna, M.A., Lopez, A., Puente, I, Yong, D.J., 'Equivalent Uniform Moment factors for Lateral Torsional Buckling of Steel Members', *Journal of the Constructional Steel Research*, Vol. 62, No. 6, Jun 2006, pp. 566–580.

Subramanian, N., 'Recent Progress in the Field of Lateral Instability of Steel Beams', *The Bridge and Structural Engineer*, Vol. 13, No. 1, Mar 1983, pp. 17–47 and No. 2, Jun 1983, pp. 1–13.

Thandavamoorthy, T.S., *Analysis of Structures – Strength and Behaviour, Oxford University Press,* New Delhi, 2005.

Timoshenko, S.P. and Gere, J.M., *Theory of Elastic Stability*, Second edition, McGraw-Hill, New York, 1961.

Trahair, N.S., Bradford, M.A., and Nethercot, D.A., *The Behaviour and Design of Steel Structures to BS 5950*, Spon Press, London/New York, 2001.

Trahair, N.S., *Flexural–Torsional Buckling of Structures*, CRC Press, Boca Raton, FL, 1993.

Trahair, N.S., Moment Capacities Of Steel Angle Sections, Research Report No. R809, Dept. Of Civil Engineering, The University Of Sydney, Nov 2001, (Http://Www.Civil.Usyd.Edu.Au/)

Vinnakota, S., 'Inelastic stability of Laterally Unsupported Beams', *Computers and Structures,* Vol. 7, No. 3, 1977.

Vogel, U. and Lindner, J., Kommentar zu DIN: 18800, Teil 2 (Gelbdruck), Stabilitatsfalle im Stahlbau, Knicken von Staben und Stabwerken, DASt. H.11, 1980, *Stahlbau Verlag*, Köln, 1980.

Wang, C.M., and Kitipornchai, S., 'Buckling Capacities of Mono-symmetric I-beams', *Journal of the Structural Div.*, ASCE, Vol. 112, No. 11, Nov 1986, pp. 2373–2391.

West, M.A., Fisher, J.M., and Griffis, L.A., 'Serviceability Design Considerations for Steel Buildings', *Steel Design Guide No. 3*, Second edition, American Institute of Steel Construction, Chicago, IL, 2003.

Winter, G., 'Lateral Bracing of Columns and Beams', *Transactions*, ASCE, Vol. 125, Part 1, 1960, pp. 809–825.

www.corusconstruction.com, last accessed on 20 July 2015.

Yura, J., 'Fundamental of Beam Bracing', *Engineering Journal*, AISC, Vol. 38, No.1, 2001, pp. 11–26.

Yura, J.A., Galambos, T.V., and Ravindra, M.K., 'The Bending Resistance of Steel Beams', *Journal of the Structural Div.*, ASCE, Vol. 104, No. ST9, Sep 1978, pp. 1355–1370.

Zirakian, T. and Showkati, H., 'Distortional Buckling of Castellated Beams', *Journal of the Constructional Steel Research*, Vol. 62, No. 8, Aug 2006.

# 11

# DESIGN OF PLATE GIRDERS

## INTRODUCTION

A plate girder is basically an I-beam built up from plates using riveting or welding. It is a deep flexural member used to carry loads that cannot be economically carried by rolled beams. Standard rolled sections may be adequate for many of the usual structures; but in situations where the load is heavier and the span is also large, the designer has the following choices (Fig. 11.1).

1. Use two or more regularly available sections, side-by-side.
2. Use a cover-plated beam, i.e., weld a plate of adequate thickness to increase the bending resistance of the flange.
3. Use a fabricated plate girder, which provides the freedom (within limits) to choose the size of web and flanges.
4. Use a steel truss or a steel-concrete-composite truss.

Among the options available, the first is usually uneconomical and does not satisfy deflection limitation. The second option is advantageous where the rolled section is marginally inadequate. Therefore, the real choice lies between the plate girder and the truss girder. The moment-resisting capacity of plate girders lies between rolled I-sections and truss girders. Truss girders involve higher costs of fabrication and erection, problems of vibration and impact, and require higher vertical clearance. For short spans (<10 m), plate girders are uneconomical due to higher connection costs and rolled I-sections are the preferred choice.

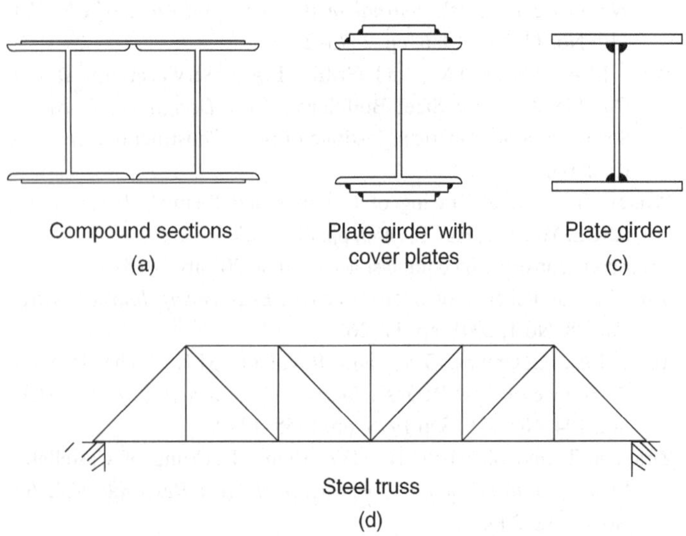


**FIG. 11.1** Options for beams of long spans to carry heavy loads

## 11.1 PLATE GIRDERS

As stated earlier, plate girders are deep flexural members used to carry loads that cannot be carried economically by rolled beams. Plate girders provide maximum flexibility and economy. In the design of a plate girder, the designer has the freedom to choose components of convenient size, but has to provide connections between the web and the flanges. Instances where large loads and large spans occur are: gantry girders provided in industrial buildings to carry the rails for large capacity overhead travelling crane, deep girders provided in power plant buildings to support bunkers, railway bridges and many other applications in industrial and residential buildings. Plate girders offer a unique flexibility in fabrication and the cross section can be uniform or non-uniform along the length. It is possible for putting the exact amount of steel required at each section along the length of the girder by changing the flange areas and keeping the same depth of the girder. In other words, it can be shaped to match the bending moment curve itself. Thus, a plate girder offers limitless possibilities to the creativity of the engineer.

It is a normal practice to fabricate plate girders by welding together three plates. In the past, plate girders were constructed by riveting or bolting, necessitating the use of angles to connect the web to the flanges (see Section 11.1.4). Since the introduction of highly automated workshops in the recent years, fabrication costs of plate girders have come down significantly. In the case of trusses and box girders, fabrication is still by manual means and hence their fabrication costs are high. Since nearly all plate girders constructed nowadays are welded (although they may use bolted field splices), in this chapter we will devote our attention to welded girders only. Moreover, as shown in Fig. 6.8 of Chapter 6, welded plate girders provide a more aesthetic appearance than bolted or riveted girders.

The material in a plate girder is used optimally as the designer has greater freedom in varying the section as per the requirements. This has resulted in the extensive use of variable depth plate girders nowadays. Thus it is possible to have tapered, cranked, and haunched girders (see Fig. 11.2). We can also have holes in the webs to accommodate the services (see Fig. 11.3). The designer may choose to reduce flange thickness (or breadth) in the zone of low applied moment, especially when a field-splice facilitates the change. Similarly, in the zone of high shear, the designer might choose a thicker web plate (see Fig. 11.4). Alternatively, higher grade steel may be employed in zones of high applied moment and shear, while standard grade steel (Fe 410) could be used elsewhere. *Hybrid girders*, with different strength material in the flanges and web, offer a method of closely matching resistance of the section to the requirements (Fan et al. 2006; Veljkovic & Johansson 2004; Schilling 1968).

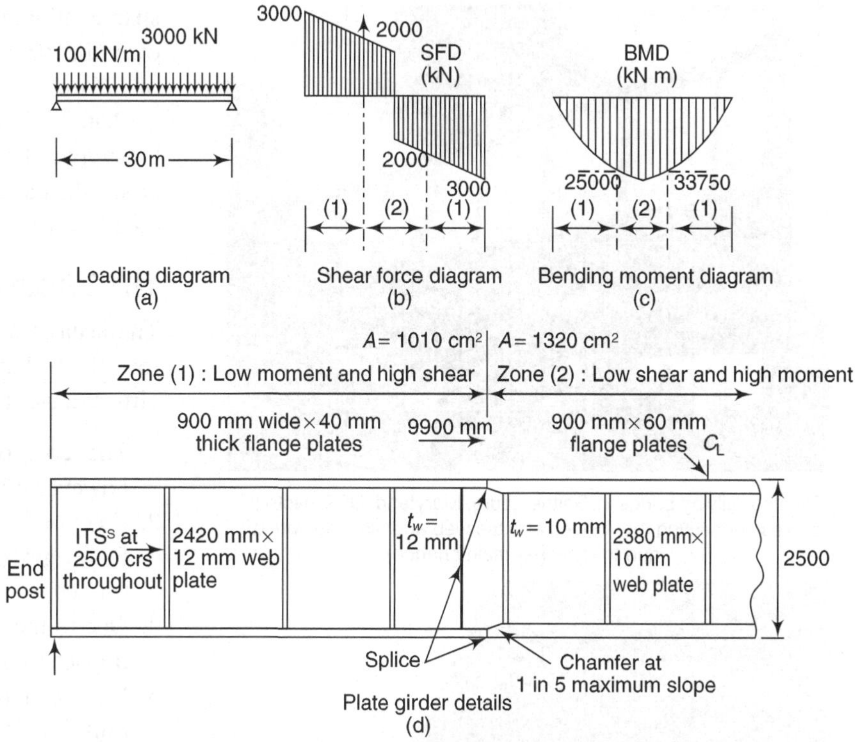


**FIG. 11.4** Plate girder with splice and variable cross section (ESDEP 2006)

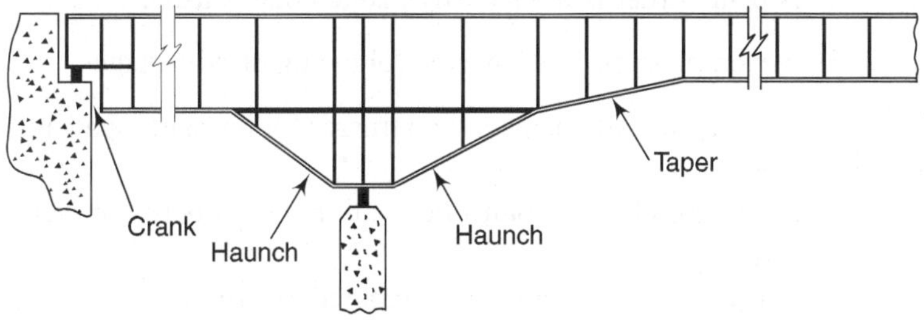


*Note* Splices, camber, variation in flanges, web and material strength are not shown

**FIG. 11.2** Plate girder with haunches, tapers, and cranks

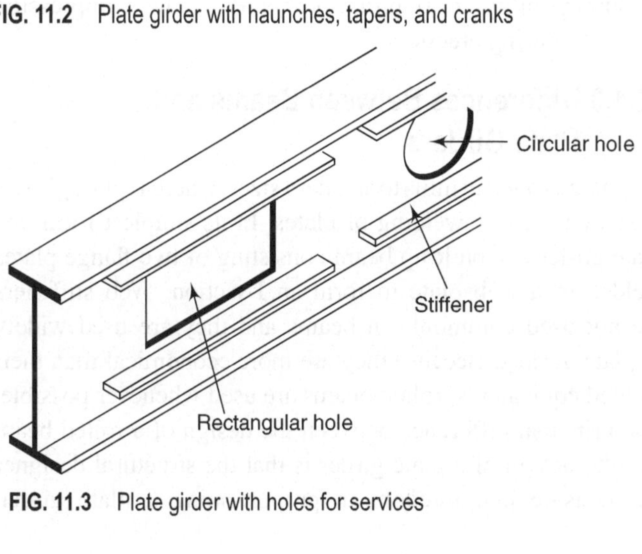


**FIG. 11.3** Plate girder with holes for services

## 11.1.1 Examples of Plate Girders

Most steel highway bridges in USA for spans less than about 24 m are steel rolled beam bridges. For longer spans, plate girders compete very well economically. Similarly, when loads are extremely large, as in railway bridges, plate girders may prove to be economical even for smaller spans. The upper economical limit of plate girder spans depend on the following factors: (a) whether the bridge is simple or continuous, (b) whether it is a highway or a railway bridge, and (c) the length of the section which can be transported in one piece. In general, plate girders are economical for railway bridges of spans 15–40 m and for highway bridges of spans 24–46 m. They may be very competitive for much longer spans, when they are continuous. They have been used in continuous bridge girders of spans in excess of 120 m. Early railway bridges (constructed during 1870–1900) were riveted plate girders, composed of angles connected to a web plate, with or without cover plates and having spans in the range of 15–45 m. Figure 11.5 shows examples of plate girders. Most of the plate girder bridges constructed after 1960s are shop welded using two flange plates and one web plate to form an I-shaped cross section. The longest plate girder bridge in the world is a three-span continuous bridge over the Save

(a) Riveted railway bridge in Gaithersburg, Maryland, USA (Note the angles connecting the flanges and the web and also the corrosion due to improper maintenance)

(b) Plate griders bridge in Pittsburg, USA (Note the lateral bracings between the griders)

(c) Two span, 107-m long,curved plate girder for a viaduct, Denver, USA (Note the lateral bracings)

**FIG. 11.5** Examples of plate girders

River at Belgrade, Yugoslavia with spans of 75–260–75 m. It has a double-box girder cross section varying in depth from 4.5 m at mid-span to 9.6 m at the pier. Another notable long span continuous plate girder bridge is the Bonn-Beuel bridge over the Rhine river in Germany which has a main span of 196 m.

Plate girders are also used in buildings when it is required to support heavy concentrated loads. Such situations may arise when a large hall with no interfering columns is desired on a lower floor of a multi-storey building.

### 11.1.2 Plate Girders Vs Trusses

The usual practical alternatives to plate girders, in the spans for which they are economical, are trusses. However, plate girders have the following advantages over trusses.

1. The cost of fabrication is lower compared to trusses; however it is higher compared to rolled steel sections.
2. Erection is faster and cheaper than trusses.
3. Plate girders require smaller vertical clearances than trusses.
4. Due to the compactness of the plate girders, vibration and impact are not serious problems.
5. Plate girders are safer than trusses. For example, if a truck runs into a bridge composed of plate girders, it would probably bend the steel plates a little; however a similar accident could cause the breaking of a member in a truss, which may even lead to the failure of the truss.
6. Points where stresses may be critical are fewer in plate girders compared to trusses.
7. A plate girder is more easily painted than a truss.

However, plate girders have the following disadvantages.

1. They are usually heavier than trusses for the same span and loads.
2. Larger number of connections are required between webs and flanges.
3. Larger exposed wind area compared to a truss.
4. Low torsional stiffness (box girders provide better torsional stiffness).
5. Susceptibility to stability problems of the compression flange during erection.

### 11.1.3 Differences Between Beams and Plate Girders

Beams are rolled in mills to standard sizes, whereas plate girders are assembled by welding of plates. In its simplest form, the plate girder is a built-up beam consisting of two flange plates welded to a web plate to form an I-section. Web stiffeners are not used commonly in beams and they are used widely in plate girders. Because they are more economical than their welded equivalents, rolled beams are used whenever possible. The principal difference between the design of a rolled beam and the design of a plate girder is that the structural designer has considerable freedom in proportioning a plate girder.

Therefore, the designer might face several structural problems, which may not arise when rolled sections are used. The most important of these are the local buckling of the compression flange and the shear buckling of the web. In a plate girder, the primary function of the flange plates is to resist bending moments by developing axial compressive and tensile stresses. The web plate resists the shear. For making the cross section efficient in resisting the in-plane bending, it is required that maximum material is placed as far away from the neutral axis as possible. From this point of view it is economical to keep the flanges as far apart as possible. The axial forces in the flanges decrease, as the depth of the girder increases. Thus, a smaller area of cross section would suffice than would be the case if a smaller depth were chosen. However, this would also mean that the web would be thin and deep. In such a situation, premature failure of the girder due to web buckling in shear might occur. Here there is a choice between thin web provided with vertical and horizontal stiffeners, and a thicker web requiring no stiffening (and therefore avoiding costly fabrication). The choice between the two depends upon a careful examination of the full costs of both forms of construction.

### 11.1.4 Types of Sections

Several possible plate girder arrangements are shown in Fig. 11.6. Figure 11.6(a) shows the simplest type of plate girder in which the flange plates are made of a pair of angles and they are connected to a solid web plate to form the plate girder. For larger moments, flange area can be increased by riveting/bolting additional plates, also known as cover plates, as shown in Fig. 11.6(b).When there are head room restrictions in buildings, and larger depths of plate girders are not possible, box girders may be an option as shown in Fig. 11.6(c). Box girders also provide greater lateral stability. If the number of cover plates is to be reduced, the flange area can be compensated by providing plates inserted between the web and flange angles [Fig. 11.6(d)]. These arrangements are sometimes adopted to keep the rivets/bolts both connecting flange angles and cover plates shorter. Figures 11.6(e) and 11.6(f) show typical welded plate girders. The form shown in Fig. 11.6(e) is the most commonly used type of welded plate girder. Here, flange angles are not required and instead of using a number of cover plates, a thick plate is used as the flange. As discussed earlier, a thinner cover plate can be used where lesser flange area is desired such as at regions away from the maximum bending moment and the plates of different thicknesses can be joined by butt welding.

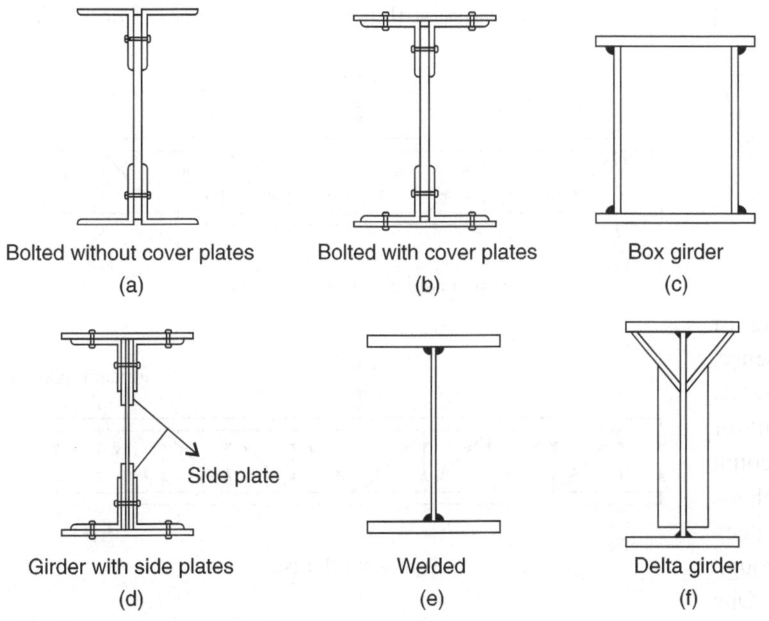


**FIG. 11.6** Types of plate girders

### 11.1.5 Elements of a Plate Girder

The various components of welded stiffened and unstiffened plate girders, shown in Fig. 11.7, are as follows:

1. Web plate
2. Flange plate with or without cover plates (with curtailment of cover plate)
3. Bearing stiffeners or end post (EP)
4. Intermediate transverse stiffeners (ITS)
5. Longitudinal stiffeners (LS)
6. Web splices
7. Flange splices
8. Connection between flange and the web
9. End bearing or end connections

To reduce the self-weight (and the corresponding self-weight bending moment), the web thickness $t$ should be limited to slender proportions (the web proportions are normally expressed in terms of the web slenderness ratio $d/t$). Slender webs (with large $d/t$ values) would buckle at relatively low values of applied shear loading. (It is also important to note that webs having a span/depth ratio smaller than 12 would result in a 'deep' beam, wherein the structural behaviour can no longer be described by conventional simple beam theories.)

Efficient and economical design usually results in slender members. Hence advantage must be taken of the post buckling capacity of the web, i.e., the ability of the girder to withstand transverse loads considerably in excess of the load at which the web buckles under shear. A girder of high strength to weight ratio can be designed by incorporating the post buckling strength of the web in the design method employed. This would be particularly advantageous where the reduction of self-weight is of prime importance. Examples of such situations arise in long span bridges, ship girders, transfer girders in buildings, etc.

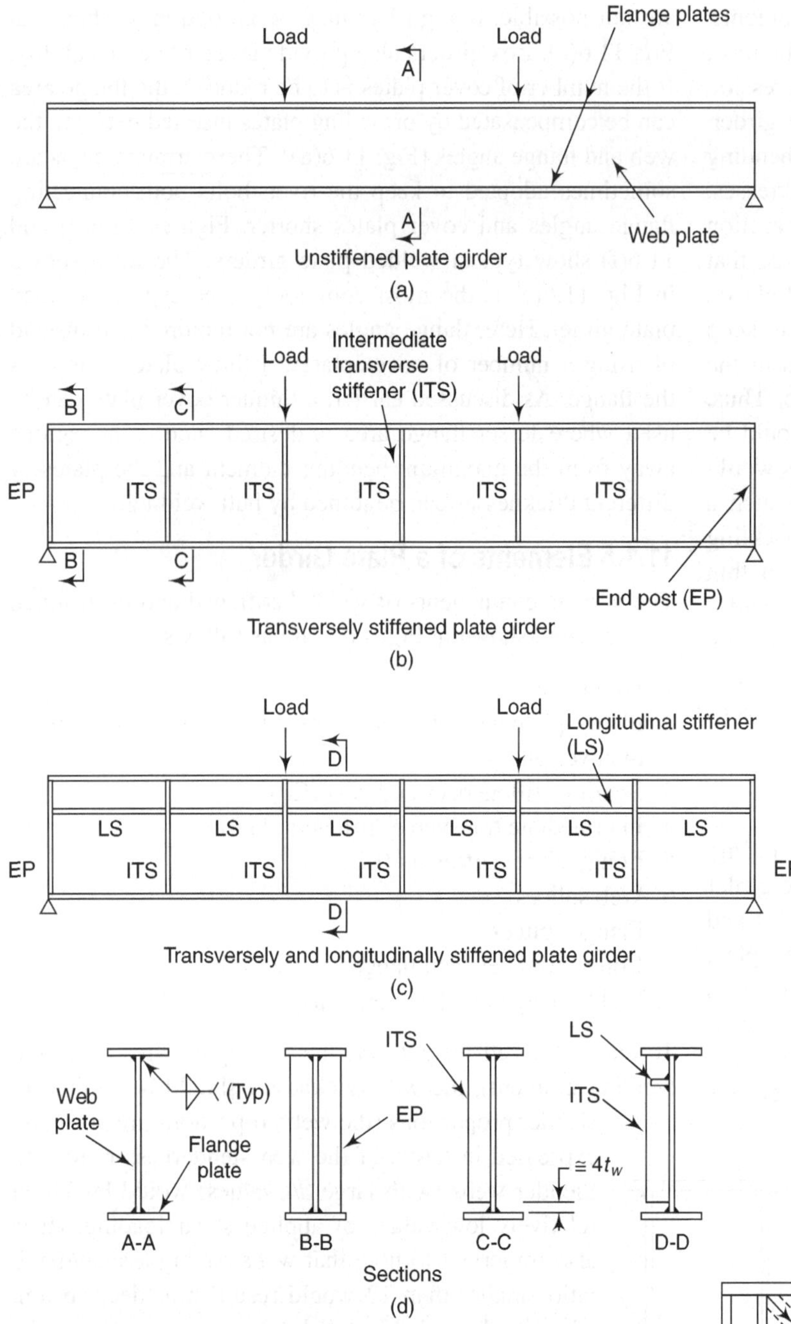


**FIG. 11.7** Stiffened and unstiffened plate girders

## 11.2 GENERAL CONSIDERATIONS

Design of structural steel members is mostly a matter of providing stability both locally and in an overall sense. For most of the hot rolled sections, the elements are so proportioned that local stability is not a consideration. But for a plate girder design, many factors are to be taken into account that would not be required with a rolled section. Problems arise in plate girders because of the deep thin webs. A deep understanding of the problem would require a thorough knowledge in stability theory, especially plate stability. One way of improving the load carrying capacity of a slender plate is to provide stiffeners; selection of the right kind of stiffening is an important consideration in a plate girder design.

The modes of failure of a plate girder are by yielding of the tension flange and buckling of the compression flange. The compression flange buckling can take place in various ways, such as vertical buckling into the web, flange local buckling, or lateral torsional buckling.

Plate girders depend on the post-buckling strength of the webs. At high shear locations in the girder web, normally near the supports and the neutral axis, the principal planes would be inclined to the longitudinal axis of the member. Along the principal planes, the principal stresses would be diagonal tension and diagonal compression. Though diagonal tension does not cause any problem, diagonal compression causes the web to buckle in a direction perpendicular to its action. This problem can be solved by any of the following three ways:

1. Reduce the depth to thickness ratio of the web such that the problem is eliminated
2. Provide web stiffeners to form panels that would enhance the shear strength of the web
3. Provide web stiffeners to form panels that would develop tension field action to resist diagonal compression

Figure. 11.8(a) shows the tension field action in a panel. As the web begins to buckle, the web loses its ability to resist the diagonal compression. The diagonal compression is then transferred to the transverse stiffeners and the flanges. The vertical

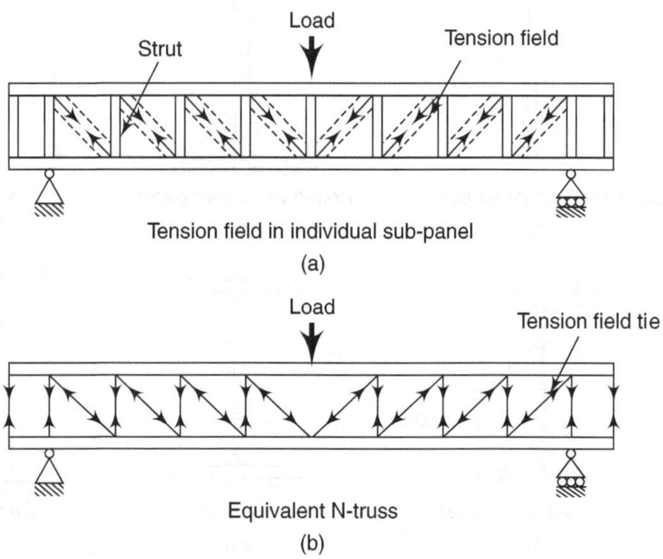


**FIG. 11.8** Tension field action and the equivalent N-truss

component of the diagonal compression is supported by the stiffeners and the flanges resist the horizontal component. The web resists only the diagonal tension and this behaviour of the web is called *tension field action*. The behaviour is very similar to a Pratt truss, in which the vertical members carry compression and the tension is carried by the diagonals, as shown in Fig. 11.8(b). The contribution of the tension field action is realized only after the web starts to buckle. The total strength of the web is made up of the web strength before buckling and the post buckling strength developed due to tension field action. Thus, the provision of suitably spaced stiffeners, called *intermediate stiffeners,* can be used to develop tension field action and improve the shear capacity of the web. The main purpose of these stiffeners is to provide stiffness to the web rather than to resist the applied loads.

Additional stiffeners, called *bearing stiffeners*, are provided at points of concentrated loads, to protect the web from the direct compressive loads. They can simultaneously act as intermediate stiffeners also.

Application of concentrated loads on the top flange may produce web yielding, web crippling, and side-sway web buckling. This form of buckling occurs when the compression in the web causes the tension flange to buckle laterally. In order to prevent this, the relative movements of the flanges should be restrained by lateral bracing. General welding procedures are followed for connecting the components during the design of welds.

## 11.3 DISTRIBUTION OF STRESS IN A PLATE GIRDER

The plate girder is generally analysed as a beam member. It would be of interest to consider here the shear stress and bending stress distributions in a rectangular section and I-section of same cross-sectional area. The shear stress distribution in a rectangular section is shown in Fig. 11.9(a). The shear stress intensity in the section may be determined from the strength of materials approach as $v = VQ/(It)$, where $V$ is the shear on the section, $Q$ is the statical moment of the area about the point where the stress is to be determined, $I$ is the moment of inertia of the section, and $t$ is the thickness of the section. For a given loading and rectangular section, $V$, $I$, and $t$ are constants.

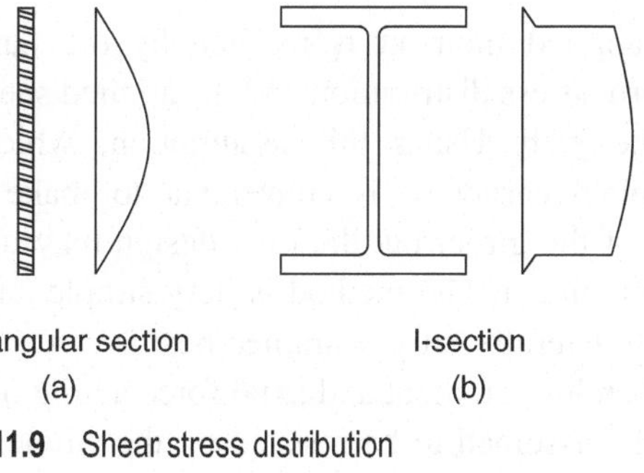


**FIG. 11.9** Shear stress distribution

Thus it can be seen that $Q$ is zero for extreme fibres of the section and maximum at the neutral axis. The shear stress distribution across the section is parabolic in shape with the maximum intensity of 3/2 of the average shear intensity.

$$\text{Unit shear } v \text{ in the section} = VQ/It$$

$$\text{Therefore, shear force per unit length} = tVQ/It = VQ/I$$

This shows that shear is proportional to $Q$, $V$ and $I$ being constant for given conditions.

The nature of shear distribution in an I-beam is shown in Fig. 11.9(b). In an I-section the value of $t$ varies across the section, with flange width as the value in the flange area and web thickness in the web portion. It may be observed from the figure that the value of $Q$ which is zero at the extreme fibres increases to a high value at top flange-web intersection and attains maximum value at the neutral axis.

The bending stress distributions in the two cross sections are shown in Fig. 11.10. It can be seen that in the I-section, the flanges carry most of the bending stresses. From the comparison of the shear and flexural stress distributions in the two sections, it may be observed that the flanges carry a major portion of the flexural load, whereas the web carries most of the shear load. Based on this behaviour, an approximate method, known as *flange area method*, is often used for a quick estimation of a trial section. The flange area method assumes that the stress distribution over the flange is uniform and the centre of gravity of the resultant force is at the centre of the respective flange.

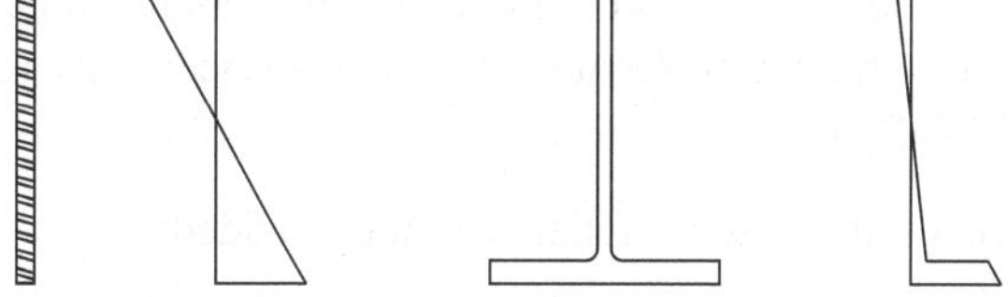


**FIG. 11.10** Bending stress distribution

## 11.4 PRELIMINARY DESIGN PROCEDURE

General dimensions of a plate girder are fixed based on its optimum behaviour. Some guidelines for proportioning the flanges and webs are given in Section 11.12. One condition that may limit the proportions of the girder is the largest size that can be fabricated in shop and transported to the site. There may be transportation problems such as clearance requirements (of overhead bridges) that limit maximum depths to about 3–3.6 m. A plate girder design begins with values of live load shears and bending moments and estimated dead loads. After the selection of the girder depth, the web may be designed to resist the shear, taking into account the minimum thickness requirements. Flanges in riveted or bolted design are made up of plates and angles. Economy in design may be achieved by using cover plates for the flanges, which may be curtailed as per requirement. Generally the cover plates constitute 40–60% of the total flange. While selecting sections for flange plates, it may be ensured that the section is not too thick so that it cannot be punched or too thin so that it will buckle locally. The web to flange fillet welds must be designed to transmit the longitudinal shear at the flange/web interface.

Figure 11.10(b) shows that a major portion of the moment of resistance in a I-section plate girder is provided by the flanges; the contribution of the web being very small. On the tension side, provision of bolt holes may reduce the effective flange area, which may be compensated by adding more flange area. In the design of a plate girder, a trial section needs to be assumed initially and the amount of load to be carried includes the self weight of the girder also. A preliminary estimate of the girder weight can be obtained using the simplified flange area method.

The design requirements for plate girders are addressed in Section 8 of IS 800 : 2007 specification. The identification of a laterally supported member as a beam or a plate girder is decided based on the $d/t_w$ ratio, where $d$ is the depth of the beam and $t_w$ is the thickness of the web. If $d/t_w$ is less than $67\varepsilon$, the member is classified as a beam and the respective clauses applicable to beams are used, irrespective of whether the member is made up of plates or a rolled section. If the ratio is greater than $67\varepsilon$, the clauses relating to plate girders are invoked.

### 11.4.1 Minimum Web Thickness

The thickness of the web in a section should satisfy the following requirements with respect to serviceability (clause 8.6.1.1 of IS 800):

(a) When transverse stiffeners are not provided:

$\frac{d}{t_w} \leq 200\varepsilon_w$ (web connection to flanges along both longitudinal edges)

$\frac{d}{t_w} \leq 90\varepsilon_w$ (web connection to flanges along one longitudinal edge only)

(b) When only transverse stiffeners are provided:

(i) When $3d \geq c \geq d$

$$\frac{d}{t_w} \leq 200\varepsilon_w$$

(ii) When $0.74d \leq c < d$

$$\frac{c}{t_w} \leq 200\varepsilon_w$$

(iii) When $c < 0.74d$

$$\frac{d}{t_w} \leq 270\varepsilon_w$$

(iv) When $c > 3d$, the web is considered as unstiffened.

(c) When transverse stiffeners and longitudinal stiffeners are provided at one level only, at $0.2d$ from the compression flange:

(i) When $2.4d \geq c \geq d$

$$\frac{d}{t_w} \leq 250\varepsilon_w$$

(ii) When $0.74\,d \leq c \leq d$

$$\frac{c}{t_w} \leq 250\varepsilon_w$$

(iii) When $c < 0.74\,d$

$$\frac{d}{t_w} \leq 340\varepsilon_w$$

(d) When there is a second longitudinal stiffener (provided at neutral axis):

$$\frac{d}{t_w} \leq 400\varepsilon_w$$

where $d$ is the depth of the web, $t_w$ is the thickness of the web, $c$ is the spacing of the transverse stiffener, $\varepsilon_w = \sqrt{250/f_{yw}}$, and $f_{yw}$ is the yield stress of the web.

These criteria are to ensure that the web will not buckle under normal service conditions. It should also be ensured that the web is strong enough so that the flange will not buckle into the web. It is also evident that a thinner web can be used if the $c/d$ ratio is less than one and that such a web panel has a higher strength.

### 11.4.2 Flexural Strength

The method by which the calculation of the bending capacity of a plate girder is determined depends upon the classification of the flanges and the web as calculated from $b/t_f$ and $d/t_w$ ratios. The limiting ratios are presented in the chapter on beams. As mentioned earlier, if the web slenderness ratio is less than $67\varepsilon$, then the member should be designed as a rolled beam section. The flexural strength of a plate girder is based on the tension flange yielding or otherwise on the compression flange buckling. The compression flange buckling strength will be decided by either flange local buckling (see Example 8.23 of Chapter 8) or lateral torsional buckling (see Section 10.8 of Chapter 10).

For sections with slender webs, when their flanges are plastic, compact, or semi-compact (i.e., $d/t_w > 67\varepsilon$), the design bending strength is calculated using one of the following methods:

1. The applied moment is resisted by the flanges with a uniform stress distribution and the applied shear is resisted by the web. Using this assumption, which has been explained earlier, it is convenient to make preliminary sizes of the girder but the final design may turn out to be uneconomical. The method is very simple and applicable only to laterally fully restrained beams.
2. The bending moment and axial force acting on the section may be assumed to be resisted by the whole section, and

the web designed for combined shear and normal stresses, by using simple elastic theory in case of semi-compact flanges and simple plastic theory in the case of compact and plastic flanges.

For sections with slender flanges, moment capacity is calculated based on the approach similar to that of a slender rolled section. In the design of plate girders, a minimum weight solution adopting thin webs may not be economic if we consider the fabrication cost of the stiffeners.

### 11.4.3 Lateral Torsional Buckling of Plate Girders

The same procedure as adopted for rolled sections (Section 10.8) may be used except that different values of the imperfection coefficient $\alpha_{LT}$ have to be used to account for higher residual stresses in welded plate girders. Approximate formulae may be used to arrive at the sections properties of the members.

### 11.4.4 Shear Strength

The shear strength of a plate girder depends upon the depth to thickness ratio of the web and the spacing of the intermediate stiffeners provided. Webs of plate girders are usually stiffened transversely as shown in Fig. 11.7. This helps to increase the ultimate shear resistance of the webs. The shear capacity of the web has two components, namely, strength before the onset of buckling and strength after post-buckling. As the shear load is increased on a stiffened web panel, the web panel buckles (see Fig. 11.11). This load does not indicate the maximum shear capacity of the web. The load can be still increased and the web panel continues to carry further load relying on the tension field action. Part of the buckled web takes the load in tension. This tension member action is across the web panel in an inclined direction to the web panel diagonal as shown in Fig. 11.12.

At this stage, the girder acts like an N-truss (Fig. 11.8) with the compression forces being carried by the flanges and the intermediate stiffeners and the buckled web resisting the tension. This additional reserve strength is termed as *tension field action*. It may be noted here that if no intermediate

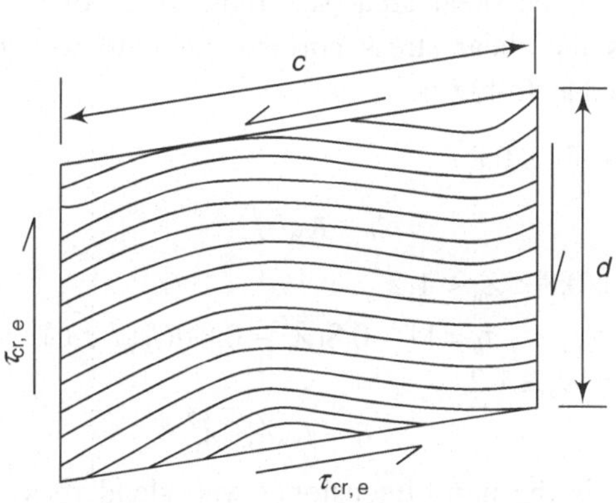


**FIG. 11.11** Buckling of the web panel

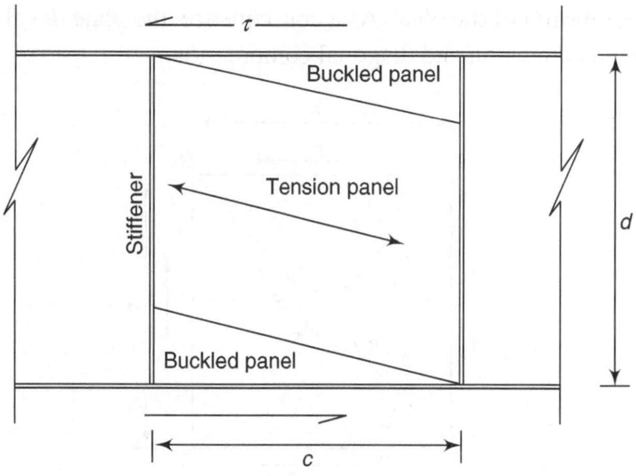


(a)

(b)

**FIG. 11.12** (a) Panel showing tension field and (b) Tension field in a plate girder (ESDEP 2006)

stiffeners are present or their spacing is large, it is not possible for tension field action to take place and the shear capacity is restricted to the strength before buckling. The stiffener spacing $c$ influences both buckling and post-buckling behaviour of the web under shear. The various stages of shear resistance are explained in the following section.

A long span plate girder may have various web panels, each panel having different combinations of bending moments and shear forces. Panels, which are closer to the support will be predominantly under shear, and those close to the centre will be under predominant bending moments. The effect of the shear alone will be considered first, followed by the case where the effect of combined bending moment and shear forces would be treated.

## 11.5 WEB PANEL SUBJECTED TO SHEAR

### 11.5.1 Shear Resistance Before Buckling (Stage 1)

When a square web plate is subjected to vertical shear, complementary shear stresses are developed to satisfy equilibrium of the plate. As a consequence, the plate develops diagonal tension and diagonal compression.

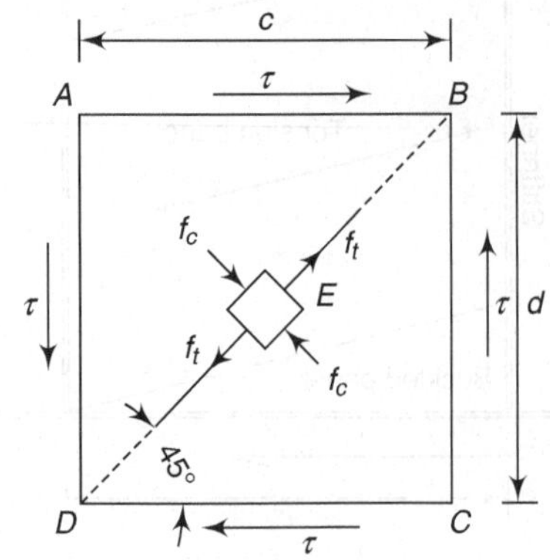


**FIG. 11.13** Unbuckled shear panel

Consider a small element $E$ in equilibrium inside the web plate subject to a shear stress $\tau$ (see Fig. 11.13). The element is subjected to principal compression along the direction AC and principal tension along the direction BD. As the applied loading is gradually increased, $\tau$ in turn will increase and the plate will buckle along the direction of the compressive diagonal AC. The plate cannot take any further increase in compressive stress. The corresponding shear stress $\tau$ in the plate is called the 'elastic critical shear stress' $\tau_{cr,e}$ of the panel. The value of $\tau_{cr,e}$ can be easily determined from the classical stability theory, if the boundary conditions of the plate can be established. However, it is very difficult to know the true boundary conditions of the plate provided by the flanges and stiffeners. Therefore, conservatively, simply supported conditions are assumed at the boundaries. The critical shear stress in such a case is given by

$$\tau_{cr,e} = k_v \frac{\pi^2 E}{12(1-\mu^2)}\left(\frac{t_w}{d}\right)^2 \qquad (11.1)$$

where $k_v$ is a coefficient which depends upon the support conditions and $\mu$ is Poisson's ratio.

$k_v = 5.35$, when transverse stiffeners are provided only at support

$= 4.0 + 5.35/(c/d)^2$ for $c/d < 1.0$

$= 5.35 + 4.0/(c/d)^2$ for $c/d \geq 1.0$

where $c$ and $d$ are the spacing of transverse stiffeners and the depth of the web, respectively.

As shown in Section 8.12.5, the limiting value of $dt_w$ is given by

$$\frac{d}{t_w}\sqrt{\frac{f_y}{250}} = 82.0$$

Thus, it is seen that for $f_y = 250$ MPa, elastic buckling due to shear will occur when the $d/t$ ratio of the web is greater than 82.

As already indicated in Section 8.12.5, elastic buckling stress may be significantly increased either by using intermediate transverse stiffeners to decrease the aspect ratio $c/d$ (thereby increasing the value of buckling coefficient $k_v$), or by using longitudinal stiffeners to decrease the depth-to-thickness ratio $d/t_w$. The aspect ratio of each panel in the range of 0.5 to 2 is found to be more efficient.

**IS 800: 2007 provisions for Resistance to Shear Buckling** Resistance to shear buckling shall be verified when (clause 8.4.2.1 of IS 800)

$$\frac{d}{t_w} > 67\varepsilon \text{ for an unstiffened web}$$

$$> 67\varepsilon\sqrt{\frac{k_v}{5.35}} \text{ for a stiffened web}$$

where $k_v$ is the shear buckling coefficient and $\varepsilon = \sqrt{250/f_y}$.

## 11.5.2 Shear Buckling Design

The nominal shear strength $V_n$ of webs with or without intermediate stiffeners as governed by buckling may be evaluated using one of the following methods:

1. Simple post-critical method
2. Tensile field theory

These two methods are discussed in the following sections.

### *Simple Post-critical Method*

The simple post-critical method is a general method and can be applied to the design of all girders. The simple post critical method can be used for webs of I-section girders, with or without intermediate transverse stiffener, provided that the web has transverse stiffeners at the supports. The nominal shear strength is given by (Clause 8.4.2.2 of IS 800)

$$V_n = V_{cr} \qquad (11.2a)$$

where $V_{cr}$ is the shear force corresponding to web buckling.

$$V_{cr} = A_v \tau_b \qquad (11.2b)$$

where $A_v$ is the shear area (see Table 10.12 of Chapter 10) and $\tau_b$ is the shear stress corresponding to web buckling, determined as follows:

(a) When $\lambda_w \leq 0.8$

$$\tau_b = f_{yw}/\sqrt{3} \qquad (11.3a)$$

(b) When $0.8 < \lambda_w < 1.2$

$$\tau_b = [1 - 0.8(\lambda_w - 0.8)](f_{yw}/\sqrt{3}) \qquad (11.3b)$$

(c) When $\lambda_w \geq 1.2$

$$\tau_b = f_{yw}/\sqrt{3}\lambda_w^2 \qquad (11.3c)$$

where $\lambda_w$ is the non-dimensional web slenderness ratio for shear buckling stress, given by

$$\lambda_w = \sqrt{f_{yw} / \left(\sqrt{3}\tau_{cr,e}\right)} \tag{11.4}$$

The elastic critical shear stress of the web $\tau_{cr,\,e}$ is given by Eqn (11.1).

A plot of Eqn (11.3) is given in Fig. 11.14. Table 11.1 provides the values of shear buckling strength $\tau_b$ for various values of stiffener spacing ratio $c/d$ and depth-to-thickness ratio $d/t_w$.

The concept of using a linear transformation between stocky webs where yielding governs and slender webs where buckling governs was first suggested by Basler (1961) and then by Rockey and Skaloud (1971). It can be observed from Fig. 11.14 that the equation suggested by the Indian code [Eqn (11.3)] disregards the post buckling strength of the web.

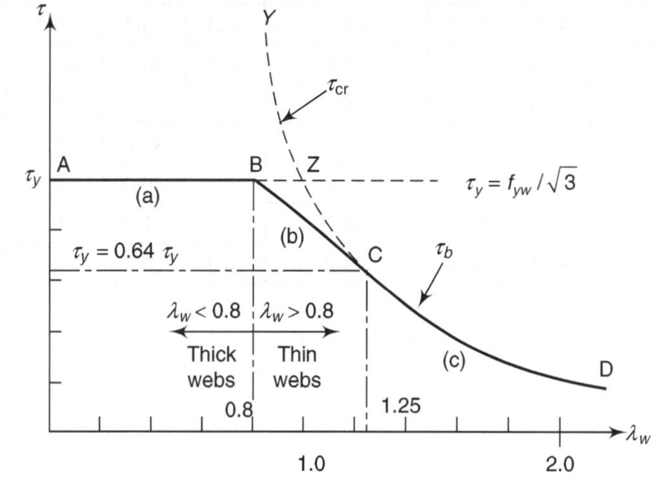


**FIG. 11.14** Simple post-critical shear strength

### 11.5.2.2 Web Strength using Tension Field Theory

As stated earlier, a plate girder with intermediate stiffeners ($1.0 \le c/d \le 3.0$) derives considerable post-buckling shear strength due to the tension field action. Tension field action can be applied to a certain range of girders only and will produce efficient designs of girders due to its taking advantage of the post-buckling reserve of resistance. According to the theory of tension field action, there are three components contributing to the predicted shear strength: the first is the primary buckling strength of the plate, the second is due to the tension field action of the web, and the third arises from the plastic moment capacity of the flanges. Basler (1961) was the first to formulate a successful model for tension-field action for plate girders of the type used in civil engineering structures. The first physical explanation for the very high shearing capacities of slender plate girder webs was provided by Rode as early as 1916 and a pure (complete tension field theory) for aircraft structures was developed by Wagner in 1929. Basler's approach (1961, 1963) disregarded the flexural rigidity of the flanges and hence did not reflect the true anchorage of the tension field along the flanges. Tension-field action has also been verified by a number of researchers (e.g., Clark & Sharp 1971; Steinhardt & Schröter 1971).

Following Basler, many different theories for the ultimate shear capacity of plate girders were developed (Ballio & Mazzolani 1983; Galambos 1998). Most notable among them were those by Takeuchi (1964), Chern and Ostapenko (1971), Fujii (1971), Komatsu (1971), Rockey and Skaloud (1971), Sharp and Clark (1971) and Steinhardt and Schröter (1971), Höglund (1973), Herzog, (1974); Porter et al. (1975).

**TABLE 11.1** Shear buckling strength $\tau_b$ (N/mm$^2$) of webs for $f_y$ = 250 MPa

| | Stiffener spacing ratio $c/d$ | | | | | | | | | | | | | |
|---|---|---|---|---|---|---|---|---|---|---|---|---|---|---|
| d/t | 0.4 | 0.5 | 0.6 | 0.7 | 0.8 | 0.9 | 1.0 | 1.2 | 1.4 | 1.6 | 1.8 | 2.0 | 2.5 | 3.0 | Infinity |
| 55 | 144.3 | 144.3 | 144.3 | 144.3 | 144.3 | 144.3 | 144.3 | 144.3 | 144.3 | 144.3 | 144.3 | 144.3 | 144.3 | 144.3 | 144.3 |
| 60 | 144.3 | 144.3 | 144.3 | 144.3 | 144.3 | 144.3 | 144.3 | 144.3 | 144.3 | 144.3 | 144.3 | 144.3 | 144.3 | 144.3 | 144.3 |
| 65 | 144.3 | 144.3 | 144.3 | 144.3 | 144.3 | 144.3 | 144.3 | 144.3 | 144.3 | 144.3 | 144.3 | 144.3 | 144.3 | 144.3 | 144.3 |
| 70 | 144.3 | 144.3 | 144.3 | 144.3 | 144.3 | 144.3 | 144.3 | 144.3 | 144.3 | 144.3 | 144.3 | 144.3 | 143.4 | 141.9 | 138.0 |
| 75 | 144.3 | 144.3 | 144.3 | 144.3 | 144.3 | 144.3 | 144.3 | 144.3 | 144.3 | 143.7 | 141.4 | 139.6 | 136.8 | 135.1 | 131.0 |
| 80 | 144.3 | 144.3 | 144.3 | 144.3 | 144.3 | 144.3 | 144.3 | 144.3 | 140.7 | 137.5 | 135.0 | 133.2 | 130.1 | 128.3 | 123.9 |
| 85 | 144.3 | 144.3 | 144.3 | 144.3 | 144.3 | 144.3 | 144.3 | 139.5 | 134.7 | 131.3 | 128.7 | 126.7 | 123.4 | 121.5 | 116.9 |
| 90 | 144.3 | 144.3 | 144.3 | 144.3 | 144.3 | 144.3 | 140.7 | 133.7 | 128.7 | 125.1 | 122.3 | 120.2 | 116.8 | 114.8 | 109.8 |
| 95 | 144.3 | 144.3 | 144.3 | 144.3 | 144.3 | 141.6 | 135.4 | 128.0 | 122.7 | 118.9 | 116.0 | 113.8 | 110.1 | 108.0 | 102.8 |
| 100 | 144.3 | 144.3 | 144.3 | 144.3 | 143.9 | 136.6 | 130.0 | 122.3 | 116.7 | 112.7 | 109.6 | 107.3 | 103.4 | 101.2 | 95.7 |
| 105 | 144.3 | 144.3 | 144.3 | 144.3 | 139.3 | 131.5 | 124.7 | 116.6 | 110.7 | 106.5 | 103.3 | 100.8 | 96.8 | 94.4 | 87.8 |
| 110 | 144.3 | 144.3 | 144.3 | 143.8 | 134.7 | 126.5 | 119.4 | 110.9 | 104.7 | 100.3 | 98.4 | 94.9 | 89.6 | 86.6 | 80.0 |
| 115 | 144.3 | 144.3 | 144.3 | 139.6 | 130.0 | 121.5 | 114.0 | 105.1 | 98.7 | 94.6 | 90.1 | 86.9 | 81.9 | 79.3 | 73.2 |

(Contd)

**TABLE 11.1** *(Contd)*

| | Stiffener spacing ratio *c/d* | | | | | | | | | | | | | |
|---|---|---|---|---|---|---|---|---|---|---|---|---|---|---|
| **d/t** | **0.4** | **0.5** | **0.6** | **0.7** | **0.8** | **0.9** | **1.0** | **1.2** | **1.4** | **1.6** | **1.8** | **2.0** | **2.5** | **3.0** | **Infinity** |
| 120 | 144.3 | 144.3 | 144.3 | 135.4 | 125.4 | 116.5 | 108.7 | 99.4 | 92.9 | 86.8 | 82.7 | 79.8 | 75.3 | 72.8 | 67.2 |
| 125 | 144.3 | 144.3 | 142.8 | 131.2 | 120.7 | 111.5 | 103.4 | 94.1 | 85.6 | 80.0 | 76.2 | 73.5 | 69.4 | 67.1 | 61.9 |
| 130 | 144.3 | 144.3 | 139.1 | 126.9 | 116.1 | 106.5 | 100.1 | 87.0 | 79.1 | 74.0 | 70.5 | 68.0 | 64.1 | 62.0 | 57.3 |
| 135 | 144.3 | 144.3 | 135.3 | 122.7 | 111.5 | 101.5 | 92.8 | 80.7 | 73.4 | 68.6 | 65.4 | 63.0 | 59.5 | 57.5 | 53.1 |
| 140 | 144.3 | 144.3 | 131.6 | 118.5 | 106.8 | 97.9 | 86.3 | 75.0 | 68.2 | 63.8 | 60.8 | 58.6 | 55.3 | 53.5 | 49.4 |
| 145 | 144.3 | 142.9 | 127.8 | 114.3 | 102.2 | 91.2 | 80.5 | 69.9 | 63.6 | 59.5 | 56.7 | 54.6 | 51.5 | 49.9 | 46.0 |
| 150 | 144.3 | 139.6 | 124.1 | 110.0 | 99.4 | 85.3 | 75.2 | 65.4 | 59.4 | 55.6 | 52.9 | 51.1 | 48.2 | 46.6 | 43.0 |
| 155 | 144.3 | 136.4 | 120.3 | 105.8 | 93.1 | 79.9 | 70.4 | 61.2 | 55.7 | 52.1 | 49.6 | 47.8 | 45.1 | 43.6 | 40.3 |
| 160 | 144.3 | 133.2 | 116.6 | 101.6 | 87.3 | 74.9 | 66.1 | 57.4 | 52.2 | 48.8 | 46.5 | 44.9 | 42.3 | 40.9 | 37.8 |
| 165 | 144.3 | 129.9 | 112.8 | 99.1 | 82.1 | 70.5 | 62.1 | 54.0 | 49.1 | 45.9 | 43.8 | 42.2 | 39.8 | 38.5 | 35.6 |
| 170 | 144.3 | 126.7 | 109.0 | 93.4 | 77.4 | 66.4 | 58.5 | 50.9 | 46.3 | 43.3 | 41.2 | 39.7 | 37.5 | 36.3 | 33.5 |
| 175 | 143.4 | 123.5 | 105.3 | 88.1 | 73.0 | 62.6 | 55.2 | 48.0 | 43.7 | 40.8 | 38.9 | 37.5 | 35.4 | 34.2 | 31.6 |
| 180 | 140.8 | 120.2 | 101.5 | 83.3 | 69.0 | 59.2 | 52.2 | 45.4 | 41.3 | 38.6 | 36.8 | 35.5 | 33.4 | 32.4 | 29.9 |
| 185 | 138.1 | 117.0 | 99.7 | 78.9 | 65.3 | 56.1 | 49.4 | 43.0 | 39.1 | 36.5 | 34.8 | 33.6 | 31.7 | 30.6 | 28.3 |
| 190 | 135.4 | 113.8 | 94.5 | 74.8 | 61.9 | 53.1 | 46.9 | 40.7 | 37.0 | 34.6 | 33.0 | 31.8 | 30.0 | 29.0 | 26.8 |
| 195 | 132.8 | 110.5 | 89.7 | 71.0 | 58.8 | 50.5 | 44.5 | 38.7 | 35.2 | 32.9 | 31.3 | 30.2 | 28.5 | 27.6 | 25.5 |
| 200 | 130.1 | 107.3 | 85.3 | 67.5 | 55.9 | 48.0 | 42.3 | 36.8 | 33.4 | 31.3 | 29.8 | 28.7 | 27.1 | 26.2 | 24.2 |
| 205 | 127.4 | 104.0 | 81.2 | 64.2 | 53.2 | 45.7 | 40.2 | | | | | | | | |
| 210 | 124.8 | 100.8 | 77.4 | 61.2 | 50.7 | 43.5 | 38.4 | | | | | Not applicable | | | |
| 215 | 122.1 | 99.4 | 73.8 | 58.4 | 48.4 | 41.5 | 36.6 | | | | | | | | |
| 220 | 119.4 | 94.9 | 70.5 | 55.8 | 46.2 | 39.6 | 34.9 | | | | | | | | |
| 225 | 116.8 | 90.8 | 67.4 | 53.3 | 44.2 | 37.9 | 33.4 | | | | | | | | |
| 230 | 114.1 | 86.9 | 64.5 | 51.0 | 42.3 | 36.3 | 32.0 | | | | | | | | |
| 235 | 111.4 | 83.2 | 61.8 | 48.9 | 40.5 | 34.7 | 30.6 | | | | | | | | |
| 240 | 108.8 | 79.8 | 59.2 | 46.9 | 38.8 | 33.3 | 29.4 | | | | | | | | |
| 245 | 106.1 | 76.6 | 56.8 | 45.0 | 37.2 | 32.0 | 28.2 | | | | | | | | |
| 250 | 103.4 | 73.5 | 54.6 | 43.2 | 35.8 | 30.7 | 27.1 | | | | | | | | |

*Note*: Intermediate values may be obtained by linear interpolation.

**TABLE 11.2** Various tension field theories for plate girders (Galambos 1998; Ballio & Mazzolani 1983)

| Investigator | Mechanism | Edge support | Unequal flanges | Longitudinal stiffener | Shear and moment |
|---|---|---|---|---|---|
| Basler (1961) | 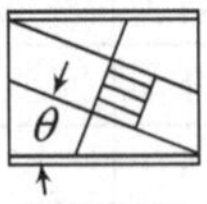 | 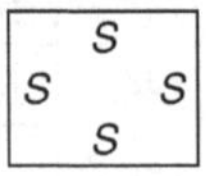 | Immaterial | Yes, Cooper (1967) | Yes |
| Takeuchi (1964) | 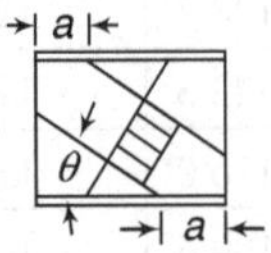 | 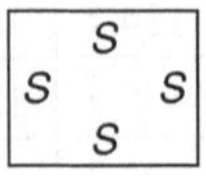 | Yes | No | No |
| Fujii (1971) | 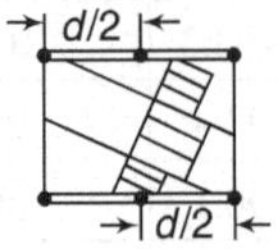 | 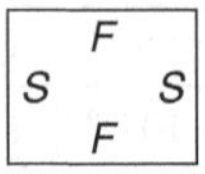 | Yes | Yes | Yes |

*(Contd)*

**TABLE 11.2** *(Contd)*

| Investigator | Mechanism | Edge support | Unequal flanges | Longitudinal stiffener | Shear and moment |
|---|---|---|---|---|---|
| Komatsu (1971) | 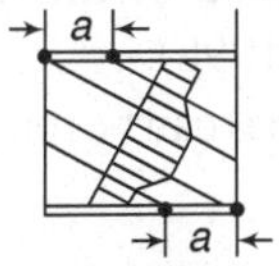 | F; S S; F | No | Yes, at mid-depth | No |
| Chern and Ostapenko (1971) | 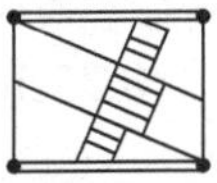 | F; S S; F | Yes | Yes | Yes |
| Porter et al. (1975) | 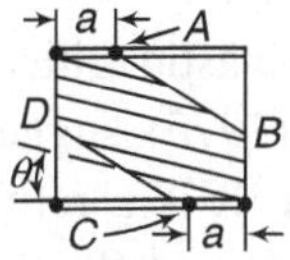 | S; S S; S | Yes | Yes | Yes |
| Höglund (1973) | 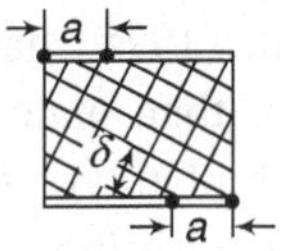 | S; S S; S | No | No | Yes |
| Herzog (1974) | 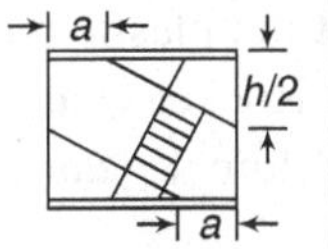 | Web buckling component neglected | Yes, in evaluating *a* | Yes | Yes |
| Sharp and Clark (1971) | 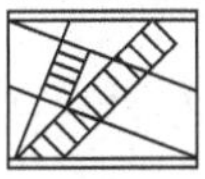 | F/2; S S; F/2 | No | No | No |
| Steinhardt and Schröter (1971) | 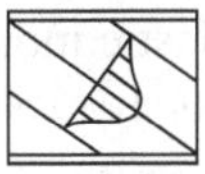 | S; S S; S | Yes | Yes | Yes |

*Note:* *S* denotes simply supported boundary and *F* denotes fixed boundary

The resulting ultimate shear strength models differed from the Basler's model in various ways (see also Table 11.2): (a) the boundary conditions for the web, (b) the inclination and the width of the tension field, and (c) the tensile stress distribution within the tension field. Detailed information about these models can be found in Ballio and Mazzolani (1983), Narayanan (1983), Dubas and Gehri, (1986), and Galambos (1998). In the following sub-section, the model as suggested by Porter et al. (1975) is only considered. This model is often referred to as the *Cardiff model*. In this model (unlike Basler's model), account is taken of the effect of the bending stiffness of the flanges on the width of the diagonal tension band. Thus the diagonal tension field is considered to be composed of a central part that anchors on the transverse stiffeners, and two additional parts anchored on the flanges (Fig. 11.15). The ultimate shear load is obtained as a sum of the critical shear buckling strength and the vertical component of the force in the band at collapse. It has been found that the shear model as proposed by Porter et al. (1975) provides a fairly good correlation between theory and test results (Maquoi 1992; Valitnat 1982).

**Tension field as per Cardiff Model** The first stage of shear resistance by the web plate till it attains its elastic critical stress has been explained in the preceding section. After the web buckles along the direction of the principal compressive stress, a new load carrying mechanism is developed along the principal tensile direction. This new mechanism, called the *tension field action*, resists any further increase in the shear load. The tensile field is constituted by the portion of the plate

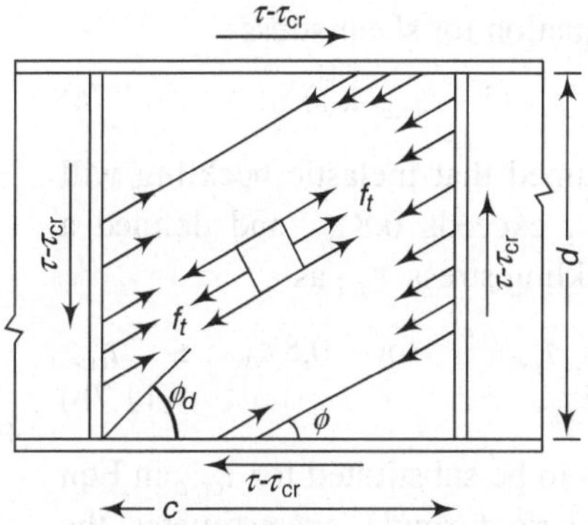


**FIG. 11.15** Post-buckled behaviour

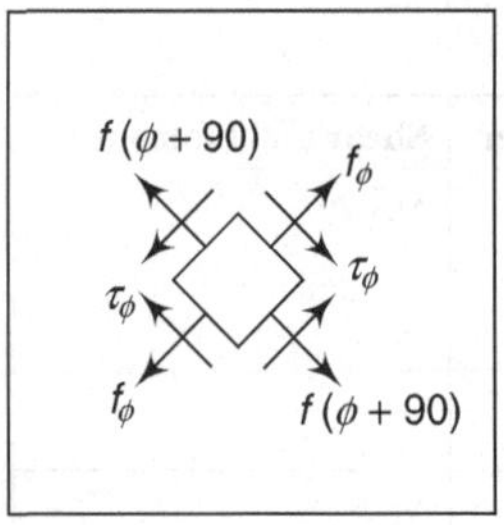


**FIG. 11.16** State of stress in the web in the post-buckled stage

in the principal tensile direction and anchored at the boundaries along the top and bottom flanges and the stiffener members on either side of the web (Fig. 11.15). At this stage, the total stress in the web plate is composed of the applied critical shear stress, $\tau_{cr,e}$ when the web plate buckled, and the post-buckled membrane tensile stress ($f_t$) due to the tension field action. The state of stress at this stage is shown in Fig. 11.16 and can be obtained by resolving the stresses in the direction along and perpendicular to the direction to the angle of inclination, $\phi$.

$$f_\phi = \tau_{cr,\,e} \sin 2\phi + f_t \tag{11.5a}$$

$$f_{(\phi+90)} = -\tau_{cr,\,e} \sin 2\phi \tag{11.5b}$$

$$\tau_\phi = -\tau_{cr,\,e} \cos 2\phi \tag{11.5c}$$

The pull exerted by the tensile membrane stresses in the web on the flexible flanges will cause the flanges to bend inwards.

On further increase of loads, the tensile membrane stress in the web plate applies greater pull on the flanges. The resultant of the plate buckling stress and the tensile membrane stress act on the flanges till the stress in the web reaches the yield value of the web. This value of stress at yield of the web can be determined by the von Misses yield criterion.

$$f_\phi^2 + f_{(\phi+90)}^2 - f_\phi \cdot f_{\phi+90} + 3\tau_\phi^2 = f_{yw}^2 \tag{11.6a}$$

where $f_{yw}$ is the tensile yield stress of the web. After substituting values for the parameters from Eqn (11.5) in Eqn (11.6a), we get (Galambos 1998 )

$$f_{yt} = -1.5\,\tau_{cr,\,e} \sin 2\phi + [f_{yw}^2 + (2.25 \sin 2\phi - 3)\ \tau_{cr,\,e}{}^2]^{0.5} \tag{11.6b}$$

Basler (1961) assumed that $f_{yt}$ and $\tau_{cr,\,e}$ are additive (as they would be if $f_{yt}$ acted at 45°) and used the resulting combination of principal stresses in a linear approximation of the von Mises yield criterion and obtained the following equation for shear stress

$$\tau_s = \tau_{cr,e} + 0.5 f_{yw} (1 - \tau_{cr,e}/\tau_{yw}) \sin \phi \tag{11.7a}$$

He also assumed that inelastic buckling will occur if $\tau_{cr,\,e}$ exceeds $0.8\,\tau_{yw}$ and defined a inelastic buckling stress $\tau_{cr,\,i}$ as

$$\tau_{cr,\,i} = (0.8\,\tau_{cr,e}\,\tau_{yw})^{0.5} \quad \text{for} \quad 0.8\,\tau_{yw} \le \tau_{cr,e} \le 1.25\,\tau_{yw} \tag{11.7b}$$

This value is to be substituted for $\tau_{cr,\,e}$ in Eqn (11.7a). Basler's formula overestimates the shear strength of a plate girder, whose flanges are incapable of supporting lateral load from the tension field. It is because, it gives the shear strength for a complete tension field instead of the limited band as observed in the tests (Galambos 1998).

Equation (11.6a) can be represented in a non-dimensionalied form as follows

$$\frac{f_{yt}}{f_{yw}} = \sqrt{1 + \left(\frac{\tau_{cr,e}}{\tau_{yw}}\right)^2 (0.75 \sin^2 2\phi - 1)} - \frac{\sqrt{3}}{2} \cdot \frac{\tau_{cr,e}}{\tau_{yw}} \sin 2\phi \tag{11.7c}$$

where $\tau_{cr,\,e}$ is the elastic critical shear stress and $\tau_{yw} = f_{yw}/\sqrt{3}$.

Porter et al. (1975) suggested a tension field consisting of a single band. It is a refinement of the model suggested earlier by Rockey and Skaloud (1971). Porter et al. (1975) and Rockey et al. (1978) have shown, both experimentally and theoretically, that plate girders fail in shear when a section of the web plate yields and plastic hinges form in the flanges, thereby permitting a shear mechanism to occur (see Fig. 11.17). Let the plastic moment capacity of the flange be denoted by $M_{pf}$. Rockey et al. (1978) have proved that the total shear load at failure $V_s$ can be obtained from the following equation.

$$V_s = \tau_{cr,\,e}\, d t_w + f_t t_w \sin^2 \phi\, (d \cot \phi - c + s_t) + 4 \frac{M_{pf}}{s_t} \tag{11.8}$$

where $M_{pf}$ is the plastic moment capacity of the flange $= f_{yf}\, b_f\, t_f^2/4$, $s_t$ is the distance between adjacent plastic hinges, and $f_t$ is the tension field strength.

It can be observed that Eqn (11.8) has been arrived at by the summation of three components, namely, the web buckling strength, post-buckling membrane yield strength of the web plate, and the plastic moment capacity of the flanges. A non-dimensionalised form of the equation can be obtained using the yield strength of the web plate $V_{yw}$ given by $f_{yw} d t_w$.

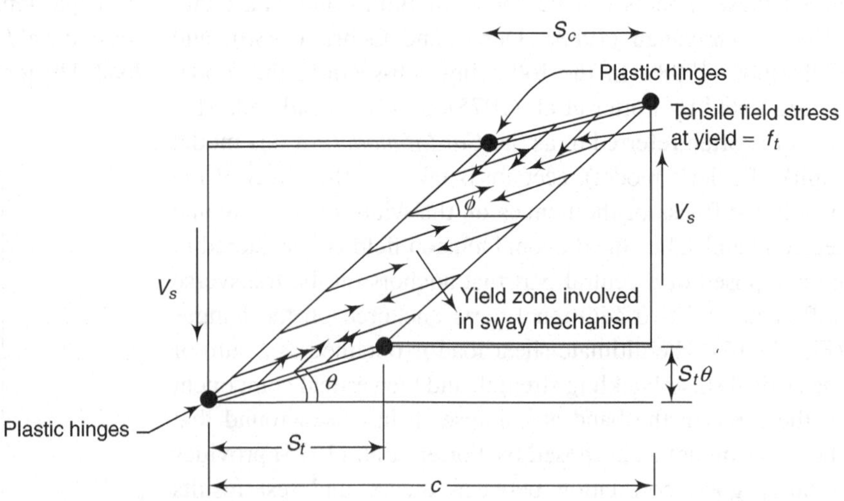


**FIG. 11.17** Collapse of the panel

The position of the plastic hinges in the flanges, $s_t$ or $s_c$, may be determined by considering the equilibrium of the flange and a free-body of a portion of web as (Rockey et al. 1973)

$$s_t = \frac{2}{\sin\phi}\sqrt{\frac{M_{\text{pf}}}{f_1 t_w}} \tag{11.9}$$

Substituting the value of $s_t$ in Eqn (11.8) and dividing throughout by $V_{\text{yw}}$, we get,

$$\frac{V_s}{V_{\text{yw}}} = \frac{\tau_{\text{cr},e}}{\tau_{\text{yw}}} + \sqrt{3}\sin^2\phi\left(\cot\phi - \frac{c}{d}\right)\frac{f_1}{f_{\text{yw}}} \tag{11.10a}$$

$$+ 4\sqrt{3}\sin\phi\sqrt{\frac{f_t}{f_{\text{yw}}}\frac{M_{\text{pf}}}{d^2 t_w f_{\text{yw}}}}$$

This equation can be solved if the value of $\phi$ is known. By a series of parametric studies, Evans (1983) has given an approximate value of $\phi$ as

$$\phi \cong \frac{2}{3}\tan^{-1}\left(\frac{d}{c}\right) \tag{11.10b}$$

It is of interest to note here that for the flanges to be able to form plastic hinges, they need to be classified as plastic. The method of analysis and the equation for the total shear load cannot be applied if the section classification is compact, semi-compact, or slender. We can study several cases of behaviour depending upon the rigidity offered by the flanges and the webs.

If the flanges of the plate girder are weak, the plastic moment capacity of the flanges would be small and in this case, the tension field is supported entirely by the vertical stiffeners on the web(Fig. 11.18) and the flanges do not support the tension field. Hence, we get

$$\frac{V_s}{V_{\text{yw}}} = \frac{\tau_{\text{cr},e}}{\tau_{\text{yw}}} + \sqrt{3}\sin^2\phi\left(\cot\phi - \frac{c}{d}\right)\frac{f_1}{f_{\text{yw}}} \tag{11.11}$$

On the other hand, if the flanges are very strong, it is observed that the distance between the plastic hinges $s_t$ increases. For the case $s_t = c$, the hinges will form at the four corners of the panel, forming a 'picture frame' type failure mechanism (Fig. 11.19). In this case, the tension field angle $\phi = 45°$ and the failure shear is given by (Narayanan 1983).

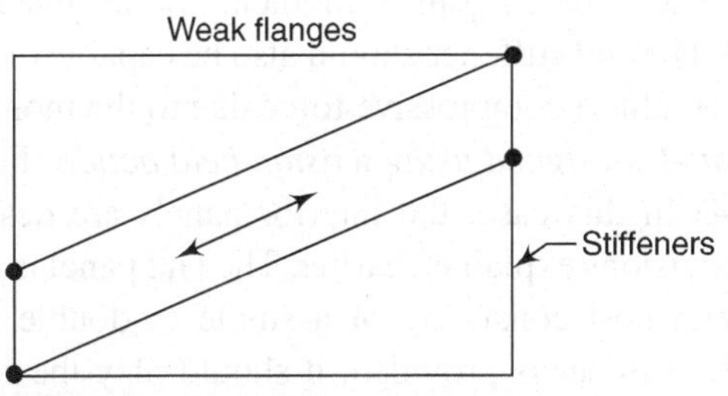

FIG. 11.18 Tension field supported fully by stiffeners

$$\frac{V_s}{V_{\text{yw}}} = \frac{1}{4}\frac{\tau_{\text{cr},e}}{\tau_{\text{yw}}} + \frac{\sqrt{3}}{2}\sqrt{\left(1 - \frac{1}{4}\left(\frac{\tau_{\text{cr},e}}{\tau_{\text{yw}}}\right)^2\right)} + 4\sqrt{3}\frac{d}{c}\frac{M_{\text{pf}}}{d^2 t_w f_{\text{yw}}} \tag{11.12}$$

When the plate girder web is very thick, yielding will take place before any buckling can occur, with $f_{\text{cr}} = f_{\text{yw}}$. The failure shear in this case is given by (Narayanan 1983)

$$\frac{V_s}{V_{\text{yw}}} = 1 + 4\sqrt{3}\left(\frac{d}{c}\right) - \frac{M_{\text{pf}}}{d^2 t_w f_{\text{yw}}} \tag{11.13}$$

When the plate girder web is very slender, the elastic buckling stress is very small and the contribution of tension field action will be significant. In this case, the ultimate shear can be obtained from the general equation [Eqn (11.10)] derived earlier. Note that the method of design including tension field action cannot be applied to gantry girders.

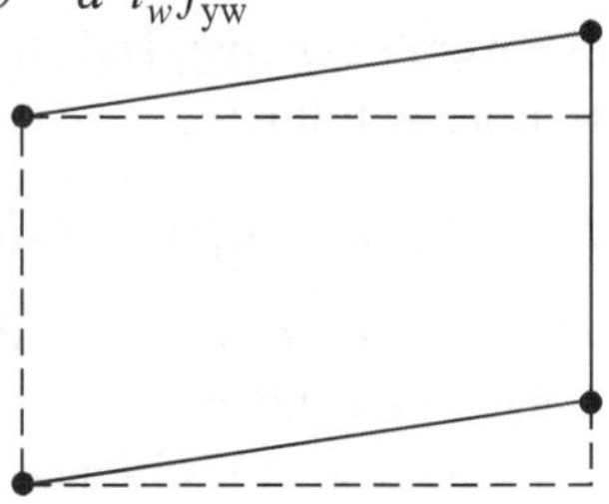
FIG. 11.19 Picture frame mechanism of strong flange

### 11.5.3 Provisions of IS 800 : 2007

The tension field method may be used for girders having transverse stiffeners at the supports and at intermediate points. In the case of end panels, the end posts provide anchorage for the tension fields. It is also necessary that $c/d \geq 1.0$, where $c$ and $d$ are the spacing of transverse stiffeners and the depth of the web, respectively.

In the tension field method, the nominal shear resistance $V_n$ is obtained from $V_n = V_{\text{tf}}$ with [clause 8.4.2.2(b) of IS 800]

$$V_{\text{tf}} = [A_v \tau_b + 0.9 w_{\text{tf}} f_v \sin\phi] \leq V_p \tag{11.14}$$

where $\tau_b$ is the buckling strength, as obtained by Eqn (11.3), $A_v$ is the shear area (see Table 10.12 of Chapter 10), $f_v$ is the yield strength of the tension field obtained from

$$f_v = \left[f_{\text{yw}}^2 - 3\tau_b^2 + \psi^2\right]^{0.5} - \psi \tag{11.15}$$

In Eqn (11.15),

$$\psi = 1.5\,\tau_b \sin 2\phi$$

$$\phi = \text{inclination of the tension field} = \tan^{-1}\left(\frac{d}{c}\right)$$

The above equation is valid for $\theta$ (slope of the panel diagonal) only. The value of $\phi$ ranges between $\theta/2$ to $\theta$. Eurocode suggests a conservative value of $\theta/1.5$.

The width of the tension field $w_{\text{tf}}$ is given by

$$w_{tf} = d\cos\phi - (c - s_c - s_t)\sin\phi$$

and $f_{yw}$ is the yield stress of the web, $d$ is the depth of the web, $c$ is the spacing of stiffeners in the web, and $s_c$ and $s_t$ are the anchorage lengths of the tension field along the compression and tension flange respectively, obtained from

$$s = \frac{2}{\sin\phi}\left[\frac{M_{fr}}{f_y t_w}\right]^{0.5} \le c \qquad (11.16)$$

In this equation, $M_{fr}$ is the reduced plastic moment of the respective flange plate (disregarding any edge stiffener), after accounting for the axial force $N_f$ in the flange, due to the overall bending, and any external axial force in the cross section, as follows

$$M_{fr} = 0.25 b_f t_f^2 f_{yf}[1 - \{N_f/(b_f t_f f_{yf}/\gamma_{mo})\}^2] \qquad (11.17)$$

where $b_f$ and $t_f$ are the width and thickness of the relevant flange, respectively.

In the case of a stiffened web of a plate girder, it would be of interest to consider the roles played by the stiffeners in the behaviour of the girders. The code provides various methods of designing the end panel depending upon whether tension field action can be developed in the end panel or not. These methods are explained in the next section.

The British code (BS 5950-1:2000) presents equations similar to Eqns (11.2) and (11.3) for calculating the post-critical shear buckling resistance in Annex H.1 of the code (simple post critical method). For calculating the elastic buckling strength, simplified equations based on Eqn (11.1) are presented in Annex H.1. However, it has to be noted that Eqn (11.3b) and Eqn (11.3c) are given as

$$\tau_b = [1 - 0.625(\lambda_w - 0.8)]\left(f_{yw}/\sqrt{3}\right) \text{ for } 0.8 < \lambda_w < 1.25$$

and

$$\tau_b = 0.9 f_{yw}/\left(\sqrt{3}\lambda_w^2\right) \quad \text{for } \lambda_w \ge 1.25$$

Hence, it allows for a higher value of shear stress for thin webs as shown in Fig. 11.20. Thus, it represents the post-buckling behaviour correctly. Equations (11.3b) and (11.3c) are used for calculating the value of $\tau_b$ in Eqn (11.14).

The American code (ANSI/AISC 360-2005) equations are also similar but are based on research done by Basler (1961) and Basler and Thürlimann (1963).

## 11.5.4 End Panel Design Without Tension Field Action

A tension field cannot normally be fully developed in an end panel. This is evident by looking at the horizontal components of the tension fields shown in Fig. 11.21. The vertical components are resisted by the stiffeners. Tension field in panel

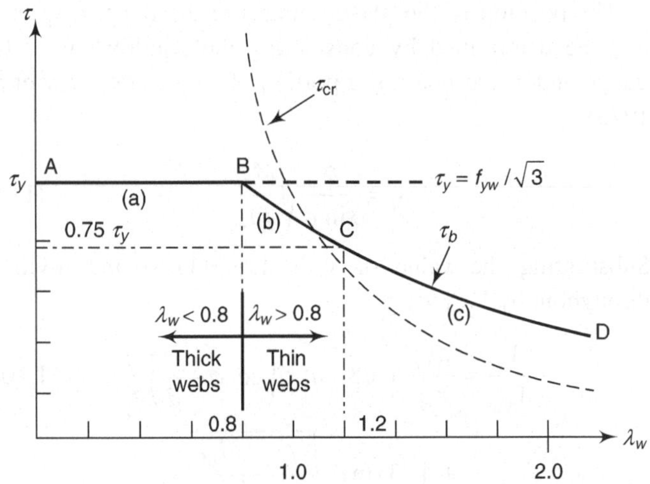


FIG. 11.20 Simple post-critical shear strength as per BS: 5950

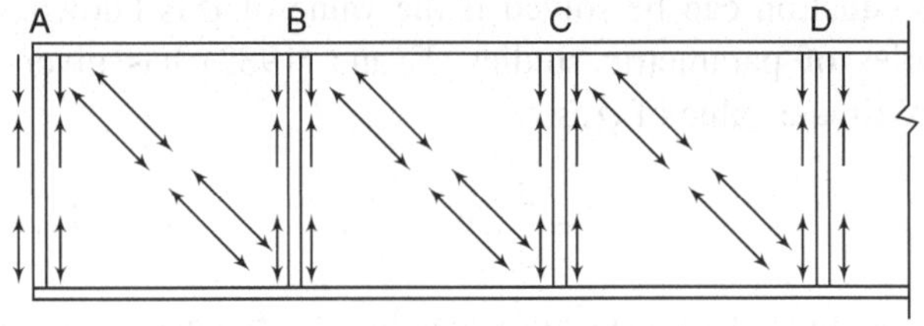


FIG. 11.21 Tension fields in a plate girder

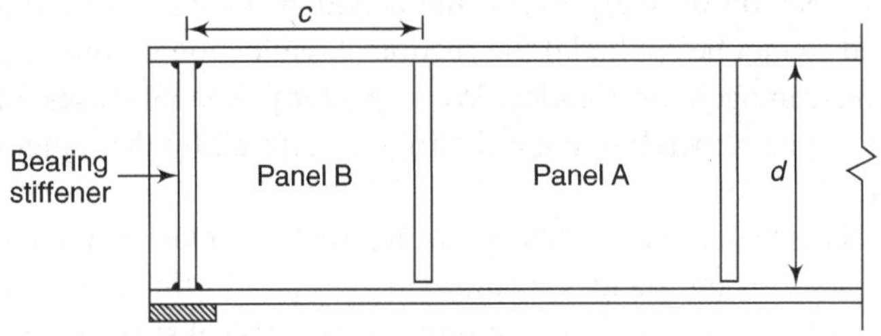


FIG. 11.22 End panel design without tension field action

CD is balanced by the tension field in panel BC. Therefore the interior panels are anchored by the neighboring panels. The end panel AB has no anchorage on the left side. Anchorage can be provided by an end stiffener that is specially designed to resist the bending induced by the tension field.

The end panel (panel *B* in Fig. 11.22) may be designed according to the post-critical method, explained earlier. Additionally, the end panel along with the stiffeners should be checked as a beam spanning between the flanges of the girder to resist a shear force $R_{tf}$ and a moment $M_{tf}$ [as given by Eqn. (11.18a)]. The end stiffener should also be capable of resisting the reaction plus a compressive force due to the moment $M_{tf}$.

*End panels designed using tension field action* (Figs 11.23 and 11.24) In this case, the interior panels are designed as per the provisions explained earlier. The end panel is provided with an end post consisting of a single or double stiffener. If a single stiffener is provided, it should play the roles of a bearing stiffener and end post (Fig. 11.23). These stiffeners need to satisfy the following criteria.

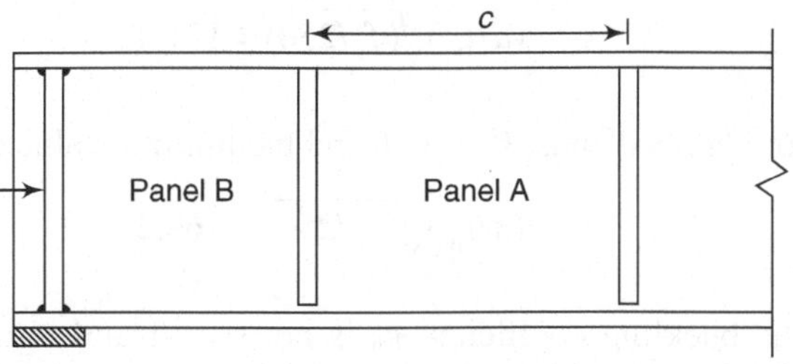


**FIG. 11.23** End panel designed with tension field action (single stiffener)

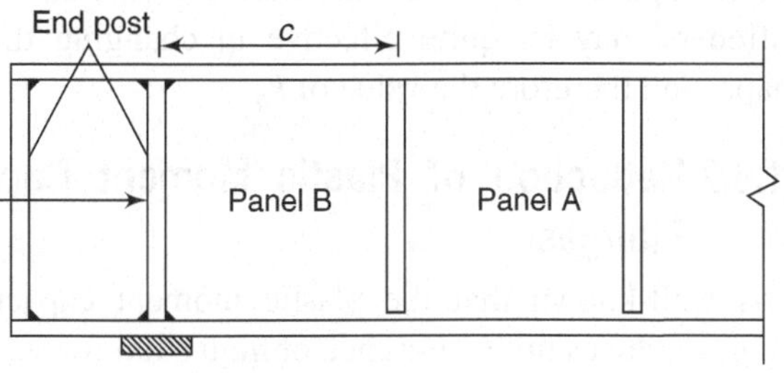


**FIG. 11.24** End panel with double stiffener

In the case of a single stiffener, the top of the end post should be rigidly connected to the flange using full strength welds. The end post should be capable of resisting the reaction plus a moment from the anchor forces equal to $2/3M_{\text{tf}}$, where $M_{\text{tf}}$ is obtained by Eqn (11.18a). The width and thickness of the end post are not to exceed the width and thickness of the flange.

The anchor field forces developed in the panel A creates a resultant longitudinal shear, $R_{\text{tf}}$, plus a moment $M_{\text{tf}}$ on the panel B. They are evaluated as follows (clause 8.5.3 of IS 800).

$$R_{\text{tf}} = \frac{H_q}{2} \quad \text{and} \quad M_{\text{tf}} = \frac{H_q d}{10} \qquad (11.18\text{a})$$

where

$$H_q = 1.25V_{\text{dp}}\left(1 - \frac{V_{\text{cr}}}{V_{\text{dp}}}\right)^{1/2} \qquad (11.18\text{b})$$

$$V_{\text{dp}} = \frac{dt_w f_y}{\sqrt{3}} \qquad (11.18\text{c})$$

If the actual factored shear force $V$ in the panel designed using tension field approach is less than the nominal shear strength $V_{\text{nt}}$, as determined earlier, then the values of $H_q$ may be reduced by the ratio

$$\frac{V - V_{\text{cr}}}{V_{\text{tf}} - V_{\text{cr}}} \qquad (11.18\text{d})$$

where $d$ is the web depth, $V_{\text{tf}}$ is the the basic shear strength for the panel utilizing tension field action [see Eqn (11.14)], and $V_{\text{cr}}$ is the critical shear strength for the panel designed without utilizing tension field action [see Eqn (11.2b)].

Alternatively, a double stiffener can be provided for the end panel as shown in Fig. 11.24. Panels A and B are designed using tension field action. The bearing stiffener is designed for compressive stresses due to the external reaction. In this case, the two stiffeners and the portion of the web projecting beyond the end support form a rigid end post to provide necessary anchorage for the tension field in the end panel. Adequate space must be available to allow the girder to project beyond its support. The end post needs to act like a beam spanning between the flanges of the girder that should resist the shear force $R_{\text{tf}}$ and bending moment $M_{\text{tf}}$ due to anchor forces (Martin & Purkiss 1992).

## 11.6 WEBS SUBJECTED TO COMBINED BENDING AND SHEAR

Till now, the focus has been on webs subjected to predominant shear. When a plate girder experiences combined bending moment and shear, its capacity is determined by the interaction between the effects of the bending moment and shear. The combination of bending and shear makes the stress conditions in the girder web more complex. Firstly, the stresses from the bending moment will combine with shear stresses to give a lower buckling load. Secondly, in the post-buckling range, the bending stresses will influence the magnitude of the tension field membrane stresses needed for developing yield in the web. Lastly, the axial forces due to bending moment will reduce the plastic moment of resistance of the flanges. The proper evaluation of all the effects is complex, but with few assumptions on the interaction of moment and shear, a simple and effective design procedure can be developed.

A typical interaction diagram between shear capacity using non-dimensionalised $V_{\text{yw}}$ (yield shear of the web) and the bending capacity non-dimensionalised using $M_p$ (plastic moment capacity of the section) is shown in Fig. 11.25. The point A in the ordinate represents the full shear capacity of the section in the absence of bending moment and given by Eqn (11.10). As seen in the figure, shear capacity reduces in the presence of bending moment till the point C. Therefore, in the region AC, as shown in the figure, the failure takes place by

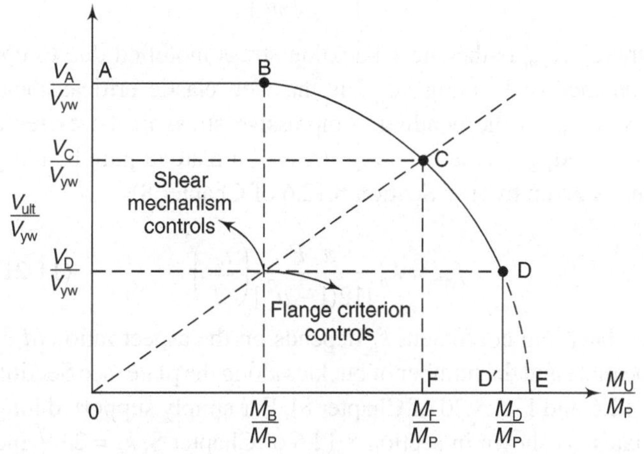


**FIG. 11.25** Interaction between bending and shear effects

the shear mechanism (see Fig. 11.17), explained earlier. In the region beyond point C, the failure may take place in any of the following ways:

1. Yielding of the flange
2. Inward buckling of the compression flange
3. Lateral buckling of the flange

The line OC represents the line of demarcation between shear failure and flexural failure. Normally, the flange yielding will occur as soon as the applied moment equals the plastic moment capacity of the flanges $M_{\text{tf}}$. Here, the contribution of the web is ignored. The expression for the flange moment capacity is given by

$$M_f = b_f t_f f_{\text{yf}}(d + t_f) \tag{11.19}$$

where $b_f$ is the breadth of the flange, $t_f$ is the thickness of the flange, $f_{\text{yf}}$ is the design stress of the flange, and $d$ is the depth of the web.

In Fig. 11.25, the point when the flange yield occurs, is denoted by F, the horizontal coordinate of the point C. Further, in the region ABC, the co-existence of the bending moment with shear leads to the following additional considerations:

1. Reduced buckling stress of the web in the presence of bending stresses
2. Reduced plastic moment capacity of the flanges due to the axial stresses, caused by the bending moment
3. The consequence of bending stresses on the web yielding capacity

These influences are discussed in the following sections.

### 11.6.1 Modification of Web Buckling Stress

The modification of web buckling stress in the presence of bending stresses is obtained by the following interaction equation (Teaching resources 2000; Gaylord et al. 1992):

$$\left[\frac{\tau_{\text{cr},\,m}}{\tau_{\text{cr},\,e}}\right]^2 + \left[\frac{f_{\text{mb}}}{f_{\text{crb}}}\right] = 1 \tag{11.20}$$

where, $\tau_{\text{cr},\,m}$ is the shear buckling stress modified due to co-existence of bending, $\tau_{\text{cr},\,e}$ is the web elastic critical shear stress, $f_{\text{mb}}$ is the bending compressive stress in the extreme fibre, and $f_{\text{crb}}$ is the buckling stress of a plate in pure bending and is given by (see Section 8.12.6 of Chapter 8)

$$f_{\text{crb}} = k_b \frac{\pi^2 E}{[12(1-\mu^2)]}\left(\frac{t_w}{d}\right)^3 \tag{11.21}$$

the buckling coefficient $k_b$ depends on the aspect ratio $c/d$ of the plate and the number of buckles along the plate (see Section 8.12.6 and Fig. 8.20 of Chapter 8). For simply supported long plates, as shown in section 8.12.6 of Chapter 8, $k_b = 23.9$ and the limiting value of $d/t_w$ is

$$(d/t_w)\sqrt{(f_y/250)} = 131.4$$

For clamped ends, $k_b = 39.6$ and the limiting value of $d/t_w$ is

$$(d/t_w)\sqrt{(f_y/250)} = 169.2$$

The buckling coefficient $k_b$ is not significantly greater than 23.9 except when the buckle length is reduced below $2d/3$. For this reason, transverse stiffeners are ineffective unless more closely spaced than $2d/3$. On the other hand, longitudinal stiffeners may be quite effective in changing the buckled shape, and therefore the value of $k_b$.

### 11.6.2 Reduction of Plastic Moment Capacity of Flanges

It is well-known that the plastic moment capacity of the flanges reduces in the presence of high axial forces due to the bending moments. The reduced plastic moment capacity may be evaluated as follows (Teaching resources, 2000);

$$M'_{\text{pf}} = M_{\text{pf}}\left[1 - \frac{f_f}{f_{\text{yt}}}\right]^2 \tag{11.22}$$

where $f_f$ is the average axial stress in the flanges between the hinges and $f_{\text{yt}}$ is the capacity of the flanges in yielding.

### 11.6.3 Modification of Membrane Stress for Web Yielding

The tensile membrane stress ($f_{\text{yt}}$) that is added to the critical web shear stress ($\tau_{\text{cr},\,e}$) is modified in the presence of bending stresses. The modified membrane stress is given by (Teaching resources, 2000)

$$f_{\text{ytm}} = -\frac{1}{2}A + \frac{1}{2}\sqrt{\left[A^2 - 4(f_b^2 + 3\tau_{\text{crm}}^2 - f_{\text{yw}}^2\right]} \tag{11.23}$$

where $A = 3\tau_{\text{cr},\,m} \sin 2\phi + f_b\sin^2\phi - 2f_b\cos^2\phi$ and $f_b$ is the bending stress in the web panel.

The value of $f_{\text{ytm}}$ depends on the value of $f_b$ and hence may have to be computed at various locations to get the resultant stress.

### 11.6.4 Webs Subjected to Pure Bending

Having considered the cases of predominant shear and combined bending and shear, we will now see the case of webs subjected to pure bending moment. In the interaction diagram, the region beyond C is the region of high bending moment with failure taking place by bending. In a plate girder with thin web, the web would buckle under compressive bending stress and therefore the compression flange would carry almost all the compressive stresses. Thus the girder cannot attain its full plastic moment of resistance $M_p$ of the cross section (point E in

Fig. 11.25). In the case of laterally supported girder, failure occurs by the inward collapse of the compression flange at an applied moment $M_u$. This moment is approximately equal to the moment required to produce first yield in the extreme fibres of the compression flange. This moment may be reduced by the influence of web buckling. Cooper (1971) has suggested an empirical formula to arrive at the ultimate moment capacity as follows:

$$\frac{M_u}{M_y} = 1 - 0.0005\frac{A_w}{A_f}\left[\frac{d}{t} - 5.7\sqrt{\frac{E}{f_{yf}}}\right] \geq \frac{M_f}{M_y} \quad (11.24)$$

where $M_y$ is the bending moment required to produce yield in the extreme fibre of the flange neglecting web buckling and $M_u$ is the moment required to produce yield in the extreme fibre of the flange. At point D, the stresses in the web will be below yield value. The ordinate of point D in the curve is given by

$$\frac{V_D}{V_C} = \sqrt{\frac{M_p - M_u}{M_{pw}}} \quad (11.25)$$

where $M_p$ is the full plastic moment capacity of the cross section and $M_{pw}$ is the plastic moment resistance of the web alone and is given by

$$M_{pw} = 0.25 t_w d^2 f_{yw} \quad (11.26)$$

Knowing all the vertical and horizontal coordinates of the critical points on the interaction curve, the complete interaction diagram can now be constructed.

### 11.6.5 Behaviour of Transverse Web Stiffeners Under Extreme Loads

As observed from the previous sections, the tension field developed in the web is supported by the flanges and the transverse web stiffeners. Thus, the web stiffeners play a very critical role in achieving the ultimate capacity of the girder. It increases the web buckling stresses, supports the tension field in the web in the post-buckling stages, and finally prevents the flanges from being pulled towards each other. Prior to buckling, stiffeners on initially plane webs are not subjected to any axial loading. But after the plate buckles the axial loads applied to the transverse stiffeners steadily increase, as the web plate develops a membrane tension field (Rockey et al. 1981). Thus, the stiffeners must have sufficient rigidity to remain straight and to contain the web buckling to individual web panels.

Now consider a typical web stiffener CD flanked on the sides by adjacent web panels, both of which have attained shear failure mechanism (see Fig. 11.26). For an intermediate stiffener, this loading represents a critical case of loading. In this figure to identify the loads which act on the flange

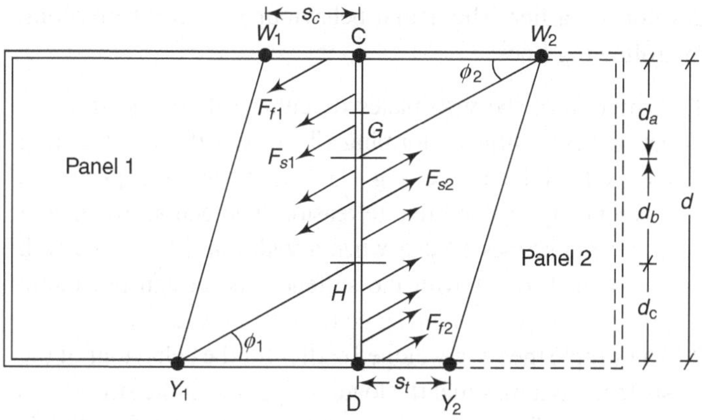


**FIG. 11.26** Force imposed on transverse stiffeners by tension field

and stiffeners, subscript $f$ is used to denote flange force and subscript $s$ to identify the forces acting on the stiffener.

The resultant load acting on the portion $W_1C$ on the left top side of the stiffener $F_{f1}$ is inclined at angle $\phi_1$. Similarly $F_{f2}$ is the resultant load acting on the bottom right side $DY_2$, inclined at an angle $\phi_2$. The vertical components of these two forces, $V_c$ and $V_d$ tend to pull the flanges towards each other. From Fig. 11.26, it is seen that the force $V_c$ applied by the flange $W_1C$ to stiffener at C is given by

$$V_c = F_{f1}\sin\phi_1 = -f_{yt}\, t s_c \sin^2\phi_1 \quad (11.27a)$$

The minus sign indicates that this load is applied in the direction from C to D.

The corresponding load at point D is given by

$$V_d = F_{f2}\sin\phi_2 = +f_{yt}\, t s_t \sin^2\phi_2 \quad (11.27b)$$

Over the range CG, the tension field in panel 1 applies a vertical load $V_1$ as follows (Rockey et al. 1981):

$$V_1 = -f_{yt} t d_a \sin\phi_1 \cos\phi_1 \quad (11.27c)$$

Similarly, panel 2 applies a vertical force $V_2$, which is given by

$$V_2 = -f_{yt} t d_c \sin\phi_2 \cos\phi_2 \quad (11.27d)$$

The horizontal components of the tension field loads act directly on the stiffener. They can be divided into three regions: (a) The top portion CG pulled to the left by the tension field in the panel on the left side of the stiffener, (b) the bottom portion HD pulled to the right by the tension field in panel on the right hand side of the stiffener. Since these two forces are more or less equal, the mid portion of the stiffener (GH) experiences no load due to the tension field action (Rockey et al. 1981).

The forces on the stiffeners can be calculated from the preceding relationships, after the ultimate shear loading and geometry of failure mechanism have been determined. However, from experimental studies, it has been observed that the actual behaviour of stiffeners is different from the model developed earlier. The actual behaviour is somewhere along the following lines:

1. A portion of the web plates on either side of the stiffener acts with the stiffener, forming a T or a cruciform in resisting the vertical load (see Fig. 11.27), though the plate has theoretically buckled due to tension field action. Rockey et al. (1981) suggested that when a width of $40t_w$ of the web is assumed to act with the stiffener, the design procedure proposed by them are in agreement with the test data.
2. The simple theory developed earlier for the behaviour of the stiffener assumes that the loading applied on the stiffener is concentric. But, in practical cases, eccentricity of loading is unavoidable. This gives rise to additional bending moment. Further, any imperfections in the stiffener also result in bending moments.

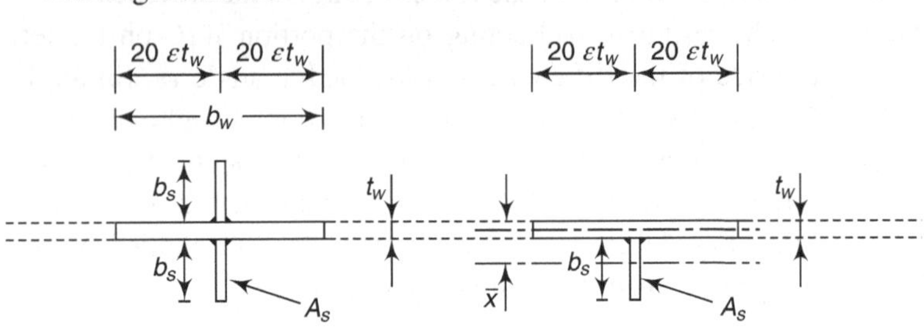


**FIG. 11.27** Effective cross section of stiffeners

Horne (1979) has suggested an approximate expression to determine the capacity of the stiffener consiering the above loadings.

$$\frac{M}{M_{ps}} = 1.0 - \frac{f_{ys}t_s(b_s - \bar{x} + 0.5t_w)^2}{M_{ps}}\frac{P}{P_s} \qquad (11.28a)$$

where

$$\bar{x} = (b_s t_s - b_w t_w)/2t_s + (t_w/2) \qquad (11.28b)$$

and $M_{ps}$ is the plastic moment capacity in the absence of axial loads, $M$ is the applied moment, $P$ is the applied load, $P_s$ is full axial yield load (squash load), $b_s$ is width of stiffener, $t_s$ is thickness of the stiffener, $f_{ys}$ is design stress of the stiffener, $t_w$ is the thickness of the web, and $b_w$ is the width of the web (may be assumed as $40t_w$).

A girder subjected to axial loads and accompanying moments which are less than the limit specified by Eqn (11.28) can be designed using the expressions given in Eqn (11.27). Design codes simplify this procedure for enabling quick sizing of the stiffeners, by assuming that the compressive force in the stiffener is constant and considering the whole depth of the web as a compression member.

### 11.6.6 Patch Loading and Web Crippling

Plate girders often support loads on the top flange which produce compression on the longitudinal edge of the web. The term *patch loading* or edge loading is used when such loads are distributed over relatively small lengths, in which case they can be regarded as concentrated loads. Transverse stiffeners are required at positions of large concentrated loads when the web cannot support these loads. The use of bearing stiffeners are not recommended in gantry girders, which are subject to moving loads, and are not realistic for bridge girders erected by launching (Maquoi 1992). In such cases, a thicker flange and/or thicker web plate may be used.

When a patch load is applied, it distributes through the bearing pad, the flange and either the web to flange fillet weld (in the case of plate girders) or the root fillet (in the case of rolled beams). For plate girders with slender webs, the ultimate patch load resistance is usually governed by web instability. Collapse occurs either by *web buckling* or web crippling (see Fig. 10.32 of Chapter 10).Web crippling consists of local buckling in the zone close to the load; it results in appreciable displacements (transverse to the web plane) in a small area of the web, without significant deformation of the remainder of the cross section. Web buckling occurs over the full depth with pronounced cross section distortion.

The change to web crippling from buckling occurs at web slenderness $d/t_w$ in the range of 60–80 (Maquoi 1991). Web crippling governs for very slender webs for short load lengths, whereas webs of medium slenderness usually collapse by web buckling. Patch load is very complex to model experimentally or analytically. A comprehensive review of experimental and theoretical studies has been presented by Roberts (see Narayanan 1983; Elgaaly 1983). A diversity of design formulae have been developed and included in different codes and standards. For example the AISC Code has adopted the equation suggested by Roberts (1981) for checking web crippling [see Section 14.8.2, Eqn (14.66)]. Roberts and Newark (1997) suggested the following equations for routine practical applications, based on a plastic collapse mechanism which results in the formation of plastic yield lines in the web plate and plastic hinges in the flange. Thus, for interior patch loading

$$F_w = \{1.1\,t_w^2\,(Ef_w)^{0.5}(t_f/t_w)^{0.25}[1 + (C_e t_w)/(d_w t_f)]\}(1/F) \qquad (11.29)$$

and for end patch loading

$$F_w = (0.8/F)\{(3.3/8)t_w^2(Ef_w)^{0.5}\,(t_f/t_w)^{0.25}[1 + (8C_{ee}t_w)/(3d_w t_f)]\} \qquad (11.30)$$

where $F_w$ is the theoretical ultimate web crippling load (in N), $t_w$ is the thickness of the web (in mm),$t_f$ is the thickness of rectangular flange (in mm), $E$ is Young's modulus (in N/mm$^2$),$f_w$ is the yield stress of the web (in N/mm$^2$), $C_e$ is the effective width of the interior patch load (in mm), $C_{ee}$ is the effective width of

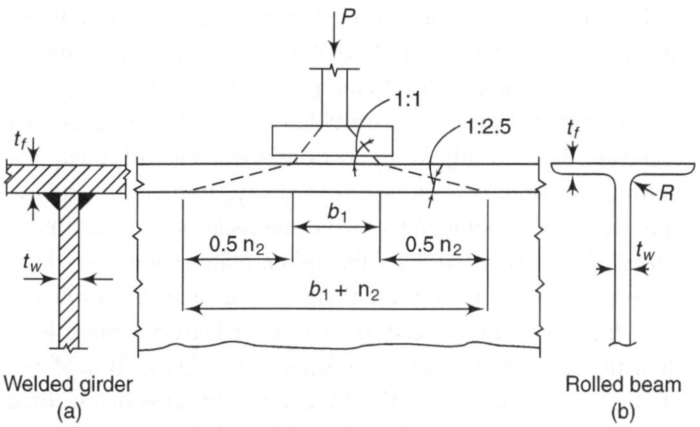


**FIG. 11.28** Dispersion of a concentrated load

end patch load (in mm), $d_w$ is the depth of the web (in mm), and $F$ is the factor of safety = 1.45 for full mechanism solution.

Roberts and Newark (1997) have also suggested equations for the case of web yielding. As per the Indian code, the length of the web relevant to the yield check is shown in Fig. 11.28.

Thus, the length of the web for yield check consists of

(a) The distance $b_1$ over which the load is applied
(b) The distance $n_2$, the dispersion of applied load through the flange thickness $t_f$, and the root fillet radius $R$

The distance $b_1$ is called the *stiff bearing length.* It is determined by assuming a dispersion of the load at a slope 1:1 through any solid steel material (such as bearing plate, flange plate, etc.). The dispersion length results from the dispersion of the load through the depth $t_f + R$ (in the case of rolled sections) or $t_f$ (in the case of plate girders) and is empirically assumed to occur at a slope of 1:2.5. Thus the ultimate patch load for a stocky web is given by

$$F_w = (b_1 + n_2)\, t_w f_{yw}/\gamma_{m0} \tag{11.31}$$

where $\gamma_{m0}$ is the partial safety factor for the material = 1.1.

An empirical value for $n_2$ is $5t_f$ or $5(t_f + R)$ for rolled sections. When the patch load is applied near the end of a girder, appropriate changes should be made to account for the limited dispersion; a conservative approach is to use half the value of $n_2$.

A more refined value for $n_2$ based on a four-hinge plastic mechanism is (Maquoi 1991)

$$n_2 = 2t_f(b_f/t_w)^{0.5}\,(f_{yf}/f_{yw})^{0.5} \tag{11.32}$$

It has to be noted that the equations proposed by Roberts and Newark properly include the effect of all the parameters that have influence on the capacity and are found to correlate closely with experimental results. Because web crippling is a local failure mode, the interaction between bending stresses and shear stresses is generally ignored.

Transverse stiffeners in the vicinity of the patch loaded zone increase the patch load resistance of girders with thick flanges. No such increases are found in girders with very thin flanges, because of the very local effect of the load.

Benefits occur only when the aspect ratio of the web panel ($c/d$) is lower than 1.5, becoming significant for aspect ratios smaller than unity. The provision of relatively small longitudinal stiffeners, placed within a distance of $0.25d$ (or 50–80 $t_w$) from the top flange, were found to improve the ultimate web crippling load (Maquoi 1992).

Eccentricities in loading with respect to the plane of the web are unavoidable, and it was found that there is a reduction in the web capacity due to the presence of an eccentricity; for example, in one case, an eccentricity of 12.5 mm reduced the web ultimate capacity to about half its capacity under in-plane load. However, the effect of load eccentricity in reducing the ultimate capacity decreases as the ratio of the flange-to-web thickness increases. Eccentricities can also arise due to moments applied to the top flange in addition to vertical load. An example would be a beam resting on the top flange of another beam and the two flanges are welded together. Rotation of the supported beam will impose a twisting moment in the flange of the supporting girder and bending of its web, which will reduce its crippling load. Currently, the codes do not address the effect of eccentricity on the reduction of the web crippling loads. More information on this topic is provided by Elgaaly (1997) and Johansson et al. (2007).

## 11.7 DESIGN OF PLATE GIRDERS USING IS 800 : 2007 PROVISIONS

In the design of plate girders, required sections are proportioned for various elements. In general, bending moments are assumed to be carried by the flanges and the shear by the web. The depth of the section is chosen in such a way that the flanges are able to carry the design bending moment.

For arriving at a good plate girder design, IS 800 : 2007 has made several provisions with respect to its sectional requirements. Thus, the code specifies minimum web thickness requirement from a serviceability point of view for different situations with regard to provision of stiffeners (see Section 11.4.1). Similarly, the code specifies web thickness requirement in order to avoid buckling of the compression flange into the web as follows (clause 8.6.1.2 of IS 800):

1. When transverse stiffeners are not provided

$$\frac{d}{t_w} \le 345\varepsilon_f$$

2. When transverse stiffeners are provided and
   (a) When $c \ge 1.5d$:

$$\frac{d}{t_w} \le 345\varepsilon_f$$

(b) When $c < 1.5d$:

$$\frac{d}{t_w} \leq 345\varepsilon_f$$

where $d$ is the depth of the web, $t_w$ is the thickness of the web, $c$ is the spacing of the transverse stiffener, $\varepsilon_f = \sqrt{250/f_{yf}}$, and $f_{yf}$ is the yield stress of the compression flange.

It should be noted that when these two sets of criteria are used, the serviceability criteria will be more critical for Fe 410 grade steel and the buckling criteria becomes more important for higher grade steels. This is because buckling strengths are dependent on Young's modulus, which is the same for all grades of steel.

### 11.7.1 Sectional Properties

In the design of a plate girder, the gross cross-sectional area may not be available for resistance in several cases due to various reasons. In some situations, the plates may be too wide and may be classified as slender. In such cases, the excess width of plates has to be deducted to get the effective area (see Examples 8.23 of Chapter 8 and 11.1). For some reasons, if open holes are to be provided in the plane perpendicular to the direction of stress, they also have to be deducted for arriving at the effective area. In a similar manner, the effective area of cross section in the tensile flange will have to be determined after deducting the area for holes. This aspect has been elaborately treated in Chapter 10. The calculation of the effective sectional area for shear is also dealt in detail in Chapter 10.

### 11.7.2 Flanges

As a rule, it is a good practice in the case of riveted or bolted constructions to provide flange angles that would form a very large part of the flange, preferably not less than one third. It is also found that keeping the requirement of flange plate to be a minimum is a common practice. In many instances, the flange angles along with a very well-machined web edge in line with the flange angles form the flange. Nevertheless, invariably at least a single plate is provided along the full length of the girder. When more than one plate is used in the flange, they are to be stitched together by tacking rivets or other means to act as a single piece. Enough number of flange plates is to be provided to resist the force/moment at any section of the girder. At regions where bending moment is less than that at the critical sections, the number of plates can be reduced or curtailed. While curtailing the plates, the curtailed plates have to be extended slightly beyond the theoretical cut-off point and the extended plate has to be sufficiently connected to develop the required resistance at the cut-off point. It is also essential that the outstand of the flange plates in a plate girder satisfies the minimum section classification requirement, as described in Chapter 10.

In instances of a plate girder design where the sections are available only in small lengths or where sectional requirements are changing along the span, provision of flange splices are permitted. Splices are provided by a plate element that connects the two adjoining sections together with adequate bolting, riveting, or welding. It is always prudent to provide splicing only at points away from the locations of maximum stress. The cross section of the splice plate is normally kept 5% more than the area of the flange elements that are spliced. Also the connections need to develop at least 5% more load than the load at the spliced portion or at least 50% of the effective strength of the spliced section. In the case of a welded construction, the flange plate should be welded together using full penetration butt weld that is capable of developing full strength of the plates connected.

In the fabrication of the plate girder, the web and flange plate are joined together with adequate rivets, bolts, or welds such that they can transmit horizontal shear arising out of the bending moments that develop in the girder and also due to any applied vertical load on the flange. As mentioned earlier, when the plate girders are exposed, viz., they facilitate inspection, the flange plates need not extend to the full length of the girder unless warranted by design requirements. Where flange plates are not provided, the web should be made flush with the flange angles or protrude a little below the angles both at the top and bottom flanges. When fillet welding is adopted for the flange-web connection, the web and flange plates are brought as close as possible to each other before welding and the maximum gap between them is limited to 1 mm. The effective sectional area for the web of a plate girder is arrived by considering the full depth of web plate and its uniform thickness. If the thickness of the web is varying in the depth of the section due to provision of tongue plates or in cases where the web contributes 25% or more of its depth to the flange area, the above provision is not applicable. In such cases the exact area has to be calculated on a case-to-case basis.

Similar to splices in flanges, webs also may need splicing or cut outs may become necessary in webs for routing of services. Such web splices and cut outs are best provided at locations away from high shear and large concentrated loads. Splices in the web are designed to withstand the shears and moments at the spliced section. Splice plates are often provided on both sides of the web in equal thickness and they are connected together. If welding is resorted to, complete full penetration butt welding is recommended. In locations where extra plate is added to the web, to increase its strength, they may be provided equally on both sides of the web plate. These plates also extend beyond the length of the requirement as arrived by calculations. The amount of shear force that these reinforcing plates can carry is restricted to the extent of horizontal shear that they transmit to the flanges by way of the fastening provided.

### 11.7.3 Stiffeners

In a plate girder, the web is often made very thin to derive maximum economy in weight. The webs, when they are inadequate to carry the load, are made strong and stable by the provision of a wide variety of stiffeners. Sometimes double stiffeners are adopted near the bearing [see Fig. 11.24] and in such cases the over hangs beyond the support should be limited to 1/8 of the depth of the girder. The stiffeners are classified based on the role they play in strengthening the web and they are mentioned as follows. It may be noted here that the stiffener provided for one function can also provide for other functions, if the design is adequate with the requirement of the other functions.

1. The function of a *bearing stiffener* is to preclude any crushing of the web at locations of heavy concentrated loads. Thus, bearing stiffeners transfer heavy reactions or concentrated loads to the full depth of the web. They are placed in pairs on the web of plate girders at unframed girder ends and where required for concentrated loads. They should fit tightly against the flanges being loaded and extend out towards the edges of flange plates as far as possible. If the load normal to the girder flange is tensile, the stiffeners must be welded to the loaded flange. If the load is compressive, it is necessary to obtain a snug fit. To accomplish this goal, the stiffeners may be welded to the flange or the outstanding leg of the stiffener may be milled.
2. A *load carrying stiffener* prevents local buckling of the web due to any concentrated load.
3. An *intermediate transverse web stiffener* mainly improves the buckling strength of the web due to shear. They continue to remain effective after the web buckles to provide anchorage for the tension field and finally they prevent the flanges from moving towards one another. They are also called as non-bearing stiffeners or stability stiffeners.
4. *Torsional stiffeners* are provided at supports to restrain the girders against torsional effects.
5. Local strengthening of the web under the combination of shear and bending is provided by *diagonal stiffener*.
6. The tensile forces from the flange are transmitted to the web through the *tension stiffener.*
7. A *longitudinal stiffener* increases the buckling resistance of the web.

The same stiffener may perform more than one function as mentioned earlier, and in such cases, their design should comply with the requirement of those functions.

In the design of various stiffeners, it is important to know the effective dimensions of the stiffener that needs to be considered for the design. The outstand of the stiffener from the face of the web is restricted to a value of $20t_q\varepsilon$. If the outstand value is anywhere between $14t_q\varepsilon$ and $20t_q\varepsilon$, a core value of $14t_q\varepsilon$ is chosen as the outstand value. In the above expressions, $t_q$ refers to the thickness of the stiffener plate. Another important parameter of the stiffener is its stiff bearing length $b_1$. It is defined as that length that does not deform appreciably in bending. The stiff bearing length $b_1$ is determined by considering the dispersion of the load at 45° through a solid material and a steel bearing element such as the bearing or flange plates as shown in Fig. 11.29.

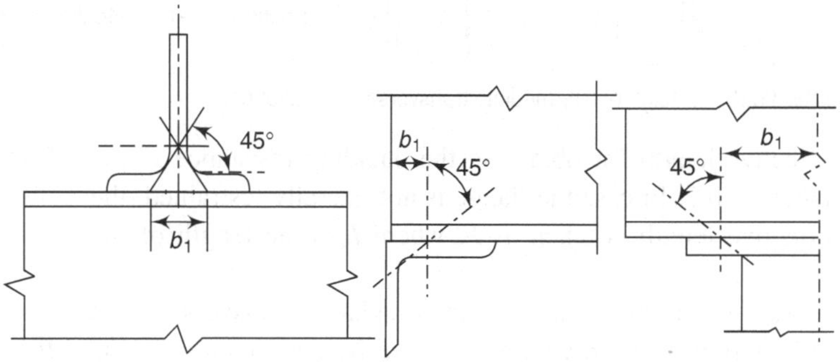

**FIG. 11.29** Stiff bearing length ($b_1$)

The eccentricity of loading with respect to the stiffener centroid is to be considered in the design of a stiffener. The eccentricity arises when the load or reaction is eccentric to the centre line of the web or when the stiffener centroid does not coincide with the centre line of the web. Buckling of the stiffeners is also a matter of serious concern and hence adequate buckling resistance needs to be provided for stiffeners. In order to evaluate the buckling resistance of the stiffener, the stiffener is considered as a strut with its radius of gyration taken about the axis parallel to the web. The effective sectional area of the stiffener is the area of the core section as determined below. The effective width of web acting together with the stiffener is taken as 20 times the web thickness on each side of the centre line of the stiffener, i.e., $40t_w$ (see Fig. 11.30), for interior stiffeners and as $20t_w$ for end stiffeners. (Note that in the American code it is taken as $25t_w$ and $12t_w$ for interior and end stiffeners respectively.) The buckling resistance of the stiffener is obtained on the basis of the design compressive stress, $f_{cd}$ of the strut using buckling curve *c* (see Section 9.7.1 of Chapter 9). In arriving at the buckling resistance the lower value between the design strength of the web material and the stiffener material is to be adopted.

The effective length of a intermediate transverse web stiffener, which is provided mainly to improve the shear buckling resistance of the web $F_{qd}$ is taken as 0.7 times the length of the stiffener. (The American code allows 0.75 times the length of the stiffener.) For other load carrying stiffeners, with the assumptions that the flange, through which the load or reaction is applied, is effectively laterally restrained, the

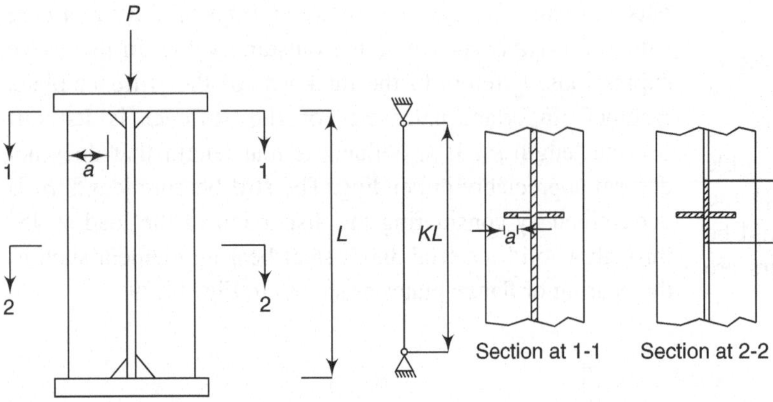


**FIG. 11.30** Design of intermediate transverse web stiffener

effective length for obtaining the buckling resistance $F_{xd}$ is taken as 0.7. In case the flange is not laterally restrained, the effective length is taken as $L$, where $L$ is the length of the stiffener.

As mentioned earlier, intermediate transverse web stiffeners mainly provide buckling resistance to the web due to shear. Sometimes intermediate stiffeners are alternated on each side of the web to gain better economy or they are placed all on one side to improve aesthetics. The spacing of these stiffeners is based on the provisions, as explained earlier depending upon the thickness of web (see Sections 11.4.1 and 11.7). The calculation of the outstand is also explained earlier in this section. The minimum requirement for intermediate transverse stiffener, when they are subjected to external loads or moments is considered in terms of a second moment of area $I_s$ about the centre line of the web (Rockey, et al. 1981, clause 8.7.2.4 of IS 800).

$$\text{if } \frac{c}{d} \geq \sqrt{2}; I_s \geq 0.75 d t_w^3 \tag{11.33a}$$

$$\text{if } \frac{c}{d} \geq \sqrt{2}; I_s \geq \frac{1.5 d^3 t_w^3}{c^2} \tag{11.33b}$$

where $d$ is the depth of the web, $t_w$ is the minimum thickness required for tension field action, and $c$ is the stiffener spacing. Equation (11.33a) is very conservative when compared to experimental results as no allowance is made for the effect of the panel aspect ratio. Equation (11.33a) is obtained by substituting the value of $c = \sqrt{2}\,d$ in Eqn (11.33b). It is of interest to note that the American code (ANSI/AISC 360-10) specifies a minimum area for the transverse stiffeners subjected to tension field action. The experimental study conducted by Lee et al. (2003), showed that this area requirement is conservative and can be eliminated. Their study also showed that the flexural rigidity of stiffeners is more important and that it should be increased several times than that specified in the American code. Earlier AISC codes specified that $I_s \geq (h/50)^4$ and AASHO specified $I_s > 0.344\, h^4\, (t_w/c)^3$

Transverse web stiffeners are also checked for buckling resistance for a stiffener force (clause 8.7.2.5 of IS 800)

$$F_q = \frac{(V - V_{cr})}{\gamma_{m0}} \leq F_{qd} \tag{11.34}$$

where $F_{qd}$ is the design buckling resistance of the intermediate stiffener, $V$ is the factored shear force adjacent to the stiffener, and $V_{cr}$ is the shear buckling resistance of the web panel without considering tension field action [Eqn (11.2b)].

Transverse web stiffeners that are subjected to external load and moments have to satisfy requirements of load carrying stiffeners. They have to also meet the interaction equation

$$\frac{F_q - F_x}{F_{qd}} + \frac{F_x}{F_{xd}} + \frac{M_q}{M_{yq}} \leq 1 \tag{11.35}$$

If $F_q < F_x$, then

$$F_q - F_x = 0$$

where $F_q$ is the stiffener force [in N, see Eqn (11.34)], $F_{qd}$ is the design buckling resistance of intermediate web stiffener about an axis parallel to the web (in N),$F_x$ is the external load or reaction at the stiffener (in N), $F_{xd}$ is the design resistance of load carrying stiffener, buckling about an axis parallel to the web (in N), $M_q$ is the moment at the stiffener due to eccentric load (in N mm), and $M_{yq}$ is the yield moment capacity of the stiffener about an axis parallel to the web (in N mm).

An important aspect of the intermediate transverse stiffener is with respect to their connection with the web. If the stiffeners are not subjected to external load, they may be connected to the web in such a way that the withstand a shear of not less than $t_w^2/5b_s$, where $t_w$ is the web thickness and $b_s$ is the stiffener outstand width.

If the stiffeners are subjected to external loading, the resulting shear to be resisted is the total shear due to loads and shear specified earlier. If the stiffeners are not subjected to external load, they may be provided in such a way as to be clear of the tension flange and the gap provided is $4t_w$ (see Fig. 11.7d). With such a gap, the fabrication problems associated with close fit can be avoided.

### 11.7.4 Design of Load Carrying Stiffeners

As defined previously, these stiffeners are provided at locations where the compressive forces that are applied through the flange by an external load or reactions exceed the buckling resistance $F_{cdw}$ of the web alone. The buckling resistance of the web alone may be calculated by using the effective length and the sectional area of the stiffener. The sectional area of the stiffener is taken as $(b_1 + n_1)t_w$, where $b_1$ is the stiff bearing length and $n_1$ is the dispersion length of the load through the web at 45° to the level of half the depth of the cross section. The buckling strength of the web is calculated about an axis parallel to the web thickness using buckling curve *c*. Although the web can be proportioned to resist any applied concentrated loads, bearing stiffeners are generally provided. If stiffeners are used at each concentrated load, the limit states of web yielding, web crippling, and side sway web buckling need not be checked.

Load carrying stiffeners are checked for their buckling and bearing resistances. For buckling check

$$F_x \leq F_{xd} \tag{11.36}$$

where $F_x$ is the external load or reaction and $F_{xd}$ is the buckling resistance.

In the event that the load carrying stiffeners also act as intermediate transverse web stiffeners, the bearing strength should be checked for the combined loads as given in Eqn (11.35). The bearing strength of the stiffener $F_{bsd}$ is obtained as follows (clause 8.7.5.2 of IS 800):

$$F_{bsd} = \frac{A_q f_{yq}}{0.8\gamma_{m0}} \tag{11.37}$$

where $A_q$ is the area of the stiffener in contact with the flange in mm$^2$ {the actual area *A* is generally less than the full cross-sectional area of the stiffener as the corners of the stiffener are often coped to clear the web-to flange fillet weld (see section 1-1 of Fig. 11.30)} and $f_{yq}$ is the yield stress of the stiffener in N/mm$^2$.

The bearing strength of the web stiffener $F_{bsd}$ as calculated above should be greater than or equal to $F_x$, the load transferred.

### 11.7.5 Design of Bearing Stiffeners

Bearing stiffeners are provided where forces, applied through a flange by loads or reaction, are in excess of the local capacity of the web at its connection to the flange, $F_w$ calculated as follows (clause 8.7.4 of IS 800):

$$F_w = \frac{(b_1 + n_2)t_w f_{yw}}{\gamma_{m0}} \tag{11.38}$$

where $b_1$ is the stiff bearing length, $n_2$ is the length obtained by dispersion through the flange to the web junction at a slope of 1:2.5 to the plane of the flange, $t_w$ is the thickness of the web, and $f_{yw}$ is the yield stress of the web.

Bearing stiffeners are designed for the difference in force between the applied load or reaction and the local capacity of the web calculated in Eqn (11.38). In all cases, if the web and stiffener are of materials with different strength, the lower of the two has to be used for the strength equation. A bearing stiffener should be as wide as the overhang of the flange through which the load is applied. The weld connecting the stiffener to the web should have the capacity to transfer the unbalanced shear force. Conversely, the weld can be designed to carry the entire concentrated load.

### 11.7.6 Design of Diagonal and Tension Stiffeners

As in the case of bearing stiffeners, the diagonal stiffeners are designed for the portion of the combination of applied shear and bearing that exceeds the capacity of the web. Tension stiffeners are designed to carry the portion of the applied load or reaction that exceeds the capacity of the web.

### 11.7.7 Design of Torsional Stiffener

Torsional stiffeners provide torsional restraint at the supports and they should satisfy the following:

1. the local capacity of the web is exceeded as calculated in Eqn (11.38)
2. second moment of area of the stiffener section about the centre line of the web $I_s$ should satisfy the following relation (clause 8.7.9 of IS 800)

$$I_s \geq 0.34\alpha_s D^3 t_{cf} \tag{11.39}$$

where $\alpha_s = 0.006$ for $KL/r_y \leq 50$

$\alpha_s = 0.3/(KL/r_y)$ for $50 < KL/r_y \leq 100$

$\alpha_s = 30/(KL/r_y)^2$ for $KL/r_y > 100$

and *D* is the depth of the beam at support (in mm), $t_{cf}$ is the maximum thickness of the compression flange of the girder (in mm), *KL* is the effective length of the laterally unsupported compression flange of the girder (in mm), and $r_y$ is the radius of gyration about minor axis (in mm).

### 11.7.8 Connection of Load Carrying Stiffeners and Bearing Stiffeners to Web

Stiffeners which are subjected to load or reaction applied through a flange have to be connected to the web adequately. This connection is designed to transmit a force equal to the lesser of

1. the tension capacity of the stiffener or
2. the sum of the forces applied at the two ends of the stiffener when they act in the same direction or the larger of the forces when they act in different directions.

It should also be ensured that the shear stress in the web due to the design force transferred by the stiffener is less than the shear strength of the web. The stiffeners which resist tension should be connected to the flange with continuous welds or non-slip fasteners. Stiffeners resisting compression can either be fitted against the loaded flange or be connected by continuous welds or non-slip fasteners. The stiffener has to be fitted or connected to both flanges if

1. the load is applied directly over the support
2. the stiffener is an end stiffener of a stiffened web
3. the stiffener also acts as torsional stiffener

## 11.8 BEHAVIOUR OF LONGITUDINALLY STIFFENED PLATE GIRDERS

In order to obtain greater economy and efficiency in the design of plate girders, slender webs are often stiffened both longitudinally and transversely. The longitudinal stiffeners are generally located in the compression zone of the girder. The main function of the longitudinal stiffeners is to increase the buckling resistance of the web. The longitudinal stiffener remains straight, thereby, subdividing the web and limiting the web buckling to smaller web panels (see Fig. 11.31). In the past it was usually thought that the resulting increase in the ultimate strength could be significant, but recent studies have shown that this is not always the case, as the additional cost of welding the longitudinal stiffeners invariably offsets any economy resulting in their use. Longitudinal stiffeners, thus, are not as effective as the transverse ones, and are frequently used in bridge girders, because many designers feel that they are more attractive.

**FIG. 11.31** Tension field in a plate girder with longitudinal stiffener (ESDEP 2006)

Studies have shown that a longitudinally stiffened plate girder subjected predominantly to shear would develop a collapse mechanism, similar to the one described previously, provided the stiffeners remained rigid up to failure (Narayanan 1983; Evans & Tang 1986). Once a tension field develops, it extends over the complete depth of the girder. Thus, the equations established previously are valid, keeping in mind that $\tau_{cr,e}$ values are enhanced due to the smaller panel dimensions.

A longitudinal stiffener is most efficient when it is placed in the compression region at about $d/5$ mm from the compression edge. This is close to the position of the crests in the buckles of an unstiffened plate. Such stiffeners may increase the buckling coefficient $k_b$ significantly, and a value of 129.4 may be achieved when the stiffener acts as if rigid. A hypothetical stiffener of zero area $A_s$ will produce this effect provided its second moment of area $I_s$ is equal to $4dt_w^3$ (Allen & Bulson 1980); in the code it is given as $4ct_w^3$. However, this value should be increased to allow for the compressive load acting on a real stiffener. The Australian code (AS 4100-1998) suggests the following minimum value of $I_s$.

$$I_s \geq 4dt_w^3\left[1+\frac{4A_s}{dt_w}\left(1+\frac{A_s}{dt_w}\right)\right] \tag{11.40}$$

where $A_s$ is the area of the stiffener in mm$^2$. A second longitudinal stiffener may be placed at the neutral axis of the girder with $I_s$ not less than $d_2t_w^3$, where $d_2$ is twice the clear distance from the compression flange to the neutral axis.

In the design of longitudinal stiffeners, linear buckling theories are often used to establish the minimum value of stiffener rigidity required to ensure that the longitudinal stiffeners remain straight at first buckling. This would ensure that the buckling is limited to individual sub-panels of the web. This invariably involves the provision of stiffeners of substantial sizes, so that they would remain straight without themselves buckling. The heavy stiffeners thus designed, though adequate, will naturally increase the weight of steel used, and therefore the cost. It seems that it is more sensible to increase the web thickness in these cases. Longitudinal stiffeners are rarely used in buildings. Fig. 11.32 shows a typical plate girder with longitudinal and transverse stiffeners. The ASCE-AASHTO Task Committee (1978) and Cooper (1967) provide a review of the theory and design of longitudinally stiffened plate girders.

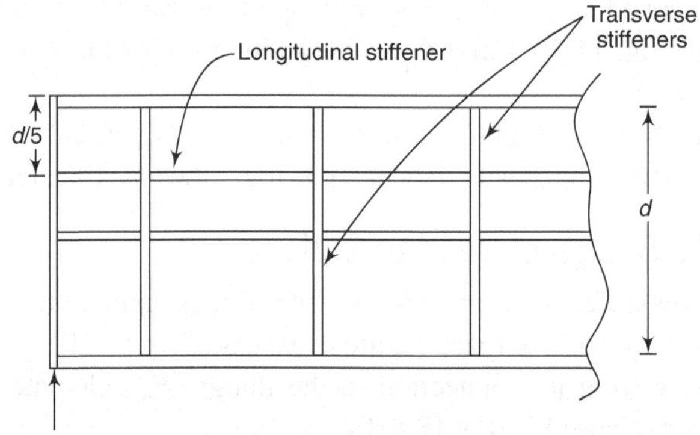


**FIG. 11.32** Longitudinal and transverse stiffeners

Bridge with *longitudinal* stiffener near Philadelphia, USA

## 11.9 GIRDERS WITH OPENINGS IN SLENDER WEBS

Holes are often provided in the webs of plate girders used in building construction to provide access to service ducts, etc. Such holes have specific influence on the behaviour of slender webs because the hole interrupts the tension field (see Fig. 11.33). Research carried out by Narayanan (1983) has shown that slender web girders with openings do have post-buckling resistance. However, some codes adopt a conservative approach and do not consider the reserve of resistance in girders with holes. To provide a simple procedure, the shear resistance of the panel with holes is taken as the buckling resistance. The drawback in such a procedure is that the shear resistance of the panel with holes is much lower than the one without a hole. One option to overcome the disadvantage is to reduce the stiffener spacing on either side of the web opening such that the buckling resistance of the panel with hole is nearly equal to the full post-buckling resistance of the adjacent panels.

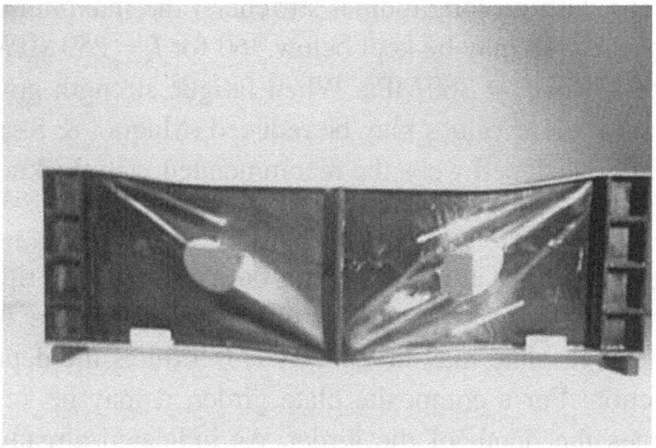

**FIG. 11.33** Interruption of tension field by web holes (ESDEP 2006)

## 11.10 HYBRID PLATE GIRDERS

In the design of plate girders, there is considerable scope for variation of cross section along the span of the girder considering the bending moment and shear force variations. One designer may reduce the flange thickness or breadth in areas of low bending moment by providing for spliced connection (see Fig. 11.4). In a zone of high shear, the designer might like to thicken the web. Another option is to employ higher grade steel in zones of high applied moment and shear and standard grade elsewhere. Hybrid girders employ different strength steel in the flanges and the web, to offer another possible way of closely matching resistance to requirements. Thus, in hybrid beams, stronger steels are used in flanges than in the web as the resistance of the web depends mainly on its slenderness ratio. The bending stress distributions of a hybrid beam at various stages of loading are shown in Fig. 11.34. Based on experimental studies by Barker et al. (2002), it has been recommended that webs of hybrid beams be designed using tension field theory with reduced web depth and that stiffener spacing be determined accordingly. One example of bridge with hybrid plate girder is the Discovery Bridge over the Missouri River (Subramanian 2011). More information about hybrid girders is provided by Barker et al. (2002), Veljkovic and Johansson (2004), and Gaylord et al. (1992).

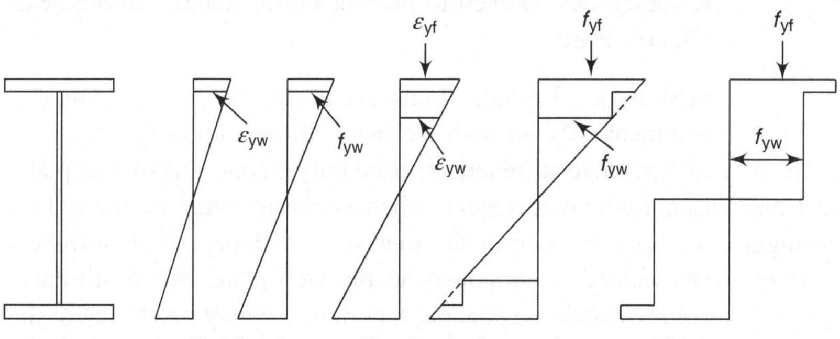


**FIG. 11.34** Bending stress and strain distribution of a hybrid plate girder at various stages of loading

## 11.11 WELDING OF GIRDER COMPONENTS

The fillet weld between the flange and the web is generally kept as small as possible depending upon the thickness of the flange and the allowable stresses. Continuous welds are desirable preferably with automatic welding equipment. Bearing stiffeners must fit tightly against the flange that transmits the load and must be attached to the flange by full penetration groove welds. The connection of the stiffener to the web may be by intermittent or continuous welds. Appropriate small openings are provided at the weld junctions to avoid interaction of different fillet welds. Transverse stiffeners are solely provided to increase the shear buckling resistance of the web and hence are fitted tightly with the compression flange but welded only to the web. But one-sided transverse stiffeners must be welded to the compression flanges also. Transverse intermediate stiffeners are stopped short of the tension flange by a distance of about $4t_w$. Welding transverse stiffeners to tension flanges may reduce the fatigue strength and may lead to brittle fracture.

## 11.12 PROPORTIONING OF THE SECTION

The cross section of the girder must be selected such that it adequately performs its functions and requires minimum cost. The functional requirements are as follows:

1. Strength to carry bending moment (adequate $Z_p$)
2. Vertical stiffness to satisfy any deflection limitation (adequate moment of inertia $I_z$)
3. Lateral stiffness to prevent lateral-torsional buckling of compression flange (adequate lateral bracing or low values of $(L_b/r_t)$)
4. Strength to carry shear (adequate web area)

5. Stiffness to improve buckling or post-buckling strength of the web (related to $d/t_w$ and $c/d$ ratios)

One method of determining the initial dimensions is to use a minimum weight analysis (Schilling 1974). However, it has to be emphasized that a minimum weight design may not result in minimum cost, unless fabrication cost is also included in the optimization process.

### 11.12.1 Optimum Girder Depth

If the moment $M$ is assumed to be resisted entirely by the flanges, then for a I-section beam,

$$M = f_y b_f t_f d \tag{11.41}$$

where $f_y$ is the design strength of the flanges, $b_f$ and $t_f$ are the width and thickness, respectively of the flanges, and $d$ is the distance between centre of flanges.

The gross-sectional area of the beam is given by

$$A = 2\, b_f t_f + d t_w \tag{11.42}$$

where $t_w$ is the thickness of web. Eliminating $b_f t_f$ using Eqn (11.41), we get

$$A = 2M/(d f_y) + d t_w \tag{11.43}$$

Defining the web slenderness as $k = d/t_w$, we get

$$A = 2M/\,(k t_w f_y) + k t_w^2 \tag{11.44}$$

Differentiating the above equation with respect to $t_w$ and setting the result equal to zero, we get the optimum value of $t_w$ as

$$t_w = [M/(f_y k^2)]^{0.33} \tag{11.45}$$

The optimum value of the depth is

$$d = (Mk/f_y)^{0.33} \tag{11.46}$$

Thus the optimum values produce a beam which has the area of single flange and the web as equal (Martin & Purkiss 1992). Extended treatment of this subject has been provided by Salmon and Johnson (1996), who also propose the following simple expression for the required area of one flange plate

$$A_f = M/(f_y d) - A_w/6 \tag{11.47}$$

where $A_w$ is the area of web plate. The above equation may be used for preliminary design purposes.

### 11.12.2 Preliminary Sizing

**Depth of Plate Girder** The depth of plate girder is of paramount importance because it governs the stiffness and bending resistance. The design has to comply with both strength and deflection requirements. The full girder depth $D$, in terms of the equivalent span $L$, may be chosen in the following range:

$D/L$ = 1/15 to 1/25 for girders in buildings,
$D/L$ = 1/12 to 1/18 for girders in highway bridges, and
$D/L$ = 1/10 to 1/15 for girders in railway bridges.

The equivalent span $L$ may be taken as the distance between the adjacent points of contraflexure under a uniformly distributed load.

The preceding values may be reduced when the top flange of the steel girder is attached to a reinforced concrete deck (composite girder) or to an orthotropic steel deck plate. A constant depth girder is chosen when a launching procedure is used for erection. However, when unequal spans are present in a continuous plate girder, it may be economical to provide variable depth plate girder.

**Web thickness** From a corrosion standpoint, the usual practice is to choose a minimum thickness of 6–8 mm. For large depth bridge girders, a minimum web thickness of 8–10 mm may be required because of handling considerations. Though thin stiffened webs have been used in the past, the tendency now is to avoid stiffeners and use thick webs, which reduce the fabrication time and cost. The web thickness may be kept constant over a length of about 4–6 m from the supports. For webs with no longitudinal stiffeners, the maximum $d/t_w$ ratio of the web may be kept below 360 for $f_y$ = 250 MPa and below 240 for $f_y$ = 360 MPa. When fatigue strength governs the design, these values may be reduced (Maquoi & Skaloud 2000). For stiffened webs the recommended web thickness is given in Sections 11.4.1 and 11.7.

**Flanges** Flanges are designed for both strength and rigidity requirements. For a non-composite plate girder, the width of the flange plate may be chosen as 0.3 times the depth of the section. For a composite plate girder, it may be kept as 0.2 times the depth of the girder. As indicated already the $b_f/t_f$ ratio of the flange should be such that the flange is either plastic or compact or semi-compact. This is to avoid local buckling before reaching the yield stress. For preliminary sizing, the flange width to thickness ratio may be limited to 24 for grade 410 steel. The thickness and width of the flange rarely exceeds 80–100 mm and 1 m respectively. The width of tension flange (i.e., bottom flange of a simply supported beam) may be increased by about 30%. Erection requirements may govern the size of flanges. Pre-heating before welding is sometimes adopted to prevent brittle fracture in the heat-affected zone.

**Stiffeners** Though transverse stiffeners are provided symmetrically on both the faces of the web at the supports, intermediate stiffeners are used only on one side of web plate. Longitudinal stiffeners, when necessary, may be located on the opposite side of the transverse stiffeners. Web stiffeners are welded continuously to the web plate. When stiffeners are provided, the panel aspect ratio $c/d$ may be chosen in the range of 1.2 to 1.6. When end panel is designed near support, without using the tension field action, the aspect ratio should be in the range 0.6–1.0.

### 11.12.3 Steps Involved in the Design of Plate Girders

The following steps are involved in the design of plate girders.

1. Assume the self weight of the girder. (The self weight $w$ may be assumed as equal to $W/200$ where $w$ is in kN/m and $W$ is the total factored load applied to the girder in kN.) Estimate the live loads. Calculate the maximum bending moment and shear force in the plate girder.
2. Using Eqns (11.45) and (11.46), assume depth and thickness of the web. Check the web thickness as per clause 8.6.1.1 and 8.6.1.2 of the code and adopt a suitable web thickness.
3. Select suitable flange plate thickness and width (approximately 0.3 times the depth of web). The flange plates should be so proportioned that it should satisfy the requirements of plastic/compact/semi-compact section. Note that due to the slender web, the plate girder cross section will always fall in the semi-compact category.
4. Check for moment capacity as per clause 8.2.1 or 8.2.2 depending on whether the plate girder is laterally supported or unsupported. It is generally advantageous to provide lateral support at sufficiently close intervals such that lateral-torsional buckling will not govern the design.
5. Check for shear resistance of the web. One may use either clause 8.4.2.2 (a) simple post–critical method or clause 8.4.2.2 (b) by considering tension field method.
6. Design of the connection between the flange plate and the web plate.
7. Design of bearing stiffeners and their connections (clauses 8.7.4, 8.7.5, and 8.7.9 of the code)
8. Design of load carrying stiffeners; if required (clause 8.7.5) and their connections.
9. Design of intermediate stiffeners, if required (clauses 8.7.2 and 8.7.1.2) and their connections (clause 8.7.2.6)
10. Design of web splice and its connections.
11. Design of flange splice and its connections.

The design of web and flange splices is covered in Sections 5.9.1 of Chapter 5 and 6.12.1 of Chapter 6.

## 11.13 PLATE GIRDERS WITH CORRUGATED WEBS

To increase the shear capacity of large steel plate girders, shaped webs may be used; in this situation, the web plate is first cold formed with waves or corrugations usually parallel to the web depth and then welded to the flanges. Although many types of corrugations are possible, trapezoidal and sinusoidal corrugations, as shown in Fig. 11.35 have received the most attention. Corrugated webs, though not yet commonly used for highway bridges in India and North America, have been incorporated into several highway bridges constructed in Europe and Japan. Girders fitted with corrugated webs tend to be most used in the construction of industrial buildings in Sweden, Germany, and other European countries (Subramanian 2011). Similar webs with reinforced and/or pre-stressed concrete flanges have been used for bridges in France. The first pre-stressed hybrid box girder bridge was the Cognac Bridge, constructed in France during 1986 (Fan et al. 2006). More recently the Shinkai and Hondani bridges have been constructed in Japan (Fan et al. 2006). A summary of the research and development in beams and girders with corrugated webs was reported by Elgaaly and Dagher (1990) and Johansson et al. (2007).

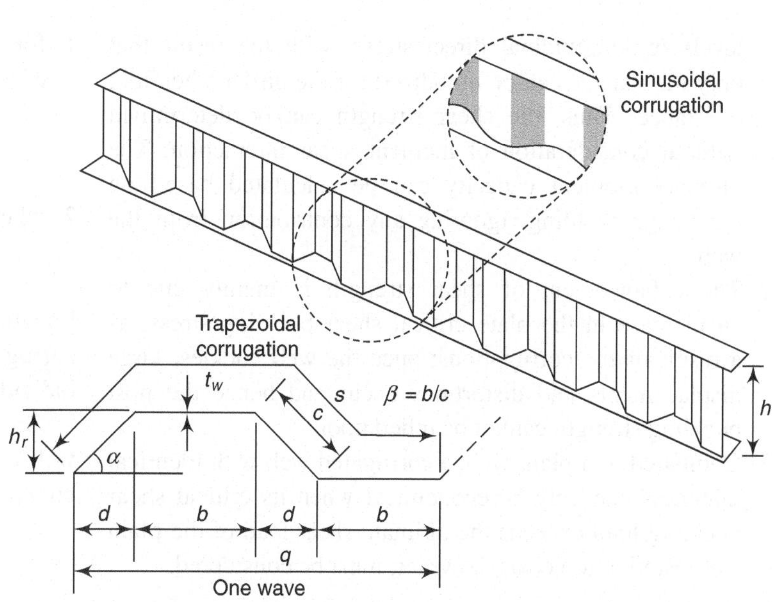


**FIG. 11.35** Profile and geometric notation of a girder with corrugated web

The results of studies on girders with corrugated web indicate that the fatigue strength of these girders can be 50% higher compared to girders with flat stiffened webs (Machacek & Tuma 2006). In addition to the improved fatigue life, these girders could be 30 to 60% lighter than the girders with flat webs and have the same capacity (Elgaaly & Dagher 1990). Thus, larger spans can be achieved with less weight. Moreover, corrugated webs improve the aesthetics of the structure. Beams used in Germany for buildings have web thicknesses that vary between 2 and 5 mm and the corresponding web height-to-thickness ratios varying between 150 and 260 mm. The corrugated webs of two bridges built in France were 8-mm thick and the web height-to-thickness ratio was in the range of 220 to 375 mm.

While designing girders with corrugated webs, the following points should be kept in mind (Maquoi 1992).

1. When the axis of the corrugations is perpendicular to the girder's longitudinal axis, the web cannot sustain significant

levels of longitudinal direct stress, with the result that only the flanges can be mobilized for the girder's bending resistance. Thus, the shear strength can be determined without consideration of moment–shear interaction. The ultimate moment capacity can be calculated based on the flange yielding, ignoring any contribution from the web.

2. The enhancement of shear strength is mainly due to an increase in the plate critical shear buckling stress, as a result of the corrugations; once the web buckles, large displacements and distortions occur and hence the post buckling strength cannot be relied upon.
3. Compared to a plain web, a corrugated web with identical thickness can only be economical when its critical shear buckling load exceeds the ultimate shear load of the plain web. Fabrication costs, however, must be considered.

The corrugations increase the out-of-plane stiffness of the web and eliminate the use of vertical stiffeners. From experimental results, it was found that failure in shear is usually due to the buckling of the web and the failure in bending is due to the yielding of the compression flange and vertical buckling into the web (Elgaaly et al. 1996; Elgaaly et al. 1997). The failure is sudden with no appreciable residual strength.

Corrugated webs have been found to fail in shear by instability, and both local and global buckling modes have been observed experimentally. In theory, local buckling involves a single flat panel or 'fold', whereas global buckling involves multiple folds, with buckles that extend diagonally over the entire depth of the web. Global buckling may be predominant for dense corrugation and local buckling for course corrugation. In experiments, however, failure modes that appear to have characteristics of both local and global buckling have been observed (Driver et al. 2006).

Plate stability theory can be used to predict the local shear buckling of corrugated webs (Lindner & Huang 1995; Lindner 1988). A given fold (longitudinal or inclined) is assumed to be supported by adjacent folds along its vertical edges and by the flanges along its horizontal edges. The elastic shear buckling stress is given by

$$(\tau_{cr,\,L})_{el} = k_L\{\pi^2 E/[12(1-\mu^2)(w/t_w)^2]\} \qquad (11.48a)$$

where $k_L$ is the buckling coefficient, which depends upon the boundary conditions and the fold aspect ratio ($w/h$), E is the Young's modulus of elasticity (in N/mm$^2$), $\mu$ is the the Poisson's ratio, $w$ is the maximum fold width (in mm, maximum of the longitudinal fold width $b$ and the inclined fold width $c$, shown in Fig. 11.35), $h$ is the web depth (in mm), and $t_w$ is the web thickness (in mm).

A small aspect ratio ($w/h$) minimizes the buckling coefficient $k_L$ and is given by the following:

1. for the longer edges—simply supported and shorter edges—fixed condition,

$$k_L = 5.34 + 2.31(w/h) - 3.44(w/h)^2 + 8.39(w/h)^3 \qquad (11.48b)$$

2. when all edges are fixed,

$$k_L = 8.98 + 5.6(w/h)^2 \qquad (11.48c)$$

Equation (11.48c) has been used by various investigators of corrugated web shear behaviour, to describe local buckling of individual fold.

Global buckling can be predicted by treating the corrugated web as an orthotropic flat web for which the elastic shear buckling stress is given by (Driver et al. 2006)

$$(\tau_{cr,\,G})_{el} = k_G\,[(Et_w^{0.5}b^{1.5})/(12h^2)]\,F(\alpha,\beta) \qquad (11.49)$$

where $k_G$ is the global shear buckling coefficient that depends on the boundary conditions. It can be taken as 31.6 for simply supported boundaries and 59.2 for fixed boundaries (Elgaaly et al. 1996).

$F(\alpha, \beta)$ is the non-dimensional coefficient characterizing the web corrugation geometry

$$F(\alpha,\beta)=\sqrt{[(1+\beta)\sin^3\alpha/(\beta+\cos\alpha)]}\cdot\{(3\beta+1)/[\beta^2(\beta+1)]\}^{0.75} \qquad (11.50)$$

where $\alpha$ = corrugation angle (see Fig. 11.35) and $\beta$ is the ratio of the longitudinal fold width $b$ to the inclined fold width $c$.

A small value of $\beta$ may lead to uneconomical design due to the large amount of web material required when the corrugations are deep. A large value of $\beta$ may result in a low global buckling strength. Values of $\beta$ normally range between 1 and 2. The value of $\alpha$ is normally taken as 30° or 45°. Lindner and Huang (1995) suggest that the value of $\alpha$ should not be less than 30° for the corrugation folds so as to provide adequate support to one another along the fold lines in order to mobilize the full shear capacity. The parameter $\alpha$ not only influences the shear strength, but also fabrication and fatigue life.

Galambos (1998) suggests the use of the following equation for calculating the global elastic buckling stress

$$(\tau_{cr,\,G})_{el} = [(k_G(D_x)^{0.25}(D_y)^{0.75})]/(t_w h^2) \qquad (11.51)$$

where

$$D_x = (q/s)Et^3_w/12$$

$$D_y = EI_y/q$$

$$I_y = 2bt_w(h_r/2)^2 + \{th_r^3/6\sin\alpha\}$$

The variables $b$, $h_r$, $q$, $s$, and $\alpha$ are as shown in Fig. 11.35.

When the elastic shear buckling stress $(\tau_{cr})_{el}$ exceeds 80% of the shear yield stress $\tau_y$, the following inelastic equation can be used for both local buckling (with $(\tau_{cr})_{el}$ taken as $(\tau_{cr,L})_{el}$ from Eqn (11.48) and global buckling (with $(\tau_{cr})_{el}$ taken as $(\tau_{cr,G})_{el}$ from Eqn (11.49) or (11.51).

$$(\tau_{cr})_{inel} = \sqrt{[0.8\tau_y(\tau_{cr})_e]} \le \tau_y \tag{11.52}$$

where $\tau_y$ is determined using the von Mises yield criterion as

$$\tau_y = f_y/\sqrt{3} \tag{11.53}$$

where $f_y$ is the uniaxial yield stress of the web material (in N/mm$^2$).

Driver et al. (2006) proposed the following equation, which considers the effect of both local and global buckling in a single interaction formula

$$\tau_n = \sqrt{[(\tau_{cr.L}\tau_{cr.G})^2/(\tau_{cr.L}^2 + \tau_{cr.G}^2)]} \tag{11.54}$$

Equation (11.54) with lower bound values for $k_L$ and $k_G$ was found to provide a reasonable lower bound for the available test results.

Under static loading, the common practice of fillet welding, the web to the flanges from one side only, was found to be adequate. Elgaaly et al. (1997) discuss the lateral-torsional buckling of girders with corrugated web and Elgaaly and Seshadri (1997) provide information on the behaviour of these girders under partial compressive edge loading. Based on eight large scale tests, Sause et al. (2006) found that plate girders with corrugated webs {having HPS 485 W steel and welded by gas metal arc (GMAW) fillet welding} exhibit a fatigue life that is longer than that of conventional plate girders with transverse stiffeners. Design rules for plate girders with corrugated webs are provided in Annex D of Eurocode 3 (EN 1993-1-5:2006). See also the commentary and worked examples by Johansson et al. (2007).

## 11.14 BOX GIRDERS

Box girders are extensively used for bridges, offshore platforms, and other structures where large loads are frequently encountered. They have considerable torsional stiffness compared to plate girders; hence they are ideally suited for girders curved in plan. Box girders may have a variety of cross-sectional shapes ranging from deep narrow box to a wide shallow box, with many webs (cells). The webs may be straight as shown in Fig.11.6(c) or sloping. Most of the box girder bridges have flanges which are wider and more slender than in plate girders. The extensive use of slender plates for flanges as well as webs makes stability considerations very important in the case of box girders.

Due to these slender plates, the effect of shear lag and lateral loading on the buckling of stiffened compression plates has to be considered. In addition, the provision and design of diaphragms, which not only retain the box shape but also distribute the forces to the support bearings, have to be considered. The information on web stability as discussed for plate girders is also applicable to box girders. The three major collapses that occurred during the erection of box girder bridges at Milford Haven (1969), West Gate (1970), and Koblenz (1971) resulted in much research on the behaviour of box girders. More details about the design aspects of box girders are provided by Galambos (1998), Paik and Thayamballi (2003) and Dubas and Gehri (1986).

Horizontally curved girders have undergone substantial development over the past years, due to the increased demand for curved roadway alignments and of the availability of computers to carryout complex analysis. The buckling stability and ultimate strength of the web plates of curved girders are affected by the horizontal curvature. More discussion on this aspect may be found in Paik and Thayamballi (2003) and Shanmugam and Wang (2006).

## EXAMPLES

**Example 11.1:** The plate girder of Example 8.23 is required to carry a factored shear of 2800 kN (see Fig.11.36).

(a) Assuming that tension field action is not utilized in the design, determine whether intermediate stiffeners are necessary. Assume $f_y = 250$ N/mm$^2$.

(b) How thick must the web be in order that this same load may be carried without the need for intermediate stiffeners?

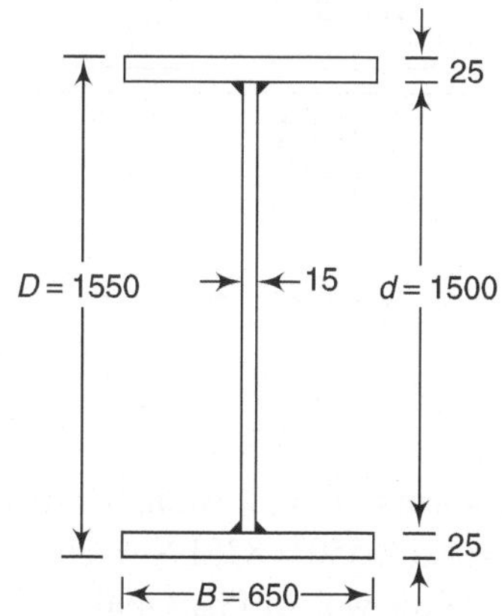


**FIG. 11.36** Cross section of the girder

**Solution:**

(a) *Determination of required stiffeners*

Using $d/t$ = 1500/15 = 100, from Table 11.1, for no stiffeners ($c/d = \infty$)

$$\tau_b = 95.7 \text{ N/mm}^2$$

Hence

$$V_n = V_{cr} = dt_w\tau_b = 1500 \times 15 \times 95.7/10^3$$
$$= 2153.2 \text{ kN} < 2800 \text{ kN}$$

Hence stiffening is required.

Required $\tau_b = 2800 \times 10^3/(1500 \times 15) = 124.4$ N/mm²

From Table 11.1, for $d/t = 100$, max $c/d$ corresponding to this strength = 1.0.

Hence stiffeners at 1500 mm (= 1.0 × 1500) intervals should be provided.

(b) *Thickness of web without stiffeners*

A trial and error approach is necessary to determine the web thickness which will support the load without any intermediate stiffener. Obviously it must be greater than 15 mm.

Try $t = 18$ mm; $d/t = 1500/18 = 83.33$

Hence $\tau_b$ (for $c/d = \infty$) from Table 11.1 = 119.2 N/mm²

Hence $V_n = 1500 \times 18 \times 119.2/10^3 = 3218.4$ kN > 2800 kN

*Note:* The web in this example is not very slender ($d/t < 100$). Hence it may be advisable to increase the thickness of the web instead of providing stiffeners, which may increase the cost of fabrication. However, inspection of Table 11.1 shows that for deeper girders (e.g., $d/t = 200$), stiffeners at a spacing of $0.9d$ double the shear strength of girders without stiffeners while stiffeners at $0.4d$ produce about 5.3 fold improvement.

**Example 11.2:** Design a load carrying stiffener for a load of 550 kN for the section ISLB 400 (see Fig. 11.37).

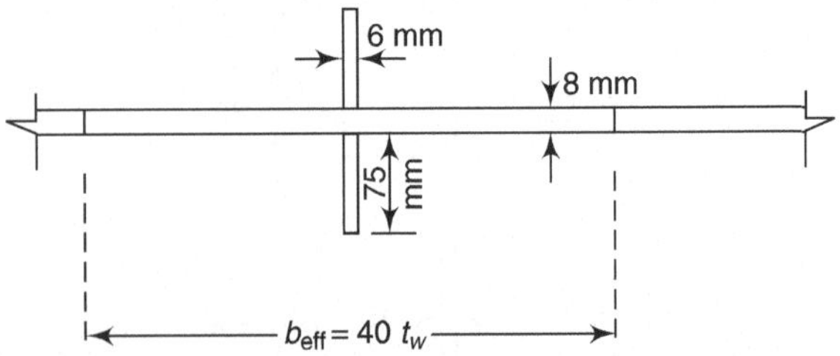


**FIG. 11.37**

**Solution:**

For ISLB 400, $t_w = 8$ mm, $t_f = 12.5$ mm, $R = 16$ mm

Factored load = 1.5 × 550 = 825 kN

Shear capacity of the web of ISLB 400

$$= \frac{A_v f_{yw}}{\gamma_{m0}\sqrt{3}} = \frac{400 \times 8 \times 250}{1.1 \times 1000 \times \sqrt{3}} = 420\ \text{kN}$$

Load to be resisted by the stiffener = 825 − 420 = 405 kN

Assume the thickness of the stiffener plate as 6 mm. The maximum allowable breadth (outstand) of stiffener (clause 8.7.1.2 of code) = $20 \times t_q \varepsilon = 20 \times 6 = 120$ mm (since $\varepsilon = 1$, for $f_y = 250$ MPa)

Take the breadth of the stiffener plate as 12.5 times the thickness of the plate (12.5 × 6 = 75 mm). Two plates of size 75 mm × 6 mm are required on either side of the web.

As per clause 8.7.1.5 of the code,

$$b_{eff} = (40 \times 8) = 320\ \text{mm}$$

Effective area = $(b_{eff} \times t_w) + (2 \times b_s \times t_s) = (320 \times 8) + (2 \times 75 \times 6) = 3460$ mm²

Moment of inertia of stiffener = $\dfrac{b_s \times d_s^3}{12} = 6 \times (2 \times 75)^3/12$

$= 1.6875 \times 10^6$ mm⁴

Assuming $c/d = 1.5$ (clause 8.7.2.4),

Minimum $I_s = 0.75 d t_w^3 = 0.75 \times 400 \times 8^3$

$= 0.154 \times 10^6\ \text{mm}^4 < 1.6875 \times 10^6\ \text{mm}^4$

Hence the adopted size of the stiffener is fine.

**Check for buckling resistance**

$$I = \frac{6 \times 150^3}{12} + \frac{320 \times 8 \times 8^3}{12} = 1.796 \times 10^6\ \text{mm}^4$$

$$r = \sqrt{\frac{I}{A}} = \sqrt{\frac{1.796 \times 10^6}{3460}} = 22.79\ \text{mm}$$

Depth of web = $400 - 2(t_f + R) = 400 - 2(12.5 + 16) = 343$ mm

Effective length of member = $0.7 \times d = 0.7 \times 343 = 240.1$ mm

$$\lambda = \frac{l}{r} = \frac{240.1}{22.79} = 10.53$$

Therefore, compressive stress, $f_{cd} = 226.75$ N/mm² (from Table 9c of the code).

Buckling resistance of the stiffener = $226.75 \times 3460/10^3$ = 784.56 kN > 405 kN.

Hence the stiffener is safe.

**Example 11.3:** Assuming stiffener spacing equal to panel depth, determine the shear capacity of the girder given in Example 11.1, including the tension field action, assuming a factored bending moment of 4500 kN m.

**Solution:**

As per clause 8.4.2.2b of the code,

$$V_n = V_{tf}$$

$$V_{tf} = [A_v \tau_b + 0.9 w_{tf} t_w f_v \sin\phi] \le V_{np}$$

From Table 11.1, for $d/t = 100$ and $c/d = 1.0$, $\tau_b = 130$ N/mm²

Reduced plastic moment of the flange plate

$$M_{fr} = 0.25 b_f t_f^2 f_{yf} [1 - \{N_f/(b_f t_f f_{yf}/\gamma_{m0})\}^2]$$

$$= 0.25 \times 650 \times 25^2 \times 250\ [1 - \{(4500 \times 10^6/1525)/(650 \times 25 \times 250/1.1)\}^2]$$

$$= 0.25 \times 650 \times 25^2 \times 250\ [1 - 0.638]$$

$$= 9.191 \times 10^6\ \text{N mm}$$

Inclination of the tension field

$$\phi = \tan^{-1}(d/c) = \tan^{-1}(1500/1500) = 45°$$

Anchorage length of the tension field in compression and tension flange

$$s = 2/\sin\phi\ [M_{fr}/(f_y\, t_w)]^{0.5} \le c$$

$= 2/\sin 45°[(9.191 \times 10^6)/(250 \times 15)]^{0.5}$
$= 140\,\text{mm} < 1500\,\text{mm}$

The width of the tension field

$w_{tf} = d \cos\phi - (c - s_c - s_t)\sin\phi$
$= 1500 \times \cos 45° - (1500 - 2 \times 140)\sin 45°$
$= 198\,\text{mm}$

Yield strength of the tension field

$f_v = [f_{yw}^2 - 3\tau_b^2 + \psi^2]^{0.5} - \psi$
$\psi = 1.5\tau_b \sin 2\phi$
$= 1.5 \times 130 \times \sin 90°$
$= 195\ \text{N/mm}^2$
$f_v = [250^2 - 3 \times 130^2 + 195^2]^{0.5} - 195$
$= 28.21\ \text{N/mm}^2$

The shear resistance

$V_{tf} = [1500 \times 15 \times 130 + 0.9 \times 198 \times 15 \times 28.21 \times \sin 45°]/1000$
$= 2925 + 53.3 = 2978.3\,\text{kN}$

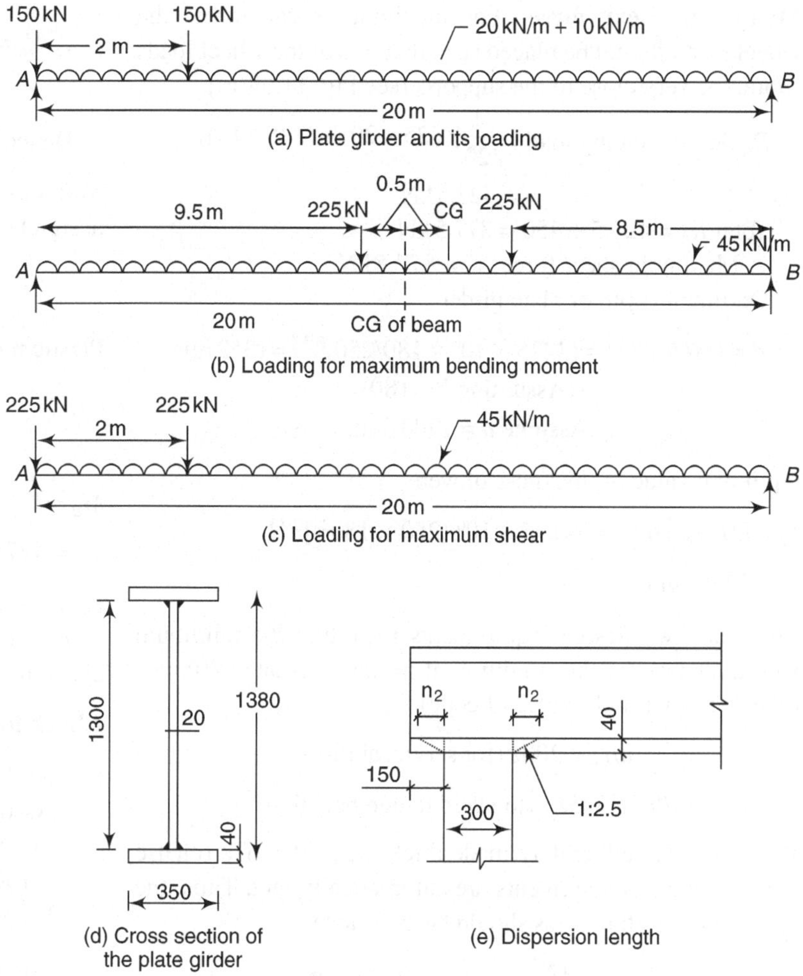

(a) Plate girder and its loading
(b) Loading for maximum bending moment
(c) Loading for maximum shear
(d) Cross section of the plate girder
(e) Dispersion length

**FIG. 11.38**

**EXAMPLE 11.4:** Design a welded plate girder for a simply supported bridge deck beam with clear span of 20 m, subjected to the following:

Dead load including self weight = 20 kN/m
Imposed load = 10 kN/m
Two moving loads = 150 kN each spaced 2 m apart

Assume that the top compression flange of the plate girder is restrained laterally and prevented from rotating. Use mild steel with $f_y = 250\,\text{MPa}$. Design as an unstiffened plate girder with thick webs.

**SOLUTION:**

**Loading**

Total factored udl on girder = 1.5 × (20 + 10) = 45 kN/m
Factored moving loads = 1.5 × 150 = 225 kN

**Maximum bending moment**

$$R_A \text{ and } R_B \text{ due to UDL} = \left(\frac{45 \times 20}{2}\right) = 450\,\text{kN}$$

The maximum bending moment due to UDL occurs at mid span; whereas the maximum bending moment due to the two moving loads will occur under one of the loads, when one of the loads is placed such that the middle of the beam divides the distance between this load (under which maximum bending moment occurs) and CG of the moving loads equally [see Fig. 11.38(b)].

CG of the loads = $a/2$(from the loads under consideration) where $a$ is the distance between the moving loads. The distance of the loads from middle of the beam = $a/4$. Hence,

Distance of the CG of loads from right support = $L/2 - a/4$
$= 20/2 - 2/4 = 9.5\,\text{m}$

Since the maximum BM under UDL is at a different location from that of maximum BM under moving loads, we have to compute the bending moment in both cases and choose the higher of the two. However, in our case the bending moment occurred under one of the moving loads.

$$R_A \text{ due to moving load} = \frac{1}{20}[(225 \times 10.5) + (225 \times 8.5)]$$
$$= 213.75\,\text{kN}$$

$\therefore R_B$ due to moving load = (2 × 225) − 213.75 = 236.25 kN
$\therefore$ Total $R_B$ = 450 + 236.25 = 686.25 kN
Total $R_A$ = 450 + 213.75 = 663.75 kN

$$\text{Maximum BM} = (663.75 \times 9.5) - \left(\frac{45 \times 9.5^2}{2}\right) = 4275\,\text{kNm}$$

$\therefore$ Maximum BM, $M_z = 4275\,\text{kNm}$

*Note*: Impact due to moving load should be considered in practice.

**Maximum shear force** For maximum shear force, the wheel loads should be placed such that one of the wheel loads should be very close to the supports [see Fig. 11.38(c)].

$$R_A \text{ due to moving load} = \frac{1}{20}[(225\times 20)+(225\times 18)]$$
$$= 427.5\,\text{kN}$$

Total $R_A = 427.5 + 450 = 877.5\,\text{kN}$

∴ Maximum shear force, $V_z = 877.5\,\text{kN}$

Optimum depth of plate girder

$$d = (Mk/f_{yf})^{0.33} = (4275 \times 10^6 \times 180/250)^{0.33} = 1352\,\text{mm}$$

(Assuming $k = 180$)

Assume $d = 1300\,\text{mm}$.

Optimum value of thickness of web

$$t_w = [M/(f_y k^2)]^{0.33} = [4275 \times 10^6/(250 \times 180^2)]^{0.33}$$
$$= 22.97\,\text{mm}$$

There are two design requirements regarding the minimum web thickness for the condition of no intermediate stiffeners (clause 8.6.1.1 and 8.6.1.2 of code)

$$d/t_w < 200\varepsilon \text{ (for serviceability)}$$
$$d/t_w \le 345\varepsilon_f^2 \text{ (to avoid flange buckling)}$$

If the web is deliberately made thick, i.e., $d/t < 67\varepsilon$ (clause 8.4.2.1), these requirements are automatically met. Thus, the minimum web thickness should be as follows:

$$t_w > \frac{1300}{67\times 1} \text{ or } t_w > 19.403\,\text{mm}$$

Hence, provide $t_w = 20\,\text{mm}$

In order to maximize the moment capacity, the cross section of the plate girder should be so proportioned that it satisfies the requirements of plastic/compact section. Thus $b_f/t_f$ should be less than $8.4\varepsilon$ or $9.4\varepsilon$ for plastic and compact sections respectively (see Table 2 of the code). Assuming $b_f$ as 0.3 times the depth of web $b_f = 0.3 \times 1300 = 390\,\text{mm}$. Provide $b_f = 350\,\text{mm}$; $t_f > b_f/(2 \times 8.4) = 20.8\,\text{mm}$. Provide

$$t_f = 40\,\text{mm}.$$

Use 1300 mm × 20 mm web plates with flange plates of 350 mm × 40 mm.

As discussed earlier, it is a *plastic section.*

**Shear capacity** As per clause 8.4 of the code

$$V \le V_d$$
$$V_d = \frac{V_n}{\gamma_{m0}}$$

Nominal plastic shear resistance, $V_n = V_p = \dfrac{A_v f_{yw}}{\sqrt{3}}$

As per clause 8.4.1.1 of the code, for welded section, $A_v = dt_w$

$$\therefore\ V_n = \frac{dt_w f_{yw}}{\sqrt{3}} = \frac{1300\times 20\times 250}{\sqrt{3}\times 1.1\times 10^3} = 3411.6\,\text{kN} > V_z$$
$$= 877.5\,\text{kN}$$

Hence safe in shear.

**Moment capacity** According to clause 8.2.1.2 of the code, design bending strength,

$$M_d = \beta_b Z_p \frac{f_y}{\gamma_{m0}} \le 1.2 Z_e \frac{f_y}{\gamma_{m0}}$$

Plastic modulus, $Z_p = 2b_f t_f (D - t_f)/2 + t_w d^2/4$

$$= 2 \times 350 \times 40 \times 1340/2 + 20 \times 1300^2/4$$
$$= 27.21 \times 10^6\,\text{mm}^3$$

$$M_d = 1.0 \times 27.21 \times 10^6 \times \frac{250}{1.10\times 10^6} = 6184.09\,\text{kNm} > M_z$$
$$= 4275\,\text{kNm}$$

Hence safe to carry the applied moment.

As the compression flange is restrained, there is no need to check for lateral-torsional buckling.

**Check for bearing stiffeners**

(a) At the supports [see Fig. 11.38(e)]

Assume that the width of support is 300 mm and that the minimum stiff bearing provided by the support $b_1 = 300/2 = 150\,\text{mm}$,

Dispersion length (1:2.5), $n_2 = 2.5 \times 40 = 100\,\text{mm}$

Local capacity of the web, $F_w = (b_1 + n_2)t_w \dfrac{f_{yw}}{\gamma_{m0}}$

$$F_w = (100 + 150) \times 20 \times \frac{250}{1.10\times 10^3} = 1136.35\,\text{kN}$$

$>$ support reaction $= 877.5\,\text{kN}$

(b) At position of moving wheel loads

$$F_w = (b_1 + n_2)t_w \frac{f_{yw}}{\gamma_{m0}}$$

Assuming, $b_1 = 0$

$$F_w = (0 + 2.5 \times 2 \times 40) \times 20 \times \frac{250}{1.10\times 10^3} = 909.09\,\text{kN}$$

$> 225\,\text{kN}$

The associated buckling resistance $F_{qd}$ is dependent on the slenderness of the unstiffened web (clause 8.7.1.5).

Slenderness ratio of the web $= L_e/r_y = 0.7L/r_y$ [with $r_y = t/2\sqrt{3}$, see Eqn (10.52)]

$$= 2.5d/t = 2.5 \times 1300/20 = 162.5$$

From Table 9c of the code $f_{cd} = 52\ \text{N/mm}^2$ and as per clause 8.7.1.3 of the code,

Stiff bearing length = 0 (assumed)

45° dispersion length (to the level of half the depth of beam) = 1380/2 = 690 mm

$$F_{qd} = (0 + 690) \times 20 \times 52/10^3 = 717.6\,\text{kN} > 225\,\text{kN}$$

The web is adequate at both supports and positions of concentrated loads. Hence, there is no need to provide bearing stiffeners.

Thus by using thick webs, the use of load bearing stiffeners may be eliminated which will minimize the fabrication.

**EXAMPLE 11.5:** Redesign the plate girder given in Example 11.4 using thin web plates.

**SOLUTION:**
From Example 11.4,

$$M_z = 4275\,\text{kNm}$$
$$F_z = 877.5\,\text{kNm}$$

As shown in Example 11.4,
optimum depth = 1352 mm and $t_w = 22.97$ mm

Let us assume web plate of 1200 mm × 12 mm size. Assuming that the bending moment is carried by the flanges,

$$A_f = M_z \times \gamma_{mo}/(f_y \times D) = 4275 \times 10^6 \times 1.1/(250 \times 1200)$$
$$= 15675\,\text{mm}^2$$

Limiting $8.4\varepsilon$ for the plastic section, the approximate flange thickness is

$$t_f = \sqrt{15675/(2 \times 8.4)} = 30.5\,\text{mm}.$$

Provide $t_f = 35$ mm.

$$b_f = 15675/35 = 447.86\,\text{mm}$$

Provide $b_f = 450$ mm.

The cross section of the plate girder is shown in Fig. 11.39.

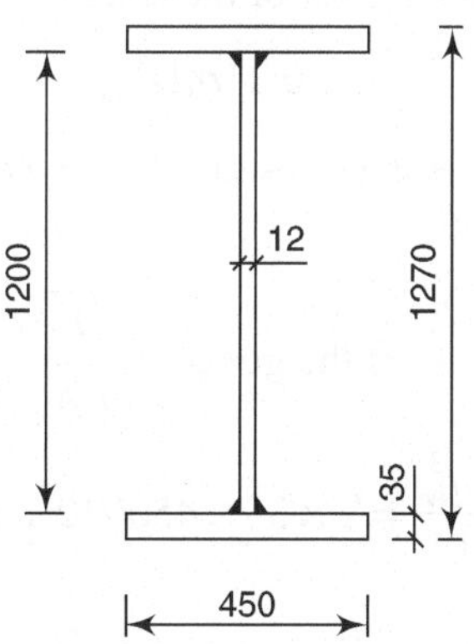


**FIG. 11.39** Cross section of plate girder

According to clause 8.2.1 of the code,

$$\frac{d}{t_w} = \frac{1200}{12} = 100 > 67\varepsilon$$

**Step 1:** *Moment capacity check*
As discussed earlier, the flanges are *plastic*

$$\text{Moment of resistance (clause 8.2.1.2)} = \frac{\beta_b Z_p f_y}{\gamma_{m0}}$$

Considering that the flanges only resist the bending moment,

$$Z_p = 2b_f t_f (D - t_f)/2 = 2 \times 450 \times 35(1270 - 35)/2$$
$$= 19.45 \times 10^6\,\text{mm}^3$$
$$M_d = 1 \times 19.45 \times 10^6 \times \frac{250}{1.10 \times 10^6}$$
$$= 4420\,\text{kN m} > 4275\,\text{kNm}$$

**Step 2:** *Shear resistance of the web*
As per clause 8.6.1.1a of the code, for serviceability

$$\frac{d}{t_w} = \frac{1200}{12} = 100 < 200\varepsilon$$

As per clause 8.6.1.2a of the code to avoid flange buckling

$$\frac{d}{t_w} \le 345\varepsilon_f^2$$

$$\frac{1200}{12} = 100 < 345\varepsilon_f^2$$

Hence the minimum web thickness requirements are met.

Let us consider the simple post-critical method. According to clause 8.4.2.2(a) of the code,

Nominal shear strength $V_n = V_{cr}$

$k_v = 5.35$, when transverse stiffeners are provided only at supports.

The elastic critical shear stress of the web

$$\tau_{cr} = \frac{k_p \pi^2 E}{12(1-\mu^2)\left(\dfrac{d}{t_w}\right)^2}$$

With $\mu = 0.3$,

$$\tau_{cr} = \frac{5.35 \times \pi^2 \times 2 \times 10^5}{12(1 - 0.3^2) \times 100^2} = 96.7\,\text{N/mm}^2$$

The non-dimensional web slenderness ratio for shear buckling stress,

$$\lambda_w = \sqrt{[f_{yw}/(\sqrt{3}\tau_{cr})]} = \sqrt{\frac{250}{(\sqrt{3} \times 96.7)}} = 1.22 > 1.20$$

Shear stress corresponding to buckling $\tau_b = \sqrt{[f_{yw}/(\sqrt{3}\lambda_w^2)]}$

$$\tau_b = 250/(\sqrt{3} \times 1.22^2) = 96.97\,\text{N/mm}^2$$

Shear force corresponding to web buckling,

$$V_{cr} = d t_w \tau_b$$
$$= 1200 \times 12 \times 96.97/10^3 = 1396.3\,\text{kN} > 877.5\,\text{kN}$$

Hence the shear strength is greater than the applied shear.

**Step 3:** *Lateral torsional buckling*

As the compression flange is restrained, there is no need to check for lateral torsional buckling.

**Step 4:** *Design of load carrying stiffeners*

(a) At the position of applied wheel load:

As per clause 8.7.4 of the code, the local capacity of web

$$F_w = (b_1 + n_2)t_w \frac{f_{yw}}{\gamma_{m0}}$$

$$n_2 = 2 \times 35 \times 2.5 = 175\,\text{mm}$$

$$b_1 = 0$$

$$F_w = [0 + 175] \times 12 \times \frac{250}{1.10 \times 10^3}$$

$$= 477.2\,\text{kN} > 225\,\text{kN}$$

The buckling resistance depends on the slenderness ratio of the web.

According to clause 8.7.1.5 of the code,

$$\lambda = 2.5\,\frac{d}{t} = 2.5 \times \frac{1200}{12} = 250$$

$$n_1 = D/2 = 1270/2 = 635$$

From Table 9c of the code, $f_{cd} = 24.3\,\text{N/mm}^2$

$\therefore$ Buckling resistance $= (0 + 2 \times 635) \times 12 \times 24.3/10^3$

$= 370.3\,\text{kN} > 225\,\text{kN}$

No stiffeners are necessary to prevent local buckling failure of the web at the point load position.

(b) At the supports:

The reaction is 877.5 kN, which is much greater than the point load of 225 kN. Hence stiffeners are necessary for both the supports.

Try a pair of 220 × 12 mm flats.

As per clause 8.7.1.2 of the code,

Maximum outstand of stiffener $= 20t_q\varepsilon = 20 \times 12 \times 1$

$= 240\,\text{mm} > 220\,\text{mm}$

$= 14 \times 12 \times 1 = 168\,\text{mm} < 220\,\text{mm}$

This means that the core area of the stiffener should be reduced to 2 × 168 mm × 12 mm.

(c) Check stiffener for buckling

As per clause 8.7.1.5, the effective section is the core area of the stiffener together with an effective length of web, on either side of the centre line of the stiffener of $20 \times t_w$. At the support, assuming that the web is available only on one side of the edge stiffener,

Local buckling resistance,

$$\text{Area} = 2 \times 168 \times 12 + 20 \times 12 \times 12 = 6912\,\text{mm}^2$$

$$I_x = \frac{12 \times 336^3}{12} + \frac{(20 \times 12)12^3}{12} = 37.967 \times 10^6\,\text{mm}^4$$

$$r = \sqrt{I_x/A} = \sqrt{\frac{37.967 \times 10^6}{6912}} = 74.11\,\text{mm}$$

$$\lambda = 0.7 \times \frac{1200}{74.11} = 11.33$$

From Table 9c of the code, $f_{cd} = 226\,\text{N/mm}^2$

Buckling resistance $= 226 \times 6912/10^3$

$= 1562.1\,\text{kN} > 877.5\,\text{kN}$

Hence the stiffener is safe.

Provide a pair of 220 × 12 mm stiffener at both the ends of the girder. Now calculate the bearing capacity of the end stiffener.

$$F_{psd} = A_q f_{yq}/(0.8 \times \gamma_{m0})$$

$$= 2(220 - 15) \times 12 \times 250/[(0.8 \times 1.1) \times 1000]$$

$$= 1397.7 > 877.5\,\text{kN}$$

Hence the stiffener is safe.

In the preceding equation, $A_q$ is the area of the stiffener in contact with the flanges. Normally, the flanges and web will be welded together, before the stiffeners are fitted. This means that the inside corners of the stiffener need to be coped/chamfered at the junction of the web and flanges, so that they will not foul with the web/flange weld. The chamfered width is taken as 15 mm.

(b) Torsional restraint provided by end stiffener

However, it might be deemed necessary that the ends of the plate girder are torsionally restrained during transportation and erection. This can be accomplished by checking the second moment of area of the end bearing stiffeners at the supports as per clause 8.7.9 of the code.

$$I_s \geq 0.34\alpha_s D^3 t_{cf}$$

Assuming that there is no restraint to the compression flange for the above situation,

$$r_y \text{ of the girder} = \sqrt{\frac{I_y}{A}}$$

$$I_y = \frac{2t_f b_f^3}{12} + \frac{dt_w^3}{12} = 2 \times 35 \times 450^3/12 + 1200 \times 12^3/12$$

$$= 531.73 \times 10^6\,\text{mm}^4$$

$$A = 2 \times 450 \times 35 + 1200 \times 12 = 45900\,\text{mm}^2$$

$$r_y = \sqrt{(531.73 \times 10^6/45900)} = 107.6\,\text{mm}$$

$$\lambda = 20 \times 1000/107.6 = 185.8 > 100$$

Hence,

$$\alpha_s = 30/\lambda^2 = 30/185.8^2 = 8.69 \times 10^{-4}$$

$$I_s \geq 0.34 \times 8.69 \times 10^{-4} \times 1270^3 \times 35 = 21.183 \times 10^6\,\text{mm}^4$$

Now $I_s = 12 \times (2 \times 220)^3/12 = 85.184 \times 10^6\,mm^4$
$> 21.183 \times 10^6\,mm^4$

Hence the provided stiffener has the necessary torsional restraint.

(c) Weld at web flange junction

Assuming fillet weld on each side of the web,

$$q_w = VA_f\bar{y}/2I_z$$

$$I_z = b_f D^3/12 - (b_f - t_w)d^3/12$$

$$= 450 \times 1270^3/12 - (450 - 12)1200^3/12$$

$$= 13742.3 \times 10^6\,mm^4$$

$$q_w = 877.5(450 \times 35) \times 635/(2 \times 13742.3 \times 10^6)$$

$$= 0.319\,kN/mm$$

From Table 6.6, provide 4 mm fillet weld (0.442 kN/mm)

(d) Weld for end stiffener

The minimum weld size required for connecting the stiffeners to the web, assuming a weld on each side of the stiffener is (clause 8.7.2.6 of the code)

$$q_1 = t_w^2/5b_s = 12^2/(5 \times 220) = 0.13\,kN/mm$$

when the stiffener also is subjected to external loading, then shear due to such loading must be added to the above value. The stiffeners have to resist the difference between the applied load and the minimum load that can be carried safely by the unstiffened web (370.3 kN in this case).

$$\text{Length of weld} = 1200 - 2 \times 15 = 1170\,mm$$

$$q_2 = (877.5 - 370.3)/1170 = 0.43\,kN/mm$$

$$q_w = q_1 + q_2 = 0.13 + 0.43 = 0.56\,kN/mm$$

$$\text{Force on each weld} = 0.56/2 = 0.28\,kN/mm$$

From Table 6.6, provide a pair of 4 mm fillet welds (0.442 kN/mm).

**EXAMPLE 11.6:** Redesign the plate girder in Example 11.5 with intermediate stiffeners and not using tension field action.

**SOLUTION:**

From Example 11.5,

$$M_z = 4275\,kNm$$

$$F_z = 877.5\,kNm$$

Assume a web plate of size 1400 × 8 mm and a flange plate of size 450 × 35 mm for each of the flanges. Since the flanges are of the same size as in Example 11.5, the calculations for moment capacity check are not repeated here.

**Step 1:** *Check for shear*

According to clause 8.4.2.2 (a) of the code,

Nominal shear strength $V_n = V_{cr}$

Assume $c/d = 1.4$; Hence stiffener spacing = 1.4 × 1400 = 1960 mm

Adopt a stiffener spacing of 2000 mm. Provide 10 panels @ 2000 mm c/c (see Fig. 11.40).

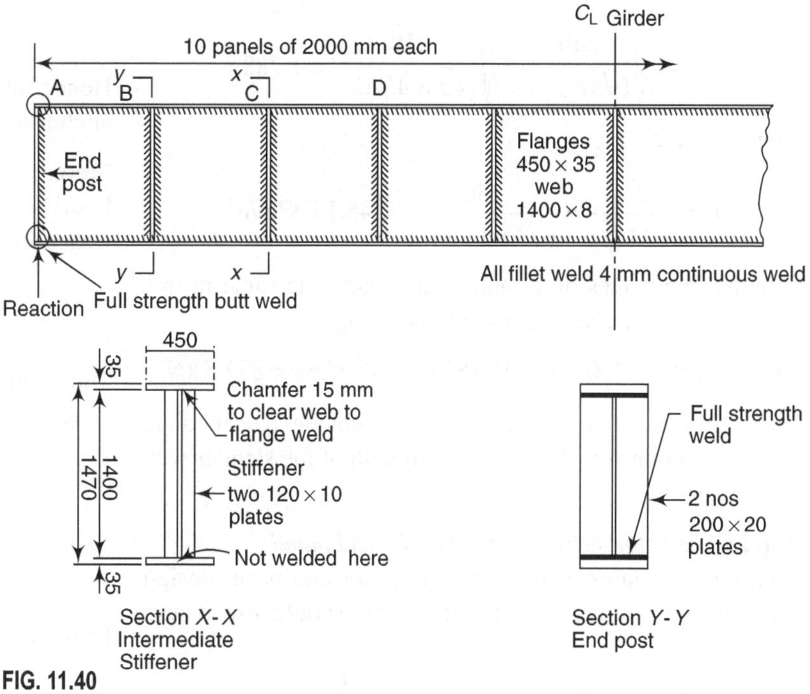


**FIG. 11.40**

Panel AB is the most critical panel for maximum shear. So design checks for the web are made for panel AB only. For the panel AB,

$$\frac{d}{t} = \frac{1400}{8} = 175$$

Since $c$ (2000 mm) > $d$ (1400 mm) and $d/t_w < 200$, the girder satisfies serviceability requirements.

$$\frac{c}{d} = \frac{2000}{1400} = 1.4286$$

$$d/t_w = 175 < 345\varepsilon_f$$

Hence as per clause 8.6.1.2 buckling of compression flange into the web will be avoided.

$$\text{For } \frac{c}{d} > 1.0;\ k_v = 5.35 + \frac{4.0}{\left(\frac{c}{d}\right)^2} = 5.35 + \frac{4.0}{(1.4286)^2}$$

$$= 7.3099$$

with $\mu = 0.3$

Average shear stress

$$q_v = \frac{877.5 \times 10^3}{1400 \times 8} = 78.35\ N/mm^2$$

The elastic critical shear stress of the web

$$\tau_{cr} = \frac{k_v \pi^2 E}{12(1-\mu^2)\left(\frac{d}{t_w}\right)^2} = \frac{7.3099 \times \pi^2 \times 2.0 \times 10^5}{12(1-0.3^2)\left(\frac{1400}{8}\right)^2}$$

$= 45.15$ N/mm$^2$

The non-dimensional web slenderness ratio for shear buckling stress,

$$\lambda_w = \sqrt{\frac{250}{(\sqrt{3}\tau_{cr,e})}} = \sqrt{\frac{250}{(\sqrt{3} \times 45.15)}} = 1.788$$

when $\lambda_w > 1.2$

$$\tau_b = \frac{f_{yw}}{(\sqrt{3}\lambda_w^2)} = \frac{250}{(\sqrt{3} \times 1.788^2)} = 45.15 \text{ N/mm}^2$$

Note that the values of $\tau_{cr}$ and $\tau_b$ are nearly the same in this case. Shear force corresponding to buckling

$$V_{cr} = dt_w\tau_b = 1400 \times 8 \times 45.15/1000 = 505.68 < 877.5 \text{ kN}$$

As per clause 8.7.1.1(a) of the code, provide intermediate stiffeners to improve the buckling strength of the slender web due to shear.

**Step 2:** *Check for shear capacity of the end panel*
According to clause 8.5.1 of the code, for end panel design (without using tension field action), we should use clause 8.5.3,

$$H_q = 1.25V_{dp}\left(1 - \frac{V_{cr}}{V_{dp}}\right)^{1/2}$$

where

$$V_{dp} = dtf_{yw}/\sqrt{3} \quad \frac{(1400 \times 8)250}{\sqrt{3}} = 1616.58 \text{ kN}$$

$$V_{cr} = 505.68 \text{ kN}$$

∴ Longitudinal shear

$$H_q = 1.25 \times 1616.58\left(1 - \frac{505.68}{1616.58}\right)^{1/2} = 1675.12 \text{ kN}$$

$$R_{tf} = \frac{H_q}{2} = \frac{1675.12}{2} = 837.56 \text{ kN}$$

$$A_v = t_w d = 8 \times 1400 = 11200 \text{ mm}^2$$

$$V_n = \frac{A_v f_{yw}}{\sqrt{3}\gamma_{m0}} = \frac{250}{\sqrt{3} \times 1.10} \times \frac{11200}{1000} = 1469.6 \text{ kN} > 837.56 \text{ kN}$$

The end panel is safe to carry the shear due to anchoring forces. Check for the moment capacity of end panel:

$$M_{tf} = \frac{H_q d}{10} = \frac{1675.12 \times 1400 \times 10^3}{10 \times 10^6} = 234.5 \text{ kN m}$$

$$y = \frac{c}{2} = \frac{2000}{2} = 1000 \text{ mm}$$

$$I = \frac{1}{12} t_w \times c^3 = \frac{1}{12} \times 8 \times 2000^3 = 533.33 \times 10^7 \text{ mm}^4$$

$$M_q = \frac{If_y}{y\gamma_{m0}} = \frac{533.33 \times 10^7}{1000 \times 10^6} \times \frac{250}{1.10} = 1212.12 \text{ kNm}$$

$$> M_{tf} = 234.5 \text{ kN m}$$

$$M_{tf} < M_q$$

Hence, the end panel can carry the bending moment due to anchor forces.

**Step 3:** *Design of stiffeners*
Load bearing stiffener at point A (see Fig.11.40),
Reaction at A = 877.5 kN

$$\text{Force } F_m \text{ due to } M_{tf} = \frac{M_{tf}}{c} = \frac{234.5 \times 10^6}{2000 \times 10^3} = 117.25 \text{ kN}$$

Total compression $F_c = 877.5 + 117.25 = 994.75$ kN
According to clause 8.7.5.2 of the code,

$$\text{Area of stiffener } A_q > \frac{0.8 \times F_c \times \gamma_{m0}}{f_{yq}}$$

$$A_q > \frac{0.8 \times 994.75 \times 1.1 \times 10^3}{250} \text{ or } A_q > 3501 \text{ mm}^2$$

Provide stiffener of two flats of size 200 × 20 mm.

$$\text{Area} = 8000 \text{ mm}^2 > 3501 \text{ mm}^2$$

(a) Check for outstand: (clause 8.7.1.2 of the code)
Outstand $b_s = 200$ m $< 20t_q\varepsilon = 20 \times 20 \times 1 = 400$ mm

$$14t_q\varepsilon = 14 \times 20 \times 1 = 280 \text{ mm}$$

$$b_s = 200 \text{ mm} < 14t_q\varepsilon$$

Hence, the criterion for the outstand has been satisfied.

(b) Buckling check: (clause 8.7.1.5 of the code)
Neglecting the buckling resistance of the web for simplicity,

$$I_x = \left(\frac{20 \times 408^3}{12}\right) - \left(\frac{1}{12} \times 20 \times 8^3\right)$$

$$= 113.19 \times 10^6 \text{ mm}^4$$

Effective area $= 200 \times 20 \times 2 = 8000$ mm$^2$

$$\text{Radius of gyration} = \sqrt{\frac{I_x}{A}} = \sqrt{\frac{113.19 \times 10^6}{8000}} = 118.951$$

Flange is restrained against rotation and lateral deflection.

$$L_e = 0.7L = 0.7 \times 1400 = 980 \text{ mm}$$

$$\lambda = \frac{L_e}{r_x} = \frac{980}{118.951} = 8.238 < 10$$

From Table 9c, for $f_y = 250$ N/mm$^2$

$$f_{cd} = 227 \text{ N/mm}^2$$

Buckling resistance of the stiffener

$$P_d = f_{cd}A_e = 227 \times 2 \times 200 \times 20 = 1816\,\text{kN} > 994.75\,\text{kN}$$

Hence, the stiffener is safe against buckling.

(c) Check stiffener as load bearing stiffener

Assume stiff bearing length,

$$b_1 = 0$$

$$n_2 = 2.5 \times 35 = 87.5\,\text{mm}$$

According to clause 8.7.4 of the code,

$$\text{Local capacity of web } F_w = (b_1 + n_2)t_w \frac{f_{yw}}{\gamma_{m0}}$$

$$F_w = (87.5) \times 8 \times \frac{250}{1.10 \times 10^3} = 159.09\,\text{kN}$$

Bearing stiffener is designed for $(F_c - F_w) = (994.75 - 159.09) = 835.66$ kN

Bearing capacity of stiffener alone

$$= \frac{250}{1.10} \times \frac{200 \times 20 \times 2}{1000} = 1818.18\,\text{kN} > 835.66\,\text{kN}$$

Hence the stiffener is safe as a load bearing stiffener.

At the location of concentrated loads, local capacity of the web

$$= (87.5 \times 2) \times 8 \times 250/(1.10 \times 10^3) = 318.16\,\text{kN} > 225\,\text{kN}$$

Hence, the stiffener is safe.

**Step 4:** *Design of intermediate stiffener at 'B' (clause 8.7.2.4 of the code)*

Stiffener B is the most critical intermediate stiffener.

(a) Minimum stiffeners

$$\frac{c}{d} = \frac{2000}{1400} = 1.429 > \sqrt{2}$$

$$I_s \geq 0.75\,dt_w^3$$

$$I_s \geq 0.75 \times 1400 \times 8^3 = 0.5376 \times 10^6\,\text{mm}^4$$

Try intermediate stiffener of two flats of 120 × 10 mm.

$$\text{Provided } I_s = \left(\frac{10 \times 248^3}{12}\right) - \left(\frac{10 \times 8^3}{12}\right) = 12.71 \times 10^6\,\text{mm}^4$$

Hence, the stiffeners have more than the minimum required stiffeness.

(b) Check for outstand (clause 8.7.1.2):

$$\text{Outstand of the stiffener} = b_s = 120\,\text{mm}$$

$$14t_q\varepsilon = 14 \times 10 = 140\,\text{mm}$$

$$b_s < 14t_q\varepsilon$$

Hence the outstand provisions of the code are satisfied.

(c) Buckling check (clause 8.7.2.5):

$$F_q = \left[\frac{V - V_{cr}}{\gamma_{m0}}\right]$$

where $V$ is the factored shear force and $V_{cr}$ is the shear buckling resistance of the web panel designed without using tension field action = 505.68 kN

Shear force @ B,

$$V_B = 877.5 - 45 \times 2 = 787.5\,\text{kN}$$

$$F_q = \frac{(787.5 - 505.68)}{1.10} = 256.2\,\text{kN}$$

Buckling resistance of intermediate stiffener at B (clause 8.7.1.5):

As per clause 8.7.1.5, an effective length of web equal to $20 \times t_w$ on each side of the centre line of stiffener can be considered along with the stiffener.

$$20t_w = 20 \times 8 = 160\,\text{mm}$$

$$I_x = \left(\frac{1}{12} \times 10 \times 248^3\right) + \left(\frac{320 \times 8^3}{12}\right) - \left(\frac{10 \times 8^3}{12}\right)$$

$$= 12.724 \times 10^6\,\text{mm}^4$$

$$\text{Area} = (240 \times 10) + (320 \times 8) = 4960\,\text{mm}^2$$

$$r_x = \sqrt{\frac{12.724 \times 10^6}{4960}} = 50.649\text{mm}$$

$$L_e = 0.7L = 0.7 \times 1400 = 980\,\text{mm}$$

$$\lambda = \frac{L_e}{r_x} = \frac{980}{50.649} = 19.349$$

From Table 9c of the code, for $f_y = 250$ MPa (buckling curve c) and $\lambda = 19.349$,

$$f_{cd} = 224.195 \text{ N/mm}^2$$

$$\text{Buckling resistance} = 224.195 \times 4960 = 1112.01\,\text{kN} > 256.2\,\text{kN}$$

Hence, the stiffener is safe against buckling.

According to clause 8.7.2.5 of the code:

Intermediate stiffener @ D subjected to external load should satisfy the following interaction equation.

$$\left(\frac{F_q - F_x}{F_{qd}}\right) + \left(\frac{F_x}{F_{xd}}\right) + \left(\frac{M_q}{M_{yq}}\right) \leq 1.0$$

where $F_q$ is the stiffener force = 256.2 kN, $F_{qd}$ is the design resistance of an intermediate web stiffener corresponding to buckling about an axis parallel to web = 1112.01 kN, $F_x$ is the external load or reaction at the stiffener = 225 kN, $F_{xd}$ is the design resistance of a load carrying stiffener corresponding to buckling about axis parallel to web at D = 1112.01 kN, and $M_q$ is the moment on the stiffener due to eccentrically applied load = 0.

$$(F_q - F_x) = (256.2 - 225) = 31.2\,\text{kN}$$

Thus, the above check reduces to

$$\frac{31.2}{1112.01} + \left(\frac{225}{1112.01}\right) = 0.23 < 1.0$$

Hence, the stiffener at D is safe.

*Note:* Intermediate stiffener should extend to the compression flange, but not necessarily be connected to it. Stiffeners not subjected to external load or moment can be terminated at a distance of about $4t$ ($4 \times 10 = 40$ mm) from the tension flange. However in our case, since we have a moving load, we may connect the intermediate stiffener to the top (compression) flange and just extend it up to the bottom flange. The calculation for weld at web-flange junction and weld for end and intermediate stiffener are similar to those given in Example 11.5.

**EXAMPLE 11.7:** Redesign the plate girder in Example 11.6 with intermediate stiffeners utilizing tension field action.

**SOLUTION:**

From Example 11.6,

$$M_z = 4275\,\text{kN m}$$

$$F_z = 877.5\,\text{kN}$$

Assume web plate of $1400 \times 8$ mm size and flange size $450 \times 35$ mm size as in Example 11.6.

Assuming that only the flanges resist the moment, the design of flanges is identical to that of Example 11.5.

**Step 1:** *Check for shear*

(For critical panel AB) with the same arrangement of stiffeners, 10 panels of 2000 mm each,

$$c = 2000\,\text{mm}$$

According to 8.4.2.2 (b) (using tension field method),

$$\frac{c}{d} = \frac{2000}{1400} = 1.4285 > 1.0$$

From Example 11.6, $\tau_{cr} = 43.15$ N/mm$^2$; $\tau_b = 45.15$ N/mm$^2$ and $V_{cr} = 505.68\,\text{kN} < 877.5\,\text{kN}$

Nominal shear resistance $V_n = V_{tf}$

$$V_{tf} = [dt_w\tau_b + 0.9w_{tf}t_w f_v \sin\phi] \le V_{np}$$

Yield strength of the tension field

$$f_v = [f_{yw}^2 - 3\tau_b^2 + \psi^2]^{0.5} - \psi$$

$\phi$ = inclination of the tension field

$$= \tan^{-1}\left(\frac{d}{c}\right) = \tan^{-1}\left(\frac{1400}{2000}\right) = 34.992$$

$$\psi = 1.5\tau_b \sin 2\phi = 1.5 \times 45.15 \times \sin(2 \times 34.992) = 63.63$$

$N_f$ = axial force in the flange, due to overall bending =

$$\frac{4275 \times 10^3}{(1400 + 35)} = 2979\,\text{kN}$$

Reduced plastic moment of the respective flange plate,

$$M_{fr} = 0.25 b_f t_f^2 f_{yf}\left[1 - \left(\frac{N_f}{\left(\dfrac{b_f t_f f_{yf}}{\gamma_{m0}}\right)}\right)^2\right]$$

$$M_{fr} = (0.25 \times 450 \times 35^2 \times 250)\left[1 - \left(\frac{2979 \times 10^3}{\left(450 \times 35 \times \dfrac{250}{1.10}\right)}\right)^2\right]$$

$$= 10.59 \times 10^6\,\text{N mm}$$

$$s = \frac{2}{\sin\phi}\left[\frac{M_{fr}}{f_y t_w}\right]^{0.5} \le c$$

$$s = \frac{2}{\sin 34.992}\left[\frac{10.59 \times 10^6}{250 \times 8}\right]^{0.5} = 253.78 < 2000\,\text{mm}$$

The width of the tension field

$$w_{ft} = d\cos\phi - (c - s_c - s_t)\sin\phi$$

$$= (1400 \times \cos 34.992) - (2000 - 253.78 - 253.78) \times \sin 34.992$$

$$= 291.07\,\text{mm}$$

$$f_v = [250^2 - 3 \times 45.15^2 + 63.63^2]^{0.5} - 63.63$$

$$= 182.2\,\text{N/mm}^2$$

$$V_{tf} = [dt_w\tau_b + 0.9w_{tf}t_w f_v \sin\phi] \le V_{np}$$

$$= [(1400 \times 8 \times 45.15) + (0.9 \times 291.07 \times 8 \times 182.2 \times \sin 34.992)]/1000$$

$$= 505.68 + 218.97 = 724.65\,\text{kN}$$

$$< 1400 \times 8 \times 250/(\sqrt{3} \times 1.1 \times 1000)$$

$$= 1469.6\,\text{kN}$$

Note that by considering the tension field action, the shear strength is increased from 505.68 kN (see Example 11.6) to 724.65 kN (43.3% increase in this example).

Nominal shear strength $V_n = V_{tf}$

$$V_n = 724.65\,\text{kN} < 877.5\,\text{kN}$$

By providing end stiffeners, the shear strength may be enhanced.

**Step 2:** *Check for shear capacity of end panel AB*

According to clause 8.5.2, end panel should be provided with an end post consisting of a single or double stiffener.

Assume double stiffeners.

(a) Anchor forces (clause 8.5.3):

$$\text{Resultant longitudinal shear } R_{\text{tf}} = \left(\frac{H_q}{2}\right)$$

$$H_q = 1.25\, V_{\text{dp}}\left[1 - \frac{V_{\text{cr}}}{V_{\text{dp}}}\right]^{1/2}$$

From clause 8.4.1,

$$V_{\text{dp}} = dtf_y/\sqrt{3} = \frac{(1400 \times 8 \times 250)}{(\sqrt{3} \times 1000)} = 1616.58\,\text{kN}$$

Actual factored shear force = 877.5 kN

If the actual shear force in the panel designed using tension field approach is less than the nominal shear stress $V_{\text{nt}}$ (724.65 kN), then the values of $H_q$ may be reduced by the ratio,

$$\left(\frac{V - V_{\text{cr}}}{V_{\text{tf}} V_{\text{cr}}}\right)$$

But in our case the shear force is greater than $V_{\text{nt}}$. From Example 11.6, $H_q = 1675.12$ kN, $R_{\text{tf}} = 837.56$ kN

$$V_n = 1469.6\,\text{kN} > 837.56\,\text{kN}$$

Hence, the end panel is safe in shear.

(b) Check for moment capacity of end panel AB: (clause 8.5.3)

From Example 11.6,

$$M_q = 1212\,\text{kN m} > M_{\text{tf}} = 234.5\,\text{kN}$$

The end panel is safe in carrying the moment.

**Step 3:** *Design of stiffeners*

Loading bearing stiffener at A

Reaction at A = 877.5 kN

From Example 11.6,

$$F_m = 117.25\,\text{kN}$$

$$\text{Required } A_q > 3501\,\text{mm}^2$$

Provide stiffener of two flats of size 150 × 15 mm with area = 4500 mm² > 3501 mm²

(a) *Check for outstand (clause 8.7.1.2 of the code):*

Outstand $b_s = 150$ mm

$$14 t_q \varepsilon = 14 \times 15 \times 1.0 = 210$$

$$20 t_q \varepsilon = 20 \times 15 \times 1.0 = 300$$

$b_s = 150$ mm, hence less than $14t_q\varepsilon$ and $20t_q\varepsilon$

Hence the criterion for the outstand has been satisfied.

(b) Buckling check (clause 8.7.1.5 of the code)

$$I_x = \left(\frac{15 \times 308^3}{12}\right) - \left(\frac{1}{12} \times 15 \times 8^3\right)$$

$$I_x = 36.522 \times 10^6\,\text{mm}^4$$

Effective area = 150 × 15 × 2 = 4500 mm²

$$\text{Radius of gyration} = \sqrt{\frac{I_x}{A}} = \sqrt{\frac{36.522 \times 10^6}{4500}} = 90.08\,\text{mm}$$

Flange restrained against lateral deflection and rotation.

From Example 11.6, $L_e = 980$ mm

$$\lambda = L_e/r_x = 980/90.08 = 10.87$$

From Table 9c of the code (buckling curve *c*), for $f_y = 250$ MPa and $\lambda = 10.87$,

$f_{\text{cd}} = 226.7$ N/mm²

$P_d = f_{\text{cd}} \times A_e = 226.7 \times 150 \times 15 \times 2/10^3 = 1020.15$ kN
$> 877.5 + M_{\text{tf}}/c$
$= 994.75$ kN

Hence, the stiffener is safe against buckling.

**Step 4:** *Check load bearing stiffener as a bearing stiffener*

Assume stiff bearing length $b_1 = 0$

From Example 11.6,

Local capacity of web $F_w = 159.09$ kN

Bearing stiffener is designed for $(F_c - F_w) = 724.65 - 159.09 = 565.56$ kN

Bearing capacity of the stiffener alone

$$= \frac{250}{1.10} \times \frac{150 \times 15 \times 2}{1000} = 1022.72\,\text{kN} > 565.56\,\text{kN}$$

Hence, the stiffener is safe against buckling.

**Step 5:** *Design of intermediate stiffener at B (clause 8.7.2.4)*

Stiffener B is the most critical intermediate stiffener.

(a) Minimum stiffness

From Example 11.6, Required $I_s \geq 0.5376 \times 10^6$ mm⁴

Try intermediate stiffener of two flats of 100 × 10 mm.

$$I_s = \left(\frac{10 \times 208^3}{12}\right) - \left(\frac{10 \times 8^3}{12}\right) = 7.49 \times 10^6\,\text{mm}^4$$

Provided $I_s >$ Required $I_s$

Hence safe.

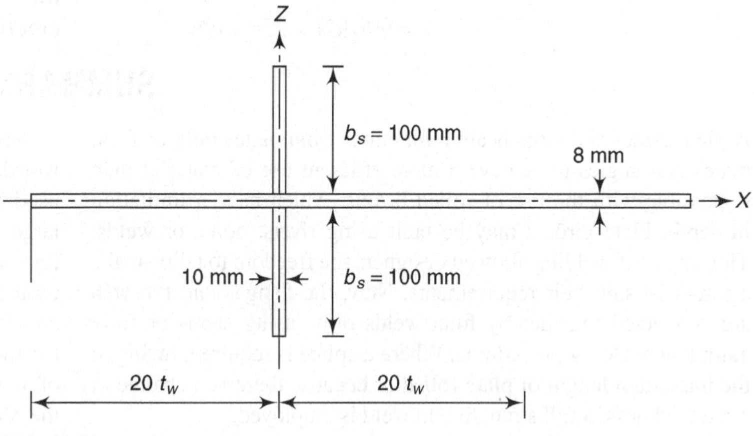


**FIG. 11.41**

(b) Check for outstand (clause 8.7.1.2):

Outstand of the stiffener $b_s = 100\,\text{mm}$

$$14t_q\varepsilon = 14 \times 10 \times 1.0 = 140\,\text{mm}$$

$$20t_q\varepsilon = 200\,\text{mm}$$

$$b_s < 14t_q\varepsilon$$

Hence, the outstand requirement of the code is satisfied.

(c) Buckling Check (clause 8.7.2.5):

$$F_q = \frac{(V - V_{cr})}{\gamma_{m0}}$$

$V$ = Factored shear force

$V_{cr}$ = Shear buckling resistance of the web panel designed without using tension field action.

$$V_{cr} = 505.68\,\text{kN}$$

Shear force at B = 877.5 – 45 × 2 = 787.5 kN

$$F_q = \frac{(787.5 - 505.68)}{1.10} = 256.2\,\text{kN}$$

Buckling resistance of intermediate stiffener at B as per clause 8.7.1.5,

$$I_x = \left(\frac{1}{12} \times 10 \times 208^3\right) + \left(\frac{40 \times 8 \times 8^3}{12}\right) - \left(\frac{10 \times 8^3}{12}\right)$$

$$= 7.512 \times 10^6\,\text{mm}^4$$

$$\text{Area} = (200 \times 10) + (40 \times 8 \times 8) = 4560\,\text{mm}^2$$

$$r_x = \sqrt{\frac{7.152 \times 10^6}{4560}} = 40.58\,\text{mm}$$

$$L_e = 0.7L = 0.7 \times 1400 = 980\,\text{mm}$$

$$\lambda = \frac{L_e}{r_x} = \frac{980}{40.58} = 24.14$$

From Table 9c of the code (buckling curve $c$), for $f_y = 250\,\text{MPa}$ and $\lambda = 24.14$,

$$f_{cd} = 218.6\,\text{N/mm}^2$$

$$\text{Buckling resistance} = 218.6 \times 4560/10^3$$

$$= 996\,\text{kN} > 256.2\,\text{kN}$$

Hence, the stiffener is safe in buckling.

According to clause 8.7.2.5 of the code, intermediate stiffener subjected to external load should satisfy the following interaction equation

$$\left(\frac{F_q - F_w}{F_{qd}}\right) + \left(\frac{F_x}{F_{xd}}\right) + \left(\frac{M_q}{M_{yq}}\right) \le 1.0$$

$$F_q = 256.2\,\text{kN}$$

$$F_{qd} = 996\,\text{kN}$$

$$F_x = 225\,\text{kN}$$

$$F_{xd} = 996\,\text{kN}$$

Since $F_q - F_w = 256.2 - 225 = 31.2\,\text{kN}$, also $M_q = 0$.

Hence the interaction equation reduces to

$$\frac{31.2}{996} + \left(\frac{225}{996}\right) = 0.257 < 1.0$$

Hence, the intermediate stiffener is safe. The details of the plate girder are similar to the girder shown in Fig. 11.40.

*Note*: The stiffeners should be connected to the web and to the compression (top) flange by welds. The calculations are similar to those given for Example 11.5.

In the preceding designs, it has been assumed that adequate lateral bracing will be provided for the compression flange of the plate girder. The calculations for location, size, and connection of lateral bracing are not included here (see Section 10.4.3.2 for the details). Normally, these bracings will be attached to the intermediate stiffeners by bolting at site. In this case, a weld between the stiffener and the compression flange would be provided to transfer the force in the bracing to the girder (Kulak & Grondin 2002). It is also usual to reduce the cross section of a welded girder at least once as the moment decreases. A transition as shown in Fig. 6.18 of the Chapter 6 will be adopted which may allow for the gradual flow of stress from one section to another. In practice, the design process would be repeated for several assumed cross sections and the best cross section, in terms of material, fabrication, and erection costs, will be chosen.

## SUMMARY

A plate girder is a large beam built either from plates only or from plates and angles to achieve a more efficient use of material than is possible with the use of rolled beams, which have a limitation in depth. Plate girders may be built using rivets, bolts, or welds. The advent of welding allowed designers the freedom to tailor-make a girder to suit their requirements. Now, the flanges and the web are connected together by fillet welds only, using semi- or fully automatic welding procedures. Where a splice is required, owing to the maximum length of plate rolled or because there is a change in plate thickness, a full strength butt weld is employed.

Several long span plate girders have been built all over the world. Box girders and hybrid girders are being increasingly used for long spans and heavy loads. Box girders, due to their large torsional rigidity, compared to plate girders, are adopted for curved bridges. In hybrid girders, different grades of steel are used for flanges (high strength) and webs (low strength) to obtain an efficient cross section. Sometimes, notch ductile steel is used for the tension flange of a road bridge to eliminate the possibility of low cycle brittle fracture. Since plate girders are quite deep, the steel need to be distributed economically between the flanges and the web so as to minimize the weight of the girder in a given situation.

Since thin webs are often adopted, the usual assumption made for small and medium plate girders is that the flanges resist most of the bending moment and the web carries the shear force. The flanges are assumed to be plastic, compact, or semi-compact. It is also possible to design assuming that that the bending moment and axial force are resisted by the whole section and the web resists the combined shear and normal stresses. The design of flanges in plate girders is similar to the design of flanges of rolled sections.

The webs are often thin and are likely to buckle under compression or shear. To avoid buckling of web due to shear and bending, and buckling or crippling of web at points of concentrated loads, the web has to be stiffened by intermediate stiffeners, horizontal stiffeners, and bearing stiffeners and their combinations. Experiments conducted in the past have shown that the web will be capable of withstanding loads in excess of the elastic buckling load, provided sufficiently heavy stiffeners are employed. This occurs as a result of *tension field action* in which the diagonal web tensile stresses act with the transverse stiffeners and the flanges to transfer the additional load by means of a truss type of action. Ultimate load is reached when the tension field yields. The code has adopted the tension field action as suggested by Rockey, Evans, and Porter of England. The details of this method, which affects the end panel of the plate girder, are explained. Various types of transverse stiffeners and longitudinal stiffeners have also been considered in depth. The provisions for the design of plate girders, as given in IS 800 : 2007 have been discussed at great length.

Plate girders with corrugated webs are being increasingly used, since they provide an aesthetic appearance as well as increase the shear strength of the web, without the use of stiffeners. Hence, the design provisions of these types of girders are discussed. Box girders are used mainly in bridge structures and an introduction about these girders is also given in this chapter.

The steps involved in the design of plate girders and some thumb rules for the preliminary proportioning of plate girders has also been discussed. The examples provided illustrate the use of equations given in this chapter.

## MULTIPLE-CHOICE QUESTIONS

11.1 For large spans and heavy loads, which of the following will be economical?
(a) I-section with plates
(b) Two I-beams placed side by side
(c) Truss
(d) Plate girder

11.2 Plate girders are often chosen in lieu of trusses when there are long spans, because they
(a) require smaller vertical clearances
(b) have lower cost for fabrication
(c) are safer than trusses
(d) all of these

11.3 The disadvantage of plate girder is
(a) it has low torsional stiffness
(b) it has more exposed wind area than truss
(c) susceptible to stability problems during erection
(d) all of these

11.4 As per IS 800, with respect to serviceability and when transverse stiffeners are not provided, the $d/t_w$ ratio of the web, should be less than
(a) $150\varepsilon_w$ (b) $200\varepsilon_w$ (c) $250\varepsilon_w$ (d) $400\varepsilon_w$

11.5 The non-dimensional web slenderness ratio for shear buckling stress, $\tau_w$, is given by
(a) $\sqrt{f_{yf}/\left(\sqrt{3}\tau_{cr,e}\right)}$
(b) $\sqrt{f_{yw}/\left(\tau_{cr,e}\right)}$
(c) $\sqrt{f_{yw}/\left(\sqrt{3}\tau_{cr,e}\right)}$
(d) $\sqrt{f_{yw}/\left(3\tau_{cr,e}\right)}$

11.6 When $\lambda_w \le 0.8$, the nominal shear strength of the web, $V_n$, is given by
(a) $A_v f_{yw}/\sqrt{3}$
(b) $A_v f_{yw}/\left(\sqrt{3}\lambda_w^2\right)$
(c) $\left(A_v f_{yw}/\sqrt{3}\right)[1-0.8\left(\lambda_w-0.8\right)]$
(d) $A_v f_{yw}/\left(\sqrt{3}\lambda_w\right)$

11.7 In the tension field method, the nominal shear strength of the web, $V_{tf}$, is given by
(a) $A_v\tau_b + 0.9w_{tf}t_w f_v$
(b) $A_v\tau_b + w_{tf}t_w f_v \sin\phi$
(c) $A_v\tau_b + 0.9w_{tf}t_w f_v \sin\phi$
(d) $A_v\tau_{cr,e} + 0.9w_{tf}t_w f_v \sin\phi$

11.8 A length of the web, on both sides of the stiffener, acting as part of the bearing stiffener in compression is
(a) $10t_w$ (b) $12t_w$ (c) $16t_w$ (d) $20t_w$

11.9 The outstand of web stiffener from the face of the web is restricted to
(a) $12t_q\varepsilon$ (b) $14t_q\varepsilon$ (c) $16t_q\varepsilon$ (d) $20t_q\varepsilon$

11.10 Intermediate vertical stiffeners are provided in plate girders to
(a) prevent local buckling
(b) prevent web buckling
(c) prevent excessive deflection
(d) prevent flange buckling

11.11 The effective length of intermediate transverse web stiffener is taken as ____________ times the length of the stiffener
(a) 0.6 (b) 0.7 (c) 0.75 (d) 0.8

11.12 When intermediate transverse web stiffeners are not subjected to external load, a gap is provided between the tension flange and the stiffener which equals
(a) $4t_w$ (b) $5t_w$ (c) $6t_w$ (d) $7.5t_w$

11.13 The connection of vertical stiffener to the web of plate girder are designed for a shear force of
(a) $t_w^3/5b_s$ (b) $t_w^2/5b_s$ (c) $t_w/5b_s$ (d) $t_w^2/b_s$

11.14 A longitudinal stiffener is efficient when it is placed in the compression region at a distance within a distance equal to
(a) $d/3$ from compression edge
(b) $d/4$ from compression edge
(c) $d/5$ from compression edge
(d) none of these

11.15 The optimum depth of plate girder is
(a) $\left(\frac{Mk}{f_y}\right)^{0.33}$
(c) $\left(\frac{Mk^2}{f_y}\right)^{0.33}$
(b) $\left(\frac{M}{f_y k^2}\right)^{0.33}$
(d) $\left(\frac{M}{f_y}\right)^{0.33}$

11.16 Why/where are bearing stiffeners provided?
(a) At the end of the plate girder
(b) At the ends of the plate girder and on both faces of the eb
(c) At the ends of the plate girder and only on one face of the web
(d) Where concentrated loads are acting and on both faces of the web

11.17 Why/where are intermediate transverse stiffeners provided?
(a) Where concentrated loads are acting
(b) In order to reduce the panel aspect ratio
(c) For aesthetic purposes
(d) On one face of the web only for aesthetic reasons

11.18 The portion of the web that acts integrally with the vertical stiffener in resisting the load is taken as
(a) 20 times the thickness of the web on either side of the stiffener
(b) 40 times the thickness of the web on either side of the stiffener
(c) 12.5 times the thickness of the web on either side of the stiffener
(d) None of these

11.19 Why/where longitudinal stiffeners provided?
(a) At d/5 distance from the compression flange, where d is the depth of the plate girder
(b) in order to increase the shear buckling strength
(c) for aesthetical purposes
(d) all of these
(e) none of these

## EXERCISES

11.1 Determine the buckling resistance moment for a welded plate girder consisting of 500 × 30 mm flange plates and a 1250 × 12 mm web plate in grade 410 steel. Assume a laterally braced span of 5.5 m.

11.2 Compute the design flexural strength of a plate girder which consists of a 8 × 1800 mm web and 30 × 450 mm flanges. Use grade 410 steel and assume that the compression flange is continuously supported.

11.3 A welded plate girder is fabricated from two 600 × 35 mm flange plates and 1500 × 12 mm web plate of grade 410 steel. What is the moment capacity of the girder? If the section is required to carry full plastic moment, what would be the changes required in the plate thickness?

11.4 Determine the shear buckling resistance of the plate girder web given in Exercise 11.1, if intermediate transverse stiffeners are spaced at 1600 mm.

11.5 A plate girder of grade 410 steel has a 12 × 1800 mm web and 75 × 550 mm flanges.
(a) Determine the design flexural strength, if the compression flange is continuously laterally supported.
(b) Determine the flexural strength, if the unrestrained length is 10 m.
(c) What is the design shear strength if no intermediate stiffeners are used.
(d) Determine the shear strength, if the stiffeners are provided with spacing of 2.5 m.
(e) Find the shear strength of the end panel, if the first intermediate stiffener is provided at 1.5 m from the end support.

11.6 Check the capacity of a pair of intermediate transverse web stiffeners of size 100 × 12 mm for the plate girder mentioned in Exercise 11.3.

11.7 A plate girder web is to be fabricated by a plate 1200 mm deep and 12 mm thick. The girder should be capable of carrying a shear load of 1400 kN without using tension field action. Assuming grade 410 steel, determine the spacing of vertical stiffeners required. What would be the increase in load carrying capacity, if tension field action is permitted?

11.8 A plate girder of grade 410 steel is composed of a 10 × 2000 mm web and 30 × 500 mm flanges. The girder span is 15 m (see Fig. 11.42). Stiffeners are placed at 1 m, 3 m, and 5 m from both ends. Determine the shear strength of each of the panels.

11.9 Check the adequacy of a pair of load bearing stiffeners 100 × 12 mm, which are placed at the support of the plate girder mentioned in Exercise 11.2. Assume that the flanges of the girder are not restrained by other structural elements against rotation. The girder is supported on a stiff bearing 250-mm long, the end panel width is 1000 mm and the reaction(factored) is 800 kN.

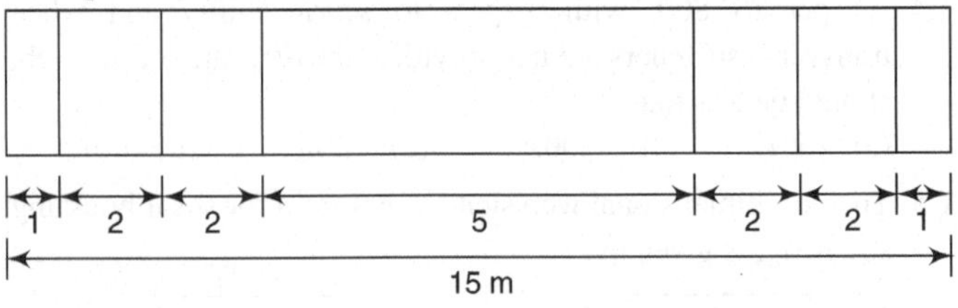


**FIG. 11.42**

11.10 Design an 18-m long simply supported welded plate girder carrying a uniformly distributed load of 50 kN/m excluding self weight, and two concentrated loads of 350 kN each at quarter points of the span. Assume that the girder is laterally supported throughout.

11.11 A 20-m long plate girder has to support a UDL and concentrated loads at one-third points. The uniform load consists of 18 kN/m dead load and 30 kN/m live load.

Each concentrated load consists of a 125 kN dead load and a 225 kN live load. There is lateral support at the ends and at the points of concentrated load. Using grade 410 steel determine the following.

(a) Design the cross section.
(b) Determine the location and size of intermediate stiffeners.
(c) Design suitable bearing stiffeners at the supports and beneath the loading points.
(d) Design welds for all the elements.

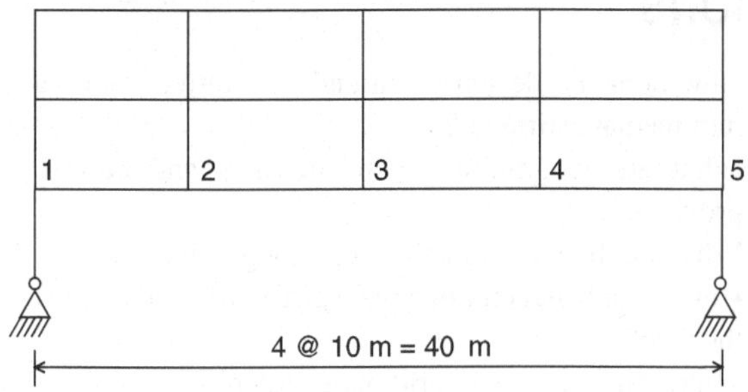


**FIG. 11.43**

11.12 A plate girder 12345 is to be provided in a building in order to create a large column-free area as shown in Fig. 11.43. The uniform load on the girder consists of 22 kN/m dead load (not including the weight of the girder) and 30 kN/m live load. In addition, the girder must support the column loads at points 2, 3, and 4 consisting of 500 kN dead load and 750 kN live load at each location. The girder is simply supported and the column loads are concentrated. The compression flange of the girder is laterally supported at 3 m intervals.
Using grade 410 steel determine the following.
(a) Design the cross section.
(b) Determine the location and size of intermediate stiffeners.
(c) Design the required bearing stiffeners at points 1, 2, 3, 4, and 5.
(d) Design welds for the components.

11.13 For the plate girder shown in Fig. 11.44, calculate the adequacy for moment, shear, and stiffener requirements. The steel is of grade 410 and stiffeners are on one side only. The unbraced length is 12 m. Spacing of stiffeners is 1900 mm.
Intermediate stiffener: 22 mm × 130 mm
Bearing stiffener: 25 mm × 180 mm

11.14 Design a welded plate girder using 410 grade steel for the codal specifications for the loading shown in Fig. 11.45. Lateral bracings are located at 3 m intervals. Try one cross section without and another with transverse stiffeners. Compare the two designs.

11.15 Design a welded plate girder to support a 10 kN UDL and a 560 kN concentrated load at mid span. The girder is simply supported with a span of 16 m and has sufficient lateral support for the compressive flange (see Fig. 11.46). Use 410 grade steel.

11.16 Determine the shear strength for a girder with transverse stiffeners located at each end and 500 mm from each end along with stiffeners every 1000 mm within the span. The web plate is 1000 × 8 mm and flange is 200 × 25 mm. Determine the moment capacity for the plate girder and the factored UDL that the girder may carry on a 25 m span.

11.17 A plate girder that spans 18 m between two columns in a multi-storey building is to be designed. The girder is one of a series over an assembly room where the columns were deleted within the assembly room. The overall girder depth cannot exceed 1800 mm. Service loads on the girder are as follows: live load = 15 kN/m and dead load = 30 kN/m (includes an estimate for the girder weight). Concentrated loads (from the columns above the girder) at 6 m from each end of the girder are as follows: live load = 500 kN and dead load = 140 kN. Use conventional design method without any intermediate web stiffeners for 410 grade steel. Solve the problem using $f_y = 330$ MPa for the flanges and $f_y = 250$ MPa for webs and stiffeners. Solve this using tension field design method with a web plate thickness of 8 mm.

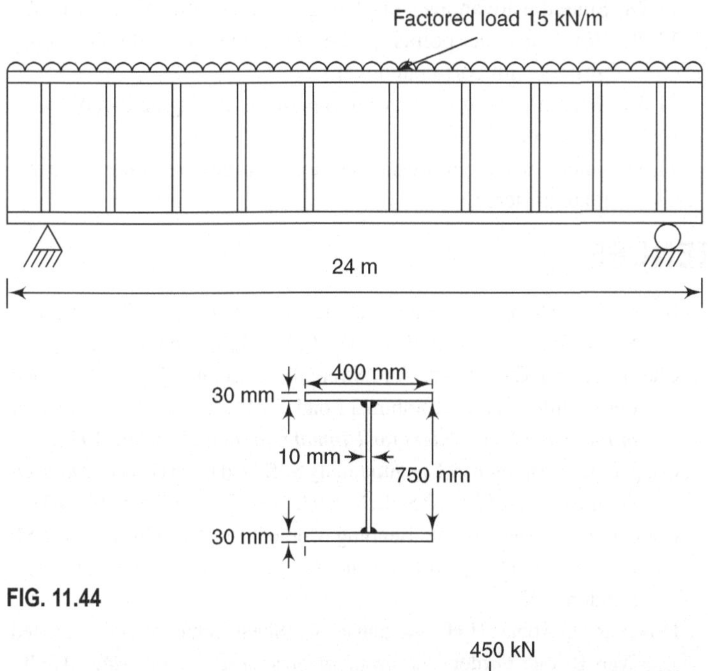


**FIG. 11.44**

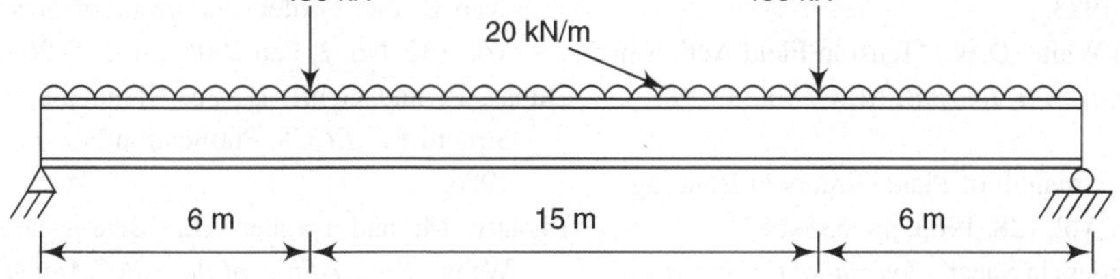


**FIG. 11.45**

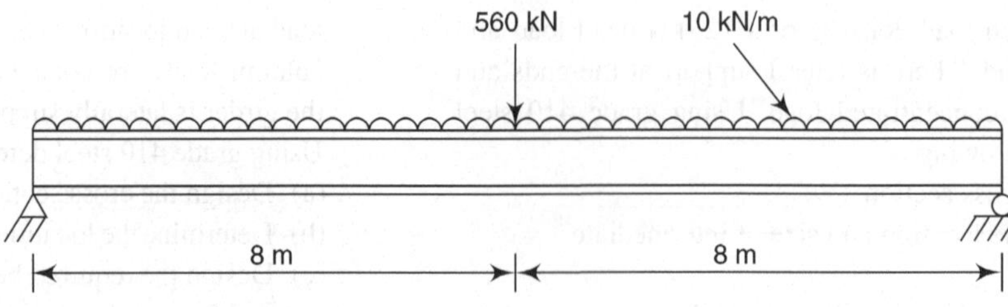


FIG. 11.46

## REVIEW QUESTIONS

11.1 What is the difference between a beam and a plate girder?
11.2 Where are plate girders used?
11.3 What are the main characteristics of a plate girder?
11.4 State some advantages and disadvantages of plate girders over trusses.
11.5 Sketch the different types of cross sections used as plate and box girders.
11.6 Why have bolted and riveted plate girders become obsolete?
11.7 What are the different modes of failures of a plate girder?
11.8 List the different elements of a welded plate girder.
11.9 Explain the tension field action that is developed in the thin webs of plate girders?
11.10 What are the various types of stiffeners?
11.11 State the minimum web thickness provisions of a IS 800 : 2007.
11.12 What are the design concepts of a plate girder?
11.13 When do we need to consider the shear buckling of the web?
11.14 How does plate girder derive post-buckling strength?
11.15 When do we provide splicing of webs and flanges?
11.16 What is meant by curtailment of plates and how this done in plate girders?
11.17 What is meant by patch loading?
11.18 What is an end post?
11.19 How does the design of an end post differ from that of an intermediate stiffener?
11.20 What are longitudinal web stiffeners and how are they provided?
11.21 What are the main functions of a longitudinal stiffener?
11.22 How is the behaviour of a plate girder affected by the holes in the web?
11.23 What is meant by a hybrid plate girder?
11.24 What are the important points that should be remembered while connecting the stiffeners to the web and flanges of a plate girder?
11.25 State the equations for the optimum value of depth and web thickness of a plate girder subjected to a moment $M$.
11.26 What is the range of the minimum thickness of the web that is usually adopted in practice?
11.27 What are the steps involved in the design of plate girders?
11.28 State the advantages of plate girders with corrugated webs?
11.29 What are the points to be remembered while designing girders with corrugated webs?
11.30 What are the modes of failures of the corrugated web due to shear buckling?
11.31 Under what circumstances are box girders preferred over plate girders?

## REFERENCES

Alinia, M.M. and Dastfan, M., Behaviour of Thin Plate Shear Walls regarding Frame Members, *Journal of Constructional Steel Research*, Vol. 62, 2006, pp. 730–738.

Allen, H.G. and Bulson, P.S., *Background to Buckling*, McGraw-Hill, UK, 1980.

ANSI/AISC:341-10, *Seismic Provisions for Structural Steel Buildings*, American Institute of Steel Construction, Chicago, IL, 2005, p. 333.

Astaneh-Asl, A. and Zhao, Q., *Cyclic Test of Steel Shear Walls—Final Report*, Dept of Civil and Environmental Engg., University of California at Berkeley, Aug 2002.

Ballio, G. and Mazzolani, F.M., *Theory and Design of Steel Structures*, Chapman & Hall, London, 1983.

Barker, M.G., Hurst, A.M., and White, D.W., 'Tension Field Action in Hybrid Steel Girders', *Engineering Journal*, AISC, Vol. 39, No. 1, 2002, pp.52–62.

Basler, K. and Thürlimann, B., 'Strength of Plate Girders in Bending', Part II, *Transactions, ASCE*, Vol. 128, 1963, pp. 655–682.

Basler, K., 'Strength of Plate Girders in Shear', *Journal of the Structural Div., Proc.* ASCE, 87, No. ST 7, Oct 1961, pp. 151–180.

Basler, K., 'Strength of Plate Girders under Combined Bending and Shear', Part II, *Transactions*, ASCE, Vol. 128, 1963, pp. 720–735.

Chern, C. and Ostapenko, A., 'Strength of Longitudinally Stiffened Plate Girders under Combined Loads', *IABSE Colloquium, Design of Plate and Box Girders for Ultimate Strength*, London, 1971.

Cooper, P.B., 'Strength of Longitudinally Stiffened Plate Girders', *Journal of the Structural Div.*, ASCE, Vol. 93, No. ST2, 1967, pp. 419–451.

Cooper, P.B., 'The Ultimate Bending Moment for Plate Girders', *IABSE Colloquium, Design of Plate and Box Girders for Ultimate Strength*, London, 1971.

Driver, R.G., Abbas, H.H., and Sause, R., 'Shear Behavior of Corrugated Web Bridge Girders', *Journal of Structural Engineering*, ASCE, Vol. 132, No. 2, Feb 2006, pp. 195–203.

Dubas, P., and Gehri, E., ed. 'Behaviour and Design of Steel Plated Structures', *ECCS*, Publication No. 44, ETH Hönggerberg, Zürich, 1986.

Elgaaly, M. and Dagher, H., 'Beams and Girders with Corrugated Webs', *Proceedings of the SSRC Annual Technical Session*, Lehigh University, Bethlehem, PA, 1990, pp. 37–53.

Elgaaly, M. and Seshadri, A., 'Girders with Corrugated Webs under Partial Compressive Edge Loading', *Journal of Structural Engineering*, ASCE, Vol. 123, No. 6, Jun 1997, pp. 783–791.

Elgaaly, M., Hamilton, R.W., and Seshadri, A., 'Shear Strength of Beams with Corrugated Webs', *Journal of Structural Engineering*, ASCE, Vol. 122, No. 4, Apr 1996, pp. 390–398.

Elgaaly, M., 'Plate and Box Girders', in *Handbook of Structural Engineering*, W.F. Chen, ed., CRC Press, Boca Raton, FL, 1997, pp. 19–1 to 19–23.

Elgaaly, M., Seshadri, A., and Hamilton, R.W., 'Bending Strength of Steel Beams with Corrugated Webs', *Journal of Structural Engineering*, ASCE, Vol. 123, No. 6, Jun 1997, pp. 772–782.

Elgaaly, M., Web Design under Compressive Edge Loads, *Engineering Journal*, AISC, Vol. 20, No. 4, 1983, pp. 153–171.

ESDEP—The European Steel Design Education Programme—Lectures, http://www.fgg.uni-lj.si/~/pmoze/ESDEP/master/wg08/l0410.htm and http://www.fgg.uni-lj.si/~/pmoze/ESDEP/master/wg08/l0420.htm, Katholieke Universiteit, Department of Building Materials and Building Technology, Nederlands, 15 Apr 2006.

Evans, H. R., Longitudinally and transversely reinforced plate girders, Chapter 1. Plated structures, stability and strength, Narayanan. R, Ed., *Elsevier applied Science Publishers*, London, 1983, pp.1–38.

Evans, H.R. and Tang, K.H., 'Longitudinal Stiffeners for Girder Webs: Their Behaviour and Design', *Journal of Constructional Steel Research*, Vol. 2, 1986, pp. 173–198.

Fan, Y-L, Mo., Y.L. and Herman, R.S., Hybrid Bridge Girders—A Preliminary Design Example, *Concrete International*, ACI, Vol. 28, No. 1, Jan 2006, pp. 65–69.

Fujii, T., 'A Comparison between Theoretical Values and Experimental Results for the Ultimate Shear Strength of Plate Girders', *IABSE Colloquium, Design of Plate and Box Girders for Ultimate Strength,* London, 1971.

Galambos, T.V., ed., *Guide to Stability Design Criteria for Metal Structures*, Fifth edition, John Wiley, New York, 1998.

Gaylord, Jr, E.H., Gaylord, C.N., and Stallmeyer, J.E., *Design of Steel Structures*, Third edition, McGraw-Hill, New York, Tokyo, 1992.

Herzog, M., 'Ultimate Strength of Plate Girders from Tests', *Journal of the Structural Div.*, ASCE, Vol. 100, No. ST5, May 1974, pp. 849–864.

Höglund, T., 'Design of Thin Plate I-girders in Shear and Bending with Special Reference to Web Buckling', Bulletin No. 94, *Div. Building Statics and Structural Engineering, Royal Inst. of Tech.*, Stockholm, 1973, p. 43.

Johansson, B., Maquoi, R., Sedlacek, G., Müller, C., and Beg, D., 'Commentary and Worked Examples to EN 1993-1-5' *Plated Structural Elements*, Joint Report, European Commission(ECCS-CECM-EKS), Aachen, Germany, Oct 2007, p. 242.

Komatsu, S., 'Ultimate Strength of Stiffened Plate Girders Subjected to Shear', *IABSE Colloquium, Design of Plate and Box Girders for Ultimate Strength*, London, 1971.

Kulak, G.L. and Grondin, G.Y., *Limit States Design in Structural Steel*, Seventh edition, Canadian Institute of Steel Construction, Toronto, Ontario, 2002.

Lee, S.C., Yoo, C.H., and Yoon, D.Y., 'New Design Rules for Intermediate Transverse Stiffeners Attached on Web Panels', *Journal of Structural Engineering*, ASCE, Vol. 129, No. 12, Dec. 2003, pp. 1607–1614, (Errata, Vol. 130, 2004, p. 693).

Lindner, J. and Huang, B., Beulwerte für trapezförmig profilierte Bleche under Schubbeanspruchung, *Der Stahlbau*, Vol. 61, No. 12, Dec 1995, pp. 370–374..

Lindner, J., Grenzschubtragfähigheit von I trägern mit trapaezförmig profilierten stegen, *Der Stahlbau*, Vol.57, No. 12, 1988, pp. 377–380.

Loov. R. and Parthasarathi, N., 'Study of Shear Provisions for Stiffened Plate Girders', *Canadian Journal of Civil Engg.*, Vol. 31, 2004, pp. 160–167.

Machacek, J. and Tuma, M., 'Fatigue Life of Girders with Undulating Web', *Journal of Constructional Steel Research*, Vol. 62, 2006, pp. 168–177.

Maquoi, R. and Skaloud, M., 'Stability of Plates and Plated Structures: General Report', *Journal of Constructional Steel Research*, Vol. 55, Nos 1–3, Jul 2000, pp. 45–68.

Maquoi, R., *Plate Girders* in *Constructional Steel Design— An International Guide*, Dowling, P.J., Harding, J.E. and Bjorhovde, R., eds, Elsevier Applied Science Publishers, London, 1992, pp. 133–173.

Martin, L.H and Purkiss, J.A., *Structural Design of Steelwork to BS 5950*, Edward Arnold, A division of Hodder and Stoughton, London, 1992.

Narayanan, R. and Roberts, T.M., ed., *Structures Subjected to Repeated Loading: Stability and Strength*, Spon Press, UK, 1990, p. 291.

Narayanan, R., ed. *Plated Structures: Stability and Strength,* Elsevier Applied Science Publishers, London, 1983.

Paik, J.K and Thayamballi, A.K., *Ultimate Limit State Design of Steel Plated Structures*, John Wiley & Sons Ltd, England, 2003, p. 544.

'Plate Girders I and II', Chapters 15, 16, and 28, *Teaching resources for structural steel design, INSDAG*, Kolkata, 2000.

Porter, D.M., Rockey, K.C., and Evans, H.R., 'The Collapse of Plate Girders Loaded in Shear', *The Structural Engineer*, Vol. 53, No. 8, 1975, pp. 313–325.

Roberts, T.M. and Newark, A.C.B., 'Strength of Webs Subjected to Compressive Edge Loading', *Journal of Structural Engineering*, ASCE, Vol. 123, No. 2, Feb 1997, pp. 176–183.

Roberts, T.M., 'Slender Plate Girders Subjected to Edge Loading', *Proceedings, Institution of Civil Engineers,* Part 2, Vol. 71, Sep 1981, pp. 805–819.

Rockey, K.C. and Skaloud, M., 'The Ultimate Load Behaviour of Plate Girders Loaded in Shear', *IABSE Colloquium, Design of Plate and Box Girders for Ultimate Strength,* Vol. II, London, 1971, pp. 1–19.

Rockey, K.C., Evans, H.R., and Porter, D.M., 'A Design Method for Predicting Collapse Behaviour of Plate Girders', *Proceedings of the Institution of Civil Engineers,* Vol. 65, 1978, pp. 85–112.

Rockey, K.C., Valtinat, G., and Tang, K.H., 'The Design of Transverse Stiffeners on Webs Loaded in Shear—An Ultimate Load Approach', *Proceedings Institution of Civil Engineers,* Part 2, Vol. 71, No. 2, Dec 1981, pp. 1069–1099.

Salmon, C.G. and Johnson, J.E., *Steel Structures: Design and Behavior*, Fourth edition, Harper Collins College Publishers, 1996, pp. 610–695.

Schilling, C.G., Chairman, 'Design of Hybrid Steel Beams, Report of the Subcommittee 1 on Beams and Girders, Joint ASCE-AASHO

Committee on Flexural Members', *Journal of the Structural Div.*, ASCE, Vol. 94, No. ST6, Jun 1968, pp. 1397–1426.

Schilling, C.G., 'Optimum Proportions for I-shaped Beams', *Journal of the Structural Div.,* ASCE, Vol. 100, No. ST12, Dec 1974, pp. 2385–2401.

Segui, W.T., *LRFD Steel Design*, Second edition, *PWS Publishing*, New York, 1999.

Sharp, M.L. and Clark, J.W., 'Thin Aluminum, Shear Webs', *Journal of the Structural Div.,* ASCE, Vol. 97, No. ST4, Apr 1971, pp. 1021–1038.

Steinhardt, O. and Schröter, W., 'Post Critical Behaviour of Aluminium Plate Girders with Transverse Stiffeners', *IABSE Colloquium, Design of Plate and Box Girders for Ultimate Strength,* London, 1971.

Subramanian, N., 'Innovations in Plate Girder Design', *New Building Materials & Construction Review*, Jul 2011, pp. 254–263.

Takeuchi, T., 'Investigation of the Load Carrying Capacity of Plate Girders', *MS Thesis*, University of Kyoto, Japan, (In Japanese), 1964.

Task committee on longitudinally stiffened plate girders of the ASCE—AASHTO Committee on Flexural Members of the Committee on Metals of the Structural Div., 'Theory and Design of Longitudinally Stiffened Plate Girders', *Journal of the Structural Div.*, ASCE, Vol. 104, No. ST4, Apr 1978, pp. 697–716.

Valitnat, G., Tragsicherheitsnachweise für spezielle trägerformen-vollwandträger mit schlanken stegen und vertikalsteifen, Stahlbau Handbuch für Studium und Praxis, Band 1, *Stahlbau-Verlags GMBH*, Köln, 1982, pp. 531–542.

Veljkovic, M. and Johansson, B., 'Design of Hybrid Steel Girders', *Journal of Constructional Steel Research*, Vol. 60, 2004, pp. 535–547.

Zahn, C.J., 'Plate Girder Design using LRFD', *Engineering Journal*, AISC, Vol. 24, No. 1, 1987, pp. 1–20.

# 12

# DESIGN OF GANTRY GIRDERS

## INTRODUCTION

In mills and heavy industrial buildings such as factories and workshops, gantry girders supported by columns and carrying cranes are used to handle and transport heavy goods, equipment, etc. There are several types of cranes; overhead travelling, under-slung, jib, gantry, and monorail are among the most common. A building may have one or several of these, either singly or in combinations. Hand-operated overhead cranes have lifting capacities of up to 50 kN and electrically operated overhead travelling cranes, called EOT cranes, can have capacities in the range of 10–3000 kN. The overhead travelling crane runway system consists of the following components (see Fig. 3.2 of Chapter 3, Figs 12.1, and 12.2):

1. The crane, comprising the crane girder (crane frame), crab or trolley, hoist, power transmitting devices, and a cab which houses the controls and operator
2. The crane rails and their attachments
3. The gantry girder
4. The gantry girder supporting columns or brackets
5. The crane stops

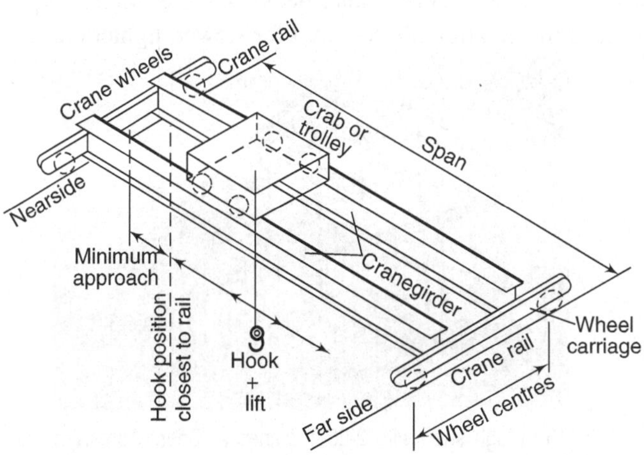


FIG. 12.1 Components of an overhead crane

The loads are lifted using a hook and moved longitudinally and transversely anywhere in the building through the movement of a crab car or trolley on the crane frame (girder) or truss and the crane wheels on the crane rails (see Figs 12.1 and 12.2). The gantry girder is supported by brackets attached to the main columns of the building or by stepped columns.

The rating of the cranes is based on the hoisting capacity, which in turn depends on the size of the crab, wheel spacing, roof clearance, etc. Usually this data is supplied by the manufacturer of cranes. Power and other requirements of

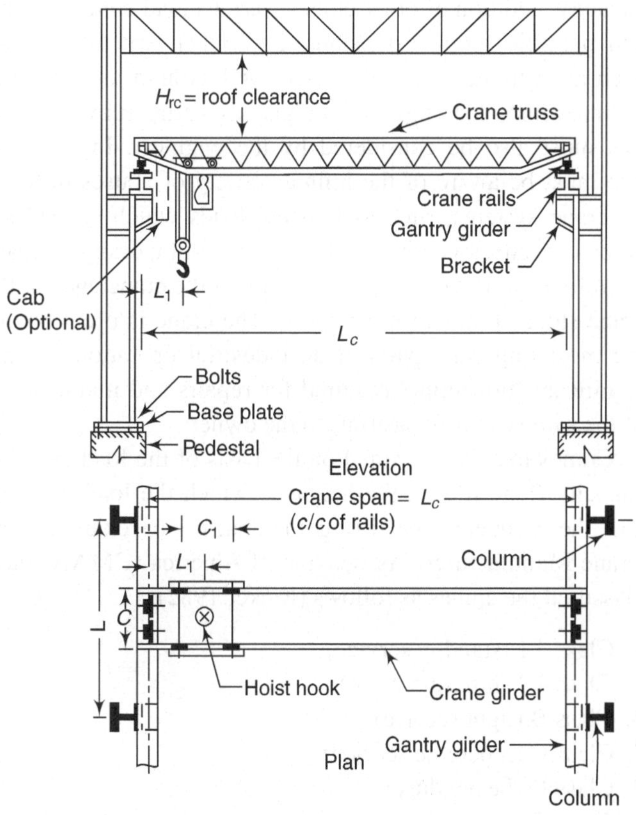


FIG. 12.2 Typical crane notations

**TABLE 12.1** Typical data for cranes (see Fig. 12.2)

| Capacity of crane (kN) | Auxiliary weight (kN) | Span $L_c$ (m) | Wheel base c (m) | Minimum hook distance of main load, $L_1$ (m) | Weight of crane bridge (kN) | Vertical clearance (m) | Weight of trolley/crab (kN) | Crane rail | | |
|---|---|---|---|---|---|---|---|---|---|---|
| | | | | | | | | Head width (mm) | Weight (kN/m) | Base width (mm) |
| 50 | — | 10.5–22.5 | 3.0–4.8 | 0.65–1.00 | 50–150 | 1.83 | 15 | CR50 | 298 | 90 |
| 100 | 30 | 10.5–22.5 | 3.2–4.8 | 0.65–1.10 | 80–210 | 1.83 | 35 | CR50 | 298 | 90 |
| 150 | 30 | 10.5–31.5 | 3.2–5.3 | 0.80–1.10 | 210–250 | 2.13 | 60 | CR60 | 400 | 105 |
| 200 | 50 | 10.5–31.5 | 3.5–5.3 | 0.80–1.10 | 160–275 | 2.13 | 75 | CR60 | 400 | 105 |
| 250 | 50 | 10.5–31.5 | 3.5–5.3 | 0.80–1.10 | 275–320 | 2.44 | 85 | CR80 | 642 | 130 |
| 300 | 80 | 10.5–31.5 | 3.8–5.3 | 0.80–1.15 | 300–360 | 2.44 | 100 | CR80 | 642 | 130 |
| 400 | 80 | 10.5–31.5 | 3.8–5.3 | 0.85–1.15 | 350–400 | 2.74 | 120 | CR80 | 642 | 130 |
| 500 | 125 | 10.5–31.5 | 4.0–5.3 | 0.85–1.20 | 400–470 | 2.74 | 135 | CR100 | 890 | 150 |
| 600 | 125 | 10.5–31.5 | 4.5–5.3 | 1.00–1.20 | 600–750 | 2.74 | 250 | CR120 | 1180 | 170 |

*Notes:* 1. Exact crane data must be obtained from the manufacturer.
2. The auxiliary load need not be considered in the design of the girder.

the crane system are also related to the capacity of the crane. Typical data for some cranes is given in Table 12.1 for guidance. The designer needs to check out this data from the manufacturer before carrying out the design of gantry girders.

While deciding on the choice of the crane, one must consider the load ca pacity, space limitations, and the class of service required. While designing the crane supporting structures, the engineer should take into account these requirements and other factors such as the likely future changes in the load capacity, addition of other cranes, various load combinations, and possible future extensions. Few other structures suffer such an extreme range of stresses and as high an incidence of maximum loadings and fatigue as crane runways, and this must also be considered by the engineer. In addition, one must be aware of the infinite variety of abuses inflicted on crane systems, such as hoisting loads which exceed the crane capacity, swinging loads as a pendulum, dragging loads laterally or longitudinally, and ramming the crane against the crane stops at an excessive speed. The crane is often one of the most important parts of an industrial operation, so any significant 'downtime' required for repairs and maintenance of the same can be disastrous to the owner.

Cranes may be classified on the basis of the load carrying capacity (load lifted), the height to which the load is lifted, and the frequency of lifting the loads. For example, the Crane Manufacturers Association of America (CMAA) have classified the cranes as follows (Ricker 1982):

1. Class A1 (standby service)
2. Class A2 (infrequent use)
3. Class B (light service)
4. Class C (moderate service)
5. Class D (heavy duty)
6. Class E (severe duty cycle service)
7. Class F (Steel Mill)

**TABLE 12.2** Suggested bridge operating speeds of overhead cranes, m/min (Extracted from CMAA Specification No. 70-2000)

| Capacity (kN) | Slow | Medium | Fast |
|---|---|---|---|
| 30 | 15 | 34.5 | 52.5 |
| 50 | 15 | 34.5 | 52.5 |
| 100 | 15 | 34.5 | 52.5 |
| 150 | 15 | 34.5 | 52.5 |
| 200 | 15 | 34.5 | 52.5 |
| 250 | 15 | 34.5 | 52.5 |
| 400 | 12 | 30 | 45 |

*Source*: Based on CMAA Specification No. 70-2000, 'Specifications for Top Running Bridge and Gantry Type Multiple Girder Electric Overhead Travelling Cranes', *Crane Manufacturers Association of America*, Charlotte, NC, 2000, p. 90.

Table 12.2 gives typical bridge operating speeds of overhead cranes.

It should be remembered that the design aspects for light cranes may be different from those for heavy cranes: cranes with long spans (over 15 m) should be treated differently than those with shorter spans; fast, heavy service cranes require special consideration not required for slower, lighter cranes.

16 m high shed with 2-25 T cranes at Orient Abrasives Ltd Porbandar
*Photo Courtesy:* Er Jayant Lakhlani

## 12.1 LOADING CONSIDERATIONS

The following important quantities need to be considered by the designer while estimating the loads on the gantry girder (see Fig. 12.3).

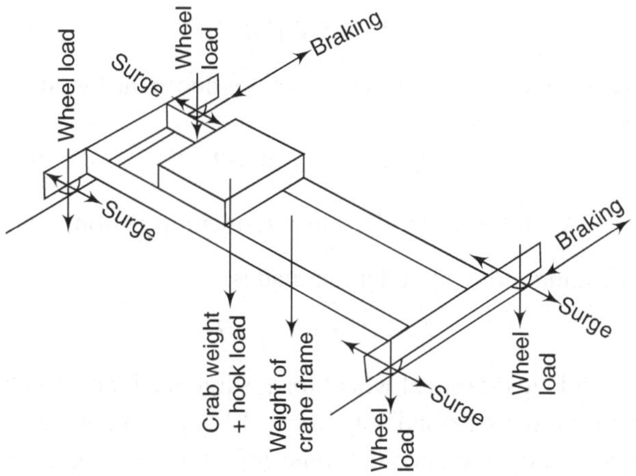


**FIG. 12.3** Loads to be considered on gantry girders

**Weight of the trolley or crab car ($W_t$)** Since the trolley moves on the crane girder (see Figs 12.1 and 12.3) along the span of the truss, its weight is transferred to the crane wheels as the axle load and finally to the gantry girder. The load transferred to the gantry will be maximum when the trolley wheels are closest to the gantry girder. The wheel load that is transferred from the trolley to the gantry girder is

$$W_1 = [W_t(L_c - L_1)]/(2L_c) \quad (12.1)$$

where $W_1$ is load of each wheel on the gantry girder (note that there are 2 wheels), $W_t$ is the weight of the trolley or crab car, $L_c$ is the distance between the gantry girders (span of truss /crane span), and $L_1$ is the distance between the CGs of the trolley and gantry.

**Weight of the crane girder ($W_c$)** The crane manufacturers also supply a pair of crane girders, which may be in the form of open web trusses or solid I-beams on which the trolley moves. These I-beams will be mounted on a crane rail using wheel carriages having four wheels, two wheels on either end (see Figs 12.1–12.3). These wheels move on rails mounted on the gantry girders. The crane rails are attached to the gantry girders with bolted clamps or hook bolts spaced about 500–1000 mm apart, depending on the lifting capacity of the crane. Rails are generally not welded to the gantry girder because of the difficulty in the readjustment of the alignment and replacement of defective or worn out rails. The hook bolts (used with slow-moving cranes having a capacity of 50 kN and bridge span of 15 m) and clamps are shown in Fig. 12.4. More details about rails and the methods of attaching them to gantry girders may be found in Ricker (1982). The weight of the crane girder is equally distributed on all the four wheels. Hence, the weight of the crane girder on each wheel is

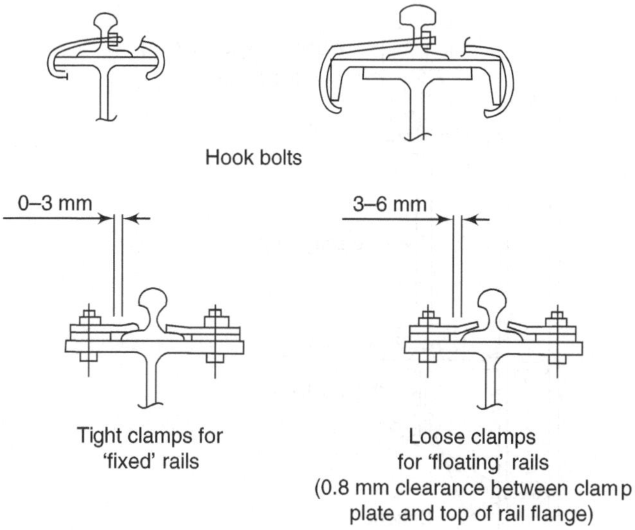


**FIG. 12.4** Methods of attaching rails to gantry girders (Ricker 1982)

*Source:* Ricker 1982, Reprinted with permission from AISC. All rights reserved.

$$W_2 = W_c/4 \quad (12.2)$$

where $W_2$ is the load due to the weight of the crane truss/girder and $W_c$ is the weight of the crane truss/girder.

**Impact load** As the load is lifted using the crane hook and moved from one place to another, and released at the required place, an impact is felt on the gantry girder. It is due to the sudden application of brakes (frictional force), moving loaded crane's acceleration, retardation, vibration, or possible slip of slings, etc. To account for these, suitable impact factors should be applied to the loads. As per IS 875 and IS 807, additional impact loads as listed in Table 12.3 should be considered (see also Table 3.9 of Chapter 3).

**TABLE 12.3** Additional impact loads on cranes

| Type of load | Impact allowance (percentage) |
|---|---|
| Vertical forces transferred to wheels | |
| (a) For EOT cranes | 25% of the maximum static wheel load |
| (b) For hand-operated cranes | 10% of the maximum static wheel load |
| Horizontal forces transverse to the rails | |
| (a) For EOT cranes | 10% of the weight of the crab and the weight lifted on the crane |
| (b) For hand-operated cranes | 5% of the weight of the crab and the weight lifted on the crane |
| Horizontal forces along the rails | 5% of the static wheel loads for EOT or hand-operated cranes |

**Lateral load (surge load)** As the crane moves with the load, a lateral load (transverse to the rail) is developed, as shown in Fig. 12.5, due to the application of brakes or the sudden acceleration of the trolley. Suppose that the total weight including the lifted weight and the trolley weight is $W$, the coefficient of friction is 0.1, the number of wheels is 4, and number of breaking wheels is 2. Then,

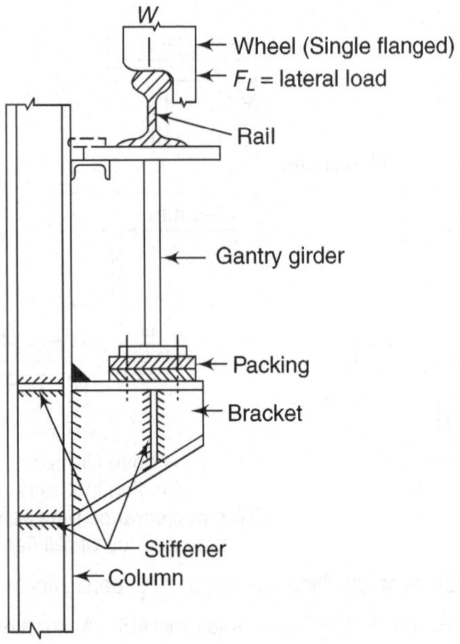


**FIG. 12.5** Application of lateral load and typical bracket

$$\text{Horizontal force} = W \times 0.1 \times (2/4) = 0.05W.$$

However, the Indian code IS 875 recommends 10%$W$ for EOT cranes (see Table 12.3). These horizontal loads are also called *surge loads*. In the design of the girder, it is usually assumed that the lateral or longitudinal surge is resisted by the compression flange alone. Note that it is customary to assume that the entire horizontal force acts on one gantry girder at the position of the wheels. But if the wheels have guides on both sides of the flange of the wheel, the force is resisted by the two opposite gantry girders and only half the horizontal load as specified in Table 12.3 acts on one girder. Thus,

$$W_L = 10/100(W_t + W_k) \text{ for EOT crane} \quad (12.3a)$$
$$W_L = 5/100(W_t + W_k) \text{ for hand-operated crane} \quad (12.3b)$$

where $W_k$ is the hook load (the maximum capacity of the crane is denoted by this load) and $W_t$ is the weight of the trolley.

**Longitudinal load (drag load)** As the crane moves longitudinally, loads parallel to the rails are caused due to the braking (stopping) or acceleration nd swing (starting) of the crane. This load is called the longitudinal load or *drag load* and is transferred at the rail level. Figure 12.6 illustrates the action of the longitudinal load and its reactions. Due to the vertical loads, the reactions $R_{a1}$ and $R_{b1}$ are developed. These reactions can be calculated easily.

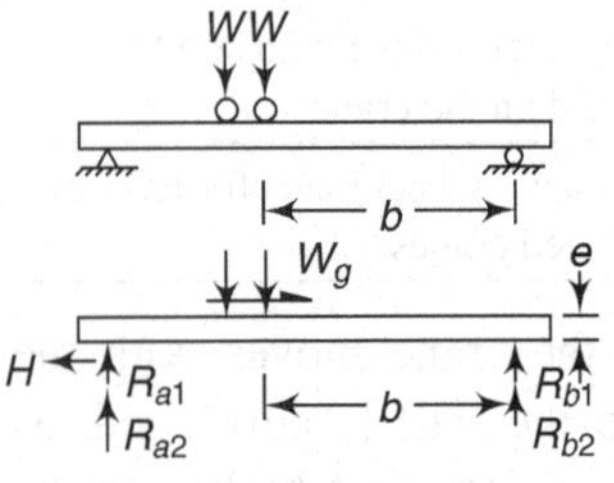


**FIG. 12.6** Braking force

Due to the braking action, a longitudinal load is developed at the top of the rails, which acts at a height $e$ from the centre of gravity of the beam. This eccentric load, in turn, generates equal and opposite reactions. Thus, the gantry girder is subjected to bending moments in addition to the axial force. The axial force could be tensile or compressive depending on the direction of motion in relation to the hinge support. The longitudinal load per wheel as per Table 12.3 is

$$W_g = 5W/100 \quad (12.4)$$

where $W$ is the wheel load. The reactions due to the longitudinal loads are

$$R_{b2} = -R_{a2} = W_g e/L \quad (12.5)$$

$$\text{Axial force } P = W_g \text{ (tension or compression)} \quad (12.6)$$

The bending moment under the load is

$$M = R_{b2}b = W_g eb/L \quad (12.7)$$

All these loads (vertical wheel load, surge load, and breaking load are shown schematically in Fig. 12.3. During preliminary design studies, the specific crane information necessary for the final design may not be available, and it becomes often necessary to estimate the loadings. For such studies, the information given in Table 12.1 or the data found in Weaver (1979) and Gaylord et al. (1996) may be useful. When the crane information does become available, it should be carefully compared with the estimated loadings and necessary adjustments should be made to the preliminary design.

## 12.2 MAXIMUM LOAD EFFECTS

Moving loads, such as crane wheels, result in bending moments and shear forces, which vary as the load travels along the supporting girder. In simply supported beams the maximum shear force will occur immediately adjacent to the support, while the maximum bending moment will occur in the region of mid-span. In general, influence lines should be used to find the load positions that produce maximum values of shear forces and bending moments (see Marshall & Nelson 1990; Coates et al. 1988 for a discussion and calculation for influence lines).

For a simply supported beam with two moving loads, the load positions which produce maximum shear force and bending moments are shown in Fig. 12.7. (The wheel load bending moment is maximum when the two loads are in such a position that the centres of gravity of the wheel loads and one of the wheel loads are equidistant from the centre of gravity of the girder. The shear due to the wheel load is maximum when one of the wheels is at the support.) From these, the maximum shear force and bending moments are (Marshall & Nelson 1990)

$$\text{Shear force (max)} = W(2 - c/L) \quad (12.8)$$

$$\text{Bending moment (max)} = WL/4 \text{ or } 2W(L/2 - c/4)^2/L \quad (12.9)$$

The greater of the bending moment values should be used.

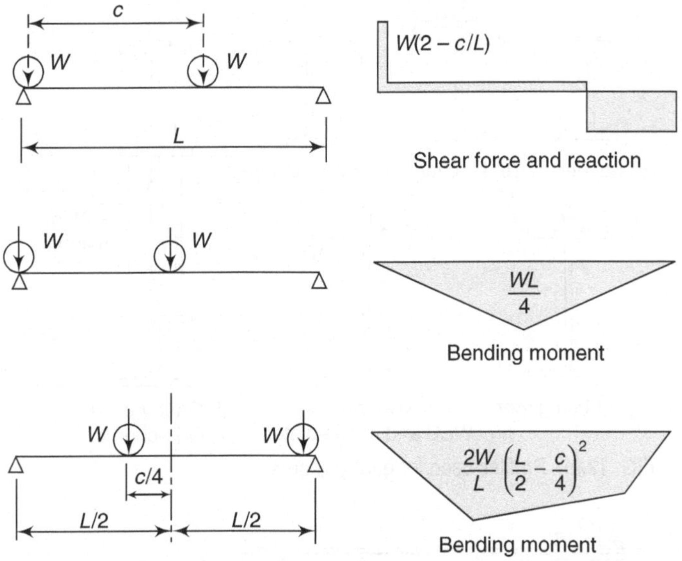


**FIG. 12.7** Maximum BM, SF, and reaction for two moving loads

The design of the bracket supporting a crane girder uses the values of maximum reaction from the adjacent simply supported beams as in Fig. 12.7. When the adjacent spans are equal, the reaction is equal to the shear force, i.e,

$$\text{Reaction (max)} = W(2 - c/L) \tag{12.10}$$

Note that in all the cases the effect of the self-weight of the gantry girder must also be considered as a uniformly distributed load.

The deflection at the mid-span due to placing the two loads symmetric with respect to it is given by

$$\Delta_c = W_c L^3[(3a/4L) - (a^3/L^3)]/(6EI) \tag{12.11}$$

where $a = (L - c)/2$, $E$ is Young's modulus of rigidity, and $I$ is the moment of inertia of the cross section.

When two cranes are operating in tandem on the same span, these have to be located closest to each other towards the mid-span to produce maximum bending moment. In such a situation, there will be four wheels, two wheels from each crane. These four wheels must be adjusted on the span such that the centroid of the load and the nearest wheel load to the CG from the centre line of the beam are at the same distance (see Fig. 12.8). Let $x$ be the distance from left to the load under which maximum bending moment occurs.

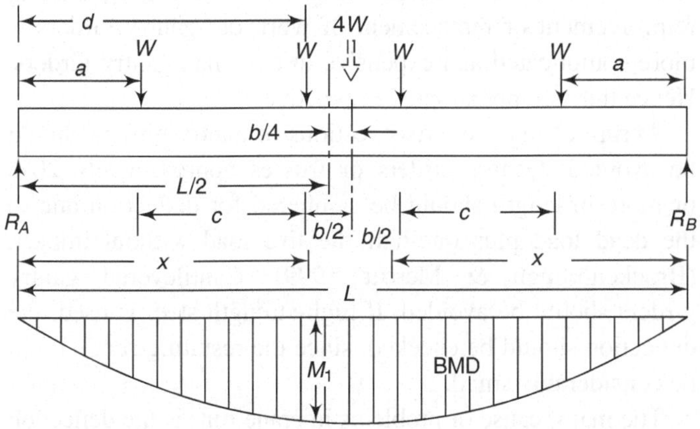


**FIG. 12.8** Maximum bending moment for four moving loads (two cranes)

$$x = (L/2) - (b/4) \tag{12.12}$$

The reaction and bending moment are

$$R_A = 4W(x/L) \tag{12.13}$$

$$\begin{aligned} M_1 &= R_A x - W_c \\ &= (4Wx^2/L) - W_c \\ &= W/L[4x^2 - cL] \end{aligned} \tag{12.14}$$

The deflection at mid span may be obtained approximately as (see Fig. 12.8)

$$\Delta = W/(24EI)[a(3L^2 - 4a^2) + b(3L^2 - 4b^2)] \tag{12.15}$$

These deflections at working loads should be less than the allowable deflections as specified in Table 12.4 (see Table 6 of IS 800 : 2007).

**TABLE 12.4** Limiting deflection of gantry girders

| Category | Maximum deflection |
|---|---|
| *Vertical deflection* | |
| Manually operated cranes | $L/500$ |
| EOT cranes with a capacity of less than 500 kN | $L/750$ |
| EOT cranes with a capacity of greater than 500 kN | $L/1000$ |
| *Lateral deflection* | |
| Relative between rails | 10 mm or $L/400$ |

## 12.3 FATIGUE EFFECTS

Gantry girders are subjected to fatigue effects due to the moving loads. Normally, light- and medium-duty cranes are not checked for fatigue effects if the number of cycles of load is less than $5 \times 10^6$. For heavy-duty cranes, the gantry girders are to be checked for fatigue loads (see also IS 1024 and IS 807). The fatigue strength is governed by the following factors.

1. Number of repetitions of load (In most cases $2 \times 10^6$ cycles of repetitions are considered to be the limit of endurance.)
2. The ratio of minimum stress to maximum stress at a section due to repetitions:

$$R = f_{min}/f_{max} \tag{12.16}$$

where $f_{max}$ is the maximum stress and $f_{min}$ is the minimum stress.

Simply supported beams and most statically determinate beams are subjected to variable stresses and hence the value of $R$ ranges from

$$0 \le R \le 1 \tag{12.17}$$

Whereas, continuous beams and frames are subjected to the reversal of stresses and hence the value of $R$ ranges from

$$-1 \leq R \leq 1 \quad (12.18)$$

3. Stress concentration (which may be caused by lack of fit), presence of notches, and sharp change of directions (at the toes of groove welds and at the toes and roots of fillet welds).

*Fatigue* is associated with the failure of a component or structure due to several cycles of repeated or cyclic loads and the level of failure stress is usually less than yield stress. Hence the fatigue strength is checked at working loads. *Fatigue limit* is defined as the stress for which the material can withstand an infinite number of cycles. Since it is difficult to obtain an infinite number of cycles of testing, $2 \times 10^6$ cycles are considered to be infinite for civil engineering structures. Tests indicate that the fatigue limit decreases with an increase in the range of $R$; the ratio of the minimum stress to the maximum stress (Gurney 1979).

The fatigue strength in IS 800 : 2007 is defined by a series of $\log f_f - \log N_{sc}$ curves, each applying to a typical category of connection (where $f_f$ is the uncorrected fatigue strength in megapascal and $N_{sc}$ is the number of stress cycles). These curves are shown in Fig. 22 of the code. Each category (see IS 800 : 2007, Table 26) is designated by a number which represents the reference value $f_c$ of the fatigue strength at 2 million cycles, i.e., $N_c = 2 \times 10^6$. In addition, Fig. 22 of the code also shows the constant-amplitude fatigue limit $f_D$, which is the limiting stress range value above which a fatigue assessment is necessary. The number of cycles corresponding to the constant-amplitude fatigue limit is 5 million cycles, i.e., $N_D = 5 \times 10^6$. The other is the cut-off limit, $f_L$, which is the limit below which stress ranges of the design spectrum do not contribute to the calculated cumulative damage. The number of cycles corresponding to this value is $10^8$ cycles, i.e., $N_L = 10^8$.

## 12.4 SELECTION OF GANTRY GIRDER

The gantry girder is subjected to vertical loading including impact loading, lateral loading, and longitudinal loading from traction, braking, and impact on crane stops. In addition, the crane beams must withstand local buckling under wheel loads and at the bottom flange over the column (in the common case where the beam bears on a column cap plate). Figure 12.9 illustrates typical profiles used for gantry girders.

1. Rolled beams with or without plates, channels or angles (for spans up to 8 m and usually for 50 kN cranes), as shown in Figs 12.9(a–c)
2. Plate girders (for spans from 6 m to 10 m—see Fig. 12.9d)
3. Plate girder with channels, angles, etc. (for spans more than 10 m—see Fig. 12.9e)
4. Box girders with angles (for spans more than 12 m)
5. Crane truss using I-sections as chord members and angles as web members (Fig. 12.9f)

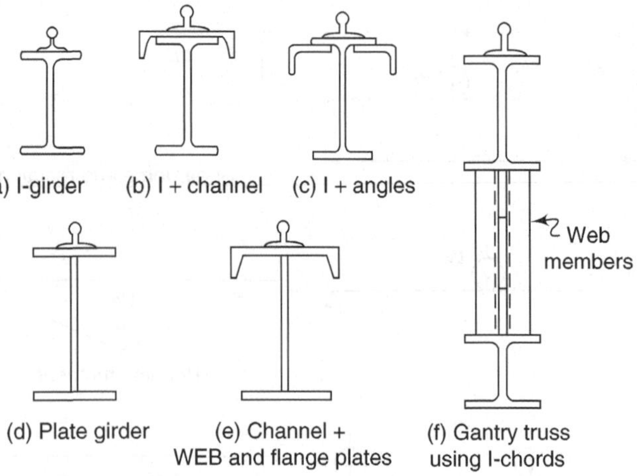


**FIG. 12.9** Profiles used for gantry girders

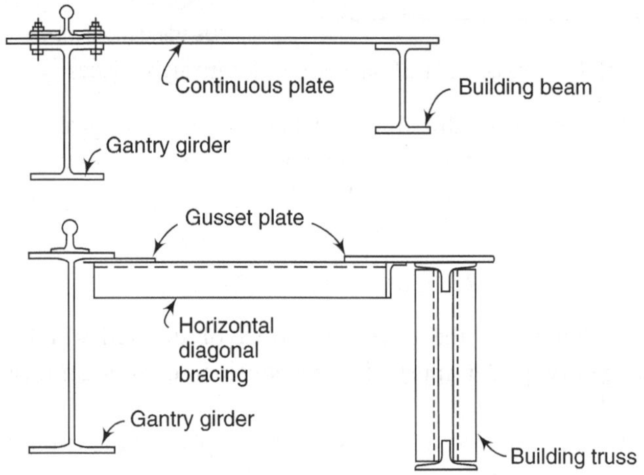


**FIG. 12.10** Gantry girder connected to building beams or trusses with plates or horizontal trusses

*Source:* Ricker 1982, Reprinted with permission from AISC. All rights reserved.

The profiles shown in Fig. 12.9 can be combined with other shapes to form horizontal trusses and walkways, as shown in Fig. 12.10.

Single-span gantry girders are desirable. Two-span gantry girders can result in uplift in columns at certain loading positions, and differential settlement of columns may result in undesirable additional stresses. Moreover, making reinforcement or replacement of worn out gantry girders is more complicated and expensive in two-span gantry girders. Hence they are not adopted in practice.

Abrupt changes in cross sections of gantry girders should be avoided. Gantry girders or trusses approximately 20 m or more in length should be cambered for deflection due to the dead load plus one-half the live load without impact. (Brockenbrough & Merritt 1999). Cantilevered gantry girders should be avoided. If high-strength steel is used, the deflection should be checked, since the resulting section will be considerably small.

The major cause of problems in crane runs is the deflection of gantry girders and the accompanying end rotation. Stretching

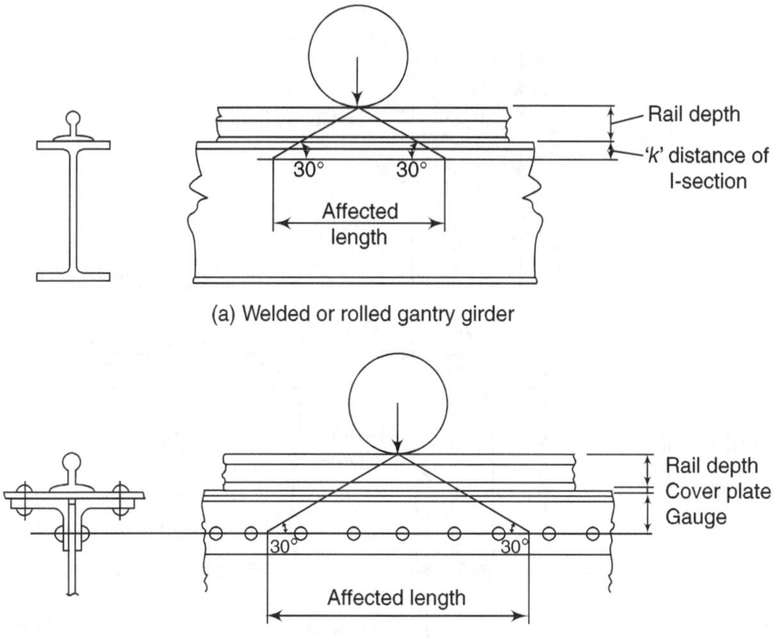


(b) Bolted or riveted gantry girder

**FIG. 12.11** Angle of dispersion for the concentrated load

*Source:* Ricker 1982, Reprinted with permission from AISC. 

of rails, opening of splice joints, column bending, skewing of the crane girders, and undulating crane motion are among the problems created by excessive deflection. Hence it is important to limit the vertical deflection of the gantry girder as per the limits given in Table 12.4. In general, it is a good idea to keep the spans as short as possible and beam depths as large as possible.

Vertical loads are delivered to the gantry girder via the crane rail. (The depth of crane rails varies from 70 mm to 150 mm and their bottom width from 70 mm to 150 mm.) The beam must be capable of withstanding this localized load. It is recommended to use a full penetration groove weld between the web and the top flange of welded plate gantry girders in order to maximize the fatigue life of the member. The length of the web that is affected by the concentrated wheel load is taken by considering a dispersion angle of 30° as shown in Fig. 12.11(a). The angle of 30° is a logical average between the 45° pure shear angle and the 22° angle used in the column stiffener analysis (Ricker 1982). Using this value of 30°, the affected length is given by

$$\text{Affected length} = 3.5 \times (\text{rail depth} + k) \qquad (12.19)$$

In the case of a plate girder, it becomes

$$\text{Affected length} = 3.5 \times (\text{rail depth} + \text{flange thickness}) \qquad (12.20)$$

In the case of riveted or bolted member (see Fig. 12.11b)

$$\text{Affected length} = 3.5 \times (\text{rail depth} + \text{cover plate thickness} + \text{gauge distance})$$

However, note that a dispersion angle of 45 degrees has been assumed in the Indian code and the dispersion is taken upto the mid-depth of the girder (see Section 10.11 of Chapter 10).

It has been found that web crushing is not critical in most of the cases (Ricker 1982).

The effects of an off-centre crane rail must be considered. Excessive rail eccentricity must be avoided, because it causes local flange bending and subjects the gantry girder to torsional moments. A limit of $0.75t_w$ is often specified for this eccentricity (see Fig. 12.12) for both wide flange beams and plate girders (Ricker 1982). Intermediate stiffeners may be welded to the underside of the top flange and down the web with a continuous weld to counteract the effect of rail eccentricity.

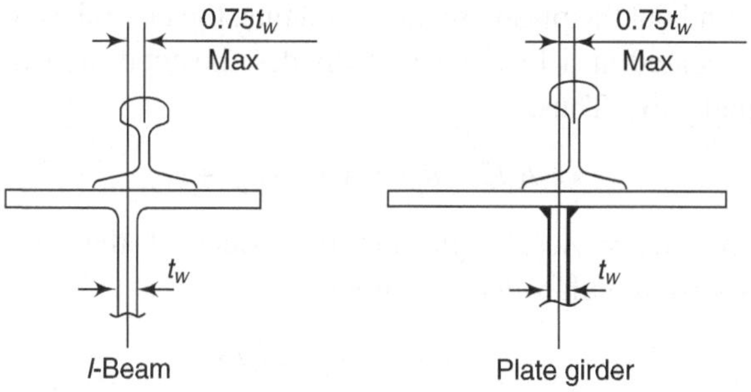


**FIG. 12.12** Maximum allowable eccentricity for rails

While designing the gantry girder to resist lateral loading, the strength of the top flange alone should be considered. If this strength is inadequate, it may be reinforced by adding a channel, plate or angles, or by making a horizontal truss or girder in the case of large lateral loads (see Figs 12.9 and 12.10). These reinforcing members are often attached by welding. Due to the fatigue factor associated with intermittent welds, it is advisable to use continuous welds. In designing gantry girders which require channels, plates, or angles to resist lateral loads, a simplified design method which considers that the beam resists the vertical forces and the lateral forces are resisted only by the channel (or plate or angle) may be used; but this will result in uneconomical designs. Hence most designers assume that the lateral load is resisted by the channel (or plates or angles) plus the top flange of the beam and that the vertical load is resisted by both the beam and the channel (or plate or angle). If clamps are used to fasten the rails above the girder, it is necessary to select member sizes which will accept the required hole spacing (see Fig. 12.13).

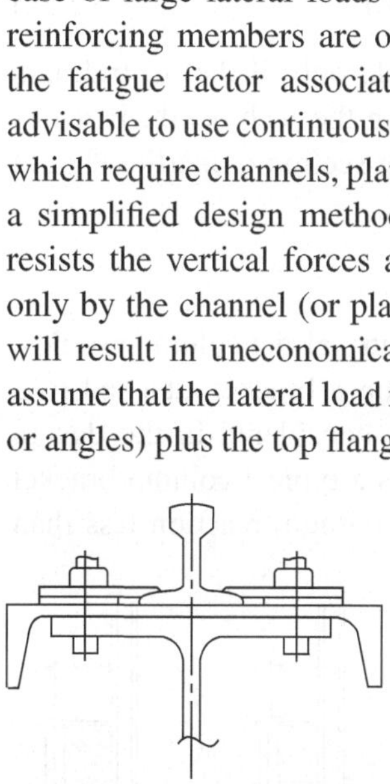

**FIG. 12.13** Clamping of rails with bolts

### 12.4.1 Section Properties

With reference to Fig. 12.14, if $A_p$ is the area of the top plate and $A_s$ is the area of the rolled I-section, then the total area

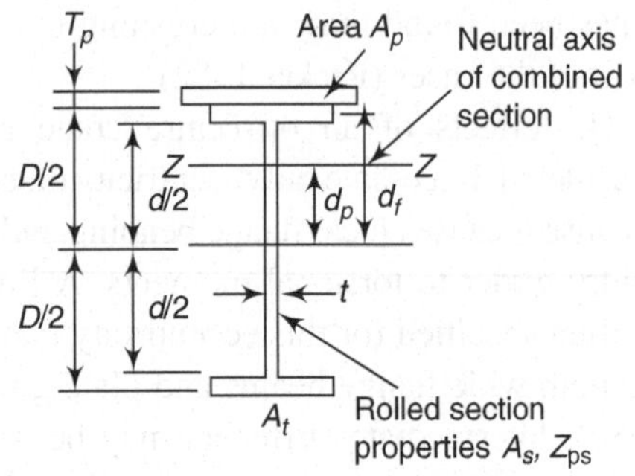


**FIG. 12.14** Rolled I-beam with top plate

$$A = A_s + A_p$$

To find out the plastic section modulus, the neutral axis should be located at a location that divides the total area into two equal parts. Thus,

$$A_s/2 + d_p t = A_s/2 - d_p t + A_p$$

Hence $d_p = A_p/2t$. Ignoring the effect of the root fillets associated with rolled I-sections,

$$Z_{ps} = 2A_f d_f + td^2/4 \tag{12.21}$$

where $A_f$ is the area of the flange and $d_f$ is the distance of the flange from the neutral axis:

$$Z_{pz} = A_f(d_f - d_p) + A_f(d_f + d_p) + t(d/2 - d_p)^2/2 + t(d/2 + d_p)^2/2 + A_p(D/2 + T_p/2 - d_p)$$

$$= 2A_f d_f + td^2/4 + td_p^2 + A_p(D/2 + T_p/2 - d_p)$$

$$= Z_{ps} + td_p^2 + A_p(D/2 + T_p/2 - d_p) \tag{12.22}$$

Note that this formula is applicable only if the neutral axis of the combined section lies within the web depth. If it lies within the flange of the section, the section properties should be determined from first principles.

## 12.4.2 Columns

Figure 12.15 shows the various gantry girder column profiles. If the gantry girder is supported on a bracket attached to a column, then the impact must be considered in the design of the bracket. Figure 12.16 shows a typical column bracket for relatively light crane loads (maximum reaction less than about 220 kN). Slots are provided in the bracket seat plate for lateral adjustment. Stiffeners are placed at the end of the beam to prevent web buckling. The bolts connecting the beam to the bracket must be strong enough to resist the longitudinal forces. Note that stiffeners are not provided in the bracket web plate directly below the gantry girder web. Omission of these stiffeners allows the seat plate to flex with the beam end as the crane passes along the runway. But stiffeners may be required for certain other bracket configurations.

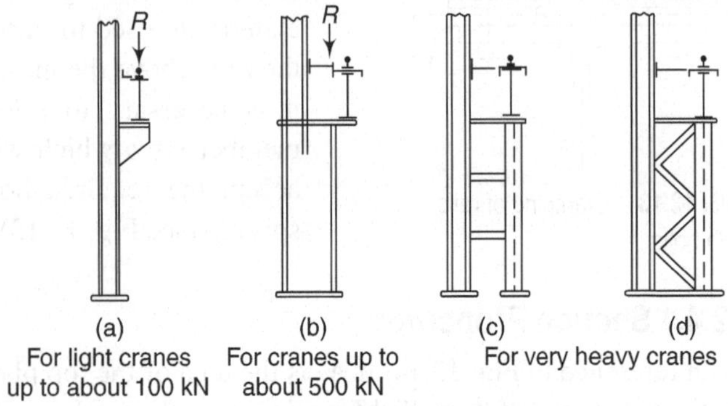


**FIG. 12.15** Typical profiles of columns to support gantry girders

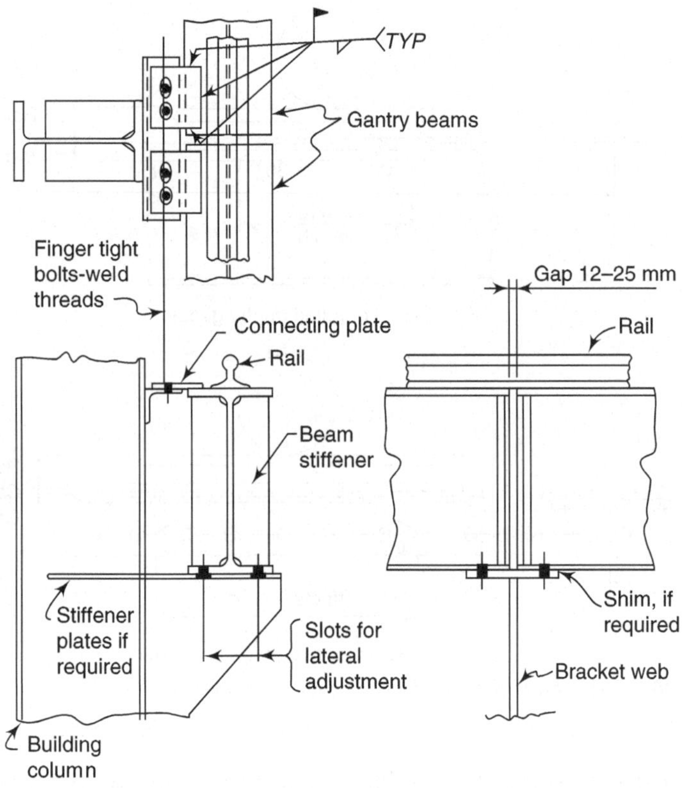


**FIG. 12.16** Details at column bracket for light cranes

*Source:* Ricker 1982, Reprinted with permission from AISC. 

Figure 12.16 shows an angle shop welded to the column flange (it could be bolted) containing longitudinal slots. The connecting plates must be individually fastened to each beam. The plates have holes and may be bolted or welded to the beams either in the shop or at the location. Note that the slots are placed in the lower of the two members, so that they will not be filled up with dirt. To determine the number of bolts required, the bearing of the bolt against the side of the slot must be checked. If it is necessary to raise the gantry girder to re-level the runway (due to settlement of the building), shims could be inserted between the bracket and the bottom flange of the beam and between the angle and the plate at the top flange. Figure 12.17 shows the other combinations of lateral-load-resisting connections.

Longitudinal forces cause torque on columns ($T = dP$) which have gantry brackets [Fig. 12.18(a)]. If the torsion is high, its effects can be significantly reduced by adding

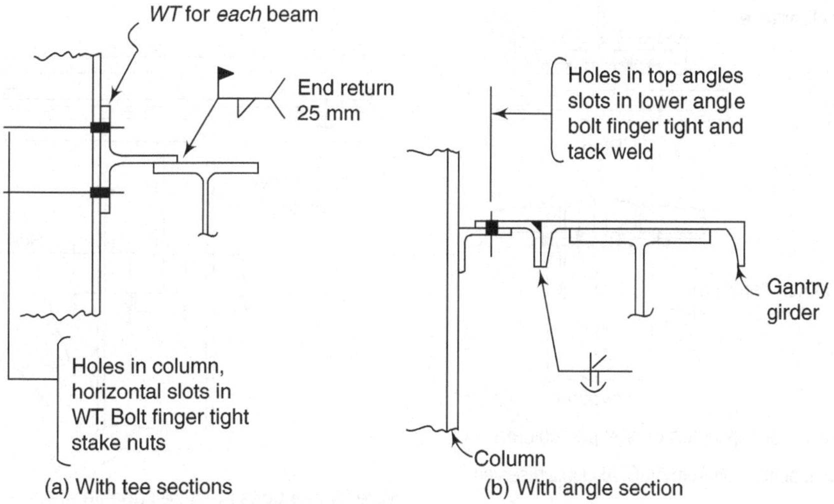


**FIG. 12.17** Lateral load resisting connections at the top flange of a gantry girder

*Source:* Ricker 1982, Reprinted with permission from AISC. All rights reserved.

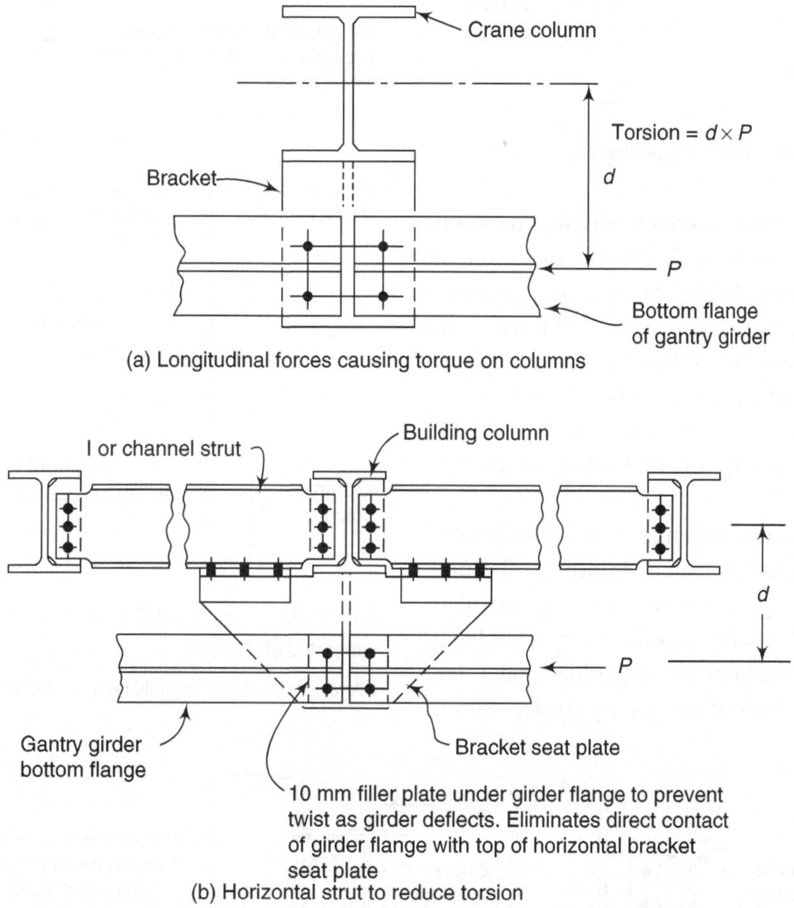


**FIG. 12.18** Torsion on columns due to longitudinal forces on the gantry girder

*Source:* Ricker 1982, Reprinted with permission from AISC. All rights reserved.

horizontal struts [Fig. 12.18(b)]. These struts are not required at every column, but may be provided near the centre of the runway in as many columns as the loads dictate.

If it becomes uneconomical to use a bracket to support the gantry girder, then stepped columns as shown in Fig. 12.15(b) can be adopted. Figure 12.19 shows a typical detail of a gantry girder supported on a stepped column.

Gantry girders supporting heavy cranes are best supported by their own columns [Figs 12.15(c) and (d)]. The usual and logical orientation of these columns is shown in Fig. 12.20.

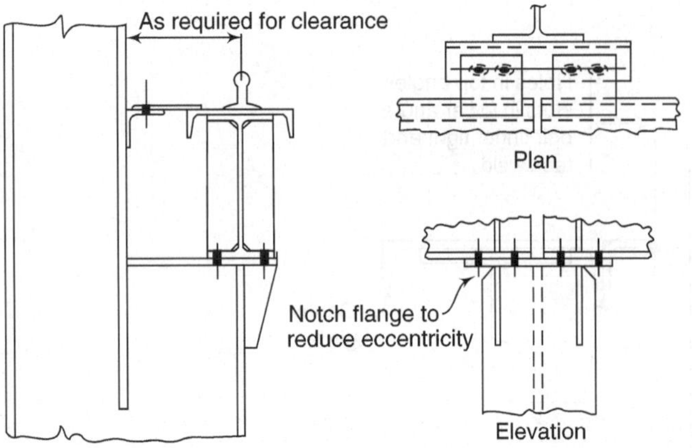


**FIG. 12.19** Detail of a gantry girder supported on stepped column

*Source:* Ricker 1982, Reprinted with permission from AISC. All rights reserved.

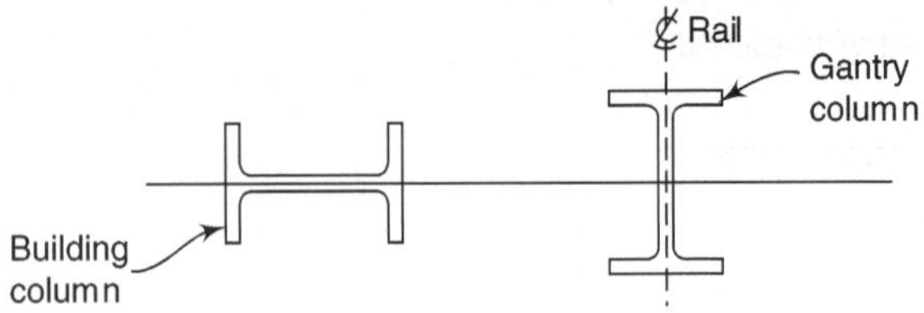


**FIG. 12.20** Orientation of columns supporting heavy cranes

The building column is oriented in such a way that its strong axis withstands the lateral wind and seismic loads on the building and the lateral crane loads. The gantry column is oriented in such a way that its strong axis is able to resist the longitudinal forces on the runway. When the gantry girder is supported on its own column over a cap plate, it is advisable not to put stiffeners below the cap plate (see Fig. 12.21). This will increase the eccentricity of the load delivered by a deflecting beam.

It is very important to note that the ends of two neighbouring gantry girders should not be connected. Similarly, the ends of two adjacent horizontal gantry girder bracing members should not be connected (tied) together, but should be attached separately to the column by individual plates (see Fig. 12.22). Similarly, the web of the gantry girders should

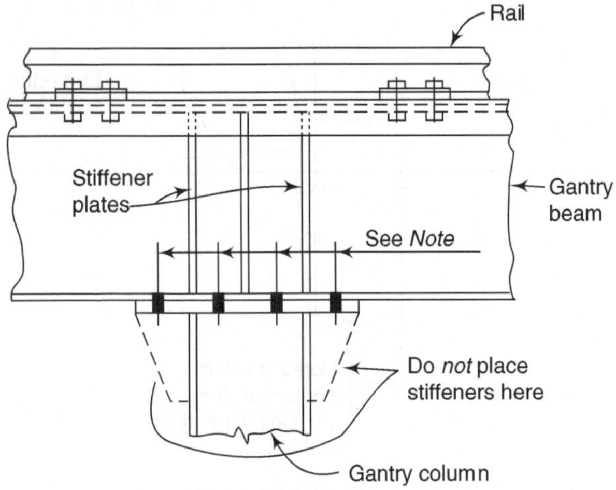


*Note:* These bolts must be capable of withstanding the longitudinal forces

**FIG. 12.21** Typical details of cap plate on columns

*Source:* Ricker 1982, Reprinted with permission from AISC. All rights reserved.

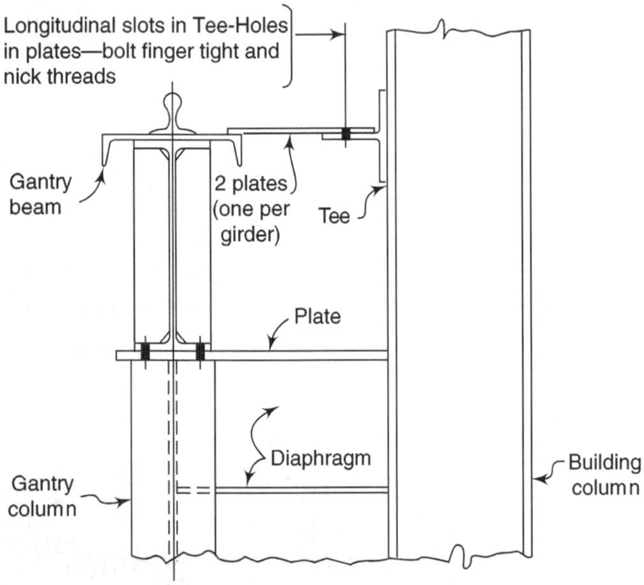


**FIG. 12.22** Typical detail when two adjacent gantry girders are to be supported on column

*Source:* Ricker 1982, Reprinted with permission from AISC. All rights reserved.

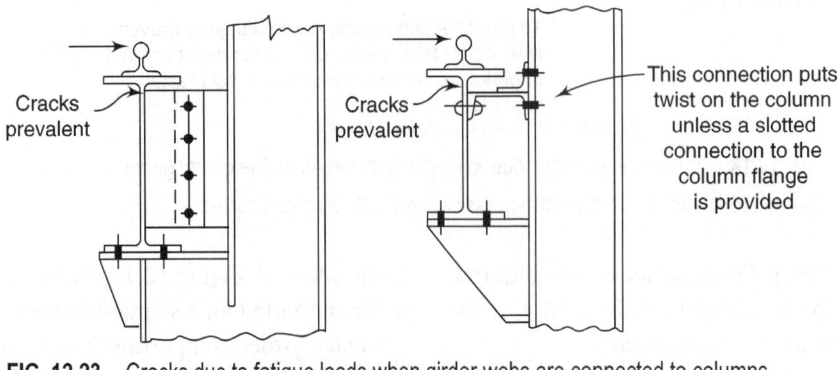


**FIG. 12.23** Cracks due to fatigue loads when girder webs are connected to columns

*Source:* Ricker 1982, Reprinted with permission from AISC. All rights reserved.

not be connected to the columns by diaphragms—which will lead to fatigue failures (see Fig. 12.23).

On very heavy crane runs, the top flange lateral bracing cannot be made of adequate strength, hence a diaphragm can be added. A single diaphragm should not serve two crane beams. Hence separate diaphragms should be provided for each beam end as shown in Fig. 12.24. Note that the diaphragms have been provided for the full depth of the gantry girder and the plate thickness should be thin to maintain flexibility (about 8–10 mm).

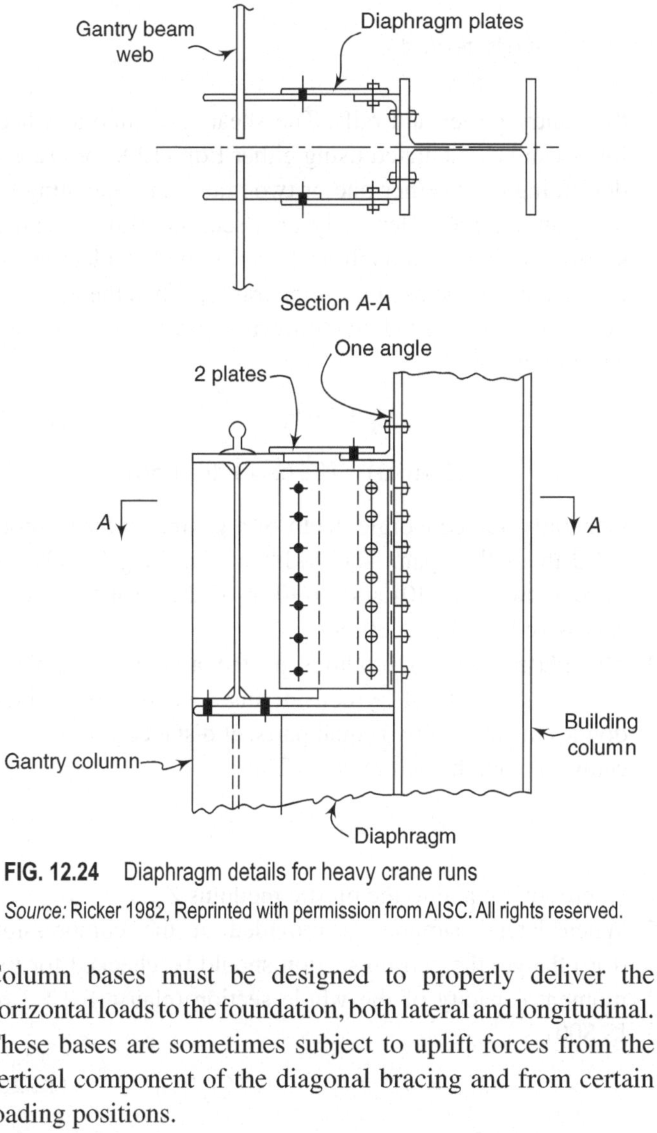


**FIG. 12.24** Diaphragm details for heavy crane runs

*Source:* Ricker 1982, Reprinted with permission from AISC. All rights reserved.

Column bases must be designed to properly deliver the horizontal loads to the foundation, both lateral and longitudinal. These bases are sometimes subject to uplift forces from the vertical component of the diagonal bracing and from certain loading positions.

## 12.4.3 Bracings

Columns supporting gantry girders should be braced laterally and longitudinally. The simplest and most effective bracing is the X-bracing system. It is better to limit the $L/r$ ratio of such bracings to 200, due to the abrupt reversal of stresses in crane runways. Bracing members should be made of double angles, wide flange beams, tubes, or pipe sections; these should never be made of rods. Single-angle bracings may be used on light cranes. Bracings should never be connected directly to the underside of gantry girders.

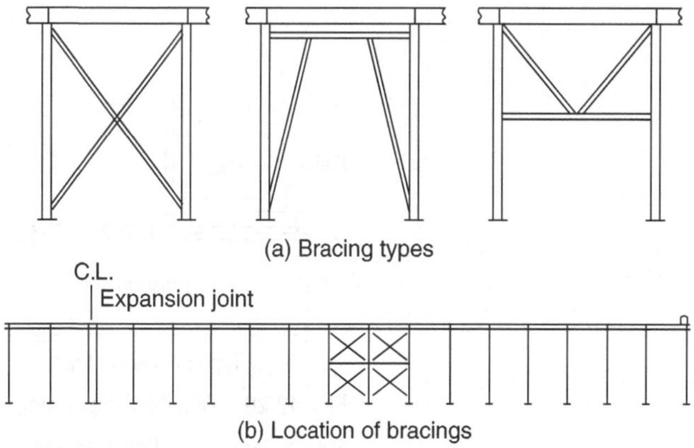


**FIG. 12.25** Bracings for columns supporting gantry girders

It is better to locate the braces near the centre of the runway, since it will allow thermal expansion and contraction to advance or retreat from a centrally 'anchored' area of the runway towards the ends. It may be often necessary to provide crane run bracing in adjacent bays, near the centrally anchored area as shown in Fig. 12.25. Knee braces should never be used, since these are the source of many crane run problems (causing undesirable restraint, column bending, and secondary stresses).

## 12.4.4 Crane Stops

Crane stops prevent the crane from moving past the end of a runway, thereby preventing the crane from falling off. Stops can also be mounted at any location along the runway to restrict the crane movement beyond that point. Two typical crane stops for light cranes and heavy-duty cranes are shown in Fig. 12.26. (Crane stops should not be confused with trolley stops (or wheel stops) which are attached directly to the crane rail by bolting through the web of the rail. Trolley stops are used for light- and slow-moving cranes only, as they are prone to damage the wheel bearing.) Note that a gap is provided between the end of the rail and the face of the stop to accommodate expansion and 'creep' (a gap of about 25 mm for every 30 m of the rail). The height of the crane stop depends on the height of the crane bumper and varies between 450 mm and 750 mm above the top of the rail. Most of the modern cranes have some energy-absorbing features such as rubber padding or hydraulic or pneumatic cylinders. The crane stop must be designed to resist the impact force $F$ as shown in Fig. 12.26, which is given by (Ricker 1982)

$$F = WV^2/(2gT) \tag{12.23}$$

where $W$ is the total weight of the crane (in kN), excluding the lifted load; $V$ is the speed of the crane (in m/s) (may be taken as 40%–50% of the full load rated speed); $g$ is the acceleration due to gravity and is equal to 9.81 m/s$^2$; and $T$ is the distance

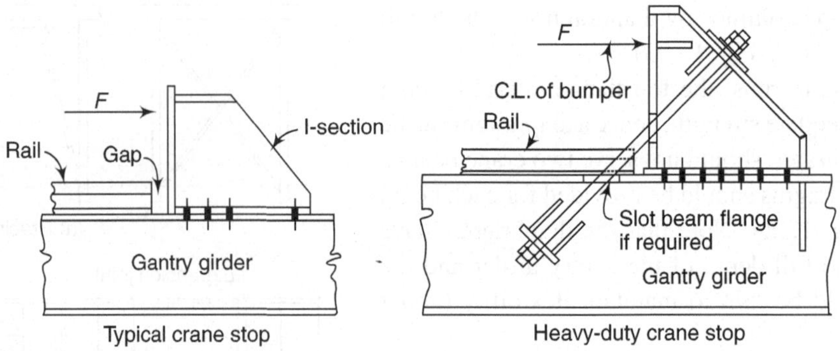


**FIG. 12.26** Details of crane stops

*Source:* Ricker 1982, Reprinted with permission from AISC. 

(in mm) the spring moves to stop the crane, usually about 40–50 mm.

Note that this formula assumes each stop to share the load equally (that the resultant inertia force due to the weight of the bridge and trolley is located at mid-span). The real problem in this formula is the determination of the value $T$, which has to be provided by the manufacturer. It should be cautioned that the cranes exposed to wind loads may also cause some problems (Weaver 1979).

## 12.5 DESIGN OF GANTRY GIRDER

The design of the gantry girder subjected to lateral loads is a trial-and-error process. As already mentioned, it is assumed that the lateral load is resisted entirely by the compression top flange of the beam and any reinforcing plates, channels, etc. and that the vertical load is resisted by the combined beam. The various steps involved in the design are given below.

1. The first step is to find the maximum wheel load. As discussed in Section 12.1, this load is maximum when the trolley is closest to the gantry girder. It can be calculated by using Eqn (12.1) and increased for the impact as specified in Table 12.3.
2. The maximum bending moment in the gantry girder due to vertical loads needs to be computed. This consists of the bending moment due to the maximum wheel loads (including impact) and the bending moment due to the dead load of the girder and rails. Equation (12.9) gives the maximum bending moment due to wheel loads when only one crane is running over the girder and Eqn (12.14) gives the maximum bending moment when two cranes are placed on one gantry girder. The bending moment due to dead loads is maximum at the centre of the girder, whereas the bending moment due to the wheel load is maximum below one of the wheels. However, for simplifying the calculations, the maximum bending moment due to the dead load is directly added to the maximum wheel load moment.
3. Next the maximum shear force is calculated. This consists of the shear force due to wheel loads and dead loads from the gantry girder and rails. The shear force due to wheel loads can be calculated using either Eqn (12.8) or (12.13) depending on whether one or two cranes are operating on the gantry girder. Generally an I-section with a channel section is chosen, though an I-section with a plate at the top flange may be used for light cranes. When the gantry is not laterally supported, the following may be used to select a trail section:

$$Z_p = M_u/f_y \tag{12.24}$$

$$Z_p\ (\text{trial}) = kZ_p\ (k = 1.40\text{–}1.50)$$

Generally, the economic depth of a gantry girder is about (1/12)th of the span. The width of the flange is chosen to be between (1/40) and (1/30)th of the span to prevent excessive lateral deflection.
4. The plastic section modulus of the assumed combined section is found out by considering a neutral axis which divides the area in two equal parts, at distance $y$ to the area centroid from the neutral axis. Thus

$$M_p = 2f_y(A/2)\bar{y} = A\bar{y}f_y \tag{12.25}$$

where $Ay$ is equal to the plastic modulus $Z_p$.
5. When lateral support is provided at the compression (top) flange, the chosen section should be checked for the moment capacity of the whole section (clause 8.2.1.2 of IS 800):

$$M_{dz} = \beta_b Z_p f_y/\gamma_{m0} \le 1.2 Z_e f_y/\gamma_{m0} \tag{12.26}$$

This value should be greater than the applied bending moment. The top flange should be checked for bending in both the axes using the interaction equation

$$(M_y/M_{ndy}) + (M_z/M_{ndz}) \le 1.0 \tag{12.27}$$

6. If the top (compression) flange is not supported, then the buckling resistance is to be checked in the same way as in step 4 but replacing $f_y$ with the design bending compressive stress $f_{bd}$ (calculated using Section 8.2.2 of the code).

7. At points of concentrated load (wheel load or reactions) the web of the girder must be checked for local buckling and, if necessary, load-carrying stiffeners must be introduced to prevent local buckling of the web.
8. At points of concentrated load (wheel load or reactions) the web of the girder must be checked for local crushing. If necessary, bearing stiffeners should be introduced to prevent local crushing of the web.
9. The maximum deflection under working loads has to be checked.

## EXAMPLES

**EXAMPLE 12.1:** Determine the moments and forces due to the vertical and horizontal loads acting on a simply-supported gantry girder given the following data:

(a) Simply-supported span = 6 m
(b) Crane's wheel centres = 3.6 m
(c) Self-weight of the girder (say) = 1.5 kN/m
(d) Maximum crane wheel load (static) = 220 kN
(e) Weight of crab/trolley = 60 kN
(f) Maximum hook load = 200 kN

Calculate also the serviceability deflection (working load).

**SOLUTION:**

**Step 1:** *Moments and forces due to self-weight*
Factored self-weight $W_d = 1.5 \times 1.5 \times 6 = 13.5$ kN
Ultimate mid-span BM, $M_1 = W_d L/8 = 13.5 \times 6/8$
$= 10.125$ kNm
Ultimate reaction $= R_{A1} = R_{B1} = W/2 = 13.5/2 = 6.75$ kN

**Step 2:** *Moments and forces due to the vertical wheel load*
Wheel load (including $\gamma_f$ and 25% impact)
$W_c = 1.5 \times 1.25 \times 220 = 412.5$ kN
Ultimate BM under wheel (case 1) as per Eqn (12.9)
$= 2W_c(L/2 - c/4)^2/L$
$= 2 \times 412.5(6/2 - 3.6/4)^2/6$
$= 606.375$ kNm
Ultimate BM under wheel (case 2)
$= W_c L/4 = 412.5 \times 6/4 = 618.75$ kNm
Hence, the maximum ultimate BM,
$M_2 = 618.75$ kNm
Ultimate reaction [Eqn (12.10)]
$R_{A2} = W_c(2 - c/L) = 412.5(2 - 3.6/6.0) = 577.5$ kN

**Step 3:** *Moment and forces due to horizontal wheel loads*
Horizontal surge load (including $\gamma_f$)
$W_{hc} = 1.5 \times 0.10(200 + 60) = 39$ kN
This is divided among the 4 wheels (assuming double-flanged wheels).
Horizontal wheel load
$W_{hc} = 39/4 = 9.75$ kN

Using calculations similar to those for vertical moments and forces, ultimate horizontal BM (case 2)
$= 2W_{hc}L/4 = 9.75 \times 6.0/4 = 14.625$ kNm
Ultimate horizontal BM (case 1)
$= 2W_{hc}(L/2 - c/4)^2/L = 2 \times 9.75(6/2 - 3.6/4)^2/6$
$= 14.33$ kNm

**Step 4:** *BM and reaction due to drag force*
Assuming that $e$ is 0.15 m and the depth of the girder is 0.6 m,
$R_{A3} = W_g e/L = 1.5 \times (0.05 \times 220 \times 1.25)(0.3 + 0.15)/6$
$= 1.55$ kN
Ultimate BM due to drag force
$M_3 = R_a(L/2 - c/4) = 1.55(3 - 0.9) = 3.255$ kNm
Hence, the maximum ultimate design BM (vertical)
$= M_1 + M_2 + M_3 = 10.125 + 618.75 + 3.255 = 632.13$ kNm
Maximum design reaction (vertical)
$R_{A1} + R_{A2} + R_{A3} = 6.75 + 577.5 + 1.55 = 585.8$ kN

**Step 5:** *Serviceability deflection due to vertical wheel load excluding impact*
$W_c = 220$ kN
$\Delta_c = W_c L^3[(3a/4L) - (a^3/L^3)/(6EI)$ with $a = (L - C)/2$
Assuming ISMB 600 with $I_z = 91{,}800 \times 10^4$ mm$^4$
$a = (6000 - 3600)/2 = 1200$ mm
$\Delta_c = 220 \times 1000 \times 6000^3[3 \times 1200/(4 \times 6000)$
$- 1200^3/6000^3]/[6 \times 2 \times 10^5 \times 91{,}800 \times 10^4]$
$= 6.125$ mm $< L/750 = 8$ mm

**EXAMPLE 12.2:** Design a gantry girder, without lateral restraint along its span, to be used in an industrial building carrying an overhead travelling crane for the following data (see Fig. 12.27).

Centre-to-centre distance between columns (i.e., span of the gantry girder) = 7.5 m

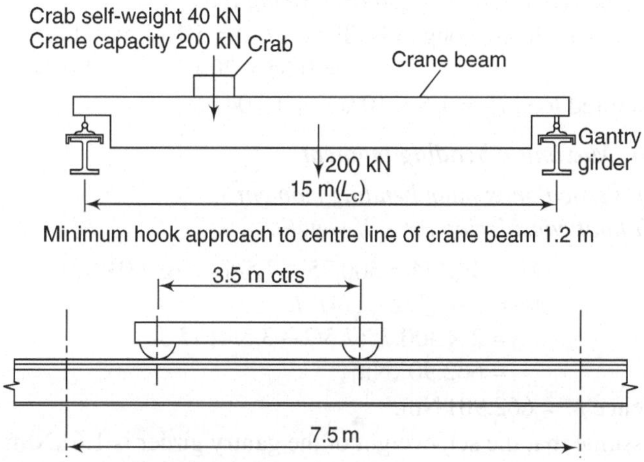


**FIG. 12.27** Design data for Example 12.2

Crane capacity = 200 k*N*
Self-weight of the crane girder excluding trolley = 200 kN
Self-weight of trolley, electrical motor, hook, etc. = 40 kN

Minimum hook approach = 1.2 m
Distance between wheel centres = 3.5 m
Centre-to-centre distance between gantry rails (i.e., span of the crane) = 15 m
Self-weight of the rail section = 300 N/m
Yield stress of steel = 250 MPa

**SOLUTION:**

**Step 1:** *Load and bending moment calculations*

*(a)* ***Load***

*(i) Vertical loading*

*Calculation of maximum static wheel load*

Maximum static wheel load due to the weight of the crane = 200/4 = 50 kN

Maximum static wheel load due to crane load

$$W_1 = [W_t(L_c - L_1)]/(2L_c)$$
$$= (200 + 40)(15 - 1.2)/(2 \times 15)$$
$$= 110.4\text{ kN}$$

Total load due to the weight of the crane and the crane load = 50 + 110.4 = 160.4 kN

To allow for impact, etc., this load should be multiplied by 25% (see Table 12.3).

Design load = 160.4 × 1.25 = 200.5 kN

∴ Factored wheel load on each wheel,

$$W_c = 200.5 \times 1.5 = 300.75\text{ kN}$$

*(ii) Lateral (horizontal) surge load*

Lateral load (per wheel) = 10%(hook + crab load)/4
= 0.1 × (200 + 40)/4
= 6 kN

Factored lateral load = 1.5 × 6 = 9 kN

*Note:* However, as per IS 875, 10% of the weight acting on one side of the rail track has to be taken for EOT cranes. Hence, the factored lateral load (per wheel) should be taken as 18 kN.

*(iii) Longitudinal (horizontal) braking load*

Horizontal force along rails (Table 12.3) = 5% of wheel load
= 0.05 × 200.5 = 10.025 kN

Factored load $P_g$ = 1.5 × 10.025 = 15.04 kN

*(b)* ***Maximum bending moment***

*(i) Vertical maximum bending moment*

*Without considering the self-weight,*

$$M_1 = W_cL/4 = 300.75 \times 7.5/4 = 563.90\text{ kNm}$$
$$M_2 = 2W_c(L/2 - c/4)^2/L$$
$$= 2 \times 300.75(7.5/2 - 3.5/4)^2/7.5$$
$$= 662.90\text{ kNm}$$

Hence $M$ = 662.90 kNm.

Assume that the self-weight of the gantry girder is 1.6 kN/m.

Total dead load = 1600 + 300 (self-weight of rail) = 1.9 kN/m

Factored DL = (1.9 × 1.5) = 2.85 kN/m

$$\text{BM due to dead load} = wl^2/8 = \left(2.85 \times \frac{7.5^2}{8}\right) = 20.04\text{ kNm}$$

*(ii) Horizontal bending moment*

Moment due to surge load = 2 × 9(7.5/2 − 3.5/4)²/7.5

$$M_y = 19.84\text{ kNm}$$

*(iii) Bending moment due to drag (assuming the rail height as 0.15 m and depth of girder as 0.6 m)*

Reaction due to drag force = $P_ge/L$ = 15.04(0.3 + 0.15)/7.5
= 0.903 kN

$$M_3 = R(L/2 - c/4) = 0.903(7.5/2 - 3.5/4)$$
$$= 2.59\text{ kNm}$$

Total design bending moment $M_z$ = 662.9 + 20.04 + 2.59
= 685.53 kNm

***(c) Shear force (For maximum shear force, one of the wheel loads should be very close to the supports)***

*(i) Vertical shear force*

Shear force due to wheel load

$$W_L(2 - c/L) = 300.75(2 - 3.5/7.5)$$
$$= 461.15\text{ kN}$$

$$\text{Shear force due to DL} = \frac{wl}{2} = 2.85 \times \frac{7.5}{2} = 10.69\text{ kN}$$

Maximum ultimate shear force

$$V_z = 10.94 + 461.15$$
$$= 472.09\text{ kN}$$

*(ii) Lateral shear force due to surge load* $V_y$ = 9(2 − 3.5/7.5)
= 13.8 kN

Reactions due to drag force = 0.903 kN
and maximum ultimate reaction

$$R_Z = 472.09\text{ kN} + 0.903$$
$$= 472.99 \text{ or } 473\text{ kN}$$

**Step 2:** *Preliminary selection of the girder*

Since $L/12$ = 7500/12 = 625, we choose the depth as 600 mm.
Therefore,
Approximate width of beam = $L/30$ = 250 mm
Since deflection governs the design, choose $I$, using the deflection limit of $L/750$,

$$I = \frac{15.6W(L-c)}{LE}[2L^2 + 2Lc - c^2]$$

= 15.6 × 200.5 × 10³ × (7500 − 3500) × [2 × 7500² + 2 × 7500 × 3500 − 3500²]/(2 × 10⁵ × 7500)
= 1.274 × 10⁹ mm⁴

Required $Z_p = 1.4 \times M/f_y$
= 1.4 × 685.53 × 10⁶/250
= 3.83 × 10⁶ mm³

Choose ISMB 600 @ 1230 N/m with a channel ISMC 300 @ 363 N/m @ the top (see Fig. 12.28).

***(a) Properties of the sections***

| ISMB 600 @ 1.23 kN/m | ISMC 300 @ 0.363 kN/m |
|---|---|
| $A = 15600\,\text{mm}^2$<br>$t_f = 20.3\,\text{mm}$<br>$t_w = 12\,\text{mm}$<br>$B = 210\,\text{mm}$<br>$I_{zz} = 91800 \times 10^4\,\text{mm}^4$<br>$I_{yy} = 2650 \times 10^4\,\text{mm}^4$<br>$R = 20\,\text{mm}$ | $A = 4630\,\text{mm}^2$<br>$t_f = 13.6\,\text{mm}$<br>$t_w = 7.8\,\text{mm}$<br>$B = 90\,\text{mm}$<br>$I_{zz} = 6420 \times 10^4\,\text{mm}^4$<br>$I_{yy} = 313 \times 10^4\,\text{mm}^4$<br>$C_y = 23.5\,\text{mm}$ |

*(i) Elastic properties of the combined section*

Total area $A = A_B + A_{ch} = 15,600 + 4630 = 20,230\,\text{mm}^2$

The distance of NA of the built-up section from the extreme fibre of tension flange,

$\bar{y} = [15,600 \times 600/2 + 4630 \times (600 + 7.8 - 23.5)]/20,230 = 365.07\,\text{mm}$

$h_1 = \bar{y} - h_B/2 = 365.07 - 600/2 = 65.07\,\text{mm}$

$h_2 = (h_B + t_{ch}) - \bar{y} - C_y = (600 + 7.8) - 365.07 - 23.5 = 219.23\,\text{mm}$

$h_3 = 607.8 - 365.07 - 7.8 = 234.93\,\text{mm}$

$I_Z = I_{ZB} + A_B h_1^2 + (I_y)_{ch} + A_{ch} \times h_2^2$

$= 91,800 \times 10^4 + 15,600 \times 65.07^2 + 313 \times 10^4 + 4630 \times 219.23^2$

$= 1.2097 \times 10^9\,\text{mm}^4 \approx 1.274 \times 10^9$ (required for deflection control)

$Z_{Zb} = 1.2097 \times 10^9/(365.07) = 3.31 \times 10^6\,\text{mm}^3$

$Z_{zt} = 1.2097 \times 10^9/(242.73) = 4.98 \times 10^6\,\text{mm}^3$

$I_{yy}$ combined $= 2650 \times 10^4 + 6420 \times 10^4 = 9070 \times 10^4\,\text{mm}^4$

$I_y$ for tension flange about the *y-y* axis

$I_{tf} = 20.3 \times 210^3/12 = 1566.6 \times 10^4\,\text{mm}^4$

For compression flange about the *y-y* axis

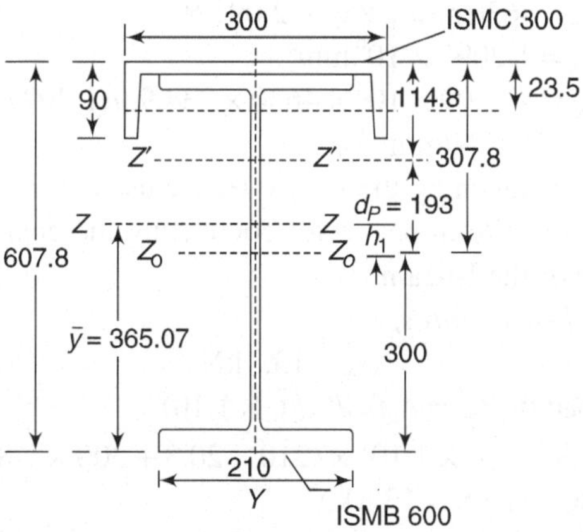


**FIG. 12.28**

$I_{cf} = 1566.6 \times 10^4 + 6420 \times 10^4 = 7986.6 \times 10^4\,\text{mm}^4$

$Z_y$ (for top flange alone) $= 7986.6 \times 10^4/150 = 532,443\,\text{mm}^3$

***(b) Calculation of plastic modulus***

The plastic neutral axis divides the area into two equal areas, i.e., $10,115\,\text{mm}^2$.

$$d_p = 4630/(2 \times 12) = 193\,\text{mm}$$

Ignoring fillets, the plastic section modulus below the equal-area axis is

$\Sigma A\bar{y} = 20.3 \times 210 \times (493 - 20.3/2) + (493 - 20.3) \times 12 \times (493 - 20.3)/2$

$= 3399.1 \times 10^3\,\text{mm}^3$

Above the equal-axea axis

$\Sigma A\bar{y} = 4630 \times (114.8 - 23.5) + 210 \times 20.3 \times (114.8 - 7.8 - 10.15) + 86.7 \times 12 \times 86.7/2$

$= 880.692 \times 10^3\,\text{mm}^3$

$Z_{pz} = 3399.1 \times 10^3 + 880.692 \times 10^3 = 4279.792 \times 10^3\,\text{mm}^3$

For the top flange only

$Z_{py} = 20.3 \times 210^2/4 + (300 - 2 \times 13.6)^2 \times 7.8/4 + 2 \times 90 \times 13.6 \times (150 - 13.6/2) = 719,479.8\,\text{mm}^3$

**Step 3:** *Check for moment capacity*

Check for plastic section

$b/t$ of the flange of the I-beam $= [(210 - 12)/2]/20.3 = 4.87 < 9.4$

$b/t$ of the flange of the channel $= (90 - 7.8)/13.6 = 6.04 < 9.4$

$d/t$ of the web of the I-section $= (600 - 2 \times 20.3)/12 = 46.6 < 84$

Hence the section is plastic.

***(a) Local moment capacity***

$1.2Z_e f_y/1.1 = 1.2 \times 3.31 \times 10^6 \times (250/1.1) \times 10^{-6} = 902.72\,\text{kNm}$

$M_{dz} = f_y Z_p/1.1 = (250/1.1) \times 4279.792 \times 10^{-3} = 972.68\,\text{kNm} > 902.72\,\text{kNm}$

Hence take $M_{dz} = 902.72\,\text{kNm}$

$M_{dy} = (f_y/1.1) \times Z_p(\text{top flange}) = (250/1.1) \times 719,479.8 \times 10^{-6} = 163\,\text{kNm}$

$1.2Z_e f_y/1.1 = 1.2 \times 532,443 \times (250/1.1) \times 10^{-6} = 145.2\,\text{kNm} < 163\,\text{kNm}$

Hence take $M_{dy} = 145.2\,\text{kNm}$.

***(b) Combined local capacity check***

$685.53/902.72 + 19.84/145.2 = 0.759 + 0.137 = 0.896 < 1$

Hence the chosen section is the right choice.

**Step 4:** *Check for buckling resistance*

As per IS 800 (Clause 8.2.2), the design bending strength

$$M_d = \beta_b Z_p f_{bd}$$

We have

$\beta_b = 1.0$

$h = 600 + 7.8 = 607.8\,\text{mm}$

$KL = 7500\,\text{mm}$

$E = 2 \times 10^5\,\text{N/mm}^2$

$t_f = 20.3 + 7.8 = 28.1\,\text{mm}$

$r_y = \sqrt{I_{yy}/A}$

$I_{yy} = (2650 \times 10^4) + (6420 \times 10^4) = 9070 \times 10^4\,\text{mm}^4$

$A = 15,600 + 4630 = 20,230\,\text{mm}^2$

$$r_y = \sqrt{\frac{9070 \times 10^4}{20,230}} = 66.96\,\text{mm}$$

According to clause 8.2.2.1 of (IS 800), elastic lateral buckling moment

$$M_{cr} = C_1 \frac{\pi^2 EI_y h}{2(KL)^2}\left[1+\frac{1}{20}\left[\frac{KL/r_y}{h/t_f}\right]^2\right]^{0.5}$$

$C_1 = 1.132$ (from Table 42 of IS 800 : 2007)

$$M_{cr} = 1.132 \times \left(\frac{\pi^2 \times 2\times 10^5 \times 9070\times 10^4 \times 607.8}{2\times 7500^2}\right)$$

$$\left[1+\frac{1}{20}\left(\frac{7500/66.96}{607.8/28.1}\right)^2\right]^{0.5} = 1675.22\times 10^6 \text{ N/mm}$$

*Note:* This formula is valid for I-sections only. For more accurate values of $M_{cr}$ for a compound section, the formula given in E1.2 (Annex E) of the code may be used.

Non-dimensional slenderness ratio

$$\lambda_{LTZ} = \sqrt{\frac{\beta_b Z_{pz} f_y}{M_{cr}}}$$

$$= \sqrt{\frac{1.0\times 4279.792\times 10^3 \times 250}{1675.22\times 10^6}} = 0.7992$$

*Along the z-direction*

$\phi_{LTZ} = 0.5[1 + \alpha_{LT}(\lambda_{LTZ} - 0.2) + \lambda_{LTZ}^2]$

$= 0.5[1 + 0.21(0.7992 - 0.2) + 0.7992^2] = 0.882$

*Note:* Since the channel will normally be connected by intermittent welds to the I-section, $\alpha_{LT}$ value has been taken as 0.21. If heavy welding is involved, take $\alpha_{LT}$ as 0.49.

$$\chi_{LTZ} = \frac{1}{[\phi_{LTZ} + (\phi_{LTZ}^2 - \lambda_{LTZ}^2)^{0.5}]} \le 1.0$$

$$= \frac{1}{0.882 + (0.882^2 - 0.7992^2)^{0.5}}$$

$= 0.7967$

$f_{bd} = f_y \chi_{LT}/\gamma_{m0}$

$\gamma_{m0} = 1.10$ (from Table 5 of the code)

$\therefore$ $f_{bd} = 0.7967 \times 250/1.1 = 181.07 \text{ N/mm}^2$

$\therefore$ $M_{dz} = \beta_b Z_{pz} f_{bd}$

$= 1.0 \times 181.07 \times 4279.792 \times 10^{-3}$

$M_{dz} = 774.9 \text{ kNm} > 685.53 \text{ kNm}$

Thus the beam is satisfactory under vertical loading. Now it is necessary to check it under biaxial bending.

For top flange only

$M_{dy} = (f_y/1.1) \times Z_{yt}$

$= (250/1.1) \times 719{,}479 \times 10^{-6} = 163.5 \text{ kNm} > 1.2 \times 532{,}433 \times 250/(1.1 \times 10^6) = 145.2 \text{ kNm}$

Hence $M_{dy} = 145.2$ kNm

***(a) Check for biaxial bending***

In order to check for biaxial bending, we substitute the terms with their values in the following equation:

$$\frac{M_z}{M_{dz}} + \frac{M_y}{M_{dy}} \not> 1.0$$

We have

$$\frac{685.53}{774.9} + \frac{19.84}{145.2} = 0.884 + 0.137 = 1.021 \cong 1.0$$

Slightly bigger size of top channel may be selected. Hence the beam is safe.

**Step 5:** *Check for shear capacity*

*For vertical load,*

$$V_z = 472.09 \text{ kN}$$

Shear capacity $= A_v f_{yw}/(\sqrt{3} \times 1.10)$

$= (600 \times 12) \times 250/(\sqrt{3} \times 1.10) \times 10^{-3}$

$= 944.75 \text{ kN} > 472.09 \text{ kN}$

The maximum shear force is 472.09 kN, which is less than 0.6 times the shear capacity, i.e.,

$$0.6 \times 944.75 = 566.85 \text{ kN}$$

Hence it is safe in vertical shear and there is no reduction in the moment capacity.

***(a) Weld design***

The required shear capacity of the weld is given by

$q = VA\bar{y}/I_Z$

$\bar{y} = h_3 = 234.93$

$A = 4630 \text{ mm}^2$, $V = 472.09$ kN

$I_Z = 1.2097 \times 10^9 \text{ mm}^4$

$q = 472.09 \times 10^3 \times 4630 \times 234.93/(1.2097 \times 10^9)$

$= 424$ N/mm

This shear is taken by the welds. Hence use a minimum weld of 4 mm (442 N/mm per weld) connecting the channel to the top flange of the I-beam.

*For lateral shear force,*

$$F_y = 13.8 \text{ kN}$$

Shear capacity $V_{ny} = A_v f_{yw}/(\sqrt{3} \times 1.10)$

$= 250/(\sqrt{3} \times 1.10) \times (210 \times 20.3 + 300 \times 7.8) \times 10^{-3}$

$= 866.41 \text{ kN} > 13.8 \text{ kN}$

Hence it is safe for resisting lateral shear.

**Step 6:** *Web buckling*

At points of concentrated loads (wheel loads or reaction) the web of the girder must be checked for local buckling (see Section 10.11)

The dispersion length under wheel (assuming it to be 150 mm and assuming an angle of dispersion of 45°). Also, assume stiff bearing length at support = 150 mm.

$$b_1 = 150 \text{ mm}$$

$$n_1 = 600/2 + 7.8 = 307.8\,\text{mm}$$

Web slenderness $\lambda = 2.5d/t$

$= 2.5 \times [600 - 2(20.3 + 20)]/12$

$= 108.2$

Stress reduction factor (from Table 8 of IS 800 : 2007) = 0.426

$$f_{cd} = 0.426 \times 250/1.1 = 96.8\,\text{MPa}$$

Buckling resistance $= (b_1 + n_1)tf_{cd}$

$= (150 + 307.8)12 \times 96.8 \times 10^{-3}$

$= 531.78\,\text{kN}$

Maximum reaction = 473 kN < 531.78 kN

Hence buckling resistance is satisfactory.

Under concentrated loads, the dispersion will be satisfactory. In addition web crippling due to wheel load at inner points is to be checked. This is left as an exercise to students.

**Step 7:** *Web bearing (see Section 10.11)*

Load dispersion at support with 1: 2.5 dispersion

Minimum stiff bearing $= R_x/(tf_{yw}/1.1) - n_2$

$n_2 = (20.3 + 20) \times 2.5 = 100.75\,\text{mm}$

$R_x = 473\,\text{kN}$ (support reaction)

$b_1 = 473 \times 10^3/(12 \times 250/1.1) - 100.75 = 72.68\,\text{mm}$

Web bearing at support requires a minimum stiff bearing of 73 mm < 150 mm.

**Step 8:** *Check for deflection at working load*

Serviceability vertical wheel load excluding impact = 160.4 kN

Deflection at mid-span

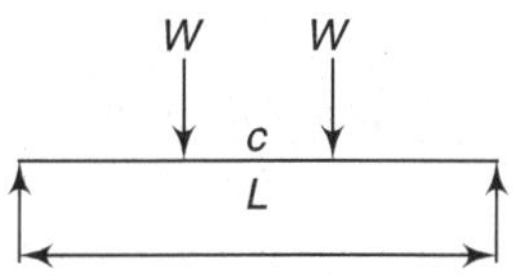


$$\Delta = WL^3[(3a/4L) - (a^3/L^3)/(6EI)$$

Where

$$a = (L - c)/2 = (7500 - 3500)/2 = 2000$$

*(i) Vertical*

Combined $I_{zz} = 1.2097 \times 10^9\,\text{mm}^4$

$$\Delta = \frac{160.4 \times 10^3 \times 7500^3}{6 \times 2 \times 10^5 \times 1.2097 \times 10^9} [3 \times 2000/(4 \times 7500) - 2000^3/7500^3]$$

$= 8.43\,\text{mm} < L/750 = 10\,\text{mm}$

*(ii) Lateral*

Only the compound top flange will be assumed to resist the applied surge load as in the bending check.

$$I = (I_z)_{ch} + I_F = 7986.6 \times 10^4\,\text{mm}^4$$

$$\Delta = \frac{6 \times 10^3 \times 7500^3}{6 \times 2 \times 10^5 \times 7986.6 \times 10^4} [3 \times 2000/(4 \times 7500) - 2000^3/7500^3]$$

$= 4.78 < 10\,\text{mm}$ (Table 6 of IS 800 : 2007)

**Step 9:** *Fatigue strength*

The gantry girder has to be checked for fatigue strength.

Assuming 2 bracket plates,

Reaction on each bracket plate = 473/2 = 236.5 kN

Horizontal reaction = 9 kN

For the calculations required for the bracket connection see Example 14.4 in Chapter 14. But since the bracket will also be subjected to fatigue loads, it will be better to use HSFG bolts rather than ordinary black bolts.

**EXAMPLE 12.3:** Repeat the design of the gantry girder given in Example 12.2, but including a lattice restraint to the compression flange. In practice, such a lattice girder is often provided in heavy industrial buildings, specifically for this purpose or to support access platforms or walkways. Note that it may not be economical to include such a restraint in order to reduce the beam size, especially in gantry girders carrying relatively small loads.

**SOLUTION:**

The loading and BM calculations are the same as in Example 12.2, hence not repeated here.

From Example 12.2,

Static wheel load = 160.4 kN

Factored lateral load = 9 kN

$M_z = 685.53\,\text{kNm}$, Shear force $V_z = 472.09\,\text{kN}$, $R_z = 473\,\text{kN}$

$M_y = 19.84\,\text{kNm}$, Shear force $V_y = 13.8\,\text{kN}$

The lattice is added to the compression flange of the gantry girder in order to reduce the beam size; a beam without the top channel section is chosen.

Assume ISWB: 600 and its properties are given below.

ISWB: 600

$A = 15600\,\text{mm}^2$

$h = 600\,\text{mm}$

$B = 210\,\text{mm}$

$t_f = 20.3\,\text{mm}$

$t_w = 12\,\text{mm}$

$I_{zz} = 91800 \times 10^4\,\text{mm}^4$

$Z_z = 3060 \times 10^3\,\text{mm}^3$

$R = 20\,\text{mm}$

**Step 1:** *Check for shear capacity*

Shear capacity $= A_v f_{yw}/(\sqrt{3} \times 1.10)$

$= (600 \times 12) \times 250/(\sqrt{3} \times 1.10) \times 10^{-3}$

$= 944.75\,\text{kN} > 472.09\,\text{kN}$

The maximum shear force is 472.09 kN, which is less than 0.6 times the shear capacity:

$$0.6 \times 944.75 = 566.85\,\text{kN}$$

Hence there is no reduction in the moment capacity.

**Step 2:** *Check for moment capacity*

As already seen in Example 12.2, the chosen section is a plastic section. Hence the moment capacity

$$M_d = \beta_b Z_p f_y / \gamma_{m0}$$

$Z_p$ of ISMB 600

$$\Sigma A\bar{y} = 2[20.3 \times 210 \times (300 - 20.3/2) + (300 - 20.3)^2 \times 12/2]$$
$$= 3410.05 \times 10^3 \text{ mm}^3$$
$$M_d = 1 \times 3410.05 \times 10^3 \times (250/1.1) \times 10^{-6}$$
$$= 775.0 \text{ kNm}$$
$$1.2 Z_e f_y / \gamma_{m0} = 1.2 \times 3060 \times 10^3 \times (250/1.1) \times 10^{-6}$$
$$= 834.55 > 775.0 \text{ kNm}$$
$$M_d > 685.53 \text{ kNm}$$

Hence the section is safe.

The effect of lateral moment $M_y$ is considered in combination with the restraint force in the later part of the calculation (see lattice girder design).

$$\text{Flange force} = 685.53/(0.600 - 0.0203) = 1182.6 \text{ kN}$$

As per Clause 8.3.1, the restraint shall be capable of resisting a force equal to 2.5% of the flange force.

$$\text{Restraint force} = \frac{2.5}{100} \times 1182.6$$
$$= 29.6 \text{ kN}$$

**Step 3:** *Check for web buckling/bearing*

As shown in Example 12.2, buckling resistance = 540.8 kN > 300.75 kN. Similarly, a stiff bearing length of 75 mm is required at support.

**Step 4:** *Check for deflection*

Calculations in Example 12.2 give a deflection of 8.43 mm for $I = 1.2097 \times 10^9 \text{ mm}^4$.

Now $I = 91{,}800 \times 10^4 \text{ mm}^4$. Hence,

$$\text{Deflection} = 8.43 \times 1.2097 \times 10^9/91800 \times 10^4 = 11.11 \text{ mm} > 10 \text{ mm}$$

Hence revise the section to ISWB 600 which has $I_{zz} = 116{,}000 \times 10^4 \text{ mm}^4$

$$\therefore \text{Deflection} = 8.43 \times 1.2097 \times 10^9/1.16 \times 10^9 = 8.79 < 10 \text{ mm}$$

Hence for ISWB 600, the deflection is within allowable limits.

**Step 5:** Connection

The lattice transmits horizontal forces to the support. Hence a separate diaphragm is not required to connect the top flange of the gantry girder to the column. The bracket connection is similar to that needed for beam in Example 12.2.

**Step 6:** *Lattice girder*

The following loads are assumed to act on the gantry girder (see the sketch):

(a) The restraint force 29.60 kN is considered to be distributed between the nodes of lattice.

(b) The surge load of 9 kN per wheel.

The forces acting on various members of the lattice girder have been found out using Truss Analysis Program (a computer program) and are given in the following table.

| Panel | Top chord | Bottom chord | Diagonal | Verticals |
|---|---|---|---|---|
| 1-2 | 0 | 33.79 | 8.73 | 0 |
| 2-3 | 37.33 | 3.78 | 17.19 | −17.32 |
| 3-4 | 67.34 | −18.9 | 25.65 | −13.10 |
| 4-5 | 90.02 | −18.67 | 52.11 | 0.13 |
| 5-6 | 82.23 | −18.67 | −0.27 | 0 |
| 6-7 | 67.34 | −11.11 | 26.19 | −4.37 |
| 7-8 | 45.13 | 3.78 | 34.65 | −8.59 |
| 8-9 | 0 | 25.99 | 43.11 | −12.83 |

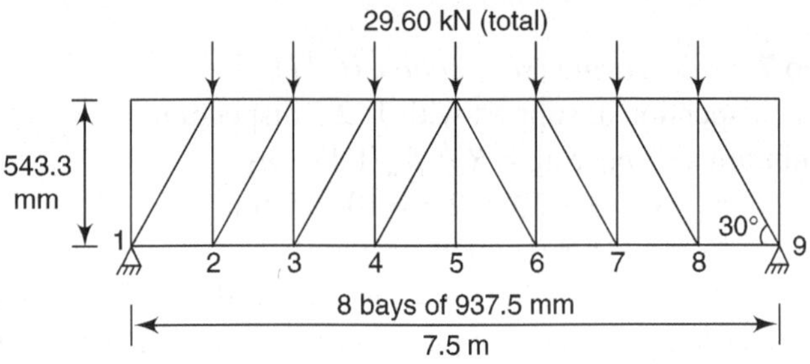


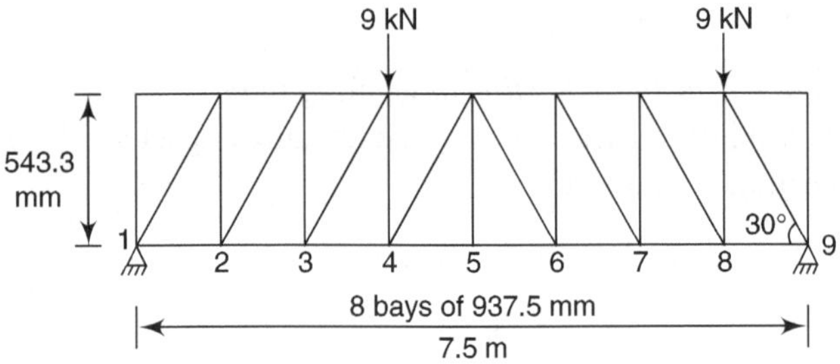


*Designing the chord members for a compression of 90.02 kN*

Use 60 × 60 × 6 angle.

$$r_{vv} = 11.5 \text{ mm}$$
$$\lambda = L_e/r_{vv} \text{ (assuming } k = 1, L_e = L)$$
$$= (1.0 \times 937.5)/11.5 = 81.52 \text{ mm}$$

From Table 8c of IS 800 : 2007, $\chi = 0.5898$.

(For curve c)

$$f_{cd} = \chi f_y / \gamma_{m0} = 0.5898 \times 250/1.10 = 134 \text{ N/mm}^2$$

***(a) Compression resistance***

Assuming the members to be welded $A_e = A_g$

$$P_d = f_{cd} A_e$$
$$= 134 \times 684 \times 10^{-3}$$
$$= 91.66 \text{ kN} > 90.02 \text{ kN}$$

*Hence safe.*

Designing the diagonals for a compression of 52.11 kN

Choose 60 × 60 × 6

$$r_{vv} = 11.5 \text{ mm}$$

$$L = \sqrt{(937.5^2 + 543.3)^2} = 1083.5\,\text{mm}$$

$$\lambda = L_e/r_{vv} = 1.0 \times 1083.5/11.5 = 94.2$$

From Table 8c of IS 800 : 2007 (For curve c)

$$\chi = 0.507$$

$$f_{cd} = \chi f_y/\gamma_{m0} = 0.507 \times 250/1.10 = 115\,\text{N/mm}^2$$

Compression resistance $= P_d = f_{cd}A_e$

$= 115 \times 684 \times 10^{-3}$

$= 78.8\,\text{kN} > 52.11$

Hence safe.

**EXAMPLE 12.4:** Check the gantry girder section designed in Example 12.2 for fatigue strength for the following data:

(a) The crane operates for 250 days per year
(b) The carne is worked daily from 9 a.m. to 4 p.m.
(c) Maximum number of trips per hour : 3
(d) Design life : 50 years

**SOLUTION:**

From Example 12.2,
Maximum bending moment, $M_u = 685.53\,\text{kNm}$
Maximum shear force, $V_{zu} = 473\,\text{kN}$
Section modulus, $Z_{ez} = 3.31 \times 10^6\,\text{mm}^3$
Partial safety factor for fatigue strength (Table 25 of IS 800), $\gamma_{mft} = 1.15$
Actual number of cycles, $N_{sc} = 3 \times 7 \times 250 \times 50 = 262,500$

**Step 1:** *Determine detail category*
From Table 26(b) of IS 800, detail category = 59
*Note:* This category is chosen because there may be intermittent longitudinal welds connecting the channel to the I-girder at the top flange.

**Step 2:** *Determine stress ranges*
As per clause 13.4, normal stress range when $N_{sc} \leq 5 \times 10^6$,

$$f_f = f_{fn}\sqrt[3]{\frac{5 \times 10^6}{N_{sc}}} = 59 \times \sqrt[3]{\frac{5 \times 10^6}{262,500}} = 157.57\,\text{N/mm}^2$$

Design fatigue strength = 157.57/1.15 = 137 N/mm²

*Actual stress range:* Minimum stress range $f_{min} = 0$
The maximum moment of 685.53 kNm has a load factor of 1.50. Since fatigue is checked at the working loads with a load factor of 1.0,
Maximum moment = 685.53/1.5 = 457.02 kNm

$$\text{Maximum stress} = \frac{M}{Z_{ez}} = \frac{457.02 \times 10^6}{3.31 \times 10^6} = 138.07\,\text{N/mm}^2 < f_y$$

$$= 250\,\text{N/mm}^2$$

Hence the stress in the gantry girder ranges from 0 to 138.07 N/mm²

Since the actual stress range (138.07 N/mm²) is more or less equal to the design fatigue strength (137 N/mm²) for this detail, the gantry girder may be considered safe.

**Step 3:** *Check for shear*
Maximum shear force with a load factor of 1.5 = 473 kN
For fatigue calculations, shear force = 473/1.5 = 315.33 kN

$$\text{Shear stress} = \frac{V}{dt_w} = \frac{1.0 \times 315.33 \times 10^3}{607.8 \times 12} = 43.23\,\text{N/mm}^2$$

$$< \tau_y = \frac{250}{\sqrt{3}} = 144.3\,\text{N/mm}^2$$

*Note*: Load factor of 1.0 has been taken as per Table 4 of IS 800.
As per clause 13.2.2.3 of IS 800

$$f \leq \frac{27}{\gamma_{mft}} = \frac{27}{1.15} = 23.48\,\text{N/mm}^2$$

Hence fatigue assessment is required at the support.

As per clause 13.4, shear stress range, $\tau_f = \tau_{fn}\sqrt[5]{\frac{5 \times 10^6}{N_{sc}}}$

$$= 59\sqrt[5]{\frac{5 \times 10^6}{262,500}} = 106.37\,\text{N/mm}^2$$

Design fatigue strength in shear $= 106.37/\gamma_{mft} = \dfrac{106.37}{1.15}$ $= 92.5\,\text{N/mm}^2 > 43.23\,\text{N/mm}^2$

Hence the gantry girder is safe in shear stress range also.

## SUMMARY

Cranes are often employed in industrial buildings to move stock, finished goods, or new materials from one place to another for processing. These cranes are supported on gantry girders, which are supported on separate columns or brackets attached to steel columns. The design of gantry girders is often complicated since they support moving load and are, hence, subjected to fatigue. Before attempting to design a gantry girder, the engineer should know the loading and clearance details from the crane manufacturer. In this chapter, some guidelines have been provided in the form of tables to do preliminary calculations and arrive at the size of gantry girders. These calculations should be checked, during the final design, with the exact details obtained from the manufacturers. The various loadings that may occur due to the moving loads have been discussed and expressions for calculating the maximum bending moment, shear, and deflection of gantry girders provided. A brief discussion on fatigue effects has been included. The various factors that may affect the choice of girders and many practical suggestions for connection with columns and bracings have been also given. The steps in the design of gantry girders have been explained with the use of practical examples.

## MULTIPLE-CHOICE QUESTIONS

12.1 Which of the following need not be considered in the design of a gantry girder?
(a) Gravity loads (c) Longitudinal loads
(b) Lateral loads (d) Wind loads

12.2 For gantry girders carrying EOT cranes, the vertical loads are to be increased to consider impact by __________ of maximum static wheel load
(a) 10% (b) 20% (c) 25% (d) 15%

12.3 For gantry girders carrying EOT cranes, the horizontal load transverse to the rails is to be increased to consider impact by
(a) 15% of maximum static wheel load
(b) 25% of maximum static wheel load
(c) 5% of the weight of crab and the weight lifted by the crane
(d) 10% of the weight of crab and the weight lifted by the crane

12.4 Limiting vertical deflection of gantry girders carrying EOT cranes with a capacity less than 500 kN is
(a) *L*/500 (b) *L*/600 (c) *L*/750 (d) *L*/1000

12.5 Maximum shear force in a gantry girder carrying a over-head travelling crane occurs when
(a) one of the wheel loads is at the support
(b) the center of the span coincides with center of gravity of wheel loads
(c) one of the wheel loads is near mid-span and the other on adjacent span
(d) none of these

12.6 Which of the following are correct?
(a) Single span gantry girders are desirable
(b) Gantry girders are always designed as laterally supported beams.
(c) When high strength girders are used it is not necessary to check deflections
(d) It is a good idea to keep short spans and deep girders

12.7 Typically a gantry girder cross section consists of an I-section over which a channel is placed as in Fig. 12.9(b). The function of the top channel is to (chose all options that are applicable)
(a) increase moment of inertia about *y*-*y* axis
(b) decrease moment of inertia about *z*-*z* axis
(c) increase torsional resistance and lateral buckling strength
(d) both (a) and (c)

12.8 For light and medium duty cranes fatigue effects need not be considered if the number of cycles of load is less than
(a) $2 \times 10^6$ (b) $3 \times 10^6$ (c) $4 \times 10^6$ (d) $5 \times 10^6$

12.9 The impact allowance in percentage to be applied to the vertical forces transferred to the wheels of an EOT crane is
(a) 10% of the maximum static wheel load
(b) 10% of the weight of the crab and the weight lifted on the crane
(c) 25% of the maximum static wheel load

12.10 The impact allowance for drag forces along the rail for an EOT crane is
(a) 10% of the static wheel load
(b) 5% of the static wheel load
(c) 10% of the weight of crab and the weight lifted on the crane.
(d) 5% of the weight of the crab and the weight lifted on the crane.

12.11 The surge load is assumed to be resisted by the
(a) whole cross section
(b) compression flange alone
(c) compression and tension flanges
(d) cross section above the neutral axis

12.12 Which of the following gantry girders are to be checked for fatigue effects?
(a) Supporting heavy-duty cranes
(b) Supporting light-duty cranes with the number of cycles of load less than $5 \times 10^6$
(c) Supporting medium-duty cranes with the number of cycles of load greater than $5 \times 10^6$
(d) Supporting light-duty cranes with the number of cycles of load greater than $5 \times 10^6$

## EXERCISES

12.1 A 50 kN hand-operated crane is provided in a building and has the following data:

| | |
|---|---|
| Centre-to-centre distance of the gantry beam (width of the building) | 16 m |
| Longitudinal spacing of columns (span of gantry) | 7.5 m |
| Weight of the crane | 40 kN |
| Wheel spacing | 3 m |
| Weight of the crab | 10 kN |
| Minimum hook approach | 1 m |
| Yield stress of steel | 250 MPa |

Design a simply supported gantry girder assuming lateral support to it.

12.2 Design a gantry girder, without lateral restraint along its span, to be used in an industrial building carrying overhead travelling crane for the following data:

| | |
|---|---|
| Centre-to-centre distance between columns (span of the gantry girder) | 6 m |
| Crane capacity | 50 kN |
| Self-weight of the crane girder excluding trolley | 40 kN |
| Self-weight of the trolley, electric motor, hook, etc. | 10 kN |
| Minimum hook approach | 1.0 m |
| Wheel centres | 3 m |
| Centre-to-centre distance between gantry rails (span of crane) | 12 m |
| Self-weight of rail section | 100 N/m |
| Yield stress of steel | 250 N/mm$^2$ |

12.3 Design a simply supported crane girder to carry an electric overhead travelling crane for the following data:

| | |
|---|---|
| Crane capacity | 300 kN |
| Weight of the crane and crab | 300 kN |
| Weight of the crane | 200 kN |

| | |
|---|---|
| Minimum hook approach | 1.2 m |
| Centre-to-centre distance between wheels | 3.2 m |
| Span of the gantry girder | 5 m |
| Centre-to-centre distance between gantries | 15 m |
| Weight of rail | 300 N/m |
| Height of rails | 75 mm |
| Yield stress of steel | 250 MPa |

12.4 Rework Exercise 12.3, including a lattice restraint to the compression flange.

12.5 Two cranes move on a gantry girder and the data of the building and crane is

| | |
|---|---|
| Bay width | 15 m |
| Column spacing | 8 m |
| Capacity of each crane | 200 kN |
| Weight of each crane girder | 140 kN |
| Weight of each crab | 15 kN |
| Spacing of wheels | 3.2 m |
| Minimum hook approach | 1.2 m |
| Minimum distance between cranes | 2 m |
| Yield stress of steel | 250 MPa |

Design a simply supported gantry girder.

12.6 Rework Exercise 12.5 assuming that the compression flange of the gantry girder is laterally supported.

12.7 Rework Exercise 12.5 assuming that the gantry girder is made of welded plates (plate girder). Note that the use of bearing stiffeners are not recommended in gantry girders.

## REVIEW QUESTIONS

12.1 What is the main purpose of a gantry girder?

12.2 List a few types of cranes.

12.3 What are the components of a crane runway system?

12.4 What are the requirements to be considered by the designer while selecting a crane and designing a crane supporting structure?

12.5 List the loads that should be considered while designing a gantry girder.

12.6 Why should rails not be welded to the gantry girder?

12.7 What is the difference between surge load and drag load of cranes?

12.8 When is surge load assumed to act on one gantry girder and when can it be assumed to be resisted by two opposite gantry girders?

12.9 Write down the expressions for maximum shear force, bending moment, and deflection at mid-span for a simply supported beam with two moving loads, each with a value *W*.

12.10 Write down the expressions for maximum shear force, bending moment, and deflection at mid-span for a simply supported beam with four moving loads, each with a value *W* (two cranes running in tandem).

12.11 List the different profiles of cross sections which are used for gantry girders.

12.12 Why are simply supported girders preferred to two-span gantry girders?

12.13 What is the limiting vertical deflection of the gantry girder for
(a) manually operated cranes
(b) EOT cranes with a capacity less than 500 kN
(c) EOT cranes with a capacity greater than 500 kN

12.14 What are the problems created by excessive deflection?

12.15 How can the effect of rail eccentricity be counteracted?

12.16 What are the possible ways to reinforce the compression flange of an I-beam to resist lateral loads?

12.17 Write short notes on the following:
(a) Columns to support gantry girders
(b) Bracings of columns supporting gantry girders
(c) Crane stops

12.18 List the various steps involved in the design of a gantry girder.

## REFERENCES


Brockenbrough, R.L. and Merrit, F.S., *Structural Steel Designers' Handbook*, Third edition, McGraw-Hill, New York, 1999, p. 1208.

Coates, R.C., Coutie, M.G., and Kong, F.K., *Structural Analysis*, Van Nostrand Reinhold, 1988, pp. 127–131.

Fisher, J.M. and Buettner D.R., *Light and Heavy Industrial Buildings*, American Institute of Steel Construction, Chicago, Illinois, 1979.

Gaylord, E.H., Gaylord, C.N., and Stallmeyer, J.E., *Structural Engineering Handbook*, Fourth edition, McGraw-Hill, New York, 1996, p. 1248.

Gurney, T.R., *Fatigue of Welded Structures*, Second edition, Cambridge University Press, England, 1979.

IS: 1024-1999 *Code of Practice for use of Welding in Bridges and Structures Subjected to Dynamic Loading (Second revision)*, Bureau of Indian Standards, New Delhi.

IS: 807-1976 *Code of Practice for Design, Manufacture, Erection and Testing (Structural Portion) of Cranes and Hoists*, Bureau of Indian Standards, New Delhi.

IS: 875-1987 *Code of Practice for Design Loads (other than Earthquake) for Buildings and Structures*, Parts 1-5, Bureau of Indian Standards, New Delhi.

IS: 3177-1977 *Code of Practice for Electrical Overhead Travelling Cranes and Gantry Girders other than Steel Works Cranes*, Bureau of Indian Standards, New Delhi.

IS: 4137-1967 *Code of Practice for Heavy Duty Electrical Overhead Travelling Cranes for Use in Steel works (with Amendment No. 1)*, Bureau of Indian Standards, New Delhi.

Marshall, W.T. and Nelson, H.M., *Structures*, Longman, 1990, pp. 79–106.

Ricker, D.T., Tips for Avoiding Crane Runway Problems, *Engineering Journal*, AISC, Vol. 19, No. 4, 1982, pp. 181–205.

Weaver, W.M., *Crane Handbook: Design, Data and Engineering Information Used in the Manufacture and Application of Overhead and Gantry Cranes*, Fourth edition, Whiting Corporation, Harvey, Illinois, 1979.

# 13

# DESIGN OF BEAM-COLUMNS

## INTRODUCTION

Most columns are subjected to bending in addition to the axial load; considerable care should be taken in a practical situation to load a column under axial load only. When significant bending is present in addition to an axial load in a member, the member is termed as a *beam-column.* The bending moments on a column may be due to any of the following effects.

**Eccentricity of axial force** Figure 13.1 shows the disposition of beams and girders at interior, corner, and exterior columns.

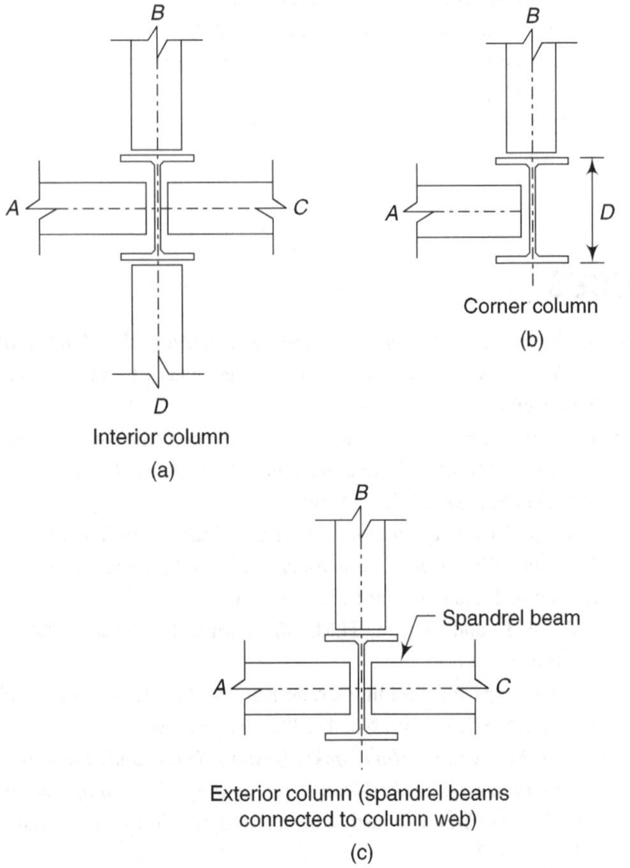


**FIG. 13.1** Eccentric loads on columns

The simple shear connection is generally used to connect beams or girders to columns in braced frames (see Fig. 13.2) and is assumed to transmit only reactions from the beams/girders to the columns. Thus, the beam B given in Fig. 13.1(b) transfers its reaction at the face of the column, that is, with an eccentricity of $e = D/2$, where $D$ is the depth of the column section. However, since the eccentricity of the reaction of beam A given in Fig. 13.1(b), framing into the web of the column, is small, its effect is often neglected. Thus, when beams having equal reactions frame into the column opposite each other as in Fig. 13.1(a) (interior column), the column loads may be assumed to be applied along the axis of the column. However even in an interior column, if the reactions from the beams {say beam B and D given in Fig. 13.1(a)} are not equal, then the difference in the value of the reactions will create a moment on the column section. The column shown in Fig. 13.1(b) will be subjected to biaxial bending and axial load whereas the column shown in Fig. 13.1(c) will be subjected to bending moment about the major axis, in addition to axial load (provided the spans of beam *A* and *C* are equal).

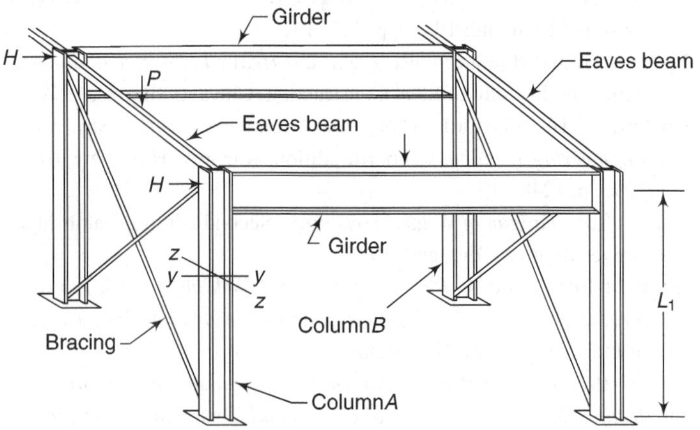


**FIG. 13.2** Beam-columns as a part of the braced frame in the *z-z* plane and a part of an unbraced frame in *y-y* plane

**Building frame action** In a multi-storey building, usually columns support beams which have similar identical connection eccentricities at each floor level. In a rigid frame building construction {see Fig. 2.8(a) of Chapter 2} the columns carry the building load axially as well as end moments from the girders that frame into them. Most building frames are braced against sway (see Fig. 2.6 of Chapter 2) by bracings or core walls, but the horizontal wind forces have to be resisted by bending actions in the columns. Due to the wind forces, the columns may bend in double-curvature bending (similar to the shape of the letter 'S') in contrast to the single curvature bending of columns due to moment, created by gravity loads. Single curvature bending is often the most critical design condition than reverse or double curvature bending. These are shown schematically in Fig. 13.3. Wind loads can also produce lateral loading on a column, giving beam-type bending moment distribution.

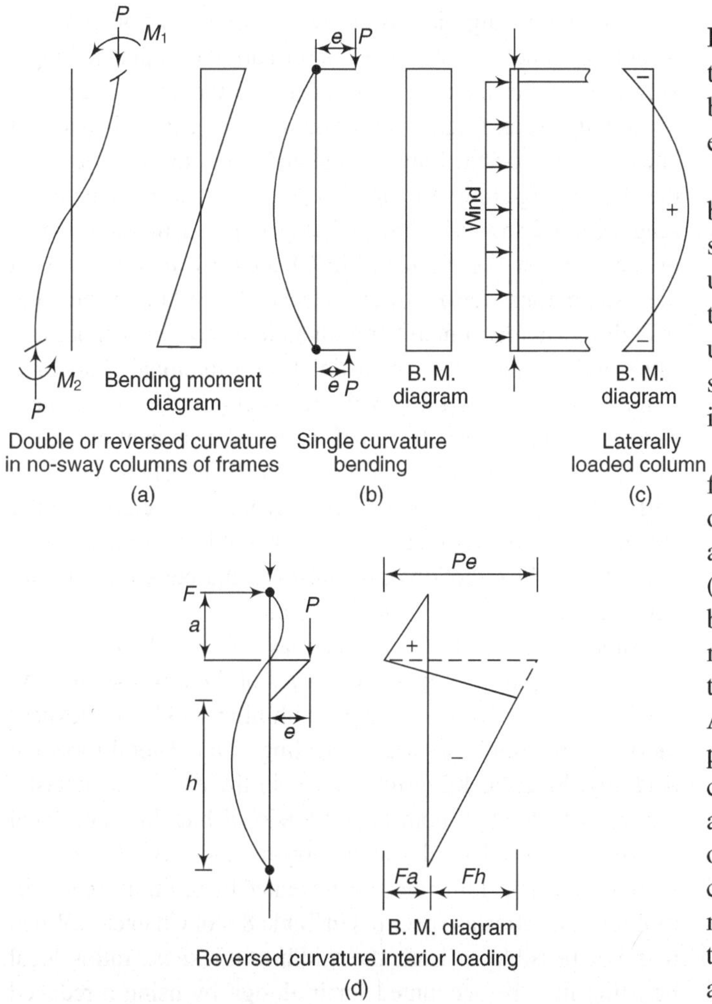


**FIG. 13.3** Single and double curvature bending in beam-columns

**Portal or gable frame action** Another common example of a column with bending moments occurs in a portal frame [See Fig.2.4(a) of Chapter 2] where the columns and rafters are subjected to relatively light axial loads combined with bending.

**Load from brackets** In industrial buildings, column brackets may be used to carry gantry girders on which the cranes move. The resulting eccentricity produces bending moments in addition to the axial loads in the columns. In this case, the column moment is not at the column ends {see Fig. 13.3(d)}.

**Transverse loads** As already discussed, wind pressure on long vertical members may produce bending moments. Similarly earthquakes also produce bending in the columns. Other design conditions also produce bending in addition to axial forces. For example, the top chords of roof and bridge trusses are often considered as pin jointed compression members, but the self weight of these members will produce bending as well. Purlins placed between panel joints of a rafter of roof trusses (in order to reduce the size of purlin or to accommodate maximum size of roofing sheets) will produce bending in rafters (see also Chapter 16).

**Fixed base condition** If the base of the column is fixed due to piles, rafts, or grillage foundation, bending moments will be present at the base of the columns, even though their top ends may be hinged.

*Beam-columns* in steel structures are often subjected to biaxial bending moments, acting in two principal planes, due to the space action of the framing system. The column cross section is usually oriented in such a way to resist significant bending about the major axis of the member. When I-shaped cross sections are used for the columns, the minor axis bending may also become significant, since the minor axis bending resistance of I-section is small compared to the major axis bending resistance.

Thus, in general, beam-columns are subjected to axial forces and bending moments. The moments may be at the ends of the member, as in rigid framed buildings, or developed at an interior point from a bracket, local beam, cable attachment (as in guyed mast or pylon of cable stayed bridges). As the bending moment on a beam-column approaches zero, the member tends to become a centrally loaded column, a problem that has been treated in Chapter 9 on compression members. As the axial forces on a beam-column approaches zero, the problem becomes that of a beam, which has been adequately covered in Chapter 10 on beams. All of the parameters that affect the behaviour of a beam or a column (such as length of the member, geometry and material properties, support conditions, magnitude and distribution of transverse loads and moments, presence or absence of lateral bracing, and whether the member is a part of an unbraced or braced frame), will also affect the behaviour, strength, and design of beam-columns (Vinnakota 2006).

Note that bending may also be produced in tension members such as the bottom chords of bridge trusses, when floor beams frame into them. Bottom chords of roof trusses may support hoisting devices or other temporary loads, thus producing bending moments in addition to the axial loads present.

The moments that are present in truss members due to end fixity are collectively termed as *secondary moments* and are often neglected in the analysis and the members are analysed as pin jointed members. (The secondary bending moments can be found out by analysing the truss as a rigid jointed truss.) This may be a reasonable procedure when the bending stresses produced by self weight of the chord member or by the intermediate purlins are relatively small and the resulting bending stresses are less than about 10% of the stresses produced by the axial loads. The effects of actual joint bending (when more than one bolt is used in the joint) can be minimized by keeping the length of the joint as short as practical compared to the total length of the member. Moreover, truss members will have relatively small $EI/L$ ratios (i.e., long members with small moment of inertia), and the moment gradient is sharp. Hence much of the length of the member will not be subjected to bending moments.

## 13.1 GENERAL BEHAVIOUR OF BEAM-COLUMNS

Beam-columns are aptly named, as sometimes they can behave essentially like restrained beams, forming plastic hinges, and under other conditions fail by buckling in a similar way to axially loaded columns or by lateral-torsional buckling similar to unrestrained beams. Under both bending moment ($M$) and axial load ($P$) the response of a typical beam-column for lateral deflection ($\delta$) or end-joint rotation ($\theta$) would be as shown in Fig. 13.4. However, both the strength attained and the form of curve is dependent on which features dominate the behaviour of the member.

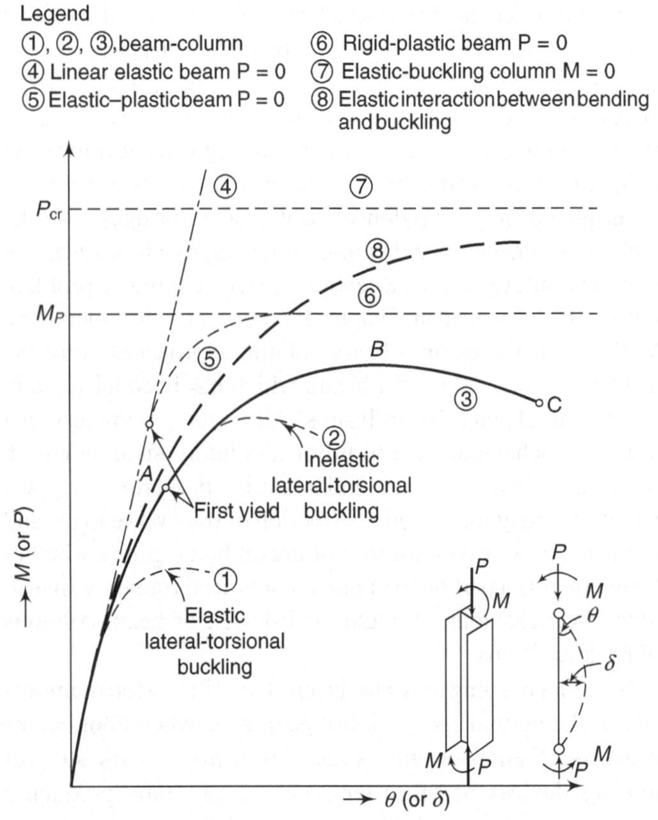


**FIG. 13.4** Behaviour of a beam-column compared with beams and columns

It may be observed that the curve is non-linear almost from the start because of the $P$-$\delta$ effect (see Section 4.2.1 and Fig. 4.2 of Chapter 4). The $P$-$\delta$ effect becomes more and more significant as the applied end moments increase. At point $A$, due to the combined effect of the primary moment $M$ and the secondary $P$-$\delta$ moment, the most severely stressed fibres of the cross sections may yield. This yielding reduces the stiffness of the member and hence the slope of the $M$-$\theta$ curve reduces beyond point $A$. As the deformation increases, the $P$-$\delta$ moment also increases. Now, this secondary moment will share a proportionally larger portion of the moment capacity of the cross section. Under increasing moment, the plasticity would spread into the section and a local hinge rotation would be developed (point $B$ on the $M$-$\theta$ curve). The hinge would now spread a short distance down the column, which causes the slight downward slope on the $M$–$\theta$ curve and at point $C$, the moment carrying capacity of the cross section is finally exhausted.

In the preceding discussion, we assumed that other forms of failure do not occur before the formation of a plastic hinge. However, if the member is slender and the cross section is torsionally weak (e.g., open cross sections), lateral-torsional buckling may occur. Lateral-torsional buckling may occur in the elastic range (curve 1 shown in Fig. 13.4) or in the inelastic range (curve 2 shown in Fig. 13.4) depending on the slenderness of the member. As discussed in Chapter 10, member with a high slenderness ratio will experience elastic lateral-torsional buckling, whereas a member with an intermediate slenderness ratio will experience inelastic lateral-torsional buckling. Lateral-torsional buckling will not occur if the slenderness ratio of the member is low or if the member is bent about the minor principal axis of the cross section. Similarly in members with cross sections having axis symmetry (e.g., circular sections) or equal moment of inertia about both principal axes (e.g., square box sections), lateral-torsional buckling will not occur, regardless of the slenderness ratio.

Another form of failure which may occur in the member is local buckling of component elements of the cross section. As seen in Chapter 10, the element with high width-to-thickness ratio is susceptible to local buckling. Like lateral-torsional buckling, local buckling may occur in the elastic or inelastic range. The effect of both lateral-torsional buckling and local buckling is to reduce the load carrying capacity of the cross section. Local buckling may be prevented by limiting the width-to-thickness ratios as specified in Table 8.6 of Chapter 8. When it is not possible to limit the width-to-thickness ratios local buckling may be accounted for in design by using a reduced width for the buckled element (see Example 9.1b of Chapter 9).

Based on the earlier discussions, the behaviour of the beam-column may be classified into the following five cases (MacGinley & Ang 1992):

1. A short beam-column subjected to axial load and uniaxial bending about either axis or biaxial bending. Failure occurs

when the plastic capacity of the section is reached, with the limitations set in the second case.

2. A slender beam-column subjected to axial load and uniaxial bending about the major axis *z-z*. If the beam-column is supported laterally against buckling about the minor axis *y-y* out of the plane bending, the beam-column fails by buckling about the *z-z* axis. It represents an interaction between column buckling and simple uniaxial bending. If the beam-column is not very slender a plastic hinge forms at the end or point of maximum moment {see Fig. 13.5(a)}. Note that this is not a common case.
3. A slender beam-column subjected to axial load and uniaxial bending about the minor axis *y-y*. Now there is no need for lateral support and no buckling out of the plane of bending. The beam-column fails by buckling about the *y-y* axis. This also represents an interaction between column buckling and simple uniaxial bending. At very low axial loads, the beam-column will attain the bending capacity about *y-y* axis {see Fig. 13.5(b)}.
4. A slender beam-column subjected to axial load and bending about the major axis *z-z*, and not restrained out of the plane of bending. The beam-column fails due to a combination of column buckling about the *y-y* axis and lateral-torsional buckling. The beam-column fails by twisting as well as deflecting in the *z-z* and *y-y* planes {see Fig. 13.5(c)}. Thus it represents an interaction between column buckling and beam buckling.
5. A slender beam-column subjected to axial load and biaxial bending and not having any lateral support. The ultimate behaviour of the beam-column is complicated by the effect of plastification, moment magnification, and lateral-torsional buckling. The failure will be similar to the fourth case but minor axis buckling will dominate.

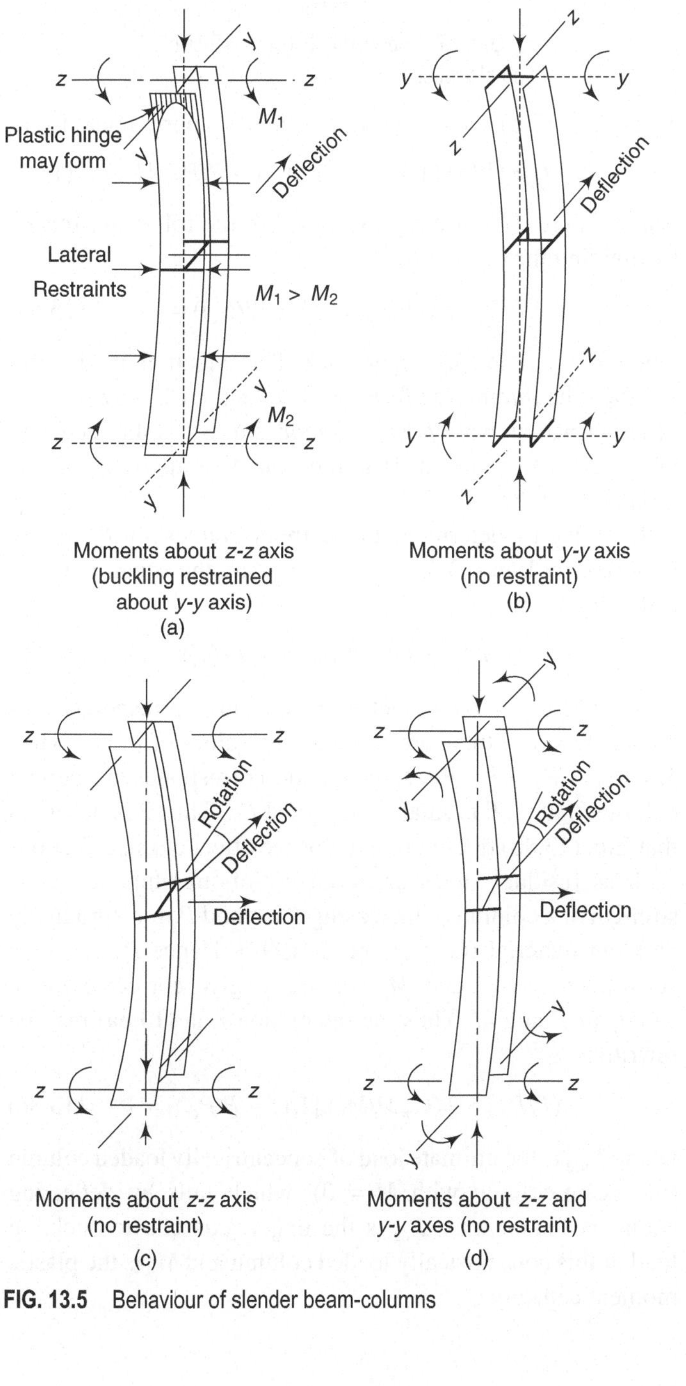


**FIG. 13.5** Behaviour of slender beam-columns

## 13.2 SECOND-ORDER MOMENTS IN BEAM-COLUMNS

The second-order moments in beam-columns may be due to the following effects:

1. $P$-$\delta$ effects, also called member effects
2. $P$-$\Delta$ effects, also called structure effects

We will discuss these briefly in the following sections.

### 13.2.1 $P$-$\delta$ Effects or Member Effects

Consider a pin-ended beam-column *AB* of length *L* and flexural rigidity *EI* as shown in Fig. 13.6. The beam-column is of uniform section and is subjected to end moments $M_A$ and $M_B$,

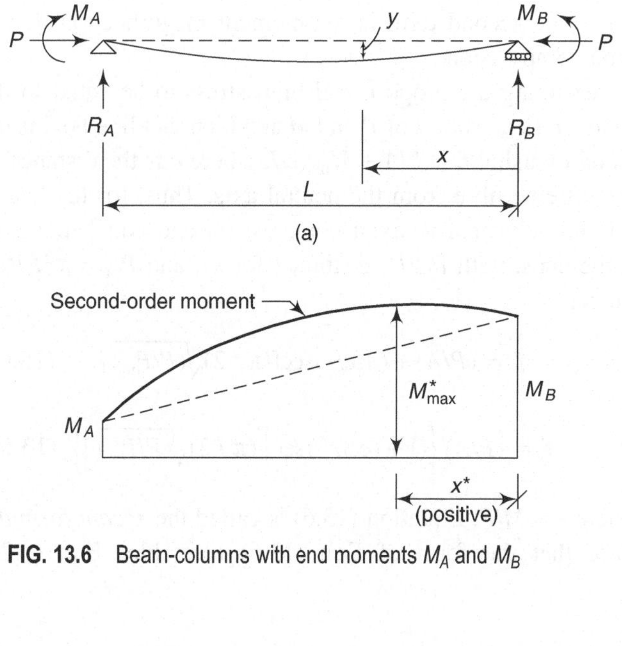


**FIG. 13.6** Beam-columns with end moments $M_A$ and $M_B$

with $M_B > M_A$, in addition to an axial compressive force $P$. The differential equation for this beam-column is given by (Bleich 1952; Timoshenko & Gere 1961; Chen & Atsuta 1976, 1977)

$$(d^2y/dx^2) + k^2y = (k^2/P)\ (M_A - M_B)\ (x/L) - (k^2/P)M_A \quad (13.1)$$

with $k^2 = P/EI$

The solution to this equation is

$$y = A\sin kx + B\cos kx + [(M_A - M_B)/P]\ (x/L) - (M_A/P) \quad (13.2)$$

substituting the boundary conditions $y = 0$ at $x = 0$ and $x = L$, and solving for $y$ and determining $M$ using $M = -EId^2y/dx^2$, we get the maximum value of bending moment as

$$M_{max} = M_B\ \text{cosec}kL\ \sqrt{[(M_A/M_B)^2 - 2(M_A/M_B)\cos kL + 1]} \quad (13.3)$$

Note that this equation is valid only in the elastic range.

When $M = M_A = M_B$, Eqn (13.3) reduces to

$$M_{max} = M\sec(kL/2) \quad (13.4)$$

Loading a beam-column with a constant moment over its length causes a constant compression along one entire flange and constitutes the most severe loading on the beam-column. Hence, it is always conservative to multiply the maximum primary moment due to any loading by sec($kL/2$), which is excessively conservative in most cases (Salmon & Johnson 1996).

In the same way, the solution for a beam-column subjected to axial load $P$ at the ends and a transverse uniformly distributed load $w$, may be obtained as (Chen & Atsuta 1976)

$$M_{max} = (wL^2/8)\{[8/(kL)^2]\ [\sec(kL/2) - 1]\} \quad (13.5)$$

where the second term is the moment magnification due to axial compression.

Assuming the proportional limit stress to be equal to the yield stress $f_y$, values of $P$ and $M$ at which yielding begins are found by using $f_y = P/A + M_{max}c/I$, where $c$ is the distance of the extreme fibre from the neutral axis. Thus, for the beam-column subjected to axial compression and equal moments at the ends, with $P/EI$ substituted for $k^2$, and $P_{cr} = \pi^2EI/L^2$, we get

$$f_y = (P/A) + (Mc/I)\sec[(\pi/2)\sqrt{(P/P_{cr})}] \quad (13.6a)$$

or

$$f_y = (P/A)\left\{(1 + (ec/r^2)\sec\left[(\pi/2)\sqrt{(P/P_{cr})}\right]\right\} \quad (13.6b)$$

where $e = M/P$. Equation (13.6) is called the *secant formula*. Note that the form of Eqn (13.6) resembles Eqn (9.29) of Chapter 9 for the initially crooked column. We may approximate the term sec[($\pi$/2) $\sqrt{(P/P_{cr})}$ ] by $1/[1 - (P/P_{cr})]$, as shown by Fig. 13.7.

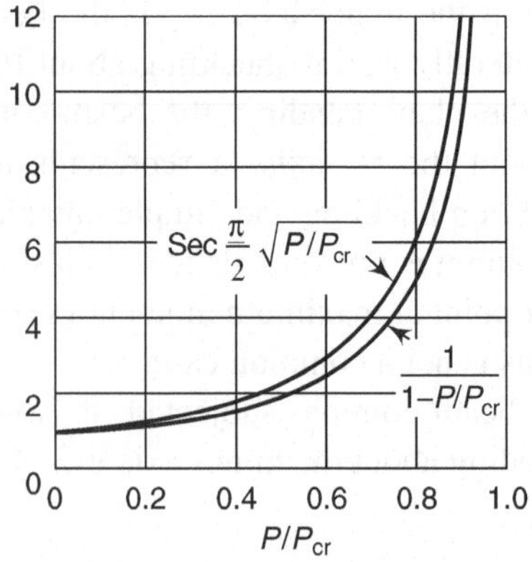


**FIG. 13.7** Approximating the effect of amplification

Thus Eqn (13.6) can be put in the simpler, approximate form

$$f_y = (P/A)\ [1 + (ec/r^2)\ \{1/(1 - P/P_{cr})\}] \quad (13.7)$$

Equation (13.7) is usually expressed in the following format by dividing throughout by $f_y$

$$(P/P_y) + (M/M_y)\ (1/(1 - P/P_{cr})) = 1 \quad (13.8a)$$

where $P_y = f_yA$ and $M_y = f_yI/c = f_yZ$. This equation in this form is called the *interaction formula*. The factor $1/[1 - (P/P_{cr})]$ is often termed as *amplification factor*, since it is the factor by which the end moment $M$ is multiplied to approximate the moment $M + Py$.

In order to determine the extreme values of $P$ and $M$ indicated by Eqn (13.8a), we may rewrite the equation in the following form

$$M/M_y = (1 - P/P_y)\ (1 - P/P_{cr}) \quad (13.8b)$$

Thus, when $P = 0$, we get $M = M_y$; this corresponds to a beam. However, there are two extreme values of $P$, i.e., when $M = 0$, $P = P_y$ or $P = P_{cr}$. Note that this corresponds to a perfect column; that is, $P$ is smaller of $P_y$ and $P_{cr}$. Thus it is observed that Eqn (13.8b) does not allow for the effect of imperfections such as residual stress or initial out-of-straightness on the strength of a column or increasing $P$ beyond $P_y$ for a gradually yielding material (Gaylord et al. 1992). Hence $P_{u,z}$ may be substituted for $P_y$ and $M_p$ for $M_y$ to give correct extreme values of $P$ and $M$. Thus the interaction Eqn (13.8a) may be rewritten as

$$(P/P_{u,z}) + (C_mM/M_p)\ [1/(1 - P/P_{cr})] = 1 \quad (13.8c)$$

where $P_{u,z}$ is the ultimate load of concentrically loaded column (the beam-column with $M = 0$), which fails by deflecting about the major axis, $P_{cr}$ is the major axis elastic buckling load of this concentrically loaded column and $M_p$ is the plastic moment capacity.

Note that in Eqn (13.8c), the coefficient $C_m$ has been included to consider the more general case of unequal end moments {see Section 13.2.2 and Chen and Zhou (1987) for the discussions on the factor $C_m$}.

It is interesting to note that prior to 1961, the following simple beam-column interaction equation was used in the codes, following the allowable stress design.

$$(f_{ac,cal}/f_{ac}) + (C_{mx}f_{bcz,cal}/f_{bcz}) + (C_{my}f_{bcy,cal}/f_{bcy}) < 1.0 \quad (13.9)$$

where $f_{ac,cal}$ is the calculated average axial compressive stress, $f_{ac}$ is the allowable axial compressive stress, $f_{bc,cal}$ is the calculated bending stress in the extreme fibre, $f_{bc}$ is the allowable bending stress, and $z$ and $y$ represent $z$-$z$ and $y$-$y$ axis.

In the 1984 version of the Indian code (IS 800 : 1984), Eqn (13.9) was permitted to be used when the ratio $f_{ac,cal}/f_{ac}$ was less than 0.15. If the ratio exceeds 0.15, the allowable bending stresses was multiplied by a factor $[1 - (f_{ac,cal}/0.6 f_{cr})]$, where $f_{cr}$ is the elastic critical stress in compression = $[\pi^2E/(KL/r)^2]$. The *equivalent moment factor* $C_m$ was taken as 0.85 for members in frames in which side sway is not prevented and as $C_m = 0.6 - 0.4\ \psi \geq 0.4$ in frames where side sway is prevented and are not subjected to transverse loading; where $\psi$ is the ratio for smaller to larger moments at the ends of the unbraced members in the plane of bending under consideration. $\psi$ is positive when the member is bent in reverse curvature and negative when bent in single curvature. For members in frames, where side sway is prevented and subjected to transverse loading, the value of $C_m$ was taken as 0.85 when the ends of the members are restrained against rotation and 1.0 when the ends of members are unrestrained against rotation. These provisions were based on the AISC code provisions.

## 13.2.2 Equivalent Moment Factor $C_m$

Member sizes of beam-columns are generally based on the magnitude of the maximum moment and the location of this maximum moment is not important in the design. Hence, the concept of equivalent moment, schematically shown in Fig. 13.8 is usually adopted in design specifications. Thus, it is assumed that the maximum second-order moment of a beam-column subjected to an axial load $P$ and end moments $M_A$ and $M_B$ with $|M_B| > |M_A|$ as shown in Fig. 13.8(a) is numerically equal to the maximum second-order moment of the same member under axial load $P$ and a pair of equal and opposite moments $M_{eq}$, as shown in Fig. 13.8(b). Note that the axial load $P$ and second-order moment $M_{max}$ are the same for both the members. The value of $M_{eq}$ is given by (Salvadori 1956; Vinnakota 2006)

$$M_{eq} = C_m|M_B| \quad (13.10)$$

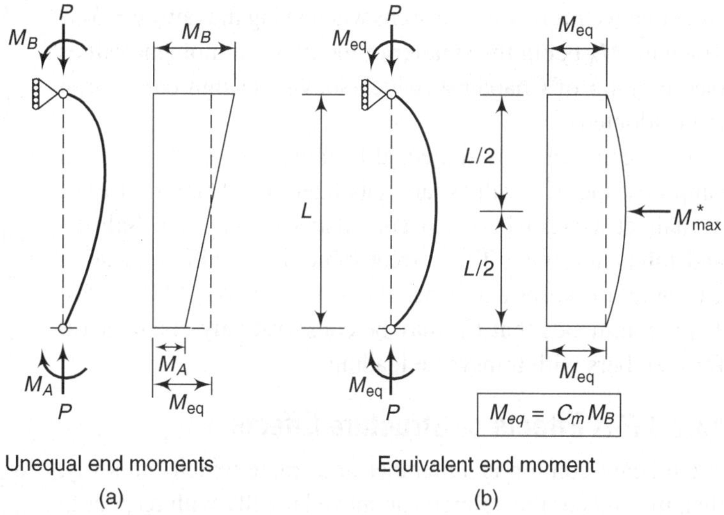


**FIG. 13.8** Concept of equivalent uniform moment

where $C_m$ is called the *equivalent moment factor* or *moment reduction factor*. This factor is a function of the moment ratio $\psi = M_A/M_B$ and also the axial load ratio $P/P_{cr}$. As discussed in Section 4.2.1.1 of Chapter 4, many simplified expressions for $C_m$ have been proposed, as a function of the moment ratio only. The following expression given by Austin (1961) has been adopted in the ANSI/AISC code for beam-columns subjected to end moments (Chen & Zhou 1987; Galambos 1998)

$$C_m = 0.6 + 0.4\ \psi \text{ where } -1.0 < \psi < 1.0 \quad (13.11)$$

For beam-columns with transverse loadings, the second-order moments can be approximated by (ANSI/AISC: 360-10)

$$C_m = 1 + \beta(\alpha P/P_{cr}) \quad (13.12)$$

where $\beta = \pi^2\ \delta_o\ EI/ML^2 - 1$, $\alpha = 1.0$, $\delta_o$ is the maximum deflection due to transverse loading (in mm), $M$ is the maximum first-order moment within the member due to transverse loading (N mm).

Thus, for a pin-ended beam-column of length $L$ with a uniformly distributed load $W$,

$$M = WL/8,\ \delta = 5WL^3/384EI,\ P_{cr} = \pi^2EI/L^2$$

$$\beta = [(\pi^2EI)/L^2]\,[(5WL^3)/(384EI)]\,(8/WL) - 1 = 1.028 - 1 = 0.028$$

Hence $C_m = 1 + 0.028P/P_{cr}$.

Similarly for a beam-column of length $L$ with a transverse load $P$ at mid length,

$$M = PL/4;\ \delta = PL^3/48EI;\ P_{cr} = \pi^2EI/L^2$$

Thus $\beta = [(\pi^2EI)/L^2]\ (PL^3/48EI)\ (4/PL) - 1 = 0.822 - 1 = -0.178$

and

$$C_m = 1 - 0.178\ P/P_{cr}$$

Note that we must consider signs when using the ratio $\psi = M_A/M_B$, with $M_A$ being the smaller of the two end moment values (see Fig. 4.5 of Chapter 4, which shows the sign convention to be adopted).

For beam-columns subjected to transverse loads between supports, the $C_m$ values are tabulated by Trahair (1993), Trahair et al. (2001), Chen and Atsuta (1977), and Salmon and Johnson (1996). The correct value of $C_m$ will be close to 1.0 for all the cases and hence the American code (AISC 360-10) recommends that $C_m$ may be conservatively taken as 1.0 for members with transverse loading.

### 13.2.3 *P*-Δ Effects or Structure Effects

If the beam-column is a member of a frame which can sway, then the ends of the column can move laterally with respect to each other and hence additional secondary moments will result {see Fig. 4.6(b) of Chapter 4}. In Fig. 4.6(b), the secondary moment produced due to side sway is equal to $P\Delta$. To study the effect of such side sway, consider a cantilever column *AB* as shown in Fig. 13.9, which is subjected to a transverse load *H* and an axial force *P*. Let us assume that the member is laterally braced and the loads are applied in a plane of symmetry.

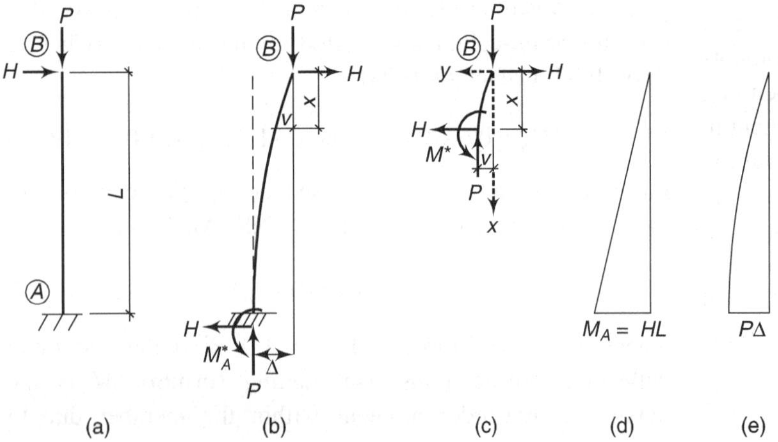


**FIG. 13.9** Beam-column in a sway frame

The differential equation for this beam-column may be written as (Vinnakota 2006)

$$EI(d^2v/dx^2) + Pv = -Hx \tag{13.13}$$

The general solution of this equation is

$$v = A_1 \sin \alpha x + A_2 \cos \alpha x - Hx/P \tag{13.14}$$

where $\alpha^2 = P/EI$.

Applying the boundary conditions $v = 0$ at $x = 0$ and $dv/dx = 0$ at $x = L$ and substituting the values of $A_1$ and $A_2$ we get,

$$v = H/P\alpha[(\sin \alpha x/\cos \alpha L) - \alpha x] \tag{13.15}$$

Differentiating this expression twice and multiplying by $-EI$, we obtain the second-order moment at any point *x* as (Vinnakota 2006)

$$M^* = H(EI/P)\alpha(\sin \alpha x/\cos \alpha L) = HL[\sin \alpha x/(\alpha L\cos \alpha L)] \tag{13.16}$$

The maximum bending moment occurring at point *A* may be obtained by substituting $x = L$ in Eqn (13.16) as

$$M_A^* = M_A(\tan\alpha L/\alpha L) \tag{13.17}$$

where $M_A = HL$ is the first-order bending moment at the base of the cantilever beam under transverse load *H* {Fig. 13.9(d)}. The expression for the maximum moment may be simplified by using the power series expansion for $\tan\alpha L$, namely

$$\tan \alpha L = \alpha L + 1/3(\alpha L)^3 + 2/15(\alpha L)^5 + 17/315(\alpha L)^7 + \ldots \tag{13.18}$$

Substituting Eqn (13.18) into Eqn (13.17) and simplifying we get (Vinnakta 2006)

$$M_A^* \approx [\{1 - 0.18(P/P_{cr})\}/\{1 - (P/P_{cr})\}]M_A \tag{13.19}$$

Equation (13.19) may be written in the form

$$M_{II} = B_2 M_I \tag{13.20}$$

where $B_2$ is the *moment amplification factor* for beam-columns in unbraced multi-storey buildings, $M_{II}$ is the second-order bending moment, and $M_I$ is the first-order bending moment. It is a measure of the second-order moment that develops in a beam-column due to the *P*-Δ effect.

The amplification factor $B_2$ for beam-columns in unbraced multi-storey frames may also be derived based on storey stiffness concept and in such a case its value is given by Eqn (4.4b) of Chapter 4.

## 13.3 ELASTIC LATERAL-TORSIONAL BUCKLING OF BEAM-COLUMNS

The differential equations of equilibrium for the lateral-torsional buckling of beams (as given in Section 10.3.1 of Chapter 10) can be extended to include the effect of a centrally applied compression force *P* by adding to the components of the end moments $M_z$, the moments produced by *P*. Figure 13.10 shows that the load *P* produces a moment $Pv$ about the beam's *z*-axis and a moment $Pu$ about its *y*-axis. Thus, the differential equations are written as (Gaylord et al. 1992)

$$-EI_z(d^2v/dx^2) = M_z + Pv \tag{13.21a}$$

$$-EI_y(d^2u/dx^2) = M_z\phi + Pu \tag{13.21b}$$

$$-EI_w d^3\phi/dx^3 + GI_t(d\phi/dx) = M_z(du/dx) + (P/A)I_p(d\phi/dx) \tag{13.21c}$$

where $EI_z$ is the bending stiffness of the section about the *z*-axis; $EI_y$ is the bending stiffness about the *y*-axis; $EI_w$ is the warping stiffness and $GI_t$ is the St Venant torsional stiffness.

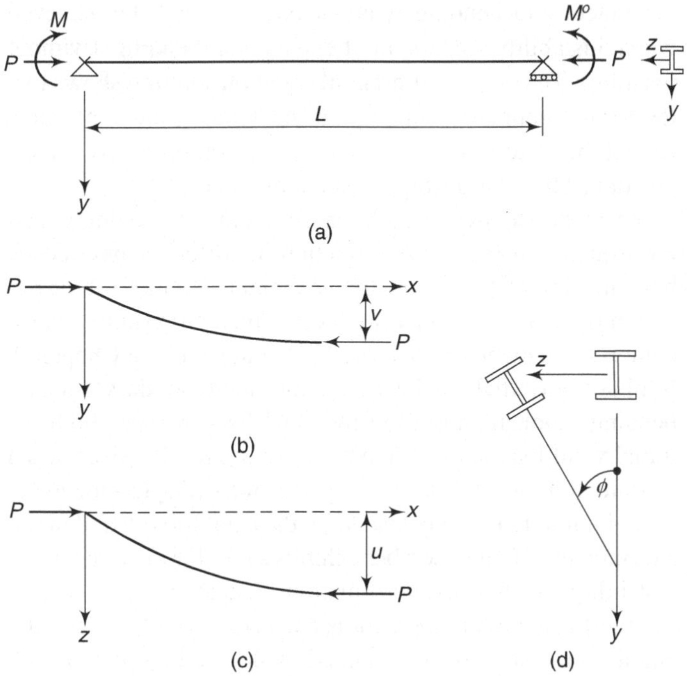


**FIG. 13.10** Lateral-torsional buckling of a *I*-shaped beam-column

For the simply supported boundary conditions, $u = 0$ and $\phi = 0$ and $d^2u/dx^2 = 0$ and $d^2\phi/dx^2 = 0$ at each end, the solution to the differential equation is given by

$$u = A\sin(\pi x/L) \text{ and } \phi = B\sin(\pi x/L)$$

Substituting these deflection functions and their derivatives into the governing differential equations, we get the elastic critical moment as (Galambos 1968)

$$M_z^2 = (I_p/A)(P_{\mathrm{cr},\,y} - P)\,(P_{\mathrm{cr},\,x} - P) \qquad (13.22)$$

where $I_p$ is the polar moment of inertia $= (I_z + I_y)$, $P_{\mathrm{cr},\,y}$ is the flexural buckling load of the member about the weak axis $= \pi^2 EI_y/L^2$, $P_{\mathrm{cr},\,x}$ is the torsional buckling load of the member $= \dfrac{1}{r_o^2}\left(\dfrac{\pi^2 EI_w}{L^2} + GI_t\right), r_o^2 = \left(I_x + I_y\right)/A$, $A$ is the area of the cross section, and $L$ is the length of the member.

The elastic critical moment for a flexurally and torsionally simply supported *I*-shaped beam under uniform moment $M$, given by Eqn (10.1a), may now be rewritten with the help of Eqn 13.22) as

$$M_{\mathrm{cr}}^2 = [\pi^2 EI_y/L^2(GI_t + (\pi^2/L^2)EI_w)] = P_{\mathrm{cr},\,y}\,P_{\mathrm{cr},\,x}\,(I_z + I_y)/A \qquad (13.23)$$

Dividing Eqn (13.22) by Eqn (13.23), we get the non-dimensional equation

$$(M_z/M_{z,\mathrm{cr}})^2 = \{1 - (P/P_{\mathrm{cr},y})\}\ \{1 - (P/P_{\mathrm{cr},x})\} \qquad (13.24)$$

Note that Eqn (13.24) is not used in design specifications. Instead, the interaction formulae derived for the beam which does not buckle laterally are used. These interaction formulae are based on Eqn (13.8c), which is the solution for Eqn (13.21a). Equation (13.8c) is arbitrarily extended to cover lateral-torsional buckling by redefining $P_{\mathrm{cr}}$ as the smaller of $P_{\mathrm{cr},z}$ and $P_{\mathrm{cr},y}$ and $M_{\mathrm{cr}}$ as the critical moment for lateral-torsional buckling. Thus, Eqn (13.8c) becomes

$$(P/P_{\mathrm{cr}\,(z\,\mathrm{or}\,y)}) + (M/M_{\mathrm{cr}})\ [C_m/(1 - P/P_{\mathrm{cr},z})] \le 1 \qquad (13.25)$$

A comparison of this formula with 36 aluminium beam-columns that failed by lateral-torsional buckling showed that the actual strengths were 4–22 % more than those predicted by Eqn (13.25) (Gaylord et al. 1992).

Inelastic lateral-buckling of beam-column has been covered by Bradford et al. (1987), Vinnakota and Aysto (1974), and Vinnakota and Aoshima (1974).

## 13.4 INTERACTION BETWEEN BEAM-COLUMN AND STRUCTURE

The buckling of beam-columns is controlled by the behaviour of the structure in which they are a part, and this has to be taken into account in the design. The end-restraints offered by other parts of the structure are often allowed for by using the effective length concept (explained in Chapter 9). However, this concept may lead to difficulties and contradictions if the designer thinks in terms of linear elastic structural behaviour, when columns fail by elastic buckling.

To explain this, consider the no-sway column shown in Fig. 13.11(a) which is bent in single curvature about its minor axis by the loading on adjacent beams. These moments, as per linear analysis, will be considered positive. Paradoxically, the same beams can restrain the column against minor axis buckling, i.e., giving the column an effective length less than $L$; however, for this to occur at failure, the end moments of the columns must be of opposite sign {as shown in Fig. 13.11(b)}. Both conditions cannot occur at the same time. Dowling et al. (1988) describe what actually happens as follows.

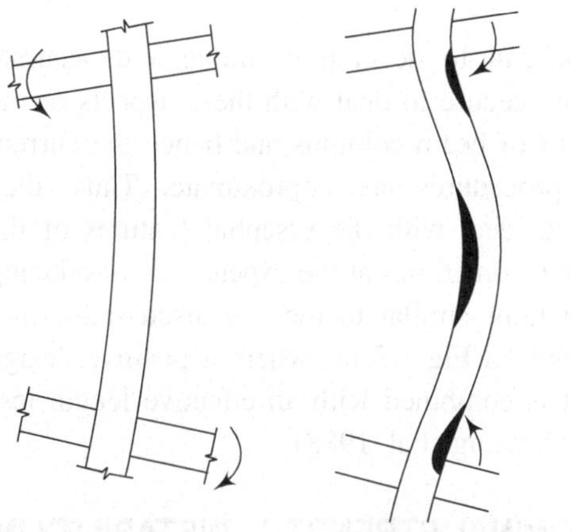


**FIG. 13.11** Interactions between beam-column and surrounding frame

When only a part of the axial load is applied, the column restraints the beams as predicted by linear elastic analysis, with positive end moments as shown inFig. 13.11(a). As the axial load approaches failure, the column loses stiffness ($EI_y$) by a factor $\alpha$. The Euler load ($P_{cr} = \pi^2 EI/L^2$) for the equivalent pin-ended column of length $L$ becomes approximately $\pi^2 EI_y \alpha/L^2$; when this is equal to the applied load, the column is incapable of restraining the beams and the end-moments must have been relaxed to zero. Under further increase of axial load the column must be supported by the beams, with negative end moments and an effective length of less than $L$, until failure occurs in the way as shown in Fig. 13.11(b).

Beam-column stiffness can also alter directly, without stability effects, as a result of elasticity; thus beam-columns can often significantly reduce their major axis applied end-moments before failure, by deforming plastically at the ends, as shown in Fig. 13.12. Most practical beam-columns behave in this way (Dowling et al. 1988).

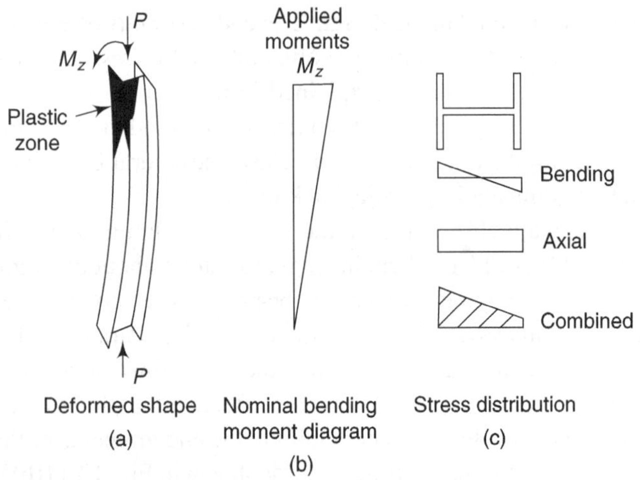


**FIG. 13.12** Simple beam-column subject to axial load and major axis moment at one end only

It is difficult to develop a simple and accurate design procedure to deal with these aspects of the behaviour of beam-columns and hence the current design procedures are approximate. Thus, they attempt to deal with the essential features of the behaviour, sometimes at the expense of introducing contradictions similar to the one discussed earlier and shown in Fig. 13.11, where a positive design moment is combined with an effective length less than $L$ (Dowling et al. 1988).

## 13.5 NOMINAL STRENGTH—INSTABILITY IN THE PLANE OF BENDING

The basic strength of beam-columns, where lateral-torsional buckling and local buckling are adequately prevented, and bending is about one axis, will be achieved when instability occurs in the plane of bending (without twisting). The elastic differential equation solution shows that the axial compression effect and the bending moment effect cannot be determined separately and combined by superposition. The relationship is also non-linear.

Residual stresses, which cause premature yielding and consequently a premature reduction in stiffness, may reduce both the initial yield and the maximum strength of beam-columns. This is similar to their effect on axially loaded compression members discussed in Section 9.6.3 of Chapter 9. Similarly, an initial crookedness, which increases the secondary bending moment caused by the axial load, reduces both the initial yield load and the maximum strength. Moreover, when a beam-column with initial imperfections (displacements $u_i$, $v_i$, and initial twist $\phi_i$) is loaded by the axial force $P$ and major axis moment $M_z$, the member exhibits a non-bifurcation type of instability, in which the deformations increase (from $u_i$, $v_i$, and $\phi_i$), until a maximum moment is reached, beyond which static equilibrium can only be sustained by decreasing the moment. The maximum strength based on the spatial behaviour of such initially crooked beam-columns $M_{cs,max}$ could be lower than the lateral-buckling load $M_{cr}$ of the corresponding initially straight beam-column as shown in Fig. 13.13 (Vinnakota 2006).

The inelastic analysis to determine the strength interaction between axial compression $P$ and bending moment $M$ for a beam-column is complicated. To trace a load deflection curve similar to that shown in Fig. 13.4, one should use some type of approximation or numerical technique. This is because the differential equations governing the inelastic behaviour of a beam-column are highly non-linear even for the simplest loading case (Chen & Atsuta 1976, 1977).

The analysis of the inelastic behaviour normally proceeds in two steps.

1. Cross section analysis
2. Member analysis

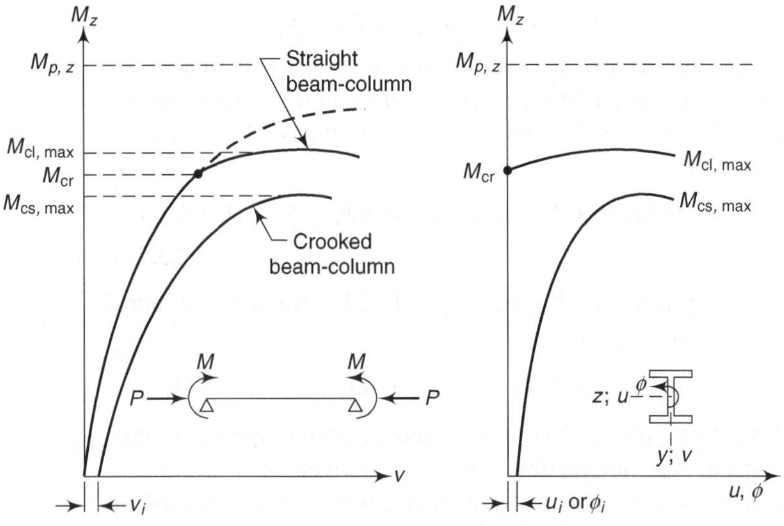


**FIG. 13.13** Inelastic lateral-torsional buckling of beam-columns

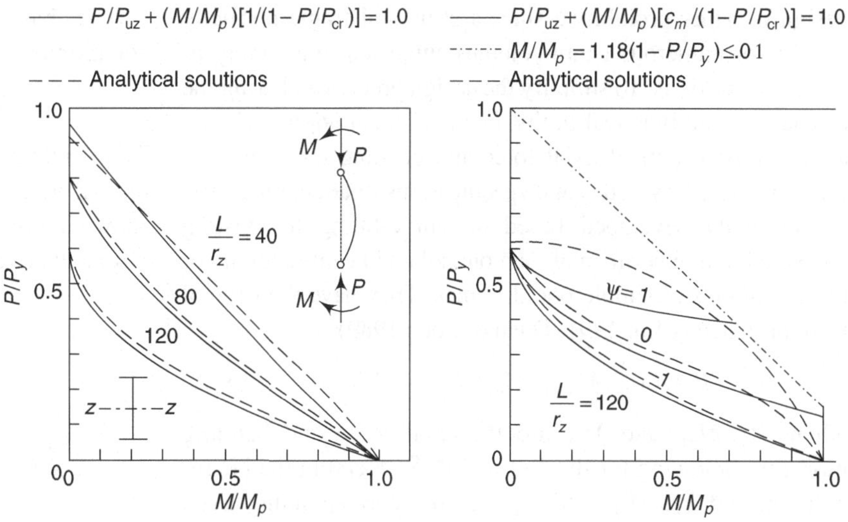


Effect of slenderness (a)

Effect of end moment ratio (b)

**FIG. 13.15** Comparison of interaction formula with the analytical solutions

In the cross section analysis, the behaviour of a cross section subjected to the combined action of axial force and bending moment is investigated. The result is expressed as a set of $M$–$\theta$–$P$ (moment–curvature–axial compression) relationship. Once $M$–$\theta$–$P$ relationship is established, member analysis can be done.

In the member analysis, the member is divided into a number of segments whereby equilibrium and compatibility conditions along the length of the member at each division point are enforced for a given set of loadings or deflections. The analysis thus consists of finding successive solutions as the applied load or deflection of the member is increased in steps. When enough of these analyses have been performed, the load-deflection relationship of the beam-column can be traced on a pointwise basis. The maximum load-carrying capacity of the member is obtained as the peak point of this load-deflection curve. The method developed by Newmark in 1943 may be used for the analysis of the in-plane bending response of beam-columns and the method by Cranston (1983) for the analysis of the biaxial bending behaviour of beam-columns (Chen & Lui 1991). The various aspects of these analyses procedures may be found in Galambos and Ketter (1961), Chen and Atsuta (1976), Chen and Lui (1991), and Pi and Trahair (1994). A typical strength interaction curve based on the study by Ketter is shown in Fig. 13.14. The analytical solutions are compared with the interaction equation {Eqn (13.8c) with $C_m$ as defined by Eqn (13.11)} in Fig. 13.15. Also shown in Fig. 13.15(b) is Eqn (8.28) of Chapter 8 (based on the plastic design approach). It is seen that the interaction equations provide a reasonable estimate of the ultimate strength of beam-columns in most of the cases. Instability in the plane of bending including the effects of transverse loading has been studied by Ketter (1961); Lu and Kamalvand (1968); Pi and Trahair (1994); English and Adams (1973); Chen (1971), and Chong–Siat–Moy (1974).

## 13.5.1 Failure by Lateral-torsional Buckling

Tests and computer studies conducted by several researchers, including inelastic effects, indicate that interaction equations similar to Eqn (13.8c) could be used to include the effect of lateral-torsional buckling also (Fukumoto and Galambos 1966; Lindner 1997, 2003). The main difference is that $P/P_u$ will be lower when based on $KL/r_y$ (weak axis) rather than $KL/r_z$ (strong axis). Similarly $M_n$ may also be less than $M_p$ because of lateral-torsional buckling. A summary of this topic is presented by Massonet (1976) which includes discussion of plastic effects.

## 13.5.2 Nominal strength—Interaction Equations

As observed from the earlier discussions the behaviour of beam-columns is affected by a number of parameters. Moreover a real beam-column may receive end moments and axial load from its connections to other members of a structure, such as a rigid frame. Hence the relation of the beam-column to the other elements of the structure is important in determining both the applied forces and the resistance of the member.

The elasto–plastic analysis involving numerical procedures (Chen and Atsuta 1976, 1977) are quite laborious and their direct use in

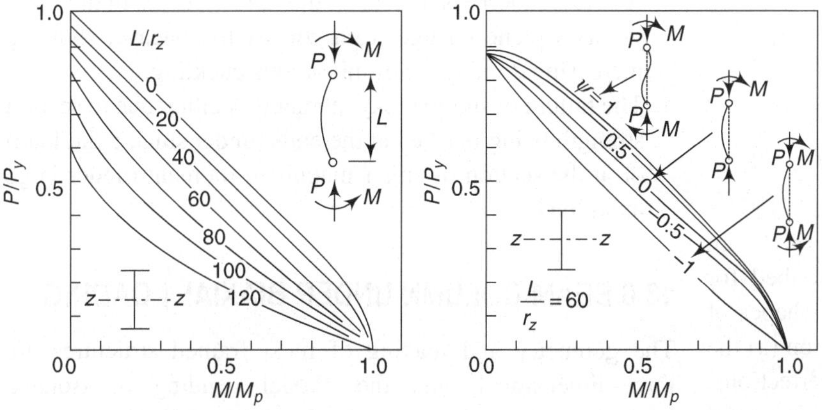


Effect of slenderness (a)

Effect of end moment ratio (b)

**FIG. 13.14** Ultimate strength interaction curves for a wide flange section for braced frame member

*Source:* Ketter 1961. Reprinted with permission from ASCE.

design is rather prohibitive since a structural frame normally consists of numerous beam-columns subjected to a variety of loading conditions. To simplify the design process and bring the problem within practical limits, *interaction equations*, relating a safe combination of axial force and bending moments, are often suggested by codes and specifications. Interaction curves are normally developed based on curve-fitting to existing analytical and experimental data on isolated beam-columns or beam-columns of simple portal frames. They have the general form of (Chen & Lui 1991; Duan & Chen 1989)

$$f\{(P_u/P_n), (M_{uz}/M_{nz}), (M_{uy}/M_{ny})\} \leq 1.0 \qquad (13.26)$$

where $P_u$, $M_{uz}$, and $M_{uy}$ are the axial force and bending moments (allowing for the $P$-$\Delta$ and $P$-$\delta$ effects) in the beam-column, and $P_n$, $M_{nz}$, and $M_{ny}$ are the corresponding axial and bending moment capacities of the member {determined as discussed in Chapter 9 (Section 9.7.1) on compression members and Chapter 10 (Sections 10.7 and 10.8) on beams}.

The three-dimensional graphical representation of Eqn (13.26) is shown in Fig. 13.16. In this figure each axis represents the capacity of the member when it is subjected to loading of one type only, while the curves represent the combination of two types of loading. The surface formed by connecting the three curves represents the interaction of axial load and biaxial bending. It is this interaction surface that is of interest to the designer.

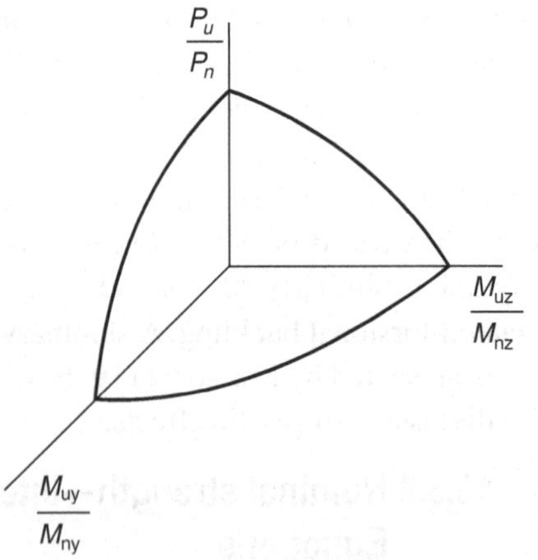


**FIG. 13.16** Ultimate interaction surface for beam-columns

The end points of the curves shown in Fig. 13.16 are dependent on the capacities of the members described for columns (Chapter 9) and beams (Chapter 10). The shapes of these curves between these end points will depend on (a) the cross-sectional shape and the beam-column imperfections, (b) the variation of moments along the beam-column, and (c) the end restraint conditions of the beam-column. All these variables can only be dealt with on an approximate basis and hence various formulae are given in the codes, which attempt to allow for the effects mentioned earlier.

The basic form of the three-dimensional interaction equation is

$$(P/P_n) + (M_z/M_{nz}) + (M_y/M_{ny}) \leq 1.0 \qquad (13.27)$$

This interaction equation results in a straight-line representation of the interaction between any two components shown in Fig. 13.17. This simplified interaction equation gives a conservative design.

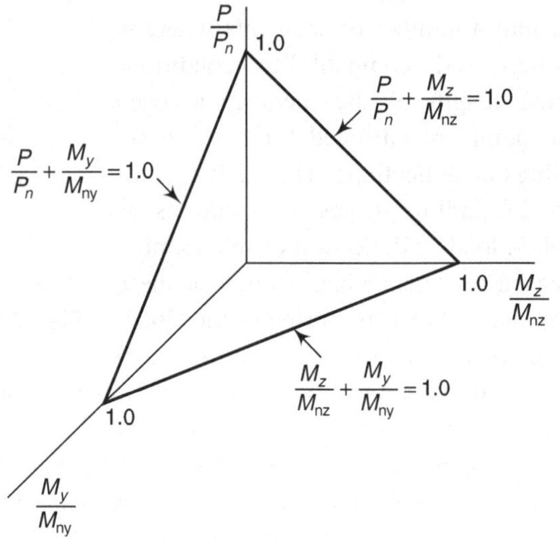


**FIG. 13.17** Simplified interaction surface

The behaviour of beam-columns subject to bending moment about minor axis is similar to that subjected to major axis bending but for the following differences.

1. In the case of slender members under small axial load, there is very little reduction of moment capacity below $M_p$, since lateral-torsional buckling is not a problem in weak axis bending.
2. The moment magnification is larger in the case of the beam-columns bending about their weak axis.
3. As the slenderness increases, the failure curves in the $P/P_n$, $y$-$y$ axis plane change from convex to concave, showing increasing dominance of minor axis buckling.
4. The failure of short/stocky members is either due to section strength being reached at the ends (under small axial load) or at the section of larger magnified moment (under large axial load).

## 13.6 BEAM-COLUMN UNDER BIAXIAL LOADING

The geometry and loading of most framed structures are three-dimensional, and the biaxial bending of isolated beam-columns with test results has been studied by several researchers (e.g., Culver 1966; Klöppel & Winkelmann 1962) and they have shown that the elastic biaxial bending of beam-column is similar to its in-plane behaviour. Thus, in this case, the major and minor axis deflections and twist begin immediately after applying the load and increases rapidly as the elastic buckling load is approached. Although the first-yield prediction of strengths of slender beam-columns, based on these methods, are of sufficient accuracy, they are found to be conservative for stocky members in which considerable yielding occurs before failure.

Sophisticated numerical analyses have also been made by researchers for the biaxial bending of inelastic beam-columns and good agreements with test results have been obtained (Vinnakota & Aoshima 1974; Vinnakota & Aysto 1974; Chen & Atsuta 1977; Pi & Trahair 1994). Such analyses require a specialized computer program, which cannot be used as a design tool, but can assist only in research.

Hence a number of attempts have been made to develop approximate methods of predicting the strength of inelastic beam-columns. One of the simplest is to extend the linear interaction equation for in-plane bending and flexural-torsional buckling {Eqn (13.25)} to the biaxial case also. This equation is given by

$$(P/P_u) + (M_z/M_{uz})\,[C_{mz}/(1 - P/P_{cr,\,z})] + (M_y/M_{uy})\,[C_{my}/(1 - P/P_{cr,y})] \leq 1 \qquad (13.28)$$

In this case $M_{uz}$ is the ultimate moment which the beam-column can support when $P = M_y = 0$ for the case of equal end moments ($\psi = -1$), while $M_{uy}$ is similarly defined.

Under biaxial bending and axial load the interaction equations should give convex failure surface as shown in Fig. 13.18. It is observed that the increase in the slenderness ratio of the member tends to reduce the strength of the member, except in the case of nearly pure bending about the weak axis ($y$-axis). For very small axial force ranges, further increases in the axial compression tends to decrease the bending strength about both axes.

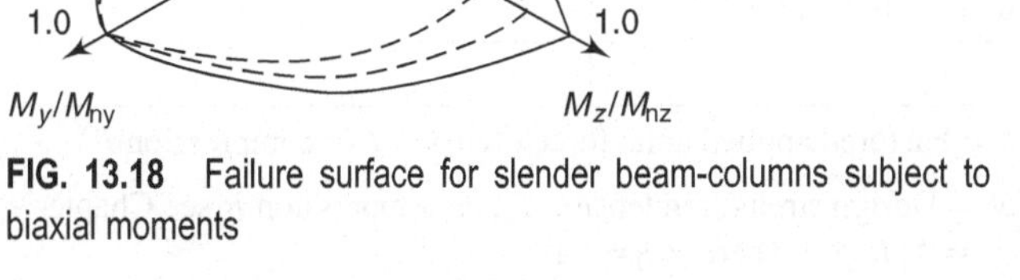


**FIG. 13.18** Failure surface for slender beam-columns subject to biaxial moments

Note that there is no allowance in Eqn (13.28) for the amplification of the minor axis moment $M_y$ by the major axis moment $M_z$, since the term $[1 - (P/P_{cr,y})]$ only allows for amplification caused by the axial load $P$. Several researchers (Chen & Tebedge 1974; Chen & Atsuta 1977; Pi & Trahair 1994) have shown that the linear interaction equation leads to very conservative results. Chen and Tebedge (1974) proposed the following non-linear interaction equations which are valid for I-shaped members in braced frames:

$$[(C_{mz}M_z)/(\phi_b\, M^*_{nz})]^\eta + [(C_{my}M_y)/(\phi_b\, M^*_{ny})]^\eta \leq 1.0 \qquad (13.29a)$$

where

$$\eta = 0.4 + (P/P_y) + (b_f/d) > 1.0 \ \text{ for } b_f/d \geq 0.3 \qquad (13.29b)$$

$$\eta = 1.0 \text{ for } b_f/d < 0.3 \qquad (13.29c)$$

in which $b_f$ is the flange width (in mm) and $d$ is the member deph (in mm), and

$$M^*_{nz} = M_{nz}\,[1 - (P/\phi_c\, P_n)]\,[1 - (P/P_{c,z})] \qquad (13.29d)$$

$$M^*_{ny} = M_{ny}\,[1 - (P/\phi_c P_n)]\,[1 - (P/P_{cr,y})] \qquad (13.29e)$$

where $P_{cr,z}$ and $P_{cr,y}$ are the Euler buckling loads about the strong and weak axis respectively, $\phi_b$ is the beam resistance factor (= 0.90), $\phi_c$ = column resistance factor (= 0.85), and $P_n$, $M_{nz}$, and $M_{ny}$ are the corresponding axial and bending moment capacities of the member {determined as discussed in Chapter 9 (Section 9.7.1) and Chapter 10 (Sections 10.7 and 10.8)}.

For square box section $\eta$ should be taken as (Galambos 1998)

$$\eta = 1.3 + [1000/(L/r)^2]\,(P/P_n) \geq 1.4 \qquad (13.30)$$

The use of Eqns (13.29) and (13.30) with $\eta$ values greater than unity recognises the convexity of the moment interaction as illustrated in Fig. 13.18. [Galambos (1998) provides the comparison of Eqn (13.28) and (13.29).]

More discussions about the behaviour and design of beam-columns subjected to biaxial bending and axial force may be found in Chen and Santathadaporn (1968), Lindner and Gietzelt (1985), Kennedy et al. (1990), and Lindner and Glitsch (2004).

## 13.7 INTERACTION EQUATIONS FOR LOCAL CAPACITY CHECK

The interaction equations discussed till now are for overall buckling check. The beam-column should also be checked for local capacity at the point of the greatest bending moment and axial load. This is usually checked at the ends of the member, but it could be within the length of the beam-column, if lateral loads are also applied. The capacity in these cases is controlled by yielding or local buckling (if it is not prevented by limiting the width-to-thickness ratios specified in the codes). The linear interaction equation for semi-compact and slender cross section is given by

$$(P/P_y) + (M_z/M_{pz}) + (M_y/M_{py}) \leq 1 \qquad (13.31)$$

where $P$ is the applied axial load, $P_y$ is the yield load = $A_g\,f_y$,

$M_z$ is the applied moment about the major axis $z$-$z$, and $M_{pz}$ is the moment capacity about the major axis $z$-$z$ in the absence of the axial load

$= M_{pz}$ if $P/P_y \le 0.15$ and
$= 1.18M_{pz}(1 - P/P_y)$ for $0.15 \le P/P_y < 1.0$

$M_y$ is the applied moment about the minor axis $y$-$y$ and $M_{py}$ is the moment capacity about the minor axis $y$-$y$ in the absence of the axial load

$= 1.19M_{py}[1 - (P/P_y)^2] \le 1.0.$

More accurate interaction equations are available for compact cross sections, which are based on the convex failure surface discussed in the previous section, which result in greater economy in design (see also Section 8.10). Chen and Atsuta (1977) and Tebedge and Chen (1974) provide the following non-linear interaction equation for compact I-shapes in which the flange width is not less than 0.5 times the depth of the section

$$(M_z/M_{pz})^{\alpha} + (M_y/M_{py})^{\alpha} \le 1 \qquad (13.32a)$$

in which $M_{pz}$, $M_{py}$, $M_z$, and $M_y$ are as defined earlier.
The value of the exponent is given by

$$\alpha = 1.6 - [(P/P_y)/2\{\ln(P/P_y)\}] \text{ for } 0.5 \le b_f/d \le 1.0 \qquad (13.32b)$$

where ln is the natural logarithm, $b_f$ is flange width (in mm), and $d$ is the member depth (in mm).

A comprehensive assessment of the accuracy of the non-linear interaction equations in predicting the load carrying capacities of biaxially loaded I-sections has been made by Pillai (1981), who found that these equations predict the capacity reasonably well compared to the experimental results.

Interaction equations for a number of sections, including circular tubes, box sections, and unsymmetrical sections such as angles are available in Chen and Lui (1971), Chen and Atsuta (1977), and Shanmugam et al. (1993).

## 13.8 CODE DESIGN PROCEDURES

Modern structural design specifications around the world have retained the generic form of the interaction formula given in Eqn (13.26). In every specification the moment $M$ is always specified as the second-order (amplified) moment obtained either from a second-order structural analysis, where equilibrium is formulated in the deformed configuration of the structure, or from an approximation of using the moments from a first-order elastic analysis, which is then multiplied by an amplification factor. Depending upon whether $P_{cr}$, the elastic critical load, is evaluated for the member length or the storey effective length, the amplification factor accounts empirically for the member or the frame stability (Galambos 1998). This versatility of the interaction equations approach makes it very useful in design.

Most limit-states design codes use a set of load and resistance factors that are based on probabilistic principles (Bjorhovde et al. 1978).

### 13.8.1 Indian Code (IS 800 : 2007) Provisions

The Indian code (IS 800 : 2007) provisions are based on the Eurocode provisions and the code requires the following two checks to be performed—local capacity check and overall buckling check.

**Local capacity check** For beam-columns subjected to combined axial force (tension or compression) and bending moment, the following interaction equation should be satisfied.

$$(M_y/M_{ndy})^{\alpha 1} + (M_z/M_{ndz})^{\alpha 2} \le 1.0 \qquad (13.33)$$

where $M_y$ and $M_z$ are the factored applied moments about the minor and major axis of the cross section, respectively and $M_{ndy}$ and $M_{ndz}$ are the design reduced flexural strength under combined axial force and the respective uniaxial moment acting alone. The value of $M_{ndy}$ and $M_{ndz}$ for plastic and compact section is approximately given in Table 13.1.

**TABLE 13.1** Approximate value of reduced flexural strength for plastic and compact sections

| Element | Equation |
|---|---|
| Plates | $M_{nd} = M_d(1 - n^2)$ |
| Welded I- or H-sections | $M_{ndz} = M_{dz}(1 - n)/(1 - 05a) \le M_{dz}$<br>$M_{ndy} = M_{dy}\left[1 - \left(\frac{n-a}{1-a}\right)^2\right] \le M_{dy}$<br>$a = (A - 2\,bt_f)/A \le 0.5;\ n \ge a$ |
| Rolled I- or H-sections without bolt holes | $M_{ndz} = 1.11M_{dz}(1 - n) \le M_{dz}$<br>$M_{ndy} = M_{dy}$ for $n \le 0.2$<br>$M_{ndy} = 1.56\,M_{dy}(1 - n)(n + 0.6)$ for $n > 0.2$ |
| Rectangular hollow sections and welded box sections | $M_{ndz} = M_{dz}(1 - n)/(1 - 0.5a_w) \le M_{dz}$<br>$M_{ndy} = M_{dy}(1 - n)/(1 - 0.5a_f) \le M_{dy}$<br>where $a_w = \frac{(A - 2bt_f)}{A} \le 0.5$ *and*<br>$a_f = \frac{(A - 2ht_w)}{A} \le 0.5$ |
| Circular hollow tubes without bolt holes | $M_{nd} = 1.04M_d(1 - n^{1.7}) \le M_d$ |
| $n = N/N_d$ | |
| $N$ = Factored applied axial force (Tension $T$ or compression $P$)<br>$N_d$ = Design strength in tension $T_d$, or compression $P$(see Chapter 9) $= A_g f_y/\gamma_{m0}$ where $\gamma_{m0} = 1.1$<br>$M_{dy}$, $M_{dz}$ = Design strength under corresponding moment acting alone (see Chapter 10)<br>$A_g$ = Gross area of cross section | |

The constants $\alpha_1$ and $\alpha_2$ are given in Table 13.2.

**TABLE 13.2** Constants $\alpha_1$ and $\alpha_2$

| Section | $\alpha_1$ | $\alpha_2$ |
|---|---|---|
| *I* and channel | $5n \geq 1$ | 2 |
| Circular tubes | 2 | 2 |
| Rectangular tubes | $1.66/(1 - 1.13n^2) \leq 6$ | $1.66/(1 - 1.13n^2) \leq 6$ |
| Solid rectangles | $1.73 + 1.8n^3$ | $1.73 + 1.8n^3$ |

where $n = N/N_d$

For semi-compact sections, without bolt holes, the code (IS 800 : 2007) suggests the following linear equation, when shear force is low

$$(N/N_d) + (M_y/M_{dy}) + (M_z/M_{dz}) \leq 1.0 \qquad (13.34)$$

**Overall buckling check** The interaction equation for overall buckling check is given by clause 9.3.2.2 the code as

$$(P/P_{dy}) + (K_y C_{my} M_y/M_{dy}) + (K_{LT} M_z/M_{dz}) \leq 1.0 \qquad (13.35a)$$

$$(P/P_{dz}) + (0.6 K_y C_{my} M_y/M_{dy}) + (K_z C_{mz} M_z/M_{dz}) \leq 1.0 \qquad (13.35b)$$

where

$C_{my}, C_{mz}$ = Equivalent uniform moment factor obtained from Table 13.3, which depends on the shape of the bending moment diagram between lateral bracing points in the appropriate plane of bending

$P$ = Factored applied axial compressive load

$P_{dy}, P_{dz}$ = Design compressive strength under axial compression as governed by buckling about minor and major axis respectively (see Chapter 9, Section 9.7.1)

$M_y, M_z$ = Maximum factored applied bending moments about minor and major axis of the member, respectively

$M_{dy}, M_{dz}$ = Design bending strength about minor and major axis considering laterally unsupported length of the cross section (see Chapter 10, Section 10.8).

$K_y, K_z, K_{LT}$ = Moment amplification factors as defined below

$$K_y = 1 + (\lambda_y - 0.2)n_y \leq 1 + 0.8n_y \qquad (13.35c)$$

$$K_z = 1 + (\lambda_z - 0.2)n_z \leq 1 + 0.8n_z \qquad 13.35d)$$

$$K_{LT} = 1 - \frac{0.1\lambda_{LT} n_y}{(C_{mLT} - 0.25)} \geq 1 - \frac{0.1 n_y}{(C_{mLT} - 0.25)} \qquad (13.35e)$$

where

$n_y, n_z$ = Ratio of actual applied axial-force to the design axial strength for buckling about minor and major axis respectively $= (P/P_{dy})$ or $(P/P_{dz})$

$C_{mLT}$ = Equivalent uniform moment factor for lateral-torsional buckling as per Table 13.3, which depends on the shape of the bending moment diagram between lateral bracing points

$\lambda_y, \lambda_z$ = Non-dimensional slenderness ratio about the minor and major axis respectively, for example, $\lambda_y = (f_y/f_{cr})^{0.5}$, where $f_{cr} = \pi^2 E/(KL/r)^2$ (see Section 9.7.1)

$\lambda_{LT}$ = Non-dimensional slenderness ratio in lateral buckling $= (f_y/f_{cr,b})^{0.5}$ and $f_{cr,b}$ is the extreme fibre bending compressive stress corresponding to elastic lateral buckling moment which may be determined as per Table 14 of the code.

The above Indian code provisions are based on the Eurocode 3 provisions and the improved interaction equations suggested by Greiner and Lindner (2006), who derived Eqns (13.35) after extensive statistical evaluations and calibration with available buckling results.

**TABLE 13.3** Equivalent uniform moment factor (Greiner & Lindner 2006)

| Bending moment diagram | Range | | $C_{my}, C_{mz}, C_{mLT}$ | |
|---|---|---|---|---|
| | | | Uniform loading | Concentrated load |
| 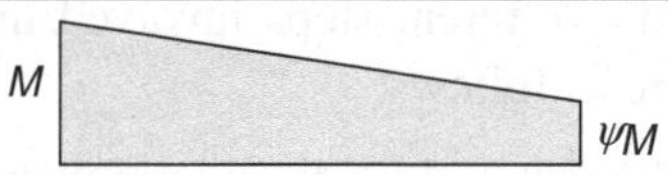  | $-1 \leq \psi \leq 1$ | | $0.6 + 0.4\psi \geq 0.4$ | |
| 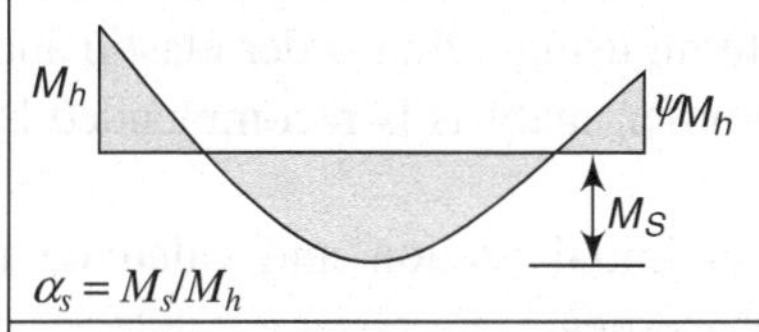  | $0 \leq \alpha_s \leq 1$ | $-1 \leq \psi \leq 1$ | $0.2 + 0.8\alpha_s \geq 0.4$ | $0.2 + 0.8\alpha_s \geq 0.4$ |
| | $-1 \leq \alpha_s \leq 0$ | $0 \leq \psi \leq 1$ | $0.1 - 0.8\alpha_s \geq 0.4$ | $-0.8\alpha_s \geq 0.4$ |
| $\alpha_s = M_s/M_h$ | | $-1 \leq \psi \leq 0$ | $0.1(1 - \psi) - 0.8\alpha_s \geq 0.4$ | $0.2(-\psi) - 0.8\alpha_s \geq 0.4$ |
| 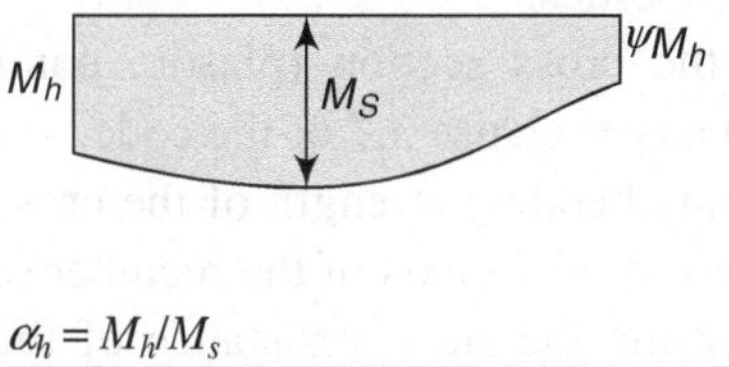  | $0 \leq \alpha_h \leq 1$ | $-1 \leq \psi \leq 1$ | $0.95 + 0.05\alpha_h$ | $0.90 + 0.10\alpha_h$ |
| | $-1 \leq \alpha_h \leq 0$ | $0 \leq \psi \leq 1$ | $0.95 + 0.05\alpha_h$ | $0.90 + 0.10\alpha_h$ |
| $\alpha_h = M_h/M_s$ | | $-1 \leq \psi \leq 0$ | $0.95 + 0.05\alpha_h (1 + 2\psi)$ | $0.90 + 0.10\alpha_h (1 + 2\psi)$ |
| For members with sway buckling mode the equivalent uniform moment factor $C_{my} = C_{mz} = 0.90$ | | | | |

*(Contd)*

**TABLE 13.3** *(Contd)*

$C_{my}$, $C_{mz}$, and $C_{mLT}$ shall be obtained according to the bending moment diagram between the relevant brace points as below:

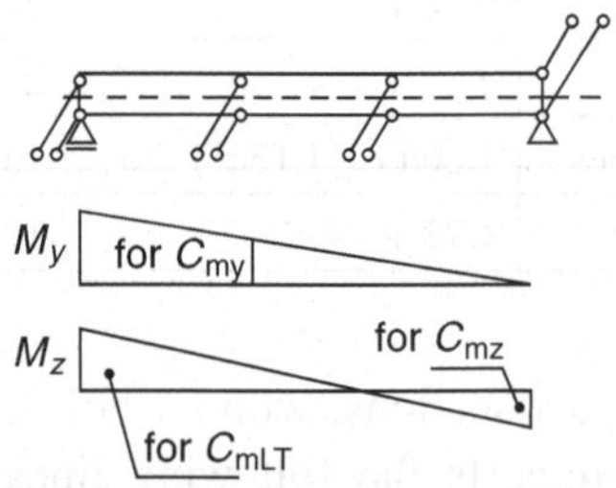


| Moment factor | Bending axis | Points braced in direction |
|---|---|---|
| $C_{my}$ | z-z | y-y |
| $C_{mz}$ | y-y | z-z |
| $C_{mLT}$ | z-z | z-z |

## 13.8.2 American Code Provisions

The American code (ANSI/AISC 360-05) gives the following linear interaction formulae for biaxially loaded beam-columns. It employs only a single equation to access all the possible modes of failure (cross section, member in-plane, and lateral-torsional strength).

1. For $P/(\phi_c P_c) \geq 0.2$

$$P/(\phi_c P_c) + (8/9)\,[M_{rz}/(\phi_b M_{cz}) + M_{ry}/(\phi_b M_{cy})] \leq 1.0 \quad (13.36a)$$

2. For $P/(\phi_c P_c) < 0.2$

$$P/(2\,\phi_c P_c) + M_{rz}/(\phi_b M_{cz}) + M_{ry}/(\phi_b M_{cy}) \leq 1.0 \quad (13.36b)$$

where $P$ is the required axial compressive strength (in $N$), $P_c$ is the available axial compressive strength (in $N$), $M_r$ is the required flexural strength calculated from first-order elastic analysis using the amplification factors $B_1$ and $B_2$ (in $N$mm), $M_c$ is the available flexural strength (in $N$mm), $z$ and $y$ are the subscripts relating to major and minor axis bending, $\phi_c$ is the resistance factor for compression = 0.90, and $\phi_b$ is the resistance factor for flexure = 0.90.

Although the shape of the 'true' beam-column strengths can be different in major and minor axis uniaxial bending, the AISC code does not distinguish the two cases. The preceding equations were developed, based on extensive calibration to 'exact' plastic-zone solutions for a wide range of non-redundant, stability critical benchmark frames (see Section 4.7.1 of Chapter 4). The American code also gives separate formulae for doubly symmetric members in single axis flexure and compression, unsymmetric, and other members subjected to flexure and axial force, and HSS subject to combined torsion, shear, flexure, and axial force.

A comparison of the provisions of American AISC-LRFD-1993, Canadian CAN/CSA-S 16.1-1994, Eurocode 3-1992, and Australian AS 4100-1990 codes is provided by White and Clarke (1997). A review of the American code provisions is provided by Chen and Lui (1985) and Sputo (1993). De Luca et al. (1991) provide a comparison of the American, Japanese, and European codes. An excellent summary of research in the field of beam-columns is provided by Massonnet (1976) and Massonnet and Maquoi (1978). Design methods used in Eurocode 3 and DIN 18800-Part 2 are given by Lindner (1997). It is interesting to note that two separate proposals for beam-column interaction formulae are given as an alternative to those proposed in Eurocode 3 by Boissonnade et al. (2002, 2004) and Greiner and Lindner (2006).

## 13.9 DESIGN OF BEAM-COLUMNS

The design of beam-columns involves a trail-and-error procedure. A trail section is selected by some process and is then checked with the appropriate interaction formula. If the section does not satisfy the equation (LHS > 1.0) or if it is too much on the safer side, indicated by LHS much less than 1.0 (that is, if it is over designed), a different section is selected and the calculations are repeated till a satisfactory section is found. Thus, the different steps involved in the design of beam-columns are as follows.

1. Determine the factored loads and moments acting on the beam-column using a first-order elastic analysis (though a second-order analysis is recommended by most of the codes)
2. Choose an initial section and calculate the necessary section properties.
3. Classify the cross section (plastic, compact, or semi-compact) as per clause 3.7 of the code.
4. Find out the bending strength of the cross section about the major and minor axis of the member (clause 8.2.1.2).
5. (a) Determine the shear resistance of the cross section (clause 8.4.1). When the design shear force exceeds $0.6V_d$, then the design bending strength must be reduced as given in clause 9.2.2 of the code.

(b) Check whether shear buckling has to be taken into account (clause 8.4.2).

6. Calculate the reduced plastic flexural strength (clause 9.3.1.2), if the section is plastic or compact.
7. Check the interaction equation for cross section resistance for biaxial bending (clause 9.3.1.1 for plastic and compact section and clause 9.3.1.3 for semi-compact section). If not satisfied go to step 2.
8. Calculate the design compressive strength $P_{dz}$ and $P_{dy}$ (clause 7.1.2) due to axial force.
9. Calculate the design bending strength governed by lateral-torsional buckling (clause 8.2.2).
10. Calculate the moment amplification factors (clause 9.3.2.2).
11. Check with the interaction equation for buckling resistance (clause 9.3.2.2). If the interaction equation is not satisfied (LHS > 1.0) or when it is over design (LHS << 1.0), go to step 2.

### 13.9.1 Selection of Initial Section

A common method used for selecting sections to resist both moments and axial loads is the *equivalent axial load or effective axial load method.* In this method, the axial load P and the bending moments $M_z$ and/or $M_y$ are replaced with fictitious concentric loads $P_{eff}$, equivalent to the actual design axial load plus the design moment. This fictitious load is called the equivalent axial load or the effective axial load.

Uang et al. (1990), suggested the following formula (based on the work of Burgett 1973) for calculating the equivalent axial load, based on the American sections.

$$P_{eff} = P + M_z m + M_y m U \quad (13.37)$$

Values of $m$ and $U$ for different values of $KL$ and for several wide flange shapes used as beam-columns are provided by Uang et al. (1990). Yura (1988) suggested a simpler approach for the initial sizing compared to Eqn (13.37) and suggested the following equation.

$$P_{eff} = P + 2M_z/d + 7.5M_y/b \quad (13.38)$$

where $d$ and $b$ are the depth and breadth of the selected beam-column.

The equivalent axial load equations were found to yield sections that are generally on the conservative side. The equivalent axial load approaches given by Eqns (13.37) and (13.38) are useful in preliminary sizing of beam-columns under gravity load combinations. However, when bending moment is predominant, those equations may result in considerable error. In this case, the *equivalent moment approach*, rather than equivalent axial force approach is preferable. Yura (1988), proposed an equation for estimating the equivalent moment as

$$M_{eq} = M_z + P_u d/2 \quad (13.39)$$

where $d$ is the depth of the section. This equation is applicable for initial beam-column sizing in unbraced frames under lateral-load combinations. A new method for the preliminary member selection of beam-column cross section (based on the interaction equations of the American code) is presented by Aminmansour (2000).

### 13.9.2 Direct Analysis for *P*-Δ Effects

It is important to realize that the moment magnification method using *K*-factors (effective length factors $K_z$ and $K_y$), as used in the interaction equation [Eqn (13.35)] is an approximate method for evaluating the *P*-Δ effects. With the availability of computer programs, which can directly account for *P*-Δ effects (by considering the equilibrium at the deformed configuration of the structure), it is not necessary to use the approximate moment magnification method, for evaluating the effect of axial loads *P* acting through the lateral deflection Δ of the structure. With the results of such a program, the structural engineer need not worry about calculating the effective length factor *K* by using Wood's curves/alignment charts or complicated equations. All columns/beam-columns whether they are gravity or frame columns can be designed by using an effective length factor $K = 1$. Note that not all programs that are advertised as programs for second-order analysis consider the *P*-$\delta$ moments. Therefore, if designers want to take account of both the *P*-$\delta$ and *P*-Δ moments, they should make certain that these moments are included in the program. The available computer programs are discussed in Chen and Lui (1991). More details about these analysis methods are given in Chapter 4.

Gaylord et al. (1992, pp. 446–451) provide examples of frames analysed using both first and second-order analysis programs and show that the second-order moments are equal to the first-order moments multiplied by the magnification factors. Thus, once the second-order moments are obtained by using such computer programs, the buckling check may be simplified by eliminating the $K_y$ and $K_z$ factors in Eqn (13.35a).

Since the design of beam-columns is an iterative process, it is time consuming to find the lightest section which will support the applied loads and moments. Hence a number of programs have been written [e.g., H. Adeli (1988); Roufaiel & Monasa (1989)], for the design of structural members (and also frames), which will aid the engineer in the design. Since there are many variables, it may be difficult to develop charts for the design of beam-columns. However, spreadsheets may be used for their design.

## 13.10 BEAM-COLUMNS SUBJECTED TO TENSION AND BENDING

Bending moments may occur in tension members due to connection eccentricity, self weight of the member or transverse loads such as wind acting along the length of the member. The effect of tension load will always reduce the primary bending moment. Hence, secondary bending effects can be conservatively ignored in the design of members subjected to an axial tensile force and bending.

**Local capacity check** The following simplified interaction equation is specified in the code for the beam-column subjected to combined axial force and bending moment (Clause 9.3.1.1 of the code)

$$(N/N_d) + (M_y/M_{dy}) + (M_z/M_{dz}) \leq 1.0 \quad (13.40)$$

where $M_{dy}$ and $M_{dz}$ are the design reduced flexural strength under combined axial force and the respective uniaxial moment acting alone, $M_y$ and $M_z$ are the factored applied moments along minor and major axis of the cross section, respectively, $N_d$ is the design strength in tension obtained from Section 6 of the code, and $N$ is the factored applied axial tensile force.

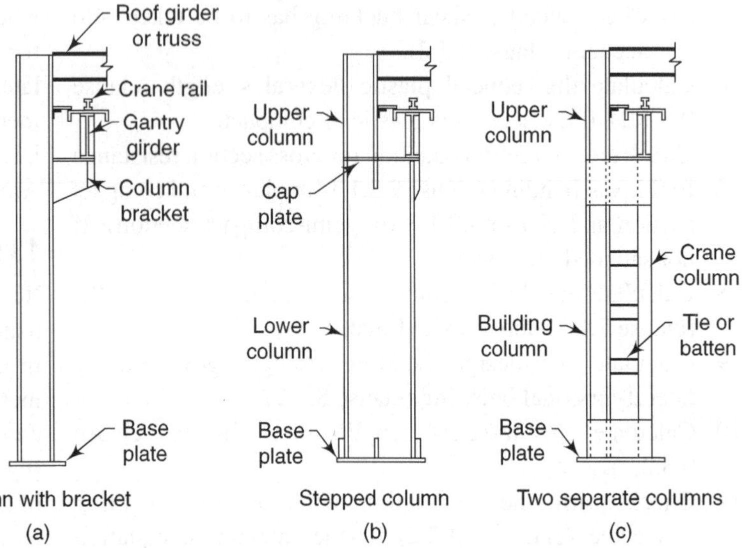


**FIG. 13.19** Crane columns

**Overall buckling check** The code stipulates that the member should be checked for lateral-torsional buckling under reduced effective moment $M_{eff}$ due to tension and bending. The reduced effective moment is given by clause 9.3.2.1 of the code as

$$M_{eff} = [M - \psi T Z_{ec}/A] \leq M_d \quad (13.41)$$

where $M$ and $T$ are the factored applied moment and tension, respectively, $A$ is the area of cross section, $Z_{ec}$ is the elastic section modulus of the section with respect to extreme compression fibre, $\psi = 0.8$, if $T$ and $M$ vary independently and 1.0 otherwise, and $M_d$ is the bending strength due to lateral-torsional buckling.

## 13.11 CRANE COLUMNS

Three common types of crane columns, used in single-storey industrial buildings, are shown in Fig. 13.19. They are as follows:

1. A column of uniform section for the entire length, with the gantry girder supported by column brackets [see Fig. 13.19(a)].
2. A stepped column, with a heavy wide flange section at the bottom and a lighter rolled *I*-beam or wide flange section at the top, supporting the roof structure [see Fig. 13.19(b)]. The gantry girder is seated on the top of the lower column. A small bracket (usually stiffened) is used when the flange of the gantry girder is overhanging, as shown in Fig. 13.19(b). To transmit the bending moments from the upper column, the junction of the top and lower column should be strongly spliced and stiffened. The effective length of such stepped columns can be determined by using Annex D2 of IS 800 : 2007.
3. For very heavy loads, a laced or battened column as shown in Fig. 13.19(c) is used, which contains two separate columns, one supporting the load from the crane and the other supporting the roof load. The two sections may be designed independently of each other. The building column may be proportioned to take the gravity and wind loads from the building and the lateral thrust from the crane. The crane column may be designed to support the vertical loads from the gantry girder and a horizontal longitudinal load at the top in the direction of the gantry girder. The two sections are separated sufficiently to provide necessary crane clearance. The advantage of having two separate columns is as follows. The crane loads can be applied concentrically on the crane column and may be oriented to obtain maximum resistance to longitudinal forces (e.g., column web can be placed parallel to the plane of the gantry girder). Similarly, the building column can be oriented independently, so that its strong axis can resist wind loads on the building and lateral forces from the crane.

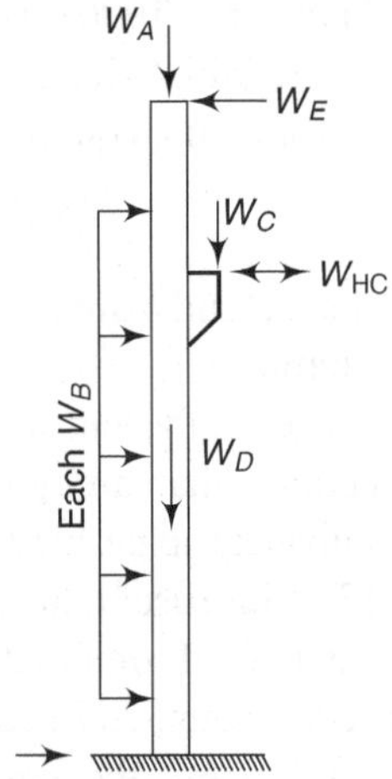


**FIG. 13.20** Loads acting on a crane column

**Loading** The different loads that act on a crane column are as follows (see Fig. 13.20):

$W_A$ = vertical load from roof truss, taken as applied concentrically

$W_B$ = horizontal loads due to wind, applied by side rails

$W_C$ and $W_{HC}$ = crane gantry loads applied through a bracket at a known eccentricity

$W_D$ = self weight of the column/sheeting

$W_E$ = resultant force carried through the truss bottom chord

In single-storey industrial buildings, the load $W_E$ occurs whenever the columns carry unequal horizontal loads or unequal moments. Values of the resultant force for different arrangement of horizontal loading are shown in Fig. 13.21 (Morris & Plum 1996).

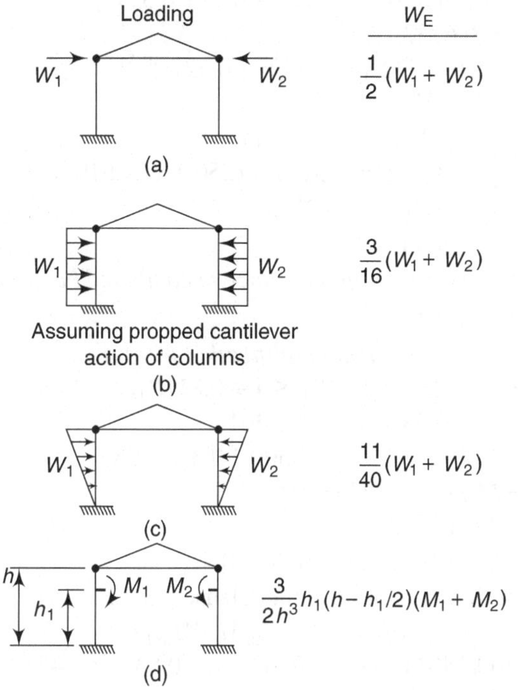


**FIG. 13.21** Resultant force carried through the bottom chord of roof truss

For the design of a column under a load system such as that shown in Fig. 13.20, load factors $\gamma_f$ must be included in the calculation. Unlike simple load cases, it will no longer be clear which combination produces the highest axial forces and moments. It may be necessary to examine all the possible combinations, although with experience three or four worst combinations may be selected. The best way of examining the effects of all the combinations is to use matrices. Matrix manipulation should be arranged in following manner:

Unfactored load matrix = $[W]$

Load factor/combination matrix = $[\gamma_f]$

Factored load matrix $[W_\gamma] = [W]\,[\gamma_f]$

Axial force and moment coefficient matrix = $[\alpha]$

Axial force and moment matrix $[F, M] = [W_\gamma]T\,[\alpha]$

The use of these matrix manipulations is shown in Example 13.3.

## EXAMPLES

**Example 13.1:** A non-sway column in a building frame with flexible joints is 4-m high and subjected to the following load and moment:

Factored axial load = 500 kN

Factored moment $M_z$:

at top of column = 27.0 Nm

at bottom of column = 45.0 kNm

Design a suitable beam-column assuming $f_y = 250$ N/mm². Take the effective length of the column as 0.8L along both the axes.

**Solution:**

**Step 1:** *Trial section*

Select ISHB 200 with $A = 4750\,\text{mm}^2$, $r_y = 45.1$ mm, $H = 200$ mm, $b = 200$ mm, and $t_f = 9.0$ mm.

$$KL = 0.8 \times 4000 = 3200\,\text{mm}$$

$$KL/r_y = 3200/45.1 = 70.95$$

$$P_e = P + 2M_z/d = 500 + 2 \times 45000/200 = 950\,\text{kN}$$

From Table 10 of the code for $h/b < 1.2$ and $t_f < 40$ mm, using curve 'c' for minor axis buckling, we have

For $KL/r_y = 70.95$ and $f_y = 250$ MPa, from Table 9c of the code, $f_{cd} = 150$ N/mm²

Hence capacity = $150 \times 4750/1000 = 712.5\,\text{kN} < 950\,\text{kN}$

Hence use ISHB 250 with $A = 6500\,\text{mm}^2$, $H = 250$ mm, and $r_y = 54.9$ mm.

$$KL/r_y = 3200/54.9 = 58.29$$

$$P_{eq} = 500 + 2 \times 45{,}000/250 = 860\,\text{kN}$$

For $KL/r_y = 58.29$ and $f_y = 250$ MPa,

From Table 9c of the code, $f_{cd} = 170.56$ N/mm²

Thus capacity = $170.56 \times 6500/1000$

= 1108 kN > 860 kN

Hence, choose ISHB 250 as the trial section.

*Section Properties*

ISHB 250 has the following cross sectional properties:

$H = 250$ mm   $A = 6500\,\text{mm}^2$   $I_z = 7740 \times 10^4\,\text{mm}^4$

$b_f = 250$ mm   $r_z = 109$ mm   $I_y = 1960 \times 10^4\,\text{mm}^4$

$t_f = 9.7$ mm   $r_y = 54.9$ mm

$t_w = 6.9$ mm   $Z_z = 619 \times 10^3\,\text{mm}^3$

$R = 10$ mm   $Z_y = 156 \times 10^3\,\text{mm}^3$

$$Z_{pz} = 2b_f t_f (H - t_f)/2 + t_w(H - 2t_f)^2/4$$

$$= 2 \times 250 \times 9.7\,(250 - 9.7)/2 + 6.9(250 - 2 \times 9.7)^2/4$$

$$= 674.46 \times 10^3\,\text{mm}^3$$

$$Z_{py} = 2t_f\, b_f^2/4 + (H - 2t_f)\, t_w^2/4$$

$$= 2 \times 9.7 \times 250^2/4 + (250 - 2 \times 9.7)6.9^2/4$$

$$= 305.87 \times 10^3\,\text{mm}^3$$

**Step 2:** *Cross section classifcaton*

$$\varepsilon = \sqrt{(250/f_y)} = \sqrt{(250/250)} = 1.0$$

*Outstand flanges* (Table 2 of the code)

$$b/t_f = (250/2)/9.7 = 12.88 < 15.7\varepsilon$$

Hence, the flange is semi-compact.

*Web*

$$d = H - 2t_f - 2R = 250 - 2 \times 9.7 - 2 \times 10 = 210.6\,\text{mm}$$

$$d/t_w = 210.6/6.9 = 30.5 < 42\varepsilon$$

Hence, the cross section is semi-compact.

**Step 3:** *Check for resistance of cross section to the combined effects (clause 9.3.1.3)*

The interaction equation is

$$(N/N_d) + (M_z/M_{dz}) \leq 1.0$$

$$N_d = A_g f_y/\gamma_{m0} = 6500 \times 250/(1.1 \times 1000) = 1477.27\,\text{kN}$$

$$M_{dz} = \beta_b Z_p f_y/\gamma_{m0}$$

where $\beta_b = Z_e/Z_p$ for a semi-compact section. Hence,

$$M_{dz} = Z_e f_y/\gamma_{m0} = 619 \times 10^3 \times 250/(1.1 \times 10^6) = 140.68\,\text{kN m}$$

Thus,

$$(500/1477.27) + (45/140.68) = 0.338 + 0.320 = 0.658 < 1.0$$

Hence, the section is safe.

**Step 4:** *Member buckling resistance in compression (clause 7.1.2)*

Effective length = $0.8L = 0.8 \times 4000 = 3200$ mm

$$KL_z/r_z = 3200/109 = 29.35$$

$$KL_y/r_y = 3200/54.9 = 58.29$$

From Table 10 of the code,

$$h/b = 250/250 = 1.0 \text{ and } t_f < 40\,\text{mm}$$

Major axis buckling, use curve *b*

Minor axis buckling, use curve *c*

$$f_{cr,z} = \pi^2 \times 2 \times 10^5/(29.35)^2 = 2291.5\ \text{Nmm}^2$$

$$\lambda_z = \sqrt{(250/2291.5)} = 0.33$$

From Table 9c of the code, for $KL/r = 58.29$ and $f_y = 250$ N/mm$^2$,

$$f_{cd} = 170.56\ \text{N/mm}^2 \text{ and}$$

$$P_{d,y} = 170.56 \times 6500/1000 = 1108\,\text{kN} > 500\,\text{kN}.$$

From Table 9b of the code, for $KL/r = 29.35$ and $f_y = 250$ N/mm$^2$,

$$f_{cd} = 215.6\ \text{N/mm}^2,$$

$$P_{d,z} = 215.6 \times 6500/1000 = 1401\ \text{k5N} > 500\,\text{kN}.$$

Hence, the section is safe.

**Step 5:** *Member buckling resistance in bending (clause 8.2.2)*

$M_d = \beta_b Z_p f_{bd}$

$\beta_b = Z_e/Z_p$ for semi-compact section

Hence $M_d = Z_e f_{bd}$

From Table 42 of the code (assuming $k = 1$)

For $\psi = 0.75$, $C_1 = 1.141$ and for $\psi = 0.5$, $C_1 = 1.323$. Hence, for $\psi = 0.6$, $C_1 = 1.25$, $L_y = 4$ m, $h/t_f = 250/9.7 = 25.77$

*Determination of $M_{cr}$*

$$f_{cr,b} = C_1\,[1473.5/(KL/r_y)]^2\,\{1 + (1/20)[(KL/r_y)/(h/t_f)]^2\}^{0.5}$$

$$= 1.25(1473.5/58.29)^2\,[1 + (1/20)(58.29/25.77)^2]^{0.5}$$

$$= 895.1\ \text{N/mm}^2$$

$$M_{cr} = 895.1 \times 619 \times 10^3/10^6 = 554\,\text{kNm}$$

Non-dimensional lateral-torsional slenderness ratio

$$\lambda_{LT} = \sqrt{(\beta_b Z_{pz} f_y/M_{cr})}$$

$$= \sqrt{0.918 \times 674.46 \times 10^3 \times 250/(554 \times 10^6)} = 0.527$$

$$\alpha_{LT} = 0.21 \text{ for rolled sections}$$

Reduction factor for lateral-torsional buckling

$$\chi_{LT} = 1/[\phi_{LT} + (\phi_{LT}^2 - \lambda_{LT}^2)^{0.5}]$$

where $\phi_{LT} = 0.5[1 + \alpha_{LT}(\lambda_T - 0.2) + \lambda_{LT}^2]$

$$= 0.5[1 + 0.21(0.527 - 0.2) + 0.527^2]$$

$$= 0.6732$$

Thus $\chi_{LT} = 1/[0.6732 + (0.6732^2 - 0.527^2)^{0.5}]$

$$= 0.9157$$

Lateral-torsional buckling resistance

$$= \chi_{LT}(f_y/\gamma_{m0})Z_e = 0.9157 \times (250/1.1) \times 619 \times 10^3/10^6$$

$$= 128.82\,\text{kNm} > 45\,\text{kNm}$$

Hence, the section is safe.

**Step 6:** *Member buckling resistance in combined bending and axial compression*

*Determination of moment amplification factors*

$$K_z = 1 + (\lambda_z - 0.2)P/P_{dz} \leq 1 + 0.8P/P_{dz}$$

$$K_z = 1 + (0.33 - 0.2)500/1401$$

$$= 1.0463 < 1 + 0.8 \times 500/1401 = 1.285$$

$$\psi_z = M_2/M_1 = 27/45 = 0.6,$$

$$C_{mz} = 0.6 + 0.4\psi$$

$$= 0.6 + 0.4 \times 0.6 = 0.84 > 0.4$$

Check with interaction formula (Clause 9.3.2.2)

$$P/P_d + [(K_z C_{mz} M_z)/M_{dz}] < 1$$

Thus, $(500/1401) + (1.0463 \times 0.84 \times 45)/128.82 = 0.357 + 0.307 = 0.664 < 1.0$

Hence section is safe against combined axial force and bending moment.

**EXAMPLE 13.2:** An I-section beam-column of length 4 m has to be designed as a ground floor column in a multi-storey building. The frame is moment-resisting in-plane and pinned out-of plane, with diagonal bracing provided in both directions. The column is subjected to major axis bending due to horizontal forces and minor axis bending due to eccentricity of loading from the floor beams. The design action effects for this column from a linear analysis program are as follows (see also Fig. 13.22).

$N = 2500\,\text{kN}$

At the base of column: $M_z = -350\,\text{kNm}$, $M_y = 0$

At the top of column: $M_z = 350\,\text{kNm}$, $M_y = 100\,\text{kNm}$

Determine whether a hot-rolled wide flange section W 310 × 310 × 226 will be suitable to resist the design action effects. Use Fe 410 grade steel.

**SOLUTION:**

**Step 1:** *Section properties*

The section properties of W 310 × 310 × 226 are as follows:

$b_f = 317$ mm

$H = 348$ mm

$t_f = 35.6$ mm     $I_z = 59{,}560 \times 10^4\ \text{mm}^4$

$t_w = 22.1$ mm     $I_y = 18{,}930 \times 10^4\ \text{mm}^4$

$R = 15.0$ mm     $r_z = 143.6$ mm

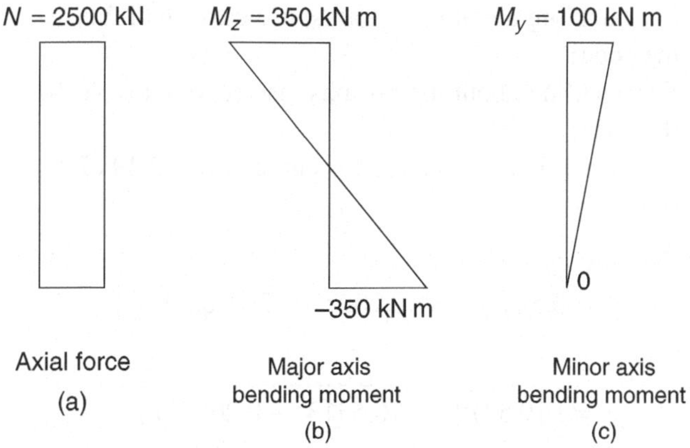


**FIG. 13.22** Design action effects on the beam-column

| | |
|---|---|
| $A = 28{,}880\,\text{mm}^2$ | $r_y = 81.0\,\text{mm}$ |
| $Z_{ez} = 3423 \times 10^3\,\text{mm}^3$ | $Z_{pz} = 39{,}48{,}812\,\text{mm}^3$ |
| $Z_{ey} = 1194 \times 10^3\,\text{mm}^3$ | $Z_{py} = 18{,}22{,}502\,\text{mm}^3$ |

The calculations for $I_t$, $I_w$, $Z_{pz}$, and $Z_{py}$ are as follows:

$$Z_{pz} = 2b_f t_f(H - t_f)/2 + t_w(H - 2t_f)^2/4$$

$$= 2 \times 317 \times 35.6\,(348 - 35.6)/2 + 22.1 \times (348 - 2 \times 35.6)^2/4$$

$$= 39{,}48{,}812\,\text{mm}^3$$

$$Z_{py} = 2 \times t_f b_f^2/4 + (H - 2t_f)t_w^2/4$$

$$= 2 \times 35.6 \times 317^2/4 + (348 - 2 \times 35.6)\,22.1^2/4$$

$$= 1{,}822{,}502\,\text{mm}^3$$

$$E = 2 \times 10^5\ \text{N/mm}^2$$

$$G = 76923\ \text{N/mm}^2$$

**Step 2:** *Cross section classification (clause 3.7)*
As the thickness is greater than 20 mm as per Table 1 of the code, $f_y = 240$ MPa. However, in the following calculations, $f_y$ has been taken as 250 MPa.

$$\varepsilon = \sqrt{(250/f_y)} = \sqrt{(250/250)} = 1$$

*Outstand flanges (Table 2 of the code)*

$$b/t_f = (317/2)35.6 = 4.45$$

Limit for class 1 flange = $9.4\varepsilon = 9.4 > 4.45$
Hence flanges are class 1 (plastic)
*Web*

$$d = H - 2t_f - 2R = 348 - 2 \times 35.6 - 2 \times 15 = 246.8\,\text{mm}$$

$$d/t_w = 246.8/22.1 = 11.1 < 42\varepsilon = 42$$

Hence web is class 1.

The overall cross section classification is class 1 (plastic).

**Step 3:** *Compression resistance of the cross section*
The design compression resistance of the cross section

$$N_d = A_g f_y/\gamma_{m0} = 28880 \times 250/(1.1 \times 1000) = 6563\,\text{kN} > 2500\,\text{kN}$$

Hence the design compression resistance is alright.

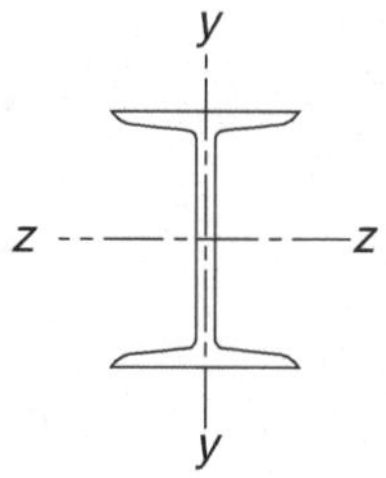


**Step 4:** *Bending resistance of the cross section (clause 8.2.1.2)*
*Major z-z axis*

Maximum bending moment = 350 kNm

The design major axis bending resistance of the cross section

$$M_{dz} = \beta_b Z_p f_y/\gamma_{m0} = (1 \times 3{,}948{,}812 \times 250/1.1) \times 10^{-6}$$

$$= 897.45\,\text{kNm} > 350\,\text{kNm}$$

*Minor y-y axis*
Maximum bending moment = 100 kNm
The design minor axis bending resistance of the cross section

$$M_{dy} = (1 \times 1822{,}502 \times 250/1.1) \times 10^{-6} = 414.2\,\text{kNm} > 100\,\text{kNm}$$

Hence the bending resistance is fine along both major *z-z* axis and minor *y-y* axis.

**Step 5:** *Shear resistance of the cross section (clause 8.4.1)*
The design plastic shear resistance of the cross section

$$V_p = A_v(f_{yw}/\sqrt{3})/\gamma_{m0}$$

*Load parallel to web*
Maximum shear force $V = [350 - (-350)]/4.0 = 175$ kN
For a rolled section, loaded parallel to the web the shear are

$$A_v = Ht_w = 348 \times 22.1 = 7690.8\,\text{mm}^2$$

$$V_p = 7690.8 \times (250/\sqrt{3})/(1.1 \times 1000) = 1009.1\,\text{kN} > 175\,\text{kN}$$

Hence the shear resistance of the cross section is alright.

*Load parallel to flanges*
Maximum shear force $V = 100/4 = 25$ kN

$$A_v = 2b_f t_f = 2 \times 317 \times 35.6 = 22{,}570.4\,\text{mm}^2$$

$$V_p = 22{,}570.4 \times (250/\sqrt{3})/(1.1 \times 1000) = 2961.5\,\text{kN} > 25\,\text{kN}$$

Hence the shear resistance is alright.

*Shear buckling (clause 8.4.2)*
Shear buckling need not be considered, provided

$$d/t_w < 67\varepsilon, \text{ for unstiffened webs}$$

$$d/t_w = 246.8/22.1 = 11.1 < 67\ (\varepsilon = 1)$$

Hence no shear buckling check is required.

**Step 6:** *Cross section resistance (clause 9.3.1)*
Provided the shear force is less than 60% of the design plastic shear resistance and provided shear buckling is not a concern, the cross section needs only satisfy the requirements of bending and axial force (clause 9.2.1). In this case shear force is less than 60% of design plastic shear resistance and hence the cross section needs to be checked for bending and axial force only.

*Reduced plastic moment resistances*
*Major z-z axis*
For rolled I- or H-sections,

$$M_{ndz} = 1.11M_{dz}(1 - n) \leq M_{dz}$$

where $n = N/N_d = 2500/6563 = 0.381$

Hence $M_{ndz} = 1.11 \times 897.45\ (1 - 0.381)$
$= 616.6\,\text{kNm} > 350\,\text{kNm}$

*Minor y-y axis*
For $n > 0.2$, $M_{ndy} = 1.56M_{ndy}(1 - n)(n + 0.6)$
$= 1.56 \times 414.2(1 - 0.381)(0.381 + 0.6)$
$= 392.4\,\text{kNm} > 100\,\text{kNm}$

Hence the moment resistances for major z-z axis and minor y-y axis are alright.

*Cross section check for biaxial bending (with reduced moment resistances)*

$$(M_y/M_{ndy})^{\alpha_1} + (M_z/M_{ndz})^{\alpha_2} \leq 1.0$$

For I- and H-sections

$$\alpha_1 = 5n \geq 1 \text{ and } \alpha_2 = 2$$
$$= 5 \times 0.381 = 1.905$$

Thus, $(100/392.4)^{1.905} + (350/616.6)^2 = 0.396 < 1.0$

**Step 7:** *Member buckling resistance in compression (clause 7.1.2)*

$$N_d = A_e f_{cd}$$
$$f_{cd} = \{(f_y/\gamma_{m0})/(\phi + [\phi^2 - \lambda^2]^{0.5}\} \leq f_y/\gamma_{m0}$$

where $\phi = 0.5[1 + \alpha(\lambda - 0.2) + \lambda^2)]$

$$\lambda = \sqrt{(f_y/f_{cr})}$$
$$f_{cr} = \pi^2 E/(KL/r)^2$$

For buckling about the major z-z axis (Table 11 of the code)

$KL_z = 0.65L = 0.65 \times 4 = 2.6$ m; $KL_z/r_z = 2600/143.6 = 18.1$

*Note*: If the stiffness of the beams are known, the effective length should be calculated by using Wood's curves. For buckling about the minor y-y axis,

$KL_y = 1.0L = 1 \times 4 = 4$ m; $KL_y/r_y = 4000/81 = 49.38$

$f_{cr,z} = \pi^2 E/(KL_z/r_z)^2 = \pi^2 \times 2 \times 10^5/(18.1)^2 = 6025$ N/mm$^2$

$\lambda_z = \sqrt{(250/6025)} = 0.2037$

$f_{cr,y} = \pi^2 \times 2 \times 10^5/(49.38)^2 = 810$ N/mm$^2$

$\lambda_y = \sqrt{(250/810)} = 0.5557$

*Selection of buckling curve and imperfection factor $\alpha$*
For hot rolled H-section (with $h/b = 348/317 = 1.09 < 1.2$ and $t_f < 100$ mm):

- For buckling about the z-z axis, use curve b (Table 10 of the code)
- For buckling about the y-y axis, use curve c (Table 10 of the code)
- For curve b, $\alpha = 0.34$ and for curve c, $\alpha = 0.49$ (Table 7 of the code)

*Buckling resistance*

$$\phi_z = 0.5\ [1 + 0.34(0.2037 - 0.2) + 0.2037^2]$$
$$= 0.5213$$
$$\chi_y = 1/[0.5213 + \sqrt{(0.5213^2 - 0.2037^2)}\ ]$$
$$= 0.9988$$

$P_{d,z} = 0.9988 \times (250/1.1) \times 28{,}880/1000 = 6556\,\text{kN} > 2500\,\text{kN}$

Buckling resistance is, thus, fine.

*Buckling resistance: Minor y-y axis*

$\phi_y = 0.5\ [1 + 0.49(0.5557 - 0.2) + 0.5557^2] = 0.7415$

$\chi_y = 1/[0.7415 + \sqrt{(0.7415^2 - 0.5557^2)}\ ] = 0.8113$

$P_{d,y} = 0.8113 \times (250/1.1) \times 28{,}880/1000 = 5325.4\,\text{kN} > 2500\,\text{kN}$

Hence the buckling resistance about the minor axis is fine.

**Step 8:** *Member buckling resistance in bending (clause 8.2.2)*
The 4-m long column is unsupported along its length with no-torsional or lateral restraint. Equal and opposite design end moments of 350 kNm are applied about the major axis. Hence the full length of the column must be checked for lateral-torsional buckling.

$$M = 350\,\text{kNm}$$
$$M_d = \beta_b Z_p f_{bd}$$

where $\beta_b = 1.0$ for plastic and compact sections.

*Determination of $M_{cr}$ ($L_y = 4000$ mm)*

$M_{cr} = C_1[(\pi^2 EI_y h)/2(KL_y)^2]\ \{1 + (1/20)\ [(KL_y/r_y)/(h/t_f)]^2\}^{0.5}$

For equal and opposite end moments ($\psi = -1$) and $K = 1.0$, $C_1 = 2.752$ from Table 42 of the code, $h/t_f = 348/35.6 = 9.78$.

$M_{cr} = 2.752[(\pi^2 \times 2.0 \times 10^5 \times 18930 \times 10^4 \times 348)/(2 \times 4000^2)]$
$\{1 + (1/20)\ [49.38/9.78]^2\}^{0.5}$
$= 16{,}866 \times 10^6$ Nmm $= 16866$ kNm

Non-dimensional lateral-torsional slenderness ratio,

$$\lambda_{Lt} = \sqrt{(\beta_b Z_{pz} f_y/M_{cr})}$$
$$= \sqrt{[1 \times 3948{,}812 \times 250/(16866 \times 10^6)]} = 0.2419$$

$\alpha_{LT} = 0.21$ for rolled section
Reduction factor for lateral-torsional buckling

$$\chi_{LT} = 1/[\ \phi_{LT} + (\phi_{LT}^2 - \lambda_{LT}^2)^{0.5}]$$

where $\phi_{LT} = 0.5\,[1 + \alpha_{LT}(\lambda_{LT} - 0.2) + \lambda_{LT}^2\,]$

$= 0.5\,[1 + 0.21(0.2419 - 0.2) + 0.2419^2]$

$= 0.5337$

Thus, $\chi_{LT} = 1/[0.5337 + (0.5337^2 - 0.2419^2)^{0.5}]$

$= 0.9907$

Lateral-torsional buckling resistance

$= X_{LT}(f_y/\gamma_{m0})\,\beta_b\,Z_{pz}$

$= 0.9907 \times 250/1.1 \times 1 \times 39{,}48{,}812$

$= 889.1 \times 10^6$ Nmm $= 889.1$ kNm

$M/M_d = 350/889.1 = 0.394 \le 1.0$

Hence this is OK.

**Step 9:** *Member buckling resistance in combined bending and axial compression*

*Determination of moment amplification factors (clause 9.3.2.2.)*

$K_z = 1 + (\lambda_z - 0.2)P/P_{dz} \le 1 + 0.8\,P/P_{dz}$

$K_z = 1 + (0.2037 - 0.2)2500/6556 = 1.0015 < 1 + 0.8 \times 2500/6556 = 1.305$

$\psi_z = M_2/M_1 = -350/350 = -1,$

$C_{mz} = 0.6 + 0.4\psi = 0.6 + 0.4 \times (-1) = 0.2 < 0.4$; Hence $C_{mz} = 0.4$

$K_y = 1 + (\lambda_y - 0.2)P/P_{dy} \le 1 + 0.8P/P_{dy}$

$K_y = 1 + (0.5557 - 0.2)2500/5325.4 = 1.167 < 1 + 0.8 \times 2500/5325.4 = 1.375$

$\psi_y = M_2/M_1 = 0,$

$C_{my} = 0.6 + 0.4\psi = 0.6 + 0.4 \times (0) = 0.6 > 0.4$

$C_{mLT} = 0.4$

$n_y = P/P_{dy} = 2500/5325.4 = 0.4694$; $\lambda_{LT} = 0.2419$

$$K_{LT} = 1 - \frac{0.1\lambda_{LT}n_y}{(C_{mLT} - 0.25)} \ge 1 - \frac{0.1n_y}{(C_{mLT} - 0.25)}$$

$K_{LT} = 1 - 0.1 \times 0.2419 \times 0.4694/(0.4 - 0.25)$

$= 0.9243 \ge 1 - 0.1 \times 0.2419/(0.4 - 0.25) = 0.84$

*Check with interaction formula (clause 9.3.2.2)*

$(P/P_{dy}) + (K_yC_{my}M_y/M_{dy}) + (K_{LT}M_z/M_{dz}) \le 1.0$

$(2500/5325.4) + (1.167 \times 0.6 \times 100)/414.2 + (0.9243 \times 350)/897.45$

$= 0.469 + 0.169 + 0.360 = 0.998 \approx 1.0$

$(P/P_{dz}) + (0.6K_yC_{my}M_y/M_{dy}) + (K_zC_{mz}M_z/M_{dz}) \le 1.0$

$(2500/6556) + (0.6 \times 1.167 \times 0.6 \times 100)/414.2 + (1.002 \times 0.4 \times 350)/897.45$

$= 0.381 + 0.101 + 0.156 = 0.638 < 1.0$

Hence the section is suitable to resist the design action effects.

**Example 13.3:** Design the crane column of an industrial building with the following data (see Fig.13.23).

**(a) Dimensions**

Overall height of the column = 10.5 m

Height of crane rail = 8 m

Free standing brick wall = 2.5 m

Crane rail eccentricity = 0.5 m

Cladding eccentricity = 0.25 m

**(b) Loading (unfactored)**

Roof truss reactions $W_A$:

Dead load = 40 kN

Imposed load = 65 kN

Wind load (suction) = 70 kN

Gantry girder reactions (vertical) $W_C$:

Dead load (girder self weight) = 20 kN

Crane load including impact (near side) = 180 kN

Crane load (far side) = 120 kN

Gantry girder reaction (horizontal) $W_{HC}$:

Crane surge load = 5 kN

Wind load on side of building $W_B$:

Wind load (UDL above brick work) = 50 kN

Self weight (column) $W_D$:

Dead load = 15 kN

Cladding $W_{DC}$:

Dead load = 12.5 kN

**Solution:**

**Step 1:** *Unfactored load matrix*

The resultant force ($W_E$ for wind) depends on the difference between wind loading on the column being designed and that of similar column on the far side of the building (as shown in Fig. 13.23). Assuming that the wind load on the far side is equal,

$W_E = 3(W_1 + W_2)/16$

$= 3(50 + 50)/16 = 18.75$ kN

The resultant force ($W_E$ for crane) depends on the difference in the crane (vertical) loading on the near and far side columns (Fig. 13.23).

$W_E = 3\,(M_1 + M_2)/(4h)$

$= 3(180 + 120) \times 0.5/(4 \times 10.5) = 10.71$ kN

The resultant force ($W_E$ for dead load) depends on the gantry girder dead load producing moments in opposite directions (see Fig. 13.23):

$W_E = 3\,(20 + 20) \times 0.5/(4 \times 10.5) = 1.43$ kN

Note that the crane surge load (horizontal) produces no resultant force in the truss as it induces equal loads in both columns in the same direction.

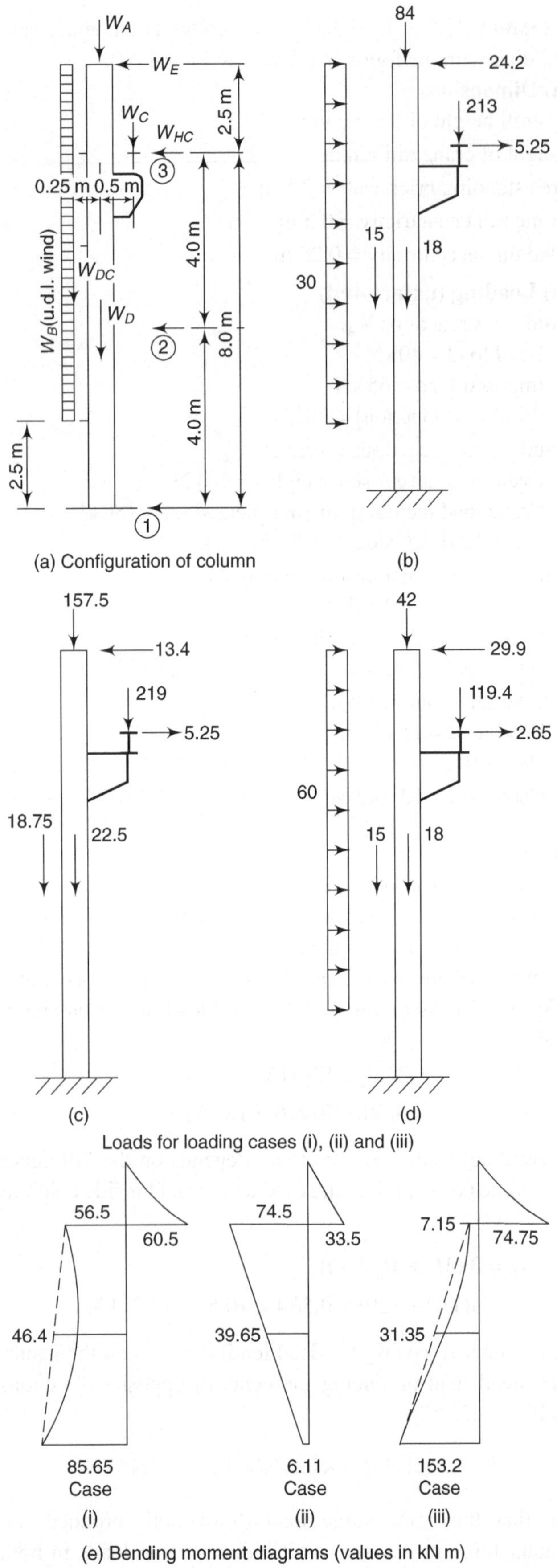

FIG. 13.23

All the unfactored loads [*W*] are shown in a matrix form as follows:

| | Dead load, $W_d$ | Imposed load, $W_I$ | Crane load vertical, $W_{cv}$ | Crane load horizontal, $W_{ch}$ | Wind load, $W_w$ |
|---|---|---|---|---|---|
| $W_A$ | 40 | 65 | 0 | 0 | –70 |
| $W_B$ | 0 | 0 | 0 | 0 | 50 |
| $W_C$ | 20 | 0 | 180 | 0 | 0 |
| $W_{HC}$ | 0 | 0 | 0 | 5 | 0 |
| $W_D$ | 15 | 0 | 0 | 0 | 0 |
| $W_{DC}$ | 12.5 | 0 | 0 | 0 | 0 |
| $W_E$ | 1.43 | 0 | 10.71 | 0 | 18.75 |

**Step 2:** *Load factors and combinations*

Possible combinations of different loads are

$$1.5W_d + 1.5W_I$$

$$1.5W_d + 1.5W_w$$

$$1.2W_d + 1.2W_I + 1.05W_{cv} + 0.6W_w + 1.05W_{ch} \quad \text{(Case i)}$$

$$1.5W_d + 1.5W_I + 1.05W_{cv} + 1.05W_{ch} \quad \text{(Case ii)}$$

$$1.2W_d + 1.2W_I + 0.53W_{cv} + 1.2W_w + 0.53W_{ch} \quad \text{(Case iii)}$$

Of these combinations, three cases are selected as shown and are given in matrix form [$r_f$] as follows:

| | Case (i) | Case (ii) | Case (iii) |
|---|---|---|---|
| $W_d$ | 1.2 | 1.5 | 1.2 |
| $W_I$ | 1.2 | 1.5 | 1.2 |
| $W_{cv}$ | 1.05 | 1.05 | 0.53 |
| $W_{ch}$ | 1.05 | 1.05 | 0.53 |
| $W_w$ | 0.6 | 0 | 1.2 |

**Step 3:** *Factored loading*

| | Case (i) | Case (ii) | Case (iii) |
|---|---|---|---|
| $W_A$ | 84 | 157.5 | 42 |
| $W_B$ | 30 | 0 | 60 |
| $W_C$ | 213 | 219 | 119.4 |
| $W_{HC}$ | 5.25 | 5.25 | 2.65 |
| $W_D$ | 18 | 22.5 | 18 |
| $W_{DC}$ | 15 | 18.75 | 15 |
| $W_E$ | 24.2 | 13.40 | 29.90 |

The loads for each loading case is shown in Figs 13.23(b), (c), and (d).

**Step 4:** *Coefficients*

To calculate the axial force, all vertical loads above the cross section being checked should be added together. To calculate the bending moments, the products of force and distance should be added together. This is shown in a matrix [$\alpha$] for axial force $F$ and bending moment $M$. The bending moments are calculated at three positions as shown in Fig. 13.23(a). The coefficients for BM are in metres.

| | $F_1$ | $M_1$ | $M_2$ | $M_3$ |
|---|---|---|---|---|
| $W_A$ | 1 | 0 | 0 | 0 |
| $W_B$ | 0 | 6.5 | 2.64* | 0.39* |
| $W_C$ | 1 | 0.5 | 0.5 | 0.5 |
| $W_{HC}$ | 0 | 8.0 | 4.0 | 0 |
| $W_D$ | 1 | 0 | 0 | 0 |
| $W_{DC}$ | 1 | −0.25 | −0.20* | 0.08* |
| $W_E$ | 0 | −10.5 | −6.5 | −2.5 |

The coefficients for bending moment in relation to the load $W_B$ allow for a reduced value of $W_B$ as well as reduced distance in calculating $\alpha$ for positions 2 and 3. For example,

$$M_2 = W_B(6.5/8) \times (6.5/2) = 2.64W_B$$

Similarly the value of $W_{DC}$ decreases for points 2 and 3 and this is reflected in the value of $\alpha$. For example,

$$M_2 = W_{DC}(6.5/8) \times (-0.25) = 0.20W_{DC}$$

**Step 5:** *BM and axial force*

As discussed already, the BM and axial force matrix is

$$[F, M] = [W\gamma]^T\,[\alpha]$$

| | $F_1$ | $M_1$ | $M_2$ | $M_3$ |
|---|---|---|---|---|
| Case (i) | 330 | 85.65 | 46.4 | 56.5 |
| Case (ii) | 417.75 | 6.11 | 39.65 | 74.5 |
| Case (iii) | 194.4 | 153.2 | 31.35 | 7.15 |

The bending moment diagram for each load case is shown in Fig. 13.23 (e). From the forces and bending moment diagram, it is clear that Case (iii) is the governing case.

**Step 6:** *Selection of trial section*

Try wide flange W 310 × 310 × 97 section

The relevant properties are

$H$ = 308 mm, $b_f$ = 305 mm, $t_f$ = 15.4 mm, $t_w$ = 9.9 mm, $R$ = 15 mm

$A = 12330\,\text{mm}^2$, $Z_z = 1444 \times 10^3\,\text{mm}^3$, $Z_y = 477.8 \times 10^3\,\text{mm}^3$

$r_z = 134.3\,\text{mm}$, $r_y = 76.9\,\text{mm}$.

From Fig. 9.37,

$KL_z = 1.5 \times 10.5 = 15.75$ m

$KL_y = 0.85 \times 8 = 6.8$ m

$KL_z/r_z = 15750/134.3 = 117.3$

$KL_y/r_y = 6800/76.9 = 88.4$

Hence $KL_z/r_z$ is critical.

$P_{eq} = P + 2M_z/h = 194.4 + 2 \times 153.2 \times 1000/308 = 1189\,\text{kN}$

$h/b = 308/305 = 1.01 < 1.2$

$t_f < 40\,\text{mm}$.

Hence use curve $b$ for major axis buckling.

For $KL/r$ = 117.3 and $f_y$ = 250 N/mm², from Table 9b of the code,

$$f_{cd} = 95\ \text{N/mm}^2$$

Thus axial capacity of the section = 95 × 12,330/1000 = 1171 kN < 1189 kN

Hence, choose a W 310 × 310 × 107 section.

**Step 7:** *Properties of the section*

Relevant properties of W 310 × 310 × 107 are

$H$ = 311 mm, $b_f$ = 306 mm, $t_w$ = 10.9 mm, $t_f$ = 17 mm, $R$ = 15 mm

$A$ = 13620 mm², $r_z$ = 134.9 mm, $r_y$ = 77.2 mm, $Z_z$ = 1594 × 10³ mm³,

$Z_y = 530.9 \times 10^3\,\text{mm}^3$

$Z_{pz} = 2b_ft_f(H - t_f)/2 + t_w(H - 2t_f)^2/4$

$= 2 \times 306 \times 17\,(311 - 17)/2 + 10.9\,(311 - 2 \times 17)^2/4$

$= 1738.5 \times 10^3\,\text{mm}^3$

$$Z_{py} = \frac{2t_fb_f^2}{4} + \left(H - 2t_f\right)t_w^2/4$$

$= 2 \times 17 \times 306^2/4 + (311 - 2 \times 17) \times 10.9^2/4$

$= 804.1 \times 10^3\,\text{mm}^3$

**Step 8:** *Check for local capacity of the section*

$b/t_f = (306/2)/17 = 9 < 9.4\varepsilon\ [\varepsilon = \sqrt{(250/250)} = 1]$

$d = H - 2(t_f + R) = 311 - 2(17 + 15) = 247\,\text{mm}$

$d/t_w = 247/10.9 = 22.67 < 42\varepsilon$

Hence the cross section is plastic.

$N_d = A_gf_y/\gamma_{m0}$

$= 13{,}620 \times 250/(1.1 \times 1000) = 3095\,\text{kN}$

$n = N/N_d = 194.4/3095 = 0.06$

$M_{dz} = \beta_bZ_pf_y/\gamma_{m0} = 1 \times 1738.5 \times 10^3 \times 250/(1.1 \times 10^6)$

$= 395.1\,\text{kNm}$

*Reduced plastic resistance*

For rolled sections,

$$M_{ndz} = 1.1 \times M_{dz}\,(1 - n) \le M_{dz}$$

$$= 1.1 \times 395.1\,(1 - 0.06) = 408.5 > 395.1\,\text{kNm}$$

Hence $M_{ndz}$ = 395.1 kNm

Interaction equation for cross section strength is

$(M_z/M_{ndz})^{\alpha_2} \le 1.0$ where $\alpha_2 = 2$ (Table 17 of the code)

Thus, $(153.2/395.1)^2 = 0.15 << 1.0$

Hence safe.

**Step 9:** *Member buckling resistance in bending (clause 8.2.2)*

$$f_{cr,b} = C_1\,[1473.5/(KL/r_y)]^2\,[1 + (1/20)\,\{(KL/r_y)/(h/t_f)\}^2]^{0.5}$$

$$\psi_z = M_2/M_1 = 7.15/153.2 = 0.047$$

From Table 42 of the code, take $C_1 = 1.563$

$h/t_f = 311/17 = 18.29$; $KL/r_y = 6800/77.2 = 88$

Hence, $f_{cr,b} = 1.563(1473.5/88)^2[1 + (1/20)(88/18.29)^2]^{0.5}$
$= 643.6$ N/mm$^2$
$M_{cr} = 643.6 \times 1594 \times 10^3/10^6 = 1025.9$ kNm

Non-dimensional lateral-torsional slenderness ratio

$$\lambda_{LT} = \sqrt{(\beta_b Z_{pz} f_y / M_{cr})}$$

$$= \sqrt{[1 \times 1738.5 \times 10^3 \times 250/(1025.9 \times 10^6)]}$$

$$= 0.65$$

$$\chi_{LT} = 1/[\phi_{LT} + (\phi_{LT}^2 - \lambda_{LT}^2)^{0.5}]$$

where $\phi_{LT} = 0.5[1 + \alpha_{LT}(\lambda_{LT} - 0.2) + \lambda_{LT}^2]$,

$\alpha_{LT} = 0.21$ for rolled sections

$= 0.5[1 + 0.21\ (0.65 - 0.2) + 0.65^2]$
$= 0.7585$

$\chi_{LT} = 1/[0.7585 + (0.7585^2 - 0.65^2)^{0.5}]$
$= 0.87$

Lateral-torsional buckling resistance

$= \chi_{LT}\,(f_y/\gamma_{m0})\,Z_p = 0.87 \times (250/1.1) \times 1738.5 \times 10^3/10^6$

$= 343.75$ kNm $> 153.2$ kNm

**Step 10:** *Check for overall buckling*

$KL/r_z = 15750/134.9 = 116.75$
$f_{cr,z} = \pi^2 E/(KL/r)^2 = \pi^2 \times 2 \times 10^5/(116.75)^2 = 144.8$ N/mm$^2$

$$\lambda_z = \sqrt{(250/144.8)} = 1.31$$

From Table 9b of the code, for $KL/r = 116.75$ and $f_y = 250$ N/mm$^2$, $f_{cd} = 95.7$ N/mm$^2$

$$P_{dz} = 95.7 \times 13{,}620/1000 = 1303\text{ kN}$$

Amplification factor

$K_z = 1 + (\lambda_z - 0.2)P/P_{dz} \le 1 + 0.8P/P_{dz}$

$K_z = 1 + (1.31 - 0.2)194.4/1303 = 1.165 > 1 + 0.8 \times 194.4/1303$
$= 1.119$

Hence $K_z = 1.119$

$\psi_z = M_2/M_1 = 7.15/153.2 = 0.046$,

$C_{mz} = 0.6 + 0.4\psi = 0.6 + 0.4 \times 0.046 = 0.6184 > 0.4$

Check with interaction formula (clause 9.3.2.2)

$$P/P_d + [(K_z C_{mz} M_z)/M_{dz}) \le 1$$

Thus, $(194.4/1303) + (1.119 \times 0.6184 \times 153.2)/343.75$
$= 0.149 + 0.308 = 0.457 < 1.0$

Hence the cross section is safe.

However, since there is some extra safety margin, the next smaller section W 310 × 310 × 97 may be adopted. The calculation for this section is left as an exercise to the reader.

In addition to the above strength checks, the column has to be checked for limit states serviceability (deflection), which may control the column size in some cases. The actual deflection may be calculated by using a computer program or by hand methods given by MacGinley and Ang (1992) and checked against the permissible deflection given in Table 6 of the code.

**EXAMPLE 13.4:** Redesign the column given in Example 13.3 as a laced column.

**SOLUTION:**

The combination of loads may be considered as before (including $\gamma_f$ factors) using the same values of $W_A$ to $W_E$. Considering case (iii), which was critical, i.e.,

$$1.2W_d + 1.2W_l + 0.53W_{cv} + 1.2W_w + 0.53W_{ch}$$

The maximum value of *BM* and force will slightly change due to the revised cross section (see Fig. 13.24)

Max. $BM = 119.4 \times 0.4 + 2.65 \times 8$
$+ 60 \times 6.5 - 29.9 \times 10.5$
$- 15 \times 0.65 - 42 \times 0.4$
$= 118.46$ kNm

Axial force $= 42 + 119.4 + 18 + 15$
$= 194.4$ kN

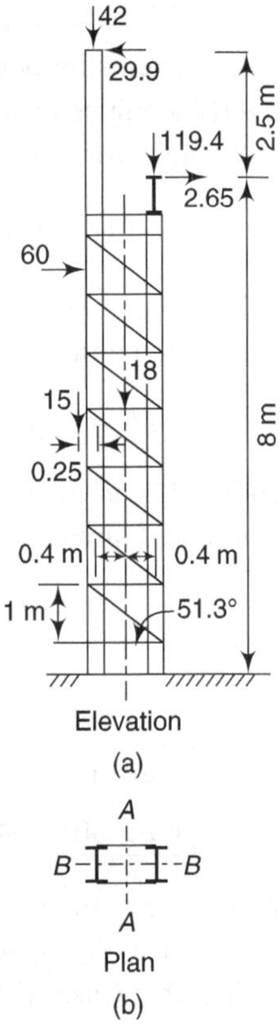


FIG. 13.24

The design of the columns may also be done independently. However, we will carry out the design assuming the column as a single integral compound column, which will result in some savings.

Use two ISHB 150 sections.

**Step 1:** *Properties of the cross section*

The required properties are

$H = 150$ mm, $B = 150$ mm, $t_f = 9.0$ mm, $t_w = 5.4$ mm, $R = 8$ mm, $A = 3450$ mm$^2$, $Z_z = 194 \times 10^3$ mm$^3$, $Z_y = 57.6 \times 10^3$ mm$^3$, $I_y = 432 \times 10^4$ mm$^4$, $I_z = 1460 \times 10^4$ mm$^4$, $r_z = 65$ mm, $r_y = 35.4$ mm

Moment of inertia of combined section

$I_a = 2\,(432 \times 10^4 + 3450 \times 400^2)$
$= 1112.64 \times 10^6$ mm$^4$

$$Z_a = 1112.64 \times 10^6/475 = 2342.4 \times 10^3\text{ mm}^3$$

$$r_a = \sqrt{[1112.64 \times 10^6/(2 \times 3450)]} = 401.5\text{ mm}$$

**Step 2:** *Section classification*

$$b/t_f = (150/2)/9 = 8.33 < 9.4\ (\varepsilon = \sqrt{(250/250)} = 1)$$

$$d = H - 2(t_f + R) = 150 - 2(9 + 8) = 116\text{ mm}$$

$$d/t_w = 116/5.4 = 21.48 < 42\varepsilon$$

Hence, the section is plastic.

**Step 3:** *Check for local capacity of the section*

$N_d = A_g f_y / \gamma_{m0}$

$= 2 \times 3450 \times 250/(1.1 \times 1000) = 1568\,\text{kN}$

$n = N/N_d = 194.4/1568 = 0.124$

$M_{dy} = \beta_b Z_p f_y / \gamma_{m0}$

$= 1 \times (2 \times 3450 \times 400) \times 250/(1.1 \times 10^6)$

$= 627.27\,\text{kNm}$

Reduced plastic moment for rolled section (from clause 9.3.1.8(c))

$M_{ndy} = M_{dy}$ when $n \leq 0.2$

$= 627.27\,\text{kNm}$

Interaction equation for cross section strength is $(M_y/M_{ndy})^{\alpha_2} \leq 1.0$ where $\alpha = 5n \geq 1$ (Table 17 of the code); hence $\alpha_1 = 1$.

Thus $(118.46/627.27) = 0.19 << 1.0$.

Hence safe.

**Step 4:** *Compressive strength*

For compressive strength the overall slenderness is based on an effective length,

From Fig. 9.38 of Chapter 9,

$KL_a = 1.5 \times 8 = 12$ m

$KL_b = 0.85 \times 8 = 6.8$ m

$r_a = 401.5\,\text{mm}$

$KL_a/r_a = 12000/401.5 = 29.89$

$KL_b/r_b = 6800/65 = 104.6$

Local slenderness $= L/r_y = 1000/35.4 = 28.2 < 50$ (clause 7.6.6)

Hence the slenderness is alright.

$L/r$ of the compound column $= 1.05 \times L/r = 1.05 \times 104.6$

$= 110$ (clause 7.6.1.5)

Local slenderness should be less than overall slenderness $\times$ 0.7

$0.7 \times 110 = 77 > 28.2$ (clause 7.6.6)

From Table 10 and Table 9c of the code, for $L/r = 110$ and $f_y = 250\ \text{N/mm}^2$,

$f_{cd} = 94.6\ \text{N/mm}^2$

**Step 5:** *Check for overall buckling*

Bending about the *A-A* axis (refer Fig. 13.24) may be assumed to produce axial forces in the two I-sections of the compound column.

Axial force = moment/centroidal distance between the two I-sections

$= 118.46/0.8 = 148\,\text{kN}$ (tension or compression)

Maximum compression in one I-section

$= 148 + 194.4/2 = 245.2\,\text{kN}$

Compression resistance of the section $= A_g f_y / \gamma_{m0}$

$= 94.6 \times 3450/(1.1 \times 1000) = 296.7\,\text{kN} > 245.2\,\text{kN}$

Hence the section is safe.

**Step 6:** *Design of lacings*

Maximum force in lowest diagonal

$= (60 + 2.65 - 29.9)/\cos 51.3°$

$= 52.4\,\text{kN}$ (compression)

In addition, the lacing should carry a transverse shear force equal to 2.5% of the axial column load, i.e., $0.025 \times 245.2 = 6.13\,\text{kN}$

Force in the lacing $= 6.13/\cos 51.3° = 9.8\,\text{kN}$

Total force to be resisted by lacing $= 52.4 + 9.8 = 62.2\,\text{kN}$ i.e., 32 kN on each side of the column.

Use a $65 \times 65 \times 6$ angle with area $= 744\,\text{mm}^2$ and $r_v = 12.6\,\text{mm}$.

Effective length of lacing $= \sqrt{[0.80^2 + 1.0^2]} = 1.28$ m

$KL/r = 1280/12.6 = 101.6 < 145$

Compressive stress from Table 9c of the code,

$f_{cd} = 105.0\ \text{N/mm}^2$

$P_d = A_e f_{cd} = 105 \times 744/1000 = 78.12\,\text{kN} > 62.2\,\text{kN}$

Hence the section is safe. In addition, provide tie-plates at the top and bottom of the column. Effective depth $> 2 \times b_f = 2 \times 150 = 300\,\text{mm}$ or 800 mm (distance between C.G. of the sections). Provide 16 mm bolts. Hence overall depth $= 800 + 2 \times 25 = 850\,\text{mm}$

Length of the plate = 800 mm

Thickness of the plate = 1/50(800) = 16 mm

Hence, $(800 \times 850 \times 16)$-mm thick tie plates are required.

**EXAMPLE 13.5:** Design the beam-column given in Example 13.1, assuming that the axial load is tensile.

Factored axial load = 500 kN (tension)

*Factored moment $M_z$:*

At top of the column: 27.0 kNm

At bottom of the column: 45 kNm

Effective length of the column = 3.2 m

**SOLUTION:**

Try ISHB 250 section. From Example 13.1, the section is semi-compact.

$M_{dz} = Z_e f_y / \gamma_{m0} = 619 \times 10^3 \times 250/(1.1 \times 10^6) = 140.68\,\text{kN}$

Assuming that the column is connected to the base plate by welding, the design strength under axial tension,

$T_{dg} = f_y A_g / \gamma_{m0} = 6500 \times 250/(1.1 \times 1000) = 1477.27\,\text{kN}$

Design strength due to rupture of critical section,

$T_{dn} = 0.9 f_u A_n / \gamma_{m1} = 0.9 \times 410 \times 6500/(1.25 \times 1000)$

$= 1918.8\,\text{kN}$

***Note*** The design strength due to block shear should also be calculated and the least value should be taken as the design strength.
For simplicity, we assume

$$T_d = 1477.27\,\text{kN}$$

*Check for section strength (clause 9.3.1.1)*

$N/N_d + (M_y/M_{dy}) + (M_z/M_{dz}) \leq 1.0$

$(500/1477.27) + (45/140.68) = 0.338 + 0.32 = 0.658 < 1.0$

Hence the section is safe.

From Example 13.1,
Lateral-torsional buckling resistance = 128.82 kNm
Reduced effective moment (clause 9.3.2.1)

$$M_{\text{eff}} = [M - \psi T Z_{ec}/A] \leq M_d$$

$$= 45 \times 10^6 - 0.8 \times 500 \times 10^3 \times 619 \times 10^3/6500$$

$$= 45 \times 10^6 - 38 \times 10^6 = 7 \times 10^6\ \text{Nmm} < 128.82\,\text{kNm}$$

*Check for overall buckling strength (clause 9.3.2.2)*

$(P/P_d) + (M_{\text{eff}}/M_{dz}) \leq 1.0$

$(500/1477.27) + (7.0/128.82) = 0.338 + 0.054 = 0.392 < 1.0$

## SUMMARY

Bending moments will be present in many practical situations in columns, in addition to the axial loads. These bending moments may be due to (a) eccentricity of axial force, (b) building frame action, (c) portal or gable frame action, (d) load from brackets, (e) transverse loads, and (f) fixed base condition. When a member is subjected to bending moment and axial force, it is called as a beam-column. The cross section of such beam-columns is oriented in such a way to resist significant bending in the major axis of the member. In general, beam-columns may be subjected to axial forces and biaxial bending moments.

All of the parameters that affect the behaviour of a beam or a column (such as the length of the member, geometry and material properties, support conditions, magnitude and distribution of transverse loads and moments, presence or absence of lateral bracing, and whether the member is a part of an unbraced or braced frame) also affect the behaviour and strength of a beam-column. The general behaviour of beam-columns is described. Thus a beam-column may fail by local buckling, overall buckling (similar to axially loaded columns), lateral-torsional buckling (similar to beams), plastic failure (short beam-columns), and by a combination of column buckling and lateral-torsional buckling. Moreover, the ultimate behaviour of beam-columns subjected to axial load and biaxial bending moments is complicated by the effect of plastification, moment magnification, and lateral-torsional buckling.

The moment magnification due to $P$-$\delta$ effects (member effects) and $P$-$\Delta$ effects (structure effects) are explained based on the differential equation approach. The simplified approach adopted in the codes by using an equivalent moment factor is also explained. Based on the differential equation approach, the elastic lateral-torsional buckling moment of beam-columns is derived.

The inelastic analysis to determine the strength of beam-column is complicated and is carried out in two steps: cross section analysis and member analysis. These methods require extensive numerical analysis procedures and hence are not suitable for design office use. However, these methods are useful in developing simplified design equations.

Hence codes and specifications of many countries suggest the use of interaction equations, which are developed based on curve-fitting the existing analytical and experimental data on isolated beam-columns or beam-columns of simple portal frames.

The elastic biaxial bending of isolated beam-columns with equal and opposite end moments acting about each principal axis is similar to its in-plane behaviour. Similar to uniaxial bending, sophisticated numerical analyses have been conducted on beam-columns subjected to axial load and biaxial moments. Based on these studies, linear and non-linear interaction equations, which are extensions of in-plane bending and flexural-torsional buckling, have been suggested. (It is important to realize that these are simplified design formula applicable to 'isolated' beam-columns. In reality the beam-column will be a member of a frame and hence there will be strong interaction between the instability of the beam-column and the general elasto-plastic behaviour of the remainder of the frame. Such effects can be taken into the design, only when advance analysis methods are employed.) Such interaction equations have been developed for beam-columns with *I*-box, and other cross sections such as angles. Similar interaction equations have also been developed for checking the local capacity of the section at the point of greatest bending moment and axial load.

The generic form of the interaction equations have been retained in the codes of practices of several countries. The interaction equations given in the Indian code (which are based on the Eurocode provisions) and ANSI/AISC code are discussed.

The various steps involved in the design of beam-columns, are given. The equivalent axial load method is described, which will be useful while selecting the initial cross section of the iterative design process. Though the codal equations are based on the first-order elastic analysis and hence include the moment amplification term, it is possible to use a second-order analysis and get the second-order moments directly. In such a case, the amplification factors should not be used in the interaction equations. Similarly the advanced analysis methods described in Chapter 4 may eliminate the determination of *K* factors. (It is of interest to note that an error of 20% in the determination of the effective length factor may result in an error of 50% in the load carrying capacity of a slender column.)

Several researchers have criticized the use of effective length factors (e.g., Cheong–Sait–Moy 1974). Hence, it is emphasized that the concept of effective length is only a simple and convenient short-cut and is used by many designers since no other better option is available. Hence, this concept must be used with some intelligence and critical spirit. The designers are not obliged to use it if they can use a better and yet sufficiently simple solution (Massonnet & Maquoi 1978).

Bending moments may occur in members subjected to tension; however, the effect of tension load will always reduce the primary bending moments. Hence in the code, a reduced effective bending moment is specified to be used with the interaction equations. Discussions on the design of crane columns are also included. All the design concepts are explained with illustrative examples.

## MULTIPLE-CHOICE QUESTIONS

13.1 Beam-columns are structural members which are subjected to
(a) axial compressive force and bending moment
(b) bending moment and shear
(c) torsional moment, bending moment, and shear
(d) all of these

13.2 The vertical members of a multi-storey frame should be designed as a
(a) column
(b) beam-column
(c) tension member
(d) beam

13.3 Second-order moments are created in beam-columns due to
(a) P-$\delta$ effects
(b) P-$\Delta$ effects
(c) Both (a) and (b)
(d) None of these

13.4 The equivalent moment factor, $C_m$, in frames where side sway is prevented is
(a) $0.6\psi - 0.4 \geq 0.4$, where $\psi$ is the ratio of the smaller to bigger moment
(b) 0.85
(c) $0.6 + 0.4\psi \geq 0.4$, where $\psi$ is the ratio of the smaller to bigger moment
(d) 1.0

13.5 A beam-column bent in single curvature as compared to that bent in double curvature will cause ____________ secondary moments.
(a) larger
(b) smaller
(c) equal
(d) negligent

13.6 The failure mode of beam-column is by
(a) local buckling
(b) flexural-torsional buckling
(c) flexural yielding
(d) all of these

13.7 The local capacity check for beam-columns as per IS 800 for circular tubes is

(a) $\left(\dfrac{M_y}{M_{ndy}}\right)+\left(\dfrac{M_z}{M_{ndz}}\right) \leq 1$

(b) $\left(\dfrac{M_y}{M_{ndy}}\right)^2+\left(\dfrac{M_z}{M_{ndz}}\right)^2 \leq 1$

(c) $\left(\dfrac{M_y}{M_{ndy}}\right)+\left(\dfrac{M_z}{M_{ndz}}\right) \leq 1$

(d) $\left(\dfrac{M_y}{M_{ndy}}\right)^2+\left(\dfrac{M_z}{M_{ndz}}\right)^2 \leq 1$

13.8 When there is low shear force, the local capacity check for semi-compact sections is

(a) $\left(\dfrac{N}{N_d}\right)+\left(\dfrac{M_y}{M_{dy}}\right)+\left(\dfrac{M_z}{M_{dz}}\right) \leq 1$

(b) $\left(\dfrac{M_y}{M_{dy}}\right)+\left(\dfrac{M_z}{M_{dz}}\right) \leq 1$

(c) $\left(\dfrac{N}{N_d}\right)^2+\left(\dfrac{M_y}{M_{dy}}\right)+\left(\dfrac{M_z}{M_{dz}}\right) \leq 1$

(d) $\left(\dfrac{N}{N_d}\right)+\left(\dfrac{M_y}{M_{dy}}\right)^2+\left(\dfrac{M_z}{M_{dz}}\right)^2 \leq 1$

13.9 One of the overall buckling checks is given by

(a) $\left(\dfrac{P}{P_{dy}}\right)+\left(\dfrac{K_y C_{my} M_y}{M_{dy}}\right)+\left(\dfrac{K_{LT} M_z}{M_{dz}}\right) \leq 1$

(b) $\left(\dfrac{P}{P_{dy}}\right)+\left(\dfrac{K_y C_{my} M_y}{M_{dy}}\right)+\left(\dfrac{K_z C_{mz} M_z}{M_{dz}}\right) \leq 1$

(c) $\left(\dfrac{P}{P_{dy}}\right)+\left(\dfrac{0.6 K_y C_{my} M_y}{M_{dy}}\right)+\left(\dfrac{K_{LT} M_z}{M_{dz}}\right) \leq 1$

(d) $\left(\dfrac{P}{P_{dy}}\right)+\left(\dfrac{K_y C_{my} M_y}{M_{dy}}\right)+\left(\dfrac{0.6 K_{LT} M_z}{M_{dz}}\right) \leq 1$

13.10 The behaviour of beam-column cross section analysis is usually expressed as a set of
(a) moment-curvature relationship
(b) moment-curvature-axial compression relationship
(c) moment-axial compression relationship
(d) axial compression-curvature relationship

## EXERCISES

13.1 A beam-column of length 5 m is subjected to a compression of 800 kN and a major axis moment of 4.5 kN m. The weaker plane of the column is strengthened by bracing. If the effective length factor is 0.8, design the beam-column, assuming Fe 410 grade steel.

13.2 Redesign the beam-column given in Example 13.2 using two channels welded together to form a box section.

13.3 A wide flange W 310 × 310 × 143 beam-column has a height of 3.54 m and is pinned at both ends. Check whether it can support a design axial load of 600 kN together with a major axis bending moment of 300 kN m applied at the top of the beam-column. Assume grade Fe 410 steel.

13.4 Design a beam-column of length 3.75 m if it carries a compressive load of 500 kN, a major axis moment of 5 kN m and a minor axis moment of 2 kN m. Assume that the effective length factor is 1.2 and the column is free to buckle in any plane.

13.5 Design a crane column (with bracket) as shown in Fig. 13.23(a) with the following data:

(a) *Dimensions*
- Overall height = 8 m
- Height of crane rail = 5.5 m
- Crane rail eccentricity = 0.4 m
- Cladding eccentricity = 0.2 m

(b) *Loading (unfactored)*
- Roof truss reactions $W_A$:
  - Dead load = 40 kN
  - Imposed load = 60 kN
  - Wind load (suction) = 50 kN
- Gantry girder reactions (vertical) $W_c$:
  - Dead load (girder self weight) = 20 kN
  - Crane load including impact (near side) = 150 kN
  - Crane load (far side) = 100 kN
- Gantry girder reaction (horizontal) $W_{HC}$:
  - Crane surge load = 4 kN
- Wind load on side of building: $W_B$; udl = 40 kN
- Self weight (column) $W_D$:
  - Dead load = 10 kN
  - Cladding $W_{DC}$ = 10 kN

13.6 Design the crane column, with the data given in Exercise 13.5 as a laced column.

13.7 Redesign the beam-column given in Exercise 13.1 assuming that the axial load is tensile.

13.8 A 6-m long tension member consisting of 2L 95 × 95 × 6 is under a tension of 80 kN. The workers desire to lift a weight by placing a pulley at the centre of the tension member. Determine the maximum load that could be lifted, assuming simple supports at the ends of the tension member.

13.9 Design the corner column of the building given in Example 9.11 (Fig. 9.59), by considering the moment due to the eccentric floor beam reaction. Assume that the moments at the floor levels can be divided between the upper and lower column lengths in proportion to the stiffness and that the stiffness of roof to second floor, second floor to first floor, and first floor to base are in the ratio 1:2.5:2.5.

## REVIEW QUESTIONS

13.1 What are beam-columns?

13.2 How are bending moments introduced in columns?

13.3 When bending moments are acting on a column in addition to axial loads, how are the columns oriented?

13.4 What are the parameters that affect the behaviour of beam-columns?

13.5 List the situations where bending moments may be introduced in tension members.

13.6 Under what situations can we neglect secondary bending in truss members?

13.7 Describe the general behaviour of a beam-column.

13.8 Under which five cases can the behaviour of a beam-column be classified?

13.9 How can $P$-$\delta$ effects be considered in the design of beam-columns?

13.10 State the general form of amplification factor and interaction formula to account for the $P$-$\delta$ effect.

13.11 What is equivalent moment factor?

13.12 What is the $P$-$\Delta$ effect?

13.13 State the two steps employed in the inelastic analysis of beam-columns.

13.14 How are the interaction curves developed?

13.15 What is the general form of the interaction equation?

13.16 Identify the difference in behaviour of beam-columns subject to bending moment about minor axis compared to applied bending moment in major axis.

13.17 What is the form of interaction equation for biaxially loaded beam-column (state the linear format only)?

13.18 What is the use of non-linear interaction equations?

13.19 State the general form of interaction equation specified in the code (IS 800) for
(a) overall buckling check
(b) local capacity check

13.20 What are the different steps to be followed while designing a beam-column?

13.21 Describe the equivalent axial load method for selecting the initial section of a beam-column.

13.22 When equivalent moment method has to be used to select the initial section of a beam-column?

13.23 What is the advantage of using a second-order analysis method over first-order elastic method, in the interaction equations?

13.24 What is the advantage of using the advanced analysis methods?

13.25 State the interaction equation used for the local capacity check of beam-columns subject to axial tension and bending moment.

13.26 What is the equation for calculating the reduced effective moment for beam-columns subjected to tension and bending moments?

13.27 What are the three types of columns used to support gantry girders in a single-storey industrial building?

## REFERENCES

Adeli, H., *Interactive Microcomputer-aided Structural Steel Design*, Prentice-Hall, Englewood Cliffs, NJ, 1988.

Aminmansour, A., 'Steel Beam-column Design: New, Fast and Practical Approach', *Practice Periodical on Structural Design and Construction, ASCE*, Vol. 9, No. 2, May 2004, pp. 93–97.

Austin, W.J., 'Strength and Design of Metal Beam-columns', *Journal of the Structural Div., ASCE,* Vol. 87, No. ST4, Apr 1961, pp. 1–34.

Bjorhovde, R., Galambos, T.V., and Ravindra, M.K., 'LRFD Criteria for Steel Beam- columns', *Journal of the Structural Div., ASCE,* Vol. 104, No. ST9, Sep 1978, pp. 1371–1387.

Bleich, F., *Buckling Strength of Metal Structures,* McGraw – Hill, New York, 1952.

Boissonnade, N, Jaspart, J-P., Muzeau, J-P., and Villette, M., 'Improvement of the Interaction Formulae for Beam-columns in Eurocode 3', *Computers and Structures,* Vol. 80, 2002, pp. 2375–2385.

Boissonnade, N., Jaspart, J-P., Muzeau, J-P, and Villette, M., 'New Interaction Formulae for Beam-columns in Eurocode 3: The French–Belgian Approach', *Journal of Constructional Steel Research,* Vol. 60, 2004, pp. 421–431.

Bradford, M.A., Cuk, P.E., Gizejowski, and Trahair, N.S., 'Inelastic Lateral Buckling of Beam-columns', *Journal of Structural Engineering, ASCE,* Vol. 113, No. 11, Nov 1987, pp. 2259–2277.

Burgett, L.B. 'Selection of a trial column section', *Engineering Journal, AISC,* Vol. 10, No. 2, 1973, pp. 54–59.

Chen, W.F., and Atsuta, T., *Theory of Beam-columns, Vol. 1: In-plane Behavior and Design,* McGraw-Hill, New York, 1976.

Chen, W.F. and Atsuta, T., *Theory of Beam-columns, Vol. 2: Space Behavior and Design,* McGraw-Hill, New York, 1977.

Chen, W.F., and Lui, E.M, *Stability Design of Steel Frames,* CRC Press, Boca Raton, Florida, 1991, p. 380.

Chen, W.F., and Lui, E.M., 'Stability Design Criteria for Steel Members and Frames in the United States', *Journal of Constructional Steel Research,* Vol. 5, 1985, pp. 31–74.

Chen, W.F., and Santathadaporn, S., 'Review of Column Behavior under Biaxial Loading', *Journal of the Structural Div. ASCE,* Vol. 94, No. ST12, Dec 1968, pp. 2999–3021.

Chen, W.F., and Zhou, S., '$C_m$ Factor in Load and Resistance Factor Design', *Journal of Structural Engineering, ASCE,* Vol. 113, No. 8, Aug 1987, pp. 1738–1754.

Chen, W.F., 'Further Studies of Inelastic Beam-column-problem', *Journal of the Structural Div., ASCE,* Vol. 97, No. ST2, Feb 1971, pp. 529–544.

Cheong–Siat-Moy, 'General Analysis of Laterally Loaded Beam-columns', *Journal of the Structural Div., ASCE,* Vol. 100, No. ST6, June 1974, pp. 1263–1278 (Also May 1974, pp. 953–970).

Culver, C.G., 'Exact Solution of Biaxial Bending Equations', *Journal of the Structural Div., ASCE,* Vol. 92, No. ST2, April 1966, pp. 63–83.

DeLuca, A., Mazzolani, F.M., and Piluso, V., 'Stability of Columns and Beam-columns: Eurocode 3 vs Other Codes', *International Conf. on Steel and Aluminium Structures,* Singapore, May 1991, pp. 288–297.

Dowling, P.J., Knowles, P., and Owens, G.W., *Structural Steel Design,* The Steel Construction Institute, UK and Butterworths, London, 1988.

Duan, L. and Chen, W.F., 'Design Interaction Equation for Steel Beam-columns', *Journal of Structural Engineering, ASCE,* Vol. 115, No. 5, May 1989, pp. 1225–1243.

English, G.W. and Adams, P.F., Experiments on Laterally Loaded Steel Beam-columns', *Journal of the Structural Div.,* ASCE, Vol. 99, No. ST7, Jul 1973, pp. 1457–1470.

Fukumoto, Y. and Galambos, T.V., 'Inelastic Lateral-torsional Buckling of Beam-columns', *Journal of the Structural Div., ASCE,* Vol. 92, No. ST2, Apr 1966, pp. 41–61.

Galambos, T.V. and Ketter, R.L., 'Columns Under Combined Bending and Thrust', *Transactions, ASCE,* Vol. 126, Part I, 1961, pp. 1–25.

Galambos, T.V., ed., *Guide to Stability Design Criteria for Metal Structures,* Fifth edition, John Wiley & Sons Inc., New York, 1998, pp. 319–364.

Galambos, T.V., *Structural Members and Frames,* Prentice Hall, New York, 1968.

Gaylord, E.H., Gaylord, C.N., and Stallmeyer, J.E., *Design of Steel Structures,* Third edition, McGraw-Hill International, New York, 1992, pp. 410–454.

Greiner, R. and Lindner, J., 'Interaction Formulae for Members Subjected to Bending and Axial Compression in Eurocode 3—The Method 2 Approach', *Journal of Constructional Steel Research,* Vol. 62, No. 8, Aug 2006, pp. 757–770.

Herzog, M.A.M., 'Unified Design Approach to Steel Beam-columns', *Journal of Constructional Steel Research,* Vol. 20, No. 4, 1991, pp. 291–303.

Kennedy, D.J.L., Picard, A., and Beaulieu, D., 'Limit States Design of Beam-columns: The Canadian Approach and Some Comparisons', *Journal of Constructional Steel Research,* Vol. 25, No. 1–2, 1993, pp. 141–164.

Ketter, R.L., 'Further Studies of the Strength of Beam-columns', *Journal of the Structural Div., ASCE,* Vol. 87, No. ST6, Aug 1961, pp. 135–152.

Klöppel, K. and Winkelmann, E., 'Experimentalle und Theoretische Untersuchungen Über die Traglast von Zweiachsig Außermittig Gedrückten Stahlstäben', *Stahlbau,* 1962, Vol. 31, pp. 33–47, 78–87, 109–119.

Lindner, J., 'Design of Steel Beams and Beam-columns', *Engineering Structures,* Vol. 19, No. 5, 1997, pp. 378–384.

Lindner, J. and Glitsch, T., 'Vereinfachter Nachweis für I-und U–Träger-beansprucht Durch Dopplete Biegung und Torsion', *Stahlbau,* Vol. 73, No. 9, Sep 2004, pp. 704–715.

Lindner, J. and Gietzelt, R., 'Zweiachsige Biegung and Längskraft-ein Ergänzter Bemessungsvorschlag', *Stahlbau,* Vol. 54, 1985, pp. 265–271.

Lindner, J., 'Design of Beams and Beam-columns', *Progress in Structural Engineering and Materials,* Vol. 5, No. 1, Jan/Mar 2003, pp. 38–47.

Lu, L-W, Kamalvand, H., 'Ultimate Strength of Laterally Loaded Columns', *Journal of the Structural Div., ASCE,* Vol. 94, No. ST6, Jun 1968, pp. 1505–1524.

MacGinley, T.J. and Ang, P.T.C., *Structural Steelwork—Design to Limit State Theory,* Butterworths/Heinemann, Oxford, UK, 1992.

Massonnet, Ch. and Maquoi, R., 'Recent Progress in the Field of Structural Stability of Steel Structures', IABSE Surveys S-5/78, *IABSE Periodica,* May 1978, pp. 1–40.

Massonnet, Ch., 'Forty Years of Research on Beam-columns in Steel', *SM Archives,* Vol. 1, North Holland Publishing Co., Sep 1976, pp. 27–157.

Morris, L.J. and Plum, D.R., *Structural Steelwork Design to BS 5950,* Second edition, Addison Wesley Longman, Essex, 1996.

Pi, Y.L. and Trahair, N.S., 'Non-linear Inelastic Analysis of Steel Beam-columns, I: Theory', *Journal of Structural Engineering, ASCE,* Vol.

120, No. 7, Jul 1994, pp. 2041–2061, and 'II: Applications', pp. 2062–2085.

Pillai, U., 'An Assessment of CSA Standard Equation for Beam-column Design', *Canadian Journal of Civil Engineering*, Vol. 8, 1981.

Roufaiel, M.S.L. and Monasa, F.F., 'Microcomputer-aided Analysis and Design of Steel Frames', *Journal of Computing in Civil Engineering, ASCE*, Vol. 3, No. 2, 1989, pp. 127–142.

Salmon, C.G. and Johnson, J.E., *Steel Structures: Design and Behavior*, Fourth edition, Harper Collins College Publishers, New York, 1996, pp. 702–787.

Salvadori, M., 'Lateral Buckling of Eccentrically Loaded *I*-columns', *Transactions, ASCE*, Vol. 121, 1956, pp. 1163–1178.

Shanmugam, N.E., Liew, J.Y.R., and Lee, S.L., 'Ultimate Strength Design of Biaxially Loaded Steel Box Beam-columns', *Journal of Constructional Steel Research*, Vol. 26, No. 2–3, 1993, pp. 99–123.

Sputo, T., 'History of Steel Beam-column Design', *Journal of Structural Engineering, ASCE*, Vol. 119, No. 2, Feb 1993, pp. 547–557.

Tebedge, N. and Chen, W.F., 'Design Criteria for *H*-columns under Biaxial Loading', *Journal of the Structural Div., ASCE*, Vol. 100, No. ST3, Mar 1974, pp. 579–598.

Timoshenko, S.P. and Gere, J.M., *Theory of Elastic Stability*, Second edition, McGraw-Hill, New York, 1961.

Trahair, N.S., Bradford, M.A., and Nethercot, D.A., *The Behaviour and Design of Steel Structures to BS 5950*, Third edition, Spon Press, London and New York, 2001, Chapter 7.

Trahair, N.S., *Flexural-torsional Buckling of Structures*, E&FN Spon, London, 1993, Chapter 11.

Uang, C.W., Wattar, S.W., and Leet, K.M., 'Proposed Revision of the Equivalent Axial Load Method for LRFD Steel and Composite Beam-column Design', *Engineering Journal, AISC*, Vol. 27, No. 4, 1990, pp. 150–157.

Vinnakota, S. and Aoshima, Y., 'Spatial Behaviour of Rotationally and Directionally Restrained Beam-columns', *Publications of IABSE*, Vol. 34-II, 1974, pp. 169–194.

Vinnakota, S. and Aysto, P., 'Inelastic Spatial Stability of Restrained Beam-columns', *Journal of the Structural Div., ASCE*, Vol. 100, No. ST11, Nov 1974, pp. 2235–2254.

Vinnakota, S., *Steel Structures: Behavior and LRFD,* McGraw – Hill Higher Education, New York, 2006, pp. 593–688.

White, D.W. and Clarke, M.J., 'Design of Beam-columns in Steel Frames, I: Philosophies and Procedures', *Journal of Structural Engineering, ASCE,* Vol. 123, No. 12, Dec 1997, pp. 1556–1564, and 'II: Comparison of Standards', pp. 1565–1575.

Yura, J.A., *Elements for Teaching Load and Resistance Factor Design*, University of Texas at Austin, TX, 1998, pp. 55–71.

# 14

# MOMENT-RESISTANT CONNECTIONS

## INTRODUCTION

As discussed in Chapter 5, simple connections are assumed to transfer only shear at some nominal eccentricity. Hence, they are used only in non-sway frames where the lateral load resistance is provided by bracings or shear walls. *Moment-resistant connections* on the other hand, are capable of transferring moment, axial force, and shear from one member to another. These kinds of connections are used in framed structures, where the joints are considered rigid. Thus these connections, where the members of a frame meet, have sufficient rigidity to prevent rotation of individual members. Another kind of situation is one in which the eccentricity of load from the centroid of the fastener group may cause forces and moments on the joint. These kinds of connections in which the connections are designed to transfer bending moments and shear or a combination of bending moment, shear, and axial force may be grouped into the following:

1. Eccentrically loaded connections
2. End plate connections
3. T-stub connections
4. Flange angle connections

Moment-resistant connections may exist in bolted as well as welded joints. Moment-resistant bolted joints are considered first in this chapter, followed by moment-resistant welded connections.

## 14.1 MOMENT-RESISTANT BOLTED CONNECTIONS

As mentioned in the introduction, moment-resistant bolted connections may be classified as (a) eccentrically loaded connection, which is again subdivided into eccentric load causing twisting moments (type I connections) and eccentric load causing bending moments (type II connections); (b) bolted moment end plate connection; (c) flange angle connections; and (d) split-beam T-stub connections. These connections briefly described in the following subsections.

### 14.1.1 Eccentrically Loaded Connections

Depending on how the bolts are placed in a connection, they are subjected to twisting or bending moments by the eccentrically applied loads. The effect of twisting moment and the shear force on the bolt group is to cause shear forces along two directions of the bolt, whereas the bending moment and shear cause tension and shear in the bolts. Hence, they are considered separately in the following sections.

### 14.1.2 Eccentric Load Causing Twisting Moments (Type I Connections)

Generally, when eccentricity of a load on a bolt group is less than approximately 60 mm, it is neglected. Joints such as the simple frame connection of Fig. 5.2(c) or 5.2(d) of Chapter 5, which is widely used, belong to this category (Bowles 1980). However, in some connections, such as the bracket connection as shown in Fig. 5.1(a) of Chapter 5, the eccentricity is too large and hence cannot be neglected. In the framed beam connections, such as those shown in Figs 5.1(b) and (e) of Chapter 5, also the resulting eccentricity may be too large to be neglected and hence taken into account in the design.

When bolt groups are subjected to shear and moment in their shear plane, the load that is eccentric with respect to the centroid of the bolt group can be replaced with a force acting through the centroid of the bolt group and a moment, with the magnitude $M = Pe$ (where $e$ is the eccentricity of the load), as shown in Fig. 14.1. Since both the moment and the concentric load result in shear effects in the bolts of the group, it is termed as *eccentric shear.* These eccentric shear forces tend to rotate the connections (i.e., a twisting moment is applied on the bolts).

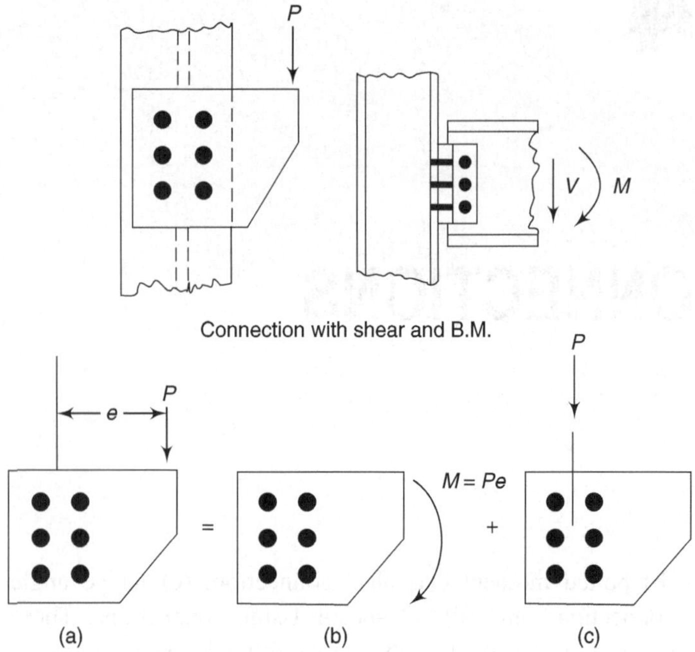


**FIG. 14.1** Replacement of eccentric shear with a moment and direct shear

Essentially two approaches are available to the designer (a) traditional elastic (vector) analysis and (b) strength (plastic) analysis.

**Elastic (Vector) analysis** This method of analysis assumes the following:

1. Deformation of the connected parts may be ignored
2. The relative movement of the connected parts are considered as the relative rigid body rotation of the two parts about some centre of rotation
3. There is friction between the 'rigid' plates and the elastic fasteners
4. This movement defines the deformations of the individual bolts, which are tangential to this centre of rotation
5. This deformation induces reactive bolt forces, which are also tangential to the centre of rotation

This elastic method of analysis has been used in the past several years, since it makes use of simple mechanics of materials concepts and has been found to yield conservative results.

Consider a connection acted upon by a moment $M$ as shown in Fig. 14.2. Neglecting friction between the plates for rotational equilibrium, the moment equals the sum of the forces multiplied by their respective distances to the centroid of the bolt group. Thus,

$$M = R_1d_1 + R_2d_2 + \ldots + R_6d_6 = \Sigma Rd \tag{14.1}$$

If we assume that the value of $R_i$ is proportional to the distance from the centroid of the fastener group, we have

$$R_1/d_1 = R_2/d_2 = \ldots = R_6/d_6 \tag{14.2}$$

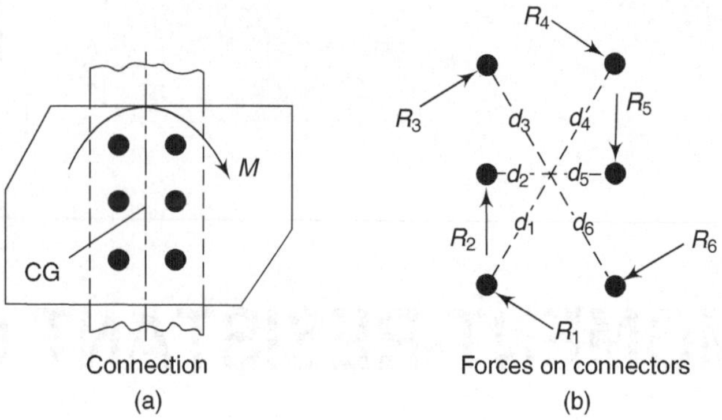


**FIG. 14.2** Pure moment (type I) connection

In terms of $R_1$, we have

$$R_1 = R_1\frac{d_1}{d_1} \quad R_2 = R_1\frac{d_2}{d_1} \quad R_3 = R_1\frac{d_3}{d_1} \quad \ldots \quad R_6 = R_1\frac{d_6}{d_1} \tag{14.3}$$

Substituting the values of Eqn (14.3) in to Eqn (14.1), we get

$$M = R_1\frac{d_1^2}{d_1} + R_1\frac{d_2^2}{d_1} + \ldots + R_1\frac{d_6^2}{d_1} \tag{14.4}$$

Simplifying the Eq. (14.4), we get

$$M = \frac{R_1}{d_1}(d_1^2 + d_2^2 + \cdots + d_6^2) = \frac{R_1}{d_1}\Sigma d_i^2 \tag{14.5}$$

Thus, the force in the fastener 1 is

$$R_1 = \frac{Md_1}{\sum d_i^2} \tag{14.6}$$

By similar reasoning, the forces in other fasteners are

$$R_2 = \frac{Md_2}{\sum d_i^2}, \cdots, R_6 = \frac{Md_6}{\sum d_i^2} \tag{14.7}$$

Or, in general

$$R = \frac{Md_i}{\sum d_i^2} \tag{14.8}$$

where $R$ is the force on the fastener at the distance $d_i$ from the centre of rotation. Note that if we want to obtain stress, we may divide $R$ by $A_b$ and the denominator in Eqn (14.8) will now become $\Sigma Ad_i^2$, which is the polar moment of inertia $J$ about the centre of rotation for the cross-sectional areas of the bolts in the group.

It is usually convenient to work with horizontal and vertical components of $R$-$R_x$ and $R_y$ respectively, which are associated with the horizontal and vertical distances $x$ and $y$ of the distance $d$.

Thus,

$$R_x = yR/d \text{ and } R_y = xR/d \tag{14.9}$$

Substituting the values of Eqn (14.9) into Eqn (14.8) and using $d^2 = x^2 + y^2$, we get

$$R_x = My/\Sigma(x^2 + y^2) \text{ and } R_y = Mx/\Sigma(x^2 + y^2) \quad (14.10)$$

Note that the $x$ and $y$ coordinates should reflect the positive and negative values of the bolt location as appropriate.

The direct shear force

$$R_v = P/n$$

where $n$ is the number of fasteners in the group. The total resultant force $R$ is given by

$$R = \sqrt{(R_y + R_v)^2 + R_x^2} \quad (14.11)$$

If there is horizontal force ($H$) and vertical force ($V$) acting eccentrically to the centroid, (with eccentricities $e_v$ and $e_h$ respectively) of the bolt group, then Eqn (14.1) can be extended as

$$R_x = H/n + (Ve_h + He_v)y/\Sigma(x^2 + y^2) \quad (14.12)$$

$$R_y = V/n + (Ve_h + He_v)x/\Sigma(x^2 + y^2) \quad (14.13)$$

This method is used in practice for ordinary bolts, high-strength friction grip bolts, and welds.

A critical bolt is the one which is subjected to the maximum resultant force. A close examination reveals that the bolt which is farthest from the centre of gravity of the bolt group and nearest to the applied load line is the most critical.

## Determination of Required Number of Bolts

From these discussions, it is clear that the forces acting on the bolts of an eccentric connection can be found if the number, pitch, and arrangement of bolts are known. However, in design scenarios, the situation is reverse. The bolt diameter is normally assumed and hence the bolt value will be known. Assuming a suitable pitch, the problem will be to find the number and arrangement of bolts in the connection. A direct procedure to determine the parameters will be involved and hence the following approximate procedure may be used.

Consider a vertical line of bolts consisting of $n$ bolts at a pitch $p$. It can be shown that whether $n$ is odd or even, the value of $\sum y^2$ of the bolts about their centre of gravity is given by [Arya and Ajmani 1989]

$$\sum y^2 = \frac{p^2 n(n-1)(n+1)}{12} \quad (14.14a)$$

Hence for $m$ rows of bolts as shown in Fig. 14.3 in a symmetrical arrangement,

$$\sum y^2 = \frac{p^2 mn(n-1)(n+1)}{12} \quad (14.14b)$$

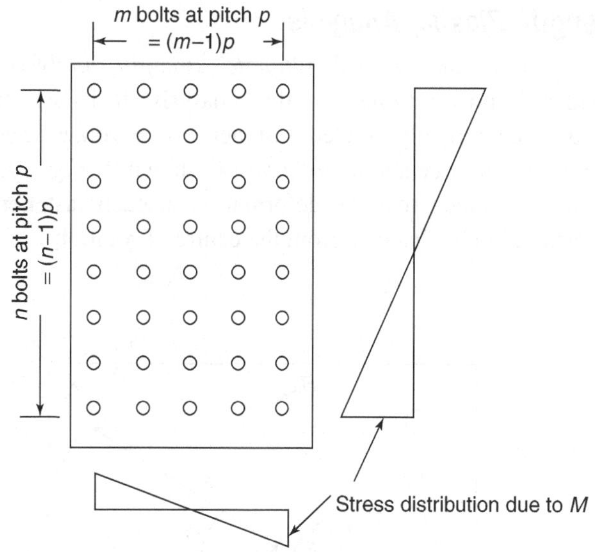


FIG. 14.3 Design of bolt group

Similarly for the group shown in Fig. 14.3 we have,

$$\sum x^2 = \frac{p^2 nm(m-1)(m+1)}{12} \quad (14.14c)$$

$$\text{Hence } \sum r^2 = \sum x^2 + \sum y^2 = \frac{p^2 nm\left(m^2 + n^2 - 1\right)}{12} \quad (14.14d)$$

For the extreme bolt,

$$r = \sqrt{\left(\frac{n-1}{2}p\right)^2 + \left(\left(\frac{m-1}{2}p\right)^2\right)} = \frac{p}{2}\sqrt{(n-1)^2 + (m-1)^2} \quad (14.14e)$$

Neglecting the effect of direct load, the force in the extreme bolt, should not exceed the bolt value $R$.

Thus we have,

$$R = \frac{Mr}{\sum r^2} = \frac{M\frac{p}{2}\sqrt{(n-1)^2 + (m-1)^2}}{\dfrac{p^2 nm\left(m^2 + n^2 - 1\right)}{12}} = \frac{6M}{pmn}\frac{\sqrt{(n-1)^2 + (m-1)^2}}{\left(m^2 + n^2 - 1\right)} \quad (14.15)$$

If the number of bolts per line, $n$, is large as compared with the number of rows of bolts, $m$, this formula may be simplified as

$$R = \frac{6M}{pmn^2} \text{ or } n = \sqrt{\frac{6M}{pmR}} \quad (14.16)$$

This equation gives a conservative estimate of the required number of bolts in a group of bolts. This formula can be used to estimate the number of rows of bolts, $m$ if $n$ is known or to find $n$ if the number of rows of bolts $m$ is known. The number so found should be checked according to Eqns 14.12 and 14.13. Examples 14.1–14.3 show the design of bracket connections.

### Strength (Plastic) Analysis

This method, also called *ultimate strength analysis* is considered most rational. In this analysis, it is assumed that the eccentrically loaded fastener group rotates about an *instantaneous centre of rotation* (as shown in Fig. 14.4). It is also assumed that the deformation at each fastener is proportional to its distance from the centre of rotation.

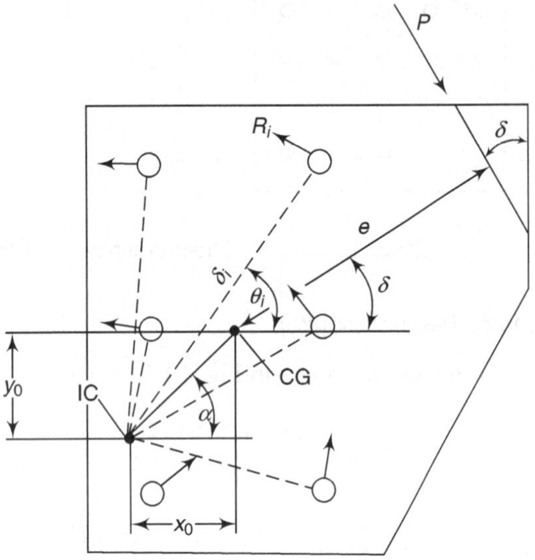


**FIG. 14.4** Instantaneous centre of rotation

With reference to Fig. 14.4, for equilibrium

$$\Sigma F_H = 0;\ \Sigma R_i \sin \theta_i - P \sin \delta = 0 \quad (14.17)$$

$$\Sigma F_v = 0;\ \Sigma R_i \cos \theta_i - P \cos \delta = 0 \quad (14.18)$$

$$\Sigma M = 0;\ \Sigma R_i d_i - P(e + x_0 \cos \delta + y_0 \sin \delta) = 0 \quad (14.19)$$

where $x_0$ and $y_0$ are the locations of the instantaneous centre from the centre of gravity of the bolt group and $P$ is the magnitude of the applied force.

When either the resistance $R_i$ is proportional to the deformation, or when the angle $\delta$ is equal to zero or 90°, the angle $\alpha$ is equal to angle $\delta$ and Eqn (14.19) reduces to

$$\Sigma M = 0;\ \Sigma R_i d_i - P(e + r_0) = 0 \quad (14.20)$$

where $r_0$ is the distance between the instantaneous centre and the centroid of the bolt group.

In fact, the concept of instantaneous centre is identical to the elastic (vector) method when the resistance $R_i$ is considered proportional to the deformation. For both elastic or strength method, the deformation is proportional to the distance $d_i$ from the instantaneous centre of rotation.

For bearing type connection, slip is neglected so that the deformation of each fastener is proportional to its distance from the instantaneous centre. The solution by this method requires the force–deformation relationship. A relation has been proposed based on experimental results by Crawford and Kulak (1971).

The procedure starts with the assumption of a location for the instantaneous centre of rotation of the fastener group. A check of the equation of equilibrium is done to find out whether or not the trial location of the instantaneous centre of rotation is the correct one. If the equations are not satisfied, a new location of the instantaneous centre is chosen and the process is repeated.

The ultimate strength method may also be used for slip-critical conditions. In this case $R_i$ is the same for all fasteners and it is not necessary to calculate the deformations (Gaylord et al. 1992). Examples of calculations based on the experimental results obtained by Crawford & Kulak (1971) are also provided by Gaylord et al. (1992) and Salmon & Johnson (1996). Crawford and Kulak have also developed a formula for joints with one line and two lines of fasteners, based on the analysis of a large number of joints. A comparison of the values obtained by the elastic (vector) analysis and the strength (plastic) analysis shows that the values obtained by the elastic analysis are conservative.

In actual construction, the distribution of shear between the bolts in a bolt group is largely influenced by the fit of the connection. Unless the components were clamped together prior to drilling there will be significant mismatch between the holes. Bolts will not start to carry any load until they are bearing against both set of plies, irrespective of their position in relation to the theoretical centre of rotation (Owens & Cheal 1989). Hence, it is suggested to use the conservative elastic method for such connections, where there will be adverse effect of lack of fit on strength.

It may be important to note that friction connections are not subject to any such uncertainity and it may be reasonable to use the strength (plastic) analysis. However, due to the difficulties encountered (interactive nature of analysis, availability of experimental load-deflection curve, etc.) in the strength analysis, designers often use the elastic (vector) analysis method for these types of friction connection groups also, which yields conservative results.

For the design of bolt groups subjected to torque, the following procedure may be adopted.

1. The number and diameter of bolts for the connection are assumed based on Eqn (14.16) and these are placed in two or more vertical rows at a suitable pitch and edge distance.
2. The resultant force on the critical bolt is worked out by using Eqn (14.11).
3. The design strength of bolt is computed. It should be more than the resultant force on the bolt. Otherwise, the number of bolts are increased and steps 1 to 3 are repeated.

### 14.1.3 Eccentric Load Causing Bending Moments (Type II Connections)

In connections where the brackets are placed in the plane of the column web (Figs 5.1(b) and 5.1(c) of Chapter 5 and

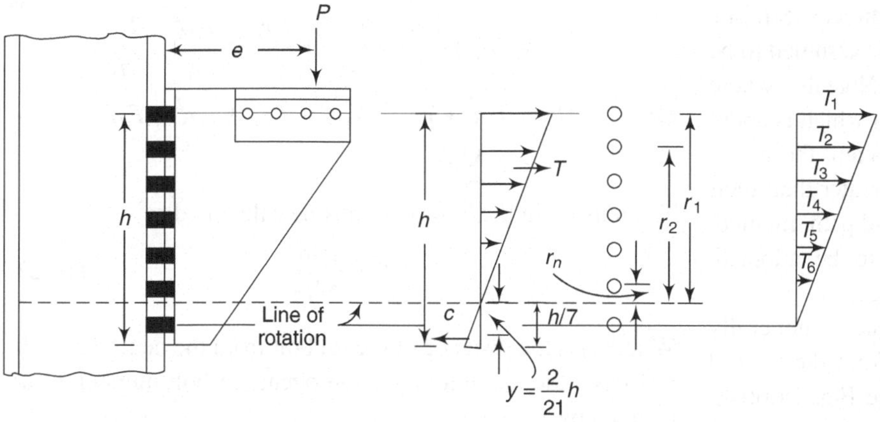


(a) Bracket (b) Distribution as per American practice (c) Distribution as per British practice

**FIG. 14.5** Bracket type II connection

Fig. 14.5), bolt groups are subject to loading eccentric to the shear plane. That is, the line action of load does not lie in the plane of the group of bolts and the line of rotation does not pass through the centre of gravity of the bolt group. Thus, the bolts are subjected to direct shear along with tension due to the moment. Extra angle seats are often provided to the bracket at the point where the reactions are transferred.

There are different approaches and assumptions to the distribution of shear between the bolts. The more conservative one is to assume all the bolts to carry equal shear. On this basis, the critical bolts are always those at the top of the connection, and those must be checked under combined shear and tension. Thus, according to this approach the shear in bolt is given by

$$V = P/n \tag{14.21}$$

where $P$ is applied load and $n$ is the total number of bolts in the bolt group. It would be more optimistic to assume that all bolts can be working at their design capacity under varying ratios of tension and shear. On this basis the residual shear capacities of all the bolts are determined, taking account of coincident bolt tension. Adequacy in shear is then checked by ensuring that the sum of those residual shear capacities is greater than applied shear. However, Owens & Cheal (1989) suggest that without matched drilling, the usual situation for general fabrication, the more conservative approach of an equal distribution of bolt shears, should be used. When HSFG bolts are used or matched drilling is ensured the latter method could be adopted. Hence, we will adopt the conservative approach only.

To estimate the amount of tension in the bolts, the position of neutral axis (line of rotation) has to be located. The bracket section below the line of rotation presses against the flange of the column and hence stiffeners are often provided to the column web opposite the lower bracket flanges, especially in connection involving end plates (see Fig. 14.6). In such cases, it is logical to assume that the compression acts at the mid depth of the stiffener as shown in Fig. 14.6 (which is referred to as 'hard spot' on the load path).

Without such a well-defined load path (hard spot), it is difficult to define the position of neutral axis, since it would be a function of stiffness of both tensile and compressive regions. In the case of end plate connections, the deformation of the column flange and end plate flexure would have significant influence on the behaviour. In British practice, the centre line of the bottom most bolt is taken as the line of rotation. However, in American practice, the neutral axis position is assumed at one-sixth to one-seventh of the depth of the connection from the bottom of the bracket as shown in Fig. 14.5 (Owens & Cheal 1989). The depth is measured from the bottom of the bracket to the topmost bolt in the connection (see Fig. 14.5). The bolt

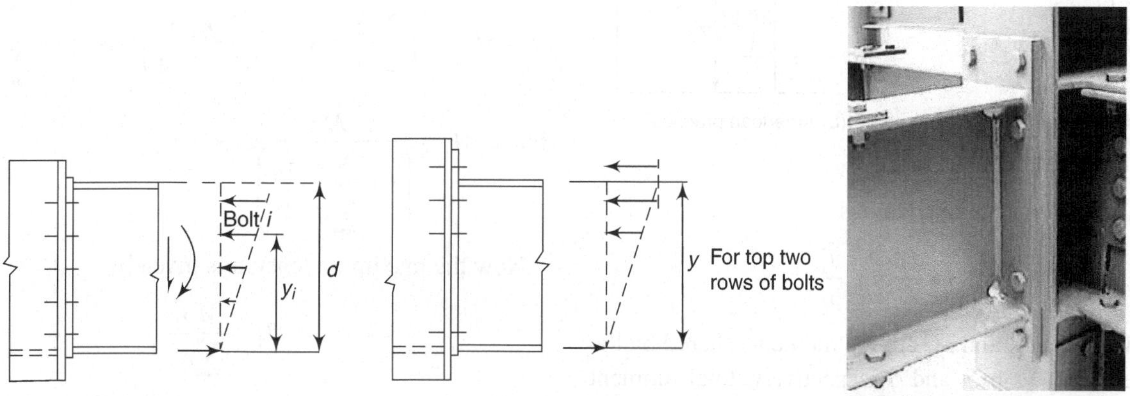


Short end plate (a) Extended end plate (b) Extended end plate connection (Note the small opening in the top and bottom of the web) (c)

**FIG. 14.6** Bolt and compressive force distribution for end-plate connection

*Source:* AISC Connections Teaching Toolkit—A Teaching Guide for Structural Steel Connections. Reprinted with permission. 

above the neutral axis will be in tension and in direct shear. The necessary compression to balance this tension is assumed to be provided by the bracket below the neutral axis. Note that where the connection is required to develop a moment which is about 80% or more than the yield moment of the beam, the above procedure will lead to flange forces that are greater than their design capacity. In such circumstances, the 'load path' method, described by Owens & Cheal (1989), has to be adopted, mobilizing the bending capacity of the web directly.

The distribution of nominal bolt tensile forces is generally assumed to be a linear function of distance from the neutral axis, where extended end plates are used [see Fig. 14.6(b)]. It has been found that the top portion of the plate behaves as a T-stub that is symmetric about the tension flange. In such cases, the forces in the top two rows of bolts may be assumed to be constant and are calculated based on the distance from the top flange centroid to the neutral axis or line of action of compressive force, as appropriate (Owens & Cheal 1989).

## *Formulae based on British and American Practices*

Let us investigate the maximum tension developed in the bolt based on the British practice as well as the American practice.

**British practice** As per British practice the bottom most bolt forms the centre of rotation (Kazimi and Jindal 1988). Let $T_1$, $T_2$, $T_3$, $T_4$, $T_5$, and $T_6$ be the tensions in the bolt number 1 to 6 from top (see Fig. 14.7a) and their corresponding distances are $r_1$, $r_2$, …, $r_6$ from the neutral axis.

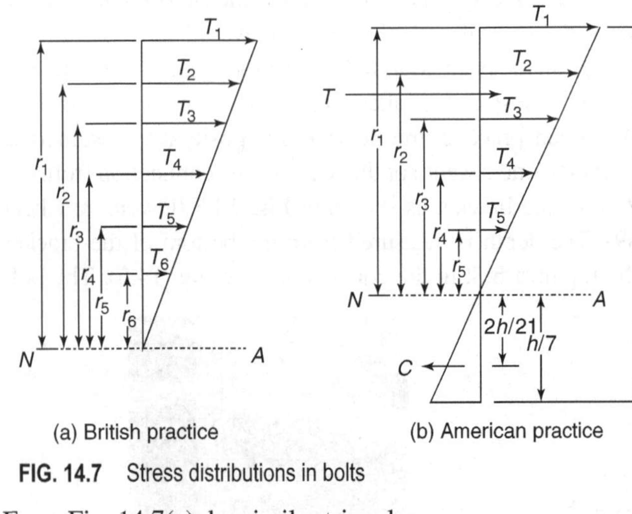


(a) British practice (b) American practice

**FIG. 14.7** Stress distributions in bolts

From Fig. 14.7(a), by similar triangles

$$\frac{T_1}{r_1} = \frac{T_2}{r_2} = \frac{T_3}{r_3} = \ldots \frac{T_6}{r_6}$$

If $M_1$, $M_2$, $M_3$, $M_4$, $M_5$, and $M_6$ are the moments shared by the bolts marked 1, 2, 3, 4, 5, and 6 respectively, total moment resisted is given by

$$M = M_1 + M_2 + M_3 + M_4 + M_5 + M_6$$

$$= T_1 r_1 + T_2 r_2 + T_3 r_3 + T_4 r_4 + T_5 r_5 + T_6 r_6$$

$$= T_1 r_1 + \frac{T_1 r_2^2}{r_1} + \frac{T_1 r_3^2}{r_1} + \frac{T_1 r_4^2}{r_1} + \frac{T_1 r_5^2}{r_1} + \frac{T_1 r_6^2}{r_1}$$

Or $$M r_1 = T_1 r_1^2 + T_1 r_2^2 + T_1 r_3^2 + T_1 r_4^2 + T_1 r_5^2 + T_1 r_6^2$$

$$= T_1 \sum r_i^2$$

The tension in the bolts can thus be calculated as

$$T_i = \frac{M r_i}{\sum r_i^2} \quad (14.22)$$

Where $r_i$ is the distance of the $i$th bolt from the neutral axis. In this case, the maximum tension occurs in bolt number 1 and is given by

$$T_1 = \frac{M r_1}{\sum r_i^2} \quad (14.22a)$$

**American practice** As per American practice, the neutral axis is assumed at a distance of $h/7$ from the bottom of the bracket, with $h$ being the depth of bracket, that is, the distance of the centre line of top bolt from the bottom (Kazimi and Jindal 1988). Let $T_1$ be the maximum tension produced in the bolt 1. Let the total moment be $M$ and the moments taken by the bolts in tension be $M'$. Let $T$ be the total tension and $C$ be the total compression in the connection.

Thus we have (see Fig. 14.7b)

$$T_1 = \frac{M' r_1}{\sum r_i^2}, \; T_2 = \frac{M' r_2}{\sum r_i^2}, \text{ and so on.}$$

$$T = T_1 + T_2 + \ldots + T_5 = \frac{M'(r_1 + r_2 + r_3 + r_4 + r_5)}{\sum r_i^2} = \frac{M' \sum r_i}{\sum r_i^2}$$

Hence $$T = C = \frac{M' \sum r_i}{\sum r_i^2}$$

Moment of compressive force $C$ about neutral axis $= C\left(\frac{2h}{21}\right)$

$$M = M' + C\left(\frac{2h}{21}\right) = M' + \frac{M' \sum r_i}{\sum r_i^2}\left(\frac{2h}{21}\right) = M'\left(1 + \frac{\sum r_i}{\sum r_i^2}\left(\frac{2h}{21}\right)\right)$$

Hence $$M' = \frac{M}{\left(1 + \frac{\sum r_i}{\sum r_i^2}\frac{2h}{21}\right)} \quad (14.23)$$

Now the maximum tension is given by

$$T_1 = \frac{M' r_1}{\sum r_i^2} \quad (14.24)$$

where $M'$ is given by Eqn (14.23).

Note that this analysis only determines the nominal bolt tension and hence appropriate allowance should be made for prying action, if any.

The design of the bracket connection as shown in Fig. 14.5 involves the following steps:

1. Assume the diameter, pitch, and edge distance for bolts
2. Compute the design strength of the bolt in shear and tension
3. Compute the required number of bolts using Eqn (14.16)
4. Compute the shear force $V$ in the extreme critical bolt
5. Calculate the tensile force $T_i$ in the extreme critical bolt using Eqn (14.22) or Eqn (14.24)
6. Check the connection for combined tension and shear by using Eqn (5.20) of Chapter 5

that is,

$$(V/V_{sd})^2 + (T_e/T_{nd})^2 \leq 1.0.$$

The various steps involved in the design of a bracket (Type II) connections are shown in Examples 14.6 and 14.7.

## 14.2 BOLTED MOMENT END PLATE CONNECTION

A bolted moment end plate connection (Fig. 14.8) is used to transfer moment, axial force, and shear force from one member to another. The end plate is fillet welded to the web and flange of beam and bolted to the column flange or web. Due to the concentration of forces in the beam, the web of columns may require additional strengthening in the form of web stiffeners, diagonal stiffeners, or web plate (see Fig.14.8).

The end plates may be extended beyond the face of the beam to carry more moments [see Fig. 14.8(b)]. In the case of deep beams connected to slender columns, haunched connections as shown in Fig. 14.8(c) may be adopted. As discussed in the previous section, by using the extended end plate connections, it is possible to transfer about 80% of the yield moment capacity of the beam. However, this connection is less stiff than welded moment connection, due to the flexure of end plate. However, they provide ease of use, since the end plates may be welded to the beams in shop and bolted at site to the column. They provide economic alternatives to connections using angle or T-sections, which require large number of bolts. (Note: The cost of bolts are expensive than steel sections.)

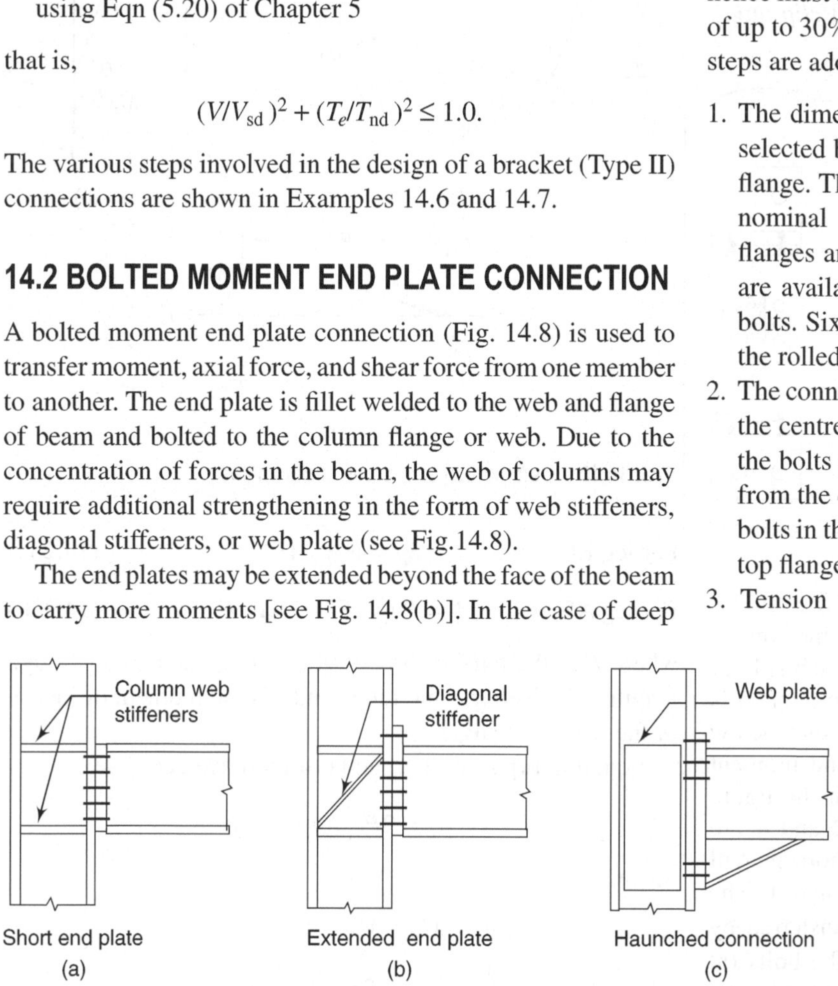


**FIG. 14.8** Rigid beam-to-column connections

In these connections, the bending moment, axial force, and shear are transferred by tension and compression or shear through the flange welds and by shear through the web welds to the end plate. Then they are transferred from the end plates to the bolts by bending and shear. Finally, they are transferred to the supporting member by bolt shear and tension and also by horizontal plate reaction (Trahair et al. 2001).

Due to the bending of the end plate, the bolts in the extended region of the end plate are subjected to prying actions, and hence must be considered on the design. Increases in bolt load of up to 30% have been observed. In the design the following steps are adopted.

1. The dimensions (length and breadth) of the end plate are selected based on the dimensions of the beam and column flange. The layout of the bolts are also fixed, by providing nominal minimum distance (of 50–60 mm) from the flanges and web of the beam, so that sufficient clearances are available for the use of power tools to tighten HSFG bolts. Six bolts are sufficient to carry the loads in most of the rolled beams.
2. The connection is assumed to 'pivot' about the 'hot spot' (at the centre of the bottom flange thickness) and the loads in the bolts are assumed to be proportional to their distances from the centre of the bottom flange. It is assumed that the bolts in the cantilever end of the plate and the bolts near the top flange of beam carry equal loads.
3. Tension and shear capacity of the bolt are computed. An approximate value of the prying force is assumed: $Q$ = tension capacity – bolt load. Using the equations for calculating the prying forces (Section 5.4.4 of Chapter 5), the moment at the toe of the weld is calculated. Equating it with the moment capacity of the plate, the thickness of end plate may be obtained. Using this thickness, the prying force is computed and checked with the assumed prying force. If it is more than the assumed prying force, the thickness has to be changed. For ductility, the thickness of the end-plate should be less than half the bolt diameter for grade 8.8 bolts and one-third bolt diameter for grade 4.6 bolts (Narayanan, et al. 2001)
4. A check is made for combined shear and tension in the bolt.
5. The capacity of the end plate to carry the reaction due to bolt forces has to be checked.
6. The welds connecting the beam to end plate are to be checked.
7. The capacity of column web to support the flange reaction has to be checked. If the column web is not able to support the reaction, the web has to be strengthened by means of web plates, web stiffeners, etc.

Example 14.7 shows the calculations involved in the preceding steps.

## 14.3 FLANGE ANGLE CONNECTIONS AND SPLIT-BEAM T-STUB CONNECTIONS

If we use an additional pair of angles in the angle seat connections to connect the web of the beam to the flange of the column, as shown in Fig. 14.9 the connection can be designed to transfer small end moments in addition to large end shears. This kind of connection is also called *clip angle connection* or *light moment connection.*

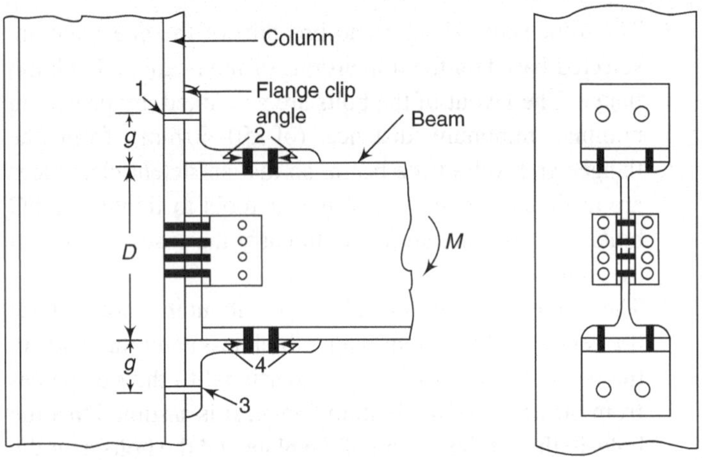


**FIG. 14.9** Flange-angle connection

As shown in Fig. 14.9, this connection consists of four angles (two pairs). One pair of angle is used to connect the web of the beam with the column flange (one angle on each side of the web) and the other pair of angles (called *clip angles*) is placed one on top of the beam flange and the other below the bottom flange of the beam. If a clockwise end moment acts on the connection, the bolts marked as 1 in the figure are subjected to tension and those marked as 2 and 4 are subjected to shear. Similarly, an anticlockwise moment will produce tension in bolts 3 and shear in bolts 2 and 4. The end shear force in the beam is assumed to be transferred by the web angle connection. The force acting on the bolts (at the top and bottom clip angles) is the moment divided by the lever arm of the bolts [$M/(D + 2g)$], where $g$ is the length as shown in Fig. 14.9. In the simple beam column connection designed for shear only, the connecting angles at the bottom and top were flexible (so that they can deform sufficiently) and hence permit the rotation necessary for simple support condition at the ends of the beam. Unless this condition is fulfilled, end moments will be induced, which will overstress the connecting bolts. On the other hand, if the connection is considered rigid, the deformation of the connecting angle should be nil or very small. Hence, it is necessary to analyse the forces and deformation in the top and bottom angles of clip-angle connection. The flange angles are deformed due to eccentric forces and moments. The thickness of the angles should be adequate to resist the internal moments imposed on them, as shown here.

Consider the free body diagram of legs AB and BC (Fig. 14.10).

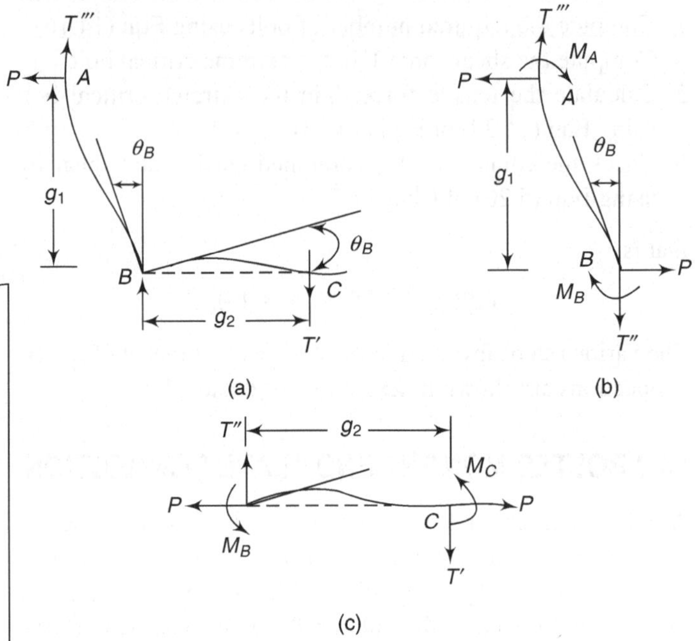


**FIG. 14.10** Deformation and forces in clip angles

For leg BC, $\theta_B = M_B g_2/4EI$ (14.25a)

For leg AB, $\theta_B = Pg_1^2/2EI - M_B g_1/EI$ (14.25b)

where $P$ is the tensile force in the bolt, $g_1$ and $g_2$ are gauge lengths, $E$ is Young's modulus, and $I$ is the moment of inertia of the leg of the angle.

Equating Eqns (14.25a) and (14.25b), we get

$$M_B = 2Pg_1^2/(4g_1 + g_2)$$

Since,

$$Pg_1 = M_B + M_A$$

$$M_A = Pg_1\,(2g_1 + g_2)/(4g_1 + g_2)$$

If we consider an equal angle, $g_1 = g_2 = g$, we get

$$M_A = 0.6Pg \quad (14.26a)$$

$$M_B = 0.4Pg \quad (14.26b)$$

$$M_C = 0.2\,Pg \quad (14.26c)$$

Hence, the maximum moment in leg AB is $0.6Pg$, where $g$ is the gauge length for the angle if only two bolts are used in the connecting leg of the column. Thus, the angle should be designed to have sufficient thickness to resist the maximum moment.

The deflection is given by Arya & Ajmani (1989)

$$\Delta = 0.133Pg^3/EI \quad (14.27)$$

Example 14.8 shows the calculations required for the design of a light moment connection using clip and flange angles.

Instead of an angle, a T-stub cut from an I-section can be used to clip the top flange of an I-beam as shown in Fig. 14.11, the deformation of the T is symmetrical with respect to point *B* so that the tangent at *B* remains vertical [Fig. 14.11(b)]. Then if *P* is the tensile force in the bolt, the moments in the T will be

$$M_A = M_B = 0.5Pg \tag{14.28}$$

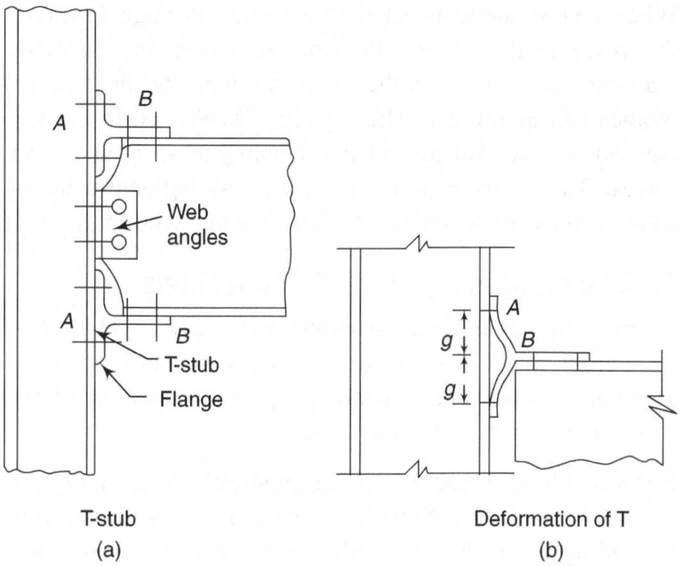


**FIG. 14.11** T-stub connection

In *split-beam T-connections* (also called *T-stub connection*), the prying action should be considered and the bolts connecting the T with the column flange should be capable of resisting the additional tension due to prying action.

Clip angle connection has the following limitations: (a) Only two bolts can be provided in one gauge line to connect the outstanding leg with the column flange. If more bolts are provided in one row (gauge line), the bolts will be subjected to non-uniform forces [see Fig. 14.12(a)] and (b). If more than one gauge line is provided, then the bolts will be subjected to non-uniform forces [see Fig. 14.12(b)]. Due to these, the number of bolts to connect the clip angle to the column flange is limited to two only and hence the connection can transfer only limited moments. Example 14.10 shows the design of a light moment connection using a T-stub and flange angles.

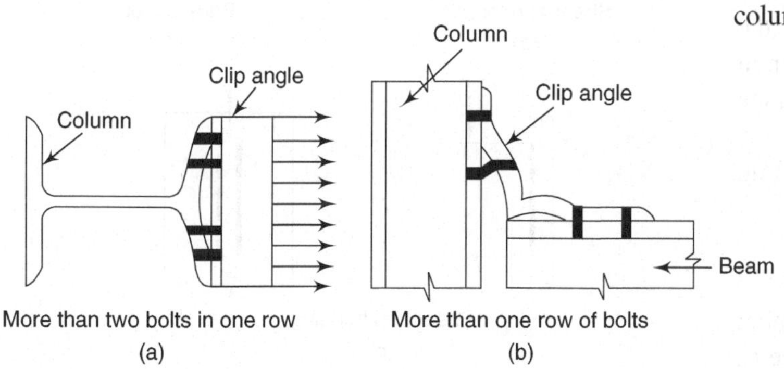


**FIG. 14.12** Non-uniform forces in bolts

**Bracket connection** If the lever arm is to be extended to accommodate more bolts, a bracket type connection is sometimes used (Fig. 14.13). These bracket type connections are more rigid compared to any type of moment connection discussed so far. But the fabrication cost is very high and hence they are not adopted in practice, except in some special cases. The connection shown in Fig. 14.13 is made up of a bracket plate and three pairs of angles. Two pairs (one pair on each flange) are connected to the beam flanges and with the bracket plate. The third pair is kept continuous across the beam and is connected to the flange of the column and to the bracket plate (see Fig. 14.13). The pitch of the bolts connecting the column flange with the bracket plate cannot be kept uniform, as bolts cannot be provided opposite the beam flange. To facilitate fabrication, the thickness of the bracket plate is kept equal to the thickness of the beam web. The column is provided with stiffeners opposite the beam flanges.

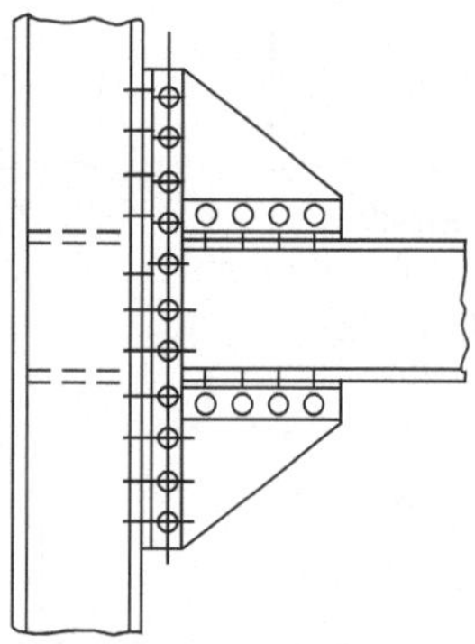

**FIG. 14.13** Bracket type connection

In all these connections, the beam reaction is transferred by shear and bearing from the web to the web bolts and to the web angle cleats. These actions are transferred by the cleats to the bolts connected to the column and to the column by shear and bearing. The moments are transferred by shear and bearing to the bolts and to the connecting beam flanges and the clip angles or T-stubs. From these elements, they are transferred by bending of the clip angle or T-stub and tension in the bolts connected to column flanges. From the bolts they are transferred to the column by horizontal forces in the connectors.

## 14.4 PLATE SUBJECTED TO HORIZONTAL PULL

No recommendations are given in IS 800 for column flange subjected to a single bolt load as shown in Fig. 14.4. Assuming 45° dispersion from the edge of the bolt hole and applying simple theory of bending, Martin & Purkins (1992) derived the following equation for the thickness of plate [see Fig. 14.14(a)].

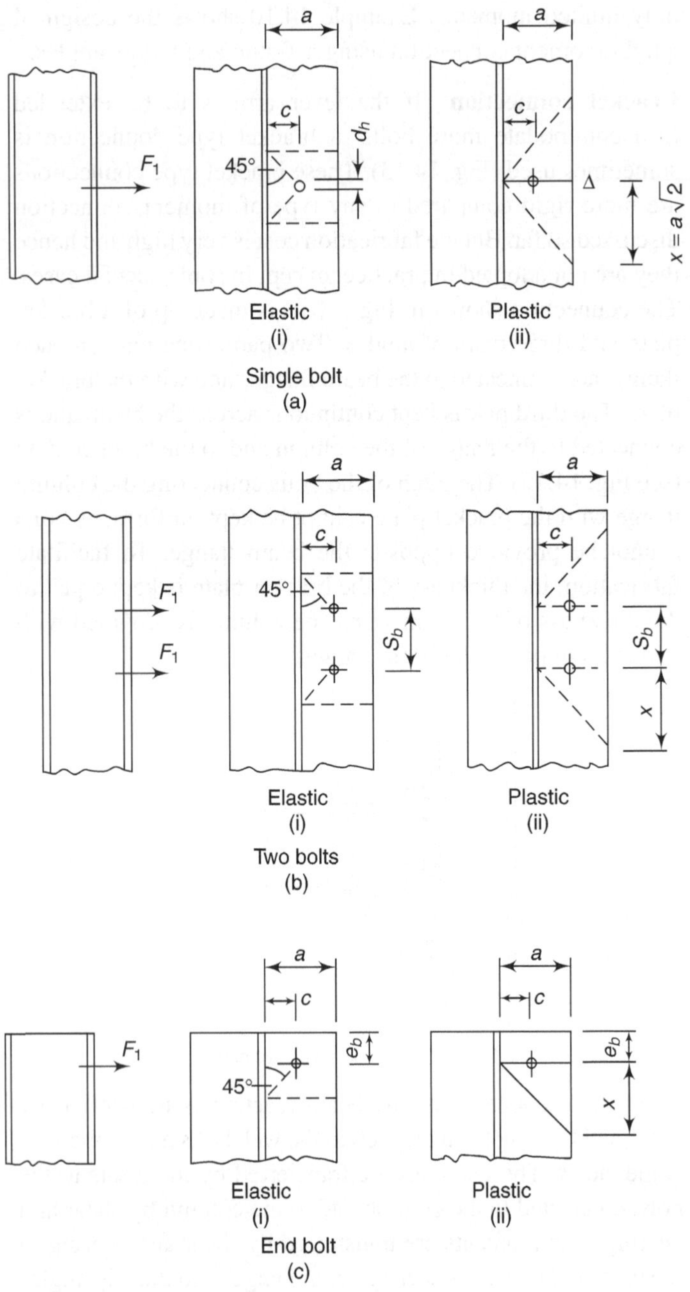


**FIG. 14.14** Concentrated bolt force in plates

$$t_p = \sqrt{T_b c/(\sqrt{2} a f_y)} \qquad (14.29)$$

where $T_b$ is the tensile force in bolt, $a$ is the length of the outstand of column flange, $c$ is the edge distance as shown in Fig. 14.14(a), and $f_y$ is the yield stress of the column flange plate.

Similarly for two bolts, where $s_b < d_h + 2c$ as shown in Fig. 14.14(b)

$$t_p = \sqrt{[4nT_b c/(af_y)]/[4\sqrt{2} + (n-1)s_b/a]} \qquad (14.30)$$

where $T_b$, $a$, $c$, and $f_y$ are as defined earlier, $n$ is the number of bolts in line and $s_b$ is the spacing of the bolts. To reduce $t_p$, reduce $T_b$ and make $s_b/a$ as large as is practical.

For a single bolt near the end of a member as shown in Fig. 14.14(c),

$$t_p = \sqrt{[\{4nT_b c/(af_y)\}/(2 + e_b/a)]} \qquad (14.31)$$

where $e_b$ is the edge distance.

## 14.5 MOMENT-RESISTANT WELDED CONNECTIONS

When beams are connected to columns through brackets, depending on the way in which they are connected, the welds may be subjected to either twisting moment or bending moment, in addition to shear forces. Welded stiffened seat connections are also subjected to bending moment and shear forces. These connections are termed as moment-resistant welded connections and are discussed in this section.

### 14.5.1 Eccentrically Loaded Connections

Loads acting eccentrically from the centroid of a weld line or weld group may cause either a twisting moment or a bending moment on the weld, depending upon the location of the welds, in addition to the direct shear forces.

**Eccentric load causing twisting moment** Since no initial tension is involved with welded connections, the eccentricity of loading, even though small, has to be considered. Also, there are situations where the loading of fillet welds is neither parallel not transverse to the axis of fillet welds, as shown in Fig. 14.15. Analysis of such eccentric loading is complicated since the load–deformation behaviour is a function of the angle q between the direction of applied load and axis of the fillet weld (see Fig. 6.39 of Chapter 6).

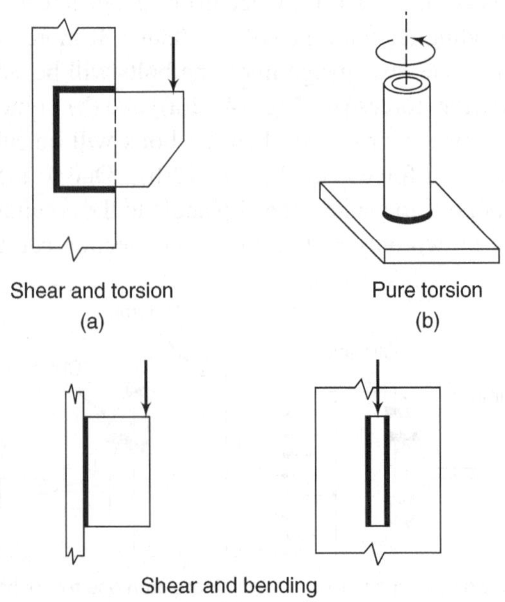


**FIG. 14.15** Types of eccentric loading

As discussed in Section 14.1.1, the strength of an eccentrically loaded fillet weld can also be determined by locating an *instantaneous centre of rotation*, using the load–deformation relationship of the fillet weld. The resistance $R_i$ of the weld segment at any distance from the instantaneous centre is proportional to such distance and acts in a direction perpendicular to the radial distance to the segment as shown in Fig. 14.16 (Salmon, et al. 2008). The more rational strength analysis, using a trial and error procedure (suitable for computer application) developed by Butler et al. (1972) and Lesik and Kennedy (1990) and adopted by AISC code is described by Salmon, et al. 2008. This procedure is based on experimental results conducted in Canada.

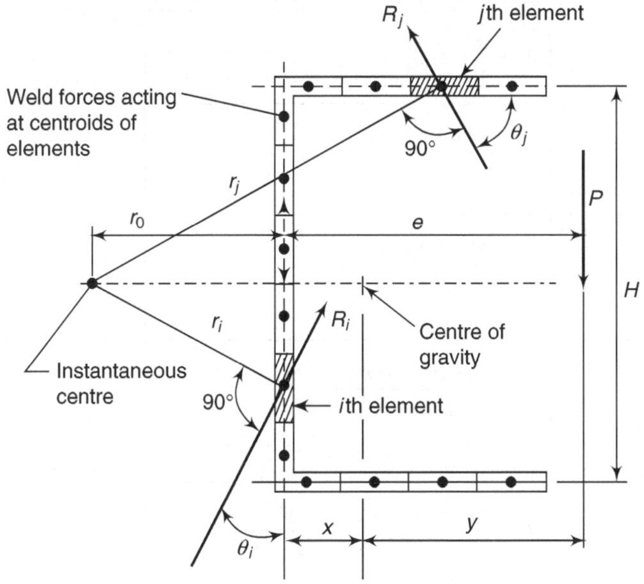


**FIG. 14.16** Principles of instantaneous centre method

Here the more conservative traditional elastic vector analysis which is easier than the strength method is described. The following assumptions are made in the elastic method.

1. Each segment of weld (of the same size) resists the concentrically applied load with an equal force.
2. The rotation caused by the torsional moment is assumed to occur about the centroid of weld configuration.
3. The load on a weld segment caused by the torsional moment is assumed to be proportional to the distance from the centroid of the weld configuration.
4. The components of the forces caused by the direct load and by torsion are combined vectorially to obtain a resultant force.

For a general case as shown in Fig. 14.17, the components of stress caused by direct stress are

$$f_x' = P_x/A \tag{14.32}$$

$$f_y' = P_y/A \tag{14.33}$$

The $x$ and $y$ components of stress resulting from torsion are

$$f_x'' = Ty/I_p = (P_x e_y + P_y e_x)y/I_p \tag{14.34}$$

$$f_y'' = Tx/I_p = (P_x e_y + P_y e_x)x/I_p \tag{14.35}$$

where $I_p$ is the polar moment of inertia

$$= I_x + I_y = \sum I_{xx} + \sum A\bar{y}^2 + \sum I_{yy} + \sum A\bar{x}^2 \tag{14.36}$$

In Eqn (14.36) $\bar{x}$ and $\bar{y}$ are the distances from the centre of gravity of the weld group to the centre of gravity of individual weld segments. $I_{xx}$ and $I_{yy}$ are the moments of inertia of the individual segments with respect to their own centroidal axes.

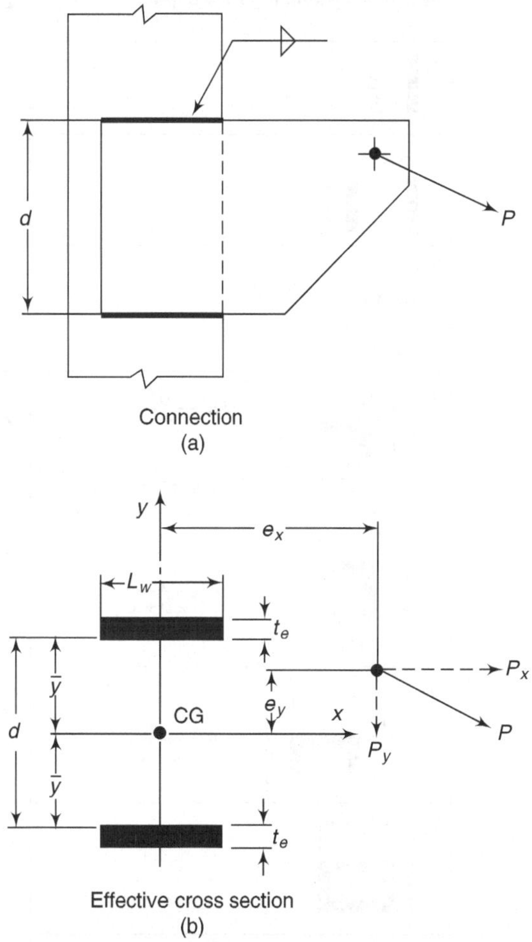


**FIG. 14.17** Eccentric bracket connection

For the weld configuration shown in Fig. 14.17 (b)

$$I_p = 2(L_w t_e^3/12) + 2[L_w t_e \bar{y}^2] + 2[t_e L_w^3/12]$$

$$= (t_e/6)[L_w t_e^2 + 12 L_w \bar{y}^2 + L_w^3] \tag{14.37a}$$

The first term is neglected since $t_e$ will be small compared to $L_w$. Hence, we get

$$I_p = (t_e/6)\,[12 L_w \bar{y}^2 + L_w^3] \tag{14.37b}$$

Treating the welds as line welds (i.e. with $t_e = 1$), and using the general terms $b$ and $d$, Blodgett (1966) has given the properties (section modulus and polar moment of inertia) for various weld configurations. They are shown in Table 14.1. Using this table, the value of $I_p$ can be written as

$$I_p = L_w(3d^2 + L_w^2)/6 \qquad (14.38)$$

The steps involved in checking the adequacy of the weld are as follows.

1. The centroid of the weld line is calculated. The twisting moment and the forces at the centroid are determined

$$T = P_x e_x + P_y e_y \qquad (14.39)$$

where $P_x$ and $P_y$ are the $x$ and $y$ components of the eccentric load and $e_x$ and $e_y$ are the eccentricities of $P_x$ and $P_y$ with respect to the centroid of weld line.
2. The critical weld points are located.
3. The force components due to twisting moment and maximum shear force for critical weld point are determined as

$$F_x^T = Ty/I_p \text{ and } F_y^T = Tx/I_p \qquad (14.40)$$

**TABLE 14.1** Properties of welds treated as lines

| Section $b$ = width; $d = Lw$ = depth | Section modulus $I_x/\bar{y}$ | Polar moment of interia, $I_p$ about centre of gravity |
|---|---|---|
| 1. 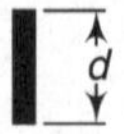 | $Z = \frac{d^2}{6}$ | $I_p = \frac{d^3}{12}$ |
| 2. 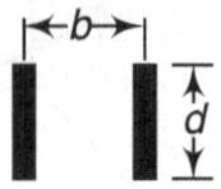 | $Z = \frac{d^2}{3}$ | $I_p = \frac{d(3b^2 + d^2)}{6}$ |
| 3. 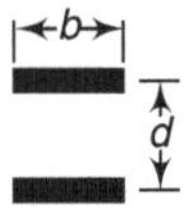 | $Z = bd$ | $I_p = \frac{b(3d^2 + b^2)}{6}$ |
| 4. 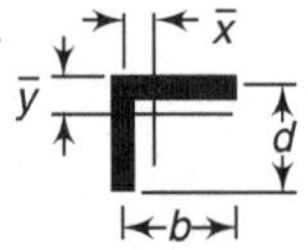 $\bar{y} = \frac{d^2}{2(b+d)}$ $\bar{x} = \frac{b^2}{2(b+d)}$ | $Z = \frac{4bd + d^2}{6}$ | $I_p = \frac{(b+d)^4 - 6b^2d^2}{12(b+d)}$ |
| 5. 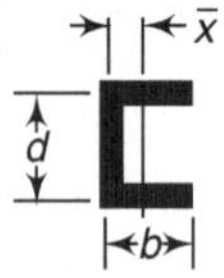 $\bar{x} = \frac{b^2}{2b+d}$ | $Z = bd + \frac{d^2}{6}$ | $I_p = \frac{8b^3 + 6bd^2 + d^3}{12} - \frac{b^4}{2b+d}$ |
| 6. 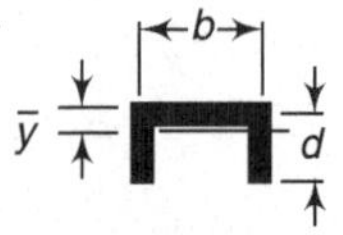 $\bar{y} = \frac{d^2}{b+2d}$ | $Z = \frac{2bd + d^2}{3}$ | $I_p = \frac{b^3 + 6b^2d + 8d^3}{12} - \frac{d^4}{2d+b}$ |
| 7. 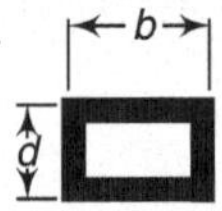 | $Z = bd + \frac{d^2}{3}$ | $I_p = \frac{(b+d)^3}{6}$ |
| 8. 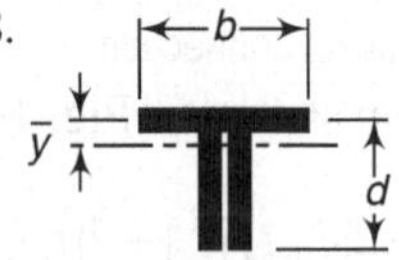 $\bar{y} = \frac{d^2}{b+2d}$ | $Z = \frac{2bd + d^2}{3}$ | $I_p = \frac{b^3 + 8d^3}{12} - \frac{d^4}{b+2d}$ |
| 9. 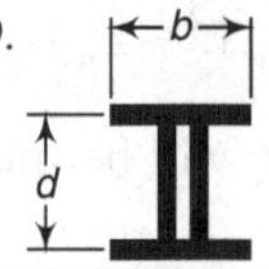 | $Z = bd + \frac{d^2}{3}$ | $I_p = \frac{b^3 + 3bd^2 + d^3}{6}$ |
| 10. 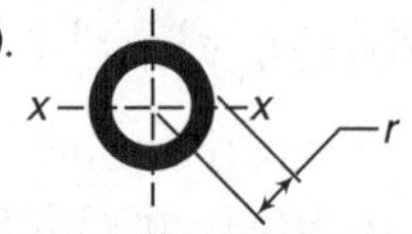 | $Z = \pi r^2$ | $I_p = 2\pi r^3$ |

where $x$ and $y$ are the coordinates of critical weld point and $I_p$ is the polar moment of inertia of the weld line about the centroid and

$$F_x^P = P_x/L_w \text{ and } F_y^P = P_y/L_w \quad (14.41)$$

where $L_w$ is the total length of the weld.

4. The resultant shear force is calculated as

$$F_R = [(F_x^P + F_x^T)^2 + (F_y^P + F_y^T)^2]^{0.5} \quad (14.42)$$

5. The maximum shear force should be less than the capacity of weld

$$F_R < R_w \text{ (weld strength)} \quad (14.43)$$

## 14.5.2 Eccentric Load Causing Bending Moment

When the applied load is eccentric to the plane of the weld configuration, as shown in Fig. 14.18, the strength method of analysis (by locating the instantaneous centre of rotation) can be used (Dawe & Kulak 1974). However, we will consider only the elastic (vector) analysis which is conservative and relatively easy to use for loading resulting in shear and tension.

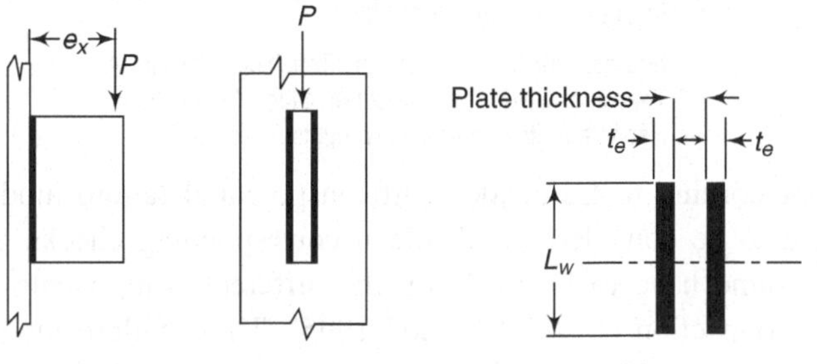


**FIG. 14.18** Loads applied eccentric to the plane of weld

The effect of eccentric load on the weld group about its centroid is equivalent to a bending moment and direct force at the centroid. The bending moments ($Pe$) cause bending tensile and compressive stresses (i.e. normal stresses) at the throat section of the fillet weld, while the direct forces cause direct or shear stresses. Either a groove weld or a fillet weld can be used in such connections. Thus, the welds must carry the loads in the same manner as the members being connected carry them. The stresses are shown in Fig. 14.19. Thus, the direct stresses in the weld = load/effective area of weld

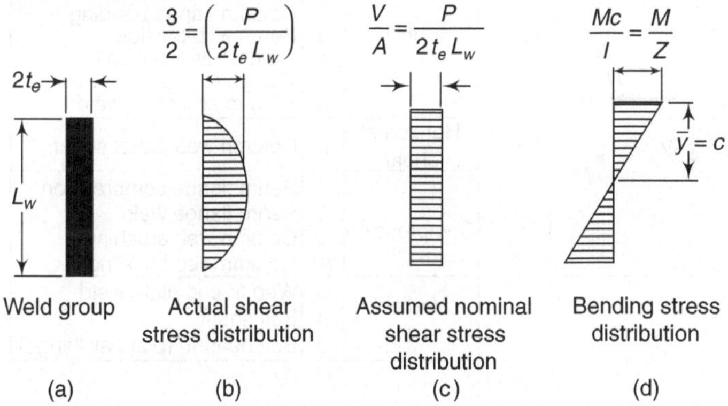


**FIG. 14.19** Stresses in weld subject to eccentric load causing bending moment

(a) In the case of a fillet weld

$$\tau_{vf,\,cal} = P/(2L_w t_e) \quad (14.44)$$

(b) For groove welds

$$\tau_{vf,\,cal} = P/dt \quad (14.45)$$

where $t$ is the thickness of plate.

The bending stress in the weld = moment/section modules

1. For fillet weld the failure is due to critical stress in the throat of the weld (at 45° to the weld leg length)

   Hence,

   $$f_b = M/Z = [Pe(L_w/2)]/[2 \times (L_w^3 t_e/12)] = 3Pe/(t_e L_w^2) \quad (14.46)$$

   The throat stress is treated as shear since a 45° line of failure is assumed. This shear stress is assumed linearly varying from zero at mid depth to the maximum value at the extreme fibres. Hence,

   $$\tau_{vf1,\,cal} = 3Pe/(t_e L_w^2) \quad (14.47)$$

2. For groove welds

   $$f_b = 6Pe/(tL_w^2) \quad (14.48)$$

   The combined stress in the fillet weld is given by the following equation

   $$f_e = \sqrt{(\tau_{vf,\,cal}^2 + \tau_{vfl,\,cal}^2)} < \text{weld strength} = f_u/(\sqrt{3}\gamma_{mw}) \quad (14.49)$$

   The combined bending and shear stress in the groove weld is checked by using the interaction formula (clause 10.5.10.1.1 of code)

   $$f_e = \sqrt{(f_{b,cal}^2 + 3\tau^2)} \quad (14.50)$$

   and $f_e$ should not exceed the values allowed for the parent metal. The code also states that the check for the combination of stresses need not be done:

   (a) for side fillet welds joining cover plates and flange plates and
   (b) for fillet welds, where sum of normal and shear stresses does not exceed

   $$f_{wd} = f_u/(\sqrt{3}\gamma_{mw}) \quad (14.51)$$

   Similarly a check for the combination of stresses in groove welds need not be done if:

   1. groove welds are axially loaded and
   2. in single and double bevel welds, where the sum of normal and shear stresses does not exceed the design normal stress and the shear stress does not exceed 50% of the design shear stress.

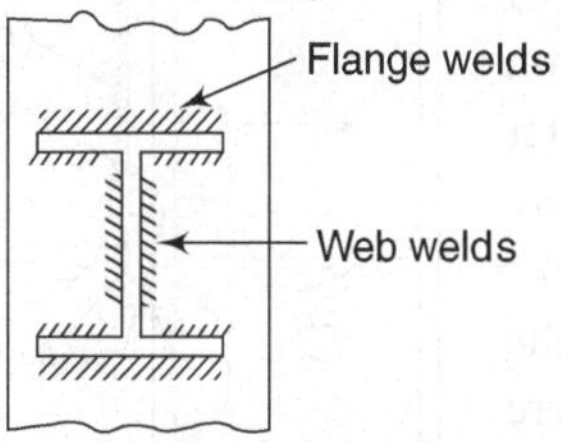


**FIG. 14.20** Welding of brackets for carrying shear and moment

Note that the locations of maximum bending and shearing stresses are not the same (see

Fig. 14.19). Hence, if the welds are used as shown in Fig. 14.20, it can be safely assumed that the web welds would carry the entire shear force and the flange welds would carry the entire bending moment.

### Design of Brackets

The design of brackets subject to combined bending and shear is done using the following steps.

1. Assume the size of the weld and compute the throat thickness, design strength, and capacity of weld ($R_{nw}$).
2. Calculate the depth of bracket (length of weld) using the following appropriate equations.
   (a) In the case of groove welds

$$d = [6M/(tf_b)]^{1/2} \tag{14.52}$$

   Where $f_b = f_y/\gamma_{m0}$ with $\gamma_{m0} = 1.10$

   (b) In the case of fillet welds

$$d = [6M/\{2t_e R_{nw}\text{ (est.)}\}]^{1/2} \tag{14.53}$$

   A reduced value of $R_{nw}$ is used to account for the direct shear effect also.
3. The direct shear stress is computed using Eqn (14.44) or (14.45), as appropriate.
4. The stress due to bending moment is computed from Eqn (14.46) or (14.48), as appropriate.
5. The equivalent stress is computed from Eqn (14.49) or (14.50), as appropriate.
6. If the equivalent stress exceeds the weld strength (fillet welds) or the design stress of the parent metal (groove weld), the length of the bracket (weld length) may be increased and the process repeated till the checks are satisfied.

## 14.6 WELDED END PLATE CONNECTIONS

End plate connection has been discussed in Section 14.2. The end plates are shop welded to the beam and connected to column flanges/web by means of HSFG bolts. The flange of the beam can be groove welded or fillet welded to the plate. The web will usually be fillet welded. A conservative approach to end plate connection design is to use the prying action concept discussed in Section 5.4.4 of Chapter 5. The region near the tension flange of the beam is designed similar to that of a split-beam T-connection. This fastener group is designed for shear and tension, including the effect of prying action. The welds are designed for the resultant force using the elastic vector analysis as below.

$$\text{Resultant force} = \sqrt{(P/A)^2 + (My/I)^2} \le \text{design strength of the weld} \tag{14.54}$$

where $P/A$ is the vertical stress due to shear force and $My/I$ is the tension component (horizontal) due to the bending moment. Several researches have studied the behaviour of end plate connections (e.g., Murray 1990; Krishnamurthy 1978).

End plate shear connection; the plate in this connection is welded to the web only and bolted to the other member at site

*Source:* AISC Connections Teaching Toolkit—A Teaching Guide for Structural Steel Connections. Reprinted with permission. All rights reserved.

According to Eurocode 3, fifteen potential failure modes have to be considered and fifteen corresponding checks for the same have to be made on the different components of the connection (BCSA/SCI 207 1995). These failure modes and corresponding checks are shown in Fig. 14.21. Most of these checks are straightforward, but the determination of the tension force in each bolt row may pose some challenges as it is linked to the ability of adjacent column flange and beam end plate on each side of the connection to transmit the bolt force. As discussed in Section 14.2, the bolts adjacent to the beam flange are considered to contribute much to the moment capacity of the connection and the bottom rows of bolts

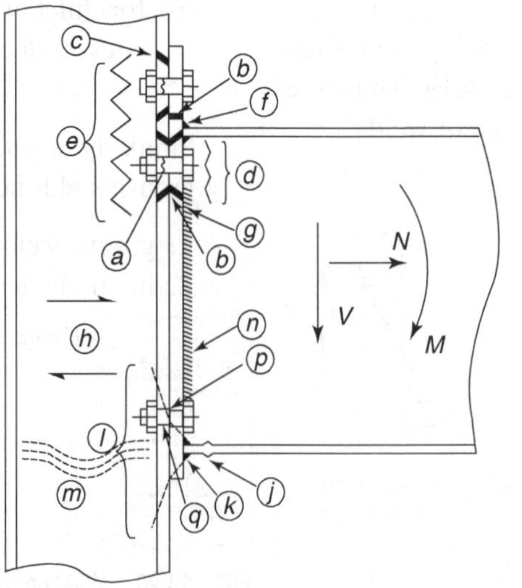


| Zone | Ref | Checklist item |
|---|---|---|
| Tension | a | Bolt tension |
| | b | End plate bending |
| | c | Column flange bending |
| | d | Beam web tension |
| | e | Column web tension |
| | f | Flange to end plate weld |
| | g | Web to end plate weld |
| Horizontal shear | h | Column web panel shear |
| Compression | j | Beam flange compression |
| | k | Beam flange weld |
| | l | Column web crushing |
| | m | Column web buckling |
| Vertical shear | n | Web to end plate weld |
| | p | Bolt shear |
| | q | Bolt bearing (plate ar flange) |

**FIG. 14.21** Potential failure modes and the corresponding design checks

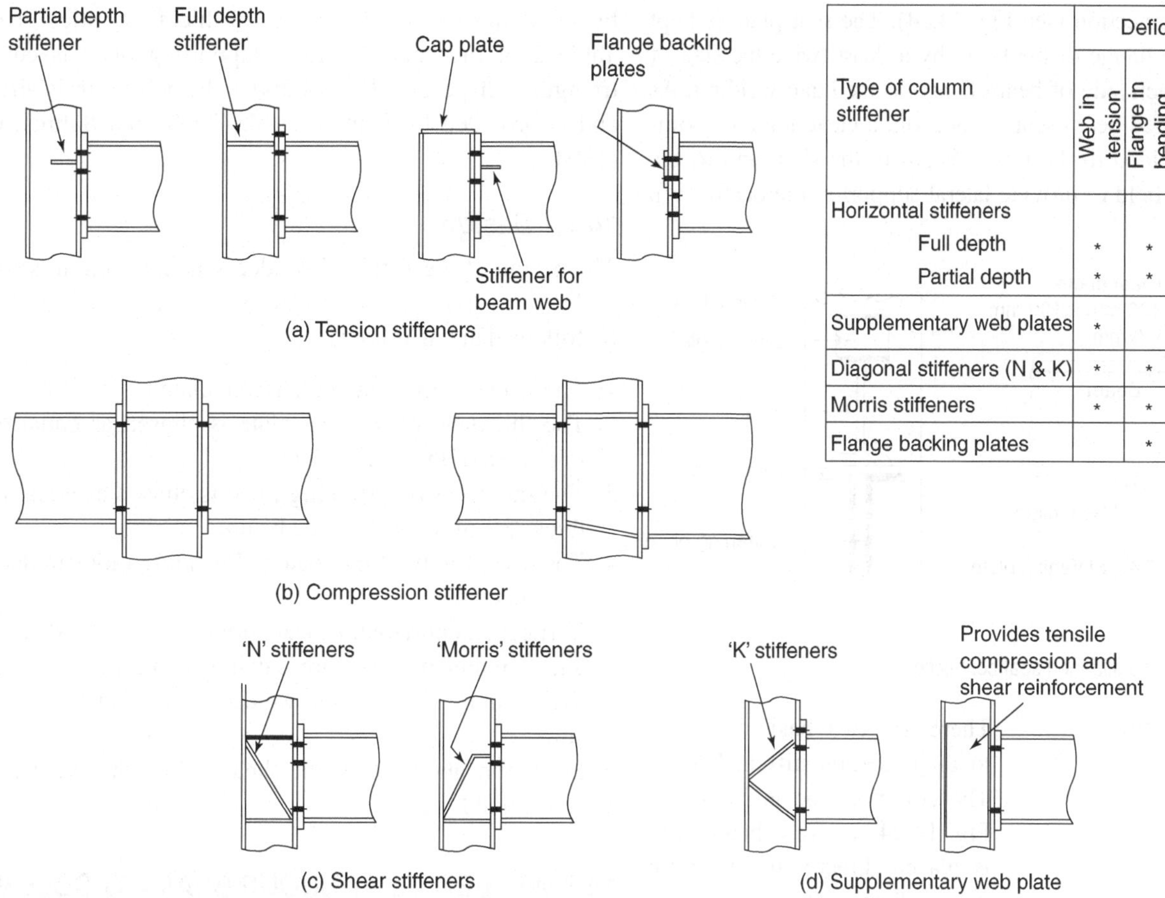


| Type of column stiffener | Deficiency | | | |
|---|---|---|---|---|
| | Web in tension | Flange in bending | Web in compression | Web in shear |
| Horizontal stiffeners | | | | |
| Full depth | * | * | * | |
| Partial depth | * | * | * | |
| Supplementary web plates | * | | * | * |
| Diagonal stiffeners (N & K) | * | * | | * |
| Morris stiffeners | * | * | | * |
| Flange backing plates | | * | | |

**FIG. 14.22** Methods of strengthening

are considered to transmit only vertical shear forces. When determining bolt row resistances the concept of *equivalent tee-stub* is used to deal with the interaction between bolts and the adjacent plating. In its full form this method is only suitable for implementation using computer software and a few standard packages also exist. The interested readers may consult BCSA/SCI 207, 1995 and BCSA/SCI P398, 2013, where a simpler approximate version of the method is presented.

While checking the components, if it is found that some components are deficient, they can be strengthened by several means. These deficiencies and the corresponding methods of strengthening are shown in Fig. 14.22 (BCSA/SCI P398, 2013). In situations where the moment at the joint exceeds the capacity of the beam section, a haunched connection may be used and a common example of such connection may be found at the eaves of portal frames.

## 14.7 WELDED STIFFENED BEAM SEAT CONNECTION

A welded stiffened seat connection for a beam is much simpler than the one with a bolted connection. It consists of two plates forming a T or a split I-section used as a seat (see Fig. 14.23). The thickness of the stem of the T should not be less than the web thickness of the beam it supports. Similarly, the thickness of seat plate should not be less than the thickness of the flange of the beam. The length of bearing is governed by the strength as well as by the web crippling requirement of the beam (as in the case of unstiffened seat angle). The depth of the stem should be short enough to avoid local buckling and it depends upon the length of the vertical weld required. The under side of the flange of the T is also welded to increase the torsional stiff-

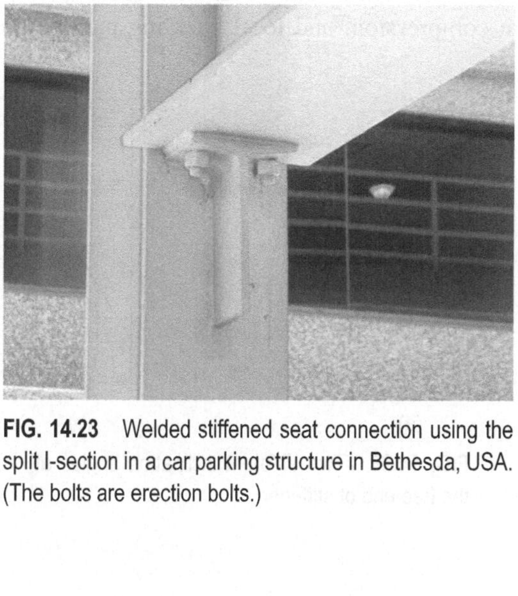

**FIG. 14.23** Welded stiffened seat connection using the split I-section in a car parking structure in Bethesda, USA. (The bolts are erection bolts.)

ness of the connection (see Fig. 14.24). The seat plate is kept wider than the flange of the beam by at least twice the size of the weld on each side of beam flange to facilitate welding. As in the unstiffened beam seat connection, a cleat angle of nominal size is welded to the top of beam in the shop and to the column at the field to provide lateral support to the beam's top flange.

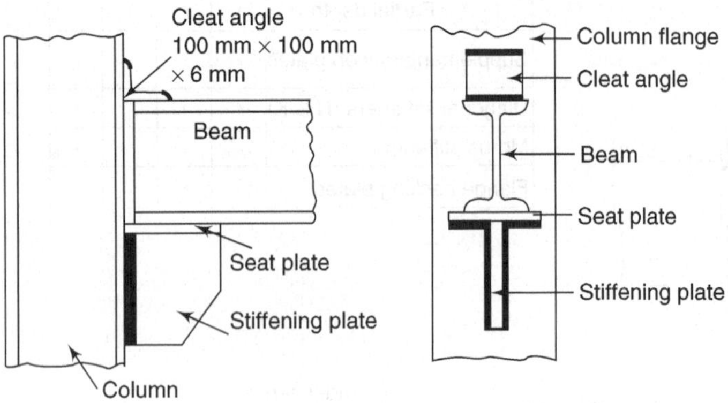


**FIG. 14.24** Welded stiffened seat connection

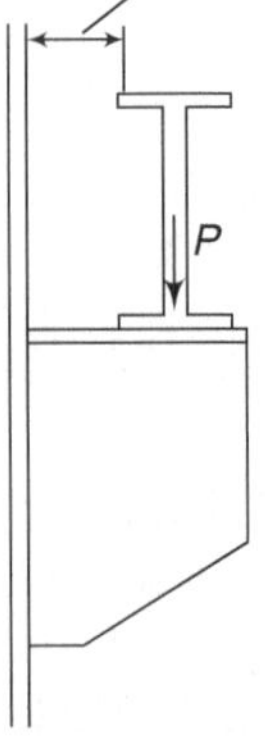


**FIG. 14.25** Bracket supporting concentrated load

There are two basic types of loading used on stiffened seats. The common one is shown in Fig. 14.24, where a beam web is placed directly in line with the stiffener. The other type which occurs when supporting gantry girders, is shown in Fig. 14.25, in this case the beam is oriented in such a way that the plane of the web is at 90° to the plane of the stiffener. A difference in behaviour of the stiffener may also result depending on the angle at which the stiffener may be cut as shown in Fig. 14.26. In Fig. 14.26(a), where the angle is 90°, the stiffener behaves similar to an unstiffened element under uniform compression, and local buckling may be prevented by satisfying the limiting width to thickness ratios given in Table 2 of the code. When a supporting plate is cut to a triangular shape as in Fig. 14.26(b), the behaviour is different and is provided by Salmon et al. (2008) and Salmon et al. (1964).

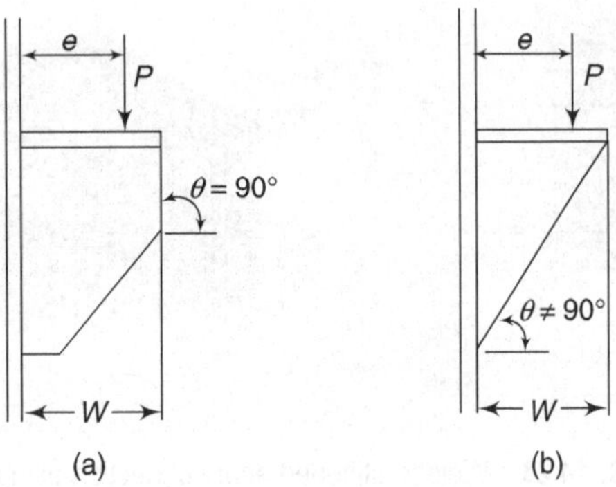


**FIG. 14.26** Two different inclination angles at the free end of stiffener

### 14.7.1 Design

The design of the stiffened welded seat connection is similar to that of the unstiffened welded seat connection. The steps to be followed are as follows.

1. The width of the seat angle is calculated.
2. The thickness of the seat plate is chosen as equal to the thickness of the flange plate.
3. The thickness of stiffening plate is chosen as equal to the thickness of the web of the beam.
4. The eccentricity of the load and bending moment due to it are calculated.
5. Vertical and horizontal shear stresses due to the shear force and bending moment are calculated. From these values, the resultant shear of the weld is computed as the vector sum.
6. From the above resultant shear value, the weld size is determined.

## 14.8 WELDED CONTINUOUS BEAM-TO-COLUMN CONNECTIONS

In continuous beam-to-column connections, it is assumed that there is full transfer of moment and little or no relative rotation of members within the joint. It is obvious that the connection must be of adequate strength to accomplish this. Local elements of the connection must be so proportioned that their premature failure will not result in a lowering of overall connection strength.

Rigid frame connections (except in pre-fabricated buildings) are always welded. A variety of arrangements are possible and some of the common welded connections to column flanges and column webs are shown in Figs 14.27 and 14.28 respectively (also see photos given here).

Most of these connections are partly shop welded and then completed at site using site welding. Splices in beams/columns are made away from this region, and at location of lower shear and moment.

For connecting the tension flange of the beam to the column, a cover plate is used Fig. 14.27(d). This plate is connected to the column flange by means of full penetration groove welds and to the beam flange by fillet welds. This welding requires a minimum gap of 4–5 mm between the plate and the column. Hence, a backing strip may be used to install the weld efficiently. The length of the plate is kept sufficiently long (about 1.5 times the width of the beam

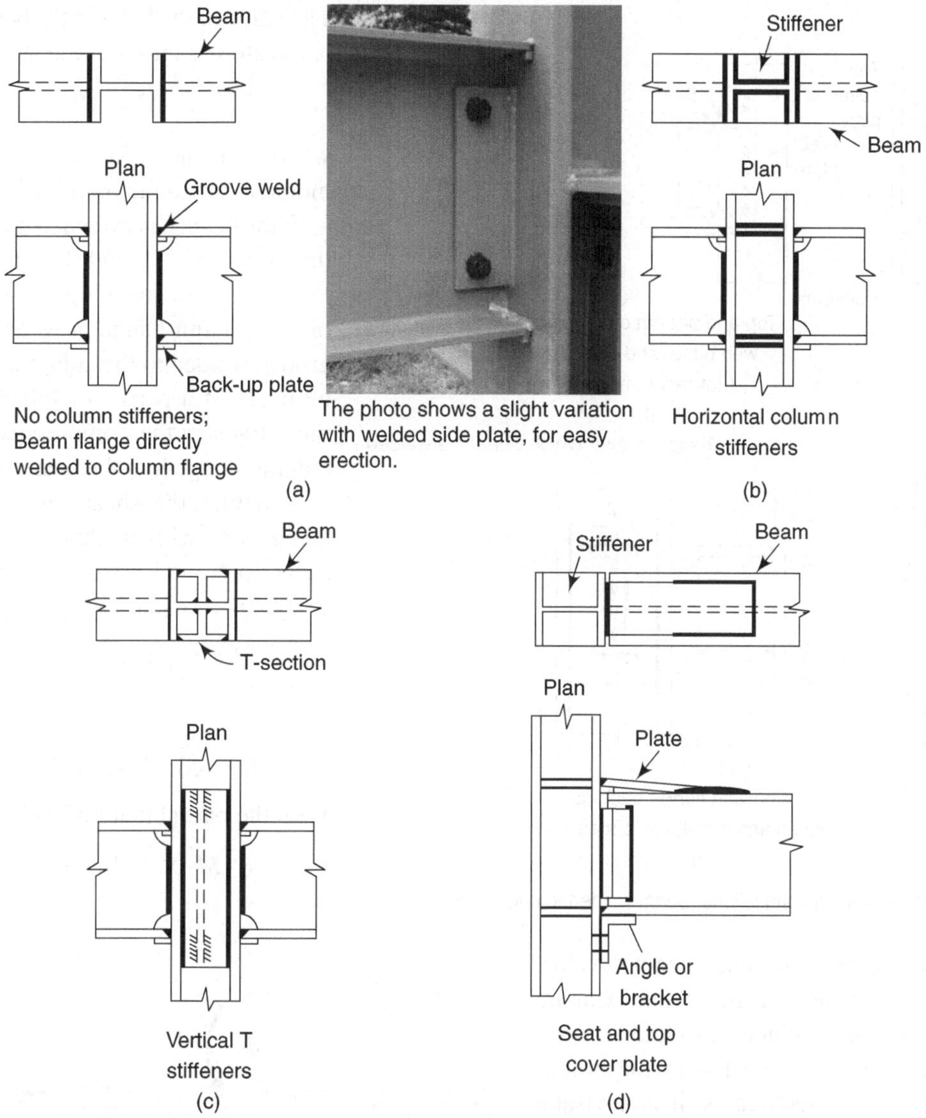


**FIG. 14.27** Continuous beam-to-column connections: Beam welded to column flange

*Source:* AISC Connections Teaching Toolkit—A Teaching Guide for Structural Steel Connections. Reprinted with permission. 

flange) so that a length at least equal to its own width is left unwelded. This unwelded length imparts ductility to the connection, since this length will yield and elongate sufficiently before the connection fails. Also the welded length is kept at least equal to the width of the plate so that full stresses are transferred to the cover plate from the flange by shear lag. (In proportioning the connection, it is usual to assume that the bending moments and thrust are carried by the flanges and the shear is carried entirely by the web.) If a long length of fillet weld is required to connect the cover plate with the beam, U-shaped slots may be cut at the ends of the plate to increase the length of the weld. It is also tapered to facilitate fillet welding. When the beam is connected to the web of the column, the top plate is cut and connected with the column flange by side welds [see Fig. 14.28(a)].

The connecting plate has little resistance to shear. Hence, when the shear force is high, a seat angle or brackets are provided to resist it. The compression flange is so connected to the column that the welds of the seat are not overstressed. Under heavy moments, the column flange may get deformed due to the loads acting on the beam flanges (tension or compression), and its moment carrying capacity is reduced. To control this, additional plates, called *stiffeners* are welded between column flanges opposite the beam flanges as shown in Figs 14.27(b) and (d). Generally these plates are provided for the full depth, though some codes allow the length of stiffener plates to be equal to half the depth of the column (Salmon et al. 2008). Similarly, when a beam frames on the web of the column [see Figs 14.28(a) and (b)], the top and bottom cover plates should be made wide enough to fit between the web and the flanges of the column and be securely welded to them.

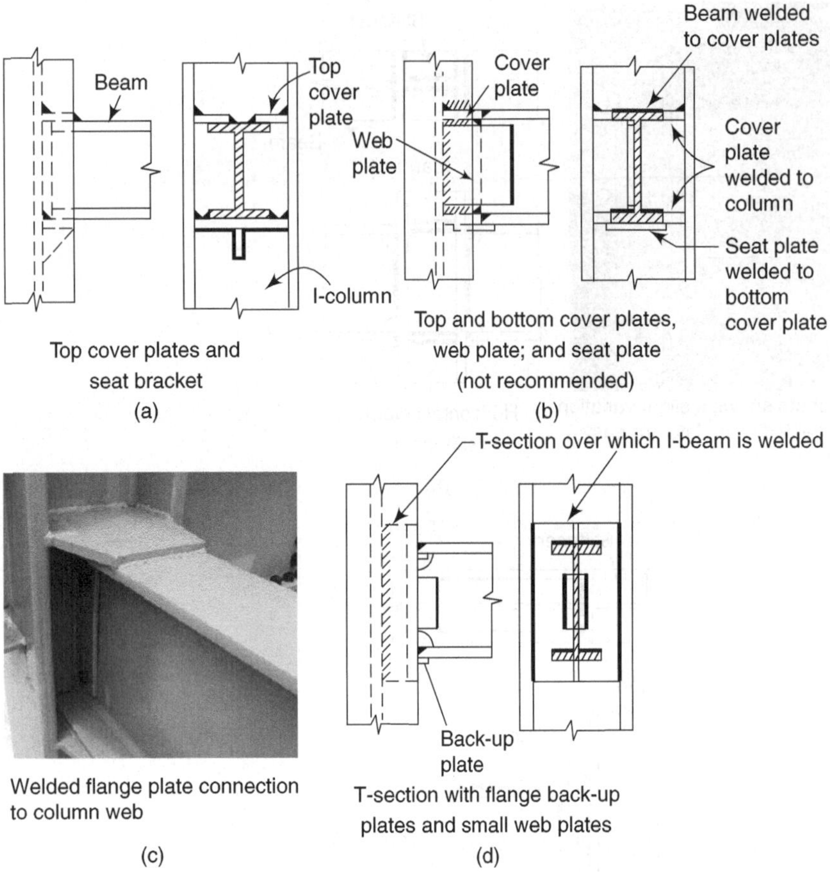


**FIG. 14.28** Continuous beam-to-column connections: Beam welded to column web

The value of the force in the beam flange, which is transferred into the web as shear is taken as

$$V_b \approx M/d_b \tag{14.55}$$

where $d_b$ is the depth of the beam and $M$ is the moment due to factored loads.

If the beam has extended into the connection, the portions AB and CD will be adequate. Portion AC is made either as a continuation of the column flange or as an equivalent plate groove welded to the column. Member BD can be made as a partial or full depth stiffener. It must transfer the compressive force from the column flange into the beam web.

Equating the shear resistance of the web along AB (with thickness $t_w$) to the applied shear force and assuming that yield in shear $\tau_y = f_y/\sqrt{3}$, we get

$$d_c t_w f_y/\sqrt{3} = M/d_b$$

or

$$t_w = \sqrt{3}\ M/(d_c d_b f_y)$$

With the partial material safety factor, we get

$$t_w \geq \sqrt{3}M \times 1.1/(d_c d_b f_y) = 1.9M/(d_c d_b f_y) \tag{14.56}$$

Normally, in continuous construction, the beam develops tension at the top and compression at the bottom near the supports. But in some tall frames subjected to lateral forces (wind or earthquake), tension may develop in the bottom flange, due to stress reversal. In such cases, it is advisable to provide a bottom plate similar to the top plate given in Fig. 14.27(d), capable of yielding at bottom also.

## 14.8.1 Stiffeners in Edge (Corner) Connection

The principal design concern of stiffeners is the transmission of concentrate loads caused by flange forces on beams to the adjacent column. When the forces in beam flanges are transferred to column flanges as compression and tension, horizontal stiffeners are required as shown in Figs 14.27 (b) and (d). Such stiffeners prevent local flange bending from the tensile load, local web yielding, web crippling, and compression buckling caused by the compressive force.

A straight corner connection formed by the intersection of a beam and column is shown in Fig. 14.29. In this arrangement, the beam is assumed to run through the connection.

The force system in the joint as per the assumption that the bending moment and thrust are carried entirely by the flanges and the shear by the web, is shown in Fig. 14.29(b).

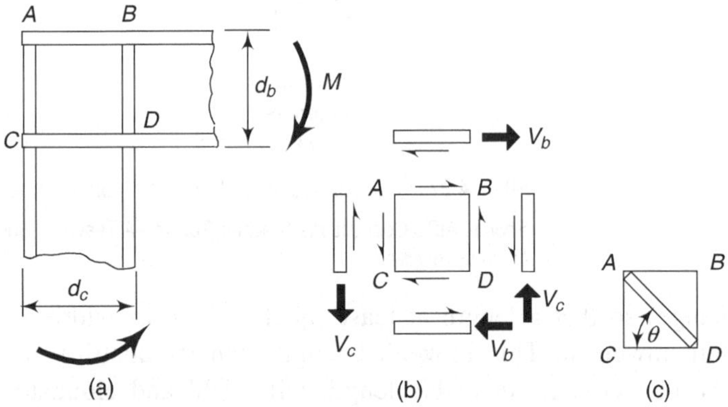


**FIG. 14.29** Beam-to-column corner connection

The stiffener is proportioned by first considering the equilibrium conditions at point A. The total force to be transmitted ($V_b$) is assumed to be shared by the stiffener and the web as

$$V_b = M/d_b = d_c t_w f_y/\sqrt{3} + f_y A_{st} \cos\theta \tag{14.57}$$

where $A_{st}$ is the total stiffener area required. Assuming that all the parts have the same yield stress, and with partial material safety factor

$$A_{st} = \{1.1/(f_y \cos\theta)\} [(M/d_b) - f_y t_w d_c/(\sqrt{3} \times 1.1)] \quad (14.58)$$

Since this element will be under compression, the ratio of its width to thickness($b/t$) should be chosen so as to avoid possibility of premature local buckling.

## 14.8.2 Interior Beam-column Joint

An interior beam-column joint may fail due to the following reasons.

1. Buckling of the column web due to the compressive forces transferred by the beam flange.
2. Failure of the groove weld in the stiff region at the beam tension flange (see Fig. 14.30).

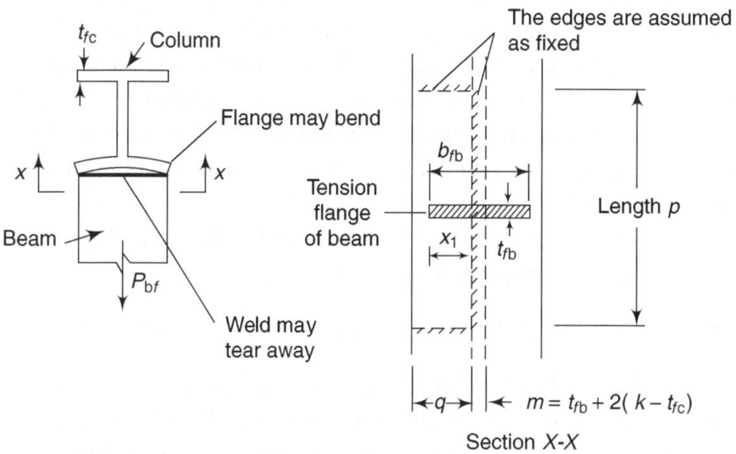


**FIG. 14.30** Failure of the groove weld at the beam-tension flange (Graham et al., 1969)

### *Horizontal Stiffener in Compression Region*

The horizontal stiffener in compression region has to be checked for

1. local web yielding (formally called web crippling),
2. compression buckling of the web, and
3. web crippling.

These criteria are discussed briefly in this section.

**Local web yielding** On the compression side, it may be assumed that the force from the beam flange is distributed out on a 2.5 to 1 slope (see clause 8.7.4), as shown in Fig. 14.31, until the least cross section of the column web is reached (i.e. at the toe of the flange-to-web fillet). The total compressive force from the flange is assumed to be counteracted by the web resistance at this location. Equating these forces, we get

$$P_{bf} = M_f/d_b \text{ or } M_f/d_b = f_{yw} t_{wc}(5k + t_{fb}) \quad (14.59)$$

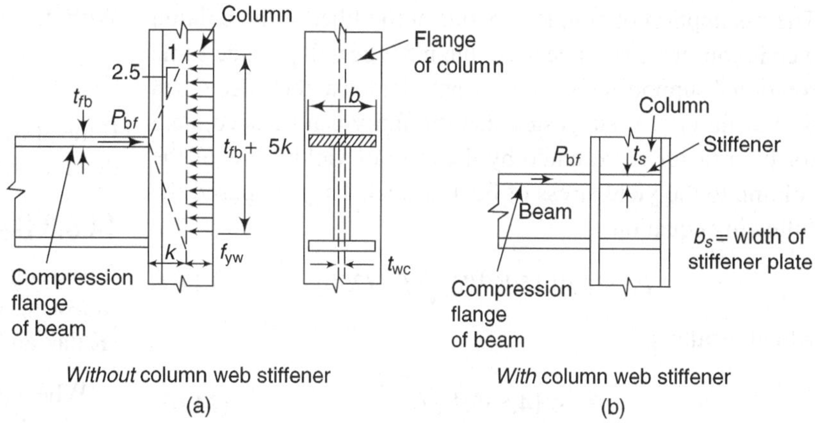


**FIG. 14.31** Column web in compression region of connection

where $P_{bf}$ is the design strength of the column web to resist the force transmitted by the beam flange under factored loads, $f_{yw}$ is the yield stress of the column web, $t_{wc}$ is the thickness of column web, $t_{fb}$ is the thickness of beam flange, $d_b$ is the depth of beam, and $k$ is the distance from extreme fibre to the toe of the fillet (see Fig. 14.31).

With the partial material safety factor, it is written as

$$P_{bf} = f_{yw} t_{wc}(5k + t_{fb})/\gamma_{mo} \geq M_f/d_b \quad (14.60)$$

If this equation is not satisfied, stiffeners are required opposite the beam compression flange. Horizontally placed stiffeners are usual and they should be used in pairs, one on each side of the column web. The area of stiffeners to resist the excess of the factored force $P_{bf}$ over the design resistance is given by

$$A_{st} = P_{bf} - (f_{yw}/1.1)(t_{fb} + 5k)\, t_{wc}/(f_{yst}/1.1) \quad (14.61)$$

where $f_{yst}$ is the design yield stress of the stiffener.

**Compression buckling of the web** The following approximate solution for the elastic buckling strength of a rectangular plate with four edges simply supported and subjected to a concentrated compressive load $P$ at mid-length between the two opposite sides is given by Timoshenko & Gere (1961-pp. 387),

$$P_{cr} = 4\pi E t^3/[12(1 - \mu^2)b] \quad (14.62)$$

where $b$ is the width of the plate in the direction of $P$. [Note that the aspect ratio $a/b$ of the plate does not appear in these equations, as it does—in the form of the coefficient $k$—in the formula for plate buckling (see Chapter 8). It is because this is a simplified form of the derived equation, which includes the aspect ratio.] Denoting the compressive force in the beam flange by $P_{bf}$ and the width $b$ of the plate by $h$, where $h$ is the column-web depth between fillets, the above equation yields the formula (after substituting for $E$, $\pi$, and $\mu$)

$$P_{bf} = 230{,}150\ (t^3/h) \quad (14.63)$$

The assumption of simple supports at the fillet of the column web is conservative since the column flange will provide some rotational support if the web has not yielded at the fillet. Chen & Newlin (1973) suggested that the fixity can be accounted for by a factor represented by the ratio of yield stress of the column to the yield stress of Fe 410 steel. They proposed the following equation:

$$P_{bf} \leq 230{,}150(t^3/h)\sqrt{(f_{yw}/250)} \quad (14.64)$$

which results in

$$P_{bf} \leq 14{,}555t^3\sqrt{f_{yw}}/h \quad (14.65)$$

This formula may be adjusted to represent a lower bound of all the test results to

$$P_{bf} \leq 10{,}10t^3\sqrt{f_{yw}}/h \quad (14.66a)$$

Using material safety factor and denoting $t$ by the web thickness of column, $t_{wc}$, we get,

$$P_{bf} \leq 10{,}710\ (t_{wc}^3/h)\sqrt{(f_{yw}/1.1)} \quad (14.66b)$$

**Web crippling** In addition to the above checks on local web yielding and compression buckling of the web, the following equation for web crippling may also be checked. (This equation is provided in the American Code.)

$$P_{bf} = (300\ t_{wc}^2/\gamma_{ml})[1 + 3(t_{fb}/d_c)(t_{wc}/t_{fc})^{1.5}]\ \sqrt{(f_{yw}t_{fc}/t_{wc})} \quad (14.67)$$

where $P_{bf}$ is the factored force from beam flange, $t_{wc}$ is the column web thickness, $\gamma_{m1}$ = partial safety factor = 1.25, $t_{fb}$ is the beam flange thickness, $t_{fc}$ is the column flange thickness, and $d_c$ is the overall depth of the column.

The above equation is based on the work of Roberts (1981).

### Horizontal Stiffener in Tension Region

From a yield line analysis performed by Graham et al. (1959) the nominal strength $P_n$ of the flange of the column can be approximated by (see Fig. 14.30)

$$P_n = 7t_{fc}^2 f_{yf} + t_{fb}mf_{yw} \quad (14.68)$$

where the first term gives the bending resistance of the column flange as two plate elements (one on each side of the web) and the second term represents the portion of load that is supported directly by column web. Using a factor of 0.8 to the local web yielding, we get

$$t_{fc} = \sqrt{[(P_n/7f_{yf})(1.25 - t_{fb}mf_{yw}/P_n)]} \quad (14.69)$$

Assuming a value of 0.15 for the second term (which was obtained from test results), we get

$$t_{fc} = \sqrt{(P_n/f_{yf})} = 0.4\sqrt{A_f} \quad (14.70)$$

With the partial safety factor

$$t_{fc} = 0.4\ \sqrt{[P_n/(f_{yf}/1.1)]} = 0.4\sqrt{[A_g/(1.1)]} \quad (14.71)$$

If the above equation is not satisfied, stiffeners are required at this location also.

## 14.8.3 Design of Stiffeners

Based on the discussions in the preceding sections, the following procedure is suggested for the design of stiffeners (Kulak and Grondin 2002):

1. When the web crippling or compression buckling of the web limit states indicates the need for stiffeners, they are designed as axially loaded compression members. Alternatively, doubler plates (vertical stiffeners) may be used. However, they are less effective than horizontal stiffeners.
2. When compression buckling of the web controls, the stiffener must extend the entire depth of the column. (They need only be fitted against the column flanges.) When local web yielding controls and the concentrated load $P_{bf}$ is applied at only one column flange, the stiffeners need not extend more than half the depth of column web. (see also Fig. 14.22)
3. When local web yielding controls, the area $A_{st}$ of the stiffener (in pairs) is given by

$$A_{st} \geq [P_{bf} - (f_{yw}/\gamma_{mo})\ (t_{fb} + 5k)t_{wc}]/(f_{yst}/\gamma_{m0}) \quad (14.72)$$

4. The ends of column web stiffeners opposite the beam tension flange must be fully welded to the column flanges.
5. The stiffeners must be proportioned for local buckling. The same slenderness limit for outstanding element of compression flange (9.4$\varepsilon$) may be used.
6. The weld joining stiffeners to the column web should be sized to carry the force in the stiffener caused by unbalanced moments on opposite sides of the column.

## 14.8.4 Shear Reinforcement

Beam column moment connections sometimes require reinforcement of the column web in shear. If the moments in the members at an interior connection differ significantly in magnitude, they may produce large shear stresses in the column web, as in the corner connections. Considering Fig. 14.32, and neglecting direct column or beam shears, since they act opposite to shear produced by the flange forces, we have

$$V = (M_1 - M_2)/d_b \quad (14.73)$$

Considering the algebraic difference of the factored moments, the shear stress can be written as

$$\tau = \Delta M_f/(d_b d_c t_w) \quad (14.74)$$

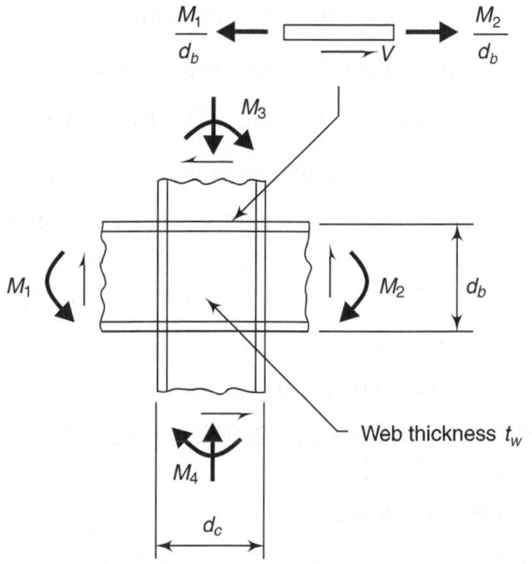


FIG. 14.32 Forces in an interior connection

setting the above shear stress equal to shear yield stress $f_{yc}/\sqrt{3}$, introducing the partial safety factor, and solving for the web thickness $t_w$, we get

$$t_w = 1.9\Delta M_f/d_b d_c f_{yc} \tag{14.75}$$

If the actual web thickness is less than that given by the above equation, a diagonal stiffener or doubler plates should be provided, as was discussed for corner connections.

The ability of beam-column rigid frame connections to develop the plastic hinge moment, as well as exhibit adequate rotation capacity (ductility) has been studied and verified by several researchers (Parfitt & Chen 1976; Grundy et al. 1980; Chen & Patel 1981; Witteveen et al. 1982; Rentschler et al. 1982; Kulak et al. 1987). Discussion and examples on vertical plate and T-stiffeners, which provide a four-way system of connection is provided by Salmon et al. (2008) and Blodgett (1966).

## 14.9 WELDED CONTINUOUS BEAM-TO-BEAM CONNECTIONS

When beams frame transversely to other beams, they may be attached to either or both sides of the web of the supporting beam, using simple seats in combination with a framed beam connection. The principal objective is to provide a means of allowing the tensile force developed in one beam flange to be carried across to the adjacent beam framing opposite the girder web. If the flanges are at the same elevation, a top plate crossing the connection as shown in Fig. 14.33(a) is used. If the flanges are at different levels, the top plates may be groove welded to the web of the girder, as shown in Fig. 14.33(b). If the difference in elevation is small, an additional plate, as shown in Fig. 14.33(c) may be used. The seat may be an angle [as in Fig. 14.33(a)] or a T (which is used when the difference between the bottom of the two beams, is too small for a sufficient length of weld on the seat angle) as shown in Fig. 14.33(b). When using a T, the heel of the seat may be groove welded in the shop and the fillet welds may be done in the field.

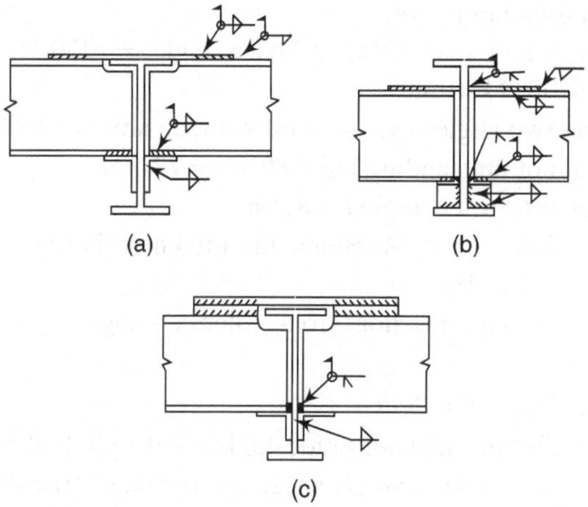


FIG. 14.33 Continuous beam-to-beam connections

When the intersecting tension flanges are not rigidly attached as shown in Fig. 14.33, the connection design is essentially a tension member design at the tension flange along with a shear connection. Salmon et al. (2008) caution that if the intersecting tension flanges are rigidly attached as shown in Fig. 14.34, the design should be checked for biaxial stresses instead of uniaxial stresses. Biaxial stress also increases possibility of brittle fracture and hence it is better to avoid the rigid connection as shown in Fig. 14.34.

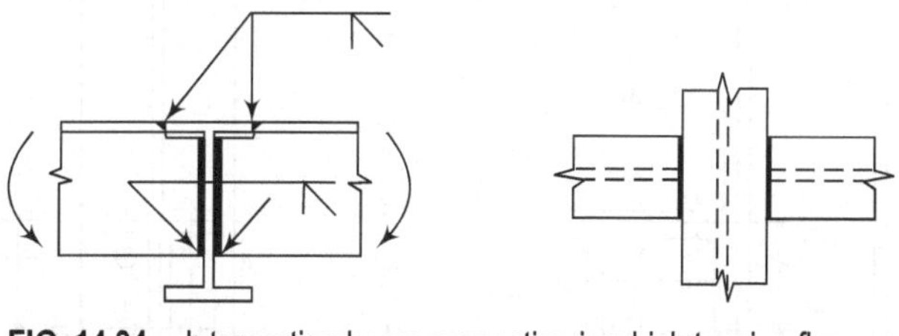

FIG. 14.34 Intersecting beam connection in which tension flanges are attached to each other

## EXAMPLES

**EXAMPLE 14.1:** Design a stiffened seat angle for a reaction of 250 kN from a beam of ISMB 400 using M20 bolts of grade 4.6. The beam has to be connected to ISSC 200 column. Assume Fe 410 grade steel ($f_y = 250$ MPa).

**SOLUTION:**

**Step 1:** *Seat angle*

Length of bearing required at root line of beam

$$= R/(t_w f_{yw}/\gamma_{m0})$$

$$= (250 \times 1000)/(8.9 \times 250/1.1) = 124 \text{ mm}$$

Assuming a clearance including tolerance of 6 mm,

Required length of the outstanding leg = 124 + 6 = 130 mm

Width of the beam = 140 mm

Provide a seat angle of 130 × 130 × 8 mm of length 140 mm connected to the beam by two M20 bolts.

**Step 2:** *Stiffener angle*

Bearing area required for stiffener angle = (250 × 1000)/(250/1.10) = 1100 mm$^2$

Select two angles ISA 80 × 80 × 8 with area = 1220 mm$^2$

Length of outstanding leg = 80 – 8 = 72 mm

Thickness of the angle $t$ = 8 mm

$B/t$ = 72/8 = 9 < 14. Hence the thickness is fine. (clause 8.7.1.2 of IS code)

Distance of end reaction from column flange, $e_x$ = 130/2 = 65 mm

**Step 3:** *Connecting bolts*

Provide 20-mm-diameter grade 4.6 bolts at a pitch of 55 mm.

Strength of bolt in single shear (from Table 5.9) = 45.3 kN

Strength of bolt in bearing = 2.5 × 0.5 × 20 × 8 × 410/(1.25 × 1000) = 65.6 kN

Hence,

Strength of bolt = 45.3 kN

Let us provide bolts in two rows with a pitch of 55 m [see Fig. 14.35]

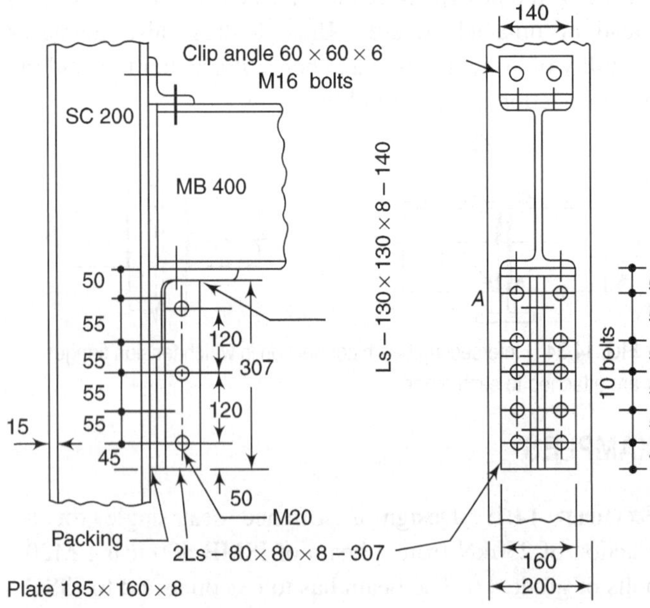


**FIG. 14.35**

$$n = \sqrt{[6M/(pn'V_{db})]}$$

$$n = \sqrt{[(6 \times 250 \times 65)/(55 \times 2 \times 45.3)]}$$

$$= 4.42$$

Provide five bolts in each row with an edge distance of 45 mm.

Depth of stiffener angle = 4 × 55 + 2 × 45 = 310 mm

$h$ = 310 – 45 = 265 mm

The neutral axis lies at $h/7$ = 265/7 = 37.86 mm

Shear force in each bolt = 250/10 = 25 kN

The critical bolt will be at the top of the connection.

$$\Sigma y = 2[(45 - 37.86) + (100 - 37.86) + (155 - 37.86) + (210 - 37.86) + (265 - 37.86)] = 1171.4 \text{ mm}$$

$$\Sigma y^2 = 2[7.14^2 + 62.14^2 + 117.14^2 + 172.14^2 + 227.14^2]$$

$$= 197{,}717.8 \text{ mm}^2$$

$$M^1 = M/[1 + (2h/21)(\Sigma y/\Sigma y^2)]$$

$$= 250 \times 10^3 \times 65/[1 + (2 \times 265/21)(1171.4/197{,}717.8)]$$

$$= 16.25 \times 10^6/1.15$$

$$= 14{,}136 \times 10^3 \text{ Nmm}$$

Tensile force = $M^1 y_n/\Sigma y^2$

$$= 14{,}136 \times 10^3 \times 227.14/197{,}717.8 = 16{,}240 \text{ N}$$

Tensile strength of 20-mm bolt (Table 5.9) = 68.5 kN

As per clause 10.3.6 of IS code,

$$(25/45.3)^2 + (16.24/68.5)^2 = 0.305 + 0.056 = 0.361 < 1.0$$

Hence the connection is safe.

Provide a top clip angle of 60 × 60 × 6 mm with four M16 bolts and a packing plate of size 185 × 160 × 8 mm as shown in Fig. 14.35.

**EXAMPLE 14.2:** The web plate of a plate girder is 10 mm thick and 1000 mm deep. Using 16 mm diameter grade 8.8 HSFG bolts, find their number and arrangement to splice the web at a section where the web carries a shear force of 200 kN and a bending moment of 900 kNm.

**SOLUTION:**

Since the HSFG bolts are in double shear,

Strength of 16 mm bolt in double shear = 2 × 33.7 = 67.4 kN (Table 5.11)

Number of bolts that can be accommodated in one vertical line at 60 mm pitch = 1000/60 = 16

Required number of rows of bolts =

$$m = \sqrt{\frac{6M}{pn^2R}} = \sqrt{\frac{6 \times 900 \times 10^3}{60 \times 16^2 \times 67.4}} = 5.22, \text{ say 5 (assumed on the lower side)}$$

*Check:*

Total number of bolts, N = 16 × 5 = 80

$$\sum r^2 = \frac{p^2 nm\left(m^2 + n^2 - 1\right)}{12} = \frac{60^2 \times 16 \times 5(5^2 + 16^2 - 1)}{12}$$

$$= 6.72 \times 10^6 \text{ mm}^2$$

For the extreme bolt, $x$ = 120 mm and $y$ = 450 mm

$$V = \frac{V_n}{N} + \frac{Mx}{\sum r^2} = \frac{200 \times 10^3}{80} + \frac{900 \times 10^6}{6.72 \times 10^6} \times 120 = 2500 + 16071 = 18571\,\text{N} = 18.57\,\text{kN}$$

$$V = \frac{H_n}{N} + \frac{My}{\sum r^2} = 0 + \frac{900 \times 10^6}{6.72 \times 10^6} \times 450 = 60268\,\text{N}$$

$$= 60.27\,\text{kN}$$

Resultant force = $\sqrt{18.57^2 + 60.27^2} = 63\,\text{kN} < 67.4\,\text{kN}$

Hence provide 5 rows with each row containing 16 numbers 16 mm bolts at 60 mm pitch.

*Note*: These details should be provided on either side of the splice joint of the web.

**Example 14.3:** Given the bracket connection shown in Fig. 14.36. With 24-mm-diameter grade 4.6 bolts and plate of Fe 410 steel, is the bolt pattern and plate adequate for the given load in a bearing type connection assuming threads in the shear plane?

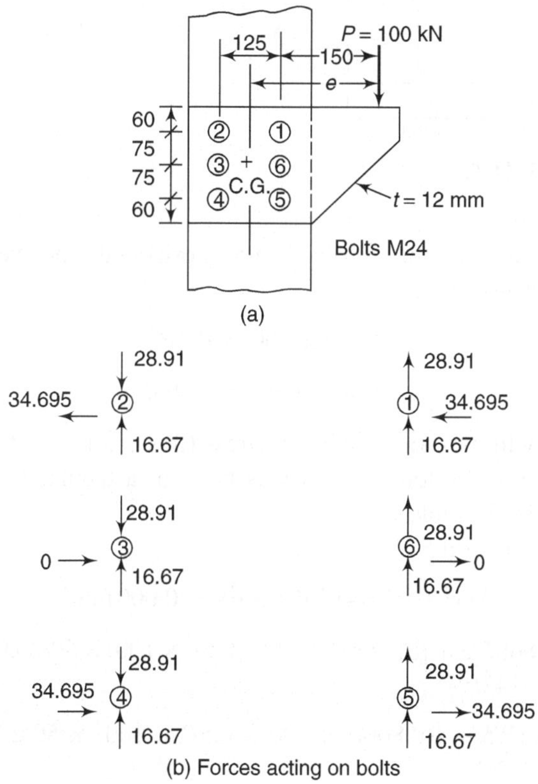


**FIG. 14.36**

**Solution:**

**Step 1:** *Determination of bolt forces*

Since the bolt pattern is symmetrical (as in most practical problems), the centroid is readily located and marked as *CG* in Fig. 14.36.

$$x = 125/2 = 62.5\,\text{mm}$$

$$\Sigma x^2 = 6 \times 62.5^2 = 23{,}437.5\,\text{mm}^2$$

$$\Sigma y^2 = 2(2 \times 75^2) = 22{,}500\,\text{mm}^2$$

$$\Sigma(x^2 + y^2) = 45{,}937.5\,\text{mm}^2$$

$$e = 150 + 125/2 = 212.5\,\text{mm}$$

Hence,

$$R_{x_i} = [212.5(100)y_i]/45{,}937.5 = 0.4626y_i$$

$$R_{y_i} = 0.4626x_i$$

The forces acting on the bolts are calculated as shown in the table below.

| Bolt | $x$ (mm) | $y$ (mm) | $R_x$ (kN) | $R_y$ (kN) |
|---|---|---|---|---|
| 1 | 62.5 | 75 | 34.695 | 28.91 |
| 2 | 62.5 | 75 | 34.695 | 28.91 |
| 3 | 62.5 | 0 | 0 | 28.91 |
| 4 | 62.5 | 75 | 34.695 | 28.91 |
| 5 | 62.5 | 75 | 34.695 | 28.91 |
| 6 | 62.5 | 0 | 0 | 28.91 |

Direct shear due to the external load is

$$R_v = P/N = 100/6 = 16.67\,\text{kN}$$

The forces acting on the bolts are shown in Fig. 14.36. From this figure, it is easy to see that bolts 1 and 5 are highly stressed and are critical, while bolt 3 is the least stressed.

The resisting force on bolt 1 is computed as

$$R = \sqrt{(28.91 + 16.67)^2 + 34.695^2} = \sqrt{3281.28} = 57.28\,\text{kN}$$

**Step 2:** *Check for stresses*

$f_v$ = Stress in the bolt = $57.28 \times 1000/353 = 162.3\,\text{MPa} <$ $400/(\sqrt{3} \times 1.25) = 185\,\text{MPa}$

Check for plate bearing:

$f_b = 57.28/(0.024 \times 12) = 198.88\,\text{MPa} < 410\,\text{MPa}$, Hence ok.

Check for possible tension rupture of the plate along the forward bolt line:

Moment of inertia $I = 12(2 \times 60 + 2 \times 75)^3/12 - 2(12 \times 24)(75)^2 = 16{,}443{,}000\,\text{mm}^4$

Section modulus $Z = 16{,}443{,}000 \times (2/270) = 121{,}800\,\text{mm}^3$

Bending moment at forward bolt line = $100 \times 150 = 15{,}000$ kNmm

$$f_b = M/Z = 15{,}000 \times 1{,}000/121{,}800$$

$$= 123.2\,\text{MPa} < 250/1.1\,\text{MPa} = 227\,\text{MPa}$$

Check plate local buckling b/t = 150/12 = 12.5 < 29.3 (Table 2 of IS 800)

**Step 3:** *Check for shear strength of plate*

Shear strength of place $= f_y A_v/(\sqrt{3} \times 1.1)$

$$= 250 \times (270 - 3 \times 27) \times 12/(\sqrt{3} \times 1.1)$$

$$= 297\,\text{kN} < 140\,\text{kN}$$

Thus, the joint is adequate for bolt shear, plate bearing on bolt, and bending.

**EXAMPLE 14.4:** Determine the size of the bolts required to connect the bracket (supporting a gantry girder) to the column shown in Fig. 14.37.

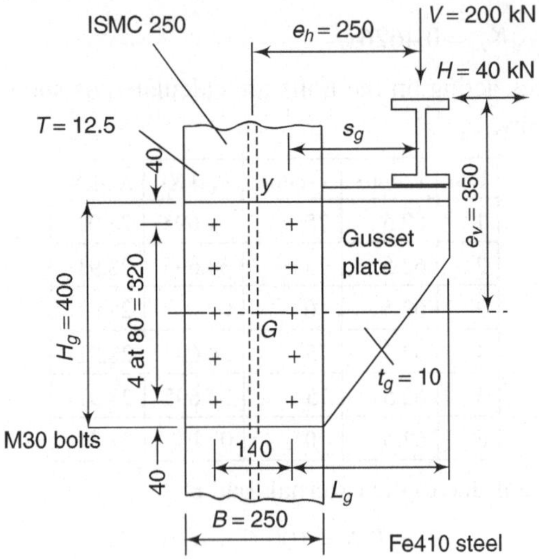


FIG. 14.37

**SOLUTION:**

$$x = 140/2 = 70\,\text{mm}$$

$$\Sigma x^2 = 10 \times 70^2 = 49{,}000\,\text{mm}^4$$

$$\Sigma y^2 = 4(80^2 + 160^2) = 128{,}000\,\text{mm}^4$$

$$\Sigma(x^2 + y^2) = 177{,}000\,\text{mm}^4$$

$$e_h = 250\,\text{mm}; \quad e_v = 350\,\text{mm}$$

The maximum vector force in the direction of *y-y* axis is on a bolt farthest from the centroid of the bolt group.

$$R_y = V/n + [(Ve_h + He_v)\, x_n/\Sigma(x^2 + y^2)]$$

$$= 200/10 + [(200 \times 250 + 40 \times 350)70/177{,}000]$$
$$= 45.3\,\text{kN}$$

The maximum vector force in the direction of *x-x* axis is on the same bolt

$$R_x = H/n + [(Ve_h + He_v)\, y_n/\Sigma(x^2 + y^2)]$$

$$= 40/10 + [(200 \times 250 + 40 \times 350)160]/177{,}000$$
$$= 61.8\,\text{kN}$$

Resultant vector force on the bolt

$$R = \sqrt{45.3^2 + 61.8^2} = 76.62\,\text{kN}$$

Assuming ten 24-mm-diameter grade 4.6 bolts, stress in the bolt

$$f_v = 76.62 \times 1000/353 = 217\,\text{MPa} > 185\,\text{MPa}$$

Hence, provide ten 30-mm-diameter grade 4.6 bolts, then stress in the bolt,

$$f_v = 76.62 \times 1000/561 = 136.5\,\text{MPa} < 185\,\text{MPa}$$

Note that some codes restrict the bolt diameter to 24 mm for a flange width of 250 mm. As in the previous example, checks should be made for plate bearing, tension rupture of the plate, and plate buckling. This is left as an exercise for the reader.

**EXAMPLE 14.5:** Find the maximum load inclined at 60° to the horizontal, which the bracket shown in Fig. 14.38 can transmit if five grade 8.8 bolts with a diameter of 20 mm are used and plates connected are 10-mm thick. Determine the load (i) if the joint is considered a slip joint and (ii) if the joint is considered as a non-slip joint.

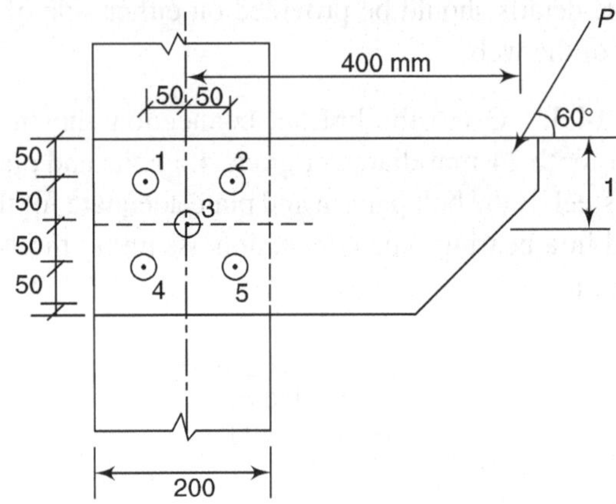


FIG. 14.38

**SOLUTION:**

The force can be resolved into horizontal and vertical components as

$$P_h = P\cos 60° = 0.5P$$

$$P_v = P\sin 60° = 0.866P$$

Bolt 5 will have the maximum stress (this can be verified by calculating the forces in various bolts in a tabular form as shown in Example 14.3)

Force in bolt 5 in *y-y* direction

$$\Sigma(x^2 + y^2) = 4(50^2 + 50^2) = 20{,}000\,\text{mm}^4$$

$$R_y = 0.866P/5 + [(0.866P \times 400 - 0.5P \times 100) \times 50/20{,}000]$$
$$= 0.9142P$$

$$R_x = 0.5P/5 + [(0.866P \times 400 - 0.5P \times 100) \times 50/20{,}000]$$
$$= 0.841P$$

$$\text{Resultant force in the bolt} = P\sqrt{0.9142^2 + 0.841^2} = 1.242P$$

(a) Joint is considered as slip joint

Strength of a 20-mm-diameter bolt (grade 8.8) from Table 5.7:

Bearing = 410 × 10 × 20/1000 = 82 kN

Shear = 370 × 245/1000 = 90.65 kN

Therefore,

Bolt strength = 82 kN

Equating bolt strength with the maximum force, we get

$$1.242P = 82\,\text{kN}$$

or

$$P = 82/1.242 = 66\,\text{kN}$$

(b) Joint is considered as non-slip joint
Slip resistance of a 20-mm-diameter bolt from Table 5.11 = 52.6 kN
Therefore,

$$1.242P = 52.6\,\text{kN}$$

Thus,

$$P = 42.35\,\text{kN}$$

Hence, if the joint is a non-slip joint, the load carrying capacity of the bolt group is reduced by 36% (from 66 kN to 42.35 kN), for the same grade and diameter of the bolt.

**Example 14.6:** Design a bracket connection to transfer a factored load of 120 kN as shown in Fig. 14.39.

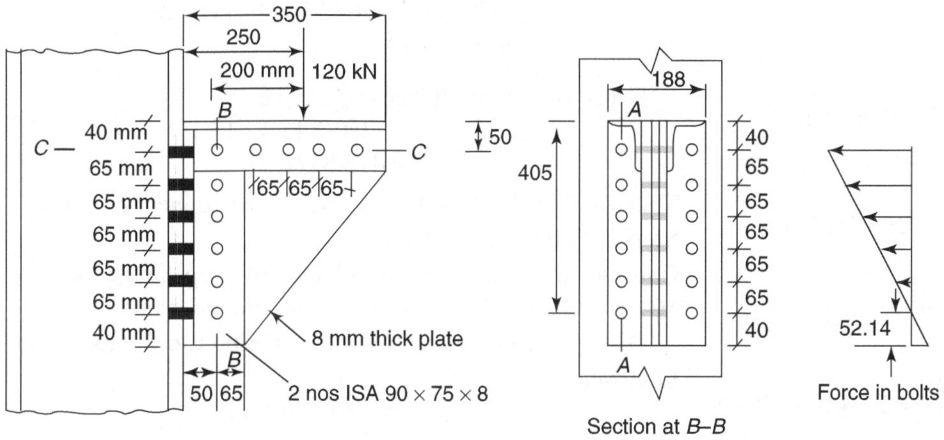


FIG. 14.39

**Solution:**

**Step 1:** *Bolts along A–A*

Bolts connecting the bracket to column flange are subject to a bending moment

$$= 120 \times 250 = 30{,}000\,\text{kNmm}$$

Provide two rows of bolts of size M20 and grade 4.6.

Capacity of an M20 bolt in single shear (Table 5.9) = 45.3 kN

Capacity of a bolt in tension = 68.5 kN

Adopt a pitch = 2.5 × 20 = 50 mm, adopt 65 mm

$$\text{Required number of bolts} = \sqrt{(6M/n'pV_{sd})}$$

$$= \sqrt{[6 \times 30{,}000/(2 \times 65 \times 45.3)]} = 5.53$$

Adopt six bolts in a row; edge distance for a 20-mm bolt is 33 mm. Adopt 40-mm edge distance.

Hence,

$$\text{Depth of bracket plate} = 5 \times 65 + 2 \times 40 = 405\,\text{mm}$$

$$h = 405 - 40 = 365\,\text{mm}$$

Assuming that the neutral axis is at $h/7$ from the bottom of connections, that is,

$$365/7 = 52.14\,\text{mm}$$

Thus, except the bottom bolt, all the other bolts will be in tension.

$$\Sigma y = 2[(65 + 40 - 52.14) + (130 + 40 - 52.14) + (195 + 40 - 52.14) + (260 + 40 - 52.14) + (325 + 40 - 52.14)]$$

$$= 2(52.86 + 117.86 + 182.86 + 247.86 + 312.86)$$

$$= 1828.6\,\text{mm}$$

$$\Sigma y^2 = 2(52.86^2 + 117.86^2 + 182.86^2 + 247.86^2 + 312.86^2)$$

$$= 418{,}878\,\text{mm}^2$$

Tensile force in the critical bolt at the top = $My/\Sigma y^2$ = 30,000 × $10^3$ × 312.86/418,878 = 22407 N
= 22.41 kN

*Note*: We can also use the *M'* parameter (US practice) to calculate tensile force in the critical bolt, as shown in Example 14.1.

Shear force in the critical bolt = 120/12 = 10 kN

*Check* $(10/45.3)^2 + (22.41/68.5)^2 = 0.048 + 0.107 = 0.155 < 1$

*Note*: Number of bolts in the line can be reduced from 6 to 5.

**Step 2:** *Bolts along B–B*

Moment on the bolts = 120 × 200 = 24,000 kNmm

These bolts will be in double shear.

Strength of M20 bolts in double shear = 90.6 kN

Assuming thickness of angles and gusset as 8 mm.

$$\text{Strength of the bolt in bearing} = 2.5 \times 0.5 \times 20 \times 8 \times 400/(1.25 \times 1000) = 64\,\text{kN}$$

Hence,

Strength of the bolt = 64 kN

Assume pitch = 65 mm

$$n = \sqrt{[6 \times 24{,}000 \times 10^3/(65 \times 64{,}000)]} = 5.88$$

Provide 6 bolts at a pitch of 65 mm and edge distance of 40 mm.

Shear in bolt = 120/6 = 20 kN

Neutral axis in assumed at the centre of 405 mm.

Tension in bolt = $Mr/\Sigma r^2$

$= 24{,}000 \times 162.5/[2(32.5^2 + 97.5^2 + 162.5^2)]$

$= 52.75\,\text{kN}$

Resultant force in the bolt

$$= \sqrt{(20^2 + 52.75^2)}$$

$$= 56.41\,\text{kN} < 64\,\text{kN}$$

Hence the connection is safe.

**Step 3:** *Bolts along C–C*

These bolts are subjected to an eccentric load of 120 kN.

Again assuming four 20-mm-diameter bolts with a pitch of 65 mm

Distance of centre of gravity of these four bolts from the face of column

$$= 350 - 40 - 3 \times 65/2 = 212.5\,\text{mm}.$$

Hence eccentricity

$$e = 250 - 212.5 = 37.5$$

$$M = 120 \times 37.5 = 4{,}500\,\text{kNmm}$$

Shear in bolt = 120/4 = 30 kN

Tension in bolt = $Mr/\Sigma r^2$

$$= 4500 \times 97.5/[2(32.5^2 + 97.5^2)]$$

$$= 20.77\,\text{kN}$$

$$\text{Resultant force} = \sqrt{20.77^2 + 30^2} = 36.5\,\text{kN} < 64\,\text{kN}$$

**Step 4:** *Check for shear capacity*

Shear capacity of connected leg of 90 × 75 × 8 angle

$$= \frac{f_y A_v}{1.1 \times \sqrt{3}} = \frac{2 \times [250 \times (8 \times 405 - 6 \times 22)]}{1.1 \times \sqrt{3}} = 573\,\text{kN} > 120\,\text{kN}$$

Shear capacity of the leg connected to flange of column will be similar and hence will be adequate.

Hence connection is safe.

**Example 14.7:** Design a bolted end plate connection between an ISMB 300 beam and an ISHB 200 column, to transfer a vertical factored shear of 120 kN and a factored hogging bending moment of 120 kNm. Use HSFG bolts of diameter 20 mm. The connection is as shown in Fig. 14.40.

**Solution:**

For ISMB 300, $b_f = 140\,\text{mm}$, $t_f = 13.1\,\text{mm}$, $t_w = 7.7\,\text{mm}$

**Step 1:** *Bolt forces*

The connection is assumed to pivot about the 'hard spot' at the bottom flange and the forces in the bolt are assumed to be proportional to the distance from the centre of the bottom flange. It is also assumed that the bolt forces $F_1$ and $F_2$ are acting at the centre of the top flange.

Bolt forces:

By taking the moment about the centre of bottom flange:

$$120 \times 10^3 + 20\left(\frac{300}{2} - \frac{13.1}{2}\right)$$

$$= (2F_1 + 2F_2)(300 - 13.1) + 2F_3 \times 63.45$$

Also as per the assumption $\dfrac{2(F_1 + F_2)}{286.9} = \dfrac{2F_3}{63.45}$

we get,

$$1{,}22{,}869 = \left(\frac{2F_1}{286.9}\right)\left[2 \times 286.9^2 + 63.45^2\right]$$

$$1{,}22{,}869 = 1175.67F_1$$

$$F_2 = F_1 = 104.51\,\text{kN}$$

$$F_3 = (63.45/286.9)F_1 = 23.11\,\text{kN}$$

Reaction at bottom flange

$$F_c = 2(104.5 + 104.51 + 23.11) - 20$$

$$= 444.26\,\text{kN}$$

Capacity of beam flange = $(f_y/\gamma_{m0})A$

$= 250/1.1) \times 13.1 \times 140 \times 10^{-3}$

$= 416.8\,\text{kN}$

This is only 6.2% less than $F_c$. Hence the connection is safe.

**Step 2:** *End plate and bolts*

Consider the portion of end plate above top flange (assume 10-mm fillet welds to flange and 180-mm wide end plate)

Distance from the centre line of the bolt to the toe of the fillet weld

$$l_v = 40 - 10 = 30\,\text{mm}$$

Adopted end distance $l_e = 50\,\text{mm}$

Effective length of end plate per bolt = 180/2 = 90 mm

Tension capacity of M20 bolt (Table 5.11) = 141 kN

Allowable prying force $Q = 141 - 104.51 = 36.49\,\text{kN}$

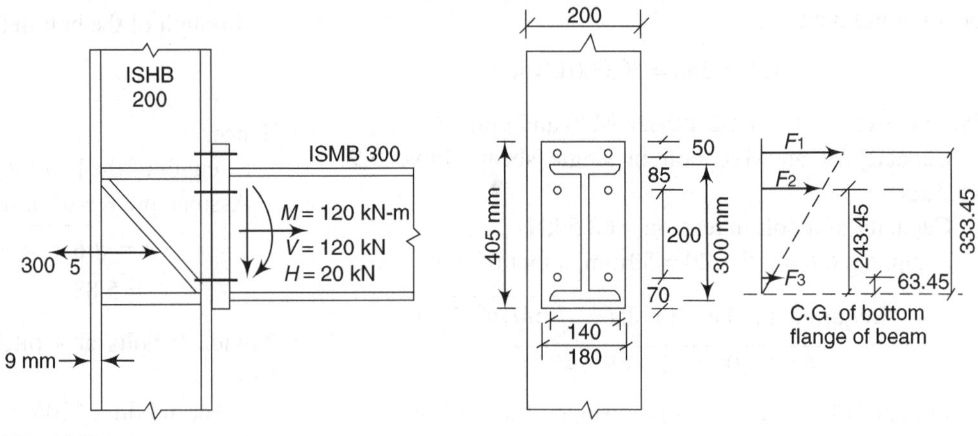


**FIG. 14.40**

Moment at the toe of the weld = $Tl_v - Ql_e$

= 104.51 × 30 – 36.49 × 50 = 1,310.8 Nm

Moment capacity of the plate = $(f_y/1.10)(wT^2/4)$

Therefore,

$$T = \sqrt{[1{,}310.8 \times 10^3 \times 1.10 \times 4/(250 \times 90)]} = 16\,\text{mm}$$

Adopt $T$ = 20 mm

Proof stress = $0.7f_{ub}$ = 0.7 × 800/1000 = 0.56 kN/mm²

**Step 3:** *Check for actual prying forces*

$Q = (l_v/2l_e)\,[T_e - \beta\gamma f_o b_e t^4/(27 l_e l_v^2)]$

= [30/(2 × 50)] [104.51 – 2 × 1.5 × 0.56 × 90 × 20⁴/(27 × 50 × 30²)]

= 0.3(104.51 – 19.91) = 25.38 kN < 36.49 kN (assumed)

(*Note*: $\beta$ = 2 for non-pre-loaded and $\gamma$ = 1.5 for factored load)

**Step 4:** *Check for combined shear and tension*

Shear capacity of M20 bolt (from Table 5.11) = 52.6 kN

Shear per bolt = 120/6 = 20 kN

Tensile capacity of the bolt = 141 kN

$(20/52.6)^2 + [(104.51 + 25.38)/141]^2 = 0.145 + 0.848 = 0.993 < 1.00$

Hence the connection is safe.

The end plate may be checked for shear capacity as done in the previous examples. As the thickness is higher, it will be safe.

In addition to the above checks, the welds connecting flange to end plate and beam web to end plate should be checked. The column flange should also be checked. Usually, stiffeners are provided in the web of the column where the beam flange is supported as shown in Fig. 14.40. These calculations are shown in Example 14.21.

**Example 14.8:** Design a flange angle connection using M16 bolts of grade 4.6 to transfer a factored moment of 12kNm and a shear of 150kN from a beam of ISMB 350 to a column of ISHB 300.

**Solution:**

For M16 bolts (from Table 5.9):

Strength of a bolt in single shear = 29 kN

Strength in double shear = 29 × 2 = 58 kN

Bearing strength on web of thickness 8.1 mm (ISMB 350), assuming $k_b$ = 0.5

= 2.5 × 0.5 × 16 × 8.1 × 410/(1.25 × 1000) = 53.1 kN

Tensile strength of bolt (from Table 5.9) = 43.8 kN

**Step 1:** *Flange clip angles*

Provide two flange clip angles of size ISA 125 × 95 × 12 with 95-mm leg, connected to the column flange.

Gauge distance for 95-mm leg (from Fig. 5.10) = 55 mm

Lever arm = 350 + 2 × 55 = 460 mm

Providing two bolts to connect flange clip angle with column flange,

Moment capacity = 2 × 42.7 × 0.46 = 39.28 > 12 kNm (applied moment)

Force on bolts connected to beam flange = Moment/depth of beam

= 12/0.35 = 34.29 kN

Required number of bolts = 34.29/29 = 1.18

Therefore, provide four bolts in 2 rows.

**Step 2:** *Flange angle thickness*

Pull $P$ in the bolt connecting column flange with clip angle

$$= 12 \times 10^3/460 = 26.09\,\text{kN}$$

$$M_A = 0.6 \times 26.09 \times (55 - 12/2) = 767\,\text{kNm}$$

Moment capacity of the 12-mm thick leg of the angle

$= 1.2 \times (f_y/\gamma_{mo})Z$

$= 1.2 \times (250/1.1) \times (140 \times 12^2/6) \times 10^{-3}$

= 916.4 kNm > 767 kNm. Hence, the thickness provided is adequate.

**Step 3:** *Web clip angle*

The bolts will be in double shear (connected to web of beam)

Required number of bolts = 150/58 = 2.58

Provide 3 bolts of M16 at a pitch of 50 mm and edge distance of 40 mm. Use thickness of web angle as 8 mm (depth of beam < 400 mm).

Length of web angle = 2 × (50 + 40) = 180 mm

Provide 180-mm long angle of size 75 × 75 × 8 mm. The bolts connecting the web angle to the column flange will be in single shear. Hence, provide six M16 bolts.

**Step 4:** *Check for prying force*

$l_v$ = edge distance = 40 mm

$\gamma$ = 1.5 (factored load)

$\beta$ = 2 (non-pre-loaded)

$f_o = 0.7f_{ub} = 0.7 \times 400 = 280$ MPa

$Q = [l_v/(2l_e)]\,[T_e - \beta\gamma f_o b_e t^4/(27 l_e l_v^2)]$

= [55/(2 × 40)] [26.09 – 2 × 1.5 × 0.28 × 140 × 12⁴/(27 × 40 × 51.5²)]

= 15.45 kN

Total load in the bolts = 26.09 + 15.45 = 41.54 kN

Capacity of two M16 bolts in tension = 2 × 43.8 = 87.6 kN > 41.54 kN

Hence the connection is safe.

**Step 5:** *Check for thickness of column flange*

Assuming an edge distance of 40 mm on the column flange, thickness required for the column flange

$$t = \sqrt{[F_t c/(\sqrt{2}af_y)]};\; a = 250/2 - 7.6/2 = 121.2\,\text{mm}$$

$= \sqrt{[41.54 \times 10^3 \times 40/(\sqrt{2} \times 121.2 \times 250)]} = 6.22\,\text{mm} < 10.6\,\text{mm}.$

Hence the thickness provided is adequate.
The designed details are shown in Fig. 14.41.

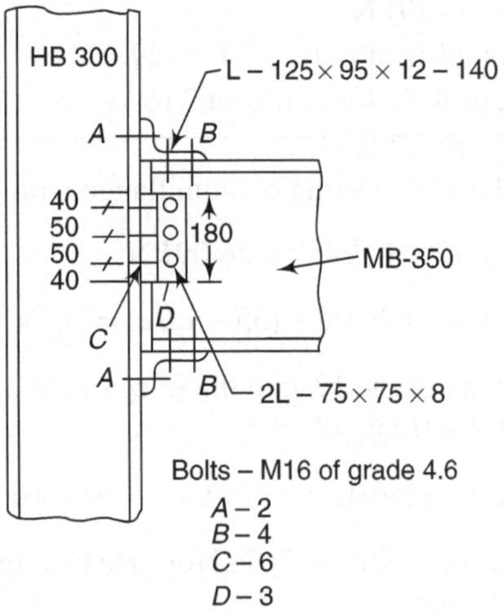


**FIG. 14.41**

**Step 6:** *Check for size of web angle*
Shear strength of web angle.

$$= \frac{2(180 - 3 \times 18) \times 8 \times 250}{(\sqrt{3} \times 1.1)}$$

$$= 264\text{kN} > 150\text{kN}$$

**EXAMPLE 14.9:** Calculate the safe loads that the bracket shown in Fig. 14.42 can carry, assuming 6 mm fillet welds.

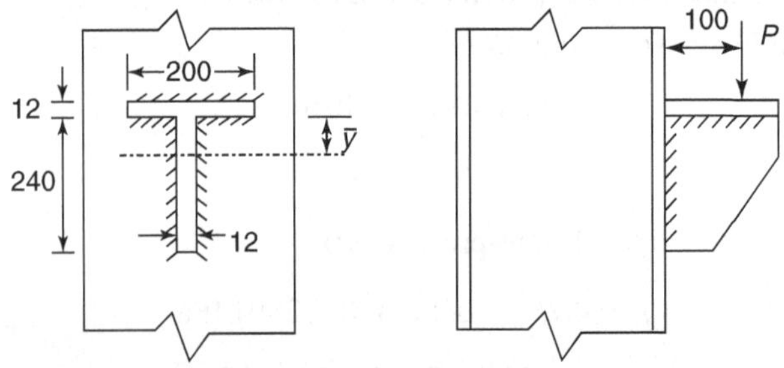


**FIG. 14.42**

**SOLUTION:**
From Table 14.1, for the welded configuration $b$ = 200 mm and $d$ = 240 mm

$$Z = \frac{2bd + d^2}{3} = \frac{2 \times 200 \times 240 + 240^2}{3} = 51,200 \text{ mm}^3$$

Polar moment of inertia

$$I_p = \frac{b^3 + 8d^3}{12} - \frac{d^4}{b + 2d} = \frac{200^3 + 8 \times 240^3}{12} - \frac{240^4}{200 + 2 \times 240}$$

$$= 5.0036 \times 10^6 \text{mm}^4$$

$$\text{Neutral axis distance, } \bar{y} = \frac{d^2}{b + 2d} = \frac{240^2}{200 + 2 \times 240} = 84.7 \text{mm}$$

Total length of weld, $L_w = 200 + (200 - 12) + 2 \times 240 = 868$ mm

Moment M = 100P

$$R_x = \frac{My}{I_p} = \frac{100P \times 84.7}{5.0036 \times 10^6} = 1.693 \times 10^{-3} P$$

$$R_y = \frac{Mx}{I_p} = \frac{100P \times 100}{5.0036 \times 10^6} = 1.999 \times 10^{-3} P$$

Stress due to direct shear, $R = P/L_w = P/868 = 1.152 \times 10^{-3} P$

$$\text{Resultant} = 10^{-3}\text{P}\sqrt{(1.693)^2 + (1.999 + 1.152)^2}$$

$$= 3.577 \times 10^{-3}\text{P N/mm}$$

$$\text{Weld resistance} = \frac{0.7 \times 6 \times 410}{\sqrt{3} \times 1.25} = 795.36 \text{N}$$

$$\text{Hence } P = \frac{795.36}{3.577} = 222.35 \text{kN}$$

**EXAMPLE 14.10:** Design a T-stub (split beam T) connection to transfer a factored shear of 150 kN and a factored moment of 50 kNm from the end of a beam ISMB 350 to a column of ISHB 300 using 16 mm bolts.

**SOLUTION:**
Try M16 bolts. For M16 bolts (from the previous example):
Strength in single shear = 29 kN
Strength in double shear = 58 kN
Bearing strength (against 8.1 thick plate) = 53.1 kN
Tensile strength = 43.8 kN

**Step 1:** *Web angle connection*
Provide similar arrangement as in the previous example, since the shear force is same.

Thus provide 75 × 75 × 8 mm angle with six M16 bolts connecting to the flanges of the column and three M16 bolts connecting to the beam web. Use a T-stub 220 mm long cut from ISMB 450.

Flange width of ISMB 450 = 150 mm
Flange thickness = 17.4 mm
Web thickness = 9.4 mm
Root radius = 15 mm

**Step 2:** *Split beam-to-beam top flange connection*
Using 20mm diameter bolts
Flange force, $P = 50 \times 10^3/350 = 142.86$ kN
Required number of bolts = 142.86/45.3 = 3.2
Provide four bolts in two rows.
Thickness of T-web (adopting a width of 140 mm)

$$t = P/(b \times f_y/\gamma_{m0})$$
$$= 142.86 \times 1000/[(140 - 2 \times 18) \times (250/1.10)]$$

= 6 mm < 9.4 mm. Hence the thickness provided is adequate.

**Step 3:** *Split beam-to-column flange connection*
For 1SMB 450, $t_f$ = 17.4 mm
$l_e$ = minimum edge distance = 1.5 × 16 = 24 mm
Provide an edge distance of 30 mm

$a$ = distance from fillet line to bolt = [(150/2) – (9.4/2) – 15 – 30]

$= 25.3\,\text{mm}$

Moment in T-stub $= 0.5P \times a$

$= 0.5 \times 142.86 \times 25.3$

$= 1807.2\,\text{kNmm}$

Moment capacity of T-stub flange

$= 1.2 \times (250/1.10) \times 140 \times 17.4^2/6 \times 10^{-3}$

$= 1926.65 > 1807.2\,\text{kNmm}$

Hence, thickness provided is as required. The force in a row of bolts connecting the T-stub to column

$= P/2 = 142.86/2 = 71.43\,\text{kN}$

The tensile force in 16-mm-diameter bolts = 43.8 × 2 = 87.6 kN > 71.43 kN
However, provide four 20-mm-diameter bolts, in two rows with a pitch of 60 mm and an edge distance of 40 mm.

Tension capacity of four M20 bolts = 4 × 68.5 = 274 kN > 142.86 kN
Hence ok.

Tensile force in bolt due to the applied moment

$= 142.86/4 = 35.72\,\text{kN}$

**Step 4:** *Check for prying force*

$l_e = 30\,\text{mm}$ (provided)

$l_v = [(150/2) - (9.4/2) - 30] = 40.3\,\text{mm}$

$\gamma = 1.5$ (factored load)

$\beta = 2$ (non-pre-loaded)

$f_o = 0.7 f_{ub} = 0.7 \times 400 = 280\,\text{MPa}$

$Q = (l_v/2l_e)\,[T_e - \beta\gamma f_o\, b_e t^4/(27 l_e l_v^2)] = [40.3/(2 \times 30)]\,[35.72 - 2 \times 1.5 \times 0.28 \times 140 \times 17.4^4/(27 \times 30 \times 40.3^2)] = 18.48\,\text{kN}$

Total tensile load in bolt = 35.72 + 18.48 = 54.2 kN < 68.5 kN

Hence the bolts are adequate. The details of the connection are given in Fig. 14.43.

**Step 5:** *Check for thickness of column flange*
Assuming an edge distance of 40 mm on the column flanges

$s_b = 150 - (2 \times 30) = 90\,\text{mm}$
$a = (250 - 7.6)/2 = 121.2$

Required thickness

$$t_b = \sqrt{[4nF_1 c/(af_y)]/(4\sqrt{2} + (n-1)s_b/a)}$$

$$= \sqrt{[4 \times 2 \times 48.48 \times 1000 \times 40/(21.2 \times 250)]/(4\sqrt{2} + 1 \times 90/121.2)}$$

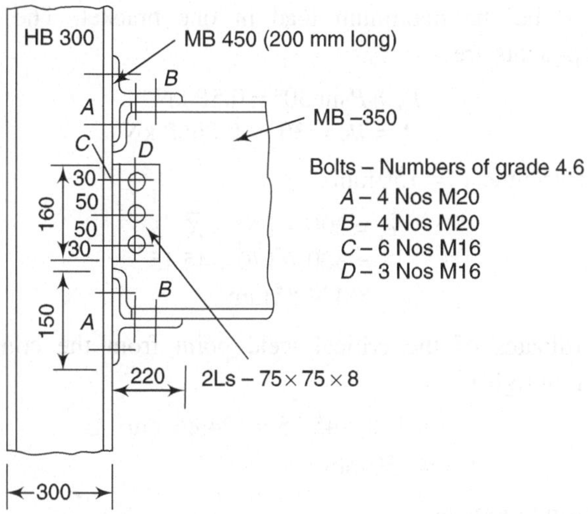


**FIG. 14.43**

$$= \sqrt{(512/6.10)} = 8.94\,\text{mm} < 10.6\,\text{mm}$$

Hence the thickness provided is adequate.

**EXAMPLE 14.11:** Determine the maximum load that could be resisted by the bracket shown in Fig. 14.44, assuming the plate thickness does not affect the result and if 6-mm shop fillet welds are used.

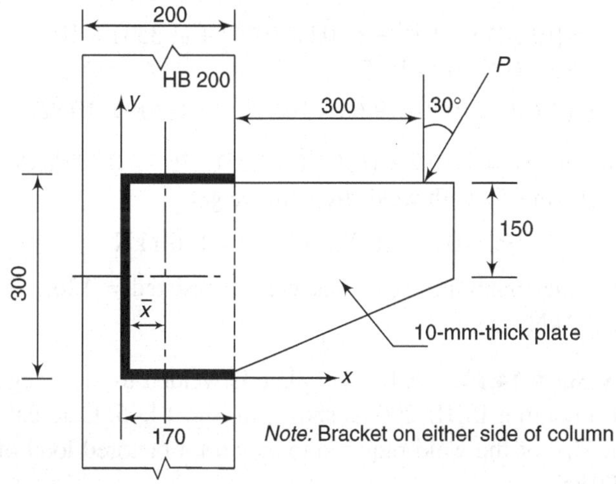


**FIG. 14.44**

**SOLUTION:**
Total length of weld = 2 × 170 + 300 = 640 mm
Polar moment of inertia (from Table 14.1)

$I_p = (8b^3 + 6bd^2 + d^3)/12 - b^4/(2b + d)$
$= (8 \times 170^3 + 6 \times 170 \times 300^2 + 300^3)/12 - 170^4/(2 \times 170 + 300)$
$= 11.87 \times 10^6\,\text{mm}^4$

C.G. of the weld line (Table 6.7)

$\bar{Y} = 300/2 = 150\,\text{mm}$
$\bar{X} = b^2/(2b + d) = 170^2/(2 \times 170 + 300)$
$= 45.15\,\text{mm}$

Let $P$ be the maximum load in one bracket. The load components are

$$P_x = P\sin 30° = 0.5P \text{ kN}$$
$$P_y = P\cos 30° = 0.866P \text{ kN}$$

Eccentricities $e_x = 150\,\text{mm}$

$$e_y = 300 + 170 - \bar{X} = 300 + 170 - 45.15 = 424.85\,\text{mm}$$

Coordinates of the critical weld point from the controid (bottom right)

$$x = 170 - 45.15 = 124.85\,\text{mm}$$
$$y = 150\,\text{mm}$$

Twisting moment

$$T = -0.5P \times 150 + 0.866P \times 424.85 = 292.92P \text{ kN mm}$$

$$F_x^P = 0.5P/640 = 0.7813 \times 10^{-3}P \leftarrow$$

$$F_y^P = 0.866P/640 = 1.3531 \times 10^{-3}P \downarrow$$

$$F_x^T = Ty/I_p = 292.92P \times 150/(11.87 \times 10^6) = 3.702 \times 10^{-3}P$$

$$F_y^T = Tx/I_p = 292.92P \times 124.85/(11.87 \times 10^6) = 3.081 \times 10^{-3}P$$

$$F_R = [(F_x^P + F_x^T)^2 + (F_y^P + F_y^T)^2]^{0.5}$$
$$= [(0.7813 \times 10^{-3} + 3.702 \times 10^{-3})^2 + (1.3531 \times 10^{-3} + 3.081 \times 10^{-3})^2]^{0.5}P$$
$$= (20.10 \times 10^{-6} + 19.66 \times 10^{-6})^{0.5}P = 6.306 \times 10^{-3}P$$

Weld strength $= 6 \times 0.7 \times 410/(\sqrt{3} \times 1.25) \times 10^{-3} = 0.795\,\text{kN/mm}$

Equating $F_R$ with weld strength, we get

$$P_u = 0.795/(6.306 \times 10^{-3}) = 126.1\,\text{kN}$$

Thus, the maximum load that can be resisted = 126.1 × 2 = 252.2 kN.

**EXAMPLE 14.12:** A bracket plate is welded to the flange of a column ISHB 200 as shown in Fig. 14.45. Calculate the size of the weld required to support a factored load of 100 kN.

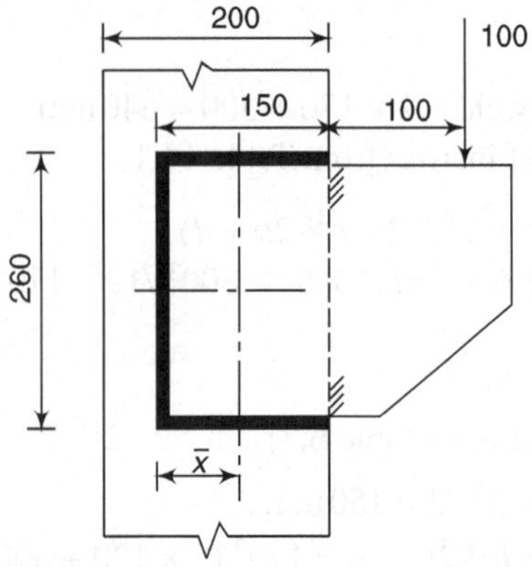


**FIG. 14.45**

**SOLUTION:**

Total length of weld (ignoring end return) = 260 + 2 × 150 = 560 mm

$\bar{X}$ = distance of centroid of weld group

$$= b^2/(2b + d) = 150^2/(2 \times 150 + 260) = 40.18\,\text{mm}$$

$$\bar{Y} = 260/2 = 130\,\text{mm}$$

Eccentricity $e_y = 100 + 150 - 40.18 = 209.82\,\text{mm}$

Polar moment inertia

$$I_p = [(8b^3 + 6bd^2 + d^3)/12] - [(b^4/(2b + d)]$$
$$= (8 \times 150^3 + 6 \times 150 \times 260^2 + 260^3)/12 - 150^4/(2 \times 150 + 260)$$
$$= 7.88 \times 10^6\,\text{mm}^4$$

Coordinates of critical weld point

$$x = 150 - \bar{X} = 150 - 40.18 = 109.82$$
$$y = 130\,\text{mm}$$

Stress due to direct shear

$$R_v = P/L_w = 100 \times 1000/560 = 178.57\,\text{N/mm}$$

From torsion $T$ about the centroid of configuration

$$R_x = Ty/I_p = (100 \times 10^3 \times 209.82) \times 130/7.88 \times 10^6 = 346.15\,\text{N/mm}$$

$$R_y = Tx/I_p = (100 \times 10^3 \times 209.82) \times 109.82/7.88 \times 10^6 = 292.42\,\text{N/mm}$$

$$\text{Resultant force } R = \sqrt{[346.15^2 + (292.42 + 178.57)^2]} = 584.5\,\text{N/mm}$$

Weld resistance $= 0.7 \times a \times 410/(\sqrt{3} \times 1.25) = 132.56a$

Therefore,

$$132.56\,a = 584.5$$

or $a = 4.4\,\text{mm}$, say 5 mm

Hence, use a 5 mm E43 weld.

**EXAMPLE 14.13:** Determine the load carrying capacity of the joint shown in Fig. 14.46 for 6-mm, 8-mm, and 10-mm welds. Assume shop welding.

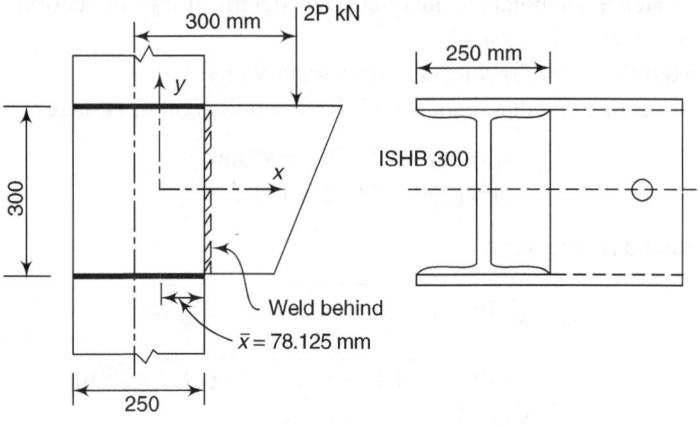


**FIG. 14.46**

**SOLUTION:**

Assume unit weld throat

$\bar{X} = b^2/(2b + d) = 250^2/(2 \times 250 + 300) = 78.125$ mm

Length of weld = 2 × 250 + 300 = 800 mm

Eccentricity of load = 300 – (250/2) + 78.125 = 253.125 mm

Polar moment of inertia (Table 14.1)

$$I_p = (8b^3 + 6bd^2 + d^3)/12 - b^4/(2b + d)$$

$$= (8 \times 250^3 + 6 \times 250 \times 300^2 + 300^3)/12 - 250^4/800$$

$$= 19.03 \times 10^6 \text{ mm}^4$$

$$F_x^P = P/800 = 1.25 \times 10^{-3} P \text{ kN/mm}$$

Coordinates of critical weld point (bottom left)

$$y = 150 \text{ mm}, x = 250 - 78.125 = 171.875 \text{ mm}$$

$$F_x^T = Ty/I_p = (P \times 253.125 \times 150)/(19.03 \times 10^6) = 1.995 \times 10^{-3} \text{ kN/mm}$$

$$F_y^T = Tx/I_p = (P \times 253.125 \times 171.875)/(19.03 \times 10^6) = 2.286 \times 10^{-3} \text{ kN/mm}$$

$$F_R = [(1.25 \times 10^{-3} + 2.286 \times 10^{-3})^2 + (1.995 \times 10^{-3})^2]^{0.5} P = 4.06 \times 10^{-3} P \text{ kN/mm}$$

For weld length $s$, throat size

$$t_e = 0.7s$$

For grade Fe 410 steel

$$R_w = 410/(\sqrt{3} \times 1.25) = 189.37 \text{ N/mm}^2$$

The following table gives the maximum load capacity of the joints.

| Weld leg, mm | Throat size, mm | P kN |
|---|---|---|
| 6 | 4.2 | $189.37 \times 10^{-3} \times 4.2/(4.06 \times 10^{-3}) = 195.9$ |
| 8 | 5.6 | $189.37 \times 10^{-3} \times 5.6/(4.06 \times 10^{-3}) = 261.2$ |
| 10 | 7.0 | $189.37 \times 10^{-3} \times 7.0/(4.06 \times 10^{-3}) = 326.5$ |

**EXAMPLE 14.14:** Determine the length of a fillet weld required to carry the load indicated by Fig. 14.47. Assume an 8-mm shop weld.

**SOLUTION:**

Factored load = 1.5 × 175 = 262.5 kN

Required length of the weld = $L_w = \sqrt{6M/R_{nw}}$

Falnge thickness of ISHB 300 = 10.6 mm

Maximum size of weld = 10.6 – 1.5 = 9.1 mm

Minimum size of weld = 5 mm (Table 21 of code)

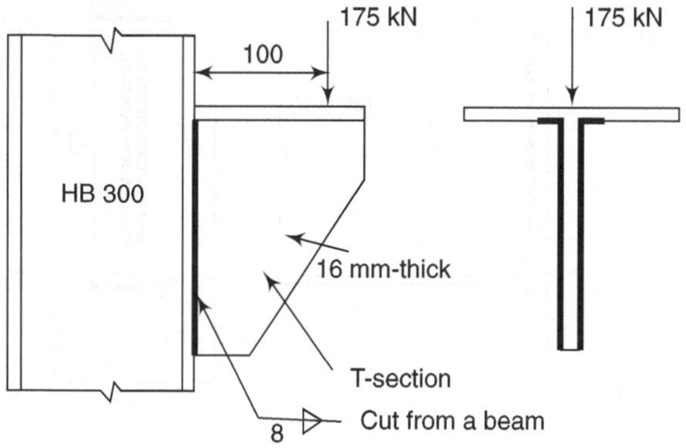


**FIG. 14.47**

Adopt 8-mm weld

Design strength of the weld = $f_{wd} = 410/(\sqrt{3} \times 1.25) = 189$ N/mm$^2$

$$R_{nw} = 189 \times 0.7 \times 8 = 1058.4 \text{ N/mm}$$

$$M_u = 262.5 \times 100 = 26250 \text{ kN/mm per two lines of weld}$$

$$L_w = \sqrt{[(6 \times 26250 \times 1000/2)/900(\text{est})]} = 295.8 \text{ mm}$$

A reduced value of $R_{nw}$ has been used to account for the direct shear effect also.

Try $L_w = 300$ mm

Neglecting the end returns, stress due to direct shear

$$\tau_v = P/2L_w t_e$$

$$= (262.5 \times 1000)/(2 \times 300 \times 0.7 \times 8)$$

$$= 78.125 \text{ N/mm}^2$$

Stress due to bending

$$f_b = My/I = 262.5 \times 1000 \times 100 \times 150/[(2 \times (300^3/12) \times 0.7 \times 8)] = 156.25 \text{ N/mm}^2$$

$$\text{Resultant stress} = \sqrt{(78.125)^2 + (156.25)^2}$$

$$= 174.69 \text{ N/mm}^2 < 189 \text{ N/mm}^2$$

Hence with a length of 300 mm, the fillet weld will be able to carry the required load.

**EXAMPLE 14.15:** Determine the size of weld for a connection as shown in Fig. 14.48. Assume site welding. The joint is subjected to a factored shear force of 180 kN and factored bending moment of 25 kNm.

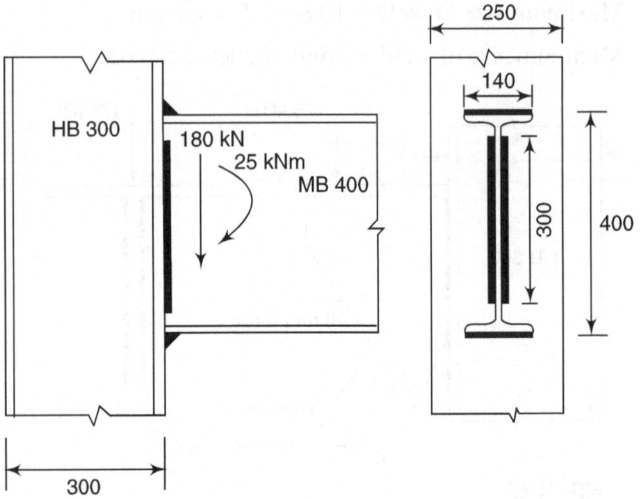

**FIG. 14.48**

**Solution:**

Assuming unit weld throat

Area = 2 × 140 + 2 × 300 = 880 mm²

$I_{zz} = 2 \times 140 \times (400/2)^2 + 2 \times 300^3/12 = 15.7 \times 10^6\,\text{mm}^4$

$Z$ for the extreme fibre = $15.7 \times 10^6/(400/2) = 78.5 \times 10^3\,\text{mm}^3$

Direct stress = 180/880 = 0.205 kN/mm

Bending stress = $25 \times 10^3/78.5 \times 10^3 = 0.318$ kN/mm

Vector sum = $(0.318^2 + 0.205^2)^{0.5} = 0.378$ kN/mm

Weld size = $0.378 \times 10^3/[410/(\sqrt{3} \times 1.50) \times 0.7] = 3.43$ mm

Thickness of flange and web of MB 400 are 16.0 mm and 8.9 mm, respectively.

Hence,

Minimum thickness = 3 mm (Table 21 of code)

Maximum thickness = 8.9 – 1.5 = 7.4 mm

Adopt a 5-mm fillet weld of grade E 43.

**Example 14.16:** A 10-mm thick bracket plate is used to transmit a reaction of 80 kN (factored load) at an eccentricity of 80 mm from the column flange. Design the butt weld for grade Fe 410 steel and E 43 electrode. See Fig. 14.49 for details.

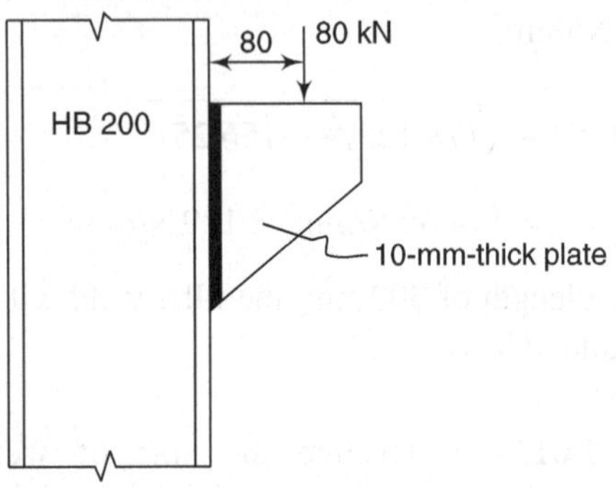

**FIG. 14.49**

**Solution:**

Adopting a double-V groove weld,

Effective throat thickness = 10 mm

Depth of bracket plate from Eqn (14.51)

$$d = \sqrt{6M/(tf_b)}$$

$$= \sqrt{[6 \times 80 \times 10^3 \times 80/(10 \times 250/1.1)]}$$

$$= 130\,\text{mm}$$

Adopt a depth of 150 mm. Direct shear stress,

$$\tau_{vf} = 80 \times 10^3/(150 \times 10)$$

$$= 53.33\,\text{N/mm}^2$$

Bending stress = $f_b = 6M/td^2$

$$= 6 \times 80 \times 10^3 \times 80/(10 \times 150^2)$$

$$= 170.67\,\text{N/mm}^2$$

*Check*

$$f_e = \sqrt{(170.67^2 + 3 \times 53.33^2)} = 194.1\,\text{N/mm}^2$$

$$< 250/1.1 = 227\,\text{N/mm}^2$$

Hence the joint is safe.

**Example 14.17:** Design stiffened seat angle connection for a reaction of 250 kN from beam of ISMB 400. This beam has to be connected to a column of size ISHB 200. Assume Fe 410 grade steel and shop welding.

**Solution:**

From Example 14.1,

Length of bearing required = 124 mm

Provide a clearance of 5 mm (including tolerance).

Required width of seating plate = 124 + 5 = 129 mm

Width of flange of beam = 140 mm

Thickness of beam flange = 16 mm

Adopt a seating plate of size 130 × 160 mm and thickness 16 mm. Thickness of stiffening plate should be equal or greater than thickness of web of beam = 8.9 mm

Provide a 12-mm thick stiffening plate.

Distance of end reaction from the outer edge of the seat plate

$$= 129 - 124/2 = 67\,\text{mm}$$

Hence,

Bending moment = 250 × 67 = 16,750 kN/mm

The vertical stem should not buckle. As per Table 2 of IS 800,

$$d/t < 18.9$$

Therefore,

$$d < 18.9 \times 12 = 226.8\,\text{mm}$$

Adopt a depth of 220 mm. Taking moments from the bottom of the seat plate, C.G. of the weld group

$\bar{y}_1 = (2 \times 220 \times 110 + 0)/[(2 \times 220 + (160 - 12)] = 82.3$ mm

$\bar{y}_2 = 220 - 82.3 = 137.7$ mm

Assuming unit weld thickness,

$I_{xx}$ of the weld $= [(160 - 12) \times 82.3^2 + 2 \times (220^3/12) + 2 \times 220 \times (137.7 - 110)^2]$

$= 3.114 \times 10^6 \text{ mm}^4$

Vertical shear $\tau_{vf1} = 250 \times 10^3/(2 \times 220 + 148)$

$= 425.2$ N/mm

Horizontal shear $\tau_{vf2} = [16750 \times 10^3 \times 82.3/(3.114 \times 10^6)]$

$= 442.7$ N/mm

Resultant shear stress/mm = Strength of weld/mm

$$\sqrt{(425.2^2 + 442.7^2)} = 0.7 \times s \times 410/(\sqrt{3} \times 1.25)$$

$$613.8 = 132.56s$$

Therefore, $s = 4.63$. Provide 5-mm weld.

Also provide a nominal size of top cleat of size 75 × 75 × 6 mm and connect it by a 4-mm fillet weld as shown in Fig. 14.50.

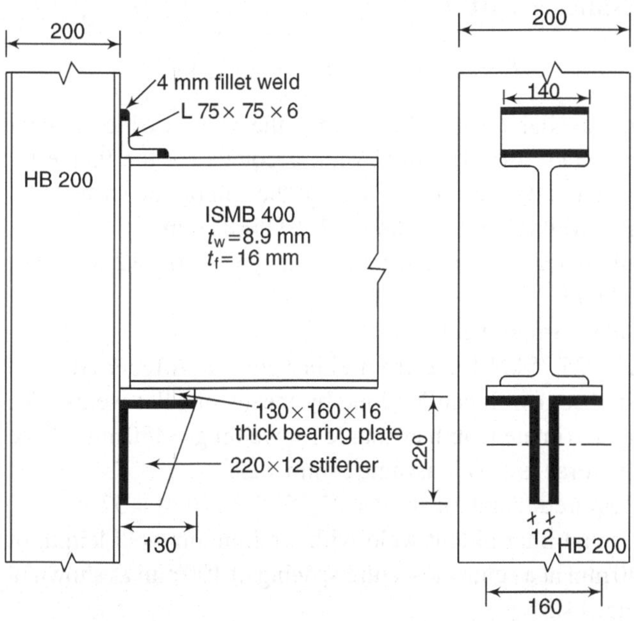


FIG. 14.50

**EXAMPLE 14.18:** An ISA 125 × 75 × 8 is welded with the flange of a column ISHB 300. The bracket carries a factored load of 75 kN at a distance of 38 mm as shown in Fig. 14.51. Design the connection. Use Fe 410 grade steel.

**SOLUTION:**

Flange width of ISHB 300 = 250 mm

Provide welding at the top and bottom of the seating angle, which acts as a bracket. The top weld will be in tension and the bottom will be in compression, in addition to sharing the shear force. Provide 250 mm length of weld at the top and bottom of the angle, and denoting $t$ as the effective throat thickness.

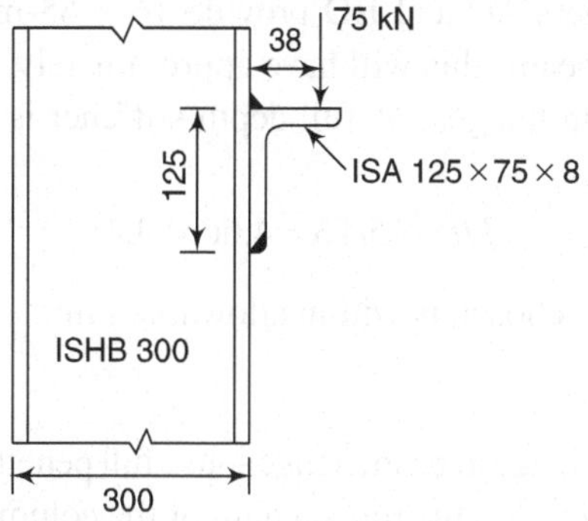


FIG. 14.51

Strength of the weld $= t \times 0.7 \times 410/(\sqrt{3} \times 1.5) = 110.4t$

Shear force in 1 mm length of weld $= 75{,}000/(2 \times 250)$

$= 150$ N

Force in 1-mm length of weld due to bending moment (see Table 14.1)

$= (75{,}000 \times 38)/(125 \times 250) = 91.2$ N

Resultant force $= \sqrt{(150^2 + 91.2^2)} = 175.6$ N

Thus,

$$110.4t \geq 175.6$$

or

$$t \geq 1.59$$

Hence, a weld size of 3 mm should be provided.

**EXAMPLE 14.19:** Design the corner connection between a column and a beam both having ISMB 400 size with Fe 410 steel. The factored moment and shear acting on the connection are 200 kNm and 450 kN, respectively.

**SOLUTION:**

Required properties of ISMB 400, $d = 400$ mm, $t_f = 16.0$ mm, $t_w = 8.9$ mm, $b_f = 140$ mm, $r = 14$ mm, and $k = 16 + 14 = 30$ mm

**Step 1:** *Check for web thickness*

The required web thickness

$t_w = 1.9M/(d_c d_b f_y)$

$= 1.9 \times 200 \times 10^6/(400 \times 400 \times 250) = 9.5 \text{ mm} > 8.9 \text{ mm}$

Hence, provide diagonal stiffeners.

**Step 2:** *Area of diagonal stiffeners*

$$A_{st} = [1.1/(f_y \cos\theta)]\,[(M/d_b) - f_y t_w d_c/(1.1 \times \sqrt{3})]$$

$$= [1.1/(250 \times 0.707)]\,[200 \times 10^6/400 - 250 \times 8.9 \times 400/(1.1 \times \sqrt{3})]$$

$$= 205 \text{ mm}^2$$

Hence, provide two stiffeners on each side of the beam web with an area of 103 mm² each.

Choose 55 × 6 mm plates, area = 330 mm² (for each plate)
Allowable $b/t \leq 9.4$, actual $b/t = 55/6 = 9.17$
Hence, the $b/t$ chosen is within allowable limit.

For stiffeners AC and BD provide 16 × 65-mm plates on each side of beam (this will have approximately same area as that of column flanges). A full depth stiffener is used for AC and BD.

$$b/t = 65/16 = 4.06 < 9.4$$

Hence, the $b/t$ chosen is within allowable limit.

**Step 3:** *Welds*

(a) Column flange to beam flange—use full penetration groove weld to develop the full strength of the column flanges
(b) Column web to beam flange—since it has been assumed that the column flanges carry all the moment, the web is assumed to carry entire shear.
Strength of weld of leg size 1 mm = $1 \times 0.7 \times 410/(\sqrt{3} \times 1.5)$
= 110.4 N/mm
Available weld length = 2[400 – 2(16 + 14)] = 680 mm
Required leg size = $450 \times 10^3/(680 \times 110.4) = 5.99$ mm
Hence, provide a 6-mm weld.
(c) Stiffeners: The stiffeners are attached by means of full penetration groove welds at C and D as shown in Fig. 14.52.
At either stiffener, the force to be transferred,

$$F = 200 \times 10^6/400 = 500 \times 10^3\,\text{N}$$

Weld length available for both stiffeners at AC

$$= 2[400 - 2(16 + 14)] = 680\,\text{mm}$$

Required leg size of the weld = $500 \times 10^3/(680 \times 110.4)$ = 6.66 mm

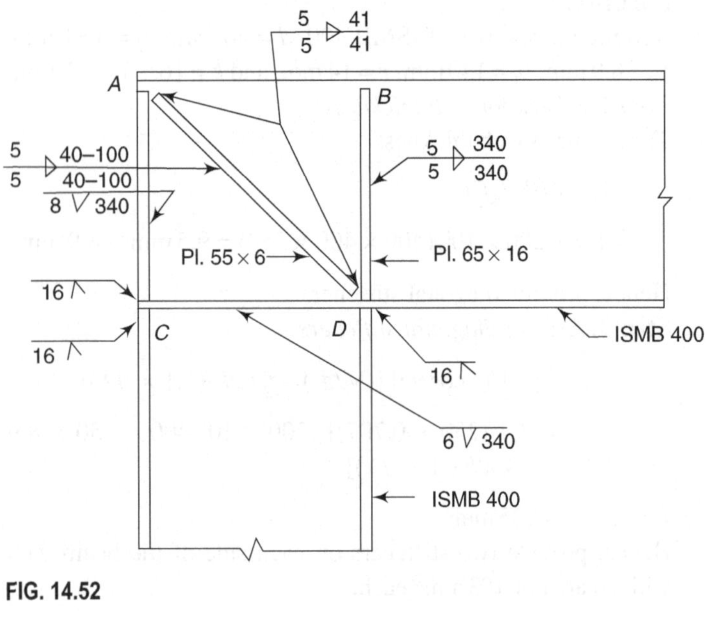


FIG. 14.52

Hence, provide 8-mm fillet welds.
Stiffener BD must carry the difference between the flange force, 500 kN, and the resistance provided by the beam web. The resistance provided by beam web is given by

$$F_{wb} = t_{wb}(t_{fc} + 5k)f_y/1.1$$
$$= 8.9\,(16 + 5 \times 30) \times (250/1.1) \times 10^{-3} = 335.77\,\text{kN}$$

Thus,
Force in the stiffener = 500 – 335.77 = 164.2 kN
Length available for welding for stiffener BD = 4 × 340 = 1360 mm
Required leg size of the weld = $164.2 \times 10^3/(1360 \times 110.4)$ = 1.09 mm
Use a full size stiffener with a 5-mm weld.
*Note:* Stiffener AC and BD should not be extended to meet the underside of the top flange of the beam, since the stiffener force has already been transferred into the beam web. Hence, they are stopped at the point of the web-to-flange fillet, that is, 30 mm below the top flange. Fitting the stiffeners to the underside of the beam flange is expensive and not required.

Using the actual stiffener area (excess area was provided in order to meet the local buckling requirement), the force in each stiffener at AD is

$$P = 103 \times 250 \times 10^{-3} = 25.75\,\text{kN}$$

Length available for welding along the 55-mm wide stiffener plate is 55 – 14= 41 mm (14 mm accounts for the flange-to-web fillet at this location). Thus, at the end of one stiffener,
Length available for welding = 2 × 41 = 82 mm
Required leg size of the weld = $25.75 \times 10^3/(82 \times 110.4)$ = 2.84 mm
Provide a 5-mm weld.

The 25.75 kN force from either end of stiffener AD into the beam web can be done by means of fillet welds. The approximate length available for welding = 480 mm. Since the force is small, assume 5-mm weld.
Required length of weld = 25.75/(5 × 110.4) = 52 mm
Use an intermittent weld with a minimum weld length of 40 mm at a centre-to-centre spacing of 100 mm as shown in Fig. 14.52.

**Example 14.20:** Design a welded moment connection between an ISMB 500 which transmits an end reaction of 300 kN (factored) and a factored bending moment of 150 kN, and the flange of a column of ISHB 300 as shown in Fig. 14.53. Assume steel grade Fe 410.

**Solution:**
Beam ISMB 500

$$t_{wb} = 10.2\,\text{mm and } t_{fb} = 17.2\,\text{mm}$$

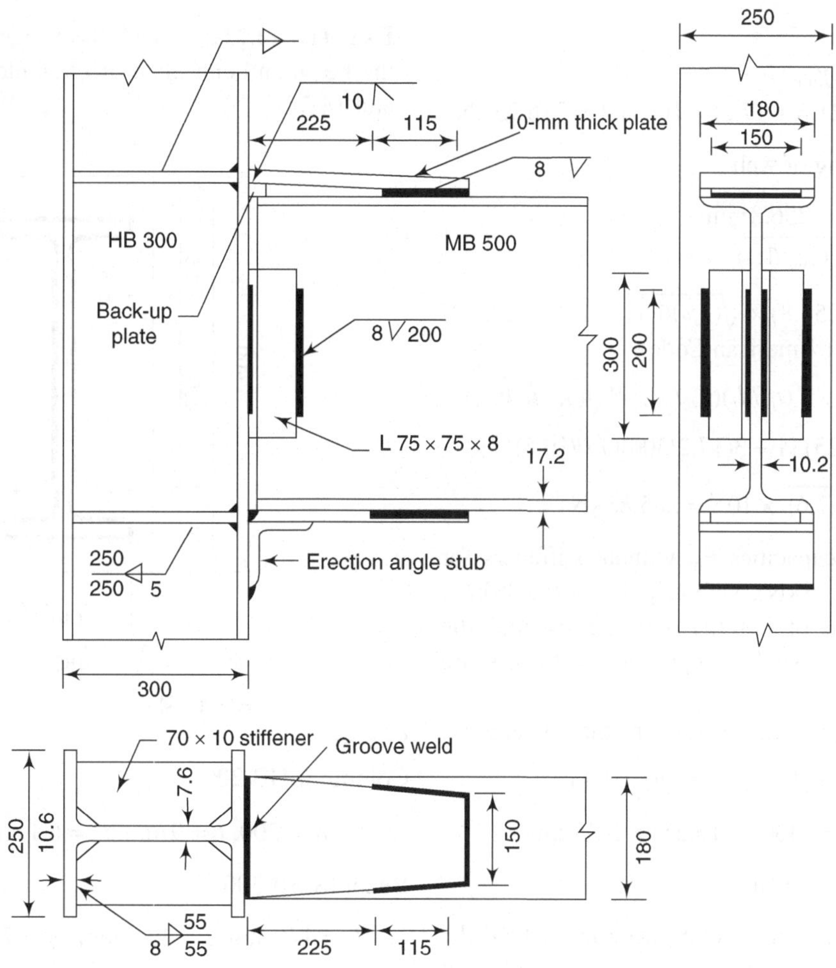


FIG. 14.53

Column ISHB 300

$$t_{wc} = 7.6\,\text{mm and } t_{fc} = 10.6\,\text{mm}$$
$$k = 10.6 + 11.0$$
$$= 21.6\,\text{mm}$$

Shear = 300 kN
Moment = 150 kNm
Flange force = $T = C = M/d_b = 150 \times 10^3/500 = 300$ kN

**Step 1:** *Topplate*
Use a fillet weld of size 8 mm.
Required area of the top plate = $300 \times 10^3/(250/1.1) = 1320\,\text{mm}^2$
Maximum width of the plate = $180 - 4 \times 8 = 148$ mm
Thickness of the plate = 1320/148 = 8.92 mm, say 10 mm
Use top and bottom plate of size 150 × 10 mm and assume site welding.
Strength of the weld = $8 \times 0.7 \times 410/(\sqrt{3} \times 1.5) = 883$ N/mm
Required length of the fillet weld = $300 \times 10^3/883 = 340$ mm
Length of the weld along the plate = (340 – 150)/2 = 95 mm
Provide side length of the weld = 95 + 2 × 8 = 111 mm, say 115 mm

Required length of plate = 150 × 1.5 + 115 = 340 mm
Use a top plate of size 150 × 10 mm and 340 mm long.

Use full strength single-bevel groove weld at column with a backing strip of size 40 × 8 mm. For compression, the erection stub may be used as backing strip.
The end reaction is 300 kN.
Length of weld = $300 \times 10^3/(883 \times 2) = 170$ mm
Length of angle = 0.6 × 500 = 300 mm
Use two angles 75 × 75 × 8 – 300 mm
Use weld length/side = 200 mm

**Step 2:** *Shear strength of web angle*
Shear strength of connected leg of angle

$$= f_y A_v/(\sqrt{3} \times 1.1)$$
$$= 2[250 \times (8 \times 300)]/(\sqrt{3} \times 1.1)$$
$$= 629\,\text{kN} > 300\,\text{kN}$$

**Step 3:** *Stiffeners*
Column-flange stiffener at beam tension (top) flange

$$= 0.4\sqrt{(A_f/1.1)} = 0.4\sqrt{(150 \times 10/1.1)} = 14.77\,\text{mm} > 10.6\,\text{mm}$$

Hence stiffener is required.
Column-web stiffener at beam compression (bottom flange)

(a) Local web yielding

$$P_{bf} = f_{wc} t_{wc}(5k + t_{fb})/\gamma_{mo}$$
$$= 250 \times 7.6(5 \times 21.6 + 17.2) \times 10^{-3}/1.1 = 216.25\,\text{kN}$$

(b) Compression buckling of web

$$h_c = 300 - 2 \times 21.6 = 256.8\,\text{mm}$$
$$P_{bf} = 10{,}710(t_{wc}^3/h)\sqrt{(f_{yw}/1.1)}$$
$$= 10710 \times (7.6^3/256.8) \times \sqrt{(250/1.1)} \times 10^{-3} = 276\,\text{kN}$$

(c) Web crippling (as per American Code)

$$P_{bf} = (300 t_{wc}^2/\gamma_{m1})[1 + 3\,(t_{fb}/d_c)(t_{wc}/t_{fc})^{1.5}]\sqrt{(f_{yw} t_{fc}/t_{wc})}$$
$$= (300 \times 7.6^2/1.25)\,[1 + 3(17.2/300)(7.6/10.6)^{1.5}]$$
$$\sqrt{(250 \times 10.6/7.6)} \times 10^{-3} = 285.88\,\text{kN}$$

Since the factored load capacities $P_{bf}$ without stiffeners for local web yielding (216.25 kN), web crippling (285.88 kN), and compression buckling of web (276 kN) are less than the 300 kN required to be resisted, compression stiffeners are required.

Taking the lowest ($P_{bf}$ = 216.25 kN) of the three strengths,

$$\text{Required } A_{st} = (P_u - 216.25)/(f_{yst}/1.1)$$
$$= (300 - 216.25) \times 10^3/(250/1.1)$$
$$= 369\,\text{mm}^2$$

Stiffeners to prevent local web yielding need not extend the full depth of the column web, unless the force is applied at both column webs. Stiffeners to prevent web crippling should extend full depth and be designed as columns, the same way as bearing stiffener in plate girders.

Local buckling limit $b/t$ = 9.4.

Use two (70 mm × 10 mm)-size stiffeners,

Area = 700 mm² (each) with $b/t$ = 7.0

Hence, the area is greater than required.

Local buckling limits do not apply to tension elements. However, for practical purposes, provide the same stiffener at the top also.

Maximum force to be carried by stiffener = 300 – 216.25 = 83.75

Adopting a weld size of 8 mm,

Strength of weld = 8 × 0.7 × 410/(√3 × 1.50) = 883 N/mm

Required length of weld = 83.75/(2 × 0.883)

= 47.3 mm > length available

Hence the stiffener dimensions are correct.

**Step 4:** *Column web shear*

$$t_{wc} = 1.9M/(d_c d_b f_y)$$
$$= 1.9 \times 150 \times 10^6/(500 \times 300 \times 250)$$
$$= 7.6\,\text{mm} = \text{web thickness of the column}$$

Hence, no diagonal web stiffener is required.

**Example 14.21:** Check the welded connection between the beam and end plate of Example 14.7 (See Figs 14.40 and 14.54).

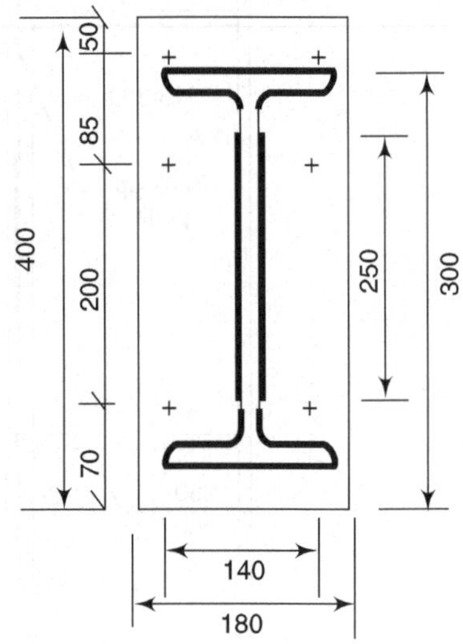


**FIG. 14.54**

**Solution:**

Column ISHB 200

$b_f$ = 200, $t_f$ = 9 mm, $t_w$ = 6.1 mm, and $R$ = 9 mm

Beam ISMB 300

$b_f$ = 140 mm, $t_f$ = 13.1 mm, $t_w$ = 7.7 mm, $R$ = 14.0 mm

Factored bending moment = 120 kNm

Factored shear force = 120 kN

**Step 1:** *Welds*

Capacity of unit weld = 1 × 0.7 × 410/(√3 × 1.25) = 132.5 N/mm

Assuming unit weld throat and 140-mm long welds at each flange (top and bottom) and 240-mm long on each face of web of the beam,

$$\text{Area} = 2 \times 140 + [2 \times (140 - 7.7)] + 2 \times 240 = 1024.6\,\text{mm}^2$$

$$I_{zz} = 2 \times 140 \times (300/2)^2 + 2 \times 240^3/12 + 2 \times 132.3 \times (300/2 - 13.1)^2$$
$$= 13.563 \times 10^6\,\text{mm}^4$$

$Z$ for the extreme fibre = 13.563 × 10⁶/(300/2)

= 90.42 × 10³ mm³

Direct stress = 120/1024.6 = 0.117 kN/mm

Bending stress = 120 × 10³/(90.42 × 10³) = 1.327 kN/mm

Vector sum = (0.117² + 1.327²)^0.5 = 1.332 kN/mm

Weld size = 1.332 × 10³/132.5 = 10.05

*Note:* The thickness of web is 7.7 mm only; hence we can adopt only 6 mm weld on web. Extra plates have to be provided to increase the weld length.

**Step 2:** *Check for stiffener*

Column flange stiffener at beam tension (top) flange

$= 0.4\sqrt{(A_f/1.1)} = 0.4 \times \sqrt{(140 \times 13.1/1.1)} = 16.33 > 9\,\text{mm}$

Hence a stiffener is required.

Column web stiffener at beam compression (bottom flange)

(a) Local web yielding

$$P_{bf} = f_{wc}\, t_{wc}\, (5k + t_{fb})/\gamma_{m0}$$
$$= 250 \times 6.1\ (5 \times 18 + 13.1) \times 10^{-3}/1.1 = 142.94\,\text{kN}$$

(b) Compression buckling of the web

$$P_{bf} = 10710\ (t_{wc}^3/h)\sqrt{(f_{yw} \times 1.1)}$$
$$= 10710 \times (6.1^3/164) \times \sqrt{(250 \times 1.1)} \times 10^{-3} = 223.46\,\text{kN}$$

Flange force = 120 × $10^3$/300 = 400 kN > $P_{bf}$ (142.94 or 223.46 kN)

Hence, stiffeners are required.

Required $A_{st} = (P_u - P_{bf})/(f_{yst}/1.1)$
$= (400 - 142.94) \times 10^3/(250/1.1) = 1131\,\text{mm}^2$

Local buckling limit, $b/t = 9.4$

Use two (120 × 14)-mm thick stiffener with

$$\text{Area} = 1680\,\text{mm}^2$$

and

$$b/t = 8.57$$

Provide the same stiffener for tension flange also.

Maximum force carried by the stiffeners = 400 – 142.94 = 257.1 kN

Length available for welding = 120 – 9 = 111 mm

Size of the weld = (257.1/2) × $10^3$/(2 × 111 × 132.5) = 4.4 mm

Provide a 5-mm weld.

**Step 3:** *Column web shear*

$$t_{wc} = 1.9M/(d_c d_b f_y)$$
$$= 1.9 \times 120 \times 10^6/(200 \times 300 \times 250) = 15.2 > 7.7\,\text{mm}$$

$$A_{st} = [1.1/(f_y \cos\theta)]\ [M/d_b - f_y t_w d_c/(\sqrt{3} \times 1.1)]$$
$$= 1.1/(250 \times 0.56)\ [120 \times 10^6/300 - 250 \times 6.1 \times 200/(\sqrt{3} \times 1.1)]$$
$$= 1885\,\text{mm}^2$$

Provide two 135 × 14 diagonal stiffener,

$$\text{Area} = 1890\,\text{mm}^2$$

Load to be taken by stiffener

$$= M/d_b - f_y t_w d_c/(\sqrt{3} \times 1.1) = 239.9\,\text{kN}$$

Approximate length available for welding = 320 mm

Required size of the weld = 239.9 × $10^3$/(2 × 320 × 132.5) = 2.8 mm

Provide a 5-mm weld.

## SUMMARY

Details of various moment-resistant connections such as eccentrically loaded connections (load causing twisting moment and bending moment), end plate connections, and beam-column moment joints have been covered. Since it is impossible to consider all the combinations of connections, only those that are frequently encountered in practice are discussed. The concepts presented have been explained with several numerical examples. Interested readers may find more details about the testing, analysis, or design of these moment-resistant connections from the references listed in this chapter.

## MULTIPLE-CHOICE QUESTIONS

14.1 Moment-resistant connections can transfer the following from one member to another:
(a) Axial force and bending moment
(b) Bending moment and shear
(c) Shear and axial force
(d) Bending moment, shear, and axial force

14.2 The moment rotation curve of a joint in a frame is shown here:

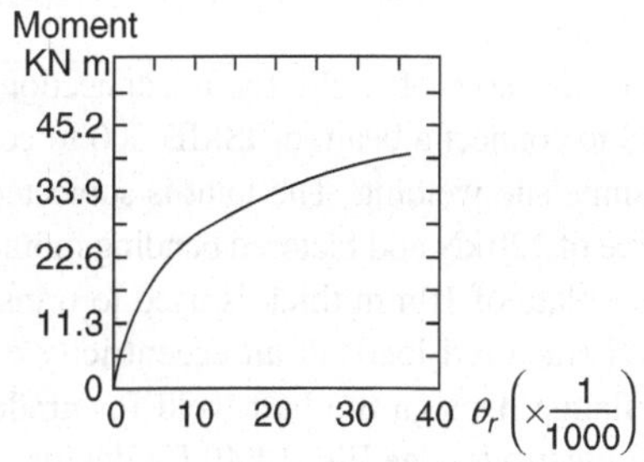


It is a ______________.
(a) pin joint
(b) flexible joint
(c) semi-rigid joint
(d) rigid joint

14.3 In the bracket connection shown here, which bolt will have maximum resultant force?

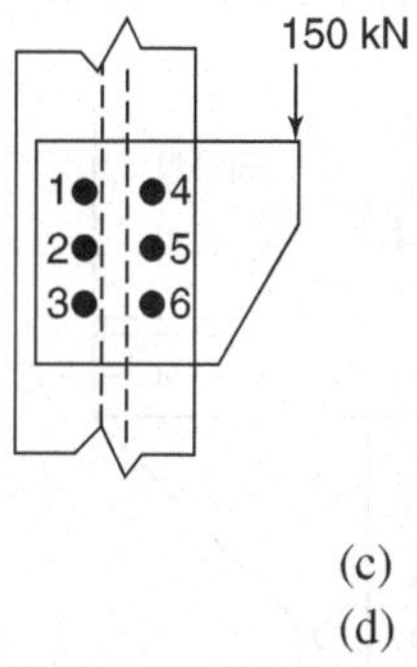


(a) Bolt 1
(b) Bolt 4
(c) Bolt 6
(d) Bolt 3

14.4 An example of light moment connection is
(a) single web angle connection
(b) end plate connection
(c) stiffened seat connection
(d) T-stub connection

14.5 In the bracket connection of question 14.3, the force in the bolt is given by the formula

(a) $\frac{Md_i}{\Sigma d_i}$ (b) $\frac{Md_i^2}{\Sigma d_i^2}$ (c) $\frac{M}{\Sigma d_i^2}$ (d) $\frac{Md_i}{\Sigma d_i^2}$

14.6 In pure moment type connection, when a horizontal force $H$ and vertical force $V$ are acting eccentrically to the centroid (with eccentricities $e_v$ and $e_h$ respectively), the component of force $R_x$ acting in a bolt (in a bolt group) in x direction is given by

(a) $\frac{H}{n}+\frac{(Ve_h+He_v)y}{\Sigma(x^2+y^2)}$ (c) $\frac{H}{n}+\frac{(Ve_h+He_v)y}{\Sigma(x+y)}$

(b) $\frac{H}{n}+\frac{(Ve_h+He_v)}{\Sigma(x^2+y^2)}$ (d) $\frac{H}{n}+\frac{(V+He_v)y}{\Sigma(x^2+y^2)}$

14.7 The required number of bolts $n$ in one row in a bolt group with bolts having pitch $p$, number of rows $m$ and strength of one bolt $R$, is given by

(a) $n=\sqrt{\frac{6M}{pmR}}$ (c) $n=\sqrt{\frac{6M}{pR}}$

(b) $n=\sqrt{\frac{M}{pmR}}$ (d) $n=\sqrt{\frac{6Mp}{mR}}$

14.8 The combined bending and shear stress in the groove weld is checked by

(a) $f_e=\sqrt{f_b^2+3\tau^2}$ (c) $f_e=\sqrt{f_b+3\tau^2}$

(b) $f_e=\sqrt{f_b^2+3\tau}$ (d) $f_e=\sqrt{f_b^2+\tau^2}$

14.9 The horizontal stiffener in compression region of a beam-column joint has to be checked for
(a) local web yielding
(b) compression buckling of the web
(c) web crippling
(d) all of these

## EXERCISES

14.1 Design a stiffened seat angle for a reaction of 160 kN from beam of ISMB 300 using M16 bolts of grade 4.6. The beam has to be connected to ISHC 200 column. Assume Fe 410 grade steel with $f_y$ = 250 MPa (see Fig.14.35 for the arrangement of connection).

14.2 Given the bracket connection shown in Fig. 14.36. With 16mm diameter grade 4.6 bolts and plate of Fe 410 steel, is the bolt pattern and plate adequate for supporting the given load of 50kN in a bearing type connection assuming threads in the shear plane?

14.3 Determine the size of the bolts required to connect the bracket (supporting a gantry girder) to the ISHB column 200 shown in Fig. 14.37, with $V$ = 100 kN, $H$ = 25 kN, $e_v$ = 250mm, $e_h$ = 200mm, $x$ = 100mm and M24 grade 4.6 bolts.

14.4 Find the maximum load inclined at 45° to the horizontal, which the bracket shown in Fig. 14.38 can transmit if 5–16mm diameter grade 8.8 bolts are used and the plates connected are 10mm thick. Determine the load for (a) if the joint is considered a slip joint and (b) if the joint is considered as a non-slip joint.

14.5 Design a bracket connection to transfer a load of 150 kN as shown here.

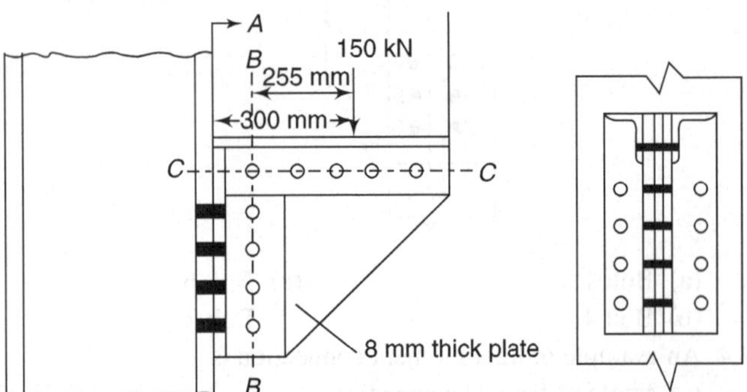


14.6 Design a bolted end plate connection between an ISMB 250 beam and an ISHB 200 column, to transfer a vertical factored shear of 90 kN and a factored hogging bending moment of 80 kNm. Use HSFG bolts of diameter 20 mm. The arrangement of connection is as shown in Fig. 14.40.

14.7 Design a flange angle connection using M20 bolts of grade 4.6 to transfer a moment of 10 kNm and a shear of 200 kN from a beam of ISMB 350 to a column of ISHB 300 (see Fig.14.41 for the arrangement of connection).

14.8 Design a T-stub connection to transfer a shear of 100 kN and a moment of 30 kNm from the end of a beam ISMB 250 to a column of ISHB 300 (See Fig.14.43 for the arrangement of connection).

14.9 Determine the maximum load that could be resisted by the bracket shown in Fig. 14.44, (with the load acting at 45°) assuming the plate thickness does not affect the result and if 8 mm shop fillet welds are used.

14.10 A bracket plate is welded to the flange of a column ISHB 300 as shown in Fig. 14.45. Calculate the size of the weld required to support a factored load of 150 kN.

14.11 Determine the load carrying capacity of the joint shown in Fig.14.46 for 3 mm, 5mm and 7mm welds. Assume field welding.

14.12 Determine the length of fillet weld required to carry the load indicated by Fig. 14.47. Assume 8mm shop weld and that the load of 200 kN is acting at 120 mm eccentricity from the face of the column.

14.13 Determine the size of weld for a connection as shown in Fig.14.48 to connect a beam of ISMB 300 to column of ISHB 200. Assume site welding. The joint is subjected to a factored shear force of 120 kN and factored bending moment of 20 kNm.

14.14 A bracket plate of 10mm thick is used to transmit a reaction of 100 kN (factored load) at an eccentricity of 60 mm from column flange. Design the butt weld for grade Fe 410 steel and E 43 electrode. See Fig. 14.49 for details.

14.15 Design stiffened seat angle connection for a reaction of 200 kN from beam of ISMB 300. This beam has to be connected to a column of size ISHB 200. Assume Fe 410 grade steel and shop welding (see Fig. 14.50 for the arrangement).

14.16 An ISA 150 × 115 × 8 is welded with the flange of a column ISHB 300. The bracket carries a factored load of 100 kN at a distance of 50 mm as shown in Fig. 14.51. Design the connection using Fe 410 grade steel.

14.17 Design the corner connection between a column and a beam both having ISMB 300 size with Fe 410 steel. The factored moment and shear acting on the connection are 120 kN m and 300 kN respectively.

14.18 Design a welded moment connection between an ISMB 400 which transmits an end reaction of 200 kN (factored) and a factored bending moment of 100 kN, and the flange of a column of ISHB 200 as shown in Fig.14.53. Assume steel grade Fe 410.

14.19 Check the welded connection between the beam of size ISMB 250 and end plate of size 180 mm × 350 mm with 6 nos 20 mm dia HSFG bolts connected to an ISHC 200 column (see Figs 14.40 and 14.54 for the arrangement of bolts).

## REVIEW QUESTIONS

14.1 What are the two main types of moment resistance connections?

14.2 Write the expression for calculating the force $R$ in a bolt subjected to a moment $M$ and located at a distance $d$ from the centre of rotation.

14.3 What is the expression used to find the number of bolts when a bolt group is subjected to applied moment or torque?

14.4 What is the design procedure adopted for the design of bolt group subjected to torque?

14.5 What are the steps that are involved in the design of a bracket connection?

14.6 Write short notes on the following:
(a) Bolted moment end plate connection
(b) Flange angle connection
(c) Split-beam tee connection
(d) Beam-to-beam connection
(e) Beam splices
(f) Column splices

14.7 Why are HSFG bolts preferred in joints subjected to fatigue?

14.8 When are backup strips used in groove welds? Under what condition should they not be used?

14.9 Under what circumstances are stiffeners provided?

14.10 Write short notes on the following:
(a) Moment-resistant connections
(b) Stiffened beam seat connections
(c) Continuous beam-to-column connections
(d) Continuous beam-to-beam connections

## REFERENCES


Arya, A.S. and J.L. Ajmani, *Design of Steel Structures*, Fourth edition, Nem Chand & Bros, Roorkee, 1989, p. 948.

Blodgett, O.W., *Design of Welded Structures*, The James F. Lincoln Arc Welding Foundation, Cleveland, 1966.

Bowles, J.E. *Structural Steel Design*, McGraw Hill, New York, 1980, p. 536.

BCSA/SCI 207/95, *Joints in Steel Construction—Moment Connections*, The Steel Construction Institute, Ascot and The British Constructional Steelwork Association, UK, 1997, p. 239.

BCSA/SCI P398/13, *Joints in Steel Construction: Moment-Resisting Joints to Eurocode 3*, The Steel Construction Institute, Ascot and The British Constructional Steelwork Association, UK, 2013, p. 172.

Butler, L.J. and G.L. Gulak, 'Strength of Fillet Welds as a Function of Direction of Load', *Welding Research Supplement, Welding Journal*, 1968, pp. 321–345.

Butler, L.J., S. Pal, and G.L. Kulak, 'Eccentrically Loaded Welded Connections', *Journal of the Structural Div.*, ASCE, Vol. 98, No. ST5, May 1972, pp. 989–1005.

Chen, W.F. and Newlin, D.E., 'Column Web Strength in Beam-to-column Connections', *Journal of the Structural Div.*, ASCE, Vol. 99, No. ST9, Sep 1973, pp. 1978–1984.

Chen, W.F. and Patel, K.V., 'Static Behaviour of Beam-to-column Moment Connections', *Journal of the Structural Div*, ASCE, Vol. 107, No. ST9, Sep 1981, pp. 1815–1838.

Crawford, S.F. and Kulak, G.L., 'Eccentrically Loaded Bolted Connections', *Journal of the Structural Div.*, ASCE, Vol. 97, No. ST3, Mar 1971, pp. 765–783.

Dawe, J.L. and Kulak, G.L., 'Welded Connections under Combined Shear and Moment', *Journal of the Structural Div.*, ASCE, Vol. 100, No. ST4, Apr 1974, pp. 727–741.

Gaylord, E.H., Gaylord, C.N., and Stallmeyer, J.E., *Design of Steel Structures*, Third edition, McGraw Hill, New York, 1992, p. 792.

Graham, J.D., Sherbourne, A.N., Khabbaz, R.N., and Jensen, C.D., *Welded Interior Beam-to-column Connections,* American Institute of Steel Constructions, New York, 1959.

Grundy, P., Thomas, I.R., and Bennetts, I.D., 'Beam-to-column Moment Connections', *Journal of the Structural Div.*, ASCE, Vol. 106, No. ST1, Jan 1980, pp. 313–330 and Disc. By Krishnamurthy N., Dec 1980, pp. 2573–2574.

Kazimi, S.M.A. and R.S. Jindal, *Design of Steel Structures*, Second edition, Prentice-Hall of India Pvt. Ltd., New Delhi, 1988, p. 674.

Krishnamurthy, N., 'A Fresh Look at Bolted End-plate Behavior and Design', *Engineering Journal*, AISC, Vol. 15, No. 2, 2nd Quarter, 1978, pp. 39–49. Also see discussions, Vol. 16, No. 2, 1979, pp. 54–64.

Kulak, G.L. and Grondin, G.Y., *Limit States Design in Structural Steel*, Seventh edition, Canadian Institute of Steel Construction, 2002.

Kulak, G.L., Fisher, J.W., and Struik, J.H.A, *Guide to Design Criteria for Bolted and Riveted Joints*, Second edition, John Wiley, New York, 1987, p. 309.

Lesik, D.F. and Kennedy D.J.L., 'Ultimate Strength of Fillet Welded Connections Loaded in Plane', *Canadian Journal of Civil Engineering*, Vol. 17, No.1, 1990, pp. 55–67.

Martin, L.H. and Purkiss, J.A., *Structural Design of Steelwork to BS 5950*, Edward Arnold, London, 1992, p. 468.

Murray, T.M., *Extended End-plate Moment Connections*, Steel Design Guide Series 4, American Institute of Steel Construction, Chicago, 1990.

Narayanan, R. et al., *Teaching Material On Structural Steel Design* (45 Chapters based on Limit State Design of Steel Structures), INSDAG, New Delhi, 2001, http://www.steel-insdag.org/TM_Contents.asp.

Owens, G.W. and Cheal, B.D., *Structural Steelwork Connections*, Butterworths, London, 1989, p. 330.

Parfitt, J. Jr. and Chen, W.F., 'Tests of Welded Steel Beam-to-column Connections', *Journal of the Structural Div.*, ASCE, Vol. 102, No. ST1, Jan 1976, pp. 189–202.

Rentschler, G.P., Chen, W.F., and Driscoll, G.C., 'Beam-to-column Web Connection Details', *Journal of the Structural Div.*, ASCE, Vol. 108, No. ST2, Feb 1982, pp. 303–409.

Roberts, T.M., 'Slender Plate Girders Subjected to Edge Loading', *Proc. of the Institution of Civil Engineers,* Part 2, Sep 1981, p. 71.

Salmon, C.G., Buettner, D.R., and O'Sheridan, T.C., 'Laboratory Investigation of Unstiffened Triangular Bracket Plates', *Journal of the Structural Division,* ASCE, Vol. 90, No. ST4, Apr 1964, pp. 257–278.

Salmon, C.G., Johnson, J.E., and Malhas, F.A., *Steel Structures, Design and Behavior*, Fifth edition, Harper Collins College Publishers, New York, 2008, p. 888.

Timoshenko S.P. and Gere, J.M., *Theory of Elastic Stability*, Second edition, McGraw Hill, New York, 1961.

Trahair, N.S., Bradford, M.A., and Nethercot, D.A., *The Behaviour and Design of Steel Structures to BS 5950*, Third edition, Spon Press, London and New York, 2001, p. 448.

Witteveen, J., Stark, J.W.B., Bijlaard, F.S.K., and Zoetemeijer, P., 'Welded and Bolted Beam-to-column Connections', *Journal of the Structural Div.*, ASCE, Vol. 108, No. ST2, Feb 1982, pp. 433–455.

# 15 DESIGN OF BASE PLATES AND CAPS

## INTRODUCTION

For transmitting the load from columns to its foundations, *base plates* are used. Base plates assist in reducing the intensity of loading and distributing it over the foundations. The area of base plate is so chosen that the intensity of load distributed is less than the bearing capacity of concrete on which it rests.

The safety of a column and thus of a structure depends mainly upon the stability of the foundations and consequently on the bases, in the case of steel columns. Hence, column base plates should be designed with great care. The design of a base plate is generally assumed to be on the condition that the distribution of load under the base is uniform and the outstanding portions of the base plate are treated as cantilevers.

## 15.1 COLUMN BASES AND CAPS

The main types of bases used are shown in Fig. 15.1. These are as follows:

1. Slab base
2. Gussetted base
3. Pocket base

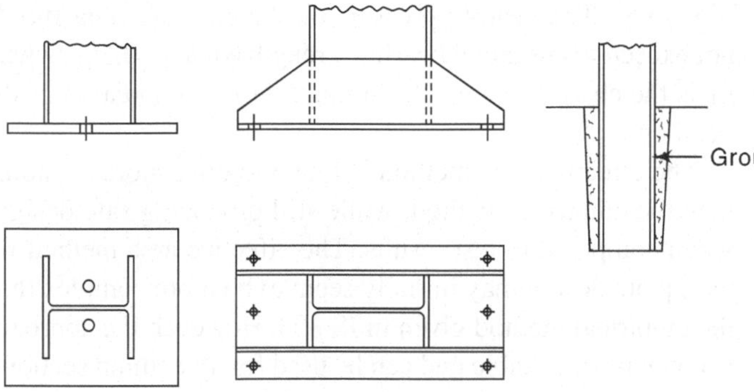


**FIG. 15.1** Column bases

With respect to slab and gussetted bases, depending on the values of axial load and moment, there may be compression over the whole base or compression over part of the base and tension in the holding-down bolts. Horizontal loads are resisted by shear in the weld between column and base plates, friction, and bond between the base and the concrete. Though the AISC code does not allow the anchor rods to transfer substantial shear, ACI-318, Appendix D gives the limit states to be checked in the anchorage, including the steel strength of the anchor in shear, as well as the various concrete limit states. If the base plate has a grout pad of any substantial thickness and the anchor rod does not bear against the base plate (the base plate holes will be larger than the anchor rod and hence in many cases the base connection will not bear against the side of the hole), then bending will be introduced in the rod in addition to shear. The bending capacity of the anchor rods is limited and hence the AISC code does not allow shear transfer through anchor rods (AISC design guide 1-2005). Hence AISC suggests the use of a shear key or lug or embedded plate with welded side plates to transfer a large horizontal shear force from the column base to the foundation (see Fig. 15.2). Note that the horizontal loads will be substantial for earthquake loading or wind loading.

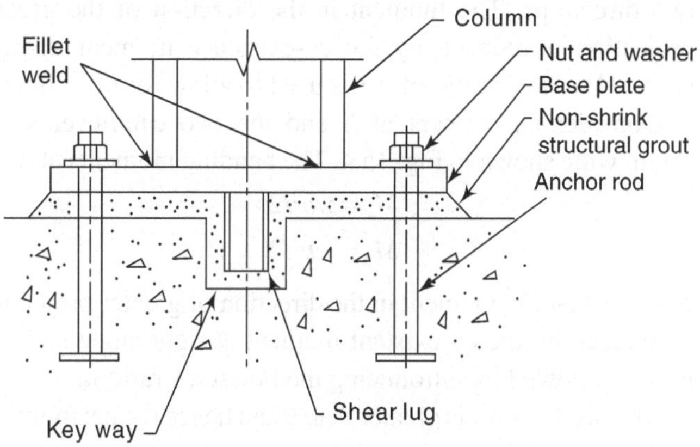


**FIG. 15.2** Use of shear lug to transfer heavy shear force

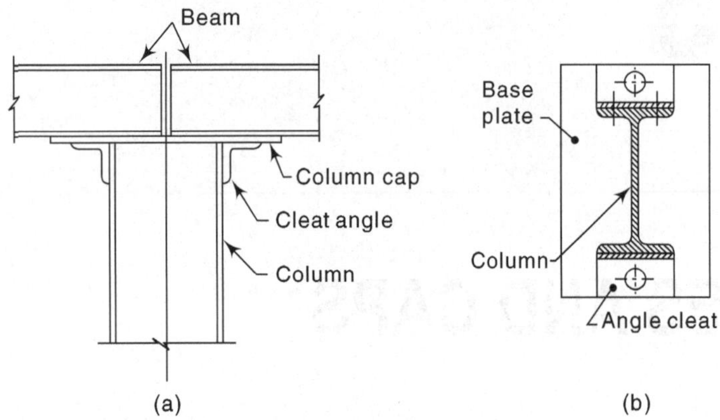


FIG. 15.3 (a) Column cap (elevation) (b) Column base (plan)

Lightly loaded columns are provided with thick slab bases. The slab base is free from pockets where corrosion may start. Base plates with especially large loads require more than a simple plate. This may result in a double layer of plates, a grillage system, or the use of stiffeners to reduce the plate thickness. The *column caps* serve similar purpose except that they act as a link between load coming on the columns and the column itself (see Fig.15.3).

## 15.1.1 Slab Bases

The design of slab bases with concentric load is covered in Section 7.4.3 of IS 800 : 2007. This states that where the rectangular plate is loaded by *I*-, *H*-, channel, box, or rectangular hollow sections, the minimum thickness of base plate $t_s$ should be

$$t_s = [2.5w(a^2 - 0.3b^2)\gamma_{mo}/f_y]^{0.5} > t_f \qquad (15.1)$$

where $w$ is the pressure on the underside of the slab base due to the factored compressive load on the column (assumed as uniformly distributed over the area of the slab base), $a$ and $b$ are the larger and smaller projections of the slab base beyond the rectangle, circumscribing the column, respectively, $f_y$ is the yield strength of the base plate, $t_f$ is the flange thickness of the compression member, and $\gamma_{m0}$ is the partial safety factor for material = 1.10.

Equation (15.1) takes into account plate bending in two directions. The moment in the direction of the greater projection is reduced by the co-existence moment at right angles. Poisson's ratio of 0.3 is used to allow for this effect.

Consider an element at $A$ and the two cantilever strips 1-mm wide shown in Fig. 15.4. The bending moments at $A$ are

$$M_x = wa^2/2$$

$$M_y = wb^2/2$$

Assume that the moment in the direction of greater projection is reduced by the co-existant moment at right angle and this may be allowed by introducing the Poisson's ratio $\mu$.

The projection $a$ is greater than $b$ and hence the net moment, with $\mu = 0.3$, is

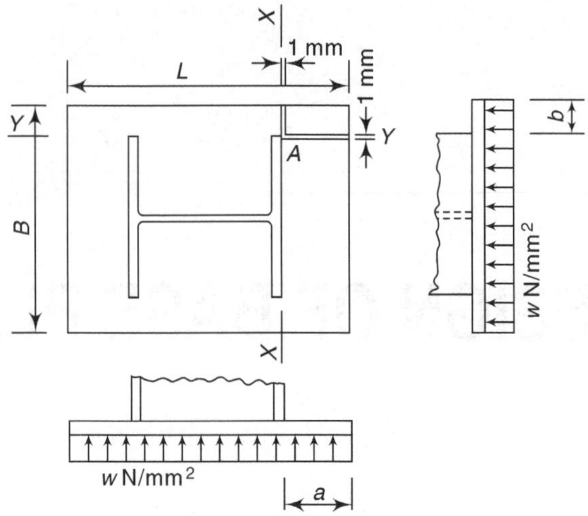


FIG. 15.4 Moment in column base plate

$$M_x = wa^2/2 - 0.3wb^2/2$$
$$= w/2(a^2 - 0.3b^2) \qquad (15.2a)$$

The moment capacity of the plate is given by

$$M_p = 1.2f_yZ_e$$

The elastic modulus for the cantilever strip is $t^2/6$. Thus,

$$M_p = 1.2f_yt^2/6 \qquad (15.2b)$$

Equating Eqns (15.2a) and (15.2b), solving for $t$, and applying the partial factor of safety for material we get Eqn (15.1), given in the code.

## 15.1.2 Effective Area of Base Plate

If the size of the base plate is larger than that required to limit the bearing pressure of the surface below the base plate, clause 7.4.1.1 of IS 800 assumes that an equal projection $c$ of the base plate from the base of the column and gusset may be taken as effective in transferring the column load, as shown in Fig. 15.5. It has to be noted that the actual distribution of pressure beneath a base plate is extremely complex. However, IS 800:2007 assumes a uniform distribution of pressure beneath an effective area of the base plate (as shown in Fig. 15.5). The bearing pressure on the effective area should not exceed the nominal bearing strength equal to $0.6f_{ck}$, where $f_{ck}$ is the characteristic cube strength of the foundation or the bedding material.

The effective area method is found to offer more economy than the empirical method, while still producing safe designs when compared to test results. The effective area method for base plate design may initially seem to be more complex than the empirical method given in IS 800. However, the approach is much more reliable and can be used for all column sections. The basic design procedure is as follows:

1. Calculate the required area = $P_u/(0.6f_{ck})$, where $P_u$ is the applied axial compressive load.

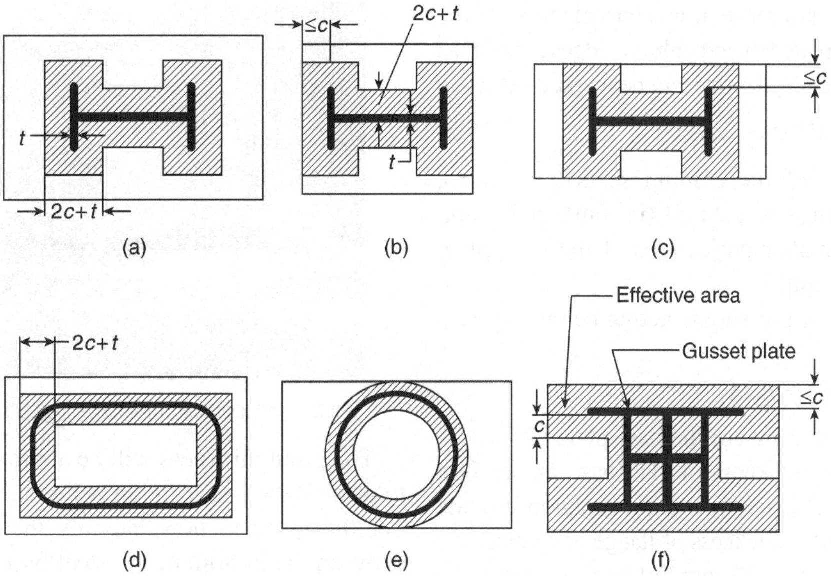


**FIG. 15.5** Effective area of base plate

2. Calculate outstand $c$ by equating required area to actual area expressed as a function of $c$. The expression for the actual effective area of an I- or H-section (see Fig. 15.6a) is

$$(b_f + 2c)(h + 2c) - [\{h - 2(t_f + c)\}(b_f - t_w)] \quad (15.3)$$

which may be approximately taken as $4c^2 + 2c(2b_f + h)$, where $b_f$ is the breadth of flange, and $h$ is the height of the section

3. Check that there is no overlap of effective area between flanges (see Fig. 15.6b). This will occur if $2c >$ the distance between the inner faces of the flanges. If an overlap exists, modify the expression for effective area and recalculate $c$.
4. Check the effective area fits on the size of base plate selected (see Fig.15.6c). If effective area does not fit on the base plate, modify the expression for effective area to allow for the limitations of the plate size and recalculate $c$, or select a larger base plate. For the case shown in Fig.15.6(c), the modified expression for the effective area can be written as:

$$4c^2 + 2c(3b_f + h) - b_f h - 2 \times (b_f + 2c) \times (c - a) \quad (15.4)$$

5. If $c$ has been recalculated step 3 should be repeated.
6. Calculate required plate thickness $t_p$ using the following expression:

$$t_p = c[2.5w\,\gamma_{m0}/f_{yp}]^{0.5} \quad (15.5)$$

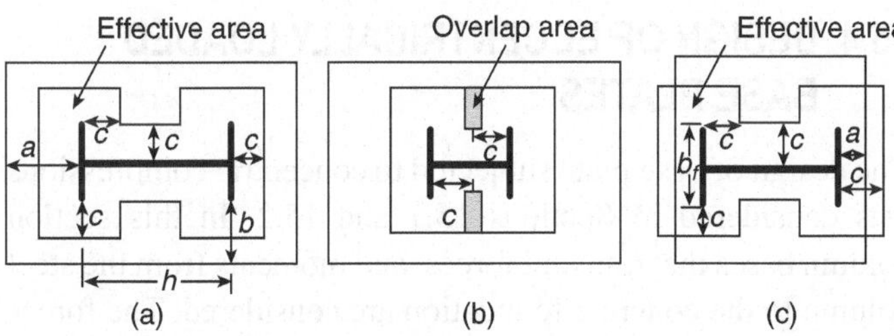


**FIG. 15.6** Calculation of effective area

where $w$ is the pressure under the base plate based on an assumed uniform distribution of pressure under the effective area ($\le 0.6 f_{ck}$), $f_{yp}$ is the design strength of the base plate. This expression for the plate thickness has been derived from equating the moment produced by the uniform load $w$ to the elastic moment capacity of the base plate (both per unit length).

### 15.1.3 Weld: Column to Slab Base

The code states in clause 7.4.3.4 that where the cap or slab base is fillet welded directly to the column, the contact surfaces should be machined to give a perfect bearing and the welding shall be sufficient to transmit the forces. When full strength butt welds are provided, machining of contact surfaces is not required. Also, where the end of the column is connected directly to the base plate, by means of full penetration butt welds, the connection is deemed to transmit to the base all the forces and moments to which the column is subjected.

## 15.2 DESIGN OF BASE PLATE

The design of base plate consists of finding out its size and thickness. The following are the design steps to be followed:

1. Assuming the grade of concrete, calculate the bearing strength of concrete, which is given by $0.45 f_{ck}$ as per clause 34.4 of IS 456:2000 (Note that 1S 800, in clause 7.4.1 suggests a value of $0.6 f_{ck}$)
2. Required area of slab base may be computed by

$$\text{Required } A = P_u/\text{bearing strength of concrete} \quad (15.6)$$

where $P_u$ is the factored concentric load on the column

3. From the above, choose a size for the base plate $L \times B$, so that $L \times B$ is greater than the required area. Though a few designers

prefer to have a square base plate, it is advisable to keep the projections of the base plate beyond column edges $a$ and $b$ as equal (see Fig. 15.4). Hence, the size can be worked out as

$$(D + 2b)\,(b_f + 2a) = A \qquad (15.7)$$

where $D$ is the depth of the column section (in mm), $b_f$ is the breadth of flange of column (in mm), and $a$ and $b$ are the larger and smaller projections of the base plate beyond the column (in mm).

4. Calculate the intensity of pressure $w$ acting below the base plate using

$$w = P_u / A_1$$

where $A_1$ is the provided area of the base plate ($L \times B$)

5. Calculate the minimum thickness of slab base as per Eqn (15.1). If it is less than $t_f$, thickness of flange of column, provide the thickness of slab base = thickness of flange of column
6. Provide nominal two or four 20-mm holding-down bolts
7. Check the weld length connecting the base plate with the column (this check is required only for fillet welds)

## 15.3 SOME PRACTICAL ASPECTS

Anchor bolts vary in size from approximately 20-mm diameter to 60-mm diameter. Anchor rod layouts for base plates should provide ample clearance distance for the washer from the column shaft and its weld, as well as reasonable edge distance. When the hole edge is not subjected to lateral force, even an edge distance that provides a clear dimension as small as 12 mm of material from the edge of the hole to the edge of the plate will normally suffice, though field issues with anchor rod placement may necessitate a larger dimension to allow some slotting of the base plate holes. When the hole's edge is subjected to lateral force, the edge distance must be large enough for the necessary force transfer. Anchor bolts, subject to corrosive conditions may be galvanized.

When ordering galvanized bolts, the threads should be *chased* so the nuts will work freely. Most anchor bolts are made from grade 410 steel though high-strength bolts may also be used. Most anchor bolts come with hexagonal nuts. Because the base plates will be provided with oversize holes, it is necessary to provide thick erection washers under the nuts.

The following three methods can be used effectively to prepare a landing site for the erection of a column (Collins and Schlafly 2010):

1. Leveling plates that may also serve as anchor rod templates (for small to medium sized base plates, say up to 560 mm)
2. Leveling nuts and washers on anchor rods beneath the base plate
3. Pre-set base plates

The details of these methods may be found in AISC design guide 1-2005.

Column under erection with leveling nuts below base plate

The grout thickness will be approximately between 20 mm to 40 mm, the lesser figure being common for smaller plates. For the grout to flow laterally there must be ample space between the bottom of the steel base plate and the top of the concrete. About 40 mm would be a minimum space. For large base plates (say over 900 mm it may be necessary to drill a hole in the base plate near the centre but not so as to foul the column section. The grout is fed through this hole and this hole should be approximately 75 mm in diameter. For very large plates or long rectangular base plates, two grout holes may be required (Ricker 1989). Many good quality grouts are available in the market with varying strengths. They should be of the non-shrink variety (Warner 2004). More grout space should be allowed for a inexperienced foundation contractor.

Often anchor bolts are dislocated at sites. In addition to being out of place, the bolts may be tilted, have too much or too little projection above the top of the concrete, or be at the wrong elevation. Since most of the column bases will be having oversized holes, small dislocation of anchor bolts can usually be tolerated. Bolts which are not vertical (tilted) may be straightened with a rod bending device. For anchor bolts dislocated up to about 20 mm, the concrete may be chipped away to a depth of few centimeters and the bending device may be used to bend the bolt into proper position (Ricker 1989). Anchor bolts dislocated over 20 mm usually require that the base plate be slotted. More remedies for large errors are discussed by Ricker (1989). It has to be noted that anchor rods cannot be pretensioned like HSFG bolts in bolted joints. Anchor rods will relax and lose any pretension as the bond along the length of the rod releases and creep occurs after few weeks of tightening (Collins and Schlafly 2010).

## 15.4 DESIGN OF ECCENTRICALLY LOADED BASE PLATES

The design of base plate subjected to concentric compressions was considered in Sections 15.1 and 15.2. In this section column bases that transmit forces, and moments from the steel column to the concrete foundation are considered. The forces may be axial loads, shear forces, and moment about either

(a) (b) (c)

Base plate connections (a) Welded slab base plate (b) Gusseted base plate (c) Moment-resisting base plate

*Source:* AISC Connections Teaching Toolkit—A Teaching Guide for Structural Steel Connections. Reprinted with permission from AISC. 

axis or any combinations of them. The main function of the base plate is to distribute the loads to the weaker material.

The common design deals with axial load and moment about major axis. With respect to slab and *gusseted base* (see Fig. 15.1), there may be two separate cases (a) compression over the whole base or (b) compression over part of the base and tension in the *holding-down bolts* (also called *anchor bolts*). The relative values of moment and axial load determine which case will occur in a given instance. Horizontal loads are resisted by shear in the weld between column and base plates, friction and bond between the base plate and concrete, and shear in the holding down bolts. As mentioned already, though ANSI/AISC code does not allow the anchor rods to transfer substantial shear, whereas ACI-318 code, Appendix D gives the limit states to be checked in the anchor bolts, including the shear strength of anchor bolts. Hence ANSI/AISC code suggests the use of a shear key or lug to transfer a large horizontal force from the column to the foundation (see Fig. 15.2). Note that the horizontal loads are generally small except when earthquake loads are considered. See the photos at the top of the page for the various types of base plates.

### 15.4.1 Compression Over the Whole Base Plate

A column base and loading are shown in Fig. 15.7. If the ratio of moment/axial load is less than $L/6$, where $L$ is the base length, then a positive pressure exists over the whole base and may be calculated from equilibrium alone. In this case, nominal holding down bolts (2 to 4) are provided to locate the base plate accurately.

$$\text{Eccentricity of the load } e = M/P$$

where $P$ is the total load on the base plate and $M_z$ is the major axis moment on the base. The area of the base is given by

$$A = BL$$

where $B$ is the breadth of the base plate, and $L$ is the length of the base plate. The maximum pressure on the concrete foundation is

$$p_{max} = (P/A) + (M_z/Z_z) \tag{15.8}$$

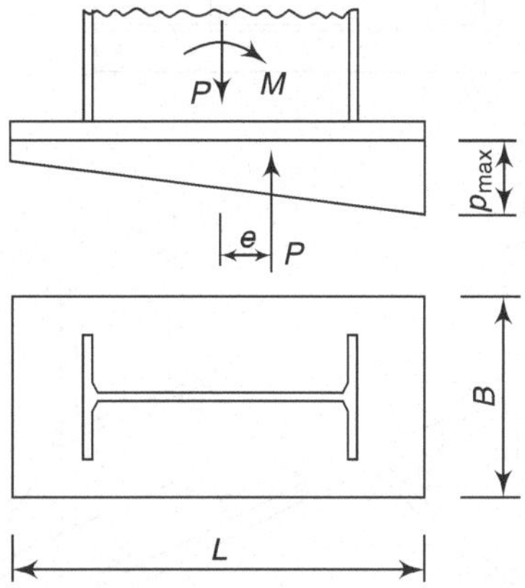


**FIG. 15.7** Compression over the whole base plate

where $Z_z$ is the modulus of the base plate about $Z$-axis $= BL^2/6$.

When there are bi-axial moments, then

$$p_{max} = (P/A) + (M_z/Z_z) + (M_y/Z_y) \tag{15.9}$$

where $M_y$ is the minor axis moment on the plate and $Z_y$ is the modulus about the $y$-$y$ axis $= LB^2/6$.

The maximum pressure must not exceed the bearing strength of the concrete, which is taken as $0.45f_{ck}$, where $f_{ck}$ is the characteristic compressive strength of concrete.

In this case, the size of the base plate is established by successive trials. If the length is fixed, the breadth may be determined so that the bearing strength of the concrete is not exceeded. The weld size between base plate and the column is determined using the same requirements that were set out for the axially loaded base plate. Example 15.3 shows the calculations required for a base plate with compression over the whole of the base plate.

## 15.5 GUSSETED BASE PLATE

When gussets are provided as shown in Fig. 15.8 they support the base plate against bending and hence the thickness of the base plate could be reduced. Now, a part of the load is transmitted from the column through the gussets to the base plate.

The gussets are subjected to bending from the upward pressure under the base as shown in Fig. 15.8. The top edge of the gusset plate will be in compression and hence must be checked for buckling. To prevent this, we should choose the dimensions

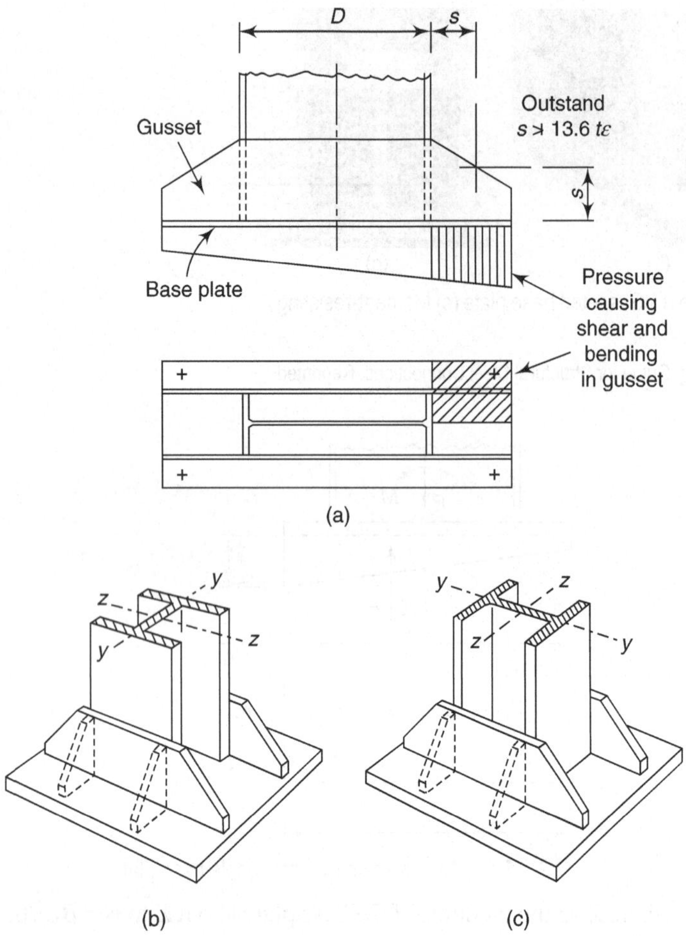


**FIG. 15.8** Gusseted base plates

of outstand in such a way that they are within the limiting width-to-thickness ratios of welded sections, as given in Table 2 of the code. Thus, for the configuration shown in Fig.15.8, the limiting width-to-thickness ratios for the gusset plate is given by

(a) For the portion of gusset plate welded to the flanges of the column

$$D \leq 29.3\varepsilon t \tag{15.10}$$

(b) Outstand of the gusset plate from the column or base plate

$$S \leq 13.6\ \varepsilon t \tag{15.11}$$

where $\varepsilon = (250/f_{yg})^{0.5}$, $t$ is the thickness of the gusset plate, and $f_{yg}$ is the design yield strength of the gusset plate.

The gusset plates are designed to resist shear and bending. The moment in the gusset should not exceed $f_{yg}Z_e/\gamma_{m0}$, where $Z_e$ is the elastic modulus of the gusset and $\gamma_{m0}$ is the partial safety factor for material = 1.1. An example of a gusseted base plate is provided in Example 15.4.

## 15.6 COMPRESSION OVER PART OF THE BASE PLATE AND TENSION IN HOLDING DOWN BOLTS

The design methods suggested in literature are based on the following assumptions (Lescouarc'h & Colson 1992).

1. The part of the foundation under the end plate is considered to be a vertical reinforced concrete beam. The anchor bolts in tension are considered as reinforcements in the beam. The other anchor bolts are ignored. The contact area between the end plate and the concrete is the cross section of the beam.
2. The end plate is very stiff and hence the reinforced concrete section under the end plate remains plane after the loads are applied.
3. The behaviour of the anchor bolts is elastic.

Two different methods of calculation are available, which are as follows.

1. *Elastic behaviour* This method is based on a linear variation of pressure on the concrete, from zero at the neutral axis and $p_{max}$ at one edge of the end plate.
2. *Plastic behaviour* In this method, the pressure on the concrete is assumed to be constant from the neutral axis to the edge of the end plate. The value of this pressure is taken as $0.45f_{ck}$, where $f_{ck}$ is the characteristic cube compressive strength of concrete.

We will adopt the plastic analysis. Consider the base plate as shown in Fig.15.9. Considering the force and moment equilibrium, the following equations could be written

$$N = 0.45f_{ck}yB - F_b \tag{15.12}$$

$$M = 0.45f_{ck}yB(L/2 - y/2) + F_b a \tag{15.13}$$

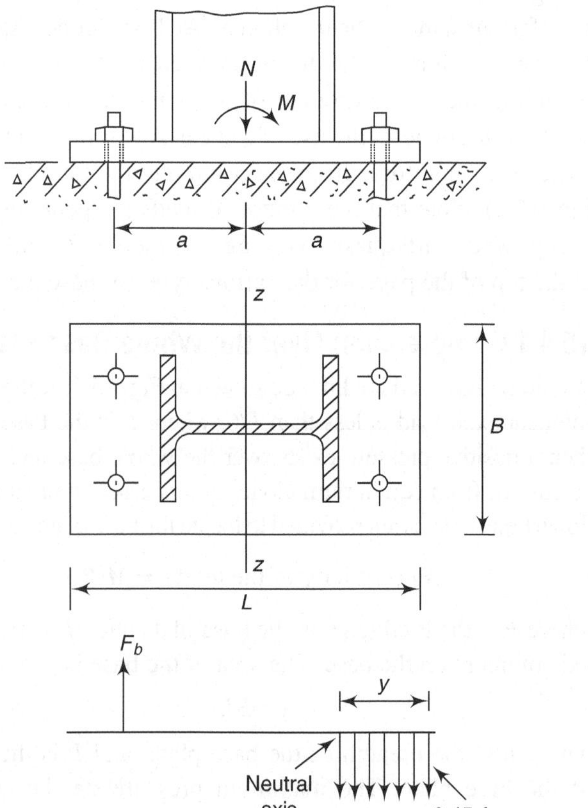


**FIG. 15.9** Compression over part of the base plate and tension in anchor bolts

From this, the value of $y$ can be expressed as

$$y = L/2 + a - \{(L/2 + a)^2 - [2(M + Na)/(0.45f_{ck}B)]\}^{0.5} \quad (15.14)$$

Then

$$p_{max} = 0.45f_{ck}$$

$$F_b = 0.45f_{ck}yB - N \quad (15.15a)$$

Note that this solution is applicable for

$$0 \leq (M + Na) \leq 0.5 \times 0.45f_{ck}B\,(L/2 + a)^2 \quad (15.15b)$$

In most cases, the elastic method leads to larger anchor bolts and a thinner base plate; hence welding may be easier and stiffeners may not be necessary (Lescouarc'h and Colson 1992).

**Effect of shear force** As mentioned earlier, friction between end plate and the concrete is generally sufficient to transmit the shear force from the column to the foundation. When the concrete is assumed to behave plastically, the magnitude of the compressive force on concrete due to normal force and bending moment is given by

$$F_c = p_{max}yB$$

A steel-steel friction coefficient of 0.3 is adopted instead of the steel-concrete coefficient of 0.45, as suggested in clause 7.4.1 of IS 800, because steel shims are often placed under the end plate. When $V < 0.3F_c$, a shear connector is not necessary (Lescouarc'h & Colson 1992).

**Base plates subject to heavy bending moments** Modest moments can be resisted by a base plate welded to the bottom of the column. However, significant bending moments may require any one of the arrangements shown in Figs 15.10(a)–(e). Though the fabrication is more complex, they have the following advantages.

1. The base plate is very stiff and may be conveniently assumed as fixed in the analysis.
2. The tension in the anchor bolts is more uniform due to the stiffness of the collar.
3. The total elongation of anchor bolts before rupture is large, because their free length is large.
4. The tightening of the nuts is easy.

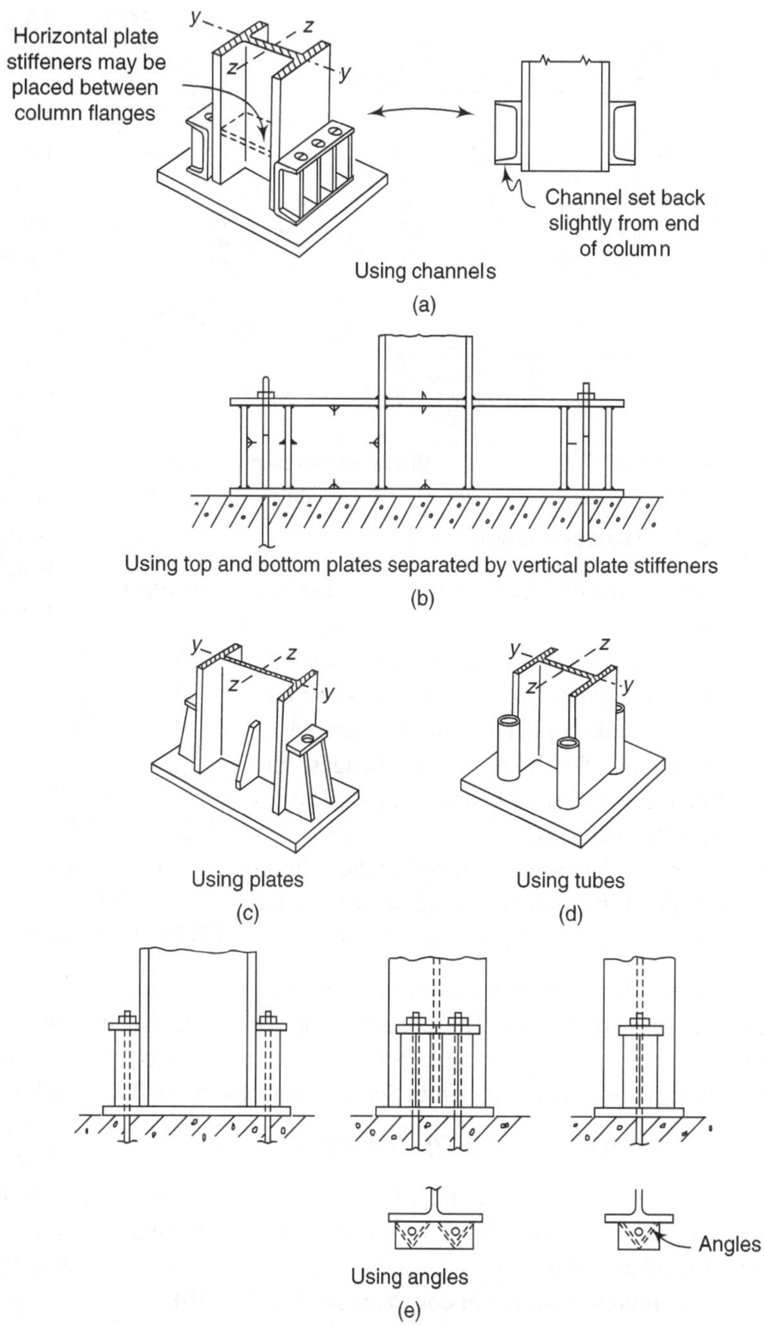


**FIG. 15.10** Base plates with collar

*Source:* Blodgett 1966. Reprinted with permission from the James F. Lincoln Arc Welding Foundation, USA. 

## 15.7 COLUMN END EMBEDDED IN CONCRETE FOUNDATION (POCKET BASES)

In this type of base, the column is grouted into a pocket in the concrete foundation as shown in Fig. 15.1. The solution appears simple, but the erection is not easy, since the column is not stable before the concrete hardens.

The following assumptions can be used in calculations (see Fig. 15.11).

1. The axial load is resisted by direct bearing and bond between steel and concrete.
2. The moment is resisted by compression forces in the concrete acting on the flanges of the column. The forces act on both faces of the flanges of the column. The action of resisting moment is shown in Fig. 15.11(b). Thus, elastic behaviour of concrete is considered and the distribution is

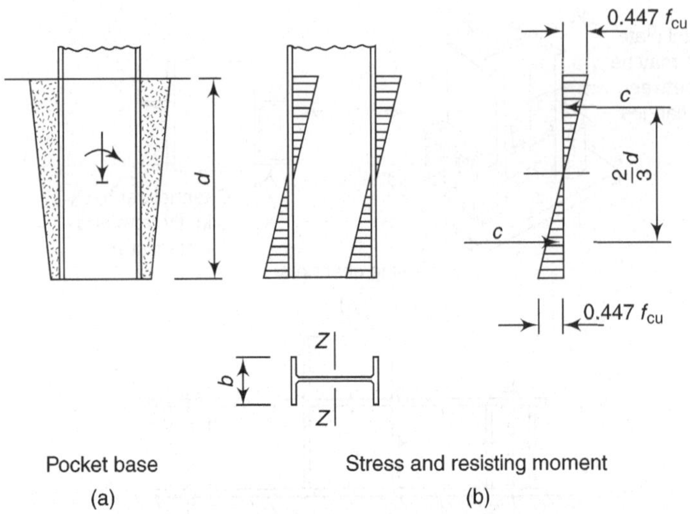


**FIG. 15.11** Pocket base design

similar to the distribution in the cross section of an elastic beam.

3. The upper part (about 50 mm) of the concrete foundation cannot support any pressure due to local cracks and hence may be ignored.
4. Due to the flexibility of the flanges, the effective width of a flange is limited to 20 times its thickness.
5. For practical reasons, the length of the column embedded in concrete should be at least 1.5 times the depth of the column section.

Additional shear connectors may be welded to the flanges to enhance the performance. These shear connectors are designed for shear only, similar to simple bolts. The design strength of the concrete in compression from Section 5 of IS 456-2000, is

$$0.67 f_{ck}/\gamma_m = 0.67 f_{ck}/1.5 = 0.447 f_{ck}$$

where $f_{ck}$ is the characteristic cube strength of concrete at 28 days. The required depth of embedment in concrete is determined as follows:

The compressive force in concrete (see Fig. 15.11b)

$$C = 0.447 f_{ck}/2 \times d/2 \times b = 0.112 f_{ck} b_d$$

The moment of resistance of the concrete

$$M = 2 \times C \times 2d/3 = 0.149 f_{ck} bd^2$$

= moment of resistance of the steel section

$= \beta f_y Z_{pz}/\gamma_{m0}$ ($\gamma_{m0} = 1.1$)

Thus, $d = [\beta f_y Z_{pz}/(1.1 \times 0.149 f_{ck} b)]^{0.5}$

For plastic or compact section

$$d = [f_y Z_{pz}/(0.164 f_{ck} b)]^{0.5} \quad (15.16)$$

More accurate design method for pocket bases is provided by Petersen (1990).

## 15.8 ANCHOR BOLTS (FOUNDATION BOLTS) AND SHEAR CONNECTORS

The specification for foundation bolts is given in IS 5624-1993. Some typical shank forms of black foundation bolts as per this code are given in Fig. 15.12. The types of anchor bolts include cast-in-place anchors and post-installed anchors [state-of-the-art report CEB (1994)]. Cast-in-place anchors, placed before concrete is cast, include a bolt with a curved end (used when there is no or low uplift forces) and a headed anchor. Figures 15.13(a) and (b) show a hooked bolt anchor, which allow clearance for their positioning. The angle section at the bottom (it can even be a circular rod) is encased in concrete, and is useful to anchor the bolt during erection, before filling the hole with concrete. The tensile load transfer is by bonding between anchor bolt and concrete filling. Bonding between filling and concrete wall is guaranteed whenever the contact surface is sufficiently rough.

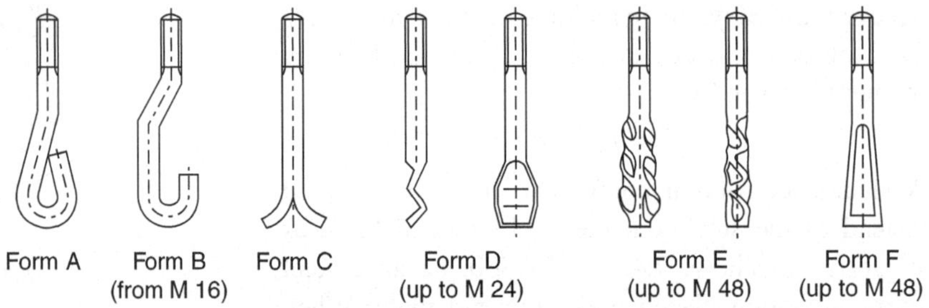


**FIG.15.12** Typical shank forms of foundation bolts

For heavy uplift forces, a plate may be fixed at the end of the bolt (as shown in Fig. 15.13d). The headed anchors transfer tensile load by mechanical bearing of the head, nut and possibly by bond between anchor shank and the surrounding concrete. As these pre-installed anchors do not allow any clearance, they need very accurate positioning. Hammer head anchor bolts are sometimes employed, which allow dimensional adjustment and transmit the load without relying on steel/concrete or concrete/concrete bonding

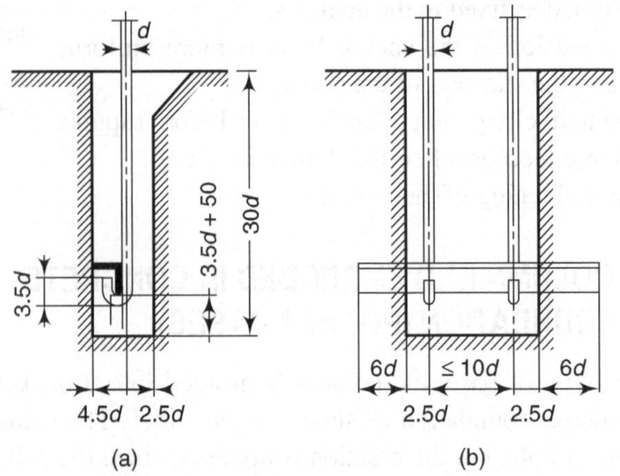


**FIG. 15.13** Hooked bolt anchors and headed anchors

| d | 1 Anchor | 2 Anchors |
|---|---|---|
| M16 | ∠60 × 60 × 6 | ∠80 × 80 × 8 |
| M20 | ∠75 × 75 × 8 | ∠90 × 90 × 10 |
| M24 | ∠90 × 90 × 10 | ∠110 × 110 × 10 |
| M27 | ∠110 × 110 × 10 | ∠130 × 130 × 12 |
| M30 | ∠130 × 130 × 12 | ∠150 × 150 × 16 |

(c)

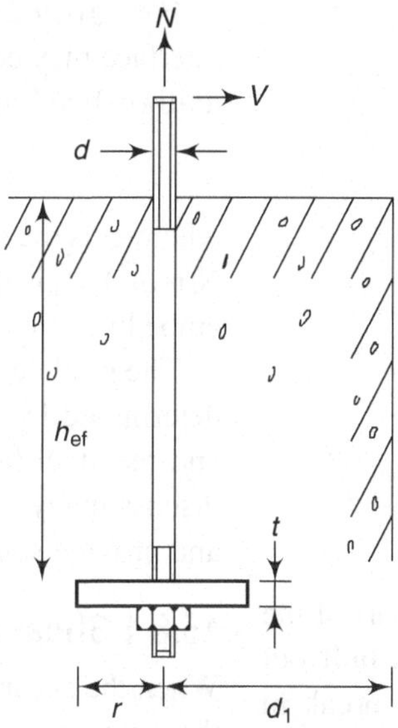


(d)

**FIG. 15.13** *(Contd)*

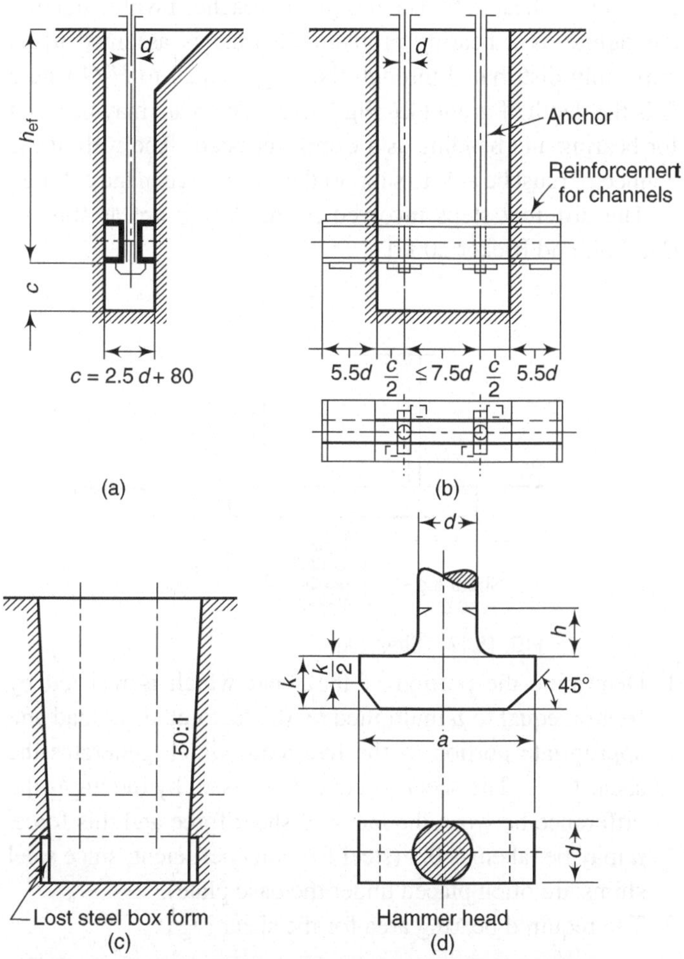


| Bolt | M24 | M30 | M36 | M42 | M48 | M56 | M64 |
|---|---|---|---|---|---|---|---|
| 1. anchor | ][75 | ][75 | ][100 | ][100 | ][125 | ][125 | ][150 |
| 2. anchor | ][100 | ][100 | ][125 | ][150 | ][175 | ][200 | ][200 |

(e)

**FIG. 15.14** Hammer head anchor bolts (Petersen 1990)

(Fig. 15.14). In this type of construction, two back-to-back rolled steel channels are placed horizontally together with reinforcing bars and the concrete is poured in the foundation, leaving holes (depth = *c*) below the two channels as shown in Fig. 15.14. These holes may be obtained by using polystyrene or steel formwork. The extension below the channels is generally obtained by a steel box, which is left behind in the concrete [see Fig. 15.14(c)]. The anchor bolt having a forged head as shown in Fig. 15.14(d) or built up by two welded plates, is inserted and rotated through 90°. The tensile force is transmitted by bearing between the hammer head and the horizontal channels, which behave as a beam on two (or more) concrete supports.

*Post-installed anchors*, installed in hardened concrete, are classified according to their load-transfer mechanisms. *Adhesive or bonded anchors* transfer load through cementitious grout or chemical adhesive. Details of installation and behaviour of these types of anchors are provided by Subramanian and Cook (2002, 2004). Epoxy is the most widely used adhesive though resins such as vinyl esters, polyesters, methacrylates, and acrylics have also been used. *Mechanical anchors* transfer load by friction or bearing and include expansion anchors and undercut anchors.

*Expansion anchors* work by expansion of a wedge or sleeve mechanism against the surrounding concrete. In torque-controlled anchors, the expansion is generated by applying a pre-determined torque (see Fig. 15.15). In displacement-controlled anchors, the expansion is generated as the anchors are driven in. *Under cut anchors* are placed in a hole that is locally widened (undercut) using a special drilling tool and are then set by anchoring elements against the sides of the undercut portion of the hole, usually by applying a torque to the anchor (see Fig. 15.16). Mechanical anchors loaded in tension apply reaction forces to the concrete at the expansion

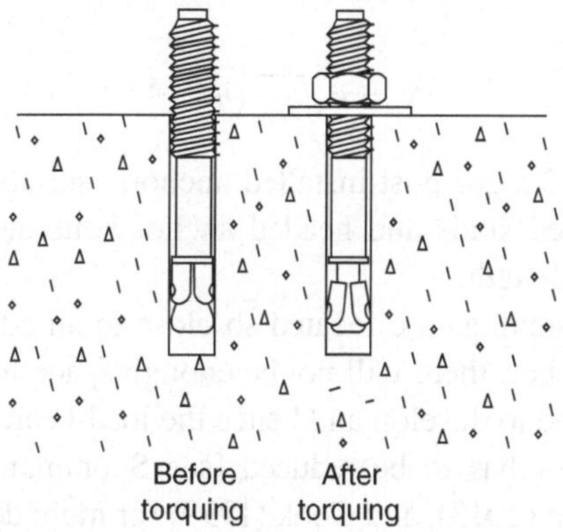


**FIG. 15.15** Torque-controller expansion anchors

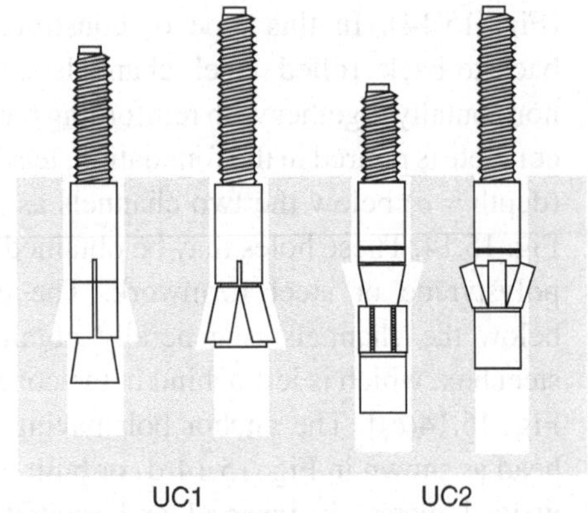


**FIG. 15.16** Undercut anchors

mechanism, usually near the end of the embedded part of the anchor. Failures typically occur when the anchors pull out a cone of concrete, slip out of the hole, or simply break in response to the failure of steel component.

Bonded anchors are often not as strong as other post-installed anchor systems because aggregate is generally not used in the holes in which these are placed and grouted. Grouted anchor failures occur most often between the grout and original concrete. Proprietary cement-based grouts can achieve a compressive strength of 27.5 MPa within 3 hours and 48 MPa within 24 hours. Average bond stress varies between 7.3–21.5 MPa for cementitious grouts and 17.8–19.4 MPa for polymer grouts (Subramanian & Cook 2004).

Failure modes of cast-in-place headed anchors are generally limited to yield and fracture of the anchor shank and concrete breakout (Subramanian & Vasanthi 1991). After conducting an exhaustive number of experiments, Eligehausen and his associates have suggested a truncated pyramid failure model (with a 35° slope of failure cone) which has been incorporated in the Eurocode (Subramanian 2000).

Fuchs et al. (1995) suggested the concrete capacity design (CCD) approach to fastenings in concrete, which has been adopted by the ACI code (Appendix D of ACI 318-2014). According to this method the tensile capacity of anchors may be determined using the following formula (Subramanian 2000)

$$N_u = k\sqrt{f_{ck}}\,(h_{ef})^{1.5} \qquad (15.17)$$

where $k$ = 13.5 for post-installed anchors and 15.5 for cast-in-situ headed studs and headed anchor bolts and $h_{ef}$ is the embedment length.

When fastenings are located so close to an edge or to an adjacent anchor, there will not be enough space for complete concrete cone to develop and hence the load-bearing capacity of the anchor has to be reduced [see Subramanian (2000), Subramanian (2013), and Cook (1999) for more details of the CCD method and worked examples].

For grouted anchors bond failure at the grout-concrete interface may occur. This failure mode is best represented by a uniform bond stress model. The strength in this case is given by

$$N_{bond} = \tau_o\,\pi\,d_o h_{ef} \qquad (15.18)$$

where $\tau_o$ is the grout-concrete bond strength of the product (in N/mm$^2$), $d_o$ is the diameter of the hole (in mm), and $h_{ef}$ is the embedment length.

The predicted mean strength of a headed grouted anchor is determined by the lower value of Eqn (15.17), Eqn (15.18), and the steel failure strength. Swiatek and Whitback (2004) discuss many practical problems connected with anchor rods and provide some easy solutions.

### 15.8.1 Shear connectors

When the shear force to be transmitted to the foundation from the column is high, a shear lug (a short length of a rolled I-section or a plate) is welded perpendicular to the bottom of the base plate (see Fig.15.17). Failure occurs when a wedge of concrete shears off. The design approaches involve treating the failure as a bearing problem. Bearing is assumed to be uniformly distributed through the height equal to *H*-*G* where *G* is the depth of grout (see Fig 15.17). The plate may be sized for bearing and bending as a cantilever beam. The web of the connector must be able to support the shear force in the column.

The different steps involved in the design are as follows (DeWolf and Ricker 2006).

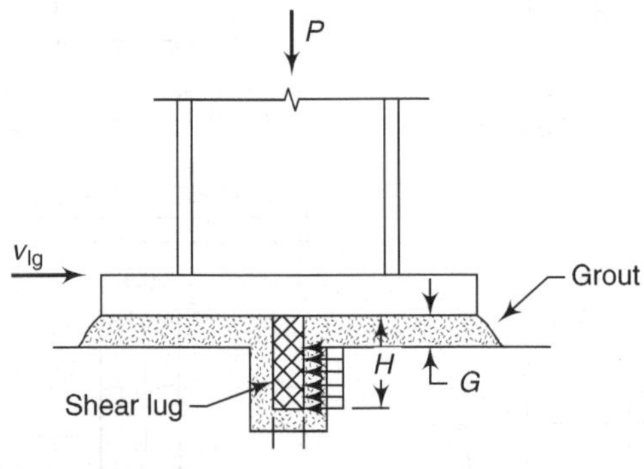


**FIG. 15.17** Shear lug

1. Determine the portion of the shear which is resisted by friction equal to $\mu$ multiplied by the factored dead load and appropriate portion of the live load, which generates the shear force. The shear force to be resisted by the lug is the difference between the factored shear force and this force. $\mu$ may be taken as 0.3 (steel friction coefficient, since steel shims are often placed under the base plate).
2. The required bearing area for the shear lug is

$$A_{lg} = V_{lg}/0.45 f_{ck} \qquad (15.19)$$

3. Determine the shear lug dimensions, assuming that bearing occurs on the portion of the lug below the concrete foundation.

4. The factored cantilever end moment $M_{lg}$ acting on a unit length of the shear lug is

$$M_{lg} = (V_{lg}/W)(H + G)/2 \quad (15.20)$$

where $W$ is the total horizontal width of the lug, $H$ is the vertical height of the lug, and $G$ is the thickness of grout (see Fig. 15.17)

5. The shear lug thickness is determined as folows:

$$t_{lg} = \sqrt{[4M_{lg}/(f_y/\gamma_{m0})]} \leq t_b \quad (15.21)$$

where $t_b$ is the thickness of the base plate. Example 15.6 illustrates the use of these steps.

DeWolf and Ricker (2006) provide more details about the design of base plates. Drake and Elkin (1999) present a step-by-step methodology for the design of base plates and anchor rods using factored loads and rectangular pressure distribution. DeWolf and Sarisley (1980) and Thambiratnam and Paramasivam (1986) conducted experiments to study the behaviour of base plates under the action of axial loads and moments. They found that the behaviour at failure was not always consistent with the assumption used in the design approaches.

## 15.9 GRILLAGE FOUNDATION

The grillage foundation may be used under very heavily loaded columns. The base plate of such columns may be large and hence may not be economical. Grillage foundation may also be adopted in places where the safe bearing capacity of soil may be low, thus resulting in the large size of foundations and base plates. Steel grillage footings were extensively used during the latter part of the last century, but have become obsolete due to the development and use of reinforced concrete foundations. They are, however, still used for underpinning purposes.

A typical grillage footing consists of two or more tiers of steel beams (usually I-sections) with each tier placed at right angles to the one below it (see Fig. 15.18). The beams in each tier are usually held together by a system of bolts and pipe spacers. The beams should be clean and unpainted, and the whole system is completely filled and encased in concrete with at least 150 mm of cover on the bottom and 100 mm at all other points. Sometimes, in lieu of concrete (other than the encasement), the spaces between the steel beams may be filled with cement grout. We will only consider grillage subjected to axial load in this chapter.

As the grillage beams are encased in concrete, most of the provisions of cased beams are applicable to these beams also. It has to be noted that IS 800: 2007 does not contain provisions for encased beams. The earlier version of IS 800:1984 had provisions for encased columns and beams in section 10.

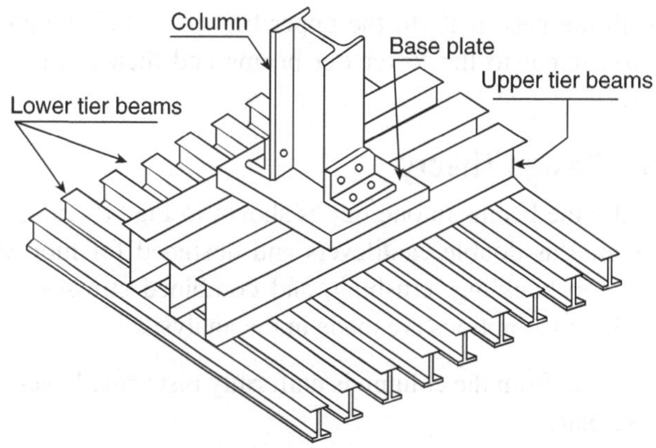


**FIG. 15.18** Typical grillage foundation

Provisions for cased sections may be found in clause 4.14 of BS 5950-1: 2000. As per these provisions, the following conditions must be fulfilled:

1. The surface of the beams should be unpainted and free from oil, grease, dirt, and loose rust. The beams should be solidly encased in ordinary dense concrete with a 28 day cube strength of at least 25 N/mm$^2$.
2. The concrete casing should extend the full length of the member and its connections. The concrete should be thoroughly compacted, especially below cleats, cap plates, and beam soffits. There should be sufficient clearance at all points so that concrete flows around the steel elements and can be compacted properly. It should be noted that the concrete filling does not carry any load; it simply keeps the beams in position and prevents their corrosion.
3. The surfaces and edges of the flanges of the beam should have sufficient cover of not less than 100 mm.
4. The beams in each tier should be spaced in such a way that the minimum distance between the flanges of two consecutive beams is greater than 75 mm. This condition is specified such that concrete can be placed and compacted properly.
5. Pipe separators are provided between the beams to maintain proper spacing between beams.
6. The overall dimensions of the steel section should not exceed 1000 mm by 500 mm, the dimension of 1000 mm being measured parallel to the web or webs.

It has to be noted that the size of base plate is independent of the bearing capacity of the concrete block, since the base plate will rest directly on the upper tier of steel beams. The steel column is connected to the upper tier with the help of a base plate, side angles, and gusset plate. These connecting elements are also embedded in the concrete so that the joint acts as a rigid joint. The thickness of the base plate should be sufficient to prevent punching and its dimensions should cover the transverse direction of upper tier beams completely. The load on the column is transmitted through the base plate to the upper tier of beams, which in turn transfer it to the lower tier beams and then to the soil below.

### 15.9.1 Design Theory

Consider the beams in one tier as shown in Fig. 15.19. They are treated as double cantilevers and designed for moment, shear, web buckling, crushing, and combined stresses. The following assumptions are made in the analysis.

1. The load from the column is uniformly distributed over the base plate.
2. Each tier of the beams distributes the load uniformly to the lower tier.
3. The intensity of bearing pressure is uniform.

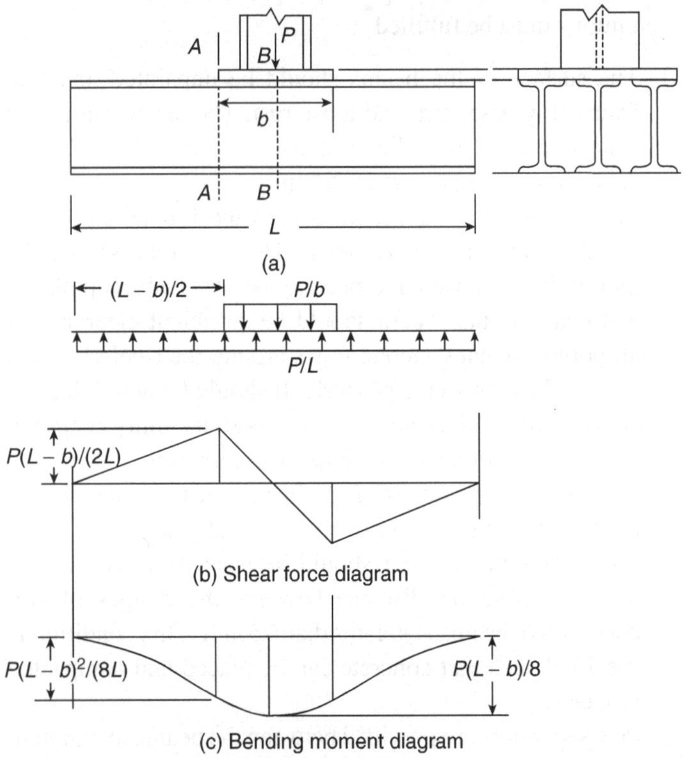


**FIG. 15.19** Shear force and bending moment diagram

Let $P$ be the total load at column base, $G$ the weight of grillage, and $p$ be the safe bearing capacity of soil. Then,

Area of grillage, $A = (P + G)/p$

Uniform load on top of beam $= P/b$

Uniform load on the underside of beam $= P/L$

$$\text{Shear at } A\text{-}A = \left(\frac{P}{L}\right)\left(\frac{L}{2}-\frac{b}{2}\right) = \frac{P}{2L}(L-b) \qquad (15.22)$$

Shear at $B$-$B = 0$

The bending moment in the beam is zero at the ends and is maximum at the center.

Bending moment at $A$-$A$

$$= \frac{P}{2L}(L-b)\frac{(L-b)}{4} = \frac{P}{8L}(L-b)^2 \qquad (15.23)$$

Bending moment at $B$-$B$

$$= \frac{P}{L}\left(\frac{L}{2}\right)\left(\frac{L}{4}\right) - \frac{P}{b}\left(\frac{b}{2}\right)\left(\frac{b}{4}\right) = \frac{P}{8}(L-b) \qquad (15.24)$$

The shear force and bending moment diagram are shown in Fig. 15.19.

### Design Procedure

The various steps in the design of grillage foundation are given as follows:

1. Compute the load on column, $P$. Add 10 percent of the column load to the load $P$ to find the total service load on the foundation.
2. Find the required area of the bottom tier by dividing the total service load by the safe bearing capacity of soil and adopt a suitable size. Square or rectangular footing may be adopted.
3. Calculate the maximum bending moment for upper tier beams from Eqn (15.24).
4. Compute the plastic section modulus for all beams. $Z_p$ per beam may be obtained by dividing this value by the number of beams in the top layer. Choose the lightest section satisfying the aforementioned criterion from the available IS beam sections.
5. Calculate the maximum shear force from Eqn (15.22).
6. Check the beam section for its adequacy in shear, web buckling, and web bearing as follows:

(a) *Shear stress*:

$$\text{Average shear stress, } \tau_b = \frac{V_u}{nht_w} \le \frac{f_y}{1.1\sqrt{3}} \qquad (15.25)$$

where $n$ is the number of beams.

(b) *Web buckling*:

The web of the beam is liable to buckle under compressive stresses, especially under point load. For design purposes, the applied load is considered to be distributed over a length equal to the length of the bearing plate plus half the depth of beam. The slenderness ratio of the web is calculated using Eqn (10.52) of Chapter 10. From Table 9c of the code, the design compressive stress, $f_{cd}$, may be found for this slenderness ratio and yield stress of steel. Then the web buckling strength (see Fig. 15.20 of Chapter 10) is checked as

$$F_{wb} = (b + 2n_1)t_w f_{cd} > \text{shear in each beam} \qquad (15.26)$$

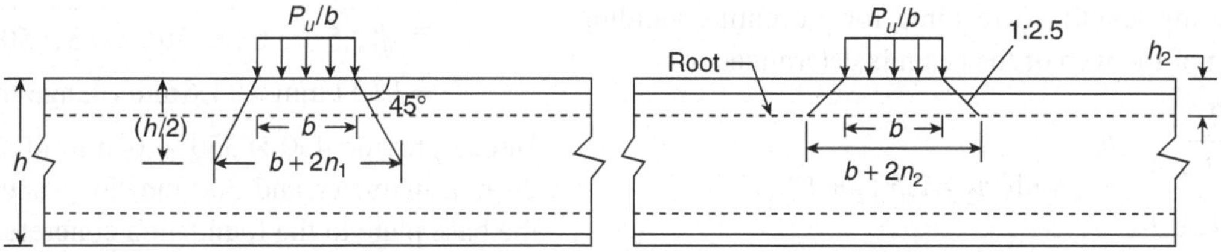


**FIG. 15.20** Load distribution for web crippling and bearing

(c) *Web crippling/bearing*:

The web bearing capacity (see Fig. 15.20) is checked as

$$F_{\text{Crip}} = (b + 2n_2)t_w f_y / \gamma_{m0} > \text{shear in each beam} \quad (15.27)$$

where $b$ is the width of base plate of column and $n_2$ is calculated as

2.5 × (root radius + flange thickness), as dispersion of load is assumed in the code at a slope of 1: 2.5.

7. Check for combined stresses at the edge of base plate. The equivalent stress is calculated as (see clause 10.5.10.2.2 of the code)

$$f_e = \sqrt{f_{\text{bc}}^2 + f_{\text{p}}^2 - f_{\text{bc}} f_{\text{p}} + 3\tau_{\text{b}}^2} \le 0.9 f_{\text{y}} \quad (15.28)$$

8. Calculate the maximum bending moment and shear force for the lower tier beams, based on the same principles as used for upper tier beams.
9. Compute the plastic section modulus and select the number of beams to be provided in the lower tier.
10. Check the beams for shear and crippling.

## 15.10 BEAM BEARING PLATES

When a beam is supported on masonry or concrete wall, the high bearing pressure developed under the beam at support may crush the supporting masonry. To avoid this and to distribute the bearing pressure over a larger area on a weaker material like masonry/concrete, a bearing plate is usually provided as shown in Fig. 15.21. The bearing plate also distributes the reaction from the beam over a sufficient length of the beam, so that web crippling and web yielding may be prevented.

### 15.10.1 Design of Bearing Plates

We normally assume that the reaction from the beam at support is spread uniformly on the masonry/concrete wall when a bearing plate is provided, and the bearing plate is subjected to a uniform upward pressure equal to the reaction divided by the area of bearing plate. It has to be noted that the actual pressure distribution under the bearing plate is non-uniform and difficult to measure; hence uniform pressure is assumed to simplify the calculations. The reaction from the beam is assumed to be distributed over the beam bearing plate with dimensions $N$ along the beam and $C$ across the beam, where $C$ equals thickness of web plus twice the radius of fillet of beam, as shown in Fig. 15.21(b).

The bearing plate is modeled as a double cantilever beam projecting from the web of the beam and subjected to upward distributed load from the bottom, as shown in Fig. 15.21(b). The maximum width $N$ of the bearing plate is limited by the wall thickness. In the case of bearing plates resting over concrete foundation, the allowable bearing pressure may be taken as 0.45 $f_{\text{ck}}$ where $f_{\text{ck}}$ is the characteristic strength of concrete. For brick masonry, the allowable bearing pressure may be taken as 4.0 MPa. The required area of the bearing plate ($N \times B$) may be obtained by dividing the reaction from the beam by the allowable bearing pressure. It has to be noted that clause 34.4 of IS 456 allows the basic bearing stress of 0.45 $f_{\text{ck}}$ be increased by a factor $\sqrt{A_1/A_2}$ with an upper limit of $\sqrt{A_1/A_2} = 2$, where $A_2$ is the area of the loaded area and $A_1$ is the area of supporting surface that is geometrically similar to and concentric with the loaded area ($A_2$ may have sides extending at two horizontal to one vertical, until they first reach the edge of the supporting wall). We may, however, assume $A_2 = A_1 = A = 1$, which will result in conservative values.

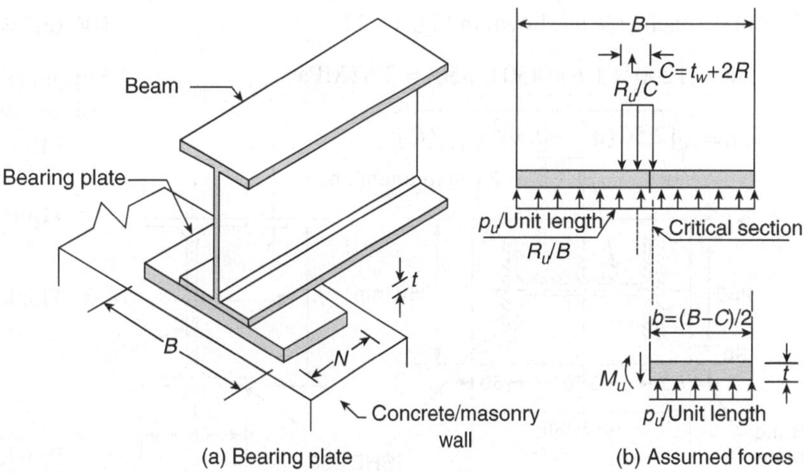


**FIG. 15.21** Beam bearing plate

The stiff bearing length, $N$, required for preventing yielding and crippling of the web of the beam is determined as

$$N \geq \frac{R_u}{t_w\left(\frac{f_{yw}}{\gamma_{m0}}\right)} - n_2 \quad ; \text{ with } n_2 = 2.5(t_f + R)$$

where $R_u$ is the factored reaction, $t_w$ and $t_f$ are the thickness of web and flange of the beam respectively, $f_{yw}$ is the yield strength of web, $\gamma_{m0}$ is the partial safety factor (=1.1), and $R$ is the radius of fillet of the beam. If the supporting width of the wall is not wide enough, then $N$ is taken as the width of the supporting wall and bearing stiffeners are provided to prevent web crippling of the web of beam.

Once the value of $N$ is determined, the length $B$ of the bearing plate can be computed easily from the required area. This length $B$ should be wide enough to be attached to the beam ($> b_f + 2\,t_f$); however should be as narrow as permitted.

The thickness of bearing plate may be found out by equating the bending moment due to the assumed cantilever action with the plastic moment capacity of the plate, which may be taken as $1.2 f_{yp} t_p^2 / 6$, where $f_{yp}$ is the yield strength and $t_p$ is the thickness of the bearing plate.

## EXAMPLES

**EXAMPLE 15.1:** Design the base plate for an ISHB 350 column to carry a factored load of 1200 kN. Assume Fe 410 grade steel and M25 concrete.

**SOLUTION:**

For ISHB 350 section,

$h = 350$ mm; $b = 250$ mm; $t_f = 11.6$ mm; $t_w = 8.3$ mm

Bearing strength of concrete $= 0.45 f_{ck} = 0.45 \times 25 = 11.25$ N/mm$^2$

*Note*: As per IS 800, we can use $0.6\,f_{ck}$.

Required area of base plate $= 1200 \times 10^3/11.25 = 1{,}06{,}667$ mm$^2$

Use a base plate of size $450 \times 350$ mm with area $= 1{,}57{,}000$ mm$^2$.

If the ISHB 350 is kept at the centre, the projection will be 50 mm on each side as shown in Fig. 15.22

$$w = (1200 \times 10^3)/(450 \times 350) = 7.62 \text{ MPa}$$

$$t_s = \sqrt{[2.5w(a^2 - 0.3b^2)\gamma_{m0}/f_y]}$$

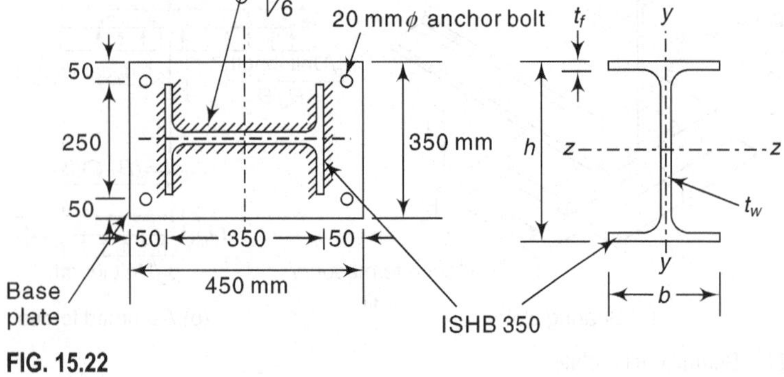


**FIG. 15.22**

$$= \sqrt{[2.5 \times 7.62 \times (50^2 - 0.3 \times 50^2) \times 1.10/250]}$$

$= 12.11$ mm $> 11.6$ mm (flange thickness)

Hence provide $450 \times 350 \times 14$ mm plate. Also provide four 20-mm diameter and 300-mm long anchor bolts to connect the base plate to the foundation concrete.

*Weld connecting base plate to column*

Use a 6-mm fillet weld all round the column section to hold the base plate in place. Note that the surfaces are to be machined for direct bearing.

Total length available for welding along the periphery of ISHB 350

$$= 2(250 + 250 - 8.3 + 350 - 11.6) = 1660.2 \text{ mm}$$

After deducting end returns of the weld, at the rate of two times the size of the weld at each end, we get,

$$L_{eff} = 1660.2 - 2(4 + 2)2a = 16660.2 - 24 \times 6 = 1516.2 \text{ mm}$$

Capacity of the weld $= 0.7 \times 6 \times 189/1000 = 0.7938$ kN/mm

Required length of the weld $= 1200/0.7938$

$= 1511$ mm $< 1516.2$ mm

Hence a 6 mm weld is adequate.

*Note*: This is a fillet weld on the edge and not on the rounded ends of the member. Hence the limitation 0.75 times thickness (clause 10.5.8.2 of the code) will not apply here.

**EXAMPLE 15.2:** Design the base plate for an ISHB 350 column to carry a factored load of 1200 kN. Assume Fe 410 grade steel and M25 concrete. Use the ffective area method of design.

**SOLUTION:**

For ISHB 350 section,

$h = 350$ mm; $b = 250$ mm; $t_f = 11.6$ mm; $t_w = 8.3$ mm

Bearing strength of concrete $= 0.45 f_{ck} = 0.45 \times 25 = 11.25$ N/mm$^2$

Required area of base plate $= 1200 \times 10^3/11.25 = 1{,}06{,}667$ mm$^2$

Effective bearing area,

$$A_{bc} = (b_f + 2c)(h + 2c) - [\{h - 2(t_f + c)\}(b_f - t_w)]$$

$$106{,}667 = (250 + 2c)(350 + 2c) - [\{350 - 2(11.6 + c)\}(250 - 8.3)]$$

Simplifying we get $4c^2 + 1683.4c - 98154.5 = 0$

Solving we get $c = 52$ mm

Minimum length of base plate $= h + 2c = 350 + 2 \times 52$
$= 454$ mm

Minimum width of base plate $= b_f + 2c = 250 + 2 \times 52$
$= 354$ mm

Thickness of plate, $t_p = c[2.5w\,\gamma_{m0}/f_{yp}]^{0.5}$

$$= 52\left(\frac{2.5 \times 0.6 \times 25 \times 1.1}{250}\right)^{0.5} = 21.12$$

Provide $460 \times 360 \times 22$ mm thick base plate (see Fig. 15.23).

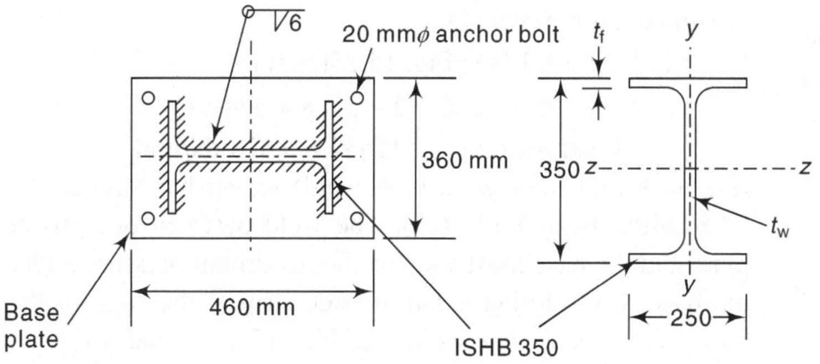


**FIG. 15.23**

**EXAMPLE 15.3:** Design the base plate for the column in Example 13.1 subjected to a factored moment of 45 kNm and a factored axial load of 500 kN. The column size is ISHB 250. The cube compressive strength of concrete in the foundation is $f_{ck}$ = 25 N/mm². Use grade 410 steel.

**SOLUTION:**

**Step 1:** *Size of the base plate*

$$e = 45 \times 10^3/500 = 90\,\text{mm}$$

If the base plate is made 6*e* in length there will be compressive pressure over the whole of the base.

$$6e = 6 \times 90 = 540\,\text{mm}$$

The required breadth to limit the bearing pressure to $0.45f_{ck}$ (= 11.25 N/mm²) is

$$B = 2P/(L \times 0.45f_{ck}) = (2 \times 500 \times 10^3)/(540 \times 0.45 \times 25)$$
$$= 164.6\,\text{mm}$$

Provide a rectangular base plate of size 540× 400 mm. The arrangement of the base plate is shown in Fig. 15.24.

$$\text{Area} = 540 \times 400 = 216 \times 10^3\,\text{mm}^2$$
$$\text{Modulus } Z = 400 \times 540^2/6 = 19.44 \times 10^6\,\text{mm}^3$$

Maximum pressure:

$$p_{max} = 500 \times 10^3/(216 \times 10^3) + 45 \times 10^6/(19.44 \times 10^6)$$
$$= 2.31 + 2.31$$
$$= 4.62\,\text{N/mm}^2$$
$$p_{min} = 2.31 - 2.31 = 0$$

**Step 2:** *Thickness of base plate*

Consider a 1-mm wide strip as shown in Fig. 15.24(b). This acts as a cantilever from the face of the column with the loading caused by pressure on the base. This method gives a conservative design for the thickness of base plate, since the plate action due to bending in two directions at right angles is not considered.

$$\text{Base pressure at section } XX = [(540 - 145)/540] \times 4.62$$
$$= 3.38\,\text{N/mm}^2$$

For the trapezoidal pressure loading on the cantilever strip as shown in the figure, the moment at *XX* is calculated as follows

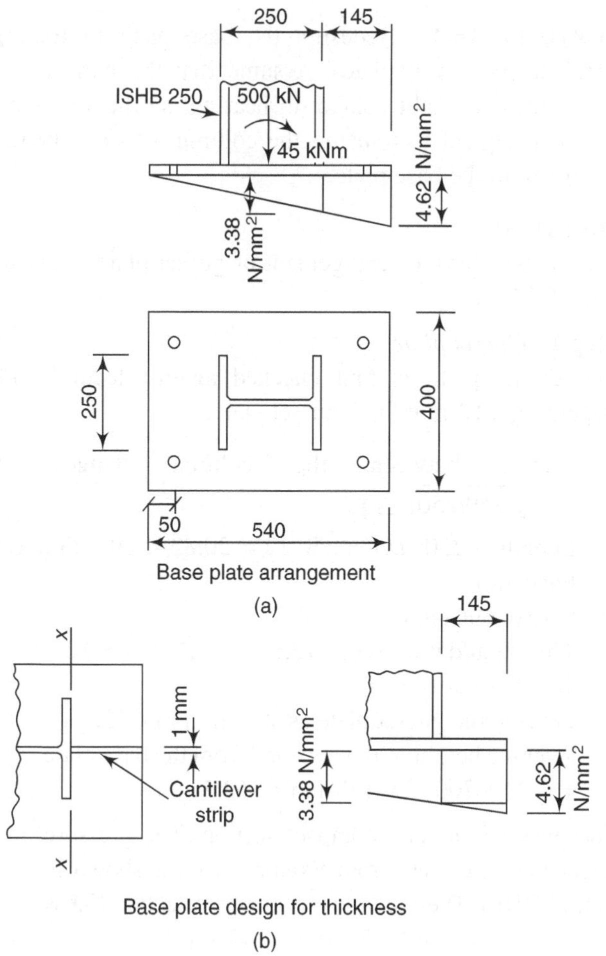


**FIG. 15.24**

$$M_x = (3.38 \times 145^2/2) + (4.62 - 3.38) \times 145/2 \times 2/3 \times 145$$
$$= 44.22 \times 10^3\,\text{Nmm}$$

Moment capacity of plate = $1.2f_y Z_e/\gamma_{m0}$

where $Z_e = t^2/6$

Hence $44.22 \times 10^3 = 1.2 \times 250 \times t^2/(6 \times 1.1)$

$$= 45.45\,t^2$$

Thickness of plate $t = \sqrt{(44.2 \times 10^3/45.5)} = 31.18\,\text{mm}$

Hence, use a 32-mm thick plate.

**Step 3:** *Weld connecting beam-column to base plate*

The base plate has been designed on the basis of linear distribution of pressure. For consistency the weld will be designed on the same basis.

Beam-column size: ISHB 250; $A$ = 6500 mm²; $Z_z = 619 \times 10^3\,\text{mm}^3$

Axial stress = $500 \times 10^3/6500 = 76.92$ N/mm²

Bending stress = $45 \times 10^6/619 \times 10^3 = 72.70$ N/mm²

On the basis of elastic stress distribution, there is compressive stress over the whole of the base. The base plate and column are to be machined for tight contact so that the weld is required only to hold the base plate in position. Use a 6 mm continuous fillet weld around the column profile.

**EXAMPLE 15.4:** Redesign the base plate in Example 15.3 using gusset plates. Assume that the parts are not machined for tight contact in bearing, so the welds have to be designed to transmit the column load and bending moment to the base plate.

**SOLUTION:**
Let us consider the arrangement of gusset plate as shown in Fig. 15.25.

**Step 1:** *Gusset plate*
The gusset plate is first checked against local buckling. Assuming a 12-mm thick gusset plate,

(a) Gusset between the column flanges, with $\varepsilon = \sqrt{(250/250)} = 1$
Length = 250; $D/t$ = 250/12 = 20.83 < 29.3 (Table 2 of the code)
(b) Gusset outstand
This should not exceed $13.6\,\varepsilon t = 13.6 \times 1 \times 12 =$ 163.2 mm > 145 mm
Provide the gusset plate as shown in Fig. 15.25.
Average height of the outstand from the base plate = (100 + 200)/2 = 150 mm < 163.2 mm

The gusset is a semi-compact section. The pressure under the base plate from Example 15.3 is shown in Fig. 15.25(b). The shear on one gusset at section *X-X* is

$$V = (4.62 + 3.38)/2 \times 145 \times 200$$
$$= 116{,}000 \text{ N} = 116 \text{ kN}$$

The bending moment at *X-X* axis is

$$M_x = 3.38 \times 145 \times 200 \times 145/2 + 1.24 \times 145 \times 200/2 \times 2/3 \times 145$$
$$= 8.84 \times 10^6 \text{ Nmm} = 8.84 \text{ kNm}$$

The shear capacity

$$V_n = V_p = A_v f_{yg}/(\sqrt{3}\,\gamma_{m0})$$
$$= 200 \times 12 \times 250/(\sqrt{3} \times 1.1 \times 1000)$$
$$= 314.9 \text{ kN} > 116 \text{ kN}$$
$$V = 116 \text{ kN} < 0.6 \times 314.9 = 188.95 \text{ kN}$$

The moment capacity is not reduced by the effect of shear.

$$M_g = Z_e f_y/\gamma_{m0} = (12 \times 200^2/6) \times 250/(1.1 \times 10^6)$$
$$= 18.18 \text{ kNm} > 8.84 \text{ kNm}$$

Hence the size of the gusset plate is satisfactory.

**Step 2:** *Gusset plate to column weld*
The welds between the column, gussets, and base plate have to transmit all the load to the base plate. These welds are shown in Fig. 15.25(c).
Weld connecting column-gusset-base plate:

$$\text{Load per weld} = 500/2 + 45/0.25$$
$$= 430 \text{ kN}$$

Assuming an 8-mm weld,
Length of the weld {see Fig. 15.25(c)(iv)}

$$= 250 + 200 \times 2 - 2 \times 8 = 634 \text{ mm}$$
$$\text{Load per mm} = 430/634 = 0.678 \text{ kN/mm}$$

Use an 8-mm fillet weld (site weld) which has strength of 0.884 MPa (from Table 6.5). The weld between one gusset plate and the base must support the maximum pressure under the base. Considering a 1-mm wide strip at the edge of the base plate, load on one weld {see Figs 15.25(a) and (b)}

$$= 4.62 \times 200/(2 \times 10^3) = 0.462 \text{ kN/mm}$$

Provide a 5-mm fillet weld with strength = 0.553 kN/mm (from Table 6.5).

**Step 3:** *Thickness of the base plate*
Consider a 1-mm wide strip, at the edge of the base plate, as shown in Fig. 15.25(a). It is assumed to act as a beam with overhanging ends as shown in Fig.15.25(c)(v). The bending moments are

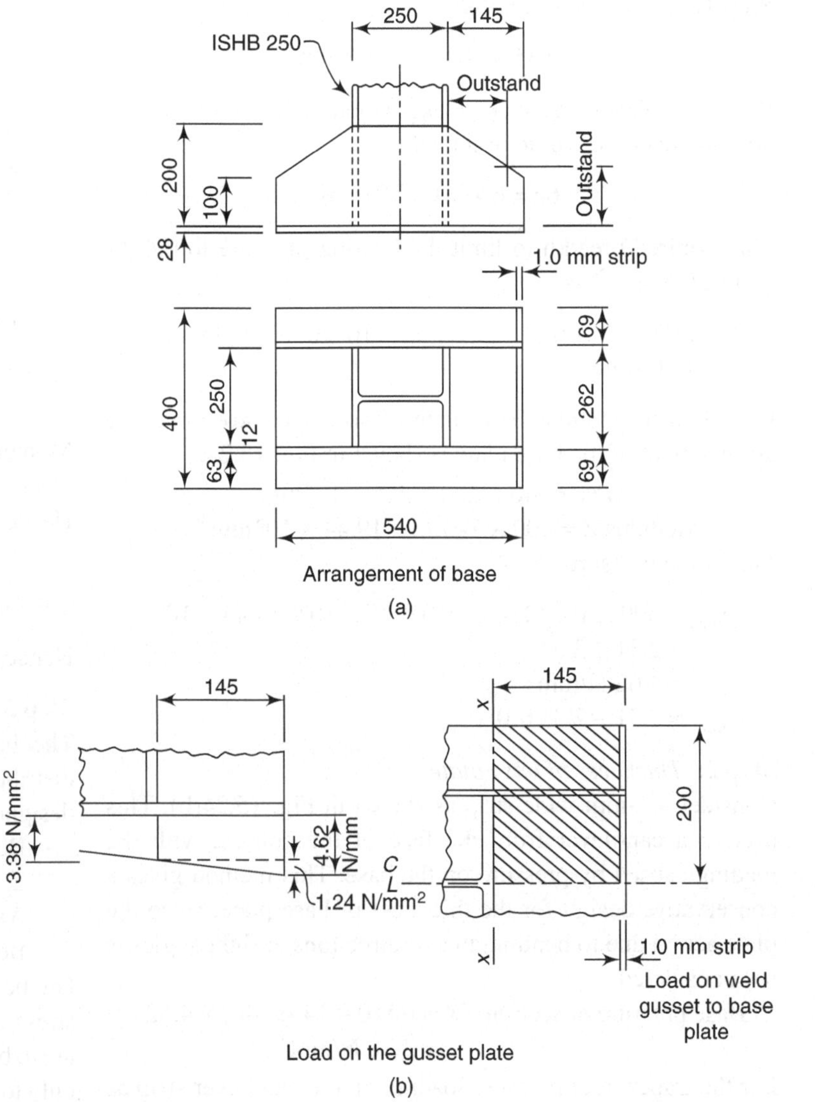


Arrangement of base
(a)

Load on the gusset plate
(b)

**FIG. 15.25**

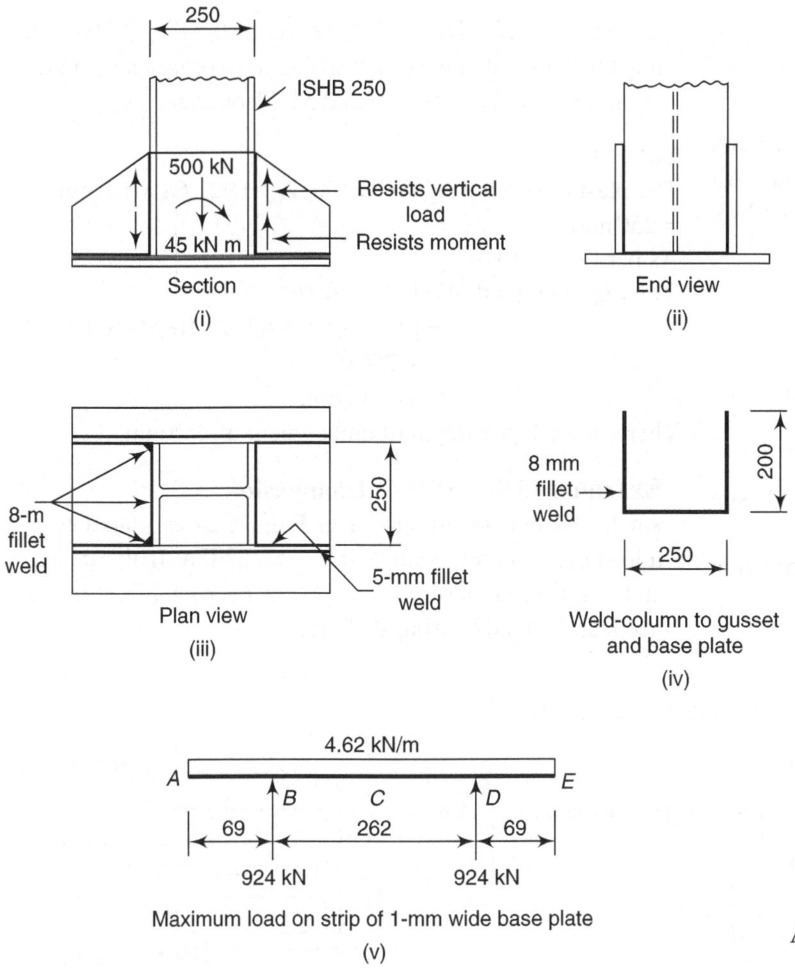


**FIG. 15.25** *(Contd)*

$$M_B = 4.62 \times 69^2/2 = 10998 \text{ Nmm}$$

$$M_C = 924 \times (262/2) - 4.62 \times 200^2/2 = 28644 \text{ Nmm}$$

Moment capacity of base plate $= 1.2 f_y Z_e/\gamma_{m0}$

$$= 1.2 \times (250/1.1) \times t^2/6$$

Hence $28644 = 1.2 \times (250/1.1) \times t^2/6$

$$t^2 = 630.2$$

or $t = 25.1$ mm

Hence, a 28-mm thick base plate is required.

**EXAMPLE 15.5:** Design the base plate for a W 310 × 310 × 143 column subjected to a factored axial load of 450 kN and bending moment of 85 kNm. Assume a M30 grade concrete for foundation.

**SOLUTION:**

Assume a base plate of size 520 × 520 mm (as shown in Fig. 15.26) and four 20-mm diameter anchor bolts of grade 4.6.

$$\text{Eccentricity} = M/F = 85 \times 10^6/(450 \times 10^3)$$

$$= 189 \text{ mm}$$

Length of the plate = 520 mm < 6 × 189 mm

Hence there will be tension in the holding down bolts.

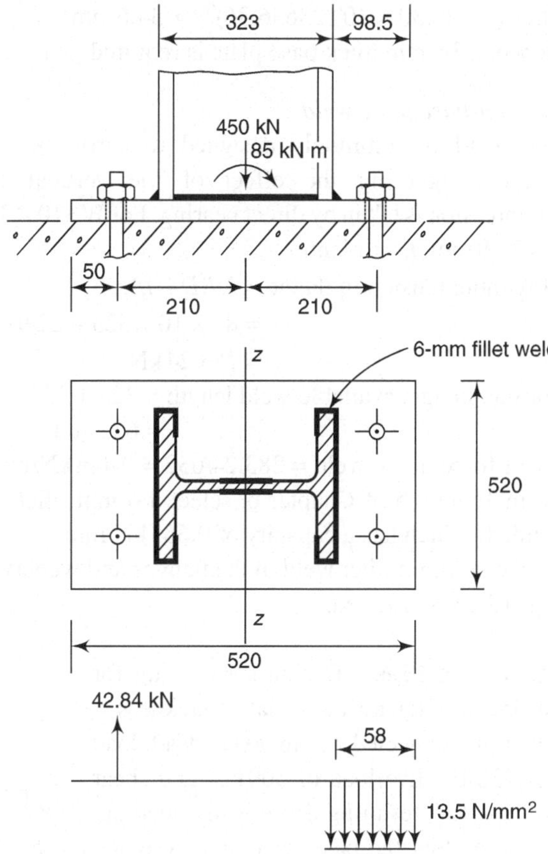


**FIG. 15.26**

As per Eqn (13.48), with $a = (520 - 100)/2 = 210$ mm

$$y = 520/2 + 210 - [(520/2 + 210)^2 - 2(85 \times 10^6 + 450 \times 10^3 \times 210)/(0.45 \times 30 \times 520)]^{0.5}$$

$$= 470 - [220{,}900 - 51{,}140]^{0.5}$$

$$= 58 \text{ mm}$$

$$F_b = 0.45 f_{ck}\, yB - N$$

$$F_b = 0.45 \times 30 \times 58 \times 520 - 450 \times 10^3$$

$$= -42840 \text{ N} = -42.84 \text{ kN (Tensile)}$$

Tensile strength of two 20-mm diameter bolts of grade 4.6 with length = 100 mm (from Table 5.9 of Chapter 5)

$$= 66.6 \text{ kN} \times 2 = 133.2 \text{ kN} > 42.84 \text{ kN}$$

Tensile capacity based on concrete breakout failure.

$$N_u = k\sqrt{f_{ck}}\; h_{ef}^{1.5}$$

$$= 15.5 \times \sqrt{30} \times 100^{1.5} = 84.9 \text{ kN} > 42.84/2 \text{ kN}$$

Hence anchor bolts are satisfactory.

*Plate thickness*

Maximum bending moment

$$= 58 \times 520 \times (0.45 \times 30) \times [98.5 - (58/2)]/10^6$$

$$= 28.30 \text{ kNm}$$

Moment capacity of base plate $= 1.2(f_y/\gamma_{m0})Z_e$

$$= 1.2 \times (250/1.1)(t^2/6) \times 520$$

$$= 23636.36t^2$$

Thus $t = (28.30 \times 10^6/23636.36)^{0.5} = 34.6\,\text{mm}$
Hence, a 36-mm thick base plate is required.

*Column-base plate weld*
The weld is commonly designed to carry the maximum moment, ignoring the effect of the vertical load. All compression is taken by direct bearing. For $W310 \times 310 \times 143$, $t_f = 22.9\,\text{mm}$, $t_w = 14\,\text{mm}$

Maximum tension in flange $= M/(D - t_f)$
$= 85 \times 10^3/(323 - 22.9)$
$= 283.24\,\text{kN}$

For one flange, available weld length $= 323 + 323 - 14$
$= 632\,\text{mm}$

Shear force in the weld $= 283.24/632 = 0.448\,\text{kN/mm}$
From Table 6.5 of Chapter 6, select a 6-mm fillet weld (site welded) which has a capacity of 0.663 kN/mm.
Hence, a 6-mm fillet weld in the flanges and web as shown in Fig. 15.26 is required.

**EXAMPLE 15.6:** Design a shear lug for a 350 mm square base plate (thickness = 30 mm) subjected to an axial dead load of 450 kN, live load of 500 kN, and shear of 240 kN resulting from wind loading. Assume that the base plate and shear lug are of Fe 410 grade steel and the foundation is M25 grade concrete.

**SOLUTION:**

1. $V_{\text{lg}1} = 1.2(240) - 0.3[1.2\,(450 + 500)]$
   $= -54\,\text{kN}$
   $V_{\text{lg}2} = 1.5(240) - 0.3[1.5(450)] = 157.5\,\text{kN}$
2. $A_{\text{lg}} = V/(0.45 f_{\text{ck}}) = 157.5 \times 10^3/(0.45 \times 25)$
   $= 14 \times 10^3\,\text{mm}^2$
3. Assume a shear lug with width $W = 200\,\text{mm}$. The height of the bearing portion is

   $$H - G = 14 \times 10^3/200 = 70\,\text{mm}$$

   Assuming a grout depth of 25 mm,
   Required depth of shear lug $= 70 + 25$
   $= 95\,\text{mm}$
4. The cantilever end moment

   $$M_{\text{lg}} = (V/W)(H + G)/2 = (157.5 \times 10^3/200)\,(95 + 25)/2$$
   $$= 47250\ \text{N mm}$$
5. The required thickness $t_{\text{lg}} = \sqrt{[4M_{\text{lg}}/(f_y/\gamma_{m0})]}$

   $$= \sqrt{[4 \times 47250/(250/1.1)}$$
   $$= 28.83\,\text{mm} < 30\,\text{mm}$$

   Use a 200-mm wide, 95-mm high, and 30-mm thick shear lug.

**EXAMPLE 15.7** Determine the depth than an ISHB 300 must be embedded in a pocket base to develop its strength. Assume grade Fe 410 steel and M25 concrete.

**SOLUTION:**
The plastic modulus of ISHB 300, $Z_{\text{pz}} = 921.68 \times 10^3\,\text{mm}^3$, $b_f = 250\,\text{mm}$
As per Eqn (15.16),
The required depth $d = [f_y Z_{\text{pz}}/(0.164 f_{\text{ck}}\, b)]^{0.5}$
$= [250 \times 921.68 \times 10^3/(0.164 \times 25 \times 250)]^{0.5}$
$= 474.13\,\text{mm}$
Therefore, adopt a depth of embedment of 480 mm.

**EXAMPLE 15.8: Stiff seat connection**
An ISMB 600 beam shown in Fig. 15.27 is seated on a reinforced concrete wall with $f_{\text{ck}} = 25\,\text{MPa}$. If the design end reaction is 300 kN, design the bearing plate and if needed, the load bearing stiffeners.

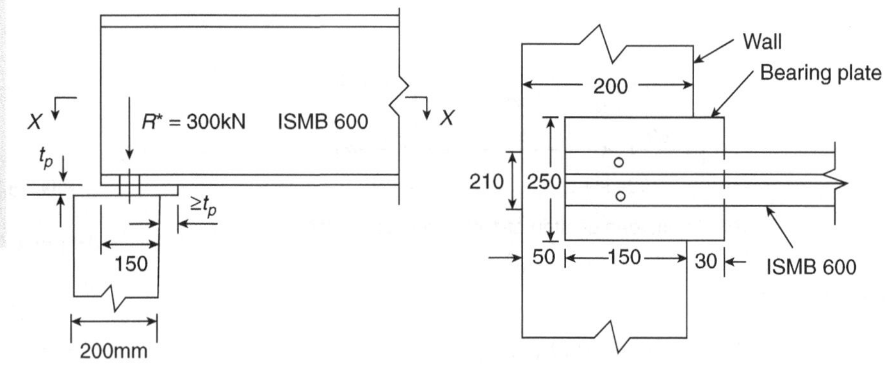

(a) Elevation (b) View X-X

**FIG. 15.27**

**SOLUTION:**
The bearing plate is assumed to distribute the beam end reaction uniformly to the area of concrete under it. The bearing area $A_1$ must be such that the design bearing strength of concrete is not exceeded.
For ISMB 600, $t_w = 12\,\text{mm}$, $t_f = 20.3\,\text{mm}$, $b_f = 210\,\text{mm}$, and $R = 20\,\text{mm}$
Bearing strength of concrete $= 0.45 f_{\text{ck}} = 0.45 \times 25 = 11.25\,\text{N/mm}^2$

$$\text{Required area of beam seat} = 300 \times \frac{10^3}{11.25} = 26{,}667\,\text{mm}^2$$

Required width of base plate $= 26{,}667/150 = 178\,\text{mm} < 210\,\text{mm}$
*Check for web bearing*

$$N = \frac{R^*}{t_w\left(\dfrac{f_{yw}}{\gamma_{m0}}\right)} = \frac{300 \times 10^3}{12 \times 250/1.1} = 110\,\text{mm} < 150\,\text{mm, Hence}$$

take $N = 150\,\text{mm}$

$$n_2 = 2.5(t_f + R) = 2.5(20.3 + 20) = 100.75$$

$$F_w = (150 + 100.75) \times 12 \times \frac{250}{1.1} = 683.8\,\text{kN} > 300\,\text{kN}$$

Hence it is safe.

Let us provide 2 numbers 20 mm diameter bolts of grade 4.6 along the width to connect the bearing plate with the bottom flange of beam and anchoring it into the concrete.

Minimum edge distance for 20 mm bolt = 33 mm (see Table 5. 6).

*Size of bearing plate*

Length of plate = 150 + 30 (assuming 30 mm thick plate) = 180 mm

Width of plate = 210 + 2 × 20 = 250 mm > 178 mm

Adopt a bearing plate of size 250 mm × 180 mm.

The bearing plate is designed as an inverted cantilever loaded with the bearing pressure

$$\text{w} = \text{P/A} = \frac{300 \times 10^3}{250 \times 150} = 8\ \text{N/mm}^2$$

The cantilever length = (250 – 210)/2 = 20 mm

Considering a 1 mm wide cantilever of length 20 mm,

Bending moment = 8 × $20^2/2$ = 1600 Nmm

$$M_p = \frac{1.2 f_y t^2}{6} = \frac{1.2 \times 250 t^2}{6} = 50t^2$$

Hence $t_p = \sqrt{1600/50} = 5.7\,\text{mm}$

Provide a 12 mm thick plate.

*Check for shear strength of web*

Shear strength of the web

$$= ht_w f_y/(1.1 \times \sqrt{3}) = 600 \times 12 \times 250/(1.1 \times \sqrt{3}) = 944.75\,\text{kN}$$

> 300 kN

Hence stiffener is not required.

*Note*: Web crippling strength should also be checked.

**EXAMPLE 15.9** Design a grillage foundation for a built-up column consisting of two ISHB 400 sections with their webs spaced at 300 mm and connected by 10 mm thick battens and subjected to a service axial load of 2500 kN. Assume SBC of soil as 250 kN/m² and gusseted base plate of size 750 mm by 750 mm and thickness 40 mm. The arrangement is shown in Fig. 15.28.

**SOLUTION:**

Let us adopt a square grillage foundation with two tiers of girders.

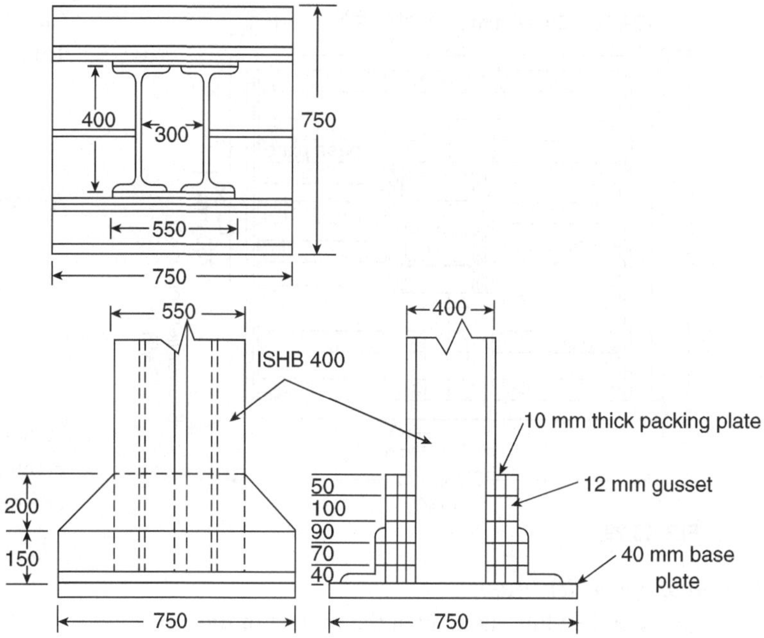


**FIG. 15.28**

Factored axial load from column = 1.5 × 2500 = 3750 kN

Assuming self-weight of foundation as 10%,

Total service load in foundation = 1.1 × 2500 = 2750 kN

$$\text{Required area of bottom tier} = \frac{2750}{250} = 11\ \text{m}^2$$

Side length of bottom tier of square grillage = $\sqrt{11} = 3.317$ m

Let us adopt a grillage of size 3.375 m × 3.375 m with a concrete cover of 125 mm on all the sides. Provide 3.5 m × 3.5 m footing block at the bottom tier (see Fig. 15.29).

*Thickness of base plate*

$$\text{Thickness of base plate, } t_p = \sqrt{\frac{2.5 w \gamma_{m0}}{f_y}(a^2 - 0.3b^2)}$$

$$\text{Where } w = \frac{3750 \times 10^3}{750 \times 750} = 6.67\ \text{MPa}$$

$$b = \frac{750 - (400 + 10 + 12)}{2} = 164\ \text{mm},$$

$$a = \frac{750 - (300 + 2 \times 9.1)}{2} = 215.9\ \text{mm}$$

$$t_p = \sqrt{\frac{2.5 \times 6.67}{\left(\frac{250}{1.1}\right)}(215.9^2 - 0.3 \times 164^2)} = 53\,\text{mm}$$

To reduce the thickness, use stiffeners as shown in Fig. 15.29.

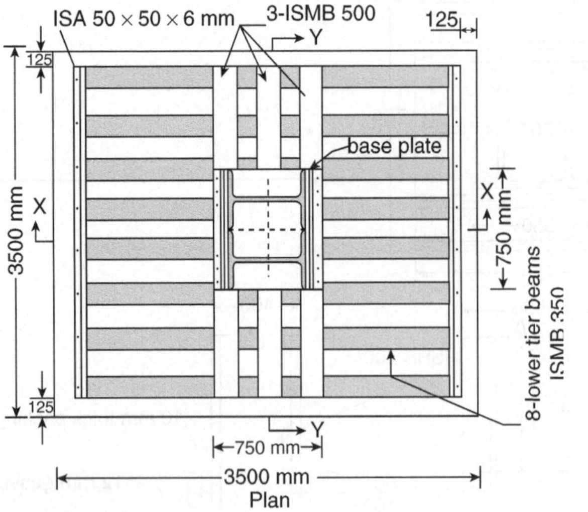


FIG. 15.29

*Design of upper tier beams*

The maximum bending moment at the middle of the tier

$$M_u = \frac{P_u(L-b)}{8} = \frac{3750\,(3.5-0.75)}{8} = 1289.06 \text{ kNm}$$

Plastic section modulus required for all beams of upper tier

$$Z_p = \frac{M_u \gamma_{m0}}{f_y} = \frac{1289.06 \times 10^6 \times 1.1}{250} = 5.67 \times 10^6 \text{ mm}^3$$

$$Z_p \text{ per beam} = 5.67 \times \frac{10^6}{3} = 1.89 \times 10^6 \text{ mm}^3$$

Adopt ISMB 500 with $Z_p = 2074.67 \times 10^3\,\text{mm}^3$ with $t_f = 17.2$ mm $t_w = 10.2$ mm, $b_f = 180$ mm, and $R = 17$ mm.

The clear distance between the flanges is usually kept greater than 75 mm.

Minimum distance between flanges

$$= \frac{750 - 3(180)}{3-1} = 105 \text{ mm} > 75 \text{ mm}$$

Hence it is adequate.

*Check for local buckling*

$$\varepsilon = \sqrt{\left(\frac{250}{f_y}\right)} = \sqrt{\left(\frac{250}{250}\right)} = 1$$

Flange local buckling $\dfrac{b}{t_f} = \dfrac{\left(\dfrac{180}{2}\right)}{17.2} = 5.23 < 9.4\varepsilon$

Web local buckling $\dfrac{d}{t_w} = \dfrac{500 - 2(17.2+17)}{10.2} = 42.3 \approx 42\varepsilon$

Hence the section is non-slender.

*Check for shear*

Total shear force in the tier,

$$V_u = \frac{P_u(L-b)}{2L} = \frac{3750\,(3.5-0.75)}{2 \times 3.5} = 1473.2 \text{ kN}$$

Average shear stress in the beam.

$$\tau_b = \frac{V_u}{nht_w} = \frac{1473.2 \times 1000}{3 \times 500 \times 10.2}$$

$$= 96.29 \text{ MPa} \le \frac{f_y}{1.1 \times \sqrt{3}} = \frac{250}{1.1 \times \sqrt{3}} = 131 \text{ MPa}$$

Hence it is safe against shear.

Provide three numbers ISMB 500 beams at a spacing of 285 mm c/c in the upper tier (see Fig. 15.30).

*Check for web buckling*

$$r = \frac{t_w}{2\sqrt{3}} = \frac{10.2}{2\sqrt{3}} = 2.94 \text{ mm}$$

Effective length of web

$= 0.7\,[500 - 2\,(17.2 + 17)] = 302.12$ mm

$L/r = 302.12/2.94 = 102.8$

From Table 9c of the code, $f_{cd} = 103.5$ N/mm$^2$

$$n_1 = \frac{500}{2} = 250 \text{ mm};\; b_1 + 2n_1 = 750 + 2 \times 250 = 1250 \text{ mm}$$

$$A_b = 1250 \times 10.2 = 12{,}750 \text{ mm}^2$$

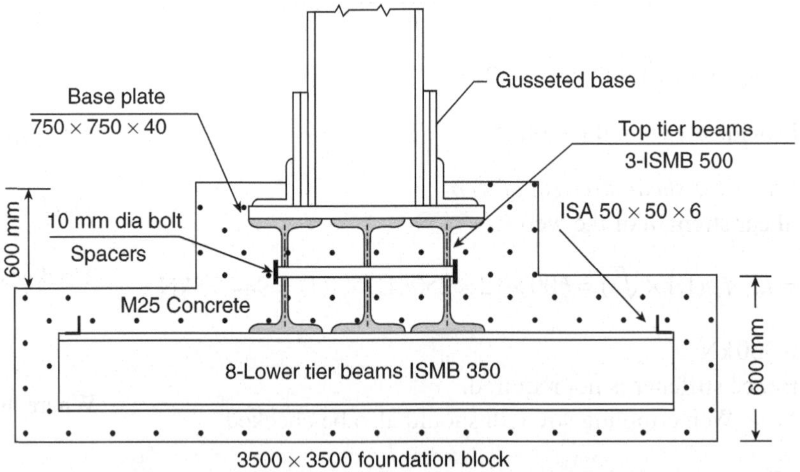


FIG. 15.30 Section Y-Y of the grillage foundation

$$\text{Buckling resistance} = A_b f_{cd} = 12{,}750 \times \frac{103.5}{1000} = 1319.6$$

kN > 3750/3 = 1250 kN

Hence it is safe.

*Check for web buckling*

$$F_w = (b_1 + 2n_2) t_w f_y / \gamma_{m0}$$

$b_1 = 750$ mm

$2n_2 = 2 \times 2.5$(root radius + flange thickness) (since angle of dispersion as per code is 1:2.5)

$2n_2 = 2 \times 2.5\,(17 + 17.2) = 171$ mm

$$F_w = \frac{(b_1 + 2n_2)t_w f_y}{\gamma_{m0}} = \frac{(750 + 2\times171)\times10.2\times250}{1.1\times1000}$$

$$= 2135\,\text{kN} > 3750/3 = 1250\,\text{kN}$$

Hence web is safe against buckling.
Bearing length at a dispersion of 1:2.5 = 750 + 2 × 2.5 (17.2 + 17) = 921 mm
Bearing stress

$$f_b = 3750 \times \frac{1000}{3\times921\times10.2} = 133 \text{ MPa}$$

*Check for combined stress*
The combined stress is checked at the edge of base plate.
Shear = 1473.2kN

$$\text{Moment} = 1473.2\left[\frac{(3.5-0.75)}{4}\right] = 1012.8\,\text{kNm}$$

Bending stress in compression at this location of the section (at the root of top fillet)

$$f_{bc} = \frac{M}{nZ_p} = \frac{1012.8\times10^6}{3\times2074.67\times10^3}\left(\frac{250-17.2-17}{250}\right)$$

$$= 140.5 \text{ N/mm}^2$$

Equivalent stress

$$f_e = \sqrt{f_{bc}^2 + f_p^2 - f_{bc}f_p + 3\tau_b^2}$$

$$= \sqrt{140.5^2 + 133^2 - 140.5\times133 + 3\times96.29^2}$$

$$= 215.8 \text{ N/mm}^2 < 0.9f_y = 0.9\times250 = 225\,\text{N/mm}^2$$

*Design of lower tier beams*
In the transverse direction of bottom tier, $L$ = 3.5 m and $a$ = 750 mm, which is the same as the upper tier. Hence the bending moment and shear force will be the same as that of the upper tier.
Let us consider 8 beams in the lower layer.
Plastic section modulus $Z_p$ required for all beams of lower tier $= 5.67\times10^6\,\text{mm}^3$

$$Z_p \text{ per beam} = 5.67\times\frac{10^6}{8} = 708.75\times10^3 \text{ mm}^3$$

Let us select ISMB 350 with $Z_p = 889.57\times10^3\,\text{mm}^3$, $t_w$ = 8.1 mm, $t_f$ = 14.2 mm, $b_f$ = 140 mm, and $R$ = 14 mm (see Fig. 15.31)

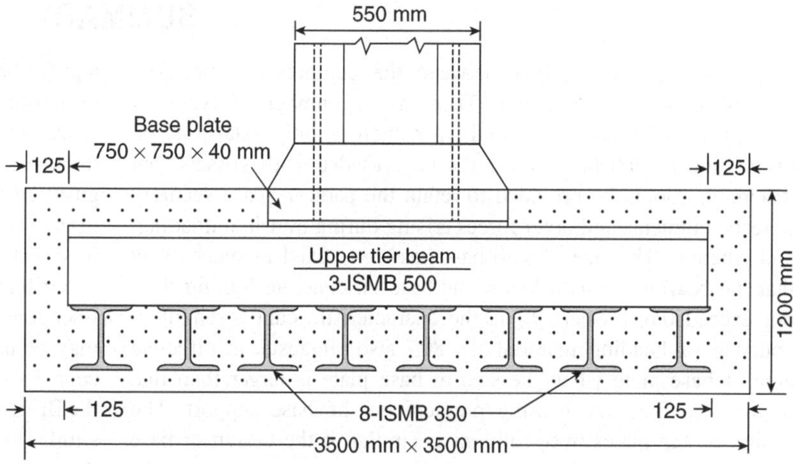


**FIG. 15.31** Section X-X of the grillage foundation

*Local buckling check*

$$\text{Flange local buckling } \frac{b}{t_f} = \frac{\left(\frac{140}{2}\right)}{14.2} = 4.93 < 9.4\varepsilon$$

Web local buckling

$$\frac{d}{t_w} = \frac{350 - 2(14.2+14)}{8.1} = 36.25 < 42\varepsilon$$

Hence the section is non-slender.
Minimum distance between beams

$$= \frac{3500 - (2\times125 + 8\times140)}{(8-1)} = 304.3\,\text{mm} > 75\,\text{mm}$$

*Check for shear*
Total shear force in the lower tier, $V_u = 1473.2$ kN
Average shear stress in the beam

$$\tau_b = \frac{V_u}{nht_w} = \frac{1473.2\times1000}{8\times350\times8.1}$$

$$= 64.95 \text{ MPa} \le \frac{f_y}{1.1\times\sqrt{3}} = \frac{250}{1.1\times\sqrt{3}} = 131\,\text{MPa}$$

Hence it is safe against shear.

*Separators*

1. Adopt 10 mm diameter spacers to connect the top tier beams.
2. Separators made of ISA 50 × 50 × 6 angles and 3.25 m long, welded or bolted with 12 mm diameter bolts to the flanges of the lower-tier girders at the two ends to prevent the girders from moving.

## SUMMARY

Base plates provide a link between the compression members and the concrete foundation. There are a number of types of base plates. Slab bases are used when there is only axial load on the columns. Slab bases need not be provided with gussets, but fastenings should be provided to retain the parts in place securely to resist moments and other forces arising during transit, unloading, and erection. The size of slab base has to be fixed in such a way that the bearing strength below the base on concrete footing does not exceed $0.6f_{ck}$ where $f_{ck}$ is the characteristic cube strength of concrete or bedding material. IS 800 also suggests an effective area of base plate when the size of base plate is larger than that required to limit the bearing pressure on the base support. The design of cap plates over columns is similar to the design of base plates.

In the design of base plates subjected to bending and axial load two cases arise: (a) base plate with compression over the whole area and (b) base plate with compression over part of the area resulting in tension in the holding down bolts (anchor bolts). Both the cases are covered. The provision of gussets will result in reduced thickness of base plates, though increasing the fabrication cost. Such gusseted base plates are also discussed. Information on the design of anchor bolts and shear lugs (to resist heavy shear forces) is also included.

Grillage footing consists of two or more tiers of steel beams (usually I-sections), with each tier placed at right angles to the one below it, may be used under heavily loaded columns, or where safe bearing capacity of soil is low. An approximate method of design based on the British code is discussed. Design of beam bearing plates, which is similar to the design of column base plates, is also discussed. All the design concepts explained in this chapter all detailed with illustrative examples.

## MULTIPLE-CHOICE QUESTIONS

15.1 The bearing strength of concrete is taken in IS 800 as
(a) $0.4f_{ck}$ (b) $0.45f_{ck}$ (c) $0.5f_{ck}$ (d) $0.6f_{ck}$

15.2 The grout below the base plate may have a thickness of
(a) 10 mm to 20 mm (b) 20 mm to 30 mm
(b) 20 mm to 40 mm (d) 10 mm to 30 mm

15.3 The thickness of base plate is determined from the
(a) flexural strength of plate
(b) bearing strength of supporting concrete
(c) shear strength of the plate
(d) punching criteria

15.4 The minimum thickness of rectangular slab base is calculated as
(a) $t_p = \sqrt{[2.5w(0.3a^2 - b^2)\gamma_{m0} / f_y]}$
(b) $t_p = \sqrt{[2.5w(a^2 - 0.3b^2)\gamma_{m0} / f_y]}$
(c) $t_p = \sqrt{[2.5wa^2\gamma_{m0} / f_y]}$
(d) $t_p = \sqrt{[3w(a^2 - 0.3b^2)\gamma_{m0} / f_y]}$

15.5 When the intensity of bearing pressure due to axial force and the maximum intensity of bearing pressure due to moment are equal, then the resulting bearing pressure below the base plate will have
(a) tension at one end and compression at the other end
(b) zero at one end and compression at the other end
(c) uniform compression throughout
(d) uniform tension throughout

15.6 To prevent buckling of the top edge of gusset plate due to bending compression, the outstand of the gusset plate from the edge of column should be limited to
(a) $9.4\varepsilon t$ (b) $13.6\varepsilon t$ (c) $29.3\varepsilon t$ (d) $8.4\varepsilon t$

15.7 In pocket bases the length of column embedded in concrete should be greater than __________ times the depth of the column section.
(a) 1.2 (b) 1.3 (c) 1.4 (d) 1.5

15.8 Anchor bolts are provided in column bases to
(a) resist tensile forces due to moments
(b) fix column in correct position during erection
(c) act as reinforcement in the pedestal below the base plate
(d) all of these

15.9 Failure modes of cast-in-place headed anchors include
(a) yielding of anchor bolt
(b) concrete breakout failure
(c) failure of bond between anchor and surrounding concrete
(d) (a) and (b)
(e) (a), (b), and (c)

15.10 As per concrete capacity design tensile capacity of cast-in-situ headed anchors is calculated as
(a) $15.5\sqrt{f_{ck}}\sqrt{h_{ef}}$
(b) $15.5\sqrt{f_{ck}}\,(h_{ef})^{1.5}$
(c) $15.5\sqrt{f_{ck}}\,h_{ef}$
(d) $15.5f_{ck}(h_{ef})^{1.5}$

15.11 When high shear is transmitted at the base of a column, it is best resisted by
(a) pocket base (c) shear lug
(b) anchor bolts (d) gusseted base

15.12 In grillage foundation, the minimum distance between the flanges of two consecutive beams should be greater than
(a) 50 mm (b) 75 mm (c) 100 mm (d) 150 mm

15.13 In grillage foundation, the surfaces and edges of the flanges of the beams should be embedded in concrete with a cover greater than
(a) 50 mm (b) 75 mm (c) 100 mm (d) 150 mm

# EXERCISES

15.1 Design a base plate for a column of ISHB 200 carrying a factored load of 600 kN. Assume that the column is supported on a concrete of grade M25.

15.2 Design the base plate for an ISHB 400 column to carry a factored load of 1500 kN. Assume Fe 410 grade steel and M25 concrete. Use the effective area method of design.

15.3 Design a suitable column and its base plate to support a factored load of 3500 kN. Take the effective length as 5m and the grade of concrete as M35.

15.4 Design the base plate for an ISHB 300 column subjected to a factored axial load of 800 kN and a factored moment of 40 kNm in the major axis. Assume M25 concrete for the foundation and Grade Fe 410 steel.

15.5 Redesign the base plate in Exercise 15.4 using gusset plates.

15.6 Design the base plate for a ISHB 300 column subjected to a factored axial load of 350 kN and a bending moment of 52.5 kNm. Assume M30 concrete for foundation and Grade Fe 410 steel.

15.7 Design a base plate for a ISHB 300 column subjected to a factored axial load of 300 kN, bending moment of 60 kNm and a shear force of 300 kN. Assume that the base plate and shear lug are of Fe 410 grade steel and the foundation is M 25 concrete.

15.8 An ISMB 500 beam is seated on a reinforced concrete wall with $f_{ck}$ =25 MPa, as shown in Fig. 15.27. If the design end reaction is 250 kN, design the bearing plate and if needed, the load bearing stiffeners.

15.9 Design a grillage foundation for a built-up column consisting of two ISHB 350 sections with their webs spaced at 300 mm and connected by 10 mm thick battens and subjected to a service axial load of 2000 kN. Assume SBC of soil as 250 kN/m$^2$ and gusseted base plate of size 650 mm by 650 mm and thickness 30 mm. The arrangement will be similar to that shown in Fig. 15.28.

# REVIEW QUESTIONS

15.1 What are the types of base plates used in practice?

15.2 State the different steps to be followed while designing a slab base.

15.3 How is the base plate designed when the size of the base plate is larger than that required to limit bearing pressure?

15.4 Why are anchor bolts normally not considered to transfer shear forces?

15.5 What is the purpose of anchor bolts in a base plate having compression over the whole area?

15.6 What are the criteria for design in the case of a base plate having compression over the whole area?

15.7 How are the gusset plates sized in the gusseted base plate?

15.8 State the assumptions that are normally made in the design of base plates having compression over part of the plate and tension in holding down bolts.

15.9 What are the advantages of base plates with collar?

15.10 State the assumptions made in the design of pocket bases.

15.11 What are the two basic types of anchor bolts?

15.12 What is the advantage of using hooked bolt over cast-in-place headed anchor, for carrying small uplift forces?

15.13 How do the headed anchors transfer tensile load to the foundation?

15.14 State the advantage of using post-installed anchors.

15.15 List the different types of post-installed anchors.

15.16 Why are bonded anchors not as strong as other post-installed anchors?

15.17 Explain the difference in the failure modes of cast-in-place headed anchors and grouted anchors.

15.18 What is a shear lug? When is it used?

15.19 What is the suggested value of $\mu$ for base plates? Why is this value suggested?

15.20 List the different design steps for sizing the shear lug.

15.21 What are grillage foundations? How are they designed?

15.22 What are bearing plates? How are they designed?

# REFERENCES


ACI 318-2014, *Building Code Requirements for Structural Concrete and Commentary*, American Concrete Institute, Farmington Hills, MI, 2014I.

*AISC Design Guide 1: Column Base Plates,* Second edition, American Institute of Steel Construction, Chicago, IL, 2005.

ANSI/AISC 360-10, *Specification for Structural Steel Buildings*, American Institute of Steel Construction, Chicago, IL, 2010 (www.aisc.org).

Collins, J. and Schlafly, T.J., 'Base Plates and Anchor Rods', *Modern Steel Construction*, AISC, Jun 2010, p. 3.

Cook, R.A., *Strength Design of Anchorage to Concrete*, Portland Cement Association, Skokie, IL, 1999, p. 86.

DeWolf, J.T. and Ricker, D.T., *Column Base Plates, Steel Design Guide Series* No. 1, Second edition, AISC, Chicago, IL, 2006.

DeWolf, J.T. and Sarisley, E.F., 'Column Base Plates with Axial Loads and Moments', *Journal of the Structural Div., ASCE*, Vol. 106, Nov 1980, pp. 2167–2184.

Drake, R.M. and Elkin, S.J., 'Beam-column Base Plate Design, LRFD Method', *Engineering Journal, AISC*, Vol. 36, No. 1, 1999, pp. 29–38, Discussion by Doyle, J.M. and Fisher, J.M., Vol. 42, No. 4, 2005, pp. 273–274.

Fuchs, W., Eligehausen, R., and Breen, J.E., 'Concrete Capacity Design (CCD) Approach for Fastening to Concrete', *ACI Structural Journal*, Jan–Feb 1995, Vol. 92, No. 1, pp. 73–94.

IS 5624-1993, *Foundation Bolts – Specification*, First Revision, Bureau of Indian Standards, New Delhi, Reaffirmed 1998, p. 5.

Lee, D.-Y., Goel, S.C., and Stojadinovic, B., 'Exposed Column-Base Plate Connections Bending about Weak Axis: I. Numerical Parametric

Study', *International Journal of Steel Structures*, Vol. 8, 2008, pp. 11–27 and 'II. Experimental Study', pp. 29–41, Also Discussion by Subramanian, N., Vol. 8, No. 4, Oct–Dec 2008, pp. 383–387 [http://www.kssc.or.kr/wonmun/KSSC_3_2008_8_1_29(C).pdf].

Lescouarc'h, Y. and Colson, A., *Column Bases in Constructional Steel Design—An International Guide*, Dowling, P.J., Harding, J.F., and Bjorhovde, R., ed., Elsevier Applied Science, London, 1992, pp. 379–396.

Petersen, C., *Stahlbau-Grundlagen der Berechnung und Baulichen Ausbuildung von Stahlbauten*, Second edition, Vieweg Verlag, Braunschweig, 1990.

Ricker, D.T., 'Some Practical Aspects of Column Base Selection', *Engineering Journal,* AISC, Third Quarter, Vol. 26, No. 3, Third Quarter, 1989, pp. 81–89.

*State-of-the-art Report (CEB), Fastenings to Concrete and Masonry Structures*, Comite Euro-International du Beton, Thomas Telford, UK, 1994, p. 249.

Subramanian, N. and Cook, R.A., 'Behaviour of Grouted Anchors', *The Indian Concrete Journal*, Vol. 78, No. 4, Apr 2004, pp. 14–21.

Subramanian, N. and Cook, R.A., 'Installation, Behaviour and Design of Bonded Anchors', *The Indian Concrete Journal*, Vol. 76, No. 1, Jan 2002, pp. 47–56.

Subramanian, N. and Vasanthi, V., 'Design of Anchor Bolts in Concrete', *The Bridge and Structural Engineer*, Vol. XXI, No. 3, Sep 1991, pp. 47–73.

Subramanian, N., 'Recent Developments in the Design of Anchor Bolts', *The Indian Concrete Journal,* Vol. 74, No. 7, July 2000, pp. 407–414.

Subramanian, N., *Design of Reinforced Concrete Structures*, Oxford University Press, 2013, p. 880.

Swiatek, D. and Whitbeck, E., 'Anchor Rods', *Modern Steel Construction*, Vol. 44, No. 12, Dec 2004, pp. 31–33.

Thambiratnam, D.P. and Paramasivam, P., 'Base Plates Under Axial Loads and Moments', *Journal of Structural Engineering, ASCE*, Vol. 112, No. 5, May 1986, pp. 20, 1166–1181.

Warner, J., *Practical Handbook of Grouting: Soil, Rock and Structures*, John Wiley & sons Inc., New York, 2004, p. 720.

16

# DESIGN OF ROOF TRUSSES AND INDUSTRIAL BUILDINGS

## INTRODUCTION

High-rise steel buildings account for a very small percentage of the total number of structures that are built around the world. The majority of steel structures being built are low-rise buildings, which are generally of one storey only. Industrial buildings, a subset of low-rise buildings are normally used for steel plants, automobile industries, utility and process industries, thermal power stations, warehouses, assembly plants, storage, garages, small scale industries, etc. These buildings require large column free areas. Hence interior columns, walls, and partitions are often eliminated or kept to a minimum. Most of these buildings may require adequate head room for the use of an over head travelling crane. The planning and design of these units requires knowledge of the following areas (Fisher 1984):

1. Site information
2. Soil conditions
3. Plant layout and work flow
4. Preferred bay sizes
5. Future expansion plans
6. Loading docks
7. Crane types and capacity
8. Roofing, side cladding, and wall material preferences
9. Heating, ventilation, and air-conditioning (HVAC) equipments loads
10. Availability of raw materials
11. Availability of waste disposal and sanitary facilities
12. Parking facilities
13. Preferred fabricators/contractors
14. Budget and project schedule

The structural engineer has to consider the following points during the planning and design of industrial buildings (Fisher 1984):

1. Selection of roofing and wall material
2. Selection of bay width
3. Selection of structural framing system
4. Roof trusses
5. Purlins, girts, and sag rods
6. Bracing systems to resist lateral loads
7. Gantry girders, columns, base plates, and foundations

Out of the listed points, gantry girders which support cranes have been discussed in Chapter 12 and columns and base plates have been discussed in Chapters 9,13, and 15. Foundations are made with reinforced concrete and are outside the scope of this book. Hence, this chapter focusses on the rest of points in the sections to follow.

## 16.1 SELECTION OF ROOFING AND WALL MATERIAL

The type of roof deck, type of purlin used, purlin spacing, deflections of secondary structural members, roof pitch, and drainage requirements are all determined by the choice of roofing. The roof weight also affects the gravity load design of the roof system and in the case of seismic calculations, the lateral load design.

Similar considerations apply to the cladding/wall systems. In selecting the cladding/wall system, the designer should consider the following areas:

1. Cost
2. Interior surface requirements
3. Aesthetic appearance (including colour)
4. Acoustics and dust control
5. Maintenance
6. Ease and speed of erection
7. Insulating properties
8. Fire resistance

Note that *cladding* carries only its own weight and the weight of the loads imposed by wind. In the case of roofs, the sheeting supports insulation and water proofing in addition to self weight and weight of loads due to wind and/or snow. Hence, it is often termed as *roof decking*. The cladding/wall system will have an impact on the design of girts, wall bracing, eave members, and foundation.

In India, corrugated galvanized iron (GI) sheets are usually adopted as coverings for roofs and sides of industrial buildings. Now light-gauge cold-formed ribbed steel or aluminium decking (manufactured by cold drawing flat steel or aluminium strips through dies to produce the required section) is also available. Sometimes asbestos cement (AC) sheets are also provided as roof coverings owing to their superior insulating properties. Their insulating properties may be enhanced by painting them white on the top surface. These three types of sheets are discussed briefly in the following section.

### 16.1.1 Steel or Aluminium Decking/Cladding

The modern built-up roof system consists of three basic components: steel/aluminium deck, thermal insulation, and membrane. The structural deck transmits gravity, wind, and earthquake forces to the roof framing. Thermal insulation is used for reducing heating and cooling costs, increasing thermal comfort, and preventing condensation on interior building surfaces. The membrane is the water-proofing component of the roof systems. On sloping roofs, the insulation consists of the insulation board or glass wool. On flat roofs, insulation board, felt, and bitumen are laid over the steel decking as shown in Fig. 16.1.

The steel decking has a ribbed cross section, with ribs generally spaced at 150 mm (centre to centre) and 37.5 mm or 50 mm deep (see Fig. 16.2). The sloped-side ribs measure about 25 mm wide at the top for a narrow rib deck, 44 mm for an intermediate rib deck, and 62.5 mm for a wide rib deck. Wide rib decking is more popular, which can be used with 25-mm-thick insulation boards. Thinner insulation boards may require narrow deck rib opening.

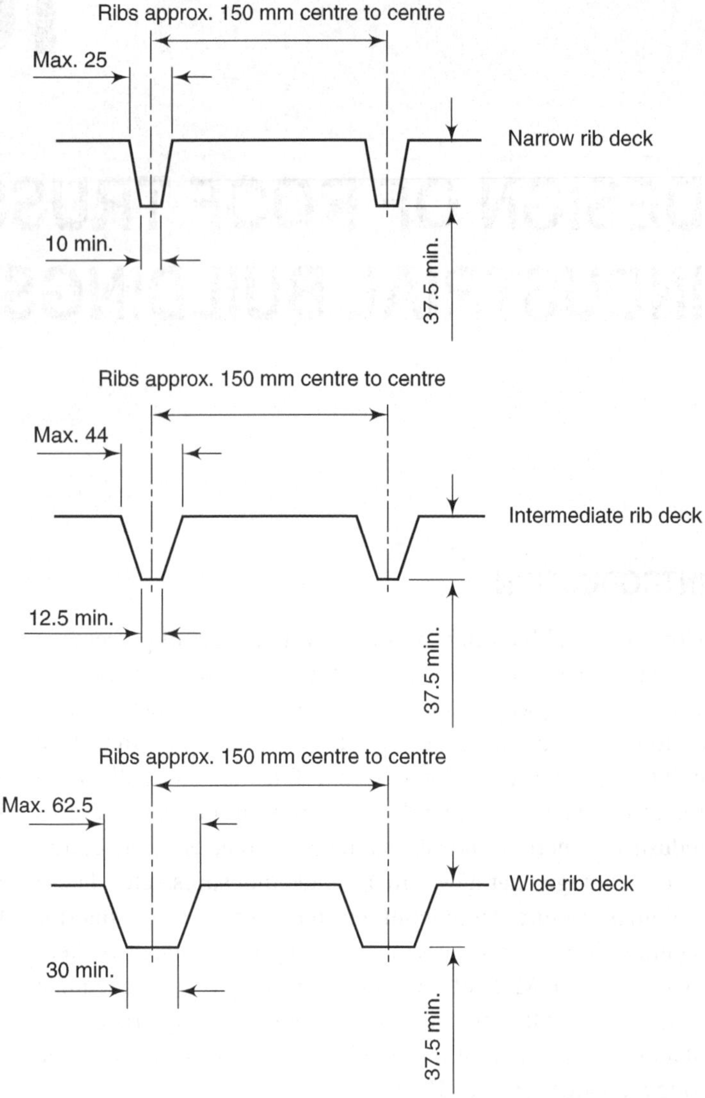


**FIG. 16.2** Typical profiles of roof deck

Wide rib decking also has higher section properties than other patterns, and hence can be used to span greater distances. These steel decks may be anchored to supporting flexural members by puddle welds by a welder, (power-activated and pneumatically driven fasteners, and self-drilling screws can also be used), as soon as the deck is placed properly on the rafters or top chord of the roof truss (Vinnakota 2006).

Steel decks are available in different thicknesses, depths, rib spacing, widths, and lengths. They are available with or without stiffening elements, with or without acoustical material, as in cellular and non-cellular forms. The cellular decks can be used to provide electrical, telephone, and cable wiring and also serve as ducts for air distribution. They are also available with different coatings and in different colours. They are easy to maintain, durable, and aesthetically pleasing.

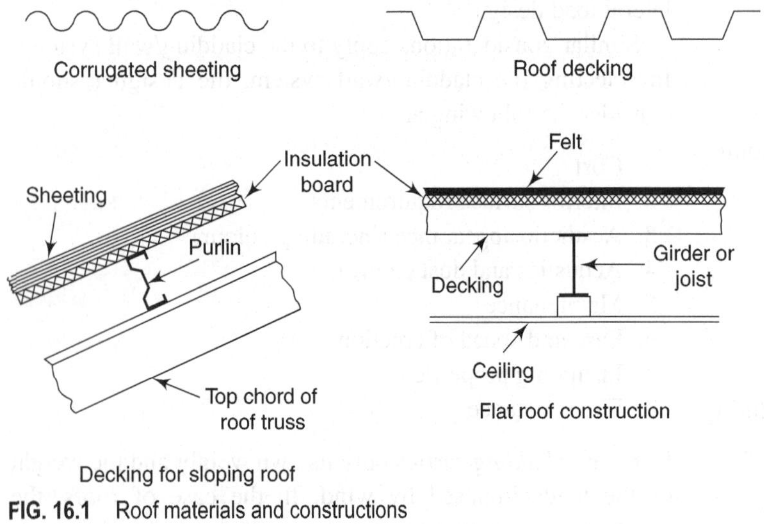


**FIG. 16.1** Roof materials and constructions

When properly anchored to supporting members, steel/aluminium decks provide lateral stability to the top flange of the structural member. They also resist the uplift forces due to wind during the construction stage. Steel decks may be considered as a simply supported or continuous one-way beam depending on the purlin and joist spacing. Aluminium sheets also offer excellent corrosion resistance. But they expand approximately twice as much as steel and are easily damaged in hailstorms. Moreover, aluminium sheeting should be separated not only from steel purlins but also from any non-aluminium roof-top framing and conduits, in order to avoid bi-metallic corrosion. The fasteners connecting aluminium sheets to steel purlins should be made of stainless steel. The aluminium alloy panels should be at least 0.8 mm thick and at least 1 mm thick for longer spans.

The load carrying capacity of the deck is influenced by the depth of the cross section, the span length, the thickness of metal, and whether it is simply supported or continuous. The profile steel decking may be assumed to have a design strength 0.93 times the characteristic strength or the specified yield strength and Young's modulus of 210 GPa. The calculations of section properties to determine the effects of bending, shear, deflections, etc. are complex (similar in nature to those for the design of cold-formed sections and are thus iterative). It should be noted that the moment carrying capacity is limited to just causing yield and hence is based on the elastic section modulus (Martin & Purkiss 1992). The manufacturers provide load tables, which can be used to select the deck for the required span. Note that these decks are available in lengths of 6 m and above and have thickness in the range of 0.37 mm (28 gauges) to 1.5 mm (16 gauges). The weight of roofing varies from 0.3 kN/m$^2$ to 1.0 kN/m$^2$, including the weight of joists, and 0.05 kN/m$^2$ to 0.1 kN/m$^2$, excluding the weight of joists. (See also Section 16.5.11.) For accurate information on weights, the manufacturer's literature should be consulted. Joists carrying roof decking can be spaced at 6 m or more, depending on the thickness of the decking sheet and depth of profile.

Metal roofing can be classified by the method of attachment to supports. *Through-fastened roofs* are attached directly to purlins, usually by self-tapping screws, self-drilling, or lock rivets. *Standing-seam roofs*, on the other hand, are connected indirectly by concealed clips formed into the seams. Standing-seam roofing is often used in USA and was introduced by Butler Manufacturing Company in 1969. Though the through-fastened lapped-seam roofing is inexpensive, straightforward, and easy to erect, it is penetrated by fasteners, and hence is susceptible to leaking. The only protection against leaking is the rubber or neoprene washer provided under the head of the fastener. A properly installed screw has its neoprene washer slightly visible from under the edge of the metal washer. The fasteners also prevent the sheets from thermal expansion and contraction. Repeated expansion and contraction of the sheets may tear the metal around the connecting screws and eventually lead to leaking. Hence, the width of a building with through-fastened roofs should not exceed about 18 m. When subjected to strong fluctuating wind loading, these sheets experience metal fatigue and may fail locally, by cracking around the fasteners (Lynn & Stathopoulos 1985; Xu 1995). The size and spacing of fasteners depend upon the forces they are designed to resist. The fastener spacing may be closer in the roof areas, subjected to high wind loading (see Fig. 3.34 of Chapter 3), than in the middle of the roof. For durability against corrosion, stainless steel or aluminium fasteners should be used in exposed areas. Galvanized or cadmium-plated screws are suitable for interior applications.

In structural standing-seam metal roofing, the elevated seams of adjacent standing-seam profiles are formed at site by portable seaming machines or, less reliably, by hand tools. A factory applied sealant is normally placed in the female corrugation of the seam. To accommodate expansion and contraction, the panels are attached to purlins by concealed clips that permit the roof to move. The biggest disadvantage of the standing-seam roof is that it provides no lateral bracing to purlins and offers little diaphragm action. Hence, a separate system of purlin bracings and a separate horizontal diaphragm structure are required. Moreover, standing-seam roofing is best suited for rectangular building layouts only. Though they may be expensive initially, life-cycle cost calculations may prove them to be economical in the long run. The most popular anti-corrosive coating for steel roofing is based on metallurgically bonded zinc, aluminium, or a combination of the two. Zinc-aluminium coated roofing has displaced galvanized roofing in USA. When another layer of protection is required, acrylic, polyester, or fluorocarbon based paints are sprayed onto the metal and baked at the factory. More details about steel decking are provided by Petersen (1990), Schittich (2001), and Newman (2004).

### 16.1.2 Galvanized Iron (GI) Sheets

Most common sizes of corrugated GI sheets are as follows:

1. 8 corrugations (75 mm wide and 19 mm deep) per sheet
2. 10 or 11 corrugations (75 mm wide and 19 mm deep) per sheet

The available sizes of sheets are as follows:

1. Length—1.8, 2.2, 2.5, 2.8, and 3 m
2. Width—0.75 m and 0.9 m
3. Thickness—0.63, 0.8, 1.0, 1.25, and 1.6 mm

The weights of the sheets vary from 50–156 N/m$^2$ (see Table 3.3 of Chapter 3). When the sheets are installed, side laps and end laps should be provided to make the joint water proof. The following overlaps are normally used:

1. For roof: Side overlap—1½ to 2 corrugations
   End overlap—150 mm
2. For side cladding: Side overlap—1 corrugation
   End overlap—100 mm

The sheets are fastened to purlins or side girts by 8-mm diameter J- or L-type hook bolts with GI nuts along with GI and bituminous felt washers at a maximum pitch of 350 mm. Where laps do not occur over supports, 6-mm diameter bolts at a maximum pitch of 250 mm for roofs and 300–450 mm for sides are used. Typical fastening details are shown in Fig. 16.3.

Spacing of purlins and girts, which support the sheeting is governed by the length of the sheet, thickness of the sheet, and applied loading. The approximate section modulus of the corrugated GI sheeting may be taken as

$$Z = (4/15)bdt \quad (16.1)$$

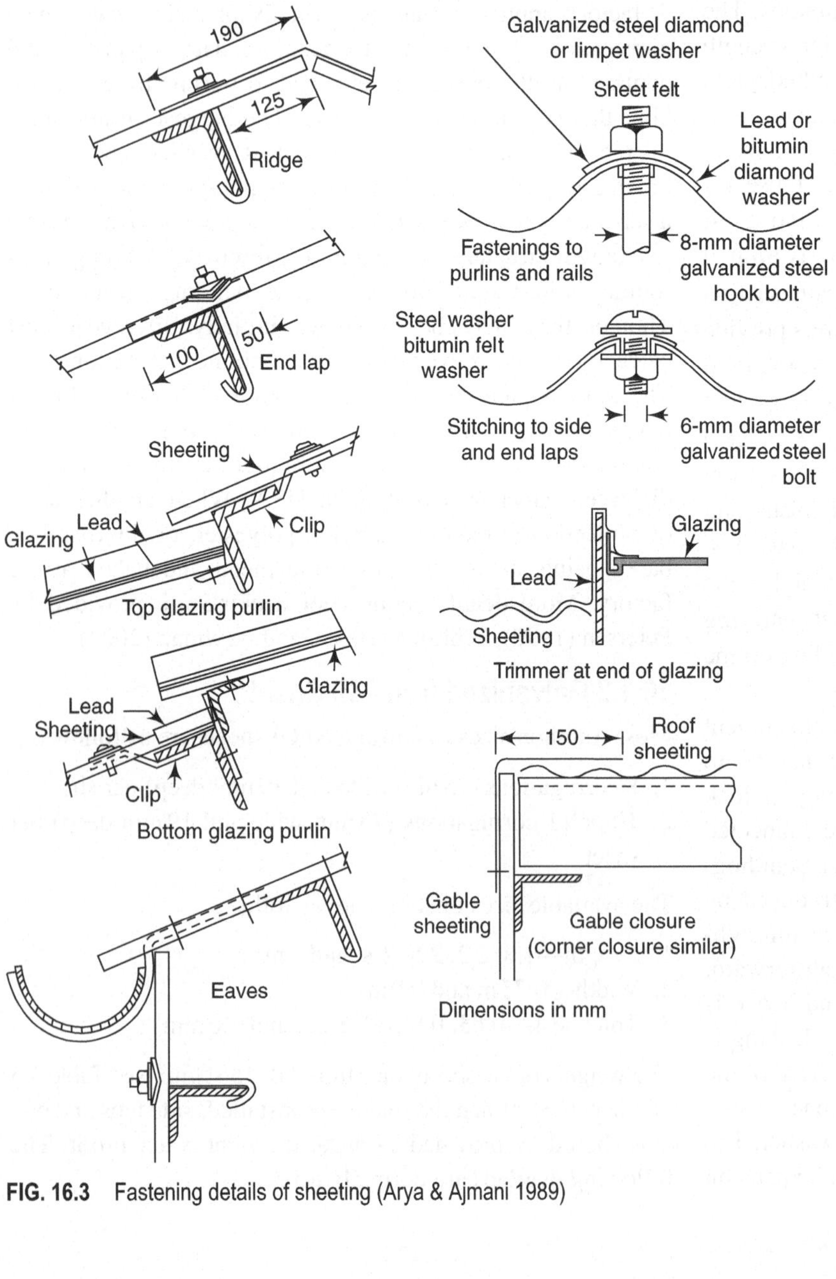


**FIG. 16.3** Fastening details of sheeting (Arya & Ajmani 1989)

where $b$ is the curvilinear width (equal to $1.13 \times$ covering width), $d$ is the depth of the corrugation, and $t$ is the thickness of the sheet.

The allowable bending stress may be taken as 0.75% of 165 N/mm$^2$. Hence for a 750-mm wide, 1-mm thick, and 18-mm deep, corrugated GI sheet subjected to a wind load of 1.6 kN/m$^2$, the maximum spacing of purlin may be worked out as follows:

$$\text{Wind load/m} = 1.60 \times 0.75 = 1.2\ \text{kN/m}$$

$$\text{Bending moment} = wl^2/8 = 1.2 \times 10^3 L^2/(10^3 \times 8) = 0.15L^2\ \text{N mm}$$

$$Z = (4/15) \times 1.13 \times 750 \times 18 \times 1.0 = 4068\ \text{mm}^3$$

Thus,

$$0.15L^2 = 4068 \times (0.75 \times 165)$$

or

$$L = 1832\ \text{mm}$$

$$\text{Required length of sheet} = 1832 + \text{end lap} = 1832 + 150 = 1981\ \text{mm}$$

Hence, adopt a purlin spacing of 1.8 m for a 2000-mm long and 750-mm wide sheet. Similarly, the maximum purlin spacing for sheets with other thicknesses may be worked out.

### 16.1.3 Asbestos Cement Sheets

Asbestos cement sheets may be used to cover the roof as an alternative to corrugated steel sheets. (These sheets are banned in many countries due to the risk of lung cancer caused by inhaling the fibres, while working with these sheets.) AC sheets are manufactured in two shapes, corrugated and Trafford, as shown in Fig. 16.4 and are available in lengths of 1.75, 2.0, 2.5, and 3 m. They are manufactured in thicknesses of 6 mm or 7 mm. The maximum permissible spacing of purlins is as follows:

1. for 6-mm sheet—1.4 m
2. for 7-mm sheet—1.6 m

For side cladding, the spacing may be increased by 300 mm.

A side overlap of one corrugation is normally given. The end lap should not be less than 150 mm for slopes less than 18° and for flatter slopes this overlap may be increased. For side covering, an overlap of 100 mm is sufficient. The purlin spacing should be so adjusted with the lengths of the sheets that the longitudinal overlaps fall on the purlins to which they are

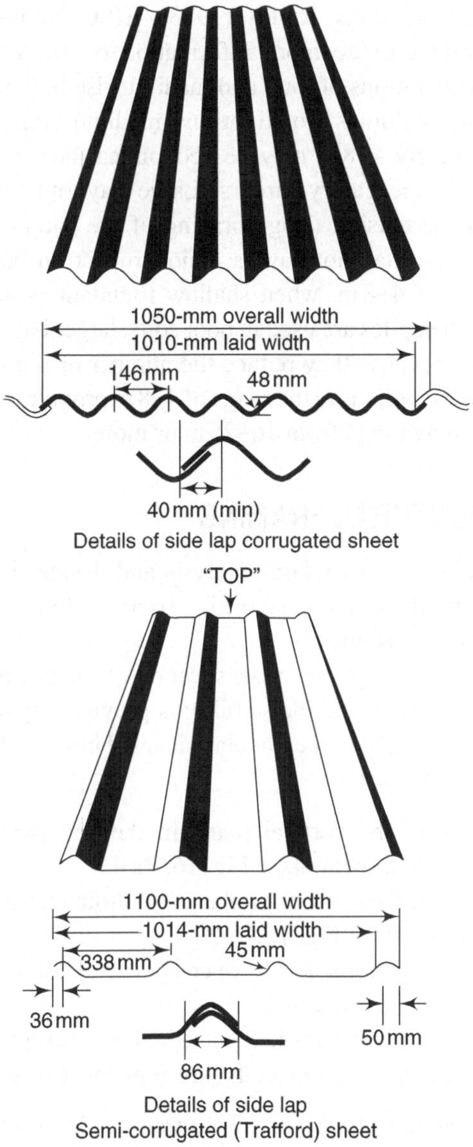


**FIG. 16.4** Asbestos cement sheets

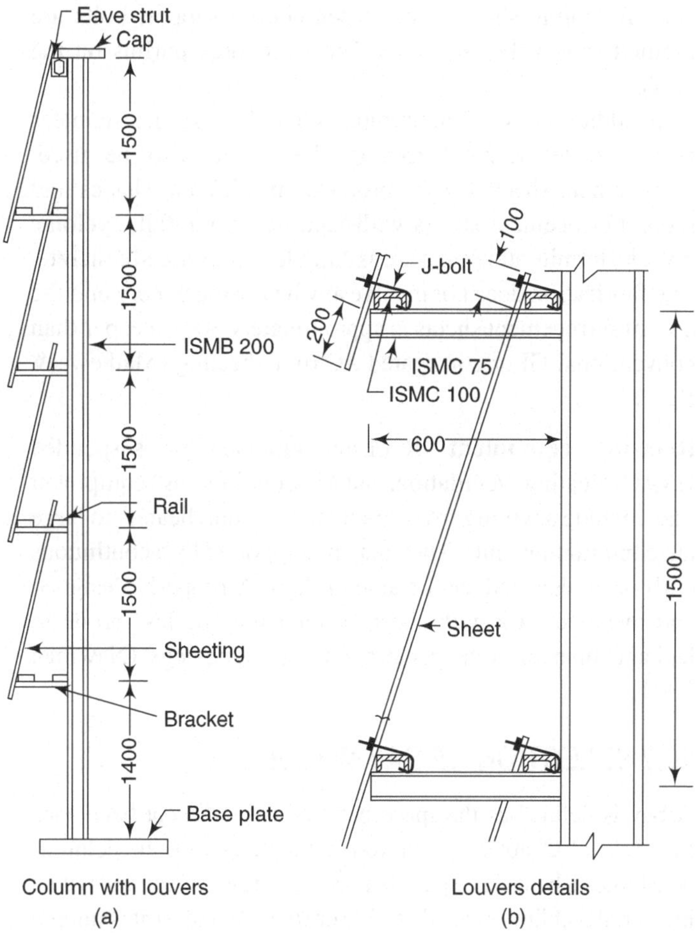


**FIG. 16.5** Typical louver arrangement

directly fastened. Spacing of purlins should be so fixed that the cutting of sheets is avoided (see also Section 16.5.6). The overhang near the eaves end should not be more than 400 mm.

The weight of asbestos sheets varies from 160 N/m$^2$ to 170 N/m$^2$ (see Table 3.3 of Chapter 3). The load per square metre of the sheet on the slope may be increased by 30% to get the load per square metre of the plan area, to account for the larger area on the slope and additional material in the side and end lapping. The sheets are fastened to purlins or girts by using 8-mm-diameter hook bolts at a maximum spacing of 350 mm.

Fixed or adjustable louvers are often provided in sheeted wall for ventilation purposes. Louvers are openings in the wall covered with a sloping sheeted cover cantilevered out from the columns for protection against rain and sun but are designed to allow air and sunlight in. A typical louver arrangement is shown in Fig. 16.5. The sheeting and the louvers are designed against wind pressure acting on the sides of the building.

It has been observed that the roofing sheets of industrial sheds were found to fail in uplift under extreme wind conditions as in a cyclone. (As discussed in Chapter 3, the local pressure coefficients given in Tables 3.31 and 3.33 have to be used for the calculation of loads on cladding and their fixtures. Note that the local pressure coefficients are much higher than the external pressure coefficients used for the design of structural frameworks.) Narayanan and Mathews (2003) recommend the following based on experimental tests carried out on 250 full-scale roofing sheet systems.

(a) Over tightening of J-bolts while fixing AC sheets should be avoided, since it causes cracks in the AC sheets leading to failure at low loads.
(b) The holes for fixing AC sheets must be drilled and not punched. Great care should be taken while drilling the bolt holes in corrugated GI sheets.
(c) Good quality metallic and bitumen washers should be used to get the required strength under uplift forces.
(d) GI sheets of thickness 0.6 mm or more should be used.

Narayanan and Mathews (2003) also found that channel and pipe purlins reduce the capacities of the roofing sheets compared to angle purlins. Large reduction in strength of AC and GI roofing sheets occur when central support bolts are omitted (when the sheets are laid over three purlins on the roof).

In addition to steel, aluminium, GI and AC sheets, stainless steel and ferrocement roofing sheets can also be used. Ferrocement sheets can be produced in different shapes and sizes. Ferrocement sheets withstand heavy rainfall, cyclone, fire, and termite attack and are as durable as reinforced concrete. The fabrication does not involve any heavy machinery and the cost of ferrocement sheets is approximately 30% cheaper than conventional GI or corrugated AC roof sheeting (Mathews & Rao 1979).

**Roof-top equipment** Roof-top mounted or suspended HVAC (Heating, Ventilation, and Air Conditioning) equipment may include anything from small fans and unit heaters to large air-conditioning units. They may be supported by a continuous curb or an elevated steel frame on legs. A properly designed and installed curb with sheet flashing may be less prone to leakage than discrete penetrations at frame legs (Newman 2004).

## 16.2 SELECTION OF BAY WIDTH

A bay is defined as the space between two adjacent bents (see Fig. 1.42 of Chapter 1). The roof truss along with the columns constitutes a bent. The space between two rows of columns of an industrial building is called an aisle or *span*. An industrial building may have a single span or multiple spans. Figure 16.6 shows industrial buildings with single, double, and multiple spans.

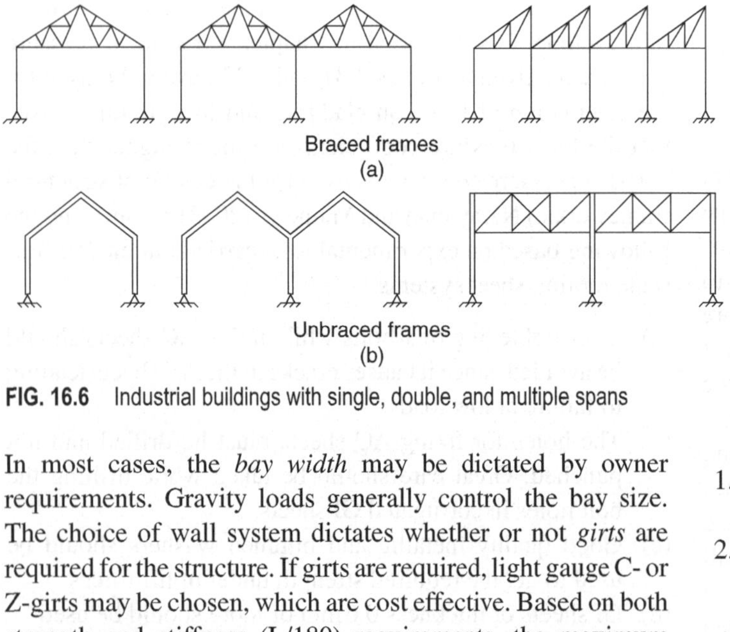


**FIG. 16.6** Industrial buildings with single, double, and multiple spans

In most cases, the *bay width* may be dictated by owner requirements. Gravity loads generally control the bay size. The choice of wall system dictates whether or not *girts* are required for the structure. If girts are required, light gauge C- or Z-girts may be chosen, which are cost effective. Based on both strength and stiffness (L/180) requirements, the maximum economical span of such girts is approximately 9 m. Hence, for buildings without cranes, a 9-m bay is the most economical choice (Fisher 1984). A 12-m bay may prove economical for large square buildings. Ruddy (1983), AISC (805-91), and Fisher (1991) provide more information on the economic design considerations of low- and medium-rise buildings.

For crane buildings (for light and medium cranes), bays of approximately 4–8 m may be economical because of the cost of the crane gantry girders. Large bays may increase the cost of the tension flange bracing of the gantry girders. Soil conditions may not have a major impact on bay width in the range of 4–8 m, when shallow foundations are used. However, when piles are used in poor soils, larger bays may be economical, because they reduce the number of foundations. Though the bay widths in the range of 4–8 m provide economy, truss spans may range from 10–25 m or more.

## 16.3 STRUCTURAL FRAMING

For the purpose of structural analysis and design, industrial buildings are classified as (see Fig. 16.6) (a) braced frames and (b) unbraced frames.

In braced buildings, the trusses rest on columns with hinge type of connections and the stability is provided by *bracings* in the three mutually perpendicular planes. These bracings are identified as follows:

1. Bracings in the vertical plane in the end bays in the longitudinal direction[see Fig. 16.7(a)]
2. Bracings in the horizontal plane at bottom chord level of the roof truss[see Fig. 16.7(c)]
3. Bracings in the plane of upper chords of the roof truss [see Figs 16.7(a)and (b)]
4. Bracings in the vertical plane in the end cross sections usually at the gable ends [see Figs 16.7(a) and (c)]

The function of a bracing is to transfer horizontal loads from the frames (such as due to wind or earthquake or horizontal surge due to acceleration and breaking of travelling cranes) to the foundation. The *longitudinal bracing* on each longitudinal end provides stability in the longitudinal direction. The *gable bracings* provide stability in the lateral direction. The *tie bracings* at the bottom chord level transfer lateral loads (due to wind or earthquake) of trusses to the end gable bracings.

Similarly, stability in the horizontal plane is provided by the following:

1. a *rafter bracing* in the end bays, which provides stability to trusses in their planes or
2. a *bracing system* [see Fig. 16.7(c)] at the level of bottom chords of trusses, which provides stability to the bottom chords of the trusses.

Purlins act as lateral bracings to the compression chords of the roof trusses, which increase the design strength of the compression chords. The lateral ties provide similar functions

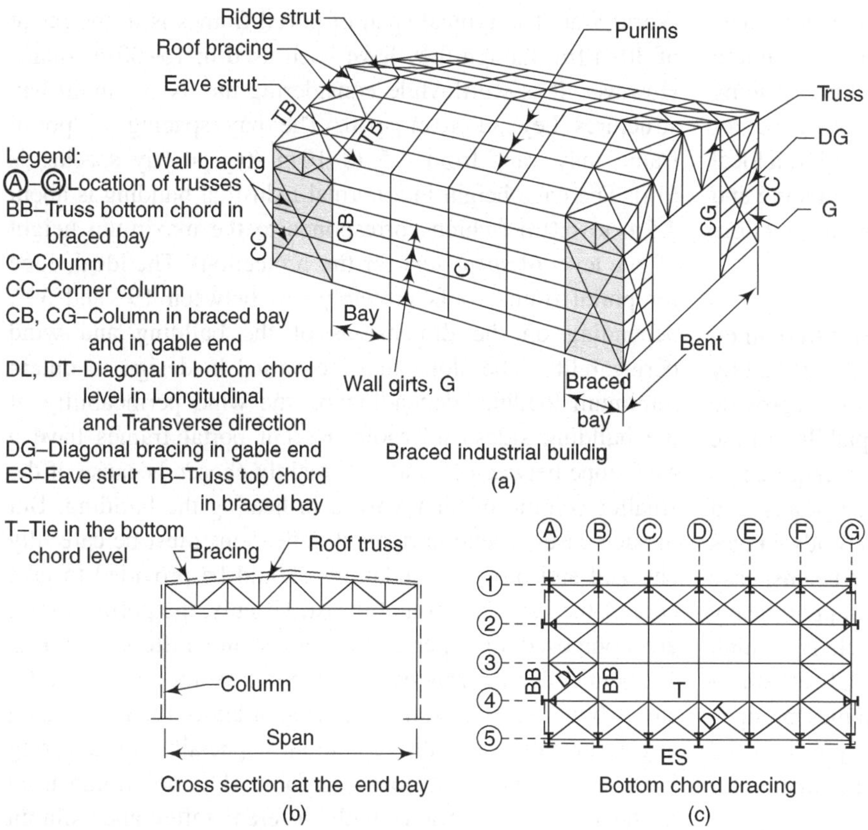


**FIG. 16.7** Structural framing for an industrial building

to the bottom chord members when they are subjected to compression due to reversal of loading (see Section 16.5.9 also). X bracings (as shown in Fig. 16.7) are the commonly used bracing systems. K-type bracing systems may also be used. If the building is lengthy, bracings in the end bays alone may not be sufficient. In these cases, every fourth or fifth bay is braced and the roof upper chord bracings are also provided in these bays.

*Braced frames* are efficient in resisting the loads and do not sway. However, the braces introduce obstructions in some bays and may cause higher forces or uplift forces in some places. Wide flange columns are often used for exterior columns of braced frames. (For interior columns of braced frames with height less than 7 m, Square Hollow Section (SHS) columns may yield most economical solution because of their high radius of gyration about both axes.)

## 16.3.1 Knee Bracing of Columns

In the braced frames, the roof truss is directly placed over the columns, which may be an I-section with a bracket or a built-up section (see Sections 9.9.5 and 9.15 of Chapter 9 for more details about built-up columns). Usually the column is assumed to be fixed at the base and hinged at the top [see Fig. 16.8(a)]. This will result in heavy bending moments (due to lateral loads) in the columns and hence are adopted when the lateral loads are small and large foundations can be provided. The moments are also small when the height of the building is small. Hence, such end conditions are suitable for wide industrial buildings of small height. To avoid this problem, trusses are sometimes braced laterally by *knee braces* as shown in Fig. 16.8(b) to the first panel point. This arrangement reduces the bending moment at the base when the column is fixed at the base and consequently reduces the size of the foundation. When the column is hinged at the base as shown in Fig. 16.8(c), no bending moment is transferred at the base and the size of foundation can be reduced further. Heavy moments in such an arrangement are present only at the junction of the knee brace and column. Since the solution with a knee brace sacrificed some interior head room, it has now lost its popularity in favour of a truss design with some depth at supports, incorporating the column into the truss-work [see Fig. 16.8(d)]. In this case, the columns are erected and braced first, followed by the erection of trusses with relatively simple truss-to-column connections. Normally these trusses are partially assembled in the shop, to the maximum permissible shipping width, about 3.6 m, and then fully assembled at site.

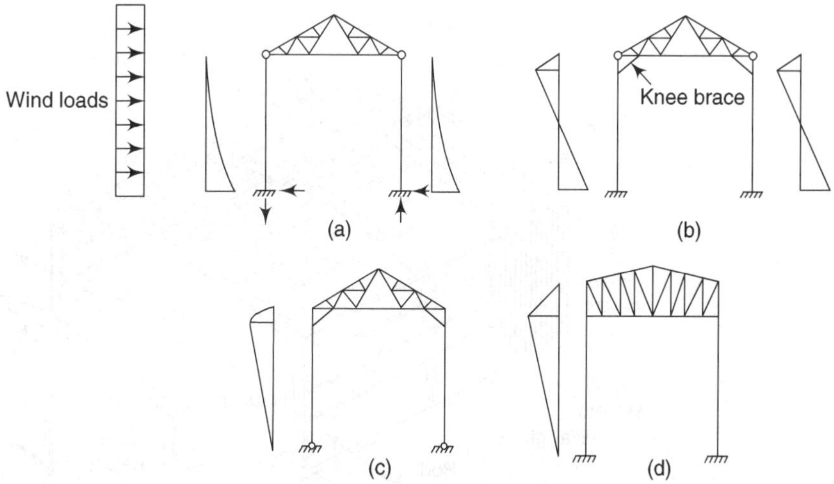


**FIG. 16.8** Methods to reduce bending moments in columns of industrial buildings

When knee braces are provided, the analysis model should include the entire system, i.e., the trusses together with the column should be analysed to get the bending moments in the columns. Due to vertical (gravity) loads; knee bracing will be under compression. However, due to lateral loads, knee bracing will be under compression or tension, depending upon the direction of the lateral load. Hence, knee bracing and its connection with the column are to be designed for compressive as well as tensile loads. Since there will be a local concentration of force in the junction of the knee bracing and the column, local strengthening may be needed at the junction.

## 16.3.2 Unbraced Frames

Unbraced frames in the form of *portal frames* is the most common form of construction for industrial buildings, distinguished by its simplicity, clean lines, and economy. The frames can provide large column free areas, offering maximum adaptability of the space inside the building. Such large span buildings require less foundation, and eliminate internal columns, valley gutters, and internal drainage. Portal frame buildings offer many advantages such as more effective use of steel than in simple beams, easy extension at any time in the future, and ability to support heavy concentrated loads. The disadvantages include relatively high material unit cost and susceptibility to differential settlement and temperature stresses. In addition, these frames produce horizontal reaction on the foundation, which may be resisted by providing a long tie beam or by designing the foundation for this horizontal reaction.

Basically, a portal frame is a rigid jointed plane frame made from hot-rolled or cold-rolled sections, supporting the roofing and side cladding via hot-rolled or cold-formed purlins and sheeting rails (see Fig. 16.9). The basic structural form of portal frames was developed during the Second World War. The typical span of portal frames is in the range of 30–40 m, though they have been used in 15–80 m, spans. However, it is worthwhile considering the use of multi-bay structures beyond 40 m span. The bay spacing of portal frame may vary from 4.5 to 10 m (typical bay spacing is 6 m). The eave height in a normal industrial building is about 4.5 m to 6.0 m (which corresponds to the maximum height of one level of *sprinklers* for fire protection). The ideal slope for portal frame roofs probably lies between 19° and 30°, depending on the dimensions of the building and wind permeability. The slope may be varied in design to obtain minimum loading, depending on the wind permeability of the building sides and roofs. Recent portal frames have a roof slope between 6° and 12°, mainly chosen because of the smaller volume of air involved in heating the building. But in such cases, frame horizontal deflections must be carefully checked and proper foundations should be provided to take care of the large horizontal thrust. Usually, the portal frames are composed of tapered stanchions and rafters (see Fig. 16.9) in order to provide an economic solution for single-storey buildings. Portal frames made of lattice members using angles or tubes are also common, especially for covering long spans. The most common form of portal frame used is the pinned-base frame with different rafter and column member size and haunches at both the eaves and apex connections (see Fig. 16.10). The simple unhaunched portal frame and its connections are shown in Fig. 16.11. Using a solid-web rafter for the portal frame has the disadvantage of placing pipes and conduits below the bottom flange.

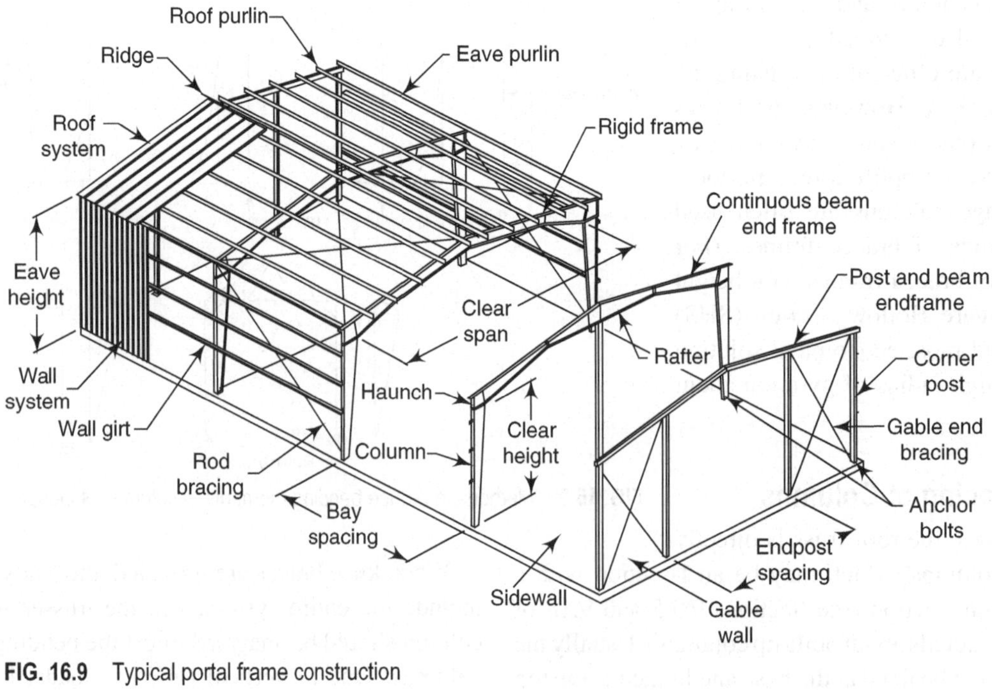


**FIG. 16.9** Typical portal frame construction

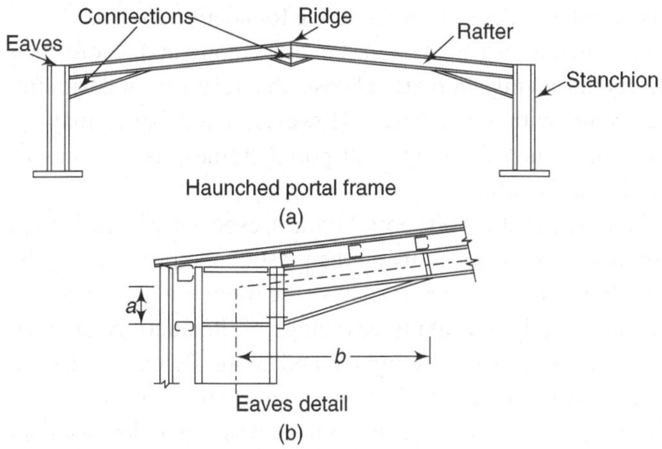


**FIG. 16.10** Typical haunched portal frame

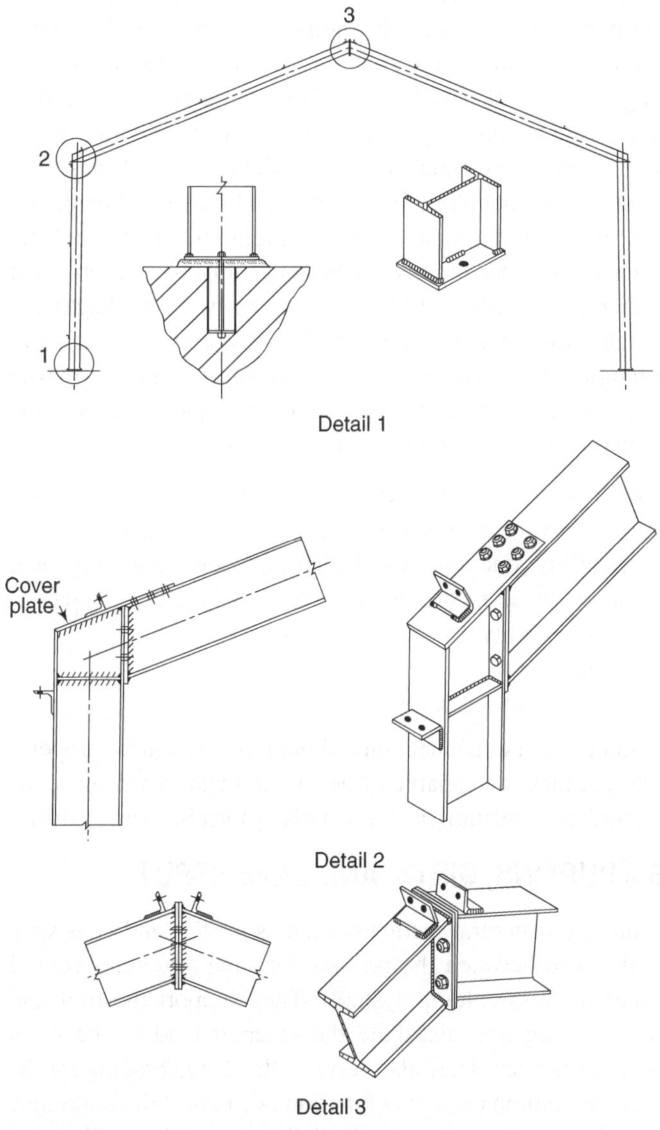


**FIG. 16.11** Simple unhaunched portal frame

This problem may be overcome by adopting castellated or *lattice rafters*, as shown in Fig. 16.12).

The lateral stability parallel to the portal frame is provided by the frame itself. However, to resist the loads perpendicular to the portal frame, wall and roof bracings have to be provided, as shown in Fig. 16.9. The roof bracings, which may be in the form of diagonal steel rods, are placed below the purlins for simplicity. These diagonal rods and the rafters of the primary

Continuous lattice portal for Kirloskar Small Engines Division at Rajkot.

*Photo Courtesy:* Er Jayant Lakhlani

Lattice portal at SRM University, Chennai (note the steel structure inside the portal)

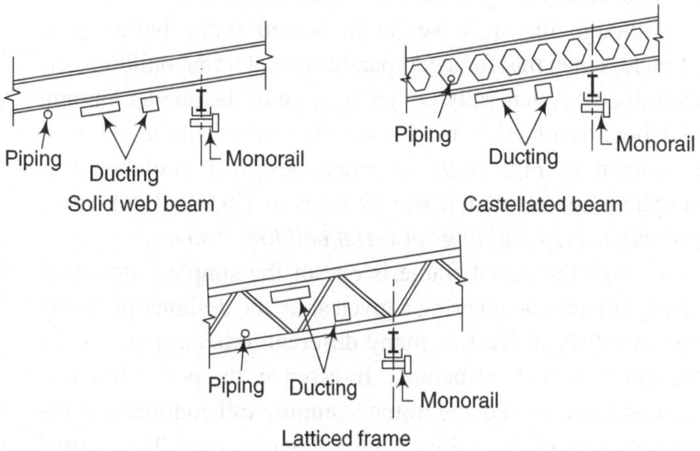


**FIG. 16.12** Solid web, castellated, and lattice rafters

portal frames constitute a horizontal truss and are some times termed as *roof diaphragm*. The secondary framing—wall girts and roof purlins—collects the load from the wall and roof covering and distributes them to the main portal frames and at the same time provides lateral restraint to the members of the portal frame.

The end frame or *gables* of a portal frame can be framed in two different ways. They can be formed using a frame identical to the intermediate portals and then framing the gable or end posts up to the underside of the portal rafter (see Fig. 16.9). Suitable arrangements must be made at the post/rafter connections to accommodate vertical movement of the portal rafter (Dowling et al. 1988). This method of gable framing may be economical, if the building is expected to be extended in the future. When there is no scope for extension, the gable rafter can be designed as a continuous member, spanning between gable or end posts (as shown in Fig. 16.9). This form of gable framing has poor sway resistance and hence needs in-plane bracing. The gable posts are generally designed as propped cantilevers. They support the gable sheeting which are subjected to wind loads; they sometimes support secondary steel work for doors, openings, etc.

For buildings without cranes, the choice of column type and orientation depends primarily on the height of the building, the wall system, and the lateral load carrying system. In general, wide-flange columns are the most economical choice for exterior columns in the buildings. This is particularly true for rigid frames, where rigid frame action is required in one direction and girts are used to brace the weak axis of these columns.

The lateral loads are transmitted along the wall from brace to brace by *eave struts*. The eave struts are designed for axial compression or for combined axial compression and biaxial bending. The diagonal wall bracing in the form of a rod may be attached to the column through the *hillside washer*, which is a cast circular element with a vertically slotted hole that allows for variable angles of rod insertion. A matching vertically slotted hole is made in the column web. More details about hillside washer are provided by Newman (2004).

Although the steel weight in braced frame buildings is often less than that for a comparable portal frame building, the overall cost is generally higher because of the greater amount of labour involved in fabrication. The portal frame systems, as shown in Fig. 16.9, are often designed, prefabricated, supplied, and erected at site by firms in USA and are called *pre-engineered buildings* or *metal building systems*.

Though the portal frame is one of the simplest structural arrangements for covering a given area, the designer probably has to satisfy at least as many different structural criteria as for more complex structures. In essence, the portal frame is analysed as a rigid plane frame assuming full continuity at the intersections of the column and rafter members. The normal practice is to assume that the columns are pinned at the bases. It is because the cost of the concrete foundation for fixed bases (owing to the effect of large bending moments), more than offsets the savings in material costs that result from designing the frame with fixed bases. However, fixed bases may be necessary when designing tall portal frames, as a means of limiting deflections.

Since the mid 1950's, portal frames, especially in the United Kingdom, are designed using the plastic methods of analysis, as it allows the engineer to analyse the frame easily and design it economically. By taking advantage of the ductility of steel, plastic design produces lighter and more slender structural members than similar rigid frames designed by elastic theory. The plastic design of portal frames has been discussed in Section 8.9 of Chapter 8. The complete design of a single storey portal frame is provided by Morris and Plum (1996), who also caution about plastic analysis programs based on the modified stiffness matrix method, which may produce false solutions. Complete information on the elastic and plastic design method of portal frames is provided by King (2005). Large scale testing of portal frames has been reported by Li and Li (2002) and Mahendran and Moor (1999). It has been found that by including the effect of end frame and cladding, in a three-dimensional analysis program, more realistic values of bending moments and deflections can be obtained (Mahendran & Moor 1999). Several software packages are available for the analysis and design of portal frames (www.structural-engineering.fsnet.co.uk). While attempting a plastic design, in addition to member strength (capacity) check, the following checks must also be carried out:

1. Reductions in the plastic moment due to the effect of axial force and shear force (see Section 8.10 of Chapter 8)
2. Instability due to local buckling, lateral buckling, and column buckling (see Sections 8.11 and 8.12 of Chapter 8)
3. Frame stability
4. Brittle fracture
5. Deflection at service loads

In addition, the connections should be designed properly such that they are capable of developing and maintaining the required moment until the frame fails by forming a mechanism.

## 16.4 PURLINS, GIRTS, AND EAVE STRUT

Secondary structural members such as *purlins* and *girts* span the distance between the primary building structures (portal frames or truss-column system). They support the roof and wall covering and distribute the external load to the main frames or trusses. They also serve as the flange bracing for the rafters or columns and may function as a part of the building's lateral load resisting system. Purlin is a part of the roof bracing system and girts form a part of the wall bracing system of the building. The behaviour and design of purlins has been discussed in Section 10.16 of Chapter 10 and the design of sag rods in Section 7.10 of Chapter 7. Examples 10.11 of Chapter 10 and 7.11 of Chapter 7 illustrate the various steps involved in the design of purlins and sag rods, respectively. When sag rods are used for bracing the purlins top flange, it is advantageous to locate the sag rods 50–75 mm below the top of the compression flange. The weight of purlins in the total weight of the steel structure could vary from 10% to 25%. The weight of purlins may be equal to or even greater than the weight of the trusses. Hence they have to be designed properly. Usual members adopted for purlins include channels, angles, tubes, and cold-formed C- and Z-sections. When cold-formed sections are used, they should be properly protected with anti-corrosive treatment, since their thickness ranges from 1.6 mm to 4 mm only. Cold-formed C-, Z-, or sigma-purlins may be economical than hot-rolled purlins for spans of 5 m to 8 m. Important aspects of cold-rolled purlins and their connection to rafter members are provided by Newman (2004) and Ghosn and Sinno (1996). Angles and channel purlins without sag bars may be economical up to 5 m and tubes up to 6 m. Open web joists (see 16.5.11) can be used as purlins, when the span of purlins exceeds 12 m.

The main function of *girts* is to transfer wind loads from wall materials to the primary frame. Girts are positioned horizontally (see Fig. 16.7) to span between the columns. When the space between primary columns is more than 9 m, *wind columns* may be provided to reduce the girt span. Wind columns are essentially intermediate vertical girts spanning from the foundation to the eave. Since typical eave strut may not be capable of resisting the lateral reaction imposed by the wind column, a system of diagonal braces should be provided to transfer the lateral reaction to the adjacent primary framing columns. Similar to purlin spacing, girt spacing is governed by the load—resisting properties of wall panels.

The third type of secondary structural members, after purlins and girts, is the *eave strut.* This member is located at the intersection of the roof and the exterior wall (see Fig. 16.7) and hence acts as both the first purlin and the last (highest) girt. The building's eave height is measured to the top of this member.

The eave strut is a relatively strong member and its functions are as follows:

1. It serves as a stiff binder beam.
2. Cladding is often hung from the eave strut; hence the total load of cladding including side girts should be carried by this beam.
3. In braced buildings, the wind bracing along the eave strut acts as a truss in the plan view [see Fig. 16.7(c)]. As already discussed, this truss transfers the horizontal loads on the roof and cladding to the gable end bracings. Therefore, the eave strut acts as a compression chord of the wind bracing truss.
4. Eave strut also supports drain gutters and other secondary elements.

Since a relatively stiff section is required, the eave girder is often composed of a built-up two channels face-to-face.

## 16.5 PLANE TRUSSES

A structure that is composed of a number of line members pin-connected at the ends to form a triangulated framework is called a *truss.* If all the members lie in a plane, the structure is a *planar truss.* In a truss, the members are so arranged that all the loads and reactions occur only at the joints (intersection points of the members). The centroidal axis of each member is straight, coincides with the line connecting the joint centres at each end of the member, and lies in a plane that also contains the lines of action of all the loads and reactions. The primary principle underlying the use of the truss as a load-carrying structure is that arranging elements into a triangular configuration results in a stable shape. Any deformations that occur in this stable structure are relatively minor and are associated with small changes in member length caused by the forces in the members by the external loads. Similarly, the angle formed between any two members remains relatively unchanged under load.

Trusses were found to have been constructed as early as 500 B.C. when Romans built a bridge using a form of timber truss across the Danube river. Though the potential of trusses were known and used in a few large public buildings in USA and Italy, the bridge builders of the early nineteenth century were responsible for the systematic use of the truss systems. The books by Emiland Gauthey in 1809 and 1813 (*Traité de la Construction des Ponts*) and Squire Whipple in 1847 (*A Work on Bridge Building*) provided the foundation for the analysis and design of truss systems (Schodek 2001). After this period several trusses have been constructed all over the world to cover large spans.

Many truss structures are three-dimensional in nature and a complete analysis would require consideration of the spatial interconnection of the members (Subramanian 2006). However, in simple roof systems the three-dimensional framework can be subdivided into planar components for analysis as planar trusses, purlins, etc., without seriously compromising the accuracy of the results. The external loads (which are applied at the joints) produce only tensile or compressive forces in the individual members of the truss. For common trusses with vertically acting loads, compressive forces are usually developed in the top chord members and tensile forces in the bottom chord members. Though the forces in the web members of a truss may be either tension or compression, there is often an alternating pattern of tensile and compressive forces present. Note that when the external loads reverse in direction (e.g., as in the case of wind loads) the top chords will be in tension and bottom chords will be in compression. Hence, it is often necessary to design the various members of a truss both for tension and compression and select the member size based on the critical force.

It is extremely important to note that when the loads are applied directly onto truss members themselves (as in the case of intermediate purlins), bending stresses will also develop in those members in addition to the basic tensile or compressive stresses. This results in complicated design procedures (they should be designed as per the provisions of beam-columns discussed in Chapter 13) and the overall efficiency of the truss is reduced (see Section 16.5.6 also).

### 16.5.1 Analysis of Trusses

The first step in the analysis of a truss is always to determine whether the truss under consideration is a stable configuration of members. There exists a relation between *m*, the number of members, *j*, the number of joints, and *r*, the reaction components. Thus, the expression

$$m = 2j - r \tag{16.2}$$

must be satisfied, if the truss is a internal statically determinate structure. The least number of reaction components required for external stability is *r* (equal to 3 for plane trusses). If *m* exceeds $(2j - r)$, then the excess members are called redundant members and the truss is said to be statically indeterminate. A similar expression can be written as follows for a three-dimensional truss work.

$$m = 3j - r \tag{16.3}$$

In the above expression $r = 6$. If fewer members, than those given by the expressions in Eqns (16.2) and (16.3), are present, then the truss will be unstable. (A tetrahedron is an example of a stable three-dimensional truss.) Note that these expressions are more correct indicators of whether the forces in the given truss can be calculated by the equations of statics alone than a predictor of stability. It is also interesting to note that a truss with a single diagonal cable member could be unstable (since they will buckle when subjected to compressive forces), a fact not predicted by Eqns (16.2) and (16.3). If cables are used, an X-type cross-cable system should be provided. (In such a case, one cable will take the tension, while the other buckles due to the compressive forces.)

For a determinate truss, when purlins are located at the nodal points, the member forces can be found by employing the laws of statics to assure internal equilibrium of the truss. The process requires the repeated use of free-body diagrams, from which individual member forces are determined. *The method of joints* is a technique of truss analysis in which the member forces are determined by the sequential isolation of joints—the unknown member forces at one end of the truss are solved, which are then used to determine the member forces at subsequent joints. The other method is known as the *method of sections* in which the equilibrium of a part of the truss is considered, and the member forces are determined by using the three equations of equilibrium. $\Sigma F_x = 0$, $\Sigma F_y = 0$, and $\Sigma M = 0$. The forces in triangulated trusses may also be found by graphical means using force diagrams. In this chapter, only plane trusses are considered and interested readers may refer to Subramanian (2006) for the analysis and design methods of three-dimensional trusses. The details of the method of joints and method of sections are provided by Schodek (2001), Thandavamoorthy (2005), and Brockenbrough & Merritt (2006).

In some situations, it may not be possible to provide ideal pin connections at the joints of trusses. When the actual joint conditions are such that the ends of the members are not free to rotate, local bending moments develop in the members in addition to axial forces. If these bending moments are too large, members have to be designed for combined axial and bending stresses. However, bending stresses induced by rigid joints are found to be about 10–20% of the axial stresses. BS: 5950 states that such secondary stresses will be insignificant if the slenderness of the chord members in the plane of the truss is greater than 50 and that of most of the web members is greater than 100. When the exact positions of point loads on the top chord are not known, the local bending moment may be taken as *PL*/6, where *P* is the point load in kN and *L* is the length of the panel of rafter in metres. Moreover, the maximum stresses from secondary moments occur at the ends of the members and are not likely to cause collapse. For preliminary design purposes, these secondary bending stresses are often ignored, though they have to be considered in the final design. It may be of interest to note that the increase in joint rigidity may reduce the defection of the trusses. Rigid jointed frameworks may be analysed by the *slope deflection* or the *moment distribution method* though a host of other methods are also available. The details of slope deflection and moment distribution methods may be found in Thandavamoorthy (2005) and Chen and Lui (2005).

In statically indeterminate trusses (which have more number of members than a determinate truss), though the principles of statics are still valid, they can not be analysed using the method of joints or the method of sections. It is because we may have more number of unknowns than the equations of equilibrium. For analysing these indeterminate trusses, *matrix methods of structural analysis* are used (Livesley 1975). In this method of structural analysis, a set of simultaneous equations that describe the load-deformation characteristics of the structure under consideration is formed. These equations are solved using matrix algebra to obtain deformations or forces. From the deformations of joints, the member forces may be determined. Matrix algebra is ideally suited for setting up and solving equations in the computer. Two methods of matrix structural analysis are available. The *flexibility or force method* assumes the forces as unknowns and the *displacement or stiffness method* assumes the deformations as unknowns. The details of stiffness and displacement methods of structural analysis may be found in Wang (1973), Harrison (1990), and Livesley (1975). Several commercial computer programs are available for the analysis of indeterminate trusses, which are based mainly on the stiffness method of structural analysis (e.g., SAP 2000 by Computers & Structures, Inc., California, STAAD Pro 2004 by Research Engineers International, RISA–2D, and RISA–3D by RISA Technologies and MIDAS/Civil by MIDASoft, Inc.).

The analysis to find the forces in a multi-member truss by hand calculation can be tedious and time-consuming. Hence in practice, computer programs are often used to analyse determinate or indeterminate trusses. The hand methods (such as the method of sections) and some hand calculations are used to verify the results obtained from the computer analysis (e.g., top chord forces in a parallel chord truss, reactions at the supports).

Using these computer programs, the input geometry for the nodes of the trusses can be specified by coordinate data or by interactive graphic interface. Support conditions may be defined through the various joint restraint options. Member and end conditions may be defined as pinned. Most of these programs require the user to specify member sizes, so that the analysis can be performed. While forces in determinate structures do not depend upon the 'initial' member sizes given by the user, forces in indeterminate structures do depend upon initial member sizes. Hence it may be necessary to repeat the analysis two or three times, such that the 'initial' member sizes and the designed member sizes are the same. The member sizes given in Tables 16.1 and 16.2 may be used for giving the 'initial' member size input for the trusses shown in Fig. 16.13. These sizes are applicable to trusses with purlins placed at truss joints, having a minimum number of longitudinal ties, as shown in Fig. 16.13. These sizes are derived based on (a) the rise to span ratio being greater than 1/6 and (b) wind permeability of the building being ≤ 20%.

**TABLE 16.1** Initial member size for roof trusses (angle sections) with $f_y = 250$ MPa

| Span (m) | Member notations (see Fig. 16.13) a | b | c, d | e | f |
|---|---|---|---|---|---|
| Truss spacing < 5 m | | | | | |
| 10.0 | 1-65 × 6* | 1-50 × 6 | 1-50 × 6 | 1-45 × 6 | 1-50 × 6 |
| 16.0 | 2-50 × 6 | 2-50 × 6 | 2-50 × 6 | 1-45 × 6 | 1-60 × 6 |
| 21.0 | 2-60 × 6** | 2-50 × 6 | 2-50 × 6 | 1-50 × 6 | 1-65 × 6 |
| 26.0 | 2-65 × 6 | 2-60 × 6 | 2-60 × 6 | 1-60 × 6 | 1-65 × 6 |

*(Contd)*

**TABLE 16.1** *(Contd)*

| Span (m) | Member notations (see Fig. 16.13) a | b | c, d | e | f |
|---|---|---|---|---|---|
| (Longitudinal ties: ISA 75 × 6) | | | | | |
| 5 m < Truss spacing < 6.5 m | | | | | |
| 10.0 | 1-75 × 6 | 1-60 × 6 | 1-60 × 6 | 1-45 × 6 | 1-60 × 6 |
| 16.0 | 2-50 × 6 | 2-50 × 6 | 2-50 × 6 | 1-50 × 6 | 1-65 × 6 |
| 21.0 | 2-65 × 6 | 2-60 × 6 | 2-60 × 6 | 1-50 × 6 | 1-75 × 6 |
| 26.0 | 2-75 × 6 | 2-65 × 6 | 2-65 × 6 | 1-65 × 6 | 1-75 × 6 |
| (Longitudinal ties: ISA 90 × 6) | | | | | |

* 1-65 × 6 means single angle ISA 65 × 65 × 6

** 2-60 × 6 means double angle ISA 60 × 60 × 6

**TABLE 16.2** Initial member sizes for roof trusses with $f_y = 220$ MPa (Tubular members)

| Span (m) | Member Notations (see Fig. 16.13) a | b | c, d | e | f |
|---|---|---|---|---|---|
| Truss spacing-5 m | | | | | |
| 10.5 | 60.3 × 3.25* | 48.3 × 3.25 | 48.3 × 3.25 | 42.4 × 3.25 | 42.5 × 3.25 |
| 16.0 | 76.1 × 3.25 | 48.3 × 3.25 | 60.3 × 3.65 | 42.4 × 3.25 | 42.5 × 3.25 |
| 21.0 | 88.9 × 4.05 | 60.3 × 3.25 | 76.1 × 3.25 | 42.4 × 3.25 | 48.3 × 3.25 |
| 26.0 | 101.6 × 4.85 | 76.1 × 3.25 | 88.9 × 4.05 | 48.3 × 3.25 | 60.3 × 3.25 |
| (Longitudinal tie = ISO 60.3 × 3.65) | | | | | |
| 5 m < truss spacing < 6.5 m | | | | | |
| 10.5 | 76.1 × 3.25 | 48.3 × 3.25 | 60.3 × 3.65 | 42.4 × 3.25 | 42.4 × 3.25 |
| 16.0 | 88.9 × 4.05 | 60.3 × 3.25 | 76.1 × 3.65 | 42.4 × 3.25 | 48.3 × 3.25 |
| 21.0 | 101.6 × 4.85 | 76.1 × 3.25 | 88.9 × 4.05 | 48.3 × 3.25 | 48.3 × 3.25 |
| 26.0 | 114.3 × 4.50 | 76.1 × 3.65 | 88.9 × 4.85 | 48.3 × 3.25 | 60.3 × 3.25 |
| (Longitudinal ties = ISO 76.1 × 3.25) | | | | | |

* 60.3 × 3.25 means tube having 60.3 mm outer diameter and 3.25 mm thickness.

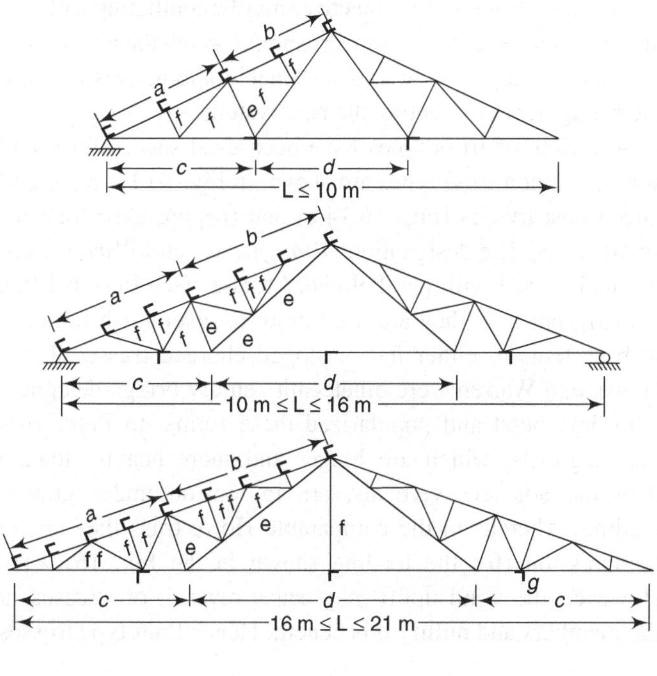

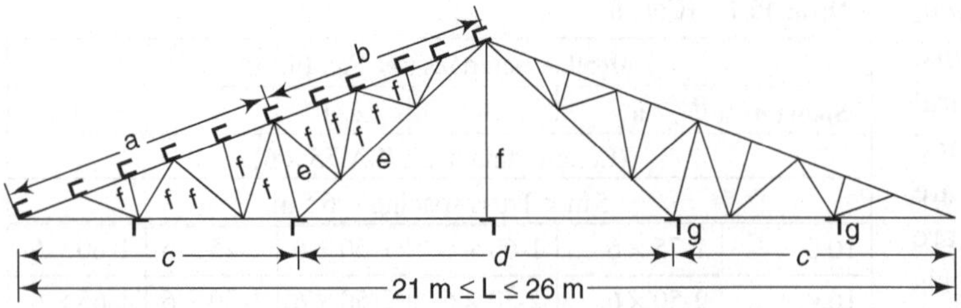


**FIG. 16.13** Member notation: guidelines to member design (Refer Tables 16.1 and 16.2)

## 16.5.2 Types of Trusses and Truss Configurations

Prior to analysing and selecting members for a roof truss, three engineering decisions are to be made: (a) the form of the chords must be determined, that is, whether they should be flat or sloping and whether they should be straight or curved; (b) the pattern of internal triangulation; (c) whether the trusses are simply supported or continuous. Important dimensional variables include the spans and depths of trusses, lengths of specific truss members (especially compression members), spacing of trusses, and transverse purlin spacing (this, in turn, dictates the way loads are applied on to the trusses and frequently, the placement of nodes within a truss).

The above decisions are governed by spatial requirements, economic advantages of a particular system, structural requirements and efficiency (efficient layouts will minimize the lengths of long compression members and maximize the length of tension members; depth of the truss is an important parameter in minimizing the weight as well as volume) and constructional efficiency. These are related to the fabrication and erection of a truss; trusses having parallel upper and lower chords simplify fabrication and erection. Similarly, reducing the number of member sizes will also result in ease of joint connections. Some of the objectives may be conflicting with one another. Hence, design engineers should weigh the advantages and disadvantages of several reasonable alternatives and use their judgment in selecting the best system.

A variety of truss types have been used successfully and some common truss types are shown in Fig. 16.14. King and queen post trusses [Figs 16.14(a) and (b)] are used for very small spans. The designations *Pratt*, *Howe*, and *Warren* were originally used with parallel chord trusses [see Figs16.14(i), (j), (h)], but now they are used more to distinguish between web systems in either flat or sloped chorded trusses. Pratt, Howe, and Warren were nineteenth century bridge designers who developed and popularized these forms. In Pratt truss the diagonals, which are longer and more heavily loaded than the adjacent verticals, are in tension under gravity loading; whereas in the comparable Howe truss they are in compression [for the loading shown in the Fig. 16.14(j)]. However, the wind uplift may cause reversal of stresses in the members and nullify this benefit. Hence Pratt type trusses are more desirable than the Howe type trusses. Note that the diagonals having a slope of 40°–50° with the horizontal have been found to be the most efficient. Also the reversal of the direction of diagonals at mid-span in such trusses is characteristic of design for symmetrical loading. In a Warren truss, approximately half the diagonals are in compression. All except the end verticals are secondary members, and hence may be eliminated, without affecting the overall stability.

The advantage of *parallel chord trusses* is that they use webs of the same lengths and thus reduce fabrication costs for very long spans. The economical span to depth ratio of parallel chord trusses is in the range of 12 m to 24 m. In the case of very deep trusses, it may be necessary to use K or diamond bracing pattern in order to reduce the effective length of members and also to achieve appropriate inclination of the web as shown in Fig. 16.14(k).

For very long span pitched roof, some depth of truss is provided at the ends, resulting in trapezoidal configuration for the trusses [see Fig. 16.14(l)]. Using this configuration results in the reduction of axial forces in the chord members adjacent to the supports. The secondary bending effects in those members are also reduced. The vertical members at supports with length about 1/10th of the truss height at mid-span are found to reduce the forces in the members adjacent to supports considerably (Sree Ramachandra Murthy et al. 2004). For very long spans (greater than 30 m), it may be economical to have trapezoidal trusses with sloping bottom and top chords. Such a configuration will reduce the length of web members and will result in uniform force in the chord members over the entire span. The slope of the bottom chord equal to about half the slope of the top chord is found to be more efficient.

The *subdivided panel truss* [Fig. 16.14(g)] is more common in bridges, but also used in long-span trusses in industrial buildings. In these types of trusses, by placing two secondary members in each panel, the unsupported lengths of long diagonal and compression chord members are reduced. This also reduces the bending moments in chord members resulting from placing the purlins between panel points.

The *pitched Fink truss* [Fig. 16.14(f)] usually proves to be economic for small spans (< 9 m), since the web members in such trusses are arranged in a fashion to obtain shorter members. As already mentioned, Pratt, Howe, and Warren trusses need not have the top chord parallel to the bottom one. Such an arrangement is used to provide a slope for drainage [Fig. 16.14(c) and 16.14(d)]. Pratt trusses with four or six panels are used for spans varying between 6 m to 15 m. The *compound Fink truss* shown in Fig. 16.14(f) may be used for a longer span. The simple *fan truss* [Fig. 16.14(e)] may be used to span 12 m and a compound *fan truss* can span up to 24 m. Fan trusses are often used when the rafter

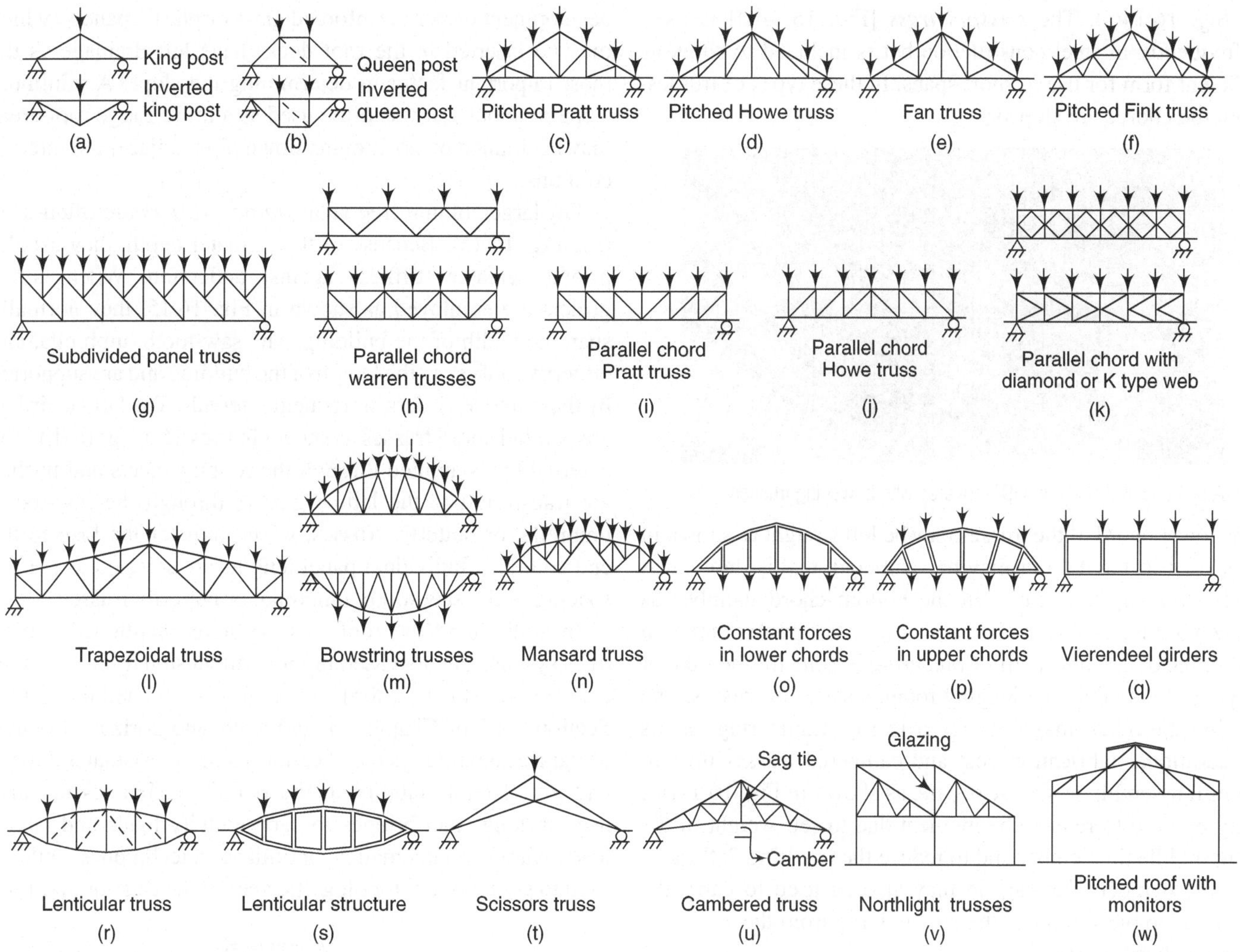


**FIG. 16.14** Common types of trusses

members of the roof trusses are to be subdivided into odd number of panels. A combination of fink and fan can also be used.

The *bowstring truss* [Fig. 16.14(m)] has an upper chord which is curved to some geometrical form, usually a circular arc but occasionally a parabola, ellipse, or multicentered curve. When straight members are used for the top chords, they are called *Mansard trusses* [Fig. 16.14(n)]. Any web system may be used, since the web stresses are normally small. If, for example, a parabolic upper chord is used and the truss is subjected to a uniform load, there will be no stress in the web members, uniform tension throughout the lower chord, and compression in the upper chord (forces in upper chord members will be slightly higher than the lower chord members due to their slope). In other words, the behaviour will be similar to that of a tied arch of the same rib form. The diagonals, though carrying zero forces, are provided for stability under varying loading conditions, if the joints are pin-connected. Both bowstring and mansard trusses are used for longer spans. The diagonals can be removed as shown in Figs 16.14(o) and 16.14(p) but in such cases, the joints have to be made rigid. Now the members will be subjected to axial forces and bending moments and hence cannot be strictly considered as trusses and are called *bowstring girders*. When the chords are parallel as shown in Fig. 16.14(q), they are called as *Vierendeel girders*. In a Vierendeel girder, the loading is carried by a combination of pure flexure and flexure due to shear induced by the relative deformation between the ends of the top and bottom chord members, similar to that found in castellated beams. Though vierendeel girders are usually fabricated using I-sections, their load carrying capacity may be enhanced by using rolled hollow sections with butt or fillet welded connections. They may be analysed using elastic analysis and the moment capacities of the members checked using effective length factors for frames not braced against side sway (Martin & Purkiss 1992).

The Lens shaped truss called as *lenticular truss,* as shown in Fig. 16.14(r) has been used in some bridges in USA. Here again the web members will carry zero forces and hence may be removed and the joints may be made rigid as shown in Fig. 16.14(s). The *scissors truss* [Fig. 16.14(t)] is used infrequently in steel construction but is included to show as efficient form for use in short spans. In these types of trusses, both the chords are sloped.

A building in Delaware, USA covered with bowstring trusses

The lower chord of the trusses may be left straight as shown in Figs 16.14(c) and (f) or may be cambered, that is, fabricated with slight upward curve in the bottom chord member as shown in Fig. 16.14(u). The camber may be approximately in the range of 0.5 m to 1.0 m. *Cambering* is done for the sake of appearance so that in a lengthy room, a series of trusses, one behind the other, may not appear to sag. Cambering results in additional fabrication cost and cambered trusses involve careful assembly at site. A *sag tie*, as shown in Fig. 16.14(u), may be used to reduce the moment due to self weight in the long middle tie member and to reduce the resulting deflection of this member. The sag tie may also be used to carry the load due to the weight of the ceiling hung from the bottom chord, if any.

In roof trusses, the drainage, lighting, and ventilation requirements are the most important elements in establishing the upper chord slope, but occasionally, structural or aesthetic reasons may also control.

In single storey industrial buildings of the type shown in Figs 16.14(v) and (w), drainage is provided toward the eaves and valleys where longitudinal gutters and downspouts are used to carry off water. Artificial lighting is supplemented by windows in the sides of the monitors or the steeper slopes of north light or saw tooth roofs. Portions of the windows can be opened for natural ventilation. (Though north light windows do not admit direct sun and allow only diffused light, they can provide reasonable light up to a distance of about twice the height of the window above the floor. Hence in larger buildings, it may be necessary to provide lighting through sky-lights, monitors, etc. The window area must be at least one eighth of the floor area to provide proper lighting.) Improvements in mechanical ventilation, artificial lighting, and development of wire-reinforced glass or plastic panels (which may be inserted in the roof deck) have left drainage as the most important factor in determining the slope. A minimum slope of about 1:100 is often used as a limit. Large roof areas may be drained by positioning down pipes adjacent to internal columns.

For large column free areas, *lattice girders* are often used (see Fig. 16.15). Because of their greater depth, they usually provide greater stiffness against deflection. When lattice girders are employed as shown in Fig. 16.15, they normally span the width of the building. The saw-tooth, umbrella, and butterfly roofs span the length of the building and are supported by these lattice girders at frequent intervals. The lattice girders are parallel chord trusses as shown in Figs 16.14(g) to (k). The external loads applied through the roofing sheets and purlins are transferred to the lattice girders through the saw-tooth, umbrella, or butterfly trusses, which transfer the load to the end columns. Individual panel lengths are selected as per the spacing of the saw-tooth, umbrella, or butterfly trusses.

In addition to the roof truss systems mentioned earlier, truss systems are also used in three-dimensional space frames (see Subramanian 2006), in multi-storey buildings (see Section 2.4.2 of Chapter 2), and walls and horizontal planes of industrial buildings (see Section 16.3) to resist lateral loads and give lateral stability to the buildings. Trusses are also used in long span bridges to carry gravity loads and lateral loads. Various mathematical optimization techniques could be used to get optimum topology (shape) of the truss or optimum

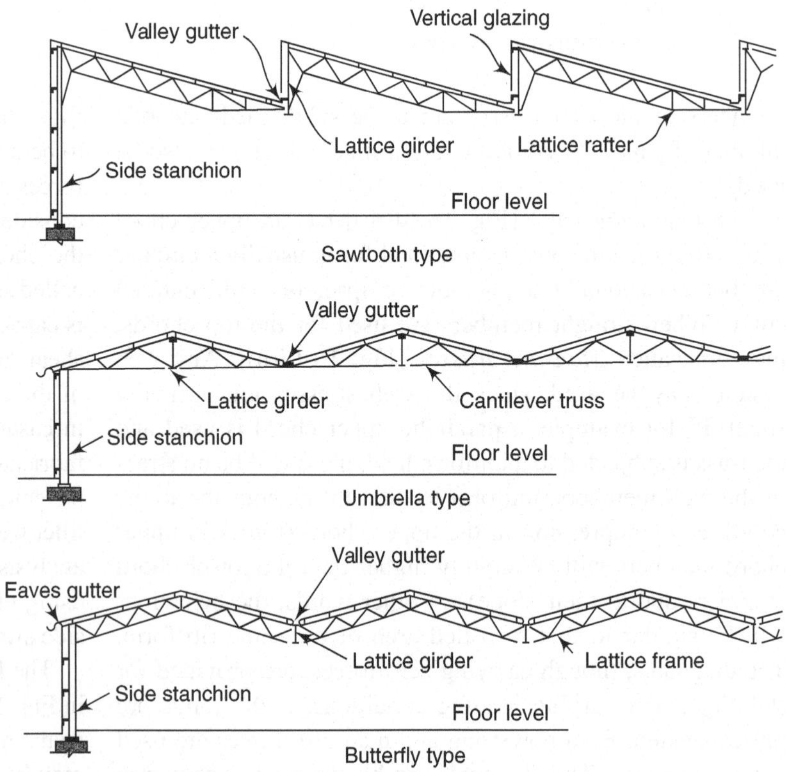


**FIG. 16.15** Sawtooth, umbrella, and butterfly roofs

member sizes (refer to Pandia Raj & Kalyanaraman 2005; Camp & Bichon 2004; Rajeev & Krishnamoorthy 1997; Ramasamy & Rajasekaran 1996).

### 16.5.3 Pitches of Trusses

As seen in Fig. 16.14, most of the trusses are pitched. This is mainly done to drain off rain water on the sheeted slopes. In addition to providing the slope, the joints in the sheetings should be effectively sealed with mastic or washers. The *pitch of a truss* is defined as the ratio of the height of the truss to its span. The pitches usually provided for various types of roof coverings are given in Table 16.3.

**TABLE 16.3** Pitch for roof trusses

| Roof covering | Pitch |
|---|---|
| Corrugated GI sheets | 1/3 to 1/6 |
| Corrugated AC sheets | 1/6 to 1/12 |
| Lapped shingles (e.g. wood, asphalt, clay, and tile) | 1/24 to 1/12 |
| Flat roof and trapezoidal trusses | 1/48 to 1/12 |

A pitch of 1/4 is found economical in cases where the roof has to carry snow loads in addition to wind loads. Where snow loads do not occur, lower pitches up to 1/6 are suitable. Lower pitches are advantageous since the wind pressure on the roof is reduced.

### 16.5.4 Depth of Truss

The depth of the truss determines its stiffness in relation to its span and also its economy. Shallow trusses have larger forces in the members than deep trusses. The length of web members and their slenderness ratio will also be more in deep trusses. Hence, the web members will have larger size in deep trusses. Some guidelines for selecting the depth of roof trusses are given in Table 16.4 (Allen & Iano 2001, *SP*-38). Space frames, decking, arches, etc., are also included for comparison. Roof depth in the range of 1/12 to 1/8$^{\text{th}}$ of span for continuous trusses will be economical. Simply supported trusses should be somewhat deeper, perhaps about 1/10$^{\text{th}}$ span as the minimum. While fixing the height, the clearance to shipping should be considered. Thus, if a truss has a span of 36 m, a depth of 3.6 m might be difficult to transport due to the bridge clearances; hence a depth of 3 m—though it is only one-twelfth of the span—may prove to be more economical.

**TABLE 16.4** Approximate span range and depth for roof systems

| System | Approximate depth | Approximate span range (m) | Typical span (m) |
|---|---|---|---|
| Decking | L/30–L/50 | 1.5–4 | 3 |
| Wide flange beams | L/18–L/28 | 3–20 | 9 |
| Plate girders | L/15–L/20 | 7.5–24 | 12.5 |
| Open-web girders | L/18–L/22 | 7–35 | 18 |

*(Contd)*

**TABLE 16.4** *(Contd)*

| System | Approximate depth | Approximate span range (m) | Typical span (m) |
|---|---|---|---|
| Fink truss | L/5–L/6 | 10–18 | 15 |
| Howe or Pratt or fan truss | L/5–L/6 | 10–15 | 12 |
| Bowstring truss | L/6–L/10 | 18–36 | 24 |
| Trapezoidal truss | L/8–L/12 | 18–36 | 24 |
| 3–D latticed truss | L/4–L/15 | 22–45 | 30 |
| Arches | L/3–L/5 | 18–55 | 36 |
| Double layer grids | L/12–L/20 | 10–35 | 25 |

The depth of the truss must also be considered in relation to the length of the panels, which in turn, depends upon the spacing of the purlins and upon the necessity of getting a favourable slope for the diagonal web members. Steep diagonals can be avoided in deep trusses with relatively close-spaced purlins by using a subdivided truss [Fig. 16.14(g)].

### 16.5.5 Spacing of Trusses

The spacing of trusses is mostly determined by the spacing of supporting columns, which in turn is determined by the functional requirements. Where there are no functional requirements, the spacing should be such that the cost of the roof is minimized. The larger the spacing, the smaller the cost of trusses, but larger is the cost of purlins and vice-versa. Roof coverings also cost more, if the spacing of the trusses is large.

Let us derive an approximate formula for arriving at the minimum cost, by considering the following variables.

$S$ is the spacing of the trusses, $C_t$ is the cost of trusses/unit area, $C_p$ is the cost of purlins/unit area, $C_r$ is the cost of roof coverings/unit area, and $C$ is the overall cost of the roof system/unit area.

Since the cost of the truss is inversely proportional to the spacing of truss,

$$C_t = k_1/S$$

where $k_1$ is a constant. Similarly, the cost of purlins is directly proportional to the square of spacing of trusses. Thus,

$$C_p = k_2S^2$$

The cost of roof coverings is directly proportional to the spacing of trusses. Thus, we have

$$C_r = k_3S$$

$$\text{Total cost } C = C_t + C_p + C_r$$
$$= (k_1/S) + k_2S^2 + k_3S$$

For the overall cost is to be minimum, $dC/dS$ should be zero. Thus,

$$-(k_1/S^2) + 2k_2S + k_3 = 0$$

or $(-k_1/S) + 2k_2S^2 + k_3S = 0$

or $-C_t + 2C_p + C_r = 0$

Thus, we get $C_t = C_r + 2C_p$ (16.4)

Equation (16.4) shows that an economic system is obtained when the cost of trusses is equal to the cost of roof covering plus twice the cost of purlins. This equation cannot be used to obtain the economic spacing of trusses. It has been found that the economic range of spacing is 1/5 to 1/3 of span. For lighter load, say, carrying no snow or superimposed load except wind, the larger spacing may be more economical. Spacing of 3–4.5 m for spans up to 15 m and 4.5–6 m for spans of 15–30 m may result in economy. The spacing of long-span trusses (longer than about 45 m) should be carefully considered. Such trusses may be more economical when spaced at 12–15 m, rather than at 6–7.5 m and may necessitate the use of open web girders in place of purlins.

### 16.5.6 Spacing of Purlins

The spacing of purlins depends largely on the maximum safe span of the roof covering and glazing sheets. Hence, they should be less than or equal to their safe spans when they are directly placed on purlins. Thus for corrugated GI sheets, the purlin spacing may vary from 1.5 to 1.75 m, and for corrugated AC sheets, it is limited to 1.4 m, for 6-mm thick sheets, and 1.6 m, for 7-mm thick sheets. For larger spans, if the configuration of the truss is such that it is not feasible to place purlins at the nodes of upper chords, the purlins are placed between the nodes, thus introducing bending moments in the upper chords, in addition to the compressive force due to truss action (see Fig. 16.16). Hence in this case, the weight of the truss may be increased by about 10–15%. Therefore, it is preferable to place purlins at the nodal point of the truss, so that the upper chord members are subjected to only direct compression.

As discussed in Chapter 3, wind loading is not uniform over the roof; for example, the loading is much higher along the roof's perimeter and sometimes along the ridge (see Fig. 3.34 of Chapter 3). Instead of using structural roofing panels of heavier gauges, in the areas of higher localized loads, it is better to space the purlins closer.

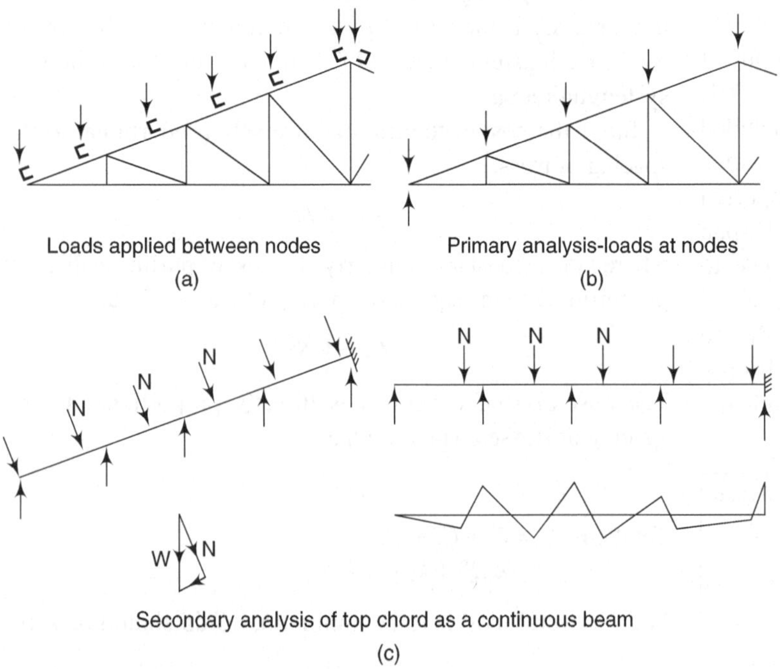


**FIG. 16.16** Loads applied between nodes of truss

### 16.5.7 Loads on Trusses

The main loads on trusses are dead, imposed, and wind loads. The dead load is due to sheeting or decking and their fixtures, insulation, felt, false ceiling (if provided), weight of purlins, and self weight. This load may range from 0.3 to 1.0 kN/m$^2$. Also the truss may be used for supporting some pipe line, fan, lighting fixtures, etc. Hence to take into account this probability, it may be worthwhile considering an occasional load of about 5 to 10 kN distributed at the lower panel points of the truss.

The weights of the purlins are known in advance as they are designed prior to the trusses. Since the weight of the truss is small compared to the total dead and imposed loads, considerable error in the assumed weight of the truss will not have a great impact on the stresses in the various members. For live load up to 2 kN/m$^2$, the following formula may be used to get an approximate estimate of the weight of the trusses:

$$w = 20 + 6.6L \quad (16.5)$$

where $w$ is the weight of the truss in N/m$^2$ and $L$ is the span of the truss in m. For welded trusses, the self weight of the truss is given by

$$w = 53.7 + 0.53A \quad (16.6)$$

where $A$ is the area of one bay.

For live loads greater than 2 kN/m$^2$, the value of $w$ may be multiplied by the ratio of actual live load in kN/m$^2$/2. The dead weight of the truss, inclusive of lateral bracing, may also be assumed to be equal to about 10% of the load it supports. Long span trusses are likely to be heavier. The weight of bracings may be assumed to be 12–15 N/m$^2$ of the plan area. The weight of the truss should be computed after it has been designed to make sure that it is within 10% of the assumed weight.

The imposed load on roofs will be as per IS 875-Part 2 (see Section 3.3 and Table 3.7 of Chapter 3). The snow loads may be computed as given in Section 3.3.2 of Chapter 3. The wind loads should be calculated as per IS 875-Part 3 (see Section 3.13 of Chapter 3). Wind loads are important in the design of light roofs where the suction can cause reversal of load in truss members. For example, a light angle member is satisfactory when used as a tie but may buckle when the reversal of load makes it to act as a strut. In the case of flat roofs with heavy decking, the wind uplift will not be greater than the dead load, and hence it need not be considered in the design.

Since earthquake load on a building depends on the mass of the building, earthquake loads usually do not govern the design of light industrial buildings. Thus wind loads usually govern the design of normal trussed roofs. However, in the case of industrial buildings with large mass located at the roof, the earthquake load may govern the design. These loads are calculated as per IS 1893 (Part 1)–2002 (see Section 3.12 of Chapter 3).

### 16.5.8 Load Combination for Design

As mentioned earlier, the earthquake loads are not critical in the design of industrial building, since the weight of the roof is not considerable. Hence, the following combinations of loads are considered when there is no crane load:

1. Dead load + imposed load (live load)
2. Dead load + snow load
3. Dead load + wind load (wind direction being normal to the ridge or parallel to ridge whichever is severe)
4. Dead load + imposed load + wind load (which may not be critical in most of the cases)

The third combination should be considered with internal positive air pressure and internal suction air pressure separately to determine the worst combination of wind load. When crane load is present, the load combinations as given in Section 3.15 should be considered. The load combinations mentioned in this section should be considered along with appropriate partial load factors as discussed in Section 3.15 of Chapter 3.

All the loads are estimated per square metre of projected area in the plan and the loads are assumed to act as concentrated loads at points where purlins are located on the upper chord. Their value depends on the roof area supported by the purlins. The weight of the truss is included in the purlin point loads. More discussion on load combinations is provided by Bakota (1977).

### 16.5.9 Design of Truss Members

The members of the trusses are made of either rolled steel sections or built-up sections depending upon the span length and intensity of loading. Rolled steel single or double angles, T-sections, hollow circular, square, or rectangular sections are used in the roof trusses of industrial buildings [see Fig. 16. 17(a)]. In long-span roof trusses and short span bridges, heavier rolled steel sections, such as channels and I-sections are used [see Fig. 16.17(b)]. Built-up I-sections, channels, angles, and plates are used in the case of long-span bridge trusses [Fig. 16.17(c)]. Access to the surface of the members for inspection, cleaning, and repainting during service are important considerations while using built-up sections. Hence in highly corrosive environments, fully closed welded box-sections or hollow sections are used, with their ends

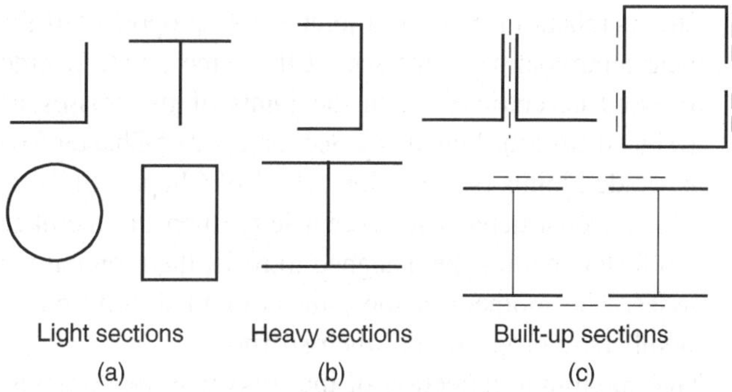


**FIG. 16.17** Cross section of truss members

fully sealed to reduce the maintenance cost and improve the durability of the trusses (Subramanian 2000).

The various steps involved in the design of truss members are as follows:

1. Depending upon the span, required lighting, and available roofing material, the type of truss is selected and a line diagram of the truss is sketched.
2. Various loads acting over the truss are calculated using IS 875 (Parts 1-5).
3. The purlins are designed and the loads acting on the truss at the purlin points are computed.
4. The roof truss is analysed for the various load combinations using the graphical method or the method of sections or joints or by a computer program and the forces acting on the members for various combinations are tabulated.
5. Each member may experience a maximum compressive or tensile force (called the design force) under a particular combination of loads. Note that a member which is under tension in one loading combination may be subjected to reversal of stresses under some other loading combination. Hence, the members have to be designed for both maximum compression and maximum tension and the size for the critical force has to be adopted. The design for compression is done as per Sections 9.9.4, 9.13.1, 9.14, and 9.16 of Chapter 9. The principal rafter is designed as a continuous strut and the other compression members are designed as discontinuous struts. The limiting slenderness ratios are discussed in Section 9.14.1 of Chapter 9. The effective length of compression members is taken as per Section 9.9.4 of Chapter 9. Similarly, the design for tension is done as per Section 7.8.2 of Chapter 7.
6. When purlins are placed at intermediate points, i.e., between the nodes of the top chord, the top chord will be subjected to bending moment in addition to axial compression. Since the rafter is a continuous member, the bending moments may be computed by any suitable method (say, moment distribution method or computer program). Then the member is designed for combined bending and axial compression as per Section 13.8.1 of Chapter 13.

7. The members meeting at a joint are so proportioned that their centroidal axes intersect at the same point, in order to avoid eccentricity. Then the joints of the trusses are designed either as bolted (see Section 5.7.2 of Chapter 5) or as welded joints (see Section 6.11.1 of Chapter 6). If the joint is constructed with eccentricity, then the members and fasteners must be designed to resist the moment that arises. The moment at the joint is divided between the members in proportion of their stiffness.
8. The maximum deflection of the truss may be computed by using either strain energy method or matrix stiffness analysis program. A computer analysis gives the value of deflection as part of the output. This deflection should be less than that specified in Table 2.8 of Chapter 2.
9. The detailed drawings and fabrication drawings are prepared and the material-take-off is worked out.
10. The lateral bracing members are then designed. When a cross braced wind girder is used, as shown in Fig. 16.18(a), it is necessary to use a computer analysis program, since the truss will be redundant. However, it is usual to neglect the compression diagonal and assume that the panel's shear is taken by the tension diagonals, as shown in Fig. 16.18(b). This idealization is useful to make the wind girder determinate and obtain the forces in various members by using method of sections or method of joints.

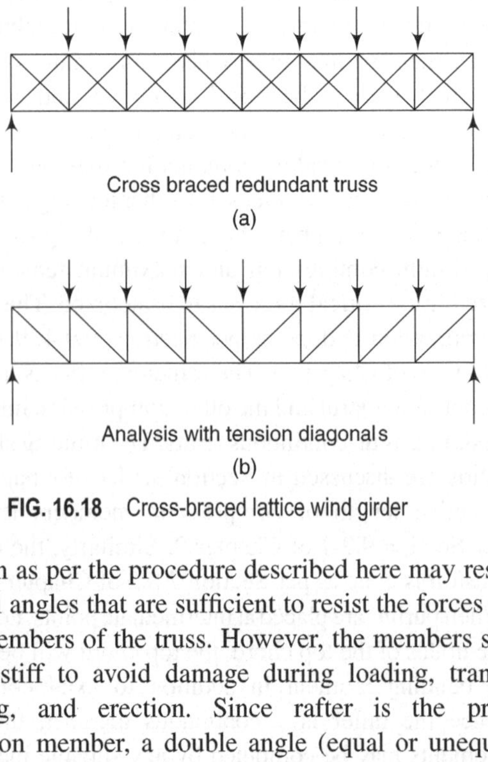


**FIG. 16.18** Cross-braced lattice wind girder

The design as per the procedure described here may result in very small angles that are sufficient to resist the forces in the various members of the truss. However, the members should be fairly stiff to avoid damage during loading, transport, off-loading, and erection. Since rafter is the primary compression member, a double angle (equal or unequal) is often preferred. Similarly, the main ties may be subjected to compression during handling or due to wind suction. Moreover, these ties often have a long unsupported length and hence double angle sections are used for these main ties also. All other web members can be designed as single angle members.

From practice, the following minimum sections are recommended for use in compound fink roof trusses.

Rafters—2 ISA 75 × 50 × 6
Main ties—2 ISA 75 × 50 × 6
Centre tie—2 ISA 65 × 45 × 6
Main sling—2 ISA 65 × 45 × 6
Main strut—ISA 65 × 45 × 6
All other members—ISA 50 × 50 × 6

The width of the members should be kept as minimum as possible, since wide members have greater secondary stresses.

While trusses are stiff in their plane, they are very weak out-of-plane. Consider the planar truss as shown in Fig. 16.19(a), in which, the top chord is braced at each panel point. The top chord when subjected to in-plane loading may buckle in the horizontal $xz$ plane or the vertical $xy$ plane. Unless we provide members having equal moment of inertia about both axes (e.g., square or round members including hollow sections), we have to calculate the buckling strength in the horizontal and vertical plane and adopt the least strength. When transverse members are provided, as shown in Fig. 16.19(b), it is still possible for the top chords to buckle in the horizontal plane. The effective lengths of the top chord members, as far as their resistance to buckling in the horizontal plane is concerned, is $2L$ and not just $L$, where $L$ is the distance between the nodal points. Note that the members in the vertical plane do nothing to prevent this type of buckling in $xz$ plane. Hence in order to make the buckling load in both the horizontal and vertical plane equal, we may have to provide members which are stiffer in the horizontal plane (such as rectangular sections, double angles, or H-sections) as shown in Fig. 16.19(b). When such lateral bracings in the form of purlins are not provided, the entire top chord of the truss may buckle laterally.

The above discussions are valid for the bottom chord member of the truss also, when it is subjected to compression due to reversal of stresses owing to wind suction. Hence lateral (longitudinal) ties are often provided at regular intervals in the bottom chords also (Fisher 1983). Depending upon the $L/r$ ratio of the top and bottom chord members in the horizontal and vertical planes, it may be advantageous to adopt unequal angles. The longitudinal ties are under tension in most load situations but may be subjected to compression under wind loading condition depending upon the bracing orientation. There must be at least two longitudinal ties to form a truss action under wind load condition. The longitudinal ties may also be used to support false ceiling. It is desirable to restrict the slenderness ratio of such ties to 250, to avoid sagging.

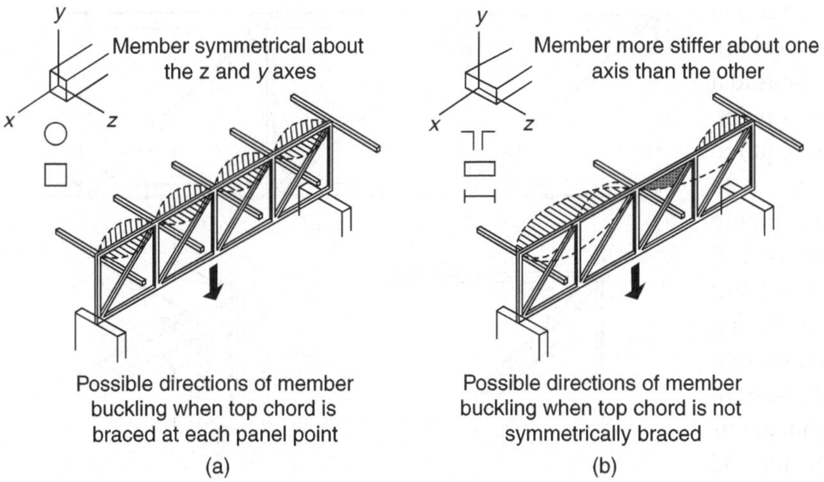


**FIG. 16.19** Lateral buckling of truss members: use of transverse members for bracing

## 16.5.10 Connections in Trusses

Members of trusses can be jointed by riveting, bolting, or welding. As explained in Chapter 5, rivets have become obsolete and for important structures high-strength friction grip (HSFG) bolts and welds are often preferred. Trusses having short span are usually fabricated in shops using welding and transported to site as one unit. Longer span trusses are prefabricated in segments by welding in shop and assembled at site by bolting or welding. For example, fink or compound fink trusses are fabricated as two halves in the workshop. The two halves are transported and assembled at site, where the centre tie is also fitted up (see detail 1 and 3 of Fig. 16.20). In such situations, the main slings will be subjected to severe handling stresses and hence are made of double angle sections.

If the rafter and tie members are made of T-sections, angle diagonals can be directly welded or bolted to the web of the T-sections. Often, it may not be possible to confine the connection within the width of the member due to inadequate space to accommodate the joint length. In such cases, gusset plates are used. The size, shape, and thickness of gusset plates depend upon the size of the members being joined, number and size of bolt or length of

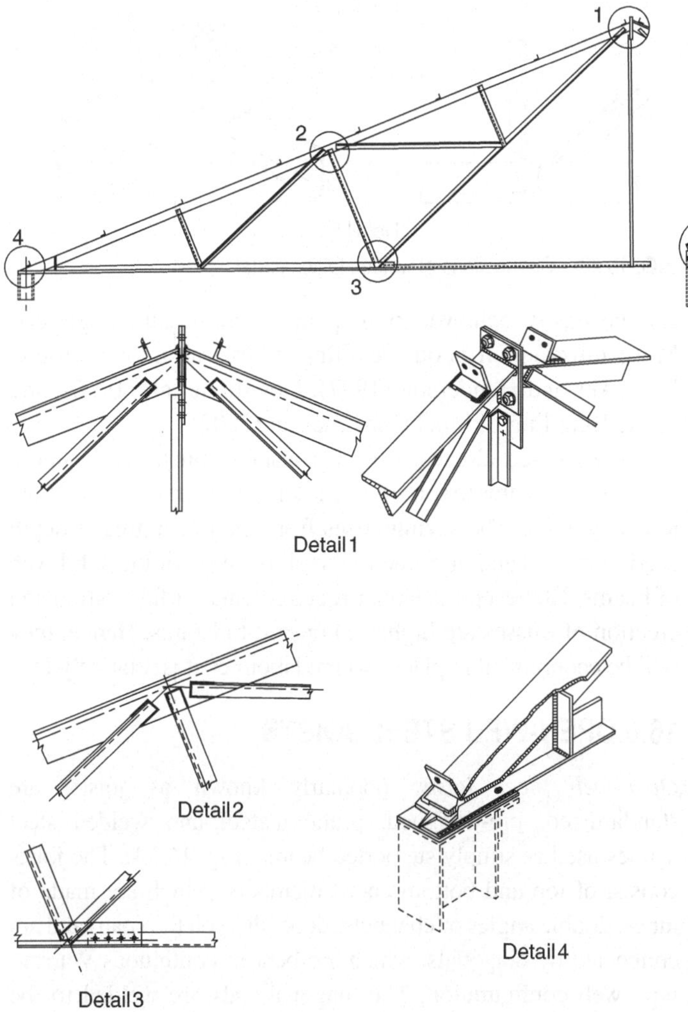


**FIG. 16.20** Connection details of welded fink roof truss

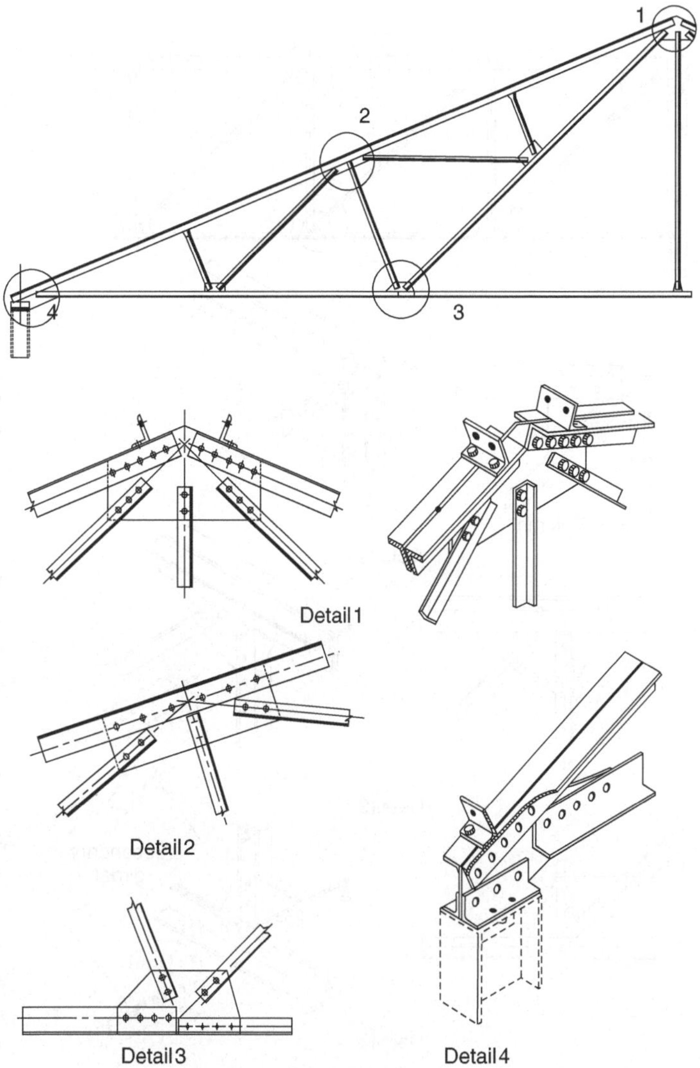


**FIG. 16.21** Connections of bolted fink roof trusses

the required weld, and the force to be transmitted (see Section 5.7.2 of Chapter 5 for more discussion on gusset plate design). The connections should be so arranged that the centroidal axes of members meeting at the connection meet at a point. Example 5.15 of Chapter 5 shows the calculations required for the design of a bolted truss joint. Figures 16.21 and 16.22 show typical bolted joints in trusses and lattice girders. Figure 16.23 shows a typical joints in a welded lattice girder. Typical welded joints using circular tubes were shown in Figs 6.58(a) and (c) and joints using hollow sections were shown in Fig. 6.60 of Chapter 6. Examples 6.9 and 6.10 of Chapter 6 showed the calculations required for the design of welded joints in trusses. The choice between welded and bolted connections depends on the equipment available with the fabricator and availability of electricity at site (if site welding is preferred). However, when large number of trusses are made, welded joints are economical. Moreover, welded joints give better appearance and are easy to maintain. Standard joints should be used with as much repetition of member shapes and sizes, end preparation, and fabrication operations as possible. This

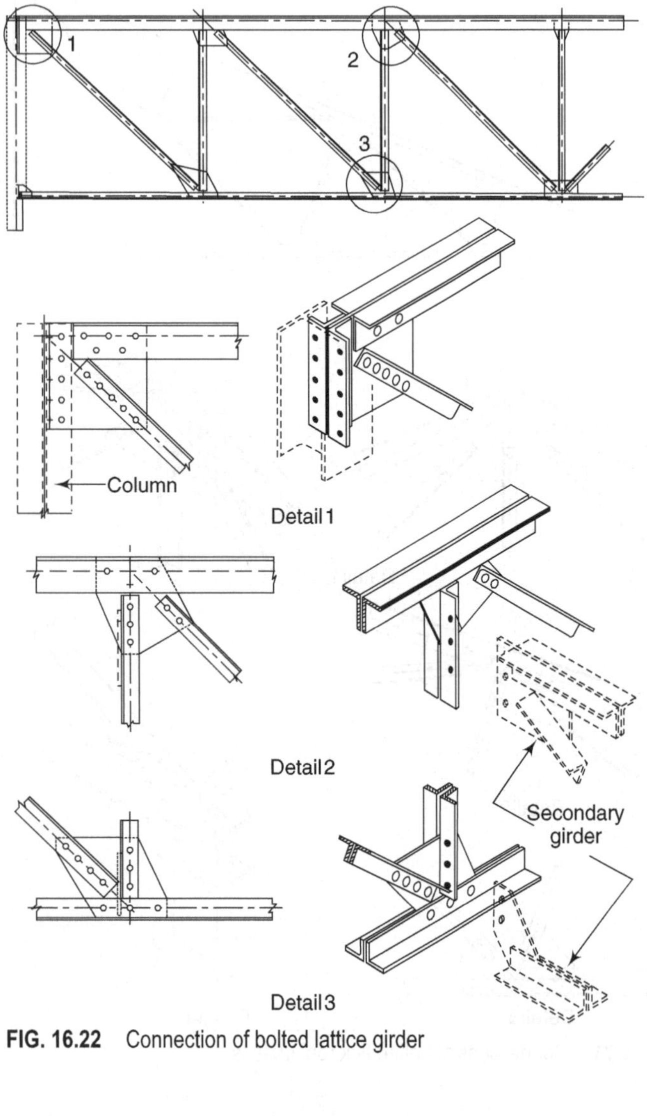

**FIG. 16.22** Connection of bolted lattice girder

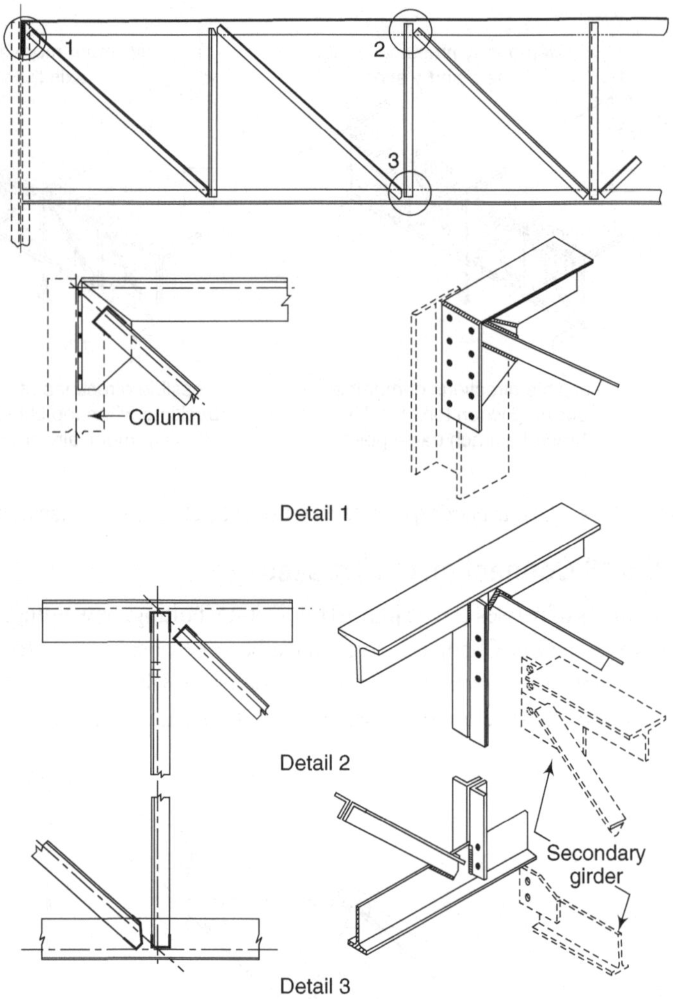

**FIG. 16.23** Connections of welded lattice girder

can be easily achieved with parallel chord lattice girders. More information about detailing of trusses may be found in Packer and Henderson (1997), Dowling et al. (1998), and AISC Detailing of Steel Construction (2002).

As discussed already, trusses consume much less material compared to beams to span the same length and transfer moderate to heavy loads. The saving arises from the much greater depth used in trusses and also due to the elimination of the solid web of beams. However, the labour requirements for fabrication and erection of trusses are higher compared to beams. Hence, they will be economical in places where labour cost is relatively low.

## 16.6 OPEN-WEB STEEL JOISTS

*Open-web steel joists,* popularly known as joists, are standardized, pre-designed, prefabricated, and welded steel trusses used as simply supported beams (Fig. 16.24). The joists consist of top and bottom chord members, which are made of tubes, double angles or channels (depending on the span) and are connected by diagonals, which are bent in continuous Warren-type web configuration. The diagonal rods are welded to the top and bottom chord members. Sometimes, the diagonals are also made of angles. They are generally hung from the supporting girders or walls and are tack welded to the flanges of the supporting girders to prevent movement during construction. Lateral bracing members, called bridging members are also provided at regular intervals. Once the bracings are properly secured and roof deck attached, the system is stable. Three-dimensional open-web girders (similar to those found in railway platforms) inherently provide strong resistance to lateral buckling. The lateral buckling resistance provided by these three-dimensional girders is largely dependent on the spacing of members in the third dimension (see Fig. 16.25); larger spacing is desirable than a very small spacing. More details about the behaviour of such three-dimensional girders are provided by Ganapathy (1972) and Varghese & Ganapathy (1974). These open-web girders are used extensively in USA, Germany, and other countries since they ensure accuracy of shop fabrication, speedy erection, and need only minimum semi-skilled labour at site. The open-web design also saves space by allowing passage of piping, conduits, and HVAC ducts. Joist spacing is governed by the roof deck's capacity. In USA, these joists are designed and built by their manufactures in accordance with Steel Joist Institute Specifications (2002). These joists are often used with relatively flat roofs, since large slopes present some difficulties in making sloped bearing seats and require careful structural analysis. (The joists can not resist any appreciable torsion, since they do not have solid webs to transfer torsional stresses; hence some system to resist the perpendicular-to-web force component must be incorporated.) More discussion about open-web joists in sloped roofs is provided by Newman (2004). The open-web joists are economical for spans ranging from 7.5 m to 15 m and may be cost effective for spans in the range of 18 m to 40 m. The system is customarily supported on wide-flange I-section or tubular columns and requires wall bracing for lateral stability (Newman 2004).

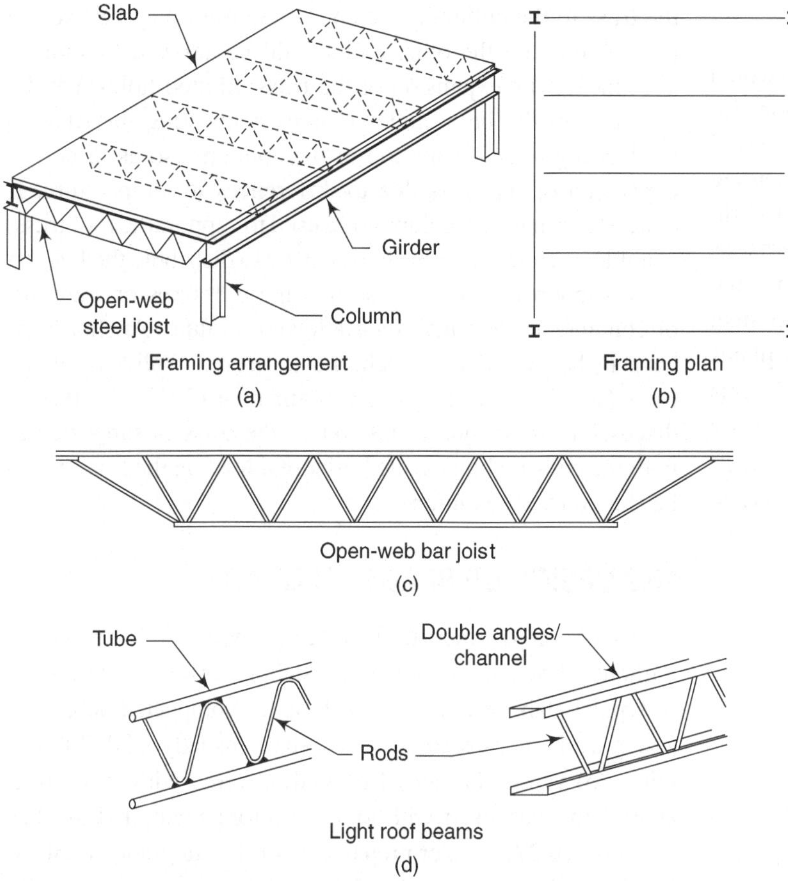


**FIG. 16.24** Open-web steel joists

A commercial building at Gaithersburg, MD, USA with open web joists

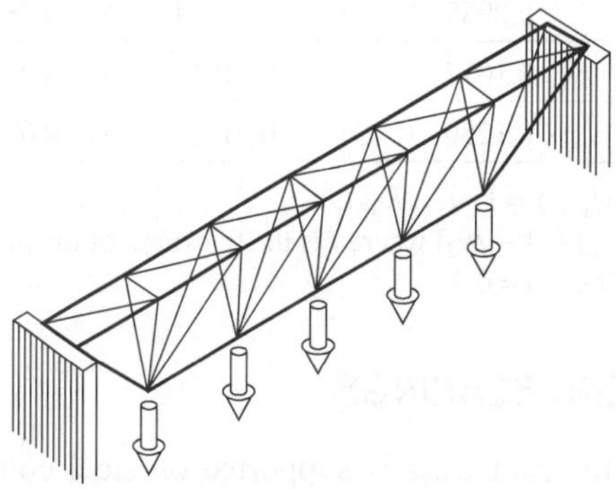

**FIG. 16.25** Three-dimensional lattice girder

## 16.7 FLOOR PLATES

Where platforms are required around some equipment, which are supported by steel frames, rolled steel plates with non-slip patterns (see Fig. 16.26) are used. Such plates, often referred to as *chequered plates*, are available in thicknesses ranging from 3 mm (approximate weight 287 N/m$^2$) to 12.0 mm (weight 990 N/m$^2$) with an increment of 1.5 mm. The patterns as shown in Fig. 16.26 project approximately 1.5 mm above the plain plate. These plates may be welded, or bolted to steel beams on all four sides acting as two-way supported plates with edges pinned or fixed. Table 16.5 gives the coefficients for maximum moment and deflection produced by uniformly distributed load in such plates of long side $l_y$ and short side $l_x$. This table may be used to design floor plates using appropriate bending stress and allowable deflection, say $l_x/100$.

**FIG. 16.26** Steel floor plates with non-slip raised patterns

**TABLE 16.5** Moments and deflections in uniformly loaded rectangular plates (Timoshenko & Krieger 1968)

| $l_y/l_x$ | Simply supported edges | | Clamped edges | |
|---|---|---|---|---|
| | $K_1$ | $K_2$ | $K_1$ | $K_2$ |
| 1.0 | 0.0479 | 0.00406 | −0.0513 | 0.00126 |
| 1.1 | 0.0553 | 0.00485 | −0.0581 | 0.00150 |
| 1.2 | 0.0626 | 0.00564 | −0.0639 | 0.00172 |
| 1.3 | 0.0693 | 0.00638 | −0.0687 | 0.00191 |
| 1.4 | 0.0753 | 0.00705 | −0.0726 | 0.00207 |
| 1.5 | 0.0812 | 0.00772 | −0.0757 | 0.00220 |
| 1.6 | 0.0862 | 0.00830 | −0.0780 | 0.00230 |
| 1.7 | 0.0908 | 0.00883 | −0.0799 | 0.00238 |
| 1.8 | 0.0948 | 0.00931 | −0.0812 | 0.00245 |
| 1.9 | 0.0985 | 0.00974 | −0.0822 | 0.00249 |
| 2.0 | 0.1017 | 0.01013 | −0.0829 | 0.00254 |
| ∞ | 0.1250 | 0.01302 | −0.0833 | 0.00260 |

$M = K_1 w l_x^2$; $\Delta = K_2 w l_x^4/K_3$; $l_y > l_x$

$K_3 = Et^3/[12(1 - \mu^2)]$ where $t$ is the thickness of the plate and $\mu$ is Poisson's ratio = 0.3.

## 16.8 END BEARINGS

When the roof truss is supported on steel columns, suitable connections should be provided to transfer the reaction from the truss to the column. One end of the truss may be fixed to the column and the other end should be allowed to slide to account for the expansion of the truss. Slotted holes may be provided in the base angles (angles connecting the truss to the bearing plates at the top of the columns) so as to permit expansion of the truss due to differences in temperature. If concrete or masonry columns are used to support the roof truss, suitable bearing plates have to be used to distribute the load on these supporting members so that the pressures on masonry or concrete are less than their allowable values. Anchor bolts have to be provided at each end to prevent uplifting of the truss (see Section 15.8 and Subramanian (2013) for further discussion on anchor bolts). When the truss is supported at both the ends by hinges, the horizontal load on the truss has to be shared by the columns.

## 16.9 DESIGN OF SLABS ON GRADE

A discussion on industrial buildings will not be complete without a brief mention of some slab-on-grade design issues. A typical slab on grade consists of (a) compacted subgrade, i.e., gravel or crushed stone sub-base usually 150–300 mm thick, (b) vapour barrier (if needed) covered with sand layer, and (c) the slab itself with joints, reinforcement, and finishes (see Fig. 16.27). Proper preparation of the subgrade is critical for the performance of the slab, since even a most carefully constructed slab may fail when placed on poorly compacted or unsuitable soil. The sub-base helps the slab to span over any poorly compacted spots by spreading concentrated loads over a larger area. It also provides effective drainage under the slab. The effectiveness of the sub-base increases with its depth. Though the use of vapour barriers (polyethylene sheets) is debatable, it prevents the dissipation of water during curing and may result in slab curling. A sand layer on the top of the vapour layer will mitigate the curling problem. The structural design of the slab is straight forward and is covered in ACI 302.

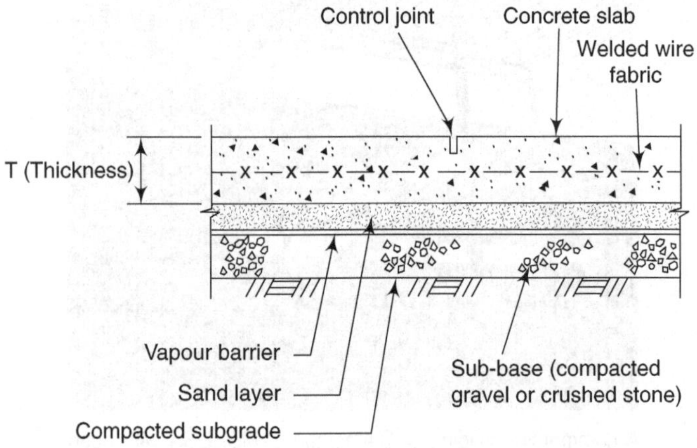


**FIG. 16.27** Components of slab on grade

## EXAMPLES

**Example 16.1:** Design a roof truss, rafter bracing, purlin, tie runner, side runner, and eave girder for an industrial building located at Guwahati with a span of 20 m and a length of 50 m. The roofing is galvanized iron sheeting. Basic wind speed is 50 m/s and the terrain is an open industrial area. Building is class B building with a clear height of 8 m at the eaves.

**Solution:**

**Step 1:** *Structural model*

A trapezoidal truss is adopted with a roof slope of 1 to 5 and end depth of 1 m. For this span range, the trapezoidal trusses would be normally efficient and economical.

Approximate span to depth ratio is about L/8 to L/12. Adopt a depth of 3 m at mid-span.

$$\text{Span/depth ratio} = 20/3.0 = 6.67$$

Hence the span/depth ratio is fine.

Truss spacing may be in the range of 1/4$^{th}$ to 1/5$^{th}$ of the span length. Hence adopt a spacing of 20/4 = 5 m. Then,

$$\text{Number of bays} = 50/5 = 10$$

The plan of the building and the elevation of the truss are shown in Figs 16.28(a) and (b).

**Step 2:** *Loading*

Calculation for dead load:

GI sheeting = 0.085
Fixings = 0.025
Services = 0.100
Total load = 0.210 kN/m$^2$

For 5 m bays,

$$\text{Roof dead load} = 0.21 \times 20 \times 5 = 21\,\text{kN}$$

$$\text{Weight of purlin (assuming } 70\,\text{N/m}^2) = 0.07 \times 5 \times 20 = 7\,\text{kN}$$

$$\text{Self-weight of one truss*} = 0.1067 \times 5 \times 20 = 10.67\,\text{kN}$$

$$\text{Total dead load} = 38.67\,\text{kN}$$

* For welded sheet roof trusses, the self-weight is given approximately by

$$w = 53.7 + 0.53\,A = 53.7 + 0.53 \times 5 \times 20 = 0.1067\,\text{kN/m}^2$$

Calculation for nodal dead loads:

Since the truss has 16 internal nodes at the top chord [see Fig.16.28(b)],

$$\text{Intermediate nodal dead load } (W_1) = 38.67/16 = 2.42\,\text{kN}$$

$$\text{Dead load at end nodes } (W_1/2) = 2.42/2 = 1.21\,\text{kN}$$

(All these loads act vertically downwards at the nodes.)

*Wind load as per IS 875 (Part 3)-1987*

Basic wind speed in Guwahati = 50 m/s

Wind load $F$ on a roof truss by static wind method is given by (clause 6.2.1 of IS 875) as follows:

$$F = (C_{pe} - C_{pi}) \times A \times P_d$$

where $C_{pe}$ and $C_{pi}$ are the force coefficients for the exterior and interior of the building.

Value of $C_{pi}$:

Assume wall openings between 5%–20% of wall area (clause 6.2.3.2 of IS 875 and Table 3.32 of Chapter 3), we have

$$C_{pi} = \pm 0.5$$

Value of $C_{pe}$:

Roof angle = $\alpha = \tan^{-1}(1/5) = 11.3°$

Height of the building to eaves $h$ = 8 m

Short dimension of the buildig in plan $w$ = 20 m

Building height to width ratio is given by

$$\frac{h}{w} = \frac{8}{20} = 0.4 < 0.5$$

*Wind angle – 0° [Table 5 of IS 875 (Part 3) and Table 3.31]*

For 10° in windward side, $C_{pe} = -1.2$ and for leeward side $C_{pe} = -0.4$

For 20° in windward side and leeward side $C_{pe} = -0.4$

Roof angle $\alpha = 11.3°$

Then by interpolation we get

$$C_{pe} = -1.1 \text{ for windward and } -0.4 \text{ for leeward}$$

*Wind angle – 90° [Table 5 of IS 875 (Part 3) and Table 3.31]*

For 10° in windward and leeward, $C_{pe} = -0.8$

For 20° in windward and leeward, $C_{pe} = -0.7$

For 11.3°, $C_{pe} = -0.79$ for windward and leeward

Risk coefficient, $k_1$ = 1.0, assuming that the industrial building is under general category and its probable life is 50 years.

*Terrain, height and structure size factor, $k_2$:*

Roof elevation: 8–11 m

Considering category 1 (exposed open terrain) and class B structure (length between 20–50 m) from Table 2 of IS 875 (Part 3)-1987, for 11 m, $k_2 = 1.038$

Assume topography factor $k_3 = 1.0$ (because of flat land)

*Wind pressure calculation*

Total height of the building = 11 m

Basic wind speed $V_b$ = 50 m/s

Design wind speed

$$V_z = k_1 \times k_2 \times k_3 \times V_b$$

$$k_1 = 1.0;\ k_2 = 1.038;\ k_3 = 1.0;$$

$$V_z = 1.038 \times 1 \times 1 \times 50 = 51.9\,\text{m/s}$$

$$\text{Design wind pressure } p_d = 0.6 v_z^2 = 0.6 \times (51.9)^2 = 1616.17\,\text{N/m}^2 = 1.616\,\text{kN/m}^2$$

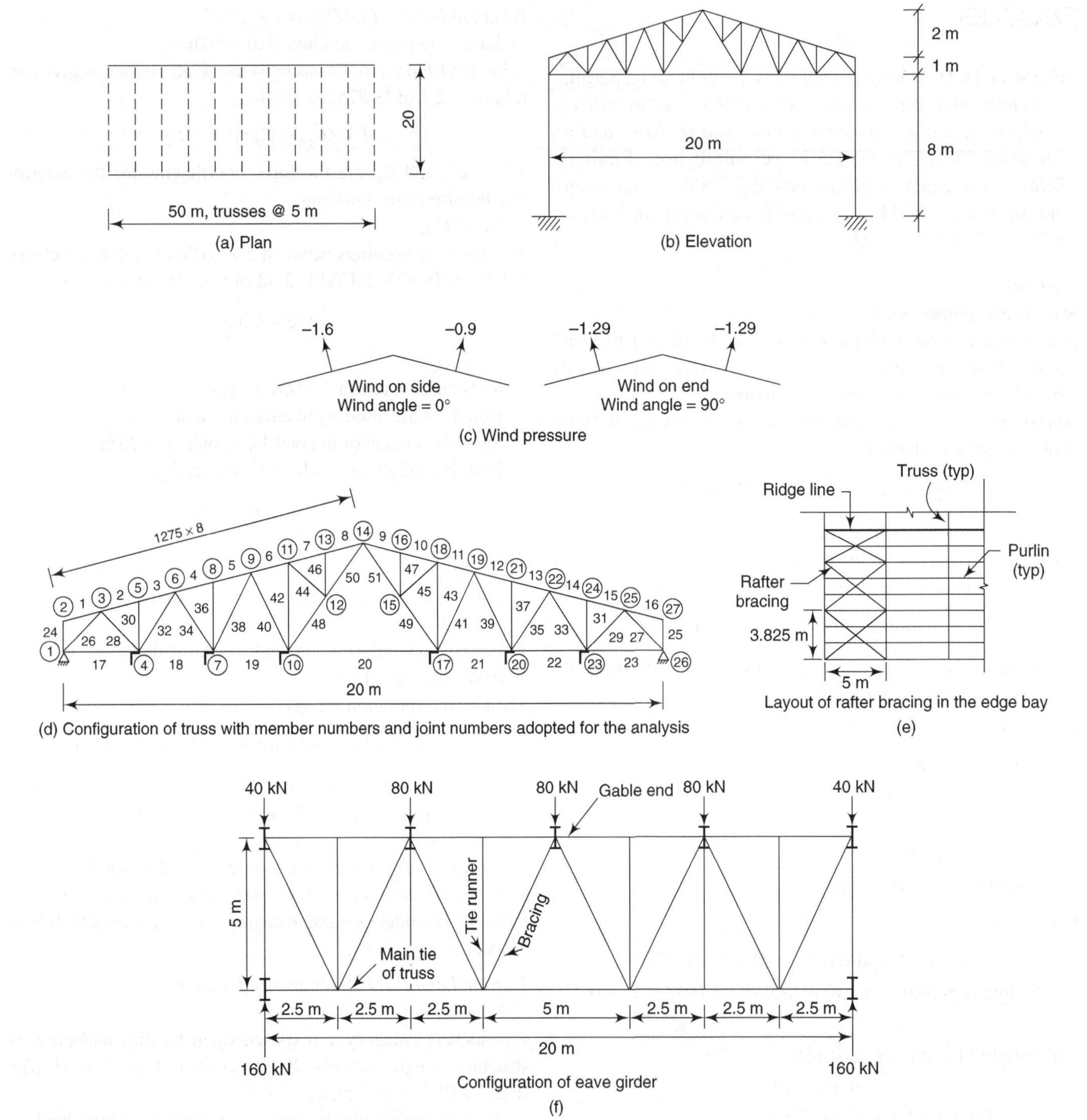


**FIG. 16.28**

*Wind load on roof truss*

| Wind angle | Pressure coefficient | | | ($C_{pe} \pm C_{pi}$) | | $A \times P_d$ (kN) | Wind load, F (kN) | |
|---|---|---|---|---|---|---|---|---|
| | $C_{pe}$ | | $C_{pi}$ | Wind-ward | Lee-ward | | Wind-ward | Lee-ward |
| | Wind-ward | Lee-ward | | | | | | |
| 0° | −1.10 | −0.4 | −0.5 | −1.6 | −0.9 | 10.3 | −16.48 | −9.27 |
| | | | 0.5 | −0.6 | 0.1 | 10.3 | −6.18 | 1.03 |
| 90° | −0.79 | −0.79 | −0.5 | −1.29 | −1.29 | 10.3 | −13.29 | −13.29 |
| | | | 0.5 | −0.29 | −0.29 | 10.3 | −2.987 | −2.987 |

The critical wind pressure is shown in Fig.16.28(c).

**Step 3:** *Design of purlin*

Span of purlin = 5 m

Spacing of purlin = 1.275 m

$\theta = 11.3°$

Load calculations:

Live load = 0.75 − (11.3 − 10)0.02
= 0.724 kN/m$^2$ > 0.4 kN/m$^2$

Dead load = 0.21 kN/m$^2$

Wind pressure = 1.616 × 1.6 = 2.586 kN/m$^2$

Load combinations:

1. DL + LL = 0.21 + 0.724 = 0.934 kN/m$^2$
2. DL + WL
   Normal to slope = −2.586 + 0.21cos11.3 = −2.38 kN/m$^2$
   Parallel to slope = 0.21 sin 11.3 = 0.041 kN/m$^2$

(a) Load combination 1: DL + LL

$$w_z = (0.934\cos 11.3) \times 1.275 = 1.168\,\text{kN/m}$$
$$w_y = (0.934\sin 11.3) \times 1.275 = 0.233\,\text{kN/m}$$

where $w_z$ is the load normal to $z$-axis, $w_y$ is the load normal to $y$-axis, and 1.275 is the spacing of the purlin. Due to continuity of purlins, factored bending moments and shear force are as follows:

$$M_z = 1.5 \times 1.168 \times 5^2/10 = 4.38\,\text{kN m}$$
$$M_y = 1.5 \times 0.233 \times 5^2/10 = 0.874\,\text{kN m}$$
$$SF_z = 1.5 \times 1.168 \times 5/2 = 4.38\,\text{kN}$$

Try MC100 for which the properties are as follows:

$$D = 100\,\text{mm};\ b_f = 50\,\text{mm};\ t_w = 5\,\text{mm};\ t_f = 7.7\,\text{mm}$$
$$I_{zz} = 192 \times 10^4\,\text{mm}^4$$
$$Z_{ez} = 37.3 \times 10^3\,\text{mm}^3,\ Z_{ey} = 7.71 \times 10^3\,\text{mm}^3$$
$$Z_{pz} = 43.83 \times 10^3\,\text{mm}^3,\ Z_{py} = 16.238 \times 10^3\,\text{mm}^3$$

Section classification:

$$b/t_f = 50/7.7 = 6.49 < 9.4$$
$$d/t_w = (100 - 2 \times 7.7)/5.0 = 16.92 < 42$$

Hence the section is plastic.

*Check for shear capacity*

As per clause 8.4 of IS 800,

$$A_v = (100 \times 5.0) = 500\,\text{mm}^2$$
$$= \frac{A_v f_{yw}}{\sqrt{3}\gamma_{m0}} = \frac{500 \times 250}{\sqrt{3} \times 1.10 \times 10^3} = 65.6\,\text{kN} > 4.38\,\text{kN}$$

Hence shear capacity is vey large compared to the shear force.

*Check for moment capacity*

$$M_{dz} = \frac{\beta_b Z_{pz} f_y}{\gamma_{m0}} = \frac{1 \times 43.83 \times 250 \times 10^3}{1.10 \times 10^6} = 9.96\ \text{KNm}$$

The above value should be less than

$$\frac{1.2 \times 37.3 \times 250 \times 10^3}{1.10 \times 10^6} = 10.17\ \text{KNm}$$

Hence $M_{dz} = 9.96\,\text{kNm} > M_z = 4.38\,\text{kNm}$

Hence the assumed section is safe.

$$M_{dy} = \frac{1 \times 16.238 \times 250 \times 10^3}{1.10 \times 10^6}$$
$$= 3.69\,\text{kNm}$$

The above value should be less than

$$\frac{1.2 \times 7.71 \times 250 \times 10^3}{1.1} = 2.10\,\text{kNm}$$

Hence $M_{dy} = 3.69\,\text{kNm} < M_y = 0.874\,\text{kNm}$

Hence the section is satisfactory.

*Check for biaxial bending*

$$\frac{M_z}{M_{dz}} + \frac{M_y}{M_{dy}} \le 1$$

Thus, $\dfrac{4.38}{9.96} + \dfrac{0.874}{3.69} = 0.68 < 1.0$

*Check for deflection*

Calculation for deflection is based on the serviceability condition, i.e, with unfactored imposed loads.

$$W = 1.168 \times 5 = 5.84\,\text{kN}$$
$$\delta = \frac{5WL^3}{384EI_z}$$
$$= \frac{5 \times 5.84 \times 1000 \times 5000^3}{384 \times 2 \times 10^5 \times 192 \times 10^4}$$
$$= 24.75\,\text{mm}$$

As per IS 800, Table 6, deflection limit is $\dfrac{L}{150}$

$$= 33.33\,\text{mm} > 24.75\,\text{mm}$$

Hence the deflection is within allowable limits.

(b) Load combination 2: DL + WL

$$w_z = 2.38 \times 1.275 = 3.035\,\text{kN/m}$$
$$w_y = 0.041 \times 1.275 = 0.052\,\text{kN/m}$$

Factored bending moments in this case are

$$M_z = 1.5 \times 3.035 \times 5^2/10 = 11.38\,\text{kNm} > M_{dz} = 9.96\,\text{kNm}$$
$$M_y = 1.5 \times 0.052 \times 5^2/10 = 0.195\,\text{kNm} < M_{dy} = 3.69\,\text{kNm}$$

Hence, the section is not safe. Let us adopt MC125, which has an

$$I_{zz} = 425 \times 10^4\,\text{mm}^2,$$
$$Z_{pz} = 77.88 \times 10^3\,\text{mm}^3 \text{ and } Z_{py} = 29.46 \times 10^3\,\text{mm}^3$$
$$M_{dz} = 1 \times 77.88 \times 250 \times 10^{-3}/1.1 = 17.7\,\text{kNm}$$
$$M_{dy} = 1 \times 29.46 \times 20 \times 10^{-3}/1.1 = 6.69\,\text{kNm}$$

Thus, the check for biaxial beding is

$$\frac{11.38}{17.7} + \frac{0.195}{6.69} = 0.67 < 1.0$$

Hence the section is safe.

*Check for deflection*

$$\delta = \frac{5 \times (3.035 \times 5) \times 1000 \times 5000^3}{384 \times 2.0 \times 10^5 \times 425 \times 10^4}$$
$$= 29.06 < 33.33\,\text{mm}$$

Note that the purlins at the edges and ridge of the building will be subjected to a local pressure of $1.4p \pm 0.5p$ [as per Table 5 of IS 875 (Part 3)-2015] instead of $1.6p$ taken in the preceding calculations. Hence, the purlins at the edges or at the ridge of the building have to be checked for this local pressure or closer spacing of purlins may be adopted at these locations.

**Step 4:** *Truss analysis and design*

Tributary area for each node of the truss:

Length of each panel along sloping roof

$$= \frac{1.25}{\cos 11.3^\circ} = 1.275 < 1.4\,\text{m}$$

Spacing of trusses = 5 m

Tributary area for each node of the truss = 5 × 1.275 = 6.375 m²

Imposed load calculations:

From IS 875 (Part 2)-1987,

Live load = 0.75 kN/m²

Reduction due to slope (see Table 3.7 and footnote 3 of Chapter 3)

$= (0.75 - 0.02 \times 1.3)2/3 = 0.483\,\text{kN/m}^2$

Load at intermediate nodes $W_2 = 0.483 \times 5 \times 1.25 = 3.02\,\text{kN}$

Load at end nodes $W_2/2 = 1.51\,\text{kN}$

(All these loads act vertically downwards.)

Maximum $C_{pe} \pm C_{pi}$ (critical wind loads to be considered for analysis):

| Wind Angle | Windward side ($W_3$) | | Leeward ($W_4$) | |
|---|---|---|---|---|
| | Intermediate nodes $W_3$ | End and apex nodes $W_3/2$ | Intermediate nodes $W_4$ | End and apex nodes $W_4/2$ |
| 0° | −16.48 | −8.24 | −9.27 | −4.64 |
| 90° | −13.29 | −6.645 | −13.29 | −6.645 |

*Loads in kN

All these loads act perpendicular to the top chord member of the truss.

*Forces in the members* The truss has been modelled as a pin jointed plane truss as shown in Fig. 16.28(d) and analysed using the software PLTRUSS developed by Computer Design Consultants, Chennai. The analysis results are tabulated in Table 16.6 [see truss configuration shown in Fig. 16.28(d) for member numbers]

Load factors and combinations (Table 4 of IS 800):

For dead + imposed load

$$1.5 \times \text{DL} + 1.5 \times \text{LL}$$

For dead + wind load

$$1.5 \times \text{DL} + 1.5 \times \text{WL}$$

Dead + imposed + wind loading case will not be critical as wind loads act in opposite direction to dead and imposed loads.

**Truss Reactions (kN)**

| Joint number | Case 1(DL + LL) | | Case 2(DL + WL(0)) | | Case 3(DL + WL(90)) | |
|---|---|---|---|---|---|---|
| | X | Y | X | Y | X | Y |
| 1 | 0 | 43.52 | 11.31 | −94.62 | 0 | −84.84 |
| 26 | 0 | 43.52 | 0 | −68.61 | 0 | −84.84 |

**TABLE 16.6** Member forces under factored loads in kN

| Member number | Dead load + Live load | Dead load + Wind load (0°) | Dead load + Wind load (90°) |
|---|---|---|---|
| 1. | 0 | 2.472 | 1.985 |
| 2. | −97.086 | 212.066 | 193.914 |
| 3. | −97.086 | 217.01 | 197.898 |
| 4. | −124.83 | 269.25 | 253.304 |
| 5. | −124.83 | 274.20 | 257.29 |
| 6. | −124.83 | 263.50 | 258.09 |
| 7. | −128.99 | 279.45 | 270.58 |
| 8. | −128.99 | 284.39 | 274.57 |
| 9. | −128.99 | 244.35 | 274.57 |
| 10. | −128.99 | 241.57 | 270.57 |
| 11. | −124.83 | 233.41 | 258.08 |
| 12. | −124.83 | 221.17 | 257.29 |
| 13. | −124.83 | 218.39 | 253.30 |
| 14. | −97.086 | 163.98 | 197.9 |
| 15. | −97.086 | 161.20 | 193.91 |
| 16. | 0 | 1.39 | 1.99 |
| 17. | 61.20 | −141.17 | −118.91 |
| 18. | 113.66 | −251.86 | −219.05 |
| 19. | 124.67 | −261.51 | −237.61 |
| 20. | 108.801 | −202.49 | −201.68 |
| 21. | 124.67 | −212.69 | −237.61 |
| 22. | 113.66 | −186.67 | −219.06 |
| 23. | 61.20 | −97.66 | −118.91 |
| 24. | −4.08 | 10.79 | 8.34 |
| 25. | −4.08 | 5.27 | 8.34 |
| 26. | −86.55 | 185.54 | 168.17 |
| 27. | −86.55 | 138.114 | 168.17 |
| 28. | 48.084 | −98.26 | −92.45 |
| 29. | 48.084 | −79.65 | −92.45 |
| 30. | −8.16 | 21.58 | 16.69 |
| 31. | −8.16 | 10.55 | 16.69 |
| 32. | −31.76 | 58.86 | 59.82 |
| 33. | −31.76 | 56.25 | 59.82 |
| 34. | 15.04 | −20.94 | −26.92 |
| 35. | 15.04 | −30.85 | −26.92 |
| 36. | −8.16 | 21.58 | 16.69 |
| 37. | −8.16 | 10.55 | 16.69 |
| 38. | −4.67 | 5.20 | 5.97 |
| 39. | −4.67 | 16.65 | 5.97 |
| 40. | −4.67 | 26.38 | 12.41 |
| 41. | −4.67 | −2.46 | 12.41 |

*(Contd)*

**TABLE 16.6** *(Contd)*

| Member number | Dead load + Live load | Dead load + Wind load (0°) | Dead load + Wind load (90°) |
|---|---|---|---|
| 42. | −12.24 | 32.38 | 25.03 |
| 43. | −12.24 | 15.82 | 25.03 |
| 44. | 5.225 | −13.82 | −10.69 |
| 45. | 5.225 | −6.75 | −10.69 |
| 46. | 8.16 | 21.58 | 16.69 |
| 47. | 8.16 | 10.55 | 16.69 |
| 48. | 21.245 | −72.17 | −46.70 |
| 49. | 21.245 | −17.793 | −46.70 |
| 50. | 27.62 | −89.024 | −59.74 |
| 51. | 27.62 | −26.03 | −59.74 |

**Step 5:** *Design of top chord member*

Factored compressive force = 128.99 kN

Factored tensile force = 284.39 kN

Trying two ISA 75 × 75 × 6 mm @ 0.136 kN/m

Sectional properties:

Area of cross section $A = 2 \times 866 = 1732\,\text{mm}^2$

Radius of gyration $r_{zz} = 23\,\text{mm}$, $I_y = 45.7 \times 10^4\,\text{mm}^4$, $c_y = 20.6\,\text{mm}$

Assuming 8-mm thick gusset plate,

$$I_y = 2[45.7 \times 10^4 + 866\,(4 + 20.6)^2] = 196.21 \times 10^4\,\text{mm}^4$$

$$r_y = \sqrt{(196.21 \times 10^4/1732)} = 33.66\,\text{mm}$$

Section classification:

$$\varepsilon = (250/f_y)^{0.5} = (250/250)^{1/2} = 1.0$$

$$b/t = 75/6 = 12.5 < 15.7\varepsilon$$

∴ the section is semi-compact.

As no member in the section is slender, the full section is effective and there is no need to adopt reduction factor.

Maximum unrestrained length = $L$ = 1275 mm

$KL = 0.85 \times L = 0.85 \times 12.75 = 1083.75\,\text{mm}$

*Note*: The effective length of top chord member may be taken as 0.7–1.0 times the distance between centres of connections as per clause 7.2.4 of IS 800. We have assumed the effective length factor as 0.85.

$$\lambda_y = 1083.75/23 = 47.12 < 180$$

Hence $\lambda_y$ is within the allowable limits. From Table 9c of the code for $KL/r = 47.12$ and $f_y = 250$ MPa,

$f_{cd} = 187.32\,\text{N/mm}^2$

Axial capacity = 187.32 × 1732/1000

= 324.4 kN > 128.99 kN

Hence, section is safe against axial compression.

Axial tension capacity of the section

= 1732 × 250/1.10

= 393.64 kN > 284.39 kN

*Note*: Design stength governed by tearing at the net section should also be checked.

Hence, section is safe in tension.

*Note*: Though a smaller section may be chosen, this section is adopted to take care of handling stresses.

**Step 6:** *Design of bottom chord member*

Factored compressive force = 202.49 kN

Factored tensile force = 108.801 kN

Try two L 100 × 100 × 8 @ 0.242 kN/m.

Sectional properties:

Area of cross section A = 2 × 1540 = 3080 mm²

Radius of gyration $r_z = 30.7\,\text{mm}$, $I_y = 145 \times 10^4\,\text{mm}^4$, $C_y = 27.6\,\text{mm}$

Assuming a 10-mm thick gusset plate

$$I_y = 2[145 \times 10^4 + 1540\,(5 + 27.6)^2] = 6.173 \times 10^6\,\text{mm}^4$$

$$r_y = \sqrt{(6.173 \times 10^6/3080} = 44.77\,\text{mm}$$

Section classification:

$$b/t = 90/6 = 15 < 15.7\varepsilon$$

∴ the section is semi-compact.

Axial tension capacity of the selected section

$= 3080 \times (250/1.10) \times 10^{-3}$

= 700 kN ≫ 108.801 kN

Hence, section is safe in tension. Providing longitudinal tie runner at every bottom node of the truss,

Maximum unrestrained length = $L$ = 5000 mm

$r_y = 44.71\,\text{mm}$

$\lambda_y = 5000 \times 0.85/44.71 = 95$

From Table 9c of IS 800 for $\lambda_y = 95$ and $f_y = 250$ MPa,

$f_{cd} = 114\,\text{N/mm}^2$

Axial capacity = 114 × 3080/1000 = 351.12 kN > 202.49 kN

Hence, section is safe against axial compression also.

*Note*: It may be economical to adopt unequal angles for the top and bottom chord members. However, many unequal angle sections are not readily available in the market. Hence, equal angle sections have been used in the truss in this example.

**Step 7:** *Design of web member*

Maximum compressive force = 98.26 kN

Maximum tensile force = 185.54 kN

Try ISA 90 × 90 × 6.0 with

$A = 1050\,\text{mm}^2$; $r_{zz} = 27.7\,\text{mm}$; $r_{vv} = 17.5\,\text{mm}$

Section classification:

$$b/t = 90/6 = 15 < 15.7\varepsilon$$

Hence, the section is not slender.

Length of member = 1768 mm

Assuming the angle is loaded through one leg, slenderness ratio $\lambda_e$ for flexural torsional buckling strength of single angle in compression through one of its legs,

$$\lambda_e = \sqrt{k_1 + k_2\lambda_{vv}^2 + k_3\lambda_\phi^2}$$

From Section 7.5.1 and Tabe 12 of IS 800,

$k_1 = 0.7$; $k_2 = 0.6$; $k_3 = 5$

$$\lambda_{vv} = \frac{(L/r_{vv})}{\varepsilon\sqrt{(\pi^2 E)/250}} \text{ and } \lambda_\phi = \frac{b_1 + b_2}{\varepsilon\sqrt{\pi^2 E/250} \times 2t}; \varepsilon = \left(\frac{250}{250}\right)^{0.5}$$

$= 1.0$

$L/r_{vv} = 0.85 \times 1768/17.5 = 85.87 < 180$

$$\lambda_{vv} = \frac{85.87}{\sqrt{\frac{\pi^2 \times 2 \times 10^5}{250}}} = 0.97$$

$b_1 = b_2 = 90$

$$\lambda_\phi = \frac{90 + 90}{\sqrt{\frac{\pi^2 \times 2 \times 10^5}{250}} \times 2 \times 6} = 0.169$$

$$\lambda_e = \sqrt{0.7 + 0.6 + 0.97^2 + 5 + 0.169^2}$$

$= 1.186$

$\alpha = 0.49$

$\phi = 0.5[1 + 0.49\,(1.186 - 0.2) + 1.186^2]$

$= 1.445$

$$\therefore f_{cd} = \frac{250/1.10}{1.445 + [1.445^2 - 1.186^2]^{0.5}} = 100.1\text{ N/mm}^2$$

Design compressive strength = 1050 × 100.1/1000

= 105.1 kN > 98.26 kN

*Note*: If we use Table 9c, for $f_y$ = 250 MPa and $L/r$ = 85.87, we get

$$f_{cd} = 127.20\text{ N/mm}^2$$

Thus by using the equations in clause 7.5.1.2, in this case, there is a reduction of 27% in the value of design compressive stresses. Hence, the section is safe in compression.

Tensile capacity of the section = (250/1.1) × 1050/1000

= 238.64 kN > 185.54 kN

Hence ISA 90 × 90 × 6 is adequate for the web member.

*Note* Web members away from the support have less axial force. However, their length will be more. If desired, these may be designed and a smaller section be adopted.

**Step 8:** *Check for deflection*

Maximum deflection from computer output = 20.07 mm

Allowable deflection as per Table 6 of IS 800 : 2007

= span/240 = 20000/240 = 83.33 mm > 20.07 mm

Hence the deflection is within allowable limits.

**Step 9:** *Design of rafter bracing members*

Considering the layout of the rafter bracing as shown in Fig. 16.28(e),

Design wind pressure = 1.616 kN/m²

Maximum force coefficient = –1.6

Factored wind load on rafter bracing

= 1.5 × 1.616× 1.6 × 3.825 × 5/2 × sec 11.3°

= 37.8 kN

Length of bracing = $\sqrt{(3825^2 + 5000^2)}$ = 6295.29 mm

Try 90 × 90 × 6, $A$ = 1050 mm², $r_{min}$ = 17.5 mm, and

$L/r$ = 6295.29 /17.5 = 359.7 < 400 (Table 3 of IS 800 : 2007)

In the X bracing system, as shown in Fig. 16.28(e), the compression bracing will buckle and only the tension bracing will be effective. Also, the bracing members are usually bolted to the trusses at site.

Axial tensile capacity:

Design strength of member due to yielding of gross section

$T_{dg} = A_g f_y/\gamma_{m0}$

= 1050 × (250/1.1)/1000

= 238.64 kN > 37.8 kN

Design strength due to rupture of critical section $T_{dn} = \alpha A_n f_u/\gamma_{m1}$

$\alpha$ = 0.6 (Assuming two bolts of 16 mm diameter at the ends)

$A_n$ = 1050 – 18 × 6 = 942 mm²

$T_{dn}$ = 0.6 × 942 × (410/1.25) × 1000 = 185 kN > 37.8 kN

Hence L 90 × 90 × 6 is safe. The member has been found to be safe for block shear failure.

*Note*: The forces in the bracing members are often small and rarely govern the design; but their slenderness limitations decide the size because of their long length.

**Step 10:** *Design of tie runner*

Portion of wind load from gable end along the ridge will be transferred as axial load to tie runners provided along the length of building at tie level. Assume three intermediate gable end columns at a spacing of 5 m.

Wind load on cladding:

$L/w$ = 50/20 = 2.5 < 4 and $h/w$ = 11/20 = 0.55 > 1/2

From Table 4 of IS 875 (Part 3)-2015, external wind pressure = +0.7, with internal pressure of ±0.5, maximum pressure = 1.2$p$

Factored wind load on intermediate runner

= 1.5 × 1.616 × 1.2 × 5 × (8/2 + 3/2)

= 80 kN (Tension)

Min $r$ required = 5000 × 0.85/250 = 17 (maximum allowable $KL/r$ = 250)

Try L 90 × 90 × 6, with $A$ = 1050 mm², $r_{min}$ = 17.5 mm,

Design strength due to yielding of gross section

$T_{dg} = A_g f_y/\gamma_{m0}$

= 1050 × (250/1.1) /1000 = 238.64 kN > 80 kN

The design strength due to rupture of critical section (see the design of rafter bracing)

= 185 kN > 80 kN

Hence L 90 × 90 × 6 is adequate.

**Step 11:** *Design of Side Runner*

Assuming that the side sheetings are provided at a spacing of 1.25 m,

Span of side runner = 5 m

*Calculation of loads*

(a) Vertical loads

Self weight of side runner = 9.56 kg/m (MC100)

Weight of side sheeting = 5 kg/m$^2$

Thus weight of GI sheeting = 5 × 1.25 = 6.25 kg/m

Total $w_y$ = 15.81 kg/m

= 0.16 kN/m

(b) Wind loads

Maximum wind force coefficient = 1.2

Wind load UDL = 1.2 × 1.616 × 1.25

= 2.424 kN/m

Factored loading

$w_y = 1.5 \times 0.16 = 0.24$ kN/m

$w_z = 1.5 \times 2.424 = 3.64$ kN/m

Try MC100, which has the following properties:

$D = 100$ mm; $b_f = 50$ mm; $t_w = 5.0$ mm; $t_f = 7.7$ mm;

$I_{zz} = 192 \times 10^4$ mm$^4$;

$Z_{ez} = 37.3 \times 10^3$ mm$^3$; $Z_{pz} = 43.83 \times 10^3$ mm$^3$;

$Z_{ey} = 7.71 \times 10^3$ mm$^3$

$Z_{py} = 16.238 \times 10^3$ mm$^3$

Assuming continuity of side runner,

$BM_z = wl^2/8 = 3.64 \times 5^2/10 = 9.1$ kNm

$BM_y = 0.24 \times 5^2/10 = 0.6$ kNm

Shear capacity:

From the design of purlin, shear capacity of the section = 65.6 kN

$V = 3.64 \times 5/2 = 9.1$ kN < 65.6 kN

Hence safe against shear.

*Check moment capacity*

From the calculations of purlin design

$M_{dz} = 9.96$ kNm > 9.1 kNm

$M_{dy} = 3.69$ kNm > 0.6 kNm

Check for biaxial bending

$$\frac{9.1}{9.96} + \frac{0.6}{3.69} = 1.08 > 1.0$$

Hence, the section may be revised to ISMC 125. The deflection has been calculated (similar to the purlin) and found to be within the codal limits. Note that for both the purlin and side runner, biaxial bending has been considered as per the code. If the sheeting is assumed to resist the in-plane loads, a smaller size of purlin and side runner may be sufficient to resist the bending moment in the $z$-direction.

**Step 12:** *Design of eave girder*

Eave girders are provided at tie level and along both longitudinal and transverse directions to carry the wind forces at the tie level (see Fig. 16.7). Here the design of an eave girder along transverse direction is only considered. The design of an eave girder along the longitudinal direction is similar to the one in the transverse direction. The eave girder is in the form of a truss by connecting the bottom chord members of the truss to the gable columns at one end and main truss bottom points at the other end. The configuration of the eave girder is given in Fig. 16.28(f). In this example we are assuming a truss at both the ends of the building.

Wind loads on eave girders:

Design wind pressure = 1.616 kN/m$^2$

Maximum force coefficient = 1.2

Eaves girder bracings are connected between gable columns which are assumed to be spaced at 5 m centre to centre and truss bottom joints.

Wind load on internal joint = [1.616 × 1.2 × 5 × (8/2 + 3/2)] × 1.5

= 80 kN

Wind load on end joints = 40 kN

Reaction = (3 × 80 + 2 × 40) /2 = 160 kN

Using the method of joints, the maximum force in the bracing at the end of the eave girder $F_{br}$ = (160 − 40)/cos 26.565 = ± 134.2 kN

Members of the girder are subjected to reversal of stresses and hence they have been checked for compression as well as tension.

Length of the bracing = $\sqrt{2.5^2 + 5^2} = 5.59$ m

Required $r_{min}$ = 5590/350 = 15.97 mm

Try section L 150 × 150 × 10 with the following properties:

$A = 2920$ mm$^2$; $r_{min} = 29.8$ mm

Section classification:

$$\varepsilon = (250/f_y)^{0.5} = 1.0$$

Flange:

$B/T = 150/10 = 15 < 15.7\varepsilon$

∴ it is semi compact.

Maximum unsupported length = 5590 mm

$$\lambda = \frac{5590 \times 0.85}{29.8} = 160$$

From Table 9c of IS 800 with $\lambda = 185$ and $f_y = 250$ MPa;

$f_{cd} = 53.3$ MPa

Axial capacity in compression = 53.3 × 2920/1000

= 155.6 kN > 134.2 kN

*Note*: The capacity in compression has to be checked as per the equations given in clause 7.5.1.2 of the code, since the connection is through one leg only. For simplicity, we have found the capacity using Table 9c. The capacity as calculated using clause 7.5.1.2 is also safe. Hence the section is safe.

Axial tensile capacity (due to yielding of gross section)

= 2920 × 250/(1.1 × 1000)

= 663.64 kN ≫ 134.2 kN

Design strength due to rupture of critical section.

$T_{dn} = \alpha A_n f_u / \gamma_{m1}$

= 0.6 × (2920 − 18 × 10) × 410 / (1.25 × 1000)

= 539 kN > 134.2 kN

The section has to be checked for block shear failure also. Note that the tension capacity is very high compared to the capacity in compression. It is due to the long length of the member. In order to reduce the member size, X bracing could be adopted and the member may be designed to be effective only in tension.

Note that the top and bottom chords of the eave girder are the bottom chord members of the main truss and carry a force of ±160 kN. The bottom chord members have been designed for a compressive force of 202.49 kN and a tensile force of 108.8 kN (tensile capacity 700 kN). Hence, the sections are safe under gable wind loads. Similarly, the vertical members are tie runners (with a maximum forceof 80 kN), which have been designed already.

Slotted holes may be provided in the base angles to permit expansion of the truss. For a change of 34°C (see Table 3.11 of Chapter 3), the maximum change in length will be 0.000012 × 34 × 20,000 = 8.16 mm. The slot for a 20-mm bolt may be 22 mm wide and 30 mm long.

**EXAMPLE 16.2:** Design a Vierendeel girder for covering the industrial building given in Example 16.1. Use Fe 410 grade steel.

**SOLUTION:**

**Step 1:** *Calculation of loads*

*Dead loads*

GI sheeting = 0.085

Fixings = 0.025

Services = 0.1

Total = 0.21 kN/m$^2$

For 5 m bays,

Roof dead load = 0.21 × 5 × 20 = 21 kN

Weight of purlin (assuming 70 N/m$^2$) = 0.07 × 5 × 20 = 7 kN

Self weight of the girder (assumed) = 0.5 kN/m × 20

= 10 kN

Therefore,

Total dead load = 38 kN

**Distribution of dead load**

Assuming the configuration of Vierendeel girder with eight panels as shown in Fig. 16.29, intermediate nodal dead load ($w_1$) = 38/8 = 4.75 kN

Dead load at end nodes = 4.75/2 = 2.375 kN

**Imposed load**

Live load = 0.75 kN/m$^2$ as per IS 875 (Part 2)-1987

For 5-m bays,

Total LL = 0.75 × 5 × 20 = 75 kN

For intermediate nodes,

LL = 75/8 = 9.375 kN

At end nodes,

LL = 9.375/2 = 4.69 kN

Note that in this example only dead and live loads have been considered, whereas in practice the effect of wind loads should also be considered. However, the wind loading may not be critical because of the flatness of the roof.

**Step 2:** *Analysis to find the forces in the members*

The Vierendeel girder has been modelled as a plane frame as shown in Fig. 16.29 (assuming all the members have the same cross section) and analysed using the plane frame program developed by Computer Design Consultants, Chennai. The bending moment, shear force, and axial force values as obtained from the computer program are shown in Fig. 16.29.

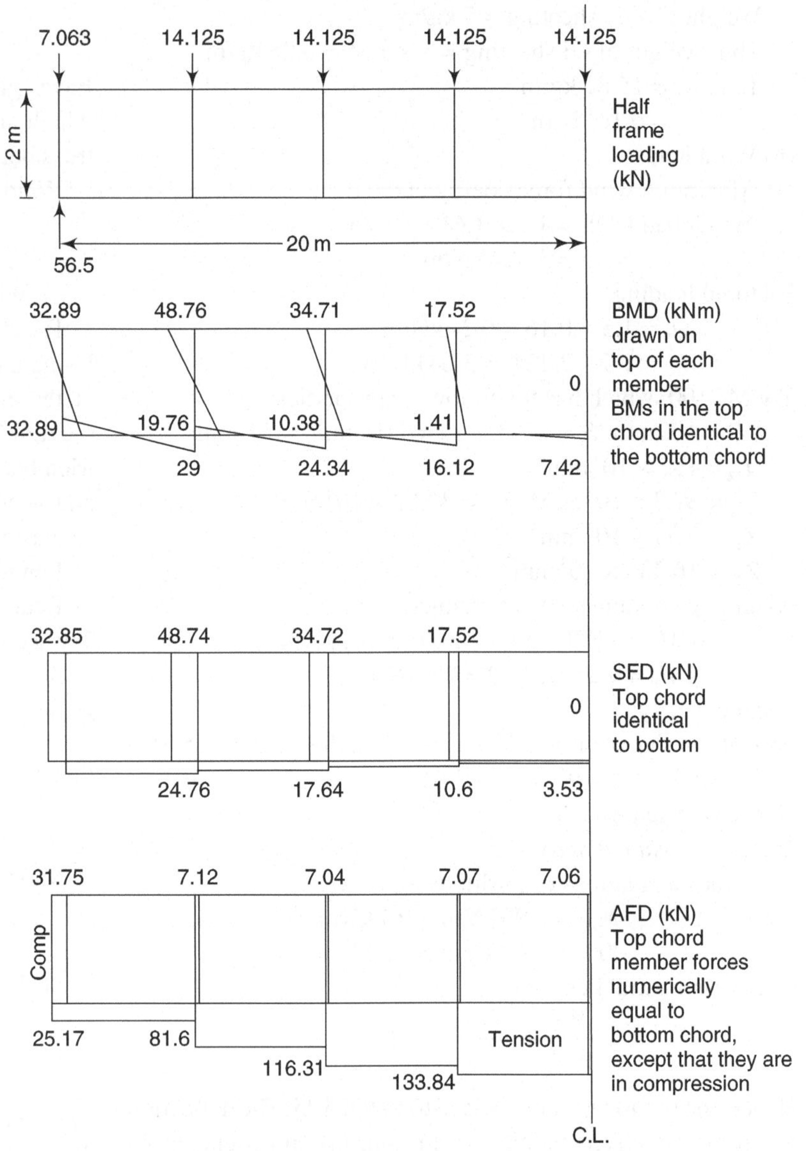


**FIG. 16.29** Analysis results of the girder

**Step 3:** *Design of members*

*Top chord members*

Maximum bending moment = 32.89 × 1.5 = 49.34 kNm

Maximum axial force = 133.84 × 1.5

= 200.76 kN (compression)

Shear force = 24.76 × 1.5 = 37.14 kN

(a) *Trial section*

Select SHS 180 × 180 × 6 mm

$A = 4083\ \text{mm}^2$, $r_{yy} = 70.6\ \text{mm}$

To obtain the effective length factor, the stiffness of the members framing into the column and the column itself has to be calculated. Since the second moments of area of all the members are assumed to be same, for the purpose of calculating relative stiffness they may be set equal to unity. Thus, $K_c = 1/2.5 = 0.4$, $K_b = 1/2 = 0.5$. Using Appendix E of the code,

$$\beta_1 = \Sigma K_c/\Sigma(K_c + K_b)$$
$$= 0.4/(0.4 + 0.5) = 0.444$$
$$\beta_2 = (0.4 + 0.4)/(0.4 + 0.4 + 0.5) = 0.615$$

Hence, efective length factor (considering the frame as sway frame)

$$K = \left[\frac{1-0.2(\beta_1+\beta_2)-0.12\beta_1\beta_2}{1-0.8(\beta_1+\beta_2)+0.6\beta_1\beta_2}\right]^{0.5}$$

$$= \left[\frac{1-0.2(0.444+0.615)-0.12\times0.444\times0.615}{1-0.8(0.444+0.615)-0.6\times0.444\times0.615}\right]^{0.5}$$

$$= \left(\frac{0.7554}{0.3166}\right)^{0.5} = 1.545$$

$KL = 1.545 \times 2500 = 3862.5$ mm

$KL/r_y = 3862.5/70.6 = 54.71$

From Section 13.9.1 of Chapter 13,

$$P_{eq} = P + 2M_z/d = 200.76 + 2 \times 49.34/0.18 = 748.98 \text{ kN}$$

From Table 10 of the code, using curve *a*, for

$$KL/r_y = 54.71 \text{ and } f_y = 250 \text{ MPa}$$

from Table 9a of the code,

$$f_{cd} = 200.29 \text{ MPa}$$

Hence capacity $P_{dz} = 200.29 \times 4083/1000 = 817.78$ kN > 748.98 kN

Hence use SHS 180 × 180 × 6 mm

*Section properties*

$A = 4083$ mm$^2$, $I_z = 2036.52 \times 10^4$ mm$^4$

$I_y = 2036.52 \times 10^4$, $t = 6$ mm

$Z_z = 226.28 \times 10^3$ mm$^3$

$Z_p = 264.35 \times 10^3$ mm$^3$

(b) *Cross section classification*

$$\varepsilon = \sqrt{250/f_y} = \sqrt{250/250} = 1.0$$

*Outstand flanges* (see Table 8.7),

$$b/t_f = 180/6 = 30 < 42\varepsilon^2$$

Hence, the flange is **plastic**.

*Web*

$$d = H - 2t_f = 180 - 2 \times 6 = 168 \text{ mm}$$
$$d/t_w = 168/6 = 28 < 67.1\varepsilon$$

Hence, the cross section is also plastic.

(c) *Member buckling resistance in compression*

As calculated in the previous section, the capacity of the member

$$P_{dz} = 817.78 \text{ kN} > 200.76 \text{ kN}$$

(d) *Check for local capacity of the cross section to the combined effects*

Nominal plastic shear resistance

$$V_p = A_v f_{yw}/(\sqrt{3}\gamma_{m0})$$

$$A_v = Ah/(b+h)$$
$$= 4083 \times 180/(180 + 180) = 2041.5 \text{ mm}^2$$
$$V_p = 2041.5 \times 250/(\sqrt{3} \times 1.1 \times 1000)$$
$$= 267.87 \text{ kN} \gg 37.14 \text{ kN}$$

Hence, the moment capacity of the section need not be reduced (clause 9.2.1 of the code). The interaction equation is

$$(N/N_d) + (M_z/M_{dz}) \le 1.0$$
$$N_d = A_g f_y/\gamma_{m0} = 4083 \times 250(1.1 \times 1000) = 927.95 \text{ kN}$$
$$M_{dz} = \beta_b Z_p f_y/\gamma_{m0}$$

where $\beta_b = 1.0$ for the plastic section.

Hence $M_{dz} = 1 \times 264.35 \times 10^3 \times 250/(1.1 \times 10^6) = 60.08$ kN.

Thus,

$$\frac{200.76}{927.95} + \frac{49.34}{60.08} = 0.22 + 0.82 = 1.04$$

Though the value is greater than 1.0, we may notice that the maximum axial force and maximum bending moment do not occur at the same section and hence the member is safe.

(e) *Member buckling resistance in bending*

$$M_d = \beta_b Z_p f_{bd}$$

$\beta_b = 1.0$ for the plastic section and $h/t_f = 180/6 = 30$

Assuming that purlins are provided at each nodal point, which provide lateral restraint at these points.

$$KL/r = 0.85 \times 2500/70.6 = 30.1$$

From Table 14 of the code for $KL/r = 30.1$ and $h/t_f = 30$,

$$f_{cr,b} = 2462 \text{ N/mm}^2$$

From Table 13a of the code, for $f_{cr,b} = 2462$ N/mm$^2$ and

$f_y = 250$ N/mm$^2$

$f_{bd} = 227.3$ N/mm$^2$

$$M_{dz} = 1 \times 264.35 \times 10^3 \times 227.3/10^6$$
$$= 60.08 \text{ kN m} < 49.34 \text{ kN m}$$

Hence, the section is safe.

(f) *Check for overall buckling*

$$f_{cr,z} = \pi^2 E/(KL/r)^2 = \pi^2 \times 2 \times 10^5/(54.71)^2 = 659.47 \text{ N/mm}^2$$
$$\lambda_y = \lambda_z = \sqrt{(250/659.47)} = 0.616$$
$$P_{dz} = 817.78 \text{ kN}$$

*Determination of moment amplification factor*

For flexural buckling failure,

$$K_z = K_y = 1 + (\lambda_z - 0.2)n_z \le 1 + 0.8n_z$$
$$n_z = P/P_{dz} = 200.76/817.78 = 0.245$$

Hence, $K_z = K_y = 1 + (0.616 - 0.2) \times 0.245 = 1.102 \le 1 + 0.8 \times 0.245 = 1.196$

$$\psi = -29/32.89 = -0.881$$
$$C_{mz} = 0.6 + 0.4\psi \ge 0.4$$

$\therefore C_{mz} = 0.6 + 0.4 \times -0.881 = 0.246 < 0.4$ Hence $C_{mz} = 0.4$

Check with interaction formula, i.e.,

$$\frac{P}{P_{dz}} + 0.6\frac{C_{my}M_y}{M_{dy}} + K_z\frac{C_{mz}M_z}{M_{dz}} \le 1.0$$

$$\frac{200.76}{817.78} + 0 + \frac{1.102 \times 0.4 \times 49.34}{60.08} = 0.245 + 0.362$$
$$= 0.607 < 1.0$$

Hence, the section is safe.

Similar calculations have been done for vertical and bottom chord members and the section is found to be safe. Hence, an 180 × 180 × 6 mm SHS is required for all the members. The bottom flange of the Vierendeel girder should be stiffened by additional plates at the end of the girder, in order to connect the truss to the steel stanchions or base plates (if the girder is supported on masonry construction) by means of bolts. Typical end details at the top and bottom of the girder are shown in Fig. 16.30.

**Step 4:** *Deflection check*

The maximum deflection from the computer output is 46.66 mm.

Allowable deflection (as per Table 6 of the code) = span/180 = 20 × 1000/180 = 111.1 mm > 46.66 mm.

Hence, the section is safe.

**Step 5:** *Self weight check*

Weight of top and bottom chords = 2 × 20 × 320.5 = 12820 N

Vertical members = 9 × 2 × 320.5 = 5769 N

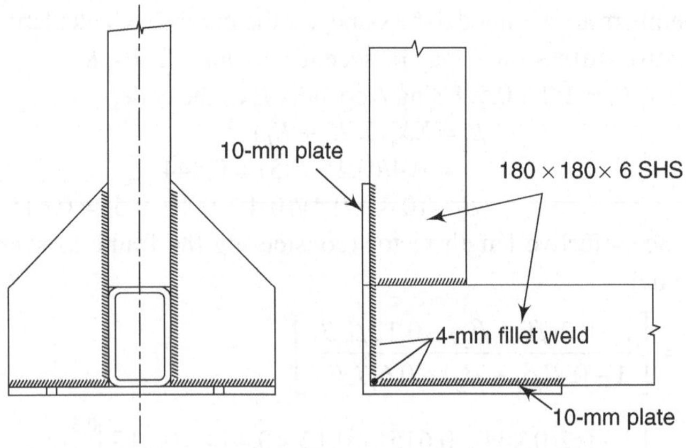


**FIG. 16.30** Typical end details of the Vierendeel girder

Total weight = 18589 N = 18589/(20 × 1000) = 0.93 kN/m

This is higher than the assumed weight of 0.5 kN/m. However, since the weight of the truss is only 8.85% of the total weight, this additional weight may not affect the design.

## SUMMARY

Structural steel is often the material of choice for the construction of single storey industrial buildings, which constitute the major percentage of the total number of steel structures built around the world. The planning and design of these buildings require the knowledge of several items such as site condition, plant layout and work flow, availability of new materials and waste disposal facilities, HVAC equipment, crane types and capacity, future expansion plans, and budget and project schedule. In particular, the structural engineer should select items such as roofing and walling material, bay width, structural framing system, and type and shape of trusses.

A number of factors have to be considered while selecting the roofing (decking) or wall (cladding) material. Steel, aluminium, galvanized iron, asbestos, stainless steel, and ferrocement sheets can be used as cladding or decking material. Metal roofing can be classified by the method of attachment to supports. Through-fastened roofs are directly attached to purlins and hence provide lateral stability to the purlins. However, standing- seam roofing (which is used extensively in USA) is connected indirectly by concealed clips formed in to the seam and requires a separate system of purlin bracings. Some details about the various types of sheeting are provided. Some guidelines for fixing the bay width of industrial buildings are also given.

Depending on the structural framing system adopted, industrial buildings may be classified as braced frames and unbraced frames. In braced frames, trusses rest on columns with hinge type connections and stability is provided by bracings in three mutually perpendicular directions. To reduce the bending moment in columns, sometimes knee braces are provided. Hence, a discussion on knee braces has also been included.

Unbraced frames in the form of portal frames offer several advantages over braced frames. These portal frames are often prefabricated in the factory and assembled at site and termed as pre-engineered buildings or metal building systems. The lateral stability of portal frames in the longitudinal direction is provided by bracings. Though these portal frames may be designed economically by using plastic methods of structural analysis, elastic methods can also be employed. It is important to design and detail the connections of portal frames properly such that they are capable of developing the required design moment resistance. Braced frames may be economical than unbraced frames in situations where the labour costs are low.

Since the weight of purlins may be equal to the weight of trusses, they should be properly designed. Channels, angles, tubes, cold-formed C-, Z- or sigma sections are employed as purlins. The functions of girts are similar to purlins except that they are used in the walls. The eave strut is located at the intersection of the roof and exterior wall and has to be designed carefully.

A triangulated framework of pin-ended members is called a truss. In most situations, the loads are applied at the nodal points of trusses by purlins. When the purlins are placed in between the nodal points, the top chord members have to be designed for the secondary bending moments. A number of hand methods are available for the analysis of trusses (e.g., method of sections, method of joints, and graphical method). With the availability of digital computers and software packages, the trusses are often analysed using these software packages. However, hand methods will be quite useful to check the results of the computer output (especially the errors made in the input data to the programs). The software packages require the member sizes of the truss to be given as input. Hence, some guidelines to assume the initial member sizes are given.

The various types of trusses and their configurations are described. It is to be noted that these configurations may change from project to project and is often selected based on aesthetics, economy, and performance. Some guidelines are also provided to select the pitch, depth, and spacing of trusses. The loads and load combination to be considered are briefly discussed and the various steps involved in the design of truss members are given. Sometimes it may be necessary to adopt a bigger section than those indicated by the actual design to take care of transportation and handling stresses. A brief discussion on the connections has been included. A few sketches showing the connection details of welded and bolted trusses are provided. A brief description of open-web steel joists which can be used for roofing industrial buildings or as replacement for purlins when the spacing of trusses is more than 8–12m is also given. Floor plates, end bearings, and slabs on grade are also discussed briefly. The design of various members of trusses is illustrated through examples.

It has to be noted that with the knowledge of the behaviour and design of plates, beams, columns, tension members, compression members, and beam-columns, different types of structures (e.g., towers, multi-storey buildings, water tanks, bridges, chimneys, etc.) can be designed, using the appropriate code of practice.

## MULTIPLE-CHOICE QUESTIONS

16.1 The usual range of bay width is
(a) 2–4 m (b) 4–8 m (c) 20–25 m (d) 10–15 m

16.2 In a long building, bracings should be provided in
(a) the end bays only
(b) every fourth or fifth bay
(c) end bays and the middle bay
(d) every 8th bay

16.3 It is economical to use the following for exterior columns of braced frames:
(a) Wide flange sections
(b) Square hollow sections
(c) ISMB sections
(d) Circular hollow sections

16.4 When knee braces are provided, the analysis model should
(a) include knee braces, columns, and trusses together
(b) include trusses only
(c) include columns and trusses only
(d) none of these

16.5 A disadvantage of unbraced portal frame is
(a) high material cost
(b) susceptibility to differential settlement
(c) provision of tie beams to take care of horizontal reaction
(d) all of these

16.6 The typical span of portal frames is
(a) 5–10 m (c) 20–30 m
(b) 10–20 m (d) 30–40 m

16.7 The typical bay spacing of portal frames is
(a) 2–4 m (c) 5–20 m
(b) 4.5–10 m (d) 10–20 m

16.8 The weight of purlins in a steel shed will be
(a) approximately equal to the weight of trusses
(b) very much greater than the weight of trusses
(c) much smaller than the weight of trusses
(d) none of these

16.9 Cold formed Z or C section purlins have thickness in the range of
(a) 4–6 mm (c) 6–8 mm
(b) 1.6–4 mm (d) 3.5–5.5 mm

16.10 The normal pitch for a truss with corrugated AC sheeting is
(a) 1/6 to 1/12 (c) 1/24 to 1/12
(b) 1/3 to 1/6 (d) 1/48 to 1/12

16.11 For an economical spacing of roof truss, if $C_t$, $C_r$, and $C_p$ are the cost of truss, roof covering, and purlin respectively, then
(a) $C_t = C_r + C_p$ (c) $C_t = 2C_r + C_p$
(b) $C_t = C_r + 2C_p$ (d) $C_t = 2C_r + 1.5C_p$

16.12 Approximate weight of roof truss (N/m$^2$) is given by
(a) $6.6L + 20$ (c) $5L - 25$
(b) $3.5L - 50$ (d) $3L + 40$

16.13 Wind load on a steel truss for an industrial building will depend on
(a) location of the structure
(b) shape of the structure
(c) height of the structure
(d) all of these

16.14 Secondary stresses in roof trusses are induced due to
(a) rigidity of joints, especially in welded trusses
(b) purlins placed at intermediate points on rafters
(c) eccentricity of bolt line with respect to centroid of the member cross section
(d) all of these

16.15 Minimum imposed load on roof truss should be greater than
(a) 0.2 kN/m$^2$ (c) 0.6 kN/m$^2$
(b) 0.4 kN/m$^2$ (d) 0.75 kN/m$^2$

16.16 The top chord of a roof truss is inclined at an angle of 25°. No access is provided for maintenance. The imposed load to be considered in the design is
(a) 0.2 kN/m$^2$ (c) 0.6 kN/m$^2$
(b) 0.45 kN/m$^2$ (d) 0.75 kN/m$^2$

## EXERCISES

16.1 Design a roof truss for a railway platform of size 30 × 12 m situated in Chennai and as shown in Fig. 16.31. Assume asbestos cement sheetings.

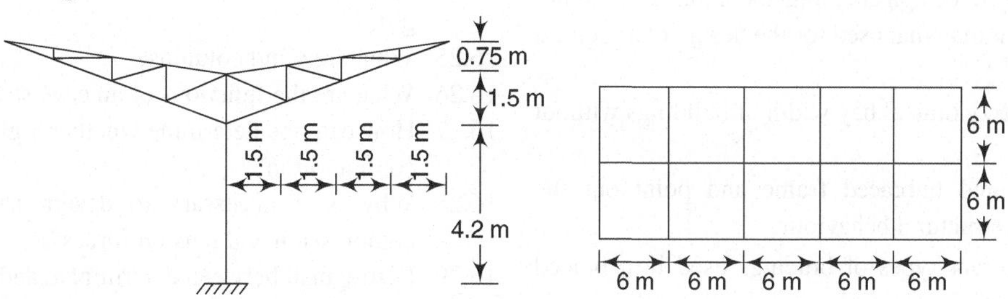


**FIG. 16.31**

16.2 An industrial building is shown in Fig. 16.32. The frames are at 5 m centres and the length of the building is 40 m. The purlin spacing of the roof is as shown in Fig. 16.32(b). The building is situated in Delhi. Assume live and wind loads as per IS 875 (part 2 and part 3) and the roof is covered with GI sheeting. Design the roof truss using angle members and gusseted joints. The truss is to be fabricated using welded joints in two parts for transport and assembled at site using bolted joints at A, B, and C as shown in Fig. 16.32.

16.3 Design the members of the truss of the previous exercise using tubular members.

16.4 A flat roof building of 18 m span has 1.5-m deep trusses at 5 m centres. The total dead load is 0.7 kN/m$^2$ and the imposed load is 0.75 kN/m$^2$. Design the truss using angle sections with welded internal joints and bolted field splices.

16.5 Design a Vierendeel girder for covering the building given in the previous exercise.

16.6 An industrial building is made of 10 portal frames spaced 6 m apart. The frame has a span of 20 m and 4 m rise with a column height of 6 m above ground level. Assuming the column bases are hinged, design the frame for dead, live, and wind loads as per IS 875.

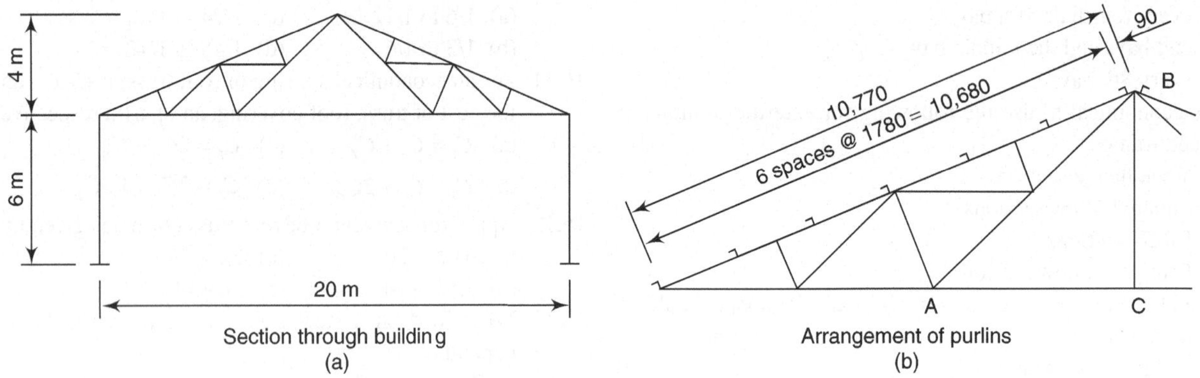


**FIG. 16.32**

## REVIEW QUESTIONS

16.1 List the items that are to be considered while planning and designing an industrial building.

16.2 List the items to be considered while selecting a cladding/decking system.

16.3 Name some of the cladding/decking materials that are used in practice.

16.4 What are the purposes of structural decking?

16.5 Under what condition will the decking provide lateral stability to the top flange of purlins?

16.6 What are the advantages and drawbacks of the following:
(a) Aluminium decking
(b) GI sheeting
(c) Asbestos sheeting
(d) Ferrocement sheeting

16.7 State the difference between through-fastened roofing and standing-seam roofing.

16.8 What are the advantages and disadvantages of standing-seam roofing?

16.9 Why is it necessary to design cladding and fixtures for higher pressure coefficient than that used for the design of structural frameworks?

16.10 What is the most economical bay width of buildings without cranes?

16.11 Sketch a braced and unbraced frame and point out the difference in their structural behaviour.

16.12 What are the different types of bracings used in a braced building?

16.13 What is the function of a bracing?

16.14 State the advantages of using a knee brace.

16.15 Why is it necessary to analyse the entire braced system, including columns, when knee braces are provided?

16.16 List the advantages and disadvantages of an unbraced frame over a braced frame.

16.17 Is it necessary to provide bracings in a portal frame? Why?

16.18 What is the ideal slope for the rafter of portal frame roofs? Why have smaller slopes been adopted in recent portal frames?

16.19 What is the disadvantage of using a solid-web rafter? How can it be overcome?

16.20 What are the two methods of framing the end frames or gables of the portal frame structure? Which one is preferred?

16.21 What are the necessary checks to be performed while using plastic design for portal frames?

16.22 State the difference between a purlin and a girt.

16.23 When are open-web girders used as purlins instead of hot-rolled or cold-rolled sections?

16.24 What are the sections that are normally used as purlins or girts?

16.25 What are wind columns?

16.26 What are the functions of an eave strut?

16.27 How can one determine whether a given truss forms a stable configuration?

16.28 Why is it necessary to design truss members for both compression and tension forces?

16.29 Distinguish between determinate and indeterminate trusses?

16.30 When are bending moments to be considered in the design of the top chord of trusses?
16.31 State the different truss configurations that are often used in practice.
16.32 Distinguish between fan and fink truss.
16.33 Why are Pratt trusses more advantageous compared to Howe trusses?
16.34 What are the advantages of parallel chord trusses?
16.35 How is the loading applied to a Vierendeel girder resisted?
16.36 What are the requirements that are considered while fixing the upper chord slope of trusses?
16.37 State the advantage of north light roof trusses over other forms of trusses?
16.38 What is the economical depth of a simply supported and continuous truss?
16.39 What is the economic range of spacing of a truss?
16.40 How is the spacing of purlins fixed?
16.41 What are the load combinations that are usually considered for truss analysis?
16.42 List the various steps involved in the design of truss members?
16.43 Why are the minimum sections recommended and adopted for truss members, even though a lighter section may be indicated by the design?
16.44 Describe the behaviour of top and bottom chord members of a truss when lateral purlins/ties are not provided at each node.
16.45 When are gusset plates used in a truss having T-section for rafters and bottom tie members?
16.46 What is an open-web joist?
16.47 What is the difference between a chequered plate and a plain plate?
16.48 Why is it necessary to provide connections that will allow movement in the supports of trusses?

## REFERENCES


ACI Committee 302, *Guide for Concrete Floor and Slab Construction*, ACI 302.1 R – 96, American Concrete Institute, Detroit, MI, 1996.

AISC 805-91, *Design of Low- and Medium-rise Steel Buildings, Design Guide 5*, American Institute of Steel Construction, Chicago, IL, 1991.

AISC 807-04, *Industrial Buildings: Roofs to Anchor Rods*, Second edition, *Design Guide 7*, American Institute of Steel Construction, Chicago, IL, 2004.

AISC 810-97, Erection Bracing of Low-rise Structural Steel Frames, *Design Guide 10*, American Institute of Steel Construction, Chicago, IL, 1997.

AISC 814-02, *Staggered Truss Framing Systems, Design Guide 14*, American Institute of Steel Construction, Chicago, IL, 2002.

Allen, E. and Iano, J., *The Architect's Studio Companion: Rules of Thumb for Preliminary Design*, Third edition, Wiley, New York, 2001, p. 416.

Bakota, J.F., 'Mill Building Design Procedure', *Engineering Journal*, AISC, Vol. 14, 4th Quarter, 1977, pp. 130–137.

Brockenbrough, R.L. and Merritt, F.S., *Structural Steel Designer's Handbook*, Fourth edition, McGraw-Hill Professional, 2006, p. 800.

Camp, C.V. and Bichon, B.J., 'Design of Space Trusses using Ant Colony Optimization', *Journal of Structural Engg., ASCE*, Vol. 130, No. 5, May 2004, pp. 741–751.

Chen, W.F and Lui, E.M., ed., *Handbook of Structural Engineering*, Second edition, CRC Press, Boca Raton, FL, 2005, p. 1768.

*Detailing for Steel Construction*, Second edition, American Institute for Steel Construction, Chicago, IL, 2002.

*Diaphragm Design Manual*, Steel Deck Institute, Fox River Grove, IL, 1995.

Dowling, P.J., Knowles, P., and Owens, G.W., *Structural Steel Design*, The Steel Construction Institute and Butterworths, London, 1988.

Epstein, H.I., Smith, E.M., and Mitchell, J.D., 'Analysis and Assumptions of Cold Formed Purlins', *Practice Periodical on Structural Design and Construction, ASCE*, Vol. 3, No. 2, May 1998, pp. 60–67.

Fisher, J.M., 'Industrial buildings—Guidelines and Criteria', *Engineering Journal, AISC*, Vol. 21, Third Quarter, 1984, pp. 149-153.

Fisher, J.M., 'Structural Details in Industrial Buildings', *Engineering Journal, AISC*, Vol. 18, 3rd Quarter, 1981, pp. 83–89, Discussions, 2nd Quarter,1982, pp. 139–140.

Fisher, J.M, 'The Importance of Tension Chord Bracing', *Engineering Journal, AISC*, Vol. 20, 3rd Quarter, 1983, pp.103–106, Discussions, 2nd Quarter, 1984, pp. 122–123.

Fisher, J.M., *Industrial Buildings in Constructional Steel Design, An International Guide*, Dowling, P.J., Harding, J.E., and Bjorhovde, R., ed., Elsevier Applied Science, London/New York, 1991, pp. 627–644.

Ganapathy, C.C., *An Investigation on the Strength and Behaviour of Latticed Steel Structures*, Thesis submitted for the award of PhD Degree, Indian Institute of Technology, Madras, 1972.

Ghosn, A.A. and Sinno, R.R., 'Load Capacity of Nested, Laterally Braced, Cold-formed Steel Z-section Beams', *Journal of Structural Engineering, ASCE*, Vol. 122, No. 8, Aug 1996, pp. 968–971; Vol. 128, No. 11, Nov 2002, pp. 1423–1428.

Harrison, H.B., *Structural Analysis and Design*, Parts 1 and 2, Second edition, Pergamon Press, Oxford, UK, 1990.

http://www.structural-engineering.fsnet.co.uk/free.htm (lists a number of free structural software packages)

King, C.M., *Design of Steel Portal frames for Europe*, Steel Construction Institute, Ascot, 2005.

LaBoube, R.A., 'Estimating Uplift Capacity of Light Steel Roof System', *Journal of Structural Engineering, ASCE*, Vol. 118, No. 3, Mar 1992, pp. 848–852.

Li, J-J. and Li, G-Q, 'Large-scale Testing of Steel Portal Frames Comprising Tapered Beams and Columns', *Advances in Structural Engineering*, Vol. 5, No. 4, Oct 2002, pp. 259–269.

Livesely, R.K., *Matrix Methods of Structural Analysis*, Second edition, Pergamon Press, Oxford, UK, 1975.

Lynn, B.A. and Stathopoulos, 'Wind-induced Fatigue on Low Metal Buildings', *Journal of Structural Engineering, ASCE*, Vol. 111, No. 4, Apr 1985, pp. 826–839.

Mahendran, M. and Moor, C., 'Three-dimensional Modeling of Steel Portal Frame Buildings', *Journal of Structural Engineering, ASCE*, Vol. 125, No. 8, Aug 1999, pp. 870–878.

Martin, L.H. and Purkiss, J.A., *Structural Design of Steelwork to BS 5950*, Edward Arnold, London, 1992.

Mathews, M.S. and Rao, P.S., *Corrugated Ferrocement Sheets*, ACI SP-61, American Concrete Institute, Detroit, MI, 1979.

Narayanan, S. and Mathews, M.S., 'Behaviour of Roofing Sheet System under Static Uplift Loads', *Journal of Institution of Engineers (India), Civil Engg. Div.*, Vol. 84, Nov 2003, pp. 216–222

Newman, A., *Metal Building Systems: Design and Specifications*, Second edition, McGraw-Hill, New York, 2004, p. 576.

Packer, J.A., and Henderson, J.E., *Hollow Structural Section: Connection and Trusses—A Design Guide*, Second edition, Canadian Institute of Steel Construction, Toronto, Ontario, June 1997, p. 464.

Pandia Raj, R., and Kalyanaraman, V., GA-based Optimal Design of Steel Truss Bridge, *6th World Congress of Structural and Multi-disciplinary Optimization*, Rio de Janeiro, 30 May–3 June 2005, Brazil.

Petersen, C., *Stahlbau*, Friedr. Vieweg & Sohn Verlag, Braunschweig, 1990, (Chapter 17, Trapezprofil-Bauweise, pp. 719–745).

Rajeev, S. and Krishnamoorthy, C.S., 'Generic Algorithms-based Methodology for Design Optimization of Trusses', *Journal of Structural Engg., ASCE*, Vol. 123, No. 3, Mar 1997, pp. 350–358.

Ramasamy, J.V. and Rajasekaran, S., 'Artificial Neural Network and Generic Algorithm for the Design Optimization of Industrial Roofs—A Comparison', *Computers and Structures*, Vol. 58, No. 4, 1996, pp. 747–755.

Ruddy, J., Economics of Low-rise Steel Framed Structures, *Engineering Journal, AISC*, Vol. 20, 3rd Quarter, 1983, pp. 107–118.

Schittich, C., *In detail: Building Skins, Concepts, Layers, Materials*, Birkhouser, Switzerland, 2001, p. 120.

Schodek, D.L., *Structures*, Fourth edition, Prentice Hall, Upper Saddle River, New Jersey, 2001, p. 581.

Sree Ramachandra Murthy, D., Bhanu Prasad, D., Ramesh K.V., and Narasimha Rao, D.L., 'Economical Aspects in the Design of Steel Trusses', *Civil Engineering and Construction Review*, Nov 2004, pp. 30–38.

*Standard Specifications for Open Web Steel Joists*, K-Series, Steel Joist Institute, 41st edition, Myrtle Beach, SC, 2002.

Subramanian, N., *Space Structures: Principles and Practice*, Multi-Science Publishing company Ltd, Essex, UK, 2006.

Subramanian, N., 'Corrosion Protection of Steel Structures', *The Master Builder*, Vol. 2, No. 1, Jan–Feb 2000, pp. 63–66 and No. 2, Apr–May 2000, pp. 14–17.

Subramanian, N., *Design of Reinforced Concrete Structures*, Oxford University Press, New Delhi, 2013, p. 880.

Thandavamoorthy, T.S., *Analysis of Structures—Strength and Behaviour*, Oxford University Press, New Delhi, 2005, p. 860.

Timoshenko, S.P. and Krieger, S.W., *Theory of Plates and Shells*, Second edition, McGraw-Hill, New York, 1968.

Vinnakota, S., *Steel Structures: Behavior & LRFD*, McGraw-Hill Higher Education, New York, 2006.

Wang, C.K., *Computer Methods in Advanced Structural Analysis*, Intext Educational Publishers, New York, 1973.

Xu, Y.L., 'Fatigue Performance of Screw-fastened Light-gauge Steel Roof Sheets', *Journal of Structural Engineering, ASCE*, Vol. 121, No. 3, Mar 1995, pp. 389–398 also Discussions, Vol. 122, No. 12, Dec 1996, pp. 1475–1483.

17

# SEISMIC-RESISTANT DESIGN

## INTRODUCTION

As discussed in Section 3.9 of Chapter 3, a huge amount of property damage and loss of life is due to natural hazards and the main culprits are the earthquakes, high speed winds, and floods [in India we do not have proper code for flood-resistant design, similar to the document available in USA (ASCE 24-05)]. We may use the same analysis procedures and software packages to arrive at the axial and shear forces and bending moments acting on structures subjected to lateral loads due to wind, earthquake, or for the combination of loads, as discussed in Section 3.15 of Chapter 3 (these lateral loads may be calculated by using the procedures described in Sections 3.12 and 3.13 of Chapter 3 and applied at the respective joints of a structure in the horizontal direction). Once the forces and moments are found out, we may proceed with the design as in the case of gravity loads.

However, we need to bear in mind that designing for earthquake is very different from designing for other loads such as dead load (DL), live load (LL), and wind load (WL). This is because, for loads other than earthquake loads, we design the structure for the actual load (though they are probabilistic, codes have specified the maximum probable loads based on measurements); however, it has to be noted that the wind loads specified in IS 875-Part 3:1987 have to be increased and rationalized as given in NICEE-GSDMA documents and ASCE 7-10. However, when we design any structure for earthquake loads, we consider about 1/6th to 1/10th of the earthquake load only [This is because, we do not consider the loads due to *maximum credible earthquake* or *maximum considered event* (MCE), but consider only the *design basis earthquake* (DBE), that is, an earthquake that the structure is required to safely withstand with repairable damage—see Section 3.12 of Chapter 3. In IS 1893 (Part 1):2002, DBE is taken as 50% of MCE. In addition, we divide the load by the *response reduction factor*]. It may be noted that as per UBC 1997 and IBC 2000, the DBE corresponds to an earthquake having 10% probability of being exceeded in 50 years, that is, a 475-year return period. [IS 1893(Part-1):2002 merely states that it is an earthquake that can reasonably be expected to occur at least once during the design life of the structure]. Even the *design-response spectrum* we consider is a smooth, broad-banded spectrum (needs to be site-specific in order to be more exact—in the design of important structures such as nuclear power stations, only *site-specific response spectrum* has to be used). The most important point to remember is that the earthquake-resistant design codes allow damage to the structure but disallow total collapse and death of occupants. Thus the codes rely on over-strength and ductility for the safety of the structure.

The combined effect of over strength, redundancy, and ductility is expressed in terms of *response reduction factor* ($R$) in the codes, which is used to arrive at the design horizontal seismic coefficient (see Section 3.12 and Table 3.22 of Chapter 3). It has to be noted that the precise estimation of *over-strength* is difficult to determine since many factors contributing to it involve uncertainties. The actual strength of materials, the contribution of non-structural elements, and the actual participation of some structural elements such as reinforced-concrete slabs are factors leading to high uncertainties (Humar and Ragozar 1996). In addition, not all factors contributing to over-strength are favourable. For example, flexural over-strength of members leads to increased shear forces when plastic hinges form, which may result in brittle shear failures. Non-structural elements, for example brick infills, could lead to shear failure due to short column effect or to soft storey failures (Park 1996). Moreover, the over-strength factor varies widely according to the period of the structure, the design intensity level, the structural system and the ductility level assumed in the design. This compounds the difficulties associated with evaluating this factor accurately (Elnashai and Mwafy 2002). Though conventional seismic design procedures in all modern seismic codes still adopt force-based design criteria (by reducing the anticipated seismic forces using a single reduction factor, to arrive at the design force level) it may be prudent to control the deformation during the design, which is the major cause of damage and collapse of structures subjected to earthquakes. Hence, a brief discussion about response reduction factor is provided.

The factors influencing seismic design are listed and briefly discussed. The survival of a building during an earthquake depends not only on structural analysis, design, and detailing, but also on architectural considerations. Proper planning is more important than design and detailing; regular structures may survive an earthquake but irregular structures may fail. There are many types of irregularities and these may be grouped as *plan irregularities* and *vertical irregularities*. These irregularities are to be avoided for better performance during earthquakes. Other aspects of planning and design in earthquake zones (for example, location of openings in walls, avoidance of long cantilevers and floating columns) and consideration of the vertical component of earthquakes (for designing horizontal projections and for structures close to the faults or with heavy mass at the top) are also discussed.

Several systems can be adopted to provide adequate resistance to seismic lateral forces. The most common are: moment-resisting frame, a combined system of moment frames and shear walls, braced frames with horizontal diaphragms, and a combination of all these systems. These systems and the concept of strong column and weak beams (which will result in the plastic hinges forming in beams, rather than in columns) are described.

Steel plate shear walls (SPSW) due to the reduced thickness, as compared with concrete shear walls, offer significant advantages in terms of cost, performance, and ease of design and erection. They are considered an alternative to braced frames and can provide equivalent strength and stiffness. National Building Code of Canada (1994), AISC 341-05, and FEMA 450-2004 introduced provisions for the design of SPSW. These provisions are described briefly along with the advantages and drawbacks of SPSW.

For better performance of the structure, the connections between columns and beams should not fail before the beams or columns. Extensive testing on these connections has resulted in three recommendations for their better performance: (a) a toughening scheme, (b) a strengthening scheme, and (c) a weakening scheme. These schemes are discussed along with the pre-qualified seismic moment connections, which are recommended in the AISC 341 code. A brief review of devices that can be used to reduce the effects of earthquakes is also provided.

## 17.1 RESPONSE REDUCTION FACTOR

As already discussed in Section 3.12 of Chapter 3, the code provisions allow the structure to be damaged in the case of severe shaking. It has to be noted that conventional seismic design procedures adopt force-based design criteria instead of a displacement-based criteria. The basic concept of displacement-based criteria is to design the structure for a target displacement rather than the specified strength (see Priestley et al. 2007 for more details of displacement-based seismic design). Hence, the deformation (which causes damage and collapse of structures during earthquakes), is controlled during the design. However, unfortunately, the traditional concept of reducing the seismic forces using a single reduction factor (called the *response reduction factor*), to arrive at the design force level, is still widely used. This is because structures designed as per this concept have performed satisfactory during full-scale tests and also during earthquakes. This reduction is justified in seismic codes based on reserve strength and ductility (see Fig. 17.1), which improve the capability of the structure to absorb and dissipate energy.

Thus, as per the code methodology, the structure is designed for a seismic force much less than that expected under strong earthquakes, if the structure were to remain linearly elastic. IS 1893(Part 1):2002 provides clauses for the calculation of the probable realistic force to be applied for an elastic structure, and then divides that force by a factor of 2R. [It has to be noted that for important structures the *Probabilistic Seismic Hazard Analysis Map* (PSHA map) of India, found at http://www.hpsdma.nic.in/ should be consulted, instead of the seismic zones map provided in IS 1893 code.] For example, if we consider a structure in zone V, as per IS 1893 (Part 1), $Z = 0.36$ gives a probable indication of the ground acceleration. For $T = 0.3$ s, $S_a/g = 2.5$ as per Fig. 3.24 of Chapter 3, which means that if the building remains elastic, it may experience a maximum horizontal force equal to 90% of its weight ($0.36 \times 2.5 = 0.90$). If we use an $R$ factor of 5 and an importance factor of 1, then we have to design the building for a horizontal force equal to 0.09 times the weight [$0.90 / (2 \times 5)$]. It is clear from this example that the designer is going to design the building for only one-tenth of the maximum elastic force and hence should provide adequate ductility and quality control for good post-yield behaviour. In other words, the term $R$ gives an indication of the level of over-strength and ductility that a structure is expected to have. The intent of the $R$ factor is to simplify the structural design process such that only linearly elastic static analysis (i.e., equivalent static method) is needed for most building design (see Fig. 17.1). It is of interest to note that the $R$ factor could be traced back to the *K factor*, which was used in the first edition of Structural Engineers Association of California's Blue book in 1959 (SEAOC Recommended Lateral Force Requirements and Commentary 1959). $R$ values generally range from 1 for systems that have no ability to provide ductile response to 5 for systems that are capable of highly ductile response.

Thus, as per IS 1893 (Part 1), the structure can be designed for a much lower force than is implied by the strong shaking by considering the following contributing factors, which will prevent the collapse of the structure: over-strength, redundancy, and ductility (see Fig. 17.1). These factors are discussed briefly in the following sections.

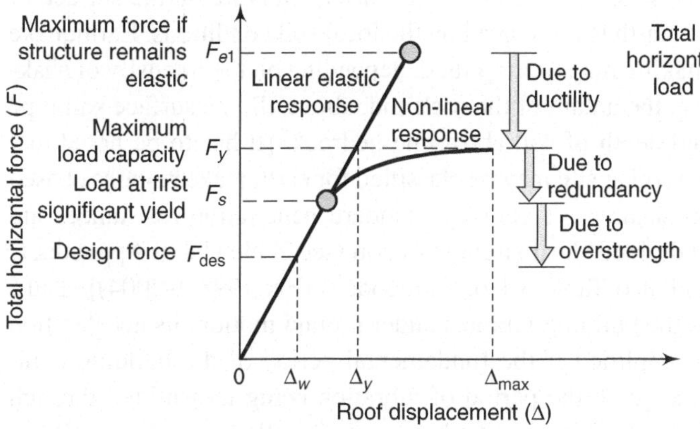


**FIG. 17.1** Concept of response reduction factor (Jain and Murty 2004)
*Source:* Reprinted with permission from NICEE.

## 17.1.1 Over-strength

The factors that account for the yielding of the structure at loads higher than the design load are as follows.

1. Partial safety factors
   (a) Partial safety factor for seismic load
   (b) Partial safety factor for gravity load
   (c) Partial safety factor for materials
2. Material properties
   (a) Provided member size larger than the required size as per design
   (b) Strain hardening in steel resulting in higher strength
   (c) Higher material strength under cyclic load (see Fig. 1.14 in Chapter 1)
3. Strength contribution of non-structural elements such as masonry infills
4. Special ductile detailing, especially at the joints, which also contributes to the increase in strength

## 17.1.2 Redundancy

It has to be noted that yielding at a location of a structure will not result in collapse. After the first yield, the forces are redistributed and the structure collapses only after forming a *mechanism* (see Section 8.4 of Chapter 8 for details). This type of action in a redundant structure provides an additional safety margin.

## 17.1.3 Ductility

As defined in Chapter 1, *ductility* is the capacity of materials/structures to absorb energy by deforming into the inelastic range. (Ductility $\mu$ may also be defined as the ratio of the ultimate deformation $\Delta_{max}$ at an assumed collapse point to the yield deformation $\Delta_y$—see Fig. 17.1). This capacity of the structure to absorb energy, with acceptable deformations and without failure, is a very desirable characteristic in any earthquake-resistant design. Ductility is often measured by the *hysteretic behaviour* of critical components such as the column-beam assembly of a moment-resisting frame. The hysteretic behaviour is usually predicted by the cyclic moment-rotation or force-deflection behaviour of the assembly as shown in Fig. 17.2. The slope of the curve represents the stiffness of the structure and the enclosed area, the dissipated energy. Perfect ductility is defined by the ideal elastic–perfectly plastic (also called elasto-plastic) curves, which are difficult to achieve in ideal elasto-plastic materials such as steel. *Hysteretic energy* is the energy dissipated by inelastic cyclic deformations and is given by the area within the load deformation curve shown in Fig. 17.2, also called the *hysteretic curve.* Under ideal conditions, *hysteresis loops* of the form shown in Fig. 17.2(a) result, where the energy absorbed will be about 70–80% of that of an equivalent elasto-plastic loop. Limited energy dissipation curves are shown in Fig. 17.2(b). The degradation of strength and stiffness under repeated inelastic cycling is called *low-cycle fatigue.*

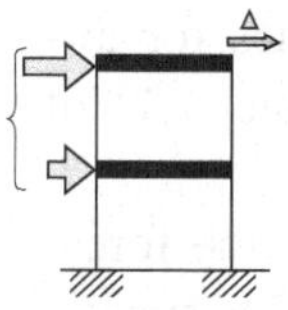


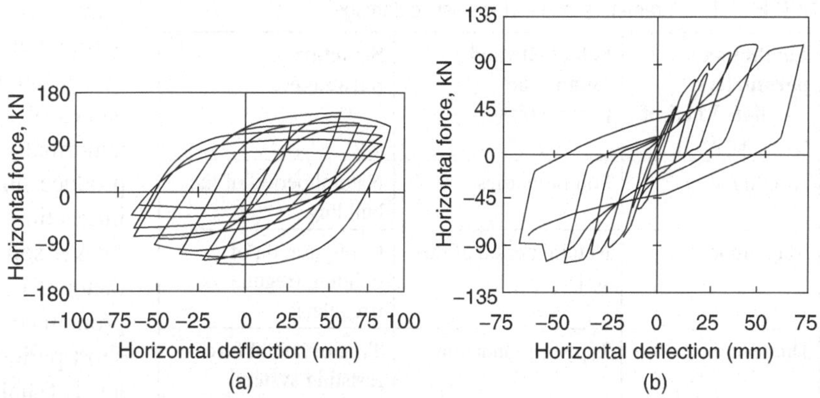


**FIG. 17.2** Hysteretic behaviour (a) Curve representing large energy dissipation (b) Curve representing limited energy dissipation

When a structure yields, the following things happen:

1. There is more energy dissipation in the structure due to hysteresis.
2. The structure becomes softer and its natural period increases; due to this it has to resist a lower seismic force (see Fig. 3.24 of Chapter 3).

Thus, higher ductility indicates that the structure can withstand stronger earthquakes without complete collapse. The values prescribed in the code (IS: 1893 and IS: 800) for

$R$ are based on the observed performance of buildings in past earthquakes, expected ductility, over-strength, practice in other countries, and include all the factors discussed in the previous paragraphs (Jain 1995). The $R$ values for different types of steel structures are given in Table 3.22 of Chapter 3.

As already indicated, a large area enclosed by the force-deformation loops indicates more dissipation of hysteretic energy. One way of ensuring high ductility and energy dissipation capacity is to use slightly thicker sections thereby avoiding local buckling (however, thick welded sections will result in brittle fracture associated with lamellar tearing). Plastic and compact sections (see Chapter 8) are more ductile than semi-compact and slender sections. Slenderness ratio and axial load ratio of the members may also control ductility. Ductility and energy dissipation capacity are important factors in resisting severe earthquakes (note that these two quantities are interrelated and a large demand on one may tend to decrease the other).

## 17.2 FACTORS INFLUENCING SEISMIC DAMAGE

The seismic damage caused at a particular site is not only dependent on the earthquake but also influenced by the soil at site and the characteristics of the foundation and structural elements. These parameters influencing seismic damage are listed in Table 17.1.

**TABLE 17.1** Parameters influencing seismic damage

| **Earthquake parameters (Section 3.11.1 of Chapter 3)** | **Soil at site and foundation parameters** | **Structural parameters** |
|---|---|---|
| Amplitude | Soil properties | Natural period of the building |
| Magnitude | Natural period of the soil | Configuration of building (regular vs irregular) |
| Duration | Type of foundation | Type of lateral force-resisting system (MRF, braced frame, shear wall, etc.) |
| Frequency | – | Detailing of joints |
| Distance of site from epicenter or fault | Geographical conditions between the epicenter and the site(a building built on the top of the ridge may be subjected to intensified shaking) | Construction material (steel, concrete, wood, masonry) and quality control at site |

The parameters concerned with the earthquake were discussed briefly in Section 3.11 of Chapter 3 (more details may be found in Bolt 2006). The other factors are discussed in the following subsections.

### 17.2.1 Influence of Soil Properties and Foundation Types

The seismic motion that reaches a structure on the surface of the earth is influenced by the local soil conditions. Earthquake shaking may be amplified, depending on the intensity of shaking, the nature of the rock and, above all, the surface soil type and depth of soil above the bedrock. [It has to be noted that the soil at site may be classified for earthquake response based on shear wave velocity, standard penetration resistance, and undrained shear strength of soil (see Table C.8 of Appendix C and also Table 3.1 of Eurocode 8 (EN 1998-1-2004)]. Thus, as the building vibrates under ground motion, its acceleration is amplified if the fundamental period of the building coincides with the period of vibration being transmitted through the soil. This amplified response is called *resonance.* [It has to be noted that normal single storey to 20 storey buildings typically have fundamental natural periods in the range of 0.1 to 2.0 s, while the natural periods of soil may be in the range of 0.4 to 2 s, depending on the nature of the soil on ground (see also Table C.7 of Appendix C). Hard ground or rock will experience short period vibration. Very soft ground may have a period of up to 2 seconds but, unlike a structure, it cannot sustain longer period motions except under certain unusual conditions (Arnold 2006). Approximate fundamental natural periods of other structures are as follows: suspension bridges: 6 s, elevated water tanks: 4 s, RC chimneys: 2 s, and large gravity dams: 0.8 s. Taller and more flexible buildings have natural periods in the range of 1–5 s or greater].

Special attention should be paid to soils that have very low values of shear wave velocity, low internal damping, and an abnormally extended range of linear behaviour, since they produce unusual seismic site amplification and soil-structure interaction effects (EN 1998-1:2004). A layer of soft and thicker soil above the bed rock may result in an amplification factor of 1.5–6 over the rock shaking. This amplification is most pronounced at longer periods, and may not be so significant at short periods. The amplification also tends to decrease as the level of shaking increases (Arnold 2006). As a result, earthquake damage tends to be more severe in areas of soft ground, as in the 1906 San Francisco earthquake, the 1989 Loma Prieta earthquake (Northern California), and the 1985 Mexico City earthquake (Mexico City's downtown area is about 400 km away from the 8.1-magnitude earthquake epicenter, and had silt and volcanic clay sediments of the bed of the historic Lake Texcoco, which were between 7–37 m deep and had a high water content. The earthquake caused the soft ground under the downtown buildings to vibrate for over 90 seconds at its long natural period of around 2 seconds. This caused buildings that were between about 6–20 storeys in height to resonate at a similar period, greatly increasing the accelerations within them. Taller buildings suffered little damage).

*Soil liquefaction* is another effect caused by earthquakes. A saturated, uniform, fine grained sand or silt, when subjected to repeated vibration, experiences an increase in pore water pressure due to a redistribution of its particles, with a consequent reduction in shear strength. This produces a 'quick sand' type condition, with a loss of bearing capacity, causing settlement and collapse of structures.

For better seismic response, proper precautions have to be taken at the planning stage itself. It is preferable to select a site where bedrock is available close to the surface, so that foundations can be laid directly on the rock. The differential movement of foundation due to seismic motions is an important cause of structural damage, especially in heavy, rigid structures that cannot accommodate such movements. Hence, if the foundation is on soft soil with spread footings, adequate plinth or tie beams should be provided to counter differential settlement. If the loads are heavy, pile foundations with strong pile caps may be provided. A raft foundation is ideal for resisting differential settlements, but may prove to be expensive. In sandy or silty soils, if the water table is near the foundation level, the following methods may be adopted to prevent liquefaction.

1. Drainage may be installed to lower the ground water table and remove the pore water (it should be checked weather the resulting settlement will not affect adjacent structures).
2. A porous overburden may be placed over the site to produce over-consolidation, which results in increased pore pressures being required before liquefaction can occur.
3. If there are no adjacent structures, pre-consolidation of the soil may be achieved by vibro-flotation techniques.
4. In order to increase the shear strength of soil, soil grouting or chemical injection may be employed.
5. All deleterious soil may be removed and replaced with better soil.
6. Pile foundation may be employed with piles extending and resting on to a sound soil layer below the unsatisfactory layer.

Appendix F of the draft IS:1893 gives a simplified procedure for evaluating the liquefaction potential of a soil deposit.

Different types of foundation in the same structure, foundations on different soil types for the same structure, and foundations at different levels (such as those found in hill slopes) should be avoided for better seismic performance.

## 17.3 RULES TO BE FOLLOWED FOR BUILDINGS IN SEISMIC AREAS

To perform well in an earthquake, a building should possess the following four main attributes: (a) simple and regular configuration, (b) adequate lateral strength, (c) adequate stiffness, and (d) adequate ductility. Thus, structural planning of steel buildings should be done in such a way that the beams yield prior to the columns, the strength of connections should be greater than the strength of beams and columns framing into the connection members, the connections should guarantee high strength, ductility, energy dissipation capacity, and excessive lateral sway is avoided.

Buildings with a simple regular geometry and uniformly distributed mass and stiffness in plan and elevation (regular structures), as shown in Fig. 17.3, have been found to suffer less damage in earthquakes than those with irregular structures. Hence, columns and walls should be arranged in grid fashion and should not be staggered in plan. The effect of asymmetry will induce torsional oscillations in structures and stress concentrations at re-entrant corners. Irregularities may be grouped as *plan irregularities* and *vertical irregularities*.

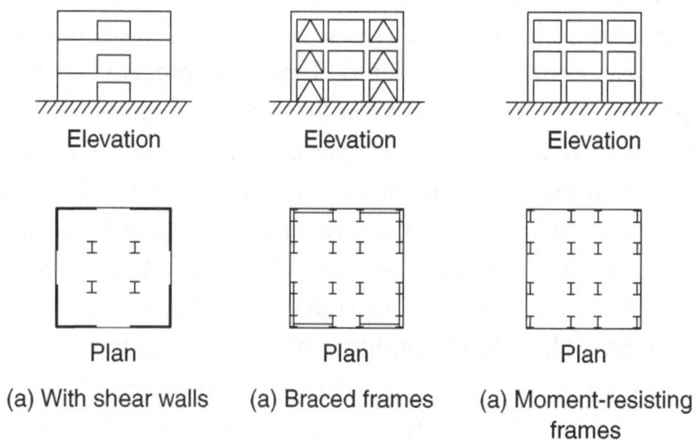


**FIG. 17.3** Buildings with regular configuration

## 17.4 PLAN IRREGULARITIES

The several plan irregularities that should be carefully considered and avoided while designing buildings in seismic areas are briefly discussed in this section.

### 17.4.1 Irregularity due to Re-entrant Corners

Buildings with re-entrant corners (i.e., plans in the shape of L, H, V, +, Y, W, or any other letter, except the O-shape) should be avoided. Projection of the structure beyond any re-entrant corner should not be greater than 15% of its plan dimension in the given direction (see Fig. 17.4). It will be advantageous to split such plans into separate rectangles by using *seismic joints* at junctions of the individual wings. (Fisher 2005 and Saunders 2005)

### 17.4.2 Torsional Irregularity

When floor diaphragms (a *diaphragm* is a horizontal system that transmits lateral forces to the vertical resisting elements; for example, reinforced concrete floors or horizontal bracing systems that transmit lateral forces to the columns)

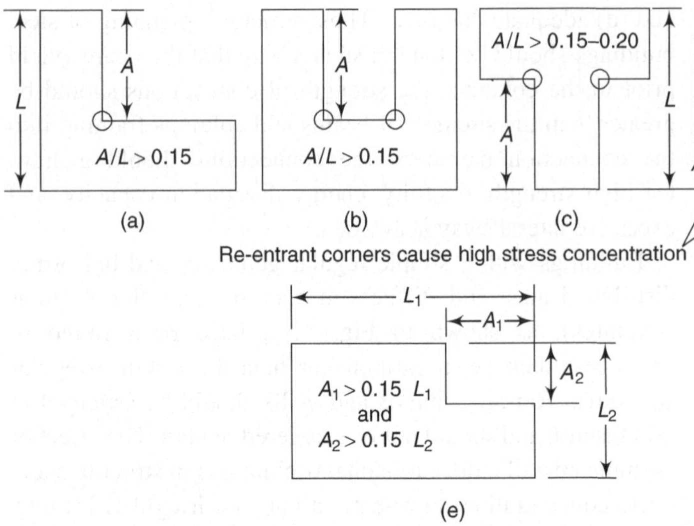


**FIG. 17.4** Plan irregularities

cause severe damage (or even collapse) of structures, is provided by Özmen et al. (2014).

As shown in Fig. 17.5(b), seismic force-resisting systems, such as shear walls should not be placed asymmetrically or the masses (for example, heavy swimming pools at the top of buildings) should not be placed eccentrically. Torsion should be minimized by making the building symmetrical and regular in geometry and stiffness, and by providing lateral load-resisting elements at the building's perimeter as shown in Fig. 17.5(a).

It has to be noted that provisions are not available in the Indian codes (IS 456:2000 or IS 13920: 1993) for the design of diaphragms. Some guidelines and discussions are provided by Subramanian (2013) [see Section 10.10 of the book]. Various researchers have identified that the commonly employed *equivalent static analysis method* for the design of diaphragms, under-estimates the acceleration of floors, particularly in the lower levels of the buildings (see for example, Gardiner, et al. 2008).

are considered rigid in their own plane in relation to the vertical structural elements that resist the lateral forces, torsional irregularity must be considered. Torsion is present whenever there is an eccentricity between the *center of mass* (CM) and the *center of rigidity* (CR), also called the center of stiffness. When the maximum storey drift, computed with design eccentricity at one end of a structure transverse to an axis, is less than 1.2 times the average of the storey drifts at the two ends of the structure, torsional irregularity need not be considered. Thus, in the building plan shown in Fig. 17.5(c), the drift $\Delta_2$ should not be greater than $1.2[(\Delta_1 + \Delta_2)/2]$. If $\Delta_2$ is greater than $1.4[(\Delta_1 + \Delta_2)/2]$, it is considered to be a case of extreme torsional irregularity (in the Canadian code the factor is 1.7 instead of 1.4). More discussions on torsional irregularity, which is one of the most important factors and may

### 17.4.3 Diaphragm Discontinuity

The roof/floor deck or slabs respond to lateral loads like a deep beam. In the American code (ACI 318:2014), it is assumed that the deck or floor slab acts as the web of a continuous beam carrying shear, and the perimeter spandrel beams or walls act as the compression and tension chords (flanges)

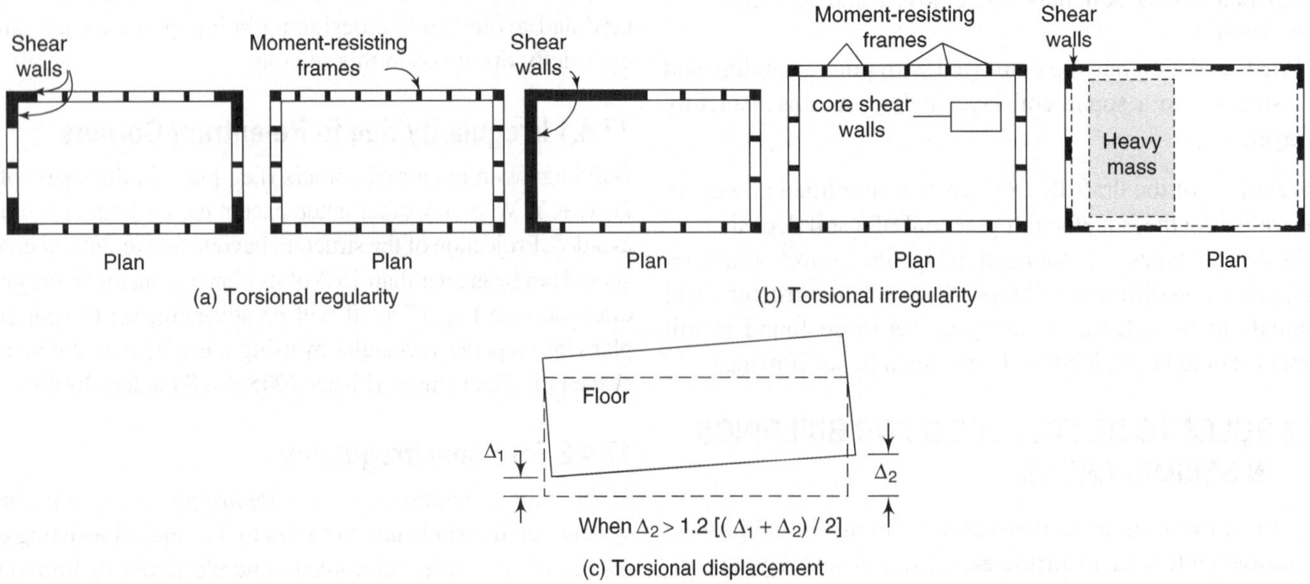


**FIG. 17.5** Torsional irregularity

of the continuous beam. Thus, inappropriate location or large size of openings (due to stair or elevator cores, atriums, skylights, etc.) creates problems similar to those related to cutting a hole in the web of a beam. Such openings reduce the ability of the diaphragm to transfer the load and may cause failure.

Hence, for diaphragms with abrupt discontinuities or variations in stiffness, the cut outs or open areas as shown in Fig. 17.6 should not be greater than 50% of the gross enclosed diaphragm area. Similarly, there should not be changes in the effective diaphragm stiffness of more than 50% from one storey to the next. This is because diaphragm discontinuity changes the lateral load distribution to different elements, in contrast to a rigid floor diaphragm.

More details about the analysis, design, and constructional aspects of diaphragms may be found in the ACI code (Section 21.11), Thomas and Sengupta (2008), and Moehle et al. (2010). Details of the analysis of chord forces for slabs with large openings may be found in Sengupta and Shetty (2011).

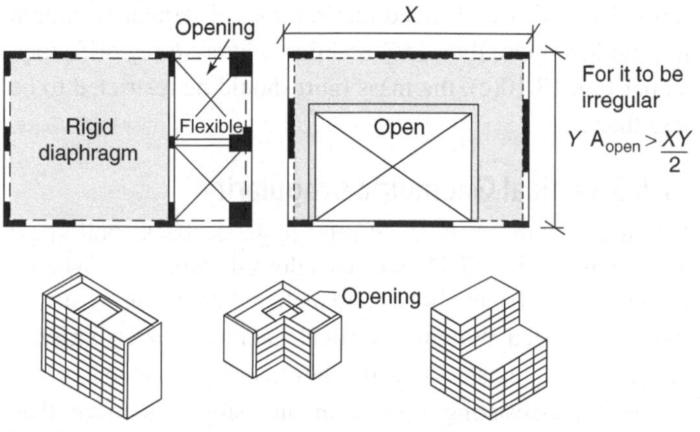


**FIG. 17.6** Diaphragm discontinuity

### 17.4.4 Out of Plane Offsets

Out-of-plane offset as shown in Fig. 17.7 is a serious irregularity, since it imposes excessive vertical and lateral load effects on horizontal elements. It is recommended to carry all shear walls down to the foundation level without any offsets, especially in places that may experience moderate to severe earthquakes. When such an arrangement is adopted, it is very important to provide adequate stiffness to the columns supporting the upper storey shear walls and also confining reinforcement throughout the reinforced concrete columns, extending beyond the beam-column joints.

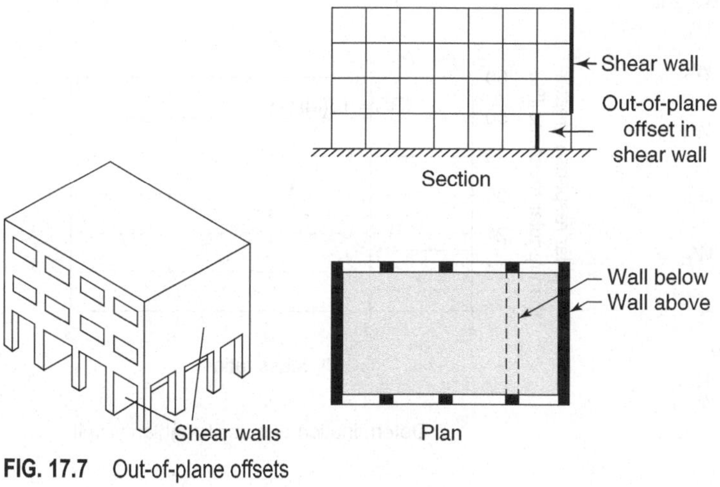


**FIG. 17.7** Out-of-plane offsets

### 17.4.5 Non-parallel System

The vertical elements resisting the lateral forces should be parallel to or symmetrical about the major orthogonal axes or the lateral force resisting elements. When they are not parallel as shown in Fig. 17.8, additional load combinations are necessary (see also Section 3.15.1 of Chapter 3).

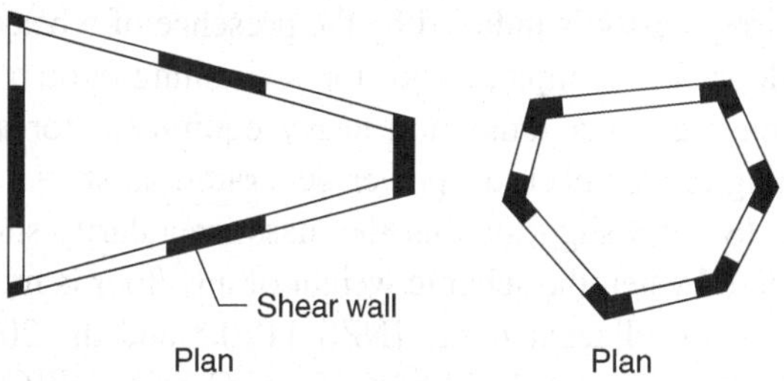


**FIG. 17.8** Non-parallel systems

## 17.5 VERTICAL IRREGULARITIES

Similar to horizontal irregularities, various kinds of vertical irregularities should also be carefully considered and avoided in seismic zones. Some of these vertical irregularities are briefly discussed in the following subsections.

### 17.5.1 Stiffness Irregularity

Omitting the exterior walls or all the walls for parking lots in the ground floor leaves all the columns at the ground level as the only elements available to resist lateral forces, thus causing an abrupt change in stiffness at that level. In general, a storey with lateral stiffness less than 70% of that in the storey above or less than 80% of the average stiffness of the three storeys above is considered a *soft storey* (see Fig. 17.9).

*Buildings on stilts* may have lateral stiffness less than 60% of that in the storey above or less than 70% of the average stiffness of the three storeys above and hence considered to have an *extreme soft storey*. Soft-storey buildings are known for their poor performance during earthquakes and many buildings have collapsed during the 2001 Bhuj earthquake. During California's Loma Prieta earthquake of 1989, nearly half of all homes became uninhabitable due to soft-storey failure.

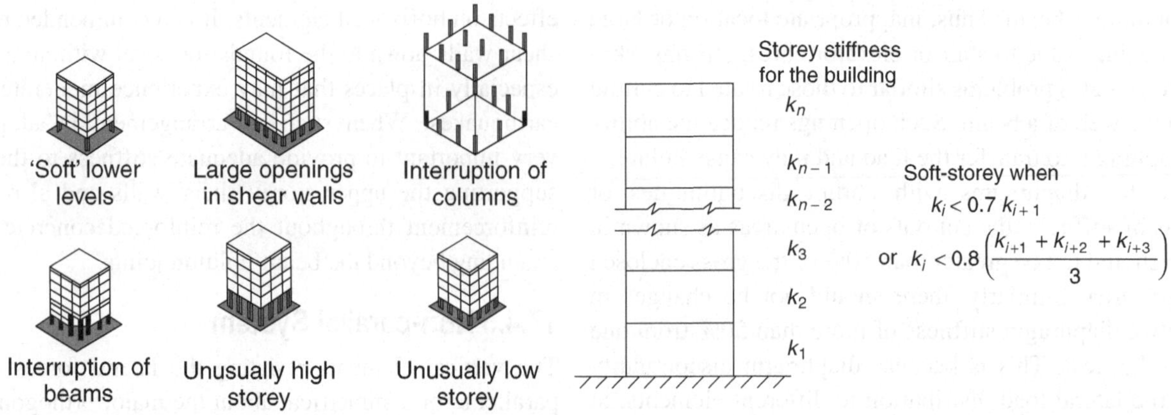


**FIG. 17.9** Stiffness irregularity

## 17.5.2 Mass Irregularity

Mass irregularity is induced by the presence of a heavy mass on a floor, for example, a roof-top swimming pool, a service floor having water tanks and heavy equipment for air conditioning and/or back-up power generator, as shown in Fig. 17.10. IS 1893 suggests that the mass irregularity should be considered when the seismic weight of any floor is more than 200% of its adjacent floors [NZS 1170.5 and the 2005 edition of the National Building Code of Canada (NBCC) consider 150% instead of 200%]. However, the code suggests that a roof that is lighter than the floor below need not be considered.

It has to be noted that this limit of 1.5–2.0 times the seismic weight for mass irregularity and also other limits for other types of irregularities have been specified in codes from engineering judgment and not based on rigorous quantitative analysis. A more meaningful comparison will be obtained if structures designed to a target drift are compared with the actual drift demand. A study by Sadashiva et al. (2009) found that the increased mass, when present at either the first floor level or at the roof, produced higher drift demands than when located at the mid-height. They also proposed a simple equation as shown in Fig. 17.10(c), where IRR represents the irregular response greater than regular response; and MR is the mass ratio. Thus, if it is decided that the mass irregularity should not produce more than 15% additional inter-storey drift, then as per Fig. 17.10(c), the mass ratio should be restricted to be less than 2.

## 17.5.3 Vertical Geometric Irregularity

Buildings with vertical offsets (e.g., setback buildings) as shown in Fig. 17.11, are not allowed, especially when a larger dimension is above the smaller dimension, as it acts like an inverted pyramid. Vertical geometric irregularity is considered to exist where the horizontal dimension of the lateral force-resisting system in any storey is more than 150% (130% in the Canadian code) of that in an adjacent storey.

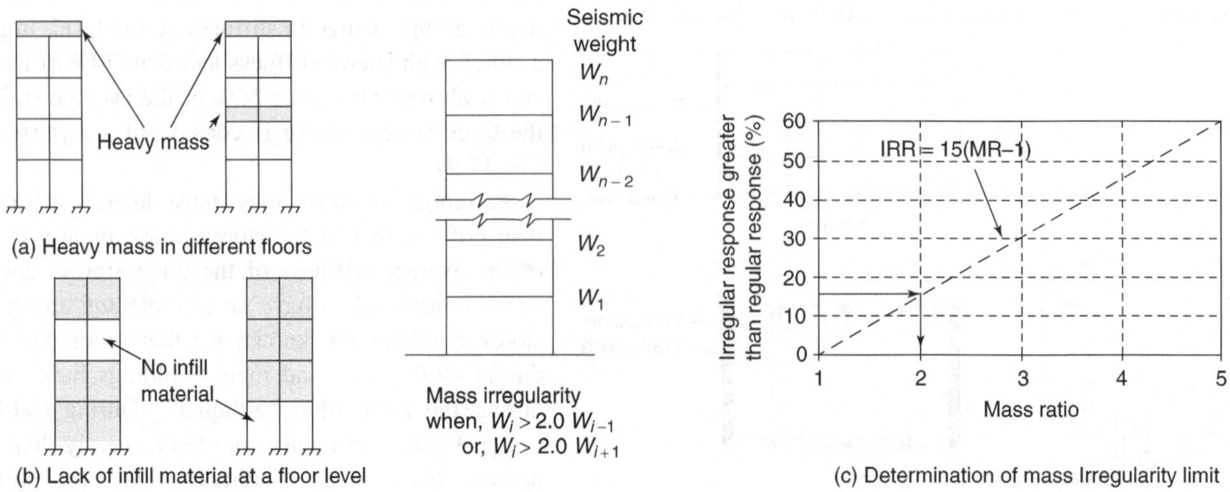


**FIG. 17.10** Mass irregularity

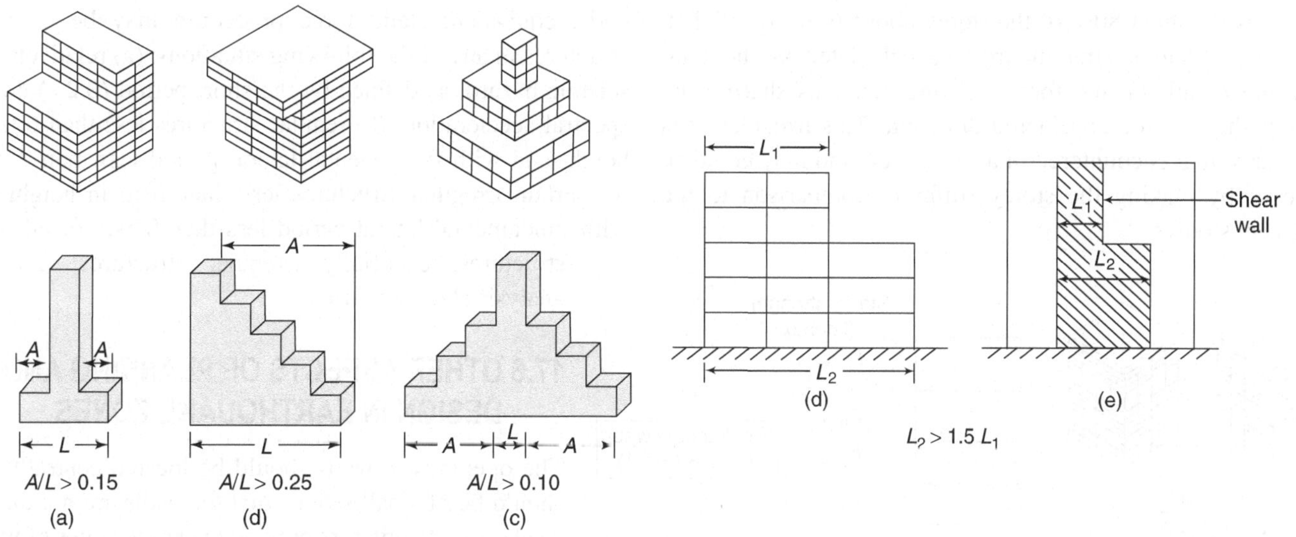


**FIG. 17.11** Vertical geometric irregularity

### 17.5.4 In-plane Discontinuity in Vertical, Lateral Force-resisting Elements

*In-plane discontinuity in vertical, lateral force-resisting elements* is considered to exist when there is an in-plane offset of a lateral load-resisting element (which is greater than the length of these elements) of the lateral force-resisting system or a reduction in lateral stiffness of the resisting element in the storey below (see Fig. 17.12).

### 17.5.5 Discontinuity in Capacity (Weak Storey)

Weak-storey configuration is often used in multi-storey apartment buildings and hospital buildings, in which not only the first floor is designed without walls (for using it as car parking), but also has a greater height than the rest of the floors. This irregularity can also be present at the first floor or at other intermediate floors (to provide meeting halls, restaurants, or machine rooms). There are numerous examples of many buildings presenting a combination of these types of irregularities, soft and weak storey, making them seismically vulnerable.

It has to be noted that the infilled brick walls in the upper storeys increase the lateral stiffness of the frame by a factor of three to four times than that of lower weak storey. This makes the much stiffer upper storeys to behave like a rigid block, and most of the horizontal displacement of the building occurs locally in the soft storey alone. Thus, the dynamic ductility demand during an earthquake gets concentrated in the soft storey and the upper storeys tend to remain elastic; the severely strained 'soft' storey causes total collapse of the building and much smaller damages occur in the upper storeys (unless the building collapses to the ground). In several past earthquakes the main cause of collapse is the ground floor soft storey (Bachmann 2002).

According to IS 1893 (Part 1):2002, a storey may be considered as a weak storey when the lateral strength of that

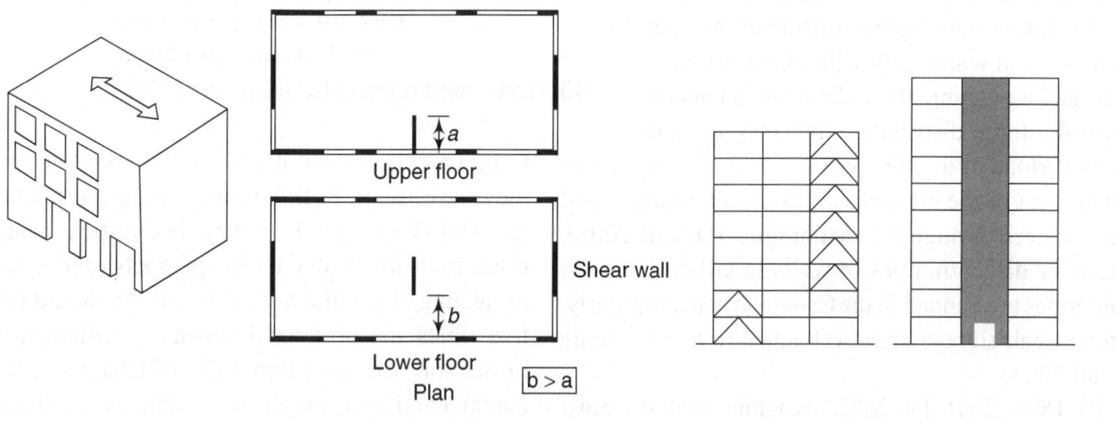


(a) In-plane offset of an element of the lateral force resisting system

(b) Reduction in lateral stiffness of an element in the storey below

**FIG. 17.12** In-plane discontinuity in vertical, lateral force-resisting elements

storey is less than 80% of the storey above (see Fig. 17.13). The storey lateral strength may be calculated as the total strength of all seismic force resisting elements sharing the storey shear in the considered direction. This irregularity is similar to that encountered in a soft storey, and may be taken care of by making the storey stiffer in comparison to the adjacent storeys.

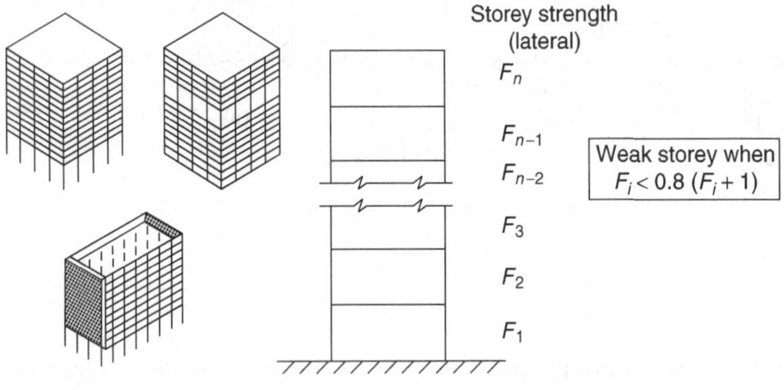


**FIG. 17.13** Weak storey

More details about the architectural planning of buildings in order to reduce earthquake damage are provided by Arnold (2006), Arnold and Raitherman (1982), Ambrose and Vergun (1997), Bachmann 2002, FEMA 454 (2006), Mathews et al. (1977), Murty (*Earthquake Tips* 2004), and Subramanian (1994).

The irregularities discussed till now are found to have a detrimental effect on building behaviour—some of them affect the behaviour considerably and may lead to complete collapse. Stiffness or weight irregularities and torsionally stiff irregular buildings can be dealt with by using a dynamic analysis—which will overcome the shortcomings of applying a static load distribution that is based on a 'first mode' shape for uniform buildings. Torsionally flexile buildings can be dealt with by analysing the structure for an additional torsional moment load case. In-plane or out-of-plane offsets of the lateral load-resisting system (especially with offset walls), are more serious and a dynamic three-dimensional analysis may improve the force distribution, but may not solve the potential overload of the supporting structures, large force transfers in and out of the elements at the discontinuities, and large force transfers through the diaphragms (DeVall 2003). Hence, offsets or discontinuities of walls in tall buildings in high seismic zones are banned in the Canadian code. Similarly, the dangerous weak storeys are also banned in most seismic zones (DeVall 2003).

As per IS 1893 (Part 1): 2002, dynamic analysis must be performed for irregular buildings and regular buildings when they exceed certain heights in different earthquake zones (see Table 3.20 of Chapter 3). As per the Canadian code, equivalent static force procedure may be used for structures in any of the following situations: (a) relatively low seismic hazard, as defined by the short-period (0.2 s) design spectral acceleration; (b) regular structures less than 60 m in height and with fundamental lateral period less than 2 s; or (c) certain irregular structures less than 20 m in height and with fundamental lateral period less than 0.5s. For all other structures, especially irregular structures, dynamic analysis should be used.

## 17.6 OTHER ASPECTS OF PLANNING AND DESIGN IN EARTHQUAKE ZONES

The openings in walls should be located centrally and should be of small size so that the walls are not unduly weakened. Ventilators provided near the edges of walls, adjacent to columns, will create a short column effect and result in the failure of the columns. There will be a similar effect if openings are provided from column to column. Since the infill masonry wall is much stiffer than the columns, column hinges form in the column at the top as well as at the top of the infill masonry wall rather than at the top and bottom of the column (see Fig. 17.14). If the column flexural capacity is $M_p$, the shear in the columns increases by the factor $H/h$, where $H$ is the height of the column and $h$ is the height of infill masonry wall, resulting in non-ductile failure of the columns; several failures in the past earthquakes may be attributed to this effect (FEMA P750:2009, Bachmann 2002).

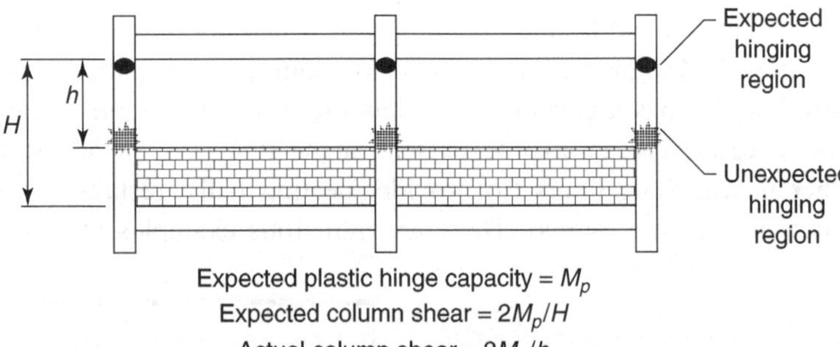


**FIG. 17.14** Short-column effect (FEMA P750:2009)

Long cantilevers and floating columns should be avoided, as they have found to fail during earthquakes. Clause 7.12.2 of IS 1893 (Part 1):2002 states that vertical cantilever projections such as water tanks, parapets, and smoke chimneys projecting above the roof of buildings should be designed for five times the horizontal seismic coefficient, $A_h$, discussed in Section 3.12 (see Eqn 3.28) of Chapter 3. Similarly horizontal cantilever projections such as sunshades (*chhajjas*) and balconies should be designed for five times the *vertical seismic coefficient*. Clause 7.12.2 of IS 1893 (Part 1):2002 suggests that this value may be taken as $10A_h/3$. An excellent introduction to the earthquake protection of non-structural elements in buildings is provided by Murty et al. 2013b.

Concrete stairways often suffer seismic damage due to their inhibition of drift between connected floors. This can be avoided by providing a slip joint at the lower end of each stairway to eliminate the bracing effect of the stairway or by tying stairways to stairway shear walls [see Chapter 17 of Subramanian (2013) for more details].

Masonry and infill (non-structural) walls should be reinforced by vertical and horizontal reinforcing bands to avoid their failure under a severe earthquake. Other non-structural elements should be carefully detailed or tied so that they may not fall under severe shaking. Murty et al. (2013b) and FEMA 454 (see Chapter 9) provide information on earthquake protection of non-structural elements in buildings.

It has to be noted that failure of a beam causes localized effect whereas that of a column may affect the stability of the whole building. Hence it makes good sense to make columns stronger than beams. This can be achieved by appropriate sizing of the member and detailing. This concept is called *strong-column, weak-beam concept* (see Section 17.8.2).

When buildings are too close to each other, they may pound on each other. Connections and bridges between buildings should be avoided and buildings with different sizes and shapes should be adequately spaced. When building heights do not match, the roof of the shorter building may pound the mid-height of the columns of the taller one, resulting in dangerous consequences. The buildings or two adjacent units of the same building should be separated by a distance equal to $R$ times the sum of the calculated storey displacements to avoid pounding (for the value of $R$, see Table 3.22 of Chapter 3). As per clause 7.11.3 of IS 1893 (Part 1): 2002, this value may be multiplied by a factor of 0.5 if the two units have same floor elevation.

## 17.6.1 Consideration of Vertical Component of Earthquake

Clause 6.1.1 of IS 1893 (Part 1):2002 suggests considering the vertical component of earthquake, in structures with large spans, those in which stability is a criterion for design or for overall stability analysis of structures. It also specifies that that special attention should be paid to the effect of vertical component of earthquake on pre-stressed or cantilever beams, girders, and slabs. As discussed in the previous section, IS 1893 suggests that the vertical component be taken as 2/3rd of the peak horizontal component. It has to be noted that the characteristics of the vertical component of earthquake are significantly different than those of the horizontal component. Observations from the damage patterns of the 1989 Loma Prieta, 1994 Northridge, 1995 Kobe, 1999 Chi Chi, 2010 Darfield, and 2011 Christchurch (where a vertical PGA of 2.21 was recorded) earthquakes emphasize the significance of vertical seismic effects, especially in near-field conditions. It was also found that the vertical-to-horizontal (V/H) spectral ratio is a strong function of natural period, source-to-site distance, and local site conditions (Dana et al. 2014). Hence, their effects cannot be ignored in design, especially for structures close to the faults or with heavy mass concentrated at the top, such as huge water tanks, bridges with heavy decks etc., where heavy damages are expected to the lower part of the structure.

The earthquake records show that the vertical component is richer in high frequency content than the horizontal component. This results in high vertical response spectral ordinates at short periods, especially at sites close to the fault. High vertical spectral acceleration at short periods can affect structural systems and components that have short vertical natural periods. In fact, based on the recorded structural response of twelve instrumented structures, Bozorgnia et al. (1998) identified that the vertical natural periods of several structural systems and components fall in the range of 0.075 to 0.26 s. Thus, a period-independent ratio of 2/3, adopted in several codes, is a grossly unconservative approximation of the V/H spectral ratio at short periods, and is a relatively conservative approximation at long periods, especially for sites with firm soil located near active faults (Bozorgnia and Campbell 2004a and 2004b).

Several researchers have concluded that certain compressive, tensile, shear, and flexural failures in structural members may be due to the high values of seismic vertical forces. In steel structures subject to near-field ground motions, the vertical ground motion may induce ultra-low cycle fatigue in connections. Furthermore, vertical frequencies in buildings are not influenced by changes in height or lateral stiffness of buildings. This can lead to resonance and significant component demands when the fundamental vertical frequencies of the structures match the vertical pulses of ground motions (Dana et al. 2014). Recent earthquakes and studies have shown that the vertical-to-horizontal (V/H) spectral ratios can reach a value of 1.7 for short periods and 0.7 for long periods. This implies that the commonly adopted code specified ratio of V/H = 2/3 is far exceeded, especially in the short-period range. This effect is pronounced in near-fields of high-frequency ground motions and in unconsolidated soil environments; hence in such situations, site specific spectra has to be used in the design for vertical ground motions (Dana et al. 2014).

Recent research has also shown that the vertical components of earthquakes have less pronounced effects in the perimeter/corner columns than in the interior columns. Perimeter/corner columns receive more forces from seismic horizontal components than those of interior columns as the perimeter/corner columns provide resistance to overturning (Dana et al., 2014). In addition, the contribution of gravity forces is larger for interior columns since the effect of overturning is negligible at interior columns. It has also been found that the vertical component of earthquake affects the forces in the upper storey columns than in the lower storey columns. A similar pattern was observed in beams. The effects due to the vertical components of earthquake are less in beams that have smaller forces due to horizontal component and having smaller tributary areas. The effects of the vertical component of earthquake were found to be more pronounced in the interior beams, upper storey beams and the long span beams. Hence care should be taken while checking the upper storey beams and columns of multi-storey buildings, especially at sites close to the fault.

## 17.7 SEISMIC FORCE RESISTING SYSTEMS

Several systems can be adopted to provide adequate resistance to seismic lateral forces. The most common systems are moment-resisting frames (though they consist of a three-dimensional space frame, for analysis purposes, they may be considered as two-dimensional in several cases); a combined system of moment frames and shear walls; braced frames with horizontal diaphragms; and a combination of these systems (see Fig. 17.15 and Table 17.2). Out of these, moment-resisting frames may be economical for buildings with only up to 30 storeys (the infill walls of non-reinforced masonry also provide some stiffness). Shear wall and braced systems (which are more rigid than moment-resisting frames) are economical up to 40 storeys (see also Fig. 2.15 of Chapter 2). When frames and shear walls are combined, the system is called *dual system.* A moment-resisting frame when provided with specified details for increasing the ductility and energy absorbing capacity of its components is called a *special moment-resisting frame (SMRF)* or *special moment frame* (*SMF*); otherwise it is called an *ordinary moment-resisting frame (OMRF)* or *ordinary moment frame (OMF).* Similarly, braced frames may be classified as *ordinary concentrically braced frames* (*OCBF*), *special concentrically braced frames* (*SCBF*), and *eccentrically braced frames* (*EBF*).

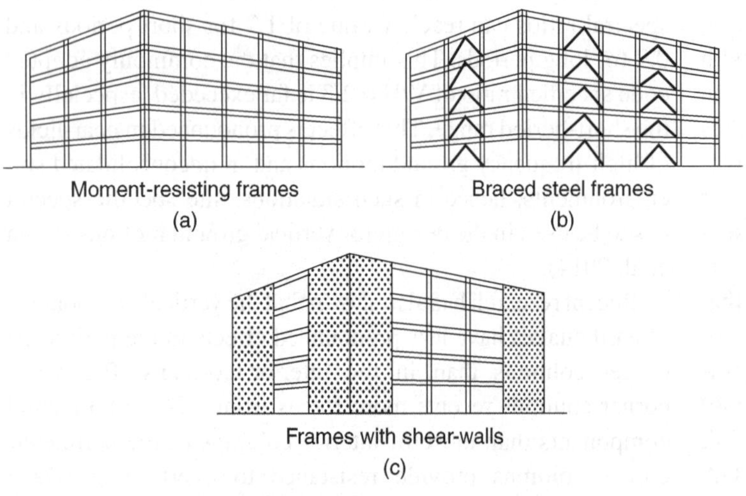


**FIG. 17.15** Lateral force-resisting systems

The design engineer should not consider the structure as composed of a summation of different parts (such as beams, columns, trusses, and walls) but as a completely integrated system, which has its own properties with respect to lateral force response. Thus, he or she should follow the flow of forces through the structure into the foundation and make sure that every connection along the path of forces is adequate to maintain the integrity of the system. It is also necessary to provide adequate redundancy in the structure. When a primary system yields or fails, the redundancy allows the lateral forces to be redistributed to a secondary system to prevent *progressive collapse*. It has to be realized that the forces due to earthquake are not static but dynamic, (cyclic and repetitive) and hence the deformations will be well beyond those determined from the elastic design.

Recently, innovative design concepts have been developed to better protect structures, together with their occupants and contents, from the damaging effects of destructive environmental forces including those due to winds, waves and earthquakes. *Base isolation* is a passive structural control system, pioneered in New Zealand by Dr Bill Robinson during the 1970s, where some isolators are used to substantially decouple the building from its foundations resting on shaking grounds, thus protecting the structural integrity of the building. There are four types of base isolation – elastomeric, sliding, rolling, and a combination of these. With these types of base isolation, the drift of the building during an earthquake is greatly reduced.

Another concept is the use of *passive energy dissipation devices* or *dampers* which absorb or consume a portion of the input energy, thereby reducing energy dissipation demand on primary structural members and minimizing possible structural damage. Devices that are commonly used for seismic protection of structures include viscous fluid dampers, viscoelastic solid dampers, friction dampers and metallic dampers [Soong and Dargush (1997) and Symans et al. (2008)]. These innovative devices are briefly discussed in Section 17.14.

An important aspect of earthquake-resistant design is to understand the impact of earthquake on the structure, and the damage it suffers. While less building damage will result in low repair costs, significant damage (or collapse) will result in higher cost of repair/replacement and also significant time for repair. Thus, a successful seismic design should result in the selection of a structure that is not only economical but also suffer minimum damage with corresponding low post-earthquake repair cost. The behavior of each structural system will differ with the type and duration of earthquake and soil types (FEMA 454:2006). Systems with stable cyclic behavior, good energy dissipation, and controlled inter-storey drift will yield low repair costs (Fig. 17.16).

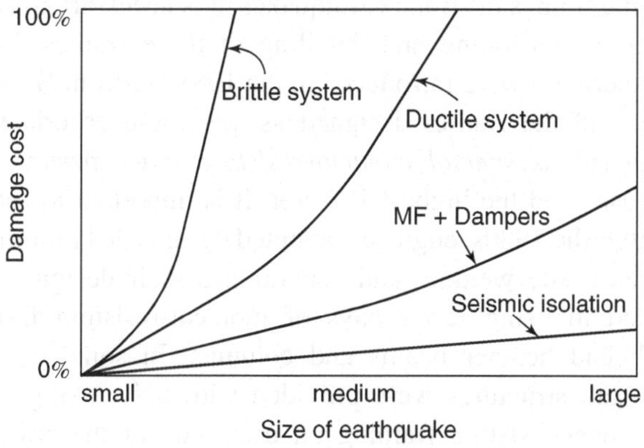


**FIG. 17.16** Performance of different systems during earthquakes

Table 17.2 compares the different seismic system strategies discussed in this section (FEMA 454:2006). Some reinforced concrete systems are also included for comparison. The response reduction factors shown in the last column of the table are from IS 1893 (Part 1): 2002 and IS 800: 2007 (The values of $R$ shown in brackets are from ASCE 7-10-They are not to be compared directly with the values given in Indian codes, as the formulation in these codes are different). It has to be noted that ASCE 7-10 provides $R$ values for 85 different building systems, whereas IS 1893 (Part 1): 2002 provides only for 14 systems.

## 17.8 MOMENT-RESISTING FRAMES (MRFs)

Although steel 'special' moment frame is a relatively new concept used in building codes, steel moment frames have been used for more than one hundred years. The 10-storey,

**TABLE 17.2** Comparison of different seismic system characteristics

| System | | Non-linear drift | Cyclic behaviour | Energy dissipation | Post EQ repair cost | Response reduction factor, R (ASCE 7-10) |
|---|---|---|---|---|---|---|
| Steel frame with unreinforced masonry wall | 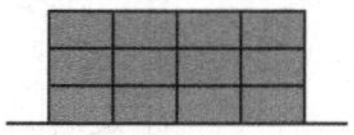 | Medium | Stable | Medium | Medium | 1.5 (3) |
| Steel frame + *RC* shear walls | 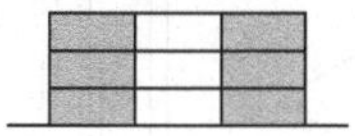 | Medium | Stable | Medium | Medium | 3–4 (6–7) |
| Non-ductile *RC* MF | 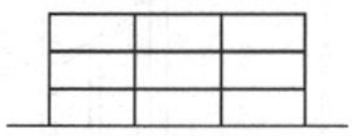 | Large to collapse | Unstable | Low | High | 3 (3) |
| Steel concentrically braced frame | 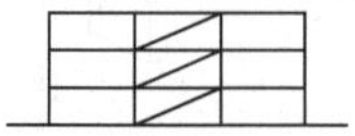 | Large to collapse | Unstable | Low | High | 4.5 (3.25–6) |
| *RC* Shear walls | 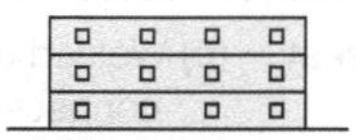 | Medium | Stable to Unstable | Medium to high | Medium to high | 3–4 (5–6) |
| Steel OMF | 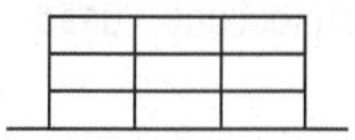 | Medium to Large | Unstable | Medium | High | 4 (3.5) |
| Steel SMF | 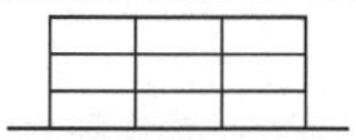 | Medium to Large | Stable to semi-stable | Medium | Medium to high | 5 (8) |
| Ductile *RC* MF | 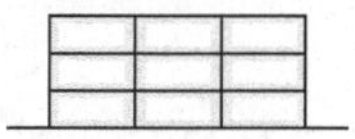 | Medium to Large | Stable to semi-stable | Medium | Medium to high | 5 (8) |
| Steel EBF | 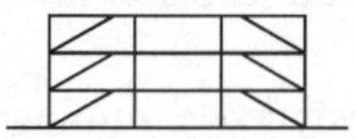 | Low to medium | Stable | Medium to high | Low to medium | 5 (8) |
| Damper + steel MF | 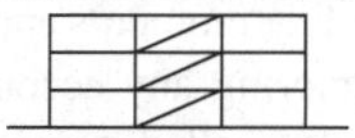 | Low to medium | Stable | Very high to high | Low | –(8.5*) |
| Buckling restrained braced frame | 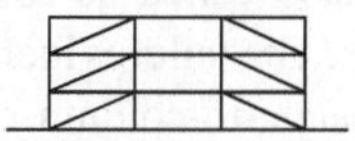 | Medium | Stable | Medium to high | Low | – (8) |
| Seismic isolation | 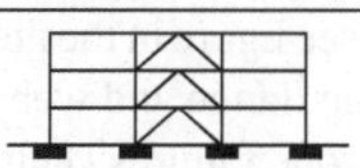 | Low inter-storey drift | Stable | Very high | Very low | – (1.6–2.0*) |
| Rocking system | 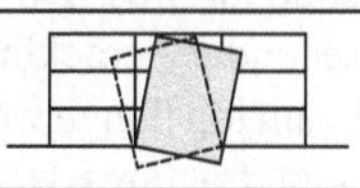 | Large rocking motion | Stable | High | Low | – (8.5*) |

*suggested in FEMA 454:2006

42 m tall Home Insurance Building in Chicago, built in 1884, is considered the first skyscraper with steel moment frames. In just 28 years, skyscraper technology progressed to the 60-storey Woolworth Building in New York. This and other tall buildings, built till the 1900s, were constructed with built up H-sections and the steel framing, at their perimeters, were completely encased by masonry, concrete, or a combination of both, to provide fire resistance. However, the designers of these early moment frame structures neglected the contribution of concrete/masonry encasements, which substantially stiffened and strengthened the steel framing and provided significant fixity at the framing connections for both gravity and lateral loads. By the early 1900s, the built-up sections were replaced by rolled H-sections (a good example being the Empire State Building in New York, USA; see Fig. 1.43(d) of Chapter 1). Most or all of the columns, interior and exterior, were part of the lateral load-resisting system. The weight of the walls, cladding, and floor systems of these buildings were so great that their column design was controlled by the gravity loads and not by the lateral loads.

The buildings built after World War II adopted modern glass and aluminum curtain wall systems instead of the infill perimeter masonry walls, for reasons of economy. These frames had connections without gussets, using angles or split tees to connect the top and bottom flanges to the steel H-columns, as shown in Fig. 17.17(a). When welding was introduced in the 1950s, the angles and split tees were replaced by flange plates that were shop welded to the column flanges, then riveted to the beam flanges (rivets were replaced by high strength bolts in the 1960s). During 1970–1994, designers used the connection as shown in 17.17(b), which had field-welded, complete joint penetration (CJP) groove welds to connect beam flanges to columns, and shop-welded, field-bolted shear plates joining beam webs to columns.

Earlier research by Professor Egor Popov at the University of California at Berkeley in the 1960s and 1970s showed that in order to obtain superior inelastic behavior of steel moment-resisting frames in strong earthquakes, it is necessary to adopt proper proportioning and detailing of these frames. These design criteria were introduced in the 1988 Uniform Building Code, and the frames designed as per these criteria were designated as *special moment-resisting space frames*, and were assigned the highest $R$ factor. It is important to realize that after the 1980s, engineers adopted designs that minimized expensive site welding and economized their designs. This resulted in using fewer bays of moment-resisting frames, which had heavier beams and columns. In some extreme cases, tall structures were provided with only a single bay of moment-resisting framing on each side of the building, resulting in less redundant structures with more concentrated lateral force resistance (Hamburger et al. 2009).

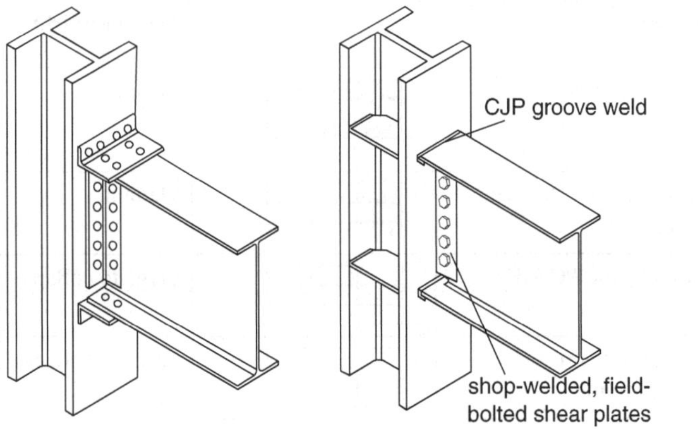


(a) Riveted, unstiffened seat angle connection (b) Welded unreinforced flange-bolted web connection used during 1970-1994

**FIG 17.17** Earlier connections used in moment-resistant frames (Hamburger et al. 2009)

## CASE STUDY

### Damages during 1994 Northridge Earthquake

After the 1994 Northridge earthquake (Los Angeles, USA), engineers found that the welded moment connections in the special moment-resisting frames of over 200 buildings experienced brittle fractures in this magnitude 6.7 earthquake [Gioncu and Mazzolani (2002), Yang and Popov (1997)]. Until this earthquake, many engineers regarded these buildings as highly resistant to earthquake damage. These beam-to-column connections experienced rotation levels well below the plastic moment capacity of framing members, but still suffered damages. (None of these steel frame buildings collapsed, but the unexpected type and severity of damage proved the inadequacy of the building code provisions available at that time). Most of the failures occurred in the upper half to two-thirds of tall buildings and at all levels in low-rise buildings (fewer than six storeys).

The joints were designed assuming that the top and bottom flanges resisted the bending moments and the shear was resisted by the web connection, ignoring any eccentricity in the connection. Horizontal stiffeners (also called as *continuity plates*) were also provided in the columns. The failures included non-ductile fractures of the beam bottom flange-to-column complete-joint-penetration (CJP) groove welding (see figure in the next page) which propagated into the adjacent column flange and web and into the beam bottom flange, and also panel zone failures. This failure was accompanied in some instances by secondary cracking of the beam web shear plate and failure of the beam top flange weld. Even though a number of these different types of fractures were observed, the majority

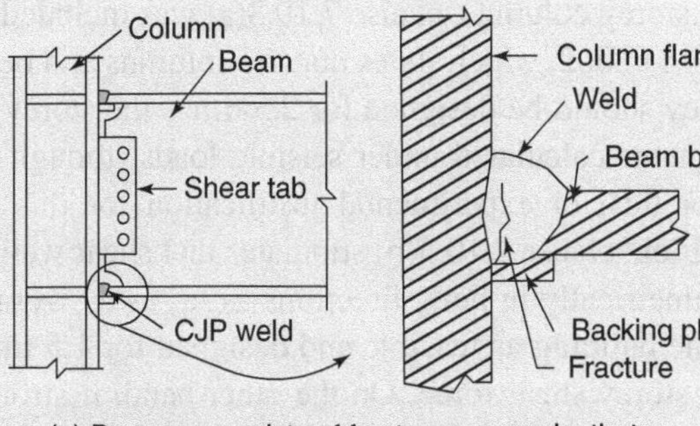


(a) Damage consists of fractures or cracks that initiate in the welded joints of the beams to columns

(b) Fracturing of W14 column at welded beam-to-column connection in Northridge earthquake

Damages to beam-column connections during 1994 Northridge earthquake, USA; Severe stress concentrations inherent in the configuration of the connection were not considered in the design

*Source:* FEMA 354, FEMA 355D

of these fractures were found near the welded bottom beam flange joint. Failures occurred in connections with and without column-flange stiffeners as well as connections with and without return welds on the shear connection plates. Both wide-flange columns and built-up box sections appear to have been affected. The damage to steel buildings during the 1995 Kobe earthquake in Japan (in which damage occurred in near-field regions due to very high velocities) was even more disturbing: 10% of the steel buildings in Kobe designed to current Japanese building standards collapsed.

The factors that contributed to the damage include the following (Williams 2004):

1. Stress concentration at the bottom flange weld, due to the notch effect produced by backing strips left in place
2. Poor welding practices, including the use of weld metal of low toughness
3. Uncontrolled deposition rates
4. The use of larger members than those previously tested or the use of higher strength girders
5. Lack of control of basic material properties (large variation of member strength from the prescribed values)
6. Less system redundancy and higher strain demands on connections
7. Inadequate quality control during construction
8. The tri-axial restraint existing at the center of beam flanges and at the beam-column interface, which inhibited yielding

The FEMA (Federal Emergency Management Agency) formed in mid-1994 the *SAC Joint Venture* [consisting of the Structural Engineers Association of California (SEAOC), the Applied Technology Council (ATC), and Consortium of Universities for Research in Earthquake Engineering (CUREE)] and funded $12 million over eight years. Their goal was to develop reliable, practical, and cost-effective guidelines and standards of practice for the repair or upgrading of damaged steel moment frame buildings, the design of new steel buildings, and the identification and rehabilitation of at-risk steel buildings. Detailed analytical studies of components, sub-assemblages, and complete structural systems and experimental testing of over 120 full-scale sub-assemblages were conducted. This resulted in the publication of several reports- FEMA 350, 351, 352, 353, and 355 series, which formed the basis for the AISC 341-05 code (Now, AISC 341-10) on *Seismic Provisions for Structural Steel Buildings* (www.sacsteel.org). In addition to this code, AISC developed another ANSI (American National Standards Institute) approved standard, AISC 358-05, which presents materials, design, detailing, fabrication, and inspection requirements for a series of pre-qualified moment connections. AISC updates and reissues this standard from time to time, as and when additional research results are available.

## 17.8.1 Design Principles for Special Moment Frame

The design base shear equations provided in Clause 7.5.3 of IS 1893 (Part 1):2002 and Clause 12.3 of IS 800:2007 incorporate a factor of $2R$ in the denominator of the equation for determining the horizontal seismic coefficient $A_h$ (see Section 3.12 and Eqn 3.28 of Chapter 3), which reflects the expected degree of inelastic response for design-basis earthquakes, and also the ductility capacity of the framing system. As per Table 23 of IS 800, steel special moment frame (SMF) is permitted to be designed using $R = 5$ [ordinary moment frame (OMF) should be designed with an $R$ value of 4 only], and is expected to sustain multiple cycles of significant inelastic response when subjected to design-level ground motion. As mentioned earlier, several steel SMF may have substantial over-strength, due to a number of factors: columns and beams having more than the designed area and moment of inertia, due to the discrete available sections, oversized columns to meet strong-column/weak-beam criteria, use of oversize sections to provide sufficient stiffness for drift control, and variability in the strength of the steel material itself. As a result, several steel SMF may remain elastic during earthquake shaking.

The proportioning and detailing requirements specified in Section 12.10 (for OMF) as well as Section 12.11(for SMF) of IS 800:2007 are intended to provide ductile inelastic response. The primary goals of these provisions are as follows (Hamburger et al. 2009):

1. Achieve a strong-column/weak-beam condition that distributes inelastic response over several stories
2. Avoid P-delta instability under gravity loads and anticipated lateral seismic drifts
3. Incorporate details that enable ductile flexural response in yielding regions

### 17.8.2 Strong-column, Weak-beam Concept

When buildings are subjected to earthquake loads, plastic hinges will be formed at the ends of the members where there are heavy bending moments (as they are often designed only for a fraction of the earthquake loads), and subsequently will fail when there are enough plastic hinges to form a mechanism. A few possible mechanisms in which a structure may fail is shown in Fig. 17.18. The distribution of damage over the height of the building depends on the distribution of lateral drift. If the building has weak columns or long columns in a particular storey, most of the inelastic portion of the structure's drift will be concentrated in these storeys, as shown in Fig. 17.18(a), resulting in very large P-$\Delta$ effects at those locations (note that in the current design practice, contribution masonry infill is considered in the mass of the building, but neglected in estimation of stiffness; thus the actual behavior of building is not captured in design). Such soft storey columns should generally be avoided for better performance of the structure. Due to the failure of several buildings having soft storey columns, clause 7.10.3(a) was included in IS 1893 (Part 1):2002, which states that the columns and beams of soft storey should be designed for 2.5 times the storey shears and moments calculated under seismic loads (though there is no theoretical or experimental justification for this clause). In addition, clause 7.10.3(b) stipulates that shear walls be placed symmetrically in both directions as far away from the center of the building as feasible and designed for 1.5 times the lateral storey shear force. On the other hand, if strong columns are provided throughout the building height, drift will be more uniformly distributed (Fig. 17.18c), and localized damage will be reduced. Buildings, with columns which have equal strength as beams, may result in an intermediate mechanism as shown in Fig. 17.18(b).

It is also important to recognize that the columns in a given storey support the weight of the entire building above those columns, whereas the beams only support the gravity loads of the particular floor; hence, failure of a column is of greater importance than the failure of a beam, as column failure will result in total collapse of the entire building. Recognizing this fact, building codes often specify that columns should be stronger than the beams that frame into them. This *strong-column/weak-beam principle* is fundamental to achieving safe behavior of frames during strong earthquake ground shaking. Buildings, following this principle, will fail in *beam-hinge mechanism* (beams yielding before the columns) and not in the *storey mechanism* (columns yielding before the beams). Storey mechanism must be avoided as it causes greater damage to the building). Therefore, column should be stronger than the beams meeting at a joint. Note that the beam-hinge

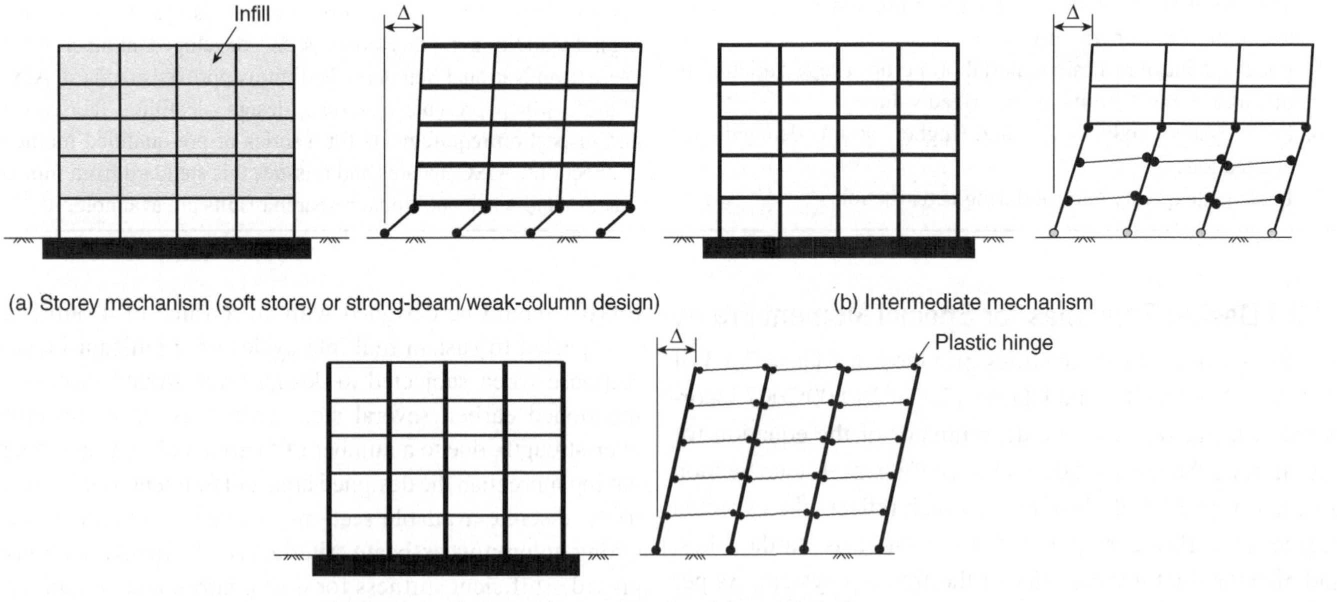


(a) Storey mechanism (soft storey or strong-beam/weak-column design)

(b) Intermediate mechanism

(c) Beam mechanism (strong-column/weak-beam design)

**FIG. 17.18** Different failure mechanisms of moment-resisting frames

mechanism also has a plastic hinge at the base of ground floor column (see Fig. 17.18c); hence it is important not to have any splicing at this location and to have ductile detailing, so that the required rotation capacity is achieved.

Murty et al. (2013a), based on parametric studies, found that an increase in the column to beam strength ratio increases (a) lateral load (base shear) capacity of the building and (b) lateral deformation and ductility capacity of the building (see Fig. 17.19). Even though this study is based on reinforced concrete frames, similar behaviour is likely in steel moment frames also. The sequence of hinge formation is critical in a building in addition to its capacity curve and the location of plastic hinges in the building. It is preferable to have hinges to form in beams before they are formed in columns. Such a possibility is found to occur when column to beam strength ratio is more than 3.6 (Murty et al. 2013a). As the column to beam strength ratio of about 3 to 4 is impractical in most practical cases, a lower strength ratio of 1.2 is adopted in clause 12.11.3.2 of IS 800: 2007 (see Eqn 17.1) at the joint, whereas Section E3 of AISC 341-10 adopts a column to beam strength ratio of 1.0 only at the joint. Due to this some column yielding associated with an intermediate mechanism, as shown in Fig. 17.18(b) is to be expected, and the columns must be detailed accordingly. Flexural yielding of columns at the base is permitted by AISC 341-10.

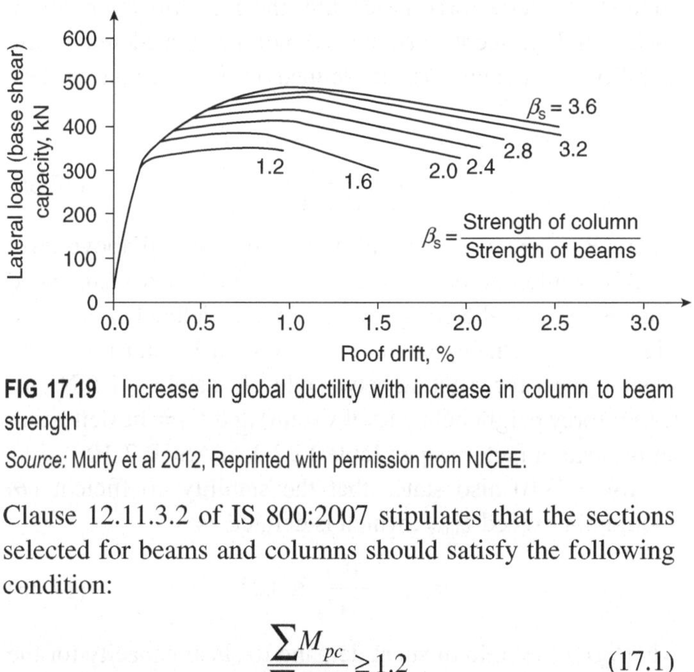


**FIG 17.19** Increase in global ductility with increase in column to beam strength

*Source:* Murty et al 2012, Reprinted with permission from NICEE.

Clause 12.11.3.2 of IS 800:2007 stipulates that the sections selected for beams and columns should satisfy the following condition:

$$\frac{\sum M_{pc}}{\sum M_{pb}} \geq 1.2 \tag{17.1}$$

where, $\sum M_{pc}$ is the sum of the nominal flexural strengths of the columns (including haunches where used) above and below the joint to the beam centerline, with a reduction for the axial force in the column (When the centerlines of opposing beams in the same joint do not coincide, the mid-line between centerlines may be used), and $\sum M_{pb}$ *is* the sum of expected flexural strengths of the beams at the intersection of the beam and column centerlines.

When determining available column flexural strength, it is important to consider the axial loads in the column, as they will reduce the moment capacity (see Section 8.10 of Chapter 8). Hence, AISC 341-10 suggests that the value of $\Sigma M_{pc}$ may be calculated using

$$\sum M_{pc} = \sum Z_{pc}(f_{yc} - f_a) \text{ with } f_a = P_{uc}/A_g \geq 0 \tag{17.1b}$$

The value of $\Sigma M_{pb}$ may be calculated using AISC 341-10 as

$$\sum M_{pc} = \sum (M_p + M_v) \tag{17.1 c}$$

where, $A_g$ is the gross area of column, mm$^2$; $f_{yc}$ is the specified minimum yield stress of column, MPa; $Z_{pc}$ is the plastic section modulus of the column about the axis of bending, mm$^3$; and $P_{uc}$ is axial force in the column, N; $M_v$ (the additional moment due to shear amplification) = $V_p S_h$, $V_p$ is the shear at plastic hinge location, N; and $S_h$ is the distance from column center line to plastic hinge location, mm.

As per AISC 341-10, this requirement of Eqn (17.1) need not be met when $f_a < 0.3f_{yc}$ for all load combinations, except for the two special factored load combinations specified in AISC 341-10 when the factored axial load on the column exceeds 40% of the nominal capacity and any of the following conditions hold:

1. The joint is at the top storey of a multi-storey frame
2. The joint is in a single-storey frame.
3. The sum of the available shear strengths of all exempted columns is less than 20% of the available shear strengths of all moment frame columns for a specific storey in the total frame and the sum of the available shear strengths of all exempted columns in a specific frame is less than 33% of the available shear strengths of all moment frame columns.

## 17.8.3 Provisions in IS 800:2007 for SMF

As per IS 800:2007, SMF can be used in any seismic zone and for any importance-factor value. According to Clause12.11of IS 800:2007, the beam and column sections used in SMF should be either plastic or compact; in addition at potential plastic hinge locations, plastic sections only should be used. It has to be noted that Table D1.1 of AISC 341 defines separate *b/t* limits for members subject to seismic loads and are parts of SMF; these limits are prescribed for highly ductile and moderately ductile members; these 'seismically compact' sections are expected to achieve a deformation ductility of at least 4. When the column-beam moment ratio as per Eqn (17.1) is above 2.0, AISC 341-10 suggests that the columns will remain elastic. Hamburger et al. (2009) cautions that the strong-column/weak-beam provisions of AISC 341-10 may not be adequate to avoid formation of storey mechanisms in all cases; hence, the designers may increase column sizes, beyond code requirements, to obtain better performance during severe earthquakes. Such an increase in sizes may reduce the need to provided expensive web stiffener or doubler plates in beam-column junctions, but will increase the total weight of steel used. Other provisions in IS 800:2007 are similar but are not as elaborate as those given in AISC 341.

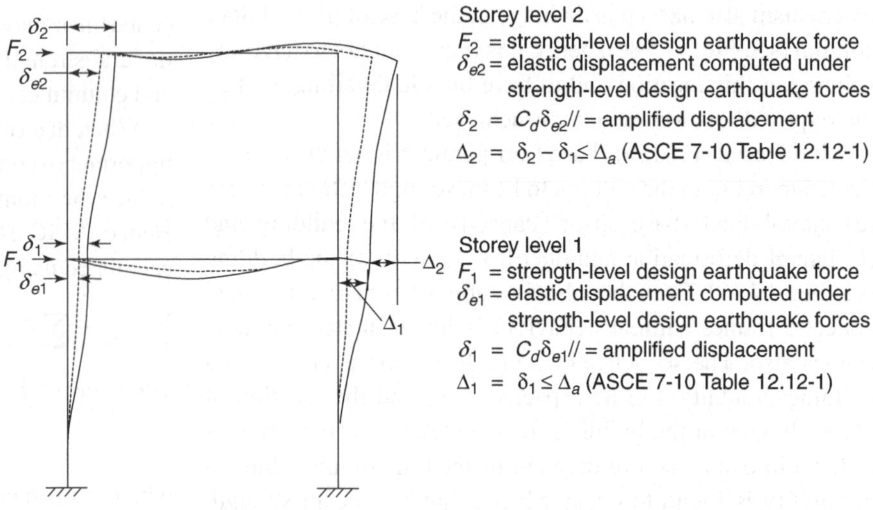


**FIG. 17.20** Storey drifts determination in special moment frames, as per ASCE 7-10

## 17.8.4 Proportioning for Drift

It is important to realize that the size of beams in a steel SMF is controlled by the consideration of drift. Consequently, the size of columns is also drift controlled, because the strong-column/weak-beam concept demands larger columns when larger beams are provided. The only exception may be the end columns in SMF, which have high axial load demands, and hence may be controlled by strength design criteria (Hamburger et al. 2009).

Members of the seismic force-resisting system (SFRS) that are anticipated to undergo inelastic deformation have been classified as either *moderately ductile members* or *highly ductile members*. As per the code, during the design earthquake, highly ductile members in special moment frames (SMF) are anticipated to undergo significant plastic rotation of 0.04 radians or more without degradation in strength and stiffness below the full yield value ($M_p$). The member rotations result from either flexure or flexural buckling.

The checking of *storey drift* (storey drift may be defined as the displacement of one level relative to the other level above or below) for lateral loads should be as per IS 1893 (Part 1):2002. This code stipulates that the storey drift in any storey (due to the minimum specified lateral loads with a partial load factor of 1.0), should not exceed 0.004 times the storey height. For buildings located in seismic zones IV and V, it should be ensured that the structural components, that are not a part of the seismic force-resisting system in the direction of consideration, do not lose their vertical load-carrying capacity under the induced moments resulting from storey deformations equal to $R$ times the storey displacements [see also Clause 7.11 of IS 1893 (Part 1):2002].

As per ASCE 7-10, the storey drift ($\Delta$) is computed as the difference of deflections at centers of mass at top and bottom of storey (see Fig. 17.20). The deflection at level $x$ at the center of mass ($\delta_x$) is determined as follows:

$$\delta_x = \frac{C_d \delta_{xe}}{I} \tag{17.2}$$

where, $C_d$ is the deflection amplification factor, as given in Table 12.2-1 of ASCE 7-10 (for SMF, $C_d = 5.5$ and for OMF $C_d = 3$; Generally $C_d$ is 1/2 to 4/5 the value of $R$), $\delta_{xe}$ is the deflection determined by elastic analysis, and $I$ is the importance factor. The factor $C_d$ is used to adjust lateral displacements for the structure determined under the influence of design seismic forces to the actual anticipated lateral displacement in response to design earthquake shaking. Generally, the more ductile a system is, the greater will be the difference between the value of $R$ and $C_d$.

It has to be noted that as per ASCE 7-10, when using equivalent lateral force procedure, the P-Δ effects on storey shears and moments drifts need not be considered if the stability coefficient ($\theta$), as defined in Eqn (17.3a) is less than 0.10:

$$\theta = \frac{P_x \Delta}{V_x h_{sx} C_d} \tag{17.3a}$$

where $P_x$ is the total vertical design load at and above level $x$ (kN); while computing $P_x$, no individual load factor need exceed 1.0, $\Delta$ is the design storey drift as defined earlier (see Fig. 17.20) occurring simultaneously with $V_x$ (mm) $V_x$ is the seismic shear force acting between levels $x$ and $x - 1$ (kN), $h_{sx}$ is the storey height below level $x$ (mm) and $C_d$ = the deflection amplification factor as per Table 12.2-1 of ASCE 7-10.

ASCE 7-10 also states that the stability coefficient ($\theta$) should not exceed $\theta_{\max}$, which is defined as:

$$\theta_{max} = \frac{0.5}{\beta C_d} \le 0.25 \tag{17.3b}$$

where $\beta$ is the ratio of shear demand to shear capacity for the storey between levels $x$ and $x - 1$. This ratio may be conservatively taken as 1.0. When the stability coefficient ($\theta$) is greater than 0.10 but less than or equal to $\theta_{max}$, the incremental factor related to P-Δ effects on displacements and member forces shall be determined by rational analysis. Alternatively, it is permitted to multiply displacements and member forces by $1.0/(1 - \theta)$, to include P-Δ effects. When $\theta$ is greater than $\theta_{max}$, the structure will become unstable and hence should be redesigned. When the P-Δ effect is included in an automated analysis, Eqn (17.3b) should still be satisfied, however, the value of $\theta$ computed from Eqn (17.3a) using the results of the P-Δ analysis is permitted to be divided by $(1 + \theta)$ before checking Eqn (17.3b). The computed drift has to be compared with the allowable drift as given in Table of 12.12-1 ASCE 7-10 [it ranges between 0.01 $h_{sx}$ to 0.02 $h_{sx}$, where $h_{sx}$ is the storey height below level $x$ (mm)]. If the computed drifts are more than allowable, the stiffness of the members have to be increased, analysis and design performed again and the drifts compared again to check whether they are within allowable limits. It has to be noted that the response spectrum method (see Section 3.12.4 of Chapter 3) will result in more economical designs for steel SMF than the equivalent static method (Hamburger et al. 2009).

### Joint Panels

It is important to realize that the storey drifts are caused by flexural and shear deformations in beams and columns and also by the shear deformations in joint *panel zones*, causing a shear (rocking) mode of drift, and by axial deformations in the columns causing flexural mode of drift (Hamburger et al. 2009). In general, the beam bending is the largest contributor, while column bending is the smallest. Panel zone shear deformations may contribute to about 15% to 30% to the total shear mode of drift. Clause 12.7.3b of ASCE 7-10 stipulates that the contribution of panel zone deformations to overall storey drift should be included, when checking drift limits. Hamburger et al. (2009) provide approximate equations to calculate storey drift due to beam flexure, column flexure and panel zone shear deformations. The effect of panel zone shear deformations can also be directly incorporated in the analytical model by using the *scissors elements* or a *panel zone parallelogram model*. Potential failure modes include shear buckling, and if doubler plates are included, fracture at the welds. Failure modes may also include column flange bending, web crippling, and web buckling. Checking of panel zones and the joint panel model proposed by Krawinkler (1978) are discussed briefly in Section 4.5.2 of Chapter 4. More details about these analytical models and the behaviour of panel zones can be found in Krawinkler (1978); FEMA-335C (2000); and Lee, et al. (2002).

### Beam-column and Other Connections

As per Clauses 12.11 of IS 800:2007, all the beam-column connections of SMF should be rigid and should be designed to withstand a moment of at least 1.2 $M_p$ of the connected beam. When reduced beam section is used (see Section 17.10), its flexural strength, determined at the column face, should be at least 0.8 $M_p$ of the unreduced section, at a storey drift angle of 0.04 rad.

The beam-to column connection of SMFs should be designed to withstand a shear resulting from the load combination of 1.2DL + 0.5LL plus the shear resulting from the application of 1.2 $M_p$ in the same direction at each end of the beam. In column connections along the strong axis, the *panel zone* has to be checked for shear buckling in accordance with clause 8.4.2 of IS 800. Column web doubler plates or diagonal stiffeners may be used to strengthen the web against shear buckling. Continuity plates (tension stiffener) should be provided in all strong axis welded connections, except in end plate connection. Depending on the type of connection used (see Section 17.9), the following failure modes have to be checked [Hamburger et al. (2009)]:

1. Net section fracture at bolts, shearing and tensile failure of bolts, bolt bearing, and block shear failures
2. Fracture in or around welds
3. Fracture in highly strained base material
4. Fracture in weld access holes

Failure modes of column splices are similar to those of beam-to-column connections. Failure modes of column bases may include anchor bolt failures or pull out, fracture in base plate, or in column-to-base plate connections, and local buckling of gusset plates.

## 17.8.5 Continuity Plates

In a rigid, fully welded connection, *continuity plates* of thickness greater than the thickness of beam flanges have to be provided and welded to the column flanges and the web. The individual thickness of column webs and doubler plates should be such that (see Fig. 17.21)

$$t \geq (d_p + b_p)/90 \qquad (17.4)$$

where $t$ is the thickness of column web or doubler plate, $d_p$ is the *panel zone* depth between the continuity plates, and $b_p$ is the panel zone width between column flanges.

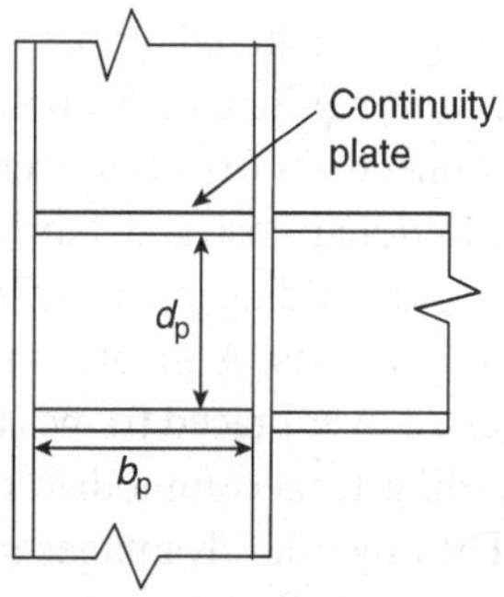


FIG. 17.21 Continuity plates

### Protected Zones

As per AISC 341-10, the plastic hinging zones at the ends of SMF beams should be treated as *protected zones*. In general,

## CASE STUDY

### Torre Latinoamericana—The skyscraper that withstood two earthquakes

Torre Latinoamericana, Mexico

*Source:* Reprinted with permission from Mr Juan Pablo Ortiz Arechiga through www.flickr.com

The Torre Latinoamericana (English: Latin–American Tower) is a skyscraper in downtown Mexico City, Mexico, built during 1956 (height 188 m with 44 storeys). It is widely recognized internationally as an engineering and architectural landmark since it was the world's first major skyscraper successfully built on highly active seismic land.

The project was designed and executed by Dr Leonardo Zeevaert and his brother Adolfo Zeevaert, Mexican civil engineers born in Veracruz. Prof. Nathan M. Newmark, of the University of Illinois was the main consultant. Its design consists of a steel-frame construction and deep-seated piles, which were necessary due to the Mexico City's frequent earthquakes and muddy soil composition. Before construction, both engineers carried out a number of soil mechanics studies at the construction site, and designed the structure accordingly. Today this is a mandatory practice, but at that time it was quite an innovation. Foundation work started in 1949 with 361 numbers of 33 m deep pylons. While it was being built, detractors said that this huge building will not withstand Mexico City's earthquakes.

The tower gained international attention when it withstood the 7.9-magnitude earthquake in 1957, thanks to its outstanding design and strength. This performance resulted in the recognition in the form of the American Institute of Steel Construction Award of Merit for 'the tallest building ever exposed to a huge seismic force' (as is attested by plaques in the building's lobby and observation deck). Interestingly, this building experienced another bigger earthquake on 19 September 1985, with a magnitude of 8.1, which destroyed many buildings in Mexico City, especially the ones built in the downtown area, in the tower's neighborhood. The Torre Latinoamericana withstood the force produced by this earthquake also without problems, and has thus become a symbol of safety in Mexico City. Today, the tower is considered as one of the safest buildings in the Mexico city despite its potentially dangerous location.

During the September 1985 earthquake, the engineer Adolfo Zeevaert was inside his office on the 25th floor. From that vantage point he was able to witness the destruction taking place; collapse of several buildings and the dust cloud that followed—all the while feeling the movements inside the tower. It could arguably be said that it was the first time that a builder and designer of a tall building witnessed first-hand its behavior during a massive earthquake.

---

for unreinforced connections, the protected zone will extend from the face of the column to one half of the beam depth beyond the plastic hinge point.

Special moment frames have the advantages of architectural simplicity and relatively low base shear. Moment-frame column and beam sizes can be significantly heavier per linear foot than in braced frames due to their means of transferring forces and resisting lateral drift. Heavier sections lead to higher overall tonnages (material costs), and possibly the need for larger erection equipment. Doubler plates required at beam-columns connections (to strengthen the column web locally) result in additional cost and fabrication time. SMF may require more field-welding than braced frames, which leads to higher erection costs. A greater number of moment frames is often required over braced frames to provide enough stiffness in the building to accommodate drift requirements for serviceability. The other disadvantages include connection cost and connection testing.

## 17.8.6 Ordinary Moment Frames (OMF)

*Ordinary moment frame* (OMF) should not be used in severe seismic zones (IV and V) and for buildings with important factor greater than unity in seismic zone III (Clause 12.10 of IS 800:2007). As per IS 800:2007, during the design earthquake, moderately ductile members in ordinary moment frame (OMFs) are anticipated to undergo moderate plastic rotation of 0.02 radians without degradation in strength and stiffness below the full yield value ($M_p$). For OMFs semi-rigid moment connection is permitted in clause 12.10; however it is desirable to have rigid joints. As per Clauses 12.10 of IS 800:2007, OMFs with rigid moment connections should be designed to withstand a moment of at least 1.2 $M_p$ of the connected beam. Similar to the case of SMF, rigid as well as semi-rigid connections should be designed to withstand a shear resulting from the load combination of 1.2DL + 0.5LL plus the shear resulting from the application of 1.2 $M_p$ in the same direction at each end of the beam. In rigid, fully welded connection, continuity plates of thickness greater than the thickness of beam flanges have to be provided and welded to the column flanges and the web.

When semi-rigid connections are used, the connection should be designed to withstand a moment of at least 0.5 $M_p$ of the connected beam or the maximum moment that can be delivered by the system, whichever is less. In semi-rigid joints, the design moment should be achieved within a rotation of 0.01 radians (The Fyre-Morris model given in Annex F of IS 800:2007 may be used to check the curvature). The stiffness and strength of semi-rigid connections should be considered in the analysis and design, in addition the overall stability of the frame should be ensured. It has to be noted that AISC 341 defines *intermediate moment frames (IMF)* also, which may fall between the requirements of SMF and OMF.

### 17.8.7 Designing Buildings with SMF/OMF

An initial choice must be made about the different members of the frames, based on past experience. After obtaining the seismic forces acting at different levels as explained in section 3.12 of Chapter 3, the forces and moments in different members can then be obtained by using any suitable standard computer program for the various load combinations specified in the code (see Section 3.15 of Chapter 3). If the software considers only elastic analysis, moments due to P-delta effects must also be determined. Then, the building should be checked for the overall drift, storey drift, and joint rotation specified in Section 12 of IS 800:2007. If they are not within the allowable limits, the member sizes have to be increased and the process repeated again till they are satisfied.

The size of columns and beams should also be checked for their required strength and strong-column weak-beam concept. The beam-column connections should also be detailed carefully and the necessary bracings should be designed and provided. The structure must also be designed to resist overturning effects caused by seismic forces. Provisions should be made for increase in shear forces on the lateral force resisting elements resulting from the horizontal torsional moment caused due to the eccentricity between the center of mass and center of rigidity [see Section 7.9 of IS 1893 (Part 1):2002 for these details]. IS 1893 also contains special provisions for buildings with soft or weak storey(s), torsional irregularity, non-structural elements, considerations of diaphragm flexibility, and modeling of brick infill panels.

An example of a twelve-storey moment-resisting steel frame as per the 2009 NEHRP Recommended Seismic provisions is provided by Charney et al. (2013). The cost of SMF can be minimized by using the following rules of thumb:

1. Avoid doubler plates by increasing column sizes.
2. Provide partial penetration welds in lieu of complete penetration welds at beam-to-column moment connections (can be used only in non/low-seismic zones).
3. Use fixed-column base connections to reduce drift. They provide more stiff frames at the first storey than pinned-column base connections do, resulting in reduced drift deflections.
4. Decrease fabrication and/or erection labor cost wherever possible, for example, by using higher sections will result in net a lower overall cost.

More details of the design, design guidance, and detailing and constructability issues of SMF may be found in AISC design guides 12 and 13 and in the NEHRP seismic design technical brief 2 (Hamburger et al. 2009).

## 17.9 SEISMIC MOMENT CONNECTIONS

After the 1994 Northridge earthquake, the AISC 341 code requires that the performance of beam-column connections used in SMF should be demonstrated through testing. Hence full-size testing has to be conducted to ensure that the connections are capable of developing 0.04 radians of inter-storey drift without excessive strength loss, when subject to cyclic loading. As this testing is expensive and only a few laboratories are equipped to conduct such large scale testing, AISC 341 permits the use of prequalified connections (see Section 17.9.4).

Extensive testing has been done in the past on beam-column connections subjected to cyclic loading, and based on these tests the following three basic philosophies have been suggested to improve the connection behaviour (FEMA 350:2000): (a) a toughening scheme, (b) a strengthening scheme, and (c) a weakening scheme. Often all of these schemes are used in combination.

### 17.9.1 Toughened Connections

In toughened connection, attention is paid to the complete penetration weld details between the beam and the column. Notch-tough electrodes are specified, which have a Charpy V-notch value of 27 J at $-18°$ C and 54 J at $21°$ C. In addition, backing strips are removed and replaced with reinforcing fillet welds to eliminate the notch effect and to remove any weld flaws (weld flaws are likely to occur at the bottom flange, where the beam web prevents continuous weld passes). At the top flange, a reinforcing fillet is added to secure the backing bar to the column flange. Test results showed that to achieve the inelastic rotation demands of SMF, the beam web to column connection should be a complete joint penetration weld. It was also realized that the size, shape, and finish of weld access holes should be detailed in such a way as to reduce stress concentrations in the region around the access holes (see Fig. 17.22).

Another important aspect is the provision of column continuity plates in all cases. The thickness of these plates should be at least equal to the thickness of beam flange (not including cover plates) or one half of the total effective flange thickness (including cover plate). Welding of continuity plates to column flanges should be with full penetration groove welds, while the plate to column-web may be with double sided

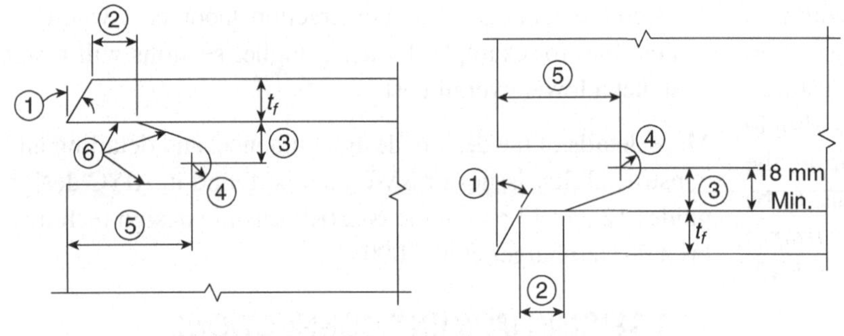


*Notes:* 1. Bevel as required for selected groove weld.
2. Large of $t_f$ or 12mm (plus 0.5 $t_f$, or minus 0.25$t_f$)
3. 0.75 $t_f$ to $t_f$, 20mm minimum (∓6mm)
4. 10mm minimum radius (plus not limited, minus 0)
5. 3 $t_f$ (∓12mm)

Tolerances shall not accumulate to the extent that the angle of the access hole cut to the flange surface exceeds 25°

**FIG. 17.22** Weld access hole details

*Source:* FEMA 350:2000.

fillet welds. Notch-tough electrodes should be used in all cases, and welding in the *k* region of the column should be avoided.

Seismic Structural Design Associates has developed a proprietary system, called DSDA *SlottedWeb$^{TM}$ connection*, which has a horizontal slot in the web of the beam, near the flanges, which separates the beam's web and flange effectively as shown in Fig. 17.23. There is also a shear plate that is shop welded to the column flange and fillet welded to the beam at the site.

The slot in the SlottedWeb$^{TM}$ connection isolates the beam section in such a way that the shear is resisted entirely by the web and the flanges resisting only the bending moment.

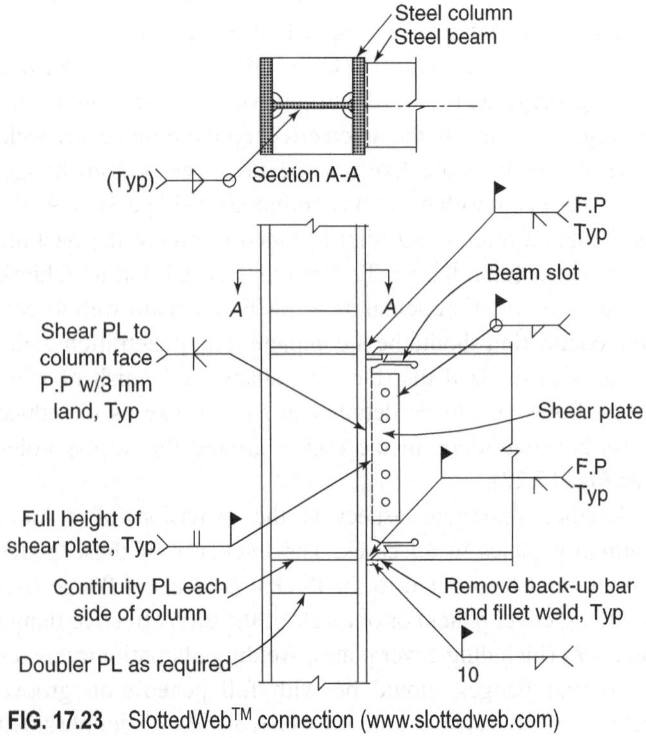


**FIG. 17.23** SlottedWeb$^{TM}$ connection (www.slottedweb.com)

According to SSDA, the SlottedWeb$^{TM}$ connection results in controlled ductile beam yielding and also prolonged fatigue life due to the following:

1. The beam web slots eliminate the seismic shear in the beam flanges, which reduces the large stress and strain gradients across and through the beam flanges and at the weld access hole by permitting the flanges to flex out of plane. Typically, the elastic stress and strain concentration factors of the pre-Northridge connections (caused by the large beam flange shear that resulted in severe local beam flange and/or column flange distortions) were reduced from 4.0 to 5.0 down to 1.2 to 1.4.
2. The lateral-torsional buckling of beams, which is prevalent in non-slotted beams, is eliminated. The separation of the beam flanges and beam web allow the flanges and web to buckle independently nd concurrently, thus eliminating a the twisting mode of buckling. Elimination of this buckling mode is important for perimeter seismic moment frames that support the exterior cladding of the building. Even a small amount of twisting of these perimeter beams will be reflected as significant visible distortion of the exterior cladding.
3. The separation of the beam web and the flange results in biaxial rather than triaxial stress and strain states in the region of the connection, thus increasing the fatigue life of the connection.
4. Residual weld stresses are significantly reduced because of the long structural separation between the vertical web and horizontal flange welds.
5. Full scale low cycle fatigue tests have shown that the fatigue life was increased considerably (more than tripled) over the non-slotted connections that subject the beam flange/welds to a large portion of the beam seismic vertical shear.

Analytical and experimental studies have shown that SlottedWeb$^{TM}$ develops the full plastic moment capacity of the beam and does not reduce its elastic stiffness. According to SSDA, SlottedWeb$^{TM}$ is more economical than reduced beam section (RBS) moment connections or Kaiser bolted brackets, since SlottedWeb$^{TM}$ involves fewer additional parts and cuts, and the work is involved in the thinner web material (Average connection cost is about $1840 per connection, as against $2360 for RBS connection). SlottedWeb$^{TM}$ has been used in the Yerba Buena Tower, San Francisco, Hyatt San Diego and GAP Embarcadero, San Francisco. More details about this connection may be found in www.slottedweb.com.

## 17.9.2 Strengthened Connections

In this design philosophy, the portion of the beam adjacent to the column, where the maximum moment occurs during seismic loading, is strengthened. This will force the plastic hinge to form away from the joint, in the un-strengthened portion of beam. The strengthening may be done using cover plates, rib plates, side plates or haunches at the beam-to-column interface (see Fig. 17.24). However, the code requirement of strong-column/weak-beam condition is to be satisfied. It is also recommended that the connections be designed in such a way to ensure plastic hinge location at a distance of half the beam depth from the column face.

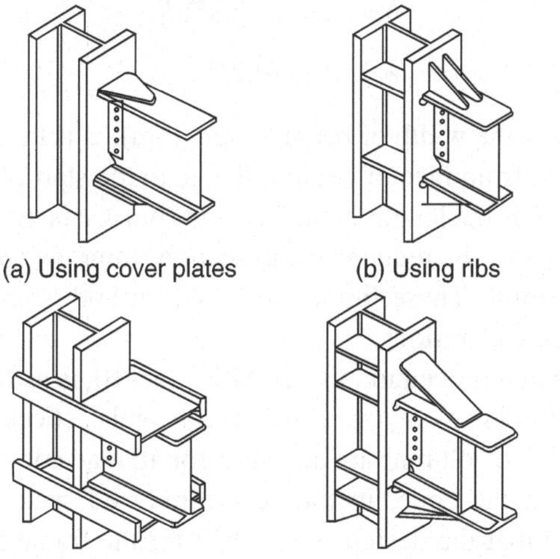

(a) Using cover plates (b) Using ribs

(c) Using side plates (d) Using haunches

**FIG. 17.24** Strengthened connections

When these connections are used, the effective section modulus of the beam at the connection is increased, decreasing the bending stress at the extreme fiber of the section, and also the total force resisted by the flange welds. The extrapolated moment near the column face, $M_f$, can now be well above the beam plastic moment, $M_p$, and hence must be considered in design (Tamboli 2009).

**Cover-plated connections** Experimental research on cover-plated connections has shown that these connections perform well in the inelastic range (FEMA 350:2000). However, when cover plates are used, proper detailing is to be followed for obtaining ductile behaviour. Typically, cover plates are fillet-welded to the beam flanges and groove-welded to the column flange. A common detail is shown in Fig. 17.25(a). Note that only the long sides of the cover plates are welded to the beam flanges. For ease of welding at site, the bottom cover plate is oversized and the top plate undersized, so that down-hand welding can be adopted at each location. When oversized top and bottom cover plates are used, the top plate is shop-welded to the beam and the bottom plate is field-welded. Cross-welds to the beam flanges at the end of the cover plates are not recommended, since they do not perform well in the inelastic region. Moreover at the column flange, both the cover plate and the beam flange should be connected by the same groove weld as shown in Fig. 17.25(b), and not by separate groove welds, since they introduce the 'notch' effect. Design of this connection with checks to satisfy FEMA requirements is provided by Tamboli (2009). See Example 14.20 of Chapter 14 for the design of cover plated connection, but the detailing should be as per Fig. 17.25 for earthquake resistance.

When haunched connections are used, the haunch may be located on the bottom flange only (since the top flange will support the floor slab). Haunches may be made from triangular cut sections of structural tee sections or by using plates, and stiffeners are provided in the column and beam webs, where the haunches end (see Fig. 17.24d). More details about this type of connection may be found in Yu et al (1997).

The action of vertical rib plate is similar to that of a cover plate or the haunch; it strengthens the section by increasing the section modulus while distributing the beam-flange force over a larger area of the column flange (see Fig. 17.24b). It is possible to use a single rib plate at the center of the beam flange, but it is better to use multiple ribs on each flange to direct the beam-flange force away from the center of the beam flange. Such a detail may also reduce the stress concentration at the center of the beam-flange

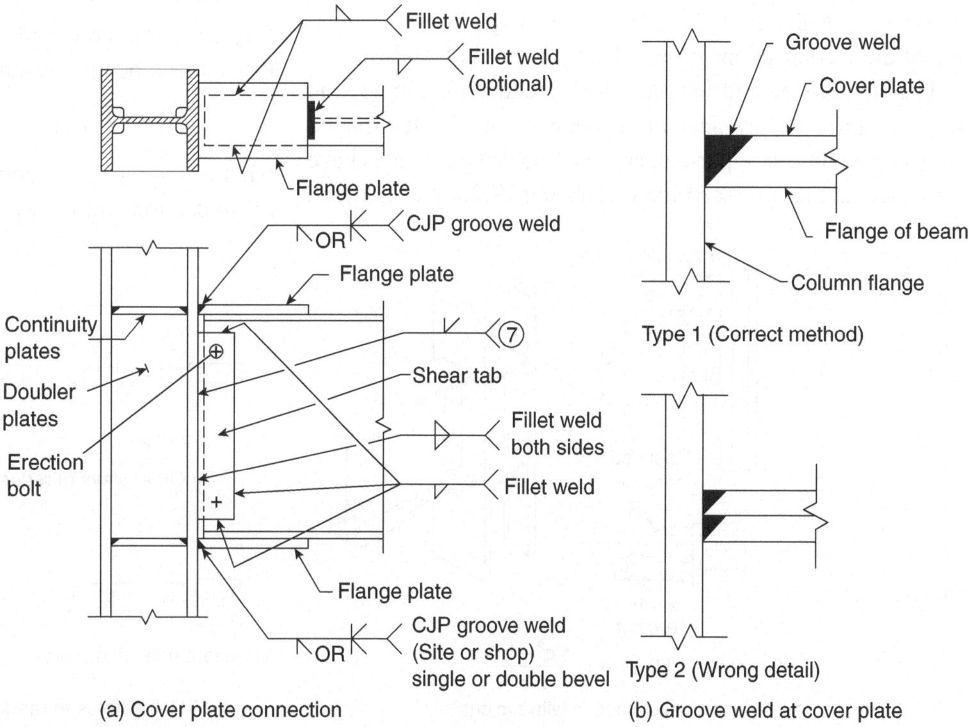


(a) Cover plate connection (b) Groove weld at cover plate

**FIG. 17.25** Connection with cover plate as per FEMA 350:2000

groove weld. The ductile behaviour of such rib strengthened connections has been validated in a few experimental results.

### 17.9.3 Weakened Connections

A *weakened connection* is achieved by removing portions of the beam flange to create a reduced beam section (see Fig. 17.26). Hence this connection is also called *reduced beam section* (RBS) *moment connection*. Thus the designer chooses a location where the plastic hinge will occur by creating a weak link, or fuse, in the moment capacity of the beam. A proprietary reduced beam connection called 'dog bone' connection is shown in Fig. 17.26(a). The effect of 'dogbone' is similar to that of cover plate connections. With cover plates the connection is made stronger than the beam by strengthening the connection. In the *dogbone*, the connection is effectively made stronger than the beam by weakening the beam. The geometry of the RBS must be selected in such a way that the factored nominal moment capacity is not exceeded, at the critical beam section adjacent to the column. This method has potential benefits where the strengthening scheme had drawbacks—that is, the strong-column weak-beam concept and panel zone strength requirements are easily achieved. While producing the same effect of cover plates, the dogbone connection can be constructed with relatively simpler details, resulting in a more reliable and economic solution. However, the reduction in flange area may reduce the stiffness and stability of beam flange, and may increase the susceptibility of lateral torsional buckling of the beam in the reduced section. Hence additional lateral bracing is recommended in these locations. The earliest application of dogbone connection was made in 1969 (Iwankiw and Carter 1996 and Engelhardt et al. 1998) and the experimental validation of these connections is provided by Chen et al. (1996).

The shape, size and location of the reduced beam section may affect the performance of the connection. Various shaped such as straight-cut, taper-cut, arc-cut, and drilled flanges have been tried and tested (see Figs 17.26b and 17.26c). In general, arc-cut reduced beam sections have provided favourable results (Iwankiw and Carter 1996 and Iwankiw 1997). The design methodology presented by FEMA 350 (2000) and AISC 358-10 is applicable to RBS with curved arc cuts.

The distance of the RBS away from the column face, *a*, and the length *b* up to which the cut is made in the beam may be chosen as follows (FEMA 350:2000, AISC 358-10):

$$a = (0.5 \text{ to } 0.75) b_{bf} \tag{17.5}$$

$$b = (0.65 \text{ to } 0.85) d_b \tag{17.6}$$

$$c = (0.1 \text{ to } 0.25) b_{bf} \tag{17.7}$$

where $b_{bf}$ is the width of beam flange (mm), $a$ is the horizontal distance from face of column flange to the start of an RBS cut (mm), $b$ is the length of an RBS cut (mm), $c$ is the depth of cut at center of the reduced beam section (mm), and $d$ = depth of beam (mm). These limits are based on both stability and strength considerations.

The design procedure, as per AISC 358-10, is as follows:

***Step 1***: After choosing the values of a, b, and c, check whether the beams and columns are adequate for all load combinations specified in the code, including the reduced section of the beam, and that the design storey drift for the frame complies with applicable limits specified by the code. Calculation of elastic drift shall consider the effect of the reduced beam section. In lieu of more detailed calculations, AISC 358 allows to calculate the effective elastic drifts by multiplying elastic drifts based on gross beam sections by 1.1 for flange reductions up to 50% of the beam flange width. Linear interpolation may be used for lesser values of beam width reduction.

***Step 2***: Compute the plastic section modulus at the center of the reduced beam section:

$$Z_{RBS} = Z_{pz} - 2ct_{bf}\left(d - t_{bf}\right) \tag{17.8}$$

where $Z_{RBS}$ = plastic section modulus at center of the reduced beam section (mm$^3$), $Z_{pz}$ = plastic section modulus about the

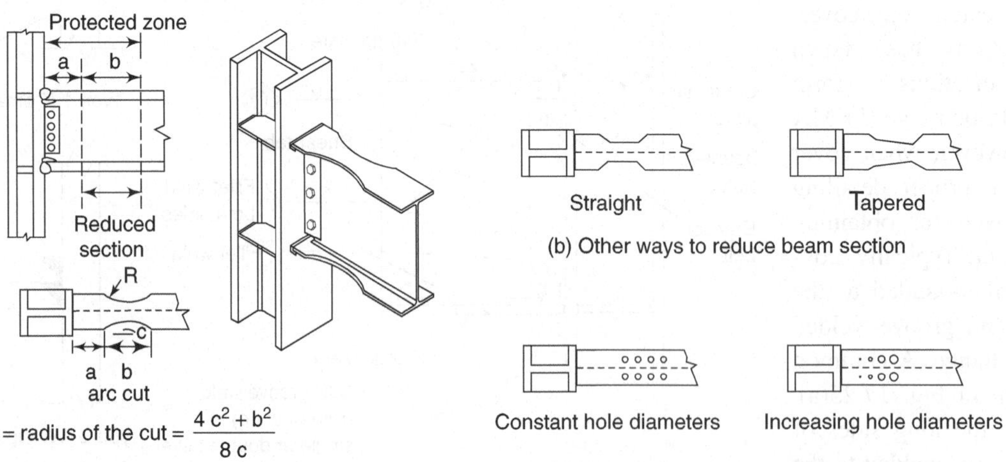


(a) Reduced beam section with arc cut

(b) Other ways to reduce beam section

(b) Using holes to reduce beam section

**FIG. 17.26** Weakened connections

$z$-axis, for full beam cross section (mm$^3$), $t_{bf}$ = thickness of beam flange (mm),

***Step 3*:** Compute the probable maximum moment, $M_p$, at the center of the reduced beam section:

$$M_p = C_p R_y f_y Z_{RBS} \quad (17.9)$$

where $f_y$ is the specified minimum yield stress (MPa), $R_y$ is the ratio of the expected yield stress to the specified minimum yield stress (see Table A3.1 of AISC 341-10), and $C_p$ is a factor to account for peak connection strength, including strain hardening, local restraint, additional reinforcement, and other connection conditions,. AISC 358-10 suggests to take the value of $C_p$ as follows:

$$C_p = \frac{f_y + f_u}{2f_y} \le 1.2 \quad (17.10)$$

where $f_y$ is the specified minimum yield stress of the member (MPa), and $f_u$ is the specified minimum tensile strength of the member (MPa).

***Step 4*:** Compute the shear force at the center of the reduced beam sections at each end of the beam.

The shear force at the center of the reduced beam sections should be determined from a free body diagram of the portion of the beam between the centers of the reduced beam sections. This calculation shall assume the moment at the center of each reduced beam section is $M_p$ and according to AISC 358 should include gravity loads acting on the beam based on the load combination of $1.2DL + 0.5IL + 0.2S$, as suggested in ASCE 7-10.

***Step 5*:** Compute the probable maximum moment at the face of the column.

The moment at the face of the column shall be computed from a free-body diagram of the segment of the beam between the center of the reduced beam section and the face of the column, as shown in Fig. 17.27.

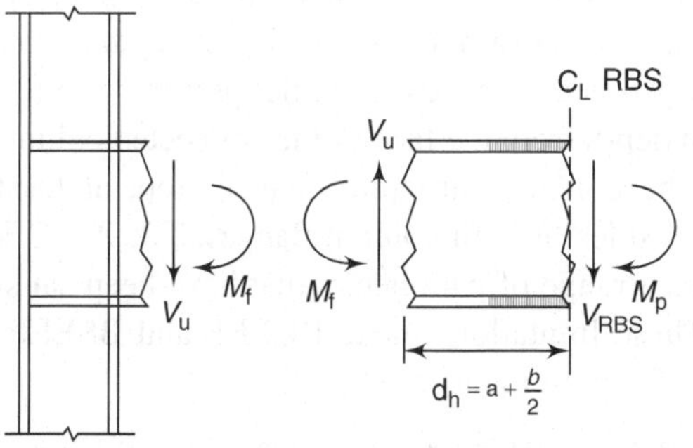


**FIG. 17.27** Free-body diagram between center of RBS and face of column

Based on this free-body diagram, the moment at the face of the column is computed as follows:

$$M_f = M_p + V_{RBS} d_h \quad (17.11)$$

where $M_f$ = probable maximum moment at face of column (Nmm), $d_h$ = distance from face of the column to the plastic hinge (mm)= $a + b/2$, and $V_{RBS}$ = larger of the two values of shear force at the center of the reduced beam section at each end of the beam (N).

It has to be noted that Eqn 17.11 neglects the gravity load on the portion of the beam between the center of the reduced beam section and the face of the column. If desired, the gravity load on this small portion of the beam may be included.

***Step 6*:** Compute $M_{pe}$, the plastic moment of the beam based on the expected yield stress:

$$M_{pe} = R_y f_y Z_{pz} \quad (17.12)$$

***Step 7*:** Check the flexural strength of the beam at the face of the column:

$$M_f = \phi_d M_{pe} \quad (17.13)$$

For ductile limit states $\phi_d = 1$

If Eqn17.13 is not satisfied, adjust the values of *c, a* and *b*, or adjust the section size, and repeat Steps 2 through 7.

***Step 8*:** Determine the required shear strength, $V_u$, of beam and beam web-to-column connection and then check design shear strength of beam. The required shear strength, $V_u$ may be calculated as

$$V_u = \frac{2M_p}{L_h} + V_{gravity} \quad (17.14)$$

where $V_u$ is the required shear strength of beam and beam web-to-column connection (N), $L_h$ is the distance between *plastic hinge locations* (mm), $V_{gravity}$ is the beam shear force resulting from $1.2D + 0.5L + 0.2S$ (N)

***Step 9*:** Design the beam web-to-column connection.

***Step 10*:** Check continuity plate requirements.

***Step 11*:** Check column-beam relationship limitations.

### 17.9.4 Pre-qualified Seismic Moment Connections

Though the recent version of the code, IS 800:2007, contains provisions for design and detailing for seismic loads, it does not suggest the type of connections which are suitable for high or intermediate seismic zones. As mentioned earlier, as the testing of full scale specimens of the beam-column joints is expensive and time consuming, AISC 341 permits the use of prequalified connections [Subramanian (2010)]. These prequalified connections have been demonstrated by extensive testing and analysis and the expert review panel of AISC has approved them to be capable of providing code specified joint rotation without degradation in strength and stiffness below the full yield value.

These connections may be adopted in India also for better performance in strong or intermediate earthquakes.

#### *Types of Pre-qualified Moment Connections*

AISC 358-2010 suggests 10 types of pre-qualified connections as given in Table 17.3 (the last two have been proposed in the 2016 version of the code):

**TABLE 17.3** Prequalified moment connections as per AISC 358-10

| Category | Connection description | Chapter in AISC 358-16 | Acronym | Permissible systems |
|---|---|---|---|---|
| Welded, fully restrained | Welded unreinforced flanges, welded web | 8 | WUF-W | OMF, SMF |
| | Reduced beam section | 5 | RBS | OMF, SMF |
| | SidePlate moment connection | 11 | SidePlate | OMF, SMF |
| Bolted, fully restrained | Bolted unstiffened extended end plate | 6 | BUEEP | OMF, SMF |
| | Bolted stiffened extended end plate | 6 | BSEEP | OMF, SMF |
| | Bolted flange plates | 7 | BFP | OMF, SMF |
| Cast iron connection | Kaiser bolted bracket | 9 | KBB | OMF, SMF |
| | ConXtech ConXL moment connection | 10 | ConXL | OMF,SMF |
| Bolted, partly restrained | Double tee moment connection | 13 | DT | OMF, SMF |
| | Simpson strong-tie strong frame moment connection | 12 | | OMF, SMF |

A brief discussion about these prequalified connections is given here. More details about them and their methods of design may be found from AISC 358-2010.

## Reduced Beam Section Connection

In reduced beam section (RBS) moment connection (also known as the 'dog bone' connection), some portions of the beam flanges are removed in a pre-determined fashion, adjacent to the beam-column connection. In such a connection, yielding and plastic hinges are forced to form away from the connection at the reduced section of the beam. It has already been discussed in the previous section.

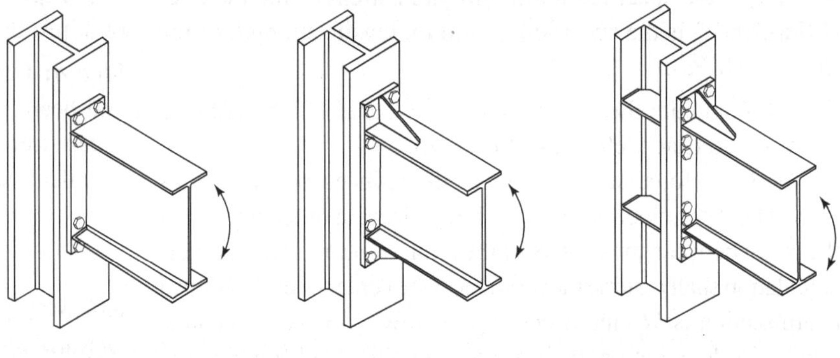


**FIG. 17.28** Bolted un-stiffened extended end-plate (BUEEP) and bolted stiffened extended end-plate (BSEEP) moment connections (From AISC 2010 *Prequalified Connections for Special and Intermediate Steel Moment Frames for Seismic Applications with Supp. No. 1 and Supp. No. 2* Reprinted with permission from AISC. All rights reserved)

## Bolted Unstiffened and Stiffened Extended End-plate Moment Connections

*Bolted unstiffened extended end plate (BUEEP) connections* and *Bolted stiffened extended end plate (BSEEP) connections* are made by welding the beam section to an end plate which is in-turn bolted to the column flange. The beam web and flanges are to be connected using either a CJP groove weld or a pair of fillet welds each having a size of 0.75 times the beam web thickness, but greater than 6 mm. Other specifications for protected zones, bolt pitch and spacing, etc. are provided in AISC 358-10.Three types of these connections are pre-qualified by AISC 358 (see Fig. 17.28). ACI 358 gives equations to check the various limit states of this type of connection such as flexural yielding of the beam section or end plates, yielding of column panel zone, shear or tension failure of the end-plate bolts, and rupture of the various welded joints. These provisions are intended to ensure inelastic deformation of the connection by beam yielding.

Extended end-plate moment connections in SMF systems with concrete structural slabs are prequalified subject to a few conditions, such as (a) the nominal beam depth is greater than 610 mm, (b) shear connectors are not present within 1.5 times the beam depth from the face of the connected column flange, and (c) the concrete structural slab is kept at least 25 mm from both sides of both column flanges. The AISC 358 code also gives a range of parameters that have been satisfactorily tested. These limitations make BUEEP and BSEEP suitable primarily for pre-engineered buildings.

## Bolted Flange Plate Moment Connection

*Bolted flange plate (BFP) moment connections* consist of plates welded to column flanges and bolted to beam flanges as shown in Fig. 17.29. Identical top and bottom plates are used. Flange plates are connected to column flange by using CJP groove welds and beam flanges are connected to the plates by using high strength friction grip bolts. The web of the beam is connected to the column flange using a bolted single-plate

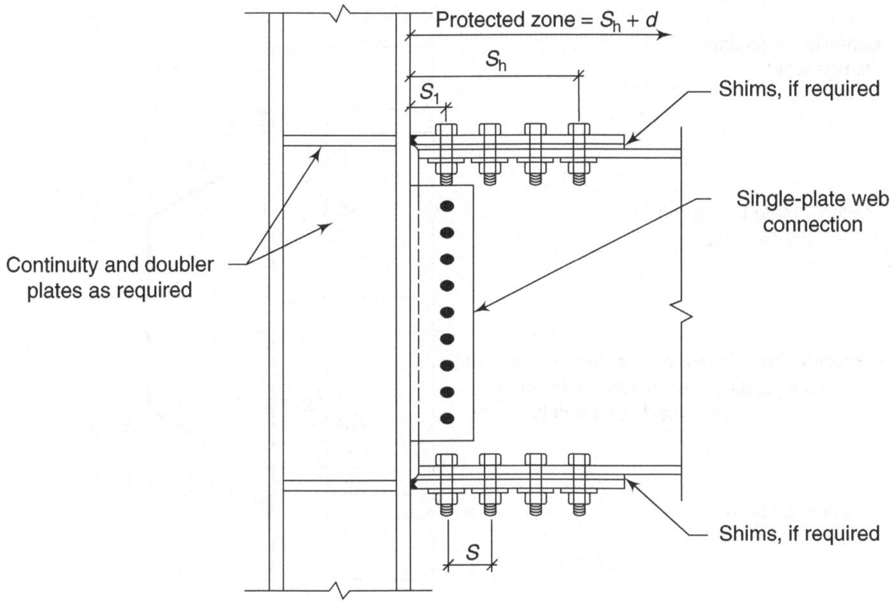


**FIG. 17.29** Bolted flange plate (BFP) moment connection

*Source:* AISC 2010 *Prequalified Connections for Special and Intermediate Steel Moment Frames for Seismic Applications with Supp. No. 1 and Supp. No. 2.* Reprinted with permission from AISC. All rights reserved

shear connection, with bolts in short-slotted holes. In this connection, yielding and plastic hinge formation are designed to occur in the beam near the end of the flange plates.

The flange plates and web shear plate are shop welded to the column flange and bolted to the beam flanges and web, respectively at site. Bolts shall be arranged symmetrically about the axes of the beam and shall be limited to two bolts per row in the flange plate connections. ASTM A490 or A490M bolts (maximum size 28 mm) with threads excluded from the shear plane should be used for the beam flange connections. The seismic behaviour of BFP momment connections were studied by Sato et al. (2008).

The experimental investigations on seismic behaviour showed the following (AISC 358-10):

1. Specimens achieved an inter-storey drift angle of 0.06 radians before failure
2. Initial yielding of the beam was at the last bolt away from the face of the column
3. Bolt slip of the flange plate bolts, which occurs at similar resistance levels to the initial yielding in the beam flange
4. Secondary yielding in the column panel zone, which occurs as the expected moment capacity and as strain hardening of beam hinge occurs
5. Limited yielding of the flange plate, which may occur at the maximum deformations

This sequence of yielding has resulted in very large inelastic deformation capacity for the BFP moment connection. However, the design procedure for this type of connection is more complex than other pre-qualified connections and is provided in AISC 358-10.

## Welded Un-reinforced Flange-welded Web Moment Connection

Unlike other pre-qualified connections, in the *welded un-reinforced flange-welded web (WUF-W) connection*, the plastic hinge location is not moved away from the column face. Rather, the design and detailing features are intended to allow it to achieve SMF performance without fracture. Inelastic rotation is intended to occur in the beam in the region adjacent to the face of the column. In this connection the beam flanges are welded directly to the column flange using CJP groove welds. The beam web is bolted to a single-plate shear connection for erection. This plate is subsequently used as a backing bar for welding the beam web directly to column flange using CJP groove weld, which extends to the full depth of the web (that is, from weld access hole to weld access hole). A fillet weld is also used to connect the shear plate to the beam web, as shown in Fig. 17.30. The single plate connection adds stiffness to the beam web connection. A special seismic weld access hole and detailing, as shown in Fig. 17.30(b), are specified for the WUF-W moment connection, to reduce stress-concentration in the region around the access hole. The design procedure for the WUF-W moment connection is available in AISC 358-10. More details about the behavior of these connections are provided by Lee et al. (2005).

## Kaiser Bolted Bracket Moment Connection

In Kaiser bolted bracket (KBB) moment connection, a cast steel (high-strength) bracket is fastened to each beam flange and bolted to the column flange as shown in Fig 17.31. The bracket can be either bolted (B-series) or welded (W-series) to the beam. The bracket is proportioned to develop the probable maximum moment strength of the beam, such that yielding and plastic hinge formation occurs in the beam at the end of bracket away from the column flange. This connection is designed to eliminate field welding and facilitate speedy frame erection. To provide erection tolerance, the bracket column bolt holes are vertically short-slotted and the column bolt holes are slightly oversized.

Several tests on this type bolted bracket connection were conducted at Lehigh University (Adan and Gibb 2009).Then it was patented with the United States Patent and Trademark Office by of Steel Cast Connection LLC. The advantage of using casting is that it will not have HAZ issues or residual stresses that would be found in welds.

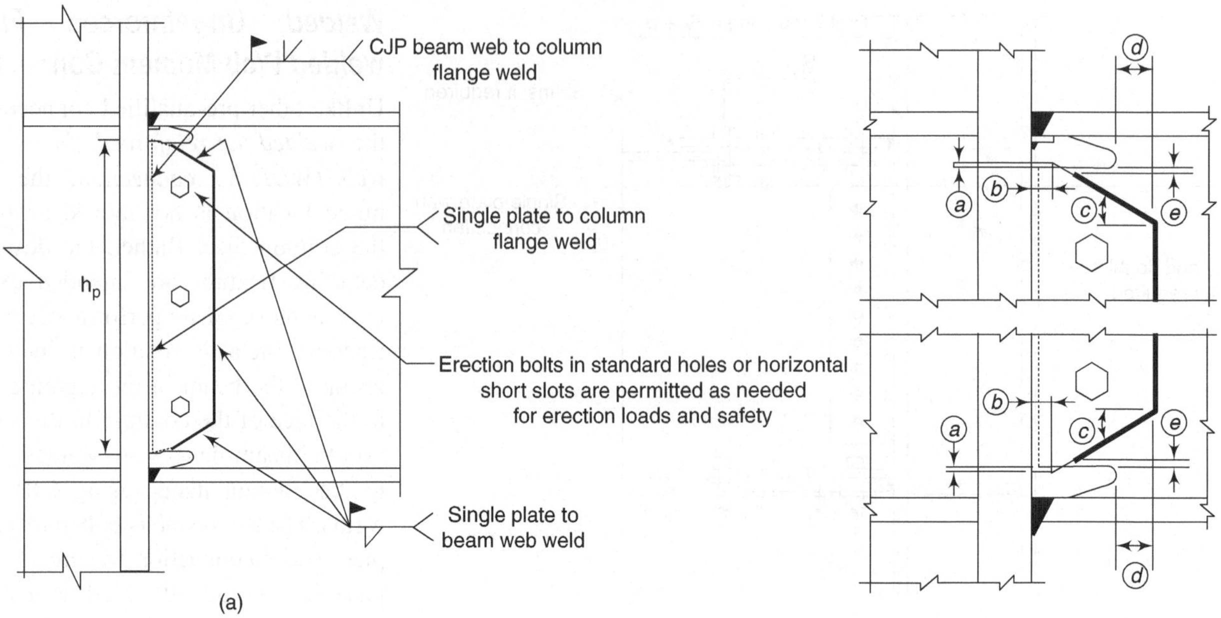


Notes:
a. 6 mm minimum, 12 mm maximum.
b. 25 mm minimum
c. 30° (±10°)
d. 50 mm minimum
e. 12 mm minimum distance, 25 mm maximum distance, from end of fillet weld to edge of weld access hole.

**FIG. 17.30** Welded un-reinforced Flange – welded Web (WUF-W) moment connection (a) connection, (b) Detailing of connection

*Source:* AISC 2010 *Prequalified Connections for Special and Intermediate Steel Moment Frames for Seismic Applications with Supp. No. 1 and Supp. No. 2.* Reprinted with permission from AISC. 

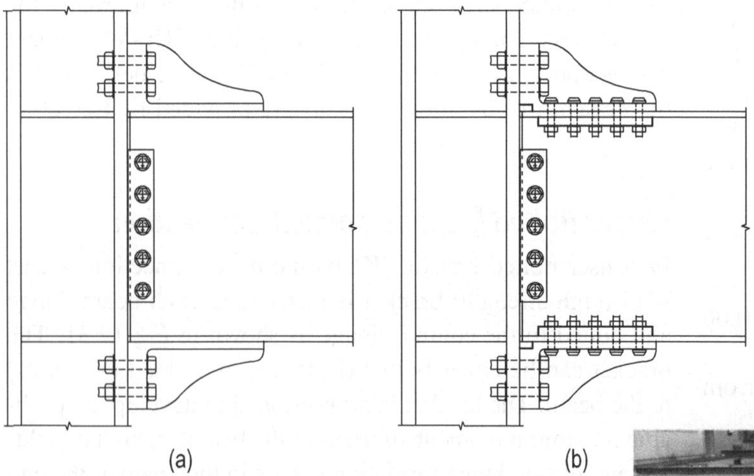


**FIG. 17.31** Kaiser bolted bracket (KBB) moment connections (a) Beam welded to bracket, (b) beam bolted to bracket

*Source:* AISC 2010 *Prequalified Connections for Special and Intermediate Steel Moment Frames for Seismic Applications with Supp. No. 1 and Supp. No. 2.* Reprinted with permission from AISC. 

This bracket is available in USA, in various sizes and bolt patterns to match the demand required for strength and ductility. The bracket is shop welded to the beam and field bolted to the column. The brackets used in Wasatch property Management Corporate Headquarters building in Utah, USA are shown in Fig. 17.32. The use of these cast steel brackets resulted in a saving of $3,000 per joint, due to the avoidance of complete penetration field welding, doubler plates, continuity plates and ultra-sonic testing and also from the reduced beam tonnage(Cartwright,2006). The design procedure and detailing requirements for these connections are given in AISC 358-2010. More information about these KBB moment connections can be had from www.steelcastconnections.com.

**FIG. 17.32** Use of Kaiser bolted brackets in Wasatch Property Management Corporate Headquarters building in Utah, USA

*Courtesy:* William C. Gibb of Steel Cast Connections LLC

## SidePlate Connections

SidePlate® is a patented, award winning SMF connection developed by SidePlate System Inc., Los Angeles, U.S.A. It can be used in moment frame and braced frames, for both uniaxial and biaxial framing applications. SidePlate™ uses a series of flange and web plates with horizontal shop and field fillet welds to create a rigid, fixed connection between wide-flange columns and beams (see Fig. 17.33). In the shop, beam flange cover plates are fillet welded to the top and bottom of the beam and erection angles are fillet welded to the web. Column side plates and horizontal shear plates are fillet welded to the column web. Figure 17.34 shows the field fillet- welded version as well as the new field bolted version of SidePlate® connection.

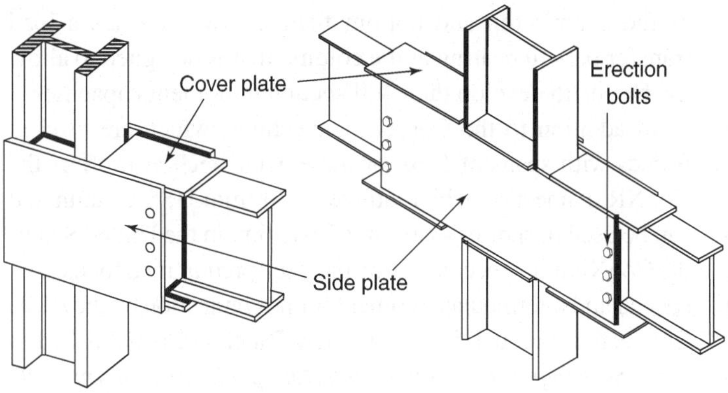


**FIG. 17.33** SidePlate® Connection

*Courtesy:* Henry Gallart, President, SidePlate Systems, Inc.

(a)

(b)

**FIG. 17.34** SidePlate® (a) Field fillet-welded version (b) Field-bolted version

*Courtesy:* Henry Gallart, President, SidePlate Systems, Inc.

In the field, column trees are erected and full-length beams are hoisted into place between the two pairs of column side plates. The beams are bolted to the column side plates (typically consists of four to six A325 bolts) and four horizontal fillet welds are applied to complete the beam column connection. Column design options include rolled or built-up 'H' wide flange sections, and tube steel or built-up box columns. Up to three storey-length of column trees can be fabricated, transported and erected and are typically transported by trucks or rail car to the job site.

In 2009, SidePlate System Inc. modified the original SidePlate® connection, which included a beam stub in a column tree assembly and required a field CJP beam splice. The new connection, referred to as the SidePlate FRAME™, has eliminated the beam stub, CJP splice, and reduced the fillet weld sizes by about half of the original connection [Cordova and Hamburger (2011)]. Thus, the new SidePlate FRAME™ is a 100% fillet welded connection and eliminates the need for CJP welds. These improvements, which resulted from extensive analytical and experimental studies, have reduced shop fabrication time and improved constructability [Cordova and Hamburger (2011)]. The cyclic rotational capacity of this moment connection system was tested at the Charles Lee Powell Structural Research Laboratories, University of California, San Diego, which showed that it exceeds all beam-to-column prequalification requirements of ANSI/AISC 341-05.

According to SidePlate System Inc., there is no limit on column or beam size; hence, deeper and lighter sections can be used reducing the total steel weight and cost. It also recommends that panel zones can be modeled as completely rigid and increased beam stiffness provided by the column side plates can be included in the analytical model; this will again result in savings in steel weight. The side plates of this system can also be extended to permit attachment of braces for dual systems, and dampers for energy dissipated structures. More information, on this patented connection can be had from www.sideplate.com.

The second supplement to ANSI/AISC 358s2-14 included the proprietary SidePlate® moment connection to the roster of prequalified connections and included the design procedure.

## ConXtech's ConXL™ and ConXR™ Moment Connections

The American inventor Robert J. Simmons introduced the ConXR™ moment connection in 2004, through his company ConXtech. Two types of systems are currently available: ConXR™ and ConXL™. The ConXL connection is pre-qualified as a fully restrained moment connection of wide flange beams to concrete-filled 400 mm HSS or built-up box columns. The ConXL connection forms a compression collar around a square column using high-strength pretensioned bolts, with no field welding. The compression collars are made up of collar corner and collar flange assemblies. The collar corner assemblies, either forged or cast steel, are attached to the column and the forged steel collar flange assembly is attached to the end of a beam. The individual interlocking collar components are manufactured with contact surfaces milled with three-dimensional tapers designed to safely align and lock together for ease of erection during field assembly.

A column can accept up to four moment-connected beams, one at each face of the column (see Fig. 17.35). Note that each compression collar at a moment node (intersection of moment beams and moment column) must be completed with *flange collars* at each face of the column. At the building perimeter or building corners, collar *flange blanks* (collars with no beam attached) are used to complete the compression collar around the column (Cordova and Hamburger, 2011). All beams connecting to a moment node can have different weights, but are required to have the same nominal depth. Beams typically use a reduced beam section (RBS) to efficiently meet strong-column/weak-beam requirements, but an RBS is not a requirement. Four collar corner assemblies are required at each moment node.

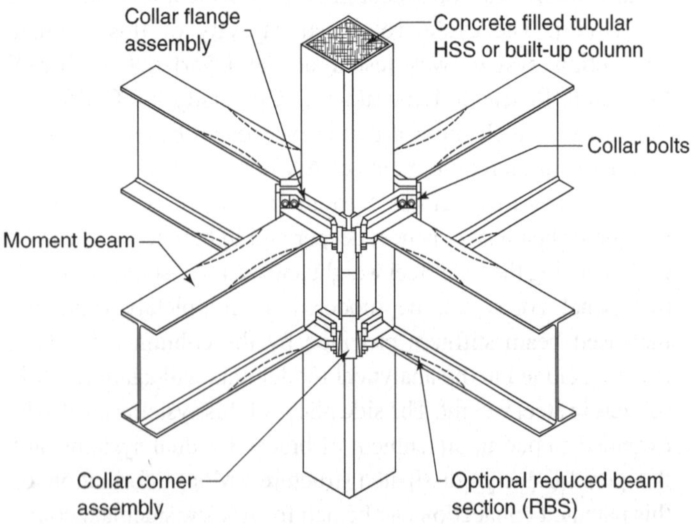


**FIG. 17.35** Assembled ConXL$^{TM}$ moment connection

*Courtesy:* ConXtech Inc.

In a shop environment, *beam collar flanges* are welded into assemblies with a *web extension* connecting the top and bottom collar flanges. The *web extension* is manufactured in different depths allowing standardized collar flanges to be paired with beams of various beam depths. The column collar corners are similarly shop welded into assemblies with a collar corner middle piece of various lengths connecting the standardized top and bottom collar corners. The collar flange assemblies are then welded to the ends of wide flange beams, and *collar corner assemblies* are welded to the corners of the column at the proper floor framing elevations.

On the job site, beams are lowered into the tapered *collar corner assemblies* on the column. Once all beams (or *collar flange blanks* used at the perimeter and corners) are erected around a column, the *collar flanges* are bolted to one another around the column using high-strength bolts. The four *collar flanges* at the column faces forms a compressive collar around the column when the bolts are pretensioned. A ConXL moment node will have compression collars at the same elevation as the beam's top and bottom flange, which creates a load transfer path in bearing at the column that is designed to allow the beams to develop their full sectional moment capacities.

In addition to the ConXL connections, which are used in frames with spans of 5.5m or more, ConXtech also offers the ConXR connection which utilizes a 200 mm HSS column and can be used in shorter spans of 2.5m to 6m. In the United States, the ConXL moment connection has been prequalified for use as a special and intermediate moment frame connection by the AISC Connection Prequalification Review Panel and is published in the *AISC Prequalified Connections for Special and Intermediate Steel Moment Frames for Seismic Applications* (AISC 358-s1-11) under Chapter 10 – ConXtech ConXL Moment Connection. The ConXR moment connection meets prequalification requirements for a SMF connection through full-scale cyclic testing as per Chapter K of the Seismic Provisions for Structural Steel Buildings (AISC 341-10). The ConXtech moment connections can also be used as an ordinary moment connection.

Both ConXtech connections can also be used in a bi-axial moment-resisting frame. They provide lateral resistance about 2 axes, resulting in reduced member sizes and foundation loads. A ConXtech design typically utilizes most of the column to beam connections as moment connections, the resulting redundancy also providing the key resistance to *progressive collapse*. The advantages of using this patented system are ease of system detailing, fabrication, erection and lack of field welding; all these factors leading to reduced time and cost. The design steps are provided in AISC 358-10 and experimental validation is provided by Seek and Murray (2005). More information on this system may be found at www.ConXtech.com.

The details of other moment connections approved for inclusion in the 2016 edition of AISC 358, i.e. *Double Tee and Simpson Strong Tie Strong Frame moment connection* may be found in AISC 358-16 and www.strongtie.com/products/strongframe/special_mf/intro.asp respectively. Another interesting all field bolted connection called the *Pin-Fuse Joint* has been developed by Skidmore, Owings & Merrill (SOM) which has started its prequalification process. Cordova and Hamburger (2011) provide details of Pin-Fuse Joint and also describe about a cast, high-strength steel connection called *Cast Connex* (www.castconnex.com).

## 17.10 BRACED FRAMES

These systems consist of steel frames with diagonal steel braces placed in selected bays. Floors are cast-in-place concrete slabs or metal deck and concrete (see Fig. 17.36). Braced frames provide resistance to lateral forces acting on a structure. The members of a braced frame act as a truss system and are subjected primarily to axial stress. Depending on the diagonal force, length, required stiffness and clearances, the diagonal members can be made of double angles, channels, tees, tubes or even wide flange shapes. Besides performance, the shape of the diagonal is often based on connection considerations. The braces are often placed around service cores and elevators, were frame diagonals may be enclosed within permanent walls. The braces can also be joined to form a closed or partially closed three dimensional cell so that torsional loads can be resisted effectively. A height-to-width ratio of 8-10 is considered to form a reasonably effective bracing system. Braced frames are most effective at the perimeter of the building, where they can control the torsional response of the building. Braced frames can also be an effective system for seismic retrofit due to their high stiffness and because they can be assembled from pieces of relatively small size and weight. Braced frames are inherently stiff. Thus, for the structures of the same height, braced frames will normally have a lower period than a moment frame.

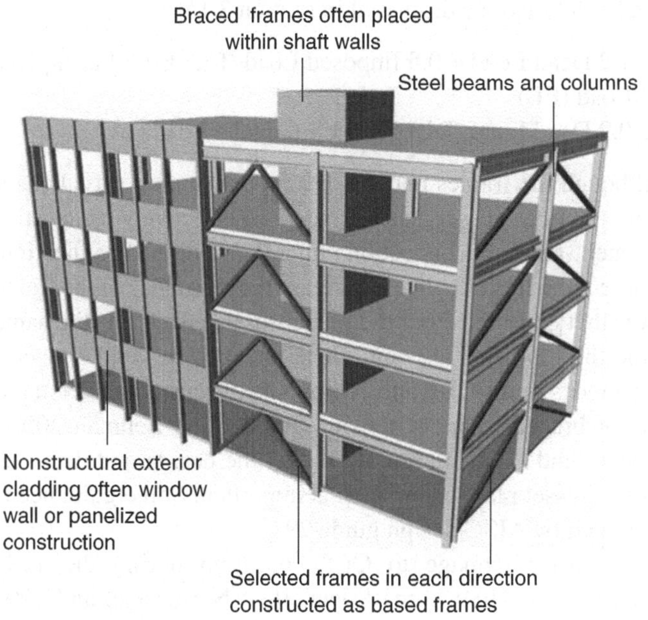


**FIG. 17.36** Braced frames

*Source:* FEMA 454-06

Braced frames may be grouped into concentrically braced frames (CBFs), and eccentrically braced frames (EBFs), depending on their ductility characteristics. In addition, CBFs are subdivided into two categories, namely, ordinary concentrically braced frames (OCBFs) and special concentrically braced frames (SCBFs).

### 17.10.1 Concentrically Braced Frames (CBFs)

*Concentrically braced frames* (CBFs) are structures resisting lateral loads through a vertical concentric truss system, the axes of the members aligning concentrically at the joints. In recent years, typical steel construction in regions of high seismic risk has shifted from moment-resisting frames to concentrically braced frames. In CBFs the axes of all members, i.e. columns, beams and braces, intersect at a common point such that the member forces are axial. The Chevron bracing, cross bracing (X-bracing), and diagonal bracing (single diagonal or K-bracing) are classified as concentrically braced and are shown in Fig. 17.37(a) – (d).

Braces buckle in compression and yield in tension. The initial compressive buckling capacity is normally smaller than tension yield force, and for subsequent buckling cycles, the bucking capacity is further reduced by prior inelastic excursion. Hence, bracing systems should be balanced so that lateral resistance in tension and compression is similar in both directions. It requires that diagonal bracing must be used in matched tensile and compressive pairs. This balancing is directly achieved in X-bracing, multi-storey X-bracing and chevron bracing. X-bracing is most commonly used with light bracing on shorter structures (Sabelli et al. 2013).

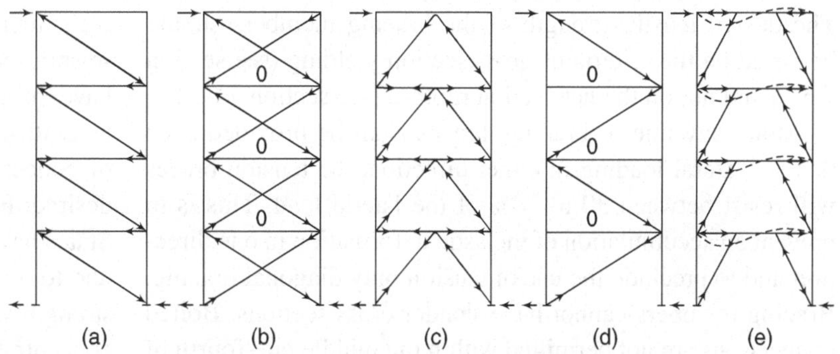


**FIG. 17.37** Types of bracings and the load path (a) Single diagonal bracing (b) X-bracing (c) Chevron (inverted V) bracing (d) Single-diagonal, alternate direction bracing (e) Knee bracing (EBF)

On the other hand, EBFs utilize axial offsets to deliberately introduce flexure and shear into framing beams to increase ductility. For example, in the knee bracing shown in Fig. 17.37(e), the end parts of the beam are in compression/tension with the entire beam subject to double curvature bending. (Note that in all the frames shown in Fig. 17.37, a reversal in the direction of horizontal load will reverse all actions and deformations in each of the members). The EBFs are discussed in detail in the next section.

The inability to provide reversible inelastic deformation is the principle disadvantage of CBFs. After buckling, an axially loaded member loses strength and does not return to its original straight configuration. To reduce the possibility of this occurring, during moderate earthquakes, more stringent design requirements are imposed on bracing members (see clause 12.7 of the IS 800:2007). Thus OBCFs are not allowed in seismic zones IV and V and for buildings with an importance factor greater than unity (I >1.0) in zone III; a K-bracing is not permitted in earthquake zones by clause 12.7.1.2 of the code (the inelastic deformation and buckling of K-bracing members may produce lateral deflection of the connected columns, causing collapse) Similarly, tension-only bracing is also prohibited, for the reasons discussed earlier.

Given the variation of design forces with period stipulated in most building codes, and the lower response modification factor, *R*, associated with conventional CBFs (*R* for OCBF is 4 and SCBF is 4.5 as against SMF of 5), the minimum lateral design load stipulated for a concentrically braced frame is generally larger than for a moment-resisting frame. In spite of these higher forces, the inherently large lateral stiffness of braced frames is generally adequate to satisfy the lateral drift requirements in current codes without further increasing the member stiffness and strength (Uriz and Mahin 2008). OBCFs should have to be designed to withstand inelastic deformation corresponding to a joint rotation of 0.02 radians without degradation in strength and stiffness, below the full yield value. The slenderness of bracing members should not exceed 120 and the required compressive strength shall not exceed 0.8 $P_d$, where $P_d$ is the design strength in axial compression. The design tensile strength of the bracing members should be based on the criteria of gross section yielding (see section 7.6.1) and not on the net section rupture (see section 7.6.2).

Along any line of bracing, braces shall be provided such that for lateral loading in either direction, the tension braces will resist between 30 to 70% of the lateral load. This is to prevent an accumulation of inelastic deformation in one direction and to preclude the use of tension only diagonal bracing. Bracing members cannot have slender cross sections. Bolted connections are not permitted within the middle one-fourth of the clear brace length, by the code. For braces consisting of built-up sections, the tack fastener spacing should be provided in such a way that the unfavorable slenderness ratio of individual element between the fasteners is less than 0.4 times the governing slenderness ratio of the brace.

According to clause 12.7.3 of the code, the end connections in bracings should be designed to withstand the minimum of the following:

1. A tensile force in the bracing equal to $1.2f_yA_g$
2. Force in the brace due to load combinations specified in clause 12.2.3 of the code
3. Maximum force that can be transferred to the brace by the system

In addition the connection should be designed for a moment of 1.2 $M_p$ of the braced section about the buckling axis. The gusset plates should also be checked for buckling out of their plane.

It is to be noted that the code in clause 12.2.3 specifies that the frame must be analyzed for the following two additional load combinations, which are different from the load combinations discussed in section 3.15.

1. 1.2 Dead Load + 0.5 Imposed Load (LL) ± 2.5 Earthquake Load (EL)
2. 0.9 Dead Load (DL) ± 2.5 Earthquake Load (EL)

Although the frames may be designed assuming truss behaviour, the large gusset plate connections effectively create stiff, moment-resistant connection rather than a pinned connection. These moments effectively increase the resistance of the frame over that expected from the plane truss analysis. At the same time the moments also introduce unexpected yield and failure modes in the CBF and complicate the current understanding of braced frame behaviour (Roeder and Lehman 2008). Roeder and Lehman also stress that the design and detailing of the gusset plate connection is important. Some guidance is provided by AISC design guide 29.

Extensive damage to CBFs occurred during the 1985 Mexico City, 1989 Loma Prieta, 1994 Northridge, and 1995 Kobe earthquakes. The most severe damage was observed in frames where braces were proportioned to resist tension only, where connections were weaker than the braces attached to them, where braces framed into columns, and where braces were inclined principally in one direction. Since then, the seismic provisions for structural steel buildings (AISC 341) have been updated to prohibit or restrict such conditions. Notwithstanding these improvements, the seismic performance of concentrically braced steel frames may still be less than desired. For example, conventional braces used in the United States have limited ductility capacity and are prone to fracture due to low-cycle fatigue. They also tend to lose compressive strength when loaded in the inelastic range, which leads to a concentration of damage in weakened stories. Prompted by these observations, seismic design requirements for braced frames changed considerably during the 1990s, and the concept of special CBFs was introduced in AISC codes. Additionally, researchers have undertaken a variety of investigations to develop ways to improve the performance of CBFs through:

1. The introduction of new structural configurations
2. The use of special bracing elements, including those utilizing
   (a) composite action
   (b) metallic yielding
   (c) high-performance materials
   (d) friction and viscous damping
3. The introduction of new behavior modes, such as uplifting foundations

Hence SCBFs are now increasingly used instead of CBFs in earthquake zones (Sabelli et al. 2013)

## 17.10.2 Special Concentrically Braced Frames (SCBFs)

As per clause 12.8 of the code, these frames should be designed to withstand inelastic deformation corresponding to a joint rotation of 0.04 radians without degradation in strength and stiffness below the full yield value. They are allowed to be used in any zone and for any building. The slenderness ratio of the bracing members should not exceed 160 and the required compressive strength should not exceed the design strength in axial compression, $P_d$. Along the line of bracing, braces should be provided such that for lateral loading in either direction, the tension braces resist 30 to 70% of the load. The bracing and column sections used in SCBFs should be plastic sections. The bracing members should be made of steel meeting the requirement of CVN impact value greater than 27J at – 30°C.

These provisions are for X-braces only. For other types of bracings such as Chevron or V-type bracings, and for eccentrically braced frames, the code does not give any guideline. More information about the design of such bracings may be found in Becker (1995), Becker and Ishler (1996), Bruneau, et al. (1997), Bozorgnia and Bertero (2004), and Williams (2004).

The connections in a braced frame may be subjected to impact loading during an earthquake and, in order to avoid brittle fracture, must be designed to withstand the minimum of the following:

1. A tensile force in the bracing equal to $1.1 f_y A_g$
2. The force in the brace due to the following load combinations
   (a) 1.2 DL + 0.5LL ± 2.5 EL
   (b) 0.9 DL ± 2.5EL
3. The maximum forces that can be transferred to the brace by the system.

The connection should be checked to withstand a moment of 1.2 times the full plastic moment of the braced section about the buckling axis and for tension rupture, and block shear under the aforementioned loading. The gusset plates should be checked for buckling out of their plane, and sufficient length should be provided for plastic hinge formation.

Recent research has shown that the current practice of providing a linear clearance of twice the thickness of gusset plates (see Fig. 17.38a), leads to thicker and larger size of gusset plates. This creates a rotationally stiff joint, which limits the rotation of the connection and leads to extensive frame yielding. Based on the recent research, Roeder and Lehman (2008) suggest providing an elliptical clearance of eight times the thickness of gusset plate (see Fig. 17.38b). This will not only result in smaller, thinner and compact gusset plates, but also greater ductility and inelastic deformation of the system. Welds joining the gusset plate to the beam and column should be sized using the plastic capacity of the gusset plate rather than the expected resistance of the brace.

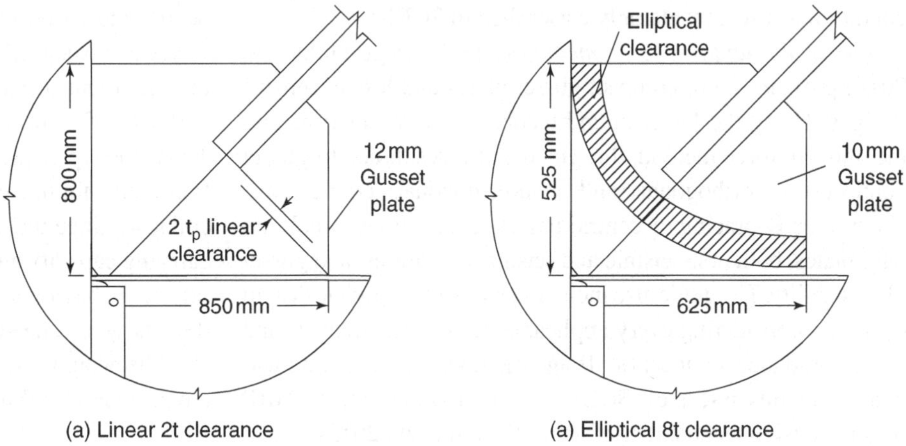


**FIG. 17.38** Improved connection detail for CBFs

*Source:* Roeder and Lehman 2008. Reprinted with permission, STRUCTURE magazine, February 2008.

Similar to OCBFs, bolted connections are not permitted within the middle one-fourth of the clear brace length, by the code. For braces consisting of built-up sections, the tack fastener spacing should be provided in such a way that the unfavorable slenderness ratio of individual element between the fasteners is less than 0.4 times the governing slenderness ratio of the brace.

The code also specifies that the columns of SCBFs should be plastic sections and the splices should be located within the middle one third height of the column. The splices should be capable of resisting at least the nominal shear strength and 50% of the nominal flexural strength of the smaller connected section.

Conventional SCBF systems have inherent problems due to the vastly different compression and tension capacities of the braces. During a major seismic event the compression brace will most likely buckle resulting in the companion tension brace resisting the majority of the demand. Recent full scale testing of a SCBF by the University of California at Berkeley (Uriz and Mahin 2008) highlighted relatively high rate of failure of traditional CBFs in several recent earthquakes. It also pointed out a variety of failure modes such as connections and details unable to develop the full tensile capacity of the braces, and local buckling and fracture of plastic hinge regions at the mid-span of buckled braces. In addition, tests and post-earthquake reconnaissance investigations suggest that CBFs with relatively robust braces may be susceptible to a number of other failure modes, including fracture of the connection of the gusset plates to the supporting beams and columns, failures in columns or base plates. However, when we use *buckling-restrained braced frame* (which is a special case of a special concentrically braced frame), the high ductility of the braces results in very small probability of collapse, and negligible loss in lateral load capacity, even for very rare events (see Section 17.13 for the details of buckling-restrained braced frame). Moreover, replacing the BRB after a major earthquake and bringing back the system to the original condition is also considerably easier than in SCBFs.

Use of the *uniform force method* (AISC Design Guide 29, 2014) provides more compact gusset plates and less expensive designs. Using the lower bound theorem of limit analysis and the uniform force method, this guide addresses: brace-to-gusset connections, orthogonal and non-orthogonal connections, chevron or K-bracing, eccentric braces, connections at column base plates, both non-seismic and seismic conditions, and gusset plate stability. The guide also includes extensive, complete design examples considering every applicable limit state. Analysis and design guidance, connection design methods, and detailing and constructability issues of SCBFs may be found in the NEHRP Seismic design technical brief no. 8 (Sabelli et al., 2013).

### 17.10.3 Eccentrically Braced Frames (EBF)

The bracing member in an eccentrically braced frames (EBF) is connected to the beam so as to form a short link beam between the braces and the column or between two opposing braces (see Fig. 17.39). Thus the eccentric bracing is a unique structural system that attempts to combine the strength and stiffness of a braced frame with the inelastic behaviour and energy dissipation characteristics of a moment frame. The link beam acts as a *structural fuse* to prevent buckling of the brace from large overloads that may occur during major earthquakes. EBFs display better architectural versatility than CBF to provide space for openings.

When an EBF is subjected to lateral load, the axial force induced in the braces is transferred in the form of high levels of shear and bending moment in the link. However, the link is usually subjected to typically low levels of axial force. Consequently, links will normally experience shear and/or flexural yielding during an earthquake and undergo formation of shear or flexural plastic hinges. Thus the links in EBF, by undergoing plastic deformations, allow dissipation of seismic energy and act as fuse to prevent damage in other parts of the frame. In addition, eccentrically braced frames may be designed to control frame deformations and minimize damage to architectural finishes during seismic loading (Williams 2004).

The web buckling is prevented by providing adequate stiffeners in the link. Links longer than twice the depth of beam tend to develop plastic hinges while shorter links tend to yield in shear. Buildings using eccentric bracing are lighter than moment-resisting frames and, while retaining the elastic stiffness of concentrically braced frames, are more ductile. Thus, they provide an economical system in seismic zones. Premature failure of the link does not cause the structure to collapse, since the structure continues to retain its vertical load carrying capacity and stiffness. Although this system has a good seismic behavior, such links are not readily disposable elements -beams would need shoring, floor slabs might require repairs, etc. The design and other details of eccentrically braced systems are provided by Williams (2004) and Bruneau, et al. (1997).

## 17.11 DUAL SYSTEMS

Dual systems consist of steel frames with concrete slabs or concrete fill over metal deck. Shear walls are provided as vertical transportation cores, isolated walls in selected bays, or as a perimeter wall system (see Fig. 17.40). In this dual system, the shear walls resist most of the lateral force (since they are stiffer than frames). Hence, there is a concern that if these walls fail, the building may suddenly collapse. This concern arises due to the fact that most buildings have only a small number of shear walls with little redundancy. Hence the code stipulates that the moment-resisting frames have to be designed to resist

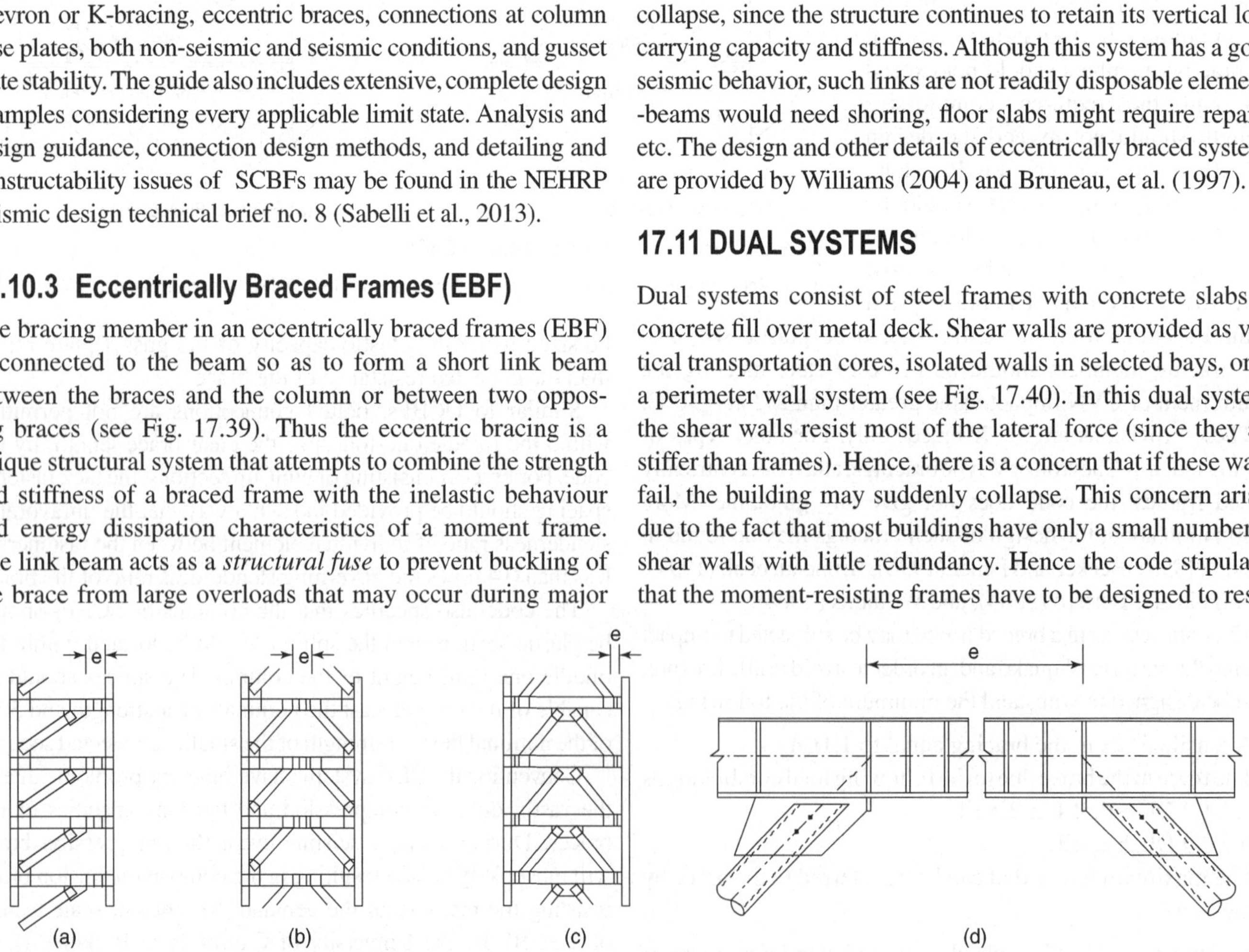


**FIG. 17.39** Eccentric bracing system (a)–(c) Common types of bracing (d) Detail

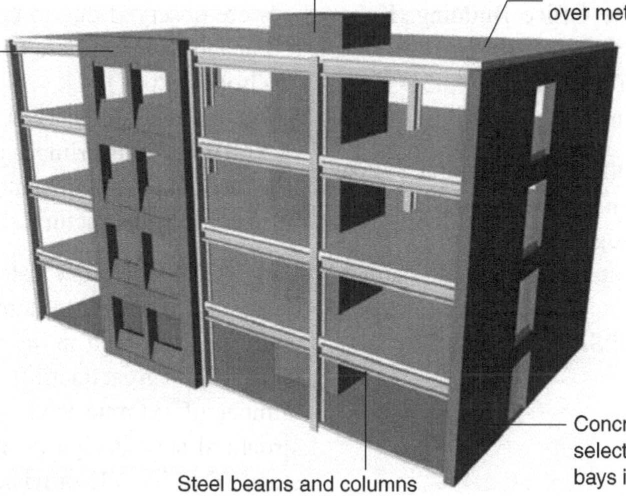


**FIG. 17.40** Dual system of steel frames and shear walls

*Source:* FEMA 454-06

at least 25% of the design base shear, if the analysis indicates that the frames are taking less than 25% of total seismic load (see also the commentary of draft IS 875 code, clause C.4.9), in order to consider the system as a dual system and apply the appropriate *R* factor [ranging from 4.0–5.0, as per Table 7 of IS 1893 (part 1)]. This provision of designing the moment frame for resisting at least 25% of the total base shear provides a 'back-up' resistance to seismic forces.

In general, a dual system has a comparably higher value of *R* since a secondary lateral support system is available to assist the primary non-bearing lateral support system. However, in this case, the storey drifts should be calculated by analyzing the frame for the actual design force and not for 25% of design force (Jain 1995).

## 17.12 STEEL PLATE SHEAR WALLS (SPSW)

*Steel plate shear walls* (SPSW) due to the reduced thickness, as compared with concrete shear walls, offer significant advantages in terms of cost, performance and ease of design and erection. They are considered as an alternative to braced frames and can provide equivalent strength and stiffness. National Building Code of Canada (1994), AISC 341-10, and FEMA 450:2004 introduced provisions for the design of SPSW.

Although the post-buckling behaviour of plates under monotonic load has been investigated by several researchers for more than half a century, post-buckling strength of plates under cyclic loading has not been investigated till now (Caccese et al. 1993; Driver et al. 1998; and Kulak et al. 2001). The results of these investigations revealed that plates can be subjected to a few reversed cycles of loading in the post-buckling domain, without damage. However, steel plate shear walls (SPSW) have been used as the primary lateral load resisting system in buildings for more than three decades in United States, Canada and Japan. Some of these buildings include (Seilie and Hooper 2005) the following:

1. United States Federal Courthouse, Seattle, WA – 23-storey building with a height of 107 m
2. Sylmar Hospital, Los Angeles, CA – 6-storey building
3. Canam Manac Headquarters Expansion, St George, Quebec, Canada, 6-storey building. The shear walls had a width of 2.6 m centre-to-centre of columns and an overall height of 23 m. The infill plates are only 4.8 mm thick and the columns are W 250s. At the floor levels, double C 200 × 17 members were used to anchor the tension field in the infill plates as well as carry gravity loads around the shafts. (Driver and Grondin 2001). Figure 17.41 shows the planar SPSW the system adopted for this building.

**FIG. 17.41** Planar SPSW system of the Canam HQ building, Canada

*Source:* Seilie and Hooper, *From Steel Plate Shear Walls: Practical Design and Construction*, Modern Steel Construction, April 2005. Reprinted with permission from AISC. All rights reserved.

4. The 50-storey (171 m tall) Hyatt Regency Hotel at Reunion, Dallas, TX
5. The 35-storey (130 m tall) Kobe Office Building, Kobe, Japan
6. Shinjuku Nomura building, Tokyo Japan – 51-storey building having a height of 211 m

Out of these buildings the Kobe Office Building and Sylmar Hospital have withstood fairly significant earthquakes (Kobe earthquake in 1995 and Northridge earthquake in 1994, respectively) and survived without any structural damage (Seilie and Hooper 2005).

There are three different types of SPSW systems:

1. Unstiffened, thin SPSW
2. Stiffened SPSW
3. Composite concrete SPSW

In North America unstiffened, thin SPSW are common while in Japan, the stiffened SPSW system is often used. In this section we will confine our discussion on unstiffened thin SPSW system only.

### 17.12.1 Advantages of SPSW

The following are the advantages of SPSW system:

1. SPSW allow for less structural wall thickness in comparison to the thickness of concrete shear walls. This results in saving of rentable floor area.
2. Steel savings as much as 50% have been achieved in structures employing a steel plate shear wall system rather than a moment-resisting frame. When compared with reinforced concrete shear walls, the steel system offers reduced foundation costs. This feature makes SPSW more amenable for upgrading the lateral load resistance of an existing structure without overstressing the foundation. However, proper connection at the interface of the steel panel and concrete frame members has to be provided to transfer the interfacial forces. We must also consider compatibility between the nominal ductility of the existing frame and the ductile nature of the steel infill panel. A connection system using HSS tube collars has been developed at the University of Alberta, Canada and tested. This system is found to have several advantages over the more common drilled expansion or adhesive anchors. In addition this system and enhances the behaviour of existing concrete frames (Driver and Grondin 2005).
3. The use of SPSW system reduces construction time. It is not only fast to erect, but also involves no curing period.
4. As mentioned earlier, the thin plates have excellent post-buckling capacity.SPSW ductility is superior to braced frames and even moment frame systems. It has been found that the system can survive up to 4 to 5% drift without experiencing significant damage (which is more than that expected in many moment frame systems). Even though some pinching and tearing close to the corners of the panel were observed due to bending, it did not reduce the plate capacity and stiffness (Astaneh and Zhao 2002; Driver and Grondin 2001).
5. As pointed out earlier, at least two buildings that use SPSW as their primary lateral force-resisting system have experienced significant earthquakes and survived with insignificant structural damage (Astaneh and Zhao 2002).

Large-scale laboratory tests have been conducted to verify the behaviour of SPSW systems (Driver et al 1998). These tests confirmed the high initial stiffness, large energy dissipation capacity and great ductility of these systems, even after a large number of extreme load cycles. The Canadian Standard for structural steel design (CAN / CSA – S16 2001) is the first standard to include provisions for unstiffened SPSW design, though it included analysis and design provisions in the 1994 edition itself in a non-mandatory appendix. Similar provisions have been incorporated in the AISC seismic provisions for steel buildings, where it is denoted as *special plate shear wall* (ANSI / AISC 341 – 10).

However, the SPSW system has the following disadvantages:

1. SPSW systems are more flexible than concrete shear walls. Hence it may be necessary to provide additional flexural stiffness to SPSW. One way of improving the flexural stiffness may be to provide large concrete composite concrete infill steel pipe columns at all corners of the core wall. This type of stiffening will also improve its overturning capacity.
2. Excessive initial compressive force in the SPSW panel may delay the development of the tension-field action. Hence, the construction sequence may be so scheduled to avoid build up of excessive compression in the panel.

### 17.12.2 Analysis and Design of SPSW

A typical SPSW system consists of steel plate panels (web element), vertical boundary element (columns) and horizontal boundary elements (beams). Thus, in a SPSW system, the building columns correspond to the flanges in a plate girder and the floor beams correspond to the girder vertical stiffeners. However, unlike girders, sizeable top and bottom horizontal bracing elements are required in SPSW to anchor the significant tension fields that develop at these ends of the structural system. Also the use of wide flange structural shapes (which have substantial in-plane bending stiffness) for the vertical and horizontal boundary elements favorably impacts orientation of the angle of development of the tension field action, and makes possible the use of very slender webs. For these reasons, the use of beam and plate girder design provisions as discussed in Chapters 10 and 11 are not appropriate for the design of SPSW (Berman and Bruneau 2004).

The key principal for design in AISC 341 is that significant inelastic deformation capacity in SPSW is provided primarily through web plate yielding and as plastic-hinge formation in the ends of *horizontal boundary elements* (HBEs). *Vertical boundary elements* (VBEs) are not expected to yield in shear; VBEs are not expected to yield in flexure except at the column base. It has been found that yielding of the web in SPSW occurs by development of tension field action at an angle of about 45° (in the range 38 - 43°) from the vertical and buckling of the plate in the orthogonal direction. The sizing of vertical and horizontal boundary elements makes it possible to develop this tension field action across the entire webs.

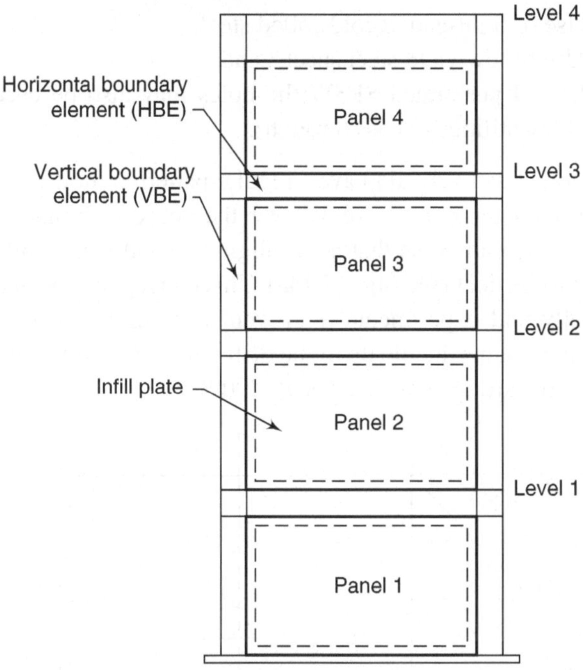


**FIG. 17.42** Typical steel plate shear wall

For the purposes of analysis each web plate can be modeled as a series of inclined pin-ended struts or strips oriented at an angle $\alpha$ (see Figs 17.42 and 17.43). A minimum of 10 struts per panel has to be used to provide realistic results. Timler and Kulak (1983) derived the equation for the inclination angle $\alpha$ (in radians) of the tension field as the angle between the direction of the strip and the vertical direction as given in Eqn (17.15) (ANSI / AISC 341 – 10):

$$\tan^4\alpha = \frac{1+\left(\dfrac{t_w L}{2A_c}\right)}{1+t_w h\left(\dfrac{1}{A_b}+\dfrac{h^3}{360 I_c L}\right)} \tag{17.15}$$

where $h$ is the distance between the centerlines of horizontal boundary elements (mm), $A_b$ is the gross-sectional area of horizontal boundary elements (mm$^2$), $A_c$ is the gross-sectional area of vertical boundary elements (mm$^2$), $I_c$ is the moment of inertia of vertical boundary element about axis perpendicular to the plane of the web plate (mm$^4$), $t_w$ is the thickness of the web, and $L$ is the distance between center line of vertical boundary elements (mm).

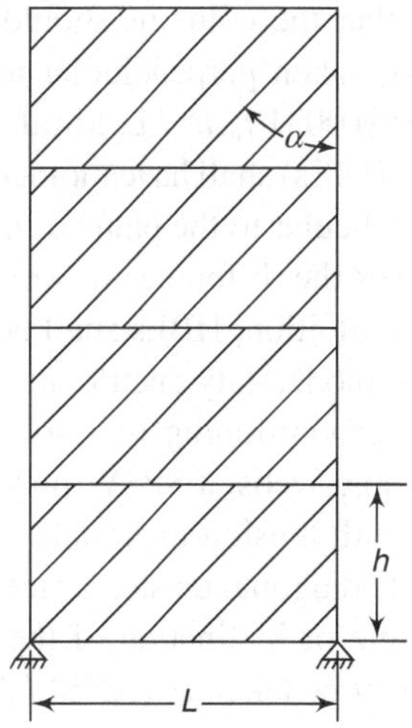


**FIG. 17.43** Strip model representation of a SPSW

The design shear strength of the panel (with a system over strength factor of 1.2) is given by (ANSI / AISC 341 – 10)

$$V_n = 0.42 f_y t_w L_{cf} \sin 2\alpha / \gamma_{m0} \tag{17.16}$$

where, $t_w$ = thickness of the web (mm), $L_{cf}$ is the clear distance between flanges of vertical boundary elements (mm), $\alpha$ is the angle of web yielding in degrees, as measured relative to the vertical. The angle of inclination, $\alpha$, is permitted to be taken as 40°, or can be calculated as per Eqn (17.15), and $\gamma_{m0}$ = partial safety factor = 1.1

The ratio of panel length to height, $L/h$ should be limited to $0.8 < L/h < 2.5$. This limit has been specified based on past research, which has not investigated the seismic behaviour of SPSW having $L/h$ greater than 2.0. Also the modeling with strips is reasonably accurate only when $L/h$ is greater than 0.8. Past research has focused on walls with $L/t_w$ ratio ranging from 300-800. Drift limits will indirectly constrain this ratio. Since thicker plates will delay the development of tension – field action, a minimum value of 180 may be adopted for $L/t_w$ ratio.

ASCE 7-10 also specifies the values for variables used in design: Seismic response modification coefficient $R = 7$; Over strength factor $\Omega_o = 2$; Deflection amplification factor $C_d = 6$. For a dual system with SMF the modifications required are: $R = 8$, $\Omega_o = 2.5$, $C_d = 6.5$. (Note these values are related to ASCE 7-10 and not with IS 800:2007, which does not include provisions for the design of SPSW).

The boundary elements are very important to the proper performance of SPSW systems. As mentioned already, the top and bottom horizontal boundary elements will have substantial size (in order to anchor the forces due to tension–field action) and the intermediate horizontal boundary elements are relatively small in size. It is important that the columns (vertical boundary elements) are strong enough such that they will not fail before the plate develops its full tension field. ANSI / AISC 341 – 10 suggests that the columns should have moment of inertia about an axis taken perpendicular to the plane of the web, $I_c$, greater than $0.0031\, t_w\, h^4 / L$. Similarly, the horizontal boundary elements (HBEs) shall have moments of inertia about an axis taken perpendicular to the plane of the web, $I_b$, greater than $0.0031L^4/h$ times the difference in web plate thicknesses above and below. In addition, HBEs shall be braced to satisfy the requirements for moderately ductile members.

For preliminary proportioning of horizontal and vertical boundary elements and webs, a SPSW may be approximated by a vertical truss with tension diagonals. Each web may be represented by single diagonal tension brace within the storey. For an assumed angle of inclination of the tension field, the web thickness, $t_w$ may be taken as (ANSI / AISC 341 – 10)

$$t_w = \frac{2A\Omega_s \sin\theta}{L\sin 2\alpha} \tag{17.17}$$

where $A$ is the area of the equivalent tension brace (mm$^2$), $\theta$ is the angle between the vertical and the longitudinal axis of the equivalent diagonal brace, $L$ is the distance between centre lines of vertical boundary elements (mm), $\alpha$ is the assumed angle of inclination of the tension field measured from vertical (see Eqn17.15), and $\Omega_s$ is the system over strength factor, which may be taken as 1.2 for SPSW. The value for the area A, is initially estimated from an equivalent brace size to meet the structure's drift requirements.

The expected tensile strength of the web strips is defined as $R_yF_yA_s$, where $A_s$ = area of a strip = $(L\cos\alpha + H\sin\alpha)/n$, (mm$^2$), $L$ = width of panel (mm), $H$ = height of panel (mm), $n$ = number of strips per panel and $n$ should be taken greater than or equal to 10.

As mentioned earlier, one of the critical factors limiting the implementation of SPSWs is the large column sizes required to resist the combined axial and flexural demands from overturning, frame action and web plate forces (Berman, 2014). Recent research by Tsai et al. (2014) has developed recommendations for design that allow the formation of the column plastic hinges at a height of 0.25 to 0.33 times the storey height above the base (instead of forming at the base as previously recommended), where the moment is typically maximum in the compression column. This reduces flexural demands significantly and does not impact performance of the system as long as the column does not form a plastic hinge at the top of the first storey. This will result in a 20% reduction in column weight with no impact on performance.

For other provisions like protected zones, demand critical welds, HBE-to-VBE Connections, Connections of Webs to Boundary Elements, column splices, etc. of SPSW reference should be made to AISC 341-10 and AISC Design Guide 20.

## Perforated Webs

It has been found that the available hot-rolled steel plates are often thicker or stronger than those required as per design for the webs of SPSW in low to medium rise buildings. In such cases, using the minimum available thickness would result in large panel force over-strength, proportionally larger design demands on the surrounding VBE and HBE, and thus the economy of the SPSW may be compromised. Several solutions are available to alleviate this concern, viz.:

(a) Use of light-gauge cold-rolled steel,
(b) Use of low yield steel sheets, and
(c) Use of perforated SPSW(the holes may also be used to allow utilities to pass through).

Perforated webs should have a regular pattern of holes of uniform diameter spaced evenly over the entire web-plate area in an array pattern so that holes align diagonally at a uniform angle to vertical (see Fig. 17.44a). This concept has been analytically and experimentally proven to be effective and the system remains ductile up to the drift demands corresponding to severe earthquakes (Vian et al. 2009).

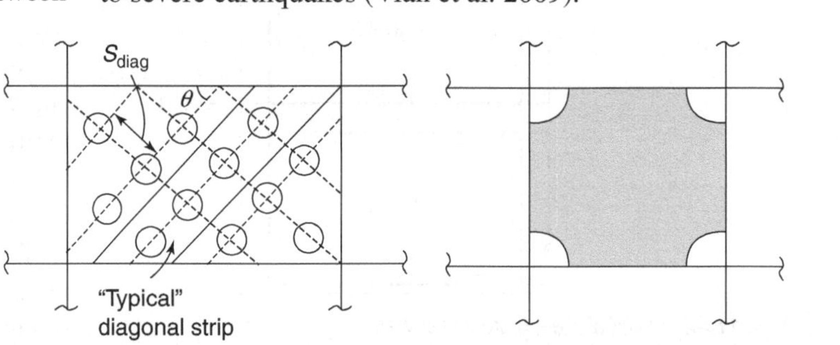


(a) SPSW with perforated web (b) SPSW with reinforced corner cut-outs

**FIG. 17.44** SPSW with openings

The spacing of holes, $S_{diag}$, should be at least $1.67D$. The distance between the first holes and web connections to the HBEs and VBEs should be at least $D$, but shall not exceed $(D + 0.7S_{diag})$, where D is the diameter of the holes. The panel design shear strength in this case is calculated as

$$V_n = 0.42 f_y t_w L_{cf}\left(1 - \frac{0.7D}{S_{diag}}\right) / \gamma_{m0} \tag{17.18}$$

The stiffness of such regularly perforated infill plates should be calculated using an effective web-plate thickness, $t_{eff}$, given by:

$$t_{eff} = \frac{1 - \dfrac{\pi}{4}\left(\dfrac{D}{S_{diag}}\right)}{1 - \dfrac{\pi}{4}\left(\dfrac{D}{S_{diag}}\right)\left(1 - \dfrac{N_r D \sin\alpha}{H_c}\right)} t_w$$

where, $H_c$ is the clear column (and web-plate) height between beam flanges (mm), $N_r$ is the number of horizontal rows of perforations, $t_w$ = web-plate thickness (mm), and $\alpha$ is the angle of the shortest center-to-center lines in the opening array to vertical(degrees).

The effective expected tension stress to be used in place of the effective tension stress for analysis is $R_y f_y (1 - 0.7\ D/S_{diag})$.

AISC 341-10 also permits quarter-circular cut-outs at the corners of the webs provided that the webs are connected to a reinforcement arching plate following the edge of the cutouts (see Fig. 17.44b). AISC 341-10 may be consulted for the provisions of webs with cut-outs at the corners.

### Coupled SPSWs

Even though *coupled SPSWs* may offer the flexibility of using SPSW systems around lift wells of tall buildings, till recently no guidance was available on design methods, steel coupling beam detailing, and general behavior. Borello and Fahnestock (2013) recently developed design concepts for coupled SPSW, and recommend target values for the degree of coupling (ratio of the overturning moment resisted by the individual walls to the total overturning moment). They also showed that significant steel weight savings can be achieved when two individual walls are coupled and by using nonlinear analysis, demonstrated that they have excellent seismic performance.

### Self-Centering SPSW

Current research in earthquake engineering research is focused on minimizing residual drift and ensuring simple post-earthquake repair strategies. In this context, *self-centering steel plate shear walls* (SC-SPSW ) has been developed. The schematic of this system is shown in Fig. 17.45. In this new system, the web plate is intended to yield under cyclic loading, whereas the boundary elements and PT connection elements remain undamaged. Thus, the post-tensioned beam-to-column connections provide recentering after earthquakes and web plate tension field action provides stiffness and energy dissipation. The damaged web plate can easily be replaced after any major earthquake. Large-scale subassemblage tests, shake table tests on systems with different connections and two-storey full-scale proof-of-concept tests were conducted by Bruneau and associates to study the effects of various design parameters on the system and connection response [Dowden et al. (2012), Clayton et al. (2012)]. These experimental results show that the SC-SPSW system has high ductility, high initial stiffness, recentering capabilities, an overall system response as anticipated, and more energy dissipation than expected.

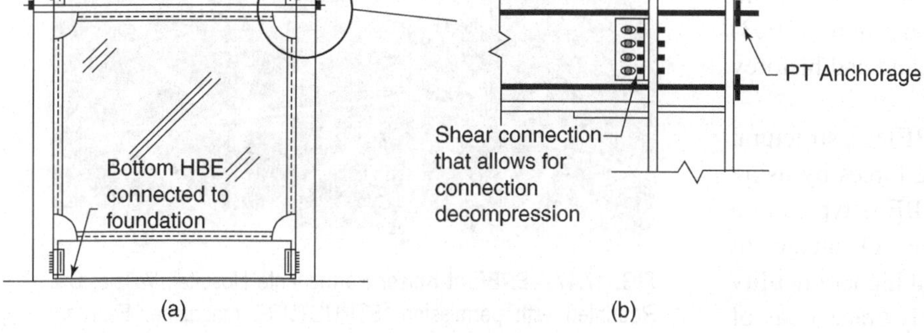


**FIG. 17.45** Schematic diagram of self-centering SPSW

## 17.13 BUCKLING-RESTRAINED BRACES (BRB)

When a concentrically braced frame (CBF) system, as discussed in section 17.10.1, is subjected to earthquake loads, the braces will be subjected to alternate compression and tension. Hence traditionally, CBFs have been treated as high strength, low ductility systems, because, the steel braces show significant strength and ductility in tension, whereas deliver only a fraction of this strength and ductility in compression, due to buckling.

Theoretically, buckling in compression members can be eliminated by providing lateral bracings at close intervals, such that the un-braced length of the member approaches a small value. In the 1980s Prof. Akira Wada of the Tokyo Institute of technology, developed a system, called *Un-bonded Brace*$^{TM}$ in collaboration with Nippon Steel Corporation; his inspiration came from the collarbone of the human body. In this system, which resembled a typical bone (bigger at the ends and a reduced section in the middle), Euler buckling of the central steel core is prevented by encasing it over its entire length in a steel tube.

The load-resisting component of a BRB is a steel core restrained against overall buckling by an outer casing filled with concrete, which is the restraining mechanism preventing buckling (see Fig. 17.46). Bonding of the steel core to the concrete is prevented during the manufacture to ensure that the BRB components remain separate to prevent composite action that would change the behavior. It has to be noted that that the steel rectangular hollow section (RHS) core is divided into five segments: the central restrained yielding segment with a reduced section within the RHS; restrained, non-yielding transition segments of larger area than the yielding segment at both ends within the RHS; and unrestrained,

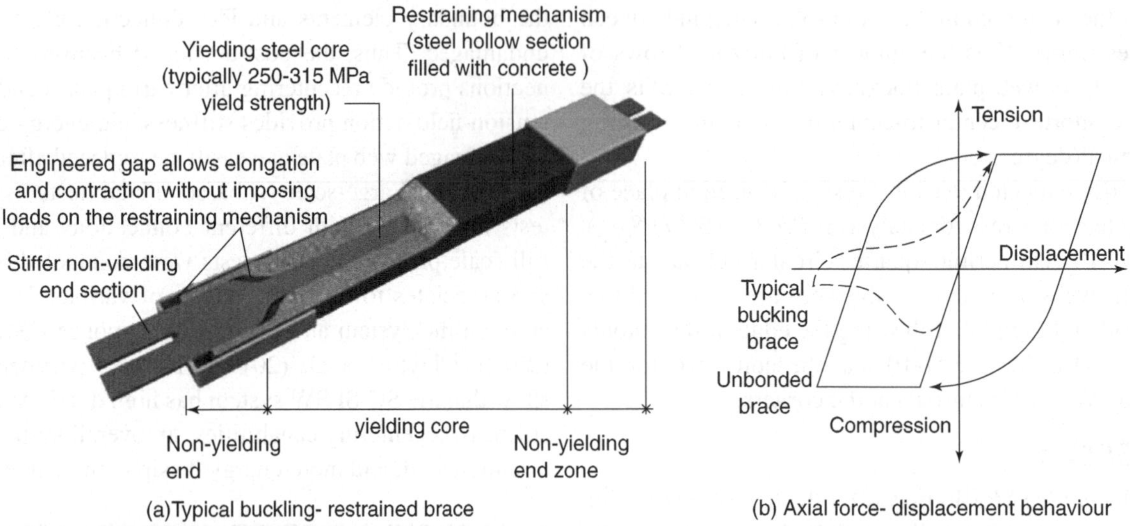


**FIG. 17.46** Typical buckling-restrained brace and its behaviour as opposed to conventional brace

*Source:* Specifying Buckling-Restrained Brace Systems, Modern Steel Construction, November 2009. Reprinted with permission from AISC. All rights reserved.

non-yielding connection segments at both ends that extend past the RHS and connect to the frame, typically by means of gusset plates. By confining the inelastic behavior to axial yielding of the steel core, great ductility can be achieved by the brace itself. The ductility of the steel material is realized over the majority of the brace length. Thus the hysteretic performance of these braces is similar to that of the steel core material (López and Sabelli 2004). Fig. 17.46(a) shows a typical WILDCAT™ BRB manufactured by Star Seismic (www.starseismic.net). Fig. 17.46(b) shows the comparison of the behaviour of a BRB and a conventional brace. Most of the BRBs developed to date are proprietary, but all of them are based on the similar concept. Some of them are Unbonded Brace™ manufactured by Nippon Steel Corporation, CoreBrace™ by CoreBrace, WILDCAT™ and Star Seismic™ Modular Systems by Star Seismic, and POWERCAT™ by PKM steel.

Interestingly a similar concept of '*core loaded sleeved strut*' was originally proposed by Er. B.N. Sridhara of Bangalore. This sleeved strut was experimentally studied for compression load carrying capacity in IIT Madras by Kalyanaraman et al. 1994, Prasad 1992 and Sridhara 1990. Sridhara also obtained an US patent for his invention and Star Seismic™ and CoreBrace™ are now manufactured in USA using his patent. Interested reader may consult Uang and Nakashima (2003) for a more detailed summary on the background and historey of the buckling-restrained braced frames.

A *buckling-restrained braced frame* (BRBF) is a structural steel frame which resists a building's lateral forces by using the buckling-restrained brace (BRB). A BRBF is typically a special case of a concentrically braced frame. Advantages of a BRBF over other CBFs are that they exhibit higher ductility and energy dissipation. BRBs most commonly brace a bay of steel frame diagonally or in a chevron pattern. The symmetrical capacity in tension and compression allows BRB braces to be used in single-diagonal configurations without penalty.

According to AISC 341-10, BRBF are expected to provide significant inelastic deformation capacity primarily through brace yielding in tension and compression.

The first BRBF system was installed in the United States at UC-Davis in 2000. Till 2008, more than 150 structures and 20,000 BRBs have been used in the USA alone (López 2008). The 56-floor Los Angeles Convention Center, the Bennet Federal Building in Salt Lake City, Utah, and the 60-storey One Rincon Hill building in San Francisco are among those fitted with the 'sleeved column' braces. BRBs have been used in a variety of applications such as bridges, horizontal diaphragm elements, buildings, high rise outrigger frames, externally anchored braces, and in wind towers. Fig. 17.47 shows BRBFs at Kaiser Permanente Hospital, Vallejo, California.

**FIG. 17.47** BRBF at Kaiser Permanente Hospital, Vallejo, CA. Reprinted with permission, STRUCTURE magazine, February 2008.

BRBF may be used in future, not only as primary lateral force resisting elements in new construction, but also as supplemental hysteretic dampers in *seismic retrofitting*, since the original motivation behind the initial development of the BRBF system was to use it for seismic retrofitting. Careful analysis and brace sizing can result in a considerable increase in damping without an intolerable decrease in building period.

BRBFs are typically designed using an equivalent-lateral-force method. As in the typical design procedure employed for other concentrically braced-frame types, a linear elastic analysis is done and the frame is subjected to a reduced seismic load in order to determine the required strength and to verify adequate stiffness of the frame (ASCE 7-10 specifies the following to be used in design: Seismic response modification coefficient $R = 8$; Over strength factor $\Omega_o = 2.5$; and deflection amplification factor $C_d = 5$). For a BRBF with braces proportioned according to this method, the difference between the elastic and inelastic deformation modes is much different than for a Special Concentrically Braced Frame (SCBF). Because of this, an inelastic dynamic analysis is not typically required, although inelastic analyses give a much better estimate of brace ductility demands than elastic analyses (López and Sabelli 2004).

For such an elastic analysis to be valid, the braces used in the analysis should correspond to tested brace behaviour, and similarly, brace tests should confirm to the strength and ductility assumed in the analysis. Accordingly, BRBF design is based on the results of successful tests. The BRBF design procedure requires the columns to have the strength to resist the vertical component of the expected yield strength of each brace in a frame. Once BRBs have been designed for adequate strength, the adjoining frame elements are designed to the adjusted BRB strengths corresponding to a storey drift of at least 2% of the storey height or two times the design storey drift, whichever is larger, in addition to brace deformations resulting from deformation of the frame due to gravity loading. This design philosophy allows the column to remain elastic during a seismic event, while the BRBs yield and absorb the seismic energy. Hence BRBF performs with a higher degree of ductility than conventional braced frames. The adjusted brace strength in compression is specified in AISC 341-10 as $\beta\omega R_y P_{ysc}$, where $\beta$ is the compression strength adjustment factor, $\omega$ is the strain hardening adjustment factor, $P_{ysc}$ is the axial yield strength of *steel core* (MPa), and $R_y$ is the ratio of the expected yield stress to the specified minimum yield stress, $f_y$.

Design storey shear has to be shared between the braces and the braced frame columns in proportion to their relative rigidities. BRBF columns resist high tensile loads. As a result, complete joint penetration welds and thick plates are normally specified at the column base, with a vertical gusset stiffening the joint. The moment generated at the column base will be resisted by a concrete-compression anchor-rod-tension couple. The shear generated at the column base will be resisted by steel elements (angles or plates) parallel to the frame and welded to the top of the base plate allowing the anchor rods to resist tension only. Because of these procedures, simple truss-force models are not sufficient, and a model that includes flexural properties is required.

Unlike the massive gusset plates required on the special concentric braced frame, the connections for the BRBF system are of smaller size and require less welding/bolting. Use of the *uniform force method* (AISC Design Guide 29, 2014) provides more compact gusset plates and less expensive designs. Recent testing has demonstrated that gusset plate connections may be critical aspect of the design of BRBF. Thornton and Muir (2009) provide detailing of the gusseted joint with beam hinges, to take into account the potential distortional forces induced by large seismic drifts. BRBFs do not need zipper columns in chevron configurations and require lighter beams. Section F4 of AISC 341-10 contains provisions for BRBF design. A design example of seven-storey office building with BRBFs is provided by López and Sabelli (2004).

It has to be noted that BRBF's cannot be spliced. Welded or bolted splices of braces are not allowed in situations where the braces are likely to be subjected to inelastic demands because that would probably result in undesirable behavior leading to possible brittle fracture (Hussain et al. 2006).

Experimental results conducted by Vargas and Bruneau (2009) indicate that the objectives of the *structural fuse* concept can be successfully achieved in the case of buckling restrained braced frames (i.e., beams and columns performed elastically, while BRBs worked as metallic fuses and dissipated the seismically induced energy). In general, analytical models reasonably predicted maximum response values for the BRB frames. They also proposed an eccentric gusset-plate detail which was found to be effective in preventing performance problems, such as local buckling and out-of-plane buckling of the plates at the connection point.

### Composite Moment Frames

Provisions for composite systems like composite ordinary/intermediate moment frames composite SMF, composite partially restrained moment frames, composite ordinary braced frames, composite special concentrically braced frames, composite eccentrically braced frames, composite ordinary shear walls (reinforced concrete walls with structural steel or composite sections serving as boundary elements) and special shear walls (coupled wall systems with steel or composite coupling beams) can also be used to resist earthquake.

## 17.14 DEVICES TO REDUCE EARTHQUAKE EFFECTS

In addition to the aforementioned guidelines for analysis and design, the structural engineer now has the option of using a variety of devices to ensure safety or serviceability of the structure under severe earthquakes. These devices either isolate the structure from ground vibration or absorb the energy provided by the earthquake to the building. They are similar to the shock absorbers provided in motor vehicles, which absorb the vibrations caused by the undulated road surfaces. But unlike shock absorbers, the vibration control needed for buildings is lateral, because the most destructive seismic motions act in the horizontal direction.

### 17.14.1 Base Isolation

The concept of *base isolation* is to introduce special bearings (called *base isolators*) between the ground and foundation of the structure, such that the building is isolated from the ground. (This concept is similar to the provision of neoprene bearings at the supports below the bridge decks). Buildings resting on such base isolators are called *base-isolated buildings*. Figure 17.48(a) shows the basic elements of a base isolation system; the supplemental *dampers* shown are optional and hence may or may not be utilized within an isolation system. These dampers absorb energy and thus increase the damping of the building.

By decoupling the structure from ground shaking, isolators reduce the level of response in the structure that would otherwise occur in a conventional, fixed-base building (see Fig. 17.48b). Conversely, base-isolated buildings may be designed with a reduced level of earthquake load to produce the same degree of seismic protection. Qualitatively, a conventional structure experiences deformations within each storey of the structure (i.e., inter-storey drifts) and amplified accelerations at upper floor levels. In contrast, base-isolated structures will experience deformation primarily at the base of the structure (i.e., within the isolation system) and the accelerations are relatively uniform over the height.

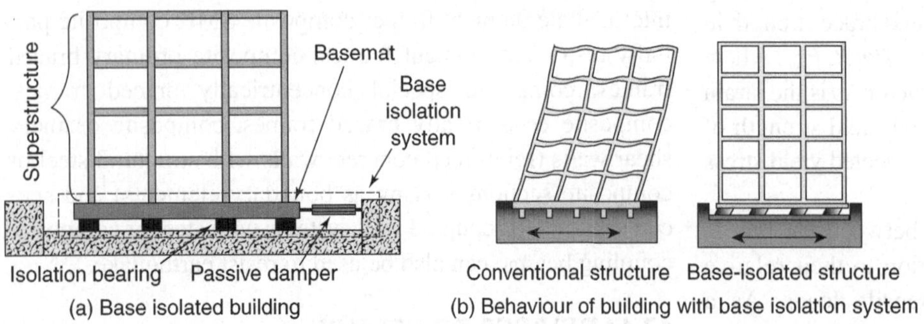


**FIG. 17.48** Concept of base isolation

Figure 17.49(a) shows typical acceleration design response spectra for three different damping levels. The major effect of seismic isolation is to increase the natural period which reduces the acceleration and thus force demand on the structure. Thus the forces induced by ground shaking will be much smaller than those experienced by 'fixed-base buildings' directly resting on the ground. In terms of energy, an isolation system shifts the fundamental period of a structure away from the strongest components in the earthquake ground motion, thus reducing the amount of energy transferred into the structure. The energy that is transmitted to the structure is largely dissipated by efficient energy dissipation mechanisms within the isolation system.

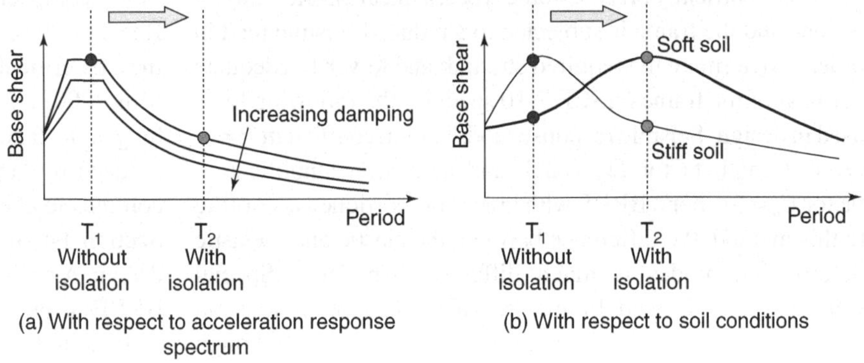


**FIG. 17.49** Effect of seismic isolation

However, as shown in Fig. 17.48(b), softer soils tend to produce ground motion at higher periods which, in-turn, amplifies the response of structures having high periods. Hence, seismic isolation systems, should not be used in sites with soft soils, for example in Mexico City (The fundamental natural period of soft soil in Mexico City is found to be approximately 2*s*). Thus, base isolation systems are most effective on structures built on stiff soil and structures with low fundamental period (low-rise building); on the other hand they are least effective on structures built on soft soil and structures with high fundamental period (high-rise building).

Though the first structure which used the principle of base isolation is believed to be the *Tomb of Cyrus* in Pasargadae, a city in ancient Persia (now Iran) in the 6th century BC, the American architect, Frank Lloyd Wright, was the first person to implement the idea of base isolation for isolating Imperial Hotel structure in Tokyo. He provided closely spaced short length piles in the top 2.5 m layer of firm soil that covered a deep deposit of shaky mud. The building survived the devastating 8.3 magnitude 1923 Tokyo earthquake, while other buildings around it collapsed (But eventually the foundation sank irrecoverably into the silt, and the structure was demolished in 1968).

The present day modern base-isolation devices started with the pioneering work done by R. Ivan Skinner, W.H. Robinson, and G.H. McVerry at the Physics and Engineering Laboratory of the Department of Scientific and Industrial Research (PEL, DSIR) in New Zealand during 1977 (they used the World's first isolator developed by them in the William Clayton Building, New Zealand) and later by Prof. James M. Kelly at the University of California at Berkeley. The first base-isolated building in the United States is the Foothill Communities Law and Justice Center, about 97 km east of downtown Los Angeles. Completed in 1985, the building is four stories high with a full basement and sub-basement for the isolation system, which consists of 98 high-damping elastomeric bearings. The superstructure of the building has a structural steel frame stiffened by braced frames in some bays. As of 2015, 500 structures in USA have been seismically isolated (Walters, 2015). During January 2002, the 300-bed district hospital in Bhuj, was the first in India to be installed with 280 lead-rubber and sliding bearings (this hospital replaced the one that collapsed tragically in the Bhuj earthquake). The first large base-isolated building in Japan was completed in 1986. Since the Kobe earthquake, more than 2000 base-isolated buildings, many of them apartment blocks, have been constructed in Japan- total number of buildings with seismic isolation in Japan till April 2015 is 7800 (Walters 2015). Base-isolation is being adopted in several buildings all over the world. As per Walters (2015) more than 4000 buildings in China have been equipped with base isolators. The application of this technology to high-rise buildings is also becoming popular, but requires very large isolators. Isolators of up to 1,600 mm diameter and around 600 mm height are currently available, a size capable of sustaining over 20 MN axial load and 800 mm shear displacement (Nishi et al. 2009). In addition to buildings, seismic isolation has been used for the seismic protection of structures such as bridges, liquefied natural gas (LNG) tanks, and offshore platforms.

The efficacy of this technology was verified during the 1994 Northridge earthquake of California, USA, the 1995 Kobe earthquake of Japan, as well as the 2008 Sichuan earthquake of China. For example, a California hospital remained operational, unlike conventionally built structures in the area (Nishi et al. 2009).

## Types of Base Isolation Systems

Many types of isolation system have been proposed and have been developed to varying stages, with some remaining only in concepts and others having installed in several projects. Fig. 17.50 shows the various types of base isolation systems; they may be broadly classified as (a) elastomeric bearings (lead-rubber bearing, high-damping natural rubber bearing, low-damping natural or synthetic rubber bearing, low damping natural rubber with lead core), (b) sliding bearings (flat sliding bearing, spherical sliding bearing, friction pendulum systems), (c) sliding/friction bearing, (d) rolling systems (using cylindrical rods or elliptical bearing), and (f) combined systems (electricide-de-France system and resilient-friction base isolators). Some of the frequently used isolator systems are shown in Fig. 17.51.

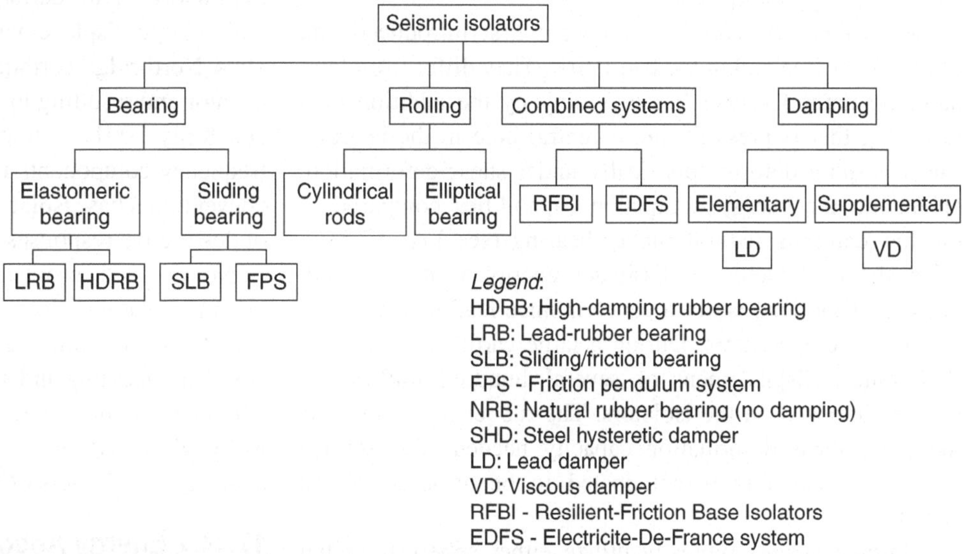


**FIG. 17.50** Types of seismic isolators

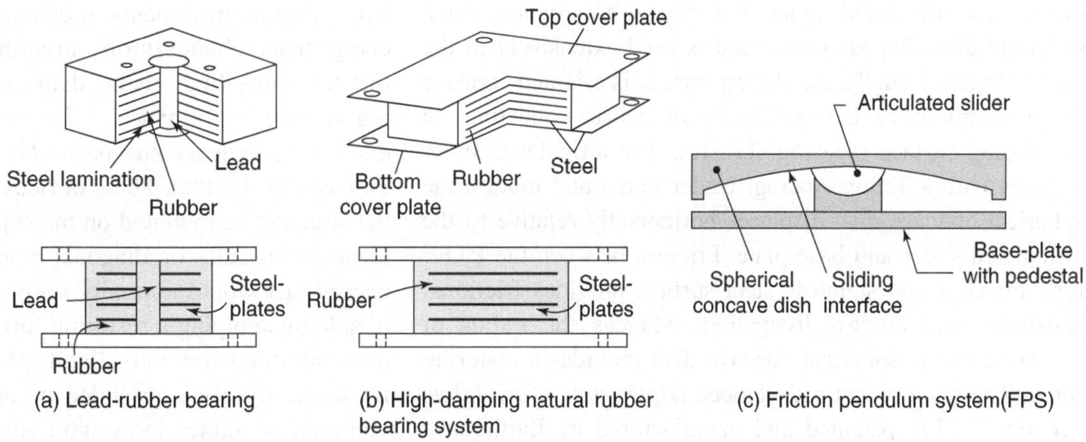


**FIG. 17.51** Some frequently used base-isolators

*Elastomeric bearing* is the most common type of base isolation device, and consist of alternating rubber and thin steel plates layers (about 3 mm thick), firmly bonded to each other (Figs 17.51a and 17.51b). The bearings are constructed by placing un-vulcanized rubber sheets and steel shims in a mold, then subjecting the mold to elevated temperature and pressure to simultaneously vulcanize and bond the rubber. The steel plate reinforcement provides a high compressive stiffness to reduce vertical deflection under the heavy weight of the structure, making the isolator stable. The rubber layers provide the very low horizontal stiffness needed to give the structure a horizontal natural frequency (typically 0.5 Hz), lower than the peak frequencies of an earthquake (Nishi et al. 2009). This decouples the structure from ground shaking, reducing the transmission of earthquake forces into the structure and protecting both the structure and its contents (50–85% reduction has been achieved). In addition, rubber cover is provided on the top, bottom, and sides of the bearing to protect the steel plates. In some cases, a lead cylinder is installed in the center of the bearing to provide high initial stiffness and a mechanism for energy dissipation.

*Lead-rubber bearings* (LRB) were first introduced and used in New Zealand in the late 1970s. They differ from low-damping natural rubber bearings only by the addition of a lead-plug that is press-fit into a central hole in the bearing. The lead-plug deforms plastically under shear deformation, enhancing the energy dissipation capabilities compared to the low-damping natural rubber bearing (see Fig. 17.51a). After the lead yields, it dissipates energy as it is cycled. Fatigue of the lead is not a concern since lead recrystallizes at normal temperatures. During a large earthquake, a shear (horizontal) displacement of several hundred millimeters may be imposed on the isolators. The rubber layers provide the large shear deformation capacity needed. The service life of the isolators is anticipated to be at least several decades.

*Sliding bearings* typically utilize either spherical or flat sliding surfaces. The *friction pendulum system* (FPS) bearing utilizes an articulated slider that moves horizontally on a spherical dish-shaped surface and is used extensively in the United States. Usually, the sliding surface is oriented concave down to minimize the possibility of debris collecting on the sliding surface (see Fig. 17.51c). The articulated slider is faced with a Teflon coating. Under horizontal motion the spherical concave dish displaces horizontally relative to the articulated slider and base-plate. Friction between the PTFE type material and stainless steel surface provides frictional resistance and energy dissipation, whereas the radius of curvature of the spherical concave dish provides a restoring force. The most recently developed *triple friction pendulum* version of FPS, patented and manufactured by Earthquake Protection Systems, Inc., contains a compound articulated slider with multiple sliding surfaces to allow control of the sliding sequence and the resulting hysteresis curve (Walters 2015).

Mayes et al. (2012) compared the cost-benefit analysis of isolated and non-isolated buildings and concluded that considering the cost of earthquake insurance premiums, using base isolation without earthquake insurance can be a more cost-effective solution than a conventional fixed based structure with insurance, despite the first cost premium for base isolation.

It has to be noted that normal base-isolation systems provide only horizontal isolation and are rigid or semi-rigid in the vertical direction. A rare exception to this rule is the full isolation (horizontal and vertical) of a building in southern California isolated by large helical coil springs and viscous dampers (Kircher 2012). The implementation of the base isolation requires optimal design, which depends on the magnitude and frequency range of the earthquake that is being considered. Recent research reveals that the base isolation system may be vulnerable for buildings situated in the near-fault and far-fault earthquakes zones. Near-fault earthquakes with a large displacement and long-period pulse, such as the 1994 Northridge earthquake, may lead to over-stretching of isolator and resulting in malfunctioning of the system (Jangid and Kelly 2001). While far-field earthquakes (with its low-frequency components falling into the resonant region of the conventional base isolation system) may result in amplification of destructive responses to the protected structures.

No detailed provisions exist in the Indian codes for seismic isolators. More information on seismic base-isolators, analytical and numerical models, code provisions for seismic isolation, buckling and stability of isolators, design examples, computer applications, and recent trends may be found in Kelly (2012), Kircher, 2012, Naeim and Kelly (1999), Skinner et al. (1993), Walters (2015), and Zhou and Xian (2001).

### 17.14.2 Energy Absorbing Devices

While a base isolator may be effective at protecting buildings from seismic movements, it cannot necessarily dissipate the energy that it obtains during an earthquake. Designers realized that for controlling seismic damage in buildings and improving seismic performance, a *damping system* (seismic energy dissipating devices) must be used in conjunction with an isolator (see Fig. 17.48a). These devices can be used at the base of the structure or mounted on the superstructure at appropriate locations (usually on diagonal braces). Dampers deplete the energy that would normally keep a building oscillating side to side by applying a restoring force to bring the superstructure and foundation into alignment. They act like the hydraulic shock absorbers provided in automobiles, absorbing the vibration of sudden jerks and transmitting only a part of the vibration above the chassis of the vehicles. *Dampers* were first used in the 1960s to absorb the vibration caused by winds in tall buildings and only since the 1990s are they being used to protect buildings against the effects of earthquakes. When the device merely absorbs the energy during vibration without any energy input from outside, it is termed a *passive device*. On the other hand, if it opposes the vibration by means of an external energy source, it is called an *active device*. We could also have semi-active and hybrid dampers. Commonly used dampers are shown in Fig. 17.52. They include the following (Patil and Reddy 2012):

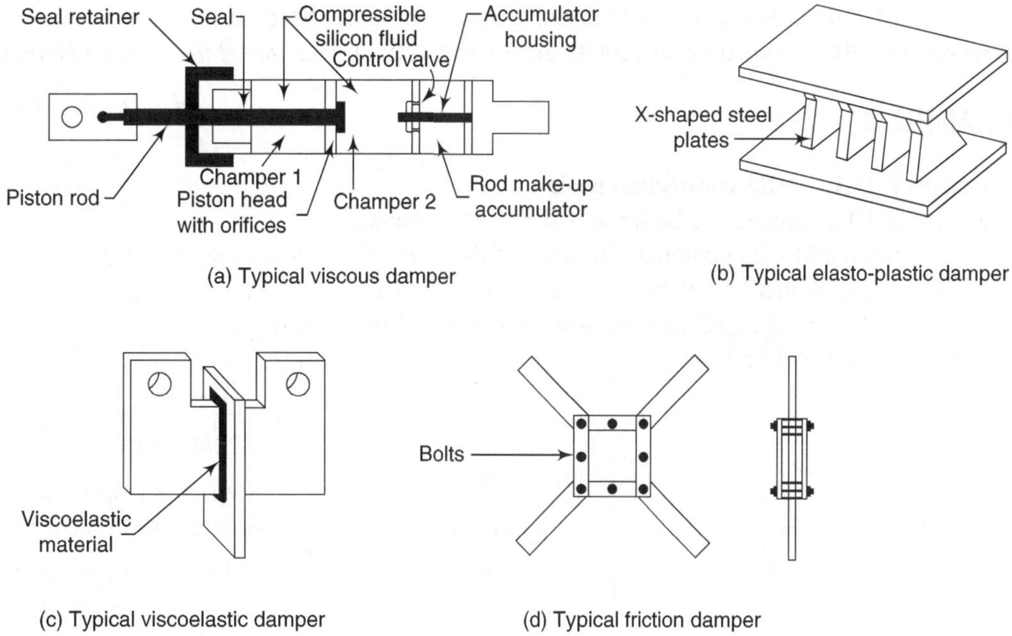


**FIG. 17.52** Various types of dampers

1. *Viscous dampers* (VDs)—They consist of a piston-cylinder arrangement filled with a viscous silicon-based fluid. As the damper piston rod and piston head are stroked, fluid is forced to flow through orifices around the piston head, which absorbs the energy.
2. *Friction dampers* (FDs)—energy is absorbed by the friction between two layers, which are made to rub against each other.
3. *Elasto-plastic dampers* (EPDs) or *added damping and stiffness* (ADAS) *dampers* are made of number of small 'X' shaped plates, which yield at small deformation thereby dissipating high amount of energy. These metallic yielding devices is similar in principle to the buckling restrained brace (BRB), discussed in Section 17.13. It has to be noted that BRBs are regarded as being part of a bracing system, rather than as part of a damping system.
4. *Viscoelastic dampers* (VEDs)—containing viscoelastic material, sandwiched between two steel plates, which undergoes shear deformation, thus dissipating energy.
5. *Shape memory alloy dampers* (SMADs) made of nickel–titanium (Ni–Ti) alloy it has an interesting pseudo-elastic property by which the alloy—regains its initial shape when external load is removed. This property is useful in putting back the structure to its original shape. It can also sustain large amount of inelastic deformation (Song et al. 2006).
6. *Lead extrusion dampers* (LEDs) work on the principle of extrusion of lead. It absorbs vibration energy by plastic deformation of the lead, during which mechanical energy is converted into heat; lead gets heated up and on being extruded and recrystallizes immediately and recovers its original mechanical properties before next extrusion.

These dampers are attached to the main structural framing system via a bracing system. The bracing system may be diagonal bracing, chevron bracing or cross-bracing. Other types of dampers are as follows:

**Tuned mass dampers (TMD)** TMD was first suggested by Frahm in 1909, to attenuate undesirable vibrations in ships. They are extra masses attached to the structure by a spring-dashpot system and designed to vibrate out of phase with the structure. Energy is dissipated by the dashpot due to the relative motion between the mass and the structure. TMDs have been successfully installed in many tall structures throughout the world, especially for dampening wind induced vibrations. These installations include the 535 m tall CN Tower in Canada, the 60 storey John Hancock Building in Boston, USA, the 305 m tall Center-Point Tower in Sydney, Australia, and the 101 storey (508 m tall) Taipei 101 Tower in Taipei [Tuan and Shang (2014)]. However, TMDs has been shown to have a varying effect on the vibrations of structures subjected to earthquake ground motions and in some cases TMDs can be detrimental to the structure.

**Tuned liquid dampers (TLD)** They are essentially water tanks mounted on structures and dissipate energy by the splashing of the water. The motion of the liquid may be hindered by orifices to get additional energy dissipation.

**Hydraulic activators** They are active vibration control devices and have a sensor to sense the vibration and activate the activator to counter it. These devices require external energy source, and are expensive.

More information on energy absorbing devices, latest developments, and applications may be found in Soong and Dargush 1997, Kareem, et al.(1999), Spencer Jr and Nagarajaiah (2003), Symans et al, (2008), and Li et al. (2013).

## EXAMPLES

**Example 17.1: RBS connection design**

Design an RBS connection between a IPE 600 M beam and a W460 × 280 × 235 column. The span of the beam is 9 m. The flange reduction will be an arc-cut shape. We will use the guidelines of AISC 358 and gravity loads will be neglected (see Fig. 17.52).

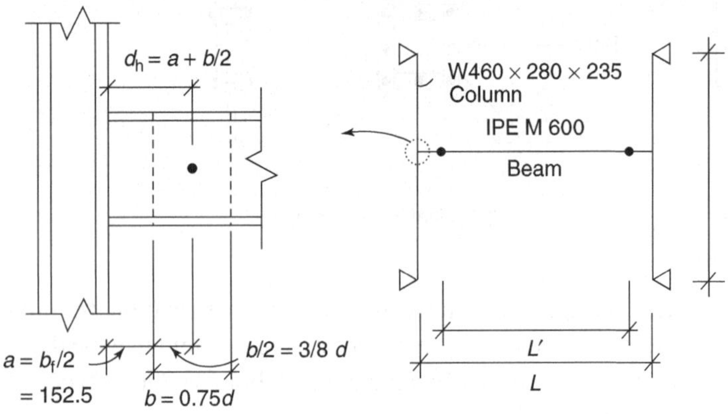


**FIG. 17.52**

**Solution:**

HE M 600, $b_{bf} = 305$ mm, $t_f = 40$ mm, $d = 620$ mm, $t_w = 21$ mm, $R = 27$ mm, $Z_{pz} = 7660 \times 10^3\ \text{mm}^3$, Since $t_f = 40$ mm, as per Table 1.9, $f_y = 240$ MPa, $f_u = 410$ MPa

W 530 × 310 × 248, $b_f = 315$ mm, $t_f = 34.5$ mm, $d = 571$ mm, $t_w = 19$ mm, $R = 14$ mm

Let us assume the dimensions of $a$, $b$, and $c$ as follows:

$a = 0.5b_{bf} = 305/2 = 152.5$ mm

$b = 0.75d = 0.75(620) = 465$ mm

$c = 0.2b_{bf} = 0.2(305) = 61$ mm.

**Step 1:** *The plastic section modulus at the center of the reduced beam*

$$Z_{RBS} = Z_{pz} - 2ct_{bf}\left(d - t_{bf}\right) = 7660 \times 10^3 - 2 \times 61 \times 40(620 - 40) = 4829.6 \times 10^3$$

**Step 2:** *Compute probable maximum moment, $M_p$, at the center of the reduced beam section*

$$M_p = C_p R_y f_y Z_{RBS}$$

$$C_p = \frac{f_y + f_u}{2f_y} = \frac{240 + 410}{2 \times 240} = 1.35 > 1.2,$$

Hence take $C_p = 1.2$

From Table A3.1 of AISC 341-10, $R_y = 1.5$ for grade Fe 250 steel

Hence

$$M_p = C_p R_y f_y Z_{RBS} = 1.2 \times 1.5 \times 240 \times 4829.6 \times 10^3 / 10^6 = 2086.38\ \text{kNm}$$

**Step 3:** *Compute the shear force at the center of the reduced beam sections at each end of the beam*

Distance of plastic hinge from column face = $d_h = a + b/2 = 152.5 + 3/8 \times 620 = 385$ mm

Distance between plastic hinges = $L' = L - d_c - 2d_h = 9000 - 571 - 2\times 385 = 7659$ mm

$$V_p = \frac{2M_p}{L'} = \frac{2 \times 2086.38}{7.659} = 544.8\ \text{kN}$$

**Step 4:** *Compute probable maximum moment at the face of the column*

$$M_f = M_p + V_{RBS} d_h = 2086.38 + 544.8 \times 0.385 = 2296.13\ \text{kNm}$$

**Step 5:** *Compute $M_{pe}$, the plastic moment of the beam based on the expected yield stress:*

$$M_{pe} = R_y f_y Z_{pz} = 1.5 \times 240 \times 7660 \times 10^3 = 2872.5\ \text{kNm}$$

**Step 6:** *Check the flexural strength of the beam at the face of the column*

$$M_f = \phi_d M_{pe} = 1 \times 2872.5 = 2872.5\,kNm > 2296.13\ \text{kNm}$$

Hence OK.

*Note:* In addition checks in steps 8–11, as per section 17.9.3 should also be done.

## SUMMARY

There appears to be an increasing earthquake activity throughout the world. The recent earthquakes have demonstrated that the damages and loss of lives will be extensive if the buildings are not designed and detailed properly. Designing for earthquake is different than designing for other loads, as earthquake codes allow us to consider only a fraction of the earthquake load in the design (as it will be expensive to design for the full earthquake loads, which are also less frequent and not predictable). The codes specify a factor called the response reduction factor ($R$) and also divide the maximum considered earthquake by a factor to arrive at the design basis earthquake. The response reduction factor thus, represents the combined effect of over strength, redundancy, and ductility. It has to be noted that the values of $R$ may change as and when more research is conducted on the behaviour of different structural systems. The parameters influencing seismic damage are briefly discussed.

Past earthquakes have demonstrated that structures having simple and regular configuration, adequate lateral strength, adequate stiffness, and adequate ductility have performed well. Hence the rules to be followed while planning buildings in high seismic zones are provided and the plan (due to re-entrant corners, torsional, diaphragm discontinuity, out-of-plane offsets, and non-parallel systems) and vertical regularities (stiffness, mass, vertical geometry, in-plane discontinuity in vertical, lateral force resisting elements, weak storey), which should be avoided are clearly indicated. Other aspects of planning such as short column effect and careful design and detailing of cantilevers and floating columns are also discussed. Importance and necessity of considering vertical component of earthquake in structures close to the faults or with heavy mass concentrated at the top are also stressed.

The most common systems adopted to resist seismic forces include: moment-resisting frames, a combined system of moment frames and shear walls, braced frames with horizontal diaphragms and a combination of these systems. Moment-resisting frames are classified as special moment-resisting frame (SMRF), ordinary moment-resisting frame (OMRF) or ordinary moment frame (OMF), depending on the details adopted at and near their joints. The proportioning and detailing requirements of these moment-resisting frames as per IS 800 as well as the strong column-weak beam concept have been explained. Proportioning for drift requirements, considering panel zone shear deformations is also discussed.

Though the recent version of steel code (IS 800) contained provisions for seismic design and detailing, designers are not given guidance to choose proper beam-to-column connections, in SMF and IMFs. After 10 years of extensive research, initiated by Federal Emergency Management Agency, USA (after the 1994 Northridge earthquake in USA and the 1995 Kobe earthquake in Japan, which revealed serious drawbacks in the welded moment connections of SMFs), AISC has developed a few pre-qualified connections, which have shown to provide the required amount of ductility. A brief description of such connections is given to aid the designers. For more details about the design methods, the readers should consult AISC 358-10 and AISC 341-10.

Similarly, braced frames may be classified as ordinary concentrically braced frames (OCBF), special concentrically braced frames (SCBF), and eccentrically braced frames (EBF). These systems are explained briefly.

While designing these seismic-resistant systems, it is necessary to provide adequate redundancy in the structure to prevent progressive collapse. It is also necessary to select a system in which the damaged element during an earthquake could be replaced effectively and easily after an earthquake (similar to a fuse in an electrical system). Two such systems, namely the steel plate shear walls and the buckling-restrained braced frames are also described.

Innovative design concepts have been developed recently to better protect structures, against the destructive effects of earthquakes. Base isolation is a passive structural control system, where isolators are used to decouple the building from its foundations resting on shaking grounds, thus protecting the structural integrity of the building, Four types of base isolation – elastomeric, sliding, rolling and a combination of these are briefly described.

Another recent concept is the use of passive energy dissipation devices called dampers which absorb part of the input energy, thereby reducing energy dissipation demand on primary structural members and minimizing possible structural damage. These devises in the form of viscous fluid dampers, visco-elastic solid dampers, friction dampers and metallic dampers are also briefly discussed.

## MULTIPLE-CHOICE QUESTIONS

17.1 Response reduction factor for OMRF is taken in IS 1893 as
(a) 3 (b) 4 (c) 5 (d) 3.5

17.2 Response reduction factor for SMRF is taken in IS 1893 as
(a) 3 (b) 4 (c) 5 (d) 3.5

17.3 When shear walls are also provided in addition to MRFs, MRFs should be designed to resist at least
(a) 30% of the total seismic load
(b) 20% of the total seismic load
(c) 25% of the total seismic load
(d) none of these

17.4 Which of the following influence seismic design?
(a) Duration of earthquake
(b) Distance of site from epicenter
(c) Amplitude of earthquake
(d) All of these

17.5 Normal buildings with up to 20 stories will have fundamental natural periods in the range of
(a) 0.5–20 s (b) 0.1–2.0 s
(c) 1–5 s (d) none of these

17.6 The fundamental period $T$ of soil layer of thickness $H$, having an average shear wave velocity of Vs is
(a) $T = 4H/Vs$
(b) $T = 2H/Vs$
(c) $T = 8H/Vs$

17.7 To perform well in an earthquake, a building should posses
(a) regular configuration and adequate strength
(b) adequate stiffness and ductility
(c) regular configuration, adequate strength, stiffness, and ductility
(d) adequate strength and ductility

17.8 Which shape of plan is better in earthquake zone?
(a) L (b) O (c) H (d) Y

17.9 Torsional irregularity need not be considered, when the maximum storey drift is less than the average of drifts at the two ends of the building by
(a) 1.2 times
(b) 1.4 times
(c) 2.0 times

17.10 The building should be considered to have extreme torsional irregularity when the maximum storey drift is greater than the average of drifts at the two ends of the building by
(a) 1.2 times
(b) 1.4 times
(c) 2.0 times

17.11 Diaphragm discontinuity should be considered when open areas are greater than
(a) 40% of the gross diaphragm area
(b) 50% of the gross diaphragm area
(c) 60% of the gross diaphragm area

17.12 A storey is considered soft storey when it's lateral stiffness is less than

(a) 50% of the stiffness in the storey above or 60% of the average stiffness of 3 stories above
(b) 60% of the stiffness in the storey above or 70% of the average stiffness of 3 stories above
(c) 70% of the stiffness in the storey above or 80% of the average stiffness of 3 stories above

17.13 Mass irregularity exists when the seismic weight of any floor is
(a) 150 % more than that of its adjacent floors
(b) 200 % more than that of its adjacent floors
(c) 250 % more than that of its adjacent floors

17.14 Vertical geometric irregularity exists when the horizontal dimension of the lateral force-resisting system in any storey is
(a) more than 130% of that in an adjacent storey
(b) more than 150% of that in an adjacent storey
(c) more than 175% of that in an adjacent storey

17.15 A storey is considered as a weak storey when the lateral strength of that storey is
(a) Less than 75% lateral strength of the storey above
(b) Less than 80% lateral strength of the storey above
(c) Less than 85% lateral strength of the storey above

17.16 Vertical cantilever projections should be designed for
(a) 3 times the horizontal seismic coefficient, $A_h$
(b) 4 times the horizontal seismic coefficient, $A_h$
(c) 5 times the horizontal seismic coefficient, $A_h$

17.17 Horizontal cantilever projections should be designed for
(a) 8/3 times the horizontal seismic coefficient, $A_h$
(b) 3 times the horizontal seismic coefficient, $A_h$
(c) 10/3 times the horizontal seismic coefficient, $A_h$

17.18 Two adjacent buildings with same floor elevation should be separated by a distance
(a) *R*/2 times the sum of the calculated storey displacements
(b) 2R/3 times the sum of the calculated storey displacements
(c) *R* times the sum of the calculated storey displacements

17.19 The vertical seismic coefficient is taken in IS 1893 as
(a) 1/2 times that of horizontal seismic coefficient
(b) 2/3 times that of horizontal seismic coefficient
(c) 1.25 times that of horizontal seismic coefficient

17.20 The common seismic force-resisting system consist of
(a) moment-resisting rigid frames
(b) shear walls
(b) braced frames
(d) all of these

17.21 As per IS 1893, the columns and beams of soft storey should be designed for
(a) 1.5 times the storey shears and moments
(b) 2.0 times the storey shears and moments
(c) 2.5 times the storey shears and moments

17.22 IS 1893 also suggests shear walls in such soft stories which should be designed for
(a) 1.5 times the lateral storey shear force
(b) 2.0 times the lateral storey shear force
(c) 2.5 times the lateral storey shear force

17.23 To have strong-column weak-beam structures, IS 800 suggests the summation strength of columns to beams ratio at a joint be greater than
(a) 1.1 (b) 1.2 (c) 1.5

17.24 Members in SMF should undergo plastic rotation of
(a) greater than 0.025 radians
(b) greater than 0.030 radians
(c) greater than 0.040 radians

17.25 Members in OMF should undergo plastic rotation of
(a) greater than 0.025 radians
(b) greater than 0.030 radians
(c) greater than 0.040 radians

17.26 Which of the following strategies are adopted to improve connection behaviour?
(a) Toughening system
(b) Strengthening system
(c) Weakening system
(d) All of these

17.27 In earthquake zones, concentrically braced frames may have
(a) X-bracing
(b) K-bracing
(c) tension-only bracing

17.28 Joints in OCBF should be designed to have joint rotation greater than
(a) 0.020 radians
(b) 0.025 radians
(c) 0.040 radians

17.30 In special concentrically braced frames, the slenderness of bracing members should not exceed
(a) 125 (b) 150 (c) 160 (d) 180

## REVIEW QUESTIONS

17.1 What is the significance of response reduction factor?
17.2 What are the three factors that are considered while arriving at the *R* values?
17.3 State the parameters that influence seismic design.
17.4 Why a layer of soft soil above bed rock has to be considered carefully while designing structures for earthquakes?
17.5 What is meant by liquefaction?
17.6 What are the methods adopted to prevent liquefaction?
17.7 What are the four main attributes a building should posses in order to perform well during earthquakes?
17.8 What are the two types of irregularities that should be considered in seismic design?
17.9 What are the shapes of buildings that are to be avoided? If such shapes are inevitable what precautions should we have to take?
17.10 When torsional irregularities should be considered? When a building is considered to have extreme torsional irregularity? How their effects could be minimized?
17.11 What is diaphragm irregularity?

17.12 How can out-of-plane offset irregularity be avoided? What detailing should be adopted when they are present?
17.13 How do we deal with non-parallel systems?
17.14 When is a building considered to have soft storey?
17.15 What constitutes mass irregularity?
17.16 When is a building considered to have vertical geometric irregularity?
17.17 When is a building considered to have in-plane discontinuity in vertical lateral force resisting elements?
17.18 Which storey is considered as a week storey?
17.19 How to tackle a building with some irregularity?
17.20 What is short column effect?
17.21 How vertical and horizontal cantilever projections are suggested to be designed in IS 1893?
17.22 How can we avoid pounding as per IS 1893?
17.23 Why and when is it important to consider vertical component of earthquake?
17.24 List four systems that can be adopted to resist seismic lateral forces.
17.25 List 4 factors that contributed to the damage of welded joints in the Northridge earthquake.
17.26 What is strong-column, weak-beam concept? How is it considered in IS 800:2007?
17.27 Write short notes on (a) concentrically braced frames, (b) eccentrically braced frames, and (c) moment-resisting frames.
17.28 What is the minimum percentage of lateral load to be resisted by a moment-resistant frame in a dual system?
17.29 What are the provisions in IS 1893 (Part 1):2002 for the design of columns, beams and shear walls in a soft storey?
17.30 What are the provisions in IS 800:2007 for the design of members in SMF?
17.31 How storey drift is checked in a SMF as per IS 1893 (Part 1):2002? How is it checked differently in ASCE 7-10?
17.32 How do panel zone shear deformations affect the total shear mode of drift?
17.33 State some of the rules specified in IS 800:2007 for beam-column joints and connections of SMF.
17.34 How the column web or doubler plate thickness is determined empirically by using IS 800:2007?
17.35 State some of the advantages and disadvantages of SMF over braced frames.
17.36 How the design of an OMF differs from an SMF as per IS 800:2007?
17.37 How can the cost of SMF be minimized?
17.38 What are the three schemes by using which the connection behaviour of SMFs be improved? Explain these schemes, by taking examples.
17.39 State the design steps given in AISC 358-10 for the design of reduced beam sections.
17.40 What are pre-qualified seismic moment connections?
17.41 Name any five pre-qualified seismic moment connections, specified in AISC 358.
17.42 Write short notes on the following:
(a) Bolted unstiffened extended end plate (BUEEP) connections and bolted stiffened extended end plate (BSEEP) connections
(b) Bolted flange plate (BFP) moment connections
(c) Welded un-reinforced flange-welded web (WUF-W) connections
(d) Kaiser bolted bracket (KBB) moment connections
(e) SidePlate connections
(f) ConXtech's ConXL™ moment connections
(g) Braced frames
17.43 Discuss about the various types of bracings that can be used with Concentrically Braced Frames.
17.44 What are the advantages of using elliptical clearance over linear clearance in gusset plates of special concentrically braced frames (SCBFs)?
17.45 What are the problems faced by using SCBFs in recent earthquakes?
17.46 How do eccentrically braced frames (EBF) behave during an earthquake and what are their advantages over SCBFs?
17.47 What are dual systems? Explain the design features of these systems.
17.48 What is the minimum percentage of lateral load to be resisted by a moment-resistant frame in a dual system?
17.49 What are the three types of steel plate shear walls?
17.50 What are the advantages of SPSW over reinforced concrete shear walls? What are its disadvantages?
17.51 Why is the behaviour of SPSW different from that of a plate girder?
17.52 Explain the model that is used in AISC code for the analysis of SPSW.
17.53 State the method used for the preliminary proportioning of SPSW.
17.54 What necessitates the use of perforated webs in SPSWs?
17.55 Write short notes on
(a) Coupled SPSW
(b) Self-centering SPSW
17.56 What are buckling restrained braces? What are the advantages of BRBF over other concentrically braced frames?
17.57 Explain the concept of base isolation. How base-isolated buildings behave during an earthquake compared to fixed base buildings?
17.58 What are the three broad types of isolators?
17.59 Write short notes on (a) elastomeric bearing, (b) lead-rubber bearing, (c) sliding bearing, and (d) energy absorbing devices.
17.60 What are the different types of dampers used to control seismic damage?

## REFERENCES


Adan, S.M., and Gibb, W., 'Experimental Evaluation of Kaiser-bolted Bracket Steel Moment-resisting Connections', *Engineering Journal, AISC*, Vol. 46, No.3, Third quarter 2009, pp. 181–195.

Ambrose, J. and Vergun, D., *Simplified Building Design for Wind and Earthquake Forces,* Third edition, Wiley–Inter science, New York, NY, 1995, p. 368.

AISC Design Guide 12—*Modification of Existing Steel Welded Moment Frame Connections for Seismic Resistance* (Authors: Gross, J.L., Engelhardt, M.D., Uang, C.-M., Kasai, K., and Iwankiw, N.*R.*), American Institute of Steel Construction, Inc., Chicago, 1999.

AISC Design Guide 13—*Stiffening of Wide-Flange Columns at Moment Connections: Wind and Seismic Applications*, (Author: Carter, C.J.), American Institute of Steel Construction, Inc., Chicago, 2003.

AISC Design Guide 20- *Steel Plate Shear Walls*, (Authors: Sabelli *R.*, and Bruneau, M.), American Institute of Steel Construction, Inc., Chicago, 2006.

AISC Design Guide 29: *Vertical Bracing Connections-Analysis and Design*, (Authors: W.A. Thornton and L.S. Muir), American Institute of Steel Construction, Inc., Chicago, 2014.

ANSI/AISC 358-10, *Pre-qualified Connections for Special and Intermediate Steel Moment Frames for Seismic Applications including Supplement* Nos 1 and 2, American Institute of Steel Construction, Inc., Chicago, IL, 2010 and Draft AISC 358:2016.

ANSI/AISC 341-10, *Seismic Provisions for Structural Steel Buildings*, American Institute of Steel Construction, Inc., Chicago, IL., 2010.

Arnold, C., 'Seismic Issues in Architectural Design', in FEMA 454 *Designing for Earthquakes- A Manual for Architects*, FEMA (Federal Emergency Management Agency, Washington, D.C.) and NEHRP (National Earthquake Hazards Reduction Program), Dec. 2006, p. 392.

Arnold, C. and Reitherman, R., *Building Configuration and Seismic Design*, John Wiley, 1982, p. 310.

Bachmann, H., *Seismic Conceptual Design of Buildings—Basic Principles for Engineers, Architects, Building Owners, and Authorities*, Swiss Federal Office for Water and Geology (FOWG), Biel, 2002, p. 81.

Becker, R., *Seismic Design of Special Concentrically Braced Frames*, Structural Steel Education Council, Morga, CA, 1995.

Becker, R. and Ishler, M., *Seismic Design Practice for Eccentrically Braced Frames*, Structural Steel education Council, Morga, CA, 1996.

Berman, J.W., 'Advances in Steel Plate Shear Walls, *STRUCTURE Magazine*, ASCE, Sep 2014, pp. 26–27.

Berman, J.W. and Bruneau, M., 'Steel Plate Shear Walls are not Plate Girders', *Engineering Journal*, AISC, Vol. 41, No. 3, 2004, pp. 95–106.

Bolt, B.A., *Earthquakes*, Fifth edition, W.H. Freeman (Macmillan Education), London, 2006, p. 320.

Borello, D. and Fahnestock, L., 'Seismic Design and Analysis of Steel Plate Shear Walls with Coupling', *Journal of Structural Engineering*, ASCE, Vol. 139, Special Issue: NEES 2: Advances in Earthquake Engineering, 2013, pp. 1263–1273.

Bozorgnia, Y. and Bertero, V.V., *Earthquake Engineering: From Engineering Seismology to Performance – Based Engineering*, CRC Press, Boca Raton, Florida, 2004, p. 1152.

Bozorgnia, Y., and Campbell, K.W., Chapter 5, Engineering Characterization of Ground Motion, In *Earthquake Engineering: From Engineering Seismology to Performance-based Engineering*, Bozorgnia and Bertero (eds), CRC Press, June 2004a, p. 74, http://www.ce.berkeley.edu/~mahin/CE227web/Bozorgni-CampbellCh_BerteroBozorgnia.pdf.

Bozorgnia, Y. and Campbell, K.W., 'The Vertical-to-horizontal Response Spectral Ratio and Tentative Procedures for Developing Simplified V/H and Vertical Design Spectra', *Journal of Earthquake Engineering*, Vol. 8, 2004b, pp. 175–207.

Bozorgnia, Y., Mahin, S.A., and Brady, A.G., 'Vertical Response of Twelve Structures Recorded During the Northridge Earthquake', *Earthquake Spectra*, Vol. 14, 1998, pp. 411–432.

Bruneau, M., Uang, C.M., and Whittaker, A., *Ductile Design of Steel Structures*, Second edition, McGraw Hill, New York, 2011, p. 921.

Caccese, V., Elgaaly, M., and Chen, R., Experimental Study of Thin Steel Plate Shear Walls under Cyclic Load, *Journal of Structural Engineering*, ASCE, Vol. 119, No. 2, Feb 1993, pp. 573-587, also Post-buckling of Steel Plate Shear Walls under Cyclic Loads, pp. 588–605.

Carter, C.J. and Grubb, K.A., 'Prequalified Moment Connections (revisited)', *Modern Steel Construction*, Vol. 50, No. 1, Jan 2010, pp. 54–57.

Cartwright, C., 'Wasatch Property Management Corporate Headquarters', *The Newsletter of the Structural Engineers Association of Utah*, Vol. XI, No. 1, Sep 2006, pp. 2–3.

Charney, F., Tola, A.T., and Atlayan, O., 'Structural Anaysis: Example 1, Twelve Storey Moment-resisting Steel Frame', 2009 *NEHRP Recommended Seismic Provisions*-FEMA P-752 CD, June 2013, www.fema.gov/media-library-data/1393888487069-c071850c3050c34da135376baa478a1b/P-752_Unit4.pdf, last accessed on 5th Aug 2015.

Chen, S.J., Yeh, C.H., and Chu, J.M., 'Ductile Steel Beam-to-Column Connections for Seismic Resistance', *Journal of Structural Engineering*, ASCE, Vol. 122, No. 11, Nov 1996, pp. 1292–1299.

Clayton, P. M., Berman, J. W., and Lowes, L. N., 'Seismic Design and Performance of Self-centering Steel Plate Shear Walls', *Journal of Structural Engineering*, ASCE, Vol. 138, No. 1, Jan 2012, pp. 22–30.

Cordova, P.P., and Hamburger, R.O., 'Steel Connections: Proprietary or Public Domain', *Modern Steel Construction*, AISC, Oct 2011, pp. 24–31.

Dana, M., Cussen, A., Chen Y.N., Davis, C., Greer, M., Houston, J., Littler, P., and Roufegarinejad, A., 'Effects of the Seismic Vertical Component on Structural Behavior - An Analytical Study of Current Code Practices and Potential Areas of Improvement', *Proceedings of the 10th National Conference in Earthquake Engineering*, Earthquake Engineering Research Institute, Anchorage, AK, 21–25 Jul 2014, p. 11, https://nees.org/resources/12050/download/10NCEE-001107.pdf, last accessed on 20 Jul 2015.

DeVall, R.H., 'Background Information for Some of the Proposed Earthquake Design Provisions for the 2005 Edition of the National Building Code of Canada', *Canadian Journal of Civil Engineering*, Vol. 30, 2003, pp. 279–286.

Dowden, D.M., Purba, *R.*, and Bruneau, M., 'Behavior of Self-Centering Steel Plate Shear Walls and Design Consideration', *Journal of Structural Engineering*, ASCE, Vol. 138, No. 1, January 1, 2012, pp.11–21.

Driver, R.G. and Grondin, G.Y., 'Steel Plate Shear Walls, now Performing on the Main Stage', *Modern Steel Construction*, AISC, Vol. 41, No. 10, Sep 2001, pp. 48–58.

Driver, R.G., Kulak, G.L., Kennedy, D.J.L., and Elwi, A.E., Cyclic Test of a Four-storey Steel Plate Shear Wall, *Journal of Structural Engineering*, ASCE, Vol. 124, No. 2, Feb 1998, pp. 111–120.

Elnashai, A.S. and Mwafy, A.M., 'Overstrength and Force Reduction Factors of Multi-storey Reinforced-Concrete Buildings', *The*

*Structural Design of Tall Buildings*, Wiley Interscience, Vol. 11, 2002, pp. 329-351.

Engelhardt, M.D., Winneberger, T., Zekany, A.J., and Potyraj, T.J., 'Experimental Investigation of Dogbone Moment Connections', *Engineering Journal,* AISC, Vol. 35, No. 4, Fourth quarter 1998, pp. 128–139.

EN 1998-1:2004, *Eurocode 8—Design of Structures for Earthquake Resistance - Part 1: General Rules, Seismic Actions and Rules for Buildings*, European Committee for Standardization, Brussels, Dec 2003, p. 229.

FEMA 350, *Recommended Seismic Design Criteria for New Steel Moment-Frame Buildings*, Federal Emergency Management Agency, Washington D.C., Jun 2000, p. 224.

FEMA-355C, *State-of-the Art Report on Systems Performance of Steel Moment Frames Subject to Earthquake Ground Shaking*, Federal Emergency Management Agency, Washington D.C., Sep 2000, p. 344.

FEMA 355D, *State-of-the art Report on Connection Performance*, prepared by the SAC Joint Venture for Federal Emergency Management Agency, Washington, D.C., 2000, http://www.sacsteel.org/, last accessed on 20 Jul 2015.

FEMA 451, *2003 NEHRP Recommended Provisions: Design Examples*, National Institute of Building Sciences, Washington, D.C., Aug 2006, p. 860.

FEMA 454, *Designing for Earthquakes—A Manual for Architects*, Federal Emergency Management Agency, Washington D.C., Dec 2006, p. 392.

FEMA P-750, *NEHRP Recommended Seismic Provisions for New Buildings and Other Structures*, Building Seismic Safety Council, Washington D.C., 2009, p. 388.

Fisher, J.M., 'Expansion Joints: Where, When, and How', *Modern Steel Construction*, AISC, Vol. 45, No. 4, Apr 2005, pp. 29–34.

Gardiner, D.R., D.K. Bull, and A.J. Carr, 'Trends in Internal Forces in Concrete Floor Diaphragms of Multi-storey Structures during Seismic Shaking', *The 14th World Conference on Earthquake Engineering,* Beijing, China, 12–17 Oct 2008, http://www.iitk.ac.in/nicee/wcee/article/14_05-06-0178.PDF, last accessed on 29 Jul 2015.

Gioncu, V. and F.M. Mazzolani, *Ductility of Seismic Resistant Steel Structures*, Spon Press, London & New York, 2002, p. 685.

Gross, J.L., Engelhardt, M.D., Uang, C.-M., Kasai, K., and Iwankiw, N.R., *Modification of Existing Welded Steel Moment Frame Connections for Seismic Resistance,* STEEL Design Guide series 12, American Institute of Steel Construction, Inc., Chicago, IL, 1999.

Hamburger, R.O., Krawinkler, H., Malley, J.O., and Adan, S.M., *Seismic Design of Steel Special Moment Frames: A Guide for Practicing Engineers,* NEHRP Seismic design technical brief no.2, National Institute of Standards and Technology, Gaithersburg, USA, June 2009, p. 33, www.nehrp.gov/pdf/nistgcr9-917-3.pdf, last accessed on 1 Aug 2015.

Haskett, T., B. Breukelman, J. Robinson, J. Kottelenberg., *Tuned Mass Dampers Under Excessive Structural Excitation,* Motioneering Inc., Ontario, Canada, 2004, p. 7, http://www.waterfordmgmt.com/school/Articles/tmd%20article.pdf, last accessed on 20 Jul 2015.

Humar, J.L. and Ragozar M.A., 'Concept of Over-strength in Seismic Design', In *Proceedings 11th World Congress on Earthquake Engineering*, International Association for Earthquake Engineering (IAEE), Acapulco, Mexico, Paper 639, 1996.

Hussain, S., Benschoten, P.V., Al Satari, M., and Lin, S., 'Buckling Restrained Braced Frame (BRBF) Structures: Analysis, Design and Approvals Issues', *Proceedings of the 75th SEAOC Annual Convention*, Long Beach, California, 13–16 Sep 2006, http://www.starseismic.net/wp-content/uploads/2013/08/brbfpaperfinal.pdf, last accessed on 17 Aug 2015.

International Network For The Design Of Earthquake Resilient Cities (INDERC): http://inderc.blogspot.com/2012/04/seismic-regulations-versus-modern.html

Iwankiw, N.R. and Carter, C.J., 'The Dogbone: A New Idea to Chew On', *Modern Steel Construction, AISC,* Vol. 36, No. 4, Apr 1996, pp. 18–23.

Ivankiw, N.R., 'Ultimate Strength Considerations for Seismic Design of the Reduced Beam Section (Internal Plastic Hinge)', *Engineering Journal*, AISC, Vol. 34, No. 1, 1997, pp. 3–16.

Jain, S.K., Explanatory Examples on Indian Seismic Code IS: 1893 (Part 1), www.iitk.ac.in/nicee/IITK-GSDMA/EQ21.pdf, last accessed on 30 Dec 2004.

Jain, S.K., 'A Proposed Draft of IS: 1893 Provisions on Seismic Design of Buildings – Part II Commentary and Examples', *Journal of Structural Engineering*, SERC, Vol. 22, No. 2, Jul 1995, pp. 73–90.

Jain, S.K. and Murty, C.V.R., *Proposed Draft Provisions and Commentary on Indian Code IS: 1893 (Part 1)*, www.iitk.ac.in/nicee/IITK-GSDMA/EQ05.pdf, last accessed on 10 Oct 2015.

Jangid, R.S. and Kelly, J. M., 'Base Isolation for Near-fault Motions', *Earthquake Engineering & Structural Dynamics,* Vol. 30, 2001, pp. 691–707.

Kalayanaraman, V., Sridhara, B.N., and Mahadevan, K., *Sleeved Column System, Proc. of the SSRC Task Group Meetings and Task Force Sessions*, Lehigh Univ., Bethlehem, Pa., 1994.

Kareem, A., Kijewski, T., and Tamura, Y., 'Mitigation of Motions of Tall Buildings with Specific Examples of Recent Applications', *Wind and Structures*, Vol. 2, No. 3, 1999, pp. 201–251.

Krawinkler, H., 1978, 'Shear in Beam-Column Joints in Seismic Design of Steel Frames', *Engineering Journal*, AISC, Vol. 5, No. 3, pp. 82–91.

Kelly, J.M., *Earthquake-resistant Design with Rubber*, Second edition, Springer-Verlag, London, 2012, p. 243.

Kircher, C.A., 'Chapter 12: Seismically Isolated Structures', in FEMA P-751 , *2009 NEHRP Recommended Seismic Provisions: Design Examples*, Federal Emergency Management Agency, Washington D.C., Sep 2012, p. 916.

Kulak, G.L., Kennedy, L., Driver, R.G., and Medhekar, M., 'Steel Plate Shear Walls – An Overview', *Engineering Journal*, AISC, Vol. 39, No. 1, 2001, pp. 50–62.

Lee, D., Cotton, S.C., Dexter, R.J., Hajjar, J.F., Ye, Y., and Ojard, S.D., *Column Stiffener Detailing and Panel Zone Behavior of Steel Moment Frame Connections*, Report, University of Minnesota, Jun 2002.

Lee, D., Cotton, S.C., Hajjar, J., Dexter, R.J. and Ye, Y. (2005a), 'Cyclic Behavior of Steel Moment-resisting Connections Reinforced by Alternative Column Stiffener Details I. Connection Performance and Continuity Plate Detailing,' *Engineering Journal*, AISC, Vol. 42, No. 4 (Fourth quarter), 2005, pp. 189–214, 'Part II. Panel Zone Behavior and Doubler Plate Detailing', Vol. 42, No. 4, 2005, pp. 215–238.

Li, J., Li, Y., Li, W., and Samali, B., 'Development of Adaptive Seismic Isolators for Ultimate Seismic Protection of Civil Structures', *Sensors and Smart Structures Technologies for Civil, Mechanical, and Aerospace Systems,* 2013, pp. 1–12.

López, W.A., 'On Designing with Buckling–Restrained Braced Frames', *STRUCTURE Magazine*, ASCE, Jul 2008, pp. 40–41.

López, W.A. and Sabelli, *R*., *Seismic Design of Buckling-restrained Restrained Braced Frames*, Steel Tips, Structural Steel Educational Council, Jul 2004, p. 66, http://www.steeltips.org/steeltips/tip_details.php?id=76, last accessed on 20 Jul 2015.

Mathews, M.S., Venugopal, M.S., and Subramanian, N., 'Planning and Design of Houses in Disaster Areas', *International Conference on Disaster Area Housing*, Turkey, Sep 1977.

Mayes, R.L., Brown, A.G., and Pietra, D., 'Using Seismic Isolation and Energy Dissipation to Create Earthquake-resilient Buildings', Paper Number 093, 2012 NZSEE Conference www.nzsee.org.nz/db/2012/Paper093.pdf, last assessed on 19 Aug 2015.

Moehle, J.P., J.D. Hooper, D.J. Kelly, T.R. Meyer, 'Seismic Design of Cast-in-place Concrete Diaphragms, Chords, and Collectors—A Guide for Practicing Engineers', *NEHRP Seismic Design Technical Brief No. 3*, National Institute of Standards and Technology, Gaithersburg, MD, Aug 2010, p. 29, http://www.nehrp.gov/pdf/nistgcr10-917-4.pdf, last accessed on 19 Sep 2015.

Murty, C.V.R., *IITK—BMTPC Earthquake Tips*, Dept of Civil Engineering, Indian Institute of Technology, Kanpur & Building Material and Technology Promotion Council, New Delhi, Sep 2005, p. 48, www.nicee.org/EQTips/IITK-BMTPC.htm, last accessed on 20 Jul 2015.

Murty, C.V.R., Goswami, R., Vijayanarayanan, A.R., and Mehta, V.V., *Some Concepts in Earthquake Behaviour of Buildings*, National Information Center of Earthquake Engineering (NICEE), Indian Institute of Technology, Kanpur, 2013a, p. 262, http://www.iitk.ac.in/nicee/IITK-GSDMA/EBB_001_30May2013.pdf, last accessed on 20 Jul 2015.

Murty, C.V.R., Goswami, R., Vijayanarayanan, A.R., Pradeep Kumar, R., and Mehta, V.V., *Introduction to Earthquake Protection of Non-structural Elements in Buildings*, National Information Center of Earthquake Engineering (NICEE), Indian Institute of Technology, Kanpur, 2013b, p. 168, http://www.iitk.ac.in/nicee/IITK-GSDMA/NSE_002_31May2013.pdf, last accessed on 20 Jul 2015.

Naeim, F., and Kelly, J.M., *Design of Seismic Isolated Structures; From Theory to Practice*, John Wiley, New York, 1999, p. 304.

Nishi,T., Kelly, J.M., and Zhou, F.L., 'Rubber Structural Mounts Save Lives During Earthquakes', *ISO Focus*, Jun 2009, pp. 26–28.

Özmen, G., Girgin, K., and Durgun, Y., 'Torsional Irregularity in Multi-storey Structures', *International Journal on Advanced Structural Engineering*, Springer, Vol. 6, No. 4, Dec 2014, pp. 121–131.

Park R., 'Explicit Incorporation of Element and Structure Over-strength in the Design Process', *Proceedings 11th World Conference on Earthquake Engineering* (WCEE), Acapulco, Mexico, Paper 2130, 1996, http://www.iitk.ac.in/nicee/wcee/article/11_2130.pdf, last accessed on 20 Jul 2015.

Patil, S.J. and Reddy. G.R. 'State Of Art Review - Base Isolation Systems For Structures', *International Journal of Emerging Technology and Advanced Engineering*, Vol. 2, No. 7, Jul 2012, pp. 438–453.

Priestley, M.J.N., Calvi, G.M., and Kowalsky, M.J., *Displacement-based Seismic Design of Structures*, IUSS Press, Pavia, Italy, 2007, p. 721.

Roeder, C.W., and Lehman, D.E., 'Seismic Design and Behavior of Concentrically Braced Steel Frames', *Structure Magazine*, Feb 2008, pp. 37–39.

Sabelli, R., 'Recommended Provisions for Buckling-restrained Braced Frames', *AISC Engineering Journal*, Fourth quarter 2004, pp 155–175.

Sabelli, R., Roeder, C.W., and Hajjar, J.F., 'Seismic Design of Steel Special Concentrically Braced Frame Systems-A Guide for Practicing Engineers', *NEHRP Seismic Design Technical Brief*, No. 8, NIST GCR 13-917-24, National Institute of Standards and Technology, Gaithersburg, USA, Jul 2013, p. 36.

Sadashiva, V.K., MacRae, G.A., and Deam, B.L., 'Determination of Structural Irregularity Limits-Mass Irregularity Example', *Bulletin of the New Zealand Society for Earthquake Engineering*, Vol. 42, No. 4, Dec. 2009, pp. 288–301.

Sato, A., Newell, J.D., Uang, C.-M., 'Cyclic Behavior and Seismic Design of Bolted Flange Plate Steel Moment Connections', *AISC Engineering Journal*, Vol. 45, Fourth quarter, 2008, pp. 221–233.

Saunders, M.C., 'Seismic Joints in Steel Frame Buildings', *Modern Steel Construction*, AISC, Vol. 45, No. 4, Apr 2005, pp. 57–62.

Seek, M.W. and Murray, T.M., *Cyclic Test of 8-Bolt Extended Stiffened Steel Moment End Plate Connection with Concrete Structural Slab*, Report submitted to the American Institute of Steel Construction, Virginia Polytechnic Institute and State University, Blacksburg, VA, 2005.

Seilie, I.F. and Hooper, J.D., 'Steel Plate Shear Walls - Practical Design and Construction, *Modern Steel Construction*, AISC, Vol. 45, No. 4, Apr 2005, pp. 37–42.

Sengupta, A.K. and A.D. Shetty, 'Analysis of Chord Forces in Floors with Large Openings or Re-entrant Corners in Reinforced Concrete Buildings', *ICI Journal*, Indian Concrete Institute, Jul–Sep, Vol. 12, No. 2, 2011, pp. 7–15.

Skinner, R.I., Robinson, W.H., and McVerry, G.H., *An Introduction to Seismic Isolation*, John Wiley & sons, New York, 1993, p. 376.

Song, G., Ma, N., and Li, H.-N., 'Applications of Shape Memory Alloys in Civil Structures', *Engineering Structures*, Vol. 28, 2006, pp. 1266–1274.

Soong, T.T. and Dargush, G.F., *Passive Energy Dissipation Systems in Structural Engineering,* John Wiley, New York, 1997, p. 368.

Spencer Jr, B.F. and Nagarajaiah, S., 'State-of-the-art of Structural Control', *Journal of Structural Engineering*, ASCE, Vol. 129, No. 7, Jul 2003, pp. 845–856.

Subramanian, N., 'How to Build Earthquake Proof Structures', *Civil Engineering and Construction Review*, Vol. 7, No. 4, Apr 1994, pp. 39–41.

Subramanian, N., *Design of Reinforced Concrete Structures*, Oxford University Press, New Delhi, 2013, p. 865.

Subramanian, N., 'Pre-qualified Seismic Moment Connections', *New Building Materials & Construction World (NBM & CW)*, Vol. 17, No. 2, Aug 2010, pp. 160–171.

Sumer, A., Fleischman, R.B., and Hoskisson, B.E., 'Development of a Cast Modular Connector for Seismic-resistant Steel Moment Frames, Part 1: Prototype development, Part 2: Experimental

Verification', *Engineering Journal, AISC*, Vol. 44, No.3, Third quarter, 2007, pp. 195–231.

Symans, M.D., Charney, F.A., Whittaker, A.S., Constantinou, M.C., Kircher, C.A., Johnson, M.W., and McNamara, R.J., 'Energy Dissipation Systems for Seismic Applications: Current Practice and Recent Developments, *Journal of Structural Engineering*, ASCE, Vol. 134, No. 1, Jan 2008, pp. 3–21.

Tamboli, A.R., *Handbook of Structural Steel Connection Design and Details*, Second edition, McGraw Hill, New York, 2009, p. 656.

Taranath, B.S., *Steel, Concrete and Composite Design of Tall Buildings*, Second edition, McGraw Hill, New York, 1998, p. 998.

Thomas, P. and A.K. Sengupta, 'Analysis of chord forces in Reinforced Concrete Buildings Subjected to Seismic Loads', *ICI Journal*, Indian Concrete Institute, Vol. 9, No. 3, Oct–Dec 2008, pp. 21–27.

Thornton, W.A. and Muir, L.S., 'Design of Vertical Bracing Connections for High-seismic Drift', *Modern Steel Construction*, AISC, Vol. 49, No. 3, Mar 2009, pp. 61–65, http://msc.aisc.org/globalassets/modern-steel/archives/2009/03/2009v03_vertical_bracing.pdf, last accessed on 17 Aug 2015.

Timler, P.A. and Kulak, G.L., *Experimental Study of Steel Plate Shear Walls,* Structural Engineering Report No. 114, Department of Civil Engineering, The University of Alberta, Canada, Nov 1983.

Tsai, K.C., Li, C.H, and Lee, H.C., 'Seismic Design and Testing of the Bottom Vertical Boundary Elements in Steel Plate Shear Walls. Part 1: Design Methodology, Part 2: Experimental Studies', *Earthquake Engineering and Structural Dynamics*, Vol. 43, Nov 2014, pp. 2155–2177 and Dec 2014, pp. 2237–2259.

Tuan, A.Y. and G.Q. Shang, 'Vibration Control in a 101-Storey Building Using a Tuned Mass Damper', *Journal of Applied Science and Engineering*, Vol. 17, No. 2, 2014, pp. 141–56.

Uang, C.M., and Nakashima, M., 'Steel Buckling-restrained Frames', in *Earthquake Engineering: Recent Advances and Application*, Chapter 16, Y. Bozorgnia and V. V. Bertero, eds, CRC Press, 2003.

Uriz, P. and Mahin, S.A., *Towards Earthquake Resistant Design of Concentrically Braced Steel-frame Structures*, Report 2008/8, Pacific Earthquake Engineering Research Center(PEER),University of California, Berkeley, Nov 2008, p. 401.

Vargas, R. and Bruneau, M., 'Analytical Response and Design of Buildings with Metallic Structural Fuses- I', *Journal of Structural Engineering*, ASCE, Vol. 135, No. 4, Apr 2009, pp. 386–393, Part II - Experimental Response of Buildings Designed with Metallic Structural Fuses, pp. 394–403.

Vian D., Bruneau, M., and Purba, R., 'Special Perforated Steel Plate Shear Walls with Reduced Beam Section Anchor Beams II: Analysis and Design Recommendations', *Journal of Structural Engineering,* ASCE, Vol. 135, No. 3, Mar 2009, pp. 221–228.

Walters, M., 'Seismic Isolation-The Gold Standard of Seismic Protection', *STRUCTURE Magazine*, ASCE, Jul 2015, p. 11–14.

Williams, A., *Seismic Design of Building and Bridges*, Third edition, Engineering Press, California, 2004.

www.hpsdma.nic.in/Development%20of%20Probablistic%20Seismic%20Hazard%20Map%20of%20India.pdf

www.starseismic.net

www.steelcastconnections.com

Yang, T.S. and E. E. Popov, 'Comparison of MRF Connection Damage due to Northridge Quake with Theory', In *Behaviour of Steel Structures in Seismic Areas, STSSA 97*, (Eds Mazzolani, F. M. and H. Akiyama), Kyoto, 2–8 Aug 1997, 10/17, Salerno, pp. 897–904.

Yu, Q.S., Noel, S., and Uang, C.M., *Experimental and Analytical Studies on Seismic Rehabilitation of Pre-Northridge Steel Moment Connections*: *RBS and Haunch Approaches*, Structural Engg. Div., Report No. SSRP-97/08, University of California, San Diego, 1997.

Zhou, F. and Xian, Q., 'Recent Development on Seismic Isolation, Energy Dissipation, Passive and Semi-active Control of Structures in P.*R.* China', in *Earthquake Engineering Frontiers in the New Millennium*,. B.F. Spencer and Y.S. Hu, (Editors) A.A. Balkema, Lisse, Netherlands, 2001, pp. 279–285.

# 18

# DESIGN PRINCIPLES FOR STEEL ENABLING STRUCTURES

## INTRODUCTION

'*Enabling works*' or '*temporary structures*' are structures that are often required at construction sites to provide access to workers and equipment at site, or to support loads due to green concrete or for other similar purposes. BS 5975:2008 defines them as follows:

*Parts of the works that allow or enable construction of, protect, support or provide access to, the permanent works and which might or might not remain in place at the completion of the works*

*An engineered solution used to support or protect either an existing structure or the permanent works during construction, or to support an item of plant or equipment, or the vertical sides or slopes of an excavation during construction operations on site or to provide access.*

Enabling structures are the major consumers of structural steel in the construction industry. This consumption varies from 10% to as high as 40% of the total project cost, depending on the type of project and possible repetitions of the said 'enabling works'. Unfortunately, the construction industry often neglects the design and construction of *enabling works*. Several catastrophic structural failures were attributed to the faulty design and detailing of steel enabling works.

In spite of being so sensitive there are very limited standards/guidelines that are available for the design of enabling structures. BS 5975:1982 was the first code under the title *Code of Practice for Falsework*. The second and third editions of this code were published in 1996 and 2008. BS EN 12812:2004 was published in 2004 and exists in parallel with this standard. The only substantive European Standards specific to temporary works are contained in the suite of codes BS EN 12810 – 12813. They consider falsework and tied access scaffolding. These standards do not stand alone but explain how the Eurocodes should be applied and modified to suit the particular application. IS 14687 was published only in 1999 (see reference list at the end of this chapter). Considering this situation, the author decided to add this present chapter to bring better awareness about this topic amongst students, teachers, and consultants. It also may bring out some key issues where there is always a difference of opinion or a debate amongst designers/contractors and clients. In certain projects even such issues had led to contractual conflicts and greater impact of project cost and time. Here the aim is not to give a detailed treatment on enabling works, but to just create awareness amongst engineers. For detailed study and further information on the design of steel enabling structures, the reference books and codes provided at the end of this chapter may be consulted.

Enabling/temporary structures widely used at construction sites are as listed here: (a) Earthworks (b) Piling platforms (c) Formwork, staging, and falsework, (d) Propping (e) Scaffolding (f) Façade retention (g) Hoardings (h) Launching girders (i) Load-out jetties and temporary bridges (j) Temporary barricades (k) access ladders, platforms, and elevators (l) Aggregate processing plants (batching plants, crushing plants, sand washing plants, chilling plants, etc.) (m) Crane mattresses (n) Floating crafts/barges and jack-up platforms (o) Tower crane bases (p) Protection from overhead electric lines

There are many other types/sub-types of enabling works; however we will consider only the important type of enabling works and associated principles and issues in this chapter.

## 18.1 MAJOR DESIGN CONSIDERATIONS FOR ENABLING WORKS

Some of the major design considerations of enabling works are discussed in this section:

**Design life and factor of safety** The life of permanent structures usually ranges from 50–100 years or even more (especially in bridges), whereas the effective useable life of enabling/temporary structures in any project ranges from a few days to a few weeks and in rare cases, only a few months (it never exceeds more than four years). Considering such a short usable life in any particular project (although the same enabling works may be used in other projects of similar nature by the construction company) the probability of the considered design load being exceeded is very less. It has to be noted that the cost of enabling works used by contractors is often not paid by the clients. Hence the contractors' designers are forced to optimize the design of enabling works. This is done by use of minimal factor of safety/load factor (1.1 to 1.3) and even allowing over-stressing of members to permissible limits (usually 25 to 33%).

**Method of design and governing criteria** Normally limit states method is preferred for the design of permanent structures (steel or concrete). However in enabling structures like formwork/scaffolding, the design is usually governed by the deflections criteria rather than the stresses. Moreover, as the structures are not loaded beyond the yield limit, the working stress method is preferred/applicable. Many times clients are not aware of these facts and insist the designers to design the enabling works using LSM and even check them for earthquake and other such rare phenomena.

**Loads acting on the enabling works** Unlike permanents structures that are designed for earthquake, wind, and their critical combinations with the dead and live loads, enabling works need not be (and can't be) designed for such extreme combinations, as these loads seldom act on these structures.

**Detailing of enabling works** From this description, it may be clear that the design of enabling structures is very critical and highly optimized area (as it is not separately paid to contractors by the clients). Hence the correct transfer of forces arising at site needs to be predefined and detailed accordingly. In case of enabling works where heavy lifting/shifting is done, the 'working points' must be strictly shown in the fabrications drawings and must be ensured during actual fabrications. Apart from working points, the load bearing stiffeners, doubler plates, and washers have to be provided at locations of stress concentration or at support points. It requires correct judgment and large experience of working at construction sites to decide such points and can't be done just by working in the office.

One more important point to be kept in mind is *degree of indeterminacy*. Unlike in permanent works, where redundancy is preferred, which helps in resisting catastrophic actions (like seismic and other events where material non-linearity/plastic behaviour/hinge formation is desired), in enabling works, simple determinate configuration and direct load transfer is preferred (it is worth noting here that even though the enabling works are extremely sensitive units at any construction sites, they are operated by unskilled workers).

**Number of repetitions, wear and tear, and handling stresses** Large numbers of reuses/repetitions is always desired (and attempted) by contractors to recover the high initial investment cost in the enabling works (as explained later); in fact large number of repetitions is the key to success of any enabling work and even timely completion of any construction work. However, in order to increase the numbers of repetitions of components like formwork and staging, the contractor should not make early deshuttering of the staging towers/H-frames below any wet concrete. Such an attempt may lead to sudden failure of the freshly cast concrete members. The formwork can be removed after the concrete gains sufficient strength as stipulated in IS 456:2000 (Clause 11.3) or IS 14687:1999 (Clause 9.5). But unlike formwork, which supports wet concrete and hence gets locked till maturity of concrete, other enabling works can be efficiently used with large numbers of repetitions. Enabling structures are often very roughly handled by workers at sites. This leads to large wear and tear and handling stresses in the components. Often based on the type of enabling structure, it is desirable to keep a provision for such secondary effects by rounding-off the estimated size of structural components (for e.g., if a designer has estimated a 4 mm thick steel skin plate for formwork, then he may use 5 mm or 6 mm thickness depending upon the number of estimated repetitions and his own past experience).

The designer should also specify the locations at which any particular enabling structure is to be lifted and also should check for the handling stresses caused due to such lifting. In certain enabling structures (like launching girders/trusses) the handling stresses may govern the design.

## 18.2 COST OF ENABLING WORKS

The cost of enabling works is one of the key issues in many construction projects. The cost depends on the method of construction and the number of repetitions expected. The cost of enabling works is usually not paid separately by the clients; hence it has to be considered in the *workmanship* while doing the rate analysis. Normally in mega projects, structural steel is the most preferred option—the total rate per kg of enabling works inclusive of its design, fabrication, treatment, transportation, and erection (for an item like bridge launching girder) could be as high as that of permanent structural steel structure. As of today, the rate per kg of mild steel varies from ₹42 to 45 per kg and the 'fabrication + transportation + erection' rate varies as per the intricacies of the structure and the distance from the fabrication shop to the site (normally varies from ₹40/kg to as high as ₹100/kg if the item is imported from outside India). If the cost turns out to be more than this, then 'future usability', the total impact on the project, and alternate methods of construction using other types of enabling work may be explored by the engineer.

In any normal project the total cost of enabling works ranges from 10% to as high as 45%. The cost of formwork and other steel units used in residential multi-storey building projects can be reused comfortably in next project as well. However, this may not happen in unique projects like construction of nuclear power plants. It must be remembered that the higher cost of enabling works have a direct impact on the tender cost. Hence, instead of merely looking at the cost of the enabling works the efficiency should be taken into account. Many times, to optimize the cost of the enabling work and lower the tender cost, the contractors also take into account the available enabling works with a reduced cost; but in such cases, it is primarily important to check whether the said enabling works in steel is in an intact condition and there is no reduction in its strength due to corrosion.

The repeated use and the wear and tear in the structural steel used for manufacturing typical enabling works may reduce its load carrying capacity. Hence before starting of any new project the available enabling structure has to be thoroughly inspected. Regular maintenance, though may add to the cost of the enabling work, should not be neglected, as corrosion or dents on plates and other components, may result in a serious reduction in strength/stiffness. Although routine maintenance helps to keep the enabling work intact, after the specified number of repetitions, the life of such enabling works may come down considerably; in such situations instead of investing more on maintenance or refurbishing, the contractor may salvage the old enabling unit and purchase/manufacture a new one.

## 18.3 REVIEW OF TYPICAL ENABLING WORKS

Typical enabling works are reviewed in this section.

### *Steel Formwork and Staging*

For concrete construction, formwork or moulds are required. It is also known as *shuttering*. The supporting props or steel towers are known as *staging*. In olden days timber was most commonly used for such formwork and staging. However, currently steel and aluminium are used most frequently.

In the construction industry, a significant quantity of structural steel is consumed in making the formwork. The choice of timber, steel, or aluminium is made based on the number of repetitions (reuses) expected at site. The cost of steel formwork is approximately three times higher than the cost of the timber formwork (but it also depends on the configuration of the main concrete structure). Normally timber formwork can give a repetition of 10–12 times, whereas steel formwork can comfortably give repetitions up to 100 uses. Thus, though the cost of steel formwork is higher, the higher number of repetitions helps to reduce the overall cost of construction. In addition, the maintenance required for steel formwork is lesser than that of timber formwork. However, it has to be noted that the on-site rectification of timber formwork is easier than the steel formwork.

Nowadays aluminium and plastic formworks are also available in the market. The advantage of these two materials over steel formwork is the lighter weight. Still, considerable quantity of steel is consumed in civil construction projects for formwork and is one of the most favoured options (see Figs 18.1–18.4).

**FIG. 18.1** Typical image of steel formwork for casting a bridge segment

*Courtesy:* Reprinted with permission from M/s Afcons Infrastructure Ltd

### *Bridge Launching Girders and Ground-based Erection System/Ground Supported Staging*

Speed of construction is more vital for modern bridge construction, as early completion of projects generates revenue and gives advantages such as cost and time saving, as well as comfort and safety to the end users. In congested cities, consideration of this fact prompts bridge designers to opt for precast concrete construction using precast concrete I-girders and segments. Such I-girders or segments are usually cast in a *casting yard* using steel formwork as shown in Fig. 18.1. After casting the concrete, it is allowed to gain sufficient strength, so that it can be lifted. The handling stresses induced in the girder are checked by the designers separately. Then, the precast segments are transported to the site on trailers. Subsequently the segments are lifted using a launching girder or ground supported staging (GSS) system as shown in Fig. 18.5. The segments are picked up one by one and then dry matching is done. Further the segments are glued to each other by using epoxy grout and then applied with *temporary prestress*.

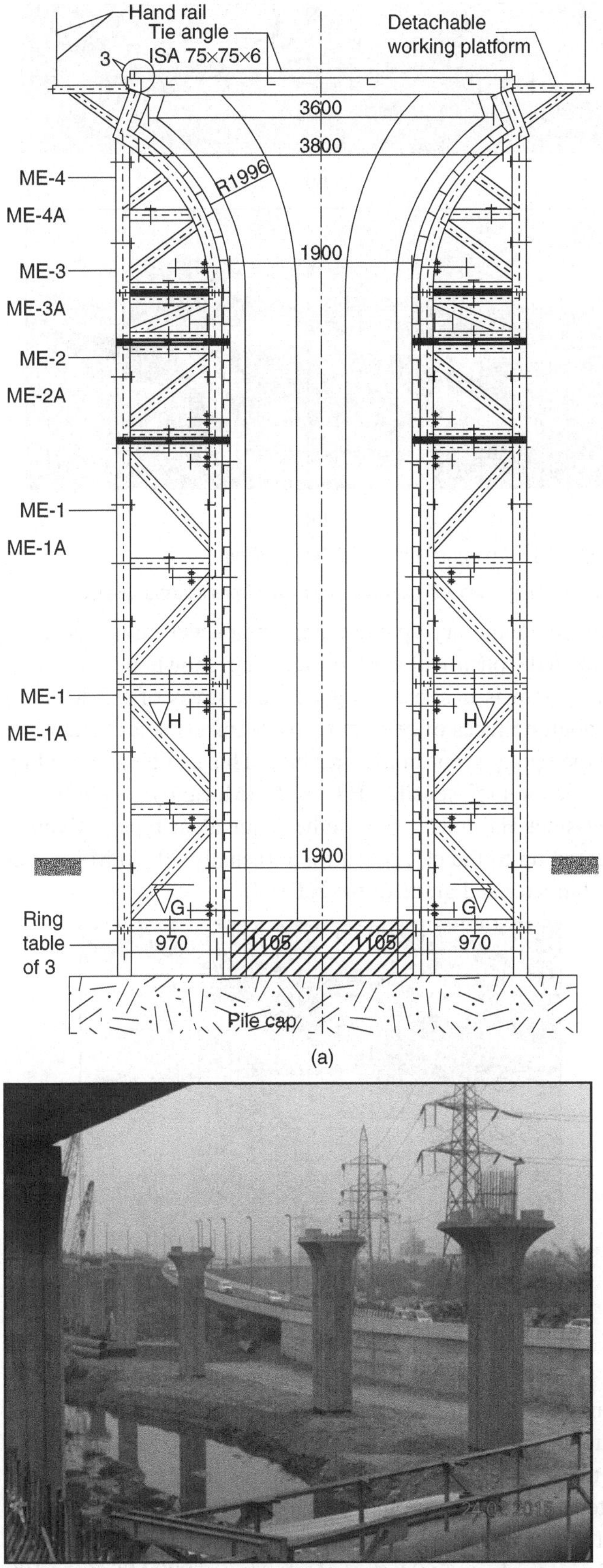


**FIG. 18.2** (a) Typical drawing issued by designer for bridge pier formwork (b) A view of piers cast using the pier formwork

*Courtesy:* Reprinted with permission from M/s Afcons Infrastructure Ltd

**FIG. 18.3** Typical image of steel formwork for a bridge pier used in a PWD project

*Courtesy:* Reprinted with permission from M/s Afcons Infrastructure Ltd

**FIG. 18.4** Steel staging erected over a live railway track for casting bridge deck in Kalamsari, Kerala

*Courtesy:* Reprinted with permission from M/s Afcons Infrastructure Ltd

**FIG. 18.5** Typical truss type bridge launching girder used in Delhi Metro rail project, manufactured and supplied by M/s SPIC, Germany

*Courtesy:* Reprinted with permission from M/s Afcons Infrastructure Ltd

After dry matching and gluing is complete, the glue is allowed to get hardened for a specified duration. Latter the final post-tensioning is done and the complete span is ready to sustain superimposed loads. Figure 18.6 shows the typical plate girder type steel launching girders and Fig. 18.7 shows the typical GSS system used in segmental construction. The steel lifter beam is shown in Fig. 18.8.

**FIG. 18.6** Typical plate girder type bridge launching girder used in Delhi Metro rail Project- Manufactured and Supplied by M/s Bridgecon Ltd, India

*Courtesy:* Reprinted with permission from M/s Afcons Infrastructure Ltd

(a)

(b)

**FIG. 18.7** Typical GSS system (a) During erection stage, before placing segments (b) Typical view after placing segments on GSS

*Courtesy:* Reprinted with permission from M/s Afcons Infrastructure Ltd

(a)

(b)

**FIG. 18.8** (a) Segment lifter (b) During lifting

*Courtesy:* Reprinted with permission from M/s Afcons Infrastructure Ltd

Typical temporary prestressing brackets/beams used in segmental construction of bridges are shown in Fig. 18.9. The system is designed to generate a stress of 0.3 MPa on the abutting faces of two adjoining segments so as to squeeze out the epoxy grout in the interface. This can be achieved by the use of high-tensile (HT) rods (Macalloy, DYWIDAG, Freyssinet HT rods, or any other equivalent type), which is stressed up to the calculated force (usually 50–70 MT as per the diameter and grade of the HT rod).

**FIG. 18.9** Typical view of steel bracket used for the application of temporary prestressing force

*Courtesy:* Reprinted with permission from M/s Afcons Infrastructure Ltd

It may be of interest to note that the typical weight of super structure (bridge proper) ranges from 300–900 MT, depending on the length of the span and width of the deck. The values stated here are generic, and give an idea about the heavy masses that are handled/erected using launching girder (300 MT is the equivalent of 60 Indian elephants and 900 MT about 180 elephants).

The weight of typical segment shown in Figs 18.6–18.8, and popularly used in Metro rail projects, ranges from 45–80 MT.

While handling such large weights much care has to be taken; otherwise catastrophic accidents may take place. It is also worth noting that the weight of LG/GSS that is used for the erection of the superstructure ranges from 100–350 MT. Hence, optimization is tried by the contractors. Still the risk is handling these enabling works cannot be neglected at all. Section 18.4 describes a few failure case studies of mega enabling works to show the importance of designing and executing them properly.

### Steel Enabling Structures in Underground Metro and Marine Projects

Structural steel is a widely used material for the manufacture of enabling structures. The reasons being, its high strength as compared to concrete, timber, aluminium, and plastic, and its easy availability in the open market. Steel can also sustain both tension and compression comfortably. High strength results in lighter member sizes, which require lighter cranes and other equipment/man power for erection of enabling works. Another reason for using structural steel in enabling works is its *weldability*. This property helps the contractors/site engineers to cater to unexpected or last minute changes at site (in principle last minute changes are not desirable but it is inevitable in the construction industry).

In underground construction projects the retention of side walls/stability of cut soil are of prime importance. The surcharge on the top of the adjoining soil and raised ground-water table (in rainy season), leads to lateral shifting/instability of the cut portions; such inward shifting of the soil leads to movement in adjoining buildings, roads etc., eventually cracking them severely. To avoid such soil movement, struts and walers are provided from the excavated side of the soil (before struts are installed, a horizontal member called waler is placed against the soil support. Intermediate struts are then installed from waler to waler across the excavation). Figure 18.10 shows the typical struts and waler beams of the Kolkata Underground Metro railway project. As per the depth of the excavation, multiple layers of the strutting and waling system may be given. It has to be noted that a large size waling and strutting creates hindrance to lowering/lifting the construction materials up/down the trench. Hence, to reduce the size of members, high-strength steel is preferred by the enabling works designers.

The force acting on the struts is due to active earth pressure (soil movement inside) and the bending action caused by the self-weight of strut. In certain cases like deep excavations with loose soil, the soil movement may be large and to control this, the designer has to apply a preload in the struts, thus designing them to overcome the passive soil resistance. The method of calculations of active/passive force, estimation of preload, and the design procedure of the strut and waler is out of the scope of this book. Expert literature may be referred to by the readers to learn more about these aspects (Souder 2015 and Ratay 2012). Figure 18.11 shows the use of steel sheet piles for a cofferdam in a mega underground excavation for pump house.

(a)

(b)

**FIG. 18.10** Steel struts and walers (a) Typical view of steel struts and walers used in an underground metro project (b) Typical heavily bolted spliced joint in a waler

*Courtesy:* Reprinted with permission from M/s Afcons Infrastructure Ltd

**FIG. 18.11** Use of steel sheet piles in the cofferdam for pump house

*Courtesy:* Reprinted with permission from M/s Afcons Infrastructure Ltd

### Other Miscellaneous Enabling Works

Figure 18.12 shows the extensive use of steel sections in a temporary shed at Kolkata underground Metro rail project, for casting the tunnel lining segments.

(a) (b)

**FIG. 18.12** Large temporary steel shed used at Kolkata underground Metro rail project for casting the tunnel lining segments (a) Inner view (b) Outer view

*Courtesy:* Reprinted with permission from M/s Afcons Infrastructure Ltd

## 18.4 FAILURE OF ENABLING WORKS

As explained earlier, enabling or the temporary structures consume a major chunk of structural steel in the construction industry. In spite of this fact, they are considered to be inferior or secondary by a few designers. Hence, the detailing/construction/maintenance of enabling works are neglected to a large extent. Such negligence often leads to minor or sometimes even catastrophic failures. As mentioned earlier, enabling works are not separately paid for, and hence the designers engaged in the design of enabling works are often not provided with sufficient resources and time to do the required magnitude of detailing/drawings.

The following are the common mistakes that lead to failure of enabling works:

1. The working points are often missed during fabrication.
2. The members are loaded at points other than those specified by the designer.
3. Stiffeners at shear force locations are not provided at sites.
4. Inferior (corroded) material is used at site.
5. Improper handling is done at site. Lifting devices (slings, wire ropes, jacks, etc.) of improper capacity are used.
6. Safety instructions are not followed.

Some failures that occurred at sites are shown in Figs 18.13–18.16. Some more failures of temporary structures and their causes may be found in http://failures.wikispaces.com.

**FIG. 18.13** Collapse of the scaffolding supporting one of the beams killing 4 persons on 19 February 1959

*Courtesy:* Reprinted with permission from HSE, ICE London

**FIG. 18.14** Failure of launching girder due to improper/excessive load during construction stage

## 18.5 CONCLUDING REMARKS

This chapter emphasizes the importance of enabling works and the large amount of steel as a construction material that is consumed by them. Unlike European or American construction industries, enabling works are highly neglected in India. Large cost cutting and unsafe practices are followed at site in several enabling works which is a severe risk. A majority of recorded construction failures are due to the failure of enabling works only. As enabling works consume large amount of structural steel, proper and safe design, detailing, and execution is desirable.

**FIG. 18.15** Failure of bridge erection truss due to improper handling during construction

However, safe use does not mean consuming more structural steel (or being over-conservative), but just proper detailing, correct load transfer, and use at site in the desired/specified way.

The aim of adding this chapter in this edition of the book is to create more awareness and interest about enabling works amongst engineering students, teachers, consultants, clients, and even site engineers. The key principles described in this chapter will facilitate/motivate students to explore this topic further. It is practically not possible to cover all types of enabling works in a single chapter, and it is also not the aim of this book. The list of references may help the students to understand temporary structures better. It has to be remembered that practice makes an engineer perfect and experience is an excellent teacher. Patient study and visit to more const-ruction sites will make the students better engineers in the future.

**FIG. 18.16** Failure of temporary bridge during well-sinking operation in Mahanadi river at Bhubaneshwar

## EXAMPLES

**EXAMPLE 18.1:** Calculate the lateral pressure acting on the vertical formwork of a pile cap of a bridge using ACI-347 and CIRIA-108 formulae. Size of the pile cap is 5.4 m × 8.6 m × 1.8 m height. Consider the ambient temperature of concrete as 25°C. Density of fresh concrete is 26 kN/cu.m and the rate of pouring of concrete inside the formwork is 0.5 m/hr. Compare the magnitude of lateral pressure acting on the formwork due to fresh concrete using both ACI-347 and CIRIA-108 report.

**SOLUTION:**

Temperature of concrete = $T_c$ = 25° C

Rate of rise of concrete (i.e., pour rate) = R = 0.5 m/hr

*Notes:*

1. It has to be noted that the temperature of concrete while placing in formwork is usually controlled between 18 °C to 25 °C. In no case can the temperature of concrete at the time of placement be more than 32 °C. Lower temperatures of concrete are preferable, unless the water content causes freezing effect. Also note that though a lower temperature of concrete is preferred from heat of hydration point of view, it leads to an increase in the lateral pressure on the formwork.
2. The rate of rise of fresh concrete inside the formwork is usually determined by the planning engineer at site; capacity of batching plant/transit mixer and the volume of formwork is to be used for estimating the rate of rise/pour rate of concrete. Faster pours lead to larger lateral pressures on formwork.

Maximum lateral pressure $P$ (in kN/m$^2$, i.e., kPa) acting on the formwork is determined using the following formulae as per American Code (ACI-347):

$$P = C_c C_w \left(7.2 + \frac{785R}{T_C + 17.8}\right) < wh$$

where, $C_c$ = multiplying coefficient to take care of the type of cement (also called as chemistry coefficient) and $C_w$ =

coefficient to take care of unit weight of the concrete. They are determined using the values in Tables 18.1 and 18.2.

**TABLE 18.1** Unit weight coefficient $C_w$ used in form pressure equations

| Unit weight (density) of concrete | $C_w$ (in lb version) | $C_w$ (SI version) |
|---|---|---|
| Less than 140 lb/ft$^3$ (<2240 kg/m$^3$) | $C_w = 0.5[1 + (w/145)]$ but not less than 0.80 | $C_w = 0.5[1 + (w/2320)]$ but not less than 0.80 |
| 140 to 150 lb/ft$^3$ (2240 to 2400 kg/m$^3$) | 1.0 | 1.0 |
| More than 150 lb/ft$^3$ (>2400 kg/m$^3$) | $C_w = w/145$ | $C_w = w/2320$ |

*Note:* $w$ = unit weight (density) of concrete, in lb/ft$^3$ (kg/m$^3$)

**TABLE 18.2** Chemistry coefficient $C_c$ used in form pressure equations

| Cement type or blend | $C_c$ |
|---|---|
| Types I, II, and III cements without retarders* | 1.0 |
| Types I, II, and III cements with a retarder | 1.2 |
| Other types or blends containing less than 70% slag or 40% fly ash without retarders* | 1.2 |
| Other types or blends containing less than 70% slag or 40% fly ash with a retarder* | 1.4 |
| Blends containing more than 70% slag or 40% fly ash | 1.4 |

*Retarders include any admixture, such as a retarder, retarding water-reducer, mid-range water-reducing admixture, or high-range water-reducing admixture, that delays setting of concrete

In the present case, $C_w = 2600/2320 = 1.12$, considering OPC cement without retarders and $C_c = 1.0$.

Thus

$$P = 1.0 \times 1.12 \times [7.2 + (785 \times 0.5)/(25 + 17.8)] = 18.4\,\text{kN/m}^2$$

The lateral load due to concrete acting on the formworks is shown in Fig. 18.17.

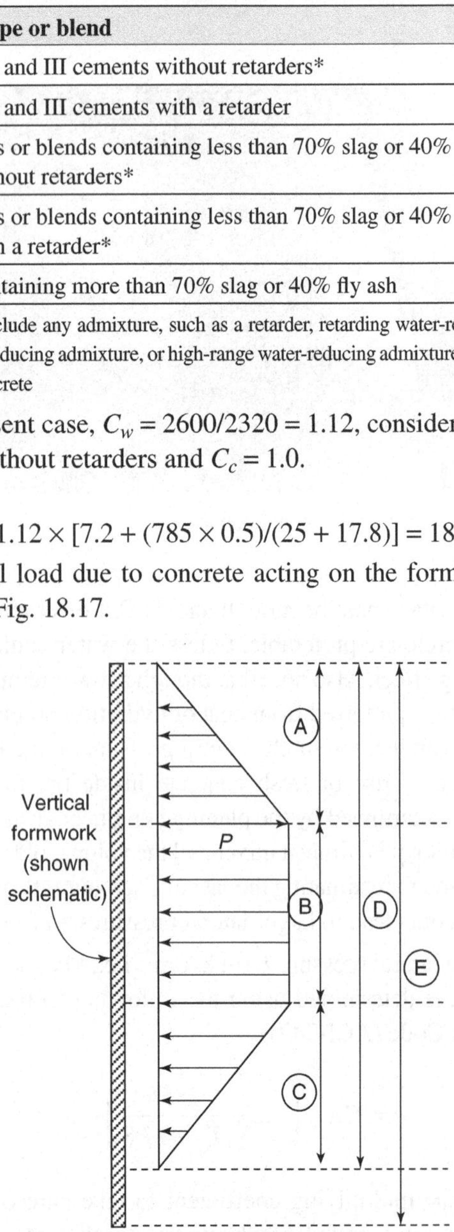


**FIG. 18.17** Pressure on formwork

In Fig. 18.17,

A = Zone in which lateral pressure due to concrete varies linearly
= Density of concrete/$P$ = 2.6/1.84 = 1.41 m

B = Zone of constant lateral pressure

C = Zone in which concrete has begun the strength gain, usually after 6 to 7 hours time (as per the final setting time of the cement), it ends at the zone where concrete has already obtained *final set*. For slow pours, height of C may be less and fast pours more.

In the present case as the height of formwork is only 1.8 m, C = 0

D = Effective zone of lateral pressure due to concrete measured from the top edge of freshly placed concrete = A + B + C

E = Total height of the formwork = 1.8 m in present example

Now, let us determine the maximum lateral pressure $P$ (in kN/m$^2$, i.e., kPa) acting on formwork using the CIRIA-108 report using

$P = D\,[\,C_1\,(R)^{1/2} + C_2.K\,(H - C_1\,(R)^{1/2})^{1/2}\,]$ or Dh, whichever is smaller

where,

$D$ = weight density of concrete, kN/m$^2$ = 26 kN/m$^2$ in the present case

$C_1$ = Coefficient depending on the size and shape of the formwork (taken as 1.0 for walls and 1.5 for columns) = taken as 1 in the present example

$C_2$ = Coefficient depending on the constituent materials of concrete as shown in Table 18.3. In the present example $C_2 = 0.3$.

$R$ = Rate of rise or pour rate = 0.5 m/h in the present example

$H$ = Vertical height of the formwork = 1.8 m in the present example

$h$ = Height of any typical pour = 1.8 m in the present case (in certain cases, the overall structure may require two or more pours due to the large size; in such cases respective $h$ value may be taken in design)

$K$ = Temperature coefficient = $\left(\dfrac{36}{T+16}\right)^2$ = 0.948 in the present example

$T$ = Concrete temperature at placing, °C = 24 in the present example

**TABLE 18.3** Coefficient $C_2$ for CIRIA-108 formula

| Type of concrete | Value for $C_2$ |
|---|---|
| OPC, RHPC, or SRPC without admixtures | 0.3 |
| OPC, RHPC, or SRPC with any admixtures, except retarder | 0.3 |
| OPC, RHPC, or SRPC with retarder | 0.45 |
| LHPBFC, PBFC, PPFAC, or blends containing less than 70% ggbfs or 40% pfa without admixture | 0.45 |

(*Contd*)

**TABLE 18.3** *(Contd)*

| Type of concrete | Value for $C_2$ |
|---|---|
| LHPBFC, PBFC, PPFAC or blends containing less than 70% ggbfs or 40% pfa with admixture, except retarder | 0.45 |
| LHPBFC, PBFC, PPFAC or blends containing less than 70% ggbfs or 40% pfa with retarder | 0.6 |
| Blend containing more than 70% ggbfs or 40% pfa | 0.6 |

OPC: Ordinary Portland cement; RHPC: Rapid hardening Portland cement; SRPC: Sulfate resisting Portland cement; LHPBFC: Low heat Portland blast furnace cement; PBFC: Portland blast furnace cement; PPFAC: Portland Pozzolana with pulverized fly ash cement; ggbfs: Ground granulated blast furnace slag; pfa: Pulverized fly ash.

Thus $P = 26\,[1.0\,(0.5)^{1/2} + 0.3 \times 0.948\,(1.8 - 1\,(0.5)^{1/2})^{1/2}]$
$= 26.114\,\text{kN/m}^2$

From this example it may be clear that in general, the CIRIA formula always gives higher lateral pressure than the ACI formula; but in low temperature concrete, say, 5°C and below, the CIRIA formula gives lesser pressure for concrete.

*Note:* Apart from the aforementioned formulae, CIRIA-108 report also provides the lateral pressure due to concrete in the form of readymade tables.

**EXAMPLE 18.2:** A 200 mm thick reinforced concrete slab of a residential building is required to be cast using steel formwork and staging with standard steel tube sections as shown in Fig. 18.18. Design the suitable formwork and staging system for the same.

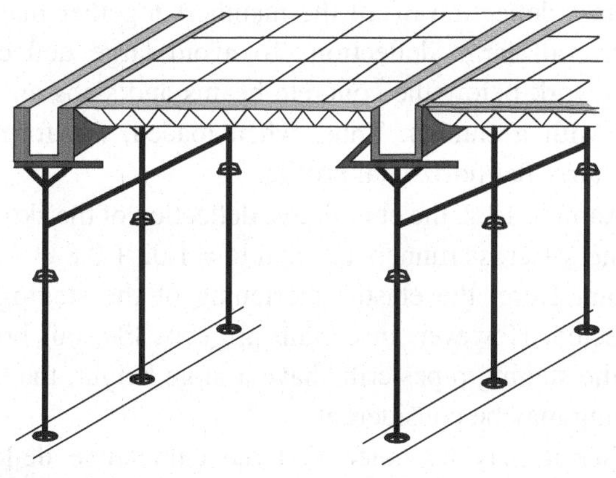

**FIG. 18.18** Prop and span assembly

**SOLUTION:**

As the fresh concrete load acts vertically (along gravity), the ACI or CIRIA formula for lateral loads is not required for estimation of pressure; the self-weight of the fresh concrete acts directly on the formwork and this is calculated as follows:

1. Weight of the fresh concrete = 0.2 m × 26 kN/m$^3$ = 5.2 kN/m$^2$
2. Construction live load (due to workers/equipment) = 3.4 kN/m$^2$ (as specified in IRC-87)
3. Self-weight of the formwork = 1 kN/m$^2$ (self-weight of formwork may be assumed suitably by the designer based on his own past experience. It generally varies between 1–2.5 kN/m$^2$)

Thus load acting on unit area of formwork = 5.2 + 3.4 + 1
= 9.6 kN/m$^2$

Assume that a 3.15 mm thick (i.e., 10 gauge) steel plate is used to support the fresh concrete (such a plate in contact with the fresh concrete is also called as a *skin plate*). The steel skin plate is supported using steel stiffeners ISA 50 × 50 × 5, spaced at 300 mm c/c distance.

Thus the forces and stresses in the plate (skin plate) are as follows (see Fig. 18.19):

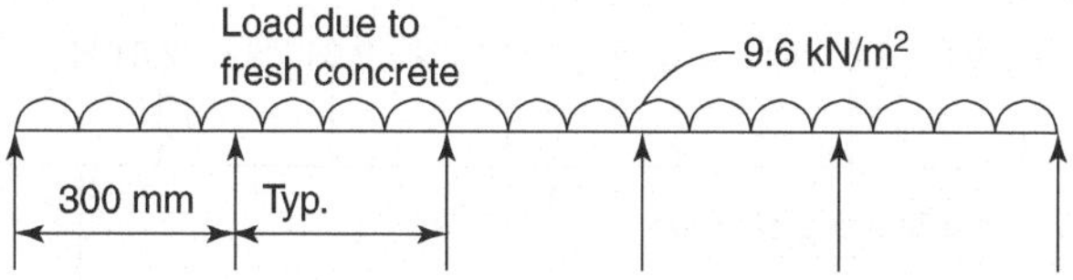


**FIG. 18.19**

$$\text{Bending moment} = M = \frac{wl^2}{10} = 0.0864\ \text{kNm/m}$$

Bending moment is estimated assuming that the plate is continuous, which is usually true at site for large size slabs/ formwork.

$$\text{Section modulus} = Z = \frac{b.t^2}{6} = \frac{1000 \times 3.15^2}{6} = 1653\ \text{mm}^3$$

$$\text{Moment of inertia} = I = \frac{bt^3}{12} = \frac{1000 \times 3.15^3}{12} = 2604\ \text{mm}^4$$

Bending stress = $M/Z = 0.0864 \times 10^6/1653$
$= 52.3$ MPa < 165 MPa

Hence within allowable limits.

Maximum deflection of the plate = $0.0065\,wL^4/EI$ (continuous span)
$= 0.0065 \times 9.6 \times 300^4/(2 \times 10^5 \times 2604)$
**= 1.02 mm**

Hence within allowable limits.

Allowable deflection in the skin plate is $L/325$ or 3 mm whichever is lesser.

The load on each stiffener welded to the skin plate is = 9.6 × 0.3 = 2.88 kN/m

Assuming that the stiffeners below the skin plate are supported at every 1 m c/c using cross beam (also called as cross runners), the forces and stresses in them are as follows:

$$\text{Bending moment} = M = \frac{wl^2}{10} = \frac{2.88 \times 1^2}{10} = 0.288\ \text{kNm}$$

$$\text{Shear force in the stiffener} = V = wl/2 = \frac{2.88 \times 1}{2} = 1.41\ \text{kN}$$

For L 50 × 50 × 5, section modulus = $Z = 3100\ \text{mm}^3$

Moment of inertia = $I$ = 1,10,000 mm$^4$

Bending stress = $M/Z = 0.288 \times 10^6/3100$
$= 92.91\text{MPa} < 165$ MPa
Hence within allowable limits.
Shear stress = $V/A_w = 1.41 \times 10^3/(50 \times 5)$
$= 5.64\text{MPa} < 100$ MPa
Hence within allowable limits.
Maximum deflection of the plate = $0.0065\, wL^4/EI$
$= 0.0065 \times 2.82 \times 1000^4/(2 \times 10^5 \times 1{,}10{,}000)$
= **0.833 mm** (Hence OK)
Allowable deflection in stiffener is $L/325 = 1000/325$
$= 3.08$ mm.

Assuming ISMB 100 as a section used for cross runners, the load on these members is = $2 \times 1.41 = 2.81$ kN concentrated load at every 300 mm c/c. Assuming that the cross runners are supported at every 1 m c/c using H-frame/pipes, as shown in Fig. 18.20.

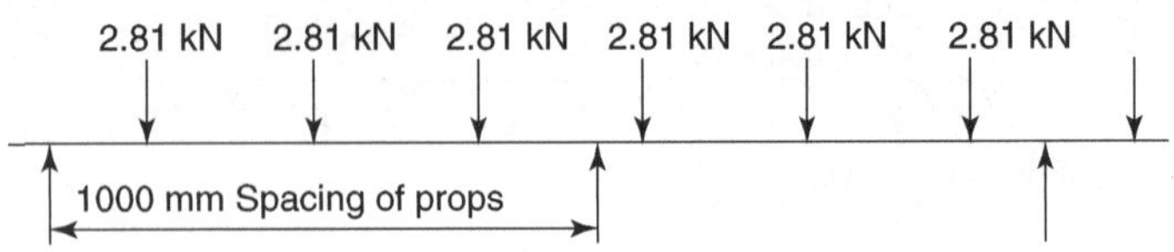


**FIG. 18.20**

Approximate bending moment
$= M = (2.81 \times 1.5) \times 0.5 - 2.81 \times 0.3 = 1.26\,\text{kNm}$
Shear force in the stiffener = $2.81 \times 1.5 = 4.22$ kN
Section modulus = $Z = 36600\,\text{mm}^3$
Moment of inertia = $I = 18{,}30{,}000\,\text{mm}^4$
Bending stress = $M/Z = 1.26 \times 10^6/36600$
$= 34.4$ MPa $< 165$ MPa
Hence within allowable limits.
Shear stress = $V/A_w = 4.22 \times 10^3/(100 \times 4.7)$
$= 8.9$ MPa $< 100$ MPa
Hence within allowable limits.
Maximum deflection of the plate = $0.045\, PL^3/EI$
$= 0.045 \times 2810 \times 1000^3/(2 \times 10^5 \times 18{,}30{,}000)$
= **0.345 mm** (Hence OK)
Allowable deflection in stiffener is L/325 = 1000/325
$= 3.08$ mm.

The cross runners are supported using standard mild steel tubes (called props). The load on the top of pipe is = $(4.221 \times 2) \times 1.2 = 10.1$Kn. Note here the multiplier 1.2 is to take care of the continuity effect, because of which the reaction on the inner props is slightly more than the outer props. Popularly, NB 40 (m) pipes of grade YSt 22 grade are used as staging pipe or for H-frames. If it is used, stresses in the pipe are as follows:

Axial load on staging pipe = 10.1 kN
Cross-sectional area of the NB40 (M) pipe (IS 1161:1998) = 453 $\text{mm}^3$
Actual compressive stress in pipe = $10.1 \times 1000/453$
$= 22.29$ MPa
Radius of gyration of the pipe = 16 mm
Assuming typical floor height as = 3000 mm
Slenderness ratio of pipe = 3000/16 = 187.5
Hence allowable compressive stress in the pipe (Table 2 of IS 806:1968) = 30 MPa > 22.29 MPa (Hence OK).

## Design Summary

1. A skin plate of thickness 3.15 mm is used to support the green concrete.
2. The skin plate will be supported using ISA 50 × 50 × 5 stiffeners placed at 300mm c/c. For ease of construction, contractors pre-fabricate panels of some defined sizes (say 1.0 × 1.0, 2.0 × 1.0, etc) by welding the plate with stiffeners in advance and directly place them over the cross runners. This saves time and also facilitates great ease while assembling and dismantling the formwork at site.
3. The panels, as explained in the example, will be supported over cross runners ISMB 100 spaced at 1 m c/c.
4. NB 40 (m), YSt 22 class pipes will be used as staging pipes at every 1 m c/c, in both the directions to support the cross runners.

In addition to these, suitable bracings (in both the directions) will be provided to the staging pipes.

**EXAMPLE 18.3:** Estimate the cumulative deflection and the *pre-camber* to be provided in the staging designed in Example 18.2.

**SOLUTION:**

As seen in Example 18.2, due to the weight of wet concrete, the formwork/staging components deflect. In long spans the cumulative deflection of all the members together may lead to an overall large deflection. To avoid large deflections, the formwork below the concrete beams and slabs are often erected with a camber. Thus, when loaded, the formwork gives a fairly flat/horizontal profile.

In Example 18.2, the cumulative deflection of the skin plate + stiffeners + cross runners is equal to = 1.02 + 0.833 + 0.345 = 2.2 mm. Here, the elastic shortening of the staging pipe is neglected. However, in certain projects, like tall bridges, where the staging pipes/cribs have a large height, the elastic shortening may be considerable.

Further it may be noted that the cumulative deflection is only 2.2 mm, which is less than the commonly followed maximum value of 3 mm. In certain projects (such as nuclear plants or LnG tanks, a very smooth concrete finish is desired; in such cases, the 'cumulative deflection' of formwork components may govern the entire design) the concrete project beyond permissible limits (usually specified by client/contact document) may have to be finished using grinding machines. Usually a parabolic pre-camber is adopted for residential slabs as shown in Fig. 18.21.

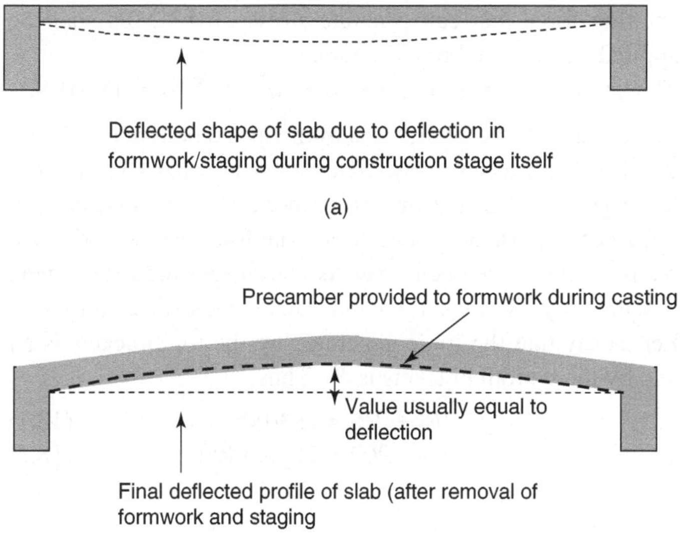


**FIG. 18.21** Formwork without and with camber (a) Final profile of slab without camber (b) Slab with a pre-camber

For special/mega projects, pre-camber may require a detailed construction-stage analysis.

**EXAMPLE 18.4:** In one of the bridge projects, a reinforced concrete beam of 9 m length is used with M-50 grade concrete and Fe 500 grade reinforcement. The width of the beam is 1200 mm and depth is 1800 mm. The beam is designed for two point loads (reaction due to live load) each of 250 kN at 1/3rd span. The concrete cube test results of approved M-50 mix at the 3rd, 7th, 14th, 21st, and 28th days are as shown in the following table. Comment on the earliest possible de-shuttering time of the formwork.

**TABLE 18.4** Concrete tube test results

| Age of concrete | Breaking compressive strength |
|---|---|
| 3 | 21.08 MPa |
| 7 | 32.22 Mpa |
| 14 | 48.03 MPa |
| 21 | 52.12 Mpa |
| 28 | 55.4 MPa |

**SOLUTION:**

Usually the formwork stripping time is determined as specified in IS:456 (Clause 11.3) or IS:14687 (Clause 9.3). See Table 18.5.

**TABLE 18.5** Stripping time for formwork as per IS 456

| Type of formwork | Minimum period before striking formwork |
|---|---|
| Vertical formwork to columns, walls, beams | 16–24 hours |
| Soffit formwork to slabs (props to be refixed immediately after removal of formwork) | 3 days |
| Soffit formwork to beams (props to refixed immediately after removal of formwork) | 7 days |
| Props to slabs<br>(a) Spanning up to 4.5m<br>(b) Spanning over 4.5m | <br>7 days<br>14 days |
| Props to slabs<br>(a) Spanning up to 6 m<br>(b) Spanning over 6 m | <br>14 days<br>21 days |

In our case, the length of the beam is 9m, that is, greater than 6m. Hence ideally the formwork striking time would be as follows:

1. Soffit formwork panels: 7 days
2. Vertical Side formwork panel: 16 to 24 hours
3. Props below soffit: 21 days

However, as stated at the beginning of IS:456 (Clause 11.3), based on the concrete strength and loading at the time of striking the formwork/props, the period of striking can be shortened as follows:

The dead load (DL) of concrete after hardening = $1.2 \times 1.8 \times 25 = 54$ kN/m

The forces induced in the beam due to DL and LL are summarized here:

| | Bending moment | % | Shear force | % |
|---|---|---|---|---|
| DL | 546 kNm | 42% | 243 kN | 49 % |
| LL | 750 kNm | 58% | 250 kN | 51 % |
| Total | 1296 kNm | 100 | 493 kN | 100 |

At the time of formwork removal, only DL will be active, which has about 42% contribution to BM and 49% contribution to shear. Hence at the time of striking the stresses generated in the section are of the same proportion. Now as per Clause11.3:

*Forms shall not be released until the concrete has achieved a strength of at least twice the stress to which the concrete may be subjected at the time of removal of formwork.*

Thus, from this clause, the formwork can be de-shuttered when the concrete has gained strength equal to 'twice that of 42%', that is, 84% (42 MPa). From Table 18.4 it is clear that the strength developed at the 14th day satisfies this codal requirement, hence formwork/props can be stripped at as early as the 14th day and need not be kept till the 28th day. However, remember that though the formwork and props are removed on the 14th day, curing has to be continued till the 28th day as it has relation to durability of concrete, as well.

**Example 18.5:** Determine the force required for the design of temporary prestressing brackets in segmental bridge construction. The effective cross-sectional area of two typical segments to be glued using epoxy grout is equal to 6.10 m$^2$. The total depth of segment is 2.45 m with neutral axis located at 960 mm from top.

**Solution:**

The temporary prestressing brackets, as seen in Fig. 18.9, are used to press the segments against each other, after application of wet epoxy grout on the abutting surfaces. The force is determined as per guidelines of IRC-SP-65-2005 code (Clause 4.0). As per this code, the temporary compressive abutting force that can generate stress of 0.3 MPa shall be applied and retained till 24 hours.

Thus, total force required = $6.10 \times 10^6 \times 0.3/10^3 = 1830$ kN

The distance of the neutral axis from top and bottom is 960 mm and 1490 mm respectively. Hence the estimated PT force has to be generated in the inverse proportion of the distances of CG so as to cause only concentric axial force and no additional moment. Typical segment cross section and desired arrangement of temporary PT brackets are indicated in Figs 18.22 and 18.23. Let us say that the force generated by the top brackets is $P_1$ and by the bottom brackets is $P_2$. Thus,

$$2P_1 + 2P_2 = 1830 \text{ kN} \quad (18.1)$$

$$2P_1 \times 960 = 2P_2 \times 1490 \quad (18.2)$$

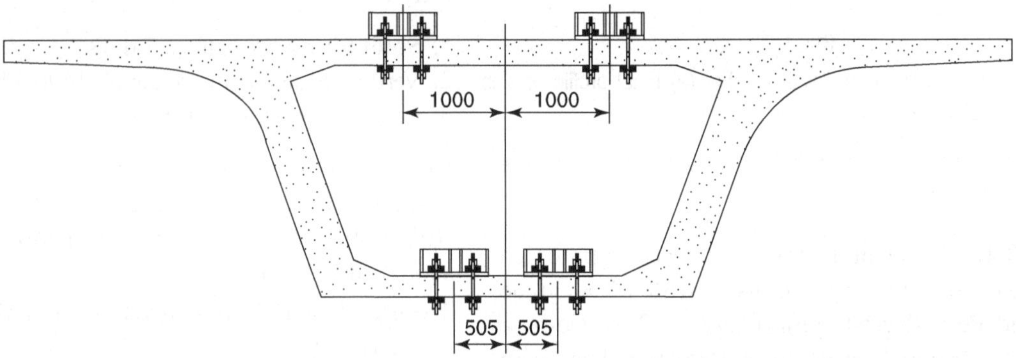


**FIG. 18.22** Typical cross section of a segment with temporary PT brackets

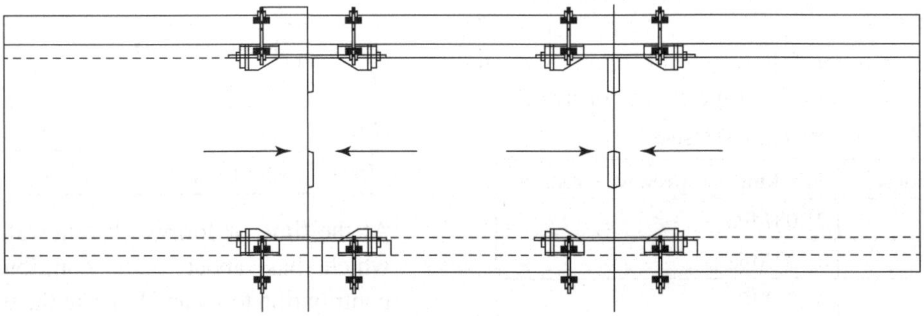

**FIG. 18.23** Typical side view showing temporary PT brackets and direction of force

Solving Eqn (18.2), we get $P_1 = 1.55 \times P_2$; substituting it back in Eqn (18.1), we get $P_2 = 358$ kN and $P_1 = 557$ kN. Now according to this force, suitable size of HT rod and PT bracket can be designed as per regular procedure followed in the design of steel structures.

**Example 18.6:** Design a lifter system (lifter beam and a cross beam) for a concrete segment mentioned in Example 18.5. The weight of the segment is 450 kN, length (along the direction of bridge) is 3 m, and width at top is 9.8 m. Consider lifting points spaced at 6.1 m c/c in section and 1.1 m c/c along length. Consider impact factor of 10%.

**Solution:**

*Design of main lifter beam*:

The design load is equal to = $450 \times 1.1 = 495$ kN

Bending moment in lifter beam = $(495/2) \times (6.1/2)$
= 754.88 KNm

Shear force in the lifter beam = 495/2 = 247.5 kN

Assuming the cross-sectional dimensions of the lifter beam as shown in Fig. 18.24.

Key structural properties estimated from these dimensions are as follows:

Cross-sectional area = $A = 2 \times (25 \times 450 + 10 \times 450)$
= 31,500 mm$^2$

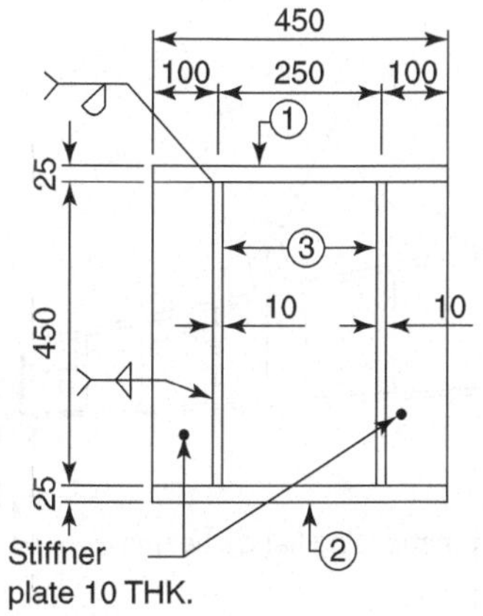


**FIG. 18.24** Cross section of lifter beam

Effective shear area = $A_w = 2 \times (10 \times 500) = 10{,}000\,\text{mm}^2$
Moment of inertia = $I_{xx} = 1/12 \times (450 \times 500^3 - 430 \times 450^3)$
$= 1.422 \times 10^9\,\text{mm}^4$
Section modulus = $Z_{xx} = I_{xx}/(D/2) = 5.689 \times 10^6\,\text{mm}^3$
Radius of gyration = $(I_{xx}/A) = 212.48\,\text{mm}$
Stresses and deflections in the lifter beam due to these factors are,
Bending stress = $M/Z_{xx} = (754.88 \times 10^6)/(5.689 \times 10^6)$
$= 132.69$ MPa < 165 MPa (Hence OK)
Average shear stress = $V/A_w = 247.5 \times 10^3/10{,}000$
$= 24.75$ MPa < 100 MPa
(Hence OK)
Central deflection = $PL^3/(48\,EI)$
$= 495 \times 10^3 \times 6100^3/(48 \times 2 \times 10^5 \times 1.422 \times 10^9)$
$= 8.23\,\text{mm} < L/325\ (18.77\,\text{mm})$ – Hence OK

*Note:* Usually a tapered section is preferred, which uses steel optimally. In such case exact deflection may be found out using conjugate beam method or any other suitable method.

*Design of cross beam:*
The design load is equal to = $(450 \times 1.1)/2 = 247.5\,\text{kN}$
Bending moment in lifter beam = $(247.5/2) \times (1.1/2)$
$= 68.06\,\text{kNm}$
Shear force in the lifter beam is = $247.5/2 = 123.75\,\text{kN}$
The cross-sectional dimensions for the cross beam are shown in Fig. 18.25.
Key structural properties estimated from the dimensions given in Fig. 18.25 are,
Cross-sectional area = $A = 2 \times (16 \times 300 + 16 \times 264)$
$= 18{,}048\,\text{mm}^2$
Effective shear area = $A_w = 2 \times (16 \times 296) = 9472\,\text{mm}^2$
Moment of inertia = $I_{xx} = 1/12 \times (300 \times 296^3 - 268 \times 264^3)$
$= 0.237 \times 10^9\,\text{mm}^4$
Section modulus = $Z_{xx} = I_{xx}/(D/2) = 1.604 \times 10^6\,\text{mm}^3$
Stresses and deflections in the lifter beam due to these are,
Bending stress = $M/Z_{xx} = (68.06 \times 10^6)/(1.604 \times 10^6)$
$= 42.43\,\text{MPa} < 165\,\text{MPa}$ (Hence OK)

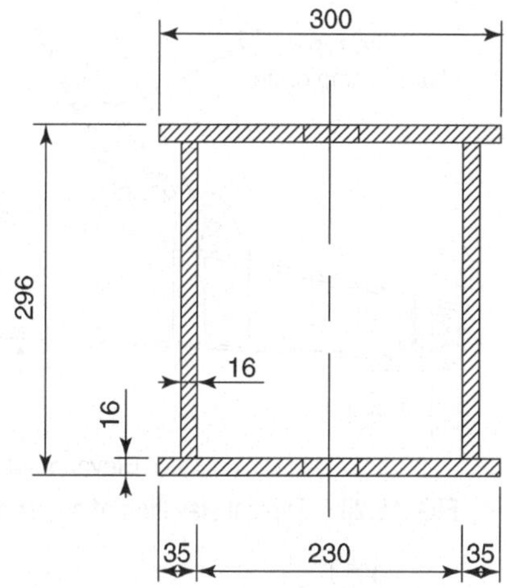


**FIG. 18.25** Cross section of cross-beam

Average shear stress = $V/A_w = 123.75 \times 10^3/9472$
$= 13.06\,\text{MPa} < 100\,\text{MPa}$
(Hence OK)
Central Deflection = $PL^3/(48\,EI)$
$= 247.5 \times 10^3 \times 1100^3/(48 \times 2 \times 10^5 \times 0.237 \times 10^9)$
$= 0.144\,\text{mm} < L/325\ (3.3\text{mm})$ (Hence OK)

*Note:* The lifter beam is engaged with the cross beam using steel pin, and the cross beam with the segment using high tension (HT) rods. Different makes of HT rods are available in market (Mcalloy, Dywidag, Freyssinet, etc.)

*Design of a pin between lifter and cross beam*:
Shear force for the pin = 247.5 kN/2
= 123.75 kN (double shear case)
Using a 80 mm diameter pin of EN8 grade,
Cross-sectional area = $5026.54\,\text{mm}^2$
Hence shear stress in the pin = 24.62 MPa > 186MPa (Hence OK)

*Notes:*

1. The yield stress on EN8 grade material is 465 MPa and allowable shear is $0.4f_y$, that is, 186 MPa. Though, in this example, it appears like there is high margin for safety in the pin, at actual sites the pins are not handled carefully by the construction workers; even the holes in which they are engaged may not be very smooth. All these cases may lead to overloading/stress concentration in the pin. In Indian codes there is no specific provision for the analysis and design of pins. Whereas BS code demands a factor of safety of 2 for all such important parts.
2. In case the pin is of a longer length, then apart from shear it is subjected to bending, and hence the same should also be checked.

Figures 18.26–18.28 show the typical details of lifter beam and cross beam as designed in this example and how they are used to lift a segment with crane.

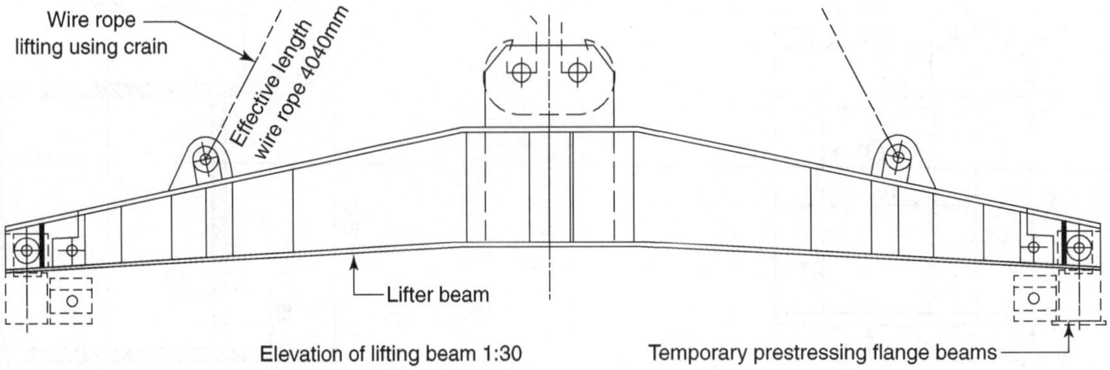


**FIG. 18.26** Typical elevation of a lifter beam and cross beams (also refer Fig. 18.8)

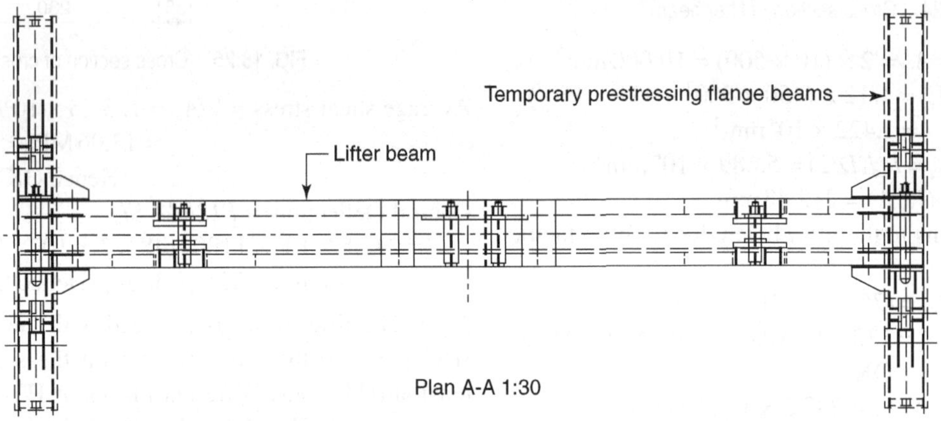


**FIG. 18.27** Typical plan of the lifter beam and cross beams

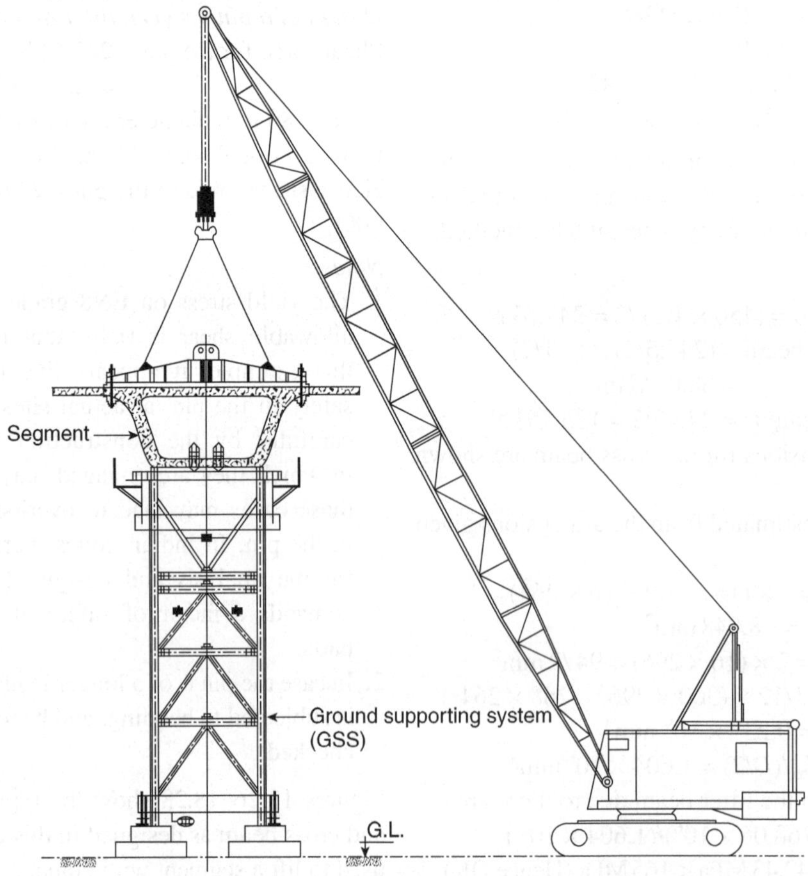


**FIG. 18.28** Typical arrangement of feeding segment on GSS using lifter beam

# MULTIPLE-CHOICE QUESTIONS

18.1 Enabling works consume major part of structural steel in construction sector ranging from
(a) 10% to 40% of the total project cost
(b) 1% to 2% of the total project cost
(c) 0.1% to 0.4% i.e. very negligible part of total project cost

18.2 A majority of the severe catastrophic structural failures have happened during construction due to
(a) lack of use of properly designed formwork/staging and other enabling works
(b) earthquakes
(c) blast loading during construction

18.3 The formwork design can be prepared as per the guidelines of
(a) IS:14687-1999
(b) ACI-347
(c) CIRIA-108 Report
(d) IRC:87
(e) Any one of these as per type of contract

18.4 Match the following:

| Type of structure | Design life (retention period) |
|---|---|
| (a) Formwork for a column | (i) 14 days |
| (b) Permanent steel building | (ii) 100+ years |
| (c) Bridge launching girder | (iii) 21 days |
| (d) Concrete dam | (iv) 24 hrs |
| (e) Tunnel formwork | (v) 50 years |
| (f) Formwork for a beam (L < 6m) | (vi) 4 years |
| (g) Formwork for a beam (L > 6m) | (vii) 2 years |

18.5 In most of the construction projects the cost incurred on enabling works is
(a) not separately paid by the client
(b) separately paid by the client
(c) paid separately as a lump-sum in 'item rate' projects

18.6 A typical formwork for a small pile cap on ground, has to be designed for
(a) temperature and shrinkage and creep effects
(b) earthquake loads and wind loads
(c) p–delta effect
(d) none of these

18.7 The formwork/staging can be de-shuttered (removed) as per the guidelines of
(a) IS:456 (Clause 11.3)
(b) IS 14687 (Clause 9.5)
(c) CIRIA report number 73 (formwork striking time by Harrison T.A.)
(d) any one of these as per applicability in project

# EXERCISES

18.1 Prepare a table of concrete lateral pressures using ACI and CIRIA formulae for different values of concrete pour rates $R$ and different temperatures $T$. Consider the variation in $R$ from 0.5 m/h to 4 m/h with an increment of 0.5 m/h. Consider temperature range from −5 °C to 30 °C with an increment of 5°. *Hint*: Refer formulae and procedure given in Example 18.1.

18.2 Plot a graph of curves from the table prepared in Exercise 18.1.

18.3 Design a formwork and staging scheme for an RCC slab of 500 mm thickness in an industrial building. Consider 5 mm thick skin plate and ISA 65 × 65 × 5 as stiffeners below the skin plate. Determine spacing as per IS:800 and IS:14687 requirements. Also design suitable cross runners and staging tubes. Consider the floor height as 4.5 m.

18.4 Determine the cumulative deflection for the problem in Exercise 18.3 and also plot the pre-camber profile for it.

18.5 Prepare a bill of material of quantity of structural steel consumed in Exercise 18.3.

18.6 A concrete pile cap of size 5.3 m × 8.3 m × 2.5 m thickness is to be constructed with M35 grade concrete. Assuming the pour rate as 1 m/h, determine the lateral pressure acting on the side formwork. Design the skin plate and stiffeners for the side formwork assuming spacing of supporting runners (often called as waler) as 1.25 m. The runners are supported with inclined pipes at every 2 m c/c.

18.7 Design the side formwork for a concrete pier of 2 m × 2 m plan area and 8 m height. Assume concrete pour rate as 4 m/h and placement temperature as 20°C.

# REVIEW QUESTIONS

18.1 What is the definition of enabling works as per BS 5975-2008 code?

18.2 List the codes that are used for the design of temporary works.

18.3 List the widely used enabling works in typical construction projects.

18.4 What is the average design life of any typical enabling work?

18.5 Why are the designers of enabling works (temporary structures) getting forced to optimize the design of enabling works? How do they manage to do this without sacrificing the structural safety and stability?

18.6 How and why is the working stress method suitable for the design of enabling works? Should the enabling works be designed for earthquake and tsunami loads?

18.7 Should the formwork/barricades at large heights of the building be designed for wind load?

18.8 What is the importance of the correct structural detailing in enabling works? Why should the following details be shown correctly on the drawings: working points, load bearing stiffeners, doubler plates, washers, etc.?

18.9 Write a short note on degree of indeterminacy in (a) permanent structures and (b) enabling structures.

18.10 Why is a larger number of repetitions desired for use of enabling works and how is it achieved (what care has to be taken)?

18.11 Why is regular maintenance required for enabling works? When is reduction in strength required to be considered in design?

18.12 List the reasons because of which steel is preferred as a construction material for enabling works in underground and marine works. Also explain in which cases high strength steel is preferred over mild steel.

18.13 What are the commonly observed reasons behind failure of enabling works?

18.14 Write a short note on the repeated use of enabling works. What factors influence the choice amongst (a) timber formwork, (b) steel formwork, and (c) plastic formwork?

## REFERENCES


ACI-347, *Guide to Formwork for Concrete*, American Concrete Institute, Farmington Hills, MI, 2004.

BS 5975:2008, *Code of Practice for Temporary Works Procedures and the Permissible Stress Design of Falsework*, British Standards Institution, London, 2008.

CIRIA Report 108, *Concrete Pressure on Formwork*, Construction Industry Research and Information Association, London, UK, 1985, p. 31.

EN 12810-1:2003 *Facade Scaffolds made of Prefabricated Components— Product Specifications,* European Committee for Standardization, Brussels.

EN 12810-2:2003 *Facade Scaffolds made of Prefabricated Components— Particular Methods of Structural Design*, European Committee for Standardization, Brussels.

EN 12811-1:2003 *Temporary Works Equipment Scaffolds—Performance Requirements and General Design*, European Committee for Standardization, Brussels.

EN 12811-2:2004 *Temporary Works Equipment—Information on Materials*, European Committee for Standardization, Brussels.

EN 12811-3:2002 *Temporary works equipment—Load Testing*, European Committee for Standardization, Brussels.

EN 12811-4:2013 *Temporary Works Equipment—Protection Fans for Scaffolds, Performance Requirements and Product Design*, European Committee for Standardization, Brussels.

EN 12812:2008 *Falsework–Performance Requirements and General Design*, European Committee for Standardization, Brussels.

EN 12813:2004 *Temporary Works Equipment—Load Bearing Towers of Prefabricated Components, Particular Methods of Structural Design*, European Committee for Standardization, Brussels.

Grant, M. and Pallett, P.F. (ed.,) *Temporary Works: Principles of Design and Construction*, ICE Publishing, Thomas Telford Ltd, London, 2012, p. 414.

http://failures.wikispaces.com/Overview+of+Temporary+Structures+Failures+in+Construction, last accessed on 4 September 2015.

https://failures.wikispaces.com/Steel+Structural+System+Collapses+%26+Failures+During+Construction, last accessed on 4 September 2015.

http://www.ice.org.uk/ICE_Web_Portal/media/northwest/presentations/ICE-29th-November-2012--Steve-Parncutt.pdf, last accessed on 4 September 2015.

Hurd, M.K. SP-4: *Formwork Work for Concrete*, Seventh edition, American Concrete Institute, Farmington Hills, MI, 2005, p. 500.

IRC 87:2011, *Guidelines for the Formwork, Falsework and Temporary Structures* (First Revision), Indian Roads Congress, New Delhi, May 2011.

IS 14687:1999, *Guidelines for Falsework for Concrete Structures*, Bureau of Indian Standards, New Delhi, Sep 1999, p. 20.

IS 1161:1998, *Steel Tubes for Structural Purposes—Specification*, Bureau of Indian Standards, New Delhi.

IS 806:1968, *Code of Practice for Use of Steel Tubes in General Building Construction*, Bureau of Indian Standards, New Delhi.

Jha, K.N., *Formwork for Concrete Structures*, Tata McGraw Hill Inc., New Delhi, 2012, p. 574.

*National Seminar on Formwork* by ACI (India chapter), Lecture notes, Held on 23–24 July 2009 at Chennai, India.

Oberlender G. and Peurifoy, R.L., *Formwork for Concrete Structures*, Fourth edition, McGraw Hill Professional, New York, 2010, p. 544.

Peurifoy, R.L. and C.J Schexnayder, *Construction Planning, Equipment and Methods*, Seventh edition, McGraw Hill Higher Education, 2005.

Ratay, R.T., *Handbook of Temporary Structures in Construction*, Third edition, McGraw-Hill, New York, 2012, p. 928.

Rosignoli, M., *Bridge Launching*, Thomas Telford Publishing, 2002, p. 352.

Souder, C., *Temporary Structure Design*, Wiley, Hoboken, NJ, 2015, p. 429.

Technical catalogues of various suppliers of formwork and launching girders


In addition to these references, safety related standards such as HSE, OSHA, and OHASIS have guidelines for the construction of various enabling (temporary) structures.

APPENDIX

# A

# PROPERTIES OF STRUCTURAL STEEL SECTIONS

The structural designer has choice of a variety of sections, which are available in the market. This appendix provides properties of structural steel sections often used in practice. For more complete details of *I*-sections, channels, equal and unequal angles, and *T*-sections refer to IS: 808-1989. Note that IS: 808 does not give values of the plastic section modulus. Hence these values for *I*-sections and channels have been provided based on IS: 800. Note that there are some small differences in the values given by IS: 808 and IS: 800. Only the values given by IS: 808 have been used in this book. However, these differences in values will not affect the design much. Also included in this appendix are the wide flange sections, which have been introduced recently (more information on these sections may be obtained from M/s. Jindal Vijayanagar Steel Limited). Properties of some castellated beams (*I*-section and channels) have also been included. Properties of circular tubes (IS: 1161-1998), square and rectangular hollow sections (IS: 4923-1997 and Tata steel catalog), and cold-formed lipped channel and zed sections (IS: 811-1987) are also included in this appendix. More details about these sections may be found in the respective Indian Standard.

## A.1 INDIAN STANDARD ROLLED STEEL PLATES

Steel plates are available in the following widths and thicknesses.

**Widths:** 160, 180, 200, 220, 250, 280, 320, 355, 400, 450, 500, 560, 630, 710, 800, 900, 1000, 1100, 1250, 1400, 1600, 1800, 2000, 2200, and 2500

**Thickness:** 5.0, 5.5, 6.0, 7.0, 8, 9, 10, 11, 12, 14, 16, 18, 20, 22, 25, 28, 32, 36, 40, 45, 50, 56, 63, 71, and 80

**TABLE A.1** Sectional properties for beams

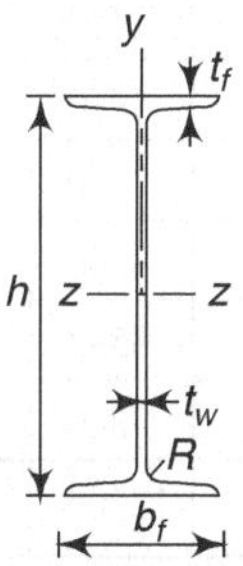


| | | Sectional dimensions | | | | | | Sectional properties | | | | | | | |
|---|---|---|---|---|---|---|---|---|---|---|---|---|---|---|---|
| Designation | Mass N/m | Area (mm²) | $h$ (mm) | $R$ (mm) | $b_f$ (mm) | $t_w$ (mm) | $t_f$ (mm) | $I_z$ (cm⁴) | $I_y$ (cm⁴) | $r_z$ (mm) | $r_y$ (mm) | $Z_z$ (cm³) | $Z_y$ (cm³) | Plastic modulus $Z_{pz}$ (cm³) | Shape factor |
| MB 100 | 89 | 1140 | 100 | 9 | 50 | 4.7 | 7 | 183 | 12.9 | 40 | 10.5 | 36.6 | 5.16 | 41.24 | 1.1268 |
| MB 125 | 133 | 1700 | 125 | 9 | 70 | 5 | 8 | 445 | 38.5 | 51.6 | 15.1 | 71.2 | 11 | 81.85 | 1.1399 |
| MB 150 | 150 | 1910 | 150 | 9 | 75 | 8 | 8 | 718 | 46.8 | 61.3 | 15.7 | 95.7 | 12.5 | 110.48 | 1.1401 |
| MB 175 | 196 | 2500 | 175 | 10 | 85 | 5.8 | 9 | 1260 | 76.7 | 71.3 | 17.6 | 144 | 18 | 166.08 | 1.1422 |
| MB 200 | 242 | 3080 | 200 | 11 | 100 | 5.7 | 10 | 2120 | 137 | 82.9 | 21.1 | 212 | 27.4 | 253.86 | 1.1358 |
| MB 225 | 311 | 3970 | 225 | 12 | 110 | 6.5 | 11.8 | 3440 | 218 | 93.1 | 23.4 | 306 | 39.7 | 348.27 | 1.1385 |
| MB 250 | 373 | 4750 | 250 | 13 | 125 | 6.9 | 12.5 | 5130 | 335 | 104 | 26.5 | 410 | 53.5 | 465.71 | 1.1345 |

*(Contd)*

**TABLE A.1** (*Contd*)

| Designation | Mass N/m | Sectional dimensions: Area (mm²) | h (mm) | R (mm) | $b_f$ (mm) | $t_w$ (mm) | $t_f$ (mm) | Sectional properties: $I_z$ (cm⁴) | $I_y$ (cm⁴) | $r_z$ (mm) | $r_y$ (mm) | $Z_z$ (cm³) | $Z_y$ (cm³) | Plastic modulus $Z_{pz}$ (cm³) | Shape factor |
|---|---|---|---|---|---|---|---|---|---|---|---|---|---|---|---|
| MB 300 | 460 | 5860 | 300 | 14 | 140 | 7.7 | 13.1 | 8990 | 486 | 124 | 28.6 | 599 | 69.5 | 651.74 | 1.1362 |
| MB 350 | 524 | 6670 | 350 | 14 | 140 | 8.1 | 14.2 | 13600 | 538 | 143 | 28.4 | 779 | 76.8 | 889.57 | 1.1421 |
| MB 400 | 615 | 7840 | 400 | 14 | 140 | 8.9 | 16 | 20500 | 622 | 162 | 28.2 | 1020 | 88.9 | 1176.18 | 1.1498 |
| MB 450 | 724 | 9220 | 450 | 15 | 150 | 9.4 | 17.4 | 30400 | 834 | 182 | 30.1 | 1350 | 111 | 1533.36 | 1.15 |
| MB 500 | 869 | 11100 | 500 | 17 | 180 | 10.2 | 17.2 | 45200 | 1370 | 202 | 35.2 | 1810 | 152 | 2074.67 | 1.1471 |
| MB 550 | 1040 | 13200 | 550 | 18 | 190 | 11.2 | 19.3 | 64900 | 1830 | 222 | 37.3 | 2360 | 193 | 2711.98 | 1.1492 |
| MB 600 | 1230 | 15600 | 600 | 20 | 210 | 12 | 20.3 | 91800 | 2650 | 242 | 41.2 | 3060 | 252 | 3510.63 | 1.1471 |

**TABLE A.2** Sectional properties of columns and heavy weight beams

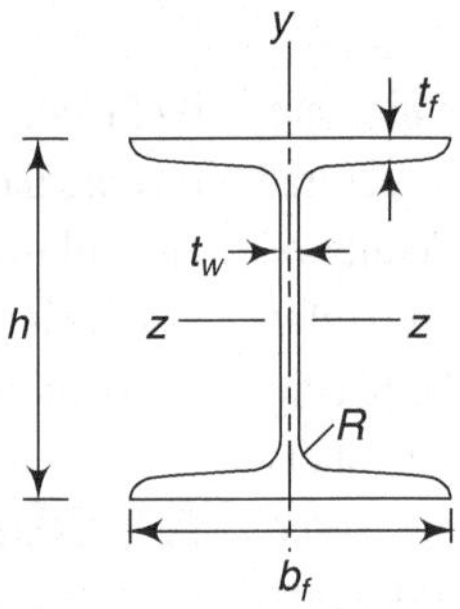


| Designation | Mass (N/m) | Sectional dimensions: Area (mm²) | h (mm) | R (mm) | $b_f$ (mm) | $t_w$ (mm) | $t_f$ (mm) | Sectional properties: $I_z$ (cm⁴) | $I_y$ (cm⁴) | $r_z$ (mm) | $r_y$ (mm) | $Z_z$ (cm³) | $Z_y$ (cm³) | Plastic modulus $Z_{pz}$ (cm³) | Shape factor |
|---|---|---|---|---|---|---|---|---|---|---|---|---|---|---|---|
| *Column sections* | | | | | | | | | | | | | | | |
| SC 100 | 200 | 2550 | 100 | 12 | 100 | 6 | 10 | 436 | 136 | 41.3 | 23.1 | 87.2 | 27.2 | 99.60 | 1.1422 |
| SC 120 | 262 | 3340 | 120 | 12 | 120 | 6.5 | 11 | 842 | 255 | 50.2 | 27.6 | 140 | 42.6 | 159.49 | 1.1392 |
| SC 140 | 333 | 4240 | 140 | 12 | 140 | 7 | 12 | 1470 | 438 | 58.9 | 32.1 | 211 | 62.5 | 238.59 | 1.1307 |
| SC 150 | 371 | 4740 | 152 | 11.7 | 152 | 7.9 | 11.9 | 1970 | 700 | 64.5 | 38.4 | 259 | 91.9 | 285.87 | 1.1038 |
| SC 160 | 419 | 5340 | 160 | 15 | 160 | 8 | 13 | 2420 | 695 | 67.4 | 36.1 | 303 | 86.8 | 341.67 | 1.1276 |
| SC 180 | 505 | 6440 | 180 | 15 | 180 | 8.5 | 14 | 3740 | 1060 | 76.2 | 40.5 | 415 | 117 | 467.42 | 1.1263 |
| SC 200 | 603 | 7680 | 200 | 18 | 200 | 9 | 15 | 5530 | 1530 | 84.8 | 44.6 | 553 | 153 | 620.03 | 1.1212 |
| SC 220 | 704 | 8980 | 220 | 18 | 220 | 9.5 | 16 | 7880 | 2160 | 93.5 | 49 | 716 | 196 | 802.02 | 1.1201 |
| SC 250 | 856 | 10900 | 250 | 23 | 250 | 10 | 17 | 12500 | 3260 | 107 | 54.6 | 997 | 260 | 1106.89 | 1.1102 |
| *Heavy weight beams/columns* | | | | | | | | | | | | | | | |
| HB 150 | 271 | 3450 | 150 | 8 | 150 | 5.4 | 9 | 1460 | 432 | 65 | 35.4 | 194 | 57.6 | 213.87 | 1.1024 |
| HB 200 | 373 | 4750 | 200 | 9 | 200 | 6.1 | 9 | 3600 | 967 | 87.1 | 45.1 | 361 | 96.7 | 394.31 | 1.0923 |
| HB 225 | 431 | 5490 | 225 | 10 | 225 | 6.5 | 9.1 | 5300 | 1350 | 98 | 49.6 | 469 | 120 | 511.55 | 1.0907 |
| HB 250 | 510 | 6500 | 250 | 10 | 250 | 6.9 | 9.7 | 7740 | 1960 | 109 | 54.9 | 619 | 156 | 674.46 | 1.0896 |
| HB 300 | 588 | 7480 | 300 | 11 | 250 | 7.6 | 10.6 | 12600 | 2200 | 130 | 54.1 | 836 | 175 | 914.60 | 1.0940 |
| HB 350 | 674 | 8590 | 350 | 12 | 250 | 8.3 | 11.6 | 19200 | 2450 | 149 | 53.4 | 1090 | 196 | 1202.97 | 1.1036 |
| HB 400 | 774 | 9870 | 400 | 14 | 250 | 9.1 | 12.7 | 28100 | 2730 | 169 | 52.6 | 1400 | 218 | 1548.92 | 1.1064 |
| HB 450 | 872 | 11100 | 450 | 15 | 250 | 9.8 | 13.7 | 39200 | 3000 | 188 | 51.8 | 1740 | 239 | 1931.87 | 1.1103 |

**TABLE A.3** Sectional properties of channel sections

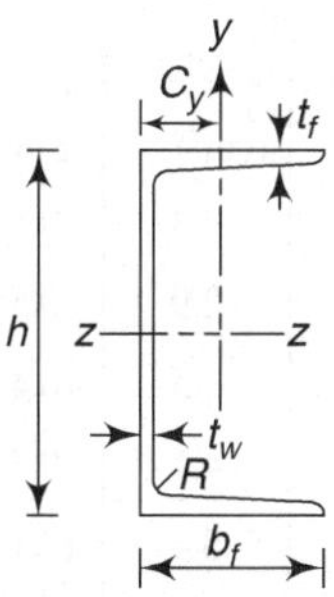


| Designation | Mass (N/m) | Sectional dimensions | | | | | | | Sectional properties | | | | | | | |
|---|---|---|---|---|---|---|---|---|---|---|---|---|---|---|---|---|
| | | Area ($mm^2$) | h (mm) | R (mm) | $b_f$ (mm) | $t_w$ (mm) | $t_f$ (mm) | $C_y$ (mm) | $I_z$ ($cm^4$) | $I_y$ ($cm^4$) | $r_z$ (mm) | $r_y$ (mm) | $Z_z$ ($cm^3$) | $Z_y$ ($cm^3$) | Plastic modulus $Z_{pz}$ ($cm^3$) | Shape factor |
| MC 75 | 71.4 | 910 | 75 | 8.5 | 40 | 4.8 | 7.5 | 13.2 | 78.5 | 12.9 | 29.4 | 11.9 | 20.9 | 4.81 | 24.57 | 1.1756 |
| MC 100 | 95.6 | 1220 | 100 | 9 | 50 | 5 | 7.7 | 15.4 | 192 | 26.7 | 39.7 | 14.8 | 37.3 | 7.71 | 44.48 | 1.1584 |
| MC 125 | 131 | 1670 | 125 | 9.5 | 65 | 5.3 | 8.2 | 19.5 | 425 | 61.1 | 50.5 | 19.1 | 68.1 | 13.4 | 77.88 | 1.1436 |
| MC 150 | 168 | 2130 | 150 | 10 | 75 | 5.7 | 9 | 22 | 788 | 103 | 60.8 | 22 | 105 | 19.5 | 120.00 | 1.1429 |
| MC 175 | 196 | 2490 | 175 | 10.5 | 75 | 6 | 10.2 | 21.9 | 1240 | 122 | 70.4 | 22.1 | 141 | 23 | 161.92 | 1.1484 |
| MC 200 | 223 | 2850 | 200 | 11 | 75 | 6.2 | 11.4 | 22 | 1830 | 141 | 80.2 | 22.2 | 181 | 26.4 | 209.92 | 1.1598 |
| MC 225 | 261 | 3330 | 225 | 12 | 80 | 6.5 | 12.4 | 23.1 | 2710 | 188 | 90.2 | 23.7 | 241 | 33 | 276.03 | 1.1453 |
| MC 250 | 306 | 3900 | 250 | 12 | 80 | 7.2 | 14.1 | 23 | 3880 | 211 | 99.2 | 23.7 | 307 | 38.5 | 354.65 | 1.1552 |
| MC 300 | 363 | 4630 | 300 | 13 | 90 | 7.8 | 13.6 | 23.5 | 6420 | 313 | 118 | 26 | 428 | 47.1 | 495.67 | 1.1581 |
| MC 350 | 427 | 5440 | 350 | 14 | 100 | 8.3 | 13.5 | 24.4 | 10000 | 434 | 136 | 28.2 | 576 | 57.3 | 670.76 | 1.1645 |
| MC 400 | 501 | 6380 | 400 | 15 | 100 | 8.8 | 15.3 | 24.2 | 15200 | 508 | 154 | 28.2 | 760 | 67 | 888.79 | 1.1695 |

**TABLE A.4** Sectional properties of equal leg angles

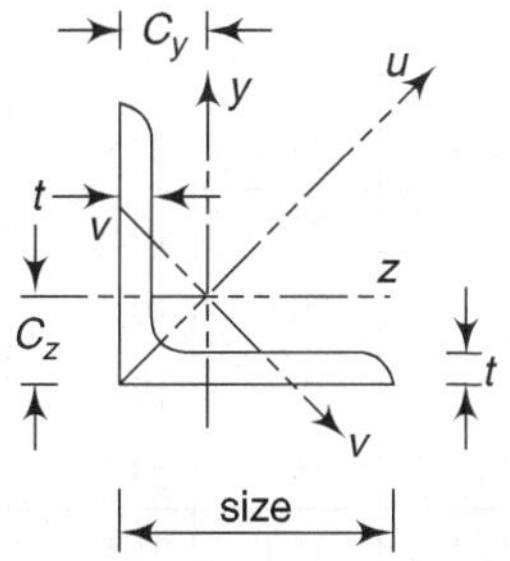


| Designation | Mass (N/m) | Sectional dimensions | | | Sectional properties | | | | | | | | |
|---|---|---|---|---|---|---|---|---|---|---|---|---|---|
| | | Area ($mm^2$) | $C_z$ (mm) | $C_y$ (mm) | $I_z$ ($cm^4$) | $I_y$ ($cm^4$) | $r_z$ (mm) | $r_y$ (mm) | $r_{u(max)}$ (mm) | $r_{v(min)}$ (mm) | $Z_z$ ($cm^3$) | $Z_y$ ($cm^3$) | $Z_{pz}$ ($cm^3$) |
| L20 20 × 3 | 9 | 112 | 5.9 | 5.9 | 0.4 | 0.4 | 5.8 | 5.8 | 7.3 | 3.7 | 0.3 | 0.3 | 0.52 |
| × 4 | 11 | 145 | 6.3 | 6.3 | 0.5 | 0.5 | 5.8 | 5.8 | 7.2 | 3.7 | 0.4 | 0.4 | 0.67 |
| L25 25 × 3 | 11 | 141 | 7.1 | 7.1 | 0.8 | 0.8 | 7.3 | 7.3 | 9.3 | 4.7 | 0.4 | 0.4 | 0.84 |
| × 4 | 14 | 184 | 7.5 | 7.5 | 1 | 1 | 7.3 | 7.3 | 9.1 | 4.7 | 0.6 | 0.6 | 1.08 |
| × 5 | 18 | 225 | 7.9 | 7.9 | 1.2 | 1.2 | 7.2 | 7.2 | 9.1 | 4.7 | 0.7 | 0.7 | 1.31 |
| L30 30 × 3 | 14 | 173 | 8.3 | 8.3 | 1.4 | 1.4 | 8.9 | 8.9 | 11.3 | 5.7 | 0.6 | 0.6 | 1.23 |
| × 4 | 18 | 226 | 8.7 | 8.7 | 1.8 | 1.8 | 8.9 | 8.9 | 11.2 | 5.7 | 0.8 | 0.8 | 1.59 |
| × 5 | 22 | 277 | 9.2 | 9.2 | 2.1 | 2.1 | 8.8 | 8.8 | 11.1 | 5.7 | 1.0 | 1.0 | 1.93 |

(*Contd*)

**TABLE A.4** *(Contd)*

| Designation | Mass | Sectional dimensions | | | Sectional properties | | | | | | | | |
|---|---|---|---|---|---|---|---|---|---|---|---|---|---|
| | (N/m) | Area $(mm^2)$ | $C_z$ (mm) | $C_y$ (mm) | $I_z$ $(cm^4)$ | $I_y$ $(cm^4)$ | $r_z$ (mm) | $r_y$ (mm) | $r_{u(max)}$ (mm) | $r_{v(min)}$ (mm) | $Z_z$ $(cm^3)$ | $Z_y$ $(cm^3)$ | $Z_{pz}$ $(cm^3)$ |
| L35 35 × 3 | 16 | 203 | 9.5 | 9.5 | 2.3 | 2.3 | 10.5 | 10.5 | 13.3 | 6.7 | 0.9 | 0.9 | 1.69 |
| × 4 | 21 | 266 | 10.0 | 10.0 | 2.9 | 2.9 | 10.5 | 10.5 | 13.2 | 6.7 | 1.2 | 1.2 | 2.20 |
| × 5 | 26 | 327 | 10.4 | 10.4 | 3.5 | 3.5 | 10.4 | 10.4 | 13.1 | 6.7 | 1.4 | 1.4 | 2.68 |
| × 6 | 30 | 386 | 10.8 | 10.8 | 4.1 | 4.1 | 10.3 | 10.3 | 12.9 | 6.7 | 1.7 | 1.7 | 3.14 |
| L40 40 × 3 | 18 | 234 | 10.8 | 10.8 | 3.4 | 3.4 | 12.1 | 12.1 | 15.4 | 7.7 | 1.2 | 1.2 | 2.23 |
| × 4 | 24 | 307 | 11.2 | 11.2 | 4.5 | 4.5 | 12.1 | 12.1 | 15.3 | 7.7 | 1.6 | 1.6 | 2.91 |
| × 5 | 30 | 378 | 11.6 | 11.6 | 5.4 | 5.4 | 12.0 | 12.0 | 15.1 | 7.7 | 1.9 | 1.9 | 3.56 |
| × 6 | 35 | 447 | 12.0 | 12.0 | 6.3 | 6.3 | 11.9 | 11.9 | 15.0 | 7.7 | 2.3 | 2.3 | 4.18 |
| L45 45 × 3 | 21 | 264 | 12.0 | 12.0 | 5 | 5 | 13.8 | 13.8 | 17.4 | 8.7 | 1.5 | 1.5 | 2.85 |
| × 4 | 27 | 347 | 12.5 | 12.5 | 6.5 | 6.5 | 13.7 | 13.7 | 17.3 | 8.7 | 2 | 2 | 3.72 |
| × 5 | 34 | 428 | 12.9 | 12.9 | 7.9 | 7.9 | 13.6 | 13.6 | 17.2 | 8.7 | 2.5 | 2.5 | 4.56 |
| × 6 | 40 | 507 | 13.3 | 13.3 | 9.2 | 9.2 | 13.5 | 13.5 | 17.0 | 8.7 | 2.9 | 2.9 | 5.37 |
| L50 50 × 3 | 23 | 295 | 13.2 | 13.2 | 6.9 | 6.9 | 15.3 | 15.3 | 19.4 | 9.7 | 1.9 | 1.9 | 3.54 |
| × 4 | 30 | 388 | 13.7 | 13.7 | 9.1 | 9.1 | 15.3 | 15.3 | 19.3 | 9.7 | 2.5 | 2.5 | 4.63 |
| × 5 | 38 | 479 | 14.1 | 14.1 | 11 | 11 | 15.2 | 15.2 | 19.2 | 9.7 | 3.1 | 3.1 | 5.68 |
| × 6 | 45 | 568 | 14.5 | 14.5 | 12.9 | 12.9 | 15.1 | 15.1 | 19.0 | 9.6 | 3.6 | 3.6 | 6.70 |
| L55 55 × 5 | 41 | 527 | 15.3 | 15.3 | 14.7 | 14.7 | 16.7 | 16.7 | 21.1 | 10.6 | 3.7 | 3.7 | 6.93 |
| × 6 | 49 | 626 | 15.7 | 15.7 | 17.3 | 17.3 | 16.6 | 16.6 | 21.0 | 10.6 | 4.4 | 4.4 | 8.19 |
| × 8 | 64 | 818 | 16.5 | 16.5 | 22 | 22 | 16.4 | 16.4 | 20.7 | 10.6 | 5.7 | 5.7 | 10.58 |
| × 10 | 79 | 1000 | 17.2 | 17.2 | 26.3 | 26.3 | 16.2 | 16.2 | 20.3 | 10.6 | 7 | 7 | 12.83 |
| L60 60 × 5 | 45 | 575 | 16.5 | 16.5 | 19.2 | 19.2 | 18.2 | 18.2 | 23.1 | 11.6 | 4.4 | 4.4 | 8.31 |
| × 6 | 54 | 684 | 16.9 | 16.9 | 22.6 | 22.6 | 18.2 | 18.2 | 22.9 | 11.5 | 5.2 | 5.2 | 9.82 |
| × 8 | 70 | 896 | 17.7 | 17.7 | 29 | 29 | 18.0 | 18.0 | 22.7 | 11.5 | 6.8 | 6.8 | 12.72 |
| × 10 | 86 | 1100 | 18.5 | 18.5 | 34.8 | 34.8 | 17.8 | 17.8 | 22.3 | 11.5 | 8.4 | 8.4 | 15.46 |
| L65 65 × 5 | 49 | 625 | 17.7 | 17.7 | 24.7 | 24.7 | 19.9 | 19.9 | 25.1 | 12.6 | 5.2 | 5.2 | 9.81 |
| × 6 | 58 | 744 | 18.1 | 18.1 | 29.1 | 29.1 | 19.8 | 19.8 | 25.0 | 12.6 | 6.2 | 6.2 | 11.61 |
| × 8 | 77 | 976 | 18.9 | 18.9 | 37.4 | 37.4 | 19.6 | 19.6 | 24.7 | 12.5 | 8.1 | 8.1 | 15.06 |
| × 10 | 94 | 1200 | 19.7 | 19.7 | 45 | 45 | 19.4 | 19.4 | 24.4 | 12.5 | 9.9 | 9.9 | 18.34 |
| L70 70 × 5 | 53 | 677 | 18.9 | 18.9 | 31.1 | 31.1 | 21.5 | 21.5 | 27.1 | 13.6 | 6.1 | 6.1 | 11.44 |
| × 6 | 63 | 806 | 19.4 | 19.4 | 36.8 | 36.8 | 21.4 | 21.4 | 27.0 | 13.6 | 7.3 | 7.3 | 13.54 |
| × 8 | 83 | 1060 | 20.2 | 20.2 | 47.4 | 47.4 | 21.2 | 21.2 | 26.7 | 13.5 | 9.5 | 9.5 | 17.60 |
| × 10 | 102 | 1300 | 21.0 | 21.0 | 57.2 | 57.2 | 21.0 | 21.0 | 26.4 | 13.5 | 11.7 | 11.7 | 21.46 |
| L75 75 × 5 | 57 | 727 | 20.2 | 20.2 | 38.7 | 38.7 | 23.1 | 23.1 | 29.2 | 14.6 | 7.1 | 7.1 | 13.19 |
| × 6 | 68 | 866 | 20.6 | 20.6 | 45.7 | 45.7 | 23.0 | 23.0 | 29.1 | 14.6 | 8.4 | 8.4 | 15.63 |
| × 8 | 89 | 1140 | 21.4 | 21.4 | 59 | 59 | 22.8 | 22.8 | 28.8 | 14.5 | 11 | 11 | 20.34 |
| × 10 | 110 | 1400 | 22.2 | 22.2 | 71.4 | 71.4 | 22.6 | 22.6 | 28.4 | 14.5 | 13.5 | 13.5 | 24.84 |
| L80 80 × 6 | 73 | 929 | 21.8 | 21.8 | 56 | 56 | 24.6 | 24.6 | 31.1 | 15.6 | 9.6 | 9.6 | 17.86 |
| × 8 | 96 | 1220 | 22.7 | 22.7 | 72.5 | 72.5 | 24.4 | 24.4 | 30.8 | 15.5 | 12.6 | 12.6 | 23.28 |
| × 10 | 118 | 1500 | 23.4 | 23.4 | 87.7 | 87.7 | 24.1 | 24.1 | 30.4 | 15.5 | 15.5 | 15.5 | 28.47 |
| × 12 | 140 | 1780 | 24.2 | 24.2 | 102 | 102 | 23.9 | 23.9 | 30.1 | 15.4 | 18.3 | 18.3 | 33.44 |

*(Contd)*

**TABLE A.4** *(Contd)*

| Designation | Mass (N/m) | Sectional dimensions | | | Sectional properties | | | | | | | | |
|---|---|---|---|---|---|---|---|---|---|---|---|---|---|
| | | Area ($mm^2$) | $C_z$ (mm) | $C_y$ (mm) | $I_z$ ($cm^4$) | $I_y$ ($cm^4$) | $r_z$ (mm) | $r_y$ (mm) | $r_{u(max)}$ (mm) | $r_{v(min)}$ (mm) | $Z_z$ ($cm^3$) | $Z_y$ ($cm^3$) | $Z_{pz}$ ($cm^3$) |
| L90 90 × 6 | 82 | 1050 | 24.2 | 24.2 | 80.1 | 80.1 | 27.7 | 27.7 | 35.0 | 17.5 | 12.2 | 12.2 | 22.78 |
| × 8 | 108 | 1380 | 25.1 | 25.1 | 104 | 104 | 27.5 | 27.5 | 34.7 | 17.5 | 16 | 16 | 29.76 |
| × 10 | 134 | 1700 | 25.9 | 25.9 | 127 | 127 | 27.3 | 27.3 | 34.4 | 17.4 | 19.8 | 19.8 | 36.47 |
| × 12 | 158 | 2020 | 26.6 | 26.6 | 148 | 148 | 27.1 | 27.1 | 34.1 | 17.4 | 23.3 | 23.3 | 42.93 |
| L100 100 × 6 | 92 | 1170 | 26.7 | 26.7 | 111 | 111 | 30.9 | 30.9 | 39.1 | 19.5 | 15.2 | 15.2 | 28.30 |
| × 8 | 121 | 1540 | 27.6 | 27.6 | 145 | 145 | 30.7 | 30.7 | 38.8 | 19.5 | 20 | 20 | 37.05 |
| × 10 | 149 | 1900 | 28.4 | 28.4 | 177 | 177 | 30.5 | 30.5 | 38.5 | 19.4 | 24.7 | 24.7 | 45.48 |
| × 12 | 177 | 2260 | 29.2 | 29.2 | 207 | 207 | 30.3 | 30.3 | 38.2 | 19.4 | 29.2 | 29.2 | 53.61 |
| L110 110 × 8 | 134 | 1710 | 30.0 | 30.0 | 197 | 197 | 34.0 | 34.0 | 42.8 | 21.8 | 24.6 | 24.6 | 45.13 |
| × 10 | 166 | 2110 | 30.9 | 30.9 | 240 | 240 | 33.7 | 33.7 | 42.5 | 21.6 | 30.4 | 30.4 | 55.48 |
| × 12 | 197 | 2510 | 31.7 | 31.7 | 281 | 281 | 33.5 | 33.5 | 42.2 | 21.5 | 35.9 | 35.9 | 65.50 |
| × 16 | 257 | 3280 | 33.2 | 33.2 | 357 | 357 | 33.0 | 33.0 | 41.5 | 21.4 | 46.5 | 46.5 | 84.62 |
| L130 130 × 8 | 159 | 2030 | 35.0 | 35.0 | 331 | 331 | 40.4 | 40.4 | 51.0 | 25.9 | 34.9 | 34.9 | 63.69 |
| × 10 | 197 | 2510 | 35.9 | 35.9 | 405 | 405 | 40.2 | 40.2 | 50.7 | 25.7 | 43.1 | 43.1 | 78.48 |
| × 12 | 235 | 2990 | 36.7 | 36.7 | 476 | 476 | 39.9 | 39.9 | 50.3 | 25.6 | 51 | 51 | 92.86 |
| × 16 | 307 | 3920 | 38.2 | 38.2 | 609 | 609 | 39.4 | 39.4 | 49.7 | 25.4 | 66.3 | 66.3 | 120.48 |
| L150 150 × 10 | 229 | 2920 | 40.8 | 40.8 | 634 | 634 | 46.6 | 46.6 | 58.7 | 29.8 | 58 | 58 | 105.48 |
| × 12 | 273 | 3480 | 41.6 | 41.6 | 746 | 746 | 46.3 | 46.3 | 58.4 | 29.7 | 68.8 | 68.8 | 125.03 |
| × 16 | 358 | 4560 | 43.1 | 43.1 | 959 | 959 | 45.8 | 45.8 | 57.7 | 29.4 | 89.7 | 89.7 | 162.74 |
| × 20 | 441 | 5620 | 44.6 | 44.6 | 1160 | 1160 | 45.3 | 45.3 | 57.1 | 29.3 | 110 | 110 | 198.73 |
| L200 200 × 12 | 369 | 4690 | 53.9 | 53.9 | 1830 | 1830 | 62.4 | 62.4 | 78.7 | 39.9 | 125 | 125 | 226.44 |
| × 16 | 485 | 6180 | 55.6 | 55.6 | 2370 | 2370 | 61.9 | 61.9 | 78.0 | 39.6 | 164 | 164 | 296.37 |
| × 20 | 600 | 7640 | 57.1 | 57.1 | 2880 | 2880 | 61.4 | 61.4 | 77.3 | 39.3 | 201 | 201 | 363.80 |
| × 25 | 739 | 9410 | 59.0 | 59.0 | 3470 | 3470 | 60.7 | 60.7 | 76.1 | 39.1 | 246 | 246 | 444.82 |

**TABLE A.5** Sectional properties of unequal leg angles

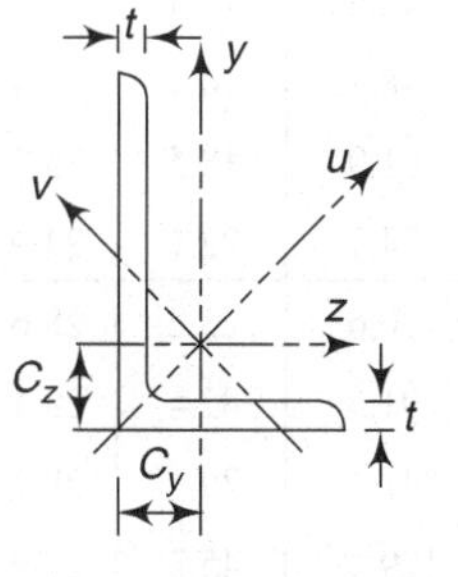


| Designation | Mass (N/m) | Sectional dimensions | | | Sectional properties | | | | | | | | |
|---|---|---|---|---|---|---|---|---|---|---|---|---|---|
| | | Area ($mm^2$) | $C_z$ (mm) | $C_y$ (mm) | $I_z$ ($cm^4$) | $I_y$ ($cm^4$) | $r_z$ (mm) | $r_y$ (mm) | $r_{u(max)}$ (mm) | $r_{v(min)}$ (mm) | $Z_z$ ($cm^3$) | $Z_y$ ($cm^3$) | $Z_{pz}$ ($cm^3$) |
| L30 20 × 3 | 11 | 141 | 9.8 | 4.9 | 1.2 | 0.4 | 9.2 | 5.4 | 9.9 | 4.1 | 0.6 | 0.3 | 1.15 |
| × 4 | 14 | 184 | 10.2 | 5.3 | 1.5 | 0.5 | 9.2 | 5.4 | 9.8 | 4.1 | 0.8 | 0.4 | 1.48 |
| × 5 | 18 | 225 | 10.6 | 5.7 | 1.9 | 0.6 | 9.1 | 5.3 | 9.7 | 4.1 | 1.0 | 0.4 | 1.78 |

*(Contd)*

**TABLE A.5** *(Contd)*

| Designation | Mass (N/m) | Sectional dimensions | | | Sectional properties | | | | | | | | |
|---|---|---|---|---|---|---|---|---|---|---|---|---|---|
| | | Area ($mm^2$) | $C_z$ (mm) | $C_y$ (mm) | $I_z$ ($cm^4$) | $I_y$ ($cm^4$) | $r_z$ (mm) | $r_y$ (mm) | $r_{u(max)}$ (mm) | $r_{v(min)}$ (mm) | $Z_z$ ($cm^3$) | $Z_y$ ($cm^3$) | $Z_{pz}$ ($cm^3$) |
| L40 25 × 3 | 15 | 188 | 13 | 5.7 | 3 | 0.9 | 12.5 | 6.8 | 13.3 | 5.2 | 1.1 | 0.5 | 2.06 |
| × 4 | 19 | 246 | 13.5 | 6.2 | 3.8 | 1.1 | 12.5 | 6.8 | 13.2 | 5.2 | 1.4 | 0.6 | 2.67 |
| × 5 | 24 | 302 | 13.9 | 6.6 | 4.6 | 1.4 | 12.4 | 6.7 | 13.1 | 5.2 | 1.8 | 0.7 | 3.25 |
| × 6 | 28 | 356 | 14.3 | 6.9 | 5.4 | 1.6 | 12.3 | 6.6 | 12.9 | 5.2 | 2.1 | 0.9 | 3.80 |
| L45 30 × 3 | 17 | 218 | 14.2 | 6.9 | 4.4 | 1.5 | 14.2 | 8.4 | 15.2 | 6.3 | 1.4 | 0.7 | 2.67 |
| × 4 | 22 | 286 | 14.7 | 7.3 | 5.7 | 2 | 14.1 | 8.4 | 15.1 | 6.3 | 1.9 | 0.9 | 3.48 |
| × 5 | 28 | 352 | 15.1 | 7.7 | 6.9 | 2.4 | 14 | 8.3 | 15 | 6.3 | 2.3 | 1.1 | 4.25 |
| × 6 | 33 | 416 | 15.5 | 8.1 | 8 | 2.8 | 13.9 | 8.2 | 14.9 | 6.3 | 2.7 | 1.3 | 4.98 |
| L50 30 × 3 | 18 | 234 | 16.3 | 6.6 | 5.9 | 1.6 | 15.9 | 8.3 | 16.7 | 6.5 | 1.7 | 0.7 | 3.23 |
| × 4 | 24 | 307 | 16.8 | 7 | 7.7 | 2.1 | 15.8 | 8.2 | 16.6 | 6.3 | 2.3 | 0.9 | 4.22 |
| × 5 | 30 | 378 | 17.2 | 7.4 | 9.3 | 2.5 | 15.7 | 8.1 | 16.5 | 6.3 | 2.8 | 1.1 | 5.16 |
| × 6 | 35 | 447 | 17.6 | 7.8 | 10.9 | 2.9 | 15.6 | 8 | 16.4 | 6.3 | 3.4 | 1.3 | 6.05 |
| L60 40 × 5 | 37 | 476 | 19.5 | 9.6 | 16.9 | 6 | 18.9 | 11.2 | 20.2 | 8.5 | 4.2 | 2 | 7.78 |
| × 6 | 44 | 565 | 19.9 | 10 | 19.9 | 7 | 18.8 | 11.1 | 20.1 | 8.5 | 5 | 2.3 | 9.17 |
| × 8 | 58 | 737 | 20.7 | 10.8 | 25.4 | 8.8 | 18.6 | 11 | 19.8 | 8.4 | 6.5 | 3 | 11.81 |
| L65 45 × 5 | 41 | 526 | 20.7 | 10.8 | 22.1 | 8.6 | 20.5 | 12.8 | 22.2 | 9.6 | 5 | 2.5 | 9.28 |
| × 6 | 49 | 625 | 21.1 | 11.2 | 26 | 10.1 | 20.4 | 12.7 | 22.1 | 9.5 | 5.9 | 3 | 10.96 |
| × 8 | 64 | 817 | 21.9 | 12 | 33.2 | 12.8 | 20.2 | 12.5 | 21.8 | 9.5 | 7.7 | 3.9 | 14.15 |
| L70 45 × 5 | 43 | 552 | 22.7 | 10.4 | 27.2 | 8.8 | 22.2 | 12.6 | 23.6 | 9.6 | 5.7 | 2.5 | 10.63 |
| × 6 | 52 | 656 | 23.2 | 10.9 | 32 | 10.3 | 22.1 | 12.5 | 23.5 | 9.6 | 6.8 | 3 | 12.56 |
| × 8 | 67 | 858 | 24 | 11.6 | 41 | 13.1 | 21.9 | 12.4 | 23.2 | 9.5 | 8.9 | 3.9 | 16.24 |
| × 10 | 83 | 1050 | 24.8 | 12.4 | 49.3 | 15.6 | 21.6 | 12.2 | 22.9 | 9.5 | 10.9 | 4.8 | 19.69 |
| L75 50 × 5 | 47 | 602 | 23.9 | 11.6 | 34.1 | 12.2 | 23.8 | 14.2 | 25.6 | 10.7 | 6.7 | 3.2 | 12.38 |
| × 6 | 56 | 716 | 24.4 | 12 | 40.3 | 14.3 | 23.7 | 14.1 | 25.5 | 10.7 | 8 | 3.8 | 14.64 |
| × 8 | 74 | 938 | 25.2 | 12.8 | 51.8 | 18.3 | 28.5 | 14 | 25.2 | 10.6 | 10.4 | 4.9 | 18.98 |
| × 10 | 90 | 1150 | 26 | 13.6 | 62.2 | 21.8 | 23.3 | 13.8 | 24.9 | 10.6 | 12.7 | 6 | 23.06 |
| L80 50 × 5 | 49 | 627 | 26 | 11.2 | 40.6 | 12.3 | 25.5 | 14 | 27 | 10.7 | 7.5 | 3.2 | 13.91 |
| × 6 | 59 | 746 | 26.4 | 11.6 | 48 | 14.4 | 25.4 | 13.9 | 26.9 | 10.7 | 9 | 3.8 | 16.46 |
| × 8 | 77 | 978 | 27.3 | 12.4 | 61.9 | 18.5 | 25.2 | 13.7 | 26.6 | 10.6 | 11.7 | 4.9 | 21.37 |
| × 10 | 94 | 1200 | 28.1 | 13.2 | 74.7 | 22.1 | 24.9 | 13.6 | 26.3 | 10.6 | 14.4 | 6 | 26.00 |
| L90 60 × 6 | 68 | 865 | 28.7 | 13.9 | 70.6 | 25.2 | 28.6 | 17.1 | 30.7 | 12.8 | 11.5 | 5.5 | 21.38 |
| × 8 | 89 | 1140 | 29.6 | 14.8 | 91.5 | 32.4 | 28.4 | 16.9 | 30.4 | 12.8 | 15.1 | 7.2 | 27.85 |
| × 10 | 110 | 1400 | 30.4 | 15.5 | 111 | 39.1 | 28.1 | 16.7 | 30.1 | 12.7 | 18.6 | 8.8 | 34.00 |
| × 12 | 130 | 1660 | 31.2 | 16.3 | 129 | 45.2 | 27.9 | 16.5 | 29.8 | 12.7 | 22 | 10.3 | 39.85 |
| L100 65 × 6 | 75 | 955 | 31.9 | 14.7 | 96.7 | 32.4 | 31.8 | 18.4 | 34 | 13.9 | 14.2 | 6.4 | 26.42 |
| × 8 | 99 | 1260 | 32.8 | 15.5 | 126 | 41.9 | 31.6 | 18.3 | 33.8 | 13.9 | 18.7 | 8.5 | 34.48 |
| × 10 | 122 | 1550 | 33.7 | 16.3 | 153 | 50.7 | 31.4 | 18.1 | 33.5 | 13.8 | 23.1 | 10.4 | 42.19 |
| L100 75 × 6 | 80 | 1010 | 30.1 | 17.8 | 101 | 48.7 | 31.5 | 21.9 | 35 | 15.9 | 14.4 | 8.5 | 27.32 |
| × 8 | 105 | 1340 | 31 | 18.7 | 132 | 63.3 | 31.4 | 21.8 | 34.8 | 15.9 | 19.1 | 11.2 | 35.68 |
| × 10 | 130 | 1650 | 31.9 | 19.5 | 160 | 76.9 | 31.2 | 21.6 | 34.5 | 15.8 | 23.6 | 13 | 43.69 |
| × 12 | 154 | 1950 | 32.7 | 20.3 | 188 | 89.5 | 31 | 21.4 | 34.2 | 15.8 | 27.9 | 16.3 | 51.36 |

*(Contd)*

**TABLE A.5** *(Contd)*

| Designation | Mass (N/m) | Sectional dimensions: Area ($mm^2$) | $C_z$ (mm) | $C_y$ (mm) | Sectional properties: $I_z$ ($cm^4$) | $I_y$ ($cm^4$) | $r_z$ (mm) | $r_y$ (mm) | $r_{u(max)}$ (mm) | $r_{v(min)}$ (mm) | $Z_z$ ($cm^3$) | $Z_y$ ($cm^3$) | $Z_{pz}$ ($cm^3$) |
|---|---|---|---|---|---|---|---|---|---|---|---|---|---|
| L125 75 × 6 | 92 | 1170 | 40.5 | 15.9 | 188 | 51.6 | 40.1 | 21 | 42.3 | 16.2 | 22.2 | 8.7 | 40.93 |
| × 8 | 121 | 1540 | 41.5 | 16.8 | 246 | 67.2 | 40 | 20.9 | 42.1 | 16.1 | 29.4 | 11.5 | 53.63 |
| × 10 | 149 | 1900 | 42.4 | 17.6 | 300 | 81.6 | 39.7 | 20.7 | 41.8 | 16.1 | 36.5 | 14.2 | 65.88 |
| L125 95 × 6 | 101 | 1290 | 37.2 | 22.4 | 205 | 103 | 39.9 | 28.3 | 44.3 | 20.7 | 23.4 | 14.3 | 43.33 |
| × 8 | 134 | 1700 | 38 | 23.2 | 268 | 135 | 39.7 | 28.1 | 44.1 | 20.5 | 30.9 | 18.8 | 56.83 |
| × 10 | 165 | 2110 | 38.9 | 24 | 328 | 164 | 39.5 | 27.9 | 43.8 | 20.4 | 38.1 | 23.1 | 69.88 |
| × 12 | 197 | 2500 | 39.7 | 24.8 | 385 | 192 | 39.2 | 27.7 | 43.5 | 20.3 | 45.1 | 27.3 | 82.48 |
| L150 75 × 8 | 137 | 1750 | 52.4 | 15.4 | 410 | 71.1 | 48.8 | 20.2 | 49.9 | 16.2 | 42 | 11.9 | 74.08 |
| × 10 | 170 | 2160 | 53.3 | 16.2 | 502 | 86.3 | 48.2 | 20 | 49.6 | 16.1 | 51.9 | 14.7 | 91.19 |
| × 12 | 202 | 2570 | 54.2 | 17 | 590 | 100 | 47.9 | 19.8 | 49.3 | 16 | 61.6 | 17.3 | 107.76 |
| L150 115 × 8 | 163 | 2070 | 44.8 | 27.6 | 474 | 244 | 47.8 | 34.3 | 53.3 | 25 | 45.1 | 28 | 82.88 |
| × 10 | 201 | 2570 | 45.7 | 28.4 | 582 | 299 | 47.6 | 34.1 | 53.1 | 24.8 | 55.8 | 34.5 | 102.19 |
| × 12 | 240 | 3050 | 46.5 | 29.2 | 685 | 351 | 47.4 | 33.9 | 52.8 | 24.7 | 66.2 | 40.8 | 120.96 |
| × 16 | 314 | 4000 | 48.1 | 30.7 | 878 | 447 | 46.9 | 33.4 | 52.1 | 24.4 | 86.2 | 53 | 177.24 |
| L200 100 × 10 | 229 | 2920 | 69.8 | 20.3 | 1230 | 215 | 64.8 | 27.1 | 66.8 | 21.7 | 94.3 | 26.9 | 165.25 |
| × 12 | 273 | 3480 | 70.7 | 21.1 | 1450 | 251 | 64.6 | 26.9 | 66.5 | 21.6 | 112 | 31.9 | 196.03 |
| × 16 | 358 | 4570 | 72.3 | 22.7 | 1870 | 320 | 64 | 26.6 | 65.9 | 21.3 | 147 | 41.3 | 255.42 |
| L200 150 × 10 | 269 | 3430 | 60.2 | 35.5 | 1410 | 689 | 64.1 | 44.8 | 71 | 32.8 | 101 | 60.2 | 184.00 |

**TABLE A.6** Sectional properties of rolled steel tee sections

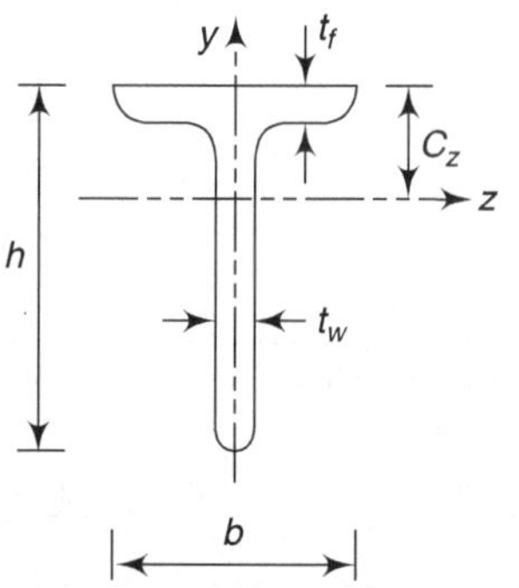


| Designation | Weight (N/m) | Sectional area ($mm^2$) | Depth of section $h$ (mm) | Width of flange $b$ (mm) | Thickness of flange $t_f$ (mm) | Thickness of web $t_w$ (mm) | Centre of gravity $C_z$ (mm) | Moment of inertia: $I_z$ ($cm^4$) | $I_y$ ($cm^4$) | Radius of gyration: $r_z$ (mm) | $r_y$ (mm) | Moduli of section: $Z_z$ ($cm^3$) | $Z_y$ ($cm^3$) |
|---|---|---|---|---|---|---|---|---|---|---|---|---|---|
| ISNT 20 | 9 | 113 | 20 | 20 | 3 | 3 | 6 | 0.4 | 0.2 | 5.9 | 3.9 | 0.3 | 0.2 |
| ISNT 30 | 14 | 175 | 30 | 30 | 3 | 3 | 8.3 | 1.4 | 0.6 | 8.9 | 5.7 | 0.6 | 0.4 |
| ISNT 40 | 35 | 448 | 40 | 40 | 6 | 6 | 12 | 6.3 | 3 | 11.8 | 8.2 | 2.2 | 1.5 |
| ISNT 50 | 45 | 570 | 50 | 50 | 6 | 6 | 14.4 | 12.7 | 5.9 | 15 | 10.2 | 3.6 | 2.4 |
| ISNT 60 | 54 | 690 | 60 | 60 | 6 | 6 | 16.7 | 22.5 | 10.1 | 18.1 | 12.1 | 5.2 | 3.4 |
| ISNT 80 | 96 | 1225 | 80 | 80 | 8 | 8 | 22.3 | 71.2 | 32.3 | 24.1 | 16.2 | 12.3 | 8.1 |
| ISNT 100 | 150 | 1910 | 100 | 100 | 10 | 10 | 27.9 | 173.8 | 79.9 | 30.2 | 20.5 | 24.1 | 16 |
| ISNT 150 | 228 | 2908 | 150 | 150 | 10 | 10 | 39.5 | 608.8 | 257.5 | 45.6 | 30.3 | 54.6 | 35.7 |

*(Contd)*

**TABLE A.6** *(Contd)*

| Designation | Weight (N/m) | Sectional area (mm²) | Depth of section h (mm) | Width of flange b (mm) | Thickness of flange $t_f$ (mm) | Thickness of web $t_w$ (mm) | Centre of gravity $C_z$ (mm) | Moment of inertia | | Radius of gyration | | Moduli of section | |
|---|---|---|---|---|---|---|---|---|---|---|---|---|---|
| | | | | | | | | $I_z$ (cm⁴) | $I_y$ (cm⁴) | $r_z$ (mm) | $r_y$ (mm) | $Z_z$ (cm³) | $Z_y$ (cm³) |
| ISHT 75 | 153 | 1949 | 75 | 150 | 9 | 8.4 | 16.2 | 96.2 | 230.2 | 22.2 | 34.4 | 16.4 | 30.1 |
| ISHT 100 | 200 | 2547 | 100 | 200 | 9 | 7.8 | 19.1 | 193.8 | 497.3 | 27.6 | 44.2 | 24 | 49.3 |
| ISHT 125 | 274 | 3485 | 125 | 250 | 9.7 | 8.8 | 23.7 | 415.4 | 1005.8 | 34.5 | 53.7 | 41 | 79.9 |
| ISHT 150 | 294 | 3742 | 150 | 250 | 10.6 | 7.6 | 26.6 | 573.7 | 1096.8 | 39.2 | 54.1 | 46.5 | 87.7 |
| ISST 100 | 81 | 1037 | 100 | 50 | 10 | 5.8 | 30.3 | 99 | 9.6 | 30.9 | 9.6 | 14.2 | 3.8 |
| ISST 150 | 157 | 1996 | 150 | 75 | 11.6 | 8 | 37.5 | 450.2 | 37 | 47.5 | 13.6 | 43.9 | 9.9 |
| ISST 200 | 284 | 3622 | 200 | 165 | 12.5 | 8 | 47.8 | 1267.8 | 358.2 | 59.2 | 31.5 | 83.3 | 43.4 |
| ISST 250 | 375 | 4775 | 250 | 180 | 12.1 | 9.2 | 64 | 2774.4 | 532 | 76.2 | 33.4 | 149.2 | 59.1 |
| ISLT 50 | 40 | 511 | 50 | 50 | 6.4 | 4 | 11.9 | 9.9 | 6.4 | 13.9 | 11.2 | 2.6 | 2.5 |
| ISLT 75 | 71 | 904 | 75 | 80 | 6.8 | 4.8 | 17.2 | 41.9 | 27.6 | 21.5 | 17.5 | 7.2 | 6.9 |
| ISLT 100 | 127 | 1616 | 100 | 100 | 10.8 | 5.7 | 21.3 | 116.6 | 75 | 26.9 | 21.5 | 14.8 | 15 |
| ISJT 75 | 35 | 450 | 75 | 50 | 4.6 | 3 | 20 | 24.8 | 4.6 | 23.5 | 10.1 | 4.5 | 1.8 |
| ISJT 87.5 | 40 | 514 | 87.5 | 50 | 4.8 | 3.2 | 25 | 39 | 4.8 | 27.5 | 9.7 | 6.2 | 1.9 |
| ISJT 100 | 50 | 632 | 100 | 60 | 5 | 3.4 | 28.1 | 63.5 | 8.6 | 31.7 | 11.7 | 8.8 | 2.9 |
| ISJT 112.5 | 64 | 814 | 112.5 | 80 | 5 | 3.7 | 30.1 | 101.6 | 20.2 | 35.3 | 15.8 | 12.3 | 5.1 |

**TABLE A.7** Sectional properties of parallel flange beams and columns

IPE - European *I*-beams
HE - European wide flange beams
W - American wide flange beams
UC - British universal columns
HD - Wide flange columns

H - Depth
B - Flange width
$t_w$ - Web thickness
$t_f$ - Flange thickness
*R* - Fillet radius

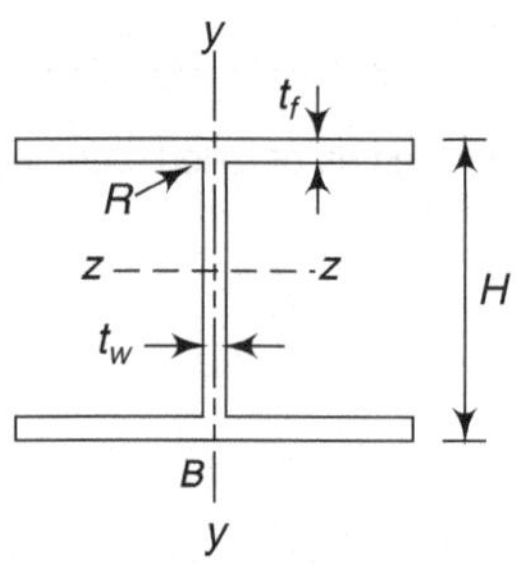


| Designation | Mass (N/m) | Sectional area (mm²) | Main dimensions (mm) | | | | | Moment of inertia (cm⁴) | | Radius of gyration(mm) | | Modulus of section (cm³) | |
|---|---|---|---|---|---|---|---|---|---|---|---|---|---|
| | | | *H* | *B* | $t_w$ | $t_f$ | *R* | $I_z$ | $I_y$ | $r_z$ | $r_y$ | $Z_z$ | $Z_y$ |
| *(1) Nominal size 200 mm* | | | | | | | | | | | | | |
| IPE 200 | 224 | 2848 | 200 | 100 | 5.6 | 8.5 | 12 | 1943 | 142.4 | 82.6 | 22.4 | 194.3 | 28.47 |
| HE 200 A | 423 | 5383 | 190 | 200 | 6.5 | 10 | 18 | 3692 | 1336 | 82.8 | 49.8 | 388.6 | 133.6 |
| HE 200 B | 613 | 7808 | 200 | 200 | 9 | 15 | 18 | 5696 | 2003 | 85.4 | 50.7 | 569.6 | 200.3 |
| HE 200 M | 1030 | 13130 | 220 | 206 | 15 | 25 | 18 | 10640 | 3651 | 90.0 | 52.7 | 967.4 | 354.5 |
| W 200 × 135 × 26.6 | 266 | 3400 | 207 | 133 | 5.8 | 8.4 | 10 | 2587 | 329.8 | 87.2 | 31.1 | 250 | 49.6 |
| W 200 × 135 × 31.3 | 313 | 3992 | 210 | 134 | 6.4 | 10.2 | 10 | 3139 | 409.6 | 88.7 | 32 | 298.9 | 61.13 |
| W 200 × 165 × 35.9 | 359 | 4575 | 201 | 165 | 6.2 | 10.2 | 10 | 3438 | 764.3 | 86.7 | 40.9 | 342.1 | 92.64 |
| W 200 × 165 × 41.7 | 417 | 5317 | 205 | 166 | 7.2 | 11.8 | 10 | 4088 | 900.5 | 87.7 | 41.2 | 398.8 | 108.5 |
| IPE 220 | 262 | 3337 | 220 | 110 | 5.9 | 9.2 | 12 | 2772 | 204.9 | 91.1 | 24.8 | 252 | 37.25 |
| HE 220 A | 505 | 6434 | 210 | 220 | 7 | 11 | 18 | 5410 | 1955 | 91.7 | 55.1 | 515.2 | 177.7 |
| HE 220 B | 715 | 9104 | 220 | 220 | 9.5 | 16 | 18 | 8091 | 2843 | 94.3 | 55.9 | 735.5 | 258.5 |
| HE 220 M | 1170 | 14940 | 240 | 226 | 15.5 | 26 | 18 | 14600 | 5012 | 98.9 | 57.9 | 1217 | 443.5 |

*(Contd)*

**TABLE A.7** *(Contd)*

| Designation | Mass (N/m) | Sectional area (mm²) | Main dimensions (mm) | | | | | Moment of inertia (cm⁴) | | Radius of gyration(mm) | | Modulus of section (cm³) | |
|---|---|---|---|---|---|---|---|---|---|---|---|---|---|
| | | | $H$ | $B$ | $t_w$ | $t_f$ | $R$ | $I_z$ | $I_y$ | $r_z$ | $r_y$ | $Z_z$ | $Z_y$ |
| *Column Sections* | | | | | | | | | | | | | |
| UC 200 × 203 × 46 | 461 | 5873 | 203.2 | 203.6 | 7.2 | 11 | 10.2 | 4568 | 1548 | 88.2 | 51.3 | 449.6 | 152.1 |
| UC 200 × 203 × 52 | 520 | 6628 | 206.2 | 204.3 | 7.9 | 12.5 | 10.2 | 5259 | 1778 | 89.1 | 51.8 | 510.1 | 174 |
| UC 200 × 203 × 60 | 600 | 7637 | 209.6 | 205.8 | 9.4 | 14.2 | 10.2 | 6125 | 2065 | 89.6 | 52 | 584.4 | 200.6 |
| UC 200 × 203 × 71 | 710 | 9043 | 215.8 | 206.4 | 10 | 17.3 | 10.2 | 7618 | 2537 | 91.8 | 53 | 706 | 245.9 |
| UC 200 × 203 × 86 | 860 | 10960 | 222.2 | 209.1 | 12.7 | 20.5 | 10.2 | 9449 | 3127 | 92.8 | 53.4 | 850.5 | 299.1 |
| UC 200 × 200 × 100 | 1000 | 12670 | 229 | 210 | 14.5 | 27.3 | 10 | 11000 | 3660 | 93.2 | 53.7 | 961 | 349 |
| *(2) Nominal size 250 mm* | | | | | | | | | | | | | |
| IPE 240 | 307 | 3912 | 240 | 120 | 6.2 | 9.8 | 15 | 3892 | 283.6 | 99.7 | 26.9 | 324.3 | 47.27 |
| HE 240 A | 603 | 7684 | 230 | 240 | 7.5 | 12 | 21 | 7763 | 2769 | 100.5 | 60 | 675.1 | 230.7 |
| HE 240 B | 832 | 10600 | 240 | 240 | 10 | 17 | 21 | 11260 | 3923 | 103.1 | 60.8 | 938.3 | 326.9 |
| HE 240 M | 1570 | 19960 | 270 | 248 | 18 | 32 | 21 | 24290 | 8153 | 110.3 | 63.9 | 1799 | 657.5 |
| W 250 × 145 × 32.7 | 327 | 4175 | 258 | 146 | 6.1 | 9.1 | 13 | 4895 | 472.6 | 108.3 | 33.6 | 379.4 | 64.74 |
| W 250 × 145 × 38.5 | 385 | 4929 | 262 | 147 | 6.6 | 11.2 | 13 | 6014 | 593.7 | 110.5 | 34.7 | 459.1 | 80.77 |
| W 250 × 145 × 44.8 | 448 | 5732 | 266 | 148 | 7.6 | 13 | 13 | 7118 | 703.5 | 111.4 | 35 | 535.2 | 95.06 |
| W250 × 200 × 49.1 | 491 | 6254 | 247 | 202 | 7.4 | 11 | 13 | 7070 | 1510 | 106 | 49 | 572 | 150 |
| W250 × 200 × 58 | 580 | 7426 | 252 | 203 | 8 | 13.5 | 13 | 8740 | 1880 | 108 | 50.3 | 694 | 185 |
| W250 × 200 × 67 | 670 | 8559 | 257 | 204 | 8.9 | 15.7 | 13 | 10400 | 2220 | 110 | 50.9 | 809 | 218 |
| HD 260 × 68.2 | 682 | 8682 | 250 | 260 | 7.5 | 12.5 | 24 | 10450 | 3668 | 109.7 | 65 | 836.4 | 282.1 |
| HD 260 × 93 | 930 | 11840 | 260 | 260 | 10 | 17.5 | 24 | 14920 | 5135 | 112.2 | 65.8 | 1148 | 395 |
| HD 260 × 114 | 1140 | 14570 | 268 | 262 | 12.5 | 21.5 | 24 | 18910 | 6456 | 113.9 | 66.6 | 1141 | 492.8 |
| HD 260 × 142 | 1420 | 18030 | 278 | 265 | 15.5 | 26.5 | 24 | 24330 | 8236 | 116.2 | 67.6 | 1750 | 621.6 |
| HD 260 × 172 | 1720 | 21960 | 290 | 268 | 18 | 32.5 | 24 | 31310 | 10450 | 119.4 | 69 | 2159 | 779.7 |
| HE 280 A | 764 | 9726 | 270 | 280 | 8 | 13 | 24 | 13670 | 4763 | 118.6 | 70 | 1013 | 340.2 |
| HE 280 B | 1030 | 13140 | 280 | 280 | 10.5 | 18 | 24 | 19270 | 6595 | 121.1 | 70.9 | 1376 | 471 |
| HE 280 M | 1890 | 24020 | 310 | 288 | 18.5 | 33 | 24 | 39550 | 13160 | 128.3 | 74 | 2551 | 914.1 |
| *Column Sections* | | | | | | | | | | | | | |
| W 250 × 250 × 73 | 730 | 9299 | 253 | 254 | 8.6 | 14.2 | 13 | 11290 | 3880 | 110.2 | 64.6 | 892.1 | 305.5 |
| W 250 × 250 × 80 | 800 | 10210 | 256 | 255 | 9.4 | 15.6 | 13 | 12570 | 4314 | 111 | 65 | 982.4 | 338.3 |
| W 250 × 250 × 89 | 890 | 11410 | 260 | 256 | 10.7 | 17.3 | 13 | 14260 | 4841 | 111.8 | 65.1 | 1097 | 378.2 |
| W 250 × 250 × 101 | 1010 | 12900 | 264 | 257 | 11.9 | 19.6 | 13 | 16380 | 5549 | 112.7 | 65.6 | 1241 | 431.9 |
| W 250 × 250 × 115 | 1150 | 14620 | 269 | 259 | 13.5 | 22.1 | 13 | 18940 | 6405 | 113.8 | 66.2 | 1408 | 494.6 |
| W 250 × 250 × 131 | 1310 | 16700 | 275 | 261 | 15.4 | 25.1 | 13 | 22150 | 7446 | 115.2 | 66.8 | 1611 | 570.6 |
| W 250 × 250 × 149 | 1490 | 18970 | 282 | 263 | 17.3 | 28.4 | 13 | 25940 | 8622 | 116.9 | 67.4 | 1840 | 655.7 |
| W 250 × 250 × 167 | 1670 | 21320 | 289 | 265 | 19.2 | 31.8 | 13 | 30020 | 9879 | 118.7 | 68.1 | 2078 | 745.6 |

*(Contd)*

**TABLE A.7** (*Contd*)

| Designation | Mass (N/m) | Sectional area (mm²) | Main dimensions (mm) | | | | | Moment of inertia (cm⁴) | | Radius of gyration(mm) | | Modulus of section (cm³) | |
|---|---|---|---|---|---|---|---|---|---|---|---|---|---|
| | | | $H$ | $B$ | $t_w$ | $t_f$ | $R$ | $I_z$ | $I_y$ | $r_z$ | $r_y$ | $Z_z$ | $Z_y$ |
| *(3) Nominal size 300 mm* | | | | | | | | | | | | | |
| IPE 300 | 422 | 5381 | 300 | 150 | 7.1 | 10.7 | 15 | 8356 | 603.8 | 124.6 | 33.5 | 557.1 | 80.5 |
| HE 300 A | 880 | 11250 | 290 | 300 | 8.5 | 14 | 27 | 18260 | 6310 | 127.4 | 74.9 | 1260 | 420.6 |
| HE 300 B | 1170 | 14910 | 300 | 300 | 11 | 19 | 27 | 25170 | 8563 | 129.9 | 75.8 | 1678 | 570.9 |
| HE 300 M | 2380 | 30310 | 340 | 310 | 21 | 39 | 27 | 59200 | 19400 | 139.8 | 80 | 3482 | 1252 |
| W 310 × 100 × 23.8 | 238 | 3038 | 305 | 101 | 5.6 | 6.7 | 8 | 4280 | 115.6 | 118.7 | 19.5 | 280.7 | 22.89 |
| W 310 × 100 × 28.3 | 283 | 3609 | 309 | 102 | 6 | 8.9 | 8 | 5431 | 158.1 | 122.7 | 20.9 | 351.5 | 30.99 |
| W 310 × 100 × 32.7 | 327 | 4181 | 313 | 102 | 6.6 | 10.8 | 8 | 6507 | 191.9 | 124.7 | 21.4 | 415.8 | 37.62 |
| W 310 × 165 × 38.7 | 387 | 4953 | 310 | 165 | 5.8 | 9.7 | 8 | 8527 | 726.8 | 131.2 | 38.3 | 550.1 | 88.1 |
| W 310 × 165 × 44.5 | 445 | 5691 | 313 | 166 | 6.6 | 11.2 | 8 | 9934 | 854.7 | 132.1 | 38.8 | 634.8 | 103 |
| W 310 × 165 × 52 | 520 | 6678 | 317 | 167 | 7.6 | 13.2 | 8 | 11851 | 1026 | 133.2 | 39.2 | 747.7 | 122.9 |
| W 310 × 200 × 60 | 600 | 7588 | 303 | 203 | 7.5 | 13.1 | 15 | 12900 | 1830 | 130 | 49.1 | 851 | 180 |
| W 310 × 200 × 67 | 670 | 8503 | 306 | 204 | 8.5 | 14.6 | 15 | 14500 | 2070 | 131 | 49.3 | 948 | 203 |
| W 310 × 200 × 74 | 740 | 9484 | 310 | 205 | 9.4 | 16.3 | 15 | 16500 | 2340 | 132 | 49.7 | 1060 | 228 |
| W 310 × 250 × 79 | 790 | 10046 | 306 | 254 | 8.8 | 14.6 | 15 | 17700 | 3990 | 133 | 63 | 1160 | 314 |
| W 310 × 250 × 86 | 860 | 10998 | 310 | 254 | 9.1 | 16.3 | 15 | 19800 | 4450 | 134 | 63.6 | 1280 | 350 |
| *Column Sections* | | | | | | | | | | | | | |
| W 310 × 310 × 97 | 970 | 12330 | 308 | 305 | 9.9 | 15.4 | 15 | 22240 | 7286 | 134.3 | 76.9 | 1444 | 477.8 |
| W 310 × 310 × 107 | 1070 | 13620 | 311 | 306 | 10.9 | 17 | 15 | 24790 | 8123 | 134.9 | 77.2 | 1594 | 530.9 |
| W 310 × 310 × 117 | 1170 | 14970 | 314 | 307 | 11.9 | 18.7 | 15 | 27510 | 9024 | 135.6 | 77.6 | 1753 | 587.9 |
| W 310 × 310 × 129 | 1290 | 16510 | 318 | 308 | 13.1 | 20.6 | 15 | 30770 | 10040 | 136.5 | 78 | 1935 | 651.9 |
| W 310 × 310 × 143 | 1430 | 18230 | 323 | 309 | 14 | 22.9 | 15 | 34760 | 11270 | 138.1 | 78.6 | 2153 | 729.4 |
| W 310 × 310 × 158 | 1580 | 20050 | 327 | 310 | 15.5 | 25.1 | 15 | 38630 | 12470 | 138.8 | 78.9 | 2363 | 804.8 |
| W 310 × 310 × 179 | 1790 | 22770 | 333 | 313 | 18 | 28.1 | 15 | 44530 | 14380 | 139.9 | 79.5 | 2675 | 918.7 |
| W 310 × 310 × 202 | 2020 | 25800 | 341 | 315 | 20.1 | 31.8 | 15 | 51982 | 16588 | 141.9 | 80.2 | 3049 | 1053 |
| W 310 × 310 × 226 | 2260 | 28880 | 348 | 317 | 22.1 | 35.6 | 15 | 59560 | 18930 | 143.6 | 81 | 3423 | 1194 |
| HE 320 A | 976 | 12440 | 310 | 300 | 9 | 15.5 | 27 | 22930 | 6985 | 135.8 | 74.9 | 1479 | 465.7 |
| HE 320 B | 1270 | 16130 | 320 | 300 | 11.5 | 20.5 | 27 | 30820 | 9239 | 138.2 | 75.7 | 1926 | 615.9 |
| HE 320 M | 2450 | 31200 | 359 | 309 | 21 | 40 | 27 | 68130 | 19710 | 147.8 | 79.5 | 3796 | 1276 |
| *(4) Nominal size 350 mm* | | | | | | | | | | | | | |
| HE 340 A | 1050 | 13350 | 330 | 300 | 9.5 | 16.5 | 27 | 27690 | 7436 | 144 | 74.6 | 1678 | 495.7 |
| HE 340 B | 1340 | 17090 | 340 | 300 | 12 | 21.5 | 27 | 36660 | 9690 | 146.5 | 75.3 | 2156 | 646 |
| HE 340 M | 2480 | 31580 | 377 | 309 | 21 | 40 | 27 | 76370 | 19710 | 155.5 | 79 | 4052 | 1276 |
| IPE 360 | 571 | 7273 | 360 | 170 | 8 | 12.7 | 18 | 16270 | 1043 | 149.5 | 37.9 | 903.6 | 122.8 |
| HE 360 A | 1120 | 14280 | 350 | 300 | 10 | 17.5 | 27 | 33090 | 7887 | 152.2 | 74.3 | 1891 | 525.8 |
| HE 360 B | 1420 | 18060 | 360 | 300 | 12.5 | 22.5 | 27 | 43190 | 10140 | 154.6 | 74.9 | 2400 | 676.1 |
| HE 360 M | 2500 | 31880 | 395 | 308 | 21 | 40 | 27 | 84870 | 19520 | 163.2 | 78.3 | 4297 | 1268 |

(*Contd*)

**TABLE A.7** *(Contd)*

| Designation | Mass (N/m) | Sectional area (mm²) | Main dimensions (mm) | | | | | Moment of inertia (cm⁴) | | Radius of gyration(mm) | | Modulus of section (cm³) | |
|---|---|---|---|---|---|---|---|---|---|---|---|---|---|
| | | | $H$ | $B$ | $t_w$ | $t_f$ | $R$ | $I_z$ | $I_y$ | $r_z$ | $r_y$ | $Z_z$ | $Z_y$ |
| *Column Sections* | | | | | | | | | | | | | |
| W 360 × 370 × 134 | 1340 | 17060 | 356 | 369 | 11.2 | 18 | 20 | 41510 | 15080 | 156 | 94 | 2332 | 817.3 |
| W 360 × 370 × 147 | 1470 | 18790 | 360 | 370 | 12.3 | 19.8 | 20 | 46290 | 16720 | 157 | 94.3 | 2572 | 903.9 |
| W 360 × 370 × 162 | 1620 | 20630 | 364 | 371 | 13.3 | 21.8 | 20 | 51540 | 18560 | 158.1 | 94.9 | 2832 | 1001 |
| W 360 × 370 × 179 | 1790 | 22830 | 368 | 373 | 15 | 23.9 | 20 | 57440 | 20680 | 158.6 | 95.2 | 3122 | 1109 |
| W 360 × 370 × 196 | 1960 | 25030 | 372 | 374 | 16.4 | 26.2 | 20 | 63630 | 22860 | 159.4 | 95.6 | 3421 | 1222 |
| W 360 × 410 × 216 | 2160 | 27550 | 375 | 394 | 17.3 | 27.7 | 20 | 71140 | 28250 | 160.7 | 101.3 | 3794 | 1434 |
| W 360 × 410 × 237 | 2370 | 30090 | 380 | 395 | 18.9 | 30.2 | 20 | 78780 | 31040 | 161.8 | 101.6 | 4146 | 1572 |
| W 360 × 410 × 262 | 2620 | 33460 | 387 | 398 | 21.1 | 33.3 | 20 | 89410 | 35020 | 163.5 | 102.3 | 4620 | 1760 |
| W 360 × 410 × 287 | 2870 | 36630 | 393 | 399 | 22.6 | 36.6 | 20 | 99710 | 38780 | 165 | 102.9 | 5074 | 1944 |
| W 360 × 410 × 314 | 3140 | 39920 | 399 | 401 | 24.9 | 39.6 | 20 | 110200 | 42600 | 166.2 | 103.3 | 5525 | 2125 |
| W 360 × 410 × 347 | 3470 | 44200 | 407 | 404 | 27.2 | 43.7 | 20 | 124900 | 48090 | 168.1 | 104.3 | 6140 | 2380 |
| *(5) Nominal size 400 mm* | | | | | | | | | | | | | |
| IPE 400 | 663 | 8446 | 400 | 180 | 8.6 | 13.5 | 21 | 23130 | 1318 | 165.5 | 39.5 | 1156 | 146.4 |
| HE 400 A | 1250 | 15900 | 390 | 300 | 11 | 19 | 27 | 45070 | 8564 | 168.4 | 73.4 | 2311 | 570.9 |
| HE 400 B | 1550 | 19780 | 400 | 300 | 13.5 | 24 | 27 | 57680 | 10820 | 170.8 | 74 | 2884 | 721.3 |
| HE 400 M | 2560 | 32580 | 432 | 307 | 21 | 40 | 27 | 104100 | 19340 | 178.8 | 77 | 4820 | 1260 |
| W 410 × 140 × 38.8 | 388 | 4970 | 399 | 140 | 6.4 | 8.8 | 11 | 12620 | 403.5 | 159.3 | 28.5 | 632.6 | 57.65 |
| W 410 × 140 × 46.1 | 461 | 5880 | 403 | 140 | 7 | 11.2 | 11 | 15550 | 513.6 | 162.6 | 29.5 | 771.9 | 73.37 |
| W 410 × 180 × 53 | 530 | 6800 | 403 | 177 | 7.5 | 10.9 | 11 | 18600 | 1009 | 165.4 | 38.5 | 922.9 | 114 |
| W 410 × 180 × 60 | 600 | 7580 | 407 | 178 | 7.7 | 12.8 | 11 | 21570 | 1205 | 168.7 | 39.9 | 1060 | 135.4 |
| W 410 × 180 × 67 | 670 | 8580 | 410 | 179 | 8.8 | 14.4 | 11 | 24530 | 1379 | 169.1 | 40.1 | 1196 | 154.1 |
| W 410 × 180 × 75 | 750 | 9520 | 413 | 180 | 9.7 | 16 | 11 | 27460 | 1559 | 169.8 | 40.5 | 1330 | 173.2 |
| W 410 × 180 × 85 | 850 | 10830 | 417 | 181 | 10.9 | 18.2 | 11 | 31530 | 1803 | 170.6 | 40.8 | 1512 | 199.3 |
| W 410 × 260 × 100 | 1000 | 12700 | 415 | 260 | 10 | 16.9 | 11 | 39800 | 4950 | 177 | 62.4 | 1920 | 381 |
| W 410 × 260 × 114 | 1140 | 14600 | 420 | 261 | 11.6 | 19.3 | 11 | 46200 | 5720 | 178 | 62.6 | 2200 | 438 |
| W 410 × 260 × 132 | 1320 | 16840 | 425 | 263 | 13.3 | 22.2 | 11 | 53900 | 6740 | 179 | 63.3 | 2540 | 513 |
| W 410 × 260 × 149 | 1490 | 19030 | 431 | 265 | 14.9 | 25 | 11 | 61900 | 7770 | 180 | 63.9 | 2870 | 586 |
| *(6) Nominal size 450 mm* | | | | | | | | | | | | | |
| IPE 450 | 776 | 9882 | 450 | 190 | 9.4 | 14.6 | 21 | 33740 | 1676 | 184.8 | 41.2 | 1500 | 176.4 |
| HE 450 A | 1400 | 17800 | 440 | 300 | 11.5 | 21 | 27 | 63720 | 9465 | 189.2 | 72.9 | 2896 | 631 |
| HE 450 B | 1710 | 21800 | 450 | 300 | 14 | 26 | 27 | 79890 | 11720 | 191.4 | 73.3 | 3551 | 781.4 |
| HE 450 M | 2630 | 33540 | 478 | 307 | 21 | 40 | 27 | 131500 | 19340 | 198 | 75.9 | 5501 | 1260 |
| W 460 × 150 × 52 | 520 | 6620 | 450 | 152 | 7.6 | 10.8 | 11 | 21200 | 634 | 178.9 | 30.9 | 942 | 83.43 |
| W 460 × 150 × 60 | 600 | 7580 | 455 | 153 | 8 | 13.3 | 10 | 25480 | 796.1 | 183.3 | 32.4 | 1120 | 104.1 |
| W 460 × 150 × 68 | 680 | 8730 | 459 | 154 | 9.1 | 15.4 | 10 | 29680 | 940.5 | 184.4 | 32.8 | 1293 | 122.1 |

*(Contd)*

**TABLE A.7** (*Contd*)

| Designation | Mass (N/m) | Sectional area (mm²) | Main dimensions (mm) | | | | | Moment of inertia (cm⁴) | | Radius of gyration(mm) | | Modulus of section (cm³) | |
|---|---|---|---|---|---|---|---|---|---|---|---|---|---|
| | | | $H$ | $B$ | $t_w$ | $t_f$ | $R$ | $I_z$ | $I_y$ | $r_z$ | $r_y$ | $Z_z$ | $Z_y$ |
| W 460 × 190 × 74 | 740 | 9460 | 457 | 190 | 9 | 14.5 | 10 | 33260 | 1661 | 187.5 | 41.9 | 1456 | 174.8 |
| W 460 × 190 × 82 | 820 | 10440 | 460 | 191 | 9.9 | 16 | 10 | 37000 | 1862 | 188.3 | 42.2 | 1608 | 195 |
| W 460 × 190 × 89 | 890 | 11390 | 463 | 192 | 10.5 | 17.7 | 10 | 40960 | 2093 | 189.6 | 42.9 | 1769 | 218 |
| W 460 × 190 × 97 | 970 | 12350 | 466 | 193 | 11.4 | 19 | 10 | 44680 | 2282 | 190.2 | 43.1 | 1917 | 237.8 |
| W 460 × 190 × 106 | 1060 | 13460 | 469 | 194 | 12.6 | 20.6 | 10 | 48790 | 2515 | 190.4 | 43.2 | 2081 | 259.2 |
| *Column Sections* | | | | | | | | | | | | | |
| W 460 × 280 × 113 | 1130 | 14400 | 463 | 280 | 10.8 | 17.3 | 18 | 55600 | 6335 | 196.5 | 66.3 | 2402 | 452.5 |
| W 460 × 280 × 128 | 1280 | 16360 | 467 | 282 | 12.2 | 19.6 | 18 | 63690 | 7333 | 197.3 | 67 | 2728 | 520.1 |
| W 460 × 280 × 144 | 1440 | 18410 | 472 | 283 | 13.6 | 22.1 | 18 | 72600 | 8358 | 198.6 | 67.4 | 3076 | 590.7 |
| W 460 × 280 × 158 | 1580 | 20080 | 476 | 284 | 15 | 23.9 | 18 | 79620 | 9137 | 199.1 | 67.5 | 3346 | 643.5 |
| W 460 × 280 × 177 | 1770 | 22600 | 482 | 286 | 16.6 | 26.9 | 18 | 91040 | 10510 | 200.7 | 68.2 | 3777 | 734.7 |
| W 460 × 280 × 193 | 1930 | 24820 | 489 | 283 | 17 | 30.5 | 18 | 103000 | 11500 | 204 | 68.1 | 4210 | 813 |
| W 460 × 280 × 213 | 2130 | 27290 | 495 | 285 | 18.5 | 33.5 | 18 | 115000 | 13000 | 205 | 69 | 4650 | 912 |
| W 460 × 280 × 235 | 2350 | 30100 | 501 | 287 | 20.6 | 36.6 | 18 | 128000 | 14500 | 206 | 69.4 | 5110 | 1010 |
| *(7) Nominal size 500 mm* | | | | | | | | | | | | | |
| IPE 500 | 907 | 11550 | 500 | 200 | 10.2 | 16 | 21 | 48200 | 2142 | 204.3 | 43.1 | 1928 | 214.2 |
| HE 500 A | 1550 | 19750 | 490 | 300 | 12 | 23 | 27 | 86970 | 10370 | 209.8 | 72.4 | 3550 | 691.1 |
| HE 500 B | 1870 | 23860 | 500 | 300 | 14.5 | 28 | 27 | 107200 | 12620 | 211.9 | 72.7 | 4287 | 841.6 |
| HE 500 M | 2700 | 34430 | 524 | 306 | 21 | 40 | 27 | 161900 | 19150 | 216.9 | 74.6 | 6180 | 1252 |
| W 530 × 210 × 92 | 920 | 11760 | 533 | 209 | 10.2 | 15.6 | 14 | 55240 | 2379 | 216.7 | 45 | 2073 | 227.7 |
| W 530 × 210 × 101 | 1010 | 12940 | 537 | 210 | 10.9 | 17.4 | 14 | 61760 | 2692 | 218.5 | 45.6 | 2300 | 256.4 |
| W 530 × 210 × 109 | 1090 | 13870 | 539 | 211 | 11.6 | 18.8 | 14 | 66730 | 2951 | 219.3 | 46.1 | 2476 | 279.7 |
| W 530 × 210 × 123 | 1230 | 15690 | 544 | 212 | 13.1 | 21.2 | 14 | 76100 | 3377 | 220.2 | 46.4 | 2798 | 318.6 |
| W 530 × 210 × 138 | 1380 | 17640 | 549 | 214 | 14.7 | 23.6 | 14 | 86160 | 3870 | 221 | 46.8 | 3139 | 361.7 |
| W 530 × 310 × 150 | 1500 | 19220 | 543 | 312 | 12.7 | 20.3 | 14 | 101000 | 10300 | 229 | 73.2 | 3720 | 660 |
| W 530 × 310 × 165 | 1650 | 21090 | 546 | 313 | 14 | 22.2 | 14 | 111000 | 11400 | 229 | 73.5 | 4070 | 728 |
| W 530 × 310 × 182 | 1820 | 23170 | 551 | 315 | 15.2 | 24.4 | 14 | 124000 | 12700 | 231 | 74 | 4500 | 806 |
| W 530 × 310 × 196 | 1960 | 25060 | 554 | 316 | 16.5 | 26.3 | 14 | 134000 | 13900 | 231 | 74.5 | 4840 | 880 |
| W 530 × 310 × 213 | 2130 | 27920 | 560 | 318 | 18.3 | 29.2 | 14 | 151000 | 15700 | 233 | 75 | 5390 | 987 |
| W 530 × 310 × 248 | 2480 | 31440 | 571 | 315 | 19 | 34.5 | 14 | 178000 | 18000 | 238 | 75.7 | 6230 | 1140 |
| *(8) Nominal size 550 mm* | | | | | | | | | | | | | |
| IPE 550 | 1060 | 13440 | 550 | 210 | 11.1 | 17.2 | 24 | 67120 | 2668 | 223.5 | 44.5 | 2441 | 254.1 |
| HE 550 A | 1660 | 21180 | 540 | 300 | 12.5 | 24 | 27 | 111900 | 10820 | 229.9 | 71.5 | 4146 | 721.3 |
| HE 550 B | 1990 | 25410 | 550 | 300 | 15 | 29 | 27 | 136700 | 13080 | 232 | 71.7 | 4971 | 871.8 |
| HE 550 M | 2780 | 35440 | 572 | 306 | 21 | 40 | 27 | 198000 | 19160 | 236.4 | 73.5 | 6923 | 1252 |

(*Contd*)

**TABLE A.7** *(Contd)*

| Designation | Mass (N/m) | Sectional area ($mm^2$) | Main dimensions (mm) | | | | | Moment of inertia ($cm^4$) | | Radius of gyration(mm) | | Modulus of section ($cm^3$) | |
|---|---|---|---|---|---|---|---|---|---|---|---|---|---|
| | | | $H$ | $B$ | $t_w$ | $t_f$ | $R$ | $I_z$ | $I_y$ | $r_z$ | $r_y$ | $Z_z$ | $Z_y$ |
| *(9) Nominal size 600 mm* | | | | | | | | | | | | | |
| IPE 600 | 1220 | 15600 | 600 | 220 | 12 | 19 | 24 | 92080 | 3387 | 243 | 46.6 | 3069 | 307.9 |
| HE 600 A | 1780 | 22650 | 590 | 300 | 13 | 25 | 27 | 141200 | 11270 | 249.7 | 70.5 | 4787 | 751.4 |
| HE 600 B | 2120 | 27000 | 600 | 300 | 15.5 | 30 | 27 | 171000 | 13530 | 251.7 | 70.8 | 5701 | 902 |
| HE 600 M | 2850 | 36370 | 620 | 305 | 21 | 40 | 27 | 237400 | 18980 | 255.5 | 72.2 | 7660 | 1244 |
| W 610 × 230 × 101 | 1010 | 12980 | 603 | 228 | 10.5 | 14.9 | 14 | 76470 | 2950 | 242.7 | 47.7 | 2536 | 258.8 |
| W 610 × 230 × 113 | 1130 | 14440 | 608 | 228 | 11.2 | 17.3 | 14 | 87570 | 3425 | 246.2 | 48.7 | 2881 | 300.5 |
| W 610 × 230 × 125 | 1250 | 15960 | 612 | 229 | 11.9 | 19.6 | 14 | 98650 | 3932 | 248.6 | 49.6 | 3224 | 343.4 |
| W 610 × 230 × 140 | 1400 | 17850 | 617 | 230 | 13.1 | 22.2 | 14 | 111990 | 4514 | 250.5 | 50.3 | 3630 | 392.5 |
| W 610 × 325 × 155 | 1550 | 19730 | 611 | 324 | 12.7 | 19 | 14 | 129000 | 10780 | 255.7 | 73.9 | 4222 | 666 |
| W 610 × 325 × 174 | 1740 | 22200 | 616 | 325 | 14 | 21.6 | 14 | 147200 | 12370 | 257.4 | 74.6 | 4778 | 761 |
| W 610 × 325 × 195 | 1950 | 24930 | 622 | 327 | 15.4 | 24.4 | 14 | 167900 | 14240 | 259.5 | 75.6 | 5398 | 871 |
| W 610 × 325 × 217 | 2170 | 27760 | 628 | 328 | 16.5 | 27.7 | 14 | 190800 | 16310 | 262.1 | 76.7 | 6076 | 995 |
| W 610 × 325 × 241 | 2410 | 30340 | 635 | 329 | 17.1 | 31 | 14 | 214200 | 18430 | 265.7 | 77.9 | 6746 | 1120 |
| W 610 × 325 × 262 | 2620 | 33270 | 641 | 327 | 19 | 34 | 14 | 235990 | 19850 | 266.3 | 77.2 | 7363 | 1214 |
| W 610 × 325 × 285 | 2850 | 36360 | 647 | 329 | 20.6 | 37.1 | 14 | 260700 | 22060 | 267.8 | 77.9 | 8059 | 1341 |
| W 610 × 325 × 341 | 3410 | 43370 | 661 | 333 | 24.4 | 43.9 | 14 | 318300 | 27090 | 270.9 | 79 | 9630 | 1627 |
| W 610 × 320 × 372 | 3720 | 47630 | 669 | 335 | 26.4 | 48 | 20 | 355000 | 30200 | 273 | 79.6 | 10600 | 1800 |
| *(10) Nominal size 650 mm* | | | | | | | | | | | | | |
| HE 650 A | 1900 | 24160 | 640 | 300 | 13.5 | 26 | 27 | 175200 | 11720 | 269.3 | 69.7 | 5474 | 781.6 |
| HE 650 B | 2250 | 28630 | 650 | 300 | 16 | 31 | 27 | 210600 | 13980 | 271.2 | 69.9 | 6480 | 932.3 |
| HE 650 M | 2930 | 37370 | 668 | 305 | 21 | 40 | 27 | 281700 | 18980 | 274.5 | 71.3 | 8433 | 1245 |
| *(11) Nominal size 700mm* | | | | | | | | | | | | | |
| HE 700 A | 2040 | 26050 | 690 | 300 | 14.5 | 27 | 27 | 215300 | 12180 | 287.5 | 68.4 | 6241 | 811.9 |
| HE 700 B | 2410 | 30640 | 700 | 300 | 17 | 32 | 27 | 256900 | 14440 | 289.6 | 68.7 | 7340 | 962.7 |
| HE 700 M | 3010 | 38300 | 716 | 304 | 21 | 40 | 27 | 329300 | 18800 | 293.2 | 70.1 | 9198 | 1237 |

**TABLE A.8** Sectional properties of castellated beams

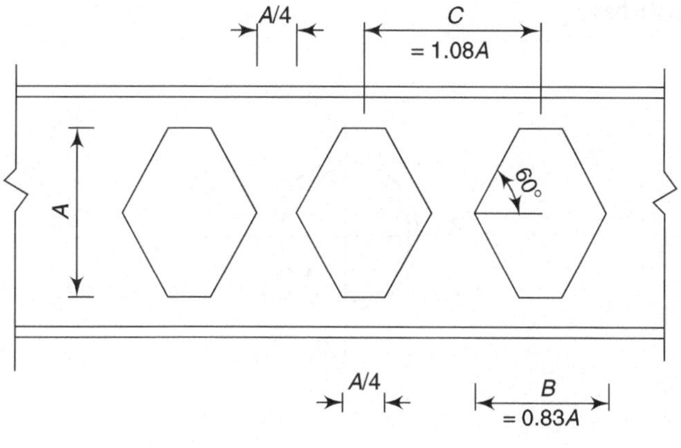

| Original | Castellated | Over size (mm) | Weight (N/m) | Size of castellated holes | | | Net Moment of inertia $I_z$ (cm$^4$) | Design radius of gyration $r_y$ (mm) | Net elastic modulus $Z_z$ (cm$^3$) |
|---|---|---|---|---|---|---|---|---|---|
| | | | | Depth (mm) $A$ | Width (mm) $B$ | Spacing (mm) $C$ | | | |
| ISMB-150 | NCB-150 | 225 × 80 | 150 | 152 | 126 | 164 | 1723 | 15.2 | 152.51 |
| ISMB-175 | NCB-175 | 263 × 90 | 196 | 177 | 146 | 191 | 3012 | 17.0 | 228.63 |
| ISMB-200 | NCB-200 | 300 × 100 | 242 | 203 | 168 | 219 | 5314 | 19.8 | 352.52 |
| ISMB-225 | NCB-225 | 338 × 110 | 311 | 228 | 189 | 246 | 8168 | 21.5 | 481.91 |
| ISMB-250 | NCB-250 | 375 × 125 | 373 | 254 | 210 | 274 | 12178 | 24.4 | 646.07 |
| ISMB-300 | NCB-300 | 450 × 140 | 460 | 304 | 252 | 328 | 20297 | 25.9 | 898.1 |
| ISMB-350 | NCB-350 | 525 × 140 | 524 | 355 | 294 | 382 | 32157 | 25.8 | 1219.22 |
| ISMB-400 | NCB-400 | 600 × 140 | 615 | 406 | 336 | 437 | 48279 | 26.9 | 1601.31 |
| ISMB-450 | NCB-450 | 675 × 150 | 724 | 457 | 378 | 492 | 71693 | 27.1 | 2113.27 |
| ISMB-500 | NCB-500 | 750 × 180 | 869 | 508 | 420 | 547 | 106483 | 31.7 | 2824.49 |
| ISMB-550 | NCB-550 | 825 × 190 | 1040 | 558 | 462 | 601 | 152750 | 33.5 | 3685.16 |
| ISMB-600 | NCB-600 | 900 × 210 | 1230 | 609 | 504 | 656 | 216098 | 37.1 | 4778.29 |

**TABLE A.9** Sectional properties of castellated channels

| Original | Castellated | Overall size (mm) | Weight (N/m) | Size of castellated holes | | | Net moment of inertia $I_Z$ (cm$^4$) | Design radius of gyration $r_y$ (mm) | Net elastic modulus $Z_z$ (cm$^3$) |
|---|---|---|---|---|---|---|---|---|---|
| | | | | Depth (mm) | Width (mm) | Spacing (mm) | | | |
| ISMC-75 | NCC-75 | 113 × 40 | 71.4 | 76 | 63 | 82 | 186 | 11.8 | 32.89 |
| ISMC-100 | NCC-100 | 150 × 50 | 95.6 | 102 | 84 | 109 | 451 | 14.4 | 59.75 |
| ISMC-125 | NCC-125 | 188 × 65 | 131 | 127 | 105 | 137 | 998 | 18.7 | 105.93 |
| ISMC-150 | NCC-150 | 225 × 75 | 168 | 152 | 126 | 164 | 1863 | 21.5 | 164.69 |
| ISMC-175 | NCC-175 | 263 × 75 | 196 | 178 | 147 | 192 | 2919 | 21.5 | 221.23 |
| ISMC-200 | NCC-200 | 300 × 75 | 223 | 203 | 168 | 219 | 4337 | 21.4 | 287.61 |
| ISMC-225 | NCC-225 | 338 × 80 | 261 | 229 | 189 | 246 | 6417 | 22.9 | 378.25 |
| ISMC-250 | NCC-250 | 375 × 80 | 306 | 254 | 210 | 274 | 9095 | 22.8 | 482.52 |
| ISMC-300 | NCC-300 | 450 × 90 | 363 | 305 | 252 | 328 | 15084 | 24.5 | 666.85 |
| ISMC-350 | NCC-350 | 525 × 100 | 427 | 356 | 294 | 383 | 23663 | 26.4 | 896.65 |
| ISMC-400 | NCC-400 | 600 × 100 | 501 | 406 | 336 | 438 | 35660 | 26.4 | 1182.35 |

**TABLE A.10** Sectional properties of circular tubes

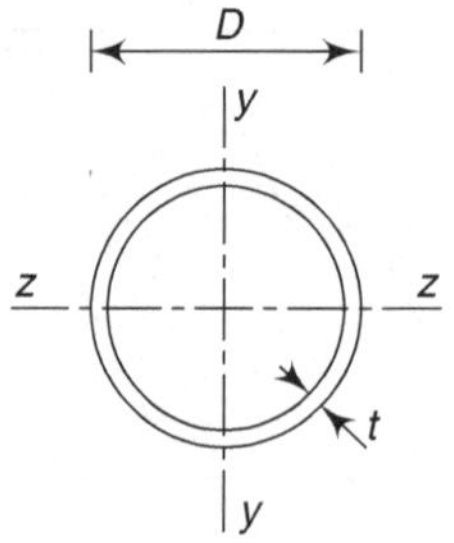

| Normal bore (mm) | Outside diameter (mm) | Class (mm) | Thickness (mm) | Weight (N/m) | Area of cross section (mm$^2$) | Surface | | Moment of inertia (cm$^4$) | Modulus section (cm$^3$) | Radius of gyration (mm) | Plastic section modulus(cm$^3$) |
|---|---|---|---|---|---|---|---|---|---|---|---|
| | | | | | | External (cm$^2$/m) | Internal | | | | |
| 15 | 21.3 | Light | 2.00 | 9.62 | 121 | 669 | 543 | 0.57 | 0.54 | 6.9 | 0.75 |
| | | Medium | 2.65 | 12.2 | 155 | | 503 | 0.69 | 0.65 | 6.7 | 0.93 |
| | | Heavy | 3.25 | 14.5 | 184 | | 465 | 0.77 | 0.73 | 6.5 | 1.07 |
| 20 | 26.9 | Light | 2.35 | 14.2 | 181 | 845 | 697 | 1.38 | 1.02 | 8.7 | 1.42 |
| | | Medium | 2.65 | 15.8 | 202 | | 679 | 1.5 | 1.12 | 8.6 | 1.57 |
| | | Heavy | 3.25 | 19.0 | 241 | | 641 | 1.72 | 1.28 | 8.4 | 1.83 |
| 25 | 33.7 | Light | 2.65 | 20.4 | 258 | 1059 | 892 | 3.14 | 1.86 | 11.0 | 2.56 |
| | | Medium | 3.25 | 24.6 | 311 | | 855 | 3.65 | 2.16 | 10.8 | 3.03 |
| | | Heavy | 4.05 | 29.9 | 377 | | 804 | 4.22 | 2.51 | 10.6 | 3.58 |
| 32 | 42.4 | Light | 2.65 | 26.1 | 331 | 1332 | 1166 | 6.57 | 3.1 | 14.1 | 4.20 |
| | | Medium | 3.25 | 31.5 | 400 | | 1128 | 7.71 | 3.64 | 13.9 | 5.00 |
| | | Heavy | 4.05 | 38.6 | 488 | | 1078 | 9.07 | 4.28 | 13.6 | 5.98 |
| 40 | 48.3 | Light | 2.9 | 32.7 | 414 | 1517 | 1335 | 10.7 | 4.43 | 16.1 | 5.99 |
| | | Medium | 3.25 | 36.1 | 460 | | 1313 | 11.73 | 4.86 | 16.0 | 6.61 |
| | | Heavy | 4.05 | 44.3 | 563 | | 1263 | 13.9 | 5.75 | 15.7 | 7.96 |
| 50 | 60.3 | Light1 | 2.9 | 41.4 | 523 | 1894 | 1712 | 21.59 | 7.16 | 20.3 | 9.57 |
| | | Light2 | 3.25 | 45.7 | 582 | | 1690 | 23.77 | 7.89 | 20.2 | 10.59 |
| | | Medium | 3.65 | 51.0 | 650 | | 1605 | 26.17 | 8.68 | 20.1 | 11.74 |
| | | Heavy | 4.5 | 61.7 | 789 | | 1612 | 30.9 | 10.2 | 19.8 | 14.05 |
| 65 | 76.1 | Light | 3.25 | 58.4 | 744 | 2391 | 2187 | 49.44 | 13 | 25.8 | 17.27 |
| | | Medium | 3.65 | 65.3 | 831 | | 2161 | 54.65 | 14.4 | 25.6 | 19.19 |
| | | Heavy | 4.5 | 79.2 | 1010 | | 2108 | 65.12 | 17.1 | 25.4 | 23.06 |
| 80 | 88.9 | Light | 3.25 | 68.6 | 874 | 2793 | 2589 | 80.31 | 18.1 | 30.3 | 23.85 |
| | | Medium | 4.05 | 84.8 | 1080 | | 2538 | 97.38 | 21.9 | 30.0 | 28.67 |
| | | Heavy | 4.85 | 101.0 | 1280 | | 2488 | 113.46 | 25.5 | 29.8 | 34.30 |
| 90 | 101.6 | Light | 3.65 | 88.2 | 1120 | 3192 | 2862 | 134.9 | 26.6 | 34.7 | 34.95 |
| | | Medium | 4.05 | 97.5 | 1240 | | 2937 | 147.9 | 29.1 | 34.5 | 38.55 |
| | | Heavy | 4.85 | 116.0 | 1470 | | 2887 | 172.9 | 34 | 34.2 | 45.33 |
| 100 | 114.3 | Light | 3.65 | 99.7 | 1270 | 3591 | 3361 | 194.4 | 34 | 39.1 | 44.77 |
| | | Medium | 4.5 | 121 | 1550 | | 3308 | 234.3 | 41 | 38.9 | 54.23 |
| | | Heavy | 5.4 | 145 | 1850 | | 3252 | 274.5 | 48 | 38.5 | 64.21 |
| 110 | 127 | Light | 4.5 | 136 | 1730 | 3990 | 3707 | 325.3 | 51.2 | 43.3 | 67.52 |
| | | Medium | 4.85 | 146 | 1860 | | 3685 | 347.7 | 54.8 | 43.2 | 72.39 |
| | | Heavy | 5.4 | 162 | 2060 | | 3651 | 382 | 60.2 | 43.0 | 79.83 |
| 125 | 139.7 | Light | 4.5 | 149 | 1910 | 4389 | 4106 | 437.2 | 62.6 | 47.8 | 82.27 |
| | | Medium | 4.85 | 162 | 2050 | | 4084 | 467.6 | 66.6 | 47.7 | 88.08 |
| | | Heavy | 5.4 | 179 | 2280 | | 4050 | 514.5 | 73.7 | 47.5 | 97.57 |
| 135 | 152.4 | Light | 4.5 | 164 | 2090 | 4788 | 4505 | 572.2 | 75.1 | 52.3 | 98.47 |
| | | Medium | 4.85 | 177 | 2250 | | 4483 | 612.5 | 80.4 | 52.2 | 105.77 |
| | | Heavy | 5.4 | 195 | 2490 | | 4449 | 674.5 | 88.5 | 52.0 | 116.62 |

(*Contd*)

**TABLE A.10** *(Contd)*

| Normal bore (mm) | Outside diameter (mm) | Class (mm) | Thickness (mm) | Weight (N/m) | Area of cross section (mm²) | Surface External (cm²/m) | Surface Internal | Moment of inertia (cm⁴) | Modulus section (cm³) | Radius of gyration (mm) | Plastic section modulus(cm³) |
|---|---|---|---|---|---|---|---|---|---|---|---|
| 150 | 165.1 | Light | 4.5 | 178 | 2270 | 5187 | 4904 | 732.6 | 88.7 | 56.8 | 116.13 |
| | | Medium | 4.85 | 192 | 2440 | | 4882 | 784.5 | 95 | 56.7 | 124.56 |
| | | Heavy | 5.4 | 212 | 2710 | | 4847 | 864.7 | 105 | 56.5 | 137.88 |
| 150 | 168.3 | Light | 4.5 | 181 | 2320 | 5287 | 5005 | 777.2 | 92.4 | 57.9 | 121.05 |
| | | Medium | 4.85 | 196 | 2490 | | 4982 | 832.4 | 98.9 | 57.8 | 129.65 |
| | | Heavy1 | 5.4 | 217 | 2760 | | 4948 | 917.7 | 109 | 57.6 | 143.24 |
| | | Heavy2 | 6.3 | 253 | 3210 | | 4891 | 1053 | 125 | 57.3 | 165.69 |
| 175 | 193.7 | Light | 4.85 | 226 | 2870 | 6085 | 5780 | 1284 | 133 | 66.8 | 172.65 |
| | | Medium | 5.4 | 250 | 3190 | | 5746 | 1417 | 146 | 66.6 | 191.35 |
| | | Heavy | 5.9 | 273 | 3480 | | 5715 | 1536 | 159 | 66.4 | 208.20 |
| 200 | 219.1 | Light | 4.85 | 257 | 3260 | 6883 | 6579 | 1874 | 171 | 75.8 | 222.48 |
| | | Medium | 5.6 | 294 | 3760 | | 6531 | 2142 | 195 | 75.5 | 255.71 |
| | | Heavy | 5.9 | 310 | 3950 | | 6513 | 2247 | 205 | 75.4 | 268.27 |
| 225 | 244.5 | Heavy | 5.9 | 342 | 4420 | 7681 | 7311 | 3149 | 258 | 84.4 | 335.93 |

**TABLE A.11** Sectional properties of square hollow sections

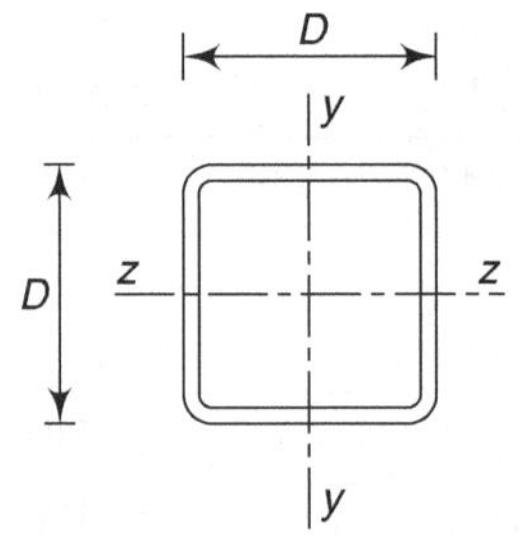


| Square hollow sections $D \times B$ (mm) | Thickness $t$ (mm) | Unit weight $w$ (N/m) | Sectional area $A$ (mm²) | Moment of inertia $I_z$ (cm⁴) | $I_y$ (cm⁴) | Radius of gyration $r_z$ (mm) | $r_y$ (mm) | Elastic modulus $Z_z$ (cm³) | $Z_y$ (cm³) | Plastic modulus $Z_{pz}$ (cm³) | $Z_{py}$ (cm³) | Torsional constant $J$(cm⁴) |
|---|---|---|---|---|---|---|---|---|---|---|---|---|
| 25 × 25 | 1.6 | 11.2 | 143 | 1.28 | 1.28 | 9.4 | 9.4 | 1.02 | 1.02 | 1.24 | 1.24 | 1.96 |
| | 2 | 13.6 | 174 | 1.48 | 1.48 | 9.2 | 9.2 | 1.19 | 1.19 | 1.48 | 1.48 | 2.29 |
| | 2.6 | 16.9 | 216 | 1.72 | 1.72 | 8.9 | 8.9 | 1.38 | 1.38 | 1.76 | 1.76 | 2.68 |
| | 3.2 | 19.8 | 253 | 1.89 | 1.89 | 8.6 | 8.6 | 1.51 | 1.51 | 2.22 | 2.22 | 2.96 |
| 32 × 32 | 2 | 18 | 230 | 3.36 | 3.36 | 12.1 | 12.1 | 2.1 | 2.1 | 2.21 | 2.21 | 5.3 |
| | 2.6 | 22.6 | 288 | 4.02 | 4.02 | 11.8 | 11.8 | 2.51 | 2.51 | 3.11 | 3.11 | 6.45 |
| | 3.2 | 26.9 | 342 | 4.54 | 4.54 | 11.5 | 11.5 | 2.84 | 2.84 | 3.59 | 3.59 | 7.41 |
| 38 × 38 | 2 | 21.8 | 278 | 5.88 | 5.88 | 14.6 | 14.6 | 3.1 | 3.1 | 3.7 | 3.7 | 9.31 |
| | 2.6 | 27.5 | 351 | 7.14 | 7.14 | 14.3 | 14.3 | 3.76 | 3.76 | 4.57 | 4.57 | 11.51 |
| | 3.2 | 32.9 | 419 | 8.18 | 8.18 | 14 | 14 | 4.3 | 4.3 | 5.34 | 5.34 | 13.45 |
| | 4 | 39.5 | 503 | 9.26 | 9.26 | 13.6 | 13.6 | 4.87 | 4.87 | 6.22 | 6.22 | 15.67 |
| 40 × 40 | 2.6 | 29.2 | 372 | 8.45 | 8.45 | 15.1 | 15.1 | 4.22 | 4.22 | 5.12 | 5.12 | 13.63 |
| | 2.9 | 32.1 | 409 | 9.11 | 9.11 | 14.9 | 14.9 | 4.56 | 4.56 | 5.58 | 5.58 | 14.85 |

*(Contd)*

**TABLE A.11** (*Contd*)

| Square hollow sections $D \times B$ (mm) | Thickness $t$ (mm) | Unit weight $w$ (N/m) | Sectional area $A$ (mm$^2$) | Moment of inertia | | Radius of gyration | | Elastic modulus | | Plastic modulus | | Torsional constant |
|---|---|---|---|---|---|---|---|---|---|---|---|---|
| | | | | $I_z$ (cm$^4$) | $I_y$ (cm$^4$) | $r_z$ (mm) | $r_y$ (mm) | $Z_z$ (cm$^3$) | $Z_y$ (cm$^3$) | $Z_{pz}$ (cm$^3$) | $Z_{py}$ (cm$^3$) | $J$(cm$^4$) |
| | 3.2 | 34.9 | 445 | 9.72 | 9.72 | 14.8 | 14.8 | 4.86 | 4.86 | 6.01 | 6.01 | 16 |
| | 4 | 42 | 535 | 11.07 | 11.07 | 14.4 | 14.4 | 5.54 | 5.54 | 7.02 | 7.02 | 18.75 |
| 49.5 × 49.5 | 2.6 | 36.9 | 470 | 16.91 | 16.91 | 19 | 19 | 6.83 | 6.83 | 8.17 | 8.17 | 27.19 |
| | 2.9 | 40.7 | 519 | 18.37 | 18.37 | 18.8 | 18.8 | 7.42 | 7.42 | 8.93 | 8.93 | 29.81 |
| | 3.6 | 49.3 | 628 | 21.42 | 21.42 | 18.5 | 18.5 | 8.66 | 8.66 | 12.81 | 12.81 | 35.54 |
| | 4.5 | 59.5 | 758 | 24.64 | 24.64 | 18 | 18 | 9.96 | 9.96 | 12.47 | 12.47 | 72.15 |
| 60 × 60 | 2.6 | 45.5 | 580 | 31.33 | 31.33 | 23.3 | 23.3 | 10.44 | 10.44 | 12.34 | 12.34 | 50.08 |
| | 2.9 | 50.3 | 641 | 34.21 | 34.21 | 23.1 | 23.1 | 11.4 | 11.4 | 13.56 | 13.56 | 56.12 |
| | 3.2 | 55 | 701 | 36.94 | 36.94 | 23 | 23 | 12.31 | 12.31 | 14.73 | 14.73 | 60.02 |
| | 4 | 67.1 | 855 | 43.55 | 43.55 | 22.6 | 22.6 | 14.52 | 14.52 | 17.64 | 17.64 | 72.41 |
| | 4.8 | 78.5 | 1001 | 49.22 | 49.22 | 22.2 | 22.2 | 16.41 | 16.41 | 20.27 | 20.27 | 83.86 |
| 72 × 72 | 3.2 | 67.1 | 854 | 66.32 | 66.32 | 27.9 | 27.9 | 18.42 | 18.42 | 21.8 | 21.8 | 106.81 |
| | 4 | 82.2 | 1047 | 79.03 | 79.03 | 27.5 | 27.5 | 21.95 | 21.95 | 26.32 | 26.32 | 129.85 |
| | 4.8 | 96.6 | 1231 | 90.31 | 90.31 | 27.1 | 27.1 | 25.09 | 25.09 | 30.49 | 30.49 | 151.55 |
| 91.5 × 91.5 | 3.6 | 96.7 | 1232 | 156.49 | 156.49 | 35.6 | 35.6 | 34.21 | 34.21 | 40.24 | 40.24 | 251.17 |
| | 4.5 | 118.8 | 1514 | 187.57 | 187.57 | 35.2 | 35.2 | 41 | 41 | 48.8 | 48.8 | 306.78 |
| | 5.4 | 140.1 | 1785 | 215.68 | 215.68 | 34.8 | 34.8 | 47.14 | 47.14 | 56.77 | 56.77 | 359.76 |
| 113.5 × 113.5 | 4.8 | 159.2 | 2028 | 393.3 | 393.3 | 44 | 44 | 69.3 | 69.3 | 81.81 | 81.81 | 637.45 |
| | 5.4 | 177.4 | 2260 | 432.58 | 432.58 | 43.8 | 43.8 | 76.23 | 76.23 | 90.55 | 90.55 | 708.69 |
| 132 × 132 | 4.8 | 187.1 | 2383 | 634.39 | 634.39 | 51.6 | 51.6 | 96.12 | 96.12 | 112.69 | 112.69 | 1018.3 |
| | 5.4 | 208.8 | 2659 | 700.11 | 700.11 | 51.3 | 51.3 | 106.08 | 106.08 | 125.02 | 125.02 | 1134.25 |
| 180 × 180 | 4 | 217.8 | 2775 | 1421.74 | 1421.74 | 71.6 | 71.6 | 157.97 | 157.97 | 182.21 | 182.21 | 2224.31 |
| | 5 | 269.7 | 3436 | 1736.87 | 1736.87 | 71.1 | 71.1 | 192.99 | 192.99 | 224.02 | 224.02 | 2747.93 |
| | 6 | 320.5 | 4083 | 2036.52 | 2036.52 | 70.6 | 70.6 | 226.28 | 226.28 | 264.35 | 264.35 | 3259.23 |
| | 7 | 370.3 | 4718 | 2321.04 | 2321.04 | 70.1 | 70.1 | 257.89 | 257.89 | 303.24 | 303.24 | 3758.53 |
| 220 × 220 | 4 | 266.1 | 3415 | 2639.14 | 2639.14 | 87.9 | 87.9 | 239.92 | 239.92 | 275.47 | 275.47 | 4099.49 |
| | 5 | 332.5 | 4236 | 3238.02 | 3238.02 | 87.4 | 87.4 | 294.37 | 294.37 | 339.73 | 339.73 | 5076.22 |
| | 6 | 395.9 | 5043 | 3813.36 | 3813.36 | 87 | 87 | 346.67 | 346.67 | 402.18 | 402.18 | 6034.53 |
| | 7 | 458.3 | 5838 | 4365.55 | 4365.55 | 86.5 | 86.5 | 396.67 | 396.67 | 462.83 | 462.83 | 6974.82 |
| 250 × 250 | 5 | 379.6 | 4836 | 4805.01 | 4805.01 | 99.7 | 99.7 | 384.4 | 384.4 | 442.26 | 442.26 | 7494.83 |
| | 6 | 452.4 | 5763 | 5672 | 5672 | 99.2 | 99.2 | 453.76 | 453.76 | 524.45 | 524.45 | 8920.44 |
| | 7 | 524.2 | 6678 | 6508.73 | 6508.73 | 98.7 | 98.7 | 520.7 | 520.7 | 604.58 | 604.58 | 10322.7 |

**TABLE A.12** Sectional properties of rectangular hollow sections

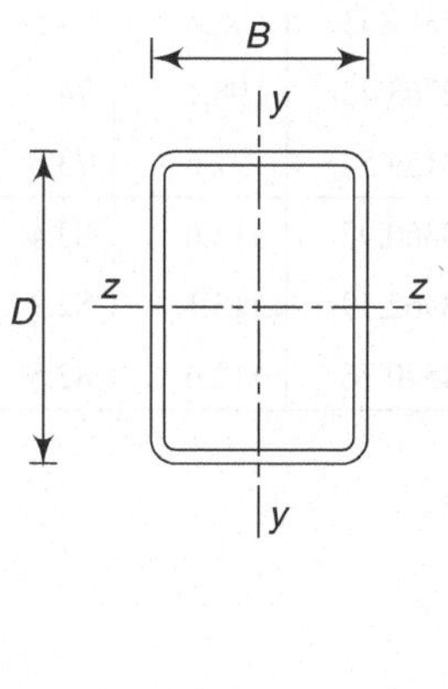

| RHS $D \times B$ (mm) | Thickness $t$ (mm) | Unit weight $w$ (N/m) | Area $A$ (mm²) | Moment of inertia | | Radius of gyration | | Elastic modulus | | Plastic modulus | | Torsional constant $J$(cm⁴) |
|---|---|---|---|---|---|---|---|---|---|---|---|---|
| | | | | $I_z$ (cm⁴) | $I_y$ (cm⁴) | $r_z$ (mm) | $r_y$ (mm) | $Z_z$ (cm³) | $Z_y$ (cm³) | $Z_{pz}$ (cm³) | $Z_{py}$ (cm³) | |
| 50 × 25 | 2 | 21.5 | 274 | 8.38 | 2.81 | 17.5 | 10.1 | 3.35 | 2.25 | 4.26 | 2.62 | 6.79 |
| | 2.6 | 27.1 | 346 | 10.16 | 3.36 | 17.1 | 9.9 | 4.06 | 2.69 | 5.26 | 3.17 | 8.27 |
| | 3.2 | 32.4 | 413 | 11.63 | 3.8 | 16.8 | 9.6 | 4.65 | 3.04 | 6.14 | 3.73 | 9.52 |
| | 4 | 38.8 | 495 | 13.13 | 4.23 | 16.3 | 9.2 | 5.25 | 3.38 | 7.13 | 4.29 | 10.86 |
| 60 × 40 | 2.6 | 37.3 | 476 | 22.76 | 12.09 | 21.9 | 15.9 | 7.59 | 6.05 | 9.36 | 7.07 | 25.59 |
| | 2.9 | 41.2 | 525 | 24.74 | 13.11 | 21.7 | 15.8 | 8.25 | 6.56 | 10.25 | 7.73 | 28.02 |
| | 3.6 | 49.8 | 635 | 28.9 | 15.23 | 21.3 | 15.5 | 9.63 | 7.62 | 12.16 | 9.15 | 33.3 |
| | 4.5 | 60.2 | 767 | 33.31 | 17.44 | 20.8 | 15.1 | 11.1 | 8.72 | 14.32 | 10.75 | 39.34 |
| 66 × 33 | 2.6 | 36.9 | 470 | 25.15 | 8.43 | 23.1 | 13.4 | 7.62 | 5.11 | 9.68 | 5.94 | 20.75 |
| | 2.9 | 40.7 | 519 | 27.33 | 9.12 | 22.9 | 13.3 | 8.28 | 5.53 | 10.59 | 6.49 | 22.65 |
| | 3.6 | 49.3 | 628 | 31.87 | 10.52 | 22.5 | 12.9 | 9.66 | 6.37 | 12.56 | 7.66 | 26.71 |
| | 4.5 | 59.5 | 758 | 36.64 | 11.93 | 22 | 12.5 | 11.1 | 7.23 | 14.77 | 8.94 | 31.21 |
| 80 × 40 | 2.6 | 45.5 | 580 | 46.58 | 15.74 | 28.4 | 16.5 | 11.65 | 7.87 | 14.64 | 9.01 | 38.5 |
| | 2.9 | 50.3 | 641 | 50.87 | 17.11 | 28.2 | 16.3 | 12.72 | 8.56 | 16.08 | 9.88 | 42.23 |
| | 3.2 | 55 | 701 | 54.94 | 18.41 | 28 | 16.2 | 13.74 | 9.21 | 17.46 | 10.72 | 45.83 |
| | 4 | 67.1 | 855 | 64.79 | 21.49 | 27.5 | 15.9 | 16.2 | 10.74 | 20.91 | 12.78 | 54.77 |
| | 4.8 | 78.5 | 1001 | 73.22 | 24.03 | 27.1 | 15.5 | 18.3 | 12.02 | 24.01 | 14.6 | 62.81 |
| 96 × 48 | 3.2 | 67.1 | 854 | 98.61 | 33.28 | 34 | 19.7 | 20.54 | 13.87 | 25.85 | 15.92 | 82.13 |
| | 4 | 82.2 | 1047 | 117.54 | 39.32 | 33.5 | 19.4 | 24.49 | 16.38 | 31.21 | 19.14 | 99.11 |
| | 4.8 | 96.6 | 1231 | 134.35 | 44.55 | 33 | 19 | 27.99 | 18.56 | 36.13 | 22.08 | 114.8 |
| 122 × 61 | 3.6 | 96.7 | 1232 | 232.61 | 78.83 | 43.4 | 25.3 | 38.13 | 25.84 | 47.71 | 29.43 | 193.91 |
| | 4.5 | 118.8 | 1514 | 278.94 | 93.78 | 42.9 | 24.9 | 45.73 | 30.75 | 57.85 | 35.56 | 235.39 |
| | 5.4 | 140.1 | 1785 | 320.83 | 107.03 | 42.4 | 24.5 | 52.6 | 35.09 | 67.29 | 41.22 | 274.29 |
| 120 × 60 | 3.2 | 85.1 | 1085 | 199.88 | 67.95 | 42.9 | 25 | 33.31 | 22.65 | 41.51 | 25.63 | 165.83 |
| 145 × 82 | 4.8 | 159.2 | 2028 | 555.16 | 228.5 | 52.3 | 33.6 | 76.57 | 55.73 | 94.93 | 63.93 | 534.27 |
| | 5.4 | 177.4 | 2260 | 610.85 | 250.59 | 52 | 33.3 | 84.26 | 61.12 | 105.07 | 70.66 | 592.7 |
| 172 × 92 | 4.8 | 187.1 | 2383 | 917.13 | 346.91 | 62 | 38.2 | 106.64 | 75.41 | 132.08 | 87.82 | 826.04 |
| | 5.4 | 208.8 | 2659 | 1012.47 | 381.74 | 61.7 | 37.9 | 117.73 | 82.99 | 146.55 | 94.86 | 918.1 |
| 220 × 140 | 4 | 217.8 | 2775 | 1892.55 | 947.64 | 82.6 | 58.4 | 172.05 | 135.38 | 206.35 | 151.68 | 2000.01 |
| | 5 | 269.7 | 3436 | 2313.36 | 1155.23 | 82.1 | 58 | 210.31 | 165.03 | 253.73 | 186.3 | 2467.63 |
| | 6 | 320.5 | 4083 | 2713.97 | 1351.66 | 81.5 | 57.5 | 246.72 | 193.09 | 299.46 | 219.65 | 2922.95 |
| | 7 | 370.3 | 4718 | 3094.76 | 1537.22 | 81 | 57.1 | 281.34 | 219.6 | 343.55 | 251.72 | 3366.29 |
| 260 × 180 | 4 | 268.1 | 3415 | 3357.53 | 1917.45 | 99.2 | 74.9 | 258.27 | 213.05 | 306 | 238.53 | 3822.78 |
| | 5 | 332.5 | 4236 | 4121.36 | 2349.53 | 98.6 | 74.5 | 317.03 | 261.06 | 377.44 | 294.02 | 4730.34 |
| | 6 | 395.9 | 5043 | 4855.87 | 2763.43 | 98.1 | 74 | 373.53 | 307.05 | 446.88 | 347.87 | 5619.5 |
| | 7 | 458.3 | 5838 | 5561.5 | 3159.5 | 97.6 | 73.6 | 427.81 | 351.06 | 514.35 | 400.12 | 6490.62 |
| 300 × 200 | 5 | 379.6 | 4836 | 6241.05 | 3360.92 | 113.6 | 83.4 | 416.07 | 336.09 | 495.65 | 376.37 | 6882.77 |
| | 6 | 452.4 | 5736 | 7370.23 | 3962.19 | 113.1 | 82.9 | 491.35 | 396.22 | 587.83 | 446.07 | 8186.02 |
| | 7 | 524.2 | 6678 | 8460.99 | 4540.76 | 112.6 | 82.5 | 564.06 | 454.09 | 677.72 | 513.93 | 9465.89 |

**TABLE A.13** Sectional properties of cold-formed lipped channel sections

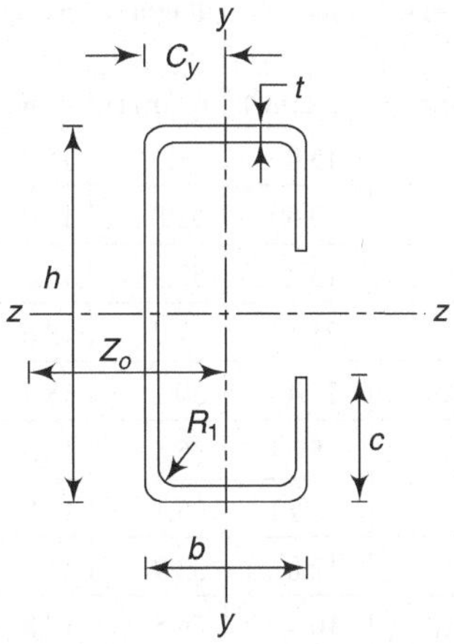


| Designation $h \times b \times c \times t$ (mm) | Weight (N/m) | Area $A$ (mm$^2$) | Centre of gravity | Moment of inertia | | Radius of gyration | | Elastic section modulus | | Shear centre $Z_o$ (mm) | Torsional constant, $J$ (cm$^4$) | Warping constant $C_w$ (cm$^6$) |
|---|---|---|---|---|---|---|---|---|---|---|---|---|
| | | | $C_y$ (mm) | $I_z$ (mm$^4$) | $I_y$ (mm$^4$) | $r_z$ (mm) | $r_y$ (mm) | $Z_z$ (cm$^3$) | $Z_y$ (cm$^3$) | | | |
| 30 × 15 × 10 × 1.25 | 6.79 | 86.6 | 6.51 | 1.05 | 0.292 | 11 | 5.81 | 0.697 | 0.344 | 16.6 | 0.004 | 1.03 |
| 30 × 15 × 10 × 1.60 | 8.32 | 106 | 6.49 | 1.22 | 0.337 | 10.7 | 5.64 | 0.815 | 0.395 | 16.8 | 0.009 | 1.14 |
| 40 × 20 × 10 × 1.25 | 8.76 | 112 | 7.9 | 2.62 | 0.657 | 15.3 | 7.67 | 1.31 | 0.543 | 20 | 0.006 | 3.06 |
| 40 × 20 × 10 × 1.60 | 10.8 | 138 | 7.87 | 3.14 | 0.773 | 15.1 | 7.48 | 1.57 | 0.637 | 20.1 | 0.011 | 3.47 |
| 50 × 25 × 10 × 1.25 | 10.7 | 137 | 9.24 | 5.23 | 1.22 | 19.6 | 9.46 | 2.09 | 0.776 | 23.3 | 0.007 | 7.59 |
| 50 × 25 × 10 × 1.60 | 13.3 | 170 | 9.2 | 6.36 | 1.46 | 19.3 | 9.26 | 2.54 | 0.923 | 23.3 | 0.014 | 8.78 |
| 50 × 25 × 15 × 2.00 | 17.7 | 226 | 10.5 | 7.79 | 2.08 | 18.6 | 9.6 | 3.12 | 1.43 | 26.7 | 0.029 | 17.8 |
| 50 × 40 × 10 × 1.25 | 13.7 | 174 | 15.8 | 7.46 | 3.81 | 20.7 | 14.8 | 2.98 | 1.57 | 37.8 | 0.009 | 23.5 |
| 50 × 40 × 10 × 1.60 | 17.1 | 218 | 15.8 | 9.17 | 4.62 | 20.5 | 14.6 | 3.67 | 1.91 | 38 | 0.018 | 28 |
| 50 × 40 × 15 × 2.00 | 22.4 | 286 | 17.3 | 11.2 | 6.45 | 19.8 | 15 | 4.5 | 2.85 | 42 | 0.037 | 55.1 |
| 50 × 40 × 15 × 3.15 | 32.9 | 419 | 17.2 | 15.4 | 8.63 | 19.2 | 14.4 | 6.16 | 3.79 | 42.8 | 0.133 | 69.7 |
| 60 × 30 × 10 × 1.60 | 15.9 | 202 | 10.5 | 11.2 | 2.44 | 23.6 | 11 | 3.73 | 1.25 | 26.5 | 0.017 | 19.4 |
| 60 × 30 × 15 × 2.00 | 20.8 | 266 | 11.8 | 13.9 | 3.48 | 22.9 | 11.4 | 4.64 | 1.92 | 30 | 0.034 | 36.1 |
| 60 × 30 × 20 × 3.15 | 32.9 | 419 | 13 | 19.4 | 5.34 | 21.5 | 11.3 | 6.46 | 3.14 | 33.5 | 0.133 | 72.6 |
| 60 × 30 × 20 × 4.00 | 39.4 | 502 | 12.9 | 21.9 | 5.92 | 20.9 | 10.9 | 7.31 | 3.47 | 34 | 0.254 | 76.4 |
| 60 × 40 × 15 × 2.00 | 24 | 306 | 16.3 | 17.3 | 6.95 | 23.8 | 15.1 | 5.76 | 2.93 | 40 | 0.04 | 72.2 |
| 60 × 40 × 20 × 3.15 | 37.8 | 482 | 17.6 | 24.5 | 10.8 | 22.6 | 15 | 8.16 | 4.83 | 44 | 0.154 | 148 |
| 60 × 40 × 20 × 4.00 | 45.7 | 582 | 17.5 | 28.2 | 12.3 | 22 | 14.5 | 9.4 | 5.46 | 44.6 | 0.296 | 162 |
| 70 × 25 × 10 × 1.60 | 15.9 | 202 | 7.87 | 14.2 | 1.65 | 26.6 | 9.03 | 4.07 | 0.962 | 20.5 | 0.017 | 16.9 |
| 70 × 25 × 15 × 2.00 | 20.8 | 266 | 9.04 | 17.9 | 2.38 | 26 | 9.48 | 5.12 | 1.49 | 23.5 | 0.034 | 30 |
| 70 × 25 × 20 × 3.15 | 32.9 | 419 | 10 | 25.2 | 3.65 | 24.6 | 9.34 | 7.21 | 2.44 | 26.4 | 0.133 | 55.4 |
| 70 × 30 × 15 × 2.00 | 22.4 | 286 | 11.1 | 20.2 | 3.7 | 26.6 | 11.4 | 5.78 | 1.95 | 28.4 | 0.037 | 46.6 |
| 70 × 30 × 20 × 3.15 | 35.3 | 450 | 12.2 | 28.7 | 5.73 | 25.3 | 11.3 | 8.21 | 3.21 | 31.7 | 0.144 | 88.3 |
| 70 × 40 × 15 × 2.00 | 25.6 | 326 | 15.3 | 24.9 | 7.39 | 27.6 | 15.1 | 7.1 | 2.99 | 38.2 | 0.043 | 93.2 |
| 70 × 40 × 20 × 3.15 | 40.3 | 513 | 16.6 | 35.8 | 11.6 | 26.4 | 15 | 10.2 | 4.95 | 42 | 0.164 | 181 |
| 70 × 40 × 25 × 4.00 | 52 | 662 | 17.8 | 42.2 | 14.9 | 25.2 | 15 | 12.1 | 6.73 | 45.5 | 0.339 | 303 |
| 80 × 40 × 10 × 1.60 | 20.9 | 266 | 13.1 | 27 | 5.51 | 31.9 | 14.4 | 6.76 | 2.04 | 32.8 | 0.022 | 70.6 |
| 80 × 40 × 20 × 3.15 | 42.8 | 545 | 15.7 | 49.7 | 12.2 | 30.2 | 15 | 12.4 | 5.05 | 40.2 | 0.175 | 221 |

*(Contd)*

**TABLE A.13** *(Contd)*

| Designation $h \times b \times c \times t$ (mm) | Weight (N/m) | Area $A$ (mm$^2$) | Centre of gravity | Moment of inertia | | Radius of gyration | | Elastic section modulus | | Shear centre $Z_0$ (mm) | Torsional constant, | Warping constant |
|---|---|---|---|---|---|---|---|---|---|---|---|---|
| | | | $C_y$ (mm) | $I_z$ (mm$^4$) | $I_y$ (mm$^4$) | $r_z$ (mm) | $r_y$ (mm) | $Z_z$ (cm$^3$) | $Z_y$ (cm$^3$) | | $J$ (cm$^4$) | $C_w$ (cm$^6$) |
| 80 × 40 × 25 × 4.00 | 55.1 | 702 | 16.9 | 59.3 | 15.9 | 29.1 | 15 | 14.8 | 6.88 | 43.6 | 0.36 | 354 |
| 80 × 50 × 10 × 1.60 | 23.4 | 298 | 17.2 | 32 | 9.59 | 32.8 | 17.9 | 7.99 | 2.92 | 42.1 | 0.025 | 123 |
| 80 × 50 × 15 × 2.00 | 30.3 | 386 | 18.8 | 40.1 | 13.3 | 32.3 | 18.6 | 10 | 4.27 | 46.1 | 0.05 | 203 |
| 80 × 50 × 20 × 3.15 | 47.7 | 608 | 20.1 | 56 | 21.1 | 31.2 | 18.6 | 14.7 | 7.07 | 50.2 | 0.196 | 382 |
| 80 × 50 × 25 × 4.00 | 61.4 | 782 | 21.5 | 70.8 | 27.4 | 30.1 | 18.7 | 17.7 | 9.61 | 54 | 0.403 | 616 |
| 90 × 40 × 10 × 1.60 | 22.1 | 282 | 12.4 | 35.6 | 5.74 | 35.5 | 14.3 | 7.9 | 2.07 | 31.5 | 0.024 | 90.8 |
| 90 × 40 × 15 × 2.00 | 28.7 | 365 | 13.8 | 45 | 8.12 | 35.1 | 14.9 | 9.99 | 3.09 | 35 | 0.048 | 148 |
| 90 × 40 × 20 × 3.15 | 45.2 | 576 | 15 | 66.3 | 12.8 | 33.9 | 14.9 | 14.72 | 5.13 | 38.5 | 0.185 | 267 |
| 90 × 50 × 10 × 1.60 | 24.6 | 314 | 16.4 | 41.8 | 10 | 36.5 | 17.8 | 9.29 | 2.97 | 40.6 | 0.026 | 158 |
| 90 × 50 × 15 × 2.00 | 31.8 | 406 | 17.9 | 52.7 | 13.9 | 36 | 18.5 | 11.7 | 4.34 | 44.4 | 0.053 | 253 |
| 90 × 50 × 20 × 3.15 | 50.2 | 639 | 19.2 | 78.1 | 22.1 | 35 | 18.6 | 17.4 | 7.2 | 48.4 | 0.206 | 463 |
| 100 × 40 × 10 × 1.60 | 23.4 | 298 | 11.8 | 45.5 | 5.94 | 39.1 | 14.1 | 9.09 | 2.1 | 30.2 | 0.025 | 114 |
| 100 × 40 × 15 × 2.00 | 30.3 | 386 | 13.1 | 57.7 | 8.43 | 38.7 | 14.8 | 11.5 | 3.13 | 33.6 | 0.05 | 182 |
| 100 × 40 × 25 × 3.15 | 50.2 | 639 | 15.5 | 88.1 | 15.1 | 37.1 | 15.4 | 17.6 | 6.17 | 39.9 | 0.206 | 438 |
| 100 × 50 × 15 × 2.00 | 33.4 | 426 | 17.1 | 67.3 | 14.5 | 39.8 | 18.4 | 13.5 | 4.4 | 42.9 | 0.056 | 312 |
| 100 × 50 × 20 × 3.15 | 52.6 | 671 | 18.4 | 101 | 23.1 | 38.7 | 18.6 | 20.1 | 7.3 | 46.6 | 0.216 | 557 |
| 100 × 50 × 25 × 4.00 | 67.7 | 862 | 19.7 | 123 | 30.2 | 37.7 | 18.7 | 24.5 | 9.95 | 50.2 | 0.446 | 847 |
| 100 × 25 × 25 × 4.00 | 52 | 662 | 9.33 | 76.5 | 5.43 | 34 | 9.05 | 15.3 | 3.46 | 24.6 | 0.339 | 144 |
| 100 × 60 × 15 × 2.00 | 36.6 | 466 | 21.3 | 76.9 | 22.6 | 40.6 | 22 | 15.4 | 5.84 | 52.3 | 0.061 | 485 |
| 100 × 60 × 20 × 3.15 | 57.6 | 734 | 22.7 | 115 | 36.1 | 39.6 | 22.3 | 23.1 | 9.66 | 56.4 | 0.237 | 872 |
| 100 × 60 × 25 × 4.00 | 74 | 942 | 24.1 | 141 | 47.1 | 38.7 | 22.4 | 28.2 | 13.1 | 60.3 | 0.488 | 1330 |
| 100 × 60 × 25 × 5.00 | 89.1 | 1130 | 23.9 | 164 | 53.9 | 38 | 21.8 | 32.9 | 15 | 60.7 | 0.912 | 1460 |
| 120 × 50 × 15 × 2.00 | 39.9 | 467 | 15.7 | 103 | 15.4 | 47 | 18.2 | 17.2 | 4.5 | 40.4 | 0.061 | 453 |
| 120 × 50 × 20 × 3.15 | 57.6 | 734 | 17 | 155 | 24.7 | 46 | 18.4 | 25.9 | 7.48 | 43.6 | 0.237 | 786 |
| 120 × 50 × 25 × 4.00 | 74 | 942 | 18.2 | 192 | 32.5 | 45.1 | 18.6 | 31.9 | 10.2 | 46.9 | 0.488 | 1150 |
| 120 × 50 × 25 × 5.00 | 89.1 | 1130 | 18.1 | 223 | 36.8 | 44.3 | 18 | 39.2 | 11.5 | 47.1 | 0.912 | 1240 |
| 120 × 60 × 20 × 3.15 | 62.5 | 797 | 21 | 177 | 38.6 | 47.1 | 22 | 29.5 | 9.91 | 53 | 0.258 | 1230 |
| 120 × 60 × 25 × 4.00 | 80.2 | 1020 | 22.3 | 216 | 50.7 | 46.2 | 22.3 | 36.4 | 13.5 | 56.7 | 0.531 | 1800 |
| 120 × 60 × 25 × 5.00 | 96.9 | 1240 | 22.2 | 256 | 58.2 | 45.5 | 21.7 | 42.7 | 15.4 | 57 | 0.995 | 1990 |
| 140 × 60 × 20 × 3.15 | 67.5 | 860 | 19.6 | 255 | 40.8 | 54.4 | 21.8 | 36.4 | 10.1 | 51 | 0.279 | 1670 |
| 140 × 60 × 25 × 4.00 | 86.5 | 1100 | 20.9 | 316 | 53.8 | 53.6 | 22.1 | 45.2 | 13.7 | 53.5 | 0.574 | 2390 |
| 140 × 60 × 25 × 5.00 | 106 | 1330 | 20.7 | 373 | 61.8 | 52.8 | 21.5 | 53.2 | 15.7 | 53.4 | 1.08 | 2640 |
| 150 × 50 × 20 × 3.15 | 65 | 828 | 15.2 | 266 | 26.7 | 56.6 | 18 | 35.4 | 7.67 | 39.7 | 0.268 | 1240 |
| 150 × 50 × 25 × 4.00 | 83.4 | 1060 | 16.3 | 331 | 35.3 | 55.8 | 18.2 | 44.1 | 10.5 | 42.7 | 0.552 | 1750 |
| 150 × 50 × 25 × 5.00 | 102 | 1280 | 16.2 | 388 | 40.1 | 55 | 17.7 | 51.8 | 11.9 | 42.7 | 1.04 | 1900 |
| 180 × 50 × 20 × 3.15 | 72.4 | 923 | 13.8 | 413 | 28.3 | 66.9 | 17.5 | 45.9 | 7.82 | 36.6 | 0.38 | 1840 |
| 180 × 50 × 25 × 4.00 | 92.8 | 1180 | 14.9 | 518 | 37.5 | 66.2 | 17.8 | 57.6 | 10.7 | 33.9 | 0.616 | 2540 |
| 180 × 50 × 25 × 5.00 | 113 | 1430 | 14.8 | 611 | 42.6 | 65.3 | 17.2 | 67.9 | 12.1 | 39.2 | 1.16 | 2790 |
| 180 × 80 × 20 × 3.15 | 87.3 | 1110 | 24.8 | 561 | 90.7 | 71 | 28.6 | 62.4 | 16.4 | 62.9 | 0.362 | 5790 |
| 180 × 80 × 25 × 4.00 | 112 | 1420 | 26.1 | 704 | 119.4 | 70.4 | 28.9 | 78.2 | 22.2 | 66.5 | 0.744 | 7990 |

*(Contd)*

**TABLE A.13** *(Contd)*

| Designation $h \times b \times c \times t$ (mm) | Weight (N/m) | Area $A$ ($mm^2$) | Centre of gravity $C_y$ (mm) | Moment of inertia | | Radius of gyration | | Elastic section modulus | | Shear centre $Z_0$ (mm) | Torsional constant, $J$ ($cm^4$) | Warping constant $C_w$ ($cm^6$) |
|---|---|---|---|---|---|---|---|---|---|---|---|---|
| | | | | $I_z$ ($mm^4$) | $I_y$ ($mm^4$) | $r_z$ (mm) | $r_y$ (mm) | $Z_z$ ($cm^3$) | $Z_y$ ($cm^3$) | | | |
| 180 × 80 × 25 × 5.00 | 136 | 1730 | 26 | 841 | 139.4 | 69.6 | 28.4 | 93.4 | 25.8 | 66.6 | 1.14 | 9090 |
| 200 × 50 × 20 × 3.15 | 77.4 | 986 | 13 | 535 | 29.2 | 73.6 | 17.2 | 53.5 | 7.89 | 34.8 | 0.321 | 2320 |
| 200 × 50 × 25 × 4.00 | 99.1 | 1260 | 14.1 | 672 | 38.8 | 73 | 17.5 | 67.2 | 10.8 | 37.4 | 0.659 | 3190 |
| 200 × 50 × 25 × 5.00 | 120 | 1530 | 14 | 795 | 44.1 | 72 | 16.9 | 79.5 | 12.2 | 37.2 | 1.24 | 3510 |
| 200 × 80 × 20 × 3.15 | 92.2 | 1170 | 23.5 | 718 | 93.9 | 78.2 | 28.3 | 71.8 | 16.6 | 60.4 | 0.383 | 7230 |
| 200 × 80 × 25 × 4.00 | 118 | 1500 | 24.8 | 903 | 124 | 77.5 | 28.7 | 90.3 | 22.4 | 63.9 | 0.787 | 9970 |
| 200 × 80 × 25 × 5.00 | 144 | 1830 | 24.7 | 1080 | 145 | 76.7 | 28.1 | 108 | 26.1 | 63.8 | 1.49 | 11900 |
| 250 × 50 × 20 × 3.15 | 89.7 | 1140 | 11.4 | 927 | 31 | 90 | 16.5 | 74.1 | 8.03 | 31 | 0.373 | 3850 |
| 250 × 50 × 25 × 4.00 | 115 | 1460 | 12.4 | 1170 | 41.3 | 89.5 | 16.8 | 93.7 | 11 | 33.3 | 0.766 | 5230 |
| 250 × 50 × 25 × 5.00 | 140 | 1780 | 12.4 | 1390 | 47 | 88.4 | 16.2 | 112 | 12 | 33 | 1.45 | 5830 |
| 250 × 80 × 20 × 3.15 | 105 | 1330 | 20.9 | 1210 | 101 | 95.5 | 27.5 | 97.2 | 17 | 55.1 | 0.435 | 11900 |
| 250 × 80 × 25 × 4.00 | 134 | 1700 | 22.1 | 1530 | 133 | 94.9 | 28 | 123 | 23 | 58.2 | 0.894 | 16200 |
| 250 × 80 × 25 × 5.00 | 164 | 2080 | 22 | 1840 | 156 | 94.1 | 27.3 | 148 | 26.8 | 58 | 1.7 | 18600 |

**TABLE A.14** Sectional properties of cold-formed zed sections

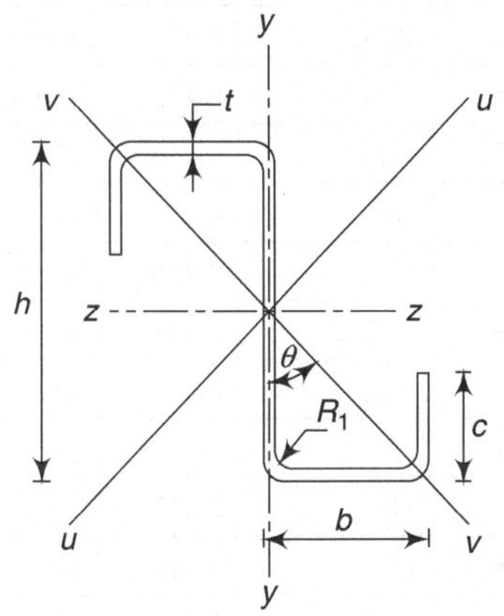


| Designation Dimensions $h \times b \times c \times t$ (mm) | Weight (N/m) | Area of section ($mm^2$) | Moment of inertia | | Radius of gyration min-$r_v$ (mm) | Angle tan $\theta$ | Section modulus | | | | Torsional constant $J$ ($cm^4$) | Warping constant $C_w$ ($cm^6$) |
|---|---|---|---|---|---|---|---|---|---|---|---|---|
| | | | $I_z$ ($cm^4$) | $I_y$ ($cm^4$) | | | $Z_z$ ($cm^3$) | $Z_y$ ($cm^3$) | $Z_u$ ($cm^3$) | $Z_v$ ($cm^3$) | | |
| 80 × 40 × 20 × 1.60 | 23.4 | 298 | 29.1 | 14.2 | 12.4 | 0.627 | 7.27 | 3.63 | 7.31 | 2.02 | 0.025 | 172 |
| 80 × 40 × 20 × 2.00 | 28.7 | 366 | 35.1 | 17 | 12.3 | 0.623 | 8.77 | 4.35 | 8.89 | 2.4 | 0.048 | 206 |
| 80 × 40 × 20 × 2.30 | 32.5 | 414 | 39.2 | 18.8 | 12.1 | 0.619 | 9.81 | 4.84 | 10 | 2.66 | 0.072 | 230 |
| 80 × 40 × 20 × 2.55 | 35.6 | 454 | 42.5 | 20.2 | 12 | 0.616 | 10.6 | 5.22 | 10.9 | 2.86 | 0.096 | 248 |
| 80 × 40 × 20 × 3.15 | 42.8 | 545 | 49.7 | 23.2 | 11.7 | 0.61 | 12.4 | 6.03 | 12.8 | 3.28 | 0.175 | 288 |
| 85 × 40 × 20 × 1.60 | 24 | 306 | 33.6 | 14.2 | 12.5 | 0.572 | 7.9 | 3.63 | 7.87 | 2.08 | 0.026 | 195 |
| 85 × 40 × 20 × 2.00 | 29.5 | 376 | 40.5 | 17 | 12.4 | 0.568 | 9.54 | 4.35 | 9.58 | 2.48 | 0.049 | 234 |
| 85 × 40 × 20 × 2.30 | 33.4 | 426 | 45.4 | 18.8 | 12.2 | 0.565 | 10.7 | 4.84 | 10.8 | 2.76 | 0.074 | 260 |
| 85 × 40 × 20 × 2.55 | 36.6 | 467 | 49.2 | 20.2 | 12.1 | 0.562 | 11.6 | 5.22 | 11.8 | 2.96 | 0.099 | 281 |
| 85 × 40 × 20 × 3.15 | 44 | 560 | 57.6 | 23.2 | 11.8 | 0.555 | 13.6 | 6.03 | 13.9 | 3.4 | 0.18 | 326 |
| 90 × 40 × 20 × 1.60 | 24.6 | 314 | 38.5 | 14.2 | 12.6 | 0.526 | 8.55 | 3.36 | 8.47 | 2.15 | 0.026 | 219 |
| 90 × 40 × 20 × 2.00 | 30.3 | 386 | 46.5 | 17 | 12.4 | 0.521 | 10.3 | 4.35 | 10.3 | 2.56 | 0.059 | 263 |

*(Contd)*

**TABLE A.14** *(Contd)*

| Designation Dimensions $h \times b \times c \times t$ (mm) | Weight (N/m) | Area of section ($mm^2$) | Moment of inertia | | Radius of gyration min-$r_v$ (mm) | Angle $\tan\theta$ | Section modulus | | | | Torsional constant $J$ ($cm^4$) | Warping constant $C_w$ ($cm^6$) |
|---|---|---|---|---|---|---|---|---|---|---|---|---|
| | | | $I_z$($cm^4$) | $I_y$($cm^4$) | | | $Z_z$ ($cm^3$) | $Z_y$ ($cm^3$) | $Z_u$ ($cm^3$) | $Z_v$ ($cm^3$) | | |
| 90 × 40 × 20 × 2.30 | 34.3 | 438 | 52.1 | 18.8 | 12.3 | 0.518 | 11.6 | 4.84 | 11.6 | 2.84 | 0.076 | 293 |
| 90 × 40 × 20 × 2.55 | 37.6 | 480 | 56.5 | 20.2 | 12.2 | 0.515 | 12.6 | 5.22 | 12.7 | 3.06 | 0.102 | 317 |
| 90 × 40 × 20 × 3.15 | 45.2 | 576 | 66.3 | 23.2 | 11.9 | 0.509 | 14.7 | 6.03 | 15 | 3.51 | 0.185 | 368 |
| 95 × 40 × 20 × 1.60 | 25.3 | 322 | 43.7 | 14.2 | 12.6 | 0.485 | 9.2 | 3.63 | 9.08 | 2.2 | 0.027 | 245 |
| 95 × 40 × 20 × 2.00 | 31.1 | 396 | 52.9 | 17 | 12.5 | 0.481 | 11.1 | 4.35 | 11.1 | 2.63 | 0.052 | 294 |
| 95 × 40 × 20 × 2.30 | 35.2 | 449 | 59.4 | 18.8 | 12.3 | 0.478 | 12.5 | 4.84 | 12.5 | 2.92 | 0.078 | 328 |
| 95 × 40 × 20 × 2.55 | 38.6 | 492 | 64.4 | 20.2 | 12.2 | 0.475 | 13.6 | 5.22 | 13.6 | 3.14 | 0.104 | 355 |
| 95 × 40 × 20 × 3.15 | 46.5 | 592 | 75.6 | 23.2 | 11.9 | 0.468 | 15.9 | 6.03 | 16.1 | 3.61 | 0.19 | 412 |
| 100 × 40 × 20 × 1.60 | 25.9 | 330 | 49.4 | 14.2 | 12.7 | 0.45 | 9.88 | 3.63 | 9.7 | 2.26 | 0.028 | 272 |
| 100 × 40 × 20 × 2.00 | 31.8 | 406 | 59.8 | 17 | 12.5 | 0.446 | 12 | 4.35 | 11.9 | 2.7 | 0.053 | 327 |
| 100 × 40 × 20 × 2.30 | 36.2 | 460 | 67.2 | 18.8 | 12.4 | 0.443 | 13.4 | 4.84 | 13.4 | 2.99 | 0.08 | 365 |
| 100 × 40 × 20 × 2.55 | 39.6 | 505 | 73 | 20.2 | 12.2 | 0.44 | 14.6 | 5.22 | 14.6 | 3.22 | 0.107 | 395 |
| 100 × 40 × 20 × 3.15 | 47.7 | 608 | 85.7 | 23.2 | 12 | 0.434 | 17.1 | 6.03 | 17.3 | 3.71 | 0.196 | 459 |
| 105 × 45 × 20 × 1.60 | 27.8 | 354 | 59.8 | 19.2 | 14 | 0.482 | 11.4 | 4.34 | 11.1 | 2.67 | 0.03 | 393 |
| 105 × 45 × 20 × 2.00 | 34.2 | 436 | 72.6 | 23 | 13.8 | 0.478 | 13.8 | 5.22 | 13.6 | 3.2 | 0.057 | 474 |
| 105 × 45 × 20 × 2.30 | 38.9 | 495 | 81.6 | 25.5 | 13.6 | 0.475 | 15.5 | 5.82 | 15.4 | 3.56 | 0.086 | 530 |
| 105 × 45 × 20 × 2.55 | 42.6 | 543 | 88.8 | 27.5 | 13.5 | 0.472 | 16.9 | 6.29 | 16.8 | 3.84 | 0.115 | 574 |
| 105 × 45 × 20 × 3.15 | 51.4 | 655 | 105 | 31.7 | 13.2 | 0.466 | 20 | 7.31 | 20 | 4.44 | 0.211 | 671 |
| 110 × 45 × 20 × 1.60 | 28.4 | 362 | 66.7 | 19.2 | 14 | 0.45 | 12.1 | 4.34 | 11.6 | 2.73 | 0.03 | 433 |
| 110 × 45 × 20 × 2.00 | 35 | 446 | 81.1 | 23 | 13.8 | 0.446 | 14.7 | 5.22 | 14.5 | 3.27 | 0.058 | 522 |
| 110 × 45 × 20 × 2.30 | 39.8 | 506 | 91.2 | 25.5 | 13.7 | 0.443 | 16.6 | 5.82 | 16.3 | 3.64 | 0.088 | 585 |
| 110 × 45 × 20 × 2.55 | 43.6 | 556 | 99.3 | 27.5 | 13.6 | 0.441 | 18 | 6.29 | 17.9 | 3.93 | 0.118 | 634 |
| 110 × 45 × 20 × 3.15 | 52.6 | 671 | 117 | 31.7 | 13.3 | 0.435 | 21.3 | 7.31 | 21.3 | 4.54 | 0.216 | 742 |
| 115 × 45 × 20 × 1.60 | 29 | 370 | 74.1 | 19.2 | 14 | 0.422 | 12.9 | 4.34 | 12.6 | 2.78 | 0.031 | 476 |
| 115 × 45 × 20 × 2.00 | 35.8 | 456 | 90.1 | 23.8 | 13.8 | 0.418 | 15.7 | 5.22 | 15.4 | 3.33 | 0.06 | 574 |
| 115 × 45 × 20 × 2.30 | 40.7 | 518 | 101 | 25.5 | 13.7 | 0.415 | 17.6 | 5.82 | 17.4 | 3.71 | 0.09 | 643 |
| 115 × 45 × 20 × 2.55 | 44.6 | 569 | 110 | 27.5 | 13.6 | 0.413 | 19.2 | 6.3 | 19 | 4.01 | 0.121 | 697 |
| 115 × 45 × 20 × 3.15 | 53.9 | 686 | 131 | 31.7 | 13.3 | 0.407 | 22.7 | 7.31 | 22.6 | 4.63 | 0.222 | 816 |
| 120 × 45 × 20 × 1.60 | 29.7 | 378 | 82 | 19.2 | 14 | 0.397 | 13.7 | 4.34 | 13.3 | 2.83 | 0.032 | 521 |
| 120 × 45 × 20 × 2.00 | 36.6 | 466 | 99.7 | 23 | 13.9 | 0.393 | 16.6 | 5.22 | 16.3 | 3.39 | 0.061 | 629 |
| 120 × 45 × 20 × 2.30 | 41.6 | 530 | 112 | 25.5 | 13.7 | 0.39 | 18.7 | 5.82 | 18.4 | 3.78 | 0.092 | 704 |
| 120 × 45 × 20 × 2.55 | 45.6 | 582 | 122 | 27.5 | 13.6 | 0.388 | 20.4 | 6.3 | 20.1 | 4.08 | 0.124 | 764 |
| 120 × 45 × 20 × 3.15 | 55.1 | 702 | 145 | 31.7 | 13.3 | 0.382 | 24.1 | 7.31 | 24 | 4.72 | 0.227 | 895 |
| 125 × 45 × 20 × 1.60 | 30.3 | 386 | 90.3 | 19.2 | 14.1 | 0.374 | 14.4 | 4.34 | 14.1 | 2.88 | 0.033 | 568 |
| 125 × 45 × 20 × 2.00 | 37.3 | 476 | 110 | 23 | 13.9 | 0.37 | 17.6 | 5.22 | 17.2 | 3.45 | 0.062 | 686 |
| 125 × 45 × 20 × 2.30 | 42.5 | 541 | 124 | 25.5 | 13.7 | 0.368 | 19.8 | 5.82 | 19.5 | 3.85 | 0.094 | 769 |
| 125 × 45 × 20 × 2.55 | 46.6 | 594 | 135 | 27.5 | 13.6 | 0.365 | 21.6 | 6.3 | 21.3 | 4.15 | 0.126 | 834 |
| 125 × 45 × 20 × 3.15 | 56.4 | 718 | 160 | 31.7 | 13.3 | 0.36 | 25.6 | 7.31 | 25.4 | 4.8 | 0.232 | 977 |
| 130 × 45 × 20 × 1.60 | 30.9 | 394 | 99.1 | 19.2 | 14.1 | 0.353 | 15.2 | 4.34 | 14.9 | 2.92 | 0.033 | 618 |
| 130 × 45 × 20 × 2.00 | 38.1 | 486 | 121 | 23 | 13.9 | 0.35 | 18.6 | 5.22 | 18.2 | 3.5 | 0.064 | 746 |
| 130 × 45 × 20 × 2.30 | 43.4 | 552 | 136 | 25.5 | 13.7 | 0.347 | 20.9 | 5.82 | 20.6 | 3.91 | 0.096 | 836 |

*(Contd)*

**TABLE A.14** *(Contd)*

| Designation Dimensions $h \times b \times c \times t$ (mm) | Weight (N/m) | Area of section (mm$^2$) | Moment of inertia | | Radius of gyration min-r$_v$ (mm) | Angle tan $\theta$ | Section modulus | | | | Torsional constant $J$ (cm$^4$) | Warping constant $C_w$ (cm$^6$) |
|---|---|---|---|---|---|---|---|---|---|---|---|---|
| | | | $I_z$(cm$^4$) | $I_y$(cm$^4$) | | | $Z_z$ (cm$^3$) | $Z_y$ (cm$^3$) | $Z_u$ (cm$^3$) | $Z_v$ (cm$^3$) | | |
| 130 × 45 × 20 × 2.55 | 47.6 | 607 | 148 | 27.5 | 13.6 | 0.345 | 22.8 | 6.3 | 22.5 | 4.22 | 0.129 | 908 |
| 130 × 45 × 20 × 3.15 | 57.6 | 734 | 176 | 31.7 | 13.3 | 0.34 | 27 | 7.31 | 26.9 | 4.88 | 0.237 | 1060 |
| 140 × 60 × 20 × 1.60 | 36 | 458 | 141 | 40.2 | 17.9 | 0.449 | 20.2 | 6.74 | 19.3 | 4.38 | 0.039 | 1400 |
| 140 × 60 × 20 × 2.00 | 44.4 | 566 | 173 | 48.5 | 17.7 | 0.445 | 24.7 | 8.22 | 23.7 | 5.28 | 0.074 | 1700 |
| 140 × 60 × 20 × 2.30 | 50.6 | 644 | 195 | 54.3 | 17.6 | 0.443 | 27.9 | 9.22 | 26.9 | 5.91 | 0.112 | 1910 |
| 140 × 60 × 20 × 2.55 | 55.7 | 709 | 213 | 58.8 | 17.4 | 0.441 | 30.5 | 10 | 29.5 | 6.41 | 0.151 | 2100 |
| 140 × 60 × 20 × 3.15 | 67.5 | 860 | 255 | 68.7 | 17.1 | 0.435 | 36.4 | 11.8 | 35.4 | 7.48 | 0.279 | 2500 |
| 150 × 60 × 20 × 1.60 | 37.2 | 474 | 166 | 40.2 | 18 | 0.406 | 22.1 | 6.79 | 21.2 | 4.5 | 0.04 | 1600 |
| 150 × 60 × 20 × 2.00 | 46 | 586 | 203 | 48.5 | 17.8 | 0.403 | 27 | 8.22 | 26 | 5.43 | 0.077 | 1970 |
| 150 × 60 × 20 × 2.30 | 52.4 | 668 | 229 | 54.3 | 17.6 | 0.401 | 30.6 | 9.22 | 29.5 | 6.08 | 0.116 | 2220 |
| 150 × 60 × 20 × 2.55 | 57.7 | 734 | 251 | 58.8 | 17.5 | 0.399 | 33.5 | 10 | 32.4 | 6.59 | 0.157 | 2420 |
| 150 × 60 × 20 × 3.15 | 70 | 891 | 300 | 68.7 | 17.2 | 0.394 | 40 | 11.8 | 38.9 | 7.7 | 0.289 | 2870 |
| 160 × 60 × 20 × 1.60 | 38.5 | 490 | 193 | 40.2 | 18 | 0.37 | 24.1 | 6.79 | 23.1 | 4.6 | 0.042 | 1870 |
| 160 × 60 × 20 × 2.00 | 47.5 | 606 | 236 | 48.5 | 17.8 | 0.367 | 29.5 | 8.22 | 28.4 | 5.56 | 0.08 | 2270 |
| 160 × 60 × 20 × 2.30 | 54.2 | 690 | 267 | 54.3 | 17.7 | 0.365 | 33.4 | 9.22 | 32.2 | 6.22 | 0.12 | 2560 |
| 160 × 60 × 20 × 2.55 | 59.7 | 760 | 292 | 58.8 | 17.6 | 0.363 | 36.5 | 10 | 35.4 | 6.75 | 0.162 | 2790 |
| 160 × 60 × 20 × 3.15 | 72.4 | 923 | 349 | 68.7 | 17.2 | 0.358 | 43.7 | 11.8 | 42.6 | 7.9 | 0.3 | 3310 |
| 170 × 60 × 20 × 1.60 | 39.7 | 506 | 222 | 40.2 | 18.1 | 0.339 | 26.1 | 6.79 | 25.1 | 4.7 | 0.043 | 2130 |
| 170 × 60 × 20 × 2.00 | 49.1 | 626 | 272 | 48.5 | 17.8 | 0.337 | 32 | 8.22 | 30.9 | 5.67 | 0.082 | 2600 |
| 170 × 60 × 20 × 2.30 | 56 | 714 | 308 | 54.3 | 17.7 | 0.334 | 36.3 | 9.22 | 35.1 | 6.36 | 0.124 | 2920 |
| 170 × 60 × 20 × 2.55 | 61.7 | 786 | 337 | 58.8 | 17.5 | 0.332 | 39.7 | 10 | 38.5 | 6.9 | 0.168 | 3190 |
| 170 × 60 × 20 × 3.15 | 74.9 | 954 | 404 | 68.7 | 17.2 | 0.328 | 47.5 | 11.8 | 46.3 | 8.07 | 0.31 | 3780 |
| 180 × 60 × 20 × 1.60 | 41 | 522 | 254 | 40.2 | 18 | 0.313 | 28.2 | 6.79 | 27.2 | 4.78 | 0.044 | 2430 |
| 180 × 60 × 20 × 2.00 | 50.7 | 646 | 311 | 48.5 | 17.8 | 0.31 | 34.6 | 8.22 | 33.5 | 5.78 | 0.085 | 2940 |
| 180 × 60 × 20 × 2.30 | 57.8 | 736 | 353 | 54.3 | 17.7 | 0.308 | 39.2 | 9.22 | 38 | 6.48 | 0.128 | 3320 |
| 180 × 60 × 20 × 2.55 | 63.7 | 811 | 386 | 58.8 | 17.6 | 0.306 | 42.9 | 10 | 41.7 | 7.03 | 0.174 | 3620 |
| 180 × 60 × 20 × 3.15 | 77.4 | 986 | 463 | 68.7 | 17.2 | 0.302 | 51.4 | 11.8 | 50.3 | 8.24 | 0.321 | 4290 |
| 190 × 60 × 20 × 1.60 | 42.2 | 538 | 289 | 40.2 | 18 | 0.289 | 30.4 | 6.79 | 29.3 | 4.84 | 0.046 | 2720 |
| 190 × 60 × 20 × 2.00 | 52.2 | 667 | 354 | 48.5 | 17.8 | 0.287 | 37.3 | 8.22 | 36.1 | 5.88 | 0.088 | 3310 |
| 190 × 60 × 20 × 2.30 | 59.6 | 760 | 401 | 55.3 | 17.6 | 0.285 | 42.2 | 9.22 | 41 | 6.59 | 0.132 | 3740 |
| 190 × 60 × 20 × 2.55 | 65.7 | 837 | 439 | 55.3 | 17.5 | 0.283 | 46.2 | 10 | 45 | 7.15 | 0.179 | 4070 |
| 190 × 60 × 20 × 3.15 | 79.8 | 1017 | 527 | 68.7 | 17.2 | 0.279 | 55.4 | 11.8 | 54.3 | 8.38 | 0.331 | 4840 |
| 200 × 60 × 20 × 1.60 | 43.5 | 554 | 326 | 40.2 | 18 | 0.269 | 32.6 | 6.79 | 31.5 | 4.94 | 0.047 | 3040 |
| 200 × 60 × 20 × 2.00 | 53.8 | 686 | 400 | 48.5 | 17.8 | 0.266 | 40 | 8.22 | 38.8 | 5.97 | 0.09 | 3710 |
| 200 × 60 × 20 × 2.30 | 61.4 | 780 | 453 | 55.3 | 17.6 | 0.265 | 45.3 | 9.22 | 44.1 | 6.7 | 0.136 | 4180 |
| 200 × 60 × 20 × 2.55 | 67.7 | 862 | 496 | 58.8 | 17.5 | 0.263 | 49.6 | 10 | 48.5 | 7.27 | 0.184 | 4560 |
| 200 × 60 × 20 × 3.15 | 82.3 | 1049 | 596 | 68.7 | 17.2 | 0.259 | 59.6 | 11.8 | 58.4 | 8.52 | 0.341 | 5420 |
| 210 × 60 × 20 × 1.60 | 44.7 | 570 | 366 | 40.2 | 17.9 | 0.251 | 34.9 | 6.79 | 33.8 | 5.01 | 0.048 | 3390 |
| 210 × 60 × 20 × 2.00 | 55.4 | 706 | 449 | 48.5 | 17.7 | 0.248 | 42.8 | 8.22 | 41.6 | 6.06 | 0.093 | 4130 |
| 210 × 60 × 20 × 2.30 | 63.2 | 805 | 509 | 54.3 | 17.6 | 0.247 | 48.5 | 9.22 | 47.3 | 6.79 | 0.14 | 4650 |
| 210 × 60 × 20 × 2.55 | 69.7 | 888 | 558 | 58.8 | 17.4 | 0.245 | 53.1 | 10 | 52 | 7.37 | 0.19 | 5080 |

*(Contd)*

**TABLE A.14** *(Contd)*

| Designation Dimensions $h \times b \times c \times t$ (mm) | Weight (N/m) | Area of section (mm$^2$) | Moment of inertia | | Radius of gyration min-r$_v$ (mm) | Angle tan $\theta$ | Section modulus | | | | Torsional constant $J$ (cm$^4$) | Warping constant $C_w$ (cm$^6$) |
|---|---|---|---|---|---|---|---|---|---|---|---|---|
| | | | $I_z$(cm$^4$) | $I_y$(cm$^4$) | | | $Z_z$ (cm$^3$) | $Z_y$ (cm$^3$) | $Z_u$ (cm$^3$) | $Z_v$ (cm$^3$) | | |
| 210 × 60 × 20 × 3.15 | 88.4 | 1080 | 667 | 68.7 | 17.1 | 0.241 | 63.8 | 11.8 | 62.7 | 8.65 | 0.352 | 6030 |
| 220 × 60 × 20 × 1.60 | 46 | 586 | 409 | 40.2 | 17.9 | 0.235 | 37.2 | 6.79 | 36.1 | 5.07 | 0.05 | 3750 |
| 220 × 60 × 20 × 2.00 | 57 | 726 | 502 | 48.5 | 17.6 | 0.232 | 45.6 | 8.22 | 44.5 | 6.13 | 0.096 | 4570 |
| 220 × 60 × 20 × 2.30 | 65 | 828 | 569 | 55.3 | 17.5 | 0.231 | 51.8 | 9.22 | 50.6 | 6.88 | 0.145 | 5160 |
| 220 × 60 × 20 × 2.55 | 71.7 | 913 | 624 | 58.8 | 17.4 | 0.229 | 56.7 | 10 | 55.6 | 7.47 | 0.196 | 5620 |
| 220 × 60 × 20 × 3.15 | 87.3 | 1110 | 750 | 68.8 | 17 | 0.226 | 68.1 | 11.8 | 67.1 | 8.76 | 0.362 | 6680 |
| 230 × 75 × 20 × 1.60 | 51 | 650 | 517 | 72.1 | 22 | 0.29 | 45 | 9.72 | 43.1 | 7.01 | 0.055 | 6990 |
| 230 × 75 × 20 × 2.00 | 63.2 | 806 | 636 | 87.5 | 21.8 | 0.287 | 55.3 | 11.8 | 53.2 | 8.52 | 0.106 | 8550 |
| 230 × 75 × 20 × 2.30 | 72.3 | 920 | 723 | 98.3 | 21.6 | 0.285 | 62.9 | 13.3 | 60.5 | 9.58 | 0.161 | 9670 |
| 230 × 75 × 20 × 2.55 | 79.7 | 1020 | 793 | 107 | 21.4 | 0.284 | 69 | 14.5 | 66.6 | 10.4 | 0.218 | 10600 |
| 230 × 75 × 20 × 3.15 | 97.2 | 1240 | 956 | 126 | 21.1 | 0.28 | 83.2 | 17.2 | 80.6 | 12.3 | 0.404 | 12600 |
| 240 × 75 × 20 × 1.60 | 52.3 | 666 | 512 | 72.1 | 21.9 | 0.272 | 47.6 | 9.72 | 45.7 | 7.1 | 0.056 | 7680 |
| 240 × 75 × 20 × 2.00 | 64.8 | 826 | 703 | 87.5 | 21.7 | 0.27 | 58.6 | 11.8 | 56.5 | 8.62 | 0.109 | 9390 |
| 240 × 75 × 20 × 2.30 | 74.1 | 944 | 799 | 98.3 | 21.6 | 0.269 | 66.6 | 13.3 | 64.3 | 9.7 | 0.165 | 10600 |
| 240 × 75 × 20 × 2.55 | 81.7 | 1040 | 878 | 107 | 21.4 | 0.267 | 73.1 | 14.5 | 70.7 | 10.6 | 0.223 | 11600 |
| 240 × 75 × 20 × 3.15 | 99.6 | 1270 | 1060 | 126 | 21.1 | 0.264 | 88.2 | 17.2 | 85.7 | 12.5 | 0.414 | 13900 |
| 250 × 75 × 20 × 1.60 | 53.5 | 682 | 629 | 72.1 | 21.9 | 0.257 | 50.3 | 9.72 | 48.5 | 7.17 | 0.058 | 8400 |
| 250 × 75 × 20 × 2.00 | 66.4 | 846 | 775 | 87.5 | 21.7 | 0.255 | 62 | 11.8 | 59.8 | 8.71 | 0.112 | 10300 |
| 250 × 75 × 20 × 2.30 | 75.9 | 966 | 881 | 98.3 | 21.5 | 0.253 | 70.5 | 13.3 | 68.2 | 9.81 | 0.169 | 11600 |
| 250 × 75 × 20 × 2.55 | 83.7 | 1070 | 967 | 107 | 21.4 | 0.252 | 77.4 | 14.5 | 75 | 10.7 | 0.229 | 12700 |
| 250 × 75 × 20 × 3.15 | 102 | 1300 | 1170 | 126 | 21 | 0.249 | 93.3 | 17.2 | 90.8 | 12.6 | 0.428 | 15200 |
| 260 × 75 × 20 × 1.60 | 54.8 | 698 | 697 | 72.1 | 21.8 | 0.234 | 53.1 | 9.72 | 51.3 | 7.24 | 0.059 | 9160 |
| 260 × 75 × 20 × 2.00 | 68 | 866 | 850 | 87.5 | 21.6 | 0.241 | 65.4 | 11.8 | 63.3 | 8.8 | 0.114 | 11200 |
| 260 × 75 × 20 × 2.30 | 77.7 | 990 | 967 | 98.3 | 21.4 | 0.279 | 74.4 | 13.3 | 72.1 | 9.91 | 0.173 | 12700 |
| 260 × 75 × 20 × 2.55 | 85.7 | 1090 | 1060 | 107 | 21.3 | 0.2338 | 81.7 | 14.5 | 79.3 | 10.8 | 0.234 | 13900 |
| 260 × 75 × 20 × 3.15 | 105 | 1330 | 1280 | 126 | 21 | 0.235 | 98.6 | 17.2 | 96.1 | 12.7 | 0.435 | 16600 |
| 270 × 75 × 20 × 1.60 | 56 | 714 | 755 | 72.1 | 21.8 | 0.23 | 55.9 | 9.72 | 54.1 | 7.31 | 0.061 | 9960 |
| 270 × 75 × 20 × 2.00 | 69.5 | 886 | 930 | 87.5 | 21.6 | 0.228 | 68.9 | 11.8 | 66.8 | 8.89 | 0.117 | 12200 |
| 270 × 75 × 20 × 2.30 | 79.5 | 1010 | 1060 | 98.3 | 21.4 | 0.227 | 78.3 | 13.3 | 76.1 | 10 | 0.177 | 13800 |
| 270 × 75 × 20 × 2.55 | 87.7 | 1120 | 1160 | 107 | 21.2 | 0.225 | 86.1 | 14.5 | 83.7 | 10.9 | 0.24 | 15100 |
| 270 × 75 × 20 × 3.15 | 107 | 1360 | 1400 | 126 | 20.9 | 0.222 | 104 | 17.2 | 102 | 12.9 | 0.446 | 18100 |
| 280 × 75 × 20 × 1.60 | 57.3 | 730 | 823 | 72.1 | 21.7 | 0.219 | 58.8 | 9.72 | 57 | 7.38 | 0.062 | 10800 |
| 280 × 75 × 20 × 2.00 | 71.1 | 906 | 1010 | 87.5 | 21.5 | 0.217 | 72.5 | 11.8 | 70.4 | 8.97 | 0.12 | 13200 |
| 280 × 75 × 20 × 2.30 | 81.3 | 1040 | 1150 | 98.3 | 21.3 | 0.215 | 82.4 | 13.3 | 80.2 | 10.1 | 0.181 | 15000 |
| 280 × 75 × 20 × 2.55 | 89.7 | 1140 | 1270 | 106.9 | 21.2 | 0.214 | 90.5 | 14.5 | 88.3 | 11 | 0.245 | 16400 |
| 280 × 75 × 20 × 3.15 | 110 | 1400 | 1530 | 126 | 20.8 | 0.211 | 109 | 17.2 | 107 | 13 | 0.456 | 19600 |
| 290 × 75 × 20 × 1.60 | 58.6 | 746 | 895 | 72.1 | 21.6 | 0.208 | 61.7 | 9.72 | 59.2 | 7.44 | 0.063 | 11700 |
| 290 × 75 × 20 × 2.00 | 72.7 | 926 | 1100 | 87.5 | 21.4 | 0.206 | 76.1 | 11.8 | 74 | 9.04 | 0.122 | 14300 |
| 290 × 75 × 20 × 2.30 | 83.7 | 1060 | 1250 | 98.3 | 21.3 | 0.205 | 86.5 | 13.3 | 84.4 | 10.2 | 0.185 | 16200 |
| 290 × 75 × 20 × 2.55 | 91.7 | 1170 | 1380 | 107 | 21.1 | 0.203 | 95.1 | 14.5 | 92.9 | 11.1 | 0.251 | 17700 |
| 290 × 75 × 20 × 3.15 | 112 | 1430 | 1670 | 126 | 20.8 | 0.2 | 115 | 17.2 | 113 | 13.1 | 0.466 | 21200 |

*(Contd)*

**TABLE A.14** *(Contd)*

| Designation Dimensions $h \times b \times c \times t$ (mm) | Weight (N/m) | Area of section (mm²) | Moment of inertia | | Radius of gyration min-$r_v$ (mm) | Angle tan $\theta$ | Section modulus | | | | Torsional constant $J$ (cm⁴) | Warping constant $C_w$ (cm⁶) |
|---|---|---|---|---|---|---|---|---|---|---|---|---|
| | | | $I_z$ (cm⁴) | $I_y$ (cm⁴) | | | $Z_z$ (cm³) | $Z_y$ (cm³) | $Z_u$ (cm³) | $Z_v$ (cm³) | | |
| 300 × 75 × 20 × 1.60 | 59.8 | 762 | 970 | 72.1 | 21 | 0.198 | 64.7 | 9.72 | 62.9 | 7.5 | 0.065 | 12600 |
| 300 × 75 × 20 × 2.00 | 74.2 | 946 | 1200 | 87.5 | 21.4 | 0.196 | 79.81 | 11.6 | 77.8 | 9.11 | 0.125 | 15400 |
| 300 × 75 × 20 × 2.30 | 84.9 | 1080 | 1360 | 98.3 | 21.2 | 0.195 | 90.8 | 13.3 | 88.7 | 10.3 | 0.189 | 17400 |
| 300 × 75 × 20 × 2.55 | 93.7 | 1190 | 1500 | 107 | 21 | 0.194 | 99.7 | 14.5 | 97.6 | 11.2 | 0.256 | 19100 |
| 300 × 75 × 20 × 3.15 | 115 | 1460 | 1810 | 126 | 20.7 | 0.191 | 121 | 17.2 | 118 | 13.2 | 0.477 | 22800 |

**TABLE A.15** Approximate radii of gyration of some commonly used sections

| Geometry of cross-section | Radius of gyration | Geometry of cross-section | Radius of gyration |
|---|---|---|---|
| 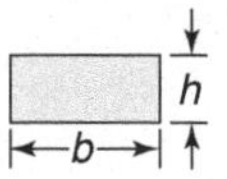 | $r_z = 0.29h$<br>$r_y = 0.29b$ | 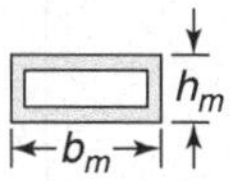 | $r_z = 0.4h_m$<br>$h_m$ = Mean $h$ |
| 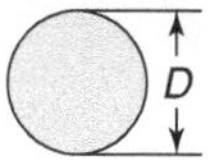 | $r = D/4$ | 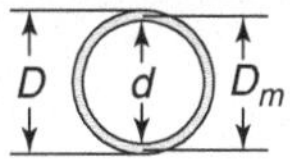 | $r = 0.35D_m$<br>$r = \sqrt{\dfrac{D^2 + d^2}{16}}$ |
| 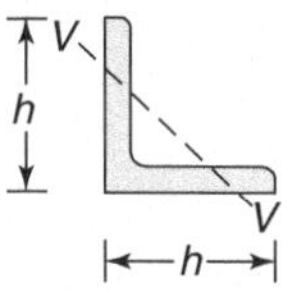 | $r_z = r_y = 0.31h$<br>$r_v = 0.197h$ | 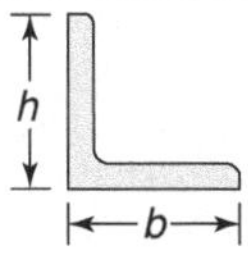 | $r_z = 0.29h$<br>$r_y = 0.32b$<br>$r_v = 0.09(h + b)$ |
| 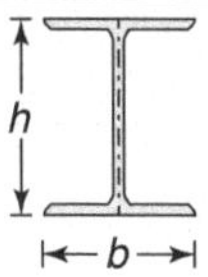 | $r_z = 0.38h$<br>$r_y = 0.22b$ | 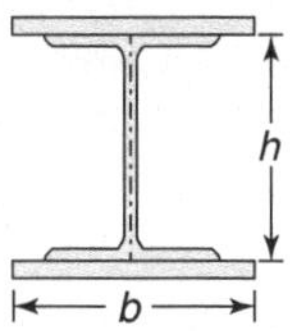 | $r_z = 0.40h$<br>$r_y = 0.21b$ |
| 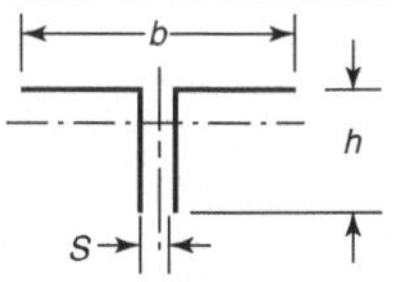 | $r_z = 0.31h$<br>$r_y = 0.215b$<br>$= b(0.21 + 0.02s)$ | 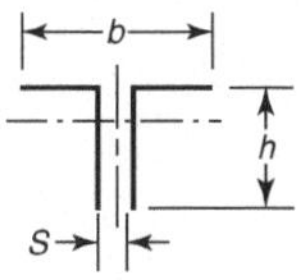 | $r_z = 0.32h$<br>$r_y = 0.21b$<br>$= b(0.19 + 0.02s)$ |
| 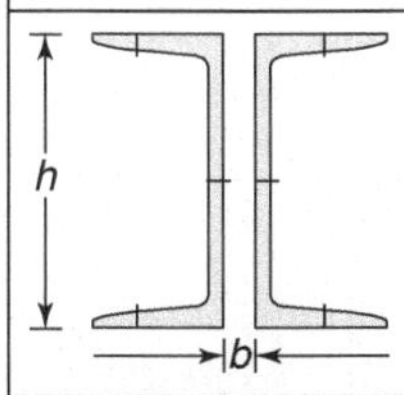 | $r_z = 0.36h$<br>$r_y = 0.60b$ | 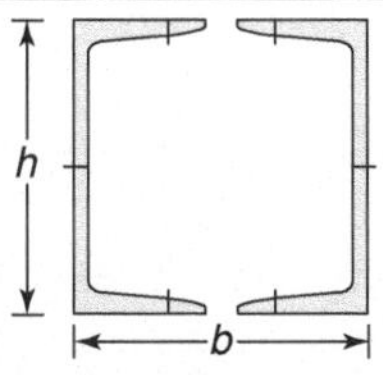 | $r_z = 0.36h$<br>$r_y = 0.45b$ |
| 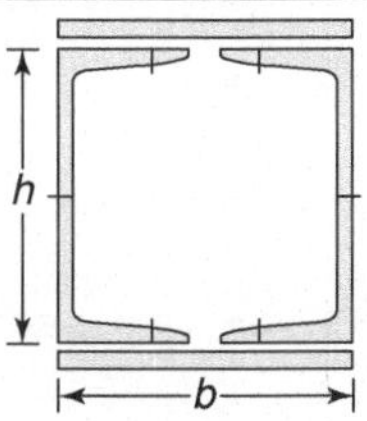 | $r_z = 0.42h$ | 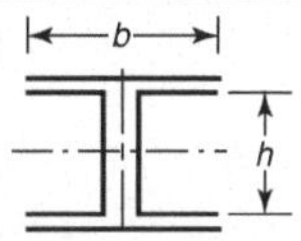 | $r_z = 0.42h$<br>$r_y = 0.32b$ |
| 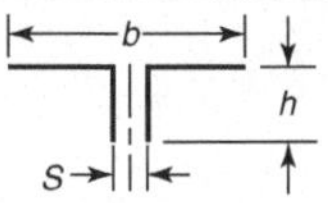 | $r_z = 0.29h$<br>$r_y = 0.24b$<br>$= b(0.23 + 0.02s)$ | 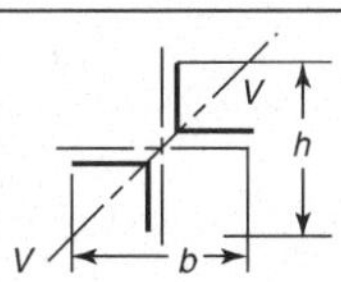 | $r_z = 0.21h$<br>$r_y = 0.21b$<br>$r_v = 0.19h$ |

*(Contd)*

**TABLE A.15** *(Contd)*

| Geometry of cross-section | Radius of gyration | Geometry of cross-section | Radius of gyration |
|---|---|---|---|
| 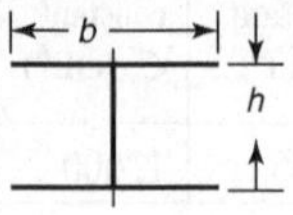 | $r_z = 0.435h$<br>$r_y = 0.25b$ | 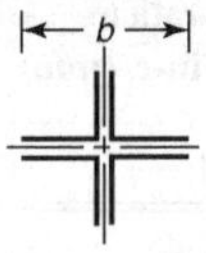 | $r_y$ = same as for 2*L* |
| 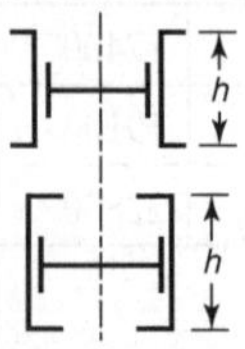 | $r_z = 0.31h$ | 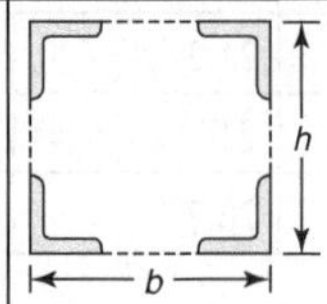 | $r_z = 0.42h$<br>$r_y = 0.42b$ |
| 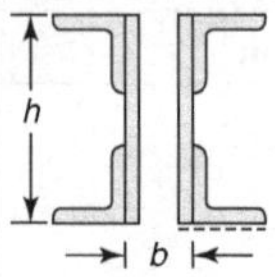 | $r_z = 0.36h$<br>$r_y = 0.53b$ | 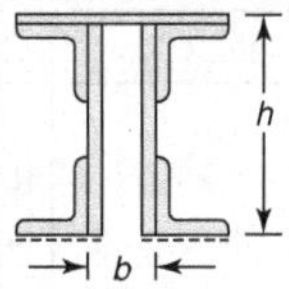 | $r_z = 0.39h$<br>$r_y = 0.55b$ |

*Note*: Horizontal axis is *z-z* and vertical axis is *y-y*

APPENDIX

B

# LOADS DUE TO TSUNAMI

## B.1 BACKGROUND

The current structural design codes provide very little guidance for loads specifically induced by tsunami effects on coastal structures; the established design codes focus mainly on loadings due to riverine floods and storm waves. There are significant differences in physical conditions between tsunami and other floods. For a typical tsunami, the water surface fluctuates near the shore with an amplitude of several metres, during a period of a few to tens of minutes. This timescale is intermediate between the typical of riverine floods stretching from hours to days, and the tens of seconds or less associated with cyclic loading of storm waves. This intermediate timescale makes the behaviour and characteristics of tsunami quite distinct from other coastal hazards, and the effects cannot be inferred from common knowledge or intuition. The timescale is long enough to allow tsunami to penetrate a great distance inland, while it is still short enough to make the tsunami flow highly transient (Yeh et al. 2005). Rapid rise and fall of tsunami inundation appears to enhance scour around a building. A relatively short lead-time for warning and evacuation likely leaves unsecured potentially hazardous objects behind, e.g., propane gas cylinders, automobiles, and boats. These objects can cause devastating effects once they become water-borne missiles. Moreover, extended run-up duration of tsunami introduces the potential to transport floating bodies such as automobiles far inland, impacting buildings in their path. Since tsunami inundation fluctuates faster, there may be greater buoyant forces exerted on buildings. Rapid water level fluctuations induce pore-pressure gradients in the soil, which may loosen the foundations, causing liquefaction of soil in some cases. Tsunami characteristics and behaviour in the run-up zone are often unpredictable since they are influenced by the type of tsunami and also the surrounding topography (Yeh et al. 2005).

## B.2 EXISTING INFORMATION IN CODES OF PRACTICES

The codes classify areas that are subject to wave heights in excess of 1 m or high- velocity wave run-up or wave-induced erosion as *coastal high-hazard zones* or *V-zones*. The following loads must be considered in the design and construction of buildings and structures in V-zones: hydrostatic (vertical and lateral), hydrodynamic, impact, surge, wave, and breaking wave loads. The effects of long-term erosion, storm-induced erosion, and local scour are to be included in the design of foundations of buildings or other structures in coastal high hazard areas. Foundation embedment should be below the depth of potential scour. All buildings and structures in potential flood zones must be designed and constructed to resist floatation from floodwater at the regulatory flood elevation. Load factors for the building materials used should be the same as the building code provisions for wind or earthquake loads combined with gravity loads. In other words, loads and stresses due to tsunamis are to be treated in the same fashion as for earthquake loading. A brief introduction of the various loads due to tsunami, as available in the existing codes, is given in the following sub-sections.

### B.2.1 Hydrostatic Force

It has been observed that hydrostatic forces are normally relatively small compared to surge and drag forces for the case of bore-like tsunamis, however for tsunamis that act as rapidly rising tide, the hydrostatic forces generally become increasingly important (Dames & Moore 1980). The following equation does not include the direct drag at the top of a wall, when the wall is less than $h$ in height.

$$F_h = \frac{1}{2}\rho g\left[h + (u_p^2/2g)\right]^2 \qquad \text{(B.1)}$$

where $F_h$ is the hydrostatic force on a wall, per unit width of wall, $\rho$ is the water density, $g$ is the gravitational acceleration, $h$ is the water depth, and $u_p$ is the velocity component normal to the wall. The resultant force will act horizontally at a distance of $h_R$ above the base of the wall where

$$h_R = \frac{1}{3}\left[h(u_p^2/2g)\right] \tag{B.2}$$

### B.2.2 Buoyant Force

All codes (ASCE 7, FEMA 55, and CCH) provide the following expression for buoyant force

$$F_b = \rho g V \tag{B.3}$$

where $V$ is the volume of water displaced by the building.

### B.2.3 Design Flood Velocity

FEMA 55 and CCH provide the following estimate of the flood velocity $u$ in the surge depth $d_s$, based on Dames and Moore (1980)

$$u = 2\sqrt{(gd_s)} \tag{B.4}$$

### B.2.4 Hydrodynamic Force

Both CCH and FEMA 55 provide the following expression for the hydrodynamic force (drag force)

$$F_d = \frac{1}{2}\rho C_d A u_p^2 \tag{B.5}$$

where $C_d$ is the drag coefficient and $A$ is the projected area of the body on the plane normal to the direction of flow. The CCH recommends $C_d = 1.0$ for circular piles, 2.0 for square piles, and 1.5 for wall sections. FEMA 55 recommends $C_d = 2.0$ for square or rectangular piles and 1.2 for round piles. In addition, Table B.1 gives drag coefficients for larger obstructions.

**TABLE B.1** Drag coefficient

| Width to depth ratio ($w/d_s$ or $w/h$) | Drag coefficient $C_d$ |
|---|---|
| From 1–12 | 1.25 |
| 13–20 | 1.3 |
| 21–32 | 1.4 |
| 33–40 | 1.5 |
| 41–80 | 1.75 |
| 81–120 | 1.8 |
| >120 | 2 |

### B.2.5 Surge Force

CCH adopted the following equation for surge force

$$F_s = 4.5\rho g h^2 \tag{B.6}$$

where $h$ is the surge height (Dames & Moore 1980). The resultant force acts at a distance approximately $h$ above the base of the wall. This equation is applicable for walls with heights equal to or greater than $3h$. Walls whose heights are less than $3h$ require surge forces to be calculated using appropriate combination of hydrostatic and hydrodynamic force equations for the given situation.

### B.2.6 Impact Force

CCH, FEMA 55, and ASCE 7 contain similar equations that resulted in the following generalized expression for impact force $F_I$ acting at the still water level:

$$F_I = m[du_b/dt] = m(u_1/\Delta t) \tag{B.7}$$

where $u_b$ is the velocity of the impacting body, $u_1$ is its approach velocity that is assumed equal to the flow velocity, $m$ is the mass of the body, $\Delta t$ is the impact duration that is equal to the time between the initial contact of the body with the building and the maximum impact force. CCH recommends $\Delta t$ values for wood construction as 1.0 second, steel construction as 0.5 second, and reinforced concrete as 0.1 second. FEMA 55 provides the $\Delta t$ values shown in Table B.2.

**TABLE B.2** Impact duration recommended by FEMA 55

| Type of construction | Duration ($t$) of impact (s) | |
|---|---|---|
| | Wall | Pile |
| Wood | 0.7–1.1 | 0.5–1.0 |
| Steel | NA | 0.2–0.4 |
| Reinforced concrete | 0.2–0.4 | 0.3–0.6 |
| Concrete masonry | 0.3–0.6 | 0.3–0.6 |

### *Breaking Waves Forces*

**Breaking wave loads on vertical piling and columns** ASCE 7 and FEMA 55 provide the following expression for the breaking wave force $F_{\text{brkp}}$:

$$F_{\text{brkp}} = \frac{1}{2}\rho g C_{\text{db}} D H_b^2 \tag{B.8}$$

where $C_{db}$ is a shape coefficient (ASCE 7 and FEMA 55 recommend $C_{db}$ values of 2.25 for square or rectangular piles and 1.75 for round piles), $D$ is the pile diameter, and $H_b$ is the breaking wave height (FEMA 55 recommends $H_b = 0.78d_s$, where $d_s$ is the design still-water flood depth).

**Breaking wave loads on vertical walls** The equations given by FEMA 55 and ASCE 7 differ slightly. FEMA 55 gives the following equations, which incorporate the lateral hydrostatic force. If these formulae are used, then the hydrostatic force should not be added.

***Case* 1** (Enclosed dry space behind the wall)

$$F_{\text{brkw}} = 1.1C_p \gamma d_s^2 + 2.41\gamma d_s^2 \tag{B.9}$$

***Case 2*** (Equal still water level on both sides of wall):

$$F_{brkw} = 1.1C_p\gamma d_s^2 + 1.9\gamma d_s^2 \quad (B.10)$$

Where $F_{brkw}$ is the total breaking wave load per unit length of wall acting at the still water level ($d_s$), $C_p$ is the dynamic pressure coefficient from Table B.3, and $\gamma$ is the specific weight of water.

**TABLE B.3** Dynamic pressure coefficient recommended by FEMA 55

| $C_p$ | Building type | Probability of exceedance |
|---|---|---|
| 1.6 | Accessory structure, low hazard to human life or property in the event of failure | 0.5 |
| 2.8 | Coastal residential building | 0.01 |
| 3.2 | High-occupancy or critical facility | 0.001 |

ASCE 7 suggests the following equations:

$$P_{max} = C_p\gamma d_s + 1.2\gamma d_s \quad (B.11)$$

$$F_t = 1.1C_p\gamma d_s^2 + 2.4\gamma d_s^2 \quad (B.12)$$

where $P_{max}$ is the maximum combined dynamic ($C_p\gamma d_s$) and static ($1.2\gamma d_s$) wave pressure, also referred to as shock pressure. $F_t$ is the total breaking wave force per unit length of the structure, and is also referred to as shock, impulse, or wave impact force acting near the still water elevation ($d_s$). $C_p$ is the dynamic pressure coefficient ($1.6 < C_p < 3.5$; see Table B.4), and $d_s$ is the still water depth at base of the building, where the wave breaks. If free water exists behind the wall, the hydrostatic component of the wave pressure and force disappears and the dynamic wave pressure and the net force are computed by

$$P_{max} = C_p\gamma d_s \quad (B.13)$$

$$F_t = 1.1\, C_p\gamma d_s^2 \quad (B.14)$$

**TABLE B.4** Dynamic pressure coefficient recommended by ASCE 7

| | Building category | $C_p$ |
|---|---|---|
| I | (low hazard) | 1.6 |
| II | (standard hazard) | 2.8 |
| III | (Substantial hazard) | 3.2 |
| IV | (essential facilities) | 3.5 |

**Loading combinations** Individual loading conditions must be applied to standard elements in appropriate load combinations. The following loading combinations are suggested by FEMA 55 (Yeh et al. 2005). The structural members considered are piles or open foundations, columns, walls, and basements. Outer edge beams at the second floor level are not mentioned because the reference codes are intended for smaller scale residential structures.

***Type 1—Columns in tsunami-prone area (required)*** The following combinations are used to calculate the force on a column from a tsunami with the additional impact force from debris.

$F_{brkp}$ (on column) + $F_I$ (on column) or $F_d$ (on column) + $F_I$ (on column)

***Type 2—Solid walls facing the shoreline in tsunami-prone areas*** Construction of non-breakaway walls or solid walls parallel to the shorelines is not recommended in structural designs. The following combinations provided are for walls, which are perpendicular to the flow of the tsunami. Tsunami effects on structural walls with additional impact force of debris.

$F_{brkw}$ (on walls facing shoreline) + $F_I$ (on one corner) or
$F_s$ (on walls facing shoreline) + $F_I$ (on one corner) or
$F_d$ (on walls facing shoreline) + $F_I$ (on one corner)

***Type 3—Vertical (buoyant) forces on structure*** This loading combination is used when there is a sudden increase in the water level. The buoyant force on the structure must also be considered with other lateral forces.

$F_b$ (for basements, swimming pools, empty above ground, and below ground tanks)

## B.3 COMMENTS

All of the equations presented earlier appear reasonable and might be applicable to tsunami cases. However, Yeh et al. (2005) have made the following comments based on recent analysis of the tsunami run-up.

1. Equation (B.1) —representation of hydrostatic force— may not be relevant to a building with—finite breadth, for which the water can flow around and quickly fill up behind the building. Hydrostatic force is usually important for a 2-D structure such as seawalls and dikes or for evaluation of an individual wall panel, where the water level outside differs substantially from the level inside.
2. The surge-speed estimate given by Eqn (B.4) was suggested by Camfield (1980) and is equivalent to a classic solution of the leading-tip of surge on a frictionless horizontal plane generated by breaking a dam with the quiescent impoundment depth of $d_s$. Hence, the computed tip velocity does not represent the velocity of flow depth $d_s$, and therefore this equation is not appropriate to represent the flow velocity for a tsunami flood passing through a structure.
3. The surge-force computation by Eqn (B.6) may result in excessively overestimated values. Note that Eqn (B.6) was derived by summing the hydrostatic force and the change in linear momentum at the impingement of a surge

front on a vertical wall. The surge impingement is treated as steady using the speed evaluated by Eqn (B.4). The estimation made by Eqn (B.6) implies that the surging force would be nine times the hydrostatic force alone. Such an excessively large surging force is contradictory to the laboratory results by Ramsden (1993). Further information about the loads to be considered for a tsunami-resistant design may be found in the references.

## REFERENCES


Camfield, F., *Tsunami Engineering*, Coastal Engineering Research Center, US Army Corps of Engineers, Special Report (SR – 6), 1980, p. 222.

Dames and Moore, *Design and Construction Standards for Residential Construction in Tsunami-prone Areas in Hawaii*, Prepared for the Federal Emergency Management Agency, USA, 1980.

Department of Planning and Permitting of Honolulu, City and County of Honolulu Building Code, Hawaii, Chapter 16 Article 11, Jul 2000.

Dudley, W.C. and Lee, M., *Tsunami*, University of Hawaii Press, Honolulu, Second edition, 1988.

FEMA 55, *Coastal Construction Manual*, Federal Emergency Management Agency, 2000.

FEMA P 646, *Guidelines for Design of Structures for Vertical Evacuation from Tsunamis*, Prepared by Applied Technology Council (ATC), California and published by Federal Emergency Management Agency (FEMA),Washington DC, Jun 2008, p. 176.

Goda, Y., *Random Seas and Design of Maritime Structures*, World Scientific Publishing Co. Pte. Ltd, Singapore, 2000, p. 443.

Ramsden, J.D., *Tsunamis: Forces on a Vertical Wall Caused by Long Waves, Bores, and Surges on a Dry Bed*, Report No. KH-R-54, W.M. Keck Laboratory, California Institute of Technology, Pasadena, CA, 1993, p. 251.

SEI/ASCE 7 – 05, *Minimum Design Loads for Buildings and Other Structures*, American Society of Civil Engineers, Reston, VA, 2006, p. 376.

Wiegel, R.L., *Oceanographical Engineering*, Prentice-Hall, Inc., Englewood Cliffs, NJ, 1964, p. 532.

Yeh, H., Robertson, I., and Preuss, J., *Development of Design Guidelines that serve as Tsunami Vertical Evacuation Sites*, Washington State Department of Natural Resources, Division of Geology and Earth Resources, Report 2005-4, Nov 2005.

APPENDIX C

# PROPERTIES OF SOILS

This appendix presents some information and tables containing properties of soils which will be of interest to the structural designer.

## C.1 SOIL TESTS

For low-rise buildings, depth of borings may be specified to be about 6 m below the anticipated foundation level, with at least one boring continuing deeper, to a lesser of 30 m, the least building dimension, or refusal. At least one soil boring should be specified for every 230 square metres of the building area for buildings over 12 m height, or having more than three storeys. For large buildings founded on poor soils, borings should be spaced at less than 15 m intervals. A minimum of five borings, one at the centre and the rest at the corners of the building, is recommended.

## C.2 ORDER OF SOIL SUITABILITY FOR FOUNDATION SUPPORT

- Best : Bed rock
- Very good : Sand and gravel
- Good : Medium to hard clay (that is kept dry)
- Poor : Silts and soft clay
- Undesirable : Organic silts and organic clay
- Unsuitable : Peat

## C.3 THE PLASTICITY INDEX (PI)

The plasticity index (PI) of the soil provides an indication of how much clay will shrink or swell. The higher the PI, the greater is the shrink-swell potential.

- PI of 0-15% : Low expansion potential
- PI of 15-25% : Medium expansion potential
- PI of 25% and above : High expansion potential

**TABLE C.1** Typical mass densities of basic soil types

| | Mass density $\rho$ (Mg/m$^3$)* | | | |
|---|---|---|---|---|
| | Poorly graded soil | | Well-graded soil | |
| Type of soil | Range | Typical value | Range | Typical value |
| Loose sand | 1.70–1.90 | 1.75 | 1.75–2.00 | 1.85 |
| Dense sand | 1.90–2.10 | 2.00 | 2.00–2.20 | 2.10 |
| Soft clay | 1.60–1.90 | 1.75 | 1.60–1.90 | 1.75 |
| Stiff clay | 1.90–2.25 | 2.00 | 1.90–2.25 | 2.07 |
| Silty soils | 1.60–2.00 | 1.75 | 1.60–2.00 | 1.75 |
| Gravelly soils | 1.90–2.25 | 2.07 | 2.00–2.30 | 2.15 |

*Values are representative of moist sand, gravel, saturated silt, and clay.

**TABLE C.2** Typical values of modulus of elasticity ($E_s$) for different types of soils

| Type of soil | $E_S$ (N/mm$^2$) |
|---|---|
| Clay | |
| Very soft | 2–15 |
| Soft | 5–25 |
| Medium | 15–50 |
| Hard | 50–100 |
| Sandy | 25–250 |
| Glacial till | |
| Loose | 10–153 |
| Dense | 144–720 |
| Very dense | 478–1, 440 |
| Loess | 14–57 |
| Sand | |
| Silty | 7–21 |
| Loose | 10–24 |
| Dense | 48–81 |

*(Contd)*

**TABLE C.2** *(Contd)*

| Type of soil | $E_S$ (N/mm²) |
|---|---|
| Sand and gravel | |
| Loose | 48–148 |
| Dense | 96–192 |
| Shale | 144–14,400 |
| Silt | 2–20 |

**TABLE C.3** Typical values of modulus of subgrade reaction ($k_s$) for different types of soils

| Type of soil | $k_s$ (kN/m³) |
|---|---|
| Loose sand | 4,800–16,000 |
| Medium dense sand | 9,600–80,000 |
| Dense sand | 64,000–1,28,000 |
| Clayey medium dense sand | 32,000–80,000 |
| Silty medium dense sand | 24,000–48,000 |
| Clayey soil: | |
| $q_u \le 200$ KN/m² | 12,000–24,000 |
| $200 < q_u \le 400$ KN/m² | 24,000–48,000 |

*(Contd)*

**TABLE C.3** *(Contd)*

| | |
|---|---|
| $q_u > 800$ KN/m² | > 48,000 |
| $q_u$ – Safe bearing capacity | |

**TABLE C.4** Typical values of Poisson's ratio ($\mu$) for soils

| Type of soil | $\mu$ |
|---|---|
| Clay (saturated) | 0.4–0.5 |
| Clay (unsaturated) | 0.1–0.3 |
| Sandy clay | 0.2–0.3 |
| Silt | 0.3–0.35 |
| Sand (dense) | 0.2–0.4 |
| Course (void ratio = 0.4 – 0.7) | 0.15 |
| Fine grained (void ratio = 0.4 – 0.7) | 0.25 |
| Rock | 0.1–0.4 (depends on type of rock) |
| Loess | 0.1–0.3 |
| Ice | 0.36 |
| Concrete | 0.15 |

**TABLE C.5** Allowable bearing pressures on soils (for preliminary design)

| Type of rock/soil | | Allowable bearing pressures (kN/m²) | | Standard penetration blow count (N) | Apparent cohesion $c_u$ (kPa) |
|---|---|---|---|---|---|
| Hard rock without lamination and defects(e.g., granite, trap, and diorite) | | 3,200 | | > 30 | |
| Laminated rocks (e.g., sand-stone and lime-stone in sound condition) | | 1,600 | | > 30 | |
| Soft or broken rock, hard shale, cemented material | | 900 | | > 30 | - |
| Soft Rock | | 450 | | > 30 | |
| Gravel | Dense | 450 | | > 30 | - |
| | Medium | 96–285 | | > 30 | - |
| Sand[1] | | *Compact and dry* | *Loose and dry* | | |
| | Coarse | 450 | 250 | 30–50 | - |
| | Medium | 250 | 48–120 | 15–30 | - |
| | Fine or silt | 150 | 100 | < 15 | - |
| Clay[2] | Very stiff | 190–450 | | 15–30 | 100–200 |
| | Medium stiff | 200–250 | | 4–15 | 25–100 |
| | Soft | 50–100 | | 0–4 | 0–25 |
| Peat, silts, made-up ground | To be determined after investigation | | | | |

*Notes*: 1. Reduce bearing pressures by half below the water table.

2. Alternatively, allow 1.2 times $c_u$ for round and square footings, and 1.0 times $c_u$ for length/width ratios of more than 4.0. Interpolate for intermediate values.

**TABLE C.6** Typical interface friction angles (NAVFAC 1982)

| | **Interface materials** | **Interface friction angle $\delta$** |
|---|---|---|
| Mass concrete against | Clean sound rock | 25 |
| | Clean gravel, gravel-sand mixtures, coarse sand | 29–31 |
| | Clean fine to medium sand, silty medium to coarse sand, silty or clayey gravel | 24–29 |
| | Clean fine sand, silty or clayey fine to medium sand | 19–24 |
| | Fine sandy silt, nonplastic silt | 17–19 |
| | Medium-stiff, stiff and silty clay | 17–19 |
| Formed concrete against | Clean gravel, gravel-sand mixture, well-graded rock fill with spalls | 22–26 |
| | Clean gravel, silty sand-gravel mixture, single-size hard rock fill | 17–22 |
| | Silty sand, gravel, or sand mixed with silt, or clay | 17 |
| | Fine sandy silt, non-plastic silt | 14 |
| Steel sheet piles against | Clean gravel, gravel-sand mixture, well- graded rock fill with spalls | 22 |
| | Clean sand, silty sand-gravel mixture, single-size hard rock fill | 17 |
| | Silty sand, gravel, or sand mixed with silt or clay | 14 |
| | Fine sandy silt, nonplastic silt | 11 |

**TABLE C.7** Typical values of fundamental period for soil deposits (for rock motions with $a_{max} = 0.4g$) (SEAOC 1980)

| **Soil depth (m)** | **Dense sand (s)** | **5 m of fill over normally consolidated clay* (s)** |
|---|---|---|
| 10 | 0.3–0.5 | 0.5–1.0 |
| 30 | 0.6–1.2 | 1.5–2.3 |
| 60 | 1.0–1.8 | 1.8–2.8 |
| 90 | 1.5–2.3 | 2.0–3.0 |
| 150 | 2.0–3.5 | — |

*Representative of San Francisco bay area

**TABLE C.8** Mean shear wave velocities (m/s) for the top 30 m of ground (Borcherdt 1994)

| **General description** | **Mean shear-wave locity** | | |
|---|---|---|---|
| | **Minimum** | **Average** | **Maximum** |
| *Firm and hard rocks* | | | |
| Hard rocks (e.g., metamorphic rocks with very widely spaced fractures) | 1400 | 1620 | — |
| *Firm to hard rocks* (e.g., granites, igneous rocks, conglomerates, sandstones, and shales with close to widely spaced fractures) | 700 | 1050 | 1400 |
| *Gravelly soils and soft to firm rocks* (e.g., soft igneous sedimentary rocks, sandstones, shales, gravels, and soils with >20% gravel) | 375 | 540 | 700 |
| *Stiff clays and sandy soils* (e.g., loose to very dense sands, silt loams, sandy clays, and medium stiff to hard clays and silty clays (N > 5 blows/ 300mm) | 200 | 290 | 375 |

(*Contd*)

| | | | |
|---|---|---|---|
| *Soft soils* (e.g., loose submerged fills and very soft (N < 5 blows/300 mm) clays and silty clays < 37 m thick) | 100 | 150 | 200 |
| *Very soft soils* (e.g., loose saturated sand, marshland, recent reclamation) | 50 | 75 | 100 |

*Note:* The fundamental time period $T$ of soil layer of thickness $H$, having average shear wave velocity $Vs$ is approximately $T = 4H/Vs$

If we assume the weighted average shear wave velocity for 30–50 m soil layer as 290 m/s, then the fundamental period of soil layer will range from 0.41 to 0.69 second. The fundamental time period of 4–6 storey buildings, including the soil- structure interaction, should fall in the above range of time period of soil layers, i.e., 0.41–0.69 sec. That is, the seismic waves in this range of time period will be allowed only to pass and filter-out the other frequencies. Therefore, there will be quasi resonance of building and the soil layer. At this point the damaging energy from the seismic waves get into the buildings having similar time period of vibration as the soil layer. If the seismic damaging energy getting into the building is more than the capacity of the structure, then the building will show distress and may collapse.

Similarly, if we assume that the weighted average shear wave velocity for 150–300 m soil layer is around 500 m/s, then the fundamental time period will range from 1.2–2.4 s. The fundamental time period of 10–15 storey building, including soil structure interaction, will fall in the above range of time period of vibrations. Therefore, there will be quasi resonance of the buildings and the soil layer and the seismic waves will affect this group of buildings which will result in damage/collapse of buildings. Hence, it is important to know the depth of soil layers above the bedrock and its properties such as the shear wave velocities, which are related in the microzonation of a region.

## REFERENCES

Borcherdt, R., 'Estimates of Site-dependent Response Spectra for Design (Methodology and Justification)', *Earthquake Spectra*, Vol. 10, No. 4, 1994, pp. 617–653.

Bowles, J.E., *Foundation Analysis and Design*, Fifth edition, McGraw Hill, New York, 1995.

Day, R.W., *Foundation Engineering Handbook*, McGraw-Hill/ASCE Press, 2006, p. 930.

Dowrick, D., *Earthquake Risk Reduction*, Wiley, Chichester, England, 2003.

NAVFAC, *Foundations and Earth Structures*, Design Manual 7.2, Naval Facilities Engineering Command, Dept. of Navy, Alexandria, Virginia, USA, 1982.

Prof D.K. Paul, IIT, Roorkee, Private Communications.

Reese, L.C., Isenhower, W.M., and Wang, S-T., *Analysis and Design of Shallow and Deep Foundations*, John Wiley, New York, 2005, p. 608.

SEAOC, *Recommended Lateral Force Requirements and Commentary*, Structural Engineers Association of California, 1980.

www.dot.ca.gov/hq/esc/construction/manuals/trenching and shoring/Ch3_soil.pdf, last accessed on 20 July 2015.

APPENDIX

# D

# BENDING MOMENT, SHEAR FORCE, AND DEFLECTION OF BEAMS AND FRAMES

The analysis methods described in Chapter 4 are to be used to find out the bending moments, shear force, and deflection of beams and frames. However, in this Appendix, formulae are provided for calculating the same for some beams and simple frames, which can be used for the design of these structures.

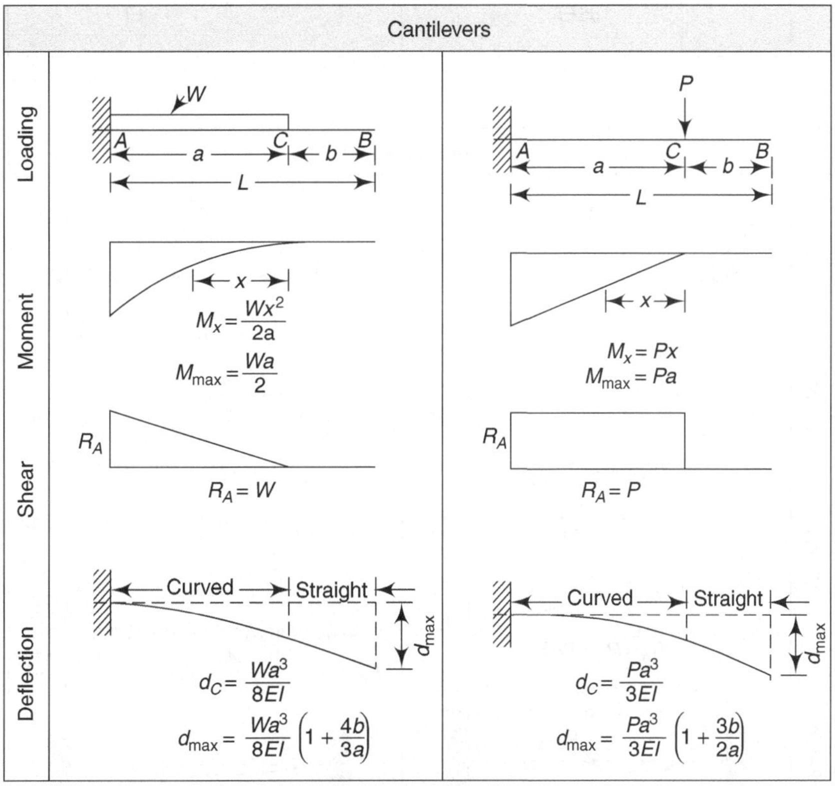

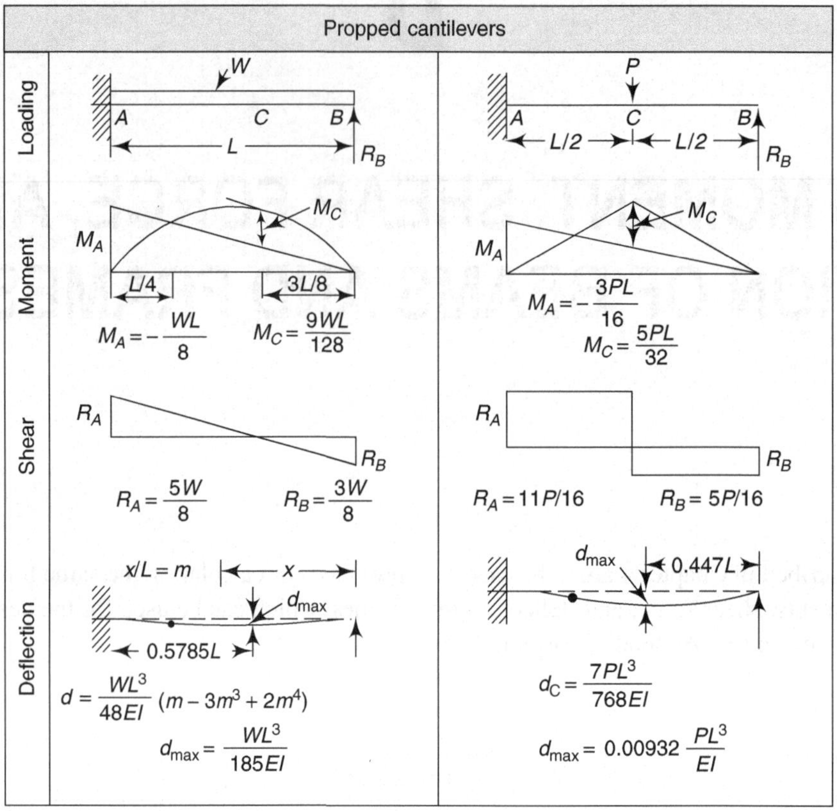
Propped cantilevers
Loading
Moment
Shear
Deflection
W
A C B
L
$R_B$
$M_C$
$M_A$
L/4
3L/8
$M_A = -\frac{WL}{8}$
$M_C = \frac{9WL}{128}$
$R_A$
$R_B$
$R_A = \frac{5W}{8}$
$R_B = \frac{3W}{8}$
x/L = m
x
$d_{max}$
0.5785L
$d = \frac{WL^3}{48EI}(m - 3m^3 + 2m^4)$
$d_{max} = \frac{WL^3}{185EI}$
P
A C B
L/2
L/2
$R_B$
$M_C$
$M_A$
$M_A = -\frac{3PL}{16}$
$M_C = \frac{5PL}{32}$
$R_A$
$R_B$
$R_A = 11P/16$
$R_B = 5P/16$
$d_{max}$
0.447L
$d_C = \frac{7PL^3}{768EI}$
$d_{max} = 0.00932\frac{PL^3}{EI}$

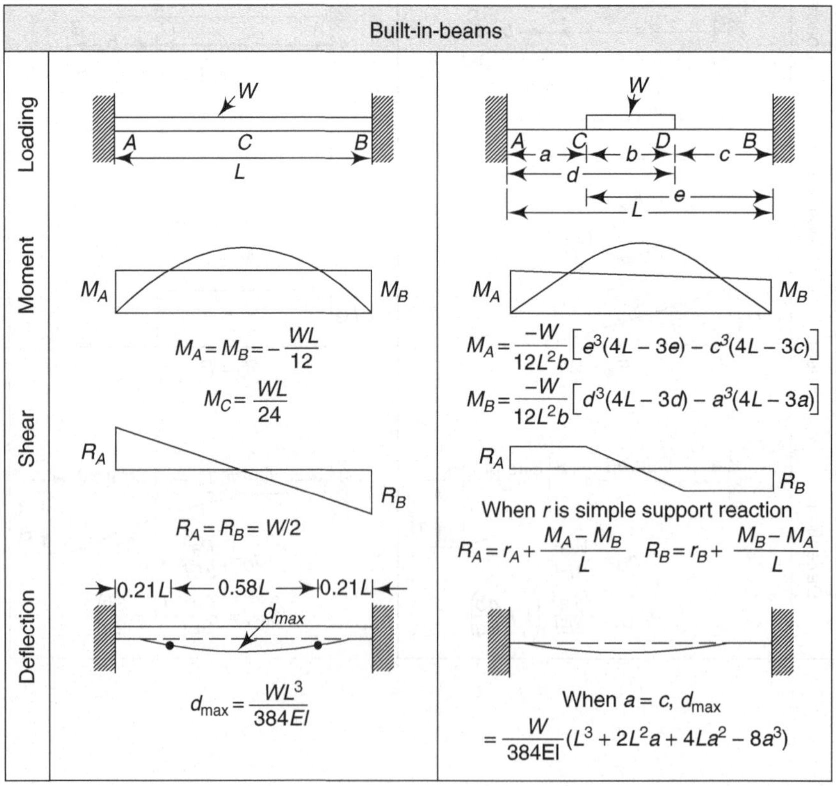
Built-in-beams
Loading
Moment
Shear
Deflection
W
A C B
L
$M_A$
$M_B$
$M_A = M_B = -\frac{WL}{12}$
$M_C = \frac{WL}{24}$
$R_A$
$R_B$
$R_A = R_B = W/2$
0.21L
0.58L
0.21L
$d_{max}$
$d_{max} = \frac{WL^3}{384EI}$
W
A C D B
a b c
d
e
L
$M_A$
$M_B$
$M_A = \frac{-W}{12L^2b}\left[e^3(4L - 3e) - c^3(4L - 3c)\right]$
$M_B = \frac{-W}{12L^2b}\left[d^3(4L - 3d) - a^3(4L - 3a)\right]$
$R_A$
$R_B$
When r is simple support reaction
$R_A = r_A + \frac{M_A - M_B}{L}$
$R_B = r_B + \frac{M_B - M_A}{L}$
When a = c, $d_{max}$
$= \frac{W}{384EI}(L^3 + 2L^2a + 4La^2 - 8a^3)$

(*Contd*)

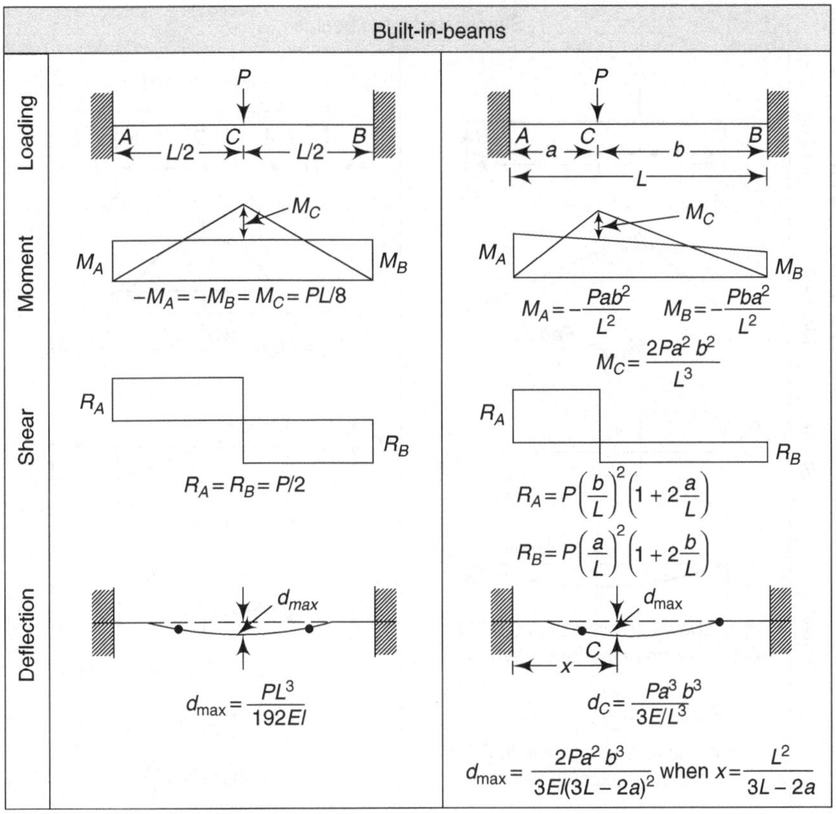
Built-in-beams
Loading
Moment
Shear
Deflection
P
A C B
L/2 L/2
$M_C$
$M_A$ $M_B$
$-M_A = -M_B = M_C = PL/8$
$R_A$ $R_B$
$R_A = R_B = P/2$
$d_{max}$
$d_{max} = \frac{PL^3}{192EI}$
P
A C B
a b
L
$M_C$
$M_A$ $M_B$
$M_A = -\frac{Pab^2}{L^2}$ $M_B = -\frac{Pba^2}{L^2}$
$M_C = \frac{2Pa^2b^2}{L^3}$
$R_A$ $R_B$
$R_A = P\left(\frac{b}{L}\right)^2\left(1 + 2\frac{a}{L}\right)$
$R_B = P\left(\frac{a}{L}\right)^2\left(1 + 2\frac{b}{L}\right)$
$d_{max}$
C
x
$d_C = \frac{Pa^3b^3}{3EIL^3}$
$d_{max} = \frac{2Pa^2b^3}{3EI(3L - 2a)^2}$ when $x = \frac{L^2}{3L - 2a}$

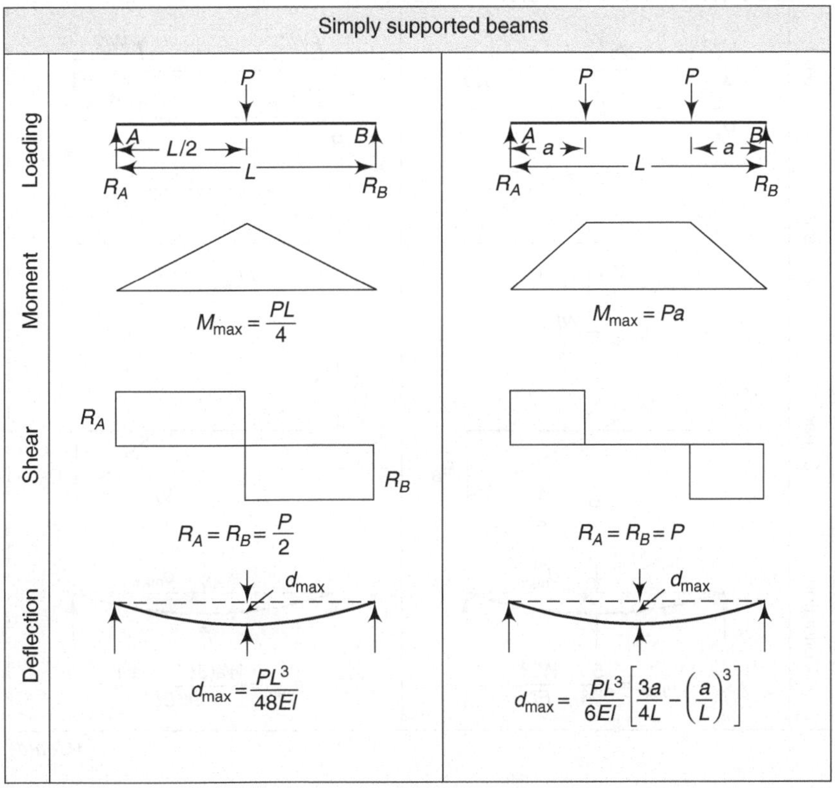
Simply supported beams
Loading
Moment
Shear
Deflection
P
A B
L/2
L
$R_A$ $R_B$
$M_{max} = \frac{PL}{4}$
$R_A$ $R_B$
$R_A = R_B = \frac{P}{2}$
$d_{max}$
$d_{max} = \frac{PL^3}{48EI}$
P P
A B
a a
L
$R_A$ $R_B$
$M_{max} = Pa$
$R_A = R_B = P$
$d_{max}$
$d_{max} = \frac{PL^3}{6EI}\left[\frac{3a}{4L} - \left(\frac{a}{L}\right)^3\right]$

(*Contd*)

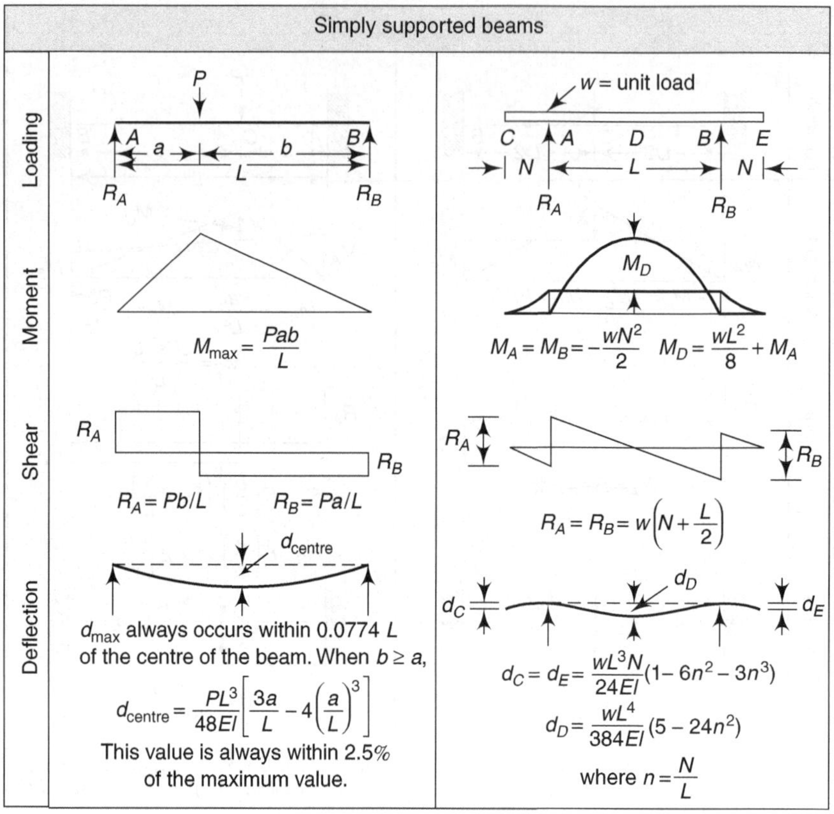
Simply supported beams
Loading
Moment
Shear
Deflection
P
A
a
b
L
B
$R_A$
$R_B$
$M_{max} = \frac{Pab}{L}$
$R_A = Pb/L$
$R_B = Pa/L$
$d_{centre}$
$d_{max}$ always occurs within 0.0774 L of the centre of the beam. When $b \geq a$,
$d_{centre} = \frac{PL^3}{48EI}\left[\frac{3a}{L} - 4\left(\frac{a}{L}\right)^3\right]$
This value is always within 2.5% of the maximum value.
w = unit load
C
A
D
B
E
N
L
N
$R_A$
$R_B$
$M_D$
$M_A = M_B = -\frac{wN^2}{2}$
$M_D = \frac{wL^2}{8} + M_A$
$R_A$
$R_B$
$R_A = R_B = w\left(N + \frac{L}{2}\right)$
$d_C$
$d_D$
$d_E$
$d_C = d_E = \frac{wL^3N}{24EI}(1 - 6n^2 - 3n^3)$
$d_D = \frac{wL^4}{384EI}(5 - 24n^2)$
where $n = \frac{N}{L}$

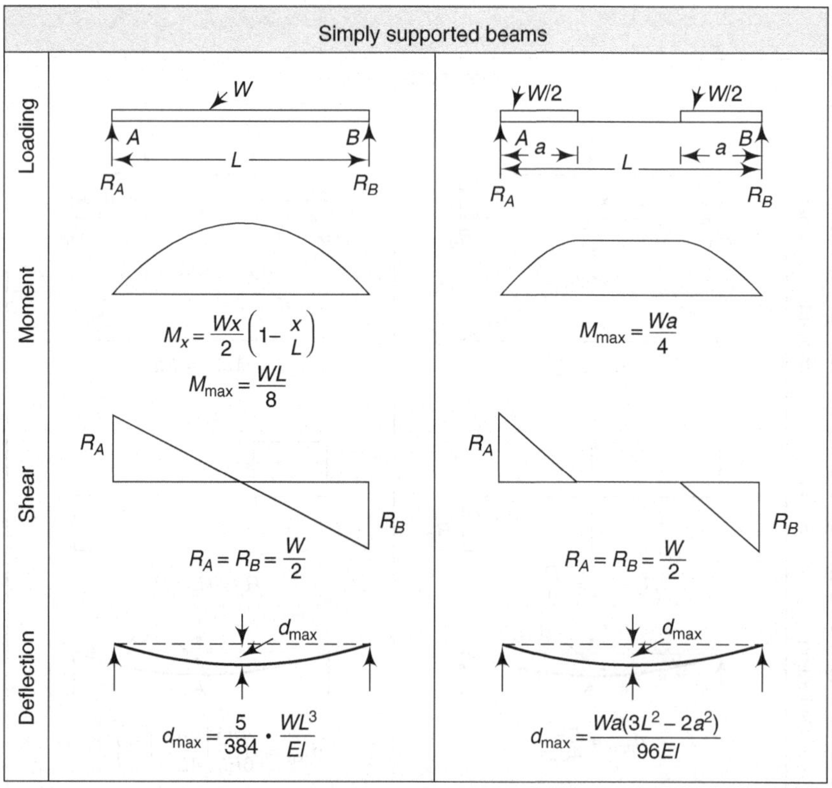
Simply supported beams
Loading
Moment
Shear
Deflection
W
A
B
L
$R_A$
$R_B$
$M_x = \frac{Wx}{2}\left(1 - \frac{x}{L}\right)$
$M_{max} = \frac{WL}{8}$
$R_A$
$R_B$
$R_A = R_B = \frac{W}{2}$
$d_{max}$
$d_{max} = \frac{5}{384} \cdot \frac{WL^3}{EI}$
W/2
W/2
A
a
L
a
B
$R_A$
$R_B$
$M_{max} = \frac{Wa}{4}$
$R_A$
$R_B$
$R_A = R_B = \frac{W}{2}$
$d_{max}$
$d_{max} = \frac{Wa(3L^2 - 2a^2)}{96EI}$

(Contd)

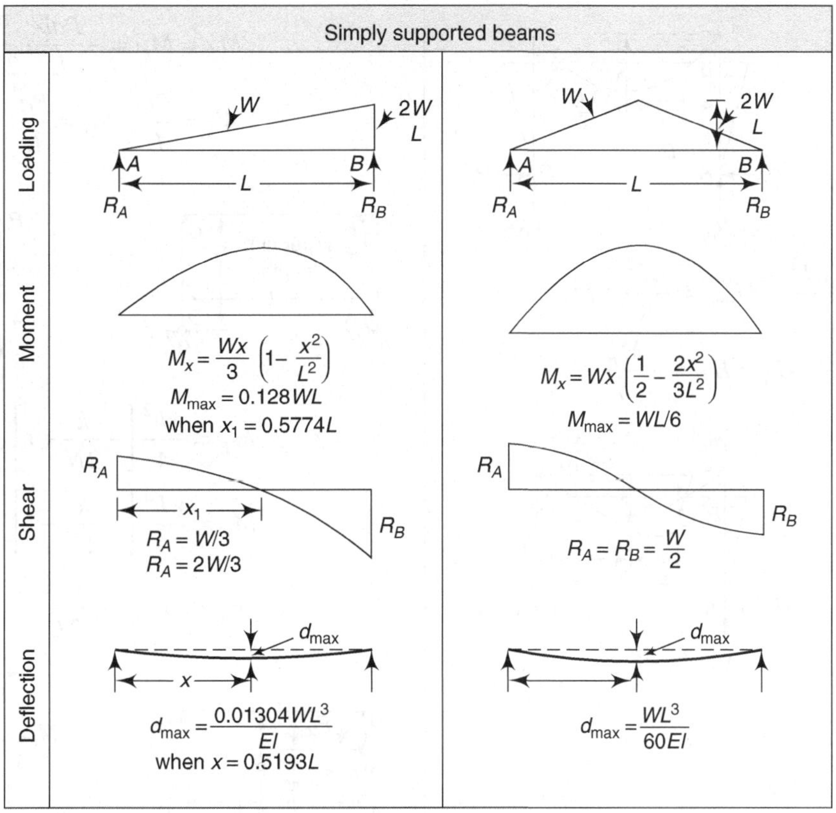


Equal span continuous beams
Uniformly distributed loads (UDL)
Moment = coefficient × $W \times L$
Reaction = coefficient × $W$
where $W$ is the UDL on one span only and $L$ is one span

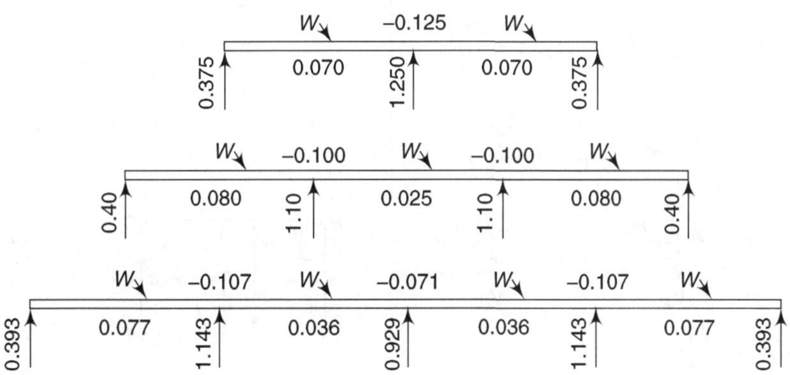


Rigid frames

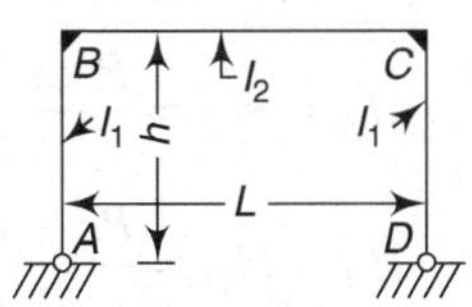


Frame data

Coefficients

$$k = \frac{I_2}{I_1} \cdot \frac{h}{L}$$

$$N = 2k + 3$$

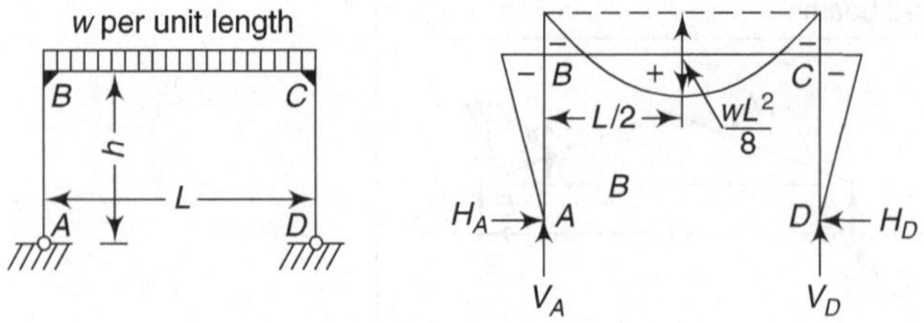


$$M_B = M_C = -\frac{wL^2}{4N} \quad M_{\max} = \frac{wL^2}{8} + M_B$$

$$V_A = V_D = \frac{wL}{2} \quad H_A = H_D = -\frac{M_B}{h}$$

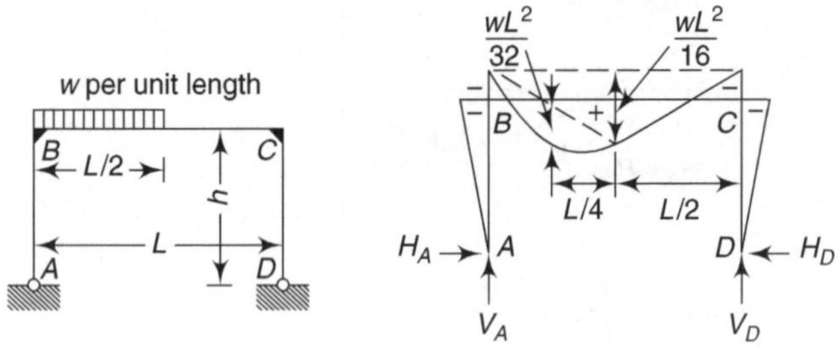


$$M_B = M_C = -\frac{wL^2}{8N}$$

$$V_A = \frac{3wL}{8} \quad V_D = \frac{wL}{8} \quad H_A = H_D = -\frac{M_B}{h}$$

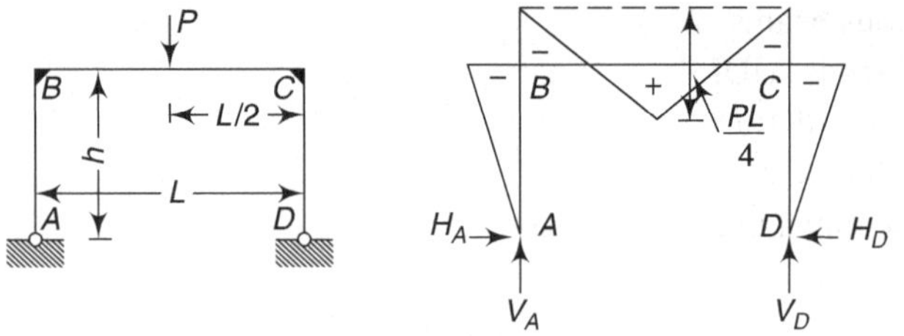


$$M_B = M_C = -\frac{3PL}{8N} \quad V_A = V_D = \frac{P}{2} \quad H_A = H_D = -\frac{M_B}{h}$$

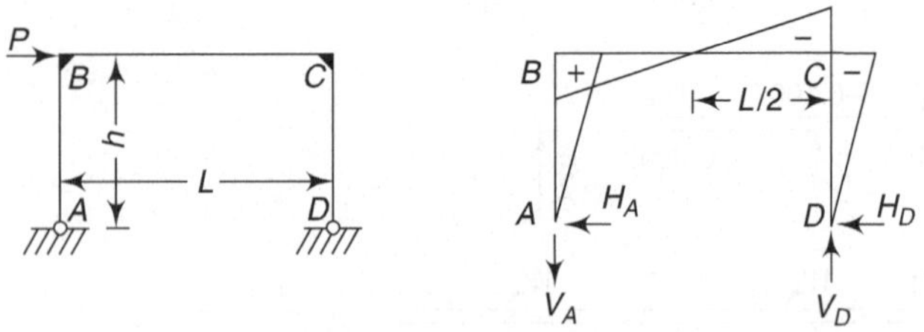


$$M_B = -M_C = +\frac{Ph}{2}$$

$$V_A = -V_D = -\frac{Ph}{L} \quad H_A = -H_D = -\frac{P}{2}$$

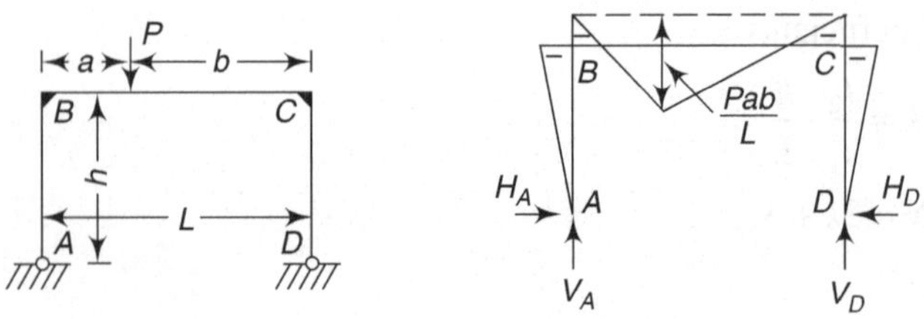


$$M_B = M_C = -\frac{Pab}{L} \cdot \frac{3}{2N}$$

$$V_A = \frac{Pb}{L} \quad V_D = \frac{Pa}{L} \quad H_A = H_D = -\frac{M_B}{h}$$

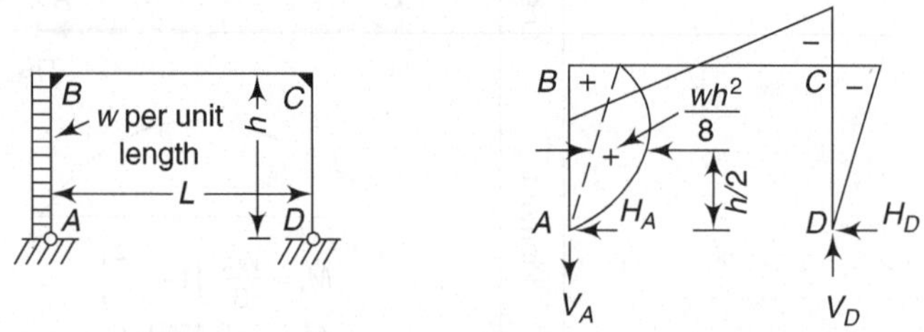


$$M_B = \frac{wh^2}{4}\left[-\frac{k}{2N} + 1\right] \quad H_D = -\frac{M_C}{h}$$

$$M_C = \frac{wh^2}{4}\left[-\frac{k}{2N} - 1\right] \quad H_A = -(wh - H_D)$$

$$V_A = -V_D = -\frac{wh^2}{2L}$$

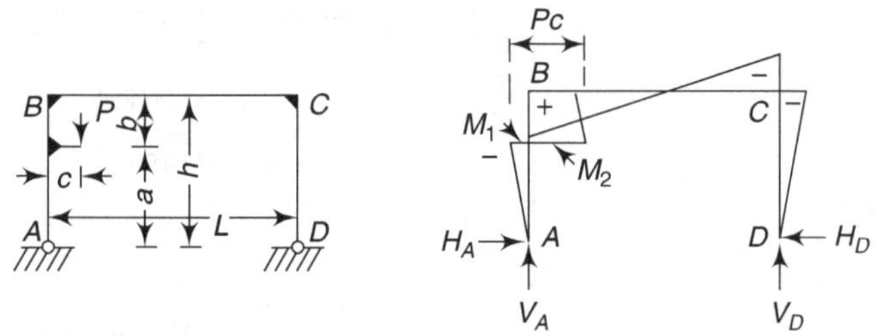


Constant: $a_1 = \frac{a}{h}$

$$M_B = \frac{Pc}{2}\left[\frac{(3a_1^2 - 1)k}{N} + 1\right] \quad H_A = H_D = -\frac{M_C}{h}$$

$$M_C = \frac{Pc}{2}\left[\frac{(3a_1^2 - 1)k}{N} - 1\right]$$

$$V_D = \frac{Pc}{L} \qquad V_A = P - V_D$$

$$M_1 = -H_A a \quad M_2 = Pc - H_A a$$

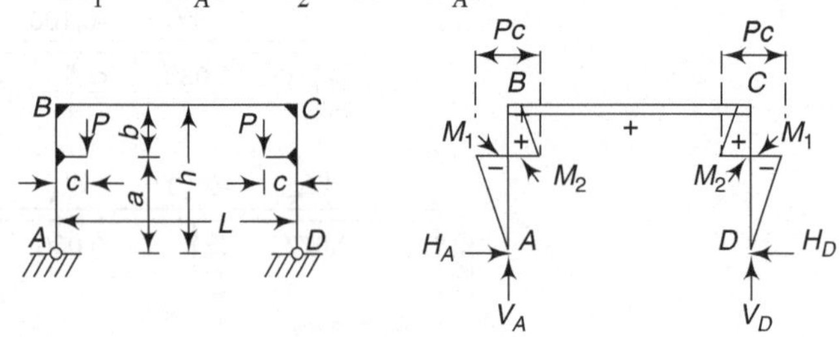


Constant: $a_1 = \frac{a}{h}$

$$M_B = M_C = \frac{Pc(3a_1^2 - 1)k}{N}$$

$$H_A = H_D = \frac{Pc - M_B}{h} \qquad V_A = V_D = P$$

$$M_1 = -H_A a \qquad M_2 = Pc - H_A a$$

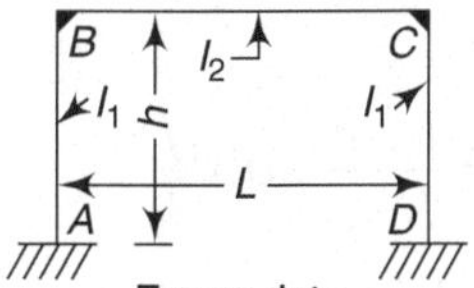


Frame data

Coefficients:

$$k = \frac{I_2}{I_1} \cdot \frac{h}{L}$$

$$N_1 = k + 2 \quad N_2 = 6k + 1$$

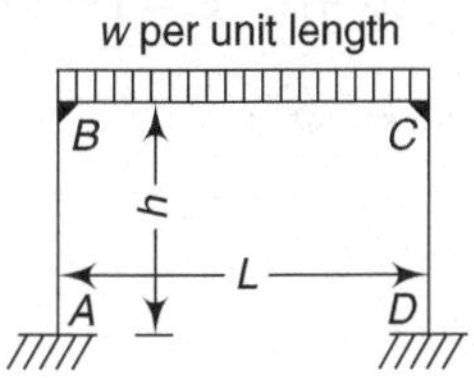


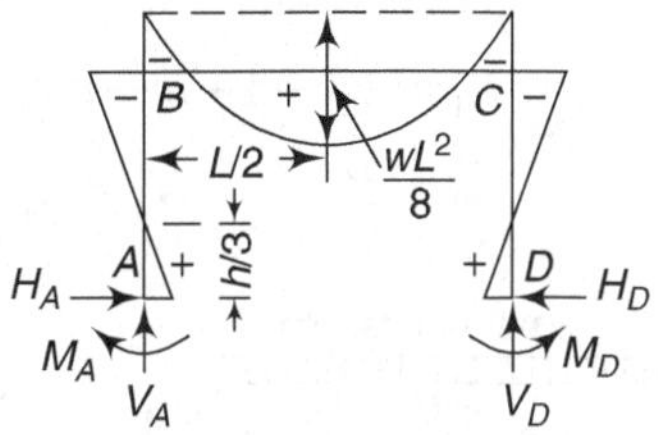


$$M_A = M_D = \frac{wL^2}{12N_1} \quad M_B = M_C = -\frac{wL^2}{6N_1} = -2M_A$$

$$M_{\max} = \frac{wL^2}{8} + M_B \quad V_A = V_D = \frac{wL}{2} \quad H_A = H_D = \frac{3M_A}{h}$$

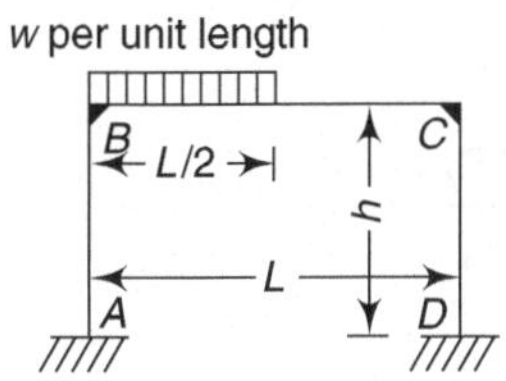


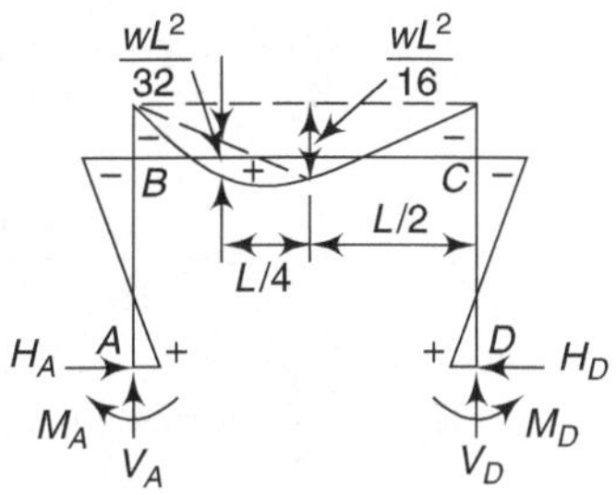


$$M_A = \frac{wL^2}{8}\left[\frac{1}{3N_1} - \frac{1}{8N_2}\right] \quad M_B = -\frac{wL^2}{8}\left[\frac{2}{3N_1} + \frac{1}{8N_2}\right]$$

$$M_D = \frac{wL^2}{8}\left[\frac{1}{3N_1} + \frac{1}{8N_2}\right] \quad M_C = -\frac{wL^2}{8}\left[\frac{2}{3N_1} - \frac{1}{8N_2}\right]$$

$$V_D = \frac{wL}{8}\left[1 - \frac{1}{4N_2}\right] \quad V_A = \frac{wL}{2} - V_D \quad H_A = H_D = \frac{wL^2}{8hN_1}$$

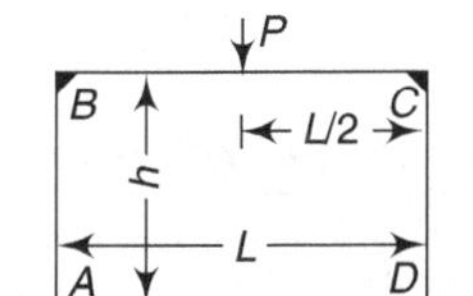


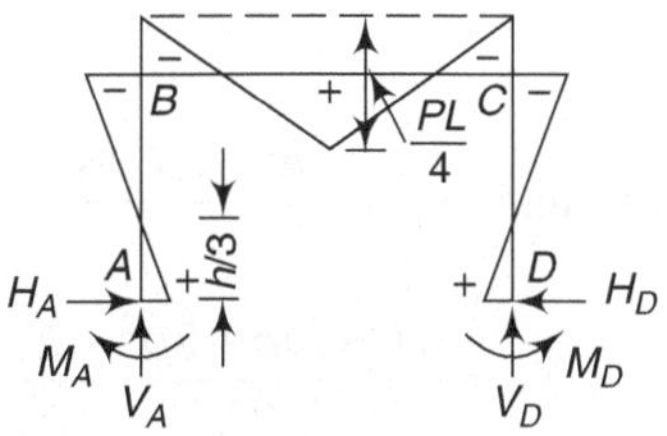


$$M_A = M_D = +\frac{PL}{8N_1} \quad M_B = M_C = -2M_A$$

$$V_A = V_D = \frac{P}{2} \quad H_A = H_D = \frac{3M_A}{h}$$

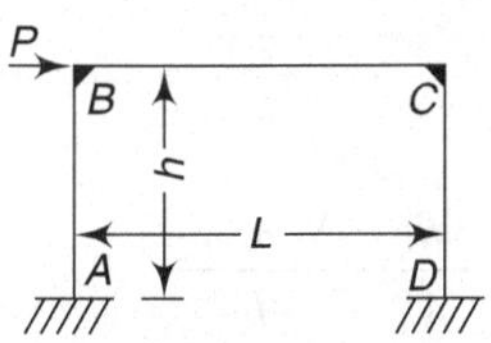


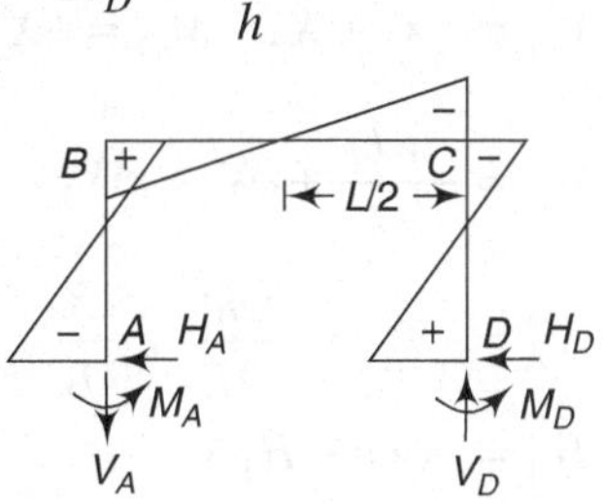


$$M_A = -\frac{Ph}{2} \cdot \frac{3k+1}{N_2} \quad M_B = +\frac{Ph}{2} \cdot \frac{3k}{N_2}$$

$$M_D = +\frac{Ph}{2} \cdot \frac{3k+1}{N_2} \quad M_C = -\frac{Ph}{2} \cdot \frac{3k}{N_2}$$

$$H_A = -H_D = -\frac{P}{2} \quad V_A = -V_D = -\frac{2M_B}{L}$$

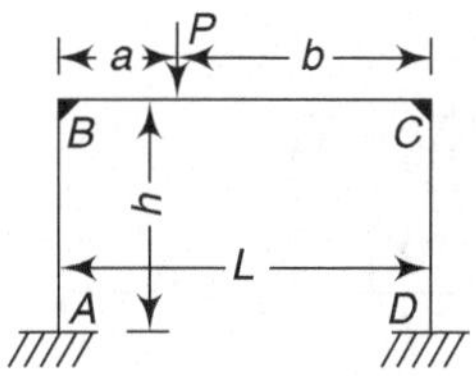


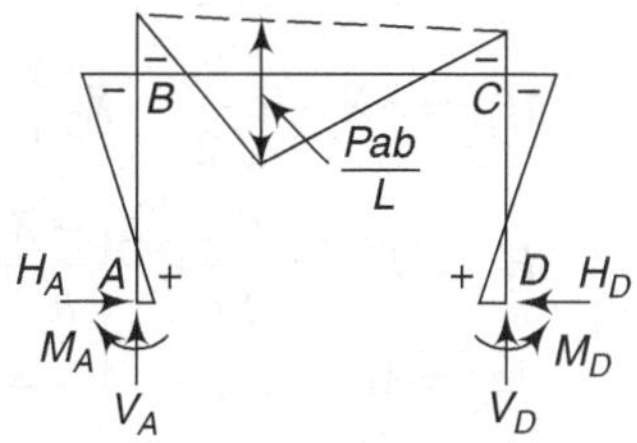


Constants: $a_1 = a/L \quad b_1 = b/L$

$$M_A = +\frac{Pab}{L}\left[\frac{1}{2N_1} - \frac{b_1 - a_1}{2N_2}\right] \quad M = -\frac{Pab}{L}\left[\frac{1}{N_1} + \frac{b_1 - a_1}{2N_2}\right]$$

$$M_D = +\frac{Pab}{L}\left[\frac{1}{2N_1} + \frac{b_1 - a_1}{2N_2}\right] \quad M_C = -\frac{Pab}{L}\left[\frac{1}{N_1} - \frac{b_1 - a_1}{2N_2}\right]$$

$$V_A = Pb_1\left[1 + \frac{a_1(b_1 - a_1)}{N_2}\right] \quad V_D = P - V_A$$

$$H_A = H_D = \frac{3Pab}{2LhN_1}$$

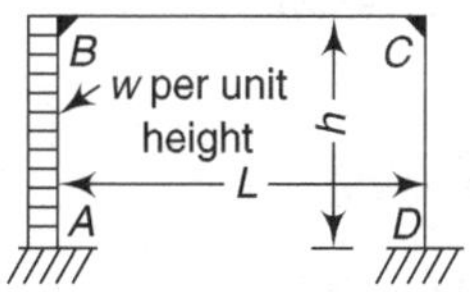


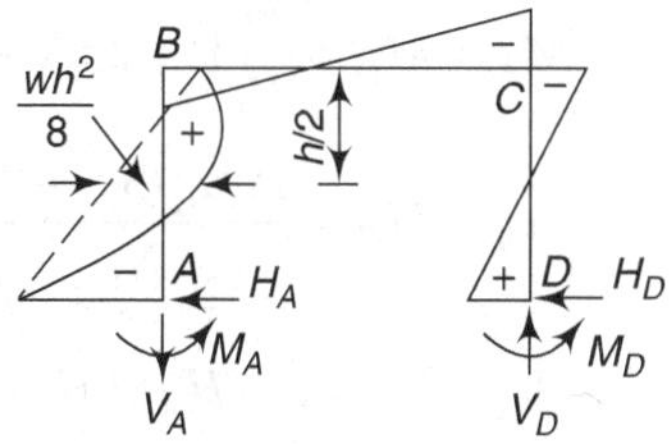


$$M_A = \frac{wh^2}{4}\left[-\frac{k+3}{6N_1} - \frac{4k+1}{N_2}\right] \quad M_B = \frac{wh^2}{4}\left[-\frac{k}{6N_1} + \frac{2k}{N_2}\right]$$

$$M_D = \frac{wh^2}{4}\left[-\frac{k+3}{6N_1} + \frac{4k+1}{N_2}\right] \quad M_C = \frac{wh^2}{4}\left[-\frac{k}{6N_1} - \frac{2k}{N_2}\right]$$

$$H_D = \frac{wh(2k+3)}{8N_1} \quad H_A = -(wh - H_D) \quad V_A = -V_D = -\frac{wh^2k}{LN_2}$$

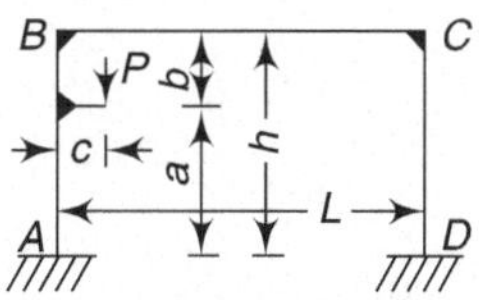


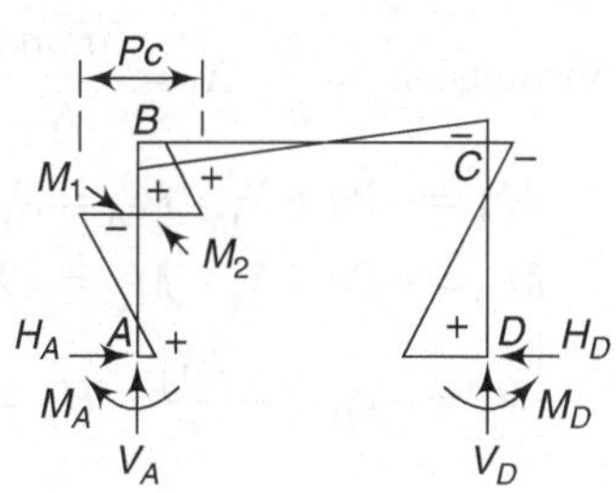

Constants: $a_1 = \dfrac{a}{h} \quad b_1 = \dfrac{b}{h}$

$$X_1 = \frac{Pc}{2N_1}[1 + 2b_1k - 3b_1^2(k+1)] \quad X_2 = \frac{Pcka_1(3a_1 - 2)}{2N_1}$$

$$X_3 = \frac{3Pcka_1}{N_2}$$

$$M_A = +X_1 - \left(\frac{Pc}{2} - X_3\right) \quad M_B = +X_2 + X_3$$

$$M_D = +X_1 + \left(\frac{Pc}{2} - X_3\right) \quad M_C = +X_2 - X_3$$

$$H_A = H_D = \frac{Pc}{2h} + \frac{X_1 - X_2}{h} \quad V_D = \frac{2X_3}{L} \quad V_A = P - V_D$$

$$M_1 = M_A - H_A a \quad M_2 = M_B + H_D b$$

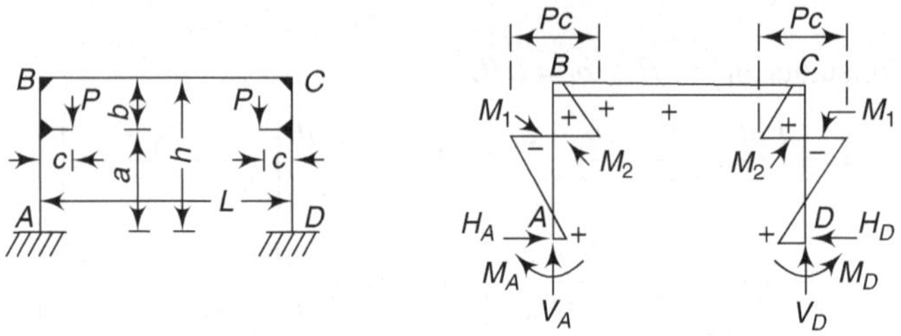


Constants: $a_1 = \dfrac{a}{h} \quad b_1 = \dfrac{b}{h}$

$$X_1 = \frac{Pc}{2N_1}[1 + 2b_1k - 3b_1^2(k+1)] \quad X_2 = \frac{Pcka_1(3a_1 - 2)}{2N_1}$$

$$M_A = M_D = \frac{Pc}{N_1}[1 + 2b_1k - 3b_1^2(k+1)] = 2X_1$$

$$M_B = M_C = \frac{Pcka_1(3a_1 - 2)}{N_1} = 2X_2$$

$$V_A = V_D = P \quad H_A = H_D = \frac{Pc + M_A - M_B}{h}$$

$$M_1 = M_A - H_A a \quad M_2 = M_B + H_D b$$

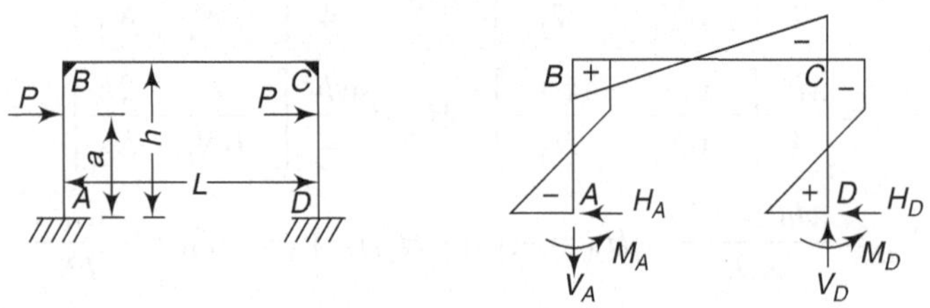


Constants: $a_1 = \dfrac{a}{h} \quad X_1 = \dfrac{3Paa_1k}{N_2}$

$$M_A = -Pa + X_1 \quad M_B = X_1$$

$$M_D = +Pa - X_1 \quad M_C = -X_1$$

$$V_A = -V_D = -\frac{2X_1}{L} \quad H_A = -H_D = -P$$

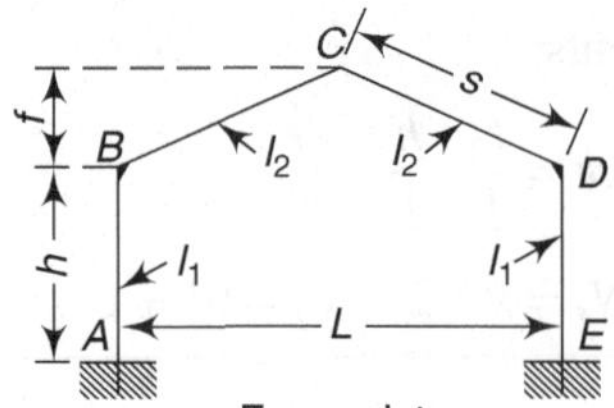


Frame data

Coefficients:

$$k = \frac{I_2}{I_1} \cdot \frac{h}{s} \quad \phi = \frac{f}{h}$$

$$m = 1 + \phi$$

$$B = 3k + 2 \quad C = 1 + 2m$$

$$K_1 = 2(k + 1 + m + m^2) \quad K_2 = 2(k + \phi^2)$$

$$R = \phi C - k \quad N_1 = K_1K_2 - R^2 \quad N_2 = 3k + B$$

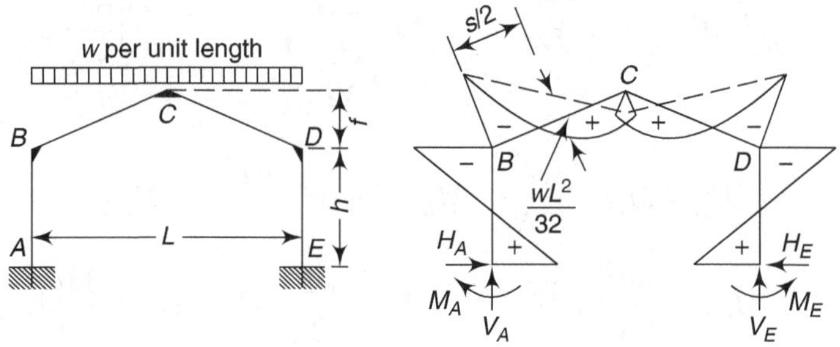


$$M_A = M_E = \frac{wL^2}{16} \cdot \frac{k(8 + 15\phi) + \phi(6 - \phi)}{N_1}$$

$$M_B = M_D = -\frac{wL^2}{16} \cdot \frac{k(16 + 15\phi) + \phi^2}{N_1}$$

$$M_C = \frac{wL^2}{8} - \phi M_A + mM_B$$

$$V_A = V_E = \frac{wL}{2} \quad H_A = H_E = \frac{M_A - M_B}{h}$$

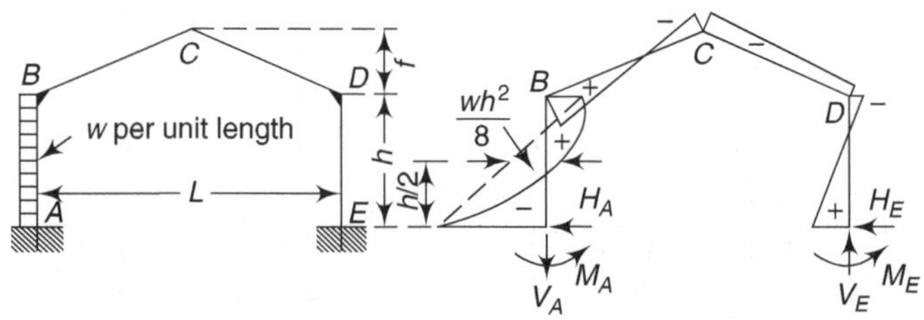


Constants: $X_1 = \dfrac{wh^2}{8} \cdot \dfrac{k(k+6) + k\phi(15 + 16\phi) + 6\phi^2}{N_1}$

$$X_2 = \frac{wh^2k(9\phi + 8\phi^2 - k)}{8N_1} \quad X_3 = \frac{wh^2(2k+1)}{2N_2}$$

$$M_A = -X_1 - X_3 \quad M_B = +X_2 + \left(\frac{wh^2}{4} - X_3\right)$$

$$M_E = -X_1 + X_3 \quad M_D = +X_2 - \left(\frac{wh^2}{4} - X_3\right)$$

$$M_C = -\frac{whf}{4} + \phi X_1 + mX_2$$

$$V_A = -V_E = -\frac{wh^2}{2L} + \frac{2X_3}{L} \quad H_E = \frac{wh}{4} - \frac{X_1 + X_2}{h}$$

$$H_A = -(wh - H_E)$$

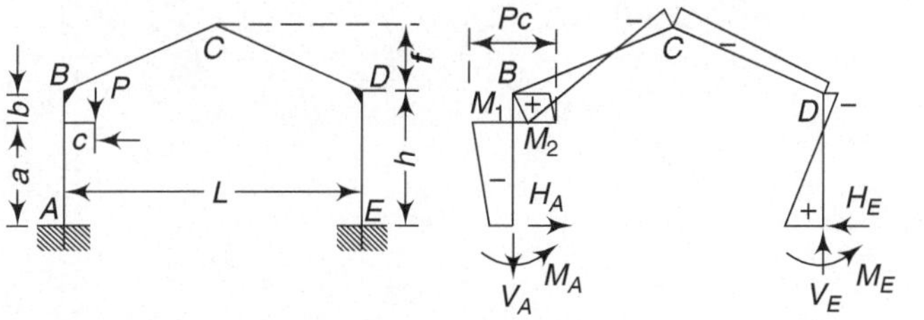


Constants: $a_1 = \dfrac{a}{h} \quad b_1 = \dfrac{b}{h}$

$$Y_1 = Pc[2\phi^2 - (1 - 3b_1^2)k] \quad Y_2 = Pc[\phi c - (3a_1^2 - 1)k]$$

$$X_1 = \frac{Y_1K_1 - Y_2R}{2N_1} \quad X_2 = \frac{Y_2K_2 - Y_1R}{2N_1} \quad X_3 = \frac{Pc}{2} \cdot \frac{b - 3(a_1 - b_1)k}{N_2}$$

$$M_A = -X_1 - X_3 \quad M_B = +X_2 + \left(\frac{Pc}{2} - X_3\right)$$

$$M_E = -X_1 + X_3 \quad M_D = +X_2 - \left(\frac{Pc}{2} - X_3\right)$$

$$M_C = -\frac{\phi Pc}{4} + \phi X_1 + mX_2$$

$$M_1 = M_A - H_A a \quad M_2 = M_B + H_E b$$

$$V_E = \frac{Pc - 2X_3}{L} \quad V_A = P - V_E \quad H_A = H_E = \frac{Pc}{2h} - \frac{X_1 + X_2}{h}$$

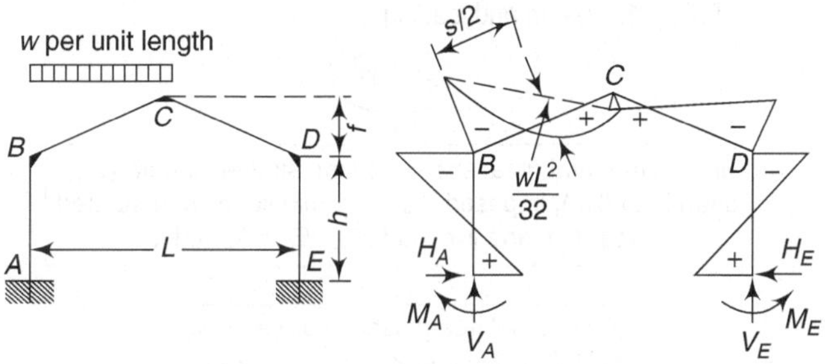


Constants: $X_1 = \dfrac{wL^2}{32} \cdot \dfrac{k(8 + 15\phi) + \phi(6 - \phi)}{N_1}$

$$X_2 = \frac{wL^2}{32} \cdot \frac{k(16 + 15\phi) + \phi^2}{N_1} \quad X_3 = \frac{wL^2}{32N_2}$$

$$M_A = +X_1 - X_3 \quad M_B = -X_2 - X_3 \quad M_E = +X_1 + X_3$$

$$M_D = -X_2 + X_3$$

$$M_C = \frac{wL^2}{16} - \phi X_1 - mX_2$$

$$V_E = \frac{wL}{8} - \frac{2X_3}{L} \quad V_A = \frac{wL}{2} - V_E \quad H_A = H_E = \frac{X_1 + X_2}{h}$$

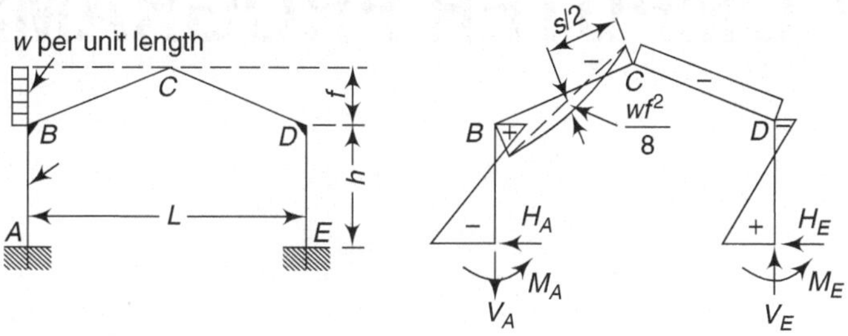


Constants: $X_1 = \dfrac{wf^2}{8} \cdot \dfrac{k(9\phi + 4) + \phi(6 + \phi)}{N_1}$

$$X_2 = \frac{wf^2}{8} \cdot \frac{k(8 + 9\phi) - \phi^2}{N_1} \quad X_3 = \frac{wfh}{8} \cdot \frac{4b + \phi}{N_2}$$

$$M_A = -X_1 - X_3 \quad M_B = +X_2 + \left(\frac{wfh}{2} - X_3\right)$$

$$M_E = -X_1 + X_3 \quad M_D = +X_2 - \left(\frac{wfh}{2} - X_3\right)$$

$$M_C = -\frac{wf^2}{4} + \phi X_1 + mX_2$$

$$V_A = -V_E = -\frac{wfh(2 + \phi)}{2L} + \frac{2X_3}{L} \quad H_E = \frac{wf}{2} - \frac{X_1 + X_2}{h}$$

$$H_A = -(wf - H_E)$$

*Source:* Kleinlogel, A., *Rahmenformeln*, Eleventh edition, Verlag von Wilhelm Ernst & Sohn., Berlin, 1949.

APPENDIX

# E

# COMPUTER PROGRAMS

## INTRODUCTION

With the advent of computers and software packages, it is now possible to analyse any structure for any geometry and given loading conditions quickly. Though several standard software packages are available for the linear, non-linear, buckling, or dynamic analysis of structures, very few design packages are available. Nonetheless, some analysis software packages do have some design routines, which are based on specific codes. Two computer programs written in Visual Basic are provided in this appendix, which can be run on any standard personal computer under the Windows environment. Visual Basic has been chosen because it uses an interactive approach to the development of computer programs. Visual Basic interprets the code as we enter it, catching and highlighting syntax or spelling errors. It also partly compiles the code as it is entered. While compiling the whole code, if the compiler finds an error, it is highlighted in the code. We can fix the error and continue compiling without having to start all over. Visual Basic also provides a variety of user interfaces in the form of text boxes, labels, radio button, pictures, etc. Visual Basic version 6.0 is used in this example, because it is simple and popular though VB.NET is the most recent version. More details about Visual Basic may be found in the Microsoft web site http://msdn.microsoft.com/library/default.asp?url=/library/en-us/vbcon98/html/vbconpart1visualbasicbasics.asp.

## E.1 DESIGN PROGRAMS

When writing design programs, it is much easier if the code of practice follows a logical sequence of requirements and checks. Also, the design stresses and factors should be given in closed-form mathematical expressions. Fortunately, the current version of IS: 800, satisfies these requirements.

### E.1.1 Beam Design Program

The simply supported steel beam design program, given in this section, designs beams of compact, semi-compact, and plastic sections. The beam arrangement is shown in Fig. E.1 and the flow chart is shown in Fig. E.2.

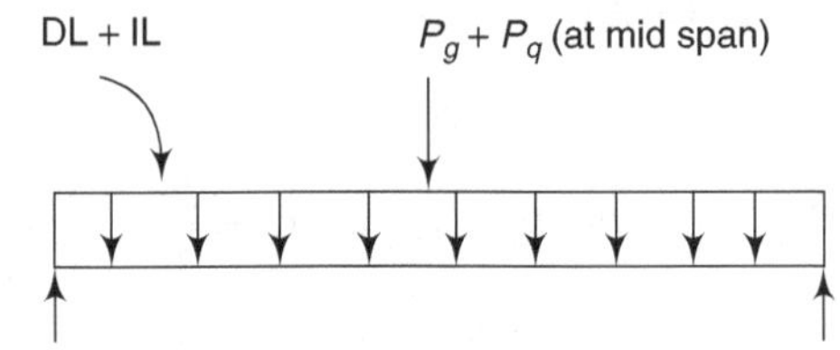


**FIG. E.1** Beam and loading

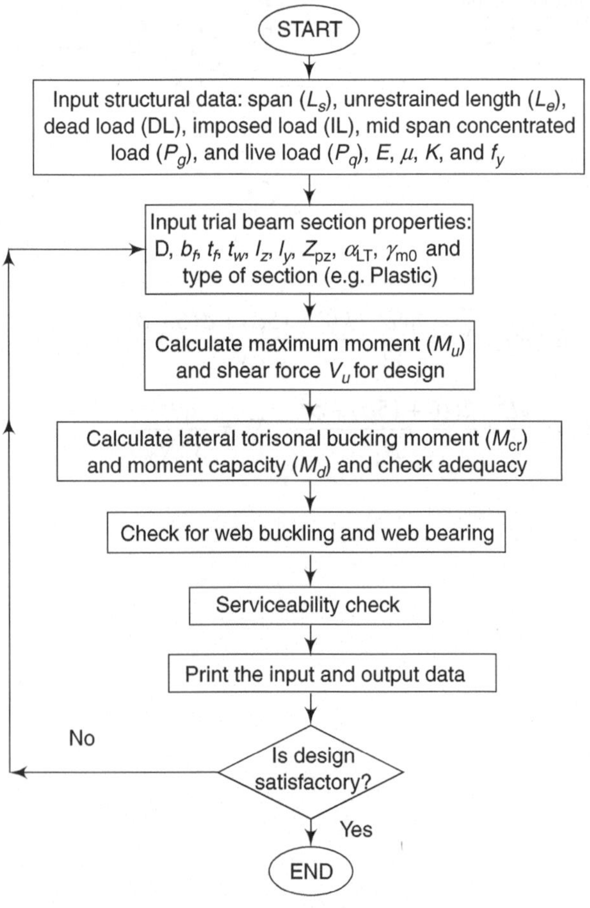


**FIG. E.2** Flowchart of beam design program

Only simple loading has been considered to reduce the length of the program (the user may modify it to suit his/her requirements). The notations used are the same as those given in IS: 800 (Section 8), and hence the program is easy to understand. The typical input screens of this program are given in Figs E.3 and E.4.

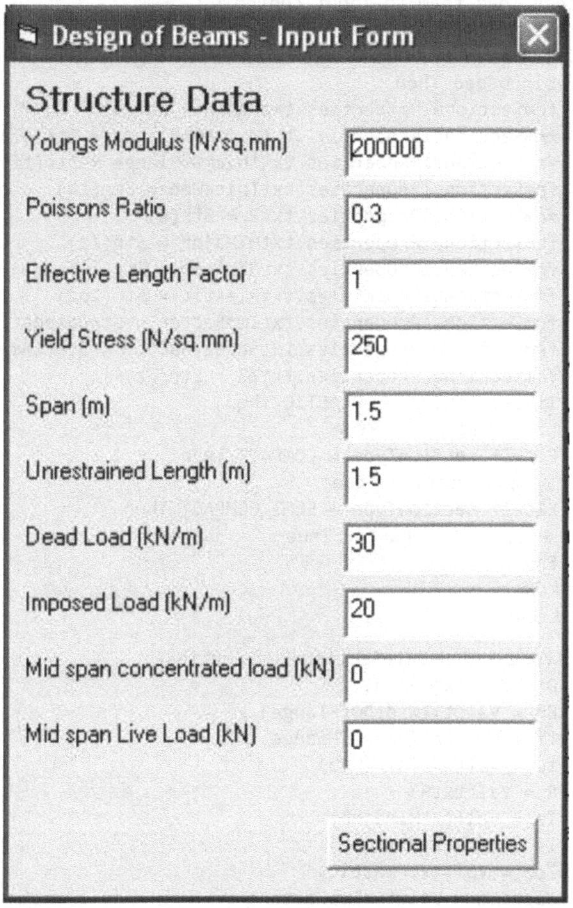


**FIG. E.3** Input screen for structure data of beam design program

The following program is written in Visual Basic for beam design.

```
‘Global Module
‘Structure properties
Public E As Double
Public EMu As Double
Public K As Double
Public Fy As Double
Public LS As Double
Public Le As Double
Public DL As Double
Public IL As Double
Public PG As Double
Public PQ As Double

‘Sectional Properties
Public D As Double
Public Bf As Double
Public tf As Double
Public tw As Double
Public R As Double
Public Iz As Double
Public Iy As Double
Public Zpz As Double
Public alphaLT As Double
Public GamaM As Double
Public Zz As Double

‘intermediate values
Public Wu As Double
Public Vu As Double
Public Pu As Double
Public Mu As Double

Public Const PI As Double = 3.14159
Public sectionType As Integer

Public Const PLASTIC As Integer = 1
Public Const COMPACT As Integer = 2
Public Const SEMI_COMPACT As Integer = 3

Public blnLoaded As Boolean
Public FileName As String
```

```
‘From Structure Data input form
Option Explicit
Private Sub cmdNext Click()
‘assign the data from input controls to
global variables
Call UpdateStructureData
Unload Me
‘Show the next input screen
frmSectionalProperties.Show
End Sub

Private Sub UpdateStructureData()
‘ transfer the data to global variables
E = Val(txtE)
EMu = Val(txtMu)
K = Val(txtK)
Fy = Val(txtFy)
LS = Val(txtSpan)
Le = Val(txtLE)
DL = Val(txtDeadLoad)
IL = Val(txtIL)
PG = Val(txtPG)
PQ = Val(txtPQ)
End Sub
```

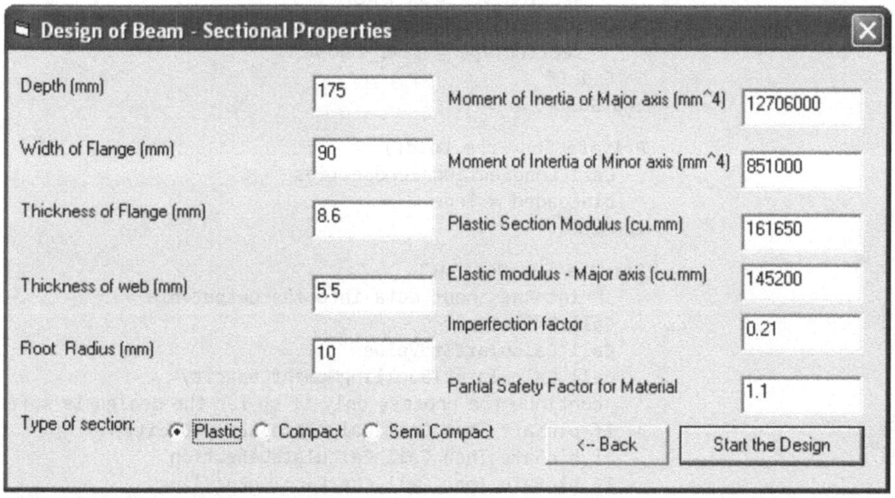


**FIG. E.4** Input screen for sectional properties of beam design program

```
'From sectional properties input form, which also has design code
Option Explicit
Dim G As Double
Dim It As Double
Dim hf As Double
Dim Iw As Double
Dim Betaf As Double
Dim Mcr As Double
Dim LamdaLT As Double
Dim LamdaT1 As Double
Dim alphaLT As Double
Dim PhiLT As Double
Dim ChiLT As Double
Dim fbd As Double

Dim Md As Double
Dim Vd As Double
Dim Deltab As Double
Dim DeltaAll As Double

Dim b1 As Double
Dim n1 As Double
Dim Ab As Double
Dim I As Double
Dim A As Double
Dim rmin As Double
Dim Lamda As Double
Dim LamdaEff As Double
Dim Phi1 As Double
Dim Phi2 As Double
Dim fcd As Double
Dim SWB As Double
Dim n2 As Double
Dim Fw As Double
Dim blnSafe As Boolean

Private Sub cmdBack Click()
   'Update the input controls with user entered Structure data
   Call LoadStructureData
   'Transfer user input sectional properties data to global variables
   Call UpdateSectionalProperties
   Unload Me

   frmBeamDesignl.Show

End Sub
Private Sub LoadStructureData()
   frmBeamDesignl.txtE = E
   frmBeamDesignl.txtMu = EMu
   frmBeamDesignl.txtK = Str(K)
   frmBeamDesignl.txtFy = Fy
   frmBeamDesignl.txtSpan = LS
   frmBeamDesignl.txtLE = Le
   frmBeamDesignl.txtDeadLoad = DL
   frmBeamDesignl.txtIL = IL
   frmBeamDesignl.txtPG = PG
   frmBeamDesignl.txtPQ = PQ
   End Sub

Private Sub cmdDesign Click()
'start the design process
'open the output file to print the results
   FileName = App.Path + "\BeamDesign.out"
   Open FileName For Output As 1
'pressume the design is safe
   blnSafe = True
   Call UpdateSectionalProperties
   Call Design
End Sub

Private Sub LoadSectionalProperties()
'if the data has been already transfered to the global variables
'then load them to input controls
'this check is necessary to sweap the default values with null values
If blnLoaded Then
   frmSectionalProperties.txtDepth = Str(D)
   frmSectionalProperties.txtWidthOfFlange = Str(Bf)
   frmSectionalProperties.txtThickOfFlange = Str(tf)
   frmSectionalProperties.txtThickWeb = Str(tw)
   frmSectionalProperties.txtR = Str(R)
   frmSectionalProperties.txtMIMajor = Str(Iz)
   frmSectionalProperties.txtMIMinor = Str(Iy)
   frmSectionalProperties.txtPlastic = Str(Zpz)
   frmSectionalProperties.txtImpFactor = Str(alphaLT)
   frmSectionalProperties.txtSafetyFactor = Str(GamaM)
   frmSectionalProperties.txtZZ = Str(Zz)
   If sectionType = PLASTIC Then
      optPlastic = True
   ElseIf sectionType = COMPACT Then
      optCompact = True
   ElseIf sectionType = SEMI_COMPACT Then
      optSemiCompact = True
   End If
End If
End Sub

Private Sub UpdateSectionalProperties()
   D = Val(txtDepth)
   Bf = Val(txtWidthOfFlange)
   tf = Val(txtThickOfFlange)
   tw = Val(txtThickWeb)
   R = Val(txtR)
   Iz = Val(txtMIMajor)
   Iy = Val(txtMIMinor)
   Zpz = Val(txtPlastic)
   alphaLT = Val(txtImpFactor)
   GamaM = Val(txtSafetyFactor)
   Zz = Val(txtZZ)
   If optPlastic Then
      sectionType = PLASTIC
   ElseIf optCompact Then
      sectionType = COMPACT
   ElseIf optSemiCompact Then
      sectionType = SEMI_COMPACT
   End If
End Sub

Private Sub Form Load()
   Call LoadSectionalProperties
   blnLoaded = True
End Sub

Private Sub Design()
   'Print the input data into the output file
   Call PrintInput
   Call CalculateStdValues
   Call CalculateLTBucklingMomentCapacity
   'continue the process only if so far the design is safe
   If blnSafe Then Call CalculateShearCapacity
   If blnSafe Then Call CalculateDeflection
   If blnSafe Then Call CheckForWebBuckling
```

```
   If blnSafe Then Call CheckForWebBearing
   Close #1
   If blnSafe Then End
End Sub

Private Sub CalculateStdValues()
   'calculate maximum Mu and Vu for design
   Dim strMsg As String

   Wu = 1.5 * (DL + IL)
   Pu = 1.5 * (PG + PQ)
   Mu = Wu * LS * LS / 8# + Pu * LS / 4#
   Vu = (Wu * LS + Pu) / 2

   Print #1, Tab(5); "Output"
   Print #1, Tab(5); "------"

   Print #1, Tab(5); "Maximum bending moment for design
   (kNm)";
   Tab(60); Format(Mu, "#######0.0##")
   Print #1, Tab(5); "Maximum shear force for design
   (kN)"; Tab(60);
   Format(Vu, "######0.0##")
End Sub

Private Sub CalculateLTBucklingMomentCapacity()
Dim Betab As Double
Dim strMsg As String

   If sectionType = SEMI_COMPACT Then
      Betab = Zz / Zpz
   Else
      Betab = 1
   End If

   Betaf = 0.5
   G = E / (2 * (1 + EMu))
   hf = D - tf
   It = (2 * Bf * tf ^ 3 + hf * tw ^ 3) / 3
   Le = K * Le
   Iw = (1 - Betaf) * Betaf * Iy * hf ^ 2
   Mcr = (((PI ^ 2 * E * Iy) / (Le * 1000) ^ 2) * (G * It
   + (PI ^ 2 * E * Iw) / (Le * 1000) ^ 2)) ^ 0.5
   LamdaLT = (Betab * Zpz * Fy / Mcr) ^ 0.5
   LamdaLT1 = (1.2 * Zz * Fy / Mcr) ^ 0.5
   If LamdaLT > LamdaLT1 then
     LamdaLT = LamdaLT1
   End If
   alphaLT = 0.21
   PhiLT = 0.5 * (1 + alphaLT * (LamdaLT - 0.2) + LamdaLT
   ^ 2)
   ChiLT = 1# / (PhiLT + (PhiLT ^ 2 - LamdaLT ^ 2) ^ 0.5)
   If (ChiLT > 1#) Then
       ChiLT = 1#
   End If
   fbd = ChiLT * Fy / GamaM

   Md = Betab * Zpz * fbd / 1000000#
   strMsg = "Torsional constant" + Str(It) + "mm^4" +
   vbCrLf
   strMsg = strMsg + "Warping constant" + Str(Iw) + "mm^6"
   + vbCrLf
   strMsg = strMsg + "Moment capacity of the section" +
   Str(Md) + "kNm" + vbCrLf
If Md < Mu Then
   MsgBox strMsg + "Moment capacity is less than" +
   Str(Mu) + "Section is unsafe. Revise the section",
   "Unsafe section"
   Unload Me
   blnSafe = False
   frmBeamDesign1.Show
   Exit Sub
End If
Print #1, Tab(5); "Torsional constant (mm^4)"; Tab(60);
Format(It, "#######0.0##")
Print #1, Tab(5); "Warping constant (mm^6)"; Tab(60);
Format(Iw, "#######0.0##")
Print #1, Tab(5); "Moment capacity of the section (kNm)";
Tab(60); Format(Md, "#######0.0##")
Print #1, Tab(5); "Moment capacity of Trial section is
adequate"
Print #1, ""

End Sub

Private Sub CalculateShearCapacity()
   Dim strMsg As String

   Vd = ((Fy / (GamaM * Sqr(3))) * D * tw) / 1000
   strMsg = "Shear capacity =" + Str(0.6 * Vd) + "kN" +
   vbCrLf + "Shear force =" + Str(Vu) + "kN" + vbCrLf
   If 0.6 * Vd < Vu Then
      MsgBox strMsg + "Shear capacity is unsafe. Revise
      the section"
      Unload Me
      blnSafe = False
      frmBeamDesign1.Show
      Exit Sub
   Else
      Print #1, Tab(5); "Shear capacity (kN)"; Tab(60);
      Format(0.6 * Vd, "#######0.0##")
      Print #1, Tab(5); "Shear force (kN)"; Tab(60);
      Format(Vu, "#######0.0##")
      Print #1, Tab(5); "Shear capacity of Trial section
      is adequate"
   End If

End Sub
Private Sub CalculateDeflection()

   Dim strMsg As String

   Deltab = (5 * (DL + IL) * (LS * 1000) ^ 4 / 384 + (PG
   + PQ) * LS ^ 3 / 48) / (E * Iz)
   DeltaAll = LS * 1000 / 300

   strMsg = "Delta allowable =" + Str(DeltaAll) + "mm" +
   vbCrLf + "Delta actual =" + Str(Deltab) + "mm" + vbCrLf
   If Deltab < DeltaAll Then
      'MsgBox strMsg + "Trial section is adequate for
      deflection check"
      Print #1, Tab(5); "Delta allowable (mm)"; Tab(60);
      Format(DeltaAll, "#######0.0##")
      Print #1, Tab(5); "Delta actual (mm)"; Tab(60);
      Format(Deltab, "#######0.0##")
      Print #1, Tab(5); "Trial section is adequate for
      deflection check"
      Print #1, ""
   Else
      MsgBox strMsg + "Revise the section for deflection
      consideration"
      Unload Me
      blnSafe = False
      frmBeamDesign1.Show
      Exit Sub
   End If
End Sub
```

```
Private Sub CheckForWebBuckling()
   Dim strMsg As String

   bl = (Bf - tw) / 2
   nl = D / 2
   Ab = (bl + nl) * tw
   I = bl * tw ^ 3 / 12
   A = bl * tw
   rmin = Sqr(I / A)
   Lamda = 0.7 * (D - 2 * (tf + R)) / rmin
   LamdaEff = (Fy * Lamda ^ 2 / (PI ^ 2 * E)) ^ 0.5
   Phil = 0.5 * (1 + 0.49 * (LamdaEff - 0.2) + LamdaEff ^ 2)
   Phi2 = Phil + (Phil ^ 2 - LamdaEff ^ 2) ^ 0.5
   fcd = Fy / (Phi2 * GamaM)
   SWB = fcd * Ab / 1000

   strMsg = "Strength against web buckling =" + Str(SWB)
   + "kN" + vbCrLf + "Shear force =" + Str(Vu) + "kN"
   If SWB > Vu Then
      'MsgBox strMsg + "Safe against web buckling"
      Print #1, Tab(5); "Strength against web buckling
      (kN)";
      Tab(60); Format(SWB, "#######0.0##")
      Print #1, Tab(5); "Shear force (kN)"; Tab(60);
      Format(Vu, "#######0.0##")
      Print #1, Tab(5); "Safe against web buckling"
   Else
      MsgBox strMsg + "Revise the section for web buckling"
      Unload Me
      blnSafe = False
      frmBeamDesignl.Show
      Exit Sub
   End If

End Sub
Private Sub CheckForWebBearing()
   Dim strMsg As String

   n2 = 2.5 * (tf + R)
   Fw = (bl + n2) * tw * Fy / (GamaM * 1000)
   strMsg = "Strength against web bearing =" + Str(Fw)
   + "kN" + vbCrLf + "Shear force =" + Str(Vu) + "kN" +
   vbCrLf
   If Fw > Vu Then
      'MsgBox strMsg + "Safe against web bearing"
      MsgBox "Design is safe and" + vbCrLf + "The output
      is saved as BeamDesign.out in the working folder"
      Print #1, Tab(5); "Strength against web bearing
      (kN)";
      Tab(60); Format(Fw, "#######0.0##")
      Print #1, Tab(5); "Shear force (kN)"; Tab(60);
      Format(Vu, "#######0.0##")
      Print #1, Tab(5); "Safe against web bearing"
      Print #1, ""
   Else
      MsgBox strMsg + "Revise the section for web bearing"
      Unload Me
      frmBeamDesignl.Show
      Exit Sub
   End If
End Sub

Private Sub optCompact Click()
   If optCompact Then
   sectionType = COMPACT
   End If

End Sub

Private Sub optPlastic Click()
   If optPlastic Then
      sectionType = PLASTIC
   End If
End Sub

Private Sub optSemiCompact Click()
   If optSemiCompact Then
      sectionType = SEMI_COMPACT
   End If
End Sub
Private Sub PrintInput()

   Print #1, Tab(10); "Beam Design"
   Print #1, Tab(10); "------"
   Print #1, ""
   Print #1, Tab(5); "Structure Data"
   Print #1, Tab(5); "-------"

   Print #1, Tab(5); "Youngs Modulus (N/sq.mm)"; Tab(60);
   Format(E, "#######0.0##")
   Print #1, Tab(5); "Poisson's Ratio"; Tab(60);
   Format(EMu, "#######0.0##")
   Print #1, Tab(5); "Eff.Length factor"; Tab(60);
   Format(K, "#######0.0##")
   Print #1, Tab(5); "Yield Stress (N/sq.mm)"; Tab(60);
   Format(Fy, "#######0.0##")
   Print #1, Tab(5); "Span (m)"; Tab(60); Format(LS,
   "#######0.0##")
   Print #1, Tab(5); "Unrestrained length (m)"; Tab(60);
   Format(Le, "#######0.0##")
   Print #1, Tab(5); "Dead load (kN/m)"; Tab(60);
   Format(DL, "#######0.0##")
   Print #1, Tab(5); "Imposed load (kN/m)"; Tab(60);
   Format(IL, "#######0.0##")
   Print #1, Tab(5); "Mid span concentrated load (kN)";
   Tab(60);
   Format(PG, "#######0.0##")
   Print #1, Tab(5); "Mid span live load (kN)"; Tab(60);
   Format(PQ, "#######0.0##")

   Print #1, Tab(5); ""
   Print #1, Tab(5); "Sectional Properties"
   Print #1, Tab(5); "------"
   Print #1, Tab(5); "Depth (mm)"; Tab(60); Format(D,
   "#######0.0##")
   Print #1, Tab(5); "Width of flange (mm)"; Tab(60);
   Format(Bf, "#######0.0##")
   Print #1, Tab(5); "Thickness of flange (mm)"; Tab(60);
   Format(tf, "#######0.0##")
   Print #1, Tab(5); "Thickness of web (mm)"; Tab(60);
   Format(tw, "#######0.0##")
   Print #1, Tab(5); "Root radius (mm)"; Tab(60);
   Format(R, "#######0.0##")
   Print #1, Tab(5); "Moment of inertia of Major axis
   (mm^4)"; Tab(60); Format(Iz, "#######0.0##")
   Print #1, Tab(5); "Moment of inertia of Minor axis
   (mm^4)"; Tab(60); Format(Iy, "#######0.0##")
   Print #1, Tab(5); "Palstic section modulus (cu.mm)";
   Tab(60); Format(Zpz, "#######0.0##")
   Print #1, Tab(5); "Elastic modulus about major axis
   (cu.mm)";
   Tab(60); Format(Zz, "#######0.0##")
   Print #1, Tab(5); "Imperfection factor"; Tab(60);
   Format(alphaLT, "#######0.0##")
   Print #1, Tab(5); "Partial safety factor for material";
   Tab(60);
   Format(GamaM, "#######0.0##")
   Print #1, ""
End Sub
```

In order to show the use of this program, the data given in Example 10.9 was used and the resulting output is as follows:

```
                         Beam Design
                         -----------
Structure Data
--------------
Youngs Modulus <N/sq. mm>                       200000.0
Poisson's Ratio                                 0.3
Eff.Length factor                               1.0
Yield Stress <N/sq.mm>                          250.0
Span <m>                                        1.5
Unrestrained length <m>                         1.5
Dead load <kN/m>                                30.0
Imposed load <kN/m>                             20.0
Mid span concentrated load <kN>                 0.0
Mid span live load <kN>                         0.0

Sectional Properties
--------------------
Depth <mm>                                      175.0
Width of flange <mm>                            90.0
Thickness of flange <mm>                        8.6
Thickness of web <mm>                           5.5
Root radius <mm>                                10.0
Moment of inertia of Major axis <mm^4>          12706000.0
Moment of inertia of Minor axis <mm^4>          851000.0
Palstic section modulus <cu.mm>                 161650.0
Elastic modulus about major axis <cu.mm>        145200.0
Imperfection factor                             0.21
Partial safety factor for material              1. 1

Output
------
Maximum bending moment for design <kNm>         21.094
Maximum shear force for design <kN>             56.25
Torsional constant <mm^4>                       4739L627
Warping constant <mm^6>                         5890826240.0
Moment capacity of the section <kNm>            31.044
Moment capacity of Trial section is adequate

Shear capacity <kN>                             75.777
Shear force <kN>                                56.25
Shear capacity of Trial section is adequate
Delta allowable <mm>                            5.0
Delta actual <mm>                               1.297
Trial section is adequate for deflection
check

Strength against web buckling<kN>               119.165
Shear force <kN>                                56.25
Safe against web buckling
Strength against web bearing <kN>               110.938
Shear force <kN>                                56.25
Safe against web bearing
```

## E.1.2 Beam-column Design Program

The design of a beam-column, using a hot-rolled *I*-section, is considered in the program given in this section. In all cases, axial compression is taken as positive. The sign convention for the moments at the top and bottom ends of the column is positive for moments applied clockwise and negative for moments applied anticlockwise.

Hot rolled I- or H-plastic, compact, or semi-compact sections can be designed using this program. There are two sets of input data in this program, as in the case of the beam program. The first is the structural data consisting of the following: factored axial force ($N$), factored bending moment at the top and bottom about the major axis ($M_{Z1}$, $M_{Z2}$), factored bending moment at the top and bottom about the minor axis ($M_{Y1}$, $M_{Y2}$), length in major and minor axis ($L_z$, $L_y$), effective length factors ($K_z$, $K_y$), Young's modulus ($E$), and Poisson's ratio ($\mu$). The second set of data is the member data consisting of the following: depth of the section ($D$), Area ($A$), breadth of flange ($b_f$), thickness of flange and web ($t_f$ and $t_w$), root radius ($R$), radius of gyration in the major and minor axis ($r_z$, $r_y$) moment of inertia about minor axis ($I_y$), elastic modulus about major and minor axis ($Z_z$ and $Z_y$), yield strength of material ($f_y$) and partial factor of safety of material ($\gamma_{m0}$) and type of section (plastic, compact, or semi-compact).

Based on the preceding data, the plastic modulus about the major and minor axis ($Z_{pz}$, $Z_{py}$) and the warping and torsional rigidity $I_w$, $I_t$) are calculated. The member buckling resistance in compression about the major axis and minor axis ($P_{dz}$ and $P_{dy}$) are calculated based on clause 7.1.2 of the code. The member buckling resistance in bending is computed based on clause 8.2.2, about both the axes ($M_{dz}$ and $M_{dy}$).

Using these values, the strength of the cross section is checked based on the interaction equations given in clause 9.3.1 of the code. Finally, the overall member strength is checked as per the interaction equation given in clause 9.3.2 of the code. If any one of the checks is not satisfied, the user has to change the trial section properties and run the program again. The program written in Visual Basic and typical input screens are shown here:

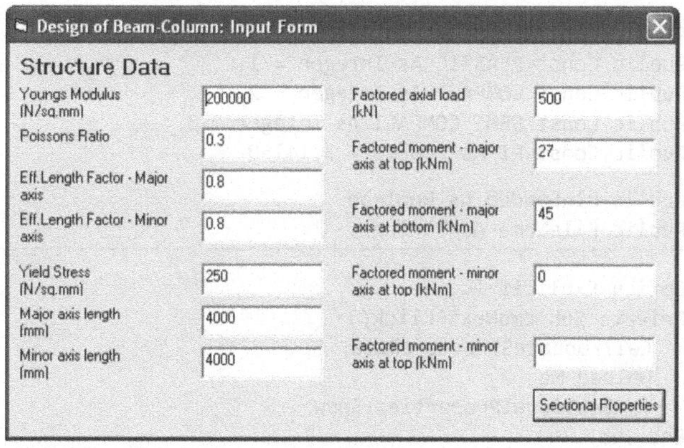


**FIG. E.5** Input screen for structure data of Beam-column program

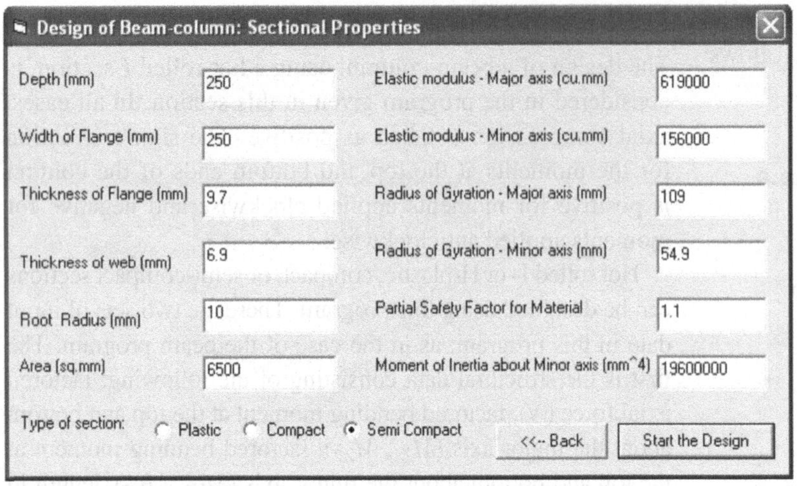


**FIG. E.6** Input screen for sectional properties of Beam-column program

```
Option Explicit
'Structure properties
Public E As Double
Public EMu As Double
Public Fy As Double
Public Lz As Double
Public Ly As Double
Public Kz As Double
Public Ky As Double
Public N As Double
Public Mz1 As Double
Public Mz2 As Double
Public My1 As Double
Public My2 As Double

'Sectional Properties
Public D As Double
Public Bf As Double
Public tf As Double
Public tw As Double
Public R As Double
Public rz As Double
Public ry As Double
Public Zz As Double
Public Zy As Double
Public Zpz As Double
Public Zpy As Double
Public GamaM As Double
Public A As Double
Public Iy As Double
Public Pd As Double

Public sectionType As Integer
Public Const PLASTIC As Integer = 1
Public Const COMPACT As Integer = 2
Public Const SEMI_COMPACT As Integer = 3
Public Const PI As Double = 3.14159

Public blnLoaded As Boolean
Public FileName As String
```

---

```
Option Explicit
Private Sub cmdNext Click()
   Call UpdateStructureData
   Unload Me
   frmSectionalProperties.Show
End Sub
```

```
Private Sub UpdateStructureData()
      E = Val(txtE)
      EMu = Val(txtMu)
      Kz = Val(txtKz)
      Ky = Val(txtKy)
      Fy = Val(txtFy)
      Ly = Val(txtLY)
      Lz = Val(txtLZ)
      My1 = Val(txtMY1)
      My2 = Val(txtMY2)
      Mz1 = Val(txtMZ1)
      Mz2 = Val(txtMZ2)
      N = Val(txtN)
End Sub
```

---

```
Option Explicit
Dim G As Double
Dim It As Double
Dim hf As Double
Dim Iw As Double
Dim Betaf As Double

Dim Mcr As Double
Dim LambdaLT As Double
Dim alphaLT As Double
Dim PhiLT As Double
Dim ChiLT As Double
Dim fbd As Double

Dim Md As Double
Dim Vd As Double
Dim Deltab As Double
Dim DeltaAl1 As Double

Dim b1 As Double
Dim n1 As Double
Dim Ab As Double
Dim I As Double
Dim A As Double
Dim rmin As Double
Dim Lamda As Double
Dim LamdaEff As Double
Dim Phi1 As Double
Dim Phi2 As Double
Dim fcd As Double
Dim SWB As Double
Dim n2 As Double
Dim Fw As Double
Dim Mz As Double
Dim Mzm As Double
Dim My As Double
Dim Mym As Double
Dim Nd As Double
Dim Mdz, Mdz1 As Double
Dim Mdy As Double
Dim StressRatio As Double
Dim StressRatio1 As Double
Dim Lrz As Double
Dim Lry As Double
Dim AlphaZ As Double
Dim AlphaY As Double
Dim Betab As Double
Dim LambdaZ As Double
Dim LambdaY As Double
Dim Pdz As Double
Dim Pdy As Double
Dim blnSafe As Boolean
```

```
Private Sub cmdBack Click()
   Call UpdateSectionalProperties
   Call LoadStructureData
   Unload Me
   frmBeamColumnDesignl.Show
End Sub
Private Sub LoadStructureData()
   frmBeamColumnDesignl.txtE = E
   frmBeamColumnDesignl.txtMu = EMu
   frmBeamColumnDesignl.txtKz = Str(Kz)
   frmBeamColumnDesignl.txtKy = Str(Ky)
   frmBeamColumnDesignl.txtFy = Fy
   frmBeamColumnDesignl.txtLY = Ly
   frmBeamColumnDesignl.txtLZ = Lz
   frmBeamColumnDesignl.txtMYl = My1
   frmBeamColumnDesignl.txtMY2 = My2
   frmBeamColumnDesignl.txtMZ1 = Mz1
   frmBeamColumnDesignl.txtMZ2 = Mz2
   frmBeamColumnDesign1.txtN = N
End Sub

Private Sub cmdDesign Click()
   FileName = App.Path + "\BeamColumn.out"
   Open FileName For Output As l
   blnSafe = True
   Call UpdateSectionalProperties
   Call Design
End Sub

Private Sub LoadSectionalProperties()
   If blnLoaded Then
      frmSectionalProperties.txtDepth = Str(D)
      frmSectionalProperties.txtWidthOfFlange = Str(Bf)
      frmSectionalProperties.txtThickOfFlange = Str(tf)
      frmSectionalProperties.txtThickWeb = Str(tw)
      frmSectionalProperties.txtR = Str(R)
      frmSectionalProperties.txtRZ = Str(rz)
      frmSectionalProperties.txtRY = Str(ry)
      frmSectionalProperties.txtA = Str(A)
      frmSectionalProperties.txtZZ = Str(Zz)
      frmSectionalProperties.txtZY = Str(Zy)
      frmSectionalProperties.txtSafetyFactor = Str(GamaM)
      frmSectionalProperties.txtIy = Str(Iy)
      If sectionType = PLASTIC Then
         optPlastic = True
      ElseIf sectionType = COMPACT Then
         optCompact = True
      ElseIf sectionType = SEMI_COMPACT Then
         optSemiCompact = True
      End If
   End If
End Sub

Private Sub UpdateSectionalProperties()
   D = Val(txtDepth)
   Bf = Val(txtWidthOfFlange)
   tf = Val(txtThickOfFlange)
   tw = Val(txtThickWeb)
   R = Val(txtR)
   rz = Val(txtRZ)
   ry = Val(txtRY)
   A = Val(txtA)
   Zz = Val(txtZZ)
   Zy = Val(txtZY)
   GamaM = Val(txtSafetyFactor)
   Iy = Val(txtIy)
   If optPlastic Then
      sectionType = PLASTIC
   ElseIf optCompact Then
      sectionType = COMPACT
   ElseIf optSemiCompact Then
      sectionType = SEMI_COMPACT
   End If
End Sub

Private Sub Form Load()
   Call LoadSectionalProperties
   blnLoaded = True
End Sub

Private Sub Design()
   blnSafe = True
   Call PrintInput
   Call ComputeInteractionEqnLocal
   If blnSafe Then Call DesignCompressionResisMajor
   If blnSafe Then Call DesignCompressionResisMinor
   If blnSafe Then Call DesignBucklingResisMajor
   If blnSafe Then Call DesignResOfCrossSection
   If blnSafe Then Call DesignShearResistance
   If blnSafe Then Call DesignComBendingAxialForce
   Close #l
   If blnSafe Then End
End Sub
Private Sub PrintInput()

   Print #l, Tab(10); "Beam-Column Design"
   Print #l, Tab(10); "----"
   Print #l, ""
   Print #l, Tab(5); "Structure Data"
   Print #l, Tab(5); "----"
   Print #l, Tab(5); "Youngs Modulus (N/sq.mm) "; Tab(60);
   Format(E, "#######0.0##")
   Print  #l,  Tab(5);  "Poisson's  Ratio  ";  Tab(60);
   Format(EMu, "#######0.0##")
   Print #l, Tab(5); "Yield Stress (N/sq.mm) "; Tab(60);
   Format(Fy, "#######0.0##")
   Print #l, Tab(5); "Major axis Length (mm) "; Tab(60);
   Format(Lz, "#######0.0##")
   Print #l, Tab(5); "Minor axis Length (mm) "; Tab(60);
   Format(Ly, "#######0.0##")
   Print #l, Tab(5); "Eff.Length factor - Major axis ";
   Tab(60); Format(Kz, "#######0.0##")
   Print #l, Tab(5); "Eff.Length factor - Minor axis ";
   Tab(60); Format(Ky, "#######0.0##")
   Print #l, Tab(5); "Factored axial load (kN) "; Tab(60);
   Format(N, "#######0.0##")
   Print #l, Tab(5); "Major axis factored moment at top
   (kNm)"; Tab(60); Format(Mzl, "#######0.0##")
   Print #l, Tab(5); "Major axis factored moment at bottom
   (kNm)"; Tab(60); Format(Mz2, "#######0.0##")
   Print #l, Tab(5); "Minor axis factored moment at top
   (kNm)"; Tab(60); Format(Myl, "#######0.0##")
   Print #l, Tab(5); "Minor axis factored moment at bottom
   (kNm)"; Tab(60); Format(My2, "#######0.0##")

   Print #l, Tab(5); ""
   Print #l, Tab(5); "Sectional Properties"
   Print #l, Tab(5); "----"
   Print  #l,  Tab(5);  "Depth  (mm)";  Tab(60);  Format(D,
   "#######0.0##")
   Print  #l,  Tab(5);  "Width  of  flange  (mm)";  Tab(60);
   Format(Bf, "#######0.0##")
   Print #l, Tab(5); "Thickness of flange (mm)"; Tab(60);
   Format(tf, "#######0.0##")
   Print #l, Tab(5); "Thickness of web (mm)"; Tab(60);
   Format(tw, "#######0.0##")
```

```
   Print #1, Tab(5); "Root radius (mm)"; Tab(60);
   Format(R, "#######0.0##")
   Print #1, Tab(5); "Radius of Gyration major axis (mm)";
   Tab(60); Format(rz, "#######0.0##")
   Print #1, Tab(5); "Radius of Gyration minor axis (mm)";
   Tab(60); Format(ry, "#######0.0##")
   Print #1, Tab(5); "Elastic modulus about major axis
   (cu.mm)"; Tab(60); Format(Zz, "#######0.0##")
   Print #1, Tab(5); "Elastic modulus about minor axis
   (cu.mm)"; Tab(60); Format(Zy, "#######0.0##")
   Print #1, Tab(5); "Partial safety factor for material ";
   Tab(60); Format(GamaM, "#######0.0##")
   Print #1, Tab(5); "Area (sq.mm)"; Tab(60); Format(A,
   "#######0.0##")
   Print #1, Tab(5); "Moment of intertia about minor axis
   (mm^4)"; Tab(60); Format(Iy, "#######0.0##")
   Print #1, ""
End Sub
Private Sub ComputeInteractionEqnLocal()
   G = E / (2 * (1 + EMu))
   If (Abs(Mzl) > Abs(Mz2)) Then
      Mz = Mz1
      Mzm = Mz2
   Else
      Mz = Mz2
      Mzm = Mz1
   End If

   If (Abs(Myl) > Abs(My2)) Then
      My = My1
      Mym = My2
   Else
      My = My2
      Mym = My1
   End If
   'Properties of the Cross-section
   Zpz = 2 * Bf * tf * (D - tf)/2 + tw * (D - 2 * tf) ^ 2/4
   Zpy = 2 * tf * Bf * Bf/4 + (D - 2 * tf) * tw * tw / 4
   It = (2 * tf ^ 3 * Bf + (D - tf) * tw ^ 3) / 3
   Iw = (D - tf) ^ 2 * Iy / 4
   Nd = A * Fy / (GamaM * 1000#)
   'Minor Axis buckling resistance in bending
   If sectionType = SEMI_COMPACT Then
      Mdy = Zy * Fy / (GamaM * 1000000#)
   Else
      Mdy = Zpy * Fy / (GamaM * 1000000#)
   End If
   Dim strMsg As String
   strMsg = "Plastic modulus about Major axis "+ Str(Zpz)
   + "cu.mm" + vbCrLf
   strMsg = strMsg + "Plastic modulus about Minor axis "+
   Str(Zpy) + "cu.mm" + vbCrLf
   strMsg = strMsg + "Torsional constant" + Str(It) +
   "mm^4" + vbCrLf
   strMsg = strMsg + "Warping constant "+ Str(Iw) + "mm^6"
   'MsgBox strMsg
   'MsgBox "Design Resistance in Bending about Minor
   Axis" + Str(Mdy) + "kNm" + vbCrLf + "Applied Moment
   about Minor Axis "+ Str(My) + "kNm"
   If sectionType = SEMI_COMPACT Then
      Betab = Zz / Zpz
   Else
      Betab = 1
   End If
   Mdz = Betab * Zpz * Fy / (GamaM * 1000000#)
   Print #1, Tab(5); "Output"
   Print #1, Tab(5); "----"
   Print #1, Tab(5); "Plastic modulus about Major axis
   (cu.mm) "; Tab(60); Format(Zpz, "#######0.0##")
   Print #1, Tab(5); "Plastic modulus about Minor axis
   (cu.mm)"; Tab(60); Format(Zpy, "#######0.0##")
   Print #1, Tab(5); "Torsional constant (mm^4)"; Tab(60);
   Format(It, "#######0.0##")
   Print #1, Tab(5); "Warping constant (mm^6)"; Tab(60);
   Format(Iw, "#######0.0##")
   Print #1, Tab(5); "Design Resistance in Bending about
   Minor Axis (kNm)"; Tab(60); Format(Mdy, "#######0.0##")
   Print #1, Tab(5); "Applied Moment about Minor Axis
   (kNm) "; Tab(60); Format(My, "#######0.0##")
   Print #1, ""
End Sub
Private Sub DesignResOfCrossSection()
Dim Alphal, Alpha2, Nr, Mndz, Mndy As Double

'Local Capacity Check
   If sectionType = SEMI_COMPACT Then
      StressRatio = (N / Nd) + (Mz / Mdz) + (My / Mdy)
   Else
      Nr = N / Nd
      Mndz = 1.11 * Mdz * (1 - Nr)
      If Mndz > Mdz Then
         Mndz = Mdz
      End If
      If Nr <= 0.2 Then
         Mndy = Mdy
      Else
         Mndy = 1.56 * Mdy * (1 - Nr) * (Nr + 0.6)
      End If
      Alpha1 = 5 * Nr
      If Alpha1 < 1 Then
         Alpha1 = 1
      End If
      Alpha2 = 2
      'section strength interaction equation
      StressRatio = (My / Mndy) ^ Alpha1 + (Mz / Mndz) ^
      Alpha2

   End If
   If StressRatio < 1# Then
      'MsgBox "Interaction equation value = "+
      Str(StressRatio) + vbCrLf + "Section is safe"
      Print #1, Tab(5); "Interaction equation (section
      strength value "; Tab(60); Format(StressRatio,
      "#######0.0##"); "
      Section is safe"
      Print #1, ""
   Else
      MsgBox "Interaction equation value = "+
      Str(StressRatio) + vbCrLf + "Section unsafe due
      to member capacity check "+ vbCrLf + "Revise the
      section"
      blnSafe = False
      Unload Me
      frmBeamColumnDesignl.Show
   End If

End Sub
Private Sub DesignCompressionResisMajor()
Dim Fcrz, Fcdz, Phiz, Phiz1 As Double

'Compression resistance - Major Axis
   Lz = Kz * Lz
   Ly = Ky * Ly
   Lrz = Lz / rz
   Lry = Ly / ry
```

```
    If (D / Bf) > 1.2 And tf <= 40 Then
       AlphaZ = 0.21
       AlphaY = 0.34
    ElseIf (D / Bf) > 1.2 And tf < 100 Then
       AlphaZ = 0.34
       AlphaY = 0.49
    ElseIf (D / Bf) <= 1.2 And tf <= 100 Then
       AlphaZ = 0.34
       AlphaY = 0.49
    Else
       AlphaZ = 0.76
       AlphaY = 0.76
    End If

    Fcrz = PI * PI * E / (Lrz * Lrz)
    LambdaZ = (Fy / Fcrz) ^ 0.5
    Phiz = 0.5 * (1 + AlphaZ * (LambdaZ - 0.2) + LambdaZ ^ 2)
    Phiz1 = Phiz + (Phiz ^ 2 - LambdaZ ^ 2) ^ 0.5
    Fcdz = Fy / (GamaM * Phiz1)
    If Fcdz > (Fy / GamaM) Then
       Fcdz = Fy / GamaM
    End If
    Pdz = Fcdz * A / 1000#
    Dim strMsg As String
    strMsg = "Compression resistance about Major Axis "+
    Str(Pdz) + "kN" + vbCrLf
    If Pdz > N Then
       'MsgBox strMsg + "Section is safe"
       Print #1, Tab(5); "Compression resistance about Major
       Axis (kN)"; Tab(60); Format(Pdz, "#######0.0##");
       "Section is safe"
       Print #1, ""
    Else
       MsgBox strMsg + "Revise the Section"
       blnSafe = False
       Unload Me
       frmBeamColumnDesign1.Show
    End If

End Sub
Private Sub DesignCompressionResisMinor()
Dim Fcry, Fcdy, Phiy, Phiy1 As Double
Dim strMsg As String

    'Compression Resistance- Minor Axis
    Fcry = PI * PI * E / (Lry * Lry)
    LambdaY = (Fy / Fcry) ^ 0.5
    Phiy = 0.5 * (1 + AlphaY * (LambdaY - 0.2) + LambdaY ^ 2)
    Phiy1 = Phiy + (Phiy ^ 2 - LambdaY ^ 2) ^ 0.5
    Fcdy = Fy / (GamaM * Phiy1)
    If Fcdy > (Fy / GamaM) Then
       Fcdy = Fy / GamaM
    End If
    Pdy = Fcdy * A / 1000#

    strMsg = "Compression resistance about Minor Axis "+
    Str(Pdy) + "kN" + vbCrLf
    If Pdy > N Then
       Print #1, Tab(5); "Compression resistance about Minor
       Axis (kNm)"; Tab(60); Format(Pdy, "#######0.0##");
       "Section is safe"
       Print #1, ""
    Else
       MsgBox strMsg + "Revise the section"
       blnSafe = False
       Unload Me
       frmBeamColumnDesign1.Show
    End If
    If Pdz > Pdy Then
       Pd = Pdy
    Else
       Pd = Pdz
    End If
       Print #1, Tab(5); "Compression Resistance (kN)";
       Tab(60); Format(Pd, "#######0.0##")
       Print #1, Tab(5); "Compressive Force (kN)"; Tab(60);
       Format(N, "#######0.0##")
       Print #1, ""
End Sub

Private Sub DesignBucklingResisMajor()
Dim PhiLT, PhiLT1, chi, C1, Chi1, Mcr1, fcrb As Double
Dim strMsg As String

'Member Buckling resistance
'Major Axis
   If Abs(Mz1) < Abs(Mz2) Then
      Chi1 = Mz1 / Mz2
   Else
      Chi1 = Mz2 / Mz1
   End If

   'Eqn suggested by Gardner and Nethercot
      C1 = 1.88 - 1.4 * Chi1 + 0.52 * Chi1 ^ 2
      If C1 > 2.7 Then C1 = 2.7

      fcrb = C1 * (1473.5 / Lry) ^ 2 * ((1 + (1 / 20) *
      (Lry / (D / tf)) ^ 2) ^ 0.5)
      Mcr = fcrb * Zz
      LambdaLT = (Betab * Zpz * Fy / Mcr) ^ 0.5
      LambdaLT1 = (1.2 * Zz * Fy / Mcr) ^ 0.5
      If LambdaLT > LambdaLT1 Then
         LambdaLT = LambdaLT1
      End If
      LambdaLT1 = (1.2 * Zz * Fy / Mcr) ^ 0.5
      If LambdaLT > LambdaLT1 Then
         LambdaLT = LambdaLT1
      End If
   alphaLT = 0.21
   PhiLT = 0.5 * (1 + alphaLT * (LambdaLT - 0.2) +
   LambdaLT ^ 2)
   PhiLT1 = PhiLT + (PhiLT ^ 2 - LambdaLT ^ 2) ^ 0.5
   chi = 1# / PhiLT1
   If chi > 1# Then
      chi = 1#
   End If
   fbd = chi * Fy / GamaM

   Mdz1 = fbd * Betab * Zpz / 1000000#
   strMsg = "Buckling Resistance in Bending about Major
   Axis" + Str(Mdz) + "kNm" + vbCrLf + "Applied Moment
   about Major axis " + Str(Mz) + "kNm" + vbCrLf
   If Mdz1 > Mz Then
      Print #1, Tab(5); "Buckling Resistance in Bending
      about Major Axis (kNm)"; Tab(60); Format(Mdzl,
      "#######0.0##")
      Print #1, Tab(5); "Applied Moment about Major axis
      (kNm)"; Tab(60); Format(Mz, "#######0.0##")
      Print #1, Tab(5); "Buckling Resistance in Bending
      about Major Axis is safe"
      Print #1, ""
   Else
      MsgBox strMsg + "Revise the Section"
      blnSafe = False
      Unload Me
      frmBeamColumnDesign1.Show
   End If
End Sub
```

```
Private Sub DesignShearResistance()

Dim MSF, Av, Vp As Double
Dim strMsg As String

  'Shear Resistance of the cross-section
  MSF = Abs(Mz1 - Mz2) / (Lz / 1000)
  Av = D * tw
  Vp = Av * Fy / (Sqr(3) * GamaM * 1000#)

  strMsg = "Shear Resistance Parallel to Web "+ Str(Vp) +
  "kN" + vbCrLf + "Max.S.F. "+ Str(MSF) + "kN" + vbCrLf
  If Vp > MSF Then
     Print #1, Tab(5); "Shear Resistance Parallel to Web
     (kN)"; Tab(60); Format(Vp, "#######0.0##")
     Print #1, Tab(5); "Maximum Shear Force (kN)";
     Tab(60); Format(MSF, "#######0.0##")
     Print #1, Tab(5); "Shear Resistance Parallel to Web
     is safe"
     Print #1, ""
  Else
     MsgBox strMsg + "Revise the Section"
     blnSafe = False
     Unload Me
     frmBeamColumnDesign1.Show
  End If

End Sub

Private Sub DesignComBendingAxialForce()
Dim Psiz, BetaMz, MulT, Mufz, Muz, Muy, KKz, Kky, psiy,
BetaMy, nz, Cmz, Cmy, CmLT, ny, klt, kltl As Double
Dim strMsg As String
'Member Bending Resistance in Combined Bending and axial
Compression
'Major Axis
  Psiz = Mzm / Mz
  nz = N / Pdz
  Kz = 1 + (LambdaZ - 0.2) * nz
  Cmz = 0.6 + 0.4 * Psiz
  If Cmz < 0.4 Then
     Cmz = 0.4
  End If
  If Kz > 1 + 0.8 * nz Then
     Kz = 1 + 0.8 * nz
  End If
  klt = 0
  Cmy = 0
  'Minor axis
  If Not (Mym = 0 And My = 0) Then
     psiy = Mym / My
     ny = N / Pdy
     Ky = 1 + (LambdaY - 0.2) * ny

     If Ky > 1 + 0.8 * ny Then
        Ky = 1 + 0.8 * ny
     End If
     Cmy = 0.6 + 0.4 * psiy
     If Cmy < 0.4 Then
        Cmy = 0.4
     End If
        CmLT = 0.6 + 0.4 * psiy
     If CmLT < 0.4 Then
        CmLT = 0.4
     End If
     klt = 1 - 0.1 * LambdaLT * ny / (CmLT - 0.25)
     kltl = 1 - 0.1 * ny / (CmLT - 0.25)
     If klt > kltl Then
        klt = kltl
     End If
  End If
  'Check with Interaction Formula for overall buckling
  StressRatiol = (N / Pdy) + (Ky * Cmy * Abs(My) / Mdy)
  + (klt * Abs(Mz) / Mdzl)
  StressRatio2 = (N / Pdz) + (0.6 * Ky * Cmy * Abs(My) /
  Mdy) + (Kz * Cmz * Abs(Mz) / Mdz1)
  strMsg = "Interaction equation value = "+
  Str(StressRatiol)
  + vbCrLf
  If StressRatiol < 1# And StressRatio2 < 1 Then
     Print #1, Tab(5); "Interaction equation value 1 ";
     Tab(60); Format(StressRatiol, "#######0.0##")
     Print #1, Tab(5); "Interaction equation value 2 ";
     Tab(60); Format(StressRatio2, "#######0.0##")
     Print #1, Tab(5); "Member is safe under combined
     Axial Force and B.M."
     Print #1, ""
     MsgBox "Member is safe" + vbCrLf + "The output is
     stored as BeamColumn.out under the working folder",
     , "Desgin Over"
  Else
     MsgBox strMsg + "Revise the section"
     blnSafe = False
     Unload Me
     frmBeamColumnDesign1.Show
  End If
End Sub

Private Sub optCompact Click()
  If optCompact Then
  sectionType = COMPACT
End If

End Sub

Private Sub optPlastic Click()
  If optPlastic Then
     sectionType = PLASTIC
  End If
End Sub

Private Sub optSemiCompact Click()
  If optSemiCompact Then
     sectionType = SEMI_COMPACT
  End If
End Sub
```

To illustrate the use of the program, a trial run of the program was conducted for Example 13.1 and the resulting output is as follows:

```
Beam-column Design
-------------

 Structure Data
 ---------------
  Youngs Modulus <N/sq.mm>             200000.0
  Poisson's Ratio                      0.3
  Yield Stress <N/sq.mm>               250.0
  Major axis Length <mm>               4000.0
  Minor axis Length <mm>               4000.0
  Eff.Length factor - Major axis       0.8
  Eff.Length factor - Minor axis       0.8
  Factored axial load <kN>             500.0
```

| | |
|---|---|
| Major axis factored moment at top <kNm> | 27.0 |
| Major axis factored moment at bottom <kNm> | 45.0 |
| Minor axis factored moment at top <kNm> | 0.0 |
| Minor axis factored moment at bottom <kNm> | 0.0 |

Sectional Properties
-------------------------

| | |
|---|---|
| Depth <mm> | 250.0 |
| Width of flange <mm> | 250.0 |
| Thickness of flange <mm> | 9.7 |
| Thickness of web <mm> | 6.9 |
| Root radius <mm> | 10.0 |
| Radius of Gyration major axis <mm> | 109.0 |
| Radius of Gyration minor axis <mm> | 54.9 |
| Elastic modulus about major axis <cu. mm> | 619000.0 |
| Elastic modulus about minor axis <cu. mm> | 156000.0 |
| Partial safety factor for material | 1.1 |
| Area <sq.mm> | 6500.0 |
| Moment of intertia about minor axis <mm^4> | 19600000.0 |

Output
-------

| | |
|---|---|
| Plastic modulus about Major axis <cu. mm> | 674456.721 |
| Plastic modulus about Minor axis <cu. mm> | 305869.717 |
| Torsional constant <mm^4> | 178425.738 |
| Warping constant <mm^6> | 282946041000.0 |
| Design Resistance in Bending about Minor Axis <kNm> | 35.455 |
| Applied Moment about Minor Axis <kNm> | 0.0 |
| Compression resistance about Major Axis <kN> | 1407.624 Section is safe |
| Compression resistance about Minor Axis <kN> | 1110.522 Section is safe |
| Compression Resistance <kN> | 1110.522 |
| Compressive Force <kN> | 500.0 |
| Buckling Resistance in Bending about | |
| Major Axis <kNm> | 128.525 |
| Applied Moment about Major axis <kNm> | 45.0 |
| Buckling Resistance in Bending about | |
| Major Axis is safe | |
| Interaction equation value | 0.658 Section is safe |
| Shear Resistance Parallel to Web <kN> | 226.348 |
| Maximum Shear Force <kN> | 5.625 |
| Shear Resistance Parallel to Web is safe | |
| Interaction equation value | 0.45 |
| Interaction equation value2 | 0.663 |
| Member is safe under combined Axial Force and B.M. | |

The inclusion of these computer programs is only to illustrate the usefulness of computers in repetitive and complex calculations, which are time consuming if done manually. However, the publisher and the author do not provide any warranty for their use to design problems. The user is advised to test the programs thoroughly, before using them.

# APPENDIX F

# DESIGN AIDS

This appendix contains design aids in the form of tables which will be useful to select compression members, beams, and tension members for the required capacity.

**TABLE F.1** Capacity of angle tension members with single row of bolts ($f_y = 250$ MPa) - Clause 6.2 to 6.4 of Code

| A (mm) | B (mm) | t (mm) | $A_g$ (mm$^2$) | No of bolts | Gauge distance (mm) | Dia. of bolt (mm) | Pitch (mm) | Edge Distance (mm) | $T_{dg}$ (kN) | $T_{dn}$ (kN) | $T_{db1}$ (KN) | $T_{db2}$ (kN) | $T_d$ (kN) |
|---|---|---|---|---|---|---|---|---|---|---|---|---|---|
| 35 | 35 | 5 | 327 | 2 | 19 | 12 | 30 | 25 | 74.3 | 60.9 | 50.1 | 48.4 | 48.4 |
| 35 | 35 | 5 | 327 | 3 | 19 | 12 | 30 | 25 | 74.3 | 70.7 | 69.8 | 62.9 | 62.9 |
| 35 | 35 | 5 | 327 | 4 | 19 | 12 | 30 | 25 | 74.3 | 74.0 | 89.5 | 77.4 | 74.0 |
| 35 | 35 | 6 | 386 | 2 | 19 | 12 | 30 | 25 | 87.7 | 75.9 | 60.1 | 58.1 | 58.1 |
| 35 | 35 | 6 | 386 | 3 | 19 | 12 | 30 | 25 | 87.7 | 85.3 | 83.7 | 75.5 | 75.5 |
| 35 | 35 | 6 | 386 | 4 | 19 | 12 | 30 | 25 | 87.7 | 88.5 | 107.4 | 92.9 | 87.7 |
| 40 | 40 | 5 | 378 | 2 | 21 | 12 | 30 | 25 | 85.9 | 66.3 | 54.5 | 51.8 | 51.8 |
| 40 | 40 | 5 | 378 | 3 | 21 | 12 | 30 | 25 | 85.9 | 81.1 | 74.2 | 66.3 | 66.3 |
| 40 | 40 | 5 | 378 | 4 | 21 | 12 | 30 | 25 | 85.9 | 86.0 | 93.9 | 80.8 | 80.8 |
| 40 | 40 | 6 | 447 | 2 | 21 | 12 | 30 | 25 | 101.6 | 84.6 | 65.4 | 62.2 | 62.2 |
| 40 | 40 | 6 | 447 | 3 | 21 | 12 | 30 | 25 | 101.6 | 98.9 | 89.1 | 79.6 | 79.6 |
| 40 | 40 | 6 | 447 | 4 | 21 | 12 | 30 | 25 | 101.6 | 103.6 | 112.7 | 97.0 | 97.0 |
| 45 | 45 | 5 | 428 | 2 | 25 | 12 | 30 | 25 | 97.3 | 77.3 | 56.0 | 53.0 | 53.0 |
| 45 | 45 | 5 | 428 | 3 | 25 | 12 | 30 | 25 | 97.3 | 89.3 | 75.7 | 67.5 | 67.5 |
| 45 | 45 | 5 | 428 | 4 | 25 | 12 | 30 | 25 | 97.3 | 96.6 | 95.4 | 82.0 | 82.0 |
| 45 | 45 | 6 | 507 | 2 | 25 | 12 | 30 | 25 | 115.2 | 91.5 | 67.2 | 63.6 | 63.6 |
| 45 | 45 | 6 | 507 | 3 | 25 | 12 | 30 | 25 | 115.2 | 110.3 | 90.8 | 81.0 | 81.0 |
| 45 | 45 | 6 | 507 | 4 | 25 | 12 | 30 | 25 | 115.2 | 117.4 | 114.5 | 98.3 | 98.3 |
| 50 | 50 | 5 | 479 | 2 | 28 | 16 | 40 | 35 | 108.9 | 81.3 | 68.4 | 65.9 | 65.9 |
| 50 | 50 | 5 | 479 | 3 | 28 | 16 | 40 | 35 | 108.9 | 96.3 | 94.6 | 84.7 | 84.7 |
| 50 | 50 | 5 | 479 | 4 | 28 | 16 | 40 | 35 | 108.9 | 103.9 | 120.9 | 103.4 | 103.4 |
| 50 | 50 | 6 | 568 | 2 | 28 | 16 | 40 | 35 | 129.1 | 96.5 | 82.1 | 79.1 | 79.1 |
| 50 | 50 | 6 | 568 | 3 | 28 | 16 | 40 | 35 | 129.1 | 118.8 | 113.6 | 101.6 | 101.6 |
| 50 | 50 | 6 | 568 | 4 | 28 | 16 | 40 | 35 | 129.1 | 126.2 | 145.1 | 124.1 | 124.1 |

*(Contd)*

**TABLE F.1** *(Contd)*

| A (mm) | B (mm) | t (mm) | $A_g$ (mm$^2$) | No of bolts | Gauge distance (mm) | Dia. of bolt (mm) | Pitch (mm) | Edge Distance (mm) | $T_{dg}$ (kN) | $T_{dn}$ (kN) | $T_{db1}$ (KN) | $T_{db2}$ (kN) | $T_d$ (kN) |
|---|---|---|---|---|---|---|---|---|---|---|---|---|---|
| 55 | 55 | 5 | 527 | 2 | 30 | 16 | 40 | 35 | 119.8 | 92.7 | 72.8 | 69.3 | 69.3 |
| 55 | 55 | 5 | 527 | 3 | 30 | 16 | 40 | 35 | 119.8 | 104.0 | 99.1 | 88.1 | 88.1 |
| 55 | 55 | 5 | 527 | 4 | 30 | 16 | 40 | 35 | 119.8 | 114.2 | 125.3 | 106.8 | 106.8 |
| 55 | 55 | 6 | 626 | 2 | 30 | 16 | 40 | 35 | 142.3 | 109.9 | 87.4 | 83.2 | 83.2 |
| 55 | 55 | 6 | 626 | 3 | 30 | 16 | 40 | 35 | 142.3 | 129.7 | 118.9 | 105.7 | 105.7 |
| 55 | 55 | 6 | 626 | 4 | 30 | 16 | 40 | 35 | 142.3 | 139.7 | 150.4 | 128.2 | 128.2 |
| 55 | 55 | 8 | 818 | 2 | 30 | 16 | 40 | 35 | 185.9 | 150.9 | 116.5 | 110.9 | 110.9 |
| 55 | 55 | 8 | 818 | 3 | 30 | 16 | 40 | 35 | 185.9 | 179.3 | 158.5 | 140.9 | 140.9 |
| 55 | 55 | 8 | 818 | 4 | 30 | 16 | 40 | 35 | 185.9 | 188.8 | 200.5 | 170.9 | 170.9 |
| 55 | 55 | 10 | 1000 | 2 | 30 | 16 | 40 | 35 | 227.3 | 199.2 | 145.6 | 138.6 | 138.6 |
| 55 | 55 | 10 | 1000 | 3 | 30 | 16 | 40 | 35 | 227.3 | 226.4 | 198.1 | 176.1 | 176.1 |
| 55 | 55 | 10 | 1000 | 4 | 30 | 16 | 40 | 35 | 227.3 | 235.5 | 250.6 | 213.6 | 213.6 |
| 60 | 60 | 5 | 575 | 2 | 35 | 16 | 40 | 35 | 130.7 | 104.0 | 72.8 | 69.3 | 69.3 |
| 60 | 60 | 5 | 575 | 3 | 35 | 16 | 40 | 35 | 130.7 | 108.9 | 99.1 | 88.1 | 88.1 |
| 60 | 60 | 5 | 575 | 4 | 35 | 16 | 40 | 35 | 130.7 | 122.5 | 125.3 | 106.8 | 106.8 |
| 60 | 60 | 6 | 684 | 2 | 35 | 16 | 40 | 35 | 155.5 | 123.5 | 87.4 | 83.2 | 83.2 |
| 60 | 60 | 6 | 684 | 3 | 35 | 16 | 40 | 35 | 155.5 | 137.8 | 118.9 | 105.7 | 105.7 |
| 60 | 60 | 6 | 684 | 4 | 35 | 16 | 40 | 35 | 155.5 | 151.2 | 150.4 | 128.2 | 128.2 |
| 60 | 60 | 8 | 896 | 2 | 35 | 16 | 40 | 35 | 203.6 | 161.0 | 116.5 | 110.9 | 110.9 |
| 60 | 60 | 8 | 896 | 3 | 35 | 16 | 40 | 35 | 203.6 | 193.8 | 158.5 | 140.9 | 140.9 |
| 60 | 60 | 8 | 896 | 4 | 35 | 16 | 40 | 35 | 203.6 | 206.6 | 200.5 | 170.9 | 170.9 |
| 60 | 60 | 10 | 1100 | 2 | 35 | 16 | 40 | 35 | 250.0 | 210.4 | 145.6 | 138.6 | 138.6 |
| 60 | 60 | 10 | 1100 | 3 | 35 | 16 | 40 | 35 | 250.0 | 247.3 | 198.1 | 176.1 | 176.1 |
| 60 | 60 | 10 | 1100 | 4 | 35 | 16 | 40 | 35 | 250.0 | 259.6 | 250.6 | 213.6 | 213.6 |
| 65 | 65 | 5 | 625 | 2 | 35 | 16 | 40 | 35 | 142.0 | 115.4 | 80.2 | 75.0 | 75.0 |
| 65 | 65 | 5 | 625 | 3 | 35 | 16 | 40 | 35 | 142.0 | 115.4 | 106.4 | 93.7 | 93.7 |
| 65 | 65 | 5 | 625 | 4 | 35 | 16 | 40 | 35 | 142.0 | 131.2 | 132.7 | 112.5 | 112.5 |
| 65 | 65 | 6 | 744 | 2 | 35 | 16 | 40 | 35 | 169.1 | 137.1 | 96.2 | 90.0 | 90.0 |
| 65 | 65 | 6 | 744 | 3 | 35 | 16 | 40 | 35 | 169.1 | 146.4 | 127.7 | 112.5 | 112.5 |
| 65 | 65 | 6 | 744 | 4 | 35 | 16 | 40 | 35 | 169.1 | 163.0 | 159.2 | 135.0 | 135.0 |
| 65 | 65 | 8 | 976 | 2 | 35 | 16 | 40 | 35 | 221.8 | 179.2 | 128.3 | 120.0 | 120.0 |
| 65 | 65 | 8 | 976 | 3 | 35 | 16 | 40 | 35 | 221.8 | 208.8 | 170.3 | 150.0 | 150.0 |
| 65 | 65 | 8 | 976 | 4 | 35 | 16 | 40 | 35 | 221.8 | 224.8 | 212.3 | 180.0 | 180.0 |
| 65 | 65 | 10 | 1200 | 2 | 35 | 16 | 40 | 35 | 272.7 | 222.5 | 160.4 | 150.0 | 150.0 |
| 65 | 65 | 10 | 1200 | 3 | 35 | 16 | 40 | 35 | 272.7 | 268.7 | 212.9 | 187.5 | 187.5 |
| 65 | 65 | 10 | 1200 | 4 | 35 | 16 | 40 | 35 | 272.7 | 284.1 | 265.4 | 225.0 | 225.0 |
| 70 | 70 | 5 | 677 | 2 | 40 | 16 | 40 | 35 | 153.9 | 126.8 | 80.2 | 75.0 | 75.0 |
| 70 | 70 | 5 | 677 | 3 | 40 | 16 | 40 | 35 | 153.9 | 126.8 | 106.4 | 93.7 | 93.7 |
| 70 | 70 | 5 | 677 | 4 | 40 | 16 | 40 | 35 | 153.9 | 136.9 | 132.7 | 112.5 | 112.5 |
| 70 | 70 | 6 | 806 | 2 | 40 | 16 | 40 | 35 | 183.2 | 150.7 | 96.2 | 90.0 | 90.0 |
| 70 | 70 | 6 | 806 | 3 | 40 | 16 | 40 | 35 | 183.2 | 150.7 | 127.7 | 112.5 | 112.5 |

*(Contd)*

**TABLE F.1** *(Contd)*

| A (mm) | B (mm) | t (mm) | $A_g$ (mm$^2$) | No of bolts | Gauge distance (mm) | Dia. of bolt (mm) | Pitch (mm) | Edge Distance (mm) | $T_{dg}$ (kN) | $T_{dn}$ (kN) | $T_{db1}$ (KN) | $T_{db2}$ (kN) | $T_d$ (kN) |
|---|---|---|---|---|---|---|---|---|---|---|---|---|---|
| 70 | 70 | 6 | 806 | 4 | 40 | 16 | 40 | 35 | 183.2 | 171.9 | 159.2 | 135.0 | 135.0 |
| 70 | 70 | 8 | 1060 | 2 | 40 | 16 | 40 | 35 | 240.9 | 197.4 | 128.3 | 120.0 | 120.0 |
| 70 | 70 | 8 | 1060 | 3 | 40 | 16 | 40 | 35 | 240.9 | 219.3 | 170.3 | 150.0 | 150.0 |
| 70 | 70 | 8 | 1060 | 4 | 40 | 16 | 40 | 35 | 240.9 | 240.0 | 212.3 | 180.0 | 180.0 |
| 70 | 70 | 10 | 1300 | 2 | 40 | 16 | 40 | 35 | 295.5 | 242.2 | 160.4 | 150.0 | 150.0 |
| 70 | 70 | 10 | 1300 | 3 | 40 | 16 | 40 | 35 | 295.5 | 285.7 | 212.9 | 187.5 | 187.5 |
| 70 | 70 | 10 | 1300 | 4 | 40 | 16 | 40 | 35 | 295.5 | 305.6 | 265.4 | 225.0 | 225.0 |
| 75 | 75 | 5 | 727 | 2 | 40 | 16 | 40 | 35 | 165.2 | 138.1 | 87.6 | 80.7 | 80.7 |
| 75 | 75 | 5 | 727 | 3 | 40 | 16 | 40 | 35 | 165.2 | 138.1 | 113.8 | 99.4 | 99.4 |
| 75 | 75 | 5 | 727 | 4 | 40 | 16 | 40 | 35 | 165.2 | 143.3 | 140.1 | 118.2 | 118.2 |
| 75 | 75 | 6 | 866 | 2 | 40 | 16 | 40 | 35 | 196.8 | 164.4 | 105.1 | 96.8 | 96.8 |
| 75 | 75 | 6 | 866 | 3 | 40 | 16 | 40 | 35 | 196.8 | 164.4 | 136.6 | 119.3 | 119.3 |
| 75 | 75 | 6 | 866 | 4 | 40 | 16 | 40 | 35 | 196.8 | 181.4 | 168.1 | 141.8 | 141.8 |
| 75 | 75 | 8 | 1140 | 2 | 40 | 16 | 40 | 35 | 259.1 | 215.5 | 140.1 | 129.1 | 129.1 |
| 75 | 75 | 8 | 1140 | 3 | 40 | 16 | 40 | 35 | 259.1 | 230.9 | 182.1 | 159.1 | 159.1 |
| 75 | 75 | 8 | 1140 | 4 | 40 | 16 | 40 | 35 | 259.1 | 255.9 | 224.1 | 189.1 | 189.1 |
| 75 | 75 | 10 | 1400 | 2 | 40 | 16 | 40 | 35 | 318.2 | 264.9 | 175.2 | 161.4 | 161.4 |
| 75 | 75 | 10 | 1400 | 3 | 40 | 16 | 40 | 35 | 318.2 | 303.7 | 227.7 | 198.9 | 198.9 |
| 75 | 75 | 10 | 1400 | 4 | 40 | 16 | 40 | 35 | 318.2 | 327.8 | 280.1 | 236.3 | 236.3 |
| 80 | 80 | 6 | 929 | 2 | 45 | 20 | 50 | 40 | 211.1 | 170.9 | 113.4 | 106.0 | 106.0 |
| 80 | 80 | 6 | 929 | 3 | 45 | 20 | 50 | 40 | 211.1 | 170.9 | 152.7 | 134.7 | 134.7 |
| 80 | 80 | 6 | 929 | 4 | 45 | 20 | 50 | 40 | 211.1 | 192.9 | 192.1 | 163.3 | 163.3 |
| 80 | 80 | 8 | 1220 | 2 | 45 | 20 | 50 | 40 | 277.3 | 224.3 | 151.2 | 141.4 | 141.4 |
| 80 | 80 | 8 | 1220 | 3 | 45 | 20 | 50 | 40 | 277.3 | 246.1 | 203.6 | 179.5 | 179.5 |
| 80 | 80 | 8 | 1220 | 4 | 45 | 20 | 50 | 40 | 277.3 | 271.0 | 256.1 | 217.7 | 217.7 |
| 80 | 80 | 10 | 1500 | 2 | 45 | 20 | 50 | 40 | 340.9 | 275.8 | 188.9 | 176.7 | 176.7 |
| 80 | 80 | 10 | 1500 | 3 | 45 | 20 | 50 | 40 | 340.9 | 322.4 | 254.6 | 224.4 | 224.4 |
| 80 | 80 | 10 | 1500 | 4 | 45 | 20 | 50 | 40 | 340.9 | 346.6 | 320.2 | 272.1 | 272.1 |
| 80 | 80 | 12 | 1780 | 2 | 45 | 20 | 50 | 40 | 404.5 | 325.8 | 226.7 | 212.0 | 212.0 |
| 80 | 80 | 12 | 1780 | 3 | 45 | 20 | 50 | 40 | 404.5 | 396.3 | 305.5 | 269.3 | 269.3 |
| 80 | 80 | 12 | 1780 | 4 | 45 | 20 | 50 | 40 | 404.5 | 419.8 | 384.2 | 326.6 | 326.6 |
| 90 | 90 | 6 | 1050 | 2 | 50 | 20 | 50 | 40 | 238.6 | 198.2 | 122.2 | 112.8 | 112.8 |
| 90 | 90 | 6 | 1050 | 3 | 50 | 20 | 50 | 40 | 238.6 | 198.2 | 161.6 | 141.5 | 141.5 |
| 90 | 90 | 6 | 1050 | 4 | 50 | 20 | 50 | 40 | 238.6 | 207.5 | 201.0 | 170.1 | 170.1 |
| 90 | 90 | 8 | 1380 | 2 | 50 | 20 | 50 | 40 | 313.6 | 260.6 | 163.0 | 150.4 | 150.4 |
| 90 | 90 | 8 | 1380 | 3 | 50 | 20 | 50 | 40 | 313.6 | 262.4 | 215.5 | 188.6 | 188.6 |
| 90 | 90 | 8 | 1380 | 4 | 50 | 20 | 50 | 40 | 313.6 | 298.3 | 267.9 | 226.8 | 226.8 |
| 90 | 90 | 10 | 1700 | 2 | 50 | 20 | 50 | 40 | 386.4 | 321.2 | 203.7 | 188.1 | 188.1 |
| 90 | 90 | 10 | 1700 | 3 | 50 | 20 | 50 | 40 | 386.4 | 351.7 | 269.3 | 235.8 | 235.8 |
| 90 | 90 | 10 | 1700 | 4 | 50 | 20 | 50 | 40 | 386.4 | 386.6 | 334.9 | 283.5 | 283.5 |
| 90 | 90 | 12 | 2020 | 2 | 50 | 20 | 50 | 40 | 459.1 | 380.0 | 244.4 | 225.7 | 225.7 |

*(Contd)*

**TABLE F.1** *(Contd)*

| *A* (mm) | *B* (mm) | *t* (mm) | $A_g$ (mm$^2$) | No of bolts | Gauge distance (mm) | Dia. of bolt (mm) | Pitch (mm) | Edge Distance (mm) | $T_{dg}$ (kN) | $T_{dn}$ (kN) | $T_{db1}$ (KN) | $T_{db2}$ (kN) | $T_d$ (kN) |
|---|---|---|---|---|---|---|---|---|---|---|---|---|---|
| 90 | 90 | 12 | 2020 | 3 | 50 | 20 | 50 | 40 | 459.1 | 438.4 | 323.2 | 282.9 | 282.9 |
| 90 | 90 | 12 | 2020 | 4 | 50 | 20 | 50 | 40 | 459.1 | 472.4 | 401.9 | 340.2 | 340.2 |
| 100 | 100 | 6 | 1170 | 2 | 60 | 20 | 50 | 40 | 265.9 | 225.4 | 122.2 | 112.8 | 112.8 |
| 100 | 100 | 6 | 1170 | 3 | 60 | 20 | 50 | 40 | 265.9 | 225.4 | 161.6 | 141.5 | 141.5 |
| 100 | 100 | 6 | 1170 | 4 | 60 | 20 | 50 | 40 | 265.9 | 225.4 | 201.0 | 170.1 | 170.1 |
| 100 | 100 | 8 | 1540 | 2 | 60 | 20 | 50 | 40 | 350.0 | 296.9 | 163.0 | 150.4 | 150.4 |
| 100 | 100 | 8 | 1540 | 3 | 60 | 20 | 50 | 40 | 350.0 | 296.9 | 215.5 | 188.6 | 188.6 |
| 100 | 100 | 8 | 1540 | 4 | 60 | 20 | 50 | 40 | 350.0 | 316.7 | 267.9 | 226.8 | 226.8 |
| 100 | 100 | 10 | 1900 | 2 | 60 | 20 | 50 | 40 | 431.8 | 366.6 | 203.7 | 188.1 | 188.1 |
| 100 | 100 | 10 | 1900 | 3 | 60 | 20 | 50 | 40 | 431.8 | 367.7 | 269.3 | 235.8 | 235.8 |
| 100 | 100 | 10 | 1900 | 4 | 60 | 20 | 50 | 40 | 431.8 | 417.7 | 334.9 | 283.5 | 283.5 |
| 100 | 100 | 12 | 2260 | 2 | 60 | 20 | 50 | 40 | 513.6 | 434.5 | 244.4 | 225.7 | 225.7 |
| 100 | 100 | 12 | 2260 | 3 | 60 | 20 | 50 | 40 | 513.6 | 467.4 | 323.2 | 282.9 | 282.9 |
| 100 | 100 | 12 | 2260 | 4 | 60 | 20 | 50 | 40 | 513.6 | 516.3 | 401.9 | 340.2 | 340.2 |
| 110 | 110 | 8 | 1710 | 2 | 65 | 20 | 50 | 40 | 388.6 | 333.3 | 174.8 | 159.5 | 159.5 |
| 110 | 110 | 8 | 1710 | 3 | 65 | 20 | 50 | 40 | 388.6 | 333.3 | 227.3 | 197.7 | 197.7 |
| 110 | 110 | 8 | 1710 | 4 | 65 | 20 | 50 | 40 | 388.6 | 333.3 | 279.7 | 235.9 | 235.9 |
| 110 | 110 | 10 | 2110 | 2 | 65 | 20 | 50 | 40 | 479.5 | 412.1 | 218.5 | 199.4 | 199.4 |
| 110 | 110 | 10 | 2110 | 3 | 65 | 20 | 50 | 40 | 479.5 | 412.1 | 284.1 | 247.1 | 247.1 |
| 110 | 110 | 10 | 2110 | 4 | 65 | 20 | 50 | 40 | 479.5 | 445.3 | 349.7 | 294.9 | 294.9 |
| 110 | 110 | 12 | 2510 | 2 | 65 | 20 | 50 | 40 | 570.5 | 489.0 | 262.2 | 239.3 | 239.3 |
| 110 | 110 | 12 | 2510 | 3 | 65 | 20 | 50 | 40 | 570.5 | 491.2 | 340.9 | 296.6 | 296.6 |
| 110 | 110 | 12 | 2510 | 4 | 65 | 20 | 50 | 40 | 570.5 | 556.6 | 419.6 | 353.8 | 353.8 |
| 110 | 110 | 16 | 3280 | 2 | 65 | 20 | 50 | 40 | 745.5 | 637.5 | 349.5 | 319.1 | 319.1 |
| 110 | 110 | 16 | 3280 | 3 | 65 | 20 | 50 | 40 | 745.5 | 709.2 | 454.5 | 395.4 | 395.4 |
| 110 | 110 | 16 | 3280 | 4 | 65 | 20 | 50 | 40 | 745.5 | 771.9 | 559.5 | 471.8 | 471.8 |
| 130 | 130 | 8 | 2030 | 2 | 80 | 20 | 50 | 40 | 461.4 | 406.0 | 186.6 | 168.6 | 168.6 |
| 130 | 130 | 8 | 2030 | 3 | 80 | 20 | 50 | 40 | 461.4 | 406.0 | 239.1 | 206.8 | 206.8 |
| 130 | 130 | 8 | 2030 | 4 | 80 | 20 | 50 | 40 | 461.4 | 406.0 | 291.6 | 245.0 | 245.0 |
| 130 | 130 | 10 | 2510 | 2 | 80 | 20 | 50 | 40 | 570.5 | 502.9 | 233.2 | 210.8 | 210.8 |
| 130 | 130 | 10 | 2510 | 3 | 80 | 20 | 50 | 40 | 570.5 | 502.9 | 298.8 | 258.5 | 258.5 |
| 130 | 130 | 10 | 2510 | 4 | 80 | 20 | 50 | 40 | 570.5 | 502.9 | 364.4 | 306.2 | 306.2 |
| 130 | 130 | 12 | 2990 | 2 | 80 | 20 | 50 | 40 | 679.5 | 598.1 | 279.9 | 252.9 | 252.9 |
| 130 | 130 | 12 | 2990 | 3 | 80 | 20 | 50 | 40 | 679.5 | 598.1 | 358.6 | 310.2 | 310.2 |
| 130 | 130 | 12 | 2990 | 4 | 80 | 20 | 50 | 40 | 679.5 | 610.7 | 437.3 | 367.5 | 367.5 |
| 130 | 130 | 16 | 3920 | 2 | 80 | 20 | 50 | 40 | 890.9 | 782.9 | 373.2 | 337.3 | 337.3 |
| 130 | 130 | 16 | 3920 | 3 | 80 | 20 | 50 | 40 | 890.9 | 782.9 | 478.1 | 413.6 | 413.6 |
| 130 | 130 | 16 | 3920 | 4 | 80 | 20 | 50 | 40 | 890.9 | 877.4 | 583.1 | 490.0 | 490.0 |
| 150 | 150 | 10 | 2920 | 2 | 90 | 20 | 50 | 40 | 663.6 | 593.8 | 262.7 | 233.5 | 233.5 |
| 150 | 150 | 10 | 2920 | 3 | 90 | 20 | 50 | 40 | 663.6 | 593.8 | 328.4 | 281.2 | 281.2 |
| 150 | 150 | 10 | 2920 | 4 | 90 | 20 | 50 | 40 | 663.6 | 593.8 | 394.0 | 329.0 | 329.0 |

*(Contd)*

**TABLE F.1** *(Contd)*

| A (mm) | B (mm) | t (mm) | $A_g$ (mm²) | No of bolts | Gauge distance (mm) | Dia. of bolt (mm) | Pitch (mm) | Edge Distance (mm) | $T_{dg}$ (kN) | $T_{dn}$ (kN) | $T_{db1}$ (KN) | $T_{db2}$ (kN) | $T_d$ (kN) |
|---|---|---|---|---|---|---|---|---|---|---|---|---|---|
| 150 | 150 | 12 | 3480 | 2 | 90 | 20 | 50 | 40 | 790.9 | 707.1 | 315.3 | 280.2 | 280.2 |
| 150 | 150 | 12 | 3480 | 3 | 90 | 20 | 50 | 40 | 790.9 | 707.1 | 394.0 | 337.5 | 337.5 |
| 150 | 150 | 12 | 3480 | 4 | 90 | 20 | 50 | 40 | 790.9 | 707.1 | 472.8 | 394.8 | 394.8 |
| 150 | 150 | 16 | 4560 | 2 | 90 | 20 | 50 | 40 | 1036.4 | 928.2 | 420.4 | 373.6 | 373.6 |
| 150 | 150 | 16 | 4560 | 3 | 90 | 20 | 50 | 40 | 1036.4 | 928.2 | 525.4 | 450.0 | 450.0 |
| 150 | 150 | 16 | 4560 | 4 | 90 | 20 | 50 | 40 | 1036.4 | 954.7 | 630.3 | 526.3 | 526.3 |
| 150 | 150 | 20 | 5620 | 2 | 90 | 20 | 50 | 40 | 1226.2 | 1124.3 | 516 | 456.1 | 456.1 |
| 150 | 150 | 20 | 5620 | 3 | 90 | 20 | 50 | 40 | 1226.2 | 1124.3 | 642 | 551.6 | 551.6 |
| 150 | 150 | 20 | 5620 | 4 | 90 | 20 | 50 | 40 | 1226.2 | 1253.0 | 768 | 647.0 | 647.0 |
| 200 | 200 | 12 | 4690 | 2 | 115 | 20 | 50 | 40 | 1065.9 | 979.7 | 403.9 | 348.4 | 348.4 |
| 200 | 200 | 12 | 4690 | 3 | 115 | 20 | 50 | 40 | 1065.9 | 979.7 | 482.6 | 405.7 | 405.7 |
| 200 | 200 | 12 | 4690 | 4 | 115 | 20 | 50 | 40 | 1065.9 | 979.7 | 561.3 | 462.9 | 462.9 |
| 200 | 200 | 16 | 6180 | 2 | 115 | 20 | 50 | 40 | 1404.5 | 1291.7 | 538.5 | 464.5 | 464.5 |
| 200 | 200 | 16 | 6180 | 3 | 115 | 20 | 50 | 40 | 1404.5 | 1291.7 | 643.4 | 540.9 | 540.9 |
| 200 | 200 | 16 | 6180 | 4 | 115 | 20 | 50 | 40 | 1404.5 | 1291.7 | 748.4 | 617.2 | 617.2 |
| 200 | 200 | 20 | 7640 | 2 | 115 | 20 | 50 | 40 | 1666.9 | 1572.2 | 663.6 | 565.2 | 565.2 |
| 200 | 200 | 20 | 7640 | 3 | 115 | 20 | 50 | 40 | 1666.9 | 1572.2 | 789.6 | 660.7 | 660.7 |
| 200 | 200 | 20 | 7640 | 4 | 115 | 20 | 50 | 40 | 1666.9 | 1572.2 | 915.6 | 756.1 | 756.1 |
| 200 | 200 | 25 | 9410 | 2 | 115 | 20 | 50 | 40 | 2053.1 | 1937.3 | 829.6 | 706.5 | 706.5 |
| 200 | 200 | 25 | 9410 | 3 | 115 | 20 | 50 | 40 | 2053.1 | 1937.3 | 987.0 | 825.8 | 825.8 |
| 200 | 200 | 25 | 9410 | 4 | 115 | 20 | 50 | 40 | 2053.1 | 1949.5 | 1145.5 | 954.1 | 945.1 |

*Note:* For thickness ≥ 20, $f_y$ has been reduced to 240 MPa as per Table 1 of IS 800:2007; in all the cases $T_{db2}$ is governing

**TABLE F.2** Capacity of concentrically loaded angles in compression ($f_y$ = 250 MPa) - clause 7.1.2 of the code

| Designation | Mass (N/m) | Sectional dimensions | | Capacity in KN for effective length KL (m) | | | | | | | | | | | | | |
|---|---|---|---|---|---|---|---|---|---|---|---|---|---|---|---|---|---|
| | | Area (mm²) | $r_v$(Min) (mm) | 0.5 | 1 | 1.5 | 2 | 2.5 | 3 | 3.5 | 4 | 4.5 | 5 | 5.5 | 6 | 6.5 | 7 |
| L35 35 × 5 | 26 | 327 | 6.7 | 47 | 20 | 10 | 6 | 4 | 3 | 2 | 2 | 1 | 1 | 1 | 1 | 1 | 1 |
| × 6 | 30 | 386 | 6.7 | 56 | 23 | 11 | 7 | 4 | 3 | 2 | 2 | 1 | 1 | 1 | 1 | 1 | 1 |
| L40 40 × 5 | 30 | 378 | 7.7 | 61 | 28 | 14 | 9 | 6 | 4 | 3 | 2 | 2 | 2 | 1 | 1 | 1 | 1 |
| × 6 | 35 | 447 | 7.7 | 72 | 33 | 17 | 10 | 7 | 5 | 4 | 3 | 2 | 2 | 1 | 1 | 1 | 1 |
| L45 45 × 5 | 34 | 428 | 8.7 | 74 | 38 | 20 | 12 | 8 | 6 | 4 | 3 | 3 | 2 | 2 | 2 | 1 | 1 |
| × 6 | 40 | 507 | 8.7 | 87 | 45 | 24 | 14 | 10 | 7 | 5 | 4 | 3 | 3 | 2 | 2 | 2 | 1 |
| L50 50 × 5 | 38 | 479 | 9.7 | 87 | 49 | 27 | 16 | 11 | 8 | 6 | 5 | 4 | 3 | 2 | 2 | 2 | 2 |
| × 6 | 45 | 568 | 9.6 | 102 | 58 | 31 | 19 | 13 | 9 | 7 | 5 | 4 | 3 | 3 | 2 | 2 | 2 |
| L55 55 × 5 | 41 | 527 | 10.6 | 99 | 60 | 34 | 21 | 14 | 10 | 8 | 6 | 5 | 4 | 3 | 3 | 2 | 2 |
| × 6 | 49 | 626 | 10.6 | 117 | 72 | 41 | 25 | 17 | 12 | 9 | 7 | 6 | 5 | 4 | 3 | 3 | 2 |
| × 8 | 64 | 818 | 10.6 | 153 | 94 | 53 | 33 | 22 | 16 | 12 | 9 | 7 | 6 | 5 | 4 | 4 | 3 |
| × 10 | 79 | 1000 | 10.6 | 188 | 115 | 65 | 40 | 27 | 19 | 15 | 11 | 9 | 7 | 6 | 5 | 4 | 4 |
| L60 60 × 5 | 45 | 575 | 11.6 | 111 | 73 | 43 | 27 | 18 | 13 | 10 | 8 | 6 | 5 | 4 | 4 | 3 | 3 |
| × 6 | 54 | 684 | 11.5 | 132 | 86 | 51 | 32 | 21 | 15 | 12 | 9 | 7 | 6 | 5 | 4 | 4 | 3 |

*(Contd)*

**TABLE F.2** *(Contd)*

| Designation | Mass (N/m) | Sectional dimensions | | Capacity in KN for effective length KL (m) | | | | | | | | | | | | | |
|---|---|---|---|---|---|---|---|---|---|---|---|---|---|---|---|---|---|
| | | Area ($mm^2$) | $r_v$(Min) (mm) | 0.5 | 1 | 1.5 | 2 | 2.5 | 3 | 3.5 | 4 | 4.5 | 5 | 5.5 | 6 | 6.5 | 7 |
| × 8 | 70 | 896 | 11.5 | 173 | 112 | 66 | 41 | 28 | 20 | 15 | 12 | 9 | 8 | 6 | 5 | 5 | 4 |
| × 10 | 86 | 1100 | 11.5 | 212 | 138 | 81 | 51 | 34 | 25 | 19 | 14 | 12 | 9 | 8 | 7 | 6 | 5 |
| *L65 65 × 5 | 49 | 625(600) | 12.6 | 119 | 82 | 51 | 32 | 22 | 16 | 12 | 9 | 8 | 6 | 5 | 4 | 4 | 3 |
| × 6 | 58 | 744 | 12.6 | 148 | 102 | 63 | 40 | 27 | 20 | 15 | 12 | 9 | 8 | 6 | 5 | 5 | 4 |
| × 8 | 77 | 976 | 12.5 | 193 | 133 | 82 | 52 | 35 | 26 | 19 | 15 | 12 | 10 | 8 | 7 | 6 | 5 |
| × 10 | 94 | 1200 | 12.5 | 237 | 164 | 100 | 64 | 44 | 31 | 24 | 18 | 15 | 12 | 10 | 9 | 7 | 6 |
| *L70 70 × 5 | 53 | 677(600) | 13.6 | 121 | 88 | 57 | 37 | 25 | 18 | 14 | 11 | 9 | 7 | 6 | 5 | 4 | 4 |
| × 6 | 63 | 806 | 13.6 | 163 | 118 | 76 | 49 | 34 | 25 | 19 | 15 | 12 | 10 | 8 | 7 | 6 | 5 |
| × 8 | 83 | 1060 | 13.5 | 214 | 154 | 99 | 64 | 44 | 32 | 24 | 19 | 15 | 12 | 10 | 9 | 8 | 7 |
| × 10 | 102 | 1300 | 13.5 | 262 | 189 | 121 | 79 | 54 | 39 | 30 | 23 | 19 | 15 | 13 | 11 | 9 | 8 |
| *L75 75 × 5 | 57 | 727(600) | 14.6 | 123 | 93 | 62 | 41 | 29 | 21 | 16 | 12 | 10 | 8 | 7 | 6 | 5 | 4 |
| × 6 | 68 | 866 | 14.6 | 178 | 134 | 90 | 59 | 41 | 30 | 23 | 18 | 14 | 12 | 10 | 8 | 7 | 6 |
| × 8 | 89 | 1140 | 14.5 | 234 | 175 | 117 | 77 | 54 | 39 | 30 | 23 | 19 | 15 | 13 | 11 | 9 | 8 |
| × 10 | 110 | 1400 | 14.5 | 288 | 215 | 144 | 95 | 66 | 48 | 36 | 28 | 23 | 19 | 16 | 13 | 11 | 10 |
| *L80 80 × 6 | 73 | 929(864) | 15.6 | 180 | 140 | 97 | 66 | 46 | 34 | 26 | 20 | 16 | 13 | 11 | 9 | 8 | 7 |
| × 8 | 96 | 1220 | 15.5 | 254 | 196 | 136 | 92 | 64 | 47 | 36 | 28 | 23 | 19 | 15 | 13 | 11 | 10 |
| × 10 | 118 | 1500 | 15.5 | 313 | 242 | 167 | 113 | 79 | 58 | 44 | 34 | 28 | 23 | 19 | 16 | 14 | 12 |
| × 12 | 140 | 1780 | 15.4 | 370 | 285 | 197 | 132 | 93 | 68 | 51 | 40 | 32 | 27 | 22 | 19 | 16 | 14 |
| *L90 90 × 6 | 82 | 1050(864) | 17.5 | 184 | 149 | 110 | 78 | 55 | 41 | 31 | 25 | 20 | 16 | 14 | 12 | 10 | 9 |
| × 8 | 108 | 1380 | 17.5 | 294 | 238 | 176 | 124 | 88 | 65 | 50 | 39 | 32 | 26 | 22 | 19 | 16 | 14 |
| × 10 | 134 | 1700 | 17.4 | 362 | 293 | 215 | 151 | 108 | 80 | 61 | 48 | 39 | 32 | 27 | 23 | 20 | 17 |
| × 12 | 158 | 2020 | 17.4 | 430 | 348 | 256 | 180 | 128 | 95 | 73 | 57 | 46 | 38 | 32 | 27 | 23 | 20 |
| *L100 100 × 6 | 92 | 1170(864) | 19.5 | 188 | 157 | 122 | 90 | 66 | 49 | 38 | 30 | 24 | 20 | 17 | 14 | 12 | 11 |
| × 8 | 121 | 1540 | 19.5 | 334 | 280 | 217 | 160 | 117 | 88 | 68 | 53 | 43 | 36 | 30 | 26 | 22 | 19 |
| × 10 | 149 | 1900 | 19.4 | 412 | 344 | 267 | 196 | 143 | 107 | 83 | 65 | 53 | 44 | 37 | 31 | 27 | 23 |
| × 12 | 177 | 2260 | 19.4 | 490 | 409 | 318 | 233 | 170 | 127 | 98 | 78 | 63 | 52 | 44 | 37 | 32 | 28 |
| *L110 110 × 8 | 134 | 1710(1536) | 21.8 | 339 | 291 | 237 | 182 | 137 | 104 | 81 | 65 | 53 | 44 | 37 | 31 | 27 | 24 |
| × 10 | 166 | 2110 | 21.6 | 465 | 399 | 323 | 247 | 186 | 141 | 110 | 88 | 71 | 59 | 50 | 42 | 36 | 32 |
| × 12 | 197 | 2510 | 21.5 | 553 | 473 | 383 | 293 | 220 | 167 | 130 | 103 | 84 | 70 | 59 | 50 | 43 | 37 |
| × 16 | 257 | 3280 | 21.4 | 722 | 618 | 499 | 381 | 285 | 217 | 168 | 134 | 109 | 90 | 76 | 65 | 56 | 48 |
| *L130 130 × 8 | 159 | 2030(1536) | 25.9 | 346 | 307 | 263 | 216 | 172 | 135 | 107 | 87 | 71 | 59 | 50 | 43 | 37 | 32 |
| *L130 130 × 10 | 197 | 2510(2400) | 25.7 | 540 | 478 | 410 | 335 | 266 | 209 | 166 | 134 | 110 | 91 | 77 | 66 | 57 | 50 |
| × 12 | 235 | 2990 | 25.6 | 673 | 595 | 509 | 416 | 329 | 259 | 205 | 166 | 136 | 113 | 96 | 82 | 71 | 62 |
| × 16 | 307 | 3920 | 25.4 | 881 | 779 | 665 | 542 | 428 | 336 | 266 | 214 | 176 | 146 | 124 | 106 | 91 | 80 |
| L150 150 × 10 | 229 | 2920(2400) | 29.8 | 549 | 496 | 439 | 376 | 313 | 255 | 207 | 170 | 141 | 118 | 100 | 86 | 75 | 65 |
| × 12 | 273 | 3480 | 29.7 | 795 | 718 | 636 | 545 | 452 | 368 | 299 | 245 | 203 | 170 | 145 | 124 | 108 | 94 |
| × 16 | 358 | 4560 | 29.4 | 1041 | 939 | 830 | 709 | 586 | 476 | 386 | 316 | 261 | 219 | 186 | 160 | 138 | 121 |
| × 20 | 441 | 5620 | 29.3 | 1234 | 1116 | 989 | 848 | 705 | 575 | 468 | 384 | 318 | 267 | 227 | 195 | 169 | 148 |
| *L200 200 × 12 | 369 | 4690(3456) | 39.9 | 809 | 753 | 695 | 633 | 567 | 497 | 430 | 369 | 316 | 271 | 234 | 204 | 179 | 158 |
| × 16 | 485 | 6180 | 39.6 | 1446 | 1344 | 1241 | 1129 | 1009 | 884 | 763 | 653 | 559 | 480 | 414 | 360 | 315 | 278 |

*(Contd)*

**TABLE F.2** *(Contd)*

| Designation | Mass (N/m) | Sectional dimensions | | Capacity in KN for effective length KL (m) | | | | | | | | | | | | | |
|---|---|---|---|---|---|---|---|---|---|---|---|---|---|---|---|---|---|
| | | Area ($mm^2$) | $r_v$(Min) (mm) | 0.5 | 1 | 1.5 | 2 | 2.5 | 3 | 3.5 | 4 | 4.5 | 5 | 5.5 | 6 | 6.5 | 7 |
| × 20 | 600 | 7640 | 39.3 | 1718 | 1599 | 1477 | 1347 | 1207 | 1060 | 918 | 788 | 675 | 580 | 502 | 437 | 382 | 337 |
| × 25 | 739 | 9410 | 39.1 | 2116 | 1968 | 1817 | 1656 | 1481 | 1300 | 1124 | 964 | 826 | 709 | 613 | 533 | 467 | 412 |

*For these sections the limiting width to thickness ratio as given in Table 2 of IS 800 are exceeded; hence only the effective width is taken in the calculation of capacity.

*Note:* For thicknesses ≥ 20, $f_y$ has been reduced to 240 MPa as per Table 1 of IS 800:2007.

**TABLE F.3** Capacity of eccentrically connected angles in compression ($f_y = 250$ MPa)-with fixed ends (Clause 7.5.1.2, $k_1 = 0.20$, $k_2 = 0.35$, and $k_3 = 20$)

| Designation | Mass (N/m) | Sectional dimensions | | Capacity in KN for effective length KL (m) | | | | | | | | | | | | | |
|---|---|---|---|---|---|---|---|---|---|---|---|---|---|---|---|---|---|
| | | Area ($mm^2$) | $r_v$(Min) (mm) | 0.5 | 1 | 1.5 | 2 | 2.5 | 3 | 3.5 | 4 | 4.5 | 5 | 5.5 | 6 | 6.5 | 7 |
| L35 35 × 5 | 26 | 327 | 6.7 | 51 | 34 | 21 | 14 | 10 | 7 | 5 | 4 | 3 | 3 | 2 | 2 | 2 | 1 |
| × 6 | 30 | 386 | 6.7 | 62 | 41 | 25 | 16 | 11 | 8 | 6 | 5 | 4 | 3 | 3 | 2 | 2 | 2 |
| L40 40 × 5 | 30 | 378 | 7.7 | 60 | 44 | 29 | 20 | 14 | 10 | 8 | 6 | 5 | 4 | 3 | 3 | 2 | 2 |
| × 6 | 35 | 447 | 7.7 | 73 | 53 | 35 | 23 | 16 | 12 | 9 | 7 | 6 | 5 | 4 | 3 | 3 | 3 |
| L45 45 × 5 | 34 | 428 | 8.7 | 68 | 53 | 37 | 26 | 19 | 14 | 11 | 9 | 7 | 6 | 5 | 4 | 4 | 3 |
| × 6 | 40 | 507 | 8.7 | 84 | 65 | 45 | 32 | 23 | 17 | 13 | 10 | 8 | 7 | 6 | 5 | 4 | 4 |
| L50 50 × 5 | 38 | 479 | 9.7 | 75 | 61 | 46 | 33 | 25 | 19 | 14 | 12 | 9 | 8 | 7 | 6 | 5 | 4 |
| × 6 | 45 | 568 | 9.6 | 93 | 75 | 56 | 40 | 29 | 22 | 17 | 14 | 11 | 9 | 8 | 7 | 6 | 5 |
| L55 55 × 5 | 41 | 527 | 10.6 | 81 | 68 | 53 | 40 | 30 | 23 | 18 | 15 | 12 | 10 | 8 | 7 | 6 | 5 |
| × 6 | 49 | 626 | 10.6 | 102 | 85 | 66 | 49 | 37 | 28 | 22 | 18 | 14 | 12 | 10 | 9 | 7 | 6 |
| × 8 | 64 | 818 | 10.6 | 141 | 118 | 90 | 66 | 49 | 38 | 29 | 23 | 19 | 16 | 13 | 11 | 10 | 9 |
| × 10 | 79 | 1000 | 10.6 | 177 | 148 | 113 | 83 | 61 | 46 | 36 | 29 | 23 | 19 | 16 | 14 | 12 | 10 |
| L60 60 × 5 | 45 | 575 | 11.6 | 86 | 75 | 60 | 47 | 36 | 29 | 23 | 18 | 15 | 13 | 11 | 9 | 8 | 7 |
| × 6 | 54 | 684 | 11.5 | 109 | 94 | 75 | 58 | 44 | 34 | 27 | 22 | 18 | 15 | 13 | 11 | 9 | 8 |
| × 8 | 70 | 896 | 11.5 | 153 | 132 | 104 | 79 | 60 | 46 | 36 | 29 | 24 | 20 | 17 | 14 | 12 | 11 |
| × 10 | 86 | 1100 | 11.5 | 195 | 167 | 132 | 100 | 75 | 58 | 45 | 36 | 30 | 25 | 21 | 18 | 15 | 13 |
| *L65 65 × 5 | 549 | 625(600) | 12.6 | 87 | 77 | 65 | 52 | 41 | 33 | 26 | 22 | 18 | 15 | 13 | 11 | 10 | 8 |
| × 6 | 58 | 744 | 12.6 | 117 | 103 | 85 | 68 | 53 | 42 | 34 | 27 | 23 | 19 | 16 | 14 | 12 | 11 |
| × 8 | 77 | 976 | 12.5 | 166 | 146 | 119 | 94 | 72 | 57 | 45 | 36 | 30 | 25 | 21 | 18 | 16 | 14 |
| × 10 | 94 | 1200 | 12.5 | 212 | 186 | 152 | 118 | 91 | 71 | 56 | 45 | 37 | 31 | 26 | 22 | 19 | 17 |
| *L70 70 × 5 | 53 | 677(600) | 13.6 | 84 | 76 | 65 | 54 | 44 | 36 | 29 | 24 | 20 | 17 | 15 | 13 | 11 | 10 |
| × 6 | 63 | 806 | 13.6 | 124 | 111 | 95 | 77 | 62 | 50 | 40 | 33 | 28 | 23 | 20 | 17 | 15 | 13 |
| × 8 | 83 | 1060 | 13.5 | 178 | 160 | 134 | 108 | 86 | 68 | 55 | 45 | 37 | 31 | 26 | 23 | 20 | 17 |
| × 10 | 102 | 1300 | 13.5 | 228 | 204 | 171 | 137 | 108 | 85 | 68 | 55 | 46 | 38 | 32 | 28 | 24 | 21 |
| *L75 75 × 5 | 557 | 727(600) | 14.6 | 81 | 75 | 65 | 55 | 46 | 38 | 31 | 26 | 22 | 19 | 16 | 14 | 12 | 11 |
| × 6 | 68 | 866 | 14.6 | 130 | 118 | 103 | 86 | 70 | 58 | 47 | 39 | 33 | 28 | 24 | 21 | 18 | 16 |
| × 8 | 89 | 1140 | 14.5 | 189 | 172 | 148 | 122 | 99 | 80 | 65 | 53 | 44 | 37 | 32 | 27 | 24 | 21 |
| × 10 | 110 | 1400 | 14.5 | 244 | 222 | 190 | 156 | 125 | 100 | 81 | 66 | 55 | 46 | 39 | 34 | 30 | 26 |
| *L80 80 × 6 | 673 | 929(864) | 15.6 | 126 | 116 | 103 | 88 | 73 | 61 | 51 | 42 | 36 | 31 | 26 | 23 | 20 | 18 |
| × 8 | 96 | 1220 | 15.5 | 200 | 184 | 161 | 136 | 112 | 91 | 75 | 62 | 52 | 44 | 38 | 33 | 29 | 25 |
| × 10 | 118 | 1500 | 15.5 | 260 | 239 | 208 | 175 | 143 | 116 | 95 | 78 | 65 | 55 | 47 | 41 | 35 | 31 |
| × 12 | 140 | 1780 | 15.4 | 318 | 291 | 254 | 212 | 172 | 139 | 113 | 93 | 78 | 66 | 56 | 48 | 42 | 37 |

*(Contd)*

**TABLE F.3** *(Contd)*

| Designation | Mass (N/m) | Sectional dimensions | | Capacity in KN for effective length KL (m) | | | | | | | | | | | | | |
|---|---|---|---|---|---|---|---|---|---|---|---|---|---|---|---|---|---|
| | | Area ($mm^2$) | $r_v$(Min) (mm) | 0.5 | 1 | 1.5 | 2 | 2.5 | 3 | 3.5 | 4 | 4.5 | 5 | 5.5 | 6 | 6.5 | 7 |
| *L90 90 × 6 | 82 | 1050(864) | 17.5 | 118 | 111 | 101 | 89 | 77 | 66 | 56 | 48 | 41 | 36 | 31 | 27 | 24 | 21 |
| × 8 | 108 | 1380 | 17.5 | 218 | 205 | 184 | 161 | 137 | 115 | 97 | 82 | 70 | 60 | 52 | 45 | 39 | 35 |
| × 10 | 134 | 1700 | 17.4 | 288 | 270 | 242 | 210 | 177 | 148 | 124 | 104 | 88 | 75 | 64 | 56 | 49 | 43 |
| × 12 | 158 | 2020 | 17.4 | 356 | 333 | 298 | 258 | 217 | 181 | 150 | 126 | 106 | 90 | 77 | 67 | 58 | 51 |
| *L100 100 × 6 | 92 | 1170(864) | 19.5 | 110 | 105 | 97 | 88 | 78 | 69 | 60 | 52 | 46 | 40 | 35 | 31 | 28 | 25 |
| × 8 | 121 | 1540 | 19.5 | 234 | 222 | 204 | 183 | 160 | 139 | 119 | 103 | 89 | 77 | 67 | 59 | 52 | 46 |
| × 10 | 149 | 1900 | 19.4 | 314 | 298 | 273 | 243 | 211 | 181 | 155 | 132 | 113 | 97 | 84 | 74 | 65 | 57 |
| × 12 | 177 | 2260 | 19.4 | 392 | 371 | 340 | 302 | 261 | 223 | 189 | 161 | 137 | 118 | 102 | 89 | 78 | 69 |
| *L110 110 × 8 | 134 | 1710(1536) | 21.8 | 223 | 214 | 200 | 183 | 165 | 146 | 129 | 113 | 99 | 87 | 77 | 68 | 60 | 54 |
| × 10 | 166 | 2110 | 21.6 | 339 | 325 | 303 | 276 | 246 | 216 | 188 | 163 | 142 | 124 | 109 | 96 | 85 | 75 |
| × 12 | 197 | 2510 | 21.5 | 427 | 409 | 380 | 345 | 306 | 267 | 232 | 200 | 173 | 150 | 131 | 115 | 102 | 90 |
| × 16 | 257 | 3280 | 21.4 | 591 | 565 | 525 | 475 | 419 | 364 | 314 | 270 | 232 | 201 | 174 | 153 | 134 | 119 |
| *L130 130 × 8 | 159 | 2030(1536) | 25.9 | 201 | 195 | 187 | 176 | 163 | 150 | 136 | 123 | 111 | 100 | 91 | 82 | 74 | 67 |
| *L130 130 × 10 | 197 | 2510(2400) | 25.7 | 360 | 350 | 333 | 312 | 287 | 261 | 235 | 211 | 189 | 168 | 150 | 135 | 121 | 109 |
| × 12 | 235 | 2990 | 25.6 | 485 | 471 | 447 | 418 | 383 | 347 | 311 | 277 | 246 | 218 | 194 | 173 | 155 | 139 |
| × 16 | 307 | 3920 | 25.4 | 689 | 667 | 633 | 589 | 538 | 485 | 432 | 382 | 337 | 297 | 263 | 233 | 207 | 185 |
| *L150 150 × 10 | 229 | 2920(2400) | 29.8 | 332 | 325 | 314 | 299 | 282 | 263 | 243 | 223 | 204 | 186 | 170 | 155 | 141 | 129 |
| × 12 | 273 | 3480 | 29.7 | 534 | 522 | 503 | 478 | 449 | 416 | 383 | 349 | 318 | 288 | 261 | 236 | 214 | 195 |
| × 16 | 358 | 4560 | 29.4 | 777 | 758 | 729 | 691 | 646 | 596 | 545 | 494 | 445 | 400 | 360 | 324 | 292 | 264 |
| × 20 | 441 | 5620 | 29.3 | 966 | 943 | 906 | 857 | 800 | 737 | 671 | 606 | 545 | 488 | 437 | 392 | 352 | 318 |
| *L200 200 × 12 | 369 | 4690(3456) | 39.9 | 445 | 440 | 432 | 421 | 407 | 391 | 374 | 355 | 336 | 317 | 298 | 280 | 262 | 246 |
| × 16 | 485 | 6180 | 39.6 | 951 | 939 | 919 | 893 | 861 | 824 | 783 | 739 | 695 | 650 | 606 | 564 | 524 | 487 |
| × 20 | 600 | 7640 | 39.3 | 1230 | 1214 | 1188 | 1152 | 1109 | 1058 | 1003 | 944 | 884 | 823 | 764 | 708 | 655 | 605 |
| × 25 | 739 | 9410 | 39.1 | 1603 | 1581 | 1546 | 1499 | 1440 | 1373 | 1299 | 1220 | 1139 | 1059 | 980 | 905 | 834 | 768 |

*For these sections the limiting width to thickness ratio as given in Table 2 of IS 800 are exceeded; hence only the effective width is taken in the calculation of capacity

*Note:* For thicknesses ≥ 20, $f_y$ has been reduced to 240 MPa as per Table 1 of IS 800:2007.

**TABLE F.4** Capacity of eccentrically connected angles in compression ($f_y = 250$ MPa)-with hinged ends (clause 7.5.1.2, $k_1 = 1.25$, $k_2 = 0.5$, and $k_3 = 60$)

| Designation | Sectional dimensions | | | Capacity in KN for effective length KL (m) | | | | | | | | | | | | | |
|---|---|---|---|---|---|---|---|---|---|---|---|---|---|---|---|---|---|
| | Mass (N/m) | Area ($mm^2$) | $r_v$(Min) (mm) | 0.5 | 1 | 1.5 | 2 | 2.5 | 3 | 3.5 | 4 | 4.5 | 5 | 5.5 | 6 | 6.5 | 7 |
| L35 35 × 5 | 26 | 327 | 6.7 | 26 | 18 | 12 | 9 | 6 | 5 | 4 | 3 | 2 | 2 | 2 | 1 | 1 | 1 |
| × 6 | 30 | 386 | 6.7 | 32 | 22 | 15 | 10 | 7 | 5 | 4 | 3 | 3 | 2 | 2 | 2 | 1 | 1 |
| L40 40 × 5 | 30 | 378 | 7.7 | 30 | 23 | 16 | 12 | 9 | 7 | 5 | 4 | 3 | 3 | 2 | 2 | 2 | 1 |
| × 6 | 35 | 447 | 7.7 | 37 | 28 | 20 | 14 | 10 | 8 | 6 | 5 | 4 | 3 | 3 | 2 | 2 | 2 |
| L45 45 × 5 | 34 | 428 | 8.7 | 33 | 26 | 20 | 15 | 11 | 9 | 7 | 6 | 5 | 4 | 3 | 3 | 2 | 2 |
| × 6 | 40 | 507 | 8.7 | 41 | 33 | 25 | 19 | 14 | 11 | 8 | 7 | 6 | 5 | 4 | 3 | 3 | 3 |
| L50 50 × 5 | 38 | 479 | 9.7 | 35 | 30 | 24 | 19 | 14 | 11 | 9 | 7 | 6 | 5 | 4 | 4 | 3 | 3 |
| × 6 | 45 | 568 | 9.6 | 45 | 38 | 30 | 23 | 18 | 14 | 11 | 9 | 7 | 6 | 5 | 4 | 4 | 3 |
| L55 55 × 5 | 41 | 527 | 10.6 | 37 | 32 | 27 | 22 | 17 | 14 | 11 | 9 | 8 | 7 | 6 | 5 | 4 | 4 |
| × 6 | 49 | 626 | 10.6 | 48 | 42 | 34 | 27 | 21 | 17 | 14 | 11 | 9 | 8 | 7 | 6 | 5 | 4 |

*(Contd)*

**TABLE F.4** *(Contd)*

| Designation | Sectional dimensions | | | Capacity in KN for effective length KL (m) | | | | | | | | | | | | | |
|---|---|---|---|---|---|---|---|---|---|---|---|---|---|---|---|---|---|
| | Mass (N/m) | Area ($mm^2$) | $r_v$(Min) (mm) | 0.5 | 1 | 1.5 | 2 | 2.5 | 3 | 3.5 | 4 | 4.5 | 5 | 5.5 | 6 | 6.5 | 7 |
| × 8 | 64 | 818 | 10.6 | 71 | 60 | 48 | 38 | 29 | 23 | 19 | 15 | 13 | 11 | 9 | 8 | 7 | 6 |
| × 10 | 79 | 1000 | 10.6 | 91 | 77 | 61 | 47 | 37 | 29 | 23 | 19 | 15 | 13 | 11 | 9 | 8 | 7 |
| L60 60 × 5 | 45 | 575 | 11.6 | 38 | 34 | 29 | 24 | 20 | 17 | 14 | 11 | 10 | 8 | 7 | 6 | 5 | 5 |
| × 6 | 54 | 684 | 11.5 | 51 | 45 | 38 | 31 | 25 | 20 | 17 | 14 | 12 | 10 | 8 | 7 | 6 | 6 |
| × 8 | 70 | 896 | 11.5 | 76 | 66 | 55 | 44 | 35 | 28 | 23 | 19 | 16 | 13 | 11 | 10 | 8 | 7 |
| × 10 | 86 | 1100 | 11.5 | 99 | 86 | 70 | 56 | 44 | 35 | 28 | 23 | 19 | 16 | 14 | 12 | 10 | 9 |
| *L65 65 × 5 | 49 | 625(600) | 12.6 | 38 | 35 | 30 | 26 | 22 | 18 | 15 | 13 | 11 | 10 | 8 | 7 | 6 | 6 |
| × 6 | 58 | 744 | 12.6 | 53 | 48 | 42 | 35 | 29 | 24 | 20 | 17 | 14 | 12 | 11 | 9 | 8 | 7 |
| × 8 | 77 | 976 | 12.5 | 81 | 72 | 61 | 50 | 41 | 33 | 28 | 23 | 19 | 16 | 14 | 12 | 11 | 9 |
| × 10 | 94 | 1200 | 12.5 | 107 | 95 | 79 | 65 | 52 | 42 | 35 | 29 | 24 | 20 | 17 | 15 | 13 | 12 |
| *L70 70 × 5 | 53 | 677(600) | 13.6 | 36 | 33 | 30 | 26 | 23 | 19 | 16 | 14 | 12 | 11 | 9 | 8 | 7 | 6 |
| × 6 | 63 | 806 | 13.6 | 55 | 51 | 45 | 39 | 33 | 28 | 23 | 20 | 17 | 15 | 13 | 11 | 10 | 9 |
| × 8 | 83 | 1060 | 13.5 | 85 | 78 | 67 | 57 | 47 | 39 | 33 | 27 | 23 | 20 | 17 | 15 | 13 | 11 |
| × 10 | 102 | 1300 | 13.5 | 114 | 103 | 88 | 74 | 61 | 50 | 41 | 35 | 29 | 25 | 21 | 18 | 16 | 14 |
| *L75 75 × 5 | 57 | 727(600) | 14.6 | 34 | 32 | 29 | 26 | 23 | 20 | 17 | 15 | 13 | 11 | 10 | 9 | 8 | 7 |
| × 6 | 68 | 866 | 14.6 | 57 | 53 | 48 | 42 | 36 | 31 | 27 | 23 | 20 | 17 | 15 | 13 | 12 | 10 |
| × 8 | 89 | 1140 | 14.5 | 89 | 82 | 73 | 63 | 53 | 45 | 38 | 32 | 27 | 24 | 20 | 18 | 16 | 14 |
| × 10 | 110 | 1400 | 14.5 | 121 | 110 | 97 | 82 | 69 | 58 | 48 | 41 | 35 | 30 | 26 | 22 | 20 | 17 |
| *L80 80 × 6 | 73 | 929(864) | 15.6 | 54 | 51 | 47 | 42 | 37 | 32 | 28 | 24 | 21 | 18 | 16 | 14 | 13 | 11 |
| × 8 | 96 | 1220 | 15.5 | 93 | 87 | 78 | 68 | 59 | 50 | 43 | 37 | 32 | 27 | 24 | 21 | 19 | 16 |
| × 10 | 118 | 1500 | 15.5 | 127 | 117 | 104 | 90 | 77 | 65 | 55 | 47 | 40 | 35 | 30 | 26 | 23 | 21 |
| × 12 | 140 | 1780 | 15.4 | 160 | 147 | 130 | 112 | 94 | 80 | 67 | 57 | 48 | 42 | 36 | 31 | 28 | 24 |
| *L90 90 × 6 | 82 | 1050(864) | 17.5 | 49 | 47 | 44 | 40 | 37 | 33 | 29 | 26 | 23 | 20 | 18 | 16 | 15 | 13 |
| × 8 | 108 | 1380 | 17.5 | 98 | 93 | 86 | 78 | 69 | 61 | 53 | 46 | 41 | 36 | 32 | 28 | 25 | 22 |
| × 10 | 134 | 1700 | 17.4 | 137 | 129 | 118 | 105 | 92 | 80 | 69 | 60 | 52 | 46 | 40 | 35 | 31 | 28 |
| × 12 | 158 | 2020 | 17.4 | 176 | 165 | 150 | 132 | 115 | 99 | 86 | 74 | 64 | 56 | 49 | 43 | 38 | 34 |
| *L100100 × 6 | 92 | 1170(864) | 19.5 | 44 | 43 | 41 | 38 | 35 | 33 | 30 | 27 | 24 | 22 | 20 | 18 | 16 | 15 |
| × 8 | 121 | 1540 | 19.5 | 102 | 98 | 92 | 85 | 77 | 70 | 62 | 56 | 49 | 44 | 39 | 35 | 32 | 29 |
| × 10 | 149 | 1900 | 19.4 | 146 | 139 | 130 | 118 | 106 | 94 | 83 | 74 | 65 | 57 | 51 | 45 | 41 | 36 |
| × 12 | 177 | 2260 | 19.4 | 189 | 180 | 167 | 151 | 135 | 119 | 104 | 92 | 80 | 71 | 63 | 55 | 49 | 44 |
| *L110 110 × 8 | 134 | 1710(1536) | 21.8 | 95 | 92 | 88 | 82 | 76 | 70 | 64 | 58 | 53 | 48 | 43 | 39 | 36 | 32 |
| × 10 | 166 | 2110 | 21.6 | 153 | 148 | 140 | 130 | 119 | 108 | 98 | 88 | 79 | 70 | 63 | 57 | 51 | 46 |
| × 12 | 197 | 2510 | 21.5 | 202 | 194 | 183 | 169 | 153 | 138 | 123 | 110 | 98 | 87 | 78 | 70 | 63 | 57 |
| × 16 | 257 | 3280 | 21.4 | 296 | 283 | 264 | 242 | 218 | 194 | 172 | 152 | 134 | 119 | 106 | 94 | 84 | 76 |
| *L130 130 × 8 | 159 | 2030(1536) | 25.9 | 81 | 80 | 77 | 74 | 71 | 67 | 63 | 59 | 55 | 51 | 47 | 44 | 40 | 37 |
| *L130 130 × 10 | 197 | 2510(2400) | 25.7 | 155 | 152 | 146 | 139 | 132 | 123 | 114 | 105 | 97 | 89 | 82 | 75 | 69 | 63 |
| × 12 | 235 | 2990 | 25.6 | 220 | 215 | 206 | 195 | 183 | 169 | 156 | 143 | 131 | 119 | 109 | 99 | 90 | 83 |
| × 16 | 307 | 3920 | 25.4 | 334 | 325 | 310 | 291 | 270 | 247 | 226 | 205 | 185 | 167 | 151 | 137 | 124 | 113 |
| *L150 150 × 10 | 229 | 2920(2400) | 29.8 | 137 | 135 | 132 | 128 | 123 | 117 | 111 | 105 | 98 | 92 | 86 | 81 | 75 | 70 |
| × 12 | 273 | 3480 | 29.7 | 232 | 228 | 222 | 214 | 204 | 193 | 182 | 170 | 159 | 148 | 137 | 127 | 118 | 109 |

*(Contd)*

**TABLE F.4** *(Contd)*

| Designation | Sectional dimensions | | | Capacity in KN for effective length KL (m) | | | | | | | | | | | | | |
|---|---|---|---|---|---|---|---|---|---|---|---|---|---|---|---|---|---|
| | Mass (N/m) | Area ($mm^2$) | $r_v$(Min) (mm) | 0.5 | 1 | 1.5 | 2 | 2.5 | 3 | 3.5 | 4 | 4.5 | 5 | 5.5 | 6 | 6.5 | 7 |
| × 16 | 358 | 4560 | 29.4 | 365 | 358 | 346 | 330 | 312 | 293 | 273 | 253 | 233 | 215 | 198 | 182 | 167 | 154 |
| × 20 | 441 | 5620 | 29.3 | 476 | 465 | 448 | 426 | 401 | 374 | 346 | 319 | 293 | 268 | 245 | 225 | 206 | 188 |
| *L200 200 × 12 | 369 | 4690(3456) | 39.9 | 178 | 177 | 175 | 172 | 168 | 164 | 159 | 154 | 149 | 143 | 137 | 132 | 126 | 120 |
| × 16 | 485 | 6180 | 39.6 | 414 | 410 | 403 | 394 | 384 | 371 | 358 | 343 | 328 | 313 | 297 | 282 | 267 | 253 |
| × 20 | 600 | 7640 | 39.3 | 569 | 563 | 552 | 538 | 521 | 502 | 481 | 459 | 436 | 413 | 390 | 368 | 346 | 326 |
| × 25 | 739 | 9410 | 39.1 | 780 | 770 | 754 | 733 | 708 | 679 | 647 | 614 | 581 | 547 | 515 | 483 | 453 | 424 |

*For these sections the limiting width to thickness ratio as given in Table 2 of IS 800 are exceeded; hence only the effective width is taken in the calculation of capacity.
*Note:* For thicknesses ≥ 20, $f_y$ has been reduced to 240 MPa as per Table 1 of IS 800:2007.

**TABLE F.5** Capacity of I-section compression members (Clause 7.1.2.1)

| Section designation | Compression resistance $P_{dz}$ and $P_{dy}$ (kN) for effective length $L_e$ (m) | | | | | | | | | | | |
|---|---|---|---|---|---|---|---|---|---|---|---|---|
| | 2 | 2.5 | 3 | 3.5 | 4 | 4.5 | 5 | 5.5 | 6 | 6.5 | 7 | |
| MB100 | $P_{dz}$ | 234 | 219 | 199 | 175 | 150 | 127 | 107 | 91 | 79 | 68 | 59 |
| | $P_{dy}$ | 48 | 32 | 23 | 17 | 13 | 10 | 8 | 7 | 6 | 5 | 4 |
| MB125 | $P_{dz}$ | 364 | 351 | 335 | 315 | 290 | 262 | 233 | 205 | 180 | 158 | 140 |
| | $P_{dy}$ | 134 | 92 | 66 | 50 | 39 | 31 | 25 | 21 | 18 | 15 | 13 |
| MB150 | $P_{dz}$ | 417 | 407 | 394 | 379 | 361 | 339 | 313 | 286 | 258 | 232 | 208 |
| | $P_{dy}$ | 160 | 110 | 80 | 60 | 47 | 38 | 31 | 26 | 22 | 19 | 16 |
| MB175 | $P_{dz}$ | 553 | 542 | 530 | 515 | 499 | 479 | 456 | 429 | 399 | 368 | 337 |
| | $P_{dy}$ | 248 | 175 | 128 | 97 | 76 | 61 | 50 | 42 | 35 | 30 | 26 |
| MB200 | $P_{dz}$ | 689 | 678 | 666 | 652 | 637 | 620 | 600 | 577 | 551 | 522 | 490 |
| | $P_{dy}$ | 389 | 288 | 216 | 166 | 131 | 105 | 87 | 73 | 62 | 53 | 46 |
| MB225 | $P_{dz}$ | 894 | 881 | 868 | 85 1 | 838 | 821 | 801 | 779 | 754 | 725 | 693 |
| | $P_{dv}$ | 561 | 430 | 329 | 255 | 202 | 164 | 135 | 114 | 97 | 83 | 72 |
| MB250 | $P_{dz}$ | 1076 | 1063 | 1049 | 1035 | 1019 | 1002 | 983 | 963 | 940 | 913 | 884 |
| | $P_{dy}$ | 749 | 603 | 475 | 375 | 301 | 245 | 203 | 171 | 146 | 125 | 109 |
| MB300 | $P_{dz}$ | 1337 | 1324 | 1310 | 1296 | 1282 | 1266 | 1249 | 1231 | 1212 | 1190 | 1167 |
| | $P_{dy}$ | 975 | 809 | 651 | 521 | 421 | 345 | 287 | 242 | 207 | 178 | 155 |
| MB350 | $P_{dz}$ | 1530 | 1517 | 1504 | 1490 | 1476 | 1462 | 1447 | 1431 | 1414 | 1396 | 1376 |
| | $P_{dy}$ | 1105 | 914 | 734 | 587 | 474 | 388 | 323 | 272 | 232 | 200 | 175 |
| MB400 | $P_{dz}$ | 1805 | 1792 | 1779 | 1765 | 1751 | 1736 | 1722 | 1706 | 1690 | 1673 | 1654 |
| | $P_{dy}$ | 1293 | 1066 | 855 | 683 | 551 | 451 | 375 | 316 | 269 | 232 | 203 |
| MB450 | $P_{dz}$ | 2130 | 2116 | 2102 | 2088 | 2073 | 2059 | 2011 | 2028 | 2012 | 1995 | 1978 |
| | $P_{dy}$ | 1584 | 1339 | 1095 | 886 | 721 | 593 | 494 | 418 | 357 | 308 | 269 |
| MB500 | $P_{dz}$ | 2571 | 2556 | 2541 | 2525 | 2510 | 2494 | 2478 | 2462 | 2445 | 2428 | 2410 |
| | $P_{dy}$ | 2060 | 1829 | 1572 | 1322 | 1103 | 923 | 778 | 662 | 569 | 493 | 432 |
| MB550 | $P_{dz}$ | 3064 | 3048 | 3031 | 3015 | 2998 | 2981 | 2964 | 2947 | 2930 | 2911 | 2893 |
| | $P_{dy}$ | 2506 | 2257 | 1974 | 1686 | 1423 | 1200 | 1017 | 869 | 749 | 651 | 570 |
| MB600 | $P_{dz}$ | 3484 | 3467 | 3450 | 3434 | 3417 | 3400 | 3383 | 3366 | 3348 | 3330 | 3312 |
| | $P_{dy}$ | 2955 | 2726 | 2460 | 2170 | 1882 | 1619 | 1392 | 1201 | 1042 | 910 | 800 |
| SC100 | $P_{dz}$ | 501 | 460 | 414 | 363 | 314 | 269 | 231 | 199 | 172 | 150 | 132 |
| | $P_{dy}$ | 322 | 246 | 190 | 149 | 119 | 97 | 80 | 68 | 58 | 50 | 43 |

*(Contd)*

**TABLE F.5** *(Contd)*

| Section designation | Compression resistance $P_{dz}$ and $P_{dy}$ (kN) for effective length $L_e$ (m) | | | | | | | | | | | |
|---|---|---|---|---|---|---|---|---|---|---|---|---|
| | 2 | 2.5 | 3 | 3.5 | 4 | 4.5 | 5 | 5.5 | 6 | 6.5 | 7 | |
| SC120 | $P_{dz}$ | 688 | 650 | 606 | 557 | 503 | 449 | 397 | 350 | 308 | 272 | 241 |
| | $P_{dy}$ | 495 | 401 | 321 | 258 | 209 | 172 | 144 | 122 | 105 | 90 | 79 |
| SC140 | $P_{dz}$ | 899 | 861 | 819 | 772 | 719 | 662 | 603 | 545 | 490 | 439 | 394 |
| | $P_{dy}$ | 698 | 592 | 492 | 406 | 336 | 280 | 236 | 201 | 173 | 151 | 132 |
| SC150 | $P_{dz}$ | 1019 | 982 | 941 | 896 | 846 | 790 | 731 | 671 | 611 | 555 | 503 |
| | $P_{dy}$ | 855 | 759 | 660 | 566 | 482 | 410 | 351 | 303 | 263 | 230 | 202 |
| SC160 | $P_{dz}$ | 1155 | 1115 | 1073 | 1025 | 973 | 915 | 853 | 788 | 723 | 660 | 601 |
| | $P_{dy}$ | 936 | 819 | 702 | 593 | 500 | 423 | 360 | 308 | 267 | 233 | 205 |
| SC180 | $P_{dz}$ | 1414 | 1373 | 1330 | 1283 | 1232 | 1176 | 1115 | 1049 | 981 | 911 | 843 |
| | $P_{dy}$ | 1187 | 1065 | 938 | 814 | 700 | 601 | 517 | 447 | 389 | 341 | 301 |
| SC200 | $P_{dz}$ | 1705 | 1662 | 1618 | 1570 | 1519 | 1463 | 1403 | 1338 | 1268 | 1196 | 1121 |
| | $P_{dy}$ | 1467 | 1339 | 1203 | 1065 | 933 | 813 | 708 | 617 | 541 | 477 | 422 |
| SC220 | $P_{dz}$ | 2011 | 1967 | 1921 | 1873 | 1821 | 1766 | 1706 | 1642 | 1574 | 1501 | 1424 |
| | $P_{dy}$ | 1766 | 1634 | 1493 | 1346 | 1201 | 1063 | 937 | 826 | 730 | 647 | 576 |
| SC250 | $P_{dz}$ | 2468 | 2422 | 2374 | 2325 | 2274 | 2219 | 2162 | 2100 | 2034 | 1964 | 1890 |
| | $P_{dy}$ | 2207 | 2067 | 1919 | 1762 | 1602 | 1444 | 1294 | 1156 | 1032 | 923 | 827 |
| HB150 | $P_{dz}$ | 742 | 716 | 686 | 654 | 618 | 578 | 536 | 492 | 449 | 408 | 370 |
| | $P_{dy}$ | 599 | 522 | 445 | 374 | 314 | 265 | 225 | 193 | 167 | 145 | 128 |
| HB200 | $P_{dz}$ | 1057 | 1032 | 1005 | 977 | 946 | 914 | 878 | 840 | 798 | 755 | 711 |
| | $P_{dy}$ | 911 | 833 | 750 | 665 | 584 | 510 | 444 | 388 | 341 | 300 | 266 |
| HB225 | $P_{dz}$ | 1235 | 1209 | 1182 | 1155 | 1125 | 1094 | 1060 | 1024 | 986 | 944 | 901 |
| | $P_{dy}$ | 1084 | 1004 | 919 | 831 | 743 | 659 | 582 | 513 | 454 | 403 | 359 |
| HB250 | $P_{dz}$ | 1474 | 1447 | 1419 | 1390 | 1360 | 1329 | 1295 | 1260 | 1222 | 1181 | 1139 |
| | $P_{dy}$ | 1318 | 1235 | 1147 | 1055 | 960 | 866 | 776 | 694 | 620 | 555 | 497 |
| HB300 | $P_{dz}$ | 1716 | 1690 | 1664 | 1637 | 1610 | 1581 | 1552 | 1521 | 1488 | 1454 | 1418 |
| | $P_{dy}$ | 1511 | 1414 | 1311 | 1202 | 1091 | 982 | 879 | 784 | 699 | 625 | 560 |
| HB350 | $P_{dz}$ | 1986 | 1960 | 1934 | 1907 | 1881 | 1853 | 1825 | 1796 | 1766 | 1734 | 1701 |
| | $P_{dy}$ | 1729 | 1616 | 1496 | 1369 | 1240 | 1114 | 994 | 886 | 789 | 704 | 630 |
| HB400 | $P_{dz}$ | 2296 | 2270 | 2243 | 2217 | 2190 | 2163 | 2135 | 2107 | 2078 | 2048 | 2016 |
| | $P_{dy}$ | 1979 | 1847 | 1705 | 1557 | 1407 | 1260 | 1123 | 998 | 888 | 791 | 707 |
| HB450 | $P_{dz}$ | 2595 | 2568 | 2541 | 2514 | 2487 | 2460 | 2433 | 2405 | 2376 | 2347 | 2317 |
| | $P_{dy}$ | 2217 | 2065 | 1903 | 1733 | 1561 | 1395 | 1240 | 1100 | 977 | 870 | 777 |
| UC 152 × 152 × 23 | $P_{dz}$ | 630 | 608 | 583 | 556 | 525 | 492 | 456 | 420 | 383 | 349 | 316 |
| | $P_{dy}$ | 519 | 457 | 394 | 335 | 283 | 240 | 205 | 176 | 152 | 133 | 117 |
| UC 152 × 152 × 30 | $P_{dz}$ | 828 | 800 | 769 | 735 | 698 | 657 | 612 | 566 | 520 | 475 | 432 |
| | $P_{dy}$ | 689 | 611 | 531 | 455 | 388 | 330 | 282 | 243 | 211 | 185 | 163 |
| UC 152 × 152 × 37 | $P_{dz}$ | 1021 | 987 | 950 | 909 | 864 | 815 | 761 | 705 | 649 | 594 | 542 |
| | $P_{dy}$ | 853 | 758 | 660 | 567 | 484 | 412 | 353 | 304 | 264 | 231 | 204 |
| UC 203 × 203 × 46 | $P_{dz}$ | 1309 | 1277 | 1245 | 1211 | 1174 | 1134 | 1091 | 1045 | 995 | 943 | 888 |
| | $P_{dy}$ | 1170 | 1089 | 1002 | 911 | 819 | 731 | 649 | 575 | 510 | 454 | 405 |
| UC 203 × 203 × 52 | $P_{dz}$ | 1478 | 1444 | 1407 | 1369 | 1328 | 1284 | 1237 | 1185 | 1130 | 1072 | 1011 |
| | $P_{dy}$ | 1324 | 1233 | 1136 | 1035 | 932 | 833 | 741 | 657 | 583 | 519 | 464 |

*(Contd)*

**TABLE F.5** *(Contd)*

| Section designation | | Compression resistance $P_{dz}$ and $P_{dy}$ (kN) for effective length $L_e$ (m) | | | | | | | | | | |
|---|---|---|---|---|---|---|---|---|---|---|---|---|
| | | 2 | 2.5 | 3 | 3.5 | 4 | 4.5 | 5 | 5.5 | 6 | 6.5 | 7 |
| UC 203 × 203 × 60 | $P_{dz}$ | 1702 | 1662 | 1621 | 1577 | 1531 | 1481 | 1426 | 1367 | 1305 | 1238 | 1169 |
| | $P_{dy}$ | 1525 | 1421 | 1310 | 1194 | 1077 | 962 | 856 | 760 | 675 | 601 | 537 |
| UC 203 × 203 × 71 | $P_{dz}$ | 2022 | 1977 | 1929 | 1879 | 1826 | 1769 | 1707 | 1640 | 1569 | 1493 | 1414 |
| | $P_{dy}$ | 1817 | 1697 | 1569 | 1434 | 1297 | 1164 | 1038 | 924 | 822 | 733 | 656 |
| UC 203 × 203 × 86 | $P_{dz}$ | 2453 | 2399 | 2342 | 2282 | 2219 | 2151 | 2077 | 1998 | 1913 | 1822 | 1728 |
| | $P_{dy}$ | 2207 | 2062 | 1908 | 1747 | 1582 | 1421 | 1269 | 1130 | 1007 | 898 | 804 |
| UC 254 × 254 × 73 | $P_{dz}$ | 2113 | 2075 | 2036 | 1996 | 1954 | 1910 | 1863 | 1813 | 1760 | 1704 | 1644 |
| | $P_{dy}$ | 1957 | 1860 | 1759 | 1652 | 1540 | 1426 | 1311 | 1198 | 1092 | 993 | 902 |
| UC 254 × 254 × 89 | $P_{dz}$ | 2574 | 2528 | 2482 | 2434 | 2383 | 2331 | 2275 | 2216 | 2153 | 2086 | 2015 |
| | $P_{dy}$ | 2386 | 2270 | 2148 | 2020 | 1886 | 1748 | 1610 | 1474 | 1345 | 1224 | 1114 |
| UC 254 × 254 × 107 | $P_{dz}$ | 3101 | 3047 | 2991 | 2934 | 2874 | 2812 | 2746 | 2676 | 2601 | 2522 | 2438 |
| | $P_{dy}$ | 2876 | 2737 | 2592 | 2439 | 2279 | 2114 | 1948 | 1785 | 1630 | 1485 | 1351 |
| UC 254 × 254 × 132 | $P_{dz}$ | 3145 | 3091 | 3036 | 2979 | 2921 | 2859 | 2795 | 2726 | 2654 | 2576 | 2494 |
| | $P_{dy}$ | 2920 | 2782 | 2638 | 2486 | 2327 | 2163 | 1997 | 1834 | 1678 | 1531 | 1395 |
| UC 254 × 254 × 167 | $P_{dz}$ | 4856 | 4775 | 4692 | 4608 | 4520 | 4429 | 4333 | 4232 | 4124 | 4010 | 3889 |
| | $P_{dy}$ | 4517 | 4308 | 4090 | 3862 | 3622 | 3375 | 3124 | 2876 | 2636 | 2410 | 2201 |
| UC 305 × 305 × 97 | $P_{dz}$ | 2837 | 2795 | 2753 | 2711 | 2667 | 2622 | 2576 | 2527 | 2476 | 2422 | 2366 |
| | $P_{dy}$ | 2672 | 2567 | 2458 | 2345 | 2227 | 2104 | 1977 | 1848 | 1720 | 1595 | 1475 |
| UC 305 × 305 × 118 | $P_{dz}$ | 3455 | 3405 | 3354 | 3303 | 3251 | 3197 | 3141 | 3083 | 3022 | 2958 | 2890 |
| | $P_{dy}$ | 3258 | 3131 | 3000 | 2864 | 2723 | 2575 | 2422 | 2268 | 2113 | 1961 | 1816 |
| UC 305 × 305 × 137 | $P_{dz}$ | 4014 | 3956 | 3899 | 3840 | 3780 | 3718 | 3655 | 3588 | 3518 | 3445 | 3368 |
| | $P_{dy}$ | 3787 | 3641 | 3491 | 3335 | 3171 | 3002 | 2826 | 2648 | 2470 | 2295 | 2126 |
| UC 305 × 305 × 158 | $P_{dz}$ | 4684 | 4618 | 4551 | 4484 | 4415 | 4344 | 4270 | 4194 | 4114 | 4030 | 3942 |
| | $P_{dy}$ | 4423 | 4254 | 4081 | 3901 | 3713 | 3517 | 3315 | 3109 | 2902 | 2699 | 2504 |
| UC 305 × 305 × 198 | $P_{dz}$ | 5821 | 5740 | 5659 | 5578 | 5494 | 5409 | 5321 | 5229 | 5133 | 5033 | 4928 |
| | $P_{dy}$ | 5503 | 5297 | 5087 | 4868 | 4640 | 4403 | 4157 | 3907 | 3655 | 3406 | 3165 |
| UC 305 × 305 × 240 | $P_{dz}$ | 7060 | 6964 | 6868 | 6772 | 6673 | 6572 | 6468 | 6360 | 6248 | 6130 | 6007 |
| | $P_{dy}$ | 6681 | 6435 | 6184 | 5923 | 5652 | 5370 | 5078 | 4779 | 4478 | 4179 | 3889 |
| UC 305 × 305 × 283 | $P_{dz}$ | 8329 | 8219 | 8109 | 7997 | 7884 | 7768 | 7648 | 7525 | 7396 | 7262 | 7122 |
| | $P_{dy}$ | 7890 | 7605 | 7314 | 7013 | 6699 | 6373 | 6035 | 5689 | 5339 | 4991 | 4652 |
| UC 356 × 368 × 129 | $P_{dz}$ | 3808 | 3761 | 3713 | 3665 | 3617 | 3567 | 3517 | 3465 | 3411 | 3355 | 3296 |
| | $P_{dy}$ | 3661 | 3547 | 3433 | 3316 | 3195 | 3070 | 2940 | 2806 | 2669 | 2529 | 2390 |
| UC 356 × 368 × 153 | $P_{dz}$ | 4517 | 4461 | 4406 | 4349 | 4292 | 4234 | 4175 | 4114 | 4051 | 3985 | 3917 |
| | $P_{dy}$ | 4343 | 4210 | 4075 | 3937 | 3795 | 3648 | 3496 | 3338 | 3177 | 3013 | 2848 |
| UC 356 × 368 × 177 | $P_{dz}$ | 5231 | 5167 | 5103 | 5038 | 4973 | 4906 | 4838 | 4768 | 4696 | 4621 | 4543 |
| | $P_{dy}$ | 5031 | 4878 | 4722 | 4564 | 4400 | 4231 | 4056 | 3875 | 3689 | 3500 | 3311 |
| UC 356 × 368 × 202 | $P_{dz}$ | 5969 | 5896 | 5824 | 5751 | 5677 | 5602 | 5525 | 5446 | 5365 | 5280 | 5192 |
| | $P_{dy}$ | 5743 | 5569 | 5393 | 5213 | 5028 | 4837 | 4639 | 4434 | 4224 | 4010 | 3795 |

*Note:* For thicknesses ≥ 20, $f_y$ has been reduced to 240 MPa as per Table 1 of IS 800:2007.

**TABLE F.6** Capacity of square hollow section compression members (Clause 7.1.2.1)

| Square hollow sections $D \times B$ (mm × mm) | Thick-ness $t$(mm) | Unit $wt$ N/m | Sec. Area $A(mm^2)$ | Radius of gyration $r_z$ (mm) | Length(m) | | | | | | | | | | | | | |
|---|---|---|---|---|---|---|---|---|---|---|---|---|---|---|---|---|---|---|
| | | | | | 0.5 | 1 | 1.5 | 2 | 2.5 | 3 | 3.5 | 4 | 4.5 | 5 | 5.5 | 6 | 6.5 | 7 |
| 25 × 25 | 1.6 | 11.2 | 143 | 9.4 | 28 | 17 | 9 | 5 | 3 | 2 | 2 | 1 | 1 | 1 | 1 | 1 | 1 | 0 |
| | 2 | 13.6 | 174 | 9.2 | 34 | 20 | 10 | 6 | 4 | 3 | 2 | 2 | 1 | 1 | 1 | 1 | 1 | 1 |
| | 2.6 | 16.9 | 216 | 8.9 | 42 | 24 | 12 | 7 | 5 | 3 | 2 | 2 | 1 | 1 | 1 | 1 | 1 | 1 |
| | 3.2 | 19.8 | 253 | 8.6 | 48 | 27 | 13 | 8 | 5 | 4 | 3 | 2 | 2 | 1 | 1 | 1 | 1 | 1 |
| 32 × 32 | 2 | 18 | 230 | 12.1 | 47 | 36 | 22 | 13 | 9 | 6 | 5 | 4 | 3 | 2 | 2 | 2 | 1 | 1 |
| | 2.6 | 22.6 | 288 | 11.8 | 59 | 45 | 26 | 16 | 10 | 7 | 6 | 4 | 3 | 3 | 2 | 2 | 2 | 1 |
| | 3.2 | 26.9 | 342 | 11.5 | 69 | 52 | 30 | 18 | 12 | 8 | 6 | 5 | 4 | 3 | 3 | 2 | 2 | 2 |
| 38 × 38 | 2 | 21.8 | 278 | 14.6 | 58 | 50 | 35 | 22 | 15 | 11 | 8 | 6 | 5 | 4 | 3 | 3 | 2 | 2 |
| | 2.6 | 27.5 | 351 | 14.3 | 73 | 62 | 43 | 27 | 18 | 13 | 10 | 8 | 6 | 5 | 4 | 3 | 3 | 3 |
| | 3.2 | 32.9 | 419 | 14 | 87 | 73 | 50 | 31 | 21 | 15 | 11 | 9 | 7 | 6 | 5 | 4 | 3 | 3 |
| | 4 | 39.5 | 503 | 13.6 | 104 | 87 | 57 | 36 | 24 | 17 | 13 | 10 | 8 | 6 | 5 | 4 | 4 | 3 |
| 40 × 40 | 2.6 | 29.2 | 372 | 15.1 | 78 | 68 | 49 | 32 | 21 | 15 | 11 | 9 | 7 | 6 | 5 | 4 | 3 | 3 |
| | 2.9 | 32.1 | 409 | 14.9 | 86 | 74 | 53 | 34 | 23 | 16 | 12 | 9 | 8 | 6 | 5 | 4 | 4 | 3 |
| | 3.2 | 34.9 | 445 | 14.8 | 93 | 80 | 57 | 37 | 25 | 18 | 13 | 10 | 8 | 7 | 5 | 5 | 4 | 3 |
| | 4 | 42 | 535 | 14.4 | 112 | 95 | 66 | 42 | 28 | 20 | 15 | 12 | 9 | 8 | 6 | 5 | 5 | 4 |
| 49.5 × 49. 5 | 2.6 | 36.9 | 470 | 19 | 100 | 92 | 77 | 57 | 40 | 29 | 22 | 17 | 14 | 11 | 9 | 8 | 7 | 6 |
| | 2.9 | 40.7 | 519 | 18.8 | 111 | 101 | 85 | 62 | 44 | 32 | 24 | 19 | 15 | 12 | 10 | 9 | 7 | 6 |
| | 3.6 | 49.3 | 628 | 18.5 | 134 | 122 | 101 | 73 | 52 | 37 | 28 | 22 | 18 | 14 | 12 | 10 | 9 | 7 |
| | 4.5 | 59.5 | 758 | 18 | 161 | 146 | 119 | 85 | 59 | 43 | 32 | 25 | 20 | 16 | 14 | 12 | 10 | 9 |
| 60 × 60 | 2.6 | 45.5 | 580 | 23.3 | 126 | 118 | 107 | 89 | 68 | 52 | 40 | 31 | 25 | 21 | 17 | 15 | 12 | 11 |
| | 2.9 | 50.3 | 641 | 23.1 | 139 | 130 | 117 | 97 | 75 | 56 | 43 | 34 | 27 | 22 | 19 | 16 | 14 | 12 |
| | 3.2 | 55 | 701 | 23 | 152 | 142 | 128 | 106 | 81 | 61 | 47 | 37 | 30 | 24 | 20 | 17 | 15 | 13 |
| | 4 | 67.1 | 855 | 22.6 | 185 | 173 | 155 | 127 | 97 | 72 | 55 | 43 | 35 | 29 | 24 | 20 | 17 | 15 |
| | 4.8 | 78.5 | 1001 | 22.2 | 216 | 202 | 180 | 146 | 110 | 82 | 63 | 49 | 40 | 32 | 27 | 23 | 20 | 17 |
| 72 × 72 | 3.2 | 67.1 | 854 | 27.9 | 186 | 178 | 166 | 149 | 126 | 100 | 79 | 63 | 51 | 42 | 36 | 30 | 26 | 23 |
| | 4 | 82.2 | 1047 | 27.5 | 228 | 218 | 203 | 182 | 152 | 121 | 95 | 76 | 61 | 51 | 42 | 36 | 31 | 27 |
| | 4.8 | 96.6 | 1231 | 27.1 | 268 | 255 | 238 | 212 | 176 | 139 | 109 | 87 | 70 | 58 | 49 | 41 | 35 | 31 |
| 91.5 × 91.5 | 3.6 | 96.7 | 1232 | 35.6 | 271 | 262 | 251 | 237 | 218 | 192 | 163 | 136 | 113 | 95 | 80 | 69 | 59 | 52 |
| | 4.5 | 118.8 | 1514 | 35.2 | 333 | 322 | 308 | 291 | 266 | 234 | 198 | 164 | 136 | 114 | 97 | 83 | 71 | 62 |
| | 5.4 | 140.1 | 1785 | 34.8 | 393 | 379 | 363 | 341 | 312 | 273 | 230 | 190 | 158 | 132 | 112 | 95 | 82 | 72 |
| 113.5 × 113.5 | 4.8 | 159.2 | 2028 | 44 | 450 | 438 | 424 | 409 | 389 | 364 | 331 | 294 | 255 | 220 | 190 | 164 | 143 | 125 |
| | 5.4 | 177.4 | 2260 | 43.8 | 501 | 487 | 473 | 455 | 433 | 404 | 368 | 326 | 283 | 244 | 210 | 181 | 158 | 138 |
| 132 × 132 | 4.8 | 187.1 | 2383 | 51.6 | 530 | 518 | 506 | 492 | 475 | 454 | 428 | 397 | 360 | 321 | 284 | 250 | 220 | 195 |
| | 5.4 | 208.8 | 2659 | 51.3 | 592 | 578 | 564 | 548 | 529 | 506 | 477 | 441 | 399 | 356 | 314 | 276 | 243 | 215 |
| 180 × 180 | 5 | 269.7 | 3436 | 71.1 | 770 | 757 | 744 | 731 | 717 | 701 | 683 | 661 | 636 | 607 | 573 | 535 | 495 | 454 |
| | 6 | 320.5 | 4083 | 70.6 | 914 | 899 | 884 | 868 | 851 | 832 | 810 | 784 | 754 | 718 | 677 | 632 | 583 | 535 |
| | 7 | 370.3 | 4718 | 70.1 | 1056 | 1039 | 1021 | 1003 | 983 | 960 | 934 | 904 | 869 | 827 | 779 | 725 | 669 | 613 |
| 220 × 220 | 6 | 395.9 | 5043 | 87 | 1133 | 1118 | 1103 | 1087 | 1071 | 1054 | 1035 | 1015 | 991 | 965 | 934 | 899 | 860 | 817 |
| | 7 | 458.3 | 5838 | 86.5 | 1311 | 1294 | 1276 | 1258 | 1239 | 1219 | 1198 | 1173 | 1146 | 1115 | 1079 | 1038 | 992 | 941 |
| 250 × 250 | 6 | 452.4 | 5763 | 99.2 | 1297 | 1282 | 1266 | 1251 | 1236 | 1219 | 1202 | 1183 | 1162 | 1139 | 1114 | 1085 | 1052 | 1016 |
| | 7 | 524.2 | 6678 | 98.7 | 1503 | 1485 | 1467 | 1449 | 1431 | 1412 | 1392 | 1370 | 1345 | 1319 | 1289 | 1255 | 1217 | 1174 |

**TABLE F.7** Capacity of circular section compression members (Clause 7.1.2.1)

| Nominal bore (mm) | Outside diameter (mm) | Thick-ness (mm) | Weight (N/m) | Area of cross section (mm²) | Radius of gyration (mm) | Length (m) | | | | | | | | | | | | | |
|---|---|---|---|---|---|---|---|---|---|---|---|---|---|---|---|---|---|---|---|
| | | | | | | 0.5 | 1 | 1.5 | 2 | 2.5 | 3 | 3.5 | 4 | 4.5 | 5 | 5.5 | 6 | 6.5 | 7 |
| 15 | 21.3 | 2.65 | 12.2 | 155 | 6.7 | 26 | 11 | 5 | 3 | 2 | 1 | 1 | 1 | 1 | 0 | 0 | 0 | 0 | 0 |
| 20 | 26.9 | 2.65 | 15.8 | 202 | 8.6 | 39 | 21 | 11 | 6 | 4 | 3 | 2 | 2 | 1 | 1 | 1 | 1 | 1 | 1 |
| 25 | 33.7 | 3.25 | 24.6 | 311 | 10.8 | 62 | 44 | 24 | 15 | 10 | 7 | 5 | 4 | 3 | 3 | 2 | 2 | 1 | 1 |
| 32 | 42.4 | 3.25 | 31.5 | 400 | 13.9 | 83 | 70 | 47 | 30 | 20 | 14 | 10 | 8 | 6 | 5 | 4 | 4 | 3 | 3 |
| 40 | 48.3 | 3.25 | 36.1 | 460 | 16 | 97 | 86 | 64 | 43 | 29 | 21 | 16 | 12 | 10 | 8 | 7 | 6 | 5 | 4 |
| 50 | 60.3 | 3.65 | 51 | 650 | 20.1 | 139 | 129 | 111 | 85 | 61 | 45 | 34 | 27 | 21 | 17 | 15 | 12 | 11 | 9 |
| 65 | 76.1 | 3.65 | 65.3 | 831 | 25.6 | 181 | 171 | 158 | 138 | 111 | 86 | 67 | 53 | 43 | 35 | 29 | 25 | 21 | 19 |
| 80 | 88.9 | 4.05 | 84.8 | 1080 | 30 | 236 | 226 | 214 | 196 | 170 | 140 | 113 | 91 | 74 | 61 | 51 | 44 | 38 | 33 |
| 90 | 101.6 | 4.05 | 97.5 | 1240 | 34.5 | 273 | 263 | 252 | 237 | 215 | 188 | 158 | 131 | 108 | 90 | 76 | 65 | 56 | 49 |
| 100 | 114.3 | 4.5 | 121 | 1550 | 38.9 | 342 | 332 | 320 | 305 | 285 | 259 | 227 | 194 | 164 | 139 | 118 | 101 | 88 | 76 |
| 110 | 127 | 4.85 | 146 | 1860 | 43.2 | 412 | 401 | 388 | 374 | 355 | 331 | 300 | 264 | 229 | 196 | 169 | 146 | 127 | 111 |
| 125 | 139.7 | 4.85 | 162 | 2050 | 47.7 | 455 | 444 | 432 | 418 | 402 | 381 | 354 | 321 | 285 | 250 | 218 | 190 | 166 | 146 |
| 135 | 152.4 | 4.85 | 177 | 2250 | 52.2 | 501 | 490 | 478 | 465 | 449 | 430 | 407 | 378 | 344 | 307 | 272 | 240 | 212 | 187 |
| 150 | 165.1 | 4.85 | 192 | 2440 | 56.7 | 544 | 533 | 521 | 509 | 494 | 477 | 456 | 430 | 399 | 365 | 328 | 293 | 261 | 233 |
| 150 | 168.3 | 4.85 | 196 | 2490 | 57.8 | 556 | 544 | 533 | 520 | 506 | 489 | 469 | 443 | 413 | 379 | 343 | 307 | 275 | 245 |
| 175 | 193.7 | 5.4 | 250 | 3190 | 66.6 | 714 | 701 | 689 | 675 | 661 | 644 | 625 | 602 | 575 | 542 | 506 | 466 | 426 | 387 |
| 200 | 219.1 | 5.6 | 294 | 3760 | 75.5 | 843 | 830 | 817 | 803 | 789 | 773 | 755 | 735 | 711 | 683 | 651 | 615 | 575 | 533 |

**TABLE F.8** Capacity of *I*-section beams (Clause 8.2.2)

| Section | Moment capacity $M_p$(kNm) | Shear capacity $V_p$(kN) | Buckling resistance moment $M_d$ (kNm) for effective length $L_e$ (m) | | | | | | | | |
|---|---|---|---|---|---|---|---|---|---|---|---|
| | | | 2 | 3 | 4 | 5 | 6 | 7 | 8 | 9 | 10 |
| MB100 | 9.473 | 61.672 | 5.792 | 4.246 | 3.310 | 2.706 | 2.288 | 1.981 | 1.748 | 1.563 | 1.414 |
| MB125 | 18.602 | 82.010 | 13.196 | 10.019 | 7.884 | 6.467 | 5.475 | 4.747 | 4.189 | 3.749 | 3.394 |
| MB150 | 25.109 | 98.412 | 17.072 | 12.377 | 9.510 | 7.703 | 6.475 | 5.587 | 4.915 | 4.389 | 3.966 |
| MB175 | 37.745 | 133.184 | 27.431 | 20.324 | 15.661 | 12.681 | 10.651 | 9.185 | 8.078 | 7.212 | 6.515 |
| MB200 | 57.695 | 149.586 | 44.646 | 33.884 | 26.103 | 21.048 | 17.617 | 15.154 | 13.303 | 11.861 | 10.705 |
| MB225 | 79.152 | 191.903 | 65.387 | 52.983 | 42.205 | 34.508 | 29.075 | 25.103 | 22.089 | 19.726 | 17.825 |
| MB250 | 105.843 | 226.348 | 90.590 | 75.709 | 61.160 | 50.110 | 42.169 | 36.347 | 31.934 | 28.484 | 25.714 |
| MB300 | 148.123 | 303.109 | 129.896 | 110.593 | 89.711 | 73.022 | 60.959 | 52.188 | 45.610 | 40.518 | 36.465 |
| MB350 | 202.175 | 371.997 | 175.290 | 145.636 | 114.790 | 91.656 | 75.616 | 64.241 | 55.850 | 49.426 | 44.356 |
| MB400 | 267.314 | 467.129 | 231.557 | 191.771 | 150.487 | 119.756 | 98.572 | 83.612 | 72.607 | 64.203 | 57.579 |
| MB450 | 348.491 | 555.044 | 307.514 | 260.805 | 207.922 | 165.844 | 136.178 | 115.145 | 99.703 | 87.952 | 78.727 |
| MB500 | 471.516 | 669.201 | 429.080 | 379.051 | 313.549 | 252.260 | 205.892 | 172.451 | 147.967 | 129.504 | 115.165 |
| MB550 | 616.359 | 808.290 | 566.823 | 508.859 | 430.880 | 352.256 | 289.488 | 243.063 | 208.692 | 182.649 | 162.381 |
| MB600 | 797.870 | 944.755 | 717.170 | 659.588 | 580.164 | 489.117 | 407.324 | 343.003 | 294.137 | 256.756 | 227.604 |
| UB 203 × 133 × 25 | 57.537 | 151.980 | 50.915 | 43.531 | 35.130 | 28.294 | 23.387 | 19.866 | 17.258 | 15.261 | 13.686 |
| UB 203 × 133 × 30 | 70.409 | 173.667 | 62.895 | 55.140 | 46.237 | 38.433 | 32.451 | 27.964 | 24.540 | 21.860 | 19.712 |
| UB 254 × 146 × 31 | 88.032 | 197.926 | 79.139 | 68.737 | 55.827 | 44.617 | 36.451 | 30.630 | 26.374 | 23.157 | 20.649 |
| UB 254 × 146 × 37 | 108.525 | 211.625 | 98.376 | 87.311 | 73.679 | 60.998 | 51.099 | 43.698 | 38.104 | 33.771 | 30.330 |
| UB 254 × 146 × 43 | 127.411 | 245.258 | 116.092 | 104.461 | 90.408 | 76.785 | 65.556 | 56.805 | 49.997 | 44.614 | 40.273 |
| UB 305 × 165 × 40 | 139.452 | 238.866 | 128.422 | 115.440 | 97.830 | 79.901 | 65.530 | 54.904 | 47.051 | 41.114 | 36.503 |

(*Contd*)

**TABLE F.8** (*Contd*)

| Section | Moment capacity $M_p$(kNm) | Shear capacity $V_p$(kN) | Buckling resistance moment $M_d$ (kNm) for effective length $L_e$ (m) | | | | | | | | |
|---|---|---|---|---|---|---|---|---|---|---|---|
| | | | 2 | 3 | 4 | 5 | 6 | 7 | 8 | 9 | 10 |
| UB 305 × 165 × 46 | 161.491 | 269.546 | 149.153 | 135.148 | 116.614 | 97.349 | 81.296 | 69.042 | 59.774 | 52.643 | 47.028 |
| UB 305 × 165 × 54 | 190.134 | 321.763 | 176.148 | 160.967 | 141.533 | 120.981 | 103.080 | 88.859 | 77.781 | 69.069 | 62.094 |
| HEA320 | 351.038 | 366.093 | 345.540 | 334.814 | 323.010 | 309.635 | 294.429 | 277.535 | 259.567 | 241.435 | 223.996 |
| HEB320 | 469.481 | 482.875 | 462.578 | 449.140 | 435.108 | 420.107 | 403.942 | 386.629 | 368.413 | 349.728 | 331.092 |
| HEM 320 | 988.978 | 989.237 | 979.867 | 958.635 | 939.444 | 921.400 | 903.889 | 886.513 | 869.018 | 851.249 | 833.127 |
| UB 356 × 171 × 45 | 172.728 | 322.765 | 158.812 | 141.837 | 118.196 | 94.453 | 75.997 | 62.702 | 53.078 | 45.924 | 40.448 |
| UB 356 × 171 × 51 | 200.314 | 344.704 | 184.916 | 166.693 | 141.723 | 115.844 | 94.881 | 79.334 | 67.850 | 59.182 | 52.466 |
| UB 356 × 171 × 57 | 226.255 | 380.500 | 209.384 | 189.960 | 163.876 | 136.382 | 113.387 | 95.884 | 82.710 | 72.624 | 64.720 |
| UB 356 × 171 × 67 | 271.873 | 433.923 | 252.686 | 231.623 | 204.398 | 175.051 | 149.116 | 128.391 | 112.229 | 99.531 | 89.381 |
| UB 406 × 178 × 54 | 235.872 | 406.772 | 217.831 | 195.782 | 164.637 | 132.225 | 106.418 | 87.654 | 74.036 | 63.916 | 56.183 |
| UB 406 × 178 × 60 | 268.788 | 421.277 | 249.204 | 225.851 | 193.381 | 158.591 | 129.759 | 108.199 | 92.256 | 80.243 | 70.960 |
| UB 406 × 178 × 67 | 302.097 | 472.734 | 280.619 | 255.595 | 221.492 | 184.557 | 153.114 | 129.054 | 110.958 | 97.144 | 86.358 |
| UB 406 × 178 × 74 | 337.274 | 514.577 | 314.016 | 287.578 | 252.342 | 213.757 | 179.899 | 153.282 | 132.864 | 117.048 | 104.558 |
| IPEA360 | 195.762 | 309.691 | 180.750 | 162.879 | 138.252 | 112.686 | 92.027 | 76.757 | 65.512 | 57.047 | 50.504 |
| IPE360 | 285.562 | 377.902 | 259.050 | 231.872 | 199.383 | 168.580 | 143.612 | 124.323 | 109.379 | 97.586 | 88.086 |
| IPEO360 | 259.244 | 439.416 | 240.185 | 218.924 | 191.098 | 161.610 | 136.307 | 116.560 | 101.403 | 89.621 | 80.275 |
| IPEA400/NPB 400 × 180 × 57.4 | 244.335 | 364.649 | 227.241 | 206.660 | 177.666 | 145.802 | 118.964 | 98.802 | 83.907 | 72.719 | 64.107 |
| IPE 400/NPB 400 × 180 × 66.3 | 281.437 | 451.383 | 261.627 | 238.390 | 206.502 | 171.742 | 142.109 | 119.474 | 102.496 | 89.572 | 79.505 |
| IPEO 400/NPB 400 × 180 × 75.7 | 325.761 | 514.209 | 303.850 | 278.903 | 245.617 | 208.747 | 175.991 | 150.051 | 130.082 | 114.592 | 102.350 |
| UB 457 × 191 × 67 | 330.026 | 505.693 | 307.960 | 281.157 | 242.965 | 199.805 | 162.695 | 134.612 | 113.855 | 98.300 | 86.368 |
| UB 457 × 191 × 74 | 371.322 | 539.691 | 347.368 | 318.792 | 278.891 | 233.235 | 192.750 | 161.319 | 137.670 | 119.714 | 105.794 |
| UB 457 × 191 × 82 | 411.904 | 597.558 | 385.908 | 355.432 | 313.726 | 265.853 | 222.471 | 188.058 | 161.744 | 141.515 | 125.679 |
| UB 457 × 191 × 89 | 453.349 | 638.458 | 425.536 | 393.501 | 350.601 | 301.191 | 255.311 | 218.002 | 188.932 | 166.275 | 148.349 |
| UB 457 × 191 × 98 | 503.068 | 698.867 | 473.090 | 439.265 | 395.104 | 344.259 | 295.885 | 255.431 | 223.207 | 197.677 | 177.223 |
| IPEA450/NPB 450 × 190 × 67.2 | 321.913 | 445.767 | 301.335 | 276.388 | 240.947 | 199.970 | 163.754 | 135.887 | 115.115 | 99.478 | 87.450 |
| IPE 450/NPB 450 × 190 × 77.6 | 369.073 | 555.044 | 345.163 | 316.810 | 277.468 | 232.635 | 192.849 | 161.865 | 138.476 | 120.663 | 106.814 |
| IPEO 450/NPB 450 × 190 × 92.4 | 447.362 | 658.179 | 419.854 | 388.338 | 346.349 | 298.129 | 253.299 | 216.729 | 188.147 | 165.809 | 148.095 |
| IPEA500/NPB 500 × 200 × 79.4 | 422.546 | 547.800 | 397.865 | 367.775 | 325.160 | 273.787 | 225.965 | 188.000 | 159.284 | 137.531 | 120.761 |
| IPE 500/NPB 500 × 200 × 90.7 | 478.934 | 669.201 | 450.553 | 416.600 | 369.455 | 313.517 | 261.455 | 219.752 | 187.872 | 163.474 | 144.486 |
| IPEO 500/NPB 500 × 200 × 107.3 | 574.131 | 796.743 | 541.570 | 503.666 | 453.039 | 393.120 | 335.313 | 287.044 | 248.922 | 219.025 | 195.310 |
| UB 533 × 210 × 82 | 460.061 | 665.485 | 433.418 | 400.480 | 353.096 | 295.295 | 241.533 | 199.169 | 167.393 | 143.515 | 125.243 |
| UB 533 × 210 × 92 | 528.576 | 706.507 | 499.445 | 463.865 | 413.947 | 352.583 | 293.459 | 245.322 | 208.402 | 180.223 | 158.403 |
| UB 533 × 210 × 101 | 585.832 | 760.575 | 554.440 | 516.515 | 464.388 | 400.417 | 337.540 | 285.157 | 244.269 | 212.657 | 187.930 |
| UB 533 × 210 × 109 | 635.017 | 821.176 | 601.600 | 561.622 | 507.588 | 441.569 | 375.854 | 320.148 | 276.043 | 241.571 | 214.379 |

(*Contd*)

**TABLE F.8** *(Contd)*

| Section | Moment capacity $M_p$(kNm) | Shear capacity $V_p$(kN) | Buckling resistance moment $M_d$ (kNm) for effective length $L_e$ (m) | | | | | | | | |
|---|---|---|---|---|---|---|---|---|---|---|---|
| | | | 2 | 3 | 4 | 5 | 6 | 7 | 8 | 9 | 10 |
| UB 533 × 210 × 122 | 718.464 | 907.378 | 681.946 | 638.948 | 582.506 | 514.149 | 444.684 | 383.951 | 334.583 | 295.223 | 263.700 |
| UB 610 × 229 × 101 | 645.876 | 830.243 | 614.443 | 574.726 | 518.677 | 446.130 | 371.585 | 308.560 | 259.583 | 222.178 | 193.360 |
| UB 610 × 229 × 113 | 736.712 | 884.968 | 702.478 | 659.412 | 600.104 | 523.660 | 443.151 | 372.939 | 317.098 | 273.760 | 239.973 |
| UB 610 × 229 × 125.1 | 826.501 | 955.932 | 789.365 | 742.933 | 680.421 | 600.601 | 515.262 | 438.936 | 376.914 | 328.011 | 289.421 |
| UB 610 × 229 × 139.9 | 932.455 | 1060.923 | 891.812 | 841.432 | 775.270 | 691.963 | 601.952 | 519.437 | 450.783 | 395.642 | 351.507 |
| IPEA600/NPB 600 × 220 × 107.6 | 682.930 | 767.692 | 649.805 | 608.476 | 551.200 | 478.021 | 402.573 | 337.980 | 287.146 | 247.900 | 217.377 |
| IPE 600/NPB 600 × 220 × 122.4 | 767.298 | 944.755 | 729.114 | 682.307 | 618.297 | 538.013 | 456.145 | 386.045 | 330.562 | 287.412 | 253.599 |
| IPEO 600/NPB 600 × 220 × 154.5 | 985.171 | 1200.626 | 939.500 | 884.801 | 814.008 | 727.185 | 635.624 | 552.540 | 483.304 | 427.301 | 382.082 |
| HEAA 600 / WPB 600 × 300 × 128.8 | 785.881 | 899.092 | 768.552 | 739.592 | 704.276 | 659.270 | 603.120 | 539.327 | 475.488 | 417.853 | 368.762 |
| HEA600/WPB 600 × 300 × 177.8 | 1178.455 | 1006.426 | 1156.580 | 1117.432 | 1072.511 | 1019.154 | 956.079 | 884.886 | 810.314 | 737.982 | 671.792 |
| HEB 600/WPB 600 × 300 × 211.9 | 1422.716 | 1220.309 | 1396.995 | 1351.225 | 1300.204 | 1241.674 | 1174.676 | 1100.424 | 1022.437 | 945.224 | 872.474 |
| HEM 600/WPB 600 × 300 × 285 | 1956.114 | 1708.432 | 1924.662 | 1867.074 | 1806.600 | 1741.597 | 1671.276 | 1595.877 | 1516.727 | 1435.976 | 1356.027 |
| HEAA 700/WPB 700 × 300 × 149.9 | 1055.662 | 1142.891 | 1030.819 | 990.553 | 940.736 | 876.359 | 795.669 | 705.015 | 616.212 | 537.701 | 471.887 |
| HEA700/WPB 700 × 300 × 204.5 | 1553.772 | 1312.816 | 1522.742 | 1469.153 | 1406.540 | 1330.683 | 1239.696 | 1136.856 | 1030.580 | 929.744 | 839.547 |
| HEB 700/WPB 700 × 300 × 240.5 | 1848.161 | 1561.470 | 1812.108 | 1750.114 | 1679.484 | 1596.501 | 1499.698 | 1391.743 | 1279.440 | 1170.573 | 1070.545 |
| HEM 700/WPB 700 × 300 × 300.7 | 2350.855 | 1972.963 | 2308.919 | 2235.044 | 2154.469 | 2064.323 | 1963.454 | 1853.196 | 1737.469 | 1621.562 | 1510.281 |
| HEAA 800/WPB 800 × 300 × 172 | 1351.465 | 1414.508 | 1317.266 | 1263.609 | 1196.131 | 1107.653 | 996.560 | 873.837 | 756.697 | 655.474 | 571.995 |
| HEA800/WPB 800 × 300 × 224 | 1913.894 | 1554.909 | 1873.312 | 1804.952 | 1723.528 | 1622.715 | 1499.858 | 1360.893 | 1219.502 | 1088.456 | 973.910 |
| HEB 800/WPB 800 × 300 × 262 | 2261.445 | 1837.024 | 2214.411 | 2135.511 | 2043.559 | 1932.751 | 1800.918 | 1653.217 | 1501.565 | 1358.050 | 1229.614 |
| HEM 800/WPB 800 × 300 × 317 | 2774.852 | 2243.006 | 2720.809 | 2628.695 | 2524.975 | 2404.825 | 2266.487 | 2113.358 | 1953.921 | 1798.146 | 1653.409 |
| HEAA 900/WPB 900 × 300 × 198 | 1746.222 | 1712.368 | 1700.188 | 1629.254 | 1539.287 | 1420.540 | 1271.687 | 1109.159 | 956.308 | 825.769 | 718.951 |
| HEA900/WPB 900 × 300 × 252 | 2385.364 | 1868.515 | 2332.190 | 2244.678 | 2139.174 | 2006.900 | 1844.607 | 1661.876 | 1478.586 | 1311.536 | 1167.614 |
| HEB 900/WPB 900 × 300 × 291 | 2788.332 | 2184.746 | 2727.263 | 2627.103 | 2508.681 | 2363.762 | 2189.599 | 1994.761 | 1797.282 | 1613.739 | 1452.291 |
| HEM 900/WPB 900 × 300 × 333 | 3445.381 | 3223.976 | 3365.226 | 3240.920 | 3097.075 | 2926.227 | 2727.255 | 2509.314 | 2289.156 | 2082.207 | 1896.813 |

Kulkarni and Jirage (2011) also developed graphs and charts for the design of steel angle tension members. Similar charts were prepared by Pasnur and Patil (2013)for steel beams. Kulkarni and Patil(2011) also developed design aids for flexure members and beam-columns made of channel sections in the form of graphs and some of these are included here with the permission of the authors (refer to the paper for ISJC and ISLC beams).

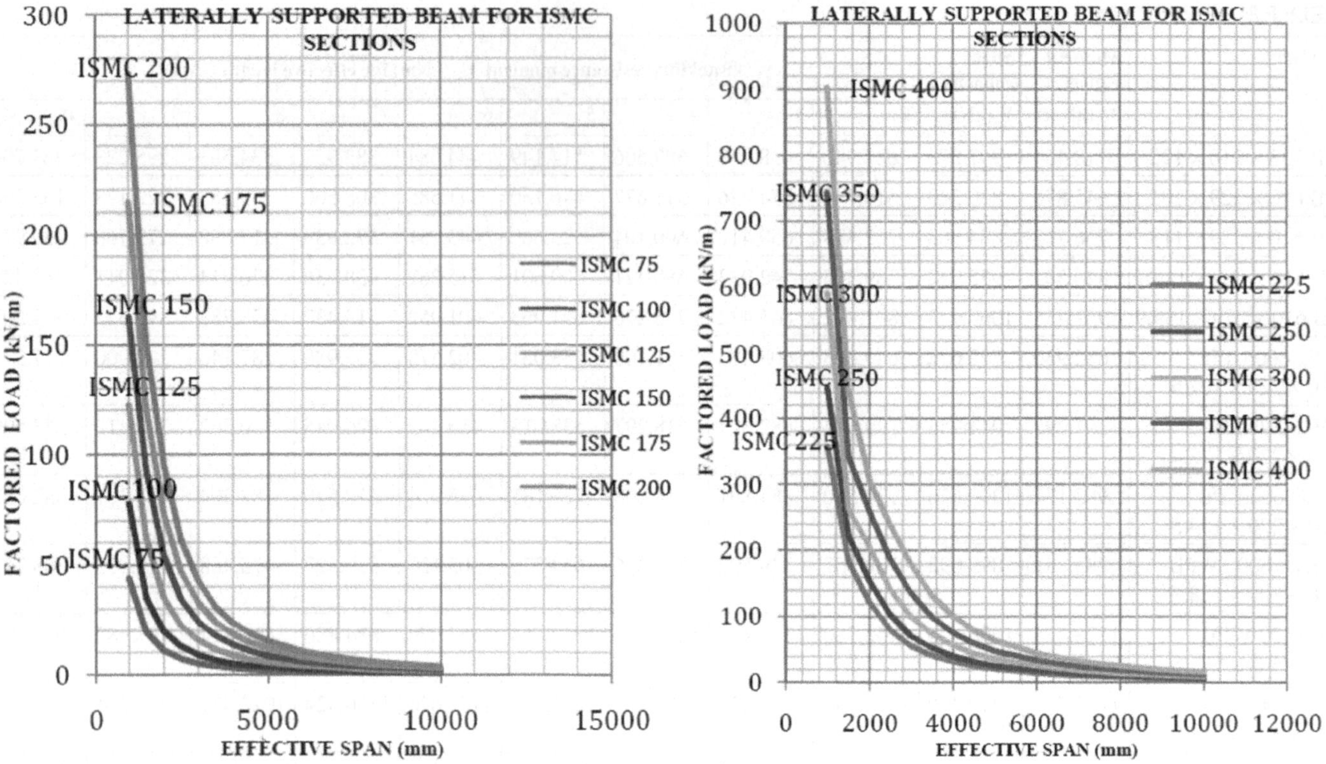


**FIG. F.1** Factored load vs effective span for laterally supported ISMC beams

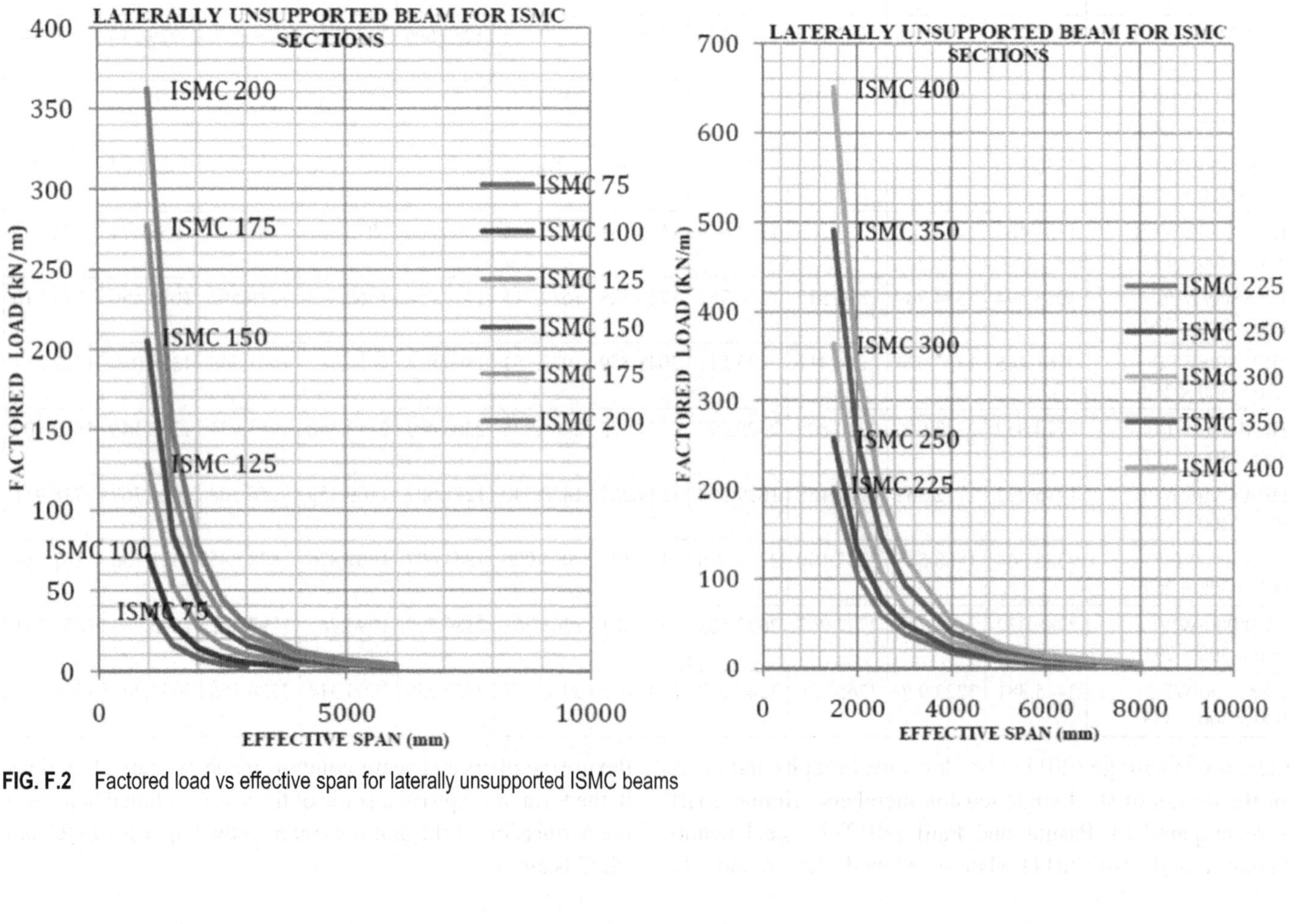


**FIG. F.2** Factored load vs effective span for laterally unsupported ISMC beams

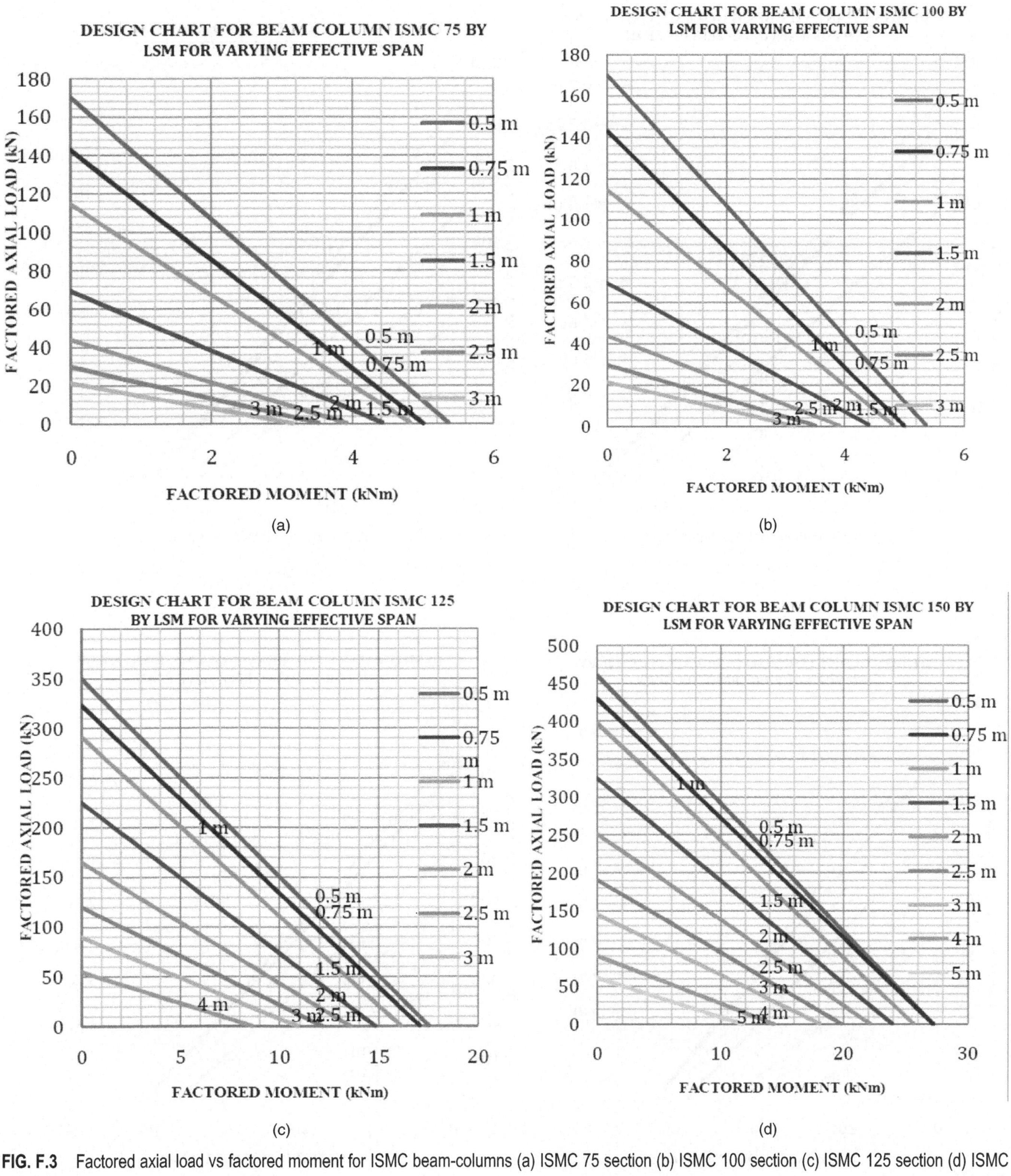


**FIG. F.3** Factored axial load vs factored moment for ISMC beam-columns (a) ISMC 75 section (b) ISMC 100 section (c) ISMC 125 section (d) ISMC 150 section *(Contd)*

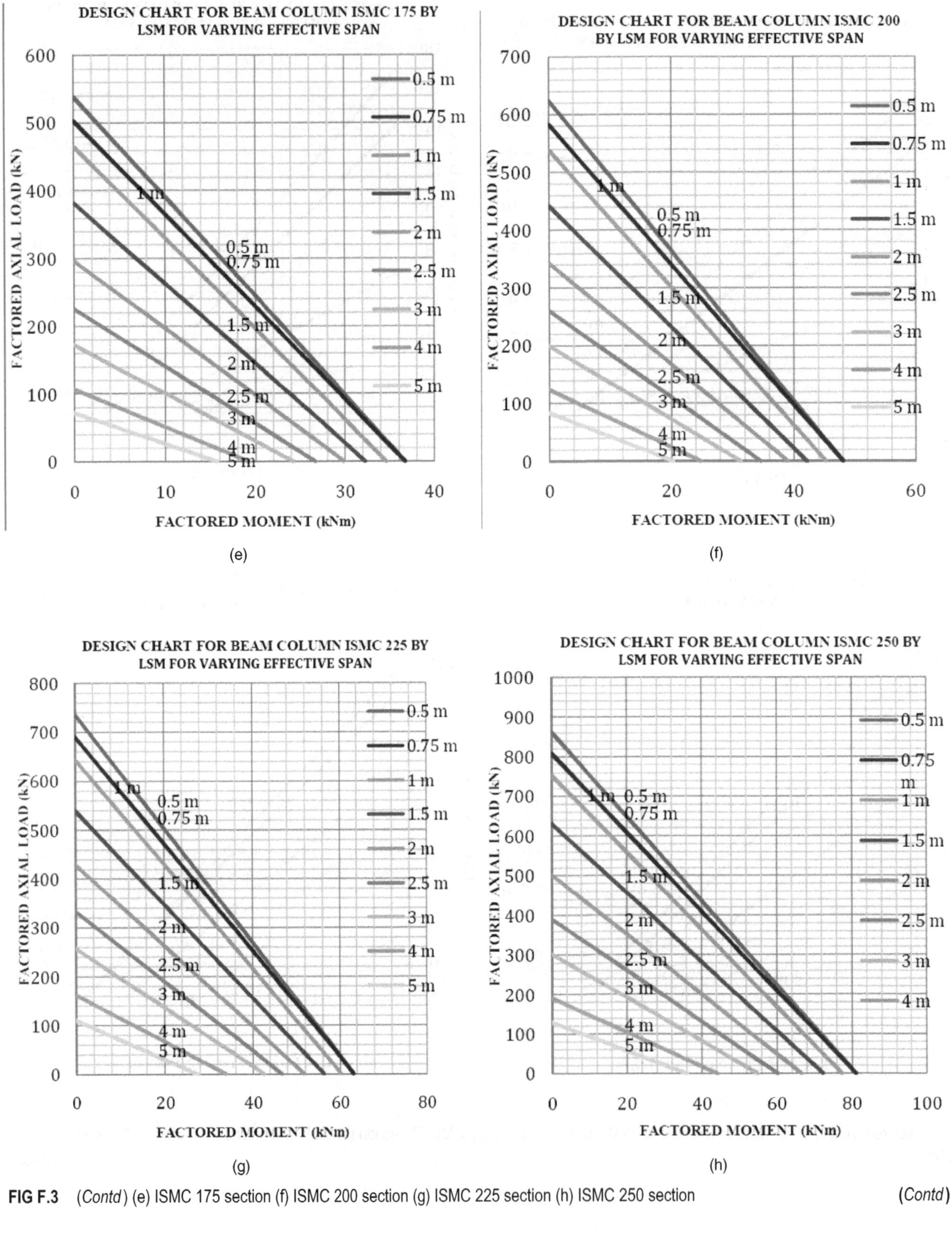


**FIG F.3** (*Contd*) (e) ISMC 175 section (f) ISMC 200 section (g) ISMC 225 section (h) ISMC 250 section (*Contd*)

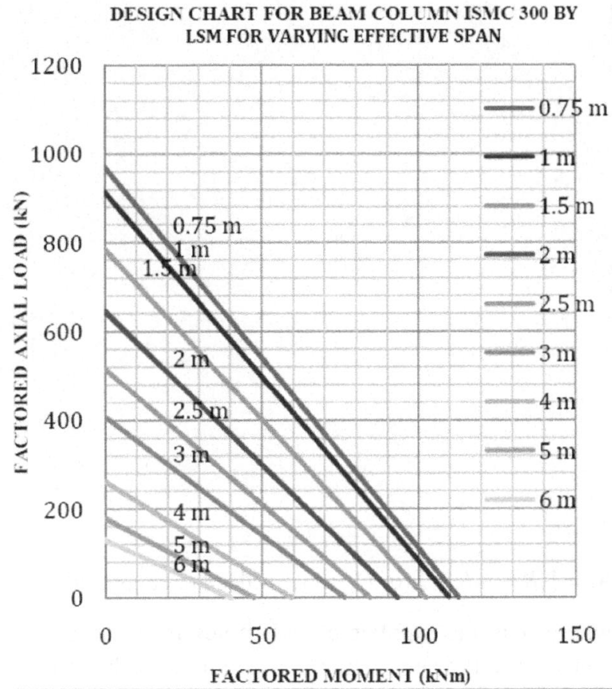


(i)

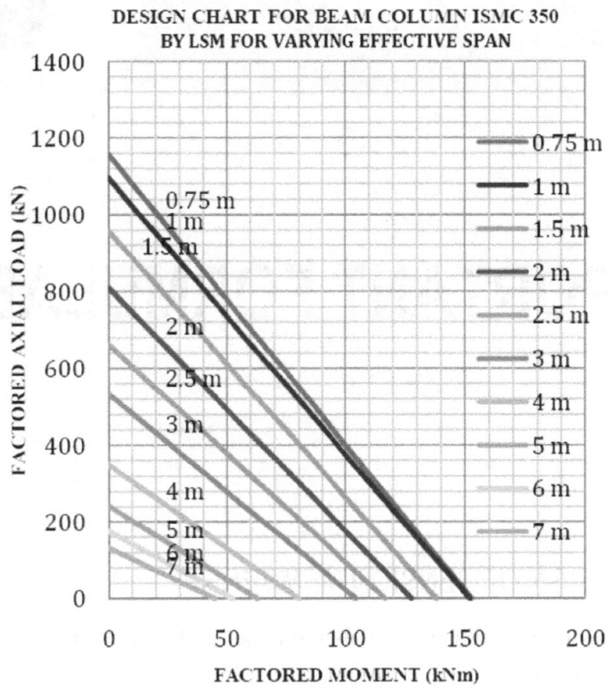


(j)

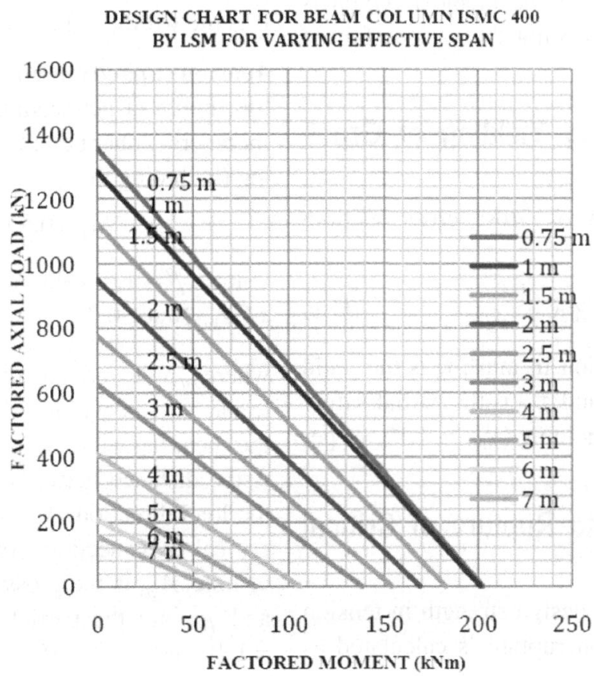


(k)

**FIG F.3** (*Contd*) (i) ISMC 300 section (j) ISMC 350 section (k) ISMC 400 section

# REFERENCES

Kulkarni, R.B. and Jirage, R.S., 'Comparative Study of Steel Angles as Tension Members Designed by Working Stress Method and Limit State Method', *International Journal of Scientific & Engineering Research*, Vol. 2, No. 11, Nov 2011, pp. 1–7.

Kulkarni, R.B. and Patil, V.A., 'Design Aids of Flexural Members and Beam-columns Based on Limit State Method', *International Journal of Scientific & Engineering Research*, Vol. 2, No. 10, Oct 2011, pp. 1–16.

Pasnur, L.A. and Patil, S.S., 'Comparative Study of Beam using IS: 800-1984 & IS: 800-2007', *International Journal of Engineering and Innovative Technology*, Vol. 2, No. 10, Apr 2013, pp. 241–243.

APPENDIX G

# IMPORTANT FORMULAE

## G.1 DESIGN OF TENSION MEMBERS

When designing a tension member, the factored design tension $T$, in the member should satisfy the following

$$T < T_d$$

Where, $T_d$ is the design strength of the member which is calculated as below talking into account the various failure modes.

### G.1.1 Design Strength due to Yielding of Gross Section

Design strength due to yielding of gross cross section, calculated as follows

$$T_{dg} = f_y A_g / \gamma_{m0} \quad \text{(G.1)}$$

where $A_g$ gross area of cross section in mm$^2$, $f_y$ is the yield strength of the material (in MPa) and $\gamma_{m0}$ is the partial safety factor for failure in tension by yielding.($\gamma_{m0}$ = 1.10 as per IS: 800).

### G.1.2 Design Strength due to Rupture of Critical Section

**Tension capacity of plates** The design strength in tension of a plate governed by net section rupture is calculated as follows:

$$T_{dn} = 0.9 f_u A_n / \gamma_{m1} \quad \text{(G.2)}$$

where $f_u$ is the ultimate strength of the material in MPa, $\gamma_{m1}$ = partial safety factor for failure due to rupture of cross-section ($\gamma_{m1}$ = 1.25 as per IS: 800), and $A_n$ is the net effective area of the cross section in mm$^2$ and is given as follows:

The net area, $A_n$, can be easily computed as

$$A_n = (B - nd_h)t \text{ (if the holes are not staggered)} \quad \text{(G.3a)}$$

$$A_n = [(B - nd_h + \Sigma(p^2/4g)]t \text{ (if holes are staggered)} \quad \text{(G.3b)}$$

where $n$ is the number of bolt holes in the critical section considered, the summation is over all the paths of the critical sections normal to the direction of tensile force, $d_h$ is the diameter of bolt hole, $p$ is the staggered pitch length between line of bolt holes, $g$ is the gauge length between the bolt holes, $B$ is the width, and $t$ is the thickness of member.

**Tension capacity of angles** The design strength as governed by rupture at net section of single and double angle tension members (either bolted or welded) and connected through one leg (including the shear lag effect) is given by

$$T_{dn} = 0.9 f_u A_{nc} / \gamma_{m1} + \beta A_{go} f_y / \gamma_{m0} \quad \text{(G.4)}$$

where, $\beta = 1.4 - 0.076\,[(b_s/L_c)\,(w/t)\,(f_y/f_u)] \le (0.9 f_u \gamma_{m0}) / (f_y \gamma_{m1})$ but $\ge 0.7$

where $f_u$ and $f_y$ are the ultimate and yield stress of the material, $w$ and $t$ are the width and thickness of the outstanding leg, $b_s$ is the shear lag width, $L_c$ is the length of the end connection, that is the distance between the outermost bolts measured along the load direction, $A_{nc}$ is the net area of the connected leg at critical cross section, computed after deducting the diameter of hole, $A_{go}$ is the gross area of the outstanding leg, and $\gamma_{m0}$ and $\gamma_{m1}$ are the partial safety factors for the material ($\gamma_{m0}$ = 1.10 and $\gamma_{m1}$ = 1.25).

### G.1.3 Design Strength due to Block Shear

The block shear strength $T_{db}$ of connection is taken as the smaller of

$$T_{db1} = A_{vg} f_y / (\sqrt{3}\gamma_{m0}) + 0.9 f_u A_{tn} / \gamma_{m1} \quad \text{(G.5a)}$$

$$T_{db2} = 0.9 f_u A_{vn} / (\sqrt{3}\,\gamma_{m1}) + f_y A_{tg} / \gamma_{m0} \quad \text{(G.5b)}$$

Where, $A_{vg}$, $A_{vn}$ = minimum gross and net area in shear along bolt line parallel to external force respectively,

$A_{tg}$, $A_{tn}$ = minimum gross and net area in tension from the bolt hole to the toe of the angle, perpendicular to the line of force respectively, and $f_u, f_y$ = ultimate and yield stress of the material respectively.

## G.2 DESIGN OF COMPRESSION MEMBERS

When designing a compression member, the factored design compression $P$, in the member should satisfy the following

$$P < P_d$$

where, $P_d = A_e f_{cd}$
where, $T_d$ is the design compression strength of the member, $A_e$ = effective sectional area.

The design stress in compression, $f_{cd}$ is given by

$$f_{cd} = [(f_y/\gamma_{m0})/\{\phi + (\phi^2 - \bar{\lambda}^2)^{0.5}\}] = \chi f_y/\gamma_{mo} \le f_y/\gamma_{m0} \quad \text{(G.6)}$$

where $\phi = 0.5[1 + \alpha(\bar{\lambda} - 0.2) + \bar{\lambda}^2]$, $\bar{\lambda}$ is a non-dimensional effective slenderness ratio = $\sqrt{(f_y/f_{cr})}$, $f_{cr}$ is the Euler buckling stress = $\pi^2 E / (KL/r)^2$, $KL/r$ is the effective slenderness ratio, $\alpha$ is the imperfection factor (See Table G.1), $\chi$ is the stress reduction factor = $1/[\phi + (\phi^2 - \lambda^2)^{0.5}]$, and $\gamma_{m0}$ is the partial safety factor for material strength = 1.10.

**TABLE G.1** Imperfection factor $\alpha$

| Buckling curve | a | b | c | D |
|---|---|---|---|---|
| $\alpha$ | 0.21 | 0.34 | 0.49 | 0.76 |

### *Effective Length of Columns in Frames*

The effective length factor $K$, based on Wood's curves is as follows:

For non-sway frames:

$$K = \frac{[1 + 0.145(\beta_1 + \beta_2) - 0.265\beta_1\beta_2]}{[2 - 0.364(\beta_1 + \beta_2) - 0.247\beta_1\beta_2]} \quad \text{(G.7a)}$$

For sway frames (Moment-resisting frames)

$$K = \left\{\frac{[1 - 0.2(\beta_1 + \beta_2) - 0.12\beta_1\beta_2]}{[1 - 0.8(\beta_1 + \beta_2) + 0.6\beta_1\beta_2]}\right\}^{0.5} \quad \text{(G.7b)}$$

Where, $$\beta_i = \Sigma K_c/(\Sigma K_c + K_b) \quad \text{(G.7c)}$$

Where, $K_c$ and $K_b$ = effective flexural stiffness of the columns or beams meeting at the joint at the ends of the columns and rigidly connected at the joints.

$$K_c \text{ or } K_b = C(I/L)$$

Where, $I$ is the moment of inertia about an axis perpendicular to the plan of the frame, $L$ is the length of the member, taken as center to center distance of the intersecting member, and $C$ is a correction factor (see Table 35 of the code).

### *Single Angle Struts Loaded through One Leg*

When the single angle is loaded eccentrically through one of its legs, the flexural torsional buckling strength may be evaluated using an equivalent slenderness ratio, $\lambda_e$ given by

$$\lambda_e = \sqrt{[k_1 + k_2\lambda_{vv}^2 + k_3\lambda_\phi^2]} \quad \text{(G.8a)}$$

where, $k_1$, $k_2$, and $k_3$ are constants depending on end conditions and given in Table 12 of the code.

$$\lambda_{vv} = (L / r_{vv})/(\varepsilon\sqrt{(\pi^2 E / 250)}) \quad \text{(G.8b)}$$

and $$\lambda_\phi = (b_1 + b_2)/(\varepsilon\sqrt{(\pi^2 E / 250)} \times 2t) \quad \text{(G.8c)}$$

Where, $L$ = unsupported length of the member, $r_{vv}$ = radius of gyration about the minor axis, $b_1$, $b_2$ = width of the two legs of the angle, t = thickness of the leg, and $\varepsilon$ = yield stress ratio $(250 / f_y)^{0.5}$

### *Minimum Thickness of Rectangular Slab Base*

Minimum thickness of rectangular slab base, $t_s$, is calculated as

$$t_s = [2.5\, w\,(a^2 - 0.3b^2)\,\gamma_{mo} / f_y]^{0.5} > t_f \quad \text{(G.9)}$$

where, $w$ = pressure on the underside of the slab base due to the factored compressive load on the column (assumed as uniformly distributed over the area of the slab base), $a$, $b$ = larger and smaller projection of the slab base beyond the rectangle circumscribing the column, respectively, $f_y$ = yield strength of the base plate, $t_f$ = flange thickness of compression member, and $\gamma_{mo}$ = partial safety factor for material = 1.10.

## G.3 DESIGN OF BEAMS

The factored design moment, $M$ at any section in a beam due to external actions should satisfy the relationship

$$M \le M_d,$$

where, $M_d$ is the design bending strength of the section.

### G.3.1 Laterally Supported Beams
### Shear Less Than $0.6V_d$

When the design shear force, $V$ is less than $0.6V_d$, the design bending strength, $M_d$, is given by

$$M_d = \beta_b Z_p f_y / \gamma_{m0} \le 1.2\, Z_e f_y / \gamma_{m0}, \quad \text{(G.10)}$$

where $\beta_b$ =1.0 for plastic and compact section, $\beta_b = Z_e/Z_p$ for semi-compact section, $Z_e$ and $Z_p$ are the elastic and plastic section modulus of the cross section respectively, and $\gamma_{m0}$ is the partial safety factor for material strength = 1.10.

### *Shear Greater Than $0.6V_d$*

When the design shear force, $V$ is greater than $0.6\ V_d$, the design bending strength, $M_d$ is calculated as follows:

**Plastic or compact section**

$$M_{\text{dv}} = M_d - \beta(M_d - M_{\text{fd}}) \leq 1.2z_e f_y/\gamma_{\text{m0}}, \text{ where} \quad (G.11)$$

$$\beta = (2V/V_d - 1)^2 \quad (G.12)$$

$M_d$ = design plastic moment of the whole section disregarding high shear force effect, but considering web buckling effects, $V$ = factored applied shear force, $V_d$ = design shear strength as governed by web yielding or web buckling = $0.6Dt_w f_y$, $M_{\text{fd}}$ = plastic design strength of the area of the cross section excluding the shear area, considering partial safety factor $\gamma_{\text{m0}}$.

**Semi-compact section**

$$M_{\text{dv}} = M_d(1-\beta) \leq \frac{Z_e f_y}{\gamma_{\text{m0}}} \quad (G.13)$$

where, $Z_e$ = elastic section modulus of the whole section.

### Design for Shear

The factored design shear force, $V$, in a beam due to external actions should satisfy

$$V \leq V_d, \quad (G.14a)$$

where, $V_d$, the design strength is given by

$$V_d = V_n/\gamma_{\text{m0}} \quad (G.14b)$$

The nominal shear strength of a cross section, $V_n$, may be governed by plastic shear resistance or strength of the web governed by shear buckling.

The nominal plastic shear resistance under pure shear is given by $V_n = V_p$, where

$$V_p = \frac{A_v f_{\text{yw}}}{\sqrt{3}} \quad (G.15)$$

$A_v$ = shear area (equals $ht_w$ for hot-rolled sections and $dt_w$ for welded sections, with $h$ and $d$ are overall depth and clear depth of the web between flanges respectively), and $f_{\text{yw}}$ = yield strength of the web.

## G.3.2 Laterally Unsupported Beams

The design bending strength of a laterally unsupported beam as governed by lateral torsional buckling is given by:

$$M_d = \beta_b Z_p f_{\text{bd}} \quad (G.16)$$

Where, $\beta_b$ = 1.0 for plastic and compact sections and equals $Z_e/Z_p$ for semi-compact section, $Z_p$, $Z_e$ = plastic section modulus and elastic section modulus with respect to extreme compression fibre, and $f_{\text{bd}}$ = design bending compressive stress.

The design bending compressive stress is given by,

$$f_{\text{bd}} = \chi_{\text{LT}} f_y/\gamma_{\text{m0}} \quad (G.17a)$$

where, $\chi_{\text{LT}}$ = reduction factor to account for lateral torsional buckling given by:

$$\chi_{\text{LT}} = \frac{1}{\phi_{\text{LT}} + \left[\phi_{\text{LT}}^2 - \lambda_{\text{LT}}^2\right]^{0.5}} \leq 1.0 \quad (G.17b)$$

in which $\phi_{\text{LT}} = 0.5\left[1 + \alpha_{\text{LT}}(\lambda_{\text{LT}} - 0.2) + \lambda_{\text{LT}}^2\right]$

The values of imperfection factor, $\alpha_{\text{LT}}$, for lateral torsional buckling of beams is given by:

$\alpha_{\text{LT}}$ = 0.21 for rolled section, $\alpha_{\text{LT}}$ = 0.49 for welded section

$\gamma_{\text{m0}}$ = partial safety factor for material = 1.10

The non-dimensional slenderness ratio, $\lambda_{\text{LT}}$, is given by

$$\lambda_{\text{LT}} = \sqrt{\beta_b Z_p f_y / M_{\text{cr}}} \quad (G.18)$$

$$= \sqrt{f_y / f_{\text{cr},b}}$$

Where $M_{\text{cr}}$ is the elastic critical moment, and $f_{\text{cr},b}$ is the extreme fibre compressive elastic lateral buckling stress, which may be calculated using

$$f_{\text{cr},b} = (1473.5/(KL/r_y)^2\left[1 + \frac{1}{20}\left(\frac{KL/r_y}{h/t_f}\right)^2\right]^{0.5} \quad (G.19)$$

The elastic critical moment corresponding lateral torsional buckling of a doubly symmetric prismatic beam subjected to uniform moment in the unsupported length and torsionally restraining lateral supports is given by

$$M_{\text{cr}} = C_1 \frac{\pi^2 EI_y}{(KL)^2}\left[\frac{I_w}{I_y} + \frac{GI_t(KL)^2}{\pi^2 EI_y}\right]^{0.5} \quad (G.20)$$

where

$I_y$, $I_w$, $I_t$ = moment of inertia about the minor axis, warping constant and St. Venant's torsion constant of the cross section, respectively

$G$ = Shear modulus of rigidity

$KL$ = effective length between points of restraint against lateral torsional buckling

$C_1$ = equivalent uniform moment factor, which may be obtained from Table 42 of IS 800.

Clause E-1.2 of Annex E of the code contains the equation for calculating the elastic critical moment of a section which is symmetrical only about the minor axis.

### Shear Buckling of Webs

The shear stress at which buckling of a perfect plate takes place is given by

$$\tau_{\text{cr},v} = \frac{k_v \pi^2 E}{12(1-\mu^2)(d/t_w)^2} \quad (G.21)$$

where, $\tau_{\text{cr},v}$ is the elastic critical shear buckling stress of the web and $k$ is the buckling coefficient. For a plate with all four edges simply supported, $k_v = 5.35$. When stiffeners are provided only at supports

$$k_v = 4 + \frac{5.35}{(L/d)^2} \quad L/d \leq 1 \tag{G.22a}$$

$$= 5.35 + \frac{4}{(L/d)^2} \quad L/d \geq 1 \tag{G.22b}$$

where $(L/d)$ is the aspect ratio of the plate, $L$ = Length/spacing between transverse stiffeners and $d$ is the depth of the web.

### Web Crippling and Web Buckling Strengths

The *web buckling strength* at the support is given by

$$P_{wb} = (b_1 + n_1) t_w f_c \tag{G.23}$$

Where, $t_w$ is the web thickness, $b_1$ is the length of bearing, $n_1$ is assumed as half the depth of beam, and $f_c$ is the allowable compressive stress corresponding to the assumed *web* strut, which is taken from Table 9c of the code for an $L/r$ ratio equal to $2.5\, d/t_w$, where, d is the depth of the *web* in between the flanges fillets.

The *crippling strength* of the web (also called as *web bearing capacity*) at supports is calculated as

$$P_{wb} = (b_1 + n_2) t_w f_{yw} / \gamma_{m0} \tag{G.24}$$

Where, $f_{yw}$ is the design yield strength of the web, $n_2$ is taken as $2.5(t_f + R)$, and $R$ is the radius of flange fillet. At an interior point where concentrated load is acting, the crippling strength is given by,

$$P_{crip} = (b_1 + 2n_2) t / \gamma_{m0} \tag{G.25}$$

If the above bearing capacity or crippling strength of the beam web is extended, stiffeners must be provided to carry the load.

## G.4 DESIGN OF PURLINS

Purlins are checked for the following interaction equations:

1. Check for local capacity by using the interaction equation

$$(M_z/M_{dz}) + (M_y / M_{dy}) \leq 1.0$$

Where $M_{dz} = Z_{pz} f_y / \gamma_{mo} \leq 1.2\, Z_{ez} f_y / \gamma_{mo}$

$$M_{dy} = Z_{py} f_y / \gamma_{mo} \leq \gamma_f Z_{ez} f_y / \gamma_{mo}$$

Where $M_z$ and $M_y$ are the applied bending moments about the major and minor axis respectively and $\gamma_f$ is the partial safety factor for load.

2. Under wind section (combined with dead load), the bottom flange of the purlin, which is laterally unsupported will be under compression. Hence under this loading case, the lateral-torsional buckling capacity of the section has to be calculated using Eqn.G.16 and an overall member buckling check is to be made by using the interaction equation.

$$(M_z/M_{z,cr}) + (M_y/M_{dy}) \leq 1.0$$

where $M_{z,cr}$ is the lateral-torsional buckling strength of the member.

In addition the deflection is checked to see whether they are under permissible limits (Table 6 of the code); shear capacity will always be adequate.

## G.5 DESIGN OF PLATE GIRDERS

The thickness of the web in a section shall satisfy the following requirements with respect to serviceability:

(a) When transverse stiffeners are not provided:

$\frac{d}{t_w} \leq 200\varepsilon$ (web connection to flanges along both longitudinal edges)

$\frac{d}{t_w} \leq 90\varepsilon$ (web connection to flanges along one longitudinal edge only)

(b) When only transverse stiffeners are provided:

(i) When $3d \geq c \geq d$

$$\frac{d}{t_w} \leq 200\varepsilon_w$$

(ii) When $0.74\,d < c < d$

$$\frac{c}{t_w} \leq 200\varepsilon_w$$

(iii) When $c < 0.74\,d$

$$\frac{d}{t_w} \leq 270\varepsilon_w$$

When $c > 3d$ the web is considered as unstiffened.

(c) When transverse stiffeners and longitudinal stiffeners at one level only are provided, at 0.2 d from compression flange:

(i) When $2.4d \geq c > d$

$$\frac{d}{t_w} \leq 250\varepsilon_w$$

(ii) When $0.74\,d < c \leq d$

$$\frac{c}{t_w} \leq 250\varepsilon_w$$

(iii) When $c < 0.74\,d$

$$\frac{d}{t_w} \leq 340\varepsilon_w$$

(d) When there is a second longitudinal stiffener (provided at neutral axis):

$$\frac{d}{t_w} \leq 400\varepsilon_w$$

These criteria are to ensure that the web will not buckle under normal service conditions.

Similarly, the code specifies web thickness requirement in order to avoid buckling of the compression flange into the web as given below:

(a) When transverse stiffeners are not provided:

$$\frac{d}{t_w} \leq 345\varepsilon_f^2$$

(b) When transverse stiffeners are provided and

(i) When $c \geq 1.5d$:

$$\frac{d}{t_w} \leq 345\varepsilon_f^2$$

(ii) When $c < 1.5d$:

$$\frac{d}{t_w} \leq 345\varepsilon_f$$

where, $d$ = depth of the web, $t_w$ = thickness of the web, $c$ = spacing of transverse stiffener, $\varepsilon_w = \sqrt{250/f_{yw}}$, and $f_{yw}$ = yield stress of the web.

### Shear Buckling Design

Resistance to shear buckling shall be verified when

$$d/t_w > 67\varepsilon \text{ for an unstiffened web}$$

$$> 67\varepsilon\sqrt{\frac{k_v}{5.35}} \text{ for a stiffened web}$$

Where, $k_v$ = shear buckling coefficient (see Eq. G.22)

### Simple Post-critical Method

The nominal shear strength is given by:

$$V_n = V_{cr} \quad \text{(G.26a)}$$

where $V_{cr}$ = shear force corresponding to web buckling

$$V_{cr} = A_v \tau_b \quad \text{(G.26b)}$$

where $A_v$ = shear area and $\tau_b$ = shear stress corresponding to web buckling, determined as follows:

(a) When $\lambda_w \leq 0.8$ $\quad \tau_b = f_{yw}/\sqrt{3}$ (G.27a)

(b) When $0.8 < \lambda_w < 1.2$ $\quad \tau_b = \left[1 - 0.8(\lambda_w - 0.8)\right]\left(f_{yw}/\sqrt{3}\right)$ (G.27b)

(c) When $\lambda_w \geq 1.2$ $\quad \tau_b = f_{yw}/\left(\sqrt{3}\,\lambda_w^2\right)$ (G.27c)

where, $\lambda_w$ = non –dimensional web slenderness ratio for shear buckling stress, given by

$$\lambda_w = \sqrt{f_{yw}/\left(\sqrt{3}\,\tau_{cr,e}\right)} \quad \text{(G.28)}$$

The elastic critical shear stress of the web, $\tau_{cr,e}$ is given by Eqn.G.21.

### Tension Field Method

In the tension field method, the nominal shear resistance, $V_n$, is obtained from $V_n = V_{tf}$

where,

$$V_{tf} = \left[A_v \tau_b + 0.9\, w_{tf}\, t_w\, f_v \sin\phi\right] \leq V_p \quad \text{(G.29)}$$

Where

$\tau_b$ = buckling strength, as obtained by Eqn G.27, $A_v$ = shear area

$f_v$ = yield strength of the tension field obtained from

$$f_v = \left[f_{yw}{}^2 - 3\tau_b^2 + \psi^2\right]^{0.5} - \psi \quad \text{(G.30)}$$

where

$\psi = 1.5\, \tau_b \sin 2\phi$

$\phi$ = inclination of the tension field = $\tan^{-1}\left(\frac{d}{c}\right)$

The width of the tension field, $w_{tf}$, is given by:

$w_{tf} = d\cos\phi - (c - s_c - s_t)\sin\phi$

$f_{yw}$ = yield stress of the web, $d$ = depth of the web, $c$ = spacing of stiffeners in the web, $\tau_b$ = shear stress corresponding to buckling of web as per Eqn.G.27, and $s_c$, $s_t$ = anchorage lengths of tension field along the compression and tension flange respectively, obtained from:

$$s = \frac{2}{\sin\phi}\left[\frac{M_{fr}}{f_y\, t_w}\right]^{0.5} \leq c \quad \text{(G.31)}$$

where

$M_{fr}$ = reduced plastic moment of the respective flange plate (disregarding any edge stiffener) after accounting for the axial force, $N_f$ in the flange, due to overall bending and any external axial force in the cross section, as given below:

$$M_{fr} = 0.25 b_f t_f^2 f_{yf}\left[1 - \left(N_f / \left(b_f t_f f_{yf}/\gamma_{m0}\right)\right)^2\right] \quad \text{(G.32)}$$

Where $b_f$, $t_f$ = width and thickness of the relevant flange respectively.

### End Panel Design Without Tension Field Action

The anchor field forces developed in the panel next to end panel creates a resultant longitudinal shear, $R_{tf}$, plus a moment $M_{tf}$ on the end panel. They are evaluated as shown below.

$$R_{tf} = \frac{H_q}{2} \quad \text{and} \quad M_{tf} = \frac{H_q d}{10} \quad \text{(G.33a)}$$

where $$H_q = 1.25 V_{dp}\left(1 - \frac{V_{cr}}{V_{dp}}\right)^{1/2} \quad \text{(G.33b)}$$

and $$V_{dp} = \frac{d t_w f_y}{\sqrt{3}} \tag{G.33c}$$

If the actual factored shear force, $V$, in the panel designed using tension field approach is less than the nominal shear strength, $V_{nt}$, as determined earlier, then the values of $H_q$ may be reduced by the ratio

$$\frac{V - V_{cr}}{V_{tf} - V_{cr}} \tag{G.33d}$$

Where $d$ = web depth, $V_{tf}$ = the basic shear strength for the panel utilizing tension field action Eqn.G.29, and $V_{cr}$ = critical shear strength for the panel designed without utilizing tension field action (see Eqn.G.26b).

## G.6 DESIGN OF STIFFENER

The outstand of the stiffener from the face of the web is restricted to a value of $20 t_q \varepsilon$. If the outstand value is anywhere between $14 t_q \varepsilon$ and $20 t_q \varepsilon$, a core value of $14 t_q \varepsilon$ is chosen as the outstand value. In the above expressions $t_q$ refers to thickness of the stiffener plate. The effective width of web acting together with the stiffener is taken as 20 times the web thickness on each side of the centre line of the stiffener, i.e., $40 t_w$, for interior stiffeners and as $20 t_w$ for end stiffeners.

The minimum requirement for intermediate transverse stiffener, when they are subjected to external loads or moments is considered in terms of a second moment of area, $I_s$, about the centerline of the web.

$$\text{if } \frac{c}{d} \geq \sqrt{2};\ I_s \geq 0.75 d t_w^3 \tag{G.34a}$$

$$\text{if } \frac{c}{d} < \sqrt{2};\ I_s \geq \frac{1.5 d^3 t_w^3}{c^2} \tag{G.34b}$$

where, $d$ = depth of the web, $t_w$ = minimum thickness required for tension field action, and $c$ = stiffener spacing.

The bearing strength of the stiffener $F_{bsd}$ is obtained as follows

$$F_{bsd} = \frac{A_q f_{yq}}{0.8 \gamma_{m0}} > F_x \tag{G.35}$$

where, $A_q$ = Area of the stiffener in contact with the flange, mm$^2$, $F_{yq}$ = yield stress of the stiffener, N/mm$^2$.

Transverse web stiffeners are also checked for buckling resistance for a stiffener force

$$F_q = V - \frac{V_{cr}}{\gamma_{m0}} \leq F_{qd} \tag{G.36}$$

Where $F_{qd}$ = Design buckling resistance of the intermediate stiffener, $V$ = Factored shear force adjacent to the stiffener and $V_{cr}$ = Shear buckling resistance of the web panel without considering tension field action (Eqn.G.26b).

Transverse web stiffeners that are subjected to external load and moments have to satisfy requirements of load carrying stiffeners. They have to also meet the interaction equation

$$\frac{F_q - F_x}{F_{qd}} + \frac{F_x}{F_{xd}} + \frac{M_q}{M_{yq}} \leq 1 \tag{G.37}$$

If $F_q < F_x$, then $F_q - F_x = 0$

Where, $F_q$ = stiffener force (Eqn.G.36), $F_{qd}$ = design buckling resistance of intermediate web stiffener about an axis parallel to the web, $F_x$ = External load or reaction at the stiffener, $F_{xd}$ = Design resistance of load carrying stiffener, buckling about an axis parallel to the web, $M_q$ = moment at the stiffener due to eccentric load, $M_{yq}$ = Yield moment capacity of the stiffener about an axis parallel to the web.

Second moment of area of the torsional stiffener section about the centre line of the web, $I_s$ should satisfy the following relation

$$I_s \geq 0.34 \alpha_s D^3 t_{cf} \tag{G.38}$$

Where $\alpha_s = 0.006$ for $KL/r_y \leq 50$

$\alpha_s = 0.3/KL/r_y$ for $50 < KL/r_y \leq 100$

$\alpha_s = 30/(KL/r_y)^2$ for $KL/r_y > 100$

$D$ = depth of the beam at support, $t_{cf}$ = Maximum thickness of the compression flange of the girder, $KL$ = Effective length of the laterally unsupported compression flange of the girder, and $r_y$ = radius of gyration about minor axis.

## G.7 DESIGN OF MEMBERS SUBJECTED TO COMBINED FORCES

### G.7.1 Combined Axial Force and Bending Moment

**Local capacity check** For beam-columns subjected to combined axial force (tension or compression) and bending moment, the following interaction equation should be satisfied.

$$\left(\frac{M_y}{M_{ndy}}\right)^{\alpha_1} + \left(\frac{M_z}{M_{ndz}}\right)^{\alpha_2} \leq 1.0 \tag{G.39}$$

Where $M_y$, $M_z$ = factored applied moments about the minor and major axis of the cross- section, respectively, $M_{ndy}$,$M_{ndz}$ = design reduced flexural strength under combined axial force and the respective uniaxial moment acting alone.

The value of $M_{ndy}$ and $M_{ndz}$ for plastic and compact section is approximately given in Table G.2.

**TABLE G.2** Approximate value of reduced flexural strength for plastic and compact sections

| Element | Equation |
|---|---|
| Plates | $M_{nd} = M_d(1 - n^2)$ |

*(Contd)*

**TABLE G.2** *(Contd)*

| Element | Equation |
|---|---|
| Welded I- or H-sections | $M_{ndz} = M_{dz}(1-n)/(1-0.5a) \le M_{dz}$<br>$M_{ndy} = M_{dy}\left[1-\left(\frac{n-a}{1-a}\right)^2\right] \le M_{dy}$<br>$a = (A - 2bt_f)/A \le 0.5$ |
| Rolled I- or H-sections without bolt holes | $M_{ndz} = 1.11 M_{dz}(1-n) \le M_{dz}$<br>$M_{ndy} = M_{dy}$ for $n \le 0.2$<br>$M_{ndy} = 1.56 M_{dy}(1-n)(n+0.6)$ |
| Rectangular hollow sections and welded box sections | $M_{ndz} = M_{dz}(1-n)/(1-0.5a_w) \le M_{dz}$<br>$M_{ndy} = M_{dy}(1-n)/(1-0.5a_f) \le M_{dy}$<br>where $a_w = (A - 2bt_f)/A \le 0.5$<br>$a_f = (A - 2ht_w)/A \le 0.5$ |
| Circular hollow tubes without bolt holes | $M_{nd} = 1.04 M_d(1-n^{1.7}) \le M_d$ |

$n = N/N_d$

$N$ = Factored applied axial force (Tension $T$ or compression $P$)

$N_d$ = Design strength in tension $T_d$, or compression $P$ (see Chapter 9)
$= A_g f_y/\gamma_{mo}$ where $\gamma_{mo} = 1.1$

$M_{dy}$, $M_{dz}$ = Design strength under corresponding moment acting alone (see chapter 10)

$A_g$ = Gross area of cross section

The constants $\alpha_1$ and $\alpha_2$ are given in Table G.3.

**TABLE G.3** Constants $\alpha_1$ and $\alpha_2$

| Section | $\alpha_1$ | $\alpha_2$ |
|---|---|---|
| I and channel | $5n \ge 1$ | 2 |
| Circular tubes | 2 | 2 |
| Rectangular tubes | $1.66/(1-1.13n^2) \le 6$ | $1.66/(1-1.13n^2) \le 6$ |
| Solid rectangles | $1.73 + 1.8n^3$ | $1.73 + 1.8n^3$ |

$N = N/N_d$

For semi compact sections without bolt holes, the code (IS: 800) suggests the following linear equation, when shear force is low

$$(N/N_d) + (M_y/M_{dy}) + (M_z/M_{dz}) \le 1.0 \quad \text{(G.40)}$$

**Overall buckling check** The interaction equation for overall buckling check is given by the code as

$$\frac{P}{P_{dy}} + K_y\frac{C_{my}M_y}{M_{dy}} + K_{LT}\frac{M_z}{M_{dz}} \le 1.0 \quad \text{(G.41a)}$$

$$\frac{P}{P_{dz}} + 0.6K_y\frac{C_{my}M_y}{M_{dy}} + K_z\frac{C_{mz}M_z}{M_{dz}} \le 1.0 \quad \text{(G.41b)}$$

where

$C_{my}$, $C_{mz}$, $C_{mLT}$ = Equivalent uniform moment factors obtained from Table 13.3, which depends on the shape of the bending moment diagram between lateral bracing points in the appropriate plane of bending

$P$ = Factored applied axial compressive load.

$P_{dy}$, $P_{dz}$ = Design compressive strength under axial compression as governed by buckling about minor and major axis respectively (See Section 9.7.1 of Chapter 9).

$M_y$, $M_z$ = Maximum factored applied bending moments about minor and major axis of the member, respectively.

$M_{dy}$, $M_{dz}$ = Design bending strength about minor and major axis considering laterally unsupported length of the cross-section (see Section 10.8 of Chapter 10).

$K_y$, $K_z$, $K_{LT}$ = Interaction factors as defined below

$$K_y = 1 + (\lambda_y - 0.2)n_y \le 1 + 0.8n_y \quad \text{(G.41c)}$$

$$K_z = 1 + (\lambda_z - 0.2)n_z \le 1 + 0.8\,n_z \quad \text{(G.41d)}$$

$$K_{LT} = 1 - \frac{0.1\lambda_{LT}n_y}{(C_{mLT} - 0.25)} \ge 1 - \frac{0.1n_y}{(C_{mLT} - 0.25)} \quad \text{(G.41e)}$$

Where

$n_y$, $n_z$ = Ratio of actual applied axial force to the design axial stength for buckling about minor and major axis respectively $= (P/P_{dy})$ or $(P/P_{dz})$

$C_{mLT}$ = equivalent uniform moment factor for lateral-torsional buckling as per Table 13.3, which depends on the shape of the bending moment diagram between lateral bracing points

$\lambda_y$, $\lambda_z$ = Non-dimensional slenderness ratio about the minor and major axis respectively, For example

$\lambda_y = (f_y/f_{cr})^{0.5}$, where $f_{cr} = \pi^2 E/(KL/r)^2$ (See Section 9.7.1)

$\lambda_{LT}$ = Non-dimensional slenderness ratio in lateral buckling $= (f_y/f_{cr,b})^{0.5}$ and $f_{cr,b}$ is the extreme fibre bending compressive stress corresponding to elastic lateral buckling moment which may be determined as per Table 14 of the code.

### G.7.2 Beam-columns Subjected to Tension and Bending

**Local capacity check** The following simplified interaction equation is specified in the code for the beam-column subjected to combined axial force and bending moment.

$$(N/N_d) + (M_y/M_{dy}) + (M_z/M_{dz}) \le 1.0 \quad \text{(G.42)}$$

where

$M_{dy}$, $M_{dz}$ = design reduced flexural strength under combined axial force and the respective uniaxial moment acting alone

$M_y$, $M_z$ = factored applied moments along minor and major axis of the cross section, respectively

$N_d$ = design strength in tension obtained from section 6 of the code

$N$ = factored applied axial tensile force.

**Overall buckling check** The code stipulates that the member should be checked for lateral-torsional buckling under reduced effective moment $M_{eff}$ due to tension and bending. The reduced effective moment is given by the code as

$$M_{eff} = [M - \psi T Z_{ec}/A] \le M_d \quad \text{(G.43)}$$

where
$M,T$ = factored applied moment and tension, respectively
$A$ = area of cross section
$Z_{ec}$ = elastic section modulus of the section with respect to extreme compression fibre.
$\psi$ = 0.8 if T and $M$ vary independently
= 1.0 otherwise
$M_d$ = bending strength due to lateral-torsional buckling.

## G.8 DESIGN OF BOLTED CONNECTIONS

We can have bearing type bolt or friction grip type bolt.

### G.8.1 Bearing Type Bolt

#### *Shear Capacity*

The factored shear force $V_{sb}$ should satisfy

$$V_{sb} \le V_{nsb}/\gamma_{mb} \quad (G.44)$$

where $\gamma_{mb}$ = partial safety factor for material of the bolt, taken as 1.25 in the code. The nominal capacity, $V_{nsb}$, of a bolt in shear is calculated as

$$V_{nsb} = (f_u/\sqrt{3})(n_n A_{nb} + n_s A_{sb})\beta_{lj}\,\beta_{lg}\,\beta_{pk} \quad (G.45)$$

where $n_n$ = number of shear planes with threads intercepting the shear plane, $n_s$ = number of shear planes without threads intercepting the shear plane, $\beta_{lj}$ = reduction factor for long connections = 1.075 − $l_j/(200\text{d})$ with $0.75 \le \beta_{lj} \le 1.0$, $\beta_{lg}$ = reduction factor for large grip length = $8d/(3d + l_g) \le \beta_{lj}$, and $\beta_{pk}$ = reduction factor to account for packing plates in excess of 6 mm = $(1 - 0.0125\,t_{pk})$.

#### *Tension Capacity*

The factored tension force $T_b$ shall satisfy

$$T_b \le T_{nb}/\gamma_{mb} \quad (G.46)$$

where $\gamma_{mb}$ = partial safety factor for material resistance of connection = 1.25

The nominal capacity of a bolt in tension, $T_{nb}$, is calculated as

$$T_{nb} = 0.90\,f_{ub}\,A_n < f_{yb}\,A_{sb}\,(\gamma_{mb}/\gamma_{m0}) \quad (G.47)$$

where $A_{sb}$ = shank area of bolt, $f_{yb}$ = yield stress of the bolt, $\gamma_{mb}$ = partial safety factor for bolt material resistance governed by ultimate stress =1.25, $\gamma_{m0}$ = partial safety factor for material resistance governed by yielding =1.10.

#### *Bearing Capacity*

The design bearing strength of bolt on any plate $V_{db}$, shall satisfy

$$V_{db} \le V_{npb}/\gamma_{mb} \quad (G.48)$$

where $\gamma_{mb}$ = partial safety factor for material resistance of connection = 1.25

The nominal bearing strength of the bolt $V_{npb}$ is calculated as

$$V_{npb} = 2.5\,k_b d t f_u \quad (G.49)$$

where $f_u$ = the ultimate tensile stress of the plate in MPa, $d$ = nominal diameter of the bolt in mm, $t$ = summation of the thicknesses of the connected plates experiencing bearing stress in the same direction, and $k_b$ is smaller of $e/(3d_o)$, $p/(3d_o) - 0.25$, $f_{ub}/f_u$ and 1.0, where $f_{ub}$ is the ultimate tensile stress of the bolt, e is the edge distance, $p$ is the pitch of the fastener along bearing direction, and $d_o$ is the diameter of the bolt hole. $V_{npb}$ should be multiplied by a factor 0.7 for over size or short slotted holes and by 0.5 for long slotted holes.

#### *Bolts Subjected to Combined Shear and Tension*

Where bolts are subjected to both tension and shear, the following circular interaction curve should be satisfied.

$$\left(\frac{V}{V_{nd}}\right)^2 + \left(\frac{T_e}{T_{nd}}\right)^2 \le 1.0 \quad (G.50)$$

where $V$ = applied shear, $V_{nd}$ = design shear capacity, $T_e$ = externally applied tension, $T_{nd}$ = design tension capacity.

### G.8.2 Friction Grip Type Bolt

#### *Slip Resistance*

The factored design force $V_{sf}$, should satisfy

$$V_{sf} \le V_{nsf}/\gamma_{mf} \quad (G.51)$$

Where $\gamma_{mf}$ = 1.10 if slip resistance is designed at service load, $\gamma_{mf}$ = 1.25 if slip resistance is designed at ultimate load. The nominal shear capacity of a bolt, $V_{nsf}$, is calculated as

$$V_{nsf} = \mu_f\,n_e\,K_h\,F_o\beta_{lj} \quad (G.52)$$

Where, $\mu_f$ = coefficient of friction (called as slip factor) as per Table 20 of code with $\mu_f \le 0.55$, $n_e$ = number of effective interfaces offering frictional resistance to slip, $K_h$ = 1.0 for fasteners in clearance holes; = 0.85 for fasteners in oversized and short slotted holes, and for fasteners in long slotted holes loaded perpendicular to the slot; = 0.7 for fasteners in long slotted holes loaded parallel to the slot, $F_o$ = minimum bolt tension (proof load) at installation = $A_{nb}f_o$, $A_{nb}$= nominal shank area of bolt, $f_o$ = proof stress = 0.7 $f_{ub}$, $f_{ub}$ = ultimate tensile stress of bolt, $\beta_{lj}$ = reduction factor for long connections = 1.075 − $l_j/(200\text{ d})$ with $0.75 \le \beta_{lj} \le 1.0$.

***Capacity after slipping*** The design capacity at ultimate load may be calculated as per bearing type bolts.

#### *Tension Resistance*

The HSFG bolt subject to a factored tension force, $T_f$, should satisfy the following

$$T_f \le T_{nf}/\gamma_{mb} \quad (G.53)$$

where $\gamma_{mb}$ = partial safety factor for the material of the bolt = 1.25, The nominal design tensile strength of bolts is similar to that of black bolts (Eq. G.47).

***Consideration of prying force*** When prying force, $Q$, is significant it is calculated as below and added to the tension in the bolt.

$$Q = \{l_v/2\, l_e\}\,[T_e - \beta\gamma f_o\, b_e\, t^4/(27 l_e\, l_v^2)] \qquad (G.54)$$

where $l_v$ = distance from the bolt center line to the toe of the fillet weld or to half the root radius of a rolled section, $l_e$ = distance between prying force and bolt center line, it taken as the maximum of {end distance or $l_e = 1.1t\sqrt{(\beta f_0/f_y)}$ }, $\beta$ = 2 for non-tensioned bolt and 1 for pre-tensioned bolt, $\gamma$ = 1.5, $b_e$ = effective width of flange per pair of bolts, $f_o$ = proof stress in consistent units, $t$ = thickness of the end plate.

### HSFG Bolts Subjected to Combined Shear and Tension

Where HSFG bolts are subjected to both tension and shear, the circular interaction curve as given below should be used

$$\left(\frac{V}{V_{nd}}\right)^2 + \left(\frac{T_e}{T_{nd}}\right)^2 \leq 1.0 \qquad (G.55)$$

where $V$ = applied shear, $V_{nd}$ = design shear capacity, $T_e$ = externally applied tension, and $T_{nd}$ = design tension capacity.

### Eccentrically Loaded Connection

When an eccentric load acts on a bolt group, producing twisting moment, the force $R$ on the maximum stressed bolt at the distance $d_i$ from the center of rotation is

$$R = \frac{Md_i}{\Sigma d_i^2} \qquad (G.56)$$

If $R_x$ and $R_y$ are the horizontal and vertical components of $R$,

$$R_x = \frac{My}{\sum(x^2+y^2)} \quad \text{and} \quad R_y = \frac{Mx}{\sum(x^2+y^2)} \qquad (G.57)$$

The $x$ and $y$ coordinates should reflect the positive and negative values of the bolt location as appropriate.

The direct shear force $R_v = \frac{P}{n}$, where $n$ = number of fasteners in the group.

The total resultant force $R$ is given by

$$R = \sqrt{(R_y + R_v)^2 + R_x^2} \qquad (G.58)$$

If the bolt group is subjected to applied moment or torque, the required number of bolts is given by

$$n = \sqrt{\frac{6M}{pmR}} \qquad (G.59)$$

Where $p$ is the pitch, $M$ is the applied moment, $R$ is the design shear strength of a single bolt and $m$ is the number of rows of bolts (e.g. if two vertical rows of bolts are provided, use $m = 2$)

## G.9 DESIGN OF WELDED JOINTS

The design strength of fillet weld is given by

$$P_{dw} = L_w\, t_e\, f_u/(\sqrt{3}\,\gamma_{mw})\, \beta_{lw} > \text{Factored load} \qquad (G.60)$$

Where $P_{dw}$ = Design strength of fillet weld, $L_w$ is the length of the weld, $t_e$ is the effective throat thickness = 0.7s, $s$ is the size of weld, $\gamma_{mw}$ = 1.25 (shop welding) = 1.5 (site welding), and $\beta_{lw}$ is the reduction factor for long connections and equals 1.0 when the $L_w \leq 150\, t_1$, and equals $1.2 - 0.2\, l_j/(150\, t_1) \leq 1.0$, when $L_w \leq 150\, t_1$.

When subjected to combined tensile and shear stresses, the equivalent stress, $f_e$, should satisfy

$$f_e = \sqrt{(f_a + 3q^2)} \leq \frac{f_u}{\sqrt{3}\gamma_{mw}}$$

Where $f_a$ = normal stress due to axial force or bending moment, and $q$ = shear stress due to shear force or tension.

When the weld group is subjected to twisting moment, the force components and maximum shear force for critical weld point are determined as

$$F_x^T = Ty/I_p \quad \text{and} \quad F_y^T = Tx/I_p \qquad (G.61)$$

Where $x$ and $y$ are the $X$ and $Y$ coordinates of critical weld point and $I_p$ = polar moment of inertia of weld line about the centroid

$$F_x^P = P_x/L_w \quad \text{and} \quad F_y^P = P_y/L_w \qquad (G.62)$$

where $L_w$ = total length of the weld

The resultant shear force is calculated and checked as

$$F_R = [(F_x^P + F_x^T)^2 + (F_y^P + F_y^T)^2]^{0.5} \leq \text{Weld strength}$$

### Weld Subjected to Combined Bearing, Bearing, and Shear

When a weld is subjected to combined bearing, bearing, and shear, the equivalent stress, $f_e$, should be checked as below

$$f_e = \sqrt{f_b^2 + f_{br}^2 - f_b f_{br} + 3q^2} \leq f_y/\gamma_{mw}$$

Where $f_e$ is the equivalent stress, $f_b$ is the stress due to bending, $f_{br}$ is the stress due to bearing and $q$ is the shear stress.

APPENDIX

# H

# CONVERSION FACTORS

## H.1 CONVERSION FACTORS

Some conversion factors useful in structural design, especially when the designer is using the American standard codes (which are still in FPS units), are provided in this appendix.

| Quantity | To convert | To | Multiply by |
|---|---|---|---|
| Length | inch (in.) | mm | 25.4 |
| | foot (ft) | m | 0.3048 |
| | metre (m) | ft | 3.2808 |
| | mile | km | 1.609 |
| Area | in.$^2$ | mm$^2$ | 645.16 |
| | ft$^2$ | m$^2$ | 0.0929 |
| | m$^2$ | ft$^2$ | 10.764 |
| Volume | in.$^3$ | mm$^3$ | 16,387 |
| | ft$^3$ | m$^3$ | 0.02832 |
| | m$^3$ | ft$^3$ | 35.315 |
| | gallon | litre | 3.7854 |
| | litre | gallon | 0.2642 |
| Mass per unit volume | lb/ft$^3$ | kg/m$^3$ | 16.0185 |
| | kg/m$^3$ | lb/ft$^3$ | 0.062428 |
| Force | kilo pound (kip) | kN | 4.448 |
| | lb | N | 4.448 |
| | ton(2000lb) | kN | 8.896 |
| | N | lb | 0.2248 |
| | kN | Kip | 0.2248 |
| Pressure,stress | psi | MPa | 0.006895 |
| | ksi | MPa | 6.895 |
| | kN/m$^2$ | kip/ft$^2$ | 0.02089 |
| | psf | N/m$^2$ | 47.88 |
| | N/m$^2$ | psf | 0.02089 |
| | MPa | ksi | 0.145 |
| | MPa | psi | 145.0 |
| Moments | in.-lb | Nm | 0.1130 |
| | kip-in. | kNm | 0.1130 |
| | kip-ft | kNm | 1.3558 |
| | kNm | ft-kip | 0.7376 |
| Uniform loading | kip/ft | kN/m | 14.59 |
| | kip/in. | kN/m | 175.2 |
| | kN/m | kip/ft | 0.06852 |
| Speed | mile/h | m/s | 0.447 |
| Acceleration | ft/s$^2$ | m/s$^2$ | 0.3048 |
| Density | lb/in.$^3$<br>lb/ft$^3$ | kg/m$^3$<br>kg/m$^3$ | 27,680<br>16.02 |
| Temperature | Degree fahrenheit (°F) | Degree celsius (°C) | (t°–32)/1.8 |
| Inertia | in.$^4$ | mm$^4$ | 416,231 |
| Energy | ft-lb | Joule (Nm/9.81) | 1.356 |

(*Contd*)

1 Pa = 1N/m$^2$, 1 MPa = $10^6$ Pa = 1 N/mm$^2$, g = 32.17 ft/s$^2$ = 9.807 m/s$^2$, 1 erg = $10^{-7}$ J, 1 Hz(Hertz) = 1 cycle/s, °K(Kelvin) = °C + 273

## H.2 BASIC SI UNITS RELATING TO STRUCTURAL STEEL DESIGN

| Quantity | Unit | Symbol |
|---|---|---|
| Length | metre | m |
| Mass | kilogram | kg |
| Time | second | s |

## H.3 DERIVED SI UNITS RELATING TO STRUCTURAL DESIGN

The SI unit of force is Newton (N). 1 Newton is the force which causes a mass of 1 kg to have an acceleration of 1 m/s$^2$. The acceleration due to gravity is 9.807 m/s$^2$ approximately, and hence the weight of a mass of 1 kg is 9.807 N.

| Quantity | Unit | Symbol | Formula |
|---|---|---|---|
| Force | Newton | N | kgm/s$^2$ |
| Pressure, Stress | Pascal | Pa | N/m$^2$ |
| Energy or work | Joule | J | Nm |

# GLOSSARY

## A

**Alloy steel** – Refers to steels made with deliberate additions of one or more alloying elements during steelmaking to enhance the properties of the steel. The most common alloying elements are Mn, Cr, Mo, and Ni.

**Accidental loads** – Loads that act occasionally and are of short duration, such as those due to impact of vehicles, explosion, fire, etc.

**Accompanying load** – Live (imposed) load acting along with leading imposed load, but causing lower actions and/or deflections

**Action effect or load effect** – Internal force, axial, shear, bending, or twisting moment, due to external actions and temperature loads

**Actions** – The primary cause for stress or deformations in a structure such as dead, live, wind, seismic, or temperature loads

**Actual length** – Length between the centre-to-centre of intersection points, with supporting members or the cantilever length in the case of a free standing member

**Allowable stress design (ASD)** – A method of proportioning structural components in which the stresses in members computed under the action of all appropriate combinations of working (service) loads, do not exceed the specified allowable stresses

**Anchor bolts** – Bolts that connect steel column base plate to the concrete foundation.

**Annealing** – Heat treatment which is used to impart changes in mechanical properties such as strength and hardness. Typically, the annealing process is used to induce softness and to relieve stress, allowing the steel to be further worked. In this process steel is heated to a temperature of about 910°C for a sufficient amount of time and then cooled slowly.

**Austenitic** – This is the largest category of stainless steel, accounting for about 70% of all production. The austenitic class offers the most resistance to corrosion in the stainless group, owing to its substantial nickel content and higher levels of chromium.

## B

**Base plate** – A steel plate welded to the base of a steel column, which distributes the loads applied on the column over a larger area of foundation concrete, in such a way that the bearing capacity of concrete is not exceeded. It is usually anchored into the concrete foundation by means of anchor bolts.

**Beam** – A straight or curved structural member, primarily supporting loads, applied at right angles to the longitudinal axis/moments. These are members subjected primarily to bending and shear actions. Beams may be simply supported, continuous, cantilevered, or have fixed ends.

**Beam-column** – A beam that carries compressive axial force in addition to bending moments; that is, it carries loads both parallel and transverse to its longitudinal axis.

**Bearing plate** – A plate provided at the support points of a beam or truss, which transmits the reaction over a larger area on masonry or concrete supports, in such a way that the bearing capacity of the support material is not exceeded.

**Bearing-type connection** – A connection using bolts in snug-tight condition and the load is transferred to the other member through bearing of bolts against plates inside the bolt hole

**Biaxial bending** – The general state of a member which is subjected to bending actions in both principal planes together with axial compression and torsion actions

**Bifurcation** – A term relating to the load-deflection behavior of a perfectly straight and perfectly centered compression element at critical load

**Blast oxygen furnace (BOF)** – A furnace in which oxygen is blown into it to convert molten pig iron (from the blast furnace) into liquid steel and to remove carbon as carbon monoxide gas

**Blind rivet** – A mechanical fastener capable of joining two members together where access to the assembly is limited to one side only

**Block shear failure** – A failure of bolted connection in which the failure occurs in shear at a row of bolt holes parallel to the applied load and by tensile rupture along the perpendicular face

**Braced frame** – A frame in which lateral loads are resisted by means of diagonal braces or other system of bracings

**Brittle cladding** – Claddings, such as asbestos cement sheets, which get damaged before undergoing considerable deformation

**Brittle fracture** – Sudden fracture without yielding of material under service condition (with negligible ductile deformation), caused by low temperature or sudden change in stress

**Buckle** – To kink, wrinkle, bulge, or otherwise lose original shape as a result of elastic or inelastic strain

**Buckling load** – The load at which an element, a member, or a structure as a whole, either collapses in service or buckles in a load test and develops excessive lateral (out of plane) deformation or instability

**Buckling strength or resistance** – The maximum load or moment a member can withstand without buckling

**Built-up section** – A member fabricated by connecting two or more elements (plates, angles, I-sections, etc.) to form a compound section, acting as a single member

**Butt joint** – A joint formed by two surfaces abutting at right angles

## C

**Camber** – An upward curvature of a beam or truss introduced during fabrication to compensate for the deflection under service loads

**Cantilever** – A beam that is fixed at one end and is free to deflect at the other end

**Capacity** – The maximum force or moment resisted by the section, without failure due to rupture, yielding, or buckling

**Castellated beam** – A rolled I-section steel beam, the web of which is first divided by a length-wise zigzag cut, then welded together so as to join the peaks of both halves, thus increasing its depth and strength

**Cast iron** – A ferrous alloy with more than 2.1% carbon content and typically has significant amounts of silicon, normally in the 1–3% range

**Characteristic load (action)** – The value of specified load (action), which is not expected to be exceeded with more than 5% probability, during the life span if the structure. Thus, the characteristic load will not be exceeded 95% of the time.

**Characteristic strength (stress)** – The value of strength (stress), below which not more than 5% of test results are expected to occur

**Cladding** – Profiled sheet for walls

**Cleat** – A short member (often an angle section) used in a connection

**Coating** – The process of covering steel with another material (tin, chrome, and zinc), primarily for corrosion resistance

**Cold-formed steel structural members** – Shapes that are manufactured by press-braking blanks sheared from sheets, cut length of coils or plates, or by roll forming cold- or hot-rolled coils or sheets. The word 'cold' refers to the forming process, not to the type of steel that is used.

**Column** – A vertical or nearly vertical member carrying predominantly compressive axial force and some moments.

**Compact section** – A cross-section capable of developing plastic moment capacity, but does not have adequate plastic rotation capacity required for the formation of plastic collapse mechanism, due to local buckling.

**Compound beam** – A beam consisting of rolled steel sections and flange plates and used in situations where the depth is limited and rolled sections alone will not be sufficient to carry the load

**Composite slab** – A slab where structural load carrying capacity is achieved by the composite action of concrete slab and steel floor decking

**Connection** – The means by which members are connected together and through which forces and moments are transferred

**Constant stress range** – The amplitude between which the stress ranges under cyclic loading is constant during the life of the structure or a structural element

**Crane load** – Horizontal and vertical loads from cranes

**Critical load** – The load at which bifurcation occurs, as determined by a theoretical stability analysis

**Coped end** – The connected end of a member, where one or both flanges are cut away to facilitate connection with other members

**Corrosion** – An electro chemical process, caused by atmosphere, moisture, or other agents, leading to oxidation of the metal

**Cover plate** – A plate welded to the top or bottom flanges of a member to increase its load carrying capacity

**Critical load** – The value of load at which a straight column becomes unstable

**Cumulative fatigue** – Total damage due to fatigue loading of varying stress ranges

**Cut-off limit** – The stress range, corresponding to the particular detail, below which cyclic loading need not be considered in cumulative fatigue damage evaluation (corresponds to $10^8$ numbers of cycles in most cases)

## D

**Dead loads** – The self-weight of all walls, partitions, floors, roofs, and other permanent fixtures acting on a member.

**Deformation capacity** – A measure of the ability of the structure to deform without failing, when the plastic collapse mechanism develops

**Design life** – Time period for which a structure or a structural element is required to perform its function without damage

**Design load** – A value of load obtained by multiplying the characteristic load with a load factor

**Design strength/resistance** – the resistance or capacity provided by the structure, element, or connection to resist the design load

**Detail category** – The designation given to a particular detail to indicate the S-N curve to be used in fatigue assessment

**Discontinuity** – A sudden change in the cross section of a loaded member, causing a stress concentration at the location

**Ductility** – Ability of the material or a structure to deform beyond the limit of yield deformation before failure or fracture. The ratio of ultimate to yield deformation is usually termed as ductility.

**Durability** – The ability of a material to resist deterioration over the design life

**Dynamic load** – A load that varies with time, for example, seismic load

## E

**Earthquake load** – The inertia forces produced in a structure due to the ground movement during an earthquake

**Eave strut** – A member at the intersection of roof truss and the exterior wall, supporting the side cladding; in braced buildings it also acts as a compression chord of the wind bracing truss

**Edge or end distance** – Distance from the centre of a fastener hole to the edge of an element measured parallel to the direction of load transfer

**Effective lateral restraint** – Restraint that produces sufficient resistance to prevent deformation in the lateral direction

**Effective length** – Actual length of a member between points of effective restraint or effective restraint and free end, multiplied by a factor to take account of the end conditions, such that it has the same elastic buckling load as the actual member

**Effective width** – Reduced flat width of a compression element in a flexure or compression members computed to account for local buckling when the flat width-to-thickness ratio exceeds a certain limit

**Elastic critical moment** – The elastic moment which initiates lateral-torsional buckling of a laterally unsupported beam

**Elastic cladding** – Claddings, such as metal sheets, that can undergo considerable deformation without damage

**Elastic design** – Design, which assumes elastic behaviour of materials throughout the service load range

**Elastic limit** – It is the stress below which the material regains its original size and shape when the load is removed. In steel design, it is taken as the yield stress.

**Electric arc furnace (EAF)** – A furnace that melts (and thus recycles) steel scrap for the production of liquid steel using electrical power

**Empirical method** – A simplified method of design based on experience or by experimental research

**Environmental loads** – The loads caused by environmental effects, such as wind, temperature, snow, earthquake, etc.

**Equivalent uniform moment factor** – A factor that takes into account the effects of end moments in a beam-column member

**Erection loads** – The actions (loads and deformations) experienced by the structure exclusively during erection

**Exposed surface area to mass ratio** – The ratio of the surface area exposed to the fire (in $mm^2$) to the mass of steel (in kg). In the case of members with tire protection material applied, the exposed surface area is to be taken as the internal surface area of the fire protection material.

## F

**Fabrication tolerance** – Amount of deviation allowed in the nominal dimensions and geometry in fabrication activities, such as cutting to length, finishing of ends, cutting of bevel angles, etc.

**Factored load** – *See* Design load

**Factor of safety** – The factor by which the yield stress of the material of a member is divided to arrive at the permissible stress in the material

**Fastener** – A bolt, pin, rivet, or weld used in a connection

**Fatigue** – Damage caused by repeated fluctuations of stress, leading to progressive cracking and fracture of a structural element

**Fatigue strength** – The stress range for a category of detail, depending upon the number of cycles the structure is required to withstand during its design life

**Fillet weld** – A type of weld that is triangular in shape and joins two surfaces at right angles to each other in a lap joint, T-joint, or corner joint; they are the most common types of welds

**Fire exposure condition**

(a) Three-sided fire exposure condition: Steel member incorporated in or in contact with a concrete or masonry floor or wall (at least against one surface)—members with more than one face in contact with a concrete or masonry floor or wall may be treated as three-sided fire exposure.

(b) Four-sided fire exposure condition: Steel member, which may be exposed to fire on all sides

**Fire protection system** – The fire protection material and its method of attachment to the steel member

**Fire resistance** – The ability of an element, component, or structure, to fulfill for a stated period of time, the required stability, integrity, thermal insulation, and/or other expected performance specified in a standard fire test

**Fire resistance level** – The fire resistance grading period for a structural element or system, in minutes, which is required to be attained in the standard fire test

**First-order elastic analysis** – An analysis in which the material is modeled as linear-elastic and equilibrium is expressed in terms of the geometry of the undeformed structure; here the moments caused by the product of the load and deflections are ignored

**Flat width** – Width of the straight portion of the element, excluding the bent portion of the section

**Flat width-to-thickness ratio** – Ratio of the flat width of an element to its thickness

**Flexural buckling** – Buckling of a column due to flexure

**Flexural member** – *See* Beam

**Flexural stiffness** – Stiffness of a member against rotation as expressed by the value of bending moment required to cause a unit rotation while all other degrees of freedom of the joints of the member except the one under consideration are assumed to be restrained

**Flexural torsional buckling** – A mode of buckling in which the member deflects and twists

**Foundation bolt** – *See* Anchor bolt

**Friction type connection** – A connection made by using pre-tensioned high-strength bolts where shear force transfer is due to mobilization of friction between the connected plates due to clamping force developed at the interface of connected plates by the bolt pre-tension

## G

**Gantry girder** – Steel rolled section (composite or single), which supports moving wheels of cranes used to transfer heavy loads in factories from one place to another

**Gauge** – The spacing between adjacent parallel lines of fasteners, transverse to the direction of load/stress

**Girts** – Horizontal members provided along the sides of industrial buildings, which are designed to resist bending due to wind loads

**Global stability** – The stability of the whole structure against buckling, overturning, sliding, and uplift

**Gravity load** – Loads arising due to gravitational effects

**Grillage foundation** – Consists of two or more tiers of steel beam layers arranged one over the other at right angles to each other and all the beams are completely encased in well compacted concrete; used to transfer heavy load from column to soil of low bearing capacity

**Groove weld** – A weld in the groove between a pair of members; used to connect members that are aligned in the same plane and often used in butt joints

**Gusset plate** – The steel plate to which the members intersecting at a joint are connected

## H

**H-section** – A H-section (UC section) is about as wide as it is deep, and flange and web thicknesses are chosen such that $I_{xx} = I_{yy}$. They are used for columns.

**High-performance steel** – A steel that has an optimized balance of properties such as strength, weldability, toughness, ductility, corrosion resistance, and formability to give maximum performance while remaining cost-effective

**High shear** – High shear condition is caused when the actual shear due to factored load is greater than about 60% of design shear resistance. When a cross section is subjected to coexisting bending and high shear, the moment resistance of the section has to be reduced to account for the presence of high shear.

**Hollow section** – A thin walled section having a square, rectangular, or circular cross-section. In square or rectangular hollow sections, the nominal outer corner radius is equal to twice the wall thickness.

**Hot rolled section** – Structural steel sections that are formed by rolling mills from molten steel; examples are I-sections, channels, angles, etc.

**Hybrid steel section** – A section with flanges of high-strength steel and web (or webs) of lower strength steel

## I

**I-section** – An I-section (UB section) is much deeper than it is wide; usually depth is greater than 1.2 times the width. It is generally used to resist bending loads as a beam.

**Imposed load** – The load is produced by the intended use or occupancy including distributed, concentrated, impact, vibration, and snow loads but excluding, dead, wind, earthquake, and temperature loads.

**Inelastic behaviour** – Deformations accompanied by yielding

**Initial imperfection** – Deviation from perfect geometry, for example, initial crookedness of a member or initial out-of-flatness of a plate

**Instability** – The phenomenon which disables an element, member, or a structure to carry further load due to excessive deflection lateral to the direction of loading and vanishing stiffness

**Interference effects** – They are caused when a tall structure is surrounded by other structures of similar size; the wind forces/pressures acting on structures upstream/downstream are modified significantly, due to the high turbulence and the formation of large size 'eddies'.

## J

**Joint** – An element that connects the members of a structure together and enables the moment and force transfer between them

**Joist** – A horizontal supporting member that runs between foundations, walls, or beams to support a ceiling or floor

## K

**Killed steel** – Steel that has its oxygen content reduced, typically through addition of aluminium; hence the term *aluminium killed.*

## L

**Lap joint** – A joint between two pieces of metal, in which the two pieces overlap without any change in form

**Lamellar tearing** – A kind of brittle fracture that may occur in the base metal beneath the weld

**Lateral restraint for a beam** – Restraint that prevents lateral movement of the compression flange of the beam (*see* also Effective lateral restraint)

**Lattice girder** – A girder with top and bottom flanges connected by a lattice (interlaced) web

**Leading imposed load** – Imposed load causing higher action and/or deflection

**Limit state** – Any limiting condition beyond which a structure would cease to fulfill its intended function

**Limit states method or design (LSM or LSD)** – A method that uses multiple safety factor format in order to provide adequate safety at ultimate loads as well as adequate serviceability at service loads

**Lintel** – A horizontal member (beam) over an opening, such as a door or window, that carries the wall load

**Live load** – gravity load (*see* Imposed load)

**Load** – An externally applied force or action (*see* also Actions)

**Load effects** – These are internal forces such as axial force, shear force, bending moment, or torsion developed due to the external actions or temperature loads

**Load factor** – a factor used to take account of the uncertainties involved in estimating the loads supported by the structure

**Load and resistance factor design (LRFD)** – A limit states design adopted by the American Institute of Steel Construction

**Local buckling** – Buckling of an element or parts of element of a section characterized by a formation of waves or ripples along the member. It modifies the capacity of cross sections.

**Lug angle** – An additional short angle used to reduce the joint length and shear lag effects; it is more effective at the beginning of the connection

## M

**Main member** – A structural member, which is primarily responsible for carrying and distributing the applied load or action

**Mechanism** – Formation of sufficient plastic hinges in an element or structure, such that the element or structure deforms indefinitely under constant load

**Mild steel** – Low carbon steel, often also referred to as *soft steel* with carbon content generally under 0.25%

**Mill tolerance** – Amount of variation allowed from the nominal dimensions and geometry, with respect to cross-sectional area, non-parallelism of flanges, and out of straightness such as sweep or camber, in a product, as manufactured in a steel mill

**Multiple stiffened element** – A cold-formed element that is stiffened between webs, or between a web and a stiffened edge, by means of intermediate stiffeners that are parallel to the direction of stress

## N

**Natural period** – It is the time period of a structure's undamped free vibration

**Nominal strength** – The capacity of a structure or component to resist the effect of loads; calculated in accordance with provisions and assumptions of the strength design method of the applicable code, and is the strength before the application of any strength-reduction factors

**Normalizing** – Heat treatment for relief of internal stress, based on heating and subsequent air cooling

**Normal stress** – Stress component acting normal to the face, plane, or section

**Notional load** – A lateral load applied on the structure to approximate the effect of member imperfections; usual magnitude is 0.5% of gravity loads in any storey

## O

**Open web steel joist** – *See* Lattice girder

## P

**P-delta effect** – The secondary effect of column axial loads and lateral deflection on the moments in the structural frames/members

**Partial safety factor** – The factor is normally greater than unity and is multiplied with the loads (actions) or divided by the resistances to obtain the design values. These factors take into account the variability in loads, section geometries and material properties.

**Pattern load** – The load that is arranged in such a way to give the most severe effect

**Period of structural adequacy under fire** – The time ($t$), in minutes, for the member to reach the limit state of structural inadequacy in a standard fire test

**Permissible stress** – The maximum stress that is permitted in elements, members, or structures under the nominal/service load (action), when a structure is designed by the working stress method.

**Pitch of fasteners** – The centre-to-centre distance between individual fasteners in a line, in the direction of load/ stress

**Pitch of a truss** – The angle the top chord of the truss makes with the horizontal

**Plastic analysis** – A structural analysis that allows for redistribution of moments in continuous member or frame, taking into account the rotation of plastic hinge

**Plastic collapse** – The failure stage at which sufficient number of plastic hinges have formed due to the applied loads (actions) in a structure leading to a failure mechanism

**Plastic design** – Design against the limit state of plastic collapse

**Plastic hinge** – A yielding zone with significant inelastic rotation, which forms in a member, when the plastic moment capacity is reached at a section

**Plastic modulus ($Z_p$)** – Static moment of tension and compression areas about equal area axis of the cross section

**Plastic moment ($M_p$)** – Moment capacity of a cross section when the entire cross section has yielded due to bending moment; its value equals yield stress times the plastic modulus of the section

**Plastic section** – A cross section, which can develop a plastic hinge and sustain plastic moment over sufficient plastic rotation required for formation of plastic failure mechanism of the member or structure

**Plate girder** – A deep I-beam built up from plates using bolts, rivets, or welds

**Poisson's ratio** – It is the absolute value of the ratio of lateral strain to longitudinal strain under uniaxial loading. For structural steel it is taken as 0.3 in the elastic range and may reach a value of 0.5 in the plastic range.

**Portal frame** – A low-rise structure, comprising columns and horizontal or pitched rafters, connected by moment-resisting connections

**Post buckling strength** – The capacity that a member or a structure can reach after initial buckling

**Pre-loaded bolt** – A bolt tightened to specified initial tension, also called *high-strength friction grip (HSFG) bolt*

**Principle of superposition** – For all linear systems, the net response at a given place and time caused by two or more actions is the sum of the responses which would have been caused by each action individually; applicable only in linear elastic analysis

**Proof stress of bolt** – The stress to which high-strength friction grip (HSFG) bolts are pre-tensioned

**Proof stress of high-strength steel** – The stress that will cause a specified small, permanent extension of the material undergoing a tensile test. Usually the stress to produce 0.2% extension is adopted for high-strength steel which will not exhibit a definite yield point.

**Proof testing** – The application of test loads to a structure, sub-structure, member, or connection to ascertain the structural characteristics of only that specific unit

**Prototype testing** – Testing of a structure, substructure, members, or connections to ascertain their structural characteristics

**Prying force** – Additional tensile force developed in a bolt as a result of the flexing of a connection component such as a beam end plate or leg of an angle

**Purlin** – A horizontal member between main beams that supports roof sheeting

## Q

**Quenching** – Rapid cooling – typically undertaken to obtain a specific property such as increased hardness of steel

## R

**Radius of gyration** – The distance from an axis at which the mass of a body may be assumed to be concentrated. It is calculated as square root of ($I/A$), where $I$ is the moment of inertia and $A$ is the area of the cross section.

**Redistribution of moment** – A process that results in successive formation of plastic hinges until the collapse load is reached

**Residual stress** – The stresses that exist in an unloaded member after it has been formed into a finished product

**Resistance** – The capacity of a structure or a structural element to resist the effect of loads

**Resistance factor** – A factor that accounts for the deviations of the actual strength from the nominal value

**Response reduction factor** – A factor by which elastic responses of the structure generated under the action of an earthquake are reduced to obtain the design values of the responses

**Response spectrum** – The representation of the maximum response of idealized single-degree freedom systems having certain period and damping, during earthquake ground motion

**Rotation** – The change in angle at a joint between the original orientation of two linear members and their final position under loading

**Rotational stiffness** – The moment required to produce unit rotation at a joint

## S

**Safety factor** – A ratio of the stress (or strength) at incipient failure to the computed stress (or strength) at design load

**Sag rods** – The steel rods that provide lateral support to the purlins and for connecting principal ties of roof trusses

**Secondary member** – A member that is provided for the overall stability and or for restraining the main member from buckling or similar modes of failure

**Second-order analysis** – An analysis in which equilibrium is formulated for the deformed structure; here the moments caused by the product of the load and deflections are included

**Seismic base isolation** – A design technique that reduces the force demand on structures by isolating them from the damaging effect of the ground motion

**Self-drilling screw** – A screw that drills its own hole and forms its mating thread in one operation

**Self-tapping screw** – A screw that taps a counter-thread in a prepared hole

**Semi-compact section** – A cross section that can attain the yield stress, but not the full plastic moment before failure, due to local plate buckling

**Serviceability limit state** – A limit state of acceptable service condition, exceeding which causes serviceability failure; acceptable service condition may include appearance, comfort, durability, and better performance during service loads.

**Service load** – The design loads appropriate for the serviceability limit state

**Shape factor** – The ratio of plastic moment ($M_p$) to the yield moment ($M_y$) of the cross section *or* the ratio of the plastic section modulus ($Z_p$) to the elastic section modulus ($Z_e$)

**Shear centre** – The point in the cross section of the beam through which the resultant transverse force must act if the beam is not to twist

**Shear connector** – A short length of rolled I-section or plate, welded perpendicular to the bottom of steel base plate, to increase the shear resistance. Also, a connector (such as a welded stud, spiral bar, or short length of channel) which resists horizontal shear between elements of a composite beam.

**Shear force** – The in-plane force at any transverse cross-section of a member of a structure

**Shear lag** – A phenomenon that occurs in thin wide flanges of beams in which shear straining causes distribution of bending normal stresses to be non-uniform

**Shear stress** – The stress component acting parallel to a face, plane, or cross-section

**Sheeting** – Profiled sheet for floors, roofs, or walls

**Sleeved column** – A column in which a solid core is loosely placed inside a sleeve and the load is applied to the solid core only

**Slender section** – A cross section in which the elements buckle locally before reaching yield moment

**Slenderness ratio** – The ratio of the effective length of a member to the radius of gyration of the cross section about the axis under consideration

**Slip critical joint** – A type of bolted structural steel connection that relies on friction between the two connected elements rather than bolt shear or bolt bearing to join two structural elements

**Slip resistance** – Limiting value of shear that can be applied in a friction grip connection before slip occurs

**S-N curve** – The curve defining the relationship between the number of cycles for fracture, and the maximum stress or stress range due to the applied cyclic load

**Snow load** – Load on a structure due to the accumulation of snow and ice on surfaces such as roof

**Snug tight** – The tightness of a bolt achieved by a few impacts of an impact wrench or by the full effort of a person using a standard spanner

**Splice** – A connection between two similar collinear members

**Stability limit state** – A limit state corresponding to the loss of static equilibrium of a structure by excessive deflection transverse to the direction of predominant loads

**Stainless** – An alloy of carbon and iron that has a minimum chromium content of 10.5%

**Stiffener** – An element, usually a flat plate or angle, connected perpendicular to the longitudinal axis of the member to strengthen it against buckling or to transfer shear or to prevent the out-of-plane deformations of plates

**Stiffened or partially stiffened element** – A flat compression element in which both edges parallel to the direction of stress are stiffened either by a web, flange, stiffening lip, or intermediate stiffener

**Strain** – Deformation per unit length or unit angle

**Strain hardening** – The phenomenon of increase in stress with increase in strain beyond yielding

**Strength** – Resistance to failure by yielding or buckling

**Strength limit state** – A limit state of collapse or loss of structural integrity

**Stress** – The internal force per unit area of the original cross section

**Stress analysis** – The analysis of the internal force and stress condition in an element, member, or structure

**Stress concentration** – Localized stresses (near edges or opening) that are greater than the average stress in the member

**Stress cycle counting** – Sum of individual stress cycles from stress history arrived at using any rational method

**Stress range** – Algebraic difference between two extremes of stresses in a cycle of loading

**Stress spectrum** – Histogram of stress cycles produced by a nominal loading event design spectrum, during design life

**Structural adequacy for fire** – The ability of the member to carry the test load when exposed to the standard fire test

**Structural analysis** – The computation of stress, strain, and deflection characteristics of a structure subjected to load effects

**Structural engineering** – The science and art of designing and making, with economy and elegance, buildings, bridges, frameworks, and other similar structures so that they can safely resist the applied loads

**Strut** – A compression member, which may be oriented in any direction

**Sway** – The lateral deflection of a frame

**Sway member** – A member in which the transverse displacement of one end, relative to the other, is not effectively prevented

## T

**Tangent modulus** – The slope of the inelastic stress–strain curve which is used to predict buckling of inelastic members under increasing load

**Tensile strength** – The maximum load applied in a tensile test divided by the original cross-sectional area of the test piece; also known as *ultimate tensile strength*

**Tensile stress** – The characteristic stress corresponding to rupture in tension, specified for the grade of steel in the appropriate Indian Standard

**Tension field action** – Post-buckling behavior of a plate girder panel under shear force, during which diagonal compressive stresses cause the web to form diagonal waves

**Thickness** – Thickness of base steel, exclusive of the coating

**Torsional buckling** – Buckling of a column by twisting

**Torsional flexural buckling** – A mode of buckling in which compression members can bend and twist simultaneously. This type of buckling mode is critical, in particular when the shear center of the section does not coincide with the centroid.

**Torsional restraint** – A restraint that prevents rotation of a member about its longitudinal axis

**Transverse** – Direction along the stronger axes of the cross-section of the member

**Truss** – A structure composed of a number of struts and tension members pin-connected at the ends to form a triangulated framework

## U

**Ultimate limit state** – The state which, if exceeded, can cause collapse of a part or the whole of the structure

**Ultimate load design** – A method of design in which the ultimate load capacity of the members is computed based on factored load

**Ultimate tensile strength** – *See* Tensile stress

**Unbraced frame** – A frame resisting lateral loads by the bending resistance of the frame members and their connections

**Unstiffened compression element** – A flat compression element that is stiffened at one of the two edges parallel to the direction of stress

## V

**Vierendeel girder** – An open-web girder without diagonal members, with rigid joints between the top and bottom chords and the verticals

## W

**Warping** – A mode of deformation in which plane cross sections do not remain plane

**Web crippling** – Local failure mode of the web of a beam in the region of concentrated load or support reaction

**Weldability** – Ability of steel to be welded without changing its basic mechanical properties

**Width-to-thickness ratio** – Ratio of the flat width of an element to its thickness, where *flat width* is the width of the straight portion of the element, excluding the bent portion of the section

**Wind loads** – Load experienced by member or structure due to wind pressure acting on the surfaces

**Working stress design (WSD)** – *See* Allowable stress design

## Y

**Yield moment ($M_y$)** – The moment of resistance of a cross section when the outermost fibre reaches the yield stress. Its value is obtained by multiplying the elastic modulus of the cross section by the yield stress.

**Yield point** – The maximum stress recorded in a tensile or compressive test of steel specimen prior to entering the plastic range

**Yield strength** – The stress at which permanent deformation occurs in a tensile test. In high-strength steels, this point is not readily distinguishable and a proof strength value is used as an alternative (*see* also Proof stress of high-strength steel).

# Suggested Reading


Allen, H.G. and Bulson, P.S., 1980, *Background to Buckling*, McGraw-Hill Book Company (UK) Ltd., London, p. 582.

Al-Nageim, H.K. and MacGinley,T.J., 2005, *Steel Structures: Practical Design Studies,* Third edition, Taylor & Francis, Oxon, UK, p. 352.

Arya, A.S., Ajmani, J.L., and Awadhesh Kumar, 2014, *Design of Steel Structures*, Sixth edition, Nem Chand & Bros, Roorkee, p. 760.

Ballio, G. and Mazzolani, F.M., 1983, *Theory and Design of Steel Structures*, Chapman and Hall, London, p. 633.

Bjorhovde, R., Colson, A., and Brozzetti, J., Nov 1990, 'Classification System for Beam-to-column Connections', *Journal of Structural Engineering*, ASCE, Vol. 116, No. 11, pp. 3059–3076.

Bleich, F., *Buckling Strength of Metal Structures,* 1952, McGraw–Hill, New York, p. 508.

Blodgett, O.W., *Design of Welded Structures*, 1966, The James P. Lincoln Arc Welding Foundation, Cleveland, Ohio.

Bowles, J.E., 1980, *Structural Steel Design*, McGraw Hill, New York, 536 pp.

Bresler, B., Lin, T.Y., and Scalzi, J.B., 1970, *Design of Steel Structures*, Second edition, Wiley Eastern Pvt. Ltd, New Delhi, p. 830.

Bruneau, M., Uang, C.M., and Whittaker, A., *Ductile Design of Steel Structures*, Second edition, McGraw Hill, New York, 2011, p. 921.

Buchanan, A.H., 2001, *Structural Design for Fire Safety*, John Wiley & Sons Ltd., Chichester, England, p. 421.

Chen, W.F. and Lui, E.M., eds, 2005, *Handbook of Structural Engineering*, Second edition, CRC Press, Boca Raton, FL, p. 1768.

Chen,W.F. and Kim, S-E., 1997, *LRFD Steel Design Using Advanced Analysis*, CRC Press, Boca Raton and New York, p. 441.

Davison, B. and Owens, G.W., 2012, *Steel Designers Manual*, Seventh edition, The Steel Construction Institute, Wiley-Blackwell, London, p. 1398.

Dayaratnam, P., 2004, *Design of Steel Structures*, Second edition, S. Chand & Co. Ltd, p. 880.

Driver, R.G., Grondin, G.Y., and Kulak, G.L., 2006, 'Unified Block Shear Equation for Achieving Consistent Reliability', *Journal of Constructional Steel Research,* Vol. 62, pp. 210–222.

Dowling, P.J., Knowles, P., and Owens, G.W., 1988, *Structural Steel Design*, The Steel Construction Institute and Butterworths, London, p. 399.

Duggal, S.K., *Limit State Design of Steel Structures,* 2014, Second edition, McGraw Hill Education (India) Private Limited, p. 888.

Fisher, J.M., 1991, Industrial Buildings, in *Constructional Steel Design, An International Guide*, Dowling, P.J., Harding, J.E. and Bjorhovde, R., eds, Elsevier Applied Science, London/New York, pp. 627–644.

Gardner, L. and Nethercot, D.A., 2005, *Designers' Guide to EN 1993-1-1, Euro code 3: Design of Steel Structures—General Rules and Rules for Buildings*, Thomas Telford, London and Steel Construction Institute.

Gaylord, Jr, E.H., Gaylord, C.N., and Stallmeyer, J.E., 1992, *Design Of Steel Structures*, Third edition, McGraw Hill International Edition, New York, p. 792.

Greiner, R. and Lindner, J., Aug 2006, 'Interaction Formulae for Members Subjected to Bending and Axial Compression in Eurocode 3—The Method 2 Approach', *Journal of Constructional Steel Research*, Vol. 62, No. 8, pp.757–770.

Hamburger, R.O., Krawinkler, H., Malley, J.O, and Adan, S.M., *Seismic Design of Steel Special Moment Frames—Guide to Practicing Engineers*, National Institute of Standards and Technology, Gaithersburg, MD, June 2009, p. 37 (http://www.nehrp.gov/pdf/nistgcr9-917-3.pdf).

Horne, M.R. and Morris, L.J., 1981, *Plastic Design of Low-rise Frames*, Granada, London, p. 238.

Johnston, B.G., Lin, F.-J., and Galambos, T.V., 1980, *Basic Steel Design*, Second edition, Prentice Hall Inc., Englewood Cliffs, NJ, p. 352.

Kirby, P.A. and Nethercot, D.A., 1979, *Design for Structural Stability*, Granada Publishing, London, p. 180.

Kulak, G.L. and Grondin, G.Y., 2010, *Limit States Design in Structural Steel*, Ninth edition, Canadian Institute of Steel Construction, Toronto, Ontario, p. 403.

Kulak, G.L., Fisher, J.W. and Struik, J.H.A, 1987, *Guide to Design Criteria for Bolted and Riveted Joints*, Second edition, John Wiley, New York, p. 309.

Lam D., Ang T-C., Chiew, S-P., 2014, *Structural Steelwork—Design to Limit State Theory*, Fourth edition, CRC Press (A Spon Press Book), p. 378.

Martin, L.H. and Purkiss, J.A., 1992, *Structural Design of Steel Work to BS 5950*, Edward Arnold, London, p. 468.

McGuire, W., *Steel Structures*, 1968, Prentice Hall Inc., Englewood Cliffs, NJ, p. 1112.

Narayanan, R., ed., 1982, *Axially Compressed Structures: Stability and Strength*, Elsevier Applied Science Publishers, Essex, p. 326.

Narayanan, R., ed., 1983, *Beams and Beam Columns: Stability and Strength*, Applied Science Publishers, Barking, Essex, England, p. 252.

Narayanan, R., et al., *Teaching Resources for Structural Steel Design*, 2 volumes, Institute for Steel Development and Growth (INSDAG), Kolkata, 2001.

Nethercot, D.A., 2001, *Limit States Design of Structural Steelwork*, Third edition, Spon Press, London, p. 270.

Owens, G.W. and Cheal, B.D., 1989, *Structural Steelwork Connections*, Butterworths, London, p. 330.

Petersen, C., 1990, *Stahlbau-Grundlagen der Berechnung und Baulichen Ausbuildung von Stahlbauten*, Second edition, Vieweg Verlag, Braunschweig.

Petersen, C., 1980, *Statik und Stabilität der Baukonstruktionen*, Friedr. Vieweg & Sohn, Braunschweig/Wiesbaden, Germany, p. 960.

Purkiss, J.A., 1996, *Fire Safety Engineering Design of Structures*, Butterworth Heinemann, Oxford, England, p. 342.

Roik, K., Carl, J., and Lindner, J., 1972, *Biege Torsions Probleme Gerader Dünnwandiger Stäbe*, Verlag von Wilhelm Ernst & Sohn, Berlin.

Salmon, C.G., Johnson, J.E., and Malhas, F.A., 2008, *Steel Structures—Design and Behaviour*, Fifth ed., Harper Collins, College Publishers, New York, 2008, p. 888.

Subramanian, N., 2007, *Space Structures: Principles and Practice*, Multi-Science Publishing Co. Ltd, Essex, UK, p. 862.

Taranath, B.S., 1998, *Steel, Concrete and Composite Design of Tall Buildings*, Second edition, McGraw Hill, New York, p. 1024.

Trahair, N.S., Bradford, M.A., Nethercot, D.A., and Gardner,L., 2008, *The Behaviour and Design of Steel Structures to EC3*, Fourth edition, Taylor & Francis, London and New York, p. 490.

Vinnakota, S., 2005, *Steel Structures: Behavior & LRFD*, McGraw-Hill Higher Education, New York, p. 928.

Ziemian, R.D., ed., 2010, *Guide to Stability Design Criteria for Metal Structures*, Sixth edition, John Wiley & Sons, Inc., New York, p. 1024.

## ONLINE RESOURCES

www.nptel.iitm.ac.in/courses/IIT-MADRAS/Design_Steel_Structures_I/index.php

www.nptel.iitm.ac.in/courses/IIT-MADRAS/Design_Steel_Structures_II/index.php

www.steel-insdag.org/TM_Contents.asp

www.steeltools.org

www.tatasteelconstruction.com/en/design-guidance/structural-design

http://www.constructalia.com/english/publications#.Va_3gPlViko

http://sections.arcelormittal.com/library/design-manuals-steel-building-in-europe.html

www.aisc.org

http://msc.aisc.org/en/modernsteel/archives/

www.steelconstruction.info/

www.thestructuralengineer.info/online-library

www.efunda.com/formulae/solid_mechanics/plates/theory.cfm

www.efunda.com/formulae/solid_mechanics/columns/intro.cfm

http://engineeringtraining.tpub.com/14071/index.htm

www.sefindia.org/

www.seaint.org/

www.youtube.com/watch?v=3KVg5kljuU0

http://bgstructuralengineering.com/

www.steelconstruct.com

www.esdep.org/

*All websites were last accessed on 7 October 2015.*

# INDEX

# About the Author

Dr N. Subramanian is a consulting engineer and mentor now living in Maryland, USA, and former Chief Executive of *Computer Design Consultants*, Chennai. A PhD and Post-Doctoral Fellow of the *Indian Institute of Technology, Madras*, he has more than 40 years of professional experience which includes teaching, research, and consultancy. He also worked in the then West Germany as *Alexander von Humboldt Fellow* with Prof. J. Lindner and Prof. Ch. Peterson. He has served as a consultant to several leading organizations in India and designed several noteworthy structures involving RCC, structural steel, and cold-formed steel. More than 600 microwave/transmission towers were designed by him, one of which earned a certificate of merit from the Consultancy Development Centre, DSIR, Govt of India, 1992.

Dr Subramanian has also developed several special-purpose software packages for analysis and design of different types of structures. Being a member/fellow of several professional bodies, he took very active part in these organizations, and has been the All India Vice President of Association of Consulting Engineers (India) and also the Indian Concrete Institute.

Dr Subramanian has authored more than 25 wide-selling books and more than 230 technical papers, published in national and international journals and conferences. He is also the author of *Steel Structures: Design and Practice* (OUP), *Design of Reinforced Concrete Structures* (OUP), *Access to Computer Education* (OUP), and Space *Structures: Principles and Practice* (Multi-Science, UK). Three of these books earned the coveted *ACCE-Nagadi Award* and are prescribed as text books.

Dr Subramanian is in the editorial/review boards of several prestigious International journals and earned several awards, including the *Tamilnadu Scientist Award* and the *Lifetime Achievement Award* given jointly by the Indian Concrete Institute and L&T.

He is active as mentor to young engineers and is one of the main contributors to the *Structural Engineering Forum of India*, which has more than 17,500 members. He has also served as the President of the Rotary Club of Madras Metro.

# Related Titles

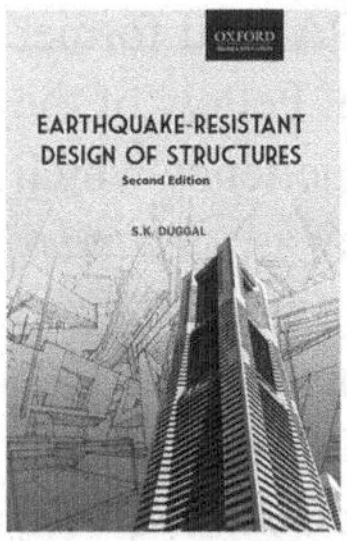


**Earthquake-resistant Design of Structures (Second Edition)** (9780198083528)

**S.K. Duggal**

Motilal Nehru National Institute of Technology, Allahabad

Beginning with an introduction to earthquakes and ground motion, the book provides a detailed coverage of structures and soil in terms of their seismic response. The need for placing emphasis on conceptual design is reinforced by enumerating factors that cause damage and by offering guidelines for efficient seismic-resistant design, with special attention to timber, masonry, concrete, and steel buildings.

**Key Features:**

- Elucidates the theoretical as well as practical aspects of earthquake-resistant design of structures
- Includes a case study on the 2001 Bhuj Earthquake to facilitate insight into the effects of seismic forces on structures
- Provides standard tables both in the text and in the appendices for ready reference
- Highlighted information on design specifications and norms

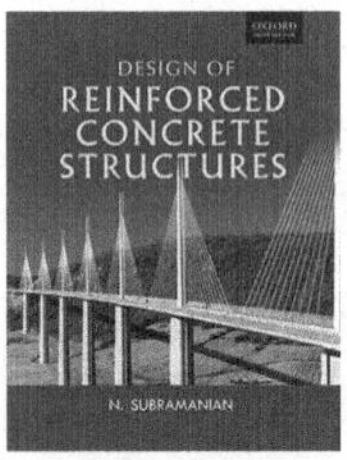


**Design of Reinforced Concrete Structures** (9780198086949)

**N. Subramanian**

Consultant engineer, Maryland, USA
Former Chief Executive of Computer Design Consultants, Chennai

This book provides an extensive coverage of the design of reinforced concrete structures in accordance with the current Indian codes of practice (IS 456:2000 and IS 13920:1993). The text is based on the modern limit states design and covers topics such as the properties of concrete, structural forms, and loadings.

**Key Features:**

- Based on the latest version of IS 456:2000 and IS 13920:1993 and includes latest developments and state-of-the-art trends
- Presents 23 case studies from actual practice that reflect on structural developments and failures
- Presents a chapter on multi-storey buildings using STAAD. Pro software

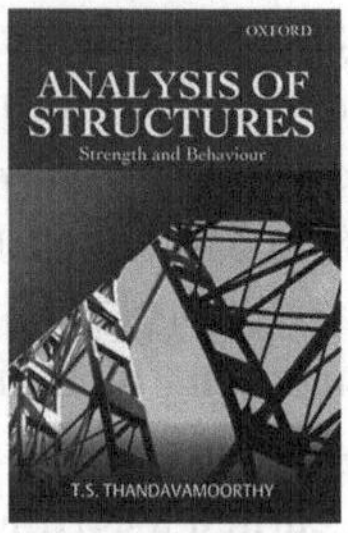


**Analysis of Structures: Strength and Behaviour** (9780195670035)

**T.S. Thandavamoorthy**

Professor, Civil Engineering,
Adhiparasakthi Engineering College, Anna University

Designed for an introductory course, this book adopts a modern and practical approach to structural analysis. The book provides a balanced coverage of concepts, basic definitions, and analytical techniques in the field of structural analysis.

**Key Features:**

- Includes topics such as offshore structures, safety auditing of structures, and experimental stress analysis
- Includes a large number of illustrations and photographs to supplement the text
- Promotes self-study by including numerous solved examples and chapter-end exercises
- Practicing professionals would find this book extremely useful for its coverage of the latest techniques.

**Steel Structures: Design and Practice** (9780198068815)

**N. Subramanian**

Consultant engineer, Maryland, USA
Former Chief Executive of Computer Design Consultants, Chennai

A perfect blend of theory and practice, the book provides an extensive coverage of the design of steel structures in accordance with the latest code of practice for general construction in steel (IS 800 : 2007). The book is based on the modern limit state approach to design and covers topics such as properties of steel, types of steel structures, loadings and load combinations, design considerations, design of tension and compression members, beams, plate and gantry girders, beam columns, and trusses.

**Key Features:**

- Features numerous figures, tables, review questions, and practice problems to supplement the text
- Contains case studies in almost every chapter and includes more than 100 worked-out examples, illustrating the different steps involved in the design of various structural elements
- Includes appendices on structural steel and soil properties and computer programs for beams, columns, and beam-columns. Provides design aids in Appendix D

**Other Related Titles**

9780198069188 Thandavamoorthy : *Structural Analysis 1/e*
9780199454723 R. Subramanian : *Fundamentals of Surveying and Levelling 1/e*
9780198061106 R. Subramanian : *Strength of Materials 2/e*